Periodic Table of the Elements

Main groups		Transition metals										Main groups					
1 1A	2 2A	3 3B	4 4B	5 5B	6 6B	7 7B	8 8B	9 8B	10 8B	11 1B	12 2B	13 3A	14 4A	15 5A	16 6A	17 7A	18 8A
1 H 1.00794																	2 He 4.002602
3 Li 6.941	4 Be 9.012182											5 B 10.811	6 C 12.0107	7 N 14.0067	8 O 15.9994	9 F 18.998403	10 Ne 20.1797
11 Na 22.989769	12 Mg 24.3050											13 Al 26.981538	14 Si 28.0855	15 P 30.973762	16 S 32.065	17 Cl 35.453	18 Ar 39.948
19 K 39.0983	20 Ca 40.078	21 Sc 44.955912	22 Ti 47.867	23 V 50.9415	24 Cr 51.9961	25 Mn 54.938045	26 Fe 55.845	27 Co 58.933195	28 Ni 58.6934	29 Cu 63.546	30 Zn 65.38	31 Ga 69.723	32 Ge 72.64	33 As 74.92160	34 Se 78.96	35 Br 79.904	36 Kr 83.798
37 Rb 85.4678	38 Sr 87.62	39 Y 88.90585	40 Zr 91.224	41 Nb 92.90638	42 Mo 95.96	43 Tc (98)	44 Ru 101.07	45 Rh 102.90550	46 Pd 106.42	47 Ag 107.8682	48 Cd 112.411	49 In 114.818	50 Sn 118.710	51 Sb 121.760	52 Te 127.60	53 I 126.90447	54 Xe 131.293
55 Cs 132.90545	56 Ba 137.327	71 Lu 174.9668	72 Hf 178.49	73 Ta 180.9479	74 W 183.84	75 Re 186.207	76 Os 190.23	77 Ir 192.217	78 Pt 195.094	79 Au 196.96657	80 Hg 200.59	81 Tl 204.3833	82 Pb 207.2	83 Bi 208.98040	84 Po (209)	85 At (210)	86 Rn (222)
87 Fr (223)	88 Ra (226)	103 Lr (262)	104 Rf (265)	105 Db (268)	106 Sg (271)	107 Bh (272)	108 Hs (270)	109 Mt (276)	110 Ds (281)	111 Rg (280)	112 Cn (285)	113 (284)	114 Fl (289)	115 (288)	116 Lv (293)	117 (292)	118 (294)

*Lanthanide series	57 *La 138.9055	58 Ce 140.116	59 Pr 140.90765	60 Nd 144.242	61 Pm (145)	62 Sm 150.36	63 Eu 151.964	64 Gd 157.25	65 Tb 158.92535	66 Dy 162.500	67 Ho 164.93032	68 Er 167.259	69 Tm 168.93421	Y 173
†Actinide series	89 †Ac (227)	90 Th 232.0381	91 Pa 231.03588	92 U 238.02891	93 Np (237)	94 Pu (244)	95 Am (243)	96 Cm (247)	97 Bk (247)	98 Cf (251)	99 Es (252)	100 Fm (257)	101 Md (258)	1 N (25

GENERAL CHEMISTRY

ATOMS FIRST

SECOND EDITION

JOHN E. McMURRY

Cornell University

ROBERT C. FAY

Cornell University

PEARSON

Boston Columbus Indianapolis New York San Francisco Upper Saddle River
Amsterdam Cape Town Dubai London Madrid Milan Munich Paris Montréal Toronto
Delhi Mexico City São Paulo Sydney Hong Kong Seoul Singapore Taipei Tokyo

Editor in Chief: *Adam Jaworski*
Senior Acquisitions Editor: *Terry Haugen*
Senior Marketing Manager: *Jonathan Cottrell*
Director of Development, Chemistry and Geosciences: *Jennifer Hart*
Project Editor: *Jessica Moro*
Assistant Editor: *Coleen Morrison*
Development Editor: *Carol Pritchard-Martinez*
Editorial Assistant/Assistant Editor: *Erin Kneuer*
Marketing Assistant: *Nicola Houston*
Senior Media Producer: *Kristin Mayo*
Managing Editor, Chemistry and Geosciences: *Gina M. Cheselka*
Senior Production Project Manager: *Beth Sweeten*
Production Management: *Greg Johnson, PreMediaGlobal*
Compositor: *PreMediaGlobal*
Illustrator: *Precision Graphics*
Image Lead: *Maya Melenchuk*
Photo Researcher: *Eric Schrader*
Text Permissions Manager: *Alison Bruckner*
Text Permissions Researcher: *Jen Roach, PreMediaGlobal*
Design Manager: *Marilyn Perry*
Interior Designer: *Cenveo*
Cover Designer: *Elise Lansdon Design*
Operations Specialist: *Jeffrey Sargent*
Cover Image Credit: *Molecular Chair, Design by Antonio Pio Saracino – www.antoniopiosaracino.com*

Credits and acknowledgments borrowed from other sources and reproduced, with permission, in this textbook appear on the appropriate page within the text or on p. C-1.

McMurry, John.
General chemistry : atoms first / John E. McMurry, Robert C. Fay.—Second edition.
pages cm
Includes index.
ISBN-13: 978-0-321-80926-1
ISBN-10: 0-321-80926-2
1. Chemistry—Textbooks. I. Fay, Robert C., 1936- II. Title.
QD33.2.M364 2014
540—dc23

2012036322

4 5 6 7 8 9 10—V011—16 15 14

PEARSON

www.pearsonhighered.com

ISBN-10: 0-321-80926-2
ISBN-13: 978-0-321-80926-1

Brief Contents

Contents

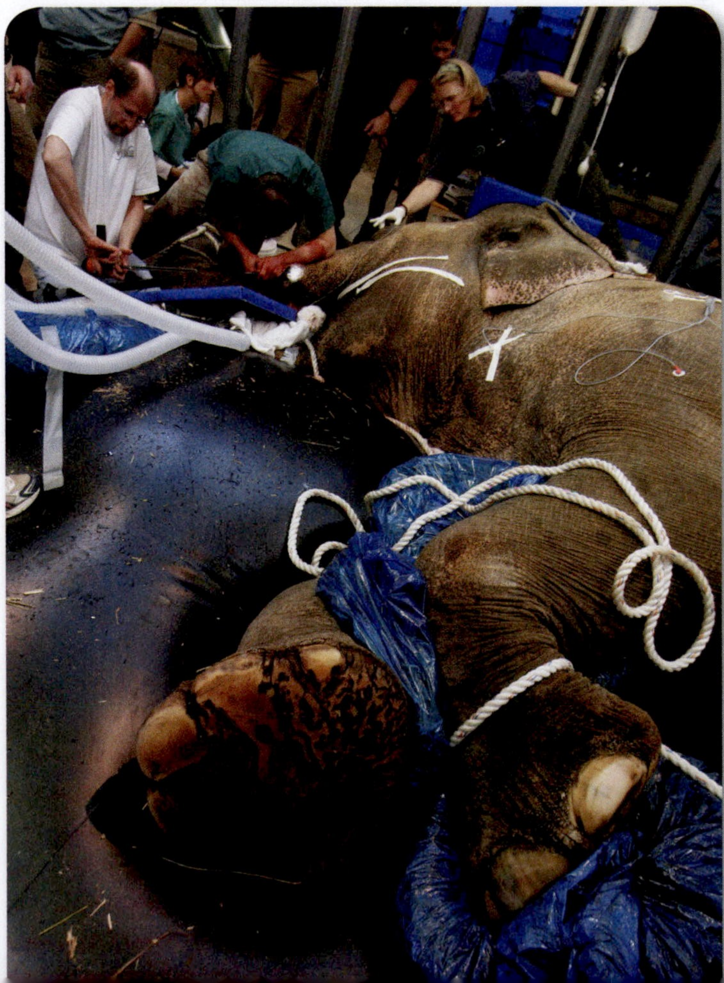

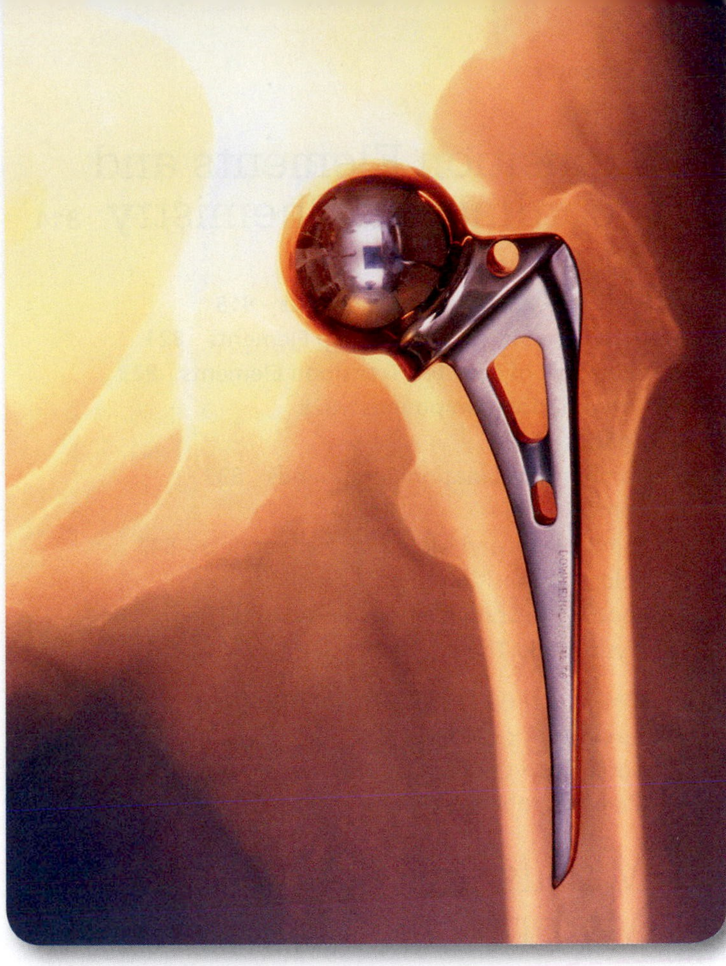

Preface

Our primary purpose in writing this book has been to fashion a clear and cohesive introduction to chemistry, covering both important principles and important facts. We write to explain chemistry to students today the way we wish it had been explained to us years ago when we were students ourselves. We can't claim that learning chemistry will always be easy, but we *can* promise that we have done our best in planning, writing, and illustrating this book to make the learning process as smooth as possible.

Perhaps the first thing you will notice about this book is that its organization is different from that of other general chemistry textbooks. Rather than follow the typical ordering of topics, in which stoichiometry and aqueous reactions come first, this book takes what has come to be called an atoms-first approach. Instead of launching immediately into stoichiometry, we start at the logical beginning of the chemical story by discussing *atoms*—their history, stability, electronic structure, and consequent periodicity. This approach makes it possible to tell a cohesive story about chemistry that follows an intuitive logic in progressing from the simplest building blocks to successively more complex concepts.

Once atoms have been fully described in Chapters 1 and 2, we proceed next to discuss how and why atoms bond together to make chemical compounds. Ion formation and ionic bonding come first, in Chapter 3, followed by covalent bonding and the structures of molecules in Chapters 4 and 5. This organization takes students immediately into real chemistry rather than making them first spend time with chemical arithmetic. Only then, when all the fundamental pieces are in place, are chemical reactions and stoichiometric mass relationships introduced in Chapters 6 and 7.

At this point in the narrative, the ordering of topics becomes more familiar, starting with thermochemistry and chemical energy (Chapter 8), moving to bulk properties of pure substances (Chapters 9–10), and continuing with the properties of solutions (Chapter 11) and with all the topics necessary for a study of chemical transformations: kinetics, equilibrium, thermodynamics, and electrochemistry (Chapters 12–17). Then, in Chapters 18–21, the concepts described in earlier chapters are applied to discussing the chemistry of main-group and transition elements, metals, and modern solid-state materials. The book ends with a chapter devoted to nuclear transformations followed by a brief look at organic and biological chemistry.

It's important to note that our atoms-first approach is not the same thing as 'theory first.' Science nearly always begins with observation and experiment, and only later are theories developed to explain observations. While we begin with atoms and move on logically to increasing complexity, we do not put theory before experiment. For instance, we do not put the kinetic-molecular theory before the gas laws or the collision theory before rate laws. Our objective throughout is to present chemistry in the way that science actually works.

To help students succeed in learning chemistry, we have put extraordinary effort into this book. Transitions between topics are smooth, explanations are lucid, and reminders of earlier material are frequent. Insofar as possible, distractions within the text are minimized. Each chapter is broken into numerous sections to provide frequent breathers, and each section has a consistent format. Sections generally begin with an explanation of their subject, move to a Worked Example that shows how to solve problems, and end with one or more Problems for the reader to work through. Each chapter ends with a brief *FYI* (including five that are new to this edition) that describes an interesting application or extension of the chapter topic. Throughout the book, every attempt has been made to explain chemistry in a visual and intuitive way so that it can be understood by all who give it an honest effort.

NEW TO THE SECOND EDITION

In preparing this second edition, we have reworked the entire book at the sentence level and made many hundreds of alterations, updates, and reviewer-requested reorganizations to make it as easy as possible for our readers to understand and learn chemistry. In addition, a number of more substantial changes and rewrites have been made. Among the more important are the following:

- *Chapter 0 — Chemical Tools; Experimentation and Measurement* One of the most commonly voiced complaints about General Chemistry texts is that they begin with a chapter on units and measurement. Many students, however, learned this material in high school and are immediately bored, while many others have to struggle. This book takes a new approach to the problem by beginning with the eye-catching number, Chapter 0. Chapter 0 is a real chapter like any other, but its numbering clearly lets people know that it covers preliminary material to bring students who need it up to speed while letting others bypass it if they choose.
- *Chapter 1 — The Structure and Stability of Atoms* This first chemical chapter in the book is completely focused on covering *atoms.* It contains introductory material about elements, the periodic table, atomic structure, and atomic weights. Nuclear Chemistry is also briefly introduced in Chapter 1 and followed with a thorough treatment in Chapter 22.
- *Chapter 4 — Atoms and Covalent Bonds* and *Chapter 5 — Covalent Bonds and Molecular Structure* Coverage of covalent bonding and molecular structure has been broken into two chapters in this new edition to make the subject less intimidating and slow the pace of the discussion.
- *Chapter 18 — Hydrogen, Oxygen and Water* The chapter has been streamlined throughout, and the former Sections 18.2 and 18.14 on hydrogen isotopes and the reactivity of water have been deleted in this second edition.
- *Chapter 19 — Main-Group Elements* The chapter has been shortened by removing the former Section 19.9 on germanium, tin, and lead and eliminating the previous coverage of polyphosphoric acids.
- *Chapter 22 — Nuclear Chemistry* This new chapter has been added to the second edition at the request of users. It provides a cohesive discussion of nuclear reactions, fission, fusion, nuclear transmutation, and applications of nuclear chemistry, building on the brief introduction to nuclear chemistry provided in Chapter 1.
- *Chapter 23 — Organic and Biological Chemistry* The former Chapter 22 on organic chemistry in the first edition has been tightened in its previous coverage and expanded to provide new coverage of the major classes of biomolecules—amino acids, peptides, proteins, carbohydrates, lipids, and nucleic acids.
- The art in this new edition has been improved in many ways to make the numbered figures more self-contained, informative, and easily read:
  - Numbered figures are placed in an unobtrusive shadow box to make them more clearly distinguished.
  - **Figure** and **Table** references in the text are called out in bold color so that it's easy to find the text corresponding to a given figure or table.
  - Internal art captions in figures are set off in a different font from art labels so that students can more readily grasp the main points of each illustration.
  - Numerous small explanations are placed directly on the relevant parts of the figures themselves instead of having long captions beneath figures. The effect is to make the text flow naturally into the figures and thereby entice readers to spend more time understanding those figures.
  - Important text within the illustrations is color-coded to focus attention on it.
- More than 250 **NEW** problems have been added throughout the text to help build students' problem-solving skills and conceptual understanding of chemistry.
- **Remember...** notes are placed in the margin to help students connect concepts from previous chapters to current material. In addition, important forward references are also called out, with the subject described in the adjacent margin under the heading **Looking Ahead...**
- Lists of Learning Outcomes keyed to problem numbers and relevant Key Equations have been added to every chapter.

- Just as important as the addition of new material and features is that the best features of the previous edition have been retained:
  - The design remains spacious, readable, and unintimidating.
  - The writing style remains clear and concise.
  - Worked problems are identified by subject and are immediately followed by a similar problem for students to solve.
  - Each chapter ends with a summary, a list of key words with accompanying page references, and a large set of end-of-chapter problems.
  - Most end-of-chapter problems are classified by text section and paired by topic. These are followed by a group of unclassified Chapter Problems and a final set of Multiconcept Problems, which draw on and connect concepts from several chapters.
  - Extensive use of visual, non-numerical, Conceptual Problems continues, both within and at the ends of chapters. These Conceptual Problems, which test the understanding of principles rather than the ability to put numbers into a formula, have been a hallmark of this text since the first edition. Don't make the mistake of thinking that these problems are simple just because they don't have numbers. Many are real challenges that will test the ability of any student.
  - Ballpark Checks follow many of the Worked Examples to reinforce students' conceptual understanding and give them confidence that they have arrived at the right answer.

We sincerely hope that this new edition will meet the goals we have set for it and that both students and faculty will find it to be friendly, accessible, and above all effective in teaching chemistry.

ACKNOWLEDGEMENTS

Our thanks go to our families and to the many talented people who helped bring this new edition into being. In addition, we are grateful to Terry Haugen, Senior Acquisitions Editor, and Coleen Morrison, Assistant Editor, for their insights and suggestions that improved the book, to Jonathan Cottrell, Senior Marketing Manager, who brought new energy to marketing the second edition, to Carol Pritchard-Martinez for her work in improving the art program and manuscript development, to Beth Sweeten and Gina Cheselka for their production efforts, and to Eric Schrader for his photo research.

We are particularly pleased to acknowledge the outstanding contributions of several colleagues who created the many important supplements that turn a textbook into a complete package:

- Lee Don Bienski at Blinn College, who updated the accompanying Test Bank
- Dennis Taylor at Clemson University, who updated the Instructor Resource Manual
- Joseph Topich at Virginia Commonwealth University, who prepared both the full and partial solutions manuals
- Mingming Xu at West Virginia University and Donald Linn at Indiana University–Purdue University Fort Wayne, who contributed valuable content for the Instructor Resource DVD
- Stephanie Dillon at Florida State University and Sandra Chimon-Peszek at DePaul University, who prepared a dedicated Atoms First Lab Manual to accompany the text.
- Amina El-Ashmawy and Dawn Richardson from Collin County Community College and Margie Haak from Oregon State University for consultation on writing the MasteringChemistry tutorials to reflect an atoms-first approach.

In addition, we are grateful to Mingming Xu, Donald Linn, and David Boatright of the University of West Georgia for error checking the entire text. And finally, we want to thank our colleagues at so many other institutions who read, criticized, and improved our work.

John McMurry
Robert C. Fay

REVIEWERS OF THE SECOND EDITION

David Boatright, *University of West Georgia*
Tabitha Chigwada, *West Virginia University*
Sandra Chimon-Peszek, *DePaul University*
Ron Duchovic, *Indiana University—Purdue University, Fort Wayne*
Amina El-Ashmawy, *Collin County Community College*
Sherman Henzel, *Monroe Community College*
Angela King, *Wake Forest University*
John Krenos, *Rutgers University*
Donald Linn, *Indiana University—Purdue University, Fort Wayne*
Charity Lovitt, *Bellevue College*
Rosemary Loza, *Ohio State University*
Mark Matthews, *Durham Technical Community College*
Jack McKenna, *St. Cloud State University*
Craig McLaughlin, *Illinois State University*
Dinty Musk, *Ohio Dominican University*
Edith Osborne, *Angelo State University*
Katherine Parks, *Motlow State Community College*
Robert Pike, *College of William and Mary*
Jimmy Reeves, *University of North Carolina, Wilmington*
Mark Schraff, *West Virginia University*
Jacob Schroeder, *Clemson University*
Kimberly Specht, *Denison University*
Lydia Tien, *Monroe Community College*
Kimberly Woznack, *California University of Pennsylvania*
David Zax, *Cornell University*

REVIEWERS OF THE FIRST EDITION

Aaron Baba, *Xavier University*
Chipley Bennett, *Spartanburg Community College*
Mark Benvenuto, *University of Detroit Mercy*
Mary Jo Bojan, *Pennsylvania State University*
Michelle Boucher, *Utica College*
Doug Chapman, *Southern Oregon University*
Holly Cymet, *Morgan State University*
Deborah Exton, *University of Oregon*
Levente Fabry-Asztalos, *Central Washington University*
Jordan Fantini, *Denison University*
Vladimir Garkov, *Mary Baldwin College*
Steven Graham, *St. John's University*
Harold H. Harris, *University of Missouri, St. Louis*
Angela King, *Wake Forest University*
Nancy E. Lowmaster, *Allegheny College*
Stephen Mezyk, *California State University, Long Beach*
Tracy Morkin, *Emory University*
Gary Mort, *Lane Community College*
Doug Mulford, *Emory University*
Richard Mullins, *Xavier University*
Robert Pike, *College of William and Mary*
John Pollard, *University of Arizona*
Clyde Webster, *University of California, Riverside*

Supplements

FOR THE STUDENT

MasteringChemistry® with Pearson eText. MasteringChemistry® from Pearson has been designed and refined with a single purpose in mind: to help educators create that moment of understanding with their students. The Mastering online homework and tutoring system delivers self-paced tutorials that provide individualized coaching, focus on your course objectives, and are responsive to each student's progress. The Mastering system helps instructors maximize class time with customizable, easy-to-assign, and automatically graded assessments that motivate students to learn outside of class and arrive prepared for lecture. By complementing their teaching with our engaging technology and content, instructors can be confident their students will arrive at that moment—the moment of true understanding.

Student Solutions Manual (0321813324) by Joseph Topich, Virginia Commonwealth University. This manual for students contains solutions to all in-chapter problems and even-numbered end-of-chapter problems.

Lab Manual (0321813375) by Stephanie Dillon, Florida State University, and Sandra Chimon-Peszek, DePaul University. This laboratory manual contains experiments written specifically to correspond with the Second Edition of General Chemistry: Atoms First by McMurry/Fay. Each experiment covers one or more topics discussed within a chapter of the textbook, with the dual goal of 1) helping students understand the underlying concepts covered in the lecture course, and 2) staying true to an atoms-first approach. This manual contains 31 experiments with a focus on real-world applications and a dedication to the atoms-first approach. Each experiment contains a set of pre-laboratory questions (also assignable in MasteringChemistry), an introduction, a background section explaining concepts that each student is expected to master for a full understanding of the experimental results, a step-by-step procedure (including safety information), and a report section featuring post-laboratory questions.

FOR THE INSTRUCTOR

MasteringChemistry® with Pearson eText. MasteringChemistry® from Pearson has been designed and refined with a single purpose in mind: to help educators create that moment of understanding with their students. The Mastering online homework and tutoring system delivers self-paced tutorials that provide individualized coaching, focus on your course objectives, and are responsive to each student's progress. The Mastering system helps instructors maximize class time with customizable, easy-to-assign, and automatically graded assessments that motivate students to learn outside of class and arrive prepared for lecture. By complementing their teaching with our engaging technology and content, instructors can be confident their students will arrive at that moment—the moment of true understanding.

MasteringChemistry® updates include:

- **NEW! 15 Pause and Predict Video Quizzes** bring chemistry to life with lab demonstrations illustrating key topics in general chemistry. Students are asked to predict the outcome of experiments as they watch the videos; a set of multiple-choice questions challenges students to apply the concepts from the video to related scenarios.
- **NEW! 10 PhET tutorials** have been developed around interactive applets that foster conceptual understanding and active learning. Topics include acid–base solutions, balancing chemical equations, and molecular polarity.
    - **Multiple-choice Reading Questions** are provided for each chapter, making it easy to hold students accountable for doing assigned readings before lecture.

- **Approximately 500 end-of-chapter questions are new or revised** and are supported by the tutorial questions in MasteringChemistry. The overall number of algorithmic and randomized problems has also been increased for the new edition.
- **All tutorials, end-of-chapter questions, reading quizzes, and test bank questions have been tagged to learning outcomes.**
- **A set of new and revised MasteringChemistry tutorials** has been developed to support the authors' atoms-first philosophy
- NEW! A dedicated atoms-first lab manual with pre-lab/post-lab question has been added to allow successful implementation of this approach with assignable questions in MasteringChemistry,

Instructor Solutions Manual (0321813057) by Joseph Topich, Virginia Commonwealth University. This solutions manual provides worked-out solutions to all in-chapter and end-of-chapter questions and problems. With instructor's permission, this manual may be made available to students.

Test Bank (download only) (0321813049) by Lee Don Bienski, Blinn College. The test bank is downloadable directly from the Intstructor Resource Center in either Microsoft Word or TestGen formats and contains nearly 4000 multiple-choice questions

Instructor Resource Manual (0321813359) by Dennis Taylor, Clemson University. This Manual contains teaching tips, a list of common misconceptions, lecture outlines, and suggested chapter learning goals for students, as well as lecture/laboratory demonstrations and literature references. It also describes various resources, such as printed test bank questions, animations, and movies that are available to instructors.

Instructor's Resource DVD (0321813332). This IRDVD provides an integrated collection of resources designed to enhance your classroom lectures. The IRDVD contains all illustrations, tables and photos from the text in JPEG and PDF format. The DVD also includes four pre-built PowerPoint™ Presentations (lecture, worked examples, images, clicker questions), the Instructor Resource Manual, and interactive animations, movies and 3D molecules. Also included are word files and the TestGen computerized software with the TestGen version of the test bank.

Instructor Manual to the Lab Manual (0321813693) by Stephanie Dillon, Florida State University, and Sandra Chimon-Peszek, DePaul University. The Instructor Manual guides the instructor through the process of planning, setting up, and teaching each laboratory exercise. For each exercise, instructors can quickly and easily reference a master list of materials for ordering supplies, an approximate time frame for completing each exercise, a series of tips on preparation, and an answer key for all of the student activities and questions within the text.

About the Authors

John McMurry (*left*), educated at Harvard and Columbia, has taught more than 20,000 students in general and organic chemistry over a 40-year period. An Emeritus Professor of Chemistry at Cornell University, Dr. McMurry previously spent 13 years on the faculty at the University of California at Santa Cruz. He has received numerous awards, including the Alfred P. Sloan Fellowship (1969–71), the National Institute of Health Career Development Award (1975–80), the Alexander von Humboldt Senior Scientist Award (1986–87), and the Max Planck Research Award (1991). With the publication of this new edition, he has authored or coauthored 36 textbooks in various fields of chemistry.

Robert C. Fay (*right*), Professor Emeritus of Chemistry at Cornell University, has taught general and inorganic chemistry at Cornell for 45 years, beginning in 1962. Known for his clear, well-organized lectures, Dr. Fay was the 1980 recipient of the Clark Distinguished Teaching Award. He has also taught as a visiting professor at Harvard University and the University of Bologna (Italy). A Phi Beta Kappa graduate of Oberlin College, Dr. Fay received his Ph.D. from the University of Illinois. He has been an NSF Science Faculty Fellow at the University of East Anglia and the University of Sussex (England) and a NATO/Heineman Senior Fellow at Oxford University.

Understanding the Cohesive Story of Chemistry:

Friends,

Our primary purpose in writing this text is to explain chemistry to students today the way we wish it had been explained to us years ago, when we were students.

The great advantage of an atoms-first organization is that **it allows us to start at the logical beginning of the chemical story by discussing atoms**—their history, stability, electronic structure, and consequent periodicity—and then continue on to bonding and reactions.

We are particularly grateful to the many instructors with experience teaching from our book who took the time to tell us how they adjusted to the organization of an atoms-first text and **helped their students appreciate connections between topics.**

- The second edition begins with Chapter 0 Chemical Tools: Experimentation and Measurement to bring students with widely diverse backgrounds up to speed on the topics of units and measurements so that Chapter 1 can focus entirely on atoms and elements.
- Chapter 2 then presents atomic structure and periodicity, which allows for a seamless transition to learning in Chapters 3–5.
- Chapters 3-5 show how atoms join to form chemical compounds. As a result, students come to see the various kinds of bonding as a continuum of atomic interactions. Our coverage of bonding models is distinguished by many well-received molecular drawings and conceptual problems.
- Chapters 6 and 7 introduce reaction chemistry in response to requests from coordinators of accompanying laboratory courses.

The changes to the Second Edition have been made with the goal of creating a unified thread of ideas that helps students build a better foundation and gain a deeper understanding of the chapters that follow.

As you strive to help your students establish an appreciation and understanding of general chemistry, we hope you will consider using this book in your endeavors.

Sincerely,
John McMurry and Bob Fay

The great advantage of an atoms-first organization is that it allows us to start at the logical beginning of the chemical story by discussing atoms—their history, stability, electronic structure, and consequent periodicity—and then continue on to bonding and reactions.

Annotated Table of Contents
Chapters 1– 7: Atoms-First Chapters

Chapters 1 and 2 start at the logical beginning of the chemical story by discussing atoms—their history, stability, electronic structure, and consequent periodicity. Nuclear Chemistry is briefly introduced in Chapter 1.

In this edition, chapters 4 and 5 have been split into two chapters to slow the pace and make these topics less intimidating.

Stoichiometric mass relationships are introduced in Chapters 6. After the foundation has been set in Chapters 1–6, Chapter 7 follows with chemical reactions in aqueous solutions.

The Atoms First Approach

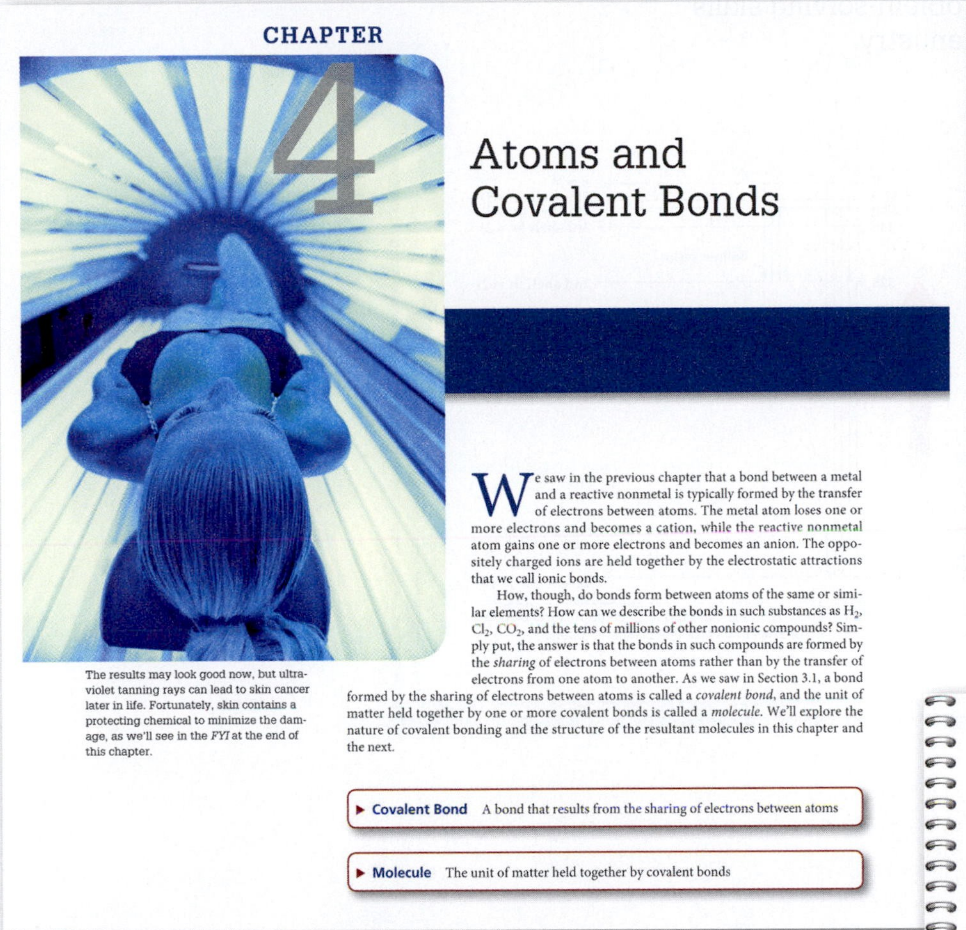

CHAPTER 4

Atoms and Covalent Bonds

We saw in the previous chapter that a bond between a metal and a reactive nonmetal is typically formed by the transfer of electrons between atoms. The metal atom loses one or more electrons and becomes a cation, while the reactive nonmetal atom gains one or more electrons and becomes an anion. The oppositely charged ions are held together by the electrostatic attractions that we call ionic bonds.

How, though, do bonds form between atoms of the same or similar elements? How can we describe the bonds in such substances as H_2, Cl_2, CO_2, and the tens of millions of other nonionic compounds? Simply put, the answer is that the bonds in such compounds are formed by the *sharing* of electrons between atoms rather than by the transfer of electrons from one atom to another. As we saw in Section 3.1, a bond formed by the sharing of electrons between atoms is called a *covalent bond*, and the unit of matter held together by one or more covalent bonds is called a *molecule*. We'll explore the nature of covalent bonding and the structure of the resultant molecules in this chapter and the next.

The results may look good now, but ultraviolet tanning rays can lead to skin cancer later in life. Fortunately, skin contains a protecting chemical to minimize the damage, as we'll see in the *FYI* at the end of this chapter.

▸ **Covalent Bond** A bond that results from the sharing of electrons between atoms

▸ **Molecule** The unit of matter held together by covalent bonds

▲ **Chapter Openers** employ a conceptual approach by telling the story of chemistry. Openers are followed by learning objectives that focus students on key concepts.

▲ **NEW! Author-written lab manual** reflects the text's atoms-first approach with 28 experiments focusing on real-world applications.

Seeing chemical concepts in contex

Numerous example types are incorporated throughout the text to help build students' problem-solving skills and a conceptual understanding of chemistry.

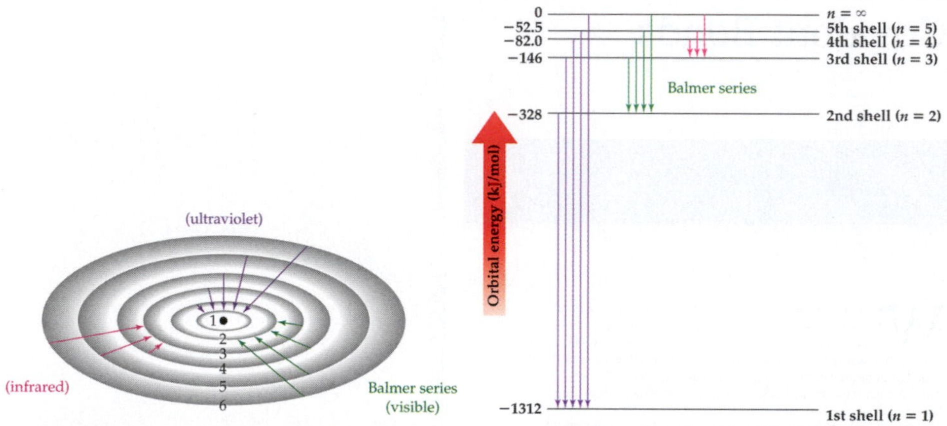

When an electron falls from a higher-energy outer-shell orbital to a lower-energy inner-shell orbital, it emits electromagnetic energy whose frequency corresponds to the energy difference between the orbitals.

The different spectral series correspond to electronic transitions from outer-shell orbitals to different inner-shell orbitals.

Clear **quantitative examples** help students make the connection between chemical reasoning and math.

REMEMBER...

Lattice energy (U) is the amount of energy that must be supplied to break an ionic solid into its individual gaseous ions and is thus a measure of the strength of the solid's ionic bonds. (Section 3.10)

LOOKING AHEAD...

The bulk materials we see in everyday life are usually solids or liquids rather than gases. In **Chapter 10**, we'll look at the forces that operate between molecules to hold them in close contact and at the resultant melting and boiling processes that happen when the forces are overcome.

Looking Ahead feature helps students more logically link key concepts, while **Remember**... margin notes jog their memories on related topics from previous chapters.

A multi-layered approach to problem solving

Worked Conceptual Examples follow many worked examples to emphasize the conceptual nature of problem solving, often using molecular illustrations.

Worked Example 6.5

Converting Mass to Moles

How many moles of sucrose are in a tablespoon of sugar containing 2.85 g? (The molar mass of sucrose, $C_{12}H_{22}O_{11}$, was calculated in Worked Example 6.4 to be 342.0 g/mol)

STRATEGY

The problem gives the mass of sucrose and asks for a mass-to-mole conversion. Use the molar mass of sucrose as a conversion factor, and set up an equation so that the unwanted unit cancels.

SOLUTION

$$2.85\ \text{g sucrose} \times \frac{1\ \text{mol sucrose}}{342.0\ \text{g sucrose}} = 0.008\,33\ \text{mol sucrose}$$
$$= 8.33 \times 10^{-3}\ \text{mol sucrose}$$

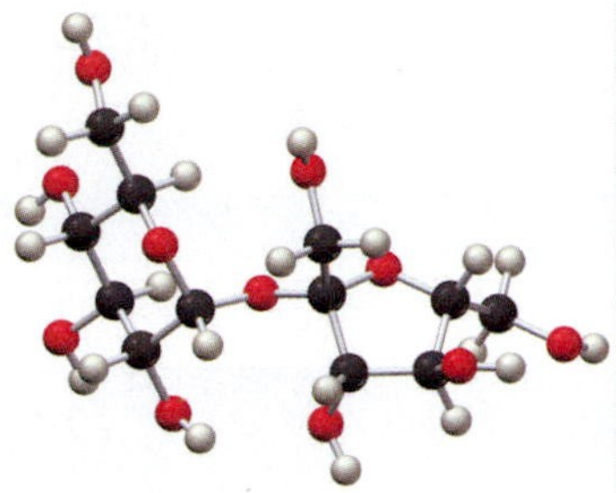

Sucrose

✔ BALLPARK CHECK

Because the molecular weight of sucrose is 342.0, 1 mol of sucrose has a mass of 342.0 g. Thus, 2.85 g of sucrose is a bit less than one-hundredth of a mole, or 0.01 mol. The estimate agrees with the detailed solution.

Ballpark Checks follow many worked examples to reinforce students' conceptual understanding of the concept just expressed numerically.

PROBLEM 3.11 Which atom or ion in each of the following pairs would you expect to be larger? Explain.
(a) O or O^{2-} **(b)** O or S **(c)** Fe or Fe^{3+} **(d)** H or H^-

CONCEPTUAL PROBLEM 3.12 Which of the following spheres represents a K^+ ion, which a K atom, and which a Cl^- ion?

r = 227 pm r = 184 pm r = 133 pm

CONCEPTUAL PROBLEM 5.11 Identify each of the following sets of hybrid orbitals:

(a) (b) (c)

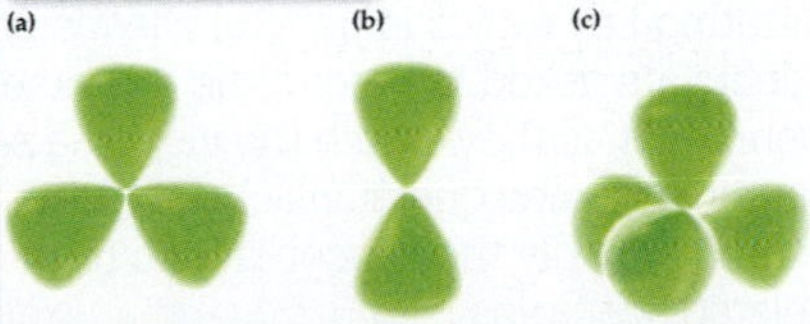

CONCEPTUAL PROBLEM 5.12 The following ball-and-stick molecular model is a representation of ibuprofen, a nonprescription drug used for minor pain and inflammation. Only the connections between atoms are shown, not whether the bonds are single, double, or triple (red = O, gray = C, ivory = H).
(a) Indicate the positions of the multiple bonds and lone pairs in ibuprofen.
(b) What is the geometry around each carbon atom?
(c) What is the hybridization of each carbon atom?

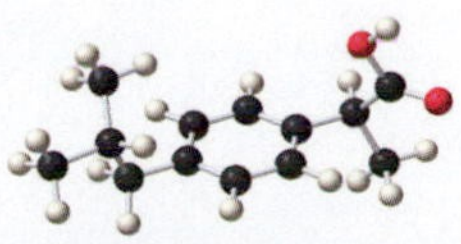

Ibuprofen

Non-numerical **Concept Problems** measure understanding of principles rather than the ability to simply plug numbers into a formula.

MasteringChemistry®

MasteringChemistry provides dynamic, engaging experiences that personalize and activate learning for each student, delivering self-paced tutorials that provide students with individualized coaching set to your course objectives. MasteringChemistry helps students arrive better prepared for lecture and lab.

Engaging Experiences

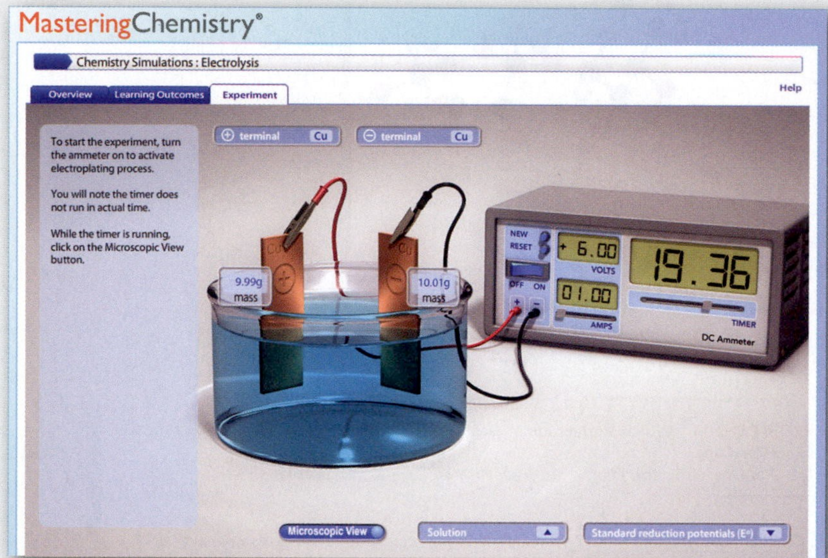

NEW! Visualizations are tutorials that enable students to make connections between real-life phenomena and the underlying chemistry that explains such phenomena. The tutorials increase students' understanding of chemistry and clearly illustrate cause-and-effect relationships and include Interactive Simulations.

NEW! 15 Pause and Predict Video Quizzes bring chemistry to life with lab demonstrations illustrating key topics in general chemistry. Students are asked to predict the outcome of experiments as they watch the videos; a set of multiple-choice questions challenges students to apply the concepts from the video to related scenarios.

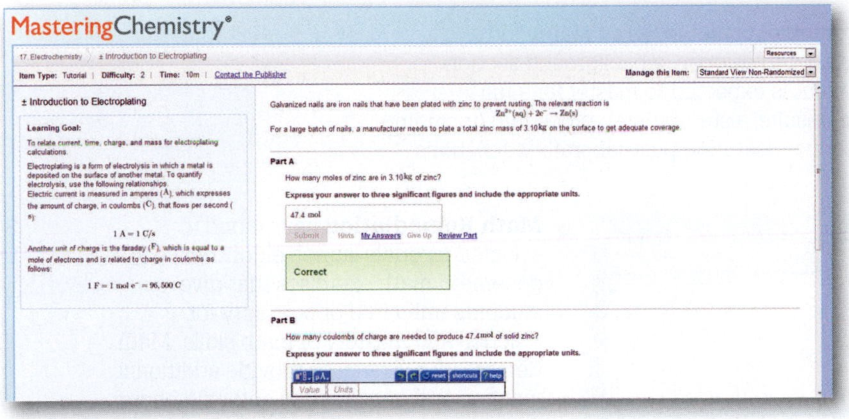

NEW! MasteringChemistry tutorials have all been evaluated and in many cases edited, revised or rewritten by an advisory board of expert chemists, all teaching with the atoms-first approach, to ensure the reinforcement of this approach.

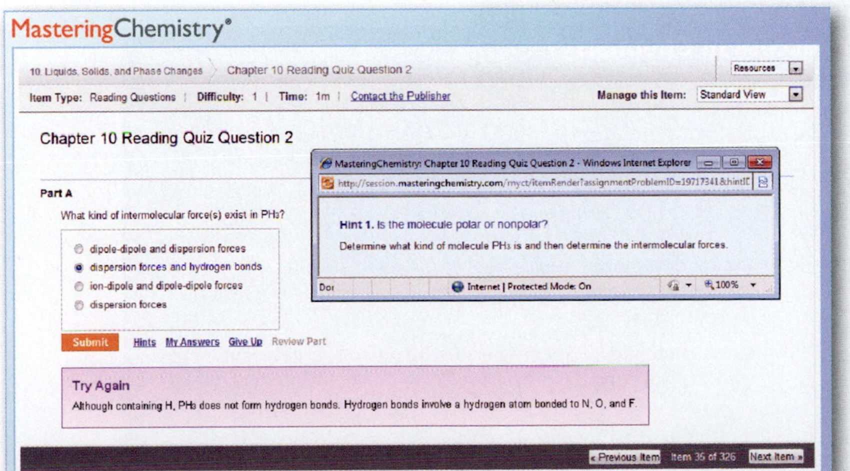

NEW! Multiple-choice Reading Questions are provided for each chapter, making it easy to hold students accountable for doing assigned readings before lecture.

MasteringChemistry®

The new atoms-first lab manual is integrated with MasteringChemistry through assignable pre-laboratory questions for each experiment.

NEW! A dedicated atoms-first lab manual written by authors teaching with this approach reflects the changes needed to successfully integrate the lab with the course. Each experiment covers one or more topics discussed within a chapter of the textbook, helping students understand the underlying concepts covered in the lecture course and staying true to an Atoms First Approach.

The lab manual features 28 experiments that focus on real-world applications. Experiments include an introduction, a background section explaining concepts that each student is expected to master for a full understanding of the experimental results, a step-by-step procedure (including safety information), and a report section featuring post-laboratory questions.

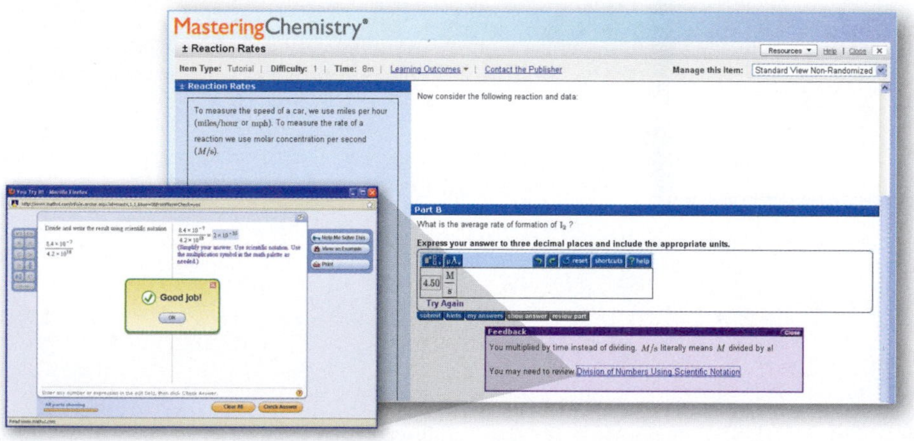

Math Remediation links found in selected tutorials launch algorithmically generated math exercises that give students unlimited opportunity for practice and mastery of math skills. Math Remediation exercises provide additional practice and free up class and office-hour time to focus on the chemistry. Exercises include guided solutions, sample problems, and learning aids for extra help, and offer helpful feedback when students enter incorrect answers.

Do your students need math remediation for success in Chemistry?

NEW! MyReadinessTest™ Math Prep for Chemistry provides a stand-alone solution for those math topics applicable to chemistry.

Powered by MathXL, MyReadinessTest offers:

- **Diagnostic test** with study plan linked to exercises covering math topics targeted for chemistry preparation
- **Self-paced remediation**, with unlimited practice on the math skills most needed for success in chemistry
- **A library of more than 1,200 problems** chosen by professors and subject matter experts
- **Pre-loaded diagnostic tests** that instructors can assign, as well as the ability to create or import their own custom tests and problems
- **Gradebook view of scores and study plans** to show where students are struggling the most

Visit **www.myreadinesstest.com** or contact your Pearson Representative for more information.

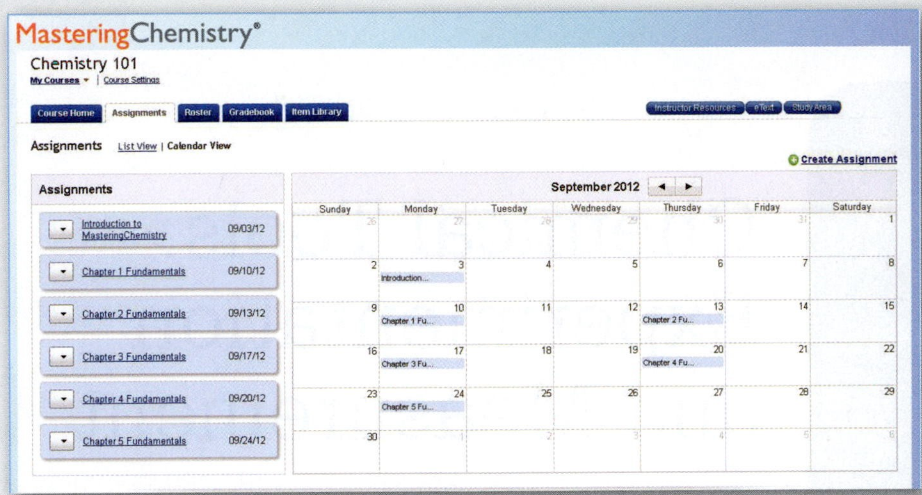

NEW! Calendar Features

The Course Home default page now features a **Calendar View** displaying upcoming assignments and due dates.

- Instructors can schedule assignments by dragging and dropping the assignment onto a date in the calendar. If the due date of an assignment needs to change, instructors can drag the assignment to the new due date and change the "available from and to dates" accordingly.
- The calendar view gives students a syllabus-style overview of due dates, making it easy to see all assignments due in a given month.

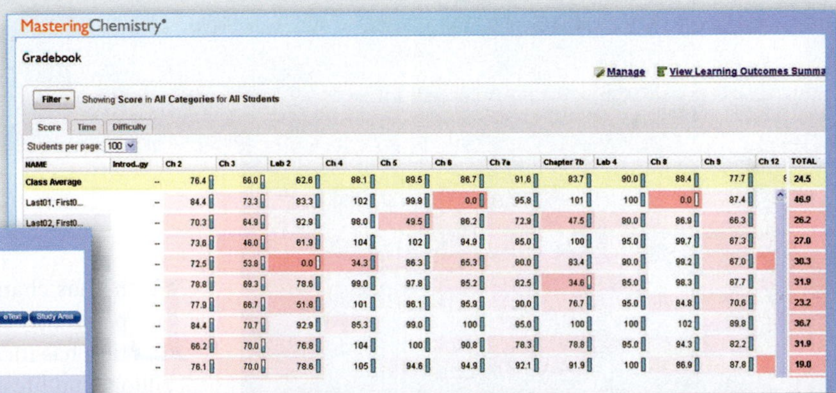

Gradebook

Every assignment is automatically graded. Shades of red highlight struggling students and challenging assignments.

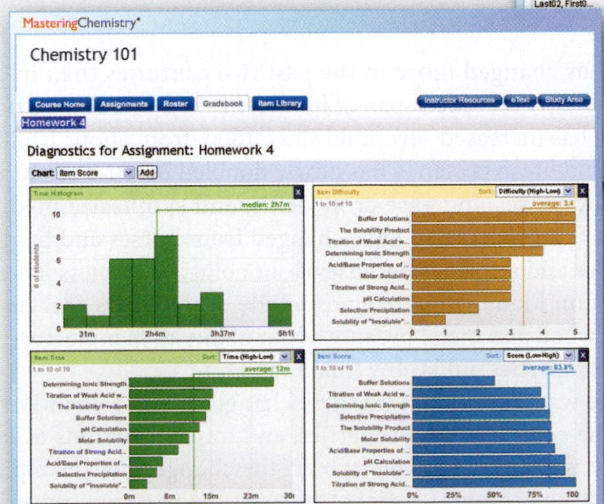

Gradebook Diagnostics

This screen provides you with your favorite diagnostics. With a single click, charts summarize the most difficult problems, vulnerable students, grade distribution, and even score improvement over the course.

STUDENTS – PLEASE READ. REALLY.

Who ever heard of beginning a book with Chapter 0? It's not that the material in this chapter is unimportant. In fact, it's *very* important because it gives you some tools that you must know how to use before beginning your study of chemistry. The problem is that you and your classmates, who typically take Introductory Chemistry in the first year of college, are so diverse in your backgrounds and preparations that many of you will have

CHAPTER

Accurate measurements, such as those obtained on this 300-year-old Persian balance, have always been a crucial part of scientific experiments.

Chemical Tools: Experimentation and Measurement

Life has changed more in the past two centuries than in all the previously recorded span of human history. The Earth's population has increased sevenfold since 1800, from about 1 billion to 7 billion, and life expectancy has nearly doubled because of our ability to control diseases, increase crop yields, and synthesize medicines. Methods of transportation have changed from horses and buggies to automobiles and airplanes because of our ability to harness the energy in petroleum. Many goods are now made of polymers and ceramics instead of wood and metal because of our ability to manufacture materials with properties unlike any found in nature.

In one way or another, all these changes involve **chemistry**, the study of the composition, properties, and transformations of matter. Chemistry is deeply involved in both the changes that take place in nature and the profound social changes of the past two centuries. In addition, chemistry is central to the current revolution in molecular biology that is now exploring the details of how life is genetically controlled. No educated person today can understand the modern world without a basic knowledge of chemistry.

already taken a chemistry Advanced Placement course in high school while many others of you will have had either no high school background in chemistry or at best a one-semester introduction. Thus, some of you will have already learned the material in this chapter and will be comfortable using it, while others of you will have to carefully brush up on it or learn it for the first time.

The choice about whether you need to learn, brush up on, or move quickly past the material in Chapter 0 is yours (or your professor's). But be careful in making your choice. You *do* need to master and feel comfortable with this chapter before going on.

CONTENTS

0.1 EXPERIMENT → HYPOTHESIS → THEORY: APPROACHING CHEMISTRY

By opening this book, you have already decided that you need to know more about chemistry. Perhaps you want to learn how medicines are made, how genes can be sequenced and manipulated, how fertilizers and pesticides work, how living organisms function, how new high-temperature ceramics are used in space vehicles, or how microelectronic circuits are etched onto silicon chips. How do you approach chemistry?

One way to approach chemistry or any other science is to look around you and try to think of logical explanations for what you see. You would certainly observe, for instance, that different substances have different forms and appearances. Some substances are gases, some are liquids, and some are solids; some are hard and shiny, but others are soft and dull. You'd also observe that different substances behave differently: Iron rusts but gold does not; copper conducts electricity but sulfur doesn't. How can these and a vast number of other observations be explained?

▶ Gold, one of the most valuable of elements, has been prized since antiquity for its beauty and resistance to corrosion.

In fact, the natural world is far too complex to be understood by looking and thinking alone, so a more active approach is needed. Specific questions must be asked and experiments must be carried out to find their answers. Only when the results of many experiments are known can we devise a tentative interpretation, or *hypothesis*, that explains the results. The hypothesis, in turn, can be used to make more predictions and to suggest more experiments until a consistent explanation, or **theory**, is finally arrived at.

Experiments and observations come first in chemistry or any other science, and theories to explain those observations come later after much hard work and a lot of thought. Nevertheless, it's important to keep in mind that scientific theories are not laws of nature and can never be absolutely proven. There's always the chance that a new experiment might give results that can't be explained by present theory. All a theory can do is to represent the best explanation that we can come up with at any given time. If new experiments uncover results that present theories can't explain, the theories will have to be modified or perhaps even replaced.

0.2 EXPERIMENTATION AND MEASUREMENT IN CHEMISTRY

Chemistry is an experimental science. But if our experiments are to be reproducible, we must be able to describe fully the substances we're working with—their amounts, volumes, temperatures, and so forth. Thus, one of the most important requirements in chemistry is that we have a way to measure things.

Under an international agreement concluded in 1960, scientists throughout the world now use the International System of Units for measurement, abbreviated **SI** for the French *Système Internationale d'Unités*. Based on the metric system, which is used in all countries except the United States, Liberia, and Myanmar, the SI system has seven fundamental units (**Table 0.1**). These seven fundamental units, along with others derived from them, suffice for all scientific measurements. We'll look at three of the most commonly used units in this chapter—those for mass, length, and temperature—and will discuss others as the need arises in later chapters.

TABLE 0.1 The Seven Fundamental SI Units of Measure

Physical Quantity	Name of Unit	Abbreviation
Mass	kilogram	kg
Length	meter	m
Temperature	kelvin	K
Amount of substance	mole	mol
Time	second	s
Electric current	ampere	A
Luminous intensity	candela	cd

One problem with any system of measurement is that the sizes of the units often turn out to be inconveniently large or small. For example, a chemist describing the diameter of a sodium atom (0.000 000 000 372 m) would find the meter (m) to be inconveniently large, but an astronomer describing the average distance from the Earth to the Sun (150,000,000,000 m) would find the meter to be inconveniently small. For this reason, SI units are modified by using prefixes when they refer to either smaller or larger quantities. For example, the prefix *milli-* means one-thousandth, so a *milli*meter (mm) is 1/1000 of 1 meter. Similarly, the prefix *kilo-* means one thousand, so a *kilo*meter (km) is 1000 meters. [Note that the SI unit for mass (kilogram) already contains the *kilo-* prefix.] A list of prefixes is shown in **Table 0.2**, with the most commonly used ones in red.

TABLE 0.2 Some Prefixes for Multiples of SI Units

Factor	Prefix	Symbol	Example
1,000,000,000,000 = 10^{12}	tera	T	1 teragram (Tg) = 10^{12} g
1,000,000,000 = 10^9	giga	G	1 gigameter (Gm) = 10^9 m
1,000,000 = 10^6	mega	M	1 megameter (Mm) = 10^6 m
1,000 = 10^3	kilo	k	1 kilogram (kg) = 10^3 g
100 = 10^2	hecto	h	1 hectogram (hg) = 100 g
10 = 10^1	deka	da	1 dekagram (dag) = 10 g
0.1 = 10^{-1}	deci	d	1 decimeter (dm) = 0.1 m
0.01 = 10^{-2}	centi	c	1 centimeter (cm) = 0.01 m
0.001 = 10^{-3}	milli	m	1 milligram (mg) = 0.001 g
*0.000 001 = 10^{-6}	micro	μ	1 micrometer (μm) = 10^{-6} m
*0.000 000 001 = 10^{-9}	nano	n	1 nanosecond (ns) = 10^{-9} s
*0.000 000 000 001 = 10^{-12}	pico	p	1 picosecond (ps) = 10^{-12} s
*0.000 000 000 000 001 = 10^{-15}	femto	f	1 femtomole (fmol) = 10^{-15} mol

*It is becoming common in scientific work to leave a thin space every three digits to the right of the decimal point in very small numbers, analogous to the comma placed every three digits to the left of the decimal point in large numbers.

Notice how numbers that are either very large or very small are indicated in Table 0.2 using an exponential format called **scientific notation**. For example, the number 55,000 is written in scientific notation as 5.5×10^4, and the number 0.003 20 as 3.20×10^{-3}. Review Appendix A if you are uncomfortable with scientific notation or if you need to brush up on how to do mathematical manipulations on numbers with exponents.

Notice also that all measurements contain both a number and a unit label. A number alone is not much good without a unit to define it. If you asked a friend how far it was to the nearest tennis court, the answer "3" alone wouldn't tell you much. 3 blocks? 3 kilometers? 3 miles?

PROBLEM 0.1 Express the following quantities in scientific notation:
(a) The diameter of a sodium atom, 0.000 000 000 372 m
(b) The distance from the Earth to the Sun, 150,000,000,000 m

PROBLEM 0.2 What units do the following abbreviations represent?
(a) μg **(b)** dm **(c)** ps **(d)** kA **(e)** mmol

0.3 FUNDAMENTAL UNITS: MEASURING MASS

Let's look in more detail at the measurement of some common quantities in chemistry, beginning with mass. **Mass** is defined as the amount of *matter* in an object. **Matter**, in turn, is a catchall term used to describe anything with a physical presence—anything you can touch, taste, or smell. (Stated more scientifically, matter is anything that has mass.) Mass is measured in SI units by the **kilogram (kg**; 1 kg = 2.205 U.S. lb). Because the kilogram is too large for many purposes in chemistry, the metric **gram (g**; 1 g = 0.001 kg), the **milligram (mg**; 1 mg = 0.001 g = 10^{-6} kg), and the **microgram (μg**; 1 μg = 0.001 mg = 10^{-6} g = 10^{-9} kg) are more commonly used. One gram is a bit less than half the mass of a new U.S. dime.

$$1 \text{ kg} = 1000 \text{ g} = 1{,}000{,}000 \text{ mg} = 1{,}000{,}000{,}000\ \mu\text{g} \quad (2.205 \text{ lb})$$
$$1 \text{ g} = 1000 \text{ mg} = 1{,}000{,}000\ \mu\text{g} \quad (0.035\,27 \text{ oz})$$
$$1 \text{ mg} = 1000\ \mu\text{g}$$

▲ The mass of a U.S. dime is approximately 2.27 g. How many milligrams is this?

The standard kilogram is defined as the mass of the *international prototype kilogram*, a cylindrical bar of platinum–iridium alloy stored in a vault in a suburb of Paris, France. There are 40 original copies of this bar distributed throughout the world, with two (Numbers K4 and K20) stored at the U.S. National Institute of Standards and Technology near Washington, D.C.

The terms "mass" and "weight," although often used interchangeably, have quite different meanings. *Mass* is a measure of the amount of matter in an object, whereas *weight* is a measure of the force that gravity exerts on an object. Mass is independent of an object's location: Your body has the same amount of matter whether you're on Earth or on the Moon. Weight, however, *does* depend on an object's location. If you weigh 140 lb on Earth, you would weigh only about 23 lb on the Moon, which has a lower gravity than Earth.

At the same location on Earth, two objects with identical masses experience an identical pull of Earth's gravity and have identical weights. Thus, the mass of an object can be measured by comparing its weight to the weight of a reference standard of known mass. Much of the confusion between mass and weight is simply due to a language problem. We speak of "weighing" when we really mean that we are measuring mass by comparing two weights. **Figure 0.1** shows two types of balances normally used for measuring mass in the laboratory.

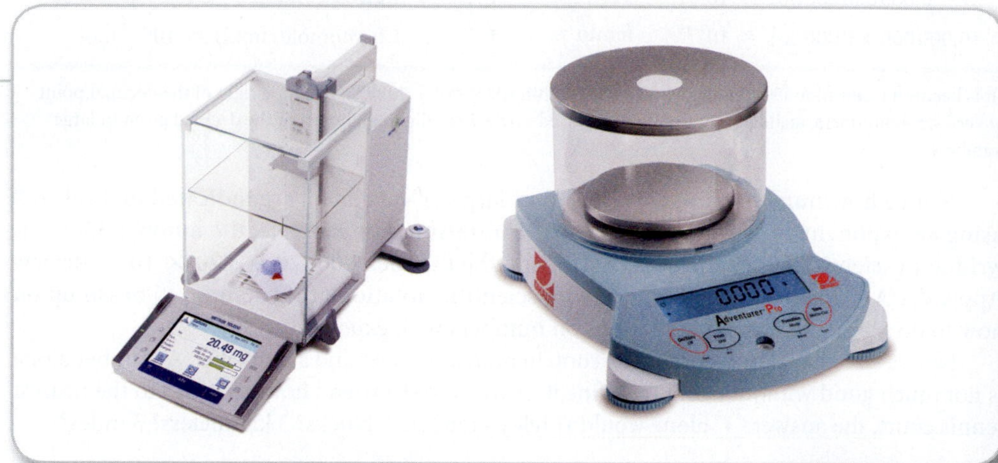

FIGURE 0.1

Some balances used for measuring mass in the laboratory.

0.4 FUNDAMENTAL UNITS: MEASURING LENGTH

The **meter (m)** is the standard unit of length in the SI system. Although originally defined in 1790 as being 1 ten-millionth of the distance from the equator to the North Pole, the meter was redefined in 1889 as the distance between two thin lines on a bar of platinum–iridium alloy stored near Paris, France. To accommodate an increasing need for precision, the meter was redefined again in 1983 as equal to the distance traveled by light through a vacuum in 1/299,792,458 second. Although this new definition isn't as easy to grasp as the distance between two scratches on a bar, it has the great advantage that it can't be lost or damaged.

One meter is 39.37 inches, about 10% longer than an English yard and much too large for most measurements in chemistry. Other, more commonly used measures of length are the **centimeter (cm**; $1 \text{ cm} = 0.01 \text{ m}$, a bit less than half an inch), the **millimeter (mm**; $1 \text{ mm} = 0.001 \text{ m}$, about the thickness of a U.S. dime), the **micrometer** (**μm**; $1 \, \mu\text{m} = 10^{-6} \text{ m}$), the **nanometer** (**nm**; $1 \text{ nm} = 10^{-9} \text{ m}$), and the **picometer** (**pm**; $1 \text{ pm} = 10^{-12} \text{ m}$). Thus, a chemist might refer to the diameter of a sodium atom as 372 pm (3.72×10^{-10} m).

$$1 \text{ m} = 100 \text{ cm} = 1000 \text{ mm} = 1{,}000{,}000 \, \mu\text{m} = 1{,}000{,}000{,}000 \text{ nm} \qquad (1.0936 \text{ yd})$$

$$1 \text{ cm} = 10 \text{ mm} = 10{,}000 \, \mu\text{m} = 10{,}000{,}000 \text{ nm} \qquad (0.3937 \text{ in.})$$

$$1 \text{ mm} = 1000 \, \mu\text{m} = 1{,}000{,}000 \text{ nm}$$

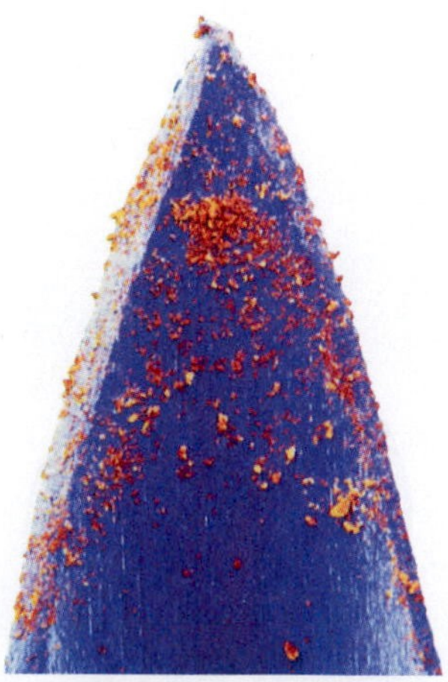

▲ The bacteria on the tip of this pin have a length of about 5×10^{-7} m. How many nanometers is this?

0.5 FUNDAMENTAL UNITS: MEASURING TEMPERATURE

Just as the kilogram and the meter are slowly replacing the pound and the yard as common units for mass and length measurement in the United States, the **degree Celsius** (°C) is slowly replacing the degree Fahrenheit (°F) as the common unit for temperature measurement. In scientific work, however, the **kelvin (K)** has replaced both. (Note that we say only "kelvin," not "degree kelvin.")

For all practical purposes, the kelvin and the degree Celsius are the same size—both are one-hundredth of the interval between the freezing point of water and the boiling point of water at standard atmospheric pressure. The only real difference between the two units is that the numbers assigned to various points on the scales differ. Whereas the Celsius scale assigns a value of 0 °C to the freezing point of water and 100 °C to the boiling point of water, the Kelvin scale assigns a value of 0 K to the coldest possible temperature, −273.15 °C, sometimes called *absolute zero*. Thus, 0 K = −273.15 °C and 273.15 K = 0 °C. For example, a warm spring day with a Celsius temperature of 25 °C has a Kelvin temperature of 25° + 273.15° = 298 K.

$$\text{Temperature in K} = \text{Temperature in °C} + 273.15°$$
$$\text{Temperature in °C} = \text{Temperature in K} - 273.15°$$

In contrast to the Kelvin and Celsius scales, the common Fahrenheit scale specifies an interval of 180° between the freezing point (32 °F) and the boiling point (212 °F) of water. Thus, it takes 180 degrees Fahrenheit to cover the same range as 100 degrees Celsius (or kelvins) so a degree Fahrenheit is only 100/180 = 5/9 as large as a degree Celsius. **Figure 0.2** compares the Fahrenheit, Celsius, and Kelvin scales.

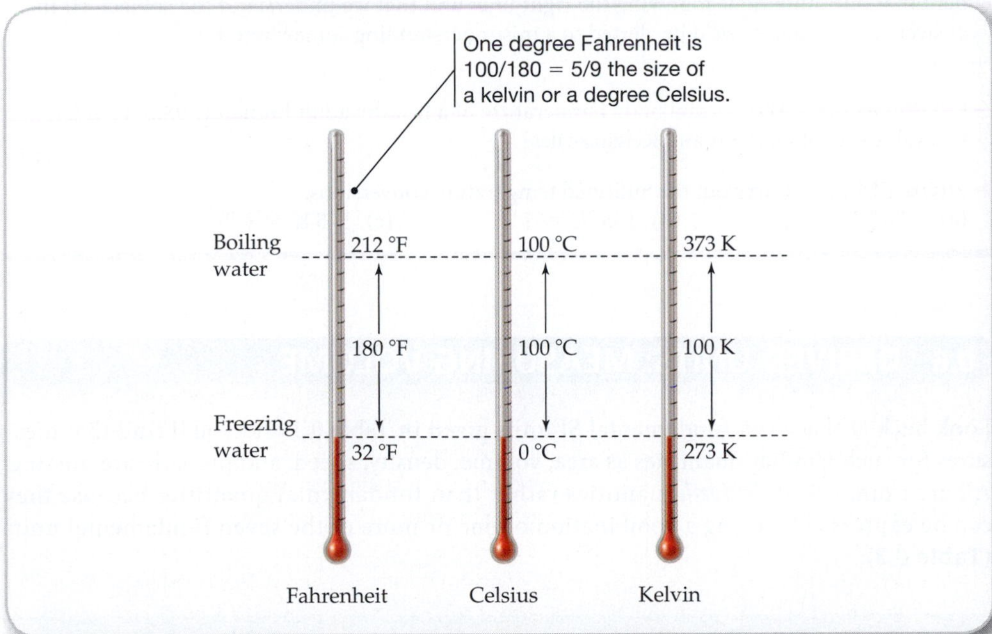

FIGURE 0.2

A comparison of the Fahrenheit, Celsius, and Kelvin temperature scales.

Two adjustments are needed to convert between Fahrenheit and Celsius scales—one to adjust for the difference in degree size and one to adjust for the difference in zero points. The size adjustment is made using the relationships 1 °C = 9/5 °F and 1 °F = 5/9 °C. The zero-point adjustment is made by remembering that the freezing point of water is higher by 32° on the Fahrenheit scale than on the Celsius scale. Thus, if you want to convert from Celsius to Fahrenheit, you first change the size of the Celsius value to that of a Fahrenheit value (multiply °C by 9/5) and then adjust the zero point of that Fahrenheit value (add 32°). If you want to convert from Fahrenheit to Celsius, you first adjust the zero-point of the Fahrenheit value (by subtracting 32°) and then change the size of that Fahrenheit value to a Celsius value (multiply by 5/9). The following formulas describe the conversions, and Worked Example 0.1 shows how to do a calculation.

Celsius to Fahrenheit

$$°\text{F} = \left(\frac{9\ °\text{F}}{5\ \cancel{°\text{C}}} \times \cancel{°\text{C}}\right) + 32\ °\text{F}$$

Fahrenheit to Celsius

$$°\text{C} = \frac{5\ °\text{C}}{9\ \cancel{°\text{F}}} \times (\cancel{°\text{F}} - 32\ \cancel{°\text{F}})$$

Worked Example 0.1

Converting from Fahrenheit to Celsius

The melting point of table salt is 1474 °F. What temperature is this on the Celsius and Kelvin scales?

SOLUTION

There are two ways to do this and every other problem in chemistry. One is to plug numbers into a formula and hope for the best; the other is to think things through to be sure you understand what's going on. The formula approach works only if you use the right equation; the thinking approach always works. Let's try both ways.

The formula approach: Set up an equation using the temperature conversion formula for changing from Fahrenheit to Celsius:

$$°C = \left(\frac{5\ °C}{9\ °F}\right)(1474\ °F - 32\ °F) = 801\ °C$$

Converting to kelvin gives a temperature of 801° + 273.15° = 1074 K.

The thinking approach: We're given a temperature in degrees Fahrenheit, and we need to convert to degrees Celsius. A temperature of 1474 °F corresponds to 1474 °F − 32 °F = 1442 °F above the freezing point of water. Because a degree Fahrenheit is only 5/9 as large as a degree Celsius, 1442 degrees Fahrenheit above freezing equals 1442 × 5/9 = 801 degrees Celsius above freezing (0 °C), or 801 °C. The same number of degrees above freezing on the Kelvin scale (273.15 K) corresponds to a temperature of 273.15 + 801 = 1074 K.

Because the answers obtained by the two approaches agree, we can feel fairly confident that our thinking is following the right lines and that we understand the subject. (If the answers did *not* agree, we'd be alerted to a misunderstanding somewhere.)

▲ The melting point of sodium chloride is 1474 °F, or 801 °C.

PROBLEM 0.3 The normal body temperature of a healthy adult human is 98.6 °F. What is this value on both Celsius and Kelvin scales?

PROBLEM 0.4 Carry out the indicated temperature conversions.
(a) −78 °C = ? K (b) 158 °C = ? °F (c) 375 K = ? °F

0.6 DERIVED UNITS: MEASURING VOLUME

Look back at the seven fundamental SI units given in Table 0.1 and you'll find that measures for such familiar quantities as area, volume, density, speed, and pressure are missing. All are examples of *derived* quantities rather than fundamental quantities because they can be expressed by using a combination of one or more of the seven fundamental units (**Table 0.3**).

TABLE 0.3 Some Derived Units and the Quantities They Measure

Quantity	Definition	Derived Unit (Name)
Area	Length times length	m^2
Volume	Area times length	m^3
Density	Mass per unit volume	kg/m^3
Speed	Distance per unit time	m/s
Acceleration	Change in speed per unit time	m/s^2
Force	Mass times acceleration	$(kg \cdot m)/s^2$ (newton, N)
Pressure	Force per unit area	$kg/(m \cdot s^2)$ (pascal, Pa)
Energy	Force times distance	$(kg \cdot m^2)/s^2$ (joule, J)

Volume, the amount of space occupied by an object, is measured in SI units by the **cubic meter** (m^3), defined as the amount of space occupied by a cube 1 meter on edge (**Figure 0.3**).

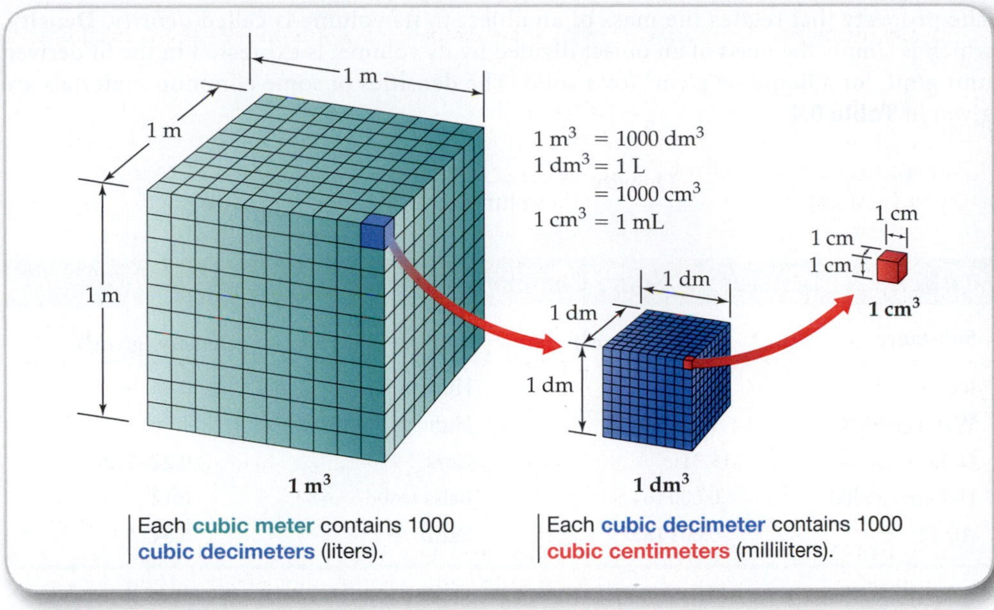

FIGURE 0.3

Units for measuring volume. A cubic meter is the volume of a cube 1 meter along each edge.

A cubic meter equals 264.2 U.S. gallons, much too large a quantity for normal use in chemistry. As a result, smaller, more convenient measures are commonly employed. Both the **cubic decimeter** ($1\ \mathbf{dm^3} = 0.001\ m^3$), equal in size to the more familiar metric **liter (L)**, and the **cubic centimeter** ($1\ \mathbf{cm^3} = 0.001\ dm^3 = 10^{-6}\ m^3$), equal in size to the metric **milliliter (mL)**, are particularly convenient. Slightly larger than 1 U.S. quart, a liter has the volume of a cube 1 dm on edge. Similarly, a milliliter has the volume of a cube 1 cm on edge (Figure 0.3).

$$1\ m^3 = 1000\ dm^3 = 1{,}000{,}000\ cm^3 \qquad (264.2\ gal)$$
$$1\ dm^3 = 1\ L = 1000\ mL \qquad (1.057\ qt)$$

Figure 0.4 shows some of the equipment frequently used in the laboratory for measuring liquid volume.

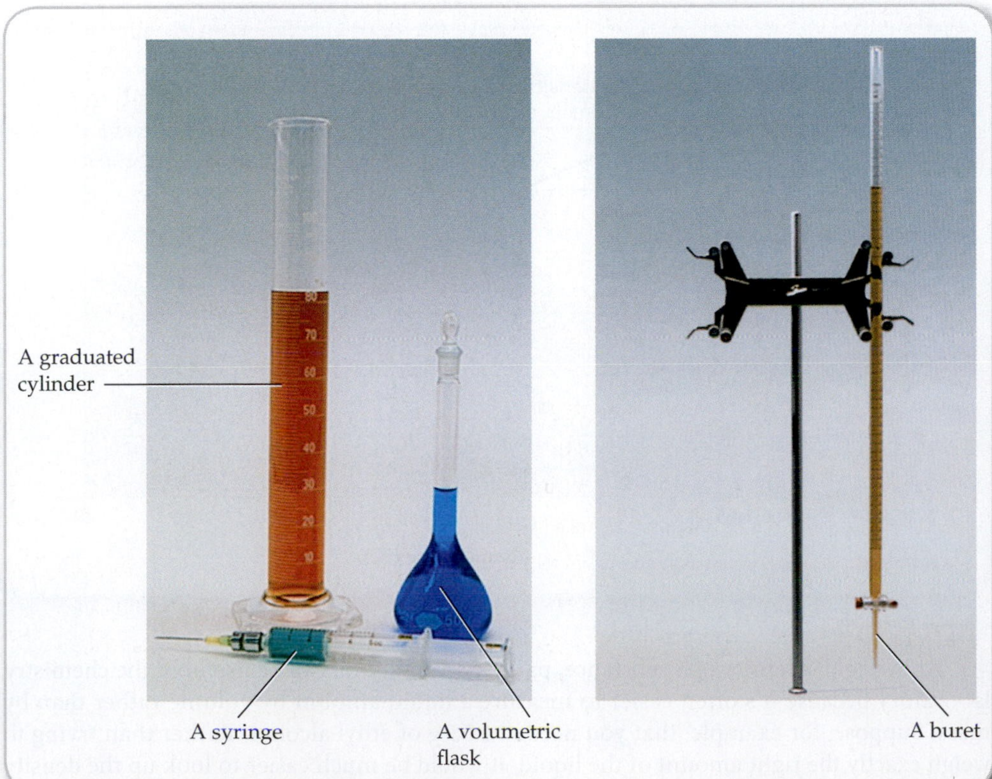

FIGURE 0.4

Common items of laboratory equipment used for measuring liquid volume.

0.7 DERIVED UNITS: MEASURING DENSITY

▲ Which weighs more, the brass weight or the pillow? Actually, both have identical masses and weights, but the brass weight has a higher density because its volume is much smaller.

The property that relates the mass of an object to its volume is called *density*. **Density**, which is simply the mass of an object divided by its volume, is expressed in the SI derived unit g/mL for a liquid or g/cm^3 for a solid. The densities of some common materials are given in **Table 0.4**.

$$\textbf{Density} = \frac{\text{Mass (g)}}{\text{Volume (mL or cm}^3\text{)}}$$

TABLE 0.4 Densities of Some Common Materials

Substance	Density (g/cm^3)	Substance	Density (g/cm^3)
Ice (0 °C)	0.917	Human fat	0.94
Water (3.98 °C)	1.0000	Human muscle	1.06
Gold	19.31	Cork	0.22–0.26
Helium (25 °C)	0.000 164	Balsa wood	0.12
Air (25 °C)	0.001 185	Earth	5.54

Because most substances change in volume when heated or cooled, densities are temperature-dependent. At 3.98 °C, for instance, a 1.0000 mL container holds exactly 1.0000 g of water (density = 1.0000 g/mL). As the temperature is raised, however, the volume occupied by the water expands so that only 0.9584 g fits in the 1.0000 mL container at 100 °C (density = 0.9584 g/mL). When reporting a density, the temperature must also be specified.

Although most substances expand when heated and contract when cooled, water behaves differently. Water contracts when cooled from 100 °C to 3.98 °C, but below this temperature it begins to expand again. Thus, the density of liquid water is at its maximum of 1.0000 g/mL at 3.98 °C but decreases to 0.999 87 g/mL at 0 °C (**Figure 0.5**). When freezing occurs, the density drops still further to a value of 0.917 g/cm^3 for ice at 0 °C. Ice and any other substance with a density less than that of water will float, but any substance with a density greater than that of water will sink.

FIGURE 0.5

The density of water at different temperatures.

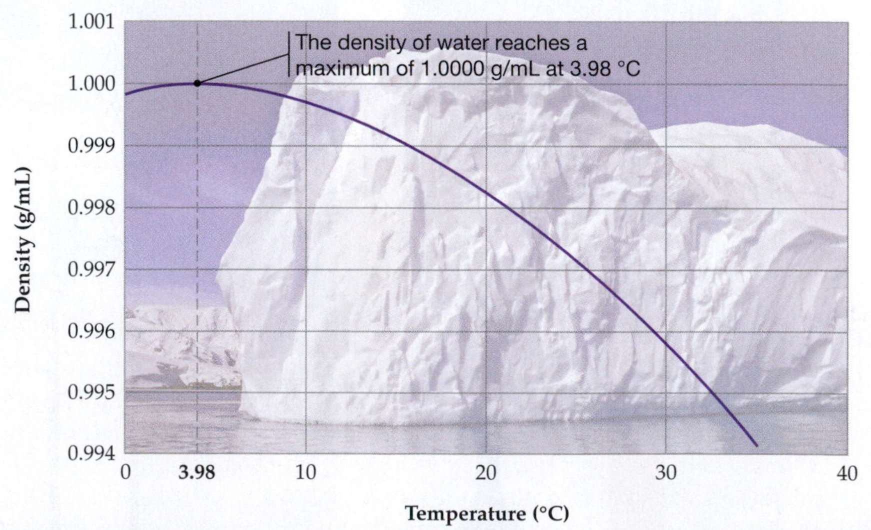

Knowing the density of a substance, particularly a liquid, can be useful in the chemistry laboratory because it's often easier to measure a liquid amount by volume rather than by mass. Suppose, for example, that you needed 1.55 g of ethyl alcohol. Rather than trying to weigh exactly the right amount of the liquid, it would be much easier to look up the density of ethyl alcohol in a standard reference book such as *The Handbook of Chemistry and Physics* (0.7893 g/mL at 20 °C) and then measure the correct volume with a syringe.

$$\text{Density} = \frac{\text{Mass}}{\text{Volume}} \quad \text{so} \quad \text{Volume} = \frac{\text{Mass}}{\text{Density}}$$

$$\text{Volume} = \frac{1.55\ \cancel{g}\ \text{ethyl alcohol}}{0.7893\ \dfrac{\cancel{g}}{\text{mL}}} = 1.96\ \text{mL ethyl alcohol}$$

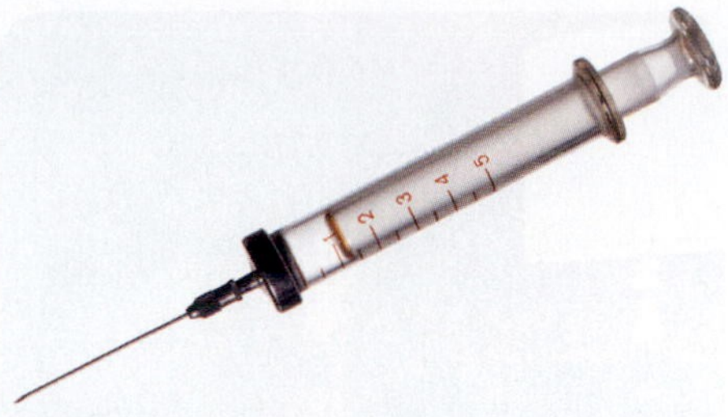

▲ The mass of a liquid is easily measured with a syringe if the density of the liquid is known.

Worked Example 0.2

Calculating a Density

What is the density of the element copper in g/cm^3 if a sample weighing 324.5 g has a volume of 36.2 cm^3?

SOLUTION

Density is mass divided by volume:

$$\text{Density} = \frac{\text{Mass}}{\text{Volume}} = \frac{324.5\ \text{g}}{36.2\ \text{cm}^3} = 8.96\ \text{g/cm}^3$$

Worked Example 0.3

Using Density to Calculate a Volume

What is the volume in cm^3 of 454 g of gold? (See Table 0.4.)

STRATEGY AND SOLUTION

As described in the text, density is mass divided by volume, so volume is mass divided by density:

$$\text{Density} = \frac{\text{Mass}}{\text{Volume}} \quad \text{so} \quad \text{Volume} = \frac{\text{Mass}}{\text{Density}}$$

$$\text{Volume} = \frac{454\ \cancel{g}\ \text{gold}}{19.31\ \cancel{g}/\text{cm}^3} = 23.5\ \text{cm}^3\ \text{gold}$$

PROBLEM 0.5 What is the density of glass in g/cm^3 if a sample weighing 27.43 g has a volume of 12.40 cm^3?

PROBLEM 0.6 Chloroform, a substance once used as an anesthetic, has a density of 1.483 g/mL at 20 °C. How many milliliters would you use if you needed 9.37 g?

0.8 DERIVED UNITS: MEASURING ENERGY

The last derived unit we'll consider at this point is **energy**, a word familiar to everyone though a bit hard to define without a lengthy explanation. We'll wait until Chapter 8 to give that explanation and discuss the topic more fully but will introduce it now because energy is so important in chemistry and shows up in so many contexts.

Energy has the unit of mass (kg) times the square of velocity (m/s)^2, or (kg · m^2)/s^2, an SI derived unit that is given the name **joule (J)** after the English physicist James Prescott Joule (1818–1889). The joule is a fairly small amount of energy—it takes roughly 100,000 J to heat a coffee cup full of water from room temperature to boiling—so kilojoules (kJ) are more frequently used in chemistry.

LOOKING AHEAD ...

The simplest definition of **energy** is to say that it is the capacity to supply heat or do work. All of Chapter 8 is devoted to explaining this topic and to seeing the importance of energy relationships in chemistry.

In addition to the SI energy unit joule, some chemists and biochemists still use the unit calorie (cal, with a lowercase c). Originally defined as the amount of energy necessary to raise

▲ The 75 watt incandescent bulb in this lamp uses energy at the rate of 75 J/s. Only about 5% of that energy appears as light, however; the remaining 95% is given off as heat.

the temperature of 1 g of water by 1 °C (specifically, from 14.5 °C to 15.5 °C), 1 calorie is now defined as exactly 4.184 J.

$$1 \text{ cal} = 4.184 \text{ J} \qquad \text{and} \qquad 1 \text{ J} = 0.2390 \text{ cal}$$

Nutritionists also use the somewhat confusing unit Calorie (Cal, with a capital C), which is equal to 1000 calories, or 1 kilocalorie (kcal), or 4.184 kilojoules (kJ).

$$1 \text{ Cal} = 1000 \text{ cal} = 1 \text{ kcal} = 4.184 \text{ kJ}$$

The energy value, or caloric content, of food is measured in Calories. Thus, the statement that a banana contains 70 Calories means that 70 Cal (70 kcal, or 290 kJ) of energy is released when the banana is used by the body for fuel.

PROBLEM 0.7 A Big Mac hamburger from McDonald's contains 540 Calories.
(a) How many kilojoules does a Big Mac contain?
(b) For how many hours could the amount of energy in a Big Mac light a 100 watt lightbulb? (1 watt = 1 J/s)

0.9 ACCURACY, PRECISION, AND SIGNIFICANT FIGURES IN MEASUREMENT

Making measurements like those discussed in the previous sections is something that most of us do every day, whether in cooking, carpentry, or chemistry. But how good are those measurements? Any measurement is only as good as the skill of the person doing the work and the reliability of the equipment being used. You've probably noticed, for instance, that you often get slightly different readings when you weigh yourself on a common bathroom scale and on a scale at the doctor's office, so there's always some uncertainty about your real weight. The same is true in chemistry—there is always some uncertainty in the value of a measurement.

In talking about the degree of uncertainty in a measurement, we use the words *accuracy* and *precision*. Although most of us use the words interchangeably in daily life, there's actually an important distinction between them. **Accuracy** refers to how close to the true value a given measurement is, whereas **precision** refers to how well a number of independent measurements agree with one another. To see the difference, imagine that you weigh a tennis ball whose true mass is 54.441 778 g. Assume that you take three independent measurements on each of three different types of balance to obtain the data shown in the following table.

▲ This tennis ball has a mass of about 54 g.

Measurement #	Bathroom Scale	Lab Balance	Analytical Balance
1	0.1 kg	54.4 g	54.4418 g
2	0.0 kg	54.5 g	54.4417 g
3	0.1 kg	54.3 g	54.4418 g
(average)	(0.07 kg)	(54.4 g)	(54.4418 g)

If you use a bathroom scale, your measurement (average = 0.07 kg) is neither accurate nor precise. Its accuracy is poor because it measures to only one digit that is far from the true value, and its precision is poor because any two measurements may differ substantially. If you now weigh the ball on an inexpensive laboratory balance, the value you get (average = 54.4 g) has three digits and is fairly accurate, but it is still not very precise because the three readings vary from 54.3 g to 54.5 g, perhaps due to air movements in the room or to a sticky mechanism. Finally, if you weigh the ball on an expensive analytical balance like those found in research laboratories, your measurement (average = 54.4418 g) is both precise and accurate. It's accurate because the measurement is very close to the true value, and it's precise because it has six digits that vary little from one reading to another.

To indicate the uncertainty in a measurement, the value you record should use all the digits you are sure of plus one additional digit that you estimate. In reading a thermometer that has a mark for each degree, for example, you could be certain about the digits of the nearest mark—say 25 °C—but you would have to estimate between two marks—say between 25 °C and 26 °C—to obtain a value of 25.3 °C.

The total number of digits recorded for a measurement is called the measurement's number of **significant figures**. For instance, the mass of the tennis ball as determined on the single-pan balance (54.4 g) has three significant figures, whereas the mass determined on the analytical balance (54.4418 g) has six significant figures. All digits but the last are certain; the final digit is an estimate, which we generally assume to have an error of plus or minus one (±1).

Finding the number of significant figures in a measurement is usually easy but can be troublesome if zeros are present. Look at the following four quantities:

4.803 cm	Four significant figures: 4, 8, 0, 3
0.006 61 g	Three significant figures: 6, 6, 1
55.220 K	Five significant figures: 5, 5, 2, 2, 0
34,200 m	Anywhere from three (3, 4, 2) to five (3, 4, 2, 0, 0) significant figures

The following rules cover the different situations that arise:

1. **Zeros in the middle of a number are like any other digit; they are always significant.** Thus, 4.803 cm has four significant figures.
2. **Zeros at the beginning of a number are not significant;** they act only to locate the decimal point. Thus, 0.006 61 g has three significant figures. (Note that 0.006 61 g can be rewritten as 6.61×10^{-3} g or as 6.61 mg.)
3. **Zeros at the end of a number and *after* the decimal point are always significant.** The assumption is that these zeros would not be shown unless they were significant. Thus, 55.220 K has five significant figures. (If the value were known to only four significant figures, we would write 55.22 K.)
4. **Zeros at the end of a number and *before* the decimal point may or may not be significant.** We can't tell whether they are part of the measurement or whether they just locate the decimal point. Thus, 34,200 m may have three, four, or five significant figures. Often, however, a little common sense is helpful. A temperature reading of 20 °C probably has two significant figures rather than one, since one significant figure would imply a temperature anywhere from 10 °C to 30 °C and would be of little use. Similarly, a volume given as 300 mL probably has three significant figures. On the other hand, a figure of 93,000,000 mi for the distance between the Earth and the Sun probably has only two or three significant figures.

The fourth rule shows why it's helpful to write numbers in scientific notation rather than ordinary notation: Doing so makes it possible to indicate the number of significant figures. Thus, writing the number 34,200 as 3.42×10^4 indicates three significant figures but writing it as 3.4200×10^4 indicates five significant figures.

One further point about significant figures: Certain numbers, such as those obtained when counting objects, are exact and have an effectively infinite number of significant figures. A week has exactly 7 days, for instance, not 6.9 or 7.0 or 7.1, and a foot has exactly 12 inches, not 11.9 or 12.0 or 12.1. In addition, the power of 10 used in scientific notation is an exact number. That is, the number 10^3 is exactly 1000, but the number 1×10^3 has one significant figure.

Worked Example 0.4

Significant Figures

How many significant figures does each of the following measurements have?

(a) 0.036 653 m (b) 7.2100×10^{-3} g (c) 72,100 km (d) $25.03

SOLUTION

(a) 5 (by rule 2) (b) 5 (by rule 3) (c) 3, 4, or 5 (by rule 4)
(d) $25.03 is an exact number

PROBLEM 0.8 A 1.000 mL sample of acetone, a common solvent sometimes used as a paint remover, was placed in a small bottle whose mass was known to be 38.0015 g. The following values were obtained when the acetone-filled bottle was weighed: 38.7798 g, 38.7795 g, and 38.7801 g. How would you characterize the precision and accuracy of these measurements if the true mass of the acetone was 0.7791 g?

PROBLEM 0.9 How many significant figures does each of the following quantities have? Explain your answers.

(a) 76.600 kJ	**(b)** 4.50200×10^3 g	**(c)** 3000 nm	**(d)** 0.00300 mL
(e) 18 students	**(f)** 3×10^{-5} g	**(g)** 47.60 mL	**(h)** 2070 mi

CONCEPTUAL PROBLEM 0.10 Characterize each of the following dartboards according to the accuracy and precision of the results.

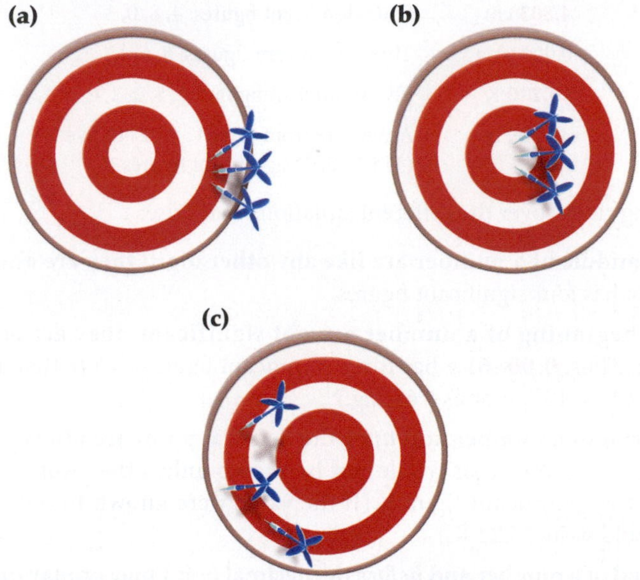

0.10 ROUNDING NUMBERS

▲ Calculators often display more figures than are justified by the precision of the data.

It often happens when dealing with measurements, particularly when doing arithmetic on a calculator, that a quantity appears to have more significant figures than are really justified. You might calculate the gas mileage of your car, for instance, by finding that it takes 11.70 gallons of gasoline to drive 278 miles:

$$\text{Mileage} = \frac{\text{Miles}}{\text{Gallons}} = \frac{278 \text{ mi}}{11.70 \text{ gal}} = 23.760684 \text{ mi/gal (mpg)}$$

Although the answer on the calculator has eight digits, your measurement is really not as precise as it appears. In fact, your answer is precise to only three significant figures and should be **rounded off** to 23.8 mi/gal by removing all nonsignificant digits.

How do you decide how many figures to keep and how many to remove? For most purposes, a simple procedure using just two rules is sufficient.

1. **In carrying out a multiplication or division, the answer can't have more significant figures than either of the original numbers.** If you think about it, this rule is just common sense. After all, if you don't know the number of miles you drove to better than three significant figures (278 could mean 277, 278, or 279), you can't calculate your mileage to more than the same number of significant figures.

Three significant figures (278 mi) → Three significant figures (23.8 mi/gal)

$$\frac{278 \text{ mi}}{11.70 \text{ gal}} = 23.8 \text{ mi/gal}$$

Four significant figures (11.70 gal)

2. **In carrying out an addition or subtraction, the answer can't have more digits to the right of the decimal point than either of the original numbers.** For example, if you have 3.18 L of water and you add 0.013 15 L more, you now have 3.19 L. Again, this rule is just common sense. If you don't know the volume you started with past the second decimal place (it could be 3.17, 3.18, or 3.19), you can't know the total of the combined volumes past the same decimal place.

3.18? ??	← Ends **two** places past decimal point
+ 0.013 15	← Ends **five** places past decimal point
3.19? ??	← Ends **two** places past decimal point

Once you decide how many digits to retain for your answer, the rules for rounding off numbers are straightforward:

1. **If the first digit you remove is less than 5, round down by dropping it and all following digits.** Thus, 3.864 becomes 3.86 when rounded to three significant figures because the first (and only) of the dropped digits is less than 5.
2. **If the first digit you remove is 5 or greater, round up by adding 1 to the digit on the left.** Thus, 3.865 becomes 3.87 when rounded to three significant figures because the first (and only) of the dropped digits is 5.

These two rules arise because of the nature of the decimal system, in which the first digit dropped in rounding can have any of ten values, 0 through 9. By rounding down in the five cases when the first dropped digit is 0, 1, 2, 3, or 4 and rounding up in the five cases when the first dropped digit is 5, 6, 7, 8, or 9, rounding up and rounding down have the same 50% chance of occurring,

Worked Example 0.5

A Calculation Using Significant Figures

It takes 9.25 hours to fly from London, England, to Chicago, Illinois, a distance of 3952 miles. What is the average speed of the airplane in miles per hour?

SOLUTION

First, set up an equation dividing the number of miles flown by the number of hours:

$$\text{Average speed} = \frac{3952 \text{ mi}}{9.25 \text{ h}} = 427.243\,24 \text{ mi/h}$$

Next, decide how many significant figures should be in your answer. Because the problem involves a division, and because one of the quantities you started with (9.25 h) has only three significant figures, the answer must also have three significant figures. Finally, round off your answer. The first digit to be dropped (2) is less than 5, so the answer 427.243 24 must be rounded off to 427 mi/h.

In doing this or any other problem, use all figures for the calculation, whether significant or not, and then round off the final answer. Don't round off at any intermediate step.

PROBLEM 0.11 Round off each of the following quantities to the number of significant figures indicated in parentheses:

(a) 3.774 499 L (4) **(b)** 255.0974 K (3)
(c) 55.265 kg (4) **(d)** 906.401 kJ (5)

PROBLEM 0.12 Carry out the following calculations, expressing each result with the correct number of significant figures:

(a) 24.567 g + 0.044 78 g = **?** g
(b) 4.6742 g ÷ 0.003 71 L = **?** g/L
(c) 0.378 mL + 42.3 mL − 1.5833 mL = **?** mL

CONCEPTUAL PROBLEM 0.13 What is the temperature reading on the following Celsius thermometer? How many significant figures do you have in your answer?

0.11 CONVERTING MEASUREMENTS FROM ONE UNIT TO ANOTHER

▲ Speed skaters have to convert from laps to meters to find out how far they have gone.

Because so many scientific activities involve numerical calculations—measuring, weighing, preparing solutions, and so forth—it's often necessary to convert a quantity from one unit to another. Converting between units isn't difficult; we all do it every day. If you run 7.5 laps around a 200 meter track, for instance, you have to convert between the distance unit *lap* and the distance unit *meter* to find that you have run 1500 m (7.5 laps times 200 meters per lap). Converting from one scientific unit to another is just as easy.

$$7.5\ \cancel{\text{laps}} \times \frac{200\ \text{meters}}{1\ \cancel{\text{lap}}} = 1500\ \text{meters}$$

The simplest way to carry out calculations that involve different units is to use the **dimensional-analysis method**. In this method, a quantity described in one unit is converted into an equivalent quantity with a different unit by using a **conversion factor** to express the relationship between units.

$$\text{Original quantity} \times \text{Conversion factor} = \text{Equivalent quantity}$$

As an example, we know from Section 0.4 that 1 meter equals 39.37 inches. Writing this relationship as a fraction restates it in the form of a conversion factor, either meters per inch or inches per meter.

Conversion factors between meters and inches $\dfrac{1\ \text{m}}{39.37\ \text{in.}}$ equals $\dfrac{39.37\ \text{in.}}{1\ \text{m}}$ equals 1

Note that this and all other conversion factors are effectively equal to 1 because the quantity above the division line (the numerator) is equal in value to the quantity below the division line (the denominator). Thus, multiplying by a conversion factor is equivalent to multiplying by 1 and so does not change the value of the quantity.

The key to the dimensional-analysis method of problem solving is that units are treated like numbers and can thus be multiplied and divided just as numbers can. The idea when solving a problem is to set up an equation so that unwanted units cancel, leaving only the desired units. Usually it's best to start by writing what you know and then manipulating that known quantity. For example, if you know your height is 69.5 inches and want to find it in meters, you can write down the height in inches and set up an equation multiplying the height by the conversion factor in meters per inch. The unit "in." cancels from the equation because it appears both above and below the division line, and the only unit that remains is "m."

$$\underset{\text{Starting quantity}}{69.5\ \cancel{\text{in.}}} \times \underset{\text{Conversion factor}}{\frac{1\ \text{m}}{39.37\ \cancel{\text{in.}}}} = \underset{\text{Equivalent quantity}}{1.77\ \text{m}}$$

The dimensional-analysis method gives the right answer only if the equation is set up so that the unwanted units cancel. If the equation is set up in any other way, the units won't cancel properly and you won't get the right answer. Thus, if you were to multiply your height in inches by the incorrect conversion factor inches per meter, you would end up with a meaningless answer using meaningless units.

Wrong! $69.5\ \text{in.} \times \frac{39.37\ \text{in.}}{1\ \text{m}} = 2740\ \text{in.}^2/\text{m}$ **??**

The main drawback to using the dimensional-analysis method is that it's easy to get the right answer without really understanding what you're doing. It's therefore best after solving a problem to think through a rough estimate, or "ballpark" solution, as a check on your work. If your ballpark check isn't close to the answer you get from the detailed solution, there's a misunderstanding somewhere and you should think through the problem again.

Even if you don't make an estimate, it's important to be sure that your calculated answer makes sense. If, for example, you were trying to calculate the volume of a human cell and you came up with the answer 5.3 cm^3, you should realize that such an answer couldn't possibly be right. Cells are too tiny to be distinguished with the naked eye, but a volume of 5.3 cm^3 is about the size of a walnut.

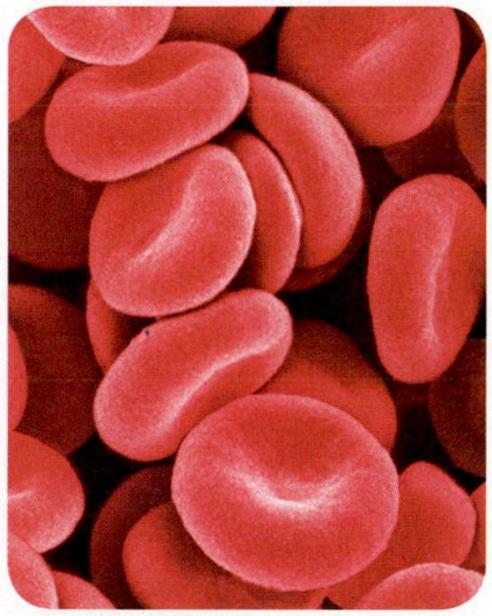

▲ What is the volume of a red blood cell?

The dimensional-analysis method and the use of ballpark checks are techniques that will help you solve problems of many kinds, not just unit conversions. Problems sometimes seem complicated, but you can usually sort out the complications by analyzing the problem properly.

- Identify the information you're given, including units.
- Identify the information you need in the answer, including units.
- Find a relationship between the known information and needed answer, and plan a strategy for getting from one to the other.
- Solve the problem.
- Make a rough estimate to be sure your calculated answer is reasonable.

Examples 0.6–0.9 show how to devise strategies and estimate answers. To conserve space, we'll use this approach routinely in only the next few chapters and more sparingly thereafter, but you should make it a standard part of all your problem solving.

Worked Example 0.6

Unit Conversions and Significant Figures

The Bugatti Veyron Super Sport is currently the fastest production car in the world, with a measured top speed of 267 miles per hour (and a price of $2.4 million). What is this speed in kilometers per hour?

▲ The Bugatti Veyron Super Sport has a top speed of 267 mph.

STRATEGY

The known information is the speed in mi/h; the unknown is the speed in km/h. Find the appropriate conversion factor inside the back cover of this book, and use the dimensional-analysis method to set up an equation so the "mi" units cancel. The answer should be rounded to three significant figures.

SOLUTION

$$\frac{267\ \cancel{\text{mi}}}{1\ \text{h}} \times \frac{1.609\ \text{km}}{1\ \cancel{\text{mi}}} = 430\ \frac{\text{km}}{\text{h}}$$

A very fast car!

continued on next page

✔ BALLPARK CHECK

The answer is certainly large, perhaps several hundred km/h. A better estimate is to realize that, because 1 mi = 1.609 km, it takes a bit more than 1 1/2 times as many kilometers as miles to measure the same distance. Thus, 267 mi is a bit more than 400 km, and 267 mi/h is over 400 km/h. The estimate agrees with the detailed solution.

Worked Example 0.7

Complex Unit Conversions

A large sport utility vehicle moving at a speed of 125 km/h might use gasoline at a rate of 16 L per 100 km. What does this correspond to in mi/gal?

STRATEGY

We are given a gasoline mileage in the units L/km (or km/L), and we need to find the mileage in the units mi/gal. Thus, two conversions are necessary, one from kilometers to miles and one from liters to gallons. It's best to do multiple conversions one step at a time until you get used to them. First, convert the distance from kilometers to miles and the amount of fuel from liters to gallons, and then divide the distance by the amount of fuel to find the mileage.

SOLUTION

$$100\ \cancel{\text{km}} \times \frac{0.6214\ \text{mi}}{1\ \cancel{\text{km}}} = 62.14\ \text{mi} \qquad 16\ \cancel{\text{L}} \times \frac{1\ \text{gal}}{3.79\ \cancel{\text{L}}} = 4.22\ \text{gal}$$

$$\frac{62.14\ \text{mi}}{4.22\ \text{gal}} = 14.73\ \frac{\text{mi}}{\text{gal}} \qquad \text{Round off to 15 mi/gal}$$

Notice that extra digits are carried through the intermediate calculations and only the final answer is rounded off.

When you become more confident in working multiple conversion problems, you can set up one large equation in which all unwanted units cancel.

$$\frac{100\ \cancel{\text{km}}}{16\ \cancel{\text{L}}} \times \frac{3.79\ \cancel{\text{L}}}{1\ \text{gal}} \times \frac{0.6214\ \text{mi}}{1\ \cancel{\text{km}}} = 14.73\ \frac{\text{mi}}{\text{gal}} \qquad \text{Round off to 15 mi/gal}$$

✔ BALLPARK CHECK

The mileage is probably low, perhaps around 15 mi/gal. This is a difficult problem to estimate, however, because it requires several different conversions. It's therefore best to think the problem through one step at a time, writing down the intermediate estimates:

- A distance of 100 km per 16 L is approximately 6 km/L.
- Because 1 km is about 0.6 mi, 6 km/L is about 4 mi/L.
- Because 1 L is approximately 1 qt, or 1/4 gal, 4 mi/L is about 16 mi/gal.

This estimate agrees with the detailed solution.

Worked Example 0.8

Complex Unit Conversions

The volcanic explosion that destroyed the Indonesian island of Krakatau on August 27, 1883, released an estimated 4.3 cubic miles (mi^3) of debris into the atmosphere and affected global weather for years. In SI units, how many cubic meters (m^3) of debris were released?

STRATEGY

We are given a volume in cubic miles and need to convert to cubic meters. It's probably simplest to convert first from mi^3 to km^3 and then from km^3 to m^3.

SOLUTION

$$4.3\ \cancel{\text{mi}^3} \times \left(\frac{1\ \text{km}}{0.6214\ \cancel{\text{mi}}}\right)^3 = 17.92\ \text{km}^3$$

$$17.92\ \cancel{\text{km}^3} \times \left(\frac{1000\ \text{m}}{1\ \cancel{\text{km}}}\right)^3 = 1.792 \times 10^{10}\ \text{m}^3$$

$$= 1.8 \times 10^{10}\ \text{m}^3 \quad \text{Rounded off}$$

✔ BALLPARK CHECK

As in the previous Worked Example, this is a difficult problem to estimate because it requires several different conversions. Take it one step at a time: One meter is much less than 1 mile, so it takes a large number of cubic meters to equal 1 mi^3, and the answer is going to be very large. Because 1 km is about 0.6 mi, 1 km^3 is about $(0.6)^3 = 0.2$ times as large as 1 mi^3. Thus, each mi^3 contains about 5 km^3, and 4.3 mi^3 contains about 20 km^3. Each km^3, in turn, contains $(1000\ m)^3 = 10^9\ m^3$. The volume of debris from the Krakatau explosion was therefore about $20 \times 10^9\ m^3$, or $2 \times 10^{10}\ m^3$. The estimate agrees with the detailed solution.

Worked Example 0.9

Complex Unit Conversions

Precious metals such as gold, silver, and platinum are measured in *troy ounces* (ozt) rather than common avoirdupois ounces (oz), where 1 ozt = 31.103 g and 1 oz = 28.35 g. Furthermore, a troy pound (lbt) contains only 12 ozt while an avoirdupois pound (lb) contains 16 oz. How many ozt are in 1 lb, and how many oz are in 1 lbt? Which would you rather receive as a gift from a rich relative, 1 avoirdupois ounce of gold or 1 troy ounce? One avoirdupois pound of gold or 1 troy pound?

▲ Gold and other precious metals are measured in troy ounces rather than common avoirdupois ounces.

STRATEGY

This problem contains a lot of information that makes it sound more difficult than it is. You can cut through the difficulties, however, by setting up equations in which unwanted units cancel and only desired units remain.

SOLUTION

$$\frac{1\ \text{ozt}}{31.103\ \cancel{\text{g}}} \times \frac{28.35\ \cancel{\text{g}}}{1\ \cancel{\text{oz}}} \times \frac{16\ \cancel{\text{oz}}}{1\ \text{lb}} = 14.58\ \frac{\text{ozt}}{\text{lb}}$$

$$\frac{12\ \cancel{\text{ozt}}}{1\ \text{lbt}} \times \frac{31.103\ \cancel{\text{g}}}{1\ \cancel{\text{ozt}}} \times \frac{1\ \text{oz}}{28.35\ \cancel{\text{g}}} = 13.17\ \frac{\text{oz}}{\text{lbt}}$$

One troy ounce of gold (31.103 g) is larger than one avoirdupois ounce (28.35 g) and makes a nicer gift, but one avoirdupois pound (16 oz) is much larger than one troy pound (13.17 oz) and makes a great gift.

✔ BALLPARK CHECK

One troy ounce (31.103 g) is slightly larger than one avoirdupois ounce (28.35 g) so there will be slightly fewer of them in an avoirdupois pound, perhaps about 15 ozt/lb rather than 16 oz/lb. Furthermore, because an avoirdupois ounce is slightly smaller than a troy ounce, there will be slightly more of them in a troy pound, perhaps about 13 oz/lbt rather than 12 ozt/lbt. These estimates agree with the detailed solution.

▶ PROBLEM 0.14 Calculate answers to the following problems, and check your solutions by making ballpark estimates.

(a) The melting point of gold is 1064 °C. What is this temperature in degrees Fahrenheit?

(b) How large, in cubic centimeters, is the volume of a red blood cell if the cell has a cylindrical shape with a diameter of 6×10^{-6} m and a height of 2×10^{-6} m?

▶ PROBLEM 0.15 Gemstones are weighed in *carats*, with 1 carat = 200 mg (exactly). What is the mass in grams of the Hope Diamond, the world's largest blue diamond at 44.4 carats? What is this mass in ounces?

▲ At 44.4 carats, the Hope Diamond is the largest blue diamond in the world.

▶ PROBLEM 0.16 A pure diamond with a mass of 0.1000 g contains 5.014×10^{21} carbon atoms and has a density of 3.52 g/cm^3. What is the volume of the Hope Diamond (Problem 0.15), and how many carbon atoms does it contain?

FYI THE RISKS AND BENEFITS OF CHEMICALS

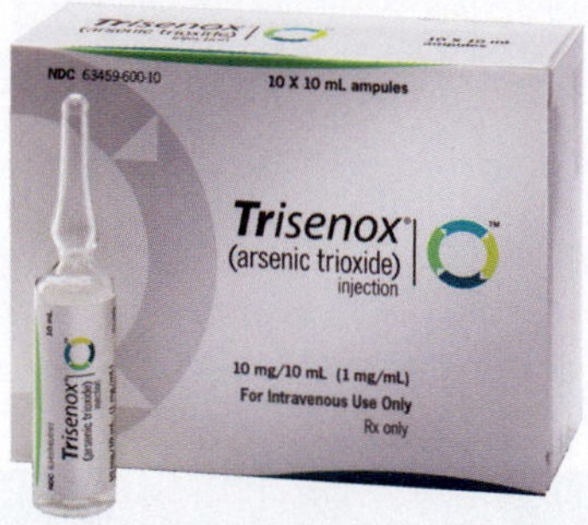

▲ Is this a poison or a treatment for leukemia?

Life is not risk-free—we all take many risks each day, often without even thinking about it. We may decide to ride a bike rather than drive, even though the likelihood per mile of being killed on a bicycle is 10 times greater than in a car. We may decide to smoke cigarettes, even though smoking kills more than 170,000 people each year in the United States.

What about risks from "chemicals"? News reports sometimes make it seem that our food is covered with pesticides and filled with dangerous additives, that our land is polluted by toxic waste dumps, and that our medicines are unsafe. How bad are the risks from chemicals, and how are the risks evaluated?

First, it's important to realize that *everything*, including your own body, is made of chemicals—that's what matter is. There is no such thing as a "chemical-free" food, cosmetic, cleanser, or anything else. Second, there is no meaningful distinction between a "natural" substance and a "synthetic" one; a chemical is a chemical. Many naturally occurring substances—snake venom and botulism toxin, for example—are extraordinarily poisonous, and many synthetic substances—polyethylene and nylon, for example—are harmless.

Risk evaluation of chemicals is carried out by exposing test animals, usually mice or rats, to a chemical and then monitoring for signs of harm. To limit the expense and time needed for testing, the amounts administered are often hundreds or thousands of times larger than those a person might normally encounter. The *acute chemical toxicity* (as opposed to chronic toxicity) observed in animal tests is reported as an *LD_{50} value*, the amount of a substance per kilogram of body weight that is a lethal dose for 50% of the test animals. Some LD_{50} values of different substances are shown in **Table 0.5**. The lower the value, the more toxic the substance.

TABLE 0.5 Some LD_{50} Values in Rats

Substance	LD_{50} (g/kg)	Substance	LD_{50} (g/kg)
Strychnine	0.005	Chloroform	1.2
Arsenic trioxide	0.015	Iron(II) sulfate	1.5
DDT	0.115	Ethyl alcohol	7.1
Aspirin	1.1	Sodium cyclamate	12.8

Even with an LD_{50} value established in test animals, the risk of human exposure to a given substance is still hard to assess. If a substance is harmful to rats, is it necessarily harmful to humans? How can a large dose for a small animal be translated into a small dose for a large human? All substances are toxic to some organisms to some extent, and the difference between help and harm is often a matter of degree. Vitamin A, for example, is necessary for vision, yet it can promote cancer at high doses. Arsenic trioxide is the most classic of poisons, yet it induces remissions in some types of leukemia and was approved in 2000 by the U.S. Food and Drug Administration for drug use under the name Trisenox.

Even water can be toxic if drunk in large amounts because it dilutes the salt in body fluids and causes a potentially life-threatening condition called *hyponatremia* that has resulted in the death of several marathon runners. Furthermore, how we evaluate risk is strongly influenced by familiarity. Many foods contain natural ingredients far more toxic than synthetic additives or pesticide residues, but the ingredients are ignored because the foods are familiar.

All decisions involve tradeoffs. Does the benefit of a pesticide that increases food production outweigh a possible health risk to 1 person in 1 million? Do the beneficial effects of a new drug outweigh a potentially dangerous side effect in a small number of users? Different people will have different opinions, but an honest evaluation of the facts is surely better than a purely emotional response.

PROBLEM 0.17 Oxalic acid, found in dark-green leafy vegetables such as spinach and rhubarb, has a reported LD_{50} of 600 mg/kg in humans. If dry rhubarb leaves (not stalks) contain 0.5% oxalic acid by weight, how many pounds of dry leaves would a 175-pound person have to consume to have a 50% chance of dying? How many pounds of undried leaves, whose mass is about 95% water?

SUMMARY

Chemistry is the study of the composition, properties, and transformations of **matter**. It is best approached by posing questions, conducting experiments, and devising **theories** to interpret the experimental results.

Accurate measurement is crucial to scientific experimentation. The units used are those of the Système Internationale (**SI units**). There are seven fundamental SI units, together with other derived units. **Mass**, the amount of matter in an object, is measured in **kilograms (kg)**; **length** is measured in **meters (m)**; **temperature** is measured in **kelvin (K)**; and **volume** is measured in **cubic meters (m^3)**. The more familiar metric **liter (L)** and **milliliter (mL)** are also still used for measuring volume, and the **Celsius degree (°C)** is still used for measuring temperature. **Density** relates mass to volume and is measured in the derived SI unit g/cm^3 or g/mL. **Energy** is a quantity used throughout chemistry and is measured in the derived SI unit $(kg \cdot m^2)/s^2$, or **joule (J)**.

Because many experiments involve numerical calculations, it's often necessary to manipulate and convert different units of measure. The simplest way to carry out such conversions is to use the **dimensional-analysis method**, in which an equation is set up so that unwanted units cancel and only the desired units remain. It's also important when measuring physical quantities or carrying out calculations to indicate the precision of the measurement by **rounding off** the result to the correct number of **significant figures**.

KEY WORDS

accuracy *12*
Celsius degree (°C) *7*
centimeter (cm) *6*
chemistry *2*
conversion factor *16*
cubic centimeter (cm^3) *9*
cubic decimeter (dm^3) *9*
cubic meter (m^3) *9*
density *10*
dimensional-analysis method *16*
energy *11*
gram (g) *5*
joule (J) *11*
kelvin (K) *7*
kilogram (kg) *5*
liter (L) *9*
mass *5*
matter *5*
meter (m) *6*
microgram (μg) *5*
micrometer (μm) *6*
milligram (mg) *5*
milliliter (mL) *9*
millimeter (mm) *6*
nanometer (nm) *6*
picometer (pm) *6*
precision *12*
rounding off *14*
scientific notation *5*
SI unit *4*
significant figure *13*
theory *4*
volume *9*

KEY EQUATIONS

- Converting between Celsius and Fahrenheit temperatures **(Section 0.5)**

CELSIUS TO FAHRENHEIT

$$^{\circ}\mathrm{F} = \left(\frac{9\,^{\circ}\mathrm{F}}{5\,^{\circ}\cancel{\mathrm{C}}} \times {}^{\circ}\cancel{\mathrm{C}}\right) + 32\,^{\circ}\mathrm{F}$$

FAHRENHEIT TO CELSIUS

$$^{\circ}\mathrm{C} = \frac{5\,^{\circ}\mathrm{C}}{9\,^{\circ}\cancel{\mathrm{F}}} \times ({}^{\circ}\cancel{\mathrm{F}} - 32\,^{\circ}\cancel{\mathrm{F}})$$

LEARNING OUTCOMES

- Express and do calculations with numbers using scientific notation. **(Section 0.2)** ***Problems 1, 34, 35, 40–45, 48–51, 67***
- Learn the fundamental SI units for physical quantities and the prefixes that modify them by size. **(Section 0.2)** ***Problems 2, 22–29, 32, 33, 39***
- Use and interconvert SI and non-SI units for mass and length. **(Sections 0.3, 0.4)** ***Problems 35, 36, 66–73, 81, 93***
- Interconvert temperatures on the Fahrenheit, Celsius, and Kelvin scales. **(Section 0.5)** ***Problems 3, 4, 26, 27, 52–57, 76, 84, 88, 92***
- Know and use the derived SI units for volume **(Section 0.6)** and density. **(Section 0.7)** ***Problems 5, 6, 20, 21, 58–63, 68, 69, 74, 75, 78–80, 83, 85, 89, 90***
- Know and use the derived SI unit for energy. **(Section 0.8)** ***Problems 7, 64, 65, 80, 82***
- Determine the number of significant figures in a measurement **(Section 0.9)**, and round the answer from a calculation to the correct number of significant figures. **(Section 0.10)** ***Problems 8, 9, 11–13, 30, 31, 40, 41, 46–51, 67***
- Use conversion factors and the dimensional-analysis method to convert measurements from one unit to another. **(Section 0.11)** ***Problems 14–16, 38, 52–57, 66–73, 76, 77, 80, 91***

CONCEPTUAL PROBLEMS

Problems at the end of each chapter begin with a section called Conceptual Problems. These problems are visual rather than numerical and are intended to probe your understanding rather than your facility with numbers and formulas. Answers to even-numbered problems can be found at the end of the book following the appendixes.

Problems 0.1–0.17 appear within the chapter.

0.18 Which block in each of the following drawings of a balance is more dense, red or green? Explain.

(a) **(b)**

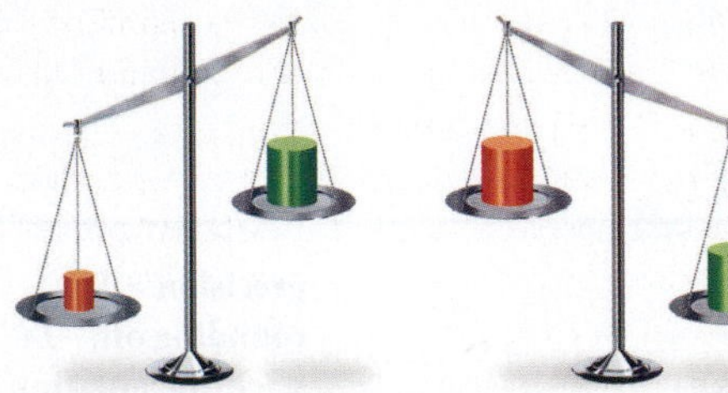

0.19 How many milliliters of water does the graduated cylinder in **(a)** contain, and how tall in centimeters is the paper clip in **(b)**? How many significant figures do you have in each answer?

(a) **(b)**

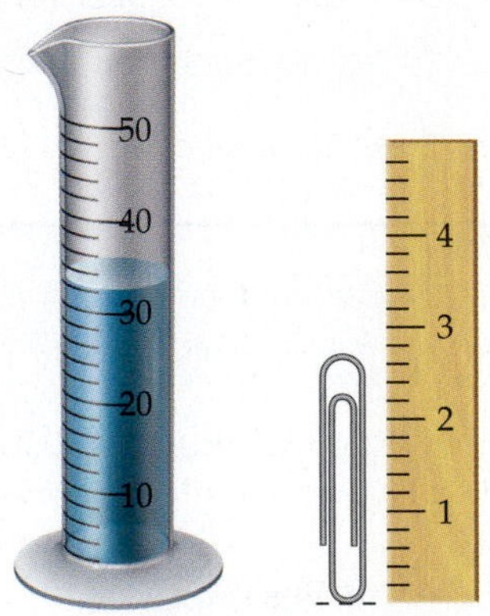

0.20 Assume that you have two graduated cylinders, one with a capacity of 5 mL **(a)** and the other with a capacity of 50 mL **(b)**. Draw a line in each, showing how much liquid you would add if you needed to measure 2.64 mL of water. Which cylinder will give the more accurate measurement? Explain.

(a) **(b)**

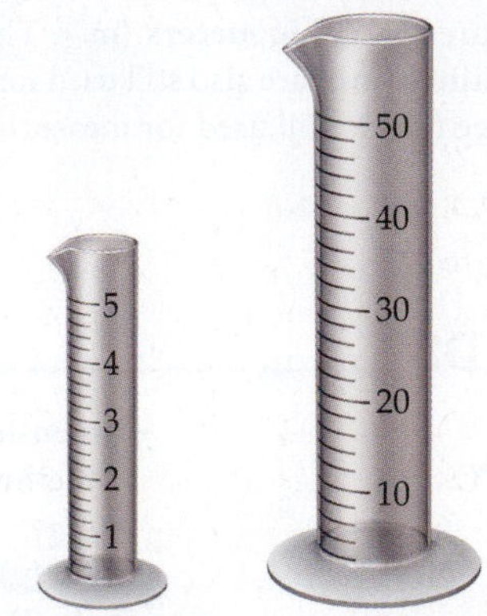

0.21 The following cylinder contains three liquids that don't mix with one another: water (density = 1.0 g/mL), vegetable oil (density = 0.93 g/mL), and mercury (density = 13.5 g/mL). Which liquid is which?

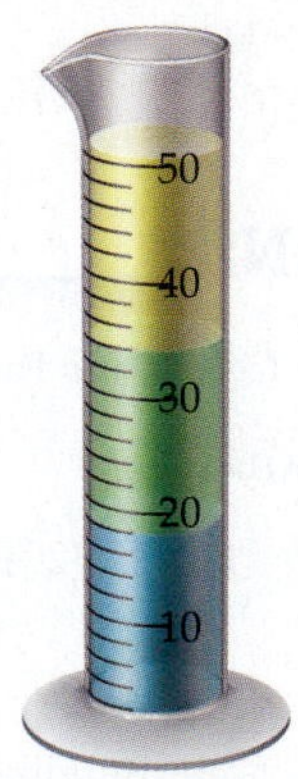

SECTION PROBLEMS

The Section Problems at the end of each chapter cover specific topics from the various sections of the chapter. These problems are presented in pairs, with each even-numbered problem followed by an odd-numbered one requiring similar skills.

Units, Significant Figures, and Rounding (Sections 0.2, 0.9, 0.10)

0.22 What is the difference between mass and weight?

0.23 What is the difference between a derived SI unit and a fundamental SI unit? Give an example of each.

0.24 What SI units are used for measuring the following quantities? For derived units, express your answers in terms of the six fundamental units.

(a) Mass **(b)** Length
(c) Temperature **(d)** Volume
(e) Energy **(f)** Density

0.25 What SI prefix corresponds to each of the following multipliers?

(a) 10^3 **(b)** 10^{-6}
(c) 10^9 **(d)** 10^{-12}
(e) 10^{-2} **(f)** 10^{-9}

0.26 Which is larger, a Fahrenheit degree or a Celsius degree? By how much?

0.27 What is the difference between a kelvin and a Celsius degree?

0.28 What is the difference between a cubic decimeter (SI) and a liter (metric)?

0.29 What is the difference between a cubic centimeter (SI) and a milliliter (metric)?

0.30 Which of the following statements use exact numbers?

(a) 1 ft = 12 in. **(b)** 1 cal = 4.184 J
(c) The height of Mt. Everest is 29,035 ft.
(d) The world record for the 1 mile run, set by Morocco's Hicham El Guerrouj in July, 1999, is 3 minutes, 43.13 seconds.

0.31 What is the difference in mass between a nickel that weighs 4.8 g and a nickel that weighs 4.8673 g?

0.32 Bottles of wine sometimes carry the notation "Volume = 75 cL." What does the unit cL mean?

0.33 What do the following abbreviations stand for?
(a) dL (b) dm
(c) µm (d) nL
(e) MJ (f) pg

0.34 Which quantity in each of the following pairs is larger?
(a) 5.63×10^6 cm or 6.02×10^1 km
(b) 46 µs or 3.2×10^{-2} ms
(c) 200,098 g or 17×10^1 kg

0.35 Which quantity in each of the following pairs is smaller?
(a) 154 pm or 7.7×10^{-9} cm
(b) 1.86×10^{11} µm or 2.02×10^2 km
(c) 2.9 GA or 3.1×10^{15} µA

0.36 How many picograms are in 1 mg? In 35 ng?

0.37 How many microliters are in 1 L? In 20 mL?

0.38 Carry out the following conversions:
(a) 5 pm = ____ cm = ____ nm
(b) 8.5 cm^3 = ____ m^3 ____ mm^3
(c) 65.2 mg = ____ g = ____ pg

0.39 Which is larger, and by approximately how much?
(a) A liter or a quart (b) A mile or a kilometer
(c) A gram or an ounce (d) A centimeter or an inch

0.40 How many significant figures are in each of the following measurements?
(a) 35.0445 g (b) 59.0001 cm
(c) 0.030 03 kg (d) 0.004 50 m
(e) 67,000 m^2 (f) 3.8200×10^3 L

0.41 How many significant figures are in each of the following measurements?
(a) $130.95 (b) 2000.003 g
(c) 5 ft 3 in. (d) 510 J
(e) 5.10×10^2 J (f) 10 students

0.42 The Vehicle Assembly Building at the John F. Kennedy Space Center in Cape Canaveral, Florida, is the largest building in the world, with a volume of 3,666,500 m^3. Express this volume in scientific notation.

0.43 The diameter of the Earth at the equator is 7926.381 mi. Round this quantity to four significant figures; to two significant figures. Express the answers in scientific notation.

0.44 Express the following measurements in scientific notation:
(a) 453.32 mg (b) 0.000 042 1 mL (c) 667,000 g

0.45 Convert the following measurements from scientific notation to standard notation:
(a) 3.221×10^{-3} mm (b) 8.940×10^5 m
(c) $1.350\,82 \times 10^{-12}$ m^3 (d) 6.4100×10^2 km

0.46 Round the following quantities to the number of significant figures indicated in parentheses:
(a) 35,670.06 m (4, 6) (b) 68.507 g (2, 3)
(c) 4.995×10^3 cm (3) (d) $2.309\,85 \times 10^{-4}$ kg (5)

0.47 Round the following quantities to the number of significant figures indicated in parentheses:
(a) 7.0001 kg (4) (b) 1.605 km (3)
(c) 13.2151 g/cm^3 (3) (d) 2,300,000.1 (7)

0.48 Express the results of the following calculations with the correct number of significant figures:
(a) 4.884×2.05 (b) $94.61 \div 3.7$
(c) $3.7 \div 94.61$ (d) $5502.3 + 24 + 0.01$
(e) $86.3 + 1.42 - 0.09$ (f) 5.7×2.31

0.49 Express the results of the following calculations with the correct number of significant figures:
(a) $\frac{3.41 - 0.23}{5.233} \times 0.205$ (b) $\frac{5.556 \times 2.3}{4.223 - 0.08}$

0.50 The world record for the women's outdoor 20,000 meter run, set in 2000 by Tegla Loroupe, is 1:05:26.6 (seconds are given to the nearest tenth). What was her average speed, expressed in miles per hour with the correct number of significant figures? Assume that the race distance is accurate to 5 significant figures.

0.51 In the U.S., the emissions limit for carbon monoxide in motorcycle engine exhaust is 12.0 g of carbon monoxide per kilometer driven. What is this limit expressed in mg per mile with the correct number of significant figures?

Temperature (Section 0.5)

0.52 The normal body temperature of a goat is 39.9 °C, and that of an Australian spiny anteater is 22.2 °C. Express these temperatures in degrees Fahrenheit.

0.53 Of the 90 or so naturally occurring elements, only four are liquid near room temperature: mercury (melting point = −38.87 °C), bromine (melting point = −7.2 °C), cesium (melting point = 28.40 °C), and gallium (melting point = 29.78 °C). Convert these melting points to degrees Fahrenheit.

0.54 Tungsten, the element used to make filaments in lightbulbs, has a melting point of 6192 °F. Convert this temperature to degrees Celsius and to kelvin.

0.55 Suppose that your oven is calibrated in degrees Fahrenheit but a recipe calls for you to bake at 175 °C. What oven setting should you use?

0.56 Suppose that you were dissatisfied with both Celsius and Fahrenheit units and wanted to design your own temperature scale based on ethyl alcohol (ethanol). On the Celsius scale, ethanol has a melting point of −117.3 °C and a boiling point of 78.5 °C, but on your new scale calibrated in units of degrees ethanol, °E, you define ethanol to melt at 0 °E and boil at 200 °E.
(a) How does your ethanol degree compare in size with a Celsius degree?
(b) How does an ethanol degree compare in size with a Fahrenheit degree?
(c) What are the melting and boiling points of water on the ethanol scale?
(d) What is normal human body temperature (98.6 °F) on the ethanol scale?
(e) If the outside thermometer reads 130 °E, how would you dress to go out?

0.57 Answer parts (a)–(d) of Problem 0.56 assuming that your new temperature scale is based on ammonia, NH_3. On the Celsius scale, ammonia has a melting point of −77.7 °C and a boiling point of −33.4 °C, but on your new scale calibrated in units of degrees ammonia, °A, you define ammonia to melt at 0 °A and boil at 100 °A.

Density (Section 0.7)

0.58 The density of silver is 10.5 g/cm^3. What is the mass (in kilograms) of a cube of silver that measures 0.62 m on each side?

0.59 A vessel contains 4.67 L of bromine, whose density is 3.10 g/cm^3. What is the mass of the bromine in the vessel (in kilograms)?

0.60 Aspirin has a density of 1.40 g/cm^3. What is the volume in cubic centimeters of an aspirin tablet weighing 250 mg? Of a tablet weighing 500 lb?

0.61 Gaseous hydrogen has a density of 0.0899 g/L at 0 °C, and gaseous chlorine has a density of 3.214 g/L at the same temperature. How many liters of each would you need if you wanted 1.0078 g of hydrogen and 35.45 g of chlorine?

0.62 What is the density of lead in g/cm^3 if a rectangular bar measuring 0.50 cm in height, 1.55 cm in width, and 25.00 cm in length has a mass of 220.9 g?

0.63 What is the density of lithium metal in g/cm^3 if a cylindrical wire with a diameter of 2.40 mm and a length of 15.0 cm has a mass of 0.3624 g?

Energy (Section 0.8)

0.64 The combustion of 45.0 g of methane (natural gas) releases 2498 kJ of heat energy. How much energy in kilocalories (kcal) would combustion of 0.450 ounces of methane release?

0.65 Sodium (Na) metal undergoes a chemical reaction with chlorine (Cl) gas to yield sodium chloride, or common table salt. If 1.00 g of sodium reacts with 1.54 g of chlorine, 2.54 g of sodium chloride is formed and 17.9 kJ of heat is released. How much sodium and how much chlorine in grams would have to react to release 171 kcal of heat?

Unit Conversions (Section 0.11)

0.66 Carry out the following conversions:

(a) How many grams of meat are in a quarter-pound hamburger (0.25 lb)?

(b) How tall in meters is the Willis Tower (formerly called the Sears Tower) in Chicago (1454 ft)?

(c) How large in square meters is the land area of Australia (2,941,526 mi^2)?

0.67 Convert the following quantities into SI units with the correct number of significant figures:

(a) 5.4 in. **(b)** 66.31 lb **(c)** 0.5521 gal

(d) 65 mi/h **(e)** 978.3 yd^3 **(f)** 2.380 mi^2

0.68 The volume of water used for crop irrigation is measured in acre-feet, where 1 acre-foot is the amount of water needed to cover 1 acre of land to a depth of 1 ft.

(a) If there are 640 acres per square mile, how many cubic feet of water are in 1 acre-foot?

(b) How many acre-feet are in Lake Erie (total volume = 116 mi^3)?

0.69 The height of a horse is usually measured in *hands* instead of in feet, where 1 hand equals 1/3 ft (exactly).

(a) How tall in centimeters is a horse of 18.6 hands?

(b) What is the volume in cubic meters of a box measuring $6 \times 2.5 \times 15$ hands?

0.70 Concentrations of substances dissolved in solution are often expressed as mass per unit volume. For example, normal human blood has a cholesterol concentration of about 200 mg/100 mL. Express this concentration in the following units:

(a) mg/L **(b)** μg/mL **(c)** g/L **(d)** ng/μL

(e) How much total blood cholesterol in grams does a person have if the normal blood volume in the body is 5 L?

0.71 Weights in England are commonly measured in *stones*, where 1 stone = 14 lb. What is the weight in pounds of a person who weighs 8.65 stones?

0.72 Among the many alternative units that might be considered as a measure of time is the *shake* rather than the second. Based on the expression "faster than a shake of a lamb's tail," we'll define 1 shake as equal to 2.5×10^{-4} s. If a car is traveling at 55 mi/h, what is its speed in cm/shake?

0.73 Administration of digitalis, a drug used to control atrial fibrillation in heart patients, must be carefully controlled because even a modest overdosage can be fatal. To take differences between patients into account, drug dosages are prescribed in terms of mg/kg body weight. Thus, a child and an adult differ greatly in weight, but both receive the same dosage per kilogram of body weight. At a dosage of 20 μg/kg body weight, how many milligrams of digitalis should a 160 lb patient receive?

CHAPTER PROBLEMS

Chapter Problems are unpaired, are not identified by subject, and can cover topics from any of the various sections of the chapter. In addition, they are generally a bit more challenging than earlier Section Problems.

0.74 When an irregularly shaped chunk of silicon weighing 8.763 g was placed in a graduated cylinder containing 25.00 mL of water, the water level in the cylinder rose to 28.76 mL. What is the density of silicon in g/cm^3?

0.75 Lignum vitae is a hard, durable, and extremely dense wood used to make ship bearings. A sphere of this wood with a diameter of 7.60 cm has a mass of 313 g.

(a) What is the density of the lignum vitae sphere?

(b) Will the sphere float or sink in water?

(c) Will the sphere float or sink in chloroform? (The density of chloroform is 1.48 g/mL.)

0.76 Sodium chloride has a melting point of 1074 K and a boiling point of 1686 K. Convert these temperatures to degrees Celsius and to degrees Fahrenheit.

0.77 A large tanker truck for carrying gasoline has a capacity of 3.4×10^4 L.

(a) What is the tanker's capacity in gallons?

(b) If the retail price of gasoline is $3.00 per gallon, what is the value of the truck's full load of gasoline?

0.78 The density of chloroform, a widely used solvent, is 1.4832 g/mL at 20 °C. How many milliliters would you use if you wanted 112.5 g of chloroform?

0.79 Sulfuric acid (density = 1.8302 g/cm^3) is produced in a larger amount than any other chemical, approximately 3.6×10^{11} lb/yr worldwide. What is the volume of this amount in liters?

0.80 Answer the following questions:

(a) An old rule of thumb in cooking says: "A pint's a pound the world around." What is the density in g/mL of a substance for which 1 pt = 1 lb exactly?

(b) There are exactly 640 acres in 1 square mile. How many square meters are in 1 acre?

(c) A certain type of wood has a density of 0.40 g/cm^3. What is the mass of 1.0 cord of this wood in kg, where 1 cord is 128 cubic feet of wood?

(d) A particular sample of crude oil has a density of 0.85 g/mL. What is the mass of 1.00 barrel of this crude oil in kg, where a barrel of oil is exactly 42 gallons?

(e) A gallon of ice cream contains exactly 32 servings, and each serving has 165 Calories, of which 30.0% are derived from fat. How many Calories derived from fat would you consume if you ate one half-gallon of ice cream?

0.81 A 1.0 ounce piece of chocolate contains 15 mg of caffeine, and a 6.0 ounce cup of regular coffee contains 105 mg of caffeine. How much chocolate would you have to consume to get as much caffeine as you would from 2.0 cups of coffee?

0.82 A bag of Hershey's Kisses contains the following information:

Serving size: 9 pieces = 41 grams

Calories per serving: 230

Total fat per serving: 13 g

(a) The bag contains 2.0 lbs of Hershey's Kisses. How many Kisses are in the bag?

(b) The density of a Hershey's Kiss is 1.4 g/mL. What is the volume of a single Hershey's Kiss?

(c) How many Calories are in one Hershey's Kiss?

(d) Each gram of fat yields 9 Calories when metabolized. What percent of the Calories in Hershey's Kisses are derived from fat?

0.83 Vinaigrette salad dressing consists mainly of olive oil and vinegar. The density of olive oil is 0.918 g/cm^3, the density of vinegar is 1.006 g/cm^3, and the two do not mix. If a certain mixture of olive oil and vinegar has a total mass of 397.8 g and a total volume of 422.8 cm^3, what is the volume of oil and what is the volume of vinegar in the mixture?

0.84 At a certain point, the Celsius and Fahrenheit scales "cross," giving the same numerical value on both. At what temperature does this crossover occur?

0.85 Imagine that you place a cork measuring 1.30 cm × 5.50 cm × 3.00 cm in a pan of water and that on top of the cork you place a small cube of lead measuring 1.15 cm on each edge. The density of cork is 0.235 g/cm^3, and the density of lead is 11.35 g/cm^3. Will the combination of cork plus lead float or sink?

0.86 The LD_{50} of aspirin in rats is given in the *FYI* at the end of this chapter. If a baby aspirin tablet contains 81 mg of aspirin, how many whole tablets would a 0.75 lb rat have to consume to have at least a 50% chance of dying from the dose?

0.87 An Eastern diamondback rattlesnake was milked until 0.134 g of venom was obtained. The venom was then administered subcutaneously in equal portions to 550 mice with an average weight of 0.70 oz, and exactly half the mice died. What is the LD_{50} (in g/kg) for the snake venom in mice? See the *FYI* at the end of this chapter.

0.88 A 125 mL sample of water at 293.2 K was heated for 8 min, 25 s so as to give a constant temperature increase of 3.0 °F/min. What is the final temperature of the water in degrees Celsius?

0.89 A calibrated flask was filled to the 25.00 mL mark with ethyl alcohol. By weighing the flask before and after adding the alcohol, it was determined that the flask contained 19.7325 g of alcohol. In a second experiment, 25.0920 g of metal beads were added to the flask, and the flask was again filled to the 25.00 mL mark with ethyl alcohol. The total mass of the metal plus alcohol in the flask was determined to be 38.4704 g. What is the density of the metal in g/mL?

0.90 Brass is an alloy, or solid mixture, of copper and zinc. What is the mass in grams of a brass cylinder having a length of 1.62 in. and a diameter of 0.514 in. if the composition of the brass is 67.0% copper and 33.0% zinc by mass? The density of copper is 8.92 g/cm^3, and the density of zinc is 7.14 g/cm^3. Assume that the density of the brass varies linearly with composition.

0.91 Ocean currents are measured in *sverdrups* (Sv) where 1 Sv = $10^9\ m^3/s$. The Gulf Stream off the tip of Florida, for instance, has a flow of 35 Sv.

(a) What is the flow of the Gulf Stream in milliliters per minute?

(b) What mass of water in the Gulf Stream flows past a given point in 24 hours? The density of seawater is 1.025 g/mL.

(c) How much time is required for 1 petaliter (PL; 1 PL = 10^{15} L) of seawater to flow past a given point?

0.92 The element gallium (Ga) has the second largest liquid range of any element, melting at 29.78 °C and boiling at 2204 °C at atmospheric pressure.

(a) What is the density of gallium in g/cm^3 at 25 °C if a 1 in. cube has a mass of 0.2133 lb?

(b) Assume that you construct a thermometer using gallium as the fluid instead of mercury, and that you define the melting point of gallium as 0 °G and the boiling point of gallium as 1000 °G. What is the melting point of sodium chloride (801 °C) on the gallium scale?

0.93 Distances over land are measured in *statute miles* (5280 ft), but distances over water are measured in *nautical miles*, where 1 nautical mile was originally defined as 1 minute of arc along an Earth meridian, or 1/21,600 of the Earth's circumference through the poles. A ship's speed through the water is measured in *knots*, where 1 knot = 1 nautical mile per hour. Historically, the unit *knot* derived from the practice of measuring a ship's speed by throwing a log tied to a knotted line over the side. The line had a knot tied in it at intervals of 47 ft 3 in., and the number of knots run out in 28 seconds was counted to determine speed.

(a) How many feet are in a nautical mile? How many meters?

(b) The northern bluefin tuna can weigh up to 1500 pounds and can swim at speeds up to 48 miles per hour. How fast is this in knots?

(c) A *league* is defined as 3 nautical miles. The Mariana Trench, with a depth of 35,798 feet, is the deepest point in the ocean. How many leagues deep is this?

(d) By international agreement, the nautical mile is now defined as exactly 1852 meters. By what percentage does this current definition differ from the original definition, and by what percentage does it differ from a statute mile?

CHAPTER 1

The Structure and Stability of Atoms

Where do atoms come from? Cosmologists believe that the atoms now found in all matter were formed billions of years ago by the gravitational collapse of massive stars. We'll learn more in the *FYI* at the end of this chapter.

People have always been fascinated by changes, particularly by changes that are dramatic or useful. In the ancient world, the change that occurred when a stick of wood burned, gave off heat, and turned into a small pile of ash was especially important. Similarly, the change that occurred when a reddish lump of rock (iron ore) was heated with charcoal (carbon) and produced a gray metal (iron) useful for making weapons, tools, and other implements was of enormous value. Observing such changes eventually caused people to think about what different materials might be composed of and led to the idea of fundamental substances that we today call *elements*.

At the same time people were pondering the question of elements, they were also thinking about related matters: What is an element made of? Is matter continuously divisible into ever smaller and smaller pieces, or is there an ultimate limit? Can you cut a piece of gold in two, take one of the pieces and cut *it* in two, and so on infinitely, or is there a point at which you must stop? Although most thinkers, including Plato and Aristotle, believed that matter is continuous, the Greek philosopher Democritus (460–370 BC) disagreed. Democritus proposed that elements are composed of tiny indestructible particles that we now call **atoms**, from the Greek word *atomos*, meaning "indivisible." Little else was learned about elements and atoms until the birth of modern experimental science some 2000 years later.

▲ Samples of mercury, silver, and sulfur (clockwise from top left).

1.1 CHEMISTRY AND THE ELEMENTS

Everything you see around you is formed from one or more of 118 presently known *elements*. An **element** is a fundamental substance that can't be chemically changed or broken down into anything simpler. Mercury, silver, and sulfur are common examples, as listed in **Table 1.1**.

CONTENTS

TABLE 1.1 Names and Symbols of Some Common Elements

Aluminum	**Al**	Chlorine	**Cl**	Manganese	**Mn**	Copper (*cuprum*)	**Cu**
Argon	**Ar**	Fluorine	**F**	Nitrogen	**N**	Iron (*ferrum*)	**Fe**
Barium	**Ba**	Helium	**He**	Oxygen	**O**	Lead (*plumbum*)	**Pb**
Boron	**B**	Hydrogen	**H**	Phosphorus	**P**	Mercury (*hydrargyrum*)	**Hg**
Bromine	**Br**	Iodine	**I**	Silicon	**Si**	Potassium (*kalium*)	**K**
Calcium	**Ca**	Lithium	**Li**	Sulfur	**S**	Silver (*argentum*)	**Ag**
Carbon	**C**	Magnesium	**Mg**	Zinc	**Zn**	Sodium (*natrium*)	**Na**

Actually, the previous statement about everything being made of one or more of 118 elements is an exaggeration because only about 90 of the 118 occur naturally. The remaining 28 have been produced artificially by nuclear chemists using high-energy particle accelerators.

Furthermore, only 83 of the 90 or so naturally occurring elements are found on Earth in any appreciable abundance. Hydrogen is thought to account for approximately 75% of the observed mass in the universe; oxygen and silicon together account for 75% of the mass of Earth's crust; and oxygen, carbon, and hydrogen make up more than 90% of the mass of the human body (**Figure 1.1**). By contrast, there are probably less than 20 grams of the element francium (Fr) dispersed over the entire Earth at any one time. Francium is an unstable radioactive element, atoms of which are continually being formed and destroyed. We'll discuss **radioactivity** briefly in Sections 1.10 and 1.11 and then again in more detail in Chapter 22.

LOOKING AHEAD ...

As we'll see in Sections 1.10 and 1.11 and in Chapter 22, **radioactivity** is the spontaneous decay of certain unstable atoms with emission of some form of radiation.

For simplicity, chemists refer to specific elements using one- or two-letter symbols. As shown by the examples in Table 1.1, the first letter of an element's symbol is always capitalized and the second letter, if any, is lowercase. Many of the symbols are just the first one or two letters of the element's English name: H = hydrogen, C = carbon, Al = aluminum, and so forth. Other symbols are derived from Latin or other languages: Na = sodium (Latin, *natrium*), Pb = lead (Latin, *plumbum*), W = tungsten (German, *Wolfram*). The names, symbols, and other information about all 118 known elements are given inside the front cover of this book, organized in a format you've undoubtedly seen before called the *periodic table*.

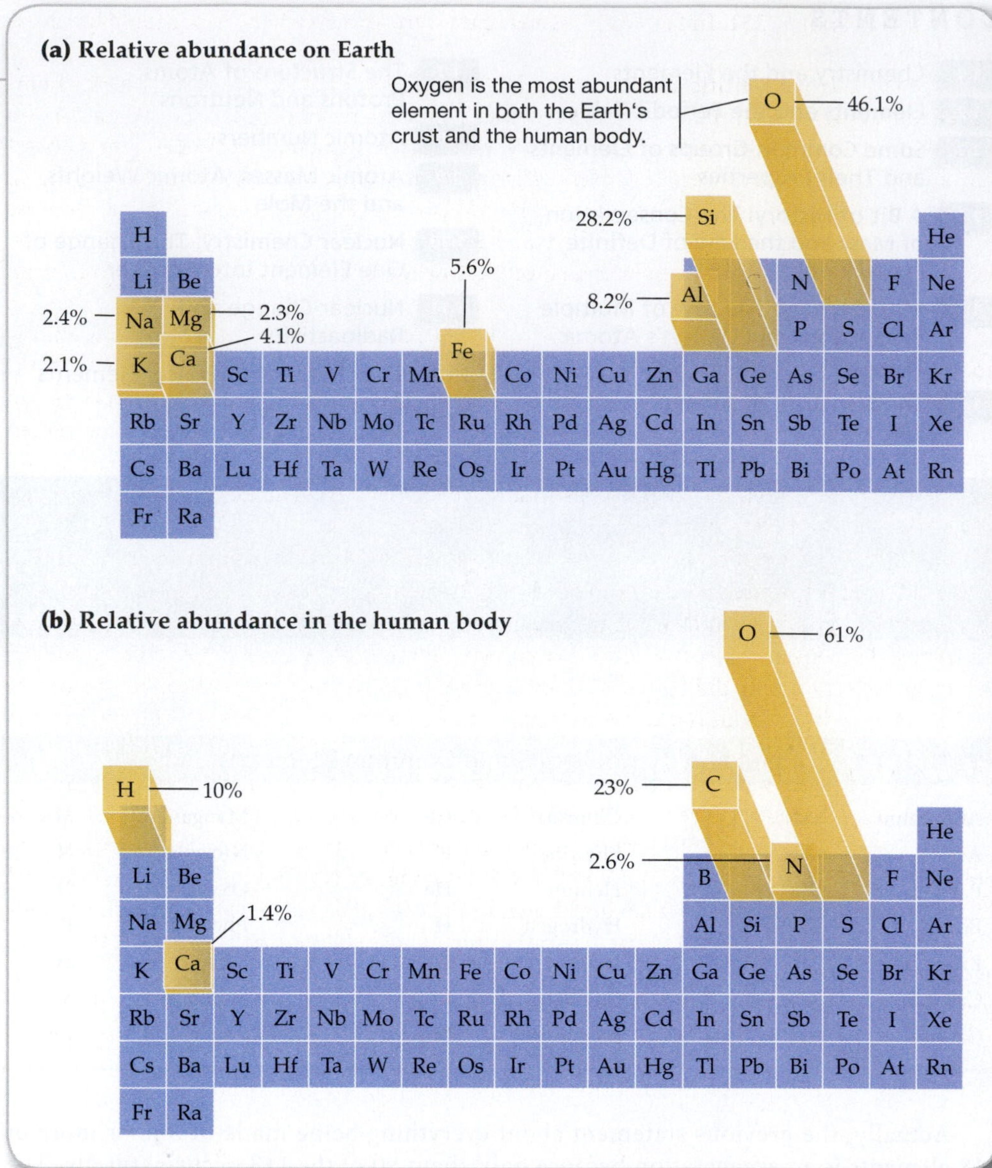

FIGURE 1.1

Estimated elemental composition by mass percent of (a) Earth's crust and (b) the human body. Only the major constituents are shown in each case; small amounts of many other elements are also present.

PROBLEM 1.1 Look at the alphabetical list of elements inside the front cover, and find the symbols for the following elements:
(a) Cadmium (used in batteries)
(b) Antimony (used in alloys with other metals)
(c) Americium (used in smoke detectors)

PROBLEM 1.2 Look at the alphabetical list of elements inside the front cover, and tell what elements the following symbols represent:

(a) Ag	**(b)** Rh	**(c)** Re
(d) Cs	**(e)** Ar	**(f)** As

1.2 ELEMENTS AND THE PERIODIC TABLE

Ten elements have been known since the beginning of recorded history: antimony (Sb), carbon (C), copper (Cu), gold (Au), iron (Fe), lead (Pb), mercury (Hg), silver (Ag), sulfur (S), and tin (Sn). The first "new" element to be found in several thousand years was arsenic (As), discovered in about 1251. In fact, only 24 elements were known when the United States was founded in 1776.

As the pace of scientific discovery quickened in the late 1700s and early 1800s, chemists began to look for similarities among elements that might allow general conclusions to be drawn. Particularly important among the early successes was Johann Döbereiner's observation in 1829 that there were several *triads*, or groups of three elements, that appeared to behave similarly. Calcium (Ca), strontium (Sr), and barium (Ba) form one such triad; chlorine (Cl), bromine (Br), and iodine (I) form another; and lithium (Li), sodium (Na), and potassium (K) form a third. By 1843, 16 such triads were known, and chemists were searching for an explanation.

▲ Left to right, samples of **chlorine**, **bromine**, and **iodine**, one of Döbereiner's triads of elements with similar chemical properties.

Numerous attempts were made in the mid-1800s to account for the similarities among groups of elements, but the breakthrough came in 1869 when the Russian chemist Dmitri Mendeleev created the forerunner of the modern **periodic table**. Mendeleev's creation is an ideal example of how a scientific theory develops. At first there is only disconnected information—a large number of elements and many observations about their properties and behavior. As more and more facts become known, people try to organize the data in ways that make sense until ultimately a consistent theory emerges.

A good theory must do two things: It must explain known facts, and it must make predictions about phenomena yet unknown. If the predictions are tested and found true, then the theory is a good one and will stand until additional facts require that it be modified or discarded. Mendeleev's theory about how known chemical information could be organized passed all tests. Not only did the periodic table arrange data in a useful and consistent way to explain known facts about chemical reactivity, it also led to several remarkable predictions that were later found to be correct.

Using the experimentally observed chemistry of the elements as his primary organizing principle, Mendeleev arranged the known elements in order of the relative masses of their constituent particles (called their *atomic weights*, as we'll see in Section 1.9) and then grouped them according to their chemical reactivity. On so doing, he realized that there were several "holes" in the table, some of which are shown in **Figure 1.2**. The chemical behavior of aluminum (relative mass ≈ 27.3) is similar to that of boron (relative mass ≈ 11), but there was no element known at the time that fit into the slot below aluminum. In the same way, silicon (relative mass ≈ 28) is similar in many respects to carbon (relative mass ≈ 12), but there was no element known that fit below silicon.

FIGURE 1.2

A portion of Mendeleev's periodic table. The table shows the relative masses of atoms as known at the time and some of the holes representing unknown elements.

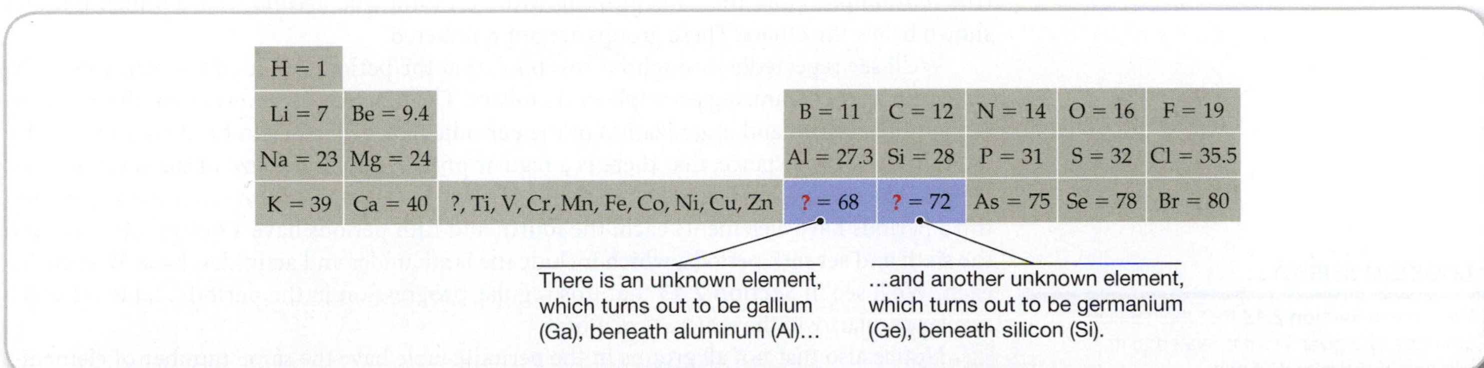

Looking at the holes in the table, Mendeleev predicted that two then-unknown elements existed and might be found at some future time. Furthermore, he predicted with remarkable accuracy what the properties of these unknown elements would be. The element immediately below aluminum, which he called *eka*-aluminum from a Sanskrit word meaning "first," should have a relative mass near 68 and should have a low melting point. Gallium, discovered in 1875, has exactly these properties. The element below silicon, which Mendeleev called *eka*-silicon, should have a relative mass near 72 and should be dark gray in color. Germanium, discovered in 1886, fits the description perfectly **(Table 1.2)**.

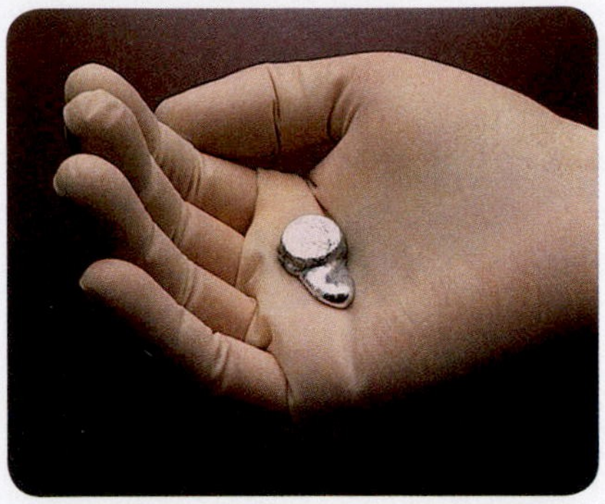

▲ Gallium is a shiny, low-melting metal.

▲ Germanium is a hard, gray semimetal.

TABLE 1.2 A Comparison of Predicted and Observed Properties for Gallium (*eka*-Aluminum) and Germanium (*eka*-Silicon)

Element	Property	Mendeleev's Prediction	Observed Property
	Relative mass	68	69.7
Gallium	Density	5.9 g/cm^3	5.91 g/cm^3
	Melting point	Low	29.8 °C
	Relative mass	72	72.6
Germanium	Density	5.5 g/cm^3	5.35 g/cm^3
	Color	Dark gray	Light gray

In the modern periodic table, shown in **Figure 1.3**, elements are placed on a grid with 7 horizontal rows, called **periods,** and 18 vertical columns, called **groups**. When organized in this way, *the elements in a given group have similar chemical properties*. Lithium, sodium, potassium, and the other metallic elements on the left side of the periodic table in group 1A behave similarly. Beryllium, magnesium, calcium, and the other elements in group 2A behave similarly. Fluorine, chlorine, bromine, and the other elements in group 7A on the right side of the table behave similarly, and so on throughout. (Mendeleev, by the way, was completely unaware of the existence of the group 8A elements—He, Ne, Ar, Kr, Xe, and Rn—because none were known when he constructed his table. All are colorless, odorless gases with little or no chemical reactivity, and none were discovered until 1894, when argon was first isolated.)

The overall form of the periodic table is well accepted, but chemists in different countries have historically used different conventions for labeling the groups. To resolve these difficulties, an international standard calls for numbering the groups from 1 to 18 going left to right. This standard has not yet found complete acceptance, however, and we'll continue to use the U.S. system of numbers and capital letters—group 3B instead of group 3 and group 7A instead of group 17, for example. Labels for the newer system are shown in Figure 1.3, above those for the U.S. system.

One further comment: There are actually 32 groups in the periodic table rather than 18, but to make the table fit manageably on a page, the 14 elements beginning with lanthanum (the *lanthanides*) and the 14 beginning with actinium (the *actinides*) are pulled out and shown below the others. These groups are not numbered.

We'll see repeatedly throughout this book that the periodic table of the elements is the most important organizing principle in chemistry. The time you take now to familiarize yourself with the layout and organization of the periodic table will pay you back later on. Notice in Figure 1.3, for instance, that there is a regular progression in the size of the seven periods (rows). The first period has only 2 elements, hydrogen (H) and helium (He); the second and third periods have 8 elements each; the fourth and fifth periods have 18 elements each; and the sixth and seventh periods, which include the lanthanides and actinides, have 32 elements each. We'll see in **Section 2.13** that this regular progression in the periodic table reflects a similar regularity in the structure of atoms.

LOOKING AHEAD ...

We'll see in **Section 2.13** that the detailed structure of a given atom is related to its position in the periodic table.

Notice also that not all groups in the periodic table have the same number of elements. The two larger groups on the left and the six larger groups on the right, labeled A in the U.S. system, are called the **main groups**. Most of the elements on which life is based—carbon, hydrogen, nitrogen, oxygen, and phosphorus, for instance—are main-group elements. The 10 smaller groups in the middle of the table, labeled B in the U.S. system, are called the **transition metal groups**. Most of the metals you're probably familiar with—iron, copper, zinc, and gold, for instance—are transition metals. And the 14 groups shown separately at the bottom of the table are called the **inner transition metal groups**.

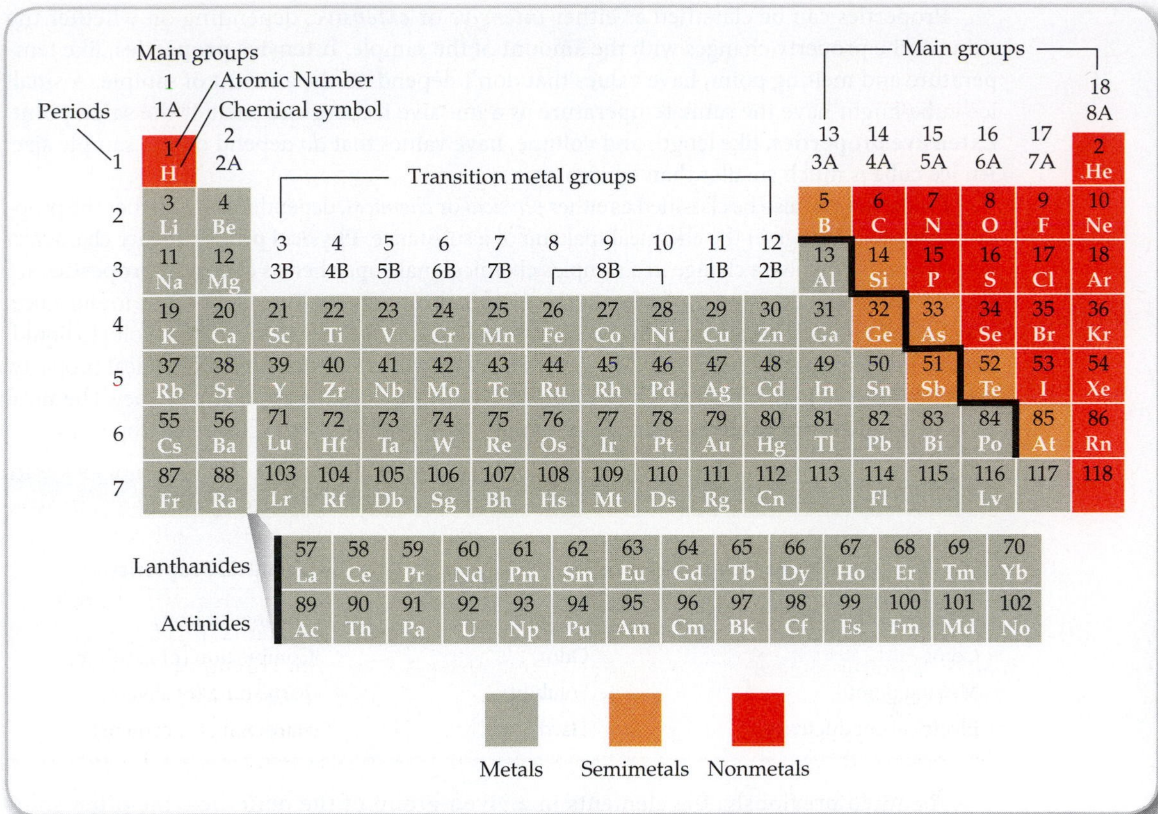

FIGURE 1.3

The periodic table. Each element is identified by a one- or two-letter symbol and is characterized by an atomic number. The table begins with hydrogen (H, atomic number 1) in the upper left-hand corner and continues to the yet unnamed element with atomic number 118. The 14 elements beginning with lanthanum (La, atomic number 57) and the 14 elements beginning with actinium (Ac, atomic number 89) are pulled out and shown below the others.

Elements are organized into 18 vertical columns, or *groups*, and 7 horizontal rows, or *periods*. The two groups on the left and the six on the right are the *main groups*; the ten in the middle are the *transition metal groups*. The 14 elements beginning with lanthanum are the *lanthanides*, and the 14 elements beginning with actinium are the *actinides*. Together, the lanthanides and actinides are known as the *inner transition metal groups*. Two systems for numbering the groups are shown above the top row and are explained in the text.

Those elements (except hydrogen) on the left side of the zigzag line running from boron (B) to astatine (At) are metals; those elements (plus hydrogen) to the right of the line are nonmetals; and seven of the nine elements abutting the line are metalloids, or semimetals.

1.3 SOME COMMON GROUPS OF ELEMENTS AND THEIR PROPERTIES

Any characteristic that can be used to describe or identify matter is called a **property**. Examples include volume, amount, odor, color, and temperature. Still other properties include such characteristics as melting point, solubility, and chemical behavior. For example, we might list some properties of sodium chloride (table salt) by saying it is a solid that melts at 1474 °F (or 801 °C), dissolves in water, and undergoes a chemical reaction to give an insoluble white precipitate when it comes into contact with a silver nitrate solution.

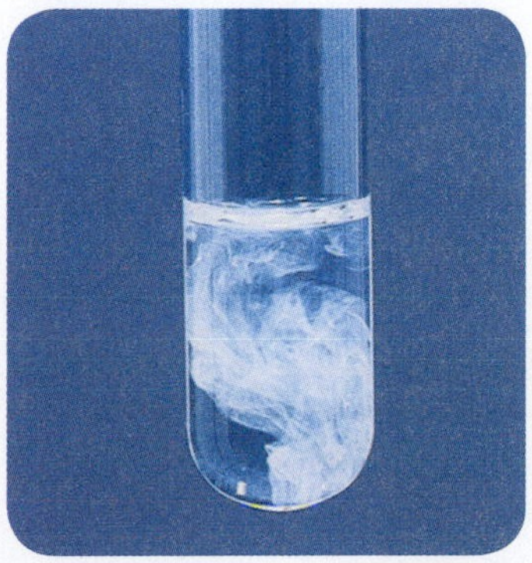

▲ Addition of a solution of silver nitrate to a solution of sodium chloride in water yields a white precipitate of solid silver chloride.

Properties can be classified as either *intensive* or *extensive*, depending on whether the value of the property changes with the amount of the sample. **Intensive properties**, like temperature and melting point, have values that don't depend on the amount of sample: A small ice cube might have the same temperature as a massive iceberg and melt at the same point. **Extensive properties,** like length and volume, have values that *do* depend on the sample size: An ice cube is much smaller than an iceberg.

Properties can also be classified as either *physical* or *chemical*, depending on whether the property involves a change in the chemical makeup of a substance. **Physical properties** are characteristics that don't involve a change in a sample's chemical makeup, whereas **chemical properties** are characteristics that do involve a change in chemical makeup. The melting point of ice, for instance, is a physical property because melting causes the water to change only in form, from solid to liquid, but not in chemical makeup. The rusting of an iron bicycle left in the rain is a chemical property, however, because iron combines with oxygen and moisture from the air to give the new chemical substance called rust. **Table 1.3** lists other examples of both physical and chemical properties.

TABLE 1.3 Some Examples of Physical and Chemical Properties

Physical Properties		Chemical Properties
Temperature	Amount	Rusting (of iron)
Color	Odor	Combustion (of gasoline)
Melting point	Solubility	Tarnishing (of silver)
Electrical conductivity	Hardness	Hardening (of cement)

As noted previously, the elements in a given group of the periodic table often show remarkable similarities in their chemical properties. Look at the following groups, for instance, to see some examples:

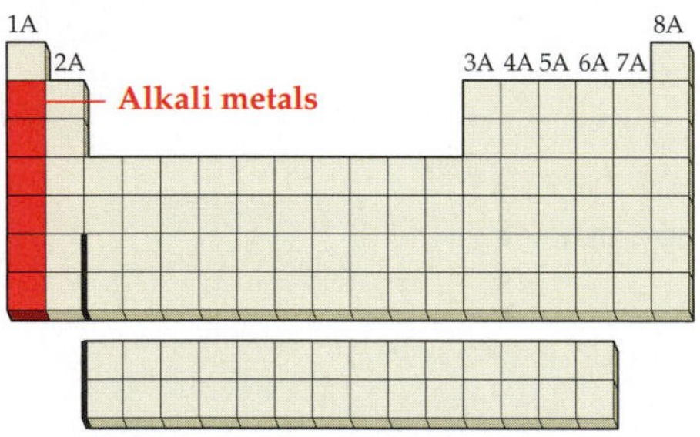

- **Group 1A—Alkali metals** Lithium (Li), sodium (Na), potassium (K), rubidium (Rb), and cesium (Cs) are soft, silvery metals. All react rapidly—often violently—with water to form products that are highly alkaline, or basic—hence the name *alkali metals*. Because of their high reactivity, the alkali metals are never found in nature in the pure state but only in combination with other elements. Francium (Fr) is also an alkali metal but, as noted previously, it is so rare that little is known about it.

  Note that group 1A also contains hydrogen (H) even though, as a colorless gas, it is completely different in appearance and behavior from the alkali metals. We'll see the reason for this classification in Section 2.13.

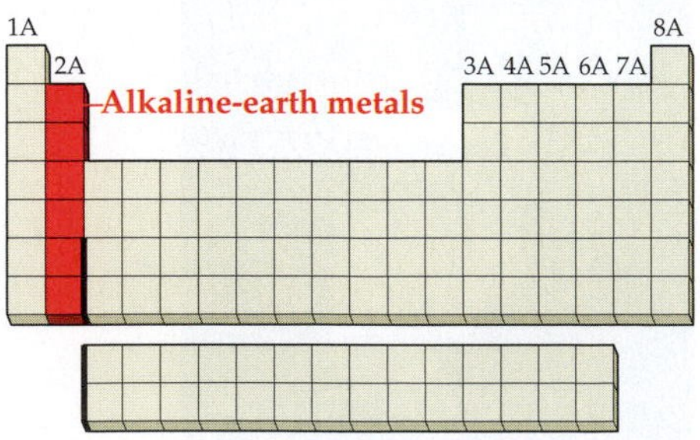

- **Group 2A—Alkaline-earth metals** Beryllium (Be), magnesium (Mg), calcium (Ca), strontium (Sr), barium (Ba), and radium (Ra) are also lustrous, silvery metals but are

▲ Sodium, one of the alkali metals, reacts violently with water to yield hydrogen gas and an alkaline solution.

▲ Magnesium, one of the alkaline-earth metals, burns in air.

less reactive than their neighbors in group 1A. Like the alkali metals, the alkaline-earth metals are never found in nature in the pure state.

- **Group 7A—Halogens** Fluorine (F), chlorine (Cl), bromine (Br), and iodine (I) are colorful, corrosive nonmetals. They are found in nature only in combination with other elements, such as with sodium in table salt (sodium chloride, NaCl). In fact, the group name *halogen* is taken from the Greek word *hals*, meaning "salt." Astatine (At) is also a halogen, but it exists in such tiny amounts that little is known about it.

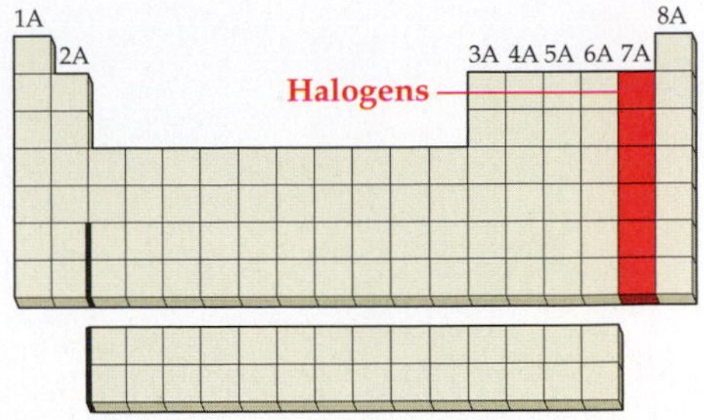

- **Group 8A—Noble gases** Helium (He), neon (Ne), argon (Ar), krypton (Kr), xenon (Xe), and radon (Rn) are colorless gases with very low chemical reactivity. Helium and neon don't combine with any other element; argon, krypton, and xenon combine with very few.

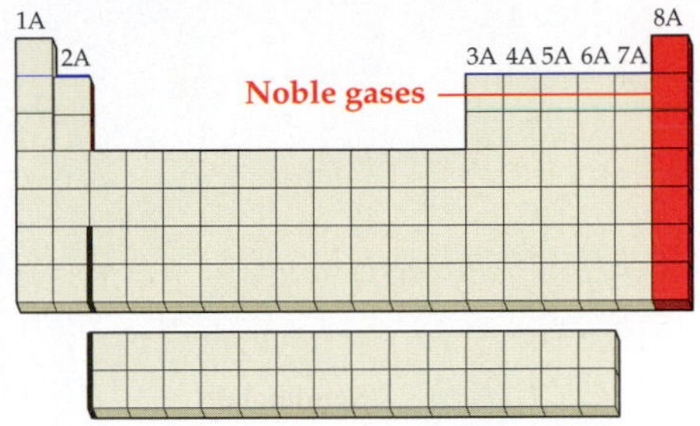

▲ Bromine, a halogen, is a corrosive dark red liquid at room temperature.

▲ Neon, one of the noble gases, is used in neon lights and signs.

As indicated previously in Figure 1.3, the elements of the periodic table are often divided into three major categories: metals, nonmetals, and semimetals.

- **Metals** Metals, the largest category of elements, are found on the left side of the periodic table, bounded on the right by a zigzag line running from boron (B) at the top to astatine (At) at the bottom. The metals are easy to characterize by their appearance. All except mercury are solid at room temperature, and most have the silvery shine we normally associate with metals. In addition, metals are generally malleable rather than brittle, can be twisted and drawn into wires without breaking, and are good conductors of heat and electricity.

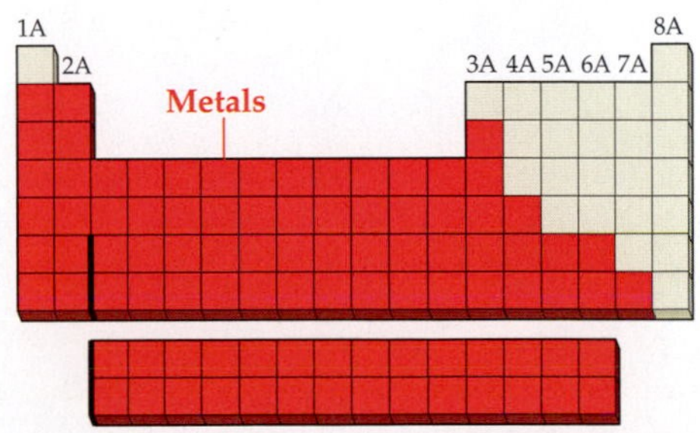

- **Nonmetals** Except for hydrogen, nonmetals are found on the right side of the periodic table and, like metals, are easy to characterize by their appearance. Eleven of the seventeen nonmetals are gases, one is a liquid (bromine), and only five are solids at room temperature (carbon, phosphorus, sulfur, selenium, and iodine). None are silvery in appearance, and several are brightly colored. The solid nonmetals are brittle rather than malleable and are poor conductors of heat and electricity.

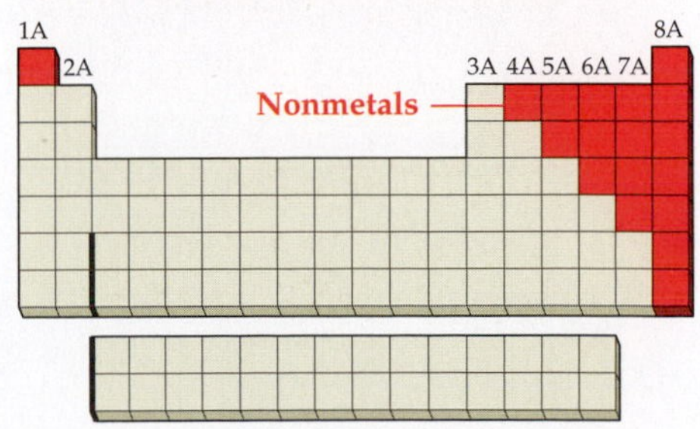

▲ Lead, aluminum, copper, gold, iron, and silver (clockwise from left) are typical metals. All conduct electricity and can be formed into wires.

▲ Bromine, carbon, phosphorus, and sulfur (clockwise from top left) are typical nonmetals. None conduct electricity or can be made into wires.

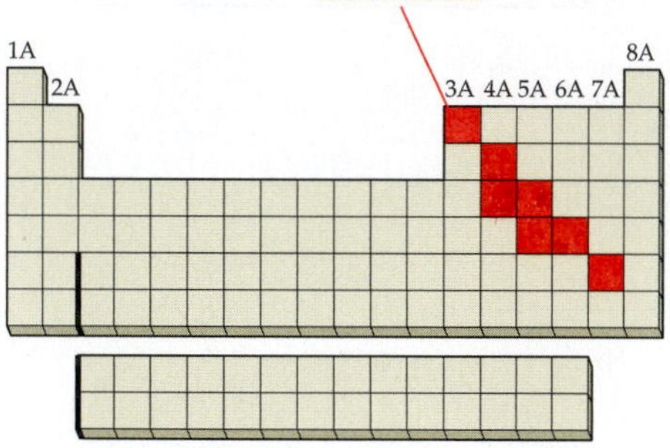

- **Semimetals** Seven of the nine elements adjacent to the zigzag boundary between metals and nonmetals—boron, silicon, germanium, arsenic, antimony, tellurium, and astatine—are called semimetals because their properties are intermediate between those of their metallic and nonmetallic neighbors. Although most are silvery in appearance and all are solid at room temperature, semimetals are brittle rather than malleable and tend to be poor conductors of heat and electricity. Silicon, for example, is a widely used *semiconductor*, a substance whose electrical conductivity is intermediate between that of a metal and an insulator.

PROBLEM 1.3 Identify the following elements as metals, nonmetals, or semimetals:
(a) Ti **(b)** Te **(c)** Se **(d)** Sc **(e)** At **(f)** Ar

CONCEPTUAL PROBLEM 1.4 The three so-called coinage metals are located near the middle of the periodic table. Identify them.

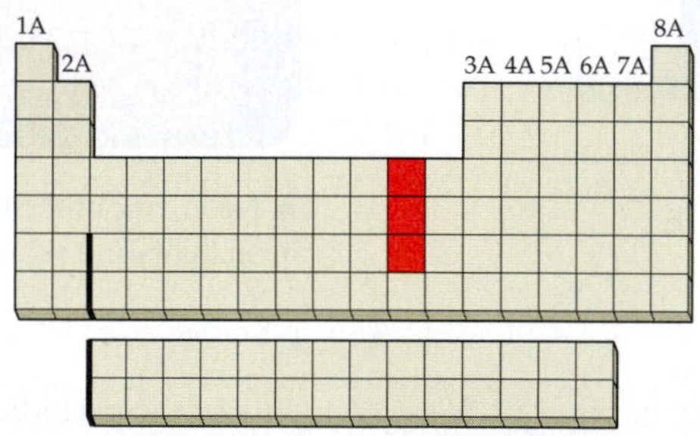

1.4 A BIT OF HISTORY: THE CONSERVATION OF MASS AND THE LAW OF DEFINITE PROPORTIONS

The Englishman Robert Boyle (1627–1691) is generally credited with being the first to study chemistry as a separate intellectual discipline and the first to carry out rigorous chemical experiments. Through a careful series of researches into the nature and behavior of gases, Boyle provided clear evidence for the atomic makeup of matter. In addition, Boyle was the first to clearly define an element as a substance that cannot be chemically broken down further and to suggest that a substantial number of different elements might exist. Atoms of these different elements, in turn, can join together in different ways to yield a vast number of different substances we call **chemical compounds**.

Progress in chemistry was slow in the decades following Boyle, and it was not until the work of Joseph Priestley (1733–1804) that the next great leap was made. Priestley prepared and isolated the gas oxygen in 1774 by heating the compound mercury oxide (HgO) according to the chemical equation we would now write as $2\ HgO \rightarrow 2\ Hg + O_2$.

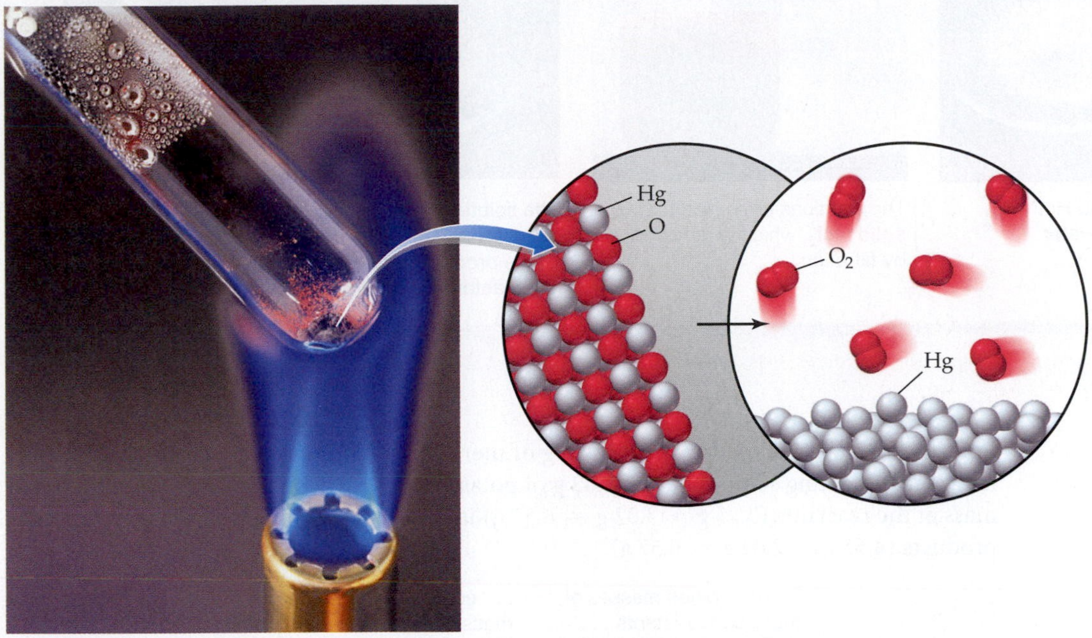

◀ Heating the red powder HgO causes it to decompose into the silvery liquid mercury and the colorless gas oxygen.

In this standard format for writing chemical transformations, each compound is described by its **chemical formula**, which lists the symbols of its constituent elements and uses subscripts to indicate the number of atoms of each. If no subscript is given, the number 1 is understood. Thus, sodium chloride (table salt) is written as NaCl, water as H_2O, and sucrose (table sugar) as $C_{12}H_{22}O_{11}$. A chemical reaction is written in a standard format called a **chemical equation**, in which the reactant substances undergoing change are written on the left, the product substances being formed are written on the right, and an arrow is drawn between them to indicate the direction of the chemical transformation.

Soon after Priestley's discovery, Antoine Lavoisier (1743–1794) showed that oxygen is the key substance involved in combustion. Furthermore, Lavoisier demonstrated with careful measurements that when combustion is carried out in a closed container, the mass of the combustion products exactly equals the mass of the starting reactants. When hydrogen gas burns and combines with oxygen to yield water (H_2O), for instance, the mass of the water formed is equal to the mass of the hydrogen and oxygen consumed. Called the **law of mass conservation**, this principle is a cornerstone of chemical science.

▶ **Law of mass conservation** Mass is neither created nor destroyed in chemical reactions.

It's easy to demonstrate the law of mass conservation by carrying out an experiment like that shown in **Figure 1.4**. If 3.25 g of mercury nitrate [$Hg(NO_3)_2$] and 3.32 g of potassium iodide (KI) are each dissolved in water and the solutions are mixed, an immediate chemical reaction occurs leading to formation of the insoluble orange solid mercury iodide (HgI_2).

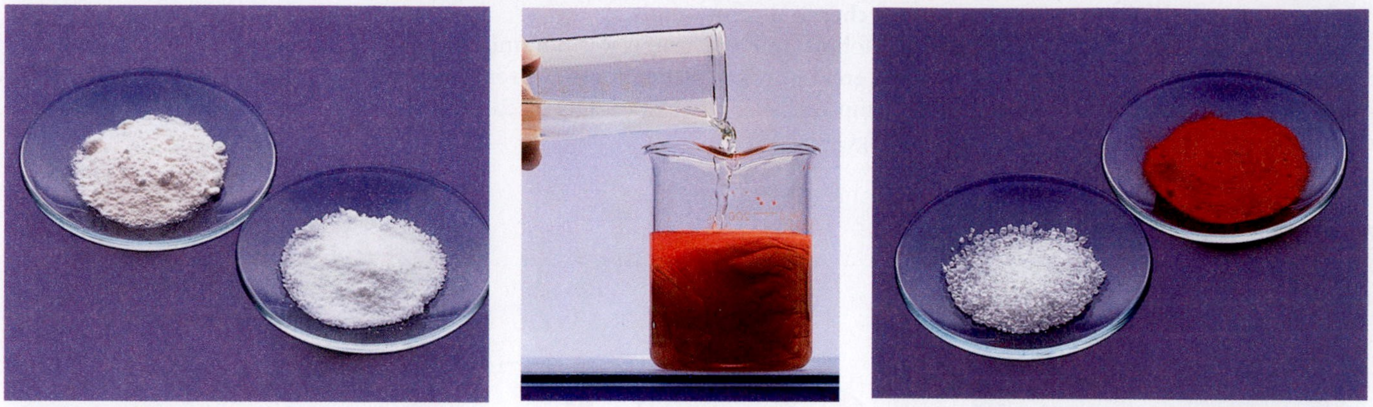

Known amounts of solid KI and solid $Hg(NO_3)_2$ are weighed and then dissolved in water.

The solutions are mixed to give **solid HgI_2**, which is removed by filtration.

The solution that remains is evaporated to give solid KNO_3. On weighing, the combined masses of the products equal the combined masses of the reactants.

FIGURE 1.4

An illustration of the law of mass conservation. In any chemical reaction, the combined masses of the final products equal the combined masses of the starting reactants.

Filtering the reaction mixture gives 4.55 g of mercury iodide, and evaporation of the water from the remaining solution leaves 2.02 g of potassium nitrate (KNO_3). Thus, the combined mass of the reactants (3.25 g + 3.32 g = 6.57 g) is exactly equal to the combined mass of the products (4.55 g + 2.02 g = 6.57 g).

The combined masses of these two reactants . . . | . . . equals the combined masses of these two products.

$$\underset{\text{Mercury nitrate}}{Hg(NO_3)_2} + \underset{\text{Potassium iodide}}{2\,KI} \longrightarrow \underset{\text{Mercury iodide}}{HgI_2} + \underset{\text{Potassium nitrate}}{2\,KNO_3}$$

Further investigations in the decades following Lavoisier led the French chemist Joseph Proust (1754–1826) to formulate a second fundamental chemical principle that we now call the **law of definite proportions**.

▶ **Law of definite proportions** Different samples of a pure chemical compound always contain the same proportion of elements by mass.

Every sample of water (H_2O) contains 1 part hydrogen and 8 parts oxygen by mass; every sample of carbon dioxide (CO_2) contains 3 parts carbon and 8 parts oxygen by mass; and so on. *Elements combine in specific proportions, not in random proportions.*

1.5 MORE HISTORY: THE LAW OF MULTIPLE PROPORTIONS AND DALTON'S ATOMIC THEORY

At the same time that Proust was formulating the law of definite proportions, the English schoolteacher John Dalton (1766–1844) was exploring along similar lines. His work led him to propose what has come to be called the **law of multiple proportions**.

▶ **Law of multiple proportions** Elements can combine in different ways to form different chemical compounds, whose mass ratios are simple whole-number multiples of each other.

▲ Copper metal reacts with nitric acid (HNO_3) to yield the brown gas NO_2.

The key to Dalton's proposition was his realization that the *same* elements sometimes combine in different ratios to give *different* chemical compounds. For example, nitrogen and oxygen can combine either in a 7:8 = 0.875 mass ratio to make the compound we know today as nitric oxide (NO) or in a 7:16 = 0.4375 mass ratio to make the compound we know as nitrogen dioxide (NO_2). The first compound contains exactly twice as much nitrogen (or half as much oxygen) as the second:

NO contains 7 g nitrogen per 8 g oxygen   N:O mass ratio = 7:8 = 0.875

NO_2 contains 7 g nitrogen per 16 g oxygen   N:O mass ratio = 7:16 = 0.4375

$$\frac{\text{N:O mass ratio in NO}}{\text{N:O mass ratio in NO}_2} = \frac{0.875}{0.4375} = 2$$

This result makes sense only if we assume that matter is composed of discrete atoms that have characteristic masses and combine with one another in specific and well-defined ways (**Figure 1.5**).

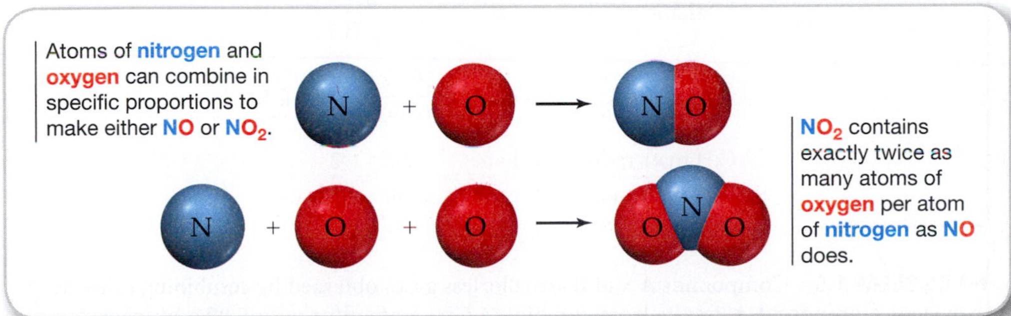

FIGURE 1.5

An illustration of Dalton's law of multiple proportions.

Taking all three laws together—the law of mass conservation, the law of definite proportions, and the law of multiple proportions—ultimately led Dalton to propose a new theory of matter that is still accepted today. He reasoned as follows:

- **Elements are made up of tiny particles called atoms.** Although Dalton didn't know what atoms were like, he nevertheless felt they were necessary to explain why there were so many different elements.
- **Each element is characterized by the mass of its atoms.** Dalton realized that there must be some feature that distinguishes the atoms of one element from those of another. Because Proust's law of definite proportions showed that elements always combine in specific mass ratios, Dalton reasoned that the distinguishing feature among atoms of different elements must be *mass*. Atoms of the same element have the same mass, but atoms of different elements have different masses.
- **The chemical combination of elements to make different chemical compounds occurs when atoms bond together in small whole-number ratios.** Only if whole numbers of atoms combine will different samples of a pure chemical compound always contain the same proportion of elements by mass (the law of definite proportions and the law of multiple proportions). Fractional parts of atoms are never involved in chemical reactions.

▲ These samples of sulfur and carbon have different masses but contain the same number of atoms.

- **Chemical reactions only rearrange how atoms are combined in chemical compounds; the atoms themselves don't change.** Dalton realized that atoms must be chemically indestructible for the law of mass conservation to be valid. If the same numbers and kinds of atoms are present in both reactants and products, then the masses of reactants and products must also be the same.

Not everything that Dalton proposed was correct. He thought, for instance, that water had the formula HO rather than H_2O. Nevertheless, his atomic theory of matter was ultimately accepted and came to form a cornerstone of modern chemical science.

▲ Sulfur burns with a bluish flame to yield colorless SO_2 gas.

Worked Example 1.1

Using the Law of Multiple Proportions

The two compounds methane and propane are both constituents of natural gas. A sample of methane contains 5.70 g of carbon atoms and 1.90 g of hydrogen atoms combined in a certain way, while a sample of propane contains 4.47 g of carbon atoms and 0.993 g of hydrogen atoms combined in a different way. Show that the two compounds obey the law of multiple proportions.

STRATEGY

Find the C:H (or H:C) mass ratio in each compound, and then compare the two ratios to see whether they are small whole-number multiples of each other.

SOLUTION

$$\text{Methane:} \quad \text{C:H mass ratio} = \frac{5.70\text{ g C}}{1.90\text{ g H}} = 3.00$$

$$\text{Propane:} \quad \text{C:H mass ratio} = \frac{4.47\text{ g C}}{0.993\text{ g H}} = 4.50$$

$$\frac{\text{C:H mass ratio in methane}}{\text{C:H mass ratio in propane}} = \frac{3.00}{4.50} = \frac{2}{3}$$

PROBLEM 1.5 Compounds **A** and **B** are colorless gases obtained by combining sulfur with oxygen. Compound **A** results from combining 6.00 g of sulfur with 5.99 g of oxygen, and compound **B** results from combining 8.60 g of sulfur with 12.88 g of oxygen. Show that the mass ratios in the two compounds are simple multiples of each other.

1.6 THE STRUCTURE OF ATOMS: ELECTRONS

Dalton's atomic theory is fine as far as it goes, but it leaves unanswered the obvious question: What is an atom made of? Dalton himself had no way of answering this question, and it was not until nearly a century later that experiments by the English physicist J. J. Thomson (1856–1940) provided some clues. Thomson's experiments involved the use of *cathode-ray tubes* (CRTs), early predecessors of the tubes still found in older televisions and computer displays.

As shown in **Figure 1.6a**, a cathode-ray tube is a sealed glass vessel from which the air has been removed and in which two thin pieces of metal, called *electrodes*, have been sealed. When a sufficiently high voltage is applied across the electrodes, an electric current flows through the tube from the negatively charged electrode (the *cathode*) to the positively charged electrode (the *anode*). If the tube is not fully evacuated but still contains a small amount of air or other gas, the flowing current is visible as a glow called a *cathode ray*. Furthermore, if the anode has a hole in it and the end of the tube is coated with a phosphorescent substance such as zinc sulfide, some of the rays emanating from the cathode pass through the hole and strike the end of the tube, where they are visible as a bright spot of light—exactly what happens in a CRT television screen or computer monitor.

Experiments by a number of physicists in the 1890s had shown that cathode rays can be deflected by bringing either a magnet or an electrically charged plate near the tube (**Figure 1.6b**). Because the beam is produced at a negative electrode and is deflected toward a positive plate, Thomson proposed that cathode rays must consist of tiny, negatively charged particles, which we now call **electrons**. Furthermore, because electrons are emitted from electrodes made of many different elements, all these different elements must contain electrons.

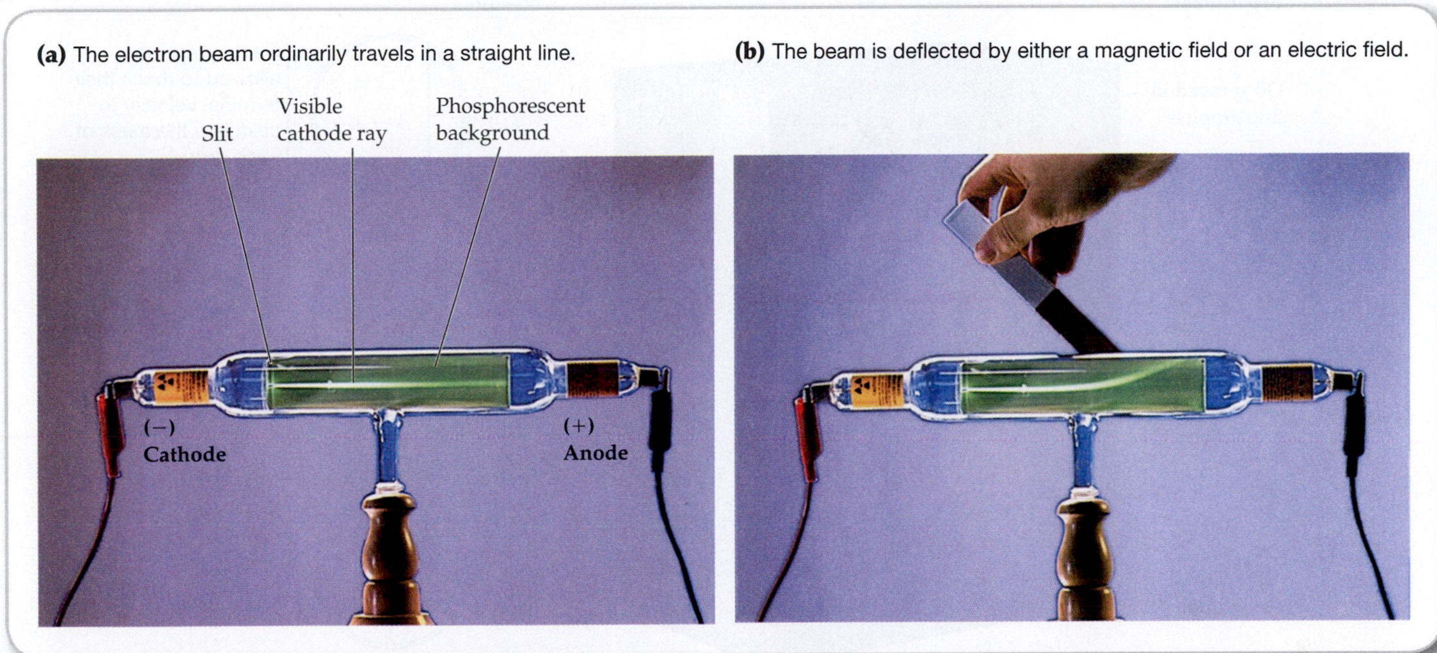

FIGURE 1.6

A cathode-ray tube. In a cathode-ray tube, a stream of electrons emitted from the negatively charged cathode passes through a slit, moves toward the positively charged anode, and is visualized by a phosphorescent strip.

Thomson reasoned that the amount of deflection of the electron beam due to a nearby magnetic or electric field should depend on three factors:

1. **The strength of the deflecting magnetic or electric field.** The stronger the magnet or the higher the voltage on the charged plate, the greater the deflection.
2. **The size of the negative charge on the electron.** The larger the charge on the particle, the greater its interaction with the magnetic or electric field and the greater the deflection.
3. **The mass of the electron.** The lighter the particle, the greater its deflection (just as a table-tennis ball is more easily deflected than a bowling ball).

By carefully measuring the amount of deflection caused by electric and magnetic fields of known strength, Thomson was able to calculate the ratio of the electron's electric charge to its mass: its *charge-to-mass ratio, e/m*. The modern value is

$$\frac{e}{m} = 1.758\,821 \times 10^{8}\ \text{C/g}$$

where e is the magnitude of the charge on the electron in the derived SI unit coulombs (C) and m is the mass of the electron in grams. (We'll say more about **coulombs** and **electrical charge** in Sections 17.3 and 17.14.) Note that because e is defined as a positive quantity, the actual (negative) charge on the electron is $-e$.

LOOKING AHEAD ...

We'll see in Sections 17.3 and 17.14 that **coulombs** and **electrical charge** are fundamental to *electrochemistry*—the area of chemistry involving batteries, fuel cells, electroplating of metals, and the like.

Thomson was able to measure only the ratio of charge to mass, not charge or mass itself, and it was left to the American R. A. Millikan (1868–1953) to devise a method for measuring the mass of an electron (**Figure 1.7**). In Millikan's experiment, a fine mist of oil was sprayed into a chamber and the tiny oil droplets were allowed to fall between two horizontal plates. Observing the spherical droplets through a telescopic eyepiece made it possible to determine how rapidly they fell through the air, which in turn allowed their masses to be calculated. The droplets were then given a negative charge by irradiating them with X rays. By applying a voltage to the plates, with the upper plate positive, it was possible to counteract the downward fall of the charged droplets and keep them suspended.

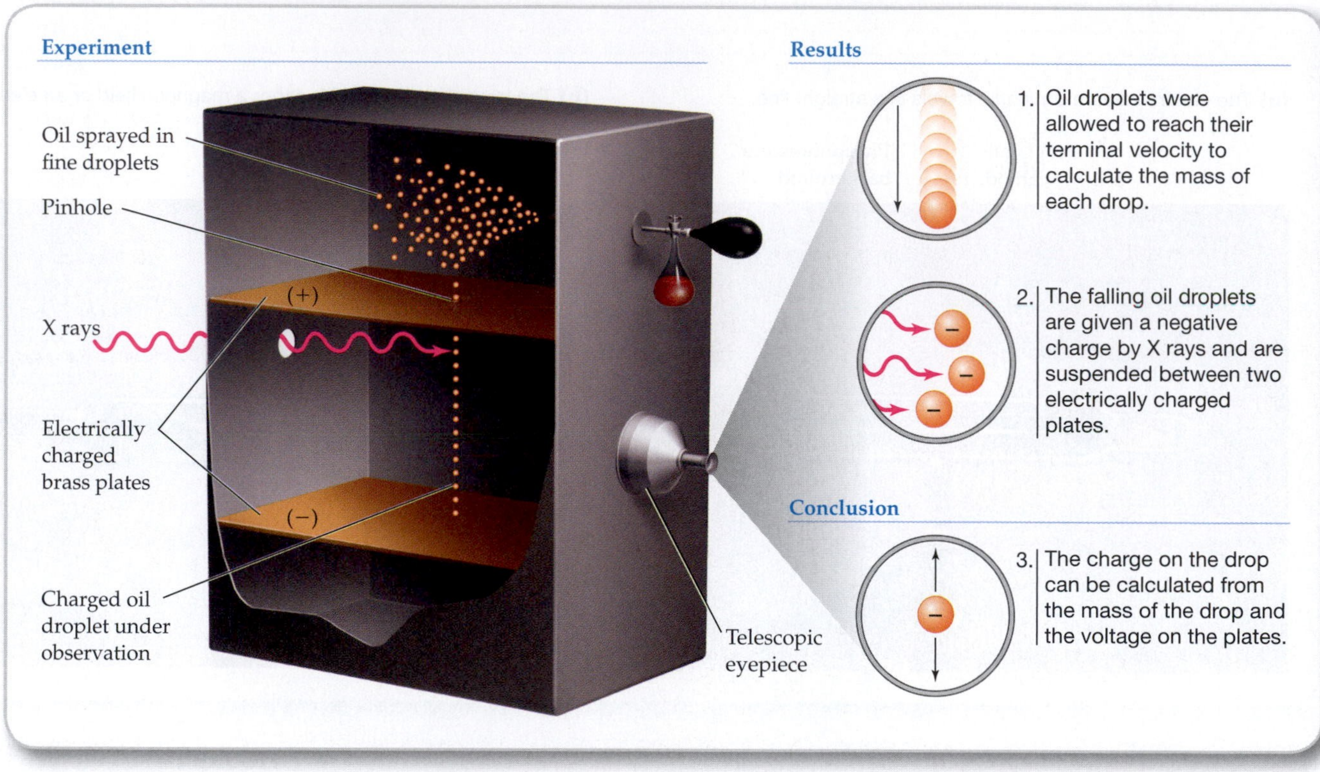

FIGURE 1.7

Millikan's oil drop experiment.

With the voltage on the plates and the mass of the droplets known, Millikan was able to show that the charge on a given droplet was always a small whole-number multiple of e, whose modern value is $1.602\,177 \times 10^{-19}$ C. Substituting the value of e into Thomson's charge-to-mass ratio then gives the mass m of the electron as $9.109\,383 \times 10^{-28}$ g:

$$\text{Because} \quad \frac{e}{m} = 1.758\,821 \times 10^{8}\ \text{C/g}$$

$$\text{then} \quad m = \frac{e}{1.758\,821 \times 10^{8}\ \text{C/g}} = \frac{1.602\,177 \times 10^{-19}\ \cancel{\text{C}}}{1.758\,821 \times 10^{8}\ \cancel{\text{C}}/\text{g}}$$

$$= 9.109\,383 \times 10^{-28}\ \text{g}$$

1.7 THE STRUCTURE OF ATOMS: PROTONS AND NEUTRONS

Think about the consequences of Thomson's cathode-ray experiments. Because matter is electrically neutral overall, the fact that the atoms in an electrode can give off *negatively* charged particles (electrons) must mean that those same atoms also contain *positively* charged particles for electrical balance. The search for those positively charged particles and for an overall picture of atomic structure led to a landmark experiment published in 1911 by the New Zealand physicist Ernest Rutherford (1871–1937).

Rutherford's work involved the use of ***alpha (α) particles***, a type of emission previously found to be given off by a number of naturally occurring radioactive elements, including radium, polonium, and radon. Rutherford knew that alpha particles are about 7000 times more massive than electrons and that they have a positive charge that is twice the magnitude of the charge on an electron, but opposite in sign.

When Rutherford directed a beam of alpha particles at a thin gold foil, he found that almost all the particles passed through the foil undeflected. A very small number, however (about 1 of every 20,000), were deflected at an angle, and a few actually bounced back toward the particle source (**Figure 1.8**).

LOOKING AHEAD ...

We'll see in Section 1.11 that **alpha (α) particles** are helium nuclei, He^{2+}, and are emitted by certain radioactive atoms.

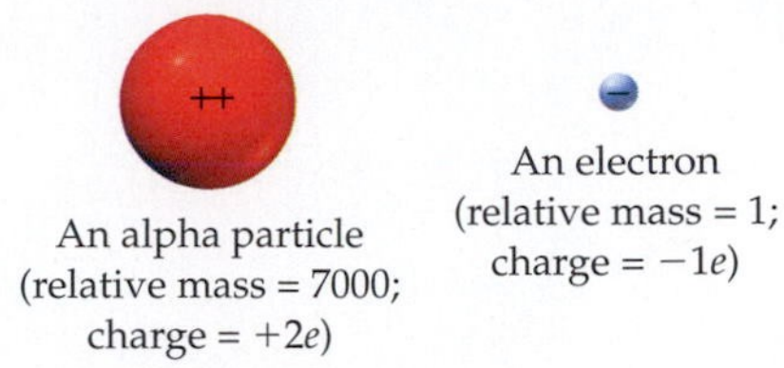

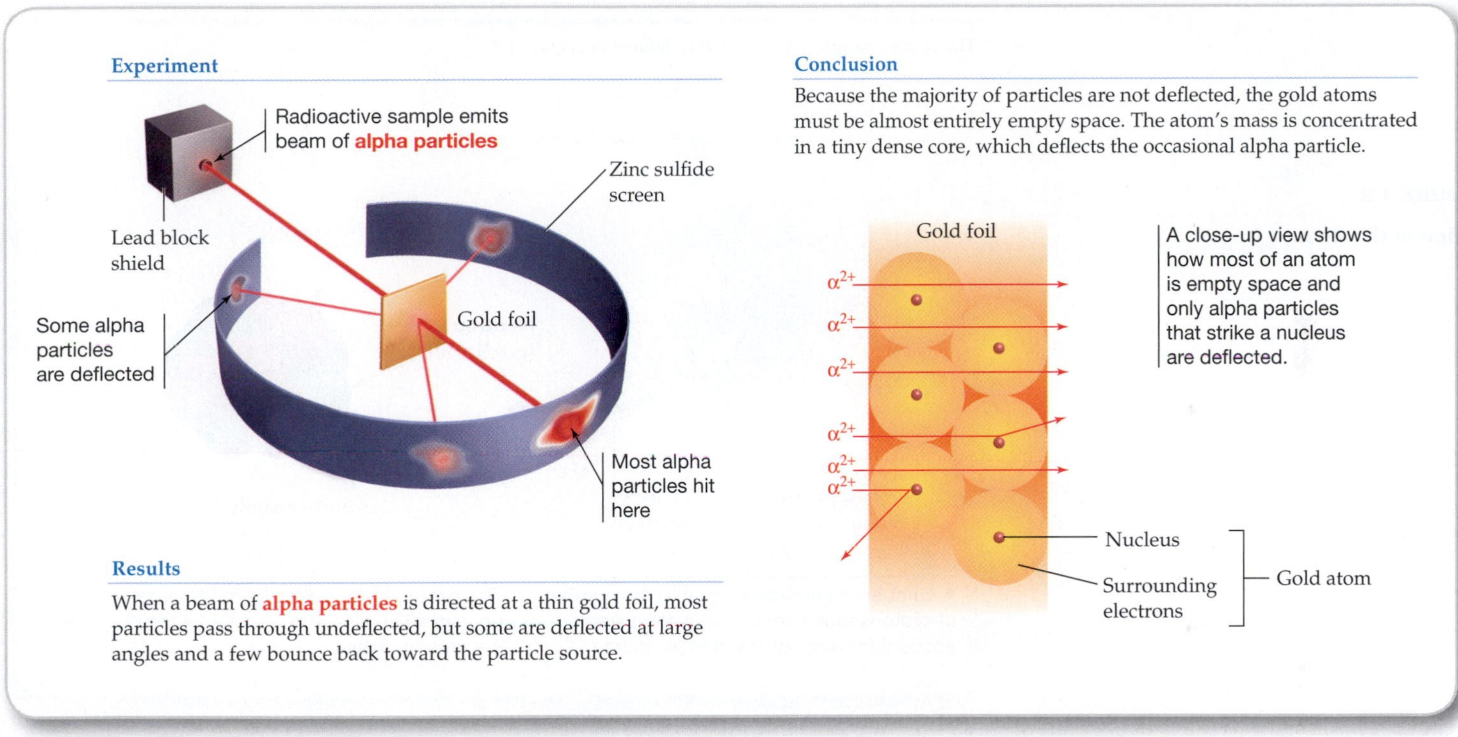

FIGURE 1.8

Rutherford's scattering experiment.

Rutherford explained his results by proposing that a metal atom must be almost entirely empty space and have its mass concentrated in a tiny central core that he called the **nucleus**. If the nucleus contains the atom's positive charges and most of its mass, and if the electrons are a relatively large distance away, then it is clear why the observed scattering results are obtained: Most alpha particles encounter empty space as they fly through the foil. Only when a positive alpha particle chances to come near a small but massive positive nucleus is it repelled strongly enough to make it bounce backward.

Modern measurements show that an atom has a diameter of roughly 10^{-10} m and that a nucleus has a diameter of about 10^{-15} m. It's difficult to imagine from these numbers alone, though, just how small a nucleus really is. For comparison purposes, if an atom were the size of a large domed stadium, the nucleus would be approximately the size of a small pea in the center of the playing field.

Further experiments by Rutherford and others between 1910 and 1930 showed that a nucleus is composed of two kinds of particles, called *protons* and *neutrons*. **Protons** have a mass of $1.672\,622 \times 10^{-24}$ g (about 1836 times that of an electron) and are positively charged. Because the charge on a proton is opposite in sign but equal in size to that on an electron, the numbers of protons and electrons in a neutral atom are equal. **Neutrons** ($1.674\,927 \times 10^{-24}$ g) are almost identical in mass to protons but carry no charge, and the number of neutrons in a nucleus is not directly related to the numbers of protons

▲ The relative size of the nucleus in an atom is roughly the same as that of a pea in the middle of this huge stadium.

and electrons. **Table 1.4** compares the three fundamental particles that make up an atom, and **Figure 1.9** gives an overall view of the atom.

TABLE 1.4 A Comparison of Subatomic Particles

	Mass		Charge	
Particle	grams	u*	coulombs	e
Electron	$9.109\,383 \times 10^{-28}$	$5.485\,798 \times 10^{-4}$	$-1.602\,177 \times 10^{-19}$	-1
Proton	$1.672\,622 \times 10^{-24}$	1.007 276	$+1.602\,177 \times 10^{-19}$	$+1$
Neutron	$1.674\,927 \times 10^{-24}$	1.008 665	0	0

* The unified atomic mass unit (u) is defined in Section 1.9.

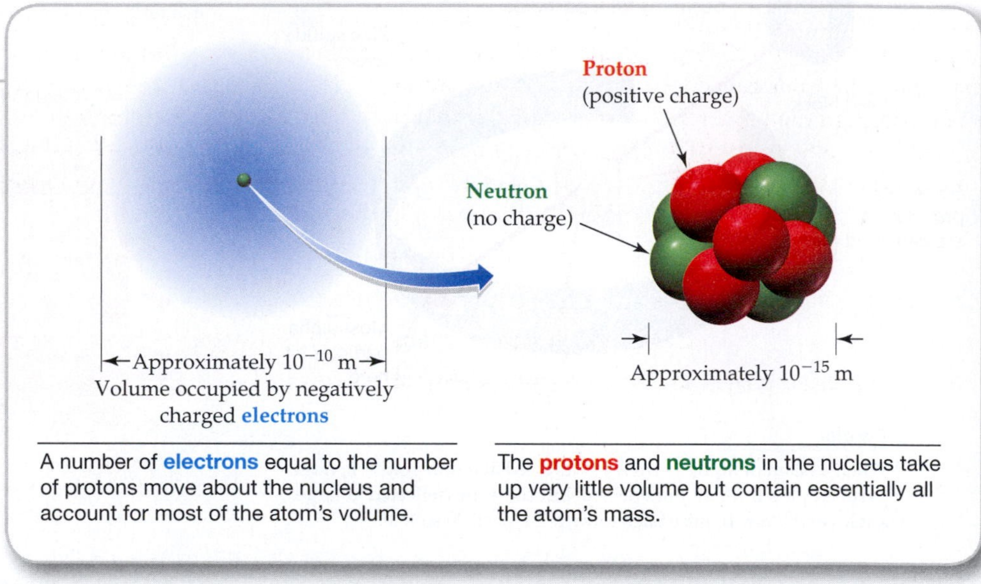

FIGURE 1.9

A view of the atom.

Worked Example 1.2

Calculations Using Atomic Size

Ordinary "lead" pencils contain no lead but instead are made of a form of carbon called graphite. If a pencil line is 0.35 mm wide and the diameter of a carbon atom is 1.5×10^{-10} m, how many atoms wide is the line?

STRATEGY

Begin with the known information, and set up an equation using appropriate conversion factors so that the unwanted units cancel. In this example, let's begin with the width of the pencil line in millimeters, convert to meters, and then divide the line width in meters by the diameter of a single atom in meters.

SOLUTION

$$\text{Atoms} = 0.35\ \text{mm} \times \frac{1\ \text{m}}{1000\ \text{mm}} \times \frac{1\ \text{atom}}{1.5 \times 10^{-10}\ \text{m}} = 2.3 \times 10^{6}\ \text{atoms}$$

✔ BALLPARK CHECK

A single carbon atom is about 10^{-10} m across, so it takes 10^{10} carbon atoms placed side by side to stretch 1 m, 10^{7} carbon atoms to stretch 1 mm, and about 0.3×10^{7} (or 3×10^{6}; 3 *million*) carbon atoms to stretch 0.35 mm. The estimate agrees with the solution.

PROBLEM 1.6 The gold foil Rutherford used in his scattering experiment had a thickness of approximately 0.005 mm. If a single gold atom has a diameter of 2.9×10^{-8} cm and the atoms are arranged so as to touch each other in a straight line, how many atoms thick was Rutherford's foil?

PROBLEM 1.7 A small speck of carbon the size of a pinhead contains about 10^{19} atoms, the diameter of a carbon atom is 1.5×10^{-10} m, and the circumference of the Earth at the equator is 40,075 km. How many times around the Earth would the atoms from this speck of carbon extend if they were laid side by side?

1.8 ATOMIC NUMBERS

Thus far, we've described atoms only in general terms and have not yet answered the most important question: What is it that makes one atom different from another? How, for example, does an atom of gold differ from an atom of carbon? The answer turns out to be quite simple: *Elements differ from one another according to the number of protons in their atoms' nuclei*, a value called the element's **atomic number (*Z*)**. That is, all atoms of a given element contain the same number of protons in their nuclei. All hydrogen atoms, atomic number 1, have 1 proton; all helium atoms, atomic number 2, have 2 protons; all carbon atoms, atomic number 6, have 6 protons; and so on. Of course, every neutral atom also contains a number of electrons equal to its number of protons.

▶ **Atomic number (*Z*)** = Number of protons in an atom's nucleus
= Number of electrons around an atom's nucleus

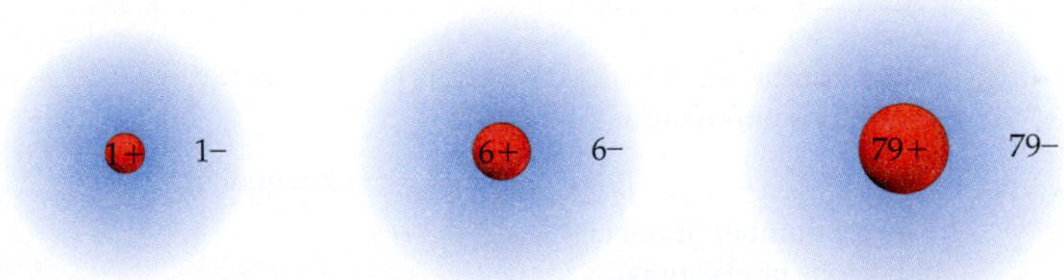

A hydrogen atom (1 proton; 1 electron) A carbon atom (6 protons; 6 electrons) A gold atom (79 protons; 79 electrons)

In addition to protons, the nuclei of all atoms other than hydrogen also contain neutrons. The sum of the numbers of protons (*Z*) and neutrons (*N*) in an atom is called the atom's **mass number (*A*)**. That is, $A = Z + N$.

▶ **Mass number (*A*)** = Number of protons (*Z*) + number of neutrons (*N*)

Most hydrogen atoms have 1 proton and no neutrons, so their mass number $A = 1 + 0 = 1$. Most helium atoms have 2 protons and 2 neutrons, so their mass number $A = 2 + 2 = 4$. Most carbon atoms have 6 protons and 6 neutrons, so their mass number $A = 6 + 6 = 12$; and so on. Except for hydrogen, atoms always contain at least as many neutrons as protons, although there is no simple way to predict how many neutrons a given atom will have.

Notice that we said *most* hydrogen atoms have mass number 1, most helium atoms have mass number 4, and most carbon atoms have mass number 12. In fact, different atoms of the same element can have different mass numbers depending on how many neutrons they have. Atoms with identical atomic numbers but different mass numbers are called **isotopes**. Hydrogen, for example, has three isotopes.

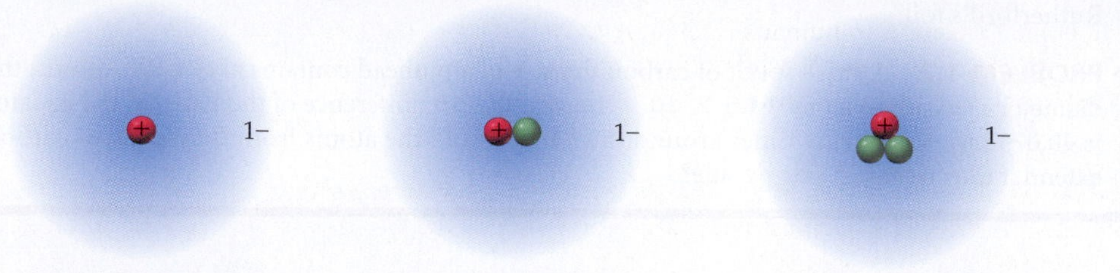

Protium—one proton (●) and no neutrons; mass number = 1

Deuterium—one proton (●) and one neutron (●); mass number = 2

Tritium—one proton (●) and two neutrons (●); mass number = 3

All hydrogen atoms have 1 proton in their nucleus (otherwise they wouldn't be hydrogen), but 99.985% of them have no neutrons. These hydrogen atoms, called *protium*, have mass number 1. In addition, 0.015% of hydrogen atoms, called *deuterium*, have 1 neutron and mass number 2. Still other hydrogen atoms, called *tritium*, have 2 neutrons and mass number 3. An unstable, radioactive isotope, tritium occurs only in trace amounts on Earth but is made artificially in nuclear reactors. As other examples, there are 15 known isotopes of nitrogen, only 2 of which occur naturally on Earth, and 25 known isotopes of uranium, only 3 of which occur naturally. In total, more than 3600 isotopes of the 118 known elements have been identified.

A specific isotope is represented by showing its element symbol, with its mass number as a left superscript and its atomic number as a left subscript. Thus, protium is represented as ${}^{1}_{1}H$, deuterium as ${}^{2}_{1}H$, and tritium as ${}^{3}_{1}H$. Similarly, the two naturally occurring isotopes of nitrogen are represented as ${}^{14}_{7}N$ (spoken as "nitrogen-14") and ${}^{15}_{7}N$ (nitrogen-15). The number of neutrons in an isotope is not given explicitly but can be calculated by subtracting the atomic number (subscript) from the mass number (superscript). For example, subtracting the atomic number 7 from the mass number 14 indicates that a ${}^{14}_{7}N$ atom has 7 neutrons.

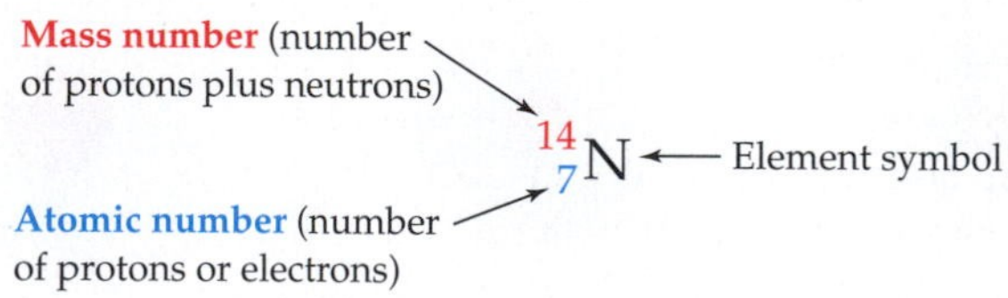

The number of neutrons in an atom has relatively little effect on the atom's chemical properties. The chemical behavior of an element is determined almost entirely by the number of electrons it has, which in turn is determined by the number of protons in its nucleus. All three isotopes of hydrogen therefore behave similarly (although not identically) in their chemical reactions.

▲ Uranium-235 is used as fuel in this nuclear-powered icebreaker.

Worked Example 1.3

Interpreting an Isotope Symbol

The isotope of uranium used to generate nuclear power is ${}^{235}_{92}U$. How many protons, neutrons, and electrons does an atom of ${}^{235}_{92}U$ have?

STRATEGY

The atomic number (subscript 92) in the symbol ${}^{235}_{92}U$ indicates the number of protons and electrons in the atom. The number of neutrons is the difference between the mass number (superscript 235) and the atomic number (92).

SOLUTION

An atom of ${}^{235}_{92}U$ has 92 protons, 92 electrons, and $235 - 92 = 143$ neutrons.

Worked Example 1.4

Writing an Isotope Symbol

Element **X** is toxic to humans in high concentration but is essential to life in low concentrations. Identify element **X**, whose atoms contain 24 protons, and write the symbol for the isotope of **X** that has 28 neutrons.

STRATEGY

The number of protons in an atom's nucleus is the element's atomic number *Z*, given in the periodic table. The mass number *A* is the sum of the atomic number and the number of neutrons.

SOLUTION

According to the periodic table, the element with atomic number 24 is chromium (Cr). The particular isotope of chromium in this instance has a mass number of 24 + 28 = 52 and is written $^{52}_{24}\text{Cr}$.

PROBLEM 1.8 The isotope $^{75}_{34}\text{Se}$ is used medically for the diagnosis of pancreatic disorders. How many protons, neutrons, and electrons does an atom of $^{75}_{34}\text{Se}$ have?

PROBLEM 1.9 Chlorine, one of the elements in common table salt (sodium chloride), has two main isotopes, with mass numbers 35 and 37. Look up the atomic number of chlorine, tell how many neutrons each isotope contains, and give the standard symbol for each.

PROBLEM 1.10 An atom of element **X** contains 47 protons and 62 neutrons. Identify the element, and write the symbol for the isotope in the standard format.

1.9 ATOMIC MASSES, ATOMIC WEIGHTS, AND THE MOLE

Pick up a pencil, and look at the small amount of tip visible. How many atoms (pencil lead is made of carbon) do you think are in the tip? One thing is certain: Atoms are so tiny that the number needed to make a visible sample is enormous. In fact, even the smallest speck of dust visible to the naked eye contains at least 10^{17} atoms. Thus, the mass in grams of a single atom is much too small a number for convenience and chemists therefore use a unit called either the *atomic mass unit (amu)* or, more correctly, the **unified atomic mass unit (u)**. Also known as a *dalton* (Da) in biological work, one unified atomic mass unit is defined as exactly 1/12 the mass of an atom of $^{12}_{6}\text{C}$ and is equal to $1.660\,538\,783 \times 10^{-24}$ g.

$$\textbf{1 u} = \frac{\text{Mass of one } ^{12}_{6}\text{C atom}}{12} = 1.660\,538\,783 \times 10^{-24}\ \text{g}$$

Because the mass of an atom's electrons is negligible compared to the mass of its protons and neutrons, defining 1 u as 1/12 the mass of a $^{12}_{6}\text{C}$ atom means that protons and neutrons both have a mass of almost exactly 1 u (Table 1.4 on page 42). Thus, the mass of a specific atom—called the atom's **atomic mass**—is numerically close to the atom's mass number. A $^{1}_{1}\text{H}$ atom, for instance, has an atomic mass of 1.007 825; a $^{235}_{92}\text{U}$ atom has an atomic mass of 235.043 930; and so forth. In practice, atomic masses are taken to be dimensionless and the unit u is understood rather than specified.

Most elements occur naturally as a mixture of different isotopes. Thus, if you look at the periodic table inside the front cover, you'll see listed below the symbol for each element a value called the element's *atomic weight.* Again, the unit u is understood but not specified.

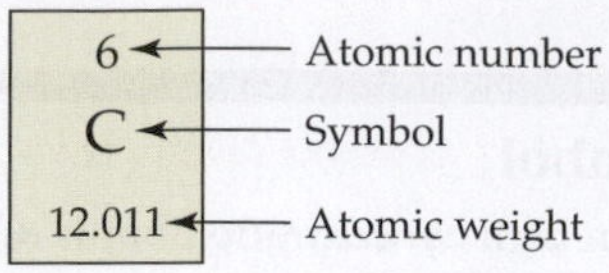

An element's **atomic weight** is the weighted average of the atomic masses of the element's naturally occurring isotopes. Carbon, for example, occurs on Earth as a mixture of two major isotopes, ${}^{12}_{6}C$ (98.89% natural abundance) and ${}^{13}_{6}C$ (1.11% natural abundance). Although the atomic mass of any *individual* carbon atom is either exactly 12 (a ${}^{12}C$ atom) or 13.0034 (a ${}^{13}C$ atom), the *average* atomic mass—that is, the atomic weight—of a large collection of carbon atoms is 12.011. A third carbon isotope, ${}^{14}_{6}C$, also exists naturally, but its abundance is so low that it can be ignored when calculating atomic weight.

$$\begin{aligned}\text{Atomic weight of C} &= (\text{Mass of } {}^{12}_{6}\text{C})(\text{Abundance of } {}^{12}_{6}\text{C}) + (\text{Mass of } {}^{13}_{6}\text{C})(\text{Abundance of } {}^{13}_{6}\text{C}) \\ &= (12)(0.9889) + (13.0034)(0.0111) \\ &= 11.867 + 0.144 = 12.011\end{aligned}$$

The real value of atomic weights is that they act as conversion factors between numbers of atoms and masses; that is, they allow us to *count* a large number of invisible atoms by *weighing* a sample of the substance. For instance, knowing that carbon has an atomic weight of 12.011 lets us calculate that a small pencil tip made of carbon and weighing 15 mg (1.5×10^{-2} g) contains 7.5×10^{20} atoms:

$$(1.5 \times 10^{-2}\ \cancel{g})\left(\frac{1\ \cancel{u}}{1.6605 \times 10^{-24}\ \cancel{g}}\right)\left(\frac{1\ \text{C atom}}{12.011\ \cancel{u}}\right) = 7.5 \times 10^{20}\ \text{C atoms}$$

As another example, the fact that the atomic weight of silver is 107.868 means that a silver ring weighing 1.872 g contains 1.045×10^{22} silver atoms:

$$(1.872\ \cancel{g})\left(\frac{1\ \cancel{u}}{1.6605 \times 10^{-24}\ \cancel{g}}\right)\left(\frac{1\ \text{Ag atom}}{107.868\ \cancel{u}}\right) = 1.045 \times 10^{22}\ \text{Ag atoms}$$

Let's think some more about what it means to be able to count atoms by using atomic weights. As we've just seen, the total mass of a given number of atoms is the atom's atomic weight times the number of atoms. Thus, if you have two samples of different elements that both contain the same number of atoms, the ratio of masses of the two samples must be the same as the ratio of their atomic weights. Take carbon and silver, for instance. The average mass ratio of *one* carbon atom (12.011) to *one* silver atom (107.868) is 12.011:107.868 = 0.111 35, so the mass ratio of *any* given number of C atoms to the same number of Ag atoms is *always* 0.111 35. Furthermore, the numerical value of the ratio is the same regardless of the mass units used since the units cancel. That is, 12.011 *mg* of carbon contains exactly the same number of atoms as 107.868 *mg* of silver, and 12.011 *kg* of carbon contains exactly the same number of atoms as 107.868 *kg* of silver.

$$\frac{12.011}{107.868} = \frac{12.011\ \cancel{mg}}{107.868\ \cancel{mg}} = \frac{12.011\ \cancel{g}}{107.868\ \cancel{g}} = \frac{12.011\ \cancel{kg}}{107.868\ \cancel{kg}} = 0.111\ 35$$

When referring to the enormous numbers of atoms that make up the visible amounts we typically deal with, chemists use the fundamental SI unit for amount called a *mole*, abbreviated *mol*. One **mole** of any element is the amount whose mass in grams, called its **molar mass,** is numerically equal to its atomic weight. One mole of carbon atoms has a mass of 12.011 g, one mole of silver atoms has a mass of 107.868 g, and so on. Molar mass thus acts as a conversion factor that lets you convert between mass in grams and number of atoms. *Whenever you have the same number of moles of different elements, you also have the same number of atoms.*

How many atoms are there in a mole? Experiments show that one mole of any element contains $6.022\ 141 \times 10^{23}$ atoms, a value called **Avogadro's number**, abbreviated N_A, after the Italian scientist who first recognized the importance of the mass/number relationship. Avogadro's number of atoms of any element—that is, one mole—has a mass in grams equal to the element's atomic weight.

▲ These samples of helium, sulfur, copper, and mercury each contain 1 mole. Do they have the same mass?

▶ **Mole** The amount of an element whose mass in grams is numerically equal to the element's atomic weight and thus contains Avogadro's number of atoms.

It's hard to grasp the magnitude of a quantity as large as Avogadro's number, but some comparisons might give you a sense of scale: The age of the universe in seconds (13.7 billion years, or 4.32×10^{17} s) is less than one millionth the size of Avogadro's number. The number of liters of water in the world's oceans (1.3×10^{21} L) is less than one hundredth the size of Avogadro's number. The mass of the Earth in kilograms (5.98×10^{24} kg) is only ten times Avogadro's number, and so on. We'll return to the **mole** throughout this book and see some of its many its uses in Chapter 6.

LOOKING AHEAD ...

We'll see in Chapter 6 that the **mole** is the fundamental unit used to calculate the relative amounts of chemical substances needed for carrying out chemical reactions.

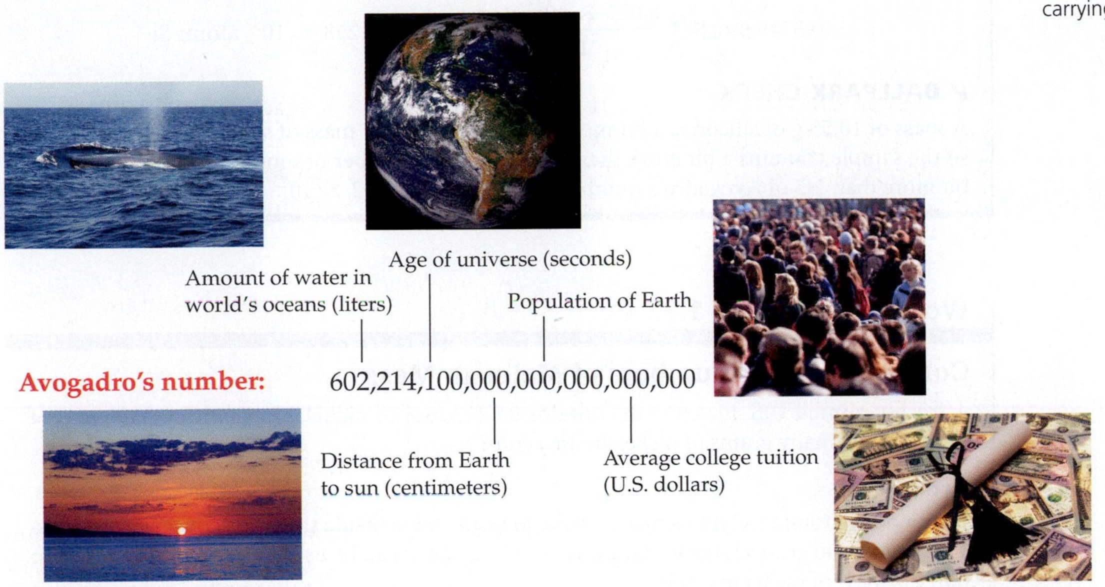

Worked Example 1.5

Calculating an Atomic Weight

Chlorine has two naturally occurring isotopes: ${}^{35}_{17}\text{Cl}$, with a natural abundance of 75.76% and an atomic mass of 34.969, and ${}^{37}_{17}\text{Cl}$, with a natural abundance of 24.24% and an atomic mass of 36.966. What is the atomic weight of chlorine?

STRATEGY

The atomic weight of an element is the weighted average of the atomic masses of the element's naturally occurring isotopes and is equal to the sum of the masses of each isotope times the natural abundance of that isotope. (To convert a percent abundance to a decimal abundance, divide by 100%, which is equivalent to dividing by 1.)

$$\text{Atomic weight} = (\text{Mass of } {}^{35}_{17}\text{Cl})(\text{Abundance of } {}^{35}_{17}\text{Cl}) + (\text{Mass of } {}^{37}_{17}\text{Cl})(\text{Abundance of } {}^{37}_{17}\text{Cl})$$

SOLUTION

$$\text{Atomic weight} = (34.969)\left(\frac{75.76\%}{100\%}\right) + (36.966)\left(\frac{24.24\%}{100\%}\right)$$

$$= (34.969)(0.7576) + (36.966)(0.2424) = 35.45$$

✔ BALLPARK CHECK

The atomic weight must be somewhere between 35 and 37, the masses of the two individual isotopes, and is closer to the mass of the more abundant isotope—perhaps about 35.5.

Worked Example 1.6

Converting from Mass to Numbers of Moles and Atoms

How many moles and how many atoms of silicon are in a sample weighing 10.53 g? The atomic weight of silicon is 28.0855.

STRATEGY

The fact that the atomic weight of silicon is 28.0855 means that 1 mol of silicon has a mass of 28.0855 g. Use this molar mass to convert between mass and number of moles, and then multiply by Avogadro's number to find the number of atoms.

SOLUTION

$$(10.53\ \cancel{\text{g Si}})\left(\frac{1\ \text{mol Si}}{28.0855\ \cancel{\text{g Si}}}\right) = 0.3749\ \text{mol Si}$$

$$(0.3749\ \cancel{\text{mol Si}})\left(\frac{6.022 \times 10^{23}\ \text{atoms Si}}{1\ \cancel{\text{mol Si}}}\right) = 2.258 \times 10^{23}\ \text{atoms Si}$$

✔ BALLPARK CHECK

A mass of 10.53 g of silicon is a bit more than 1/3 the molar mass of silicon (28.0855 g/mol), so the sample contains a bit more than 0.33 mol. This number of moles, in turns, contains a bit more than 1/3 of Avogadro's number of atoms, or about 2×10^{23} atoms.

Worked Example 1.7

Converting from Numbers of Moles to Mass

A freshly minted U.S. nickel coin contains 0.0213 mol of nickel, along with 0.0590 mol of copper. How many grams of nickel are in a coin?

STRATEGY

Look up the atomic weight of nickel in the periodic table inside the front cover of this book. That number in grams is the molar mass of nickel, which can be used as a conversion factor to find the mass of nickel in a coin.

SOLUTION

The atomic weight of nickel is 58.6934, so the mass of 1 mol of nickel (the molar mass) is 58.6934 g/mol. Use that molar mass as a conversion factor between moles and grams.

$$(0.0213\ \cancel{\text{mol Ni}})\left(58.6934\ \frac{\text{g Ni}}{\cancel{\text{mol Ni}}}\right) = 1.25\ \text{g Ni}$$

✔ BALLPARK CHECK

One mole of nickel has a mass of about 60 g, so 0.02 mol of nickel has a mass of about 0.02×60 g, or 1.2 g.

▶ PROBLEM 1.11 Copper metal has two naturally occurring isotopes: copper-63 (69.15%; atomic mass = 62.93) and copper-65 (30.85%; atomic mass = 64.93). Calculate the atomic weight of copper, and check your answer in a periodic table.

▶ PROBLEM 1.12 Based on your answer to Problem 1.11, how many atoms of copper are in an old penny made of pure copper and weighing 2.15 g?

▶ PROBLEM 1.13 What is the mass in grams of each of the following samples?
(a) 1.505 mol of Ti **(b)** 0.337 mol of Na **(c)** 2.583 mol of U

▶ PROBLEM 1.14 How many moles are in each of the following samples?
(a) 11.51 g of Ti **(b)** 29.127 g of Na **(c)** 1.477 kg of U

1.10 NUCLEAR CHEMISTRY: THE CHANGE OF ONE ELEMENT INTO ANOTHER

We saw in Section 1.4 at the beginning of this chapter, and we'll see repeatedly throughout the book, that the identities of atoms don't change when different chemical substances react with one another to give products. Yet it has been known for more than a century that atoms *can* change identity, resulting in the conversion of one element into another. Atomic weapons, nuclear power, and radioactive radon gas in our homes are all topics of societal importance, and all involve **nuclear chemistry**—the study of the properties and changes of atomic nuclei. We'll give just a brief introduction to this important subject in the concluding two sections of this chapter, but will return for a much more detailed look in Chapter 22.

▲ Nuclei of radioactive ^{235}U undergo spontaneous change in this nuclear power plant, forming different elements and releasing an enormous amount of energy in the process.

Take the element carbon, for instance. There are 15 known isotopes of carbon, two of which occur commonly (^{12}C and ^{13}C) and one of which (^{14}C) is produced in small amounts in the upper atmosphere by the action of cosmic rays on ^{14}N atoms. The remaining 12 carbon isotopes have been produced artificially. Only the two commonly occurring isotopes are stable; the other 13 undergo spontaneous changes to their nuclei. Carbon-14, for instance, slowly changes to give nitrogen-14 plus an electron, a process we can write as the following *nuclear equation* (as opposed to a chemical equation, described previously in Section 1.4).

$$^{14}_{6}C \longrightarrow {}^{14}_{7}N + {}^{0}_{-1}e$$

In a **nuclear equation**, the element symbols represent only the *nuclei* of atoms rather than the entire neutral atoms, so the subscript represents only the number of nuclear charges (protons). An emitted electron is written as ${}^{0}_{-1}e$, where the superscript 0 indicates that the mass of an electron is essentially zero when compared to that of a proton or neutron and the subscript indicates that the charge is -1.

Nuclear reactions, such as the spontaneous change of ^{14}C to ^{14}N, are distinguished from chemical reactions, such as that of sodium with chlorine, in several ways:

- A nuclear reaction involves a change in an atom's nucleus, usually producing a different element. A chemical reaction, by contrast, involves only a change in the way that different atoms are combined. A chemical reaction never changes the nuclei themselves or produces a different element.

- Different isotopes of an element have essentially the same behavior in chemical reactions but often have completely different behavior in nuclear reactions.
- The energy change accompanying a nuclear reaction is far greater than that accompanying a chemical reaction. The nuclear transformation of 1.0 g of uranium-235 ($^{235}_{92}U$) releases more than *one million* times as much energy as the chemical combustion of 1.0 g of methane.

1.11 NUCLEAR CHANGE AND RADIOACTIVITY

Scientists have known since 1896 that many nuclei are **radioactive**—they undergo a spontaneous change, or nuclear *decay*, and emit some form of *radiation*. Early studies of radioactivity by Ernest Rutherford in 1897 showed that there are three common types of radiation with markedly different properties: *alpha* (α), *beta* (β), and *gamma* (γ) *radiation*, named after the first three letters of the Greek alphabet.

Alpha (α) Radiation

Using the simple experiment shown in **Figure 1.10**, Rutherford found that α **radiation** consists of a stream of particles that are repelled by a positively charged electrode, attracted by a negatively charged electrode, and have a mass-to-charge ratio identifying them as helium nuclei, $^{4}_{2}He^{2+}$. Alpha particles thus consist of two protons and two neutrons.

FIGURE 1.10

The effect of an electric field on α, β, and γ radiation.

Experiment

The radioactive source in the shielded box emits radiation, which passes between two electrodes.

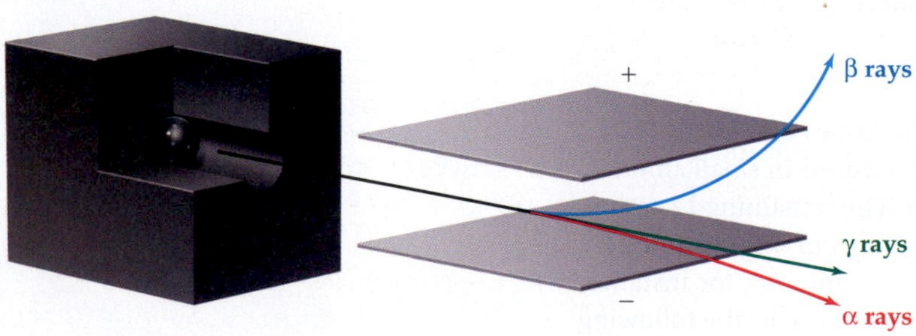

Results

Beta radiation is strongly deflected toward the positive electrode.

Gamma radiation is undeflected.

Alpha radiation is deflected toward the negative electrode.

Conclusion

Alpha particles and beta particles are both electrically charged, but they carry charges with opposite signs. Alpha particles have a positive charge, while beta particles have a negative charge. Gamma rays are not affected by an electric field and therefore have no electric charge.

Because the emission of an α particle from a nucleus results in a loss of two protons and two neutrons, it reduces the atomic number of the nucleus by 2 and reduces the mass number by 4. Alpha emission is particularly common for heavy radioactive isotopes, or **radioisotopes**. Uranium-238, for example, spontaneously emits an α particle and forms thorium-234.

$$^{238}_{92}U \longrightarrow {}^{4}_{2}He + {}^{234}_{90}Th$$

$^{238}_{92}U$: 92 protons + 146 neutrons = 238 nucleons

$^{4}_{2}He$: 2 protons + 2 neutrons = 4 nucleons

$^{234}_{90}Th$: 90 protons + 144 neutrons = 234 nucleons

Notice how the nuclear equation for the radioactive decay of uranium-238 is written. The equation is said to be *balanced* because the total number of neutrons and protons, collectively called **nucleons**, or nuclear particles, is the same on both sides of the equation and the number of charges on the nuclei and on any elementary particles (protons and electrons) is the same on both sides. In the decay of $^{238}_{92}U$ to give $^{4}_{2}He$ and $^{234}_{90}Th$, for instance, there are 238 nucleons and 92 nuclear charges on both sides of the equation.

Beta (β) Radiation

Further work by Rutherford in the late 1800s showed that β **radiation** consists of a stream of particles that are attracted to a positive electrode (Figure 1.10), repelled by a negative electrode, and have a mass-to-charge ratio identifying them as electrons, $^{0}_{-1}e$, often symbolized β^-. Beta emission occurs when a neutron in the nucleus spontaneously decays into a proton plus an electron, which is then ejected. The product nucleus has the same mass number as the starting nucleus because a neutron has turned into a proton, but it has a higher atomic number because it has the newly created proton. The reaction of iodine-131 to give xenon-131 is an example:

53 protons / 78 neutrons / 131 nucleons → $^{131}_{53}I$

54 protons / 77 neutrons / 131 nucleons → $^{131}_{54}Xe$

0 nucleons but −1 charge → $^{0}_{-1}e$

$$^{131}_{53}I \longrightarrow\ ^{131}_{54}Xe + \ ^{0}_{-1}e$$

Writing the emitted β particle as $^{0}_{-1}e$ in the nuclear equation makes clear the charge balance of the nuclear reaction. The subscript in the $^{131}_{53}I$ nucleus on the left (53) is balanced by the sum of the two subscripts on the right ($54 - 1 = 53$).

Gamma (γ) Radiation

Gamma (γ) radiation is unaffected by either electric or magnetic fields (Figure 1.10) and has no mass. Like visible light, ultraviolet rays, and X rays, γ radiation is simply **electromagnetic radiation** of very high energy, which we'll discuss in more detail in Section 2.1. Gamma radiation almost always accompanies α and β emission as a mechanism for the release of energy, but it is often not shown when writing nuclear equations because it changes neither the mass number nor the atomic number of the product nucleus.

LOOKING AHEAD ...

We'll see in Section 2.1 that a study of **electromagnetic radiation** was crucial historically to understanding the details of how electrons are distributed within an atom.

Positron Emission and Electron Capture

In addition to α, β, and γ radiation, two other types of radioactive decay processes also occur commonly: *positron emission* and *electron capture*. **Positron emission** occurs when a proton in the nucleus changes into a neutron plus an ejected *positron* ($^{0}_{1}e$ or β^+), a particle that can be thought of as a positive electron. A positron has the same mass as an electron but an opposite charge.

The result of positron emission is a decrease in the atomic number of the product nucleus but no change in the mass number. Potassium-40, for example, undergoes positron emission to yield argon-40, a nuclear reaction important in geology for dating rocks. Note once again that the sum of the two subscripts on the right of the nuclear equation ($18 + 1 = 19$) is equal to the subscript in the $^{40}_{19}K$ nucleus on the left.

19 protons / 21 neutrons / 40 nucleons → $^{40}_{19}K$

18 protons / 22 neutrons / 40 nucleons → $^{40}_{18}Ar$

0 nucleons but +1 charge → $^{0}_{1}e$

$$^{40}_{19}K \longrightarrow\ ^{40}_{18}Ar + \ ^{0}_{1}e$$

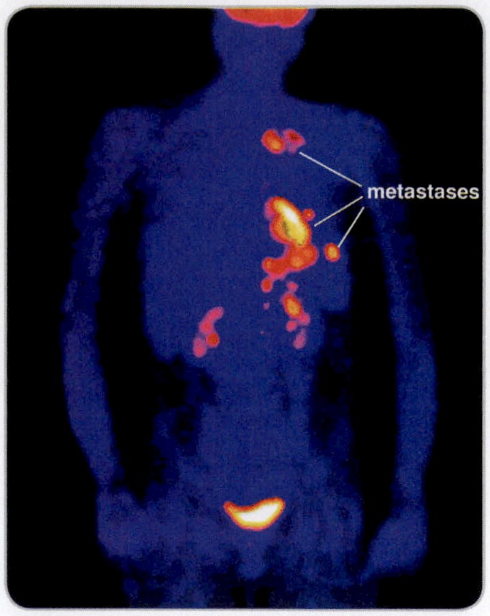

▲ The PET scan of this patient shows the presence of recurrent lung carcinoma.

You might already know that the PET scans used in medical imaging stands for *positron emission tomography*. A chemical compound containing a positron-emitting isotope, usually ^{18}F, is injected into the body and accumulates at a certain site, such as in a tumor. When decay of the ^{18}F nucleus occurs, the emitted positron reacts with a nearby electron and is instantly annihilated, releasing gamma rays whose position in the body can be detected.

Electron capture is a process in which the nucleus captures one of the surrounding electrons in an atom, thereby converting a proton into a neutron. The mass number of the product nucleus is unchanged, but the atomic number decreases by 1, just as in positron emission. The conversion of mercury-197 into gold-197 is an example:

80 protons
117 neutrons
197 nucleons

Inner-shell electron

79 protons
118 neutrons
197 nucleons

$$^{197}_{80}Hg + ^{0}_{-1}e \longrightarrow ^{197}_{79}Au$$

Characteristics of the different kinds of radioactive decay processes are summarized in **Table 1.5**.

TABLE 1.5 A Summary of Radioactive Decay Processes

Process	Symbol	Change in Atomic Number	Change in Mass Number	Change in Neutron Number
Alpha emission	$^{4}_{2}He$ or α	−2	−4	−2
Beta emission	$^{0}_{-1}e$ or β^-	+1	0	−1
Gamma emission	$^{0}_{0}\gamma$ or γ	0	0	0
Positron emission	$^{0}_{1}e$ or β^+	−1	0	+1
Electron capture	E. C.	−1	0	+1

As noted at the beginning of the previous section, we've given only a brief introduction to the subject of nuclear chemistry at this point. There's much more to come on this important topic in Chapter 22.

Worked Example 1.8

Balancing Nuclear Equations

Balance the following nuclear equations by identifying the species represented by the question marks:

(a) Alpha emission from curium-242: $^{242}_{96}Cm \rightarrow ^{4}_{2}He + ?$

(b) Beta emission from magnesium-28: $^{28}_{12}Mg \rightarrow ^{0}_{-1}e + ?$

(c) Positron emission from xenon-118: $^{118}_{54}Xe \rightarrow ^{0}_{1}e + ?$

STRATEGY

The key to balancing nuclear equations is to make sure that the number of nucleons is the same on both sides of the equation and that the number of nuclear charges plus electron or positron charges is the same.

SOLUTION

(a) In α emission, the mass number decreases by 4 and the atomic number decreases by 2, giving plutonium-238: ${}^{242}_{96}Cm \rightarrow {}^{4}_{2}He + {}^{238}_{94}Pu$

(b) In β emission, the mass number is unchanged and the atomic number increases by 1, giving aluminum-28: ${}^{28}_{12}Mg \rightarrow {}^{0}_{-1}e + {}^{28}_{13}Al$

(c) In positron emission, the mass number is unchanged and the atomic number decreases by 1, giving iodine-118: ${}^{118}_{54}Xe \rightarrow {}^{0}_{1}e + {}^{118}_{53}I$

PROBLEM 1.15 Write a balanced nuclear equation for each of the following processes:

(a) Beta emission from ruthenium-106: ${}^{106}_{44}Ru \rightarrow {}^{0}_{-1}e + ?$

(b) Alpha emission from bismuth-189: ${}^{189}_{83}Bi \rightarrow {}^{4}_{2}He + ?$

(c) Electron capture by polonium-204: ${}^{204}_{84}Po + {}^{0}_{-1}e \rightarrow ?$

PROBLEM 1.16 What particle is produced by decay of thorium-214 to radium-210?

$$ {}^{214}_{90}Th \rightarrow {}^{210}_{88}Ra + ? $$

CONCEPTUAL PROBLEM 1.17 Identify the isotopes involved, and tell what type of decay process is occurring in the following nuclear reaction:

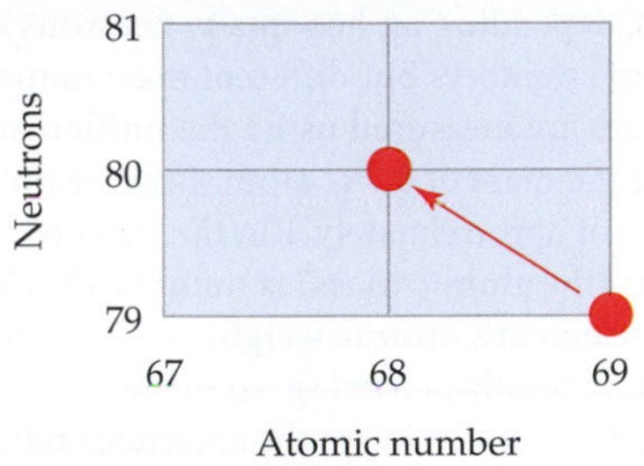

FYI THE ORIGIN OF CHEMICAL ELEMENTS

Cosmologists theorize that the universe began some 13.7 billion years ago in an extraordinary event they call the big bang. Initially, the temperature must have been inconceivably high, but after 1 second, the temperature had dropped to about 10^{10} K and elementary particles began to form: protons, neutrons, and electrons, as well as positrons and *neutrinos*—neutral particles with a mass much less than that of an electron. After 3 minutes, the temperature had dropped to 10^9 K, and protons began fusing with neutrons to form helium nuclei (${}^{4}_{2}He^{2+}$). Matter remained in this form for many millions of years until the expanding universe had cooled to about 10,000 K. Electrons were then able to bind to protons and to helium nuclei, forming stable hydrogen and helium atoms.

The attractive force of gravity acting on regions of higher-than-average density slowly produced massive local concentrations of matter and ultimately formed billions of galaxies, each with many billions of stars. As the gas clouds of hydrogen and helium condensed under gravitational attraction and stars formed, their temperatures reached 10^7 K and their densities reached 100 g/cm^3. Protons and neutrons again fused to yield helium nuclei, generating vast amounts of heat and light.

Most of these early stars probably burned out after a few billion years, but a few were so massive that, as their nuclear fuel diminished, gravitational attraction caused a rapid contraction leading to still higher core temperatures and higher densities—up to 5×10^8 K and 5×10^5 g/cm^3. Much larger nuclei were now formed, including carbon, oxygen, silicon, magnesium, and iron. Ultimately, these massive stars underwent gravitational collapse, resulting in the synthesis of still heavier elements and explosions visible throughout the universe as *supernovas*.

▲ The stars in the Milky Way galaxy condensed from gas clouds under gravitational attraction.

▲ In 1667, the instantaneous gravitational collapse of the massive star Cassiopeia A resulted in a supernova explosion, whose remnants are still visible.

Matter from exploding supernovas was blown throughout the galaxy, forming a new generation of stars and planets. Our own sun and solar system formed only about 4.5 billion years ago from matter released by former supernovas. Except for hydrogen and helium, all the atoms in our bodies, our planet, and our solar system were created more than 5 billion years ago in exploding stars. We are made of stardust.

PROBLEM 1.18 What are the first two elements thought to be formed in the big bang?

PROBLEM 1.19 The following two processes are among the nuclear reactions that power our Sun. Balance both.

(a) ${}^{1}_{1}H + {}^{1}_{1}H \rightarrow {}^{2}_{1}H + ?$

(b) ${}^{1}_{1}H + {}^{2}_{1}H \rightarrow ?$

SUMMARY

All matter is formed from atoms of one or more of 118 presently known **elements**—fundamental substances that can't be chemically broken down. Elements are symbolized by one- or two-letter abbreviations and can be organized into a **periodic table** with **groups** (columns) and **periods** (rows). Elements in the same group of the periodic table show similar chemical behavior. The two larger groups on the left and the six larger groups on the right of the table are called the **main groups,** the 10 smaller ones in the middle of the table are called the **transition metal groups,** and the 14 shown separately at the bottom of the table are called the **inner transition metal groups**. Elements are classified as **metals, nonmetals,** or **semimetals**.

The characteristics, or **properties**, that are used to describe matter can be classified in several ways. **Physical properties** are those that can be determined without changing the chemical composition of the sample, whereas **chemical properties** are those that do involve a chemical change in the sample. **Intensive properties** are those whose values do not depend on the size of the sample, whereas **extensive properties** are those whose values do depend on sample size.

Elements are made of tiny particles called **atoms**, which can combine in simple numerical ratios according to the **law of multiple proportions**. Atoms, in turn, are composed of three fundamental particles: **protons** are positively charged, **electrons** are negatively charged, and **neutrons** are neutral. According to the nuclear model of an atom, protons and neutrons are clustered into a dense core called the **nucleus**, while electrons move around the nucleus at a relatively great distance.

Elements differ from one another according to how many protons their atoms contain, a value called the **atomic number** (Z) of the element. The sum of an atom's protons and neutrons is its **mass number** (A). Although all atoms of a specific element have the same atomic number, different atoms of an element can have different mass numbers, depending on how many neutrons they have. Atoms with identical atomic numbers but different mass numbers are called **isotopes**. Atomic masses are measured using the **unified atomic mass unit** (**u**), defined as 1/12 the mass of a ${}^{12}C$ atom. Because both protons and neutrons have a mass of approximately 1 u, the mass of an atom in unified atomic mass units (the **atomic mass**) is numerically close to the atom's mass number. The element's **atomic weight** is a weighted average of the atomic masses of its naturally occurring isotopes.

When referring to the enormous numbers of atoms that make up visible amounts of matter, the fundamental SI unit called a *mole* is used. One **mole** is the amount whose mass in grams, called its **molar mass**, is numerically equal to the atomic weight. Numerically, one mole of any element contains 6.022×10^{23} atoms, a value called **Avogadro's number** (N_A)

Nuclear chemistry is the study of the properties and reactions of atomic nuclei. Nuclear reactions differ from chemical reactions in that they involve a change in an atom's nucleus, producing a different element. Nuclear reactions are written using balanced **nuclear equations**, in which the element symbols represent only the nuclei rather than neutral atoms.

Radioactivity is the spontaneous emission of radiation from an unstable nucleus. **Alpha (α) radiation** consists of helium nuclei, small particles containing two protons and two neutrons (${}^{4}_{2}He^{2+}$). **Beta (β) radiation** consists of electrons (${}^{0}_{-1}e$), and **gamma (γ) radiation** consists of high-energy electromagnetic radiation that has no mass. **Positron emission** is the conversion of a proton in the nucleus into a neutron plus an ejected *positron* (${}^{0}_{1}e$ or β^{+}), a particle that has the same mass as an electron but a positive charge. **Electron capture** is the capture of a surrounding electron by a proton in the nucleus. The process is accompanied by the emission of γ rays and results in the conversion of a proton in the nucleus into a neutron.

KEY WORDS

alkali metal *32*
alkaline-earth metal *32*
alpha (α) radiation *50*
atom *26*
atomic mass *45*
atomic number (Z) *43*
atomic weight *46*
Avogadro's number (N_A) *46*
beta (β) radiation *51*
chemical compound *35*
chemical equation *35*
chemical formula *35*
chemical property *32*
electron *39*
electron capture *52*
element *26*
extensive property *32*
gamma (γ) radiation *51*
group (periodic table) *30*
halogen *33*
inner transition metal group *30*
intensive property *32*
isotope *44*
law of definite proportions *36*
law of mass conservation *35*
law of multiple proportions *36*
main group *30*
mass number (A) *43*
metal *33*
molar mass *46*
mole *46*
neutron *41*
noble gas *33*
nonmetal *33*
nuclear chemistry *49*

LEARNING OUTCOMES

- Recognize the symbols of common elements, and describe the organization of the periodic table. **(Sections 1.1–1.3)** ***Problems 1–4, 20–23, 30–51***
- Know and apply the law of mass conservation **(Section 1.4)**, the law of definite proportions **(Section 1.5)**, and the law of multiple proportions. **(Section 1.5)** ***Problems 24–26, 52–67, 104, 105***
- Know the reasoning behind Dalton's atomic theory. **(Section 1.5)** ***Problems 52, 53***
- Explain the results of Thomson's cathode-ray experiments and Millikan's oil-drop experiment. **(Section 1.6)**
- Explain the results of Rutherford's experiment bombarding gold foil with alpha particles. **(Section 1.7)**
- Explain the difference between atomic number and mass number, and know how to write and interpret the symbol of an isotope. **(Section 1.8)** ***Problems 8–10, 28, 68, 70, 74–83, 106, 107***
- Explain the difference between the atomic weight of an element and the atomic mass of an isotope, and relate atomic weight to the isotopic composition of an element. **(Section 1.9)** ***Problems 11, 69, 71–73, 84–87, 102, 104, 108***
- Understand what a mole is, and know how to use molar mass to relate masses and numbers of atoms. **(Section 1.9)** ***Problems 12–14, 60–67, 105***
- Know how to write a balanced nuclear equation. **(Sections 1.10, 1.11)** ***Problems 15, 16, 19, 92–101, 112, 113***
- Know the different kinds of radioactive decay processes and the results of each. **(Section 1.11)** ***Problems 17, 29, 88–91***

CONCEPTUAL PROBLEMS

Problems 1.1–1.19 appear within the chapter.

1.20 Where on the following outline of a periodic table are the indicated elements or groups of elements?

(a) Alkali metals
(b) Halogens
(c) Alkaline-earth metals
(d) Transition metals
(e) Hydrogen
(f) Helium

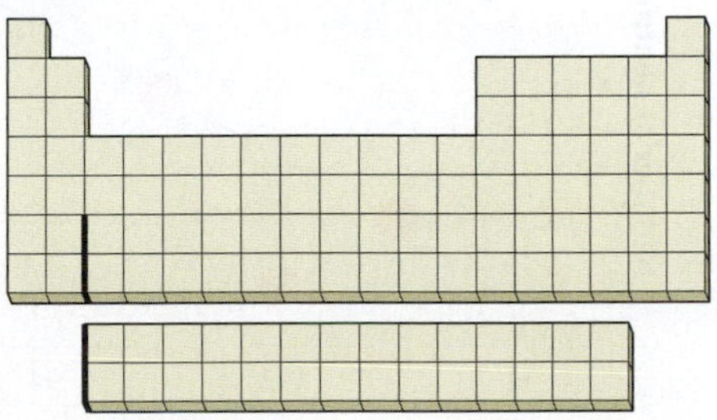

1.21 Where on the following outline of a periodic table does the dividing line between metals and nonmetals fall?

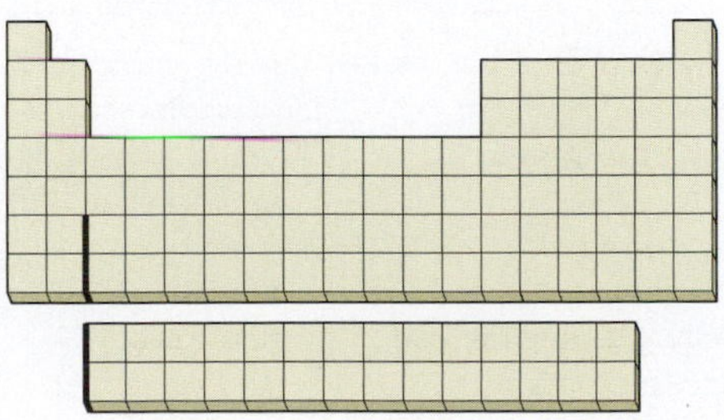

1.22 Is the red element on the following periodic table likely to be a gas, a liquid, or a solid? What is the atomic number of the blue element? Name at least one other element that is chemically similar to the green element.

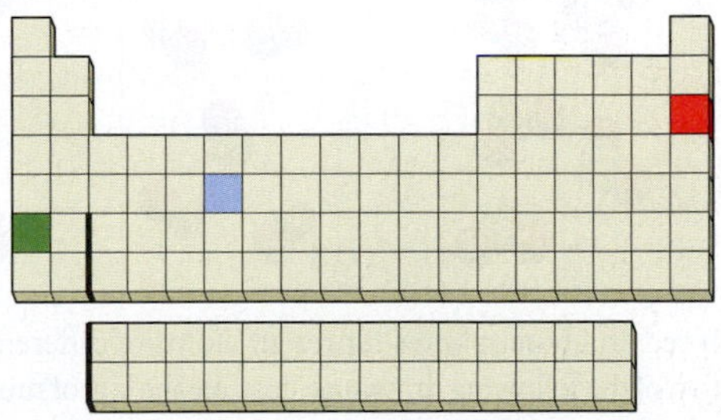

1.23 The element indicated on the following periodic table is used in smoke detectors. Identify it, give its atomic number, and tell what kind of group it's in.

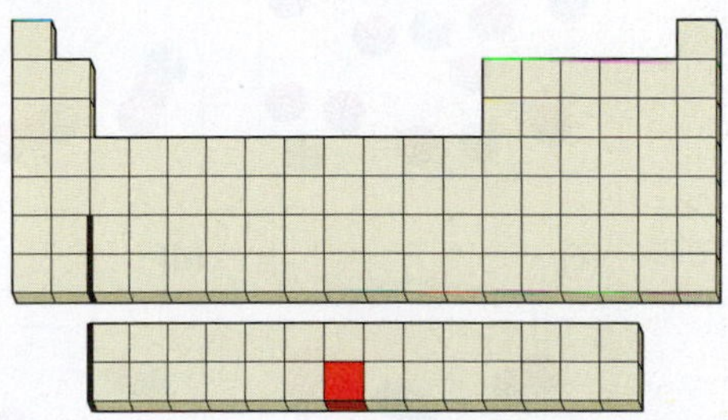

1.24 If yellow spheres represent sulfur atoms and red spheres represent oxygen atoms, which of the following drawings shows a collection of sulfur dioxide (SO_2) units?

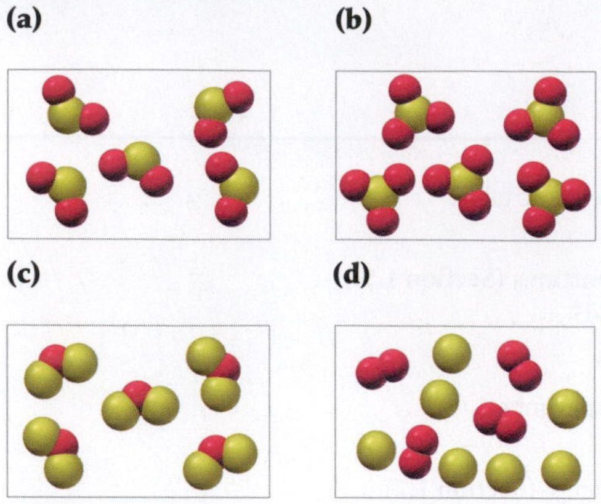

1.25 Assume that the mixture of substances in drawing (**a**) undergoes a reaction. Which of the drawings (**b**)–(**d**) represents a product mixture consistent with the law of mass conservation?

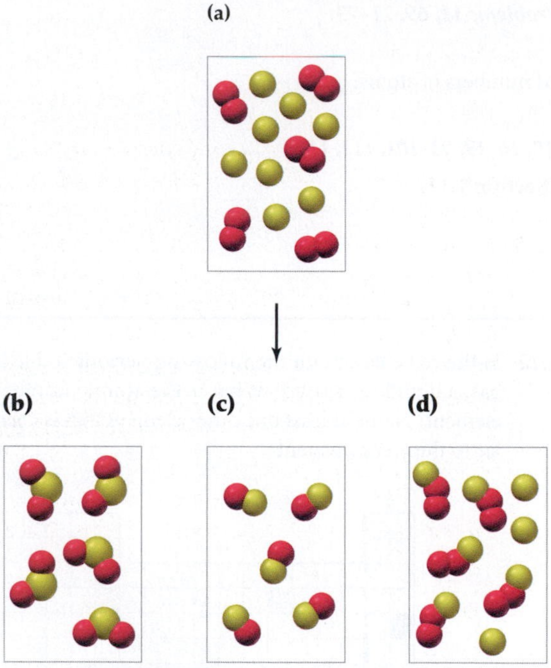

1.26 If red and blue spheres represent atoms of different elements, which two of the following drawings illustrate the law of multiple proportions?

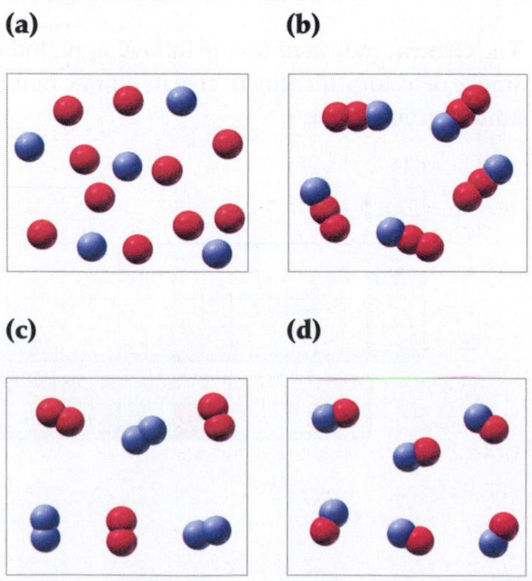

1.27 Which of the following three drawings represents a neutral Na atom, which represents a Ca atom with two positive electrical charges (Ca^{2+}), and which represents an F atom with one minus charge (F^-)?

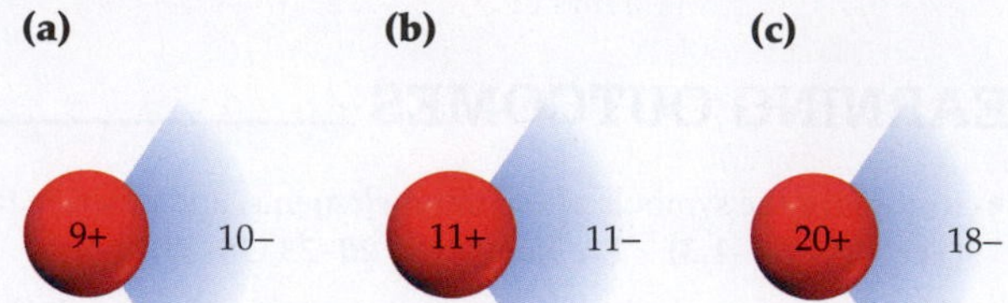

1.28 In the following drawings, red spheres represent protons and blue spheres represent neutrons. Which of the drawings represent different isotopes of the same element, and which represents a different element altogether? Explain.

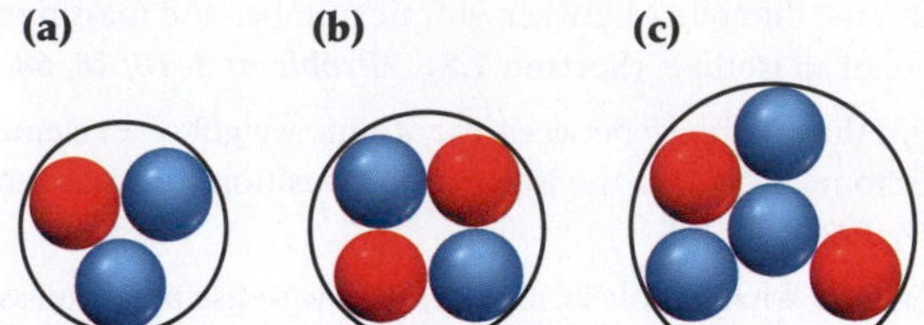

1.29 Isotope **A** decays to isotope **E** through the following series of steps, in which the products of the individual decay events are themselves radioactive and undergo further decay until a stable nucleus is ultimately reached. Two kinds of processes are represented, one by the shorter arrows pointing right and the other by the longer arrows pointing left.

(a) What kind of nuclear decay process does each kind of arrow correspond to?

(b) Identify and write the symbol ${}^{A}_{Z}X$ for each isotope in the series:

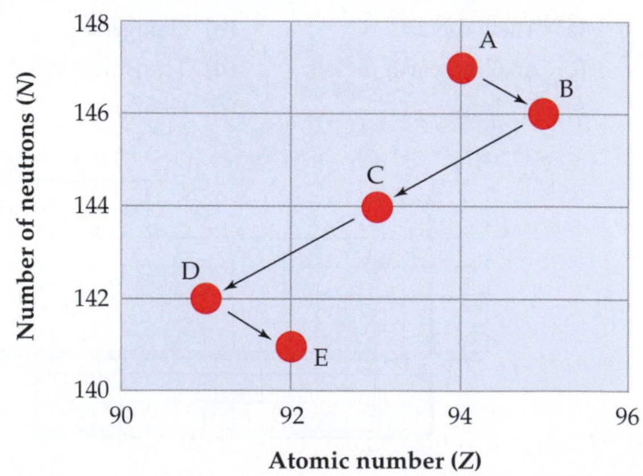

SECTION PROBLEMS

Elements and the Periodic Table (Sections 1.1–1.3)

1.30 How many elements are presently known? About how many occur naturally?

1.31 What are the rows called and what are the columns called in the periodic table?

1.32 How many groups are there in the periodic table? How are they labeled?

1.33 What common characteristics do elements within a group of the periodic table have?

1.34 Where in the periodic table are the main-group elements found? Where are the transition metal groups found?

1.35 Where in the periodic table are the metallic elements found? Where are the nonmetallic elements found?

1.36 What is a semimetal, and where in the periodic table are semimetals found?

1.37 List several general properties of the following:
(a) Alkali metals
(b) Noble gases
(c) Halogens

1.38 Without looking at a periodic table, list as many alkali metals as you can. (There are five common ones.)

1.39 Without looking at a periodic table, list as many alkaline-earth metals as you can. (There are five common ones.)

1.40 Without looking at a periodic table, list as many halogens as you can. (There are four common ones.)

1.41 Without looking at a periodic table, list as many noble gases as you can. (There are six common ones.)

1.42 At room temperature, a certain element is found to be a soft, silver-colored solid that reacts violently with water and is a good conductor of electricity. Is the element likely to be a metal, a nonmetal, or a semimetal?

1.43 At room temperature, a certain element is found to be a shiny, silver-colored solid that is a poor conductor of electricity. When a sample of the element is hit with a hammer, it shatters. Is the element likely to be a metal, a nonmetal, or a semimetal?

1.44 In which of the periodic groups 1A, 2A, 5A, and 7A do all elements have the first letter of their symbol the same as the first letter of their name?

1.45 For which elements in groups 1A, 2A, 5A, and 7A of the periodic table is the first letter of their symbol different from the first letter of their name?

1.46 What are the symbols for the following elements?
(a) Gadolinium (used in color TV screens)
(b) Germanium (used in semiconductors)
(c) Technetium (used in biomedical imaging)
(d) Arsenic (used in pesticides)

1.47 What are the symbols for the following elements?
(a) Cadmium (used in rechargeable ni-cad batteries)
(b) Iridium (used for hardening alloys)
(c) Beryllium (used in the space shuttle)
(d) Tungsten (used in lightbulbs)

1.48 Give the names corresponding to the following symbols:
(a) Te (b) Re (c) Be (d) Ar
(e) Pu (f) Sb

1.49 Give the names corresponding to the following symbols:
(a) B (b) Rh (c) Cf (d) Os
(e) Ga (f) Cs

1.50 What is wrong with each of the following statements?
(a) The symbol for tin is Ti.
(b) The symbol for manganese is Mg.
(c) The symbol for potassium is Po.
(d) The symbol for helium is HE.

1.51 What is wrong with each of the following statements?
(a) The symbol for carbon is ca.
(b) The symbol for sodium is So.
(c) The symbol for nitrogen is Ni.
(d) The symbol for chlorine is Cr.

Atomic Theory (Sections 1.4 and 1.5)

1.52 How does Dalton's atomic theory account for the law of mass conservation and the law of definite proportions?

1.53 What is the law of multiple proportions, and how does Dalton's atomic theory account for it?

1.54 A sample of mercury with a mass of 114.0 g was combined with 12.8 g of oxygen gas, and the resulting reaction gave 123.1 g of mercury oxide, HgO. How much oxygen was left over after the reaction was complete?

1.55 A sample of $CaCO_3$ was heated, causing it to form CaO and CO_2 gas. Solid CaO remained behind, while the CO_2 escaped to the atmosphere. If the $CaCO_3$ weighed 612 g and the CaO weighed 343 g, how many grams of CO_2 were formed in the reaction?

1.56 Benzene, ethane, and ethylene are just three of a large number of *hydrocarbons*—compounds that contain only carbon and hydrogen. Show how the following data are consistent with the law of multiple proportions.

Compound	Mass of carbon in 5.00 g sample	Mass of hydrogen in 5.00 g sample
Benzene	4.61 g	0.39 g
Ethane	4.00 g	1.00 g
Ethylene	4.29 g	0.71 g

1.57 In addition to carbon monoxide (CO) and carbon dioxide (CO_2), there is a third compound of carbon and oxygen called *carbon suboxide*. If a 2.500 g sample of carbon suboxide contains 1.32 g of C and 1.18 g of O, show that the law of multiple proportions is followed.

1.58 The atomic weight of carbon (12.011) is approximately 12 times that of hydrogen (1.008).
(a) Show how you can use this knowledge to calculate possible formulas for benzene, ethane, and ethylene (Problem 1.56).
(b) Show how your answer to part (a) is consistent with the actual formulas for benzene (C_6H_6), ethane (C_2H_6), and ethylene (C_2H_4).

1.59 What is a possible formula for carbon suboxide (Problem 1.57)?

1.60 If the atomic weight of an element is *x*, what is the mass in grams of 6.02×10^{23} atoms of the element?

1.61 If 6.02×10^{23} atoms of element **Y** have a mass of 83.80 g, what is the identity of **Y**?

1.62 If the atomic weight of an element is *x*, what is the mass in grams of 3.17×10^{20} atoms of the element?

1.63 If 4.61×10^{21} atoms of element **Z** have a mass of 0.815 g, what is the identity of **Z**?

1.64 A compound of zinc and sulfur contains 67.1% zinc by mass. What is the ratio of zinc and sulfur atoms in the compound?

1.65 There are two compounds of titanium and chlorine. One compound contains 31.04% titanium by mass, and the other contains 74.76% chlorine by mass. What are the ratios of titanium and chlorine atoms in the two compounds?

1.66 In methane, one part hydrogen combines with three parts carbon by mass. If a sample of a compound containing only carbon and hydrogen contains 32.0 g of carbon and 8.0 g of hydrogen, could the sample be methane? If the sample is not methane, show that the law of multiple proportions is followed for methane and this other substance.

1.67 In borane, one part hydrogen combines with 3.6 parts boron by mass. A compound containing only hydrogen and boron contains 6.0 g of hydrogen and 43.2 g of boron. Could this compound be borane? If it is not borane, show that the law of multiple proportions is followed for borane and this other substance.

Elements and Atoms (Sections 1.6–1.9)

1.68 What is the difference between an atom's atomic number and its mass number?

1.69 What is the difference between an element's atomic number and its atomic weight?

1.70 The subscript giving the atomic number of an atom is often left off when writing an isotope symbol. For example, ${}^{13}_{6}C$ is often written simply as ${}^{13}C$. Why is this allowable?

1.71 Iodine has a *lower* atomic weight than tellurium (126.90 for iodine, 127.60 for tellurium) even though it has a *higher* atomic number (53 for iodine, 52 for tellurium). Explain.

1.72 Copper has two naturally occurring isotopes, including ${}^{65}Cu$. Look at the periodic table, and tell whether the second isotope is ${}^{63}Cu$ or ${}^{66}Cu$.

1.73 Sulfur has four naturally occurring isotopes, including ${}^{33}S$, ${}^{34}S$, and ${}^{36}S$. Look at the periodic table, and tell whether the fourth isotope is ${}^{32}S$ or ${}^{35}S$.

1.74 Give the names and symbols for the following elements:
(a) An element with atomic number 6
(b) An element with 18 protons in its nucleus
(c) An element with 23 electrons

1.75 The radioactive isotopes cesium-137 and iodine-131 were released in substantial amounts after the nuclear power plant disaster at Fukushima, Japan, on March 11, 2011. Write the symbols for both isotopes in standard format.

1.76 Write symbols for the following isotopes:
(a) Radon-220
(b) Polonium-210
(c) Gold-197

1.77 Write symbols for the following isotopes:
(a) $Z = 58$ and $A = 140$
(b) $Z = 27$ and $A = 60$

1.78 How many protons, neutrons, and electrons are in each of the following atoms?
(a) ${}^{15}_{7}N$ **(b)** ${}^{60}_{27}Co$ **(c)** ${}^{131}_{53}I$ **(d)** ${}^{142}_{58}Ce$

1.79 How many protons and neutrons are in the nucleus of each of the following atoms?
(a) ${}^{27}Al$ **(b)** ${}^{32}S$ **(c)** ${}^{64}Zn$ **(d)** ${}^{207}Pb$

1.80 Identify the following elements:
(a) ${}^{24}_{12}X$ **(b)** ${}^{58}_{28}X$ **(c)** ${}^{104}_{46}X$ **(d)** ${}^{183}_{74}X$

1.81 Identify the following elements:
(a) ${}^{202}_{80}X$ **(b)** ${}^{195}_{78}X$ **(c)** ${}^{184}_{76}X$ **(d)** ${}^{209}_{83}C$

1.82 Which of the following isotope symbols can't be correct? Explain.
${}^{18}_{9}F$ ${}^{12}_{5}C$ ${}^{33}_{35}Br$ ${}^{18}_{8}O$ ${}^{11}_{5}Bo$

1.83 Which of the following isotope symbols can't be correct? Explain.
${}^{14}_{7}Ni$ ${}^{131}_{54}Xe$ ${}^{54}_{26}Fe$ ${}^{73}_{23}Ge$ ${}^{1}_{2}He$

1.84 Naturally occurring boron consists of two isotopes: ${}^{10}B$ (19.9%) with an atomic mass of 10.0129 and ${}^{11}B$ (80.1%) with an atomic mass of 11.009 31. What is the atomic weight of boron? Check your answer by looking at a periodic table.

1.85 Naturally occurring silver consists of two isotopes: ${}^{107}Ag$ (51.84%) with an atomic mass of 106.9051 and ${}^{109}Ag$ (48.16%) with an atomic mass of 108.9048. What is the atomic weight of silver? Check your answer in a periodic table.

1.86 Magnesium has three naturally occurring isotopes: ${}^{24}Mg$ (23.985) with 78.99% abundance, ${}^{25}Mg$ (24.986) with 10.00% abundance, and a third with 11.01% abundance. Look up the atomic weight of magnesium, and then calculate the mass of the third isotope.

1.87 A sample of naturally occurring silicon consists of ${}^{28}Si$ (27.9769), ${}^{29}Si$ (28.9765), and ${}^{30}Si$ (29.9738). If the atomic weight of silicon is 28.0855 and the natural abundance of ${}^{29}Si$ is 4.68%, what are the natural abundances of ${}^{28}Si$ and ${}^{30}Si$?

Nuclear Reactions and Radioactivity (Sections 1.10–1.11)

1.88 Positron emission and electron capture both give a product nucleus whose atomic number is 1 less than the starting nucleus. Explain.

1.89 What is the difference between an α particle and a helium atom?

1.90 Why does beta emission *raise* the atomic number of the product while positron emission *lowers* the atomic number?

1.91 Why does gamma emission not change the atomic number of the product?

1.92 Complete and balance the following nuclear equations:
(a) ${}^{126}_{50}Sn \rightarrow {}^{0}_{-1}e + ?$
(b) ${}^{210}_{88}Ra \rightarrow {}^{4}_{2}He + ?$
(c) ${}^{77}_{37}Rb \rightarrow {}^{0}_{1}e + ?$
(d) ${}^{76}_{36}Kr + {}^{0}_{-1}e \rightarrow ?$

1.93 Complete and balance the following nuclear equations:
(a) ${}^{90}_{38}Sr \rightarrow {}^{0}_{-1}e + ?$
(b) ${}^{247}_{100}Fm \rightarrow {}^{4}_{2}He + ?$
(c) ${}^{49}_{25}Mn \rightarrow {}^{0}_{1}e + ?$
(d) ${}^{37}_{18}Ar + {}^{0}_{-1}e \rightarrow ?$

1.94 What particle is produced in each of the following decay reactions?
(a) ${}^{188}_{80}Hg \rightarrow {}^{188}_{79}Au + ?$
(b) ${}^{218}_{85}At \rightarrow {}^{214}_{83}Bi + ?$
(c) ${}^{234}_{90}Th \rightarrow {}^{234}_{91}Pa + ?$

1.95 What particle is produced in each of the following decay reactions?

(a) $^{24}_{11}Na \rightarrow ^{24}_{12}Mg + ?$

(b) $^{135}_{60}Nd \rightarrow ^{135}_{59}Pr + ?$

(c) $^{170}_{78}Pt \rightarrow ^{166}_{76}Os + ?$

1.96 Write balanced nuclear equations for the following processes:

(a) Alpha emission of ^{162}Re

(b) Electron capture of ^{138}Sm

(c) Beta emission of ^{188}W

(d) Positron emission of ^{165}Ta

1.97 Write balanced nuclear equations for the following processes:

(a) Beta emission of ^{157}Eu

(b) Electron capture of ^{126}Ba

(c) Alpha emission of ^{146}Sm

(d) Positron emission of ^{125}Ba

1.98 Americium-241, a radioisotope used in smoke detectors, decays by a series of 12 reactions involving sequential loss of $\alpha, \alpha, \beta, \alpha, \alpha, \beta, \alpha, \alpha, \alpha, \beta, \alpha$, and β particles. Identify each intermediate nucleus and the final stable product nucleus.

1.99 Radon-222 decays by a series of three α emissions and two β emissions. What is the final stable nucleus?

1.100 Thorium-232 decays by a 10-step series, ultimately yielding lead-208. How many α particles and how many β particles are emitted?

1.101 How many α particles and how many β particles are emitted in the 11-step decay of ^{235}U into ^{207}Pb?

CHAPTER PROBLEMS

1.102 Germanium has five naturally occurring isotopes: ^{70}Ge, 20.5%, 69.924; ^{72}Ge, 27.4%, 71.922; ^{73}Ge, 7.8%, 72.923; ^{74}Ge, 36.5%, 73.921; and ^{76}Ge, 7.8%, 75.921. What is the atomic weight of germanium?

1.103 Fluorine occurs naturally as a single isotope. How many protons, neutrons, and electrons are present in deuterium fluoride (^{2}HF)? (Deuterium is ^{2}H.)

1.104 Ammonia (NH_3) and hydrazine (N_2H_4) are both compounds of nitrogen and hydrogen. Based on the law of multiple proportions, how many grams of hydrogen would you expect 2.34 g of nitrogen to combine with to yield ammonia? To yield hydrazine?

1.105 If 3.670 g of nitrogen combines with 0.5275 g of hydrogen to yield compound **X**, how many grams of nitrogen would combine with 1.575 g of hydrogen to make the same compound? Is **X** ammonia or hydrazine (Problem 1.104)?

1.106 Identify the following atoms:

(a) A halogen with 53 electrons

(b) A noble gas with $A = 84$

1.107 Hydrogen has three isotopes (^{1}H, ^{2}H, and ^{3}H), and chlorine has two isotopes (^{35}Cl and ^{37}Cl). How many isotopic kinds of HCl are there? Write the formula for each, and tell how many protons, neutrons, and electrons each contains.

1.108 Prior to 1961, the predecessor to the unified atomic mass unit was called simply the atomic mass unit (amu) and was defined as 1/16 of the atomic mass of oxygen-16; that is, the atomic mass of a ^{16}O atom was defined as exactly 16 amu. What was the mass of a ^{12}C atom prior to 1961 if the atomic mass of oxygen-16 on today's scale is 15.9994?

1.109 What was the mass in unified atomic mass units of a ^{40}Ca atom prior to 1961 if its mass on today's scale is 39.9626? (See Problem 1.108.)

1.110 The *molecular weight* of a compound is the sum of the atomic weights of all atoms in the molecule. What is the molecular weight of acetaminophen ($C_8H_9NO_2$), the active ingredient in Tylenol?

1.111 The *mass percent* of an element in a compound is the mass of the element (total mass of the element's atoms in the compound) divided by the mass of the compound (total mass of all atoms in the compound) times 100%. What is the mass percent of each element in acetaminophen? (See Problem 1.110.)

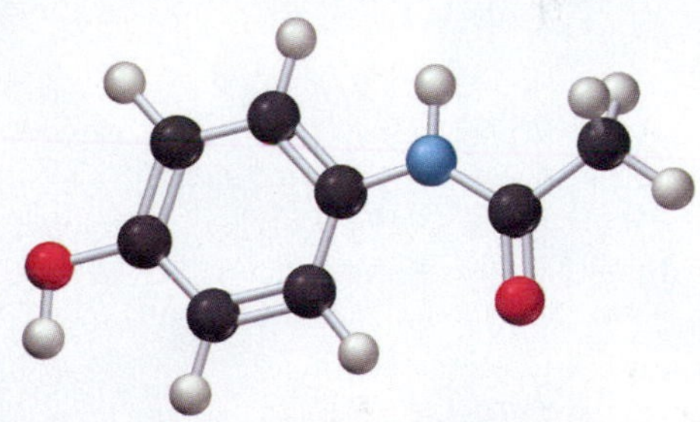

Acetaminophen ($C_8H_9NO_2$)

1.112 Radioactive ^{100}Tc decays to form ^{100}Mo. There are two possible pathways for this decay. Write balanced equations for both.

1.113 ^{226}Ac can decay by any of three different nuclear processes: α emission, β emission, or electron capture. Write a balanced nuclear equation for the decay of ^{226}Ac by each process.

1.114 In an alternate universe, the smallest negatively charged particle, analogous to our electron, is called a blorvek. To determine the charge on a single blorvek, an experiment like Millikan's with charged oil droplets was carried out and the following results were recorded:

Droplet number	Charge (C)
1	7.74×10^{-16}
2	4.42×10^{-16}
3	2.21×10^{-16}
4	4.98×10^{-16}
5	6.64×10^{-16}

(a) Based on these observations, what is the largest possible value for the charge on a blorvek?

(b) Further experiments found a droplet with a charge of 5.81×10^{-16} C. Does this new result change your answer to part **(a)**? If so, what is the new largest value for the blorvek's charge?

CHAPTER 2

Periodicity, the presence of repeating patterns, is common throughout nature, as seen here in these basalt columns at the biosphere reserve in Gangolfsberg, Germany.

Periodicity and the Electronic Structure of Atoms

The periodic table, introduced in Section 1.2, is the most important organizing principle in chemistry. If you know the properties of any one element in a group, or column, of the periodic table, you can make a good guess at the properties of every other element in the same group and even of the elements in neighboring groups. Although the periodic table was originally constructed from empirical observations, its scientific underpinnings have long been established and are well understood.

To see why it's called the *periodic* table, look at the graph of atomic radius versus atomic number in **Figure 2.1**, which shows a periodic

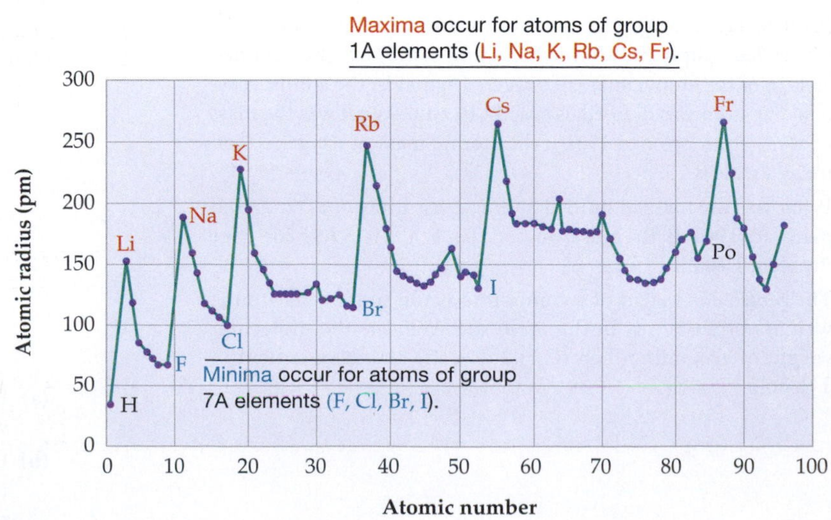

FIGURE 2.1

A graph of atomic radius in picometers (pm) versus atomic number. A clear rise-and-fall pattern of periodicity is evident. (Accurate data are not available for group 8A elements.)

CONTENTS

rise-and-fall pattern. Beginning on the left with atomic number 1 (hydrogen), the radii of the atoms increase to a maximum at atomic number 3 (lithium), then decrease to a minimum, then increase again to a maximum at atomic number 11 (sodium), then decrease, and so on. It turns out that all the maxima occur for atoms of the group 1A elements—Li, Na, K, Rb, Cs, and Fr—and the minima occur for atoms of the group 7A elements—F, Cl, Br, and I.

There's nothing unique about the periodicity of atomic radii shown in Figure 2.1. Any of several dozen other physical or chemical properties of atoms could be plotted in a similar way with similar results. We'll look at several examples of such periodicity in this chapter and the next and will see why that periodicity occurs.

2.1 THE NATURE OF RADIANT ENERGY AND THE ELECTROMAGNETIC SPECTRUM

What fundamental property of atoms is responsible for the periodic variations we observe in atomic radii and in so many other characteristics of the elements? This question occupied the thoughts of chemists for more than 50 years after Mendeleev, and it was not until well into the 1920s that the question was answered. To understand how the answer slowly emerged, it's first necessary to look at the nature of visible light and other forms of radiant energy. Historically, studies of the interaction of radiant energy with matter have provided immense insight into atomic structure.

Although they appear quite different to our senses, visible light, infrared radiation, microwaves, radio waves, X rays, and so on are all different forms of *electromagnetic radiation*, or **radiant energy**. Collectively, they make up the **electromagnetic spectrum**, shown in **Figure 2.2**.

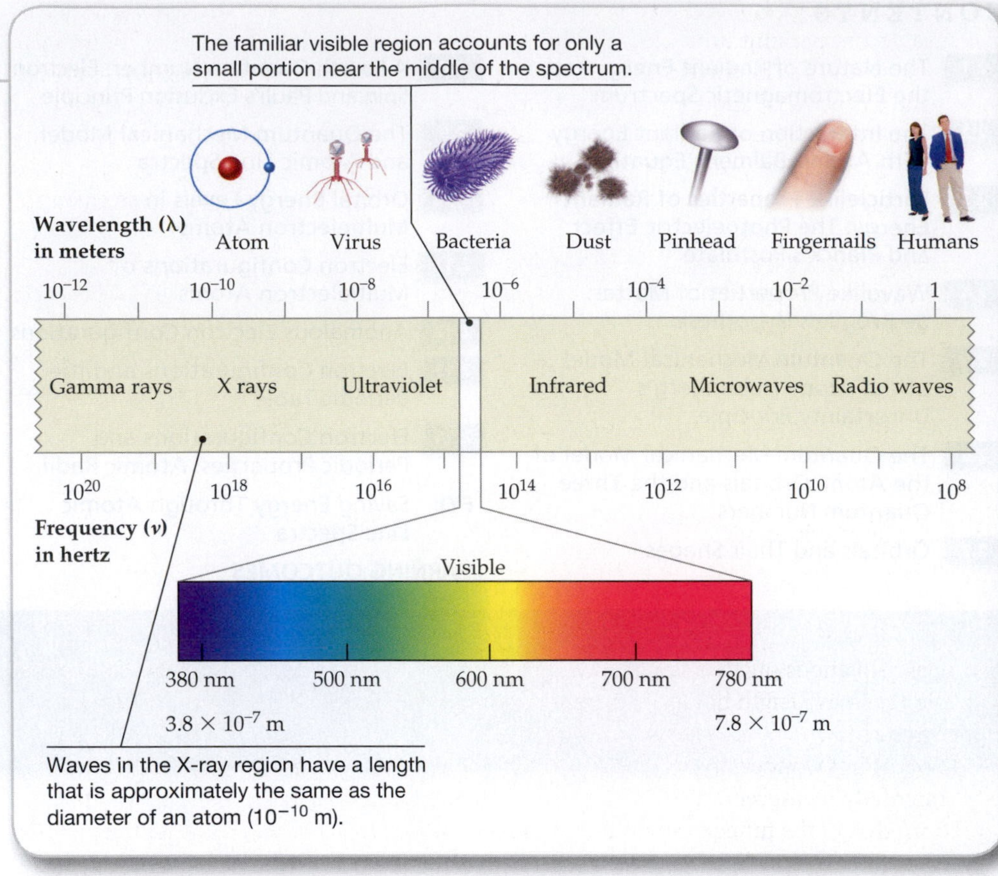

FIGURE 2.2

The electromagnetic spectrum. The spectrum consists of a continuous range of wavelengths and frequencies, from radio waves at the low-frequency end to gamma rays at the high-frequency end.

Radiant energy traveling through a vacuum behaves in some ways like ocean waves traveling through water. Like ocean waves, radiant energy is characterized by a *frequency*, a *wavelength*, and an *amplitude*. If you could stand in one place and look at a sideways, cutaway view of an ocean wave moving through the water, you would see a regular rise-and-fall pattern like that in **Figure 2.3**.

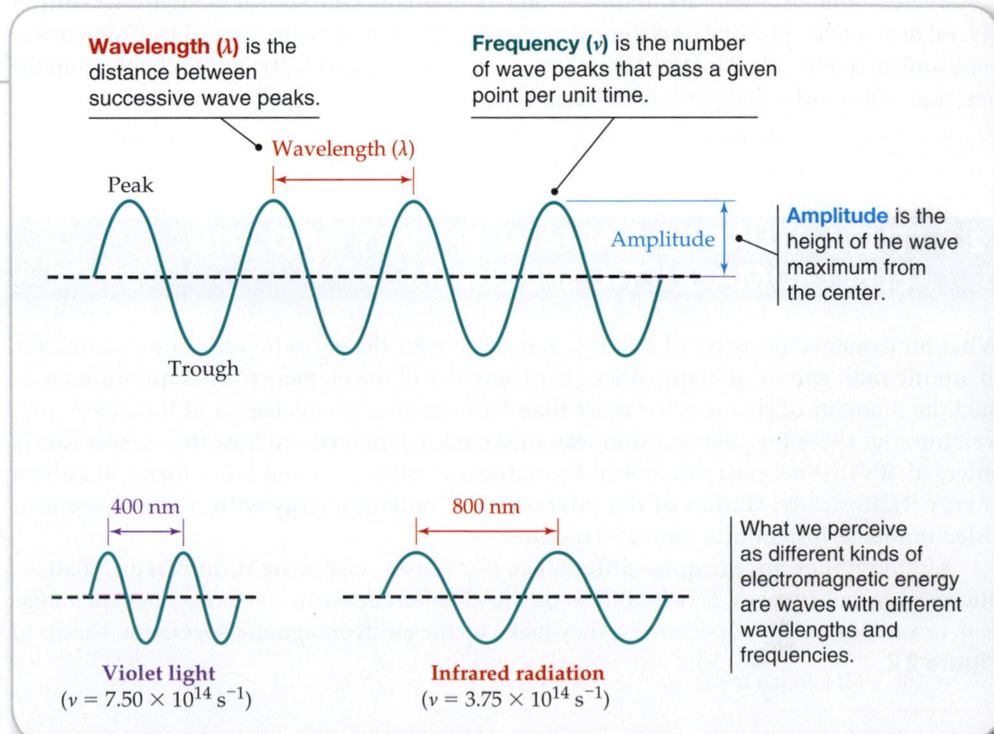

FIGURE 2.3

The nature of electromagnetic waves. The waves are characterized by a wavelength, a frequency, and an amplitude.

The **frequency** (**ν**, Greek nu) of a wave is simply the number of wave peaks that pass by a given point per unit time, usually expressed in units of waves per second (s^{-1}), or **hertz** (**Hz**; $1\text{ Hz} = 1\text{ s}^{-1}$). The **wavelength** (**λ**, Greek lambda) of the wave is the distance from one wave peak to the next, and the **amplitude** of the wave is the height of the wave, measured from the center line between peak and trough. Physically, what we perceive as the intensity of radiant energy is proportional to the square of the wave amplitude. A feeble beam and a blinding glare of light may have the same wavelength and frequency, but they differ greatly in amplitude.

Multiplying the wavelength of a wave in meters (m) by its frequency in reciprocal seconds (s^{-1}) gives the speed of the wave in meters per second (m/s). The rate of travel of all radiant energy in a vacuum is a constant value, commonly called the speed of light and abbreviated *c*. Its numerical value is defined as exactly $2.997\,924\,58 \times 10^8$ m/s, usually rounded off to 3.00×10^8 m/s:

$$\text{Wavelength} \times \text{Frequency} = \text{Speed}$$

$$\lambda\,(\text{m}) \times \nu\,(\text{s}^{-1}) = c\,(\text{m/s})$$

which can be rewritten as:

$$\lambda = \frac{c}{\nu} \quad \text{or} \quad \nu = \frac{c}{\lambda}$$

These equations say that frequency and wavelength are inversely related: Radiant energy with a longer wavelength has a lower frequency, and energy with a shorter wavelength has a higher frequency.

(You might wonder how the speed of light *c* can be a defined quantity rather than an experimentally measured one. The answer is that *c* is a **derived quantity**, whose value follows from that of the fundamental SI units for length (*m*) and time (*s*). Because the meter is defined as the distance light travels through a vacuum in 1/299,792,458 s (Section 0.4), the speed of light is 299,792,458 m/s.)

▲ Ocean waves, like electromagnetic waves, are characterized by a wavelength, a frequency, and an amplitude.

REMEMBER...

A **derived quantity** is one that can be expressed using a combination of one or more of the seven fundamental SI units shown previously in Table 0.1. (Section 0.6)

Worked Example 2.1

Calculating a Frequency from a Wavelength

The light blue glow given off by mercury streetlamps has a wavelength of 436 nm. What is its frequency in hertz?

STRATEGY

We are given a wavelength and want to find the corresponding frequency. Wavelength and frequency are inversely related by the equation $\lambda\nu = c$, which can be solved for ν. Don't forget to convert from nanometers to meters.

SOLUTION

$$\nu = \frac{c}{\lambda} = \frac{\left(3.00 \times 10^8 \dfrac{\cancel{\text{m}}}{\text{s}}\right)}{(436\ \cancel{\text{nm}})\left(\dfrac{1\ \cancel{\text{m}}}{10^9\ \cancel{\text{nm}}}\right)}$$

$$= 6.88 \times 10^{14}\ \text{s}^{-1} = 6.88 \times 10^{14}\ \text{Hz}$$

The frequency of the light is $6.88 \times 10^{14}\ \text{s}^{-1}$, or 6.88×10^{14} Hz.

▶ **PROBLEM 2.1** What is the frequency of a gamma ray with $\lambda = 3.56 \times 10^{-11}$ m? Of a radar wave with $\lambda = 10.3$ cm?

▶ **PROBLEM 2.2** What is the wavelength in meters of an FM radio wave with frequency $\nu = 102.5$ MHz? Of a medical X ray with $\nu = 9.55 \times 10^{17}$ Hz?

▲ Does the blue glow from this mercury lamp correspond to a longer or shorter wavelength than the yellow glow from a sodium lamp?

CONCEPTUAL PROBLEM 2.3 Two electromagnetic waves are represented below.

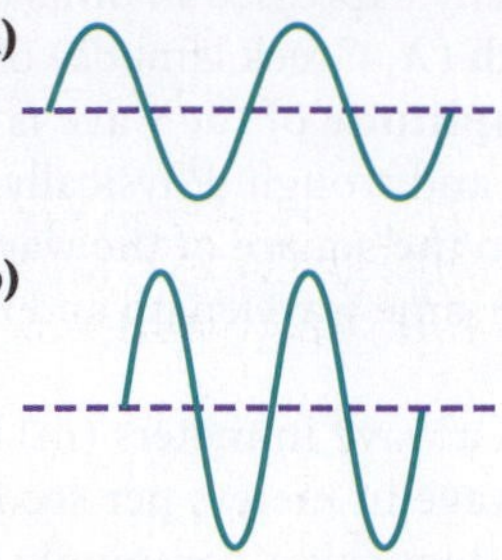

(a) Which wave has the higher frequency?
(b) Which wave represents a more intense beam of light?
(c) Which wave represents blue light, and which represents red light?

2.2 THE INTERACTION OF RADIANT ENERGY WITH ATOMS: BALMER'S EQUATION

The light that we see from the sun or from a typical incandescent lightbulb is "white" light, meaning that it consists of an essentially continuous distribution of wavelengths spanning the entire visible region of the electromagnetic spectrum. When a narrow beam of white light is passed through a glass prism, the different wavelengths travel through the glass at different rates. As a result, the white light is separated into its component colors, ranging from red at the long-wavelength end of the spectrum (780 nm) to violet at the short-wavelength end (380 nm) (**Figure 2.4a**). This separation into colors also occurs when light travels through water droplets in the air, forming a rainbow, or through oriented ice crystals in clouds, causing an unusual weather phenomenon called a *parhelion*, or sundog (**Figure 2.4b**).

FIGURE 2.4

Separation of white light into its constituent colors. (**a**) When a narrow beam of ordinary white light is passed through a glass prism, different wavelengths travel through the glass at different rates and appear as different colors. A similar effect occurs when light passes through water droplets in the air, forming a rainbow, or (**b**) through ice crystals in clouds, causing a *parhelion*, or sundog.

What do visible light and other kinds of radiant energy have to do with atomic structure? It turns out that atoms give off light when heated or otherwise energetically excited, thereby providing a clue to their atomic makeup. Unlike the white light from the sun, though, the light given off by an energetically excited atom is not a continuous distribution of wavelengths. When passed first through a narrow slit to get a thin beam and then through a prism, the light emitted by an excited atom is found to consist of only a few specific wavelengths rather than a full rainbow of colors, giving a series of discrete lines on an otherwise dark background—a so-called **line spectrum**. Excited sodium atoms, produced by heating NaCl or some other compound of sodium in the flame of a Bunsen burner, give off yellow light; hydrogen atoms give off a red light made of several different wavelengths (**Figure 2.5**); and so on. In fact, the brilliant colors of fireworks result from the light emitted by mixtures of metal atoms that have been heated by explosive powder.

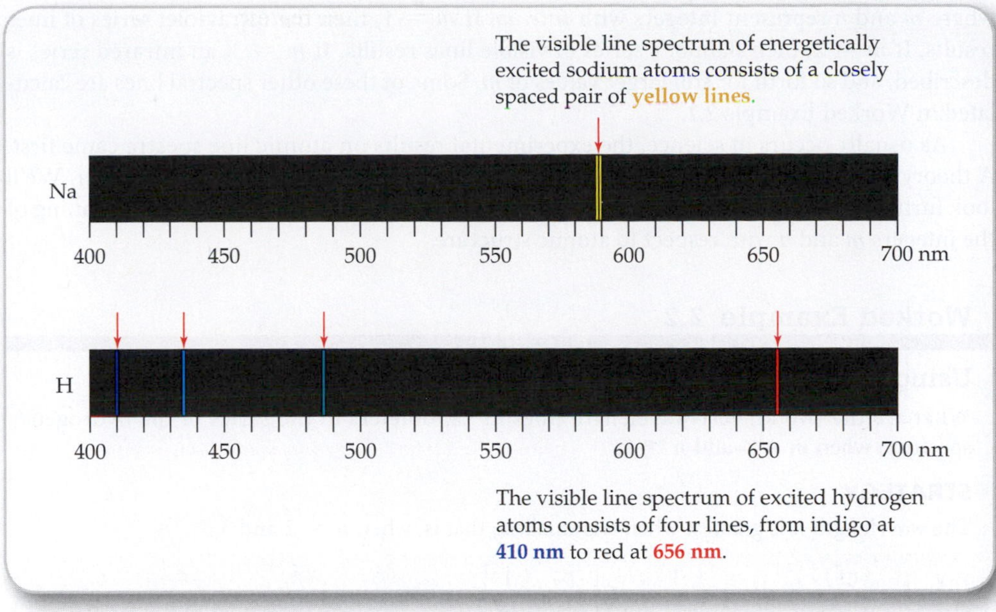

FIGURE 2.5

Atomic line spectra.

Soon after the discovery that energetic atoms emit light of specific wavelengths, chemists began cataloging the line spectra of various elements. They rapidly found that each element has its own unique spectral "signature," and they began using the results to identify the elements present in minerals and other substances. Not until the work of the Swiss schoolteacher Johann Balmer in 1885, though, was a pattern discovered in atomic line spectra. It was known at the time that hydrogen produced a spectrum with four lines, as shown in Figure 2.5. The wavelengths of the four lines are 656.3 nm (red), 486.1 nm (blue-green), 434.0 nm (blue), and 410.1 nm (indigo).

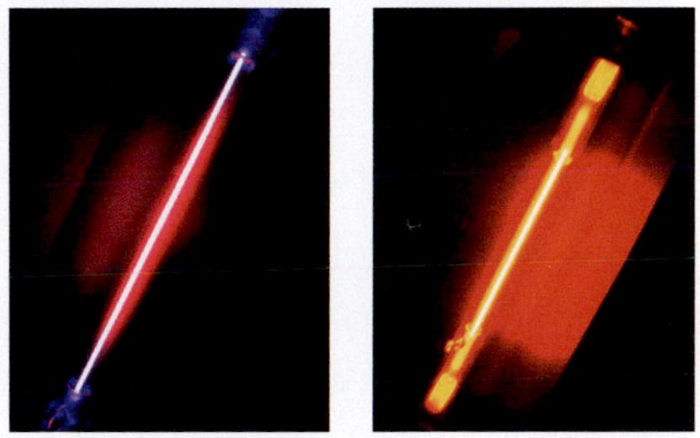

▲ Excited hydrogen atoms give off a red light, and excited neon atoms emit orange light.

Thinking about the hydrogen spectrum and trying by trial-and-error to organize the data in various ways, Balmer discovered that the wavelengths of the four lines in the hydrogen spectrum can be expressed by the equation

$$\frac{1}{\lambda} = R_\infty\left[\frac{1}{2^2} - \frac{1}{n^2}\right] \quad \text{or} \quad \nu = R_\infty \cdot c\left[\frac{1}{2^2} - \frac{1}{n^2}\right]$$

where R_∞ is a constant (now called the *Rydberg constant*) equal to $1.097 \times 10^{-2}\ \text{nm}^{-1}$ and n is an integer greater than 2. The red spectral line at 656.3 nm, for example, results from Balmer's equation when $n = 3$:

$$\frac{1}{\lambda} = [1.097 \times 10^{-2}\ \text{nm}^{-1}]\left[\frac{1}{2^2} - \frac{1}{3^2}\right] = 1.524 \times 10^{-3}\ \text{nm}^{-1}$$

$$\lambda = \frac{1}{1.524 \times 10^{-3}\ \text{nm}^{-1}} = 656.3\ \text{nm}$$

Similarly, a value of $n = 4$ gives the blue-green line at 486.1 nm, a value of $n = 5$ gives the blue line at 434.0 nm, and so on. Solve Balmer's equation yourself to make sure.

Subsequent to the discovery of the Balmer series of lines in the visible region of the electromagnetic spectrum, many other spectral lines were found to be present in *nonvisible* regions of the spectrum. Hydrogen, for instance, shows a series of spectral lines in the ultraviolet region, along with several other series in the infrared region.

By adapting Balmer's equation, the Swedish physicist Johannes Rydberg was able to show that every line in the entire spectrum of hydrogen can be fit by a generalized **Balmer–Rydberg equation**:

▶ **Balmer–Rydberg equation** $$\frac{1}{\lambda} = R_\infty\left[\frac{1}{m^2} - \frac{1}{n^2}\right] \quad \text{or} \quad \nu = R_\infty \cdot c\left[\frac{1}{m^2} - \frac{1}{n^2}\right]$$

where m and n represent integers with $n > m$. If $m = 1$, then the ultraviolet series of lines results. If $m = 2$, then Balmer's series of visible lines results. If $m = 3$, an infrared series is described, and so forth for still larger values of m. Some of these other spectral lines are calculated in Worked Example 2.2.

As usually occurs in science, the experimental results on atomic line spectra came first. A theory of atomic structure to explain those results didn't appear until 40 years later. We'll look further at the Balmer–Rydberg equation in Section 2.9 and ultimately see the meaning of the integers m and n with respect to atomic structure.

Worked Example 2.2

Using the Balmer–Rydberg Equation

What are the two longest-wavelength lines in nanometers in the series of the hydrogen spectrum when $m = 1$ and $n > 1$?

STRATEGY

The wavelength λ is greatest when n is smallest; that is, when $n = 2$ and 3.

$$\frac{1}{\lambda} = R_\infty\left[\frac{1}{m^2} - \frac{1}{n^2}\right] \qquad \text{where } m = 1$$

SOLUTION

Solving the equation first when $n = 2$ gives

$$\frac{1}{\lambda} = R_\infty\left[\frac{1}{1^2} - \frac{1}{2^2}\right] = (1.097 \times 10^{-2}\ \text{nm}^{-1})\left(1 - \frac{1}{4}\right) = 8.228 \times 10^{-3}\ \text{nm}^{-1}$$

or $$\lambda = \frac{1}{8.228 \times 10^{-3}\ \text{nm}^{-1}} = 121.5\ \text{nm}$$

Solving the equation next when $n = 3$ gives

$$\frac{1}{\lambda} = R_\infty\left[\frac{1}{1^2} - \frac{1}{3^2}\right] = (1.097 \times 10^{-2}\ \text{nm}^{-1})\left(1 - \frac{1}{9}\right) = 9.751 \times 10^{-3}\ \text{nm}^{-1}$$

or $$\lambda = \frac{1}{9.751 \times 10^{-3}\ \text{nm}^{-1}} = 102.6\ \text{nm}$$

The two longest-wavelength lines are at 121.5 nm and 102.6 nm.

Worked Example 2.3

Using the Balmer–Rydberg Equation

What is the shortest-wavelength line in nanometers in the series of the hydrogen spectrum when $m = 1$ and $n > 1$?

STRATEGY

The shortest-wavelength line occurs when n is infinitely large so that $1/n^2 = 0$. That is, if $n = \infty$, then $1/n^2 = 0$.

SOLUTION

$$\frac{1}{\lambda} = R_\infty\left[\frac{1}{1^2} - \frac{1}{\infty^2}\right] = (1.097 \times 10^{-2}\ \text{nm}^{-1})(1 - 0) = 1.097 \times 10^{-2}\ \text{nm}^{-1}$$

or $$\lambda = \frac{1}{1.097 \times 10^{-2}\ \text{nm}^{-1}} = 91.16\ \text{nm}$$

The shortest-wavelength line is at 91.16 nm.

PROBLEM 2.4 The Balmer equation can be extended beyond the visible portion of the electromagnetic spectrum to include lines in the ultraviolet. What is the wavelength in nanometers of ultraviolet light in the Balmer series when $m = 2$ and $n = 7$?

PROBLEM 2.5 What is the longest-wavelength line in nanometers in the infrared series for hydrogen where $m = 3$?

PROBLEM 2.6 What is the shortest-wavelength line in nanometers in the infrared series for hydrogen where $m = 3$?

2.3 PARTICLELIKE PROPERTIES OF RADIANT ENERGY: THE PHOTOELECTRIC EFFECT AND PLANCK'S POSTULATE

The existence of atomic line spectra and the fit of the visible hydrogen spectrum to the Balmer–Rydberg equation imply the existence of a general underlying principle about atomic structure, but it was more than 40 years before that principle was found.

One important step toward developing a model of atomic structure came in 1905, when Albert Einstein (1879–1955) proposed an explanation of the **photoelectric effect**. Scientists had known since the late 1800s that irradiating a clean metal surface with light causes electrons to be ejected from the metal **(Figure 2.6)**. Furthermore, the frequency of the light used for the irradiation must be above some threshold value, which is different for every metal. Blue light ($\nu \approx 6.5 \times 10^{14}$ Hz) causes metallic sodium to emit electrons, for instance, but red light ($\nu \approx 4.5 \times 10^{14}$ Hz) has no effect.

▲ Night-vision goggles are based on the photoelectric effect of metal alloys that emit electrons when irradiated with infrared light.

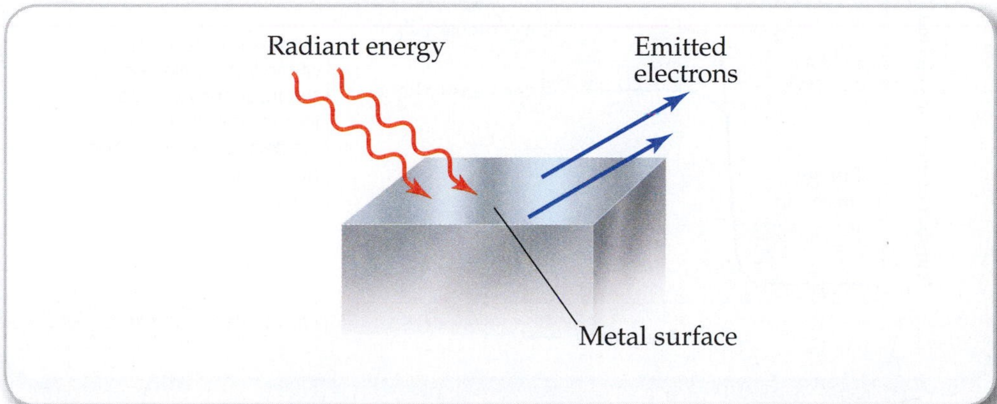

FIGURE 2.6

The photoelectric effect. Irradiation of a clean metal surface with radiant energy causes electrons to be emitted.

Einstein explained the photoelectric effect by assuming that a beam of light behaves as if it were a stream of small particles, now called **photons**, whose energy (E) is related to their frequency, ν, (or wavelength, λ) by an equation called **Planck's postulate** after the German physicist Max Planck (1858–1947).

▶ **Planck's postulate** $E = h\nu = hc/\lambda$

The proportionality constant h in the equation represents a physical constant known as *Planck's constant* and has the value $h = 6.626 \times 10^{-34}\ \text{J} \cdot \text{s}$. For example, one photon of red light with a frequency $\nu = 4.62 \times 10^{14}\ \text{s}^{-1}$ (wavelength $\lambda = 649$ nm) has an energy of 3.06×10^{-19} J. [Recall from Section 0.8 that the SI unit for energy is the *joule* (J), where $1\ \text{J} = 1\ (\text{kg} \cdot \text{m}^2)/\text{s}^2$.]

$$E = h\nu = (6.626 \times 10^{-34}\ \text{J} \cdot \text{s})(4.62 \times 10^{14}\ \text{s}^{-1}) = 3.06 \times 10^{-19}\ \text{J}$$

You might also recall from Section 1.9 that 1 *mole* (mol) of anything is the amount that contains Avogadro's number (6.022×10^{23}) of entities. Thus, it's convenient to express radiant energy on a per-mole basis rather than a per-photon basis. Multiplying the per-photon energy of 3.06×10^{-19} J by Avogadro's number gives an energy of 184 kJ/mol for red light.

$$\left(3.06 \times 10^{-19} \frac{\text{J}}{\text{photon}}\right)\left(6.022 \times 10^{23} \frac{\text{photon}}{\text{mol}}\right) = 1.84 \times 10^{5} \frac{\text{J}}{\text{mol}}$$

$$\left(1.84 \times 10^{5} \frac{\text{J}}{\text{mol}}\right)\left(\frac{1\ \text{kJ}}{1000\ \text{J}}\right) = 184\ \text{kJ/mol}$$

Higher frequencies and shorter wavelengths correspond to higher energy radiation, while lower frequencies and longer wavelengths correspond to lower energy. Blue light ($\lambda \approx 450$ nm), for instance, has a shorter wavelength and is more energetic than red light ($\lambda \approx 650$ nm). Similarly, an X ray ($\lambda \approx 1$ nm) has a shorter wavelength and is more energetic than an FM radio wave ($\lambda \approx 10^{10}$ nm, or 10 m).

If the energy (or frequency) of the photon striking a metal is below a minimum value, no electron is ejected. Above the threshold level, however, sufficient energy is transferred to an electron to overcome the attractive forces holding the electron to the metal and allow the electron to escape (**Figure 2.7**). The amount of energy necessary to eject an electron is called the **work function (Φ)** of the metal and is lowest for the group 1A and group 2A elements. That is, elements on the left side of the periodic table hold their electrons less tightly than other metals and lose them more readily (Table 2.1).

TABLE 2.1 Work Functions of Some Common Metals

Element	Work Function (Φ) (kJ/mol)
Cs	188
K	221
Na	228
Ca	277
Mg	353
Cu	437
Fe	451

FIGURE 2.7

Dependence of the photoelectric effect on frequency.

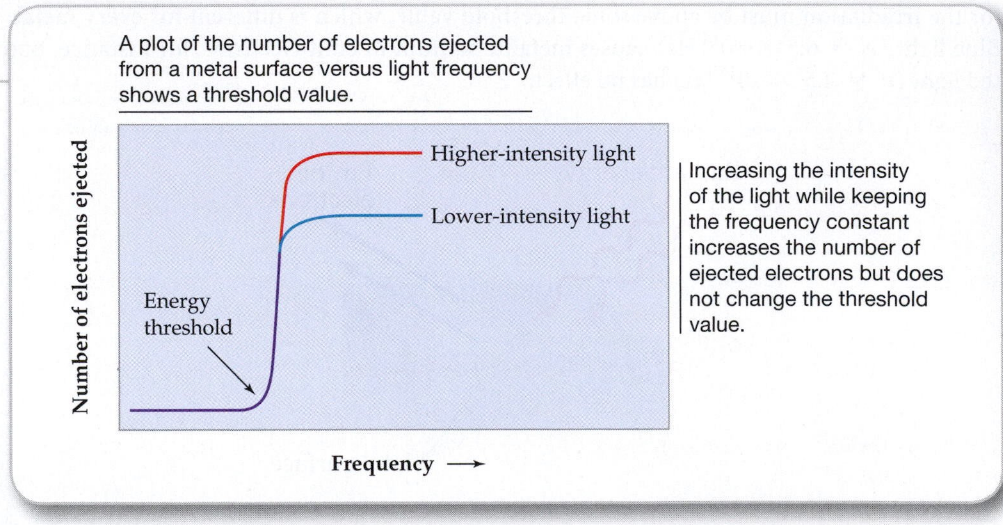

Note again that the energy of an individual photon depends only on its frequency (or wavelength), not on the intensity of the light beam. The intensity of a light beam is a measure of the *number* of photons in the beam, whereas frequency is a measure of the *energies* of those photons. A low-intensity beam of high-energy photons might easily knock a few electrons loose from a metal, but a high-intensity beam of low-energy photons might not be able to knock loose a single electron.

As a rough analogy, think of throwing balls of different masses at a glass window. A thousand table-tennis balls (lower energy) would only bounce off the window, but a single baseball (higher energy) would break the glass. In the same way, low-energy photons bounce off the metal surface, but a single photon at or above a certain threshold energy can "break" the metal and dislodge an electron.

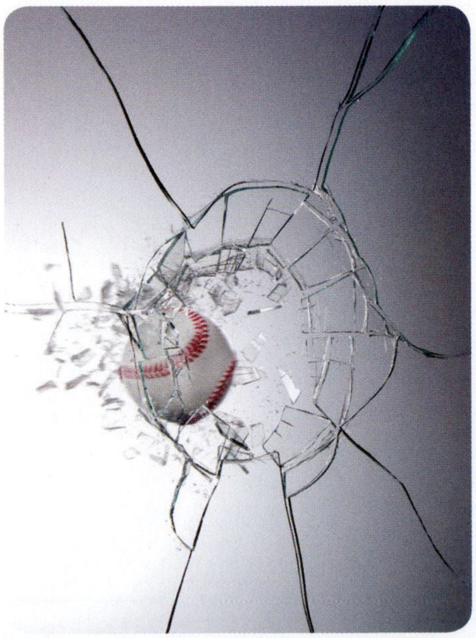

▲ A glass window can be broken by a single baseball, but a thousand table-tennis balls would only bounce off.

The main conclusion from Einstein's work was that the behavior of light and other forms of radiant energy is more complex than had been formerly believed. In addition to behaving as waves, light can also behave simultaneously as small particles. The idea might seem strange at first but becomes less so if you think of light as analogous to matter. Both radiant energy and matter are said to be *quantized*, meaning that both occur only in discrete amounts. Just as there can be either 1 or 2 hydrogen atoms but not 1.5 or 1.8, there can be 1 or 2 photons of light but not 1.5 or 1.8. The amount, or **quantum**, of energy corresponding to one photon of light is almost inconceivably small, just as the amount of matter in one atom is inconceivably small, but the idea is the same. An analogy of quantization from daily life is that of stairs versus a ramp. A ramp changes height continuously, but stairs change height only in discrete amounts and are thus quantized.

Once the particlelike nature of radiant energy is accepted, part of the puzzle of atomic line spectra can be explained. Energetically excited atoms are unable to emit light of continuously varying wavelengths and therefore don't give a continuous spectrum. Instead, the atoms are for some reason constrained to emit quanta of specific energies and therefore give a line spectrum.

Based on this insight, the Danish physicist Niels Bohr (1885–1962) proposed in 1913 a model of the atom in which electrons circle the nucleus at specific distances in a series of concentric orbits. Furthermore, said Bohr, each orbit has associated with it a specific energy level that increases with increasing distance from the nucleus. That is, the energy levels of the orbits are quantized so that only certain specific orbits corresponding to certain specific energies for the electrons are available. The observed spectral emission lines result when an electron falls from a higher-energy outer orbit to a lower-energy inner orbit, emitting a quantum of energy corresponding to the energy difference between the two orbits (**Figure 2.8**).

▲ A ramp changes height continuously, but stairs are quantized, changing height only in discrete amounts. In the same way, radiant energy is not continuous but is emitted only in discrete amounts.

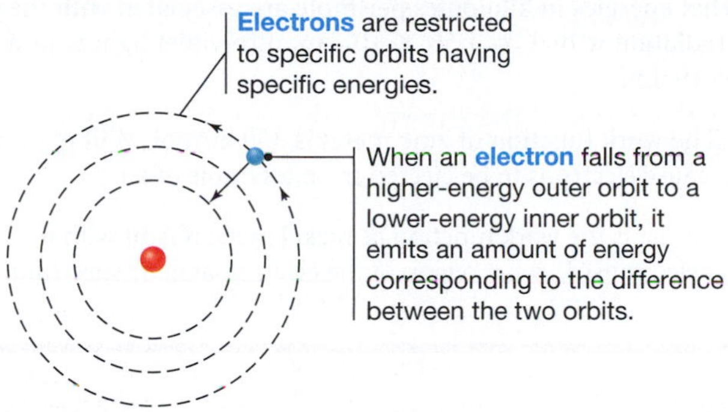

FIGURE 2.8

The Bohr model of the hydrogen atom. In this model, an electron can travel only in specific circular orbits around the nucleus.

Worked Example 2.4

Calculating the Energy of a Photon from Its Frequency

What is the energy in kilojoules per mole of photons corresponding to radar waves with $\nu = 3.35 \times 10^8$ Hz?

STRATEGY

The energy of a photon with frequency ν can be calculated with the equation $E = h\nu$. To find the energy per mole of photons, the energy of one photon must be multiplied by Avogadro's number (Section 1.9).

SOLUTION

$$E = h\nu = (6.626 \times 10^{-34}\ \text{J}\cdot\cancel{\text{s}})(3.35 \times 10^8\ \cancel{\text{s}^{-1}}) = 2.22 \times 10^{-25}\ \text{J}$$

$$\left(2.22 \times 10^{-25}\ \frac{\text{J}}{\cancel{\text{photon}}}\right)\left(6.022 \times 10^{23}\ \frac{\cancel{\text{photon}}}{\text{mol}}\right) = 0.134\ \text{J/mol}$$

$$= 1.34 \times 10^{-4}\ \text{kJ/mol}$$

Worked Example 2.5

Predicting the Frequency for a Photoelectric Effect Given a Work Function

The work function of lithium metal is $\Phi = 283$ kJ/mol. What is the minimum frequency of light needed to eject electrons from lithium?

continued on next page

STRATEGY

The work function of a metal is the minimum energy needed to eject electrons: $E = 283$ kJ/mol for lithium. Convert the work function from kJ/mol to J/photon to find the energy of a single photon, rearrange the Planck equation $E = h\nu$ to the form $\nu = E/h$, and solve for the photon frequency ν needed to eject electrons.

SOLUTION

$$E = \left(283\,\frac{\text{kJ}}{\text{mol}}\right)\left(\frac{1000\text{ J}}{1\text{ kJ}}\right)\left(\frac{1\text{ mol}}{6.022 \times 10^{23}\text{ photon}}\right) = 4.70 \times 10^{-19}\,\frac{\text{J}}{\text{photon}}$$

$$\nu = \frac{E}{h} = \frac{4.70 \times 10^{-19}\text{ J}}{6.626 \times 10^{-34}\text{ J}\cdot\text{s}} = 7.09 \times 10^{14}\text{ s}^{-1} \quad\text{or}\quad 7.09 \times 10^{14}\text{ Hz}$$

PROBLEM 2.7 What is the energy in kilojoules per mole of photons corresponding to the shortest-wavelength line in the series of the hydrogen spectrum when $m = 1$ and $n > 1$? (Worked Example 2.3)?

PROBLEM 2.8 The biological effects of a given dose of radiant energy generally become more serious as the energy of the radiation increases: Infrared radiation has a pleasant warming effect, ultraviolet radiation causes tanning and burning, and X rays can cause serious tissue damage. What energies in kilojoules per mole are associated with the following wavelengths: infrared radiation with $\lambda = 1.55 \times 10^{-6}$ m, ultraviolet light with $\lambda = 250$ nm, and X rays with $\lambda = 5.49$ nm?

PROBLEM 2.9 The work function of zinc metal is 350 kJ/mol. Will photons of violet light with $\lambda = 390$ nm cause electrons to be ejected from a sample of zinc?

PROBLEM 2.10 What is the work function of nickel metal if light with $\nu = 1.22 \times 10^{15}$ Hz is necessary to eject electrons? In what region of the electromagnetic spectrum (Figure 2.2) does this light fall?

2.4 WAVELIKE PROPERTIES OF MATTER: DE BROGLIE'S HYPOTHESIS

The analogy between matter and radiant energy developed by Einstein in the early 1900s was further extended in 1924 by the French physicist Louis de Broglie (1892–1987). De Broglie suggested that, if *light* can behave in some respects like *matter*, then perhaps *matter* can behave in some respects like *light*. That is, just as light is particlelike as well as wavelike, perhaps matter is wavelike as well as particlelike.

In developing his theory about the wavelike behavior of matter, de Broglie focused on the inverse relationship between energy and wavelength for photons:

$$\text{Since } E = \frac{hc}{\lambda} \quad \text{then} \quad \lambda = \frac{hc}{E}$$

Using the famous equation $E = mc^2$ proposed in 1905 by Einstein and substituting for E then gives

$$\lambda = \frac{hc}{E} = \frac{hc}{mc^2} = \frac{h}{mc}$$

De Broglie suggested that a similar equation might be applied to moving particles like electrons by replacing the speed of light, c, by the speed of the particle, v. The resultant **de Broglie hypothesis** thus allows calculation of a "wavelength" of an electron or of any other particle or object of mass m moving at velocity v:

▶ **de Broglie hypothesis** $\lambda = \dfrac{h}{mv}$

As an example of the de Broglie hypothesis, look at the electron in a hydrogen atom. The mass of an electron is 9.11×10^{-31} kg, and the velocity v of an electron in a hydrogen atom is 2.2×10^{6} m/s (about 1% of the speed of light). Thus, the de Broglie wavelength of an electron in a hydrogen atom is 3.3×10^{-10} m, or 330 pm. Note that Planck's constant (h), which is usually expressed in units of joule seconds (J · s), is expressed for the present purposes in units of (kg · m^2)/s [1 J = 1 (kg · m^2)/s^2].

$$\lambda = \frac{h}{mv} = \frac{6.626 \times 10^{-34}\,\frac{\cancel{\text{kg}} \cdot \text{m}^2}{\cancel{\text{s}}}}{(9.11 \times 10^{-31}\,\cancel{\text{kg}})\left(2.2 \times 10^{6}\,\frac{\cancel{\text{m}}}{\cancel{\text{s}}}\right)} = 3.3 \times 10^{-10}\ \text{m}$$

What does it mean to say that light and matter act both as waves and as particles? The answer, at least on the everyday human scale, is "not much" because our common sense isn't up to the task. Our intuition has been developed from personal experiences, using our eyes and other senses to tell us how light and matter are "supposed" to behave. We have no personal experience on the atomic scale, though, and thus have no common-sense way of dealing with the behavior of light and matter at that level. On the atomic scale, where distances and masses are so small, light and matter behave in a manner very different from what we're used to. On that tiny scale where chemistry takes place, the wave/particle behavior of light and matter is extremely important.

PROBLEM 2.11 What is the de Broglie wavelength in meters of a small car with a mass of 1150 kg traveling at a velocity of 55.0 mi/h (24.6 m/s)? Is this wavelength longer or shorter than the diameter of an atom (approximately 200 pm)?

2.5 THE QUANTUM MECHANICAL MODEL OF THE ATOM: HEISENBERG'S UNCERTAINTY PRINCIPLE

With the particlelike nature of radiant energy and the wavelike nature of matter now established, let's return to the problem of atomic structure. Several models of atomic structure were proposed in the late nineteenth and early twentieth centuries, such as the Bohr model described in Section 2.3. The Bohr model was important historically because of its conclusion that electrons have only specific energy levels available to them, but the model fails for atoms with more than one electron and must be discarded.

The breakthrough in understanding atomic structure came in 1926, when the Austrian physicist Erwin Schrödinger (1887–1961) proposed what has come to be called the **quantum mechanical model** of the atom. The fundamental idea behind the model is that it's best to abandon the notion of an electron as a small particle moving around the nucleus in a defined path and to concentrate instead on the electron's wavelike properties. In fact, it was shown in 1927 by Werner Heisenberg (1901–1976) that it is *impossible* to know both the position of an electron and what path it follows in an atom—a statement called the **Heisenberg uncertainty principle**.

The Heisenberg uncertainty principle can be understood by imagining what would happen if we tried to determine the position of an electron in an atom at a given moment. For us to "see" the electron, light photons of an appropriate frequency would have to interact with and bounce off the electron. But such an interaction would transfer energy from the photon to the electron, thereby increasing the energy of the electron and making it move faster. Thus, the very act of determining the electron's position would make that position change.

In mathematical terms, Heisenberg's principle states that the uncertainty in the electron's position, Δx, times the uncertainty in its momentum, Δmv, is equal to or greater than the quantity $h/4\pi$:

▶ **Heisenberg uncertainty principle** $(\Delta x)(\Delta mv) \geq \dfrac{h}{4\pi}$

According to this equation, we can never know both the position and the velocity of an electron (or of any other object) beyond a certain level of precision. If we know the *velocity* with a high degree of certainty (Δmv is small), then the *position* of the electron must be uncertain (Δx must be large). Conversely, if we know the position of the electron exactly (Δx is small), then we can't know its velocity (Δmv must be large). As a result, an electron will always appear as something of a blur whenever we attempt to make any physical measurements of its position and velocity.

A brief calculation can help make the importance of the uncertainty principle clearer. As mentioned in the previous section, the mass m of an electron is 9.11×10^{-31} kg and the velocity v of an electron in a hydrogen atom is approximately 2.2×10^{6} m/s. If we assume that the velocity is known to within 5%, or 0.1×10^{6} m/s, then the uncertainty in the electron's position in a hydrogen atom is greater than 1.5×10^{-10} m, or 150 pm. But since the diameter of a hydrogen atom is only 240 pm, *the uncertainty in the electron's position is similar in size to the size of the atom itself!*

▲ Even the motion of very fast objects such as bullets can be captured in daily life. On the atomic scale, however, velocity and position can't both be known precisely at the same time.

$$\text{If } (\Delta x)(\Delta mv) \geq \frac{h}{4\pi} \quad \text{then } (\Delta x) \geq \frac{h}{(4\pi)(\Delta mv)}$$

$$\Delta x \geq \frac{6.626 \times 10^{-34}\ \frac{\text{kg} \cdot \text{m}^2}{\text{s}}}{(4)(3.1416)(9.11 \times 10^{-31}\ \text{kg})\left(0.1 \times 10^{6}\ \frac{\text{m}}{\text{s}}\right)}$$

$$\Delta x \geq 1.5 \times 10^{-10}\ \text{m} \quad \text{or} \quad 150\ \text{pm}$$

When the mass m of an object is relatively large, as in daily life, then both Δx and Δv in the Heisenberg relationship are very small, so we have no problem in measuring both position and velocity for visible objects. The problem arises only on the atomic scale.

2.6 THE QUANTUM MECHANICAL MODEL OF THE ATOM: ORBITALS AND THE THREE QUANTUM NUMBERS

Schrödinger's quantum mechanical model of atomic structure is framed mathematically in what is called a *wave equation* because it is similar in form to the equation used to describe the motion of ordinary waves in fluids. The solutions to the wave equation are called *wave functions*, or **orbitals**, and are represented by the symbol ψ (Greek psi). A good way to think about an electron's orbital is to regard it as an expression whose square, ψ^2, defines the probability of finding the electron within a given region of space around the nucleus. As Heisenberg showed, we can never be completely *certain* about an electron's position in an atom, but an orbital tells where the electron will most *probably* be found.

Wave equation $\xrightarrow{\text{Solve}}$ Wave function or orbital (ψ) $\longrightarrow$ Probability of finding electron in a region of space (ψ^2)

LOOKING AHEAD...

Although the quantum mechanical model uses only three quantum numbers to describe an orbital, we'll see in Section 2.8 that a fourth quantum number, called the **spin quantum number, m_s,** is also needed to complete the entire picture of atomic structure.

An orbital is characterized by three parameters called **quantum numbers**, represented as n, l, and m_l, which describe the energy level of the orbital and the three-dimensional shape and orientation of the region in space occupied by an electron in that orbital.

- **The principal quantum number (n)** is a positive integer ($n = 1, 2, 3, 4, \ldots$) on which the size and energy level of the orbital primarily depend. For hydrogen and other one-electron atoms, such as He^{+}, the energy of an orbital depends *only* on n. For atoms with more than one electron, the energy level of an orbital depends both on n and on the l quantum number.

  As the value of n increases, the number of allowed orbitals increases and the size of those orbitals becomes larger, thus allowing an electron to be farther from the nucleus. Because it takes energy to separate a negative charge from a positive charge, this increased distance between the electron and the nucleus means that the energy of the electron in the orbital increases as the quantum number n increases.

We often speak of orbitals as being grouped according to the principal quantum number n into successive layers, or **shells**, around the nucleus. Those orbitals with $n = 3$, for example, are said to be in the third shell.

- **The angular-momentum quantum number (l)** defines the three-dimensional shape of an orbital. For an orbital whose principal quantum number is n, the angular-momentum quantum number l can have any integral value from 0 to $n - 1$. Thus, within each shell, there are n different shapes for orbitals.

If $n = 1$, then $l = 0$

If $n = 2$, then $l = 0$ or 1

If $n = 3$, then $l = 0, 1$, or 2

. . . and so forth

Just as it's convenient to think of orbitals as being grouped into shells according to the principal quantum number n, we often speak of orbitals within a shell as being further grouped into **subshells** according to the angular-momentum quantum number l. Different subshells are usually referred to by letter rather than by number, following the order *s*, *p*, *d*, *f*, *g*. (Historically, the letters *s*, *p*, *d*, and *f* arose from the use of the words *sharp*, *principal*, *diffuse*, and *fundamental* to describe various lines in atomic spectra.) After *f*, successive subshells are designated alphabetically: *g*, *h*, and so on.

Quantum number l:	0	1	2	3	4 . . .
Subshell notation:	*s*	*p*	*d*	*f*	*g* . . .

As an example, an orbital with $n = 3$ and $l = 2$ is called a 3*d* orbital: 3 to represent the third shell and *d* to represent the $l = 2$ subshell.

- **The magnetic quantum number (m_l)** defines the spatial orientation of the orbital with respect to a standard set of coordinate axes. For an orbital whose angular-momentum quantum number is l, the magnetic quantum number m_l can have any integral value from $-l$ to $+l$. Thus, within each subshell—orbitals with the same shape, or value of l—there are $2l + 1$ different spatial orientations for those orbitals. We'll explore this point further in the next section.

If $l = 0$ (or *s*), then $m_l = 0$

If $l = 1$ (or *p*), then $m_l = -1, 0$, or $+1$

If $l = 2$ (or *d*), then $m_l = -2, -1, 0, +1$, or $+2$

. . . and so forth

A summary of the allowed combinations of quantum numbers for the first four shells is given in **Table 2.2**.

TABLE 2.2 Allowed Combinations of Quantum Numbers n, l, and m_l for the First Four Shells

n	l	m_l	Orbital Notation	Number of Orbitals in Subshell	Number of Orbitals in Shell
1	0	0	1*s*	1	1
2	0	0	2*s*	1	4
	1	−1, 0, +1	2*p*	3	
3	0	0	3*s*	1	9
	1	−1, 0, +1	3*p*	3	
	2	−2, −1, 0, +1, +2	3*d*	5	
4	0	0	4*s*	1	16
	1	−1, 0, +1	4*p*	3	
	2	−2, −1, 0, +1, +2	4*d*	5	
	3	−3, −2, −1, 0, +1, +2, +3	4*f*	7	

The energy levels of various orbitals are shown in **Figure 2.9**. As noted earlier in this section, the energy levels of different orbitals in a hydrogen atom depend only on the principal quantum number n, but the energy levels of orbitals in multielectron atoms depend on both n and l. In other words, the orbitals in a given shell all have the same energy for hydrogen but have slightly different energies for other atoms, depending on their subshell. In fact, there is even some crossover of energies from one shell to another. A 3d orbital in some multielectron atoms has a higher energy than a 4s orbital, for instance.

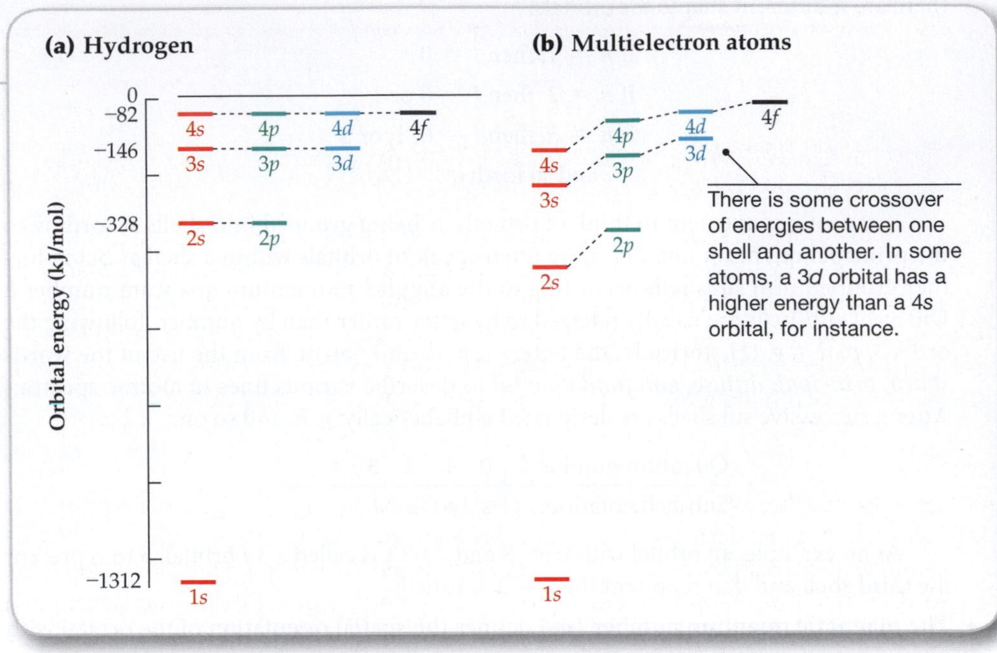

FIGURE 2.9

Energy levels of atomic orbitals: (**a**) hydrogen and (**b**) a typical multielectron atom. The differences between energies of various subshells in (**b**) are exaggerated for clarity.

Worked Example 2.6

Using Quantum Numbers to Identify an Orbital

Identify the shell and subshell of an orbital with the quantum numbers $n = 3, l = 1, m_l = 1$.

STRATEGY

The principal quantum number n gives the shell number, and the angular-momentum quantum number l gives the subshell designation. The magnetic quantum number m_l is related to the spatial orientation of the orbital within the subshell.

SOLUTION

A value of $n = 3$ indicates that the orbital is in the third shell, and a value of $l = 1$ indicates that the orbital is of the p type. Thus, the orbital has the designation 3p.

Worked Example 2.7

Assigning Quantum Numbers to an Orbital

Give the possible combinations of the three quantum numbers for a 4p orbital.

STRATEGY

The designation 4p indicates that the orbital has a principal quantum number $n = 4$ and an angular-momentum quantum number $l = 1$. The magnetic quantum number m_l can have any of the three values −1, 0, or +1.

SOLUTION

The allowable combinations are

$$n = 4, l = 1, m_l = -1 \qquad n = 4, l = 1, m_l = 0 \qquad n = 4, l = 1, m_l = +1$$

PROBLEM 2.12 Extend Table 2.2 to show allowed combinations of quantum numbers when $n = 5$. How many orbitals are in the fifth shell?

PROBLEM 2.13 Give orbital notations for electrons in orbitals with the following quantum numbers:
(a) $n = 2, l = 1, m_l = 1$ (b) $n = 4, l = 3, m_l = -2$ (c) $n = 3, l = 2, m_l = -1$

PROBLEM 2.14 Give the allowed combinations of three quantum numbers for the following orbitals:
(a) A 3*s* orbital (b) A 2*p* orbital (c) A 4*d* orbital

2.7 ORBITALS AND THEIR SHAPES

We said in the previous section that the square of a wave function, or orbital, describes the probability of finding the electron within a specific region of space within an atom. The shape of that spatial region is defined by the angular-momentum quantum number *l*, with $l = 0$ called an *s* orbital, $l = 1$ a *p* orbital, $l = 2$ a *d* orbital, and so forth. Of the various possibilities, *s*, *p*, *d*, and *f* orbitals are the most important because these are the only ones actually occupied by electrons in known elements. Let's look at each of the four individually.

s Orbitals

All *s* orbitals are spherical, meaning that the probability of finding an *s* electron depends only on distance from the nucleus, not on direction. Furthermore, because there is only one possible orientation of a sphere in space, an *s* orbital has $m_l = 0$ and there is only one *s* orbital per shell.

As shown in **Figure 2.10**, the value of ψ^2 for an *s* orbital is greatest near the nucleus and then drops off rapidly as distance from the nucleus increases, although it never goes all the way to zero, even at a large distance. As a result, there is no definite boundary to an atom

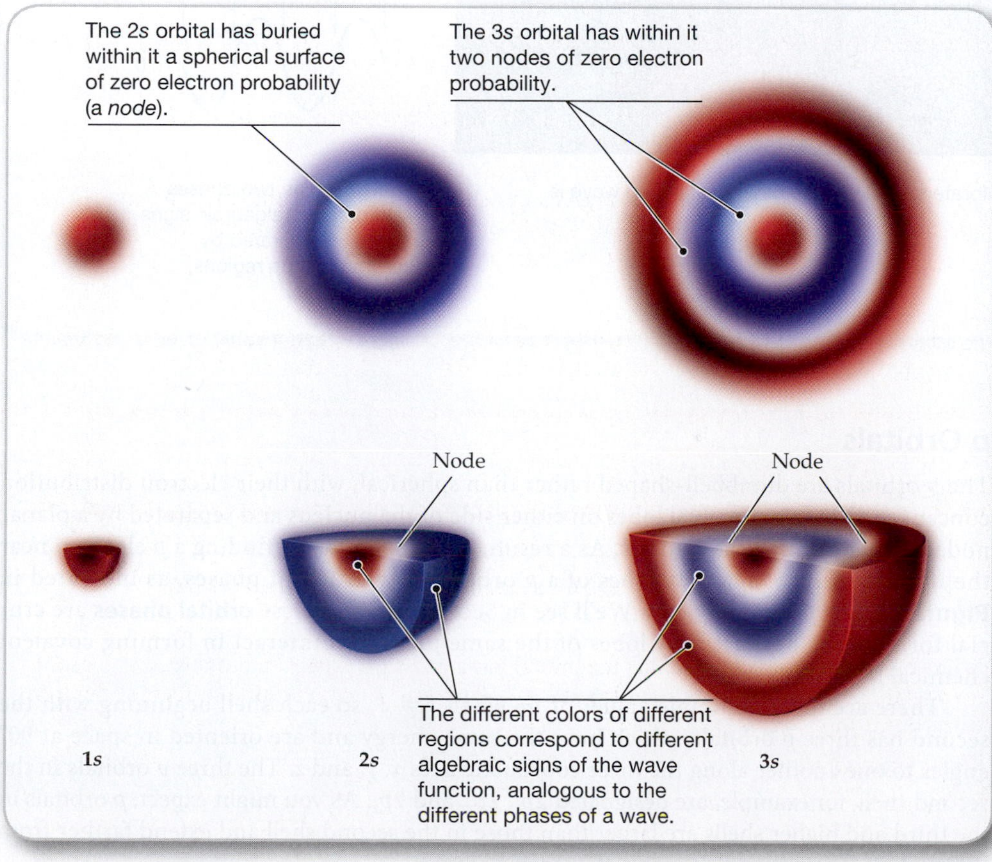

FIGURE 2.10

Representations of 1*s*, 2*s*, and 3*s* orbitals. Slices through these spherical orbitals are shown on the top and cutaway views on the bottom, with the probability of finding an electron represented by the density of the shading.

and no definite size. For purposes like that of Figure 2.10, however, we usually imagine a boundary surface enclosing the volume where an electron spends most (say, 95%) of its time.

Although all *s* orbitals are spherical, there are significant differences among the *s* orbitals in different shells. For one thing, the size of the *s* orbital increases in successively higher shells, implying that an electron in an outer-shell *s* orbital is farther from the nucleus on average than an electron in an inner-shell *s* orbital and is therefore higher in energy. For another thing, the electron distribution in an outer-shell *s* orbital has more than one region of high probability. As shown in Figure 2.10, a 2*s* orbital is essentially a sphere within a sphere and has two regions of high probability, separated by a surface of zero electron probability called a **node**. Similarly, a 3*s* orbital has three regions of high probability and two spherical nodes.

The concept of an orbital node—a surface of zero electron probability separating regions of nonzero probability—is difficult to grasp because it raises the question: How does an electron get from one region of the orbital to another if it can't be at the node? The question is misleading, though, because it assumes particlelike behavior for the electron rather than the wavelike behavior that forms the basis of the quantum mechanical model.

In fact, nodes are an intrinsic property of all waves, from moving waves of water in the ocean to the stationary, or standing, wave generated by vibrating a rope or guitar string (**Figure 2.11**). A node simply corresponds to the zero-amplitude part of the wave. On either side of the node is a nonzero wave amplitude. Note that a wave has two **phases**—peaks above the zero line and troughs below—corresponding to different algebraic signs, + and −. Similarly, the different regions of 2*s* and 3*s* orbitals have different phases, + and −, as indicated in Figure 2.10 by different colors.

When a rope is fixed at one end and vibrated rapidly at the other, a standing wave is generated.

The wave has two phases with different algebraic signs, + and −, separated by zero-amplitude regions, called *nodes*.

FIGURE 2.11

A standing wave in a vibrating rope.

p Orbitals

The *p* orbitals are dumbbell-shaped rather than spherical, with their electron distribution concentrated in two identical lobes on either side of the nucleus and separated by a planar node cutting through the nucleus. As a result, the probability of finding a *p* electron near the nucleus is zero. The two lobes of a *p* orbital have different phases, as indicated in **Figure 2.12** by different colors. We'll see in Section 5.2 that these **orbital phases** are crucial for bonding because only lobes of the same phase can interact in forming covalent chemical bonds.

LOOKING AHEAD ...

We'll see in Section 5.2 that a bond forms between two atoms only if an orbital of a given **phase** on one atom interacts with an orbital of the same phase on the other atom. No bond can form if the phases of the two interacting orbitals are not the same.

There are three allowable values of m_l when $l = 1$, so each shell beginning with the second has three *p* orbitals, which have the same energy and are oriented in space at 90° angles to one another along the three coordinate axes *x*, *y*, and *z*. The three *p* orbitals in the second shell, for example, are designated $2p_x$, $2p_y$, and $2p_z$. As you might expect, *p* orbitals in the third and higher shells are larger than those in the second shell and extend farther from the nucleus. Their shape is roughly the same, however.

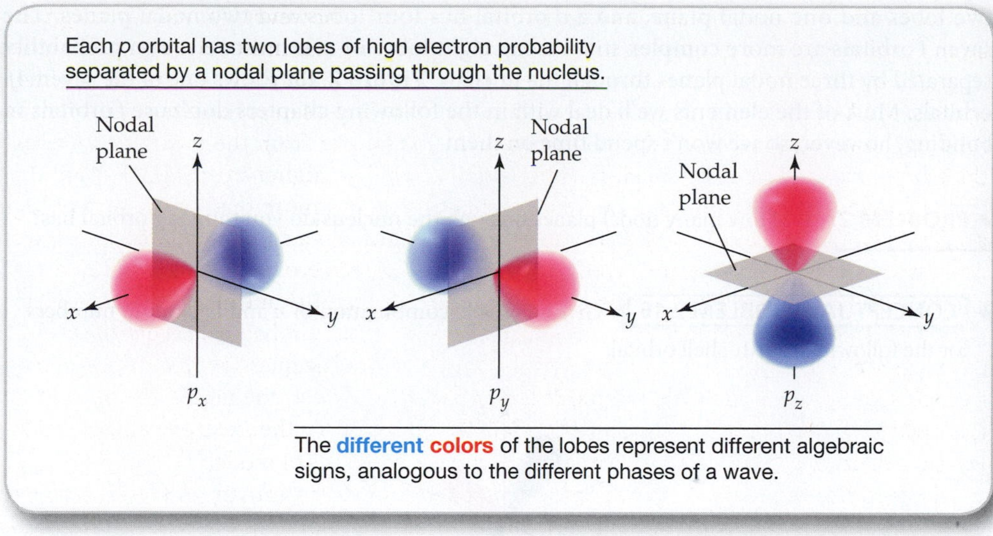

FIGURE 2.12

Representations of the three 2*p* orbitals. Each orbital is dumbbell-shaped and oriented in space along one of the three coordinate axes *x*, *y*, or *z*.

d and *f* Orbitals

The third and higher shells each contain five *d* orbitals, which differ from their *s* and *p* counterparts because they have two different shapes. Four of the five *d* orbitals are cloverleaf-shaped and have four lobes of maximum electron probability separated by two nodal planes through the nucleus (**Figure 2.13 a–d**). The fifth *d* orbital is similar in shape to a p_z orbital but has an additional donut-shaped region of electron probability centered in the *xy* plane (**Figure 2.13e**). Despite their different shapes, all five *d* orbitals in a given shell have the same energy. As with *p* orbitals, alternating lobes of the *d* orbitals have different phases.

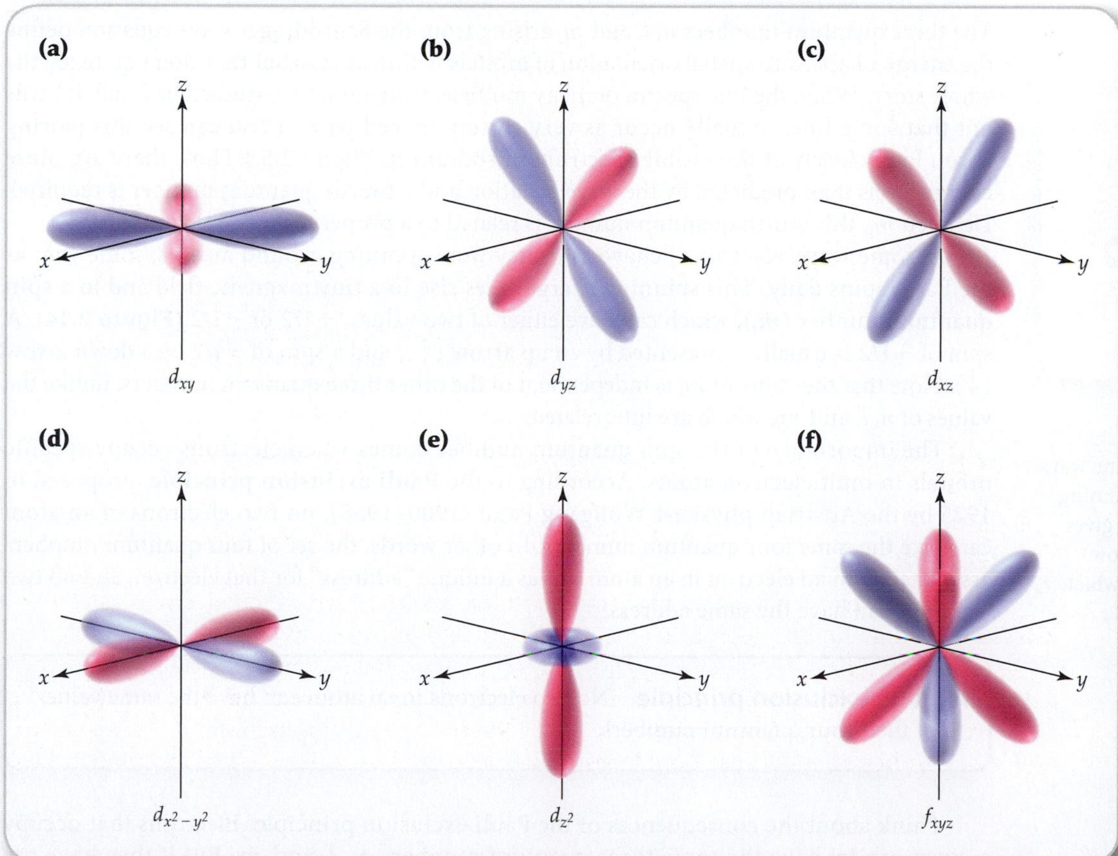

FIGURE 2.13

Representations of the five 3*d* orbitals. Four of the orbitals are shaped like a cloverleaf (**a–d**), and the fifth is shaped like an elongated dumbbell inside a donut (**e**). Also shown is one of the seven 4*f* orbitals (**f**). As with *p* orbitals in Figure 2.12, the different colors of the lobes reflect different phases.

You may have noticed that both the number of nodal planes through the nucleus and the overall geometric complexity of the orbitals increase with the *l* quantum number of the subshell: An *s* orbital has one lobe and no nodal plane through the nucleus, a *p* orbital has two lobes and one nodal plane, and a *d* orbital has four lobes and two nodal planes. The seven *f* orbitals are more complex still, having eight lobes of maximum electron probability separated by three nodal planes through the nucleus. **Figure 2.13f** shows one of the seven 4*f* orbitals. Most of the elements we'll deal with in the following chapters don't use *f* orbitals in bonding, however, so we won't spend time on them.

PROBLEM 2.15 How many nodal planes through the nucleus do you think a *g* orbital has?

CONCEPTUAL PROBLEM 2.16 Give a possible combination of *n* and *l* quantum numbers for the following fourth-shell orbital:

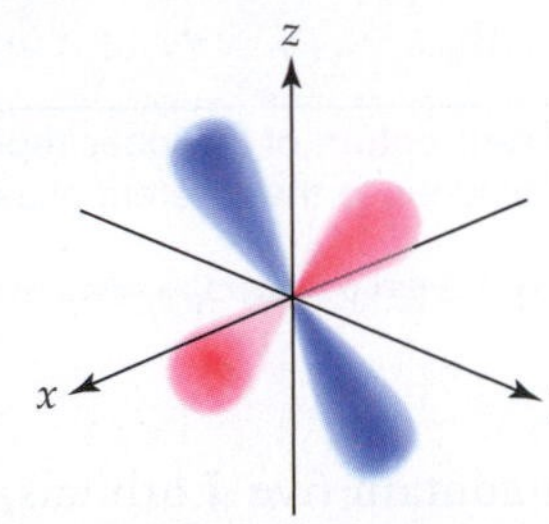

2.8 A FOURTH QUANTUM NUMBER: ELECTRON SPIN AND PAULI'S EXCLUSION PRINCIPLE

The three quantum numbers n, l, and m_l arising from the Schrödinger wave equation define the energy, shape, and spatial orientation of orbitals within atoms, but they don't quite tell the whole story. When the line spectra of many multielectron atoms are studied in detail, it turns out that some lines actually occur as very closely spaced *pairs*. (You can see this pairing if you look closely at the visible spectrum of sodium in Figure 2.5.) Thus, there are more energy levels than predicted by the wave equation and a fourth quantum number is required. Denoted m_s, this fourth quantum number is related to a property called *electron spin*.

In some ways, electrons behave as if they were spinning around an axis, somewhat as the Earth spins daily. This spinning charge gives rise to a tiny magnetic field and to a **spin quantum number (m_s),** which can have either of two values, $+1/2$ or $-1/2$ (**Figure 2.14**). A spin of $+1/2$ is usually represented by an up arrow (↑), and a spin of $-1/2$ by a down arrow (↓). Note that the value of m_s is independent of the other three quantum numbers, unlike the values of n, l, and m_l, which are interrelated.

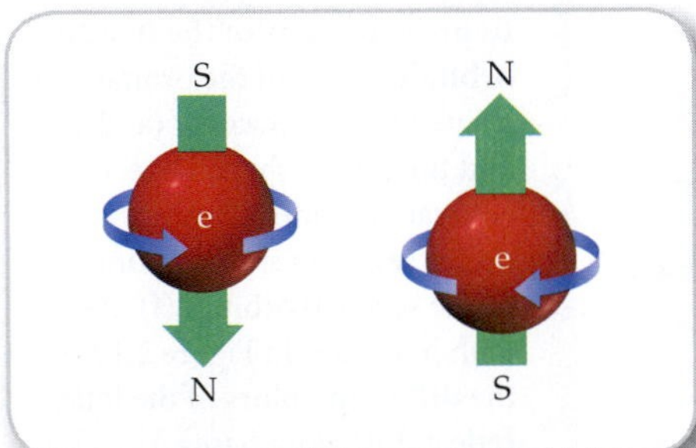

FIGURE 2.14

Electron spin. Electrons behave in some ways as if they were tiny charged spheres spinning around an axis. This **spin** (blue arrow) gives rise to a tiny **magnetic field** (green arrow) and to a fourth quantum number, m_s, which can have a value of either $+1/2$ or $-1/2$.

The importance of the spin quantum number comes when electrons occupy specific orbitals in multielectron atoms. According to the **Pauli exclusion principle**, proposed in 1925 by the Austrian physicist Wolfgang Pauli (1900–1958), no two electrons in an atom can have the same four quantum numbers. In other words, the set of four quantum numbers associated with an electron in an atom acts as a unique "address" for that electron, and no two electrons can have the same address.

▶ **Pauli exclusion principle** No two electrons in an atom can have the same values of their four quantum numbers.

Think about the consequences of the Pauli exclusion principle. Electrons that occupy a given orbital have the same three quantum numbers, n, l, and m_l. But if they have the *same* values for n, l, and m_l, they must have *different* values for the fourth quantum number: either $m_s = +1/2$ or $m_s = -1/2$. Thus, an orbital can hold only two electrons, which must have opposite spins. An atom with x number of electrons therefore has at least $x/2$ occupied orbitals (although it might have more if some of its orbitals are only half-filled).

2.9 THE QUANTUM MECHANICAL MODEL AND ATOMIC LINE SPECTRA

Now that we've seen how atomic structure is described according to the quantum mechanical model, let's return briefly to finish the subject of atomic line spectra that was first mentioned in Section 2.2. How does the quantum mechanical model of atomic structure account for the discrete wavelengths of light found in a line spectrum and for the **Balmer–Rydberg equation** that gives the values of those wavelengths?

REMEMBER...

The **Balmer–Rydberg equation** has two variables m and n, which are integers with $n > m$. (Section 2.2)

$$\frac{1}{\lambda} = R_\infty\left[\frac{1}{m^2} - \frac{1}{n^2}\right]$$

Each electron in an atom occupies an orbital, and each orbital has a specific energy level. Thus, the energy levels available to electrons are quantized and can have only the specific values associated with the orbitals they occupy. When an atom is heated in a flame or electric discharge, the heat energy causes an electron to be promoted from a lower-energy orbital to a higher-energy orbital. In a hydrogen atom, for instance, the electron might jump from the 1s orbital to a second-shell orbital, to a third-shell orbital, or to an orbital in any higher shell, depending on the amount of energy added.

But the energetically excited atom is relatively unstable, so the electron rapidly returns to a lower-energy level accompanied by *emission* of energy equal to the difference between the higher and lower orbitals. Because the energies of the orbitals are quantized, *the amount of energy emitted is also quantized.* Thus, we observe the emission of only specific frequencies of radiation (**Figure 2.15**). By measuring the frequencies emitted by excited hydrogen atoms, we can calculate the energy differences between pairs of orbitals.

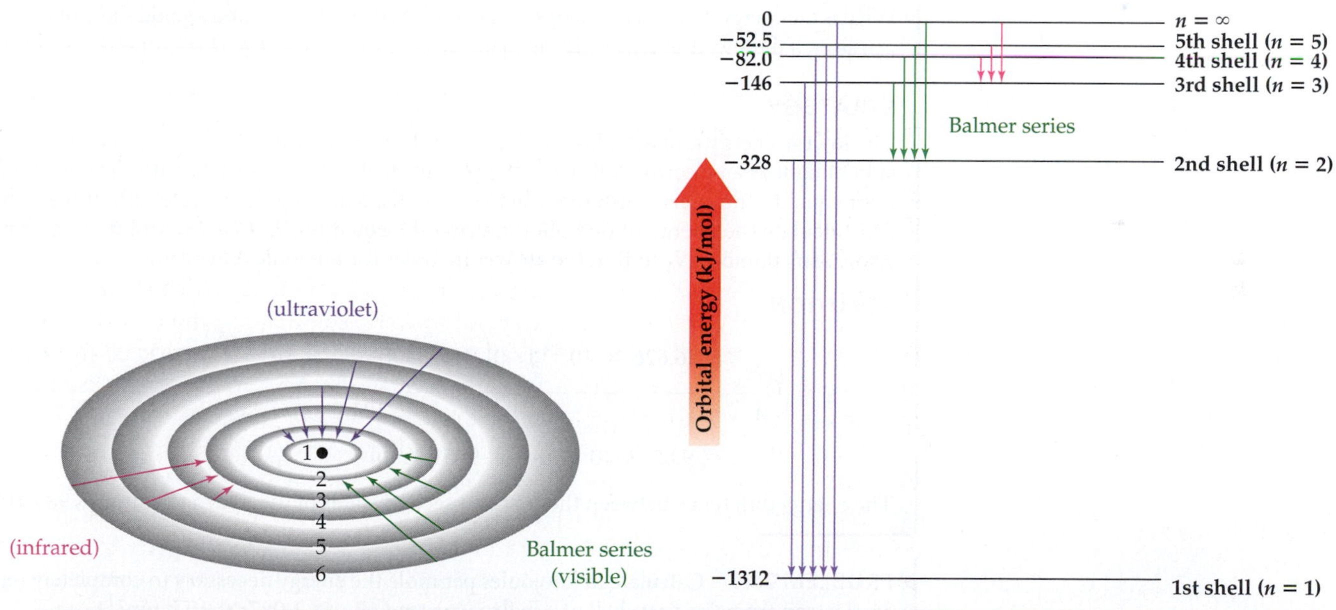

When an electron falls from a higher-energy outer-shell orbital to a lower-energy inner-shell orbital, it emits electromagnetic energy whose frequency corresponds to the energy difference between the orbitals.

The different spectral series correspond to electronic transitions from outer-shell orbitals to different inner-shell orbitals.

FIGURE 2.15

The origin of atomic line spectra.

The variables n and m in the Balmer–Rydberg equation for hydrogen represent the principal quantum numbers of the two orbitals involved in the electronic transition. The variable n corresponds to the principal quantum number of the higher-energy, outer-shell orbital that the transition is *from*, and the variable m corresponds to the principal quantum number of the lower-energy, inner-shell orbital that the transition is *to*. When $m = 1$, for example, the frequencies of emitted light correspond to energy differences between various outer-shell orbitals and the first-shell orbital. When $m = 2$ (the visible Balmer series), the frequencies correspond to energy differences between outer-shell orbitals and the second-shell orbitals, and so on.

$$\frac{1}{\lambda} = R_{\infty}\left(\frac{1}{m^2} - \frac{1}{n^2}\right)$$

Shell the transition is *to* (inner-shell) — Shell the transition is *from* (outer-shell)

Notice in Figure 2.15 that as n becomes larger and approaches infinity, the energy difference between the n shell and the first shell converges to a value of 1312 kJ/mol. That is, 1312 kJ is released when electrons come from a great distance (the "infinite" shell) and add to H^+ to give a mole of hydrogen atoms, each with an electron in its first shell:

$$H^+ + e^- \longrightarrow H + \textbf{Energy} \quad (1312\ \text{kJ/mol})$$

Because the energy *released* on adding an electron to H^+ is equal to the energy *absorbed* on removing an electron from a hydrogen atom, we can also say that 1312 kJ/mol is required to remove the electron from a hydrogen atom. We'll see in the next chapter that the amount of energy necessary to remove an electron from a given atom provides an important clue about that element's chemical reactivity.

What is true for hydrogen is also true for all other atoms: All atoms show atomic line spectra when energetically excited electrons fall from higher-energy orbitals in outer shells to lower-energy orbitals in inner shells. As you might expect, though, these spectra become very complex for multielectron atoms in which different orbitals within a shell no longer have the same energies and in which a large number of electronic transitions are possible.

Worked Example 2.8

Calculating the Energy Difference between Two Orbitals

What is the energy difference in kilojoules per mole between the first and second shells of the hydrogen atom if the lowest-energy emission in the spectral series with $m = 1$ occurs at $\lambda = 121.5$ nm?

STRATEGY

The lowest-energy emission line in the spectral series with $m = 1$ corresponds to the emission of light as an electron falls from the second shell to the first shell, with the energy of that light equal to the energy difference between shells. Knowing the wavelength of the light, we can calculate the energy of one photon using the equation, $E = hc/\lambda$, and then multiply by Avogadro's number N_A to find the answer in joules (or kilojoules) per mole:

SOLUTION

$$E = \frac{hcN_A}{\lambda} = \frac{(6.626 \times 10^{-34}\ \text{J}\cdot\text{s})\left(3.00 \times 10^{8}\ \frac{\text{m}}{\text{s}}\right)\left(10^{9}\ \frac{\text{nm}}{\text{m}}\right)(6.022 \times 10^{23}\ \text{mol}^{-1})}{121.5\ \text{nm}}$$

$$= 9.85 \times 10^{5}\ \text{J/mol} = 985\ \text{kJ/mol}$$

The energy difference between the first and second shells of the hydrogen atom is 985 kJ/mol.

PROBLEM 2.17 Calculate in kilojoules per mole the energy necessary to completely remove an electron from the first shell of a hydrogen atom ($R_{\infty} = 1.097 \times 10^{-2}\ \text{nm}^{-1}$).

2.10 ORBITAL ENERGY LEVELS IN MULTIELECTRON ATOMS

As we said in Section 2.6, the energy level of an orbital in a hydrogen atom, which has only one electron, is determined by its principal quantum number n. Within a shell, all hydrogen orbitals have the same energy, independent of their other quantum numbers. The situation is different in multielectron atoms, however, where the energy level of a given orbital depends not only on the shell but also on the subshell. The *s*, *p*, *d*, and *f* orbitals within a given shell have slightly different energies in a multielectron atom, as shown previously in Figure 2.9, and there is even some crossover of energies between orbitals in different shells.

The difference in energy between subshells in multielectron atoms results from electron–electron repulsions. In hydrogen, the only electrical interaction is the attraction of the positive nucleus for the negative electron, but in multielectron atoms there are many different interactions. Not only are there the attractions of the nucleus for each electron, there are also the repulsions between every electron and its neighbors.

The repulsion of outer-shell electrons by inner-shell electrons is particularly important because the outer-shell electrons are pushed farther away from the nucleus and are thus held less tightly. Part of the electrical attraction of the nucleus for an outer electron is thereby canceled, an effect we describe by saying that the outer electrons are *shielded* from the nucleus by the inner electrons (**Figure 2.16**). As a result, the nuclear charge actually felt by an electron, called the **effective nuclear charge, Z_{eff}**, is often substantially lower than the actual nuclear charge Z.

▶ **Effective nuclear charge** $Z_{eff} = Z_{actual} - \text{Electron shielding}$

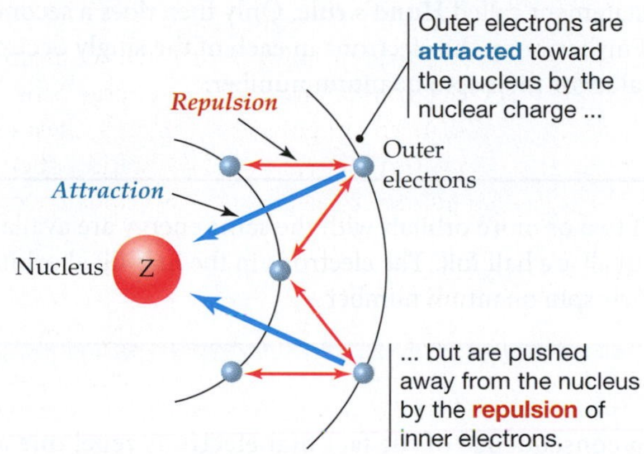

FIGURE 2.16

The origin of electron shielding and Z_{eff}. The outer electrons feel a diminished attraction to the nucleus because inner electrons shield them from feeling the full nuclear charge.

How does electron shielding lead to the observed energy differences among orbitals within a shell? The answer is a consequence of the differences in orbital shapes. Compare a 2*s* orbital with a 2*p* orbital, for instance. The 2*s* orbital is spherical and has a large probability density near the nucleus, while the 2*p* orbitals are dumbbell-shaped and have a node at the nucleus (Section 2.7). An electron in a 2*s* orbital therefore spends more time closer to the nucleus than an electron in a 2*p* orbital does and is less shielded. A 2*s* electron thus feels a higher Z_{eff}, is more tightly held by the nucleus, and is lower in energy than a 2*p* electron. In the same way, a 3*p* electron spends more time closer to the nucleus, feels a higher Z_{eff}, and has a lower energy than a 3*d* electron. More generally, within any given shell, a lower value of the angular-momentum quantum number *l* corresponds to a higher Z_{eff} and to a lower energy for the electron.

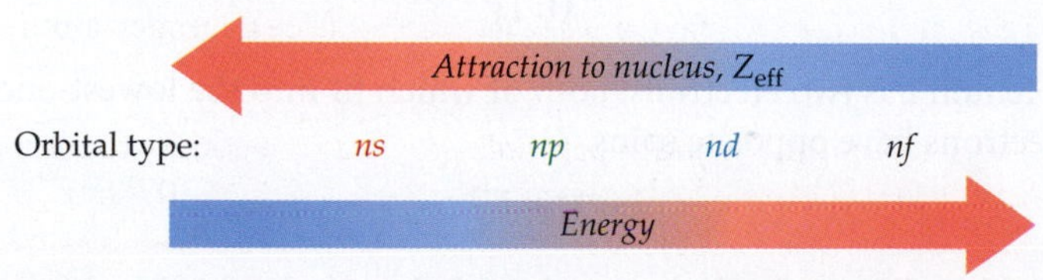

The idea that electrons in different orbitals are shielded differently and feel different values of Z_{eff} is a very useful one that we'll return to on numerous occasions to explain various chemical phenomena.

2.11 ELECTRON CONFIGURATIONS OF MULTIELECTRON ATOMS

All the parts are now in place to provide an electronic description for the atoms of every element. Knowing the relative energies of the various orbitals, we can predict for each element which orbitals are occupied by electrons—the element's **electron configuration**.

A set of three guidelines called the **aufbau principle**, from the German word for "building up," guides the filling order of orbitals. In general, each successive electron added to an atom occupies the lowest-energy orbital available. The resultant lowest-energy configuration is called the **ground-state electron configuration** of the atom. Often, several orbitals will have the same energy level—for example, the three *p* orbitals or the five *d* orbitals in a given subshell. Orbitals that have the same energy level are said to be **degenerate**.

Guidelines of the aufbau principle:

1. **Lower-energy orbitals fill before higher-energy orbitals.** The ordering of energy levels for orbitals was shown in Figure 2.9 on page 74.
2. **An orbital can hold only two electrons, which must have opposite spins.** This is just a restatement of the Pauli exclusion principle (Section 2.8), emphasizing that no two electrons in an atom can have the same set of four quantum numbers.
3. **If two or more degenerate orbitals are available, one electron goes into each until all are half-full**, a statement called **Hund's rule**. Only then does a second electron fill one of the orbitals. Furthermore, the electrons in each of the singly occupied orbitals must have the same value for their spin quantum number.

▶ **Hund's rule** If two or more orbitals with the same energy are available, one electron goes in each until all are half full. The electrons in the half-filled orbitals all have the same value of their spin quantum number.

Hund's rule is a consequence of the fact that electrons repel one another and therefore remain as far apart as possible. Not surprisingly, they can remain farther apart and be lower in energy if they are in different orbitals describing different spatial regions than if they are in the same orbital occupying the same region. It also turns out that electrons in half-filled orbitals stay farther apart on average if they have the same spin rather than opposite spins.

Electron configurations are normally represented by listing the *n* quantum number and the *s*, *p*, *d*, or *f* designation of the occupied orbitals, beginning with the lowest energy one. A superscript is added to indicate the number of electrons occupying each orbital. Let's look at some examples to see how the aufbau criteria are applied.

- **Hydrogen:** Hydrogen has only one electron, which must go into the lowest-energy, 1*s* orbital. Thus, the ground-state electron configuration of hydrogen is $1s^1$.

$$\text{H: } 1s^1$$

- **Helium:** Helium has two electrons, both of which fit into the lowest-energy, 1*s* orbital. The two electrons have opposite spins.

$$\text{He: } 1s^2$$

- **Lithium and beryllium:** With the 1*s* orbital full, both the third and fourth electrons go into the next available orbital, 2*s*.

$$\text{Li: } 1s^2\,2s^1 \qquad \text{Be: } 1s^2\,2s^2$$

- **Boron through neon:** The six elements from boron through neon have their three 2*p* orbitals filled successively. Because these three 2*p* orbitals have the same energy, they are degenerate and are thus filled according to Hund's rule. In carbon, for instance, the two 2*p* electrons are in different orbitals, which can be arbitrarily specified as $2p_x$, $2p_y$, or $2p_z$ when writing the electron configuration. The same is true of nitrogen, whose three 2*p* electrons must be in three different orbitals. Although not usually noted in the written electron configuration, the electrons in each of the singly occupied carbon and nitrogen 2*p* orbitals have the same value of the spin quantum number—either $+1/2$ or $-1/2$.

For clarity, we sometimes specify electron configurations using orbital-filling diagrams, in which electrons are represented by arrows. The two values of the spin quantum numbers are indicated by having the arrow point either up (↑) or down (↓). An up–down pair (↑↓) indicates that an orbital is filled, while a single up or down arrow indicates that an orbital is half filled. Note in the diagrams for carbon and nitrogen that the degenerate $2p$ orbitals are half filled rather than filled, according to Hund's rule, and that the electron spin is the same in each.

		1s	2s	2p		
B: $1s^2\,2s^2\,2p^1$	or	↓↑	↓↑	↑	—	—
C: $1s^2\,2s^2\,2p_x^{\,1}\,2p_y^{\,1}$	or	↓↑	↓↑	↑	↑	—
N: $1s^2\,2s^2\,2p_x^{\,1}\,2p_y^{\,1}\,2p_z^{\,1}$	or	↓↑	↓↑	↑	↑	↑

From oxygen through neon, the three $2p$ orbitals are successively filled. For fluorine and neon, it's no longer necessary to distinguish among the different $2p$ orbitals, so we can simply write $2p^5$ and $2p^6$.

		1s	2s	2p		
O: $1s^2\,2s^2\,2p_x^{\,2}\,2p_y^{\,1}\,2p_z^{\,1}$	or	↓↑	↓↑	↓↑	↑	↑
F: $1s^2\,2s^2\,2p^5$	or	↓↑	↓↑	↓↑	↓↑	↑
Ne: $1s^2\,2s^2\,2p^6$	or	↓↑	↓↑	↓↑	↓↑	↓↑

- **Sodium and magnesium:** The $3s$ orbital is filled next, giving sodium and magnesium the ground-state electron configurations shown. Note that we often write the configurations in a shorthand version by giving the symbol of the noble gas in the previous row to indicate electrons in filled shells and then specifying only those electrons in partially filled shells.

Neon configuration ($1s^2\,2s^2\,2p^6$)

Na: $1s^2\,2s^2\,2p^6\;3s^1$ or [Ne] $3s^1$

Mg: $1s^2\,2s^2\,2p^6\;3s^2$ or [Ne] $3s^2$

- **Aluminum through argon:** The $3p$ orbitals are now filled according to the same criteria used previously for filling the $2p$ orbitals of boron through neon. Rather than explicitly identify which of the degenerate $3p$ orbitals are occupied in Si, P, and S, we'll simplify the writing by giving just the total number of electrons in the subshell. For example, we'll write $3p^2$ for silicon rather than $3p_x^{\,1}3p_y^{\,1}$.

Al: [Ne] $3s^2\,3p^1$ **Si:** [Ne] $3s^2\,3p^2$ **P:** [Ne] $3s^2\,3p^3$

S: [Ne] $3s^2\,3p^4$ **Cl:** [Ne] $3s^2\,3p^5$ **Ar:** [Ne] $3s^2\,3p^6$

- **Elements past argon:** Following the filling of the $3p$ subshell in argon, the first crossover in the orbital filling order is encountered, as noted previously in Figure 2.9b. Rather than continue filling the third shell by populating the $3d$ orbitals, the next two electrons in potassium and calcium go into the $4s$ subshell. Only then does filling of the $3d$ subshell occur to give the first transition metal series from scandium through zinc.

K: [Ar] $4s^1$ **Ca:** [Ar] $4s^2$ **Sc:** [Ar] $4s^2\,3d^1$ ⟶ **Zn:** [Ar] $4s^2\,3d^{10}$

The experimentally determined ground-state electron configurations of the elements are shown in **Figure 2.17**.

1 1A	2 2A	3 3B	4 4B	5 5B	6 6B	7 7B	8 8B	9 8B	10 8B	11 1B	12 2B	13 3A	14 4A	15 5A	16 6A	17 7A	18 8A
1 H $1s^1$																	2 He $1s^2$
3 Li $2s^1$	4 Be $2s^2$											5 B $2s^22p^1$	6 C $2s^22p^2$	7 N $2s^22p^3$	8 O $2s^22p^4$	9 F $2s^22p^5$	10 Ne $2s^22p^6$
11 Na $3s^1$	12 Mg $3s^2$											13 Al $3s^23p^1$	14 Si $3s^23p^2$	15 P $3s^23p^3$	16 S $3s^23p^4$	17 Cl $3s^23p^5$	18 Ar $3s^23p^6$
19 K $4s^1$	20 Ca $4s^2$	21 Sc $4s^23d^1$	22 Ti $4s^23d^2$	23 V $4s^23d^3$	24 Cr $4s^13d^5$	25 Mn $4s^23d^5$	26 Fe $4s^23d^6$	27 Co $4s^23d^7$	28 Ni $4s^23d^8$	29 Cu $4s^13d^{10}$	30 Zn $4s^23d^{10}$	31 Ga $4s^23d^{10}4p^1$	32 Ge $4s^23d^{10}4p^2$	33 As $4s^23d^{10}4p^3$	34 Se $4s^23d^{10}4p^4$	35 Br $4s^23d^{10}4p^5$	36 Kr $4s^23d^{10}4p^6$
37 Rb $5s^1$	38 Sr $5s^2$	39 Y $5s^24d^1$	40 Zr $5s^24d^2$	41 Nb $5s^14d^4$	42 Mo $5s^14d^5$	43 Tc $5s^24d^5$	44 Ru $5s^14d^7$	45 Rh $5s^14d^8$	46 Pd $4d^{10}$	47 Ag $5s^14d^{10}$	48 Cd $5s^24d^{10}$	49 In $5s^24d^{10}5p^1$	50 Sn $5s^24d^{10}5p^2$	51 Sb $5s^24d^{10}5p^3$	52 Te $5s^24d^{10}5p^4$	53 I $5s^24d^{10}5p^5$	54 Xe $5s^24d^{10}5p^6$
55 Cs $6s^1$	56 Ba $6s^2$	71 Lu $6s^24f^{14}5d^1$	72 Hf $6s^24f^{14}5d^2$	73 Ta $6s^24f^{14}5d^3$	74 W $6s^24f^{14}5d^4$	75 Re $6s^24f^{14}5d^5$	76 Os $6s^24f^{14}5d^6$	77 Ir $6s^24f^{14}5d^7$	78 Pt $6s^14f^{14}5d^9$	79 Au $6s^14f^{14}5d^{10}$	80 Hg $6s^24f^{14}5d^{10}$	81 Tl $6s^24f^{14}5d^{10}6p^1$	82 Pb $6s^24f^{14}5d^{10}6p^2$	83 Bi $6s^24f^{14}5d^{10}6p^3$	84 Po $6s^24f^{14}5d^{10}6p^4$	85 At $6s^24f^{14}5d^{10}6p^5$	86 Rn $6s^24f^{14}5d^{10}6p^6$
87 Fr $7s^1$	88 Ra $7s^2$	103 Lr $7s^25f^{14}6d^1$	104 Rf $7s^25f^{14}6d^2$	105 Db $7s^25f^{14}6d^3$	106 Sg $7s^25f^{14}6d^4$	107 Bh $7s^25f^{14}6d^5$	108 Hs $7s^25f^{14}6d^6$	109 Mt $7s^25f^{14}6d^7$	110 Ds $7s^15f^{14}6d^9$	111 Rg $7s^15f^{14}6d^{10}$	112 Cn $7s^25f^{14}6d^{10}$	113 —	114 Fl	115 —	116 Lv	117 —	118 —

57 La $6s^25d^1$	58 Ce $6s^24f^15d^1$	59 Pr $6s^24f^3$	60 Nd $6s^24f^4$	61 Pm $6s^24f^5$	62 Sm $6s^24f^6$	63 Eu $6s^24f^7$	64 Gd $6s^24f^75d^1$	65 Tb $6s^24f^9$	66 Dy $6s^24f^{10}$	67 Ho $6s^24f^{11}$	68 Er $6s^24f^{12}$	69 Tm $6s^24f^{13}$	70 Yb $6s^24f^{14}$
89 Ac $7s^26d^1$	90 Th $7s^26d^2$	91 Pa $7s^25f^26d^1$	92 U $7s^25f^36d^1$	93 Np $7s^25f^46d^1$	94 Pu $7s^25f^6$	95 Am $7s^25f^7$	96 Cm $7s^25f^76d^1$	97 Bk $7s^25f^9$	98 Cf $7s^25f^{10}$	99 Es $7s^25f^{11}$	100 Fm $7s^25f^{12}$	101 Md $7s^25f^{13}$	102 No $7s^25f^{14}$

FIGURE 2.17

Outer-shell, ground-state electron configurations of the elements.

2.12 SOME ANOMALOUS ELECTRON CONFIGURATIONS

The guidelines discussed in the previous section for determining ground-state electron configurations work well but are not completely accurate. A careful look at Figure 2.17 shows that 91 electron configurations are correctly accounted for by the aufbau principle but that 21 of the predicted configurations are incorrect.

These 21 anomalous configurations can be explained by the unusual stability of half-filled and fully filled subshells. Chromium, for example, which we would predict to have the configuration [Ar] $4s^2\,3d^4$, actually has the configuration [Ar] $4s^1\,3d^5$. By moving an electron from the 4*s* orbital to an energetically similar 3*d* orbital, chromium trades one filled subshell ($4s^2$) for two half-filled subshells ($4s^1\,3d^5$), thereby allowing the two electrons to be farther apart and minimizing their mutual repulsion. In the same way, copper, which we would predict to have the configuration [Ar] $4s^2\,3d^9$, actually has the configuration [Ar] $4s^1\,3d^{10}$. By transferring an electron from the 4*s* orbital to a 3*d* orbital, copper trades one filled subshell ($4s^2$) for a different filled subshell ($3d^{10}$) and gains a half-filled subshell ($4s^1$).

Most of the anomalous electron configurations shown in Figure 2.17 occur in elements with atomic numbers greater than $Z = 40$, where the energy differences between subshells are small. In all cases, the transfer of an electron from one subshell to another lowers the total energy of the atom because of a decrease in electron–electron repulsions.

PROBLEM 2.18 Look at the electron configurations in Figure 2.17, and identify the 21 anomalous ones.

2.13 ELECTRON CONFIGURATIONS AND THE PERIODIC TABLE

Why are electron configurations so important, and what do they have to do with the periodic table? The answers begin to emerge when you look closely at Figure 2.17. Focusing only on the electrons in the outermost shell, called the **valence shell**, *all the elements in a given group of the periodic table have similar valence-shell electron configurations* (**Table 2.3**). The group 1A elements, for example, all have an s^1 valence-shell configuration; the group 2A elements have an s^2 valence-shell configuration; the group 3A elements have an $s^2\,p^1$ valence-shell configuration; and so on across every group of the periodic table (except for the small number of anomalies). Furthermore, because the valence-shell electrons are outermost and least tightly held, they are the most important for determining an element's properties, thus explaining why the elements in a given group of the periodic table have similar chemical behavior.

The periodic table can be divided into four regions, or blocks, of elements according to the orbitals being filled (**Figure 2.18**). The group 1A and 2A elements on the left side of the table are called the ***s*-block elements** because they result from the filling of an *s* orbital; the group 3A–8A elements on the right side of the table are the ***p*-block elements** because they result from the filling of *p* orbitals; the transition metal ***d*-block elements** in the middle of the table result from the filling of *d* orbitals; and the lanthanide/actinide ***f*-block elements** detached at the bottom of the table result from the filling of *f* orbitals.

Thinking of the periodic table as outlined in Figure 2.18 provides a useful way to remember the order of orbital filling. Beginning at the top left corner of the periodic table and going across successive rows gives the correct orbital-filling order automatically. The first row of the periodic table, for instance, contains only the two *s*-block elements H and He, so the first available *s* orbital (1*s*) is filled first. The second row begins with two *s*-block elements (Li and Be) and continues with six *p*-block elements (B through Ne), so the next available *s* orbital (2*s*) and then the first available *p* orbitals (2*p*) are filled. The third row is similar to the second row, so the 3*s* and 3*p* orbitals are filled. The fourth row again starts with two *s*-block elements (K and Ca) but is then followed by 10 *d*-block elements (Sc through Zn) and six *p*-block elements (Ga through Kr). Thus, the order of orbital filling is 4*s* followed by the first

TABLE 2.3 Valence-Shell Electron Configurations of Main-Group Elements

Group	Valence-Shell Electron Configuration	
1A	ns^1	(1 total)
2A	ns^2	(2 total)
3A	$ns^2\,np^1$	(3 total)
4A	$ns^2\,np^2$	(4 total)
5A	$ns^2\,np^3$	(5 total)
6A	$ns^2\,np^4$	(6 total)
7A	$ns^2\,np^5$	(7 total)
8A	$ns^2\,np^6$	(8 total)

FIGURE 2.18

Blocks of the periodic table. Each block corresponds to filling a different kind of orbital.

The arrangement of the periodic table provides a method for remembering the order of orbital filling. Beginning at the top left and moving across successive rows, the order is 1s → 2s → 2p → 3s → 3p → 4s → 3d → 4p and so on.

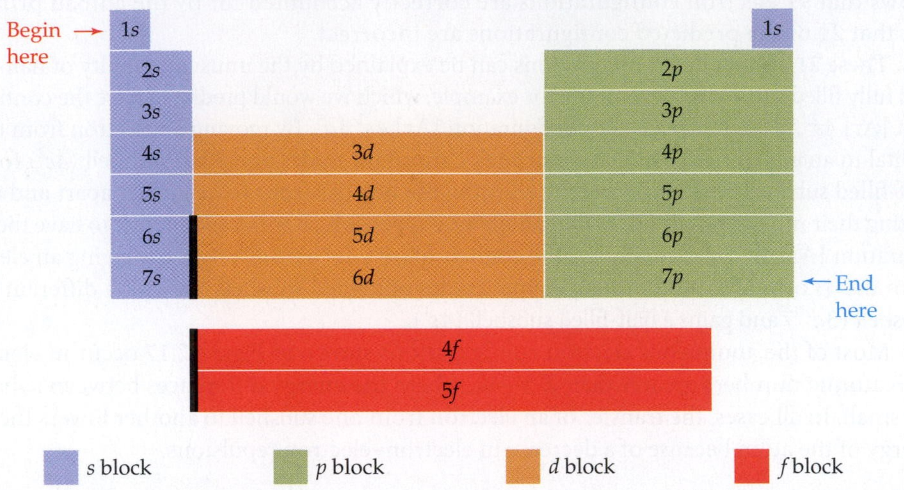

available d orbitals ($3d$) followed by $4p$. Continuing through successive rows of the periodic table gives the entire filling order:

$$1s \rightarrow 2s \rightarrow 2p \rightarrow 3s \rightarrow 3p \rightarrow 4s \rightarrow 3d \rightarrow 4p \rightarrow 5s \rightarrow 4d \rightarrow 5p \rightarrow$$
$$6s \rightarrow 4f \rightarrow 5d \rightarrow 6p \rightarrow 7s \rightarrow 5f \rightarrow 6d \rightarrow 7p$$

Worked Example 2.9

Assigning a Ground-State Electron Configuration to an Atom

Give the ground-state electron configuration of arsenic, $Z = 33$, and draw an orbital-filling diagram, indicating the electrons as up or down arrows.

STRATEGY

Think of the periodic table as having *s*, *p*, *d*, and *f* blocks of elements, as shown in Figure 2.18. Start with hydrogen at the upper left, and fill orbitals until 33 electrons have been added. Remember that only 2 electrons can go into an orbital and that each one of a set of degenerate orbitals must be half filled before any one can be completely filled.

SOLUTION

$$\textbf{As: } 1s^2\, 2s^2\, 2p^6\, 3s^2\, 3p^6\, 4s^2\, 3d^{10}\, 4p^3 \quad \text{or} \quad [\text{Ar}]\; 4s^2\, 3d^{10}\, 4p^3$$

An orbital-filling diagram indicates the electrons in each orbital as arrows. Note that the three $4p$ electrons all have the same spin:

As: [Ar] ↑↓ (4s) ↑↓ ↑↓ ↑↓ ↑↓ ↑↓ (3d) ↑ ↑ ↑ (4p)

Worked Conceptual Example 2.10

Identifying an Atom from Its Ground-State Electron Configuration

Identify the atom with the following ground-state orbital filling diagram:

[Kr] ↑↓ ↑ ↑ ↑ ↑ ↑ __ __ __

STRATEGY

One way to do this problem is to identify the electron configuration and decide which atom has that configuration. Alternatively, you can just count the electrons, thereby finding the atomic number of the atom.

SOLUTION

The atom whose ground-state electron configuration is depicted is in the fifth row because it follows krypton. It has the configuration $5s^2\,4d^5$, which identifies it as technetium. Alternatively, it has $36 + 7 = 43$ electrons and is the element with $Z = 43$.

PROBLEM 2.19 Give expected ground-state electron configurations for the following atoms, and draw orbital-filling diagrams for parts **(a)**–**(c)**.

(a) Ti ($Z = 22$) **(b)** Zn ($Z = 30$)
(c) Sn ($Z = 50$) **(d)** Pb ($Z = 82$)

PROBLEM 2.20 Take a guess. What do you think is a likely ground-state electron configuration for the sodium *ion*, Na^+, formed by loss of an electron from a neutral sodium atom? That is, which electron in a neutral sodium atom is least tightly held and easiest to remove? What is a likely ground-state electron configuration for the chloride ion, Cl^-, formed by adding an electron to a neutral chlorine atom?

CONCEPTUAL PROBLEM 2.21 Identify the atom with the following ground-state electron configuration:

[Ar] ↓↑   ↓↑ ↓↑ ↓↑ ↑ ↑   __ __ __

2.14 ELECTRON CONFIGURATIONS AND PERIODIC PROPERTIES: ATOMIC RADII

We began this chapter by saying that atomic radius is one of many elemental properties to show periodic behavior. You might wonder, though, how we can talk about a definite size for an atom, having said in Section 2.7 that the spatial regions occupied by electrons around atoms have no specific boundaries. What's usually done is to define an atom's radius as being half the distance between the nuclei of two identical atoms when they are bonded together in chemical compounds. In Cl_2, for example, the distance between the two chlorine nuclei is 198 pm; in diamond (elemental carbon), the distance between two carbon nuclei is 154 pm. Thus, we say that the atomic radius of chlorine is half the Cl–Cl distance, or 99 pm, and the atomic radius of carbon is half the C–C distance, or 77 pm.

It's possible to check the accuracy of atomic radii by making sure that the assigned values are additive. For instance, since the atomic radius of Cl is 99 pm and the atomic radius of C is 77 pm, the distance between Cl and C nuclei when those two atoms are bonded together ought to be roughly 99 pm + 77 pm, or 176 pm. In fact, the measured distance between chlorine and carbon in chloromethane (CH_3Cl) is 178 pm, remarkably close to the expected value.

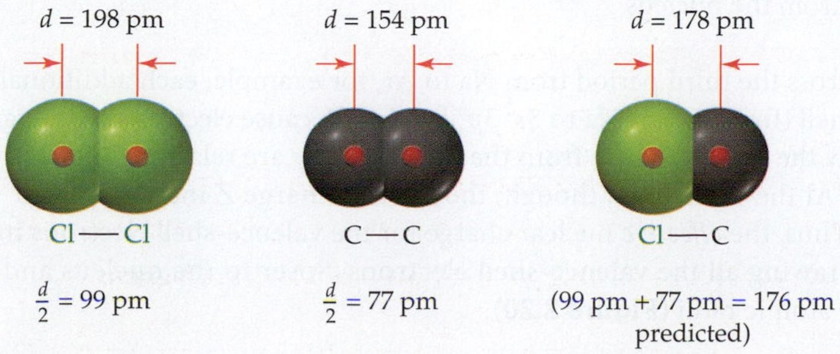

As shown pictorially in **Figure 2.19** and graphically in Figure 2.1 on page 60 at the beginning of this chapter, a comparison of atomic radius versus atomic number shows a periodic rise-and-fall pattern. Atomic radii increase going down a group of the periodic table (Li < Na < K < Rb < Cs, for instance) but decrease going across a row from left to right (Na > Mg > Al > Si > P > S > Cl, for instance). How can this behavior be explained?

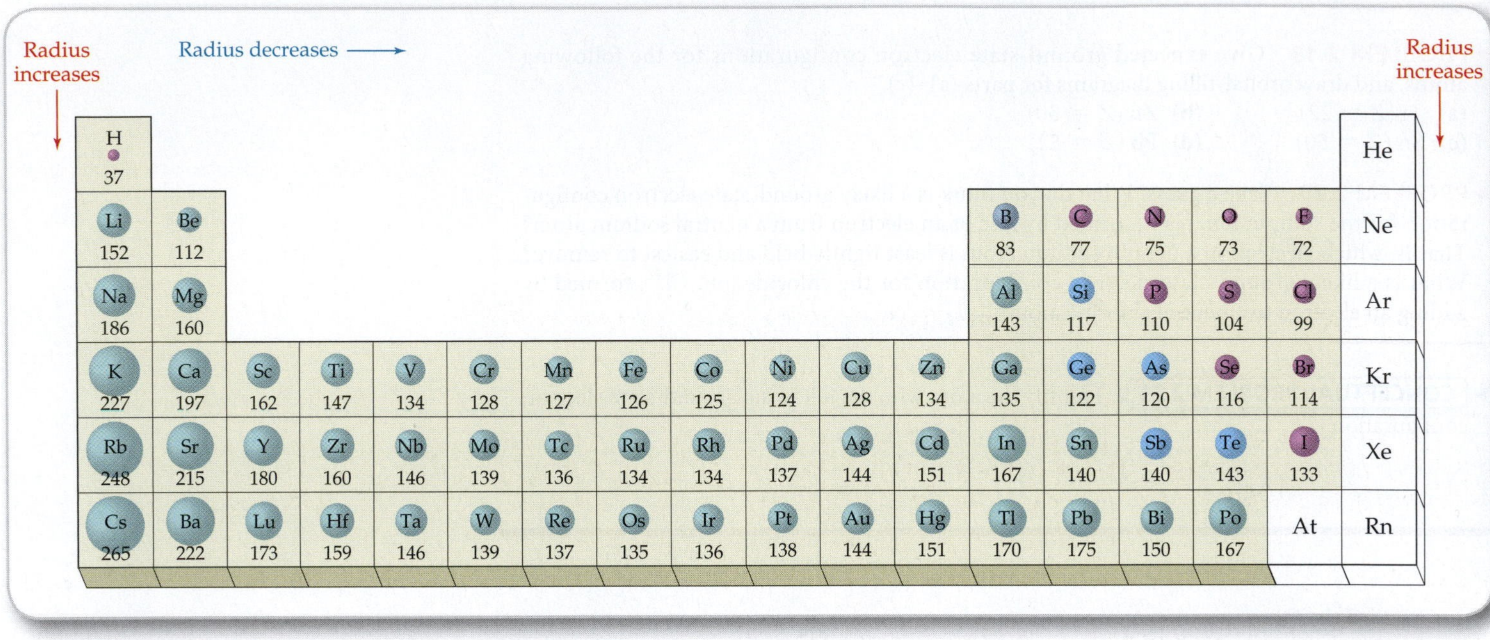

FIGURE 2.19

Atomic radii of the elements in picometers.

The increase in radius going down a group of the periodic table occurs because successively larger valence-shell orbitals are occupied. In Li, for example, the outermost occupied shell is the second one ($2s^1$); in Na it's the third one ($3s^1$); in K it's the fourth one ($4s^1$); and so on through Rb ($5s^1$), Cs ($6s^1$), and Fr ($7s^1$). Because larger shells are occupied, the atomic radii are also larger.

The decrease in radius from left to right across the periodic table occurs because of an increase in effective nuclear charge caused by the increasing number of protons in the nucleus. As we saw in Section 2.10, Z_{eff}, the effective nuclear charge actually felt by an electron, is lower than the true nuclear charge Z because of shielding by other electrons in the atom. The amount of shielding felt by an electron depends on both the shell and subshell of the other electrons with which it is interacting. As a general rule, a valence-shell electron is:

- Strongly shielded by electrons in inner shells, which are closer to the nucleus
- Less strongly shielded by other electrons in the same shell, according to the order $s > p > d > f$
- Only weakly shielded by other electrons in the same subshell, which are at the same distance from the nucleus

Going across the third period from Na to Ar, for example, each additional electron adds to the same shell (from $3s^1$ for Na to $3s^2\,3p^6$ for Ar). Because electrons in the same shell are at approximately the same distance from the nucleus, they are relatively ineffective at shielding one another. At the same time, though, the nuclear charge Z increases from +11 for Na to +18 for Ar. Thus, the *effective* nuclear charge for the valence-shell electrons increases across the period, drawing all the valence-shell electrons closer to the nucleus and progressively shrinking the atomic radii (**Figure 2.20**).

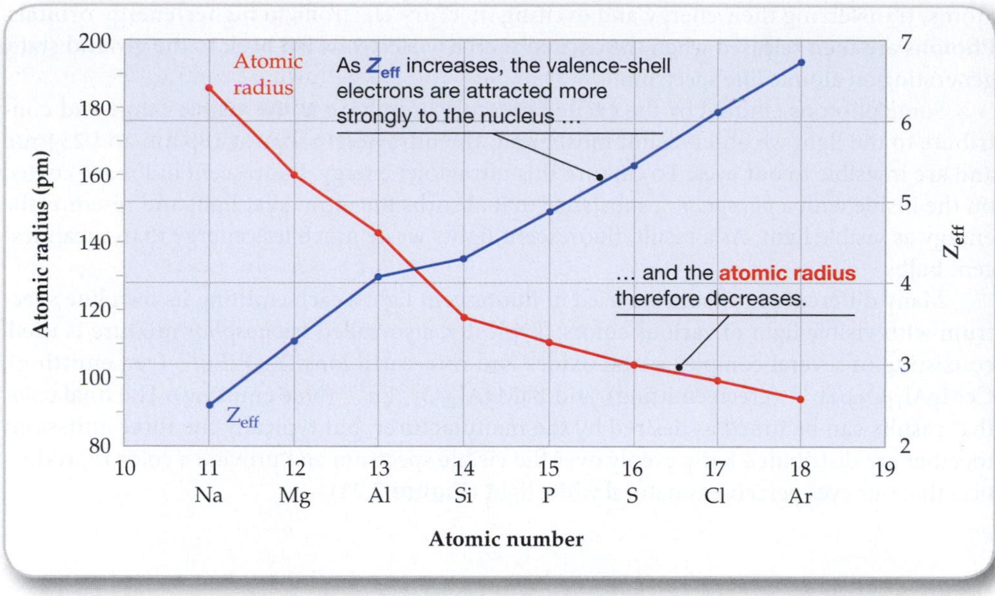

FIGURE 2.20

Plots of atomic radius and calculated Z_{eff} for the highest-energy electron versus atomic number.

What is true of atomic radius is also true of other atomic properties, whose periodicity can be explained by electron configurations. We'll see more examples in the next chapter and draw some general conclusions about chemical reactivity.

PROBLEM 2.22 Which atom in each of the following pairs would you expect to be larger? Explain.

(a) Mg or Ba **(b)** W or Hf **(c)** Si or Sn **(d)** Os or Lu

FYI SAVING ENERGY WITH ATOMIC LINE SPECTRA

In the standard incandescent lightbulb that has been used for more than a century, an electrical current passes through a thin tungsten filament, which is thereby heated and begins to glow. The wavelengths and intensity of the light emitted depend on the temperature of the glowing filament—typically about 2500 °C—and cover the range from ultraviolet (200–400 nm), through visible (400–800 nm), to infrared (800–2000 nm). The ultraviolet frequencies are blocked by the glass of the bulb, the visible frequencies pass through (the whole point of the lightbulb, after all), and the infrared frequencies warm the bulb and its surroundings.

Despite its long history, an incandescent bulb is actually an extremely inefficient device. In fact, only about 5% of the electrical energy consumed by the bulb is converted into visible light. Most of the remaining 95% is converted into heat. Thus, many households and businesses are replacing their incandescent lightbulbs with modern compact fluorescent bulbs in an effort to use less energy. (Not that fluorescent bulbs are terribly efficient themselves—only about 20% of the energy they consume is converted into light—but that still makes them about four times better than incandescents.)

▲ Compact fluorescent bulbs like those shown here offer a much more energy-efficient way to light a home than do typical incandescent lightbulbs.

A fluorescent bulb is, in essence, a variation of the cathode-ray tube described in Section 1.6. The bulb has two main parts, an argon-filled glass tube (either straight or coiled) containing a small amount of mercury vapor, and electronic circuitry that provides a controlled high-voltage current. The current passes through a filament, which heats up and emits a flow of electrons through the tube. In the tube, some of the flowing electrons collide with mercury atoms, transferring their energy and exciting mercury electrons to higher-energy orbitals. Photons are then released when the excited mercury electrons fall back to the ground state, generating an atomic line spectrum.

Some photons emitted by the excited mercury atoms are in the visible range and contribute to the light we observe, but most are in the ultraviolet range at 185 nm and 254 nm and are invisible to our eyes. To capture this ultraviolet energy, fluorescent bulbs are coated on the inside with a *phosphor*, a substance that absorbs the ultraviolet light and re-emits the energy as visible light. As a result, fluorescent lights waste much less energy than incandescent bulbs.

Many different phosphors are used in fluorescent lights, each emitting its own line spectrum with visible light of various colors. Typically, a so-called *triphosphor* mixture is used, consisting of several complex metal oxides and rare-earth ions: $Y_2O_3{:}Eu^{3+}$ (red emitting), $CeMgAl_{11}O_{19}{:}Tb^{3+}$ (green emitting), and $BaMgAl_{10}O_{17}{:}Eu^{2+}$ (blue emitting). The final color that results can be tuned as desired by the manufacturer, but typically the three emissions together are distributed fairly evenly over the visible spectrum and provide a color reproduction that our eyes perceive as natural white light (**Figure 2.21**).

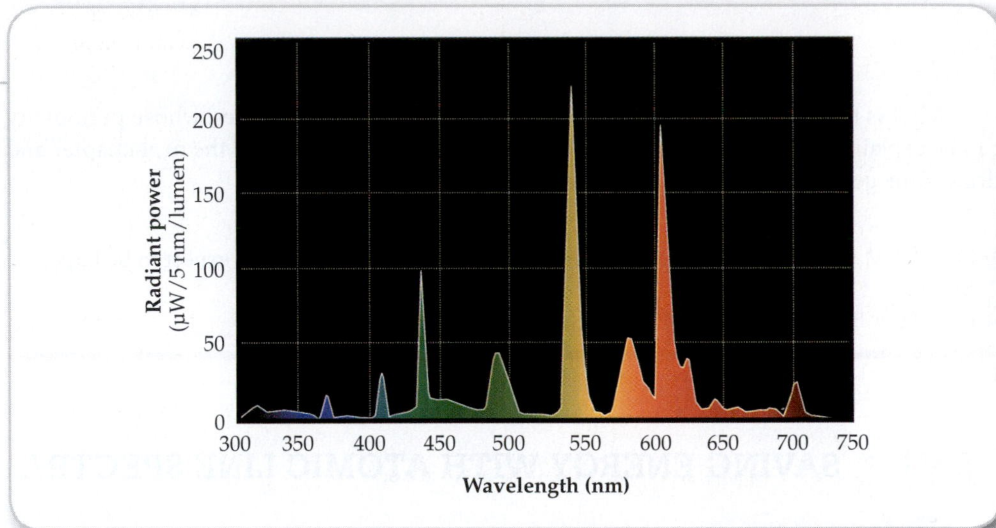

FIGURE 2.21

The triphosphor spectrum emitted from a typical fluorescent bulb. The triphosphor spectrum is distributed over the visible spectrum and is perceived by our eyes as white light.

PROBLEM 2.23 How do atomic line spectra give rise to the light emitted from fluorescent bulbs?

SUMMARY

Understanding the nature of atoms and molecules begins with an understanding of light and other kinds of **radiant energy** that make up the **electromagnetic spectrum**. An electromagnetic wave travels through a vacuum at the speed of light (c) and is characterized by its **frequency** (ν), **wavelength** (λ), and **amplitude**. Unlike the white light of the sun, which consists of a nearly continuous distribution of wavelengths, the light emitted by an excited atom consists of only a few discrete wavelengths, a so-called **line spectrum**.

Atomic line spectra arise because radiant energy is emitted only in discrete amounts, or **quanta**. Just as light behaves in some respects like a stream of small particles, called **photons**, so electrons and other tiny units of matter behave in some respects like waves. The wavelength of a particle of mass m traveling at a velocity v is given by the **de Broglie hypothesis**, $\lambda = h/mv$, where h is Planck's constant.

The **quantum mechanical model** of atomic structure, proposed in 1926 by Erwin Schrödinger, describes an atom by a mathematical equation similar to that used to describe wave motion. The behavior of each electron in an atom is characterized by a *wave function*, or **orbital**, whose square defines the probability of finding the electron in a given region of space. Each orbital has a set of three parameters called **quantum numbers**. The **principal quantum number** n defines the size of the orbital; the **angular-momentum quantum number** l defines the shape of the orbital; and the **magnetic quantum number** m_l defines the spatial orientation of the orbital. In a hydrogen atom, which contains only one electron, the energy of an orbital depends only on n. In a multielectron atom, the energy of an orbital depends on both n and l. In addition, the **spin quantum number** m_s specifies the electron spin as either $+1/2$ or $-1/2$. No two electrons in an atom can have the same values for their four quantum numbers.

Orbitals are grouped into successive layers, or **shells**, according to their principal quantum number n. Within a shell, orbitals are grouped into s, p, d, and f **subshells** according to their angular-momentum quantum numbers l. An orbital in an s subshell is spherical, an orbital in a p subshell is dumbbell-shaped, and four of the five orbitals in a d subshell are cloverleaf-shaped.

The **ground-state electron configuration** of a multielectron atom can be determined by the guidelines given in the **aufbau principle**:

1. The lowest-energy orbitals fill first.
2. Only two electrons of opposite spin go into any one orbital (the **Pauli exclusion principle**).
3. If two or more orbitals are equal in energy (**degenerate**), each is half filled before any one of them is completely filled (**Hund's rule**).

The periodic table is the most important organizing principle of chemistry. It is successful because elements in each group of the periodic table have similar **valence-shell electron configurations** and therefore have similar properties. For example, atomic radii of elements show a periodic rise-and-fall pattern according to the positions of the elements in the table. Atomic radii increase going down a group because n increases, and they decrease from left to right across a period because the **effective nuclear charge (Z_{eff})** increases.

KEY WORDS

amplitude *63*
angular-momentum quantum number (l) *73*
aufbau principle *82*
Balmer–Rydberg equation *65*
***d*-block element** *85*
de Broglie hypothesis *70*
degenerate *82*
effective nuclear charge (Z_{eff}) *81*
electromagnetic spectrum *61*
electron configuration *81*
***f*-block element** *85*
frequency (ν) *63*
ground-state electron configuration *82*
Heisenberg uncertainty principle *71*
hertz (Hz) *63*
Hund's rule *82*
line spectrum *64*
magnetic quantum number (m_l) *73*
node *76*
orbital *72*
***p*-block element** *85*
Pauli exclusion principle *78*
phase *76*
photoelectric effect *67*
photon *67*
Planck's postulate *67*
principal quantum number (n) *72*
quantum *68*
quantum mechanical model *71*
quantum number *72*
radiant energy *61*
***s*-block element** *85*
shell *73*
spin quantum number (m_s) *78*
subshell *73*
valence shell *85*
wavelength (λ) *63*
work function (Φ) *68*

KEY EQUATIONS

- The relationship between wavelength (λ) and frequency (ν) **(Section 2.1)**

$$\lambda = \frac{c}{\nu} \quad \text{or} \quad \nu = \frac{c}{\lambda}$$

- The Balmer–Rydberg equation to account for atomic line spectra **(Section 2.2)**

$$\frac{1}{\lambda} = R_\infty\left[\frac{1}{m^2} - \frac{1}{n^2}\right] \quad \text{or} \quad \nu = R_\infty \cdot c\left[\frac{1}{m^2} - \frac{1}{n^2}\right]$$

where m and n are integers with $n > m$ and R_∞ is the Rydberg constant

- Planck's postulate relating energy (E) and frequency (ν) or wavelength (λ) **(Section 2.3)**

$$E = h\nu = hc/\lambda$$

where h is Planck's constant, $6.626 \times 10^{-34}\ \text{J} \cdot \text{s}$

- de Broglie's hypothesis relating wavelength (λ), mass (m) and velocity (v) **(Section 2.4)**

$$\lambda = \frac{h}{mv}$$

- Heisenberg's uncertainty principle **(Section 2.5)**

$$(\Delta x)(\Delta mv) \geq \frac{h}{4\pi}$$

LEARNING OUTCOMES

- Know how a wave is characterized by its wavelength, frequency, and amplitude and how to interconvert wavelength and frequency. **(Section 2.1)** ***Problems 1–3, 17, 27, 32–37, 40, 41, 45, 113***
- Use the Balmer–Rydberg equation to calculate atomic line spectra and energy differences among orbitals. **(Sections 2.2, 2.9)** ***Problems 4–6, 46, 47, 70, 71, 98–102***
- Use Planck's postulate to calculate the energy of a photon of radiant energy. **(Section 2.3)** ***Problems 7–10, 38, 39, 42–46, 48, 49, 70–73, 103–105, 108, 109, 116, 117***

- Use de Broglie's hypothesis to relate the mass of a particle to its wavelength. **(Section 2.4)** ***Problems 11, 50–55***
- Describe the Heisenberg uncertainty principle. **(Section 2.5)** ***Problems 56, 57, 59***
- Know the meaning of the four quantum numbers, and be able to assign quantum numbers to an electron in a given orbital. **(Section 2.6)** ***Problems 12–14, 31, 58, 64–69, 74, 75, 107, 121***
- Describe the shapes of specific orbitals. **(Section 2.7)** ***Problems 15, 16, 31, 60, 61***
- Describe the Pauli exclusion principle. **(Section 2.8)**
- Know how effective nuclear charge and electron shielding affect orbital energy levels in multielectron atoms. **(Section 2.10)** ***Problems 62, 63, 84, 85, 114***
- Use the aufbau principle to assign ground-state electron configurations to atoms. **(Sections 2.11, 2.12)** ***Problems 18, 26, 28, 29, 76–83, 86, 87, 106, 110***
- Explain the relationship between an element's position in the periodic table and the electron configuration of its atoms. **(Section 2.13)** ***Problems 19–21, 24, 25, 74, 88, 89, 96, 97***
- Know how electron configuration affects atomic radius. **(Section 2.14)** ***Problems 22, 90–93, 111***

CONCEPTUAL PROBLEMS

Problems 2.1–2.23 appear within the chapter.

2.24 Where on the blank outline of the periodic table do elements that meet the following descriptions appear?

(a) Elements with the valence-shell ground-state electron configuration $ns^2\ np^5$

(b) An element whose fourth shell contains two *p* electrons

(c) An element with the ground-state electron configuration [Ar] $4s^2\ 3d^{10}\ 4p^5$

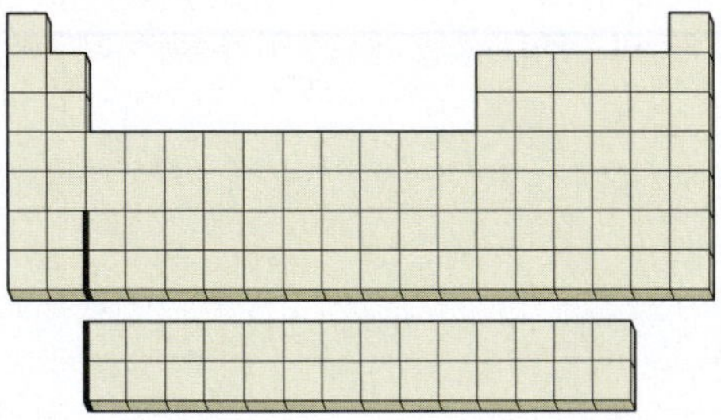

2.25 Where on the periodic table do elements that meet the following descriptions appear?

(a) Elements with electrons whose largest principal quantum number is $n = 4$

(b) Elements with the valence-shell ground-state electron configuration $ns^2\ np^3$

(c) Elements that have only one unpaired *p* electron

(d) The *d*-block elements

(e) The *p*-block elements

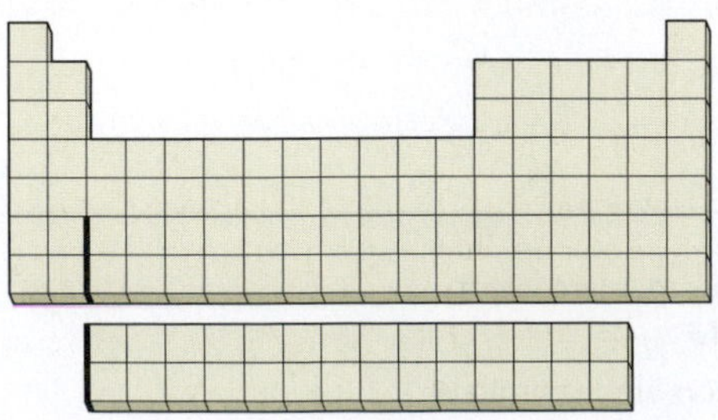

2.26 One of the elements shown on the following periodic table has an anomalous ground-state electron configuration. Which is it—red, blue, or green—and why?

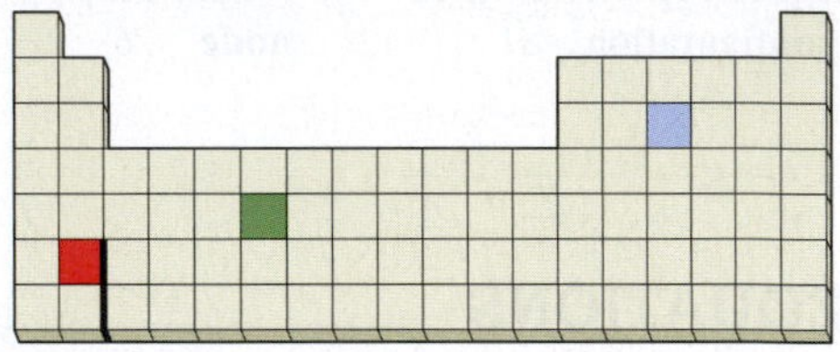

2.27 Two electromagnetic waves are represented below.

(a) Which wave has the greater intensity?

(b) Which wave corresponds to higher-energy radiation?

(c) Which wave represents yellow light, and which represents infrared radiation?

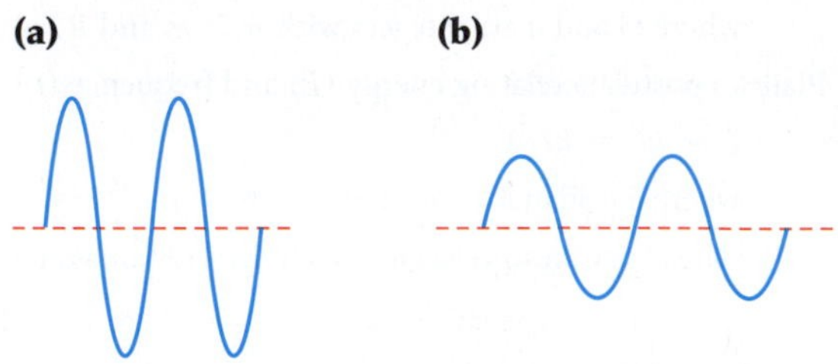

2.28 What atom has the following orbital-filling diagram?

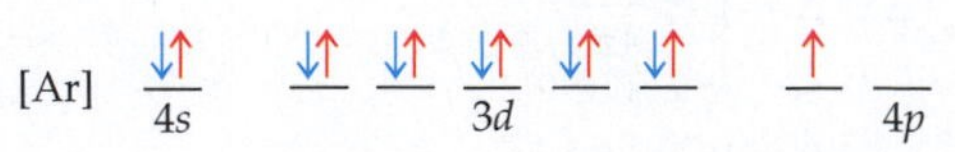

2.29 The following orbital-filling diagram represents an excited state rather than the ground state of an atom. Identify the atom, and give its ground-state electron configuration.

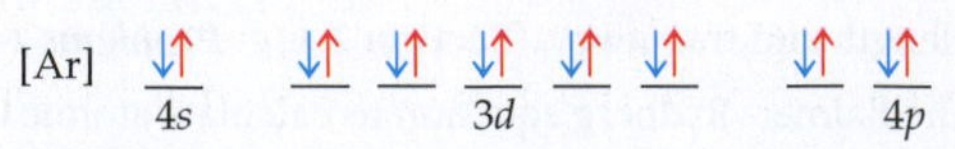

2.30 Which of the following three spheres represents a Ca atom, which an Sr atom, and which a Br atom?

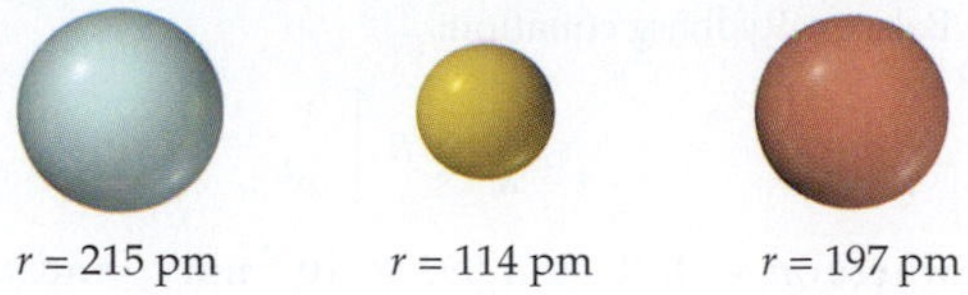

2.31 Identify each of the following orbitals, and give n and l quantum numbers for each.

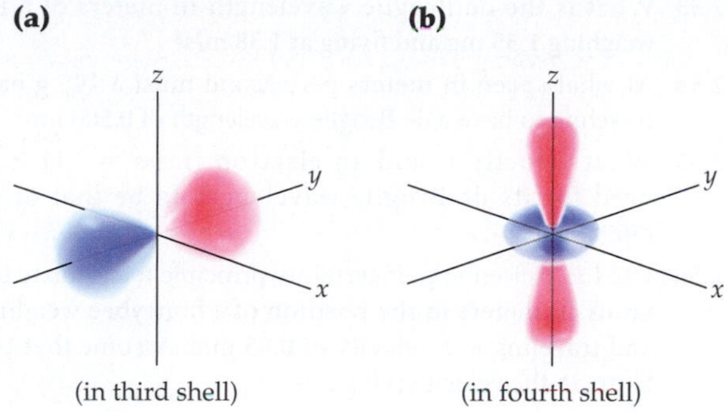

SECTION PROBLEMS

Radiant Energy and Atomic Spectra (Sections 2.1–2.3)

2.32 Which has the higher frequency, red light or violet light? Which has the longer wavelength? Which has the greater energy?

2.33 Which has the higher frequency, infrared light or ultraviolet light? Which has the longer wavelength? Which has the greater energy?

2.34 The Hubble Space Telescope detects radiant energy in the wavelength range 1.15×10^{-7} m to 2.0×10^{-6} m. What region of the electromagnetic spectrum is found completely within this range? What regions fall partially in this range?

2.35 The Green Bank Telescope in West Virginia—the world's largest steerable radio telescope—detects frequencies from 290 MHz to 90 GHz. What region or regions of the electromagnetic spectrum are found completely or partially within its detection range?

2.36 What is the wavelength in meters of ultraviolet light with $\nu = 5.5 \times 10^{15}\ s^{-1}$?

2.37 What is the frequency of a microwave with $\lambda = 4.33 \times 10^{-3}$ m?

2.38 Calculate the energies of the following waves in kilojoules per mole, and tell which member of each pair has the higher value.

(a) An FM radio wave at 99.5 MHz and an AM radio wave at 1150 kHz

(b) An X ray with $\lambda = 3.44 \times 10^{-9}$ m and a microwave with $\lambda = 6.71 \times 10^{-2}$ m

2.39 The MRI (magnetic resonance imaging) body scanners used in hospitals operate with 400 MHz radiofrequency energy. How much energy does this correspond to in kilojoules per mole?

2.40 A certain cellular telephone transmits at a frequency of 825 MHz and receives at a frequency of 875 MHz.

(a) What is the wavelength of the transmitted signal in cm?

(b) What is the wavelength of the received signal in cm?

2.41 Optical fibers allow the fast transmission of vast amounts of data. In one type of fiber, the wavelength of transmitted light is 1.3×10^3 nm.

(a) What is the frequency of the light?

(b) Fiber optic cable is available in 12 km lengths. How long will it take for a signal to travel that distance assuming that the speed of light in the cable is the same as in a vacuum?

2.42 What is the wavelength in meters of photons with the following energies? In what region of the electromagnetic spectrum does each appear?

(a) 90.5 kJ/mol (b) 8.05×10^{-4} kJ/mol

(c) 1.83×10^3 kJ/mol

2.43 What is the energy of each of the following photons in kilojoules per mole?

(a) $\nu = 5.97 \times 10^{19}\ s^{-1}$ (b) $\nu = 1.26 \times 10^{6}\ s^{-1}$

(c) $\lambda = 2.57 \times 10^{2}$ m

2.44 The data encoded on CDs, DVDs, and Blu-ray discs is read by lasers. What is the wavelength in nanometers and the energy in joules of the following lasers?

(a) CD laser, $\nu = 3.85 \times 10^{14}\ s^{-1}$

(b) DVD laser, $\nu = 4.62 \times 10^{14}\ s^{-1}$

(c) Blu-ray laser, $\nu = 7.41 \times 10^{14}\ s^{-1}$

2.45 The semimetal germanium is used as a component in photodetectors, which generate electric current when exposed to light. If a germanium photodetector responds to photons in the range $\lambda = 400-1700$ nm, will the following light sources be detected?

(a) A laser with $\nu = 4.35 \times 10^{14}\ s^{-1}$

(b) Photons with $E = 43$ kJ/mol

(c) Electromagnetic radiation with $\nu = 706$ THz

2.46 According to the equation for the Balmer line spectrum of hydrogen, a value of $n = 3$ gives a red spectral line at 656.3 nm, a value of $n = 4$ gives a green line at 486.1 nm, and a value of $n = 5$ gives a blue line at 434.0 nm. Calculate the energy in kilojoules per mole of the radiation corresponding to each of these spectral lines.

2.47 According to the values cited in Problem 2.46, the wavelength differences between lines in the Balmer series become smaller as n becomes larger. In other words, the wavelengths converge toward a minimum value as n becomes very large. At what wavelength in nanometers do the lines converge?

2.48 The work function of silver metal is 436 kJ/mol. What frequency of light is needed to eject electrons from a sample of silver?

2.49 What is the work function of gold metal in kJ/mol if light with $\lambda = 234$ nm is necessary to eject electrons?

Particles and Waves (Section 2.4–2.5)

2.50 Protons and electrons can be given very high energies in particle accelerators. What is the wavelength in meters of an electron (mass $= 9.11 \times 10^{-31}$ kg) that has been accelerated to 99% of the speed of light? In what region of the electromagnetic spectrum is this wavelength?

2.51 If a proton (mass $= 1.673 \times 10^{-24}$ g) is accelerated to 25% of the speed of light, what is its wavelength in meters? In what region of the electromagnetic spectrum is this wavelength?

2.52 What is the de Broglie wavelength in meters of a baseball weighing 145 g and traveling at 156 km/h?

2.53 What is the de Broglie wavelength in meters of a mosquito weighing 1.55 mg and flying at 1.38 m/s?

2.54 At what speed in meters per second must a 145 g baseball be traveling to have a de Broglie wavelength of 0.500 nm?

2.55 What velocity would an electron (mass $= 9.11 \times 10^{-31}$ kg) need for its de Broglie wavelength to be that of red light (750 nm)?

2.56 Use the Heisenberg uncertainty principle to calculate the uncertainty in meters in the position of a honeybee weighing 0.68 g and traveling at a velocity of 0.85 m/s. Assume that the uncertainty in the velocity is 0.1 m/s.

2.57 The mass of a helium-4 atom is 4.0026 u, and its average velocity at 25 °C is 1.36×10^3 m/s. What is the uncertainty in meters in the position of a helium atom if the uncertainty in its velocity is 1%?

Orbitals and Quantum Mechanics (Sections 2.6–2.9)

2.58 What are the four quantum numbers, and what does each specify?

2.59 What is the Heisenberg uncertainty principle, and how does it affect our description of atomic structure?

2.60 Why do we have to use an arbitrary value such as 95% to determine the spatial limitations of an orbital?

2.61 How many nodal surfaces does a 4*s* orbital have? Draw a cutaway representation of a 4*s* orbital showing the nodes and the regions of maximum electron probability.

2.62 What is meant by the term effective nuclear charge, Z_{eff}, and what is it due to?

2.63 How does electron shielding in multielectron atoms give rise to energy differences among 3*s*, 3*p*, and 3*d* orbitals?

2.64 Give the allowable combinations of quantum numbers for each of the following electrons:

(a) A 4*s* electron (b) A 3*p* electron
(c) A 5*f* electron (d) A 5*d* electron

2.65 Give the orbital designations of electrons with the following quantum numbers:

(a) $n = 3, l = 0, m_l = 0$ (b) $n = 2, l = 1, m_l = -1$
(c) $n = 4, l = 3, m_l = -2$ (d) $n = 4, l = 2, m_l = 0$

2.66 Tell which of the following combinations of quantum numbers are not allowed. Explain your answers.

(a) $n = 3, l = 0, m_l = -1$ (b) $n = 3, l = 1, m_l = 1$
(c) $n = 4, l = 4, m_l = 0$

2.67 Which of the following combinations of quantum numbers can refer to an electron in a ground-state cobalt atom ($Z = 27$)?

(a) $n = 3, l = 0, m_l = 2$ (b) $n = 4, l = 2, m_l = -2$
(c) $n = 3, l = 1, m_l = 0$

2.68 What is the maximum number of electrons in an atom whose highest-energy electrons have the principal quantum number $n = 5$?

2.69 What is the maximum number of electrons in an atom whose highest-energy electrons have the principal quantum number $n = 4$ and the angular-momentum quantum number $l = 0$?

2.70 The wavelength of light at which the Balmer series converges (Problem 2.47) corresponds to the amount of energy required to completely remove an electron from the second shell of a hydrogen atom. Calculate this energy in kilojoules per mole.

2.71 One series of lines of the hydrogen spectrum is caused by emission of energy accompanying the fall of an electron from outer shells to the fourth shell. The lines can be calculated using the Balmer–Rydberg equation:

$$\frac{1}{\lambda} = R_\infty\left[\frac{1}{m^2} - \frac{1}{n^2}\right]$$

where $m = 4$, $R_\infty = 1.097 \times 10^{-2}\ \text{nm}^{-1}$, and n is an integer greater than 4. Calculate the wavelengths in nanometers and energies in kilojoules per mole of the first two lines in the series. In what region of the electromagnetic spectrum do they fall?

2.72 Sodium atoms emit light with a wavelength of 330 nm when an electron moves from a 4*p* orbital to a 3*s* orbital. What is the energy difference between the orbitals in kilojoules per mole?

2.73 Excited rubidium atoms emit red light with $\lambda = 795$ nm. What is the energy difference in kilojoules per mole between orbitals that give rise to this emission?

Electron Configuratio ns (Sections 2.10–2.13)

2.74 Why does the number of elements in successive periods of the periodic table increase by the progression 2, 8, 18, 32?

2.75 Which two of the four quantum numbers determine the energy level of an orbital in a multielectron atom?

2.76 In a multielectron atom, which orbital in each of the following pairs is higher in energy?

(a) 5*p* or 5*d* (b) 4*s* or 3*p* (c) 6*s* or 4*d*

2.77 Order the orbitals for a multielectron atom in each of the following lists according to increasing energy:

(a) 4*d*, 3*p*, 2*p*, 5*s* (b) 2*s*, 4*s*, 3*d*, 4*p* (c) 6*s*, 5*p*, 3*d*, 4*p*

2.78 According to the aufbau principle, which orbital is filled immediately *after* each of the following in a multielectron atom?

(a) 4*s* (b) 3*d* (c) 5*f* (d) 5*p*

2.79 According to the aufbau principle, which orbital is filled immediately *before* each of the following?

(a) 3*p* (b) 4*p* (c) 4*f* (d) 5*d*

2.80 Give the expected ground-state electron configurations for the following elements:

(a) Ti (b) Ru (c) Sn (d) Sr (e) Se

2.81 Give the expected ground-state electron configurations for atoms with the following atomic numbers:

(a) $Z = 55$ (b) $Z = 40$ (c) $Z = 80$ (d) $Z = 62$

2.82 Draw orbital-filling diagrams for the following atoms. Show each electron as an up or down arrow, and use the abbreviation of the preceding noble gas to represent inner-shell electrons.

(a) Rb (b) W (c) Ge (d) Zr

2.83 Draw orbital-filling diagrams for atoms with the following atomic numbers. Show each electron as an up or down arrow, and use the abbreviation of the preceding noble gas to represent inner-shell electrons.

(a) $Z = 25$ (b) $Z = 56$ (c) $Z = 28$ (d) $Z = 47$

2.84 Order the electrons in the following orbitals according to their shielding ability: 4*s*, 4*d*, 4*f*.

2.85 Order the following elements according to increasing Z_{eff}: Ca, Se, Kr, K.

2.86 How many unpaired electrons are present in each of the following ground-state atoms?

(a) O (b) Si (c) K (d) As

2.87 Identify the following atoms:

(a) It has the ground-state electron configuration $[Ar]\ 4s^2\ 3d^{10}\ 4p^1$.

(b) It has the ground-state electron configuration $[Kr]\ 4d^{10}$.

2.88 At what atomic number is the filling of a *g* orbital likely to begin?

2.89 Assuming that *g* orbitals fill according to Hund's rule, what is the atomic number of the first element to have a filled *g* orbital?

Electron Configurations and Periodic Properties (Section 2.14)

2.90 Why do atomic radii increase going down a group of the periodic table?

2.91 Why do atomic radii decrease from left to right across a period of the periodic table?

2.92 Order the following atoms according to increasing atomic radius: S, F, O.

2.93 Which atom in each of the following pairs has a larger radius?

(a) Na or K **(b)** V or Ta **(c)** V or Zn **(d)** Li or Ba

2.94 The amount of energy that must be added to remove an electron from a neutral atom to give a positive ion is called the atom's *ionization energy*. Which would you expect to have the larger ionization energy, Na or Mg? Explain.

2.95 The amount of energy released when an electron adds to a neutral atom to give a negative ion is called the atom's *electron affinity*. Which would you expect to have the larger electron affinity, C or F? Explain.

2.96 What is the expected ground-state electron configuration of the recently discovered element with $Z = 116$?

2.97 What is the atomic number and expected ground-state electron configuration of the yet undiscovered element directly below Fr in the periodic table?

CHAPTER PROBLEMS

2.98 Orbital energies in single-electron atoms or ions, such as He^+, can be described with an equation similar to the Balmer–Rydberg equation:

$$\frac{1}{\lambda} = Z^2 R_\infty \left[\frac{1}{m^2} - \frac{1}{n^2}\right]$$

where Z is the atomic number. What wavelength of light in nm is emitted when the electron in He^+ falls from $n = 3$ to $n = 2$?

2.99 Like He^+, the Li^{2+} ion is a single-electron system (Problem 2.98). What wavelength of light in nm must be absorbed to promote the electron in Li^{2+} from $n = 1$ to $n = 4$?

2.100 Use the Balmer equation to calculate the wavelength in nanometers of the spectral line for hydrogen when $n = 6$. What is the energy in kilojoules per mole of the radiation corresponding to this line?

2.101 Lines in a certain series of the hydrogen spectrum are caused by emission of energy accompanying the fall of an electron from outer shells to the fifth shell. Use the Balmer–Rydberg equation to calculate the wavelengths in nanometers and energies in kilojoules per mole of the two longest-wavelength lines in the series. In what region of the electromagnetic spectrum do they fall?

2.102 What is the shortest wavelength in nanometers in the series you calculated in Problem 2.101?

2.103 What is the wavelength in meters of photons with the following energies? In what region of the electromagnetic spectrum does each appear?

(a) 142 kJ/mol **(b)** 4.55×10^{-2} kJ/mol

(c) 4.81×10^4 kJ/mol

2.104 What is the energy of each of the following photons in kilojoules per mole?

(a) $\nu = 3.79 \times 10^{11}\ s^{-1}$ **(b)** $\nu = 5.45 \times 10^4\ s^{-1}$

(c) $\lambda = 4.11 \times 10^{-5}$ m

2.105 The *second* in the SI system is defined as the duration of 9,192,631,770 periods of radiation corresponding to the transition between two energy levels of a cesium-133 atom. What is the energy difference between the two levels in kilojoules per mole?

2.106 Write the symbol, give the ground-state electron configuration, and draw an orbital-filling diagram for each of the following atoms. Use the abbreviation of the preceding noble gas to represent the inner-shell electrons.

(a) The heaviest alkaline-earth metal

(b) The lightest transition metal

(c) The heaviest actinide metal

(d) The lightest semimetal

(e) The group 6A element in the fifth period

2.107 Imagine a universe in which the four quantum numbers can have the same possible values as in our universe except that the angular-momentum quantum number l can have integral values of $0, 1, 2, \ldots, n + 1$ (instead of $0, 1, 2, \ldots, n - 1$).

(a) How many elements would be in the first two rows of the periodic table in this universe?

(b) What would be the atomic number of the element in the second row and fifth column?

(c) Draw an orbital-filling diagram for the element with atomic number 12.

2.108 Cesium metal is often used in photoelectric sensors used to turn lights on and off automatically because the amount of energy necessary to eject electrons from a cesium surface is relatively small—only 188 kJ/mol. What wavelength of light in nanometers does this correspond to?

2.109 The laser light used in compact disc players has $\lambda = 780$ nm. In what region of the electromagnetic spectrum does this light appear? What is the energy of this light in kilojoules per mole?

2.110 Draw orbital-filling diagrams for the following atoms. Show each electron as an up or down arrow, and use the abbreviation of the preceding noble gas to represent inner-shell electrons.

(a) Sr **(b)** Cd **(c)** Has $Z = 22$ **(d)** Has $Z = 34$

2.111 The atomic radii of Y (180 pm) and La (187 pm) are significantly different, but the radii of Zr (160 pm) and Hf (159 pm) are essentially identical. Explain.

2.112 You're probably familiar with using Scotch Tape for wrapping presents but may not know that it can also generate electromagnetic radiation. When Scotch Tape is unrolled in a vacuum (but not in air), photons with a range of frequencies around $\nu = 2.9 \times 10^{18}\ s^{-1}$ are emitted in nanosecond bursts.

(a) What is the wavelength in meters of photons with $\nu = 2.9 \times 10^{18}\ s^{-1}$?

(b) What is the energy in kJ/mol of photons with $\nu = 2.9 \times 10^{18}\ s^{-1}$?

(c) What type of electromagnetic radiation are these photons?

2.113 Hard wintergreen-flavored candies are *triboluminescent*, meaning that they emit flashes of light when crushed. (You can see it for yourself if you look in a mirror while crunching a wintergreen Life Saver in your mouth in a dark room.) The strongest emission is around $\lambda = 450$ nm.

(a) What is the frequency in s^{-1} of photons with $\lambda = 450$ nm?

(b) What is the energy in kJ/mol of photons with $\lambda = 450$ nm?

(c) What is the color of the light with $\lambda = 450$ nm?

2.114 One method for calculating Z_{eff} is to use the equation

$$Z_{eff} = \sqrt{\frac{(E)(n^2)}{1312\ \text{kJ/mol}}}$$

where E is the energy necessary to remove an electron from an atom and n is the principal quantum number of the electron. Use this equation to calculate Z_{eff} values for the highest-energy electrons in potassium ($E = 418.8$ kJ/mol) and krypton ($E = 1350.7$ kJ/mol).

2.115 One watt (W) is equal to 1 J/s. Assuming that 5.0% of the energy output of a 75 W lightbulb is visible light and that the average wavelength of the light is 550 nm, how many photons of visible light are emitted by the lightbulb each second?

2.116 Microwave ovens work by irradiating food with microwave radiation, which is absorbed and converted into heat. Assuming that radiation with $\lambda = 15.0$ cm is used, that all the energy is converted to heat, and that 4.184 J is needed to raise the temperature of 1.00 g of water by 1.00 °C, how many photons are necessary to raise the temperature of a 350 mL cup of water from 20 °C to 95 °C?

2.117 Photochromic sunglasses, which darken when exposed to light, contain a small amount of colorless AgCl embedded in the glass. When irradiated with light, metallic silver atoms are produced and the glass darkens: $AgCl \rightarrow Ag + Cl$. Escape of the chlorine atoms is prevented by the rigid structure of the glass, and the reaction therefore reverses as soon as the light is removed. If 310 kJ/mol of energy is required to make the reaction proceed, what wavelength of light is necessary?

2.118 The amount of energy necessary to remove an electron from an atom is a quantity called the *ionization energy*, E_i. This energy can be measured by a technique called *photoelectron spectroscopy*, in which light of wavelength λ is directed at an atom, causing an electron to be ejected by the photoelectric effect. The kinetic energy of the ejected electron (E_K) is measured by determining its velocity, v ($E_K = mv^2/2$), and E_i is then calculated using the conservation of energy principle. That is, the energy of the incident light equals E_i plus E_K. What is the ionization energy of selenium atoms in kilojoules per mole if light with $\lambda = 48.2$ nm produces electrons with a velocity of 2.371×10^6 m/s? The mass, m, of an electron is 9.109×10^{-31} kg.

2.119 X rays with a wavelength of 1.54×10^{-10} m are produced when a copper metal target is bombarded with high-energy electrons that have been accelerated by a voltage difference of 30,000 V. The kinetic energy of the electrons (see Problem 2.118) equals the product of the voltage difference and the electronic charge in coulombs, where 1 volt-coulomb = 1 J.

(a) What is the kinetic energy in joules and the de Broglie wavelength in meters of an electron that has been accelerated by a voltage difference of 30,000 V?

(b) What is the energy in joules of the X rays emitted by the copper target?

2.120 In the Bohr model of atomic structure, electrons are constrained to orbit a nucleus at specific distances, given by the equation

$$r = \frac{n^2 a_0}{Z}$$

where r is the radius of the orbit, Z is the charge on the nucleus, a_0 is the *Bohr radius* and has a value of 5.292×10^{-11} m, and n is a positive integer ($n = 1, 2, 3, \ldots$) like a principal quantum number. Furthermore, Bohr concluded that the energy level E of an electron in a given orbit is

$$E = \frac{-Ze^2}{2r}$$

where e is the charge on an electron. Derive an equation that will let you calculate the difference ΔE between any two energy levels. What relation does your equation have to the Balmer–Rydberg equation?

2.121 Assume that the rules for quantum numbers are different and that the spin quantum number m_s can have any of three values, $m_s = -1/2, 0, +1/2$, while all other rules remain the same.

(a) Draw an orbital-filling diagram for the element with $Z = 25$, showing the individual electrons in the outermost subshell as up arrows, down arrows, or 0. How many partially filled orbitals does the element have?

(b) What is the atomic number of the element in the 3rd column of the 4th row under these new rules? What block does it belong to (s, p, d, or f)?

2.122 Given the subshells $1s$, $2s$, $2p$, $3s$, $3p$, and $3d$, identify those that meet the following descriptions:

(a) Has $l = 2$

(b) Can have $m_l = -1$

(c) Is empty in a nitrogen atom

(d) Is full in a carbon atom

(e) Contains the outermost electrons in a beryllium atom

(f) Can contain two electrons, both with spin $m_s = +1/2$

2.123 A hydrogen atom with an electron in the first shell ($n = 1$) absorbs ultraviolet light with a wavelength of 1.03×10^{-7} m. To what shell is the electron promoted?

MULTICONCEPT PROBLEMS

2.124 A minimum energy of 7.21×10^{-19} J is required to produce the photoelectric effect in chromium metal.

(a) What is the minimum frequency of light needed to remove an electron from chromium?

(b) Light with a wavelength of 2.50×10^{-7} m falls on a piece of chromium in an evacuated glass tube. What is the minimum de Broglie wavelength of the emitted electrons? (Note that the energy of the incident light must be conserved; that is, the photon's energy must equal the sum of the energy needed to eject the electron plus the kinetic energy of the electron. See Problem 2.118.)

2.125 A photon produced by an X-ray machine has an energy of 4.70×10^{-16} J.

(a) What is the frequency of the photon?

(b) What is the wavelength of radiation of frequency **(a)**?

(c) What is the velocity of an electron with a de Broglie wavelength equal to **(b)**?

(d) What is the kinetic energy of an electron traveling at velocity **(c)**? (See Problem 2.118.)

2.126 An energetically excited hydrogen atom has its electron in a 5*f* subshell. The electron drops down to the 3*d* subshell, releasing a photon in the process.

(a) Give the n and l quantum numbers for both subshells, and give the range of possible m_l quantum numbers.

(b) What wavelength of light is emitted by the process?

(c) The hydrogen atom now has a single electron in the 3*d* subshell. What is the energy in kJ/mol required to remove this electron?

2.127 Consider the noble gas xenon.

(a) Write the electron configuration of xenon using the abbreviation of the previous noble gas to represent inner-shell electrons.

(b) When xenon absorbs 801 kJ/mol of energy, it is excited into a higher-energy state in which the outermost electron has been promoted to the next available subshell. Write the electron configuration for this excited xenon.

(c) The energy required to completely remove the outermost electron from the excited xenon atom is 369 kJ/mol, almost identical to that of cesium (376 kJ/mol). Explain.

CHAPTER 3

Atoms and Ionic Bonds

This statue at the 700-year-old Wieliczka salt mine in Poland was carved by miners entirely out of salt. We'll learn more about this remarkable substance in the *FYI* at the end of this chapter.

Now that we have a good understanding of the structure of isolated atoms, let's extend that understanding by seeing how and why atoms join together to form chemical compounds. The forces that hold atoms together in chemical compounds are called *chemical bonds* and are of two types: *ionic bonds* and *covalent bonds*. As a general rule, ionic bonds form primarily between a metal atom and a nonmetal atom, while covalent bonds form primarily between two nonmetal atoms. We'll begin this study of chemical bonding by first seeing a brief introduction to both kinds of bonds and then focusing on ionic bonds in the remainder of this chapter. We'll then cover covalent bonds in Chapters 4 and 5.

3.1 IONS, MOLECULES, AND CHEMICAL BONDS

Imagine what must happen when two atoms approach each other at the beginning of a chemical reaction. Because the electrons of an atom occupy a much greater volume than the nucleus, it's the electrons that actually make the contact when atoms collide. Thus, it's the electrons that form the connections, or **chemical bonds,** that join atoms together in compounds.

Covalent Bonds: Molecules

A **covalent bond**, the most common kind of chemical bond, results when two atoms share several (usually two) electrons. A simple way to think about a covalent bond is to imagine it as a tug-of-war. If two people pull on the same rope, they are effectively joined together: Neither person can escape from the other as long as both hold on to the rope. Similarly with atoms: When two atoms both hold on to the same electrons, the atoms are bonded together (**Figure 3.1**).

CONTENTS

The two teams are joined together because both are tugging on the same rope.

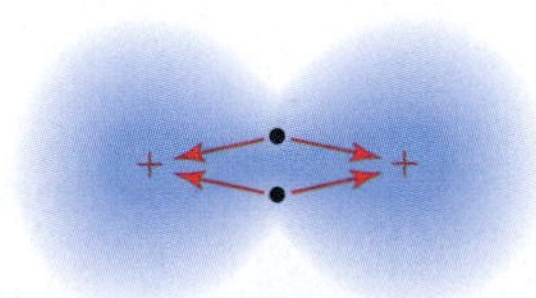

Similarly, two atoms are joined together when both **nuclei (+)** tug on the same **electrons (dots)**.

FIGURE 3.1

A covalent bond between atoms is analogous to a tug-of-war.

The unit of matter that results when two or more atoms are joined by covalent bonds is called a **molecule**. A hydrogen chloride molecule (HCl), for instance, results when a hydrogen atom and a chlorine atom share two electrons in a bond. A water molecule (H_2O) results when an oxygen atom shares two electrons with each of two hydrogen atoms. An ammonia molecule (NH_3) results when a nitrogen atom shares two electrons with each of three hydrogen atoms, and so on. To visualize these and other molecules, it helps to imagine the individual atoms as spheres joined together in specific three-dimensional shapes, as shown in **Figure 3.2.** *Ball-and-stick* models specifically indicate the covalent bonds between atoms, while *space-filling* models accurately portray overall molecular shape but don't explicitly show the bonds.

Even some *elements* exist as molecules rather than as individual atoms. Hydrogen, nitrogen, oxygen, fluorine, chlorine, bromine, and iodine, for instance, all exist as *diatomic* (two-atom) molecules whose two identical atoms are held together by covalent bonds. We therefore have to write them as such—H_2, N_2, O_2, F_2, Cl_2, Br_2, and I_2—when using any

FIGURE 3.2

Molecular models. Drawings such as these help in visualizing molecules.

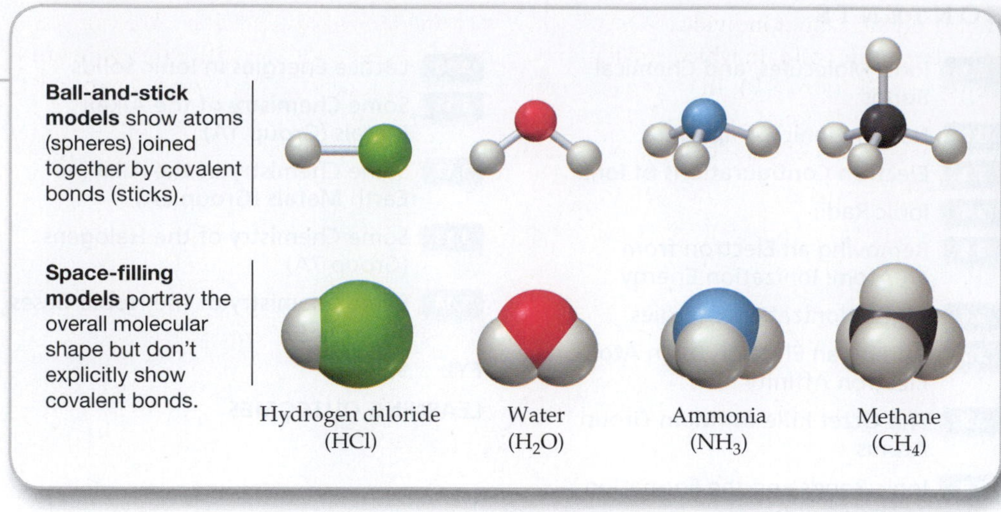

of these elements in a chemical equation. Notice that all these diatomic elements except hydrogen cluster toward the far right side of the periodic table.

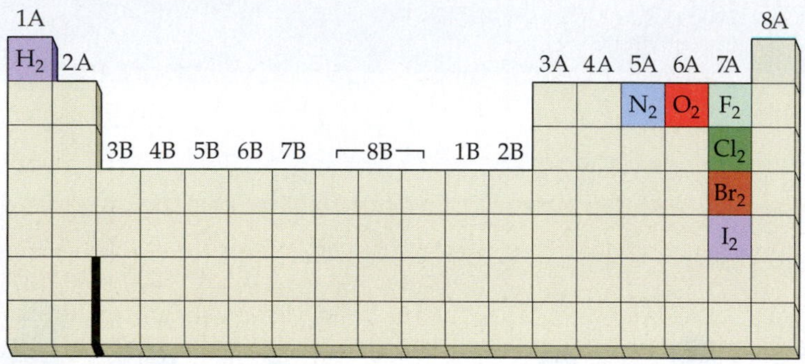

Ionic Bonds

In contrast to a covalent bond, an **ionic bond** results when a complete transfer of one or more electrons from one atom to another occurs, resulting in the formation of two charged particles, called **ions**. When sodium metal comes in contact with chlorine gas, for example, a neutral sodium atom transfers an electron to a neutral chlorine atom, giving two ions. Because the Na atom *loses* one electron, it loses one negative charge and becomes an Na^+ ion with a charge of +1. Conversely, because the Cl atom *gains* an electron, it gains a negative charge and becomes a Cl^- ion with a charge of −1. Positive ions such as Na^+ are called **cations** (**cat**-ions), and negative ions such as Cl^- are called **anions** (**an**-ions).

A sodium atom → Na; A chlorine molecule → Cl_2; A sodium cation → Na^+; A chloride anion → Cl^-

$$Na + \frac{1}{2}\,Cl_2 \longrightarrow Na^+ + Cl^-$$

A similar reaction takes place when magnesium atoms and chlorine molecules (Cl_2) come in contact to form $MgCl_2$. A magnesium atom transfers two electrons, one to each of two chlorine atoms, yielding the doubly charged Mg^{2+} cation and two singly charged Cl^- anions.

$$Mg + Cl_2 \longrightarrow Mg^{2+} + Cl^- + Cl^- \quad (MgCl_2)$$

Because opposite charges attract each other, positively charged cations like Na^+ and Mg^{2+} feel a strong electrical attraction to negatively charged anions like Cl^-, an attraction that we call an ionic bond. Unlike what happens when covalent bonds are formed, though, we

▲ Chlorine is a toxic green gas, sodium is a reactive metal, and sodium chloride is a harmless white solid.

can't really talk about individual Na^+Cl^- *molecules* under normal conditions. We can speak only of an **ionic solid**, in which equal numbers of Na^+ and Cl^- ions are packed together in a regular way **(Figure 3.3).** In a crystal of table salt, for instance, each Na^+ ion is surrounded by six nearby Cl^- ions, and each Cl^- ion is surrounded by six nearby Na^+ ions, but we can't specify what pairs of ions "belong" to each other as we can with atoms in covalently bonded molecules.

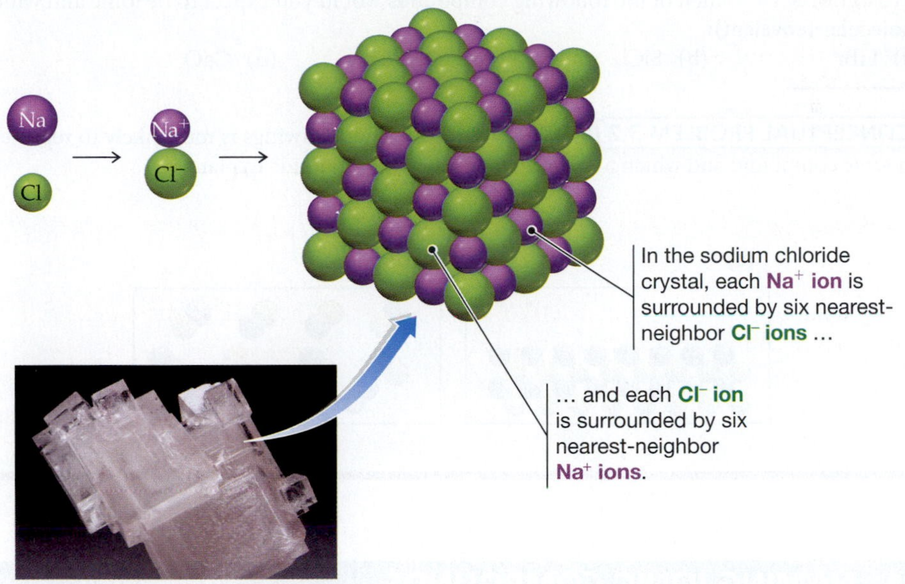

FIGURE 3.3

The arrangement of Na^+ ions and Cl^- ions in a crystal of sodium chloride. There are no discrete "molecules" of NaCl. Instead, the entire crystal is an ionic solid.

Charged, covalently bonded *groups* of atoms, called **polyatomic ions**, are also common—ammonium ion (NH_4^+), hydroxide ion (OH^-), nitrate ion (NO_3^-), and the doubly charged sulfate ion (SO_4^{2-}) are examples. You can think of these polyatomic ions as charged molecules because they consist of specific numbers and kinds of atoms joined together by covalent bonds, with the overall unit having a positive or negative charge (**Figure 3.4**). When writing the formulas of substances that contain more than one of these ions, parentheses are placed around the entire polyatomic unit to show that the atoms belong together. The formula $Ba(NO_3)_2$, for instance, indicates a substance made of Ba^{2+} cations and NO_3^- polyatomic anions in a 1:2 ratio.

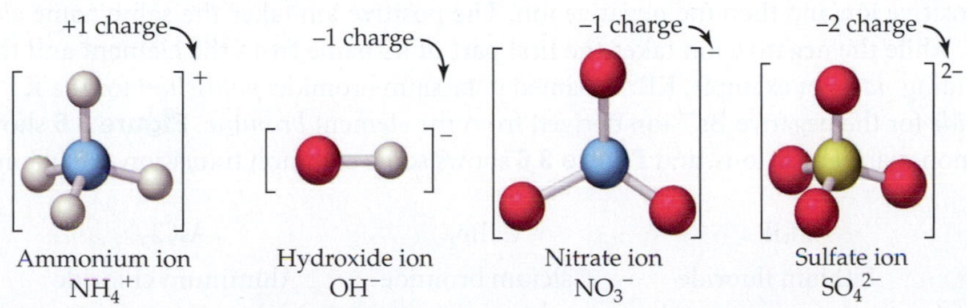

FIGURE 3.4

Molecular models of some polyatomic ions.

Worked Example 3.1

Identifying Ionic and Molecular Compounds

Which of the following compounds would you expect to be ionic and which molecular (covalent)?

(a) BaF_2 **(b)** SF_4 **(c)** PH_3 **(d)** CH_3OH

continued on next page

STRATEGY

Remember that covalent bonds generally form between nonmetal atoms, yielding molecular compounds, while ionic bonds form between metal and nonmetal atoms.

SOLUTION

Compound **(a)** is composed of a metal (barium) and a nonmetal (fluorine) and is likely to be ionic. Compounds **(b)**–**(d)** are composed entirely of nonmetals and are likely to be molecular.

PROBLEM 3.1 Which of the following compounds would you expect to be ionic and which molecular (covalent)?

(a) LiBr (b) $SiCl_4$ (c) BF_3 (d) CaO

CONCEPTUAL PROBLEM 3.2 Which of the following drawings is most likely to represent an ionic compound and which a molecular (covalent) compound? Explain.

(a) **(b)**

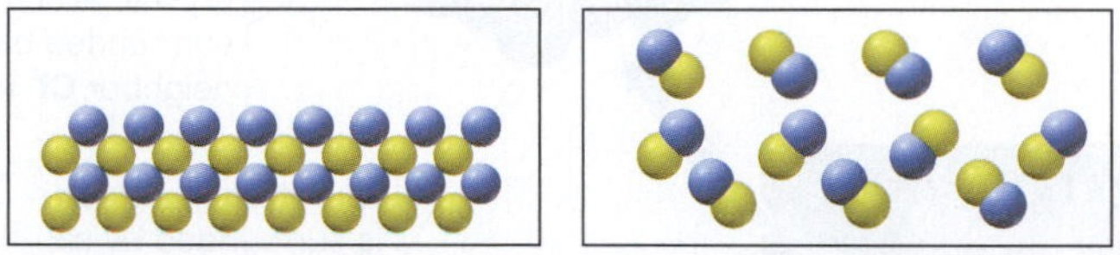

3.2 NAMING IONIC COMPOUNDS

▲ Morphine, a pain-killing agent found in the opium poppy, was named after Morpheus, the Greek god of dreams.

In the early days of chemistry, when few pure substances were known, newly discovered compounds were often given fanciful names—morphine, quicklime, potash, and barbituric acid (said to be named by its discoverer in honor of his friend Barbara) to cite a few. Today, with more than 67 million pure compounds known, there would be chaos without a systematic method for naming compounds. Every chemical compound must have a name that not only defines it uniquely but also allows chemists (and computers) to know its chemical structure. Let's begin by seeing how simple ionic compounds are named and then introduce additional rules in later chapters as the need arises.

Naming Binary Ionic Compounds

Binary ionic compounds—those made of only two elements—are named by identifying first the positive ion and then the negative ion. The positive ion takes the same name as the element, while the negative ion takes the first part of its name from the element and then adds the ending *-ide*. For example, KBr is named potassium bromide: *potassium* for the K^+ ion, and *bromide* for the negative Br^- ion derived from the element *brom*ine. **Figure 3.5** shows some common main-group ions, and **Figure 3.6** shows some common transition-metal ions.

LiF	$CaBr_2$	$AlCl_3$
Lithium fluoride	Calcium bromide	Aluminum chloride

There are several interesting points about Figure 3.5. Notice, for instance, that metal atoms tend to form cations and nonmetal atoms tend to form anions. Notice also that elements within a given group of the periodic table form similar kinds of ions and that the charge on the ion is related to the group number. Main-group metals usually form cations whose charge is equal to the group number. Group 1A elements form singly positive ions (M^+, where M is a metal), group 2A elements form doubly positive ions (M^{2+}), and group 3A elements form triply positive ions (M^{3+}). Main-group nonmetals usually form anions whose charge is equal to the group number in the U.S. system minus eight. Thus, group 6A

FIGURE 3.5

Main-group cations (blue) and anions (red). A **cation** bears the same name as the element it is derived from; an **anion** name has an *-ide* ending.

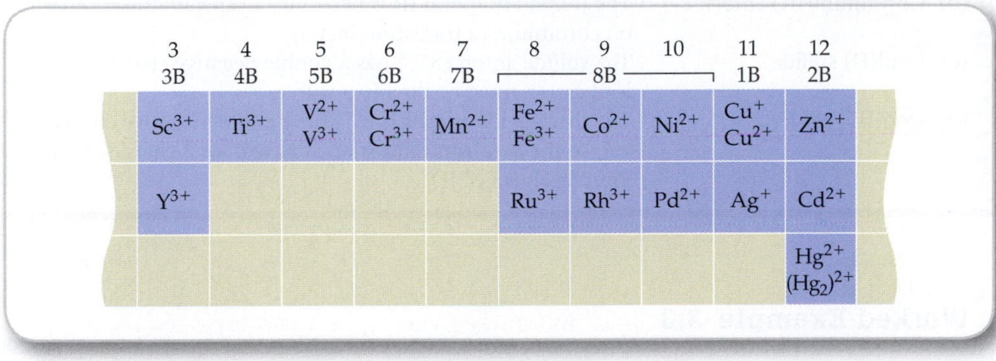

FIGURE 3.6

Common transition-metal ions. Only ions that exist in aqueous solution are shown.

elements form doubly negative ions ($6 - 8 = -2$), group 7A elements form singly negative ions ($7 - 8 = -1$), and group 8A elements form no ions at all ($8 - 8 = 0$). We'll see the reason for this behavior when we discuss the **octet rule** in Section 3.8.

Notice also in both Figures 3.5 and 3.6 that some metals form more than one kind of cation. Iron, for instance, forms both the doubly charged Fe^{2+} ion and the triply charged Fe^{3+} ion. In naming different ions of a given element, we distinguish between them by using a Roman numeral in parentheses to indicate the number of charges on each. Thus, $FeCl_2$ is named iron(II) chloride and $FeCl_3$ is iron(III) chloride. Alternatively, an older method distinguishes between the ions by using the Latin name of the element (*ferrum* in the case of iron) together with the ending *-ous* for the ion with lower charge and *-ic* for the ion with higher charge. Thus, you might see $FeCl_2$ called ferrous chloride and $FeCl_3$ called ferric chloride. Although still in use, particularly on the labels that list ingredients of products, this older system is being phased out and we'll rarely use it in this book.

	Fe^{2+}	Fe^{3+}	Sn^{2+}	Sn^{4+}
	Iron(II) ion	Iron(III) ion	Tin(II) ion	Tin(IV) ion
or	Ferrous ion	Ferric ion	Stannous ion	Stannic ion
	(From the Latin *ferrum* = iron)		(From the Latin *stannum* = tin)	

In any neutral compound, the total number of positive charges must equal the total number of negative charges. Thus, you can always figure out the number of positive charges

LOOKING AHEAD ...

As we'll see in Section 3.8, the **octet rule** summarizes the chemical behavior of main-group elements by saying that they tend to undergo reactions that leave them with eight outer-shell electrons.

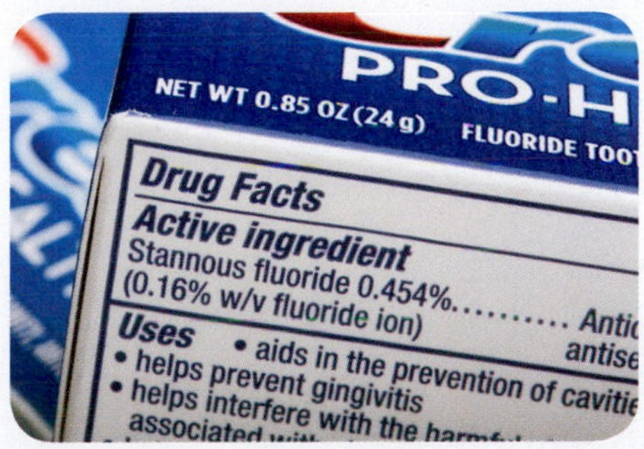

▲ Many product labels use the older system of naming with *-ous* and *-ic* endings.

▲ Crystals of iron(II) chloride tetrahydrate are greenish, and crystals of iron(III) chloride hexahydrate are brownish yellow.

on a metal cation by counting the number of negative charges on the associated anion(s). In $FeCl_2$, for instance, the iron ion must be Fe(II) because there are two Cl^- ions associated with it. Similarly, in $TiCl_3$ the titanium ion is Ti(III) because there are three Cl^- anions associated with it. As a general rule, a Roman numeral is needed for transition-metal compounds to avoid ambiguity. In addition, the main-group metals tin (Sn), thallium (Tl), and lead (Pb) can form more than one kind of ion and need Roman numerals for naming their compounds. Metals in group 1A and group 2A form only one cation, however, so Roman numerals are not needed.

Worked Example 3.2

Naming Binary Ionic Compounds

Give systematic names for the following compounds:

(a) $BaCl_2$ **(b)** $CrCl_3$ **(c)** PbS **(d)** Fe_2O_3

STRATEGY

Try to figure out the number of positive charges on each cation by counting the number of negative charges on the associated anion(s). Refer to Figures 3.5 and 3.6 as necessary.

SOLUTION

(a) Barium chloride — No Roman numeral is necessary because barium, a group 2A element, forms only Ba^{2+}.

(b) Chromium(III) chloride — The Roman numeral III is necessary to specify the +3 charge on chromium (a transition metal).

(c) Lead(II) sulfide — The sulfide anion (S^{2-}) has a double negative charge, so the lead cation must be doubly positive.

(d) Iron(III) oxide — The three oxide anions (O^{2-}) have a total negative charge of −6, so the two iron cations must have a total charge of +6. Thus, each is Fe(III).

Worked Example 3.3

Converting Names into Formulas

Write formulas for the following compounds:

(a) Magnesium fluoride **(b)** Tin(IV) oxide **(c)** Iron(III) sulfide

STRATEGY

For transition-metal compounds, the charge on the cation is indicated by the Roman numeral in the name. Knowing the number of positive charges, you can then figure out the number of necessary negative charges for the associated anions.

SOLUTION

(a) MgF_2 — Magnesium (group 2A) forms only a 2+ cation, so there must be two fluoride ions (F^-) to balance the charge.

(b) SnO_2 — Tin(IV) has a +4 charge, so there must be two oxide ions (O^{2-}) to balance the charge.

(c) Fe_2S_3 — Iron(III) has a +3 charge and sulfide ion a −2 charge (S^{2-}), so there must be two irons and three sulfurs.

▶ **PROBLEM 3.3** Give systematic names for the following compounds:

(a) CsF **(b)** K_2O **(c)** CuO **(d)** BaS **(e)** $BeBr_2$

▶ **PROBLEM 3.4** Write formulas for the following compounds:

(a) Vanadium(III) chloride **(b)** Manganese(IV) oxide

(c) Copper(II) sulfide **(d)** Aluminum oxide

CONCEPTUAL PROBLEM 3.5 Three binary ionic compounds are represented on the following periodic table: red with red, green with green, and blue with blue. Name each, and tell its likely formula.

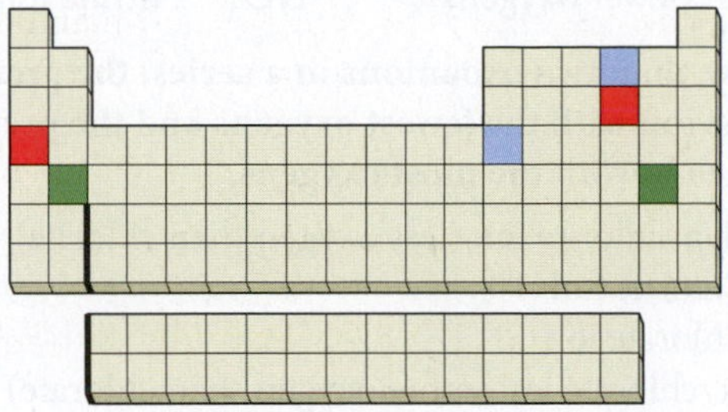

Naming Compounds with Polyatomic Ions

Ionic compounds that contain a polyatomic ion (Section 3.1) are named in the same way as binary ionic compounds: First the cation is identified and then the anion. For example, $Ba(NO_3)_2$ is called *barium nitrate* because Ba^{2+} is the cation and the NO_3^- polyatomic anion has the name *nitrate*. Unfortunately, there is no simple systematic way for naming the polyatomic ions themselves, so it's necessary to memorize the names, formulas, and charges of the most common ones, listed in **Table 3.1**. The ammonium ion (NH_4^+) is the only cation on the list; all the others are anions.

TABLE 3.1 Some Common Polyatomic Ions

Formula	Name	Formula	Name
Cation		NO_2^-	Nitrite
NH_4^+	Ammonium	NO_3^-	Nitrate
Singly charged anions		**Doubly charged anions**	
$CH_3CO_2^-$	Acetate	CO_3^{2-}	Carbonate
CN^-	Cyanide	CrO_4^{2-}	Chromate
ClO^-	Hypochlorite	$Cr_2O_7^{2-}$	Dichromate
ClO_2^-	Chlorite	O_2^{2-}	Peroxide
ClO_3^-	Chlorate	HPO_4^{2-}	Hydrogen phosphate
ClO_4^-	Perchlorate	SO_3^{2-}	Sulfite
$H_2PO_4^-$	Dihydrogen phosphate	SO_4^{2-}	Sulfate
HCO_3^-	Hydrogen carbonate (or bicarbonate)	$S_2O_3^{2-}$	Thiosulfate
HSO_4^-	Hydrogen sulfate (or bisulfate)	**Triply charged anion**	
OH^-	Hydroxide	PO_4^{3-}	Phosphate
MnO_4^-	Permanganate		

Several points about the ions in Table 3.1 should be noted:

- **The names of most polyatomic anions end in *-ite* or *-ate*.** Only hydroxide (OH^-), cyanide (CN^-), and peroxide (O_2^{2-}) have the *-ide* ending.
- **Several of the ions form a series of oxoanions**, binary polyatomic anions in which an atom of a given element is combined with different numbers of oxygen atoms—hypochlorite (ClO^-), chlorite (ClO_2^-), chlorate (ClO_3^-), and perchlorate (ClO_4^-), for instance.

- **When there are only two oxoanions in a series, as with sulfite (SO_3^{2-}) and sulfate (SO_4^{2-}), the ion with fewer oxygens takes the *-ite* ending and the ion with more oxygens takes the *-ate* ending.**

SO_3^{2-}	Sulf*ite* ion (fewer oxygens)	SO_4^{2-}	Sulf*ate* ion (more oxygens)
NO_2^-	Nitr*ite* ion (fewer oxygens)	NO_3^-	Nitr*ate* ion (more oxygens)

- **When there are more than two oxoanions in a series, the prefix *hypo-* (meaning "less than") is used for the ion with the fewest oxygens and the prefix *per-* (meaning "more than") is used for the ion with the most oxygens.**

ClO^-	*Hypo*chlorite ion (less oxygen than chlorite)
ClO_2^-	Chlorite ion
ClO_3^-	Chlorate ion
ClO_4^-	*Per*chlorate ion (more oxygen than chlorate)

- **Several pairs of ions are related by the presence or absence of a hydrogen.** The hydrogen carbonate anion (HCO_3^-) differs from the carbonate anion (CO_3^{2-}) by the presence of H^+, and the hydrogen sulfate anion (HSO_4^-) differs from the sulfate anion (SO_4^{2-}) by the presence of H^+. The ion that has the additional hydrogen is sometimes referred to using the prefix *bi-*, although this usage is now discouraged; for example, $NaHCO_3$ is sometimes called sodium bicarbonate.

HCO_3^-	Hydrogen carbonate (*bi*carbonate) ion	CO_3^{2-}	Carbonate ion
HSO_4^-	Hydrogen sulfate (*bi*sulfate) ion	SO_4^{2-}	Sulfate ion

Worked Example 3.4

Naming Compounds with Polyatomic Ions

Give systematic names for the following compounds:
(a) $LiNO_3$ **(b)** $KHSO_4$ **(c)** $CuCO_3$ **(d)** $Fe(ClO_4)_3$

STRATEGY

Unfortunately, there is no alternative: The names and charges of the common polyatomic ions must be memorized. Refer to Table 3.1 if you need help.

SOLUTION

(a) Lithium nitrate	Lithium (group 1A) forms only the Li^+ ion and does not need a Roman numeral.
(b) Potassium hydrogen sulfate	Potassium (group 1A) forms only the K^+ ion.
(c) Copper(II) carbonate	The carbonate ion has a −2 charge, so copper must be +2. A Roman numeral is needed because copper, a transition metal, can form more than one ion.
(d) Iron(III) perchlorate	There are three perchlorate ions, each with a −1 charge, so the iron must have a +3 charge.

Worked Example 3.5

Writing Formulas of Compounds with Polyatomic Ions

Write formulas for the following compounds:
(a) Potassium hypochlorite
(b) Silver(I) chromate
(c) Iron(III) carbonate

SOLUTION

(a) $KClO$	Potassium forms only the K^+ ion, so only one ClO^- is needed.
(b) Ag_2CrO_4	The polyatomic chromate ion has a −2 charge, so two Ag^+ ions are needed.
(c) $Fe_2(CO_3)_3$	Iron(III) has a +3 charge and the polyatomic carbonate ion has a −2 charge, so there must be two iron ions and three carbonate ions. The polyatomic carbonate ion is set off in parentheses to indicate that there are three of them.

PROBLEM 3.6 Give systematic names for the following compounds:
(a) $Ca(ClO)_2$ (b) $Ag_2S_2O_3$ (c) NaH_2PO_4 (d) $Sn(NO_3)_2$ (e) $Pb(CH_3CO_2)_4$ (f) $(NH_4)_2SO_4$

PROBLEM 3.7 Write formulas for the following compounds:
(a) Lithium phosphate (b) Magnesium hydrogen sulfate (c) Manganese(II) nitrate (d) Chromium(III) sulfate

CONCEPTUAL PROBLEM 3.8 The following drawings are those of solid ionic compounds, with red spheres representing the cations and blue spheres representing the anions in each.

(1)

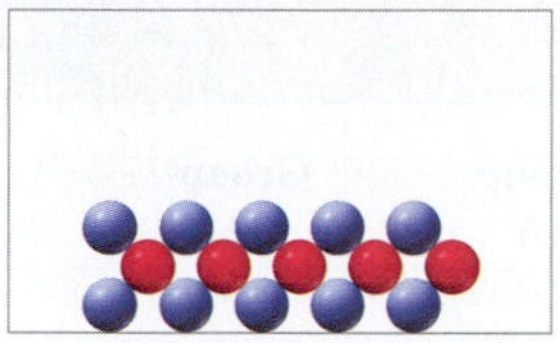

(2)

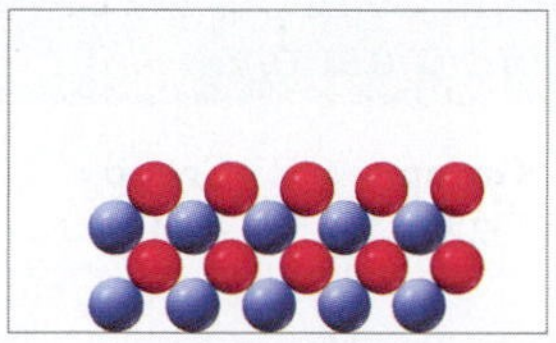

Which of the following formulas are consistent with each drawing?
(a) LiBr (b) $NaNO_2$ (c) $CaCl_2$ (d) K_2CO_3 (e) $Fe_2(SO_4)_3$

3.3 ELECTRON CONFIGURATIONS OF IONS

As we've now seen, metals, on the left side of the periodic table, tend to give up electrons in their chemical reactions and form cations. Conversely, halogens and other reactive nonmetals, on the right side of the table, tend to accept electrons in their chemical reactions and form anions. What are the **ground-state electron configurations** of the resultant ions?

For main-group elements, the electrons given up by a metal in forming a cation come from the highest-energy occupied orbital, while the electrons that are accepted by a nonmetal in forming an anion go into the lowest-energy unoccupied orbital according to the **aufbau principle** (Section 2.11). When a sodium atom ($1s^2\,2s^2\,2p^6\,3s^1$) reacts with a chlorine atom and gives up an electron, for example, the valence-shell 3*s* electron of sodium is lost, giving an Na^+ ion with the noble-gas electron configuration of neon ($1s^2\,2s^2\,2p^6$). At the same time, when the chlorine atom ($1s^2\,2s^2\,2p^6\,3s^2\,3p^5$) accepts an electron from sodium, the electron fills the remaining vacancy in the 3*p* subshell to give a Cl^- ion with the noble-gas electron configuration of argon ($1s^2\,2s^2\,2p^6\,3s^2\,3p^6$).

$$\textbf{Na: } 1s^2\,2s^2\,2p^6\,3s^1 \xrightarrow{-e^-} \textbf{Na}^+\textbf{: } 1s^2\,2s^2\,2p^6$$

$$\textbf{Cl: } 1s^2\,2s^2\,2p^6\,3s^2\,3p^5 \xrightarrow{+e^-} \textbf{Cl}^-\textbf{: } 1s^2\,2s^2\,2p^6\,3s^2\,3p^6$$

What is true for sodium is also true for the other elements in group 1A: All form positive ions by losing their valence-shell *s* electron when they undergo reaction, and all the resultant ions have noble-gas electron configurations. Similarly for the elements in group 2A: All form a doubly positive ion when they react, losing both their valence-shell *s* electrons. An Mg atom ($1s^2\,2s^2\,2p^6\,3s^2$), for example, goes to an Mg^{2+} ion with the neon configuration $1s^2\,2s^2\,2p^6$ by loss of its two 3*s* electrons.

$$\textbf{Group 1A atom: } [\text{Noble gas}]\ ns^1 \xrightarrow{-e^-} \textbf{Group 1A ion}^+\textbf{: } [\text{Noble gas}]$$

$$\textbf{Group 2A atom: } [\text{Noble gas}]\ ns^2 \xrightarrow{-2e^-} \textbf{Group 2A ion}^{2+}\textbf{: } [\text{Noble gas}]$$

REMEMBER...
The **ground-state electron configuration** of an atom or ion is a description of the atomic orbitals that are occupied in the lowest-energy state of the atom or ion. (Section 2.11)

REMEMBER...
According to the **aufbau principle**, lower-energy orbitals fill before higher-energy ones and an orbital can hold only two electrons, which have opposite spins. (Section 2.11)

Just as the group 1A and 2A metals *lose* the appropriate number of electrons to yield ions with noble-gas configurations, the group 6A and group 7A nonmetals *gain* the appropriate number of electrons when they react with metals. The halogens in group 7A gain one electron to form singly charged anions with noble-gas configurations, and the elements in group 6A gain two electrons to form doubly charged anions with noble-gas configurations. Oxygen ($1s^2\ 2s^2\ 2p^4$), for example, becomes the O^{2-} ion with the neon configuration ($1s^2\ 2s^2\ 2p^6$) when it reacts with a metal:

$$\textbf{Group 6A atom: } [\text{Noble gas}]\ ns^2\ np^4 \xrightarrow{+2e^-} \textbf{Group 6A ion}^{2-}\textbf{: } [\text{Noble gas}]\ ns^2\ np^6$$
$$\textbf{Group 7A atom: } [\text{Noble gas}]\ ns^2\ np^5 \xrightarrow{+e^-} \textbf{Group 7A ion}^{-}\textbf{: } [\text{Noble gas}]\ ns^2\ np^6$$

The electron configurations of common main-group ions are listed in **Table 3.2**.

TABLE 3.2 Some Common Main-Group Ions and Their Noble-Gas Electron Configurations

Group 1A	Group 2A	Group 3A	Group 6A	Group 7A	Electron Configuration
H^+					[None]
H^-					[He]
Li^+	Be^{2+}				[He]
Na^+	Mg^{2+}	Al^{3+}	O^{2-}	F^-	[Ne]
K^+	Ca^{2+}	*Ga^{3+}	S^{2-}	Cl^-	[Ar]
Rb^+	Sr^{2+}	*In^{3+}	Se^{2-}	Br^-	[Kr]
Cs^+	Ba^{2+}	*Tl^{3+}	Te^{2-}	I^-	[Xe]

*These ions don't have a true noble-gas electron configuration because they have an additional filled *d* subshell.

The situation is a bit different for ion formation from the transition-metal elements in groups 1B–8B than it is for the main-group elements. Transition metals react with nonmetals to form cations by first losing their valence-shell *s* electrons and then losing one or more *d* electrons. As a result, the remaining valence electrons in transition-metal cations occupy *d* orbitals. Iron, for instance, forms the Fe^{2+} ion by losing its two 4*s* electrons and forms the Fe^{3+} ion by losing two 4*s* electrons and one 3*d* electron:

$$\textbf{Fe: } [\text{Ar}]\ 4s^2\ 3d^6 \xrightarrow{-2e^-} \textbf{Fe}^{2+}\textbf{: } [\text{Ar}]\ 3d^6$$
$$\textbf{Fe: } [\text{Ar}]\ 4s^2\ 3d^6 \xrightarrow{-3e^-} \textbf{Fe}^{3+}\textbf{: } [\text{Ar}]\ 3d^5$$

It may seem strange that building up the periodic table adds the 3*d* electrons *after* the 4*s* electrons, whereas forming an ion from a transition metal removes the 4*s* electrons *before* the 3*d* electrons. Note, though, that the two processes are not the reverse of one another, so they can't be compared directly. Building up the periodic table adds one electron to the valence shell and also adds one positive charge to the nucleus, but forming an ion removes an electron from the valence shell without altering the nucleus.

PROBLEM 3.9 Predict the ground-state electron configuration for each of the following ions, and explain your answers.
(a) Ra^{2+} **(b)** Y^{3+} **(c)** Ti^{4+} **(d)** N^{3-}

PROBLEM 3.10 What doubly positive ion has the following ground-state electron configuration? $1s^2\ 2s^2\ 2p^6\ 3s^2\ 3p^6\ 3d^{10}$

3.4 IONIC RADII

Just as there are systematic differences in the **radii of atoms** (Section 2.14), there are also systematic differences in the radii of ions. As shown in **Figure 3.7** for the elements of groups 1A and 2A, atoms shrink dramatically when an electron is removed to form a cation. The radius of an Na atom, for example, is 186 pm, but that of an Na^+ cation is 102 pm. Similarly, the radius of an Mg atom is 160 pm and that of an Mg^{2+} cation is 72 pm.

The cation that results when an electron is removed from a neutral atom is smaller than the original atom both because the electron is removed from a large, valence-shell orbital and because there is an increase in the **effective nuclear charge, Z_{eff}** for the remaining electrons (Section 2.10). On going from a neutral Na atom to a charged Na^+ cation, for example, the electron configuration changes from $1s^2\,2s^2\,2p^6\,3s^1$ to $1s^2\,2s^2\,2p^6$. The valence shell of the Na *atom* is the *third* shell, but the valence shell of the Na^+ *cation* is the *second* shell. Thus, the Na^+ ion has a smaller valence shell than the Na atom and therefore a smaller size. In addition, the effective nuclear charge felt by the valence-shell electrons is greater in the Na^+ cation than in the neutral atom. The Na atom has 11 protons and 11 electrons, but the Na^+ cation has 11 protons and only 10 electrons. The smaller number of electrons in the cation means that they shield one another to a lesser extent and therefore are pulled in more strongly toward the nucleus.

REMEMBER...

Atomic radii increase down a column of the periodic table because successively larger valence-shell orbitals are occupied, and radii decrease from left to right across a row of the periodic table because the effective nuclear charge increases across the row. (Section 2.14)

REMEMBER...

The effective nuclear charge (Z_{eff}) felt by an electron is due to electron shielding of outer electrons by inner electrons and may be substantially lower than the actual n clear charge: $Z_{eff} = Z_{actual} -$ electron shielding. (Section 2.10)

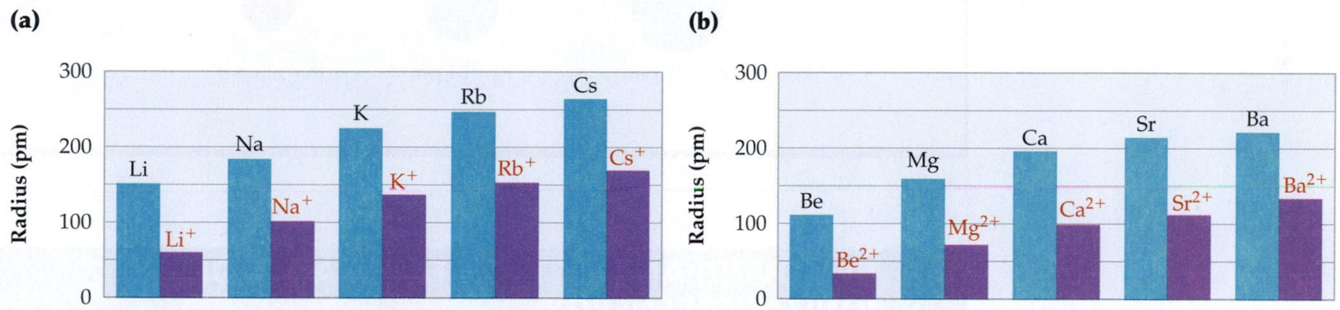

Cations are smaller than the corresponding **neutral atoms**, both because the principal quantum number of the valence-shell electrons is smaller for the cations than it is for the neutral atoms and because Z_{eff} is larger.

FIGURE 3.7

Radii of (a) group 1A atoms and their cations; (b) group 2A atoms and their cations.

The same effects felt by the group 1A elements when a single electron is lost are felt by the group 2A elements when two electrons are lost. For instance, loss of two valence-shell electrons from an Mg atom ($1s^2\,2s^2\,2p^6\,3s^2$) gives the Mg^{2+} cation ($1s^2\,2s^2\,2p^6$). The smaller valence shell of the Mg^{2+} cation and the increase in effective nuclear charge combine to cause a dramatic shrinkage. A similar shrinkage occurs whenever any metal atom on the left-hand two-thirds of the periodic table is converted into a cation.

Just as neutral atoms *shrink* when converted to cations by loss of one or more electrons, they *expand* when converted to anions by gain of one or more electrons. As shown in **Figure 3.8**

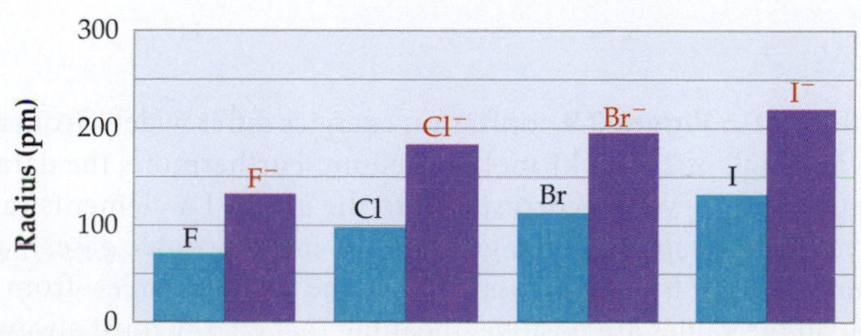

Anions are larger than their **neutral atoms** because of additional electron–electron repulsions and a decrease in Z_{eff}.

FIGURE 3.8

Radii of the group 7A atoms (halogens) and their anions.

for the group 7A elements (halogens), the expansion is dramatic. Chlorine, for example, nearly doubles in radius, from 99 pm for the neutral atom to 184 pm for the chloride anion.

The expansion that occurs when a group 7A atom gains an electron to yield an anion can't be accounted for by a change in the quantum number of the valence shell, because the added electron simply completes an already occupied p subshell: [Ne] $3s^2$ $3p^5$ for a Cl atom becomes [Ne] $3s^2$ $3p^6$ for a Cl^- anion, for instance. Thus, the expansion is due entirely to the decrease in effective nuclear charge and the increase in electron–electron repulsions that occurs when an extra electron is added.

PROBLEM 3.11 Which atom or ion in each of the following pairs would you expect to be larger? Explain.
(a) O or O^{2-} **(b)** O or S **(c)** Fe or Fe^{3+} **(d)** H or H^-

CONCEPTUAL PROBLEM 3.12 Which of the following spheres represents a K^+ ion, which a K atom, and which a Cl^- ion?

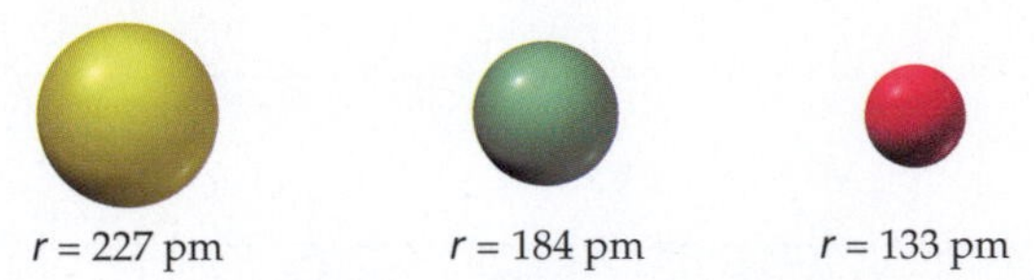

3.5 REMOVING AN ELECTRON FROM AN ATOM: IONIZATION ENERGY

We saw in the previous chapter that absorption of radiant energy by an atom leads to a change in electron configuration. When energy is added, a valence-shell electron is promoted from a lower-energy orbital to a higher-energy one with a larger principal quantum number n. If enough energy is absorbed, the electron can even be removed completely from the atom, leaving behind a cation. The amount of energy necessary to remove the highest-energy electron from an isolated neutral atom in the gaseous state is called the atom's **ionization energy** (**E_i**). For hydrogen, $E_i = 1312.0$ kJ/mol.

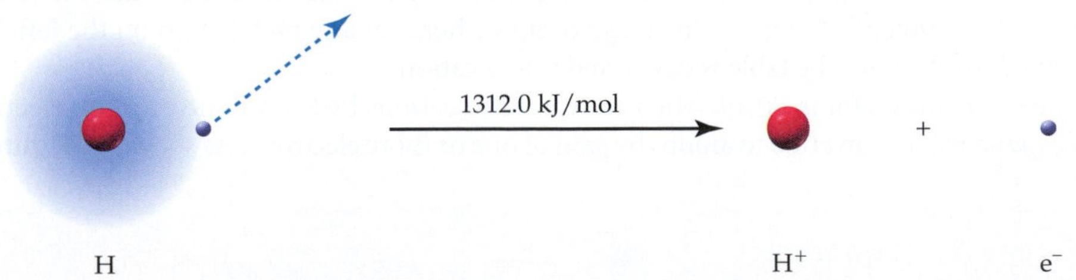

As shown by the plot in **Figure 3.9**, ionization energies differ widely, from a low of 375.7 kJ/mol for cesium to a high of 2372.3 kJ/mol for helium. Furthermore, the data show a clear periodicity. The minimum E_i values correspond to the group 1A elements (alkali metals), the maximum E_i values correspond to the group 8A elements (noble gases), and a gradual increase in E_i occurs from left to right across a row of the periodic table—from Na to Ar, for example. Note that all the values are positive, meaning that energy must always be added to remove an electron from an atom.

The periodicity evident in Figure 3.9 can be explained by looking at electron configurations. Atoms of the group 8A elements have filled valence subshells, either s for helium or both s and p for the other noble gases. As we saw in Section 2.14, an electron in a filled valence subshell feels a relatively high Z_{eff} because electrons in the same subshell don't **shield** one another very strongly. As a result, the electrons are held tightly to the nucleus, the radius of the atom is small, and the energy necessary to remove an electron is relatively large. Atoms of group 1A elements, by contrast, have only a single *s* electron in their valence shell. This single valence electron is shielded from the nucleus by all the inner-shell electrons, called the **core electrons**, resulting in a low Z_{eff}. The valence electron is thus held loosely, and the energy necessary to remove it is relatively small.

REMEMBER...

Valence electrons are strongly **shielded** by inner-shell (core) electrons, are less strongly shielded by other electrons in the same shell according to the order $s > p > d > f$, and are only weakly shielded by other electrons in the same subshell. (Section 2.14)

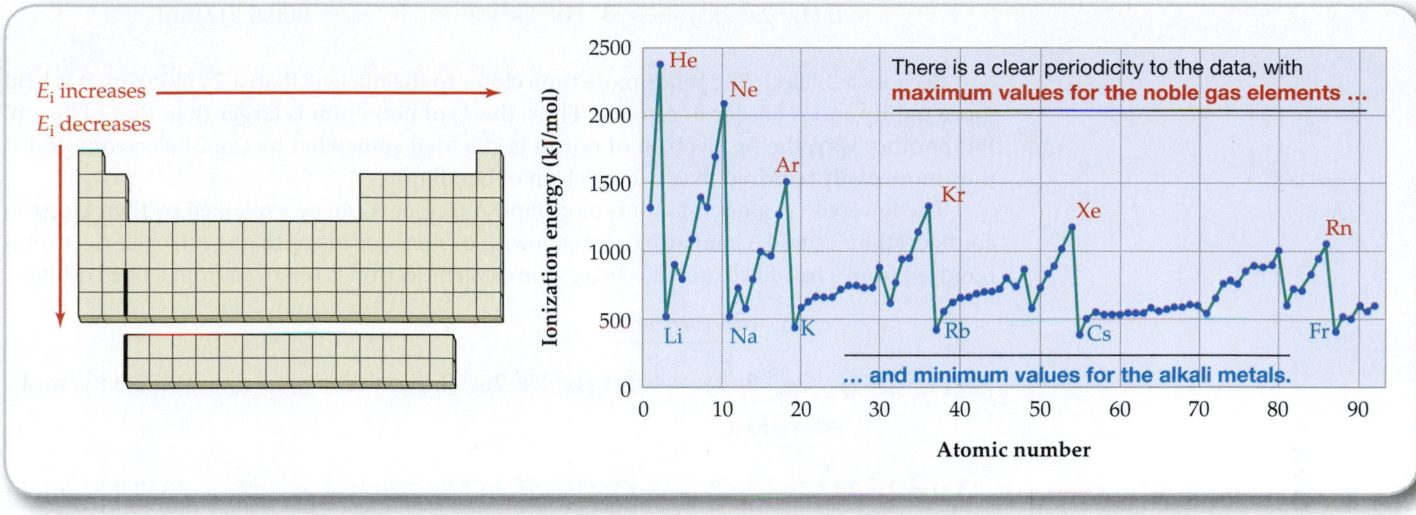

FIGURE 3.9

Ionization energies of the first 92 elements.

The plot of ionization energies in Figure 3.9 also shows other trends in the data beyond the obvious periodicity. One such trend is that ionization energies gradually decrease going down a group in the periodic table, from He to Rn and from Li to Fr, for instance. As the atomic number increases going down a group, both the principal quantum number of the valence-shell electrons and their average distance from the nucleus also increase. As a result, the valence-shell electrons are less tightly held and E_i is smaller.

Yet another point about the E_i data is that minor irregularities occur across a row of the periodic table. A close look at E_i values of the first 20 elements (**Figure 3.10**) shows that the E_i of beryllium is larger than that of its neighbor boron and the E_i of nitrogen is larger than that of its neighbor oxygen. Similarly, magnesium has a larger E_i than aluminum and phosphorus has a slightly larger E_i than sulfur.

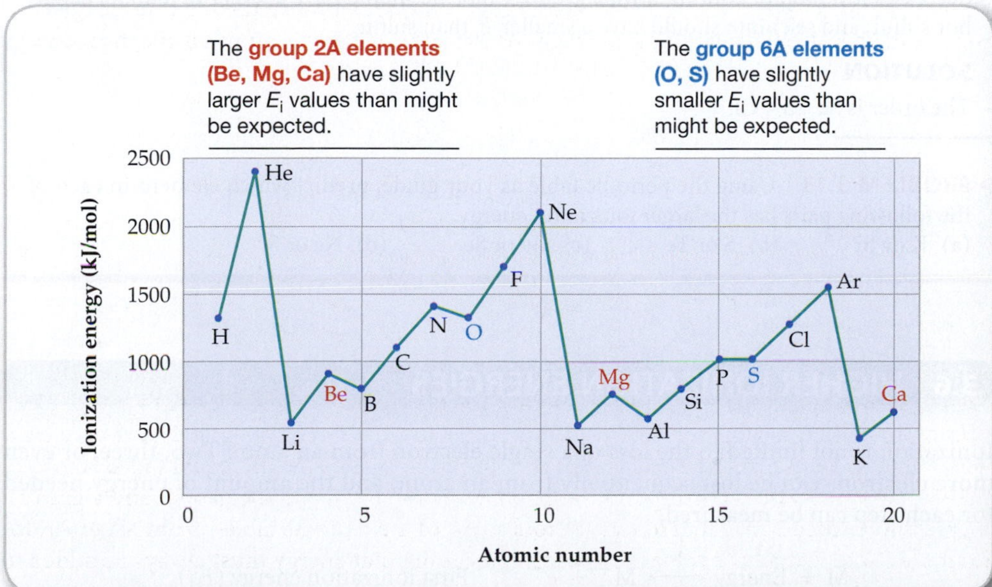

FIGURE 3.10

Ionization energies of the first 20 elements.

The slightly enlarged E_i values for the group 2A elements Be, Mg, and others can be explained by their electron configurations. Compare beryllium with boron, for instance. A 2*s* electron is removed on ionization of beryllium, but a $2p$ electron is removed on ionization of boron:

s electron removed

$$\text{Be}\ (1s^2\ 2s^2) \longrightarrow \text{Be}^+\ (1s^2\ 2s^1) + e^- \qquad E_i = 899.4\ \text{kJ/mol}$$

p electron removed

$$\text{B}\ (1s^2\ 2s^2\ 2p^1) \longrightarrow \text{B}^+\ (1s^2\ 2s^2) + e^- \qquad E_i = 800.6\ \text{kJ/mol}$$

Because a $2s$ electron spends more time closer to the nucleus than a $2p$ electron, it is held more tightly and is harder to remove. Thus, the E_i of beryllium is larger than that of boron. Put another way, the $2p$ electron of boron is shielded somewhat by the $2s$ electrons and is thus more easily removed than a $2s$ electron of beryllium.

The lowered E_i values for atoms of group 6A elements can be explained by their electron configurations as well. Comparing nitrogen with oxygen, for instance, the nitrogen electron is removed from a half-filled orbital, whereas the oxygen electron is removed from a filled orbital:

Half-filled orbital

$$\text{N}\ (1s^2\ 2s^2\ 2p_x^{\,1}\ 2p_y^{\,1}\ 2p_z^{\,1}) \longrightarrow \text{N}^+\ (1s^2\ 2s^2\ 2p_x^{\,1}\ 2p_y^{\,1}) + e^- \qquad E_i = 1402.3\ \text{kJ/mol}$$

Filled orbital

$$\text{O}\ (1s^2\ 2s^2\ 2p_x^{\,2}\ 2p_y^{\,1}\ 2p_z^{\,1}) \longrightarrow \text{O}^+\ (1s^2\ 2s^2\ 2p_x^{\,1}\ 2p_y^{\,1}\ 2p_z^{\,1}) + e^- \qquad E_i = 1313.9\ \text{kJ/mol}$$

Because electrons repel one another and tend to stay as far apart as possible, electrons that are forced together in a filled orbital are slightly higher in energy than those in a half-filled orbital, so removing one is slightly easier. Thus, oxygen has a smaller E_i than nitrogen.

Worked Example 3.6

Ionization Energies

Arrange the elements Se, Cl, and S in order of increasing ionization energy.

STRATEGY

Ionization energy generally increases from left to right across a row of the periodic table and decreases from top to bottom down a group. Chlorine should have a larger E_i than its neighbor sulfur, and selenium should have a smaller E_i than sulfur.

SOLUTION

The order is Se < S < Cl.

PROBLEM 3.13 Using the periodic table as your guide, predict which element in each of the following pairs has the larger ionization energy:
(a) K or Br **(b)** S or Te **(c)** Ga or Se **(d)** Ne or Sr

3.6 HIGHER IONIZATION ENERGIES

Ionization is not limited to the loss of a single electron from an atom. Two, three, or even more electrons can be lost sequentially from an atom, and the amount of energy needed for each step can be measured.

$$\text{M} + \text{Energy} \longrightarrow \text{M}^+ + e^- \qquad \text{First ionization energy } (E_{i1})$$

$$\text{M}^+ + \text{Energy} \longrightarrow \text{M}^{2+} + e^- \qquad \text{Second ionization energy } (E_{i2})$$

$$\text{M}^{2+} + \text{Energy} \longrightarrow \text{M}^{3+} + e^- \qquad \text{Third ionization energy } (E_{i3})$$

... and so forth

Successively larger amounts of energy are required for each ionization step because it is much harder to pull a negatively charged electron away from a positively charged ion than from a neutral atom. Interestingly, though, the energy differences between successive steps vary dramatically from one element to another. Removing the second electron from sodium takes nearly 10 times as much energy as removing the first one (4562 versus 496 kJ/mol), but removing the second electron from magnesium takes only twice as much energy as removing the first one (1451 versus 738 kJ/mol).

Large jumps in successive ionization energies are also found for other elements, as can be seen by following the zigzag line in **Table 3.3**. Magnesium has a large jump between its second and third ionization energies, aluminum has a large jump between its third and fourth ionization energies, silicon has a large jump between its fourth and fifth ionization energies, and so on.

TABLE 3.3 Higher Ionization Energies (kJ/mol) for Main-Group Third-Row Elements

Group	1A	2A	3A	4A	5A	6A	7A	8A
E_i Number	Na	Mg	Al	Si	P	S	Cl	Ar
E_{i1}	496	738	578	787	1,012	1,000	1,251	1,520
E_{i2}	4,562	1,451	1,817	1,577	1,903	2,251	2,297	2,665
E_{i3}	6,912	7,733	2,745	3,231	2,912	3,361	3,822	3,931
E_{i4}	9,543	10,540	11,575	4,356	4,956	4,564	5,158	5,770
E_{i5}	13,353	13,630	14,830	16,091	6,273	7,013	6,540	7,238
E_{i6}	16,610	17,995	18,376	19,784	22,233	8,495	9,458	8,781
E_{i7}	20,114	21,703	23,293	23,783	25,397	27,106	11,020	11,995

The zigzag line marks the large jumps in ionization energies.

The large increases in ionization energies that follow the zigzag line in Table 3.3 are yet another consequence of electron configurations. It's relatively easier to remove an electron from a *partially* filled valence shell because Z_{eff} is lower, but it's relatively harder to remove an electron from a *filled* valence shell because Z_{eff} is higher. In other words, ions formed by reaction of main-group elements usually have filled *s* and *p* subshells (a noble-gas electron configuration), which corresponds to having eight electrons (an *octet*) in the valence shell of an atom or ion. Sodium ([Ne] $3s^1$) loses only one electron easily, magnesium ([Ne] $3s^2$) loses only two electrons easily, aluminum ([Ne] $3s^2\,3p^1$) loses only three electrons easily, and so on across the row. We'll explore the stability of valence-shell electron octets further in Section 3.8.

8 electrons in outer (2nd) shell

$$\mathrm{Na}\ (1s^2\,2s^2\,2p^6\,3s^1) \longrightarrow \mathrm{Na}^+\ (1s^2\,2s^2\,2p^6) + e^-$$

$$\mathrm{Mg}\ (1s^2\,2s^2\,2p^6\,3s^2) \longrightarrow \mathrm{Mg}^{2+}\ (1s^2\,2s^2\,2p^6) + 2\,e^-$$

$$\mathrm{Al}\ (1s^2\,2s^2\,2p^6\,3s^2\,3p^1) \longrightarrow \mathrm{Al}^{3+}\ (1s^2\,2s^2\,2p^6) + 3\,e^-$$

$$\vdots \qquad\qquad \vdots$$

$$\mathrm{Cl}\ (1s^2\,2s^2\,2p^6\,3s^2\,3p^5) \longrightarrow \mathrm{Cl}^{7+}\ (1s^2\,2s^2\,2p^6) + 7\,e^-$$

Worked Example 3.7

Higher Ionization Energies

Which has the larger fifth ionization energy, Ge or As?

STRATEGY

Look at their positions in the periodic table. The group 4A element germanium has four valence-shell electrons and thus has four relatively low ionization energies, whereas the group 5A element arsenic has five valence-shell electrons and has five low ionization energies.

SOLUTION

Germanium $(E_{i5} = 9021 \text{ kJ/mol})$ has a larger E_{i5} than arsenic $(E_{i5} = 6043 \text{ kJ/mol})$.

PROBLEM 3.14

(a) Which has the larger third ionization energy, Be or N?

(b) Which has the larger fourth ionization energy, Ga or Ge?

PROBLEM 3.15 Three atoms have the following electron configurations:

(a) $1s^2\ 2s^2\ 2p^6\ 3s^2\ 3p^1$ (b) $1s^2\ 2s^2\ 2p^6\ 3s^2\ 3p^5$ (c) $1s^2\ 2s^2\ 2p^6\ 3s^2\ 3p^6\ 4s^1$

Which of the three has the largest E_{i1}? Which has the smallest E_{i4}?

CONCEPTUAL PROBLEM 3.16 Order the indicated three elements according to the ease with which each is likely to lose its third electron:

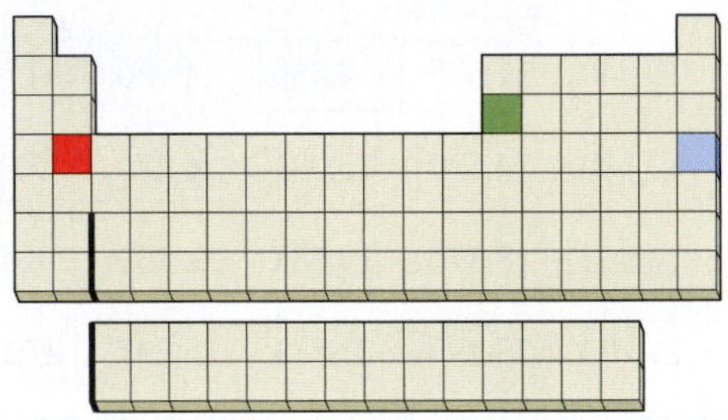

3.7 ADDING AN ELECTRON TO AN ATOM: ELECTRON AFFINITY

Just as it's possible to measure the energy change on *removing* an electron from an atom to form a *cation*, it's also possible to measure the energy change on *adding* an electron to an atom to form an *anion*. An element's **electron affinity (E_{ea})** is the energy change that occurs when an electron is added to an isolated atom in the gaseous state.

Ionization energies (Section 3.5) are always positive because energy must always be added to separate a negatively charged electron from the resultant positively charged cation. Electron affinities, however, are generally negative because energy is usually released when a neutral atom adds an additional electron.* We'll see in Section 8.2 that this same convention is used throughout chemistry: A positive energy change means that energy is added, and a negative energy change means that energy is released.

*We have defined E_{ea} as the energy *released* when a neutral atom *gains* an electron to form an anion and have given it a negative sign. Some books and reference sources adopt the opposite point of view, defining E_{ea} as the energy *gained* when an anion *loses* an electron to form a neutral atom and giving it a positive value. The two definitions are simply the reverse of one another, so the sign of the energy change is also reversed.

The more negative the E_{ea}, the greater the tendency of the atom to accept an electron and the more stable the anion that results. In contrast, an atom that forms an *unstable* anion by addition of an electron has, in principle, a positive value of E_{ea}, but no experimental measurement can be made because the process does not take place. All we can say is that the E_{ea} for such an atom is greater than zero. The E_{ea} of hydrogen, for instance, is –72.8 kJ/mol, meaning that energy is released and the H^- anion is stable. The E_{ea} of neon, however, is greater than 0 kJ/mol, meaning that Ne does not add an electron and the Ne^- anion is not stable.

$$H\,(1s^1) + e^- \longrightarrow H^-(1s^2) + 72.8\text{ kJ/mol} \qquad E_{ea} = -72.8\text{ kJ/mol}$$
$$Ne\,(1s^2\,2s^2\,2p^6) + e^- + \text{Energy} \longrightarrow Ne^-(1s^2\,2s^2\,2p^6\,3s^1) \qquad E_{ea} > 0\text{ kJ/mol}$$

As with ionization energies, electron affinities show a periodicity that is related to the electron configurations of the elements. The data in **Figure 3.11** indicate that group 7A elements have the most negative electron affinities, corresponding to the largest release of energy, while group 2A and group 8A elements have near-zero or positive electron affinities, corresponding to a small release or even an absorption of energy.

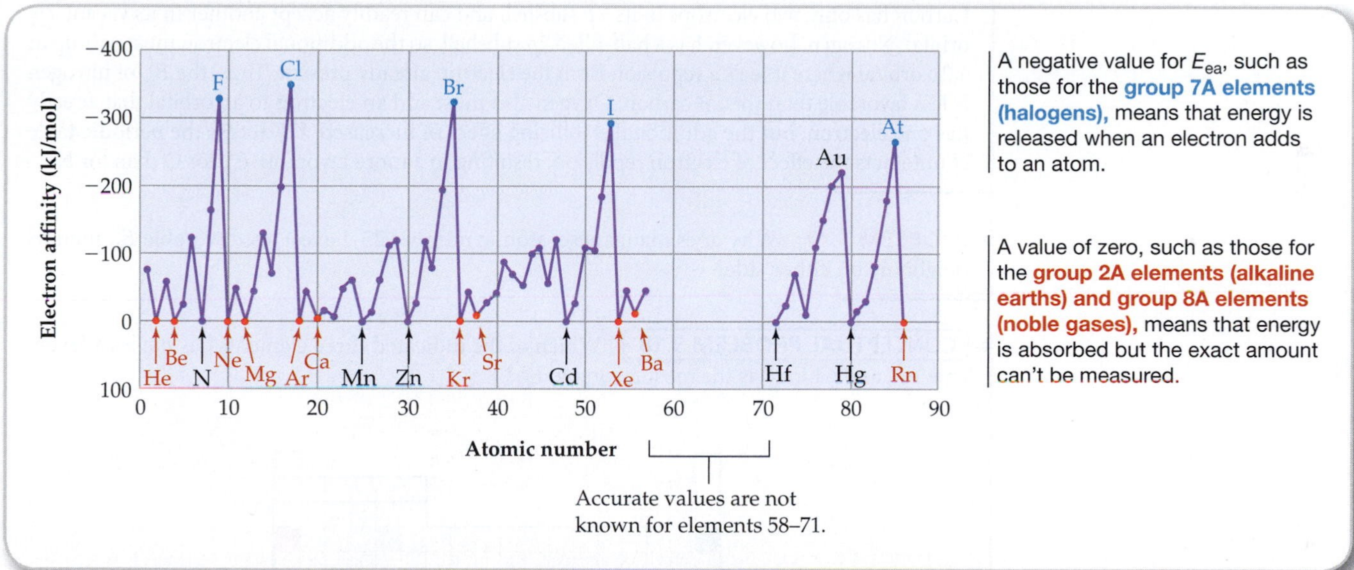

FIGURE 3.11

Electron affinities for elements 1–57 and 72–86.

The exact value of an element's electron affinity is due to an interplay of offsetting factors. Attraction between the additional electron and the nucleus favors a negative E_{ea}, but the increase in electron–electron repulsions that results from addition of the extra electron favors a positive E_{ea}.

Large negative E_{ea}'s are found for the halogens (F, Cl, Br, I) because each of these elements has both a high Z_{eff} and room in its valence shell for an additional electron. Halide anions have a noble-gas electron configuration with filled *s* and *p* subshells, and the attraction between the additional electron and the atomic nucleus is high. Positive E_{ea}'s are found for the noble-gas elements (He, Ne, Ar, Kr, Xe), however, because the *s* and *p* subshells in these elements are already full, so the additional electron must go into the next higher shell, where it is shielded from the nucleus and feels a relatively low Z_{eff}. The attraction of the nucleus for the added electron is therefore small and is outweighed by the additional electron–electron repulsions.

$$\textbf{A halogen: } Cl\,(\ldots\,3s^2\,3p^5) + e^- \longrightarrow Cl^-\,(\ldots\,3s^2\,3p^6) \qquad E_{ea} = -348.6\text{ kJ/mol}$$
$$\textbf{A noble gas: } Ar\,(\ldots\,3s^2\,3p^6) + e^- \longrightarrow Ar^-(\ldots\,3s^2\,3p^6 4s^1) \qquad E_{ea} > 0\text{ kJ/mol}$$

In looking for other trends in the data of Figure 3.11, the near-zero E_{ea}'s of the alkaline-earth metals (Be, Mg, Ca, Sr, Ba) are particularly striking. Atoms of these elements have filled *s* subshells, which means that the additional electron must go into a *p* subshell. The higher energy of the *p* subshell, together with a relatively low Z_{eff} for elements on the left side of the periodic table, means that alkaline-earth atoms accept an electron reluctantly and have E_{ea} values near zero.

$$\textbf{An alkaline earth: } Mg\,(\ldots\,3s^2) + e^- \longrightarrow Mg^-\,(\ldots\,3s^2\,3p^1) \quad E_{ea} \approx 0 \text{ kJ/mol}$$

Worked Example 3.8

Electron Affinities

Why does nitrogen have a less favorable (less negative) E_{ea} than its neighbors on either side, C and O?

STRATEGY AND SOLUTION

The magnitude of an element's E_{ea} depends on the element's valence-shell electron configuration. The electron configurations of C, N, and O are

$$\text{Carbon: } 1s^2\,2s^2\,2p_x^{\,1}\,2p_y^{\,1}$$
$$\text{Nitrogen: } 1s^2\,2s^2\,2p_x^{\,1}\,2p_y^{\,1}\,2p_z^{\,1}$$
$$\text{Oxygen: } 1s^2\,2s^2\,2p_x^{\,2}\,2p_y^{\,1}\,2p_z^{\,1}$$

Carbon has only two electrons in its $2p$ subshell and can readily accept another in its vacant $2p_z$ orbital. Nitrogen, however, has a half-filled $2p$ subshell, so the additional electron must pair up in a $2p$ orbital where it feels a repulsion from the electron already present. Thus, the E_{ea} of nitrogen is less favorable than that of carbon. Oxygen also must add an electron to an orbital that already has one electron, but the additional stabilizing effect of increased Z_{eff} across the periodic table counteracts the effect of electron repulsion, resulting in a more favorable E_{ea} for O than for N.

PROBLEM 3.17 Why does manganese, atomic number 25, have a less favorable E_{ea} than its neighbors on either side?

CONCEPTUAL PROBLEM 3.18 Which of the indicated three elements has the least favorable E_{ea}, and which has the most favorable E_{ea}?

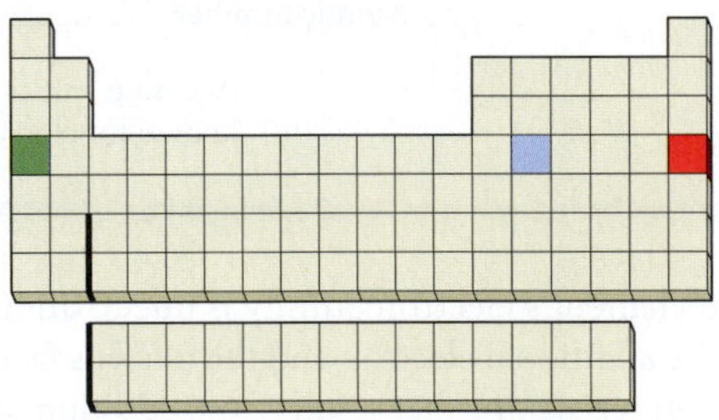

3.8 THE OCTET RULE FOR MAIN-GROUP ATOMS

Now that we've seen the experimental results about ionization energies and electron affinities, let's list the important points about ion formation and try to draw some general conclusions:

- **Group 1A (alkali-metal) elements** have a relatively low E_{i1}, so they tend to lose their ns^1 valence-shell electron easily when they react, thereby adopting the electron configuration of the noble gas in the previous row of the periodic table.
- **Group 2A (alkaline-earth) elements** have relatively low E_{i1} and E_{i2}, so they tend to lose both their ns^2 valence-shell electrons easily when they react and adopt a noble-gas electron configuration.

- **Group 7A (halogen) elements** have a relatively large negative E_{ea}, so they tend to gain one electron easily when they react, changing from $ns^2\ np^5$ to $ns^2\ np^6$ and thereby adopting the configuration of the neighboring noble gas in the same row.
- **Group 8A (noble-gas) elements** are essentially inert and undergo very few reactions. They neither gain nor lose electrons easily.

All these observations can be gathered into a single statement called the **octet rule**:

▶ **Octet rule** Main-group elements tend to undergo reactions that leave them with eight outer-shell electrons. That is, main-group elements react so that they attain a noble-gas electron configuration with filled *s* and *p* subshells in their valence electron shell.

As we'll see in the next chapter, there are exceptions to the octet rule, particularly for elements in the third and lower rows of the periodic table. Nevertheless, the rule is extremely useful for making predictions and for providing insights about chemical bonding.

Why does the octet rule work? What factors determine how many electrons an atom is likely to gain or lose? Electrons are most likely to be lost if they are held loosely in the first place—that is, if they feel a relatively low effective nuclear charge, Z_{eff}, and have relatively low ionization energies. Valence-shell electrons in the group 1A, 2A, and 3A metals, for instance, are shielded from the nucleus by core electrons, feel a low Z_{eff}, and are therefore lost relatively easily. Once the next lower noble-gas configuration is reached, though, loss of an additional electron suddenly becomes much more difficult because it must come from an inner shell, where it feels a much higher Z_{eff}.

Conversely, electrons are most likely to be gained if they can be held tightly by a high Z_{eff}. Valence-shell electrons in the group 6A and 7A elements, for example, are poorly shielded, feel high values of Z_{eff}, and aren't lost easily. The high Z_{eff} thus makes possible the gain of one or more additional electrons into vacant valence-shell orbitals. Once the noble-gas configuration is reached, though, lower-energy orbitals are no longer available. Any additional electron would have to be placed in a higher-energy orbital, where it would feel only a low Z_{eff}.

Eight is therefore the optimum number for valence-shell electrons. Taking electrons *from* a filled octet is difficult because they are tightly held by a high Z_{eff}; adding more electrons *to* a filled octet is difficult because, with *s* and *p* subshells full, no low-energy orbital is available.

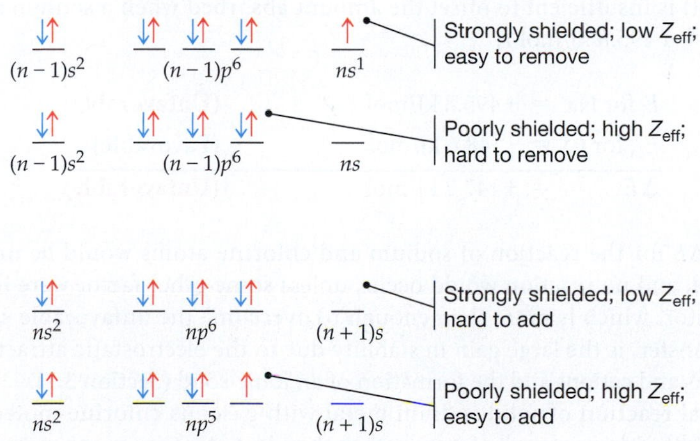

Worked Example 3.9

Chemical Reactions and the Octet Rule

Lithium metal reacts with nitrogen to yield Li_3N. What noble-gas configuration does the nitrogen atom in Li_3N have?

continued on next page

STRATEGY AND SOLUTION

The compound Li_3N contains three Li^+ ions, each formed by loss of a $2s$ electron from lithium metal (group 1A). The nitrogen atom in Li_3N must therefore gain three electrons over the neutral atom, making it triply negative (N^{3-}) and giving it a valence-shell octet with the neon configuration. The electrons that are added to N completely fill the $2p$ orbitals:

$$\textbf{N configuration: } (1s^2\,2s^2\,2p^3) \qquad \textbf{N}^{3-}\textbf{ configuration: } (1s^2\,2s^2\,2p^6)$$

PROBLEM 3.19 What noble-gas configurations are the following elements likely to adopt in reactions when they form ions?

(a) Rb **(b)** Ba **(c)** Ga **(d)** F

PROBLEM 3.20 What are group 6A elements likely to do when they form ions—gain electrons or lose them? How many?

3.9 IONIC BONDS AND THE FORMATION OF IONIC SOLIDS

When an element that gives up an electron relatively easily (that is, has a small positive ionization energy) comes in contact with an element that accepts an electron easily (that is, has a large negative electron affinity), the element with the small E_i can transfer an electron to the element with the negative E_{ea}, yielding a cation and an anion. Sodium, for instance, reacts with chlorine to give sodium chloride, Na^+Cl^-:

$$Na + Cl \longrightarrow Na^+\,Cl^-$$

$$1s^2\,2s^2\,2p^6\,3s^1 \quad 1s^2\,2s^2\,2p^6\,3s^2\,3p^5 \qquad\qquad 1s^2\,2s^2\,2p^6 \quad 1s^2\,2s^2\,2p^6\,3s^2\,3p^6$$

What about the overall energy change, ΔE, for the reaction of sodium with chlorine to yield sodium chloride? (The Greek capital letter delta, Δ, is used to represent a change in the value of the indicated quantity, in this case a change in energy ΔE.) It's apparent from E_i and E_{ea} values that the amount of energy released when a chlorine atom accepts an electron ($E_{ea} = -348.6$ kJ/mol) is insufficient to offset the amount absorbed when a sodium atom loses an electron ($E_i = +495.8$ kJ/mol):

E_i for Na	$= +495.8$ kJ/mol	**(Unfavorable)**
E_{ea} for Cl	$= -348.6$ kJ/mol	**(Favorable)**
ΔE	$= +147.2$ kJ/mol	**(Unfavorable)**

The net ΔE for the reaction of sodium and chlorine atoms would be unfavorable by +147.2 kJ/mol, and no reaction would occur, unless some other factor were involved. This additional factor, which is more than enough to overcome the unfavorable energy change of electron transfer, is the large gain in stability due to the electrostatic attractions between product anions and cations and the formation of an ionic solid (Section 3.1).

The actual reaction of solid sodium metal with gaseous chlorine molecules to form solid sodium chloride occurs all at once rather than in a stepwise manner, but it's easier to make an overall energy assessment if we imagine a series of hypothetical steps for which energy changes can be measured experimentally. There are five steps to take into account to find the overall energy change. [In writing the equations, the designations (*s*) for solid and (*g*) for gas are appended to the symbols of the reactants and products to show their physical state. In the same way, the designation (*l*) is often used for liquids, and (*aq*) is used for aqueous solutions.]

Step 1. Solid Na metal is converted into isolated, gaseous Na atoms, a process called *sublimation*. Because energy must be added to disrupt the forces holding atoms together in a solid, the heat of sublimation has a positive value: +107.3 kJ/mol for Na.

$Na(s) \longrightarrow Na(g)$ **+107.3 kJ/mol**

Step 2. Gaseous Cl_2 molecules are split into individual Cl atoms. Energy must be added to break molecules apart, and the energy required for bond breaking therefore has a positive value: +243 kJ/mol for Cl_2 (or 122 kJ/mol for 1/2 Cl_2). We'll learn more about the strengths of covalent bonds in Section 4.3.

$1/2\ Cl_2(g) \longrightarrow Cl(g)$ **+122 kJ/mol**

Step 3. Gaseous Na atoms from Step 1 are ionized into Na^+ plus electrons. The energy required is the first ionization energy of sodium (E_{i1}) and has a positive value: +495.8 kJ/mol.

$Na(g) \longrightarrow Na^+(g) + e^-$ **+495.8 kJ/mol**

Step 4. Cl^- ions are formed from Cl atoms by addition of an electron. The energy released is the electron affinity of chlorine (E_{ea}) and has a negative value: −348.6 kJ/mol.

$Cl(g) + e^- \longrightarrow Cl^-(g)$ **−348.6 kJ/mol**

Step 5. Lastly, solid NaCl is formed from Na^+ and Cl^- ions. The energy change is a measure of the overall electrostatic interactions between ions in the solid. It is the amount of energy released when isolated ions condense to form a solid, and it has a negative value: −787 kJ/mol for NaCl.

$Na^+(g) + Cl^-(g) \longrightarrow NaCl(s)$ **−787 kJ/mol**

Net reaction: $Na(s) + 1/2\ Cl_2(g) \longrightarrow NaCl(s)$

Net energy change: **−411 kJ/mol**

The five hypothetical steps in the reaction between sodium metal and gaseous chlorine are represented in **Figure 3.12.** Called a **Born–Haber cycle**, the figure shows how each step contributes to the overall energy change and how the net process is the sum of the individual steps.

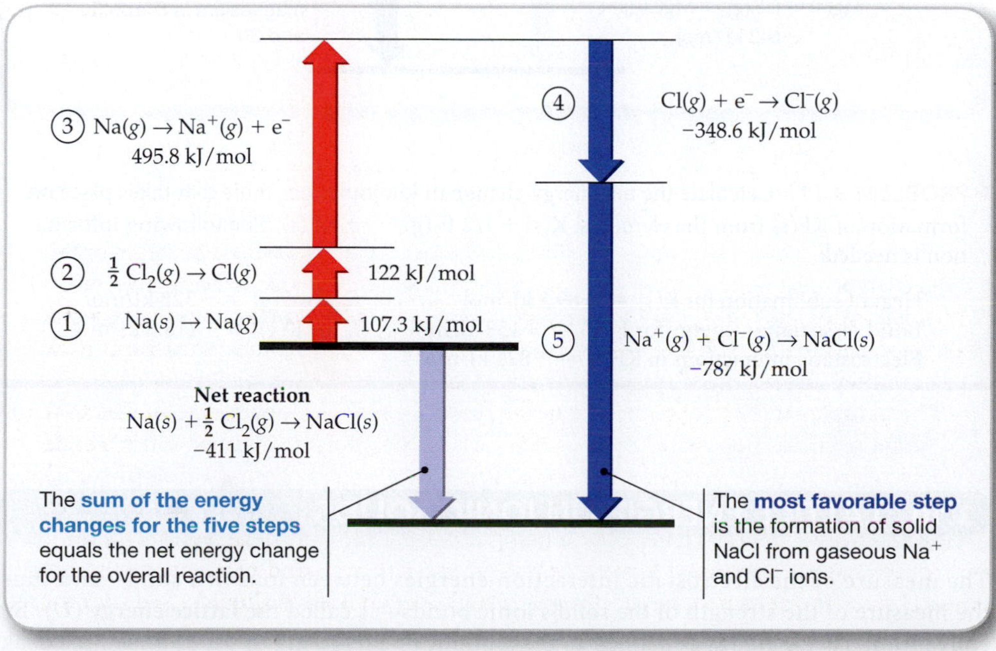

FIGURE 3.12
A Born–Haber cycle for the formation of NaCl(*s*) from Na(*s*) and Cl_2(*g*).

As indicated, Steps 1, 2, and 3 have positive values and absorb energy, while Steps 4 and 5 have negative values and release energy. The largest contribution is Step 5, which measures the electrostatic forces between ions in the solid product—that is, the strength of the ionic bonding. Were it not for this large amount of stabilization of the solid due to ionic bonding, no reaction could take place.

A similar Born–Haber cycle for the reaction of magnesium with chlorine to form magnesium chloride shows the hypothetical energy changes involved in the reaction of an alkaline-earth element (**Figure 3.13**). As in the reaction of sodium with chlorine, there are five contributions to the overall energy change in forming solid magnesium chloride. First, solid magnesium metal must be converted into gaseous magnesium atoms (sublimation). Second, the bond in Cl_2 molecules must be broken to yield chlorine atoms. Third, the magnesium atoms must lose two electrons to form the dipositive Mg^{2+} ions. Fourth, the chlorine atoms formed in Step 2 must accept electrons to form Cl^- ions. Fifth, the gaseous ions must combine to form the ionic solid, $MgCl_2$. As the Born–Haber cycle indicates, it is the large contribution from ionic bonding that releases enough energy to drive the entire process.

FIGURE 3.13

A Born–Haber cycle for the formation of $MgCl_2$ from its elements.

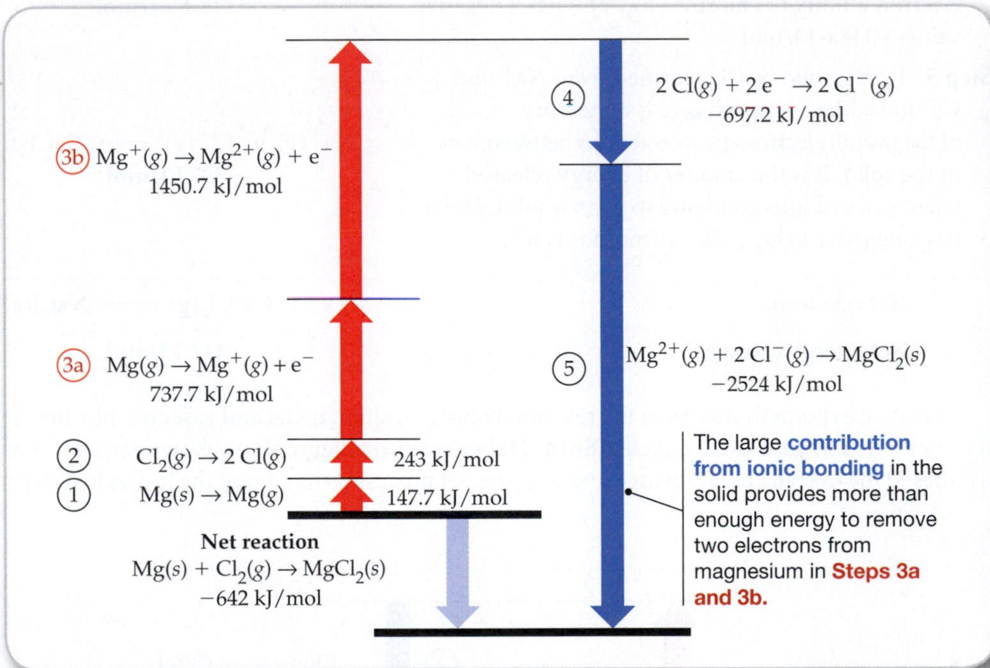

PROBLEM 3.21 Calculate the net energy change in kilojoules per mole that takes place on formation of KF(*s*) from the elements: $K(s) + 1/2\ F_2(g) \longrightarrow KF(s)$. The following information is needed:

Heat of sublimation for K(*s*) = +89.2 kJ/mol
Bond dissociation energy for $F_2(g)$ = +158 kJ/mol
Electrostatic interactions in KF(*s*) = −821 kJ/mol
E_{ea} for F(*g*) = −328 kJ/mol
E_i for K(*g*) = +418.8 kJ/mol

3.10 LATTICE ENERGIES IN IONIC SOLIDS

The measure of the electrostatic interaction energies between ions in a solid—and thus the measure of the strength of the solid's ionic bonds—is called the **lattice energy** (*U*). By convention, lattice energy is defined as the amount of energy that must be added to break up an ionic solid into its individual gaseous ions, so it has a positive value. Since the formation of a solid from ions is the reverse of the breakup, Step 5 in the Born–Haber cycle of Figure 3.12 has a negative value, $-U$.

$$NaCl(s) \longrightarrow Na^+(g) + Cl^-(g) \quad U = +787 \text{ kJ/mol} \quad \text{(Energy absorbed)}$$
$$Na^+(g) + Cl^-(g) \longrightarrow NaCl(s) \quad -U = -787 \text{ kJ/mol} \quad \text{(Energy released)}$$

The force F that results from the interaction of electric charges is described by **Coulomb's law** and is equal to a constant k times the product of the charges on the ions, z_1 and z_2, divided by the square of the distance d between their centers (nuclei):

▶ **Coulomb's law** $F = k \times \dfrac{z_1 z_2}{d^2}$

Because energy is defined by physics as being equal to force times distance, the negative of the lattice energy is

$$-U = F \times d = k \times \frac{z_1 z_2}{d}$$

where the value of the constant k in the equation depends on the arrangement of the ions in the specific compound and is different for different substances.

Lattice energies are large when the distance d between ions is small and when the charges z_1 and z_2 are large. A small distance d means that the ions are close together, which implies that they have small ionic radii. Thus, if z_1 and z_2 are held constant, the largest lattice energies belong to compounds formed from the smallest ions, as listed in **Table 3.4**.

TABLE 3.4 Lattice Energies of Some Ionic Solids (kJ/mol)

	Anion				
Cation	F^-	Cl^-	Br^-	I^-	O^{2-}
Li^+	1036	853	807	757	2925
Na^+	923	787	747	704	2695
K^+	821	715	682	649	2360
Be^{2+}	3505	3020	2914	2800	4443
Mg^{2+}	2957	2524	2440	2327	3791
Ca^{2+}	2630	2258	2176	2074	3401
Al^{3+}	5215	5492	5361	5218	15,916

Within a series of compounds that have the same anion but different cations, lattice energy increases as the cation becomes smaller. Comparing LiF, NaF, and KF, for example, cation size follows the order $K^+ > Na^+ > Li^+$, so lattice energies follow the order LiF > NaF > KF. Similarly, within a series of compounds that have the same cation but different anions, lattice energy increases as anion size decreases. Comparing LiF, LiCl, LiBr, and LiI, for example, anion size follows the order $I^- > Br^- > Cl^- > F^-$, so lattice energies follow the reverse order LiF > LiCl > LiBr > LiI.

Table 3.4 also shows that compounds of ions with higher charges have larger lattice energies than compounds of ions with lower charges. In comparing NaI, MgI_2, and AlI_3, for example, the order of charges on the cations is $Al^{3+} > Mg^{2+} > Na^+$, and the order of lattice energies is $AlI_3 > MgI_2 > NaI$.

Worked Example 3.10

Predicting Lattice Energies

Which has the larger lattice energy, NaCl or CsI?

STRATEGY

The magnitude of a substance's lattice energy is affected both by the charges on its constituent ions and by the sizes of those ions. The higher the charges on the ions and the smaller the sizes of the ions, the larger the lattice energy. In this case, all four ions—Na^+, Cs^+, Cl^-, and I^-— are singly charged and differ only in size.

SOLUTION

Because Na^+ is smaller than Cs^+ and Cl^- is smaller than I^-, the distance between ions is smaller in NaCl than in CsI. Thus, NaCl has the larger lattice energy.

WORKED CONCEPTUAL EXAMPLE 3.11

Lattice Energies

Which of the following compounds of an alkali-metal cation with a halogen anion has the larger lattice energy, and which has the smaller lattice energy? Explain.

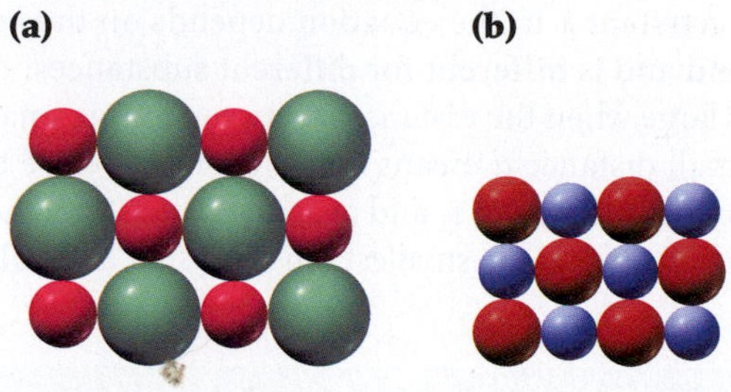

STRATEGY

The magnitude of a lattice energy depends directly on the charge on the ions and inversely on the distance between ions (that is, on the radii of the ions). In this instance, all the ions in both drawings are singly charged, so only the size of the ions is important.

SOLUTION

The ions in drawing **(b)** are smaller than those in drawing **(a)**, so **(b)** has the larger lattice energy.

PROBLEM 3.22 Which substance in each of the following pairs has the larger lattice energy?
(a) KCl or RbCl **(b)** CaF_2 or BaF_2 **(c)** CaO or KI

CONCEPTUAL PROBLEM 3.23 One of the following pictures represents NaCl and one represents MgO. Which is which, and which has the larger lattice energy?

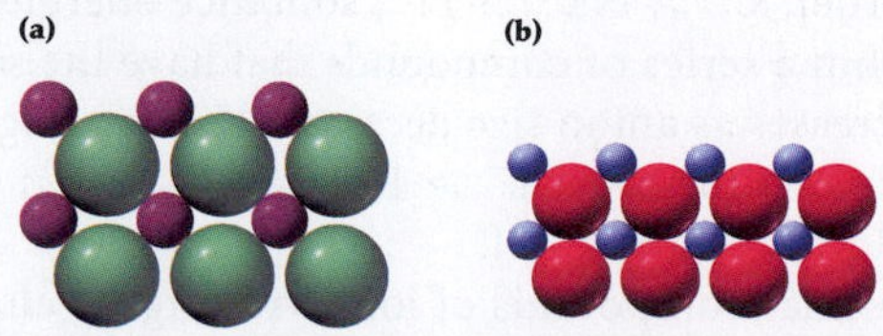

3.11 SOME CHEMISTRY OF THE ALKALI METALS (GROUP 1A)

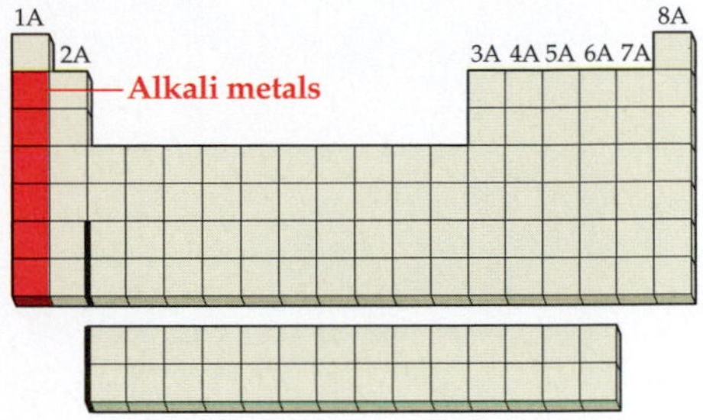

Now that we know something about ionization energies, electron affinities, lattice energies, and the octet rule, let's end this chapter with a brief look at the chemistry of some elements that form ionic bonds. The alkali metals of group 1A—Li, Na, K, Rb, Cs, and Fr—have the smallest ionization energies of all the elements because of their valence-shell ns^1 electron configurations (Figure 3.9). They therefore lose this ns^1 electron easily in chemical reactions to yield 1+ ions.

As their group name implies, the alkali metals are *metallic*. They have a bright, silvery appearance, are malleable, and are good conductors of electricity. Unlike the more common metals such as iron, though, the alkali metals are all soft enough to cut with a dull knife, have low melting points and densities, and are so reactive that they must be stored under oil to prevent their instantaneous reaction with oxygen and moisture. None are found in the elemental state in nature; they occur only in salts. Their properties are summarized in **Table 3.5**.

TABLE 3.5 Properties of Alkali Metals

Name	Melting Point (°C)	Boiling Point (°C)	Density (g/cm^3)	First Ionization Energy (kJ/mol)	Abundance on Earth (%)	Atomic Radius (pm)	Ionic (M^+) Radius (pm)
Lithium	180.5	1342	0.534	520.2	0.0020	152	68
Sodium	97.7	883	0.971	495.8	2.36	186	102
Potassium	63.3	759	0.862	418.8	2.09	227	138
Rubidium	39.3	688	1.532	403.0	0.0090	248	152
Cesium	28.4	671	1.873	375.7	0.000 10	265	167
Francium	—	—	—	~400	Trace	—	—

Alkali metals react rapidly with group 7A elements (halogens) to yield colorless, crystalline, ionic compounds called *halides*. The reactivity of an alkali metal increases as its ionization energy decreases, giving a reactivity order Cs > Rb > K > Na > Li. Cesium is the most reactive, combining almost explosively with the halogens.

$$2\,M(s) + X_2 \longrightarrow \underset{\text{A metal halide}}{2\,MX(s)}$$

where M = Alkali metal (Li, Na, K, Rb, or Cs)
X = Halogen (F, Cl, Br, or I)

▲ A sample of sodium metal

All the alkali metals also react rapidly with oxygen, but different metals give different kinds of products. Lithium reacts with O_2 to yield the *oxide*, Li_2O; sodium reacts to yield the *peroxide*, Na_2O_2; and the remaining alkali metals, K, Rb, and Cs, form either peroxides or *superoxides*, MO_2, depending on the reaction conditions and on how much oxygen is present. The reasons for the differences have to do largely with the differences in stability of the various products and with how the ions pack together in ionic solids. The alkali-metal cations have a +1 charge in all cases, but the oxygen anions might be either O^{2-}, $O_2{}^{2-}$, or $O_2{}^-$.

$$4\,Li(s) + O_2(g) \longrightarrow 2\,Li_2O(s) \qquad \text{An } \textit{oxide}\text{; the anion is } O^{2-}$$

$$2\,Na(s) + O_2(g) \longrightarrow Na_2O_2(s) \qquad \text{A } \textit{peroxide}\text{; the anion is } O_2{}^{2-}$$

$$K(s) + O_2(g) \longrightarrow KO_2(s) \qquad \text{A } \textit{superoxide}\text{; the anion is } O_2{}^{-}$$

Potassium superoxide, KO_2, is a particularly valuable compound because of its use in spacecraft and in self-contained breathing devices to remove moisture and CO_2 from exhaled air, generating oxygen in the process:

$$2\,KO_2(s) + H_2O(g) \longrightarrow KOH(s) + KO_2H(s) + O_2(g)$$
$$4\,KO_2(s) + 2\,CO_2(g) \longrightarrow 2\,K_2CO_3(s) + 3\,O_2(g)$$

Perhaps the most well-known and dramatic reaction of the alkali metals is with water to yield hydrogen gas and an alkali-metal hydroxide, MOH. In fact, it's this reaction that gives the elements their group name because the solution of metal hydroxide that results from adding an alkali metal to water is *alkaline*, or basic. Lithium undergoes the reaction with vigorous bubbling as hydrogen is released, sodium reacts rapidly with evolution of heat, and potassium reacts so violently that the hydrogen produced bursts instantly into flame. Rubidium and cesium react almost explosively.

$$2\,M(s) + 2\,H_2O(l) \longrightarrow 2\,M^+(aq) + 2\,OH^-(aq) + H_2(g)$$
$$\text{where} \quad M = \text{Li, Na, K, Rb, or Cs}$$

Lithium reacts vigorously, with bubbling.

Sodium reacts violently.

Potassium reacts almost explosively.

▲ All the alkali metals react with water to generate H_2 gas.

Finally, the alkali metals also react with ammonia, NH_3, to yield H_2 gas plus a metal *amide*, MNH_2, a reaction analogous to that between an alkali metal and water.

$$2\,M(s) + 2\,NH_3(l) \longrightarrow 2\,M^+(soln) + 2\,NH_2^-(soln) + H_2(g)$$
$$\text{where} \quad M = \text{Li, Na, K, Rb, or Cs}$$

The reaction is sufficiently slow at lower temperatures that it's possible for the alkali metals to dissolve in liquid ammonia at −33 °C, forming deep blue solutions of metal cations and dissolved electrons.

$$M(s) \xrightarrow[\text{solvent}]{\text{Liquid NH}_3} M^+(soln) + e^-(soln)$$
$$\text{where} \quad M = \text{Li, Na, K, Rb, or Cs}$$

▲ Sodium metal dissolves in liquid ammonia at low temperature to yield a blue solution of Na^+ ions and solvent-surrounded electrons.

PROBLEM 3.24 Assign charges to the oxygen-containing anions in the following compounds:
(a) Li_2O (b) K_2O_2 (c) CsO_2

PROBLEM 3.25 Complete the following equations so that the same numbers and kinds of atoms appear on both sides of the reaction arrow. If no reaction takes place, write N.R.
(a) $Cs(s) + H_2O(l) \longrightarrow$? (b) $Rb(s) + O_2(g) \longrightarrow$? (c) $K(s) + NH_3(g) \longrightarrow$?

3.12 SOME CHEMISTRY OF THE ALKALINE-EARTH METALS (GROUP 2A)

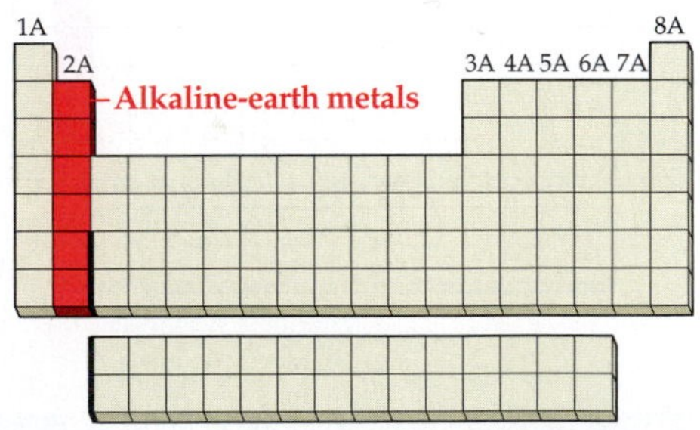

The alkaline-earth elements of group 2A—Be, Mg, Ca, Sr, Ba, and Ra—are similar to the alkali metals in many respects. They differ, however, in that they have ns^2 valence-shell electron configurations and therefore lose two electrons to give doubly positive ions, M^{2+}. Because their first ionization energy is larger than that of alkali metals (Figure 3.9), the group 2A metals are somewhat less reactive than alkali metals. The general reactivity trend is Ba > Sr > Ca > Mg > Be.

Although harder than their neighbors in group 1A, the alkaline-earth elements are still relatively soft, silvery metals. They tend, however, to have higher melting points and densities than alkali metals, as listed in **Table 3.6**. Alkaline-earth elements are less reactive toward oxygen and water than alkali metals but are nevertheless found in nature only in salts, not in the elemental state.

TABLE 3.6 Properties of Alkaline-Earth Metals

Name	Melting Point (°C)	Boiling Point (°C)	Density (g/cm^3)	First Ionization Energy (kJ/mol)	Abundance on Earth (%)	Atomic Radius (pm)	Ionic (M^{2+}) Radius (pm)
Beryllium	1287	2471	1.848	899.4	0.000 28	112	44
Magnesium	650	1090	1.738	737.7	2.33	160	66
Calcium	842	1484	1.55	589.8	4.15	197	99
Strontium	777	1382	2.54	549.5	0.038	215	112
Barium	727	1897	3.62	502.9	0.042	222	134
Radium	700	1140	~5.0	509.3	Trace	223	143

▲ Calcium metal reacts very slowly with water at room temperature.

Alkaline-earth metals react with halogens to yield ionic halides, MX_2, and with oxygen to form oxides, MO:

$$M + X_2 \rightarrow MX_2 \qquad \text{where} \qquad M = \text{Be, Mg, Ca, Sr, or Ba}$$
$$2\,M + O_2 \rightarrow 2\,MO \qquad\qquad X = \text{F, Cl, Br, or I}$$

Beryllium and magnesium are relatively unreactive toward oxygen at room temperature, but both burn with a brilliant white glare when ignited by a flame. Calcium, strontium, and barium are so reactive that they are best stored under oil to keep them from contact with air. Like the heavier alkali metals, strontium and barium form peroxides, MO_2.

With the exception of beryllium, the alkaline-earth elements react with water to yield metal hydroxides, $M(OH)_2$. Magnesium undergoes reaction only at temperatures above 100 °C, while calcium and strontium react slowly with liquid water at room temperature. Only barium reacts vigorously at room temperature.

$$M(s) + 2\,H_2O(l) \rightarrow M(OH)_2 + H_2(g) \qquad \text{where} \qquad M = \text{Mg, Ca, Sr, or Ba}$$

PROBLEM 3.26 Predict the products of the following reactions, and balance the equations so that the numbers and kinds of atoms are the same on both sides of the reaction arrows:
(a) $Be(s) + Br_2(l) \longrightarrow$ **?** **(b)** $Sr(s) + H_2O(l) \longrightarrow$ **?** **(c)** $Mg(s) + O_2(g) \longrightarrow$ **?**

PROBLEM 3.27 What product do you think is formed by reaction of magnesium with sulfur, a group 6A element? What is the charge on the sulfide ion in the product?

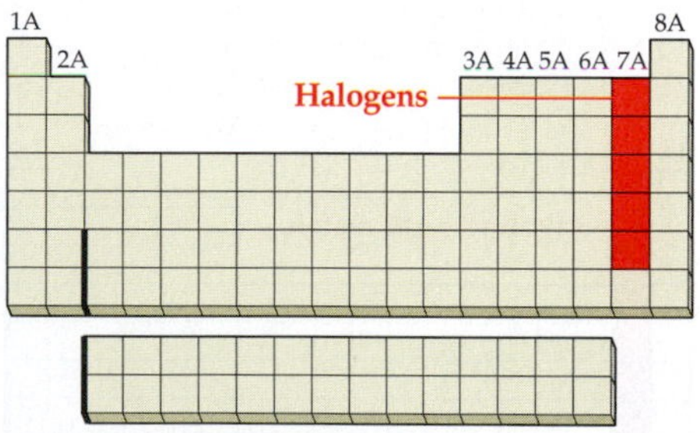

3.13 SOME CHEMISTRY OF THE HALOGENS (GROUP 7A)

The halogens of group 7A—F, Cl, Br, I, and At—are completely different from the main-group metals. The halogens are nonmetals rather than metals, they exist as diatomic molecules rather than as individual atoms, and they tend to gain rather than lose electrons when they enter into reactions because of their $ns^2\,np^5$ electron configurations. In other words, the halogens are characterized by large negative electron affinities and large positive ionization energies. Some of their properties are listed in **Table 3.7**.

TABLE 3.7 Properties of Halogens

Name	Melting Point (°C)	Boiling Point (°C)	Density (g/cm^3)	Electron Affinity (kJ/mol)	Abundance on Earth (%)	Atomic Radius (pm)	Ionic (X^-) Radius (pm)
Fluorine	−220	−188	1.50 (*l*)	−328	0.062	72	133
Chlorine	−101	−34	2.03 (*l*)	−349	0.013	99	181
Bromine	−7	59	3.12 (*l*)	−325	0.0003	114	196
Iodine	114	184	4.930 (*s*)	−295	0.000 05	133	220
Astatine	—	—	—	−270	Trace	—	—

Halogens are too reactive to occur in nature as free elements. Instead, they are found only as their anions in various minerals. Even the name *halogen* implies reactivity, since it comes from the Greek words *hals* (salt) and *gennan* (to form). Thus, a halogen is literally a salt-former.

Halogens are among the most reactive elements in the periodic table. Fluorine, in fact, forms compounds with every other element except the three noble gases He, Ne, and Ar.

As noted previously, their large negative electron affinities allow halogens to accept electrons from other atoms to yield halide anions, X^-.

Halogens react with every metal in the periodic table to yield metal halides. With alkali and alkaline-earth metals, the formula of the halide product is easily predictable. With transition metals, though, more than one product can sometimes form depending on the reaction conditions and the amounts of reactants present. Iron, for instance, can react with Cl_2 to form either $FeCl_2$ or $FeCl_3$. Without knowing a good deal more about transition-metal chemistry, it's not possible to make predictions at this point. The reaction can be generalized as

$$2\,M + n\,X_2 \longrightarrow 2\,MX_n \quad \text{where} \quad M = \text{Metal}, \; X = \text{F, Cl, Br, or I}$$

Unlike the metallic elements, halogens become less rather than more reactive going down the periodic table because of their generally decreasing electron affinity. Thus, their reactivity order is $F_2 > Cl_2 > Br_2 > I_2$. Fluorine often reacts violently, chlorine and bromine somewhat less so, and iodine often sluggishly.

In addition to their reaction with metals, halogens also react with hydrogen gas to yield hydrogen halides, HX. Fluorine reacts explosively with hydrogen as soon as the two gases come in contact. Chlorine also reacts explosively once the reaction is initiated by a spark or by ultraviolet light, but the mixture of gases is stable in the dark. Bromine and iodine react more slowly.

$$H_2(g) + X_2 \longrightarrow 2\,HX(g) \quad \text{where} \quad X = \text{F, Cl, Br, or I}$$

As we'll see in Section 7.5, hydrogen halides are useful because they behave as ***acids***—compounds that produce H^+ ions when dissolved in water. An aqueous solution of HCl, for instance, is used throughout the chemical industry in a vast number of processes, from pickling steel (removing its iron oxide coating) to dissolving animal bones for producing gelatin.

$$HX \xrightarrow[\text{in } H_2O]{\text{Dissolve}} H^+(aq) + X^-(aq)$$

Hydrogen fluoride (HF) is used frequently for etching glass because it is one of the few substances that reacts with glass.

▲ Gaseous HF is one of the few substances that reacts with and etches glass (SiO_2), according to the equation $SiO_2(s) + 4\,HF(g) \longrightarrow SiF_4(g) + 2\,H_2O(l)$.

LOOKING AHEAD ...

Acids—compounds that separate into H^+ ions and a corresponding anion when dissolved in water—are frequently encountered in chemistry and will be discussed at numerous points throughout this text, particularly in Section 7.5 and Chapter 14.

3.14 SOME CHEMISTRY OF THE NOBLE GASES (GROUP 8A)

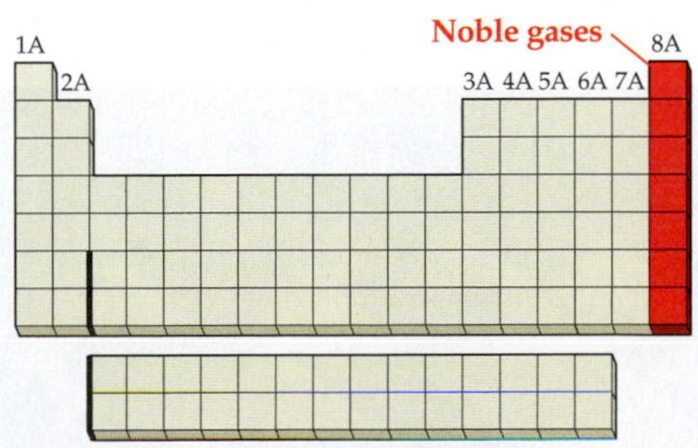

The noble gases of group 8A—He, Ne, Ar, Kr, Xe, and Rn—are neither metals nor reactive nonmetals. Rather, they are colorless, odorless, unreactive gases. Helium has a $1s^2$ valence-shell electron configuration, and the $ns^2\,np^6$ configurations for the others already contain octets, thus making it difficult for the noble gases to either gain or lose electrons.

Although sometimes referred to as rare gases or inert gases, these older names are not really accurate because the group 8A elements are neither rare nor completely inert. Argon, for instance, makes up nearly 1% by volume of dry air, and there are several dozen known compounds of krypton and xenon, although none occur naturally. Some properties of the noble gases are listed in **Table 3.8**.

Helium and neon undergo no chemical reactions and form no known compounds. Argon forms only HArF, and krypton and xenon react only with fluorine. Depending on the reaction conditions and on the amounts of reactants present, xenon can form three different fluorides: XeF_2, XeF_4, and XeF_6. All the products of noble-gas reactions are molecular rather than ionic compounds.

$$Xe(g) + F_2(g) \longrightarrow XeF_2(s)$$
$$Xe(g) + 2\,F_2(g) \longrightarrow XeF_4(s)$$
$$Xe(g) + 3\,F_2(g) \longrightarrow XeF_6(s)$$

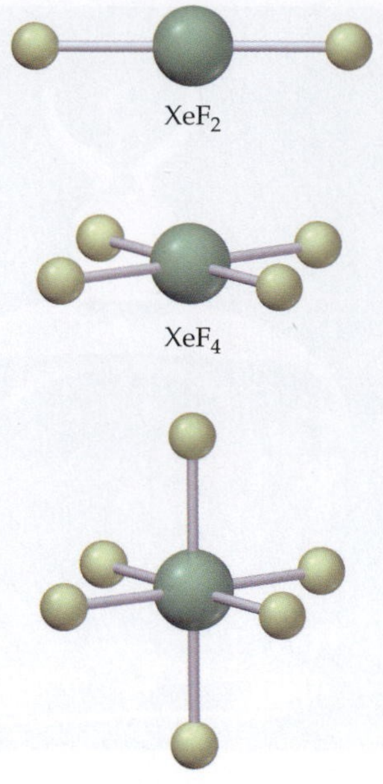

TABLE 3.8 Properties of Noble Gases

Name	Melting Point (°C)	Boiling Point (°C)	First Ionization Energy (kJ/mol)	Abundance in Dry Air (vol %)
Helium	−272.2	−268.9	2372.3	5.2×10^{-4}
Neon	−248.6	−246.1	2080.6	1.8×10^{-3}
Argon	−189.3	−185.9	1520.4	0.93
Krypton	−157.4	−153.2	1350.7	1.1×10^{-4}
Xenon	−111.8	−108.0	1170.4	9×10^{-6}
Radon	−71	−61.7	1037	Trace

The lack of reactivity of the noble gases is a consequence of their unusually large ionization energies (Figure 3.9) and their unusually small electron affinities (Figure 3.11), which result from their valence-shell electron configurations.

FYI SALT

If you're like most people, you probably feel a little guilty about reaching for the saltshaker at mealtime. The fact that high salt intake and high blood pressure often go hand in hand is surely among the most highly publicized pieces of nutritional advice to appear in recent decades.

Salt has not always been held in such disrepute. Historically, salt has been prized since the earliest recorded times as a seasoning and a food preservative. Words and phrases in many languages reflect the importance of salt as a life-giving and life-sustaining substance. We refer to a kind and generous person as "the salt of the earth," for instance, and we speak of being "worth one's salt." In Roman times, soldiers were paid in salt. The English word salary is derived from the Latin word for paying salt wages (*salarium*).

Salt is perhaps the easiest of all minerals to obtain and purify. The simplest method, used for thousands of years throughout the world in coastal climates where sunshine is abundant and rainfall is scarce, is to evaporate seawater. Although the exact amount varies depending on the source, seawater contains an average of about 3.5% by mass of dissolved substances, most of which is sodium chloride. It has been estimated that evaporation of all the world's oceans would yield approximately *4.5 million cubic miles* of NaCl.

▲ Harvesting salt from evaporated sea water

Only about 10% of current world salt production comes from evaporation of seawater. Most salt is obtained by mining the vast deposits of halite, or rock salt, formed by evaporation of ancient inland seas. These salt beds vary in thickness up to hundreds of meters and vary in depth from a few meters to thousands of meters below the Earth's surface. Salt mining has gone on for at least 3400 years, and the Wieliczka mine in Galicia, Poland, was worked continuously for more than 700 years until its closure in 1996.

Let's get back now to the dinner table. What about the link between dietary salt intake and high blood pressure? There's no doubt that most people in industrialized nations have a relatively high salt intake, and there's no doubt that high blood pressure among industrialized populations is on the rise. What's not so clear is exactly how the two observations are related.

The case against salt has been made largely by comparing widely diverse populations with different dietary salt intakes—by comparing the health of modern Americans with that of inhabitants of the Amazon rain forest, for example. Obviously, though, industrialization brings with it far more changes than simply an increase in dietary salt intake, and many of these other changes may be more important than salt in contributing to hypertension.

In a study called the DASH-Sodium study, published in 2001, a strong correlation was found between a change in salt intake and a change in blood pressure. When volunteers cut back their salt intake from 8.3 g per day—roughly what Americans typically consume—to 3.8 g per day, significant drops in blood pressure were found for most people. The largest reduction in blood pressure was seen in people already diagnosed with hypertension, but even subjects with normal blood pressure lowered their readings by several percentage points.

What should an individual do? The best answer, as in so many things, is to use moderation and common sense. People with hypertension should make a strong effort to lower their sodium intake; others might be well advised to choose unsalted snacks, use less salt in preparing food, and read nutrition labels for sodium content. But one further point: Nutritional advice is never one-size-fits-all. Long-distance runners, construction workers, farm workers, and others who sweat profusely, may well need an *increased* salt intake to avoid problems with low sodium and potassium levels.

PROBLEM 3.28 How is salt obtained commercially?

SUMMARY

The forces that hold atoms together in compounds are called **chemical bonds** and are of two types. **Covalent bonds** are formed by the sharing of electrons between atoms; **ionic bonds** are formed by the transfer of one or more electrons from one atom to another. The unit of matter held together by covalent bonds is called a **molecule**, and the charged particles formed by the gain or loss of one or more electrons are called **ions**. A **cation** is a positively charged particle, and an **anion** is a negatively charged particle. A **polyatomic ion** is a group of atoms joined by covalent bonds but containing an overall positive or negative charge.

Metallic elements on the left side of the periodic table tend to give up electrons to form cations, while the halogens and a few other nonmetallic elements on the right side of the table tend to accept electrons to form anions. The electrons given up by a main-group metal in forming a cation come from the highest-energy occupied orbital, while the electrons that are accepted by a nonmetal in forming an anion go into the lowest-energy unoccupied orbital. Sodium metal, for instance, loses its valence-shell 3*s* electron to form an Na^+ ion with the electron configuration of neon, while chlorine gains a 3*p* electron to form a Cl^- anion with the electron configuration of argon.

The amount of energy necessary to remove a valence electron from an isolated neutral atom is called the atom's **ionization energy (E_i)**. Ionization energies are smallest for metallic elements on the left side of the periodic table and largest for nonmetallic elements on the right side. As a result, metals usually give up electrons in chemical reactions.

Ionization is not limited to the removal of a single electron from an atom. Two, three, or even more electrons can be removed sequentially from an atom, although larger amounts of energy are required for each successive ionization step. In general, valence-shell electrons are much more easily removed than **core electrons.**

The amount of energy released or absorbed when an electron adds to an isolated neutral atom is called the atom's **electron affinity (E_{ea})**. By convention, a negative E_{ea} corresponds to a release of energy and a positive E_{ea} corresponds to an absorption of energy. Electron affinities are most negative for group 7A elements and most positive for group 2A and 8A elements. As a result, the group 7A elements can act as electron acceptors in chemical reactions.

In general, reactions of main-group elements can be described by the **octet rule**, which states that these elements tend to undergo reactions so as to attain a noble-gas electron configuration with filled *s* and *p* subshells in their valence shell. Elements on the left side of the periodic table tend to give up electrons until a noble-gas configuration is reached; elements on the right side of the table tend to accept electrons until a noble-gas configuration is reached; and the noble gases themselves are essentially unreactive.

Main-group metallic elements in groups 1A and 2A react with halogens in group 7A, during which the metal loses one or more electrons to the halogen. The product, a metal halide such as NaCl, is an **ionic solid** that consists of metal cations and halide anions electrostatically attracted to one another by ionic bonds. The sum of the interaction energies among all ions in a crystal is called the crystal's **lattice energy (*U*)**.

KEY WORDS

anion *100*
Born–Haber cycle *119*
cation *100*
chemical bond *98*
core electron *111*
Coulomb's law *121*
covalent bond *98*
electron affinity (E_{ea}) *114*
ion *100*
ionic bond *100*
ionic solid *101*
ionization energy (E_i) *110*
lattice energy (*U*) *120*
molecule *99*
octet rule *117*
oxoanion *105*
polyatomic ion *101*

LEARNING OUTCOMES

- Describe the difference between an ionic bond and a covalent bond. (Section 3.1) ***Problems 1, 2, 30, 50***
- Know how to name binary ionic compounds and write the formula of an ionic compound, given its name. (Section 3.2) ***Problems 3–5, 38, 44, 45, 49, 108, 111***
- Identify common polyatomic ions, and name compounds that contain them. (Section 3.2) ***Problems 6–8, 39–43, 46–48***
- Predict the likely ground-state electron configuration of an element's ion. (Section 3.3) ***Problems 9, 10, 32, 54–59***
- Explain the difference in size between an atom and its anion or cation. (Section 3.4) ***Problems 11, 12, 33, 100, 101, 114***
- Describe the periodic trends observed in ionization energy. (Sections 3.5, 3.6) ***Problems 13–16, 31, 61–71, 113***
- Describe the periodic trends observed in electron affinity. (Section 3.7) ***Problems 17, 18, 31, 70–77, 112, 114***
- Describe the octet rule, and explain why main-group ions follow it. (Section 3.8) ***Problems 19, 20, 96–98, 106, 107***
- Use a Born–Haber cycle to describe the energy changes involved in formation of an ionic solid and to calculate lattice energies. (Sections 3.9, 3.10) ***Problems 21–23, 35, 37, 78–91, 102–105, 115, 116***
- Describe the observed trends in chemical reactivity for elements in groups 1A, 2A, 7A, and 8A. (Sections 3.11–3.14) ***Problems 24–27, 92–99, 106–109***

CONCEPTUAL PROBLEMS

Problems 3.1–3.28 appear within the chapter.

3.29 In the following drawings, red spheres represent cations and blue spheres represent anions. Match each of the drawings (a)–(d) with the following ionic compounds:

(i) $Ca_3(PO_4)_2$ **(ii)** Li_2CO_3 **(iii)** $FeCl_2$ **(iv)** $MgSO_4$

(a)

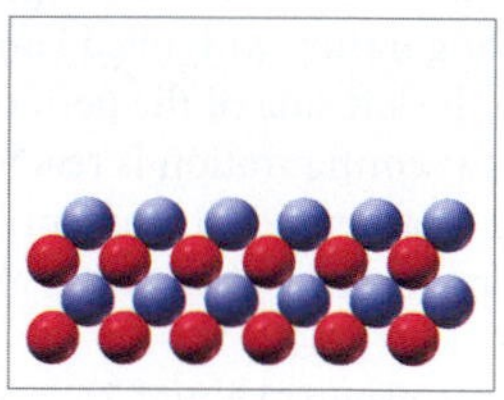

(b)

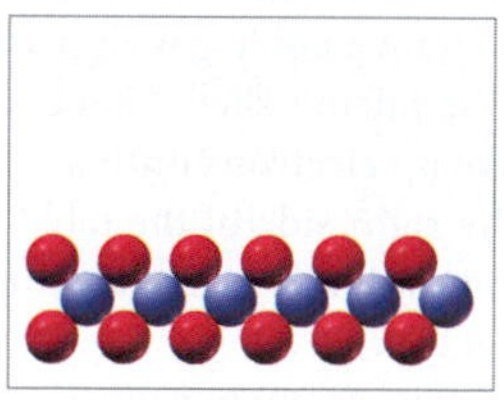

(c)

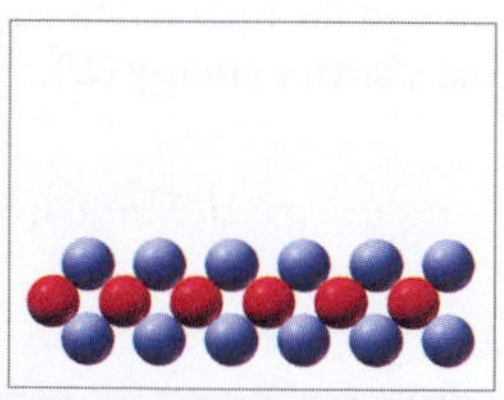

(d)

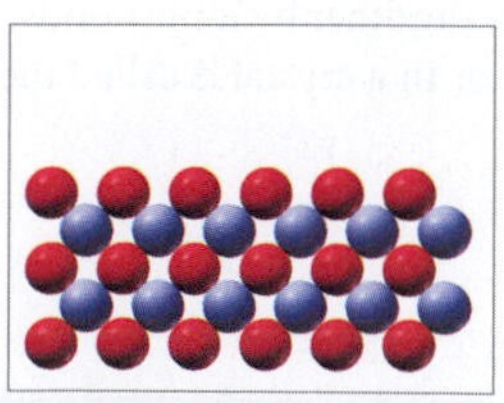

3.30 Which of the following drawings is more likely to represent an ionic compound, and which a covalent compound?

(a)

(b)

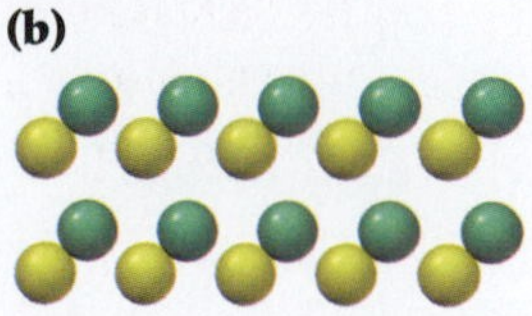

3.31 Circle the approximate part or parts of the periodic table where the following elements appear:

(a) Elements with the smallest values of E_{i1}

(b) Elements with the largest atomic radii

(c) Elements with the most negative values of E_{ea}

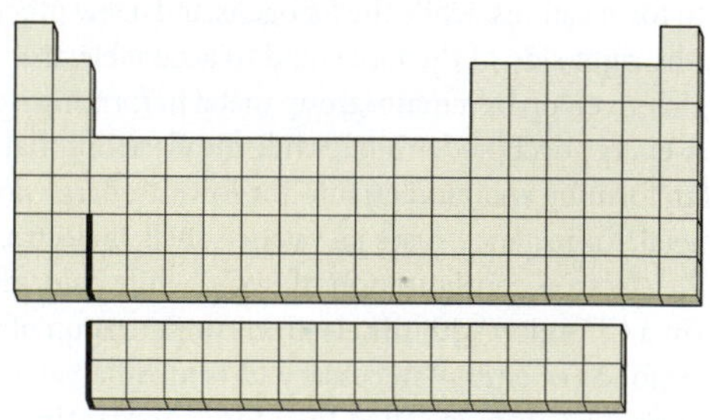

3.32 Where on the periodic table would you find the element that has an ion with each of the following electron configurations? Identify each ion.

(a) 3+ ion: $1s^2\,2s^2\,2p^6$ **(b)** 3+ ion: [Ar] $3d^3$

(c) 2+ ion: [Kr] $5s^2\,4d^{10}$ **(d)** 1+ ion: [Kr] $4d^{10}$

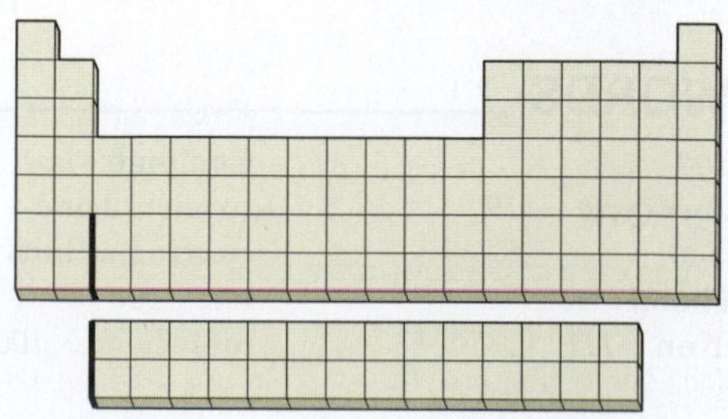

3.33 Which of the following spheres is likely to represent a metal atom, and which a nonmetal atom? Which sphere in the products represents a cation, and which an anion?

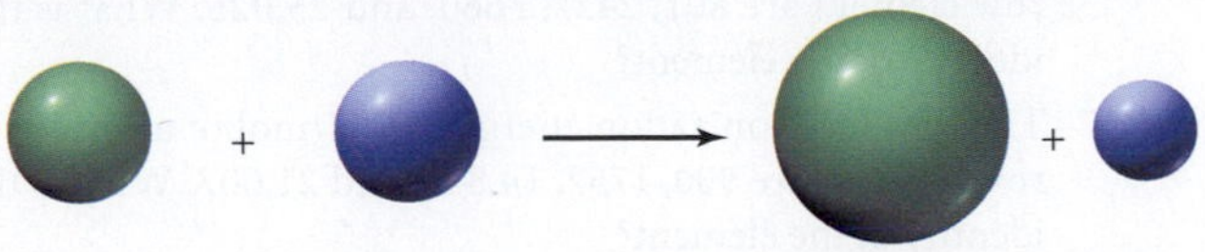

3.34 Each of the pictures **(a)**–**(d)** represents one of the following substances at 25 °C: sodium, chlorine, iodine, sodium chloride. Which picture corresponds to which substance?

(a) **(b)** **(c)** **(d)**

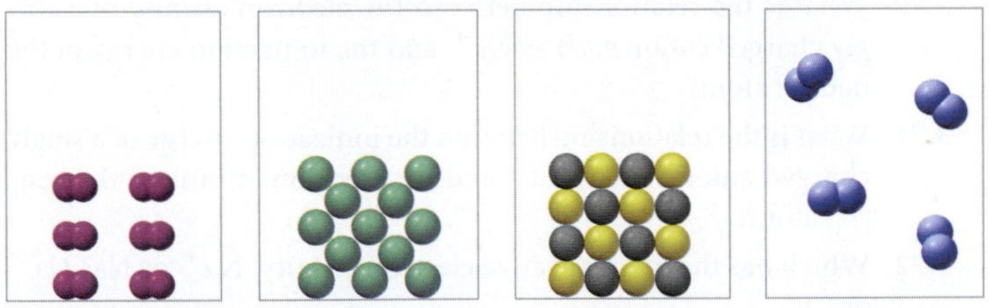

3.35 Which of the following alkali-metal halides has the largest lattice energy, and which has the smallest lattice energy? Explain.

(a) **(b)** **(c)**

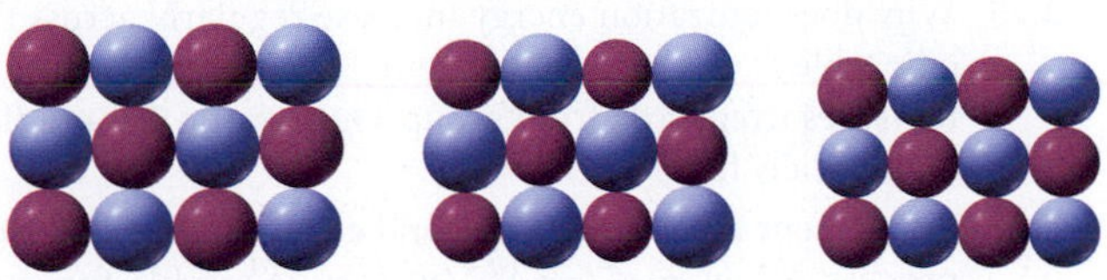

3.36 Three binary compounds are represented on the following drawing: red with red, blue with blue, and green with green. Give a likely formula for each compound.

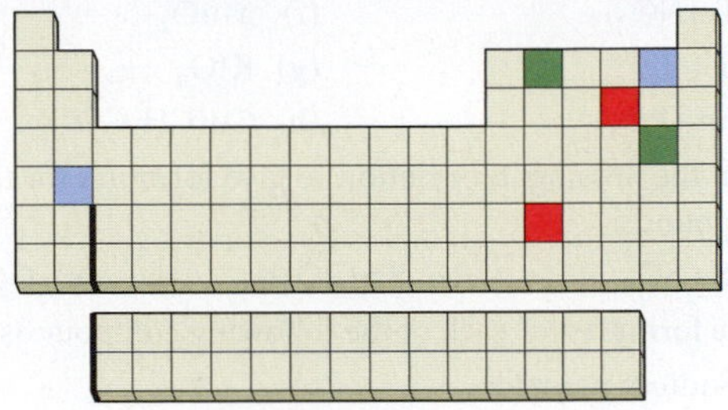

3.37 Given the following values for the formation of LiCl(*s*) from its elements, draw a Born–Haber cycle similar to that shown in Figure 3.12 on page 119.

E_{ea} for Cl(*g*) = −348.6 kJ/mol

Heat of sublimation for Li(*s*) = +159.4 kJ/mol

E_{i1} for Li(*g*) = +520 kJ/mol

Bond dissociation energy for $Cl_2(g)$ = +243 kJ/mol

Lattice energy for LiCl(*s*) = +853 kJ/mol

SECTION PROBLEMS

Naming Ionic Compounds (Section 3.2)

3.38 Write formulas for the following binary compounds:
(a) Potassium chloride
(b) Tin(II) bromide
(c) Calcium oxide
(d) Barium chloride
(e) Aluminum hydride

3.39 Write formulas for the following compounds:
(a) Calcium acetate
(b) Iron(II) cyanide
(c) Sodium dichromate
(d) Chromium(III) sulfate
(e) Mercury(II) perchlorate

3.40 Name the following ions:
(a) Ba^{2+}
(b) Cs^+
(c) V^{3+}
(d) HCO_3^-
(e) NH_4^+
(f) Ni^{2+}
(g) NO_2^-
(h) ClO_2^-
(i) Mn^{2+}
(j) ClO_4^-

3.41 Name each of the following compounds:
(a) $MgSO_4$
(b) $ZnCrO_4$
(c) $(NH_4)_2CO_3$
(d) $KClO_4$
(e) $Ca_3(PO_4)_2$
(f) $KMnO_4$

3.42 What are the formulas of the compounds formed from the following ions?
(a) Ca^{2+} and Br^-
(b) Ca^{2+} and SO_4^{2-}
(c) Al^{3+} and SO_4^{2-}

3.43 What are the formulas of the compounds formed from the following ions?
(a) Na^+ and NO_3^-
(b) K^+ and SO_4^{2-}
(c) Sr^{2+} and Cl^-

3.44 Write formulas for compounds of calcium with each of the following:
(a) Chlorine **(b)** Oxygen **(c)** Sulfur

3.45 Write formulas for compounds of rubidium with each of the following:
(a) Bromine **(b)** Nitrogen **(c)** Selenium

3.46 Give the formulas and charges of the following ions:
(a) Sulfite ion
(b) Phosphate ion
(c) Zirconium(IV) ion
(d) Chromate ion
(e) Acetate ion
(f) Thiosulfate ion

3.47 What are the charges on the positive ions in the following compounds?

(a) $Zn(CN)_2$ (e) Hg_2S
(b) $Fe(NO_2)_3$ (f) MnO_2
(c) $Ti(SO_4)_2$ (g) KIO_4
(d) $Sn_3(PO_4)_2$ (h) $Cu(CH_3CO_2)_2$

3.48 Fill in the missing information to give formulas for the following compounds:

(a) $Na_?SO_4$ (b) $Ba_?(PO_4)_?$ (c) $Ga_?(SO_4)_?$

3.49 Write formulas for each of the following compounds:

(a) Sodium peroxide
(b) Aluminum bromide
(c) Chromium(III) sulfate

Ions, Ionization Energy, and Electron Affinity (Sections 3.1, 3.3–3.7)

3.50 What is the difference between a covalent bond and an ionic bond? Give an example of each.

3.51 What is the difference between a molecule and an ion? Give an example of each.

3.52 How many protons and electrons are in each of the following ions?

(a) Be^{2+} (c) Se^{2-}
(b) Rb^+ (d) Au^{3+}

3.53 What is the identity of the element X in the following ions?

(a) X^{2+}, a cation that has 36 electrons
(b) X^-, an anion that has 36 electrons

3.54 What are the likely ground-state electron configurations of the following cations?

(a) La^{3+} (b) Ag^+ (c) Sn^{2+}

3.55 What are the likely ground-state electron configurations of the following anions?

(a) Se^{2-} (b) N^{3-}

3.56 What is the electron configuration of Ca^{2+}? What is the electron configuration of Ti^{2+}?

3.57 What tripositive ion has the electron configuration [Kr] $4d^3$? What neutral atom has the electron configuration [Kr] $5s^2\,4d^2$?

3.58 There are two elements in the transition-metal series Sc through Zn that have four unpaired electrons in their 2+ ions. Identify them.

3.59 Identify the element whose 2+ ion has the ground-state electron configuration [Ar] $3d^{10}$.

3.60 Which group of elements in the periodic table has the largest E_{i1}, and which group has the smallest? Explain.

3.61 Which element in the periodic table has the smallest ionization energy? Which has the largest?

3.62 (a) Which has the smaller second ionization energy, K or Ca?
(b) Which has the larger third ionization energy, Ga or Ca?

3.63 (a) Which has the smaller fourth ionization energy, Sn or Sb?
(b) Which has the larger sixth ionization energy, Se or Br?

3.64 Three atoms have the following electron configurations:

(a) $1s^2\,2s^2\,2p^6\,3s^2\,3p^3$
(b) $1s^2\,2s^2\,2p^6\,3s^2\,3p^6$
(c) $1s^2\,2s^2\,2p^6\,3s^2\,3p^6\,4s^2$

Which of the three has the largest E_{i2}? Which has the smallest E_{i7}?

3.65 Write the electron configuration of the atom in the third row of the periodic table that has the smallest E_{i4}.

3.66 The first four ionization energies in kJ/mol of a certain second-row element are 801, 2427, 3660, and 25,025. What is the likely identity of the element?

3.67 The first four ionization energies in kJ/mol of a certain second-row element are 900, 1757, 14,849, and 21,007. What is the likely identity of the element?

3.68 Which element in each of the following sets has the smallest first ionization energy, and which has the largest?

(a) Li, Ba, K (b) B, Be, Cl (c) Ca, C, Cl

3.69 What elements meet the following descriptions?

(a) Has the largest E_{i3} (b) Has the largest E_{i7}

3.70 What is the relationship between the electron affinity of a singly charged cation such as Na^+ and the ionization energy of the neutral atom?

3.71 What is the relationship between the ionization energy of a singly charged anion such as Cl^- and the electron affinity of the neutral atom?

3.72 Which has the more negative electron affinity, Na^+ or Na? Na^+ or Cl?

3.73 Which has the more negative electron affinity, Br or Br^-?

3.74 Why is energy usually released when an electron is added to a neutral atom but absorbed when an electron is removed from a neutral atom?

3.75 Why does ionization energy increase regularly across the periodic table from group 1A to group 8A, whereas electron affinity increases irregularly from group 1A to group 7A and then falls dramatically for group 8A?

3.76 No element has a negative second electron affinity. That is, the process $A^-(g) + e^- \rightarrow A^{2-}(g)$ is unfavorable for every element. Suggest a reason.

3.77 Why does phosphorus have a less-negative electron affinity than its neighbors silicon and sulfur?

Ionic Bonds and Lattice Energy (Sections 3.9 and 3.10)

3.78 Order the following compounds according to their expected lattice energies: LiCl, KCl, KBr, $MgCl_2$.

3.79 Order the following compounds according to their expected lattice energies: $AlBr_3$, $MgBr_2$, LiBr, CaO.

3.80 Calculate the energy change in kilojoules per mole when lithium atoms lose an electron to bromine atoms to form isolated Li^+ and Br^- ions. [The E_i for Li(g) is +520 kJ/mol; the E_{ea} for Br(g) is −325 kJ/mol.]

3.81 Cesium has the smallest ionization energy of all elements (+376 kJ/mol), and chlorine has the most negative electron affinity (−349 kJ/mol). Will a cesium atom transfer an electron to a chlorine atom to form isolated $Cs^+(g)$ and $Cl^-(g)$ ions? Explain.

3.82 Find the lattice energy of LiBr(s) in Table 3.4, and calculate the energy change in kilojoules per mole for the formation of solid LiBr from the elements. [The sublimation energy for Li(s) is +159.4 kJ/mol, the bond dissociation energy of $Br_2(g)$ is +224 kJ/mol, and the energy necessary to convert $Br_2(l)$ to $Br_2(g)$ is +30.9 kJ/mol.]

3.83 Look up the lattice energies in Table 3.4, and calculate the energy change in kilojoules per mole for the formation of the following substances from their elements:

(a) LiF(*s*) [The sublimation energy for Li(*s*) is +159.4 kJ/mol, the E_i for Li(*g*) is +520 kJ/mol, the E_{ea} for F(*g*) is −328 kJ/mol, and the bond dissociation energy of $F_2(g)$ is +158 kJ/mol.]

(b) $CaF_2(s)$ [The sublimation energy for Ca(*s*) is +178.2 kJ/mol, E_{i1} = +589.8 kJ/mol, and E_{i2} = +1145 kJ/mol.]

3.84 Born–Haber cycles, such as those shown in Figures 3.12 and 3.13, are called *cycles* because they form closed loops. If any five of the six energy changes in the cycle are known, the value of the sixth can be calculated. Use the following five values to calculate the lattice energy in kilojoules per mole for sodium hydride, NaH(*s*):

E_{ea} for H(*g*) = −72.8 kJ/mol

E_{i1} for Na(*g*) = +495.8 kJ/mol

Heat of sublimation for Na(*s*) = +107.3 kJ/mol

Bond dissociation energy for $H_2(g)$ = +435.9 kJ/mol

Net energy change for the formation of NaH(*s*) from its elements = −60 kJ/mol

3.85 Calculate a lattice energy for $CaH_2(s)$ in kilojoules per mole using the following information:

E_{ea} for H(*g*) = −72.8 kJ/mol

E_{i1} for Ca(*g*) = +589.8 kJ/mol

E_{i2} for Ca(*g*) = +1145 kJ/mol

Heat of sublimation for Ca(*s*) = +178.2 kJ/mol

Bond dissociation energy for $H_2(g)$ = +435.9 kJ/mol

Net energy change for the formation of $CaH_2(s)$ from its elements = −186.2 kJ/mol

3.86 Calculate the overall energy change in kilojoules per mole for the formation of CsF(*s*) from its elements using the following data:

E_{ea} for F(*g*) = −328 kJ/mol

E_{i1} for Cs(*g*) = +375.7 kJ/mol

E_{i2} for Cs(*g*) = +2422 kJ/mol

Heat of sublimation for Cs(*s*) = +76.1 kJ/mol

Bond dissociation energy for $F_2(g)$ = +158 kJ/mol

Lattice energy for CsF(*s*) = +740 kJ/mol

3.87 The estimated lattice energy for $CsF_2(s)$ is +2347 kJ/mol. Use the data given in Problem 3.86 to calculate an overall energy change in kilojoules per mole for the formation of $CsF_2(s)$ from its elements. Does the overall reaction absorb energy or release it? In light of your answer to Problem 3.86, which compound is more likely to form in the reaction of cesium with fluorine, CsF or CsF_2?

3.88 Calculate the overall energy change in kilojoules per mole for the formation of CaCl(*s*) from the elements. The following data are needed:

E_{ea} for Cl(*g*) = −348.6 kJ/mol

E_{i1} for Ca(*g*) = +589.8 kJ/mol

E_{i2} for Ca(*g*) = +1145 kJ/mol

Heat of sublimation for Ca(*s*) = +178.2 kJ/mol

Bond dissociation energy for $Cl_2(g)$ = +243 kJ/mol

Lattice energy for $CaCl_2(s)$ = +2258 kJ/mol

Lattice energy for CaCl(*s*) = +717 kJ/mol (estimated)

3.89 Use the data in Problem 3.88 to calculate an overall energy change for the formation of $CaCl_2(s)$ from the elements. Which is more likely to form, CaCl or $CaCl_2$?

3.90 Use the data and the result in Problem 3.84 to draw a Born–Haber cycle for the formation of NaH(*s*) from its elements.

3.91 Use the data and the result in Problem 3.83(a) to draw a Born–Haber cycle for the formation of LiF(*s*) from its elements.

Main-Group Chemistry (Sections 3.11–3.14)

3.92 Little is known about the chemistry of astatine (At) from direct observation, but reasonable predictions can be made.

(a) Is astatine likely to be a gas, a liquid, or a solid?

(b) Is astatine likely to react with sodium? If so, what is the formula of the product?

3.93 Look at the properties of the alkali metals summarized in Table 3.5, and predict reasonable values for the melting point, boiling point, density, and atomic radius of francium.

3.94 Why does chemical reactivity increase from top to bottom in groups 1A and 2A?

3.95 Why does chemical reactivity decrease from top to bottom in group 7A?

3.96 Write chemical equations for the reaction of potassium with the following substances, making sure that the numbers and kinds of atoms are the same on both sides of the equations. If no reaction occurs, write N.R.

(a) H_2O **(b)** NH_3 **(c)** Br_2 **(d)** O_2

3.97 Write chemical equations for the reaction of calcium with the following substances, making sure that the numbers and kinds of atoms are the same on both sides of the equations. If no reaction occurs, write N.R.

(a) H_2O **(b)** He **(c)** Br_2 **(d)** O_2

3.98 Write chemical equations for the reaction of chlorine with the following substances, making sure that the numbers and kinds of atoms are the same on both sides of the equations. If no reaction occurs, write N.R.

(a) H_2 **(b)** Ar **(c)** Rb

3.99 Milk of magnesia, a widely used antacid, is an aqueous suspension of $Mg(OH)_2$. How would you prepare $Mg(OH)_2$ from magnesium metal?

CHAPTER PROBLEMS

3.100 Cu^+ has an ionic radius of 77 pm, but Cu^{2+} has an ionic radius of 73 pm. Explain.

3.101 The following ions all have the same number of electrons: Ti^{4+}, Sc^{3+}, Ca^{2+}, S^{2-}. Order them according to their expected sizes, and explain your answer.

3.102 Calculate overall energy changes in kilojoules per mole for the formation of MgF(*s*) and $MgF_2(s)$ from their elements. In light of your answers, which compound is more likely to form in the reaction of magnesium with fluorine, MgF or MgF_2? The following data are needed:

E_{ea} for F(*g*) = −328 kJ/mol
E_{i1} for Mg (*g*) = +737.7 kJ/mol
E_{i2} for Mg(*g*) = +1450.7 kJ/mol
Heat of sublimation for Mg(*s*) = +147.7 kJ/mol
Bond dissociation energy for $F_2(g)$ = +158 kJ/mol
Lattice energy for $MgF_2(s)$ = +2952 kJ/mol
Lattice energy for MgF(*s*) = +930 kJ/mol (estimated)

3.103 Draw Born–Haber cycles for the formation of both MgF and MgF_2 (Problem 3.102).

3.104 We saw in Section 3.9 that the reaction of solid sodium with gaseous chlorine to yield solid sodium chloride (Na^+Cl^-) is favorable by 411 kJ/mol. Calculate the energy change for the alternative reaction that yields chlorine sodide (Cl^+Na^-), and then explain why sodium chloride formation is preferred.

$$2\,Na(s) + Cl_2(g) \longrightarrow 2\,Cl^+Na^-(s)$$

Assume that the lattice energy for Cl^+Na^- is the same as that for Na^+Cl^-. The following data are needed in addition to that found in Section 3.9:

E_{ea} for Na(*g*) = −52.9 kJ/mol E_{i1} for Cl(*g*) = +1251 kJ/mol

3.105 Draw a Born–Haber cycle for the reaction of sodium with chlorine to yield chlorine sodide (Problem 3.104).

3.106 Write chemical equations for the reaction of lithium with the following substances, making sure that the numbers and kinds of atoms are the same on both sides of the equations. If no reaction occurs, write N.R.

(a) H_2O **(b)** NH_3 **(c)** Br_2 **(d)** N_2 **(e)** O_2

3.107 Write chemical equations for the reaction of fluorine with the following substances, making sure that the numbers and kinds of atoms are the same on both sides of the equations. If no reaction occurs, write N.R.

(a) H_2 **(b)** Na **(c)** Sr

3.108 Each of the following pairs of elements will react to form a binary ionic compound. Write the formula of each compound formed, and give its name.

(a) Magnesium and chlorine
(b) Calcium and oxygen
(c) Lithium and nitrogen
(d) Aluminum and oxygen

3.109 Element **X** reacts with element **Y** to give a product containing X^{3+} ions and Y^{2-} ions.

(a) Is element **X** likely to be a metal or a nonmetal? Explain.
(b) Is element **Y** likely to be a metal or a nonmetal? Explain.
(c) What is the formula of the product?
(d) What groups of the periodic table are elements **X** and **Y** likely to be in?

3.110 Many early chemists noted a diagonal relationship among elements in the periodic table, whereby a given element is sometimes more similar to the element below and to the right than it is to the element directly below. Lithium is more similar to magnesium than to sodium, for example, and boron is more similar to silicon than to aluminum. Use your knowledge about the periodic trends of such properties as atomic radii and Z_{eff} to explain the existence of diagonal relationships.

3.111 Tellurium, a group 6A element, forms the oxoanions TeO_4^{2-} and TeO_3^{2-}. What are the likely names of these ions? To what other group 6A oxoanions are they analogous?

3.112 Use the following information plus the data given in Tables 3.3 and 3.4 to calculate the second electron affinity, E_{ea2}, of oxygen. Is the O^{2-} ion stable in the gas phase? Why is it stable in solid MgO?

Heat of sublimation for Mg(*s*) = +147.7 kJ/mol
Bond dissociation energy for $O_2(g)$ = +498.4 kJ/mol
E_{ea1} for O(*g*) = −141.0 kJ/mol
Net energy change for formation of MgO(*s*) from its elements = −601.7 kJ/mol

3.113 **(a)** Which element from each set has the largest atomic radius? Explain.

(i) Ba, Ti, Ra, Li **(ii)** F, Al, In, As

(b) Which element from each set has the smallest ionization energy? Explain.

(i) Tl, Po, Se, Ga **(ii)** Cs, Ga, Bi, Se

3.114 **(a)** Which of the elements Be, N, O, and F has the most negative electron affinity? Explain.

(b) Which of the ions Se^{2-}, F^-, O^{2-}, and Rb^+ has the largest radius? Explain.

3.115 Given the following information, construct a Born–Haber cycle to calculate the lattice energy of $CaC_2(s)$:

Net energy change for the formation of $CaC_2(s)$ = −60 kJ/mol
Heat of sublimation for Ca(*s*) = +178 kJ/mol
E_{i1} for Ca(*g*) = +590 kJ/mol
E_{i2} for Ca(*g*) = +1145 kJ/mol
Heat of sublimation for C(*s*) = +717 kJ/mol
Bond dissociation energy for $C_2(g)$ = +614 kJ/mol
E_{ea1} for $C_2(g)$ = −315 kJ/mol
E_{ea2} for $C_2(g)$ = +410 kJ/mol

3.116 Given the following information, construct a Born–Haber cycle to calculate the lattice energy of $CrCl_2I(s)$:

Net energy change for the formation of $CrCl_2I(s)$ = −420 kJ/mol
Bond dissociation energy for $Cl_2(g)$ = +243 kJ/mol
Bond dissociation energy for $I_2(g)$ = +151 kJ/mol
Heat of sublimation for $I_2(s)$ = +62 kJ/mol
Heat of sublimation for Cr(*s*) = +397 kJ/mol
E_{i1} for Cr(*g*) = +652 kJ/mol
E_{i2} for Cr(*g*) = +1588 kJ/mol
E_{i3} for Cr(*g*) = +2882 kJ/mol
E_{ea} for Cl(*g*) = −349 kJ/mol
E_{ea} for I(*g*) = −295 kJ/mol

MULTICONCEPT PROBLEMS

3.117 Consider the electronic structure of the element bismuth.

(a) The first ionization energy of bismuth is $E_{i1} = +703$ kJ/mol. What is the longest possible wavelength of light that could ionize an atom of bismuth?

(b) Write the electron configurations of neutral Bi and the Bi^+ cation.

(c) What are the n and l quantum numbers of the electron removed when Bi is ionized to Bi^+?

(d) Would you expect element 115 to have an ionization energy greater than, equal to, or less than that of bismuth? Explain.

3.118 Iron is commonly found as Fe, Fe^{2+}, and Fe^{3+}.

(a) Write electron configurations for each of the three.

(b) What are the n and l quantum numbers of the electron removed on going from Fe^{2+} to Fe^{3+}?

(c) The third ionization energy of Fe is $E_{i3} = +2952$ kJ/mol. What is the longest wavelength of light that could ionize $Fe^{2+}(g)$ to $Fe^{3+}(g)$?

(d) The third ionization energy of Ru is less than the third ionization energy of Fe. Explain.

3.119 The ionization energy of an atom can be measured by photoelectron spectroscopy, in which light of wavelength λ is directed at an atom, causing an electron to be ejected. The kinetic energy of the ejected electron (E_K) is measured by determining its velocity, v, since $E_K = 1/2mv^2$. The E_i is then calculated using the relationship that the energy of the incident light equals the sum of E_i plus E_K.

(a) What is the ionization energy of rubidium atoms in kilojoules per mole if light with $\lambda = 58.4$ nm produces electrons with a velocity of 2.450×10^6 m/s? (The mass of an electron is 9.109×10^{-31} kg.)

(b) What is the ionization energy of potassium in kilojoules per mole if light with $\lambda = 142$ nm produces electrons with a velocity of 1.240×10^6 m/s?

CHAPTER

4 Atoms and Covalent Bonds

The results may look good now, but ultraviolet tanning rays can lead to skin cancer later in life. Fortunately, skin contains a protecting chemical to minimize the damage, as we'll see in the *FYI* at the end of this chapter.

We saw in the previous chapter that a bond between a metal and a reactive nonmetal is typically formed by the transfer of electrons between atoms. The metal atom loses one or more electrons and becomes a cation, while the reactive nonmetal atom gains one or more electrons and becomes an anion. The oppositely charged ions are held together by the electrostatic attractions that we call ionic bonds.

How, though, do bonds form between atoms of the same or similar elements? How can we describe the bonds in such substances as H_2, Cl_2, CO_2, and the tens of millions of other nonionic compounds? Simply put, the answer is that the bonds in such compounds are formed by the *sharing* of electrons between atoms rather than by the transfer of electrons from one atom to another. As we saw in Section 3.1, a bond formed by the sharing of electrons between atoms is called a *covalent bond*, and the unit of matter held together by one or more covalent bonds is called a *molecule*. We'll explore the nature of covalent bonding and the structure of the resultant molecules in this chapter and the next.

▶ **Covalent Bond** A bond that results from the sharing of electrons between atoms

▶ **Molecule** The unit of matter held together by covalent bonds

CONTENTS

4.1 COMPARING IONIC AND MOLECULAR COMPOUNDS

To get an idea how ionic and molecular (covalent) compounds differ, look at the comparison between NaCl and HCl in **Table 4.1**. Sodium chloride, an ionic compound, is a white solid with a melting point of 801 °C and a boiling point of 1465 °C. Hydrogen chloride, a molecular compound, is a colorless gas with a melting point of −115 °C and a boiling point of −84.9 °C. What accounts for such large differences in properties?

TABLE 4.1 Some Physical Properties of NaCl and HCl

Property	NaCl	HCl
Physical appearance	White solid	Colorless gas
Type of bond	Ionic	Covalent
Melting point	801 °C	−115 °C
Boiling point	1465 °C	−84.9 °C

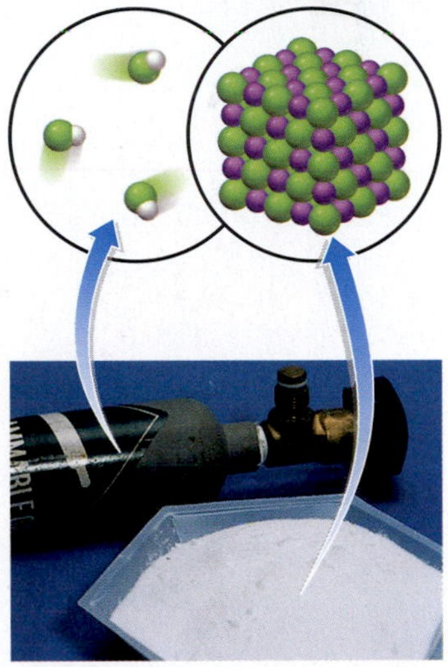

▲ Sodium chloride, an ionic compound, is a white, crystalline solid that melts at 801 °C. Hydrogen chloride, a molecular compound, is a gas at room temperature.

Ionic compounds are high-melting solids because of their ionic bonds. As discussed in Section 3.1, a visible sample of sodium chloride consists, not of individual NaCl molecules, but of a vast three-dimensional network of ions in which each Na^+ cation is attracted to many surrounding Cl^- anions and each Cl^- ion is attracted to many surrounding Na^+ ions. For sodium chloride to melt or boil so that the ions break free of one another, every ionic attraction in the entire crystal—the **lattice energy**—must be overcome, a process that requires a large input of energy.

REMEMBER...

Lattice energy (U) is the amount of energy that must be supplied to break an ionic solid into its individual gaseous ions and is thus a measure of the strength of the solid's ionic bonds. (Section 3.10)

Molecular compounds, by contrast, are low-melting solids, liquids, or even gases. A sample of a molecular compound such as hydrogen chloride consists of discrete HCl molecules. Although the covalent bond within an individual molecule may be very strong, the attractive forces between different molecules are fairly weak. As a result, relatively little energy is required to overcome the forces between molecules and cause a molecular compound to melt or boil. We'll look at the nature of the forces between molecules and at the boiling and melting processes in **Chapter 10**.

LOOKING AHEAD...

The bulk materials we see in everyday life are usually solids or liquids rather than gases. In **Chapter 10**, we'll look at the forces that operate between molecules to hold them in close contact and at the resultant melting and boiling processes that happen when the forces are overcome.

4.2 COVALENT BOND FORMATION

To see how the formation of a covalent, shared-electron bond between atoms can be described, let's look at the H–H bond in the H_2 molecule as the simplest example. When two isolated hydrogen atoms come close together, electrostatic interactions begin to develop between them. The two positively charged nuclei repel each other, and the two negatively charged electrons repel each other, but each nucleus attracts both electrons and each electron is attracted to both nuclei (**Figure 4.1**). If the attractive forces are stronger than the repulsive forces, a covalent bond is formed, with the two atoms held together and the two shared electrons occupying the region between the nuclei.

FIGURE 4.1

A covalent H–H bond. The bond is the net result of attractive and repulsive electrostatic forces.

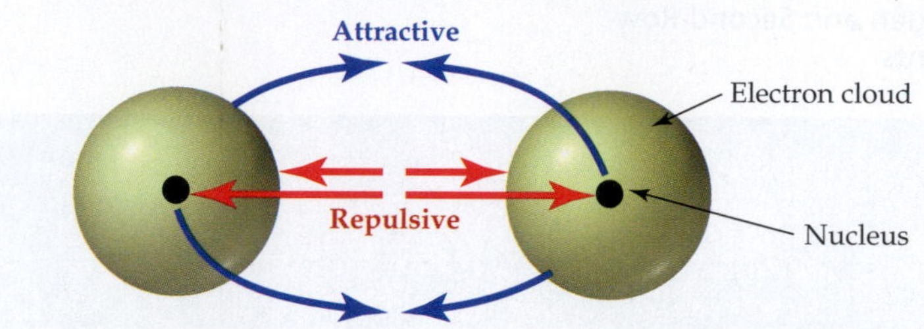

The nucleus–electron **attractions** are greater than the nucleus–nucleus and electron–electron **repulsions**, resulting in a net attractive force that binds the atoms together.

In essence, the shared electrons act as a kind of "glue" to bind the two atoms into an H_2 molecule. Because both nuclei are attracted to the same electrons, they are held together, much as two tug-of-war teams pulling on the same rope are held together.

The magnitudes of the various attractive and repulsive forces between nuclei and electrons in a covalent bond depend on how close together the atoms are. If the hydrogen atoms are too far apart, the attractive forces are small and no bond exists. If the hydrogen atoms are too close together, the repulsive interaction between the two nuclei becomes so strong that it pushes the atoms apart. Thus, there is an optimum distance between nuclei, called the **bond length**, where net attractive forces are maximized and the H–H molecule is most stable. In the H_2 molecule, the bond length is 74 pm. On a graph of energy versus internuclear distance, the bond length is the H–H distance in the minimum-energy, most stable arrangement (**Figure 4.2**).

FIGURE 4.2

A graph of potential energy versus internuclear distance for the H_2 molecule.

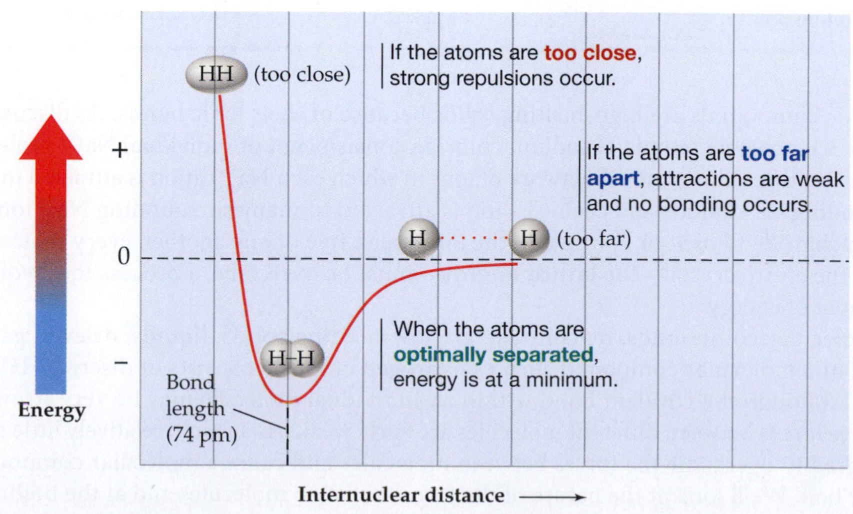

Every covalent bond has its own characteristic length that leads to maximum stability and that is roughly predictable from a knowledge of the atomic radii described in Section 2.14. For example, because the atomic radius of hydrogen is 37 pm and the atomic radius of chlorine is 99 pm, we might expect the H—Cl bond length in a hydrogen chloride molecule to be approximately 37 pm + 99 pm = 136 pm. The actual value is 127 pm.

Chemists normally represent a given molecule by giving its **structural formula**, which explicitly shows the bonds between atoms and therefore gives much more information than just a **chemical formula**. Ethyl alcohol, for instance, has the chemical formula C_2H_6O and the structural formula shown:

> **REMEMBER...**
>
> The **chemical formula** of a compound lists the symbols of the compound's constituent elements and uses subscripts to indicate the number of atoms of each element. If no subscript is given, the number 1 is understood. (Section 1.4)

[Red = O, gray = C, ivory = H]

C_2H_6O

Chemical formula

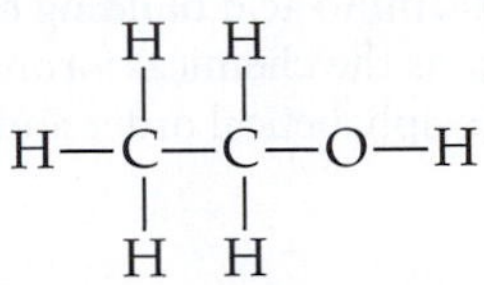

Structural formula

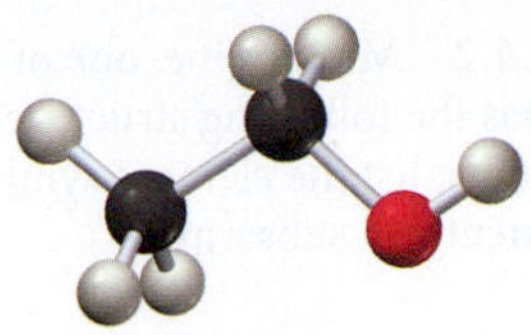

Molecular model

Ethyl alcohol

A structural formula, like that of ethanol, uses lines between atoms to indicate the covalent bonds. Thus, the two carbon atoms in ethyl alcohol are covalently bonded to each other, the oxygen atom is bonded to one of the carbon atoms, and the six hydrogen atoms are distributed three to one carbon, two to the other carbon, and one to the oxygen.

Structural formulas are particularly important in *organic chemistry*—the chemistry of carbon compounds—where the behavior of the large, complex molecules on which life is based is almost entirely governed by their structures. Take even a relatively simple substance like glucose, for instance. The molecular formula of glucose, $C_6H_{12}O_6$, tells nothing about how the atoms are connected. In fact, you could probably imagine a great many different ways. The structural formula for glucose, however, shows that five carbons and one oxygen form a ring of atoms, with the remaining five oxygens each bonded to one hydrogen and distributed onto different carbons.

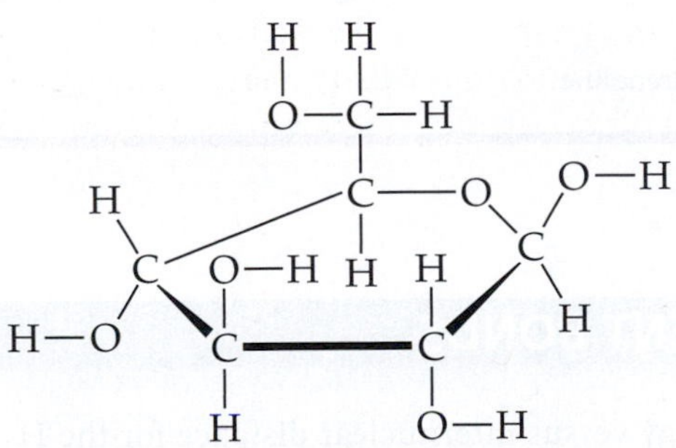

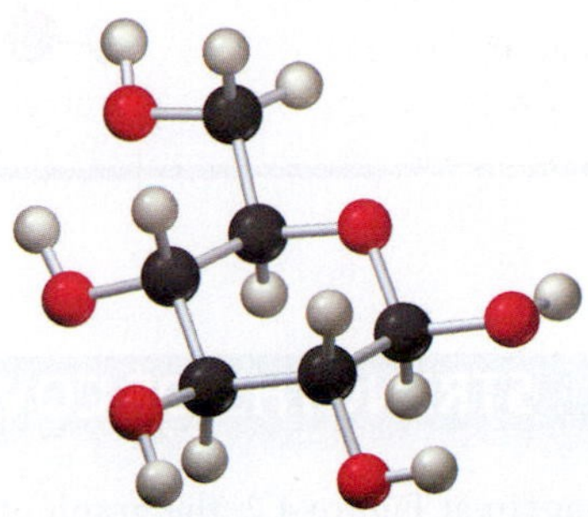

Glucose, $C_6H_{12}O_6$

Worked Example 4.1

Drawing a Structural Formula

Propane, C_3H_8, has a structure in which the three carbon atoms are bonded in a row, each end carbon is bonded to three hydrogens, and the middle carbon is bonded to two hydrogens. Draw the structural formula.

STRATEGY

Think of a compound's structural formula as a sketch that shows which atoms are bonded to one another. If bonding information is provided—perhaps by a molecular model—the structural formula is written using lines to represent covalent bonds.

continued on next page

SOLUTION

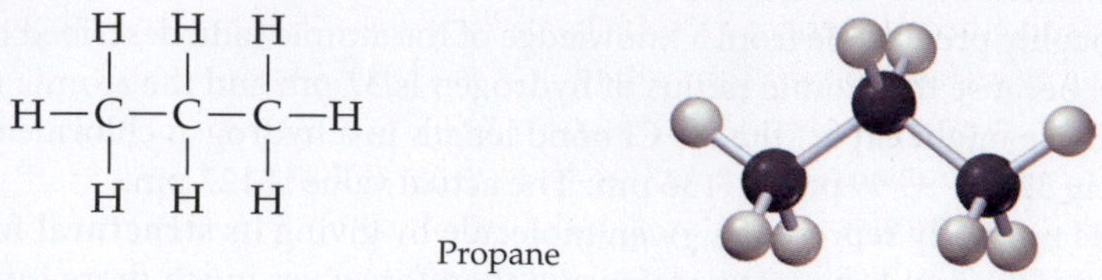

Propane

PROBLEM 4.1 Draw the structural formula of methylamine, CH_5N, a substance responsible for the odor of rotting fish. The carbon atom is bonded to the nitrogen atom and to three hydrogens. The nitrogen atom is bonded to the carbon and two hydrogens.

PROBLEM 4.2 Methionine, one of the 20 amino acid building blocks from which proteins are made, has the following structure. What is the chemical formula of methionine? In writing the formula, list the element symbols in alphabetical order and give the number of atoms of each element as a subscript.

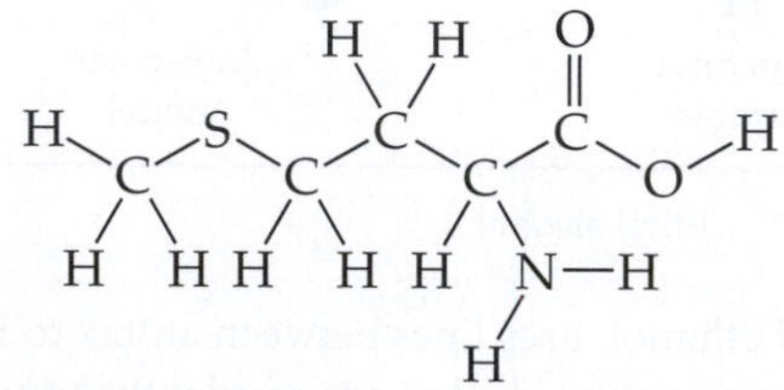

Methionine
(an amino acid)

CONCEPTUAL PROBLEM 4.3 Adrenaline, the so-called "flight or fight" hormone, can be represented by the following ball-and-stick model. What is the chemical formula of adrenaline? (gray = C, ivory = H, red = O, blue = N)

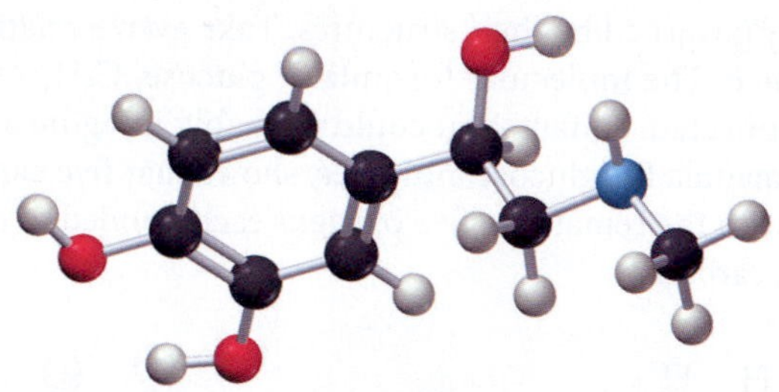

Adrenaline

4.3 STRENGTHS OF COVALENT BONDS

Look again at Figure 4.2, the graph of energy versus internuclear distance for the H_2 molecule, and notice how the H_2 molecule is lower in energy than two isolated hydrogen atoms. When pairs of hydrogen atoms bond together, they form lower-energy H_2 molecules and release 436 kJ/mol. Looked at from the other direction, 436 kJ must be *added* to split 1 mol of H_2 molecules apart into 2 mol of hydrogen atoms.

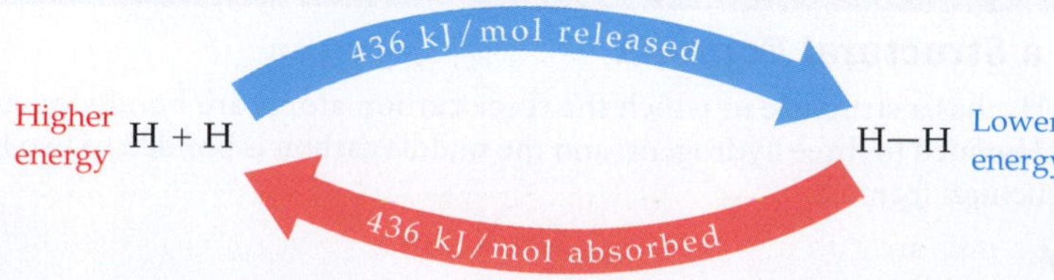

The amount of energy that must be supplied to break a covalent bond between two atoms in an isolated molecule in the gaseous state—and thus the amount of energy released when the bond forms—is called the **bond dissociation energy (*D*)**. Bond dissociation energies are always positive because energy must always be added to break a bond. Conversely, the amount of energy released on forming a bond always has a negative value.

Every bond in every molecule has its own specific bond dissociation energy. Perhaps it's not surprising, though, to find that bonds between the same pairs of atoms usually have similar dissociation energies in whatever molecules they occur. For example, carbon–carbon bonds usually have *D* values of approximately 350–380 kJ/mol regardless of the exact structure of the molecule.

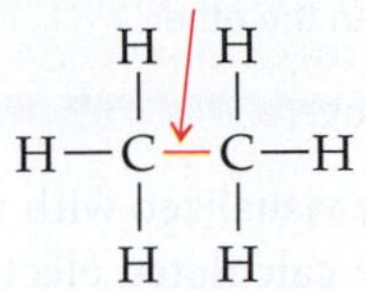
Ethane
D = 377 kJ/mol

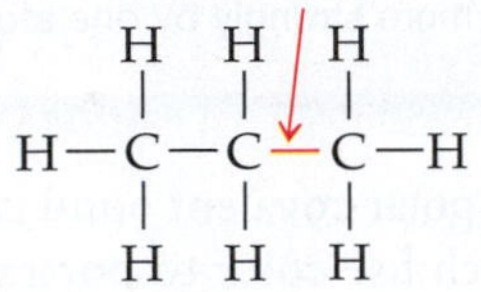
Propane
D = 370 kJ/mol

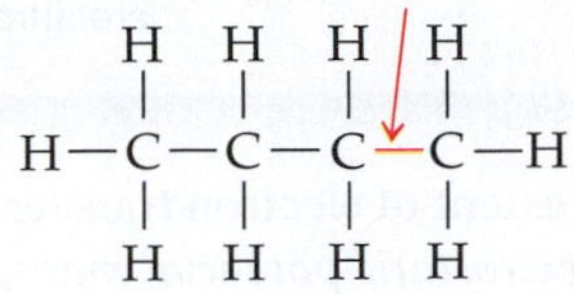
Butane
D = 372 kJ/mol

Because similar bonds have similar bond dissociation energies, it's possible to construct a table of average values to compare different kinds of bonds (**Table 4.2**). Keep in mind, though, that the actual value in a specific molecule might vary by ± 10% from the average.

TABLE 4.2 Average Bond Dissociation Energies, *D* (kJ/mol)

H–H	436[a]	C–H	410	N–H	390	O–F	180	I–I	151[a]
H–C	410	C–C	350	N–C	300	O–Cl	200	S–F	310
H–F	570[a]	C–F	450	N–F	270	O–Br	210	S–Cl	250
H–Cl	432[a]	C–Cl	330	N–Cl	200	O–I	220	S–Br	210
H–Br	366[a]	C–Br	270	N–Br	240	O–N	200	S–S	225
H–I	298[a]	C–I	240	N–N	240	O–O	180		
H–N	390	C–N	300	N–O	200	F–F	159[a]		
H–O	460	C–O	350	O–H	460	Cl–Cl	243[a]		
H–S	340	C–S	260	O–C	350	Br–Br	193[a]		
Multiple covalent bonds[b]									
C=C	728	C≡C	965	C=O	732	O=O	498[a]	N≡N	945[a]

[a]Exact value

[b]We'll discuss multiple covalent bonds in Section 4.6.

The bond dissociation energies listed in Table 4.2 cover a wide range, from a low of 151 kJ/mol for the I–I bond to a high of 570 kJ/mol for the H–F bond. As a rule of thumb, though, most of the bonds commonly encountered in naturally occurring organic molecules (C–H, C–C, C–O) have values in the range 350–400 kJ/mol.

4.4 POLAR COVALENT BONDS: ELECTRONEGATIVITY

We've given the impression up to this point that a given bond is either purely ionic, with electrons completely transferred, or purely covalent, with electrons shared equally. In fact, though, ionic and covalent bonds represent only the two extremes of a continuous range of possibilities. Between these two extremes are the large majority of bonds in which the bonding electrons are shared unequally between two atoms but are not completely transferred from one to the other. Such bonds are said to be **polar covalent bonds** (**Figure 4.3**). The lowercase Greek letter delta (δ) is used to symbolize the resultant *partial* charges on the atoms, either partial positive ($\delta+$) for the electron-poor atom that has a smaller share of the bonding electrons or partial negative ($\delta-$) for the electron-rich atom that has a larger share.

FIGURE 4.3

The bonding continuum from nonpolar covalent to ionic. [The symbol δ (Greek delta) means partial charge, either partial positive (δ+) or partial negative (δ−).]

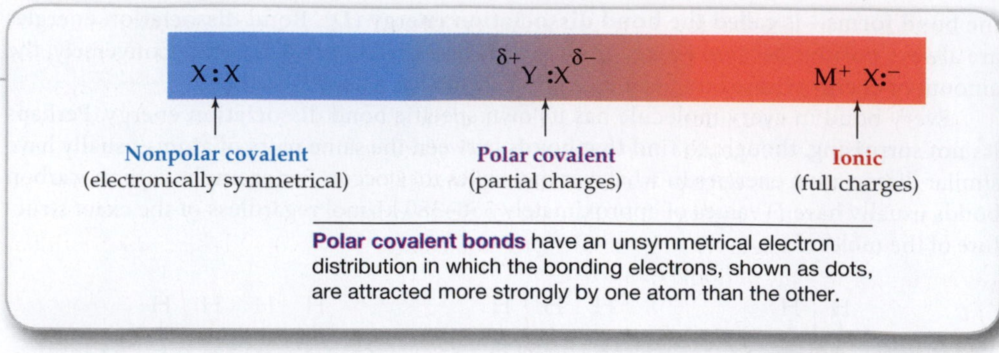

The extent of electron transfer in a polar covalent bond can be visualized with what are called *electrostatic potential maps*, which use color to portray the calculated electron distribution in an isolated, gas-phase molecule. Yellow–green represents a neutral, nonpolar atom; blue represents an electron-poor atom (partial positive charge), and red represents an electron-rich atom (partial negative charge). As examples of different points along the bonding spectrum, look at the three substances Cl_2, HCl, and NaCl.

- **Cl_2 – Nonpolar Covalent** The bond in a chlorine molecule is nonpolar covalent, with the bonding electrons shared equally by the two identical chlorine atoms. A similar situation exists in all such molecules that contain a covalent bond between two identical atoms. Thus, both chlorine atoms in Cl_2 have an identical yellow–green coloration in an electrostatic potential map, indicating their neutral, nonpolar nature.

The two bonding electrons, shown here as dots, are symmetrically distributed between the two Cl atoms.

Cl:Cl

A nonpolar covalent bond. Yellow-green represents a neutral, nonpolar atom.

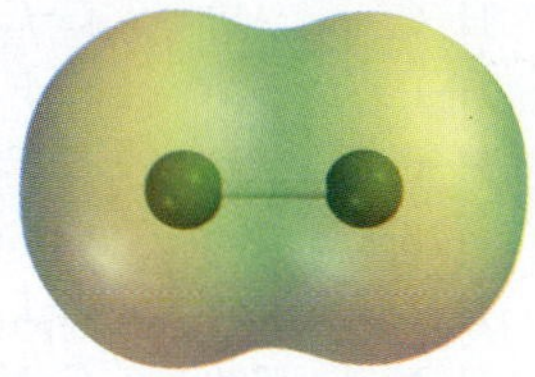

- **HCl – Polar Covalent** The bond in a hydrogen chloride molecule is polar covalent because the chlorine atom attracts the shared electron pair more strongly than the hydrogen atom does, resulting in an unsymmetrical distribution of electrons. Chlorine is thus electron-rich and has a partial negative charge (orange in the electrostatic potential map), while hydrogen is electron-poor and has a partial positive charge (blue in the electrostatic potential map). Experimentally, the H–Cl bond has been found to be about 83% covalent and 17% ionic.

$^{\delta+}H — Cl^{\delta-}$

[H :Cl]

A polar covalent bond. The two bonding electrons (dots) are attracted more strongly by Cl than by H.

- **NaCl – Ionic** The bond in solid sodium chloride is a largely ionic one between Na^+ and Cl^-. In spite of what we've said previously, though, experiments show that the NaCl bond is only about 80% ionic and that the electron transferred from Na to Cl still spends some of its time near sodium. Thus, the electron-poor sodium atom is blue in an electrostatic potential map, while the electron-rich chlorine is red.

$Na^+ Cl^-$

An ionic bond. Blue indicates a partial positive charge; red indicates a partial negative charge.

Bond polarity like that found in HCl is due to differences in the *electronegativities* of the bonded atoms, where **electronegativity (EN)** is defined as the ability of an atom in a molecule to attract the shared electrons in a covalent bond. As shown in **Figure 4.4**, electronegativities are expressed on a unitless scale, with fluorine, the most highly electronegative element, assigned a value of 4.0. Metallic elements on the left of the periodic table attract electrons only weakly and are the least electronegative elements (low EN), while halogens and other reactive nonmetals in the upper right of the table attract electrons strongly and are the most electronegative (high EN). Figure 4.4 also shows that electronegativity generally decreases within a group going down the periodic table.

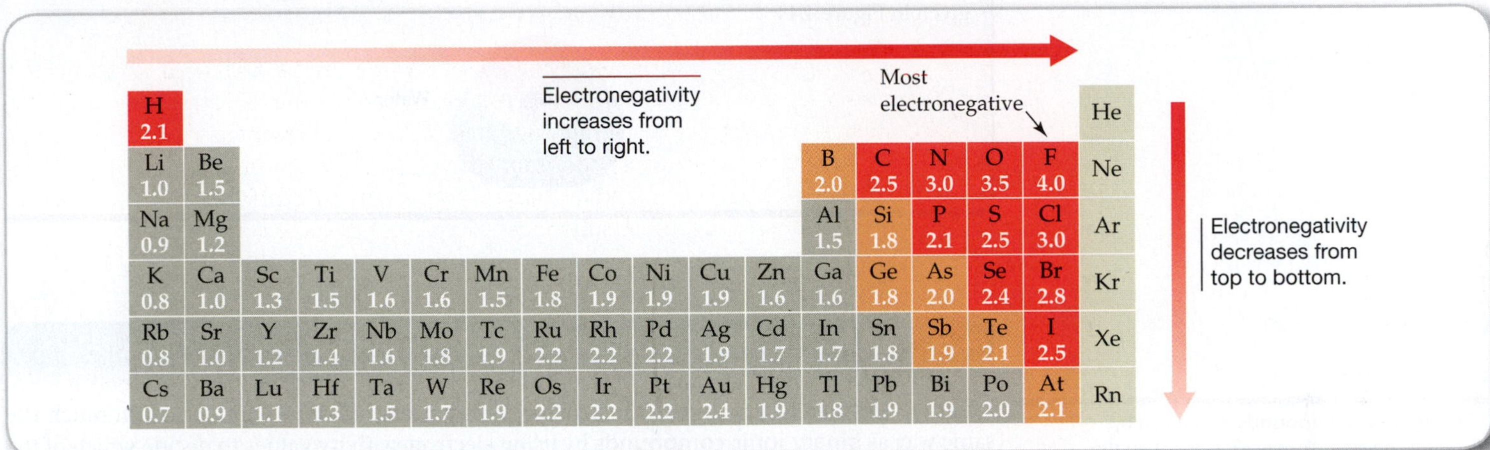

FIGURE 4.4

Electronegativity values and trends in the periodic table.

Because electronegativity measures the ability of an atom in a molecule to attract shared electrons, it seems reasonable that it should be related to **electron affinity E_{ea}** (Section 3.7) and **ionization energy E_i** (Section 3.5). Electron affinity, after all, is a measure of the tendency of an isolated atom to gain an electron, and ionization energy is a measure of the tendency of an isolated atom to lose an electron. In fact, one of the ways in which electronegativities were first calculated was by taking the average of the absolute values of E_{ea} and E_i and setting up a scale with fluorine assigned a value of 4.0.

REMEMBER...

Electron affinity (E_{ea}) is defined as the energy change that occurs when an electron is added to an isolated gaseous atom. (Section 3.7) **Ionization energy (E_i)**, in contrast, is the amount of energy needed to remove the highest-energy electron from an isolated neutral atom in the gaseous state. (Section 3.5)

How can we use electronegativities to predict bond polarity? The following guidelines generally apply:

- Bonds between atoms with the same or similar electronegativity are usually nonpolar covalent.
- Bonds between atoms whose electronegativities differ by more than about 2 units are largely ionic.
- Bonds between atoms whose electronegativities differ by less than 2 units are usually polar covalent.

Thus, we can be reasonably sure that a C–Cl bond in chloroform, $CHCl_3$, is polar covalent, while an Na^+Cl^- bond in sodium chloride is largely ionic.

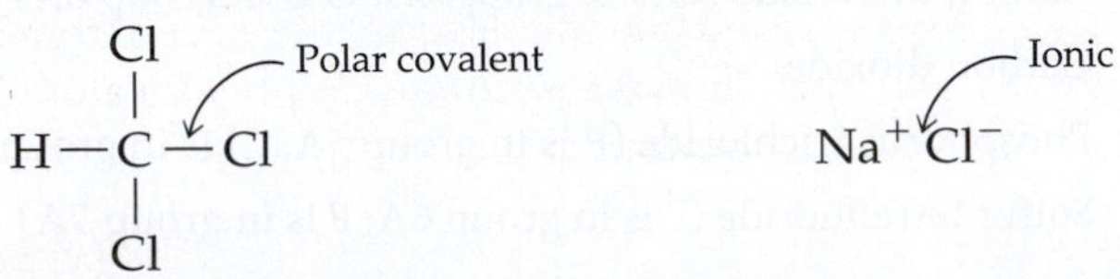

Chlorine:	EN = 3.0
Carbon:	EN = 2.5
	Difference = 0.5

Chlorine:	EN = 3.0
Sodium:	EN = 0.9
	Difference = 2.1

PROBLEM 4.4 Use the electronegativity values in Figure 4.4 to predict whether the bonds in the following compounds are polar covalent or ionic:
(a) $SiCl_4$ **(b)** CsBr **(c)** $FeBr_3$ **(d)** CH_4

PROBLEM 4.5 Order the following compounds according to the increasing ionic character of their bonds: CCl_4, $BaCl_2$, $TiCl_3$, Cl_2O.

CONCEPTUAL PROBLEM 4.6 An electrostatic potential map of water is shown below. Which atom, H or O, is positively polarized (electron-poor) and which is negatively polarized (electron-rich)? Is this polarity pattern consistent with the electronegativity values of O and H given in Figure 4.4?

4.5 NAMING MOLECULAR COMPOUNDS

REMEMBER...

Binary ionic compounds are named by identifying first the positive ion and then the negative ion. The positive ion takes the same name as the element, while the negative ion takes the first part of its name from the element and adds the ending *-ide*. (Section 3.2)

Binary molecular compounds—those made of only two elements—are named in much the same way as **binary ionic compounds** by using electronegativity values to decide which of the elements in the compound is more electron-poor, or *cationlike*, and which element is more electron-rich, or *anionlike*. As with ionic compounds, the cationlike element takes the name of the element itself, and the anionlike element takes an *-ide* ending. The compound HF, for example, is called *hydrogen fluoride*.

HF Hydrogen is more electron-poor and cationlike because it is less electronegative (EN = 2.1). Fluorine is more electron-rich and anionlike because it is more electronegative (EN = 4.0). The compound is therefore named *hydrogen fluoride*.

It's usually possible to decide which element is cationlike and which element is anionlike just by looking at their relative positions in the periodic table. The farther left and toward the bottom of the periodic table an element occurs, the more likely it is to have a low electronegativity and be cationlike. The farther right and toward the top an element occurs (except for the noble gases), the more likely it is to have a high electronegativity and be anionlike.

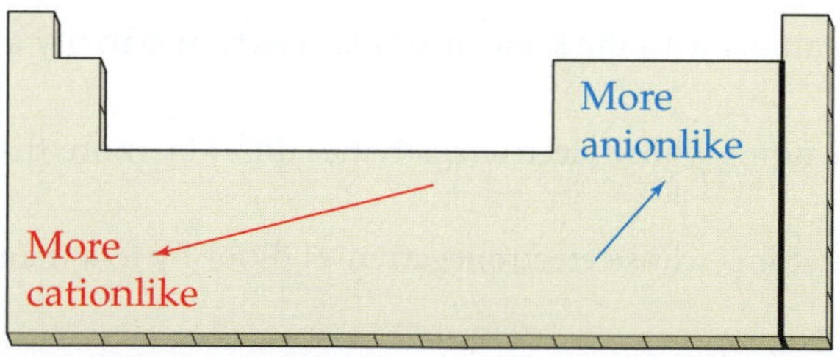

TABLE 4.3 Numerical Prefixes for Naming Compounds

Prefix	Meaning
mono-	1
di-	2
tri-	3
tetra-	4
penta-	5
hexa-	6
hepta-	7
octa-	8
nona-	9
deca-	10

The following examples show how these generalizations apply:

CO	Carbon monoxide (C is in group 4A; O is in group 6A)
CO_2	Carbon dioxide
PCl_3	Phosphorus trichloride (P is in group 5A; Cl is in group 7A)
SF_4	Sulfur tetrafluoride (S is in group 6A; F is in group 7A)
N_2O_4	Dinitrogen tetroxide (N is in group 5A; O is in group 6A)

Because nonmetals often combine with one another in different proportions to form different compounds, numerical prefixes are usually included in the names of binary molecular compounds to specify the numbers of each kind of atom present. The compound CO, for example, is called carbon *mon*oxide, and CO_2 is called carbon *di*oxide. **Table 4.3** lists the most common numerical prefixes. Note that when the prefix ends in *a* or *o* (but not *i*) and the anion name begins with a vowel (*oxide*, for instance), the *a* or *o* on the prefix is dropped to avoid having two vowels together in the name. Thus, we write carbon *mon*oxide rather than carbon *mono*oxide for CO and dinitrogen *tetr*oxide rather than dinitrogen *tetra*oxide for N_2O_4. A *mono-* prefix is not used for the atom named first: CO_2 is called carbon dioxide rather than monocarbon dioxide.

Worked Example 4.2

Naming Binary Molecular Compounds

Give systematic names for the following compounds:
(a) PCl_3 (b) N_2O_3 (c) P_4O_7 (d) BrF_3

STRATEGY

Look at a periodic table to see which element in each compound is more cationlike (less electronegative; farther to the left or lower) and which is more anionlike (more electronegative; farther to the right or higher). Then name the compound using the appropriate numerical prefix.

SOLUTION

(a) Phosphorus trichloride (b) Dinitrogen trioxide
(c) Tetraphosphorus heptoxide (d) Bromine trifluoride

PROBLEM 4.7 Give systematic names for the following compounds:
(a) NCl_3 (b) P_4O_6 (c) S_2F_2 (d) SeO_2

PROBLEM 4.8 Write formulas for compounds with the following names:
(a) Disulfur dichloride (b) Iodine monochloride
(c) Nitrogen triiodide

CONCEPTUAL PROBLEM 4.9 Give systematic names for the following compounds:

(a)

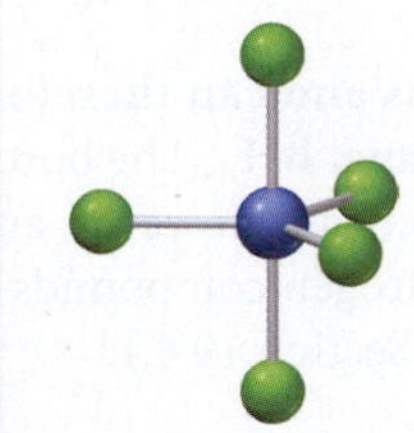

Purple = P, green = Cl

(b)

Blue = N, red = O

4.6 ELECTRON-DOT STRUCTURES

One way to picture the sharing of electrons between atoms in covalent or polar covalent bonds is to use **electron-dot structures**, often called *Lewis structures* after G. N. Lewis of the University of California at Berkeley. An electron-dot structure represents an atom's valence electrons by dots and indicates by the placement of the dots how the valence electrons are distributed in a molecule. A hydrogen molecule, for instance, is written showing a pair of dots between the hydrogen atoms, indicating that the hydrogens share the pair of electrons in a covalent bond:

An electron-pair bond

H· ·H ⟶ H:H

Two hydrogen atoms — A hydrogen molecule

By sharing two electrons in a covalent bond, each hydrogen effectively has one electron pair and the stable, $1s^2$ electron configuration of helium.

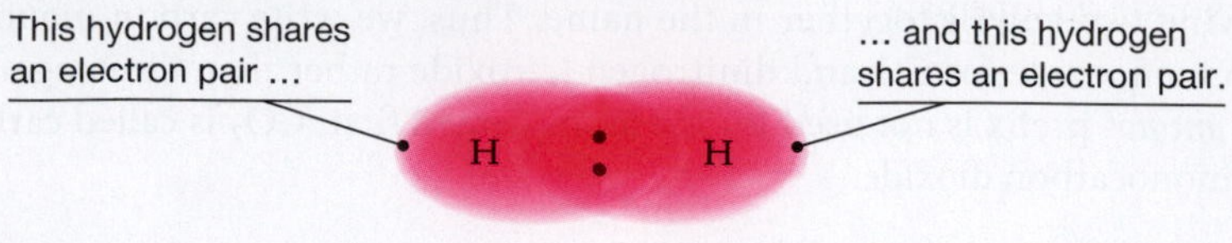

Atoms other than hydrogen also form covalent bonds by sharing electron pairs, and the electron-dot structures of the resultant molecules are drawn by assigning the correct number of valence (outer-shell) electrons to each atom. Group 3A atoms, such as boron, have three valence electrons; group 4A atoms, such as carbon, have four valence electrons; and so on across the periodic table. The group 7A element fluorine has seven valence electrons, and an electron-dot structure for the F_2 molecule shows how a covalent bond can form:

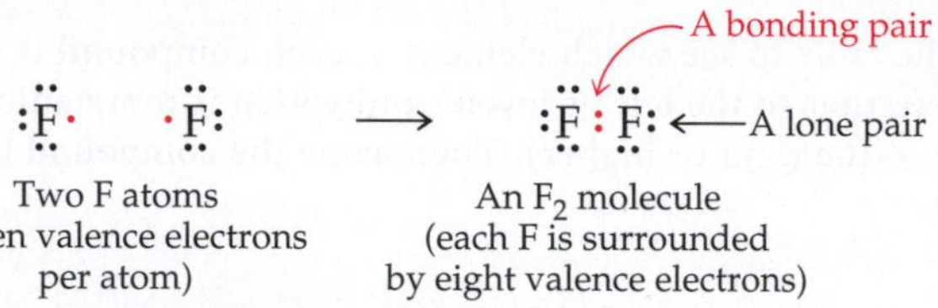

Six of the seven valence electrons in a fluorine atom are already paired in three filled atomic orbitals and thus are not shared in bonding. The seventh fluorine valence electron, however, is unpaired and can be shared in a covalent bond with another fluorine. Each atom in the resultant F_2 molecule thereby gains a noble-gas configuration with eight valence-shell electrons and thus obeys the **octet rule**, discussed in Section 3.8. The three pairs of nonbonding electrons on each fluorine atom are called **lone pairs**, or *nonbonding pairs*, and the shared electrons are called a **bonding pair**.

REMEMBER...

According to the **octet rule**, main-group elements tend to undergo reactions that give them a noble-gas electron configuration with filled *s* and *p* sublevels in their valence electron shell. (Section 3.8)

The tendency of main-group atoms to fill their *s* and *p* subshells and thereby achieve a noble-gas configuration when they form bonds is an important guiding principle that makes it possible to predict the formulas and electron-dot structures of a great many molecules. As a general rule, a main-group atom shares as many of its valence-shell electrons as possible, either until it has no more to share or until it reaches an octet configuration. The following guidelines apply for neutral molecules:

LOOKING AHEAD...

Covalent bonding in compounds of boron and hydrogen (*boron hydrides*) will be discussed in Section 19.4.

- **Group 3A elements, such as boron, have three valence electrons and can therefore form three electron-pair bonds in neutral molecules such as borane, BH_3.** The boron atom in the resultant molecule has only three bonding pairs of electrons, however, and can't reach an electron octet. (The bonding situation in boron–hydrogen compounds is actually more complicated than suggested here; we'll deal with it in Section 19.4.)

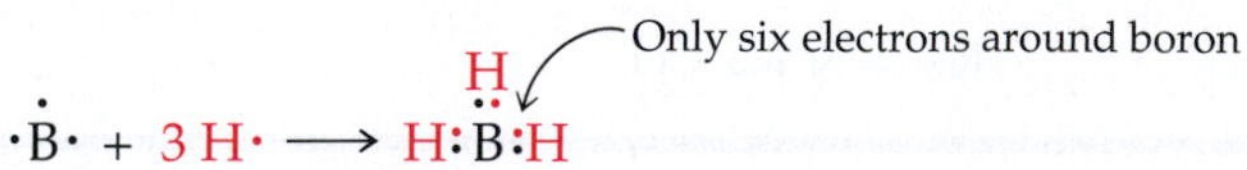

Borane

- **Group 4A elements, such as carbon, have four valence electrons and form four bonds, as in methane, CH_4.** The carbon atom in the resultant molecule has four bonding pairs of electrons.

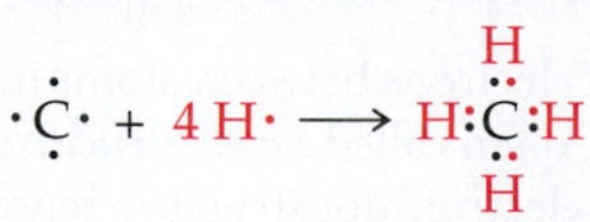

Methane

- **Group 5A elements, such as nitrogen, have five valence electrons and form three bonds, as in ammonia, NH_3.** The nitrogen atom in the resultant molecule has three bonding pairs of electrons and one lone pair.

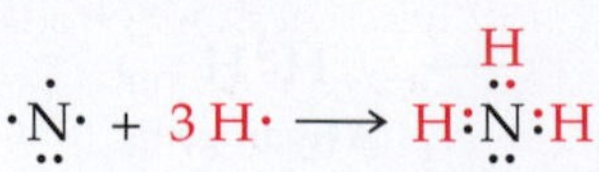

Ammonia

- **Group 6A elements, such as oxygen, have six valence electrons and form two bonds, as in water, H_2O.** The oxygen atom in the resultant molecule has two bonding pairs of electrons and two lone pairs.

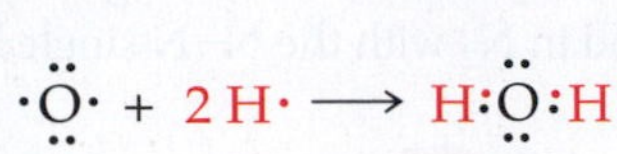

Water

- **Group 7A elements (halogens), such as fluorine, have seven valence electrons and form one bond, as in hydrogen fluoride, HF.** The fluorine atom in the resultant molecule has one bonding pair of electrons and three lone pairs.

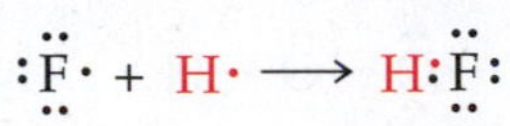

Hydrogen fluoride

- **Group 8A elements (noble gases), such as neon, rarely form covalent bonds because they already have valence-shell octets.**

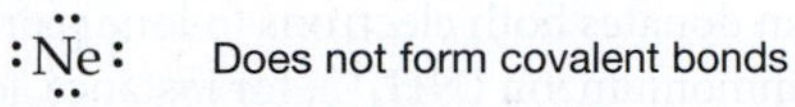

These conclusions are summarized in **Table 4.4**.

TABLE 4.4 Covalent Bonding Trends for Second-Row Elements in Neutral Molecules

Group	Number of Valence Electrons	Number of Bonds	Example
3A	3	3	BH_3
4A	4	4	CH_4
5A	5	3	NH_3
6A	6	2	H_2O
7A	7	1	HF
8A	8	0	Ne

Not all covalent bonds contain just one shared electron pair, or **single bond**, like those just discussed. In molecules such as O_2, N_2, and many others, the atoms share more than one pair of electrons, leading to the formation of *multiple* covalent bonds. The oxygen atoms in the O_2 molecule, for example, reach valence-shell octets by sharing two pairs, or four electrons, in a **double bond**. Similarly, the nitrogen atoms in the N_2 molecule share three pairs, or six electrons, in a **triple bond**. (Although the O_2 molecule does have a double bond, the following electron-dot structure is incorrect in some respects, as we'll see in Section 5.6.)

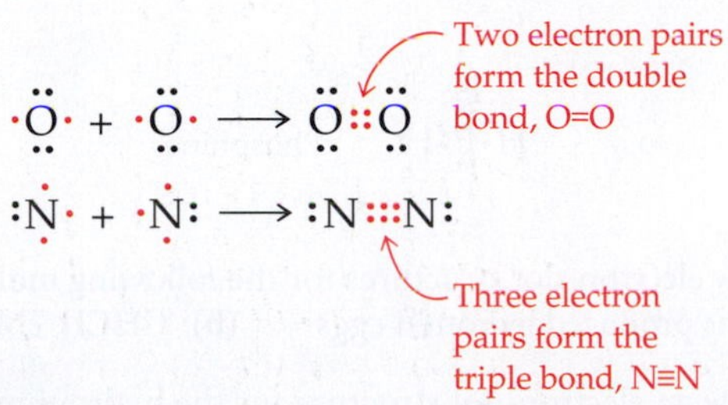

In speaking of molecules with multiple bonds, we often use the term **bond order** to refer to the number of electron pairs shared between atoms. Thus, the F—F single bond in the F_2 molecule has a bond order of 1, the O=O double bond in the O_2 molecule has a bond order of 2, and the N≡N triple bond in the N_2 molecule has a bond order of 3.

Multiple bonds are both shorter and stronger than their corresponding single-bond counterparts because there are more shared electrons holding the atoms together. Compare, for example, the O=O double bond in O_2 with the O–O single bond in H_2O_2 (hydrogen peroxide), and compare the N≡N triple bond in N_2 with the N–N single bond in N_2H_4 (hydrazine):

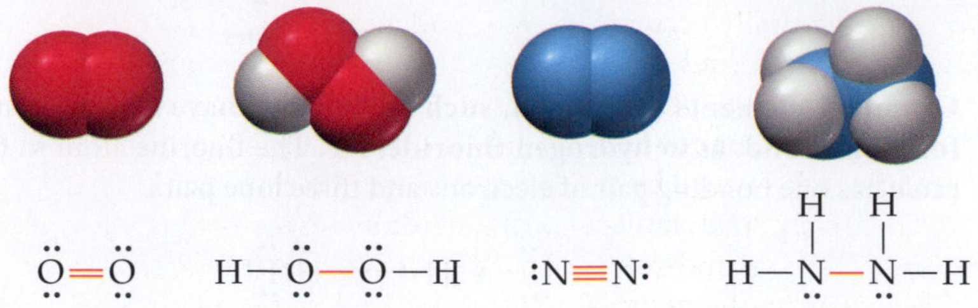

Bond length:	121 pm	148 pm	110 pm	145 pm
Bond strength:	498 kJ/mol	213 kJ/mol	945 kJ/mol	275 kJ/mol

One final point about covalent bonds involves the origin of the bonding electrons. Although most covalent bonds form when two atoms each contribute one electron, bonds can also form when one atom donates both electrons (a lone pair) to another atom that has a vacant valence orbital. The ammonium ion (NH_4^+), for instance, forms when the two lone-pair electrons from the nitrogen atom of ammonia, :NH_3, bond to H^+. Such bonds are sometimes called **coordinate covalent bonds**.

An ordinary covalent bond—each atom donates one electron.

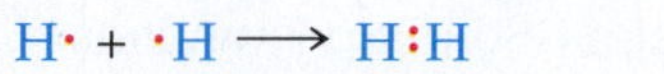

A coordinate covalent bond—the nitrogen atom donates both electrons.

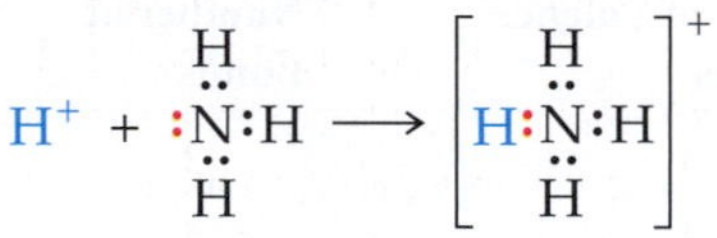

Notice that the nitrogen atom in the ammonium ion (NH_4^+), has more than the usual number of bonds—four instead of three—but that it still has an octet of valence electrons. Nitrogen, oxygen, phosphorus, and sulfur form coordinate covalent bonds frequently.

Worked Example 4.3

Drawing an Electron-Dot Structure

Draw an electron-dot structure for phosphine, PH_3.

STRATEGY

The number of covalent bonds formed by a main-group element depends on the element's group number. Phosphorus, a group 5A element, has five valence electrons and can achieve a valence-shell octet by forming three bonds and leaving one lone pair. Each hydrogen supplies one electron to the P–H bonds.

SOLUTION

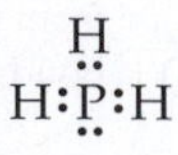

Phosphine

PROBLEM 4.10 Draw electron-dot structures for the following molecules:
(a) H_2S, a poisonous gas produced by rotten eggs **(b)** $CHCl_3$, chloroform

PROBLEM 4.11 Draw an electron-dot structure for the hydronium ion, H_3O^+, and show how a coordinate covalent bond is formed by the reaction of H_2O with H^+.

4.7 ELECTRON-DOT STRUCTURES OF COMPOUNDS CONTAINING ONLY HYDROGEN AND SECOND-ROW ELEMENTS

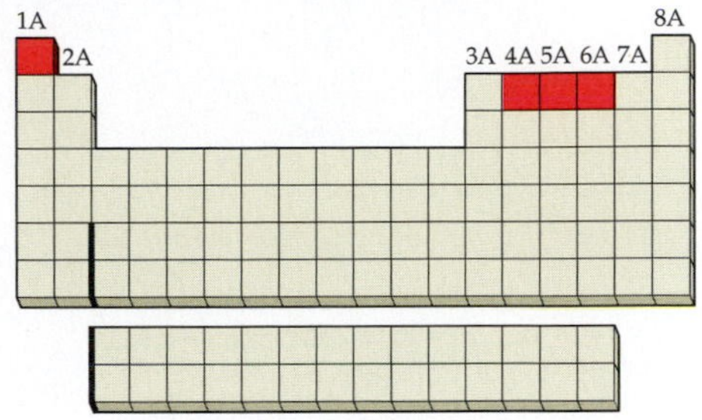

Many of the naturally occurring compounds on which life is based—proteins, fats, carbohydrates, and numerous others—contain only hydrogen and one or more of the second-row elements carbon, nitrogen, and oxygen. Electron-dot structures are relatively easy to draw for such compounds because the octet rule is almost always followed and the number of bonds formed by each element is predictable (Table 4.4).

For relatively small molecules that contain only a few second-row atoms in addition to hydrogen, the second-row atoms are always bonded to one another in a central core, with hydrogens on the periphery. In ethane (C_2H_6), for instance, two carbon atoms, each of which forms four bonds, combine with six hydrogens, each of which forms one bond. Joining the two carbon atoms and adding three hydrogens to each yields only one possible structure:

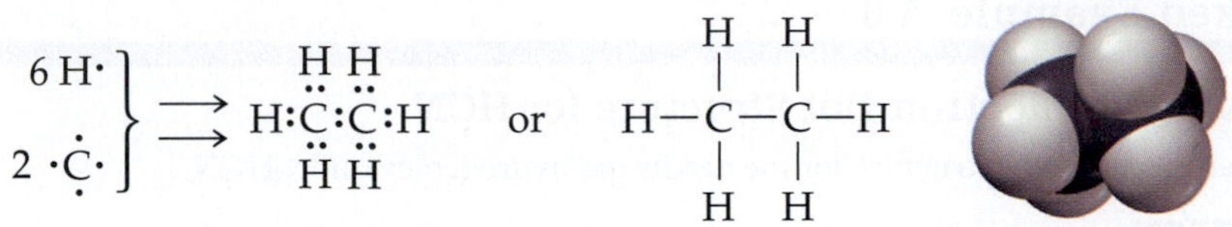

Ethane, C_2H_6

For larger molecules that contain numerous second-row atoms, there is usually more than one possible structure. In such cases, some additional knowledge about the order of connections among atoms is necessary. In drawing structures, we'll follow the usual convention of indicating a two-electron covalent bond by a line. Similarly, we'll use two lines between atoms to represent four shared electrons (two pairs) in a double bond, and three lines to represent six shared electrons (three pairs) in a triple bond. Worked Examples 4.4–4.7 and Problems 4.12–4.14 give you some practice with electron-dot structures that contain hydrogen and second-row atoms.

Worked Example 4.4

Drawing an Electron-Dot Structure for N_2H_4

Draw an electron-dot structure for hydrazine, N_2H_4.

STRATEGY

Nitrogen, a group 5A element, has five valence electrons and forms three bonds, as shown in Table 4.4. Join the two nitrogen atoms, and add two hydrogen atoms to each.

SOLUTION

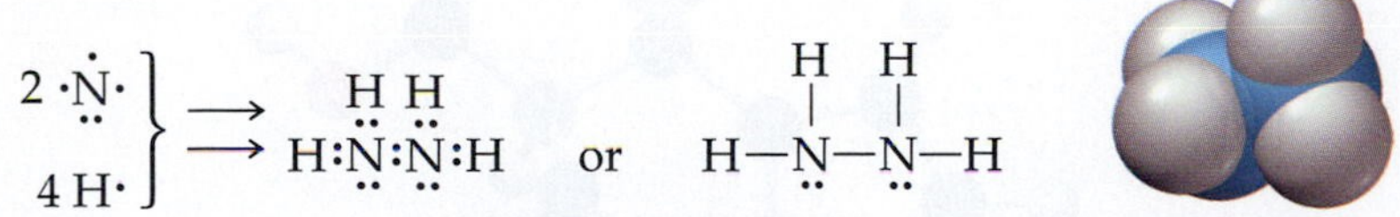

Hydrazine, N_2H_4

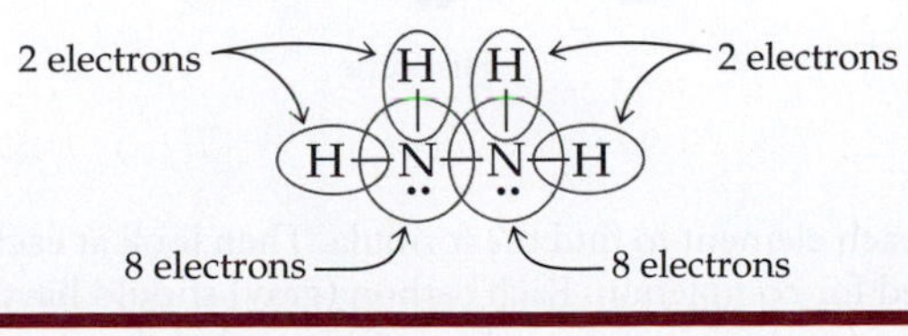

Worked Example 4.5

Drawing an Electron-Dot Structure for CO_2

Draw an electron-dot structure for carbon dioxide, CO_2.

STRATEGY

Connect the atoms so that carbon forms four bonds and each oxygen forms two bonds (Table 4.4). The only possible structure contains two carbon–oxygen double bonds.

SOLUTION

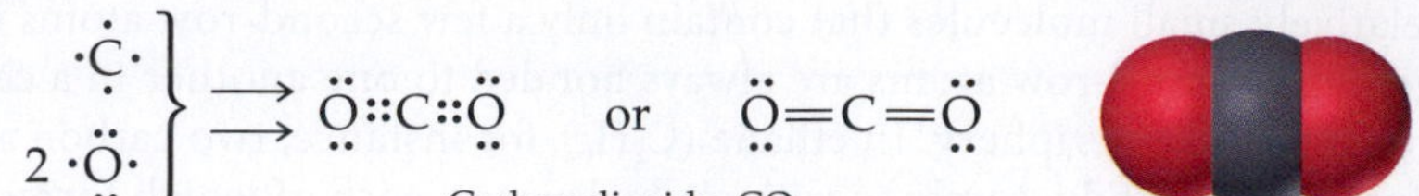

Carbon dioxide, CO_2

Worked Example 4.6

Drawing an Electron-Dot Structure for HCN

Draw an electron-dot structure for the deadly gas hydrogen cyanide, HCN.

STRATEGY

First, connect the carbon and nitrogen atoms. The only way the carbon can form four bonds and the nitrogen can form three bonds is if there is a carbon–nitrogen triple bond.

SOLUTION

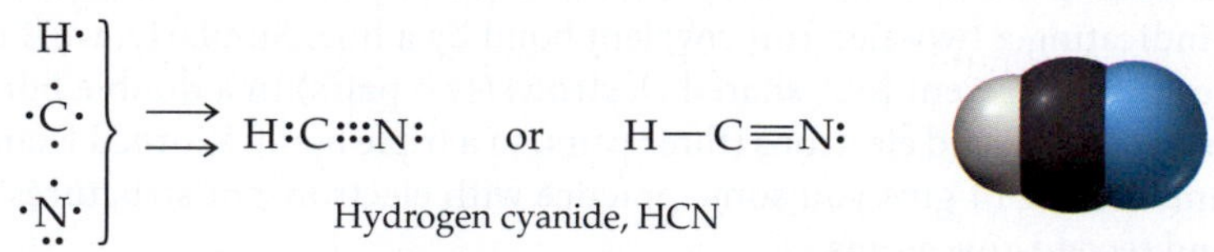

Hydrogen cyanide, HCN

Worked Conceptual Example 4.7

Identifying Multiple Bonds in Molecules

The following structure is a representation of histidine, an amino acid constituent of proteins. Only the connections between atoms are shown; multiple bonds are not indicated. Give the chemical formula of histidine, and complete the structure by showing where the multiple bonds and lone pairs are located (red = O, gray = C, blue = N, ivory = H).

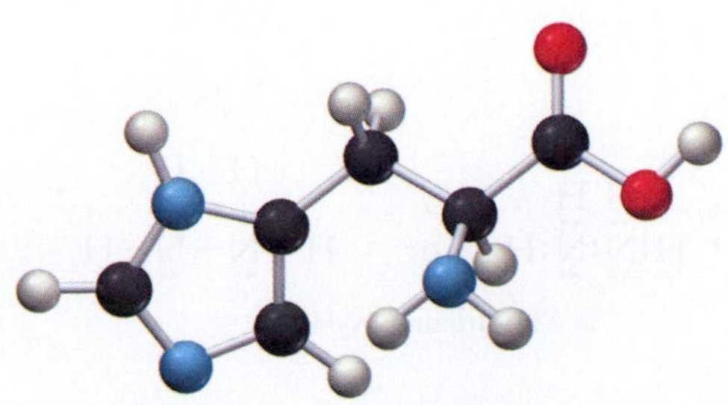

Histidine

STRATEGY

Count the atoms of each element to find the formula. Then look at each atom in the structure to find what is needed for completion. Each carbon (gray) should have four bonds, each oxygen (red) should have two bonds and two lone pairs, and each nitrogen (blue) should have three bonds and one lone pair.

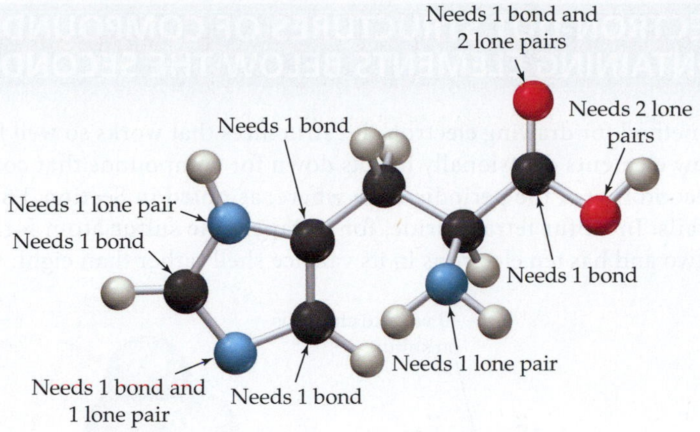

SOLUTION

Histidine has the formula $C_6H_9N_3O_2$.

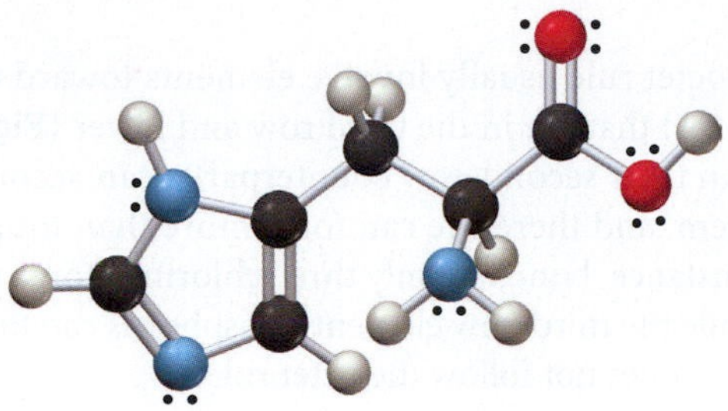

PROBLEM 4.12 Draw electron-dot structures for the following molecules:

(a) Propane, C_3H_8
(b) Hydrogen peroxide, H_2O_2
(c) Methylamine, CH_5N
(d) Ethylene, C_2H_4
(e) Acetylene, C_2H_2
(f) Phosgene, Cl_2CO

PROBLEM 4.13 There are two molecules with the formula C_2H_6O. Draw electron-dot structures for both.

CONCEPTUAL PROBLEM 4.14 The following structure is a representation of cytosine, a constituent of the DNA found in all living cells. Only the connections between atoms are shown; multiple bonds are not indicated. Give the formula of cytosine, and complete the structure by showing where the multiple bonds and lone pairs are located (red = O, gray = C, blue = N, ivory = H).

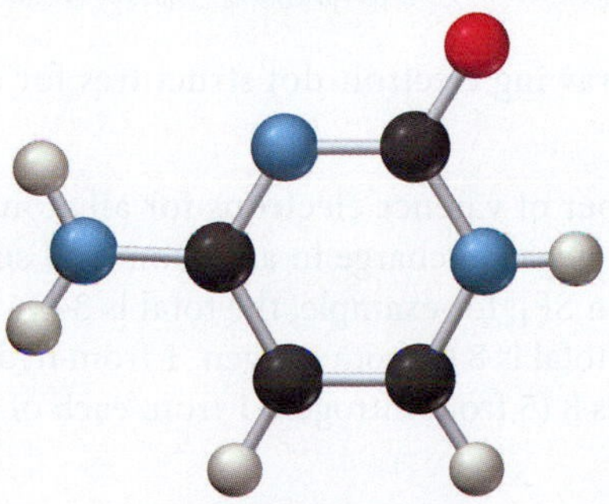

Cytosine

4.8 ELECTRON-DOT STRUCTURES OF COMPOUNDS CONTAINING ELEMENTS BELOW THE SECOND ROW

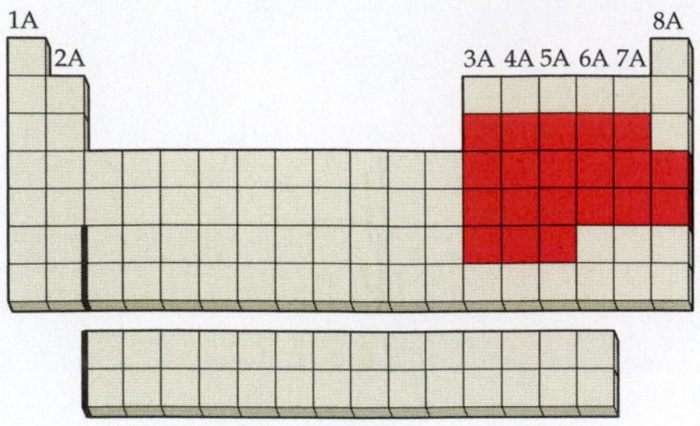

The simple method for drawing electron-dot structures that works so well for compounds of second-row elements occasionally breaks down for compounds that contain elements below the second row of the periodic table where, as noted in Section 3.8, the octet rule sometimes fails. In sulfur tetrafluoride, for instance, the sulfur atom forms four bonds rather than two and has ten electrons in its valence shell rather than eight:

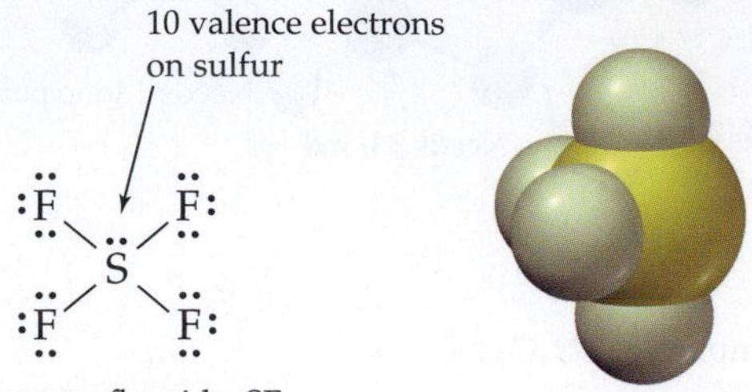

Sulfur tetrafluoride, SF_4

Exceptions to the octet rule usually involve elements toward the right side of the periodic table (groups 3A–8A) that are in the third row and lower (**Figure 4.5**). Atoms of these elements are larger than their second-row counterparts, can accommodate more than four atoms close around them, and therefore can form more than four bonds. The second-row element nitrogen, for instance, bonds to only three chlorine atoms in forming NCl_3 and thus obeys the octet rule, while the third-row element phosphorus can bond to five chlorine atoms in forming PCl_5 and thus does not follow the octet rule.

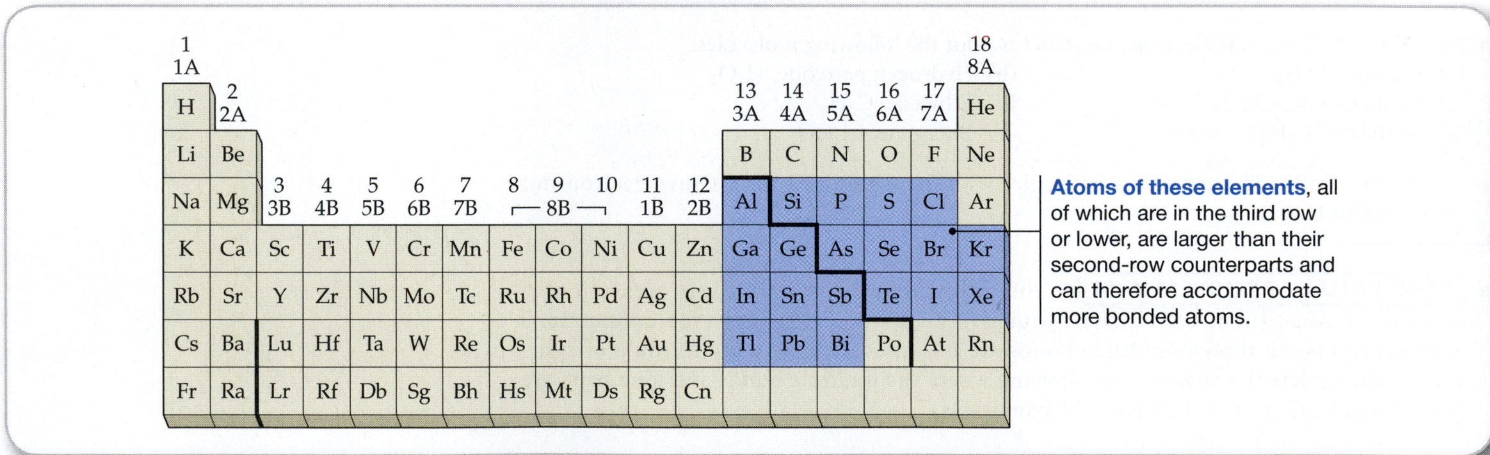

FIGURE 4.5

The octet rule occasionally fails for the main-group elements shown in blue.

A general method of drawing electron-dot structures for any compound is to use the following steps:

Step 1 Find the total number of valence electrons for all atoms in the molecule. Add one additional electron for each negative charge in an anion, and subtract one electron for each positive charge in a cation. In SF_4, for example, the total is 34 (6 from sulfur and 7 from each of 4 fluorines). In OH^-, the total is 8 (6 from oxygen, 1 from hydrogen, and 1 for the negative charge). In NH_4^+, the total is 8 (5 from nitrogen, 1 from each of 4 hydrogens, minus 1 for the positive charge).

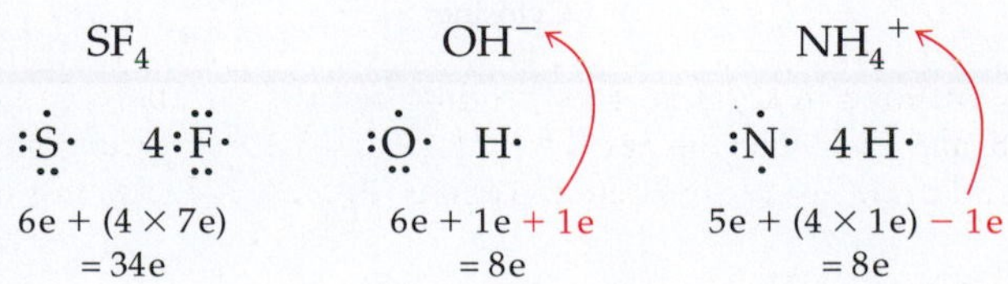

Step 2 Decide what the connections are between atoms, and draw lines to represent the bonds. Often, you'll be told the connections; other times you'll have to guess. Remember that:

- Hydrogen and the halogens usually form only one bond.
- Elements in the second row usually form the number of bonds predictable by their periodic group number and given in Table 4.4.
- Elements in the third row and lower often form more bonds than predicted by the octet rule and frequently occur in compounds as the central atom around which other atoms are grouped.

Also, it's usually the case that the central atom is the least electronegative one (except H). If, for example, you were asked to predict the connections in SF_4, a good guess would be that each fluorine forms one bond to sulfur, which occurs as the central atom and forms more bonds than are predicted by its group number.

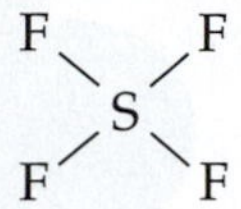

Sulfur tetrafluoride, SF_4

Step 3 Subtract the number of valence electrons used for bonding from the total number calculated in Step 1 to find the number that remain. Assign as many of these remaining electrons as necessary to the terminal atoms (other than hydrogen) so that each has an octet. In SF_4, 8 of the 34 total valence electrons are used in covalent bonding, leaving $34 - 8 = 26$. Twenty-four of the 26 are assigned to the four terminal fluorine atoms to reach an octet configuration for each:

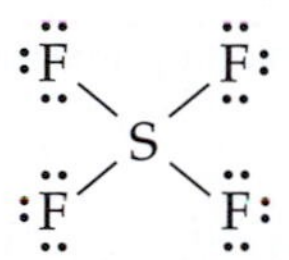

$8 + 24 = 32$ electrons distributed

Step 4 If unassigned electrons remain after Step 3, place them on the central atom. In SF_4, 32 of the 34 electrons have been assigned, leaving the final 2 to be placed on the central S atom:

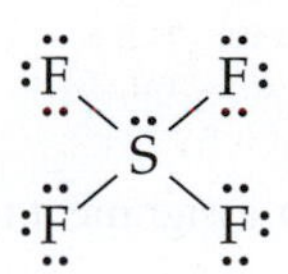

34 electrons distributed

Step 5 If no unassigned electrons remain after Step 3 but the central atom does not yet have an octet, use one or more lone pairs of electrons from a neighboring atom to form a multiple bond (either double or triple). Oxygen, carbon, nitrogen, and sulfur often form multiple bonds. Worked Example 4.9 shows how to deal with such a case.

Worked Example 4.8

Drawing an Electron-Dot Structure for PCl_5

Draw an electron-dot structure for phosphorus pentachloride, PCl_5.

STRATEGY

Follow the steps outlined in the text. First, count the total number of valence electrons. Phosphorus has 5, and each chlorine has 7, for a total of 40. Next, decide on the connections between atoms, and draw lines to indicate the bonds. Because chlorine normally forms only

continued on next page

one bond, it's likely in the case of PCl_5 that all five chlorines are bonded to a central phosphorus atom, placing 10 electrons around phosphorus rather than the usual 8.

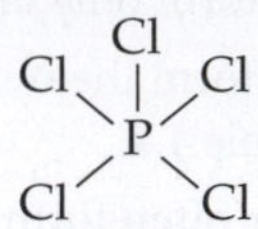

Ten of the 40 valence electrons are necessary for the five P—Cl bonds, leaving 30 to be distributed so that each chlorine has an octet. All 30 remaining valence electrons are used in this step.

SOLUTION

Phosphorus pentachloride, PCl_5

Worked Example 4.9

Drawing an Electron-Dot Structure for CH_2O

Draw an electron-dot structure for formaldehyde, CH_2O, a compound used in manufacturing the adhesives for making plywood and particle board.

STRATEGY

First, count the total number of valence electrons. Carbon has 4, each hydrogen has 1, and the oxygen has 6, for a total of 12. Next, decide on the probable connections between atoms, and draw a line to indicate each bond. In the case of formaldehyde, the less electronegative atom (carbon) is the central atom, and both hydrogens and the oxygen are bonded to carbon:

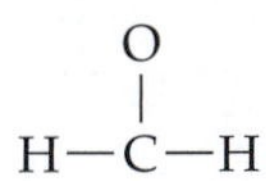

Six of the 12 valence electrons are used for bonds, leaving 6 for assignment to the terminal oxygen atom.

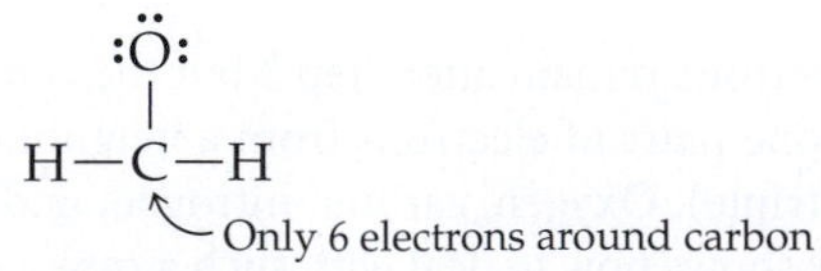

At this point, all the valence electrons are assigned, but the central carbon atom still does not have an octet. We therefore move two of the oxygen electrons from a lone pair into a bonding pair (Step 5), generating a carbon–oxygen double bond and satisfying the octet rule for both oxygen and carbon.

SOLUTION

Formaldehyde, CH_2O

Worked Example 4.10

Drawing an Electron-Dot Structure for an Ion

Draw an electron-dot structure for XeF_5^+, one of the very few noble-gas ions.

STRATEGY

Count the total number of valence electrons. Xenon has 8, each fluorine has 7, and 1 is subtracted to account for the positive charge, giving a total of 42. Then, decide on the probable connections between atoms, and draw a line for each bond. In the case of XeF_5^+, it's likely that the five fluorines are bonded to xenon, a fifth-row atom.

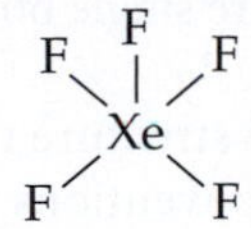

With 10 of the 42 valence electrons used in bonds, distribute as many of the remaining 32 electrons as necessary so that each of the terminal fluorine atoms has an octet. Two electrons still remain, so we assign them to xenon to give the final structure, which has a positive charge. Note that the xenon atom has a total of 12 electrons around it.

SOLUTION

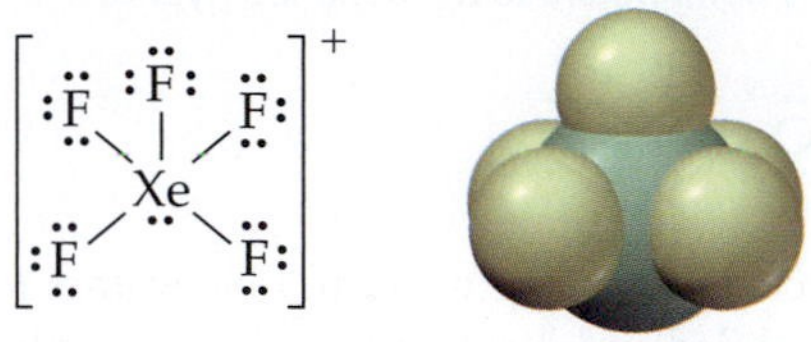

PROBLEM 4.15 Carbon monoxide, CO, is a deadly gas produced by incomplete combustion of fuels. Draw an electron-dot structure for CO.

PROBLEM 4.16 Draw an electron-dot structure for each of the following molecules:
(a) $AlCl_3$ (b) ICl_3 (c) $XeOF_4$ (d) HOBr

PROBLEM 4.17 Draw an electron-dot structure for each of the following ions:
(a) OH^- (b) H_3S^+ (c) HCO_3^- (d) ClO_4^-

4.9 ELECTRON-DOT STRUCTURES AND RESONANCE

The stepwise procedure given in the previous section for drawing electron-dot structures sometimes leads to an interesting problem. Look at ozone, O_3, for instance. Step 1 says that there are 18 valence electrons in the molecule, and Steps 2–4 let us draw the following structure:

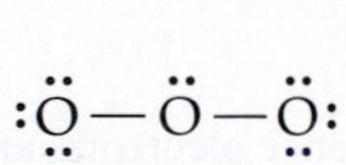

We find at this point that the central atom does not yet have an octet, and we therefore have to move one of the lone pairs of electrons from a terminal oxygen to become a bonding pair and give the central oxygen an octet. But do we take a lone pair from the "right-hand" oxygen or the "left-hand" one? Both possibilities lead to acceptable structures:

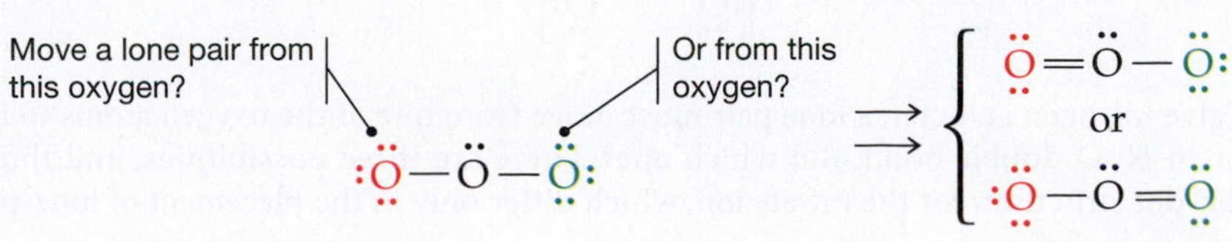

Which of the two structures for O_3 is correct? In fact, neither is correct by itself. Whenever it's possible to write more than one valid electron-dot structure for a molecule, the actual electronic structure is an *average* of the different possibilities, called a **resonance hybrid**. Note that the two individual resonance structures differ only in the placement of the valence-shell electrons (both bonding and nonbonding). The total number of valence electrons remains the same in both structures, the connections between atoms remain the same, and the relative positions of the atoms remain the same.

Ozone doesn't have one O=O double bond and one O–O single bond as the individual structures imply. Rather, ozone has two equivalent O–O bonds that we can think of as having a bond order of 1.5, midway between pure single bonds and pure double bonds. Both bonds have an identical length of 128 pm.

We can't draw a single electron-dot structure that indicates the equivalence of the two O–O bonds in O_3 because the simple conventions we use in drawings to indicate electron placement aren't good enough. Instead, the idea of resonance is indicated by drawing the two (or more) individual electron-dot structures and using a double-headed *resonance arrow* (↔) between them to show that both contribute to the resonance hybrid. *A straight, double-headed arrow always indicates resonance; it is never used for any other purpose.*

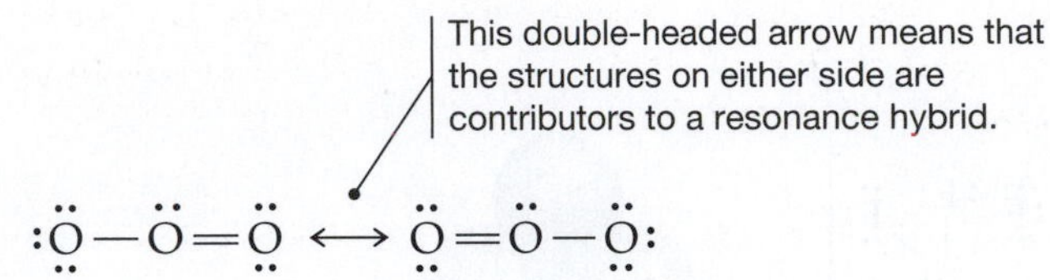

The fact that single electron-dot structures can't be written for all molecules indicates that such structures are oversimplified and don't always give an accurate representation of the electron distribution in a molecule. There's a more accurate way of describing electron distributions called *molecular orbital theory*, which we'll look at in the next chapter. This theory is more complex, however, so chemists still make routine use of simple resonance electron-dot structures.

Worked Example 4.11

Drawing Resonance Structures

The nitrate ion, NO_3^-, has three equivalent oxygen atoms, and its electronic structure is a resonance hybrid of three electron-dot structures. Draw them.

STRATEGY

Begin as you would for drawing any electron-dot structure. There are 24 valence electrons in the nitrate ion: 5 from nitrogen, 6 from each of 3 oxygens, and 1 for the negative charge. The three equivalent oxygens are all bonded to nitrogen, the less electronegative central atom:

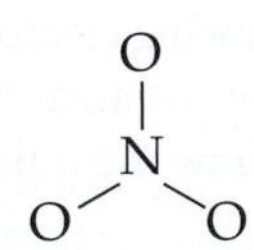

6 of 24 valence electrons assigned

Distributing the remaining 18 valence electrons among the three terminal oxygen atoms completes the octet of each oxygen but leaves nitrogen with only 6 electrons.

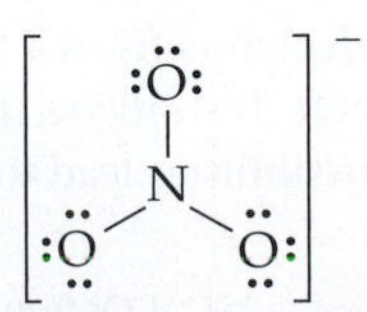

To give nitrogen an octet, a lone pair must move from one of the oxygen atoms to become part of an N–O double bond. But which one? There are three possibilities, and thus three electron-dot structures for the nitrate ion, which differ only in the placement of lone-pair and bonding electrons. The connections between atoms are the same in all three structures, and the atoms have the same positions in all structures.

SOLUTION

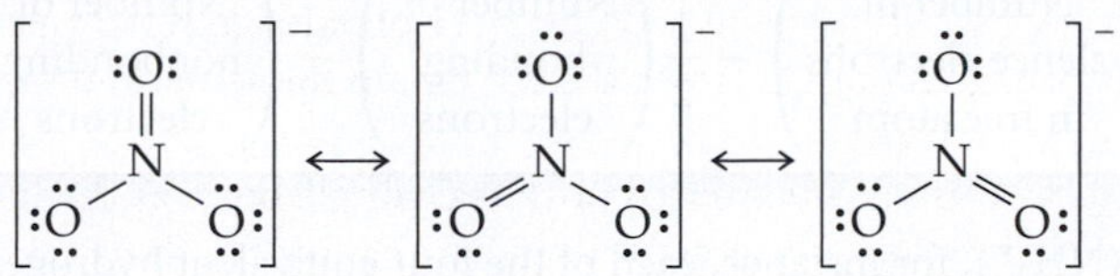

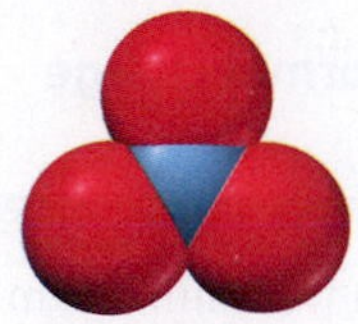

PROBLEM 4.18 Called "laughing gas," nitrous oxide (N_2O) is sometimes used by dentists as an anesthetic. Given the connections N–N–O, draw two resonance structures for N_2O.

PROBLEM 4.19 Draw as many resonance structures as possible for each of the following molecules or ions, giving all atoms (except H) octets:
(a) SO_2 **(b)** CO_3^{2-} **(c)** HCO_2^{-} **(d)** BF_3

CONCEPTUAL PROBLEM 4.20 The following structure shows the connections between atoms for anisole, a compound used in perfumery. Draw two resonance structures for anisole, showing the positions of the multiple bonds in each (red = O, gray = C, ivory = H).

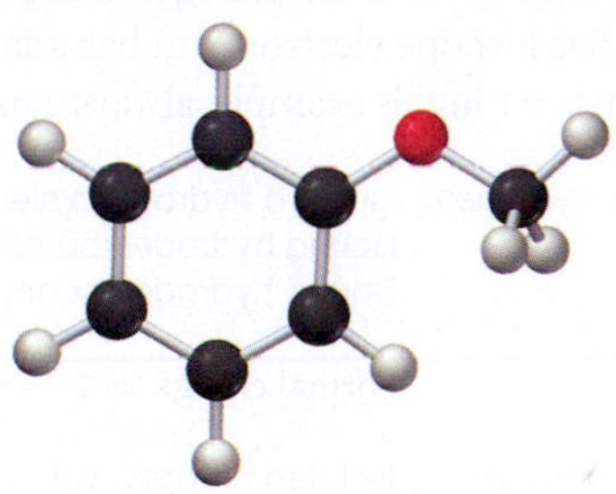

Anisole

4.10 ELECTRON-DOT STRUCTURES AND FORMAL CHARGE

Before ending this discussion of electron-dot structures, let's look at one final matter: the need in some cases to add *formal charges* to specific atoms in structures. **Formal charges** result from a kind of electron "bookkeeping" and can be calculated in the following way:

- Count the number of valence electrons around an atom in a given electron-dot structure, and compare that value with the number of valence electrons in the free atom. If the numbers aren't the same, then the atom in the molecule has either gained or lost electrons and thus has a formal charge.
- If the atom in a molecule has more electrons than the isolated atom, it has a negative formal charge; if it has fewer electrons than the isolated atom, it has a positive formal charge.

$$\textbf{Formal charge} = \begin{pmatrix}\text{Number of}\\ \text{valence electrons}\\ \text{in free atom}\end{pmatrix} - \begin{pmatrix}\text{Number of}\\ \text{valence electrons}\\ \text{in bonded atom}\end{pmatrix}$$

In counting the number of valence electrons in a bonded atom, it's necessary to make a distinction between unshared, nonbonding electrons and shared, bonding electrons. For bookkeeping purposes, an atom can be thought of as "owning" all its nonbonding electrons but only half of its bonding electrons, because the bonding electrons are shared with another atom. Thus, we can rewrite the definition of formal charge as

$$\blacktriangleright\ \textbf{Formal charge} = \begin{pmatrix}\text{Number of}\\ \text{valence electrons}\\ \text{in free atom}\end{pmatrix} - \frac{1}{2}\begin{pmatrix}\text{Number of}\\ \text{bonding}\\ \text{electrons}\end{pmatrix} - \begin{pmatrix}\text{Number of}\\ \text{nonbonding}\\ \text{electrons}\end{pmatrix}$$

In the ammonium ion (NH_4^+), for instance, each of the four equivalent hydrogen atoms has 2 valence electrons in its covalent bond to nitrogen and the nitrogen atom has 8 valence electrons, 2 from each of its four N–H bonds:

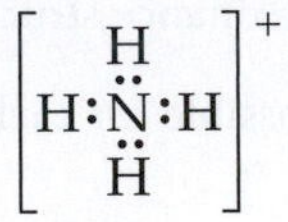

Ammonium ion
8 valence electrons around nitrogen
2 valence electrons around each hydrogen

For bookkeeping purposes, each hydrogen owns half of its 2 shared bonding electrons, or 1, while the nitrogen atom owns half of its 8 shared bonding electrons, or 4. Because an isolated hydrogen atom has 1 electron and the hydrogens in the ammonium ion each still own 1 electron, they have neither gained nor lost electrons and thus have no formal charge. An isolated nitrogen atom, however, has 5 valence electrons, while the nitrogen atom in NH_4^+ owns only 4. Thus, the nitrogen atom in the NH_4^+ ion has lost one electron and has a formal charge of +1. The sum of the formal charges on all the atoms, +1 in this example, always equals the overall charge on the ion.

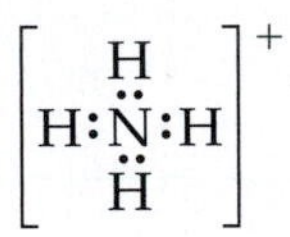

For hydrogen:	Isolated hydrogen valence electrons	1
	Bound hydrogen bonding electrons	2
	Bound hydrogen nonbonding electrons	0
	Formal charge $= 1 - \frac{1}{2}(2) - 0 = 0$	
For nitrogen:	Isolated nitrogen valence electrons	5
	Bound nitrogen bonding electrons	8
	Bound nitrogen nonbonding electrons	0
	Formal charge $= 5 - \frac{1}{2}(8) - 0 = +1$	

Formal charges have two important uses. The first is their application to organic chemistry, where the reactive parts of organic molecules can often be predicted from a knowledge of formal charges. We won't pursue the subject in this book, but you'll certainly use formal charges again if you continue in your chemical studies.

The second use of formal charges is their application to the resonance structures described in the previous section. It often happens with resonance hybrids that the two or more structures making up the hybrid are not equivalent. One of the structures may be "better" than the others, meaning that it approximates the actual electronic structure of the substance more closely than the others. The resonance hybrid in such cases is thus weighted more strongly toward the more favorable structure.

Take the organic substance called acetamide (C_2H_5NO), for instance, a compound related to proteins. We can write two valid structures for acetamide, both of which fulfill the octet rule for the C, N, and O atoms. One of the two structures has no formal charges, while the other has formal charges on the O and N atoms. (Check for yourself that the formal charges are correct.)

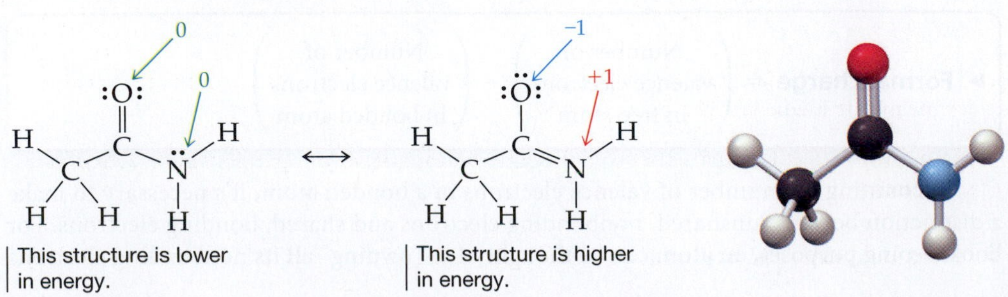

Acetamide

Which of the two nonequivalent structures is likely to give a more accurate representation of the molecule? Because energy is needed to separate + and – charges, the structure with formal charges is probably higher in energy than the structure without formal charges. Thus, the actual electronic structure of acetamide is closer to that of the more favorable, lower-energy structure.

As another example, both resonance structures for N_2O have formal charges, but the structure that places the formal negative charge on the more electronegative oxygen atom rather than on the less electronegative nitrogen atom is probably a more accurate representation of the molecule.

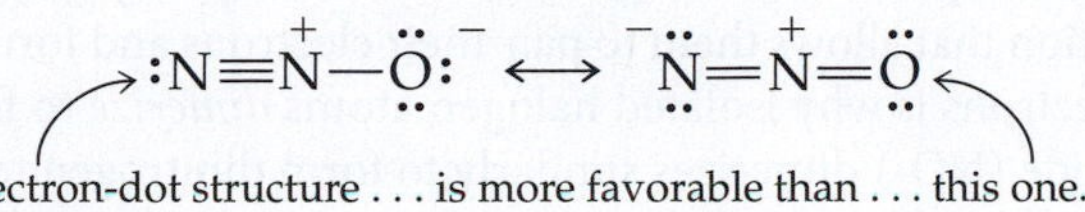

Worked Example 4.12

Calculating Formal Charges

Calculate the formal charge on each atom in the following electron-dot structure for SO_2:

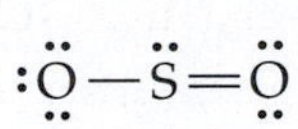

STRATEGY

Find the number of valence electrons on each atom (its periodic group number). Then subtract half the number of the atom's bonding electrons and all of its nonbonding electrons.

SOLUTION

For sulfur:	Isolated sulfur valence electrons	6
	Bound sulfur bonding electrons	6
	Bound sulfur nonbonding electrons	2
	Formal charge $= 6 - \frac{1}{2}(6) - 2 = +1$	
For singly bonded oxygen:	Isolated oxygen valence electrons	6
	Bound oxygen bonding electrons	2
	Bound oxygen nonbonding electrons	6
	Formal charge $= 6 - \frac{1}{2}(2) - 6 = -1$	
For doubly bonded oxygen:	Isolated oxygen valence electrons	6
	Bound oxygen bonding electrons	4
	Bound oxygen nonbonding electrons	4
	Formal charge $= 6 - \frac{1}{2}(4) - 4 = 0$	

The sulfur atom of SO_2 has a formal charge of +1, and the singly bonded oxygen atom has a formal charge of −1. We might therefore write the structure for SO_2 as

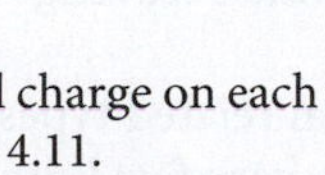

PROBLEM 4.21 Calculate the formal charge on each atom in the three resonance structures for the nitrate ion in Worked Example 4.11.

PROBLEM 4.22 Calculate the formal charge on each atom in the following electron-dot structures:

(a) Cyanate ion: 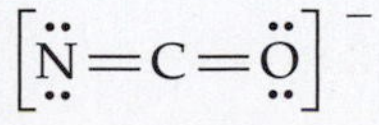 **(b)** Ozone: 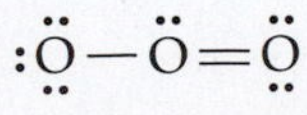

FYI FREE RADICALS AND YOUR HEALTH

▲ The skin of these sun worshipers is being protected by a free radical called melanin that absorbs the Sun's ultraviolet rays.

Except for hydrogen, almost all the atoms in almost all the 60 million known chemical compounds follow the octet rule. As a result, most compounds have an even number of electrons, which are paired up in filled orbitals. A very few substances, however, have an odd number of electrons, meaning that at least one of their electrons must be unpaired in a half-filled orbital. Such substances are called **radicals** (or sometimes *free radicals*).

Because of their unpaired electron, radicals are usually very reactive, rapidly undergoing some kind of reaction that allows them to pair their electrons and form a more stable product. This pairing of electrons is why isolated halogen atoms *dimerize* to form halogen molecules, and nitrogen dioxide (NO_2) dimerizes similarly to form dinitrogen tetroxide (N_2O_4).

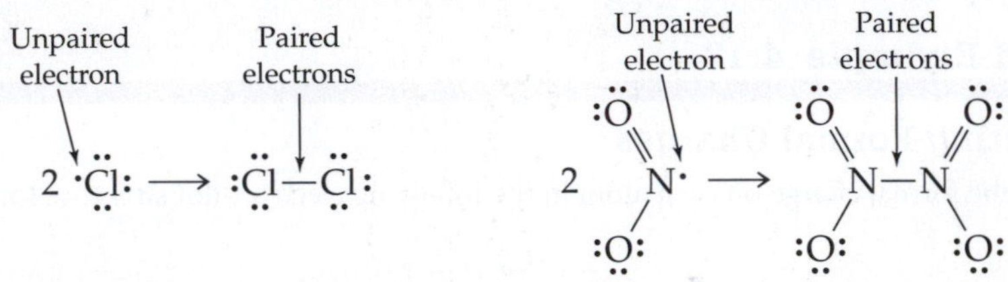

Despite their high reactivity, small amounts of radicals are involved in many common processes, including the combustion of fuels, the industrial preparation of polyethylene and other polymers, and the ongoing destruction of the Earth's ozone layer in the upper atmosphere by chlorofluorocarbons. Perhaps more surprisingly, radicals are also present in the human body, where they both mediate important life processes and act as causal factors for many diseases, including cancer, emphysema, stroke, and diabetes. Even the changes that take place as a result of normal aging are thought to involve radicals. Perhaps the most important radicals in the human body are nitric oxide (NO), superoxide ion (O_2^-), and the hydroxyl radical (OH), and melanin.

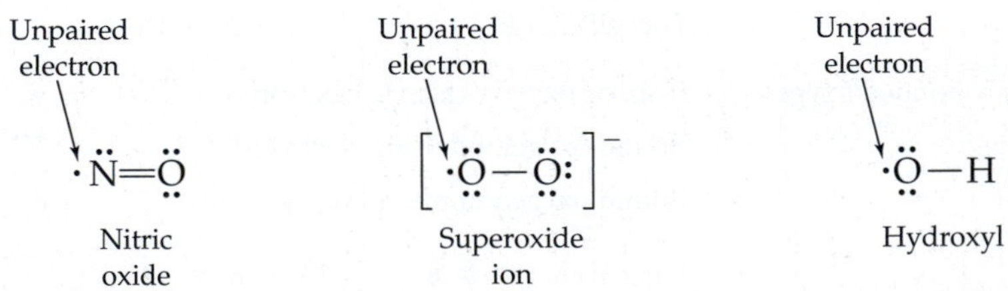

Nitric oxide, also known as an endothelium-derived relaxing factor (EDRF), is a signaling molecule that is released in the human body by cells lining the interior surface of blood vessels. This release of NO causes the muscles surrounding the vessels to relax, thereby dilating the blood vessels and increasing blood flow. Hair growth, pulmonary hypertension, and angina among other conditions are treated by drugs that affect NO release.

Superoxide ion and hydroxyl are oxygen-containing radicals that contribute both to the development of many diseases, including stroke and heart attack, and to normal human aging. The exact mechanisms by which these radicals act is not known, but they appear to damage DNA and other body chemicals by *oxidation* reactions that form carbon–oxygen bonds. Their effects are partially counteracted by substances called *antioxidants*, particularly vitamin A, vitamin C, and vitamin E, which scavenge superoxide ion and remove it from the body.

Melanin, of which there are several related types, is the biological pigment responsible for skin and hair color. It is one of the very few known stable radicals and functions in the body to absorb harmful ultraviolet rays, thereby protecting the skin from sun damage that might otherwise cause cancer. The melanin molecule is large and has a great many resonance structures. Only a part of the structure is shown:

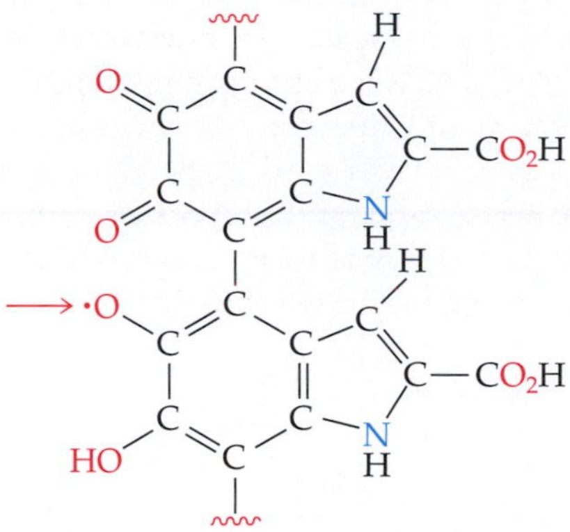
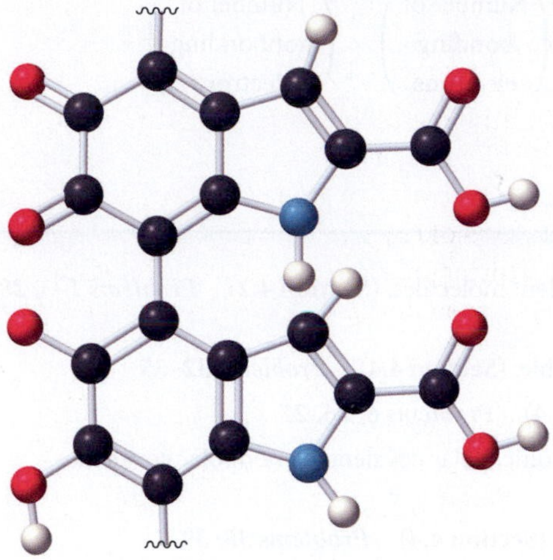

Eumelanin

PROBLEM 4.23 What is a radical, and why are they so reactive?

PROBLEM 4.24 What are the most common radicals in the human body, and what biological effects do they have?

PROBLEM 4.25 Draw an electron-dot structure for the ethyl radical, C_2H_5, which is involved in the synthesis of polyethylene polymer.

SUMMARY

A **covalent bond** results from the sharing of electrons between atoms. Every covalent bond has a specific **bond length** that leads to optimum stability and a specific **bond dissociation energy** that describes the strength of the bond. Energy is released when a bond is formed, and energy is absorbed when a bond is broken.

As a general rule, a main-group atom shares as many of its valence-shell electrons as possible, either until it has no more to share or until it reaches an octet. Atoms in the third and lower rows of the periodic table can accommodate more than the number of bonds predicted by the octet rule.

In a bond between dissimilar atoms, such as that in HCl, one atom often attracts the bonding electrons more strongly than the other, giving rise to a **polar covalent bond**. Bond polarity is due to differences in **electronegativity**, the ability of an atom in a molecule to attract shared electrons. Electronegativity increases from left to right across a row and generally decreases from top to bottom in a group of the periodic table.

An electron-dot structure represents an atom's valence electrons by dots and shows the two electrons in a **single bond** as a pair of dots or a single line shared between atoms. In the same way, a **double bond** is represented as four dots or two lines between atoms, and a **triple bond** is represented as six dots or three lines between atoms. Occasionally, a molecule can be represented by more than one electron-dot structure. In such cases, no single structure is adequate by itself and the actual electronic structure of the molecule is a **resonance hybrid** of the different individual structures. The relative importance of different resonance structures to the overall hybrid structure can be estimated by calculating **formal charges**, which tell whether atoms in the molecule have either gained or lost electrons.

KEY WORDS

bond dissociation energy (*D*) *141*
bond length *138*
bond order *147*
bonding pair *146*
coordinate covalent bond *148*
double bond *147*
electron-dot structure *145*
electronegativity (EN) *143*
formal charge *157*
lone pair *146*
polar covalent bond *141*
radical *160*
resonance hybrid *156*
single bond *147*
structural formula *139*
triple bond *147*

KEY EQUATION

- **Formal charge (Section 4.10)**

$$\text{Formal charge} = \begin{pmatrix}\text{Number of}\\ \text{valence electrons}\\ \text{in free atom}\end{pmatrix} - \frac{1}{2}\begin{pmatrix}\text{Number of}\\ \text{bonding}\\ \text{electrons}\end{pmatrix} - \begin{pmatrix}\text{Number of}\\ \text{nonbonding}\\ \text{electrons}\end{pmatrix}$$

LEARNING OUTCOMES

- Draw and interpret the structural formulas of covalent molecules. (Section 4.2) ***Problems 1–3, 28–31, 54, 55, 58–61, 79, 80***
- Describe electronegativity trends in the periodic table. (Section 4.4) ***Problems 32–35***
- Interpret an electrostatic potential map. (Section 4.4) ***Problems 6, 26, 27***
- Use electronegativity to predict whether a bond is ionic, polar covalent, or nonpolar covalent. (Section 4.4) ***Problems 4–6, 26, 36–43, 81, 84***
- Use the $\delta-/\delta+$ convention to show bond polarity. (Section 4.4) ***Problems 38, 39***
- Name a binary molecular compound, given the formula. (Section 4.5) ***Problems 7–9, 46–49, 81, 84***
- Draw electron-dot structures for polyatomic molecules and ions. (Sections 4.6–4.8) ***Problems 10–17, 52–59, 66, 72, 73, 75, 84***
- Identify the positions of multiple bonds in molecules. (Section 4.7) ***Problems 14, 28–31, 54, 55, 58–61, 79, 80***
- Draw resonance structures for molecules and polyatomic ions. (Section 4.9) ***Problems 18–20, 62–65, 68–73, 76, 77, 83***
- Calculate the formal charge on an atom in a molecule or ion. (Section 4.10) ***Problems 21, 22, 66–77***
- Use formal charges to evaluate resonance structures. (Section 4.10) ***Problems 70, 71, 73, 75***

CONCEPTUAL PROBLEMS

Problems 4.1–4.25 appear within the chapter.

4.26 Two electrostatic potential maps are shown, one of methyllithium (CH_3Li) and the other of chloromethane (CH_3Cl). Based on their polarity patterns, which do you think is which?

(a)

(b)

4.27 Electrostatic potential maps of acetaldehyde (C_2H_4O), ethane (C_2H_6), ethanol (C_2H_6O), and fluoroethane (C_2H_5F) are shown. Which do you think is which?

(a)

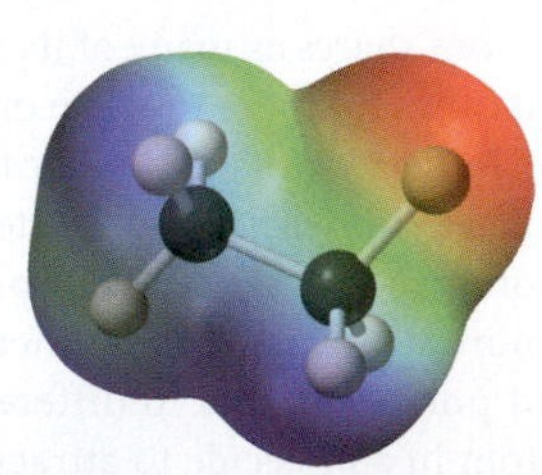

(b)

(c)

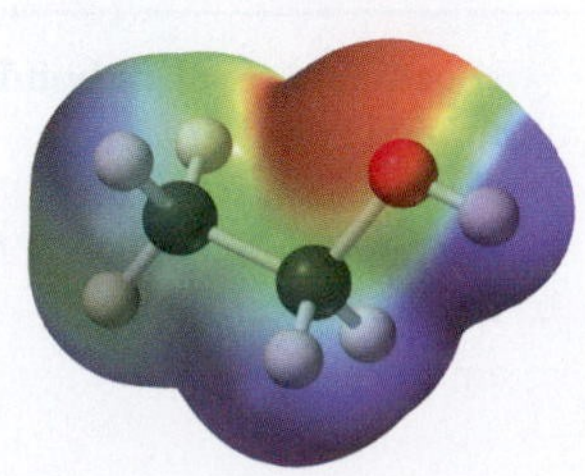

(d)

4.28 The following ball-and-stick molecular model is a representation of acetaminophen, the active ingredient in such over-the-counter pain remedies as Tylenol (red = O, gray = C, blue = N, ivory = H). Give the formula of acetaminophen, and indicate the positions of the multiple bonds.

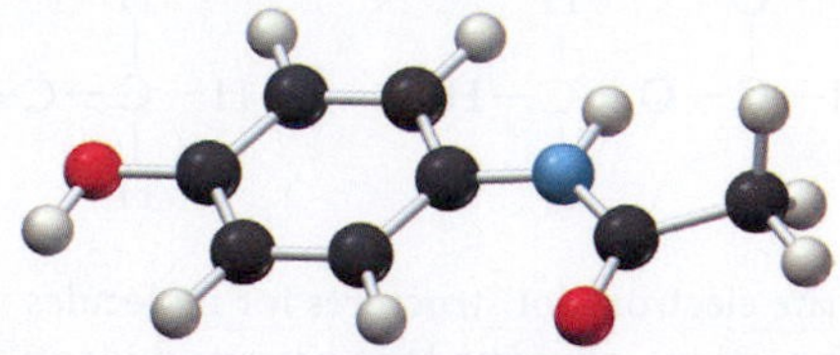

Acetaminophen

4.29 The following ball-and-stick molecular model is a representation of thalidomide, a drug that causes birth defects when taken by expectant mothers but is valuable for its use against leprosy. The lines indicate only the connections between atoms, not whether the bonds are single, double, or triple (red = O, gray = C, blue = N, ivory = H). Give the formula of thalidomide, and indicate the positions of the multiple bonds.

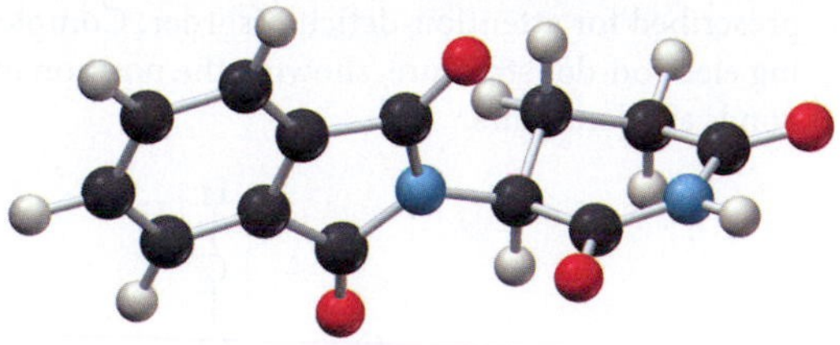

Thalidomide

4.30 Sinapaldehyde, a compound present in the toasted wood used for aging wine, has the following connections among atoms. Complete the following electron-dot structure for sinapaldehyde, and identify any multiple bonds in the molecule:

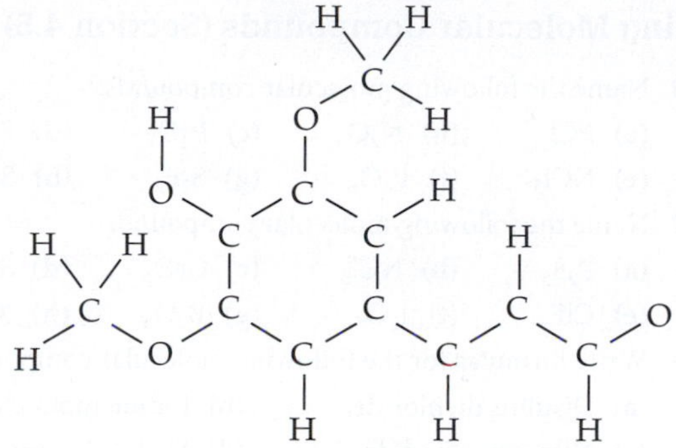

Sinapaldehyde

4.31 Vitamin C (ascorbic acid) has the following connections among atoms. Complete the electron-dot structure for vitamin C, showing lone pairs and identifying any multiple bonds:

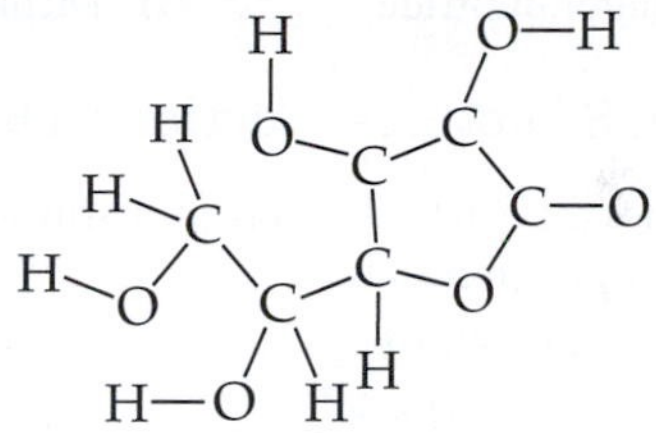

Vitamin C

SECTION PROBLEMS

Electronegativity and Polar Covalent Bonds (Section 4.4)

4.32 What general trends in electronegativity occur in the periodic table?

4.33 Predict the electronegativity of the undiscovered element with $Z = 119$.

4.34 Order the following elements according to increasing electronegativity: Li, Br, Pb, K, Mg, C.

4.35 Order the following elements according to decreasing electronegativity: C, Ca, Cs, Cl, Cu.

4.36 Which of the following substances are largely ionic, and which are covalent?

(a) HF (b) HI (c) $PdCl_2$
(d) BBr_3 (e) NaOH (f) CH_3Li

4.37 Use the electronegativity data in Figure 4.4 to predict which bond in each of the following pairs is more polar:

(a) C–H or C–Cl (b) Si–Li or Si–Cl
(c) N–Cl or N–Mg

4.38 Show the direction of polarity for each of the bonds in Problem 4.37, using the $\delta+/\delta-$ notation.

4.39 Show the direction of polarity for each of the covalent bonds in Problem 4.36, using the $\delta+/\delta-$ notation.

4.40 Which of the substances $CdBr_2$, P_4, BrF_3, MgO, NF_3, $BaCl_2$, $POCl_3$, and LiBr are:

(a) Largely ionic?
(b) Nonpolar covalent?
(c) Polar covalent?

4.41 Which of the substances S_8, $CaCl_2$, $SOCl_2$, NaF, CBr_4, BrCl, LiF, and AsH_3 are:

(a) Largely ionic?
(b) Nonpolar covalent?
(c) Polar covalent?

4.42 Using only the elements P, Cl, and Ca, give formulas for the following:

(a) An ionic compound
(b) A molecular compound with polar covalent bonds that obeys the octet rule and has no formal charges

4.43 Using only the elements Ca, Cl, Si, give formulas for the following:

(a) An ionic compound
(b) A molecular compound with polar covalent bonds that obeys the octet rule and has no formal charges

4.44 Which compound do you expect to have the stronger N–N bond, N_2H_2 or N_2H_4? Explain.

4.45 Which compound do you expect to have the stronger N–O bond, NO or NO_2? Explain.

Naming Molecular Compounds (Section 4.5)

4.46 Name the following molecular compounds:

(a) PCl_3 **(b)** N_2O_3 **(c)** P_4O_7 **(d)** BrF_3
(e) NCl_3 **(f)** P_4O_6 **(g)** S_2F_2 **(h)** SeO_2

4.47 Name the following molecular compounds:

(a) P_2S_5 **(b)** NCl_3 **(c)** CrO_3 **(d)** SiO_2
(e) ClF_3 **(f)** I_2O_5 **(g)** B_2O_3 **(h)** XeO_3

4.48 Write formulas for the following molecular compounds:

(a) Disulfur dichloride **(b)** Iodine monochloride
(c) Nitrogen triiodide **(d)** Dichlorine monoxide
(e) Chlorine trioxide **(f)** Tetrasulfur tetranitride

4.49 Write formulas for the following molecular compounds:

(a) Xenon hexafluoride **(b)** Carbon disulfide
(c) Silicon tetrachloride **(d)** Tetraiodine nonoxide
(e) Disulfur dinitride **(f)** Dichlorine hexoxide

Electron-Dot Structures (Sections 4.6–4.8)

4.50 Why does the octet rule apply primarily to main-group elements, not to transition metals?

4.51 Which of the following substances contains an atom that does not follow the octet rule?

(a) $AlCl_3$ **(b)** PCl_3 **(c)** PCl_5 **(d)** $SiCl_4$

4.52 Draw electron-dot structures for the following molecules or ions:

(a) CBr_4 **(b)** NCl_3 **(c)** C_2H_5Cl
(d) BF_4^- **(e)** O_2^{2-} **(f)** NO^+

4.53 Draw electron-dot structures for the following molecules, which contain atoms from the third row or lower:

(a) $SbCl_3$ **(b)** KrF_2 **(c)** ClO_2
(d) PF_5 **(e)** H_3PO_4 **(f)** $SeOCl_2$

4.54 Oxalic acid, $H_2C_2O_4$, is a mildly poisonous substance found in the leaves of rhubarb, spinach, and many other plants. (You'd have to eat about 15 lb or so of dry spinach leaves to ingest a lethal amount.) If oxalic acid has a C–C single bond and no C–H bond, draw its electron-dot structure showing lone pairs and identifying any multiple bonds.

4.55 Draw an electron-dot structure for carbon disulfide, CS_2, showing lone pairs and identifying any multiple bonds.

4.56 Identify the third-row elements, **X**, that form the following ions:

(a)

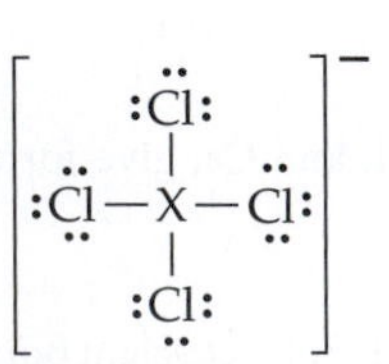

(b)

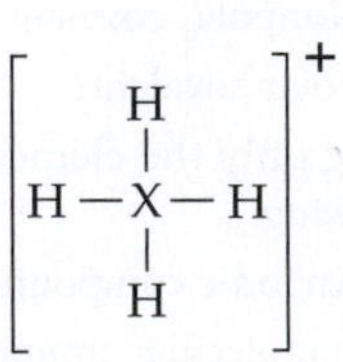

4.57 Identify the fourth-row elements, **X**, that form the following compounds:

(a)

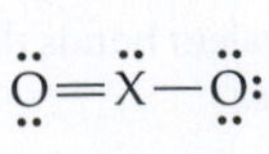

(b)

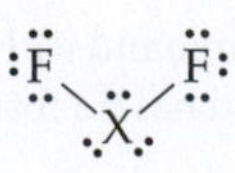

4.58 Write electron-dot structures for molecules with the following connections, showing lone pairs and identifying any multiple bonds:

(a)

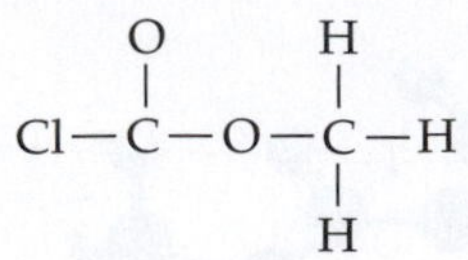

(b)

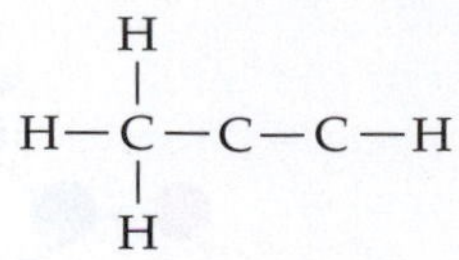

4.59 Write electron-dot structures for molecules with the following connections, showing lone pairs and identifying any multiple bonds:

(a)

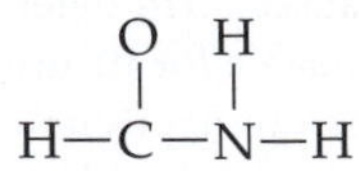

(b)

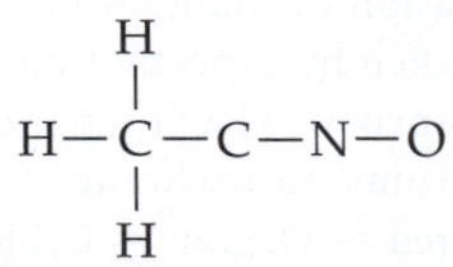

4.60 Methylphenidate ($C_{14}H_{19}NO_2$), marketed as Ritalin, is often prescribed for attention-deficit disorder. Complete the following electron-dot structure, showing the position of any multiple bonds and lone pairs.

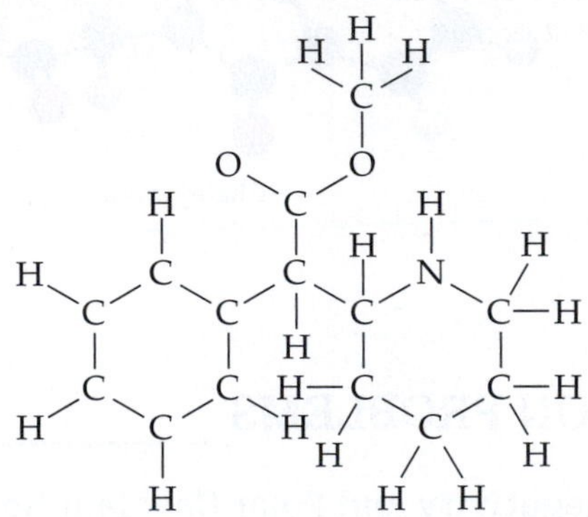

Methylphenidate
(Ritalin)

4.61 Pregabalin ($C_8H_{17}NO_2$), marketed as Lyrica, is an anticonvulsant drug prescribed for treatment of seizures. Complete the following electron-dot structure for pregabalin by adding hydrogen atoms where needed.

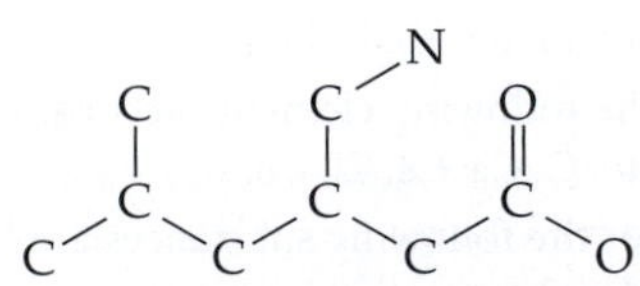

Pregabalin
(Lyrica)

Resonance and Formal Charge (Sections 4.9, 4.10)

4.62 Draw as many resonance structures as you can that obey the octet rule for each of the following molecules or ions:

(a) HN_3 **(b)** SO_3 **(c)** SCN^-

4.63 Draw as many resonance structures as you can for the following nitrogen-containing compounds. Not all will obey the octet rule.

(a) N_2O **(b)** NO **(c)** NO_2 **(d)** $N_2O_3(ONNO_2)$

4.64 Which of the following pairs of structures represent resonance forms, and which do not?

(a) 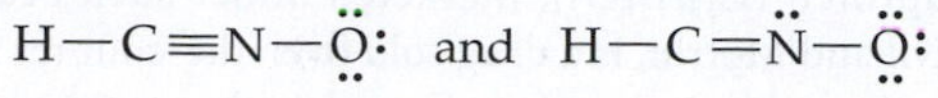

(b) 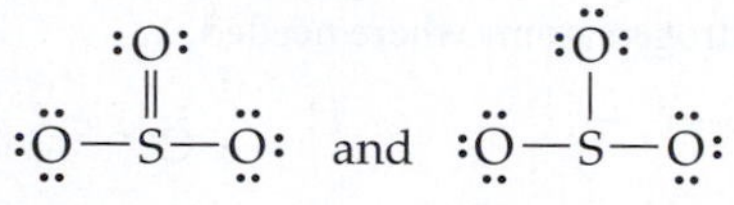

(c) 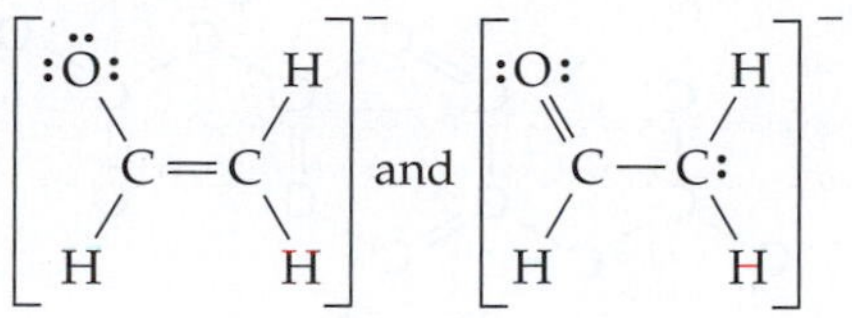

(d) 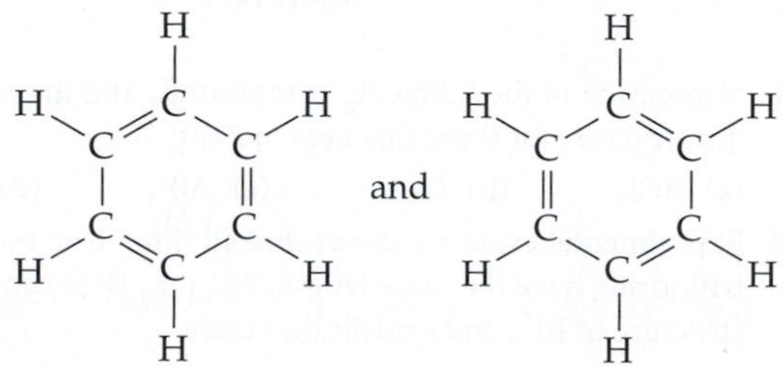

4.65 Which of the following pairs of structures represent resonance forms, and which do not?

(a) 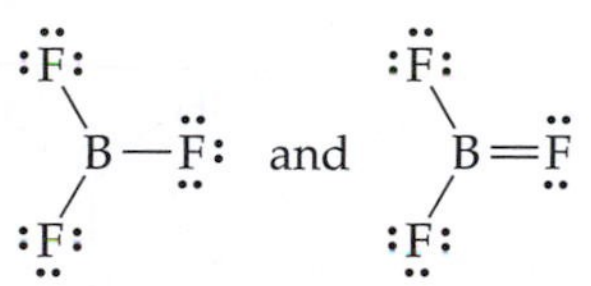

(b) 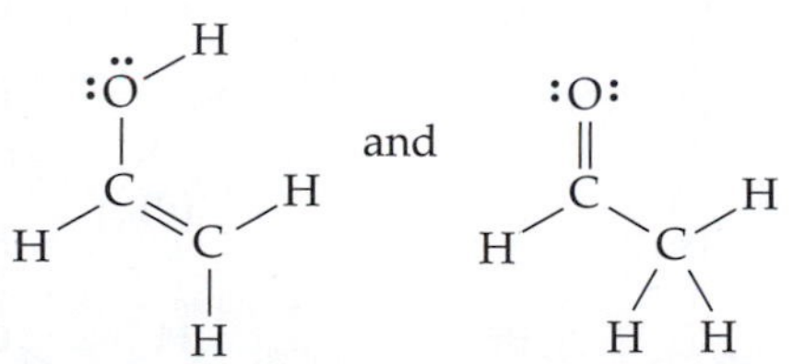

(c) 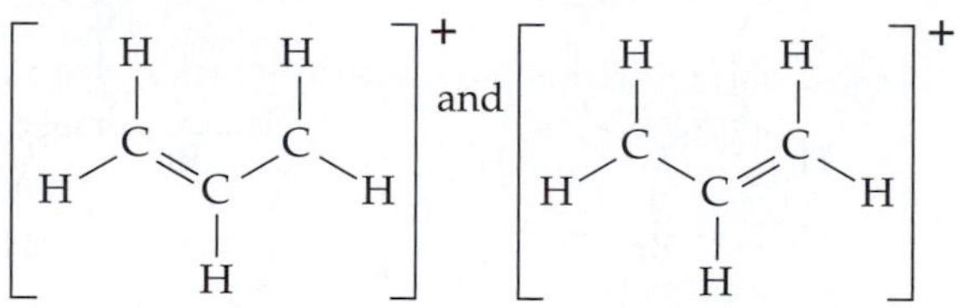

4.66 Draw an electron-dot structure for carbon monoxide, CO, and assign formal charges to both atoms.

4.67 Assign formal charges to the atoms in the following structures:

(a) 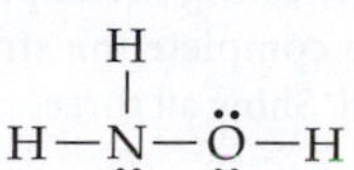

(b) 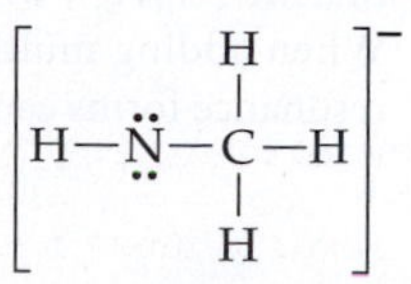

(c) 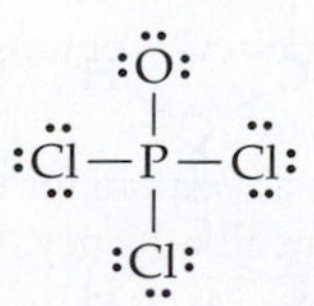

4.68 Assign formal charges to the atoms in the following resonance forms of ClO_2^-:

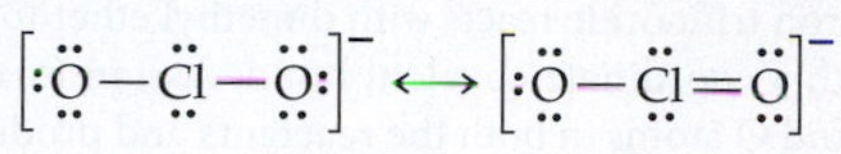

4.69 Assign formal charges to the atoms in the following resonance forms of H_2SO_3:

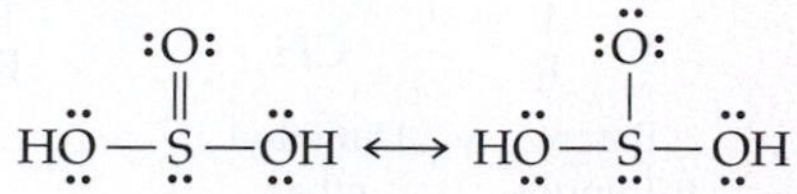

4.70 Assign formal charges to the atoms in the following structures. Which of the two do you think is the more important contributor to the resonance hybrid?

(a) 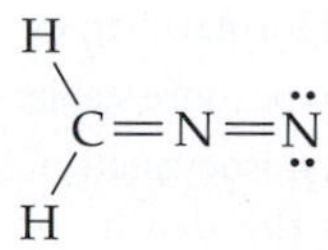

(b) 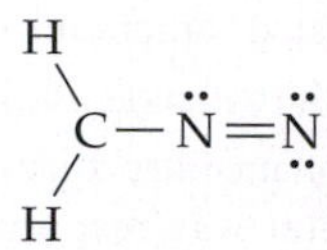

4.71 Calculate formal charges for the C and O atoms in the following two resonance structures. Which structure do you think is the more important contributor to the resonance hybrid? Explain.

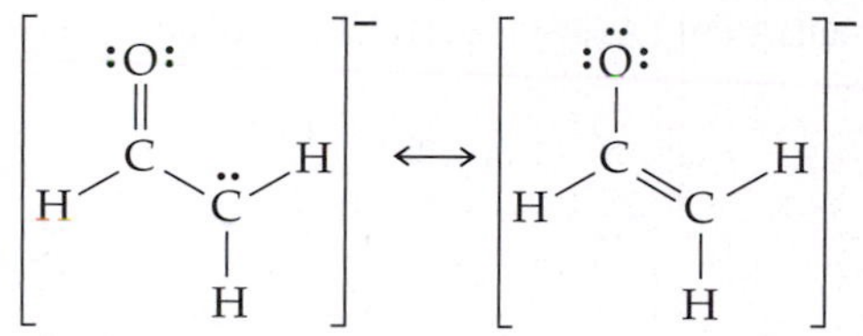

4.72 Draw two electron-dot resonance structures that obey the octet rule for trichloronitromethane, CCl_3NO_2, and show the formal charges on N and O in both structures. (Carbon is connected to the chlorines and to nitrogen; nitrogen is also connected to both oxygens.)

4.73 Draw two electron-dot resonance structures that obey the octet rule for nitrosyl chloride, NOCl (nitrogen is the central atom). Show formal charges, if present, and predict which of the two structures is a more accurate representation of the molecule.

CHAPTER PROBLEMS

4.74 Boron trifluoride reacts with dimethyl ether to form a compound with a coordinate covalent bond. Assign formal charges to the B and O atoms in both the reactants and product.

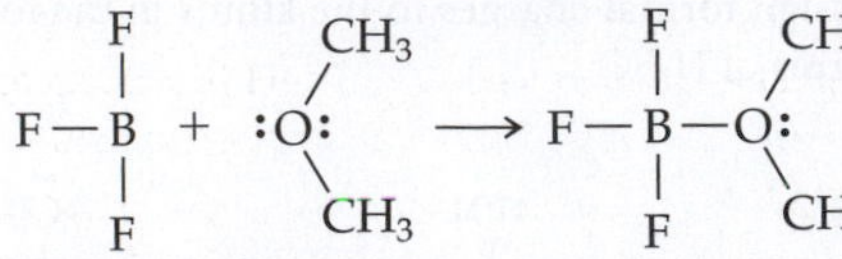

Boron trifluoride   Dimethyl ether

4.75 Thiofulminic acid, H–C≡N–S, is a highly reactive compound.

(a) Draw an electron-dot structure for thiofulminic acid, and assign formal charges.

(b) A related compound with the same formula and the connection H–N–C–S is also known. Draw an electron-dot structure for this related compound and assign formal charges.

(c) Which of the two molecules is likely to be more stable? Explain.

4.76 Draw two resonance structures for methyl isocyanate, CH_3NCO, a toxic gas that was responsible for the deaths of at least 3000 people when it was accidentally released into the atmosphere in December 1984 in Bhopal, India. Assign formal charges to the atoms in each resonance structure.

4.77 Draw three resonance structures for sulfur tetroxide, SO_4, whose connections are shown below. (This is a neutral molecule; it is not sulfate ion.) Assign formal charges to the atoms in each structure.

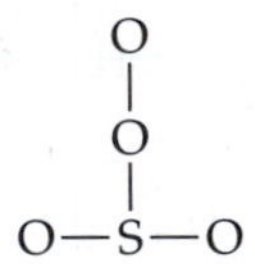

Sulfur tetroxide

4.78 The overall energy change during a chemical reaction can be calculated from a knowledge of bond dissociation energies using the following relationship:

$$\text{Energy change} = D(\text{Bonds broken}) - D(\text{Bonds formed})$$

Use the data in Table 4.2 to calculate an energy change for the reaction of methane with chlorine.

$$CH_4(g) + Cl_2(g) \longrightarrow CH_3Cl(g) + HCl(g)$$

4.79 The following molecular model is that of aspartame, $C_{14}H_{18}N_2O_5$, known commercially as NutraSweet. Only the connections between atoms are shown; multiple bonds are not indicated. Complete the structure by indicating the positions of the multiple bonds and lone pairs.

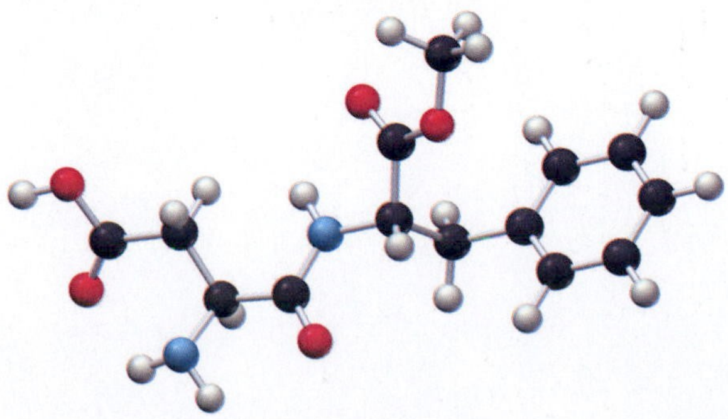

Aspartame

4.80 Ibuprofen ($C_{13}H_{18}O_2$), marketed under such brand names as Advil and Motrin, is a drug sold over-the-counter for treatment of pain and inflammation. Complete the structure of ibuprofen by adding hydrogen atoms where needed.

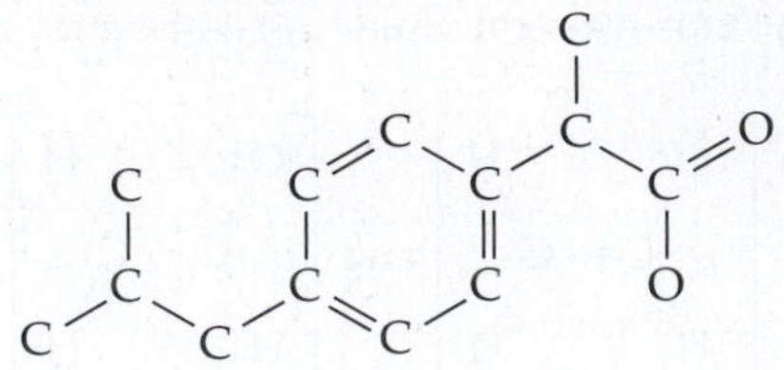

Ibuprofen

4.81 Name each of the following compounds, and draw an electron-dot structure for those that are covalent:

(a) BCl_3 **(b)** CsCl **(c)** AlF_3 **(d)** BrF_5

4.82 Experimental evidence shows that all three B–F bonds in boron trifluoride have the same length, 131 pm. Draw an electron-dot structure of BF_3, and explain the result.

4.83 Four different structures **(a)**, **(b)**, **(c)**, and **(d)** can be drawn for compounds named dibromobenzene, but only three different compounds actually exist. Explain.

(a)

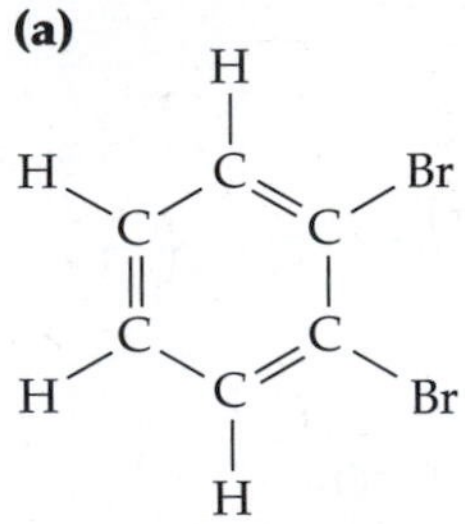

(b)

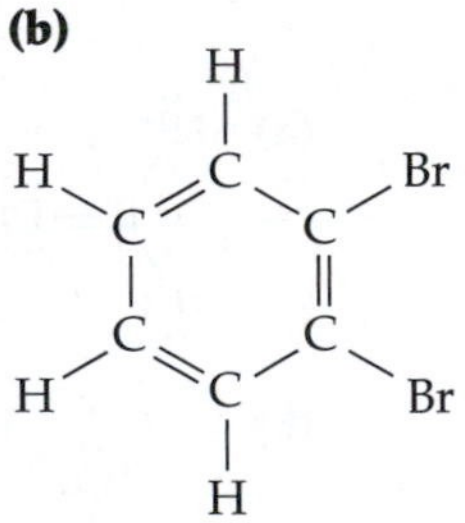

(c)

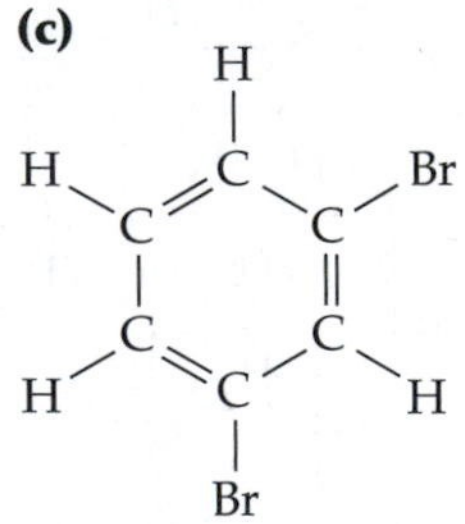

(d)

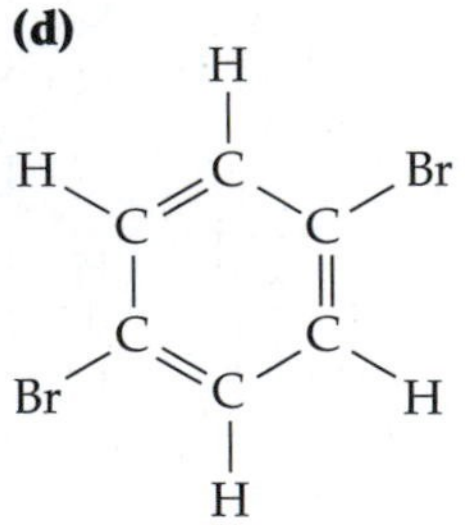

4.84 Name each of the following compounds, and draw an electron-dot structure for those that are covalent:

(a) MgF_2 **(b)** PBr_3 **(c)** $GeCl_4$ **(d)** CaO

4.85 Some mothballs used when storing clothes are made of naphthalene ($C_{10}H_8$), which has the following incomplete structure. When adding multiple bonds to complete the structure, three resonance forms can be generated. Show all three.

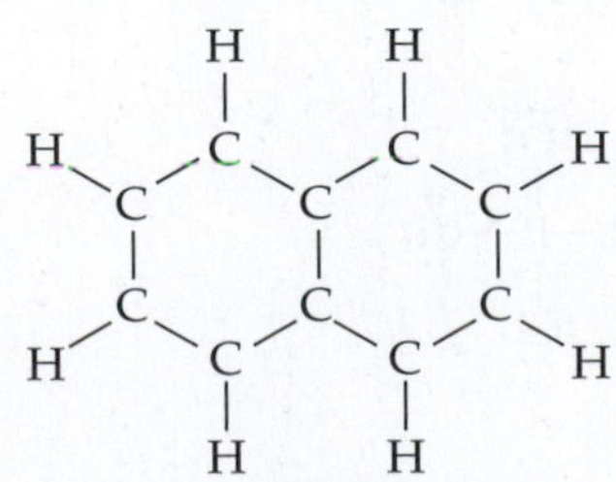

Naphthalene

MULTICONCEPT PROBLEMS

4.86 Sulfur reacts with chlorine to give a product that contains 47.5% by mass sulfur and 52.5% by mass chlorine and has no formal charges on any of its atoms. Draw the electron-dot structure of the product.

4.87 Sulfur reacts with ammonia to give a product **A** that contains 69.6% by mass sulfur and 30.4% by mass nitrogen and has a molar mass of 184.3 g.

(a) What is the formula of the product **A**?

(b) The S and N atoms in the product **A** alternate around a ring, with half of the atoms having formal charges. Draw two possible electron-dot structures for **A**.

(c) When compound **A** is heated with metallic silver at 250 °C, a new product **B** is formed. Product **B** has the same percent composition as **A** but has a molar mass of 92.2 g. Draw two possible electron-dot structures for **B** which, like **A**, also has a ring structure.

4.88 The neutral OH molecule has been implicated in certain ozone-destroying processes that take place in the upper atmosphere.

(a) Draw electron-dot structures for the OH molecule and the OH^- ion.

(b) Electron affinity can be defined for molecules just as it is defined for single atoms. Assuming that the electron added to OH is localized in a single atomic orbital on one atom, identify which atom is accepting the electron, and give the n and l quantum numbers of the atomic orbital.

(c) The electron affinity of OH is similar to but slightly more negative than that of O atoms. Explain.

4.89 Suppose that the Pauli exclusion principle were somehow changed to allow three electrons per orbital rather than two.

(a) Instead of an octet, how many outer-shell electrons would be needed for a noble-gas electron configuration?

(b) How many electrons would be shared in a covalent bond?

(c) Give the electron configuration and draw an electron-dot structure for element **X** with $Z = 12$.

(d) Draw an electron-dot structure for the molecule X_2.

CHAPTER

5

Covalent Bonds and Molecular Structure

Shape is important, both in architecture and in chemistry, as we'll see in the *FYI* at the end of this chapter.

The electron-dot structures described in the previous chapter provide a straightforward way to indicate the connections between atoms in a molecule; that is, to identify the locations of the covalent bonds. Much more information is needed, however, for a detailed picture of molecular structure. For one thing, electron-dot structures are two-dimensional while molecules are three-dimensional. In fact, knowing the exact three-dimensional shape of a molecule is often crucial to understanding the molecule's properties and behavior, a point that will be explored in more detail in the *FYI* at the end of this chapter and that is particularly important in understanding how drugs function. Furthermore, electron-dot structures explain nothing about the electronic nature of covalent bonds in quantum mechanical terms.

In this chapter, we'll look at both of these topics—molecular shape and the electronic nature of covalent bonding. Let's begin with shape.

5.1 MOLECULAR SHAPE: THE VSEPR MODEL

Look at the following ball-and-stick models of water, ammonia, and methane. Each of these molecules—and every other molecule as well—has a specific three-dimensional shape. Often, particularly for biologically important molecules, three-dimensional shape plays a crucial part in determining the molecule's chemistry and biological properties.

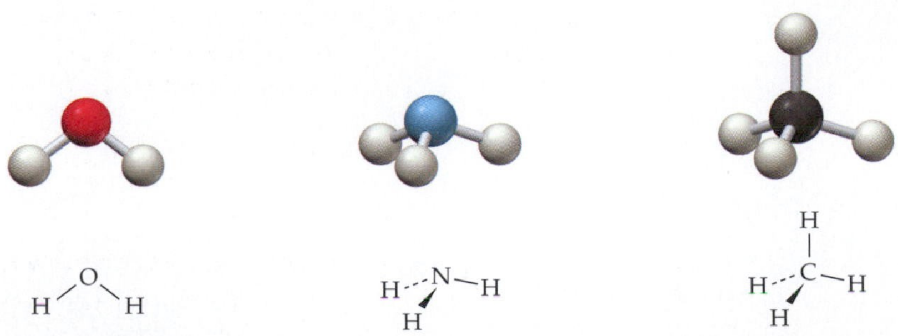

Water, H_2O Ammonia, NH_3 Methane, CH_4

CONTENTS

Like so many other properties, the shape of a molecule is determined by the electronic structure of its atoms. That shape can often be predicted using what is called the **valence-shell electron-pair repulsion (VSEPR) model**. In this model, electrons in bonds and in lone pairs can be thought of as occupying "charge clouds," or regions of electron density, that repel one another and stay as far apart as possible, thus causing molecules to assume specific shapes. There are only two steps to remember in applying the VSEPR model:

Step 1. Write an electron-dot structure for the molecule, as described in Section 4.7, and count the number of electron charge clouds surrounding the atom of interest. A charge cloud is simply the space occupied by a group of electrons, either in a bond or in a lone pair. Thus, the number of charge clouds around an atom is the total number of the atom's bonds and lone pairs. Multiple bonds count the same as single bonds because it doesn't matter how many electrons occupy each cloud.

Step 2. Predict the geometric arrangement of charge clouds around each atom by assuming that its charge clouds are oriented in space as far away from one another as possible. How they achieve this orientation depends on their number. Let's look at the possibilities.

Two Charge Clouds

When there are only two charge clouds on an atom, as occurs on the carbon atoms of CO_2 (two double bonds) and HCN (one single bond and one triple bond), the clouds are farthest apart when they point in opposite directions. Thus, CO_2 and HCN are *linear* molecules with **bond angles** of 180°.

A CO_2 molecule is linear, with a bond angle of 180°.

An HCN molecule is linear, with a bond angle of 180°.

Three Charge Clouds

When there are three charge clouds on an atom, as occurs on the carbon atom of formaldehyde (two single bonds and one double bond) and the sulfur atom of SO_2 (one single bond, one double bond, and one lone pair), the clouds are farthest apart when they lie in the same plane and point to the corners of an equilateral triangle. Thus, a formaldehyde molecule has a *trigonal planar* shape, with H–C–H and H–C=O bond angles near 120°.

Similarly, an SO_2 molecule has a trigonal planar geometric arrangement of its three charge clouds on sulfur, but one corner of the triangle is occupied by a lone pair and two corners by oxygen atoms. The molecule therefore has a *bent* shape, with an O–S=O bond angle of approximately 120° rather than 180°. (For consistency, we'll use the word *shape* to refer to the overall arrangement of atoms in the *molecule* and the word *geometry* to refer to the arrangement of charge clouds on a specific atom.)

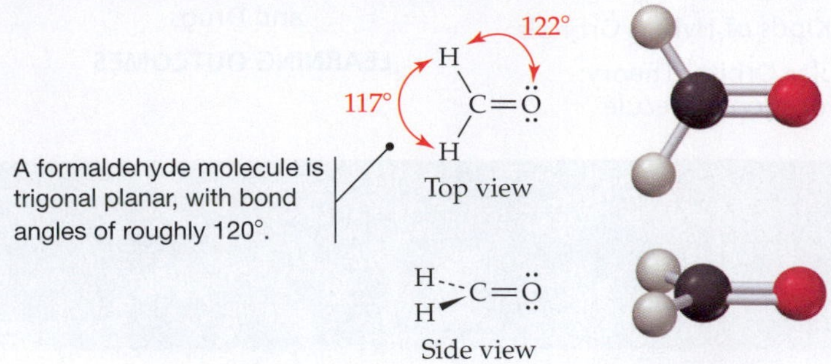

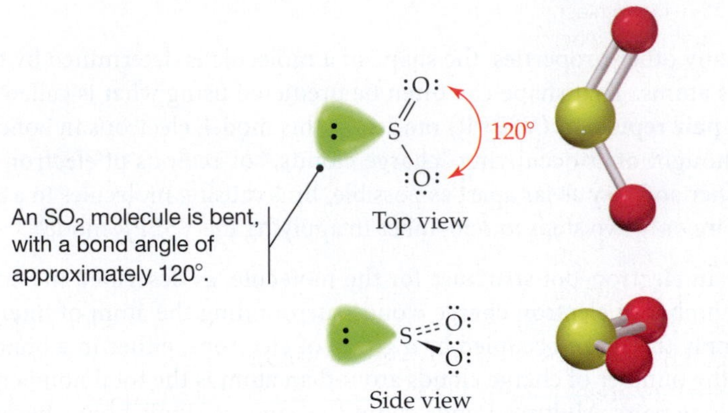

Four Charge Clouds

When there are four charge clouds on an atom, as occurs on the central atoms in CH_4 (four single bonds), NH_3 (three single bonds and one lone pair), and H_2O (two single bonds and two lone pairs), the clouds are farthest apart if they extend toward the corners of a regular tetrahedron. As shown in **Figure 5.1**, a regular tetrahedron is a geometric solid whose four identical faces are equilateral triangles. The atom lies in the center of the tetrahedron and its charge clouds point toward the four corners. In this geometric arrangement, the angle between lines drawn from the central atom to any two corners is 109.5°.

FIGURE 5.1

The tetrahedral geometry of an atom with four charge clouds.

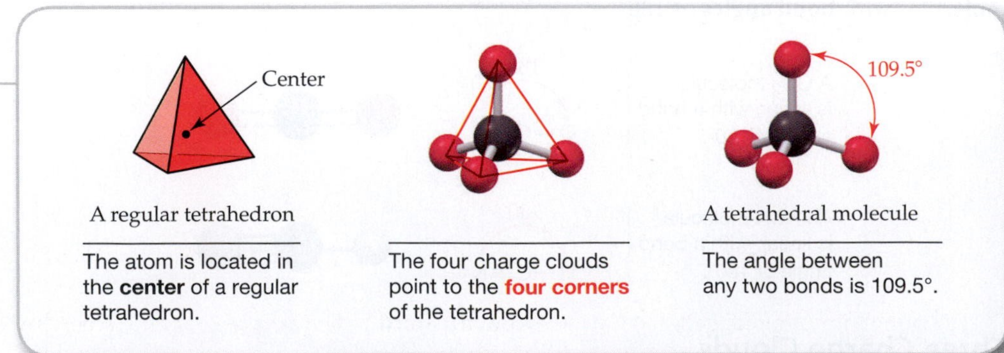

Because valence electron octets are so common, particularly for atoms of second-row elements, the atoms in a great many molecules have geometries based on the tetrahedron. In CH_4, for instance, the carbon atom has a tetrahedral arrangement of its four C–H bonds and an overall tetrahedral shape, with H–C–H bond angles of 109.5°. In NH_3, the nitrogen atom has a tetrahedral arrangement of its four charge clouds, but one corner of the tetrahedron is occupied by a lone pair, resulting in a *trigonal pyramidal* shape for the molecule. Similarly, H_2O has two corners of the tetrahedron occupied by lone pairs and thus has a *bent* shape.

A methane molecule is tetrahedral, with bond angles of 109.5°.

An ammonia molecule is trigonal pyramidal, with bond angles of 107°.

A water molecule is bent, with a bond angle of 104.5°.

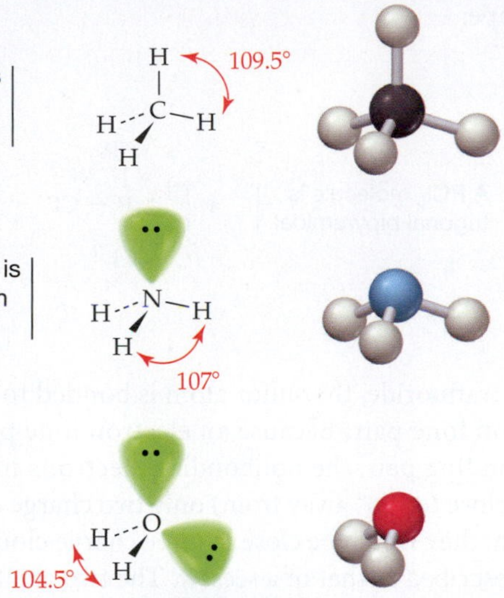

Note how the three-dimensional shapes of the molecules in the preceding structures are drawn. Solid lines are assumed to be in the plane of the paper, dashed lines recede behind the plane of the paper away from the viewer, and heavy, wedged lines protrude out of the paper toward the viewer. Note also that the H–N–H bond angles in ammonia (107°) and the H–O–H bond angle in water (104.5°) are less than the ideal 109.5° tetrahedral value. The angles are diminished somewhat from the tetrahedral value because of the presence of lone pairs. Charge clouds of lone-pair electrons spread out more than charge clouds of bonding electrons because they aren't confined to the space between two nuclei. As a result, the somewhat enlarged lone-pair charge clouds tend to compress the bond angles in the rest of the molecule.

Five Charge Clouds

Five charge clouds, such as are found on the central atoms in PCl_5, SF_4, ClF_3, and I_3^-, are oriented toward the corners of a geometric figure called a *trigonal bipyramid*. Three clouds lie in a plane and point toward the corners of an equilateral triangle, the fourth cloud points directly up, and the fifth cloud points down:

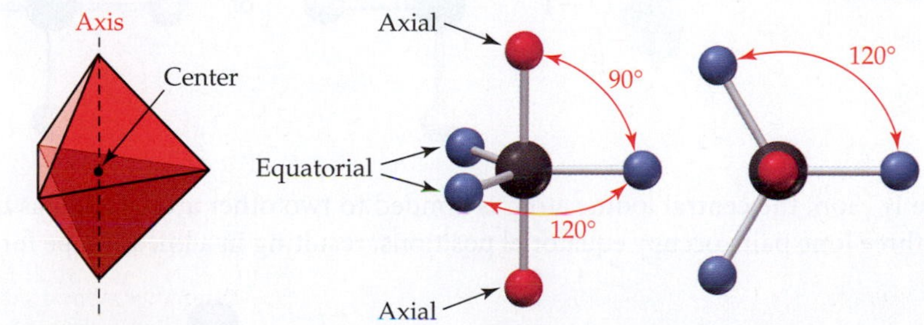

A trigonal bipyramid — Side view — Top view

Trigonal bipyramidal geometry differs from the linear, trigonal planar, and tetrahedral geometries discussed previously because it has two kinds of positions—three *equatorial* positions (around the "equator" of the bipyramid) and two *axial* positions (along the "axis" of the bipyramid). The three equatorial positions are at angles of 120° to one another and at an angle of 90° to the axial positions. The two axial positions are at angles of 180° to each other and at an angle of 90° to the equatorial positions.

Different substances containing a trigonal bipyramidal arrangement of charge clouds on an atom adopt different shapes, depending on whether the five charge clouds contain bonding or nonbonding electrons. In phosphorus pentachloride, for instance, all five positions around phosphorus are occupied by chlorine atoms, giving the molecule a trigonal bipyramidal shape:

A PCl_5 molecule is trigonal bipyramidal.

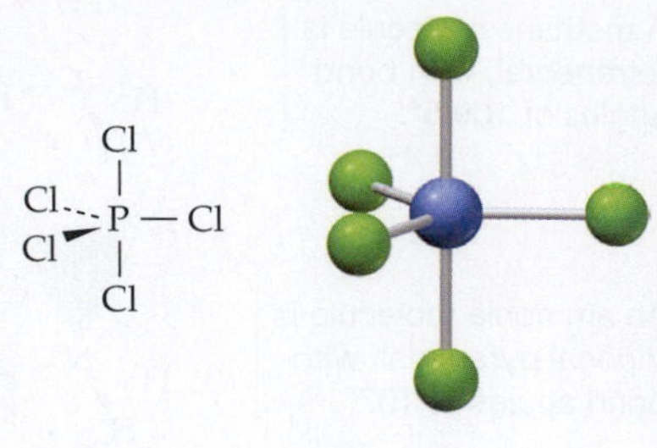

In sulfur tetrafluoride, the sulfur atom is bonded to four other atoms and has one nonbonding electron lone pair. Because an electron lone pair spreads out and occupies more space than a bonding pair, the nonbonding electrons in SF_4 occupy an equatorial position where they are close to (90° away from) only two charge clouds. Were they instead to occupy an axial position, they would be close to three charge clouds. As a result, the SF_4 molecule has a shape often described as that of a seesaw. The two axial bonds form the board, and the two equatorial bonds form the legs of the seesaw.

An SF_4 molecule is shaped like a seesaw (turn 90° to see it).

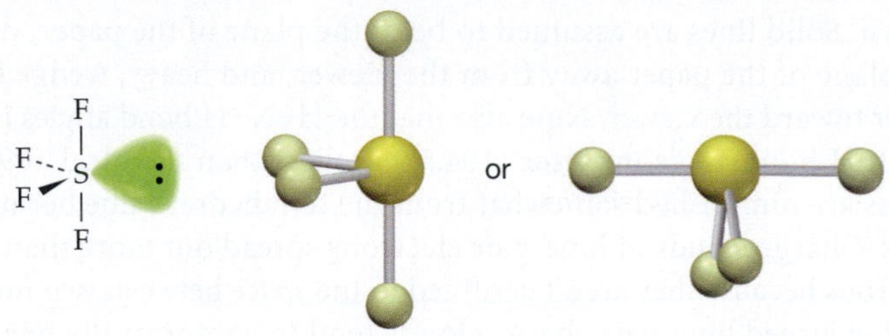

Yet another shape resulting from a trigonal bipyramidal arrangement of charge clouds occurs in chlorine trifluoride. The chlorine atom is bonded to three other atoms and has two nonbonding electron lone pairs. Both lone pairs occupy equatorial positions, giving the ClF_3 molecule a T shape.

A ClF_3 molecule is T-shaped.

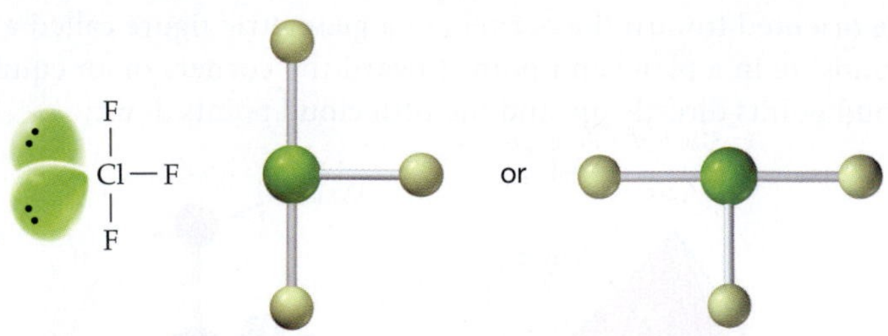

In the I_3^- ion, the central iodine atom is bonded to two other atoms and has three lone pairs. All three lone pairs occupy equatorial positions, resulting in a linear shape for I_3^-.

An I_3^- ion is linear.

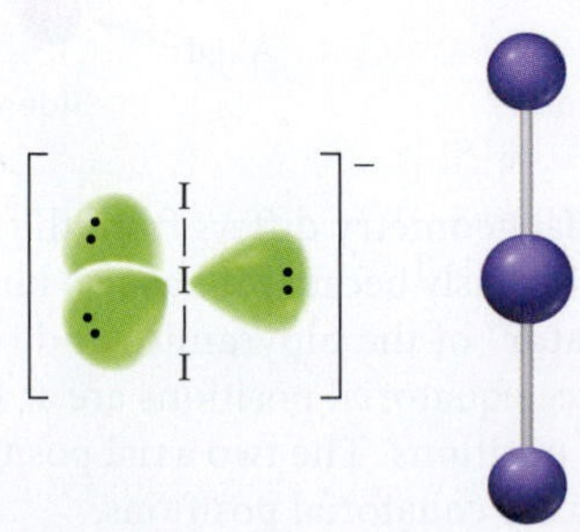

Six Charge Clouds

Six charge clouds around an atom orient toward the six corners of a regular octahedron, a geometric solid whose eight faces are equilateral triangles. All six positions are equivalent, and the angle between any two adjacent positions is 90°.

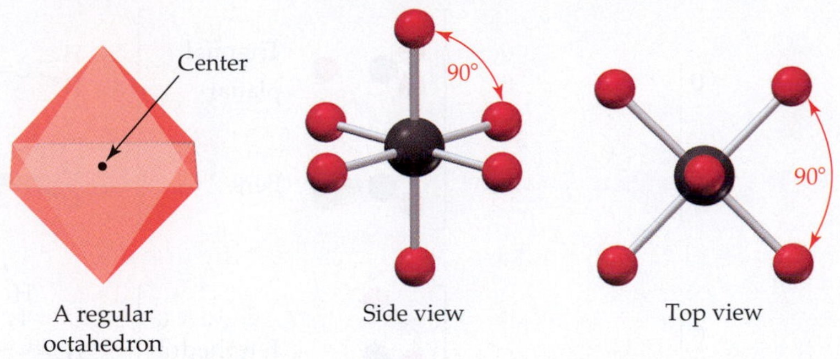

As was true in the case of five charge clouds, different shapes are possible for molecules having atoms with six charge clouds, depending on whether the clouds are of bonding or nonbonding electrons. In sulfur hexafluoride, for instance, all six positions around sulfur are occupied by fluorine atoms:

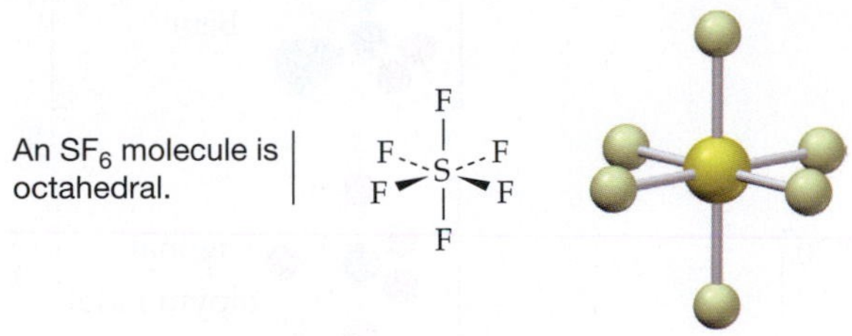

The antimony atom in the $SbCl_5{}^{2-}$ ion also has six charge clouds but is bonded to only five atoms and has one nonbonding electron lone pair. As a result, the ion has a *square pyramidal* shape—a pyramid with a square base:

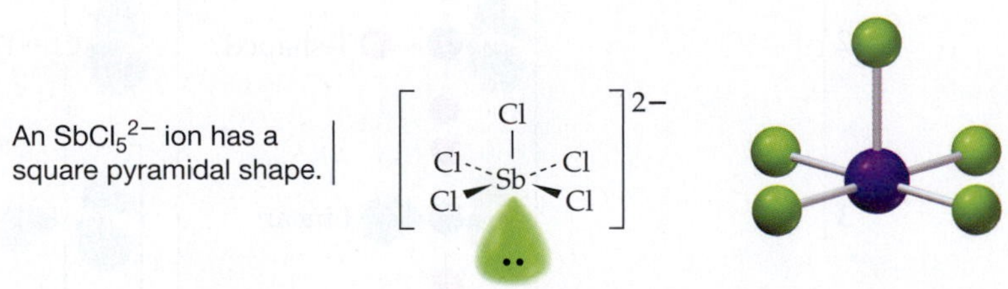

The xenon atom in XeF_4 is bonded to four atoms and has two lone pairs. The lone pairs orient as far away from each other as possible to minimize electronic repulsions, giving the molecule a *square planar* shape:

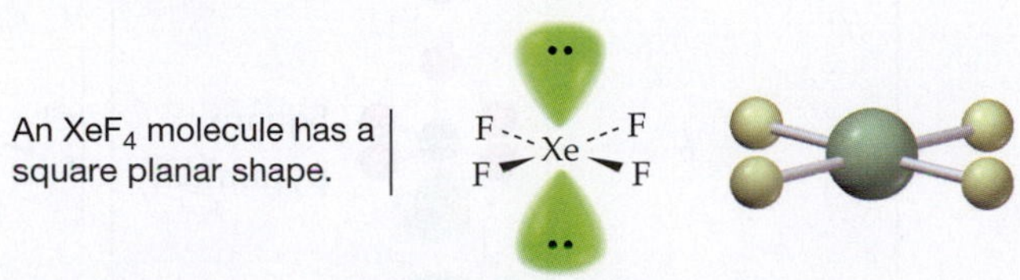

All the geometries for two to six charge clouds are summarized in **Table 5.1**.

TABLE 5.1 Geometry Around Atoms with 2, 3, 4, 5, and 6 Charge Clouds

Number of Bonds	Number of Lone Pairs	Number of Charge Clouds	Geometry and Shape	Example
2	0	2	 Linear	O=C=O
3	0	3	 Trigonal planar	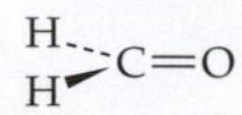
2	1	3	 Bent	
4	0	4	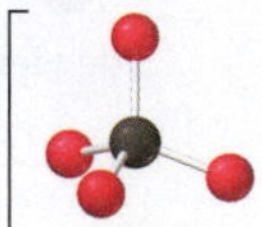 Tetrahedral	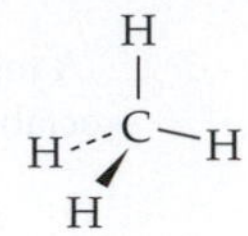
3	1	4	 Trigonal pyramidal	
2	2	4	 Bent	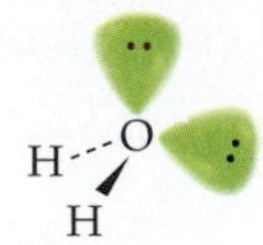
5	0	5	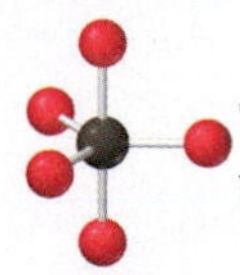 Trigonal bipyramidal	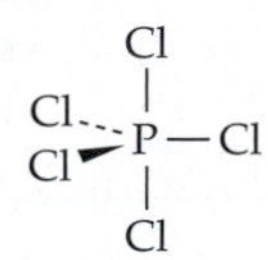
4	1	5	 Seesaw	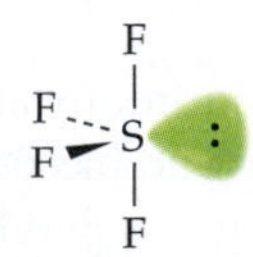
3	2	5	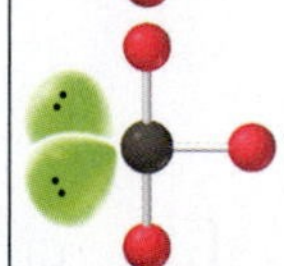 T-shaped	
2	3	5	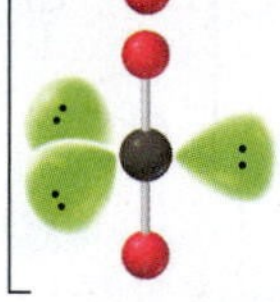 Linear	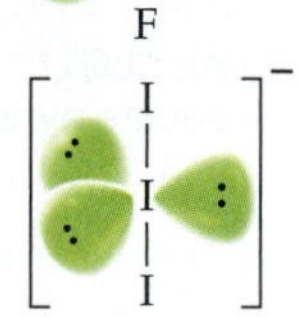
6	0	6	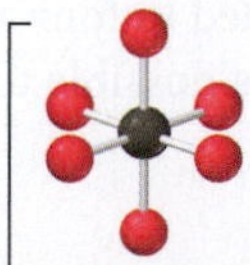 Octahedral	
5	1	6	 Square pyramidal	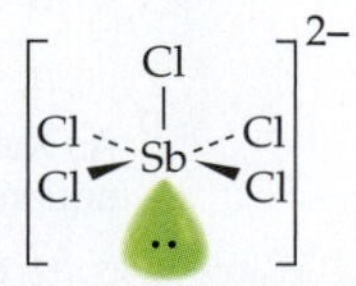
4	2	6	 Square planar	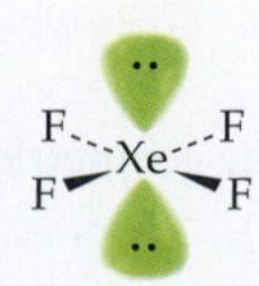

Shapes of Larger Molecules

The geometries around individual atoms in larger molecules can also be predicted from the rules summarized in Table 5.1. For example, each of the two carbon atoms in ethylene ($H_2C{=}CH_2$) has three charge clouds, giving rise to trigonal planar geometry for each carbon. The molecule as a whole has a planar shape, with H–C–C and H–C–H bond angles of approximately 120°.

Each carbon atom in ethylene has trigonal planar geometry. As a result, the entire molecule is planar, with bond angles of 120°.

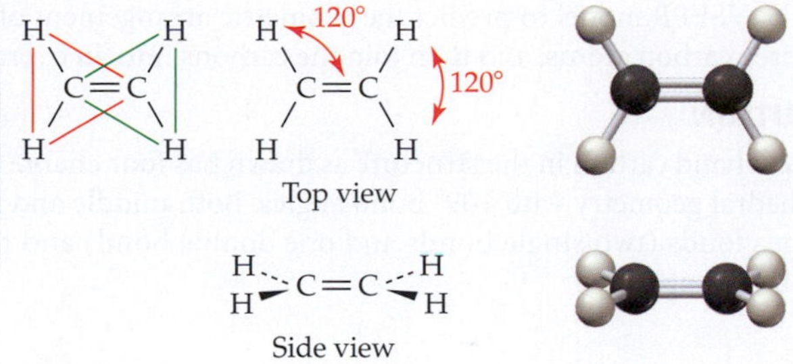

Carbon atoms bonded to four other atoms are each at the center of a tetrahedron. As shown below for ethane, $H_3C{-}CH_3$, the two tetrahedrons are joined so that the central carbon atom of one tetrahedron is a corner atom of the other.

Each carbon atom in ethane has tetrahedral geometry, with bond angles of 109.5°.

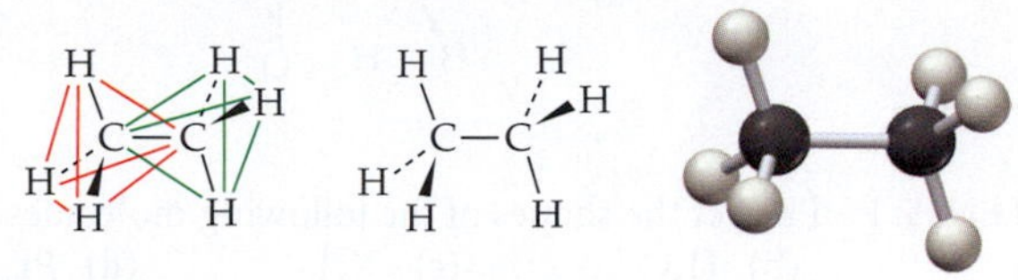

Worked Example 5.1

Using the VSEPR Model to Predict a Molecular Shape

Predict the shape of BrF_5.

STRATEGY

First, draw an electron-dot structure for BrF_5 to determine how many charge clouds surround the central bromine atom. Then use the VSEPR model to predict how the charge clouds are arranged.

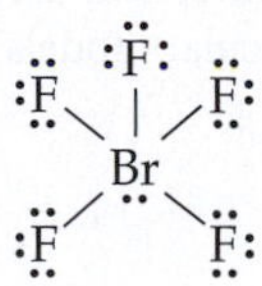

Bromine pentafluoride

SOLUTION

The central bromine atom in BrF_5 has six charge clouds—five bonds and one lone pair—implying an octahedral arrangement. The five attached atoms and one lone pair give BrF_5 a square pyramidal shape:

Worked Example 5.2

Predicting the Shape of a Larger Molecule

Pyruvic acid is a key substance in the metabolism of both carbohydrates and several amino acids. Describe the geometry around each of the carbon atoms in pyruvic acid, and draw the overall shape of the molecule.

continued on next page

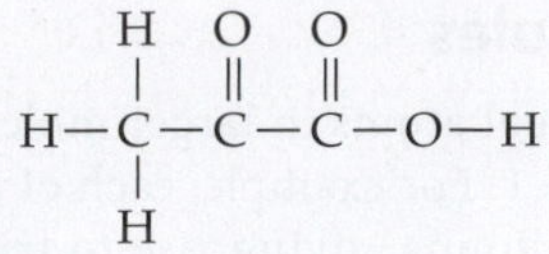

Pyruvic acid

STRATEGY

Use the VSEPR model to predict the geometric arrangement of charge clouds around each of the three carbon atoms, and then join the carbons into an overall structure.

SOLUTION

The left-hand carbon in the structure as drawn has four charge clouds (four single bonds) and tetrahedral geometry with 109° bond angles. Both middle and right-hand carbons have three charge clouds (two single bonds and one double bond) and trigonal planar geometry with 120° bond angles.

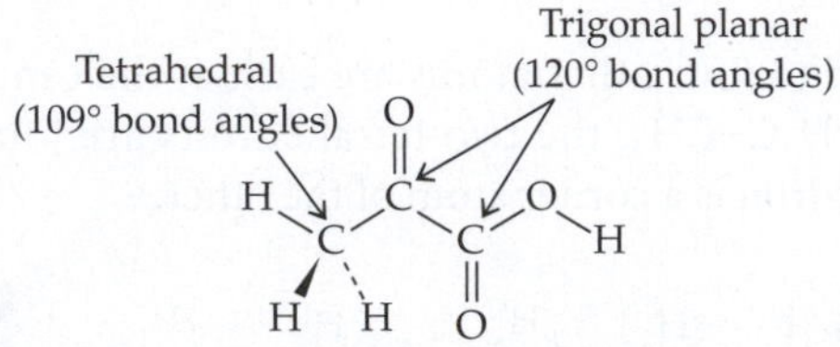

PROBLEM 5.1 Predict the shapes of the following molecules or ions:
(a) O_3 **(b)** H_3O^+ **(c)** XeF_2 **(d)** PF_6^- **(e)** $XeOF_4$
(f) AlH_4^- **(g)** BF_4^- **(h)** $SiCl_4$ **(i)** ICl_4^- **(j)** $AlCl_3$

PROBLEM 5.2 Acetic acid, CH_3CO_2H, is the main organic constituent of vinegar. Draw an electron-dot structure for acetic acid, and show its overall shape. (The two carbons are connected by a single bond, and both oxygens are connected to the same carbon.)

PROBLEM 5.3 Benzene, C_6H_6, is a *cyclic* molecule in which all six carbon atoms are joined in a ring, with each carbon also bonded to one hydrogen. Draw an electron-dot structure for benzene and predict the overall shape of the molecule.

CONCEPTUAL PROBLEM 5.4 What is the geometrical arrangement of charge clouds around the central atom in each of the following molecular models?

(a) **(b)**

5.2 VALENCE BOND THEORY

The VSEPR model discussed in the previous section provides a simple way to predict molecular shapes, but it says nothing about the electronic nature of covalent bonds. To describe bonding, two models, called *valence bond theory* and *molecular orbital theory*, have been developed. We'll look first at valence bond theory, followed by molecular orbital theory in Sections 5.5 and 5.6.

Valence bond theory provides a readily visualized orbital picture of how electron pairs are shared in a covalent bond. In this model, a covalent bond results when two atoms approach each other closely enough so that a singly occupied valence orbital on one atom spatially *overlaps* a singly occupied valence orbital on the other atom. The now-paired electrons in the overlapping orbitals are attracted to the nuclei of both atoms and thus bond the two atoms together. In the H_2 molecule, for instance, the H—H bond results from the overlap of two singly occupied hydrogen 1*s* orbitals.

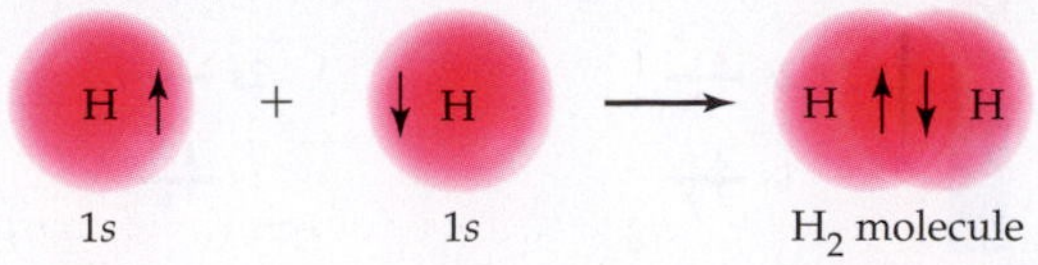

Recall that atomic orbitals arise from solutions to the **Schrödinger wave equation** and that the two lobes of a *p* atomic orbital have different **phases**, represented in this book by different colors. In the valence bond model, the two overlapping lobes must be of the same phase, and the strength of the covalent bond that forms depends on the amount of orbital overlap: the greater the overlap, the stronger the bond. This, in turn, means that bonds formed by overlap of other than *s* orbitals have a directionality to them. In the F_2 molecule, for instance, each fluorine atom has the electron configuration [He] $2s^2\,2p_x^2\,2p_y^2\,2p_z^1$ and the F—F bond results from the overlap of two singly occupied 2*p* orbitals. The two 2*p* orbitals must point directly at each other for optimum overlap to occur, and the F—F bond forms along the orbital axis. Such bonds that result from head-on orbital overlap are called **sigma (σ) bonds**.

REMEMBER...

The **Schrödinger wave equation** focuses on the wavelike properties of atoms to describe the quantum mechanical model of atomic structure. The solutions to the wave equation are called wave functions, or orbitals. (Section 2.6)

REMEMBER...

The two lobes of a *p* orbital have different mathematical signs in the wave function, corresponding to the different **phases** of a wave. (Section 3.7)

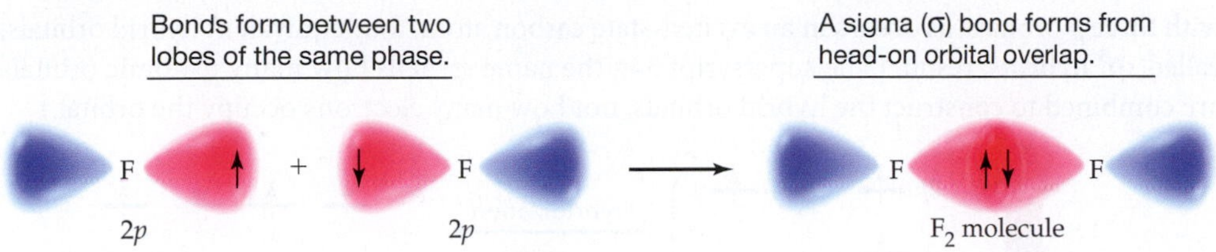

In HCl, the covalent bond involves overlap of a hydrogen 1*s* orbital with a chlorine 3*p* orbital and forms along the *p*-orbital axis:

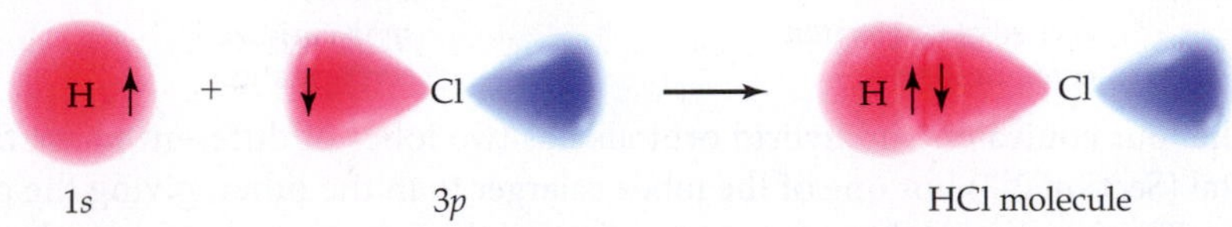

The key ideas of valence bond theory can be summarized as follows:

- Covalent bonds are formed by overlap of atomic orbitals, each of which contains one electron of opposite spin. The two overlapping lobes must be of the same phase.
- Each of the bonded atoms maintains its own atomic orbitals, but the electron pair in the overlapping orbitals is shared by both atoms.
- The greater the amount of orbital overlap, the stronger the bond. This leads to a directional character for the bond when other than *s* orbitals are involved.

5.3 HYBRIDIZATION AND sp^3 HYBRID ORBITALS

How does valence bond theory describe the electronic structure of complex polyatomic molecules, and how does it account for the observed geometries around atoms in molecules? Let's look first at a simple tetrahedral molecule such as methane, CH_4. There are several problems to be dealt with.

Carbon has the ground-state electron configuration [He] $2s^2\,2p_x^1\,2p_y^1$. It thus has four valence electrons, two of which are paired in a 2*s* orbital and two of which are unpaired in different 2*p* orbitals that we'll arbitrarily designate as $2p_x$ and $2p_y$. But how can carbon form four bonds if two of its valence electrons are already paired and only two unpaired electrons remain for sharing? The answer is that an electron must be promoted from the lower-energy

2*s* orbital to the vacant, higher-energy $2p_z$ orbital, giving an *excited-state configuration* [He] $2s^1\ 2p_x^1\ 2p_y^1\ 2p_z^1$ that has *four* unpaired electrons and can thus form four bonds.

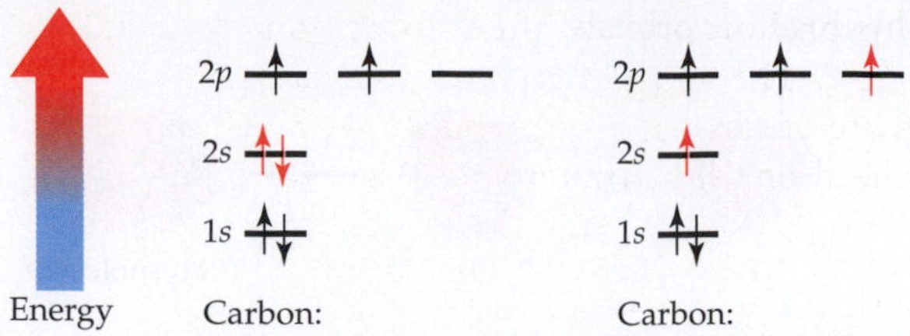

Carbon: ground-state electron configuration

Carbon: excited-state electron configuration

A second problem is more difficult to resolve: If excited-state carbon uses two kinds of orbitals for bonding, 2*s* and 2*p*, how can it form four *equivalent* bonds? Furthermore, if the three 2*p* orbitals in carbon are at angles of 90° to one another and the 2*s* orbital has no directionality, how can carbon form bonds with angles of 109.5° directed to the corners of a regular tetrahedron? The answers to these questions were provided in 1931 by Linus Pauling, who introduced the idea of *hybrid orbitals*.

Pauling showed how the quantum mechanical wave functions for *s* and *p* atomic orbitals derived from the Schrödinger wave equation can be mathematically combined to form a new set of equivalent wave functions called **hybrid atomic orbitals**. When one *s* orbital combines with three *p* orbitals, as occurs in an excited-state carbon atom, four equivalent hybrid orbitals, called sp^3 **hybrids**, result. (The superscript 3 in the name sp^3 tells how many *p* atomic orbitals are combined to construct the hybrid orbitals, not how many electrons occupy the orbital.)

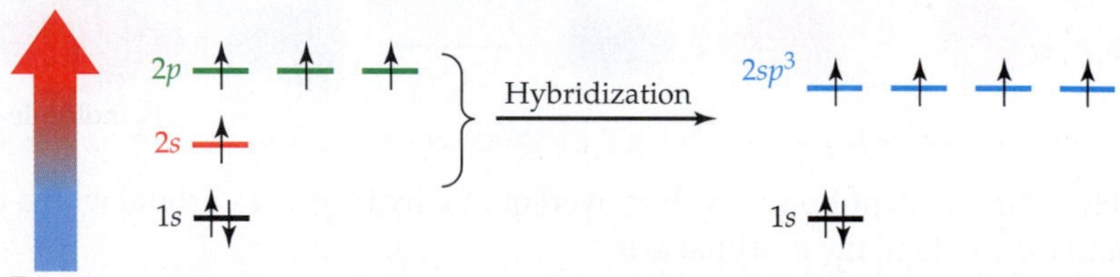

Carbon: excited-state electron configuration

Carbon: sp^3-hybridized configuration

Each of the four equivalent sp^3 hybrid orbitals has two lobes of different phase like an atomic *p* orbital (Section 2.7) but one of the lobes is larger than the other, giving the orbital a directionality. The four large lobes are oriented toward the four corners of a regular tetrahedron at angles of 109.5°, as shown in **Figure 5.2**. For consistency in the use of colors, we'll routinely show the different phases of unhybridized orbitals in red and blue and will show the large lobes of the resultant hybrid orbitals in green.

FIGURE 5.2

The formation of four sp^3 hybrid orbitals by combination of an atomic *s* orbital with three atomic *p* orbitals.

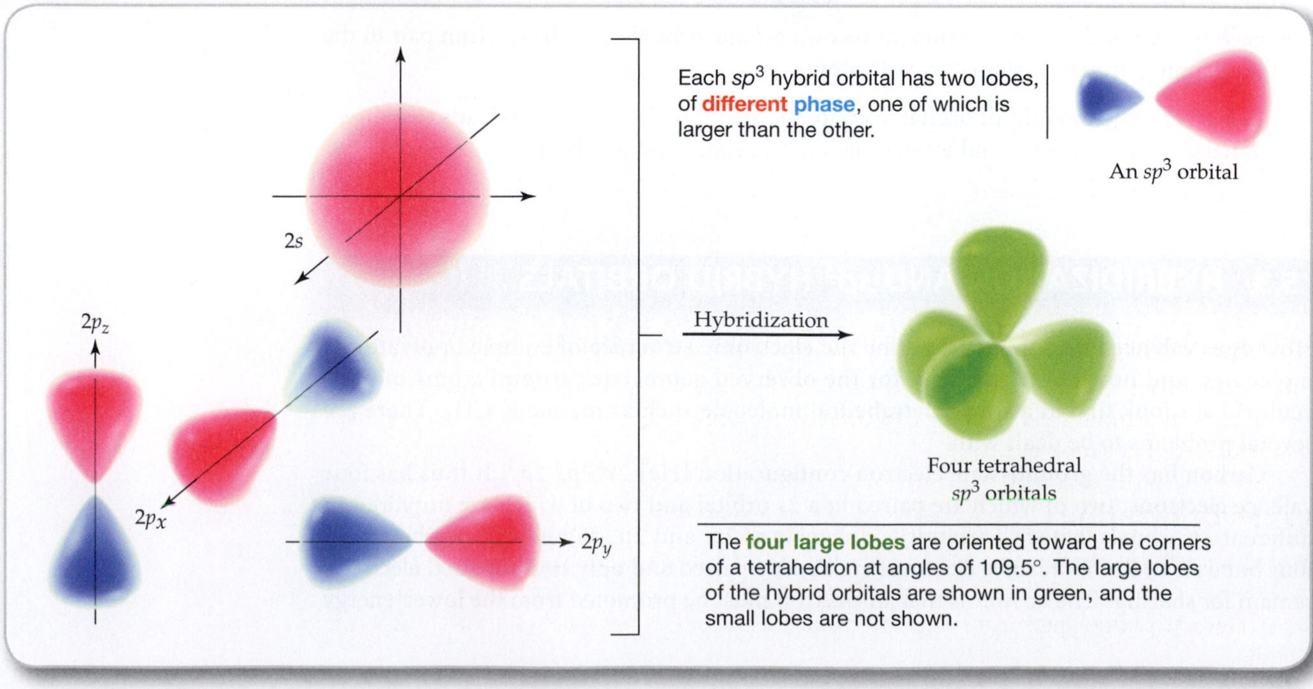

The shared electrons in a covalent bond made with a spatially directed hybrid orbital spend most of their time in the region between the two bonded nuclei. As a result, covalent bonds made with sp^3 hybrid orbitals are often strong ones. In fact, the energy released on forming the four strong C–H bonds in CH_4 more than compensates for the energy required to form the excited state of carbon. **Figure 5.3** shows how the four C–H sigma bonds in methane can form by head-on overlap of carbon sp^3 hybrid orbitals with hydrogen 1*s* orbitals.

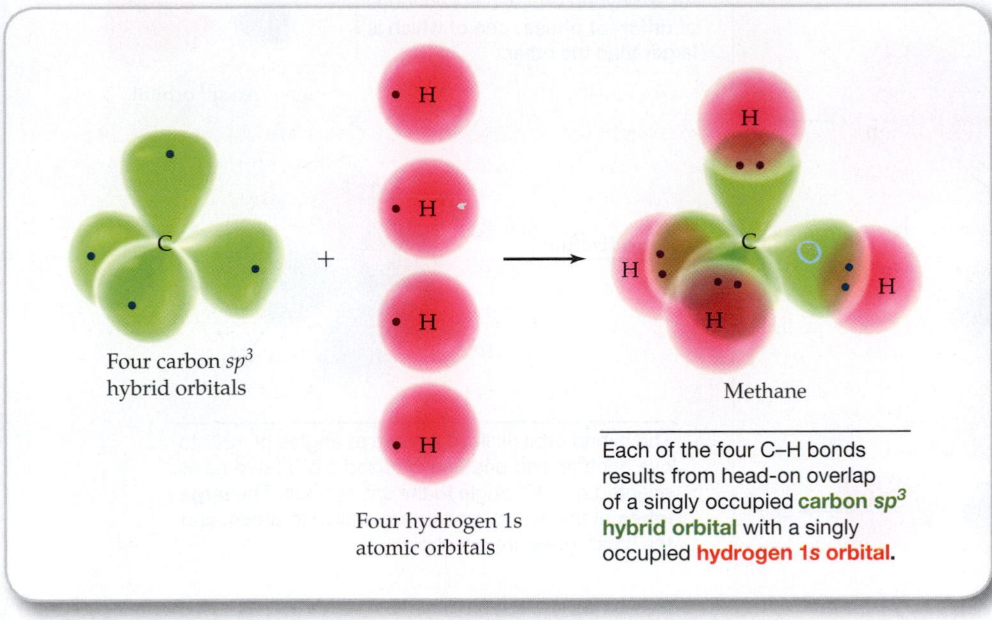

FIGURE 5.3

The bonding in methane.

The same kind of sp^3 hybridization that describes the bonds to carbon in the tetrahedral methane molecule can also be used to describe bonds to nitrogen in the trigonal pyramidal ammonia molecule, to oxygen in the bent water molecule, and to all other atoms that the VSEPR model predicts to have a tetrahedral arrangement of four charge clouds. That is, tetrahedral geometry always implies sp^3 hybridization.

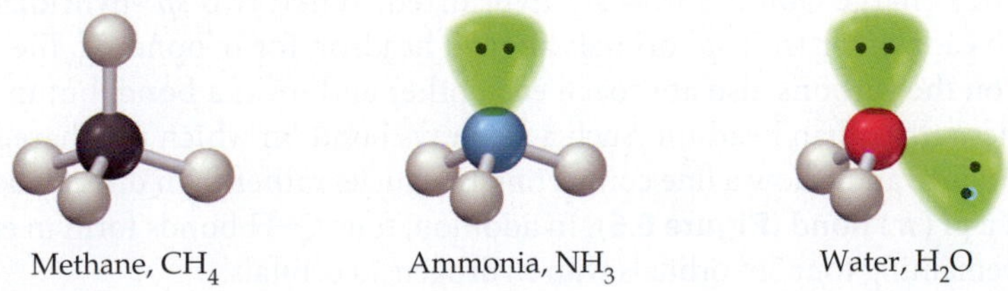

PROBLEM 5.5 Describe the bonding in propane, C_3H_8, a fuel often used in rural homes and campers, and tell what kinds of orbitals on each atom overlap to form the C–C and C–H bonds.

5.4 OTHER KINDS OF HYBRID ORBITALS

Other geometries shown in Table 5.1 can also be accounted for by specific kinds of orbital hybridization, although the situation becomes more complex for atoms with five and six charge clouds. Let's look at each.

sp^2 Hybridization

Atoms with three charge clouds undergo hybridization by combination of one atomic *s* orbital with two *p* orbitals, resulting in three **sp^2 hybrid orbitals**. These three sp^2 hybrids have trigonal planar geometry and are oriented toward the corners of an equilateral triangle at angles of 120° to one another. Just as tetrahedral geometry always implies sp^3 hybridization, trigonal planar geometry always implies sp^2 hybridization. In addition to the three sp^2 hybrid orbitals, one p orbital remains unchanged and is oriented at a 90° angle to the plane of the sp^2 hybrids, as shown in **Figure 5.4**.

FIGURE 5.4

The formation of sp^2 hybrid orbitals by combination of one s orbital and two p orbitals.

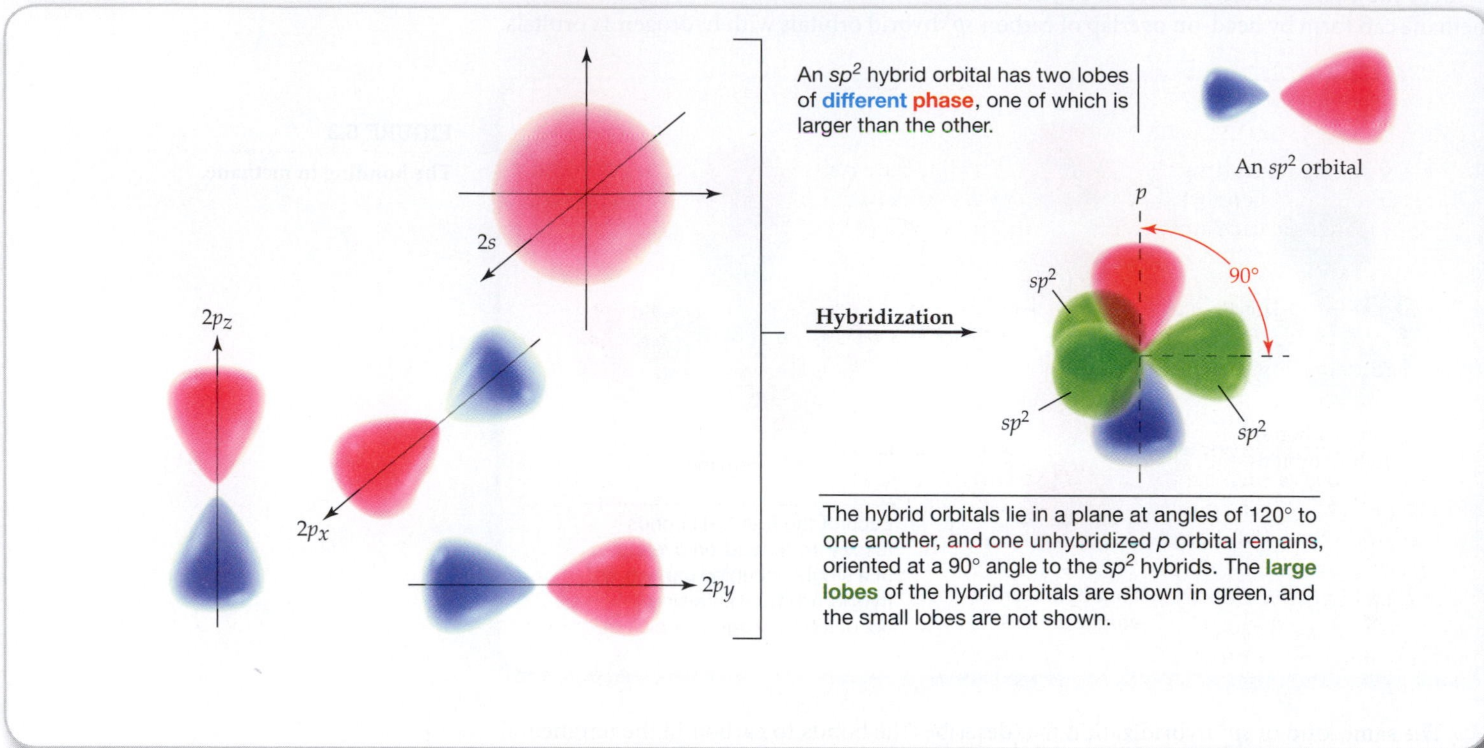

The presence of the unhybridized p orbital on an sp^2-hybridized atom has some interesting consequences. Look, for example, at ethylene, $H_2C{=}CH_2$, a colorless gas used as starting material for the industrial preparation of polyethylene. Each carbon atom in ethylene has three charge clouds and is sp^2-hybridized. When two sp^2-hybridized carbon atoms approach each other with sp^2 orbitals aligned head-on for σ bonding, the unhybridized p orbitals on the carbons also approach each other and form a bond, but in a parallel, sideways manner rather than head-on. Such a sideways bond, in which the shared electrons occupy regions above and below a line connecting the nuclei rather than directly between the nuclei, is called a **pi (π) bond** (**Figure 5.5**). In addition, four C–H bonds form in ethylene by overlap of the remaining four sp^2 orbitals with hydrogen 1s orbitals.

FIGURE 5.5

The structure of a carbon–carbon double bond.

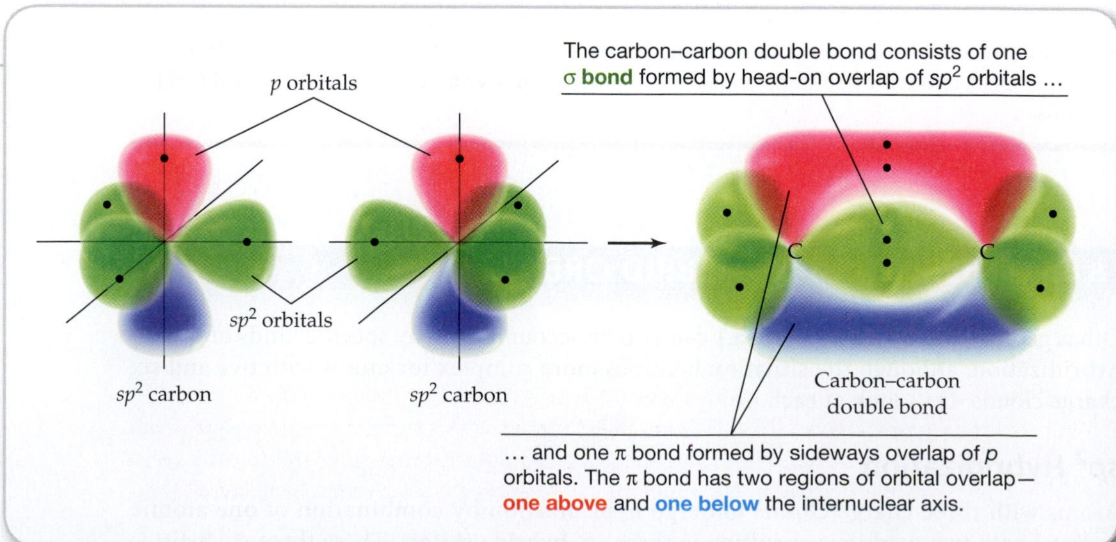

A π bond has two regions of orbital overlap, one above and one below a line drawn between the nuclei. Both regions are part of the same bond, and the two shared electrons are spread over both regions. As always, the *p* lobes must be of the same phase for overlap leading to bond formation. The net result of both σ and π overlap is the sharing of four electrons and the formation of a carbon–carbon double bond.

Worked Example 5.3

Hybrid Orbitals in Larger Molecules

Propylene, C_3H_6, a starting material used to manufacture polypropylene polymers, has its three carbon atoms connected in a row. Draw the overall shape of the molecule, and tell what kinds of orbitals on each atom overlap to form the C–C and C–H bonds.

STRATEGY

Draw an electron-dot structure of the molecule, use the VSEPR model to predict the geometry around each carbon atom, and predict the hybridization of each carbon atom based on the VSEPR bond angles.

SOLUTION

A molecule with the formula C_3H_6 and its carbons connected in a row does not have enough hydrogens to give each carbon four single bonds. Thus, propylene must contain a double bond and have two trigonal planar carbon atoms along with one tetrahedral carbon. Trigonal planar atoms have 120° bond angles and sp^2 hybridization, while tetrahedral atoms have 109° bond angles and sp^3 hybridization. Thus, the C—C single bond is formed by overlap of an sp^3 orbital and an sp^2 orbital, the C=C double bond is formed by overlap of two sp^2 orbitals and two *p* orbitals, and each C—H bond is formed by overlap of a hydrogen 1*s* orbital with the hybrid orbital on the carbon to which it is attached.

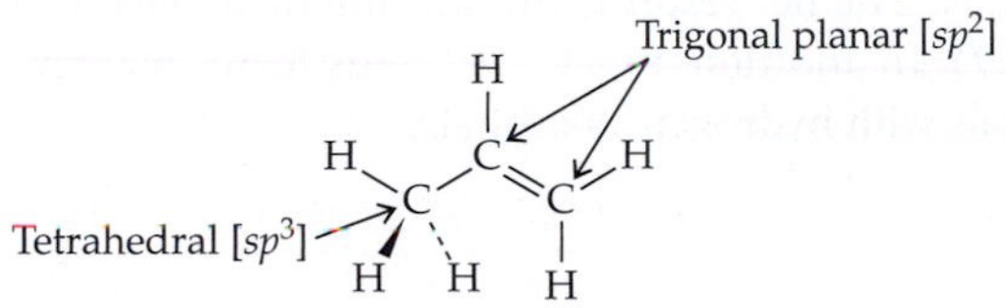

Propylene

PROBLEM 5.6 Describe the hybridization of the carbon atom in formaldehyde, $H_2C=O$, and make a rough sketch of the molecule showing the orbitals involved in bonding.

PROBLEM 5.7 Describe the hybridization of each carbon atom in lactic acid, a metabolite formed in tired muscles. Tell what kinds of orbitals on each atom overlap to form the C–C, C–O, and C–H bonds, and draw the overall shape of the molecule.

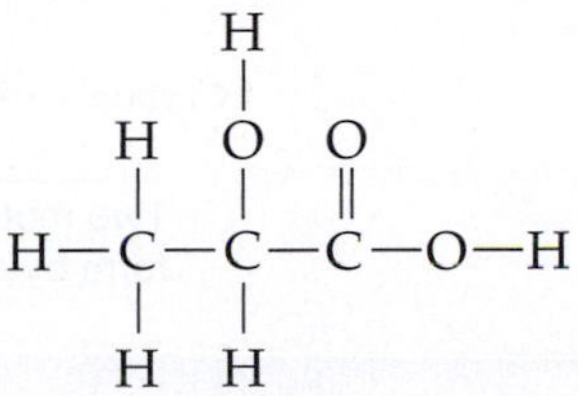

Lactic acid

sp Hybridization

Atoms with two charge clouds undergo hybridization by combination of one atomic *s* orbital with one *p* orbital, resulting in two ***sp* hybrid orbitals** that are oriented 180° from each other. Since only one *p* orbital is involved when an atom undergoes *sp* hybridization, the other two *p* orbitals are unchanged and are oriented at 90° angles to the *sp* hybrids, as shown in **Figure 5.6**.

FIGURE 5.6

***sp* Hybridization.**

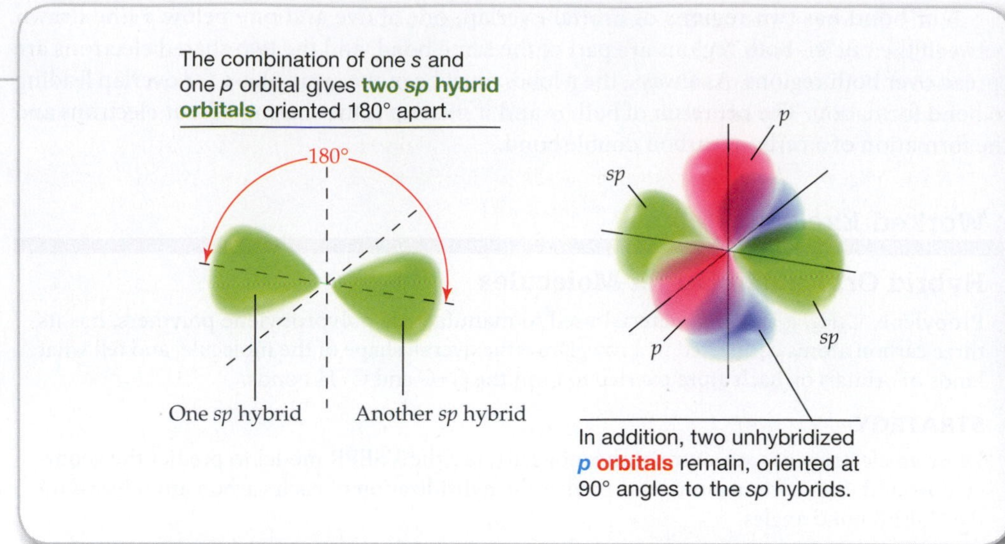

One of the simplest examples of *sp* hybridization occurs in acetylene, H—C≡C—H, a colorless gas used in welding. Both carbon atoms in the acetylene molecule have linear geometry and are *sp*-hybridized. When the two *sp*-hybridized carbon atoms approach each other with their *sp* orbitals aligned head-on for σ bonding, the unhybridized *p* orbitals on each carbon are aligned for π bonding. Two *p* orbitals are aligned in an up/down position, and two are aligned in an in/out position. Thus, there are two mutually perpendicular π bonds that form by sideways overlap of *p* orbitals, along with one σ bond that forms by head-on overlap of the *sp* orbitals. The net result is the sharing of six electrons and formation of a triple bond (**Figure 5.7**). In addition, two C—H bonds form in acetylene by overlap of the remaining two *sp* orbitals with hydrogen 1*s* orbitals.

FIGURE 5.7

Formation of a triple bond by two *sp*-hybridized atoms.

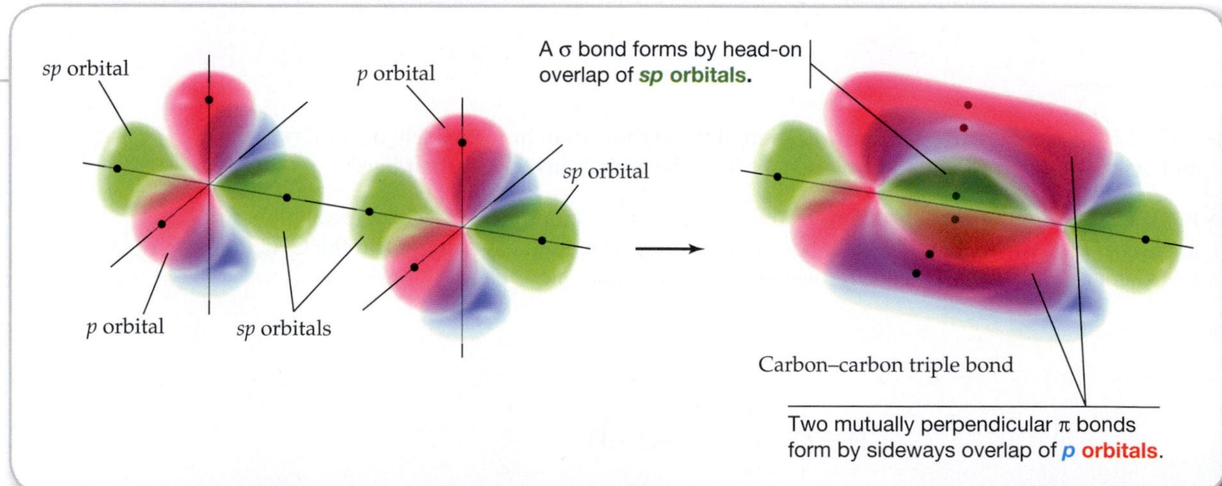

PROBLEM 5.8 Describe the hybridization of the carbon atom in the hydrogen cyanide molecule, H—C≡N, and make a rough sketch to show the hybrid orbitals it uses for bonding.

Atoms with Five and Six Charge Clouds

Main-group atoms with five or six charge clouds, such as the phosphorus in PCl_5 and the sulfur in SF_6, were at one time thought to undergo hybridization by combination of five and six atomic orbitals, respectively. Because a given shell has a total of only four *s* and *p* orbitals, however, the need to use five or six orbitals implies that *d* orbitals are involved. As we'll see in Section 20.11, hybridization involving *d* orbitals is indeed involved for many compounds of transition metals. Recent quantum mechanical studies indicate, however, that main-group compounds do not use *d* orbitals in hybridization but instead use a more complex bonding pattern that is not easily explained by valence bond theory.

A summary of the three common kinds of hybridization for main-group elements and the geometry that each corresponds to is given in **Table 5.2**.

TABLE 5.2 Hybrid Orbitals and Their Geometry

Number of Charge Clouds	Arrangement of Charge Clouds	Hybridization
2	Linear	*sp*
3	Trigonal planar	sp^2
4	Tetrahedral	sp^3

Worked Example 5.4

Predicting the Hybridization of an Atom

Describe the hybridization of the carbon atoms in allene, $H_2C{=}C{=}CH_2$, and make a rough sketch of the molecule showing its hybrid orbitals.

STRATEGY

Draw an electron-dot structure of allene to find the number of charge clouds on each atom.

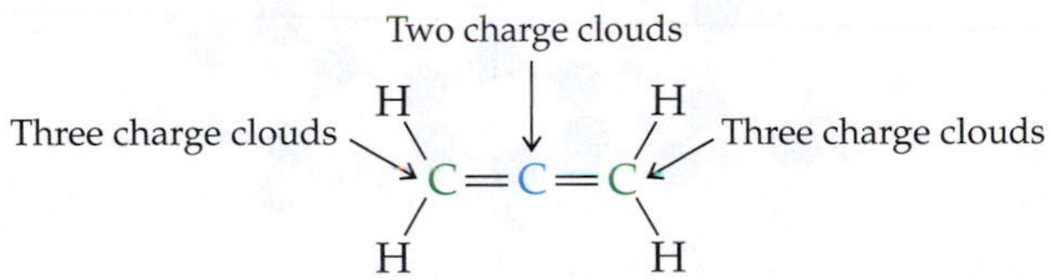

Then predict the geometry around each atom using the VSEPR model in Table 5.1.

SOLUTION

Because the central carbon atom in allene has two charge clouds (two double bonds), it has a linear geometry and is *sp*-hybridized. Because the two terminal carbon atoms have three charge clouds each (one double bond and two C–H bonds), they have trigonal planar geometry and are sp^2-hybridized. The central carbon uses its *sp* orbitals to form two σ bonds at 180° angles and uses its two unhybridized *p* orbitals to form π bonds, one to each of the terminal carbons. Each terminal carbon atom uses an sp^2 orbital for σ bonding to carbon, a *p* orbital for π bonding, and its two remaining sp^2 orbitals for C–H bonds. Note that the mutually perpendicular arrangement of the two π bonds results in a similar perpendicular arrangement of the two CH_2 groups.

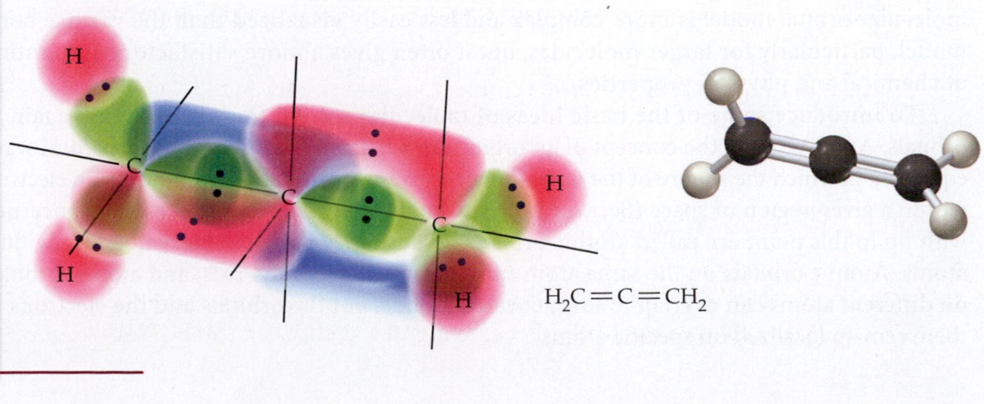

PROBLEM 5.9 Describe the hybridization of the carbon atom in carbon dioxide, and make a rough sketch of the molecule showing its hybrid orbitals and π bonds.

PROBLEM 5.10 Describe the hybridization of the carbon atom in the poisonous gas phosgene, Cl_2CO, and make a rough sketch of the molecule showing its hybrid orbitals and π bonds.

CONCEPTUAL PROBLEM 5.11 Identify each of the following sets of hybrid orbitals:

(a) (b) (c)

CONCEPTUAL PROBLEM 5.12 The following ball-and-stick molecular model is a representation of ibuprofen, a nonprescription drug used for minor pain and inflammation. Only the connections between atoms are shown, not whether the bonds are single, double, or triple (red = O, gray = C, ivory = H).

(a) Indicate the positions of the multiple bonds and lone pairs in ibuprofen.
(b) What is the geometry around each carbon atom?
(c) What is the hybridization of each carbon atom?

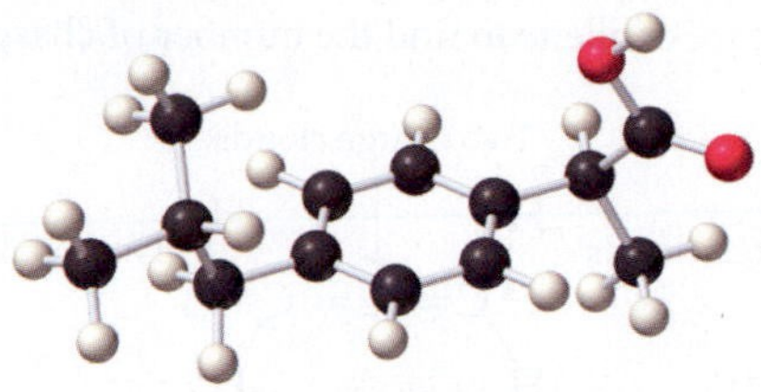

Ibuprofen

5.5 MOLECULAR ORBITAL THEORY: THE HYDROGEN MOLECULE

The valence bond model that describes covalent bonding through orbital overlap is relatively easy to visualize and leads to a satisfactory electronic description for most molecules. It does, however, have some problems. Perhaps the most serious flaw in the valence bond model is that it sometimes leads to an incorrect electronic description. For this reason, another bonding description called **molecular orbital (MO) theory** is often used. The molecular orbital model is more complex and less easily visualized than the valence bond model, particularly for larger molecules, but it often gives a more satisfactory accounting of chemical and physical properties.

To introduce some of the basic ideas of molecular orbital theory, let's look again at orbitals. As we've seen, the concept of an orbital derives from the quantum mechanical wave equation, in which the square of the wave function gives the probability of finding an electron within a given region of space (Section 2.6). The kinds of orbitals that we've been concerned with up to this point are called atomic orbitals because they are characteristic of individual atoms. Atomic orbitals on the same atom can combine to form hybrids, and atomic orbitals on different atoms can overlap to form covalent bonds, but the orbitals and the electrons in them remain localized on specific atoms.

▶ **Atomic orbital** A wave function whose square gives the probability of finding an electron within a given region of space *in an atom.*

Molecular orbital theory takes a different approach to bonding by considering the molecule as a whole rather than concentrating on individual atoms. A **molecular orbital** is to a molecule what an atomic orbital is to an atom.

▶ **Molecular orbital** A wave function whose square gives the probability of finding an electron within a given region of space *in a molecule.*

Like atomic orbitals, molecular orbitals have specific energy levels and specific shapes, and they can be occupied by a maximum of two electrons with opposite spins. The energy and shape of a molecular orbital depend on the size and complexity of the molecule and can thus be fairly complicated, but the fundamental analogy between atomic and molecular orbitals remains. Let's look at the molecular orbital description of the simple diatomic molecule H_2 to see some general features of MO theory.

Imagine what might happen when two isolated hydrogen atoms approach each other and begin to interact. The 1*s* orbitals begin to combine and spread out over both atoms. Molecular orbital theory says that there are two ways for the orbital combination to occur, which we'll refer to as additive and subtractive. The additive combination between two 1*s* atomic orbitals leads to formation of a molecular orbital that is roughly egg-shaped, whereas the subtractive combination leads to formation of a molecular orbital with a **node** between atoms (**Figure 5.8**).

REMEMBER...

A **node** is a surface of zero electron probability separating regions of nonzero probability within an orbital. (Section 2.7)

The additive combination of atomic 1*s* orbitals forms a lower-energy, **bonding molecular orbital**, σ.

The subtractive combination of atomic 1*s* orbitals forms a higher-energy, **antibonding molecular orbital**, σ*, that has a **node** between the nuclei.

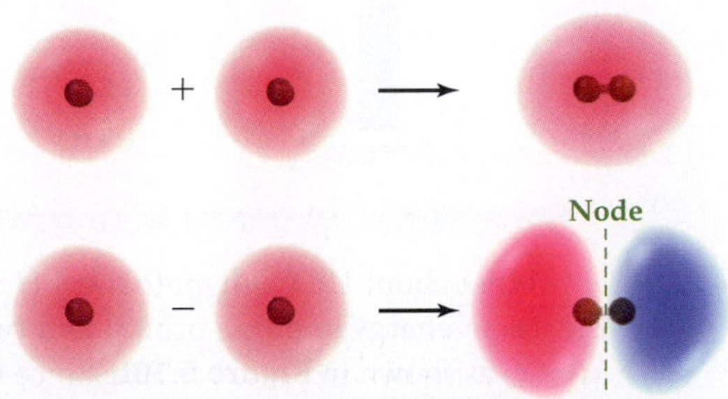

FIGURE 5.8

Formation of molecular orbitals in the H_2 molecule. What are called additive and subtractive combinations lead to two different orbitals.

The additive combination, denoted σ, is lower in energy than the two isolated 1*s* orbitals it's formed from and is called a **bonding molecular orbital** because any electrons it contains spend most of their time in the region between the two nuclei, bonding the atoms together. The subtractive combination, denoted σ^* and spoken as "sigma star," is higher in energy than the two isolated 1*s* orbitals and is called an **antibonding molecular orbital**. Any electrons it contains can't occupy the central region between the nuclei and can't contribute to bonding.

Diagrams of the sort shown in **Figure 5.9** are used to show the energy relationships of the various molecular orbitals. The two isolated H atomic orbitals are shown on either side, and the two H_2 molecular orbitals are shown in the middle. Each of the starting hydrogen atomic orbitals has one electron, which pair up and occupy the lower-energy bonding MO after covalent bond formation.

FIGURE 5.9

Energy levels of molecular orbitals for the H_2 molecule.

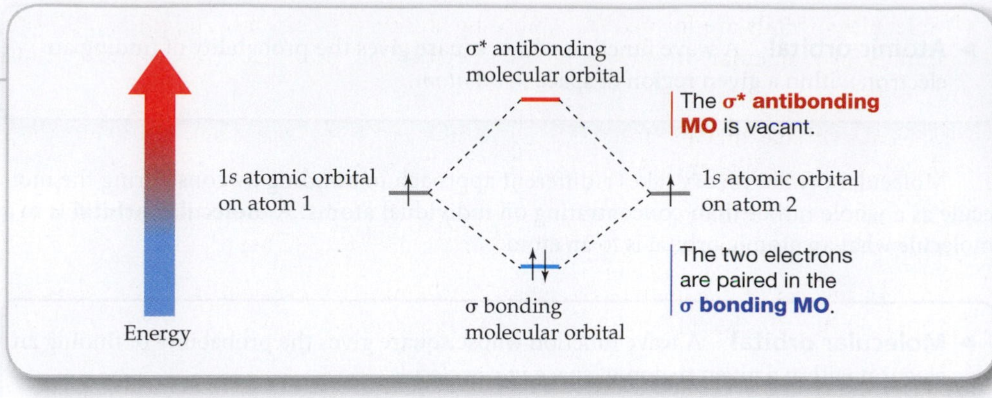

Similar MO diagrams can be drawn, and predictions about stability can be made, for related diatomic species such as H_2^- and He_2. For example, we might imagine constructing the H_2^- ion by bringing together a neutral H· atom with one electron and an H:$^-$ anion with two electrons. Since the resultant H_2^- ion has three electrons, two of them will occupy the lower-energy bonding σ MO and one will occupy the higher-energy antibonding σ^* MO as shown in **Figure 5.10a**. Two electrons are lowered in energy while only one electron is raised in energy, so a net gain in stability results. We therefore predict (and find experimentally) that the H_2^- ion is stable.

FIGURE 5.10

Energy levels of molecular orbitals for (a) the stable H_2^- ion and (b) the unstable He_2 molecule.

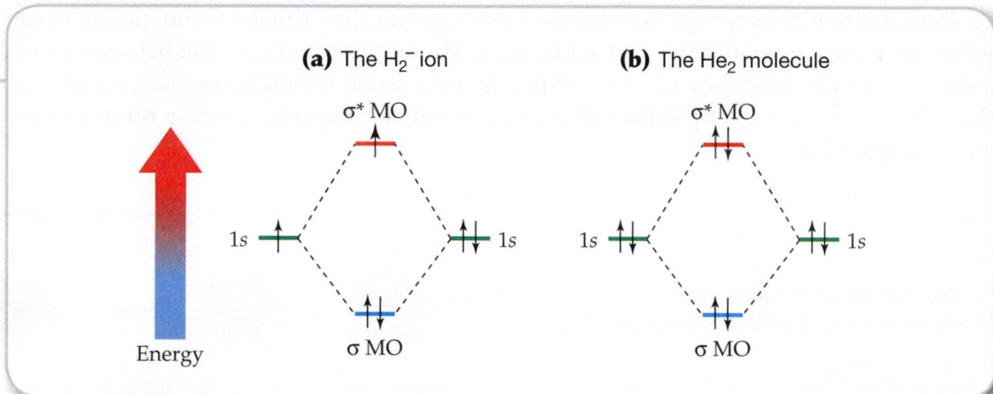

What about He_2? A hypothetical He_2 molecule has four electrons, two of which occupy the lower-energy bonding orbital and two of which occupy the higher-energy antibonding orbital, as shown in **Figure 5.10b**. Since the decrease in energy for the two bonding electrons is counteracted by the increase in energy for the two antibonding electrons, the He_2 molecule has no net bonding energy and is not stable.

Bond orders—the number of electron pairs shared between atoms (Section 4.6)—can be calculated from MO diagrams by subtracting the number of antibonding electrons from the number of bonding electrons and dividing the difference by 2:

$$\text{Bond order} = \frac{(\text{Number of bonding electrons}) - (\text{Number of antibonding electrons})}{2}$$

The H_2 molecule, for instance, has a bond order of 1 because it has two bonding electrons and no antibonding electrons. In the same way, the H_2^- ion has a bond order of 1/2, and the hypothetical He_2 molecule has a bond order of 0, which accounts for the instability of He_2.

The key ideas of the molecular orbital theory of bonding can be summarized as follows:

- Molecular orbitals are to molecules what atomic orbitals are to atoms. A molecular orbital describes a region of space in a molecule where electrons are most likely to be found, and it has a specific size, shape, and energy level.

- Molecular orbitals are formed by combining atomic orbitals on different atoms. The number of molecular orbitals formed is the same as the number of atomic orbitals combined.
- Molecular orbitals that are lower in energy than the starting atomic orbitals are bonding, and MOs that are higher in energy than the starting atomic orbitals are antibonding.
- Electrons occupy molecular orbitals beginning with the MO of lowest energy. A maximum of two electrons can occupy each orbital, and their spins are paired.
- Bond order can be calculated by subtracting the number of electrons in antibonding MOs from the number in bonding MOs and dividing the difference by 2.

PROBLEM 5.13 Construct an MO diagram for the He_2^+ ion. Is this ion likely to be stable? What is its bond order?

5.6 MOLECULAR ORBITAL THEORY: OTHER DIATOMIC MOLECULES

Having looked at bonding in the H_2 molecule, let's move up a level in complexity by looking at the bonding in several second-row diatomic molecules—N_2, O_2, and F_2. The valence bond model developed in Section 5.2 predicts that the nitrogen atoms in N_2 are triply bonded and have one lone pair each, that the oxygen atoms in O_2 are doubly bonded and have two lone pairs each, and that the fluorine atoms in F_2 are singly bonded and have three lone pairs each:

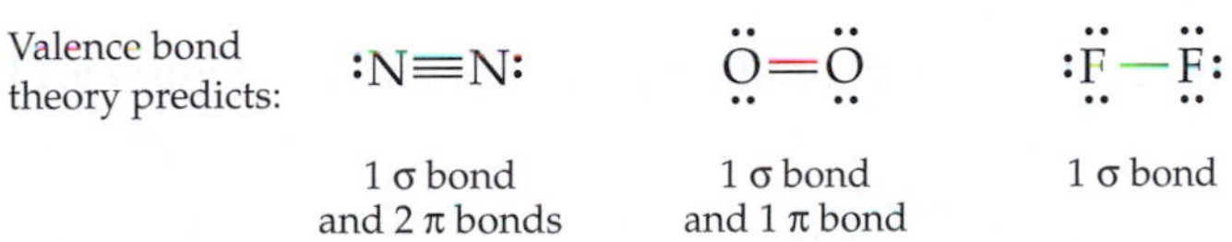

Unfortunately, this simple valence bond picture can't be right because it predicts that the electrons in all three molecules are *spin-paired*. In other words, the electron-dot structures indicate that the occupied atomic orbitals in all three molecules contain two electrons each. It can be demonstrated experimentally, however, that the O_2 molecule has two electrons that are not spin-paired and that these electrons therefore must be in different, singly occupied orbitals.

Experimental evidence for the electronic structure of O_2 rests on the observation that substances with unpaired electrons are attracted by magnetic fields and are thus said to be **paramagnetic**. The more unpaired electrons a substance has, the stronger the paramagnetic attraction. Substances whose electrons are all spin-paired, by contrast, are weakly repelled by magnetic fields and are said to be **diamagnetic**. Both N_2 and F_2 are diamagnetic, just as predicted by their electron-dot structures, but O_2 is paramagnetic. When liquid O_2 is poured over the poles of a strong magnet, the O_2 is attracted to the poles, as shown in **Figure 5.11**.

FIGURE 5.11

Paramagnetism. Why does liquid O_2 stick to the poles of a magnet?

Why is O_2 paramagnetic? Electron-dot structures and valence bond theory fail to answer this question, but MO theory explains the experimental results nicely. In a molecular orbital description of N_2, O_2, and F_2, two atoms come together and their valence-shell atomic orbitals interact to form molecular orbitals. Four orbital interactions occur in all three cases, leading to the formation of four bonding MOs and four antibonding MOs, whose relative energies are shown in **Figure 5.12**. (Note that the relative energies of the σ_{2p} and π_{2p} orbitals in N_2 are different from those in O_2 and F_2.)

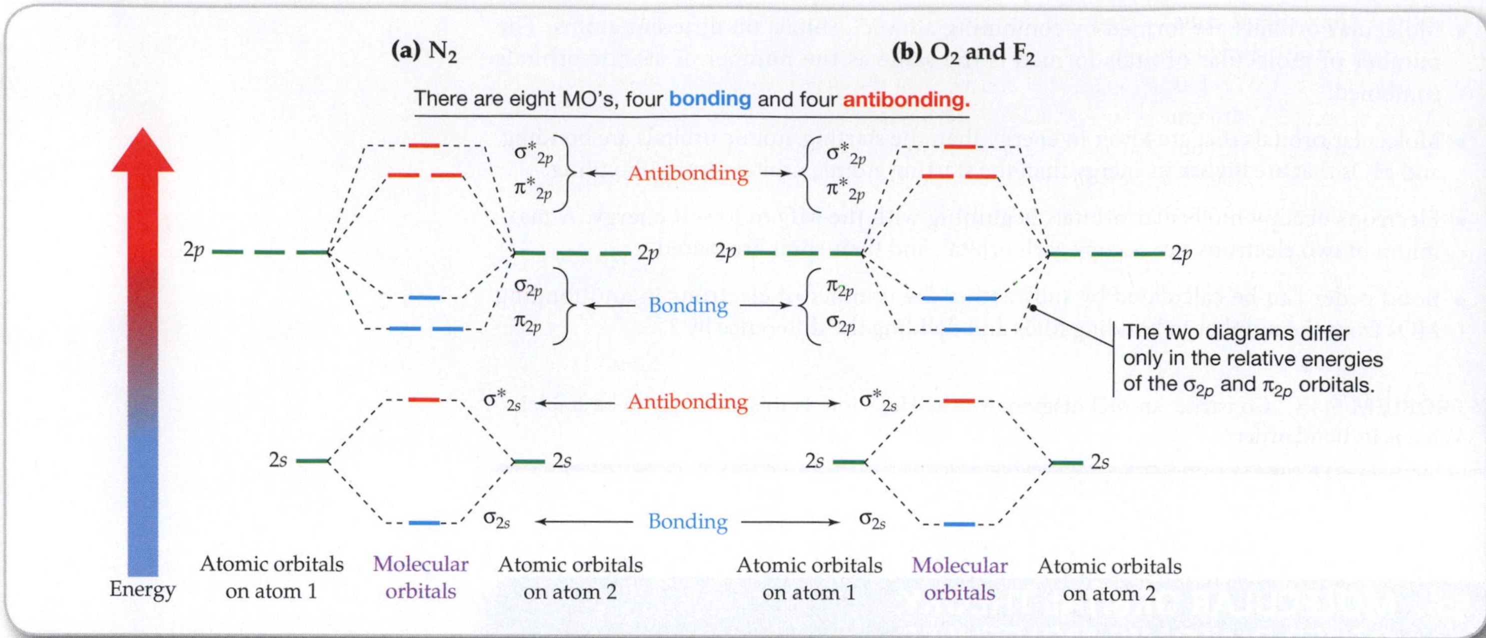

FIGURE 5.12

Energy levels of molecular orbitals for (a) N_2 and (b) O_2 and F_2.

The diagrams in Figure 5.12 show the following orbital interactions:

- The 2*s* orbitals interact to give σ_{2s} and σ^*_{2s} MOs.
- The two 2*p* orbitals that lie on the internuclear axis interact head-on to give σ_{2p} and σ^*_{2p} MOs.
- The two remaining pairs of 2*p* orbitals that are perpendicular to the internuclear axis interact in a sideways manner to give two **degenerate** π_{2p} and two degenerate π^*_{2p} MOs oriented 90° apart.

REMEMBER...

Degenerate orbitals are those that have the same energy. (Section 2.11)

The shapes of the σ_{2p}, σ^*_{2p}, π_{2p}, and π^*_{2p} MOs are shown in **Figure 5.13**.

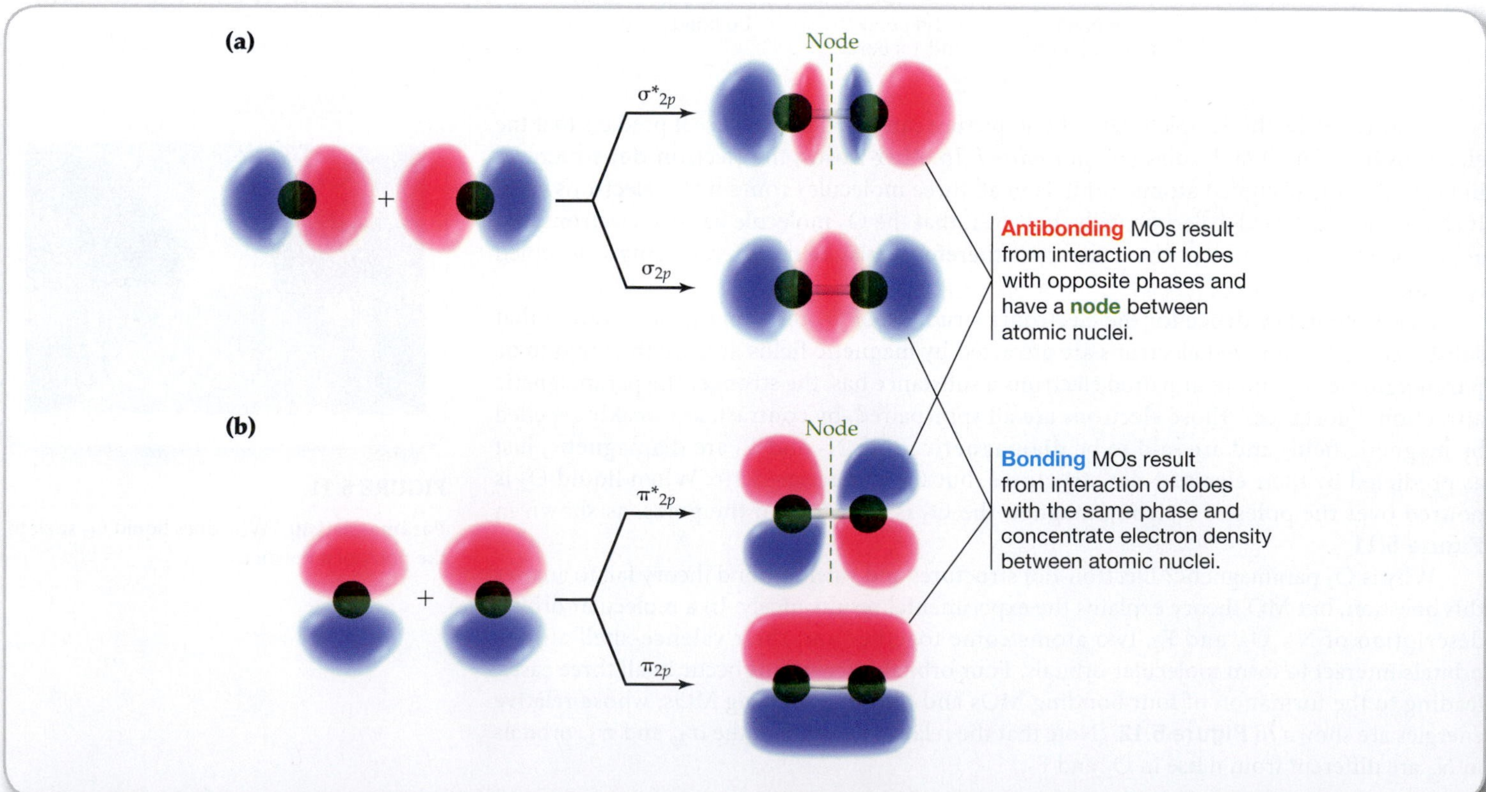

FIGURE 5.13

Formation of (a) σ_{2p} and σ^*_{2p} MOs by head-on interaction of two *p* atomic orbitals, and (b) π_{2p} and π^*_{2p} MOs by sideways interaction of two *p* orbitals.

When appropriate numbers of valence electrons are added to occupy the molecular orbitals, the results shown in **Figure 5.14** are obtained. Both N_2 and F_2 have all their electrons spin-paired, but O_2 has two unpaired electrons in the degenerate π^*_{2p} orbitals. Both N_2 and F_2 are therefore diamagnetic, whereas O_2 is paramagnetic.

We should also point out that MO diagrams like those in Figure 5.14 are usually obtained from mathematical calculations and can't necessarily be predicted. MO theory is therefore less easy to visualize and understand on an intuitive level than valence bond theory.

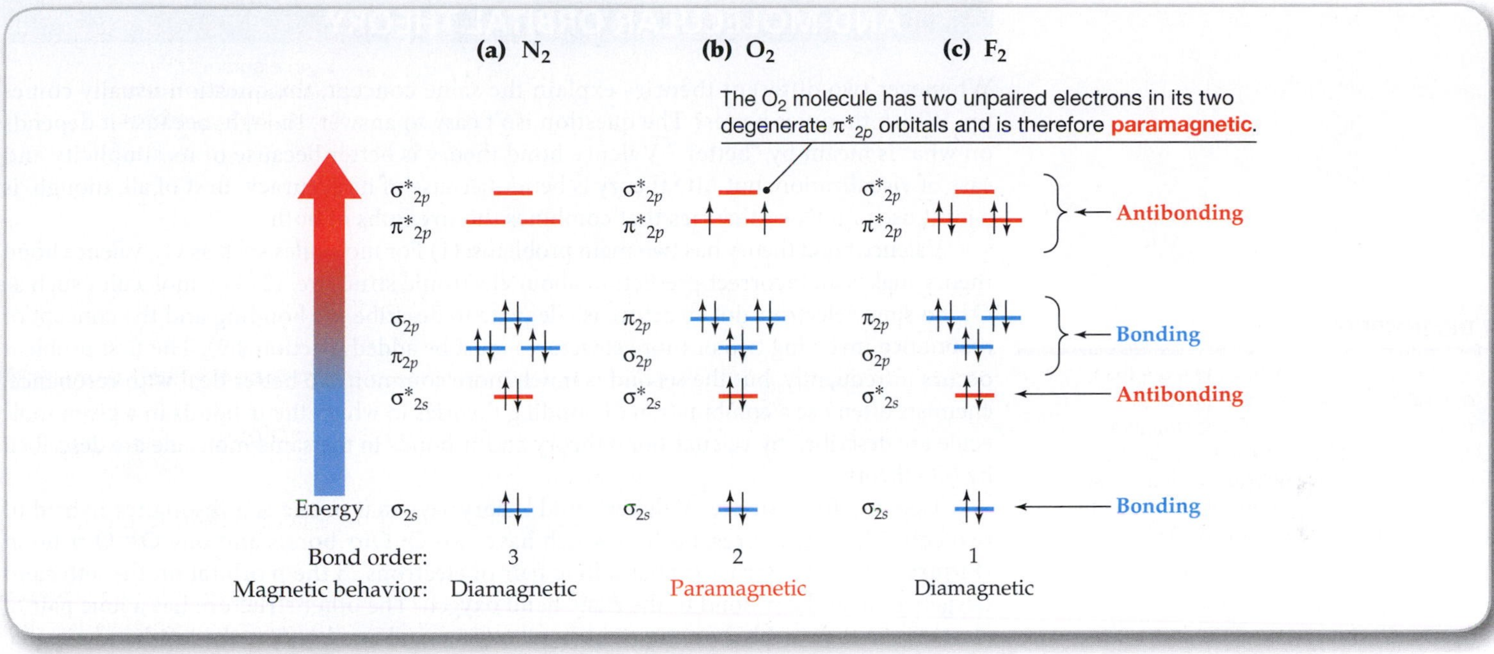

FIGURE 5.14

Energy levels of molecular orbitals for the second-row diatomic molecules (a) N_2, (b) O_2, and (c) F_2.

Worked Example 5.5

Constructing an MO Diagram for a Diatomic Molecule

Nitrogen monoxide (NO), usually called nitric oxide, is a diatomic molecule with a molecular orbital diagram similar to that of N_2 (Figure 5.12a). Show the MO diagram for nitric oxide, and predict its bond order. Is NO diamagnetic or paramagnetic?

STRATEGY

Count the number of valence electrons in NO, and assign them to the available MOs shown in Figure 5.12a, beginning with the lowest energy orbital.

SOLUTION

Nitric oxide has 5 valence electrons from N and 6 from O for a total of 11. Assigning the 11 electrons to the MOs in Figure 5.12a shows that NO has 8 bonding and 3 antibonding electrons and a bond order of $(8 - 3)/2 = 2.5$. Because it has one unpaired electron, it is paramagnetic.

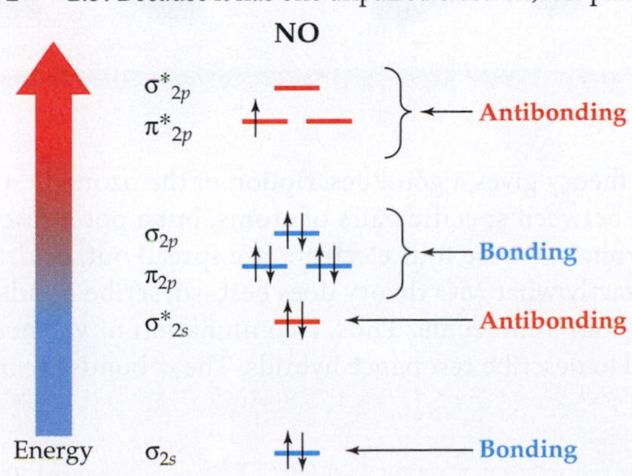

PROBLEM 5.14 The B_2 and C_2 molecules have MO diagrams similar to that of N_2 in Figure 5.12a. What MOs are occupied in B_2 and C_2, and what is the bond order in each? Would you expect either of these substances to be paramagnetic?

5.7 COMBINING VALENCE BOND THEORY AND MOLECULAR ORBITAL THEORY

Whenever two different theories explain the same concept, the question usually comes up: Which theory is better? The question isn't easy to answer, though, because it depends on what is meant by "better." Valence bond theory is better because of its simplicity and ease of visualization, but MO theory is better because of its accuracy. Best of all, though, is a joint use of the two theories that combines the strengths of both.

Valence bond theory has two main problems: (1) For molecules such as O_2, valence bond theory makes an incorrect prediction about electronic structure. (2) For molecules such as O_3, no single electron-dot structure is adequate to describe the bonding and the concept of **resonance** involving two or more structures must be added (Section 4.9). The first problem occurs infrequently, but the second is much more common. To better deal with resonance, chemists often use a combination of bonding theories in which the σ bonds in a given molecule are described by valence bond theory and π bonds in the same molecule are described by MO theory.

REMEMBER...

Whenever it's possible to write more than one valid electron-dot structure for a molecule, the actual electronic structure is an average of the different possibilities, called a **resonance hybrid**. The individual structures differ only in the placement of the valence-shell electrons. (Section 4.9)

Take O_3, for instance. Valence bond says that ozone is a resonance hybrid of two equivalent structures, both of which have two O–O σ bonds and one O=O π bond (**Figure 5.15**). One structure has a lone pair of electrons in the *p* orbital on the left-hand oxygen atom and a π bond to the right-hand oxygen. The other structure has a lone pair of electrons in the *p* orbital on the right-hand oxygen and a π bond to the left-hand oxygen. The actual structure of O_3 is an average of the two resonance forms in which four electrons occupy the entire region encompassed by the overlapping set of three *p* orbitals. The only difference between the resonance structures is in the placement of *p* electrons. The atoms themselves are in the same positions in both, and the geometries are the same in both.

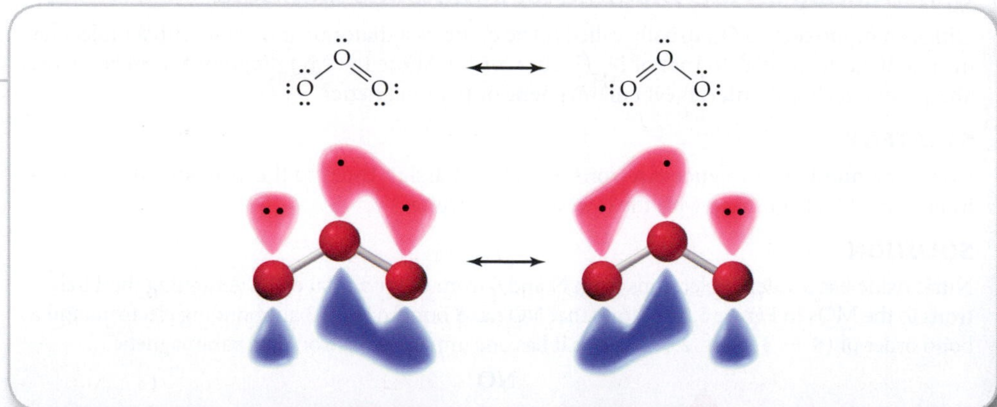

FIGURE 5.15

The structure of ozone.

Valence bond theory gives a good description of the ozone O–O σ bonds, whose electrons are localized between specific pairs of atoms, but a poor description of the π bonds among *p* atomic orbitals, whose four electrons are spread out, or *delocalized*, over the molecule. Yet this is exactly what MO theory does best—describe bonds in which electrons are delocalized throughout a molecule. Thus, a combination of valence bond theory and MO theory is often used to describe resonance hybrids. The σ bonds are best described in valence bond terminology as being localized between pairs of atoms, and the π electrons are best described by MO theory as being delocalized over the entire molecule.

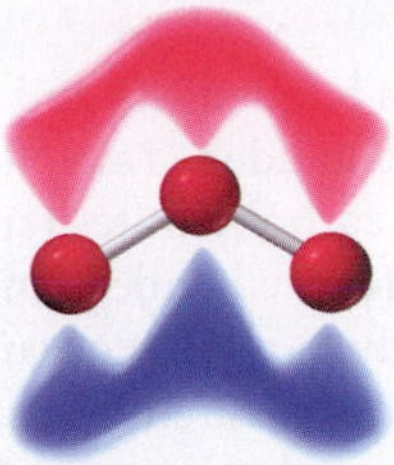

Lowest-energy π molecular orbital

Worked Example 5.6

Combining Valence Bond and MO Theory

Acetamide (C_2H_5NO), an organic compound that has been detected in interstellar space, has the following connections among atoms and can be drawn using two nonequivalent electron-dot resonance structures. Draw both, and sketch a π molecular orbital showing how the π electrons are delocalized over the oxygen and nitrogen atoms.

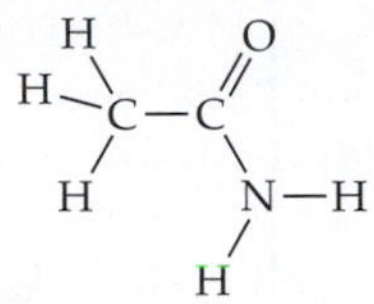

Acetamide

STRATEGY

Draw the two electron-dot resonance structures in the usual way (Section 4.7), including all nonbonding electrons. The two structures differ in that one has a C=O double bond and a lone pair on nitrogen, while the other has a C=N double bond and an additional lone pair on oxygen. Thus, a π molecular orbital corresponding to the O–C–N resonance includes all three atoms.

SOLUTION

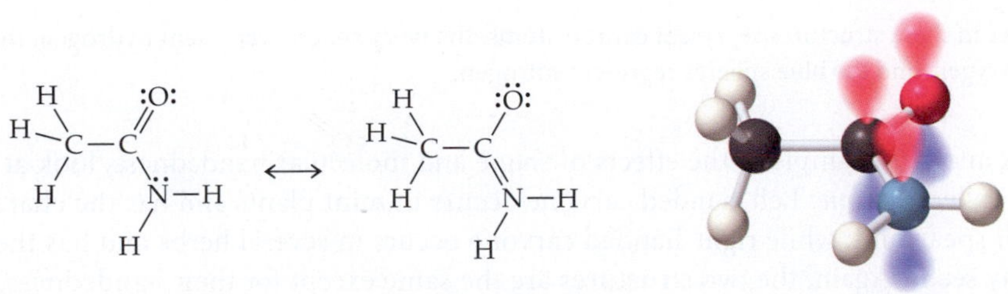

PROBLEM 5.15 Draw two electron-dot resonance structures for the formate ion, HCO_2^-, and sketch a π molecular orbital showing how the π electrons are delocalized over both oxygen atoms.

FYI MOLECULAR SHAPE, HANDEDNESS, AND DRUGS

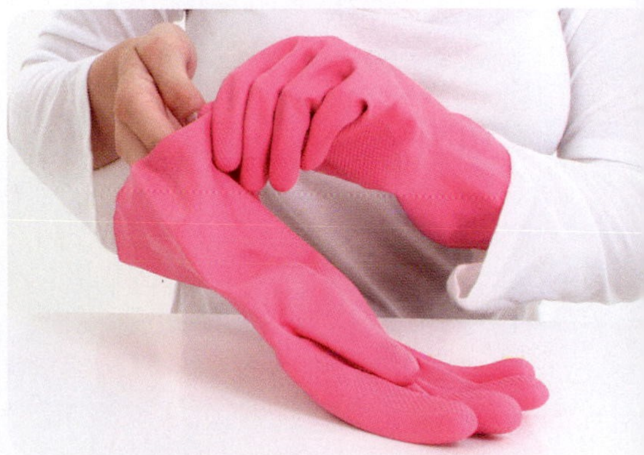

▲ A right hand fits only into a right-handed glove with a complementary shape, not into a left-handed glove.

Why does a right-handed glove fit so nicely on your right hand but not on your left hand? Why do the threads on a lightbulb twist only in one direction so that you have to turn the bulb clockwise to screw it in? The reason has to do with the shapes of the glove and the lightbulb threads and the fact that both have a *handedness* to them. When the right-handed glove is held up to a mirror, the reflected image looks like a left-handed glove. (Try it.) When the lightbulb with clockwise threads is reflected in a mirror, the threads in the mirror image twist in a counterclockwise direction.

Molecules, too, can have shapes that give them a handedness and they can thus exist in mirror-image forms, one right-handed and one left-handed. Take, for example, the main classes of biomolecules found in living organisms: carbohydrates (sugars), proteins, fats, and nucleic acids. These and most other biomolecules are handed, and usually only one of the two possible mirror-image forms occurs naturally in a given organism. The other form can often be made in the laboratory but does not occur naturally.

The biological consequences of molecular shape can be dramatic. Look at the structures of dextromethorphan and levomethorphan, for instance. (The Latin prefixes *dextro-* and *levo-* mean "right" and "left," respectively.) Dextromethorphan is a common cough suppressant found in many over-the-counter cold medicines, but its mirror image, levomethorphan, is a powerful narcotic pain-reliever similar in its effects to morphine. The two substances are chemically identical except for their handedness, yet their biological properties are very different.

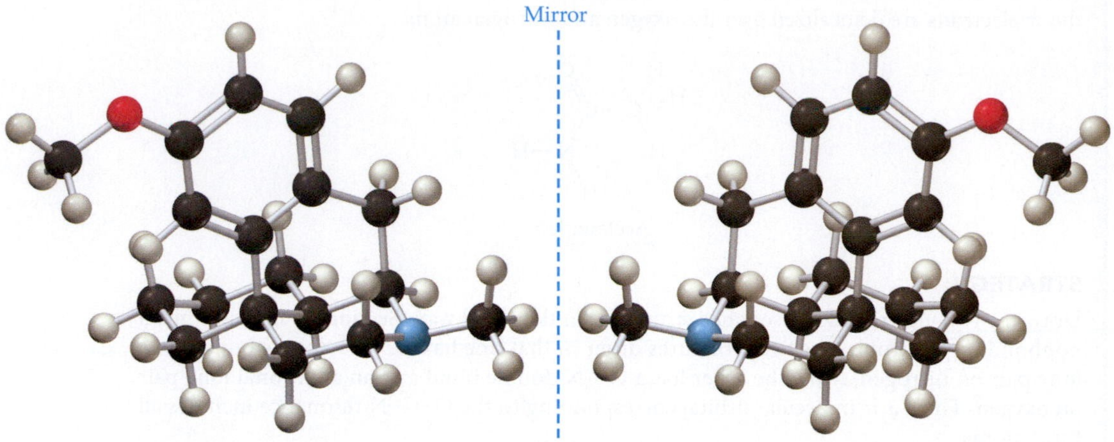

Levomethorphan
(a narcotic analgesic)

Dextromethorphan
(a cough suppressant)

▲ The black spheres in these structures represent carbon atoms, the ivory spheres represent hydrogen, the red spheres represent oxygen, and the blue spheres represent nitrogen.

As another example of the effects of shape and molecular handedness, look at the substance called *carvone*. Left-handed carvone occurs in mint plants and has the characteristic odor of spearmint, while right-handed carvone occurs in several herbs and has the odor of caraway seeds. Again, the two structures are the same except for their handedness, yet they have entirely different odors.

Why do different mirror-image forms of molecules have different biological properties? The answer goes back to the question about why a right-handed glove fits only the right hand. A right hand in a right-handed glove is a perfect match because the two shapes are complementary. Putting the right hand into a left-handed glove produces a mismatch because the two shapes are *not* complementary. In the same way, handed molecules such as dextromethorphan and carvone have specific shapes that match only complementary-shaped receptor sites in the body. The mirror-image forms of the molecules can't fit into the receptor sites and thus don't elicit the same biological response.

Precise molecular shape is critically important to every living organism. Almost every chemical interaction in living systems is governed by complementarity between handed molecules and glovelike receptor sites.

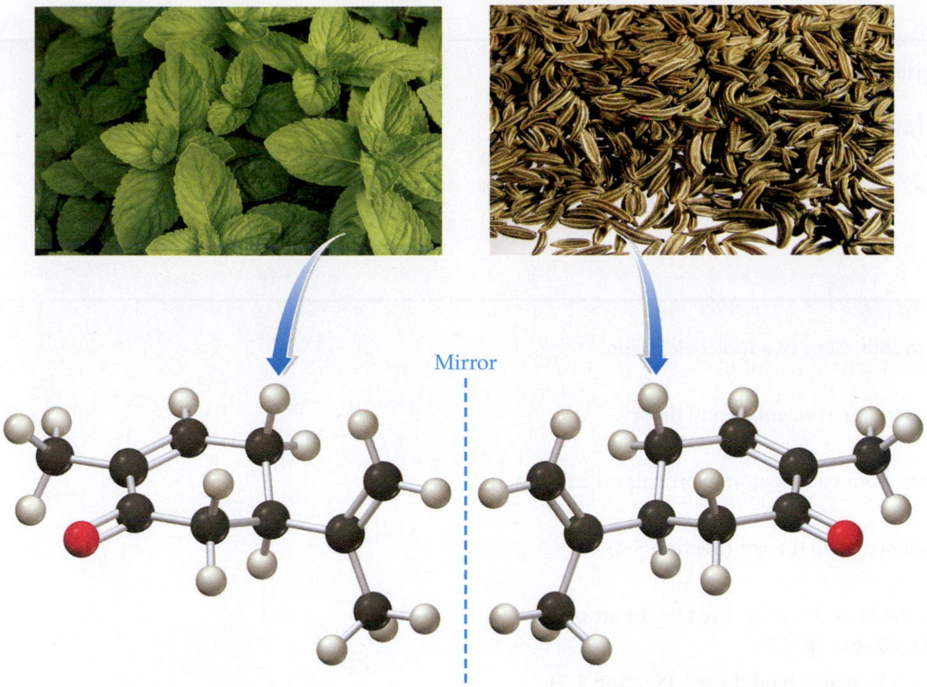

"Left-handed" carvone
(odor of spearmint)

"Right-handed" carvone
(odor of caraway)

◀ Two plants both produce carvone, but the mint leaves yield the "left-handed" form, while the caraway seeds yield the "right-handed" form.

PROBLEM 5.16 Why is molecular shape so important in biological chemistry?

CONCEPTUAL PROBLEM 5.17 One of the following two molecules has a handedness to it and can exist in two mirror-image forms; the other does not. Which is which? Why?

(a) **(b)**

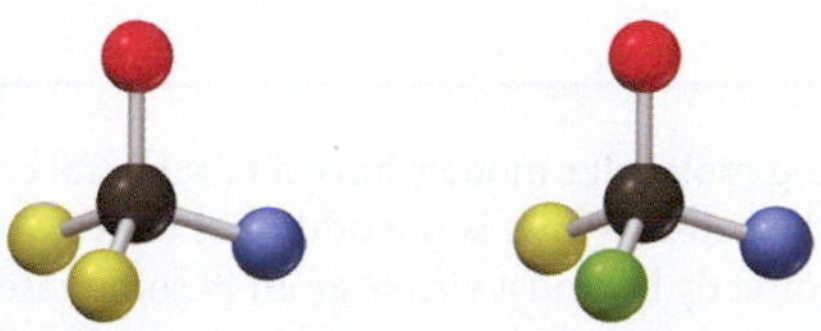

SUMMARY

Molecular shape can often be predicted using the **valence-shell electron-pair repulsion (VSEPR) model**, which treats the electrons around atoms as charge clouds that repel one another and therefore orient themselves as far away from one another as possible. Atoms with two charge clouds adopt a linear arrangement of the clouds, atoms with three charge clouds adopt a trigonal planar arrangement, and atoms with four charge clouds adopt a tetrahedral arrangement. Similarly, atoms with five charge clouds are trigonal bipyramidal and atoms with six charge clouds are octahedral.

According to **valence bond theory**, covalent bond formation occurs by the overlap of two singly occupied atomic orbital lobes of the same phase, either head-on along the internuclear axis to form a **sigma (σ) bond** or sideways above and below the internuclear axis to form a **pi (π) bond**. The observed geometry of covalent bonding in main-group compounds is described by assuming that *s* and *p* atomic orbitals combine to generate **hybrid orbitals**, which are strongly oriented in specific directions: ***sp*** **hybrid orbitals** have linear geometry, ***sp***^{2} **hybrid orbitals** have trigonal planar geometry, and ***sp***^{3} **hybrid orbitals** have tetrahedral geometry. The bonding of main-group atoms with five and six charge clouds is more complex.

Molecular orbital theory sometimes gives a more accurate picture of electronic structure than the valence bond model. A **molecular orbital** is a wave function whose square gives the probability of finding an electron in a given region of space in a molecule. Combination of two atomic orbitals gives two molecular orbitals, a **bonding MO** that is lower in energy than the starting atomic orbitals and an **antibonding MO** that is higher in energy than the starting atomic orbitals. Molecular orbital theory is particularly useful for describing delocalized π bonding in molecules.

KEY WORDS

antibonding molecular orbital *185*
bond angle *169*
bonding molecular orbital *185*
diamagnetic *187*
hybrid atomic orbital *178*
molecular orbital *185*
molecular orbital (MO) theory *184*
paramagnetic *187*
pi (π) bond *180*
sigma (σ) bond *177*
***sp* hybrid orbital** *181*
***sp*^{2} hybrid orbital** *179*
***sp*^{3} hybrid orbital** *178*
valence bond theory *176*
valence-shell electron-pair repulsion (VSEPR) model *169*

LEARNING OUTCOMES

- Use the VSEPR model to predict the bond angles and overall shape of a molecule or ion. **(Section 5.1)** ***Problems 1–4, 18–41, 73, 76–78***
- Describe how covalent sigma (σ) bonds are formed according to valence bond theory. **(Section 5.2)** ***Problems 5, 42***
- Describe the three kinds of hybrid orbitals that can form from *s* and *p* atomic orbitals on an atom. **(Sections 5.3, 5.4)** ***Problems 5–8***
- Describe how pi (π) bonds are formed according to valence bond theory. **(Section 5.4)** ***Problems 6, 7, 42***
- Relate the number of charge clouds around an atom to the hybrid orbitals used by the atom in bonding. **(Section 5.4)** ***Problems 9–12, 21–23, 44–51, 62–66, 70–73***
- Describe how covalent bonds are formed according to molecular orbital theory. **(Section 5.5)** ***Problems, 52, 53***
- Interpret the molecular orbital diagram for a second-row diatomic molecule or ion. **(Sections 5.5, 5.6)** ***Problems 13, 14, 54–59, 68, 69, 74, 75, 79***
- Use a molecular orbital diagram to determine the bond order of a diatomic molecule or ion. **(Section 5.5)** ***Problems 13, 14, 54–57, 75, 79***
- Use a molecular orbital diagram to predict whether a molecule is diamagnetic or paramagnetic. **(Section 5.6)** ***Problems 14, 58, 59, 74, 79***
- Describe how valence bond theory and molecular orbital theory can be combined to emphasize the strengths of each. **(Section 5.7)** ***Problems 15, 50, 51, 60, 61***

CONCEPTUAL PROBLEMS

Problems 5.1–5.17 appear within the chapter.

5.18 What is the geometry around the central atom in each of the following molecular models?

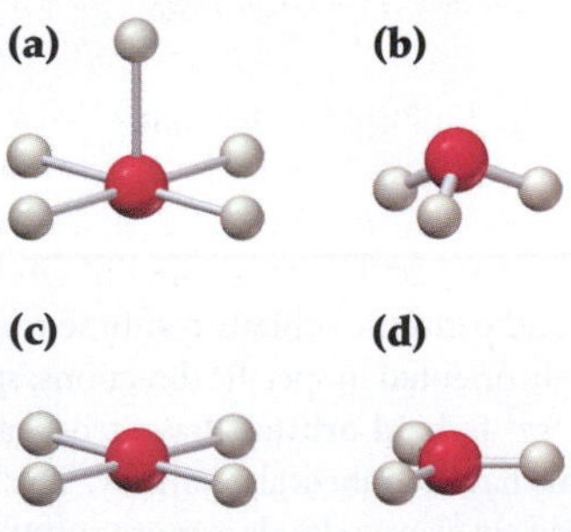

5.19 What is the geometry around the central atom in each of the following molecular models? (There may be a "hidden" atom directly behind a visible atom in some cases.)

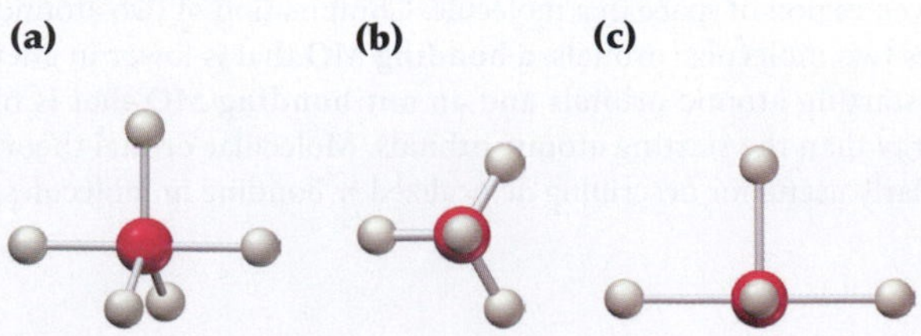

5.20 Three of the following molecular models have a tetrahedral central atom, and one does not. Which is the odd one? (There may be a "hidden" atom directly behind a visible atom in some cases.)

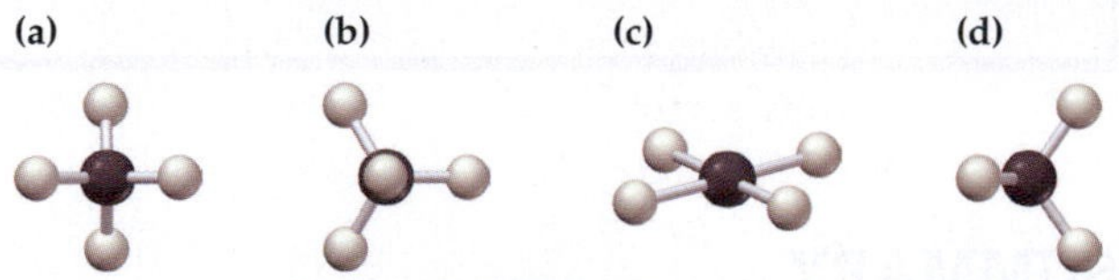

5.21 The VSEPR model is a simple predictive tool that is usually, but not always, correct. Take urea, for instance, a waste product excreted in animal urine:

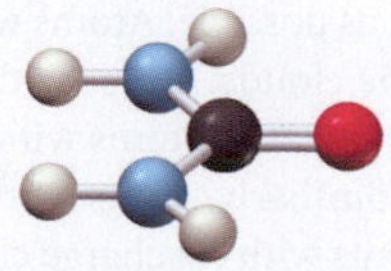
Urea

What hybridization would you expect for the C and N atoms in urea according to the VSEPR model, and what approximate values would you expect for the various bond angles? What are the actual hybridizations and bond angles based on the molecular model shown? (red = O, gray = C, blue = N, ivory = H)

5.22 The following ball-and-stick molecular model is a representation of acetaminophen, the active ingredient in several over-the-counter headache remedies (red = O, gray = C, blue = N, ivory = H):

(a) What is the geometry around each carbon in acetaminophen?
(b) What is the hybridization of each carbon?

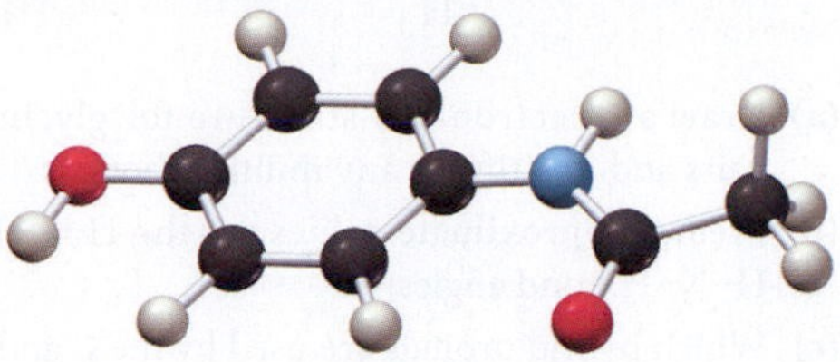

Acetaminophen

5.23 The following ball-and-stick molecular model is a representation of thalidomide, a drug that causes birth defects when taken by expectant mothers (red = O, gray = C, blue = N, ivory = H).

(a) What is the geometry around each carbon in thalidomide?
(b) What is the hybridization of each carbon?

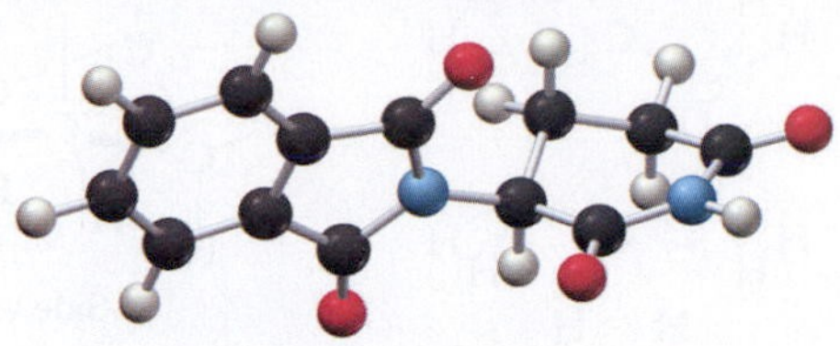

Thalidomide

SECTION PROBLEMS

The VSEPR Model (Section 5.1)

5.24 What geometric arrangement of charge clouds do you expect for atoms that have the following number of charge clouds?

(a) 3 (b) 5 (c) 2 (d) 6

5.25 What shape do you expect for molecules that meet the following descriptions?

(a) A central atom with two lone pairs and three bonds to other atoms
(b) A central atom with two lone pairs and two bonds to other atoms
(c) A central atom with two lone pairs and four bonds to other atoms

5.26 How many charge clouds are there around central atoms that have the following geometries?

(a) Tetrahedral (b) Octahedral
(c) Bent (d) Linear
(e) Square pyramidal (f) Trigonal pyramidal

5.27 How many charge clouds are there around central atoms that have the following geometries?

(a) Seesaw (b) Square planar
(c) Trigonal bipyramidal (d) T-shaped
(e) Trigonal planar (f) Linear

5.28 What shape do you expect for each of the following molecules?

(a) H_2Se (b) $TiCl_4$ (c) O_3 (d) GaH_3

5.29 What shape do you expect for each of the following molecules?

(a) XeO_4 (b) SO_2Cl_2 (c) OsO_4 (d) SeO_2

5.30 What shape do you expect for each of the following molecules or ions?

(a) SbF_5 (b) IF_4^+ (c) SeO_3^{2-} (d) CrO_4^{2-}

5.31 Predict the shape of each of the following ions:

(a) NO_3^- (b) NO_2^+ (c) NO_2^-

5.32 What shape do you expect for each of the following anions?

(a) PO_4^{3-} (b) MnO_4^- (c) SO_4^{2-} (d) SO_3^{2-}
(e) ClO_4^- (f) SCN^-

5.33 What shape do you expect for each of the following cations?

(a) XeF_3^+ (b) SF_3^+ (c) ClF_2^+ (d) CH_3^+

5.34 What bond angles do you expect for each of the following?

(a) The F–S–F angle in SF_2
(b) The H–N–N angle in N_2H_2
(c) The F–Kr–F angle in KrF_4
(d) The Cl–N–O angle in NOCl

5.35 What bond angles do you expect for each of the following?

(a) The Cl–P–Cl angle in PCl_6^-
(b) The Cl–I–Cl angle in ICl_2^-
(c) The O–S–O angle in SO_4^{2-}
(d) The O–B–O angle in BO_3^{3-}

5.36 Acrylonitrile is used as the starting material for manufacturing acrylic fibers. Predict values for all bond angles in acrylonitrile.

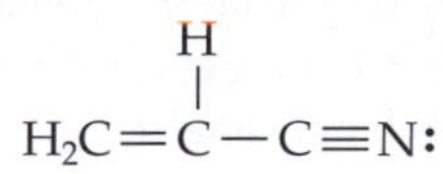

Acrylonitrile

5.37 Predict values for all bond angles in dimethyl sulfoxide, a powerful solvent used in veterinary medicine to treat inflammation.

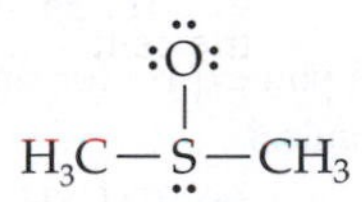

Dimethyl sulfoxide

5.38 Oceanographers study the mixing of water masses by releasing tracer molecules at a site and then detecting their presence at other places. The molecule trifluoromethylsulfur pentafluoride is one such tracer. Draw an electron-dot structure for CF_3SF_5, and predict the bond angles around both carbon and sulfur.

5.39 A potential replacement for the chlorofluorocarbon refrigerants that harm the Earth's protective ozone layer is a compound called E143a, or trifluoromethyl methyl ether, F_3COCH_3. Draw an electron-dot structure for F_3COCH_3, and predict the geometry around both the carbons and the oxygen.

5.40 Explain why cyclohexane, a substance that contains a six-membered ring of carbon atoms, is not flat but instead has a puckered, nonplanar shape. Predict the values of the C–C–C bond angles.

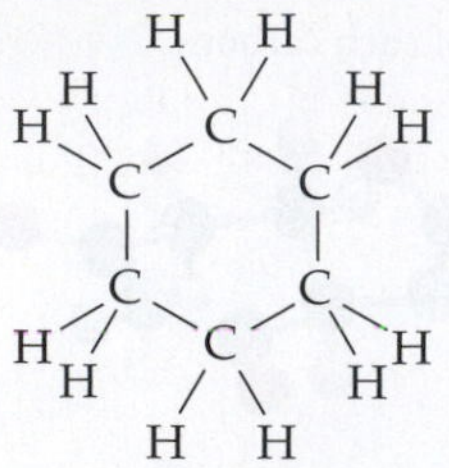

Cyclohexane

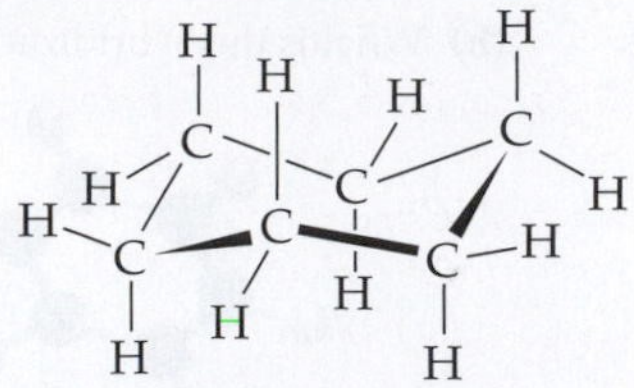

Side view

5.41 Like cyclohexane (Problem 5.40), benzene also contains a six-membered ring of carbon atoms, but it is flat rather than puckered. Explain, and predict the values of the C–C–C bond angles.

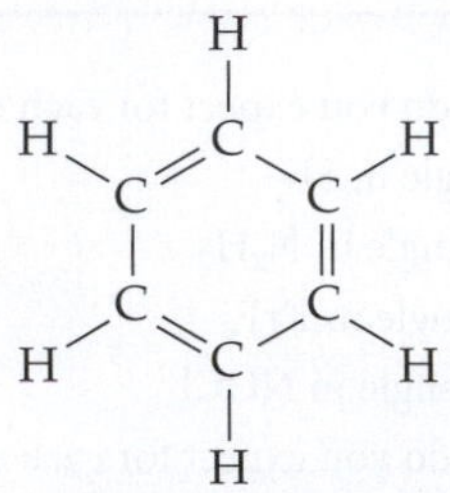

Benzene

Valence Bond Theory (Sections 5.2–5.4)

5.42 What is the difference in spatial distribution between electrons in a π bond and electrons in a σ bond?

5.43 What is the difference in spatial distribution between electrons in an sp^2 orbital and electrons in an sp^3 orbital?

5.44 What hybridization do you expect for atoms that have the following numbers of charge clouds?

(a) 2 (b) 3 (c) 4

5.45 What spatial arrangement of charge clouds corresponds to each of the following kinds of hybridization?

(a) sp^3 (b) sp^2 (c) sp

5.46 What hybridization would you expect for the indicated atom in each of the following molecules?

(a) $H_2C{=}O$ (b) BH_3 (c) CH_3SH (d) $H_2C{=}NH$

5.47 What hybridization would you expect for the indicated atom in each of the following ions?

(a) BH_4^- (b) HCO_2^- (c) CH_3^+ (d) CH_3^-

5.48 Oxaloacetic acid is an intermediate involved in the citric acid cycle of food metabolism. What are the hybridizations of the various carbon atoms in oxaloacetic acid, and what are the approximate values of the various bond angles?

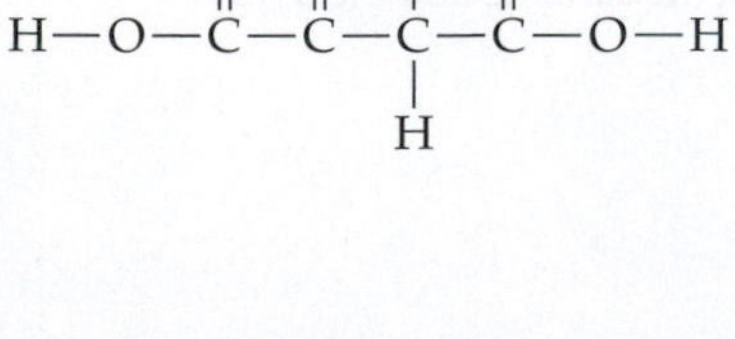

Oxaloacetic acid

5.49 The atoms in the amino acid glycine are connected as shown:

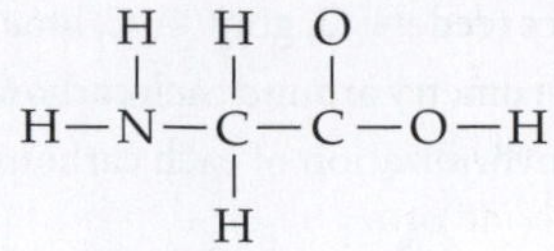

Glycine

(a) Draw an electron-dot structure for glycine, showing lone pairs and identifying any multiple bonds.

(b) Predict approximate values for the H–C–H, O–C–O, and H–N–H bond angles.

(c) Which hybrid orbitals are used by the C and N atoms?

5.50 Bupropion, marketed as Wellbutrin, is a heavily prescribed medication used in the treatment of depression. What are the hybridizations of the various carbon atoms?

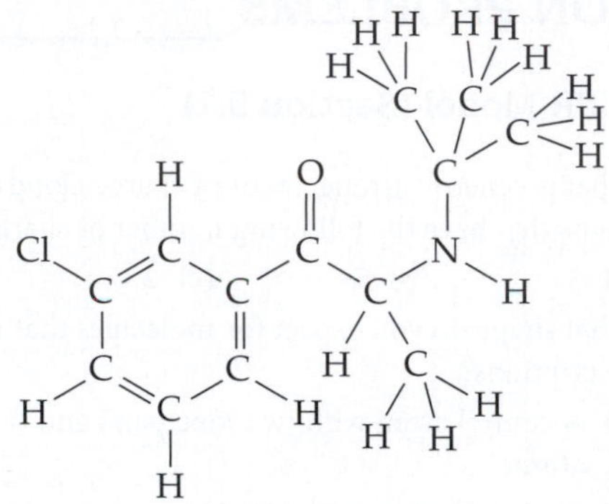

Bupropion

5.51 Efavirenz, marketed as Sustiva, is a medication used in the treatment of human immunodeficiency virus, HIV. What are the hybridizations of the various carbon atoms?

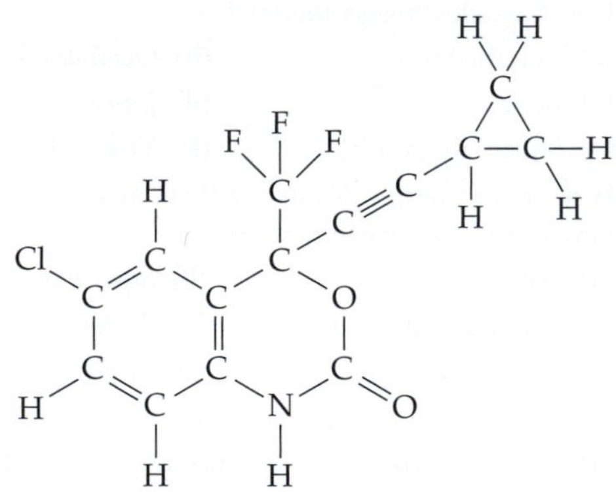

Sustiva

Molecular Orbital Theory (Sections 5.5–5.6)

5.52 What is the difference in spatial distribution between electrons in a molecular orbital and electrons in an atomic orbital?

5.53 What is the difference in spatial distribution between electrons in a bonding MO and electrons in an antibonding MO?

5.54 Use the MO energy diagram in Figure 5.12b to describe the bonding in O_2^+, O_2, and O_2^-. Which of the three is likely to be stable? What is the bond order of each? Which contain unpaired electrons?

5.55 Use the MO energy diagram in Figure 5.12a to describe the bonding in N_2^+, N_2, and N_2^-. Which of the three is likely to be stable? What is the bond order of each? Which contain unpaired electrons?

5.56 The C_2 molecule can be represented by an MO diagram similar to that in Figure 5.12a.

(a) What is the bond order of C_2?

(b) To increase the bond order of C_2, should you add or remove an electron?

(c) Give the charge and the bond order of the new species made in part **(b)**.

5.57 Look at the molecular orbital diagram for O_2 in Figure 5.14 and answer the following questions:

(a) What is the bond order of O_2?

(b) To increase the bond order of O_2, should you add or remove an electron?

(c) Give the charge and the bond order of the new species made in part **(b)**.

5.58 Look at the MO diagrams of corresponding neutral diatomic species in Figure 5.14, and predict whether each of the following ions is diamagnetic or paramagnetic. Diagrams for Li_2 and C_2 are similar to N_2; Cl_2 is similar to F_2.

(a) C_2^{2-} **(b)** C_2^{2+} **(c)** F_2^- **(d)** Cl_2 **(e)** Li_2^+

5.59 Look at the MO diagrams of corresponding neutral diatomic species in Figure 5.14, and predict whether each of the following ions is diamagnetic or paramagnetic. MO diagrams for Li_2 and C_2 are similar to N_2; Cl_2 is similar to F_2.

(a) O_2^{2+} **(b)** N_2^{2+} **(c)** C_2^+ **(d)** F_2^{2+} **(e)** Cl_2^+

5.60 Make a sketch showing the location and geometry of the *p* orbitals in the allyl cation. Describe the bonding in this cation using a localized valence bond model for σ bonding and a delocalized MO model for π bonding.

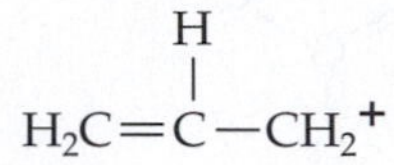

Allyl cation

5.61 Make a sketch showing the location and geometry of the *p* orbitals in the nitrite ion, NO_2^-. Describe the bonding in this ion using a localized valence bond model for σ bonding and a delocalized MO model for π bonding.

CHAPTER PROBLEMS

5.62 The following molecular model is a representation of caffeine. Identify the position(s) of the multiple bonds in caffeine, and tell the hybridization of each carbon atom (red = O, blue = N, gray = C, ivory = H).

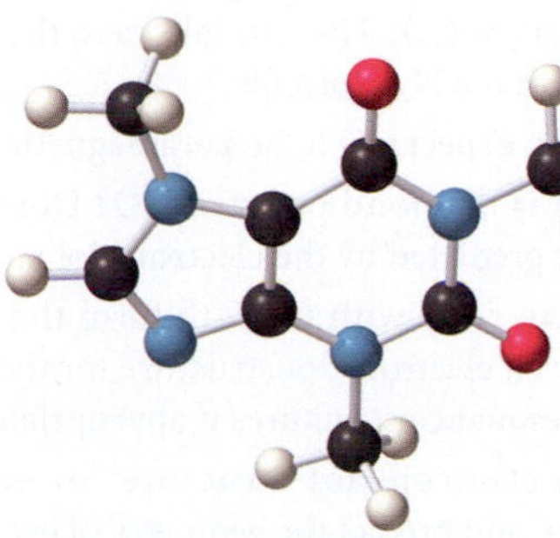

Caffeine

5.63 The following molecular model is a representation of ascorbic acid, or vitamin C. Identify the position(s) of the multiple bonds in ascorbic acid, and tell the hybridization of each carbon atom (red = O, gray = C, ivory = H).

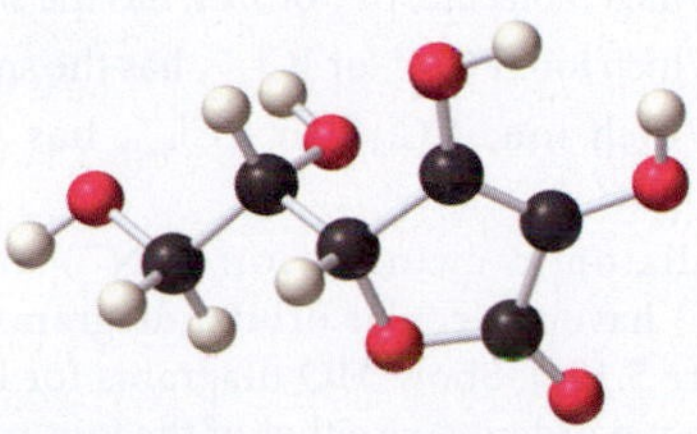

Asorbic acid

5.64 The odor of cinnamon oil is due to cinnamaldehyde, C_9H_8O. What is the hybridization of each carbon atom in cinnamaldehyde? How many σ bonds and how many π bonds does cinnamaldehyde have?

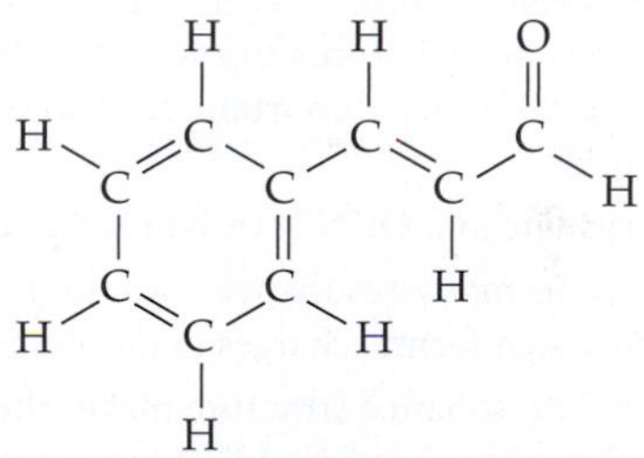

Cinnamaldehyde

5.65 What is the hybridization of the B and N atoms in borazine, what are the values of the B–N–B and N–B–N bond angles, and what is the overall shape of the molecule?

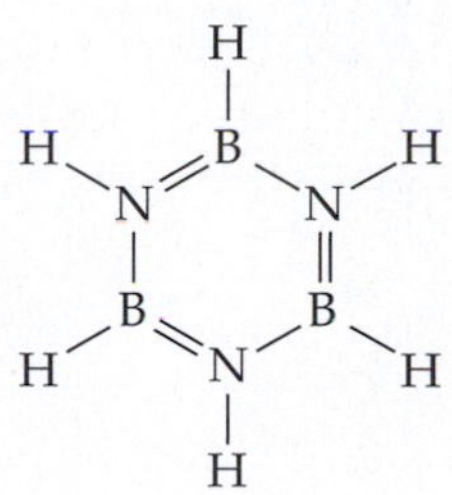

Borazine

5.66 Benzyne, C_6H_4, is a highly energetic and reactive molecule. What hybridization do you expect for the two triply bonded carbon atoms? What are the "theoretical" values for the C–C≡C bond angles? Why do you suppose benzyne is so reactive?

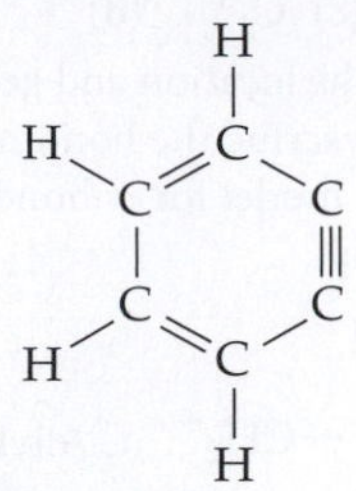

Benzyne

5.67 Propose structures for molecules that meet the following descriptions:

(a) Contains a C atom that has two π bonds and two σ bonds

(b) Contains an N atom that has one π bond and two σ bonds

(c) Contains an S atom that has a coordinate covalent bond

5.68 Calcium carbide, CaC_2, reacts with water to produce acetylene, C_2H_2, and is sometimes used as a convenient source of that substance. Use the MO energy diagram in Figure 5.12a to describe the bonding in the carbide anion, C_2^{2-}. What is its bond order?

5.69 Draw a molecular orbital energy diagram for Li_2. What is the bond order? Is the molecule likely to be stable? Explain.

5.70 The N_2O_5 molecule has six N–O σ bonds and two N–O π bonds, but has no N–N bonds and no O–O bonds.

(a) Draw an electron-dot structure of N_2O_5, and predict the shape of the molecule.

(b) What is the hybridization of each nonterminal atom in N_2O_5?

(c) Draw eight resonance structures for N_2O_5, and assign formal charges to the atoms in each. Which resonance structures make the more important contributions to the resonance hybrid?

5.71 In the cyanate ion, OCN^-, carbon is the central atom.

(a) Draw as many resonance structures as you can for OCN^-, and assign formal charges to the atoms in each.

(b) Which resonance structure makes the greatest contribution to the resonance hybrid? Which makes the least contribution? Explain.

(c) Is OCN^- linear or bent? Explain.

(d) Which hybrid orbitals are used by the C atom, and how many π bonds does the C atom form?

5.72 Aspirin has the following connections among atoms. Complete the electron-dot structure for aspirin, tell how many σ bonds and how many π bonds the molecule contains, and tell the hybridization of each carbon atom.

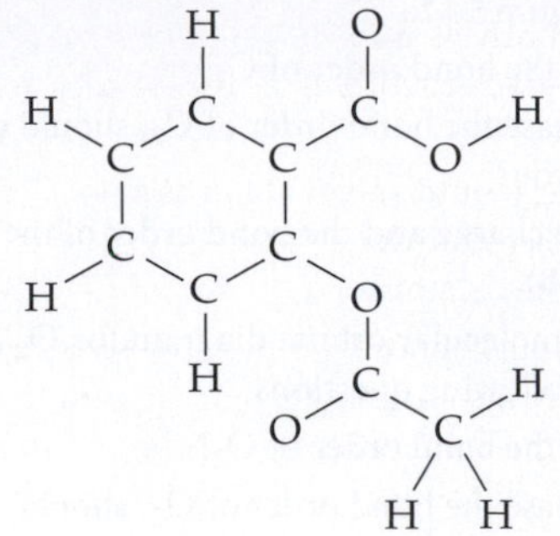

Aspirin

5.73 The cation $[H–C–N–Xe–F]^+$ is entirely linear. Draw an electron-dot structure consistent with that shape, and tell the hybridization of the C and N atoms.

5.74 At high temperatures, sulfur vapor is predominantly in the form of $S_2(g)$ molecules.

(a) Assuming that the molecular orbitals for third-row diatomic molecules are analogous to those for second-row molecules, construct an MO diagram for the valence orbitals of $S_2(g)$.

(b) Is S_2 likely to be paramagnetic or diamagnetic?

(c) What is the bond order of $S_2(g)$?

(d) When two electrons are added to S_2, the disulfide ion S_2^{2-} is formed. Is the bond length in S_2^{2-} likely to be shorter or longer than the bond length in S_2? Explain.

5.75 Carbon monoxide is produced by incomplete combustion of fossil fuels.

(a) Give the electron configuration for the valence molecular orbitals of CO. The orbitals have the same energy order as those of the N_2 molecule.

(b) Do you expect CO to be paramagnetic or diamagnetic?

(c) What is the bond order of CO? Does this match the bond order predicted by the electron-dot structure?

(d) CO can react with OH^- to form the formate ion, HCO_2^-. Draw an electron-dot structure for the formate ion, and give any resonance structures if appropriate.

5.76 Draw an electron-dot structure for each of the following substances, and predict the geometry of every nonterminal atom.

(a) $F_3S–S–F$

(b) $CH_3–C≡CCO_2^-$

5.77 The ion I_5^- is shaped like a big "V." Draw an electron-dot structure consistent with this overall shape.

5.78 Use VSEPR theory to answer the following questions:

(a) Which molecule, BF_3 or PF_3, has the smaller F–X–F angles?

(b) Which ion, PCl_4^+ or ICl_2^-, has the smaller Cl–X–Cl angles?

(c) Which ion, CCl_3^- or PCl_6^-, has the smaller Cl–X–Cl angles?

5.79 The diatomic cyanide ion (CN^-) and nitrosonium ion (NO^+) have molecular orbital diagrams similar to that of N_2 (Figure 5.12a). Show MO diagrams for both ions, and predict their bond orders. Are either of the ions paramagnetic?

MULTICONCEPT PROBLEMS

5.80 The dichromate ion, $Cr_2O_7^{2-}$, has neither Cr–Cr nor O–O bonds.

(a) Taking both 4*s* and 3*d* electrons into account, draw an electron-dot structure that minimizes the formal charges on the atoms.

(b) How many outer-shell electrons does each Cr atom have in your electron-dot structure? What is the likely geometry around the Cr atoms?

5.81 Just as individual bonds in a molecule are often polar, molecules as a whole are also often polar because of the net sum of individual bond polarities. There are three possible structures for substances with the formula $C_2H_2Cl_2$, two of which are polar overall and one of which is not.

(a) Draw the three possible structures for $C_2H_2Cl_2$, predict an overall shape for each, and explain how they differ.

(b) Which of the three structures is nonpolar, and which two are polar? Explain.

(c) Two of the three structures can be interconverted by a process called cis–trans isomerization, in which rotation around the central carbon–carbon bond takes place when the molecules are irradiated with ultraviolet light. If light with a wavelength of approximately 200 nm is required for isomerization, how much energy in kJ/mol is involved?

(d) Sketch the orbitals involved in the central carbon–carbon bond, and explain why so much energy is necessary for bond rotation to occur.

5.82 Cyclooctatetraene dianion, $C_8H_8^{2-}$, is a cyclic organic ion with the structure shown. Considering only the π bonds and not the σ bonds, cyclooctatetraene dianion can be described by the following energy diagram of its π molecular orbitals:

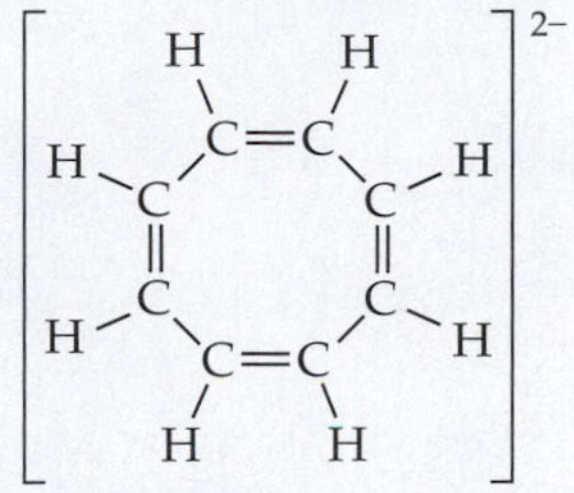

Cyclooctatetraene dianion

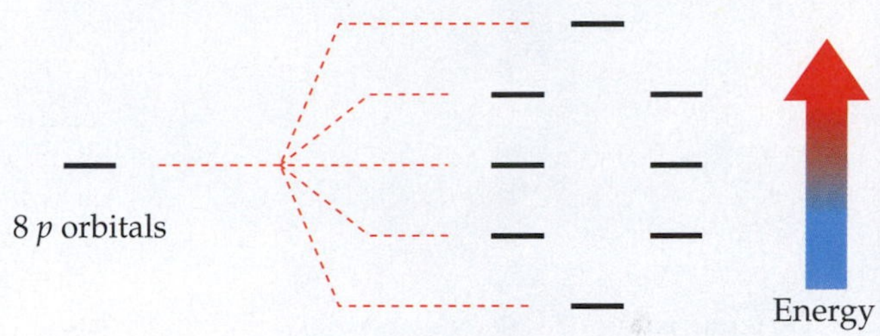

Cyclooctatetraene dianion
pi (π) molecular orbitals

(a) What is the hybridization of the 8 carbon atoms?

(b) Three of the π molecular orbitals are bonding, three are antibonding, and two are *nonbonding*, meaning that they have the same energy level as isolated *p* orbitals. Which is which?

(c) Complete the MO energy diagram by assigning the appropriate numbers of *p* electrons to the various molecular orbitals, indicating the electrons using up/down arrows (↑ ↓).

(d) Based on your MO energy diagram, is the dianion paramagnetic or diamagnetic?

CHAPTER 6

Chemical Arithmetic: Stoichiometry

How many molecules of oil are in this slick? We'll see in the *FYI* at the end of this chapter a very easy way to calculate the number.

It's sometimes easy when beginning the study of chemistry to forget that *reactions* are at the heart of the science. New words, ideas, and principles are introduced so quickly that the central concern of chemistry—the change of one substance into another—can get lost in the rush.

In previous chapters, we've concentrated on describing the structure of individual atoms and on seeing how atoms bond together to form molecules. In this chapter, we'll begin learning how to describe chemical reactions. We'll start with a look at the conventions for writing chemical equations and at the necessary mass relationships between reactants and products. With that knowledge, we'll then see how chemical formulas are determined. Finally, because most chemical reactions are carried out with solutions rather than with pure substances, we'll discuss the units for describing the concentration of substances in solution.

6.1 CHEMICAL SYMBOLS ON DIFFERENT LEVELS

Before starting the main subject of this chapter, let's first answer an apparently simple question: What does it mean when we write a chemical formula or equation? Answering this question isn't as easy as it sounds because a chemical symbol can have different meanings under different circumstances. Chemists use the same symbols to represent both the small-scale, microscopic level and the large-scale, macroscopic level, and they tend to slip back and forth between the two levels without realizing the confusion this can cause for newcomers to the field.

On the microscopic level, chemical symbols represent the behavior of individual atoms and molecules. Atoms and molecules are far too small to be seen, but we can nevertheless describe their microscopic behavior if we read the equation $2\,H_2 + O_2 \rightarrow 2\,H_2O$ to mean "Two molecules of hydrogen react with one molecule of oxygen to yield two molecules of water." It's this microscopic world that we deal with when trying to understand how reactions occur.

CONTENTS

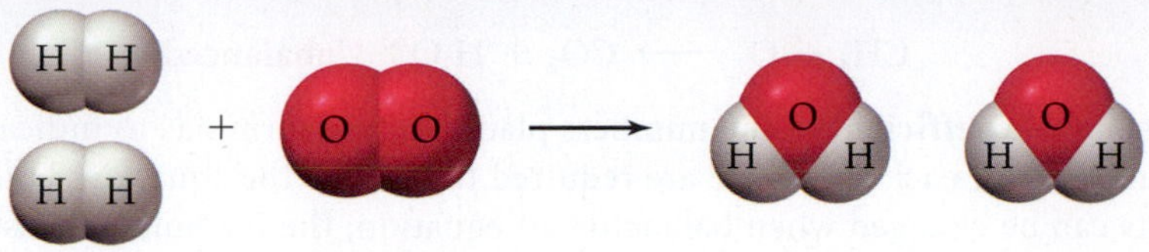

On the macroscopic level, formulas and equations represent the large-scale behavior of atoms and molecules that gives rise to visible properties. In other words, the symbols H_2, O_2, and H_2O represent not just single molecules but vast numbers of molecules that together have a set of measurable physical properties. A *single* isolated H_2O molecule is neither solid nor liquid nor gas, but a huge collection of H_2O molecules appears to us as a colorless liquid that freezes to a solid at 0 °C and boils to a gas at 100 °C.

In the same way, a single atom of copper is neither solid, liquid, nor gas, does not conduct electricity, and has no color. A large collection of copper atoms, however, appears to us as a shiny, reddish-brown metal that can be drawn into electrical wires or made into coins. It's this macroscopic behavior we deal with in the laboratory when we weigh specific amounts of reactants, place them together in a flask, and observe visible changes.

What does a chemical formula or equation mean? It means different things depending on the context. The symbol H_2O can mean either one tiny, invisible molecule or a vast collection of molecules large enough to swim in.

6.2 BALANCING CHEMICAL EQUATIONS

Previous chapters have provided several examples of reactions: hydrogen reacting with oxygen to yield water, sodium reacting with chlorine to yield sodium chloride, mercury(II) nitrate reacting with potassium iodide to yield mercury(II) iodide, and so forth.

4 H and 2 O atoms on this side | 4 H and 2 O atoms on this side

$$2\,H_2 + O_2 \longrightarrow 2\,H_2O$$

2 Na and 2 Cl atoms on this side | 2 Na and 2 Cl atoms on this side

$$2\,Na + Cl_2 \longrightarrow 2\,NaCl$$

1 Hg, 2 N, 6 O, 2 K, and 2 I atoms on this side | 1 Hg, 2 N, 6 O, 2 K, and 2 I atoms on this side

$$Hg(NO_3)_2 + 2\,KI \longrightarrow HgI_2 + 2\,KNO_3$$

Look carefully at how these equations are written. Because hydrogen, oxygen, and chlorine exist as covalent H_2, O_2, and Cl_2 diatomic molecules rather than as isolated atoms (Section 3.1), we must write them as such in chemical equations. Now look at the atoms on each side of the reaction arrow. Although we haven't explicitly stated it yet, chemical equations are always written so that they are **balanced**; that is, so that the numbers and kinds of atoms on both sides of the reaction arrow are the same. This requirement is just a consequence of the **mass conservation law**. Because atoms are neither created nor destroyed in chemical reactions, their numbers and kinds must remain the same in both products and reactants.

REMEMBER...

According to the **law of mass conservation**, mass is neither created nor destroyed in chemical reactions. (Section 1.4)

Balancing a chemical equation involves finding out how many *formula units* of each different substance take part in the reaction. A **formula unit**, as its name implies, is one unit—whether atom, ion, or molecule—corresponding to a given formula. One formula unit of NaCl is one Na^+ ion and one Cl^- ion, one formula unit of $MgBr_2$ is one Mg^{2+} ion and two Br^- ions, and one formula unit of H_2O is one H_2O molecule.

Complicated equations generally need to be balanced using a systematic method, such as we'll discuss in the next chapter, but simpler equations can often be balanced using a mixture of common sense and trial-and-error:

1. Write the unbalanced equation using the correct chemical formula for each reactant and product. In the combustion reaction of methane (CH_4; natural gas) with oxygen to yield carbon dioxide and water, for example, we begin by writing:

$$CH_4 + O_2 \longrightarrow CO_2 + H_2O \quad \text{Unbalanced}$$

2. Find suitable **coefficients**—the numbers placed before formulas to indicate how many formula units of each substance are required to balance the equation. Only these coefficients can be changed when balancing an equation; the formulas themselves can't be changed. Again using the reaction of methane with oxygen as an example, we can balance the equation by adding a coefficient of 2 to both O_2 and H_2O. By so doing, we now have 1 carbon atom, 4 hydrogen atoms, and 4 oxygen atoms on both sides of the equation:

Add these coefficients to balance the equation.

$$CH_4 + 2\,O_2 \longrightarrow CO_2 + 2\,H_2O$$

3. Reduce the coefficients to their smallest whole-number values, if necessary, by dividing them all by a common divisor.
4. Check your answer by making sure that the numbers and kinds of atoms are the same on both sides of the equation.

Let's work through some additional examples:

▲ Propane is used as a fuel in camp stoves and in rural homes.

Worked Example 6.1

Balancing a Chemical Equation

Propane, C_3H_8, is a colorless, odorless gas often used as a heating and cooking fuel in campers and rural homes. Write a balanced equation for the combustion reaction of propane with oxygen to yield carbon dioxide and water.

STRATEGY AND SOLUTION

Follow the four steps described in the text:

Step 1 Write the unbalanced equation using correct chemical formulas for all substances:

$$C_3H_8 + O_2 \longrightarrow CO_2 + H_2O \quad \text{Unbalanced}$$

Step 2 Find coefficients to balance the equation. It's usually best to begin with the most complex substance—in this case C_3H_8—and to deal with one element at a time. Look at the unbalanced equation, and note that there are 3 carbon atoms on the left side of the equation but only 1 on the right side. If we add a coefficient of 3 to CO_2 on the right, the carbons balance:

$$C_3H_8 + O_2 \longrightarrow 3\,CO_2 + H_2O \quad \text{Balanced for C}$$

Next, look at the number of hydrogen atoms. There are 8 hydrogens on the left but only 2 (in H_2O) on the right. By adding a coefficient of 4 to the H_2O on the right, the hydrogens balance:

$$C_3H_8 + O_2 \longrightarrow 3\,CO_2 + 4\,H_2O \quad \text{Balanced for C and H}$$

Finally, look at the number of oxygen atoms. There are 2 on the left but 10 on the right. By adding a coefficient of 5 to the O_2 on the left, the oxygens balance:

$$C_3H_8 + 5\,O_2 \longrightarrow 3\,CO_2 + 4\,H_2O \quad \text{Balanced for C, H, and O}$$

Step 3 Make sure the coefficients are reduced to their smallest whole-number values. In this case, our answer is already correct, but we might have arrived at a different answer through trial and error:

$$2\,C_3H_8 + 10\,O_2 \longrightarrow 6\,CO_2 + 8\,H_2O$$

Although the preceding equation is balanced, the coefficients are not the smallest whole numbers. It would be necessary to divide all coefficients by 2 to reach the final equation.

Step 4 Check your answer. Count the numbers and kinds of atoms on both sides of the equation to make sure they're the same:

3 C, 8 H, and 10 O atoms on this side | 3 C, 8 H, and 10 O atoms on this side

$$C_3H_8 + 5\,O_2 \longrightarrow 3\,CO_2 + 4\,H_2O$$

Worked Example 6.2

Balancing a Chemical Equation

The major ingredient in ordinary safety matches is potassium chlorate, $KClO_3$, a substance that can act as a source of oxygen in combustion reactions. Its reaction with ordinary table sugar (sucrose, $C_{12}H_{22}O_{11}$), for instance, occurs violently to yield potassium chloride, carbon dioxide, and water. Write a balanced equation for the reaction.

▲ Safety matches contain potassium chlorate, which acts as a source of oxygen for ignition.

STRATEGY AND SOLUTION

Step 1 Write the unbalanced equation, making sure the formulas for all substances are correct:

$$KClO_3 + C_{12}H_{22}O_{11} \longrightarrow KCl + CO_2 + H_2O \quad \text{Unbalanced}$$

Step 2 Find coefficients to balance the equation by starting with the most complex substance (sucrose) and considering one element at a time. Since there are 12 C atoms on the left and only 1 on the right, we can balance for carbon by adding a coefficient of 12 to CO_2 on the right:

$$KClO_3 + C_{12}H_{22}O_{11} \longrightarrow KCl + 12\,CO_2 + H_2O \quad \text{Balanced for C}$$

Since there are 22 H atoms on the left and only 2 on the right, we can balance for hydrogen by adding a coefficient of 11 to H_2O on the right:

$$KClO_3 + C_{12}H_{22}O_{11} \longrightarrow KCl + 12\,CO_2 + 11\,H_2O \quad \text{Balanced for C and H}$$

There are now 35 O atoms on the right but only 14 on the left (11 in sucrose and 3 in $KClO_3$). Thus, 21 oxygens must be added on the left. We can do this without disturbing the C and H balance by adding 7 more $KClO_3$, giving a coefficient of 8 for $KClO_3$ on the left:

$$8\,KClO_3 + C_{12}H_{22}O_{11} \longrightarrow KCl + 12\,CO_2 + 11\,H_2O \quad \text{Balanced for C, H, and O}$$

Potassium and chlorine can both be balanced by adding a coefficient of 8 to KCl on the right:

$$8\,KClO_3 + C_{12}H_{22}O_{11} \longrightarrow 8\,KCl + 12\,CO_2 + 11\,H_2O \quad \text{Balanced for C, H, O, K, and Cl}$$

Steps 3 and 4 The coefficients in the balanced equation are already reduced to their smallest whole-number values, and a check shows that the numbers and kinds of atoms are the same on both sides of the equation.

8 K, 8 Cl, 12 C, 22 H, and 35 O atoms | 8 K, 8 Cl, 12 C, 22 H, and 35 O atoms

$$8\,KClO_3 + C_{12}H_{22}O_{11} \longrightarrow 8\,KCl + 12\,CO_2 + 11\,H_2O$$

WORKED CONCEPTUAL EXAMPLE 6.3

Balancing a Chemical Equation

Write a balanced equation for the reaction of element **A** (red spheres) with element **B** (blue spheres) as represented below:

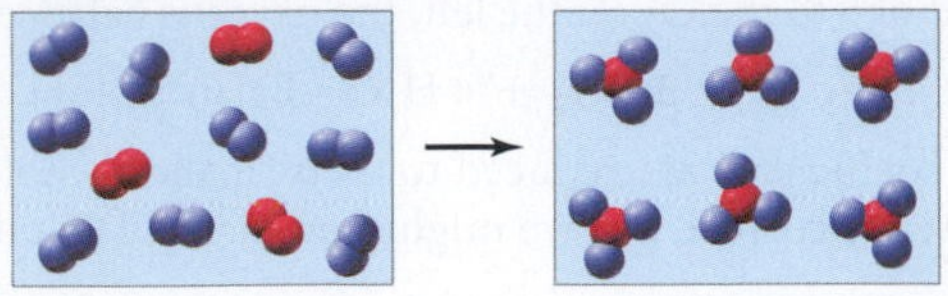

STRATEGY

Balancing the reactions shown in graphic representations of this sort is just a matter of counting the numbers of reactant and product formula units. In this example, the reactant box contains three red $\mathbf{A_2}$ molecules and nine blue $\mathbf{B_2}$ molecules, while the product box contains six $\mathbf{AB_3}$ molecules with no reactant left over.

SOLUTION

$$3\,A_2 + 9\,B_2 \longrightarrow 6\,AB_3 \quad \text{or} \quad A_2 + 3\,B_2 \longrightarrow 2\,AB_3$$

PROBLEM 6.1 Sodium chlorate, $NaClO_3$, decomposes when heated to yield sodium chloride and oxygen, a reaction used to provide oxygen for the emergency breathing masks in many airliners. Balance the equation.

PROBLEM 6.2 Balance the following equations:

(a) $C_6H_{12}O_6 \longrightarrow C_2H_6O + CO_2$ (fermentation of sugar to yield ethyl alcohol)
(b) $CO_2 + H_2O \longrightarrow C_6H_{12}O_6 + O_2$ (photosynthesis reaction in green plants)
(c) $NH_3 + Cl_2 \longrightarrow N_2H_4 + NH_4Cl$ (synthesis of hydrazine for rocket fuel)

CONCEPTUAL PROBLEM 6.3 Write a balanced equation for the reaction of element **A** (red spheres) with element **B** (green spheres) as represented below:

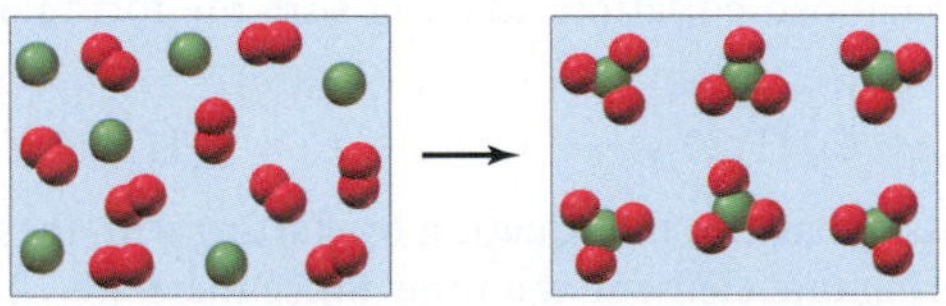

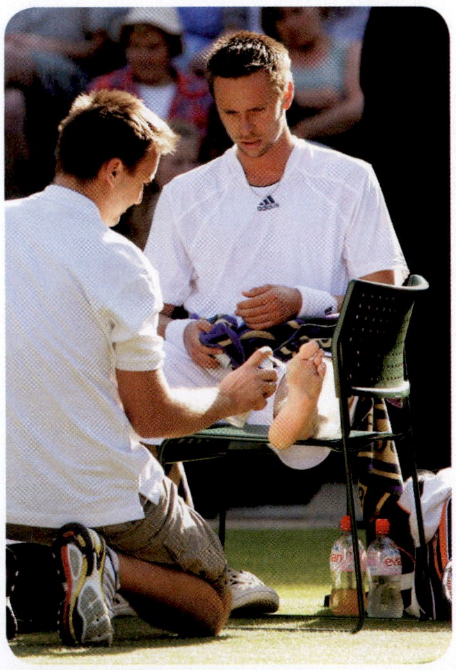

▲ Ethyl chloride is often used as a spray-on anesthetic for athletic injuries.

6.3 STOICHIOMETRY: THE ARITHMETIC OF CHEMICAL REACTIONS

Now that we know how to write balanced chemical equations, let's imagine actually carrying out a reaction in a laboratory—perhaps the reaction of ethylene, C_2H_4, with hydrogen chloride, HCl, to prepare ethyl chloride, C_2H_5Cl, a colorless, low-boiling liquid used by doctors and athletic trainers as a spray-on anesthetic for minor injuries. In writing this and other equations, the designations (*g*) for gas, (*l*) for liquid, (*s*) for solid, and (*aq*) for aqueous solutions are often appended to the symbols of reactants and products to show their physical state. We'll do this frequently from now on.

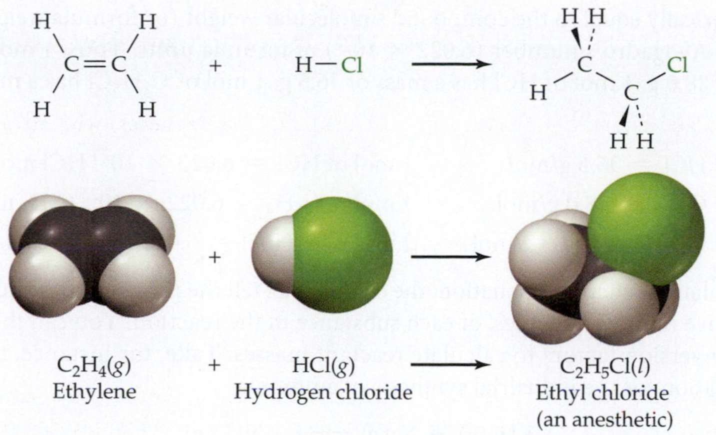

How much ethylene and how much hydrogen chloride should you use for your experiment? According to the coefficients of the balanced equation, a 1:1 numerical ratio of the two reactants is needed. But because you can't count the reactant molecules, you have to weigh them. That is, you must convert a *number* ratio of reactant molecules, as given by coefficients in the balanced equation, into a *mass* ratio to be sure you are using the right amounts.

Mass ratios are determined by using the *molecular weights* of the substances involved in a reaction. Just as the **atomic weight** of an element is the weighted average mass of the element's *atoms* in **unified atomic mass units (u)**, the **molecular weight** (mol. wt.) of a substance is the weighted average mass of the substance's *molecules*. Numerically, molecular weight or, more generally, **formula weight** (form. wt.) to include both ionic and molecular substances, equals the sum of the atomic weights of all atoms in the molecule.

REMEMBER...

The **atomic weight** of an element is the weighted average mass of the element's naturally occurring isotopes. Although atomic weight is usually written as dimensionless, the unit of atomic weight is the **unified atomic mass unit (u)**. (Section 1.9)

▶ **Molecular weight** Sum of atomic weights of all atoms in a molecule.

▶ **Formula weight** Sum of atomic weights of all atoms in a formula unit of any compound, molecular or ionic.

As examples, the molecular weight of ethylene is 28.0, the molecular weight of hydrogen chloride is 36.5, and the molecular weight of ethyl chloride is 64.5. (These numbers are rounded off to one decimal place for convenience; the actual values are known more precisely.)

For ethylene, C_2H_4:

Atomic weight of 2 C	$= 2 \times 12.0$	$=$	24.0
Atomic weight of 4 H	$= 4 \times 1.0$	$=$	4.0
Molecular weight of C_2H_4		$=$	28.0

For hydrogen chloride, HCl:

Atomic weight of H	$=$	1.0
Atomic weight of Cl	$=$	35.5
Molecular weight of HCl	$=$	36.5

For ethyl chloride, C_2H_5Cl:

Atomic weight of 2 C	$= 2 \times 12.0$	$=$	24.0
Atomic weight of 5 H	$= 5 \times 1.0$	$=$	5.0
Atomic weight of Cl		$=$	35.5
Molecular weight of C_2H_5Cl		$=$	64.5

How do we use molecular weights? We saw in Section 1.9 that one **mole** of any element is the amount whose mass, or *molar mass*, is numerically equal to the element's atomic weight in grams. In the same way, one mole of any chemical compound is the amount whose

REMEMBER...

The **mole** is the fundamental SI unit for measuring amount of matter. One mole of any substance is the amount whose mass in grams is numerically equal to the substance's atomic or formula weight. One mole contains Avogadro's number (6.022×10^{23}) of formula units. (Section 1.9)

mass is numerically equal to the compound's molecular weight (or formula weight) in grams and contains Avogadro's number (6.022×10^{23}) of formula units. Thus, 1 mol of ethylene has a mass of 28.0 g, 1 mol of HCl has a mass of 36.5 g, 1 mol of C_2H_5Cl has a mass of 64.5 g, and so on.

Mol. wt. of HCl = 36.5	Molar mass of HCl = 36.5 g/mol	1 mol of HCl = 6.022×10^{23} HCl molecules
Mol. wt. of C_2H_4 = 28.0	Molar mass of C_2H_4 = 28.0 g/mol	1 mol of C_2H_4 = 6.022×10^{23} C_2H_4 molecules
Mol. wt. of C_2H_5Cl = 64.5	Molar mass of C_2H_5Cl = 64.5 g/mol	1 mol of C_2H_5Cl = 6.022×10^{23} C_2H_5Cl molecules

In any balanced chemical equation, the coefficients tell the number of formula units, and thus the relative number of moles, of each substance in the reaction. You can then use molar masses as conversion factors to calculate reactant masses. Take, for instance, the following balanced equation for the industrial synthesis of ammonia:

$$3\,H_2(g) + N_2(g) \longrightarrow 2\,NH_3(g)$$

Knowing that one mole of any element is the amount whose mass is numerically equal to the element's atomic weight in grams, we can calculate that 3 mol of H_2 (3 mol × 2.0 g/mol = 6.0 g) is needed for reaction with 1 mol of N_2 (28.0 g) to yield 2 mol of NH_3 (2 mol × 17.0 g/mol = 34.0 g).

This number of moles of hydrogen... | ...reacts with this number of moles of nitrogen... | ...to yield this number of moles of ammonia.

$$3\,H_2(g) + 1\,N_2(g) \longrightarrow 2\,NH_3(g)$$

In referring to the chemical arithmetic needed for mole–mass conversions, we use the word **stoichiometry** (stoy-key-**ahm**-uh-tree; from the Greek *stoicheion*, "element," and *metron*, "measure"). Let's look again at the reaction of ethylene with HCl and assume that we have 15.0 g of ethylene and need to know how many grams of HCl to use in the reaction.

$$C_2H_4(g) + HCl(g) \longrightarrow C_2H_5Cl(l)$$

According to the coefficients in the balanced equation, 1 molecule of HCl reacts with 1 molecule of ethylene, so 1 mol of HCl is needed for reaction with each mole of ethylene. To find out how many grams of HCl are needed to react with 15.0 g of ethylene, we first have to find out how many moles of ethylene are in 15.0 g. We do this gram-to-mole conversion by calculating the molar mass of ethylene and using that value as a conversion factor:

$$\text{Molecular weight of } C_2H_4 = (2 \times 12.0) + (4 \times 1.0) = 28.0$$
$$\text{Molar mass of } C_2H_4 = 28.0 \text{ g/mol}$$
$$\text{Moles of } C_2H_4 = 15.0 \text{ g ethylene} \times \frac{1 \text{ mol ethylene}}{28.0 \text{ g ethylene}} = 0.536 \text{ mol ethylene}$$

Now that we know how many moles of ethylene we have (0.536 mol), we also know from the balanced equation how many moles of HCl we need (also 0.536 mol), and we have to do a mole-to-gram conversion to find the mass of HCl required. Once again, the conversion is done by calculating the molar mass of HCl and using that value as a conversion factor:

$$\text{Molecular weight of HCl} = 1.0 + 35.5 = 36.5$$
$$\text{Molar mass of HCl} = 36.5 \text{ g/mol}$$
$$\text{Grams of HCl} = 0.536 \text{ mol } C_2H_4 \times \frac{1 \text{ mol HCl}}{1 \text{ mol } C_2H_4} \times \frac{36.5 \text{ g HCl}}{1 \text{ mol HCl}} = 19.6 \text{ g HCl}$$

Thus, 19.6 g of HCl is needed to react with 15.0 g of ethylene.

Look carefully at the sequence of steps in the calculation just completed. *Moles* (numbers of molecules) are given by the coefficients in the balanced equation but *grams* are used to weigh reactants in the laboratory. Moles tell us *how many molecules* of each reactant are needed, while grams tell us *how much mass* of each reactant is needed.

$$\text{Moles} \longrightarrow \text{Numbers of molecules or formula units}$$
$$\text{Grams} \longrightarrow \text{Mass}$$

The flow chart in **Figure 6.1** illustrates the necessary conversions. You can't go directly from the number of grams of one reactant to the number of grams of another reactant. You *must* first convert to moles.

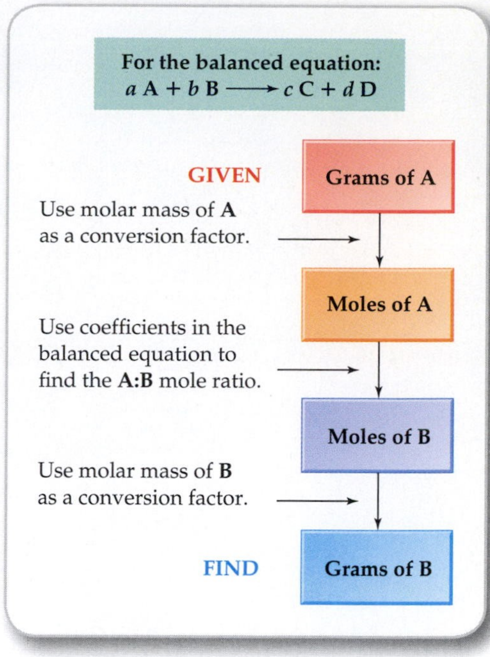

FIGURE 6.1

Conversions between moles and grams for a chemical reaction. The numbers of moles tell how many molecules of each reactant are needed, as given by the coefficients of the balanced equation; the numbers of grams tell what mass of each reactant is needed.

Worked Example 6.4

Calculating a Molecular Weight

What is the molecular weight of table sugar (sucrose, $C_{12}H_{22}O_{11}$), and what is its molar mass in g/mol?

STRATEGY

The molecular weight of a substance is the sum of the atomic weights of the constituent atoms. List the elements present in the molecule, and look up the atomic weight of each (we'll round off to one decimal place for convenience):

$$\text{C (12.0); H (1.0); O (16.0)}$$

Then, multiply the atomic weight of each element by the number of times that element appears in the chemical formula, and total the results.

SOLUTION

$$\begin{array}{rrl} C_{12} & (12 \times 12.0) = & 144.0 \\ H_{22} & (22 \times \ \ 1.0) = & 22.0 \\ O_{11} & (11 \times 16.0) = & 176.0 \\ \hline \text{Mol. wt. of } C_{12}H_{22}O_{11} & = & 342.0 \end{array}$$

Because one *molecule* of sucrose has a molecular weight of 342.0 u, 1 *mol* of sucrose has a mass of 342.0 g. Thus, the molar mass of sucrose is 342.0 g/mol.

Worked Example 6.5

Converting Mass to Moles

How many moles of sucrose are in a tablespoon of sugar containing 2.85 g? (The molar mass of sucrose, $C_{12}H_{22}O_{11}$, was calculated in Worked Example 6.4 to be 342.0 g/mol)

STRATEGY

The problem gives the mass of sucrose and asks for a mass-to-mole conversion. Use the molar mass of sucrose as a conversion factor, and set up an equation so that the unwanted unit cancels.

SOLUTION

$$2.85 \text{ g sucrose} \times \frac{1 \text{ mol sucrose}}{342.0 \text{ g sucrose}} = 0.008\,33 \text{ mol sucrose}$$
$$= 8.33 \times 10^{-3} \text{ mol sucrose}$$

✔ BALLPARK CHECK

Because the molecular weight of sucrose is 342.0, 1 mol of sucrose has a mass of 342.0 g. Thus, 2.85 g of sucrose is a bit less than one-hundredth of a mole, or 0.01 mol. The estimate agrees with the detailed solution.

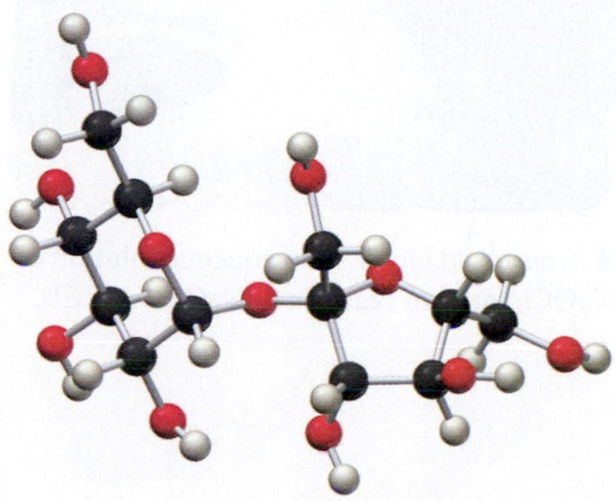

Sucrose

Worked Example 6.6

Converting Moles to Mass

How many grams are in 0.0626 mol of $NaHCO_3$, the main ingredient in Alka-Seltzer tablets?

STRATEGY

The problem gives the number of moles of $NaHCO_3$ and asks for a mole-to-mass conversion. First, calculate the molar mass of $NaHCO_3$. Then use molar mass as a conversion factor, and set up an equation so that the unwanted unit cancels.

SOLUTION

Form. wt. of $NaHCO_3 = 23.0 + 1.0 + 12.0 + (3 \times 16.0) = 84.0$

Molar mass of $NaHCO_3 = 84.0$ g/mol

$$0.0626\ \cancel{\text{mol NaHCO}_3} \times \frac{84.0\text{ g NaHCO}_3}{1\ \cancel{\text{mol NaHCO}_3}} = 5.26\text{ g NaHCO}_3$$

Worked Example 6.7

▲ Household bleach is an aqueous solution of NaOCl, made by reaction of NaOH with Cl_2.

Finding the Mass of One Reactant, Given the Mass of Another

Aqueous solutions of sodium hypochlorite (NaOCl), best known as household bleach, are prepared by reaction of sodium hydroxide with chlorine. How many grams of NaOH are needed to react with 25.0 g of Cl_2?

$$2\ \text{NaOH}(aq) + \text{Cl}_2(g) \longrightarrow \text{NaOCl}(aq) + \text{NaCl}(aq) + \text{H}_2\text{O}(l)$$

STRATEGY

Finding the relationship between numbers of reactant formula units always requires working in moles, using the general strategy outlined in Figure 6.1.

SOLUTION

First, find out how many moles of Cl_2 are in 25.0 g of Cl_2. This gram-to-mole conversion is done in the usual way, using the molar mass of Cl_2 (70.9 g/mol) as the conversion factor:

$$25.0\ \cancel{\text{g Cl}_2} \times \frac{1\text{ mol Cl}_2}{70.9\ \cancel{\text{g Cl}_2}} = 0.353\text{ mol Cl}_2$$

Next, look at the coefficients in the balanced equation. Each mole of Cl_2 reacts with 2 mol of NaOH, so 0.353 mol of Cl_2 reacts with $2 \times 0.353 = 0.706$ mol of NaOH. With the number of moles of NaOH known, carry out a mole-to-gram conversion using the molar mass of NaOH (40.0 g/mol) as a conversion factor to find that 28.2 g of NaOH is required for the reaction:

$$\text{Grams of NaOH} = 0.353\ \cancel{\text{mol Cl}_2} \times \frac{2\ \cancel{\text{mol NaOH}}}{1\ \cancel{\text{mol Cl}_2}} \times \frac{40.0\text{ g NaOH}}{1\ \cancel{\text{mol NaOH}}}$$
$$= 28.2\text{ g NaOH}$$

The problem can also be worked by combining the steps and setting up one large equation:

$$\text{Grams of NaOH} = 25.0\ \cancel{\text{g Cl}_2} \times \frac{1\ \cancel{\text{mol Cl}_2}}{70.9\ \cancel{\text{g Cl}_2}} \times \frac{2\ \cancel{\text{mol NaOH}}}{1\ \cancel{\text{mol Cl}_2}} \times \frac{40.0\text{ g NaOH}}{1\ \cancel{\text{mol NaOH}}}$$
$$= 28.2\text{ g NaOH}$$

✔ BALLPARK CHECK

The molar mass of NaOH is about half that of Cl_2, and 2 mol of NaOH is needed per 1 mol of Cl_2. Thus, the needed mass of NaOH will be similar to that of Cl_2, or about 25 g.

PROBLEM 6.4 Calculate the formula weight or molecular weight of the following substances:

(a) Fe_2O_3 (rust) (b) H_2SO_4 (sulfuric acid)
(c) $C_6H_8O_7$ (citric acid) (d) $C_{16}H_{18}N_2O_4S$ (penicillin G)

CONCEPTUAL PROBLEM 6.5 Aspirin can be represented by the adjacent ball-and-stick molecular model. Give the formula for aspirin, and calculate its molecular weight (red = O, gray = C, ivory = H). How many moles of aspirin are in a tablet weighing 500 mg? How many molecules?

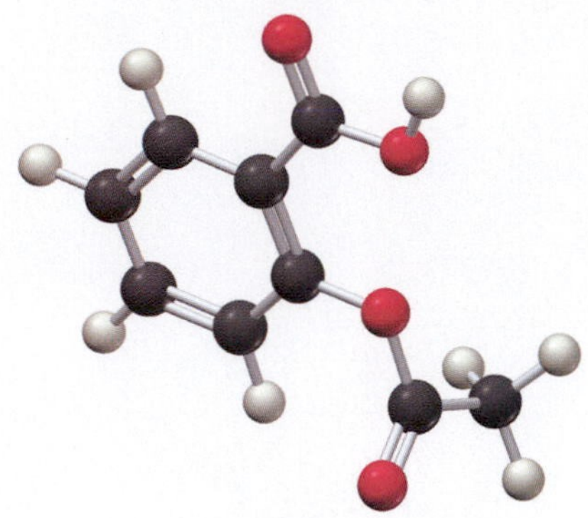

Aspirin

PROBLEM 6.6 Aspirin is prepared by reaction of salicylic acid ($C_7H_6O_3$) with acetic anhydride ($C_4H_6O_3$) according to the following equation:

$$\underset{\text{Salicylic acid}}{C_7H_6O_3} + \underset{\text{Acetic anhydride}}{C_4H_6O_3} \longrightarrow \underset{\text{Aspirin}}{C_9H_8O_4} + \underset{\text{Acetic acid}}{CH_3CO_2H}$$

(a) How many grams of acetic anhydride are needed to react with 4.50 g of salicylic acid?
(b) How many grams of aspirin will result?
(c) How many grams of acetic acid are formed as a byproduct?

6.4 YIELDS OF CHEMICAL REACTIONS

In the stoichiometry examples worked out in the preceding section, we made the unstated assumption that all reactions "go to completion." That is, we assumed that all reactant molecules are converted to products. In fact, few reactions behave so nicely. More often, a large majority of molecules react as expected, but other processes, or *side reactions*, also occur. Thus, the amount of product actually formed, called the **yield** of the reaction, is usually less than the amount predicted by calculations.

The amount of product actually formed in a reaction divided by the amount theoretically possible and multiplied by 100% is the reaction's **percent yield**. For example, if a given reaction *could* provide 6.9 g of a product according to its stoichiometry, but actually provides only 4.7 g, then its percent yield is 4.7/6.9 × 100% = 68%.

$$\text{Percent yield} = \frac{\text{Actual yield of product}}{\text{Theoretical yield of product}} \times 100\%$$

Worked Examples 6.8 and 6.9 show how to calculate and use percent yield.

Worked Example 6.8

Calculating a Percent Yield

Methyl *tert*-butyl ether (MTBE, $C_5H_{12}O$), a gasoline additive now being phased out because of health concerns, can be made by reaction of isobutylene (C_4H_8) with methanol (CH_4O). What is the percent yield of the reaction if 32.8 g of methyl *tert*-butyl ether is obtained from reaction of 26.3 g of isobutylene with sufficient methanol?

$$\underset{\text{Isobutylene}}{C_4H_8(g)} + CH_4O(l) \longrightarrow \underset{\text{Methyl tert-butyl ether (MTBE)}}{C_5H_{12}O(l)}$$

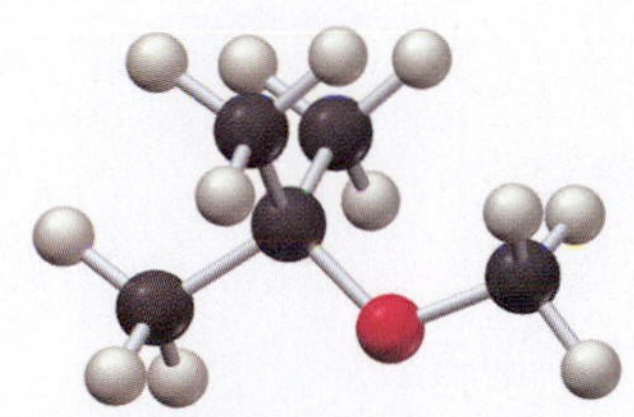

Methyl *tert*-butyl ether

continued on next page

STRATEGY

We need to calculate the amount of methyl *tert*-butyl ether that could be produced theoretically from 26.3 g of isobutylene and compare that theoretical amount to the actual amount (32.8 g). As always, stoichiometry problems begin by calculating the molar masses of reactants and products. Coefficients of the balanced equation then tell mole ratios, and molar masses act as conversion factors between moles and masses.

SOLUTION

Isobutylene, C_4H_8: Mol. wt. $= (4 \times 12.0) + (8 \times 1.0) = 56.0$

Molar mass $= 56.0$ g/mol

MTBE, $C_5H_{12}O$: Mol. wt. $= (5 \times 12.0) + (12 \times 1.0) + 16.0 = 88.0$

Molar mass of MTBE $= 88.0$ g/mol

To calculate the amount of MTBE that could theoretically be produced from 26.3 g of isobutylene, we first have to find the number of moles of reactant, using molar mass as the conversion factor:

$$26.3 \text{ g isobutylene} \times \frac{1 \text{ mol isobutylene}}{56.0 \text{ g isobutylene}} = 0.470 \text{ mol isobutylene}$$

According to the balanced equation, 1 mol of product is produced per mol of reactant, so we know that 0.470 mol of isobutylene can theoretically yield 0.470 mol of MTBE. Finding the mass of this MTBE requires a mole-to-mass conversion:

$$0.470 \text{ mol isobutylene} \times \frac{1 \text{ mol MTBE}}{1 \text{ mol isobutylene}} \times \frac{88.0 \text{ g MTBE}}{1 \text{ mol MTBE}} = 41.4 \text{ g MTBE}$$

Dividing the actual amount by the theoretical amount and multiplying by 100% gives the percent yield:

$$\frac{32.8 \text{ g MTBE}}{41.4 \text{ g MTBE}} \times 100\% = 79.2\%$$

Worked Example 6.9

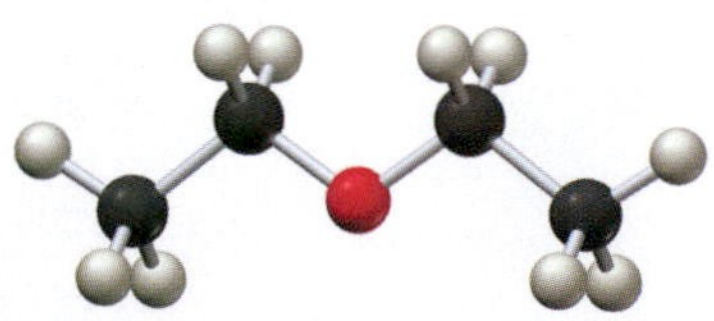

Diethyl ether

Calculating a Yield in Grams, Given a Percent Yield

Diethyl ether ($C_4H_{10}O$), the "ether" used medically as an anesthetic, is prepared commercially by treatment of ethyl alcohol (C_2H_6O) with an acid. How many grams of diethyl ether would you obtain from 40.0 g of ethyl alcohol if the percent yield of the reaction is 87%?

$$\underset{\text{Ethyl alcohol}}{2\,C_2H_6O(l)} \xrightarrow{\text{Acid}} \underset{\text{Diethyl ether}}{C_4H_{10}O(l)} + H_2O(l)$$

STRATEGY

Treat this as a typical stoichiometry problem to find the amount of diethyl ether that can theoretically be formed from 40.0 g of ethyl alcohol, and then multiply the answer by 87% to find the amount actually formed.

SOLUTION

First, calculate the molar masses of the reactant and product:

Ethyl alcohol, C_2H_6O: Mol. wt. $= (2 \times 12.0) + (6 \times 1.0) + 16.0 = 46.0$

Molar mass $= 46.0$ g/mol

Diethyl ether, $C_4H_{10}O$: Mol. wt. $= (4 \times 12.0) + (10 \times 1.0) + 16.0 = 74.0$

Molar mass $= 74.0$ g/mol

Next, find how many moles of ethyl alcohol are in 40.0 g by using molar mass as a conversion factor:

$$40.0 \text{ g ethyl alcohol} \times \frac{1 \text{ mol ethyl alcohol}}{46.0 \text{ g ethyl alcohol}} = 0.870 \text{ mol ethyl alcohol}$$

Because we started with 0.870 mol of ethyl alcohol, and because the balanced equation indicates that 2 mol of ethyl alcohol yield 1 mol of diethyl ether, we can theoretically obtain 0.435 mol of product:

$$0.870 \text{ mol ethyl alcohol} \times \frac{1 \text{ mol diethyl ether}}{2 \text{ mol ethyl alcohol}} = 0.435 \text{ mol diethyl ether}$$

We therefore need to find how many grams of diethyl ether are in 0.435 mol, using molar mass as the conversion factor:

$$0.435 \text{ mol diethyl ether} \times \frac{74.0 \text{ g diethyl ether}}{1 \text{ mol diethyl ether}} = 32.2 \text{ g diethyl ether}$$

Finally, we have to multiply the theoretical amount of product by the observed yield (87% = 0.87) to find how much diethyl ether is actually formed:

$$32.2 \text{ g diethyl ether} \times 0.87 = 28 \text{ g diethyl ether}$$

PROBLEM 6.7 Ethyl alcohol is prepared industrially by the reaction of ethylene, C_2H_4, with water. What is the percent yield of the reaction if 4.6 g of ethylene gives 4.7 g of ethyl alcohol?

$$\underset{\text{Ethylene}}{C_2H_4(g)} + H_2O(l) \longrightarrow \underset{\text{Ethyl alcohol}}{C_2H_6O(l)}$$

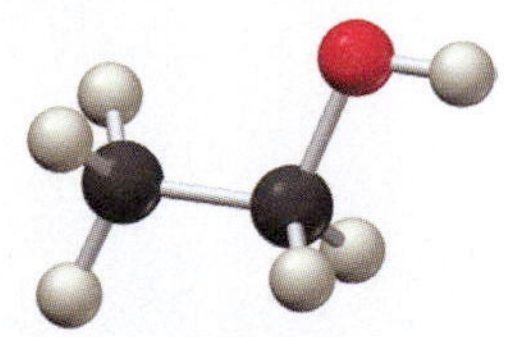

Ethyl alcohol

PROBLEM 6.8 Dichloromethane (CH_2Cl_2), used as a solvent in the decaffeination of green coffee beans, is prepared by reaction of methane (CH_4) with chlorine. How many grams of dichloromethane result from reaction of 1.85 kg of methane if the yield is 43.1%?

$$\underset{\text{Methane}}{CH_4(g)} + \underset{\text{Chlorine}}{2\,Cl_2(g)} \longrightarrow \underset{\text{Dichloromethane}}{CH_2Cl_2(l)} + 2\,HCl(g)$$

Dichloromethane

6.5 REACTIONS WITH LIMITING AMOUNTS OF REACTANTS

Because chemists usually write balanced equations, it's easy to get the impression that reactions are always carried out using exactly the right proportions of reactants. In fact, this is often not the case. Many reactions are carried out using an excess amount of one reactant—more than is actually needed according to stoichiometry. Look, for instance, at the industrial synthesis of ethylene glycol, $C_2H_6O_2$, a substance used both as automobile antifreeze and as a starting material for the preparation of polyester polymers. Approximately 18 million metric tons of ethylene glycol are prepared each year worldwide by reaction of ethylene oxide, C_2H_4O, with water at high temperature (1 metric ton = 1000 kg = 2205 lb).

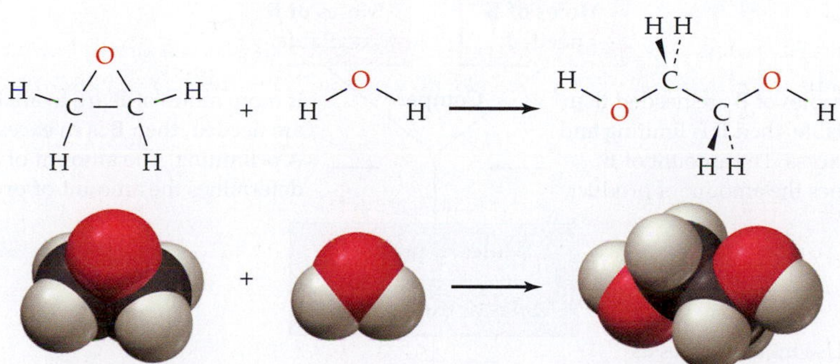

$$\underset{\text{Ethylene oxide}}{C_2H_4O} + \underset{\text{Water}}{H_2O} \xrightarrow{\text{Heat}} \underset{\text{Ethylene glycol}}{C_2H_6O_2}$$

Because water is so cheap and so abundant, it doesn't make sense to worry about using exactly 1 mol of water for each mole of ethylene oxide. Rather, it's much easier to use an excess of water to be certain that enough is present to consume entirely the more valuable ethylene oxide reactant. Of course, when an excess of water is present, only the amount required by stoichiometry undergoes reaction. The excess water does not react and remains unchanged.

Whenever the ratios of reactant molecules used in an experiment are different from those given by the coefficients of the balanced equation, a surplus of one reactant is left over after the reaction is finished. Thus, the extent to which a chemical reaction takes place depends on the reactant that is present in limiting amount—the **limiting reactant**. The other reactant is said to be the *excess reactant*.

The situation with excess reactants and limiting reactants is analogous to what sometimes happens with people and chairs. If there are five people in a room but only three chairs, then only three people can sit while the other two stand because the number of people sitting is limited by the number of available chairs. In the same way, if 5 moles of water come in contact with 3 moles of ethylene oxide, only 3 moles of water can undergo a reaction while the other 2 moles of water are unchanged.

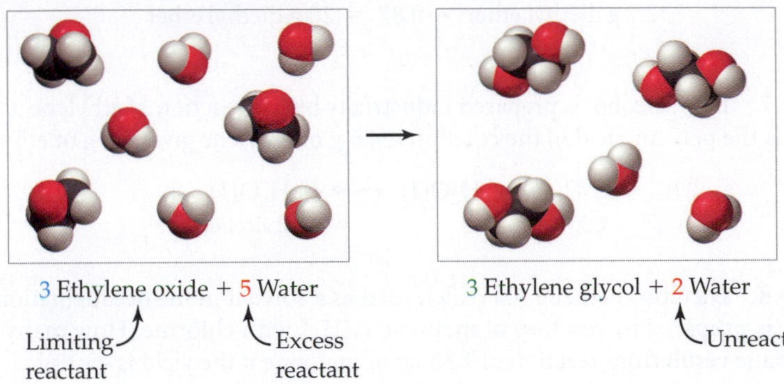

A flow chart of the steps needed for telling whether a limiting amount of one reactant is present and then calculating the amounts of excess reactant consumed and product formed is shown in **Figure 6.2**. Worked Example 6.10 illustrates the procedure.

FIGURE 6.2

Steps for identifying limiting and excess reactants and calculating amounts of products formed.

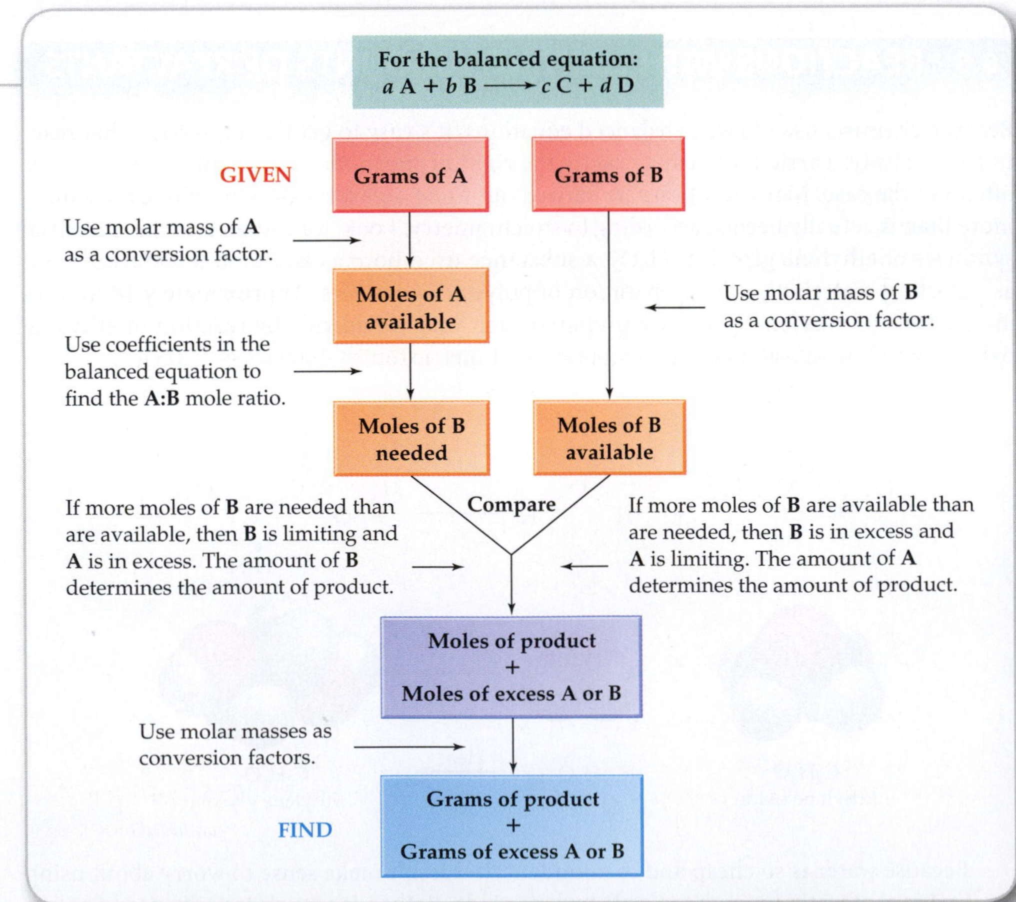

Worked Example 6.10

Calculating the Amount of an Excess Reactant and the Amounts of Products Formed

Cisplatin, an anticancer agent used for the treatment of solid tumors, is prepared by the reaction of ammonia with potassium tetrachloroplatinate, K_2PtCl_4. Assume that 10.0 g of K_2PtCl_4 and 10.0 g of NH_3 are allowed to react and that the reaction occurs in 100% yield.

$$\underset{\text{Potassium tetrachloroplatinate}}{K_2PtCl_4(aq)} + 2\,NH_3(aq) \longrightarrow \underset{\text{Cisplatin}}{Pt(NH_3)_2Cl_2(s)} + 2\,KCl(aq)$$

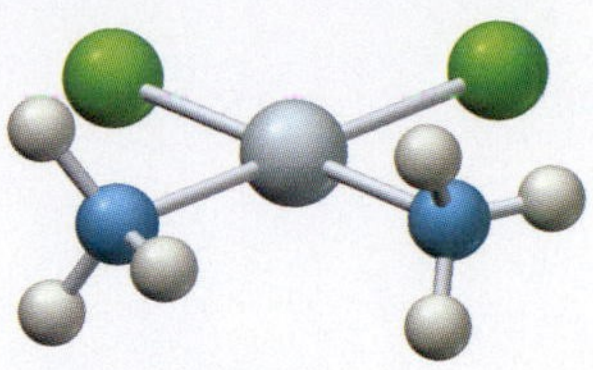

Cisplatin

(a) Which reactant is limiting, and which is in excess?
(b) How many grams of the excess reactant are consumed, and how many grams remain?
(c) How many grams of cisplatin are formed?

STRATEGY

When solving a problem that deals with limiting reactants, the idea is to find how many moles of all reactants are actually present and compare the mole ratios of those actual amounts to the mole ratios required by the balanced equation. That comparison will identify the reactant there is too much of (the excess reactant) and the reactant there is too little of (the limiting reactant). The flow chart in Figure 6.2 shows the procedure.

SOLUTION

(a) Begin by calculating the numbers of moles of reactants available. Finding the molar amounts always begins by calculating formula weights and using molar masses as conversion factors:

$$\text{Form. wt. of } K_2PtCl_4 = (2 \times 39.1) + 195.1 + (4 \times 35.5) = 415.3$$
$$\text{Molar mass of } K_2PtCl_4 = 415.3 \text{ g/mol}$$
$$\text{Moles of } K_2PtCl_4 = 10.0 \text{ g } \cancel{K_2PtCl_4} \times \frac{1 \text{ mol } K_2PtCl_4}{415.3 \text{ g } \cancel{K_2PtCl_4}} = 0.0241 \text{ mol } K_2PtCl_4$$
$$\text{Mol. wt. of } NH_3 = 14.0 + (3 \times 1.0) = 17.0$$
$$\text{Molar mass of } NH_3 = 17.0 \text{ g/mol}$$
$$\text{Moles of } NH_3 = 10.0 \text{ g } \cancel{NH_3} \times \frac{1 \text{ mol } NH_3}{17.0 \text{ g } \cancel{NH_3}} = 0.588 \text{ mol } NH_3$$

These calculations say that we have 0.588 mol of ammonia and 0.0241 mol of K_2PtCl_4, or 0.588/0.0241 = 24.4 times as much ammonia as K_2PtCl_4. The coefficients in the balanced equation, however, say that only *two* times as much ammonia as K_2PtCl_4 is needed. Thus, a large excess of NH_3 is present, and K_2PtCl_4 is the limiting reactant.

(b) With the identities of the excess reactant and limiting reactant known, we now have to find how many moles of each undergo reaction and then carry out mole-to-gram conversions to find the mass of each reactant consumed. The entire amount of the limiting reactant (K_2PtCl_4) is used up, but only the amount of the excess reactant (NH_3) required by stoichiometry undergoes reaction:

$$\text{Moles of } K_2PtCl_4 \text{ consumed} = 0.0241 \text{ mol } K_2PtCl_4$$
$$\text{Moles of } NH_3 \text{ consumed} = 0.0241 \text{ mol } \cancel{K_2PtCl_4} \times \frac{2 \text{ mol } NH_3}{1 \text{ mol } \cancel{K_2PtCl_4}} = 0.0482 \text{ mol } NH_3$$
$$\text{Grams of } NH_3 \text{ consumed} = 0.0482 \text{ } \cancel{\text{mol } NH_3} \times \frac{17.0 \text{ g } NH_3}{1 \text{ } \cancel{\text{mol } NH_3}} = 0.819 \text{ g } NH_3$$
$$\text{Grams of } NH_3 \text{ not consumed} = (10.0 \text{ g} - 0.819 \text{ g})NH_3 = 9.2 \text{ g } NH_3$$

(c) The balanced equation shows that 1 mol of cisplatin is formed for each mole of K_2PtCl_4 consumed. Thus, 0.0241 mol of cisplatin is formed from 0.0241 mol of K_2PtCl_4.

continued on next page

To determine the mass of cisplatin produced, we must calculate its molar mass and then carry out a mole-to-gram conversion:

$$\text{Form. wt. of } Pt(NH_3)_2Cl_2 = 195.1 + (2 \times 17.0) + (2 \times 35.5) = 300.1$$

$$\text{Molar mass of } Pt(NH_3)_2Cl_2 = 300.1 \text{ g/mol}$$

$$\text{Grams of } Pt(NH_3)_2Cl_2 = 0.0241 \text{ mol } Pt(NH_3)_2Cl_2 \times \frac{300.1 \text{ g } Pt(NH_3)_2Cl_2}{1 \text{ mol } Pt(NH_3)_2Cl_2}$$

$$= 7.23 \text{ g } Pt(NH_3)_2Cl_2$$

PROBLEM 6.9 Lithium oxide was used aboard the space shuttle to remove water from the air supply. If 80.0 kg of water is to be removed and 65 kg of Li_2O is available, which reactant is limiting? How many kilograms of the excess reactant remain?

$$Li_2O(s) + H_2O(g) \longrightarrow 2\,LiOH(s)$$

PROBLEM 6.10 After lithium hydroxide is produced aboard the space shuttle by reaction of Li_2O with H_2O (Problem 6.9), it is used to remove exhaled carbon dioxide from the air supply. How many grams of CO_2 can 500.0 g of LiOH absorb?

$$LiOH(s) + CO_2(g) \longrightarrow LiHCO_3(s)$$

CONCEPTUAL PROBLEM 6.11 The following diagram represents the reaction of **A** (red spheres) with $\mathbf{B_2}$ (blue spheres):

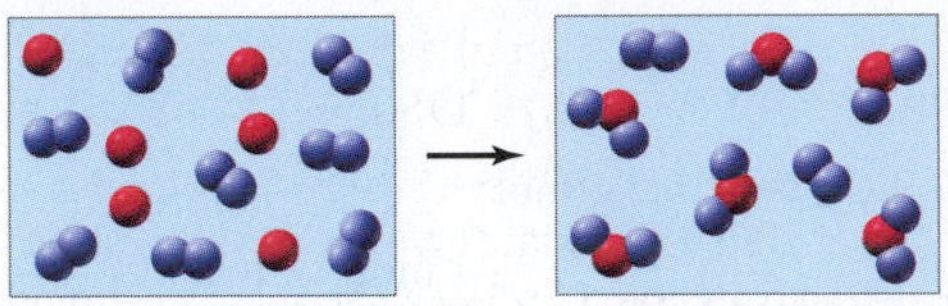

(a) Write a balanced equation for the reaction, and identify the limiting reactant.
(b) How many moles of product can be made from 1.0 mol of **A** and 1.0 mol of $\mathbf{B_2}$?

6.6 PERCENT COMPOSITION AND EMPIRICAL FORMULAS OF COMPOUNDS

All the substances we've dealt with thus far have had known formulas—water (H_2O), methane (CH_4), and ethyl alcohol (C_2H_6O), for instance. When, however, a new compound is made in the laboratory or found in nature, its formula must be experimentally determined as the first step in characterizing the compound's structure and chemical properties.

Determining the formula of a new compound begins with analyzing the substance to find what elements it contains and how much of each element is present—that is, to find its *composition*. The **percent composition** of a compound is expressed by identifying the elements present and giving the mass percent of each. For example, we might express the percent composition of a certain colorless liquid found in gasoline by saying that it contains 84.1% carbon and 15.9% hydrogen by mass. In other words, a 100.0 g sample of the compound contains 84.1 g of carbon atoms and 15.9 g of hydrogen atoms.

Knowing a compound's percent composition makes it possible to calculate the compound's chemical formula. As shown in **Figure 6.3**, the strategy is to find the relative number of moles of each element in the compound and then use those numbers to establish the *mole ratios* of the elements. The mole ratios, in turn, correspond to the subscripts in the chemical formula.

Let's use for an example the colorless liquid whose composition is 84.1% carbon and 15.9% hydrogen by mass. Arbitrarily taking 100 g of the substance to make the calculation easier, we find by using molar masses as conversion factors that the 100 g contains:

$$84.1\ \cancel{\text{g C}} \times \frac{1\ \text{mol C}}{12.01\ \cancel{\text{g C}}} = 7.00\ \text{mol C}$$

$$15.9\ \cancel{\text{g H}} \times \frac{1\ \text{mol H}}{1.008\ \cancel{\text{g H}}} = 15.8\ \text{mol H}$$

With the relative numbers of moles of C and H known, we find the mole ratio by dividing both by the smaller number (7.00):

$$C_{\left(\frac{7.00}{7.00}\right)}H_{\left(\frac{15.8}{7.00}\right)} = C_1H_{2.26}$$

The C:H mole ratio of 1:2.26 means that we can write $C_1H_{2.26}$ as a temporary formula for the liquid. Multiplying the subscripts by small integers in a trial-and-error procedure until whole numbers are found then gives the **empirical formula**, which tells the smallest whole-number ratios of atoms in the compound. In the present instance, multiplying the subscripts by 4 gives the whole-number ratio of atoms in the empirical formula C_4H_9. (The subscripts may not always be *exact* integers because of small errors in the data, but the discrepancies should be small.)

$$C_{(4\times1)}H_{(4\times2.26)} = C_4H_{9.04} = C_4H_9$$

An empirical formula determined from percent composition tells only the *ratios* of different atoms in a compound. The **molecular formula**, which tells the actual numbers of atoms in a molecule, can be either the same as the empirical formula or a multiple of it. To determine the molecular formula, it's necessary to know the *molecular weight* of the substance. In the present instance, the empirical formula weight of C_4H_9 is 57.1, but the molecular weight of the compound (a substance called octane) is 114.2, which is a multiple of 57.1.

For C_4H_9:

Atomic weight of 4 C	= (4 × 12.01)	= 12.04
Atomic weight of 9 H	= (9 × 1.01)	= 9.09
Empirical formula weight of C_4H_9		= 57.13

To find the multiple, divide the actual molecular weight (114.2) by the empirical formula weight (57.1):

$$\text{Multiple} = \frac{\text{Molecular weight}}{\text{Empirical formula weight}} = \frac{114.2}{57.1} = 2.00$$

Then multiply the subscripts in the empirical formula by this multiple to obtain the molecular formula. In our example, the molecular formula of octane is $C_{(2\times4)}H_{(2\times9)}$, or C_8H_{18}.

Just as we can find the empirical formula of a substance from its percent composition, we can also do the reverse calculation and find the percent composition of a substance from its empirical (or molecular) formula. The strategies for the two kinds of calculations are exactly opposite. Aspirin, for example, has the molecular formula $C_9H_8O_4$ and thus has a C:H:O mole ratio of 9:8:4. We can convert this mole ratio into a mass ratio, and thus into percent composition, by carrying out mole-to-gram conversions.

Let's assume we start with 1 mol of compound to simplify the calculation:

$$1\ \cancel{\text{mol aspirin}} \times \frac{9\ \cancel{\text{mol C}}}{1\ \cancel{\text{mol aspirin}}} \times \frac{12.0\ \text{g C}}{1\ \cancel{\text{mol C}}} = 108\ \text{g C}$$

$$1\ \cancel{\text{mol aspirin}} \times \frac{8\ \cancel{\text{mol H}}}{1\ \cancel{\text{mol aspirin}}} \times \frac{1.01\ \text{g H}}{1\ \cancel{\text{mol H}}} = 8.08\ \text{g H}$$

$$1\ \cancel{\text{mol aspirin}} \times \frac{4\ \cancel{\text{mol O}}}{1\ \cancel{\text{mol aspirin}}} \times \frac{16.0\ \text{g O}}{1\ \cancel{\text{mol O}}} = 64.0\ \text{g O}$$

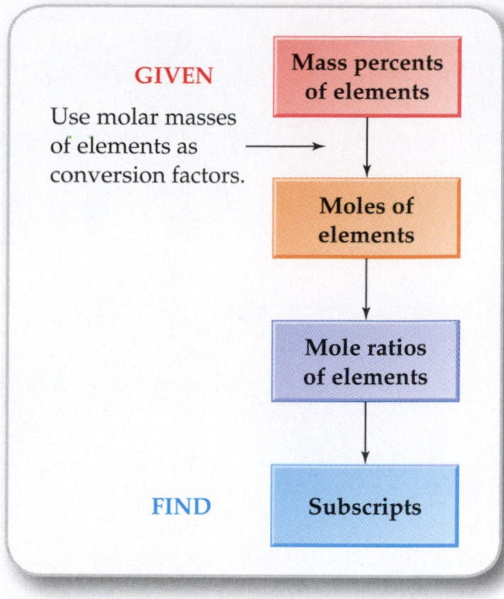

FIGURE 6.3

Calculating the formula of a compound from its percent composition.

Dividing the mass of each element by the total mass and multiplying by 100% then gives the percent composition:

$$\text{Total mass of 1 mol aspirin} = 108\text{ g} + 8.08\text{ g} + 64.0\text{ g} = 180\text{ g}$$

$$\%\text{C} = \frac{108\text{ g C}}{180\text{ g}} \times 100\% = 60.0\%$$

$$\%\text{H} = \frac{8.08\text{ g H}}{180\text{ g}} \times 100\% = 4.49\%$$

$$\%\text{O} = \frac{64.0\text{ g O}}{180\text{ g}} \times 100\% = 35.6\%$$

The answer can be checked by confirming that the sum of the mass percentages is within a rounding error of 100%: 60.0% + 4.49% + 35.6% = 100.1%.

Worked Examples 6.11 and 6.12 show further conversions between percent composition and empirical formulas.

Worked Example 6.11

Calculating an Empirical Formula from a Percent Composition

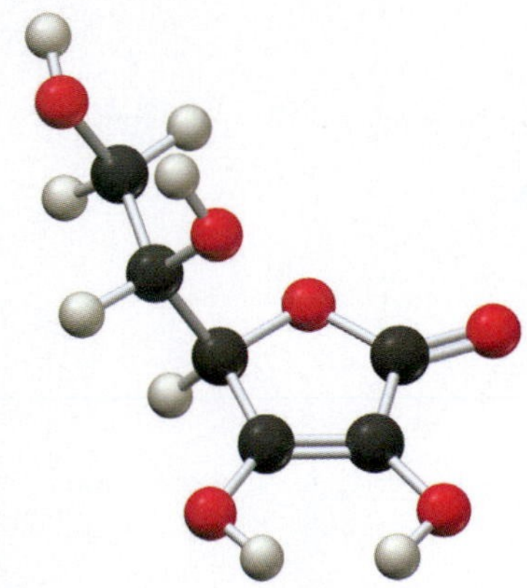

Ascorbic acid

Vitamin C (ascorbic acid) contains 40.92% C, 4.58% H, and 54.50% O by mass. What is the empirical formula of ascorbic acid?

STRATEGY

Assume that you have 100.00 g of ascorbic acid, and then carry out the procedure outlined in Figure 6.3.

SOLUTION

First, find the number of moles of each element in the sample:

$$40.92\text{ g C} \times \frac{1\text{ mol C}}{12.0\text{ g C}} = 3.41\text{ mol C}$$

$$4.58\text{ g H} \times \frac{1\text{ mol H}}{1.01\text{ g H}} = 4.53\text{ mol H}$$

$$54.50\text{ g O} \times \frac{1\text{ mol O}}{16.0\text{ g O}} = 3.41\text{ mol O}$$

Dividing each of the three numbers by the smallest (3.41 mol) gives a C:H:O mole ratio of 1:1.33:1 and a temporary formula of $C_1H_{1.33}O_1$. Multiplying the subscripts by small integers in a trial-and-error procedure until whole numbers are found then gives the empirical formula: $C_{(3\times1)}H_{(3\times1.33)}O_{(3\times1)} = C_3H_4O_3$.

Worked Example 6.12

Calculating a Percent Composition from a Formula

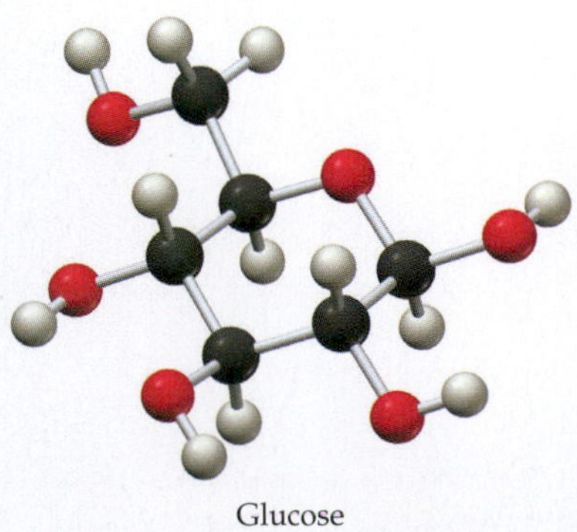

Glucose

Glucose, or blood sugar, has the molecular formula $C_6H_{12}O_6$. What is the empirical formula, and what is the percent composition of glucose?

STRATEGY AND SOLUTION

The empirical formula is found by reducing the subscripts in the molecular formula to their smallest whole-number values. In this case, dividing the subscripts by 6 reduces $C_6H_{12}O_6$ to CH_2O.

The percent composition of glucose can be calculated either from the molecular formula or from the empirical formula. Using the molecular formula, for instance, the C:H:O mole ratio of 6:12:6 can be converted into a mass ratio by assuming that we have 1 mol of compound and carrying out mole-to-gram conversions:

$$1\text{ mol glucose} \times \frac{6\text{ mol C}}{1\text{ mol glucose}} \times \frac{12.0\text{ g C}}{1\text{ mol C}} = 72.0\text{ g C}$$

$$1\text{ mol glucose} \times \frac{12\text{ mol H}}{1\text{ mol glucose}} \times \frac{1.01\text{ g H}}{1\text{ mol H}} = 12.1\text{ g H}$$

$$1\text{ mol glucose} \times \frac{6\text{ mol O}}{1\text{ mol glucose}} \times \frac{16.0\text{ g O}}{1\text{ mol O}} = 96.0\text{ g O}$$

Dividing the mass of each element by the total mass, and multiplying by 100%, gives the percent composition. Note that the sum of the mass percentages is 100%.

$$\text{Total mass of 1 mol glucose} = 72.0\text{ g} + 12.1\text{ g} + 96.0\text{ g} = 180.1\text{ g}$$

$$\%\text{ C} = \frac{72.0\text{ g C}}{180.1\text{ g}} \times 100\% = 40.0\%$$

$$\%\text{ H} = \frac{12.1\text{ g H}}{180.1\text{ g}} \times 100\% = 6.72\%$$

$$\%\text{ O} = \frac{96.0\text{ g O}}{180.1\text{ g}} \times 100\% = 53.3\%$$

PROBLEM 6.12 What is the empirical formula and what is the percent composition of dimethylhydrazine, $C_2H_8N_2$, a colorless liquid used as a rocket fuel?

PROBLEM 6.13 What is the empirical formula of the ingredient in Bufferin tablets that has the percent composition C 14.25%, O 56.93%, Mg 28.83% by mass?

PROBLEM 6.14 What is the percent composition of citric acid, an organic acid, commonly found in citrus fruits, whose structure is shown in the nearby margin? (gray = C, red = O, H = ivory)

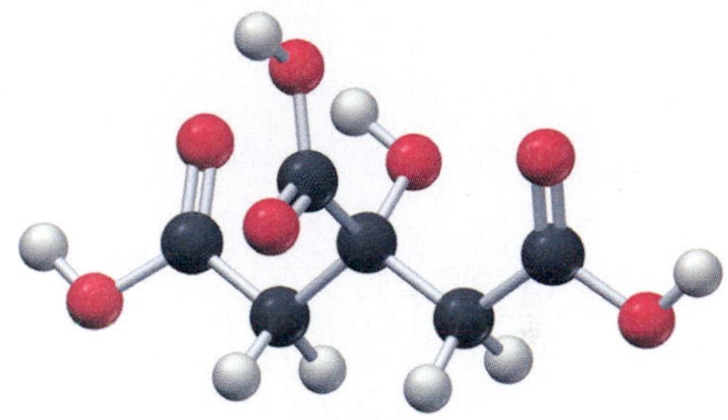

Citric acid

6.7 DETERMINING EMPIRICAL FORMULAS BY ELEMENTAL ANALYSIS

One of the most common methods used to determine percent composition and empirical formulas, particularly for organic compounds containing carbon and hydrogen, is *combustion analysis*. In this method, a compound of unknown composition is burned with oxygen to produce the volatile combustion products CO_2 and H_2O, which are separated and have their amounts determined by an automated instrument. Combustion reactions of organic compounds in an oxygen atmosphere essentially always proceed in 100% yield, so the amounts of CO_2 and H_2O formed are a direct measure of the amounts of C and H in the sample. Methane (CH_4), for instance, burns according to the balanced equation

$$CH_4(g) + 2\,O_2(g) \longrightarrow CO_2(g) + 2\,H_2O(g)$$

With the amounts of the carbon-containing product (CO_2) and hydrogen-containing product (H_2O) established, the strategy is to calculate the number of moles of carbon and hydrogen in the products, from which we can find the C:H mole ratio of the starting compound. This information, in turn, provides the chemical formula, as outlined by the flow chart in **Figure 6.4**.

As an example of how combustion analysis works, imagine that we have a sample of a pure compound—say, naphthalene, a substance often used in household moth balls. We weigh a known amount of the sample, burn it completely in pure oxygen, and then

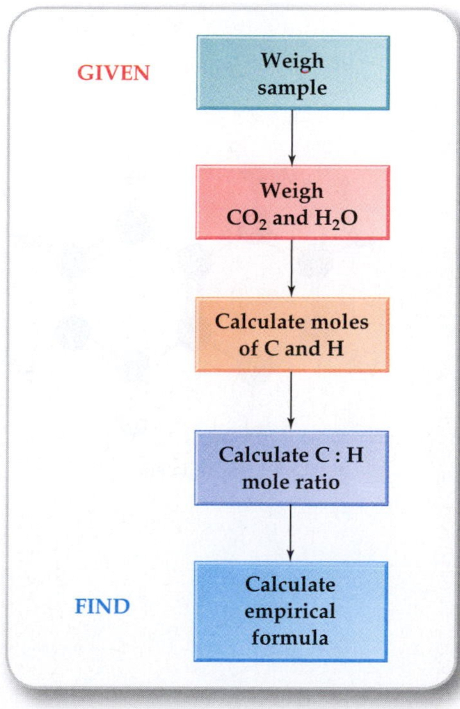

FIGURE 6.4

Determining an empirical formula from combustion analysis of a compound containing C and H.

analyze the products. Let's say that 0.330 g of naphthalene reacts with O_2 and that 1.133 g of CO_2 and 0.185 g of H_2O are formed. The first thing to find out is the number of moles of carbon and hydrogen in the CO_2 and H_2O products so that we can calculate the number of moles of each element originally present in the naphthalene sample.

$$\text{Moles of C in } 1.133 \text{ g } CO_2 = 1.133 \text{ g } CO_2 \times \frac{1 \text{ mol } CO_2}{44.01 \text{ g } CO_2} \times \frac{1 \text{ mol C}}{1 \text{ mol } CO_2}$$
$$= 0.025\,74 \text{ mol C}$$

$$\text{Moles of H in } 0.185 \text{ g } H_2O = 0.185 \text{ g } H_2O \times \frac{1 \text{ mol } H_2O}{18.02 \text{ g } H_2O} \times \frac{2 \text{ mol H}}{1 \text{ mol } H_2O}$$
$$= 0.0205 \text{ mol H}$$

Although it's not necessary in this instance since naphthalene contains only carbon and hydrogen, we can make sure that all the mass is accounted for and that no other elements are present. To do so, we carry out mole-to-gram conversions to find the number of grams of C and H in the starting sample:

$$\text{Mass of C} = 0.025\,74 \text{ mol C} \times \frac{12.01 \text{ g C}}{1 \text{ mol C}} = 0.3091 \text{ g C}$$

$$\text{Mass of H} = 0.0205 \text{ mol H} \times \frac{1.01 \text{ g H}}{1 \text{ mol H}} = 0.0207 \text{ g H}$$

$$\text{Total mass of C and H} = 0.3091 \text{ g} + 0.0207 \text{ g} = 0.3298 \text{ g}$$

Because the total mass of the C and H in the products (0.3298 g) is the same as the mass of the starting sample (0.330 g), we know that no other elements are present in naphthalene.

With the relative number of moles of C and H in naphthalene known, divide the larger number of moles by the smaller number to get the formula $C_{1.26}H_1$.

$$C_{\left(\frac{0.02574}{0.0205}\right)}H_{\left(\frac{0.0205}{0.0205}\right)} = C_{1.26}H_1$$

Then multiply the subscripts by small integers in a trial-and-error procedure until whole numbers are found to obtain the whole-number formula C_5H_4:

Multiply subscripts by 4: $C_{(4\times1.26)}H_{(4\times1)} = C_{5.04}H_4 = C_5H_4$ (Both subscripts are integers)

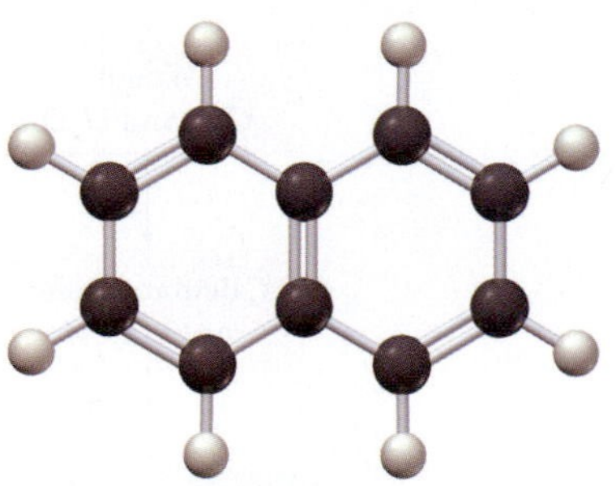

Naphthalene

Elemental analysis provides only an empirical formula. To determine the molecular formula, it's also necessary to know the substance's molecular weight. In the present instance, the molecular weight of naphthalene is 128.2, or twice the empirical formula weight of C_5H_4 (64.1). Thus, the molecular formula of naphthalene is $C_{(2\times5)}H_{(2\times4)} = C_{10}H_8$.

Worked Example 6.13 shows a combustion analysis when the sample contains oxygen in addition to carbon and hydrogen. Because oxygen yields no combustion products, its presence in a molecule can't be directly detected by this method. Rather, the presence of oxygen must be inferred by subtracting the calculated masses of C and H from the total mass of the sample.

Worked Example 6.13

Calculating an Empirical Formula from a Combustion Analysis

Caproic acid, the substance responsible for the aroma of goats, dirty socks, and old shoes, contains carbon, hydrogen, and oxygen. On combustion analysis, a 0.450 g sample of caproic acid gives 0.418 g of H_2O and 1.023 g of CO_2. What is the empirical formula of caproic acid? If the molecular weight of caproic acid is 116.2, what is the molecular formula?

STRATEGY

Use the steps outlined in Figure 6.4 to find the empirical formula of caproic acid. Then calculate a formula weight and compare it to the known molecular weight.

SOLUTION

First, find the molar amounts of C and H in the sample:

$$\text{Moles of C} = 1.023\ \text{g CO}_2 \times \frac{1\ \text{mol CO}_2}{44.01\ \text{g CO}_2} \times \frac{1\ \text{mol C}}{1\ \text{mol CO}_2} = 0.023\ 24\ \text{mol C}$$

$$\text{Moles of H} = 0.418\ \text{g H}_2\text{O} \times \frac{1\ \text{mol H}_2\text{O}}{18.02\ \text{g H}_2\text{O}} \times \frac{2\ \text{mol H}}{1\ \text{mol H}_2\text{O}} = 0.0464\ \text{mol H}$$

Next, find the number of grams of C and H in the sample:

$$\text{Mass of C} = 0.023\ 24\ \text{mol C} \times \frac{12.01\ \text{g G}}{1\ \text{mol C}} = 0.2791\ \text{g C}$$

$$\text{Mass of H} = 0.0464\ \text{mol H} \times \frac{1.01\ \text{g H}}{1\ \text{mol H}} = 0.0469\ \text{g H}$$

Subtracting the masses of C and H from the mass of the starting sample indicates that 0.124 g is unaccounted for:

$$0.450\ \text{g} - (0.2791\ \text{g} + 0.0469\ \text{g}) = 0.124\ \text{g}$$

Because we are told that oxygen is also present in the sample, the "missing" mass must be due to oxygen, which can't be detected by combustion. We therefore need to find the number of moles of oxygen in the sample:

$$\text{Moles of O} = 0.124\ \text{g O} \times \frac{1\ \text{mol O}}{16.00\ \text{g O}} = 0.007\ 75\ \text{mol O}$$

Knowing the relative numbers of moles of all three elements, C, H, and O, we divide the three numbers of moles by the smallest number (0.007 75 mol of oxygen) to arrive at a C:H:O ratio of 3:6:1.

$$\text{C}_{\left(\frac{0.02324}{0.00775}\right)}\text{H}_{\left(\frac{0.0464}{0.00775}\right)}\text{O}_{\left(\frac{0.00775}{0.00775}\right)} = \text{C}_3\text{H}_6\text{O}$$

The empirical formula of caproic acid is therefore C_3H_6O, and the empirical formula weight is 58.1. Because the molecular weight of caproic acid is 116.2, or twice the empirical formula weight, the molecular formula of caproic acid must be $C_{(2\times3)}H_{(2\times6)}O_{(2\times1)} = C_6H_{12}O_2$.

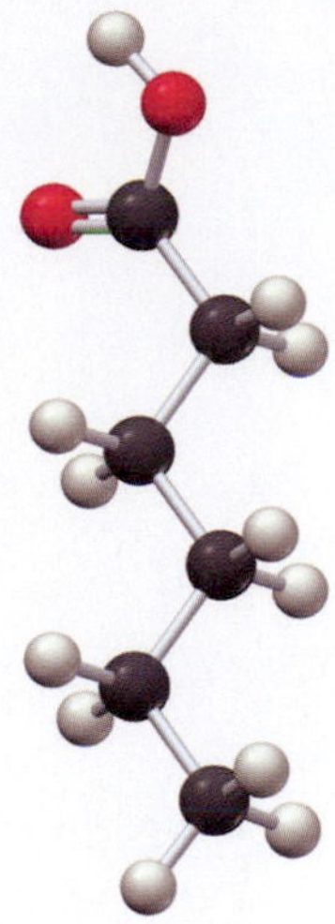

Caproic acid

PROBLEM 6.15 Menthol, a flavoring agent obtained from peppermint oil, contains carbon, hydrogen, and oxygen. On combustion analysis, 1.00 g of menthol yields 1.161 g of H_2O and 2.818 g of CO_2. What is the empirical formula of menthol?

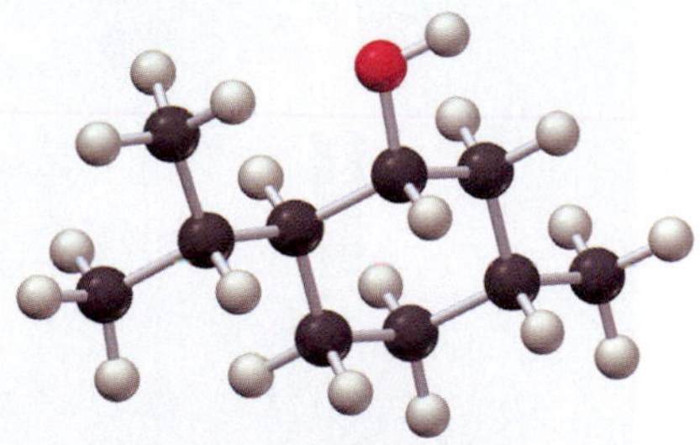

Menthol

PROBLEM 6.16 Ribose, a sugar present in the cells of all living organisms, has a molecular weight of 150 and the empirical formula CH_2O. What is the molecular formula of ribose?

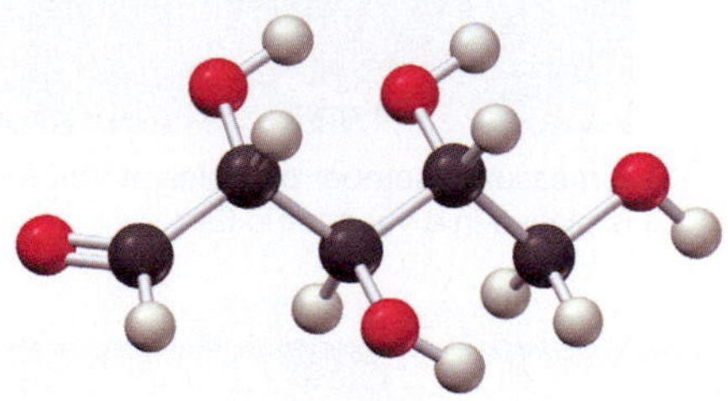

Ribose

PROBLEM 6.17 Convert the following percent compositions into molecular formulas:

(a) Diborane: H 21.86%, B 78.14%; Mol. wt. = 27.7

(b) Trioxan: C 40.00%, H 6.71%, O 53.28%; Mol. wt. = 90.08

6.8 CONCENTRATIONS OF REACTANTS IN SOLUTION: MOLARITY

Now that we know how to calculate the appropriate amounts of reactants to use in chemical reactions, one final point needs to be covered. For a chemical reaction to occur, the reacting molecules or ions must come into contact. This means that they must be mobile, which in turn means that most chemical reactions are carried out in liquid solution rather than as solids. It's therefore necessary to have a standard means to describe exact quantities of substances in solution.

As we've seen, stoichiometry calculations for chemical reactions always require working in moles. Thus, the most generally useful means of expressing a solution's concentration is **molarity (M)**—the number of moles of a substance, or **solute**, dissolved in one liter of solution. For example, a solution made by dissolving 1.00 mol (58.5 g) of NaCl in enough water to give 1.00 L of solution has a concentration of 1.00 mol/L, or 1.00 M. The molarity of any solution is found by dividing the number of moles of solute by the number of liters of solution.

$$\blacktriangleright \textbf{ Molarity (M)} = \frac{\text{Moles of solute}}{\text{Liters of solution}}$$

Note that it's the final volume of the *solution* that's important, not the starting volume of solvent used to make the solution. The final volume of the solution might be a bit larger than the volume of the solvent because of the additional volume of the solute. In practice, a solution of known molarity is prepared by weighing an appropriate amount of solute and placing it in a container called a *volumetric flask*, as shown in **Figure 6.5**. Enough solvent is added to dissolve the solute, and further solvent is added until an accurately calibrated final volume is reached. The solution is then shaken until it's uniformly mixed.

A measured number of moles of solute is placed in a volumetric flask.

Enough solvent is added to dissolve the solute by swirling.

Further solvent is added to reach the calibration mark on the neck of the flask, and the solution is shaken until uniform.

FIGURE 6.5

Preparing a solution of known molarity.

In essence, molarity is a conversion factor that relates a solution's volume to the number of moles of solute. If we know the molarity and volume of a solution, we can calculate the number of moles of solute. If we know the number of moles of solute and the molarity of the solution, we can find the solution's volume. Worked Examples 6.14 and 6.15 show how the calculations are done.

$$\text{Molarity} = \frac{\text{Moles of solute}}{\text{Volume of solution (L)}}$$

$$\text{Moles of solute} = \text{Molarity} \times \text{Volume of solution} \qquad \text{Volume of solution} = \frac{\text{Moles of solute}}{\text{Molarity}}$$

Worked Example 6.14

Calculating the Molarity of a Solution

What is the molarity of a solution made by dissolving 2.355 g of sulfuric acid (H_2SO_4) in water and diluting to a final volume of 50.0 mL?

STRATEGY

Molarity is the number of moles of solute per liter of solution. Thus, it's necessary to find the number of moles of sulfuric acid in 2.355 g and then divide by the volume of the solution.

SOLUTION

$$\text{Mol. wt. of } H_2SO_4 = (2 \times 1.0) + (32.1) + (4 \times 16.0) = 98.1$$

$$\text{Molar mass of } H_2SO_4 = 98.1 \text{ g/mol}$$

$$2.355 \text{ g } H_2SO_4 \times \frac{1 \text{ mol } H_2SO_4}{98.1 \text{ g } H_2SO_4} = 0.0240 \text{ mol } H_2SO_4$$

$$\frac{0.0240 \text{ mol } H_2SO_4}{0.0500 \text{ L}} = 0.480 \text{ M}$$

The solution has a sulfuric acid concentration of 0.480 M.

Worked Example 6.15

Calculating the Number of Moles of Solute in a Solution

Hydrochloric acid is sold commercially as a 12.0 M aqueous solution. How many moles of HCl are in 300.0 mL of 12.0 M solution?

STRATEGY

The number of moles of solute is calculated by multiplying the molarity of the solution by its volume in liters.

SOLUTION

$$\text{Moles of HCl} = (\text{Molarity of solution}) \times (\text{Volume of solution})$$

$$= \frac{12.0 \text{ mol HCl}}{1 \text{ L}} \times 0.3000 \text{ L} = 3.60 \text{ mol HCl}$$

There are 3.60 mol of HCl in 300.0 mL of 12.0 M solution.

✔ BALLPARK CHECK

One liter of 12.0 M HCl solution contains 12 mol of HCl, so 300 mL (0.3 L) of solution contains $0.3 \times 12 = 3.6$ mol.

PROBLEM 6.18 How many moles of solute are present in the following solutions?
(a) 125 mL of 0.20 M $NaHCO_3$ **(b)** 650.0 mL of 2.50 M H_2SO_4

PROBLEM 6.19 How many grams of solute would you use to prepare the following solutions?
(a) 500.0 mL of 1.25 M NaOH **(b)** 1.50 L of 0.250 M glucose ($C_6H_{12}O_6$)

PROBLEM 6.20 How many milliliters of a 0.20 M glucose ($C_6H_{12}O_6$) solution are needed to provide a total of 25.0 g of glucose?

PROBLEM 6.21 The concentration of cholesterol ($C_{27}H_{46}O$) in normal blood is approximately 0.005 M. How many grams of cholesterol are in 750 mL of blood?

Cholesterol

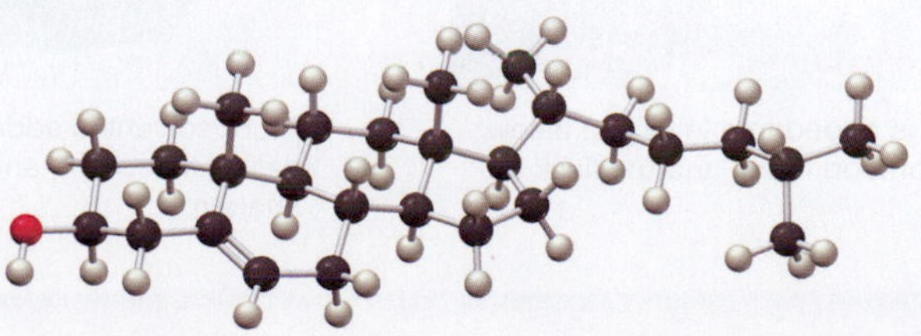

6.9 DILUTING CONCENTRATED SOLUTIONS

▲ Just as frozen orange juice concentrate must be diluted before use by adding water, many chemical solutions must also be diluted.

For convenience, chemicals are sometimes bought and stored as concentrated solutions, which are then diluted before use. Aqueous hydrochloric acid, for instance, is sold commercially as a 12.0 M solution, yet it is most commonly used in the laboratory after dilution with water to a final concentration of either 6.0 M or 1.0 M.

$$\text{Concentrated solution} + \text{Solvent} \longrightarrow \text{Dilute solution}$$

The main thing to remember when diluting a concentrated solution is that the number of moles of solute is constant. Only the volume of the solution is changed by adding more solvent. Because the number of moles of solute can be calculated by multiplying molarity times volume, we can set up the following equation:

$$\begin{aligned}\text{Moles of solute (constant)} &= \text{Molarity} \times \text{Volume}\\ &= \mathrm{M_i} \times V_\mathrm{i} = \mathrm{M_f} \times V_\mathrm{f}\end{aligned}$$

where $\mathrm{M_i}$ is the initial molarity, V_i is the initial volume, $\mathrm{M_f}$ is the final molarity, and V_f is the final volume after dilution. Rearranging this equation into a more useful form shows that the molar concentration after dilution ($\mathrm{M_f}$) can be found by multiplying the initial concentration ($\mathrm{M_i}$) by the ratio of initial and final volumes ($V_\mathrm{i}/V_\mathrm{f}$):

$$\mathrm{M_f} = \mathrm{M_i} \times \frac{V_\mathrm{i}}{V_\mathrm{f}}$$

Suppose, for example, that we dilute 50.0 mL of a solution of 2.00 M H_2SO_4 to a volume of 200.0 mL. The solution volume *increases* by a factor of four (from 50 mL to 200 mL), so the concentration of the solution must *decrease* by a factor of four (from 2.00 M to 0.500 M):

$$\mathrm{M_f} = 2.00\ \mathrm{M} \times \frac{50.0\ \cancel{\mathrm{mL}}}{200.0\ \cancel{\mathrm{mL}}} = 0.500\ \mathrm{M}$$

In practice, dilutions are usually carried out as shown in **Figure 6.6**. The volume to be diluted is withdrawn using a calibrated tube called a *pipet*, placed in an empty volumetric flask of the chosen volume, and diluted to the calibration mark on the flask. The one common exception to this order of steps is when diluting a strong acid such as H_2SO_4, where a large amount of heat is released. In such instances, it is much safer to add the acid slowly to the water rather than adding water to the acid.

FIGURE 6.6

The procedure for diluting a concentrated solution.

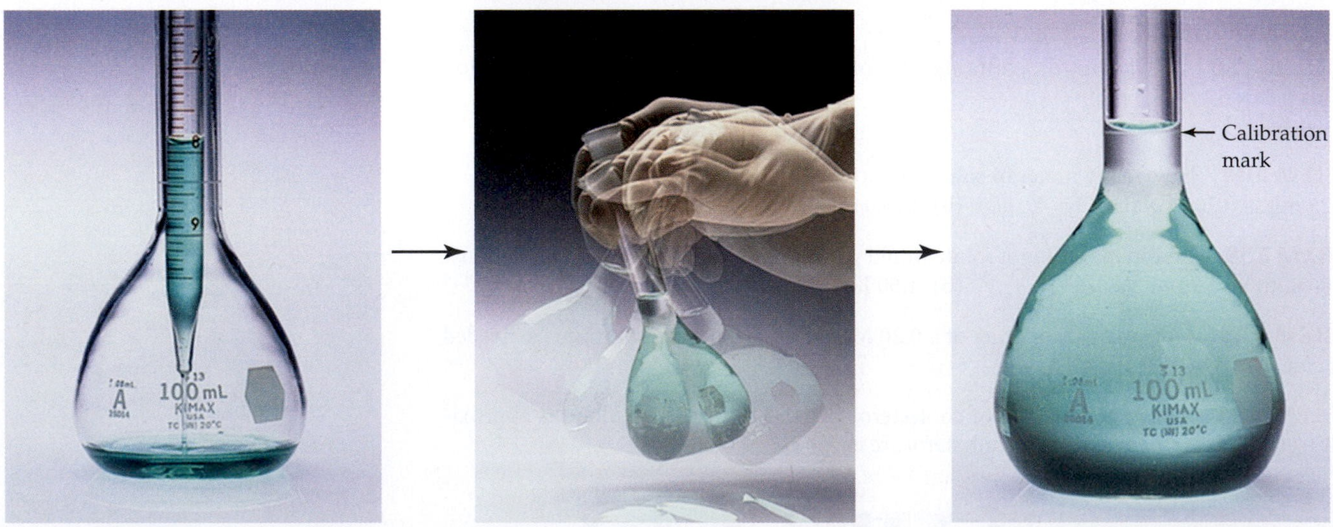

The **volume** to be diluted is placed in an empty volumetric flask.

Solvent is added to a level just below the calibration mark, and the flask is shaken.

More solvent is added to reach the calibration mark, and the flask is again shaken.

Worked Example 6.16

Diluting a Solution

How would you prepare 500.0 mL of 0.2500 M NaOH solution starting from a concentration of 1.000 M?

STRATEGY

The problem gives initial and final concentrations (M_i and M_f) and final volume (V_f) and asks for the initial volume (V_i) that we need to dilute. Rewriting the equation $M_i \times V_i = M_f \times V_f$ as $V_i = (M_f/M_i) \times V_f$ gives the answer.

SOLUTION

$$V_i = \frac{M_f}{M_i} \times V_f = \frac{0.2500\text{ M}}{1.000\text{ M}} \times 500.0\text{ mL} = 125.0\text{ mL}$$

We need to place 125.0 mL of 1.000 M NaOH solution in a 500.0 mL volumetric flask and fill to the calibration mark with water.

✔ BALLPARK CHECK

Because the concentration decreases by a factor of four after dilution (from 1.000 M to 0.2500 M), the volume must increase by a factor of four. Thus, to prepare 500.0 mL of solution, we should start with 500.0/4 = 125.0 mL.

PROBLEM 6.22 What is the final concentration if 75.0 mL of a 3.50 M glucose solution is diluted to a volume of 400.0 mL?

PROBLEM 6.23 Sulfuric acid is normally purchased at a concentration of 18.0 M. How would you prepare 250.0 mL of 0.500 M aqueous H_2SO_4? (Remember to add the acid to water rather than water to the acid.)

6.10 REACTION STOICHIOMETRY IN SOLUTIONS

We remarked in Section 6.8 that molarity is a conversion factor between numbers of moles of solute and the volume of a solution. If we know the volume and molarity of a solution, we can calculate the number of moles of solute. If we know the number of moles of solute and molarity, we can find the volume.

As indicated by the flow chart in **Figure 6.7**, using molarity is critical for carrying out stoichiometry calculations on substances in solution. Molarity makes it possible to calculate the volume of one solution needed to react with a given volume of another solution. This sort of calculation is particularly important in the chemistry of acids and bases, as shown in Worked Example 6.17.

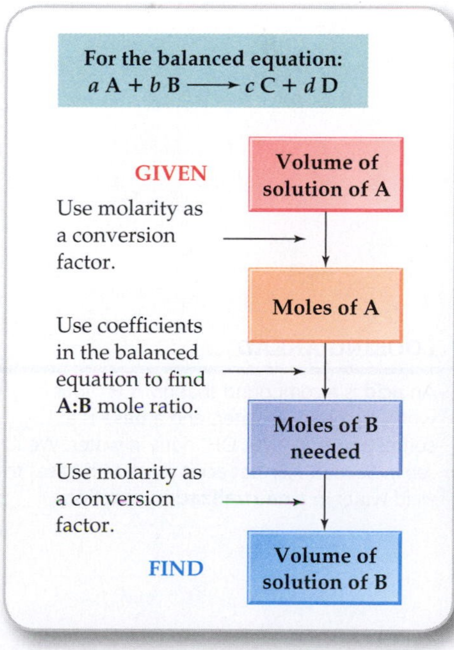

FIGURE 6.7

Using molarity as a conversion factor between moles and volume in stoichiometry calculations.

Worked Example 6.17

Reaction Stoichiometry in Solution

Stomach acid, a dilute solution of HCl in water, can be neutralized by reaction with sodium hydrogen carbonate, $NaHCO_3$, according to the equation

$$HCl(aq) + NaHCO_3(aq) \longrightarrow NaCl(aq) + H_2O(l) + CO_2(g)$$

How many milliliters of 0.125 M $NaHCO_3$ solution are needed to neutralize 18.0 mL of 0.100 M HCl?

STRATEGY

Solving stoichiometry problems always requires finding the number of moles of one reactant, using the coefficients of the balanced equation to find the number of moles of the other reactant, and then finding the amount of the other reactant. The flow chart summarizing the situation was shown in Figure 6.7.

continued on next page

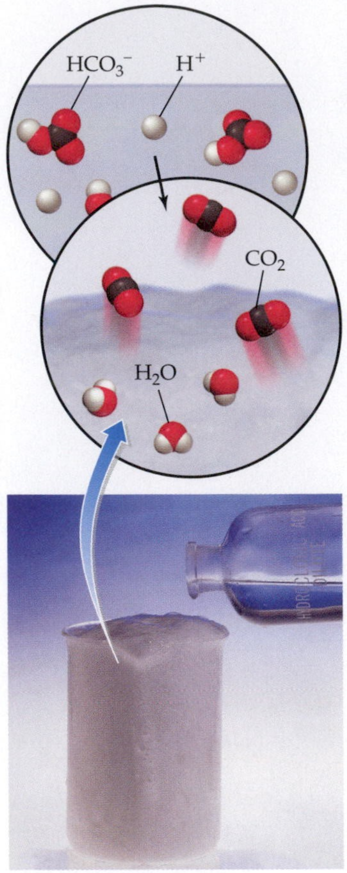

▲ Neutralization of sodium hydrogen carbonate with acid leads to release of CO_2 gas, visible in this fizzing solution.

SOLUTION

We first have to find how many moles of HCl are in 18.0 mL of a 0.100 M solution by multiplying volume times molarity:

$$\text{Moles of HCl} = 18.0\ \cancel{\text{mL}} \times \frac{1\ \cancel{\text{L}}}{1000\ \cancel{\text{mL}}} \times \frac{0.100\ \text{mol}}{1\ \cancel{\text{L}}} = 1.80 \times 10^{-3}\ \text{mol HCl}$$

Next, check the coefficients of the balanced equation to find that 1 mol of HCl reacts with 1 mol of $NaHCO_3$, and then calculate how many milliliters of a 0.125 M $NaHCO_3$ solution contains 1.80×10^{-3} mol:

$$1.80 \times 10^{-3}\ \cancel{\text{mol HCl}} \times \frac{1\ \cancel{\text{mol NaHCO}_3}}{1\ \cancel{\text{mol HCl}}} \times \frac{1\ \text{L solution}}{0.125\ \cancel{\text{mol NaHCO}_3}} = 0.0144\ \text{L solution}$$

Thus, 14.4 mL of the 0.125 M $NaHCO_3$ solution is needed to neutralize 18.0 mL of the 0.100 M HCl solution.

✔ BALLPARK CHECK

The balanced equation shows that HCl and $NaHCO_3$ react in a 1:1 molar ratio, and we are told that the concentrations of the two solutions are about the same. Thus, the volume of the $NaHCO_3$ solution must be about the same as that of the HCl solution.

▶ **PROBLEM 6.24** What volume of 0.250 M H_2SO_4 is needed to react with 50.0 mL of 0.100 M NaOH? The equation is

$$H_2SO_4(aq) + 2\,NaOH(aq) \longrightarrow Na_2SO_4(aq) + 2\,H_2O(l)$$

▶ **PROBLEM 6.25** What is the molarity of an HNO_3 solution if 68.5 mL is needed to react with 25.0 mL of 0.150 M KOH solution? The equation is

$$HNO_3(aq) + KOH(aq) \longrightarrow KNO_3(aq) + H_2O(l)$$

6.11 FINDING THE CONCENTRATION OF A SOLUTION: TITRATION

In order to use a solution of known molarity, that solution must first be made. There are two common ways to make a solution of known molarity. One way is to use the method described in Section 6.8; that is, to dissolve an accurately weighed amount of solute in enough solvent to reach an accurately calibrated volume. Often though, it's more convenient to make up a solution quickly, using an estimated amount of solute and an estimated amount of solvent, and then determine the solution's exact molarity using a procedure called *titration*.

Titration is a procedure for determining the concentration of a solution by allowing a measured volume of that solution to react with a second solution of another substance (the *standard solution*) whose concentration is known. By finding the volume of the standard solution that reacts with the measured volume of the first solution, the concentration of the first solution can be calculated. (It's necessary, though, that the reaction goes to completion and has a yield of 100%.)

LOOKING AHEAD ...

An **acid** is a compound that gives H^+ ions when dissolved in water, and a **base** is a compound that gives OH^- ions in water. We'll see in Section 7.5 that acids react with bases to yield water in a **neutralization reaction**.

To see how titration works, let's imagine that we have an HCl solution (an **acid**) whose concentration we want to find by allowing it to react with NaOH (a **base**) in what is called an **acid–base neutralization** reaction. The balanced equation is

$$NaOH(aq) + HCl(aq) \longrightarrow NaCl(aq) + H_2O(l)$$

We'll begin the titration by measuring out a known volume of the HCl solution and adding a small amount of an *indicator*, a compound that undergoes a color change during the course of the reaction. The compound phenolphthalein, for instance, is colorless in acid solution but turns red in base solution. Next, we fill a calibrated glass tube called a *buret* with an NaOH standard solution of known concentration and slowly add the NaOH to the HCl. When the phenolphthalein just begins to turn pink, the reaction is finished. That is, all the HCl has completely reacted and the solution now contains a tiny amount of excess NaOH. By then reading from the buret to find the volume of the NaOH standard solution that has been added to react with the known volume of HCl solution, we can calculate the concentration of the HCl. The strategy is summarized in **Figure 6.8**, and the procedure is shown in **Figure 6.9**.

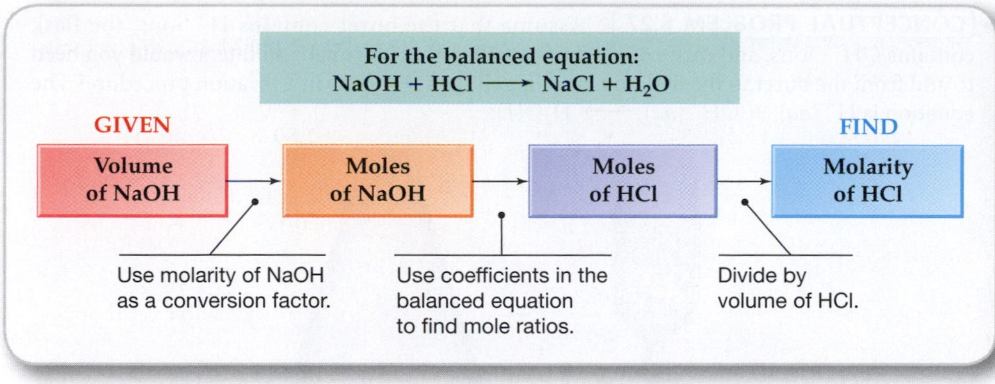

FIGURE 6.8

A flow chart for an acid–base titration. The calculations needed to determine the concentration of an HCl solution by titration with an NaOH standard solution are summarized.

Let's assume that we take 20.0 mL of the HCl solution of unknown concentration and find that we have to add 48.6 mL of 0.100 M NaOH from a buret to obtain complete reaction. Using the molarity of the NaOH standard solution as a conversion factor, we can calculate the number of moles of NaOH undergoing reaction:

$$\text{Moles of NaOH} = 0.0486\ \cancel{\text{L NaOH}} \times \frac{0.100\ \text{mol NaOH}}{1\ \cancel{\text{L NaOH}}}$$
$$= 0.004\,86\ \text{mol NaOH}$$

According to the balanced equation, the number of moles of HCl is the same as that of NaOH:

$$\text{Moles of HCl} = 0.004\,86\ \cancel{\text{mol NaOH}} \times \frac{1\ \text{mol HCl}}{1\ \cancel{\text{mol NaOH}}} = 0.004\,86\ \text{mol HCl}$$

Dividing the number of moles of HCl by the volume of the HCl solution then gives the molarity of the HCl:

$$\text{HCl molarity} = \frac{0.004\,86\ \text{mol HCl}}{0.0200\ \text{L HCl}} = 0.243\ \text{M HCl}$$

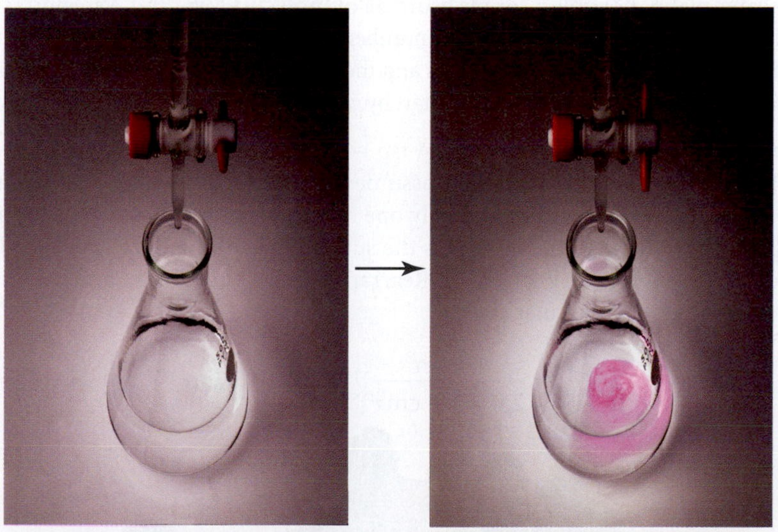

A measured volume of acid solution is placed in a flask, and phenolphthalein indicator is added.

Base solution of known concentration is added from a buret until the indicator changes color. Reading the volume of base from the buret allows calculation of the acid concentration.

FIGURE 6.9

Titration of an acid solution of unknown concentration with a base solution of known concentration.

PROBLEM 6.26 A 25.0 mL sample of vinegar (dilute acetic acid, CH_3CO_2H) is titrated and found to react with 94.7 mL of 0.200 M NaOH. What is the molarity of the acetic acid solution? The reaction is

$$NaOH(aq) + CH_3CO_2H(aq) \longrightarrow CH_3CO_2Na(aq) + H_2O(l)$$

CONCEPTUAL PROBLEM 6.27 Assume that the buret contains H^+ ions, the flask contains OH^- ions, and each has a volume of 100 mL. How many mililiters would you need to add from the buret to the flask to neutralize all the OH^- ions in a titration procedure? The equation is $H^+(aq) + OH^-(aq) \longrightarrow H_2O(l)$.

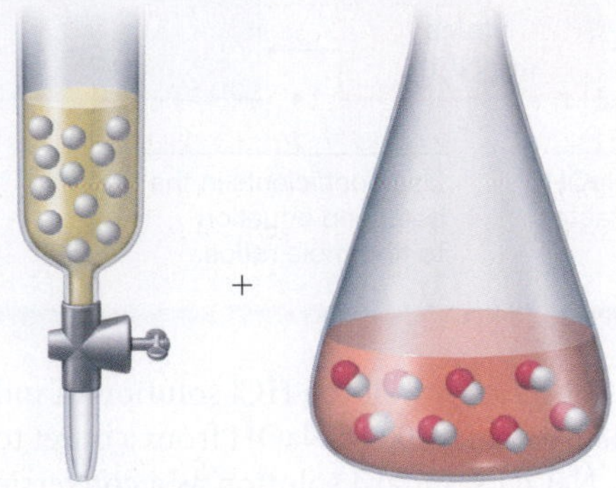

FYI DID BEN FRANKLIN HAVE AVOGADRO'S NUMBER?

▲ What did these two have in common?

At length being at Clapham, where there is on the common a large pond . . . I fetched out a cruet of oil and dropped a little of it on the water. I saw it spread itself with surprising swiftness upon the surface. The oil, although not more than a teaspoonful, produced an instant calm over a space several yards square which spread amazingly and extended itself gradually . . . making all that quarter of the pond, perhaps half an acre, as smooth as a looking glass.

Excerpt from a letter of Benjamin Franklin to William Brownrigg, 1773.

Benjamin Franklin, author and renowned statesman, was also an inventor and a scientist. Every schoolchild knows of Franklin's experiment with a kite and a key, demonstrating that lightning is electricity. Less well-known is that his measurement of the extent to which oil spreads on water makes possible a simple estimate of molecular size and Avogadro's number.

The calculation goes like this: Avogadro's number is the number of molecules in a mole. So, if we can estimate both the number of molecules and the number of moles in Franklin's teaspoon of oil, we can calculate Avogadro's number. Let's start by calculating the number of molecules in the oil.

1. The volume (V) of oil Franklin used was 1 tsp $= 4.9\ cm^3$, and the area (A) covered by the oil was 1/2 acre $= 2.0 \times 10^7\ cm^2$. Let's assume that the oil molecules are tiny cubes that pack closely together and form a layer only one molecule thick. As shown in the accompanying figure, the volume of the oil equals the surface area of the layer times the length (l) of the side of one molecule: $V = A \times l$. Rearranging this equation to find the length then gives us an estimate of molecular size:

$$l = \frac{V}{A} = \frac{4.9\ cm^3}{2.0 \times 10^7\ cm^2} = 2.4 \times 10^{-7}\ cm$$

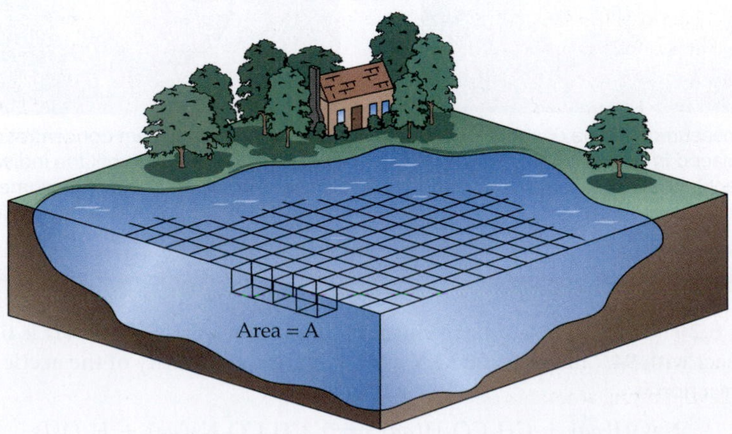

2. The area of the oil layer is the area of the side of one molecule (l^2) times the number of molecules (N) of oil: $A = l^2 \times N$. Rearranging this equation gives the number of molecules:

$$N = \frac{A}{l^2} = \frac{2.0 \times 10^7\ \text{cm}^2}{(2.4 \times 10^{-7}\ \text{cm})^2} = 3.5 \times 10^{20}\ \text{molecules}$$

3. To calculate the number of moles, we first need to know the mass (m) of the oil. This could have been determined by weighing, but Franklin neglected to do so. Let's therefore estimate the mass by multiplying the volume (V) of the oil by the density (D) of a typical oil, 0.95 g/cm^3. [Since oil floats on water, the density of oil must be a bit less than the density of water (1.00 g/cm^3).]

$$m = V \times D = 4.9\ \text{cm}^3 \times 0.95\ \frac{\text{g}}{\text{cm}^3} = 4.7\ \text{g}$$

4. We now have to make one final assumption about the molecular weight of the oil before completing the calculation. Assuming that a typical oil has mol. wt. = 900, then the mass of 1 mol of oil is 900 g. Dividing the mass of the oil by the mass of one mole gives the number of moles of oil:

$$\text{Moles of oil} = \frac{4.7\ \text{g}}{900\ \text{g/mol}} = 0.0052\ \text{mol}$$

5. Finally, the number of molecules per mole—Avogadro's number—can be obtained:

$$\text{Avogadro's number} = \frac{3.5 \times 10^{20}\ \text{molecules}}{0.0052\ \text{mol}} = 6.7 \times 10^{22}$$

The calculation is not very accurate, but Franklin wasn't really intending for us to calculate Avogadro's number when he made a rough estimate of how much his oil spread out. Nevertheless, the result isn't too bad for such a simple experiment.

PROBLEM 6.28 What do you think are the main sources of error in calculating Avogadro's number by spreading oil on a pond?

PROBLEM 6.29 Recalculate Avogadro's number assuming that the oil molecules are shaped like tall rectangular boxes rather than cubes, with two edges of equal length and the third edge four times the length of the other two. Assume also that the molecules stand on end in the water.

SUMMARY

Because mass is neither created nor destroyed in chemical reactions, all chemical equations must be **balanced**—that is, the numbers and kinds of atoms on both sides of the reaction arrow must be the same. A balanced equation tells the number ratio of reactant and product **formula units** in a reaction.

Molecular weight is the sum of the atomic weights of all the atoms in a molecule. The analogous term **formula weight** is used for ionic and other nonmolecular substances. One mole of a substance is the amount whose mass in grams—its molar mass—is numerically equal to the substance's molecular (or formula) weight. Carrying out chemical calculations using mass–mole relationships is called **stoichiometry** and is done using molar masses as conversion factors.

The amount of product actually formed in a reaction—the reaction's **yield**—is often less than the amount theoretically possible. Dividing the actual amount by the theoretical amount and multiplying by 100% gives the reaction's **percent yield**. Often, reactions are carried out with an excess of one reactant beyond that called for by the balanced equation. In such cases, the extent to which the reaction takes place depends on the reactant present in limiting amount, the **limiting reactant**.

The chemical makeup of a substance is described by its **percent composition**—the percentage of the substance's mass due to each of its constituent elements. Elemental analysis is used to calculate a substance's **empirical formula**, which gives the smallest whole-number ratio of atoms of the elements in the compound. To determine the **molecular formula**, which may be a simple multiple of the empirical formula, it's also necessary to know the substance's molecular weight.

The concentration of a substance in solution is usually expressed as **molarity (M)**, defined as the number of moles of a substance (the **solute**) dissolved per liter of solution. A solution's molarity acts as a conversion factor between solution volume and number of moles of solute, making it possible to carry out stoichiometry calculations on solutions. Often, chemicals are stored as concentrated aqueous solutions that are diluted before use. When carrying out a dilution, only the volume is changed by adding solvent; the amount of solute is unchanged. A solution's exact concentration can often be determined by **titration**.

KEY WORDS

balanced equation *202*	**formula unit** *202*	**molecular weight** *205*	**stoichiometry** *206*
coefficient *202*	**limiting reactant** *212*	**percent composition** *214*	**titration** *224*
empirical formula *215*	**molarity (M)** *220*	**percent yield** *209*	**yield** *209*
formula weight *205*	**molecular formula** *215*	**solute** *220*	

KEY EQUATIONS

- Percent yield **(Section 6.4)**

$$\textbf{Percent yield} = \frac{\text{Actual yield of product}}{\text{Theoretical yield of product}} \times 100\%$$

- Molarity **(Section 6.8)**

$$\textbf{Molarity (M)} = \frac{\text{Moles of solute}}{\text{Liters of solution}}$$

LEARNING OUTCOMES

- Balance chemical equations. **(Section 6.2)** ***Problems 1–3, 31, 33, 36, 38–43, 115, 117, 120***
- Calculate the molecular or formula weight of a substance. **(Section 6.3)** ***Problems 4, 5, 34, 44–47***
- Use molecular or formula weight to interconvert masses and moles in stoichiometry calculations. **(Section 6.3)** ***Problems 5, 6, 48–71, 126, 134, 136***
- Calculate theoretical yields and percent yields in a chemical reaction. **(Section 6.4)** ***Problems 7, 8, 72–83, 117, 120, 127***
- Determine the limiting reactant in a reaction and calculate the amount of excess reactant left over. **(Section 6.5)** ***Problems 9–11, 72–75, 78, 80–83, 117, 120, 127***
- Calculate an empirical formula from percent composition, or calculate percent composition from an empirical or molecular formula. **(Section 6.6)** ***Problems 12–14, 16, 17, 84–89, 93–97, 118, 119, 122, 129***
- Use results of combustion analysis to calculate percent composition and empirical formula. **(Section 6.7)** ***Problems 15, 90, 91, 98, 99, 125, 128, 130, 131***
- Dilute a concentrated solution. **(Section 6.9)** ***Problems 22, 23, 108, 109***
- Use molarity to convert between moles and volume in stoichiometry calculations. **(Section 6.10)** ***Problems 18, 20, 21, 24, 25, 100–111, 114, 124***
- Determine the concentration of a solution using a titration procedure. **(Section 6.11)** ***Problems 26, 27, 112, 113, 130–132, 137, 139***

CONCEPTUAL PROBLEMS

Problems 6.1–6.29 appear within the chapter.

6.30 Box **(a)** represents 1.0 mL of a solution of particles at a given concentration. Which of the boxes **(b)**–**(d)** represents 1.0 mL of the solution that results after **(a)** has been diluted by doubling the volume of its solvent?

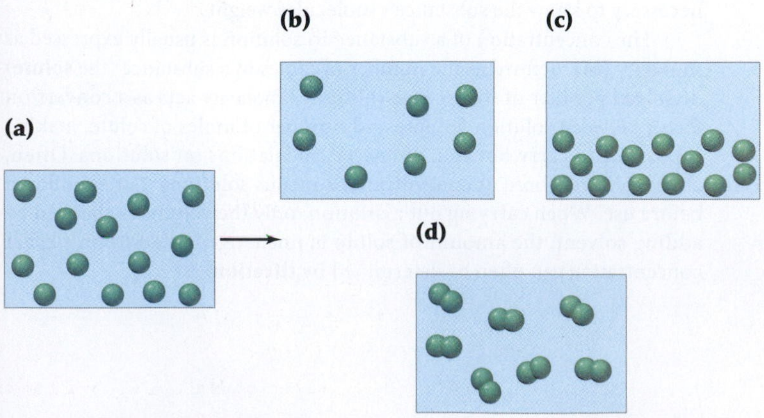

6.31 The reaction of **A** (red spheres) with **B** (blue spheres) is shown in the following diagram:

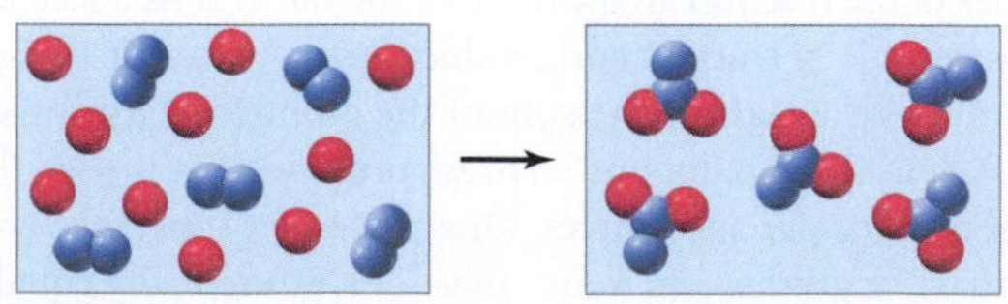

Which equation best describes the stoichiometry of the reaction?

(a) $A_2 + 2\,B \rightarrow A_2B_2$

(b) $10\,A + 5\,B_2 \rightarrow 5\,A_2B_2$

(c) $2\,A + B_2 \rightarrow A_2B_2$

(d) $5\,A + 5\,B_2 \rightarrow 5\,A_2B_2$

6.32 Cytosine, a constituent of deoxyribonucleic acid (DNA) can be represented by the following molecular model. If 0.0010 mol of cytosine is submitted to combustion analysis, how many moles of CO_2 and how many moles of H_2O would be formed? (gray = C, red = O, blue = N, ivory = H)

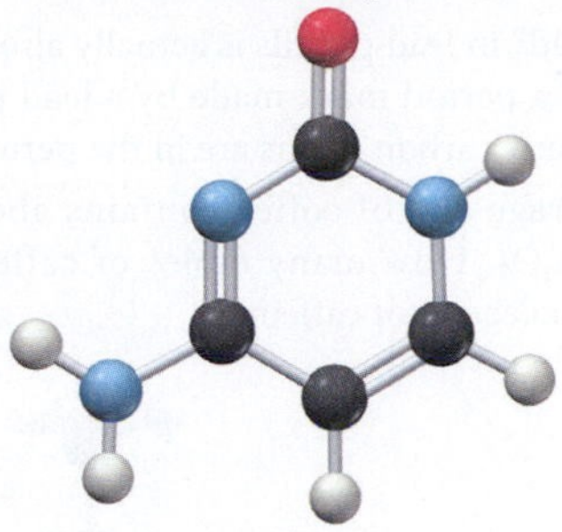

Cytosine

6.33 If blue spheres represent nitrogen atoms and red spheres represent oxygen atoms, which box represents reactants and which represents products for the reaction $2\,NO(g) + O_2(g) \longrightarrow 2\,NO_2(g)$?

(a)

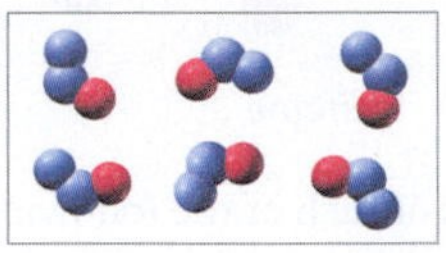

(b)

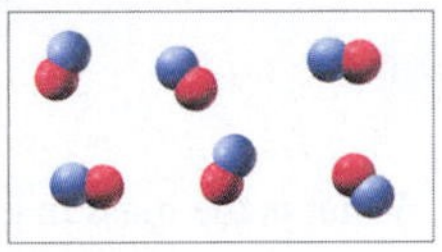

(c)

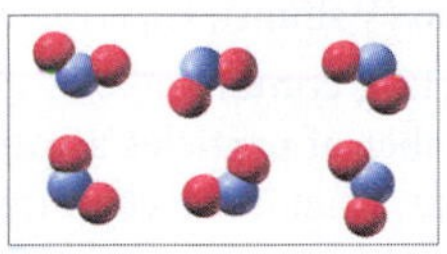

(d)

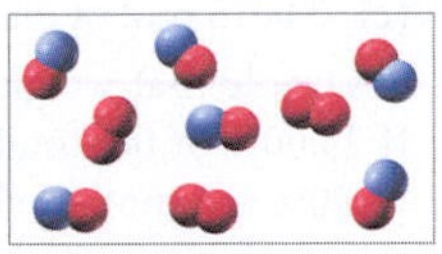

6.34 Fluoxetine, marketed as an antidepressant under the name Prozac, can be represented by the following ball-and-stick molecular model. Write the molecular formula for fluoxetine, and calculate its molecular weight (red = O, gray = C, blue = N, yellow-green = F, ivory = H).

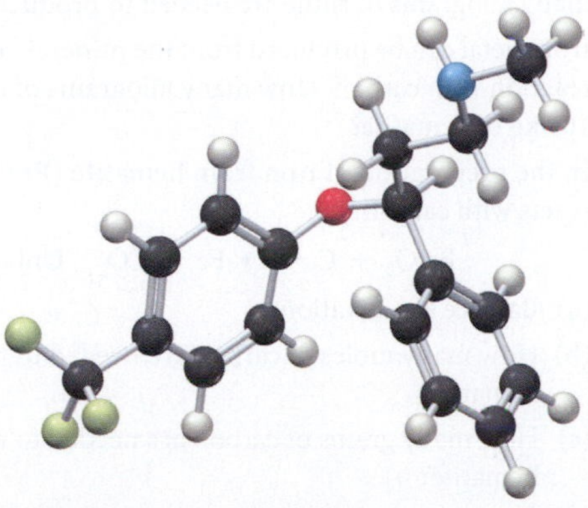

Fluoxetine

6.35 What is the percent composition of cysteine, one of the 20 amino acids commonly found in proteins? (gray = C, red = O, blue = N, yellow = S, ivory = H)

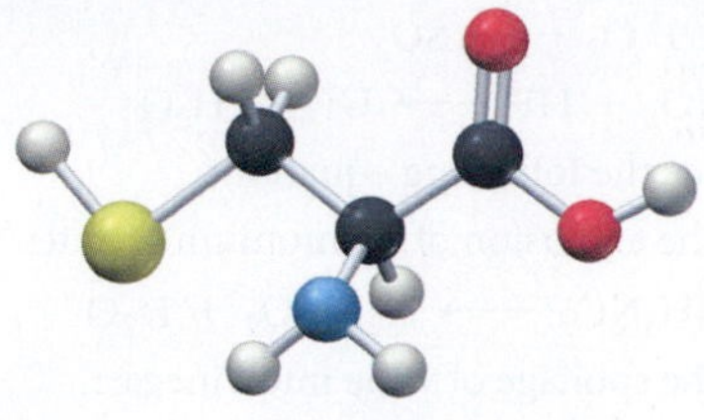

Cysteine

6.36 The following diagram represents the reaction of $\mathbf{A_2}$ (red spheres) with $\mathbf{B_2}$ (blue spheres):

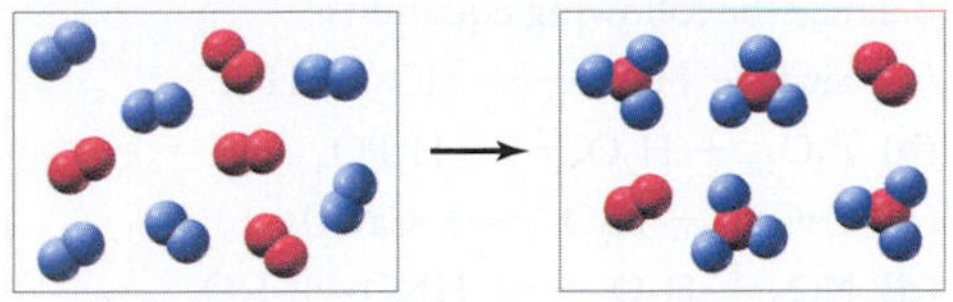

(a) Write a balanced equation for the reaction, and identify the limiting reactant.

(b) How many moles of product can be made from 1.0 mol of $\mathbf{A_2}$ and 1.0 mol of $\mathbf{B_2}$?

6.37 A hydrocarbon of unknown formula C_xH_y was submitted to combustion analysis with the following results. What is the empirical formula of the hydrocarbon?

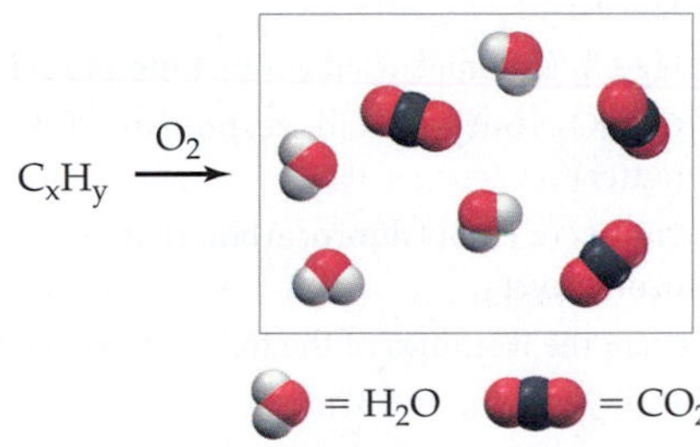

SECTION PROBLEMS

Balancing Equations (Section 6.2)

6.38 Which of the following equations are balanced?

(a) The development reaction in silver-halide photography:

$$2\,AgBr + 2\,NaOH + C_6H_6O_2 \longrightarrow 2\,Ag + H_2O + 2\,NaBr + C_6H_4O_2$$

(b) The preparation of household bleach:

$$2\,NaOH + Cl_2 \longrightarrow NaOCl + NaCl + H_2O$$

6.39 Which of the following equations are balanced? Balance any that need it.

(a) The thermite reaction, used in welding:

$$Al + Fe_2O_3 \longrightarrow Al_2O_3 + Fe$$

(b) The photosynthesis of glucose from CO_2:

$$6\,CO_2 + 6\,H_2O \longrightarrow C_6H_{12}O_6 + 6\,O_2$$

(c) The separation of gold from its ore:

$$Au + 2\,NaCN + O_2 + H_2O \longrightarrow NaAu(CN)_2 + 3\,NaOH$$

6.40 Balance the following equations:

(a) $Mg + HNO_3 \longrightarrow H_2 + Mg(NO_3)_2$

(b) $CaC_2 + H_2O \longrightarrow Ca(OH)_2 + C_2H_2$

(c) $S + O_2 \longrightarrow SO_3$

(d) $UO_2 + HF \longrightarrow UF_4 + H_2O$

6.41 Balance the following equations:

(a) The explosion of ammonium nitrate:

$NH_4NO_3 \longrightarrow N_2 + O_2 + H_2O$

(b) The spoilage of wine into vinegar:

$C_2H_6O + O_2 \longrightarrow C_2H_4O_2 + H_2O$

(c) The burning of rocket fuel:

$C_2H_8N_2 + N_2O_4 \longrightarrow N_2 + CO_2 + H_2O$

6.42 Balance the following equations:

(a) $SiCl_4 + H_2O \longrightarrow SiO_2 + HCl$

(b) $P_4O_{10} + H_2O \longrightarrow H_3PO_4$

(c) $CaCN_2 + H_2O \longrightarrow CaCO_3 + NH_3$

(d) $NO_2 + H_2O \longrightarrow HNO_3 + NO$

6.43 Balance the following equations:

(a) $VCl_3 + Na + CO \longrightarrow V(CO)_6 + NaCl$

(b) $RuI_3 + CO + Ag \longrightarrow Ru(CO)_5 + AgI$

(c) $CoS + CO + Cu \longrightarrow Co_2(CO)_8 + Cu_2S$

Molecular Weights and Stoichiometry (Section 6.3)

6.44 What are the molecular (formula) weights of the following substances?

(a) Hg_2Cl_2 (calomel, used at one time as a bowel purgative)

(b) $C_4H_8O_2$ (butyric acid, responsible for the odor of rancid butter)

(c) CF_2Cl_2 (a chlorofluorocarbon that destroys the stratospheric ozone layer)

6.45 What are the formulas of the following substances?

(a) $PCl_?$; Mol. wt. = 137.3

(b) Nicotine, $C_{10}H_{14}N_?$; Mol. wt. = 162.2

6.46 What are the molecular weights of the following pharmaceuticals?

(a) $C_{33}H_{35}FN_2O_5$ (atorvastatin, lowers blood cholesterol)

(b) $C_{22}H_{27}F_3O_4S$ (fluticasone, anti-inflammatory)

(c) $C_{16}H_{16}ClNO_2S$ (clopidogrel, inhibits blood clots)

6.47 What are the molecular weights of the following herbicides?

(a) $C_6H_6Cl_2O_3$ (2,4-dichlorophenoxyacetic acid, effective on broadleaf plants)

(b) $C_{15}H_{22}ClNO_2$ (metolachlor, pre-emergent herbicide)

(c) $C_8H_6Cl_2O_3$ (dicamba, effective on broadleaf plants)

6.48 How many grams are in a mole of each of the following substances?

(a) Ti **(b)** Br_2

(c) Hg **(d)** H_2O

6.49 How many moles are in a gram of each of the following substances?

(a) Cr **(b)** Cl_2

(c) Au **(d)** NH_3

6.50 How many moles of ions are in 27.5 g of $MgCl_2$?

6.51 How many moles of anions are in 35.6 g of AlF_3?

6.52 What is the molecular weight of chloroform if 0.0275 mol weighs 3.28 g?

6.53 What is the molecular weight of cholesterol if 0.5731 mol weighs 221.6 g?

6.54 Iron(II) sulfate, $FeSO_4$, is prescribed for the treatment of anemia. How many moles of $FeSO_4$ are present in a standard 300 mg tablet? How many iron(II) ions?

6.55 The "lead" in lead pencils is actually almost pure carbon, and the mass of a period mark made by a lead pencil is about 0.0001 g. How many carbon atoms are in the period?

6.56 An average cup of coffee contains about 125 mg of caffeine, $C_8H_{10}N_4O_2$. How many moles of caffeine are in a cup? How many molecules of caffeine?

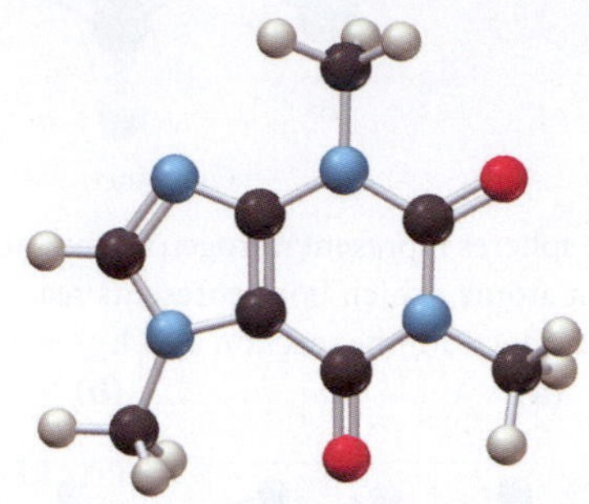

Caffeine

6.57 What is the mass in grams of each of the following samples?

(a) 0.0015 mol of sodium

(b) 0.0015 mol of lead

(c) 0.0015 mol of diazepam (Valium), $C_{16}H_{13}ClN_2O$

6.58 A sample that weighs 25.12 g contains 6.022×10^{23} particles. If 25.00% of the total number of particles are argon atoms and 75.00% are another element, what is the chemical identity of the other constituent?

6.59 A sample that weighs 107.75 g is a mixture of 30% helium atoms and 70% krypton atoms. How many particles are present in the sample?

6.60 Titanium metal is obtained from the mineral rutile, TiO_2. How many kilograms of rutile are needed to produce 100.0 kg of Ti?

6.61 Iron metal can be produced from the mineral hematite, Fe_2O_3, by reaction with carbon. How many kilograms of iron are present in 105 kg of hematite?

6.62 In the preparation of iron from hematite (Problem 6.61), Fe_2O_3 reacts with carbon:

$$Fe_2O_3 + C \longrightarrow Fe + CO_2 \quad \text{Unbalanced}$$

(a) Balance the equation.

(b) How many moles of carbon are needed to react with 525 g of hematite?

(c) How many grams of carbon are needed to react with 525 g of hematite?

6.63 An alternative method for preparing pure iron from Fe_2O_3 (Problem 6.61) is by reaction with carbon monoxide:

$$Fe_2O_3 + CO \longrightarrow Fe + CO_2 \quad \text{Unbalanced}$$

(a) Balance the equation.

(b) How many grams of CO are needed to react with 3.02 g of Fe_2O_3?

(c) How many grams of CO are needed to react with 1.68 mol of Fe_2O_3?

6.64 Magnesium metal burns in oxygen to form magnesium oxide, MgO.
(a) Write a balanced equation for the reaction.
(b) How many grams of oxygen are needed to react with 25.0 g of Mg? How many grams of MgO will result?
(c) How many grams of Mg are needed to react with 25.0 g of O_2? How many grams of MgO will result?

6.65 Ethylene gas, C_2H_4, reacts with water at high temperature to yield ethyl alcohol, C_2H_6O.
(a) How many grams of ethylene are needed to react with 0.133 mol of H_2O? How many grams of ethyl alcohol will result?
(b) How many grams of water are needed to react with 0.371 mol of ethylene? How many grams of ethyl alcohol will result?

6.66 Pure oxygen was first made by heating mercury(II) oxide:

$$HgO \xrightarrow{\text{Heat}} Hg + O_2 \quad \text{Unbalanced}$$

(a) Balance the equation.
(b) How many grams of mercury and how many grams of oxygen are formed from 45.5 g of HgO?
(c) How many grams of HgO would you need to obtain 33.3 g of O_2?

6.67 Titanium dioxide (TiO_2), the substance used as the pigment in white paint, is prepared industrially by reaction of $TiCl_4$ with O_2 at high temperature.

$$TiCl_4 + O_2 \xrightarrow{\text{Heat}} TiO_2 + 2\,Cl_2$$

How many kilograms of TiO_2 can be prepared from 5.60 kg of $TiCl_4$?

6.68 Silver metal reacts with chlorine (Cl_2) to yield silver chloride. If 2.00 g of Ag reacts with 0.657 g of Cl_2, what is the empirical formula of silver chloride?

6.69 Aluminum reacts with oxygen to yield aluminum oxide. If 5.0 g of Al reacts with 4.45 g of O_2, what is the empirical formula of aluminum oxide?

6.70 The industrial production of hydriodic acid takes place by treatment of iodine with hydrazine (N_2H_4):

$$2\,I_2 + N_2H_4 \longrightarrow 4\,HI + N_2$$

(a) How many grams of I_2 are needed to react with 36.7 g of N_2H_4?
(b) How many grams of HI are produced from the reaction of 115.7 g of N_2H_4 with excess iodine?

6.71 An alternative method for producing hydriodic acid (Problem 6.70) is the reaction of iodine with hydrogen sulfide:

$$H_2S + I_2 \longrightarrow 2\,HI + S$$

(a) How many grams of I_2 are needed to react with 49.2 g of H_2S?
(b) How many grams of HI are produced from reaction of 95.4 g of H_2S with excess I_2?

Limiting Reactants and Reaction Yield (Sections 6.4 and 6.5)

6.72 Assume that you have 1.39 mol of H_2 and 3.44 mol of N_2. How many grams of ammonia (NH_3) can you make, and how many grams of which reactant will be left over?

$$3\,H_2 + N_2 \longrightarrow 2\,NH_3$$

6.73 Hydrogen and chlorine react to yield hydrogen chloride: $H_2 + Cl_2 \longrightarrow 2\,HCl$. How many grams of HCl are formed from reaction of 3.56 g of H_2 with 8.94 g of Cl_2? Which reactant is limiting?

6.74 How many grams of the dry-cleaning solvent 1,2-dichloroethane (also called ethylene chloride), $C_2H_4Cl_2$, can be prepared by reaction of 15.4 g of ethylene, C_2H_4, with 3.74 g of Cl_2?

$$C_2H_4 + Cl_2 \longrightarrow C_2H_4Cl_2$$

1,2-Dichloroethane
(ethylene chloride)

6.75 How many grams of each product result from the following reactions, and how many grams of which reactant is left over?
(a) $(1.3\text{ g NaCl}) + (3.5\text{ g AgNO}_3) \longrightarrow (x\text{ g AgCl}) + (y\text{ g NaNO}_3)$
(b) $(2.65\text{ g BaCl}_2) + (6.78\text{ g H}_2\text{SO}_4) \longrightarrow (x\text{ g BaSO}_4) + (y\text{ g HCl})$

6.76 Nickel(II) sulfate, used for nickel plating, is prepared by treatment of nickel(II) carbonate with sulfuric acid:

$$NiCO_3 + H_2SO_4 \longrightarrow NiSO_4 + CO_2 + H_2O$$

(a) How many grams of H_2SO_4 are needed to react with 14.5 g of $NiCO_3$?
(b) How many grams of $NiSO_4$ are obtained if the yield is 78.9%?

6.77 Hydrazine, N_2H_4, once used as a rocket propellant, reacts with oxygen:

$$N_2H_4 + O_2 \longrightarrow N_2 + 2\,H_2O$$

(a) How many grams of O_2 are needed to react with 50.0 g of N_2H_4?
(b) How many grams of N_2 are obtained if the yield is 85.5%?

6.78 Limestone ($CaCO_3$) reacts with hydrochloric acid according to the equation $CaCO_3 + 2\,HCl \longrightarrow CaCl_2 + H_2O + CO_2$. If 1.00 mol of CO_2 has a volume of 22.4 L under the reaction conditions, how many liters of gas can be formed by reaction of 2.35 g of $CaCO_3$ with 2.35 g of HCl? Which reactant is limiting?

6.79 Sodium azide (NaN_3) yields N_2 gas when heated to 300 °C, a reaction used in automobile air bags. If 1.00 mol of N_2 has a volume of 47.0 L under the reaction conditions, how many liters of gas can be formed by heating 38.5 g of NaN_3? The reaction is

$$2\,NaN_3 \longrightarrow 3\,N_2(g) + 2\,Na$$

6.80 Acetic acid (CH_3CO_2H) reacts with isopentyl alcohol ($C_5H_{12}O$) to yield isopentyl acetate ($C_7H_{14}O_2$), a fragrant substance with the odor of bananas. If the yield from the reaction of acetic acid with isopentyl alcohol is 45%, how many grams of isopentyl acetate are formed from 3.58 g of acetic acid and 4.75 g of isopentyl alcohol? The reaction is

$$CH_3CO_2H + C_5H_{12}O \longrightarrow C_7H_{14}O_2 + H_2O$$

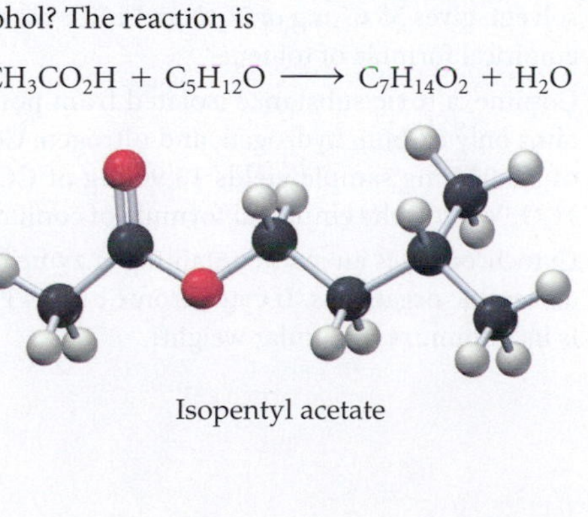

Isopentyl acetate

6.81 Cisplatin $[Pt(NH_3)_2Cl_2]$, a compound used in cancer treatment, is prepared by reaction of ammonia with potassium tetrachloroplatinate:

$$K_2PtCl_4 + 2\,NH_3 \longrightarrow 2\,KCl + Pt(NH_3)_2Cl_2$$

How many grams of cisplatin are formed from 55.8 g of K_2PtCl_4 and 35.6 g of NH_3 if the reaction takes place in 95% yield based on the limiting reactant?

6.82 If 1.87 g of acetic acid reacts with 2.31 g of isopentyl alcohol to give 2.96 g of isopentyl acetate (Problem 6.80), what is the percent yield of the reaction?

6.83 If 3.42 g of K_2PtCl_4 and 1.61 g of NH_3 give 2.08 g of cisplatin (Problem 6.81), what is the percent yield of the reaction?

Formulas and Elemental Analysis (Sections 6.6 and 6.7)

6.84 Urea, a substance commonly used as a fertilizer, has the formula CH_4N_2O. What is its percent composition by mass?

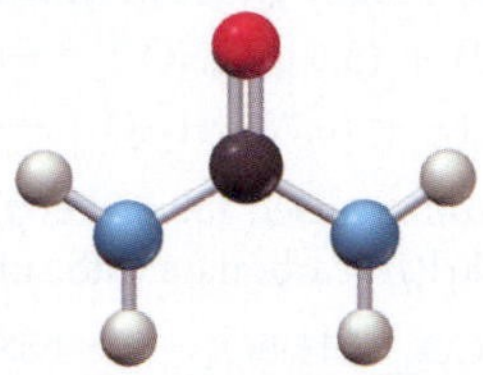

Urea

6.85 Calculate the mass percent composition of each of the following substances:

(a) Malachite, a copper-containing mineral: $Cu_2(OH)_2CO_3$

(b) Acetaminophen, a headache remedy: $C_8H_9NO_2$

(c) Prussian blue, an ink pigment: $Fe_4[Fe(CN)_6]_3$

6.86 An unknown liquid is composed of 5.57% H, 28.01% Cl, and 66.42% C. The molecular weight is 126.58. What is the molecular formula of the compound?

6.87 An unknown liquid is composed of 34.31% C, 5.28% H, and 60.41% I. The molecular weight is 210.07. What is the molecular formula of the compound?

6.88 What is the empirical formula of stannous fluoride, the first fluoride compound added to toothpaste to protect teeth against decay? Its mass percent composition is 24.25% F, 75.75% Sn.

6.89 What are the empirical formulas of each of the following substances?

(a) Ibuprofen, a headache remedy: 75.69% C, 15.51% O, 8.80% H

(b) Magnetite, a naturally occurring magnetic mineral: 72.36% Fe, 27.64% O

(c) Zircon, a mineral from which cubic zirconia is made: 34.91% O, 15.32% Si, 49.77% Zr

6.90 Combustion analysis of 45.62 mg of toluene, a commonly used solvent, gives 35.67 mg of H_2O and 152.5 mg of CO_2. What is the empirical formula of toluene?

6.91 Coniine, a toxic substance isolated from poison hemlock, contains only carbon, hydrogen, and nitrogen. Combustion analysis of a 5.024 mg sample yields 13.90 mg of CO_2 and 6.048 mg of H_2O. What is the empirical formula of coniine?

6.92 Cytochrome *c* is an iron-containing enzyme found in the cells of all aerobic organisms. If cytochrome *c* is 0.43% Fe by mass, what is its minimum molecular weight?

6.93 Nitrogen fixation in the root nodules of peas and other leguminous plants is carried out by the molybdenum-containing enzyme *nitrogenase*. What is the molecular weight of nitrogenase if the enzyme contains two molybdenum atoms and is 0.0872% Mo by mass?

6.94 Disilane, Si_2H_x, is analyzed and found to contain 90.28% silicon by mass. What is the value of *x*?

6.95 A certain metal sulfide, MS_2, is used extensively as a high-temperature lubricant. If MS_2 is 40.06% sulfur by mass, what is the identity of the metal M?

6.96 The molecular weight of an organic compound was found to be 70.042 11. Is the sample C_5H_{10}, C_4H_6O, or $C_3H_6N_2$? Exact atomic masses are: 1.007 825 (^{1}H); 12.000 00 (^{12}C); 14.003 074 (^{14}N); 15.994 915 (^{16}O).

6.97 The molecular weight of an organic compound was found to be 58.077 46. Is the sample C_4H_{10}, C_3H_6O, or $C_2H_6N_2$? Exact atomic masses are: 1.007 825 (^{1}H); 12.000 00 (^{12}C); 14.003 074 (^{14}N); 15.994 915 (^{16}O).

6.98 Combustion analysis of a 31.472 mg sample of the widely used flame retardant Decabrom gave 1.444 mg of CO_2. Is the molecular formula of Decabrom $C_{12}Br_{10}$ or $C_{12}Br_{10}O$?

6.99 The stimulant amphetamine contains only carbon, hydrogen, and nitrogen. Combustion analysis of a 42.92 mg sample of amphetamine gives 37.187 mg of H_2O and 125.75 mg of CO_2. If the molar mass of amphetamine is less than 160 g/mol, what is its molecular formula?

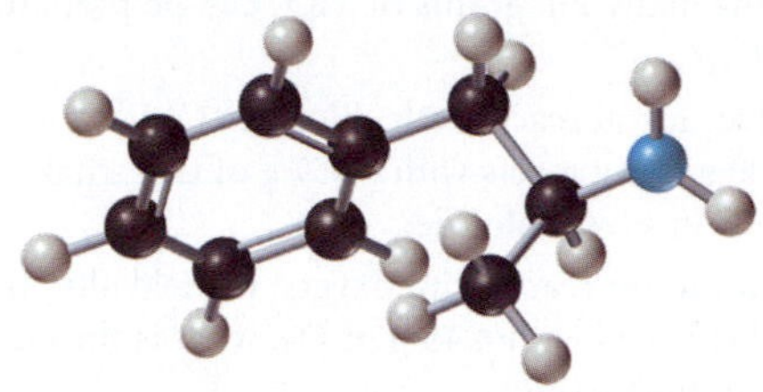

Amphetamine

Molarity, Solution Stoichiometry, Dilution, and Titration (Sections 6.8–6.11)

6.100 How many moles of solute are present in each of the following solutions?

(a) 35.0 mL of 1.200 M HNO_3

(b) 175 mL of 0.67 M glucose ($C_6H_{12}O_6$)

6.101 How many grams of solute would you use to prepare each of the following solutions?

(a) 250.0 mL of 0.600 M ethyl alcohol (C_2H_6O)

(b) 167 mL of 0.200 M boric acid (H_3BO_3)

6.102 How many milliliters of a 0.45 M $BaCl_2$ solution contain 15.0 g of $BaCl_2$?

6.103 How many milliliters of a 0.350 M KOH solution contain 0.0171 mol of KOH?

6.104 The sterile saline solution used to rinse contact lenses can be made by dissolving 400 mg of NaCl in sterile water and diluting to 100 mL. What is the molarity of the solution?

6.105 The concentration of glucose ($C_6H_{12}O_6$) in normal blood is approximately 90 mg per 100 mL. What is the molarity of the glucose?

6.106 Copper reacts with dilute nitric acid according to the following equation:

$$3\,Cu(s) + 8\,HNO_3(aq) \longrightarrow 3\,Cu(NO_3)_2(aq) + 2\,NO(g) + 4\,H_2O(l)$$

If a copper penny weighing 3.045 g is dissolved in a small amount of nitric acid and the resultant solution is diluted to 50.0 mL with water, what is the molarity of the $Cu(NO_3)_2$?

6.107 Pennies minted after 1982 are mostly zinc (97.5%) with a copper cover. If a post-1982 penny is dissolved in a small amount of nitric acid, the copper coating reacts as in Problem 6.106 and the exposed zinc reacts according to the following equation:

$$Zn(s) + 2\,HNO_3(aq) \longrightarrow Zn(NO_3)_2(aq) + H_2(g)$$

For a penny that weighs 2.482 g, what is the molarity of the $Zn(NO_3)_2$ if the resultant solution is diluted to 250.0 mL with water?

6.108 A bottle of 12.0 M hydrochloric acid has only 35.7 mL left in it. What will the HCl concentration be if the solution is diluted to 250.0 mL?

6.109 What is the volume of the solution that would result by diluting 70.00 mL of 0.0913 M NaOH to a concentration of 0.0150 M?

6.110 A flask containing 450 mL of 0.500 M HBr was accidentally knocked to the floor. How many grams of K_2CO_3 would you need to put on the spill to neutralize the acid according to the following equation?

$$2\,HBr(aq) + K_2CO_3(aq) \longrightarrow 2\,KBr(aq) + CO_2(g) + H_2O(l)$$

6.111 The odor of skunks is caused by chemical compounds called *thiols*. These compounds, of which butanethiol ($C_4H_{10}S$) is a representative example, can be deodorized by reaction with household bleach (NaOCl) according to the following equation:

$$2\,C_4H_{10}S + NaOCl(aq) \longrightarrow C_8H_{18}S_2 + NaCl(aq) + H_2O(l)$$

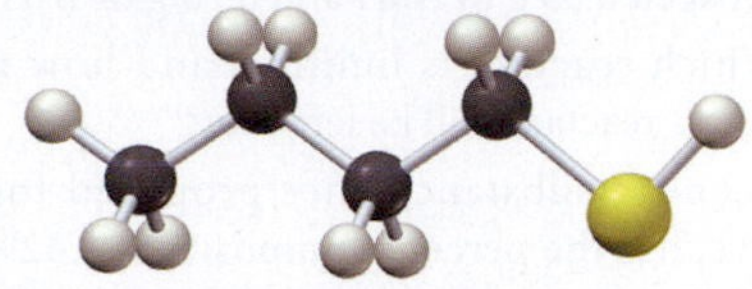

Butanethiol

How many grams of butanethiol can be deodorized by reaction with 5.00 mL of 0.0985 M NaOCl?

6.112 Potassium permanganate ($KMnO_4$) reacts with oxalic acid ($H_2C_2O_4$) in aqueous sulfuric acid according to the following equation:

$$2\,KMnO_4 + 5\,H_2C_2O_4 + 3\,H_2SO_4 \longrightarrow 2\,MnSO_4 + 10\,CO_2 + 8\,H_2O + K_2SO_4$$

How many milliliters of a 0.250 M $KMnO_4$ solution are needed to react completely with 3.225 g of oxalic acid?

6.113 Oxalic acid, $H_2C_2O_4$, is a toxic substance found in spinach leaves. What is the molarity of a solution made by dissolving 12.0 g of oxalic acid in enough water to give 400.0 mL of solution? How many milliliters of 0.100 M KOH would you need to titrate 25.0 mL of the oxalic acid solution according to the following equation?

$$H_2C_2O_4(aq) + 2\,KOH(aq) \longrightarrow K_2C_2O_4(aq) + 2\,H_2O(l)$$

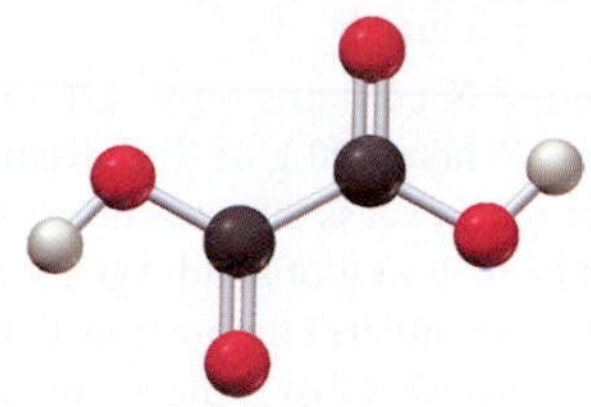

Oxalic acid

CHAPTER PROBLEMS

6.114 *Ringer's solution*, used in the treatment of burns and wounds, is prepared by dissolving 4.30 g of NaCl, 0.150 g of KCl, and 0.165 g of $CaCl_2$ in water and diluting to a volume of 500.0 mL. What is the molarity of each of the component ions in the solution?

6.115 Balance the following equations:

(a) $C_6H_5NO_2 + O_2 \longrightarrow CO_2 + H_2O + NO_2$

(b) $Au + H_2SeO_4 \longrightarrow Au_2(SeO_4)_3 + H_2SeO_3 + H_2O$

(c) $NH_4ClO_4 + Al \longrightarrow Al_2O_3 + N_2 + Cl_2 + H_2O$

6.116 The estimated concentration of gold in the oceans is 1.0×10^{-11} g/mL.

(a) Express the concentration in mol/L.

(b) Assuming that the volume of the oceans is 1.3×10^{21} L, estimate the amount of dissolved gold in grams in the oceans.

6.117 Silver sulfide, the tarnish on silverware, comes from reaction of silver metal with hydrogen sulfide (H_2S):

$$Ag + H_2S + O_2 \longrightarrow Ag_2S + H_2O \quad \text{Unbalanced}$$

(a) Balance the equation.

(b) If the reaction were used intentionally to prepare Ag_2S, how many grams would be formed from 496 g of Ag, 80.0 g of H_2S, and 40.0 g of O_2 if the reaction takes place in 95% yield based on the limiting reactant?

6.118 Give the percent composition of each of the following substances:

(a) Glucose, $C_6H_{12}O_6$

(b) Sulfuric acid, H_2SO_4

(c) Potassium permanganate, $KMnO_4$

(d) Saccharin, $C_7H_5NO_3S$

6.119 What are the empirical formulas of substances with the following mass percent compositions?

(a) Aspirin: 4.48% H, 60.00% C, 35.52% O

(b) Ilmenite (a titanium-containing ore): 31.63% O, 31.56% Ti, 36.81% Fe

(c) Sodium thiosulfate (photographic "fixer"): 30.36% O, 29.08% Na, 40.56% S

6.120 The reaction of tungsten hexachloride (WCl_6) with bismuth gives hexatungsten dodecachloride (W_6Cl_{12}).

$$WCl_6 + Bi \longrightarrow W_6Cl_{12} + BiCl_3 \quad \text{Unbalanced}$$

(a) Balance the equation.

(b) How many grams of bismuth react with 150.0 g of WCl_6?

(c) When 228 g of WCl_6 react with 175 g of Bi, how much W_6Cl_{12} is formed based on the limiting reactant?

6.121 Sodium borohydride, $NaBH_4$, a substance used in the synthesis of many pharmaceutical agents, can be prepared by reaction of NaH with B_2H_6 according to the equation $2\ NaH + B_2H_6 \longrightarrow 2\ NaBH_4$.

(a) How many grams of $NaBH_4$ can be prepared by reaction between 8.55 g of NaH and 6.75 g of B_2H_6?

(b) Which reactant is limiting, and how many grams of the excess reactant will be left over?

6.122 Ferrocene, a substance once proposed for use as a gasoline additive, has the percent composition 5.42% H, 64.56% C, and 30.02% Fe. What is the empirical formula of ferrocene?

6.123 The molar mass of HCl is 36.5 g/mol, and the average mass per HCl molecule is 36.5 u. Use the fact that $1\ u = 1.6605 \times 10^{-24}$ g to calculate Avogadro's number.

6.124 What is the molarity of each ion in a solution prepared by dissolving 0.550 g of Na_2SO_4, 1.188 g of Na_3PO_4, and 0.223 g of Li_2SO_4 in water and diluting to a volume of 100.00 mL?

6.125 Ethylene glycol, commonly used as automobile antifreeze, contains only carbon, hydrogen, and oxygen. Combustion analysis of a 23.46 mg sample yields 20.42 mg of H_2O and 33.27 mg of CO_2. What is the empirical formula of ethylene glycol? What is its molecular formula if it has a molecular weight of 62.0?

6.126 Assume that gasoline has the formula C_8H_{18} and has a density of 0.703 g/mL. How many pounds of CO_2 are produced from the complete combustion of 1.00 gal of gasoline?

6.127 A sample of $CaCO_3$ with a mass of 6.35 g is placed in 500.0 mL of 0.31 M HCl, forming $CaCl_2$, H_2O, and CO_2. What mass in grams of CO_2 is produced?

6.128 Compound **X** contains only carbon, hydrogen, nitrogen, and chlorine. When 1.00 g of **X** is dissolved in water and allowed to react with excess silver nitrate, $AgNO_3$, all the chlorine in **X** reacts and 1.95 g of solid AgCl is formed. When 1.00 g of **X** undergoes complete combustion, 0.900 g of CO_2 and 0.735 g of H_2O are formed. What is the empirical formula of **X**?

6.129 A pulverized rock sample believed to be pure calcium carbonate, $CaCO_3$, is subjected to chemical analysis and found to contain 51.3% Ca, 7.7% C, and 41.0% O by mass. Why can't this rock sample be pure $CaCO_3$?

6.130 Salicylic acid, used in the manufacture of aspirin, contains only the elements C, H, and O and has only one acidic hydrogen that reacts with NaOH. When 1.00 g of salicylic acid undergoes complete combustion, 2.23 g CO_2 and 0.39 g H_2O are obtained. When 1.00 g of salicylic acid is titrated with 0.100 M NaOH, 72.4 mL of base is needed for complete reaction. What are the empirical and molecular formulas of salicylic acid?

6.131 Compound **X** contains only the elements C, H, O, and S. A 5.00 g sample undergoes complete combustion to give 4.83 g of CO_2, 1.48 g of H_2O, and a certain amount of SO_2 that is further oxidized to SO_3 and dissolved in water to form sulfuric acid, H_2SO_4. On titration of the H_2SO_4, 109.8 mL of 1.00 M NaOH is needed for complete reaction. (Both H atoms in sulfuric acid are acidic and react with NaOH.)

(a) What is the empirical formula of **X**?

(b) When 5.00 g of **X** is titrated with NaOH, it is found that **X** has two acidic hydrogens that react with NaOH and that 54.9 mL of 1.00 M NaOH is required to completely neutralize the sample. What is the molecular formula of **X**?

6.132 Assume that you have 1.00 g of a mixture of benzoic acid (mol. wt. = 122) and gallic acid (mol. wt. = 170), both of which contain one acidic hydrogen that reacts with NaOH. On titrating the mixture with 0.500 M NaOH, 14.7 mL of base is needed to completely react with both acids. What mass in grams of each acid is present in the original mixture?

6.133 A certain alcoholic beverage contains only ethanol (C_2H_6O) and water. When a sample of this beverage undergoes combustion, the ethanol burns but the water simply evaporates and is collected along with the water produced by combustion. The combustion reaction is

$$C_2H_6O(l) + 3\ O_2(g) \longrightarrow 2\ CO_2(g) + 3\ H_2O(g)$$

When a 10.00 g sample of this beverage is burned, 11.27 g of water is collected. What is the mass in grams of ethanol, and what is the mass of water in the original sample?

6.134 A mixture of FeO and Fe_2O_3 with a mass of 10.0 g is converted to 7.43 g of pure Fe metal. What are the amounts in grams of FeO and Fe_2O_3 in the original sample?

6.135 A compound of formula XCl_3 reacts with aqueous $AgNO_3$ to yield solid AgCl according to the following equation:

$$XCl_3(aq) + 3\ AgNO_3(aq) \longrightarrow X(NO_3)_3(aq) + 3\ AgCl(s)$$

When a solution containing 0.634 g of XCl_3 was allowed to react with an excess of aqueous $AgNO_3$, 1.68 g of solid AgCl was formed. What is the identity of the atom **X**?

6.136 When eaten, dietary carbohydrates are digested to yield glucose ($C_6H_{12}O_6$), which is then metabolized to yield carbon dioxide and water:

$$C_6H_{12}O_6 + O_2 \longrightarrow CO_2 + H_2O \quad \text{Unbalanced}$$

Balance the equation, and calculate both the mass in grams and the volume in liters of the CO_2 produced from 66.3 g of glucose, assuming that 1 mol of CO_2 has a volume of 25.4 L at normal body temperature.

6.137 Other kinds of titrations are possible in addition to acid–base titrations. For example, the concentration of a solution of potassium permanganate, $KMnO_4$, can be determined by titration against a known amount of oxalic acid, $H_2C_2O_4$, according to the following equation:

$$5\ H_2C_2O_4(aq) + 2\ KMnO_4(aq) + 3\ H_2SO_4(aq) \longrightarrow 10\ CO_2(g) + 2\ MnSO_4(aq) + K_2SO_4(aq) + 8\ H_2O(l)$$

What is the concentration of a $KMnO_4$ solution if 22.35 mL reacts with 0.5170 g of oxalic acid?

6.138 A copper wire having a mass of 2.196 g was allowed to react with an excess of sulfur. The excess sulfur was then burned, yielding SO_2 gas. The mass of the copper sulfide produced was 2.748 g.

(a) What is the percent composition of copper sulfide?

(b) What is its empirical formula?

(c) Calculate the number of copper ions per cubic centimeter if the density of the copper sulfide is 5.6 g/cm^3.

6.139 A 1.268 g sample of a metal carbonate (MCO_3) was treated with 100.00 mL of 0.1083 M sulfuric acid (H_2SO_4), yielding CO_2 gas and an aqueous solution of the metal sulfate (MSO_4). The solution was boiled to remove all the dissolved CO_2 and was then titrated with 0.1241 M NaOH. A 71.02 mL volume of NaOH was required to neutralize the excess H_2SO_4.

(a) What is the identity of the metal **M**?

(b) How many liters of CO_2 gas were produced if the density of CO_2 is 1.799 g/L?

6.140 Ammonium nitrate, a potential ingredient of terrorist bombs, can be made nonexplosive by addition of diammonium hydrogen phosphate, $(NH_4)_2HPO_4$. Analysis of such a NH_4NO_3–$(NH_4)_2HPO_4$ mixture showed the mass percent of nitrogen to be 30.43%. What is the mass ratio of the two components in the mixture?

6.141 Window glass is typically made by mixing soda ash (Na_2CO_3), limestone ($CaCO_3$), and silica sand (SiO_2) and then heating to 1500 °C to drive off CO_2 from the Na_2CO_3 and $CaCO_3$. The resultant glass consists of about 12% Na_2O by mass, 13% CaO by mass, and 75% SiO_2 by mass. How much of each reactant would you start with to prepare 0.35 kg of glass?

MULTICONCEPT PROBLEMS

6.142 An unidentified metal **M** reacts with an unidentified halogen **X** to form a compound MX_2. When heated, the compound decomposes by the reaction:

$$2\,MX_2(s) \longrightarrow 2\,MX(s) + X_2(g)$$

When 1.12 g of MX_2 is heated, 0.720 g of MX is obtained, along with 56.0 mL of X_2 gas. Under the conditions used, 1.00 mol of the gas has a volume of 22.41 L.

(a) What is the atomic weight and identity of the halogen **X**?

(b) What is the atomic weight and identity of the metal **M**?

6.143 Element **X**, a member of group 5A, forms two chlorides, XCl_3 and XCl_5.

(a) If reaction of an excess of Cl_2 with 8.729 g of XCl_3 yields 13.233 g of XCl_5, what is the atomic weight and the identity of element **X**?

(b) If a mixture of XCl_3 and XCl_5 weighing 10.00 g contains 81.04% Cl by mass, how many grams of XCl_3 and how many grams of XCl_5 are present in the mixture?

6.144 A 1.005 g sample of an unknown alkaline-earth metal was allowed to react with a volume of chlorine gas that contains 1.91×10^{22} Cl_2 molecules. The resulting metal chloride was analyzed for chlorine by dissolving a 0.436 g sample in water and adding an excess of $AgNO_3(aq)$ to give a precipitate of 1.126 g of solid AgCl.

(a) What is the percent Cl in the alkaline-earth chloride?

(b) What is the identity of the alkaline-earth metal?

(c) Write balanced equations for all chemical reactions.

(d) In the reaction of the alkaline-earth metal with chlorine, which reactant is in excess and how many grams of that reactant remain unreacted?

6.145 A compound with the formula $XOCl_2$ reacts with water, yielding HCl and another acid H_2XO_3, which has two acidic hydrogens that react with NaOH. When 0.350 g of $XOCl_2$ was added to 50.0 mL of water and the resultant solution was titrated, 96.1 mL of 0.1225 M NaOH was required to react with all the acid.

(a) Write a balanced equation for the reaction of $XOCl_2$ with H_2O.

(b) What are the atomic weight and identity of element **X**?

6.146 Element **M** is prepared industrially by a two-step procedure according to the following (unbalanced) equations:

$$(1)\ M_2O_3(s) + C(s) + Cl_2(g) \longrightarrow MCl_3(l) + CO(g)$$

$$(2)\ MCl_3(l) + H_2(g) \longrightarrow M(s) + HCl(g)$$

Assume that 0.855 g of M_2O_3 is submitted to the reaction sequence. When the HCl produced in step (2) is dissolved in water and titrated with 0.511 M NaOH, 144.2 mL of the NaOH solution is required to neutralize the HCl.

(a) Balance both equations.

(b) What is the atomic weight of element **M**, and what is its identity?

(c) What mass of **M** in grams is produced in the reaction?

CHAPTER

7 Reactions in Aqueous Solution

Some chemical reactions produce toxic wastes, but green chemistry should help prevent this kind of problem in the future. We'll learn more in the *FYI* at the end of this chapter.

Ours is a world based on water. Approximately 71% of the Earth's surface is covered by water, and another 3% is covered by ice; 66% of the mass of an adult human body is water, and water is needed to sustain all living organisms. It's therefore not surprising that a large amount of important chemistry, including all those reactions that happen in our bodies, takes place in water—that is, in *aqueous solution*.

We saw in the previous chapter how chemical reactions are described and how the specific mass relationships among reactant and product substances in reactions can be calculated. In this chapter, we'll continue the study of chemical reactions by describing some ways that chemical reactions can be classified and by exploring some general ways in which reactions take place.

7.1 ELECTROLYTES IN AQUEOUS SOLUTION

Before beginning a study of reactions in water, it's helpful to look at some general properties of aqueous solutions. We all know from experience that both sugar (sucrose) and table salt (NaCl) dissolve in water. The solutions that result, though, are quite different. When sucrose, a molecular substance, dissolves in water, the resulting solution contains neutral sucrose *molecules* surrounded by water. When NaCl, an ionic substance, dissolves in water, the solution contains separate Na^+ and Cl^- *ions* surrounded by water. Because of the presence of the charged ions, the NaCl solution conducts an electric current, but the sucrose solution does not.

$$\underset{\text{Sucrose}}{C_{12}H_{22}O_{11}(s)} \xrightarrow{H_2O} C_{12}H_{22}O_{11}(aq)$$

$$NaCl(s) \xrightarrow{H_2O} Na^+(aq) + Cl^-(aq)$$

The electrical conductivity of an aqueous NaCl solution is easy to demonstrate using a battery, a light bulb, and several pieces of wire, connected as shown in **Figure 7.1**. When the wires are dipped into an aqueous NaCl solution, the positively charged Na^+ ions move through the solution toward the wire connected to the negatively charged terminal of the battery and the negatively charged Cl^- ions move toward the wire connected to the positively charged terminal of the battery. The resulting movement of electrical charges allows a current

CONTENTS

to flow, so the bulb lights. When the wires are dipped into an aqueous sucrose solution, however, there are no ions to carry the current, so the bulb remains dark.

Substances such as NaCl or KBr, which dissolve in water to produce conducting solutions of ions, are called **electrolytes**. Substances such as sucrose or ethyl alcohol, which do not produce ions in aqueous solution, are **nonelectrolytes**. Most electrolytes are ionic compounds,

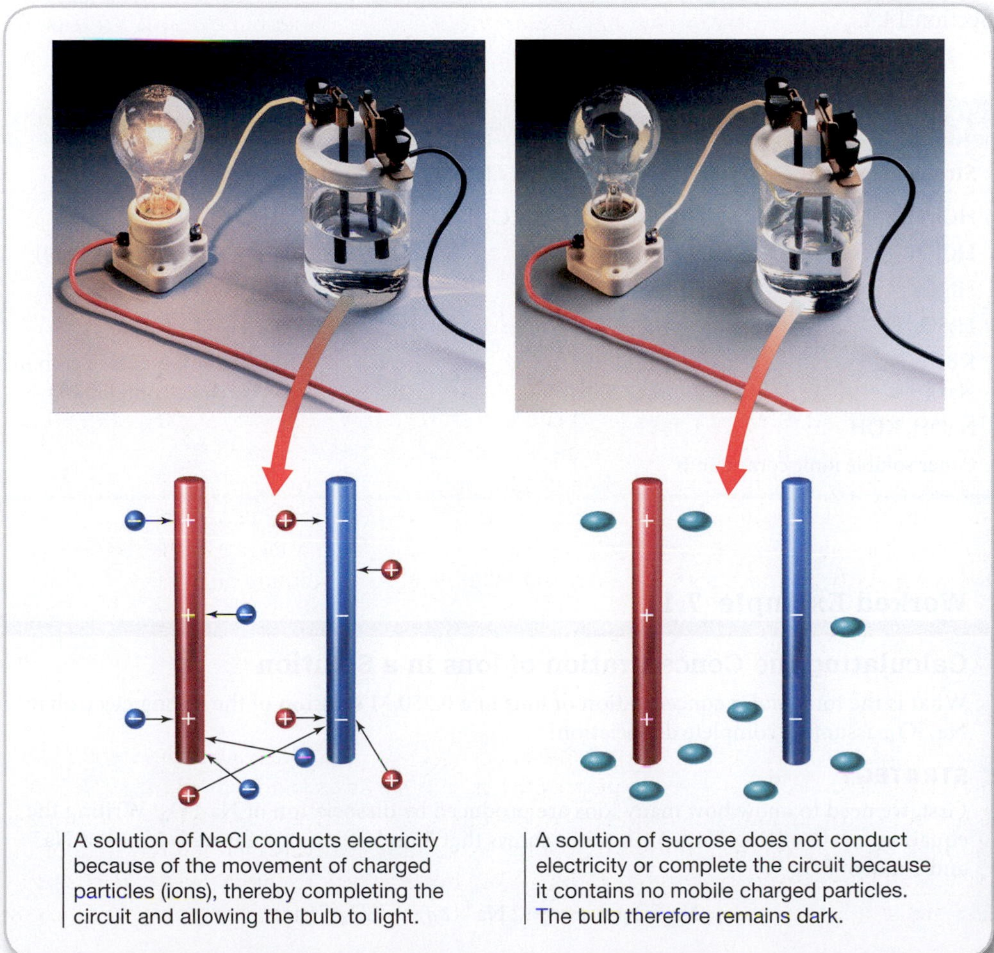

FIGURE 7.1

Testing the conductivity of aqueous solutions.

but some are molecular. Hydrogen chloride, for instance, is a gaseous molecular compound when pure but **dissociates**, or splits apart, to give H^+ and Cl^- ions when it dissolves in water.

$$HCl(g) \xrightarrow{H_2O} H^+(aq) + Cl^-(aq)$$

Compounds that dissociate to a large extent (70–100%) into ions when dissolved in water are said to be **strong electrolytes**, while compounds that dissociate to only a small extent are **weak electrolytes**. Potassium chloride and most other ionic compounds, for instance, are largely dissociated in dilute solution and are thus strong electrolytes. Acetic acid (CH_3CO_2H), by contrast, dissociates only to the extent of about 1.3% in a 0.10 M solution and is a weak electrolyte. As a result, a 0.10 M solution of acetic acid is only weakly conducting.

For 0.10 M solutions:

$$\underset{(2\%)}{KCl(aq)} \rightleftharpoons \underset{(98\%)}{K^+(aq) + Cl^-(aq)} \quad \text{Strong electrolyte}$$

$$\underset{(99\%)}{CH_3CO_2H(aq)} \rightleftharpoons \underset{(1\%)}{H^+(aq) + CH_3CO_2^-(aq)} \quad \text{Weak electrolyte}$$

Note that a forward-and-backward double arrow ($\rightleftharpoons$) is used in the dissociation equation to indicate that the reaction takes place simultaneously in both directions. That is, dissociation is a dynamic process in which an *equilibrium* is established between the forward and reverse reactions. Dissociation of acetic acid takes place in the forward direction, while recombination of H^+ and $CH_3CO_2^-$ ions takes place in the reverse direction. Ultimately, the concentrations of the reactants and products reach constant values and no longer change with time. We'll learn much more about **chemical equilibria** in Chapters 13 and 14.

LOOKING AHEAD ...

A **chemical equilibrium**, as we'll see in Chapters 13 and 14, is the state in which a reaction takes place in both forward and backward directions so that the concentrations of products and reactants remain constant over time.

A brief list of some common substances classified according to their electrolyte strength is given in **Table 7.1**. Note that pure water is a nonelectrolyte because it does not dissociate appreciably into H^+ and OH^- ions. We'll explore the dissociation of water in more detail in Section 14.4.

TABLE 7.1 Electrolyte Classification of Some Common Substances

Strong Electrolytes	Weak Electrolytes	Nonelectrolytes
HCl, HBr, HI	CH_3CO_2H	H_2O
$HClO_4$	HF	CH_3OH (methyl alcohol)
HNO_3	HCN	C_2H_5OH (ethyl alcohol)
H_2SO_4		$C_{12}H_{22}O_{11}$ (sucrose)
KBr		Most compounds of carbon (organic compounds)
NaCl		
NaOH, KOH		
Other soluble ionic compounds		

Worked Example 7.1

Calculating the Concentration of Ions in a Solution

What is the total molar concentration of ions in a 0.350 M solution of the strong electrolyte Na_2SO_4, assuming complete dissociation?

STRATEGY

First, we need to know how many ions are produced by dissociation of Na_2SO_4. Writing the equation for dissolving Na_2SO_4 in water shows that 3 mol of ions are formed: 2 mol of Na^+ and 1 mol of SO_4^{2-}.

$$Na_2SO_4(s) \xrightarrow{H_2O} 2\,Na^+(aq) + SO_4^{2-}(aq)$$

SOLUTION

Assuming complete dissociation, the total molar concentration of ions is three times the molarity of Na_2SO_4, or 1.05 M:

$$\frac{0.350 \text{ mol } Na_2SO_4}{1 \text{ L}} \times \frac{3 \text{ mol ions}}{1 \text{ mol } Na_2SO_4} = 1.05 \text{ M}$$

PROBLEM 7.1 What is the molar concentration of Br^- ions in a 0.225 M aqueous solution of $FeBr_3$, assuming complete dissociation?

CONCEPTUAL PROBLEM 7.2 Three different substances, A_2X, A_2Y, and A_2Z, are dissolved in water, with the following results. (Water molecules are omitted for clarity.) Which of the substances is the strongest electrolyte, and which is the weakest? Explain.

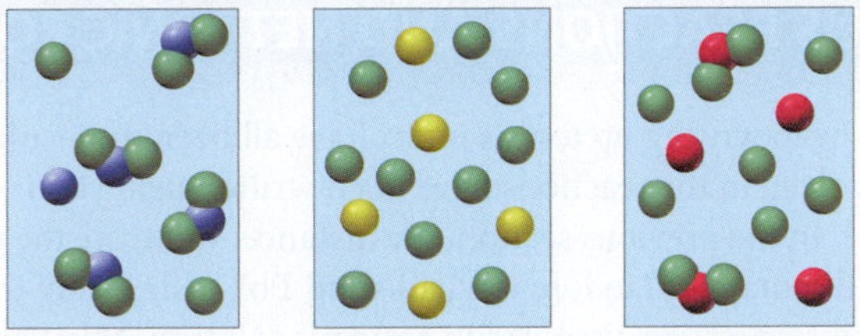

7.2 SOME WAYS THAT CHEMICAL REACTIONS OCCUR

Many common chemical reactions that take place in aqueous solution fall into one of three general categories: *precipitation reactions*, *acid–base neutralization reactions*, and *oxidation–reduction reactions*. Let's look briefly at an example of each before looking at them in more detail in subsequent sections.

▲ Reaction of aqueous $Pb(NO_3)_2$ with aqueous KI gives a yellow precipitate of PbI_2.

- **Precipitation reactions** are processes in which soluble ionic reactants (strong electrolytes) yield an insoluble solid product called a *precipitate*, which falls out of the solution thereby removing some of the dissolved ions. Most precipitations take place when the anions and cations of two ionic compounds change partners. For example, an aqueous solution of lead(II) nitrate reacts with an aqueous solution of potassium iodide to yield an aqueous solution of potassium nitrate plus an insoluble yellow precipitate of lead(II) iodide:

$$Pb(NO_3)_2(aq) + 2\,KI(aq) \longrightarrow 2\,KNO_3(aq) + PbI_2(s)$$

- **Acid–base neutralization reactions** are processes in which an acid reacts with a base to yield water plus an ionic compound called a *salt*. Acids, as we saw briefly in Section 6.11, are compounds that produce H^+ ions when dissolved in water, and bases are compounds that produce OH^- ions when dissolved in water. Thus, a neutralization reaction removes H^+ and OH^- ions from solution, just as a precipitation reaction removes metal and nonmetal ions. The reaction between hydrochloric acid and aqueous sodium hydroxide to yield water plus aqueous sodium chloride is a typical example:

$$HCl(aq) + NaOH(aq) \longrightarrow H_2O(l) + NaCl(aq)$$

- **Oxidation–reduction reactions**, or **redox reactions**, are processes in which one or more electrons are transferred between reaction partners (atoms, molecules, or ions). As a result of this electron transfer, the charges on atoms in the various reactants change. When metallic magnesium reacts with aqueous hydrochloric acid, for instance, a magnesium atom gives an electron to each of two H^+ ions, forming an Mg^{2+} ion and an H_2

molecule. The charge on the magnesium changes from 0 to +2, and the charge on each hydrogen changes from +1 to 0:

$$Mg(s) + 2\,HCl(aq) \longrightarrow MgCl_2(aq) + H_2(g)$$

PROBLEM 7.3 Classify each of the following processes as a precipitation, acid–base neutralization, or redox reaction:

(a) $AgNO_3(aq) + KCl(aq) \longrightarrow AgCl(s) + KNO_3(aq)$

(b) $Cl_2(g) + 2\,NaBr(aq) \longrightarrow Br_2(aq) + 2\,NaCl(aq)$

(c) $Ca(OH)_2(aq) + 2\,HNO_3(aq) \longrightarrow 2\,H_2O(l) + Ca(NO_3)_2(aq)$

7.3 AQUEOUS REACTIONS AND NET IONIC EQUATIONS

The equations we've been writing up to this point have all been **molecular equations**. That is, all the substances involved in the reactions have been written using their complete formulas as if they were *molecules*. In the previous section, for instance, we wrote the precipitation reaction of lead(II) nitrate with potassium iodide to yield solid PbI_2 using only the **parenthetical (*aq*)** to indicate that the substances are dissolved in aqueous solution. Nowhere in the equation was it indicated that ions are involved.

REMEMBER...
The physical state of a substance in a chemical reaction is often indicated with a **parenthetical (*s*)** for solid, **(*l*)** for liquid, **(*g*)** for gas, and **(*aq*)** for aqueous solution. (Section 3.9)

A MOLECULAR EQUATION

$$Pb(NO_3)_2(aq) + 2\,KI(aq) \longrightarrow 2\,KNO_3(aq) + PbI_2(s)$$

In fact, lead nitrate, potassium iodide, and potassium nitrate are strong electrolytes that dissolve in water to yield solutions of ions. Thus, it's more accurate to write the precipitation reaction as an **ionic equation**, in which all the ions are explicitly shown.

AN IONIC EQUATION

$$Pb^{2+}(aq) + 2\,NO_3^-(aq) + 2\,K^+(aq) + 2\,I^-(aq) \longrightarrow 2\,K^+(aq) + 2\,NO_3^-(aq) + PbI_2(s)$$

A look at this ionic equation shows that the NO_3^- and K^+ ions undergo no change during the reaction. Instead, they appear on both sides of the reaction arrow and act merely as **spectator ions**, whose only role is to balance the charge. The actual reaction, when stripped to its essentials, can be described more simply by writing a **net ionic equation**, in which the spectator ions are cancelled from both sides and only the ions undergoing change are shown—the Pb^{2+} and I^- ions in this instance.

AN IONIC EQUATION

$$Pb^{2+}(aq) + \cancel{2\,NO_3^-}(aq) + \cancel{2\,K^+}(aq) + 2\,I^-(aq) \longrightarrow \cancel{2\,K^+}(aq) + \cancel{2\,NO_3^-}(aq) + PbI_2(s)$$

A NET IONIC EQUATION

$$Pb^{2+}(aq) + 2\,I^-(aq) \longrightarrow PbI_2(s)$$

Leaving the spectator ions out of a net ionic equation doesn't mean that their presence is irrelevant. If a reaction occurs by mixing a solution of Pb^{2+} ions with a solution of I^- ions, then those solutions must also contain additional ions to balance the charge in each. That is, the Pb^{2+} solution must also contain an anion, and the I^- solution must also contain a cation. Leaving these other ions out of the net ionic equation only implies that their specific identity is not important. Any nonreactive spectator ions could fill the same role.

Worked Example 7.2

Writing a Net Ionic Equation

Aqueous hydrochloric acid reacts with zinc metal to yield hydrogen gas and aqueous zinc chloride. Write a net ionic equation for the process, and identify the spectator ions.

$$2\,HCl(aq) + Zn(s) \longrightarrow H_2(g) + ZnCl_2(aq)$$

STRATEGY

First, write the ionic equation, listing all the species present in solution. Both HCl (a molecular compound; Table 7.1) and $ZnCl_2$ (an ionic compound) are strong electrolytes that exist as ions in solution. Then find the ions that are present on both sides of the reaction arrow—the spectator ions—and cancel them to leave the net ionic equation.

SOLUTION

IONIC EQUATION

$$2\,H^+(aq) + \cancel{2\,Cl^-}(aq) + Zn(s) \longrightarrow H_2(g) + Zn^{2+}(aq) + \cancel{2\,Cl^-}(aq)$$

NET IONIC EQUATION

$$2\,H^+(aq) + Zn(s) \longrightarrow H_2(g) + Zn^{2+}(aq)$$

Chloride ion is the only spectator ion in this reaction.

▲ Zinc metal reacts with aqueous hydrochloric acid to give hydrogen gas and aqueous Zn^{2+} ions.

PROBLEM 7.4 Write net ionic equations for the following reactions:

(a) $2\,AgNO_3(aq) + Na_2CrO_4(aq) \longrightarrow Ag_2CrO_4(s) + 2\,NaNO_3(aq)$

(b) $H_2SO_4(aq) + MgCO_3(s) \longrightarrow H_2O(l) + CO_2(g) + MgSO_4(aq)$

(c) $Hg(NO_3)_2(aq) + 2\,NH_4I(aq) \longrightarrow HgI_2(s) + 2\,NH_4NO_3(aq)$

7.4 PRECIPITATION REACTIONS AND SOLUBILITY GUIDELINES

How can you predict whether a precipitation reaction will occur on mixing aqueous solutions of two substances? To do that, you must know the **solubility** of each potential product—how much of each compound will dissolve in a given amount of solvent at a given temperature. If a substance has a high solubility in water, no precipitate will form. If a substance has a low solubility in water, it's likely to precipitate from an aqueous solution.

Solubility is a complex matter, and it's not always possible to make correct predictions. Furthermore, solubilities depend on the concentrations of the reactant ions, and the very words *soluble* and *insoluble* are imprecise. Using a concentration of 0.01 M as the lower limit a substance can have to be considered soluble, a compound is soluble if it meets either (or both) of the following criteria:

1. **A compound is soluble if it contains one of the following *cations*:**
   - Li^+, Na^+, K^+, Rb^+, Cs^+ (group 1A cations)
   - NH_4^+ (ammonium ion)

   That is, essentially all ionic compounds containing an alkali metal cation or ammonium cation are soluble in water and will not precipitate, regardless of the anions present.
2. **A compound is soluble if it contains one of the following *anions*:**
   - Cl^-, Br^-, I^- (halide)

     except: Ag^+, Hg_2^{2+}, and Pb^{2+} halides
   - NO_3^- (nitrate), ClO_4^- (perchlorate), $CH_3CO_2^-$ (acetate), and SO_4^{2-} (sulfate)

     except: Sr^{2+}, Ba^{2+}, Hg_2^{2+}, and Pb^{2+} sulfates

   That is, most ionic compounds containing a halide, nitrate, perchlorate, acetate, or sulfate anion are soluble in water and will not precipitate regardless of the cations present. The exceptions that *will* precipitate are silver(I), mercury(I) and lead(II) halides, and strontium, barium, mercury(I), and lead(II) sulfates.

Looked at from the opposite perspective, a compound that does *not* contain one of the cations or anions listed above is *not* soluble. Thus, carbonates (CO_3^{2-}), sulfides (S^{2-}), phosphates (PO_4^{3-}), and hydroxides (OH^-) are generally not soluble unless they contain an alkali metal or ammonium cation. The main exceptions are the sulfides and hydroxides of Ca^{2+}, Sr^{2+}, Ba^{2+}. Solubility guidelines are summarized in **Table 7.2**.

TABLE 7.2 Solubility Guidelines for Ionic Compounds in Water

Soluble Compounds	Common Exceptions
Li^+, Na^+, K^+, Rb^+, Cs^+ (group 1A cations)	None
NH_4^+ (ammonium ion)	None
Cl^-, Br^-, I^- (halide)	Halides of Ag^+, Hg_2^{2+}, Pb^{2+}
NO_3^- (nitrate)	None
ClO_4^- (perchlorate)	None
$CH_3CO_2^-$ (acetate)	None
SO_4^{2-} (sulfate)	Sulfates of Sr^{2+}, Ba^{2+}, Hg_2^{2+}, Pb^{2+}

Insoluble Compounds	Common Exceptions
CO_3^{2-} (carbonate)	Carbonates of group 1A cations, NH_4^+
S^{2-} (sulfide)	Sulfides of group 1A cations, NH_4^+, Ca^{2+}, Sr^{2+}, and Ba^{2+}
PO_4^{3-} (phosphate)	Phosphates of group 1A cations, NH_4^+
OH^- (hydroxide)	Hydroxides of group 1A cations, NH_4^+, Ca^{2+}, Sr^{2+}, and Ba^{2+}

You might notice that most of the ions that impart solubility to compounds are singly charged—either singly positive (Li^+, Na^+, K^+, Rb^+, Cs^+, NH_4^+) or singly negative (Cl^-, Br^-, I^-, NO_3^-, ClO_4^-, $CH_3CO_2^-$). Very few doubly charged ions or triply charged ions form soluble compounds. This solubility behavior arises because of the relatively high **lattice energies** in compounds containing ions with multiple charges. The higher the lattice energy holding the ions together in a crystal, the more difficult it is to pull those ions apart during the solution process.

REMEMBER...

The **lattice energy** of an ionic compound is the amount of energy required to break up an ionic solid into its constituent ions. (Section 3.10)

Using the solubility guidelines makes it possible not only to predict whether a precipitate will form when solutions of two ionic compounds are mixed but also to prepare a specific compound by purposefully carrying out a precipitation. If, for example, you wanted to prepare a sample of solid silver carbonate, Ag_2CO_3, you could mix a solution of $AgNO_3$ with a solution of Na_2CO_3. Both starting compounds are soluble in water, as is $NaNO_3$. Silver carbonate is the only insoluble combination of ions and will therefore precipitate from solution.

$$2\,AgNO_3(aq) + Na_2CO_3(aq) \longrightarrow Ag_2CO_3(s) + 2\,NaNO_3(aq)$$

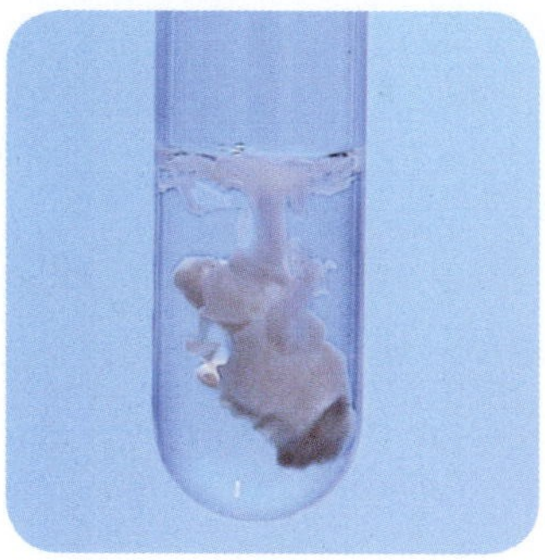

▲ Reaction of aqueous $AgNO_3$ with aqueous Na_2CO_3 gives a white precipitate of Ag_2CO_3.

Worked Example 7.3

Predicting the Product of a Precipitation Reaction

Will a precipitation reaction occur when aqueous solutions of $CdCl_2$ and $(NH_4)_2S$ are mixed? If so, write the net ionic equation.

STRATEGY

Write the possible reaction, identify the two potential products, and predict the solubility of each. In the present instance, $CdCl_2$ and $(NH_4)_2S$ might give CdS and NH_4Cl:

$$?? \; CdCl_2(aq) + (NH_4)_2S(aq) \longrightarrow CdS + 2\,NH_4Cl \; ??$$

SOLUTION

Of the two possible products, the solubility guidelines predict that CdS, a sulfide, is insoluble and that NH_4Cl, an ammonium compound and a halide, is soluble. Thus, a precipitation reaction will likely occur:

$$Cd^{2+}(aq) + S^{2-}(aq) \longrightarrow CdS(s)$$

Worked Example 7.4

Using a Precipitation Reaction to Prepare a Substance

How might you use a precipitation reaction to prepare a sample of $CuCO_3$? Write the net ionic equation.

STRATEGY

To prepare a precipitate of $CuCO_3$, a soluble Cu^{2+} compound must react with a soluble CO_3^{2-} compound.

SOLUTION

A look at the solubility guidelines suggests that a soluble copper compound, such as $Cu(NO_3)_2$, and a soluble carbonate, such as Na_2CO_3, might work. (There are many other possibilities with different spectator ions.)

$$Cu(NO_3)_2(aq) + Na_2CO_3(aq) \longrightarrow 2\,NaNO_3(aq) + CuCO_3(s)$$
$$Cu^{2+}(aq) + CO_3^{2-}(aq) \longrightarrow CuCO_3(s)$$

WORKED CONCEPTUAL EXAMPLE 7.5

Identifying Precipitation Reactions

When aqueous solutions of two ionic compounds are mixed, the following results are obtained. (Only the anion of the first compound, represented by blue spheres, and the cation of the second compound, represented by red spheres, are shown.) Which cations and anions, chosen from the following lists, are compatible with the observed results?

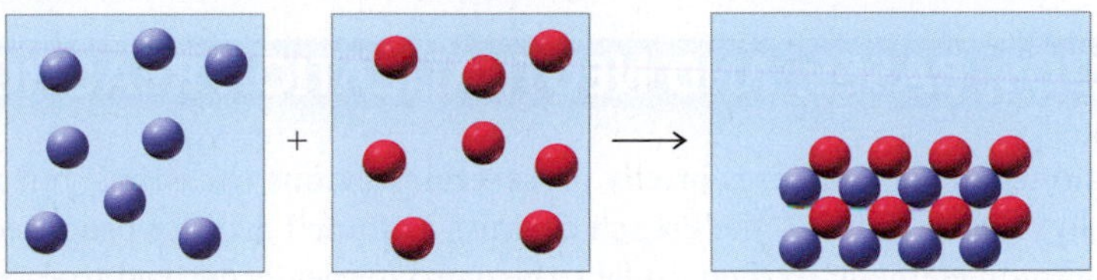

Anion	Cation
NO_3^-, Cl^-, CO_3^{2-}, PO_4^{3-}	Ca^{2+}, Ag^+, K^+, Cd^{2+}

STRATEGY

The process represented in the drawing is a precipitation reaction because ions in solution fall to the bottom of the container in an ordered arrangement. Counting the spheres shows that the cation and anion react in equal numbers (8 of each), so they must have the same number of charges—either both singly charged or both doubly charged. (There is no triply charged cation in the list.) Look at all the possible combinations, and decide which would precipitate.

SOLUTION

Possible combinations of singly charged ions: $AgNO_3$, KNO_3, AgCl, KCl
Possible combinations of doubly charged ions: $CaCO_3$, $CdCO_3$

Of the possible combinations, AgCl, $CaCO_3$, and $CdCO_3$ are insoluble, so the anion might be Cl^- or CO_3^{2-} and the cation might be Ag^+, Ca^{2+}, or Cd^{2+}.

PROBLEM 7.5 Predict whether each of the following compounds is likely to be soluble in water:

(a) $CdCO_3$ (b) MgO (c) Na_2S
(d) $PbSO_4$ (e) $(NH_4)_3PO_4$ (f) $HgCl_2$

continued on next page

PROBLEM 7.6 Predict whether a precipitation reaction will occur in each of the following situations. Write a net ionic equation for each reaction that occurs.

(a) $NiCl_2(aq) + (NH_4)_2S(aq) \longrightarrow ?$
(b) $Na_2CrO_4(aq) + Pb(NO_3)_2(aq) \longrightarrow ?$
(c) $AgClO_4(aq) + CaBr_2(aq) \longrightarrow ?$
(d) $ZnCl_2(aq) + K_2CO_3(aq) \longrightarrow ?$

PROBLEM 7.7 How might you use a precipitation reaction to prepare a sample of $Ca_3(PO_4)_2$? Write the net ionic equation.

CONCEPTUAL PROBLEM 7.8 An aqueous solution containing an anion, represented by blue spheres, is added to another solution containing a cation, represented by red spheres, and the following result is obtained. Which cations and anions, chosen from the following lists, are compatible with the observed results?

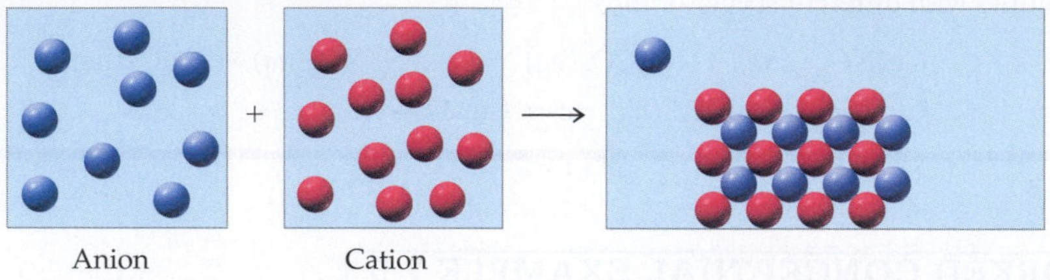

Anions: S^{2-}, PO_4^{3-}, SO_4^{2-}, ClO_4^-

Cations: Mg^{2+}, Fe^{3+}, NH_4^+, Zn^{2+}

7.5 ACIDS, BASES, AND NEUTRALIZATION REACTIONS

We've mentioned acids and bases briefly on several previous occasions, but now let's look more carefully at both. In 1777, the French chemist Antoine Lavoisier proposed that all acids contain a common element: oxygen. In fact, the word *oxygen* is derived from a Greek phrase meaning "acid former." Lavoisier's idea had to be modified, however, when the English chemist Sir Humphrey Davy (1778–1829) showed in 1810 that muriatic acid (now called hydrochloric acid) contains only hydrogen and chlorine but no oxygen. Davy's studies thus suggested that the common element in acids is *hydrogen*, not oxygen.

The relationship between acidic behavior and the presence of hydrogen in a compound was clarified in 1887 by the Swedish chemist Svante Arrhenius (1859–1927). Arrhenius proposed that an **acid** is a substance that dissociates in water to give hydrogen ions (H^+) and a **base** is a substance that dissociates in water to give hydroxide ions (OH^-):

$$\text{An acid} \quad HA(aq) \longrightarrow H^+(aq) + A^-(aq)$$
$$\text{A base} \quad MOH(aq) \longrightarrow M^+(aq) + OH^-(aq)$$

In these equations, HA is a general formula for an acid—for example, HCl or HNO_3—and MOH is a general formula for a metal hydroxide—for example, NaOH or KOH.

Although convenient to use in equations, the symbol $H^+(aq)$ does not really represent the structure of the ion present in aqueous solution. As a bare hydrogen nucleus—a proton—with no electron nearby, H^+ is much too reactive to exist by itself. Rather, the H^+ bonds to the oxygen atom of a water molecule and forms a **coordinate covalent bond**, to produce the stable **hydronium ion, H_3O^+**. We'll sometimes write $H^+(aq)$ for convenience, particularly when balancing equations, but will more often write $H_3O^+(aq)$ to represent an aqueous acid solution. Hydrogen chloride, for instance, gives $Cl^-(aq)$ and $H_3O^+(aq)$ when it dissolves in water.

REMEMBER...

A **coordinate covalent bond** is a covalent bond formed when one reactant donates both electrons (a lone pair) to another atom that has a vacant valence orbital. (Section 4.6)

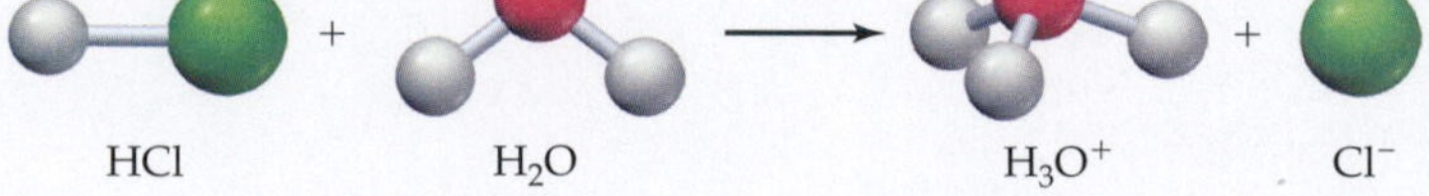

Different acids dissociate to different extents in aqueous solution. Acids that dissociate to a large extent are strong electrolytes and **strong acids,** whereas acids that dissociate to only a small extent are weak electrolytes and **weak acids**. We've already seen in Table 7.1, for instance, that HCl, $HClO_4$, HNO_3, and H_2SO_4 are strong electrolytes and therefore strong acids, while CH_3CO_2H and HF are weak electrolytes and therefore weak acids. You might note that acetic acid actually contains four hydrogens, but only the one bonded to the electronegative oxygen atom dissociates.

Acetic acid

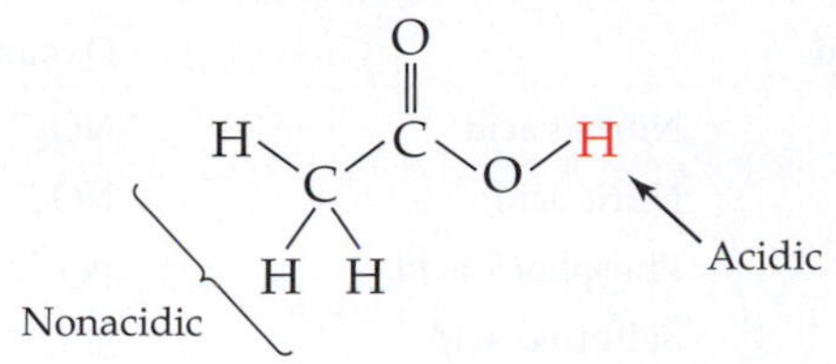

Different acids can have different numbers of acidic hydrogens and yield different numbers of H_3O^+ ions in solution. Hydrochloric acid (HCl) is said to be a **monoprotic acid** because it provides only one H^+ ion, but sulfuric acid (H_2SO_4) is a **diprotic acid** because it can provide two H^+ ions. Phosphoric acid (H_3PO_4) is a **triprotic acid** and can provide three H^+ ions. With sulfuric acid, the first dissociation of an H^+ is complete—all H_2SO_4 molecules lose one H^+—but the second dissociation is incomplete, as indicated by the double arrow in the following equation:

Sulfuric acid:
$$H_2SO_4(aq) + H_2O(l) \longrightarrow HSO_4^-(aq) + H_3O^+(aq)$$
$$HSO_4^-(aq) + H_2O(l) \rightleftharpoons SO_4^{2-}(aq) + H_3O^+(aq)$$

With phosphoric acid, none of the three dissociations is complete:

Phosphoric acid:
$$H_3PO_4(aq) + H_2O(l) \rightleftharpoons H_2PO_4^-(aq) + H_3O^+(aq)$$
$$H_2PO_4^-(aq) + H_2O(l) \rightleftharpoons HPO_4^{2-}(aq) + H_3O^+(aq)$$
$$HPO_4^{2-}(aq) + H_2O(l) \rightleftharpoons PO_4^{3-}(aq) + H_3O^+(aq)$$

Bases, like acids, can also be either strong or weak, depending on the extent to which they produce OH^- ions in aqueous solution. Most metal hydroxides, such as NaOH and $Ba(OH)_2$, are strong electrolytes and **strong bases**, but ammonia (NH_3) is a weak electrolyte and a **weak base**. Ammonia is a weak base, not because it contains OH^- ions in its formula, but because it reacts to a small extent with water to yield NH_4^+ and OH^- ions. In fact, aqueous solutions of ammonia are often called *ammonium hydroxide*, although this is really a misnomer because the concentrations of NH_4^+ and OH^- ions are low.

$$NH_3(g) + H_2O(l) \rightleftharpoons NH_4^+(aq) + OH^-(aq)$$

As with the dissociation of acetic acid, discussed in Section 7.1, the reaction of ammonia with water takes place only to a small extent (about 1%). Most of the ammonia remains unreacted, and we therefore write the reaction with a double arrow to show that a dynamic equilibrium exists between the forward and reverse reactions.

Table 7.3 summarizes the names, formulas, and classification of some common acids and bases.

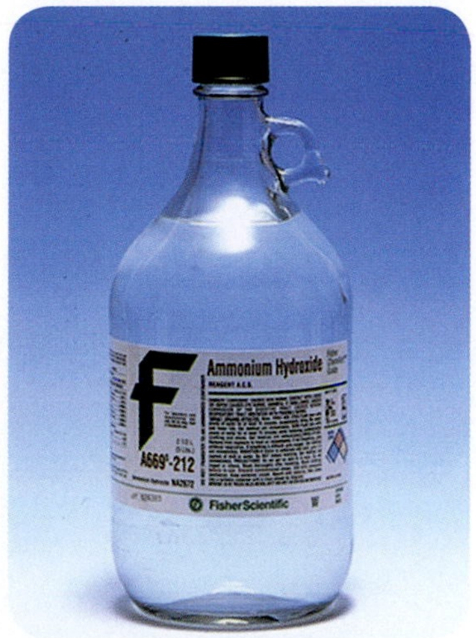

▲ Shouldn't this bottle be labeled "Aqueous Ammonia" rather than "Ammonium Hydroxide"?

TABLE 7.3 Some Common Acids and Bases

Strong acid	$HClO_4$	Perchloric acid	KOH	Potassium hydroxide	Strong base
	H_2SO_4	Sulfuric acid	NaOH	Sodium hydroxide	
	HBr	Hydrobromic acid	$Ba(OH)_2$	Barium hydroxide	
	HCl	Hydrochloric acid	$Ca(OH)_2$	Calcium hydroxide	
	HNO_3	Nitric acid			
	H_3PO_4	Phosphoric acid	NH_3	Ammonia	
	HF	Hydrofluoric acid			
	HNO_2	Nitrous acid			
Weak acid	CH_3CO_2H	Acetic acid			Weak base

Naming Acids

REMEMBER...

Oxoanions are binary polyatomic anions in which an atom of a given element is combined with different numbers of oxygen atoms. (Section 3.2)

Most acids are **oxoacids**; that is, they contain oxygen in addition to hydrogen and other elements. When dissolved in water, an oxoacid yields one or more H^+ ions and an **oxoanion** like one of those listed in **Table 7.4** and discussed previously in Section 3.2.

TABLE 7.4 Common Oxoacids and Their Anions

Oxoacid		Oxoanion	
HNO_2	Nitr*ous* acid	NO_2^-	Nitr*ite* ion
HNO_3	Nitr*ic* acid	NO_3^-	Nitr*ate* ion
H_3PO_4	Phosphor*ic* acid	PO_4^{3-}	Phosph*ate* ion
H_2SO_3	Sulfur*ous* acid	SO_3^{2-}	Sulf*ite* ion
H_2SO_4	Sulfur*ic* acid	SO_4^{2-}	Sulf*ate* ion
$HClO$	*Hypo*chlor*ous* acid	ClO^-	*Hypo*chlor*ite* ion
$HClO_2$	Chlor*ous* acid	ClO_2^-	Chlor*ite* ion
$HClO_3$	Chlor*ic* acid	ClO_3^-	Chlor*ate* ion
$HClO_4$	*Per*chlor*ic* acid	ClO_4^-	*Per*chlor*ate* ion

The names of oxoacids are related to the names of the corresponding oxoanions, with the *-ite* or *-ate* ending of the anion name replaced by *-ous acid* or *-ic acid*, respectively. In other words, the acid with fewer oxygens has an *-ous* ending, and the acid with more oxygens has an *-ic* ending. The compound HNO_2, for example, is called *nitrous acid* because it has fewer oxygens and yields the nitr*ite* ion (NO_2^-) when dissolved in water, while HNO_3 is called *nitric acid* because it has more oxygens and yields the nitr*ate* ion (NO_3^-) when dissolved in water.

NITR*OUS ACID* GIVES NITR*ITE* ION

$$HNO_2(aq) + H_2O(l) \rightleftharpoons H_3O^+(aq) + NO_2^-(aq)$$

NITR*IC ACID* GIVES NITR*ATE* ION

$$HNO_3(aq) + H_2O(l) \longrightarrow H_3O^+(aq) + NO_3^-(aq)$$

In a similar way, hypochlor*ous* acid yields the hypochlor*ite* ion, chlor*ous* acid yields the chlor*ite* ion, chlor*ic* acid yields the chlor*ate* ion, and perchlor*ic* acid yields the perchlor*ate* ion (Table 7.4).

In addition to the oxoacids, there are a small number of other common acids, such as HCl, that do not contain oxygen. For such compounds, the prefix *hydro-* and the suffix *-ic acid* are used for the aqueous solution.

HYDROGEN CHLORIDE GIVES *HYDROCHLORIC ACID*

$$HCl(g) + H_2O(l) \longrightarrow H_3O^+(aq) + Cl^-(aq)$$

HYDROGEN CYANIDE GIVES *HYDROCYANIC ACID*

$$HCN(g) + H_2O(l) \rightleftharpoons H_3O^+(aq) + CN^-(aq)$$

Worked Example 7.6

Naming Acids

Name the following acids:
(a) $HBrO(aq)$ **(b)** $H_2S(aq)$

STRATEGY

To name an acid, look at its formula and decide whether the compound is an oxoacid. If so, the name must reflect the number of oxygen atoms, according to Table 7.4. If the compound is not an oxoacid, it is named using the prefix *hydro-* and the suffix *-ic acid.*

SOLUTION

(a) This compound is an oxoacid that yields hypobromite ion (BrO^-) when dissolved in water. Its name is hypobromous acid.

(b) This compound is not an oxoacid but yields sulfide ion when dissolved in water. As a pure gas, H_2S is named hydrogen sulfide. In water solution, it is called hydrosulfuric acid.

PROBLEM 7.9 Name the following acids:
(a) HIO_4 **(b)** $HBrO_2$ **(c)** H_2CrO_4

PROBLEM 7.10 Give likely chemical formulas corresponding to the following names:
(a) Phosphorous acid **(b)** Hydroselenic acid

Neutralization Reactions

When an acid and a base are mixed in the right stoichiometric proportions, both acidic and basic properties disappear because of a neutralization reaction that produces water and an ionic **salt**. The anion of the salt (A^-) comes from the acid, and the cation of the salt (M^+) comes from the base:

A NEUTRALIZATION REACTION

$$\underset{\text{Acid}}{HA(aq)} + \underset{\text{Base}}{MOH(aq)} \longrightarrow \underset{\textbf{Water}}{H_2O(l)} + \underset{\text{A salt}}{MA(aq)}$$

Because salts are generally strong electrolytes in aqueous solution, we can write the neutralization reaction of a strong acid with a strong base as an ionic equation:

$$H^+(aq) + A^-(aq) + M^+(aq) + OH^-(aq) \longrightarrow H_2O(l) + M^+(aq) + A^-(aq)$$

Canceling the ions that appear on both sides of the ionic equation, A^- and M^+, gives the net ionic equation, which describes the reaction of any strong acid with any strong base in water:

NET IONIC EQUATION

$$H^+(aq) + OH^-(aq) \longrightarrow H_2O(l)$$
$$\text{or} \quad H_3O^+(aq) + OH^-(aq) \longrightarrow 2\,H_2O(l)$$

For the reaction of a weak acid with a strong base, a similar neutralization occurs, but we must write the molecular formula of the acid rather than simply $H^+(aq)$, because the dissociation of the acid in water is incomplete. Instead, the acid exists primarily as the neutral molecule. In the reaction of the weak acid HF with the strong base KOH, for example, we write the net ionic equation as

$$HF(aq) + OH^-(aq) \longrightarrow H_2O(l) + F^-(aq)$$

Worked Example 7.7

Writing Ionic and Net Ionic Equations for an Acid–Base Reaction

Write both an ionic equation and a net ionic equation for the neutralization reaction of aqueous HBr and aqueous $Ba(OH)_2$.

STRATEGY

Hydrogen bromide is a strong acid whose aqueous solution contains H^+ ions and Br^- ions. Barium hydroxide is a strong base whose aqueous solution contains Ba^{2+} and OH^- ions. Thus, we have a mixture of four different ions on the reactant side. Write the neutralization reaction as an ionic equation, and then cancel spectator ions to give the net ionic equation.

continued on next page

SOLUTION

IONIC EQUATION

$$2\,H^+(aq) + 2\,Br^-(aq) + Ba^{2+}(aq) + 2\,OH^-(aq) \longrightarrow 2\,H_2O(l) + 2\,Br^-(aq) + Ba^{2+}(aq)$$

NET IONIC EQUATION

$$2\,H^+(aq) + 2\,OH^-(aq) \longrightarrow 2\,H_2O(l)$$

$$\text{or}\quad H^+(aq) + OH^-(aq) \longrightarrow H_2O(l)$$

The reaction of HBr with $Ba(OH)_2$ involves the combination of a proton (H^+) from the acid with OH^- from the base to yield water and an aqueous salt ($BaBr_2$).

PROBLEM 7.11 Write a balanced ionic equation and net ionic equation for each of the following acid–base reactions:

(a) $2\,CsOH(aq) + H_2SO_4(aq) \rightarrow ?$ **(b)** $Ca(OH)_2(aq) + 2\,CH_3CO_2H(aq) \rightarrow ?$

CONCEPTUAL PROBLEM 7.12 The following pictures represent aqueous solutions of three acids HA (A = X, Y, or Z), with water molecules omitted for clarity. Which of the three is the strongest acid, and which is the weakest?

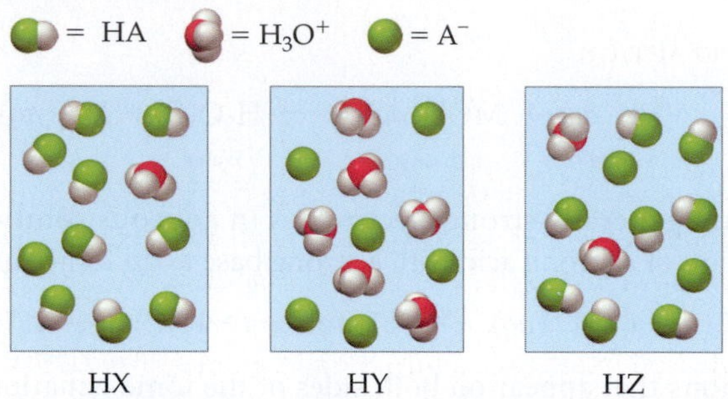

7.6 OXIDATION–REDUCTION (REDOX) REACTIONS

▲ Aqueous potassium permanganate, deep purple in color, is frequently used as an *oxidizing agent*, as described in the text.

Purple aqueous permanganate ion, MnO_4^-, reacts with aqueous Fe^{2+} ion to yield Fe^{3+} and pale pink Mn^{2+}:

$$MnO_4^-(aq) + 5\,Fe^{2+}(aq) + 8\,H^+(aq) \longrightarrow Mn^{2+}(aq) + 5\,Fe^{3+}(aq) + 4\,H_2O(l)$$

Magnesium metal burns in air with an intense white light to form solid magnesium oxide:

$$2\,Mg(s) + O_2(g) \longrightarrow 2\,MgO(s)$$

Red phosphorus reacts with liquid bromine to form liquid phosphorus tribromide:

$$2\,P(s) + 3\,Br_2(l) \longrightarrow 2\,PBr_3(l)$$

Although these and many thousands of other reactions appear unrelated, and many don't even take place in aqueous solution, all are oxidation–reduction (redox) reactions.

Historically, the word *oxidation* referred to the combination of an element with oxygen to yield an oxide, and the word *reduction* referred to the removal of oxygen from an oxide to yield the element. Such oxidation–reduction processes have been crucial to the development of human civilization and still have enormous commercial value. The oxidation (rusting) of iron metal by reaction with moist air has been known for millennia and is still a serious problem that causes enormous structural damage to buildings, boats, and bridges. The reduction of iron ore (Fe_2O_3) with charcoal (C) to make iron metal has been carried out since prehistoric times and is still used today in the initial stages of steelmaking.

$4\,Fe(s) + 3\,O_2(g) \longrightarrow 2\,Fe_2O_3(s)$ Rusting of iron: an oxidation of Fe

$2\,Fe_2O_3(s) + 3\,C(s) \longrightarrow 4\,Fe(s) + 3\,CO_2(g)$ Manufacture of iron: a reduction of Fe_2O_3

Today, the words *oxidation* and *reduction* have taken on a much broader meaning. An **oxidation** is now defined as the loss of one or more electrons by a substance, whether element, compound, or ion, and a **reduction** is the gain of one or more electrons by a substance. Thus, an oxidation–reduction, or redox, reaction is any process in which electrons are transferred from one substance to another.

Oxidation →

$A^{2-} \longrightarrow A^- + \text{electron}$

$A^- \longrightarrow A + \text{electron}$

$A \longrightarrow A^+ + \text{electron}$

$A^+ \longrightarrow A^{2+} + \text{electron}$

← *Reduction*

Reactant A might be:
a neutral atom,
a monatomic ion,
a polyatomic ion,
or a molecule.

How can you tell when a redox reaction takes place? The answer is that you assign to each atom in a compound a value called an **oxidation number** (or *oxidation state*), which indicates whether the atom is neutral, electron-rich, or electron-poor. By comparing the oxidation number of an atom before and after reaction, you can tell whether the atom has gained or lost electrons. Note that oxidation numbers don't necessarily imply ionic charges; they are just a convenient device to help keep track of electrons during redox reactions.

The rules for assigning oxidation numbers are as follows:

1. **An atom in its elemental state has an oxidation number of 0.** For example:

   Na H_2 Br_2 S Ne — Oxidation number 0

2. **An atom in a monatomic ion has an oxidation number identical to its charge.** Review Section 3.2 to see the charges on some common ions. For example:

   Na^+ (+1) Ca^{2+} (+2) Al^{3+} (+3) Cl^- (−1) O^{2-} (−2)

3. **An atom in a polyatomic ion or in a molecular compound usually has the same oxidation number it would have if it were a monatomic ion.** In the hydroxide ion (OH^-), for instance, the hydrogen atom has an oxidation number of +1, as if it were H^+, and the oxygen atom has an oxidation number of −2, as if it were a monatomic O^{2-} ion.

   H—O—H (H +1, O −2, H +1) $[O{-}H]^-$ (O −2, H +1) NH_3: H—N—H with H above N (H +1, N −3, H +1, H +1)

REMEMBER...

Electronegativity is a measure of the ability of an atom in a molecule to attract the shared electrons in a covalent bond. Metallic elements on the left of the periodic table attract electrons weakly, whereas halogens and other reactive nonmetals on the upper right of the table attract electrons strongly. (Section 4.4)

In general, the farther left an element is in the periodic table, the less **electronegative** it is and the more probable that it will be "cationlike." Metals, therefore, usually have positive oxidation numbers. The farther right an element is in the periodic table, the more electronegative it is and the more probable that it will be "anionlike." Nonmetals, such as O, N, and the halogens, usually have negative oxidation numbers.

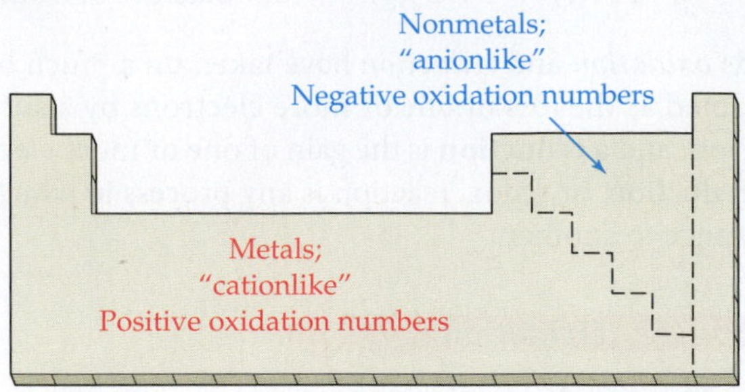

(a) **Hydrogen can be either +1 or −1.** When bonded to a metal, such as Na or Ca, hydrogen has an oxidation number of −1. When bonded to a nonmetal, such as C, N, O, or Cl, hydrogen has an oxidation number of +1.

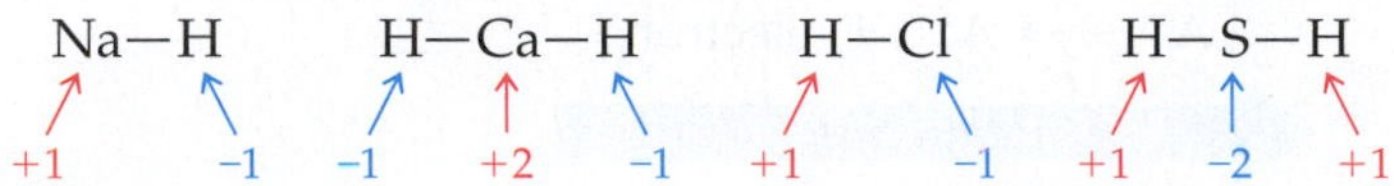

(b) **Oxygen usually has an oxidation number of −2.** The major exception is in compounds called *peroxides*, which contain either the O_2^{2-} ion or an O—O covalent bond in a molecule. Both oxygen atoms in a peroxide have an oxidation number of −1.

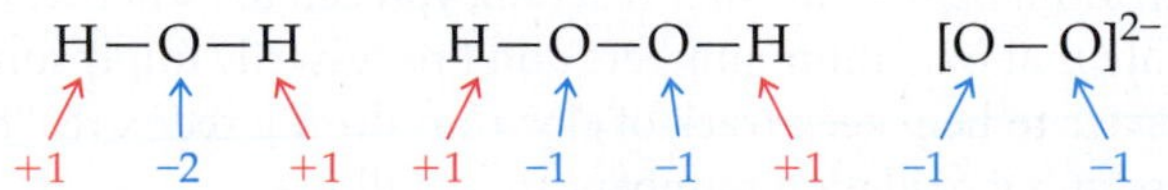

(c) **Halogens usually have an oxidation number of −1.** The major exception is in compounds of chlorine, bromine, or iodine in which the halogen atom is bonded to oxygen. In such cases, the oxygen has an oxidation number of −2, and the halogen has a positive oxidation number. In Cl_2O, for instance, the O atom has an oxidation number of −2 and each Cl atom has an oxidation number of +1.

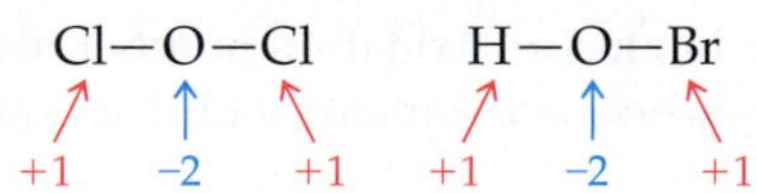

4. **The sum of the oxidation numbers is 0 for a neutral compound and is equal to the net charge for a polyatomic ion.** This rule is particularly useful for finding the oxidation number of an atom in difficult cases. The general idea is to assign oxidation numbers to the "easy" atoms first and then find the oxidation number of the "difficult" atom by subtraction. For example, suppose we want to know the oxidation number of the sulfur atom in sulfuric acid, H_2SO_4. Since each H atom is +1 and each O atom is −2, the S atom must have an oxidation number of +6 for the compound to have no net charge:

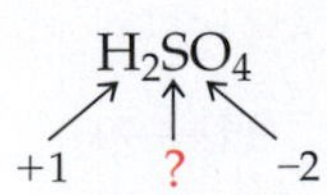

$$2(+1) + (?) + 4(-2) = 0 \text{ net charge}$$

$$? = 0 - 2(+1) - 4(-2) = +6$$

To find the oxidation number of the chlorine atom in the perchlorate anion (ClO_4^-), we know that each oxygen is −2, so the Cl atom must have an oxidation number of +7 for there to be a net charge of −1 on the ion:

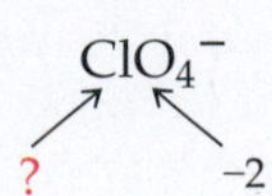

$$? + 4(-2) = -1 \text{ net charge}$$

$$? = -1 - 4(-2) = +7$$

To find the oxidation number of the nitrogen atom in the ammonium cation (NH_4^+), we know that each H atom is +1, so the N atom must have an oxidation number of −3 for the ion to have a net charge of +1:

$$\underset{?\quad +1}{NH_4^+} \qquad ? + 4(+1) = +1 \text{ net charge}$$

$$? = +1 - 4(+1) = -3$$

Worked Example 7.8

Assigning Oxidation Numbers

Assign oxidation numbers to each atom in the following substances:

(a) CdS (b) AlH_3 (c) $S_2O_3^{2-}$ (d) $Na_2Cr_2O_7$

STRATEGY

(a) The sulfur atom in S^{2-} has an oxidation number of −2, so Cd must be +2.
(b) H bonded to a metal has the oxidation number −1, so Al must be +3.
(c) O usually has the oxidation number −2, so S must be +2 for the anion to have a net charge of −2: for $(2\,S^{+2})\,(3\,O^{-2})$, $2(+2) + 3(-2) = -2$ net charge.
(d) Na is always +1, and oxygen is −2, so Cr must be +6 for the compound to be neutral: for $(2\,Na^+)\,(2\,Cr^{+6})\,(7\,O^{-2})$, $2(+1) + 2(+6) + 7(-2) = 0$ net charge.

SOLUTION

(a) CdS: Cd +2, S −2
(b) AlH_3: Al +3, H −1
(c) $S_2O_3^{2-}$: S +2, O −2
(d) $Na_2Cr_2O_7$: Na +1, Cr +6, O −2

PROBLEM 7.13 Assign an oxidation number to each atom in the following compounds:

(a) $SnCl_4$ (b) CrO_3 (c) $VOCl_3$
(d) V_2O_3 (e) HNO_3 (f) $FeSO_4$

7.7 IDENTIFYING REDOX REACTIONS

Once oxidation numbers are assigned, it's clear why all the reactions mentioned in the previous section are redox processes. Take the rusting of iron, for example. Two of the reactants, Fe and O_2, are neutral elements and have oxidation numbers of 0. In the product, however, the oxygen atoms have an oxidation number of −2 and the iron atoms have an oxidation number of +3. Thus, Fe has undergone a change from 0 to +3 (a loss of electrons, or oxidation), and O has undergone a change from 0 to −2 (a gain of electrons, or reduction). Note that the total number of electrons given up by the atoms being oxidized (4 Fe × 3 electrons/Fe = 12 electrons) is the same as the number gained by the atoms being reduced (6 O × 2 electrons/O = 12 electrons).

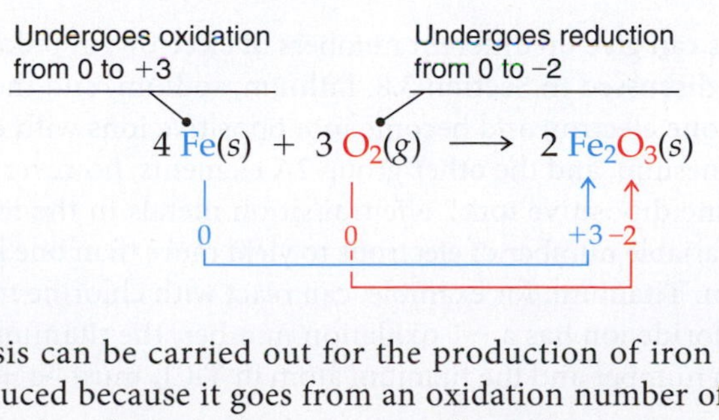

A similar analysis can be carried out for the production of iron metal from its ore. The iron atom is reduced because it goes from an oxidation number of +3 in the reactant (Fe_2O_3) to 0 in the product (Fe). At the same time, the carbon atom is oxidized because it

▲ The iron used in this prehistoric dagger handle was made by the reduction of iron ore with charcoal.

goes from an oxidation number of 0 in the reactant (C) to +4 in the product (CO_2). The oxygen atoms undergo no change because they have an oxidation number of −2 in both reactant and product. The total number of electrons given up by the atoms being oxidized (3 C × 4 electrons/C = 12 electrons) is the same as the number gained by the atoms being reduced (4 Fe × 3 electrons/Fe = 12 electrons).

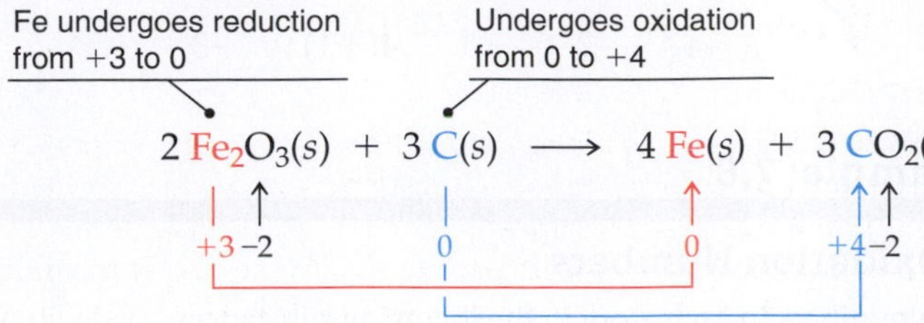

As these examples show, oxidations and reductions always occur together. Whenever one atom loses one or more electrons, another atom must gain those electrons. The substance that *causes* a reduction by giving up electrons—the iron atom in the reaction of Fe with O_2 and the carbon atom in the reaction of C with Fe_2O_3—is called a **reducing agent**. The substance that causes an oxidation by accepting electrons—the oxygen atom in the reaction of Fe with O_2 and the iron atom in the reaction of C with Fe_2O_3—is called an **oxidizing agent**. The reducing agent is itself oxidized when it gives up electrons, and the oxidizing agent is itself reduced when it accepts electrons.

REDUCING AGENT
- Causes reduction
- Loses one or more electrons
- Undergoes oxidation
- Oxidation number of atom increases

OXIDIZING AGENT
- Causes oxidation
- Gains one or more electrons
- Undergoes reduction
- Oxidation number of atom decreases

We'll see in later chapters that redox reactions are common for almost every element in the periodic table except for the noble gas elements of group 8A. In general, metals give up electrons and act as reducing agents, while reactive nonmetals such as O_2 and the halogens accept electrons and act as oxidizing agents.

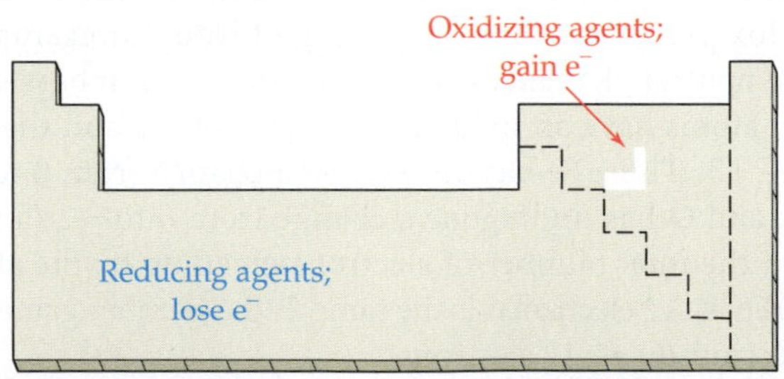

REMEMBER...

According to the **octet rule**, main-group elements tend to undergo reactions that leave them with eight outer-shell electrons; that is, with a noble gas electron configuration having filled *s* and *p* sublevels in their valence electron shell. (Section 3.8)

Different metals can give up different numbers of electrons in redox reactions, in accord with the **octet rule** discussed in Section 3.8. Lithium, sodium, and the other group 1A elements give up only one electron and become monopositive ions with oxidation numbers of +1. Beryllium, magnesium, and the other group 2A elements, however, typically give up two electrons and become dipositive ions. The transition metals in the middle of the periodic table can give up a variable number of electrons to yield more than one kind of ion depending on the exact reaction. Titanium, for example, can react with chlorine to yield either $TiCl_3$ or $TiCl_4$. Because a chloride ion has a −1 oxidation number, the titanium atom in $TiCl_3$ must have a +3 oxidation number and the titanium atom in $TiCl_4$ must be +4.

Worked Example 7.9

Identifying Oxidizing and Reducing Agents

Assign oxidation numbers to all atoms, tell in each case which substance is undergoing oxidation and which reduction, and identify the oxidizing and reducing agents.

(a) $Ca(s) + 2\,H^{+}(aq) \longrightarrow Ca^{2+}(aq) + H_2(g)$

(b) $2\,Fe^{2+}(aq) + Cl_2(aq) \longrightarrow 2\,Fe^{3+}(aq) + 2\,Cl^{-}(aq)$

STRATEGY AND SOLUTION

(a) The elements Ca and H_2 have oxidation numbers of 0; Ca^{2+} is +2 and H^+ is +1. Ca is oxidized, because its oxidation number increases from 0 to +2, and H^+ is reduced, because its oxidation number decreases from +1 to 0. The reducing agent is the substance that gives away electrons, thereby going to a higher oxidation number, and the oxidizing agent is the substance that accepts electrons, thereby going to a lower oxidation number. In the present case, calcium is the reducing agent and H^+ is the oxidizing agent.

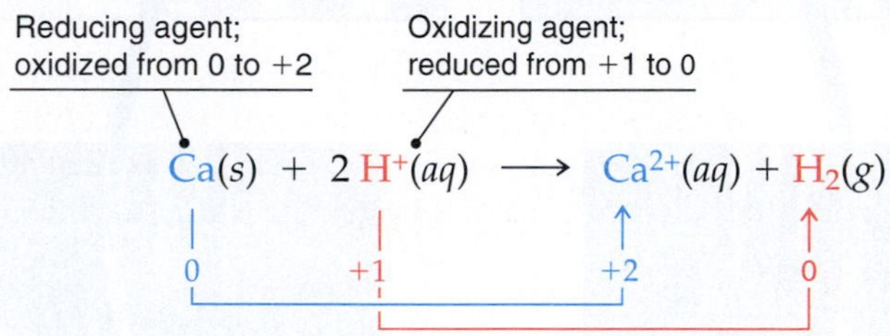

(b) Atoms of the neutral element Cl_2 have an oxidation number of 0; the monatomic ions have oxidation numbers equal to their charge:

$$\underset{+2}{\underset{\uparrow}{2\,Fe^{2+}(aq)}} + \underset{0}{\underset{\uparrow}{Cl_2(aq)}} \longrightarrow \underset{+3}{\underset{\uparrow}{2\,Fe^{3+}(aq)}} + \underset{-1}{\underset{\uparrow}{2\,Cl^{-}(aq)}}$$

Fe^{2+} is oxidized because its oxidation number increases from +2 to +3, and Cl_2 is reduced because its oxidation number decreases from 0 to −1. Fe^{2+} is the reducing agent, and Cl_2 is the oxidizing agent.

PROBLEM 7.14 Aqueous copper(II) ion reacts with aqueous iodide ion to yield solid copper(I) iodide and aqueous iodine. Write the balanced net ionic equation, assign oxidation numbers to all species present, and identify the oxidizing and reducing agents.

PROBLEM 7.15 In each of the following reactions, tell which substance is undergoing an oxidation and which a reduction, and identify the oxidizing and reducing agents.

(a) $SnO_2(s) + 2\,C(s) \longrightarrow Sn(s) + 2\,CO(g)$

(b) $Sn^{2+}(aq) + 2\,Fe^{3+}(aq) \longrightarrow Sn^{4+}(aq) + 2\,Fe^{2+}(aq)$

(c) $4\,NH_3(g) + 5\,O_2(g) \longrightarrow 4\,NO(g) + 6\,H_2O(l)$

7.8 THE ACTIVITY SERIES OF THE ELEMENTS

The reaction of an aqueous cation, usually a metal ion, with a free element to give a different cation and a different element is among the simplest of all redox processes. Aqueous copper(II) ion reacts with iron metal, for example, to give iron(II) ion and copper metal (**Figure 7.2**):

$$Fe(s) + Cu^{2+}(aq) \longrightarrow Fe^{2+}(aq) + Cu(s)$$

Similarly, aqueous acid reacts with magnesium metal to yield magnesium ion and hydrogen gas:

$$Mg(s) + 2\,H^{+}(aq) \longrightarrow Mg^{2+}(aq) + H_2(g)$$

FIGURE 7.2

The redox reaction of iron with aqueous copper(II) ion.

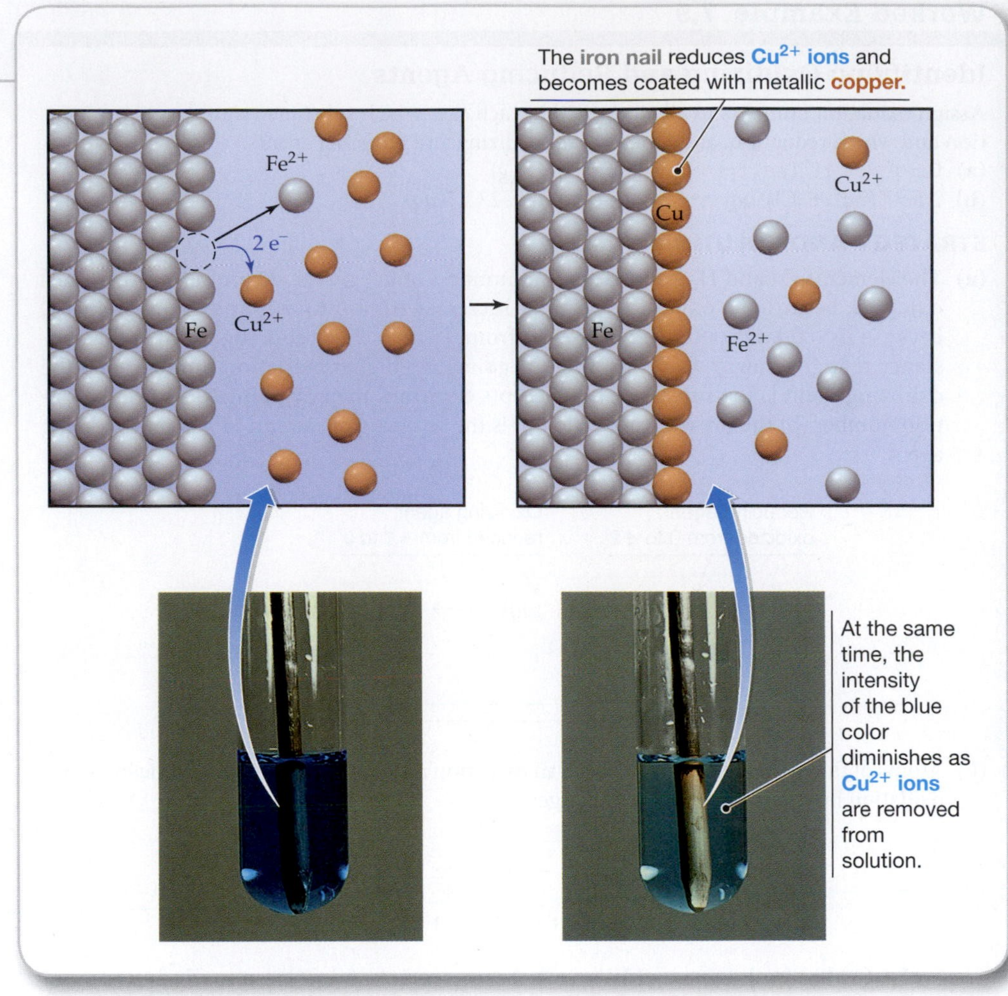

Whether a reaction occurs between a given ion and a given element depends on the relative ease with which the various substances gain or lose electrons—that is, on how easily each substance is reduced or oxidized. By noting the results from a succession of different reactions, it's possible to construct an **activity series**, which ranks the elements in order of their reducing ability in aqueous solution (**Table 7.5**).

Elements at the top of Table 7.5 give up electrons readily and are stronger reducing agents, whereas elements at the bottom give up electrons less readily and are weaker reducing agents. As a result, any element higher in the activity series will reduce the ion of any element lower in the activity series. Because copper is above silver, for example, copper metal gives electrons to Ag^+ ions (**Figure 7.3**).

$$Cu(s) + 2\,Ag^+(aq) \longrightarrow Cu^{2+}(aq) + 2\,Ag(s)$$

Conversely, because gold is below silver in the activity series, gold metal does not give electrons to Ag^+ ions.

$$Au(s) + 3\,Ag^+(aq) \not\longrightarrow Au^{3+}(aq) + 3\,Ag(s) \qquad \textit{Does not occur}$$

The position of hydrogen in the activity series is particularly important because it indicates which metals react with aqueous acid (H^+) to release H_2 gas. The metals at the top of the series—the alkali metals of group 1A and alkaline earth metals of group 2A—are such

TABLE 7.5 A Partial Activity Series of the Elements

		Oxidation Reaction
Strongly reducing	These elements react rapidly with aqueous H^+ ions (acid) or with liquid H_2O to release H_2 gas.	$Li \longrightarrow Li^+ + e^-$
		$K \longrightarrow K^+ + e^-$
		$Ba \longrightarrow Ba^{2+} + 2e^-$
		$Ca \longrightarrow Ca^{2+} + 2e^-$
		$Na \longrightarrow Na^+ + e^-$
	These elements react with aqueous H^+ ions or with steam to release H_2 gas.	$Mg \longrightarrow Mg^{2+} + 2e^-$
		$Al \longrightarrow Al^{3+} + 3e^-$
		$Mn \longrightarrow Mn^{2+} + 2e^-$
		$Zn \longrightarrow Zn^{2+} + 2e^-$
		$Cr \longrightarrow Cr^{3+} + 3e^-$
		$Fe \longrightarrow Fe^{2+} + 2e^-$
	These elements react with aqueous H^+ ions to release H_2 gas.	$Co \longrightarrow Co^{2+} + 2e^-$
		$Ni \longrightarrow Ni^{2+} + 2e^-$
		$Sn \longrightarrow Sn^{2+} + 2e^-$
		$H_2 \longrightarrow 2H^+ + 2e^-$
	These elements do not react with aqueous H^+ ions to release H_2.	$Cu \longrightarrow Cu^{2+} + 2e^-$
		$Ag \longrightarrow Ag^+ + e^-$
		$Hg \longrightarrow Hg^{2+} + 2e^-$
		$Pt \longrightarrow Pt^{2+} + 2e^-$
Weakly reducing		$Au \longrightarrow Au^{3+} + 3e^-$

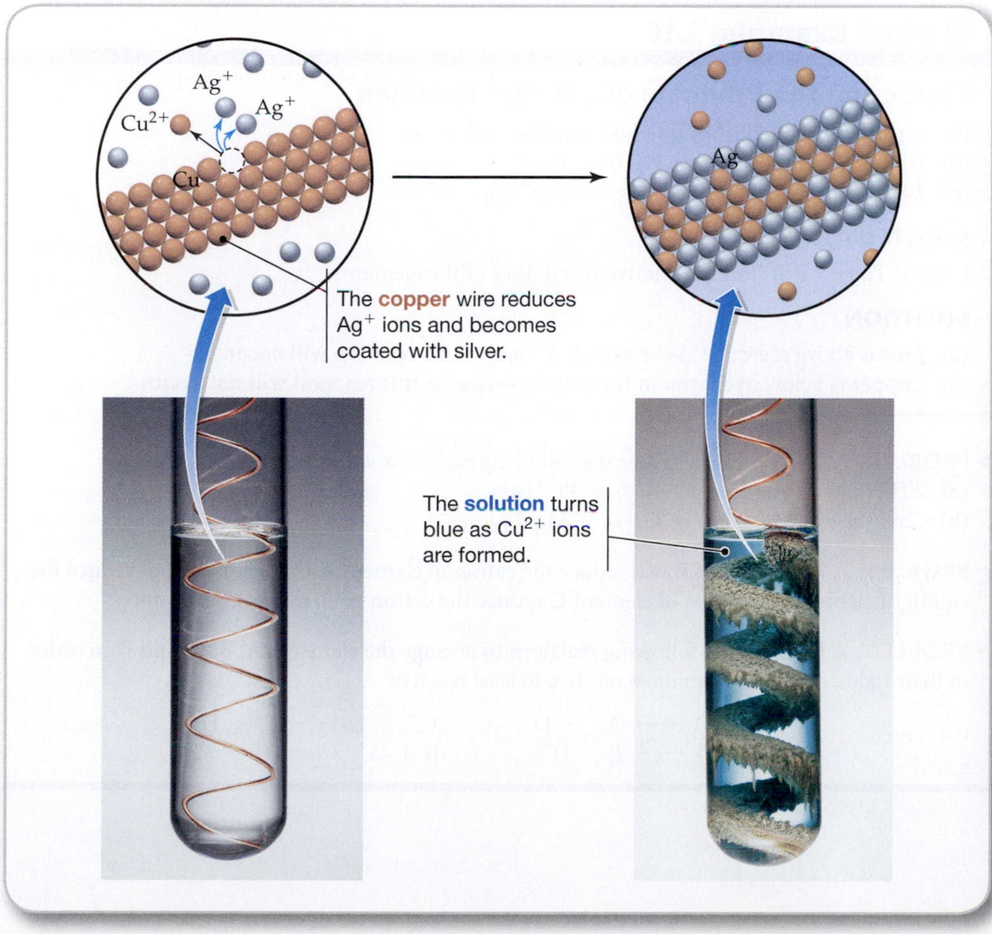

FIGURE 7.3

The redox reaction of copper with aqueous Ag^+ ion.

powerful reducing agents that they react even with pure water, in which the concentration of H^+ is very low:

Oxidized (Na) — Reduced (H of H_2O) — Unchanged (O of H_2O)

$$2\,Na(s) + 2\,H_2O(l) \longrightarrow 2\,Na^+(aq) + 2\,OH^-(aq) + H_2(g)$$

Oxidation numbers: Na(s) 0; H_2O +1, −2; Na^+ +1; OH^- −2, +1; H_2 0

In contrast, the metals in the middle of the series react with aqueous acid but not with water, and the metals at the bottom of the series react with neither aqueous acid nor water:

$$Fe(s) + 2\,H^+(aq) \longrightarrow Fe^{2+}(aq) + H_2(g)$$
$$Ag(s) + H^+(aq) \longrightarrow \text{No reaction}$$

Notice that the most easily oxidized metals—those at the top of the activity series—are on the left of the periodic table and thus have relatively low **ionization energies**. Conversely, the least easily oxidized metals—those at the bottom of the activity series—are in the transition metal groups closer to the right side of the table and thus have relatively high ionization energies.

REMEMBER...

An element's **ionization energy** is the amount of energy that must be supplied to remove an electron from an isolated neutral atom in the gas state. (Section 3.5)

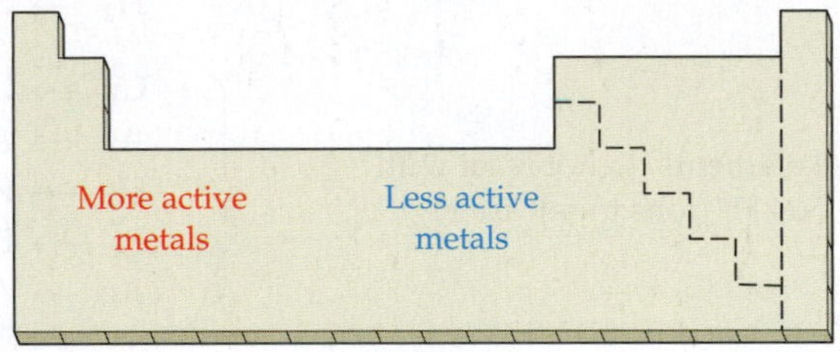

Worked Example 7.10

Predicting the Products of a Redox Reaction

Predict whether the following redox reactions will occur:

(a) $Hg^{2+}(aq) + Zn(s) \longrightarrow Hg(l) + Zn^{2+}(aq)$
(b) $2\,H^+(aq) + Cu(s) \longrightarrow H_2(g) + Cu^{2+}(aq)$

STRATEGY

Look at Table 7.5 to find the relative reactivities of the elements.

SOLUTION

(a) Zinc is above mercury in the activity series, so this reaction will occur.
(b) Copper is below hydrogen in the activity series, so this reaction will not occur.

PROBLEM 7.16 Predict whether the following reactions will occur:
(a) $2\,H^+(aq) + Pt(s) \longrightarrow H_2(g) + Pt^{2+}(aq)$
(b) $Ca^{2+}(aq) + Mg(s) \longrightarrow Ca(s) + Mg^{2+}(aq)$

PROBLEM 7.17 Element **B** will reduce the cation of element **A** ($\mathbf{A^+}$) but will not reduce the cation of element **C** ($\mathbf{C^+}$). Will element **C** reduce the cation of element **A**? Explain.

PROBLEM 7.18 Use the following reactions to arrange the elements **A**, **B**, **C**, and **D** in order of their redox reactivity from most reactive to least reactive.

$$A + D^+ \longrightarrow A^+ + D \qquad C^+ + D \longrightarrow C + D^+$$
$$B^+ + D \longrightarrow B + D^+ \qquad B + C^+ \longrightarrow B^+ + C$$

7.9 BALANCING REDOX REACTIONS BY THE HALF-REACTION METHOD

Simple redox reactions can often be balanced by the trial-and-error method described previously in Section 6.2, but many reactions are so complex that a more systematic approach is needed. Although a number of different methods are available for balancing equations, we'll look only at the **half-reaction method**, which focuses on the transfer of electrons, a subject of particular interest when discussing batteries and other aspects of electrochemistry as we'll see in Chapter 17.

The key to the half-reaction method is to realize that a redox reaction can be broken into two parts, or **half-reactions**. One half-reaction describes the oxidation part of the process, and the other describes the reduction part. Each half is balanced separately, and the two halves are then added to obtain the final equation. As an example, let's look at the reaction of aqueous potassium dichromate ($K_2Cr_2O_7$) with aqueous NaCl. The reaction occurs in acidic solution according to the unbalanced net ionic equation

$$Cr_2O_7^{2-}(aq) + Cl^-(aq) \longrightarrow Cr^{3+}(aq) + Cl_2(aq) \quad \textbf{Unbalanced}$$

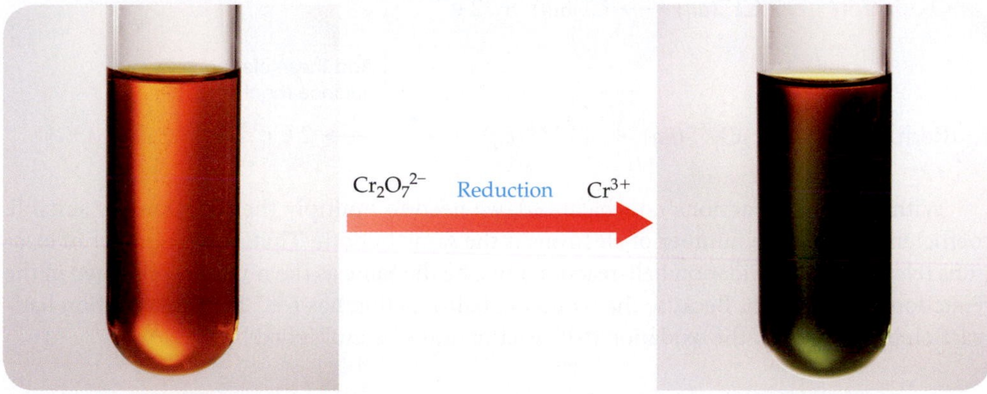

▲ The orange **dichromate ion** is reduced by addition of Cl^- to give the green **Cr^{3+} ion**.

The first step in balancing the equation is to decide which atoms have been oxidized and which have been reduced. In the present case, the chloride ion is oxidized from −1 to 0, and the chromium atom is reduced from +6 to +3. Thus, we can write two unbalanced half-reactions that show the separate parts:

Oxidation half-reaction $\quad Cl^-(aq) \longrightarrow Cl_2(aq)$

Reduction half-reaction $\quad Cr_2O_7^{2-}(aq) \longrightarrow Cr^{3+}(aq)$

With the two half-reactions identified, each is balanced separately. Begin by balancing for all atoms other than H and O. The oxidation half-reaction needs a coefficient of 2 before the Cl^-, and the reduction half-reaction needs a coefficient of 2 before the Cr^{3+}.

Add this coefficient to balance for Cl.

Oxidation $\quad 2\,Cl^-(aq) \longrightarrow Cl_2(aq)$

Add this coefficient to balance for Cr.

Reduction $\quad Cr_2O_7^{2-}(aq) \longrightarrow 2\,Cr^{3+}(aq)$

Next, balance both half-reactions for oxygen by adding H_2O to the side with less O, and then balance for hydrogen by adding H^+ to the side with less H. The oxidation half-reaction has no O or H, but the reduction half-reaction needs 7 H_2O on the product side to balance for O, and then 14 H^+ on the reactant side to balance for H:

Then, add 14 H^+ to balance for H.

First, add 7 H_2O to balance for O.

$$\text{Oxidation} \quad 2\,Cl^-(aq) \longrightarrow Cl_2(aq)$$

$$\text{Reduction} \quad Cr_2O_7^{2-}(aq) + 14\,H^+(aq) \longrightarrow 2\,Cr^{3+}(aq) + 7\,H_2O(l)$$

Now, balance both half-reactions for charge by adding electrons (e^-) to the side with the greater positive charge (or lesser negative charge). The oxidation half-reaction has 2 minus charges on the reactant side (2 Cl^-) and must therefore have 2 e^- added to the product side. The reduction half-reaction has a net of 12 positive charges on the reactant side and 6 positive charges on the product side and so must have 6 e^- added to the reactant side:

Add these electrons to balance for charge.

$$\text{Oxidation} \quad 2\,Cl^-(aq) \longrightarrow Cl_2(aq) + 2\,e^-$$

Add these electrons to balance for charge.

$$\text{Reduction} \quad Cr_2O_7^{2-}(aq) + 14\,H^+(aq) + 6\,e^- \longrightarrow 2\,Cr^{3+}(aq) + 7\,H_2O(l)$$

With both half-reactions now balanced, we need to multiply the equations by suitable coefficients so that the number of electrons is the same in both. That is, the number of electrons released in the oxidation half-reaction must be the same as the number consumed in the reduction half-reaction. Because the reduction half-reaction has 6 e^- but the oxidation half-reaction has only 2 e^-, the oxidation half-reaction must be multiplied by 3:

Multiply by this coefficient to equalize the numbers of electrons in the two half-reactions.

$$\text{Oxidation} \quad 3 \times [2\,Cl^-(aq) \longrightarrow Cl_2(aq) + 2\,e^-]$$

$$\text{or} \quad 6\,Cl^-(aq) \longrightarrow 3\,Cl_2(aq) + 6\,e^-$$

$$\text{Reduction} \quad Cr_2O_7^{2-}(aq) + 14\,H^+(aq) + 6\,e^- \longrightarrow 2\,Cr^{3+}(aq) + 7\,H_2O(l)$$

Adding the two half-reactions together and canceling the species that occur on both sides (only the electrons in this example) then gives the final balanced equation. Check the answer to make sure it is balanced both for atoms and for charge.

$$6\,Cl^-(aq) \longrightarrow 3\,Cl_2(aq) + \cancel{6\,e^-}$$

$$Cr_2O_7^{2-}(aq) + 14\,H^+(aq) + \cancel{6\,e^-} \longrightarrow 2\,Cr^{3+}(aq) + 7\,H_2O(l)$$

$$Cr_2O_7^{2-}(aq) + 14\,H^+(aq) + 6\,Cl^-(aq) \longrightarrow 3\,Cl_2(aq) + 2\,Cr^{3+}(aq) + 7\,H_2O(l)$$

Charge: $(-2) + (+14) + (6 \times -1) = +6$ Charge: $(2 \times +3) = +6$

To summarize, balancing a redox reaction in acidic solution by the half-reaction method is a six-step process, followed by a check of the answer **(Figure 7.4.)**

Worked Example 7.12 shows how to use the method for balancing an equation for a reaction that takes place in basic solution. The procedure is the same as that used for balancing a reaction in acidic solution, but OH^- ions are added as a final step to neutralize any H^+ ions that appear in the equation. This simply reflects the fact that basic solutions contain negligibly small amounts of H^+ but relatively large amounts of OH^-.

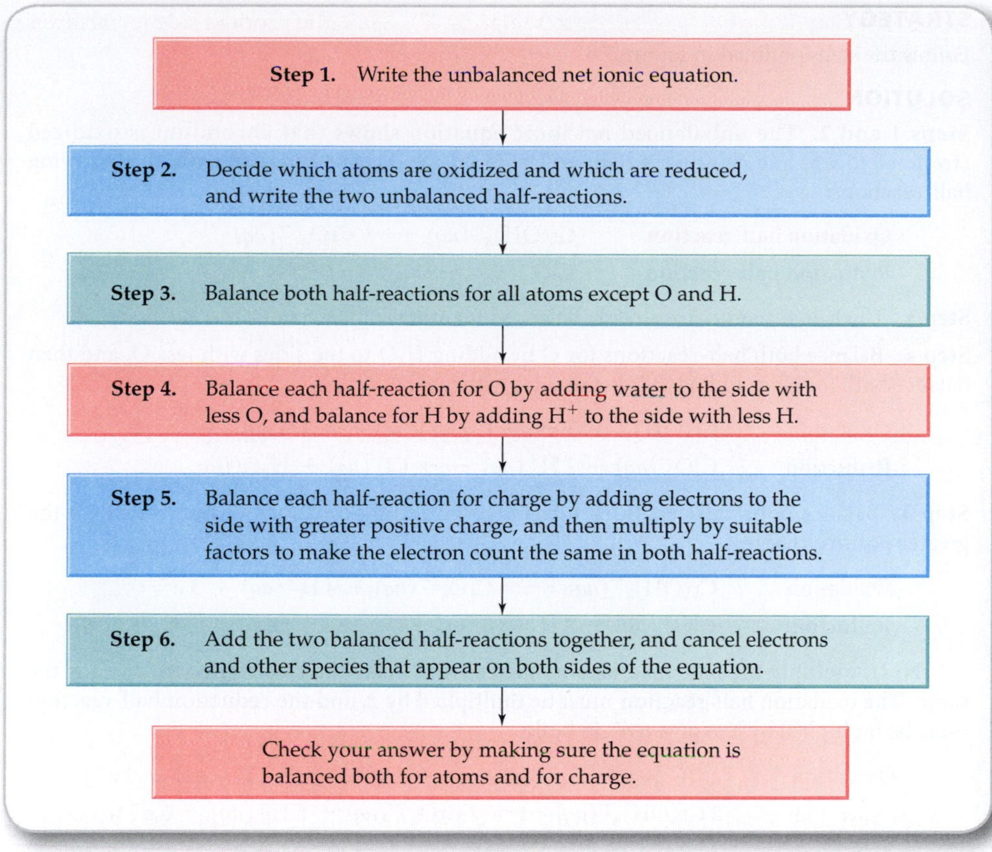

FIGURE 7.4

Using the half-reaction method to balance redox equations for reactions in acidic solution.

Worked Example 7.11

Writing Half-Reactions

Write unbalanced half-reactions for the following unbalanced net ionic equations:

(a) $Mn^{2+}(aq) + ClO_3^-(aq) \longrightarrow MnO_2(s) + ClO_2(aq)$

(b) $Cr_2O_7^{2-}(aq) + Fe^{2+}(aq) \longrightarrow Cr^{3+}(aq) + Fe^{3+}(aq)$

STRATEGY

Look at each equation to see which atoms are being oxidized (increasing in oxidation number) and which are being reduced (decreasing in oxidation number).

SOLUTION

(a)	Oxidation	$Mn^{2+}(aq) \longrightarrow MnO_2(s)$	Mn goes from +2 to +4
	Reduction	$ClO_3^-(aq) \longrightarrow ClO_2(aq)$	Cl goes from +5 to +4
(b)	Oxidation	$Fe^{2+}(aq) \longrightarrow Fe^{3+}(aq)$	Fe goes from +2 to +3
	Reduction	$Cr_2O_7^{2-}(aq) \longrightarrow Cr^{3+}(aq)$	Cr goes from +6 to +3

Worked Example 7.12

Balancing an Equation for a Reaction in Base

Aqueous sodium hypochlorite (NaOCl; household bleach) is a strong oxidizing agent that reacts with chromite ion $[Cr(OH)_4^-]$ in basic solution to yield chromate ion (CrO_4^{2-}) and chloride ion. The net ionic equation is

$$ClO^-(aq) + Cr(OH)_4^-(aq) \longrightarrow CrO_4^{2-}(aq) + Cl^-(aq) \quad \textbf{Unbalanced}$$

Balance the equation using the half-reaction method.

continued on next page

STRATEGY

Follow the steps outlined in Figure 7.4.

SOLUTION

Steps 1 and 2. The unbalanced net ionic equation shows that chromium is oxidized (from +3 to +6) and chlorine is reduced (from +1 to −1). Thus, we can write the following half-reactions:

Oxidation half-reaction $Cr(OH)_4^-(aq) \longrightarrow CrO_4^{2-}(aq)$

Reduction half-reaction $ClO^-(aq) \longrightarrow Cl^-(aq)$

Step 3. The half-reactions are already balanced for atoms other than O and H.

Step 4. Balance both half-reactions for O by adding H_2O to the sides with less O, and then balance both for H by adding H^+ to the sides with less H:

Oxidation $Cr(OH)_4^-(aq) \longrightarrow CrO_4^{2-}(aq) + 4\,H^+(aq)$

Reduction $ClO^-(aq) + 2\,H^+(aq) \longrightarrow Cl^-(aq) + H_2O(l)$

Step 5. Balance both half-reactions for charge by adding electrons to the sides with the greater positive charge:

Oxidation $Cr(OH)_4^-(aq) \longrightarrow CrO_4^{2-}(aq) + 4\,H^+(aq) + 3\,e^-$

Reduction $ClO^-(aq) + 2\,H^+(aq) + 2\,e^- \longrightarrow Cl^-(aq) + H_2O(l)$

Next, multiply the half-reactions by factors that make the electron count in each the same. The oxidation half-reaction must be multiplied by 2, and the reduction half-reaction must be multiplied by 3 to give 6 e^- in both:

Oxidation $2 \times [Cr(OH)_4^-(aq) \longrightarrow CrO_4^{2-}(aq) + 4\,H^+(aq) + 3\,e^-]$

or $2\,Cr(OH)_4^-(aq) \longrightarrow 2\,CrO_4^{2-}(aq) + 8\,H^+(aq) + 6\,e^-$

Reduction $3 \times [ClO^-(aq) + 2\,H^+(aq) + 2\,e^- \longrightarrow Cl^-(aq) + H_2O(l)]$

or $3\,ClO^-(aq) + 6\,H^+(aq) + 6\,e^- \longrightarrow 3\,Cl^-(aq) + 3\,H_2O(l)$

Step 6. Add the balanced half-reactions:

$$2\,Cr(OH)_4^-(aq) \longrightarrow 2\,CrO_4^{2-}(aq) + 8\,H^+(aq) + 6\,e^-$$
$$3\,ClO^-(aq) + 6\,H^+(aq) + 6\,e^- \longrightarrow 3\,Cl^-(aq) + 3\,H_2O(l)$$
$$2\,Cr(OH)_4^-(aq) + 3\,ClO^-(aq) + 6\,H^+(aq) + 6\,e^- \longrightarrow 2\,CrO_4^{2-}(aq) + 3\,Cl^-(aq) + 3\,H_2O(l) + 8\,H^+(aq) + 6\,e^-$$

Now, cancel the species that appear on both sides of the equation:

$$2\,Cr(OH)_4^-(aq) + 3\,ClO^-(aq) \longrightarrow 2\,CrO_4^{2-}(aq) + 3\,Cl^-(aq) + 3\,H_2O(l) + 2\,H^+(aq)$$

Finally, since we know that the reaction takes place in basic solution, we must add 2 OH^- ions to both sides of the equation to neutralize the 2 H^+ ions on the right giving 2 additional H_2O. The final net ionic equation, balanced for both atoms and charge, is

$$2\,Cr(OH)_4^-(aq) + 3\,ClO^-(aq) + 2\,OH^-(aq) \longrightarrow 2\,CrO_4^{2-}(aq) + 3\,Cl^-(aq) + 5\,H_2O(l)$$

Charge: $(2 \times -1) + (3 \times -1) + (2 \times -1) = -7$ Charge: $(2 \times -2) + (3 \times -1) = -7$

PROBLEM 7.19 Write unbalanced half-reactions for the following unbalanced net ionic equations:

(a) $MnO_4^-(aq) + IO_3^-(aq) \longrightarrow MnO_2(s) + IO_4^-(aq)$

(b) $NO_3^-(aq) + SO_2(aq) \longrightarrow SO_4^{2-}(aq) + NO_2(g)$

PROBLEM 7.20 Balance the following net ionic equation by the half-reaction method. The reaction takes place in acidic solution.

$$NO_3^-(aq) + Cu(s) \longrightarrow NO(g) + Cu^{2+}(aq) \quad \textbf{Unbalanced}$$

PROBLEM 7.21 Balance the following equation by the half-reaction method. The reaction takes place in basic solution.

$$Fe(OH)_2(s) + O_2(g) \longrightarrow Fe(OH)_3(s) \quad \textbf{Unbalanced}$$

7.10 REDOX TITRATIONS

REMEMBER...

The reaction used for a **titration** must go to completion and have a yield of 100%. (Section 6.11)

We saw in Section 6.11 that the concentration of an acid or base solution can be determined by **titration**. A measured volume of the acid or base solution of unknown concentration is placed in a flask, and a base or acid solution of known concentration is slowly added from a buret. By measuring the volume of the added solution necessary for a complete reaction, as signaled by the color change of an indicator, the unknown concentration can be calculated.

A similar procedure can be carried out to determine the concentration of many oxidizing or reducing agents using a *redox titration*. All that's necessary is that the substance whose concentration you want to determine undergo an oxidation or reduction reaction in 100% yield and that there be some means, such as a color change, to indicate when the reaction is complete. The color change might be due to one of the substances undergoing reaction or to some added indicator.

Let's imagine that we have a potassium permanganate solution whose concentration we want to find. Aqueous $KMnO_4$ reacts with oxalic acid, $H_2C_2O_4$, in acidic solution according to the following net ionic equation (K^+ is a spectator ion):

$$5\,H_2C_2O_4(aq) + 2\,MnO_4^-(aq) + 6\,H^+(aq) \longrightarrow 10\,CO_2(g) + 2\,Mn^{2+}(aq) + 8\,H_2O(l)$$

The reaction takes place in 100% yield and is accompanied by a sharp color change when the intense purple color of the MnO_4^- ion disappears.

The strategy used is outlined in **Figure 7.5**. As with acid–base titrations, the general idea is to measure a known amount of one substance—in this case, $H_2C_2O_4$—and use mole ratios from the balanced equation to find the number of moles of the second substance—in this case, $KMnO_4$—necessary for complete reaction. With the molar amount of $KMnO_4$ thus known, titration gives the volume of solution containing that amount. Dividing the number of moles by the volume gives the concentration.

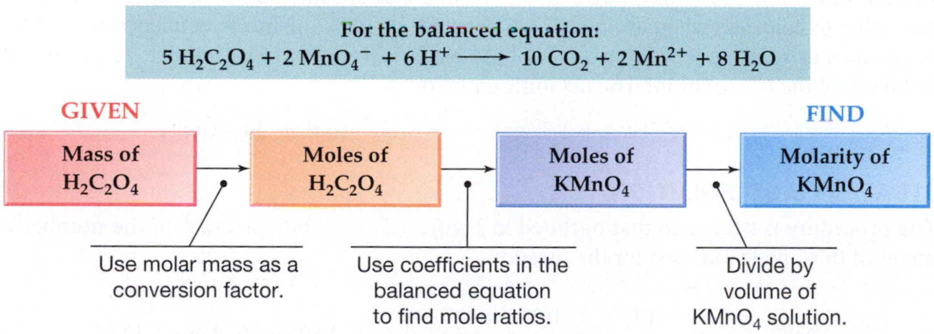

FIGURE 7.5

A summary of calculations for determining the concentration of a $KMnO_4$ solution by redox titration of $H_2C_2O_4$.

As an example of how the procedure works, let's carefully weigh some amount of $H_2C_2O_4$—say, 0.2585 g—and dissolve it in approximately 100 mL of 0.5 M H_2SO_4. The exact volume isn't important because we're concerned only with the amount of dissolved $H_2C_2O_4$, not with its concentration. Next, we place an aqueous $KMnO_4$ solution of unknown concentration in a buret and slowly add it to the $H_2C_2O_4$ solution. The purple color of the added MnO_4^- initially disappears as reaction occurs, but we continue the addition until a faint color persists, indicating that all the $H_2C_2O_4$ has reacted and that MnO_4^- ion is no longer being reduced. At this equivalence point, or *end point*, of the titration, we might find that 22.35 mL of the $KMnO_4$ solution has been added (**Figure 7.6**).

A precise amount of oxalic acid is weighed and dissolved in aqueous H_2SO_4.

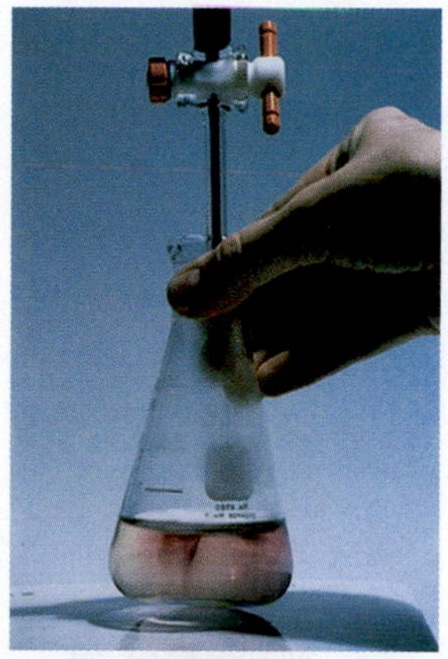

Aqueous **$KMnO_4$** of unknown concentration is added from a buret until ...

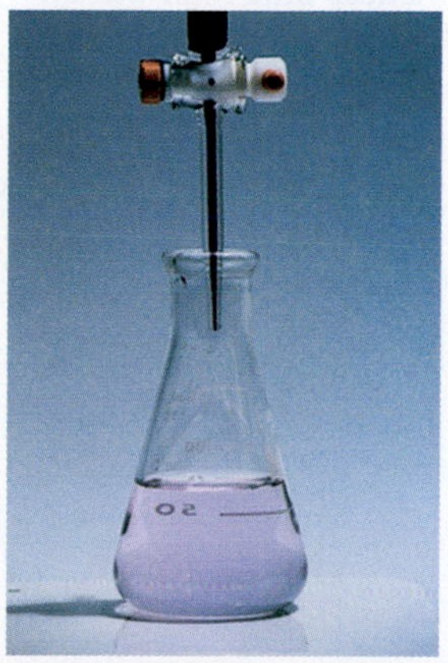

... the purple color persists, indicating that all of the oxalic acid has reacted.

FIGURE 7.6

The redox titration of oxalic acid, $H_2C_2O_4$, with $KMnO_4$.

To calculate the molarity of the $KMnO_4$ solution, we need to find the number of moles of $KMnO_4$ present in the 22.35 mL of solution used for titration. We do this by first calculating the number of moles of oxalic acid that react with the permanganate ion, using a gram-to-mole conversion with the molar mass of $H_2C_2O_4$ as the conversion factor:

$$\text{Moles of } H_2C_2O_4 = 0.2585\ \cancel{\text{g } H_2C_2O_4} \times \frac{1\ \text{mol } H_2C_2O_4}{90.04\ \cancel{\text{g } H_2C_2O_4}}$$
$$= 2.871 \times 10^{-3}\ \text{mol } H_2C_2O_4$$

According to the balanced equation, 5 mol of oxalic acid react with 2 mol of permanganate ion. Thus, we can calculate the number of moles of $KMnO_4$ that react with 2.871×10^{-3} mol of $H_2C_2O_4$:

$$\text{Moles of } KMnO_4 = 2.871 \times 10^{-3}\ \cancel{\text{mol } H_2C_2O_4} \times \frac{2\ \text{mol } KMnO_4}{5\ \cancel{\text{mol } H_2C_2O_4}}$$
$$= 1.148 \times 10^{-3}\ \text{mol } KMnO_4$$

Knowing both the number of moles of $KMnO_4$ that react (1.148×10^{-3} mol) and the volume of the $KMnO_4$ solution (22.35 mL), we can calculate the molarity:

$$\text{Molarity} = \frac{1.148 \times 10^{-3}\ \text{mol } KMnO_4}{22.35\ \cancel{\text{mL}}} \times \frac{1000\ \cancel{\text{mL}}}{1\ \text{L}} = 0.051\,36\ \text{M}$$

The molarity of the $KMnO_4$ solution is 0.051 36 M.

Worked Example 7.13

Using a Redox Reaction to Determine a Solution's Concentration

The concentration of an aqueous I_3^- solution can be determined by titration with aqueous sodium thiosulfate, $Na_2S_2O_3$, in the presence of a starch indicator. The starch turns from deep blue to colorless when all the I_3^- has reacted. What is the molar concentration of I_3^- in an aqueous solution if 24.55 mL of 0.102 M $Na_2S_2O_3$ is needed for complete reaction with 10.00 mL of the I_3^- solution? The net ionic equation is

$$2\ S_2O_3^{2-}(aq) + I_3^-(aq) \longrightarrow S_4O_6^{2-}(aq) + 3\ I^-(aq)$$

STRATEGY AND SOLUTION

The procedure is similar to that outlined in Figure 7.5. We first need to find the number of moles of thiosulfate ion used for the titration:

$$24.55\ \cancel{\text{mL}} \times \frac{1\ \cancel{\text{L}}}{1000\ \cancel{\text{mL}}} \times \frac{0.102\ \text{mol } S_2O_3^{2-}}{1\ \cancel{\text{L}}} = 2.50 \times 10^{-3}\ \text{mol } S_2O_3^{2-}$$

According to the balanced equation, 2 mol of $S_2O_3^{2-}$ ion react with 1 mol of I_3^- ion. Thus, we can find the number of moles of I_3^- ion:

$$2.50 \times 10^{-3}\ \cancel{\text{mol } S_2O_3^{2-}} \times \frac{1\ \text{mol } I_3^-}{2\ \cancel{\text{mol } S_2O_3^{2-}}} = 1.25 \times 10^{-3}\ \text{mol } I_3^-$$

Knowing both the number of moles of I_3^- (1.25×10^{-3} mol) and the volume of the I_3^- solution (10.00 mL) then lets us calculate molarity:

$$\frac{1.25 \times 10^{-3}\ \text{mol } I_3^-}{10.00\ \cancel{\text{mL}}} \times \frac{10^3\ \cancel{\text{mL}}}{1\ \text{L}} = 0.125\ \text{M}$$

The molarity of the I_3^- solution is 0.125 M.

✔ BALLPARK CHECK

According to the balanced equation, the amount of $S_2O_3^{2-}$ needed for the reaction (2 mol) is twice the amount of I_3^- (1 mol). The titration results indicate that the volume of the $S_2O_3^{2-}$ solution (24.55 mL) is a little over twice the volume of the I_3^- solution (10.00 mL). Thus, the concentrations of the two solutions must be about the same—approximately 0.1 M.

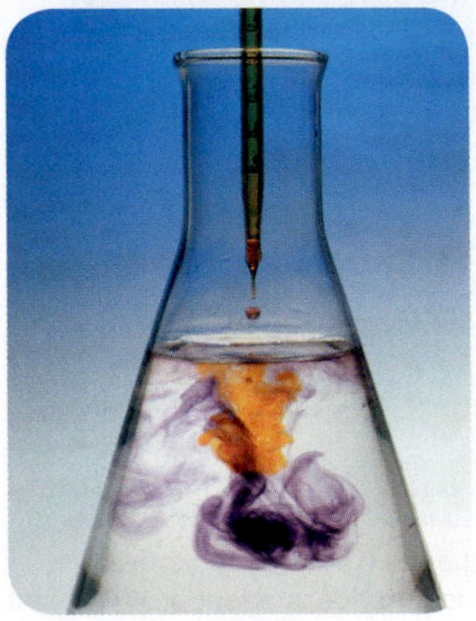

▲ The reddish I_3^- **solution** turns a **deep blue** color when it is added to a solution containing a small amount of starch.

PROBLEM 7.22 What is the molar concentration of Fe^{2+} ion in an aqueous solution if 31.50 mL of 0.105 M $KBrO_3$ is required for complete reaction with 10.00 mL of the Fe^{2+} solution? The net ionic equation is:

$$6\,Fe^{2+}(aq) + BrO_3^-(aq) + 6\,H^+(aq) \longrightarrow 6\,Fe^{3+}(aq) + Br^-(aq) + 3\,H_2O(l)$$

7.11 SOME APPLICATIONS OF REDOX REACTIONS

Redox reactions take place with every element in the periodic table except helium and neon and occur in a vast number of processes throughout nature, biology, and industry. Here are just a few examples:

- **Combustion** is the burning of a fuel by oxidation with oxygen in air. Gasoline, fuel oil, natural gas, wood, paper, and other organic substances of carbon and hydrogen are the most common fuels. Even some metals, such as magnesium and calcium, will burn in air.

$$\underset{\substack{\text{Methane}\\\text{(natural Gas)}}}{CH_4(g)} + 2\,O_2(g) \longrightarrow CO_2(g) + 2\,H_2O(l)$$

- **Bleaching** uses redox reactions to decolorize or lighten colored materials. Dark hair is bleached to turn it blond, clothes are bleached to remove stains, wood pulp is bleached to make white paper, and so on. The exact oxidizing agent used depends on the situation—hydrogen peroxide (H_2O_2) is used for hair, sodium hypochlorite (NaOCl) is used for clothes, and ozone or chlorine dioxide is used for wood pulp—but the principle is always the same. In all cases, colored impurities are destroyed by reaction with a strong oxidizing agent.

▲ Dark hair can be bleached by a redox reaction with hydrogen peroxide.

- **Batteries**, although they come in many types and sizes, are all based on redox reactions. In a typical redox reaction carried out in the laboratory—say, the reaction of zinc metal with Ag^+ to yield Zn^{2+} and silver metal—the reactants are simply mixed in a flask and electrons are transferred by direct contact between them. In a battery, however, the two reactants are kept in separate compartments and the electrons are transferred through a wire running between them.

  The inexpensive alkaline battery commonly used in flashlights and other small household items uses a thin steel can containing zinc powder and a paste of potassium hydroxide as one reactant, separated by paper from a paste of powdered carbon and manganese dioxide as the other reactant. A graphite rod with a metal cap sticks into the MnO_2 to provide electrical contact. When the can and the graphite rod are connected by a wire, zinc sends electrons flowing through the wire toward the MnO_2 in a redox reaction. The resultant electrical current can be used to light a bulb or power a small electronic device. The reaction is

$$Zn(s) + 2\,MnO_2(s) \longrightarrow ZnO(s) + Mn_2O_3(s)$$

  We'll look at the chemistry of batteries in more detail in Section 17.9.

- **Metallurgy**, the extraction and purification of metals from their ores, makes use of numerous redox processes. We'll see in Section 21.2, for example, that metallic zinc is prepared by reduction of ZnO with *coke*, a form of carbon:

$$ZnO(s) + C(s) \longrightarrow Zn(s) + CO(g)$$

- **Corrosion** is the deterioration of a metal by oxidation, such as the rusting of iron in moist air. The economic consequences of rusting are enormous: It has been estimated that up to one-fourth of the iron produced in the United States is used to replace bridges, buildings, and other structures that have been destroyed by corrosion. (The raised dot in the formula $Fe_2O_3 \cdot H_2O$ for rust indicates that one water molecule is associated with each Fe_2O_3 in an unspecified way.)

$$4\,Fe(s) + 3\,O_2(g) \xrightarrow{H_2O} \underset{\text{Rust}}{2\,Fe_2O_3 \cdot H_2O(s)}$$

- **Respiration** is the process of breathing and using oxygen for the many biological redox reactions that provide the energy needed by living organisms. The energy is released from food molecules slowly and in complex, multistep pathways, but the overall result of respiration is similar to that of a combustion reaction. For example, the simple sugar glucose ($C_6H_{12}O_6$) reacts with O_2 to give CO_2 and H_2O according to the following equation:

$$\underset{\substack{\text{Glucose}\\\text{(a carbohydrate)}}}{C_6H_{12}O_6} + 6\,O_2 \longrightarrow 6\,CO_2 + 6\,H_2O + \text{energy}$$

PROBLEM 7.23 In the common lead storage battery used in automobiles, the following two half-reactions occur in acid solution:

Oxidation $Pb(s) + HSO_4^-(aq) \longrightarrow PbSO_4(s)$

Reduction $PbO_2(s) + HSO_4^-(aq) \longrightarrow PbSO_4(s)$

Write a balanced equation for the overall battery reaction.

FYI CAN CHEMICAL REACTIONS BE GREEN?

▲ The world will always need chemical plants, like this BASF facility in Ludswigshafen, Germany—the largest chemical plant in the world. But much more attention will need to be paid in the future to minimizing the chemical wastes they generate.

Chemistry has made our lives longer, safer, and far more comfortable than they would otherwise be. The medicines, fertilizers, insecticides, adhesives, textiles, dyes, building materials, and polymers that we take for granted are all products of the chemical industry. But these benefits have not come without a price. Many chemical processes produce hazardous wastes that must be dealt with—reaction solvents and byproducts that might evaporate into the air or leach into groundwater if not disposed of properly.

It's unlikely that all chemical processes will ever be completely benign, but awareness of the environmental problems caused by chemistry has grown dramatically in recent years, giving rise to a movement called *green chemistry*. Green chemistry is the design and implementation of chemical processes that reduce waste and minimize or eliminate the generation of hazardous substances. Twelve principles form the foundation of green chemistry:

Prevent waste. Waste should be prevented rather than cleaned up after it has been created.

Maximize atom economy. Processes should maximize the incorporation of all reactant atoms into the final product to minimize leftover waste.

Use less hazardous processes. Methods of chemical synthesis should use nontoxic reactants and generate nontoxic wastes.

Design safer chemicals. Chemical products should be designed from the beginning to have minimal toxicity.

Use safer solvents. Solvents and other auxiliary substances used in reactions should be safe and used sparingly.

Design for energy efficiency. Energy usage in chemical processes should be minimized, with reactions carried out at room temperature if possible.

Use renewable feedstocks. Raw materials should come from renewable sources when feasible.

Minimize derivatives. Syntheses should be designed with minimal use of protecting groups to avoid extra steps and reduce waste.

Use catalysis. Reactions should be catalytic rather than stoichiometric.

Design for degradation. Products should be designed to be biodegradable at the end of their useful lifetimes.

Monitor pollution in real time. Processes should be monitored in real time for the formation of hazardous substances.

Prevent accidents. Chemical substances and processes should minimize the potential for fires, explosions, or other accidents.

The twelve principles of green chemistry won't all be met in most applications, but they provide a goal to aim for and they can make chemists think more carefully about the environmental implications of their work. Among the successes already recorded is a new process for converting glycerin, a byproduct of producing biodiesel fuel from vegetable oil, into propylene glycol, which is used in automobile antifreeze. The process occurs in a single step and gives only water as the byproduct.

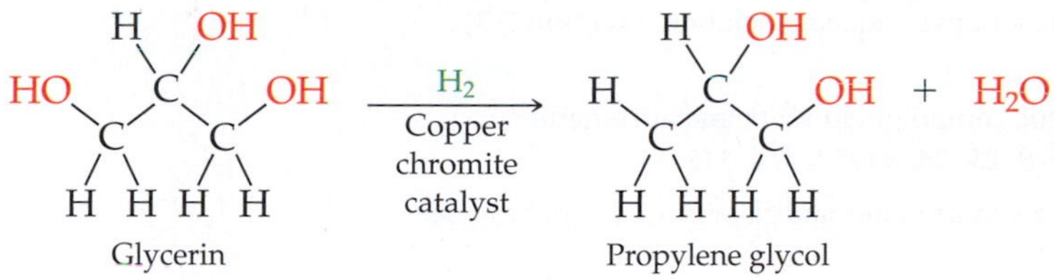

PROBLEM 7.24 Many dozens of different solvents are used in various chemical processes. If you were designing a green process that required a solvent, what properties would you look for, and what solvent might you choose?

SUMMARY

Many reactions, particularly those that involve ionic compounds, take place in aqueous solution. Substances whose aqueous solutions contain ions conduct electricity and are called **electrolytes**. Ionic compounds, such as NaCl, and molecular compounds that **dissociate** substantially into ions when dissolved in water are **strong electrolytes**. Substances that dissociate to only a small extent are **weak electrolytes**, and substances that do not produce ions in aqueous solution are **nonelectrolytes**.

Acids are compounds that dissociate in aqueous solutions to yield an anion and a **hydronium ion, H_3O^+**. Acids that dissociate to a large extent are **strong acids**; acids that dissociate to a small extent are **weak acids**. Similarly, **bases** are compounds that dissociate in aqueous solution to yield a cation and hydroxide ion, OH^-.

Aqueous reactions can be classified into three major groups. **Precipitation reactions** occur when solutions of two ionic substances are mixed and a precipitate falls from solution. To predict whether a precipitate will form, you must know the **solubility** of each potential product. **Acid–base neutralization reactions** occur when an acid is mixed with a base, yielding water and an ionic **salt**. The neutralization of a strong acid with a strong base can be written as a **net ionic equation**, in which nonparticipating, **spectator ions** are not specified:

$$H^+(aq) + OH^-(aq) \longrightarrow H_2O(l)$$

Oxidation–reduction reactions, or **redox reactions**, are processes in which one or more electrons are transferred between reaction partners. An **oxidation** is the loss of one or more electrons; a **reduction** is the gain of one or more electrons. Redox reactions can be identified by assigning to each atom in a substance an **oxidation number**, which provides a measure of whether the atom is neutral, electron rich, or electron poor. Comparing the oxidation numbers of an atom before and after reaction shows whether the atom has gained or lost electrons.

Oxidations and reductions must occur together. Whenever one substance loses one or more electrons (is oxidized), another substance gains the electrons (is reduced). The substance that causes a reduction by giving up electrons is called a **reducing agent**. The substance that causes an oxidation by accepting electrons is called an **oxidizing agent**. The reducing agent is itself oxidized when it gives up electrons, and the oxidizing agent is itself reduced when it accepts electrons.

Among the simplest of redox processes is the reaction of an aqueous cation, usually a metal ion, with a free element to give a different ion and a different element. Noting the results from a succession of different reactions makes it possible to organize an **activity series**, which ranks the elements in order of their reducing ability in aqueous solution.

Redox reactions can be balanced using the **half-reaction method**, which divides a reaction into oxidation and reduction parts and focuses on equalizing the transfer of electrons between the parts. The concentration of an oxidizing agent or a reducing agent in solution can be determined by a redox titration.

KEY WORDS

acid *244*
acid–base neutralization reaction *239*
activity series *254*
base *244*
diprotic acid *245*
dissociate *238*
electrolyte *237*
half-reaction *257*
half-reaction method *257*
hydronium ion, H_3O^+ *244*
ionic equation *240*
molecular equation *240*
monoprotic acid *245*
net ionic equation *240*
nonelectrolyte *237*
oxidation *249*
oxidation number *249*
oxidation–reduction reaction *239*
oxidizing agent *252*
oxoacid *246*
precipitation reaction *239*
redox reaction *239*
reducing agent *252*
reduction *249*
salt *247*
solubility *241*
spectator ion *240*
strong acid *245*
strong base *245*
strong electrolyte *238*
triprotic acid *245*
weak acid *245*
weak base *245*
weak electrolyte *238*

LEARNING OUTCOMES

- Classify a substance as a strong, weak, or nonelectrolyte. **(Section 7.1)** ***Problems 2, 29, 36, 37, 40, 41***
- Classify a reaction as a precipitation, acid–base neutralization, or oxidation–reduction (redox) reaction. **(Section 7.2)** ***Problems 3, 32, 33***
- Write molecular, ionic, and net ionic equations to represent reactions in aqueous solution. **(Section 7.3)** ***Problems 4, 34, 35, 60–63, 90–93, 118–119***
- Use the solubility guidelines to predict the solubility of an ionic compound in water and tell whether a precipitation reaction will occur. **(Section 7.4)** ***Problems 5–8, 25, 26, 44–57, 114, 115***
- Name an acid given its formula, write the formula of an acid given its name, and classify acids and bases as strong or weak. **(Section 7.5)** ***Problems 9, 10, 12, 27, 60, 61***
- Write the ionic equation and net ionic equation for an acid–base neutralization reaction. **(Section 7.5)** ***Problems 11, 27, 60–67, 113***
- Assign oxidation numbers to atoms in a compound. **(Section 7.6)** ***Problems 13, 14, 72–75, 106***
- Identify redox reactions, oxidizing agents, and reducing agents. **(Section 7.7)** ***Problems 14, 15, 70, 71, 76, 77***
- Use the activity series of the elements to predict if a redox reaction will occur. **(Section 7.8)** ***Problems 16–18, 30, 31, 78–81, 108***
- Use the half-reaction method to balance the equations for redox reactions. **(Section 7.9)** ***Problems 19–21, 82–93, 104, 107, 116, 118, 119, 124***
- Use a redox titration to determine the concentration of an oxidizing or reducing agent in solution. **(Section 7.10)** ***Problems 22, 94–103, 105, 110, 125, 126***

CONCEPTUAL PROBLEMS

Problems 7.1–7.24 appear within the chapter.

7.25 Assume that an aqueous solution of a cation, represented as a red sphere, is allowed to mix with a solution of an anion, represented as a yellow sphere. Three possible outcomes are represented by boxes (1)–(3):

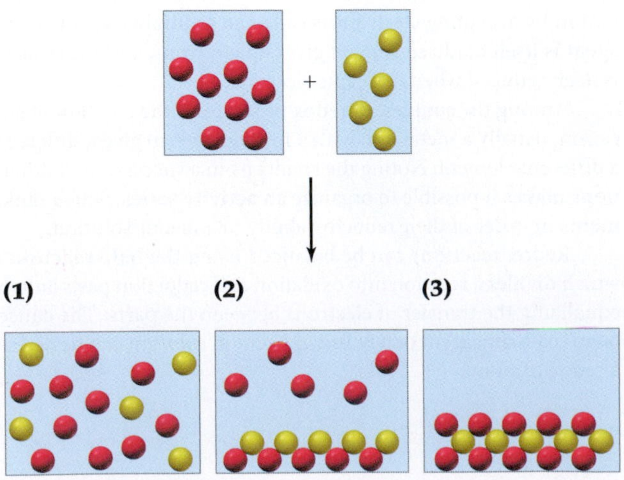

Which outcome corresponds to each of the following reactions?

(a) $2\,Na^+(aq) + CO_3^{2-}(aq) \rightarrow$
(b) $Ba^{2+}(aq) + CrO_4^{2-}(aq) \rightarrow$
(c) $2\,Ag^+(aq) + SO_3^{2-}(aq) \rightarrow$

7.26 Assume that an aqueous solution of a cation, represented as a blue sphere, is allowed to mix with a solution of an anion, represented as a red sphere, and that the following result is obtained:

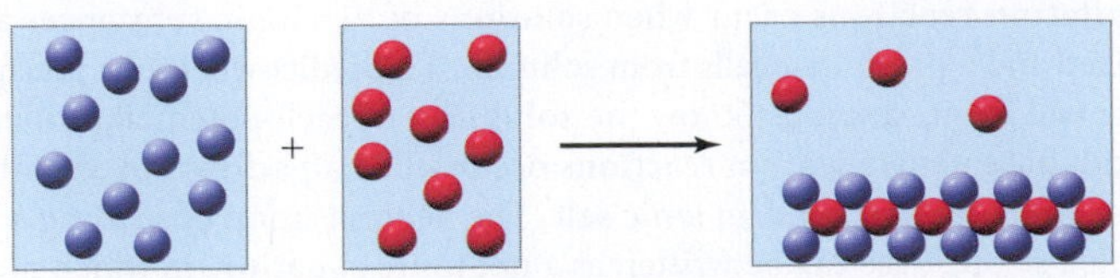

Which combinations of cation and anion, chosen from the following lists, are compatible with the observed results? Explain.

Cations: $Na^+, Ca^{2+}, Ag^+, Ni^{2+}$
Anions: $Cl^-, CO_3^{2-}, CrO_4^{2-}, NO_3^-$

7.27 Assume that an aqueous solution of OH^-, represented as a blue sphere, is allowed to mix with a solution of an acid H_nA, represented as a red sphere. Three possible outcomes are depicted by boxes (1)–(3), where the green spheres represent A^{n-}, the anion of the acid:

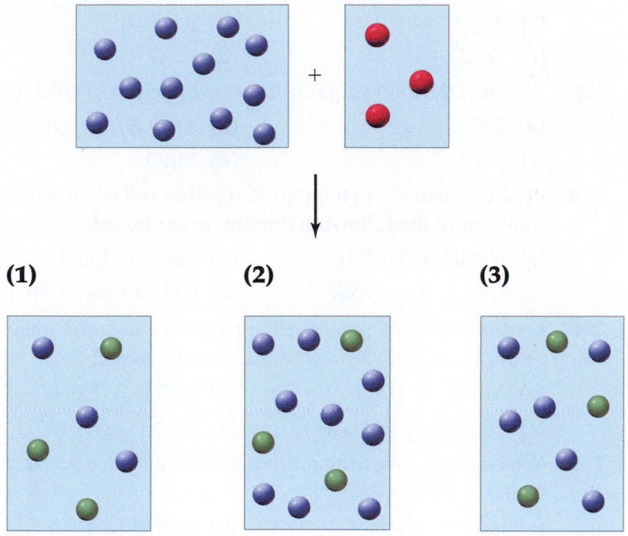

Which outcome corresponds to each of the following reactions?

(a) $HF + OH^- \rightarrow H_2O + F^-$

(b) $H_2SO_3 + 2\,OH^- \rightarrow 2\,H_2O + SO_3^{2-}$

(c) $H_3PO_4 + 3\,OH^- \rightarrow 3\,H_2O + PO_4^{3-}$

7.28 The concentration of an aqueous solution of NaOCl (sodium hypochlorite; the active ingredient in household bleach) can be determined by a redox titration with iodide ion in acidic solution:

$$OCl^-(aq) + 2\,I^-(aq) + 2\,H^+(aq) \longrightarrow Cl^-(aq) + I_2(aq) + H_2O(l)$$

Assume that the blue spheres in the buret represent I^- ions, the red spheres in the flask represent OCl^- ions, the concentration of the I^- ions in the buret is 0.120 M, and the volumes in the buret and the flask are identical. What is the concentration of NaOCl in the flask? What percentage of the I^- solution in the buret must be added to the flask to react with all the OCl^- ions?

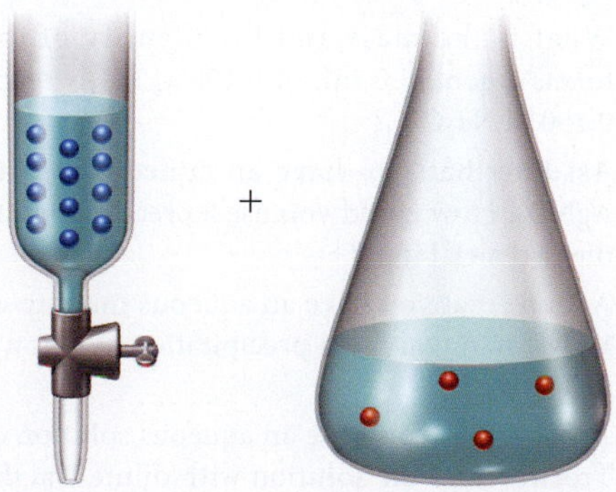

7.29 Assume that the electrical conductivity of a solution depends on the total concentration of dissolved ions and that you measure the conductivity of three different solutions while carrying out titration procedures:

(a) Begin with 1.00 L of 0.100 M KCl, and titrate by adding 0.100 M $AgNO_3$.

(b) Begin with 1.00 L of 0.100 M HF, and titrate by adding 0.100 M KOH.

(c) Begin with 1.00 L of 0.100 M $BaCl_2$, and titrate by adding 0.100 M Na_2SO_4.

Which of the following graphs corresponds to which titration?

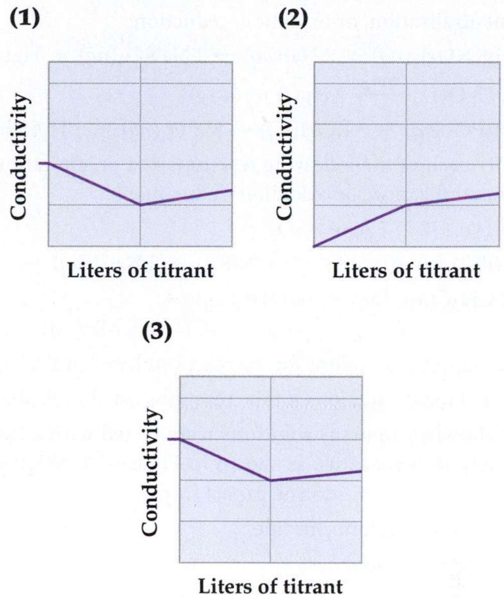

7.30 Based on the positions in the periodic table, which of the following reactions would you expect to occur?

(a) $\text{Red}^+ + \text{Green} \rightarrow \text{Red} + \text{Green}^+$

(b) $\text{Blue} + \text{Green}^+ \rightarrow \text{Blue}^+ + \text{Green}$

(c) $\text{Red} + \text{Blue}^+ \rightarrow \text{Red}^+ + \text{Blue}$

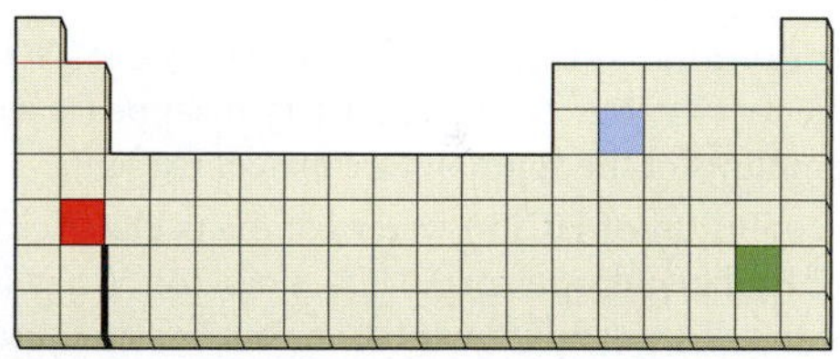

7.31 The following two redox reactions occur between aqueous cations and solid metals. Will a solution of green cations react with solid blue metal? Explain.

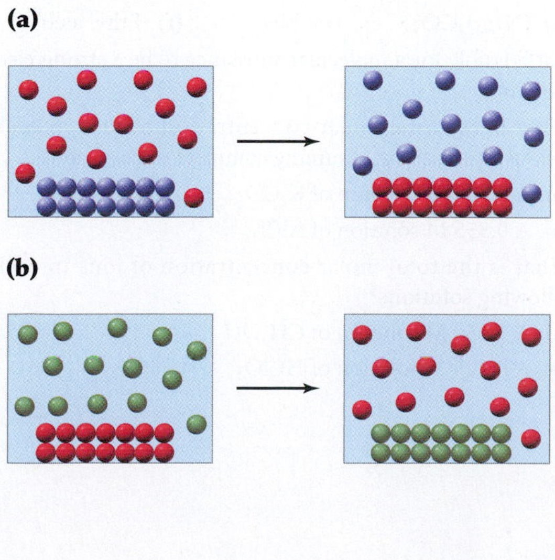

SECTION PROBLEMS

Aqueous Reactions, Net Ionic Equations, and Electrolytes (Sections 7.1–7.3)

7.32 Classify each of the following reactions as a precipitation, acid–base neutralization, or oxidation–reduction:

(a) $Hg(NO_3)_2(aq) + 2\ NaI(aq) \rightarrow 2\ NaNO_3(aq) + HgI_2(s)$

(b) $2\ HgO(s) \xrightarrow{\text{heat}} 2\ Hg(l) + O_2(g)$

(c) $H_3PO_4(aq) + 3\ KOH(aq) \rightarrow K_3PO_4(aq) + 3\ H_2O(l)$

7.33 Classify each of the following reactions as a precipitation, acid–base neutralization, or oxidation–reduction:

(a) $S_8(s) + 8\ O_2(g) \rightarrow 8\ SO_2(g)$

(b) $NiCl_2(aq) + Na_2S(aq) \rightarrow NiS(s) + 2\ NaCl(aq)$

(c) $2\ CH_3CO_2H(aq) + Ba(OH)_2(aq) \rightarrow (CH_3CO_2)_2Ba(aq) + 2\ H_2O(l)$

7.34 Write net ionic equations for the reactions listed in Problem 7.32.

7.35 Write net ionic equations for the reactions listed in Problem 7.33.

7.36 The following aqueous solutions were tested with a light bulb conductivity apparatus, as shown in Figure 7.1. What result—dark, dim, or bright—do you expect from each?

(a) 0.10 M potassium chloride

(b) 0.10 M methanol

(c) 0.10 M acetic acid

7.37 The following aqueous solutions were tested with a light bulb conductivity apparatus, as shown in Figure 7.1. What result—dark, dim, or bright—do you expect from each?

(a) 0.10 M hydrofluoric acid

(b) 0.10 M sodium chloride

(c) 0.10 M glucose ($C_6H_{12}O_6$)

7.38 Individual solutions of $Ba(OH)_2$ and H_2SO_4 both conduct electricity, but the conductivity disappears when equal molar amounts of the solutions are mixed. Explain.

7.39 A solution of HCl in water conducts electricity, but a solution of HCl in chloroform, $CHCl_3$, does not. What does this observation tell you about how HCl exists in water and how it exists in chloroform?

7.40 Classify each of the following substances as either a strong electrolyte, weak electrolyte, or nonelectrolyte:

(a) HBr (b) HF (c) $NaClO_4$

(d) $(NH_4)_2CO_3$ (e) NH_3 (f) Ethyl alcohol

7.41 Is it possible for a molecular substance to be a strong electrolyte? Explain.

7.42 What is the total molar concentration of ions in each of the following solutions, assuming complete dissociation?

(a) A 0.750 M solution of K_2CO_3

(b) A 0.355 M solution of $AlCl_3$

7.43 What is the total molar concentration of ions in each of the following solutions?

(a) A 1.250 M solution of CH_3OH

(b) A 0.225 M solution of $HClO_4$

Precipitation Reactions and Solubility Guidelines (Section 7.4)

7.44 Which of the following substances are likely to be soluble in water?

(a) Ag_2O (b) $Ba(NO_3)_2$

(c) $SnCO_3$ (d) Fe_2O_3

7.45 Which of the following substances are likely to be soluble in water?

(a) ZnS (b) $Au_2(CO_3)_3$

(c) $PbCl_2$ (d) MnO_2

7.46 Predict whether a precipitation reaction will occur when aqueous solutions of the following substances are mixed:

(a) $NaOH + HClO_4$ (b) $FeCl_2 + KOH$

(c) $(NH_4)_2SO_4 + NiCl_2$ (d) $CH_3CO_2Na + HCl$

7.47 Predict whether a precipitation reaction will occur when aqueous solutions of the following substances are mixed:

(a) $MnCl_2 + Na_2S$ (b) $HNO_3 + CuSO_4$

(c) $Hg(NO_3)_2 + Na_3PO_4$ (d) $Ba(NO_3)_2 + KOH$

7.48 Which of the following solutions will not form a precipitate when added to 0.10 M $BaCl_2$?

(a) 0.10 M $LiNO_3$ (b) 0.10 M K_2SO_4

(c) 0.10 M $AgNO_3$

7.49 Which of the following solutions will not form a precipitate when added to 0.10 M NaOH?

(a) 0.10 M $MgBr_2$ (b) 0.10 M NH_4Br

(c) 0.10 M $FeCl_2$

7.50 How would you prepare the following substances by a precipitation reaction?

(a) $PbSO_4$ (b) $Mg_3(PO_4)_2$

(c) $ZnCrO_4$

7.51 How would you prepare the following substances by a precipitation reaction?

(a) $Al(OH)_3$ (b) FeS

(c) $CoCO_3$

7.52 What is the mass and the identity of the precipitate that forms when 30.0 mL of 0.150 M HCl reacts with 25.0 mL of 0.200 M $AgNO_3$?

7.53 What is the mass and the identity of the precipitate that forms when 55.0 mL of 0.100 M $BaCl_2$ reacts with 40.0 mL of 0.150 M Na_2CO_3?

7.54 Assume that you have an aqueous mixture of $NaNO_3$ and $AgNO_3$. How could you use a precipitation reaction to separate the two metal ions?

7.55 Assume that you have an aqueous mixture of $BaCl_2$ and $CuCl_2$. How could you use a precipitation reaction to separate the two metal ions?

7.56 Assume that you have an aqueous solution of an unknown salt. Treatment of the solution with dilute NaOH, Na_2SO_4, and KCl produces no precipitate. Which of the following cations might the solution contain?

(a) Ag^+ (b) Cs^+

(c) Ba^{2+} (d) NH_4^+

7.57 Assume that you have an aqueous solution of an unknown salt. Treatment of the solution with dilute $BaCl_2$, $AgNO_3$, and $Cu(NO_3)_2$ produces no precipitate. Which of the following anions might the solution contain?

(a) Cl^- (b) NO_3^-
(c) OH^- (d) SO_4^{2-}

Acids, Bases, and Neutralization Reactions (Section 7.5)

7.58 Assume that you are given a solution of an unknown acid or base. How can you tell whether the unknown substance is acidic or basic?

7.59 Why do we use a double arrow ($\rightleftharpoons$) to show the dissociation of a weak acid or weak base in aqueous solution?

7.60 Write balanced ionic equations for the following reactions:

(a) Aqueous perchloric acid is neutralized by aqueous calcium hydroxide.
(b) Aqueous sodium hydroxide is neutralized by aqueous acetic acid.

7.61 Write balanced ionic equations for the following reactions:

(a) Aqueous hydrobromic acid is neutralized by aqueous calcium hydroxide.
(b) Aqueous barium hydroxide is neutralized by aqueous nitric acid.

7.62 Write balanced net ionic equations for the following reactions:

(a) $LiOH(aq) + HI(aq) \rightarrow ?$
(b) $HBr(aq) + Ca(OH)_2(aq) \rightarrow ?$

7.63 Write balanced net ionic equations for the following reactions. Note that $HClO_3$ is a strong acid.

(a) $Fe(OH)_3(s) + H_2SO_4(aq) \rightarrow ?$
(b) $HClO_3(aq) + NaOH(aq) \rightarrow ?$

7.64 If the following solutions are mixed, is the resulting solution acidic, basic, or neutral?

(a) 50.0 mL of 0.100 M HBr and 30.0 mL of 0.200 M KOH
(b) 100.0 mL of 0.0750 M HCl and 75.0 mL of 0.100 M $Ba(OH)_2$

7.65 If the following solutions are mixed, is the resulting solution acidic, basic, or neutral?

(a) 65.0 mL of 0.0500 M $HClO_4$ and 40.0 mL of 0.0750 M NaOH
(b) 125.0 mL of 0.100 M HNO_3 and 90.0mL of 0.0750 M $Ca(OH)_2$

7.66 How many milliliters of 1.00 M KOH must be added to neutralize the following solutions?

(a) A mixture of 0.240 M LiOH (25.0 mL) and 0.200 M HBr (75.0 mL)
(b) A mixture of 0.300 M HCl (45.0 mL) and 0.250 M NaOH (10.0 mL)

7.67 How many milliliters of 2.00 M HCl must be added to neutralize the following solutions?

(a) A mixture of 0.160 M HNO_3 (100.0 mL) and 0.100 M KOH (400.0 mL)
(b) A mixture of 0.120 M NaOH (350.0 mL) and 0.190 M HBr (150.0 mL)

Redox Reactions and Oxidation Numbers (Sections 7.6–7.8)

7.68 Where in the periodic table are the best reducing agents found? The best oxidizing agents?

7.69 Where in the periodic table are the most easily reduced elements found? The most easily oxidized?

7.70 In each of the following instances, tell whether the substance gains electrons or loses electrons in a redox reaction:

(a) An oxidizing agent
(b) A reducing agent
(c) A substance undergoing oxidation
(d) A substance undergoing reduction

7.71 Tell for each of the following substances whether the oxidation number increases or decreases in a redox reaction:

(a) An oxidizing agent
(b) A reducing agent
(c) A substance undergoing oxidation
(d) A substance undergoing reduction

7.72 Assign oxidation numbers to each element in the following compounds:

(a) NO_2 (b) SO_3
(c) $COCl_2$ (d) CH_2Cl_2
(e) $KClO_3$ (f) HNO_3

7.73 Assign oxidation numbers to each element in the following compounds:

(a) $VOCl_3$ (b) $CuSO_4$
(c) CH_2O (d) Mn_2O_7
(e) OsO_4 (f) H_2PtCl_6

7.74 Assign oxidation numbers to each element in the following ions:

(a) ClO_3^- (b) SO_3^{2-}
(c) $C_2O_4^{2-}$ (d) NO_2^-
(e) BrO^- (f) AsO_4^{3-}

7.75 Assign oxidation numbers to each element in the following ions:

(a) $Cr(OH)_4^-$ (b) $S_2O_3^{2-}$
(c) NO_3^- (d) MnO_4^{2-}
(e) HPO_4^{2-} (f) $V_2O_7^{4-}$

7.76 Which element is oxidized and which is reduced in each of the following reactions?

(a) $Ca(s) + Sn^{2+}(aq) \rightarrow Ca^{2+}(aq) + Sn(s)$
(b) $ICl(s) + H_2O(l) \rightarrow HCl(aq) + HOI(aq)$

7.77 Which element is oxidized and which is reduced in each of the following reactions?

(a) $Si(s) + 2\,Cl_2(g) \rightarrow SiCl_4(l)$
(b) $Cl_2(g) + 2\,NaBr(aq) \rightarrow Br_2(aq) + 2\,NaCl(aq)$

7.78 Use the activity series of metals (Table 7.5; page 255) to predict the outcome of each of the following reactions. If no reaction occurs, write N.R.

(a) $Na^+(aq) + Zn(s) \rightarrow ?$ (b) $HCl(aq) + Pt(s) \rightarrow ?$
(c) $Ag^+(aq) + Au(s) \rightarrow ?$ (d) $Au^{3+}(aq) + Ag(s) \rightarrow ?$

7.79 Neither strontium (Sr) nor antimony (Sb) is shown in the activity series of Table 7.5. Based on their positions in the periodic table, which would you expect to be the better reducing agent? Will the following reaction occur? Explain.

$$2\,Sb^{3+}(aq) + 3\,Sr(s) \longrightarrow 2\,Sb(s) + 3\,Sr^{2+}(aq)$$

7.80 **(a)** Use the following reactions to arrange the elements **A**, **B**, **C**, and **D** in order of their decreasing ability as reducing agents:

$A + B^+ \longrightarrow A^+ + B$ $\qquad C^+ + D \longrightarrow$ no reaction

$B + D^+ \longrightarrow B^+ + D$ $\qquad B + C^+ \longrightarrow B^+ + C$

(b) Which of the following reactions would you expect to occur according to the activity series you established in part **(a)**?

(1) $A^+ + C \longrightarrow A + C^+$

(2) $A^+ + D \longrightarrow A + D^+$

7.81 **(a)** Use the following reactions to arrange the elements **A**, **B**, **C**, and **D** in order of their decreasing ability as reducing agents:

$2\,A + B^{2+} \longrightarrow 2\,A^+ + B$ $\qquad B + D^{2+} \longrightarrow B^{2+} + D$

$A^+ + C \longrightarrow$ no reaction $\qquad 2C + B^{2+} \longrightarrow 2\,C^+ + B$

(b) Which of the following reactions would you expect to occur according to the act ivity series you established in part **(a)**?

(1) $2\,A^+ + D \longrightarrow 2\,A + D^{2+}$

(2) $D^{2+} + 2\,C \longrightarrow D + 2\,C^+$

Balancing Redox Reactions (Section 7.9)

7.82 Classify each of the following unbalanced half-reactions as either an oxidation or a reduction:

(a) $NO_3^-(aq) \rightarrow NO(g)$ $\qquad$ **(b)** $Zn(s) \rightarrow Zn^{2+}(aq)$

(c) $Ti^{3+}(aq) \rightarrow TiO_2(s)$ $\qquad$ **(d)** $Sn^{4+}(aq) \rightarrow Sn^{2+}(aq)$

7.83 Classify each of the following unbalanced half-reactions as either an oxidation or a reduction:

(a) $O_2(g) \rightarrow OH^-(aq)$

(b) $H_2O_2(aq) \rightarrow O_2(g)$

(c) $MnO_4^-(aq) \rightarrow MnO_4^{2-}(aq)$

(d) $CH_3OH(aq) \rightarrow CH_2O(aq)$

7.84 Balance the half-reactions in Problem 7.82, assuming that they occur in acidic solution.

7.85 Balance the half-reactions in Problem 7.83, assuming that they occur in basic solution.

7.86 Write unbalanced oxidation and reduction half-reactions for the following processes:

(a) $Te(s) + NO_3^-(aq) \rightarrow TeO_2(s) + NO(g)$

(b) $H_2O_2(aq) + Fe^{2+}(aq) \rightarrow Fe^{3+}(aq) + H_2O(l)$

7.87 Write unbalanced oxidation and reduction half-reactions for the following processes:

(a) $Mn(s) + NO_3^-(aq) \rightarrow Mn^{2+}(aq) + NO_2(g)$

(b) $Mn^{3+}(aq) \rightarrow MnO_2(s) + Mn^{2+}(aq)$

7.88 Balance the following half-reactions:

(a) (acidic) $Cr_2O_7^{2-}(aq) \rightarrow Cr^{3+}(aq)$

(b) (basic) $CrO_4^{2-}(aq) \rightarrow Cr(OH)_4^-(aq)$

(c) (basic) $Bi^{3+}(aq) \rightarrow BiO_3^-(aq)$

(d) (basic) $ClO^-(aq) \rightarrow Cl^-(aq)$

7.89 Balance the following half-reactions:

(a) (acidic) $VO^{2+}(aq) \rightarrow V^{3+}(aq)$

(b) (basic) $Ni(OH)_2(s) \rightarrow Ni_2O_3(s)$

(c) (acidic) $NO_3^-(aq) \rightarrow NO_2(aq)$

(d) (basic) $Br_2(aq) \rightarrow BrO_3^-(aq)$

7.90 Write balanced net ionic equations for the following reactions in basic solution:

(a) $MnO_4^-(aq) + IO_3^-(aq) \rightarrow MnO_2(s) + IO_4^-(aq)$

(b) $Cu(OH)_2(s) + N_2H_4(aq) \rightarrow Cu(s) + N_2(g)$

(c) $Fe(OH)_2(s) + CrO_4^{2-}(aq) \rightarrow Fe(OH)_3(s) + Cr(OH)_4^-(aq)$

(d) $H_2O_2(aq) + ClO_4^-(aq) \rightarrow ClO_2^-(aq) + O_2(g)$

7.91 Write balanced net ionic equations for the following reactions in basic solution:

(a) $S_2O_3^{2-}(aq) + I_2(aq) \rightarrow S_4O_6^{2-}(aq) + I^-(aq)$

(b) $Mn^{2+}(aq) + H_2O_2(aq) \rightarrow MnO_2(s) + H_2O(l)$

(c) $Zn(s) + NO_3^-(aq) \rightarrow NH_3(aq) + Zn(OH)_4^{2-}(aq)$

(d) $Bi(OH)_3(s) + Sn(OH)_3^-(aq) \rightarrow Bi(s) + Sn(OH)_6^{2-}(aq)$

7.92 Write balanced net ionic equations for the following reactions in acidic solution:

(a) $Zn(s) + VO^{2+}(aq) \rightarrow Zn^{2+}(aq) + V^{3+}(aq)$

(b) $Ag(s) + NO_3^-(aq) \rightarrow Ag^+(aq) + NO_2(g)$

(c) $Mg(s) + VO_4^{3-}(aq) \rightarrow Mg^{2+}(aq) + V^{2+}(aq)$

(d) $I^-(aq) + IO_3^-(aq) \rightarrow I_3^-(aq)$

7.93 Write balanced net ionic equations for the following reactions in acidic solution:

(a) $MnO_4^-(aq) + C_2H_5OH(aq) \rightarrow Mn^{2+}(aq) + CH_3CO_2H(aq)$

(b) $H_2O_2(aq) + Cr_2O_7^{2-}(aq) \rightarrow O_2(g) + Cr^{3+}(aq)$

(c) $Sn^{2+}(aq) + IO_4^-(aq) \rightarrow Sn^{4+}(aq) + I^-(aq)$

(d) $PbO_2(s) + Cl^-(aq) \rightarrow PbCl_2(s) + O_2(g)$

Redox Titrations (Section 7.10)

7.94 Iodine, I_2, reacts with aqueous thiosulfate ion in neutral solution according to the balanced equation

$$I_2(aq) + 2\,S_2O_3^{2-}(aq) \longrightarrow S_4O_6^{2-}(aq) + 2\,I^-(aq)$$

How many grams of I_2 are present in a solution if 35.20 mL of 0.150 M $Na_2S_2O_3$ solution is needed to titrate the I_2 solution?

7.95 How many milliliters of 0.250 M $Na_2S_2O_3$ solution is needed for complete reaction with 2.486 g of I_2 according to the equation in Problem 7.94?

7.96 Dichromate ion, $Cr_2O_7^{2-}$, reacts with aqueous iron(II) ion in acidic solution according to the balanced equation

$$Cr_2O_7^{2-}(aq) + 6\,Fe^{2+}(aq) + 14\,H^+(aq) \longrightarrow 2\,Cr^{3+}(aq) + 6\,Fe^{3+}(aq) + 7\,H_2O(l)$$

What is the concentration of Fe^{2+} if 46.99 mL of 0.2004 M $K_2Cr_2O_7$ is needed to titrate 50.00 mL of the Fe^{2+} solution?

7.97 A volume of 18.72 mL of 0.1500 M $K_2Cr_2O_7$ solution was required to titrate a sample of $FeSO_4$ according to the equation in Problem 7.96. What is the mass of the sample?

7.98 What is the molar concentration of As(III) in a solution if 22.35 mL of 0.100 M $KBrO_3$ is needed for complete reaction with 50.00 mL of the As(III) solution? The balanced equation is:

$$3\,H_3AsO_3(aq) + BrO_3^-(aq) \longrightarrow Br^-(aq) + 3\,H_3AsO_4(aq)$$

7.99 Standardized solutions of $KBrO_3$ are frequently used in redox titrations. The necessary solution can be made by dissolving $KBrO_3$ in water and then titrating it with an As(III) solution. What is the molar concentration of a $KBrO_3$ solution if 28.55 mL of the solution is needed to titrate 1.550 g of As_2O_3? See Problem 7.98 for the balanced equation. (As_2O_3 dissolves in aqueous acid solution to yield H_3AsO_3: $As_2O_3 + 3\,H_2O \rightarrow 2\,H_3AsO_3$.)

7.100 The metal content of iron in ores can be determined by a redox procedure in which the sample is first oxidized with Br_2 to convert all the iron to Fe^{3+} and then titrated with Sn^{2+} to reduce the Fe^{3+} to Fe^{2+}. The balanced equation is:

$$2\,Fe^{3+}(aq) + Sn^{2+}(aq) \longrightarrow 2\,Fe^{2+}(aq) + Sn^{4+}(aq)$$

What is the mass percent Fe in a 0.1875 g sample of ore if 13.28 mL of a 0.1015 M Sn^{2+} solution is needed to titrate the Fe^{3+}?

7.101 The concentration of the Sn^{2+} solution used in Problem 7.100 can be found by letting it react with a known amount of Fe^{3+}. What is the molar concentration of an Sn^{2+} solution if 23.84 mL is required for complete reaction with 1.4855 g of Fe_2O_3?

7.102 Alcohol levels in blood can be determined by a redox reaction with potassium dichromate according to the balanced equation

$$C_2H_5OH(aq) + 2\,Cr_2O_7^{2-}(aq) + 16\,H^+(aq) \longrightarrow 2\,CO_2(g) + 4\,Cr^{3+}(aq) + 11\,H_2O(l)$$

What is the blood alcohol level in mass percent if 8.76 mL of 0.049 88 M $K_2Cr_2O_7$ is required for complete reaction with a 10.002 g sample of blood?

7.103 Calcium levels in blood can be determined by adding oxalate ion to precipitate calcium oxalate, CaC_2O_4, followed by dissolving the precipitate in aqueous acid and titrating the resulting oxalic acid ($H_2C_2O_4$) with $KMnO_4$:

$$5\,H_2C_2O_4(aq) + 2\,MnO_4^-(aq) + 6\,H^+(aq) \longrightarrow 10\,CO_2(g) + 2\,Mn^{2+}(aq) + 8\,H_2O(l)$$

How many milligrams of Ca^{2+} are present in 10.0 mL of blood if 21.08 mL of 0.000 988 M $KMnO_4$ solution is needed for the titration?

CHAPTER PROBLEMS

7.104 Balance the equations for the following reactions in basic solution:

(a) $[Fe(CN)_6]^{3-}(aq) + N_2H_4(aq) \rightarrow [Fe(CN)_6]^{4-}(aq) + N_2(g)$

(b) $SeO_3^{2-}(aq) + Cl_2(g) \rightarrow SeO_4^{2-}(aq) + Cl^-(aq)$

(c) $Co^{2+}(aq) + HO_2^-(aq) \rightarrow Co(OH)_3(s)$

7.105 An alternative procedure to that given in Problem 7.100 for determining the amount of iron in a sample is to convert the iron to Fe^{2+} and then titrate it with a solution of $Ce(NH_4)_2(NO_3)_6$:

$$Fe^{2+}(aq) + Ce^{4+}(aq) \longrightarrow Fe^{3+}(aq) + Ce^{3+}(aq)$$

What is the mass percent of iron in a sample if 1.2284 g of the sample requires 57.91 mL of 0.1018 M $Ce(NH_4)_2(NO_3)_6$ for complete reaction?

7.106 Assign oxidation numbers to each atom in the following substances:

(a) Ethane, C_2H_6, a constituent of natural gas

(b) Borax, $Na_2B_4O_7$, a mineral used in laundry detergents

(c) $Mg_2Si_2O_6$, a silicate mineral

7.107 Balance the equations for the following reactions in acidic solution:

(a) $PbO_2(s) + Mn^{2+}(aq) \rightarrow Pb^{2+}(aq) + MnO_4^-(aq)$

(b) $As_2O_3(s) + NO_3^-(aq) \rightarrow H_3AsO_4(aq) + HNO_2(aq)$

(c) $Br_2(aq) + SO_2(g) \rightarrow Br^-(aq) + HSO_4^-(aq)$

(d) $NO_2^-(aq) + I^-(aq) \rightarrow I_2(s) + NO(g)$

7.108 **(a)** Use the following reactions to arrange the elements **A**, **B**, **C**, and **D** in order of their decreasing ability as reducing agents:

$C + B^+ \longrightarrow C^+ + B$ $\qquad$ $A^+ + D \longrightarrow$ No reaction

$C^+ + A \longrightarrow$ No reaction $\qquad$ $D + B^+ \longrightarrow D^+ + B$

(b) Which of the following reactions would you expect to occur according to the activity series you established in part **(a)**?

(1) $A^+ + C \longrightarrow A + C^+$

(2) $A^+ + B \longrightarrow A + B^+$

7.109 Some metals occur naturally in their elemental state while others occur as compounds in ores. Gold, for instance, is found as the free metal; mercury is obtained by heating mercury(II) sulfide ore in oxygen; and zinc is obtained by heating zinc(II) oxide ore with coke (carbon). Judging from their positions in the activity series, which of the metals silver, platinum, and chromium would probably be obtained by

(a) finding it in its elemental state?

(b) heating its sulfide with oxygen?

(c) heating its oxide with coke?

7.110 A sample weighing 14.98 g and containing a small amount of copper was treated to give a solution containing aqueous Cu^{2+} ions. Sodium iodide was then added to yield solid copper(I) iodide plus I_3^- ion, and the I_3^- was titrated with thiosulfate, $S_2O_3^{2-}$. The titration required 10.49 mL of 0.100 M $Na_2S_2O_3$ for complete reaction. What is the mass percent copper in the sample? The balanced equations are

$$2\,Cu^{2+}(aq) + 5\,I^-(aq) \longrightarrow 2\,CuI(s) + I_3^-(aq)$$

$$I_3^-(aq) + 2\,S_2O_3^{2-}(aq) \longrightarrow 3\,I^-(aq) + S_4O_6^{2-}(aq)$$

7.111 The solubility of an ionic compound can be described quantitatively by a value called the *solubility product constant*, K_{sp}. For the general solubility process $A_aB_b \rightleftharpoons a\,A^{n+} + b\,B^{m-}$, $K_{sp} = [A^{n+}]^a[B^{m-}]^b$. The brackets refer to concentrations in moles per liter.

(a) Write the expression for the solubility product constant of Ag_2CrO_4.

(b) If $K_{sp} = 1.1 \times 10^{-12}$ for Ag_2CrO_4, what are the molar concentrations of Ag^+ and CrO_4^{2-} in solution?

7.112 Write the expression for the solubility product constant of MgF_2 (see Problem 7.111). If $[Mg^{2+}] = 2.6 \times 10^{-4}$ mol/L in a solution, what is the value of K_{sp}?

7.113 Succinic acid, an intermediate in the metabolism of food molecules, has molecular weight = 118.1. When 1.926 g of succinic acid was dissolved in water and titrated, 65.20 mL of 0.5000 M NaOH solution was required to neutralize the acid. How many acidic hydrogens are there in a molecule of succinic acid?

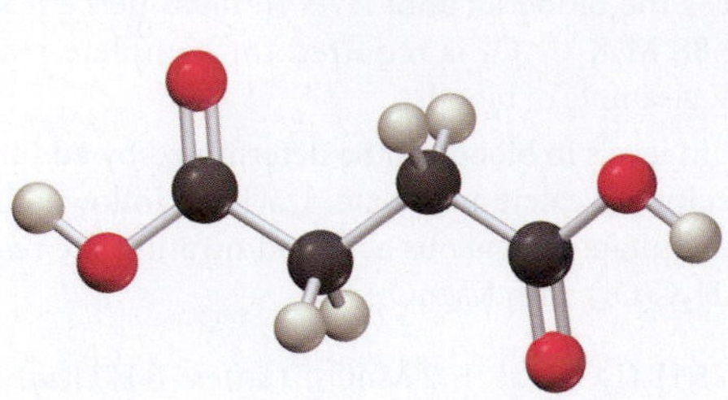

Succinic acid

7.114 How could you use a precipitation reaction to separate each of the following pairs of cations? Write the formula for each reactant you would add, and write a balanced net ionic equation for each reaction.

(a) K^+ and Hg_2^{2+} **(b)** Pb^{2+} and Ni^{2+}

(c) Ca^{2+} and NH_4^+ **(d)** Fe^{2+} and Ba^{2+}

7.115 How could you use a precipitation reaction to separate each of the following pairs of anions? Write the formula for each reactant you would add, and write a balanced net ionic equation for each reaction.

(a) Cl^- and NO_3^- **(b)** S^{2-} and SO_4^{2-}

(c) SO_4^{2-} and CO_3^{2-} **(d)** OH^- and ClO_4^-

7.116 Write a balanced net ionic equation for each of the following reactions:

(a) $Mn(OH)_2(s) + H_2O_2(aq) \xrightarrow{\text{Base}} Mn(OH)_3(s)$

(b) $MnO_4^{2-}(aq) \xrightarrow{\text{Acid}} MnO_2(s) + MnO_4^-(aq)$

(c) $IO_3^-(aq) + I^-(aq) \xrightarrow{\text{Acid}} I_3^-(aq)$

(d) $P(s) + PO_4^{3-}(aq) \xrightarrow{\text{Base}} HPO_3^{2-}(aq)$

7.117 A 100.0 mL solution containing aqueous HCl and HBr was titrated with 0.1235 M NaOH. The volume of base required to neutralize the acid was 47.14 mL. Aqueous $AgNO_3$ was then added to precipitate the Cl^- and Br^- ions as AgCl and AgBr. The mass of the silver halides obtained was 0.9974 g. What are the molarities of the HCl and HBr in the original solution?

7.118 Write balanced net ionic equations for the following reactions in acidic solution:

(a) $S_4O_6^{2-}(aq) + Al(s) \rightarrow H_2S(aq) + Al^{3+}(aq)$

(b) $S_2O_3^{2-}(aq) + Cr_2O_7^{2-}(aq) \rightarrow S_4O_6^{2-}(aq) + Cr^{3+}(aq)$

(c) $ClO_3^-(aq) + As_2S_3(s) \rightarrow Cl^-(aq) + H_2AsO_4^-(aq) + HSO_4^-(aq)$

(d) $IO_3^-(aq) + Re(s) \rightarrow ReO_4^-(aq) + I^-(aq)$

(e) $HSO_4^-(aq) + As_4(s) + Pb_3O_4(s) \rightarrow PbSO_4(s) + H_2AsO_4^-(aq)$

(f) $HNO_2(aq) \rightarrow NO_3^-(aq) + NO(g)$

7.119 Write balanced net ionic equations for the following reactions in basic solution:

(a) $C_4H_4O_6^{2-}(aq) + ClO_3^-(aq) \rightarrow CO_3^{2-}(aq) + Cl^-(aq)$

(b) $Al(s) + BiONO_3(s) \rightarrow Bi(s) + NH_3(aq) + Al(OH)_4^-(aq)$

(c) $H_2O_2(aq) + Cl_2O_7(aq) \rightarrow ClO_2^-(aq) + O_2(g)$

(d) $Tl_2O_3(s) + NH_2OH(aq) \rightarrow TlOH(s) + N_2(g)$

(e) $Cu(NH_3)_4^{2+}(aq) + S_2O_4^{2-}(aq) \rightarrow SO_3^{2-}(aq) + Cu(s) + NH_3(aq)$

(f) $Mn(OH)_2(s) + MnO_4^-(aq) \rightarrow MnO_2(s)$

7.120 A mixture of CuO and Cu_2O with a mass of 10.50 g is reduced to give 8.66 g of pure Cu metal. What are the amounts in grams of CuO and Cu_2O in the original mixture?

7.121 When 75.0 mL of a 0.100 M lead(II) nitrate solution is mixed with 100.0 mL of a 0.190 M potassium iodide solution, a yellow-orange precipitate of lead(II) iodide is formed.

(a) What mass in grams of lead(II) iodide is formed, assuming the reaction goes to completion?

(b) What is the molarity of each of the ions Pb^{2+}, K^+, NO_3^-, and I^- in the resulting solution?

7.122 A sample of metal (**M**) reacted with both steam and aqueous HCl to release H_2, but did not react with water at room temperature. When 1.000 g of the metal was burned in oxygen, it formed 1.890 g of a metal oxide, M_2O_3. What is the identity of the metal?

7.123 An unknown metal (**M**) was found not to react with either water or steam, but its reactivity with aqueous acid was not investigated. When a 1.000 g sample of the metal was burned in oxygen and the resulting metal oxide converted to a metal sulfide, 1.504 g of sulfide was obtained. What is the identity of the metal?

7.124 Hydrogen peroxide can be either an oxidizing or reducing agent depending on the circumstances. Write balanced net ionic equations for each of the following reactions, and determine in each case whether hydrogen peroxide is oxidized (acts as a reducing agent) or reduced (acts as an oxidizing agent).

(a) $HCl(aq) + H_2O_2(aq) \rightarrow Cl_2(aq) + H_2O(l)$

(b) (acidic) $MnO_4^-(aq) + H_2O_2(aq) \rightarrow Mn^{2+}(aq) + O_2(g)$

(c) $H_2O_2(aq) + Cl_2(aq) \rightarrow HCl(aq) + O_2(g)$

7.125 A mixture of acetic acid (CH_3CO_2H; monoprotic) and oxalic acid ($H_2C_2O_4$; diprotic) requires 27.15 mL of 0.100 M NaOH to neutralize it. When an identical amount of the mixture is titrated, 15.05 mL of 0.0247 M $KMnO_4$ is needed for complete reaction. What is the mass percent of each acid in the mixture? (Acetic acid does not react with MnO_4^-. The equation for the reaction of oxalic acid with MnO_4^- was given in Problem 7.103.)

7.126 Iron content in ores can be determined by a redox procedure in which the sample is first reduced with Sn^{2+}, as in Problem 7.100, and then titrated with $KMnO_4$ to oxidize the Fe^{2+} to Fe^{3+}. The balanced equation is

$$MnO_4^-(aq) + 5\,Fe^{2+}(aq) + 8\,H^+(aq) \longrightarrow Mn^{2+}(aq) + 5\,Fe^{3+}(aq) + 4\,H_2O(l)$$

What is the mass percent Fe in a 2.368 g sample if 48.39 mL of a 0.1116 M $KMnO_4$ solution is needed to titrate the Fe^{3+}?

7.127 A mixture of $FeCl_2$ and NaCl is dissolved in water, and addition of aqueous silver nitrate then yields 7.0149 g of a precipitate. When an identical amount of the mixture is titrated with MnO_4^-, 14.28 mL of 0.198 M $KMnO_4$ is needed for complete reaction. What are the mass percents of the two compounds in the mixture? (Na^+ and Cl^- do not react with MnO_4^-. The equation for the reaction of Fe^{2+} with MnO_4^- was given in Problem 7.126.)

MULTICONCEPT PROBLEMS

7.128 Assume that you dissolve 10.0 g of a mixture of NaOH and $Ba(OH)_2$ in 250.0 mL of water and titrate with 1.50 M hydrochloric acid. The titration is complete after 108.9 mL of the acid has been added. What is the mass in grams of each substance in the mixture?

7.129 The following three solutions are mixed: 100.0 mL of 0.100 M Na_2SO_4, 50.0 mL of 0.300 M $ZnCl_2$, and 100.0 mL of 0.200 M $Ba(CN)_2$.

(a) What ionic compounds will precipitate out of solution?

(b) What is the molarity of each ion remaining in the solution assuming complete precipitation of all insoluble compounds?

7.130 A 250.0 g sample of a white solid is known to be a mixture of KNO_3, $BaCl_2$, and NaCl. When 100.0 g of this mixture is dissolved in water and allowed to react with excess H_2SO_4, 67.3 g of a white precipitate is collected. When the remaining 150.0 g of the mixture is dissolved in water and allowed to react with excess $AgNO_3$, 197.6 g of a second precipitate is collected.

(a) What are the formulas of the two precipitates?

(b) What is the mass of each substance in the original 250 g mixture?

7.131 Four solutions are prepared and mixed in the following order:

(1) Start with 100.0 mL of 0.100 M $BaCl_2$

(2) Add 50.0 mL of 0.100 M $AgNO_3$

(3) Add 50.0 mL of 0.100 M H_2SO_4

(4) Add 250.0 mL of 0.100 M NH_3

Write an equation for any reaction that occurs after each step, and calculate the concentrations of Ba^{2+}, Cl^-, NO_3^-, NH_3, and NH_4^+ in the final solution, assuming that all reactions go to completion.

7.132 To 100.0 mL of a solution that contains 0.120 M $Cr(NO_3)_2$ and 0.500 M HNO_3 is added 20.0 mL of 0.250 M $K_2Cr_2O_7$. The dichromate and chromium(II) ions react to give chromium(III) ions.

(a) Write a balanced net ionic equation for the reaction.

(b) Calculate the concentrations of all ions in the solution after reaction. Check your concentrations to make sure that the solution is electrically neutral.

7.133 Sodium nitrite, $NaNO_2$, is frequently added to processed meats as a preservative. The amount of nitrite ion in a sample can be determined by acidifying to form nitrous acid (HNO_2), letting the nitrous acid react with an excess of iodide ion, and then titrating the I_3^- ion that results with thiosulfate solution in the presence of a starch indicator. The unbalanced equations are

(1) $HNO_2 + I^- \longrightarrow NO + I_3^-$ (in acidic solution)

(2) $I_3^- + S_2O_3^{2-} \longrightarrow I^- + S_4O_6^{2-}$

(a) Balance the two redox equations.

(b) When a nitrite-containing sample with a mass of 2.935 g was analyzed, 18.77 mL of 0.1500 M $Na_2S_2O_3$ solution was needed for the reaction. What is the mass percent of NO_2^- ion in the sample?

7.134 Brass is an approximately 4:1 alloy of copper and zinc, along with small amounts of tin, lead, and iron. The mass percents of copper and zinc can be determined by a procedure that begins with dissolving the brass in hot nitric acid. The resulting solution of Cu^{2+} and Zn^{2+} ions is then treated with aqueous ammonia to lower its acidity, followed by addition of sodium thiocyanate (NaSCN) and sulfurous acid (H_2SO_3) to precipitate copper(I) thiocyanate (CuSCN). The solid CuSCN is collected, dissolved in aqueous acid, and treated with potassium iodate (KIO_3) to give iodine, which is then titrated with aqueous sodium thiosulfate ($Na_2S_2O_3$). The filtrate remaining after CuSCN has been removed is neutralized by addition of aqueous ammonia, and a solution of diammonium hydrogen phosphate [$(NH_4)_2HPO_4$] is added to yield a precipitate of zinc ammonium phosphate ($ZnNH_4PO_4$). Heating the precipitate to 900 °C converts it to zinc pyrophosphate ($Zn_2P_2O_7$), which is weighed. The equations are

(1) $Cu(s) + NO_3^-(aq) \longrightarrow Cu^{2+}(aq) + NO(g)$ (in acid)

(2) $Cu^{2+}(aq) + SCN^-(aq) + HSO_3^-(aq) \longrightarrow CuSCN(s) + HSO_4^-(aq)$ (in acid)

(3) $Cu^+(aq) + IO_3^-(aq) \longrightarrow Cu^{2+}(aq) + I_2(aq)$ (in acid)

(4) $I_2(aq) + S_2O_3^{2-}(aq) \longrightarrow I^-(aq) + S_4O_6^{2-}(aq)$ (in acid)

(5) $ZnNH_4PO_4(s) \longrightarrow Zn_2P_2O_7(s) + H_2O(g) + NH_3(g)$

(a) Balance all equations.

(b) When a brass sample with a mass of 0.544 g was subjected to the preceding analysis, 10.82 mL of 0.1220 M sodium thiosulfate was required for the reaction with iodine. What is the mass percent copper in the brass?

(c) The brass sample in part **(b)** yielded 0.246 g of $Zn_2P_2O_7$. What is the mass percent zinc in the brass?

7.135 A certain metal sulfide, MS_n (where *n* is a small integer), is widely used as a high-temperature lubricant. The substance is prepared by reaction of the metal pentachloride (MCl_5) with sodium sulfide (Na_2S). Heating the metal sulfide to 700 °C in air gives the metal trioxide (MO_3) and sulfur dioxide (SO_2), which reacts with Fe^{3+} ion under aqueous acidic conditions to give sulfate ion (SO_4^{2-}). Addition of aqueous $BaCl_2$ then forms a precipitate of $BaSO_4$. The unbalanced equations are:

(1) $MCl_5(s) + Na_2S(s) \longrightarrow MS_n(s) + S(l) + NaCl(s)$

(2) $MS_n(s) + O_2(g) \longrightarrow MO_3(s) + SO_2(g)$

(3) $SO_2(g) + Fe^{3+}(aq) \longrightarrow Fe^{2+}(aq) + SO_4^{2-}(aq)$ (in acid)

(4) $SO_4^{2-}(aq) + Ba^{2+}(aq) \longrightarrow BaSO_4(s)$

Assume that you begin with 4.61 g of MCl_5 and that reaction (1) proceeds in 91.3% yield. After oxidation of the MS_n product, oxidation of SO_2, and precipitation of sulfate ion, 7.19 g of $BaSO_4(s)$ is obtained.

(a) How many moles of sulfur are present in the MS_n sample?

(b) Assuming several possible values for $n\,(n = 1, 2, 3\ldots)$, what is the atomic weight of **M** in each case?

(c) What is the likely identity of the metal **M**, and what is the formula of the metal sulfide MS_n?

(d) Balance all equations.

7.136 On heating a 0.200 g sample of a certain semimetal **M** in air, the corresponding oxide M_2O_3 was obtained. When the oxide was dissolved in aqueous acid and titrated with $KMnO_4$, 10.7 mL of 0.100 M MnO_4^- was required for complete reaction. The unbalanced equation is

$$H_3MO_3(aq) + MnO_4^-(aq) \longrightarrow H_3MO_4(aq) + Mn^{2+}(aq) \text{ (in acid)}$$

(a) Balance the equation.

(b) How many moles of oxide were formed, and how many moles of semimetal were in the initial 0.200 g sample?

(c) What is the identity of the semimetal **M**?

CHAPTER

8 Thermochemistry: Chemical Energy

Vegetable oil from the bright yellow rapeseed plant is a leading candidate for large-scale production of biodiesel fuel, as we'll see in the *FYI* at the end of this chapter.

Why do chemical reactions occur? Stated simply, the answer involves *stability*. For a reaction to take place spontaneously, the products of the reaction must be more stable than the reactants.

But what is "stability," and what does it mean to say that one substance is more stable than another? The key factor in determining the stability of a substance is *energy*. Less stable substances have higher energy and are converted during chemical reactions into more stable substances with lower energy. But what, in turn, does it mean for a substance to have higher or lower energy? We'll explore some different forms of energy in this chapter and look at the subject of **thermochemistry**, the absorption or release of heat that accompanies chemical reactions.

8.1 ENERGY AND ITS CONSERVATION

The word *energy*, though familiar to everyone, is surprisingly hard to define in simple, nontechnical terms. A good working definition, however, is to say that **energy** is the capacity to do work or supply heat. The water falling over a dam, for instance, contains energy that can be used to turn a turbine and generate electricity. A tank of propane gas contains energy that, when released in the chemical process of combustion, can heat a house or barbecue a hamburger.

Energy is classified as either *kinetic* or *potential*. **Kinetic energy (E_K)** is the energy of motion. The amount of kinetic energy in a moving object with mass *m* and velocity *v* is given by the equation

▶ **Kinetic energy** $E_K = \frac{1}{2}mv^2$ where m = mass and v = velocity

CONTENTS

The derived SI unit for energy, $(\text{kg} \cdot \text{m}^2)/\text{s}^2$, follows from the expression for kinetic energy, $E_K = (1/2)mv^2$, and is given the name **joule (J)**.

$$1\ (\text{kg} \cdot \text{m}^2)/\text{s}^2 = 1\ \text{J}$$

Potential energy (E_P), by contrast, is stored energy—perhaps stored in an object because of its height or in a molecule because of chemical reactions it can undergo. The water sitting in a reservoir behind a dam contains potential energy because of its height above the stream at the bottom of the dam. When the water is allowed to fall, its potential energy is converted into kinetic energy. Propane and other substances used as fuels contain potential energy because they can undergo a reaction with oxygen—a *combustion*—that releases heat.

Let's pursue the relationship between potential energy and kinetic energy a bit further. According to the **conservation of energy law**, energy can be neither created nor destroyed; it can only be converted from one form into another.

> ▶ **Conservation of energy law** Energy cannot be created or destroyed; it can only be converted from one form into another.

To take an example, think about a hydroelectric dam. The water sitting motionless in the reservoir behind the dam has potential energy because of its height above the outlet stream, but it has no kinetic energy because it isn't moving ($v = 0$). When the water falls through the penstocks of the dam, however, its height and potential energy decrease while its velocity and kinetic energy increase. The moving water then spins the turbine of a generator, converting its kinetic energy into electrical energy **(Figure 8.1)**.

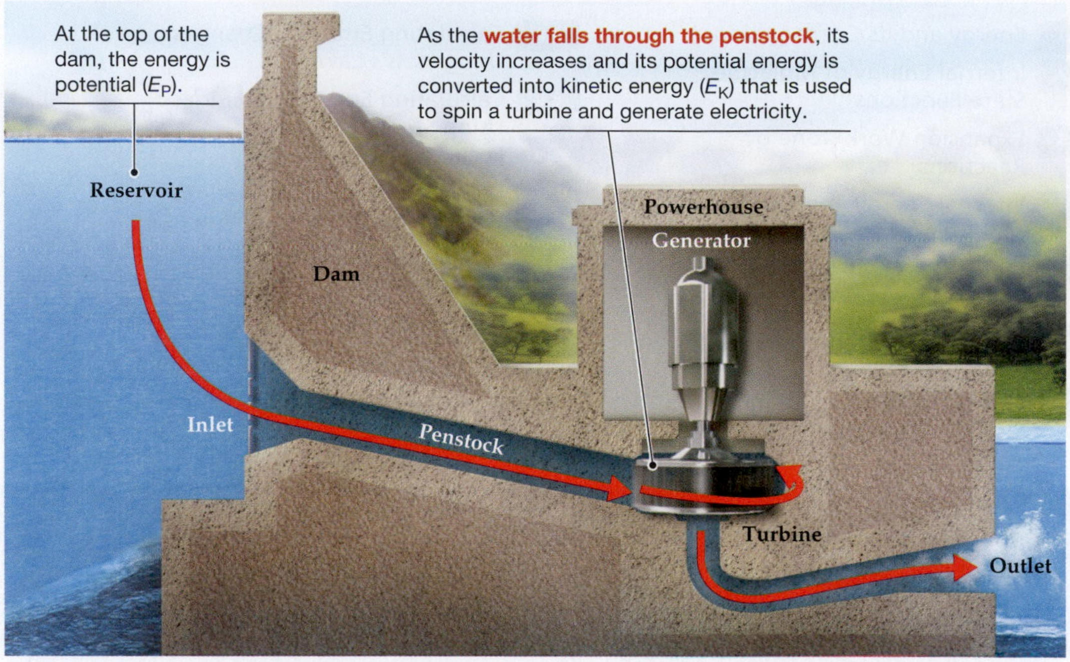

FIGURE 8.1

Conservation of energy. The total amount of energy contained by the water in the reservoir is constant.

The conversion of the kinetic energy in falling water into electrical energy illustrates several other important points. One point is that energy has many forms. Thermal energy contained in hot objects, for example, might seem very different from the kinetic energy of falling water, yet it's really quite similar. *Thermal energy* is just the kinetic energy of molecular motion, which we measure by finding the **temperature** of an object. An object has a low temperature and we perceive it as cold if its atoms or molecules are moving slowly. Conversely, an object has a high temperature and we perceive it as hot if its atoms or molecules are moving rapidly and are colliding forcefully with a thermometer or other measuring device.

Heat, in turn, is the amount of thermal energy transferred from one object to another as the result of a temperature difference between the two. Rapidly moving molecules in a hotter object collide with more slowly moving molecules in a colder object, transferring kinetic energy and causing the slower moving molecules to speed up.

Chemical energy is another kind of energy that seems different from that of the water in a reservoir, yet again is really quite similar. Chemical energy is a kind of potential energy in which chemical bonds act as the storage medium. Just as the water in a reservoir releases its potential energy when it falls to a more stable position, chemicals can release their potential energy in the form of heat or light when they undergo reactions and form more stable products. We'll explore this topic shortly.

A second point illustrated by falling water involves the conservation of energy law. To keep track of all the energy involved, it's necessary to take into account the entire chain of events that ensue from the falling water: the sound of the crashing water, the heating of the rocks at the bottom of the dam, the driving of turbines and electrical generators, the transmission of electrical power, the appliances powered by the electricity, and so on. Carrying the process to its logical extreme, it's necessary to take the entire universe into account when keeping track of all the energy in the water because the energy lost in one form *always* shows up elsewhere in another form. So important is the conservation of energy law that it's also known as the **first law of thermodynamics**.

▶ **First law of thermodynamics** Energy cannot be created or destroyed; it can only be converted from one form into another.

8.2 INTERNAL ENERGY OF MOLECULES: STATE FUNCTIONS

When keeping track of the energy changes in a chemical reaction, it's helpful to think of the reaction as being isolated from the world around it. The substances we focus on in an experiment—the starting reactants and the final products—are collectively called the *system*, while everything else—the reaction flask, the solvent, the room, the building, and so on—is called the *surroundings*. If the system could be truly isolated from its surroundings so that no energy transfer could occur between them, then the total **internal energy** (***E***) of the system, defined as the sum of all the kinetic and potential energies for every molecule or ion in the system, would be conserved and remain constant throughout the reaction. In fact, this assertion is just a restatement of the first law of thermodynamics:

▶ **First law of thermodynamics (restated)** The total internal energy E of an isolated system is constant.

In practice, of course, it's not possible to truly isolate a chemical reaction from its surroundings. In any real situation, the chemicals are in physical contact with the walls of a flask or container and the container itself is in contact with the surrounding air or laboratory bench. What's important, however, is not that the system be isolated but that we be able to measure accurately any energy that enters the system from the surroundings or leaves the system and flows to the surroundings (**Figure 8.2**). That is, we must be able to measure any *change* in the internal energy of the system. The energy change, symbolized ΔE, represents the difference in internal energy between the final state of the system after reaction and the initial state of the system before reaction:

$$\Delta E = E_{\text{final}} - E_{\text{initial}}$$

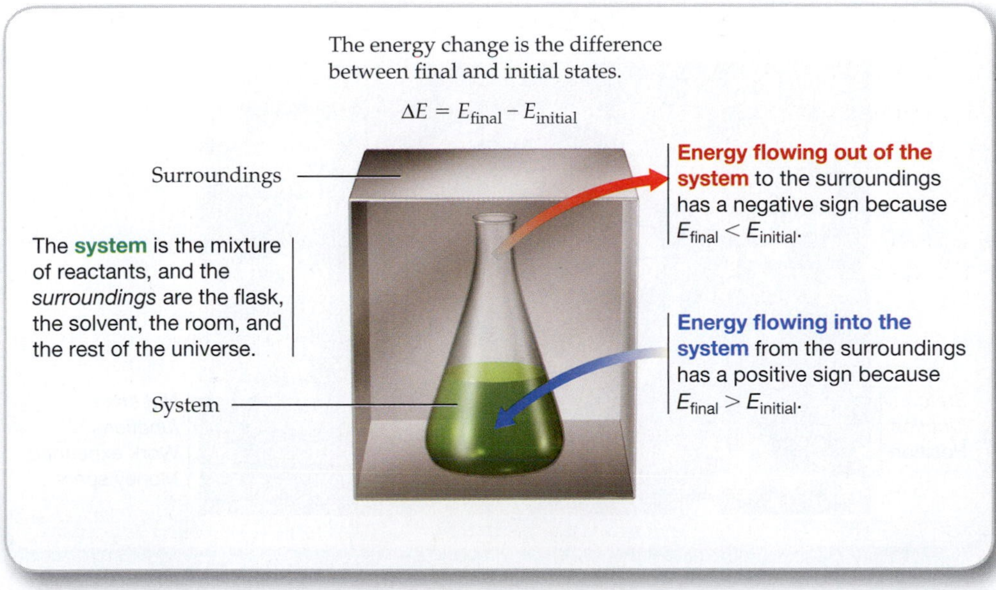

FIGURE 8.2

Energy changes in a chemical reaction.

By convention, energy changes are measured from the point of view of the system. Any energy that flows *from* the system *to* the surroundings has a negative sign because the system has lost it (that is, E_{final} is smaller than E_{initial}). Any energy that flows *to* the system *from* the surroundings has a positive sign because the system has gained it (E_{final} is larger than E_{initial}). If, for instance, we were to burn 1.00 mol of methane (CH_4) in the presence of 2.00 mol of oxygen, 802 kJ would be released as heat and transferred from the system to the surroundings. The system now has 802 kJ less energy than it started with, so $\Delta E = -802$ kJ. This energy flow can be detected and measured by placing the reaction vessel in a water bath and noting the temperature rise of the bath during the reaction.

$$CH_4(g) + 2\,O_2(g) \longrightarrow CO_2(g) + 2\,H_2O(g) + \textbf{802 kJ energy} \qquad \Delta E = -802 \text{ kJ}$$

The methane combustion experiment tells us that the products of the reaction, $CO_2(g)$ and $2\,H_2O(g)$, have 802 kJ less internal energy than the reactants, $CH_4(g)$ and $2\,O_2(g)$, even though we don't know the exact values at either the beginning ($E_{initial}$) or end (E_{final}) of the reaction. Note that the value $\Delta E = -802$ kJ for the reaction refers to the energy released when reactants are converted to products *in the molar amounts represented by coefficients in the balanced equation.* That is, 802 kJ is released when 1 mol of gaseous methane reacts with 2 mol of gaseous oxygen to give 1 mol of gaseous carbon dioxide and 2 mol of gaseous water vapor.

The internal energy of a system depends on many things: chemical identity, sample size, temperature, pressure, physical state (gas, liquid, or solid), and so forth. What the internal energy does not depend on is the system's past history. It doesn't matter what the system's temperature or physical state was an hour ago, and it doesn't matter how the chemicals were obtained. All that matters is the present condition of the system. Thus, internal energy is said to be a **state function**, meaning that its value depends only on the present state of the system. Pressure, volume, and temperature are other examples of state functions, but work and heat are not.

▶ **State function** A function or property whose value depends only on the present state, or condition, of the system, not on the path used to arrive at that state.

We can illustrate the idea of a state function by imagining a cross-country trip, say from the Artichoke Capitol of the World (Castroville, California), to the Hub of the Universe (Boston, Massachusetts). You are the system, and your position is a state function because how you got to wherever you are is irrelevant. Because your position is a state function, the *change* in your position after you complete your travel (Castroville and Boston are about 2720 miles apart) is independent of the path you take, whether through North Dakota or Louisiana (**Figure 8.3**).

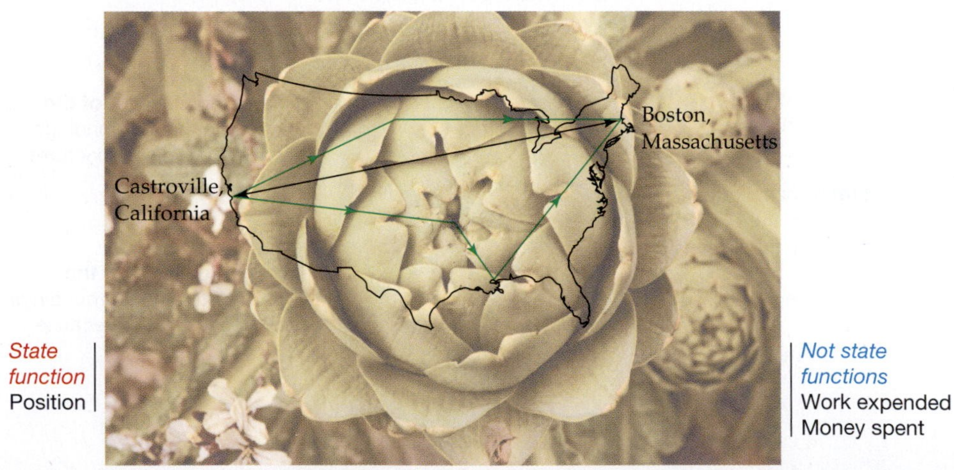

FIGURE 8.3

State functions. Because your position is a state function, the change in your position on going from Castroville, California, to Boston, Massachusetts, is independent of the path you take.

The cross-country trip shown in Figure 8.3 illustrates an important point about state functions: their reversibility. Imagine that, after traveling from Castroville to Boston, you turn around and go back. Because your final position is now identical to your initial position, the change in your position is zero. In fact, the overall change in *any* state function is zero when the system returns to its original condition. For a nonstate function, however, the overall change is not zero when the system returns to its original condition. Any work you do in making the trip, for instance, is not recovered when you return to your initial position, and any money or time you spend does not reappear.

PROBLEM 8.1 Which of the following are state functions, and which are not?
(a) The temperature of an ice cube
(b) The volume of an aerosol can
(c) The amount of time required for Paula Radcliffe to run her world-record marathon: 2:15:25

8.3 EXPANSION WORK DONE BY REACTIONS

Just as energy comes in many forms, so too does *work*. In physics, **work (w)** is defined as the force (F) that produces the movement of an object times the distance moved (d):

▶ **Work** = Force × Distance
$$w = F \times d$$

▲ This runner going uphill is doing a lot of work to overcome gravity.

When you run up stairs, for instance, your leg muscles provide a force sufficient to overcome gravity and lift you higher. When you swim, you provide a force sufficient to push water out of the way and move yourself forward.

The most common type of work encountered in chemical systems is the *expansion work* (also called *pressure–volume*, or *PV, work*) done as the result of a volume change in the system. In the combustion reaction of propane (C_3H_8) with oxygen, for instance, the balanced equation says that 7 moles of products come from 6 moles of reactants:

$$\underbrace{C_3H_8(g) + 5\,O_2(g)}_{\text{6 mol of gas}} \longrightarrow \underbrace{3\,CO_2(g) + 4\,H_2O(g)}_{\text{7 mol of gas}}$$

If the reaction takes place inside a container outfitted with a movable piston, the greater volume of gas in the product will force the piston outward against the pressure of the atmosphere (P), moving air molecules aside and thereby doing work **(Figure 8.4)**.

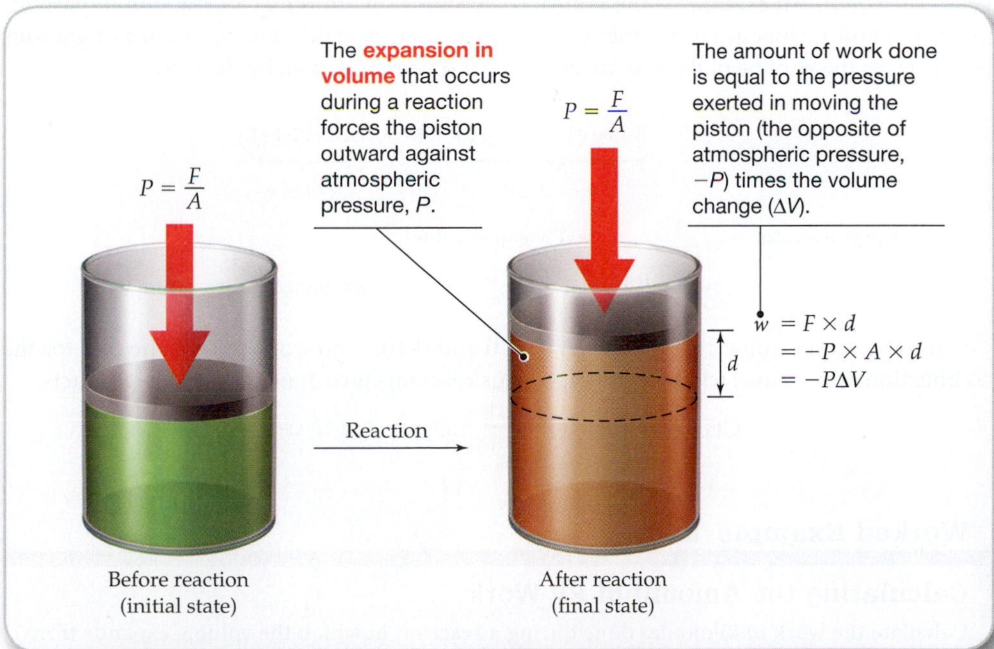

FIGURE 8.4
Expansion work in a chemical reaction.

A short calculation gives the exact amount of work done during the expansion. We know from physics that force (F) is area (A) times pressure (P). In Figure 8.4, the force that the expanding gas exerts is the area of the piston times the pressure that the gas exerts against the piston. This internal pressure is equal in magnitude but opposite in sign to the external atmospheric pressure that opposes the movement, so if external atmospheric pressure is called P, then the internal pressure has the value $-P$.

$$F = -P \times A \qquad \text{where } P \text{ is the external atmospheric pressure}$$

If the piston is pushed out a distance d, then the amount of work done is equal to force times distance, or pressure times area times distance:

$$w = F \times d = -P \times A \times d$$

This equation can be simplified by noticing that the area of the piston times the distance the piston moves equals the volume change in the system: $\Delta V = A \times d$. Thus, the amount of work done is equal to the pressure the gas exerts against the piston times the volume change, hence the name PV work:

A negative value ↘ ↙ A positive value

$$w = -P\Delta V \qquad \text{Work done during expansion}$$

What about the sign of the work done during the expansion? Because the work is done by the system to move air molecules aside as the piston rises, work energy must be leaving the system. Thus, the negative sign of the work in the preceding equation is consistent with the convention previously established for ΔE (Section 8.2), whereby we always adopt the point of view of the system. Any energy that flows out of the system has a negative sign because the system has lost it ($E_{\text{final}} < E_{\text{initial}}$).

If the pressure is given in the unit atmospheres (atm) and the volume change is given in liters, then the amount of work done has the unit liter atmosphere (L · atm), where 1 atm = 101×10^3 kg/(m · s^2) in SI units. Thus, 1 L · atm = 101 J:

$$1\ \text{L}\cdot\text{atm} = (1\ \text{L})\left(\frac{10^{-3}\ \text{m}^3}{1\ \text{L}}\right)\left(101 \times 10^3\ \frac{\text{kg}}{\text{m}\cdot\text{s}^2}\right) = 101\ \frac{\text{kg}\cdot\text{m}^2}{\text{s}^2} = 101\ \text{J}$$

When a reaction takes place with a contraction in volume rather than an expansion, the ΔV term has a negative sign so the work has a positive sign. This is again consistent with adopting the point of view of the system because the system has now gained work energy ($E_{\text{final}} > E_{\text{initial}}$). An example is the industrial synthesis of ammonia by reaction of gaseous hydrogen and nitrogen. Four moles of gaseous reactants yield only two moles of gaseous products, so the volume of the system contracts and work is gained by the system.

$$\underbrace{3\ H_2(g) + N_2(g)}_{\text{4 mol of gas}} \longrightarrow \underbrace{2\ NH_3(g)}_{\text{2 mol of gas}}$$

A positive value ↘ ↙ A negative value

$$w = -P\Delta V \qquad \text{Work gained during contraction}$$

If there is no volume change, then $\Delta V = 0$ and there is no work. Such is the case for the combustion of methane, where 3 mol of gaseous reactants give 3 mol of gaseous products:

$$CH_4(g) + 2\ O_2(g) \longrightarrow CO_2(g) + 2\ H_2O(g)$$

Worked Example 8.1

Calculating the Amount of *PV* Work

Calculate the work in kilojoules done during a reaction in which the volume expands from 12.0 L to 14.5 L against an external pressure of 5.0 atm.

STRATEGY

Expansion work done during a chemical reaction is calculated with the formula $w = -P\Delta V$, where P is the external pressure opposing the change in volume. In this instance,

$P = 5.0$ atm and $\Delta V = (14.5 - 12.0)$ L $= 2.5$ L. Remember that an expanding system loses work energy, which thus has a negative sign.

SOLUTION

$$w = -(5.0\ \text{atm})(2.5\ \text{L}) = -12.5\ \text{L}\cdot\text{atm}$$

$$(-12.5\ \text{L}\cdot\text{atm})\left(101\,\frac{\text{J}}{\text{L}\cdot\text{atm}}\right) = -1.3 \times 10^3\ \text{J} = -1.3\ \text{kJ}$$

PROBLEM 8.2 Calculate the work in kilojoules done during a synthesis of ammonia in which the volume contracts from 8.6 L to 4.3 L at a constant external pressure of 44 atm. In which direction does the work energy flow? What is the sign of the energy change?

CONCEPTUAL PROBLEM 8.3 How much work is done in kilojoules, and in which direction, as a result of the following reaction?

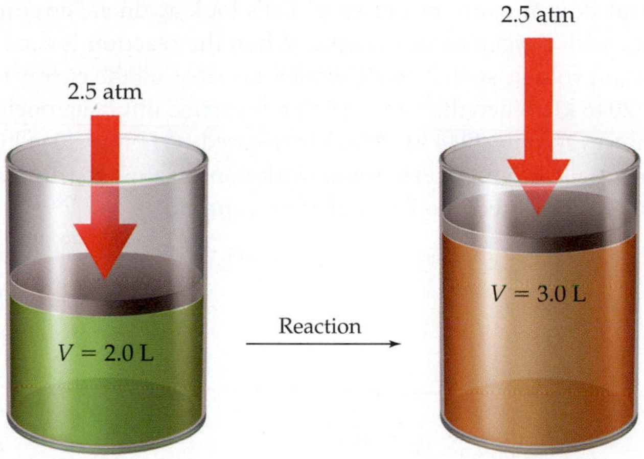

8.4 INTERNAL ENERGY AND ENTHALPY OF MOLECULES

We've seen up to this point that a system can exchange energy with its surroundings either by transferring heat or by doing work. Using the symbol q to represent transferred heat and the formula $w = -P\Delta V$ for work, we can represent the total change in the internal energy of a system, ΔE, as

$$\Delta E = q + w = q - P\Delta V$$

where q has a positive sign if the system gains heat and a negative sign if the system loses heat. Rearranging the equation to solve for q gives the amount of heat transferred:

$$q = \Delta E + P\Delta V$$

Let's look at two ways in which a chemical reaction might be carried out. On the one hand, a reaction might be carried out in a closed container with a constant volume, so that $\Delta V = 0$. In such a case, no PV work can be done so the energy change in the system is due entirely to heat transfer. We indicate this heat transfer at constant volume by the symbol q_v.

$$q_v = \Delta E \qquad \text{At constant volume; } \Delta V = 0$$

Alternatively, a reaction might be carried out in an open flask or other apparatus that keeps the pressure constant and allows the volume of the system to change freely. In such a case, $\Delta V \neq 0$ and the energy change in the system is due to both heat transfer and PV work. We indicate the heat transfer at constant pressure by the symbol q_p:

$$q_p = \Delta E + P\Delta V \qquad \text{At constant pressure}$$

Because reactions carried out at constant pressure in open containers are so common in chemistry, the heat change q_p for such a process is given a special name and symbol and is

▲ Chemical reactions are often carried out in open vessels at constant atmospheric pressure.

called the **heat of reaction,** or **enthalpy change** (**ΔH**). The **enthalpy** (***H***) of a system is the name given to the quantity $E + PV$.

$$q_p = \Delta E + P\Delta V = \underbrace{\Delta H}_{\text{Enthalpy change}}$$

Note that only the enthalpy *change* during a reaction is important. As with internal energy, enthalpy is a state function whose value depends only on the current state of the system, not on the path taken to arrive at that state. Thus, we don't need to know the exact value of the system's enthalpy before and after a reaction. All we need to know is the difference between final and initial states:

$$\Delta H = H_{\text{final}} - H_{\text{initial}}$$
$$= H_{\text{products}} - H_{\text{reactants}}$$

How big a difference is there between $q_v = \Delta E$, the heat flow at constant volume, and $q_p = \Delta H$, the heat flow at constant pressure? Let's look again at the combustion reaction of propane, C_3H_8, with oxygen as an example. When the reaction is carried out in a closed container at constant volume so that no *PV* work is possible, all the energy released is released as heat: $\Delta E = -2046$ kJ. When the same reaction is carried out in an open container at constant pressure, however, only 2044 kJ of heat is released ($\Delta H = -2044$ kJ). The difference, 2 kJ, is due to the small amount of expansion work done against the atmosphere as 6 mol of gaseous reactants are converted into 7 mol of gaseous products.

$$\underset{\text{Propane}}{C_3H_8(g)} + 5\,O_2(g) \longrightarrow 3\,CO_2(g) + 4\,H_2O(g) \qquad \begin{aligned}\Delta E &= -2046\text{ kJ}\\ \Delta H &= -2044\text{ kJ}\\ P\Delta V &= +2\text{ kJ}\end{aligned}$$

That is:

$$\left[\begin{aligned} q_v &= q_p + w \\ \Delta E &= \Delta H - P\Delta V \\ -2046\text{ kJ} &= -2044\text{ kJ} - (+2\text{ kJ}) \end{aligned}\right]$$

What is true of the propane + oxygen reaction is also true of most other reactions: The difference between ΔH and ΔE is usually small, so the two quantities are nearly equal. Of course, if no volume change occurs, then no work is done so ΔH and ΔE are the same. An example of such a process is the combustion of methane, in which 3 mol of gaseous reactants give 3 mol of gaseous products:

$$CH_4(g) + 2\,O_2(g) \longrightarrow CO_2(g) + 2\,H_2O(g) \qquad \Delta E = \Delta H = -802\text{ kJ}$$

CONCEPTUAL PROBLEM 8.4 The following reaction has $\Delta E = -186$ kJ/mol.

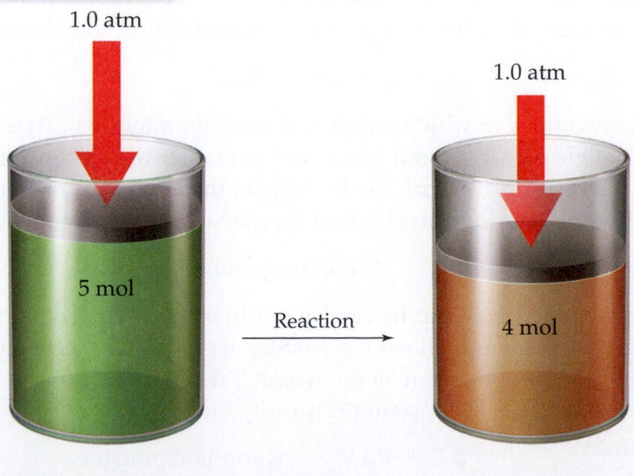

(a) Is the sign of $P\Delta V$ positive or negative? Explain.
(b) What is the sign and approximate magnitude of ΔH? Explain.

8.5 MEASURING HEAT TRANSFER: CALORIMETRY AND HEAT CAPACITY

How is the amount of heat transferred during a reaction measured? The amount of heat transferred during a reaction can be measured with a device called a *calorimeter*, shown schematically in **Figure 8.5**. At its simplest, a calorimeter is just an insulated vessel with a stirrer, a thermometer, and a loose-fitting lid to keep the contents at atmospheric pressure. The reaction is carried out inside the vessel, and the heat evolved or absorbed is calculated from the temperature change. Because the pressure inside the calorimeter is constant (atmospheric pressure), the temperature measurement makes it possible to calculate the enthalpy change ΔH during a reaction.

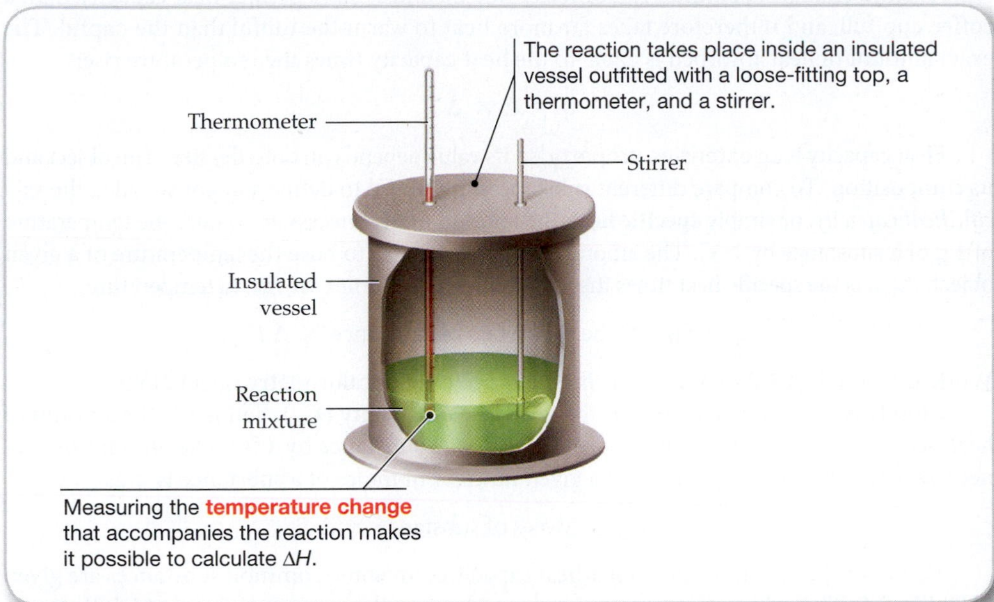

FIGURE 8.5

A calorimeter for measuring the heat flow in a reaction at constant pressure (ΔH).

A somewhat more complicated device called a *bomb calorimeter* is used to measure the heat released during a combustion reaction, or burning of a flammable substance. (More generally, a *combustion* reaction is any reaction that produces a flame.) The sample is placed in a small cup and sealed in an oxygen atmosphere inside a steel bomb that is itself placed in an insulated, water-filled container (**Figure 8.6**). The reactants are ignited electrically, and the evolved heat is calculated from the temperature change of the surrounding water. Because the reaction takes place at constant volume rather than constant pressure, the measurement provides a value for ΔE rather than ΔH.

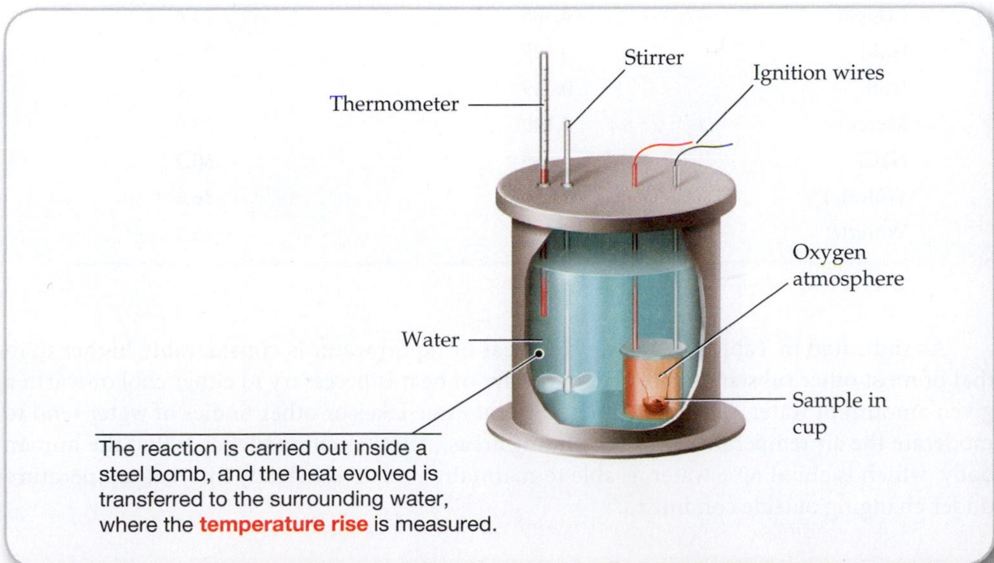

FIGURE 8.6

A bomb calorimeter for measuring the heat evolved at constant volume (ΔE) in a combustion reaction.

How can the temperature change inside a calorimeter be used to calculate ΔH (or ΔE) for a reaction? When a calorimeter and its contents absorb a given amount of heat, the temperature rise that results depends on the calorimeter's *heat capacity.* **Heat capacity (*C*)** is the amount of heat required to raise the temperature of an object or substance by a given amount, a relationship that can be expressed by the equation

$$C = \frac{q}{\Delta T}$$

where q is the quantity of heat transferred and ΔT is the temperature change that results ($\Delta T = T_{final} - T_{initial}$).

The greater the heat capacity, the greater the amount of heat needed to produce a given temperature change. A bathtub full of water, for instance, has a greater heat capacity than a coffee cup full, and it therefore takes far more heat to warm the tubful than the cupful. The exact amount of heat absorbed is equal to the heat capacity times the temperature rise:

$$q = C \times \Delta T$$

REMEMBER...

Extensive properties, such as length and volume, have values that depend on the sample size. Intensive properties, such as temperature and melting point, have values that do not depend on the amount of the sample. (Section 1.3)

Heat capacity is an **extensive property**, so its value depends on both the size of an object and its composition. To compare different substances, it's useful to define a quantity called the *specific heat capacity*, or simply **specific heat**, the amount of heat necessary to raise the temperature of 1 g of a substance by 1 °C. The amount of heat necessary to raise the temperature of a given object, then, is the specific heat times the mass of the object times the rise in temperature:

$$q = \text{Specific heat} \times \text{Mass of substance} \times \Delta T$$

Worked Example 8.3 shows how specific heats are used in calorimetry calculations.

Closely related to specific heat is the **molar heat capacity (C_m)**, defined as the amount of heat necessary to raise the temperature of 1 mol of a substance by 1 °C. The amount of heat necessary to raise the temperature of a given number of moles of a substance is thus

$$q = C_m \times \text{Moles of substance} \times \Delta T$$

Values of specific heats and molar heat capacities for some common substances are given in **Table 8.1**. The values are temperature dependent, so the temperatures at which the measurements are taken must be specified.

TABLE 8.1 Specific Heats and Molar Heat Capacities for Some Common Substances at 25 °C

Substance	Specific Heat J/(g · °C)	Molar Heat Capacity J/(mol · °C)
Air (dry)	1.01	29.1
Aluminum	0.897	24.2
Copper	0.385	24.4
Gold	0.129	25.4
Iron	0.449	25.1
Mercury	0.140	28.0
NaCl	0.859	50.2
Water(*s*)*	2.03	36.6
Water(*l*)	4.179	75.3

*At −11 °C

As indicated in Table 8.1, the specific heat of liquid water is considerably higher than that of most other substances, so a large transfer of heat is necessary to either cool or warm a given amount of water. One consequence is that large lakes or other bodies of water tend to moderate the air temperature in surrounding areas. Another consequence is that the human body, which is about 60% water, is able to maintain a relatively steady internal temperature under changing outside conditions.

Worked Example 8.2

Calculating a Specific Heat

What is the specific heat of silicon if it takes 192 J to raise the temperature of 45.0 g of Si by 6.0 °C?

STRATEGY

To find a specific heat of a substance, calculate the amount of energy necessary to raise the temperature of 1 g of the substance by 1 °C.

SOLUTION

$$\text{Specific heat of Si} = \frac{192\ \text{J}}{(45.0\ \text{g})(6.0\ °\text{C})} = 0.71\ \text{J}/(\text{g} \cdot °\text{C})$$

▲ Lake Chelan in the North Cascades of Washington State is the third deepest lake in the United States at 1486 ft. Such large masses of water moderate the temperature of the surroundings because of their high heat capacity.

Worked Example 8.3

Calculating ΔH in a Calorimetry Experiment

Aqueous silver ion reacts with aqueous chloride ion to yield a white precipitate of solid silver chloride:

$$Ag^+(aq) + Cl^-(aq) \longrightarrow AgCl(s)$$

When 10.0 mL of 1.00 M $AgNO_3$ solution is added to 10.0 mL of 1.00 M NaCl solution at 25.0 °C in a calorimeter, a white precipitate of AgCl forms and the temperature of the aqueous mixture increases to 32.6 °C. Assuming that the specific heat of the aqueous mixture is 4.18 J/(g · °C), that the density of the mixture is 1.00 g/mL, and that the calorimeter itself absorbs a negligible amount of heat, calculate ΔH in kilojoules for the reaction.

STRATEGY

Because the temperature rises during the reaction, heat must be released and ΔH must be negative. The amount of heat evolved during the reaction is equal to the amount of heat absorbed by the mixture:

$$\text{Heat evolved} = \text{Specific heat} \times \text{Mass of mixture} \times \text{Temperature change}$$

Calculating the heat evolved on a per-mole basis then gives the enthalpy change ΔH.

SOLUTION

$$\text{Specific heat} = 4.18\ \text{J}/(\text{g} \cdot °\text{C})$$

$$\text{Mass} = (20.0\ \text{mL})\left(1.00\ \frac{\text{g}}{\text{mL}}\right) = 20.0\ \text{g}$$

$$\text{Temperature change} = 32.6\ °\text{C} - 25.0\ °\text{C} = 7.6\ °\text{C}$$

$$\text{Heat evolved} = \left(4.18\ \frac{\text{J}}{\text{g} \cdot °\text{C}}\right)(20.0\ \text{g})(7.6\ °\text{C}) = 6.4 \times 10^2\ \text{J}$$

According to the balanced equation, the number of moles of AgCl produced equals the number of moles of Ag^+ (or Cl^-) reacted:

$$\text{Moles of } Ag^+ = (10.0\ \text{mL})\left(\frac{1.00\ \text{mol } Ag^+}{1000\ \text{mL}}\right) = 1.00 \times 10^{-2}\ \text{mol } Ag^+$$

$$\text{Moles of AgCl} = 1.00 \times 10^{-2}\ \text{mol AgCl}$$

$$\text{Heat evolved per mole of AgCl} = \frac{6.4 \times 10^2\ \text{J}}{1.00 \times 10^{-2}\ \text{mol AgCl}} = 64\ \text{kJ/mol AgCl}$$

Therefore, $\Delta H = -64$ kJ (negative because heat is released)

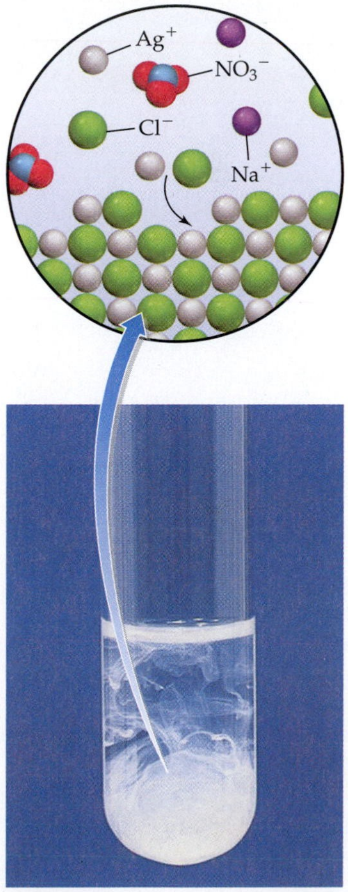

▲ The reaction of aqueous $AgNO_3$ with aqueous NaCl to yield solid AgCl is an exothermic process.

▶ **PROBLEM 8.5** Assuming that Coca Cola has the same specific heat as water [4.18 J/(g · °C)], calculate the amount of heat in kilojoules transferred when one can (about 350 g) is cooled from 25 °C to 3 °C.

▶ **PROBLEM 8.6** What is the specific heat of lead if it takes 97.2 J to raise the temperature of a 75.0 g block by 10.0 °C?

▶ **PROBLEM 8.7** When 25.0 mL of 1.0 M H_2SO_4 is added to 50.0 mL of 1.0 M NaOH at 25.0 °C in a calorimeter, the temperature of the aqueous solution increases to 33.9 °C. Assuming that the specific heat of the solution is 4.18 J/(g · °C), that its density is 1.00 g/mL, and that the calorimeter itself absorbs a negligible amount of heat, calculate ΔH in kilojoules for the reaction.

$$H_2SO_4(aq) + 2\,NaOH(aq) \longrightarrow 2\,H_2O(l) + Na_2SO_4(aq)$$

8.6 THE THERMODYNAMIC STANDARD STATE

The value of the enthalpy change reported for a reaction is the amount of heat released or absorbed when reactants are converted to products at the same temperature and pressure and in the molar amounts represented by coefficients in the balanced chemical equation. In the combustion reaction of propane, for instance, the reaction of 1 mol of propane gas with 5 mol of oxygen gas to give 3 mol of CO_2 gas and 4 mol of water vapor releases 2044 kJ. The actual amount of heat released in a specific reaction, however, depends on the specific amounts of reactants. Thus, reaction of 0.5000 mol of propane with 2.500 mol of O_2 releases $0.5000 \times 2044\text{ kJ} = 1022\text{ kJ}$.

Note that the physical states of reactants and products must be specified as solid (*s*), liquid (*l*), gaseous (*g*), or aqueous (*aq*) when enthalpy changes are reported. The enthalpy change for the reaction of propane with oxygen is $\Delta H = -2044$ kJ if water is produced as a gas but $\Delta H = -2220$ kJ if water is produced as a liquid.

$$C_3H_8(g) + 5\,O_2(g) \longrightarrow 3\,CO_2(g) + 4\,H_2O(g) \quad \Delta H = -2044\text{ kJ}$$
$$C_3H_8(g) + 5\,O_2(g) \longrightarrow 3\,CO_2(g) + 4\,H_2O(l) \quad \Delta H = -2220\text{ kJ}$$

The difference of 176 kJ between the values of ΔH for the two reactions arises because the conversion of liquid water to gaseous water requires energy. If liquid water is produced, ΔH is larger (more negative), but if gaseous water is produced, ΔH is smaller (less negative) because 44.0 kJ/mol is needed for the vaporization.

$$H_2O(l) \longrightarrow H_2O(g) \qquad \Delta H = +44.0\text{ kJ}$$
$$\text{or} \quad 4\,H_2O(l) \longrightarrow 4\,H_2O(g) \qquad \Delta H = +176\text{ kJ}$$

In addition to specifying the physical state of reactants and products when reporting an enthalpy change, it's also necessary to specify the pressure and temperature. To ensure that all measurements are reported in the same way so that different reactions can be compared, a set of conditions called the **thermodynamic standard state** has been defined.

▶ **Thermodynamic standard state** The most stable form of a substance at 1 atm pressure[1] and at a specified temperature, usually 25 °C; 1 M concentration for all substances in solution.

[1]The standard pressure, listed here and in most other books as 1 atmosphere (atm), has been redefined to be 1 *bar*, which is equal to 0.986 923 atm. The difference is small, however.

Measurements made under these standard conditions are indicated by addition of the superscript ° to the symbol of the quantity reported. Thus, an enthalpy change measured under standard conditions is called a **standard enthalpy of reaction** and is indicated by the symbol $\Delta H°$. The reaction of propane with oxygen, for example, might be written

$$C_3H_8(g) + 5\,O_2(g) \longrightarrow 3\,CO_2(g) + 4\,H_2O(g) \qquad \Delta H° = -2044 \text{ kJ}$$

Worked Example 8.4

Calculating ΔE for a Reaction

The reaction of nitrogen with hydrogen to make ammonia has $\Delta H° = -92.2$ kJ. What is the value of ΔE in kilojoules if the reaction is carried out at a constant pressure of 40.0 atm and the volume change is -1.12 L?

$$N_2(g) + 3\,H_2(g) \longrightarrow 2\,NH_3(g) \qquad \Delta H° = -92.2 \text{ kJ}$$

STRATEGY

We are given an enthalpy change ΔH, a volume change ΔV, and a pressure P and asked to find an energy change ΔE. Rearrange the equation $\Delta H = \Delta E + P\Delta V$ to the form $\Delta E = \Delta H - P\Delta V$, and substitute the appropriate values for ΔH, P, and ΔV:

SOLUTION

$$\Delta E = \Delta H - P\Delta V \qquad \text{where } \Delta H = -92.2 \text{ kJ}$$

$$P\Delta V = (40.0 \text{ atm})(-1.12 \text{ L}) = -44.8 \text{ L}\cdot\text{atm}$$

$$= (-44.8 \text{ L}\cdot\text{atm})\left(101 \frac{\text{J}}{\text{L}\cdot\text{atm}}\right) = -4520 \text{ J} = -4.52 \text{ kJ}$$

$$\Delta E = (-92.2 \text{ kJ}) - (-4.52 \text{ kJ}) = -87.7 \text{ kJ}$$

Note that ΔE is smaller (less negative) than ΔH for this reaction because the volume change is negative. The products have less volume than the reactants, so a contraction occurs and a small amount of PV work is gained by the system.

PROBLEM 8.8 The reaction between hydrogen and oxygen to yield water vapor has $\Delta H° = -484$ kJ. How much PV work is done, and what is the value of ΔE in kilojoules for the reaction of 0.50 mol of H_2 with 0.25 mol of O_2 at atmospheric pressure if the volume change is -5.6 L?

$$2\,H_2(g) + O_2(g) \longrightarrow 2\,H_2O(g) \qquad \Delta H° = -484 \text{ kJ}$$

PROBLEM 8.9 The explosion of 2.00 mol of solid trinitrotoluene (TNT; $C_7H_5N_3O_6$) with a volume of approximately 274 mL produces gases with a volume of 448 L at room temperature and 1.0 atm pressure. How much PV work in kilojoules is done during the explosion?

$$2\,C_7H_5N_3O_6(s) \longrightarrow 12\,CO(g) + 5\,H_2(g) + 3\,N_2(g) + 2\,C(s)$$

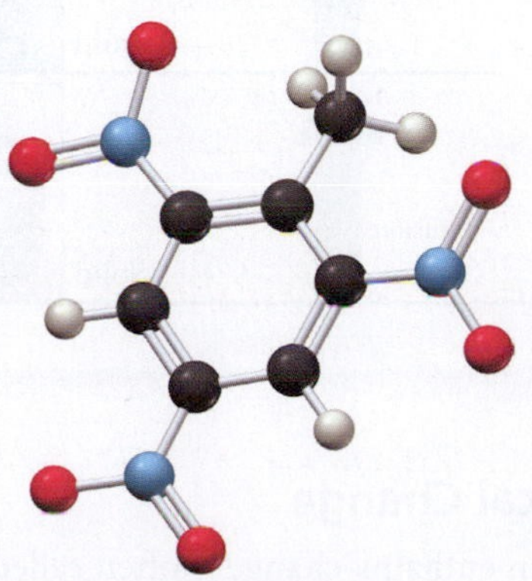

Trinitrotoluene

8.7 ENTHALPIES OF PHYSICAL AND CHEMICAL CHANGE

Almost every change in a system involves either a gain or a loss of enthalpy. The change can be either physical, such as the melting of a solid to a liquid, or chemical, such as the burning of propane. Let's look at examples of both kinds.

Enthalpies of Physical Change

What would happen if you started with a block of ice at a low temperature, say −10 °C, and slowly increased its enthalpy by adding heat? The initial input of heat would cause the temperature of the ice to rise until it reached 0 °C. Additional heat would then cause the ice to melt without raising its temperature as the added energy is expended in overcoming the forces that hold H_2O molecules together in the ice crystal. The amount of heat necessary to melt a substance without changing its temperature is called the *enthalpy of fusion*, or *heat of fusion* (ΔH_{fusion}). For H_2O, $\Delta H_{fusion} = +6.01$ kJ/mol at 0 °C.

Once the ice has melted, further input of heat raises the temperature of the liquid water until it reaches 100 °C, and adding still more heat then causes the water to boil. Once again, energy is necessary to overcome the forces holding molecules together in the liquid, so the temperature does not rise again until all the liquid has been converted entirely into vapor. The amount of heat required to vaporize a substance without changing its temperature is called the *enthalpy of vaporization*, or *heat of vaporization* (ΔH_{vap}). For H_2O, $\Delta H_{vap} = +40.7$ kJ/mol at 100 °C.

Another kind of physical change in addition to melting and boiling is **sublimation**, the direct conversion of a solid to a vapor without passing through a liquid state. Solid CO_2 (dry ice), for example, changes directly from solid to vapor at atmospheric pressure without first melting to a liquid. Because enthalpy is a state function, the enthalpy change on going from solid to vapor must be constant regardless of the path taken. Thus, at a given temperature, a substance's *enthalpy of sublimation*, or *heat of sublimation* (ΔH_{subl}), equals the sum of the heat of fusion and heat of vaporization (**Figure 8.7**).

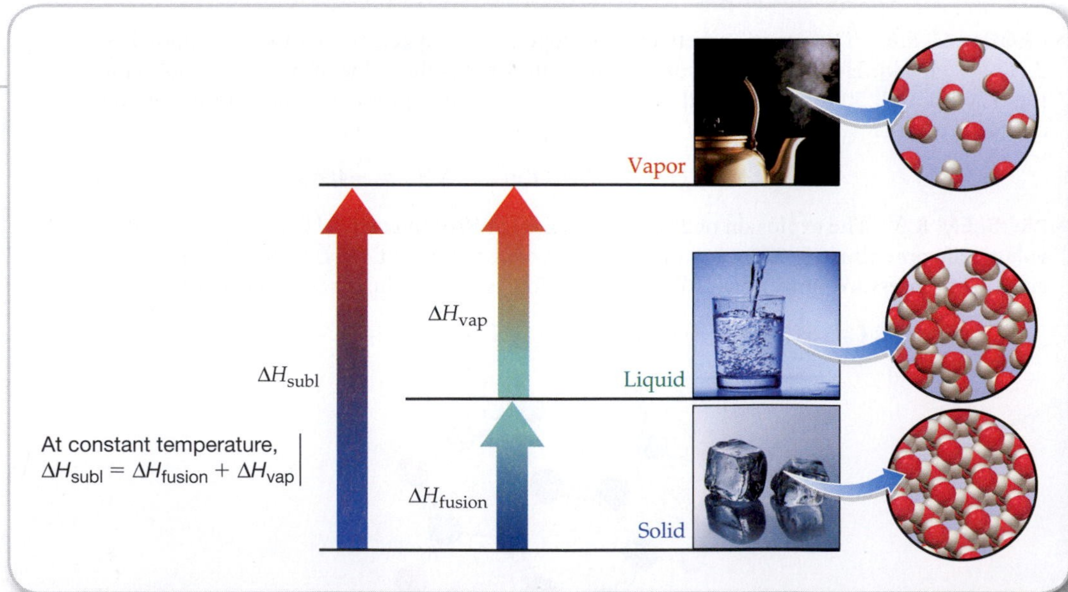

FIGURE 8.7

Enthalpy as a state function. Because enthalpy is a state function, the value of the enthalpy change from solid to vapor does not depend on the path taken between the two states.

Enthalpies of Chemical Change

We saw in Section 8.4 that an enthalpy change is often called a *heat of reaction* because it is a measure of the heat transfer into or out of a system at constant pressure. If the products of a reaction have more enthalpy than the reactants, then heat has flowed into the system from the surroundings and ΔH has a positive sign. Such reactions are said to be **endothermic** (*endo* means "within," so heat flows in). The reaction of 1 mol of barium hydroxide octahydrate[2] with ammonium chloride, for example, absorbs 80.3 kJ from the surroundings ($\Delta H° = +80.3$ kJ). The surroundings, having lost heat, become cold—so cold, in fact, that the temperature drops below freezing (**Figure 8.8**).

$$Ba(OH)_2 \cdot 8\,H_2O(s) + 2\,NH_4Cl(s) \longrightarrow BaCl_2(aq) + 2\,NH_3(aq) + 10\,H_2O(l) \qquad \Delta H° = +80.3\text{ kJ}$$

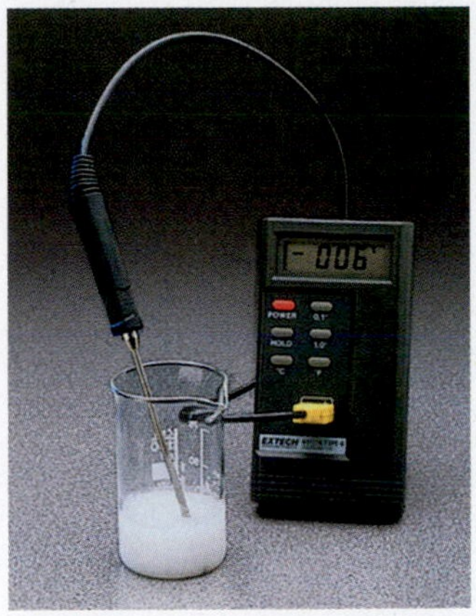

FIGURE 8.8

The endothermic reaction of barium hydroxide octahydrate with ammonium chloride. The reaction draws so much heat from the surroundings that the temperature falls below 0 °C.

If the products of a reaction have less enthalpy than the reactants, then heat has flowed out of the system to the surroundings and ΔH has a negative sign. Such reactions are said to be **exothermic** (*exo* means "out," so heat flows out). The so-called thermite reaction of aluminum with iron(III) oxide, for instance, releases so much heat ($\Delta H° = -852$ kJ), and the surroundings get so hot, that the reaction is used in construction work to weld iron.

$$2\,Al(s) + Fe_2O_3(s) \longrightarrow 2\,Fe(s) + Al_2O_3(s) \qquad \Delta H° = -852\text{ kJ}$$

As noted previously, the value of $\Delta H°$ given for an equation assumes that the equation is balanced to represent the numbers of moles of reactants and products, that all substances are in their standard states, and that the physical state of each substance is as specified. The actual amount of heat released in a specific reaction depends on the amounts of reactants, as illustrated in Worked Example 8.5.

It should also be emphasized that $\Delta H°$ values refer to the reaction going in the direction written. For the reverse reaction, the sign of $\Delta H°$ must be changed. Because of the reversibility of state functions (Section 8.2), the enthalpy change for a reverse reaction is equal in magnitude but opposite in sign to that for the corresponding forward reaction. For example, the reaction of iron with aluminum oxide to yield aluminum and iron(III) oxide—the reverse of the thermite reaction—would be endothermic and have $\Delta H° = +852$ kJ:

$$2\,Fe(s) + Al_2O_3(s) \longrightarrow 2\,Al(s) + Fe_2O_3(s) \qquad \Delta H° = +852\text{ kJ}$$
$$2\,Al(s) + Fe_2O_3(s) \longrightarrow 2\,Fe(s) + Al_2O_3(s) \qquad \Delta H° = -852\text{ kJ}$$

Worked Example 8.5

Calculating the Amount of Heat Released in a Reaction

How much heat in kilojoules is evolved when 5.00 g of aluminum reacts with a stoichiometric amount of Fe_2O_3?

$$2\,Al(s) + Fe_2O_3(s) \longrightarrow 2\,Fe(s) + Al_2O_3(s) \qquad \Delta H° = -852\text{ kJ}$$

STRATEGY

According to the balanced equation, 852 kJ of heat is evolved from the reaction of 2 mol of Al. To find out how much heat is evolved from the reaction of 5.00 g of Al, we have to find out how many moles of aluminum are in 5.00 g.

SOLUTION

The molar mass of Al is 26.98 g/mol, so 5.00 g of Al equals 0.185 mol:

$$5.00\text{ g Al} \times \frac{1\text{ mol Al}}{26.98\text{ g Al}} = 0.185\text{ mol Al}$$

Because 2 mol of Al releases 852 kJ of heat, 0.185 mol of Al releases 78.8 kJ of heat:

$$0.185\text{ mol Al} \times \frac{852\text{ kJ}}{2\text{ mol Al}} = 78.8\text{ kJ}$$

✔ BALLPARK CHECK

Since the molar mass of Al is about 27 g, 5 g of aluminum is roughly 0.2 mol, and the heat evolved is about (852 kJ/2 mol)(0.2 mol), or approximately 85 kJ.

[2]Barium hydroxide octahydrate, $Ba(OH)_2 \cdot 8\,H_2O$, is a crystalline compound that contains eight water molecules clustered around the barium ion. We'll learn more about hydrates in Section 18.13.

PROBLEM 8.10 How much heat in kilojoules is evolved or absorbed in each of the following reactions?

(a) Burning of 15.5 g of propane:

$$C_3H_8(g) + 5\,O_2(g) \longrightarrow 3\,CO_2(g) + 4\,H_2O(l) \quad \Delta H° = -2220 \text{ kJ}$$

(b) Reaction of 4.88 g of barium hydroxide octahydrate with ammonium chloride:

$$Ba(OH)_2 \cdot 8\,H_2O(s) + 2\,NH_4Cl(s) \longrightarrow BaCl_2(aq) + 2\,NH_3(aq) + 10\,H_2O(l) \quad \Delta H° = +80.3 \text{ kJ}$$

PROBLEM 8.11 Nitromethane (CH_3NO_2), sometimes used as a fuel in drag racers, burns according to the following equation. How much heat is released by burning 100.0 g of nitromethane?

$$4\,CH_3NO_2(l) + 7\,O_2(g) \longrightarrow 4\,CO_2(g) + 6\,H_2O(g) + 4\,NO_2(g) \quad \Delta H° = -2441.6 \text{ kJ}$$

8.8 CALCULATING ENTHALPY CHANGES USING HESS'S LAW

Now that we've discussed in general terms the energy changes that occur during chemical reactions, let's look at a specific example in detail. In particular, let's look at the *Haber process*, the industrial method by which more than 130 million metric tons of ammonia is produced each year worldwide, primarily for use as fertilizer (1 metric ton = 1000 kg). The reaction of hydrogen with nitrogen to make ammonia is exothermic, with $\Delta H° = -92.2$ kJ.

$$3\,H_2(g) + N_2(g) \longrightarrow 2\,NH_3(g) \qquad \Delta H° = -92.2 \text{ kJ}$$

If we dig into the details of the reaction, we find that it's not as simple as it looks. In fact, the overall reaction occurs in a series of steps, with hydrazine (N_2H_4) produced at an intermediate stage:

$$2\,H_2(g) + N_2(g) \longrightarrow \underset{\text{Hydrazine}}{N_2H_4(g)} \xrightarrow{H_2} \underset{\text{Ammonia}}{2\,NH_3(g)}$$

The enthalpy change for the conversion of hydrazine to ammonia can be measured as $\Delta H° = -187.6$ kJ, but if we wanted to measure $\Delta H°$ for the formation of hydrazine from hydrogen and nitrogen, we would have difficulty because the reaction doesn't go cleanly. Some of the hydrazine that's formed is converted into ammonia, while some of the starting nitrogen remains unreacted.

Fortunately, there's a way around the difficulty—a way that makes it possible to measure an energy change indirectly when a direct measurement can't be made. The trick is to realize that, because enthalpy is a state function, ΔH is the same no matter what path is taken between two states. Thus, the sum of the enthalpy changes for the individual steps in a sequence must equal the enthalpy change for the overall reaction, a statement known as **Hess's law:**

> **Hess's law** The overall enthalpy change for a reaction is equal to the sum of the enthalpy changes for the individual steps in the reaction.

Reactants and products in the individual steps can be added and subtracted like algebraic quantities in determining the overall equation. In the synthesis of ammonia, for instance, the sum of Steps 1 and 2 is equal to the overall reaction. Thus, the sum of the enthalpy changes for Steps 1 and 2 is equal to the enthalpy change for the overall reaction. With this knowledge, we can calculate the enthalpy change for Step 1. **Figure 8.9** shows the situation pictorially, and Worked Examples 8.6 and 8.7 give additional examples of Hess's law calculations.

Step 1.	$2\,H_2(g) + N_2(g) \longrightarrow \cancel{N_2H_4(g)}$	$\Delta H°_1 = ?$
Step 2.	$\cancel{N_2H_4(g)} + H_2(g) \longrightarrow 2\,NH_3(g)$	$\Delta H°_2 = -187.6$ kJ
Overall reaction	$3\,H_2(g) + N_2(g) \longrightarrow 2\,NH_3(g)$	$\Delta H°_{\text{reaction}} = -92.2$ kJ

Since $\Delta H^\circ_1 + \Delta H^\circ_2 = \Delta H^\circ_{\text{reaction}}$

then $\Delta H^\circ_1 = \Delta H^\circ_{\text{reaction}} - \Delta H^\circ_2$

$= (-92.2 \text{ kJ}) - (-187.6 \text{ kJ}) = +95.4 \text{ kJ}$

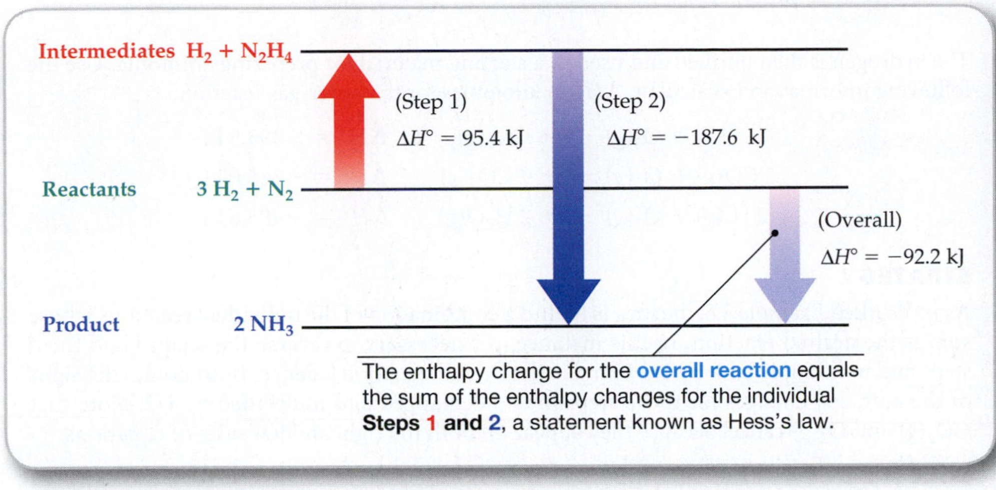

FIGURE 8.9

Enthalpy changes for steps in the synthesis of ammonia from nitrogen and hydrogen. If ΔH° values for Step 2 and for the overall reaction are known, then ΔH° for Step 1 can be calculated.

Worked Example 8.6

Using Hess's Law to Calculate ΔH°

Methane, the main constituent of natural gas, burns in oxygen to yield carbon dioxide and water:

$$CH_4(g) + 2\,O_2(g) \longrightarrow CO_2(g) + 2\,H_2O(l)$$

Use the following information to calculate ΔH° in kilojoules for the combustion of methane:

$$CH_4(g) + O_2(g) \longrightarrow CH_2O(g) + H_2O(g) \qquad \Delta H^\circ = -275.6 \text{ kJ}$$
$$CH_2O(g) + O_2(g) \longrightarrow CO_2(g) + H_2O(g) \qquad \Delta H^\circ = -526.7 \text{ kJ}$$
$$H_2O(l) \longrightarrow H_2O(g) \qquad \Delta H^\circ = +44.0 \text{ kJ}$$

STRATEGY

It often takes some trial and error, but the idea is to combine the individual reactions so that their sum is the desired reaction. The important points are that:

- All the reactants [$CH_4(g)$ and $O_2(g)$] must appear on the left.
- All the products [$CO_2(g)$ and $H_2O(l)$] must appear on the right.
- All intermediate products [$CH_2O(g)$ and $H_2O(g)$] must occur on *both* the left and the right so that they cancel.
- A reaction written in the reverse of the direction given [$H_2O(g) \rightarrow H_2O(l)$] must have the sign of its ΔH° reversed (Section 8.7).
- If a reaction is multiplied by a coefficient [$H_2O(g) \rightarrow H_2O(l)$ is multiplied by 2], then ΔH° for the reaction must be multiplied by that same coefficient.

SOLUTION

$$CH_4(g) + O_2(g) \longrightarrow \cancel{CH_2O(g)} + \cancel{H_2O(g)} \qquad \Delta H^\circ = -275.6 \text{ kJ}$$
$$\cancel{CH_2O(g)} + O_2(g) \longrightarrow CO_2(g) + \cancel{H_2O(g)} \qquad \Delta H^\circ = -526.7 \text{ kJ}$$
$$2\,[\cancel{H_2O(g)} \longrightarrow H_2O(l)] \qquad 2[\Delta H^\circ = -44.0 \text{ kJ}] = -88.0 \text{ kJ}$$

$$CH_4(g) + 2\,O_2(g) \longrightarrow CO_2(g) + 2\,H_2O(l) \qquad \Delta H^\circ = -890.3 \text{ kJ}$$

Worked Example 8.7

Using Hess's Law to Calculate $\Delta H°$

Water gas is the name for the mixture of CO and H_2 prepared by reaction of steam with carbon at 1000 °C:

$$C(s) + H_2O(g) \longrightarrow \underbrace{CO(g) + H_2(g)}_{\text{“Water gas”}}$$

The hydrogen is then purified and used as a starting material for preparing ammonia. Use the following information to calculate $\Delta H°$ in kilojoules for the water-gas reaction:

$$C(s) + O_2(g) \longrightarrow CO_2(g) \qquad \Delta H° = -393.5 \text{ kJ}$$
$$2\,CO(g) + O_2(g) \longrightarrow 2\,CO_2(g) \qquad \Delta H° = -566.0 \text{ kJ}$$
$$2\,H_2(g) + O_2(g) \longrightarrow 2\,H_2O(g) \qquad \Delta H° = -483.6 \text{ kJ}$$

STRATEGY

As in Worked Example 8.6, the idea is to find a combination of the individual reactions whose sum is the desired reaction. In this instance, it's necessary to reverse the second and third steps and to multiply both by 1/2 to make the overall equation balance. In so doing, the signs of the enthalpy changes for those steps must be changed and multiplied by 1/2. Note that $CO_2(g)$ and $O_2(g)$ cancel because they appear on both the right and left sides of equations.

SOLUTION

$$C(s) + O_2(g) \longrightarrow CO_2(g) \qquad \Delta H° = -393.5 \text{ kJ}$$
$$1/2\,[2\,CO_2(g) \longrightarrow 2\,CO(g) + O_2(g)] \qquad 1/2[\Delta H° = 566.0 \text{ kJ}] = +283.0 \text{ kJ}$$
$$1/2\,[2\,H_2O(g) \longrightarrow 2\,H_2(g) + O_2(g)] \qquad 1/2[\Delta H° = 483.6 \text{ kJ}] = +241.8 \text{ kJ}$$
$$C(s) + H_2O(g) \longrightarrow CO(g) + H_2(g) \qquad \Delta H° = +131.3 \text{ kJ}$$

The water-gas reaction is endothermic by 131.3 kJ.

Methylene chloride

PROBLEM 8.12 The industrial degreasing solvent methylene chloride, CH_2Cl_2, is prepared from methane by reaction with chlorine:

$$CH_4(g) + 2\,Cl_2(g) \longrightarrow CH_2Cl_2(g) + 2\,HCl(g)$$

Use the following data to calculate $\Delta H°$ in kilojoules for the reaction:

$$CH_4(g) + Cl_2(g) \longrightarrow CH_3Cl(g) + HCl(g) \qquad \Delta H° = -98.3 \text{ kJ}$$
$$CH_3Cl(g) + Cl_2(g) \longrightarrow CH_2Cl_2(g) + HCl(g) \qquad \Delta H° = -104 \text{ kJ}$$

CONCEPTUAL PROBLEM 8.13 The reaction of **A** with **B** to give **D** proceeds in two steps and can be represented by the following Hess's law diagram.

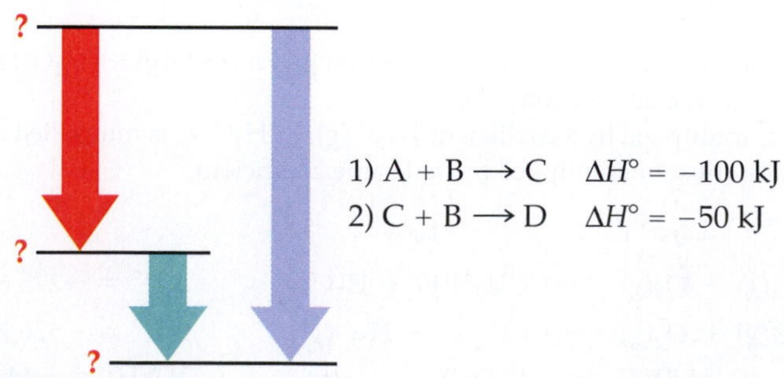

(a) What is the equation and $\Delta H°$ for the net reaction?
(b) Which arrow on the diagram corresponds to which step, and which arrow corresponds to the net reaction?
(c) The diagram shows three energy levels. The energies of which substances are represented by each?

CONCEPTUAL PROBLEM 8.14 Draw a Hess's law diagram similar to that in Problem 8.13 depicting the energy changes for the reaction in Problem 8.12.

8.9 CALCULATING ENTHALPY CHANGES USING STANDARD HEATS OF FORMATION

Where do the $\Delta H°$ values we've been using in previous sections come from? There are so many chemical reactions—hundreds of millions are known—that it's impossible to measure $\Delta H°$ for all of them. A better way is needed.

The most efficient way to manage with the smallest number of experimental measurements is to use what are called **standard heats of formation ($\Delta H°_f$)**.

▶ **Standard heat of formation** The enthalpy change $\Delta H°_f$ for the formation of 1 mol of a substance in its standard state from its constituent elements in their standard states.

Note several points about this definition. First, the "reaction" to form a substance from its constituent elements can be (and usually is) hypothetical. We can't combine carbon and hydrogen in the laboratory to make methane, for instance, yet the heat of formation for methane is $\Delta H°_f = -74.8$ kJ/mol, which corresponds to the standard enthalpy change for the hypothetical reaction

$$C(s) + 2\,H_2(g) \longrightarrow CH_4(g) \qquad \Delta H° = -74.8 \text{ kJ}$$

Second, each substance in the reaction must be in its most stable, standard form at 1 atm pressure and the specified temperature, usually 25 °C. Carbon, for instance, is most stable as solid graphite rather than as diamond under these conditions, and hydrogen is most stable as gaseous H_2 molecules rather than as H atoms. **Table 8.2** gives standard heats of formation for some common substances, and Appendix B gives a more detailed list.

TABLE 8.2 Standard Heats of Formation for Some Common Substances at 25 °C

Substance	Formula	$\Delta H°_f$ (kJ/mol)	Substance	Formula	$\Delta H°_f$ (kJ/mol)
Acetylene	$C_2H_2(g)$	+227.4	Hydrogen chloride	$HCl(g)$	−92.3
Ammonia	$NH_3(g)$	−46.1	Iron(III) oxide	$Fe_2O_3(s)$	−824.2
Carbon dioxide	$CO_2(g)$	−393.5	Magnesium carbonate	$MgCO_3(s)$	−1095.8
Carbon monoxide	$CO(g)$	−110.5	Methane	$CH_4(g)$	−74.8
Ethanol	$C_2H_5OH(l)$	−277.7	Nitric oxide	$NO(g)$	+91.3
Ethylene	$C_2H_4(g)$	+52.3	Water (*g*)	$H_2O(g)$	−241.8
Glucose	$C_6H_{12}O_6(s)$	−1273.3	Water (*l*)	$H_2O(l)$	−285.8

No elements are listed in Table 8.2 because, by definition, the most stable form of any element in its standard state has $\Delta H°_f = 0$ kJ; that is, the enthalpy change for the formation of an element from itself is zero. Defining $\Delta H°_f$ as zero for all elements thus establishes a thermochemical "sea level," or reference point, from which other enthalpy changes are measured.

How can standard heats of formation be used for thermochemical calculations? The standard enthalpy change for any chemical reaction is found by subtracting the sum of the heats of formation of all reactants from the sum of the heats of formation of all products, with each heat of formation multiplied by the coefficient of that substance in the balanced equation.

$$\Delta H°_{\text{reaction}} = \Delta H°_f(\text{Products}) - \Delta H°_f(\text{Reactants})$$

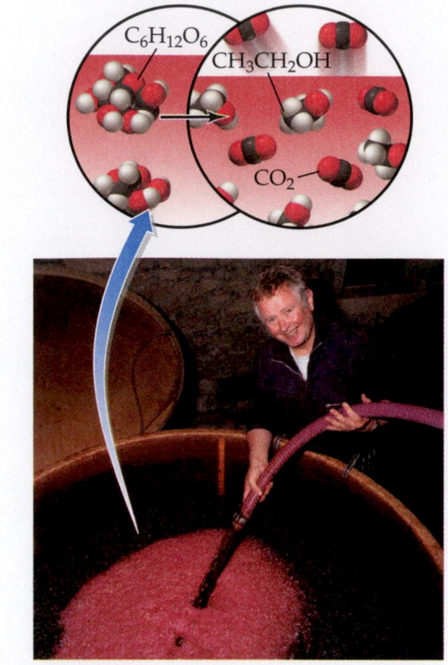

▲ Fermentation of the sugar from grapes yields the ethyl alcohol in wine.

To find $\Delta H°$ for the reaction

$$a\,\text{A} + b\,\text{B} + \cdots \longrightarrow c\,\text{C} + d\,\text{D} + \cdots$$

Subtract the sum of the heats of formation for these reactants . . .

. . . from the sum of the heats of formation for these products.

$$\Delta H°_{\text{reaction}} = [c\,\Delta H°_f(\text{C}) + d\,\Delta H°_f(\text{D}) + \cdots] - [a\,\Delta H°_f(\text{A}) + b\,\Delta H°_f(\text{B}) + \cdots]$$

As an example, let's calculate $\Delta H°$ for the fermentation of glucose to make ethyl alcohol (ethanol), the reaction that occurs during the production of alcoholic beverages:

$$C_6H_{12}O_6(s) \longrightarrow 2\,C_2H_5OH(l) + 2\,CO_2(g) \qquad \Delta H° = ?$$

Using the data in Table 8.2 gives the following answer:

$$\begin{aligned}\Delta H° &= [2\,\Delta H°_f(\text{Ethanol}) + 2\,\Delta H°_f(CO_2)] - [\Delta H°_f(\text{Glucose})]\\ &= (2\text{ mol})(-277.7\text{ kl/mol}) + (2\text{ mol})(-393.5\text{ kJ/mol}) - (1\text{ mol})(-1273.3\text{ kJ/mol})\\ &= -69.1\text{ kJ}\end{aligned}$$

The fermentation reaction is exothermic by 69.1 kJ.

Why does this calculation "work"? It works because enthalpy is a state function and the calculation is really just an application of Hess's law. That is, the sum of the individual equations corresponding to the heat of formation for each substance in the reaction equals the enthalpy change for the overall reaction:

(1) $C_6H_{12}O_6(s) \longrightarrow 6\,C(s) + 6\,H_2(g) + 3\,O_2(g)$		$-\Delta H°_f = +1273.3$ kJ
(2) $2\,[2\,C(s) + 3\,H_2(g) + 1/2\,O_2(g) \longrightarrow C_2H_5OH(l)]$	$2\,[\Delta H°_f = -277.7\text{ kJ}] =$	-555.4 kJ
(3) $2\,[C(s) + O_2(g) \longrightarrow CO_2(g)]$	$2\,[\Delta H°_f = -393.5\text{ kJ}] =$	-787.0 kJ
Net $C_6H_{12}O_6(s) \longrightarrow 2\,C_2H_5OH(l) + 2\,CO_2(g)$	$\Delta H° =$	-69.1 kJ

Note that reaction (1) represents the formation of glucose from its elements written in reverse, so the sign of $\Delta H°_f$ is reversed. Note also that reactions (2) and (3), which represent the formation of ethyl alcohol and carbon dioxide, respectively, are multiplied by 2 to arrive at the balanced equation for the overall reaction.

When we use heats of formation to calculate standard reaction enthalpies, what we're really doing is referencing the enthalpies of both products and reactants to the same point: their constituent elements. By thus referencing product and reactant enthalpies to the same point, they are referenced to each other and the difference between them is the reaction enthalpy (**Figure 8.10**). Worked Examples 8.8 and 8.9 further illustrate how to use standard heats of formation.

FIGURE 8.10

The standard reaction enthalpy, $\Delta H°$, for the generalized reaction A → B. The value of $\Delta H°$ for the reaction is the difference between the standard heats of formation of product **B** and reactant **A**.

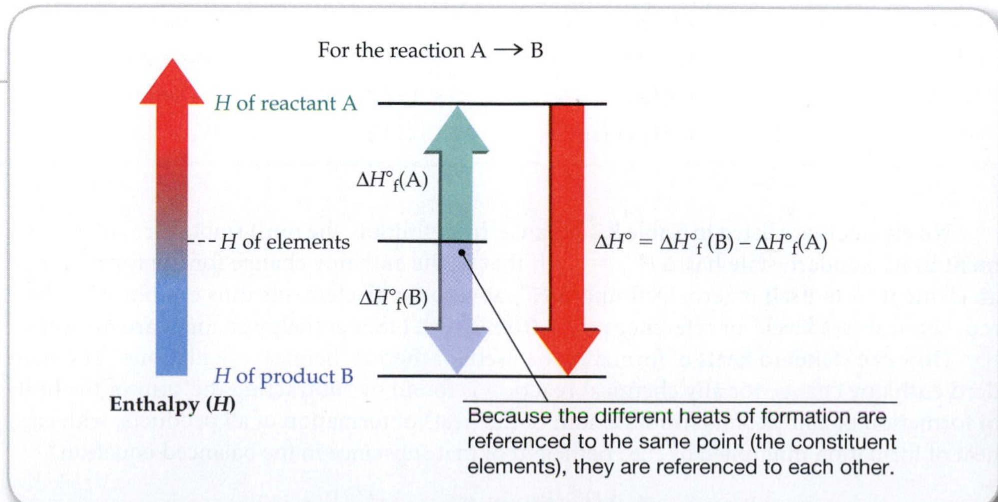

Worked Example 8.8

Using Standard Heats of Formation to Calculate $\Delta H°$

Calculate $\Delta H°$ in kilojoules for the synthesis of lime (CaO) from limestone ($CaCO_3$), the key step in the manufacture of cement.

$$CaCO_3(s) \longrightarrow CaO(s) + CO_2(g) \qquad \Delta H°_f[CaCO_3(s)] = -1207.6 \text{ kJ/mol}$$
$$\Delta H°_f[CaO(s)] = -634.9 \text{ kJ/mol}$$
$$\Delta H°_f[CO_2(g)] = -393.5 \text{ kJ/mol}$$

STRATEGY

Subtract the heat of formation of the reactant from the sum of the heats of formation of the products.

SOLUTION

$$\begin{aligned}\Delta H° &= [\Delta H°_f(CaO) + \Delta H°_f(CO_2)] - [\Delta H°_f(CaCO_3)] \\ &= (1 \text{ mol})(-634.9 \text{ kJ/mol}) + (1 \text{ mol})(-393.5 \text{ kJ/mol}) - (1 \text{ mol})(-1207.6 \text{ kJ/mol}) \\ &= +179.2 \text{ kJ}\end{aligned}$$

The reaction is endothermic by 179.2 kJ.

Worked Example 8.9

Using Standard Heats of Formation to Calculate $\Delta H°$

Oxyacetylene welding torches burn acetylene gas, $C_2H_2(g)$. Use the information in Table 8.2 to calculate $\Delta H°$ in kilojoules for the combustion reaction of acetylene to yield $CO_2(g)$ and $H_2O(g)$.

▲ Acetylene burns at a high temperature, making the reaction useful for cutting and welding iron

STRATEGY

Write the balanced equation, look up the appropriate heats of formation for each reactant and product in Table 8.2, and then carry out the calculation, making sure to multiply each $\Delta H°_f$ by the coefficient given in the balanced equation. Remember also that $\Delta H°_f(O_2) = 0$ kJ/mol.

SOLUTION

The balanced equation is

$$2\,C_2H_2(g) + 5\,O_2(g) \longrightarrow 4\,CO_2(g) + 2\,H_2O(g)$$

The necessary heats of formation from Table 8.2 are

$$\Delta H°_f[C_2H_2(g)] = +227.4 \text{ kJ/mol} \qquad \Delta H°_f[H_2O(g)] = -241.8 \text{ kJ/mol}$$
$$\Delta H°_f[CO_2(g)] = -393.5 \text{ kJ/mol}$$

The standard enthalpy change for the reaction is

$$\begin{aligned}\Delta H° &= [4\,\Delta H°_f(CO_2) + 2\,\Delta H°_f(H_2O)] - [2\,\Delta H°_f(C_2H_2)] \\ &= (4 \text{ mol})(-393.5 \text{ kJ/mol}) + (2 \text{ mol})(-241.8 \text{ kJ/mol}) - (2 \text{ mol})(227.4 \text{ kJ/mol}) \\ &= -2512.4 \text{ kJ}\end{aligned}$$

▶ **PROBLEM 8.15** Use the information in Table 8.2 to calculate $\Delta H°$ in kilojoules for the reaction of ammonia with O_2 to yield nitric oxide (NO) and $H_2O(g)$, a step in the Ostwald process for the commercial production of nitric acid.

▶ **PROBLEM 8.16** Use the information in Table 8.2 to calculate $\Delta H°$ in kilojoules for the photosynthesis of glucose ($C_6H_{12}O_6$) and O_2 from CO_2 and liquid H_2O, a reaction carried out by all green plants.

8.10 CALCULATING ENTHALPY CHANGES USING BOND DISSOCIATION ENERGIES

The procedure described in the previous section for determining overall heats of reaction from heats of formation is extremely useful, but it still presents a problem. To use the method, it's necessary to know $\Delta H°_f$ for every substance in a reaction. This implies, in turn, that vast numbers of measurements are needed because there are over 67 million known chemical compounds. In practice, though, only a few thousand $\Delta H°_f$ values have been determined.

For those reactions where insufficient $\Delta H°_f$ data are available to allow an exact calculation of $\Delta H°$, it's often possible to find an approximate $\Delta H°$ by using the average **bond dissociation energies (*D*)** discussed in Section 4.3. Although we didn't identify it as such at the time, a bond dissociation energy is really just a standard enthalpy change for the corresponding bond-breaking reaction.

REMEMBER...

Bond dissociation energy is the amount of energy that must be supplied to break a chemical bond in an isolated molecule in the gaseous state and is thus the amount of energy released when the bond forms. (Section 4.3)

$$\text{For the reaction X–Y} \longrightarrow \text{X} + \text{Y} \qquad \Delta H° = D = \text{Bond dissociation energy}$$

When we say, for example, that the bond dissociation energy of Cl_2 is 243 kJ/mol, we mean that the standard enthalpy change for the reaction $Cl_2(g) \longrightarrow 2\,Cl(g)$ is $\Delta H° = +243$ kJ. Bond dissociation energies are always positive because energy must always be put into bonds to break them.

Applying Hess's law, we can calculate an enthalpy change for any reaction by subtracting the sum of the bond dissociation energies in the products from the sum of the bond dissociation energies in the reactants.

$$\Delta H° = D(\text{Reactant bonds}) - D(\text{Product bonds})$$

In the reaction of H_2 with Cl_2 to yield HCl, for example, the reactants have one Cl–Cl bond and one H–H bond, while the product has two H–Cl bonds.

$$H_2(g) + Cl_2(g) \longrightarrow 2\,HCl(g)$$

According to the data shown previously in Table 4.2 and reproduced in **Table 8.3** for convenience, the bond dissociation energy of Cl_2 is 243 kJ/mol, that of H_2 is 436 kJ/mol, and that of HCl is 432 kJ/mol. We can thus calculate that the standard enthalpy change for the reaction is $\Delta H° = -185$ kJ.

$$\begin{aligned}\Delta H° &= D(\text{Reactant bonds}) - D(\text{Product bonds})\\ &= (D_{\text{Cl–Cl}} + D_{\text{H–H}}) - (2\,D_{\text{H–Cl}})\\ &= [(1\text{ mol})(243\text{ kJ/mol}) + (1\text{ mol})(436\text{ kJ/mol})] - (2\text{ mol})(432\text{ kJ/mol})\\ &= -185\text{ kJ}\end{aligned}$$

TABLE 8.3 Average Bond Dissociation Energies, *D* (kJ/mol)

H–H	436[a]	C–H	410	N–H	390	O–F	180	I–I	151[a]
H–C	410	C–C	350	N–C	300	O–Cl	200	S–F	310
H–F	570[a]	C–F	450	N–F	270	O–Br	210	S–Cl	250
H–Cl	432[a]	C–Cl	330	N–Cl	200	O–I	220	S–Br	210
H–Br	366[a]	C–Br	270	N–Br	240	O–N	200	S–S	225
H–I	298[a]	C–I	240	N–N	240	O–O	180		
H–N	390	C–N	300	N–O	200	F–F	159[a]		
H–O	460	C–O	350	O–H	460	Cl–Cl	243[a]		
H–S	340	C–S	260	O–C	350	Br–Br	193[a]		
Multiple covalent bonds									
C=C	728	C≡C	965	C=O	732	O=O	498[a]	N≡N	945[a]

[a]Exact value

One word of warning: Accurate calculations only result when the bond dissociation energies are known accurately. Using approximate bond-dissociation values like many of those in Table 8.3 gives only an approximate answer.

Worked Example 8.10

Using Bond Dissociation Energies to Calculate $\Delta H°$

Use the data in Table 8.3 to find an approximate $\Delta H°$ in kilojoules for the industrial synthesis of chloroform by reaction of methane with Cl_2.

$$CH_4(g) + 3\,Cl_2(g) \longrightarrow CHCl_3(g) + 3\,HCl(g)$$

STRATEGY

Identify all the bonds in the reactants and products, and look up the appropriate bond dissociation energies in Table 8.3. Then subtract the sum of the bond dissociation energies in the products from the sum of the bond dissociation energies in the reactants to find the enthalpy change for the reaction.

SOLUTION

The reactants have four C–H bonds and three Cl–Cl bonds; the products have one C–H bond, three C–Cl bonds, and three H–Cl bonds. The approximate bond dissociation energies from Table 8.3 are:

C–H	$D = 410$ kJ/mol	Cl–Cl	$D = 243$ kJ/mol
C–Cl	$D = 330$ kJ/mol	H–Cl	$D = 432$ kJ/mol

Subtracting the product bond dissociation energies from the reactant bond dissociation energies gives the approximate enthalpy change for the reaction:

$$\begin{aligned}\Delta H° &= [3\,D_{Cl-Cl} + 4\,D_{C-H}] - [D_{C-H} + 3\,D_{H-Cl} + 3\,D_{C-Cl}]\\ &= [(3\text{ mol})(243\text{ kJ/mol}) + (4\text{ mol})(410\text{ kJ/mol})] - [(1\text{ mol})(410\text{ kJ/mol})\\ &\qquad + (3\text{ mol})(432\text{ kJ/mol}) + (3\text{ mol})(330\text{ kJ/mol})]\\ &= -327\text{ kJ}\end{aligned}$$

The reaction is exothermic by approximately 330 kJ.

PROBLEM 8.17 Use the data in Table 8.3 to calculate an approximate $\Delta H°$ in kilojoules for the industrial synthesis of ethyl alcohol from ethylene: $C_2H_4(g) + H_2O(g) \rightarrow C_2H_5OH(g)$.

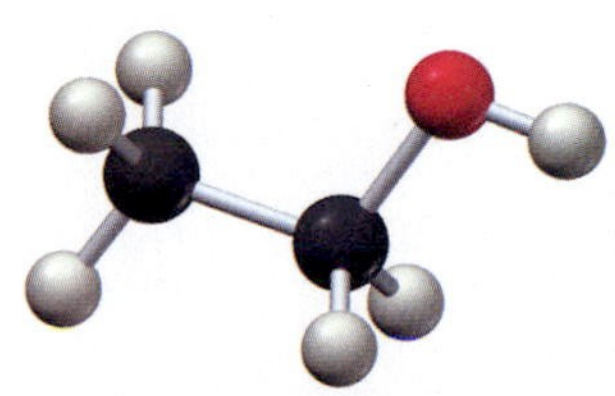
Ethyl alcohol

PROBLEM 8.18 Use the data in Table 8.3 to calculate an approximate $\Delta H°$ in kilojoules for the synthesis of hydrazine from ammonia: $2\,NH_3(g) + Cl_2(g) \longrightarrow N_2H_4(g) + 2\,HCl(g)$.

8.11 FOSSIL FUELS AND HEATS OF COMBUSTION

Surely the most familiar of all exothermic reactions is the one that takes place every time you turn up a thermostat, drive a car, or light a match: the combustion of a carbon-based fuel by reaction with oxygen to yield H_2O, CO_2, and heat. The amount of energy released on burning a substance is called its **heat of combustion**, or *combustion enthalpy*, $\Delta H°_c$, and is simply the standard enthalpy change for the reaction of 1 mol of the substance with oxygen. Hydrogen, for instance, has $\Delta H°_c = -285.8$ kJ/mol, and methane has $\Delta H°_c = -890.3$ kJ/mol. Note that the H_2O product in reporting heats of combustion is $H_2O(l)$ rather than $H_2O(g)$ because water at 25 °C in its standard state is liquid.

$$H_2(g) + 1/2\,O_2(g) \longrightarrow H_2O(l) \qquad \Delta H°_c = -285.8\text{ kJ/mol}$$
$$CH_4(g) + 2\,O_2(g) \longrightarrow CO_2(g) + 2\,H_2O(l) \qquad \Delta H°_c = -890.3\text{ kJ/mol}$$

To compare the efficiency of different fuels, it's more useful to calculate combustion enthalpies per gram or per milliliter of substance rather than per mole (**Table 8.4**). For applications where weight is important, as in rocket engines, hydrogen is ideal because its combustion enthalpy per gram is the highest of any known fuel. For applications where volume is important, as in automobiles, a mixture of *hydrocarbons*—compounds of carbon and hydrogen—such as those in gasoline is most efficient because hydrocarbon combustion enthalpies per milliliter are relatively high. Octane and toluene are representative examples.

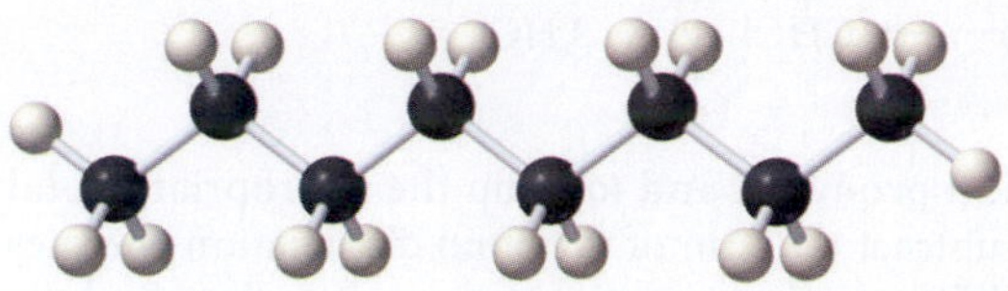

Octane

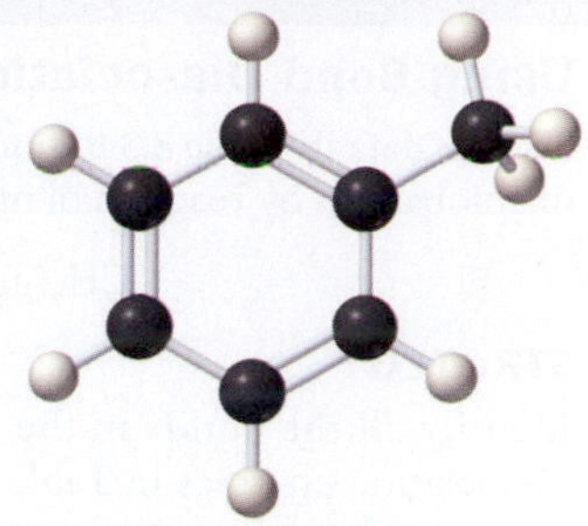

Toluene

TABLE 8.4 Thermochemical Properties of Some Fuels

	Combustion Enthalpy		
Fuel	**kJ/mol**	**kJ/g**	**kJ/mL**
Hydrogen, $H_2(l)$	−285.8	−141.8	−9.9*
Ethanol, $C_2H_5OH(l)$	−1366.8	−29.7	−23.4
Graphite, $C(s)$	−393.5	−32.8	−73.8
Methane, $CH_4(g)$	−890.3	−55.5	−30.8*
Methanol, $CH_3OH(l)$	−725.9	−22.7	−17.9
Octane, $C_8H_{18}(l)$	−5470	−47.9	−33.6
Toluene, $C_7H_8(l)$	−3910	−42.3	−36.7

*Calculated for the compressed liquid at 0 °C

You might also note in Table 8.4 that combustion enthalpies per milliliter for the commonly used alternative fuels methanol and ethanol are substantially lower than those for hydrocarbons. Methanol is particularly inefficient, with a combustion enthalpy per milliliter approximately half that of gasoline.

With the exception of hydrogen, all common fuels are organic compounds, whose energy is derived ultimately from the sun through the photosynthesis of carbohydrates in green plants. Although the details are complex, the net result of the photosynthesis reaction is the conversion of carbon dioxide and water into glucose, $C_6H_{12}O_6$, plus oxygen. Glucose, once formed, is converted into cellulose and starch, which in turn act as structural materials for plants and as food sources for animals. The conversion is highly endothermic and therefore requires a large input of solar energy. It has been estimated that the total annual amount of solar energy absorbed by the Earth's vegetation is approximately 10^{19} kJ, an amount sufficient to synthesize 5×10^{14} kg of glucose per year.

$$6\,CO_2(g) + 6\,H_2O(l) \longrightarrow C_6H_{12}O_6(s) + 6\,O_2(g) \quad \Delta H° = +2803 \text{ kJ}$$

▲ Much coal lies near the surface of the Earth and is obtained by strip mining.

The fossil fuels we use most—coal, petroleum, and natural gas—are derived from the decayed remains of organisms from previous geologic eras. Both coal and petroleum are enormously complex mixtures of compounds. Coal is primarily of vegetable origin, and many of the compounds it contains are structurally similar to graphite (pure carbon). Petroleum is a viscous liquid mixture of hydrocarbons that are primarily of marine origin, and natural gas is primarily methane, CH_4.

Coal is burned just as it comes from the mine, but petroleum must be *refined* before use. Refining begins with *distillation*, the separation of crude liquid oil into fractions on the basis of their boiling points (bp). So-called straight-run gasoline (bp 30–200 °C) consists of compounds with 5–11 carbon atoms per molecule; kerosene (bp 175–300 °C) contains compounds in the C_{11}–C_{14} range; gas oil (bp 275–400 °C) contains C_{14}–C_{25} substances; and lubricating oils contain whatever remaining compounds will distill. Left over is a tarry residue of asphalt (**Figure 8.11**).

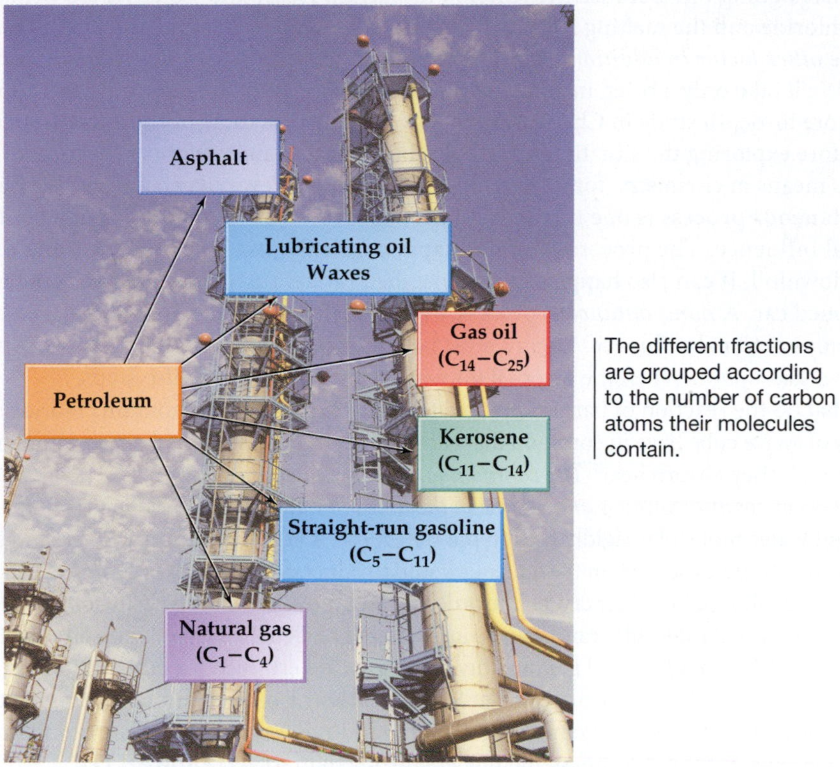

FIGURE 8.11

The products of petroleum refining.

As the world's petroleum deposits become more scarce, other sources of energy will have to be found to replace them. Hydrogen, although it burns cleanly and is relatively nonpolluting, has numerous drawbacks: low current availability, difficulty in transport and storage, and low combustion enthalpy per milliliter, to name a few. Ethanol is a potential choice as an alternative fuel but its combustion enthalpy is substantially lower than that of petroleum. At present, ethanol is produced largely by fermentation of corn or cane sugar, but attempts are being made to produce it from waste wood by the breakdown of cellulose to glucose and subsequent fermentation.

PROBLEM 8.19 Liquid butane (C_4H_{10}), the fuel used in many disposable lighters, has $\Delta H°_f = -147.5$ kJ/mol and a density of 0.579 g/mL. Write a balanced equation for the combustion of butane, and use Hess's law to calculate the enthalpy of combustion in kJ/mol, kJ/g, and kJ/mL.

Butane

8.12 AN INTRODUCTION TO ENTROPY

Before ending this chapter, let's go back to a statement made in the introduction where we said that chemical reactions and physical processes occur when the final state is more stable than the initial state. Because less stable substances generally have higher internal energy and are converted into more stable substances with lower internal energy, energy is generally released in chemical reactions. At the same time, though, we've seen that some reactions and processes occur even though they *absorb* rather than release energy. The endothermic reaction of barium hydroxide octahydrate with ammonium chloride shown previously in Figure 8.8, for example, absorbs 80.3 kJ of heat from the surroundings ($\Delta H° = +80.3$ kJ) and leaves the surroundings so cold that the temperature drops below 0 °C.

$$Ba(OH)_2 \cdot 8\,H_2O(s) + 2\,NH_4Cl(s) \longrightarrow BaCl_2(aq) + 2\,NH_3(aq) + 10\,H_2O(l) \qquad \Delta H° = +80.3 \text{ kJ}$$

An example of a physical process that takes place spontaneously yet absorbs energy takes place every time an ice cube melts. At a temperature of 0 °C, ice spontaneously absorbs heat from the surroundings to turn from solid into liquid water.

What's going on? Because the reaction of barium hydroxide octahydrate with ammonium chloride and the melting ice cube absorb heat yet still occur spontaneously, *there must be some other factor in addition to energy that determines whether a reaction or process will occur.* We'll take only a brief, introductory look at this additional factor now and will return for a more in-depth study in Chapter 16 when we take up the topic of **thermodynamics**.

LOOKING AHEAD...

The entire subject of **thermodynamics**, including a more detailed discussion of entropy and free energy, is the subject of Chapter 16.

Before exploring the situation further, it's important to understand what the word *spontaneous* means in chemistry, for it's not quite the same as in everyday language. In chemistry, a **spontaneous process** is one that, once started, proceeds on its own without a continuous external influence. The process need not happen quickly, like a spring uncoiling or a sled going downhill. It can also happen slowly, like the gradual rusting away of an iron bridge or abandoned car. A *nonspontaneous* process, by contrast, takes place only in the presence of a continuous external influence. Energy must be continuously expended to re-coil a spring or to push a sled uphill. When the external influence stops, the process also stops.

▲ Sledding downhill is a spontaneous process: Once started, it continues on its own. Dragging the sled back uphill, however, is a nonspontaneous process that requires a continuous input of energy.

What do the reaction of barium hydroxide octahydrate with ammonium chloride and the melting of an ice cube have in common that allows the two processes to take place spontaneously even though they absorb heat? *The common feature of these and all other processes that absorb heat yet occur spontaneously is an increase in the amount of molecular randomness of the system.* The eight water molecules rigidly held in the $Ba(OH)_2 \cdot 8\,H_2O$ crystal break loose and become free to move about randomly in the aqueous liquid product. Similarly, the rigidly held H_2O molecules in the solid ice lose their crystalline ordering and move around freely in liquid water.

The amount of molecular randomness in a system is called the system's **entropy (*S*)**. Entropy has the units J/K (not kJ/K) and is a quantity that can be determined for pure substances, as we'll see in Section 16.5. The larger the value of *S*, the greater the molecular randomness of the particles in the system. Gases, for example, have more randomness and higher entropy than liquids, and liquids have more randomness and higher entropy than solids (**Figure 8.12**).

FIGURE 8.12

Entropy is a measure of molecular randomness.

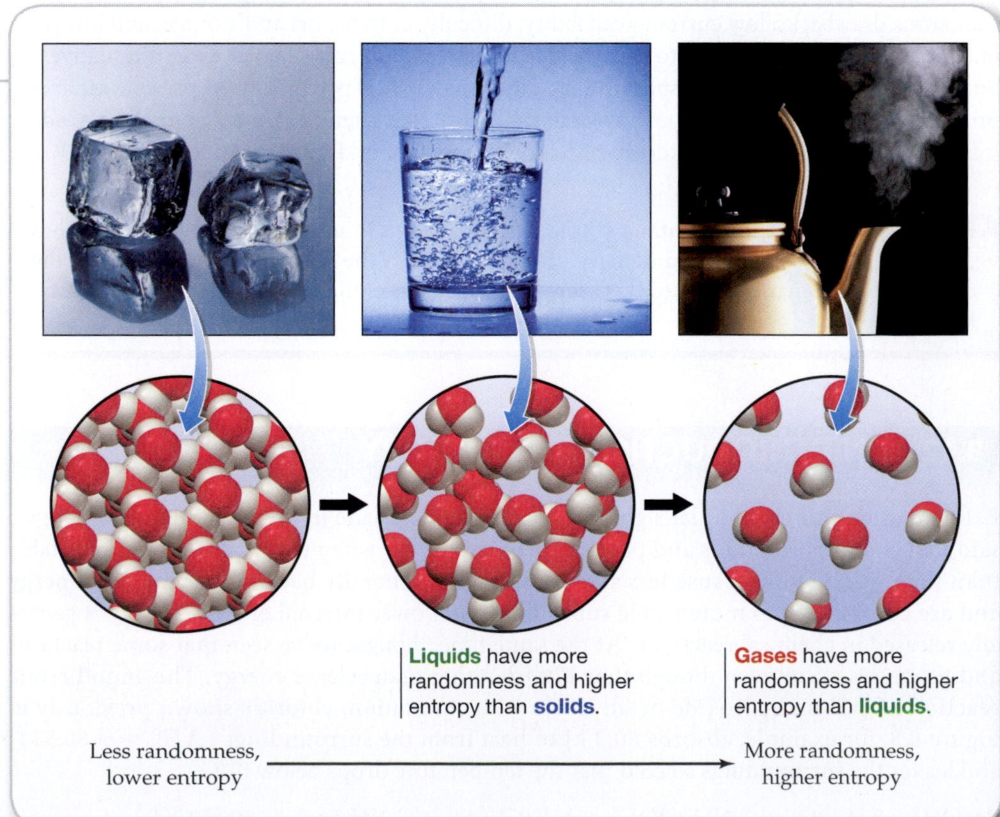

A change in entropy is represented as $\Delta S = S_{\text{final}} - S_{\text{initial}}$. When randomness increases, as it does when barium hydroxide octahydrate reacts or ice melts, ΔS has a positive value because $S_{\text{final}} > S_{\text{initial}}$. The reaction of $Ba(OH)_2 \cdot 8\,H_2O(s)$ with $NH_4Cl(s)$ has $\Delta S° = +428$ J/K, and the melting of ice has $\Delta S° = +22.0$ J/(K · mol). When randomness decreases, ΔS is negative because $S_{\text{final}} < S_{\text{initial}}$. The freezing of water, for example, has $\Delta S° = -22.0$ J/(K · mol). (As with $\Delta H°$, the superscript ° is used in $\Delta S°$ to refer to the standard entropy change in a reaction where products and reactants are in their standard states.)

Thus, two factors determine the spontaneity of a chemical or physical change in a system: a release or absorption of heat (ΔH) and an increase or decrease in molecular randomness (ΔS). To decide whether a process is spontaneous, both enthalpy and entropy changes must be taken into account:

Spontaneous process:
- Favored by **decrease in *H*** (negative ΔH)
- Favored by **increase in *S*** (positive ΔS)

Nonspontaneous process:
- Favored by **increase in *H*** (positive ΔH)
- Favored by **decrease in *S*** (negative ΔS)

Note that the two factors don't have to operate in the same direction. Thus, it's possible for a process to be *disfavored* by enthalpy (endothermic, positive ΔH) yet still be spontaneous because it is strongly *favored* by entropy (positive ΔS). The melting of ice [$\Delta H° = +6.01$ kJ/mol; $\Delta S° = +22.0$ J/(K · mol)] is just such a process, as is the reaction of barium hydroxide octahydrate with ammonium chloride ($\Delta H° = +80.3$ kJ; $\Delta S° = +428$ J/K). In the latter case, 3 mol of solid reactants produce 10 mol of liquid water, 2 mol of dissolved ammonia, and 3 mol of dissolved ions (1 mol of Ba^{2+} and 2 mol of Cl^-), with a consequently large increase in molecular randomness:

$$\underbrace{Ba(OH)_2 \cdot 8\,H_2O(s) + 2\,NH_4Cl(s)}_{\text{3 mol solid reactants}} \longrightarrow \underset{\substack{\text{3 mol}\\ \text{dissolved ions}}}{BaCl_2(aq)} + \underset{\substack{\text{2 mol dissolved}\\ \text{molecules}}}{2\,NH_3(aq)} + \underset{\substack{\text{10 mol liquid}\\ \text{water molecules}}}{10\,H_2O(l)}$$

$$\Delta H° = +80.3\text{ kJ} \longleftarrow \text{Unfavorable}$$

$$\Delta S° = +428\text{ J/K} \longleftarrow \text{Favorable}$$

Conversely, it's also possible for a process to be favored by enthalpy (exothermic, negative ΔH) yet be nonspontaneous because it is strongly disfavored by entropy (negative ΔS). The conversion of liquid water to ice is nonspontaneous above 0 °C, for example, because the process is disfavored by entropy [$\Delta S° = -22.0$ J/(K · mol)] even though it is favored by enthalpy ($\Delta H° = -6.01$ kJ/mol).

Worked Example 8.11

Predicting the Sign of ΔS for a Reaction

Predict whether $\Delta S°$ is likely to be positive or negative for each of the following reactions:

(a) $H_2C{=}CH_2(g) + Br_2(g) \longrightarrow BrCH_2CH_2Br(l)$

(b) $2\,C_2H_6(g) + 7\,O_2(g) \longrightarrow 4\,CO_2(g) + 6\,H_2O(g)$

STRATEGY

Look at each reaction, and try to decide whether molecular randomness increases or decreases. Reactions that increase the number of gaseous molecules generally have a positive ΔS, while reactions that decrease the number of gaseous molecules have a negative ΔS.

continued on next page

SOLUTION

(a) The amount of molecular randomness in the system decreases when 2 mol of gaseous reactants combine to give 1 mol of liquid product, so the reaction has a negative $\Delta S°$.

(b) The amount of molecular randomness in the system increases when 9 mol of gaseous reactants give 10 mol of gaseous products, so the reaction has a positive $\Delta S°$.

PROBLEM 8.20 Ethane, C_2H_6, can be prepared by the reaction of acetylene, C_2H_2, with hydrogen. Is $\Delta S°$ for the reaction likely to be positive or negative? Explain.

$$C_2H_2(g) + 2\,H_2(g) \longrightarrow C_2H_6(g)$$

CONCEPTUAL PROBLEM 8.21 Is the reaction represented in the following drawing likely to have a positive or a negative value of $\Delta S°$? Explain.

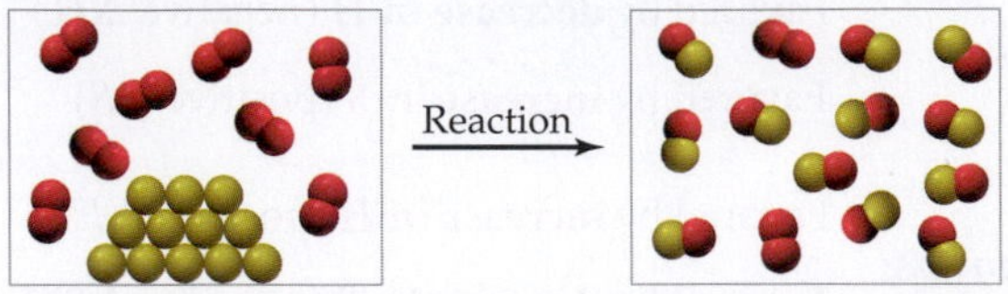

8.13 AN INTRODUCTION TO FREE ENERGY

How do we weigh the relative contributions of enthalpy changes (ΔH) and entropy changes (ΔS) to the overall spontaneity of a process? To take both factors into account when deciding the spontaneity of a chemical reaction or other process, we define a quantity called the **Gibbs free-energy change (Δ*G*)**, which is related to ΔH and ΔS by the equation $\Delta G = \Delta H - T\Delta S$.

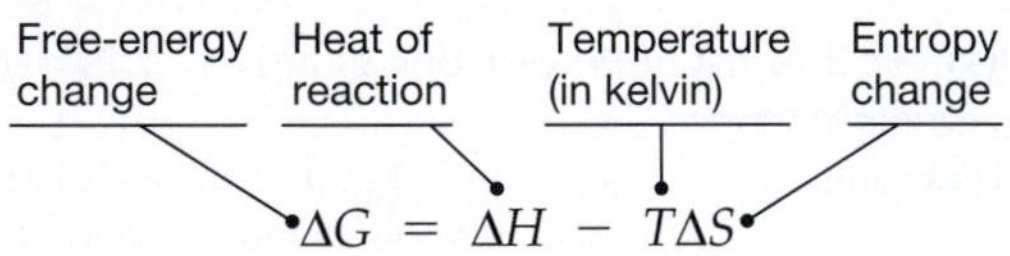

The value of the free-energy change ΔG determines whether a chemical or physical process occurs spontaneously. If ΔG has a negative value, free energy is released and the process is spontaneous. If ΔG has a value of 0, the process is neither spontaneous nor nonspontaneous but is instead at an equilibrium. And if ΔG has a positive value, free energy is absorbed and the process is nonspontaneous.

$\Delta G < 0$ Process is spontaneous

$\Delta G = 0$ Process is at equilibrium—neither spontaneous nor nonspontaneous

$\Delta G > 0$ Process is nonspontaneous

The fact that the $T\Delta S$ term in the free-energy equation is temperature dependent implies that some processes might be either spontaneous or nonspontaneous, depending on the temperature. At a low temperature, for instance, an unfavorable (positive) ΔH term might be larger than a favorable (positive) $T\Delta S$ term, but at higher temperature, the $T\Delta S$ term might be larger. Thus, an endothermic process that is nonspontaneous at low temperature can become spontaneous at higher temperature. This, in fact, is exactly what happens in the ice–water transition. At a temperature below 0 °C, the melting of ice is nonspontaneous because the unfavorable ΔH term outweighs the favorable $T\Delta S$ term. At a temperature above 0 °C, however, the melting of ice is spontaneous because the favorable $T\Delta S$ term outweighs the unfavorable ΔH term (**Figure 8.13**). At exactly 0 °C, the two terms are balanced.

$$\Delta G^\circ = \Delta H^\circ - T\Delta S^\circ$$

$$\text{At } -10\ ^\circ\text{C (263 K):}\quad \Delta G^\circ = 6.01\,\frac{\text{kJ}}{\text{mol}} - (263\text{ K})\left(0.0220\,\frac{\text{kJ}}{\text{K}\cdot\text{mol}}\right) = +0.22\text{ kJ/mol}$$

$$\text{At } 0\ ^\circ\text{C (273 K):}\quad \Delta G^\circ = 6.01\,\frac{\text{kJ}}{\text{mol}} - (273\text{ K})\left(0.0220\,\frac{\text{kJ}}{\text{K}\cdot\text{mol}}\right) = 0.00\text{ kJ/mol}$$

$$\text{At } +10\ ^\circ\text{C (283 K):}\quad \Delta G^\circ = 6.01\,\frac{\text{kJ}}{\text{mol}} - (283\text{ K})\left(0.0220\,\frac{\text{kJ}}{\text{K}\cdot\text{mol}}\right) = -0.22\text{ kJ/mol}$$

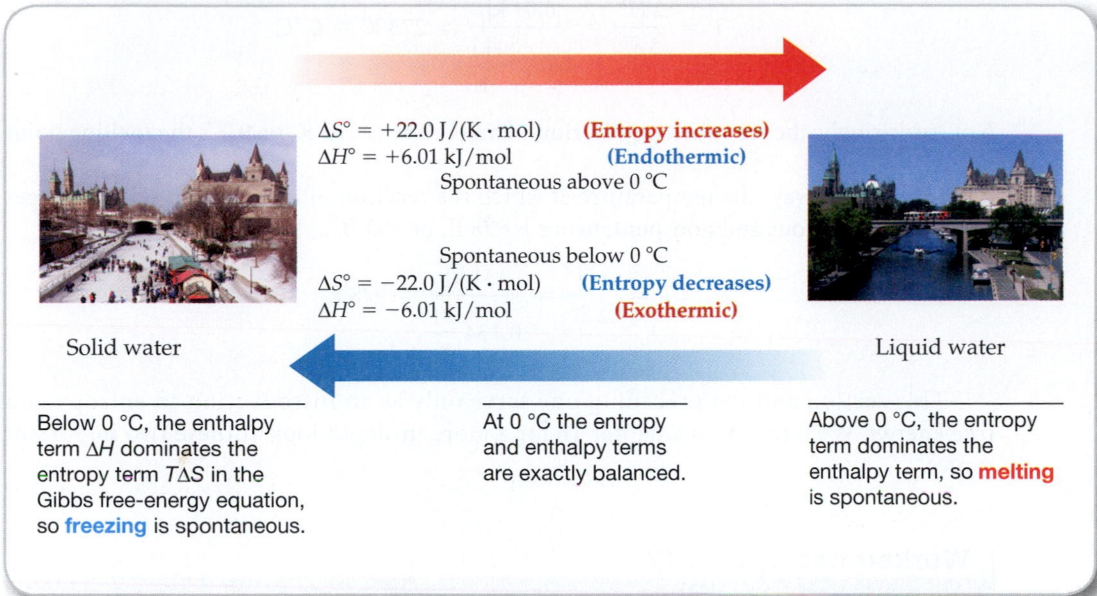

FIGURE 8.13

Melting and freezing. The melting of ice is disfavored by enthalpy ($\Delta H > 0$) but favored by entropy ($\Delta S > 0$). The freezing of water is favored by enthalpy ($\Delta H < 0$) but disfavored by entropy ($\Delta S < 0$).

An example of a chemical reaction in which temperature controls spontaneity is that of carbon with water to yield carbon monoxide and hydrogen. The reaction has an unfavorable ΔH term (positive) but a favorable $T\Delta S$ term (positive) because randomness increases when 1 mol of solid and 1 mol of gas are converted into 2 mol of gas:

$$C(s) + H_2O(g) \longrightarrow CO(g) + H_2(g) \qquad \Delta H^\circ = +131\text{ kJ} \quad \textbf{Unfavorable}$$

$$\Delta S^\circ = +134\text{ J/K} \quad \textbf{Favorable}$$

No reaction occurs if carbon and water are mixed at room temperature because the unfavorable ΔH term outweighs the favorable $T\Delta S$ term. At approximately 978 K (705 °C), however, the reaction becomes spontaneous because the favorable $T\Delta S$ term becomes larger than the unfavorable ΔH term. Below 978 K, ΔG has a positive value; at 978 K, $\Delta G = 0$; and above 978 K, ΔG has a negative value. (The calculation is not exact because values of ΔH and ΔS themselves vary somewhat with temperature.)

$$\Delta G^\circ = \Delta H^\circ - T\Delta S^\circ$$

$$\text{At } 695\ ^\circ\text{C (968 K):}\quad \Delta G^\circ = 131\text{ kJ} - (968\text{ K})\left(0.134\,\frac{\text{kJ}}{\text{K}}\right) = +1\text{ kJ}$$

$$\text{At } 705\ ^\circ\text{C (978 K):}\quad \Delta G^\circ = 131\text{ kJ} - (978\text{ K})\left(0.134\,\frac{\text{kJ}}{\text{K}}\right) = 0\text{ kJ}$$

$$\text{At } 715\ ^\circ\text{C (988 K):}\quad \Delta G^\circ = 131\text{ kJ} - (988\text{ K})\left(0.134\,\frac{\text{kJ}}{\text{K}}\right) = -1\text{ kJ}$$

The reaction of carbon with water is, in fact, the first step of an industrial process for manufacturing methanol (CH_3OH). As supplies of natural gas and oil diminish, this reaction may become useful for the manufacture of synthetic fuels.

A process is at equilibrium when it is balanced between spontaneous and nonspontaneous—that is, when $\Delta G = 0$ and it is energetically unfavorable to go either from reactants to products or from products to reactants. Thus, at the equilibrium point, we can set up the equation

$$\Delta G = \Delta H - T\Delta S = 0 \qquad \textbf{At equilibrium}$$

Solving this equation for T gives

$$T = \frac{\Delta H}{\Delta S}$$

which makes it possible to calculate the temperature at which a changeover in behavior between spontaneous and nonspontaneous occurs. Using the known values of $\Delta H°$ and $\Delta S°$ for the melting of ice, for instance, we find that the point at which liquid water and solid ice are in equilibrium is

$$T = \frac{\Delta H°}{\Delta S°} = \frac{6.01\ \text{kJ}}{0.0220\ \dfrac{\text{kJ}}{\text{K}}} = 273\ \text{K} = 0\ °\text{C}$$

Not surprisingly, the ice–water equilibrium temperature is 273 K, or 0 °C, the melting point of ice.

In the same way, the temperature at which the reaction of carbon with water changes between spontaneous and nonspontaneous is 978 K, or 705 °C:

$$T = \frac{\Delta H°}{\Delta S°} = \frac{131\ \text{kJ}}{0.134\ \dfrac{\text{kJ}}{\text{K}}} = 978\ \text{K}$$

This section and the preceding one serve only as an introduction to entropy and free energy. We'll return in Chapter 16 for a more in-depth look at these two important topics.

Worked Example 8.12

Using the Free-Energy Equation to Calculate Equilibrium Temperature

Lime (CaO) is produced by heating limestone ($CaCO_3$) to drive off CO_2 gas, a reaction used to make Portland cement. Is the reaction spontaneous under standard conditions at 25 °C? Calculate the temperature at which the reaction becomes spontaneous.

$$CaCO_3(s) \longrightarrow CaO(s) + CO_2(g) \qquad \Delta H° = +179.2\ \text{kJ};\ \Delta S° = +160.0\ \text{J/K}$$

STRATEGY

The spontaneity of the reaction at a given temperature can be found by determining whether ΔG is positive or negative at that temperature. The changeover point between spontaneous and nonspontaneous can be found by setting $\Delta G = 0$ and solving for T.

SOLUTION

At 25 °C (298 K), we have

$$\Delta G = \Delta H - T\Delta S = 179.2\ \text{kJ} - (298\ \text{K})\left(0.1600\ \frac{\text{kJ}}{\text{K}}\right) = +131.5\ \text{kJ}$$

Because ΔG is positive at this temperature, the reaction is nonspontaneous. The changeover point between spontaneous and nonspontaneous is approximately

$$T = \frac{\Delta H}{\Delta S} = \frac{179.2\ \text{kJ}}{0.1600\ \dfrac{\text{kJ}}{\text{K}}} = 1120\ \text{K}$$

The reaction becomes spontaneous above approximately 1120 K (847 °C).

Worked Conceptual Example 8.13

Predicting the Signs of ΔH, ΔS, and ΔG for a Reaction

What are the signs of ΔH, ΔS, and ΔG for the following nonspontaneous transformation?

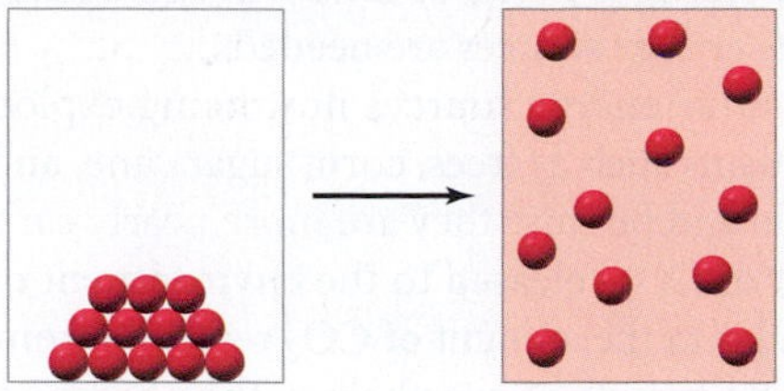

STRATEGY

First, decide what kind of process is represented in the drawing. Then decide whether the process increases or decreases the entropy (molecular randomness) of the system and whether it is exothermic or endothermic.

SOLUTION

The drawing shows ordered particles in a solid subliming to give a gas. Formation of a gas from a solid increases molecular randomness, so ΔS is positive. Furthermore, because we're told that the process is nonspontaneous, ΔG is also positive. Because the process is favored by ΔS (positive) yet still nonspontaneous, ΔH must be unfavorable (positive). This makes sense, because conversion of a solid to a liquid or gas requires energy and is always endothermic.

PROBLEM 8.22 Which of the following reactions are spontaneous under standard conditions at 25 °C, and which are nonspontaneous?

(a) $AgNO_3(aq) + NaCl(aq) \longrightarrow AgCl(s) + NaNO_3(aq)$ $\Delta G° = -55.7\ kJ$

(b) $2\,C(s) + 2\,H_2(g) \longrightarrow C_2H_4(g)$ $\Delta G° = +68.1\ kJ$

PROBLEM 8.23 Is the Haber process for the industrial synthesis of ammonia spontaneous or nonspontaneous under standard conditions at 25 °C? At what temperature (°C) does the changeover occur?

$$N_2(g) + 3\,H_2(g) \longrightarrow 2\,NH_3(g) \qquad \Delta H° = -92.2\ kJ;\ \Delta S° = -199\ J/K$$

CONCEPTUAL PROBLEM 8.24 The following reaction is exothermic:

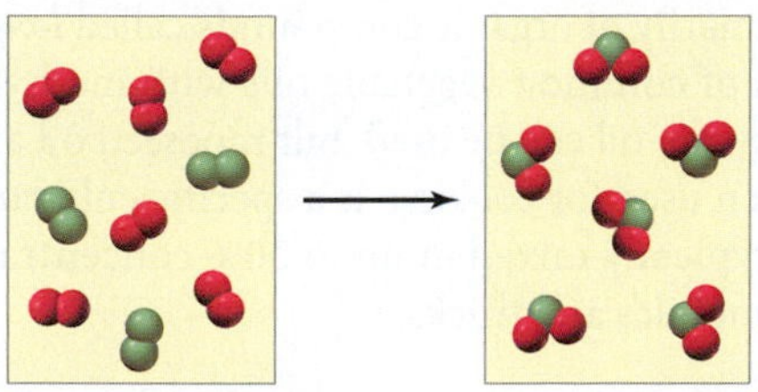

(a) Write a balanced equation for the reaction.

(b) What are the signs of ΔH and ΔS for the reaction?

(c) Is the reaction likely to be spontaneous at low temperatures only, at high temperatures only, or at all temperatures? Explain.

FYI BIOFUELS

The petroleum era began in August 1859, when the world's first oil well was drilled near Titusville, Pennsylvania. Since that time, approximately 1.2×10^{12} barrels of petroleum have been used throughout the world, primarily as fuel for automobiles (1 barrel = 42 gallons).

No one really knows how much petroleum remains on Earth. Current world consumption is approximately 3.1×10^{10} barrels per year, and currently known recoverable reserves

▲ Fermentation of cane sugar is one method for producing ethanol for use as an alternative fuel.

are estimated at 1.4×10^{12} barrels. Thus, the world's known petroleum reserves will be exhausted in approximately 45 years at the current rate of consumption. Additional petroleum reserves are still being found, but consumption is also increasing, making any prediction of the amount of time remaining highly inaccurate. Only two things are certain: The amount of petroleum remaining *is* finite, and we *will* run out at some point, whenever that might be. Thus, alternative energy sources are needed.

Of the various alternative energy sources now being explored, *biofuels*—fuels derived from recently living organisms such as trees, corn, sugar cane, and rapeseed—look promising because they are renewable and because they are more nearly carbon neutral than fossil fuels, meaning that the amount of CO_2 released to the environment during the manufacture and burning of a biofuel is similar to the amount of CO_2 removed from the environment by photosynthesis during the plant's growth. Nevertheless, the phrase *carbon neutral* doesn't mean that biofuels don't release CO_2 when burned; they release just as much CO_2 as any other carbon-based fuel.

The two biofuels receiving the most attention at present are ethanol and biodiesel. Ethanol, sometimes called bioethanol to make it seem more attractive, is simply ethyl alcohol, the same substance found in alcoholic drinks and produced in the same way: by yeast-catalyzed fermentation of carbohydrate.

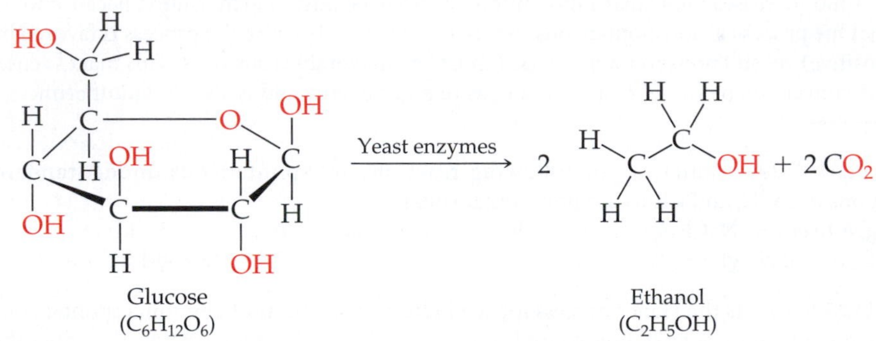

The only difference between beverage ethanol and fuel ethanol is the source of the sugar. Beverage ethanol comes primarily from fermentation of sugar in grapes (for wine) or grains (for distilled liquors), while fuel ethanol comes primarily from fermentation of cane sugar or corn. Much current work is being done, however, on developing economical methods of converting cheap cellulose-based agricultural and logging wastes into fermentable sugars.

Biodiesel consists primarily of organic compounds called *long-chain methyl esters*, which are produced by reaction of common vegetable oils with methyl alcohol in the presence of an acid catalyst. Any vegetable oil can be used, but rapeseed oil and soybean oil are the most common. (*Canola* oil, often used for cooking, is a specific cultivar of generic rapeseed.) Once formed, the biodiesel is typically mixed in up to 30% concentration with petroleum-based diesel fuel for use in automobiles and trucks.

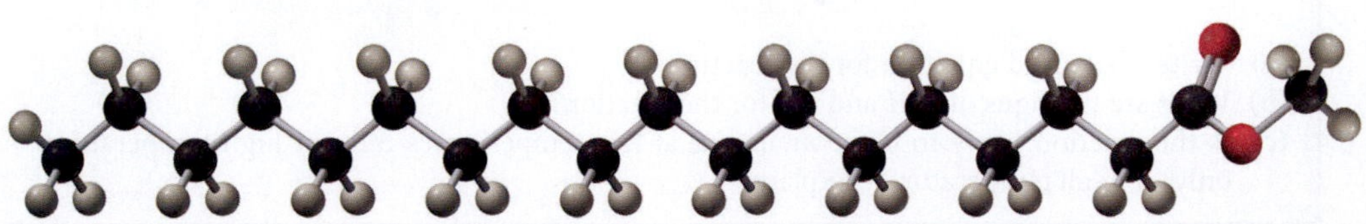

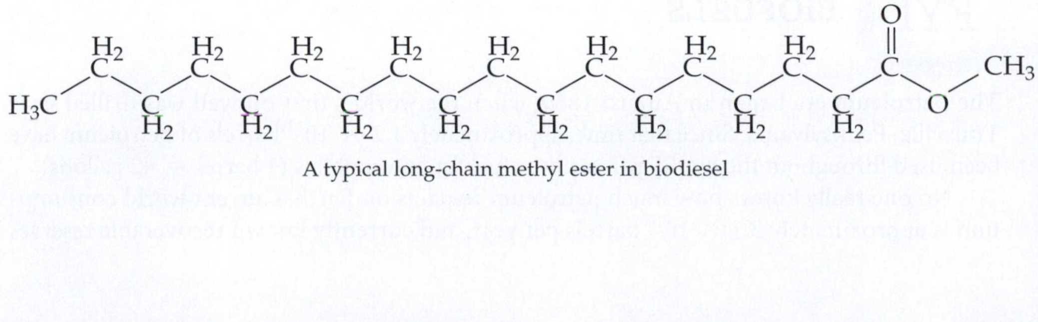

A typical long-chain methyl ester in biodiesel

PROBLEM 8.25 Write balanced equations for the combustion reactions of ethanol (C_2H_6O) and biodiesel ($C_{19}H_{38}O_2$) with oxygen to give CO_2 and H_2O.

PROBLEM 8.26 Biodiesel has a more favorable (more negative) combustion enthalpy per gram than ethanol. Explain, in light of your answer to Problem 8.25.

SUMMARY

Energy is either *kinetic* or *potential*. Kinetic energy (E_K) is the energy of motion. Its value depends on both the mass m and velocity v of an object according to the equation $E_K = (1/2)mv^2$. Potential energy (E_P) is the energy stored in an object because of its position or in a chemical substance because of its composition. **Heat** is the thermal energy transferred between two objects as the result of a temperature difference, whereas **temperature** is a measure of the kinetic energy of molecular motion.

According to the **conservation of energy law**, also known as the **first law of thermodynamics**, energy can be neither created nor destroyed. Thus, the total energy of an isolated system is constant. The total **internal energy (*E*)** of a system—the sum of all kinetic and potential energies for each particle in the system—is a **state function** because its value depends only on the present condition of the system, not on how that condition was reached.

Work (*w*) is defined as the distance moved times the force that produces the motion. In chemistry, most work is expansion work (PV work) done as the result of a volume change during a reaction when air molecules are pushed aside. The amount of work done by an expanding gas is given by the equation $w = -P\Delta V$, where P is the external pressure against which the system must push and ΔV is the change in volume of the system.

The total internal energy change that takes place during a reaction is the sum of the heat transferred (q) and the work done ($-P\Delta V$). The equation

$$\Delta E = q + (-P\Delta V) \quad \text{or} \quad q = \Delta E + P\Delta V = \Delta H$$

where ΔH is the **enthalpy change** of the system, is a fundamental equation of thermochemistry. In general, the $P\Delta V$ term is much smaller than the ΔE term, so that the total internal energy change of a reacting system is approximately equal to ΔH, also called the **heat of reaction**. Reactions that have a negative ΔH are **exothermic** because heat is lost by the system, and reactions that have a positive ΔH are **endothermic** because heat is absorbed by the system.

Because enthalpy is a state function, ΔH is the same regardless of the path taken between reactants and products. Thus, the sum of the enthalpy changes for the individual steps in a reaction is equal to the overall enthalpy change for the entire reaction, a relationship known as **Hess's law**. Using this law, it is possible to calculate overall enthalpy changes for individual steps that can't be measured directly. Hess's law also makes it possible to calculate the enthalpy change of any reaction if the standard heats of formation (ΔH°_f) are known for the reactants and products. The **standard heat of formation** is the enthalpy change for the hypothetical formation of 1 mol of a substance in its **thermodynamic standard state** from the most stable forms of the constituent elements in their standard states (1 atm pressure and a specified temperature, usually 25 °C).

In addition to enthalpy, **entropy (*S*)**—a measure of the amount of molecular randomness in a system—is also important in determining whether a process will occur spontaneously. Together, changes in enthalpy and entropy define a quantity called the **Gibbs free-energy change (ΔG)** according to the equation $\Delta G = \Delta H - T\Delta S$. If ΔG is negative, the reaction is **spontaneous**; if ΔG is positive, the reaction is nonspontaneous.

KEY WORDS

conservation of energy law *275*
endothermic *288*
energy *274*
enthalpy (*H*) *282*
enthalpy change (ΔH) *282*
entropy (*S*) *300*
exothermic *289*
first law of thermodynamics *276*
Gibbs free-energy change (ΔG) *302*
heat *276*
heat capacity (*C*) *284*
heat of combustion (ΔH°_c) *297*
heat of reaction (ΔH) *282*
Hess's law *290*
internal energy (*E*) *277*
joule (*J*) *275*
kinetic energy (E_K) *274*
molar heat capacity (C_m) *284*
potential energy (E_P) *275*
specific heat *284*
spontaneous process *300*
standard enthalpy of reaction (ΔH°) *287*
standard heat of formation (ΔH°_f) *293*
state function *278*
sublimation *288*
temperature *276*
thermochemistry *274*
thermodynamic standard state *286*
work (*w*) *279*

KEY EQUATIONS

- Kinetic energy **(Section 8.1)**

$$E_K = \frac{1}{2}mv^2 \qquad \text{where } m = \text{mass and } v = \text{velocity}$$

- Work **(Section 8.3)**

$$\text{Work } (w) = \text{Force } (F) \times \text{Distance } (d) = -P\Delta V$$

- Internal energy change **(Section 8.4)**

$$\Delta E = q + w = q - P\Delta V \qquad \text{where } q = \text{heat, } P = \text{pressure, and } \Delta V = \text{volume change}$$

- Enthalpy change **(Section 8.4)**

$$\Delta H = q_p = \Delta E + P\Delta V$$ where q_p = heat at constant pressure, P = pressure, ΔE = internal energy change, and ΔV = volume change

- Heat capacity **(Section 8.5)**

$$C = \frac{q}{\Delta T}$$ where q = heat and ΔT = temperature change

- Heat of reaction **(Section 8.9)**

For the reaction $a\text{A} + b\text{B} + \ldots \longrightarrow c\text{C} + d\text{D} + \ldots$

$$\Delta H^\circ_{\text{reaction}} = [c\,\Delta H^\circ_f(\text{C}) + d\,\Delta H^\circ_f(\text{D}) + \ldots] - [a\,\Delta H^\circ_f(\text{A}) + b\,\Delta H^\circ_f(\text{B}) + \ldots]$$

- Free-energy change **(Section 8.13)**

$$\Delta G = \Delta H - T\Delta S$$ where ΔH = enthalpy change, T = temperature, and ΔS = entropy change

LEARNING OUTCOMES

- Calculate the sign and amount of expansion work that occurs during a reaction. **(Section 8.3)** ***Problems 2–4, 8, 9, 28, 29, 32, 40, 41, 44, 45, 60–63***
- Calculate the specific heat and enthalpy change from a calorimetry experiment. **(Section 8.5)** ***Problems 5–7, 46–53, 110, 123–127***
- Use Hess's Law to calculate the standard enthalpy change in a reaction. **(Section 8.8)** ***Problems 12–14, 19, 30, 31, 78–80, 112, 117, 121, 122, 128, 133, 134***
- Use standard heats of formation to calculate the standard enthalpy change in a reaction. **(Section 8.9)** ***Problems 15, 16, 76, 77, 81–87, 111, 118, 130–132***
- Use bond dissociation energies to estimate the standard enthalpy change in a reaction. **(Section 8.10)** ***Problems 17, 18, 88–91, 116, 130***
- Predict the sign of the standard entropy change in a reaction. **(Section 8.12)** ***Problems 20, 21, 27, 34, 35, 96, 97, 99–101***
- Calculate the free-energy change for a reaction, given standard enthalpy and entropy changes, and determine if a process is favorable at a given temperature. **(Section 8.13)** ***Problems 22–24, 33, 93–95, 102–108, 113–115, 117***

CONCEPTUAL PROBLEMS

Problems 8.1–8.26 appear within the chapter.

8.27 The following reaction is exothermic:

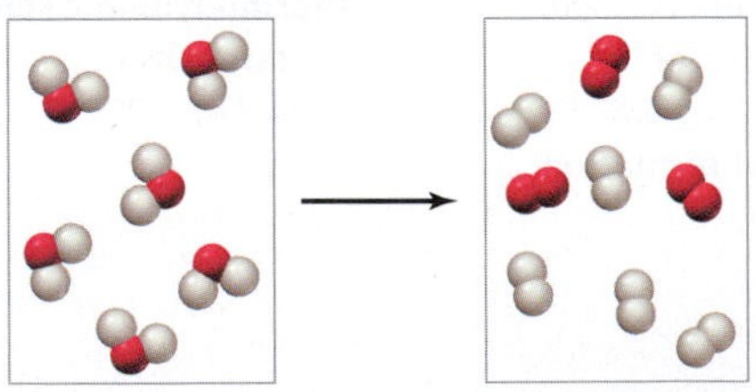

(a) Write a balanced equation for the reaction (red spheres represent **A** atoms and ivory spheres represent **B** atoms).

(b) What are the signs (+ or −) of ΔH and ΔS for the reaction?

(c) Is the reaction likely to be spontaneous at lower temperatures only, at higher temperatures only, or at all temperatures?

8.28 Imagine a reaction that results in a change in both volume and temperature:

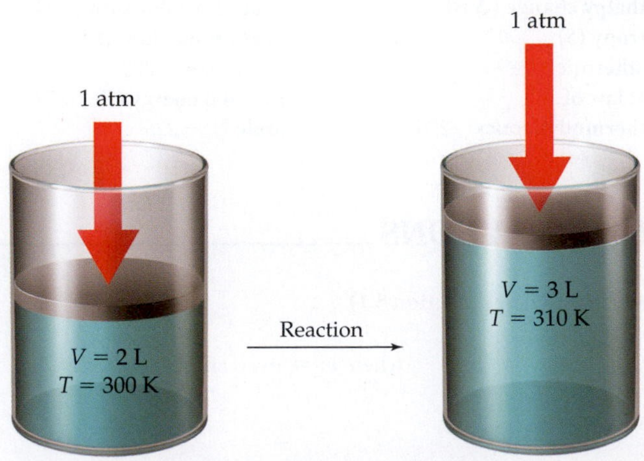

(a) Has any work been done? If so, is its sign positive or negative?

(b) Has there been an enthalpy change? If so, what is the sign of ΔH? Is the reaction exothermic or endothermic?

8.29 Redraw the following diagram to represent the situation **(a)** when work has been gained by the system and **(b)** when work has been lost by the system:

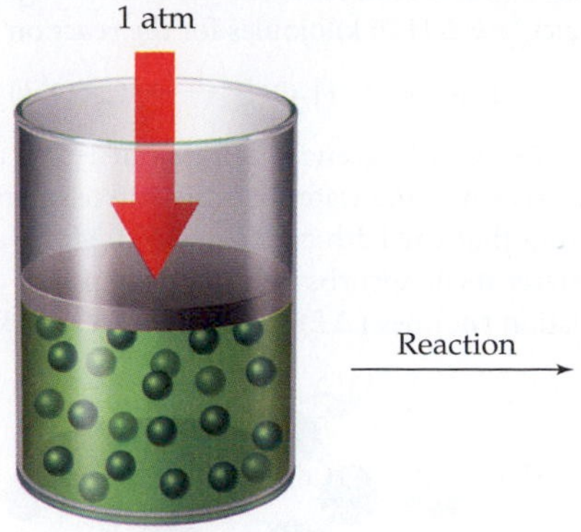

8.30 Acetylene, C_2H_2, reacts with H_2 in two steps to yield ethane, CH_3CH_3:

(1)	$HC{\equiv}CH + H_2 \longrightarrow H_2C{=}CH_2$	$\Delta H° = -175.1$ kJ
(2)	$H_2C{=}CH_2 + H_2 \longrightarrow CH_3CH_3$	$\Delta H° = -136.3$ kJ
Net	$HC{\equiv}CH + 2\,H_2 \longrightarrow CH_3CH_3$	$\Delta H° = -311.4$ kJ

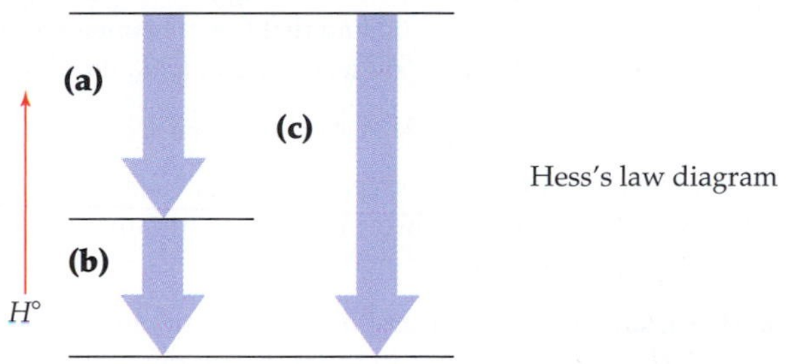

Which arrow (**a**, **b**, **c**) in the Hess's law diagram corresponds to which step, and which arrow corresponds to the net reaction? Where are the reactants located on the diagram, and where are the products located?

8.31 Draw a Hess's law diagram similar to the one in Problem 8.30 for the reaction of ethyl alcohol (CH_3CH_2OH) with oxygen to yield acetic acid (CH_3CO_2H).

(1) $CH_3CH_2OH(l) + 1/2\,O_2(g) \longrightarrow CH_3CHO(g) + H_2O(l)$ $\quad \Delta H° = -174.2$ kJ

(2) $CH_3CHO(g) + 1/2\,O_2(g) \longrightarrow CH_3CO_2H(l)$ $\quad \Delta H° = -318.4$ kJ

Net $CH_3CH_2OH(l) + O_2(g) \longrightarrow CH_3CO_2H(l) + H_2O(l)$ $\quad \Delta H° = -492.6$ kJ

8.32 A reaction is carried out in a cylinder fitted with a movable piston. The starting volume is $V = 5.00$ L, and the apparatus is held at constant temperature and pressure. Assuming that $\Delta H = -35.0$ kJ and $\Delta E = -34.8$ kJ, redraw the piston to show its position after reaction. Does V increase, decrease, or remain the same?

8.33 The following drawing portrays a reaction of the type $\mathbf{A} \longrightarrow \mathbf{B} + \mathbf{C}$, where the different colored spheres represent different molecular structures. Assume that the reaction has $\Delta H° = +55$ kJ. Is the reaction likely to be spontaneous at all temperatures, nonspontaneous at all temperatures, or spontaneous at some but nonspontaneous at others? Explain.

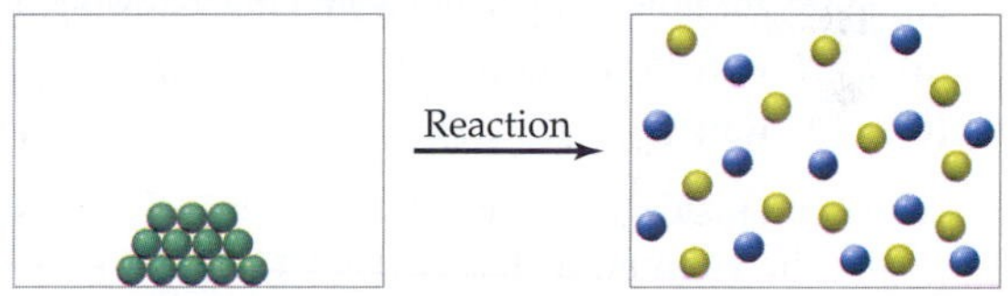

8.34 What are the signs of ΔH, ΔS, and ΔG for the following spontaneous change? Explain.

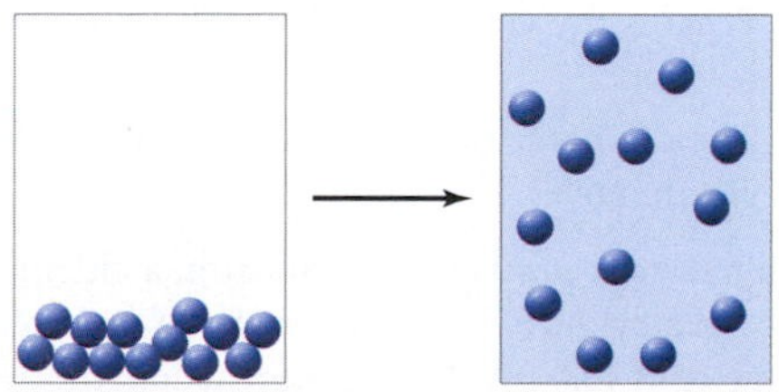

8.35 The following reaction of A_3 molecules is spontaneous:

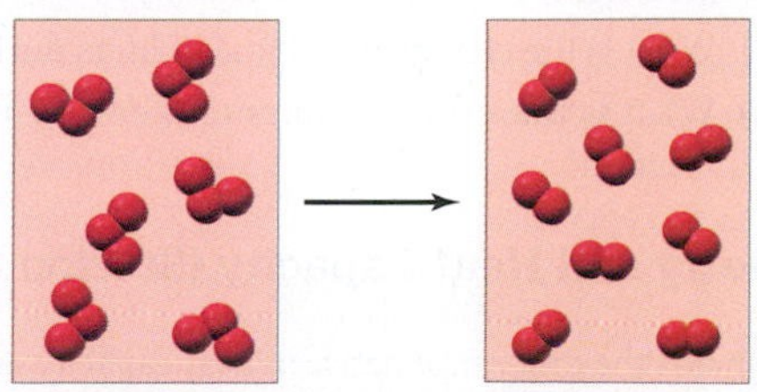

(a) Write a balanced equation for the reaction.

(b) What are the signs of ΔH, ΔS, and ΔG for the reaction? Explain.

SECTION PROBLEMS

Heat, Work, and Energy (Sections 8.1–8.3)

8.36 What is the difference between heat and temperature? Between work and energy? Between kinetic energy and potential energy?

8.37 What is internal energy?

8.38 Which has more kinetic energy, a 1400 kg car moving at 115 km/h or a 12,000 kg truck moving at 38 km/h?

8.39 Assume that the kinetic energy of a 1400 kg car moving at 115 km/h (Problem 8.38) could be converted entirely into heat. What amount of water could be heated from 20 °C to 50 °C by the car's energy?

8.40 Calculate the work done in joules by a chemical reaction if the volume increases from 3.2 L to 3.4 L against a constant external pressure of 3.6 atm. What is the sign of the energy change?

8.41 The addition of H_2 to C=C double bonds is an important reaction used in the preparation of margarine from vegetable oils. If 50.0 mL of H_2 and 50.0 mL of ethylene (C_2H_4) are allowed to react at 1.5 atm, the product ethane (C_2H_6) has a volume of 50.0 mL. Calculate the amount of *PV* work done, and tell the direction of the energy flow.

$$C_2H_4(g) + H_2(g) \longrightarrow C_2H_6(g)$$

8.42 Assume that the nutritional content of an apple—say 50 Cal (1 Cal = 1000 cal)—could be used to light a lightbulb. For how many minutes would there be light from each of the following?

(a) a 100 watt incandescent bulb (1 W = 1 J/s)

(b) a 23 watt compact fluorescent bulb, which provides a similar amount of light

8.43 A double cheeseburger has a caloric content of 440 Cal (1 Cal = 1000 cal). If this energy could be used to operate a television set, for how many hours would the following sets run?

(a) a 275-watt 46-in. plasma TV (1 W = 1 J/s)

(b) a 175-watt 46-in. LCD TV

8.44 A reaction inside a cylindrical container with a movable piston causes the volume to change from 12.0 L to 18.0 L while the pressure outside the container remains constant at 0.975 atm. (The volume of a cylinder is $V = \pi r^2 h$, where h is the height; 1 L·atm = 101.325 J.)

(a) What is the value in joules of the work w done during the reaction?

(b) The diameter of the piston is 17.0 cm. How far does the piston move?

8.45 At a constant pressure of 0.905 atm, a chemical reaction takes place in a cylindrical container with a movable piston having a diameter of 40.0 cm. During the reaction, the height of the piston drops by 65.0 cm. (The volume of a cylinder is $V = \pi r^2 h$, where h is the height; 1 L·atm = 101.325 J.)

(a) What is the change in volume in liters during the reaction?

(b) What is the value in joules of the work w done during the reaction?

Calorimetry and Heat Capacity (Section 8.5)

8.46 What is the difference between heat capacity and specific heat?

8.47 Does a measurement carried out in a bomb calorimeter give a value for ΔH or ΔE? Explain.

8.48 Sodium metal is sometimes used as a cooling agent in heat-exchange units because of its relatively high molar heat capacity of 28.2 J/(mol · °C). What is the specific heat of sodium in J/(g · °C)?

8.49 Titanium metal is used as a structural material in many high-tech applications, such as in jet engines. What is the specific heat of titanium in J/(g · °C) if it takes 89.7 J to raise the temperature of a 33.0 g block by 5.20 °C? What is the molar heat capacity of titanium in J/(mol · °C)?

8.50 When 1.045 g of CaO is added to 50.0 mL of water at 25.0 °C in a calorimeter, the temperature of the water increases to 32.3 °C. Assuming that the specific heat of the solution is 4.18 J/(g · °C) and that the calorimeter itself absorbs a negligible amount of heat, calculate ΔH in kilojoules for the reaction

$$CaO(s) + H_2O(l) \longrightarrow Ca(OH)_2(aq)$$

8.51 When 0.187 g of benzene, C_6H_6, is burned in a bomb calorimeter, the surrounding water bath rises in temperature by 7.48 °C. Assuming that the bath contains 250.0 g of water and that the calorimeter itself absorbs a negligible amount of heat, calculate combustion energies (ΔE) for benzene in both kJ/g and kJ/mol.

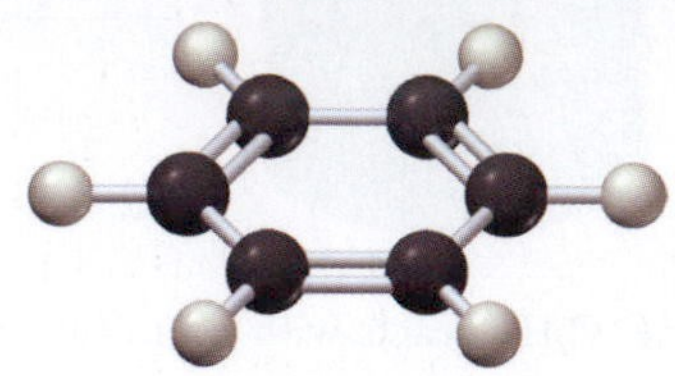

Benzene

8.52 When a solution containing 8.00 g of NaOH in 50.0 g of water at 25.0 °C is added to a solution of 8.00 g of HCl in 250.0 g of water at 25.0 °C in a calorimeter, the temperature of the solution increases to 33.5 °C. Assuming that the specific heat of the solution is 4.18 J/(g · °C) and that the calorimeter absorbs a negligible amount of heat, calculate ΔH in kilojoules for the reaction

$$NaOH(aq) + HCl(aq) \longrightarrow NaCl(aq) + H_2O(l)$$

When the experiment is repeated using a solution of 10.00 g of HCl in 248.0 g of water, the same temperature increase is observed. Explain.

8.53 Instant cold packs used to treat athletic injuries contain solid NH_4NO_3 and a pouch of water. When the pack is squeezed, the pouch breaks and the solid dissolves, lowering the temperature because of the endothermic reaction

$$NH_4NO_3(s) \xrightarrow{H_2O} NH_4NO_3(aq) \quad \Delta H = +25.7 \text{ kJ}$$

What is the final temperature in a squeezed cold pack that contains 50.0 g of NH_4NO_3 dissolved in 125 mL of water? Assume a specific heat of 4.18 J/(g · °C) for the solution, an initial temperature of 25.0 °C, and no heat transfer between the cold pack and the environment.

Energy and Enthalpy (Sections 8.4, 8.6–8.7)

8.54 What is the difference between the internal energy change ΔE and the enthalpy change ΔH? Which of the two is measured at constant pressure, and which at constant volume?

8.55 What is the sign of ΔH for an exothermic reaction? For an endothermic reaction?

8.56 Under what circumstances are ΔE and ΔH essentially equal?

8.57 Which of the following has the highest enthalpy content, and which the lowest at a given temperature: $H_2O(s)$, $H_2O(l)$, or $H_2O(g)$? Explain.

8.58 The enthalpy change for the reaction of 50.0 mL of ethylene with 50.0 mL of H_2 at 1.5 atm pressure (Problem 8.41) is $\Delta H = -0.31$ kJ. What is the value of ΔE?

8.59 Assume that a particular reaction evolves 244 kJ of heat and that 35 kJ of *PV* work is gained by the system. What are the values of ΔE and ΔH for the system? For the surroundings?

8.60 What is the enthalpy change (ΔH) for a reaction at a constant pressure of 1.00 atm if the internal energy change (ΔE) is +44.0 kJ and the volume increase is 14.0 L? ($1\ \text{L}\cdot\text{atm} = 101.325\ \text{J}$.)

8.61 A reaction takes place at a constant pressure of 1.10 atm with an internal energy change (ΔE) of +71.5 kJ and a volume decrease of 13.6 L. What is the enthalpy change (ΔH) for the reaction? ($1\ \text{L}\cdot\text{atm} = 101.325\ \text{J}$.)

8.62 Used in welding metals, the reaction of acetylene with oxygen has $\Delta H° = -1256.2\ \text{kJ}$:

$$C_2H_2(g) + 5/2\ O_2(g) \longrightarrow H_2O(g) + 2\ CO_2(g) \quad \Delta H° = -1256.2\ \text{kJ}$$

How much *PV* work is done in kilojoules and what is the value of ΔE in kilojoules for the reaction of 6.50 g of acetylene at atmospheric pressure if the volume change is −2.80 L?

8.63 Ethyl chloride (C_2H_5Cl), a substance used as a topical anesthetic, is prepared by reaction of ethylene with hydrogen chloride:

$$C_2H_4(g) + HCl(g) \longrightarrow C_2H_5Cl(g) \quad \Delta H° = -72.3\ \text{kJ}$$

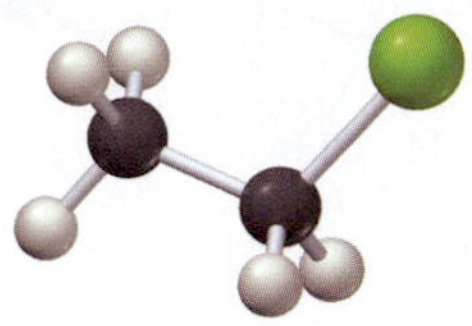

Ethyl chloride

How much *PV* work is done in kilojoules and what is the value of ΔE in kilojoules if 89.5 g of ethylene and 125 g of HCl are allowed to react at atmospheric pressure and the volume change is −71.5 L?

8.64 The familiar "ether" used as an anesthetic agent is diethyl ether, $C_4H_{10}O$. Its heat of vaporization is +26.5 kJ/mol at its boiling point. How much energy in kilojoules is required to convert 100 mL of diethyl ether at its boiling point from liquid to vapor if its density is 0.7138 g/mL?

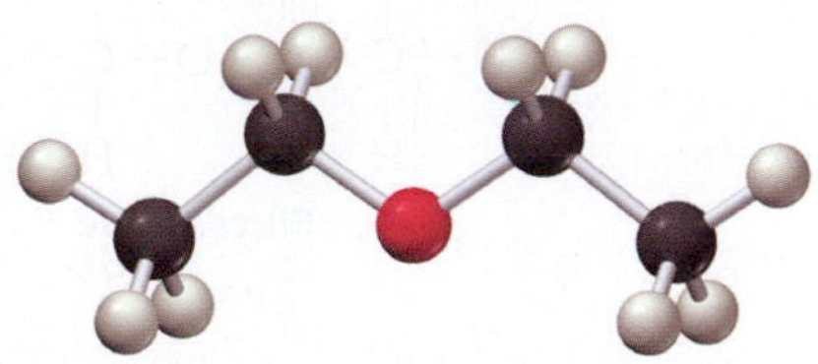

Diethyl ether

8.65 How much energy in kilojoules is required to convert 100 mL of water at its boiling point from liquid to vapor, and how does this compare with the result calculated in Problem 8.64 for diethyl ether? [$\Delta H_{vap}(H_2O) = +40.7\ \text{kJ/mol}$]

8.66 Aluminum metal reacts with chlorine with a spectacular display of sparks:

$$2\ Al(s) + 3\ Cl_2(g) \longrightarrow 2\ AlCl_3(s) \quad \Delta H° = -1408.4\ \text{kJ}$$

How much heat in kilojoules is released on reaction of 5.00 g of Al?

8.67 How much heat in kilojoules is evolved or absorbed in the reaction of 1.00 g of Na with H_2O? Is the reaction exothermic or endothermic?

$$2\ Na(s) + 2\ H_2O(l) \longrightarrow 2\ NaOH(aq) + H_2(g) \quad \Delta H° = -368.4\ \text{kJ}$$

8.68 How much heat in kilojoules is evolved or absorbed in the reaction of 2.50 g of Fe_2O_3 with enough carbon monoxide to produce iron metal? Is the process exothermic or endothermic?

$$Fe_2O_3(s) + 3\ CO(g) \longrightarrow 2\ Fe(s) + 3\ CO_2(g) \quad \Delta H° = -24.8\ \text{kJ}$$

8.69 How much heat in kilojoules is evolved or absorbed in the reaction of 233.0 g of calcium oxide with enough carbon to produce calcium carbide? Is the process exothermic or endothermic?

$$CaO(s) + 3\ C(s) \longrightarrow CaC_2(s) + CO(g) \quad \Delta H° = +464.6\ \text{kJ}$$

Hess's Law and Heats of Formation (Sections 8.8 and 8.9)

8.70 How is the standard state of an element defined?

8.71 What is a compound's standard heat of formation?

8.72 What is Hess's law, and why does it "work"?

8.73 Why do elements always have $\Delta H°_f = 0$?

8.74 What phase of matter is associated with the standard states of the following elements and compounds?

(a) Cl_2 **(b)** Hg **(c)** CO_2 **(d)** Ga

8.75 What is the phase of the standard states of the following elements and compounds?

(a) NH_3 **(b)** Fe **(c)** N_2 **(d)** Br_2

8.76 Write balanced equations for the formation of the following compounds from their elements:

(a) Iron(III) oxide
(b) Sucrose (table sugar, $C_{12}H_{22}O_{11}$)
(c) Uranium hexafluoride (a solid at 25 °C)

8.77 Write balanced equations for the formation of the following compounds from their elements:

(a) Ethanol (C_2H_6O)
(b) Sodium sulfate
(c) Dichloromethane (a liquid, CH_2Cl_2)

8.78 Sulfuric acid (H_2SO_4), the most widely produced chemical in the world, is made by a two-step oxidation of sulfur to sulfur trioxide, SO_3, followed by reaction with water. Calculate $\Delta H°_f$ for SO_3 in kJ/mol, given the following data:

$$S(s) + O_2(g) \longrightarrow SO_2(g) \quad \Delta H° = -296.8\ \text{kJ}$$
$$SO_2(g) + 1/2\ O_2(g) \longrightarrow SO_3(g) \quad \Delta H° = -98.9\ \text{kJ}$$

8.79 Calculate $\Delta H°_f$ in kJ/mol for benzene, C_6H_6, from the following data:

$$2\ C_6H_6(l) + 15\ O_2(g) \longrightarrow 12\ CO_2(g) + 6\ H_2O(l) \quad \Delta H° = -6534\ \text{kJ}$$

$$\Delta H°_f(CO_2) = -393.5\ \text{kJ/mol} \quad \Delta H°_f(H_2O) = -285.8\ \text{kJ/mol}$$

8.80 The standard enthalpy change for the reaction of $SO_3(g)$ with $H_2O(l)$ to yield $H_2SO_4(aq)$ is $\Delta H° = -227.8\ \text{kJ}$. Use the information in Problem 8.78 to calculate $\Delta H°_f$ for $H_2SO_4(aq)$ in kJ/mol. [For $H_2O(l)$, $\Delta H°_f = -285.8\ \text{kJ/mol}$.]

8.81 Acetic acid (CH_3CO_2H), whose aqueous solutions are known as *vinegar*, is prepared by reaction of ethyl alcohol (CH_3CH_2OH) with oxygen:

$$CH_3CH_2OH(l) + O_2(g) \longrightarrow CH_3CO_2H(l) + H_2O(l)$$

Use the following data to calculate $\Delta H°$ in kilojoules for the reaction:

$$\Delta H°_f[CH_3CH_2OH(l)] = -277.7 \text{ kJ/mol}$$
$$\Delta H°_f[CH_3CO_2H(l)] = -484.5 \text{ kJ/mol}$$
$$\Delta H°_f[H_2O(l)] = -285.8 \text{ kJ/mol}$$

8.82 Styrene (C_8H_8), the precursor of polystyrene polymers, has a standard heat of combustion of −4395 kJ/mol. Write a balanced equation for the combustion reaction, and calculate $\Delta H°_f$ for styrene in kJ/mol.

$$\Delta H°_f[CO_2(g)] = -393.5 \text{ kJ/mol}$$
$$\Delta H°_f[H_2O(l)] = -285.8 \text{ kJ/mol}$$

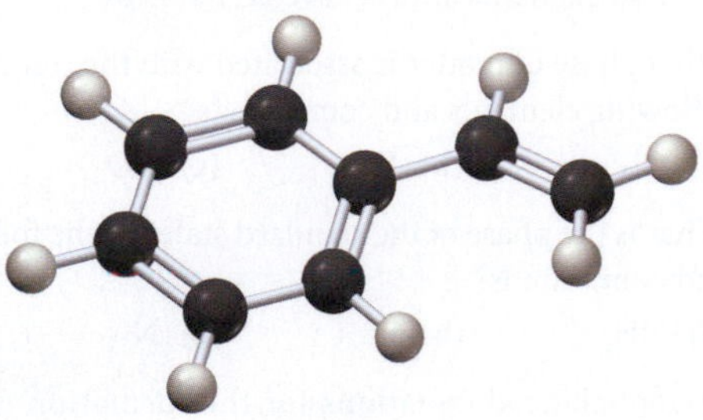

Styrene

8.83 Methyl *tert*-butyl ether (MTBE), $C_5H_{12}O$, a gasoline additive used to boost octane ratings, has $\Delta H°_f = -313.6$ kJ/mol. Write a balanced equation for its combustion reaction, and calculate its standard heat of combustion in kilojoules.

8.84 Methyl *tert*-butyl ether (Problem 8.83) is prepared by reaction of methanol(l) ($\Delta H°_f = -239.2$ kJ/mol) with 2-methylpropene(g), according to the equation

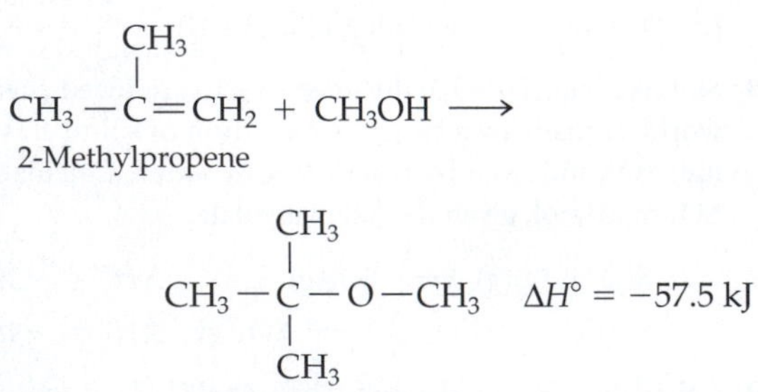

Calculate $\Delta H°_f$ in kJ/mol for 2-methylpropene.

8.85 One possible use for the cooking fat left over after making french fries is to burn it as fuel. Write a balanced equation, and use the following data to calculate the amount of energy released in kJ/mL from the combustion of cooking fat:

Formula = $C_{51}H_{88}O_6$
Density = 0.94 g/mL $\quad \Delta H°_f = -1310$ kJ/mol

8.86 Given the standard heats of formation shown in Appendix B, what is $\Delta H°$ in kilojoules for the reaction

$$CaCO_3(s) \longrightarrow CaO(s) + CO_2(g)?$$

8.87 Given the standard heats of formation shown in Appendix B, what is $\Delta H°$ in kilojoules for the reaction $3\,N_2O_4(g) + 2\,H_2O(l) \longrightarrow 4\,HNO_3(aq) + 2\,NO(g)$?

Bond Dissociation Energies (Section 8.10)

8.88 Use the bond dissociation energies in Table 8.3 on page 296 to calculate an approximate $\Delta H°$ in kilojoules for the reaction of ethylene with hydrogen to yield ethane.

$$H_2C{=}CH_2(g) + H_2(g) \longrightarrow CH_3CH_3(g)$$

8.89 Use the bond dissociation energies in Table 8.3 to calculate an approximate $\Delta H°$ in kilojoules for the industrial synthesis of isopropyl alcohol (rubbing alcohol) by reaction of water with propene.

$$\underset{\text{Propene}}{CH_3CH{=}CH_2} + H_2O \longrightarrow \underset{\text{Isopropyl alcohol}}{CH_3\overset{\overset{\displaystyle OH}{|}}{C}HCH_3}$$

8.90 Calculate an approximate heat of combustion for butane in kilojoules by using the bond dissociation energies in Table 8.3. (The strength of the O=O bond is 498 kJ/mol, and that of a C=O bond in CO_2 is 804 kJ/mol.)

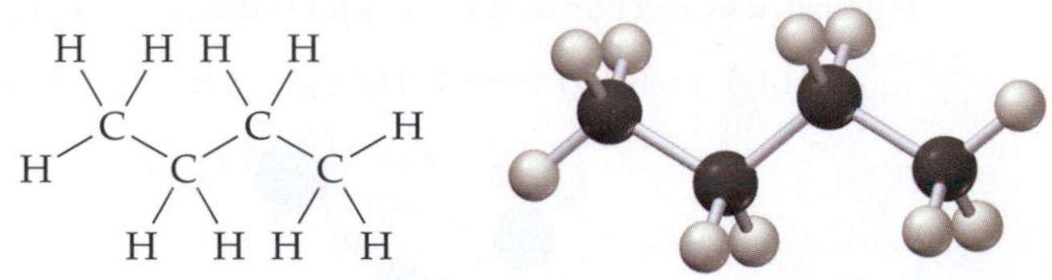

Butane

8.91 Use the bond dissociation energies in Table 8.3 to calculate an approximate heat of reaction, $\Delta H°$ in kilojoules, for the industrial reaction of ethanol with acetic acid to yield ethyl acetate (used as nail-polish remover) and water.

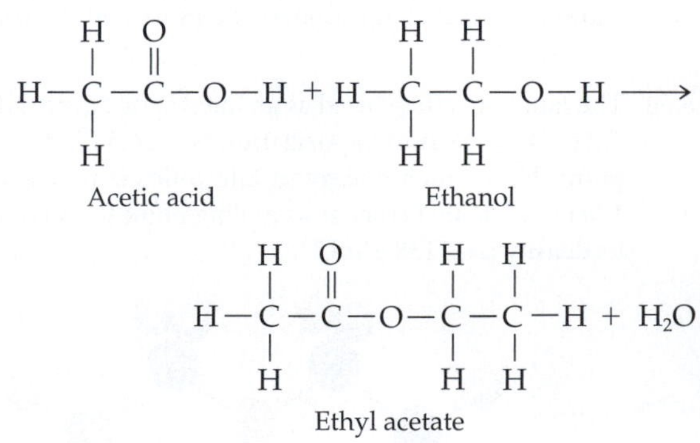

Free Energy and Entropy (Sections 8.12 and 8.13)

8.92 What does entropy measure?

8.93 What are the two terms that make up the free-energy change for a reaction, ΔG, and which of the two is usually more important?

8.94 How is it possible for a reaction to be spontaneous yet endothermic?

8.95 Is it possible for a reaction to be nonspontaneous yet exothermic? Explain.

8.96 Tell whether the entropy changes for the following processes are likely to be positive or negative:

(a) The fizzing of a newly opened can of soda

(b) The growth of a plant from seed

8.97 Tell whether the entropy changes, ΔS, for the following processes are likely to be positive or negative:

(a) The conversion of liquid water to water vapor at 100 °C

(b) The freezing of liquid water to ice at 0 °C

(c) The eroding of a mountain by a glacier

8.98 Tell whether the free-energy changes, ΔG, for the processes listed in Problem 8.97 are likely to be positive, negative, or zero.

8.99 When a bottle of perfume is opened, odorous molecules mix with air and slowly diffuse throughout the entire room. Is ΔG for the diffusion process positive, negative, or zero? What about ΔH and ΔS for the diffusion?

8.100 One of the steps in the cracking of petroleum into gasoline involves the thermal breakdown of large hydrocarbon molecules into smaller ones. For example, the following reaction might occur:

$$C_{11}H_{24} \longrightarrow C_4H_{10} + C_4H_8 + C_3H_6$$

Is ΔS for this reaction likely to be positive or negative? Explain.

8.101 The commercial production of 1,2-dichloroethane, a solvent used in dry cleaning, involves the reaction of ethylene with chlorine:

$$C_2H_4(g) + Cl_2(g) \longrightarrow C_2H_4Cl_2(l)$$

Is ΔS for this reaction likely to be positive or negative? Explain.

8.102 Tell whether reactions with the following values of ΔH and ΔS are spontaneous or nonspontaneous and whether they are exothermic or endothermic:

(a) $\Delta H = -48$ kJ; $\Delta S = +135$ J/K at 400 K

(b) $\Delta H = -48$ kJ; $\Delta S = -135$ J/K at 400 K

(c) $\Delta H = +48$ kJ; $\Delta S = +135$ J/K at 400 K

(d) $\Delta H = +48$ kJ; $\Delta S = -135$ J/K at 400 K

8.103 Tell whether reactions with the following values of ΔH and ΔS are spontaneous or nonspontaneous and whether they are exothermic or endothermic:

(a) $\Delta H = -128$ kJ; $\Delta S = +35$ J/K at 500 K

(b) $\Delta H = +67$ kJ; $\Delta S = -140$ J/K at 250 K

(c) $\Delta H = +75$ kJ; $\Delta S = +95$ J/K at 800 K

8.104 Suppose that a reaction has $\Delta H = -33$ kJ and $\Delta S = -58$ J/K. At what temperature will it change from spontaneous to nonspontaneous?

8.105 Suppose that a reaction has $\Delta H = +41$ kJ and $\Delta S = -27$ J/K. At what temperature, if any, will it change between spontaneous and nonspontaneous?

8.106 Which of the reactions **(a)–(d)** in Problem 8.102 are spontaneous at all temperatures, which are nonspontaneous at all temperatures, and which have an equilibrium temperature?

8.107 Vinyl chloride ($H_2C{=}CHCl$), the starting material used in the industrial preparation of poly(vinyl chloride), is prepared by a two-step process that begins with the reaction of Cl_2 with ethylene to yield 1,2-dichloroethane:

$$Cl_2(g) + H_2C{=}CH_2(g) \longrightarrow ClCH_2CH_2Cl(l) \quad \Delta H° = -217.5 \text{ kJ}$$
$$\Delta S° = -233.9 \text{ J/K}$$

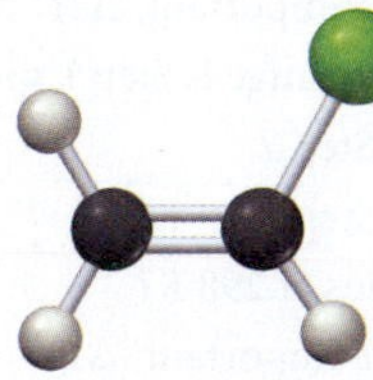

Vinyl chloride

(a) Tell whether the reaction is favored by entropy, by enthalpy, by both, or by neither, and then calculate $\Delta G°$ at 298 K.

(b) Tell whether the reaction has an equilibrium temperature between spontaneous and nonspontaneous. If yes, calculate the equilibrium temperature.

8.108 Ethyl alcohol has $\Delta H_{fusion} = +5.02$ kJ/mol and melts at -114.1 °C. What is the value of ΔS_{fusion} for ethyl alcohol?

8.109 Chloroform has $\Delta H_{vap} = +29.2$ kJ/mol and boils at 61.2 °C. What is the value of ΔS_{vap} for chloroform?

CHAPTER PROBLEMS

8.110 When 1.50 g of magnesium metal is allowed to react with 200 mL of 6.00 M aqueous HCl, the temperature rises from 25.0 °C to 42.9 °C. Calculate ΔH in kilojoules for the reaction, assuming that the heat capacity of the calorimeter is 776 J/°C, that the specific heat of the final solution is the same as that of water [4.18 J/(g · °C)], and that the density of the solution is 1.00 g/mL.

8.111 Use the data in Appendix B to find standard enthalpies of reaction in kilojoules for the following processes:

(a) $C(s) + CO_2(g) \longrightarrow 2\,CO(g)$

(b) $2\,H_2O_2(aq) \longrightarrow 2\,H_2O(l) + O_2(g)$

(c) $Fe_2O_3(s) + 3\,CO(g) \longrightarrow 2\,Fe(s) + 3\,CO_2(g)$

8.112 Find $\Delta H°$ in kilojoules for the reaction of nitric oxide with oxygen, $2\,NO(g) + O_2(g) \rightarrow N_2O_4(g)$, given the following data:

$$N_2O_4(g) \longrightarrow 2\,NO_2(g) \qquad \Delta H° = +55.3 \text{ kJ}$$
$$NO(g) + 1/2\,O_2(g) \longrightarrow NO_2(g) \qquad \Delta H° = -58.1 \text{ kJ}$$

8.113 The boiling point of a substance is defined as the temperature at which liquid and vapor coexist in equilibrium. Use the heat of vaporization ($\Delta H_{vap} = +30.91$ kJ/mol) and the entropy of vaporization [$\Delta S_{vap} = +93.2$ J/(K · mol)] to calculate the boiling point (°C) of liquid bromine.

8.114 What is the melting point of benzene in kelvin if $\Delta H_{fusion} = +9.95$ kJ/mol and $\Delta S_{fusion} = +35.7$ J/(K · mol)?

8.115 Metallic mercury is obtained by heating the mineral cinnabar (HgS) in air:

$$HgS(s) + O_2(g) \longrightarrow Hg(l) + SO_2(g)$$

(a) Use the data in Appendix B to calculate $\Delta H°$ in kilojoules for the reaction.

(b) The entropy change for the reaction is $\Delta S° = +36.7$ J/K. Is the reaction spontaneous at 25 °C?

(c) Under what conditions, if any, is the reaction nonspontaneous? Explain.

8.116 Use the average bond dissociation energies in Table 8.3 to calculate approximate reaction enthalpies in kilojoules for the following processes:

(a) $2\,CH_4(g) \longrightarrow C_2H_6(g) + H_2(g)$

(b) $C_2H_6(g) + F_2(g) \longrightarrow C_2H_5F(g) + HF(g)$

(c) $N_2(g) + 3\,H_2(g) \longrightarrow 2\,NH_3(g)$

8.117 Methanol (CH_3OH) is made industrially in two steps from CO and H_2. It is so cheap to make that it is being considered for use as a precursor to hydrocarbon fuels, such as methane (CH_4):

Step 1. $CO(g) + 2\,H_2(g) \longrightarrow CH_3OH(l) \quad \Delta S° = -332\ J/K$

Step 2. $CH_3OH(l) \longrightarrow CH_4(g) + 1/2\,O_2(g) \quad \Delta S° = +162\ J/K$

(a) Calculate $\Delta H°$ in kilojoules for Step 1.

(b) Calculate $\Delta G°$ in kilojoules for Step 1.

(c) Is Step 1 spontaneous at 298 K?

(d) Which term is more important, $\Delta H°$ or $\Delta S°$?

(e) In what temperature range is Step 1 spontaneous?

(f) Calculate $\Delta H°$ for Step 2.

(g) Calculate $\Delta G°$ for Step 2.

(h) Is Step 2 spontaneous at 298 K?

(i) Which term is more important, $\Delta H°$ or $\Delta S°$?

(j) In what temperature range is Step 2 spontaneous?

(k) Calculate an overall $\Delta G°$, $\Delta H°$, and $\Delta S°$ for the formation of CH_4 from CO and H_2.

(l) Is the overall reaction spontaneous at 298 K?

(m) If you were designing a production facility, would you plan on carrying out the reactions in separate steps or together? Explain.

8.118 Isooctane, C_8H_{18}, is the component of gasoline from which the term *octane rating* derives.

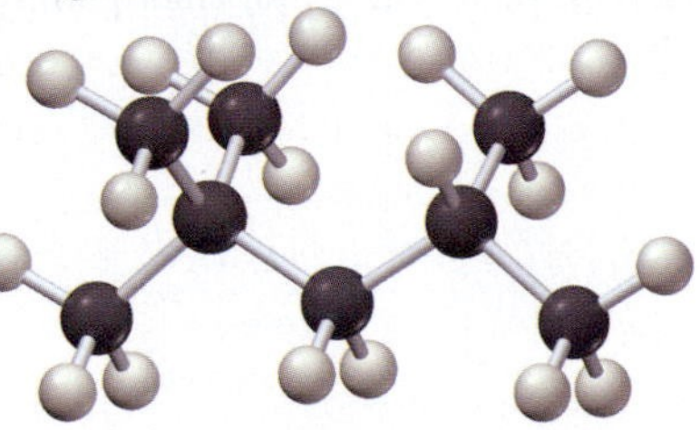

Isooctane

(a) Write a balanced equation for the combustion of isooctane(l) with O_2 to yield $CO_2(g)$ and $H_2O(l)$.

(b) The standard molar heat of combustion for isooctane(l) is −5461 kJ/mol. Calculate $\Delta H°_f$ for isooctane(l).

8.119 We said in Section 8.1 that the potential energy of water at the top of a dam or waterfall is converted into heat when the water dashes against rocks at the bottom. The potential energy of the water at the top is equal to $E_P = mgh$, where m is the mass of the water, g is the acceleration of the falling water due to gravity ($g = 9.81\ m/s^2$), and h is the height of the water. Assuming that all the energy is converted to heat, calculate the temperature rise of the water in degrees Celsius after falling over California's Yosemite Falls, a distance of 739 m. The specific heat of water is 4.18 J/(g·K).

8.120 For a process to be spontaneous, the total entropy of the system *and its surroundings* must increase; that is

$$\Delta S_{total} = \Delta S_{system} + \Delta S_{surr} > 0 \quad \text{For a spontaneous process}$$

Furthermore, the entropy change in the surroundings, ΔS_{surr}, is related to the enthalpy change for the process by the equation $\Delta S_{surr} = -\Delta H/T$.

(a) Since both ΔG and ΔS_{total} offer criteria for spontaneity, they must be related. Derive a relationship between them.

(b) What is the value of ΔS_{surr} for the photosynthesis of glucose from CO_2 at 298 K?

$$6\,CO_2(g) + 6\,H_2O(l) \longrightarrow C_6H_{12}O_6(s) + 6\,O_2(g) \quad \Delta G° = +2879\ kJ \quad \Delta S° = -262\ J/K$$

8.121 Set up a Hess's law cycle, and use the following information to calculate $\Delta H°_f$ for aqueous nitric acid, $HNO_3(aq)$. You will need to use fractional coefficients for some equations.

$3\,NO_2(g) + H_2O(l) \longrightarrow 2\,HNO_3(aq) + NO(g) \quad \Delta H° = -137.3\ kJ$

$2\,NO(g) + O_2(g) \longrightarrow 2\,NO_2(g) \quad \Delta H° = -116.2\ kJ$

$4\,NH_3(g) + 5\,O_2(g) \longrightarrow 4\,NO(g) + 6\,H_2O(l) \quad \Delta H° = -1165.2\ kJ$

$NH_3(g) \quad \Delta H°_f = -46.1\ kJ/mol$

$H_2O(l) \quad \Delta H°_f = -285.8\ kJ/mol$

8.122 Hess's law can be used to calculate reaction enthalpies for hypothetical processes that can't be carried out in the laboratory. Set up a Hess's law cycle that will let you calculate $\Delta H°$ for the conversion of methane to ethylene:

$$2\,CH_4(g) \longrightarrow C_2H_4(g) + 2\,H_2(g)$$

You can use the following information:

$2\,C_2H_6(g) + 7\,O_2(g) \longrightarrow 4\,CO_2(g) + 6\,H_2O(l) \quad \Delta H° = -3120.8\ kJ$

$CH_4(g) + 2\,O_2(g) \longrightarrow CO_2(g) + 2\,H_2O(l) \quad \Delta H° = -890.3\ kJ$

$C_2H_4(g) + H_2(g) \longrightarrow C_2H_6(g) \quad \Delta H° = -136.3\ kJ$

$H_2O(l) \quad \Delta H°_f = -285.8\ kJ/mol$

8.123 A 110.0 g piece of molybdenum metal is heated to 100.0 °C and placed in a calorimeter that contains 150.0 g of water at 24.6 °C. The system reaches equilibrium at a final temperature of 28.0 °C. Calculate the specific heat of molybdenum metal in J/(g·°C). The specific heat of water is 4.184 J/(g·°C).

8.124 Given 400.0 g of hot tea at 80.0 °C, what mass of ice at 0 °C must be added to obtain iced tea at 10.0 °C? The specific heat of the tea is 4.18 J/(g·°C), and ΔH_{fusion} for ice is +6.01 kJ/mol.

8.125 Citric acid has three dissociable hydrogens. When 5.00 mL of 0.64 M citric acid and 45.00 mL of 0.77 M NaOH are mixed at an initial temperature of 26.0 °C, the temperature rises to 27.9 °C as the citric acid is neutralized. The combined mixture has a mass of 51.6 g and a specific heat of 4.0 J/(g·°C). Assuming that no heat is transferred to the surroundings, calculate the enthalpy change for the reaction of 1.00 mol of citric acid in kJ. Is the reaction exothermic or endothermic?

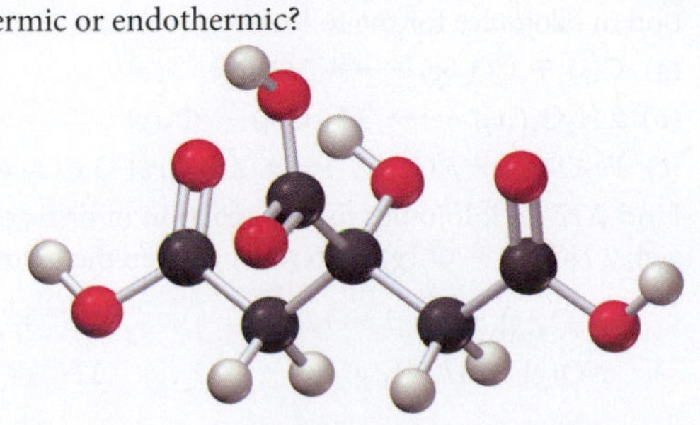

Citric acid

8.126 Assume that 100.0 mL of 0.200 M CsOH and 50.0 mL of 0.400 M HCl are mixed in a calorimeter. The solutions start out at 22.50 °C, and the final temperature after reaction is 24.28 °C. The densities of the solutions are all 1.00 g/mL, and the specific heat of the mixture is 4.2 J/(g · °C). What is the enthalpy change for the neutralization reaction of 1.00 mol of CsOH in kJ?

8.127 Imagine that you dissolve 10.0 g of a mixture of $NaNO_3$ and KF in 100.0 g of water and find that the temperature rises by 2.22 °C. Using the following data, calculate the mass of each compound in the original mixture. Assume that the specific heat of the solution is 4.18 J/(g · °C).

$$NaNO_3(s) \longrightarrow NaNO_3(aq) \quad \Delta H = +20.4 \text{ kJ/mol}$$
$$KF(s) \longrightarrow KF(aq) \quad \Delta H = -17.7 \text{ kJ/mol}$$

8.128 Consider the reaction: $4\,CO(g) + 2\,NO_2(g) \longrightarrow 4\,CO_2(g) + N_2(g)$. Using the following information, determine $\Delta H°$ for the reaction at 25 °C.

$$NO(g) \quad \Delta H°_f = +91.3 \text{ kJ/mol}$$
$$CO_2(g) \quad \Delta H°_f = -393.5 \text{ kJ/mol}$$
$$2\,NO(g) + O_2(g) \longrightarrow 2\,NO_2(g) \quad \Delta H° = -116.2 \text{ kJ}$$
$$2\,CO(g) + O_2(g) \longrightarrow 2\,CO_2(g) \quad \Delta H° = -566.0 \text{ kJ}$$

MULTICONCEPT PROBLEMS

8.129 The reaction $S_8(g) \longrightarrow 4\,S_2(g)$ has $\Delta H° = +237$ kJ.

(a) The S_8 molecule has eight sulfur atoms arranged in a ring. What is the hybridization and geometry around each sulfur atom in S_8?

(b) The average S–S bond dissociation energy is 225 kJ/mol. Using the value of $\Delta H°$ given above, what is the S=S double bond energy in $S_2(g)$?

(c) Assuming that the bonding in S_2 is similar to the bonding in O_2, give a molecular orbital description of the bonding in S_2. Is S_2 likely to be paramagnetic or diamagnetic?

8.130 Phosgene, $COCl_2(g)$, is a toxic gas used as an agent of warfare in World War I.

(a) Draw an electron-dot structure for phosgene.

(b) Using the table of bond dissociation energies (Table 8.3) and the value $\Delta H°_f = +716.7$ kJ/mol for C(g), estimate $\Delta H°_f$ for $COCl_2(g)$ at 25 °C. Compare your answer to the actual $\Delta H°_f$ given in Appendix B, and explain why your calculation is only an estimate.

8.131 Acid spills are often neutralized with sodium carbonate or sodium hydrogen carbonate. For neutralization of acetic acid, the unbalanced equations are

$$(1)\; CH_3CO_2H(l) + Na_2CO_3(s) \longrightarrow CH_3CO_2Na(aq) + CO_2(g) + H_2O(l)$$
$$(2)\; CH_3CO_2H(l) + NaHCO_3(s) \longrightarrow CH_3CO_2Na(aq) + CO_2(g) + H_2O(l)$$

(a) Balance both equations.

(b) How many kilograms of each substance is needed to neutralize a 1.000 gallon spill of pure acetic acid (density = 1.049 g/mL)?

(c) How much heat in kilojoules is absorbed or liberated in each reaction? See Appendix B for standard heats of formation; $\Delta H°_f = -726.1$ kJ/mol for $CH_3CO_2Na(aq)$.

8.132 **(a)** Write a balanced equation for the reaction of potassium metal with water.

(b) Use the data in Appendix B to calculate $\Delta H°$ for the reaction of potassium metal with water.

(c) Assume that a chunk of potassium weighing 7.55 g is dropped into 400.0 g of water at 25.0 °C. What is the final temperature of the water if all the heat released is used to warm the water?

(d) What is the molarity of the KOH solution prepared in part (c), and how many milliliters of 0.554 M H_2SO_4 are required to neutralize it?

8.133 Hydrazine, a component of rocket fuel, undergoes combustion to yield N_2 and H_2O:

$$N_2H_4(l) + O_2(g) \longrightarrow N_2(g) + 2\,H_2O(l)$$

(a) Draw an electron-dot structure for hydrazine, predict the geometry about each nitrogen atom, and tell the hybridization of each nitrogen.

(b) Use the following information to set up a Hess's law cycle, and then calculate $\Delta H°$ for the combustion reaction. You will need to use fractional coefficients for some equations.

$$2\,NH_3(g) + 3\,N_2O(g) \longrightarrow 4\,N_2(g) + 3\,H_2O(l) \quad \Delta H° = -1011.2 \text{ kJ}$$
$$N_2O(g) + 3\,H_2(g) \longrightarrow N_2H_4(l) + H_2O(l) \quad \Delta H° = -317.2 \text{ kJ}$$
$$4\,NH_3(g) + O_2(g) \longrightarrow 2\,N_2H_4(l) + 2\,H_2O(l) \quad \Delta H° = -286.0 \text{ kJ}$$
$$H_2O(l) \quad \Delta H°_f = -285.8 \text{ kJ/mol}$$

(c) How much heat is released on combustion of 100.0 g of hydrazine?

8.134 Reaction of gaseous fluorine with compound **X** yields a single product **Y**, whose mass percent composition is 61.7% F and 38.3% Cl.

(a) What is a probable molecular formula for product **Y**, and what is a probable formula for **X**?

(b) Draw an electron-dot structure for **Y**, and predict the geometry around the central atom.

(c) Calculate $\Delta H°$ for the synthesis of **Y** using the following information:

$$2\,ClF(g) + O_2(g) \longrightarrow Cl_2O(g) + OF_2(g) \quad \Delta H° = +205.4 \text{ kJ}$$
$$2\,ClF_3(l) + 2\,O_2(g) \longrightarrow Cl_2O(g) + 3\,OF_2(g) \quad \Delta H° = +532.8 \text{ kJ}$$
$$OF_2(g) \quad \Delta H°_f = +24.5 \text{ kJ/mol}$$

(d) How much heat in kilojoules is released or absorbed in the reaction of 25.0 g of **X** with a stoichiometric amount of F_2, assuming 87.5% yield for the reaction?

CHAPTER 9

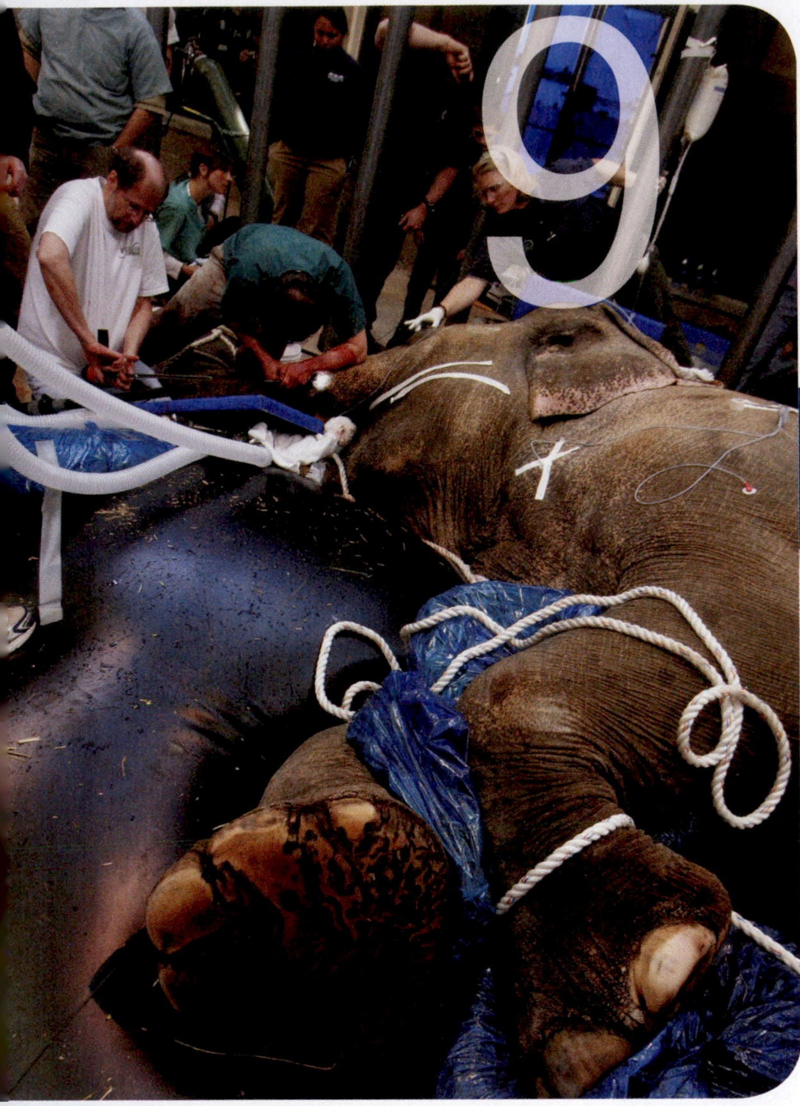

Gases: Their Properties and Behavior

Inhaled anesthetic gases have been used for more than 160 years to relieve pain during surgery. Even elephants appreciate it. We'll learn more in the *FYI* at the end of this chapter.

A quick look around tells you that matter takes many forms. Most of the things around you are *solids*, substances whose constituent atoms, molecules, or ions are held rigidly together in a definite way, giving the solid a definite volume and shape. Other substances are *liquids*, whose constituent atoms or molecules are held together less strongly, giving the liquid a definite volume but a changeable and indefinite shape. Still other substances are *gases*, whose constituent atoms or molecules have little attraction for one another and are therefore free to move about in whatever volume is available.

Although gases are few in number—only about a hundred substances are gases at room temperature and atmospheric pressure—experimental studies of their properties were enormously important in the historical development of atomic theories. We'll look briefly at this historical development in the present chapter, and we'll see how the behavior of gases can be described.

9.1 GASES AND GAS PRESSURE

We live surrounded by a blanket of air—the mixture of gases that make up the Earth's atmosphere. As shown in **Table 9.1**, nitrogen and oxygen account for more than 99% by volume of dry air. The remaining 1% is largely argon, with trace amounts of several other substances also present. Carbon dioxide, about which there is so much current concern because of its relationship to global warming, is present in air only to the extent of about 0.0390%, or 390 parts per million (ppm). Although small, this value has risen in the past 160 years from an estimated 290 ppm in 1850, as the burning of fossil fuels and the deforestation of tropical rain forests have increased.

CONTENTS

TABLE 9.1 Composition of Dry Air at Sea Level

Constituent	% Volume	% Mass
N_2	78.08	75.52
O_2	20.95	23.14
Ar	0.93	1.29
CO_2	0.0390	0.060
Ne	1.82×10^{-3}	1.27×10^{-3}
He	5.24×10^{-4}	7.24×10^{-5}
CH_4	1.7×10^{-4}	9.4×10^{-5}
Kr	1.14×10^{-4}	3.3×10^{-4}

Air is typical of gases in many respects, and its behavior illustrates several important points about gases. For instance, gas mixtures are always *homogeneous*, meaning that they are uniform in composition. Unlike liquids, which often fail to mix with one another and which may separate into distinct layers—oil and water, for example—gases always mix completely. Furthermore, gases are *compressible*. When pressure is applied, the volume of a gas contracts proportionately. Solids and liquids, however, are nearly incompressible, and even the application of great pressure changes their volume only slightly.

Homogeneous mixing and compressibility of gases both occur because the constituent particles—whether atoms or molecules—are far apart (**Figure 9.1**). Mixing occurs because individual gas particles have little interaction with their neighbors so the chemical identity of those neighbors is irrelevant. In solids and liquids, by contrast, the constituent particles are packed closely together, where they are affected by various attractive and repulsive forces that can inhibit their mixing. Compressibility is possible in gases because less than 0.1% of the volume of a typical gas is taken up by the particles themselves under normal circumstances; the remaining 99.9% is empty space. By contrast, approximately 70% of a solid's or liquid's volume is taken up by the particles.

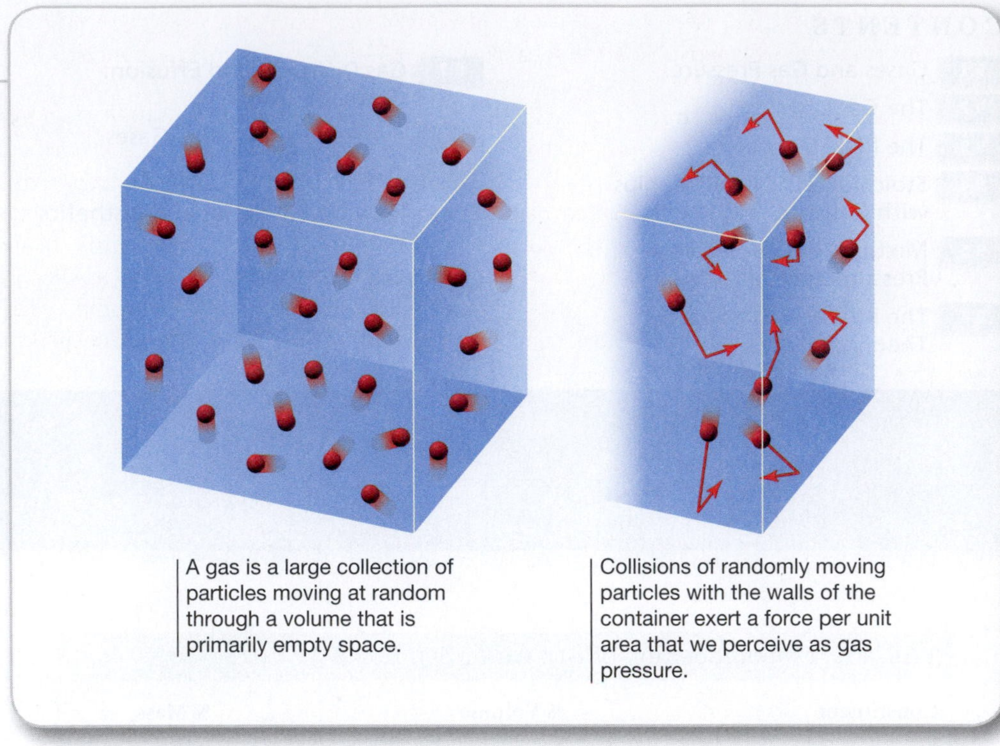

FIGURE 9.1

Molecular view of a gas.

One of the most obvious characteristics of gases is that they exert a measurable *pressure* on the walls of their container (Figure 9.1). We're all familiar with inflating a balloon or pumping up a bicycle tire and feeling the hardness that results from the pressure inside. In scientific terms, **pressure (*P*)** is defined as a force (*F*) exerted per unit area (*A*). Force, in turn, is defined as mass (*m*) times acceleration (*a*), which, on Earth, is usually the acceleration due to gravity, $a = 9.81\ \text{m/s}^2$.

$$\text{Pressure } (P) = \frac{F}{A} = \frac{m \times a}{A}$$

The SI unit for force is the **newton (N)**, where $1\ \text{N} = 1\ (\text{kg}\cdot\text{m})/\text{s}^2$, and the SI unit for pressure is the **pascal (Pa)**, where $1\ \text{Pa} = 1\ \text{N/m}^2 = 1\ \text{kg/(m}\cdot\text{s}^2)$. Expressed in more familiar units, a pascal is actually a very small amount—the pressure exerted by a mass of 10.2 mg resting on an area of $1.00\ \text{cm}^2$. In rough terms, a penny sitting on the tip of your finger exerts a pressure of about 250 Pa.

$$P = \frac{m \times a}{A} = \frac{(10.2\ \text{mg})\left(\frac{1\ \text{kg}}{10^6\ \text{mg}}\right)\left(9.81\ \frac{\text{m}}{\text{s}^2}\right)}{(1.00\ \text{cm}^2)\left(\frac{1\ \text{m}^2}{10^4\ \text{cm}^2}\right)} = \frac{1.00 \times 10^{-4}\ \frac{\text{kg}\cdot\text{m}}{\text{s}^2}}{1.00 \times 10^{-4}\ \text{m}^2}$$

$$= 1.00\ \frac{\text{kg}}{\text{m}\cdot\text{s}^2} = 1.00\ \text{Pa}$$

FIGURE 9.2

Atmospheric pressure. A column of air 1 m^2 in cross-sectional area extending from the Earth's surface through the upper atmosphere has a mass of about 10,300 kg, producing an atmospheric pressure of approximately 101,000 Pa.

Just as the air in a tire and a penny on your finger exert pressure, the mass of air in the atmosphere pressing down on the Earth's surface exerts what we call *atmospheric pressure*. In fact, a 1 m^2 column of air extending from the Earth's surface through the upper atmosphere has a mass of about 10,300 kg, producing an atmospheric pressure of approximately 101,000 Pa, or 101 kPa (**Figure 9.2**).

$$P = \frac{m \times a}{A} = \frac{10{,}300\ \text{kg} \times 9.81\ \frac{\text{m}}{\text{s}^2}}{1.00\ \text{m}^2} = 101{,}000\ \text{Pa} = 101\ \text{kPa}$$

As is frequently the case with SI units, which must serve many disciplines, the pascal is an inconvenient size for most chemical measurements. Thus, the alternative pressure units *millimeter of mercury (mm Hg), atmosphere (atm),* and *bar* are more often used.

The **millimeter of mercury**, also called a *torr* after the 17th-century Italian scientist Evangelista Torricelli (1608–1647), is based on atmospheric pressure measurements using a mercury *barometer*. As shown in **Figure 9.3**, a barometer consists of a long, thin tube that is sealed at one end, filled with mercury, and then inverted into a dish of mercury. Some mercury runs from the tube into the dish until the downward pressure of the mercury inside the column is exactly balanced by the outside atmospheric pressure, which presses on the mercury in the dish and pushes it up the column. The height of the mercury column varies slightly from day to day depending on the altitude and weather conditions, but atmospheric pressure at sea level is defined as exactly 760 mm Hg.

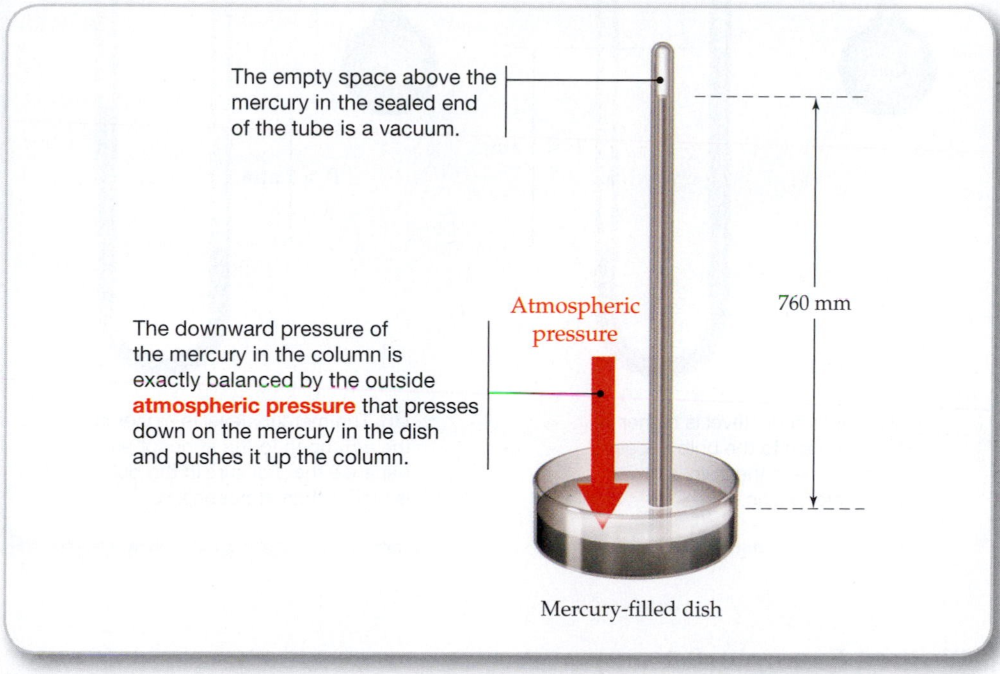

FIGURE 9.3

A mercury barometer. The barometer measures atmospheric pressure by determining the height of a mercury column supported in a sealed glass tube.

Knowing the density of mercury (1.35951×10^4 kg/m^3 at 0 °C) and the acceleration due to gravity (9.806 65 m/s^2), it's possible to calculate the pressure exerted by the column of mercury 760 mm (0.760 m) in height. Thus, 1 standard **atmosphere** of pressure (1 atm) is now defined as exactly 101,325 Pa:

$$P = (0.760\ \text{m})\left(1.359\,51 \times 10^4\ \frac{\text{kg}}{\text{m}^3}\right)\left(9.806\,65\ \frac{\text{m}}{\text{s}^2}\right) = 101{,}325\ \text{Pa}$$

$$\mathbf{1\ atm = 760\ mm\ Hg = 101{,}325\ Pa}$$

Although not strictly an SI unit, the **bar** is quickly gaining popularity as a unit of pressure because it is a convenient power of 10 of the SI unit pascal and because it differs from 1 atm by only about 1%:

$$1\ \text{bar} = 100{,}000\ \text{Pa} = 100\ \text{kPa} = 0.986\,923\ \text{atm}$$

Gas pressure inside a container is often measured using an open-end **manometer**, a simple instrument similar in principle to the mercury barometer. As shown in **Figure 9.4**, an open-end manometer consists of a U-tube filled with mercury, with one end connected to a gas-filled container and the other end open to the atmosphere. The difference between the pressure of the gas in the container and the pressure of the atmosphere is equal to the difference between the heights of the mercury levels in the two arms of the U-tube. If the gas pressure inside the container is less than atmospheric, the mercury level is higher in the arm connected to the container (**Figure 9.4a**). If the gas pressure inside the container is greater than atmospheric, the mercury level is higher in the arm open to the atmosphere (**Figure 9.4b**).

FIGURE 9.4

Open-end manometers for measuring pressure in a gas-filled bulb.

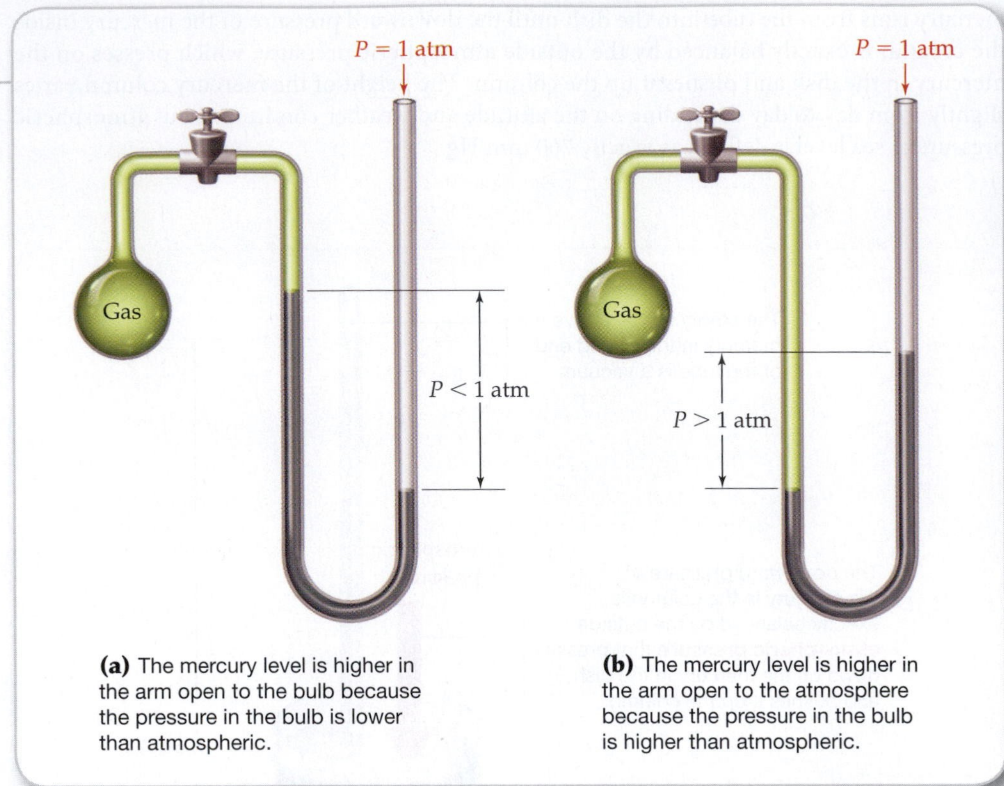

Worked Example 9.1

Converting between Different Units of Pressure

Typical atmospheric pressure on top of Mt. Everest, whose official altitude is 8848 m, is 265 mm Hg. Convert this value to pascals, atmospheres, and bars.

STRATEGY

Use the conversion factors 101,325 Pa/760 mm Hg, 1 atm/760 mm Hg, and 1 bar/10^5 Pa to carry out the necessary calculations.

SOLUTION

$$(265 \text{ mm Hg})\left(\frac{101{,}325 \text{ Pa}}{760 \text{ mm Hg}}\right) = 3.53 \times 10^4 \text{ Pa}$$

$$(265 \text{ mm Hg})\left(\frac{1 \text{ atm}}{760 \text{ mm Hg}}\right) = 0.349 \text{ atm}$$

$$(3.53 \times 10^4 \text{ Pa})\left(\frac{1 \text{ bar}}{10^5 \text{ Pa}}\right) = 0.353 \text{ bar}$$

✓ BALLPARK CHECK

One atmosphere equals 760 mm Hg pressure. Since 265 mm Hg is about 1/3 of 760 mm Hg, the air pressure on Mt. Everest is about 1/3 of standard atmospheric pressure—approximately 30,000 Pa, 0.3 atm, or 0.3 bar.

▲ Atmospheric pressure decreases as altitude increases. On the top of Mt. Everest, typical atmospheric pressure is 265 mm Hg.

Worked Example 9.2

Using an Open-End Manometer

Assume that you are using an open-end manometer (Figure 9.4) filled with mineral oil rather than mercury. What is the gas pressure in the bulb in millimeters of mercury if the level of mineral oil in the arm connected to the bulb is 237 mm higher than the level in the arm connected to the atmosphere and atmospheric pressure is 746 mm Hg? The density of mercury is 13.6 g/mL, and the density of mineral oil is 0.822 g/mL.

STRATEGY

The gas pressure in the bulb equals the difference between the outside pressure and the manometer reading. The manometer reading indicates that the pressure of the gas in the bulb is less than atmospheric because the liquid level is higher on the side connected to the sample. Because mercury is more dense than mineral oil by a factor of 13.6/0.822, or 16.5, a given pressure will hold a column of mercury only 1/16.5 times the height of a column of mineral oil.

SOLUTION

$$P_{\text{manometer}} = 237 \text{ mm mineral oil} \times \frac{0.822 \text{ mm Hg}}{13.6 \text{ mm mineral oil}} = 14.3 \text{ mm Hg}$$

$$P_{\text{bulb}} = P_{\text{outside}} - P_{\text{manometer}} = 746 \text{ mm Hg} - 14.3 \text{ mm Hg} = 732 \text{ mm Hg}$$

PROBLEM 9.1 Yet another common measure of pressure is the unit pounds per square inch (psi). How many pounds per square inch correspond to 1.00 atm? To 1.00 mm Hg?

PROBLEM 9.2 If the density of water is 1.00 g/mL and the density of mercury is 13.6 g/mL, how high a column of water in meters can be supported by standard atmospheric pressure? By 1 bar?

PROBLEM 9.3 What is the pressure in atmospheres in a container of gas connected to a mercury-filled, open-end manometer if the level in the arm connected to the container is 24.7 cm higher than in the arm open to the atmosphere and atmospheric pressure is 0.975 atm?

CONCEPTUAL PROBLEM 9.4 What is the pressure of the gas inside the apparatus shown on the right in mm Hg if the outside pressure is 750 mm Hg?

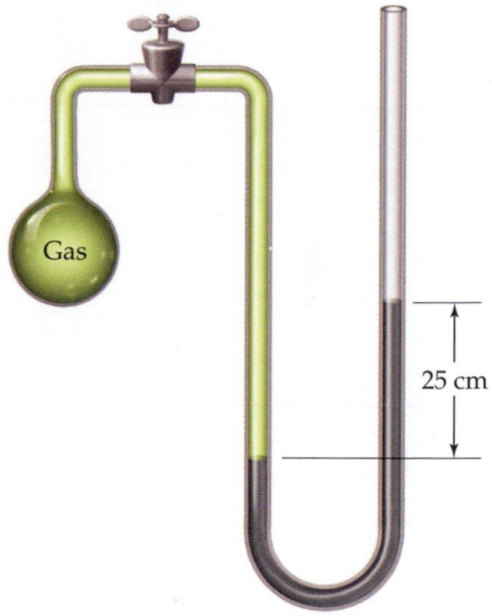

9.2 THE GAS LAWS

Unlike solids and liquids, different gases show remarkably similar physical behavior regardless of their chemical makeup. Helium and fluorine, for example, are vastly different in their chemical properties yet are almost identical in much of their physical behavior. Numerous observations made in the late 1600s showed that the physical properties of any gas can be defined by four variables: pressure (P), temperature (T), volume (V), and amount, or number of moles (n). The specific relationships among these four variables are called the **gas laws**, and a gas whose behavior follows the laws exactly is called an **ideal gas**.

Boyle's Law: The Relationship between Gas Volume and Pressure

Imagine that you have a sample of gas inside a cylinder with a movable piston at one end (**Figure 9.5**). What would happen if you were to increase the pressure on the gas by pushing down on the piston? Experience probably tells you that the volume of gas in the cylinder would decrease as you increase the pressure. According to **Boyle's law**, the volume of a fixed amount of gas at a constant temperature varies inversely with its pressure. If the gas pressure is doubled, the volume is halved; if the pressure is halved, the gas volume doubles.

▶ **Boyle's law** $V \propto 1/P$ or $PV = k$ at constant n and T

The volume of an ideal gas varies inversely with pressure. That is, P times V is constant when n and T are kept constant. (The symbol $\propto$ means "is proportional to," and k denotes a constant.)

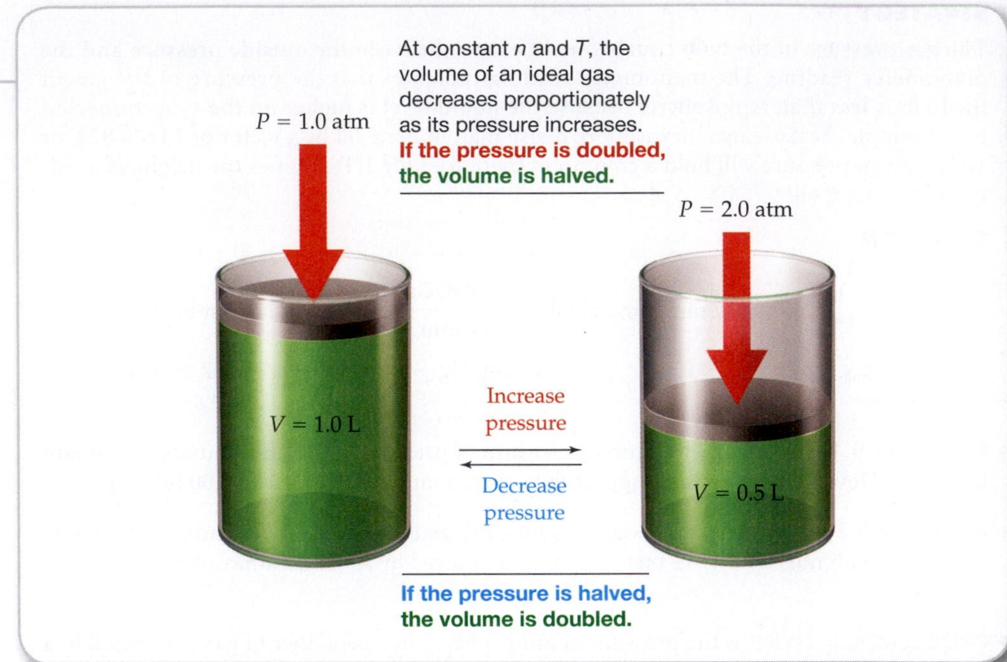

FIGURE 9.5

Boyle's law.

The validity of Boyle's law can be demonstrated by making a simple series of pressure–volume measurements on a gas sample **(Table 9.2)** and plotting them as in **Figure 9.6**. When V is plotted versus P, the result is a curve in the form of a hyperbola. When V is plotted versus $1/P$, however, the result is a straight line. Such graphical behavior is characteristic of mathematical equations of the form $y = mx + b$. In this case, $y = V$, $m =$ the slope of the line

TABLE 9.2 Pressure–Volume Measurements on a Gas Sample at Constant n, T

Pressure (mm Hg)	Volume (L)
760	1
380	2
253	3
190	4
152	5
127	6
109	7
95	8
84	9
76	10

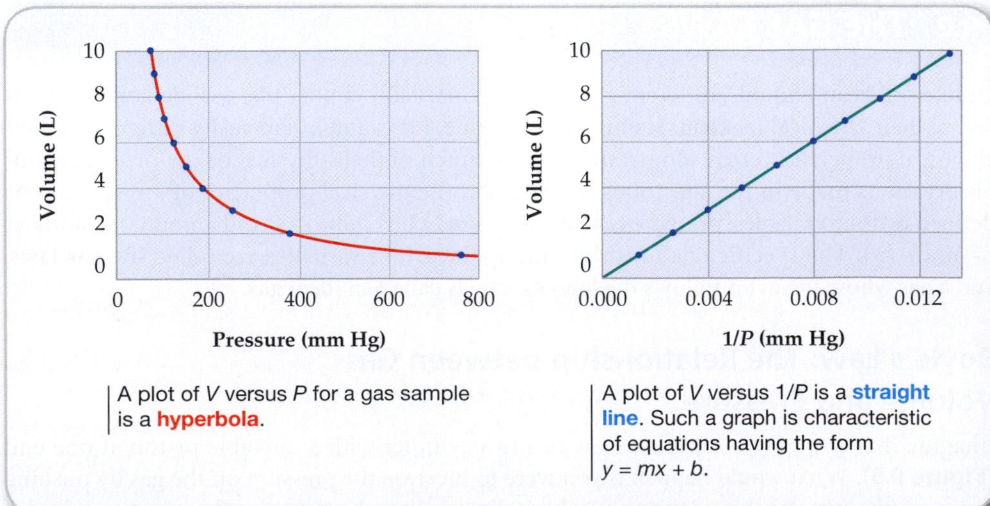

FIGURE 9.6

Boyle's law plot.

(the constant k in the present instance), $x = 1/P$, and $b =$ the y-intercept (a constant; 0 in the present instance). (See Appendix A.3 for a review of linear equations.)

$$\underset{y}{\underset{\uparrow}{V}} = \underset{m}{\underset{\uparrow}{k}}\underset{x}{\underset{\uparrow}{\left(\frac{1}{P}\right)}} + \underset{b}{\underset{\uparrow}{0}} \quad (\text{or } PV = k)$$

Charles's Law: The Relationship between Gas Volume and Temperature

Imagine again that you have a gas sample inside a cylinder with a movable piston at one end (**Figure 9.7**). What would happen if you were to raise the temperature of the sample while letting the piston move freely to keep the pressure constant? Experience tells you that the piston would move up because the volume of the gas in the cylinder would expand. According to **Charles's law**, the volume of a fixed amount of an ideal gas at a constant pressure varies directly with its absolute temperature. If the gas temperature in kelvins is doubled, the volume is doubled; if the gas temperature is halved, the volume is halved.

▶ **Charles's law** $V \propto T$ or $V/T = k$ at constant n and P

The volume of an ideal gas varies directly with absolute temperature. That is, V divided by T is constant when n and P are held constant.

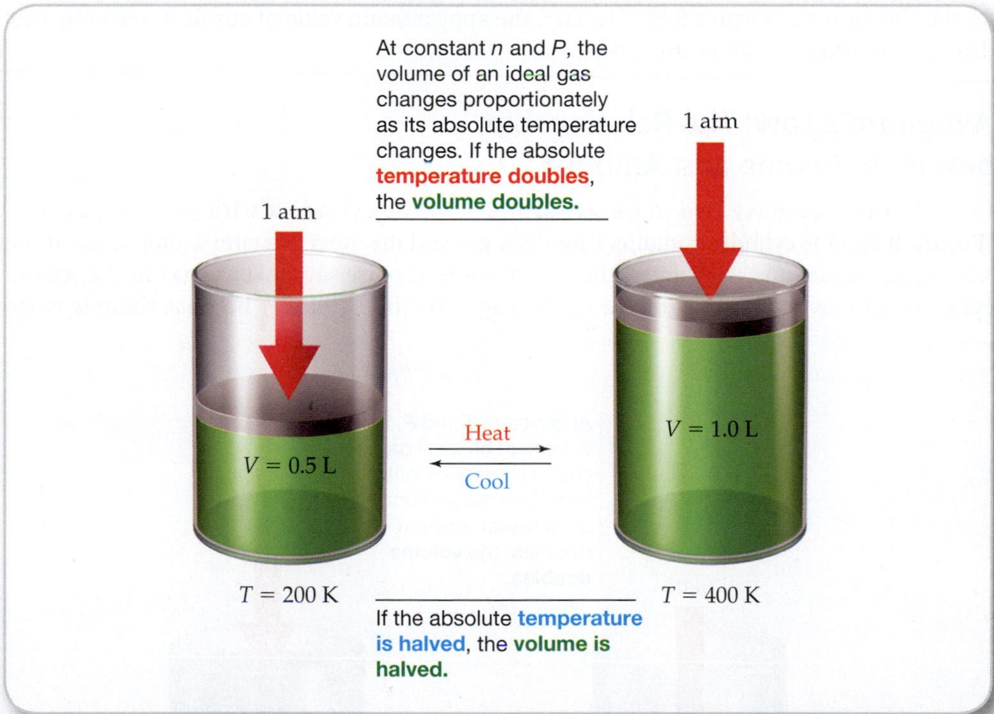

FIGURE 9.7

Charles's law.

The validity of Charles's law can be demonstrated by making a series of temperature–volume measurements on a gas sample, giving the results listed in **Table 9.3**. Like Boyle's law, Charles's law takes the mathematical form $y = mx + b$, where $y = V$, $m =$ the slope of the line (the constant k in the present instance), $x = T$, and $b =$ the y-intercept (0 in the present instance). A plot of V versus T is therefore a straight line whose slope is the constant k (**Figure 9.8**).

$$\underset{y}{\underset{\uparrow}{V}} = \underset{m}{\underset{\uparrow}{k}}\underset{x}{\underset{\uparrow}{T}} + \underset{b}{\underset{\uparrow}{0}} \quad \left(\text{or } \frac{V}{T} = k\right)$$

TABLE 9.3 Temperature–Volume Measurements on a Gas Sample at Constant n and P

Temperature (K)	Volume (L)
123	0.45
173	0.63
223	0.82
273	1.00
323	1.18
373	1.37

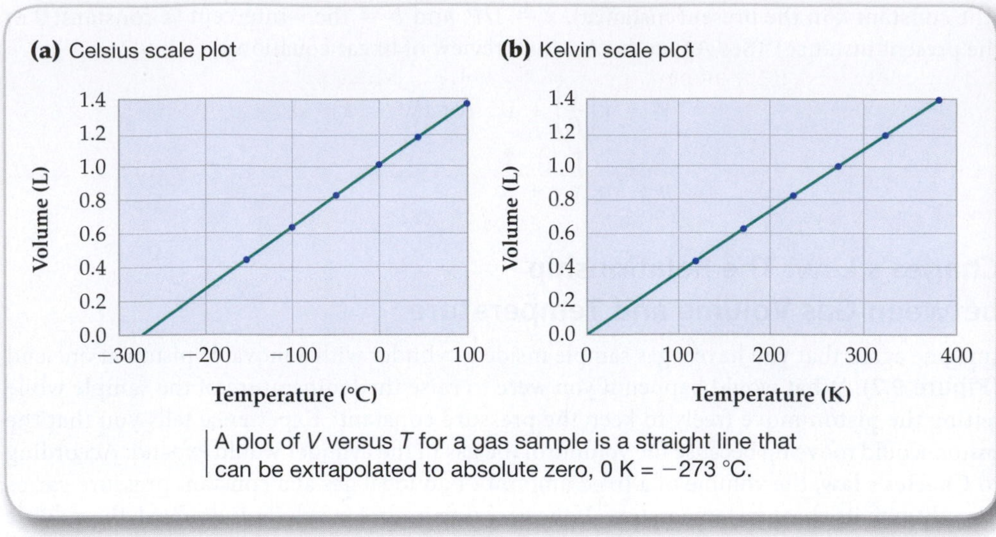

FIGURE 9.8

Charles's law plot.

The plots of volume versus temperature in Figure 9.8 demonstrate an interesting point. When temperature is plotted on the Celsius scale, the straight line can be extrapolated to $V = 0$ at $T = -273$ °C (**Figure 9.8a**). But because matter can't have a negative volume, this extrapolation suggests that −273 °C must be the lowest possible temperature, or *absolute zero* on the Kelvin scale (**Figure 9.8b**). In fact, the approximate value of absolute zero was first determined using this simple method.

Avogadro's Law: The Relationship between Volume and Amount

Imagine that you have two more gas samples inside cylinders with movable pistons (**Figure 9.9**). One cylinder contains 1 mol of a gas and the other contains 2 mol of gas at the same temperature and pressure as the first. Common sense says that the gas in the second cylinder will have twice the volume of the gas in the first cylinder because there is twice

FIGURE 9.9

Avogadro's law.

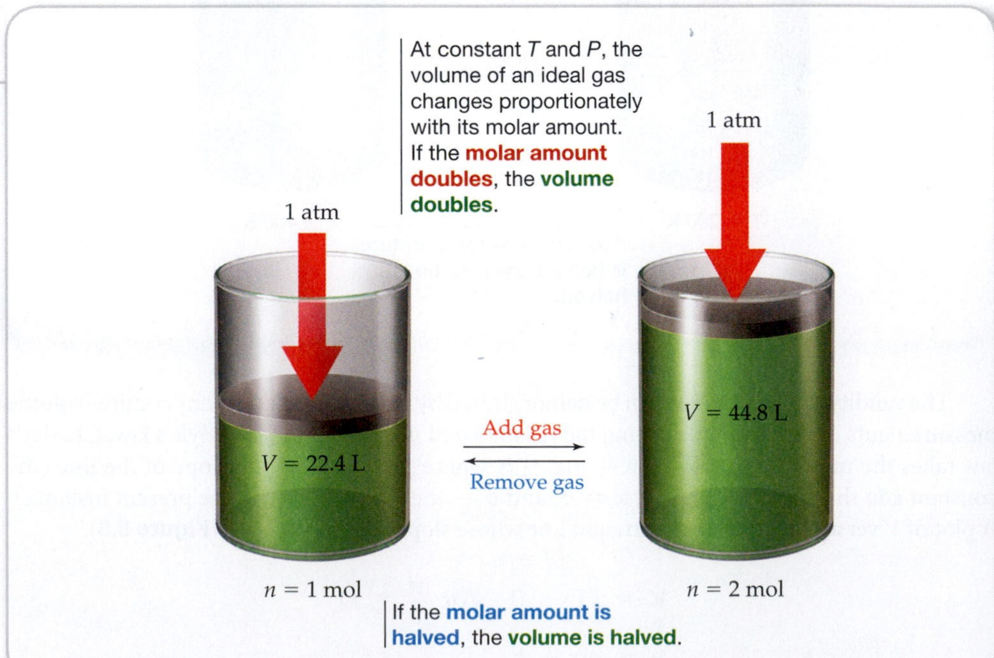

as much of it. According to **Avogadro's law**, the volume of an ideal gas at a fixed pressure and temperature depends on its molar amount. If the amount of the gas is doubled, the gas volume is doubled; if the amount is halved, the volume is halved.

> ▶ **Avogadro's law** $V \propto n$ or $V/n = k$ at constant T and P
>
> The volume of an ideal gas varies directly with its molar amount. That is, V divided by n is constant when T and P are held constant.

Put another way, Avogadro's law also says that equal volumes of different gases at the same temperature and pressure contain the same molar amounts. A 1 L container of oxygen contains the same number of moles as a 1 L container of helium, fluorine, argon, or any other gas at the same T and P. Furthermore, 1 mol of an ideal gas occupies a volume, called the **standard molar volume**, of 22.414 L at 0 °C and exactly 1 atm pressure. For comparison, the standard molar volume is nearly identical to the volume of three basketballs (or 16 cantaloupes).

WORKED CONCEPTUAL EXAMPLE 9.3

Visual Representations of Gas Laws

Show the approximate level of the movable piston in drawings **(a)** and **(b)** after the indicated changes have been made to the initial gas sample.

(a) **(b)**

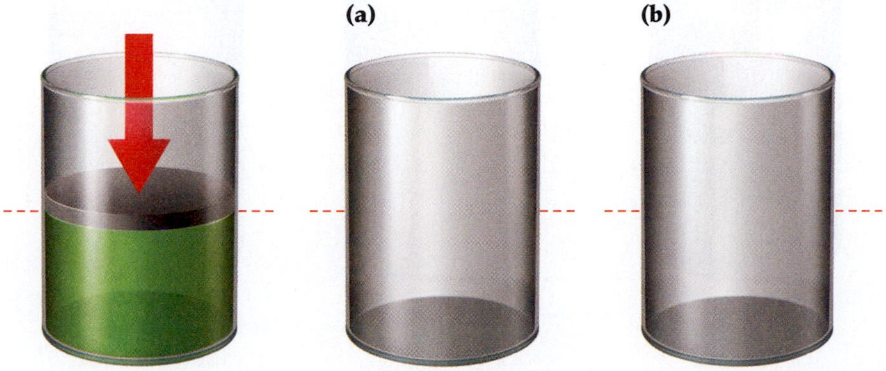

(initial)	(a)	(b)
$T = 250$ K	$T = 375$ K	$T = 250$ K
$n = 0.140$ mol	$n = 0.140$ mol	$n = 0.070$ mol
$P = 1.0$ atm	$P = 1.0$ atm	$P = 0.50$ atm

STRATEGY

Identify which of the variables P, n, and T have changed, and calculate the effect of each change on the volume according to the appropriate gas law.

SOLUTION

(a) The temperature T has increased by a factor of $375/250 = 1.5$, while the molar amount n and the pressure P are unchanged. According to Charles's law, the volume will increase by a factor of 1.5.

continued on next page

(b) The temperature T is unchanged, while both the molar amount n and the pressure P are halved. According to Avogadro's law, halving the molar amount will halve the volume, and according to Boyle's law, halving the pressure will double the volume. The two changes cancel, so the volume is unchanged.

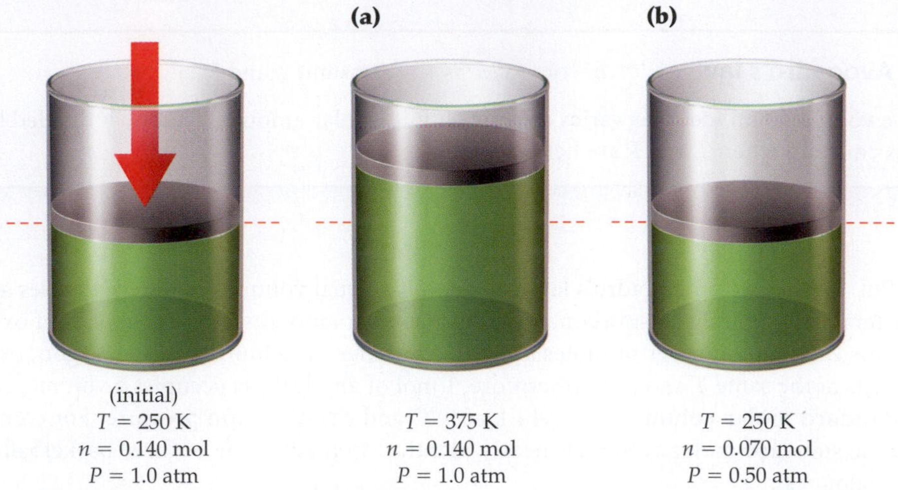

CONCEPTUAL PROBLEM 9.5 Show the approximate level of the movable piston in drawings **(a)** and **(b)** after the indicated changes have been made to the initial gas sample at a constant pressure of 1.0 atm.

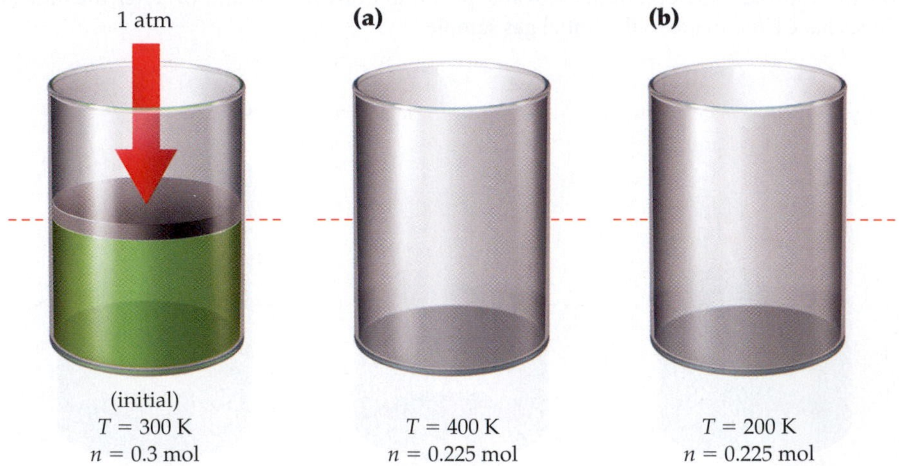

9.3 THE IDEAL GAS LAW

All three gas laws discussed in the previous section can be combined into a single statement called the **ideal gas law**, which describes how the volume of a gas is affected by changes in pressure, temperature, and amount. When the values of any three of the variables P, V, T, and n are known, the value of the fourth can be calculated using the ideal gas law. The proportionality constant R in the equation is called the **gas constant** and has the same value for all gases.

▶ **Ideal gas law** $V = \frac{nRT}{P}$ or $PV = nRT$

The ideal gas law can be rearranged in different ways to take the form of Boyle's law, Charles's law, or Avogadro's law.

Boyle's law: $PV = nRT = k$ (When n and T are constant)

Charles's law: $\frac{V}{T} = \frac{nR}{P} = k$ (When n and P are constant)

Avogadro's law: $\frac{V}{n} = \frac{RT}{P} = k$ (When T and P are constant)

The value of the gas constant R can be calculated from a knowledge of the standard molar volume of a gas. Since 1 mol of a gas occupies a volume of 22.414 L at 0 °C (273.15 K) and 1 atm pressure, the gas constant R is equal to 0.082 058 (L · atm)/(K · mol), or 8.3145 J/(K · mol) in SI units:

$$R = \frac{P \cdot V}{n \cdot T} = \frac{(1\ \text{atm})(22.414\ \text{L})}{(1\ \text{mol})(273.15\ \text{K})} = 0.082\,058\ \frac{\text{L} \cdot \text{atm}}{\text{K} \cdot \text{mol}}$$

$$= 8.3145\ \text{J/(K} \cdot \text{mol)} \quad \text{(When } P \text{ is in pascals and } V \text{ is in cubic meters)}$$

The specific conditions used in the calculation—0 °C (273.15 K) and 1 atm pressure—are said to represent **standard temperature and pressure**, abbreviated **STP**. These standard conditions are generally used when reporting measurements on gases. Note that the standard temperature for gas measurements (0 °C, or 273.15 K) is different from that usually assumed for thermodynamic measurements (25 °C, or 298.15 K; Section 8.6).

▶ **Standard temperature and pressure (STP) for gases** $T = 0\ °\text{C}$ $P = 1\ \text{atm}$

We should also point out that the standard pressure for gas measurements, still listed here and in most other books as 1 atm (101,325 Pa), has actually been redefined to be 1 bar, or 100,000 Pa. This new standard pressure is now 0.986 923 atm, making the newly defined standard molar volume 22.711 L rather than 22.414 L. Like most other books, we'll continue for the present using 1 atm as the standard pressure.

The name *ideal* gas law implies that there must be some gases whose behavior is *nonideal*. In fact, there is no such thing as an ideal gas that obeys the equation perfectly under all circumstances. All real gases are nonideal to some extent and deviate slightly from the behavior predicted by the gas laws. As **Table 9.4** shows, for example, the actual molar volume of a real gas often differs slightly from the 22.414 L ideal value. Under most conditions, though, the deviations from ideal behavior are so slight as to make little difference. We'll discuss circumstances in Section 9.8 where the deviations are greater.

TABLE 9.4 Molar Volumes of Some Real Gases at 0 °C and 1 atm

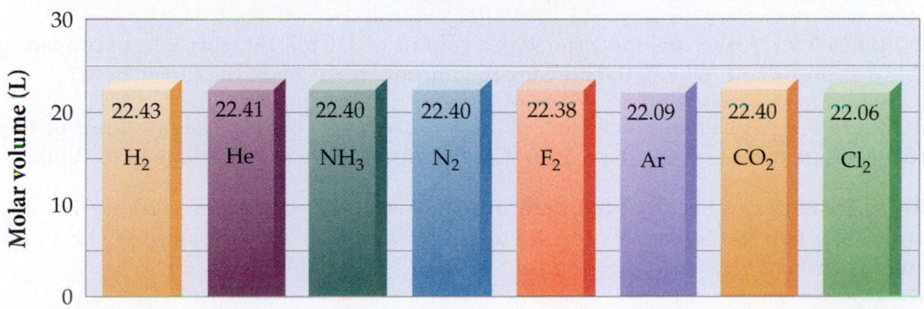

Worked Example 9.4

Gas Law Calculations

How many moles of gas (air) are in the lungs of an average adult with a lung capacity of 3.8 L? Assume that the lungs are at 1.00 atm pressure and a normal body temperature of 37 °C.

STRATEGY

This problem asks for a value of n when V, P, and T are given. Rearrange the ideal gas law to the form $n = PV/RT$, convert the temperature from degrees Celsius to kelvin, and substitute the given values of P, V, and T into the equation.

SOLUTION

$$n = \frac{PV}{RT} = \frac{(1.00\ \text{atm})(3.8\ \text{L})}{\left(0.082\ 06\ \dfrac{\text{L} \cdot \text{atm}}{\text{K} \cdot \text{mol}}\right)(310\ \text{K})} = 0.15\ \text{mol}$$

The lungs of an average adult hold 0.15 mol of air.

✓ BALLPARK CHECK

A lung volume of 4 L is about 1/6 of 22.4 L, the standard molar volume of an ideal gas. Thus, the lungs have a capacity of about 1/6 mol, or 0.17 mol.

Worked Example 9.5

Gas Law Calculations

In a typical automobile engine, the mixture of gasoline and air in a cylinder is compressed from 1.0 atm to 9.5 atm prior to ignition. If the uncompressed volume of the cylinder is 410 mL, what is the volume in milliliters when the mixture is fully compressed?

STRATEGY

This is a Boyle's law problem because only P and V are changing, while n and T remain fixed. We can therefore set up the following equation and solve for V_{final}.

$$(PV)_{\text{initial}} = (PV)_{\text{final}} = nRT$$

SOLUTION

$$V_{\text{final}} = \frac{(PV)_{\text{initial}}}{P_{\text{final}}} = \frac{(1.0\ \text{atm})(410\ \text{mL})}{9.5\ \text{atm}} = 43\ \text{mL}$$

✓ BALLPARK CHECK

Because the pressure in the cylinder increases about 10-fold, the volume must decrease about 10-fold according to Boyle's law, from approximately 400 mL to 40 mL.

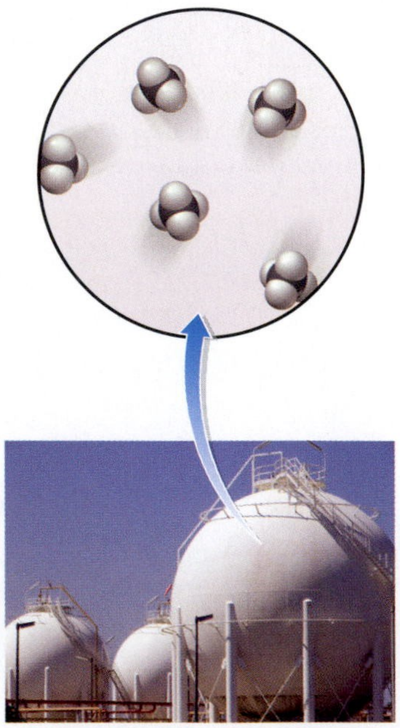

▲ How many moles of methane are in these tanks?

▶ PROBLEM 9.6 How many moles of methane gas, CH_4, are in a storage tank with a volume of 1.000×10^5 L at STP? How many grams?

▶ PROBLEM 9.7 An aerosol spray can with a volume of 350 mL contains 3.2 g of propane gas (C_3H_8) as propellant. What is the pressure in atmospheres of gas in the can at 20 °C?

▶ PROBLEM 9.8 A helium gas cylinder of the sort used to fill balloons has a volume of 43.8 L and a pressure of 1.51×10^4 kPa at 25.0 °C. How many moles of helium are in the tank?

▶ PROBLEM 9.9 What final temperature (°C) is required for the pressure inside an automobile tire to increase from 2.15 atm at 0 °C to 2.37 atm, assuming the volume remains constant?

CONCEPTUAL PROBLEM 9.10 Show the approximate level of the movable piston in drawings **(a)**, **(b)**, and **(c)** after the indicated changes have been made to the gas.

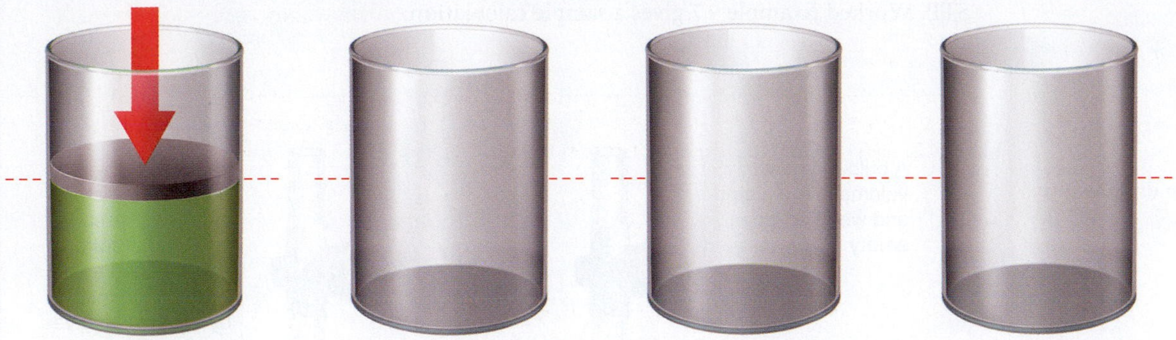

9.4 STOICHIOMETRIC RELATIONSHIPS WITH GASES

Many chemical reactions, including some of the most important processes in the chemical industry, involve gases. Approximately 130 million metric tons of ammonia, for instance, is manufactured each year worldwide by the reaction of hydrogen with nitrogen according to the equation $3\ H_2 + N_2 \rightarrow 2\ NH_3$. Thus, it's necessary to be able to calculate amounts of gaseous reactants just as it's necessary to calculate amounts of solids, liquids, and solutions.

Most gas calculations are just applications of the ideal gas law in which three of the variables *P*, *V*, *T*, and *n* are known and the fourth variable must be calculated. The reaction used in the deployment of automobile air bags, for instance, is the high-temperature decomposition of sodium azide, NaN_3, to produce N_2 gas. (The sodium is then removed by a subsequent reaction.) How many liters of N_2 gas at 1.15 atm and 30 °C are produced by decomposition of 45.0 g of NaN_3?

▲ Automobile air bags are inflated with N_2 gas produced by decomposition of sodium azide.

$$2\ NaN_3(s) \longrightarrow 2\ Na(s) + 3\ N_2(g)$$

Values for *P* and *T* are given, the value of *n* can be calculated, and the ideal gas law will then let us find *V*. To find *n*, the number of moles of N_2 gas produced, we first need to find how many moles of NaN_3 are in 45.0 g:

$$\text{Molar mass of NaN}_3 = 65.0\ \text{g/mol}$$

$$\text{Moles of NaN}_3 = (45.0\ \text{g NaN}_3)\left(\frac{1\ \text{mol NaN}_3}{65.0\ \text{g NaN}_3}\right) = 0.692\ \text{mol NaN}_3$$

Next, find how many moles of N_2 are produced in the decomposition reaction. According to the balanced equation, 2 mol of NaN_3 yields 3 mol of N_2, so 0.692 mol of NaN_3 yields 1.04 mol of N_2:

$$\text{Moles of N}_2 = (0.692\ \text{mol NaN}_3)\left(\frac{3\ \text{mol N}_2}{2\ \text{mol NaN}_3}\right) = 1.04\ \text{mol N}_2$$

Finally, use the ideal gas law to calculate the volume of N_2. Remember to use the Kelvin temperature (303 K) rather than the Celsius temperature (30 °C) in the calculation.

$$V = \frac{nRT}{P} = \frac{(1.04\ \text{mol N}_2)\left(0.08206\ \frac{\text{L}\cdot\text{atm}}{\text{K}\cdot\text{mol}}\right)(303\ \text{K})}{1.15\ \text{atm}} = 22.5\ \text{L}$$

Worked Example 9.6 illustrates another gas stoichiometry calculation.

Still other applications of the ideal gas law make it possible to calculate such properties as density and molar mass. Densities are calculated by weighing a known volume of a gas at a known temperature and pressure, as shown in **Figure 9.10.** Using the ideal gas law to find the volume at STP and then dividing the measured mass by the volume gives the density at STP. Worked Example 9.7 gives a sample calculation.

FIGURE 9.10

Determining the density of an unknown gas.

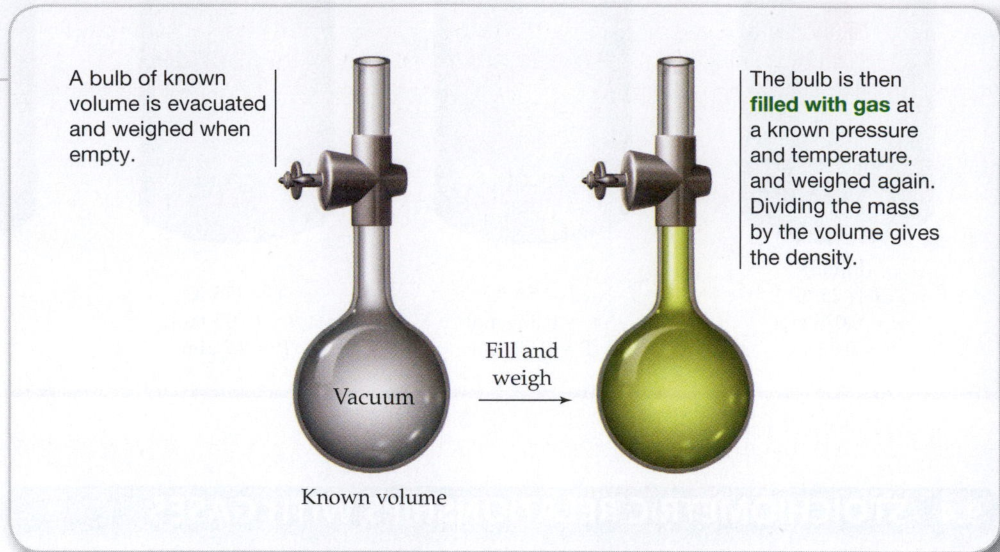

Molar masses, and therefore molecular weights, can also be calculated using the ideal gas law. Imagine, for instance, that an unknown gas found bubbling up in a swamp is collected, placed in a glass bulb, and found to have a density of 0.714 g/L at STP. What is the molecular weight of the gas?

Let's assume that we have 1.00 L of sample, which has a mass of 0.714 g. Since the density is measured at STP, we know *T*, *P*, and *V*, and we need to find *n*, the molar amount of gas that has a mass of 0.714 g:

$$n = \frac{PV}{RT} = \frac{(1.00\ \text{atm})(1.00\ \text{L})}{\left(0.082\ 06\ \dfrac{\text{L}\cdot\text{atm}}{\text{k}\cdot\text{mol}}\right)(273\ \text{K})} = 0.0446\ \text{mol}$$

Dividing the mass of the sample by the number of moles then gives the molar mass:

$$\text{Molar mass} = \frac{0.714\ \text{g}}{0.0446\ \text{mol}} = 16.0\ \text{g/mol}$$

Thus, the molar mass of the unknown gas (actually methane, CH_4) is 16.0 g/mol, and the molecular weight is 16.0.

It's often true in chemistry, particularly in gas-law calculations, that a problem can be solved in more than one way. As an alternative method for calculating the molar mass of the unknown swamp gas, you might recognize that 1 mol of an ideal gas has a volume of 22.4 L at STP. Since 1.00 L of the unknown gas has a mass of 0.714 g, 22.4 L of the gas (1 mol) has a mass of 16.0 g:

$$\text{Molar mass} = \left(0.714\ \frac{\text{g}}{\text{L}}\right)\left(22.4\ \frac{\text{L}}{\text{mol}}\right) = 16.0\ \text{g/mol}$$

Worked Example 9.8 illustrates another calculation of the molar mass of an unknown gas.

Worked Example 9.6

Finding a Mass Using Gas Law Calculations

A typical high-pressure tire on a bicycle might have a volume of 365 mL and a pressure of 7.80 atm at 25 °C. Suppose the rider filled the tire with helium to minimize weight. What is the mass of the helium in the tire?

STRATEGY

We are given *V*, *P*, and *T*, and we need to use the ideal gas law to calculate *n*, the number of moles of helium in the tire. With *n* known, we then do a mole-to-mass conversion.

SOLUTION

$$n = \frac{PV}{RT} = \frac{(7.80\ \text{atm})(0.365\ \text{L})}{\left(0.082\,06\ \dfrac{\text{L} \cdot \text{atm}}{\text{K} \cdot \text{mol}}\right)(298\ \text{K})} = 0.116\ \text{mol}$$

$$\text{Grams of helium} = 0.116\ \text{mol He} \times \frac{4.00\ \text{g He}}{1\ \text{mol He}} = 0.464\ \text{g}$$

Worked Example 9.7

Finding a Density Using Gas Law Calculations

What is the density in g/L of ammonia at STP if the gas in a 1.000 L bulb weighs 0.672 g at 25 °C and 733.4 mm Hg pressure?

STRATEGY

The density of any substance is mass divided by volume. For the ammonia sample, the mass is 0.672 g but the volume of the gas is given under nonstandard conditions and must first be converted to STP. Because the amount of sample *n* is constant, we can set the quantity *PV*/*RT* measured under nonstandard conditions equal to *PV*/*RT* at STP and then solve for *V* at STP.

SOLUTION

$$n = \left(\frac{PV}{RT}\right)_{\text{measured}} = \left(\frac{PV}{RT}\right)_{\text{STP}} \quad \text{or} \quad V_{\text{STP}} = \left(\frac{PV}{RT}\right)_{\text{measured}}\left(\frac{RT}{P}\right)_{\text{STP}}$$

$$V_{\text{STP}} = \left(\frac{733.4\ \text{mm Hg} \times 1.000\ \text{L}}{298\ \text{K}}\right)\left(\frac{273\ \text{K}}{760\ \text{mm Hg}}\right) = 0.884\ \text{L}$$

The amount of gas in the 1.000 L bulb under the measured nonstandard conditions would have a volume of only 0.884 L at STP. Dividing the given mass by this volume gives the density of ammonia at STP:

$$\text{Density} = \frac{\text{Mass}}{\text{Volume}} = \frac{0.672\ \text{g}}{0.884\ \text{L}} = 0.760\ \text{g/L}$$

Worked Example 9.8

Identifying an Unknown Using Gas Law Calculations

To identify the contents of an unlabeled cylinder of gas, a sample was collected and found to have a density of 5.380 g/L at 15 °C and 736 mm Hg pressure. What is the molar mass of the gas?

STRATEGY

Let's assume we have a 1.000 L sample of the gas weighing 5.380 g. We know the temperature, volume, and pressure of the gas and can therefore use the ideal gas law to find *n*, the number of moles in the sample. Dividing the mass by the number of moles gives the molar mass.

continued on next page

SOLUTION

$$PV = nRT \quad \text{or} \quad n = \frac{PV}{RT}$$

$$n = \frac{\left(736\text{ mm Hg} \times \dfrac{1\text{ atm}}{760\text{ mm Hg}}\right)(1.000\text{ L})}{\left(0.08206\ \dfrac{\text{L}\cdot\text{atm}}{\text{K}\cdot\text{mol}}\right)(288\text{ K})} = 0.0410\text{ mol}$$

$$\text{Molar mass} = \frac{5.380\text{ g}}{0.0410\text{ mol}} = 131\text{ g/mol}$$

The gas is probably xenon (atomic weight = 131.3).

PROBLEM 9.11 Carbonate-bearing rocks like limestone ($CaCO_3$) react with dilute acids such as HCl to produce carbon dioxide, according to the equation

$$CaCO_3(s) + 2\,HCl(aq) \longrightarrow CaCl_2(aq) + CO_2(g) + H_2O(l)$$

How many grams of CO_2 are formed by complete reaction of 33.7 g of limestone? What is the volume in liters of this CO_2 at STP?

▲ Carbonate-bearing rocks like limestone ($CaCO_3$) react with dilute acids such as HCl to produce bubbles of carbon dioxide.

PROBLEM 9.12 Propane gas (C_3H_8) is often used as a fuel in rural areas. How many liters of CO_2 are formed at STP by the complete combustion of the propane in a container with a volume of 15.0 L and a pressure of 4.5 atm at 25 °C? The unbalanced equation is

$$C_3H_8(g) + O_2(g) \longrightarrow CO_2(g) + H_2O(l)$$

PROBLEM 9.13 A foul-smelling gas produced by the reaction of HCl with Na_2S was collected, and a 1.00 L sample was found to have a mass of 1.52 g at STP. What is the molecular weight of the gas? What is its likely formula and name?

9.5 MIXTURES OF GASES: PARTIAL PRESSURE AND DALTON'S LAW

Just as the gas laws apply to all pure gases, regardless of chemical identity, they also apply to *mixtures* of gases, such as air. The pressure, volume, temperature, and amount of a gas mixture are all related by the ideal gas law.

What is responsible for the pressure in a gas mixture? Because the pressure of a pure gas at constant temperature and volume is proportional to its amount ($P = nRT/V$), the pressure contribution from each individual gas in a mixture is also proportional to *its* amount in the mixture. In other words, the total pressure exerted by a mixture of gases in a container at constant V and T is equal to the sum of the pressures of each individual gas in the container, a statement known as **Dalton's law of partial pressures.**

▶ **Dalton's law of partial pressures** $P_{total} = P_1 + P_2 + P_3 + \ldots$ at constant V and T, where $P_1, P_2, \ldots$ refer to the pressures each individual gas would have if it were alone.

The individual pressure contributions of the various gases in the mixture, P_1, P_2, and so forth, are called *partial pressures* and refer to the pressure each individual gas would exert if it were alone in the container. That is,

$$P_1 = n_1\left(\frac{RT}{V}\right) \qquad P_2 = n_2\left(\frac{RT}{V}\right) \qquad P_3 = n_3\left(\frac{RT}{V}\right) \ldots \text{ and so forth}$$

But because all the gases in the mixture have the same temperature and volume, we can rewrite Dalton's law to indicate that the total pressure depends only on the total molar amount of gas present and not on the chemical identities of the individual gases:

$$P_{total} = (n_1 + n_2 + n_3 + \ldots)\left(\frac{RT}{V}\right)$$

The concentration of any individual component in a gas mixture is usually expressed as a **mole fraction (X)**, which is defined simply as the number of moles of the component divided by the total number of moles in the mixture:

▶ **Mole fraction (X)** $= \dfrac{\text{Moles of component}}{\text{Total moles in mixture}}$

The mole fraction of component 1, for example, is

$$X_1 = \frac{n_1}{n_1 + n_2 + n_3 + \ldots} = \frac{n_1}{n_{total}}$$

But because $n = PV/RT$, we can also write

$$X_1 = \frac{P_1\left(\frac{V}{RT}\right)}{P_{total}\left(\frac{V}{RT}\right)} = \frac{P_1}{P_{total}}$$

which can be rearranged to solve for P_1, the partial pressure of component 1:

$$P_1 = X_1 \cdot P_{total}$$

This equation says that the partial pressure exerted by each component in a gas mixture is equal to the mole fraction of that component times the total pressure. In air, for example, the mole fractions of N_2, O_2, Ar, and CO_2 are 0.7808, 0.2095, 0.0093, and 0.000 39, respectively (Table 9.1), and the total pressure of the air is the sum of the individual partial pressures:

$$P_{air} = P_{N_2} + P_{O_2} + P_{Ar} + P_{CO_2} + \ldots$$

Thus, at a total air pressure of 1 atm (760 mm Hg), the partial pressures of the individual components are

$$\begin{aligned}
P_{N_2} &= 0.7808 \text{ atm } N_2 &&= 593.4 \text{ mm Hg}\\
P_{O_2} &= 0.2095 \text{ atm } O_2 &&= 159.2 \text{ mm Hg}\\
P_{Ar} &= 0.0093 \text{ atm Ar} &&= 7.1 \text{ mm Hg}\\
P_{CO_2} &= 0.000\,39 \text{ atm } CO_2 &&= 0.3 \text{ mm Hg}\\
\hline
P_{air} &= 1.0000 \text{ atm air} &&= 760.0 \text{ mm Hg}
\end{aligned}$$

There are numerous practical applications of Dalton's law, ranging from the use of anesthetic agents in hospital operating rooms, where partial pressures of both oxygen and anesthetic in the patient's lungs must be constantly monitored, to the composition of diving gases used for underwater exploration. Worked Example 9.9 gives an illustration.

Worked Example 9.9

Using Partial Pressures and Mole Fractions

At an underwater depth of 250 ft, the pressure is 8.38 atm. What should the mole percent of oxygen in the diving gas be for the partial pressure of oxygen in the gas to be 0.21 atm, the same as in air at 1.0 atm?

▲ The partial pressure of oxygen in the scuba tanks must be the same underwater as in air at atmospheric pressure.

STRATEGY

The partial pressure of a gas in a mixture is equal to the mole fraction of the gas times the total pressure. Rearranging this equation lets us solve for mole fraction of O_2.

SOLUTION

$$\text{Since} \quad P_{O_2} = X_{O_2} \cdot P_{\text{total}}, \quad \text{then} \quad X_{O_2} = \frac{P_{O_2}}{P_{\text{total}}}$$

$$X_{O_2} = \frac{0.21 \text{ atm}}{8.38 \text{ atm}} = 0.025$$

$$\text{Percent } O_2 = 0.025 \times 100\% = 2.5\% \ O_2$$

The diving gas should contain 2.5% O_2 for the partial pressure of O_2 to be the same at 8.38 atm as it is in air at 1.0 atm.

✓ BALLPARK CHECK

Because the pressure underwater is about 8 times atmospheric pressure, the percentage of O_2 in the diving gas should be about 1/8 times the percentage of O_2 in air (20%), or about 2.5%.

PROBLEM 9.14 What is the mole fraction of each component in a mixture of 12.45 g of H_2, 60.67 g of N_2, and 2.38 g of NH_3?

PROBLEM 9.15 What is the total pressure in atmospheres and what is the partial pressure of each component if the gas mixture in Problem 9.14 is in a 10.00 L steel container at 90 °C?

PROBLEM 9.16 On a humid day in summer, the mole fraction of gaseous H_2O (water vapor) in the air at 25 °C can be as high as 0.0287. Assuming a total pressure of 0.977 atm, what is the partial pressure in atmospheres of H_2O in the air?

CONCEPTUAL PROBLEM 9.17 What is the partial pressure of each gas—red, yellow, and green—if the total pressure inside the following container is 600 mm Hg?

9.6 THE KINETIC–MOLECULAR THEORY OF GASES

Thus far, we've concentrated on just describing the behavior of gases rather than on understanding the reasons for that behavior. Actually, the reasons are straightforward and were explained more than a century ago using a model called the **kinetic–molecular theory**. The kinetic–molecular theory is based on the following assumptions:

1. A gas consists of tiny particles, either atoms or molecules, moving about at random.
2. The volume of the particles themselves is negligible compared with the total volume of the gas. Most of the volume of a gas is empty space.
3. The gas particles act independently of one another; there are no attractive or repulsive forces between particles.
4. Collisions of the gas particles, either with other particles or with the walls of a container, are elastic. That is, they bounce off the walls at the same speed and energy they hit with, so that the total kinetic energy of the gas particles is constant at constant T.
5. The average kinetic energy of the gas particles is proportional to the Kelvin temperature of the sample.

Beginning with these assumptions, it's possible not only to understand the behavior of gases but also to derive quantitatively the ideal gas law (though we'll not do so here). For example, look at how the individual gas laws follow from the five postulates of kinetic–molecular theory:

- **Boyle's law ($P \propto 1/V$):** Gas pressure is a measure of the number and forcefulness of collisions between gas particles and the walls of their container. The smaller the volume at constant n and T, the more crowded together the particles are and the greater the frequency of collisions. Thus, pressure increases as volume decreases (**Figure 9.11a**).
- **Charles's law ($V \propto T$):** Temperature is a measure of the average kinetic energy of the gas particles. The higher the temperature at constant n and P, the faster the gas particles move and the more room they need to move around in to avoid increasing their collisions with the walls of the container. Thus, volume increases as temperature increases (**Figure 9.11b**).
- **Avogadro's law ($V \propto n$):** The more particles there are in a gas sample, the more volume the particles need at constant P and T to avoid increasing their collisions with the walls of the container. Thus, volume increases as amount increases (**Figure 9.11c**).
- **Dalton's law ($P_{total} = P_1 + P_2 + \ldots$):** Because gas particles are far apart and act independently of one another, the chemical identity of the particles is irrelevant. Total pressure of a fixed volume of gas depends only on the temperature T and the total number of moles of gas n. The pressure exerted by a specific kind of particle thus depends on the mole fraction of that kind of particle in the mixture (**Figure 9.11d**).

One of the more important conclusions from kinetic–molecular theory comes from assumption 5—the relationship between temperature and E_K, the kinetic energy of molecular motion. Although we won't do so in this book, it can be shown that the total kinetic energy

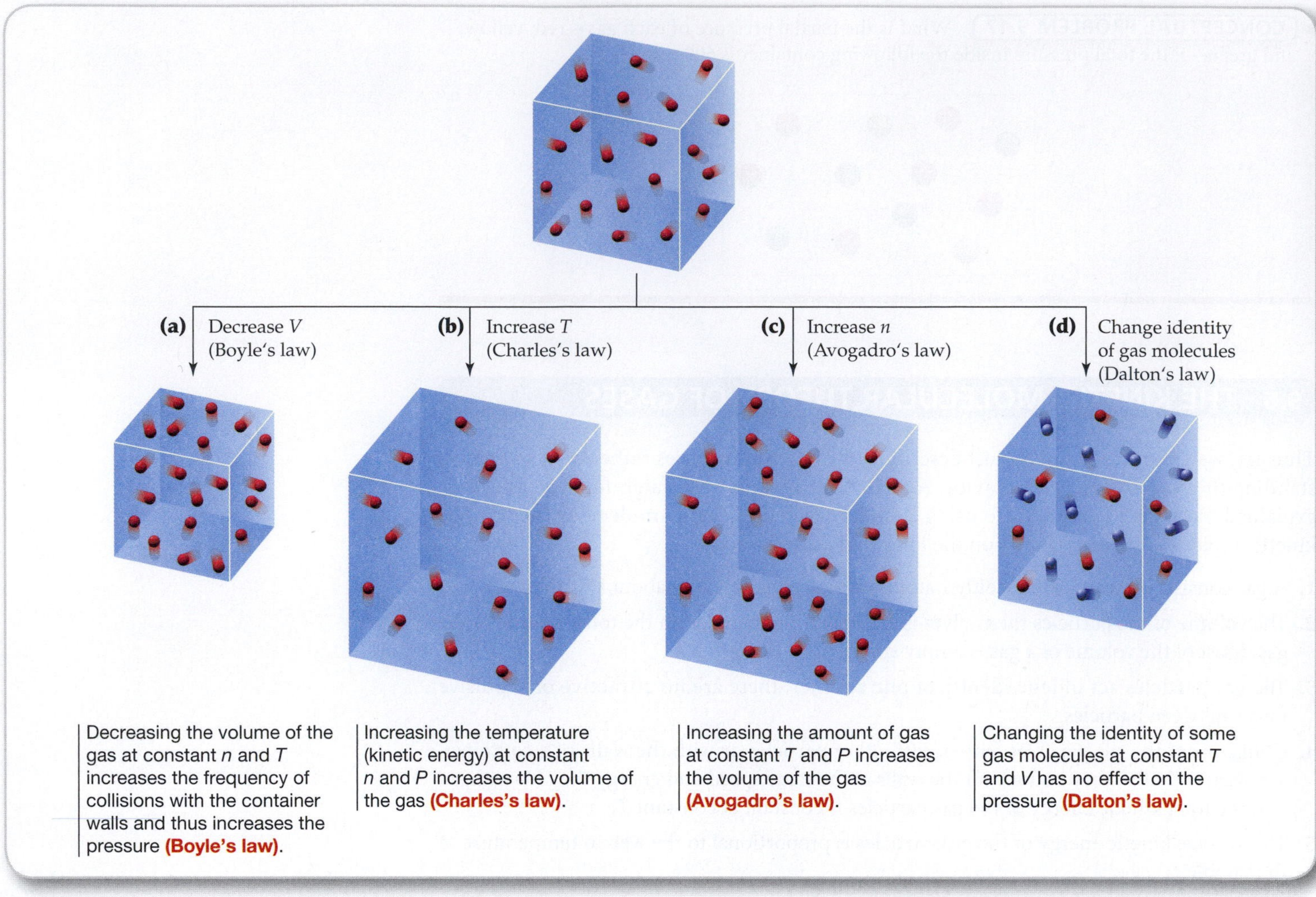

FIGURE 9.11

A kinetic-molecular view of the gas laws.

of a mole of gas particles equals $3RT/2$ and that the average kinetic energy per particle is thus $3RT/2N_A$ where N_A is Avogadro's number. Knowing this relationship makes it possible to calculate the average speed u of a gas particle at a given temperature. To take a helium atom at room temperature (298 K), for instance, we can write

$$E_K = \frac{3\,RT}{2\,N_A} = \frac{1}{2}mu^2$$

which can be rearranged to give

$$u^2 = \frac{3RT}{mN_A}$$

$$\text{or} \quad u = \sqrt{\frac{3RT}{mN_A}} = \sqrt{\frac{3RT}{M}} \quad \text{where } M \text{ is the molar mass}$$

Substituting appropriate values for R [8.314 J/(K · mol)] and for M, the molar mass of helium (4.00×10^{-3} kg/mol), we have

$$u = \sqrt{\frac{(3)\left(8.314\,\dfrac{\text{J}}{\text{K}\cdot\text{mol}}\right)(298\ \text{K})}{400 \times 10^{-3}\,\dfrac{\text{kg}}{\text{mol}}}} = \sqrt{1.86 \times 10^6\,\frac{\text{J}}{\text{kg}}}$$

$$= \sqrt{1.86 \times 10^6\,\frac{\dfrac{\text{kg}\cdot\text{m}^2}{\text{s}^2}}{\text{kg}}} = 1.36 \times 10^3\ \text{m/s}$$

Thus, the average speed of a helium atom at room temperature is more than 1.3 km/s, or about 3000 mi/h! Average speeds of some other molecules at 25 °C are given in **Table 9.5**. The heavier the molecule, the slower the average speed.

TABLE 9.5 Average Speeds (m/s) of Some Gas Molecules at 25 °C

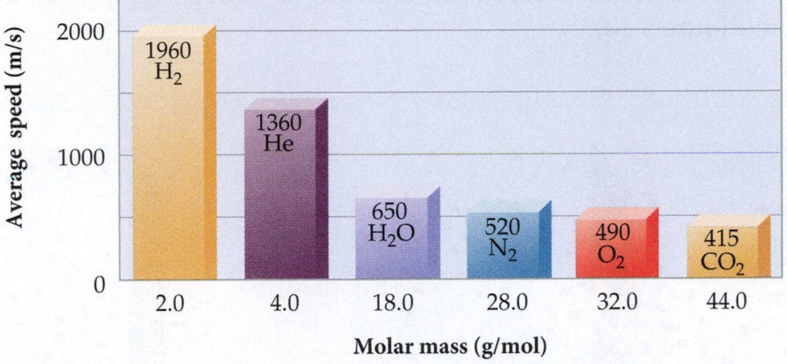

Just because the average speed of helium atoms at 298 K is 1.36 k m/s doesn't mean that all helium atoms are moving at that speed or that a given atom will travel from Maine to California in one hour. As shown in **Figure 9.12**, there is a broad distribution of speeds among particles in a gas, a distribution that flattens out and moves higher as the temperature increases. Furthermore, an individual gas particle is likely to travel only a very short distance before it collides with another particle and bounces off in a different direction. Thus, the actual path followed by a gas particle is a random zigzag.

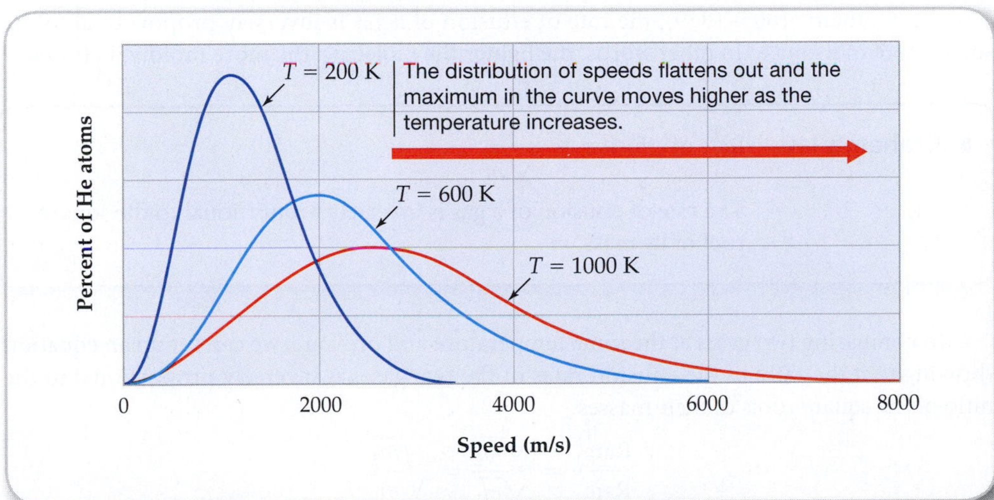

FIGURE 9.12

The distribution of speeds for helium atoms at different temperatures.

For helium at room temperature and 1 atm pressure, the average distance between collisions, called the *mean free path*, is only about 2×10^{-7} m, or 1000 atomic diameters, and there are approximately 10^{10} collisions per second. For a larger O_2 molecule, the mean free path is about 6×10^{-8} m.

PROBLEM 9.18 Calculate the average speed of a nitrogen molecule in m/s on a hot day in summer ($T = 37$ °C) and on a cold day in winter ($T = -25$ °C).

PROBLEM 9.19 At what temperature does the average speed of an oxygen molecule equal that of an airplane moving at 580 mph?

9.7 GAS DIFFUSION AND EFFUSION: GRAHAM'S LAW

The constant motion and high velocities of gas particles have some important practical consequences. One such consequence is that gases mix rapidly when they come in contact. Take the stopper off a bottle of perfume, for instance, and the odor will spread rapidly through a room as perfume molecules mix with the molecules in the air. This mixing of different molecules by random molecular motion with frequent collisions is called **diffusion**. A similar process in which gas molecules escape without collisions through a tiny hole into a vacuum is called **effusion** (**Figure 9.13**).

FIGURE 9.13

Diffusion and effusion of gases.

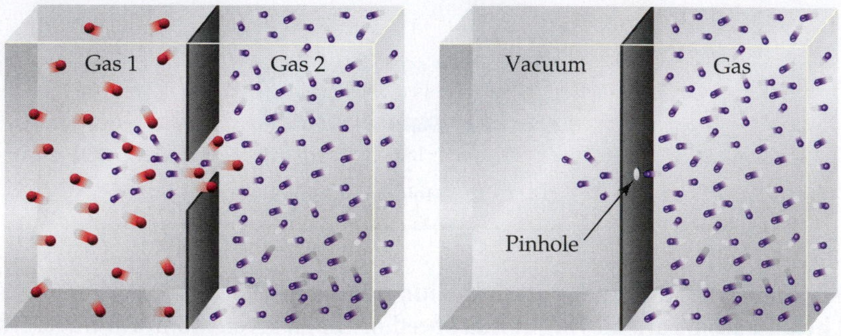

According to **Graham's law**, formulated in the mid-1800s by the Scottish chemist Thomas Graham (1805–1869), the rate of effusion of a gas is inversely proportional to the square root of its mass. In other words, the lighter the molecule, the more rapidly it effuses.

▶ **Graham's law** Rate of effusion $\propto \dfrac{1}{\sqrt{m}}$

The rate of effusion of a gas is inversely proportional to the square root of its mass, *m*.

In comparing two gases at the same temperature and pressure, we can set up an equation showing that the ratio of the effusion rates of the two gases is inversely proportional to the ratio of the square roots of their masses:

$$\frac{\text{Rate}_1}{\text{Rate}_2} = \frac{\sqrt{m_2}}{\sqrt{m_1}} = \sqrt{\frac{m_2}{m_1}}$$

The inverse relationship between the rate of effusion and the square root of the mass follows directly from the connection between temperature and kinetic energy described in the previous section. Because temperature is a measure of average kinetic energy and is independent of the gas's chemical identity, different gases at the same temperature have the same average kinetic energy:

$$\text{Since} \quad \frac{1}{2}mu^2 = \frac{3}{2}\frac{RT}{N_A} \quad \text{for any gas}$$

$$\text{then} \quad \left(\frac{1}{2}mu^2\right)_{\text{gas 1}} = \left(\frac{1}{2}mu^2\right)_{\text{gas 2}} \quad \text{at the same } T$$

Canceling the factor of 1/2 from both sides and rearranging, we find that the average speeds of the molecules in two gases vary as the inverse ratio of the square roots of their masses:

$$\text{Since} \quad \left(\frac{1}{2}mu^2\right)_{\text{gas 1}} = \left(\frac{1}{2}mu^2\right)_{\text{gas 2}}$$

$$\text{then} \quad (mu^2)_{\text{gas 1}} = (mu^2)_{\text{gas 2}} \quad \text{and} \quad \frac{(u_{\text{gas 1}})^2}{(u_{\text{gas 2}})^2} = \frac{m_2}{m_1}$$

$$\text{so} \quad \frac{u_{\text{gas 1}}}{u_{\text{gas 2}}} = \frac{\sqrt{m_2}}{\sqrt{m_1}} = \sqrt{\frac{m_2}{m_1}}$$

If, as seems reasonable, the rate of effusion of a gas is proportional to the average speed of the gas molecules, then Graham's law results.

Diffusion is more complex than effusion because of the molecular collisions that occur, but Graham's law usually works as a good approximation. One of the most important practical consequences is that mixtures of gases can be separated into their pure components by taking advantage of the different rates of diffusion of the components. For example, naturally occurring uranium is a mixture of isotopes, primarily ^{235}U (0.72%) and ^{238}U (99.28%). In uranium enrichment plants that purify the fissionable uranium-235 used for fuel in nuclear reactors, elemental uranium is converted into volatile uranium hexafluoride (bp 56 °C), and the UF_6 gas is allowed to diffuse from one chamber to another through a permeable membrane. The $^{235}UF_6$ and $^{238}UF_6$ molecules diffuse through the membrane at slightly different rates according to the square root of the ratio of their masses:

$$\text{For } ^{235}UF_6,\ m = 349.03$$

$$\text{For } ^{238}UF_6,\ m = 352.04$$

$$\text{so} \quad \frac{\text{Rate of } ^{235}UF_6 \text{ diffusion}}{\text{Rate of } ^{238}UF_6 \text{ diffusion}} = \sqrt{\frac{352.04}{349.03}} = 1.0043$$

The UF_6 gas that passes through the membrane is thus very slightly enriched in the lighter, faster-moving isotope. After repeating the process many thousands of times, a separation of isotopes can be achieved. Approximately 30% of the Western world's nuclear fuel supply—some 5000 tons per year—is produced by this gas diffusion method, although the percentage is dropping because better methods are now available.

▲ Much of the uranium-235 used as a fuel in nuclear reactors is obtained by gas diffusion of UF_6 in these cylinders.

Worked Example 9.10

Using Graham's Law to Calculate Diffusion Rates

Assume that you have a sample of hydrogen gas containing H_2, HD, and D_2 that you want to separate into pure components (H = ^1H and D = ^2H). What are the relative rates of diffusion of the three molecules according to Graham's law?

STRATEGY

First, find the masses of the three molecules: for H_2, $m = 2.016$; for HD, $m = 3.022$; for D_2, $m = 4.028$. Then apply Graham's law to different pairs of gas molecules.

SOLUTION

Because D_2 is the heaviest of the three molecules, it will diffuse most slowly. If we call its relative rate 1.000, we can then compare HD and H_2 with D_2:

Comparing HD with D_2, we have

$$\frac{\text{Rate of HD diffusion}}{\text{Rate of } D_2 \text{ diffusion}} = \sqrt{\frac{\text{mass of } D_2}{\text{mass of HD}}} = \sqrt{\frac{4.028}{3.022}} = 1.155$$

continued on next page

Comparing H_2 with D_2, we have

$$\frac{\text{Rate of } H_2 \text{ diffusion}}{\text{Rate of } D_2 \text{ diffusion}} = \sqrt{\frac{\text{mass of } D_2}{\text{mass of } H_2}} = \sqrt{\frac{4.028}{2.016}} = 1.414$$

Thus, the relative rates of diffusion are H_2 (1.414) > HD (1.155) > D_2 (1.000).

PROBLEM 9.20 Which gas in each of the following pairs diffuses more rapidly, and what are the relative rates of diffusion?
(a) Kr and O_2 **(b)** N_2 and acetylene (C_2H_2)

PROBLEM 9.21 What are the relative rates of diffusion of the three naturally occurring isotopes of neon, ^{20}Ne, ^{21}Ne, and ^{22}Ne?

9.8 THE BEHAVIOR OF REAL GASES

Before ending this discussion of gases, it's worthwhile expanding on a point made earlier: The behavior of a real gas is often a bit different from that of an ideal gas. For instance, kinetic–molecular theory assumes that the volume of the gas particles themselves is negligible compared with the total gas volume. The assumption is valid at STP, where the volume taken up by molecules of a typical gas is only about 0.05% of the total volume, but the assumption is not valid at 500 atm and 0 °C, where the volume of the molecules is about 20% of the total volume (**Figure 9.14**). As a result, the volume of a real gas at high pressure is larger than predicted by the ideal gas law.

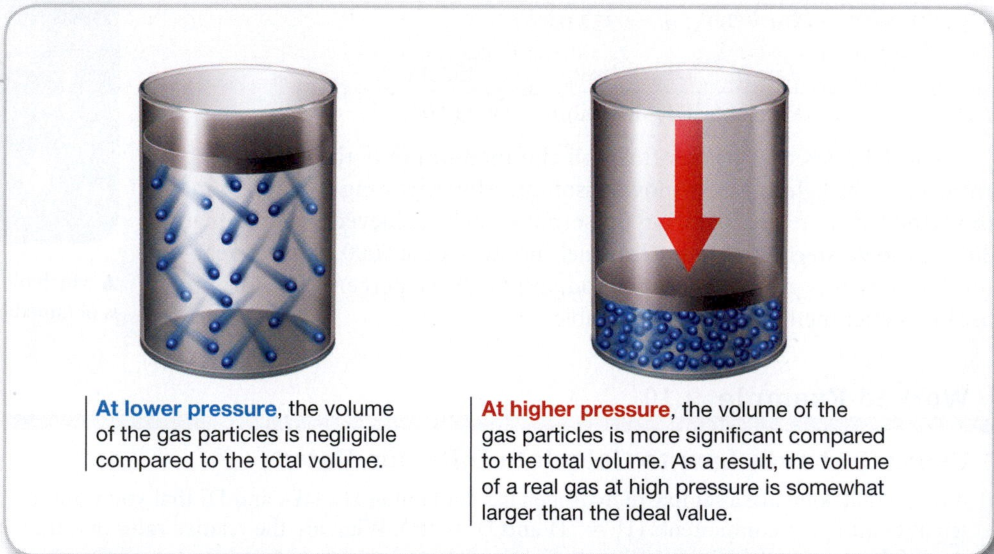

FIGURE 9.14
The volume of a real gas.

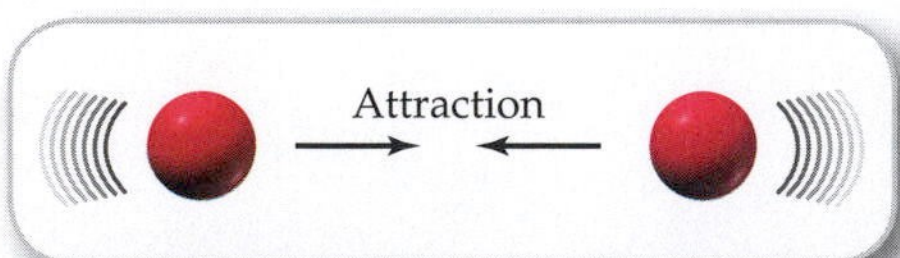

FIGURE 9.15
Molecules attract one another at distances up to about 10 molecular diameters. The result is a decrease in the actual volume of most real gases when compared with ideal gases at pressures up to 300 atm.

A second issue arising with real gases is the assumption that there are no attractive forces between particles. At lower pressures, this assumption is reasonable because the gas particles are so far apart. At higher pressures, however, the particles are much closer together and the attractive forces between them become more important. In general, intermolecular attractions become significant at a distance of about 10 molecular diameters and increase rapidly as the distance diminishes (**Figure 9.15**). The result is to draw the molecules of real gases together slightly, decreasing the volume at a given pressure (or decreasing the pressure for a given volume).

Note that the effect of molecular volume—to increase V—is opposite that of intermolecular attractions—to decrease V. The two factors therefore tend to cancel at intermediate pressures, but the effect of molecular volume dominates above about 350 atm.

Both deviations in the behavior of real gases can be dealt with mathematically by a modification of the ideal gas law called the **van der Waals equation**, which uses two correction factors, called *a* and *b*. The increase in *V*, caused by the effect of molecular volume, is corrected by subtracting an amount *nb* from the observed volume, where *n* is the number of moles of gas. The decrease in *V* at constant *P* (or, equivalently, the decrease in *P* at constant *V*), caused by the effect of intermolecular attractions, is best corrected by adding an amount an^2/V^2 to the pressure.

▶ **van der Waals equation** Correction for intermolecular attractions ↘ ↙ Correction for molecular volume

$$\left(P + \frac{an^2}{V^2}\right)(V - nb) = nRT$$

$$\text{or } P = \frac{nRT}{V - nb} - \frac{an^2}{V^2}$$

▶ **PROBLEM 9.22** Assume that you have 0.500 mol of N_2 in a volume of 0.600 L at 300 K. Calculate the pressure in atmospheres using both the ideal gas law and the van der Waals equation. For N_2, $a = 1.35\ (\text{L}^2 \cdot \text{atm})/\text{mol}^2$, and $b = 0.0387$ L/mol.

9.9 THE EARTH'S ATMOSPHERE

The mantle of gases surrounding the Earth is far from the uniform mixture you might expect. Although atmospheric pressure decreases in a regular way at higher altitudes, the profile of temperature versus altitude is much more complex (**Figure 9.16**). Four regions of the atmosphere have been defined based on this temperature curve. The temperature in the *troposphere*, the region nearest the Earth's surface, decreases regularly up to about 12 km altitude, where it reaches a minimum value, and then increases in the *stratosphere*, up to about 50 km. Above the stratosphere, in the *mesosphere*, (50–85 km), the temperature again decreases but then again increases in the *thermosphere*. To give you a feeling for these altitudes, passenger jets normally fly near the top of the troposphere at altitudes of 10 to 12 km, and the world altitude record for jet aircraft is 37.65 km—roughly in the middle of the stratosphere.

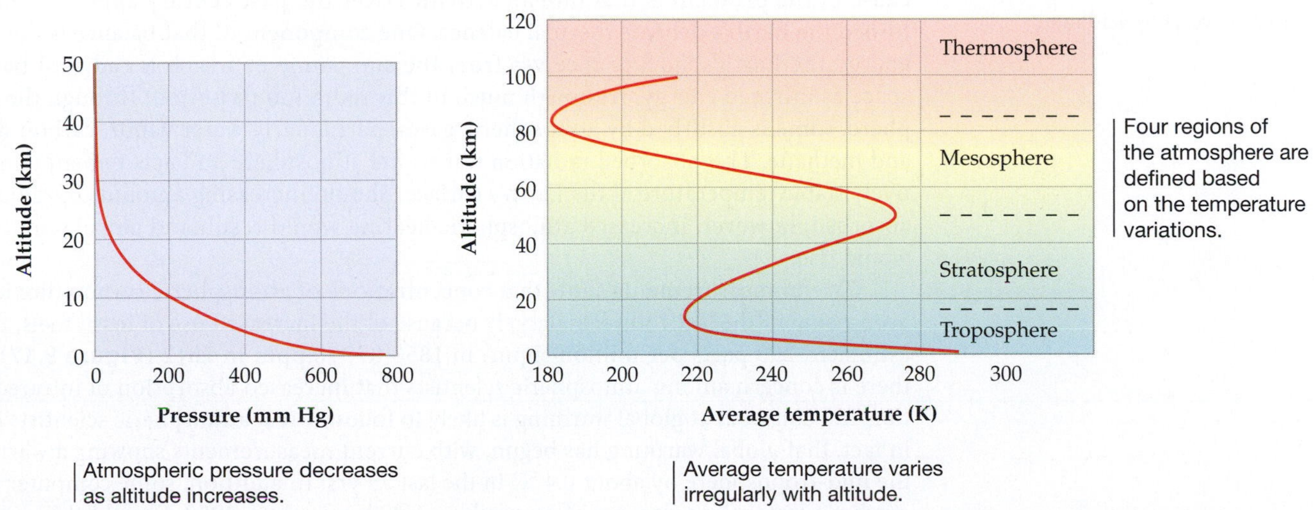

FIGURE 9.16

Variations of atmospheric pressure and average temperature with altitude.

Chemistry of the Troposphere

Not surprisingly, it's the layer nearest the Earth's surface—the troposphere—that is the most easily disturbed by human activities and has the greatest effect on the Earth's surface conditions. Among those effects, air pollution, acid rain, and global warming are particularly important.

▲ The photochemical smog over many cities is the end result of pollution from automobile exhausts.

▲ The details on this marble statue have been eaten away over the years by acid rain.

Air Pollution Air pollution has appeared in the last two centuries as an unwanted byproduct of industrialized societies. Its causes are relatively straightforward; its control is difficult. The main causes of air pollution are the release of unburned hydrocarbon molecules and the production of nitric oxide, NO, during combustion of petroleum products in older industrial plants and the more than 800 million automobile and truck engines presently in use worldwide. The NO is further oxidized by reaction with air to yield nitrogen dioxide, NO_2, which splits into NO plus free oxygen atoms in the presence of sunlight (symbolized by $h\nu$). Reaction of the oxygen atoms with O_2 molecules then yields ozone, O_3, a highly reactive substance that can further combine with unburned hydrocarbons in the air. The end result is the production of so-called photochemical smog, the hazy, brownish layer lying over many cities.

$$NO_2(g) + h\nu \longrightarrow NO(g) + O(g)$$

$$O(g) + O_2(g) \longrightarrow O_3(g)$$

Acid Rain Acid rain, a second major environmental problem, results primarily from the production of sulfur dioxide, SO_2, that accompanies the burning of sulfur-containing coal in power-generating plants. Sulfur dioxide is slowly converted to SO_3 by reaction with oxygen in air, and SO_3 dissolves in rainwater to yield dilute sulfuric acid, H_2SO_4.

$$S(\text{in coal}) + O_2(g) \longrightarrow SO_2(g)$$

$$2\,SO_2(g) + O_2(g) \longrightarrow 2\,SO_3(g)$$

$$SO_3(g) + H_2O(l) \longrightarrow H_2SO_4(aq)$$

Among the many dramatic effects of acid rain are the extinction of fish in acidic lakes throughout parts of the northeastern United States, Canada, and Scandinavia, the damage to forests throughout much of central and eastern Europe, and the deterioration everywhere of marble buildings and statuary. Marble is a form of calcium carbonate, $CaCO_3$, and, like all metal carbonates, reacts with acid to produce CO_2. The result is a slow eating away of the stone.

$$CaCO_3(s) + H_2SO_4(aq) \longrightarrow CaSO_4(aq) + H_2O(l) + CO_2(g)$$

Global Warming The third major atmospheric problem, global warming, is more complicated and less well understood than either air pollution or acid rain. The fundamental cause of the problem is that human activities over the past century appear to have disturbed the Earth's delicate thermal balance. One component of that balance is the radiant energy the Earth's surface receives from the sun, some of which is radiated back into space as infrared energy. Although much of this radiation passes out through the atmosphere, some is absorbed by atmospheric gases, particularly water vapor, carbon dioxide, and methane. This absorbed radiation warms the atmosphere and acts to maintain a relatively stable temperature at the Earth's surface. Should increasing amounts of radiation be absorbed, however, increased atmospheric heating would result and global temperatures would rise.

Careful measurements show that concentrations of atmospheric carbon dioxide have been rising in the last 160 years, largely because of the increased use of fossil fuels, from an estimated 290 parts per million (ppm) in 1850 to 390 ppm in 2012 (**Figure 9.17**). Thus, there is concern among atmospheric scientists that increased absorption of infrared radiation and consequent global warming is likely to follow. Most atmospheric scientists believe, in fact, that global warming has begun, with current measurements showing a warming of the mid-troposphere by about 0.4 °C in the last 25 yrs. In addition, some computer models predict a potential warming of as much as 3 °C by the year 2050. Should this occur, the result would be a significant increase in melting of glacial ice and a consequent rise in ocean levels.

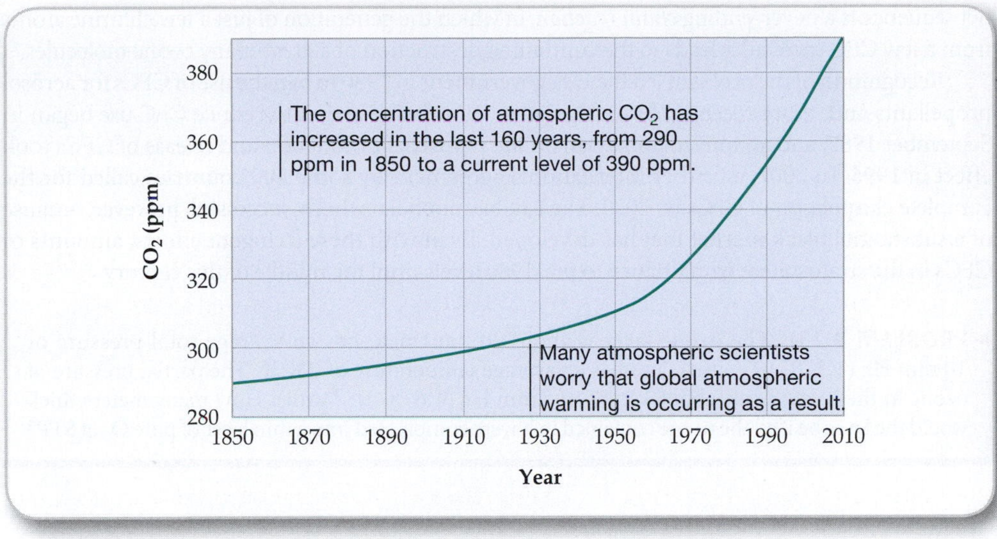

FIGURE 9.17

Annual concentration of atmospheric CO_2 since 1850.

Chemistry of the Upper Atmosphere

Relatively little of the atmosphere's mass is located above the troposphere, but the chemistry that occurs there is nonetheless crucial to maintaining life on Earth. Particularly important is what takes place in the *ozone layer*, an atmospheric band stretching from about 20 to 40 km above the Earth's surface. Ozone (O_3) is a severe pollutant at low altitudes but is critically important in the upper atmosphere because it absorbs intense ultraviolet radiation from the sun. Even though it is present in very small amounts in the stratosphere, ozone acts as a shield to prevent high-energy solar radiation from reaching the Earth's surface, where it can cause such health problems as eye cataracts and skin cancer.

Around 1976, a disturbing decrease in the amount of ozone present over the South Pole began showing up (**Figure 9.18**), and more recently a similar phenomenon has been found over the North Pole. Ozone levels drop to below 50% of normal in the polar spring before returning to near normal in the autumn.

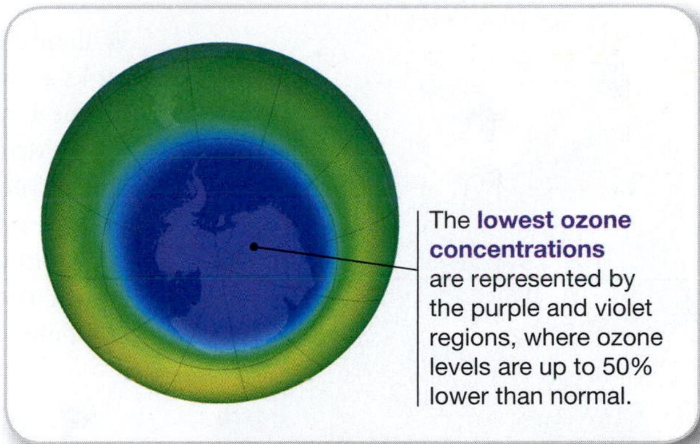

FIGURE 9.18

A false-color satellite image of the ozone hole over Antarctica in September, 2011.

The principal cause of ozone depletion is the presence in the stratosphere of *chlorofluorocarbons* (CFCs), such as CF_2Cl_2 and $CFCl_3$. Because they are inexpensive and stable, yet not toxic, flammable, or corrosive, CFCs are ideal as propellants in aerosol cans, as refrigerants, as solvents, and as fire-extinguishing agents. In addition, they are used for blowing bubbles into foamed plastic insulation. Unfortunately, the chemical stability that makes CFCs so useful also causes them to persist in the environment. Molecules released at ground level slowly diffuse into the stratosphere, where they undergo a complex series of reactions that ultimately result in ozone destruction.

Several different mechanisms of ozone destruction predominate under different stratospheric conditions. All are multistep processes that begin when ultraviolet light ($h\nu$) strikes a CFC molecule, breaking a carbon–chlorine bond and generating a chlorine atom:

$$CFCl_3 + h\nu \longrightarrow CFCl_2 + Cl$$

The resultant chlorine atom reacts with ozone to yield O_2 and ClO, and two ClO molecules then give Cl_2O_2. Further reaction occurs when Cl_2O_2 is struck by more ultraviolet light to generate O_2 and two more chlorine atoms.

$$\begin{array}{ll}
(1) & 2\,[Cl + O_3 \longrightarrow O_2 + ClO] \\
(2) & 2\,ClO \longrightarrow Cl_2O_2 \\
(3) & Cl_2O_2 + h\nu \longrightarrow 2\,Cl + O_2 \\
\hline
\text{Net:} & 2\,O_3 + h\nu \longrightarrow 3\,O_2
\end{array}$$

Look at the overall result of the above reaction sequence. Chlorine atoms are used up in the first step but are regenerated in the third step, so they don't appear in the net equation. Thus, the net sequence is a never-ending *chain reaction*, in which the generation of just a few chlorine atoms from a few CFC molecules leads to the continuing destruction of a great many ozone molecules.

Recognition of the problem led the U.S. government in 1980 to ban the use of CFCs for aerosol propellants and, more recently, for refrigerants. Worldwide action to reduce CFC use began in September 1987, and an international ban on the industrial production and release of CFCs took effect in 1996. In 2007, a further international agreement by some 200 countries called for the complete elimination of CFCs by 2020. The ban has not been wholly successful, however, because of a substantial black market that has developed. Even with these stringent efforts, amounts of CFCs in the stratosphere won't return to pre-1980 levels until the middle of this century.

PROBLEM 9.23 The ozone layer is about 20 km thick, has an average total pressure of 10 mm Hg (1.3×10^{-2} atm), and has an average temperature of 230 K. The partial pressure of ozone in the layer is only about 1.2×10^{-6} mm Hg (1.6×10^{-9} atm). How many meters thick would the layer be if all the ozone contained in it were compressed into a thin layer of pure O_3 at STP?

FYI HOW DO INHALATIONAL ANESTHETICS WORK?

▲ On April 7, 1853, Queen Victoria of England gave birth to her eighth child, while anesthetized by chloroform, changing forever the practice of obstetrics.

William Morton's demonstration in 1846 of ether-induced anesthesia during dental surgery ranks as one of the most important medical breakthroughs of all time. Before then, all surgery had been carried out with the patient conscious. Use of chloroform as an anesthetic quickly followed Morton's work, made popular by Queen Victoria of England, who in 1853 gave birth to a child while anesthetized by chloroform.

Literally hundreds of substances in addition to ether and chloroform have subsequently been shown to act as inhalational anesthetics. Isoflurane, sevoflurane, and desflurane are among the most commonly used agents at present. All three are nontoxic, nonflammable, and potent at relatively low doses.

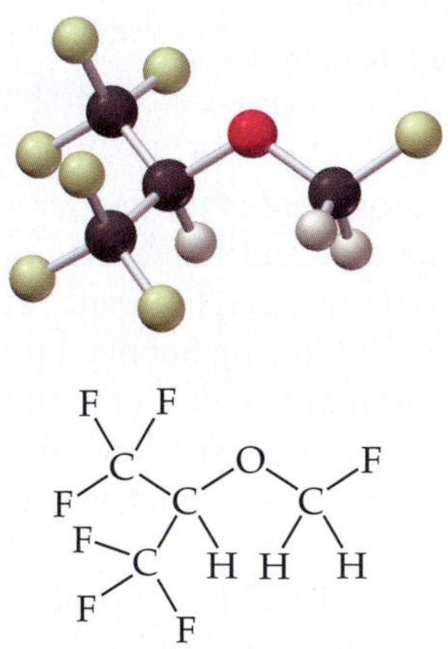

Sevoflurane

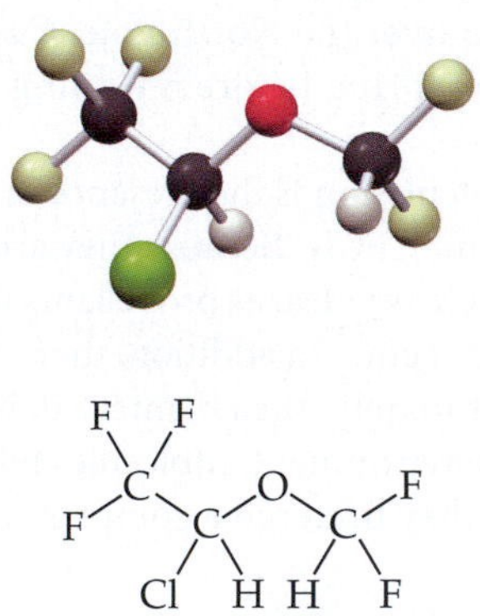

Isoflurane

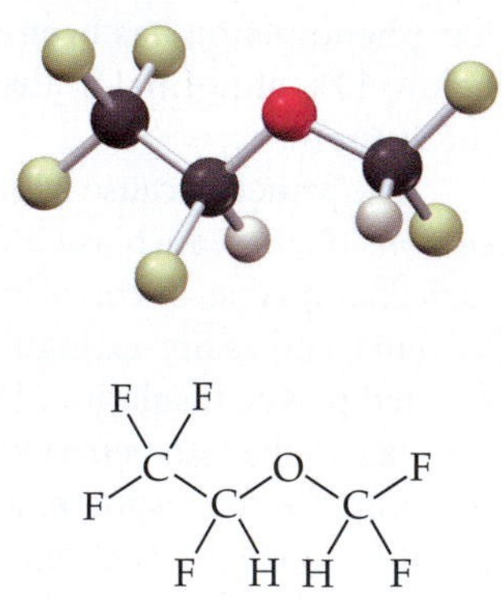

Desflurane

Despite their importance, surprisingly little is known about how inhalational anesthetics work in the body. Even the definition of anesthesia as a behavioral state is imprecise, and the nature of the changes in brain function leading to anesthesia are unknown. Remarkably, the potency of different anesthetics correlates well with their solubility in olive oil: the more soluble in olive oil, the more potent as an anesthetic. This unusual observation has led many scientists to believe that anesthetics act by dissolving in the fatty membranes surrounding nerve cells. The resultant changes in the fluidity and shape of the membranes apparently decrease the ability of sodium ions to pass into the nerve cells, thereby blocking the firing of nerve impulses.

Depth of anesthesia is determined by the concentration of anesthetic agent that reaches the brain. Brain concentration, in turn, depends on the solubility and transport of the anesthetic agent in the bloodstream and on its partial pressure in inhaled air. Anesthetic potency is usually expressed as a *minimum alveolar concentration (MAC)*, defined as the percent concentration of anesthetic in inhaled air that results in anesthesia in 50% of patients. As shown in **Table 9.6**, nitrous oxide, N_2O, is the least potent of the common anesthetics. Fewer than 50% of patients are immobilized by breathing an 80:20 mix of nitrous oxide and oxygen. Isoflurane is the most potent of the commonly used agents; a partial pressure of only 11 mm Hg is sufficient to anesthetize 50% of patients.

TABLE 9.6 Relative Potency of Inhalational Anesthetics

Anesthetic	MAC (%)	MAC (partial pressure, mm Hg)
Nitrous oxide	—	>760
Desflurane	6.2	47
Sevoflurane	2.5	19
Isoflurane	1.4	11

PROBLEM 9.24 For ether, a partial pressure of 15 mm Hg results in anesthesia in 50% of patients. What is the MAC for ether?

PROBLEM 9.25 Chloroform has a MAC of 0.77%.
(a) What partial pressure of chloroform is required to anesthetize 50% of patients?
(b) What mass of chloroform in 10.0 L of air at STP will produce the appropriate MAC?

SUMMARY

A gas is a collection of atoms or molecules moving independently through a volume that is largely empty space. Collisions of the randomly moving particles with the walls of their container exert a force per unit area that we perceive as **pressure**. The SI unit for pressure is the **pascal**, but the **atmosphere**, the **millimeter of mercury**, and the **bar** are more commonly used.

The physical condition of any gas is defined by four variables: pressure (P), temperature (T), volume (V), and molar amount (n). The specific relationships among these variables are called the **gas laws:**

Boyle's law: The volume of a gas varies inversely with its pressure. That is, $V \propto 1/P$ or $PV = k$ at constant n, T.

Charles's law: The volume of a gas varies directly with its Kelvin temperature. That is, $V \propto T$ or $V/T = k$ at constant n, P.

Avogadro's law: The volume of a gas varies directly with its molar amount. That is, $V \propto n$ or $V/n = k$ at constant T, P.

The three individual gas laws can be combined into a single **ideal gas law**, $PV = nRT$. If any three of the four variables P, V, T, and n are known, the fourth can be calculated. The constant R in the equation is called the **gas constant** and has the same value for all gases. At **standard temperature and pressure (STP**; 1 atm and 0 °C), the **standard molar volume** of an ideal gas is 22.414 L.

The gas laws apply to mixtures of gases as well as to pure gases. According to **Dalton's law of partial pressures**, the total pressure exerted by a mixture of gases in a container is equal to the sum of the pressures each individual gas would exert alone. The behavior of gases can be accounted for using a model called the **kinetic–molecular theory**, a group of five postulates:

1. A gas consists of tiny particles moving at random.
2. The volume of the gas particles is negligible compared with the total volume.
3. There are no forces between particles, either attractive or repulsive.
4. Collisions of gas particles are elastic.
5. The average kinetic energy of gas particles is proportional to their absolute temperature.

The connection between temperature and kinetic energy obtained from the kinetic–molecular theory makes it possible to calculate the average speed of a gas particle at any temperature. An important practical consequence of this relationship is **Graham's law**, which states that the rate of a gas's **effusion**, or spontaneous passage through a pinhole in a membrane, depends inversely on the square root of the gas's mass.

Real gases differ in their behavior from that predicted by the ideal gas law, particularly at high pressure, where gas particles are forced close together and intermolecular attractions become significant. The deviations from ideal behavior can be dealt with mathematically by the **van der Waals equation**.

KEY WORDS

KEY EQUATIONS

- Boyle's law (Section 9.2)

  $V \propto 1/P$ or $PV = k$ at constant n and T

- Charles's law (Section 9.2)

  $V \propto T$ or $V/T = k$ at constant n and P

- Avogadro's law (Section 9.2)

  $V \propto n$ or $V/n = k$ at constant T and P

- Ideal gas law (Section 9.3)

  $$V = \frac{nRT}{P} \quad \text{or} \quad PV = nRT$$

- Dalton's law of partial pressures (Section 9.5)

  $P_{total} = P_1 + P_2 + P_3 + \ldots$

  where $P_1, P_2, \ldots$ are the pressures each individual gas would have if it were alone.

- Average speed of gas particle at temperature T (Section 9.6)

  $$u = \sqrt{\frac{3RT}{mN_A}} = \sqrt{\frac{3RT}{M}} \quad \text{where } M \text{ is the molar mass.}$$

- van der Waals equation (Section 9.8)

  $$\left(P + \frac{an^2}{V^2}\right)(V - nb) = nRT \quad \text{or} \quad P = \frac{nRT}{V - nb} - \frac{an^2}{V^2}$$

  where a and b are correction factors to the ideal gas law.

LEARNING OUTCOMES

- Use a manometer to obtain a pressure reading. (Section 9.1) ***Problems 3, 4, 32, 33, 35, 40–43, 101***
- Use the individual gas laws and the ideal gas law to calculate a pressure, volume, molar amount, or temperature for a gas sample. (Sections 9.2 and 9.3) ***Problems 5–13, 26, 28, 29, 46–73, 102, 103, 105, 108, 110–113, 115, 116, 122, 124***
- Calculate the mole fractions and partial pressures of components in a gas mixture. (Section 9.5) ***Problems 14–17, 27, 74–83, 109, 117–120, 123, 126***
- Explain the assumptions of kinetic–molecular theory. (Section 9.6) ***Problems 29, 31, 37, 84, 106, 107***
- Calculate the speed of a gas particle at a given temperature. (Section 9.6) ***Problems 18, 19, 88–91, 96, 97***
- Calculate the relative rates of diffusion of different gas particles. (Section 9.7) ***Problems 20, 21, 34, 85, 87, 92–95, 100***
- Use the van der Waals equation to calculate the properties of real gases. (Section 9.8) ***Problems 22, 104, 114, 125***

CONCEPTUAL PROBLEMS

Problems 9.1–9.25 appear within the chapter.

9.26 A 1:1 mixture of helium (red) and argon (blue) at 300 K is portrayed below on the left. Draw the same mixture when the temperature is lowered to 150 K.

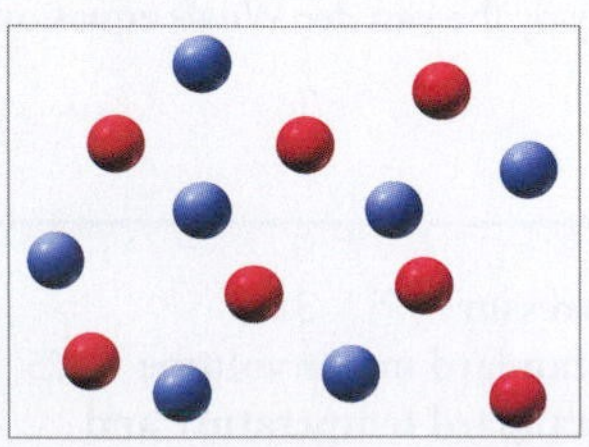

300 K

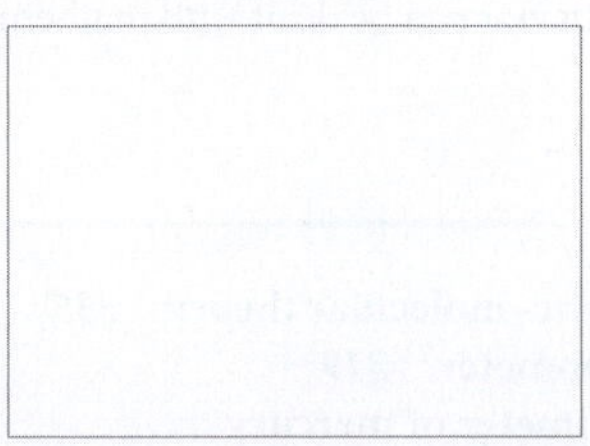

150 K

9.27 The following drawing represents a container holding a mixture of four gases, red, blue, green, and black. If the total pressure inside the container is 420 mm Hg, what is the partial pressure of each individual component?

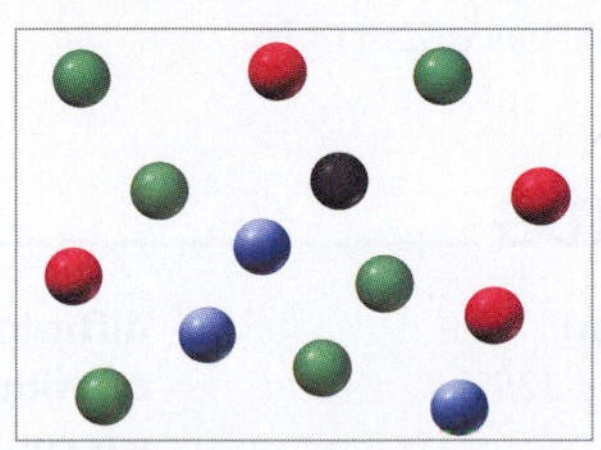

9.28 Assume that you have a sample of gas in a cylinder with a movable piston, as shown in the following drawing:

Redraw the apparatus to show what the sample will look like after **(a)** the temperature is increased from 300 K to 450 K at constant pressure, **(b)** the pressure is increased from 1 atm to 2 atm at constant temperature, and **(c)** the temperature is decreased from 300 K to 200 K and the pressure is decreased from 3 atm to 2 atm.

9.29 Assume that you have a sample of gas at 350 K in a sealed container, as represented in **(a)**. Which of the drawings **(b)**–**(d)** represents the gas after the temperature is lowered from 350 K to 150 K?

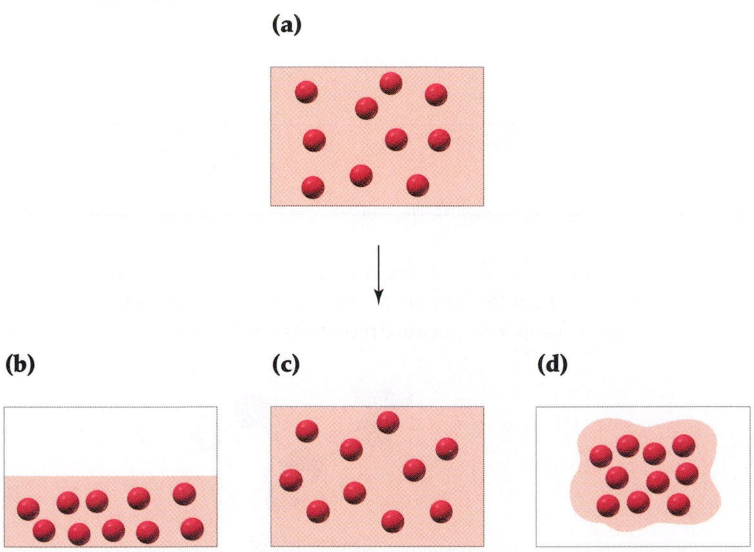

9.30 Assume that you have a mixture of He (atomic weight = 4) and Xe (atomic weight = 131) at 300 K. Which of the drawings best represents the mixture (blue = He; red = Xe)?

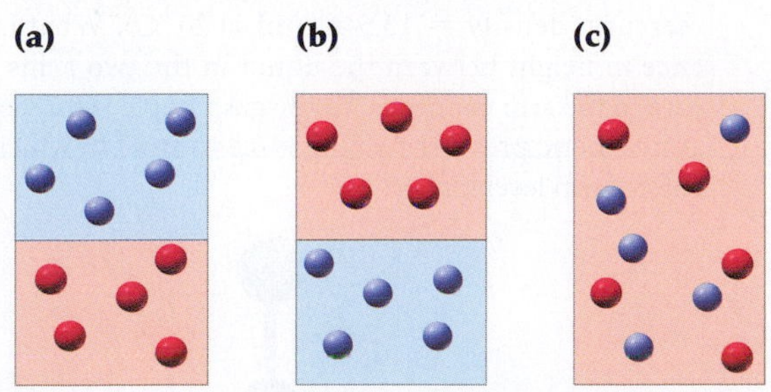

9.31 Three bulbs, two of which contain different gases and one of which is empty, are connected as shown in the following drawing. Redraw the apparatus to represent the gases after the stopcocks are opened and the system is allowed to come to equilibrium.

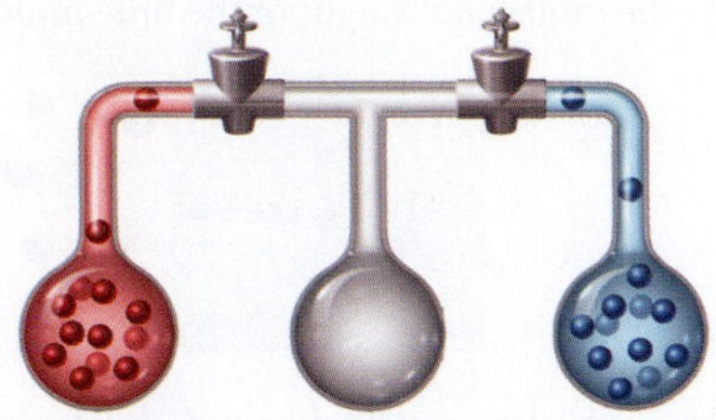

9.32 The apparatus shown is called a *closed-end* manometer because the arm not connected to the gas sample is closed to the atmosphere and is under vacuum. Explain how you can read the gas pressure in the bulb.

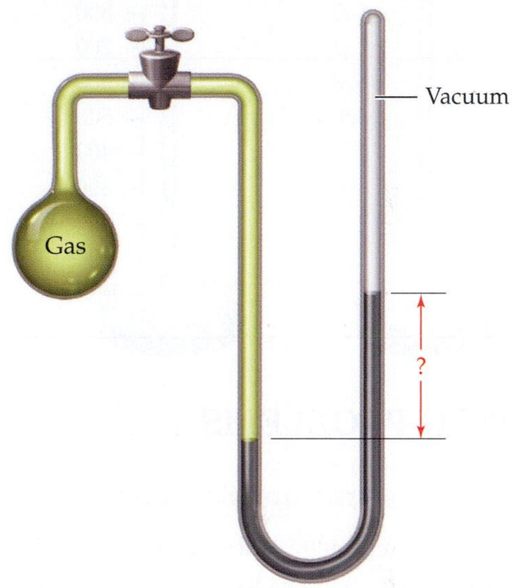

9.33 Redraw the following open-end manometer to show what it would look like when stopcock A is opened.

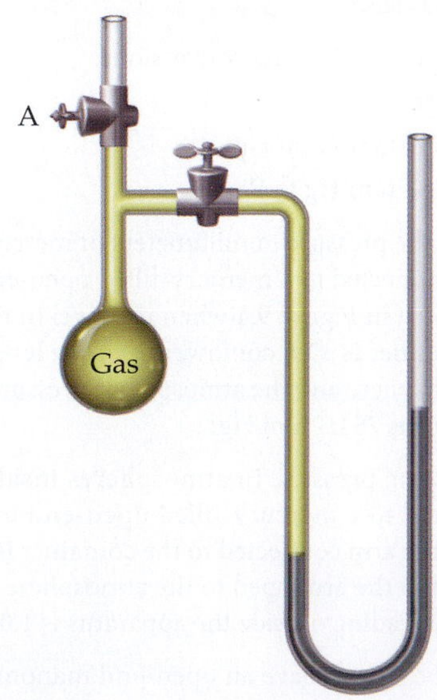

9.34 Effusion of a 1:1 mixture of two gases through a small pinhole produces the results shown below.

(a) Which gas molecules—yellow or blue—have a higher average speed?

(b) If the yellow molecules have a molecular weight of 25, what is the molecular weight of the blue molecules?

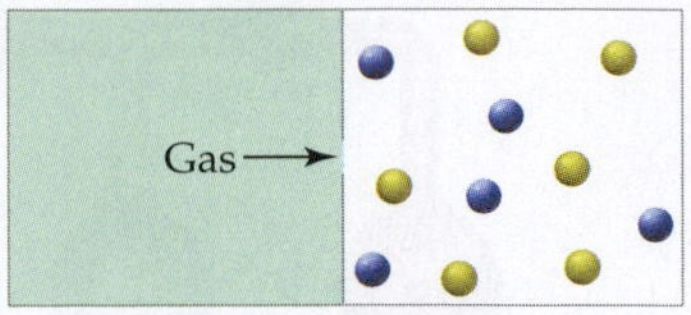

9.35 A glass tube has one end in a dish of mercury and the other end closed by a stopcock. The distance from the surface of the mercury to the bottom of the stopcock is 850 mm. The apparatus is at 25 °C, and the mercury level in the tube is the same as that in the dish.

(a) Show on drawing **(1)** what the approximate level of mercury in the tube will be when the temperature of the entire apparatus is lowered from +25 °C to −25 °C.

(b) Show on drawing **(2)** what the approximate level of mercury in the tube will be when a vacuum pump is connected to the top of the tube, the stopcock is opened, the tube is evacuated, the stopcock is closed, and the pump is removed.

(c) Show on drawing **(3)** what the approximate level of mercury in the tube will be when the stopcock in drawing **(2)** is opened.

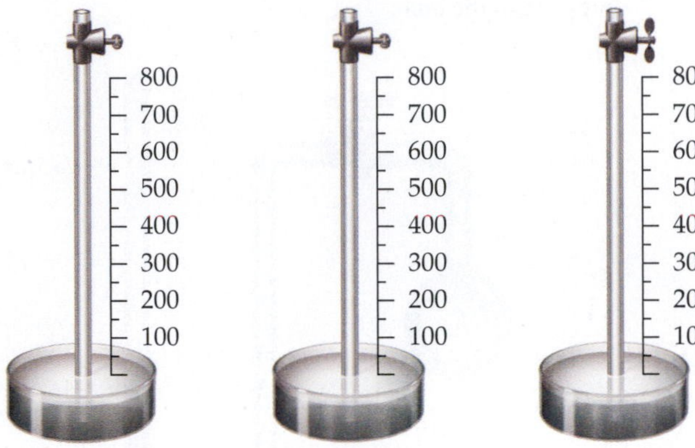

SECTION PROBLEMS

Gases and Gas Pressure (Section 9.1)

9.36 What is temperature a measure of?

9.37 Why are gases so much more compressible than solids or liquids?

9.38 Atmospheric pressure at the top of Pikes Peak in Colorado is approximately 480 mm Hg. Convert this value to atmospheres and to pascals.

9.39 Carry out the following conversions:

(a) 352 torr to kPa

(b) 0.255 atm to mm Hg

(c) 0.0382 mm Hg to Pa

9.40 What is the pressure in millimeters of mercury inside a container of gas connected to a mercury-filled open-end manometer of the sort shown in Figure 9.4 when the level in the arm connected to the container is 17.6 cm lower than the level in the arm open to the atmosphere and the atmospheric pressure reading outside the apparatus is 754.3 mm Hg?

9.41 What is the pressure in atmospheres inside a container of gas connected to a mercury-filled open-end manometer when the level in the arm connected to the container is 28.3 cm higher than the level in the arm open to the atmosphere and the atmospheric pressure reading outside the apparatus is 1.021 atm?

9.42 Assume that you have an open-end manometer filled with ethyl alcohol (density = 0.7893 g/mL at 20 °C) rather than mercury (density = 13.546 g/mL at 20 °C). What is the pressure in pascals if the level in the arm open to the atmosphere is 55.1 cm higher than the level in the arm connected to the gas sample and the atmospheric pressure reading is 752.3 mm Hg?

Ethyl alcohol

9.43 Assume that you have an open-end manometer filled with chloroform (density = 1.4832 g/mL at 20 °C) rather than mercury (density = 13.546 g/mL at 20 °C). What is the difference in height between the liquid in the two arms if the pressure in the arm connected to the gas sample is 0.788 atm and the atmospheric pressure reading is 0.849 atm? In which arm is the chloroform level higher?

Chloroform

9.44 Calculate the average molecular weight of air from the data given in Table 9.1.

9.45 What is the average molecular weight of a diving-gas mixture that contains 2.0% by volume O_2 and 98.0% by volume He?

The Gas Laws (Sections 9.2 and 9.3)

9.46 Assume that you have a cylinder with a movable piston. What would happen to the gas pressure inside the cylinder if you were to do the following?

(a) Triple the Kelvin temperature while holding the volume constant

(b) Reduce the amount of gas by 1/3 while holding the temperature and volume constant

(c) Decrease the volume by 45% at constant T

(d) Halve the Kelvin temperature and triple the volume

9.47 Assume that you have a cylinder with a movable piston. What would happen to the gas volume of the cylinder if you were to do the following?

(a) Halve the Kelvin temperature while holding the pressure constant

(b) Increase the amount of gas by 1/4 while holding the temperature and pressure constant

(c) Decrease the pressure by 75% at constant T

(d) Double the Kelvin temperature and double the pressure

9.48 Which sample contains more molecules: 1.00 L of O_2 at STP, 1.00 L of air at STP or 1.00 L of H_2 at STP?

9.49 Which sample contains more molecules: 2.50 L of air at 50 °C and 750 mm Hg pressure or 2.16 L of CO_2 at −10 °C and 765 mm Hg pressure?

9.50 Oxygen gas is commonly sold in 49.0 L steel containers at a pressure of 150 atm. What volume in liters would the gas occupy at a pressure of 1.02 atm if its temperature remained unchanged? If its temperature was raised from 20.0 °C to 35.0 °C at constant $P = 150$ atm?

9.51 A compressed air tank carried by scuba divers has a volume of 8.0 L and a pressure of 140 atm at 20 °C. What is the volume of air in the tank in liters at STP?

9.52 If 15.0 g of CO_2 gas has a volume of 0.30 L at 300 K, what is its pressure in millimeters of mercury?

9.53 If 2.00 g of N_2 gas has a volume of 0.40 L and a pressure of 6.0 atm, what is its Kelvin temperature?

9.54 The matter in interstellar space consists almost entirely of hydrogen atoms at a temperature of 100 K and a density of approximately 1 atom/cm^3. What is the gas pressure in millimeters of mercury?

9.55 Methane gas, CH_4, is sold in a 43.8 L cylinder containing 5.54 kg. What is the pressure inside the cylinder in kilopascals at 20 °C?

9.56 Many laboratory gases are sold in steel cylinders with a volume of 43.8 L. What mass in grams of argon is inside a cylinder whose pressure is 17,180 kPa at 20 °C?

9.57 A small cylinder of helium gas used for filling balloons has a volume of 2.30 L and a pressure of 13,800 kPa at 25 °C. How many balloons can you fill if each one has a volume of 1.5 L and a pressure of 1.25 atm at 25 °C?

Gas Stoichiometry (Section 9.4)

9.58 Which sample contains more molecules, 15.0 L of steam (gaseous H_2O) at 123.0 °C and 0.93 atm pressure or a 10.5 g ice cube at −5.0 °C?

9.59 Which sample contains more molecules, 3.14 L of Ar at 85.0 °C and 1111 mm Hg pressure or 11.07 g of Cl_2?

9.60 Imagine that you have two identical flasks, one containing hydrogen at STP and the other containing oxygen at STP. How can you tell which is which without opening them?

9.61 Imagine that you have two identical flasks, one containing chlorine gas and the other containing argon at the same temperature and pressure. How can you tell which is which without opening them?

9.62 What is the total mass in grams of oxygen in a room measuring 4.0 m by 5.0 m by 2.5 m? Assume that the gas is at STP and that air contains 20.95% oxygen by volume.

9.63 The average oxygen content of arterial blood is approximately 0.25 g of O_2 per liter. Assuming a body temperature of 37 °C, how many moles of oxygen are transported by each liter of arterial blood? How many milliliters?

9.64 One mole of an ideal gas has a volume of 22.414 L at STP. Assuming ideal behavior, what are the densities of the following gases in g/L at STP?

(a) CH_4

(b) CO_2

(c) O_2

9.65 What is the density in g/L of a gas mixture that contains 27.0% F_2 and 73.0% He by volume at 714 mm Hg and 27.5 °C?

9.66 An unknown gas is placed in a 1.500 L bulb at a pressure of 356 mm Hg and a temperature of 22.5 °C, and is found to weigh 0.9847 g. What is the molecular weight of the gas?

9.67 What are the molecular weights of the gases with the following densities:

(a) 1.342 g/L at STP

(b) 1.053 g/L at 25 °C and 752 mm Hg

9.68 Pure oxygen gas was first prepared by heating mercury(II) oxide, HgO:

$$2\,HgO(s) \longrightarrow 2\,Hg(l) + O_2(g)$$

What volume in liters of oxygen at STP is released by heating 10.57 g of HgO?

9.69 How many grams of HgO would you need to heat if you wanted to prepare 0.0155 mol of O_2 according to the equation in Problem 9.68?

9.70 Hydrogen gas can be prepared by reaction of zinc metal with aqueous HCl:

$$Zn(s) + 2\,HCl(aq) \longrightarrow ZnCl_2(aq) + H_2(g)$$

(a) How many liters of H_2 would be formed at 742 mm Hg and 15 °C if 25.5 g of zinc was allowed to react?

(b) How many grams of zinc would you start with if you wanted to prepare 5.00 L of H_2 at 350 mm Hg and 30.0 °C?

9.71 Ammonium nitrate can decompose explosively when heated according to the equation

$$2\,NH_4NO_3(s) \longrightarrow 2\,N_2(g) + 4\,H_2O(g) + O_2(g)$$

How many liters of gas would be formed at 450 °C and 1.00 atm pressure by explosion of 450 g of NH_4NO_3?

9.72 The reaction of sodium peroxide (Na_2O_2) with CO_2 is used in space vehicles to remove CO_2 from the air and generate O_2 for breathing:

$$2\ Na_2O_2(s) + 2\ CO_2(g) \longrightarrow 2\ Na_2CO_3(s) + O_2(g)$$

(a) Assuming that air is breathed at an average rate of 4.50 L/min (25 °C; 735 mm Hg) and that the concentration of CO_2 in expelled air is 3.4% by volume, how many grams of CO_2 are produced in 24 h?

(b) How many days would a 3.65 kg supply of Na_2O_2 last?

9.73 Titanium(III) chloride, a substance used in catalysts for preparing polyethylene, is made by high-temperature reaction of $TiCl_4$ vapor with H_2:

$$2\ TiCl_4(g) + H_2(g) \longrightarrow 2\ TiCl_3(s) + 2\ HCl(g)$$

(a) How many grams of $TiCl_4$ are needed for complete reaction with 155 L of H_2 at 435 °C and 795 mm Hg pressure?

(b) How many liters of HCl gas at STP will result from the reaction described in part (a)?

Dalton's Law and Mole Fraction (Section 9.5)

9.74 Use the information in Table 9.1 to calculate the partial pressure in atmospheres of each gas in dry air at STP.

9.75 Natural gas is a mixture of many substances, primarily CH_4, C_2H_6, C_3H_8, and C_4H_{10}. Assuming that the total pressure of the gases is 1.48 atm and that their mole ratio is 94:4.0:1.5:0.50, calculate the partial pressure in atmospheres of each gas.

9.76 A special gas mixture used in bacterial growth chambers contains 1.00% by weight CO_2 and 99.0% O_2. What is the partial pressure in atmospheres of each gas at a total pressure of 0.977 atm?

9.77 A gas mixture for use in some lasers contains 5.00% by weight HCl, 1.00% H_2, and 94% Ne. The mixture is sold in cylinders that have a volume of 49.0 L and a pressure of 13,800 kPa at 21.0 °C. What is the partial pressure in kilopascals of each gas in the mixture?

9.78 What is the mole fraction of each gas in the mixture described in Problem 9.77?

9.79 A mixture of Ar and N_2 gases has a density of 1.413 g/L at STP. What is the mole fraction of each gas?

9.80 A mixture of 14.2 g of H_2 and 36.7 g of Ar is placed in a 100.0 L container at 290 K.

(a) What is the partial pressure of H_2 in atmospheres?

(b) What is the partial pressure of Ar in atmospheres?

9.81 A 20.0 L flask contains 0.776 g of He and 3.61 g of CO_2 at 300 K.

(a) What is the partial pressure of He in mm Hg?

(b) What is the partial pressure of CO_2 in mm Hg?

9.82 A sample of magnesium metal reacts with aqueous HCl to yield H_2 gas:

$$Mg(s) + 2\ HCl(aq) \longrightarrow MgCl_2(aq) + H_2(g)$$

The gas that forms is found to have a volume of 3.557 L at 25 °C and a pressure of 747 mm Hg. Assuming that the gas is saturated with water vapor at a partial pressure of 23.8 mm Hg, what is the partial pressure in millimeters of mercury of the H_2? How many grams of magnesium metal were used in the reaction?

9.83 Chlorine gas was first prepared in 1774 by the oxidation of NaCl with MnO_2:

$$2\ NaCl(s) + 2\ H_2SO_4(l) + MnO_2(s) \longrightarrow Na_2SO_4(s) + MnSO_4(s) + 2\ H_2O(g) + Cl_2(g)$$

Assume that the gas produced is saturated with water vapor at a partial pressure of 28.7 mm Hg and that it has a volume of 0.597 L at 27 °C and 755 mm Hg pressure.

(a) What is the mole fraction of Cl_2 in the gas?

(b) How many grams of NaCl were used in the experiment, assuming complete reaction?

Kinetic–Molecular Theory and Graham's Law (Sections 9.6 and 9.7)

9.84 What are the basic assumptions of the kinetic–molecular theory?

9.85 What is the difference between effusion and diffusion?

9.86 What is the difference between heat and temperature?

9.87 Why does a helium-filled balloon lose pressure faster than an air-filled balloon?

9.88 The average temperature at an altitude of 20 km is 220 K. What is the average speed in m/s of an N_2 molecule at this altitude?

9.89 At what temperature (°C) will xenon atoms have the same average speed that Br_2 molecules have at 20 °C?

9.90 Which has a higher average speed, H_2 at 150 K or He at 375 °C?

9.91 Which has a higher average speed, a Ferrari at 145 mph or a gaseous UF_6 molecule at 145 °C?

9.92 An unknown gas is found to diffuse through a porous membrane 2.92 times more slowly than H_2. What is the molecular weight of the gas?

9.93 What is the molecular weight of a gas that diffuses through a porous membrane 1.86 times faster than Xe? What might the gas be?

9.94 Rank the following gases in order of their speed of diffusion through a membrane, and calculate the ratio of their diffusion rates: HCl, F_2, Ar.

9.95 Which will diffuse through a membrane more rapidly, CO or N_2? Assume that the samples contain only the most abundant isotopes of each element, ^{12}C, ^{16}O, and ^{14}N.

9.96 A big-league fastball travels at about 45 m/s. At what temperature (°C) do helium atoms have this same average speed?

9.97 Traffic on the German autobahns reaches speeds of up to 230 km/h. At what temperature (°C) do oxygen molecules have this same average speed?

CHAPTER PROBLEMS

9.98 What is the greatest atmospheric concern for each of the following: air pollution, acid rain, global warming, or ozone depletion?

(a) $CFCl_3$ (b) S (in coal)

(c) CO_2 (d) NO

9.99 What is the role or effect of ozone in the troposphere and in the stratosphere?

9.100 Chlorine occurs as a mixture of two isotopes, ^{35}Cl and ^{37}Cl. What is the ratio of the diffusion rates of the three species $(^{35}Cl)_2$, $^{35}Cl^{37}Cl$, and $(^{37}Cl)_2$?

9.101 What would the atmospheric pressure be in millimeters of mercury if our atmosphere were composed of pure CO_2 gas?

9.102 The surface temperature of Venus is about 1050 K, and the pressure is about 75 Earth atmospheres. Assuming that these conditions represent a Venusian "STP", what is the standard molar volume in liters of a gas on Venus?

9.103 When you look directly up at the sky, you are actually looking through a very tall, transparent column of air that extends from the surface of the Earth thousands of kilometers into space. If the air in this column were liquefied, how tall would it be? The density of liquid air is 0.89 g/mL.

9.104 Uranium hexafluoride, a molecular solid used for purification of the uranium isotope needed to fuel nuclear power plants, sublimes at 56.5 °C. Assume that you have a 22.9 L vessel that contains 512.9 g of UF_6 at 70.0 °C.

(a) What is the pressure in the vessel calculated using the ideal gas law?

(b) What is the pressure in the vessel calculated using the van der Waals equation? (For UF_6, $a = 15.80$ $(L^2 \cdot atm)/mol^2$; $b = 0.1128$ L/mol.)

9.105 A driver with a nearly empty fuel tank may say she is "running on fumes." If a 15.0 gallon automobile gas tank had only gasoline vapor remaining in it, what is the farthest the vehicle could travel if it gets 20.0 miles per gallon on liquid gasoline? Assume the average molar mass of molecules in gasoline is 105 g/mol, the density of liquid gasoline is 0.75 g/mL, the pressure is 743 mm Hg, and the temperature is 25 °C.

9.106 Two 112 L tanks are filled with gas at 330 K. One contains 5.00 mol of Kr, and the other contains 5.00 mol of O_2. Considering the assumptions of kinetic–molecular theory, rank the gases from low to high for each of the following properties:

(a) collision frequency **(b)** density (g/L)
(c) average speed **(d)** pressure

9.107 Two identical 732.0 L tanks each contain 212.0 g of gas at 293 K, with neon in one tank and nitrogen in the other. Based on the assumptions of kinetic–molecular theory, rank the gases from low to high for each of the following properties:

(a) average speed **(b)** pressure
(c) collision frequency **(d)** density (g/L)

9.108 Pakistan's K2 is the world's second tallest mountain, with an altitude of 28,251 ft. Its base camp, where climbers stop to acclimate, is located about 16,400 ft above sea level.

(a) Approximate atmospheric pressure P at different altitudes is given by the equation $P = e^{-h/7000}$, where P is in atmospheres and h is the altitude in meters. What is the approximate atmospheric pressure in mm Hg at K2 base camp?

(b) What is the atmospheric pressure in mm Hg at the summit of K2?

(c) Assuming the mole fraction of oxygen in air is 0.2095, what is the partial pressure of oxygen in mm Hg at the summit of K2?

9.109 When a 10.00 L vessel containing 42.189 g of I_2 is heated to 1173 K, some I_2 dissociates: $I_2(g) \longrightarrow 2\,I(g)$. If the final pressure in the vessel is 1.733 atm, what are the mole fractions of the two components $I_2(g)$ and $I(g)$ after the reaction?

9.110 Assume that you take a flask, evacuate it to remove all the air, and find its mass to be 478.1 g. You then fill the flask with argon to a pressure of 2.15 atm and reweigh it. What would the balance read in grams if the flask has a volume of 7.35 L and the temperature is 20.0 °C?

9.111 The apparatus shown consists of three bulbs connected by stopcocks. What is the pressure inside the system when the stopcocks are opened? Assume that the lines connecting the bulbs have zero volume and that the temperature remains constant.

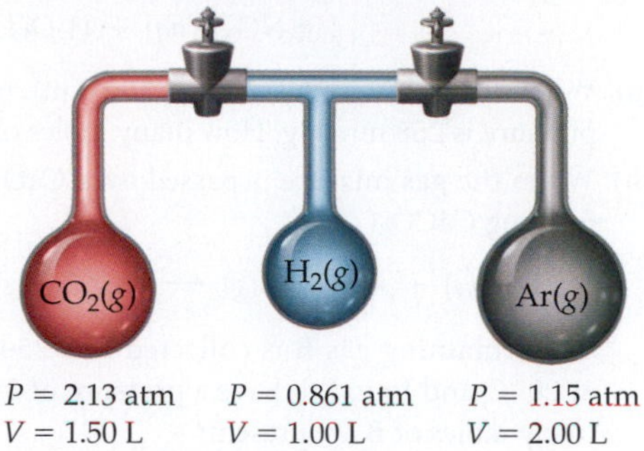

9.112 The apparatus shown consists of three temperature-jacketed 1.000 L bulbs connected by stopcocks. Bulb **A** contains a mixture of $H_2O(g)$, $CO_2(g)$, and $N_2(g)$ at 25 °C and a total pressure of 564 mm Hg. Bulb **B** is empty and is held at a temperature of −70 °C. Bulb **C** is also empty and is held at a temperature of −190 °C. The stopcocks are closed, and the volume of the lines connecting the bulbs is zero. CO_2 sublimes at −78 °C, and N_2 boils at −196 °C.

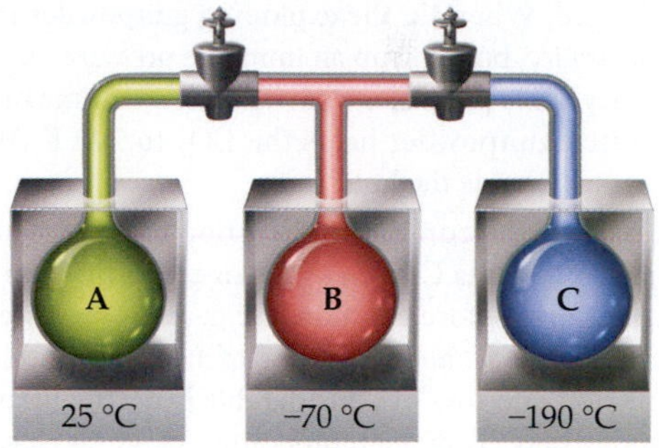

(a) The stopcock between **A** and **B** is opened, and the system is allowed to come to equilibrium. The pressure in **A** and **B** is now 219 mm Hg. What do bulbs **A** and **B** contain?

(b) How many moles of H_2O are in the system?

(c) Both stopcocks are opened, and the system is again allowed to come to equilibrium. The pressure throughout the system is 33.5 mm Hg. What do bulbs **A**, **B**, and **C** contain?

(d) How many moles of N_2 are in the system?

(e) How many moles of CO_2 are in the system?

9.113 Assume that you have 1.00 g of nitroglycerin in a 500.0 mL steel container at 20.0 °C and 1.00 atm pressure. An explosion occurs, raising the temperature of the container and its contents to 425 °C. The balanced equation is

$$4\,C_3H_5N_3O_9(l) \longrightarrow 12\,CO_2(g) + 10\,H_2O(g) + 6\,N_2(g) + O_2(g)$$

(a) How many moles of nitroglycerin and how many moles of gas (air) were in the container originally?

(b) How many moles of gas are in the container after the explosion?

(c) What is the pressure in atmospheres inside the container after the explosion according to the ideal gas law?

9.114 Use both the ideal gas law and the van der Waals equation to calculate the pressure in atmospheres of 45.0 g of NH_3 gas in a 1.000 L container at 0 °C, 50 °C, and 100 °C. For NH_3, $a = 4.17$ $(L^2 \cdot atm)/mol^2$ and $b = 0.0371$ L/mol.

9.115 When solid mercury(I) carbonate, Hg_2CO_3, is added to nitric acid, HNO_3, a reaction occurs to give mercury(II) nitrate, $Hg(NO_3)_2$, water, and two gases **A** and **B**:

$$Hg_2CO_3(s) + HNO_3(aq) \longrightarrow Hg(NO_3)_2(aq) + H_2O(l) + A(g) + B(g)$$

(a) When the gases are placed in a 500.0 mL bulb at 20 °C, the pressure is 258 mm Hg. How many moles of gas are present?

(b) When the gas mixture is passed over CaO(*s*), gas **A** reacts, forming $CaCO_3(s)$:

$$CaO(s) + A(g) + B(g) \longrightarrow CaCO_3(s) + B(g)$$

The remaining gas **B** is collected in a 250.0 mL container at 20 °C and found to have a pressure of 344 mm Hg. How many moles of **B** are present?

(c) The mass of gas **B** collected in part **(b)** was found to be 0.218 g. What is the density of **B** in g/L?

(d) What is the molecular weight of **B**, and what is its formula?

(e) Write a balanced equation for the reaction of mercury(I) carbonate with nitric acid.

9.116 Dry ice (solid CO_2) has occasionally been used as an "explosive" in mining. A hole is drilled, dry ice and a small amount of gunpowder are placed in the hole, a fuse is added, and the hole is plugged. When lit, the exploding gunpowder rapidly vaporizes the dry ice, building up an immense pressure. Assume that 500.0 g of dry ice is placed in a cavity with a volume of 0.800 L and the ignited gunpowder heats the CO_2 to 700 K. What is the final pressure inside the hole?

9.117 Consider the combustion reaction of 0.148 g of a hydrocarbon having formula C_nH_{2n+2} with an excess of O_2 in a 400.0 mL steel container. Before reaction, the gaseous mixture had a temperature of 25.0 °C and a pressure of 2.000 atm. After complete combustion and loss of considerable heat, the mixture of products and excess O_2 had a temperature of 125.0 °C and a pressure of 2.983 atm.

(a) What is the formula and molar mass of the hydrocarbon?

(b) What are the partial pressures in atmospheres of the reactants?

(c) What are the partial pressures in atmospheres of the products and the excess O_2?

9.118 Natural gas is a mixture of hydrocarbons, primarily methane (CH_4) and ethane (C_2H_6). A typical mixture might have $X_{methane} = 0.915$ and $X_{ethane} = 0.085$. Let's assume that we have a 15.50 g sample of natural gas in a volume of 15.00 L at a temperature of 20.00 °C.

(a) How many total moles of gas are in the sample?

(b) What is the pressure of the sample in atmospheres?

(c) What is the partial pressure of each component in the sample in atmospheres?

(d) When the sample is burned in an excess of oxygen, how much heat in kilojoules is liberated?

9.119 A mixture of $CS_2(g)$ and excess $O_2(g)$ is placed in a 10.0 L reaction vessel at 100.0 °C and a pressure of 3.00 atm. A spark causes the CS_2 to ignite, burning it completely, according to the equation

$$CS_2(g) + 3\,O_2(g) \longrightarrow CO_2(g) + 2\,SO_2(g)$$

After reaction, the temperature returns to 100.0 °C, and the mixture of product gases (CO_2, SO_2, and unreacted O_2) is found to have a pressure of 2.40 atm. What is the partial pressure of each gas in the product mixture?

9.120 Gaseous compound **Q** contains only xenon and oxygen. When 0.100 g of **Q** is placed in a 50.0 mL steel vessel at 0 °C, the pressure is 0.229 atm.

(a) What is the molar mass of **Q**, and what is a likely formula?

(b) When the vessel and its contents are warmed to 100 °C, **Q** decomposes into its constituent elements. What is the total pressure, and what are the partial pressures of xenon and oxygen in the container?

9.121 When 10.0 g of a mixture of $Ca(ClO_3)_2$ and $Ca(ClO)_2$ is heated to 700 °C in a 10.0 L vessel, both compounds decompose, forming $O_2(g)$ and $CaCl_2(s)$. The final pressure inside the vessel is 1.00 atm.

(a) Write balanced equations for the decomposition reactions.

(b) What is the mass of each compound in the original mixture?

9.122 A 5.00 L vessel contains 25.0 g of PCl_3 and 3.00 g of O_2 at 15 °C. The vessel is heated to 200.0 °C, and the contents react to give $POCl_3$. What is the final pressure in the vessel, assuming that the reaction goes to completion and that all reactants and products are in the gas phase?

9.123 When 2.00 mol of NOCl(*g*) was heated to 225 °C in a 400.0 L steel reaction vessel, the NOCl partially decomposed according to the equation $2\,NOCl(g) \longrightarrow 2\,NO(g) + Cl_2(g)$. The pressure in the vessel after reaction is 0.246 atm.

(a) What is the partial pressure of each gas in the vessel after reaction?

(b) What percent of the NOCl decomposed?

9.124 Ozone (O_3) can be prepared in the laboratory by passing an electrical discharge through oxygen gas: $3\,O_2(g) \longrightarrow 2\,O_3(g)$. Assume that an evacuated steel vessel with a volume of 10.00 L is filled with 32.00 atm of O_2 at 25 °C and an electric discharge is passed through the vessel, causing some of the oxygen to be converted into ozone. As a result, the pressure inside the vessel drops to 30.64 atm at 25.0 °C. What is the final mass percent of ozone in the vessel?

9.125 A steel container with a volume of 500.0 mL is evacuated, and 25.0 g of $CaCO_3$ is added. The container and contents are then heated to 1500 K, causing the $CaCO_3$ to decompose completely, according to the equation $CaCO_3(s) \longrightarrow CaO(s) + CO_2(g)$.

(a) Using the ideal gas law and ignoring the volume of any solids remaining in the container, calculate the pressure inside the container at 1500 K.

(b) Now make a more accurate calculation of the pressure inside the container. Take into account the volume of solid CaO (density = 3.34 g/mL) in the container, and use the van der Waals equation to calculate the pressure. The van der Waals constants for $CO_2(g)$ are: $a = 3.59\ (L^2 \cdot atm)/mol^2$, and $b = 0.0427$ L/mol.

9.126 Nitrogen dioxide dimerizes to give dinitrogen tetroxide: $2\,NO_2(g) \longrightarrow N_2O_4(g)$. At 298 K, 9.66 g of an NO_2/N_2O_4 mixture exerts a pressure of 0.487 atm in a volume of 6.51 L. What are the mole fractions of the two gases in the mixture?

9.127 A certain nonmetal reacts with hydrogen at 440 °C to form a poisonous, foul-smelling gas. The density of the gas at 25 °C and 1.00 atm is 3.309 g/L. What is the formula of the gas?

MULTICONCEPT PROBLEMS

9.128 An empty 4.00 L steel vessel is filled with 1.00 atm of $CH_4(g)$ and 4.00 atm of $O_2(g)$ at 300 °C. A spark causes the CH_4 to burn completely, according to the equation:

$$CH_4(g) + 2\,O_2(g) \longrightarrow CO_2(g) + 2\,H_2O(g) \quad \Delta H° = -802 \text{ kJ}$$

(a) What mass of $CO_2(g)$ is produced in the reaction?

(b) What is the final temperature inside the vessel after combustion, assuming that the steel vessel has a mass of 14.500 kg, the mixture of gases has an average molar heat capacity of 21 J/(mol · °C), and the heat capacity of steel is 0.449 J/(g · °C)?

(c) What is the partial pressure of $CO_2(g)$ in the vessel after combustion?

9.129 When a gaseous compound **X** containing only C, H, and O is burned in O_2, 1 volume of the unknown gas reacts with 3 volumes of O_2 to give 2 volumes of CO_2 and 3 volumes of gaseous H_2O. Assume all volumes are measured at the same temperature and pressure.

(a) Calculate a formula for the unknown gas, and write a balanced equation for the combustion reaction.

(b) Is the formula you calculated an empirical formula or a molecular formula? Explain.

(c) Draw two different possible electron-dot structures for the compound **X**.

(d) Combustion of 5.000 g of **X** releases 144.2 kJ heat. Look up $\Delta H°_f$ values for $CO_2(g)$ and $H_2O(g)$ in Appendix B, and calculate $\Delta H°_f$ for compound **X**.

9.130 Isooctane, C_8H_{18}, is the component of gasoline from which the term *octane rating* derives.

(a) Write a balanced equation for the combustion of isooctane to yield CO_2 and H_2O.

(b) Assuming that gasoline is 100% isooctane, that isooctane burns to produce only CO_2 and H_2O, and that the density of isooctane is 0.792 g/mL, what mass of CO_2 in kilograms is produced each year by the annual U.S. gasoline consumption of 4.6×10^{10} L?

(c) What is the volume in liters of this CO_2 at STP?

(d) How many moles of air are necessary for the combustion of 1 mol of isooctane, assuming that air is 21.0% O_2 by volume? What is the volume in liters of this air at STP?

9.131 The *Rankine* temperature scale used in engineering is to the Fahrenheit scale as the Kelvin scale is to the Celsius scale. That is, 1 Rankine degree is the same size as 1 Fahrenheit degree, and 0 °R = absolute zero.

(a) What temperature corresponds to the freezing point of water on the Rankine scale?

(b) What is the value of the gas constant R on the Rankine scale in (L · atm)/(°R · mol)?

(c) Use the van der Waals equation to determine the pressure inside a 400.0 mL vessel that contains 2.50 mol of CH_4 at a temperature of 525 °R. For CH_4, $a = 2.253$ $(L^2 \cdot atm)/mol^2$; $b = 0.04278$ L/mol.

9.132 Chemical explosions are characterized by the instantaneous release of large quantities of hot gases, which set up a shock wave of enormous pressure (up to 700,000 atm) and velocity (up to 20,000 mi/h). For example, explosion of nitroglycerin ($C_3H_5N_3O_9$) releases four gases, **A**, **B**, **C**, and **D**:

$$n\ C_3H_5N_3O_9(l) \longrightarrow a\ A(g) + b\ B(g) + c\ C(g) + d\ D(g)$$

Assume that the explosion of 1 mol (227 g) of nitroglycerin releases gases with a temperature of 1950 °C and a volume of 1323 L at 1.00 atm pressure.

(a) How many moles of hot gas are released by the explosion of 0.004 00 mol of nitroglycerin?

(b) When the products released by explosion of 0.004 00 mol of nitroglycerin were placed in a 500.0 mL flask and the flask was cooled to −10 °C, product **A** solidified and the pressure inside the flask was 623 mm Hg. How many moles of **A** were present, and what is its likely identity?

(c) When gases **B**, **C**, and **D** were passed through a tube of powdered Li_2O, gas **B** reacted to form Li_2CO_3. The remaining gases, **C** and **D**, were collected in another 500.0 mL flask and found to have a pressure of 260 mm Hg at 25 °C. How many moles of **B** were present, and what is its likely identity?

(d) When gases **C** and **D** were passed through a hot tube of powdered copper, gas **C** reacted to form CuO. The remaining gas, **D**, was collected in a third 500.0 mL flask and found to have a mass of 0.168 g and a pressure of 223 mm Hg at 25 °C. How many moles each of **C** and **D** were present, and what are their likely identities?

(e) Write a balanced equation for the explosion of nitroglycerin.

9.133 Combustion analysis of 0.1500 g of methyl *tert*-butyl ether, an octane booster used in gasoline, gave 0.3744 g of CO_2 and 0.1838 g of H_2O. When a flask having a volume of 1.00 L was evacuated and then filled with methyl *tert*-butyl ether vapor at a pressure of 100.0 kPa and a temperature of 54.8 °C, the mass of the flask increased by 3.233 g.

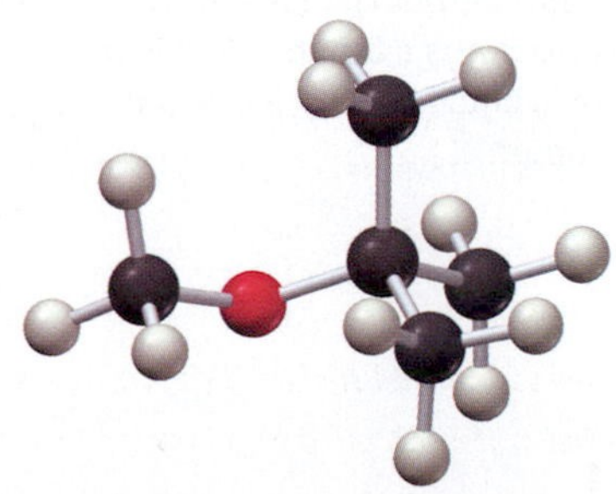

Methyl *tert*-butyl ether

(a) What is the empirical formula of methyl *tert*-butyl ether?

(b) What is the molecular weight and molecular formula of methyl *tert*-butyl ether?

(c) Write a balanced equation for the combustion reaction.

(d) The enthalpy of combustion for methyl *tert*-butyl ether is $\Delta H°_{combustion} = -3368.7$ kJ/mol. What is its standard enthalpy of formation, $\Delta H°_f$?

CHAPTER 10

Liquids, Solids, and Phase Changes

Most ionic compounds are crystalline solids at room temperature, but the ionic compound in this reactor is a liquid. We'll learn more about ionic liquids in the *FYI* at the end of this chapter.

The kinetic–molecular theory developed in the previous chapter accounts for the properties of gases by assuming that gas particles act independently of one another. Because the attractive forces between them are so weak, the particles in gases are free to move about at random and occupy whatever volume is available. The same is not true in liquids and solids, however. Liquids and solids are distinguished from gases by the presence of substantial attractive forces between particles. In liquids, these attractive forces are strong enough to hold the particles in close contact while still letting them slip and slide over one another. In solids, the forces are so strong that they hold the particles rigidly in place and prevent their movement (**Figure 10.1**).

In this chapter, we'll examine the nature of the forces responsible for the properties of liquids and solids, paying particular attention to the ordering of particles in solids and to the

FIGURE 10.1

A molecular comparison of gases, liquids, and solids.

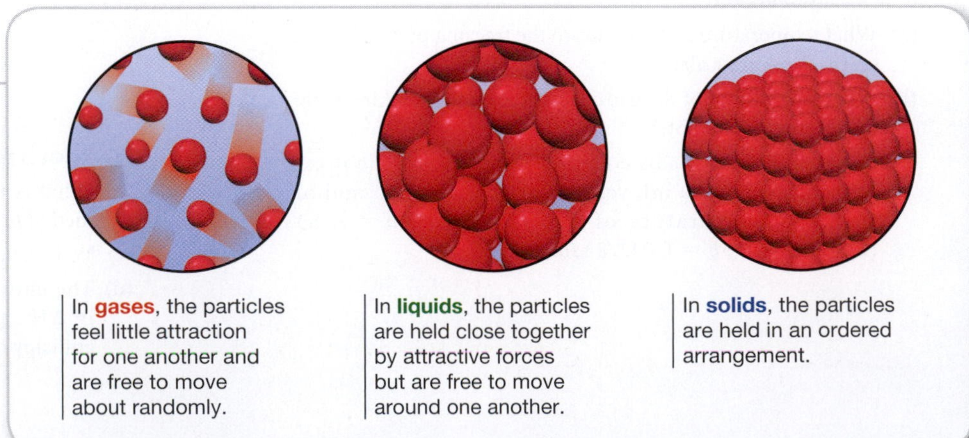

CONTENTS

different kinds of solids that result. In addition, we'll look at what happens during transitions between solid, liquid, and gaseous states and at the effects of temperature and pressure on these transitions.

10.1 POLAR COVALENT BONDS AND DIPOLE MOMENTS

Before looking at the forces between molecules, it's first necessary to develop the ideas of *bond dipoles* and *dipole moments.* We saw in Section 4.4 that **polar covalent bonds** form between atoms of different **electronegativity**. Chlorine is more electronegative than carbon, for instance, so the chlorine atom in chloromethane (CH_3Cl) attracts the electrons in the C–Cl bond more strongly than carbon does. The C–Cl bond is therefore polarized so that the chlorine atom is slightly electron-rich ($\delta-$) and the carbon atom is slightly electron-poor ($\delta+$).

Because the polar C–Cl bond in chloromethane has two polar ends—a positive end and a negative end—we describe it as being a bond **dipole**, and we often represent the dipole using an arrow with a cross at one end(↦)to indicate the direction of electron displacement. The point of the arrow represents the negative end of the dipole ($\delta-$), and the crossed end (which looks like a plus sign) represents the positive end ($\delta+$). This polarity is clearly visible in an **electrostatic potential map**, which shows the electron-rich chlorine atom as red and the electron-poor remainder of the molecule as blue-green.

REMEMBER...

Polar covalent bonds are those in which the bonding electrons are shared unequally between two atoms but are not completely transferred. Thus, they are intermediate between nonpolar covalent bonds and ionic bonds. (Section 4.4)

REMEMBER...

Electronegativity is the ability of an atom in a molecule to attract the shared electrons in a covalent bond. Metals on the left of the periodic table are the least electronegative, while halogens and other reactive nonmetals in the upper right of the table are the most electronegative. (Section 4.4)

REMEMBER...

An **electrostatic potential map** uses color to portray the calculated electron distribution in a molecule. Electron-rich regions are red, and electron-poor regions are blue. (Section 4.4)

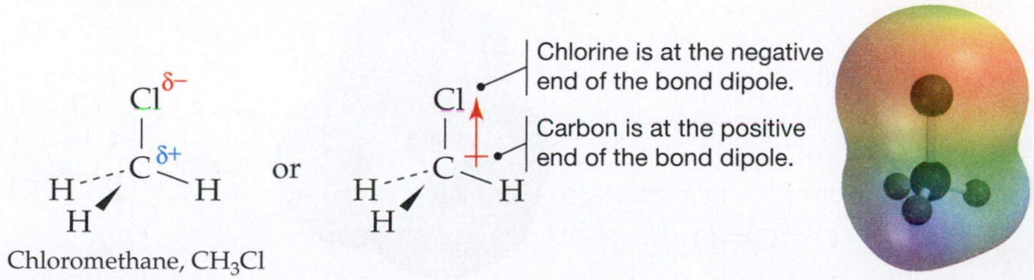

Chloromethane, CH_3Cl

Just as individual bonds in molecules are often polar, molecules as a whole are also often polar because of the net sum of individual bond polarities and lone-pair contributions. The resultant *molecular dipoles* can be looked at in the following way: Assume that there is a center of mass of all positive charges (nuclei) in a molecule and a center of mass of all negative charges (electrons). If these two centers don't coincide, the molecule has a net polarity.

The measure of net molecular polarity is a quantity called the **dipole moment,** μ (Greek mu), which is defined as the magnitude of the charge Q at either end of the molecular dipole times the distance r between the charges: $\mu = Q \times r$. Dipole moments are expressed in *debyes* (D), where $1\text{ D} = 3.336 \times 10^{-30}$ coulomb meters ($\text{C} \cdot \text{m}$) in SI units. To calibrate your thinking, the charge on an electron is 1.60×10^{-19} C. Thus, if a proton and an electron were separated by 100 pm (a bit less than the length of a typical covalent bond), then the dipole moment would be $1.60 \times 10^{-29}\text{ C} \cdot \text{m}$, or 4.80 D:

$$\mu = Q \times r$$

$$\mu = (1.60 \times 10^{-19}\text{ C})(100 \times 10^{-12}\text{ m})\left(\frac{1\text{ D}}{3.336 \times 10^{-30}\text{ C} \cdot \text{m}}\right) = 4.80\text{ D}$$

TABLE 10.1 Dipole Moments of Some Common Compounds

Compound	Dipole Moment (D)
NaCl*	9.0
CH_3Cl	1.90
H_2O	1.85
NH_3	1.47
HCl	1.11
CO_2	0
CCl_4	0

*Measured in the gas phase

It's relatively easy to measure dipole moments experimentally, and values for some common substances are given in **Table 10.1**. Once the dipole moment is known, it's then possible to get an idea of the amount of charge separation in a molecule. In chloromethane, for example, the experimentally measured dipole moment is $\mu = 1.90$ D. If we assume that the contributions of the nonpolar C–H bonds are small, then most of the chloromethane dipole moment is due to the C–Cl bond. Since the C–Cl bond distance is 179 pm, we can calculate that the dipole moment of chloromethane would be $1.79 \times 4.80\text{ D} = 8.59$ D if the C–Cl bond were ionic—that is, if a full negative charge on chlorine were separated from a full positive charge on carbon by a distance of 179 pm. But because the measured dipole moment of chloromethane is only 1.90 D, we can conclude that the C–Cl bond is only about $(1.90/8.59)(100\%) = 22\%$ ionic. Thus, the chlorine atom in chloromethane has an excess of about 0.2 electron, and the carbon atom has a deficiency of about 0.2 electron.

Perhaps it's not surprising that by far the largest dipole moment listed in Table 10.1 belongs to the ionic compound NaCl, which exists in the gas phase as a pair of Na^+ and Cl^- ions held together by a strong ionic bond. The molecular compounds water and ammonia also have substantial dipole moments because both oxygen and nitrogen are electronegative relative to hydrogen and because both O and N have lone pairs of electrons that make substantial contributions to net molecular polarity:

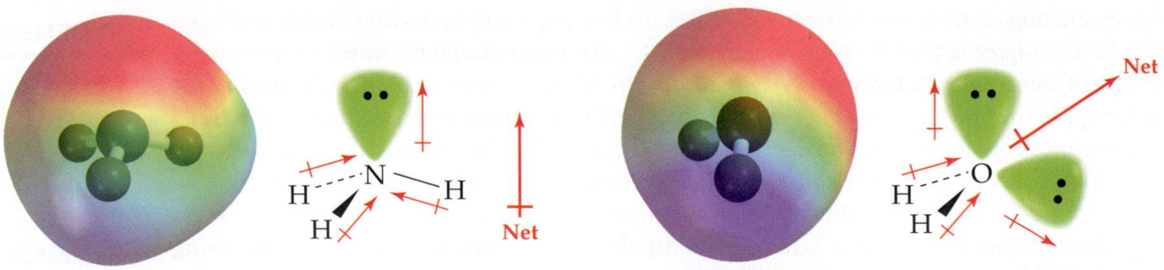

Ammonia (μ = 1.47 D) Water (μ = 1.85 D)

In contrast with water and ammonia, carbon dioxide and tetrachloromethane (CCl_4) have zero dipole moments. Molecules of both substances contain *individual* polar covalent bonds, but because of the symmetry of their structures, the individual bond polarities exactly cancel.

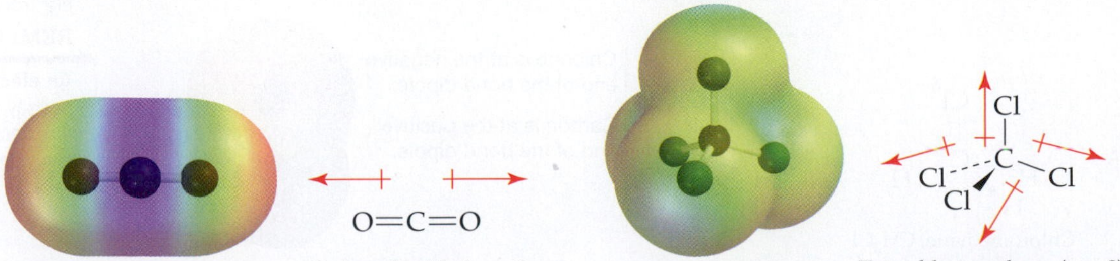

Carbon dioxide (μ = 0) Tetrachloromethane (μ = 0)

Worked Example 10.1

Calculating Percent Ionic Character from a Dipole Moment

The dipole moment of HCl is 1.11 D, and the distance between atoms is 127 pm. What is the percent ionic character of the HCl bond?

STRATEGY

If HCl were 100% ionic, a negative charge (Cl^-) would be separated from a positive charge (H^+) by 127 pm. Calculate the expected dipole moment, and compare that calculated value to the actual value.

SOLUTION

The calculated dipole moment is

$$\mu = Q \times r$$

$$\mu = (1.60 \times 10^{-19}\ \text{C})(127 \times 10^{-12}\ \text{m})\left(\frac{1\ \text{D}}{3.336 \times 10^{-30}\ \text{C}\cdot\text{m}}\right) = 6.09\ \text{D}$$

The observed dipole moment of 1.11 D for HCl implies that the H—Cl bond is only about 18% ionic:

$$\frac{1.11\ \text{D}}{6.09\ \text{D}} \times 100\% = 18.2\%$$

Worked Example 10.2

Predicting the Presence of a Dipole Moment

Would you expect vinyl chloride ($H_2C{=}CHCl$), the starting material used for preparation of poly(vinyl chloride) polymer, to have a dipole moment? If so, indicate the direction.

STRATEGY

First, use the VSEPR model described in Section 5.1 to predict the molecular shape of vinyl chloride. Then, assign polarities to the individual bonds according to the differences in electronegativity of the bonded atoms (Figure 4.4 on page 143), and make a reasonable guess about the overall polarity that would result by summing the individual contributions.

SOLUTION

Because both carbon atoms have three charge clouds, each has trigonal planar geometry. Because the unhybridized carbon 2*p* orbitals in the pi part of the C=C double bond overlap in a sideways manner, the molecule as a whole is planar:

Vinyl chloride

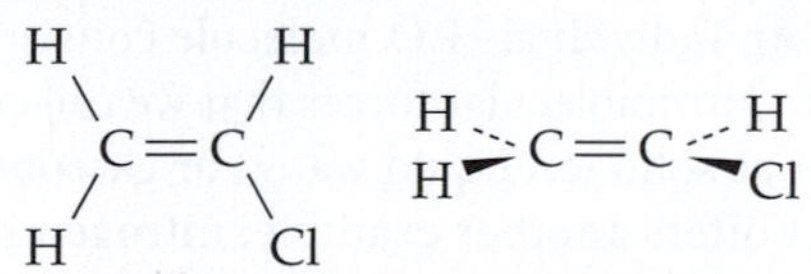

Top view Side view

Only the C—Cl bond has a substantial polarity, giving the molecule a net polarity:

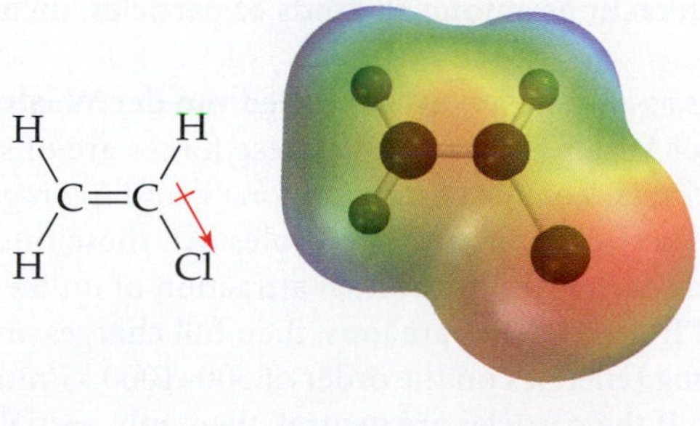

PROBLEM 10.1 The dipole moment of HF is $\mu = 1.83$ D, and the bond length is 92 pm. Calculate the percent ionic character of the H—F bond. Is HF more ionic or less ionic than HCl (Worked Example 10.1)?

PROBLEM 10.2 Tell which of the following compounds is likely to have a dipole moment, and show the direction of each.

(a) SF_6 (b) $H_2C{=}CH_2$
(c) $CHCl_3$ (d) CH_2Cl_2

CONCEPTUAL PROBLEM 10.3 The dipole moment of methanol is $\mu = 1.70$ D. Use arrows to indicate the direction in which electrons are displaced.

CONCEPTUAL PROBLEM 10.4 Methylamine, CH_3NH_2, is responsible for the odor of rotting fish. Look at the following electrostatic potential map of methylamine, and explain the observed polarity.

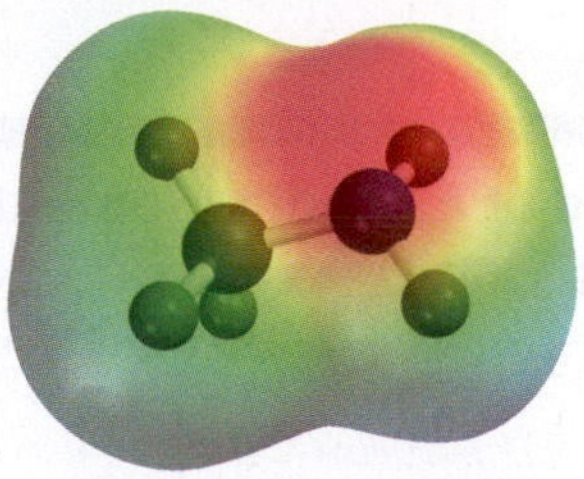

10.2 KINDS OF INTERMOLECULAR FORCES

Now that we know a bit about molecular polarities, let's see how they give rise to some of the forces that occur between molecules. The existence of such forces is easy to show. Take H_2O, for example. An individual H_2O molecule consists of three atoms joined together in a specific way by the *intra*molecular forces that we call covalent bonds. But a visible sample of H_2O exists either as solid ice, liquid water, or gaseous vapor, depending on its temperature. Nitrogen (N_2) offers another example; nitrogen is a gas at higher temperatures but becomes a liquid at low temperature (**Figure 10.2**). Clearly, there must be some ***inter*molecular forces** that act between molecules to hold them close together at certain temperatures. (Strictly speaking, the term *intermolecular* refers only to molecular substances, but we'll use it generally to refer to interactions among all kinds of particles, including molecules, ions, and atoms.)

Intermolecular forces as a whole are usually called **van der Waals forces** after the Dutch scientist Johannes van der Waals (1837–1923). These forces are of several different types, including *dipole–dipole forces*, *London dispersion forces*, and *hydrogen bonds*. In addition, *ion–dipole forces* operate between ions and molecules. All these intermolecular forces are electrostatic in origin and result from the mutual attraction of unlike charges or the mutual repulsion of like charges. If the particles are ions, then full charges are present and the ion–ion attractions are so strong (energies on the order of 500–1000 kJ/mol) that they give rise to what we call **ionic bonds**. If the particles are neutral, then only partial charges are present at best, but even so the attractive forces can be substantial.

REMEMBER...

Ionic bonds generally form between the cation of a metal and the anion of a reactive nonmetal. (Section 3.9)

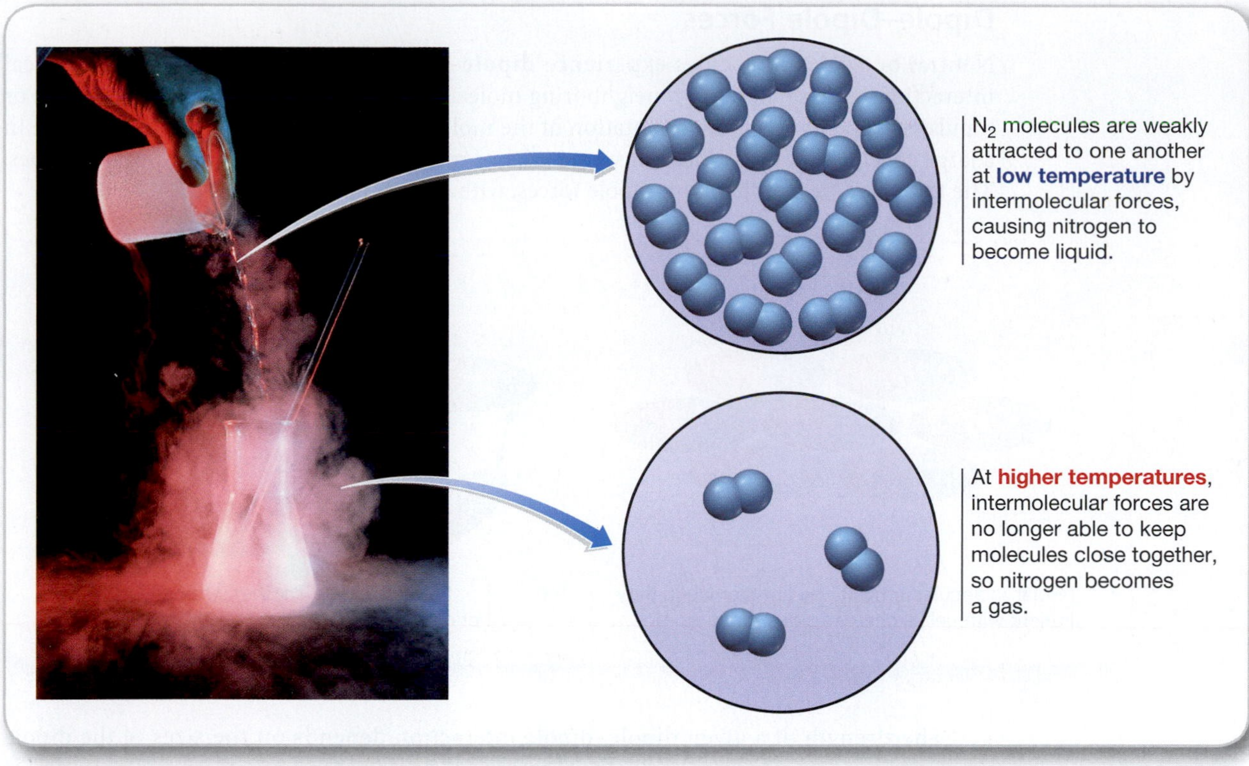

FIGURE 10.2

Intermolecular forces.

Ion–Dipole Forces

We said in the previous section that a molecule has a net polarity and an overall dipole moment if the sum of its individual bond dipoles is nonzero. One side of the molecule has a net excess of electrons and a partial negative charge ($\delta-$), while the other side has a net deficiency of electrons and a partial positive charge ($\delta+$). An **ion–dipole force** is the result of electrical interactions between an ion and the partial charges on a polar molecule **(Figure 10.3)**.

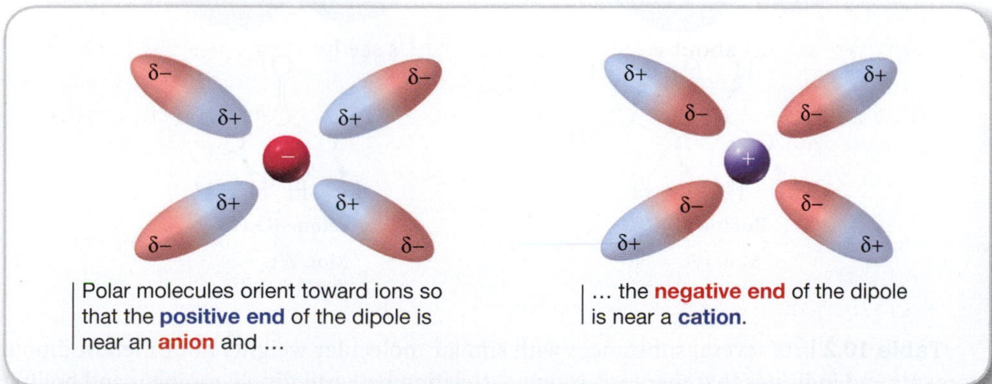

FIGURE 10.3

Ion–dipole forces.

The favored orientation of a polar molecule in the presence of ions is one where the positive end of the molecular dipole is near an anion and the negative end of the dipole is near a cation. The magnitude of the interaction energy E depends on the charge on the ion z, the strength of the dipole as measured by its dipole moment μ, and the inverse square of the distance r from the ion to the dipole: $E \propto z\mu/r^2$. Ion–dipole forces are highly variable in strength and are particularly important in aqueous solutions of ionic substances such as NaCl, in which polar water molecules surround the ions. We'll explore the **formation of solutions** in more detail in the next chapter.

LOOKING AHEAD ...

In Section 11.2, we'll look at the energy changes accompanying the **formation of solutions** and at how those energy changes are affected by different kinds of intermolecular forces.

Dipole–Dipole Forces

Neutral but polar molecules experience **dipole–dipole forces** as the result of electrical interactions among dipoles on neighboring molecules. The forces can be either attractive or repulsive, depending on the orientation of the molecules (**Figure 10.4**), and the net force in a large collection of molecules is a summation of many individual interactions of both types. The forces are weaker than ion–dipole forces, with energies on the order of 3–4 kJ/mol.

FIGURE 10.4

Dipole–dipole forces.

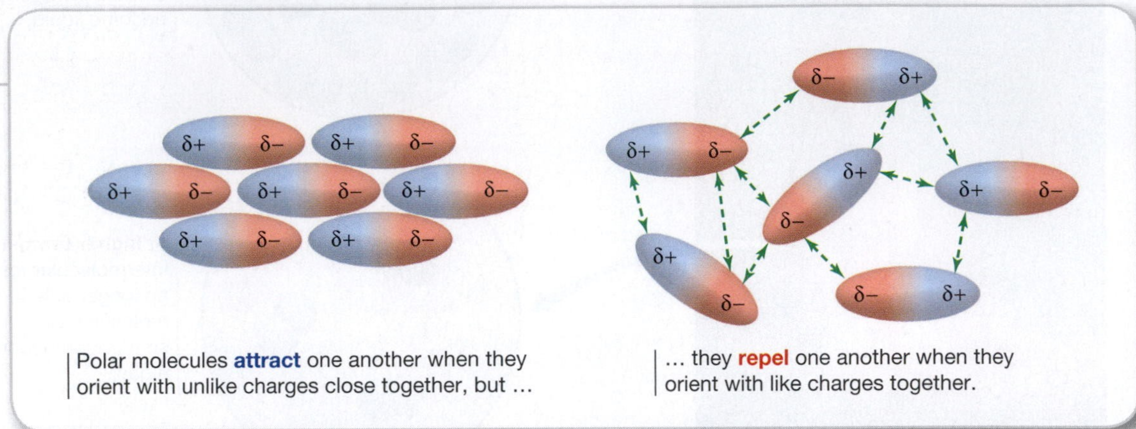

The strength of a given dipole–dipole interaction depends on the sizes of the dipole moments involved. The more polar the substance, the greater the strength of its dipole–dipole interactions. Butane, for instance, is a nonpolar molecule with a molecular weight of 58 and a boiling point of −0.5 °C, while acetone has the same molecular weight yet boils 57 °C higher because it is polar.

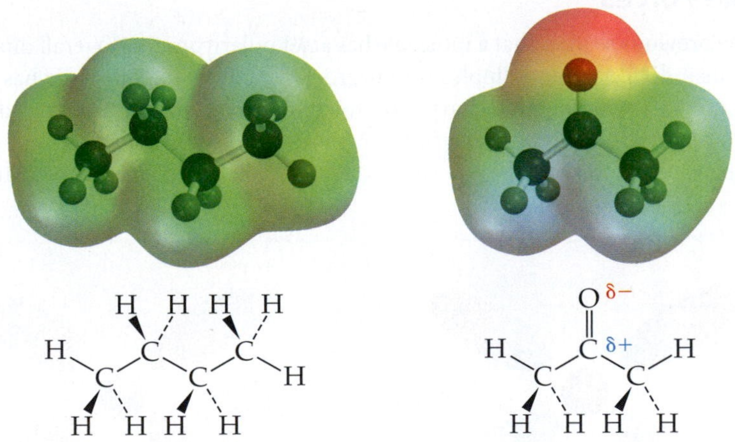

Butane (C_4H_{10})
Mol. Wt. = 58
bp = −0.5 °C

Acetone (C_3H_6O)
Mol. Wt. = 58
bp = 56.1 °C

Table 10.2 lists several substances with similar molecular weights but different dipole moments and indicates that there is a rough correlation between dipole moment and boiling

TABLE 10.2 Comparison of Molecular Weights, Dipole Moments, and Boiling Points

Substance	Mol. Wt.	Dipole Moment (D)	bp (K)
$CH_3CH_2CH_3$	44.10	0.08	231
CH_3OCH_3	46.07	1.30	248
CH_3CN	41.05	3.93	355

point. The larger the dipole moment, the stronger the intermolecular forces and the greater the amount of heat that must be added to overcome those forces. Thus, substances with higher dipole moments generally have higher boiling points.

London Dispersion Forces

The causes of intermolecular forces among charged and polar particles are relatively easy to understand, but it's less obvious how attractive forces arise among nonpolar molecules or among the individual atoms of a noble gas. Benzene (C_6H_6), for instance, has zero dipole moment and therefore experiences no dipole–dipole forces. Nevertheless, there must be *some* intermolecular forces present among benzene molecules because the substance is a liquid rather than a gas at room temperature, with a melting point of 5.5 °C and a boiling point of 80.1 °C.

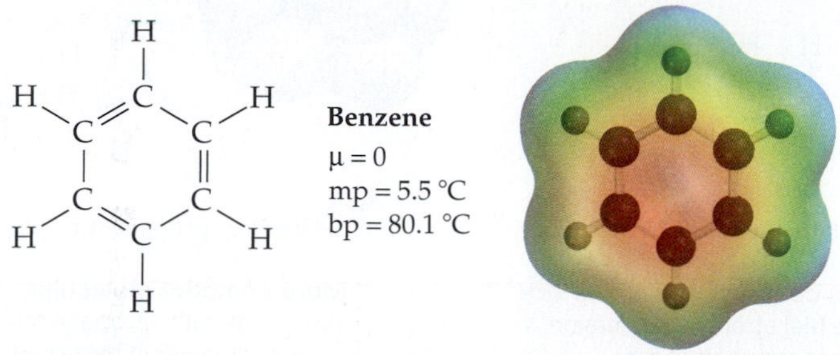

All atoms and molecules, regardless of structure, experience **London dispersion forces**, which result from the motion of electrons. Take even a simple nonpolar molecule like Br_2, for instance. Averaged over time, the distribution of electrons throughout the molecule is symmetrical, but at any given instant there may be more electrons at one end of the molecule than at the other, giving the molecule a short-lived dipole moment. This instantaneous dipole on one molecule can affect the electron distributions in neighboring molecules and *induce* temporary dipoles in those neighbors (**Figure 10.5**). As a result, weak attractive forces develop and Br_2 is a liquid at room temperature rather than a gas.

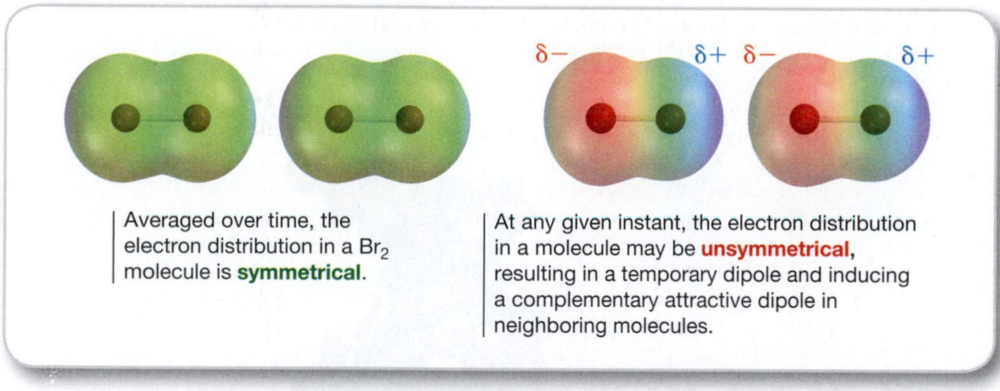

FIGURE 10.5

London dispersion forces.

Dispersion forces are generally small, with energies in the range 1–10 kJ/mol, and their exact magnitude depends on the ease with which a molecule's electron cloud can be distorted by a nearby electric field, a property referred to as *polarizability*. A smaller molecule or lighter atom is less polarizable and has smaller dispersion forces because it has only a few, tightly held electrons. A larger molecule or heavier atom, however, is more polarizable and has larger dispersion forces because it has many electrons, some of which are less tightly held and are farther from the nucleus. Among the halogens, for instance, the F_2 molecule is small and less polarizable, while I_2 is larger and more polarizable. As a result, F_2 has smaller dispersion forces and is a gas at room temperature, while I_2 has larger dispersion forces and is a solid (**Table 10.3**).

Shape is also important in determining the magnitude of the dispersion forces affecting a molecule. More spread-out shapes, which maximize molecular surface area, allow greater

TABLE 10.3 Melting Points and Boiling Points of the Halogens

Halogen	mp (K)	bp (K)
F_2	53.5	85.0
Cl_2	171.6	239.1
Br_2	265.9	331.9
I_2	386.8	457.5

contact between molecules and give rise to higher dispersion forces than do more compact shapes, which minimize molecular contact. Pentane, for example, boils at 309.2 K, whereas 2,2-dimethylpropane boils at 282.6 K. Both substances have the same molecular formula, C_5H_{12}, but pentane is longer and somewhat spread out, whereas 2,2-dimethylpropane is more spherical and compact (**Figure 10.6**).

FIGURE 10.6

The effect of molecular shape on London dispersion forces.

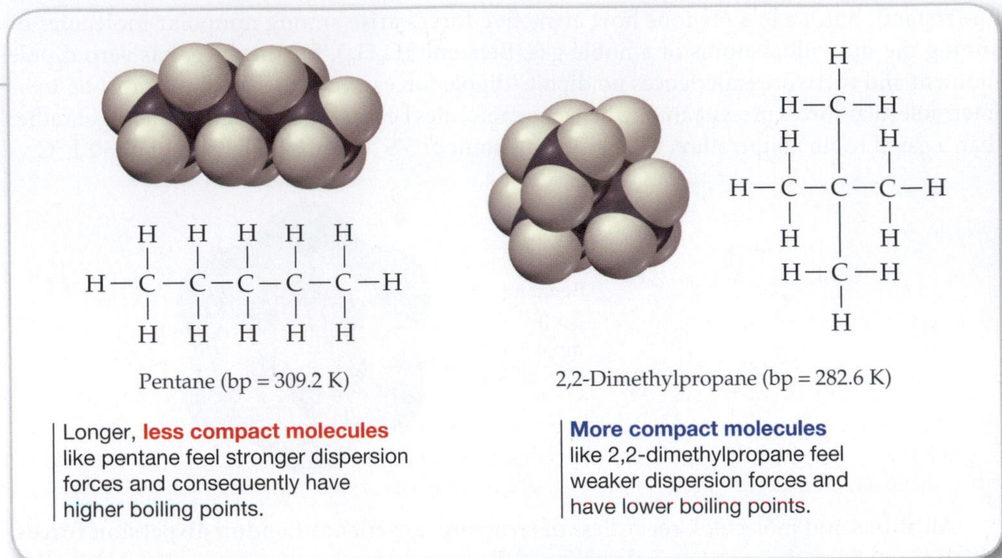

Hydrogen Bonds

In many ways, *hydrogen bonds* are responsible for life on Earth. They cause water to be a liquid rather than a gas at ordinary temperatures, and they are the primary intermolecular force that holds huge biomolecules in the shapes needed to play their essential roles in biochemistry. Deoxyribonucleic acid (DNA), for instance, contains two enormously long molecular strands that are coiled around each other and held together by hydrogen bonds.

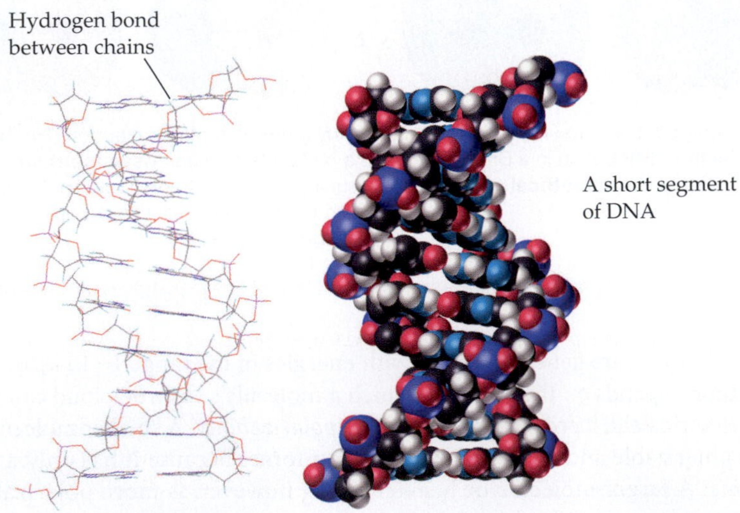

A **hydrogen bond** is an attractive interaction between a hydrogen atom bonded to a very electronegative atom (O, N, or F) and an electron-rich region elsewhere in the same molecule or in a different molecule. Most often, this electron-rich region is an unshared electron pair on another electronegative atom. For example, hydrogen bonds occur in both water and ammonia:

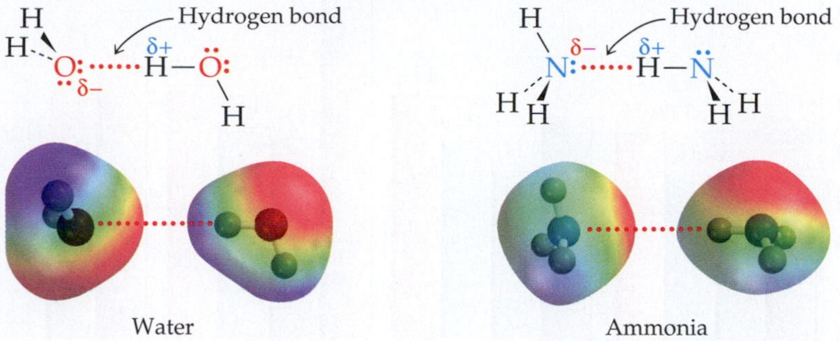

Hydrogen bonds arise because O–H, N–H, and F–H bonds are highly polar, with a partial positive charge on the hydrogen and a partial negative charge on the electronegative atom. In addition, the hydrogen atom has no core electrons to shield its nucleus and it has a small size so it can be approached closely by other molecules. As a result, the dipole–dipole attraction between the hydrogen and an unshared electron pair on a nearby atom is unusually strong, giving rise to a hydrogen bond. Water, in particular, is able to form a vast three-dimensional network of hydrogen bonds because each H_2O molecule has two hydrogens and two electron pairs (**Figure 10.7**).

FIGURE 10.7

Hydrogen bonding in water.

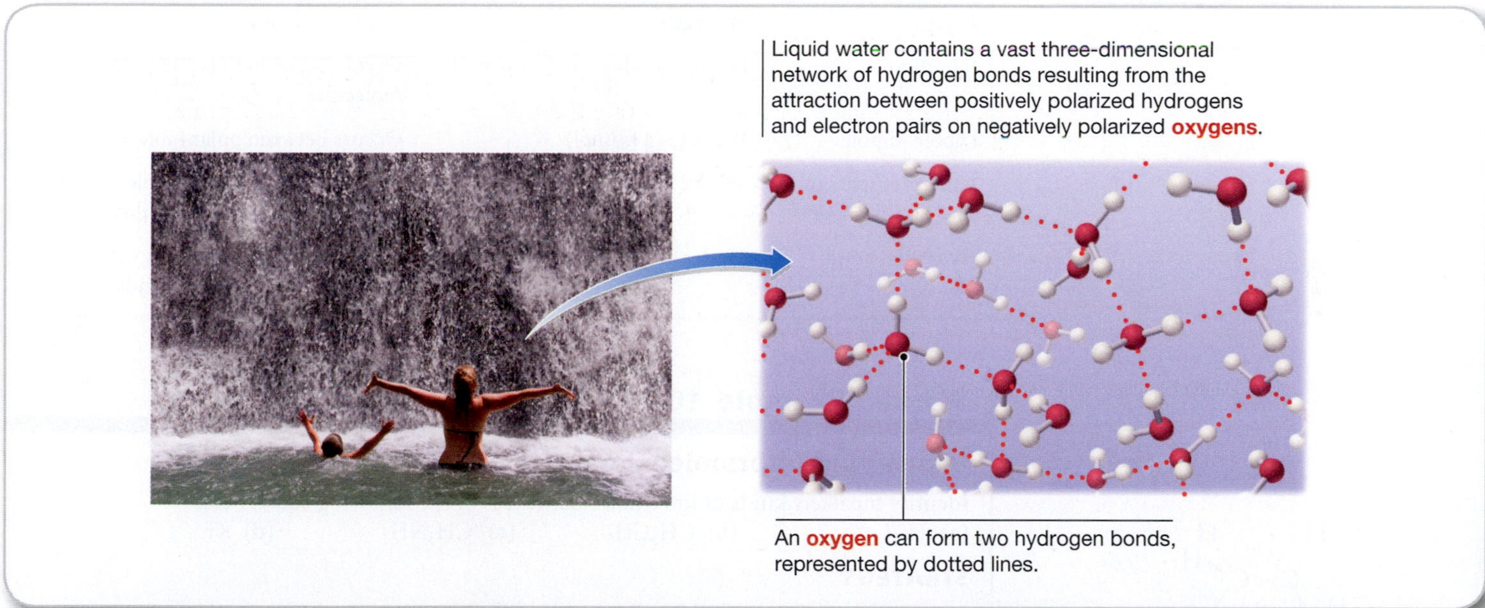

Hydrogen bonds can be quite strong, with energies up to 40 kJ/mol. To see one effect of hydrogen bonding, look at **Table 10.4**, which plots the boiling points of the covalent binary hydrides for the group 4A–7A elements. As you might expect, the boiling points generally increase with molecular weight down a group of the periodic table as a result of increased London dispersion forces—for example, $CH_4 < SiH_4 < GeH_4 < SnH_4$. Three substances, however, are clearly anomalous: NH_3, H_2O, and HF. All three have higher boiling points than might be expected because of the hydrogen bonds they contain.

TABLE 10.4 Boiling Points of the Covalent Binary Hydrides of Groups 4A, 5A, 6A, and 7A

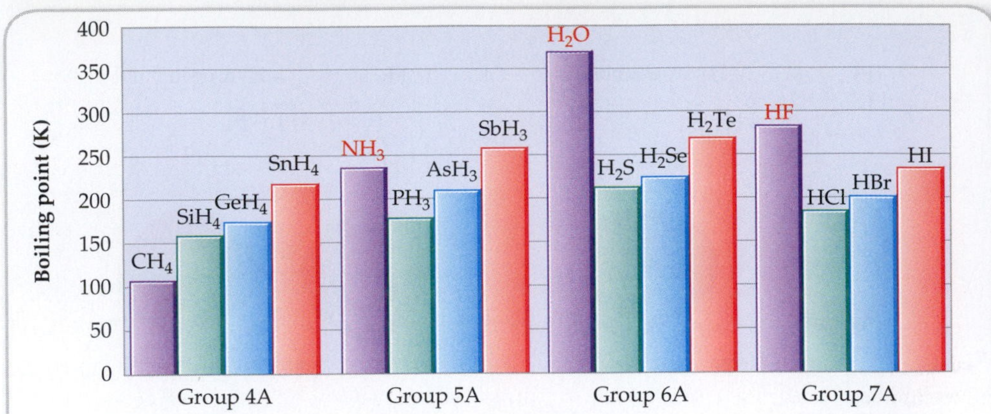

The boiling points generally increase with increasing molecular weight down a group of the periodic table, but the hydrides of nitrogen **(NH_3)**, oxygen **(H_2O)**, and fluorine **(HF)** have abnormally high boiling points because these molecules form hydrogen bonds.

A comparison of the various kinds of intermolecular forces is shown in **Table 10.5**.

TABLE 10.5 A Comparison of Intermolecular Forces

Force	Strength	Characteristics
Ion–dipole	Highly variable (10–70 kJ/mol)	Occurs between ions and polar molecules
Dipole–dipole	Weak (3–4 kJ/mol)	Occurs between polar molecules
London dispersion	Weak (1–10 kJ/mol)	Occurs between all molecules; strength depends on size, polarizability
Hydrogen bond	Moderate (10–40 kJ/mol)	Occurs between molecules with O–H, N–H, and F–H bonds

Worked Example 10.3

Identifying Intermolecular Forces

Identify the likely kinds of intermolecular forces in the following substances:

(a) HCl **(b)** CH_3CH_3 **(c)** CH_3NH_2 **(d)** Kr

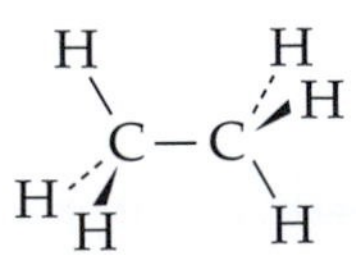

Ethane

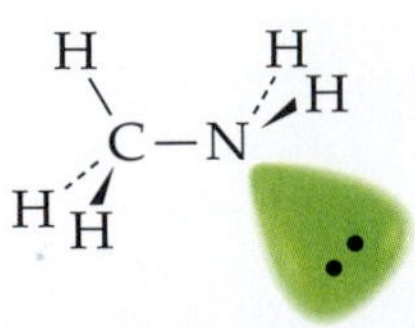

Methylamine

STRATEGY

Look at the structure of each substance, and decide what intermolecular forces are present. All molecules have dispersion forces; polar molecules have dipole–dipole forces; and molecules with O–H, N–H, or F–H bonds have hydrogen bonds.

SOLUTION

(a) HCl is a polar molecule but can't form hydrogen bonds. It has dipole–dipole forces and dispersion forces.

(b) CH_3CH_3 is a nonpolar molecule and has only dispersion forces.

(c) CH_3NH_2 is a polar molecule that can form hydrogen bonds. In addition, it has dipole–dipole forces and dispersion forces.

(d) Kr is nonpolar and has only dispersion forces.

PROBLEM 10.5 Of the substances Ar, Cl_2, CCl_4, and HNO_3, which has:
(a) The largest dipole–dipole forces?
(b) The largest hydrogen-bond forces?
(c) The smallest dispersion forces?

PROBLEM 10.6 Consider the kinds of intermolecular forces present in the following compounds, and rank the substances in likely order of increasing boiling point: H_2S (mol. wt. = 34), CH_3OH (mol. wt. = 32), C_2H_6 (mol. wt. = 30), Ar (mol. wt. = 40).

10.3 SOME PROPERTIES OF LIQUIDS

Many familiar and observable properties of liquids can be explained by the intermolecular forces just discussed. We all know, for instance, that some liquids, such as water or gasoline, flow easily when poured, whereas others, such as motor oil or maple syrup, flow sluggishly.

The measure of a liquid's resistance to flow is called its **viscosity**. Viscosity is related to the ease with which individual molecules move around in the liquid and thus to the intermolecular forces present. Substances with small, nonpolar molecules, such as pentane and benzene, experience only weak intermolecular forces and have relatively low viscosities, whereas larger, more polar substances, such as glycerol [$C_3H_5(OH)_3$], experience stronger intermolecular forces and so have higher viscosities.

Another familiar property of liquids is **surface tension**, the resistance of a liquid to spread out and increase its surface area. Surface tension is caused by the difference in intermolecular forces experienced by molecules at the surface of a liquid and those experienced by molecules in the interior. Molecules at the surface feel attractive forces on only one side and are thus pulled in toward the liquid, while molecules in the interior are surrounded and are pulled equally in all directions **(Figure 10.8)**. The ability of a water strider to walk on water and the beading up of water on a newly waxed car are both due to surface tension.

▲ Surface tension allows a water strider to walk on a pond without penetrating the surface.

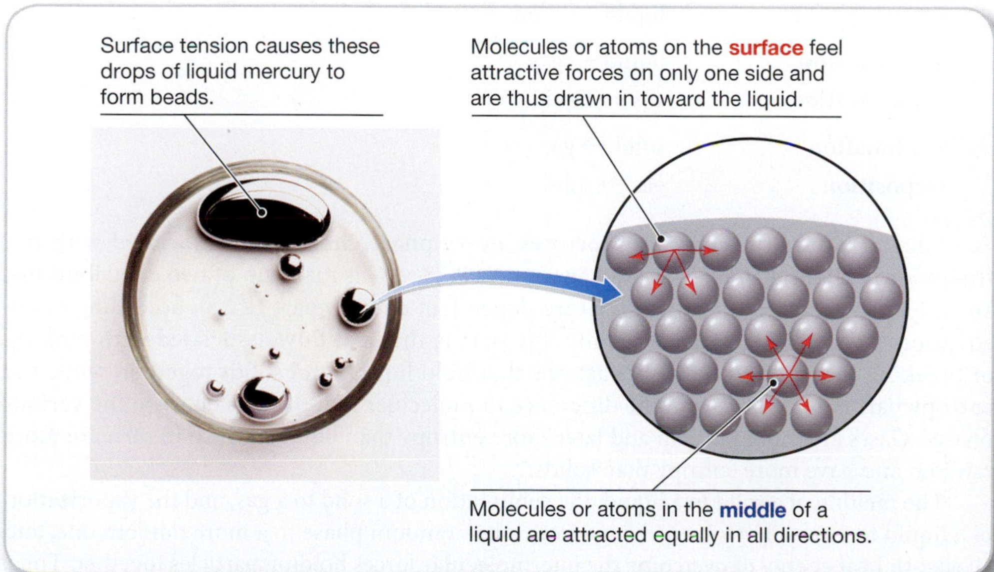

FIGURE 10.8

Surface tension. Surface tension is the resistance of a liquid to spread out and increase its surface area.

Surface tension, like viscosity, is generally higher in liquids that have stronger intermolecular forces. Both properties are also temperature-dependent because molecules at higher temperatures have more kinetic energy to counteract the attractive forces holding them together. Data for some common substances are given in **Table 10.6**. Note that mercury has a particularly large surface tension, causing droplets to form beads (Figure 10.8) and giving the top of the mercury column in a barometer a rounded shape called a *meniscus*.

TABLE 10.6 Viscosities and Surface Tensions of Some Common Substances at 20 °C

Name	Formula	Viscosity $(N \cdot s/m^2)$	Surface Tension (J/m^2)
Pentane	C_5H_{12}	2.4×10^{-4}	1.61×10^{-2}
Benzene	C_6H_6	6.5×10^{-4}	2.89×10^{-2}
Water	H_2O	1.00×10^{-3}	7.29×10^{-2}
Ethanol	C_2H_5OH	1.20×10^{-3}	2.23×10^{-2}
Mercury	Hg	1.55×10^{-3}	4.6×10^{-1}
Glycerol	$C_3H_5(OH)_3$	1.49	6.34×10^{-2}

10.4 PHASE CHANGES BETWEEN SOLIDS, LIQUIDS, AND GASES

We're all familiar with seeing solid ice melt to liquid water, liquid water freeze to solid ice or evaporate to gaseous steam, and gaseous steam condense to liquid water. Such processes, in which the physical form but not the chemical identity of a substance changes, are called **phase changes**, or *changes of state*. Matter in any one state, or **phase**, can change into either of the other two. Solids can even change directly into gases, as occurs when dry ice (solid CO_2) undergoes *sublimation*. The names of the various phase changes are:

Fusion (melting)	solid $\rightarrow$ liquid
Freezing	liquid $\rightarrow$ solid
Vaporization	liquid $\rightarrow$ gas
Condensation	gas $\rightarrow$ liquid
Sublimation	solid $\rightarrow$ gas
Deposition	gas $\rightarrow$ solid

REMEMBER...

The value of the **free-energy change (ΔG)** is a general criterion for the spontaneity of a chemical or physical process. If $\Delta G < 0$, the process is spontaneous; if $\Delta G = 0$, the process is at equilibrium; and if $\Delta G > 0$, the process is nonspontaneous. (Section 8.13)

Like all naturally occurring processes, every phase change has associated with it a **free-energy change, ΔG**. As we saw in Section 8.13, ΔG is made up of two contributions, an enthalpy part (ΔH) and a temperature-dependent entropy part ($T\Delta S$), according to the equation $\Delta G = \Delta H - T\Delta S$. The enthalpy part is the heat flow associated with making or breaking the intermolecular attractions that hold liquids and solids together, while the entropy part is associated with the difference in molecular randomness between the various phases. Gases are more random and have more entropy than liquids, which in turn are more random and have more entropy than solids.

The melting of a solid to a liquid, the sublimation of a solid to a gas, and the vaporization of a liquid to a gas all involve a change from a less random phase to a more random one, and all absorb heat energy to overcome the intermolecular forces holding particles together. Thus, both ΔS and ΔH are positive for these phase changes. By contrast, the freezing of a liquid to a solid, the deposition of a gas to a solid, and the condensation of a gas to a liquid all involve a change from a more random phase to a less random one, and all release heat energy as intermolecular attractions increase to hold particles more tightly together. Thus, both ΔS and ΔH have negative values for these phase changes. The situations are summarized in **Figure 10.9**.

Let's look at the transitions of solid ice to liquid water and liquid water to gaseous steam to see examples of energy relationships during phase changes. For the melting, or fusion, of ice to water, $\Delta H = +6.01$ kJ/mol and $\Delta S = +22.0$ J/(K · mol); for the vaporization of water

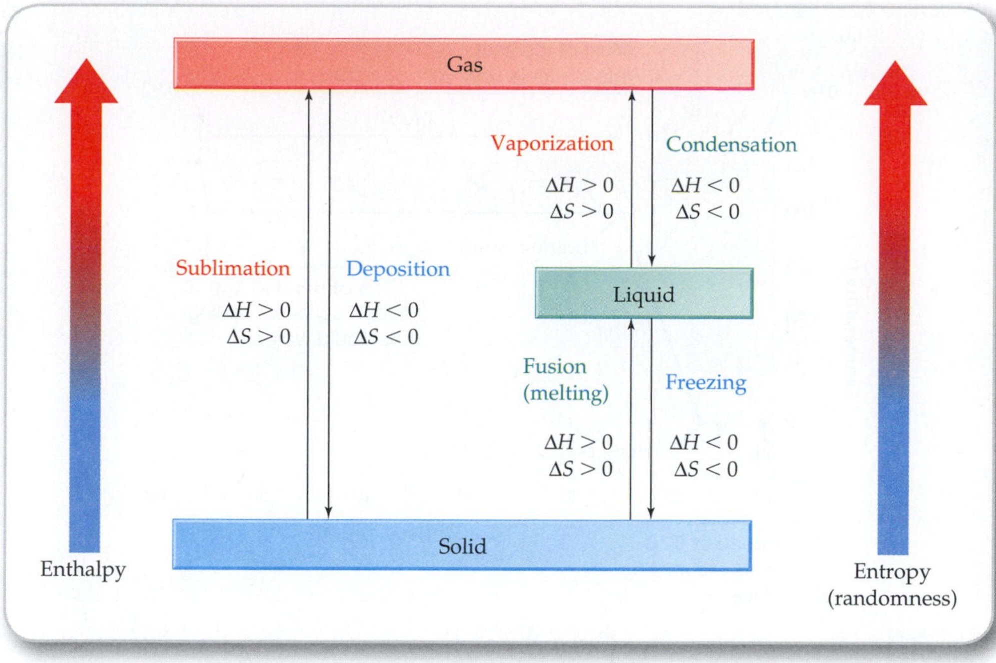

FIGURE 10.9

Phase changes and randomness. Changes from a less random phase to a more random one (up arrows) have positive values of ΔH and ΔS. Changes from a more random phase to a less random one (down arrows) have negative values of ΔH and ΔS.

to steam, $\Delta H = +40.67$ kJ/mol and $\Delta S = +109$ J/(K · mol). Both ΔH and ΔS are larger for the liquid $\rightarrow$ vapor change than for the solid $\rightarrow$ liquid change because many more intermolecular attractions need to be overcome and much more randomness is gained in the change of liquid to vapor. This highly endothermic conversion of liquid water to gaseous water vapor is used by many organisms as a cooling mechanism. When our bodies perspire on a warm day, evaporation of the perspiration absorbs heat and leaves the skin feeling cooler.

For phase changes in the opposite direction, the numbers have the same absolute values but opposite signs. That is, $\Delta H = -6.01$ kJ/mol and $\Delta S = -22.0$ J/(K · mol) for the freezing of liquid water to ice, and $\Delta H = -40.67$ kJ/mol and $\Delta S = -109$ J/(K · mol) for the condensation of water vapor to liquid water. Citrus growers take advantage of the exothermic freezing of water when they spray their trees with water on cold nights to prevent frost damage. As water freezes on the leaves, it releases heat that protects the tree.

Knowing the values of ΔH and ΔS for a phase transition makes it possible to calculate the temperature at which the change occurs. Recall from Section 8.13 that ΔG is negative for a spontaneous process, positive for a nonspontaneous process, and zero for a process at equilibrium. Thus, by setting $\Delta G = 0$ and solving for T in the free-energy equation, we can calculate the temperature at which two phases are in equilibrium. For the solid $\rightarrow$ liquid phase change in water, for instance, we have

$$\Delta G = \Delta H - T\Delta S = 0 \qquad \text{at equilibrium}$$

$$\text{or} \qquad T = \Delta H / \Delta S$$

where $\Delta H = +6.01$ kJ/mol and $\Delta S = +22.0$ J/(K · mol)

$$\text{so} \qquad T = \frac{6.01 \dfrac{\text{kJ}}{\text{mol}}}{0.0220 \dfrac{\text{kJ}}{\text{K} \cdot \text{mol}}} = 273 \text{ K}$$

In other words, ice turns into liquid water, and liquid water turns into ice, at 273 K, or 0 °C, at 1 atm pressure—hardly a surprise. In practice, the calculation is more useful in the opposite direction. That is, the temperature at which a phase change occurs is measured and then used to calculate $\Delta S (= \Delta H/T)$.

The results of continuously adding heat to a substance can be displayed on a *heating curve* like that shown in **Figure 10.10** for H_2O. Beginning with solid H_2O at an arbitrary temperature, say -25.0 °C, addition of heat raises the ice's temperature until it reaches 0 °C.

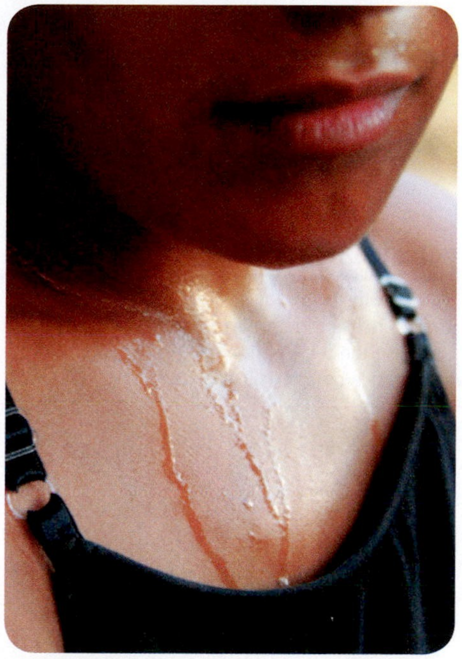

▲ Evaporation of perspiration carries away heat and cools the body after exertion.

▲ Why do citrus growers spray their trees with water on cold nights?

FIGURE 10.10

A heating curve for 1 mol of H_2O, showing the temperature changes and phase transitions that occur when heat is added.

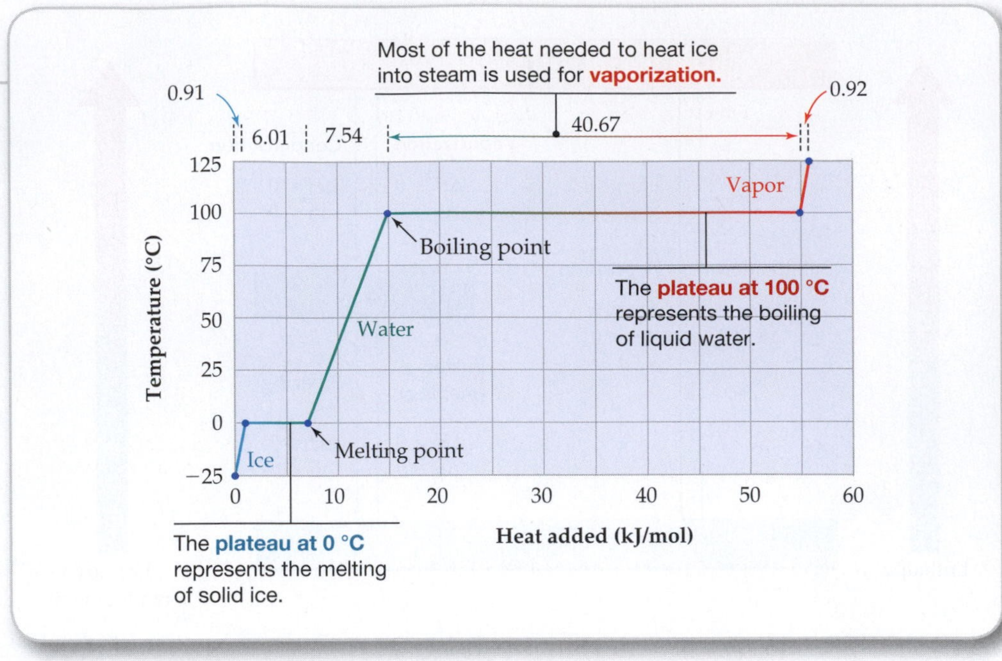

REMEMBER...

The **molar heat capacity (C_m)** of a substance is the amount of heat necessary to raise the temperature of 1 mol of the substance by 1 °C. (Section 8.5)

Because the **molar heat capacity** of ice is 36.57 J/(mol · °C), and because we need to raise the temperature 25.0 °C, 914 J/mol is required:

$$\text{Energy to heat ice from } -25\text{ °C to } 0\text{ °C} = \left(36.57\,\frac{\text{J}}{\text{mol}\cdot\text{°C}}\right)(25.0\text{ °C}) = 914\text{ J/mol}$$

Once the temperature of the ice reaches 0 °C, addition of further heat goes into disrupting hydrogen bonds and other intermolecular forces rather than into increasing the temperature, as indicated by the plateau at 0 °C on the heating curve in Figure 10.10. At this temperature—the *melting point*—solid and liquid coexist in equilibrium as molecules break free from their positions in the ice crystals and enter the liquid phase. Not until the solid turns completely to liquid does the temperature again rise. The amount of energy required to overcome enough intermolecular forces to convert a solid into a liquid is the *enthalpy of fusion*, or **heat of fusion, ΔH_{fusion}**. For ice, $\Delta H_{\text{fusion}} = +6.01$ kJ/mol.

Continued addition of heat to liquid water raises the temperature until it reaches 100 °C. Because the molar heat capacity of liquid water is 75.4 J/(mol · °C), 7.54 kJ/mol is required:

$$\text{Energy to heat water from } 0\text{ °C to } 100\text{ °C} = \left(75.4\,\frac{\text{J}}{\text{mol}\cdot\text{°C}}\right)(100\text{ °C}) = 7.54 \times 10^3\text{ J/mol}$$

Once the temperature of the water reaches 100 °C, addition of further heat again goes into overcoming intermolecular forces rather than into increasing the temperature, as indicated by the second plateau at 100 °C on the heating curve. At this temperature—the *boiling point*—liquid and vapor coexist in equilibrium as molecules break free from the surface of the liquid and enter the gas phase. The amount of energy necessary to convert a liquid into a gas is called the *enthalpy of vaporization*, or **heat of vaporization, ΔH_{vap}**. For water, $\Delta H_{\text{vap}} = +40.67$ kJ/mol. Only after the liquid has been completely vaporized does the temperature again rise.

Notice that the largest part (40.67 kJ/mol) of the 56.05 kJ/mol required to convert solid ice at −25 °C to gaseous steam at 125 °C is used for vaporization. The heat of vaporization for water is larger than the heat of fusion because only a relatively small number of hydrogen bonds must be broken to convert the solid to the liquid, but *all* hydrogen bonds must be broken to convert the liquid to the vapor.

Table 10.7 gives further data on both heat of fusion and heat of vaporization for some common compounds. What is true for water is also true for other compounds: The heat of vaporization of a compound is always larger than its heat of fusion because all intermolecular

TABLE 10.7 Heats of Fusion and Heats of Vaporization for Some Common Compounds

Name	Formula	ΔH_{fusion} (kJ/mol)	ΔH_{vap} (kJ/mol)
Ammonia	NH_3	5.66	23.33
Benzene	C_6H_6	9.87	30.72
Ethanol	C_2H_5OH	4.93	38.56
Helium	He	0.02	0.08
Mercury	Hg	2.30	59.11
Water	H_2O	6.01	40.67

forces must be overcome before vaporization can occur, but relatively fewer intermolecular forces must be overcome for a solid to change to a liquid.

Worked Example 10.4

Calculating an Entropy of Vaporization

The boiling point of water is 100 °C, and the enthalpy change for the conversion of water to steam is $\Delta H_{vap} = 40.67$ kJ/mol. What is the entropy change for vaporization, ΔS_{vap}, in J/(K · mol)?

STRATEGY

At the temperature where a phase change occurs, the two phases coexist in equilibrium and ΔG, the free-energy difference between the phases, is zero: $\Delta G = \Delta H - T\Delta S = 0$. Rearranging this equation gives $\Delta S = \Delta H/T$, where both ΔH and T are known. Remember that T must be expressed in kelvin.

SOLUTION

$$\Delta S_{vap} = \frac{\Delta H_{vap}}{T} = \frac{40.67\ \frac{\text{kJ}}{\text{mol}}}{373.15\ \text{K}} = 0.1090\ \text{kJ/(K} \cdot \text{mol)} = 109.0\ \text{J/(K} \cdot \text{mol)}$$

As you might expect, there is a large positive entropy change, corresponding to a large increase in randomness, on converting water from a liquid to a gas.

PROBLEM 10.7 Which of the following processes would you expect to have a positive value of ΔS, and which a negative value?

(a) Sublimation of dry ice
(b) Formation of dew on a cold morning
(c) Mixing of cigarette smoke with air in a closed room

PROBLEM 10.8 Chloroform ($CHCl_3$) has $\Delta H_{vap} = 29.2$ kJ/mol and $\Delta S_{vap} = 87.5$ J/(K · mol). What is the boiling point of chloroform in kelvin?

10.5 EVAPORATION, VAPOR PRESSURE, AND BOILING POINT

The conversion of a liquid to a vapor is visible when the liquid boils, but it occurs under other conditions as well. Let's imagine the two experiments illustrated in **Figure 10.11**. In one experiment, we place a liquid in an open container; in the other experiment, we place the liquid in a closed container connected to a mercury **manometer**. After a certain amount of time has passed, the liquid in the first container has evaporated, while the liquid in the second container remains but the pressure has risen. At equilibrium and at a constant temperature, the pressure increase has a constant value called the **vapor pressure** of the liquid.

REMEMBER...

The gas pressure inside a container can be measured using an open-end **manometer**, which consists of a U-tube filled with mercury. The difference between the pressure of the gas and the pressure of the atmosphere is equal to the difference between the heights of the mercury levels in the two arms of the U-tube. (Section 9.1)

FIGURE 10.11

The origin of vapor pressure.

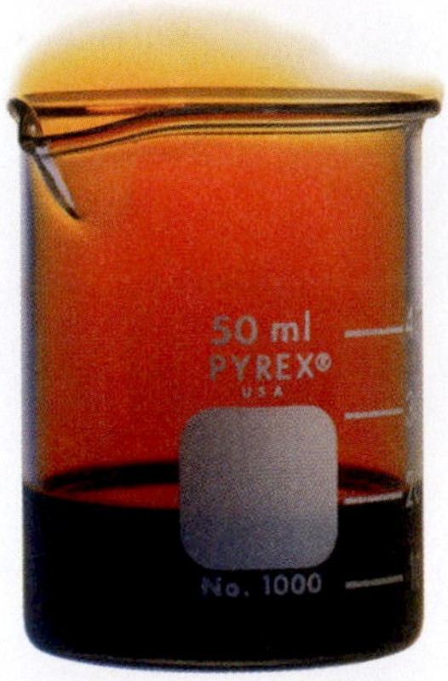

▲ Because bromine is colored, it's possible to see its reddish vapor above the liquid.

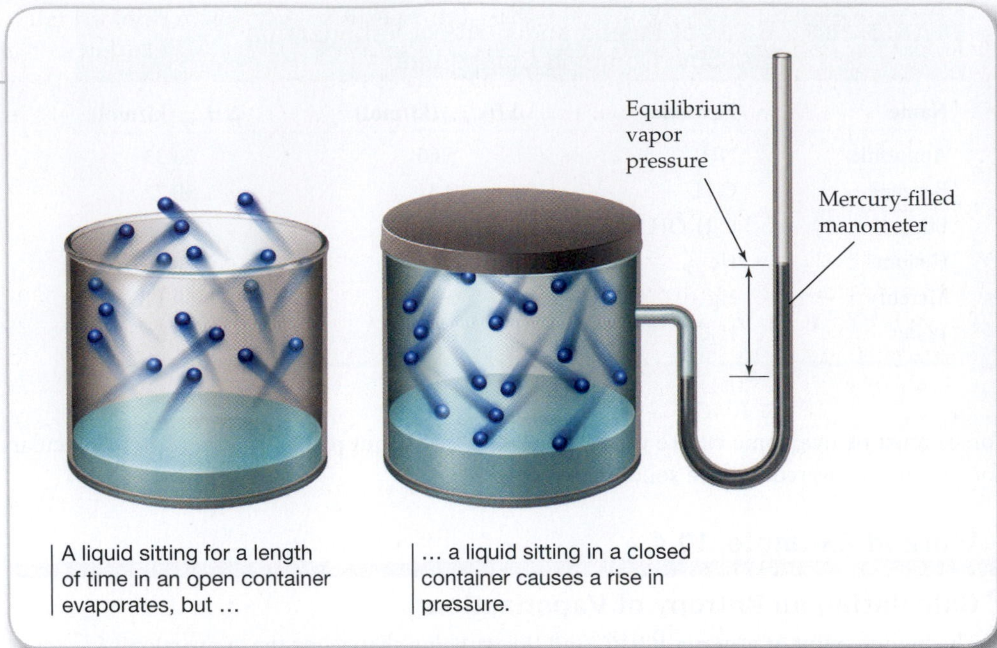

REMEMBER…

The **kinetic–molecular theory** is a group of five postulates that can be used to account for the behavior of gases and to derive the ideal gas law. Temperature and kinetic energy are related according to the equation $E_K = (3/2)RT$. (Section 9.6)

Evaporation and vapor pressure are both explained on a molecular level by the **kinetic–molecular theory**, developed in Section 9.6 to account for the behavior of gases. The molecules in a liquid are in constant motion but at a variety of speeds depending on the amount of kinetic energy they have. In considering a large sample, molecular kinetic energies follow a distribution curve like that shown in **Figure 10.12**, with the exact shape of the curve dependent on the temperature. The higher the temperature and the lower the boiling point of the substance, the greater the fraction of molecules in the sample that have sufficient kinetic energy to break free from the surface of the liquid and escape into the vapor.

FIGURE 10.12

The distribution of molecular kinetic energies in a liquid.

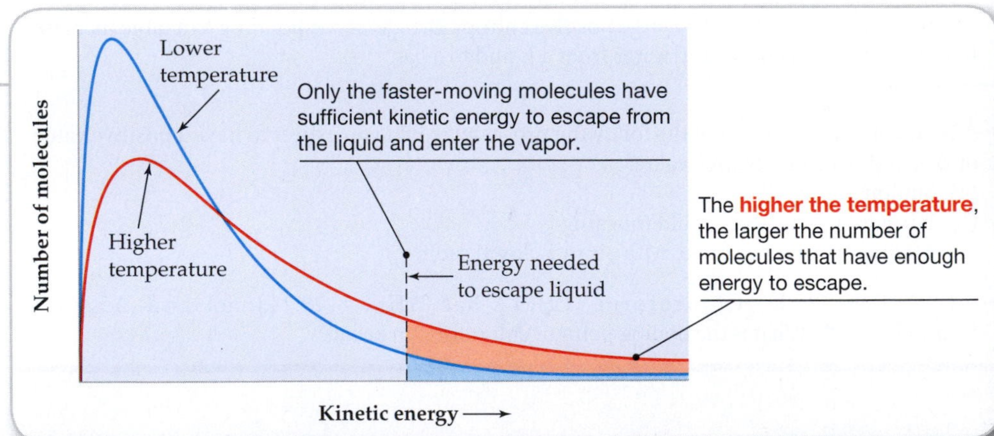

Molecules that enter the vapor phase in an open container can escape from the liquid and drift away until the liquid evaporates entirely, but molecules in a closed container are trapped. As more and more molecules pass from the liquid to the vapor, the chances increase that random motion will cause some of them to return occasionally to the liquid. Ultimately, the number of molecules returning to the liquid becomes equal to the number escaping, at which point a dynamic equilibrium exists. Although *individual* molecules are constantly passing back and forth from one phase to the other, the *total numbers* of molecules in both liquid and vapor phases remain constant.

The numerical value of a liquid's vapor pressure depends on the magnitude of the intermolecular forces present and on the temperature. The smaller the intermolecular forces, the higher the vapor pressure, because loosely held molecules escape more easily. The higher the temperature, the higher the vapor pressure, because a larger fraction of molecules have sufficient kinetic energy to escape.

The Clausius–Clapeyron Equation

As indicated in **Figure 10.13**, the vapor pressure of a liquid rises with temperature in a nonlinear way. A linear relationship *is* found, however, when the natural logarithm of the vapor pressure, $\ln P_{vap}$, is plotted against the inverse of the Kelvin temperature, $1/T$. **Table 10.8** gives the appropriate data for water, and Figure 10.13 shows the plot. As noted in Section 9.2, a linear graph is characteristic of mathematical equations of the form $y = mx + b$. In the present instance, $y = \ln P_{vap}$, $x = 1/T$, m is the slope of the line $(-\Delta H_{vap}/R)$, and b is the y-intercept (a constant, C). Thus, the data fit an expression known as the **Clausius–Clapeyron equation**:

▶ **Clausius–Clapeyron equation**

$$\ln P_{vap} = \left(-\frac{\Delta H_{vap}}{R}\right)\frac{1}{T} + C$$
$$\uparrow \quad \uparrow \quad \uparrow \quad \uparrow$$
$$y = m \quad x + b$$

where ΔH_{vap} is the heat of vaporization of the liquid, R is the gas constant (Section 9.3), and C is a constant characteristic of each specific substance.

TABLE 10.8 Vapor Pressure of Water at Various Temperatures

Temp (K)	P_{vap} (mm Hg)	$\ln P_{vap}$	$1/T$	Temp (K)	P_{vap} (mm Hg)	$\ln P_{vap}$	$1/T$
273	4.58	1.522	0.003 66	333	149.4	5.007	0.003 00
283	9.21	2.220	0.003 53	343	233.7	5.454	0.002 92
293	17.5	2.862	0.003 41	353	355.1	5.872	0.002 83
303	31.8	3.459	0.003 30	363	525.9	6.265	0.002 75
313	55.3	4.013	0.003 19	373	760.0	6.633	0.002 68
323	92.5	4.527	0.003 10	378	906.0	6.809	0.002 65

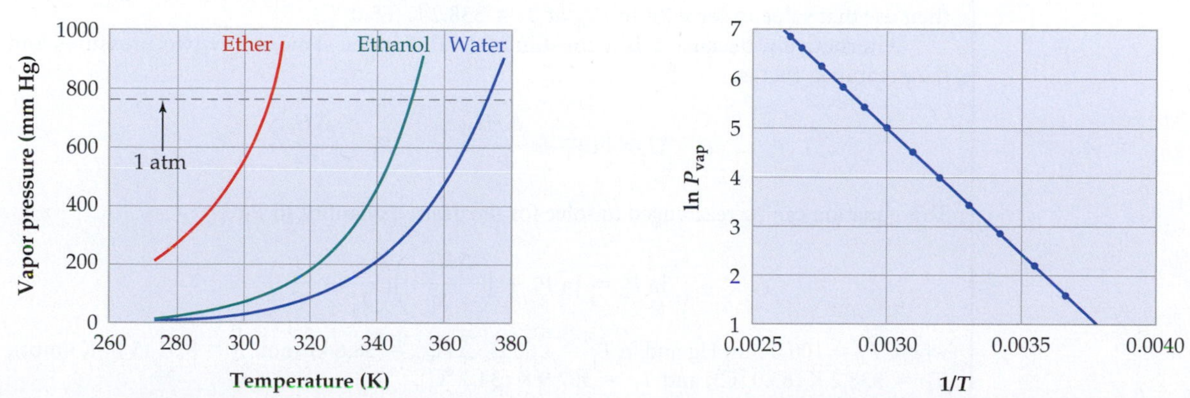

The vapor pressures of **ether**, **ethanol**, and **water** show a nonlinear rise when plotted as a function of temperature.

A plot of ln P_{vap} versus $1/T$ (kelvin) for water, prepared from the data in Table 10.8, shows a linear relationship.

FIGURE 10.13
Vapor pressure of liquids at different temperatures.

▲ What is the vapor pressure of the liquid at its boiling point

The Clausius–Clapeyron equation makes it possible to calculate the heat of vaporization of a liquid by measuring its vapor pressure at several temperatures and then plotting the results to obtain the slope of the line. Alternatively, once the heat of vaporization and the vapor pressure at one temperature are known, the vapor pressure of the liquid at any other temperature can be calculated, as shown in Worked Example 10.5.

When the vapor pressure of a liquid rises to the point where it becomes equal to the external pressure, the liquid reaches its boiling point and changes into vapor. On a molecular level, you might picture boiling in the following way: Imagine that a few molecules in the interior of the liquid momentarily break free from their neighbors and form a microscopic bubble. If the external pressure from the atmosphere is greater than the vapor pressure inside the bubble, the bubble is immediately crushed. At the temperature where the external pressure and the vapor pressure in the bubble are the same, however, the bubble is not crushed. Instead, it rises through the denser liquid, grows larger as more molecules join it, and appears as part of the vigorous action we associate with boiling.

The temperature at which a liquid boils when the external pressure is exactly 1 atm (760 mm Hg) is called the **normal boiling point**. On the plots in Figure 10.13, the normal boiling points of the three liquids are reached when the curves cross the dashed line representing 760 mm Hg—for ether, 34.6 °C (307.8 K); for ethanol, 78.3 °C (351.5 K); and for water, 100.0 °C (373.15 K).

If the external pressure is less than 1 atm, then the vapor pressure necessary for boiling is reached earlier than 1 atm and the liquid boils at a lower than normal temperature. On top of Mt. Everest, for example, where the atmospheric pressure is only about 260 mm Hg, water boils at approximately 71 °C rather than 100 °C. Conversely, if the external pressure on a liquid is greater than 1 atm, the vapor pressure necessary for boiling is reached later and the liquid boils at a greater than normal temperature. Pressure cookers take advantage of this effect by raising the boiling point of water, thereby allowing food to cook more rapidly.

Worked Example 10.5

Calculating a Vapor Pressure Using the Clausius–Clapeyron Equation

The vapor pressure of ethanol at 34.7 °C is 100.0 mm Hg, and the heat of vaporization of ethanol is 38.6 kJ/mol. What is the vapor pressure of ethanol in millimeters of mercury at 65.0 °C?

STRATEGY

There are several ways to do this problem. One way is to use the vapor pressure at $T = 307.9$ K (34.7 °C) to find a value for C in the Clausius–Clapeyron equation. You could then use that value to solve for $\ln P_{vap}$ at $T = 338.2$ K (65.0 °C).

Alternatively, because C is a constant, its value is the same at any two pressures and temperatures. That is:

$$C = \ln P_1 + \frac{\Delta H_{vap}}{RT_1} = \ln P_2 + \frac{\Delta H_{vap}}{RT_2}$$

This equation can be rearranged to solve for the desired quantity, $\ln P_2$:

$$\ln P_2 = \ln P_1 + \left(\frac{\Delta H_{vap}}{R}\right)\left(\frac{1}{T_1} - \frac{1}{T_2}\right)$$

where $P_1 = 100.0$ mm Hg and $\ln P_1 = 4.6052$, $\Delta H_{vap} = 38.6$ kJ/mol, $R = 8.3145$ J/(K · mol), $T_2 = 338.2$ K (65.0 °C), and $T_1 = 307.9$ K (34.7 °C).

SOLUTION

$$\ln P_2 = 4.6052 + \left(\frac{38{,}600 \dfrac{\text{J}}{\text{mol}}}{8.3145 \dfrac{\text{J}}{\text{K} \cdot \text{mol}}}\right)\left(\frac{1}{307.9\text{ K}} - \frac{1}{338.2\text{ K}}\right)$$

$$\ln P_2 = 4.6052 + 1.3509 = 5.9561$$

$$P_2 = \text{antiln}\,(5.9561) = 386.1\text{ mm Hg}$$

Antilogarithms are reviewed in Appendix A.2.

Worked Example 10.6

Calculating a Heat of Vaporization Using the Clausius–Clapeyron Equation

Ether has $P_{vap} = 400$ mm Hg at 17.9 °C and a normal boiling point of 34.6 °C. What is the heat of vaporization, ΔH_{vap}, for ether in kJ/mol?

STRATEGY

The heat of vaporization, ΔH_{vap}, of a liquid can be obtained either graphically from the slope of a plot of ln P_{vap} versus $1/T$ or algebraically from the Clausius–Clapeyron equation. As derived in Worked Example 10.5,

$$\ln P_2 = \ln P_1 + \left(\frac{\Delta H_{vap}}{R}\right)\left(\frac{1}{T_1} - \frac{1}{T_2}\right)$$

which can be solved for ΔH_{vap}:

$$\Delta H_{vap} = \frac{(\ln P_2 - \ln P_1)(R)}{\left(\dfrac{1}{T_1} - \dfrac{1}{T_2}\right)}$$

where $P_1 = 400$ mm Hg and $\ln P_1 = 5.991$, $P_2 = 760$ mm Hg at the normal boiling point and $\ln P_2 = 6.633$, $R = 8.3145$ J/(K·mol), $T_1 = 291.1$ K (17.9 °C), and $T_2 = 307.8$ K (34.6 °C).

SOLUTION

$$\Delta H_{vap} = \frac{(6.633 - 5.991)\left(8.3145 \dfrac{\text{J}}{\text{K} \cdot \text{mol}}\right)}{\dfrac{1}{291.1\text{ K}} - \dfrac{1}{307.8\text{ K}}} = 28{,}600\text{ J/mol} = 28.6\text{ kJ/mol}$$

PROBLEM 10.9 The normal boiling point of benzene is 80.1 °C, and the heat of vaporization is $\Delta H_{vap} = 30.7$ kJ/mol. What is the boiling point of benzene in °C on top of Mt. Everest, where $P = 260$ mm Hg?

PROBLEM 10.10 Bromine has $P_{vap} = 400$ mm Hg at 41.0 °C and a normal boiling point of 331.9 K. What is the heat of vaporization, ΔH_{vap}, of bromine in kJ/mol?

10.6 KINDS OF SOLIDS

A brief look around tells you that most substances are solids rather than liquids or gases at room temperature. That brief look also shows that there are many different kinds of solids. Some solids, such as iron and aluminum, are hard and metallic. Others, such as sugar and table salt, are crystalline and easily broken. And still others, such as rubber and many plastics, are soft and amorphous.

The most fundamental distinction between kinds of solids is that some are crystalline and others are amorphous. **Crystalline solids** are those whose constituent particles—atoms, ions, or molecules—have an ordered arrangement extending over a long range. This order on the atomic level is also seen on the visible level because crystalline solids usually have flat faces and distinct angles (**Figure 10.14a**). **Amorphous solids**, by contrast, are those whose constituent particles are randomly arranged and have no ordered long-range structure (**Figure 10.14b**). Rubber is an example.

FIGURE 10.14

Some different kinds of solids.

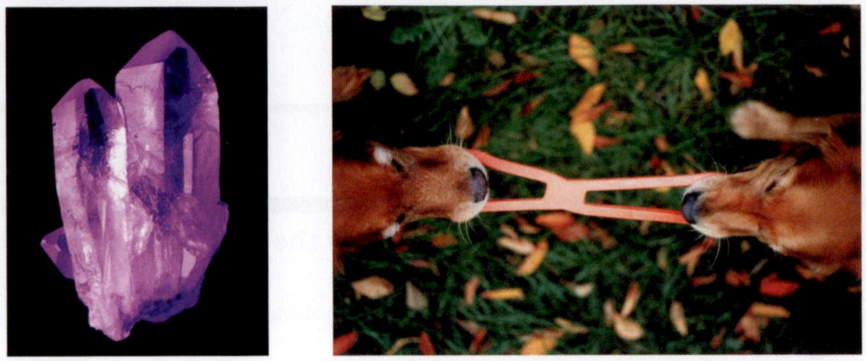

(a) A crystalline solid, such as this amethyst, has flat faces and distinct angles. These regular macroscopic features reflect a similarly ordered arrangement of particles at the atomic level.

(b) An amorphous solid like rubber has a disordered arrangement of its constituent particles.

Crystalline solids can be further categorized as *ionic*, *molecular*, *covalent network*, or *metallic*.

Ionic solids are those like sodium chloride, whose constituent particles are ions. A crystal of sodium chloride is composed of alternating Na^+ and Cl^- ions ordered in a regular three-dimensional arrangement and held together by ionic bonds, as discussed in Sections 3.1 and 3.9.

Molecular solids are those like sucrose or ice, whose constituent particles are molecules held together by the intermolecular forces discussed in Section 10.2. A crystal of ice, for instance, is composed of H_2O molecules held together in a regular way by hydrogen bonding (**Figure 10.15a**).

Covalent network solids are those like quartz (**Figure 10.15b**) or diamond, whose atoms are linked together by covalent bonds into a giant three-dimensional array. In effect, a covalent network solid is one *very* large molecule.

FIGURE 10.15

Crystal structures of ice, a molecular solid, and quartz, a covalent network solid.

(a) Ice consists of individual H_2O molecules held together in a regular manner by hydrogen bonds.

(b) Quartz (SiO_2) is essentially one very large molecule with Si–O covalent bonds. Each silicon atom has tetrahedral geometry and is bonded to four oxygens; each oxygen has approximately linear geometry and is bonded to two silicons.

(c) This shorthand representation shows how SiO_4 tetrahedra join at their corners to share oxygen atoms.

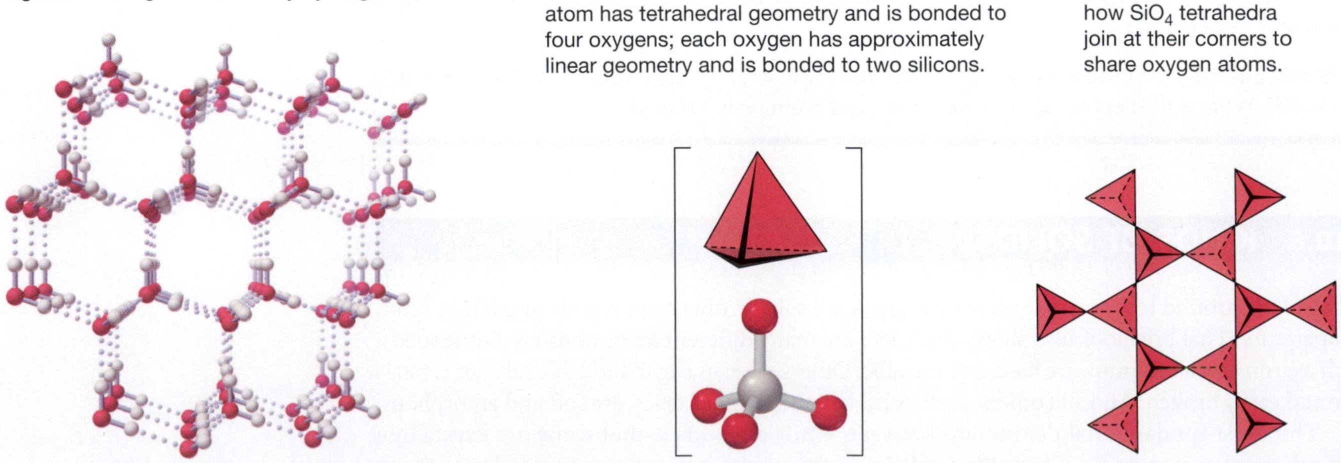

Metallic solids, such as silver or iron, also consist of large arrays of atoms, but their crystals have metallic properties such as electrical conductivity. We'll discuss metals in Chapter 21.

A summary of the different types of crystalline solids and their characteristics is given in **Table 10.9**.

TABLE 10.9 Types of Crystalline Solids and Their Characteristics

Type of Solid	Intermolecular Forces	Properties	Examples
Ionic	Ion–ion forces	Brittle, hard, high-melting	NaCl, KBr, $MgCl_2$
Molecular	Dispersion forces, dipole–dipole forces, hydrogen bonds	Soft, low-melting, nonconducting	H_2O, Br_2, CO_2, CH_4
Covalent network	Covalent bonds	Hard, high-melting	C (diamond), SiO_2
Metallic	Metallic bonds	Variable hardness and melting point, conducting	Na, Zn, Cu, Fe

10.7 PROBING THE STRUCTURE OF SOLIDS: X-RAY CRYSTALLOGRAPHY

How can the structure of a solid be found experimentally? According to a principle of optics, the wavelength of light used to observe an object must be less than twice the length of the object itself. Since atoms have diameters of around 2×10^{-10} m and the visible light detected by our eyes has wavelengths of $4–7 \times 10^{-7}$ m, it's impossible to see atoms using even the finest optical microscope. To "see" atoms, we must use "light" with a wavelength of approximately 10^{-10} m, which is in the X-ray region of the **electromagnetic spectrum**.

The origins of X-ray crystallography go back to the work of Max von Laue in 1912. On passing X rays through a crystal of sodium chloride and letting them strike a photographic plate, Laue noticed that a pattern of spots was produced on the plate, indicating that the X rays were being *diffracted* by the atoms in the crystal. A typical diffraction pattern is shown in **Figure 10.16**.

REMEMBER...

The **electromagnetic spectrum** is made up of all wavelengths of radiant energy, including visible light, infrared radiation, microwaves, radio waves, X rays, and so on. (Section 2.1)

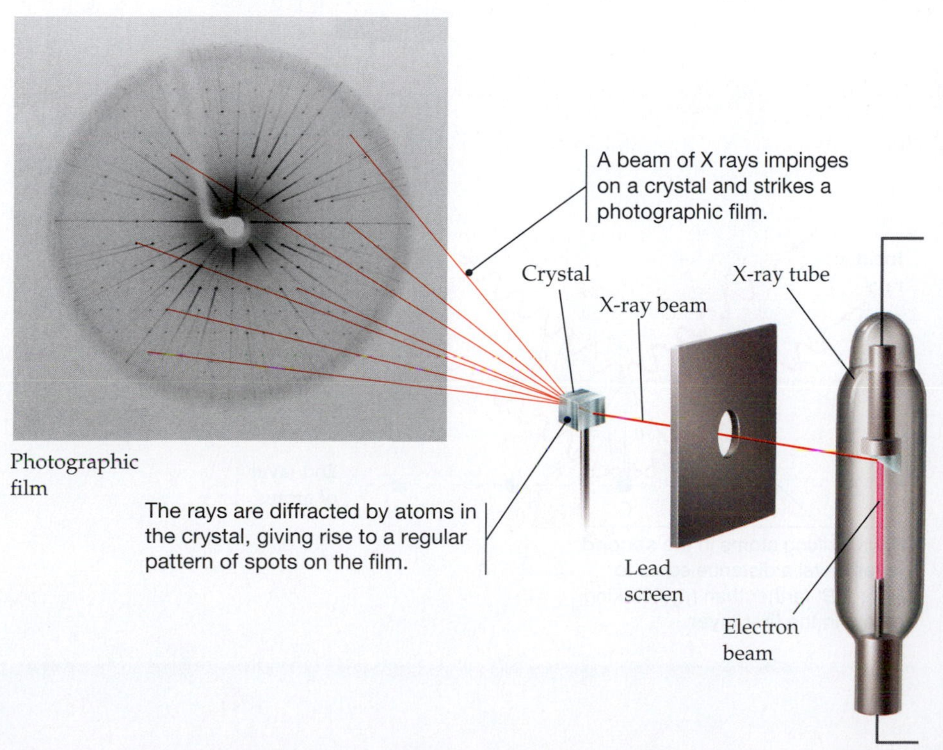

FIGURE 10.16

An X-ray diffraction experiment.

Diffraction of electromagnetic radiation occurs when a beam is scattered by an object containing regularly spaced lines (such as those in a diffraction grating) or points (such as the atoms in a crystal). This scattering happens only if the spacing between the lines or points is comparable to the wavelength of the radiation.

As shown schematically in **Figure 10.17**, diffraction is due to *interference* between two waves passing through the same region of space at the same time. If the waves are in-phase, peak to peak and trough to trough, the interference is constructive and the combined wave is increased in intensity. If the waves are out-of-phase, however, the interference is destructive and the wave is canceled. Constructive interference gives rise to the intense spots observed on Laue's photographic plate, while destructive interference causes the surrounding light areas.

FIGURE 10.17

Interference of electromagnetic waves.

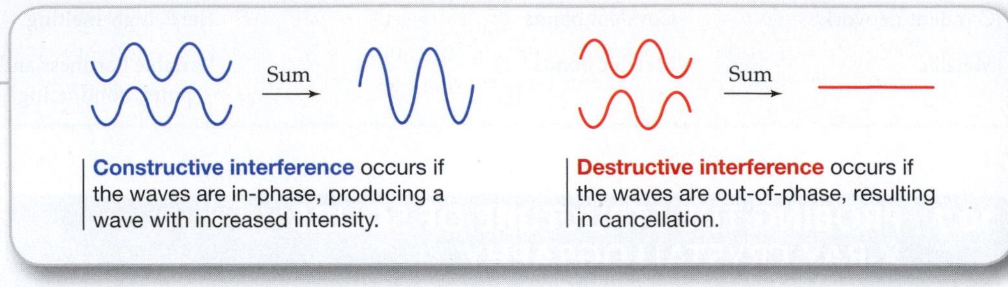

How does the diffraction of X rays by atoms in a crystal give rise to the observed pattern of spots on a photographic plate? According to an explanation advanced in 1913 by the English physicist William H. Bragg and his 22-year-old son, William L. Bragg, the X rays are diffracted by different layers of atoms in the crystal, leading to constructive interference in some instances but destructive interference in others.

To understand the Bragg analysis, imagine that incoming X rays with wavelength λ strike a crystal face at an angle θ and then bounce off at the same angle, just as light bounces off a mirror (**Figure 10.18**). Those rays that strike an atom in the top layer are all reflected at the same angle θ, and those rays that strike an atom in the second layer are also reflected at the angle θ. But because the second layer of atoms is farther from the X-ray source, the distance that the X rays have to travel to reach the second layer is farther than the distance they

FIGURE 10.18

Diffraction of X rays of wavelength λ from atoms in the top two layers of a crystal.

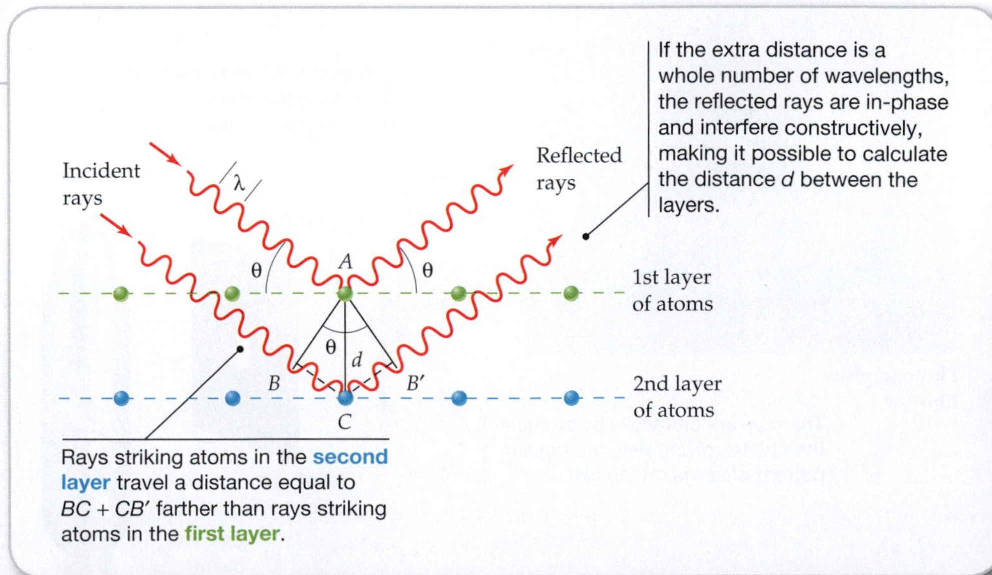

have to travel to reach the first layer by an amount indicated as BC in Figure 10.18. Using trigonometry, you can show that the extra distance BC is equal to the distance between atomic layers $d\,(=AC)$ times the sine of the angle θ:

$$\sin\theta = \frac{BC}{d} \quad \text{so} \quad BC = d\sin\theta$$

The extra distance $BC = CB'$ must also be traveled again by the *reflected* rays as they exit the crystal, making the total extra distance traveled equal to $2d\sin\theta$.

$$BC + CB' = 2d\sin\theta$$

The key to the Bragg analysis is the realization that the different rays striking the two layers of atoms are in-phase initially but can be in-phase after reflection only if the extra distance $BC + CB'$ is equal to a whole number of wavelengths $n\lambda$, where n is an integer (1, 2, 3, ...). If the extra distance is not a whole number of wavelengths, then the reflected rays will be out-of-phase and will cancel. Setting the extra distance $2d\sin\theta = n\lambda$ and rearranging to solve for d gives the **Bragg equation**:

$$BC + CB' = 2d\sin\theta = n\lambda$$

▶ **Bragg equation** $d = \dfrac{n\lambda}{2\sin\theta}$

Of the variables in the Bragg equation, the value of the wavelength λ is known, the value of $\sin\theta$ can be measured, and the value of n is a small integer, usually 1. Thus, the distance d between layers of atoms in a crystal can be calculated. For their work, the Braggs shared the 1915 Nobel Prize in Physics. The younger Bragg was 25 years old at the time.

Computer-controlled X-ray diffractometers are now available that automatically rotate a crystal and measure the diffraction from all angles. Analysis of the X-ray diffraction pattern then makes it possible to measure the interatomic distance between any two nearby atoms in a crystal. For molecular substances, this knowledge of interatomic distances indicates which atoms are close enough to form a bond. X-ray analysis thus provides a means for determining the structures of molecules (**Figure 10.19**).

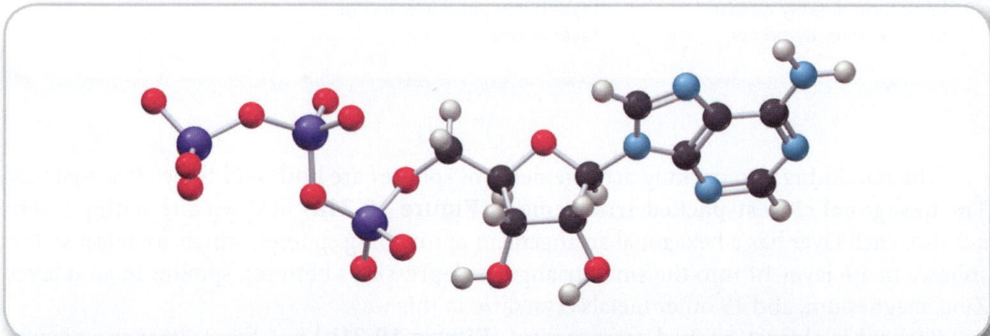

FIGURE 10.19

A computer-generated structure of adenosine triphosphate (ATP) as determined by X-ray crystallography. ATP has been called "the energy currency of the living cell," because it fuels many metabolic processes. (Black = C, red = O, blue = N, purple = P, gray = H)

10.8 THE PACKING OF SPHERES IN CRYSTALLINE SOLIDS: UNIT CELLS

How do particles—whether atoms, ions, or molecules—pack together in crystals? Let's look at metals, which are the simplest examples of crystal packing because the individual metal atoms are spheres. Not surprisingly, metal atoms (and other kinds of particles as well) generally pack together in crystals so that they can be as close as possible and maximize intermolecular attractions.

If you were to take a large number of uniformly sized marbles and arrange them in a box in some orderly way, there are four possibilities you might come up with. One way to arrange the marbles is in orderly rows and stacks, with the spheres in one layer sitting directly on top of those in the previous layer so that all layers are identical (**Figure 10.20a**). In this arrangement, called **simple cubic packing**, each sphere is touched by six neighbors—four in its own layer, one above, and one below—and is thus said to have a **coordination number** of 6. Only 52% of the available volume is occupied by the spheres in simple cubic packing, making inefficient use of space and minimizing attractive forces. Of all the metals in the periodic table, only polonium crystallizes in this way.

Alternatively, space could be used more efficiently if, instead of stacking the spheres directly on top of one another, you slightly separate the spheres in a given layer and offset the alternating layers in an *a-b-a-b* arrangement so that the spheres in the *b* layers fit into the depressions between spheres in the *a* layers, and vice versa (**Figure 10.20b**). In this arrangement, called **body-centered cubic packing**, each sphere has a coordination number of 8—four neighbors above and four below—and space is used quite efficiently: 68% of the available volume is occupied. Iron, sodium, and 14 other metals crystallize in this way.

FIGURE 10.20

Simple cubic packing and body-centered cubic packing.

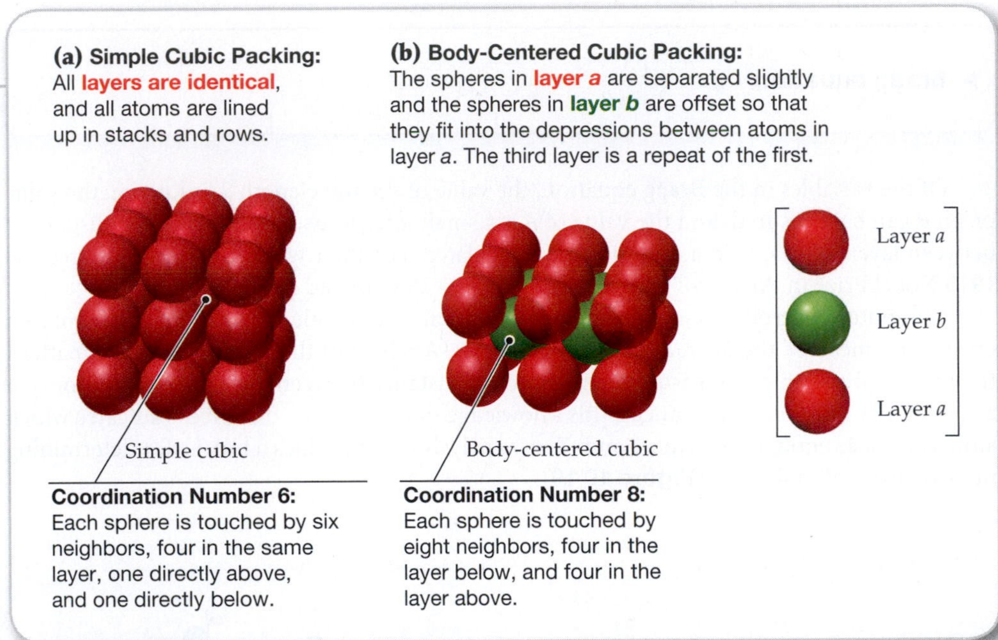

The remaining two packing arrangements of spheres are both said to be *closest-packed*. The **hexagonal closest-packed** arrangement (**Figure 10.21a**) has two alternating layers, *a-b-a-b*. Each layer has a hexagonal arrangement of touching spheres, which are offset so that spheres in a *b* layer fit into the small triangular depressions between spheres in an *a* layer. Zinc, magnesium, and 19 other metals crystallize in this way.

The **cubic closest-packed** arrangement (**Figure 10.21b**) has *three* alternating layers, *a-b-c-a-b-c*. The *a-b* layers are identical to those in the hexagonal closest-packed arrangement, but the third layer is offset from both *a* and *b* layers. Silver, copper, and 16 other metals crystallize with this arrangement.

▲ What kind of packing arrangement do these oranges have?

In both kinds of closest-packed arrangements, each sphere has a coordination number of 12—six neighbors in the same layer, three above, and three below—and 74% of the available volume is filled. The next time you're in a grocery store, look to see how the oranges or apples are stacked in their display box. They'll almost certainly have a closest-packed arrangement.

Having just taken a bulk view of how spheres can pack in a crystal, let's now take a close-up view. Just as a large brick wall is made up of many identical bricks stacked together in a repeating pattern, a crystal is made up of many small repeat units, called **unit cells**, stacked together in three dimensions.

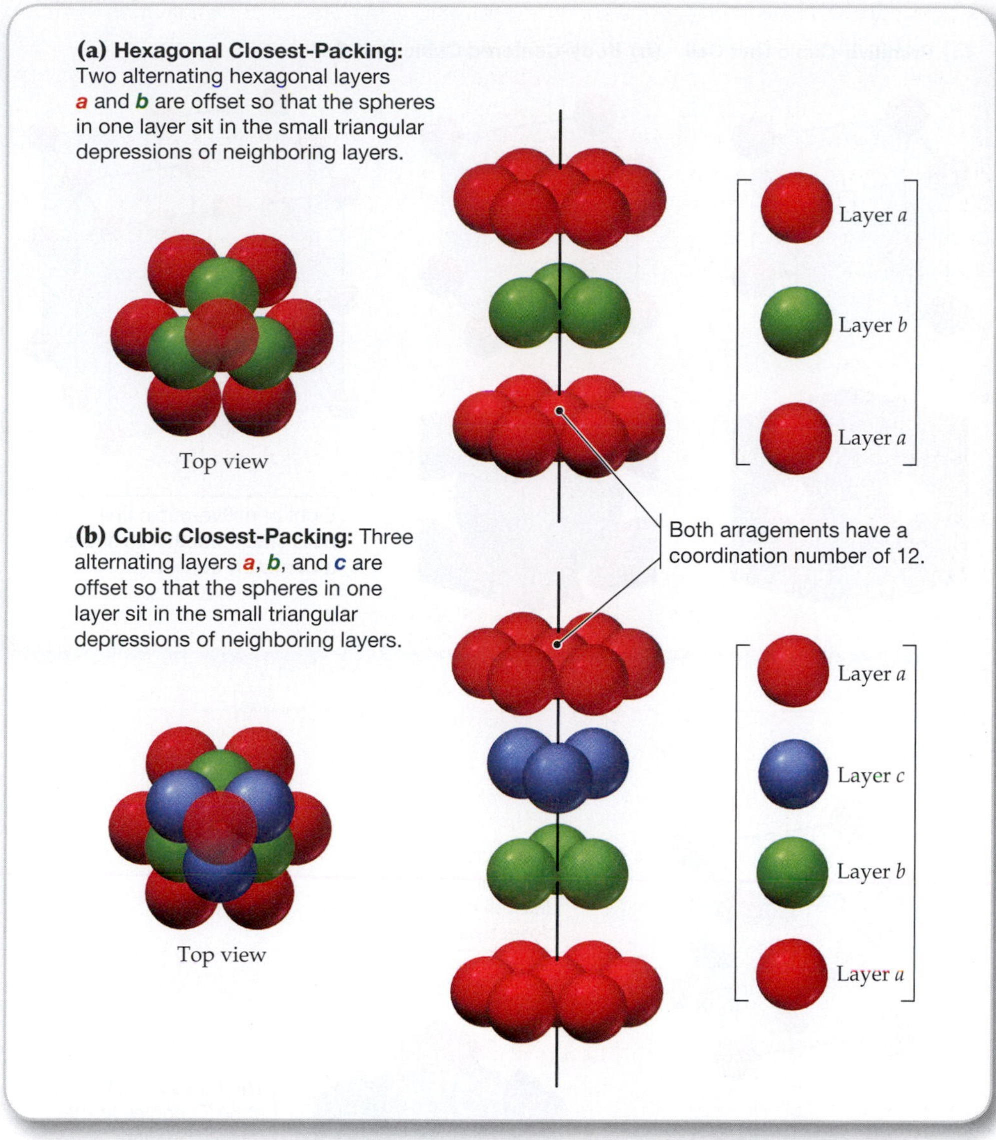

FIGURE 10.21

Hexagonal closest-packing and cubic closest-packing. In both kinds of packing, each sphere is touched by 12 neighbors, 6 in the same layer, 3 in the layer above, and 3 in the layer below.

Fourteen different unit-cell geometries occur in crystalline solids. All are parallelepipeds—six-sided geometric solids whose faces are parallelograms. We'll be concerned here only with those unit cells that have cubic symmetry; that is, cells whose edges are equal in length and whose angles are 90°. The other 11 unit cells have noncubic geometry.

There are three kinds of cubic unit cells: *primitive-cubic*, *body-centered cubic*, and *face-centered cubic*. As shown in **Figure 10.22a**, a **primitive-cubic unit cell** for a metal has an atom at each of its eight corners, where it is shared with seven neighboring cubes that come together at the same point. As a result, only 1/8 of each corner atom "belongs to" a given cubic unit. This primitive-cubic unit cell, with all atoms arranged in orderly rows and stacks, is the repeat unit found in simple cubic packing.

A **body-centered cubic unit cell** has eight corner atoms plus an additional atom in the center of the cube (**Figure 10.22b**). This body-centered cubic unit cell, with two repeating offset layers and with the spheres in a given layer slightly separated, is the repeat unit found in body-centered cubic packing.

A **face-centered cubic unit cell** has eight corner atoms plus an additional atom in the center of each of its six faces, where it is shared with one other neighboring cube (**Figure 10.23a**). Thus, 1/2 of each face atom belongs to a given unit cell. This face-centered cubic unit cell is the repeat unit found in cubic closest-packing, as can be seen by looking down the body diagonal of a unit cell (**Figure 10.23b**). The faces of the unit-cell cube are at 54.7° angles to the layers of the atoms.

▲ Just as these bricks are stacked together in a regular way on the pallet, a crystal is made of many small repeating units called unit cells that stack together in a regular way.

FIGURE 10.22

Geometries of (a) primitive-cubic and (b) body-centered cubic unit cells. Both skeletal (top) and space-filling views (bottom) are shown.

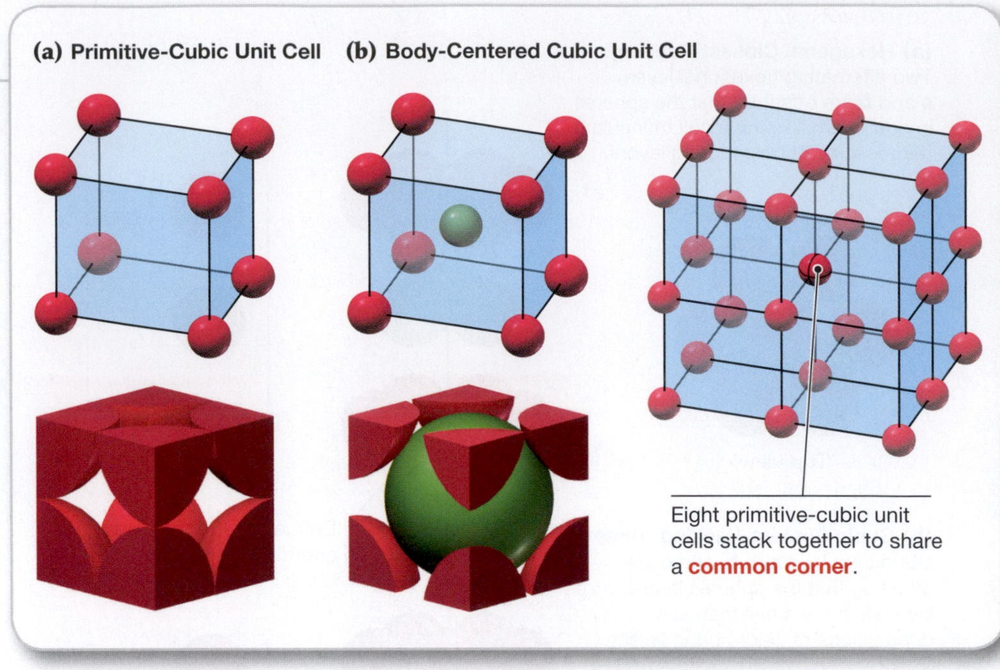

FIGURE 10.23

Geometry of a face-centered cubic unit cell.

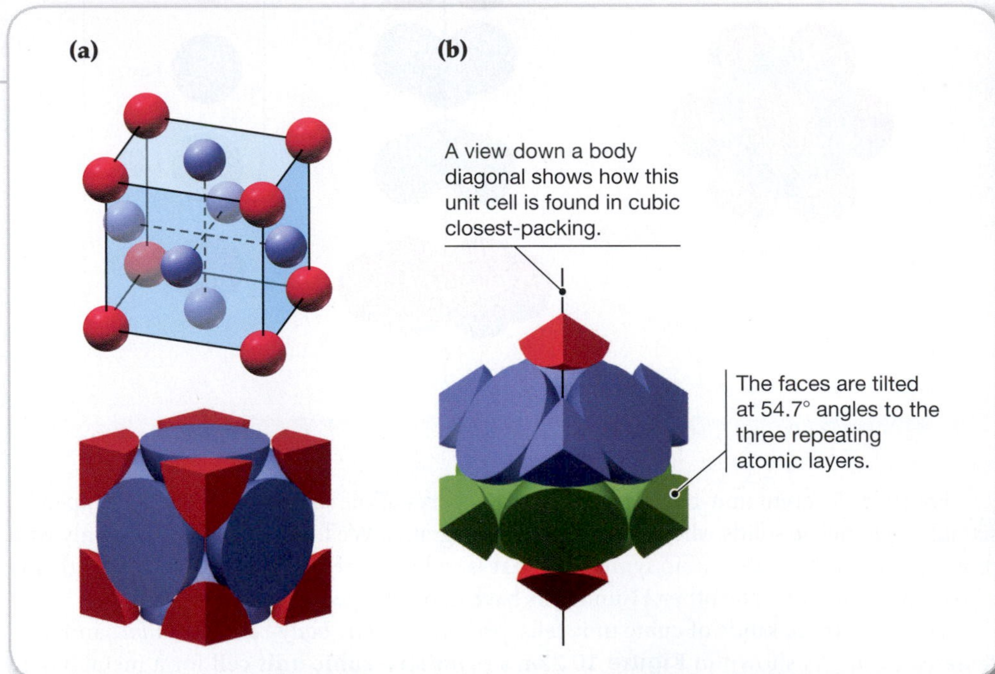

A summary of stacking patterns, coordination numbers, amount of space used, and unit cells for the four kinds of packing of spheres is given in **Table 10.10**. Hexagonal closest-packing is the only one of the four that has a noncubic unit cell.

TABLE 10.10 Summary of the Four Kinds of Packing for Spheres

Structure	Stacking Pattern	Coordination Number	Space Used (%)	Unit Cell
Simple cubic	*a-a-a-a-*	6	52	Primitive-cubic
Body-centered cubic	*a-b-a-b-*	8	68	Body-centered cubic
Hexagonal closest-packed	*a-b-a-b-*	12	74	(Noncubic)
Cubic closest-packed	*a-b-c-a-b-c-*	12	74	Face-centered cubic

Worked Example 10.7

Calculating the Number of Atoms in a Unit Cell

How many atoms are in one primitive-cubic unit cell of a metal?

STRATEGY AND SOLUTION

As shown in Figure 10.22a, a primitive-cubic unit cell has an atom at each of its eight corners. When unit cells are stacked together, each corner atom is shared by eight cubes, so that only 1/8 of each atom "belongs" to a given unit cell. Thus there is $1/8 \times 8 = 1$ atom per unit cell.

Worked Example 10.8

Using Unit-Cell Dimensions to Calculate the Radius of an Atom

Silver metal crystallizes in a cubic closest-packed arrangement with the edge of the unit cell having a length $d = 407$ pm. What is the radius in picometers of a silver atom?

STRATEGY AND SOLUTION

Cubic closest-packing uses a face-centered cubic unit cell. Looking at any one face of the cube head-on shows that the face atoms touch the corner atoms along the diagonal of the face but that corner atoms do not touch one another along the edges. Each diagonal is therefore equal to four atomic radii, $4r$:

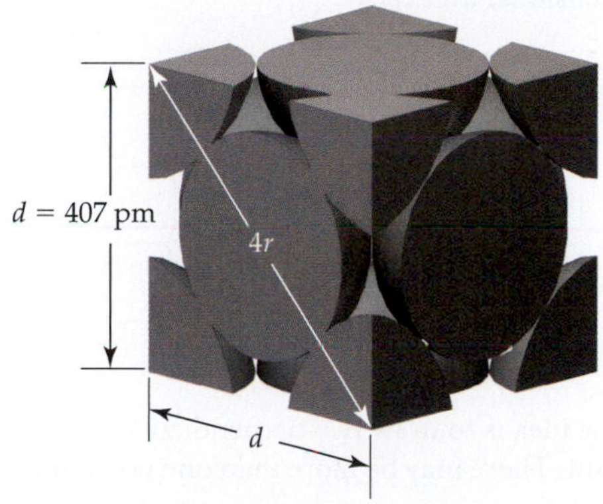

Because the diagonal and two edges of the cube form a right triangle, we can use the Pythagorean theorem to set the sum of the squares of the two edges equal to the square of the diagonal, $d^2 + d^2 = (4r)^2$ and then solve for r, the radius of one atom:

$$d^2 + d^2 = (4r)^2$$

$$2d^2 = 16r^2 \quad \text{and} \quad r^2 = \frac{d^2}{8}$$

$$\text{thus} \quad r = \sqrt{\frac{d^2}{8}} = \sqrt{\frac{(407\ \text{pm})^2}{8}} = 144\ \text{pm}$$

The radius of a silver atom is 144 pm.

Worked Example 10.9

Using Unit-Cell Dimensions to Calculate the Density of a Metal

Nickel has a face-centered cubic unit cell with a length of 352.4 pm along an edge. What is the density of nickel in g/cm^3?

STRATEGY

Density is mass divided by volume. The mass of a single unit cell can be calculated by counting the number of atoms in the cell and multiplying by the mass of a single atom. The volume of a single cubic unit cell with edge d is $d^3 = (3.524 \times 10^{-8}\ \text{cm})^3 = 4.376 \times 10^{-23}\ \text{cm}^3$.

continued on next page

SOLUTION

Each of the eight corner atoms in a face-centered cubic unit cell is shared by eight unit cells, so that only $1/8 \times 8 = 1$ atom belongs to a single cell. In addition, each of the six face atoms is shared by two unit cells, so that $1/2 \times 6 = 3$ atoms belong to a single cell. Thus, a single cell has 1 corner atom and 3 face atoms, for a total of 4, and each atom has a mass equal to the molar mass of nickel (58.69 g/mol) divided by Avogadro's number (6.022×10^{23} atoms/mol). We can now calculate the density:

$$\text{Density} = \frac{\text{Mass}}{\text{Volume}} = \frac{(4\text{ atoms})\left(\dfrac{58.69\,\dfrac{\text{g}}{\text{mol}}}{6.022 \times 10^{23}\,\dfrac{\text{atoms}}{\text{mol}}}\right)}{4.376 \times 10^{-23}\text{ cm}^3} = 8.909\text{ g/cm}^3$$

The calculated density of nickel is 8.909 g/cm^3. (The measured value is 8.90 g/cm^3.)

WORKED CONCEPTUAL EXAMPLE 10.10

Identifying a Unit Cell

Imagine a tiled floor in the following pattern. Identify the smallest repeating rectangular unit, analogous to a two-dimensional unit cell.

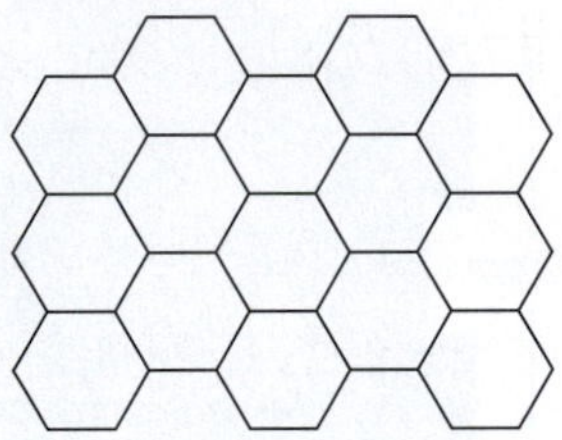

STRATEGY

Using trial and error, the idea is to draw two perpendicular sets of parallel lines that define a repeating rectangular unit. There may be more than one possibility.

SOLUTION

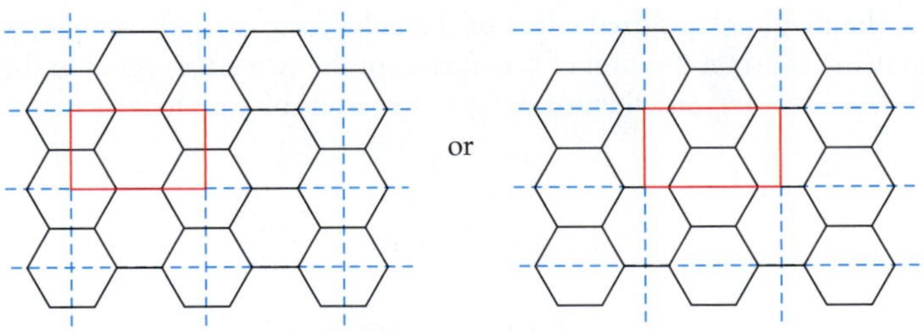

PROBLEM 10.11 How many atoms are in the following:

(a) One body-centered cubic unit cell of a metal

(b) One face-centered cubic unit cell of a metal

PROBLEM 10.12 Polonium metal crystallizes in a simple cubic arrangement, with the edge of a unit cell having a length $d = 334$ pm. What is the radius in picometers of a polonium atom?

PROBLEM 10.13 What is the density of polonium (Problem 10.12) in g/cm^3?

CONCEPTUAL PROBLEM 10.14 Imagine a tiled floor in the following pattern. Identify the smallest repeating unit, analogous to a two-dimensional unit cell.

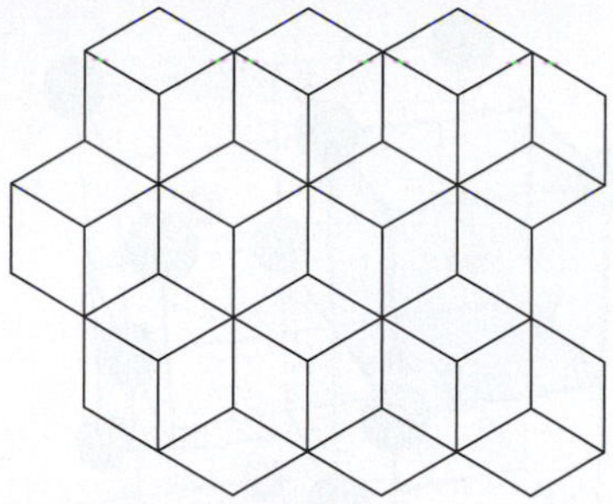

10.9 STRUCTURES OF SOME IONIC SOLIDS

Simple ionic solids such as NaCl and KBr are like metals in that the individual ions are spheres that pack together in a regular way. They differ from metals, however, in that the spheres are not all the same size—anions generally have larger **ionic radii** than cations. As a result, ionic solids adopt a variety of different unit cells, depending on the size and charge of the ions. NaCl, KCl, and a number of other salts have a face-centered cubic unit cell in which the larger Cl^- anions occupy corners and faces while the smaller Na^+ cations fit into the holes between adjacent anions (**Figure 10.24**).

REMEMBER...

Atomic radius decreases when an atom is converted to a cation by loss of an electron and increases when the atom is converted to an anion by gain of an electron. (Section 3.4)

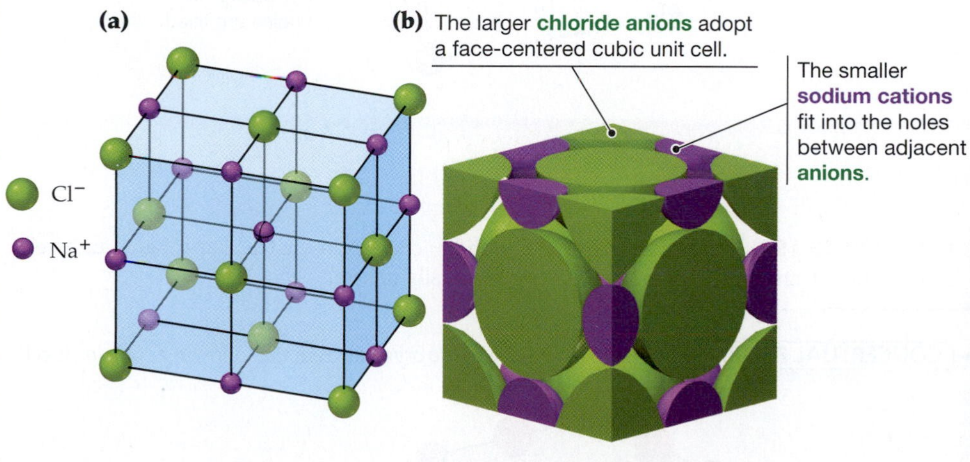

FIGURE 10.24

The unit cell of NaCl. Both a skeletal view **(a)** and a space-filling view **(b)** in which the unit cell is viewed edge-on are shown.

It's necessary, of course, that the unit cell of an ionic substance be electrically neutral, with equal numbers of positive and negative charges. In the NaCl unit cell, for instance, there are four Cl^- anions ($1/8 \times 8 = 1$ corner atom, plus $1/2 \times 6 = 3$ face atoms) and also four Na^+ cations ($1/4 \times 12 = 3$ edge atoms, plus 1 center atom). (Remember that each corner atom in a cubic unit cell is shared by eight cells, each face atom is shared by two cells, and each edge atom is shared by four cells.)

Two other common ionic unit cells are shown in **Figure 10.25**. Copper(I) chloride has a face-centered cubic arrangement of the larger Cl^- anions, with the smaller Cu^+ cations in holes so that each is surrounded by a tetrahedron of four anions. Barium chloride, by contrast, has a face-centered cubic arrangement of the smaller Ba^{2+} *cations*, with the larger Cl^- anions surrounded tetrahedrally. As required for charge neutrality, there are twice as many Cl^- anions as Ba^{2+} cations.

FIGURE 10.25

Unit cells of (a) CuCl and (b) $BaCl_2$.

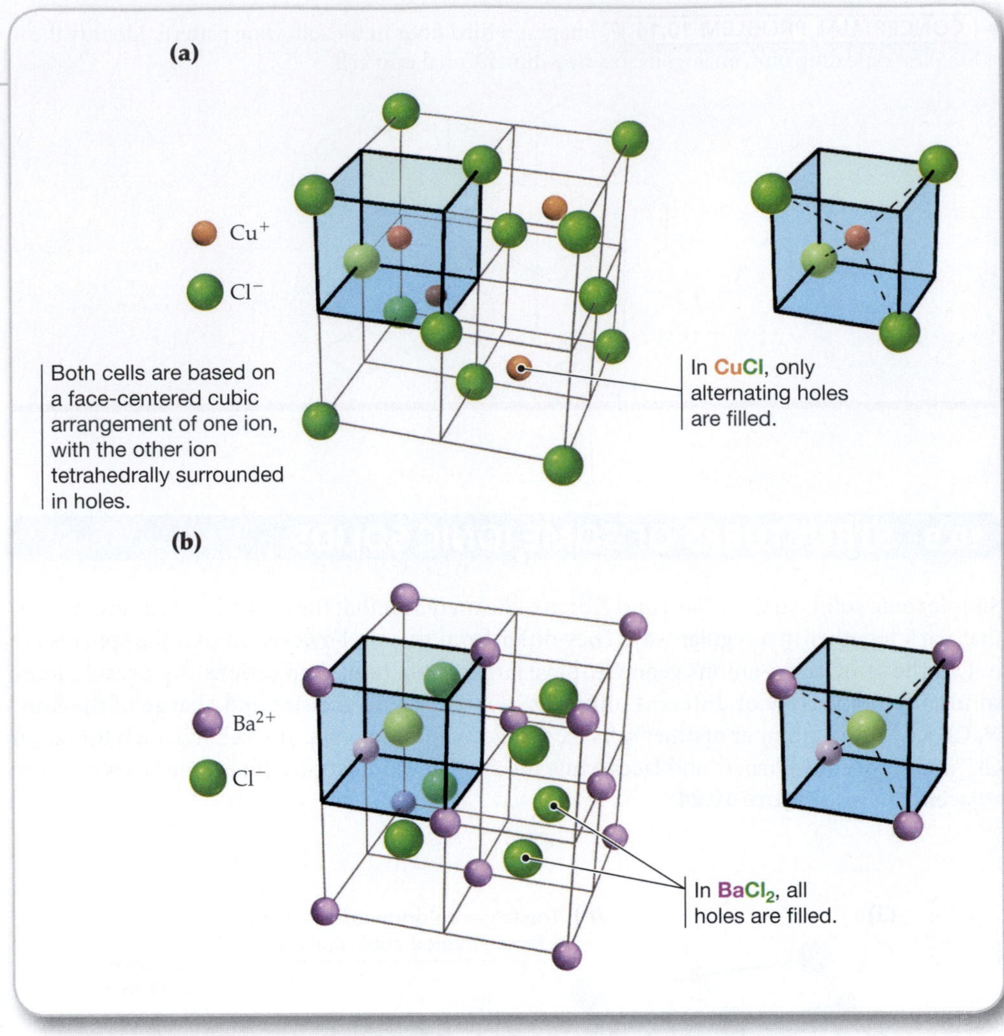

PROBLEM 10.15 Count the numbers of + and − charges in the CuCl and $BaCl_2$ unit cells (Figure 10.25), and show that both cells are electrically neutral.

CONCEPTUAL PROBLEM 10.16 Rhenium oxide crystallizes in the following cubic unit cell:

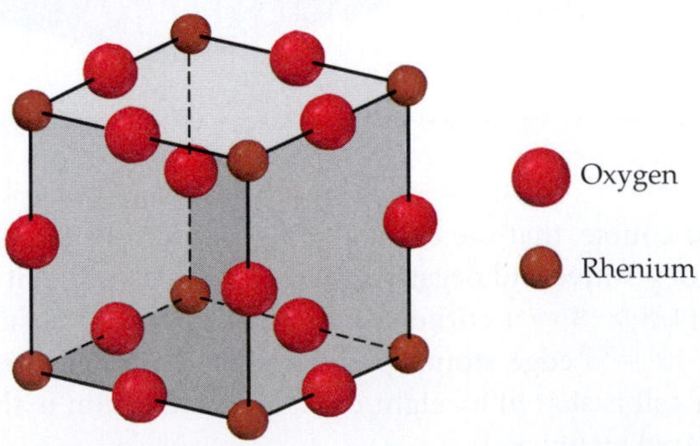

(a) How many rhenium atoms and how many oxygen atoms are in each unit cell?
(b) What is the formula of rhenium oxide?
(c) What is the oxidation state of rhenium?
(d) What is the geometry around each oxygen atom?
(e) What is the geometry around each rhenium atom?

10.10 STRUCTURES OF SOME COVALENT NETWORK SOLIDS

Carbon

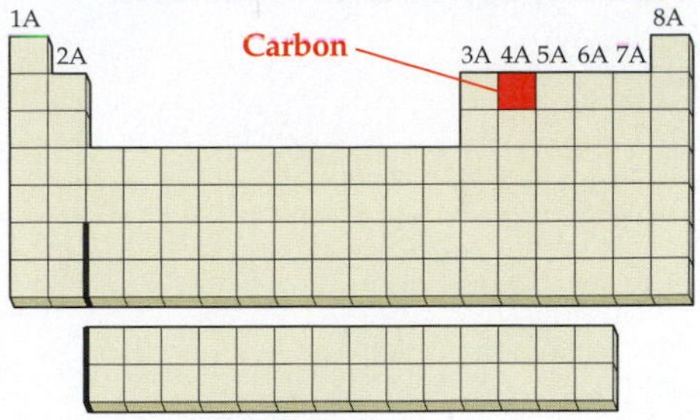

Carbon exists in more than 40 known structural forms, or **allotropes**, several of which are crystalline but most of which are amorphous. Graphite, the most common allotrope of carbon and the most stable under normal conditions, is a crystalline covalent network solid that consists of two-dimensional sheets of fused six-membered rings (**Figure 10.26a**). Each carbon atom is sp^2-hybridized and is bonded with trigonal planar geometry to three other carbons. The diamond form of elemental carbon is a covalent network solid in which each carbon atom is sp^3-hybridized and is bonded with tetrahedral geometry to four other carbons (**Figure 10.26b**).

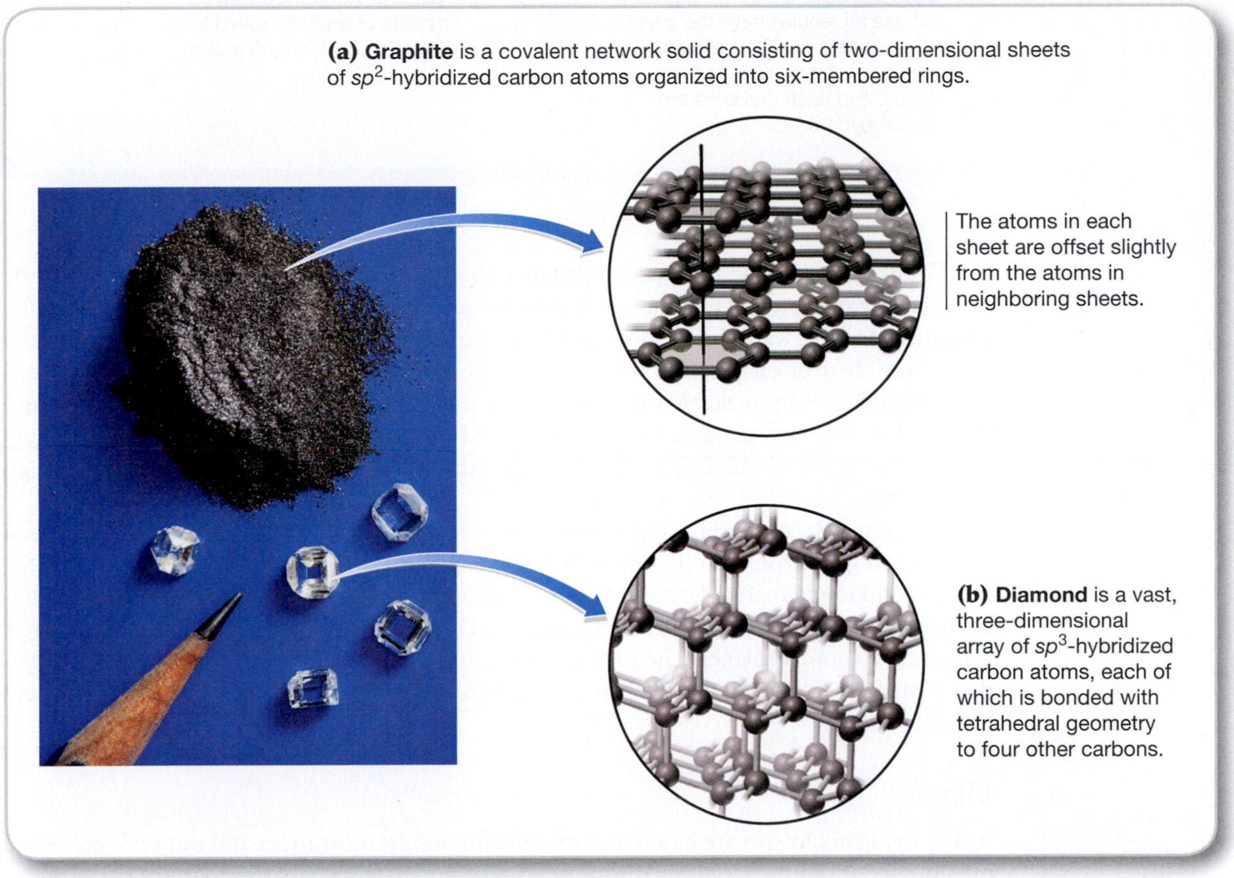

FIGURE 10.26

Two crystalline allotropes of carbon, (a) graphite and (b) diamond.

In addition to graphite and diamond, a third crystalline allotrope of carbon called *fullerene* was discovered in 1985 as a constituent of soot. Fullerene consists of spherical C_{60} molecules with the extraordinary shape of a soccer ball. The C_{60} ball has 12 pentagonal and 20 hexagonal faces, with each atom sp^2-hybridized and bonded to three other atoms (**Figure 10.27a**). Closely related to both graphite and fullerene are a group of carbon allotropes called *nanotubes*—tubular structures made of repeating six-membered carbon rings, as if a sheet of graphite were rolled up (**Figure 10.27b**). Typically, the tubes have a diameter of about 2–30 nm and a length of up to 1 mm.

The different structures of the carbon allotropes lead to widely different properties. Because of its three-dimensional network of strong single bonds that tie all atoms in a crystal together, diamond is the hardest known substance. In addition to its use in jewelry, diamond is widely used industrially for the tips of saw blades and drilling bits. It is an electrical insulator and has a melting point of about 8700 °C at a pressure of 6–10 million atm. Clear, colorless, and highly crystalline, diamonds are very rare and are found in only a few places in the world, particularly in central and southern Africa.

FIGURE 10.27

Fullerene, C_{60}, and carbon nanotubes.

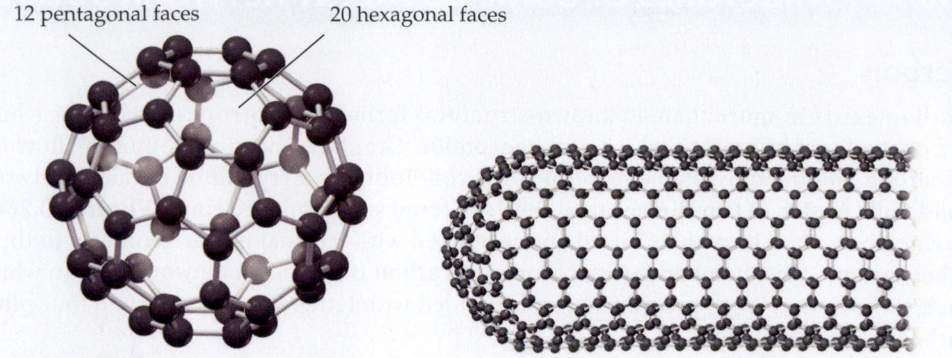

(a) Fullerene is a molecular solid whose molecules have the shape of a soccer ball. The ball has 12 pentagonal and 20 hexagonal faces, and each carbon atom is sp^2-hybridized.

(b) Carbon nanotubes consist of sheets of graphite rolled into tubes of 2–30 nm diameter.

Graphite is the black, slippery substance used as the "lead" in pencils, as an electrode material in batteries, and as a lubricant in locks. All these properties result from its sheetlike structure. Air and water molecules can adsorb onto the flat faces of the sheets, allowing the sheets to slide over one another and giving graphite its greasy feeling and lubricating properties. Graphite is more stable than diamond at normal pressures but can be converted into diamond at very high pressure and temperature. In fact, approximately 120,000 kg of industrial diamonds are synthesized annually by applying 150,000 atm pressure to graphite at high temperature.

Fullerene, black and shiny like graphite, is the subject of much current research because of its interesting electronic properties. When fullerene is allowed to react with rubidium metal, a superconducting material called rubidium fulleride, Rb_3C_{60}, is formed. Carbon nanotubes are being studied for use as fibers in the structural composites used to make golf clubs, bicycle frames, boats, and airplanes. Their tensile strength is approximately 50–60 times greater than that of steel. We'll look at the chemistry of carbon and some of its compounds in more detail in Section 19.7.

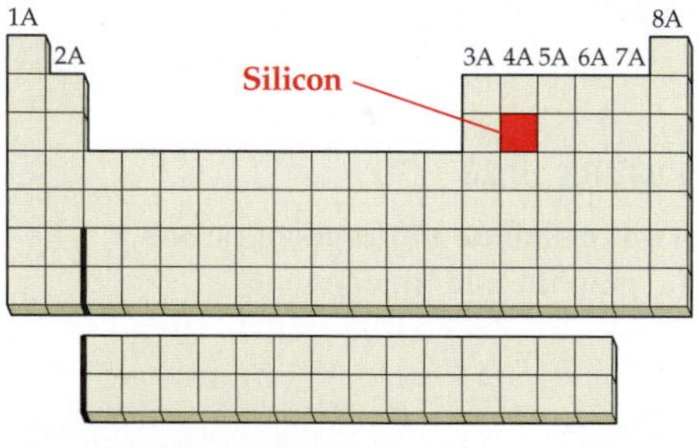

Silica

Just as living organisms are based on carbon compounds, most rocks and minerals are based on silicon compounds. Quartz and much sand, for instance, are nearly pure *silica*, SiO_2. Silicon and oxygen together, in fact, make up nearly 75% of the mass of the Earth's crust. Considering that silicon and carbon are both in group 4A of the periodic table, you might expect SiO_2 to be similar in its properties to CO_2. In fact, though, CO_2 is a molecular substance and a gas at room temperature, whereas SiO_2 (Figure 10.15b on page 374) is a covalent network solid with a melting point over 1600 °C.

The dramatic difference in properties between CO_2 and SiO_2 is due primarily to the difference in electronic structure between carbon and silicon. The π part of a *carbon*–oxygen **double bond** is formed by sideways overlap of a carbon 2*p* orbital with an oxygen 2*p* orbital (Section 5.4). If a similar *silicon*–oxygen double bond were to form, it would require overlap of an oxygen 2*p* orbital and a silicon 3*p* orbital. But because the Si–O bond distance is longer than the C–O distance and a 3*p* orbital is larger than a 2*p* orbital, overlap between the silicon and oxygen *p* orbitals is not as favorable. As a result, silicon forms four single bonds to four oxygens in a covalent network structure rather than two double bonds to two oxygens in a molecular structure.

REMEMBER...

Two atoms form a **double bond** when they approach each other with their hybrid orbitals aligned head-on for σ bonding and with their unhybridized *p* orbitals aligned in a parallel, sideways manner to form a π bond. (Section 5.4)

Heating silica above about 1600 °C breaks many of its Si–O bonds and turns it from a crystalline solid into a viscous liquid. When this fluid is cooled, some of the Si–O bonds re-form in a random arrangement and a noncrystalline, amorphous solid called *quartz glass* is formed. If additives are mixed in before cooling, a wide variety of glasses can be prepared. Common window glass, for instance, is prepared by adding $CaCO_3$ and Na_2CO_3. Addition of various transition metal ions results in the preparation of colored glasses, and addition of B_2O_3 produces a high-melting *borosilicate glass* that is sold under the trade name Pyrex. Borosilicate glass is particularly useful for cooking utensils and laboratory glassware because it expands very little when heated and is thus unlikely to crack. We'll look further at the chemistry of silicon and silicon-containing minerals in Section 19.8.

▲ Colored glasses contain transition metal ions.

10.11 PHASE DIAGRAMS

Now that we've looked at the three phases of matter individually, let's draw everything together by taking an overall view. As noted previously, any one phase of matter can change spontaneously into either of the other two, depending on the temperature and pressure. A convenient way to picture this pressure–temperature dependency of a pure substance in a closed system without air present is to use what is called a **phase diagram**. As illustrated for water in **Figure 10.28**, a typical phase diagram shows which phase is stable at different combinations of pressure and temperature. When a boundary line between phases is crossed by changing either the temperature or the pressure, a phase change occurs.

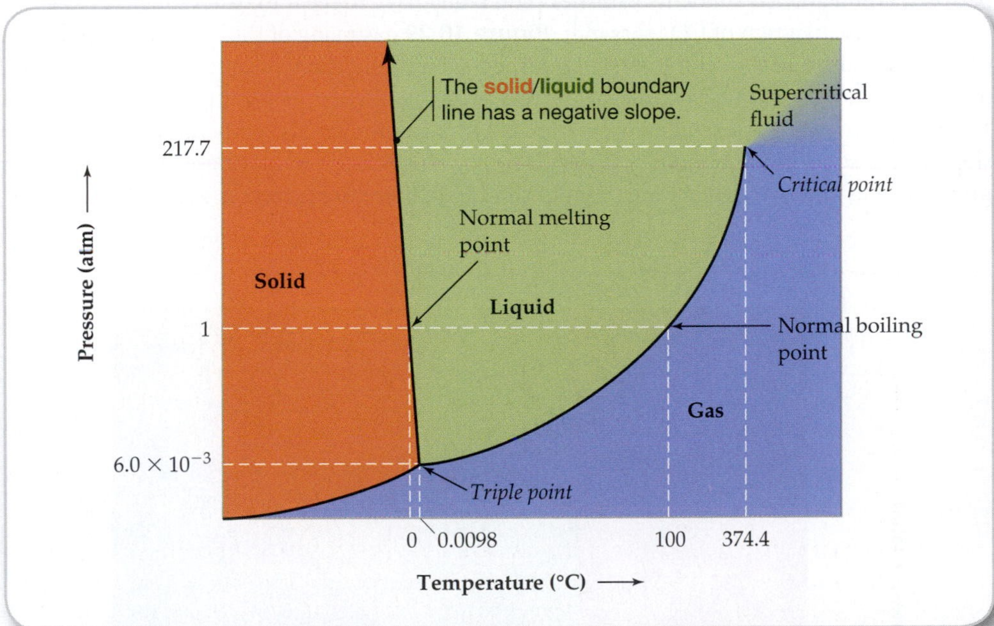

FIGURE 10.28

A phase diagram for H_2O. Various features of the diagram are discussed in the text. Note that the pressure and temperature axes are not drawn to scale.

The simplest way to understand a phase diagram is to begin at the origin in the lower left corner of Figure 10.28 and travel up and right along the boundary line between solid on the left and gas on the right. Points on this line represent pressure/temperature combinations at which the two phases are in equilibrium in a closed system and a direct phase transition between solid ice and gaseous water vapor occurs. At some point along the solid/gas line, an intersection is reached where two lines diverge to form the bounds of the liquid region. The solid/liquid boundary for H_2O goes up and slightly left, while the liquid/gas boundary continues curving up and to the right. Called the **triple point**, this three-way intersection represents a unique combination of pressure and temperature at which all three phases coexist in equilibrium. For water, the triple-point temperature T_t is 0.0098 °C, and the triple-point pressure P_t is 6.0×10^{-3} atm.

Continuing up and slightly left from the triple point, the solid/liquid boundary line represents the melting point of solid ice (or the freezing point of liquid water) at various pressures.

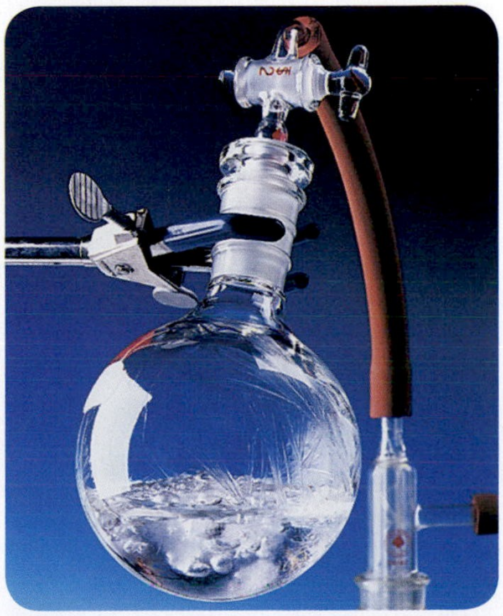

▲ At the triple point, solid exists in the boiling liquid. That is, solid, liquid, and gas coexist in equilibrium.

When the pressure is 1 atm, the melting point—called the **normal melting point**—is exactly 0 °C. There is a slight negative slope to the line, indicating that the melting point of ice decreases as pressure increases. Water is unusual in this respect, because most substances have a positive slope to their solid/liquid line, indicating that their melting points *increase* with pressure. We'll say more about this behavior shortly.

Continuing up and right from the triple point, the liquid/gas boundary line represents the pressure/temperature combinations at which liquid and gas coexist and water vaporizes (or steam condenses). In fact, the part of the curve up to 1 atm pressure is simply the vapor pressure curve we saw previously in Figure 10.13. When the pressure is 1 atm, water is at its normal boiling point of 100 °C. Continuing along the liquid/gas boundary line, we suddenly reach the **critical point**, where the line abruptly ends. The critical temperature T_c is the temperature beyond which a gas cannot be liquefied, no matter how great the pressure, and the critical pressure P_c is the pressure beyond which a liquid cannot be vaporized, no matter how high the temperature. For water, $T_c = 374.4$ °C and $P_c = 217.7$ atm.

We're all used to seeing solid/liquid and liquid/gas phase transitions, but behavior at the critical point lies so far outside our normal experiences that it's hard to imagine. Think of it this way: A *gas* at the critical point is under such high pressure, and its molecules are pressed so close together, that it becomes almost like a liquid. A *liquid* at the critical point is at such a high temperature, and its molecules are so relatively far apart, that it becomes almost like a gas. Thus, the two phases simply become the same and form a **supercritical fluid** that is neither true liquid nor true gas. No distinct physical phase change occurs on going beyond the critical point. Rather, a whitish, pearly sheen momentarily appears, and the visible boundary between liquid and gas suddenly vanishes. You really have to see it to believe it.

The phase diagram of CO_2 shown in **Figure 10.29** has many of the same features as that of water but differs in several interesting respects. First, the triple point is at $P_t = 5.11$ atm, meaning that CO_2 can't be a liquid below this pressure, no matter what the temperature. At 1 atm pressure, CO_2 is a solid below −78.5 °C but a gas above this temperature. Second, the slope of the solid/liquid boundary is positive, meaning that the solid phase is favored as the pressure rises and that the melting point of solid CO_2 therefore increases with pressure.

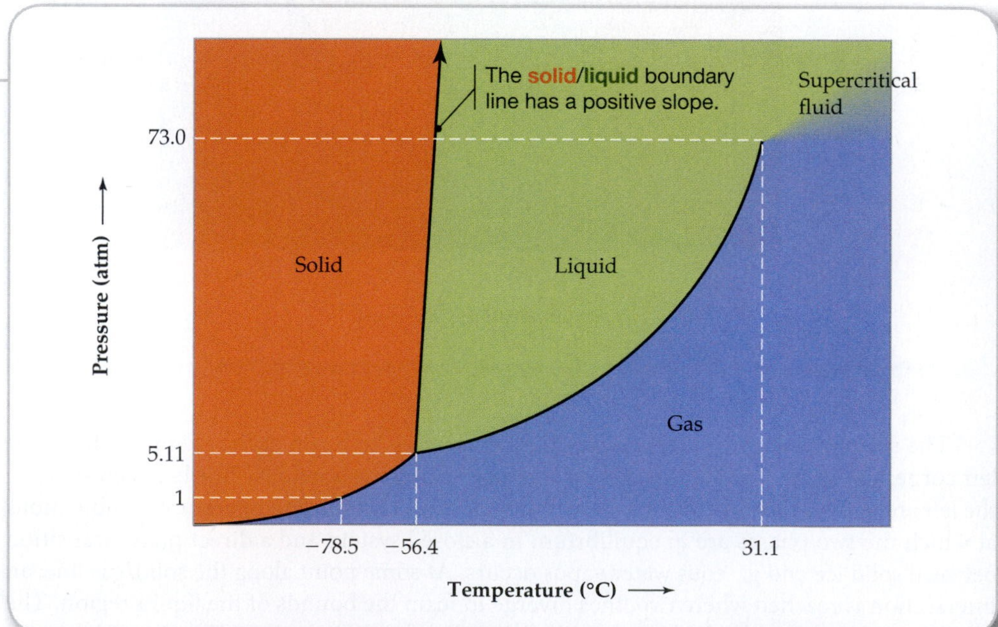

FIGURE 10.29

A phase diagram for CO_2. The pressure and temperature axes are not to scale.

The effect of pressure on the slope of the solid/liquid boundary line—negative for H_2O but positive for CO_2 and most other substances—depends on the relative densities of the solid and liquid phases. For CO_2 and most other substances, the solid phase is denser than the liquid because particles are packed closer together in the solid. Increasing the pressure pushes the molecules even closer together, thereby favoring the solid phase even more and giving the solid/liquid boundary line a positive slope. Water, however, becomes less dense when it freezes to a solid because large empty spaces are left between molecules due to the ordered three-dimensional network of hydrogen bonds in ice (Figure 10.15a). As a result, increasing the pressure favors the liquid phase, giving the solid/liquid boundary a negative slope.

Figure 10.30 shows a simple demonstration of the effect of pressure on melting point. If a thin wire with heavy weights at each end is draped over a block of ice near 0 °C, the wire rapidly cuts through the block because the increased pressure lowers the melting point of the ice under the wire, causing the ice to liquefy. If the same experiment is tried with a block of dry ice (solid CO_2), however, nothing happens. The dry ice is unaffected because the increased pressure under the wire makes melting more difficult rather than less difficult.

FIGURE 10.30

Pressure and melting point. Why does the weighted wire cut through this block of ice?

Worked Example 10.11

Interpreting a Phase Diagram

Freeze-dried foods are prepared by freezing the food and removing water by subliming the ice at low pressure. Look at the phase diagram of water in Figure 10.28, and tell the maximum pressure in mm Hg at which ice and water vapor are in equilibrium.

STRATEGY

Solid and vapor are in equilibrium only below the triple-point pressure, $P_t = 6.0 \times 10^{-3}$ atm, which needs to be converted to millimeters of mercury.

SOLUTION

$$6.0 \times 10^{-3}\ \text{atm} \times \frac{760\ \text{mm Hg}}{1\ \text{atm}} = 4.6\ \text{mm Hg}$$

PROBLEM 10.17 Look at the phase diagram of CO_2 in Figure 10.29, and tell the minimum pressure in atmospheres at which liquid CO_2 can exist.

PROBLEM 10.18 Look at the phase diagram of CO_2 in Figure 10.29, and describe what happens to a CO_2 sample when the following changes are made:

(a) The temperature is increased from −100 °C to 0 °C at a constant pressure of 2 atm.
(b) The pressure is reduced from 72 atm to 5.0 atm at a constant temperature of 30 °C.
(c) The pressure is first increased from 3.5 atm to 76 atm at −10 °C, and the temperature is then increased from −10 °C to 45 °C.

CONCEPTUAL PROBLEM 10.19 Gallium metal has the following phase diagram (the pressure axis is not to scale). In the region shown, gallium has two different solid phases.

(a) Where on the diagram are the solid, liquid, and vapor regions?
(b) How many triple points does gallium have? Circle each on the diagram.
(c) At 1 atm pressure, which phase is more dense, solid or liquid? Explain.

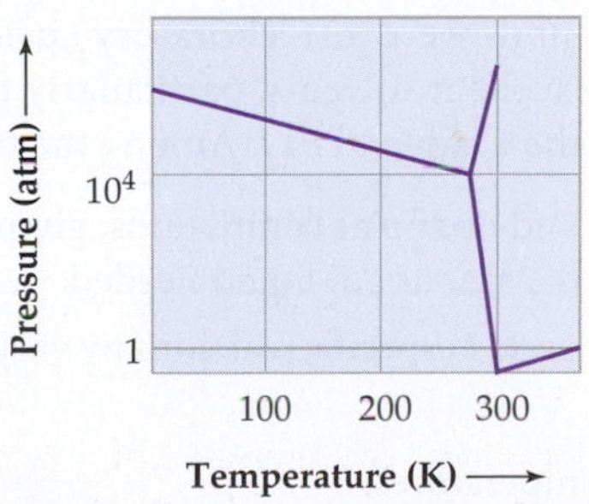

FYI LIQUIDS MADE OF IONS?

▲ Several hundred different ionic liquids, such as these from the German company BASF, are manufactured and sold commercially for use in a wide variety of chemical processes.

When you think of ionic compounds, you probably think of crystalline, high-melting solids: sodium chloride (mp = 801 °C), magnesium oxide (mp = 2825 °C), lithium carbonate (mp = 732 °C), and so on. It's certainly true that many ionic compounds fit that description, but not all. Some ionic compounds are actually liquid at room temperature. Ionic liquids, in fact, have been known for nearly a century—the first such compound to be discovered was ethylammonium nitrate, $CH_3CH_2NH_3{}^+\ NO_3{}^-$, with a melting point of just 12 °C.

Generally speaking, the ionic liquids used today are salts in which the cation has an irregular shape and in which one or both of the ions are large and bulky so that the charges are dispersed over a large volume. Both factors minimize the crystal lattice energy, thereby making the solid less stable and favoring the liquid. Typical cations are derived from nitrogen-containing organic compounds called *amines*, either 1,3-dialkylimidazolium ions or *N*-alkylpyridinium ions.

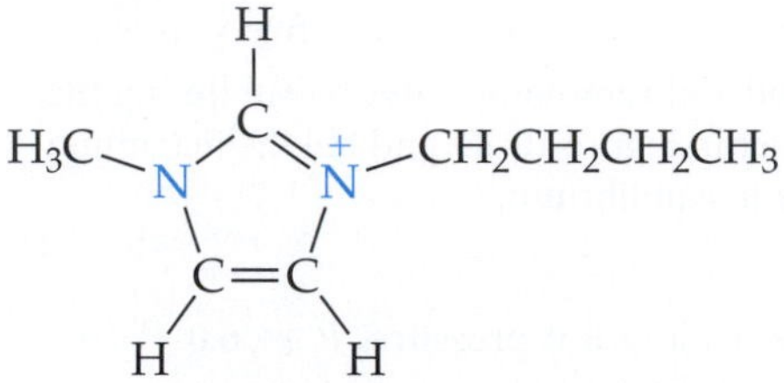

A 1,3-dialkylimidazolium ion

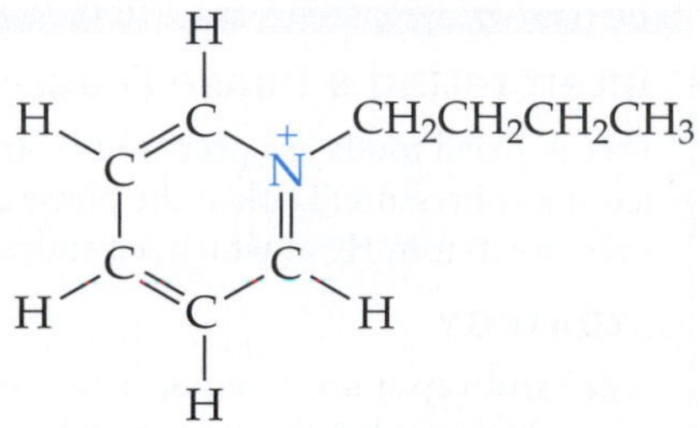

An *N*-alkylpyridinium ion

Anions are just as varied as the cations, and more than 500 different ionic liquids with different anion/cation combinations are commercially available. Hexafluorophosphate, tetrafluoroborate, alkyl sulfates, trifluoromethanesulfonate, and halides are typical anions.

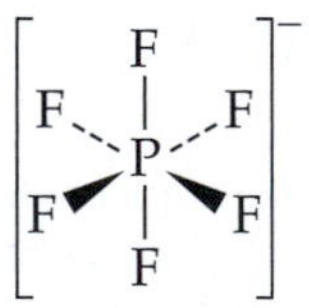

Hexafluoro-
phosphate

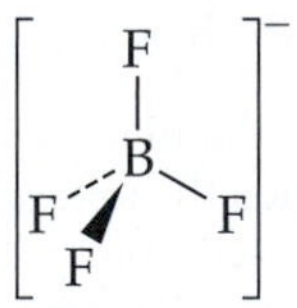

Tetrafluoro-
borate

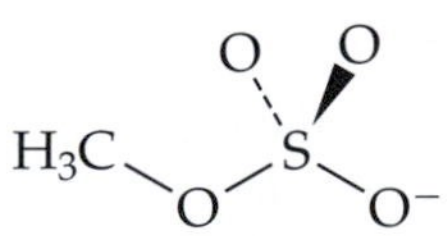

Methyl
sulfate

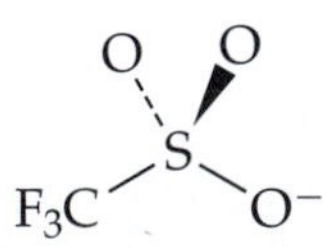

Trifluoromethane-
sulfonate

Cl^-, Br^-, I^-

Halide

For many years, ionic liquids were just laboratory curiosities. More recently, though, they have been found to be excellent solvents, particularly for use in green chemistry processes like those described in the Chapter 8 *FYI*. Among the properties of ionic liquids:

- They dissolve both polar and nonpolar compounds, giving highly concentrated solutions and thereby minimizing the amount of liquid needed.
- They can be fine-tuned for use in specific reactions by varying cation and anion structures.
- They are nonflammable.
- They are stable at high temperatures.
- They have negligible vapor pressures and don't evaporate.
- They are generally recoverable and can be reused many times.

Among their potential applications, ionic liquids are now being explored for use as electrolytes in high-temperature batteries, as solvents for the extraction of heavy organic materials from oil shale, and as replacements for toxic or flammable organic solvents in many industrial processes. We'll be hearing much more about ionic liquids in the coming years.

PROBLEM 10.20 How does an ionic liquid differ from a typical molecular liquid, such as water?

PROBLEM 10.21 What structural features do ionic liquids have that prevent them from forming solids easily?

SUMMARY

The presence of polar covalent bonds in a molecule can cause the molecule to have a net polarity, a property measured by the **dipole moment**.

Intermolecular forces, known collectively as **van der Waals forces**, are the attractions responsible for holding particles together in the liquid and solid phases. There are several kinds of intermolecular forces, all of which arise from electrical attractions. **Dipole–dipole forces** occur between two polar molecules. **London dispersion forces** are characteristic of all molecules and result from the presence of temporary dipole moments caused by momentarily unsymmetrical electron distributions. A **hydrogen bond** is the attraction between a positively polarized hydrogen atom bonded to O, N, or F and a lone pair of electrons on an O, N, or F atom of another molecule. In addition, **ion–dipole forces** occur between an ion and a polar molecule.

Matter in any one **phase**—solid, liquid, or gas—can undergo a **phase change** to either of the other two phases. Like all naturally occurring processes, a phase change has associated with it a free-energy change, $\Delta G = \Delta H - T\Delta S$. The enthalpy component, ΔH, is a measure of the change in intermolecular forces; the entropy component, ΔS, is a measure of the change in molecular randomness accompanying the phase transition. The enthalpy change for the solid–liquid transition is called the **heat of fusion**, and the enthalpy change for the liquid–vapor transition is the **heat of vaporization**.

The effects of temperature and pressure on phase changes can be displayed graphically on a **phase diagram**. A typical phase diagram has three regions—solid, liquid, and gas—separated by three boundary lines that represent pressure/temperature combinations at which two phases are in equilibrium and phase changes occur. At exactly 1 atm pressure, the temperature at the solid/liquid boundary corresponds to the **normal melting point** of the substance, and the temperature at the liquid/gas boundary corresponds to the **normal boiling point**. The three lines meet at the **triple point**, a unique combination of temperature and pressure at which all three phases coexist in equilibrium. The liquid/gas line runs from the triple point to the **critical point**, a pressure/temperature combination beyond which liquid and gas phases become a **supercritical fluid** that is neither a true liquid nor a true gas.

Solids can be characterized as **amorphous** if their particles are randomly arranged or **crystalline** if their particles are ordered. Crystalline solids can be further characterized as **ionic solids** if their particles are ions, **molecular solids** if their particles are molecules, **covalent network solids** if they consist of a covalently bonded array of atoms without discrete molecules, or **metallic solids** if their particles are metal atoms.

The regular three-dimensional network of particles in a crystal is made up of small repeating units called **unit cells**. There are 14 kinds of unit cells, 3 of which have cubic symmetry. **Simple cubic packing** uses a **primitive-cubic** unit cell, with an atom at each corner of the cube. **Body-centered cubic packing** uses a **body-centered cubic** unit cell, with an atom at the center and at each corner of the cube. **Cubic closest-packing** uses a **face-centered cubic** unit cell, with an atom at the center of each face and at each corner of the cube. A fourth kind of packing, called **hexagonal closest-packing**, uses a noncubic unit cell.

KEY WORDS

KEY EQUATIONS

- Clausius–Clapeyron equation **(Section 10.5)**

$$\ln P_{vap} = \left(-\frac{\Delta H_{vap}}{R}\right)\frac{1}{T} + C$$ where ΔH_{vap} is the heat of vaporization, R is the gas constant and C is a constant characteristic of the specific substance.

- Bragg equation **(Section 10.7)**

$$d = \frac{n\lambda}{2 \sin \theta}$$ where d is the distance between layers of atoms, λ is the wavelength of the X rays, and θ is the angle at which X rays strike the atomic layers.

LEARNING OUTCOMES

- Predict whether a given molecule has a dipole moment and its direction. **(Section 10.1)** ***Problems 1–4, 22, 32, 38–45, 108, 113***
- Identify the likely intermolecular forces experienced by a molecule. **(Section 10.2)** ***Problems 5, 6, 33–37, 46, 47, 130***
- Explain the energy changes that occur during a phase change, and use the equilibrium free-energy equation $\Delta G = \Delta H - T\Delta S = 0$ to calculate the temperature at which a phase change takes place. **(Section 10.4)** ***Problems 7, 8, 50–65, 110, 116, 117***
- Use the Clausius–Clapeyron equation to calculate the vapor pressure of a liquid at a given temperature. **(Section 10.5)** ***Problems 9, 10, 31, 66–73, 112, 118–121***
- Use the Bragg equation to determine the spacing between atomic layers in a crystalline solid. **(Section 10.7)** ***Problems 79–81, 114, 115***
- Identify the different kinds of crystalline solids. **(Section 10.6)** ***Problems 74–79, 111***
- Identify the four kinds of spherical packing arrangements in crystalline solids and the three kinds of cubic unit cells. **(Section 10.8)** ***Problems 11, 14–16, 23–25, 28, 82, 83, 124***
- Calculate the density of a substance and the atomic radii of its atoms, given its unit cell dimensions. **(Section 10.8)** ***Problems 12, 13, 84–95, 109, 125–129, 131***
- Use a phase diagram to interpret the behavior of a substance at a given temperature and pressure. **(Section 10.11)** ***Problems 17–19, 26, 29, 96, 97, 103–107, 123***
- Sketch a phase diagram given the appropriate data. **(Section 10.11)** ***Problems 98–102, 122, 132***

CONCEPTUAL PROBLEMS

Problems 10.1–10.21 appear within the chapter.

10.22 Ethyl acetate, $CH_3CO_2CH_2CH_3$, is commonly used as a solvent and nail-polish remover. Look at the following electrostatic potential map of ethyl acetate, and explain the observed polarity.

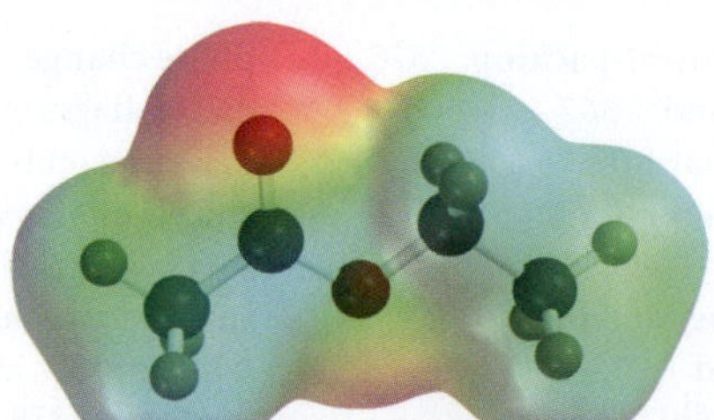

10.23 Identify each of the following kinds of packing:

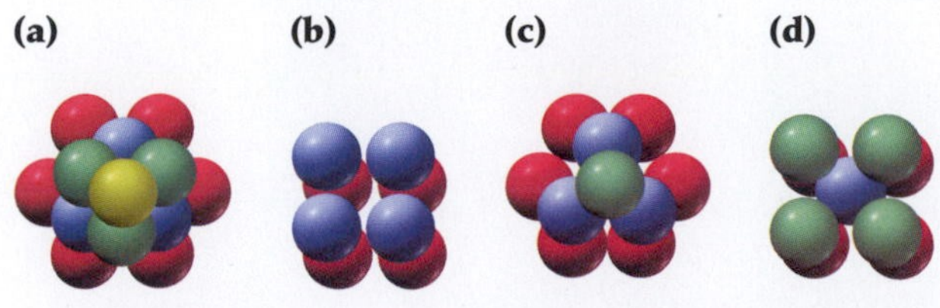

10.24 Zinc sulfide, or sphalerite, crystallizes in the following cubic unit cell:

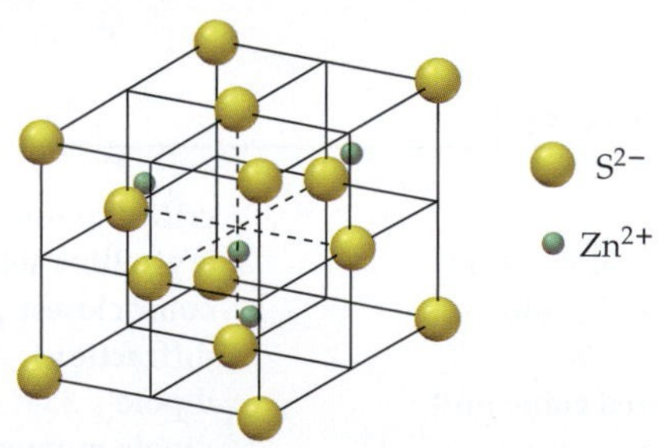

(a) What kind of packing do the sulfide ions adopt?

(b) How many S^{2-} ions and how many Zn^{2+} ions are in the unit cell?

10.25 Perovskite, a mineral containing calcium, oxygen, and titanium, crystallizes in the following cubic unit cell:

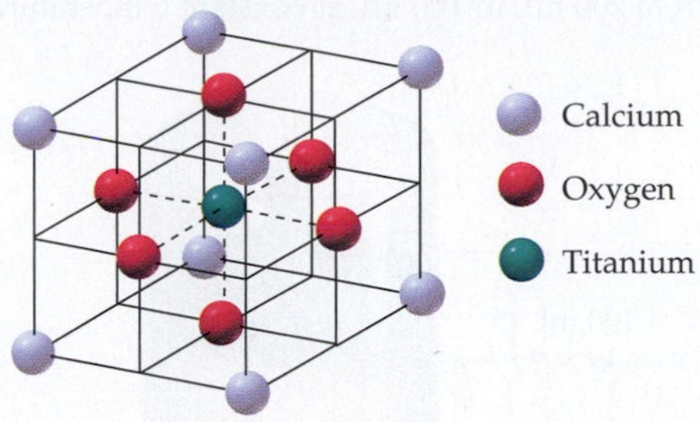

(a) What is the formula of perovskite?

(b) What is the oxidation number of the titanium atom in perovskite?

10.26 The phase diagram of a substance is shown below.

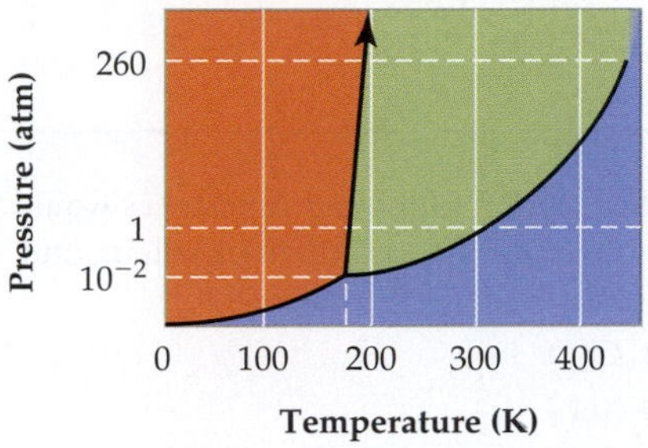

(a) Approximately what is the normal boiling point and what is the normal melting point of the substance?

(b) What is the physical state of the substance under the following conditions?

(i) $T = 150$ K, $P = 0.5$ atm

(ii) $T = 325$ K, $P = 0.9$ atm

(iii) $T = 450$ K, $P = 265$ atm

10.27 Boron nitride, BN, is a covalent network solid with a structure similar to that of graphite. Sketch a small portion of the boron nitride structure.

10.28 Imagine a tiled floor made of square and octagonal tiles in the following pattern. Identify the smallest repeating unit, analogous to a unit cell.

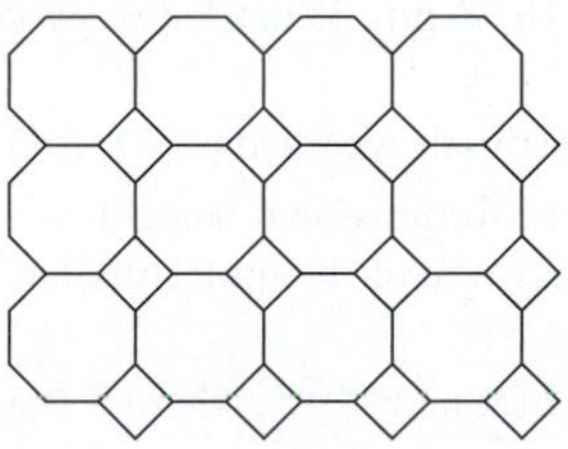

10.29 The following phase diagram of elemental carbon has three different solid phases in the region shown.

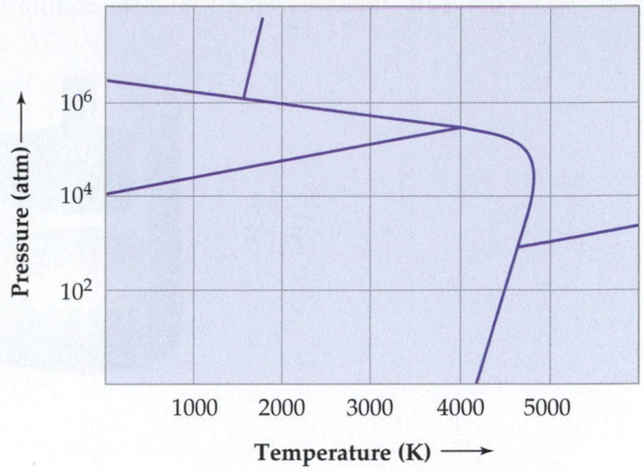

(a) Show where the solid, liquid, and vapor regions are on the diagram.

(b) How many triple points does carbon have? Circle each on the diagram.

(c) Graphite is the most stable solid phase under normal conditions. Identify the graphite phase on the diagram.

(d) On heating graphite to 2500 K at a pressure of 100,000 atm, it can be converted into diamond. Identify the diamond phase on the graph.

(e) Which phase is more dense, graphite or diamond? Explain.

10.30 Acetic acid, the principal nonaqueous constituent of vinegar, exists as a *dimer* in the liquid phase, with two acetic acid molecules joined together by two hydrogen bonds. Sketch the structure you would expect this dimer to have.

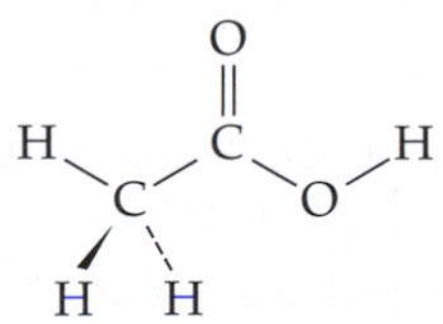

Acetic acid

10.31 Assume that you have a liquid in a cylinder equipped with a movable piston. There is no air in the cylinder, the volume of space above the liquid is 200 mL, and the equilibrium vapor pressure above the liquid is 28.0 mm Hg. What is the equilibrium pressure above the liquid when the volume of space is decreased from 200 mL to 100 mL at constant temperature?

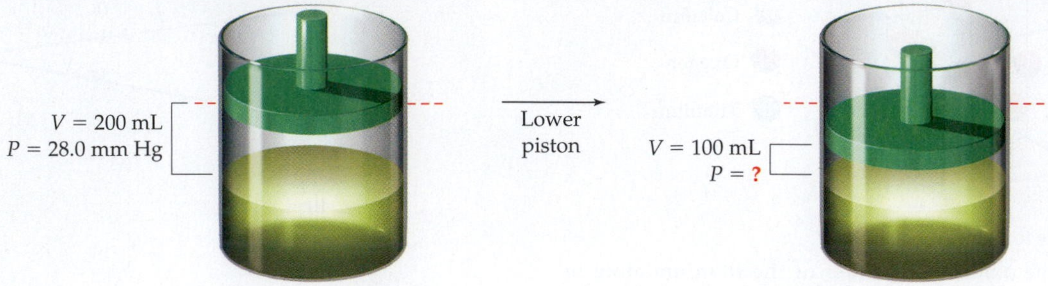

SECTION PROBLEMS

Dipole Moments and Intermolecular Forces (Sections 10.1–10.3)

10.32 Why don't all molecules with polar covalent bonds have dipole moments?

10.33 What is the difference between London dispersion forces and dipole–dipole forces?

10.34 What are the most important kinds of intermolecular forces present in each of the following substances?

(a) Chloroform, $CHCl_3$

(b) Oxygen, O_2

(c) Polyethylene, C_nH_{2n+2}

(d) Methanol, CH_3OH

10.35 Of the substances Xe, CH_3Cl, HF, which has:

(a) The smallest dipole–dipole forces?

(b) The largest hydrogen bond forces?

(c) The largest dispersion forces?

10.36 Methanol (CH_3OH; bp = 65 °C) boils nearly 230 °C higher than methane (CH_4; bp = −164 °C), but 1-decanol ($C_{10}H_{21}OH$; bp = 231 °C) boils only 57 °C higher than decane ($C_{10}H_{22}$; bp = 174 °C). Explain.

Methanol

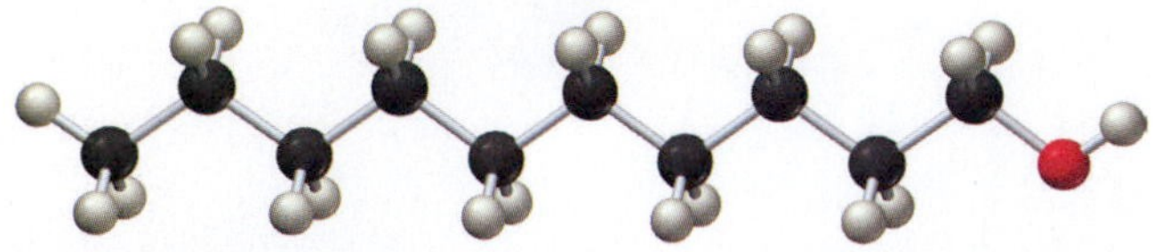

1-Decanol

10.37 Which substance in each of the following pairs would you expect to have larger dispersion forces?

(a) Ethane, C_2H_6, or octane, C_8H_{18}

(b) HCl or HI

(c) H_2O or H_2Se

10.38 Which of the following substances would you expect to have a nonzero dipole moment? Explain, and show the direction of each.

(a) Cl_2O

(b) XeF_4

(c) Chloroethane, CH_3CH_2Cl

(d) BF_3

10.39 Which of the following substances would you expect to have a nonzero dipole moment? Explain, and show the direction of each.

(a) NF_3

(b) CH_3NH_2

(c) XeF_2

(d) PCl_5

10.40 The dipole moment of BrCl is 0.518 D and the distance between atoms is 213.9 pm. What is the percent ionic character of the BrCl bond?

10.41 The dipole moment of ClF is 0.887 D and the distance between atoms is 162.8 pm. What is the percent ionic character of the ClF bond?

10.42 Why is the dipole moment of SO_2 1.63 D, but that of CO_2 is zero?

10.43 Draw three-dimensional structures of PCl_3 and PCl_5, and then explain why one of the molecules has a dipole moment and one does not.

10.44 The class of ions PtX_4^{2-}, where X is a halogen, has a square planar geometry.

(a) Draw a structure for a $PtBr_2Cl_2^{2-}$ ion that has no dipole moment.

(b) Draw a structure for a $PtBr_2Cl_2^{2-}$ ion that has a dipole moment.

10.45 Of the two compounds SiF_4 and SF_4, which is polar and which is nonpolar?

10.46 Draw a picture showing how hydrogen bonding takes place between two ammonia molecules.

10.47 1,3-Propanediol can form *intra*molecular as well as intermolecular hydrogen bonds. Draw a structure of 1,3-propanediol showing an intramolecular hydrogen bond.

1,3-Propanediol

10.48 A magnetized needle gently placed on the surface of a glass of water acts like a makeshift compass. Is it water's viscosity or its surface tension that keeps the needle on top?

10.49 Water flows quickly through the narrow neck of a bottle, but maple syrup flows sluggishly. Is this different behavior due to a difference in viscosity or in surface tension for the liquids?

Vapor Pressure and Phase Changes (Sections 10.4 and 10.5)

10.50 Why is ΔH_{vap} usually larger than ΔH_{fusion}?

10.51 Why is the heat of sublimation, ΔH_{subl}, equal to the sum of ΔH_{vap} and ΔH_{fusion} at the same temperature?

10.52 Mercury has mp $= -38.8$ °C and bp $= 356.6$ °C. What, if any, phase changes take place under the following conditions at 1.0 atm pressure?

(a) The temperature of a sample is raised from -30 °C to 365 °C.

(b) The temperature of a sample is lowered from 291 K to 238 K.

(c) The temperature of a sample is lowered from 638 K to 231 K.

10.53 Iodine has mp $= 113.7$ °C and bp $= 184.4$ °C. What, if any, phase changes take place under the following conditions at 1.0 atm pressure?

(a) The temperature of a solid sample is held at 113.7 °C while heat is added.

(b) The temperature of a sample is lowered from 452 K to 389 K.

10.54 Water at room temperature is placed in a flask connected by rubber tubing to a vacuum pump, and the pump is turned on. After several minutes, the volume of the water has decreased and what remains has turned to ice. Explain.

10.55 Ether at room temperature is placed in a flask connected by a rubber tube to a vacuum pump, the pump is turned on, and the ether begins boiling. Explain.

10.56 How much energy in kilojoules is needed to heat 5.00 g of ice from -11.0 °C to 30.0 °C? The heat of fusion of water is 6.01 kJ/mol, and the molar heat capacity is 36.6 J/(K · mol) for ice and 75.3 J/(K · mol) for liquid water.

10.57 How much energy in kilojoules is released when 15.3 g of steam at 115.0 °C is condensed to give liquid water at 75.0 °C? The heat of vaporization of liquid water is 40.67 kJ/mol, and the molar heat capacity is 75.3 J/(K · mol) for the liquid and 33.6 J/(K · mol) for the vapor.

10.58 How much energy in kilojoules is released when 7.55 g of water at 33.5 °C is cooled to -11.0 °C? (See Problem 10.56 for the necessary data.)

10.59 How much energy in kilojoules is released when 25.0 g of ethanol vapor at 93.0 °C is cooled to -11.0 °C? Ethanol has mp $= -114.1$ °C, bp $= 78.3$ °C, $\Delta H_{vap} = 38.56$ kJ/mol, and $\Delta H_{fusion} = 4.93$ kJ/mol. The molar heat capacity is 112.3 J/(K · mol) for the liquid and 65.6 J/(K · mol) for the vapor.

10.60 Draw a molar heating curve for ethanol, C_2H_5OH, similar to that shown for water in Figure 10.10 on page 368. Begin with solid ethanol at its melting point, and raise the temperature to 100 °C. The necessary data are given in Problem 10.59.

10.61 Draw a molar heating curve for sodium similar to that shown for water in Figure 10.10. Begin with solid sodium at its melting point, and raise the temperature to 1000 °C. The necessary data are mp $= 97.8$ °C, bp $= 883$ °C, $\Delta H_{vap} = 89.6$ kJ/mol, and $\Delta H_{fusion} = 2.64$ kJ/mol. Assume that the molar heat capacity is 20.8 J/(K · mol) for both liquid and vapor phases and does not change with temperature.

10.62 Naphthalene, better known as "mothballs," has bp $= 218$ °C and $\Delta H_{vap} = 43.3$ kJ/mol. What is the entropy of vaporization, ΔS_{vap} in J/(K · mol) for naphthalene?

10.63 What is the entropy of fusion, ΔS_{fusion} in J/(K · mol) for sodium? The necessary data are given in Problem 10.61.

10.64 Carbon disulfide, CS_2, has $P_{vap} = 100$ mm Hg at -5.1 °C and a normal boiling point of 46.5 °C. What is ΔH_{vap} for carbon disulfide in kJ/mol?

Carbon disulfide

10.65 The vapor pressure of $SiCl_4$ is 100 mm Hg at 5.4 °C, and the normal boiling point is 57.7 °C. What is ΔH_{vap} for $SiCl_4$ in kJ/mol?

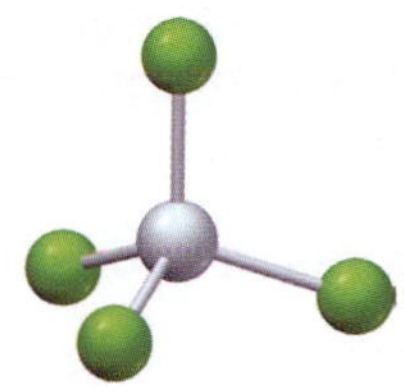
Silicon tetrachloride

10.66 What is the vapor pressure of CS_2 in mm Hg at 20.0 °C? (See Problem 10.64.)

10.67 What is the vapor pressure of $SiCl_4$ in mm Hg at 30.0 °C? (See Problem 10.65.)

10.68 Dichloromethane, CH_2Cl_2, is an organic solvent used for removing caffeine from coffee beans. The following table gives the vapor pressure of dichloromethane at various temperatures. Fill in the rest of the table, and use the data to plot curves of P_{vap} versus T and ln P_{vap} versus $1/T$.

Temp (K)	P_{vap} (mm Hg)	ln P_{vap}	$1/T$
263	80.1	?	?
273	133.6	?	?
283	213.3	?	?
293	329.6	?	?
303	495.4	?	?
313	724.4	?	?

10.69 The following table gives the vapor pressure of mercury at various temperatures. Fill in the rest of the table, and use the data to plot curves of P_{vap} versus T and $\ln P_{vap}$ versus $1/T$.

Temp (K)	P_{vap} (mm Hg)	$\ln P_{vap}$	$1/T$
500	39.3	?	?
520	68.5	?	?
540	114.4	?	?
560	191.6	?	?
580	286.4	?	?
600	432.3	?	?

10.70 Use the plot you made in Problem 10.68 to find a value in kJ/mol for ΔH_{vap} for dichloromethane.

10.71 Use the plot you made in Problem 10.69 to find a value in kJ/mol for ΔH_{vap} for mercury. The normal boiling point of mercury is 630 K.

10.72 Choose any two temperatures and corresponding vapor pressures in the table given in Problem 10.68, and use those values to calculate ΔH_{vap} for dichloromethane in kJ/mol. How does the value you calculated compare to the value you read from your plot in Problem 10.70?

10.73 Choose any two temperatures and corresponding vapor pressures in the table given in Problem 10.69, and use those values to calculate ΔH_{vap} for mercury in kJ/mol. How does the value you calculated compare to the value you read from your plot in Problem 10.71?

Kinds and Structures of Solids (Sections 10.6–10.10)

10.74 List the four main classes of crystalline solids, and give a specific example of each.

10.75 What kinds of particles are present in each of the four main classes of crystalline solids?

10.76 Which of the substances Na_3PO_4, CBr_4, rubber, Au, and quartz best fits each of the following descriptions?

(a) Amorphous solid
(b) Ionic solid
(c) Molecular solid
(d) Covalent network solid
(e) Metallic solid

10.77 Which of the substances diamond, Hg, Cl_2, glass, and KCl best fits each of the following descriptions?

(a) Amorphous solid
(b) Ionic solid
(c) Molecular solid
(d) Covalent network solid
(e) Metallic solid

10.78 Silicon carbide is very hard, has no known melting point, and diffracts X rays. What type of solid is it: amorphous, ionic, molecular, covalent network, or metallic?

10.79 Arsenic tribromide melts at 31.1 °C, diffracts X rays, and does not conduct electricity in either the solid or liquid phase. What type of solid is it: amorphous, ionic, molecular, covalent network, or metallic?

10.80 Diffraction of X rays with $\lambda = 154.2$ pm occurred at an angle $\theta = 22.5°$ from a metal surface. What is the spacing (in pm) between the layers of atoms that diffracted the X rays?

10.81 Diffraction of X rays with $\lambda = 154.2$ pm occurred at an angle $\theta = 76.84°$ from a metal surface. What is the spacing (in pm) between layers of atoms that diffracted the X rays?

10.82 What is a unit cell?

10.83 Which of the four kinds of packing used by metals makes the most efficient use of space, and which makes the least efficient use?

10.84 Copper crystallizes in a face-centered cubic unit cell with an edge length of 362 pm. What is the radius of a copper atom in picometers? What is the density of copper in g/cm^3?

10.85 Lead crystallizes in a face-centered cubic unit cell with an edge length of 495 pm. What is the radius of a lead atom in picometers? What is the density of lead in g/cm^3?

10.86 Aluminum has a density of 2.699 g/cm^3 and crystallizes with a face-centered cubic unit cell. What is the edge length of a unit cell in picometers?

10.87 Tungsten crystallizes in a body-centered cubic unit cell with an edge length of 317 pm. What is the length in picometers of a unit-cell diagonal that passes through the center atom?

10.88 In light of your answer to Problem 10.87, what is the radius in picometers of a tungsten atom?

10.89 Sodium has a density of 0.971 g/cm^3 and crystallizes with a body-centered cubic unit cell. What is the radius of a sodium atom, and what is the edge length of the cell in picometers?

10.90 Titanium metal has a density of 4.506 g/cm^3 and an atomic radius of 144.8 pm. In what cubic unit cell does titanium crystallize?

10.91 Calcium metal has a density of 1.55 g/cm^3 and crystallizes in a cubic unit cell with an edge length of 558.2 pm.

(a) How many Ca atoms are in one unit cell?
(b) In which of the three cubic unit cells does calcium crystallize?

10.92 Sodium hydride, NaH, crystallizes in a face-centered cubic unit cell similar to that of NaCl (Figure 10.24; page 383). How many Na^+ ions touch each H^- ion, and how many H^- ions touch each Na^+ ion?

10.93 Cesium chloride crystallizes in a cubic unit cell with Cl^- ions at the corners and a Cs^+ ion in the center. Count the numbers of + and − charges, and show that the unit cell is electrically neutral.

10.94 If the edge length of an NaH unit cell is 488 pm, what is the length in picometers of an Na–H bond? (See Problem 10.92.)

10.95 The edge length of a CsCl unit cell (Problem 10.93) is 412.3 pm. What is the length in picometers of the Cs–Cl bond? If the ionic radius of a Cl^- ion is 181 pm, what is the ionic radius in picometers of a Cs^+ ion?

Phase Diagrams (Section 10.11)

10.96 Look at the phase diagram of CO_2 in Figure 10.29, and tell what phases are present under the following conditions:

(a) $T = -60\,°C$, $P = 0.75$ atm
(b) $T = -35\,°C$, $P = 18.6$ atm
(c) $T = -80\,°C$, $P = 5.42$ atm

10.97 Look at the phase diagram of H_2O in Figure 10.28, and tell what happens to an H_2O sample when the following changes are made:

(a) The temperature is reduced from 48 °C to −4.4 °C at a constant pressure of 6.5 atm.

(b) The pressure is increased from 85 atm to 226 atm at a constant temperature of 380 °C.

10.98 Bromine has $T_t = -7.3$ °C, $P_t = 44$ mm Hg, $T_c = 315$ °C, and $P_c = 102$ atm. The density of the liquid is 3.1 g/cm^3, and the density of the solid is 3.4 g/cm^3. Sketch a phase diagram for bromine, and label all points of interest.

10.99 Oxygen has $T_t = 54.3$ K, $P_t = 1.14$ mm Hg, $T_c = 154.6$ K, and $P_c = 49.77$ atm. The density of the liquid is 1.14 g/cm^3, and the density of the solid is 1.33 g/cm^3. Sketch a phase diagram for oxygen, and label all points of interest.

10.100 Refer to the bromine phase diagram you sketched in Problem 10.98, and tell what phases are present under the following conditions:

(a) $T = -10$ °C, $P = 0.0075$ atm

(b) $T = 25$ °C, $P = 16$ atm

10.101 Refer to the oxygen phase diagram you sketched in Problem 10.99, and tell what phases are present under the following conditions:

(a) $T = -210$ °C, $P = 1.5$ atm

(b) $T = -100$ °C, $P = 66$ atm

10.102 Does solid oxygen (Problem 10.99) melt when pressure is applied as water does? Explain.

10.103 Assume that you have samples of the following three gases at 25 °C. Which of the three can be liquefied by applying pressure, and which cannot? Explain.

Ammonia: $T_c = 132.5$ °C and $P_c = 112.5$ atm
Methane: $T_c = -82.1$ °C and $P_c = 45.8$ atm
Sulfur dioxide: $T_c = 157.8$ °C and $P_c = 77.7$ atm

10.104 Benzene has a melting point of 5.53 °C and a boiling point of 80.09 °C at atmospheric pressure. Its density is 0.8787 g/cm^3 when liquid and 0.899 g/cm^3 when solid; it has $T_c = 289.01$ °C, $P_c = 48.34$ atm, $T_t = 5.52$ °C, and $P_t = 0.0473$ atm. Starting from a point at 200 K and 66.5 atm, trace the following path on a phase diagram:

(1) First, increase T to 585 K while keeping P constant.
(2) Next, decrease P to 38.5 atm while keeping T constant.
(3) Then, decrease T to 278.66 K while keeping P constant.
(4) Finally, decrease P to 0.0025 atm while keeping T constant.

What is your starting phase, and what is your final phase?

10.105 Refer to the oxygen phase diagram you drew in Problem 10.99, and trace the following path starting from a point at 0.0011 atm and −225 °C:

(1) First, increase P to 35 atm while keeping T constant.
(2) Next, increase T to −150 °C while keeping P constant.
(3) Then, decrease P to 1.0 atm while keeping T constant.
(4) Finally, decrease T to −215 °C while keeping P constant.

What is your starting phase, and what is your final phase?

10.106 How many phase transitions did you pass through in Problem 10.104, and what are they?

10.107 What phase transitions did you pass through in Problem 10.105?

CHAPTER PROBLEMS

10.108 Fluorine is more electronegative than chlorine (Figure 4.4), yet fluoromethane (CH_3F; $\mu = 1.86$ D) has a smaller dipole moment than chloromethane (CH_3Cl; $\mu = 1.90$ D). Explain.

10.109 What is the atomic radius in picometers of an argon atom if solid argon has a density of 1.623 g/cm^3 and crystallizes at low temperature in a face-centered cubic unit cell?

10.110 Mercury has mp = −38.8 °C, a molar heat capacity of 27.9 J/(K · mol) for the liquid and 28.2 J/(K · mol) for the solid, and $\Delta H_{fusion} = 2.33$ kJ/mol. Assuming that the heat capacities don't change with temperature, how much energy in kilojoules is needed to heat 7.50 g of Hg from a temperature of −50.0 °C to +50.0 °C?

10.111 Silicon carbide, SiC, is a covalent network solid with a structure similar to that of diamond. Sketch a small portion of the SiC structure.

10.112 In Denver, the Mile-High City, water boils at 95 °C. What is atmospheric pressure in atmospheres in Denver? ΔH_{vap} for H_2O is 40.67 kJ/mol.

10.113 There are three compounds with the formula $C_2H_2Br_2$. Two of the three have dipole moments, and one does not. Draw the structures of all three compounds, and tell which has no dipole moment.

10.114 If a protein can be induced to crystallize, its molecular structure can be determined by X-ray crystallography. Protein crystals, though solid, contain a large amount of water molecules along with the protein. The protein chicken egg-white lysozyme, for instance, crystallizes with a unit cell having angles of 90° and with edge lengths of 7.9×10^3 pm, 7.9×10^3 pm, and 3.8×10^3 pm. There are eight molecules in the unit cell. If the lysozyme molecule has a molecular weight of 1.44×10^4 and a density of 1.35 g/cm^3, what percent of the unit cell is occupied by the protein?

10.115 The molecular structure of a scorpion toxin, a small protein, was determined by X-ray crystallography. The unit cell has angles of 90°, contains 16 molecules, and has a volume of 1.019×10^2 nm^3. If the molecular weight of the toxin is 3336 and the density is about 1.35 g/cm^3, what percent of the unit cell is occupied by protein?

10.116 Magnesium metal has $\Delta H_{fusion} = 9.037$ kJ/mol and $\Delta S_{fusion} =$ 9.79 J/(K · mol). What is the melting point in °C of magnesium?

10.117 Titanium tetrachloride, $TiCl_4$, has a melting point of −23.2 °C and has $\Delta H_{fusion} = 9.37$ kJ/mol. What is the entropy of fusion, ΔS_{fusion} in J/(K · mol), for $TiCl_4$?

10.118 Dichlorodifluoromethane, CCl_2F_2, one of the chlorofluorocarbon refrigerants responsible for destroying part of the Earth's ozone layer, has $P_{vap} = 40.0$ mm Hg at −81.6 °C and $P_{vap} = 400$ mm Hg at −43.9 °C. What is the normal boiling point of CCl_2F_2 in °C?

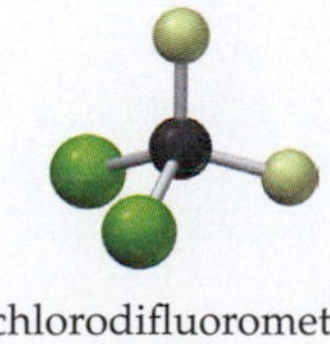

Dichlorodifluoromethane

10.119 The chlorofluorocarbon refrigerant trichlorofluoromethane, CCl_3F, has $P_{vap} = 100.0$ mm Hg at $-23\,°C$ and $\Delta H_{vap} = 24.77$ kJ/mol.

(a) What is the normal boiling point of trichlorofluoromethane in °C?

(b) What is ΔS_{vap} for trichlorofluoromethane?

10.120 Nitrous oxide, N_2O, occasionally used as an anesthetic by dentists under the name "laughing gas," has $P_{vap} = 100$ mm Hg at $-111.3\,°C$ and a normal boiling point of $-88.5\,°C$. What is the heat of vaporization of nitrous oxide in kJ/mol?

10.121 Acetone, a common laboratory solvent, has $\Delta H_{vap} = 29.1$ kJ/mol and a normal boiling point of 56.1 °C. At what temperature in °C does acetone have $P_{vap} = 105$ mm Hg?

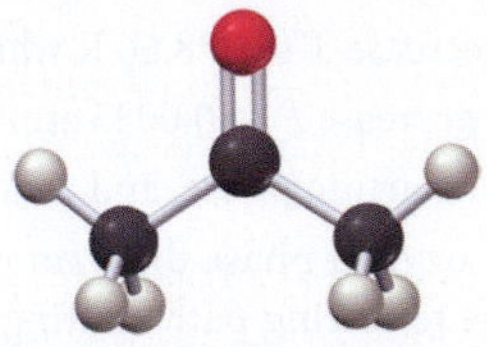

Acetone

10.122 Use the following data to sketch a phase diagram for krypton: $T_t = -169\,°C$, $P_t = 133$ mm Hg, $T_c = -63\,°C$, $P_c = 54$ atm, mp $= -156.6\,°C$, bp $= -152.3\,°C$. The density of solid krypton is 2.8 g/cm^3, and the density of the liquid is 2.4 g/cm^3. Can a sample of gaseous krypton at room temperature be liquefied by raising the pressure?

10.123 What is the physical phase of krypton (Problem 10.122) under the following conditions:

(a) $P = 5.3$ atm, $T = -153\,°C$

(b) $P = 65$ atm, $T = 250\,K$

10.124 Calculate the percent volume occupied by the spheres in a body-centered cubic unit cell.

10.125 Iron crystallizes in a body-centered cubic unit cell with an edge length of 287 pm. What is the radius of an iron atom in picometers?

10.126 Iron metal has a density of 7.86 g/cm^3 and a molar mass of 55.85 g. Use this information together with the data in Problem 10.125 to calculate a value for Avogadro's number.

10.127 Silver metal crystallizes in a face-centered cubic unit cell with an edge length of 408 pm. The molar mass of silver is 107.9 g/mol, and its density is 11.50 g/cm^3. Use these data to calculate a value for Avogadro's number.

10.128 A drawing of the NaCl unit cell is shown in Figure 10.24.

(a) What is the edge length in picometers of the NaCl unit cell? The ionic radius of Na^+ is 97 pm, and the ionic radius of Cl^- is 181 pm.

(b) What is the density of NaCl in g/cm^3?

10.129 Niobium oxide crystallizes in the following cubic unit cell:

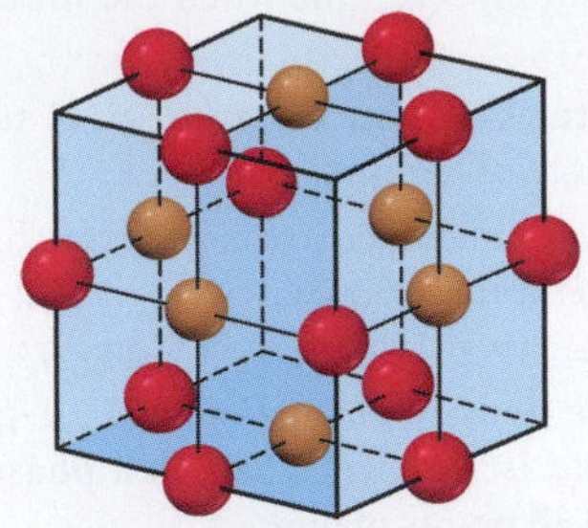

(a) How many niobium atoms and how many oxygen atoms are in each unit cell?

(b) What is the formula of niobium oxide?

(c) What is the oxidation state of niobium?

10.130 For each of the following substances, identify the intermolecular force or forces that predominate. Using your knowledge of the relative strengths of the various forces, rank the substances in order of their normal boiling points.

Al_2O_3, F_2, H_2O, Br_2, ICl, NaCl

10.131 One form of silver telluride (Ag_2Te) crystallizes with a cubic unit cell and a density of 7.70 g/cm^3. X-ray crystallography shows that the edge of the cubic unit cell has a length of 529 pm. How many Ag atoms are in the unit cell?

10.132 Substance **X** has a vapor pressure of 100 mm Hg at its triple point (48 °C). When 1 mol of **X** is heated at 1 atm pressure with a constant rate of heat input, the following heating curve is obtained:

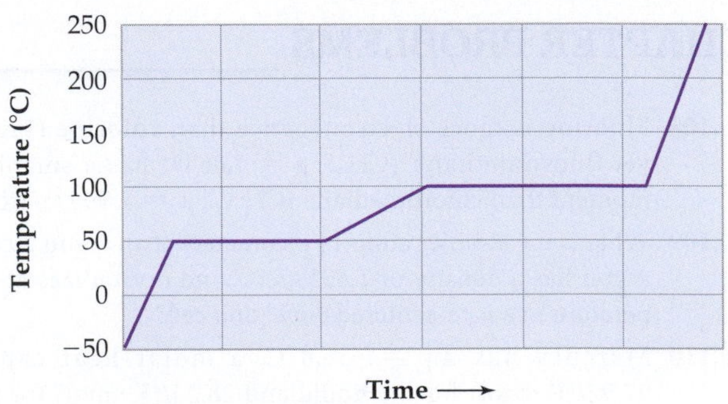

(a) Sketch the phase diagram for **X**, including labels for different phases, triple point, melting point, and boiling point.

(b) For each of the following, choose which phase of **X** (solid, liquid, or gas) fits the description:

(i) Is the most dense at 50 °C

(ii) Is the least dense at 50 °C

(iii) Has the greatest specific heat

(iv) Predominates at 80 °C and 1 atm

(v) Can have a vapor pressure of 20 mm Hg

MULTICONCEPT PROBLEMS

10.133 Look up thermodynamic data for ethanol (C_2H_5OH) in Appendix B, estimate the normal boiling point of ethanol, and calculate the vapor pressure of ethanol at 25 °C.

10.134 The mineral *magnetite* is an iron oxide ore that has a density of 5.20 g/cm^3. At high temperature, magnetite reacts with carbon monoxide to yield iron metal and carbon dioxide. When 2.660 g of magnetite is allowed to react with sufficient carbon monoxide, the CO_2 product is found to have a volume of 1.136 L at 298 K and 751 mm Hg pressure.

(a) What mass of iron in grams is formed in the reaction?

(b) What is the formula of magnetite?

(c) Magnetite has a somewhat complicated cubic unit cell with an edge length of 839 pm. How many Fe and O atoms are present in each unit cell?

10.135 A group 3A metal has a density of 2.70 g/cm^3 and a cubic unit cell with an edge length of 404 pm. Reaction of a 1.07 cm^3 chunk of the metal with an excess of hydrochloric acid gives a colorless gas that occupies 4.00 L at 23.0 °C and a pressure of 740 mm Hg.

(a) Identify the metal.

(b) Is the unit cell primitive, body-centered, or face-centered?

(c) What is the atomic radius of the metal atom in picometers?

10.136 A cube-shaped crystal of an alkali metal, 1.62 mm on an edge, was vaporized in a 500.0 mL evacuated flask. The resulting vapor pressure was 12.5 mm Hg at 802 °C. The structure of the solid metal is known to be body-centered cubic.

(a) What is the atomic radius of the metal atom in picometers?

(b) Use the data in Figure 2.19 on page 88 to identify the alkali metal.

(c) What are the densities of the solid and the vapor in g/cm^3?

10.137 Assume that 1.588 g of an alkali metal undergoes complete reaction with the amount of gaseous halogen contained in a 0.500 L flask at 298 K and 755 mm Hg pressure. In the reaction, 22.83 kJ is released ($\Delta H = -22.83$ kJ). The product, a binary ionic compound, crystallizes in a unit cell with anions in a face-centered cubic arrangement and with cations centered along each edge between anions. In addition, there is a cation in the center of the cube.

(a) What is the identity of the alkali metal?

(b) The edge length of the unit cell is 535 pm. Find the radius of the alkali metal cation from the data in Figure 3.7 on page 109, and then calculate the radius of the halide anion. Identify the anion from the data in Figure 3.8 on page 109.

(c) Sketch a space-filling, head-on view of the unit cell, labeling the ions. Are the anions in contact with one another?

(d) What is the density of the compound in g/cm^3?

(e) What is the standard heat of formation for the compound?

CHAPTER

11 Solutions and Their Properties

Coastal California Redwoods are the tallest trees in the world, with heights up to 379 ft. The rise of sap in these giant trees is due to osmotic pressure, one of the properties of solutions that we'll discuss in this chapter.

Thus far, we've been concerned only with pure substances, both elements and compounds. If you look around, though, most of the substances you see in day-to-day life are *mixtures*. Air is a gaseous mixture of (primarily) oxygen and nitrogen, gasoline is a liquid mixture of many different organic compounds, and rocks are solid mixtures of different minerals.

A **mixture** is any combination of two or more pure substances blended together in some arbitrary proportion without chemically changing the individual substances themselves. Mixtures can be classified as either *heterogeneous* or *homogeneous*, depending on their appearance. A heterogeneous mixture is one in which the mixing of components is visually nonuniform and that therefore has regions of different composition. Sugar with salt and oil with water are examples. A homogeneous mixture is one in which the mixing *is* uniform, at least to the naked eye, and that therefore has a constant composition throughout. Seawater (sodium chloride with water) and brass (copper with zinc) are examples. We'll explore the properties of some homogeneous mixtures in this chapter, with particular emphasis on the mixtures we call *solutions*.

11.1 SOLUTIONS

Homogeneous mixtures can be classified according to the size of their constituent particles as either *solutions* or *colloids*. **Solutions**, the most common class of homogeneous mixtures, contain particles with diameters in the range 0.1–2 nm, the size of a typical ion or small molecule. They are transparent, although they may be colored, and they don't separate on standing. **Colloids**, such as milk and fog, contain larger particles, with diameters in the range 2–500 nm. Although they are often murky or opaque to light, colloids don't separate on standing. Mixtures called *suspensions* also exist, having even larger particles than colloids. These are not truly homogeneous, however, because their particles separate out on standing and are visible with a low-power microscope. Blood, paint, and aerosol sprays are examples.

We usually think of a solution as a solid dissolved in a liquid or as a mixture of liquids, but there are many other kinds of solutions as well. In fact, any one state of matter can form a solution with any other state, and seven different kinds of solutions are possible (**Table 11.1**).

CONTENTS

Even solutions of one solid with another and solutions of a gas in a solid are well known. Metal alloys, such as stainless steel (4–30% chromium in iron) and brass (10–40% zinc in copper), are solid/solid solutions, and hydrogen in palladium is a gas/solid solution. Metallic palladium, in fact, is able to absorb up to 935 times its own volume of H_2 gas!

TABLE 11.1 Some Different Kinds of Solutions

Kind of Solution	Example
Gas in gas	Air (O_2, N_2, Ar, and other gases)
Gas in liquid	Carbonated water (CO_2 in water)
Gas in solid	H_2 in palladium metal
Liquid in liquid	Gasoline (mixture of hydrocarbons)
Liquid in solid	Dental amalgam (mercury in silver)
Solid in liquid	Seawater (NaCl and other salts in water)
Solid in solid	Metal alloys, such as sterling silver (92.5% Ag, 7.5% Cu)

▲ The pewter in this 200-year-old jug is a solid/solid solution of approximately 96% tin and 4% copper.

For solutions in which a gas or solid is dissolved in a liquid, the dissolved substance is called the **solute** and the liquid is called the **solvent**. When one liquid is dissolved in another, the minor component is usually considered the solute and the major component is the solvent. Thus, ethyl alcohol is the solute and water the solvent in a mixture of 10% ethyl alcohol and 90% water, but water is the solute and ethyl alcohol the solvent in a mixture of 90% ethyl alcohol and 10% water.

11.2 ENERGY CHANGES AND THE SOLUTION PROCESS

With the exception of gas/gas mixtures, such as air, the different kinds of solutions listed in Table 11.1 involve *condensed phases*, either liquid or solid. Thus, all the **intermolecular forces** described in Section 10.2 to explain the properties of pure liquids and solids are also important for explaining the properties of solutions. The situation is more complex for solutions than for pure substances, though, because three types of interactions among particles have to be taken into account: solvent–solvent interactions, solvent–solute interactions, and solute–solute interactions.

REMEMBER...

Intermolecular forces include ion–dipole, dipole–dipole, and London dispersion forces, as well as hydrogen bonds. (Section 10.2)

A good rule of thumb, often summarized in the phrase "like dissolves like," is that solutions will form when the three types of interactions are similar in kind and in magnitude. Thus, ionic solids like NaCl dissolve in polar solvents like water because the strong ion–dipole attractions between Na^+ and Cl^- ions and polar H_2O molecules are similar in magnitude to the strong hydrogen bonding attractions between water molecules and to the strong ion–ion attractions between Na^+ and Cl^- ions. In the same way, nonpolar organic substances like cholesterol, $C_{27}H_{46}O$, dissolve in nonpolar organic solvents like benzene, C_6H_6, because of the similar London dispersion forces present among both kinds of molecules. Oil, however, does not dissolve appreciably in water because the two liquids have different kinds of intermolecular forces.

▲ Why don't oil and water mix?

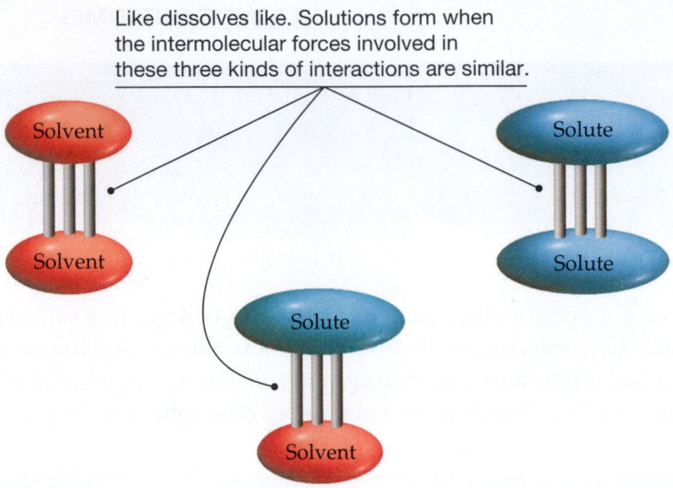

The dissolution of a solid in a liquid can be visualized as shown in **Figure 11.1** for NaCl. When solid NaCl is placed in water, those ions that are less tightly held because of their position at a corner or an edge of the crystal are exposed to water molecules, which collide with them until an ion happens to break free. More water molecules then cluster around the ion, stabilizing it by means of ion–dipole attractions. A new edge or corner

FIGURE 11.1

Dissolution of NaCl crystals in water.

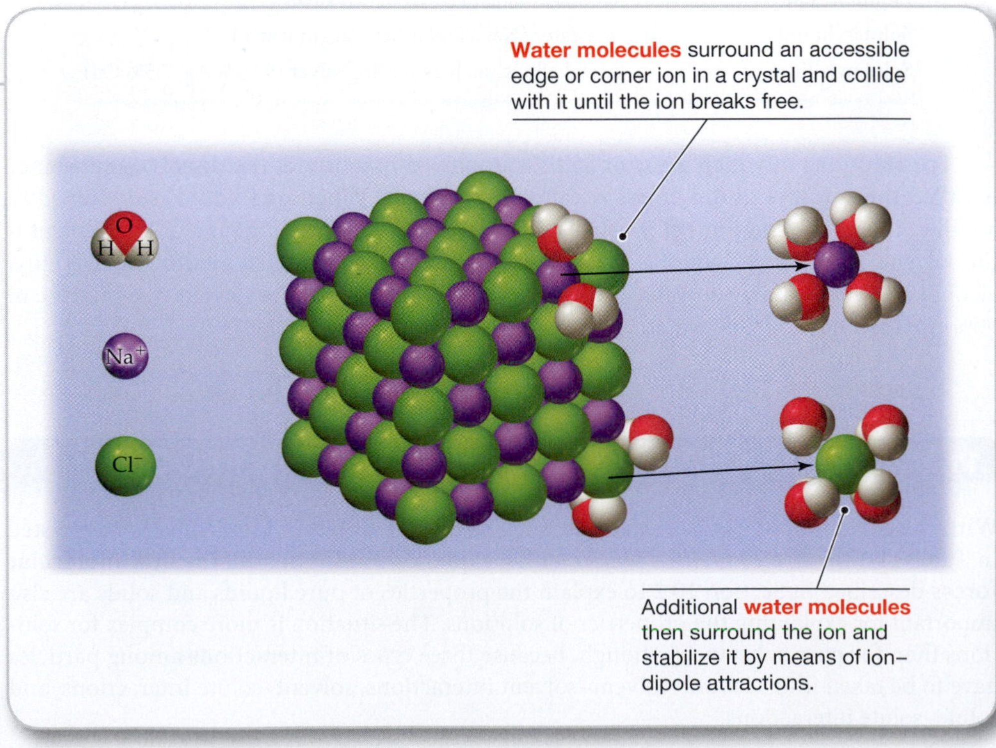

is thereby exposed on the crystal, and the process continues until the entire crystal has dissolved. The ions in solution are said to be *solvated*—more specifically, *hydrated*, when water is the solvent—meaning that they are surrounded and stabilized by an ordered shell of solvent molecules.

Like all chemical and physical processes, the dissolution of a solute in a solvent has associated with it a free-energy change, $\Delta G = \Delta H - T\Delta S$, whose value describes its spontaneity (Section 8.13). If ΔG is negative, the process is spontaneous and the substance dissolves; if ΔG is positive, the process is nonspontaneous and the substance does not dissolve. The enthalpy term ΔH measures the heat flow into or out of the system during dissolution, and the temperature-dependent entropy term $T\Delta S$ measures the change in the amount of molecular randomness in the system. The enthalpy change is called the *heat of solution*, or **enthalpy of solution (ΔH_{soln})**, and the entropy change is called the **entropy of solution (ΔS_{soln})**.

What values might we expect for ΔH_{soln} and ΔS_{soln}? Let's take the entropy change first. Entropies of solution are usually positive because molecular randomness usually increases during dissolution: +43.4 J/(K · mol) for NaCl in water, for example. When a solid dissolves in a liquid, randomness increases upon going from a well-ordered crystal to a less-ordered state in which solvated ions or molecules are able to move freely in solution. When one liquid dissolves in another, randomness increases as the different molecules intermingle (**Figure 11.2**). **Table 11.2** lists values of ΔS_{soln} for some common ionic substances.

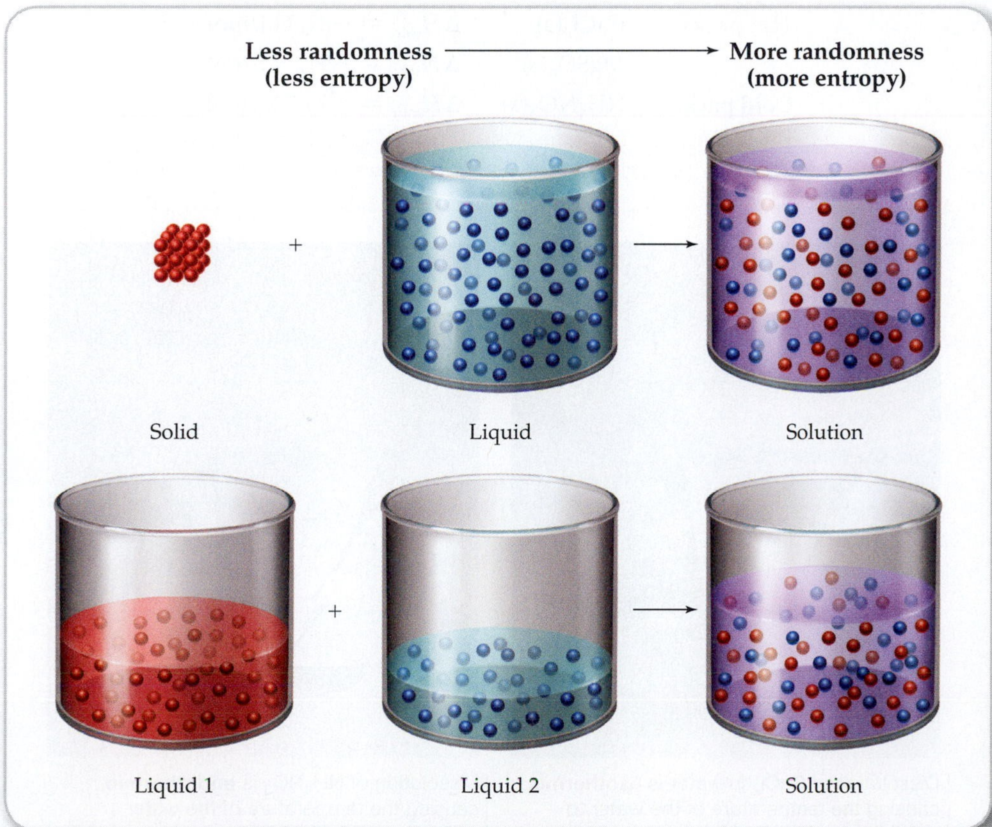

FIGURE 11.2

Entropy of solution. Entropies of solution are usually positive because molecular randomness usually increases when a solid dissolves in a liquid or one liquid dissolves in another.

TABLE 11.2 Some Enthalpies and Entropies of Solution in Water at 25 °C

Substance	ΔH_{soln} (kJ/mol)	ΔS_{soln} [J/(K · mol)]
LiCl	−37.0	10.5
NaCl	3.9	43.4
KCl	17.2	75.0
LiBr	−48.8	21.5
NaBr	−0.6	54.6
KBr	19.9	89.0
KOH	−57.6	12.9

Values for enthalpies of solution, ΔH_{soln}, are more difficult to predict (Table 11.2). Some solids dissolve exothermically and have a negative ΔH_{soln} (−37.0 kJ/mol for LiCl in water), but others dissolve endothermically and have a positive ΔH_{soln} (+17.2 kJ/mol for KCl in water). Athletes benefit from both situations when they use instant hot packs or cold packs to treat injuries. Both kinds of instant packs consist of a pouch of water and a dry chemical, either $CaCl_2$ or $MgSO_4$ for hot packs, and NH_4NO_3 for cold packs. When the pack is squeezed, the pouch breaks and the solid dissolves, either raising or lowering the temperature (**Figure 11.3**).

Hot packs:	$CaCl_2(s)$	$\Delta H_{soln} = -81.3$ kJ/mol
	$MgSO_4(s)$	$\Delta H_{soln} = -91.2$ kJ/mol
Cold pack:	$NH_4NO_3(s)$	$\Delta H_{soln} = +25.7$ kJ/mol

FIGURE 11.3

Enthalpy of solution. Enthalpies of solution can be either negative (exothermic) or positive (endothermic).

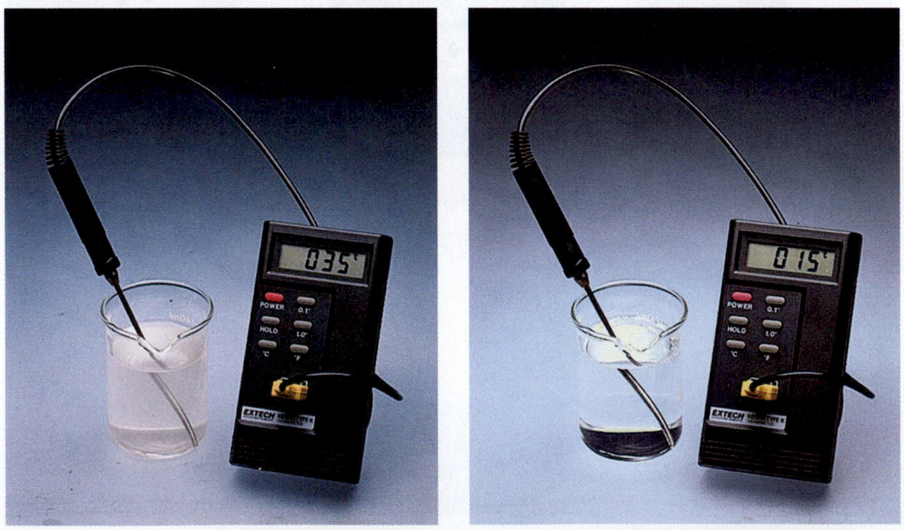

Dissolution of $CaCl_2$ in water is **exothermic**, causing the temperature of the water to rise from its initial 25 °C value.

Dissolution of NH_4NO_3 is **endothermic**, causing the temperature of the water to fall from its initial 25 °C value.

The exact value of ΔH_{soln} for a given substance results from an interplay of the three kinds of interactions mentioned at the beginning of this section:

- **Solvent–solvent interactions**: Energy is absorbed (positive ΔH) to overcome intermolecular forces between solvent molecules because the molecules must be separated and pushed apart to make room for solute particles.
- **Solute–solute interactions**: Energy is absorbed (positive ΔH) to overcome intermolecular forces holding solute particles together in a crystal. For an ionic solid, this is the **lattice energy**. As a result, substances with higher lattice energies tend to be less soluble than substances with lower lattice energies. Compounds with singly charged ions are thus more soluble than compounds with doubly or triply charged ions, as we saw when discussing solubility guidelines in Section 7.4.
- **Solvent–solute interactions**: Energy is released (negative ΔH) when solvent molecules cluster around solute particles and solvate them. For ionic substances in water, the amount of hydration energy released is generally greater for smaller cations than for larger ones because water molecules can approach the positive nuclei of smaller ions more closely and thus bind more tightly. In addition, hydration energy generally increases as the charge on the ion increases.

REMEMBER...

Lattice energy (*U*) is the sum of the electrostatic interaction energies between ions in a crystal. It is the amount of energy that must be supplied to break up an ionic solid into its individual gaseous ions. (Section 3.10)

The first two kinds of interactions are endothermic, requiring an input of energy to spread apart solvent molecules and to break apart crystals. Only the third interaction is exothermic, as attractive intermolecular forces develop between solvent and solute particles. The sum of the three interactions determines whether ΔH_{soln} is endothermic or exothermic. For some substances, the one exothermic interaction is sufficiently large to outweigh the two endothermic interactions, but for other substances, the opposite is true (**Figure 11.4**).

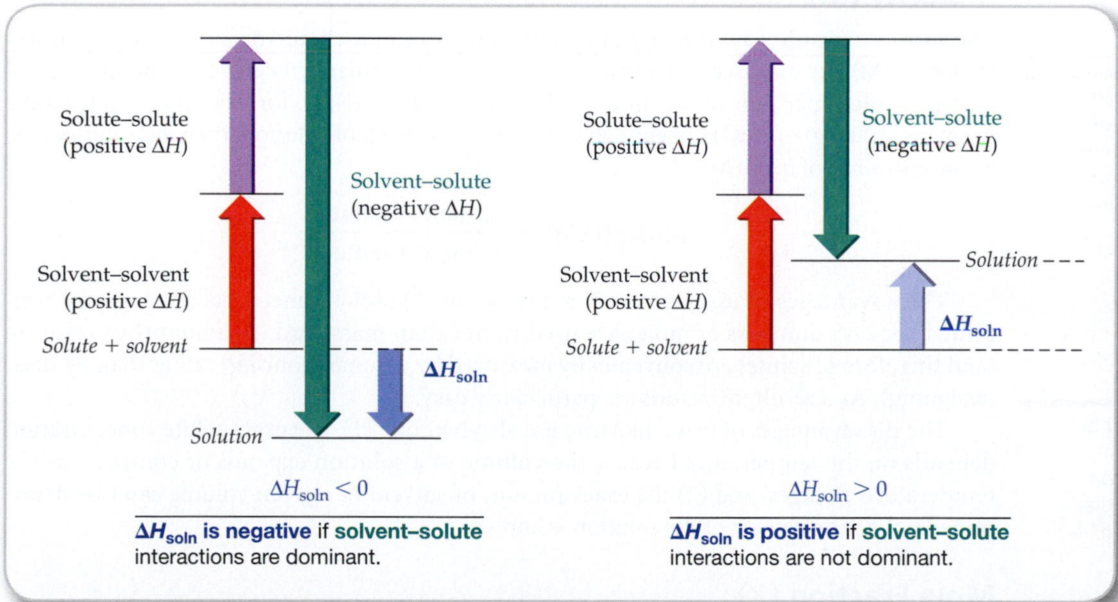

FIGURE 11.4

Components of ΔH_{soln}. The value of ΔH_{soln} is the sum of three terms: solvent–solvent, solute–solute, and solvent–solute.

Worked Example 11.1

Correlating Solubility and Chemical Structure

Pentane (C_5H_{12}) and 1-butanol (C_4H_9OH) are organic liquids with similar molecular weights but substantially different solubility behavior. Which of the two would you expect to be more soluble in water? Explain.

STRATEGY

Look at the two structures and decide on the kinds of intermolecular forces present between molecules in each case. The substance with intermolecular forces more like those in water will probably be more soluble in water.

continued on next page

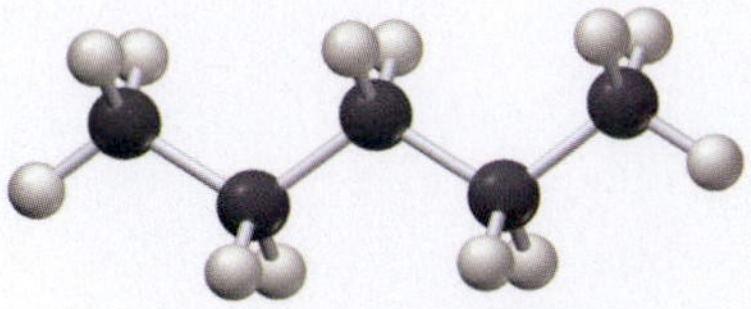

Pentane

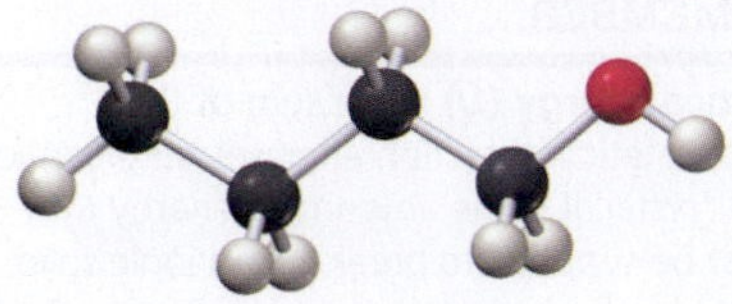

1-Butanol

SOLUTION

Pentane is a nonpolar molecule and is unlikely to have strong intermolecular interactions with polar water molecules. 1-Butanol, however, has an $-OH$ part just as water does and can form hydrogen bonds with water. As a result, 1-butanol is more soluble in water.

PROBLEM 11.1 Arrange the following compounds in order of their expected increasing solubility in water: Br_2, KBr, toluene (C_7H_8, a constituent of gasoline).

PROBLEM 11.2 Which would you expect to have the larger (more negative) hydration energy?

(a) Na^+ or Cs^+ (b) K^+ or Ba^{2+}

11.3 CONCENTRATION UNITS FOR SOLUTIONS

In daily life, it's often sufficient to describe a solution as either *dilute* or *concentrated*. In scientific work, though, it's usually necessary to know the exact concentration of a solution—that is, to know the exact amount of solute dissolved in a certain amount of solvent. There are many ways of expressing concentration, each of which has its own advantages and disadvantages. We'll look briefly at four of the most common methods: *molarity*, *mole fraction*, *mass percent*, and *molality*.

Molarity (M)

REMEMBER...

A solution of a given **molarity** is prepared by dissolving a solute in a small amount of solvent and then diluting with more solvent to the desired volume. The solution is not made by dissolving the solute in the desired volume of solvent. (Section 6.8)

The most common way of expressing concentration in a chemistry laboratory is to use **molarity (M)**. As discussed in Section 6.8, a solution's molarity is given by the number of moles of solute per liter of solution (mol/L, abbreviated M). If, for example, you dissolve 0.500 mol (20.0 g) of NaOH in enough water to give 1.000 L of solution, then the solution has a concentration of 0.500 M.

$$\text{Molarity (M)} = \frac{\text{Moles of solute}}{\text{Liters of solution}}$$

The advantages of using molarity are twofold: (1) Stoichiometry calculations are simplified because numbers of moles are used rather than mass, and (2) amounts of solution (and therefore of solute) are conveniently measured by volume (pouring) rather than by mass (weighing). As a result, **titrations** are particularly easy.

REMEMBER...

Titration is a procedure for determining the concentration of a solution by allowing a carefully measured amount of the solution to react with a standard solution of another substance whose concentration is known. (Section 6.11)

The disadvantages of using molarity are also twofold: (1) The exact solute concentration depends on the temperature because the volume of a solution expands or contracts as the temperature changes, and (2) the exact amount of solvent in a given volume can't be determined unless the density of the solution is known.

Mole Fraction (*X*)

As discussed in Section 9.5, the mole fraction (*X*) of any component in a solution is given by the number of moles of the component divided by the total number of moles making up the solution (including solvent):

$$\text{Mole fraction } (X) = \frac{\text{Moles of component}}{\text{Total moles making up the solution}}$$

For example, a solution prepared by dissolving 1.00 mol (32.0 g) of methyl alcohol (CH_3OH) in 5.00 mol (90.0 g) of water has a methyl alcohol concentration $X = 1.00\text{ mol}/(1.00\text{ mol} + 5.00\text{ mol}) = 0.167$. Note that mole fractions are dimensionless because the units cancel.

Mole fractions are independent of temperature and are particularly useful for calculations involving gas mixtures. Except in special situations, mole fractions are not often used for liquid solutions because other units are generally more convenient.

Mass Percent (Mass %)

As the name suggests, the **mass percent** of any component in a solution is the mass of that component divided by the total mass of the solution times 100%:

$$\text{Mass percent} = \frac{\text{Mass of component}}{\text{Total mass of solution}} \times 100\%$$

For example, a solution prepared by dissolving 10.0 g of glucose in 100.0 g of water has a glucose concentration of 9.09 mass %:

$$\text{Mass \% glucose} = \frac{10.0\text{ g}}{10.0\text{ g} + 100.0\text{ g}} \times 100\% = 9.09\text{ mass \%}$$

Closely related to mass percent, and particularly useful for very dilute solutions, are the concentration units **parts per million (ppm)** and **parts per billion (ppb)**:

$$\text{Parts per million (ppm)} = \frac{\text{Mass of component}}{\text{Total mass of solution}} \times 10^6$$

$$\text{Parts per billion (ppb)} = \frac{\text{Mass of component}}{\text{Total mass of solution}} \times 10^9$$

A concentration of 1 ppm for a substance means that each kilogram (1 million mg) of solution contains 1 mg of solute. For dilute aqueous solutions near room temperature, where 1 kg has a volume of 1 L, 1 ppm also means that each liter of solution contains 1 mg of solute. In the same way, a concentration of 1 ppb means that each liter of an aqueous solution contains 0.001 mg of solute.

Values in ppm and ppb are frequently used for expressing the concentrations of trace amounts of impurities in air or water. Thus, you might express the maximum allowable concentration of lead in drinking water as 15 ppb, or about 1 g per 67,000 L.

The advantage of using mass percent (or ppm) for expressing concentration is that the values are independent of temperature because masses don't change when substances are heated or cooled. The disadvantage of using mass percent is that it is generally less convenient when working with liquid solutions to measure amounts by mass rather than by volume. Furthermore, the density of a solution must be known before a concentration in mass percent can be converted into molarity. Worked Example 11.3 shows how to make the conversion.

▲ Dissolving a quarter teaspoon of table sugar in this backyard swimming pool would give a sugar concentration of about 1 ppb.

Worked Example 11.2

Using a Mass Percent Concentration

Assume that you have a 5.75 mass % solution of LiCl in water. What mass of solution in grams contains 1.60 g of LiCl?

STRATEGY

Describing a concentration as 5.75 mass % means that 100.0 g of aqueous solution contains 5.75 g of LiCl (and 94.25 g of H_2O), a relationship that can be used as a conversion factor.

SOLUTION

$$\text{Mass of soln} = 1.60\text{ g LiCl} \times \frac{100\text{ g soln}}{5.75\text{ g LiCl}} = 27.8\text{ g soln}$$

Worked Example 11.3

Using Density to Convert Mass Percent Concentration to Molarity

The density of a 25.0 mass % solution of sulfuric acid (H_2SO_4) in water is 1.1783 g/mL at 25.0 °C. What is the molarity of the solution?

STRATEGY

Describing a solution as 25.0 mass % sulfuric acid in water means that 100.0 g of solution contains 25.0 g of H_2SO_4 and 75.0 g of water. Since we want to calculate the concentration in molarity, we first need to find the number of moles of sulfuric acid dissolved in a specific mass of solution. We next use density as a conversion factor to find the volume of that solution and then calculate molarity by dividing the number of moles by the volume.

SOLUTION

First, convert the 25.0 g of H_2SO_4 into moles:

$$\frac{\text{Moles } H_2SO_4}{100.0 \text{ g solution}} = \frac{25.0 \text{ g } H_2SO_4}{100.0 \text{ g solution}} \times \frac{1 \text{ mol } H_2SO_4}{98.1 \text{ g } H_2SO_4} = \frac{0.255 \text{ mol } H_2SO_4}{100.0 \text{ g solution}}$$

Next, find the volume of 100.0 g of solution, using density as the conversion factor:

$$\text{Volume} = 100.0 \text{ g soln} \times \frac{1 \text{ mL}}{1.1783 \text{ g soln}} = 84.87 \text{ mL} = 0.084\,87 \text{ L}$$

Then, calculate the molarity of the solution:

$$\text{Molarity} = \frac{\text{moles } H_2SO_4}{\text{liters of solution}} = \frac{0.255 \text{ mol } H_2SO_4}{0.084\,87 \text{ L}} = 3.00 \text{ M}$$

The molarity of the 25.0 mass % sulfuric acid solution is 3.00 M.

▶ **PROBLEM 11.3** What is the mass percent concentration of a saline solution prepared by dissolving 1.00 mol of NaCl in 1.00 L of water?

▶ **PROBLEM 11.4** The legal limit for human exposure to carbon monoxide in the workplace is 35 ppm. Assuming that the density of air is 1.3 g/L, how many grams of carbon monoxide are in 1.0 L of air at the maximum allowable concentration?

▶ **PROBLEM 11.5** Assuming that seawater is an aqueous solution of NaCl, what is its molarity? The density of seawater is 1.025 g/mL at 20 °C, and the NaCl concentration is 3.50 mass %.

Molality (*m*)

The **molality** (*m*) of a solution is defined as the number of moles of solute per kilogram of solvent (mol/kg):

$$\text{Molality } (m) = \frac{\text{Moles of solute}}{\text{Mass of solvent (kg)}}$$

To prepare a 1.000 *m* solution of KBr in water, for example, you might dissolve 1.000 mol of KBr (119.0 g) in 1.000 kg (1000 mL) of water. You can't say for sure what the final volume of the solution will be, although it will probably be a bit larger than 1000 mL. Although the names sound similar, note the differences between molarity and molality. Molarity is the number of moles of solute per *volume* (liter) of *solution*, whereas molality is the number of moles of solute per *mass* (kilogram) of *solvent*.

The main advantage of using molality is that it is temperature-independent because masses don't change when substances are heated or cooled. Thus, it is well suited for calculating certain properties of solutions that we'll discuss later in this chapter. The disadvantages of using molality are that amounts of solution must be measured by mass rather than by volume and that the density of the solution must be known to convert molality into molarity (see Worked Example 11.5).

A summary of the four methods of expressing concentration, together with a comparison of their relative advantages and disadvantages, is given in **Table 11.3**.

TABLE 11.3 A Comparison of Various Concentration Units

Name	Units	Advantages	Disadvantages
Molarity (M)	$\frac{\text{mol solute}}{\text{L solution}}$	Useful in stoichiometry; by volume	Temperature-dependent; must know density to find solvent mass
Mole fraction (X)	none	Temperature-independent; useful in special applications	Measure by mass; must know density to convert to molarity
Mass %	%	Temperature-independent; useful for small amounts	Measure by mass; must know density to convert to molarity
Molality (m)	$\frac{\text{mol solute}}{\text{kg solvent}}$	Temperature-independent; useful in special applications	Measure by mass; must know density to convert to molarity

Worked Example 11.4

Calculating the Molality of a Solution

What is the molality of a solution made by dissolving 1.45 g of table sugar (sucrose, $C_{12}H_{22}O_{11}$) in 30.0 mL of water? The molar mass of sucrose is 342.3 g/mol.

STRATEGY

Molality is the number of moles of solute per kilogram of solvent. Thus, we need to find how many moles are in 1.45 g of sucrose and how many kilograms are in 30.0 mL of water.

SOLUTION

The number of moles of sucrose is

$$1.45 \text{ g sucrose} \times \frac{1 \text{ mol sucrose}}{342.3 \text{ g sucrose}} = 4.24 \times 10^{-3} \text{ mol sucrose}$$

Since the density of water is 1.00 g/mL, 30.0 mL of water has a mass of 30.0 g, or 0.0300 kg. Thus, the molality of the solution is

$$\text{Molality} = \frac{4.24 \times 10^{-3} \text{ mol}}{0.0300 \text{ kg}} = 0.141\ m$$

Worked Example 11.5

Using Density to Convert from Molality to Molarity

Ethylene glycol, $C_2H_4(OH)_2$, is a colorless liquid used as automobile antifreeze. If the density at 20 °C of a 4.028 m solution of ethylene glycol in water is 1.0241 g/mL, what is the molarity of the solution? The molar mass of ethylene glycol is 62.07 g/mol.

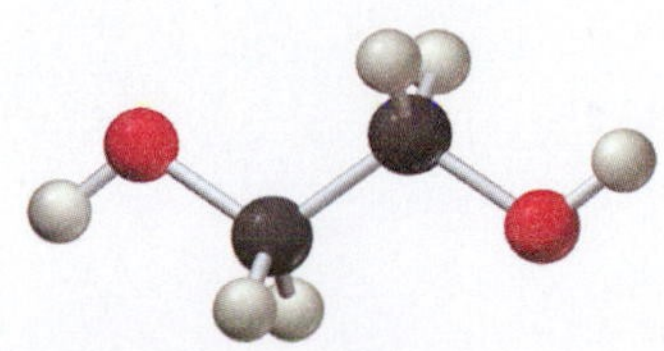
Ethylene glycol

continued on next page

STRATEGY

A 4.028 *m* solution of ethylene glycol in water contains 4.028 mol of ethylene glycol per kilogram of water. To find the solution's molarity, we need to find the number of moles of solute per volume (liter) of solution. The volume, in turn, can be found from the mass of the solution by using density as a conversion factor.

SOLUTION

The mass of the solution is the sum of the masses of solute and solvent. Assuming that 1.000 kg of solvent is used to dissolve 4.028 mol of ethylene glycol, the mass of the ethylene glycol is

$$\text{Mass of ethylene glycol} = 4.028\text{ mol} \times 62.07\,\frac{\text{g}}{\text{mol}} = 250.0\text{ g}$$

Dissolving this 250.0 g of ethylene glycol in 1.000 kg (or 1000 g) of water gives the total mass of the solution:

$$\text{Mass of solution} = 250.0\text{ g} + 1000\text{ g} = 1250\text{ g}$$

The volume of the solution is obtained from its mass by using density as a conversion factor:

$$\text{Volume of solution} = 1250\text{ g} \times \frac{1\text{ mL}}{1.0241\text{ g}} = 1221\text{ mL} = 1.221\text{ L}$$

The molarity of the solution is the number of moles of solute divided by the volume of solution:

$$\text{Molarity of solution} = \frac{4.028\text{ mol}}{1.221\text{ L}} = 3.299\text{ M}$$

Worked Example 11.6

Using Density to Convert Molarity to Other Measures of Concentration

A 0.750 M solution of H_2SO_4 in water has a density of 1.046 g/mL at 20 °C. What is the concentration of this solution in **(a)** mole fraction, **(b)** mass percent, and **(c)** molality? The molar mass of H_2SO_4 is 98.1 g/mol.

STRATEGY AND SOLUTION

(a) Let's pick an arbitrary amount of the solution that will make the calculations easier, say 1.00 L. Since the concentration of the solution is 0.750 mol/L and the density is 1.046 g/mL (or 1.046 kg/L), 1.00 L of the solution contains 0.750 mol (73.6 g) of H_2SO_4 and has a mass of 1.046 kg:

$$\text{Moles of } H_2SO_4 \text{ in 1.00 L soln} = 0.750\,\frac{\text{mol}}{\text{L}} \times 1.00\text{ L} = 0.750\text{ mol}$$

$$\text{Mass of } H_2SO_4 \text{ in 1.00 L soln} = 0.750\text{ mol} \times 98.1\,\frac{\text{g}}{\text{mol}} = 73.6\text{ g}$$

$$\text{Mass of 1.00 L soln} = 1.00\text{ L} \times 1.046\,\frac{\text{kg}}{\text{L}} = 1.046\text{ kg}$$

Subtracting the mass of H_2SO_4 from the total mass of the solution gives 0.972 kg, or 54.0 mol, of water in 1.00 L of solution:

$$\text{Mass of } H_2O \text{ in 1.00 L soln} = (1.046\text{ kg}) - (0.0736\text{ kg}) = 0.972\text{ kg } H_2O$$

$$\text{Moles of } H_2O \text{ in 1.00 L of soln} = 972\text{ g} \times \frac{1\text{ mol}}{18.0\text{ g}} = 54.0\text{ mol } H_2O$$

Thus, the mole fraction of H_2SO_4 is

$$X_{H_2SO_4} = \frac{0.750 \text{ mol } H_2SO_4}{0.750 \text{ mol } H_2SO_4 + 54.0 \text{ mol } H_2O} = 0.0137$$

(b) The mass percent concentration can be determined from the calculations in part **(a)**:

$$\text{Mass \% of } H_2SO_4 = \frac{0.0736 \text{ kg } H_2SO_4}{1.046 \text{ kg total}} \times 100\% = 7.04\%$$

(c) The molality of the solution can also be determined from the calculations in part **(a)**. Since 0.972 kg of water has 0.750 mol of H_2SO_4 dissolved in it, 1.00 kg of water would have 0.772 mol of H_2SO_4 dissolved in it:

$$1.00 \text{ kg } H_2O \times \frac{0.750 \text{ mol } H_2SO_4}{0.972 \text{ kg } H_2O} = 0.772 \text{ mol } H_2SO_4$$

Thus, the molality of the sulfuric acid solution is 0.772 *m*.

PROBLEM 11.6 What is the molality of a solution prepared by dissolving 0.385 g of cholesterol, $C_{27}H_{46}O$, in 40.0 g of chloroform, $CHCl_3$? What is the mole fraction of cholesterol in the solution?

PROBLEM 11.7 What mass in grams of a 0.500 *m* solution of sodium acetate, CH_3CO_2Na, in water would you use to obtain 0.150 mol of sodium acetate?

PROBLEM 11.8 The density at 20 °C of a 0.258 *m* solution of glucose in water is 1.0173 g/mL, and the molar mass of glucose is 180.2 g/mol. What is the molarity of the solution?

PROBLEM 11.9 The density at 20 °C of a 0.500 M solution of acetic acid in water is 1.0042 g/mL. What is the molality of the solution? The molar mass of acetic acid, CH_3CO_2H, is 60.05 g/mol.

PROBLEM 11.10 Assuming that seawater is a 3.50 mass % aqueous solution of NaCl, what is the molality of seawater?

11.4 SOME FACTORS THAT AFFECT SOLUBILITY

If you take solid NaCl and add it to water, dissolution occurs rapidly at first but then slows down as more and more NaCl is added. Eventually the dissolution stops because a dynamic equilibrium is reached where the number of Na^+ and Cl^- ions leaving a crystal to go into solution is equal to the number of ions returning from the solution to the crystal. At this point, the maximum possible amount of NaCl has dissolved and the solution is said to be **saturated** in that solute.

$$\text{Solute} + \text{Solvent} \underset{\text{crystallize}}{\overset{\text{dissolve}}{\rightleftharpoons}} \text{Solution}$$

Note that this definition requires a saturated solution to be at *equilibrium* with undissolved solid at a given temperature. Some substances, however, can form what are called **supersaturated** solutions, which contain a greater-than-equilibrium amount of solute. For example, when a saturated solution of sodium acetate is prepared at high temperature and then cooled slowly, a supersaturated solution results, as shown in **Figure 11.5**. Such a solution is unstable, however, and precipitation occurs when a tiny seed crystal of sodium acetate is added to initiate crystallization.

FIGURE 11.5

Precipitation from a supersaturated solution.

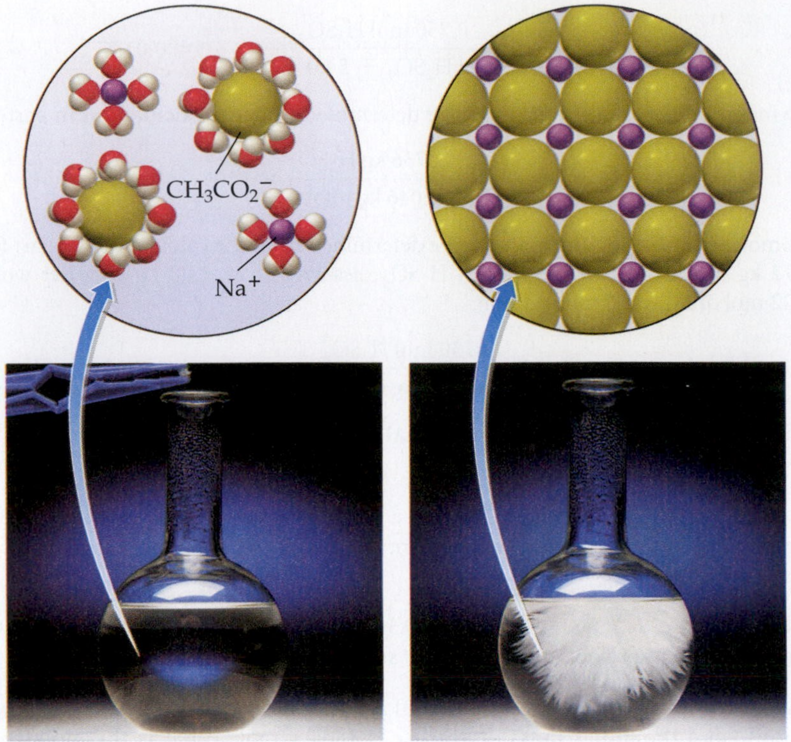

A supersaturated solution of sodium acetate in water.

When a tiny seed crystal is added, larger crystals begin to grow and precipitate from the solution until equilibrium is reached.

Effect of Temperature on Solubility

The solubility of a substance—that is, the amount of the substance per unit volume of solvent needed to form a saturated solution at a given temperature—is a physical property characteristic of that substance. Different substances can have greatly different solubilities, as shown in **Figure 11.6**. Sodium chloride, for instance, has a solubility of 35.9 g/100 mL

FIGURE 11.6

Solubilities of some common solids in water as a function of temperature.

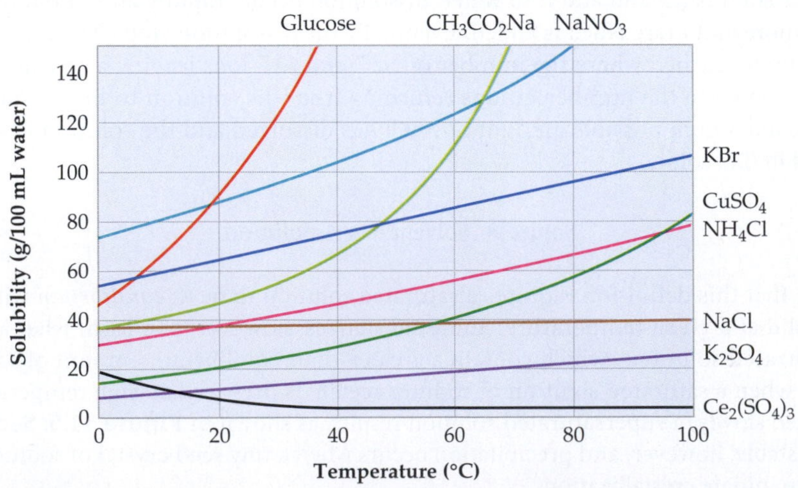

Most substances become more soluble as temperature rises, although the exact relationship is often complex and nonlinear.

of water at 20 °C, and sodium nitrate has a solubility of 87.3 g/100 mL of water at 20 °C. Sometimes, particularly when two liquids are involved, the solvent and solute are **miscible**, meaning that they are mutually soluble in all proportions. A solution of ethyl alcohol and water is an example.

Solubilities are temperature-dependent, so the temperature at which a specific measurement is made must be reported. As Figure 11.6 shows, there is no obvious correlation between structure and solubility or between solubility and temperature. The solubilities of most molecular and ionic solids increase with increasing temperature, but the solubilities of some (NaCl) are almost unchanged and the solubilities of others [$Ce_2(SO_4)_3$] actually decrease.

The effect of temperature on the solubility of gases is more predictable than its effect on the solubility of solids: Gases become less soluble in water as the temperature increases (**Figure 11.7**). One consequence of this decreased solubility is that carbonated drinks bubble continuously as they warm up to room temperature after being refrigerated. Soon, they lose so much dissolved CO_2 that they go flat. A much more important consequence is the damage to aquatic life that can result from the decrease in concentration of dissolved oxygen in lakes and rivers when hot water is discharged from industrial plants, an effect known as thermal pollution.

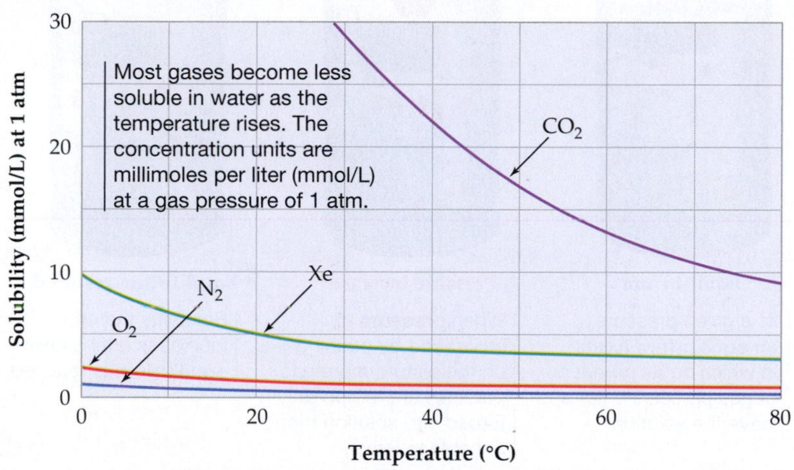

FIGURE 11.7

Solubilities of some gases in water as a function of temperature.

Effect of Pressure on Solubility

Pressure has practically no effect on the solubility of liquids and solids but has a profound effect on the solubility of gases. According to **Henry's law**, the solubility of a gas in a liquid at a given temperature is directly proportional to the partial pressure of the gas over the solution:

▶ **Henry's law:** Solubility $= k \cdot P$

The constant k in this expression is characteristic of a specific gas, and P is the partial pressure of the gas over the solution. Doubling the partial pressure doubles the solubility, tripling the partial pressure triples the solubility, and so forth. Henry's law constants are usually given in units of mol/(L · atm), and measurements are reported at 25 °C. Note that when the gas partial pressure P is 1 atm, the Henry's law constant k is numerically equal to the solubility of the gas in moles per liter.

▲ Divers who ascend too quickly can develop the bends, a condition caused by the formation of nitrogen bubbles in the blood. Treatment involves placing the diver into this decompression chamber and slowly changing the pressure.

Perhaps the most common example of Henry's law behavior occurs when you open a can of soda or other carbonated drink. Bubbles of gas immediately come fizzing out of solution because the pressure of CO_2 in the can drops and CO_2 suddenly becomes less soluble. A more serious example of Henry's law behavior occurs when a deep-sea diver surfaces too quickly and develops a painful and life-threatening condition called decompression sickness (DCS) or the *bends*. The bends occur because large amounts of nitrogen dissolve in the blood at high underwater pressures. When the diver ascends and pressure decreases too rapidly, bubbles of nitrogen form in the blood, blocking capillaries and inhibiting blood flow. The condition can be prevented by using an oxygen/helium mixture for breathing rather than air (oxygen/nitrogen), because helium has a much lower solubility in blood than nitrogen.

On a molecular level, the increase in gas solubility with increasing pressure occurs because of a change in the position of the equilibrium between dissolved and undissolved gas. At a given pressure, an equilibrium is established in which equal numbers of gas particles enter and leave the solution. When the pressure is increased, however, more particles are forced into solution than leave it, so gas solubility increases until a new equilibrium is established (**Figure 11.8**).

FIGURE 11.8

A molecular view of Henry's law.

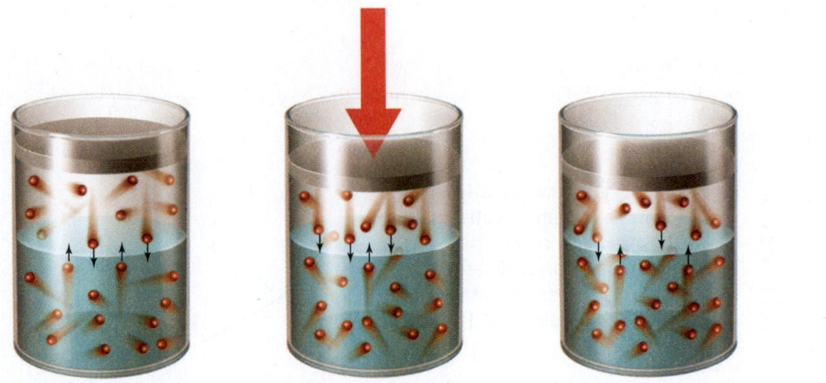

Worked Example 11.7

Using Henry's Law to Calculate Gas Solubility

The Henry's law constant of methyl bromide (CH_3Br), a gas used as a soil fumigating agent, is $k = 0.159$ mol/(L · atm) at 25 °C. What is the solubility in mol/L of methyl bromide in water at 25 °C and a partial pressure of 125 mm Hg?

STRATEGY

According to Henry's law, solubility $= k \cdot P$.

SOLUTION

$$k = 0.159 \text{ mol/(L} \cdot \text{atm)}$$

$$P = 125 \text{ mm Hg} \times \frac{1 \text{ atm}}{760 \text{ mm Hg}} = 0.164 \text{ atm}$$

$$\text{Solubility} = k \cdot P = 0.159 \frac{\text{mol}}{\text{L} \cdot \text{atm}} \times 0.164 \text{ atm} = 0.0261 \text{ M}$$

The solubility of methyl bromide in water at a partial pressure of 125 mm Hg is 0.0261 M.

PROBLEM 11.11 The solubility of CO_2 in water is 3.2×10^{-2} M at 25 °C and 1 atm pressure. What is the Henry's law constant for CO_2 in mol/(L · atm)?

PROBLEM 11.12 The partial pressure of CO_2 in air is approximately 4.0×10^{-4} atm. Use the Henry's law constant you calculated in Problem 11.11 to find the concentration of CO_2 in:

(a) A can of soda under a CO_2 pressure of 2.5 atm at 25 °C

(b) A can of soda open to the atmosphere at 25 °C

11.5 PHYSICAL BEHAVIOR OF SOLUTIONS: COLLIGATIVE PROPERTIES

The behavior of solutions is qualitatively similar to that of pure solvents but is quantitatively different. Pure water boils at 100.0 °C and freezes at 0.0 °C, for instance, but a 1.00 *m* (molal) solution of NaCl in water boils at 101.0 °C and freezes at −3.7 °C.

The higher boiling point and lower freezing point observed for a solution compared to a pure solvent are examples of **colligative properties**, which depend only on the amount of dissolved solute but not on the solute's chemical identity. The word *colligative* means "bound together in a collection" and is used because a collection of solute particles is responsible for the observed effects. Other colligative properties are a lower vapor pressure for a solution compared with the pure solvent and *osmosis*, the migration of solvent and other small molecules through a semipermeable membrane.

▶ Colligative Properties In comparing the properties of a pure solvent with those of a solution:

- The vapor pressure of the solution is lower.
- The boiling point of the solution is higher.
- The freezing (or melting) point of the solution is lower.
- The solution gives rise to *osmosis*, the migration of solvent molecules through a semipermeable membrane.

We'll look at each of the four colligative properties in more detail in Sections 11.6–11.8.

11.6 VAPOR-PRESSURE LOWERING OF SOLUTIONS: RAOULT'S LAW

Recall from Section 10.5 that a liquid in a closed container is in equilibrium with its vapor and that the amount of pressure exerted by the vapor is called the **vapor pressure**. When you compare the vapor pressure of a pure solvent with that of a solution at the same temperature, however, you find that the two values are different. If the solute is nonvolatile and has no appreciable vapor pressure of its own, as occurs when a solid is dissolved, then the vapor pressure of the solution is always lower than that of the pure solvent. If the solute *is* volatile and has a significant vapor pressure of its own, as occurs in a mixture of two liquids, then the vapor pressure of the mixture is intermediate between the vapor pressures of the two pure liquids.

REMEMBER...

At equilibrium, the rate of evaporation from the liquid to the vapor is equal to the rate of condensation from the vapor back to the liquid. The resulting **vapor pressure** is the partial pressure of the gas in the equilibrium. (Section 10.5)

Solutions with a Nonvolatile Solute

REMEMBER...

A **manometer** is a mercury-filled U-tube used for reading pressure. One end of the tube is connected to the sample container and the other end is open to the atmosphere. (Section 9.1)

It's easy to demonstrate with **manometers** that a solution of a nonvolatile solute has a lower vapor pressure than the pure solvent (**Figure 11.9**). Alternatively, you can show the same effect by comparing the evaporation rate of a pure solvent with the evaporation rate of a solution. A solution always evaporates more slowly than a pure solvent does because its vapor pressure is lower and its molecules therefore escape less readily.

FIGURE 11.9

The equilibrium vapor pressure of solutions.

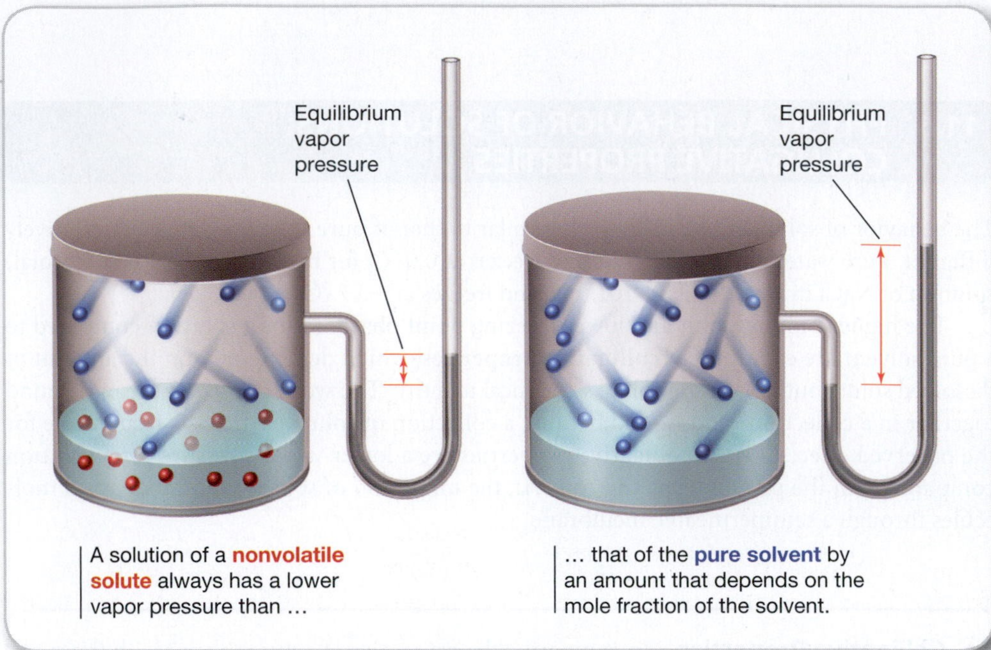

According to **Raoult's law**, the vapor pressure of a solution containing a nonvolatile solute is equal to the vapor pressure of the pure solvent times the mole fraction of the solvent. That is,

▶ **Raoult's law:** $P_{soln} = P_{solv} \times X_{solv}$

where P_{soln} is the vapor pressure of the solution, P_{solv} is the vapor pressure of pure solvent at the same temperature, and X_{solv} is the mole fraction of the solvent in the solution.

Take a solution of 1.00 mol of glucose in 15.0 mol of water at 25 °C, for instance. The vapor pressure of pure water at 25 °C is 23.76 mm Hg, and the mole fraction of water in the solution is 15.0 mol/(1.00 mol + 15.0 mol) = 0.938. Thus, Raoult's law predicts a vapor pressure for the solution of 23.76 mm Hg × 0.938 = 22.3 mm Hg, which corresponds to a vapor-pressure lowering, ΔP_{solve}, of 1.5 mm Hg:

$$P_{soln} = P_{solv} \times X_{solv} = 23.76 \text{ mm Hg} \times \frac{15.0 \text{ mol}}{1.00 \text{ mol} + 15.0 \text{ mol}} = 22.3 \text{ mm Hg}$$

$$\Delta P_{soln} = P_{solv} - P_{soln} = 23.76 \text{ mm Hg} - 22.3 \text{ mm Hg} = 1.5 \text{ mm Hg}$$

Alternatively, the amount of vapor-pressure lowering can be calculated directly by multiplying the mole fraction of the *solute* times the vapor pressure of the pure solvent. That is,

$$\Delta P_{soln} = P_{solv} \times X_{solute} = 23.76 \text{ mm Hg} \times \frac{1.00 \text{ mol}}{1.00 \text{ mol} + 15.0 \text{ mol}}$$

$$= 1.49 \text{ mm Hg}$$

If an ionic substance such as NaCl is the solute rather than a molecular substance, we have to calculate mole fractions based on the total concentration of solute *particles* (ions) rather than NaCl formula units. A solution of 1.00 mol of NaCl in 15.0 mol of water at 25 °C, for example, contains 2.00 mol of dissolved particles, assuming complete dissociation, resulting in a mole fraction for water of 0.882 and a solution vapor pressure of 21.0 mm Hg.

$$X_{water} = \frac{15.0 \text{ mol } H_2O}{1.00 \text{ mol } Na^+ + 1.00 \text{ mol } Cl^- + 15.0 \text{ mol } H_2O} = 0.882$$

$$P_{soln} = P_{solv} \times X_{solv} = 23.76 \text{ mm Hg} \times 0.882 = 21.0 \text{ mm Hg}$$

Because the mole fraction of water is smaller in the NaCl solution than in the glucose solution, the vapor pressure of the NaCl solution is lower: 21.0 mm Hg for NaCl versus 22.3 mm Hg for glucose at 25 °C.

Just as the ideal gas law discussed in Section 9.3 applies only to "ideal" gases, Raoult's law applies only to ideal solutions. Raoult's law approximates the behavior of most real solutions, but significant deviations from ideality occur as the solute concentration increases. The law works best when solute concentrations are low and when solute and solvent particles have similar intermolecular forces.

If the intermolecular forces between solute particles and solvent molecules are weaker than the forces between solvent molecules alone, then the solvent molecules are less tightly held in the solution and the vapor pressure is higher than Raoult's law predicts. Conversely, if the intermolecular forces between solute and solvent molecules are stronger than the forces between solvent molecules alone, then the solvent molecules are more tightly held in the solution and the vapor pressure is lower than predicted. Solutions of ionic substances, in particular, often have a vapor pressure significantly lower than predicted, because the ion–dipole forces between dissolved ions and polar water molecules are so strong.

A further complication is that ionic substances rarely dissociate completely, so a solution of an ionic compound usually contains fewer particles than the formula of the compound would suggest. The actual extent of dissociation can be expressed as a **van't Hoff factor (*i*)**:

▶ **van't Hoff factor** $i = \frac{\text{Moles of particles in solution}}{\text{Moles of solute dissolved}}$

Rearranging this equation shows that the number of moles of particles dissolved in a solution is equal to number of moles of solute dissolved times the van't Hoff factor.

$$\text{Moles of particles in solution} = i \times \text{Moles of solute dissolved}$$

To take a solution of NaCl as an example, the experimentally determined van't Hoff factor for 0.05 *m* NaCl is 1.9, meaning that each mole of NaCl gives only 1.9 mol of particles rather than the 2.0 mol expected for complete dissociation. Of the 1.9 mol of particles, 0.9 mol is Cl^-, 0.9 mol is Na^+, and 0.1 mol is undissociated NaCl. Thus, NaCl is only $(0.9/1.0) \times 100\% = 90\%$ dissociated in a 0.05 *m* solution and the amount of vapor pressure lowering is less than expected.

What accounts for the lowering of the vapor pressure when a nonvolatile solute is dissolved in a solvent? As we've noted on many prior occasions, a physical process such as the vaporization of a liquid to a gas is accompanied by a free-energy change, $\Delta G_{vap} = \Delta H_{vap} - T\Delta S_{vap}$. The more negative the value of ΔG_{vap}, the more favored the vaporization process. Thus, if we want to compare the ease of vaporization of a pure solvent with that of the solvent in a solution, we have to compare the signs and relative magnitudes of the ΔH_{vap} and ΔS_{vap} terms in the two cases.

The vaporization of a liquid to a gas is disfavored by enthalpy (positive ΔH_{vap}) because energy is required to overcome intermolecular attractions in the liquid. At the same time, however, vaporization is favored by entropy (positive ΔS_{vap}) because randomness increases when molecules go from a liquid state to a gaseous state.

Comparing a pure solvent with a solution, the *enthalpies* of vaporization for the pure solvent and solvent in a solution are similar because similar intermolecular forces must be overcome in both cases for solvent molecules to escape from the liquid. The *entropies* of vaporization for a pure solvent and a solvent in a solution are *not* similar, however. Because a solvent in a solution has more molecular randomness and higher entropy than a pure solvent does, the entropy *change* on going from liquid to vapor is smaller for the solvent in a solution than for the pure solvent. Subtracting a smaller $T\Delta S_{vap}$ from ΔH_{vap} thus results in a larger (less negative) ΔG_{vap} for the solution. As a result, vaporization is less favored for the solution and the vapor pressure of the solution at equilibrium is lower (**Figure 11.10**).

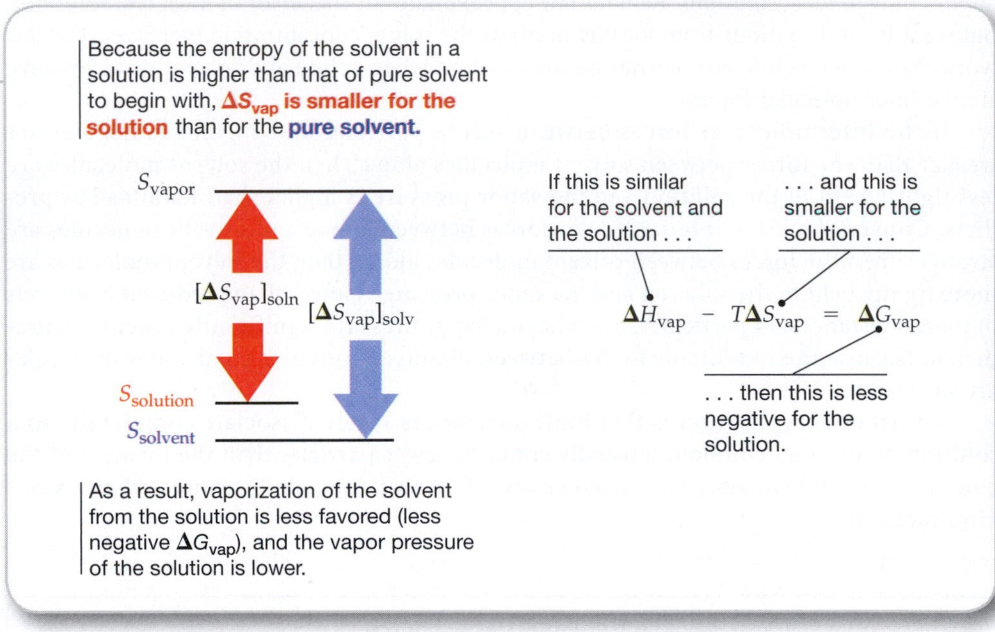

FIGURE 11.10

Vapor-pressure lowering. The lower vapor pressure of a solution relative to that of a pure solvent is due to the difference in their entropies of vaporization, ΔS_{vap}.

Worked Example 11.8

Calculating the Vapor Pressure of a Solution

What is the vapor pressure in mm Hg of a solution made by dissolving 18.3 g of NaCl in 500.0 g of H_2O at 70 °C, assuming a van't Hoff factor of 1.9? The vapor pressure of pure water at 70 °C is 233.7 mm Hg.

STRATEGY

According to Raoult's law, the vapor pressure of the solution equals the vapor pressure of pure solvent times the mole fraction of the solvent in the solution. Thus, we have to find the numbers of moles of solvent and solute and then calculate the mole fraction of solvent.

SOLUTION

First, use molar mass to calculate the number of moles of NaCl and H_2O.

$$\text{Moles of NaCl} = 18.3 \text{ g NaCl} \times \frac{1 \text{ mol NaCl}}{58.44 \text{ g NaCl}} = 0.313 \text{ mol NaCl}$$

$$\text{Moles of } H_2O = 500.0 \text{ g } H_2O \times \frac{1 \text{ mol } H_2O}{18.02 \text{ g } H_2O} = 27.75 \text{ mol } H_2O$$

Next, calculate the mole fraction of water in the solution. A van't Hoff factor of 1.9 means that the NaCl dissociates incompletely and gives only 1.9 particles per formula unit. Thus, the solution contains $1.9 \times 0.313\ \text{mol} = 0.59\ \text{mol}$ of dissolved particles and the mole fraction of water is

$$\text{Mole fraction of } H_2O = \frac{27.75\ \text{mol}}{0.59\ \text{mol} + 27.75\ \text{mol}} = 0.9792$$

From Raoult's law, the vapor pressure of the solution is

$$P_{\text{soln}} = P_{\text{solv}} \times X_{\text{solv}} = 233.7\ \text{mm Hg} \times 0.9792 = 228.8\ \text{mm Hg}$$

Worked Example 11.9

Calculating Vapor-Pressure Lowering

How many grams of sucrose must be added to 320 g of water to lower the vapor pressure by 1.5 mm Hg at 25 °C? The vapor pressure of water at 25 °C is 23.8 mm Hg, and the molar mass of sucrose is 342.3 g/mol.

STRATEGY

According to Raoult's law, $P_{\text{soln}} = P_{\text{solv}} \times X_{\text{solv}}$, which can be rearranged to the form $X_{\text{solv}} = P_{\text{soln}}/P_{\text{solv}}$. This equation can then be solved to find the number of moles of sucrose and hence the number of grams.

SOLUTION

First, calculate the vapor pressure of the solution, P_{soln}, by subtracting the amount of vapor-pressure lowering from the vapor pressure of the pure solvent, P_{solv}:

$$P_{\text{soln}} = 23.8\ \text{mm Hg} - 1.5\ \text{mm Hg} = 22.3\ \text{mm Hg}$$

Now calculate the mole fraction of water, X_{solv}.

$$\text{Since} \quad P_{\text{soln}} = P_{\text{solv}} \times X_{\text{solv}}$$

$$\text{then} \quad X_{\text{solv}} = \frac{P_{\text{soln}}}{P_{\text{solv}}} = \frac{22.3\ \text{mm Hg}}{23.8\ \text{mm Hg}} = 0.937$$

This mole fraction of water is the number of moles of water divided by the total number of moles of sucrose plus water:

$$X_{\text{solv}} = \frac{\text{Moles of water}}{\text{Total moles}}$$

Since the number of moles of water is

$$\text{Moles of water} = 320\ \text{g} \times \frac{1\ \text{mol}}{18.0\ \text{g}} = 17.8\ \text{mol}$$

then the total number of moles of sucrose plus water is

$$\text{Total moles} = \frac{\text{moles of water}}{X_{\text{solv}}} = \frac{17.8\ \text{mol}}{0.937} = 19.0\ \text{mol}$$

Subtracting the number of moles of water from the total number of moles gives the number of moles of sucrose needed:

$$\text{Moles of sucrose} = 19.0\ \text{mol} - 17.8\ \text{mol} = 1.2\ \text{mol}$$

Converting moles into grams then gives the mass of sucrose needed:

$$\text{Grams of sucrose} = 1.2\ \text{mol} \times 342.3\ \frac{\text{g}}{\text{mol}} = 4.1 \times 10^2\ \text{g}$$

PROBLEM 11.13 What is the vapor pressure in mm Hg of a solution prepared by dissolving 5.00 g of benzoic acid ($C_7H_6O_2$) in 100.00 g of ethyl alcohol (C_2H_6O) at 35 °C? The vapor pressure of pure ethyl alcohol at 35 °C is 100.5 mm Hg.

PROBLEM 11.14 How many grams of NaBr must be added to 250 g of water to lower the vapor pressure by 1.30 mm Hg at 40 °C assuming complete dissociation? The vapor pressure of water at 40 °C is 55.3 mm Hg.

CONCEPTUAL PROBLEM 11.15 The following diagram shows a close-up view of part of the vapor-pressure curve for a pure solvent and a solution of a nonvolatile solute. Which curve represents the pure solvent, and which the solution?

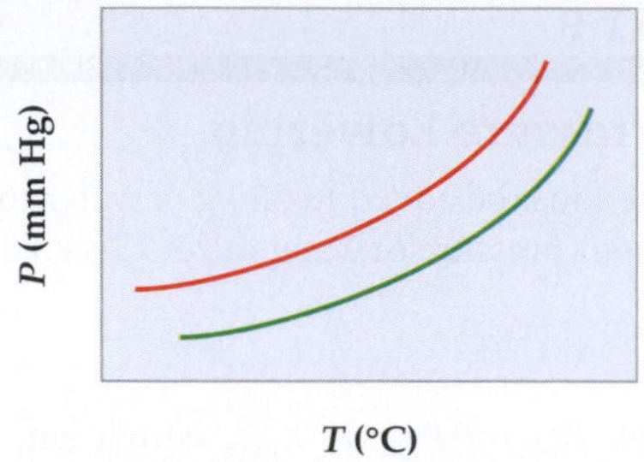

Solutions with a Volatile Solute

REMEMBER...

Dalton's law of partial pressures states that the total pressure exerted by a mixture of gases in a container at constant volume and temperature is equal to the sum of the pressures of each individual gas in the container. (Section 9.5)

According to **Dalton's law of partial pressures**, the overall vapor pressure P_{total} of a mixture of two volatile liquids, A and B, is the sum of the vapor-pressure contributions of the individual components, P_A and P_B:

$$P_{total} = P_A + P_B$$

The individual vapor pressures P_A and P_B are calculated by Raoult's law. That is, the vapor pressure of A is equal to the mole fraction of A (X_A) times the vapor pressure of pure A ($P°_A$), and the vapor pressure of B is equal to the mole fraction of B (X_B) times the vapor pressure of pure B ($P°_B$). Thus, the total vapor pressure of the solution is

$$P_{total} = P_A + P_B = (P°_A \cdot X_A) + (P°_B \cdot X_B)$$

Take a mixture of the two similar organic liquids benzene (C_6H_6, bp = 80.1 °C) and toluene (C_7H_8, bp = 110.6 °C), as an example. Pure benzene has a vapor pressure $P° = 96.0$ mm Hg at 25 °C, and pure toluene has $P° = 30.3$ mm Hg at the same temperature. In a 1:1 molar mixture of the two, where the mole fraction of each is $X = 0.500$, the vapor pressure of the solution is 63.1 mm Hg:

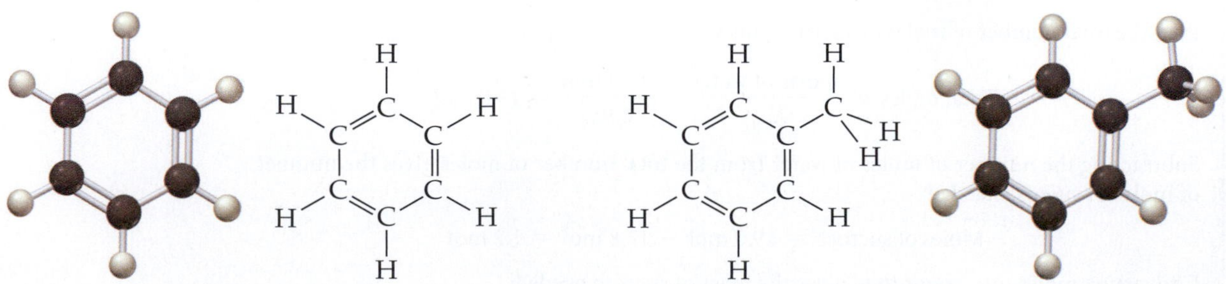

Benzene Toluene

$$\begin{aligned} P_{total} &= (P°_{benzene})(X_{benzene}) + (P°_{toluene})(X_{toluene}) \\ &= (96.0 \text{ mm Hg} \times 0.500) + (30.3 \text{ mm Hg} \times 0.500) \\ &= 48.0 \text{ mm Hg} + 15.2 \text{ mm Hg} = 63.2 \text{ mm Hg} \end{aligned}$$

Note that the vapor pressure of the mixture is intermediate between the vapor pressures of the two pure liquids (**Figure 11.11**).

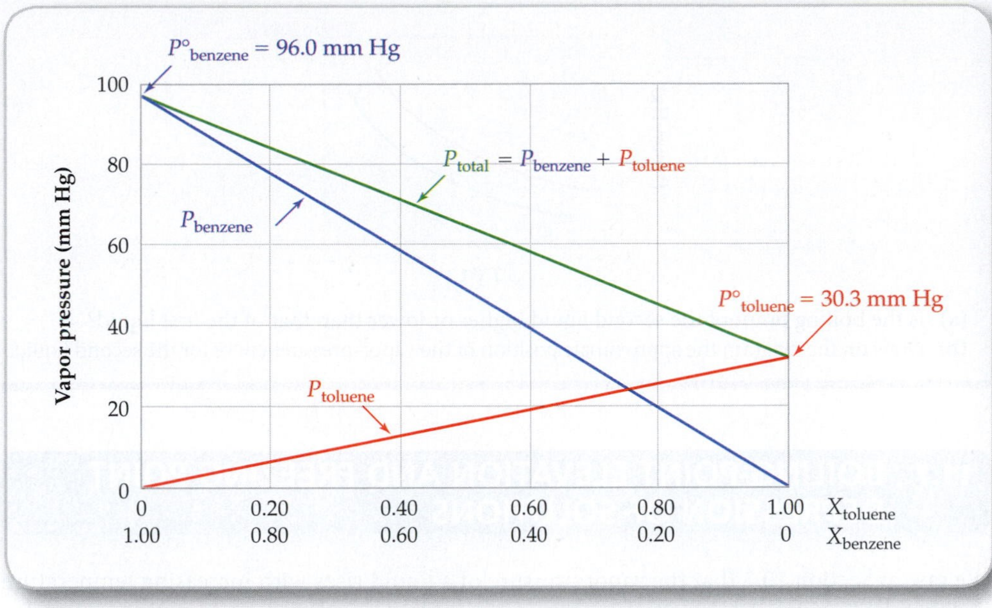

FIGURE 11.11

Raoult's law for a mixture of volatile liquids. The vapor pressure of a solution of the two volatile liquids benzene and toluene at 25 °C is the sum of the two individual contributions, each calculated by Raoult's law.

As with nonvolatile solutes, Raoult's law for a mixture of volatile liquids applies only to ideal solutions. Most real solutions show behaviors that deviate slightly from the ideal in either a positive or negative way, depending on the kinds and strengths of intermolecular forces present in the solution.

WORKED CONCEPTUAL EXAMPLE 11.10

Interpreting Vapor Pressure Curves

The following diagram shows a close-up view of part of the vapor-pressure curves for two pure liquids and a mixture of the two. Which curves represent pure liquids, and which represents the mixture?

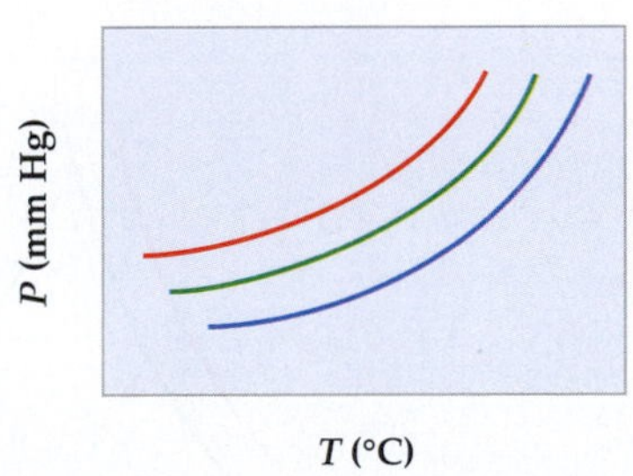

STRATEGY AND SOLUTION

The vapor pressure of a mixture of two volatile liquids is always intermediate between the vapor pressures of the two pure liquids. Thus, the top (red) and bottom (blue) curves represent pure liquids, and the middle curve (green) represents the mixture.

PROBLEM 11.16

(a) What is the vapor pressure in mm Hg of a solution prepared by dissolving 25.0 g of ethyl alcohol (C_2H_5OH) in 100.0 g of water at 25 °C? The vapor pressure of pure water is 23.8 mm Hg, and the vapor pressure of ethyl alcohol is 61.2 mm Hg at 25 °C.

(b) What is the vapor pressure of the solution if 25.0 g of water is dissolved in 100.0 g of ethyl alcohol at 25 °C?

CONCEPTUAL PROBLEM 11.17 The following phase diagram shows part of the vapor-pressure curves for a pure liquid (green curve) and a solution of the first liquid with a second volatile liquid (red curve).

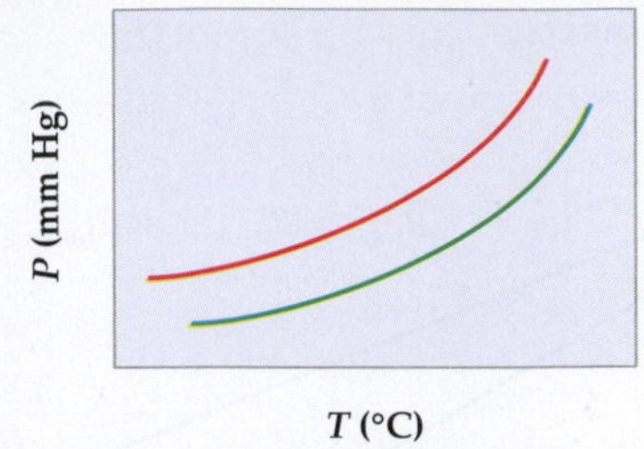

(a) Is the boiling point of the second liquid higher or lower than that of the first liquid?
(b) Draw on the diagram the approximate position of the vapor-pressure curve for the second liquid.

11.7 BOILING-POINT ELEVATION AND FREEZING-POINT DEPRESSION OF SOLUTIONS

REMEMBER... The **phase diagram** of a substance is a plot of pressure versus temperature, showing which phase is stable at any given combination of P and T. (Section 10.11)

We saw in Section 10.5 that the vapor pressure of a liquid rises with increasing temperature and that the liquid boils when its vapor pressure equals atmospheric pressure. Because a solution of a nonvolatile solute has a lower vapor pressure than a pure solvent has at a given temperature, the solution must be heated to a higher temperature to cause it to boil. Furthermore, the lower vapor pressure of the solution means that the liquid/vapor phase transition line on a **phase diagram** is always lower for the solution than for the pure solvent. As a result, the triple-point temperature T_t is lower for the solution, the solid/liquid phase transition line is shifted to a lower temperature for the solution, and the solution must be cooled to a lower temperature to freeze. **Figure 11.12** shows the situation.

The boiling point of a solution depends on the concentration of dissolved particles, just as vapor pressure does. Thus, a 1.00 *m* solution of glucose in water boils at approximately 100.51 °C at 1 atm pressure (0.51 °C above normal), but a 1.00 *m* solution of NaCl in water boils at approximately 101.02 °C (1.02 °C above normal) because there are twice as many particles (ions) dissolved in the NaCl solution as there are in the glucose solution.

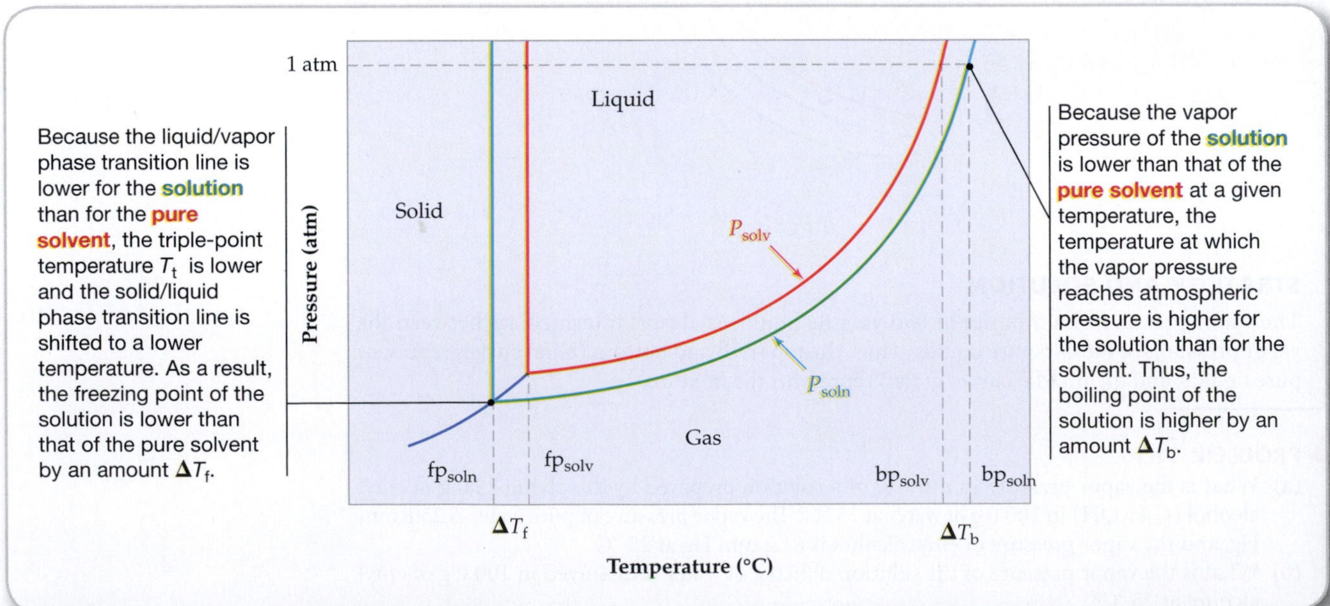

FIGURE 11.12
Phase diagrams for a pure solvent (red) and a solution of a nonvolatile solute (green).

The change in boiling point ΔT_b for a solution is

$$\Delta T_b = K_b \cdot m$$

where m is the *molal* (not molar) concentration of solute particles and K_b is the **molal boiling-point-elevation constant** characteristic of a given liquid. The concentration must be expressed in molality—the number of moles of solute particles per kilogram of solvent—rather than molarity so that the solute concentration is independent of temperature. Molal boiling-point-elevation constants are given in **Table 11.4** for some common substances.

TABLE 11.4 Molal Boiling-Point-Elevation Constants (K_b) and Molal Freezing-Point-Depression Constants (K_f) for Some Common Substances

Substance	K_b [(°C · kg)/mol]	K_f [(°C · kg)/mol]
Benzene (C_6H_6)	2.64	5.07
Camphor ($C_{10}H_{16}O$)	5.95	37.8
Chloroform ($CHCl_3$)	3.63	4.70
Diethyl ether ($C_4H_{10}O$)	2.02	1.79
Ethyl alcohol (C_2H_6O)	1.22	1.99
Water (H_2O)	0.51	1.86

The freezing point of a solution depends on the concentration of solute particles, just as the boiling point does. For example, a 1.00 m solution of glucose in water freezes at −1.86 °C, and a 1.00 m solution of NaCl in water freezes at approximately −3.72 °C. The change in freezing point ΔT_f for a solution is

$$\Delta T_f = -K_f \cdot m$$

where m is the molal concentration of solute particles and K_f is the **molal freezing-point-depression constant** characteristic of a given solvent. Some molal freezing-point-depression constants are also given in Table 11.4.

As with vapor-pressure lowering described in the previous Section, the actual amount of boiling-point elevation and freezing-point depression observed for a solution of an ionic substance depends on the extent of dissociation, as given by a van't Hoff factor. The formulas for both boiling-point elevation and freezing-point depression can be modified to take dissociation into account:

$$\Delta T_b = K_b \cdot m \cdot i \qquad \text{where } m \text{ is the molal concentration of formula units}$$
$$\Delta T_f = -K_f \cdot m \cdot i \qquad \text{and } i \text{ is the van't Hoff factor}$$

The fundamental cause of boiling-point elevation and freezing-point depression in solutions is the same as the cause of vapor-pressure lowering: the entropy difference between the pure solvent and the solvent in a solution. Let's take boiling-point elevations first. We know that liquid and vapor phases are in equilibrium at the boiling point (T_b) and that the free-energy difference between the two phases (ΔG_{vap}) is therefore zero.

$$\text{Since} \quad \Delta G_{vap} = \Delta H_{vap} - T_b \Delta S_{vap} = 0$$

$$\text{then} \quad \Delta H_{vap} = T_b \Delta S_{vap} \quad \text{and} \quad T_b = \frac{\Delta H_{vap}}{\Delta S_{vap}}$$

In comparing the enthalpies of vaporization (ΔH_{vap}) for the pure solvent and for the solvent in a solution, the values are similar because similar intermolecular forces holding the solvent molecules must be overcome in both cases. In comparing the entropies of vaporization, however, the values are not similar. Because the solvent in a solution has more molecular randomness than the pure solvent has, the entropy change between solution and vapor is smaller than the entropy change between pure solvent and vapor. But if ΔS_{vap} is smaller for the solution, then T_b must be correspondingly larger. In other words, the boiling point of the solution (T_b) is higher than that of the pure solvent (**Figure 11.13**).

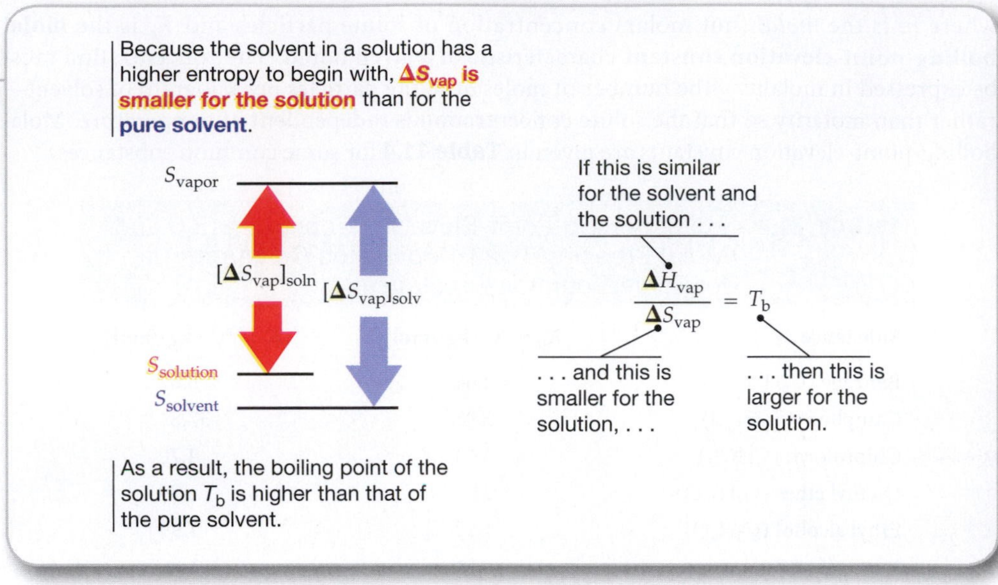

FIGURE 11.13

Boiling-point elevation. The higher boiling point of a solution relative to that of a pure solvent is due to a difference in their entropies of vaporization, ΔS_{vap}.

A similar explanation accounts for freezing-point depression. Because liquid and solid phases are in equilibrium at the freezing point, the free-energy difference between the phases (ΔG_{fusion}) is zero:

$$\text{Since} \quad \Delta G_{fusion} = \Delta H_{fusion} - T_f \Delta S_{fusion} = 0$$

$$\text{then} \quad \Delta H_{fusion} = T_f \Delta S_{fusion} \quad \text{and} \quad T_f = \frac{\Delta H_{fusion}}{\Delta S_{fusion}}$$

In comparing the solvent in solution with pure solvent, the enthalpies of fusion (ΔH_{fusion}) are similar because similar intermolecular forces between solvent molecules are involved. The entropies of fusion (ΔS_{fusion}), however, are not similar. Because the solvent in solution has more molecular randomness than the pure solvent has, the entropy change between the solvent in the solution and the solid is larger than the entropy change between pure solvent and the solid. With ΔS_{fusion} larger for the solution, T_f must be correspondingly smaller, meaning that the freezing point of the solution (T_f) is lower than that of the pure solvent (**Figure 11.14**).

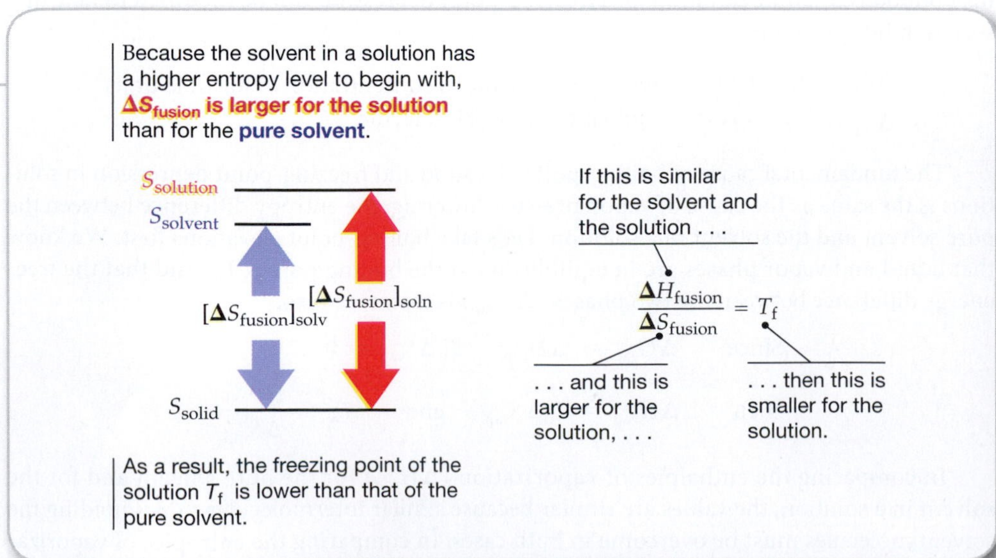

FIGURE 11.14

Freezing-point lowering. The lower freezing point of a solution relative to that of a pure solvent is due to a difference in their entropies of fusion, ΔS_{fusion}.

Worked Example 11.11

Using Boiling-Point Elevation to Calculate the Molality of a Solution

What is the molality of an aqueous glucose solution if the boiling point of the solution at 1 atm pressure is 101.27 °C? The molal boiling-point-elevation constant for water is given in Table 11.4.

STRATEGY

Rearrange the equation for molal boiling-point elevation to solve for m:

$$\Delta T_b = K_b \cdot m \quad \text{so} \quad m = \frac{\Delta T_b}{K_b}$$

where $K_b = 0.51\ (°\text{C} \cdot \text{kg})/\text{mol}$ and $\Delta T_b = 101.27\ °\text{C} - 100.00\ °\text{C} = 1.27\ °\text{C}$.

SOLUTION

$$m = \frac{1.27\ °\text{C}}{0.51\ \dfrac{°\text{C} \cdot \text{kg}}{\text{mol}}} = 2.5\ \frac{\text{mol}}{\text{kg}} = 2.5\ m$$

The molality of the solution is 2.5 m.

PROBLEM 11.18 What is the normal boiling point in °C of a solution prepared by dissolving 1.50 g of aspirin (acetylsalicylic acid, $C_9H_8O_4$) in 75.00 g of chloroform ($CHCl_3$)? The normal boiling point of chloroform is 61.7 °C, and K_b for chloroform is given in Table 11.4.

PROBLEM 11.19 What is the freezing point in °C of a solution prepared by dissolving 7.40 g of $MgCl_2$ in 110 g of water? The value of K_f for water is given in Table 11.4, and the van't Hoff factor for $MgCl_2$ is $i = 2.7$.

PROBLEM 11.20 Assuming complete dissociation, what is the molality of an aqueous solution of KBr whose freezing point is −2.95 °C? The molal freezing-point-depression constant of water is given in Table 11.4.

PROBLEM 11.21 When 9.12 g of HCl was dissolved in 190 g of water, the freezing point of the solution was −4.65 °C. What is the value of the van't Hoff factor for HCl?

CONCEPTUAL PROBLEM 11.22 The following phase diagram shows a close-up view of the liquid/vapor phase transition boundaries for pure chloroform and a solution of a nonvolatile solute in chloroform.

(a) What is the approximate boiling point of pure chloroform?

(b) What is the approximate molal concentration of the nonvolatile solute? See Table 11.4 to find K_b for chloroform.

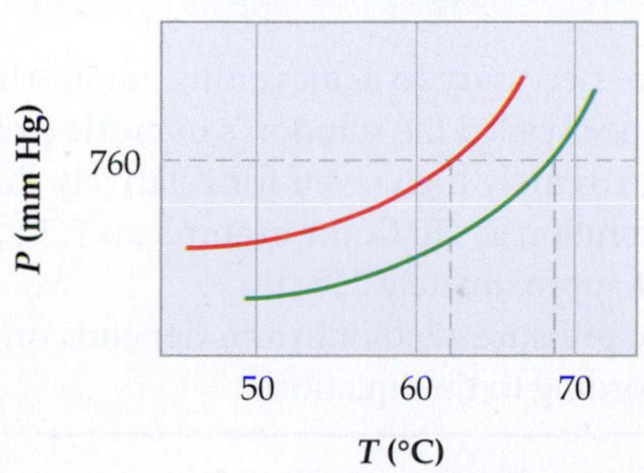

11.8 OSMOSIS AND OSMOTIC PRESSURE

Certain materials, including those that make up the membranes around living cells, are *semipermeable*. That is, they allow water or other smaller molecules to pass through, but they block the passage of larger solute molecules or solvated ions. When a solution and a pure solvent, or two solutions of different concentration, are separated by the right kind of semipermeable membrane, solvent molecules pass through the membrane in a process called **osmosis**. Although the passage of solvent through the membrane takes place in both directions, passage from the pure solvent side to the solution side is more favored and occurs at a greater rate. As a result, the amount of liquid on the pure solvent side of the membrane decreases, the amount of liquid on the solution side increases, and the concentration of the solution decreases.

Osmosis can be demonstrated with the experimental setup shown in **Figure 11.15**, in which a solution inside the bulb is separated from pure solvent in the beaker by a semipermeable membrane. Solvent passes through the membrane from the beaker to the bulb, causing the liquid level in the attached tube to rise. The increased weight of liquid in the tube creates an increased pressure that pushes solvent back through the membrane until the rates of forward and reverse passage become equal and the liquid level stops rising.

FIGURE 11.15

The phenomenon of osmosis.

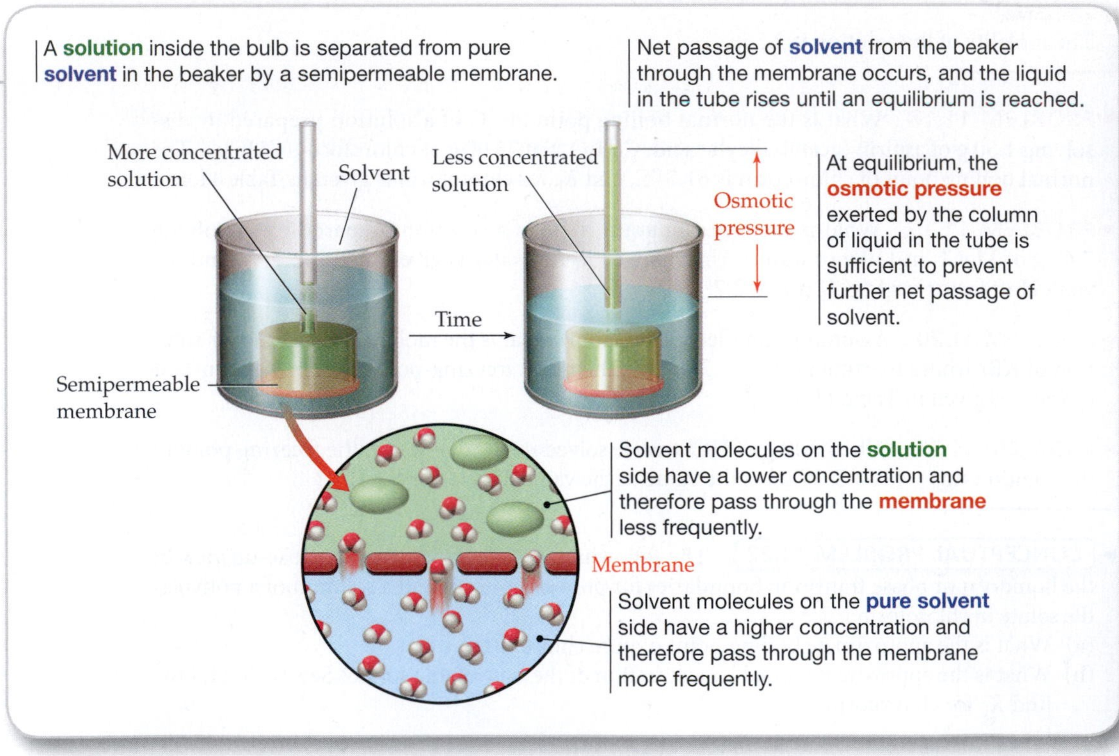

The amount of pressure necessary to achieve this equilibrium passage of solvent molecules through the membrane is called the solution's **osmotic pressure,** Π (Greek capital pi). Osmotic pressures can be extremely high, even for relatively dilute solutions. The osmotic pressure of a 0.15 M NaCl solution at 25 °C, for example, is 7.3 atm, a value that will support a difference in water level of approximately 250 ft!

The amount of osmotic pressure at equilibrium depends on the concentration of solute particles in the solution according to the equation

$$\Pi = MRT$$

where M is the molar concentration of solute particles, R is the gas constant [0.082 06 (L · atm)/(K · mol)], and T is the temperature in kelvins. For example, a 1.00 M solution of glucose in water at 300 K has an osmotic pressure of 24.6 atm:

$$\Pi = MRT = \left(1.00\ \frac{\text{mol}}{\text{L}}\right)\left(0.082\ 06\ \frac{\text{L}\cdot\text{atm}}{\text{K}\cdot\text{mol}}\right)(300\ \text{K}) = 24.6\ \text{atm}$$

Note that the solute concentration is given in *molarity* when calculating osmotic pressure rather than in molality as for other colligative properties. Because osmotic-pressure measurements are made at the specific temperature given in the equation $\Pi = MRT$, it's not necessary to express concentration in a temperature-independent unit like molality.

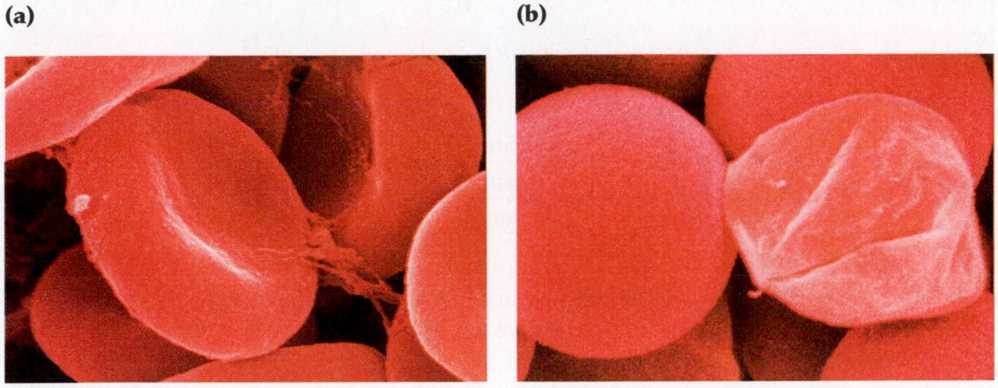

▲ **(a)** Normal red blood cells in plasma. **(b)** Rupture of the cells occurs in pure water because osmotic pressure drives water into the cells and increases the internal pressure.

Osmosis, like all colligative properties, results from a favorable increase in entropy of the pure solvent as it passes through the membrane and mixes with the solution. We can also explain osmosis on a molecular level by noting that molecules on the solvent side of the membrane, because of their greater concentration, approach the membrane a bit more frequently than molecules on the solution side, thereby passing through more often (Figure 11.15).

Worked Example 11.12

Calculating the Osmotic Pressure of a Solution

The total concentration of dissolved particles inside red blood cells and in the surrounding plasma is approximately 0.30 M, and the membrane surrounding the cells is semipermeable. What would the maximum osmotic pressure in atmospheres inside the cells be if the cells were removed from blood plasma and placed in pure water at 298 K?

STRATEGY

If red blood cells were removed from plasma and placed in pure water, water would pass through the cell membrane, causing a pressure increase inside the cells. The maximum amount of this pressure would be

$$\Pi = MRT$$

where $M = 0.30\text{ mol/L}$, $R = 0.082\,06\ (\text{L}\cdot\text{atm})/(\text{K}\cdot\text{mol})$, $T = 298\text{ K}$.

SOLUTION

$$\Pi = \left(0.30\ \frac{\text{mol}}{\text{L}}\right)\left(0.082\,06\ \frac{\text{L}\cdot\text{atm}}{\text{K}\cdot\text{mol}}\right)(298\text{ K}) = 7.3\text{ atm}$$

In fact, red blood cells rupture if the pressure differential across their membrane is greater than 5 atm, so the cells would burst if placed in pure water.

Worked Example 11.13

Using Osmotic Pressure to Calculate the Molarity of a Solution

A solution of an unknown substance in water at 293 K gives rise to an osmotic pressure of 5.66 atm. What is the molarity of the solution?

continued on next page

STRATEGY

We are given values for Π and T, and we need to solve for M in the equation $\Pi = MRT$. Rearranging this equation, we get

$$M = \frac{\Pi}{RT}$$

where $\Pi = 5.66$ atm, $R = 0.082\,06\,(\text{L}\cdot\text{atm})/(\text{K}\cdot\text{mol})$, and $T = 293$ K.

SOLUTION

$$M = \frac{5.66\text{ atm}}{\left(0.082\,06\,\dfrac{\text{L}\cdot\text{atm}}{\text{K}\cdot\text{mol}}\right)(293\text{ K})} = 0.235\text{ M}$$

PROBLEM 11.23 What osmotic pressure in atmospheres would you expect for a solution of 0.125 M $CaCl_2$ that is separated from pure water by a semipermeable membrane at 310 K? Assume 100% dissociation for $CaCl_2$.

PROBLEM 11.24 A solution of an unknown substance in water at 300 K gives rise to an osmotic pressure of 3.85 atm. What is the molarity of the solution?

11.9 SOME USES OF COLLIGATIVE PROPERTIES

▲ The Hadera seawater reverse-osmosis (SWRO) desalination plant in Israel is the world's largest, producing approximately 4.4 *billion* gallons per year of desalinated water.

Colligative properties may seem somewhat obscure, but in fact they have many practical uses, both in the chemical laboratory and in everyday life. Motorists in winter, for instance, take advantage of freezing-point lowering when they drive on streets where the snow has been melted by a sprinkling of salt. The antifreeze added to automobile radiators and the de-icer solution sprayed on airplane wings also work by lowering the freezing point of water. That same automobile antifreeze keeps radiator water from boiling over in summer by raising its boiling point.

One of the more interesting uses of colligative properties is the desalination of seawater by *reverse osmosis*. When pure water and seawater are separated by a suitable membrane, the passage of water molecules from the pure side to the solution side is faster than passage in the reverse direction. As osmotic pressure builds up, though, the rates of forward and reverse water passage eventually become equal at an osmotic pressure of about 30 atm at 25 °C. If, however, a pressure even *greater* than 30 atm is applied to the solution side, then the reverse passage of water becomes favored. As a result, pure water can be obtained from seawater (**Figure 11.16**).

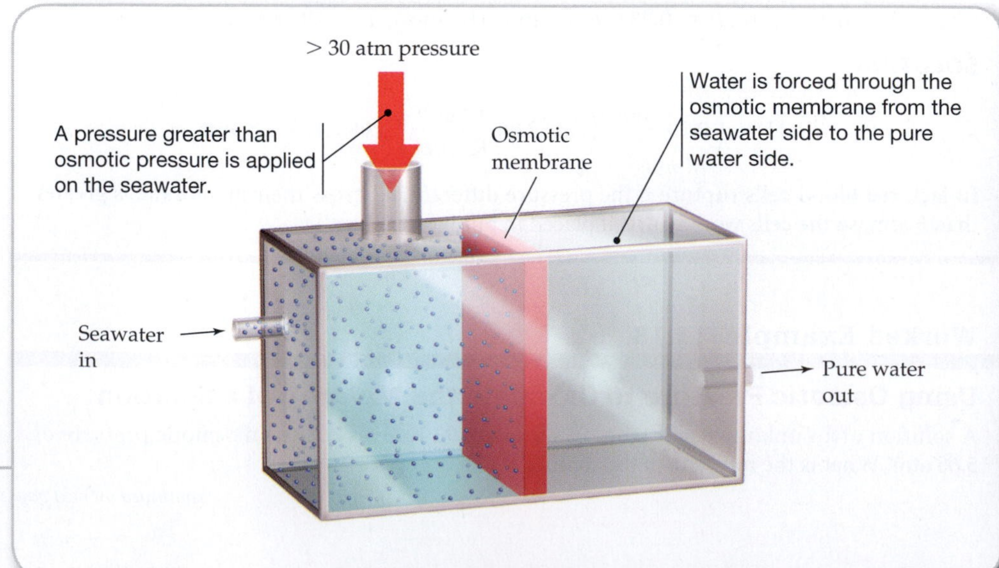

FIGURE 11.16

Desalination of seawater by reverse osmosis at high pressure.

Desalination of seawater has many applications, from use on cruise ships, to use by troops on battlefields, to use for providing drinking water to entire countries. Israel, for instance, now obtains more than 35% of its domestic consumer water demand by reverse osmosis.

In the laboratory, colligative properties are sometimes used for determining the molecular weight of an unknown substance. Any of the four colligative properties we've discussed can be used, but the most accurate values are obtained from osmotic-pressure measurements because the magnitude of the osmosis effect is so great. For example, a solution of 0.0200 M glucose in water at 300 K will give an osmotic-pressure reading of 374.2 mm Hg, a value that can easily be read to four significant figures. The same solution, however, will lower the freezing point by only 0.04 °C, a value that can be read to only one significant figure. Worked Example 11.14 shows how osmotic pressure can be used to find molecular weight.

Worked Example 11.14

Using Osmotic Pressure to Calculate the Molecular Weight of a Solute

A solution prepared by dissolving 20.0 mg of insulin in water and diluting to a volume of 5.00 mL gives an osmotic pressure of 12.5 mm Hg at 300 K. What is the molecular weight of insulin?

STRATEGY

To determine molecular weight, we need to know the number of moles of insulin represented by the 20.0 mg sample. We can do this by first rearranging the equation for osmotic pressure to find the molar concentration of the insulin solution and then multiplying by the volume of the solution to obtain the number of moles of insulin.

SOLUTION

$$\text{Since} \quad \Pi = MRT, \quad \text{then} \quad M = \frac{\Pi}{RT}$$

$$M = \frac{12.5\ \text{mm Hg} \times \dfrac{1\ \text{atm}}{760\ \text{mm Hg}}}{0.082\,06\ \dfrac{\text{L}\cdot\text{atm}}{\text{K}\cdot\text{mol}} \times 300\ \text{K}} = 6.68 \times 10^{-4}\ \text{M}$$

Since the volume of the solution is 5.00 mL, the number of moles of insulin is

$$\text{Moles insulin} = 6.68 \times 10^{-4}\ \frac{\text{mol}}{\text{L}} \times \frac{1\ \text{L}}{1000\ \text{mL}} \times 5.00\ \text{mL} = 3.34 \times 10^{-6}\ \text{mol}$$

Knowing both the mass and the number of moles of insulin, we can calculate the molar mass and thus the molecular weight:

$$\text{Molar mass} = \frac{\text{mass of insulin}}{\text{moles of insulin}} = \frac{0.0200\ \text{g insulin}}{3.34 \times 10^{-6}\ \text{mol insulin}} = 5990\ \text{g/mol}$$

The molecular weight of insulin is 5990.

PROBLEM 11.25 A solution of 0.250 g of naphthalene (mothballs) in 35.00 g of camphor lowers the freezing point by 2.10 °C. What is the molar mass of naphthalene? The freezing-point-depression constant for camphor is 37.7 (°C · kg)/mol.

PROBLEM 11.26 What is the molar mass of sucrose (table sugar) if a solution prepared by dissolving 0.822 g of sucrose in 300.0 mL of water has an osmotic pressure of 149 mm Hg at 298 K?

11.10 FRACTIONAL DISTILLATION OF LIQUID MIXTURES

Undoubtedly the most important of all commercial applications of colligative properties is in the refining of petroleum to make gasoline. Although petroleum refineries appear as a vast maze of pipes, tanks, and towers, the pipes are just for transferring the petroleum or its products

▲ Distillation of petroleum into fractions according to boiling point is carried out in the large towers of this refinery.

and the tanks are just for storage. It's in the towers that the separation of crude petroleum into usable fractions takes place. As we saw in Section 8.11, petroleum is a complex mixture of hydrocarbon molecules that are refined by distillation into different fractions: straight-run gasoline (bp 30–200 °C), kerosene (bp 175–300 °C), and gas oil (bp 275–400 °C).

Called **fractional distillation**, the separation of a mixture of different liquids into fractions with different boiling points occurs when the mixture is boiled and the vapors are condensed. Because the vapor is enriched in the component with the higher vapor pressure according to Raoult's law (Section 11.6), the condensed vapors are also enriched in that component and a partial purification can be effected. If the boil/condense cycle is repeated a large number of times, complete purification of the more volatile liquid component can be achieved.

As an example, let's look at the separation by fractional distillation of a 1:1 molar mixture of benzene and toluene ($X_{benzene}$ and $X_{toluene}$ are both 0.500). If we begin by heating the mixture, boiling occurs when the sum of the vapor pressures equals atmospheric pressure—that is, when $X \cdot P°_{benzene} + X \cdot P°_{toluene} = 760$ mm Hg. Reading from the vapor-pressure curves in **Figure 11.17** (or calculating values with the **Clausius–Clapeyron equation** as discussed in Section 10.5), we find that boiling occurs at 365.3 K (92.2 °C), where $P°_{benzene} = 1084$ mm Hg and $P°_{toluene} = 436$ mm Hg:

$$\begin{aligned} P_{mixt} &= X \cdot P°_{benzene} + X \cdot P°_{toluene} \\ &= (0.500)(1084 \text{ mm Hg}) + (0.500)(436 \text{ mm Hg}) \\ &= 542 \text{ mm Hg} + 218 \text{ mm Hg} \\ &= 760 \text{ mm Hg} \end{aligned}$$

REMEMBER...

The **Clausius–Clapeyron equation** describes the relationship between vapor pressure and heat of vaporization, according to the expression $\ln P_{vap} = (-\Delta H_{vap}/R)(1/T) + C$. (Section 10.5)

Although the starting *liquid* mixture of benzene and toluene has a 1:1 molar composition, the composition of the *vapor* is not 1:1. Of the 760 mm Hg total vapor pressure for the boiling mixture, 542/760 = 71.3% is due to benzene and 218/760 = 28.7% is due to toluene. If we now condense the vapor, the liquid we get has this same 71.3:28.7 composition. On boiling this new liquid mixture, the composition of the vapor now becomes 86.4% benzene and 13.6% toluene. A third condense/boil cycle brings the composition of the vapor to 94.4% benzene/5.6% toluene, and so on through further cycles until the desired level of purity is reached.

FIGURE 11.17

Vapor-pressure curves for pure benzene (blue), pure toluene (red), and a 1:1 mixture of the two (green).

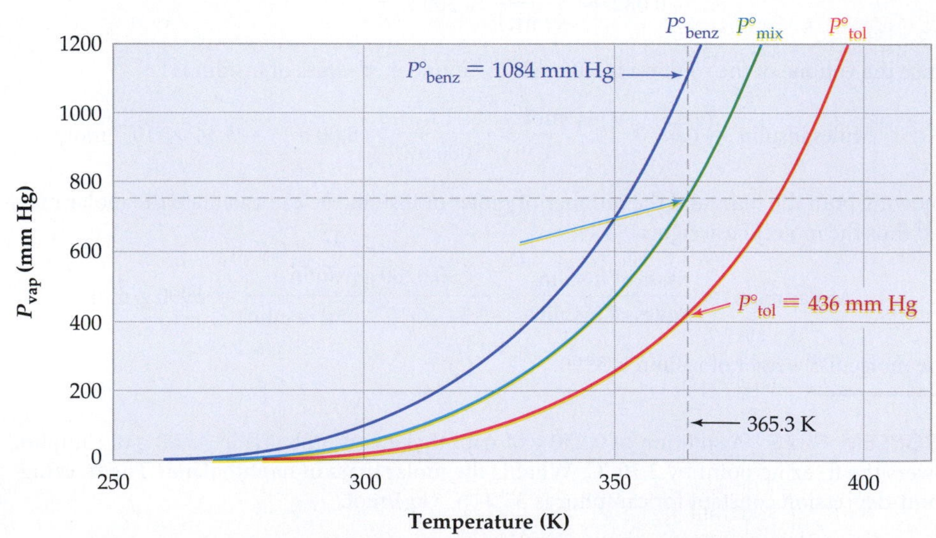

The mixture of benzene and toluene boils at 92.2 °C (365.3 K) at atmospheric pressure, intermediate between the boiling points of the two pure liquids.

Fractional distillation can be represented on a liquid/vapor phase diagram by plotting temperature versus composition, as shown in **Figure 11.18**. The lower region of the diagram represents the liquid phase, and the upper region represents the vapor phase. Between the two is a thin equilibrium region where liquid and vapor coexist.

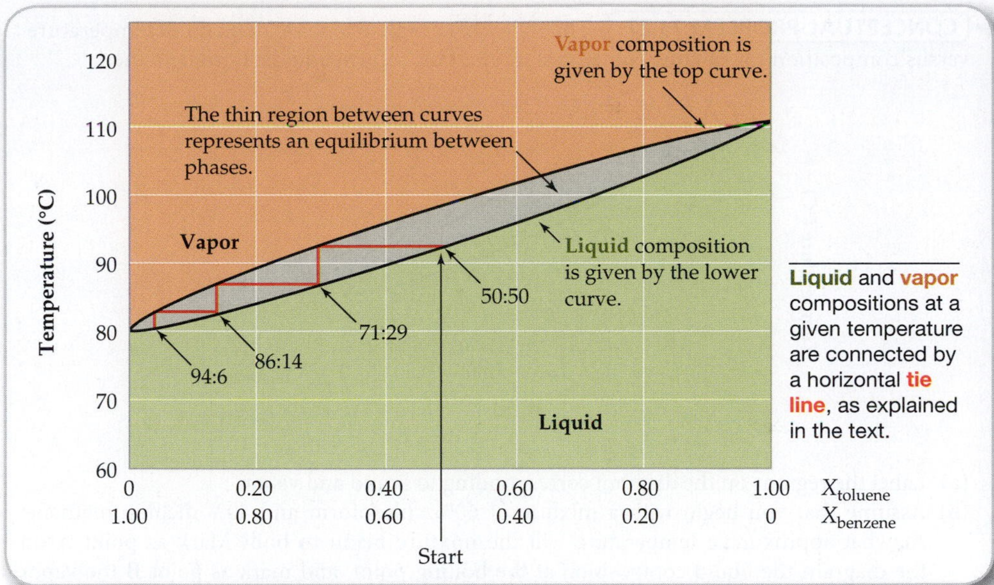

FIGURE 11.18

A phase diagram of temperature versus composition (mole fraction) for a mixture of benzene and toluene.

To understand how the diagram works, let's imagine starting in the middle of Figure 11.18 with the 50:50 benzene/toluene mixture and heating it to its boiling point (92.2 °C on the diagram). The lower curve represents the liquid composition (50:50), but the upper curve represents the vapor composition (approximately 71:29 at 92.2 °C). The two points are connected by a short horizontal line called a *tie line* to indicate that the temperature is the same at both points. Condensing the 71:29 vapor mixture by lowering the temperature gives a 71:29 liquid mixture that, when heated to *its* boiling point (86.6 °C), has an 86:14 vapor composition, as represented by another tie line. In essence, fractional distillation is simply a stairstep walk across successive tie lines in whatever number of steps is necessary to reach the desired purity.

In practice, the successive boil/condense cycles occur naturally in the distillation column, and there is no need to isolate liquid mixtures at intermediate stages of purification. Fractional distillation is therefore relatively simple to carry out and is used on a daily basis in refineries, chemical plants, and laboratories throughout the world (**Figure 11.19**).

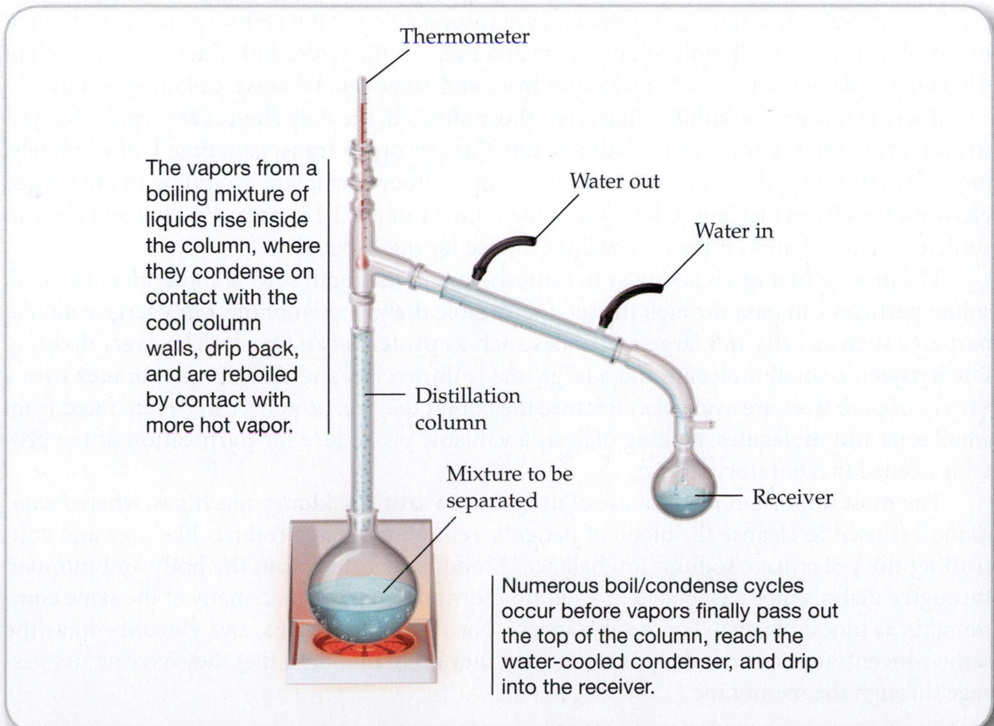

FIGURE 11.19

A simple fractional distillation column used in a chemistry laboratory.

CONCEPTUAL PROBLEM 11.27 The following graph is a phase diagram of temperature versus composition for mixtures of the two liquids chloroform and dichloromethane.

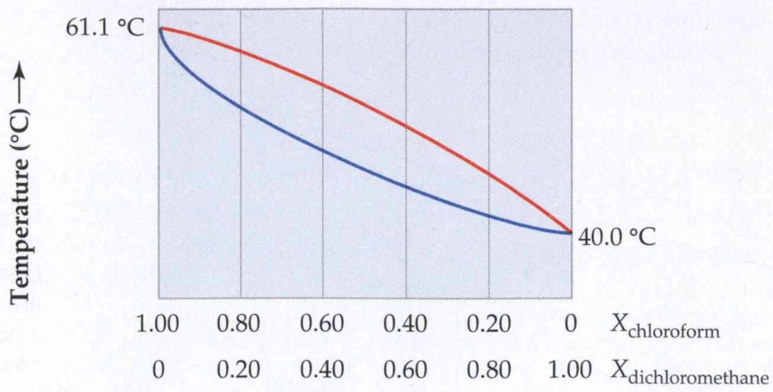

(a) Label the regions on the diagram corresponding to liquid and vapor.
(b) Assume that you begin with a mixture of 60% chloroform and 40% dichloromethane. At what approximate temperature will the mixture begin to boil? Mark as point **A** on the diagram the liquid composition at the boiling point, and mark as point **B** the vapor composition at the boiling point.
(c) Assume that the vapor at point **B** condenses and is reboiled. Mark as point **C** on the diagram the liquid composition of the condensed vapor and as point **D** on the diagram the vapor composition of the reboiled material.
(d) What will the approximate composition of the liquid be after carrying out two cycles of boiling and condensing?

FYI KIDNEY FAILURE, HEMODIALYSIS, AND COLLIGATIVE PROPERTIES

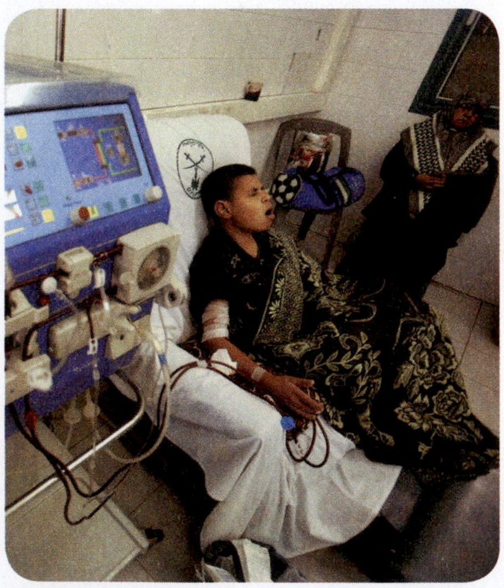

▲ Hemodialysis on this artificial kidney machine cleanses the blood of individuals whose kidneys no longer function.

The primary function of the kidneys is to filter the blood and remove wastes by producing urine. But kidneys also have a number of other important functions on which life depends, including regulation of blood pressure, body pH, and electrolyte balance. In addition, they produce several important hormones and are responsible for the reabsorption from filtered blood of ions and small molecules such as glucose, amino acids, and water. Yet more than 330,000 people in the U.S. currently suffer from end-stage renal disease, or kidney failure.

There is no cure for kidney failure. For those afflicted, the only choices are regular dialysis treatments to replace some of the kidney's functions or organ transplantation. Unfortunately, more than 91,000 patients are currently awaiting a kidney transplant according to the United Network for Organ Sharing (UNOS), and the number of available organs is nowhere close to sufficient. Thus, dialysis is the only realistic choice for many people.

The process of *dialysis* is similar to osmosis, except that both solvent molecules and small solute particles can pass through the semipermeable dialysis membrane. Only large colloidal particles such as cells and large molecules such as proteins can't pass. (The exact dividing line between a small molecule and a large one is imprecise, and dialysis membranes with a variety of pore sizes are available.) Because they don't dialyze, proteins can be separated from small ions and molecules, making dialysis a valuable procedure for purification of the proteins needed in laboratory studies.

The most important medical use of dialysis is in artificial kidney machines, where *hemodialysis* is used to cleanse the blood of patients, removing waste products like urea and controlling the potassium/sodium ion balance. Blood is diverted from the body and pumped through a dialysis tube suspended in a solution formulated to contain many of the same components as blood plasma. These substances—NaCl, $NaHCO_3$, KCl, and glucose—have the same concentrations in the dialysis solution as they do in blood, so that they have no net passage through the membrane.

Small waste products pass through the dialysis membrane from the blood to the solution side, where they are washed away, but cells, proteins, and other important blood components are prevented by their size from passing through the membrane. A typical hemodialysis treatment lasts for approximately 3–4 hours, and treatments are usually repeated 3–4 times per week.

And by the way: Please consider signing the necessary authorization card indicating your willingness to become an organ donor. In most states, all that's needed is to sign the back of your driver's license.

PROBLEM 11.28 What is the difference between a dialysis membrane and the typical semipermeable membrane used for osmosis?

SUMMARY

Solutions are homogeneous mixtures that contain particles the size of a typical ion or small molecule. Any one state of matter can mix with any other state, leading to seven possible kinds of solutions. For solutions in which a gas or solid is dissolved in a liquid, the dissolved substance is called the **solute** and the liquid is called the **solvent**.

The dissolution of a solute in a solvent has associated with it a free-energy change, $\Delta G = \Delta H - T\Delta S$. The enthalpy change is the **enthalpy of solution (ΔH_{soln})**, and the entropy change is the **entropy of solution (ΔS_{soln})**. Enthalpies of solution can be either positive or negative, depending on the relative strengths of solvent–solvent, solute–solute, and solvent–solute intermolecular forces. Entropies of solution are usually positive because molecular randomness increases when a pure solute dissolves in a pure solvent.

The concentration of a solution can be expressed in many ways, including **molarity** (moles of solute per liter of solution), **mole fraction** (moles of solute per mole of solution), **mass percent** (mass of solute per mass of solution times 100%), and **molality** (moles of solute per kilogram of solvent). When equilibrium is reached and no further solute dissolves in a given amount of solvent, a solution is said to be **saturated**. Solubilities are usually temperature-dependent, although often not in a simple way. Gas solubilities usually decrease with increasing temperature, but the solubilities of solids can either increase or decrease. The solubilities of gases also depend on pressure. According to **Henry's law**, the solubility of a gas in a liquid at a given temperature is proportional to the partial pressure of the gas over the solution.

In comparison with a pure solvent, a solution has a lower vapor pressure at a given temperature, a lower freezing point, and a higher boiling point. In addition, a solution that is separated from solvent by a semipermeable membrane gives rise to the phenomenon of **osmosis**. All four of these properties of solutions depend only on the concentration of dissolved solute particles rather than on the chemical identity of those particles and are therefore called **colligative properties**. The fundamental cause of all colligative properties is the same: the higher entropy of the solvent in a solution relative to that of the pure solvent.

Colligative properties have many practical uses, including the melting of snow by salt, the desalination of seawater by reverse osmosis, the separation and purification of volatile liquids by fractional distillation, and the determination of molecular weight by osmotic pressure measurement.

KEY WORDS

KEY EQUATIONS

- Henry's law **(Section 11.4)**

  Solubility $= k \cdot P$ where the constant k is characteristic of a specific gas and P is the partial pressure of the gas over the solution.

- Raoult's law and vapor pressure lowering **(Section 11.6)**

$$P_{soln} = P_{solv} \times X_{solv}$$
$$\Delta P_{soln} = P_{solv} \times X_{solute}$$

- Molal boiling-point elevation **(Section 11.7)**

  $\Delta T_b = K_b \cdot m$ where K_b is the molal boiling-point-elevation constant.
- Molal freezing-point depression **(Section 11.7)**

  $\Delta T_f = -K_f \cdot m$ where K_f is the molal freezing-point-depression constant.
- Osmotic pressure **(Section 11.8)**

  $\Pi = MRT$

LEARNING OUTCOMES

- Explain the energy changes that take place when a solute dissolves in a solvent to form a solution. **(Section 11.2)** ***Problems 1, 2, 38–45, 110***
- Know how to use and interconvert the concentration units molarity, mole fraction, mass percent, ppm, ppb, and molality. **(Section 11. 3)** ***Problems 3–10, 46–67, 90, 91, 117, 118, 126–128***
- Use Henry's Law to calculate the solubility of a gas at different pressures. **(Section 11.4)** ***Problems 11, 12, 68–75***
- Calculate the vapor pressure of a solution containing a nonvolatile solute. **(Section 11.6)** ***Problems 13–15, 34, 76, 80, 81, 122, 123***
- Calculate the vapor pressure of a solution containing a volatile solute. **(Section 11.6)** ***Problems 16–17, 31, 76, 77, 86–89, 115, 132***
- Calculate the amount of boiling-point elevation and freezing-point depression for a solution. **(Section 11.7)** ***Problems 18–22, 79, 82–85, 92–95, 100, 101, 104, 105, 108, 109, 114, 116, 120, 121, 125, 132–134, 136***
- Calculate osmotic pressure for a solution. **(Section 11.8)** ***Problems 23, 24, 32, 79, 96, 97, 111–113, 117, 129, 135***
- Use osmotic pressure or other colligative property to calculate the molecular weight of a solute. **(Section 11.8)** ***Problems 25, 26, 98, 99, 102, 103, 106, 107, 119, 124, 130, 131***
- Interpret the liquid/liquid phase diagram of a solution and apply the diagram to fractional distillation. **(Section 11.10)** ***Problems 27, 29, 33, 35–37, 78***

CONCEPTUAL PROBLEMS

Problems 11.1–11.28 appear within the chapter.

11.29 The following phase diagram shows part of the liquid/vapor phase-transition boundaries for pure ether and a solution of a nonvolatile solute in ether.

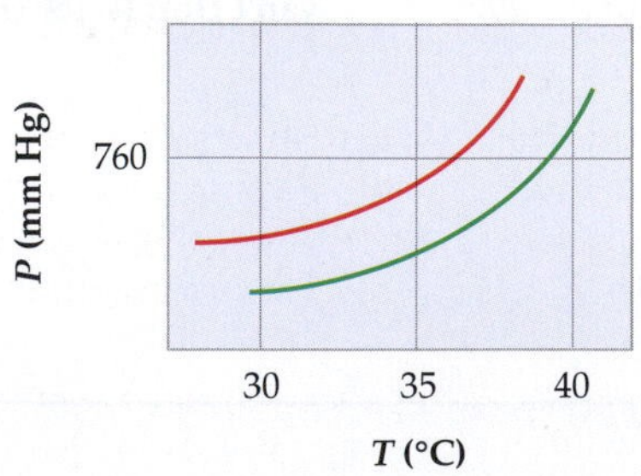

(a) What is the approximate normal boiling point of pure ether?

(b) What is the approximate molal concentration of the solute? [K_b for ether is 2.02 (°C · kg)/mol.]

11.30 Rank the situations represented by the following drawings according to increasing entropy.

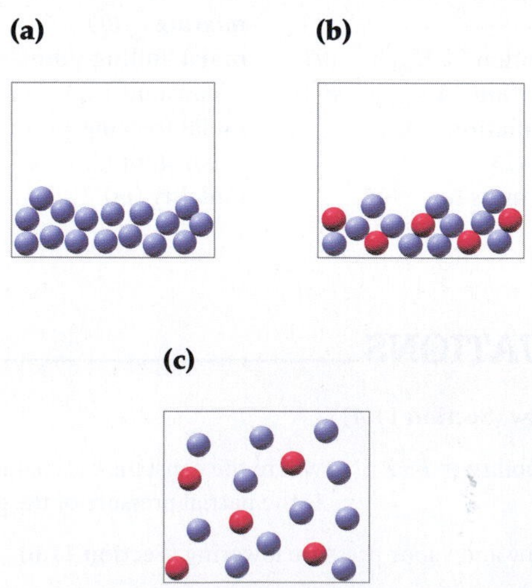

11.31 The following diagram shows a close-up view of part of the vapor-pressure curves for a solvent (red curve) and a solution of the solvent with a second liquid (green curve). Is the second liquid more volatile or less volatile than the solvent?

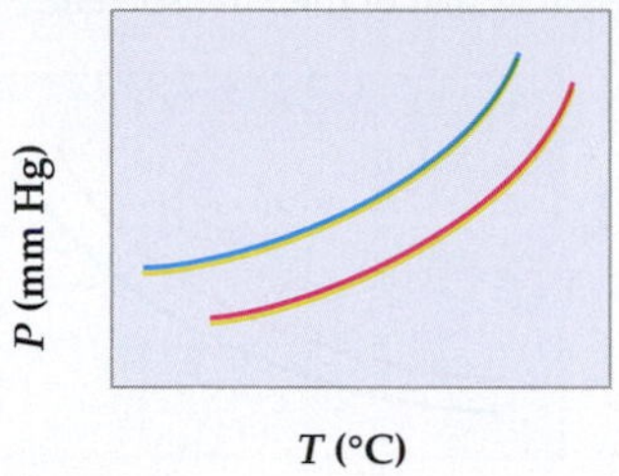

11.32 Assume that two liquids are separated by a semipermeable membrane. Make a drawing that shows the situation after equilibrium is reached.

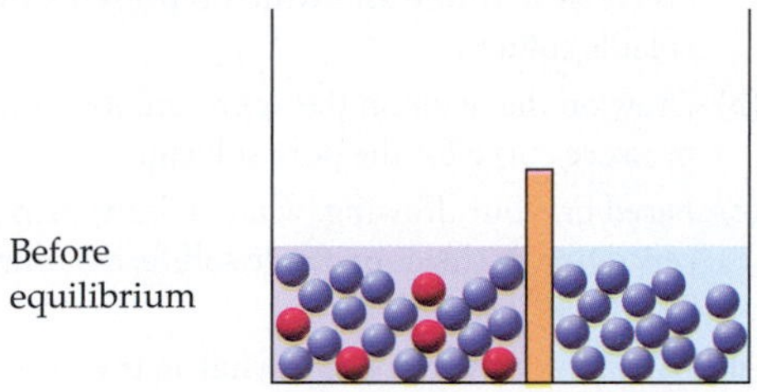

11.33 Assume that you are distilling a 1:1 molar mixture of chloroform ($CHCl_3$; bp = 61.7 °C) and tetrachloromethane (CCl_4; bp = 76.5 °C). After the mixture has been heated to its boiling point and vapor has reached the top of the column, predict the approximate temperatures at the three points indicated.

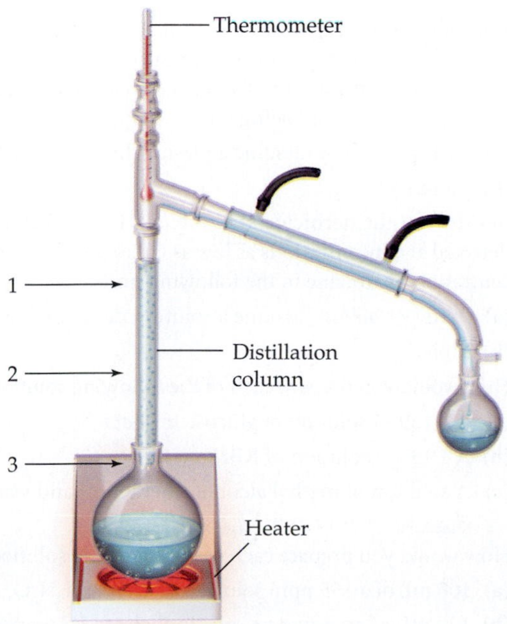

11.34 Two beakers, one with pure water (blue) and the other with a solution of NaCl in water (green), are placed in a closed container as represented by drawing **(a)**. Which of the drawings **(b)–(d)** represents what the beakers will look like after a substantial amount of time has passed?

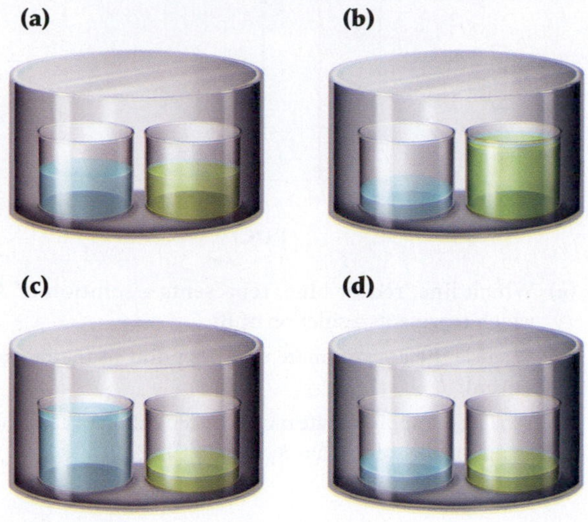

11.35 A phase diagram of temperature versus composition for a mixture of the two volatile liquids hexane (bp = 69 °C) and decane (bp = 126 °C) is shown. Assume that you begin with a mixture containing 0.60 mol of decane and 0.40 mol of hexane.

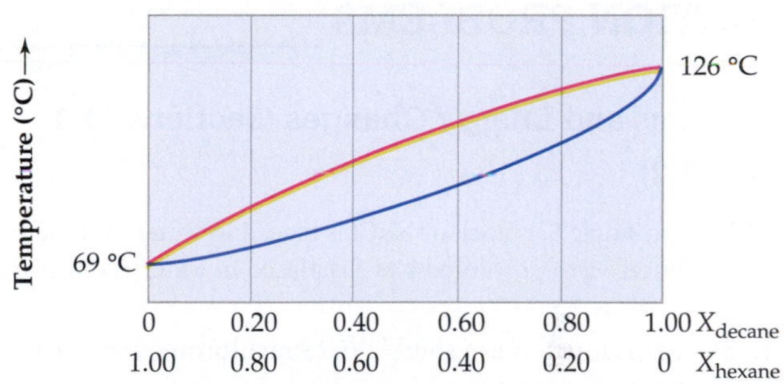

(a) What region on the diagram corresponds to vapor, and what region corresponds to liquid?

(b) At what approximate temperature will the mixture begin to boil? Mark as point **A** on the diagram the liquid composition at the boiling point, and mark as point **B** the vapor composition at the boiling point.

(c) Assume that the vapor at point **B** condenses and is reboiled. Mark as point **C** on the diagram the liquid composition of the condensed vapor and as point **D** on the diagram the vapor composition of the reboiled material.

11.36 The following phase diagram shows a very small part of the solid–liquid phase-transition boundaries for two solutions of equal concentration. Substance **A** has $i = 1$, and substance **B** has $i = 3$.

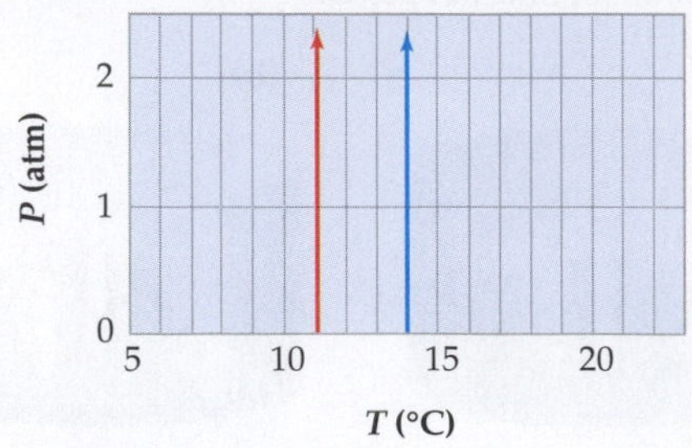

(a) Which line, red or blue, represents a solution of **A**, and which represents a solution of **B**?

(b) What is the approximate melting point of the pure liquid solvent?

(c) What is the approximate molal concentration of each solution, assuming the solvent has $K_f = 3.0\ °C/m$?

11.37 The following phase diagram shows part of the liquid–vapor phase-transition boundaries for two solutions of equal concentration, one containing a nonvolatile solute and the other containing a volatile solute whose vapor pressure at a given temperature is approximately half that of the pure solvent.

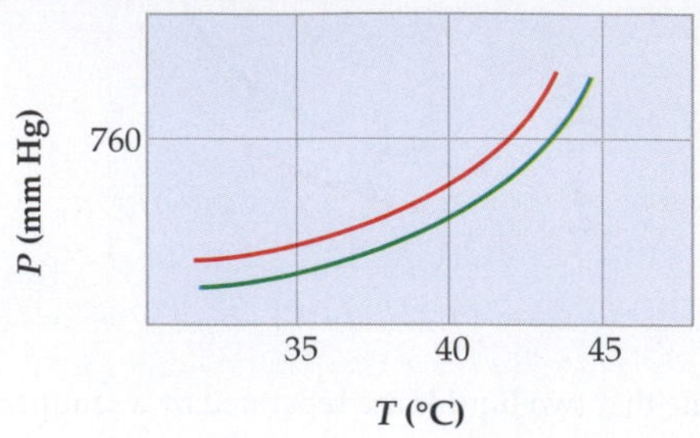

(a) Which curve, red or green, represents the solution of the nonvolatile solute and which represents the solution of the volatile solute?

(b) Draw on the diagram the approximate position of the vapor-pressure curve for the pure solvent.

(c) Based on your drawing, what is the approximate molal concentration of the nonvolatile solute, assuming the solvent has $K_b = 2.0\ °C/m$?

(d) Based on your drawing, what is the approximate normal boiling point of the pure solvent?

SECTION PROBLEMS

Solutions and Energy Changes (Sections 11.1 and 11.2)

11.38 If a single 5 g block of NaCl is placed in water, it dissolves slowly, but if 5 g of powdered NaCl is placed in water, it dissolves rapidly. Explain.

11.39 Br_2 is much more soluble in tetrachloromethane, CCl_4, than in water. Explain.

11.40 Why do ionic substances with higher lattice energies tend to be less soluble in water than substances with lower lattice energies?

11.41 Which would you expect to have the larger hydration energy, SO_4^{2-} or ClO_4^-? Explain.

11.42 Ethyl alcohol, CH_3CH_2OH, is miscible with water at 20°C, but pentyl alcohol, $CH_3CH_2CH_2CH_2CH_2OH$, is soluble in water only to the extent of 2.7 g/100 mL. Explain.

11.43 Pentyl alcohol (Problem 11.42) is miscible with octane, C_8H_{18}, but methyl alcohol, CH_3OH, is insoluble in octane. Explain.

11.44 The dissolution of $CaCl_2(s)$ in water is exothermic, with $\Delta H_{soln} = -81.3$ kJ/mol. If you were to prepare a 1.00 m solution of $CaCl_2$ beginning with water at 25.0 °C, what would the final temperature of the solution be in °C? Assume that the specific heats of both pure H_2O and the solution are the same, 4.18 J/(K · g).

11.45 The dissolution of $NH_4ClO_4(s)$ in water is endothermic, with $\Delta H_{soln} = +33.5$ kJ/mol. If you prepare a 1.00 m solution of NH_4ClO_4 beginning with water at 25.0 °C, what is the final temperature of the solution in °C? Assume that the specific heats of both pure H_2O and the solution are the same, 4.18 J/(K · g).

Units of Concentration (Section 11.3)

11.46 Propranolol ($C_{16}H_{21}NO_2$), a so-called beta-blocker that is used for treatment of high blood pressure, is effective at a blood plasma concentration of 50 ng/L. Express this concentration of propranolol in the following units:

(a) Parts per billion (assume a plasma density of 1.025 g/mL)

(b) Molarity

11.47 Residues of the herbicide atrazine ($C_8H_{14}ClN_5$) in water can be detected at concentrations as low as 0.050 μg/L. Express this concentration of atrazine in the following units:

(a) Parts per billion (assume a solution density of 1.00 g/mL)

(b) Molarity

11.48 How would you prepare each of the following solutions?

(a) A 0.150 M solution of glucose in water

(b) A 1.135 m solution of KBr in water

(c) A solution of methyl alcohol (methanol) and water in which $X_{methanol} = 0.15$ and $X_{water} = 0.85$

11.49 How would you prepare each of the following solutions?

(a) 100 mL of a 155 ppm solution of urea, CH_4N_2O, in water

(b) 100 mL of an aqueous solution whose K^+ concentration is 0.075 M

11.50 How would you prepare 165 mL of a 0.0268 M solution of benzoic acid ($C_7H_6O_2$) in chloroform ($CHCl_3$)?

11.51 How would you prepare 165 mL of a 0.0268 *m* solution of benzoic acid ($C_7H_6O_2$) in chloroform ($CHCl_3$)?

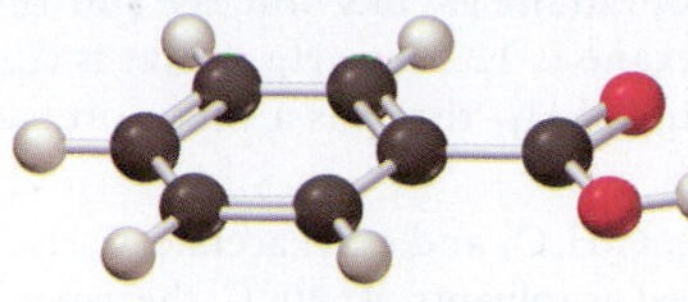

Benzoic acid

11.52 Which of the following solutions is more concentrated?

(a) 0.500 M KCl or 0.500 mass % KCl in water

(b) 1.75 M glucose or 1.75 *m* glucose in water

11.53 Which of the following solutions has the higher molarity?

(a) 10 ppm KI in water or 10,000 ppb KBr in water

(b) 0.25 mass % KCl in water or 0.25 mass % citric acid ($C_6H_8O_7$) in water

11.54 What is the mass percent concentration of the following solutions?

(a) Dissolve 0.655 mol of citric acid, $C_6H_8O_7$, in 1.00 kg of water.

(b) Dissolve 0.135 mg of KBr in 5.00 mL of water.

(c) Dissolve 5.50 g of aspirin, $C_9H_8O_4$, in 145 g of dichloromethane, CH_2Cl_2.

11.55 What is the molality of each solution prepared in Problem 11.54?

11.56 The so-called ozone layer in the Earth's stratosphere has an average total pressure of 10 mm Hg (1.3×10^{-2} atm). The partial pressure of ozone in the layer is about 1.2×10^{-6} mm Hg (1.6×10^{-9} atm). What is the concentration of ozone in parts per million, assuming that the average molar mass of air is 29 g/mol?

11.57 Persons are medically considered to have lead poisoning if they have a concentration of greater than 10 micrograms of lead per deciliter of blood. What is this concentration in parts per billion? Assume that the density of blood is the same as that of water.

11.58 What is the concentration of each of the following solutions?

(a) The molality of a solution prepared by dissolving 25.0 g of H_2SO_4 in 1.30 L of water

(b) The mole fraction of each component of a solution prepared by dissolving 2.25 g of nicotine, $C_{10}H_{14}N_2$, in 80.0 g of CH_2Cl_2

11.59 Household bleach is a 5.0 mass % aqueous solution of sodium hypochlorite, NaOCl. What is the molality of the bleach? What is the mole fraction of NaOCl in the bleach?

11.60 The density of a 16.0 mass % solution of sulfuric acid in water is 1.1094 g/mL at 25.0 °C. What is the molarity of the solution?

11.61 Ethylene glycol, $C_2H_6O_2$, is the principal constituent of automobile antifreeze. If the density of a 40.0 mass % solution of ethylene glycol in water is 1.0514 g/mL at 20 °C, what is the molarity?

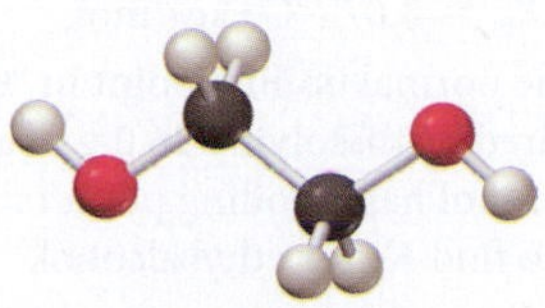

Ethylene glycol

11.62 What is the molality of the 40.0 mass % ethylene glycol solution used for automobile antifreeze (Problem 11.61)?

11.63 What is the molality of the 16.0 mass % solution of sulfuric acid in Problem 11.60?

11.64 Nalorphine ($C_{19}H_{21}NO_3$), a relative of morphine, is used to combat withdrawal symptoms in narcotics users. How many grams of a 1.3×10^{-3} *m* aqueous solution of nalorphine are needed to obtain a dose of 1.5 mg?

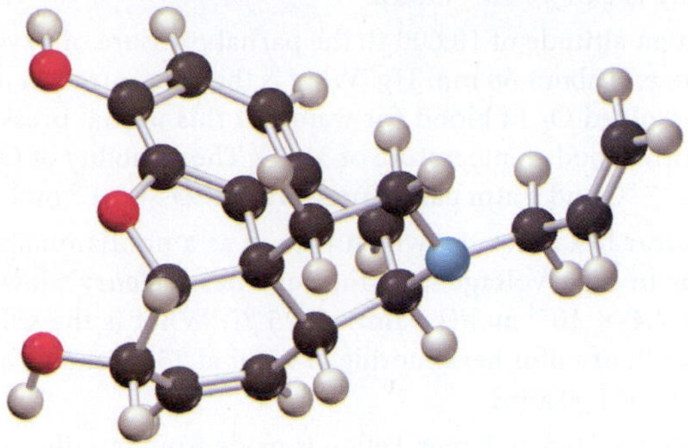

Nalorphine

11.65 How many grams of water should you add to 32.5 g of sucrose, $C_{12}H_{22}O_{11}$, to get a 0.850 *m* solution?

11.66 A 0.944 M solution of glucose, $C_6H_{12}O_6$, in water has a density of 1.0624 g/mL at 20 °C. What is the concentration of this solution in the following units?

(a) Mole fraction

(b) Mass percent

(c) Molality

11.67 Lactose, $C_{12}H_{22}O_{11}$, is a naturally occurring sugar found in mammalian milk. A 0.335 M solution of lactose in water has a density of 1.0432 g/mL at 20 °C. What is the concentration of this solution in the following units?

(a) Mole fraction

(b) Mass percent

(c) Molality

Solubility and Henry's Law (Section 11.4)

11.68 Look at the solubility graph in Figure 11.6 on page 412, and estimate which member of each of the following pairs has the higher molar solubility at the given temperature:

(a) $CuSO_4$ or NH_4Cl at 60 °C

(b) CH_3CO_2Na or glucose at 20 °C

11.69 Look at the solubility graph in Figure 11.6, and estimate which member of each of the following pairs has the higher molar solubility at the given temperature:

(a) NaCl or NH_4Cl at 40 °C

(b) K_2SO_4 or $CuSO_4$ at 20 °C

11.70 Vinyl chloride ($H_2C{=}CHCl$), the starting material from which PVC polymer is made, has a Henry's law constant of 0.091 mol/(L·atm) at 25 °C. What is the solubility of vinyl chloride in water in mol/L at 25 °C and a partial pressure of 0.75 atm?

11.71 Hydrogen sulfide, H_2S, is a toxic gas responsible for the odor of rotten eggs. The solubility of $H_2S(g)$ in water at STP is 0.195 M. What is the Henry's law constant of H_2S at 0 °C? What is the solubility of H_2S in water at 0 °C and a partial pressure of 25.5 mm Hg?

11.72 Fish generally need an O_2 concentration in water of at least 4 mg/L for survival. What partial pressure of oxygen above the water in atmospheres at 0 °C is needed to obtain this concentration? The solubility of O_2 in water at 0 °C and 1 atm partial pressure is 2.21×10^{-3} mol/L.

11.73 At an altitude of 10,000 ft, the partial pressure of oxygen in the lungs is about 68 mm Hg. What is the concentration in mg/L of dissolved O_2 in blood (or water) at this partial pressure and a normal body temperature of 37 °C? The solubility of O_2 in water at 37 °C and 1 atm partial pressure is 1.93×10^{-3} mol/L.

11.74 Sulfur hexafluoride, which is used as a nonflammable insulator in high-voltage transformers, has a Henry's law constant of 2.4×10^{-4} mol/(L · atm) at 25 °C. What is the solubility in mol/L of sulfur hexafluoride in water at 25 °C and a partial pressure of 2.00 atm?

11.75 The nonstick polymer Teflon is made from tetrafluoroethylene, C_2F_4. If C_2F_4 is a gas that dissolves in water at 298 K to the extent of 1.01×10^{-3} M with a partial pressure of 0.63 atm, what is its Henry's law constant at 298 K?

Colligative Properties (Section 11.5–11.8)

11.76 When 1 mol of NaCl is added to 1 L of water, the boiling point increases. When 1 mol of methyl alcohol is added to 1 L of water, the boiling point decreases. Explain.

11.77 When 100 mL of 9 M H_2SO_4 at 0 °C is added to 100 mL of liquid water at 0 °C, the temperature rises to 12 °C. When 100 mL of 9 M H_2SO_4 at 0 °C is added to 100 g of solid ice at 0 °C, the temperature falls to −12 °C. Explain the difference in behavior.

11.78 Draw a phase diagram showing how the phase boundaries differ for a pure solvent compared with a solution.

11.79 A solution concentration must be expressed in molality when considering boiling-point elevation or freezing-point depression but can be expressed in molarity when considering osmotic pressure. Why?

11.80 What is the vapor pressure in mm Hg of the following solutions, each of which contains a nonvolatile solute? The vapor pressure of water at 45.0 °C is 71.93 mm Hg.

(a) A solution of 10.0 g of urea, CH_4N_2O, in 150.0 g of water at 45.0 °C

(b) A solution of 10.0 g of LiCl in 150.0 g of water at 45.0 °C, assuming complete dissociation

11.81 What is the vapor pressure in mm Hg of a solution of 16.0 g of glucose ($C_6H_{12}O_6$) in 80.0 g of methanol (CH_3OH) at 27 °C? The vapor pressure of pure methanol at 27 °C is 140 mm Hg.

11.82 What is the boiling point in °C of each of the solutions in Problem 11.80? For water, $K_b = 0.51$ (°C · kg)/mol.

11.83 What is the freezing point in °C of each of the solutions in Problem 11.80? For water, $K_f = 1.86$ (°C · kg)/mol.

11.84 A 1.0 *m* solution of K_2SO_4 in water has a freezing point of −4.3 °C. What is the value of the van't Hoff factor *i* for K_2SO_4?

11.85 The van't Hoff factor for KCl is $i = 1.85$. What is the boiling point of a 0.75 *m* solution of KCl in water? For water, $K_b = 0.51$ (°C · kg)/mol.

11.86 Heptane (C_7H_{16}) and octane (C_8H_{18}) are constituents of gasoline. At 80.0 °C, the vapor pressure of heptane is 428 mm Hg and the vapor pressure of octane is 175 mm Hg. What is $X_{heptane}$ in a mixture of heptane and octane that has a vapor pressure of 305 mm Hg at 80.0 °C?

11.87 Cyclopentane (C_5H_{10}) and cyclohexane (C_6H_{12}) are volatile, nonpolar hydrocarbons. At 30.0 °C, the vapor pressure of cyclopentane is 385 mm Hg and the vapor pressure of cyclohexane is 122 mm Hg. What is $X_{pentane}$ in a mixture of C_5H_{10} and C_6H_{12} that has a vapor pressure of 212 mm Hg at 30.0 °C?

11.88 Acetone, C_3H_6O, and ethyl acetate, $C_4H_8O_2$, are organic liquids often used as solvents. At 30 °C, the vapor pressure of acetone is 285 mm Hg and the vapor pressure of ethyl acetate is 118 mm Hg. What is the vapor pressure in mm Hg at 30 °C of a solution prepared by dissolving 25.0 g of acetone in 25.0 g of ethyl acetate?

Acetone   Ethyl acetate

11.89 The industrial solvents chloroform, $CHCl_3$, and dichloromethane, CH_2Cl_2, are prepared commercially by reaction of methane with chlorine, followed by fractional distillation of the product mixture. At 25 °C, the vapor pressure of $CHCl_3$ is 205 mm Hg and the vapor pressure of CH_2Cl_2 is 415 mm Hg. What is the vapor pressure in mm Hg at 25 °C of a mixture of 15.0 g of $CHCl_3$ and 37.5 g of CH_2Cl_2?

11.90 What is the mole fraction of each component in the liquid mixture in Problem 11.88, and what is the mole fraction of each component in the vapor at 30 °C?

11.91 What is the mole fraction of each component in the liquid mixture in Problem 11.89, and what is the mole fraction of each component in the vapor at 25 °C?

11.92 A solution prepared by dissolving 5.00 g of aspirin, $C_9H_8O_4$, in 215 g of chloroform has a normal boiling point that is elevated by $\Delta T = 0.47$ °C over that of pure chloroform. What is the value of the molal boiling-point-elevation constant for chloroform?

11.93 A solution prepared by dissolving 3.00 g of ascorbic acid (vitamin C, $C_6H_8O_6$) in 50.0 g of acetic acid has a freezing point that is depressed by $\Delta T = 1.33$ °C below that of pure acetic acid. What is the value of the molal freezing-point-depression constant for acetic acid?

11.94 A solution of citric acid, $C_6H_8O_7$, in 50.0 g of acetic acid has a boiling point elevation of $\Delta T = 1.76$ °C. What is the molality of the solution if the molal boiling-point-elevation constant for acetic acid is $K_b = 3.07$ (°C · kg)/mol.

11.95 What is the normal boiling point in °C of ethyl alcohol if a solution prepared by dissolving 26.0 g of glucose ($C_6H_{12}O_6$) in 285 g of ethyl alcohol has a boiling point of 79.1 °C? See Table 11.4 on page 423 to find K_b for ethyl alcohol.

11.96 What osmotic pressure in atmospheres would you expect for each of the following solutions?

(a) 5.00 g of NaCl in 350.0 mL of aqueous solution at 50 °C

(b) 6.33 g of sodium acetate, CH_3CO_2Na, in 55.0 mL of aqueous solution at 10 °C

11.97 What osmotic pressure in mm Hg would you expect for an aqueous solution of 11.5 mg of insulin (mol. weight = 5990) in 6.60 mL of solution at 298 K? What would the height of the water column be in meters? The density of mercury is 13.534 g/mL at 298 K.

11.98 A solution of an unknown molecular substance in water at 300 K gives rise to an osmotic pressure of 4.85 atm. What is the molarity of the solution?

11.99 Human blood gives rise to an osmotic pressure of approximately 7.7 atm at body temperature, 37.0 °C. What must the molarity of an intravenous glucose solution be to give rise to the same osmotic pressure as blood?

Uses of Colligative Properties (Section 11.9)

11.100 When salt is spread on snow-covered roads at −2 °C, the snow melts. When salt is spread on snow-covered roads at −30 °C, nothing happens. Explain.

11.101 If cost per gram were not a concern, which of the following substances would be the most efficient per unit mass for melting snow from sidewalks and roads: glucose ($C_6H_{12}O_6$), LiCl, NaCl, or $CaCl_2$? Explain.

11.102 Cellobiose is a sugar obtained by degradation of cellulose. If 200.0 mL of an aqueous solution containing 1.500 g of cellobiose at 25.0 °C gives rise to an osmotic pressure of 407.2 mm Hg, what is the molecular weight of cellobiose?

11.103 Met-enkephalin is one of the so-called endorphins, a class of naturally occurring morphinelike chemicals in the brain. What is the molecular weight of met-enkephalin if 20.0 mL of an aqueous solution containing 15.0 mg of met-enkephalin at 298 K supports a column of water 32.9 cm high? The density of mercury at 298 K is 13.534 g/mL.

11.104 The freezing point of a solution prepared by dissolving 1.00 mol of hydrogen fluoride, HF, in 500 g of water is −3.8 °C, but the freezing point of a solution prepared by dissolving 1.00 mol of hydrogen chloride, HCl, in 500 g of water is −7.4 °C. Explain.

11.105 The boiling point of a solution prepared by dissolving 71 g of Na_2SO_4 in 1.00 kg of water is 100.8 °C. Explain.

11.106 Elemental analysis of β-carotene, a dietary source of vitamin A, shows that it contains 10.51% H and 89.49% C. Dissolving 0.0250 g of β-carotene in 1.50 g of camphor gives a freezing-point depression of 1.17 °C. What are the molecular weight and formula of β-carotene? [K_f for camphor is 37.7 (°C · kg)/mol.]

11.107 Lysine, one of the amino acid building blocks found in proteins, contains 49.29% C, 9.65% H, 19.16% N, and 21.89% O by elemental analysis. A solution prepared by dissolving 30.0 mg of lysine in 1.200 g of the organic solvent biphenyl gives a freezing-point depression of 1.37 °C. What are the molecular weight and formula of lysine? [K_f for biphenyl is 8.00 (°C · kg)/mol.]

CHAPTER PROBLEMS

11.108 How many grams of ethylene glycol (automobile antifreeze, $C_2H_6O_2$) dissolved in 3.55 kg of water is needed to lower the freezing point of water in an automobile radiator to −22.0 °C?

11.109 When 1 mL of toluene is added to 100 mL of benzene (bp 80.1 °C), the boiling point of the benzene solution rises, but when 1 mL of benzene is added to 100 mL of toluene (bp 110.6 °C), the boiling point of the toluene solution falls. Explain.

11.110 When solid $CaCl_2$ is added to liquid water, the temperature rises. When solid $CaCl_2$ is added to ice at 0 °C, the temperature falls. Explain.

11.111 Silver chloride has a solubility of 0.007 mg/mL in water at 5 °C. What is the osmotic pressure in atmospheres of a saturated solution of AgCl at 5 °C?

11.112 When a 2.850 g mixture of the sugars sucrose ($C_{12}H_{22}O_{11}$) and fructose ($C_6H_{12}O_6$) was dissolved in water to a volume of 1.50 L, the resultant solution gave an osmotic pressure of 0.1843 atm at 298.0 K. What is $X_{sucrose}$ of the mixture?

11.113 Glycerol ($C_3H_8O_3$) and diethylformamide ($C_5H_{11}NO$) are nonvolatile, miscible liquids. If the volume of a solution made by dissolving 10.208 g of a glycerol–diethylformamide mixture in water is 1.75 L and the solution has an osmotic pressure of 1.466 atm at 298.0 K, what is $X_{glycerol}$ of the mixture?

11.114 How many grams of naphthalene, $C_{10}H_8$ (commonly used as household mothballs), should be added to 150.0 g of benzene to depress its freezing point by 0.35 °C? See Table 11.4 on page 423 to find K_f for benzene.

11.115 Bromine is sometimes used as a solution in tetrachloromethane, CCl_4. What is the vapor pressure in mm Hg of a solution of 1.50 g of Br_2 in 145.0 g of CCl_4 at 300 K? The vapor pressure of pure bromine at 300 K is 30.5 kPa, and the vapor pressure of CCl_4 is 16.5 kPa.

11.116 Assuming that seawater is a 3.5 mass % solution of NaCl and that its density is 1.00 g/mL, calculate both its boiling point and its freezing point in °C.

11.117 There's actually much more in seawater than just dissolved NaCl. Major ions present include 19,000 ppm Cl^-, 10,500 ppm Na^+, 2650 ppm SO_4^{2-}, 1350 ppm Mg^{2+}, 400 ppm Ca^{2+}, 380 ppm K^+, 140 ppm HCO_3^-, and 65 ppm Br^-.

(a) What is the total molality of all ions present in seawater?

(b) Assuming molality and molarity to be equal, what amount of osmotic pressure in atmospheres would seawater give rise to at 300 K?

11.118 Rubbing alcohol is a 90 mass % solution of isopropyl alcohol, C_3H_8O, in water.

(a) How many grams of rubbing alcohol contains 10.5 g of isopropyl alcohol?

(b) How many moles of isopropyl alcohol are in 50.0 g of rubbing alcohol?

11.119 Although not particularly convenient, it's possible to use osmotic pressure to measure temperature. What is the Kelvin temperature if a solution prepared by dissolving 17.5 mg of glucose ($C_6H_{12}O_6$) in 50.0 mL of aqueous solution gives rise to an osmotic pressure of 37.8 mm Hg?

11.120 The van't Hoff factor for $CaCl_2$ is 2.71. What is its mass % in an aqueous solution that has T_f = −1.14 °C?

11.121 What is the van't Hoff factor for K_2SO_4 in an aqueous solution that is 5.00% K_2SO_4 by mass and freezes at −1.21 °C?

11.122 If the van't Hoff factor for LiCl in a 0.62 *m* solution is 1.96, what is the vapor pressure depression in mm Hg of the solution at 298 K? (The vapor pressure of water at 298 K is 23.76 mm Hg.)

11.123 What is the value of the van't Hoff factor for KCl if a 1.00 *m* aqueous solution shows a vapor pressure depression of 0.734 mm Hg at 298 °C? (The vapor pressure of water at 298 K is 23.76 mm Hg.)

11.124 The steroid hormone *estradiol* contains only C, H, and O; combustion analysis of a 3.47 mg sample yields 10.10 mg CO_2 and 2.76 mg H_2O. On dissolving 7.55 mg of estradiol in 0.500 g of camphor, the melting point of camphor is depressed by 2.10 °C. What is the molecular weight of estradiol, and what is a probable formula? [For camphor, $K_f = 37.7\ (°C \cdot kg)/mol$.]

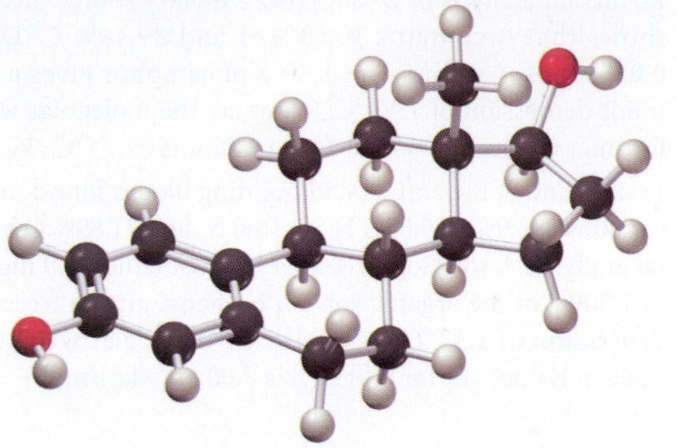

Estradiol

11.125 Many acids are partially dissociated into ions in aqueous solution. Trichloroacetic acid (CCl_3CO_2H), for instance, is partially dissociated in water according to the equation

$$CCl_3CO_2H(aq) \rightleftharpoons H^+(aq) + CCl_3CO_2^-(aq)$$

What is the percentage of molecules dissociated if the freezing point of a 1.00 *m* solution of trichloroacetic acid in water is −2.53 °C?

11.126 Addition of 50.00 mL of 2.238 *m* H_2SO_4 (solution density = 1.1243 g/mL) to 50.00 mL of 2.238 M $BaCl_2$ gives a white precipitate.

(a) What is the mass of the precipitate in grams?

(b) If you filter the mixture and add more H_2SO_4 solution to the filtrate, would you obtain more precipitate? Explain.

11.127 A solid mixture of KCl, KNO_3, and $Ba(NO_3)_2$ is 20.92 mass % chlorine, and a 1.000 g sample of the mixture in 500.0 mL of aqueous solution at 25 °C has an osmotic pressure of 744.7 mm Hg. What are the mass percents of KCl, KNO_3, and $Ba(NO_3)_2$ in the mixture?

11.128 A solution of LiCl in a mixture of water and methanol (CH_3OH) has a vapor pressure of 39.4 mm Hg at 17 °C and 68.2 mm Hg at 27 °C. The vapor pressure of pure water is 14.5 mm Hg at 17 °C and 26.8 mm Hg at 27 °C, and the vapor pressure of pure methanol is 82.5 mm Hg at 17 °C and 140.3 mm Hg at 27 °C. What is the composition of the solution in mass percent?

11.129 An aqueous solution of KI has a freezing point of −1.95 °C and an osmotic pressure of 25.0 atm at 25.0 °C. Assuming that the KI completely dissociates in water, what is the density of the solution?

11.130 An aqueous solution of a certain organic compound has a density of 1.063 g/mL, an osmotic pressure of 12.16 atm at 25.0 °C, and a freezing point of −1.03 °C. The compound is known not to dissociate in water. What is the molar mass of the compound?

11.131 At 60 °C, compound **X** has a vapor pressure of 96 mm Hg, benzene (C_6H_6) has a vapor pressure of 395 mm Hg, and a 50:50 mixture by mass of benzene and **X** has a vapor pressure of 299 mm Hg. What is the molar mass of **X**?

11.132 An aqueous solution containing 100.0 g of NaCl and 100.0 g of $CaCl_2$ has a volume of 1.00 L and a density of 1.15 g/mL. The vapor pressure of pure water at 25 °C is 23.8 mm Hg, and you can assume complete dissociation for both solutes.

(a) What is the boiling point of the solution?

(b) What is the vapor pressure of the solution at 25 °C?

11.133 Iodic acid, HIO_3, is a weak acid that undergoes only partial dissociation in water. If a 1.00 M solution of HIO_3 has a density of 1.07 g/mL and a freezing point of −2.78 °C, what percent of the HIO_3 is dissociated?

11.134 A 1.24 M solution of KI has a density of 1.15 g/cm^3.

(a) What is the molality of the solution?

(b) What is the freezing point of the solution, assuming complete dissociation of KI?

(c) The actual freezing point of the solution is −4.46 °C. What percent of the KI is dissociated?

11.135 Desert countries like Saudi Arabia have built reverse osmosis plants to produce freshwater from seawater. Assume that seawater has the composition 0.470 M NaCl and 0.068 M $MgCl_2$ and that both compounds are completely dissociated.

(a) What is the osmotic pressure of seawater at 25 °C?

(b) If the reverse osmosis equipment can exert a maximum pressure of 100.0 atm at 25.0 °C, what is the maximum volume of freshwater that can be obtained from 1.00 L of seawater?

11.136 A solution prepared by dissolving 100.0 g of a mixture of sugar ($C_{12}H_{22}O_{11}$) and table salt (NaCl) in 500.0 g of water has a freezing point of −2.25 °C. What is the mass of each individual solute? Assume that NaCl is completely dissociated.

MULTICONCEPT PROBLEMS

11.137 Treatment of 1.385 g of an unknown metal **M** with an excess of aqueous HCl evolved a gas that was found to have a volume of 382.6 mL at 20.0 °C and 755 mm Hg pressure. Heating the reaction mixture to evaporate the water and remaining HCl then gave a white crystalline compound, MCl_x. After dissolving the compound in 25.0 g of water, the melting point of the resulting solution was −3.53 °C.

(a) How many moles of H_2 gas are evolved?

(b) What mass of MCl_x is formed?

(c) What is the molality of particles (ions) in the solution of MCl_x?

(d) How many moles of ions are in solution?

(e) What are the formula and molecular weight of MCl_x?

(f) What is the identity of the metal **M**?

11.138 A compound that contains only C and H was burned in excess O_2 to give CO_2 and H_2O. When 0.270 g of the compound was burned, the amount of CO_2 formed reacted completely with 20.0 mL of 2.00 M NaOH solution according to the equation

$$2\,OH^-(aq) + CO_2(g) \longrightarrow CO_3^{2-}(aq) + H_2O(l)$$

When 0.270 g of the compound was dissolved in 50.0 g of camphor, the resulting solution had a freezing point of 177.9 °C. [Pure camphor freezes at 179.8 °C and has $K_f = 37.7$ (°C · kg)/mol.]

(a) What is the empirical formula of the compound?

(b) What is the molecular weight of the compound?

(c) What is the molecular formula of the compound?

11.139 Combustion analysis of a 36.72 mg sample of the male hormone testosterone gave 106.43 mg CO_2 and 32.100 mg H_2O as the only combustion products. When 5.00 mg of testosterone was dissolved in 15.0 mL of a suitable solvent at 25 °C, an osmotic pressure of 21.5 mm Hg was measured. What is the molecular formula of testosterone?

11.140 When 8.900 g of a mixture of an alkali metal chloride (**X**Cl) and an alkaline earth chloride (**Y**Cl_2) was dissolved in 150.0 g of water, the freezing point of the resultant solution was −4.42 °C. Addition of an excess of aqueous $AgNO_3$ to the solution yielded a white precipitate with a mass of 27.575 g. How much of each metal chloride was present in the original mixture, and what are the identities of the two metals **X** and **Y**?

11.141 Combustion analysis of a 3.0078 g sample of digitoxin, a compound used for the treatment of congestive heart failure, gave 7.0950 g of CO_2 and 2.2668 g of H_2O. When 0.6617 g of digitoxin was dissolved in water to a total volume of 0.800 L, the osmotic pressure of the solution at 298 K was 0.026 44 atm. What is the molecular formula of digitoxin, which contains only C, H, and O?

CHAPTER 12

The Rates and Mechanisms of Chemical Reactions

Chemical reactions occur as a result of energetic collisions between atoms and molecules.

Chemists ask three fundamental questions when they study chemical reactions: What happens? To what extent does it happen? How fast and by what mechanism, or sequence of reaction steps, does it happen? The answer to the first question is given by the balanced chemical equation, which identifies the reactants, the products, and the stoichiometry of the reaction. The answer to the second question is addressed in Chapter 13, which deals with **chemical equilibrium**. In this chapter, we'll look at the answer to the third question—the speeds, or rates, and the mechanisms of chemical reactions. The area of chemistry concerned with reaction rates and the sequence of steps by which reactions occur is called **chemical kinetics**.

Chemical kinetics is a subject of crucial environmental, biological, and economic importance. In the upper atmosphere, for example, maintenance or depletion of the ozone layer, which protects us from the sun's harmful ultraviolet radiation, depends on the relative rates of reactions that produce and destroy O_3 molecules. In our bodies, large protein molecules called *enzymes* increase the rates of numerous reactions essential to life processes. In the chemical industry, the profitability of many processes requires fast reaction rates. For example, the economical synthesis of ammonia, used as a fertilizer, depends on the rate at which gaseous N_2 and H_2 can be converted to NH_3. Like the biochemical reactions occurring within the body, industrial reactions can also be hastened by **catalysts**. Enzymes and other types of catalysts increase reaction rates without being consumed in the process (a topic we'll explore in Section 12.14).

Chemical reactions occur as a result of collisions between atoms and molecules. In Section 12.9 we'll look at the **collision theory** model of chemical reactions, which describes how the energy and geometry of molecular collisions affect the rates of reactions. But because science nearly always begins with observation and experiment, and rarely with theory, we'll first examine some of the experimental data that led to the collision theory.

LOOKING AHEAD...

We'll see in Chapter 13 that **chemical equilibrium** is the state reached when the concentrations of reactants and products remain constant over time.

LOOKING AHEAD...

We'll see in Section 12.14 why **catalysts** speed up chemical reactions.

LOOKING AHEAD...

We'll see in Section 12.9 that **collision theory** provides a picture of how chemical reactions occur.

CONTENTS

12.1 RATES OF CHEMICAL REACTIONS

The rates of chemical reactions differ greatly. Some reactions, such as the explosion of a gaseous mixture of H_2 and O_2, occur in a fraction of a second. Other reactions, such as the rusting of iron, are imperceptibly slow.

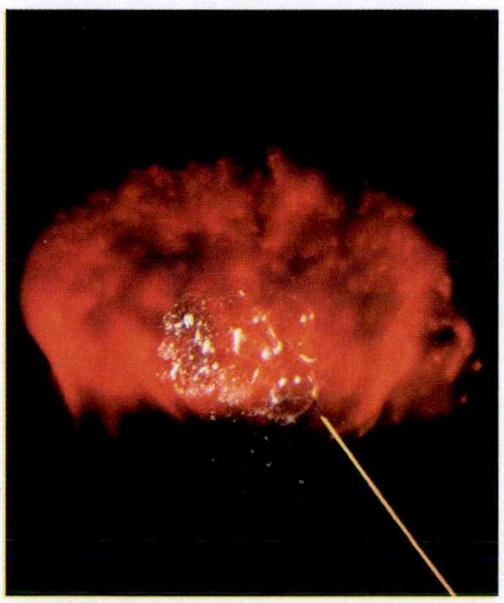

The explosion of gaseous H_2 and O_2, or

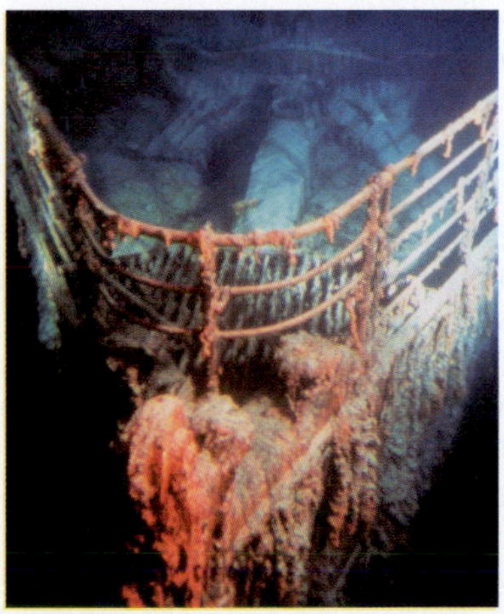

The rusting of iron?

▲ Which of these reactions is faster?

To describe a reaction rate quantitatively, we must specify how fast the concentration of a reactant or a product changes per unit time.

$$\text{Rate} = \frac{\text{Concentration change}}{\text{Time change}}$$

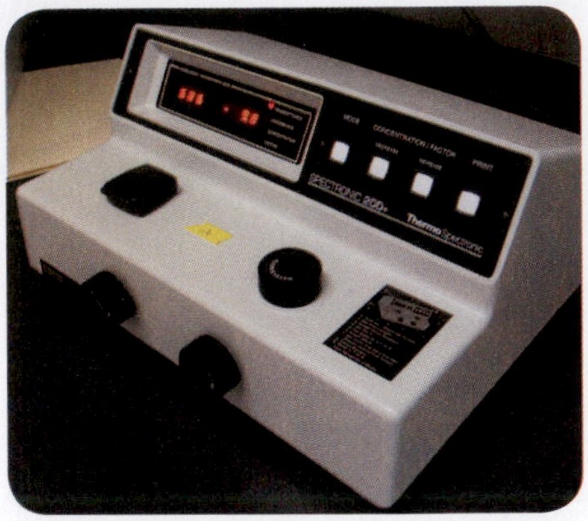

▲ The absorption of visible light, and thus the intensity of color, can be measured with a spectrophotometer.

Look, for example, at the thermal decomposition of gaseous dinitrogen pentoxide, N_2O_5, to give the brown gas nitrogen dioxide, a common air pollutant, and molecular oxygen:

$$\underset{\text{Colorless}}{2\,N_2O_5(g)} \longrightarrow \underset{\text{Brown}}{4\,NO_2(g)} + \underset{\text{Colorless}}{O_2(g)}$$

Changes in concentration as a function of time can be determined by measuring the increase in pressure as 2 gas molecules are converted to 5 gas molecules. Alternatively, concentration changes can be monitored by measuring the intensity of the brown color due to NO_2 with a spectrophotometer. Reactant and product concentrations as a function of time at 55 °C are listed in **Table 12.1**.

TABLE 12.1 Concentrations as a Function of Time at 55 °C for the Reaction $2\,N_2O_5(g) \longrightarrow 4\,NO_2(g) + O_2(g)$.

	Concentration (M)		
Time (s)	**N_2O_5**	**NO_2**	**O_2**
0	0.0200	0	0
100	0.0169	0.0063	0.0016
200	0.0142	0.0115	0.0029
300	0.0120	0.0160	0.0040
400	0.0101	0.0197	0.0049
500	0.0086	0.0229	0.0057
600	0.0072	0.0256	0.0064
700	0.0061	0.0278	0.0070

Note that the concentrations of NO_2 and O_2 increase as the concentration of N_2O_5 decreases.

The **reaction rate** is defined either as the *increase* in the concentration of a *product* per unit time or as the *decrease* in the concentration of a *reactant* per unit time. Let's look first at product formation. In the decomposition of N_2O_5, the rate of formation of O_2 is given by the equation

$$\text{Rate of Formation of } O_2 = \frac{\Delta[O_2]}{\Delta t} = \frac{\text{Conc. of } O_2 \text{ at time } t_2 - \text{Conc. of } O_2 \text{ at time } t_1}{t_2 - t_1}$$

where the square brackets surrounding O_2 denote its molar concentration; $\Delta[O_2]$ is the change in the molar concentration of O_2; Δt is the change in the time; and $\Delta[O_2]/\Delta t$ is the average rate of change in the molar concentration of O_2 during the interval from time t_1 to t_2. During the time period 300 to 400 s, for example, the average rate of formation of O_2 is 9×10^{-6} M/s:

$$\text{Rate of formation of } O_2 = \frac{\Delta[O_2]}{\Delta t} = \frac{0.0049\text{ M} - 0.0040\text{ M}}{400\text{ s} - 300\text{ s}} = 9 \times 10^{-6}\text{ M/s}$$

The most common units of reaction rate are molar per second, M/s or, equivalently, moles per liter second, mol/(L · s). We define reaction rate in terms of concentration (moles per liter) rather than amount (moles) because we want the rate to be independent of the scale of the reaction. When twice as much 0.0200 M N_2O_5 decomposes in a vessel of twice the volume, twice the number of moles of O_2 form per second, but the number of moles of O_2 *per liter* that form per second is unchanged.

Plotting the data of Table 12.1 to give the three curves in **Figure 12.1** affords additional insight into the concept of a reaction rate. Looking at the time period 300–400 s on the O_2 curve, $\Delta[O_2]$ and Δt are represented, respectively, by the vertical and horizontal sides of a right triangle. The slope of the third side, the hypotenuse of the triangle, is $\Delta[O_2]/\Delta t$, the average rate of O_2 formation during that time period. The steeper the slope of the hypotenuse, the faster the rate. Look, for example, at the triangle defined by $\Delta[NO_2]$ and Δt. The average

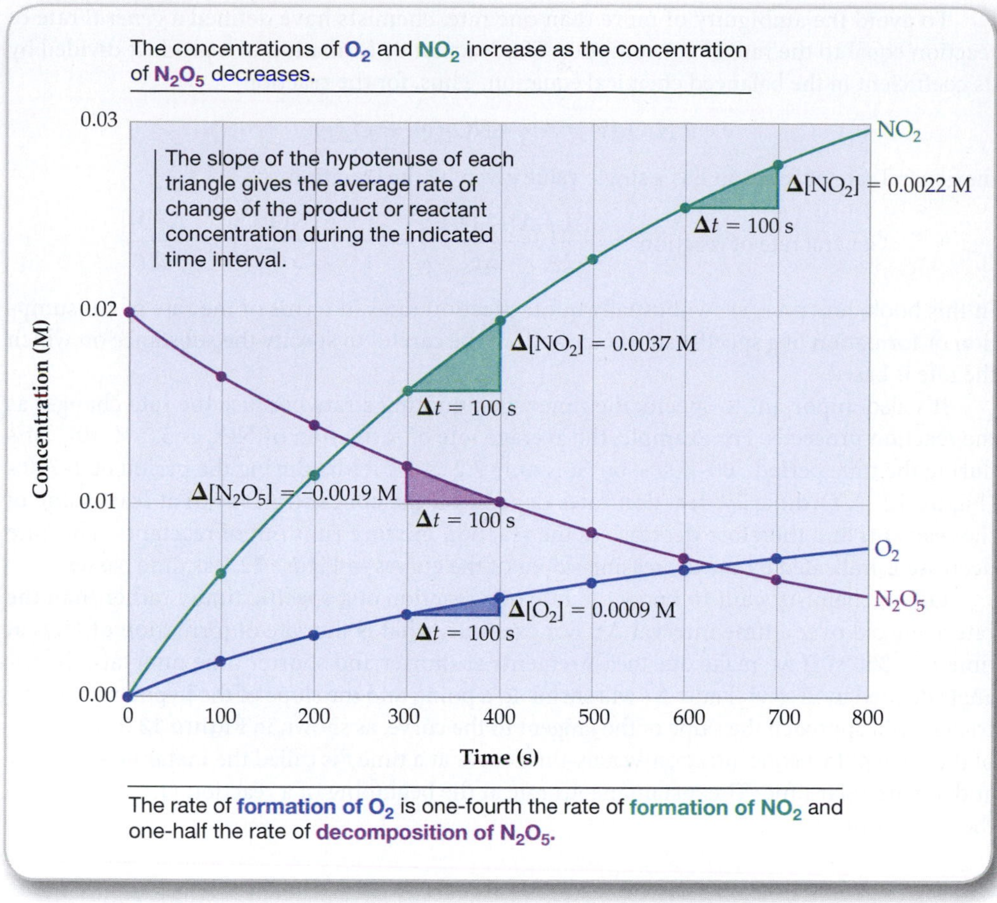

FIGURE 12.1

Concentrations as a function of time when gaseous N_2O_5 decomposes to gaseous NO_2 and O_2 at 55 °C.

rate of formation of NO_2 during the time period 300–400 s is 3.7×10^{-5} M/s, which is four times the rate of formation of O_2, in accord with the 4:1 ratio of the coefficients of NO_2 and O_2 in the chemical equation for the decomposition of N_2O_5.

$$\text{Rate of formation of } NO_2 = \frac{\Delta[NO_2]}{\Delta t} = \frac{0.0197\text{ M} - 0.0160\text{ M}}{400\text{ s} - 300\text{ s}} = 3.7 \times 10^{-5}\text{ M/s}$$

As O_2 and NO_2 form, N_2O_5 disappears. Consequently, $\Delta[N_2O_5]/\Delta t$ is negative, in accord with the negative slope of the hypotenuse of the triangle defined by $\Delta[N_2O_5]$ and Δt in Figure 12.1. Because *reaction rate is defined as a positive quantity*, we must always introduce a minus sign when calculating the rate of disappearance of a reactant. During the time period 300–400 s, for example, the average rate of decomposition of N_2O_5 is 1.9×10^{-5} M/s:

$$\text{Rate of decomposition of } N_2O_5 = \frac{-\Delta[N_2O_5]}{\Delta t} = \frac{-(0.0101\text{ M} - 0.0120\text{ M})}{400\text{ s} - 300\text{ s}} = 1.9 \times 10^{-5}\text{ M/s}$$

When quoting a reaction rate, it's important to specify the reactant or product on which the rate is based because the rates of product formation and reactant consumption may differ, depending on the coefficients in the balanced chemical equation. For the decomposition of N_2O_5, 4 mol of NO_2 form and 2 mol of N_2O_5 disappear for each mole of O_2 that forms. Therefore, the rate of formation of O_2 is 1/4 the rate of formation of NO_2 and 1/2 the rate of decomposition of N_2O_5:

$$\left(\begin{array}{c}\text{Rate of formation}\\ \text{of } O_2\end{array}\right) = \frac{1}{4}\left(\begin{array}{c}\text{Rate of formation}\\ \text{of } NO_2\end{array}\right) = \frac{1}{2}\left(\begin{array}{c}\text{Rate of decomposition}\\ \text{of } N_2O_5\end{array}\right)$$

$$\text{or} \quad \frac{\Delta[O_2]}{\Delta t} = \frac{1}{4}\left(\frac{\Delta[NO_2]}{\Delta t}\right) = -\frac{1}{2}\left(\frac{\Delta[N_2O_5]}{\Delta t}\right)$$

To avoid the ambiguity of more than one rate, chemists have defined a *general* rate of reaction equal to the rate of consumption of a reactant or formation of a product divided by its coefficient in the balanced chemical equation. Thus, for the reaction

$$2\,N_2O_5(g) \longrightarrow 4\,NO_2(g) + O_2(g)$$

the general rate of reaction has a single value given by the equation

$$\text{General rate of reaction} = -\frac{1}{2}\left(\frac{\Delta[N_2O_5]}{\Delta t}\right) = \frac{1}{4}\left(\frac{\Delta[NO_2]}{\Delta t}\right) = \frac{\Delta[O_2]}{\Delta t}$$

In this book, however, we will usually quote reaction rates in terms of the rate of consumption or formation of a specific substance and will be careful to specify the substance on which the rate is based.

It's also important to specify the time when quoting a rate because the rate changes as the reaction proceeds. For example, the average rate of formation of NO_2 is 3.7×10^{-5} M/s during the time period 300–400 s, but it is only 2.2×10^{-5} M/s during the period 600–700 s (Figure 12.1). Ordinarily, reaction rates depend on the concentrations of at least some of the reactants and therefore decrease as the reaction mixture runs out of reactants. This rate decrease is indicated by the decreasing slopes of the curves in Figure 12.1 as time passes.

Often, chemists want to know the rate of a reaction at a specific time t rather than the rate averaged over a time interval Δt. For example, what is the rate of formation of NO_2 at time $t = 350$ s? If we make our measurements at shorter and shorter time intervals, the triangle defined by $\Delta[NO_2]$ and Δt will shrink to a point, and the slope of the hypotenuse of the triangle will approach the slope of the tangent to the curve, as shown in **Figure 12.2**. The slope of the tangent to a concentration-versus-time curve at a time t is called the **instantaneous rate** at that particular time. The instantaneous rate at the beginning of a reaction ($t = 0$) is called the **initial rate**.

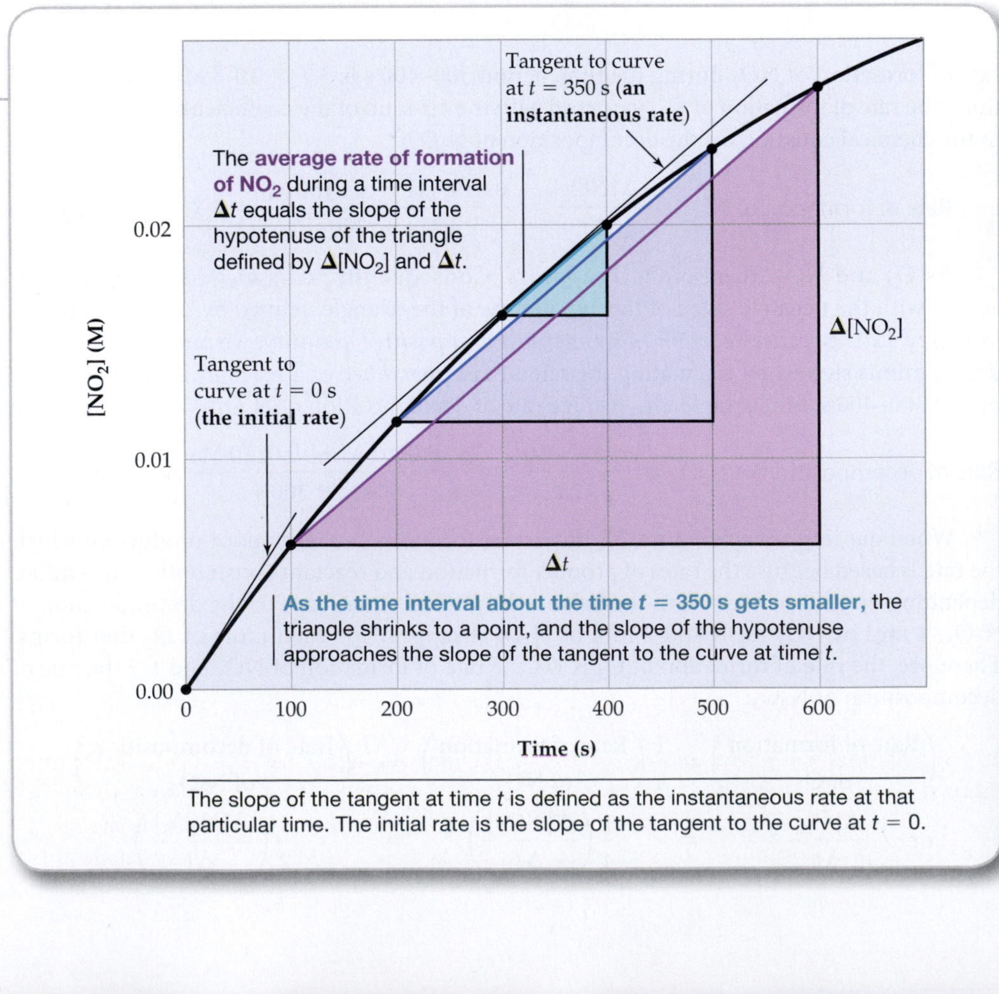

FIGURE 12.2

Concentration of NO_2 versus time when N_2O_5 decomposes at 55 °C. As measurements are made at shorter and shorter time intervals, the triangle defined by $\Delta[NO_2]$ and Δt becomes smaller and smaller, and the slope of the hypotenuse of the triangle approaches the slope of the tangent to the curve.

Worked Example 12.1

Finding Relative Rates of Product Formation and Reactant Consumption

Ethanol (C_2H_5OH), the active ingredient in alcoholic beverages and an octane booster in gasoline, is produced by the fermentation of glucose. The balanced equation is

$$C_6H_{12}O_6(aq) \longrightarrow 2\,C_2H_5OH(aq) + 2\,CO_2(g)$$

(a) How is the rate of formation of ethanol related to the rate of consumption of glucose?
(b) Write this relationship in terms of $\Delta[C_2H_5OH]/\Delta t$ and $\Delta[C_6H_{12}O_6]/\Delta t$.

STRATEGY

To find the relative rates, look at the coefficients in the balanced chemical equation.

SOLUTION

(a) According to the balanced equation, 2 mol of ethanol is produced for each mole of glucose that reacts. Therefore, the rate of formation of ethanol is twice the rate of consumption of glucose.

(b) Since the rate of formation of ethanol is $\Delta[C_2H_5OH]/\Delta t$ and the rate of consumption of glucose is $-\Delta[C_6H_{12}O_6]/\Delta t$ (note the minus sign), we can write

$$\frac{\Delta[C_2H_5OH]}{\Delta t} = -2\frac{\Delta[C_6H_{12}O_6]}{\Delta t}$$

PROBLEM 12.1 The oxidation of iodide ion by arsenic acid, H_3AsO_4, is described by the balanced equation

$$3\,I^-(aq) + H_3AsO_4(aq) + 2\,H^+(aq) \longrightarrow I_3^-(aq) + H_3AsO_3(aq) + H_2O(l)$$

(a) If $-\Delta[I^-]/\Delta t = 4.8 \times 10^{-4}$ M/s, what is the value of $\Delta[I_3^-]/\Delta t$ during the same time interval?

(b) What is the average rate of consumption of H^+ during that time interval?

PROBLEM 12.2 Use the data in Table 12.1 to calculate the average rate of decomposition of N_2O_5 and the average rate of formation of O_2 during the time interval 200–300 s.

12.2 RATE LAWS AND REACTION ORDER

We noted in Section 12.1 that the rate of decomposition of N_2O_5 depends on its concentration, slowing down as the N_2O_5 concentration decreases. To explore further how reaction rates depend on concentrations, let's consider the general reaction

$$a\,A + b\,B \longrightarrow \text{Products}$$

where A and B are the reactants and *a* and *b* are stoichiometric coefficients in the balanced chemical equation. The dependence of the reaction rate on the concentration of each reactant is given by an equation called the **rate law**:

▶ **Rate law** $\text{Rate} = -\dfrac{\Delta[A]}{\Delta t} = k[A]^m[B]^n$

In this equation, *k* is a proportionality constant called the **rate constant**. We've arbitrarily expressed the rate as the rate of disappearance of A ($-\Delta[A]/\Delta t$), but we could equally well have written it as the rate of disappearance of any reactant (say, $-\Delta[B]/\Delta t$) or as the rate of appearance of any product. The exponents *m* and *n* in the rate law indicate how sensitive the rate is to changes in the concentrations of A and B, and the exponents are generally unrelated to the coefficients a and b in the balanced chemical equation. For the simple reactions discussed in this book, the exponents are usually small positive integers. For more complex reactions, however, the exponents can be negative, zero, or even fractions.

Figure 12.3 shows how the rate changes when the concentration of A is doubled for various values of the exponent m. An exponent of 1 means that the rate depends linearly on the concentration of the corresponding reactant: if $m = 1$ and [A] is doubled, the rate doubles. If $m = 2$ and [A] is doubled, $[A]^2$ quadruples and the rate increases by a factor of 4. If m is zero, the rate is independent of the concentration of A because any number raised to the zeroth power equals one ($[A]^0 = 1$). If m is negative, the rate *decreases* as [A] increases. For example, if $m = -1$ and [A] is doubled, $[A]^{-1}$ is halved and the rate decreases by a factor of 2. In general, the rate is proportional to $[A]^m$.

FIGURE 12.3

Change in reaction rate when the concentration of reactant A is doubled for different values of the exponent m in the rate law, rate $= k[A]^m[B]^n$.

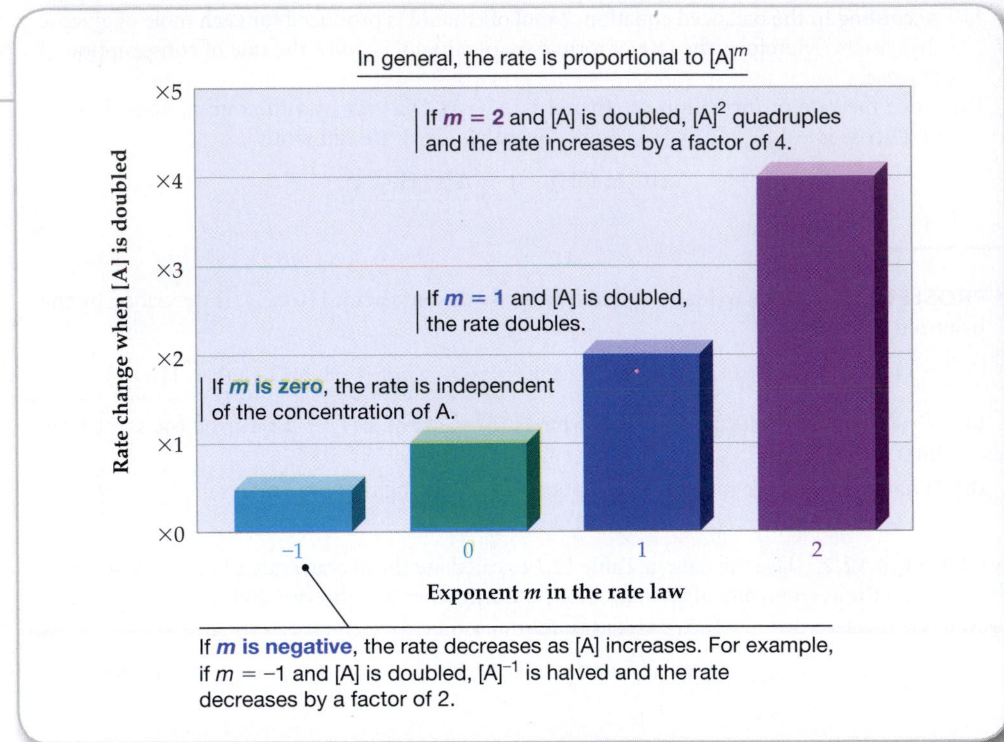

The values of the exponents m and n determine the **reaction order** with respect to A and B, respectively. The sum of the exponents $(m + n)$ defines the *overall* reaction order. Thus, if the rate law is

$$\text{Rate} = k[A]^2[B] \qquad m = 2; n = 1; \text{and } m + n = 3$$

we say that the reaction is second order in A, first order in B, and third order overall.

The values of the exponents in a rate law must be determined by experiment; they cannot be deduced from the stoichiometry of the reaction. As **Table 12.2** shows, there is no general relationship between the stoichiometric coefficients in the balanced chemical equation and the exponents in the rate law. In the first reaction in Table 12.2, for example, the coefficients of $(CH_3)_3CBr$ and H_2O in the balanced equation are both 1, but the exponents in the rate law are 1 for $(CH_3)_3CBr$ and 0 for H_2O:

$$\text{Rate} = k[(CH_3)_3CBr]^1[H_2O]^0 = k[(CH_3)_3CBr]$$

In Section 12.13, we'll see that the **exponents in a rate law** depend on exactly how the reaction occurs, that is, on the reaction mechanism.

LOOKING AHEAD...

We'll see in Section 12.13 that the **exponents in a rate law** are determined by the sequence of reaction steps and their relative rates.

TABLE 12.2 Balanced Chemical Equations and Experimentally Determined Rate Laws for Some Reactions.

Reaction*	Rate Law
$(CH_3)_3CBr(soln) + H_2O(soln) \longrightarrow (CH_3)_3COH(soln) + H^+(soln) + Br^-(soln)$	Rate $= k[(CH_3)_3CBr]$
$HCO_2H(aq) + Br_2(aq) \longrightarrow 2\,H^+(aq) + 2\,Br^-(aq) + CO_2(g)$	Rate $= k[Br_2]$
$BrO_3^-(aq) + 5\,Br^-(aq) + 6\,H^+(aq) \longrightarrow 3\,Br_2(aq) + 3\,H_2O(l)$	Rate $= k[BrO_3^-][Br^-][H^+]^2$
$H_2(g) + I_2(g) \longrightarrow 2\,HI(g)$	Rate $= k[H_2][I_2]$

*In the first reaction "(*soln*)" denotes a nonaqueous solution.

In general, the exponents in the rate law are not the same as the stoichiometric coefficients in the balanced chemical equation.

Worked Example 12.2

Finding Reaction Order from a Rate Law

The second reaction in Table 12.2, shown in progress in **Figure 12.4**, is

$$HCO_2H(aq) + Br_2(aq) \longrightarrow 2\,H^+(aq) + 2\,Br^-(aq) + CO_2(g)$$

Colorless Red Colorless

What is the order of the reaction with respect to each of the reactants? What is the overall reaction order?

FIGURE 12.4

The reaction of formic acid (HCO_2H) and bromine (Br_2). As time passes (left to right), the red color of bromine disappears because Br_2 is reduced to the colorless Br^- ion. The concentration of Br_2 as a function of time, and thus the reaction rate, can be determined by measuring the intensity of the color.

STRATEGY

To find the reaction order with respect to each reactant, look at the exponents in the rate law, not the coefficients in the balanced chemical equation. Then sum the exponents to obtain the overall reaction order.

SOLUTION

The rate law for the second reaction in Table 12.2 is

$$\text{Rate} = k[Br_2]$$

Because HCO_2H (formic acid) does not appear in the rate law, the rate is independent of the HCO_2H concentration, and so the reaction is zeroth order in HCO_2H. Because the exponent on $[Br_2]$ is 1 (it is understood to be 1 when no exponent is given), the reaction is first order in Br_2. The reaction is first order overall because the sum of the exponents is 1.

PROBLEM 12.3 Consider the last two reactions in Table 12.2. What is the reaction order with respect to each reactant in these two reactions? What is the overall reaction order for each reaction?

12.3 DETERMINING A RATE LAW: THE METHOD OF INITIAL RATES

One method of determining the values of the exponents in a rate law—the *method of initial rates*—is to carry out a series of experiments in which the initial rate of a reaction is measured as a function of different sets of initial concentrations. Take, for example, the oxidation of nitric oxide in air, one of the reactions that contributes to the formation of acid rain:

$$2\,NO(g) + O_2(g) \longrightarrow 2\,NO_2(g)$$

Some initial rate data are listed in **Table 12.3**.

TABLE 12.3 Initial Concentration and Rate Data for the Reaction $2\,NO(g) + O_2(g) \longrightarrow 2\,NO_2(g)$

Experiment	Initial [NO]	Initial $[O_2]$	Initial Rate of Formation of NO_2 (M/s)
1	0.015	0.015	0.048
2	0.030	0.015	0.192
3	0.015	0.030	0.096
4	0.030	0.030	0.384

Note that pairs of experiments are designed to investigate the effect on the initial rate of a change in the initial concentration of a single reactant. In the first two experiments, for example, the concentration of NO is doubled from 0.015 M to 0.030 M while the concentration of O_2 is held constant. The initial rate increases by a factor of 4, from 0.048 M/s to 0.192 M/s, indicating that the rate depends on the concentration of NO squared, $[NO]^2$. When [NO] is held constant and $[O_2]$ is doubled (Experiments 1 and 3), the initial rate doubles from 0.048 M/s to 0.096 M/s, indicating that the rate depends on the concentration of O_2 to the first power, $[O_2]^1$. Therefore, the rate law for the formation of NO_2 is

$$\text{Rate} = \frac{\Delta[NO_2]}{\Delta t} = k[NO]^2[O_2]$$

In accord with this rate law, which is second order in NO, first order in O_2, and third order overall, the initial rate increases by a factor of 8 when the concentrations of both NO and O_2 are doubled (Experiments 1 and 4).

The preceding method uses initial rates rather than rates at a later stage of the reaction because chemical reactions are reversible and we want to avoid complications from the reverse reaction: reactants $\leftarrow$ products. As the product concentrations build up, the rate of the reverse reaction increases and the measured rate is affected by the concentrations of both reactants and products. At the beginning of the reaction, however, the product concentrations are zero, and therefore the products can't affect the measured rate. When we measure an initial rate, we are measuring the rate of only the forward reaction, so only reactants (and catalysts; see Section 12.14) can appear in the rate law.

One aspect of determining a rate law, as we've just seen, is to establish the reaction order. Another is to evaluate the numerical value of the rate constant k. Each reaction has its own characteristic value of the rate constant, which depends on temperature but does not depend on concentrations. To evaluate k for the formation of NO_2, for instance, we can use the data from any one of the experiments in Table 12.3. Solving the rate law for k and substituting the initial rate and concentrations from the first experiment, we obtain

$$k = \frac{\text{Rate}}{[NO]^2[O_2]} = \frac{0.048\,\dfrac{M}{s}}{(0.015\,M)^2(0.015\,M)} = 1.4 \times 10^4/(M^2 \cdot s)$$

Try repeating the calculation for Experiments 2–4, and show that you get the same value of k. Note that the units of k in this example are $1/(M^2 \cdot s)$, read as "one over molar squared second." In general, the units of k depend on the number of concentration terms in the rate law and on the values of their exponents. Units for some common cases are given as follows:

Rate Law	Overall Reaction Order	Units for k
Rate $= k$	Zeroth order	M/s or $M\ s^{-1}$
Rate $= k[A]$	First order	1/s or s^{-1}
Rate $= k[A][B]$	Second order	$1/(M \cdot s)$ or $M^{-1}\ s^{-1}$
Rate $= k[A][B]^2$	Third order	$1/(M^2 \cdot s)$ or $M^{-2}\ s^{-1}$

Be careful not to confuse the rate of a reaction and its rate constant. The *rate* depends on concentrations, whereas the rate *constant* does not (it is a constant). The rate is usually expressed in units of M/s, whereas the units of the rate constant depend on the overall reaction order.

Worked Example 12.3 gives another instance of how a rate law can be determined from initial rates.

Worked Example 12.3

Determining a Rate Law from Initial Rates

Initial rate data at 25 °C are listed in the table for the reaction

$$NH_4^+(aq) + NO_2^-(aq) \longrightarrow N_2(g) + 2\,H_2O(l)$$

Experiment	Initial $[NH_4^+]$	Initial $[NO_2^-]$	Initial Rate of Consumption of NH_4^+ (M/s)
1	0.12	0.10	3.6×10^{-6}
2	0.24	0.10	7.2×10^{-6}
3	0.12	0.15	5.4×10^{-6}

(a) What is the rate law?
(b) What is the value of the rate constant including the proper units?
(c) What is the initial rate when the initial concentrations are $[NH_4^+] = 0.39$ M and $[NO_2^-] = 0.052$ M?

STRATEGY

(a) The rate law for the reaction can be written as

$$\text{Rate} = -\frac{\Delta[NH_4^+]}{\Delta t} = k[NH_4^+]^m[NO_2^-]^n$$

where m is the order of the reaction in NH_4^+ and n is the order of the reaction in NO_2^-. To find the values of the exponents m and n, compare the change in the initial rate with the change in the initial concentration of one reactant in pairs of experiments in which the concentration of the other reactant is held constant.

(b) To find the value of the rate constant k, solve the rate law for k and then substitute in the data from any one of the three experiments.

(c) To calculate the initial rate, substitute into the rate law the rate constant found in part **(b)** and the given initial concentrations of NH_4^+ (0.39 M) and NO_2^- (0.052 M).

SOLUTION

(a) Comparing Experiments 1 and 2 shows that doubling the initial concentration of NH_4^+ from 0.12 M to 0.24 M while holding the NO_2^- concentration constant increases the initial rate by a factor of 2, from 3.6×10^{-6} M/s to 7.2×10^{-6} M/s. Therefore, the initial rate is proportional to $[NH_4^+]$ and $m = 1$.

continued on next page

Comparing Experiments 1 and 3 shows that increasing the initial concentration of NO_2^- by a factor of 1.5 from 0.10 M to 0.15 M while holding the NH_4^+ concentration constant increases the initial rate by a factor of 1.5, from 3.6×10^{-6} M/s to 5.4×10^{-6} M/s. And so the initial rate is proportional to $[NO_2^-]$ and $n = 1$.

Because the reaction is first order in both NH_4^+ and NO_2^-, the rate law is

$$\text{Rate} = -\frac{\Delta[NH_4^+]}{\Delta t} = k[NH_4^+][NO_2^-]$$

A more formal way to approach this problem begins with the rate law for two experiments in which the concentration of only one reactant varies. We'll illustrate this approach for Experiments 1 and 2, which yields the value of m. (A similar procedure applied to Experiments 1 and 3 would give the value of n.) The rate laws for Experiments 1 and 2 are

$$(\text{Rate})_1 = k[NH_4^+]^m_1[NO_2^-]^n_1 = k(0.12\text{ M})^m(0.10\text{ M})^n$$

$$(\text{Rate})_2 = k[NH_4^+]^m_2[NO_2^-]^n_2 = k(0.24\text{ M})^m(0.10\text{ M})^n$$

If we then divide the second equation by the first, we obtain

$$\frac{(\text{Rate})_2}{(\text{Rate})_1} = \frac{\cancel{k}(0.24\text{ M})^m\cancel{(0.10)^n}}{\cancel{k}(0.12\text{ M})^m\cancel{(0.10)^n}} = \left(\frac{0.24}{0.12}\right)^m$$

The right-hand side of this equation equals the ratio of the experimental rates:

$$\left(\frac{0.24}{0.12}\right)^m = \frac{7.2 \times 10^{-6}\text{ M/s}}{3.6 \times 10^{-6}\text{ M/s}}$$

Finally, we can solve for m if we take the logarithm of both sides of the last equation and remember that $\log x^m = m \log x$ (see Appendix A.2 for a review of logarithms):

$$\log\left(\frac{0.24}{0.12}\right)^m = m\log\left(\frac{0.24}{0.12}\right) = \log\left(\frac{7.2 \times 10^{-6}\text{ M/s}}{3.6 \times 10^{-6}\text{ M/s}}\right)$$

$$m = \frac{\log\left(\dfrac{7.2 \times 10^{-6}\text{ M/s}}{3.6 \times 10^{-6}\text{ M/s}}\right)}{\log\left(\dfrac{0.24}{0.12}\right)} = \frac{\log 2}{\log 2} = 1$$

A similar procedure applied to Experiments 1 and 3 yields $n = 1$, and therefore the rate law is

$$\text{Rate} = -\frac{\Delta[NH_4^+]}{\Delta t} = k[NH_4^+][NO_2^-]$$

(b) Solving the rate law for k and substituting in the data from the first experiment gives

$$k = \frac{\text{Rate}}{[NH_4^+][NO_2^-]} = \frac{3.6 \times 10^{-6}\,\dfrac{\text{M}}{\text{s}}}{(0.12\text{ M})(0.10\text{ M})} = 3.0 \times 10^{-4}\text{ M}^{-1}\text{ s}^{-1}$$

(c) Substituting the initial concentrations ($[NH_4^+] = 0.39$ M; $[NO_2^-] = 0.052$ M) and the rate constant from part **(b)** ($3.0 \times 10^{-4}\text{ M}^{-1}\text{ s}^{-1}$) into the rate law gives

$$\text{Rate} = -\frac{\Delta[NH_4^+]}{\Delta t} = k[NH_4^+][NO_2^-] = (3.0 \times 10^{-4}\text{ M}^{-1}\text{ s}^{-1})(0.39\text{ M})(0.052\text{ M})$$

$$= 6.1 \times 10^{-6}\text{ M/s}$$

✔ BALLPARK CHECK

(b) It's a good idea to check the units of the rate constant. The units of k, $\text{M}^{-1}\text{ s}^{-1}$, are the expected units for a reaction that is second order overall.

(c) Because the initial $[NH_4^+]$ is about 3 times that in Experiment 1 and the initial $[NO_2^-]$ is about 1/2 times that in Experiment 1, the reaction rate is estimated to be about 3/2 times that in Experiment 1, or about 5.4×10^{-6} M/s, which agrees with the solution.

WORKED CONCEPTUAL EXAMPLE 12.4

Using the Method of Initial Rates

The relative rates of the reaction A + 2 B ⟶ products in vessels (1)–(4) are 1:2:2:4. Red spheres represent A molecules, and blue spheres represent B molecules.

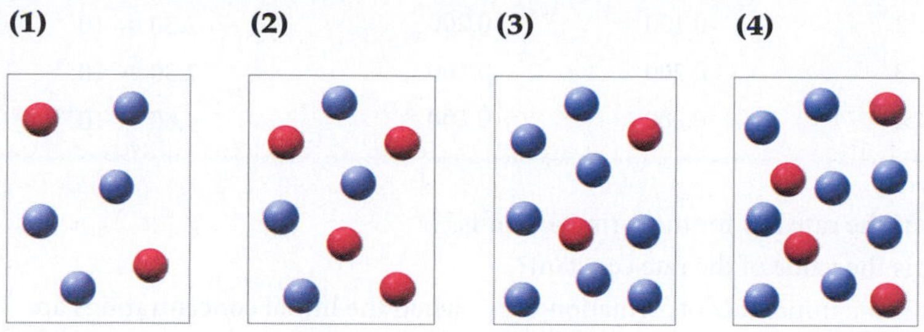

(a) What is the order of the reaction in A and B? What is the overall reaction order?
(b) Write the rate law.

STRATEGY

(a) To find the reaction order, apply the method of initial rates. Count the number of A and B molecules in vessels (1)–(4), and compare the relative rates with the relative number of molecules of each type. Assume that all four vessels have the same volume, so the concentrations are proportional to the number of molecules.
(b) The rate law can be written as rate $= k[A]^m[B]^n$, where the exponents m and n are the orders of the reaction in A and B, respectively.

SOLUTION

(a) Compare pairs of vessels in which the concentration of one reactant varies while the concentration of the other reactant remains constant. The concentration of A molecules in vessel (2) is twice that in vessel (1) while the concentration of B remains constant. Because the reaction rate in vessel (2) is twice that in vessel (1), the rate is proportional to [A], and therefore the reaction is first order in A. When the concentration of B is doubled while the concentration of A remains constant [compare vessels (1) and (3)], the rate doubles, so the reaction is first order in B. When the concentrations of both A and B are doubled, the rate increases by a factor of 4 [compare vessels (1) and (4)], in accord with a reaction that is first order in A and first order in B. The overall reaction order is the sum of the orders in A and B, or 1 + 1 = 2.
(b) Since the reaction is first order in A and first order in B, the rate law is rate $= k[A][B]$. Note that the exponents in the rate law differ from the coefficients in the balanced chemical equation, A + 2 B ⟶ products.

PROBLEM 12.4 The oxidation of iodide ion by hydrogen peroxide in an acidic solution is described by the balanced equation

$$H_2O_2(aq) + 3\,I^-(aq) + 2\,H^+(aq) \longrightarrow I_3^-(aq) + 2\,H_2O(l)$$

The rate of formation of the red triiodide ion, $\Delta[I_3^-]/\Delta t$, can be determined by measuring the rate of appearance of the color (**Figure 12.5**).

FIGURE 12.5

A sequence of photographs showing the progress of the reaction of hydrogen peroxide (H_2O_2) and iodide ion (I^-). As time passes (left to right), the red color due to the triiodide ion (I_3^-) increases in intensity.

continued on next page

Initial rate data at 25 °C and a constant H^+ ion concentration are as follows:

Experiment	Initial $[H_2O_2]$	Initial $[I^-]$	Initial Rate of Formation of I_3^- (M/s)
1	0.100	0.100	1.15×10^{-4}
2	0.100	0.200	2.30×10^{-4}
3	0.200	0.100	2.30×10^{-4}
4	0.200	0.200	4.60×10^{-4}

(a) What is the rate law for the formation of I_3^-?

(b) What is the value of the rate constant?

(c) What is the initial rate of formation of I_3^- when the initial concentrations are $[H_2O_2] = 0.300$ M and $[I^-] = 0.400$ M?

PROBLEM 12.5 What are the units of the rate constant for each of the reactions in Table 12.2?

CONCEPTUAL PROBLEM 12.6 The relative rates of the reaction A + B ⟶ products in vessels (1)–(4) are 1:1:4:4. Red spheres represent A molecules, and blue spheres represent B molecules.

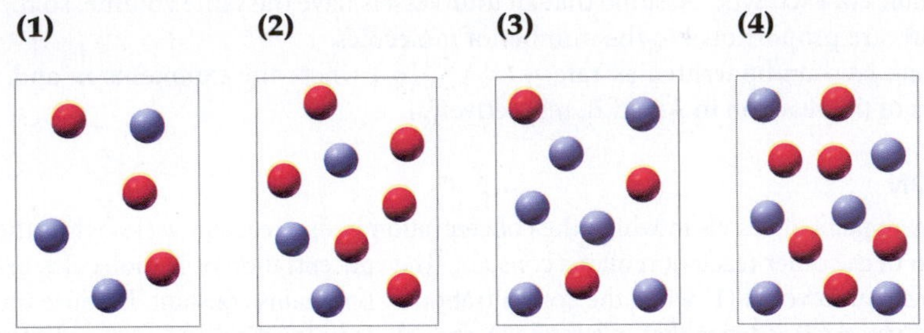

(a) What is the order of the reaction in A and B? What is the overall reaction order?

(b) Write the rate law.

12.4 FIRST-ORDER REACTIONS: THE INTEGRATED RATE LAW

Thus far we've focused on the rate law, an equation that tells how a reaction rate depends on reactant concentrations. We're also interested, however, in how reactant and product concentrations vary with time. For example, it's important not only to know the rate at which the atmospheric ozone layer is being destroyed by air pollutants, but also what the ozone concentration will be 20 years from now and how long it will take for the concentration to change by a given amount, say, 10%.

Because the kinetics of the pollution-induced decomposition of ozone is a very complicated process, let's consider instead a simple, general first-order reaction:

$$A \longrightarrow \text{Products}$$

A **first-order reaction** is one whose rate depends on the concentration of a single reactant raised to the first power.

$$\text{Rate} = -\frac{\Delta[A]}{\Delta t} = k[A]$$

An example is the decomposition of hydrogen peroxide in basic solution, which we'll discuss in Worked Example 12.5 at the end of this section:

$$2\,H_2O_2(aq) \longrightarrow 2\,H_2O(l) + O_2(g)$$

Using calculus, it's possible to convert the rate law to another form, called the **integrated rate law**:

▶ **Integrated rate law for a first-order reaction** $\ln \frac{[A]_t}{[A]_0} = -kt$

In this equation, ln denotes the natural logarithm, $[A]_0$ is the concentration of A at some initial time, arbitrarily considered to be $t = 0$, and $[A]_t$ is the concentration of A at any time t thereafter. (Appendix A.2 reviews natural logarithms.) The ratio $[A]_t/[A]_0$ is the fraction of A that remains at time t. Thus, the integrated rate law is a *concentration–time equation* that lets us calculate the concentration of A or the fraction of A that remains at any time t. It can also be used to calculate the time required for the initial concentration of A to drop to any particular value or to any particular fraction of its initial concentration (**Figure 12.6a**). Although you must be able to use the integrated rate law, as illustrated in Worked Example 12.5, you don't need to be able to derive it. Students who know some calculus, however, may be interested in the derivation shown in Appendix A.5.

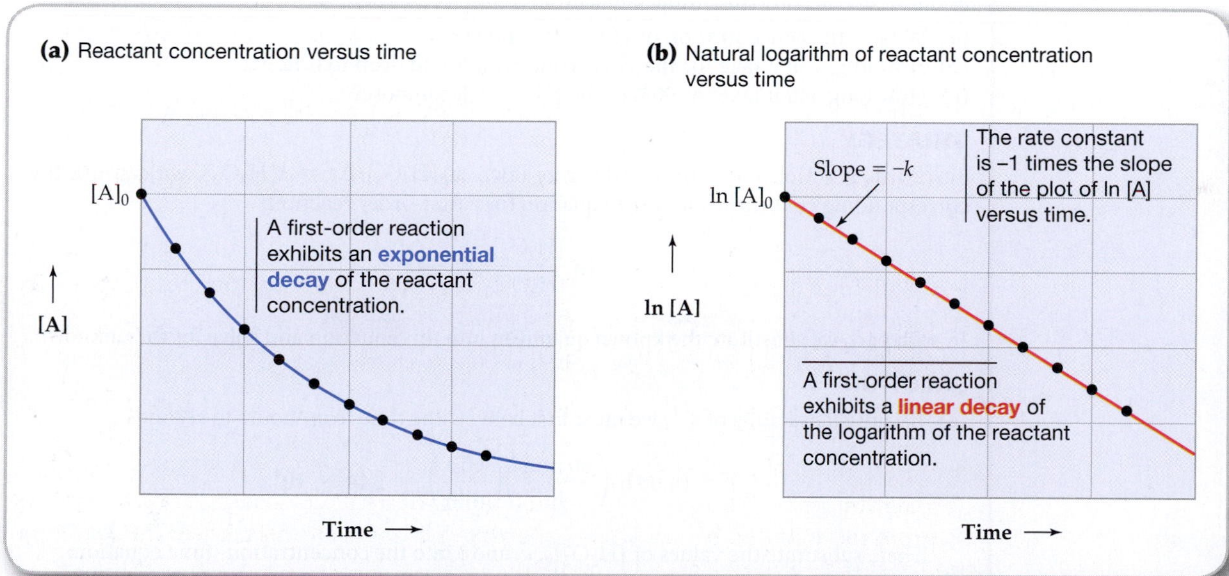

FIGURE 12.6

Plots for a first-order reaction.

Since $\ln([A]_t/[A]_0) = \ln [A]_t - \ln [A]_0$, we can rewrite the integrated rate law as

▶ **Integrated rate law for a first-order reaction** $\ln [A]_t = -kt + \ln [A]_0$

This equation is of the form $y = mx + b$, the equation for a straight line, so $\ln [A]_t$ is a linear function of time:

$$\underset{y}{\underset{\uparrow}{\ln [A]_t}} = \underset{m}{\underset{\uparrow}{(-k)}}\underset{x}{\underset{\uparrow}{t}} + \underset{b}{\underset{\uparrow}{\ln [A]_0}}$$

A graph of ln [A] versus time is therefore a straight line having a slope $m = -k$ and an intercept $b = \ln [A]_0$ **(Figure 12.6b)**. The value of the rate constant is simply equal to -1 times the slope of the straight line:

$$k = -(\text{slope})$$

This graphical method of determining a rate constant, illustrated in Worked Example 12.6, is an alternative to the method of initial rates used in Worked Example 12.3. A plot of ln [A] versus time, however, will give a straight line only if the reaction is first order in A. Indeed, a good way of testing whether a reaction is first order is to examine the appearance of such a plot.

Worked Example 12.5

Using the Integrated Rate Law for a First-Order Reaction

The decomposition of hydrogen peroxide in dilute sodium hydroxide solution is described by the equation

$$2\,H_2O_2(aq) \longrightarrow 2\,H_2O(l) + O_2(g)$$

The reaction is first order in H_2O_2, the rate constant for the consumption of H_2O_2 at 20 °C is $1.8 \times 10^{-5}\ s^{-1}$, and the initial concentration of H_2O_2 is 0.30 M.

(a) What is the concentration of H_2O_2 after 4.00 h?
(b) How long will it take for the H_2O_2 concentration to drop to 0.12 M?
(c) How long will it take for 90% of the H_2O_2 to decompose?

STRATEGY

Since this reaction has a first-order rate law, $-\Delta[H_2O_2]/\Delta t = k[H_2O_2]$, we can use the corresponding concentration–time equation for a first-order reaction:

$$\ln \frac{[H_2O_2]_t}{[H_2O_2]_0} = -kt$$

In each part, we substitute the known quantities into this equation and solve for the unknown.

SOLUTION

(a) Because k has units of s^{-1}, we must first convert the time from hours to seconds:

$$t = (4.00\ h)\left(\frac{60\ min}{h}\right)\left(\frac{60\ s}{min}\right) = 1.44 \times 10^4\ s$$

Then, substitute the values of $[H_2O_2]_0$, k, and t into the concentration–time equation:

$$\ln \frac{[H_2O_2]_t}{0.30\ M} = -(1.8 \times 10^{-5}\ s^{-1})(1.44 \times 10^4\ s) = -0.259$$

Taking the natural antilogarithm (antiln) of both sides gives

$$\frac{[H_2O_2]_t}{0.30\ M} = e^{-0.259} = 0.772$$

where the number $e = 2.718\,28\ldots$ is the base of natural logarithms (see Appendix A.2). Therefore,

$$[H_2O_2]_t = (0.772)(0.30\ M) = 0.23\ M$$

(b) First, solve the concentration–time equation for the time:

$$t = -\frac{1}{k}\ln\frac{[H_2O_2]_t}{[H_2O_2]_0}$$

Then evaluate the time by substituting the concentrations and the value of k:

$$t = -\left(\frac{1}{1.8 \times 10^{-5}\,s^{-1}}\right)\left(\ln\frac{0.12\text{ M}}{0.30\text{ M}}\right) = -\left(\frac{1}{1.8 \times 10^{-5}\,s^{-1}}\right)(-0.916) = 5.1 \times 10^4\text{ s}$$

Thus, the H_2O_2 concentration reaches 0.12 M at a time of 5.1×10^4 s (14 h).

(c) When 90% of the H_2O_2 has decomposed, 10% remains. Therefore,

$$\frac{[H_2O_2]_t}{[H_2O_2]_0} = \frac{(0.10)(0.30\text{ M})}{(0.30\text{ M})} = 0.10$$

The time required for 90% decomposition is

$$t = -\left(\frac{1}{1.8 \times 10^{-5}\,s^{-1}}\right)(\ln 0.10) = -\left(\frac{1}{1.8 \times 10^{-5}\,s^{-1}}\right)(-2.30) = 1.3 \times 10^5\text{ s (36 h)}$$

✔ BALLPARK CHECK

The concentration of H_2O_2 (0.23 M) after 4.00 h is less than the initial concentration (0.30 M). A longer period of time (14 h) is required for the concentration to drop to 0.12 M, and still more time (36 h) is needed for the concentration to fall to 0.030 M (10% of the original concentration). These results look reasonable. A plot of $[H_2O_2]$ versus time would exhibit an exponential decay of the H_2O_2 concentration, as expected for a first-order reaction.

Worked Example 12.6

Plotting Data for a First-Order Reaction

Experimental concentration-versus-time data for the decomposition of gaseous N_2O_5 at 55 °C are listed in Table 12.1 and are plotted in Figure 12.1. Use those data to confirm that the decomposition of N_2O_5 is a first-order reaction. What is the value of the rate constant for the consumption of N_2O_5 including the proper units?

STRATEGY

To confirm that the reaction is first order, check to see whether a plot of ln $[N_2O_5]$ versus time gives a straight line. The rate constant for a first-order reaction equals −1 times the slope of the straight line.

SOLUTION

Values of ln $[N_2O_5]$ are listed in the following table and are plotted versus time in the graph:

Time (s)	$[N_2O_5]$	ln $[N_2O_5]$
0	0.0200	−3.912
100	0.0169	−4.080
200	0.0142	−4.255
300	0.0120	−4.423
400	0.0101	−4.595
500	0.0086	−4.756
600	0.0072	−4.934
700	0.0061	−5.099

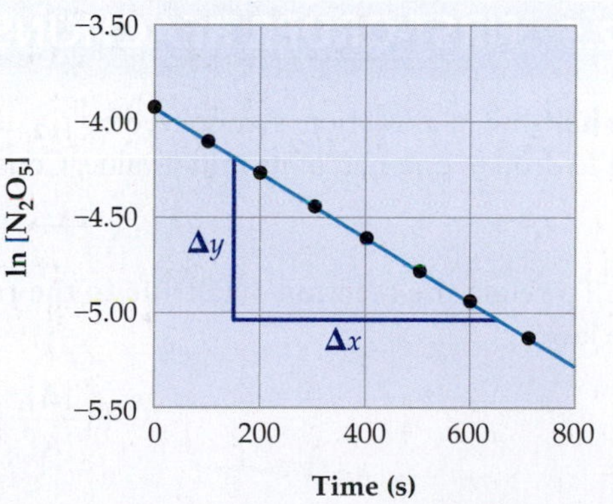

continued on next page

Because the data points lie on a straight line, the reaction is first order in N_2O_5. The slope of the line can be determined from the coordinates of any two widely separated points on the line and the rate constant k can be calculated from the slope:

$$\text{Slope} = \frac{\Delta y}{\Delta x} = \frac{(-5.02) - (-4.17)}{650 \text{ s} - 150 \text{ s}} = \frac{-0.85}{500 \text{ s}} = -1.7 \times 10^{-3} \text{ s}^{-1}$$

$$k = -(\text{Slope}) = 1.7 \times 10^{-3} \text{ s}^{-1}$$

Note that the units of k are the units expected for a first-order reaction.

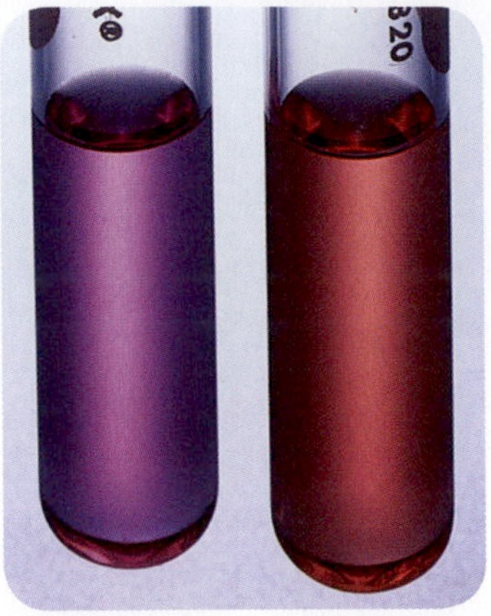

▲ Aqueous solutions of $Co(NH_3)_5Br^{2+}$ (left) and $Co(NH_3)_5(H_2O)^{3+}$ (right).

PROBLEM 12.7 In acidic aqueous solution, the purple complex ion $Co(NH_3)_5Br^{2+}$ undergoes a slow reaction in which the bromide ion is replaced by a water molecule, yielding the pinkish-orange complex ion $Co(NH_3)_5(H_2O)^{3+}$:

$$\underset{\text{Purple}}{Co(NH_3)_5Br^{2+}(aq)} + H_2O(l) \longrightarrow \underset{\text{Pinkish-orange}}{Co(NH_3)_5(H_2O)^{3+}(aq)} + Br^-(aq)$$

The reaction is first order in $Co(NH_3)_5Br^{2+}$, the rate constant at 25 °C is 6.3×10^{-6} s^{-1}, and the initial concentration of $Co(NH_3)_5Br^{2+}$ is 0.100 M.

(a) What is the molarity of $Co(NH_3)_5Br^{2+}$ after a reaction time of 10.0 h?

(b) How many hours are required for 75% of the $Co(NH_3)_5Br^{2+}$ to react?

PROBLEM 12.8 At high temperatures, cyclopropane is converted to propene, the material from which polypropylene plastics are made:

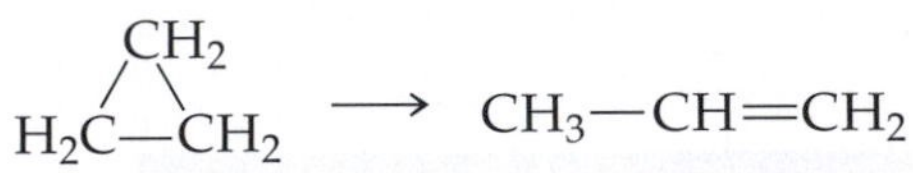

Cyclopropane Propene

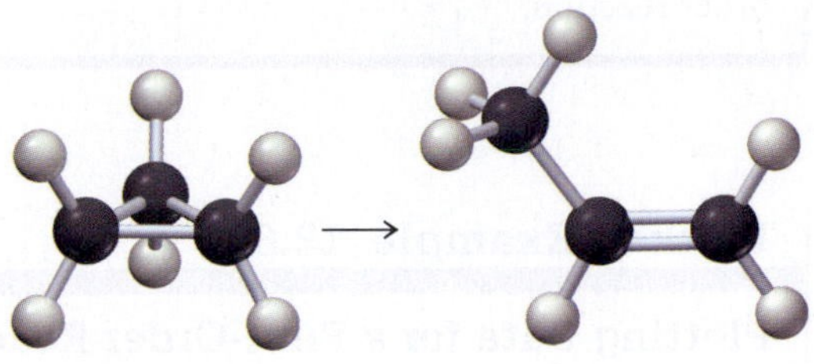

Cyclopropane Propene

Given the following concentration data, test whether the reaction is first order and calculate the value of the rate constant.

Time (min)	0	5.0	10.0	15.0	20.0
[Cyclopropane]	0.098	0.080	0.066	0.054	0.044

12.5 FIRST-ORDER REACTIONS: HALF-LIFE

The **half-life** of a reaction, symbolized by $t_{1/2}$, is the time required for the reactant concentration to drop to one-half of its initial value. Consider the first-order reaction

$$A \longrightarrow \text{Products}$$

To relate the reaction's half-life to the rate constant, let's begin with the integrated rate law:

$$\ln \frac{[A]_t}{[A]_0} = -kt$$

When $t = t_{1/2}$, the fraction of A that remains, $[A]_t/[A]_0$, is 1/2. Therefore,

$$\ln \frac{1}{2} = -kt_{1/2}$$

$$\text{so } t_{1/2} = \frac{-\ln \frac{1}{2}}{k} = \frac{\ln 2}{k} = \frac{0.693}{k}$$

▶ **Half-life of a first-order reaction** $t_{1/2} = \frac{0.693}{k}$

Thus, the half-life of a first-order reaction is readily calculated from the rate constant, and vice versa.

The half-life of a first-order reaction is a constant because it depends only on the rate constant and not on the reactant concentration. This point is worth noting because reactions that are not first order have half-lives that *do* depend on concentration; that is, the amount of time in one half-life changes as the reactant concentration changes for a non-first-order reaction.

Because the half-life of a first-order reaction is a constant, each successive half-life is an equal period of time in which the reactant concentration decreases by a factor of 2 (**Figure 12.7**).

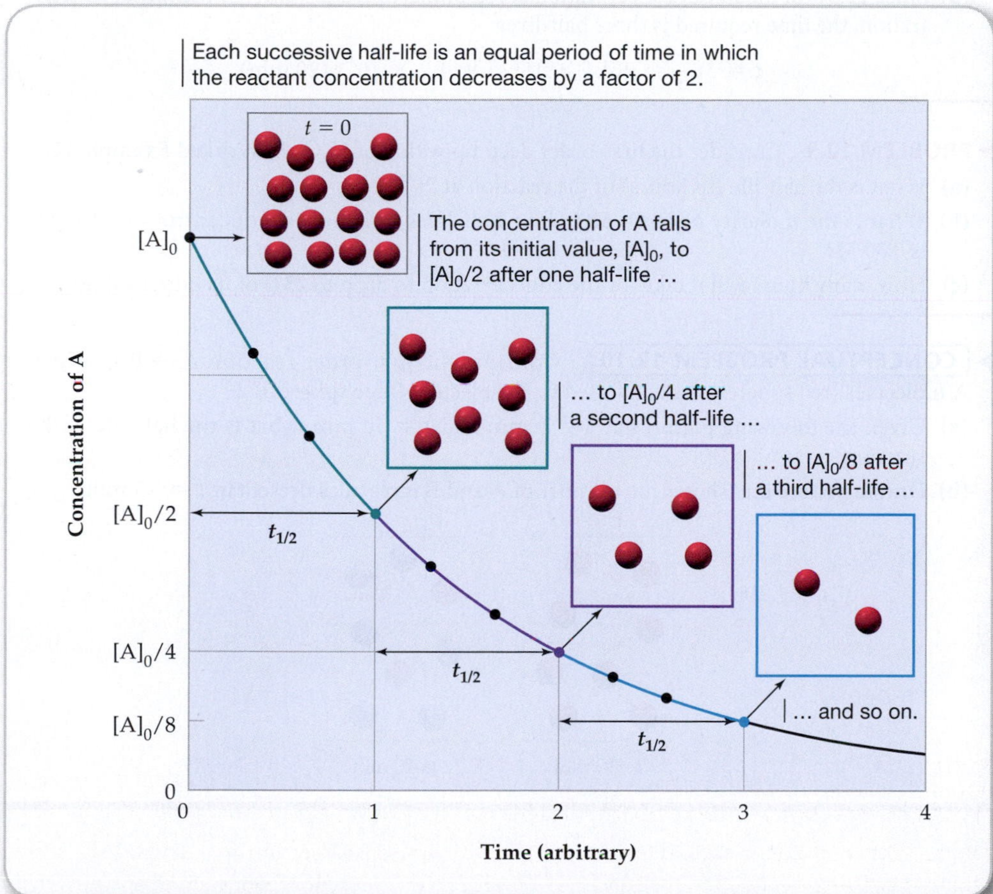

FIGURE 12.7

Concentration of a reactant A as a function of time for a first-order reaction. Each half-life represents an equal amount of time.

Worked Example 12.7

Determining the Half-Life for a First-Order Reaction

(a) Estimate the half-life for the decomposition of gaseous N_2O_5 at 55 °C from the concentration-versus-time plot in Figure 12.1.

(b) Calculate the half-life from the rate constant ($1.7 \times 10^{-3}\ s^{-1}$).

(c) If the initial concentration of N_2O_5 is 0.020 M, what is the concentration of N_2O_5 after five half-lives?

(d) How long will it take for the N_2O_5 concentration to fall to 12.5% of its initial value?

continued on next page

STRATEGY

Because the decomposition of N_2O_5 is a first-order reaction (Worked Example 12.6), we can determine its half-life either from the time required for $[N_2O_5]$ to drop to 1/2 of its initial value or from the equation $t_{1/2} = 0.693/k$. To find $[N_2O_5]$ after n half-lives, multiply its initial concentration by $(1/2)^n$ because $[N_2O_5]$ drops by a factor of 2 during each successive half-life.

SOLUTION

(a) Figure 12.1 shows that the concentration of N_2O_5 falls from 0.020 M to 0.010 M during a time period of approximately 400 s. At 800 s, $[N_2O_5]$ has decreased by another factor of 2, to 0.0050 M. Therefore, $t_{1/2} \approx 400$ s.

(b) Based on the value of the rate constant,

$$t_{1/2} = \frac{\ln 2}{k} = \frac{0.693}{1.7 \times 10^{-3}\ \text{s}^{-1}} = 4.1 \times 10^2\ \text{s (6.8 min)}$$

(c) At $5t_{1/2}$, $[N_2O_5]$ will be $(1/2)^5 = 1/32$ of its initial value. Therefore,

$$[N_2O_5] = \frac{0.020\ \text{M}}{32} = 0.000\,62\ \text{M}$$

(d) Since 12.5% of the initial concentration corresponds to 1/8 or $(1/2)^3$ of the initial concentration, the time required is three half-lives:

$$t = 3t_{1/2} = 3(4.1 \times 10^2\ \text{s}) = 1.2 \times 10^3\ \text{s (20 min)}$$

PROBLEM 12.9 Consider the first-order decomposition of H_2O_2 in Worked Example 12.5.

(a) What is the half-life (in hours) of the reaction at 20 °C?

(b) What is the molarity of H_2O_2 after four half-lives if the initial concentration of H_2O_2 is 0.30 M?

(c) How many hours will it take for the concentration to drop to 25% of its initial value?

CONCEPTUAL PROBLEM 12.10 Consider the first-order reaction $A \longrightarrow B$ in which A molecules (red spheres) are converted to B molecules (blue spheres).

(a) Given the following pictures at $t = 0$ min and $t = 10$ min, what is the half-life of the reaction?

(b) Draw a picture that shows the number of A and B molecules present at $t = 15$ min.

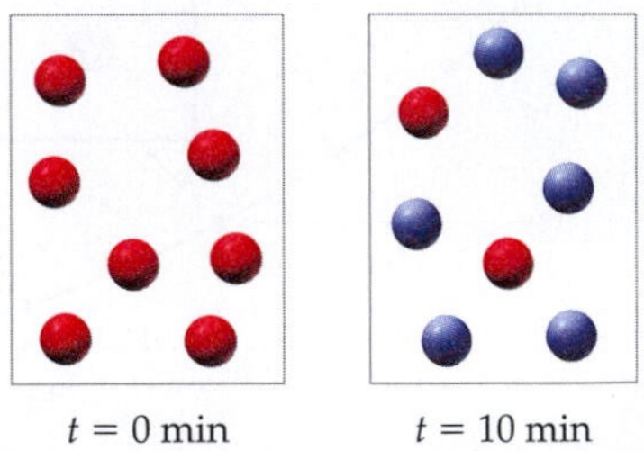

12.6 FIRST-ORDER REACTIONS: RADIOACTIVE DECAY

In Sections 1.10 and 1.11 we saw that many nuclei are radioactive. The rate at which a radioactive nucleus decays is characterized by its half-life, $t_{1/2}$. For example, carbon-14 decays to nitrogen-14 by emission of a β^- particle with a half-life of 5715 years:

$$^{14}_{6}\text{C} \longrightarrow {}^{14}_{7}\text{N} + {}^{0}_{-1}\text{e} \qquad t_{1/2} = 5715\ \text{y}$$

Radioactive decay is a first-order process, whose rate is proportional to the number of radioactive nuclei N in a sample times a first-order rate constant k, called the **decay constant**:

$$\text{Decay rate} = -\frac{\Delta N}{\Delta t} = kN$$

This rate law is strictly analogous to the rate law for a first-order chemical reaction, as can be seen by replacing the chemical reactant concentration [A] in Section 12.4 with the number of radioactive nuclei N. Consequently, the integrated rate law for radioactive decay and the relationship between the half-life and the decay constant are also analogous to the equations given in Sections 12.4 and 12.5 for a first-order chemical reaction.

The integrated rate law for radioactive decay is

$$\ln\left(\frac{N_t}{N_0}\right) = -kt$$

In this equation N_0 is the number of radioactive nuclei originally present in a sample and N_t is the number remaining at time t.

The relationship between the half-life and the decay constant is

$$t_{1/2} = \frac{\ln 2}{k} \quad \text{and} \quad k = \frac{\ln 2}{t_{1/2}}$$

Thus, if we know the value of either the decay constant k or the half-life $t_{1/2}$, we can calculate the value of the other. Furthermore, if we know the value of $t_{1/2}$, we can calculate the ratio of remaining and initial amounts of a radioactive sample N_t/N_0 at any time t by substituting the expression for k into the integrated rate law:

$$\text{Since} \quad \ln\left(\frac{N_t}{N_0}\right) = -kt \quad \text{and} \quad k = \frac{\ln 2}{t_{1/2}}$$

$$\text{then} \quad \ln\left(\frac{N_t}{N_0}\right) = (-\ln 2)\left(\frac{t}{t_{1/2}}\right)$$

Worked Example 12.8 shows how to calculate a half-life from a decay constant, and Worked Example 12.10 shows how to determine the percentage of a radioactive sample remaining at time t. Applications of **radioactive decay** will be further explored in Section 22.6.

Worked Example 12.8

Calculating a Half-Life from a Decay Constant

The decay constant for sodium-24, a **radioisotope** used medically in blood studies, is $4.63 \times 10^{-2}\ \text{h}^{-1}$. What is the half-life of ^{24}Na?

STRATEGY

Half-life can be calculated from the decay constant by using the equation

$$t_{1/2} = \frac{\ln 2}{k}$$

SOLUTION

Substituting the values $k = 4.63 \times 10^{-2}\ \text{h}^{-1}$ and $\ln 2 = 0.693$ into the equation gives

$$t_{1/2} = \frac{0.693}{4.63 \times 10^{-2}\ \text{h}^{-1}} = 15.0\ \text{h}$$

▲ The rate of radioactive decay of carbon-14 has been used to determine that the Dead Sea Scrolls are approximately 1900 years old.

LOOKING AHEAD...

As we'll see in Section 22.6, **radioactive decay** has many important medical applications.

REMEMBER ...

A **radioisotope** is an isotope that undergoes spontaneous radioactive decay. (Section 1.11)

Worked Example 12.9

Calculating a Decay Constant from a Half-Life

The half-life of radon-222, a radioactive gas of concern as a health hazard in some homes, is 3.82 days. What is the decay constant of ^{222}Rn?

STRATEGY

A decay constant can be calculated from the half-life by using the equation

$$k = \frac{\ln 2}{t_{1/2}}$$

SOLUTION

Substituting the values $t_{1/2} = 3.82$ days and $\ln 2 = 0.693$ into the equation gives

$$k = \frac{0.693}{3.82 \text{ days}} = 0.181 \text{ day}^{-1}$$

Worked Example 12.10

Using Half-Life to Calculate an Amount Remaining

Phosphorus-32, a radioisotope used in leukemia therapy, has a half-life of 14.26 days. What percent of a sample remains after 35.0 days?

STRATEGY

The ratio of remaining (N_t) and initial (N_0) amounts of a radioactive sample at time t is given by the equation

$$\ln\left(\frac{N_t}{N_0}\right) = (-\ln 2)\left(\frac{t}{t_{1/2}}\right)$$

Taking N_0 as 100%, N_t can then be obtained.

SOLUTION

Substituting values for t and for $t_{1/2}$ into the equation gives

$$\ln\left(\frac{N_t}{N_0}\right) = (-0.693)\left(\frac{35.0 \text{ days}}{14.26 \text{ days}}\right) = -1.70$$

Taking the natural antilogarithm of both sides of this equation then gives the ratio N_t/N_0:

$$\frac{N_t}{N_0} = \text{antiln}\,(-1.70) = e^{-1.70} = 0.183$$

Since the initial amount of ^{32}P was 100%, we can set $N_0 = 100\%$ and solve for N_t:

$$\frac{N_t}{100\%} = 0.183 \quad \text{so} \quad N_t = (0.183)(100\%) = 18.3\%$$

After 35.0 days, 18.3% of a ^{32}P sample remains and 100% − 18.3% = 81.7% has decayed.

▲ The flow of gases from a smokestack can be measured by releasing ^{41}Ar and monitoring its passage.

Worked Example 12.11

Using Decay Rates to Calculate a Half-Life

A sample of ^{41}Ar, a radioisotope used to measure the flow of gases from smokestacks, decays initially at a rate of 34,500 disintegrations/min, but the decay rate falls to 21,500 disintegrations/min after 75.0 min. What is the half-life of ^{41}Ar?

STRATEGY

The half-life of a radioactive decay process is obtained by finding $t_{1/2}$ in the equation

$$\ln\left(\frac{N_t}{N_0}\right) = (-\ln 2)\left(\frac{t}{t_{1/2}}\right)$$

In the present instance, though, we are given decay *rates* at two different times rather than values of N_t and N_0. Nevertheless, for a first-order process like radioactive decay, in which rate $= kN$, the ratio of the decay rate at any time t to the decay rate at time $t = 0$ is the same as the ratio of N_t to N_0:

$$\frac{\text{Decay rate at time } t}{\text{Decay rate at time } t = 0} = \frac{kN_t}{kN_0} = \frac{N_t}{N_0}$$

SOLUTION

Substituting the decay rates into the equation gives

$$\ln\left(\frac{21{,}500}{34{,}500}\right) = (-0.693)\left(\frac{75.0 \text{ min}}{t_{1/2}}\right) \quad \text{or} \quad -0.473 = \frac{-52.0 \text{ min}}{t_{1/2}}$$

$$\text{so} \quad t_{1/2} = \frac{-52.0 \text{ min}}{-0.473} = 110 \text{ min}$$

The half-life of ^{41}Ar is 110 min.

PROBLEM 12.11 The decay constant for mercury-197, a radioisotope used medically in kidney scans, is $1.08 \times 10^{-2}\ h^{-1}$. What is the half-life of mercury-197?

PROBLEM 12.12 The half-life of cesium-137, released as a result of the Fukushima Daiichi nuclear disaster, is 30.2 years. What is its decay constant?

PROBLEM 12.13 What percentage of carbon-14 ($t_{1/2}$ = 5715 years) remains in a sample estimated to be 16,230 years old?

PROBLEM 12.14 What is the half-life of iron-59, a radioisotope used medically in the diagnosis of anemia, if a sample with an initial decay rate of 16,800 disintegrations/min decays at a rate of 10,860 disintegrations/min after 28.0 days?

12.7 SECOND-ORDER REACTIONS

A **second-order reaction** is one whose rate depends either on the concentration of a single reactant raised to the second power or on the concentrations of two different reactants, each raised to the first power. For the simpler type, A → Products, the rate law is

$$\text{Rate} = -\frac{\Delta[\text{A}]}{\Delta t} = k[\text{A}]^2$$

An example is the thermal decomposition of nitrogen dioxide to yield NO and O_2:

$$2\ NO_2(g) \longrightarrow 2\ NO(g) + O_2(g)$$

Using calculus, it's possible to convert the rate law to the integrated rate law:

▶ **Integrated rate law for a second-order reaction** $\dfrac{1}{[\text{A}]_t} = kt + \dfrac{1}{[\text{A}]_0}$

You can find the derivation in Appendix A.5, but you don't have to be able to do the calculus. The integrated rate law allows us to calculate the concentration of A at any time t if the initial concentration $[\text{A}]_0$ is known.

Since the integrated rate law has the form $y = mx + b$, a graph of $1/[A]$ versus time is a straight line if the reaction is second order:

$$\underset{\substack{\uparrow \\ y}}{\frac{1}{[A]_t}} = \underset{\substack{\uparrow\uparrow \\ m\,x}}{kt} + \underset{\substack{\uparrow \\ b}}{\frac{1}{[A]_0}}$$

The slope of the straight line is the rate constant k, and the intercept is $1/[A]_0$. Thus, by plotting $1/[A]$ versus time, we can test whether the reaction is second order and can determine the value of the rate constant (see Worked Example 12.12). We can also obtain an expression for the half-life of a second-order reaction by substituting $[A]_t = [A]_0/2$ and $t = t_{1/2}$ into the integrated rate law:

$$\frac{1}{\left(\frac{[A]_0}{2}\right)} = kt_{1/2} + \frac{1}{[A]_0}$$

$$\text{so} \quad t_{1/2} = \frac{1}{k}\left(\frac{2}{[A]_0} - \frac{1}{[A]_0}\right) = \frac{1}{k[A]_0}$$

▶ **Half-life for a second-order reaction** $t_{1/2} = \dfrac{1}{k[A]_0}$

In contrast with a first-order reaction, the time required for the concentration of A to drop to one-half of its initial value in a second-order reaction depends on both the rate constant and the initial concentration. Thus, the value of $t_{1/2}$ increases as the reaction proceeds because the value of $[A]_0$ at the beginning of each successive half-life is smaller by a factor of 2. Consequently, each half-life for a second-order reaction is twice as long as the preceding one (**Figure 12.8**).

FIGUR 12.8

Concentration of a reactant A as a function of time for a second-order reaction.

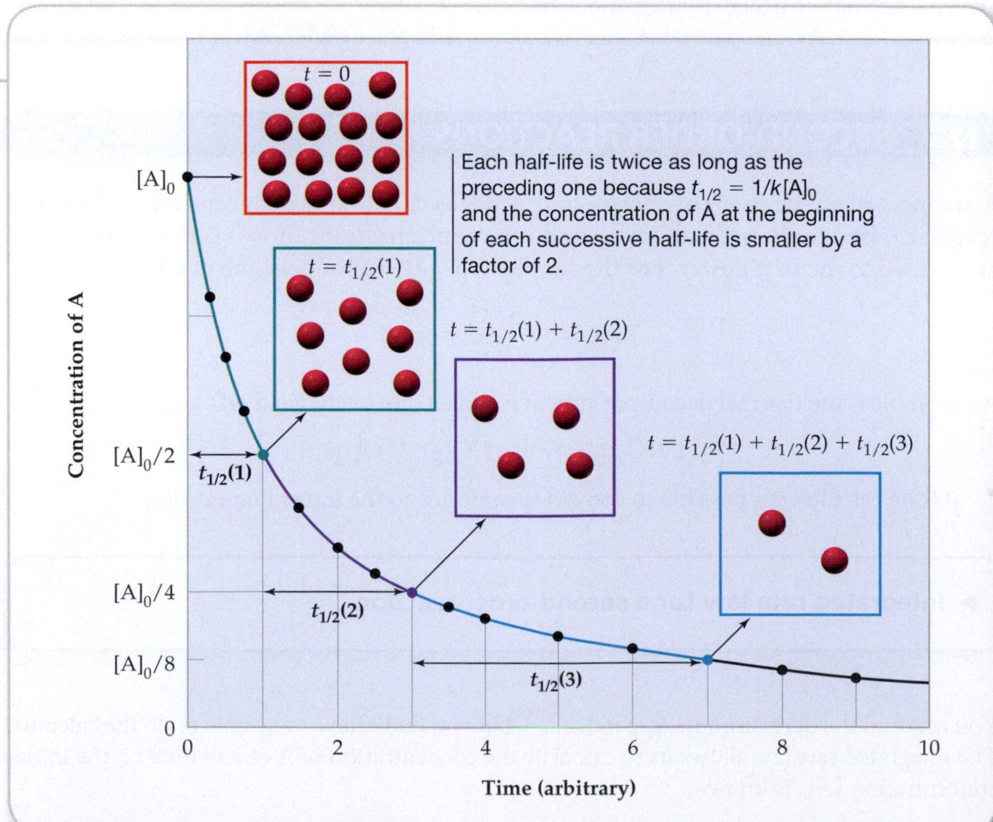

Table 12.4 summarizes some important differences between first-order and second-order reactions of the type A $\longrightarrow$ Products.

TABLE 12.4 Characteristics of First- and Second-Order Reactions of the Type A $\longrightarrow$ Products

	First-Order	Second-Order
Rate law	$-\frac{\Delta[A]}{\Delta t} = k[A]$	$-\frac{\Delta[A]}{\Delta t} = k[A]^2$
Concentration–time equation	$\ln [A] = -kt + \ln [A]_0$	$\frac{1}{[A]_t} = kt + \frac{1}{[A]_0}$
Linear graph	$\ln [A]$ versus t	$\frac{1}{[A]}$ versus t
	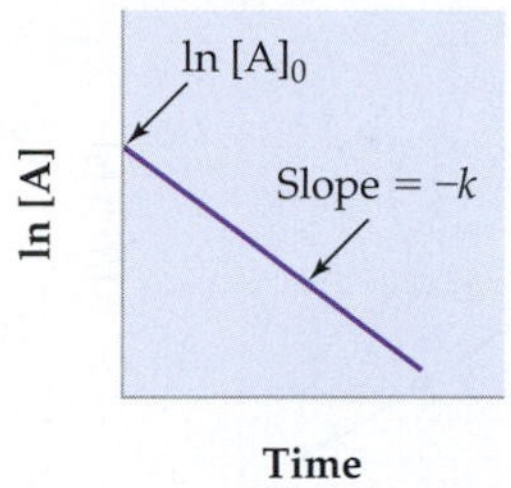	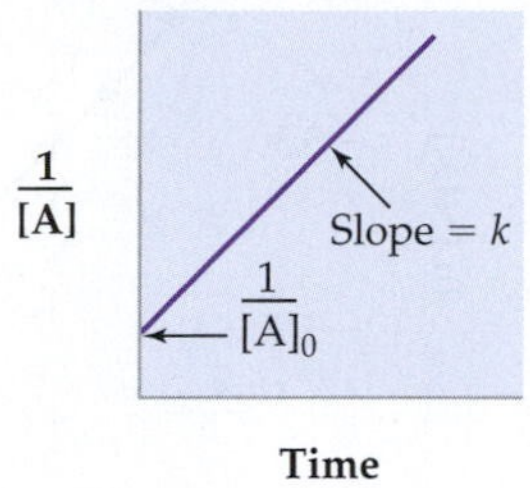
Graphical determination of k	$k = -(\text{Slope})$	$k = \text{Slope}$
Half-life	$t_{1/2} = \frac{0.693}{k}$ (constant)	$t_{1/2} = \frac{1}{k[A]_0}$ (not constant)

Worked Example 12.12

Determining Reaction Order Graphically

At elevated temperatures, nitrogen dioxide decomposes to nitric oxide and molecular oxygen:

$$2\,NO_2(g) \longrightarrow 2\,NO(g) + O_2(g)$$

Concentration–time data for the consumption of NO_2 at 300 °C are as follows:

Time (s)	$[NO_2]$	Time (s)	$[NO_2]$
0	8.00×10^{-3}	200	4.29×10^{-3}
50	6.58×10^{-3}	300	3.48×10^{-3}
100	5.59×10^{-3}	400	2.93×10^{-3}
150	4.85×10^{-3}	500	2.53×10^{-3}

(a) Is the reaction first order or second order?
(b) What is the value of the rate constant?
(c) What is the concentration of NO_2 at $t = 20.0$ min?
(d) What is the half-life of the reaction when the initial concentration of NO_2 is 6.00×10^{-3} M?
(e) What is $t_{1/2}$ when $[NO_2]_0$ is 3.00×10^{-3} M?

STRATEGY

To determine whether the reaction is first order or second order, calculate values of $\ln [NO_2]$ and $1/[NO_2]$ and then graph these values versus time. The rate constant can be obtained from the slope of the straight-line plot, and concentrations and half-lives can be calculated from the appropriate equation in Table 12.4.

continued on next page

SOLUTION

Time (s)	$[NO_2]$	$\ln [NO_2]$	$1/[NO_2]$
0	8.00×10^{-3}	−4.828	125
50	6.58×10^{-3}	−5.024	152
100	5.59×10^{-3}	−5.187	179
150	4.85×10^{-3}	−5.329	206
200	4.29×10^{-3}	−5.451	233
300	3.48×10^{-3}	−5.661	287
400	2.93×10^{-3}	−5.833	341
500	2.53×10^{-3}	−5.980	395

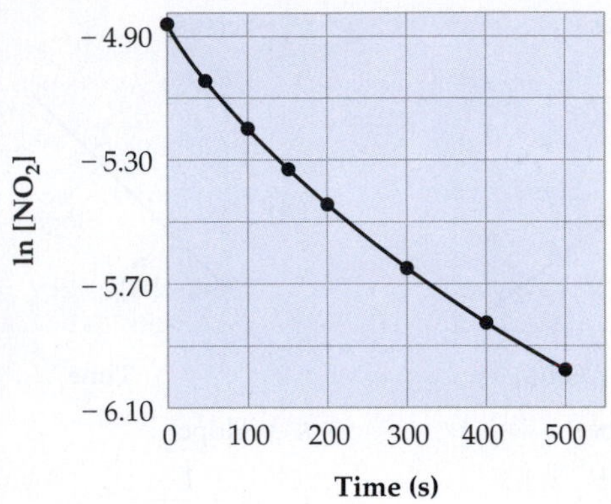

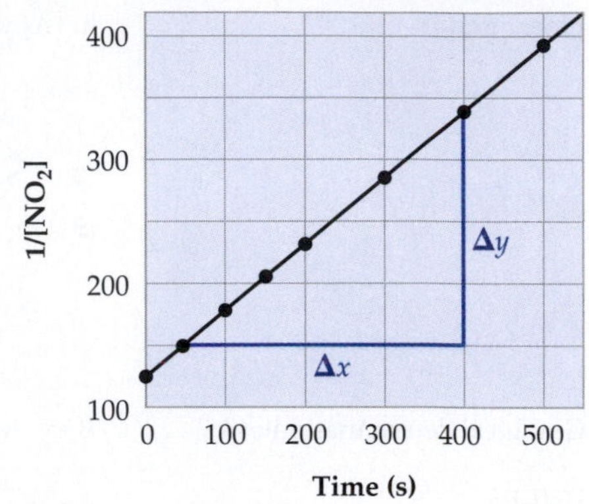

(a) The plot of $\ln [NO_2]$ versus time is curved, but the plot of $1/[NO_2]$ versus time is a straight line. The reaction is therefore second order in NO_2.

(b) The rate constant equals the slope of the straight line in the plot of $1/[NO_2]$ versus time, which we can estimate from the coordinates of two widely separated points on the line:

$$k = \text{Slope} = \frac{\Delta y}{\Delta x} = \frac{340\ \text{M}^{-1} - 150\ \text{M}^{-1}}{400\ \text{s} - 50\ \text{s}} = \frac{190\ \text{M}^{-1}}{350\ \text{s}} = 0.54/(\text{M} \cdot \text{s})$$

(c) The concentration of NO_2 at $t = 20.0$ min (1.20×10^3 s) can be calculated using the integrated rate law:

$$\frac{1}{[NO_2]_t} = kt + \frac{1}{[NO_2]_0}$$

Substituting the values of k, t, and $[NO_2]_0$ gives

$$\frac{1}{[NO_2]_t} = \left(\frac{0.54}{\text{M} \cdot \text{s}}\right)(1.20 \times 10^3\ \text{s}) + \frac{1}{8.00 \times 10^{-3}\ \text{M}}$$

$$= \frac{648}{\text{M}} + \frac{125}{\text{M}} = \frac{773}{\text{M}}$$

$$[NO_2]_t = 1.3 \times 10^{-3}\ \text{M}$$

(d) The half-life of this second-order reaction when the initial concentration of NO_2 is 6.00×10^{-3} M can be calculated from the rate constant and the initial concentration:

$$t_{1/2} = \frac{1}{k[NO_2]_0} = \frac{1}{\left(\frac{0.54}{\text{M} \cdot \text{s}}\right)(6.00 \times 10^{-3}\ \text{M})} = 3.1 \times 10^2\ \text{s}$$

(e) When $[NO_2]_0$ is 3.00×10^{-3} M, $t_{1/2} = 6.2 \times 10^2$ s (twice as long as when $[NO_2]_0$ is 6.00×10^{-3} M because $[NO_2]_0$ is now smaller by a factor of 2).

PROBLEM 12.15 Hydrogen iodide gas decomposes at 410 °C:

$$2\,HI(g) \longrightarrow H_2(g) + I_2(g)$$

The following data describe this decomposition:

Time (min)	0	20	40	60	80
[HI]	0.500	0.382	0.310	0.260	0.224

(a) Is the reaction first order or second order?
(b) What is the value of the rate constant for consumption of HI?
(c) At what time (in minutes) does the HI concentration reach 0.100 M?
(d) In how many minutes does the HI concentration drop from 0.400 M to 0.200 M?

12.8 ZEROTH-ORDER REACTIONS

A **zeroth-order reaction** of the type A ⟶ Products is one that has the rate law

$$\text{Rate} = -\frac{\Delta[\text{A}]}{\Delta t} = k[\text{A}]^0 = k(1) = k$$

Throughout the course of the reaction, the rate remains constant ($=k$), independent of the concentration of the reactant. The integrated rate law is

▶ **Integrated rate law for a zeroth-order reaction** $[\text{A}]_t = -kt + [\text{A}]_0$

This is another equation of the form $y = mx + b$, so a graph of [A] versus time is a straight line with a slope $= -k$ (**Figure 12.9**). Note that both the rate constant k and the rate of a zeroth-order reaction have a constant value equal to -1 times the slope of the [A] versus time plot.

Zeroth-order reactions are relatively uncommon, but they can occur under special circumstances. Take, for example, the decomposition of gaseous ammonia on a hot platinum surface:

$$2\,NH_3(g) \xrightarrow[\text{Pt catalyst}]{1130\text{ K}} N_2(g) + 3\,H_2(g)$$

The platinum surface is completely covered by a layer of NH_3 molecules (**Figure 12.10**), but the number of NH_3 molecules that can fit on the surface is limited by size constraints and is

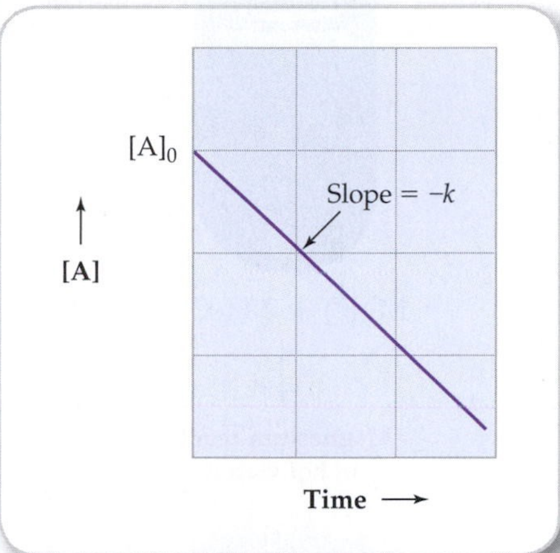

FIGURE 12.9

Concentration of a reactant A as a function of time for a zeroth-order reaction.

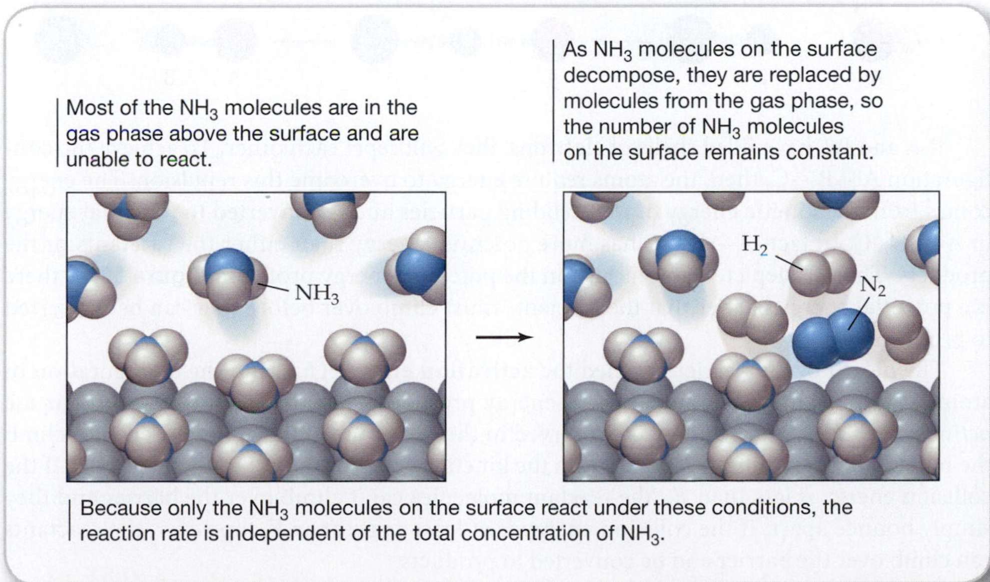

FIGURE 12.10

The constant rate of a zeroth-order reaction. The rate of decomposition of ammonia on a hot platinum surface is constant because only those NH_3 molecules attached to the surface can react.

very small compared with the total number of NH_3 molecules. Most of the NH_3 is in the gas phase above the surface. Because only the NH_3 molecules on the surface can react, the reaction rate is constant and independent of the total concentration of NH_3:

$$\text{Rate} = -\frac{\Delta[NH_3]}{\Delta t} = k[NH_3]^0 = k$$

12.9 REACTION RATES AND TEMPERATURE: COLLISION THEORY AND THE ARRHENIUS EQUATION

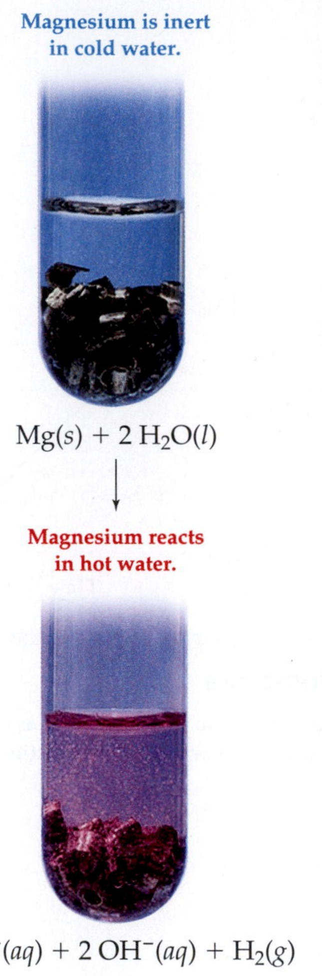

The formation of bubbles of gas and the pink color of phenolphthalein, which indicates an alkaline solution, provide evidence of the reaction.

FIGURE 12.11

Temperature affects reaction rates: Magnesium is inert in cold water (top) but reacts with hot water (bottom).

Thus far, our discussion has centered on experimental aspects of reaction rates. We've seen that the rate of a reaction usually depends on both reactant concentrations and the value of the rate constant.

Still another factor that affects reaction rates is temperature. Everyday experience tells us that the rates of chemical reactions increase with increasing temperature. Familiar fuels such as gas, oil, and coal are relatively inert at room temperature but burn rapidly at elevated temperatures. Many foods last almost indefinitely when stored in a freezer but spoil quickly at room temperature. Metallic magnesium is inert in cold water but reacts with hot water (**Figure 12.11**). As a rule of thumb, reaction rates tend to double when the temperature is increased by 10 °C.

To understand why reaction rates depend on temperature, we need a picture, or a model, of how reactions take place. According to the **collision theory** model, a bimolecular reaction occurs when two properly oriented reactant molecules come together in a sufficiently energetic collision. To be specific, let's consider one of the simplest possible reactions, the reaction of an atom A with a diatomic molecule BC to give a diatomic molecule AB and an atom C:

$$A + BC \longrightarrow AB + C$$

An example from atmospheric chemistry is the reaction of an oxygen atom with an HCl molecule to give an OH molecule and a chlorine atom:

$$O(g) + HCl(g) \longrightarrow OH(g) + Cl(g)$$

If the reaction occurs in a single step, the electron distribution about the three nuclei must change in the course of the collision such that a new bond, A–B, develops at the same time the old bond, B–C, breaks. Between the reactant and product stages, the nuclei pass through a configuration in which all three atoms are weakly linked together. We can picture the progress of the reaction as

$$A + B{-}C \longrightarrow A{-}{-}{-}B{-}{-}{-}C \longrightarrow A{-}B + C$$

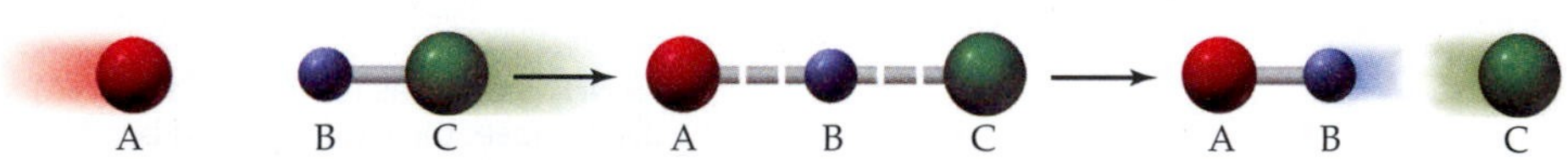

If A and BC have filled shells of electrons, they will repel each other. To achieve the configuration A---B---C, then, the atoms require energy to overcome this repulsion. The energy comes from the kinetic energy of the colliding particles and is converted to potential energy in A---B---C. In fact, A---B---C has more potential energy than either the reactants or the products. Thus, as depicted graphically on the potential energy profile in **Figure 12.12** there is a potential energy barrier that the reactants must climb over before they can be converted to products.

The height of the barrier is called the **activation energy** (E_a), and the configuration of atoms at the maximum in the potential energy profile is called the **transition state**, or the *activated complex*. Since energy is conserved in the collision, all the energy needed to climb the potential energy hill must come from the kinetic energy of the colliding molecules. If the collision energy is less than E_a, the reactant molecules can't climb over the barrier and they simply bounce apart. If the collision energy is at least as great as E_a, however, the reactants can climb over the barrier and be converted to products.

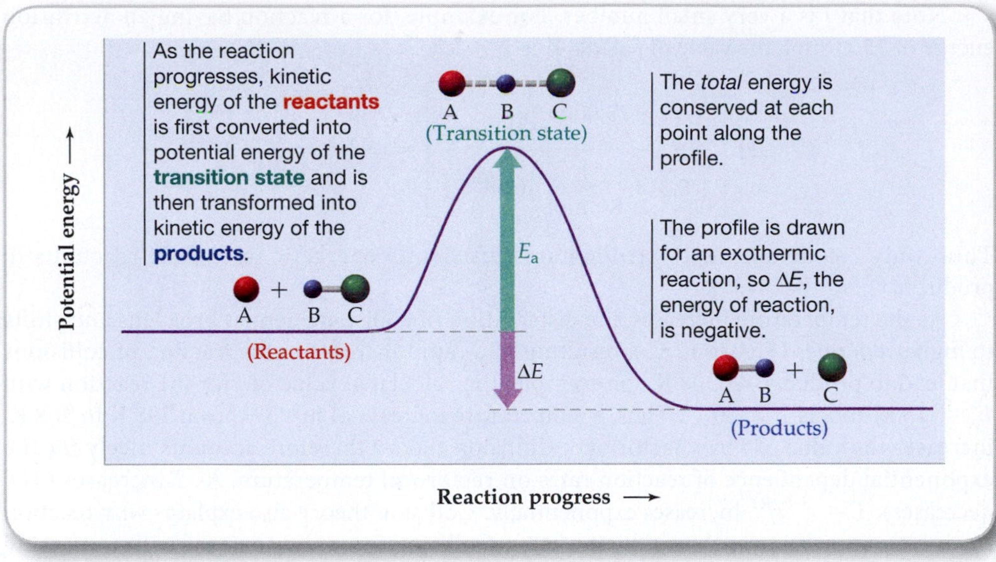

FIGURE 12.12

Potential energy profile for the reaction A + BC → AB + C.

Experimental evidence for the notion of an activation energy barrier comes from a comparison of collision rates and reaction rates. Collision rates in gases can be calculated from **kinetic–molecular theory** (Section 9.6). For a gas at room temperature (298 K) and 1 atm pressure, each molecule undergoes approximately 10^9 collisions per second, or 1 collision every 10^{-9} s. Thus, if every collision resulted in reaction, every gas-phase reaction would be complete in about 10^{-9} s. By contrast, observed reactions often have half-lives of minutes or hours, so only a tiny fraction of the collisions lead to reaction.

REMEMBER...

According to the **kinetic–molecular theory**, the average kinetic energy and thus the average speed of gas molecules depends on the Kelvin temperature. Collision rates increase with increasing concentration, diameter, and average speed of the colliding molecules. (Section 9.6)

Very few collisions are productive because very few occur with a collision energy as large as the activation energy. The fraction of collisions with an energy equal to or greater than the activation energy E_a is represented in **Figure 12.13** at two different temperatures by the areas under the curves to the right of E_a. When E_a is large compared to RT, this fraction f is approximated by the equation

$$f = e^{-E_a/RT}$$

where R is the gas constant [8.314 J/(K · mol)] and T is the absolute temperature in kelvin.

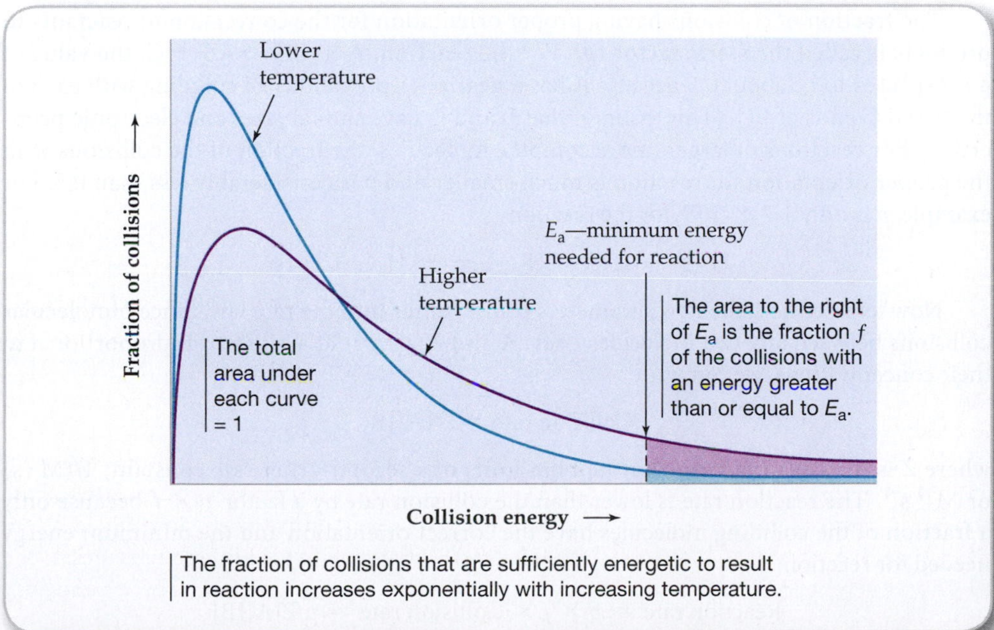

FIGURE 12.13

Plots of the fraction of collisions with a particular energy at two different temperatures.

Note that f is a very small number. For example, for a reaction having an activation energy of 75 kJ/mol, the value of f at 298 K is 7×10^{-14}:

$$f = \exp\left[\frac{-75{,}000 \dfrac{\text{J}}{\text{mol}}}{\left(8.314 \dfrac{\text{J}}{\text{K} \cdot \text{mol}}\right)(298\text{ K})}\right] = e^{-30.3} = 7 \times 10^{-14}$$

Thus, only 7 collisions in 100 trillion are sufficiently energetic to convert reactants to products.

As the temperature increases, the distribution of collision energies broadens and shifts to higher energies (Figure 12.13), resulting in a rapid increase in the fraction of collisions that lead to products. At 308 K, for example, the calculated value of f for the reaction with $E_a = 75$ kJ/mol is 2×10^{-13}. Thus, a temperature increase of just 3%, from 298 K to 308 K, increases the value of f by a factor of 3. Collision theory therefore accounts nicely for the exponential dependence of reaction rates on reciprocal temperature. As T increases ($1/T$ decreases), $f = e^{-E_a/RT}$ increases exponentially. Collision theory also explains why reaction rates are so much lower than collision rates. Collision rates also increase with increasing temperature, but only by a small amount—less than 2% on going from 298 K to 308 K.

The fraction of collisions that lead to products is further reduced by an orientation requirement. Even if the reactants collide with sufficient energy, they won't react unless the orientation of the reaction partners is correct for formation of the transition state. For example, a collision of A with the C end of the molecule BC can't result in formation of AB:

$$\text{A} + \text{C}{-}\text{B} \longrightarrow \text{A}{\cdots}\text{C}{\cdots}\text{B} \not\longrightarrow \text{A}{-}\text{B} + \text{C}$$

The reactant molecules would simply collide and then separate without reaction:

$$\text{A} + \text{C}{-}\text{B} \longrightarrow \text{A}{\cdots}\text{C}{\cdots}\text{B} \longrightarrow \text{A} + \text{C}{-}\text{B}$$

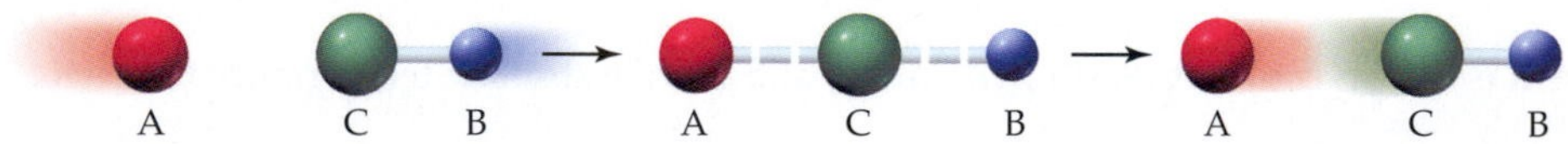

The fraction of collisions having proper orientation for the conversion of reactants to products is called the **steric factor (*p*)**. For the reaction, $A + BC \rightarrow AB + C$, the value of p is expected to be about 0.5 because A has a nearly 1:1 probability of colliding with each of the B and C ends of BC. (This assumes that B and C have similar sizes and electronic properties.) For reactions of larger, more complex molecules, the fraction of the collisions with the proper orientation for reaction is much smaller and p is considerably less than 0.5. For example, p is only 1.7×10^{-6} for the reaction

$$H_2(g) + H_2C{=}CH_2(g) \longrightarrow H_3C{-}CH_3(g)$$

Now let's see how the two parameters p and f enter into the rate law. Since bimolecular collisions between any two molecules—say, A and B—occur at a rate that is proportional to their concentrations, we can write

$$\text{Collision rate} = Z[\text{A}][\text{B}]$$

where Z is the collision frequency and has units of a second-order rate constant, $1/(\text{M} \cdot \text{s})$ or $\text{M}^{-1}\,\text{s}^{-1}$. The reaction rate is lower than the collision rate by a factor $p \times f$ because only a fraction of the colliding molecules have the correct orientation and the minimum energy needed for reaction:

$$\text{Reaction rate} = p \times f \times \text{Collision rate} = pfZ[\text{A}][\text{B}]$$

Since the rate law is

$$\text{Reaction rate} = k[\text{A}][\text{B}]$$

the rate constant predicted by collision theory is $k = pfZ$, or $k = pZe^{-E_a/RT}$:

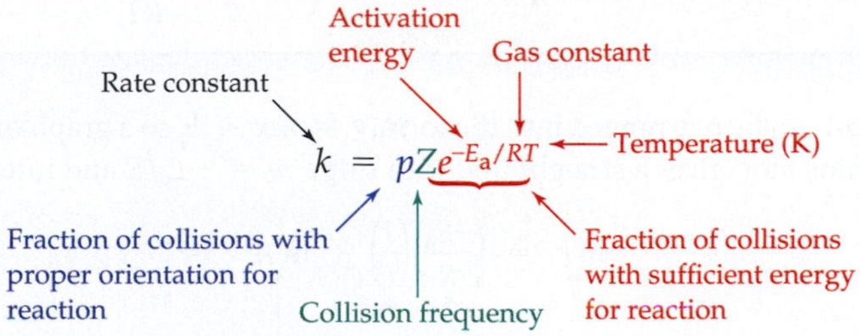

This expression is usually written in a form called the **Arrhenius equation**, named after Svante Arrhenius, the Swedish chemist who proposed it in 1889 on the basis of experimental studies of reaction rates:

▶ **Arrhenius equation** $k = Ae^{-E_a/RT}$

The parameter $A\,(=pZ)$ is called the **frequency factor**, or pre-exponential factor. In accord with the minus sign in the exponent, the rate constant decreases as E_a increases and increases as T increases.

CONCEPTUAL PROBLEM 12.16 The potential energy profile for the one-step reaction $AB + CD \rightarrow AC + BD$ follows. The energies are in kJ/mol relative to an arbitrary zero of energy.

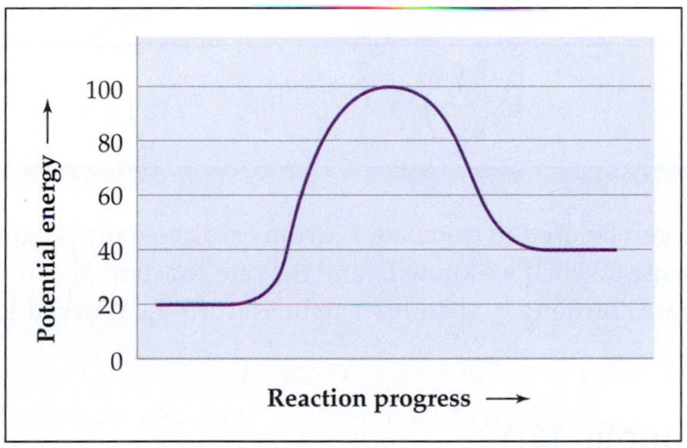

(a) What is the value of the activation energy for this reaction?
(b) Is the reaction exothermic or endothermic?
(c) Suggest a plausible structure for the transition state.
(d) Does the reaction rate increase or decrease when the temperature increases? Explain.

12.10 USING THE ARRHENIUS EQUATION

As we saw in Section 12.9, the activation energy E_a is one of the most important factors affecting the rate of a chemical reaction. Its value can be determined using the Arrhenius equation if values of the rate constant are known at different temperatures.

Taking the natural logarithm of both sides of the Arrhenius equation, we obtain the logarithmic form

▶ **Log form of the Arrhenius equation** $\ln k = \ln A - \dfrac{E_a}{RT}$

This equation can be rearranged into the form $y = mx + b$, so a graph of $\ln k$ versus $1/T$, called an Arrhenius plot, gives a straight line with slope $m = -E_a/R$ and intercept $b = \ln A$:

$$\underset{y}{\ln k} = \underset{m}{\left(\frac{-E_a}{R}\right)}\underset{x}{\left(\frac{1}{T}\right)} + \underset{b}{\ln A}$$

The experimental value of the activation energy can be determined from the slope of the straight line, as shown in Worked Example 12.13.

$$E_a = -R(\text{Slope})$$

Still another form of the Arrhenius equation can be derived that allows us to estimate the activation energy from rate constants at just two temperatures. At temperature T_1,

$$\ln k_1 = \left(\frac{-E_a}{R}\right)\left(\frac{1}{T_1}\right) + \ln A$$

and at temperature T_2,

$$\ln k_2 = \left(\frac{-E_a}{R}\right)\left(\frac{1}{T_2}\right) + \ln A$$

Subtracting the first equation from the second, and remembering that $(\ln k_2 - \ln k_1) = \ln (k_2/k_1)$, we obtain

▶ $$\ln\left(\frac{k_2}{k_1}\right) = \left(\frac{-E_a}{R}\right)\left(\frac{1}{T_2} - \frac{1}{T_1}\right)$$

This equation can be used to calculate E_a from rate constants k_1 and k_2 at temperatures T_1 and T_2. By the same token, if we know E_a and the rate constant k_1 at one temperature T_1, we can calculate the rate constant k_2 at another temperature T_2. Worked Example 12.13 shows how this is done.

Worked Example 12.13

Using the Arrhenius Equation

Rate constants for the gas-phase decomposition of hydrogen iodide, $2\ HI(g) \rightarrow H_2(g) + I_2(g)$, are listed in the following table:

Temperature (°C)	k ($M^{-1}\,s^{-1}$)
283	3.52×10^{-7}
356	3.02×10^{-5}
393	2.19×10^{-4}
427	1.16×10^{-3}
508	3.95×10^{-2}

(a) Find the activation energy (in kJ/mol) using all five data points.
(b) Calculate E_a from the rate constants at 283 °C and 508 °C.
(c) Given the rate constant at 283 °C and the value of E_a obtained in part **(b)**, what is the rate constant at 293 °C?

STRATEGY

(a) The activation energy E_a can be determined from the slope of a linear plot of ln k versus $1/T$.

(b) To calculate E_a from values of the rate constant at two temperatures, use the equation

$$\ln\left(\frac{k_2}{k_1}\right) = \left(\frac{-E_a}{R}\right)\left(\frac{1}{T_2} - \frac{1}{T_1}\right)$$

(c) Use the same equation and the known values of E_a and k_1 at T_1 to calculate k_2 at T_2.

SOLUTION

(a) Because the temperature in the Arrhenius equation is expressed in kelvin, we must first convert the Celsius temperatures to absolute temperatures. Then calculate values of $1/T$ and ln k, and plot ln k versus $1/T$. The results are shown in the following table and graph:

t (°C)	T (K)	k ($M^{-1}\,s^{-1}$)	$1/T$ (1/K)	ln k
283	556	3.52×10^{-7}	0.001 80	−14.860
356	629	3.02×10^{-5}	0.001 59	−10.408
393	666	2.19×10^{-4}	0.001 50	−8.426
427	700	1.16×10^{-3}	0.001 43	−6.759
508	781	3.95×10^{-2}	0.001 28	−3.231

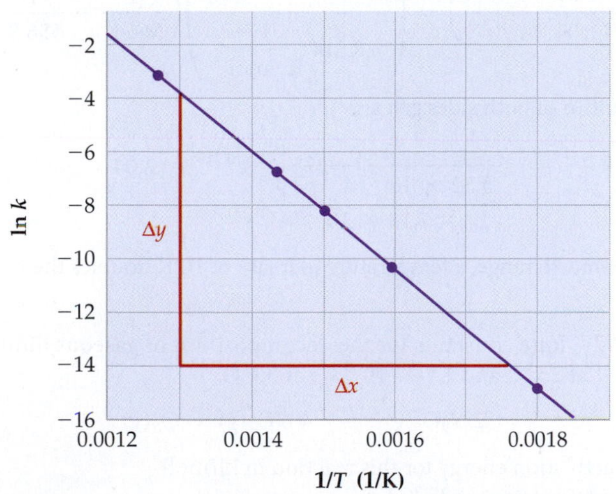

The slope of the straight-line plot can be determined from the coordinates of any two widely separated points on the line:

$$\text{Slope} = \frac{\Delta y}{\Delta x} = \frac{(-14.0) - (-3.9)}{(0.001\,75\ \text{K}^{-1}) - (0.001\,30\ \text{K}^{-1})} = \frac{-10.1}{0.000\,45\ \text{K}^{-1}} = -2.24 \times 10^4\ \text{K}$$

Finally, calculate the activation energy from the slope:

$$E_a = -R(\text{Slope}) = -\left(8.314\frac{\text{J}}{\text{K}\cdot\text{mol}}\right)(-2.24 \times 10^4\ \text{K})$$

$$= 1.9 \times 10^5\ \text{J/mol} = 190\ \text{kJ/mol}$$

Note that the slope of the Arrhenius plot is negative and the activation energy is positive. The greater the activation energy for a particular reaction, the steeper the slope of the ln k versus $1/T$ plot and the greater the increase in the rate constant for a given increase in temperature.

continued on next page

(b) Substituting the values of $k_1 = 3.52 \times 10^{-7}\ M^{-1}\,s^{-1}$ at $T_1 = 556\ K\ (283\ °C)$ and $k_2 = 3.95 \times 10^{-2}\ M^{-1}\,s^{-1}$ at $T_2 = 781\ K\ (508\ °C)$ into the equation

$$\ln\left(\frac{k_2}{k_1}\right) = \left(\frac{-E_a}{R}\right)\left(\frac{1}{T_2} - \frac{1}{T_1}\right)$$

gives

$$\ln\left(\frac{3.95 \times 10^{-2}\ M^{-1}\,s^{-1}}{3.52 \times 10^{-7}\ M^{-1}\,s^{-1}}\right) = \left(\frac{-E_a}{8.314\,\frac{J}{K \cdot mol}}\right)\left(\frac{1}{781\ K} - \frac{1}{556\ K}\right)$$

Simplifying this equation gives

$$11.628 = \left(\frac{-E_a}{8.314\,\frac{J}{K \cdot mol}}\right)\left(\frac{-5.18 \times 10^{-4}}{K}\right)$$

$$E_a = 1.87 \times 10^5\ J/mol = 187\ kJ/mol$$

(c) Use the same equation as in part **(b)**, but now the known values are

$$k_1 = 3.52 \times 10^{-7}\ M^{-1}\,s^{-1} \text{ at } T_1 = 556\ K\ (283\ °C)$$
$$E_a = 1.87 \times 10^5\ J/mol$$

and k_2 at $T_2 = 566\ K\ (293\ °C)$ is the unknown.

$$\ln\left(\frac{k_2}{3.52 \times 10^{-7}\ M^{-1}\,s^{-1}}\right) = \left(\frac{-1.87 \times 10^5\,\frac{J}{mol}}{8.314\,\frac{J}{K \cdot mol}}\right)\left(\frac{1}{566\ K} - \frac{1}{556\ K}\right) = 0.715$$

Taking the antiln of both sides gives

$$\frac{k_2}{3.52 \times 10^{-7}\ M^{-1}\,s^{-1}} = e^{0.715} = 2.04$$
$$k_2 = 7.18 \times 10^{-7}\ M^{-1}\,s^{-1}$$

In this temperature range, a temperature increase of 10 K doubles the rate constant.

PROBLEM 12.17 Rate constants for the decomposition of gaseous dinitrogen pentoxide are $3.7 \times 10^{-5}\ s^{-1}$ at 25 °C and $1.7 \times 10^{-3}\ s^{-1}$ at 55 °C.

$$2\,N_2O_5(g) \longrightarrow 4\,NO_2(g) + O_2(g)$$

(a) What is the activation energy for this reaction in kJ/mol?
(b) What is the rate constant at 35 °C?

12.11 REACTION MECHANISMS

In Section 12.9, we saw that chemical reactions occur as a result of energetic collisions between properly oriented atoms and molecules. Often, reactions occur in two or more steps. The sequence of reaction steps that describes the pathway from reactants to products is called the **reaction mechanism**. Chemists want to know the sequence in which the various reaction steps take place so that they can better control known reactions and predict new ones.

▶ **Reaction mechanism** The sequence of reaction steps that describes the pathway from reactants to products.

A single step in a reaction mechanism is called an **elementary reaction**, or **elementary step**. To clarify the crucial distinction between an elementary reaction and an overall reaction, let's consider the gas-phase reaction of nitrogen dioxide and carbon monoxide to give nitric oxide and carbon dioxide:

$$NO_2(g) + CO(g) \longrightarrow NO(g) + CO_2(g) \quad \text{Overall reaction}$$

Experimental evidence suggests that this reaction takes place by a two-step mechanism:

Step 1. $NO_2(g) + NO_2(g) \longrightarrow NO(g) + NO_3(g)$ Elementary reaction

Step 2. $NO_3(g) + CO(g) \longrightarrow NO_2(g) + CO_2(g)$ Elementary reaction

In the first elementary step, two NO_2 molecules collide with enough energy to break one N–O bond and form another, resulting in the transfer of an oxygen atom from one NO_2 molecule to the other. In the second step, the NO_3 molecule formed in the first step collides with a CO molecule, and the transfer of an oxygen atom from NO_3 to CO yields an NO_2 molecule and a CO_2 molecule (**Figure 12.14**).

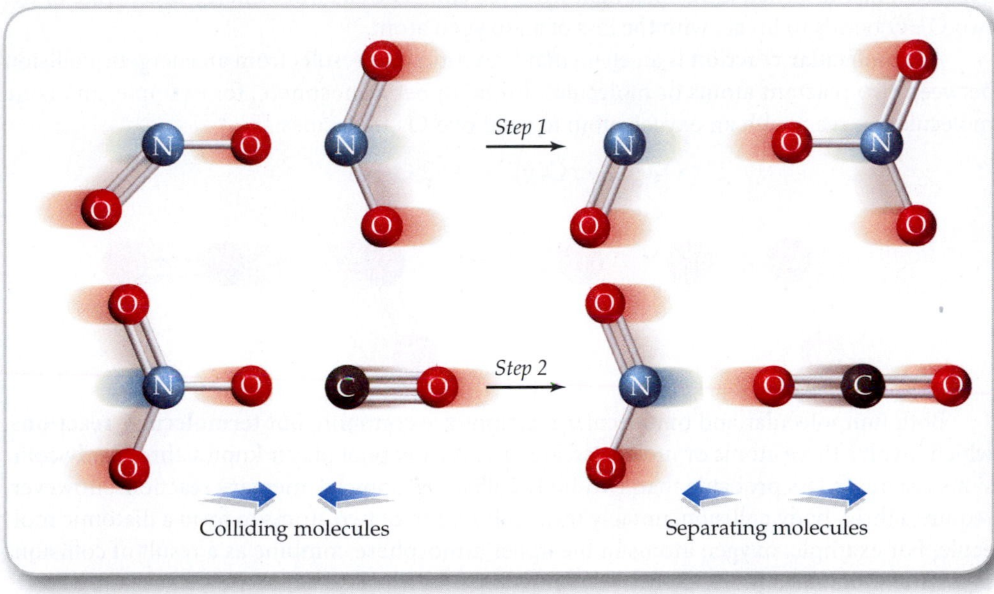

FIGURE 12.14

Elementary steps in the reaction of NO_2 with CO.

The chemical equation for an elementary reaction is a description of an individual molecular event that involves breaking and/or making chemical bonds. By contrast, the balanced equation for an overall reaction describes only the stoichiometry of the overall process but provides no information about how the reaction occurs. The equation for the overall reaction of NO_2 with CO, for example, doesn't mean that the reaction occurs by the direct transfer of an oxygen atom from an NO_2 molecule to a CO molecule.

Elementary reaction—describes an individual molecular event.

Overall reaction—describes the reaction stoichiometry.

The elementary steps in a proposed reaction mechanism must sum to give the overall reaction. When we sum the elementary steps in the reaction of NO_2 with CO and then cancel the molecules that appear on both sides of the resulting equation, for instance, we obtain the overall reaction:

Step 1. $NO_2(g) + NO_2(g) \longrightarrow NO(g) + NO_3(g)$ Elementary reaction

Step 2. $NO_3(g) + CO(g) \longrightarrow NO_2(g) + CO_2(g)$ Elementary reaction

$$NO_2(g) + \cancel{NO_2(g)} + \cancel{NO_3(g)} + CO(g) \longrightarrow NO(g) + \cancel{NO_3(g)} + \cancel{NO_2(g)} + CO_2(g)$$

$$NO_2(g) + CO(g) \longrightarrow NO(g) + CO_2(g) \quad \text{Overall reaction}$$

A species that is formed in one step of a reaction mechanism and consumed in a subsequent step, such as NO_3 in our example, is called a **reaction intermediate**. Reaction intermediates

▲ The beautiful northern lights, or aurora borealis, are often observed in the northern hemisphere at high latitudes. The light is partly produced by excited O atoms in the upper atmosphere.

don't appear in the net equation for the overall reaction, and it's only by looking at the elementary steps that their presence is noticed.

Elementary reactions are classified on the basis of their **molecularity**, the number of molecules (or atoms) on the reactant side of the chemical equation. A **unimolecular reaction**, for instance, is an elementary reaction that involves a single reactant molecule—for example, the unimolecular decomposition of ozone in the upper atmosphere:

$$O_3^*(g) \longrightarrow O_2(g) + O(g)$$

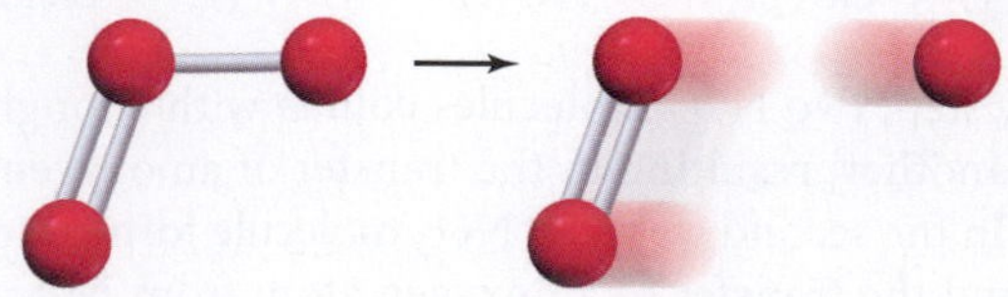

The asterisk on O_3 indicates that the ozone molecule is in an energetically excited state because it has absorbed ultraviolet light from the sun. The absorbed energy causes one of the two O—O bonds to break, with the loss of an oxygen atom.

A **bimolecular reaction** is an elementary reaction that results from an energetic collision between two reactant atoms or molecules. In the upper atmosphere, for example, an ozone molecule can react with an oxygen atom to yield two O_2 molecules:

$$O_3(g) + O(g) \longrightarrow 2\,O_2(g)$$

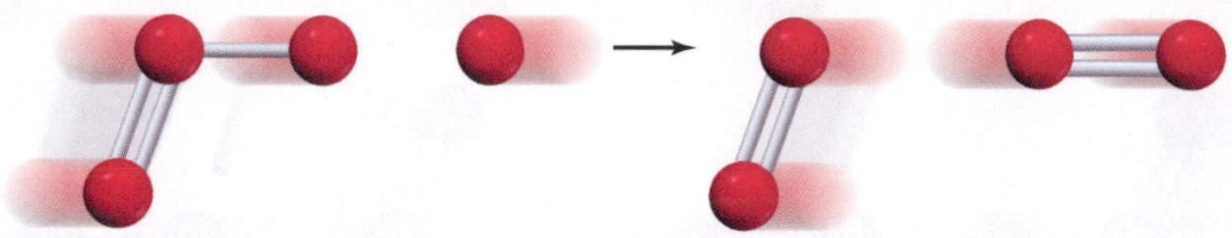

Both unimolecular and bimolecular reactions are common, but **termolecular reactions**, which involve three atoms or molecules, are rare. As any pool player knows, three-body collisions are much less probable than two-body collisions. Some elementary reactions, however, require a three-body collision, notably the combination of two atoms to form a diatomic molecule. For example, oxygen atoms in the upper atmosphere combine as a result of collisions involving some third molecule M:

$$O(g) + O(g) + M(g) \longrightarrow O_2(g) + M(g)$$

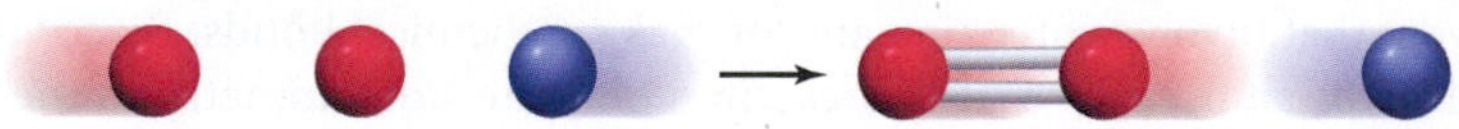

In the atmosphere, M is most likely N_2, but in principle it could be any atom or molecule. The role of M is to carry away the energy that is released when the O—O bond is formed. If M were not involved in the collision, the two oxygen atoms would simply bounce off each other and no reaction would occur.

Worked Example 12.14

Identifying Intermediates and Molecularity in a Reaction Mechanism

The following two-step mechanism has been proposed for the gas-phase decomposition of nitrous oxide (N_2O):

Step 1. $$N_2O(g) \longrightarrow N_2(g) + O(g)$$

Step 2. $$N_2O(g) + O(g) \longrightarrow N_2(g) + O_2(g)$$

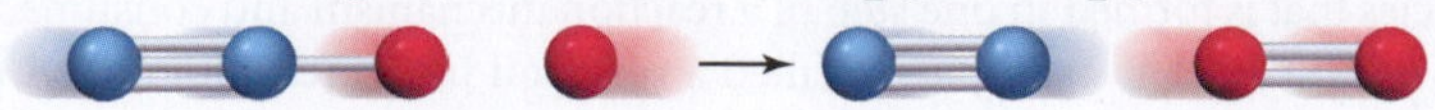

(a) Write the chemical equation for the overall reaction.
(b) Identify any reaction intermediates.
(c) What is the molecularity of each of the elementary reactions?
(d) What is the molecularity of the overall reaction?

STRATEGY

To find the overall reaction, sum the elementary steps. To identify intermediates and molecularity, look at the individual steps.

SOLUTION

(a) The overall reaction is the sum of the two elementary steps:

$$\begin{array}{lrcll} \textit{Step 1.} & N_2O(g) & \longrightarrow & N_2(g) + O(g) & \text{Elementary reaction} \\ \textit{Step 2.} & N_2O(g) + O(g) & \longrightarrow & N_2(g) + O_2(g) & \text{Elementary reaction} \\ \hline & 2\,N_2O(g) + \cancel{O(g)} & \longrightarrow & 2\,N_2(g) + \cancel{O(g)} + O_2(g) & \\ & 2\,N_2O(g) & \longrightarrow & 2\,N_2(g) + O_2(g) & \text{Overall reaction} \end{array}$$

(b) The oxygen atom is a reaction intermediate because it is formed in the first elementary step and consumed in the second step.
(c) The first elementary reaction is unimolecular because it involves a single reactant molecule. The second step is bimolecular because it involves two reactant atoms or molecules.
(d) It's inappropriate to use the word *molecularity* in connection with the overall reaction because the overall reaction does not describe an individual molecular event. Only an elementary reaction can have a molecularity.

PROBLEM 12.18 A suggested mechanism for the reaction of nitrogen dioxide and molecular fluorine is

Step 1. $NO_2(g) + F_2(g) \longrightarrow NO_2F(g) + F(g)$

Step 2. $NO_2(g) + F(g) \longrightarrow NO_2F(g)$

(a) Write the chemical equation for the overall reaction, and identify any reaction intermediates.
(b) What is the molecularity of each elementary reaction?

12.12 RATE LAWS FOR ELEMENTARY REACTIONS

Recall from Section 12.2 that the rate law for an overall chemical reaction must be determined experimentally. It can't be deduced from the stoichiometric coefficients in the balanced equation for the overall reaction. By contrast, the rate law for an elementary reaction follows directly from its molecularity because an elementary reaction is an individual molecular event. The concentration of each reactant in an elementary reaction appears in the rate law, with an exponent equal to its coefficient in the chemical equation for the elementary reaction.

Consider, for example, the unimolecular decomposition of ozone:

$$O_3(g) \longrightarrow O_2(g) + O(g)$$

The number of moles of O_3 per liter that decompose per unit time is directly proportional to the molar concentration of O_3:

$$\text{Rate} = -\frac{\Delta[O_3]}{\Delta t} = k[O_3]$$

The rate of a unimolecular reaction is always first order in the concentration of the reactant molecule.

For a bimolecular elementary reaction of the type A + B → Products, the reaction rate depends on the frequency of collisions between A and B molecules. The frequency of AB collisions involving any *particular* A molecule is proportional to the molar concentration of B, and the total frequency of AB collisions involving *all* A molecules is proportional to the molar concentration of A times the molar concentration of B (**Figure 12.15**). Therefore, the reaction obeys the second-order rate law

$$\text{Rate} = -\frac{\Delta[A]}{\Delta t} = -\frac{\Delta[B]}{\Delta t} = k[A][B]$$

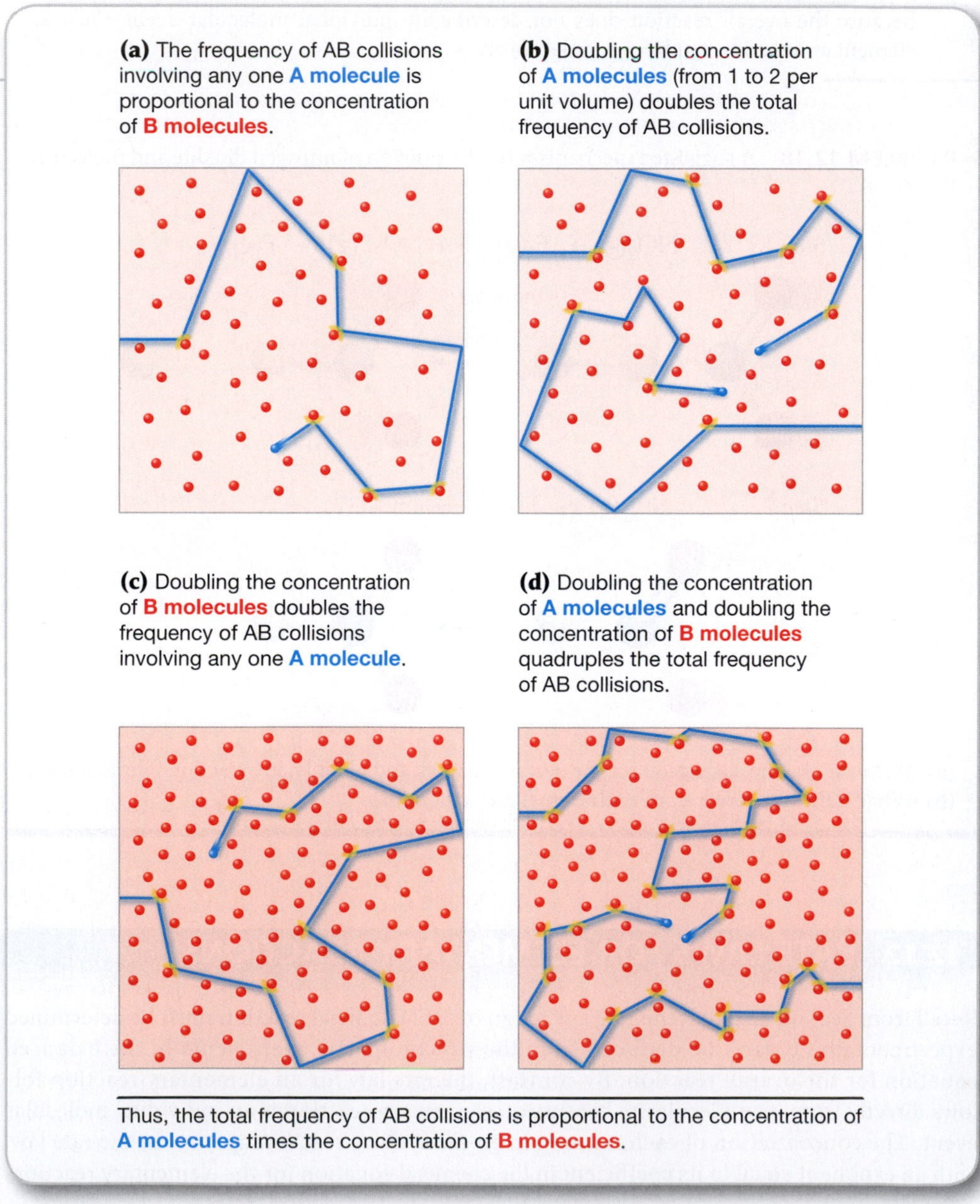

FIGURE 12.15

The effect of concentration on the frequency of collisions between A molecules (blue) and B molecules (red).

The experimental rate law for the formation of N_2 is

$$\text{Rate} = \frac{\Delta[N_2]}{\Delta t} = k[NO]^2[H_2]$$

where k is the observed rate constant for the overall reaction. To decide whether the proposed mechanism is plausible, we need to compare the experimental and predicted rate laws. The concentrations of reaction intermediates, such as N_2O_2 in the example at hand, do not appear in the experimental rate law because their concentrations are usually very small and undetermined. Only reactants and products (and catalysts, if present) appear in the rate law for an overall reaction. Therefore, we must eliminate $[N_2O_2]$ from the predicted rate law. Because the mechanism indicates that the rate of formation of N_2 equals the rate of disappearance of N_2O_2, $\Delta[N_2]/\Delta t = -\Delta[N_2O_2]/\Delta t$, we can rewrite the predicted rate law as

$$\text{Rate} = \frac{\Delta[N_2]}{\Delta t} = k_2[N_2O_2][H_2]$$

To eliminate N_2O_2 from the right side of this equation, we assume that the fast, reversible, first step in the mechanism reaches a dynamic equilibrium. (We'll have more to say about **chemical equilibrium** in Chapter 13.) The rates of the forward and reverse reactions in the first step are given by

$$\text{Rate}_{\text{forward}} = k_1[NO]^2 \qquad \text{Rate}_{\text{reverse}} = k_{-1}[N_2O_2]$$

LOOKING AHEAD...

In chapter 13, we'll see that **chemical equilibrium** is reached when the rates of the forward and reverse reactions are equal.

Just as the rates of vaporization and condensation are equal for a **liquid–vapor equilibrium** (Section 10.5), so the rates of the forward and reverse reactions are equal for a chemical equilibrium. Therefore,

REMEMBER...

In a **liquid–vapor equilibrium** the number of molecules escaping from the liquid per unit time (vaporization) equals the number returning to the liquid per unit time (condensation). (Section 10.5)

$$k_1[NO]^2 = k_{-1}[N_2O_2] \quad \text{and} \quad [N_2O_2] = \frac{k_1}{k_{-1}}[NO]^2$$

Substituting this expression for $[N_2O_2]$ into the equation for the predicted rate eliminates the intermediate and gives the final predicted rate law in terms of only reactants and products:

$$\text{Rate} = \frac{\Delta[N_2]}{\Delta t} = k_2[N_2O_2][H_2] = k_2\frac{k_1}{k_{-1}}[NO]^2[H_2]$$

The predicted and experimental rate laws now have the same form: Both are second order in NO and first order in H_2. Therefore, the proposed mechanism is consistent with experiment and is a plausible mechanism for the reaction. Comparison of the two rate laws indicates that the observed rate constant k equals k_2k_1/k_{-1}.

Procedure Used in Studies of Reaction Mechanisms

Let's summarize the procedure that chemists use in establishing a reaction mechanism. First, the rate law is determined by experiment. Then, a series of elementary steps is proposed, and the rate law predicted by the proposed mechanism is worked out. If the observed and predicted rate laws do not agree, the mechanism must be discarded and another one must be devised. If the observed and predicted rate laws *do* agree, the proposed mechanism is a plausible (though not necessarily correct) pathway for the reaction. **Figure 12.16** summarizes the procedure.

The case for a particular mechanism is strengthened considerably if a reaction intermediate can be isolated or if an unstable intermediate can be detected. It's easy to disprove a mechanism, but it's seldom possible to finally prove a mechanism because there may be an alternative reaction pathway—not yet imagined—that also fits the experimental facts. The best we can do to establish a mechanism is to accumulate a convincing body of experimental evidence that supports it. Proving a reaction mechanism is more like proving a case in a court of law than like proving a theorem in mathematics.

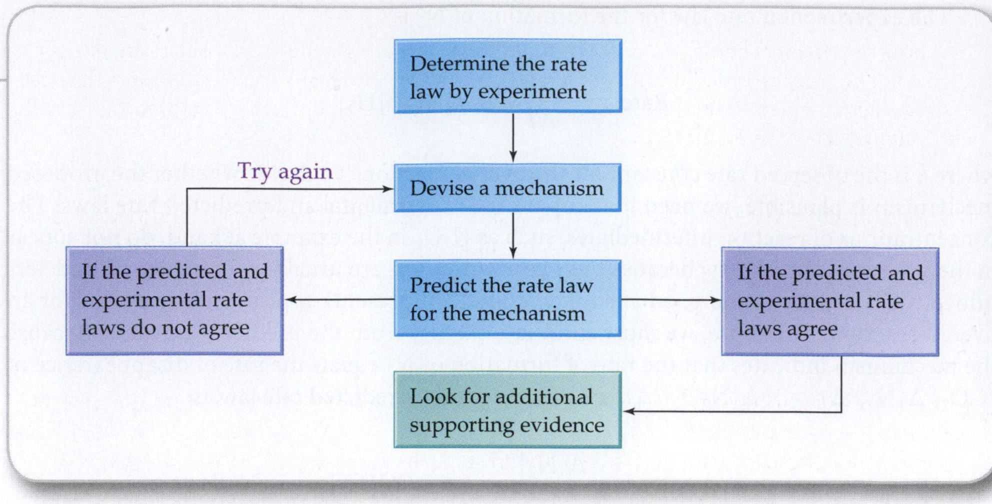

FIGURE 12.16

Flowchart illustrating the logic used in studies of reaction mechanisms.

Worked Example 12.15

Suggesting a Mechanism Given the Rate Law: Reactions with an Initial Slow Step

The following reaction has a second-order rate law:

$$H_2(g) + 2\,ICl(g) \longrightarrow I_2(g) + 2\,HCl(g) \qquad \text{Rate} = k[H_2][ICl]$$

Devise a possible reaction mechanism.

STRATEGY

The reaction doesn't occur in a single elementary step because, if it did, the rate law would be third order: Rate $= k[H_2][ICl]^2$. The observed rate law will be obtained if the rate-determining step involves the bimolecular reaction of H_2 and ICl.

SOLUTION

A plausible sequence of elementary steps is

$$H_2(g) + ICl(g) \xrightarrow{k_1} HI(g) + HCl(g) \qquad \text{Slower, rate-determining}$$
$$HI(g) + ICl(g) \xrightarrow{k_2} I_2(g) + HCl(g) \qquad \text{Faster}$$
$$H_2(g) + 2\,ICl(g) \longrightarrow I_2(g) + 2\,HCl(g) \qquad \text{Overall reaction}$$

The rate law predicted by this mechanism, rate $= k_1[H_2][ICl]$, agrees with the observed rate law.

Worked Example 12.16

Supporting a Mechanism Given the Rate Law: Reactions with an Initial Fast Step

The experimental rate law for the decomposition of ozone is second order in ozone and inverse first order in molecular oxygen:

$$2\,O_3(g) \longrightarrow 3\,O_2(g) \qquad \text{Rate} = -\frac{\Delta[O_3]}{\Delta t} = k\frac{[O_3]^2}{[O_2]}$$

Show that the following mechanism is consistent with the experimental rate law, and relate the observed rate constant k to the rate constants for the elementary reactions:

$$O_3(g) \underset{k_{-1}}{\overset{k_1}{\rightleftharpoons}} O_2(g) + O(g) \qquad \text{Faster, reversible}$$
$$O(g) + O_3(g) \xrightarrow{k_2} 2\,O_2(g) \qquad \text{Slower, rate-determining}$$
$$2\,O_3(g) \longrightarrow 3\,O_2(g) \qquad \text{Overall reaction}$$

STRATEGY

To show that the mechanism is consistent with experiment, we must derive the rate law predicted by the mechanism and compare it with the experimental rate law. If we assume that the faster, reversible step is at equilibrium, we can eliminate the concentration of the intermediate O atoms from the predicted rate law.

SOLUTION

The rate law for the rate-determining step is rate $= k_2[\text{O}][\text{O}_3]$, but the stoichiometry of the reaction indicates that the overall rate of consumption of ozone is twice the rate of the rate-determining step:

$$\text{Rate} = -\frac{\Delta[\text{O}_3]}{\Delta t} = 2k_2[\text{O}][\text{O}_3]$$

(When one O atom and one O_3 molecule react in the rate-determining step, two O_3 molecules are consumed in the overall reaction.)

The rates of the forward and reverse reactions in the faster, reversible step are given by

$$\text{Rate}_{\text{forward}} = k_1[\text{O}_3] \qquad \text{Rate}_{\text{reverse}} = k_{-1}[\text{O}_2][\text{O}]$$

Assuming that the first step is at equilibrium, we can equate the rates of the forward and reverse reactions and then solve for the concentration of the intermediate O atoms:

$$k_1[\text{O}_3] = k_{-1}[\text{O}_2][\text{O}] \qquad \text{so, } [\text{O}] = \frac{k_1[\text{O}_3]}{k_{-1}[\text{O}_2]}$$

Substituting this expression for [O] into the predicted rate law for overall consumption of ozone gives the predicted rate law in terms of only reactants and products:

$$\text{Rate} = -\frac{\Delta[\text{O}_3]}{\Delta t} = 2k_2[\text{O}][\text{O}_3] = 2k_2\frac{k_1}{k_{-1}}\frac{[\text{O}_3]^2}{[\text{O}_2]}$$

Because the predicted and experimental rate laws have the same reaction orders in O_3 and O_2, the proposed mechanism is consistent with the experimental rate law and is a plausible mechanism for the reaction. Comparison of the predicted and experimental rate laws indicates that the observed rate constant k equals $2k_2k_1/k_{-1}$.

PROBLEM 12.20 The following reaction has a first-order rate law:

$$\text{Co(CN)}_5(\text{H}_2\text{O})^{2-}(aq) + \text{I}^-(aq) \longrightarrow \text{Co(CN)}_5\text{I}^{3-}(aq) + \text{H}_2\text{O}(l)$$
$$\text{Rate} = k[\text{Co(CN)}_5(\text{H}_2\text{O})^{2-}]$$

Suggest a possible reaction mechanism, and show that your mechanism agrees with the observed rate law.

PROBLEM 12.21 The following mechanism has been proposed for the oxidation of nitric oxide to nitrogen dioxide:

$$\text{NO}(g) + \text{O}_2(g) \underset{k_{-1}}{\overset{k_1}{\rightleftharpoons}} \text{NO}_3(g) \qquad \text{Faster, reversible}$$

$$\text{NO}_3(g) + \text{NO}(g) \xrightarrow{k_2} 2\,\text{NO}_2(g) \qquad \text{Slower, rate-determining}$$

The experimental rate law for the overall reaction is

$$\text{Rate} = -\frac{\Delta[\text{NO}]}{\Delta t} = k[\text{NO}]^2[\text{O}_2]$$

(a) Write a balanced equation for the overall reaction.
(b) Show that the proposed mechanism is consistent with the experimental rate law.
(c) Relate the rate constant k to the rate constants for the elementary reactions.

12.14 CATALYSIS

▲ MnO_2 catalyzes the thermal decomposition of $KClO_3$ to KCl and O_2 gas.

Reaction rates are affected not only by reactant concentrations and temperature but also by the presence of *catalysts*. A **catalyst** is a substance that increases the rate of a transformation without itself being consumed in the process. An example is manganese dioxide, a black powder that speeds up the thermal decomposition of potassium chlorate:

$$2\,KClO_3(s) \xrightarrow[\text{Heat}]{MnO_2\text{ catalyst}} 2\,KCl(s) + 3\,O_2(g)$$

In the absence of a catalyst, $KClO_3$ decomposes very slowly, even when heated, but when a small amount of MnO_2 is mixed with the $KClO_3$ before heating, rapid evolution of oxygen ensues. The MnO_2 can be recovered unchanged after the reaction is complete.

Catalysts are enormously important, both in the chemical industry and in living organisms. Nearly all industrial processes for the manufacture of essential chemicals use catalysts to favor formation of specific products and to lower reaction temperatures, thus reducing energy costs. In environmental chemistry, catalysts such as nitric oxide play a role in the formation of air pollutants, while other catalysts, such as platinum in automobile catalytic converters, are potent weapons in the battle to control air pollution.

In living organisms, almost all the hundreds of thousands of chemical reactions that take place constantly are catalyzed by large molecules called *enzymes* (see the *FYI* at the end of this chapter). Nitrogenase, for example, an enzyme present in bacteria on the root nodules of leguminous plants such as peas and beans, catalyzes the conversion of atmospheric nitrogen to ammonia. The ammonia then serves as a fertilizer for plant growth. In the human body, the enzyme carbonic anhydrase catalyzes the reaction of carbon dioxide with water:

$$CO_2(aq) + H_2O(l) \rightleftharpoons H^+(aq) + HCO_3^-(aq)$$

The forward reaction occurs when the blood takes up CO_2 in the tissues, and the reverse reaction occurs when the blood releases CO_2 in the lungs. Remarkably, carbonic anhydrase increases the rate of these reactions by a factor of about 10^6!

How does a catalyst work? A catalyst accelerates the rate of a reaction by making available a different, lower-energy mechanism for the conversion of reactants to products. Take the decomposition of hydrogen peroxide in a basic, aqueous solution, for instance:

$$2\,H_2O_2(aq) \longrightarrow 2\,H_2O(l) + O_2(g)$$

Although unstable with respect to water and oxygen, hydrogen peroxide decomposes only very slowly at room temperature because the reaction has a high activation energy (76 kJ/mol). In the presence of iodide ion, however, the reaction is appreciably faster (**Figure 12.17**) because it can proceed by a different, lower-energy pathway:

Step 1.	$H_2O_2(aq) + I^-(aq) \longrightarrow H_2O(l) + IO^-(aq)$	Slower, rate-determining
Step 2.	$H_2O_2(aq) + IO^-(aq) \longrightarrow H_2O(l) + O_2(g) + I^-(aq)$	Faster
	$2\,H_2O_2(aq) \longrightarrow 2\,H_2O(l) + O_2(g)$	Overall reaction

The H_2O_2 first oxidizes the catalyst (I^-) to hypoiodite ion (IO^-) and then reduces the intermediate IO^- back to I^-. The catalyst does not appear in the overall reaction because it is consumed in Step 1 and regenerated in Step 2. The catalyst is, however, intimately involved in the reaction and appears in the observed rate law:

$$\text{Rate} = -\frac{\Delta[H_2O_2]}{\Delta t} = k[H_2O_2][I^-]$$

The rate law is consistent with the reaction of H_2O_2 and I^- as the rate-determining step. Notice the difference between a catalyst and a reaction intermediate: *A catalyst is consumed in one step of a reaction and is regenerated in a subsequent step, whereas an intermediate is formed in one step and is consumed in a subsequent step.*

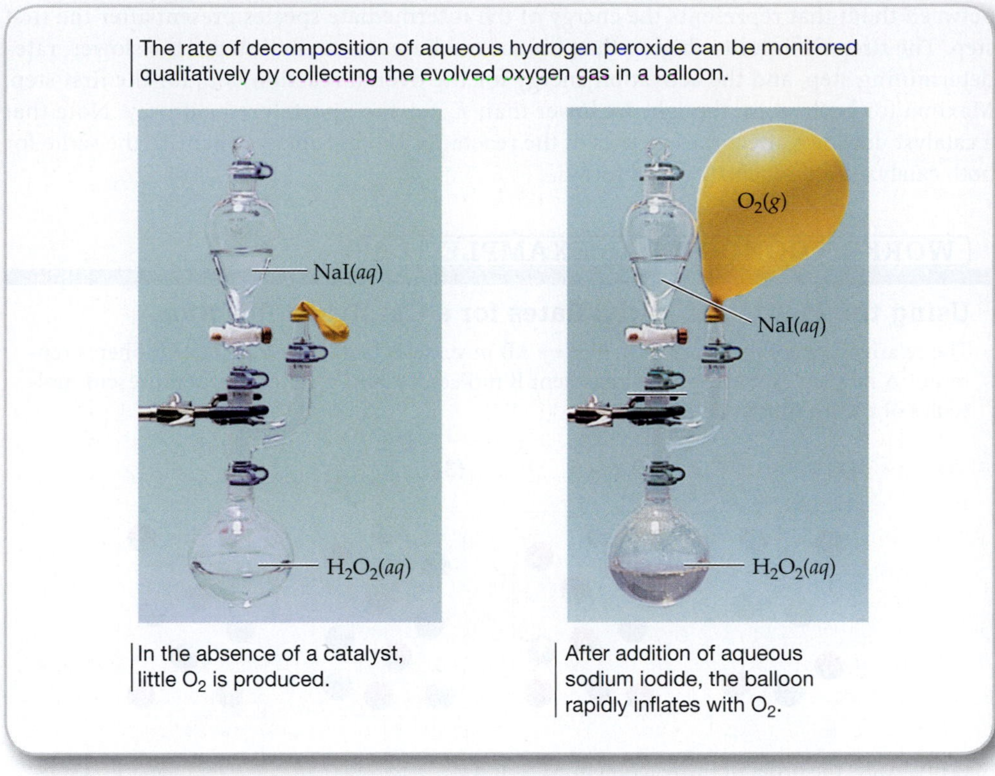

FIGURE 12.17

Iodide ion catalyzes the decomposition of aqueous hydrogen peroxide, yielding O_2 gas and water.

The catalyzed pathway for a reaction might have a faster rate than the uncatalyzed pathway either because of a larger frequency factor (A) or a smaller activation energy (E_a) in the Arrhenius equation. Usually, though, catalysts function by making available a reaction pathway with a lower activation energy (**Figure 12.18**). In the decomposition of hydrogen peroxide, for example, catalysis by I^- lowers E_a for the overall reaction by 19 kJ/mol.

Because the reaction occurs in two steps, the energy profile for the catalyzed pathway shown in **Figure 12.18(a)** exhibits two maxima (two transition states) with a minimum

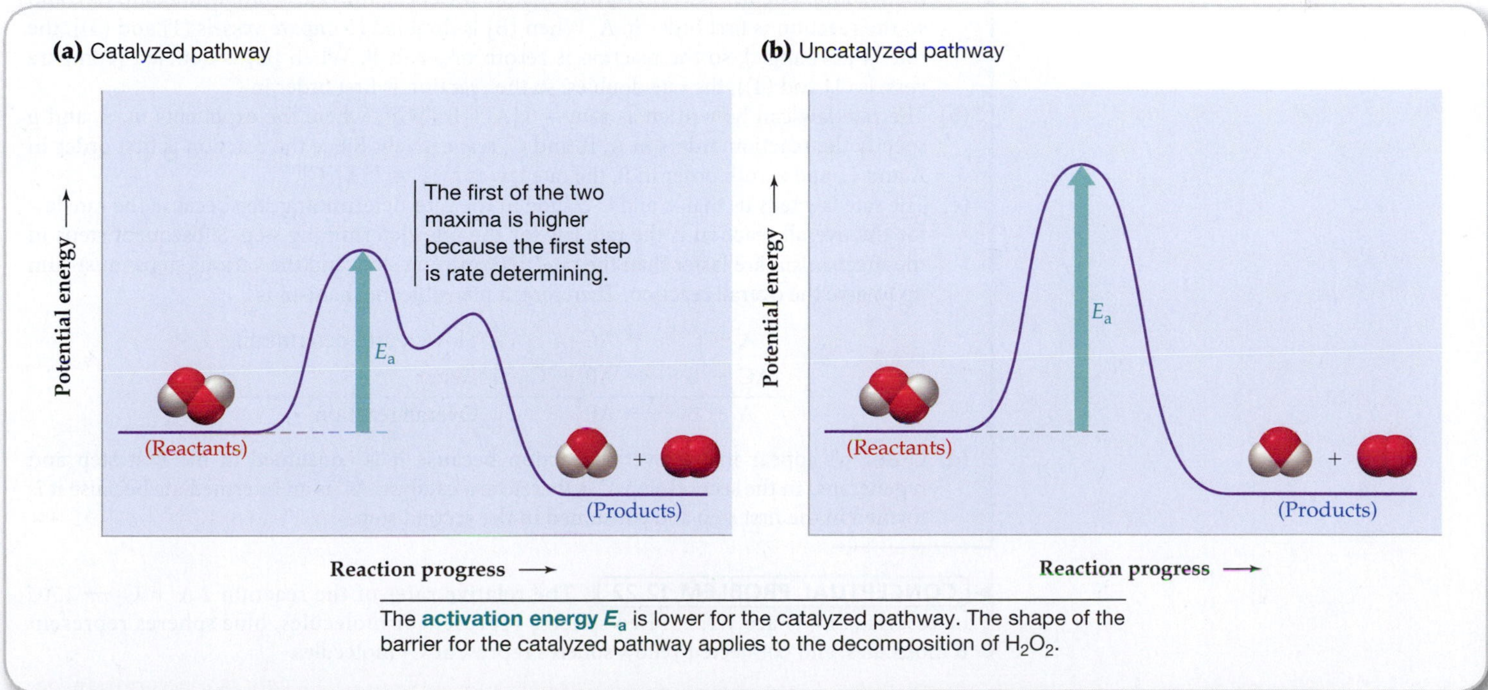

FIGURE 12.18

Typical potential energy profiles for a reaction whose activation energy is lowered by the presence of a catalyst.

between them that represents the energy of the intermediate species present after the first step. The first maximum is higher than the second because the first step is the slower, rate-determining step, and the activation energy for the overall reaction is E_a for the first step. Maxima for both steps, though, are lower than E_a for the uncatalyzed pathway. Note that a catalyst does not affect the energies of the reactants and products, which are the same for both catalyzed and uncatalyzed pathways.

WORKED CONCEPTUAL EXAMPLE 12.17

Using the Method of Initial Rates for a Catalyzed Reaction

The relative rates of the reaction A + B ⟶ AB in vessels (1)–(4) are 1:2:1:2. Red spheres represent A molecules, blue spheres represent B molecules, and yellow spheres represent molecules of a third substance C.

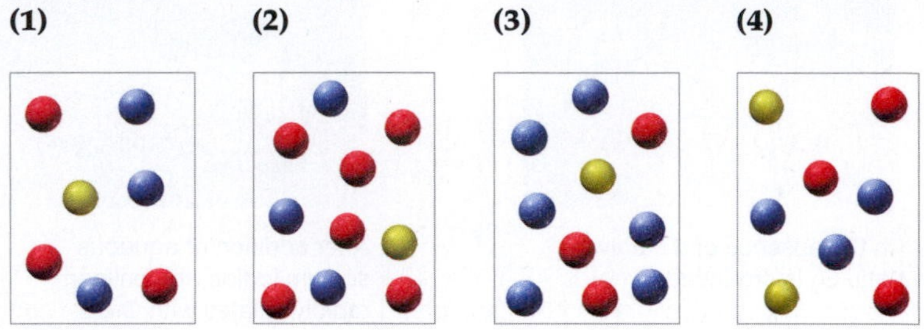

(a) What is the order of the reaction in A, B, and C?
(b) Write the rate law.
(c) Write a mechanism that agrees with the rate law.
(d) Why doesn't C appear in the equation for the overall reaction? Identify any intermediates in your mechanism.

STRATEGY AND SOLUTION

(a) Count the number of molecules of each type in vessels (1)–(4), and compare the relative rates with the relative numbers of molecules. The concentration of A molecules in vessel (2) is twice that in vessel (1), while the concentrations of B and C remain constant. Because the reaction rate in vessel (2) is twice that in vessel (1), the rate is proportional to [A], and so the reaction is first order in A. When [B] is doubled [compare vessels (1) and (3)], the rate is unchanged, so the reaction is zeroth order in B. When [C] is doubled [compare vessels (1) and (4)], the rate doubles, so the reaction is first order in C.

(b) The rate law can be written as rate = $k[A]^m[B]^n[C]^p$, where the exponents m, n, and p specify the reaction orders in A, B, and C, respectively. Since the reaction is first order in A and C, and zeroth order in B, the rate law is rate = k[A][C].

(c) The rate law tells us that A and C collide in the rate-determining step because the rate law for the overall reaction is the rate law for the rate-determining step. Subsequent steps in the mechanism are faster than the rate-determining step, and the various steps must sum up to give the overall reaction. Therefore, a plausible mechanism is

A + C ⟶ AC	Slower, rate-determining
AC + B ⟶ AB + C	Faster
A + B ⟶ AB	Overall reaction

(d) C doesn't appear in the overall reaction because it is consumed in the first step and regenerated in the second step. C is therefore a catalyst. AC is an intermediate because it is formed in the first step and consumed in the second step.

▶ **CONCEPTUAL PROBLEM 12.22** The relative rates of the reaction $2\,A + C_2 \longrightarrow 2\,AC$ in vessels (1)–(4) are 1:1:2:3. Red spheres represent A molecules, blue spheres represent B molecules, and connected yellow spheres represent C_2 molecules.

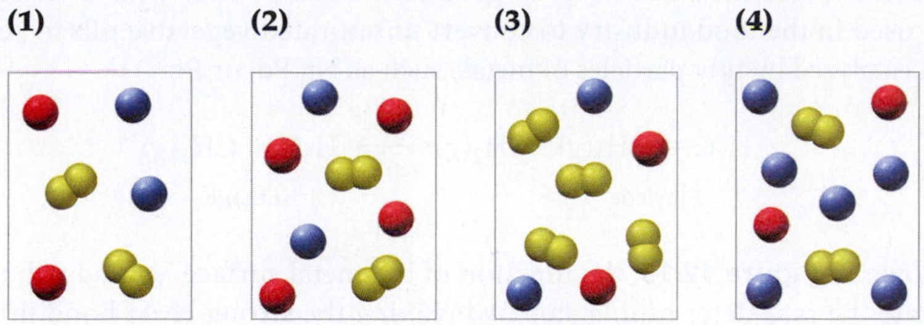

(a) What is the order of the reaction in A, B, and C_2?
(b) Write the rate law.
(c) Write a mechanism that agrees with the rate law.
(d) Identify all catalysts and intermediates in your mechanism.

12.15 HOMOGENEOUS AND HETEROGENEOUS CATALYSTS

Catalysts are commonly classified as either *homogeneous* or *heterogeneous*. A **homogeneous catalyst** is one that exists in the same phase as the reactants. For example, iodide ion is a homogeneous catalyst for the decomposition of aqueous hydrogen peroxide because both I^- and H_2O_2 are present in the same aqueous solution phase.

In the atmosphere, nitric oxide is a gas-phase homogeneous catalyst for the conversion of molecular oxygen to ozone, a process described by the following series of reactions:

$$
\begin{array}{rcl}
1/2\, O_2(g) + NO(g) & \longrightarrow & NO_2(g) \\
NO_2(g) & \xrightarrow{\text{sunlight}} & NO(g) + O(g) \\
O(g) + O_2(g) & \longrightarrow & O_3(g) \\
\hline
3/2\, O_2(g) & \longrightarrow & O_3(g) \qquad \text{Overall reaction}
\end{array}
$$

Nitric oxide first reacts with atmospheric O_2 to give nitrogen dioxide, a poisonous brown gas. Subsequently, NO_2 absorbs sunlight and dissociates to give an oxygen atom, which then reacts with O_2 to form ozone. As usual, the catalyst (NO) and the intermediates (NO_2 and O) do not appear in the chemical equation for the overall reaction.

A **heterogeneous catalyst** is one that exists in a different phase from that of the reactants. Ordinarily, the heterogeneous catalyst is a solid, and the reactants are either gases or liquids. In the Fischer–Tropsch process for manufacturing synthetic gasoline, for example, tiny particles of a metal such as iron or cobalt coated on alumina (Al_2O_3) catalyze the conversion of gaseous carbon monoxide and hydrogen to hydrocarbons such as octane (C_8H_{18}):

$$8\, CO(g) + 17\, H_2(g) \xrightarrow[\text{catalyst}]{Co/Al_2O_3} C_8H_{18}(l) + 8\, H_2O(l)$$

The mechanism of heterogeneous catalysis is often complex and not well understood. Important steps, however, frequently involve (1) attachment of reactants to the surface of the catalyst, a process called *adsorption*, (2) conversion of reactants to products on the surface, and (3) desorption of products from the surface. The adsorption step is thought to involve chemical bonding of the reactant molecules to the highly reactive metal atoms on the surface with accompanying breaking, or at least weakening, of bonds in the reactants.

To illustrate, take the catalytic hydrogenation of compounds with C=C double bonds, a reaction used in the food industry to convert unsaturated vegetable oils to solid fats. The reaction is catalyzed by tiny particles of metals such as Ni, Pd, or Pt:

$$\underset{\text{Ethylene}}{H_2C{=}CH_2(g)} + H_2(g) \xrightarrow[\text{catalyst}]{\text{Metal}} \underset{\text{Ethane}}{H_3C{-}CH_3(g)}$$

As shown in **Figure 12.19**, the function of the metal surface is to adsorb the reactants and facilitate the rate-determining step by breaking the strong H–H bond in the H_2 molecule. Because the H–H bond breaking is accompanied by the simultaneous formation of bonds from the separating H atoms to the surface metal atoms, the activation energy for the process is lowered. The H atoms then move about on the surface until they encounter the C atoms of the adsorbed C_2H_4 molecule. Subsequent stepwise formation of two new C–H bonds gives C_2H_6, which is finally desorbed from the surface.

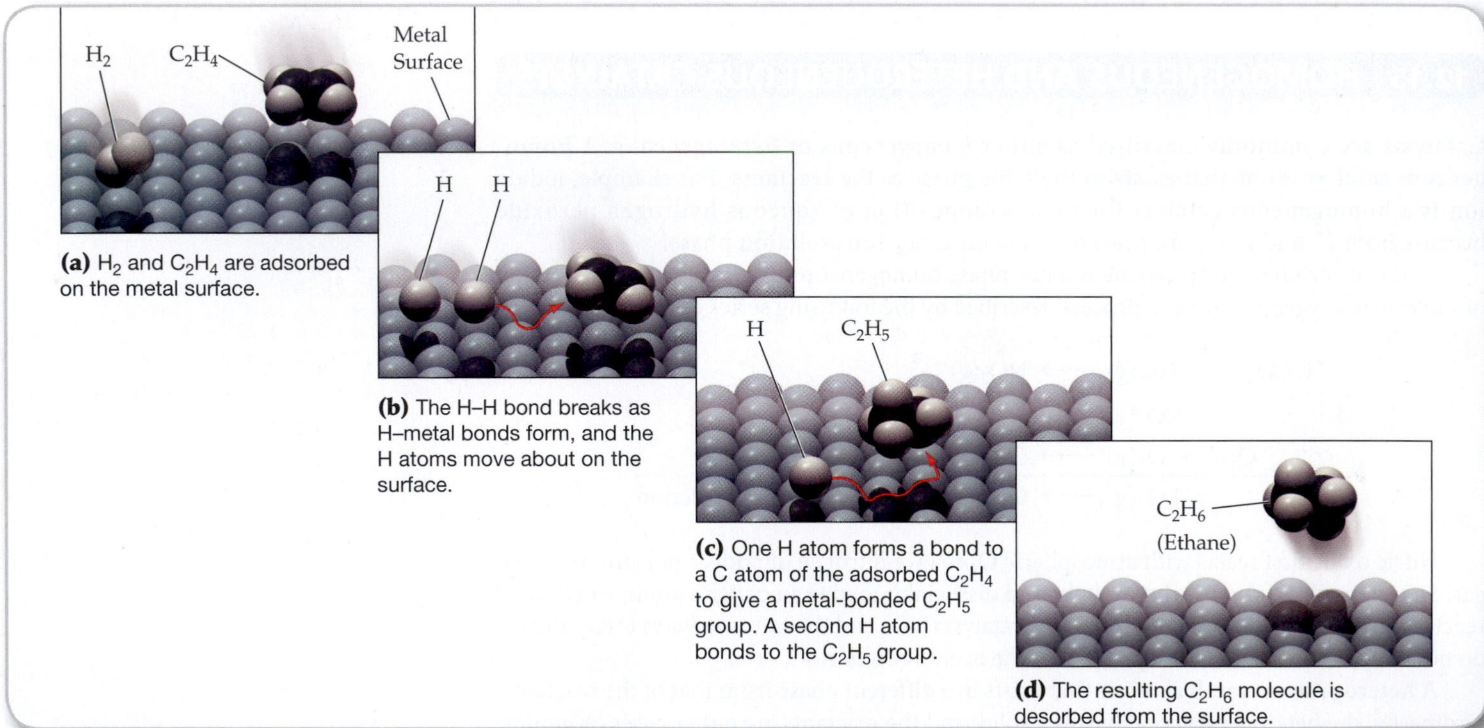

FIGURE 12.19

Proposed mechanism for the catalytic hydrogenation of ethylene (C_2H_4) on a metal surface.

Most of the catalysts used in industrial chemical processes are heterogeneous, in part because such catalysts can be easily separated from the reaction products. **Table 12.6** lists some commercial processes that employ heterogeneous catalysts.

Another important application of heterogeneous catalysts is in automobile catalytic converters. Despite much work on engine design and fuel composition, automotive exhaust emissions contain air pollutants such as unburned hydrocarbons (C_xH_y), carbon monoxide, and nitric oxide. Carbon monoxide results from the incomplete combustion of hydrocarbon fuels, and nitric oxide is produced when atmospheric nitrogen and oxygen combine at the high temperatures present in an automobile engine. Catalytic converters help convert the offending pollutants to carbon dioxide, water, nitrogen, and oxygen (**Figure 12.20**):

$$C_xH_y(g) + \left(x + \frac{y}{4}\right)O_2(g) \longrightarrow x\,CO_2(g) + \left(\frac{y}{2}\right)H_2O(g)$$
$$2\,CO(g) + O_2(g) \longrightarrow 2\,CO_2(g)$$
$$2\,NO(g) \longrightarrow N_2(g) + O_2(g)$$

TABLE 12.6 Some Heterogeneous Catalysts Used in Commercially Important Reactions

Reaction	Catalyst	Commercial Process	End Product: Commercial Uses
$2\ SO_2 + O_2 \longrightarrow 2\ SO_3$	Pt or V_2O_5	Intermediate step in the contact process for synthesis of sulfuric acid	H_2SO_4: Manufacture of fertilizers, chemicals; oil refining
$4\ NH_3 + 5\ O_2 \longrightarrow 4\ NO + 6\ H_2O$	Pt and Rh	First step in the Ostwald process for synthesis of nitric acid	HNO_3: Manufacture of fertilizers, plastics, dyes, lacquers
$N_2 + 3\ H_2 \longrightarrow 2\ NH_3$	Fe, K_2O, and Al_2O_3	Haber process for synthesis of ammonia	NH_3: Manufacture of fertilizers, nitric acid
$H_2O + CH_4 \longrightarrow CO + 3\ H_2$	Ni	Steam–hydrocarbon re-forming process for synthesis of hydrogen	H_2: Manufacture of ammonia, methanol
$CO + H_2O \longrightarrow CO_2 + H_2$	ZnO and CuO	Water-gas shift reaction to improve yield in the synthesis of H_2	H_2: Manufacture of ammonia, methanol
$CO + 2\ H_2 \longrightarrow CH_3OH$	Cu, ZnO, and Al_2O_3	Industrial synthesis of methanol	CH_3OH: Manufacture of plastics, adhesives, gasoline additives; industrial solvent
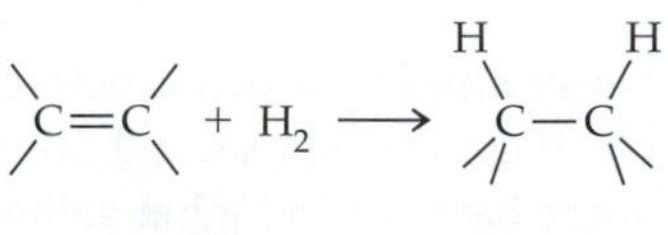	Ni, Pd, or Pt	Catalytic hydrogenation of compounds with C=C bonds, as in conversion of unsaturated vegetable oils to solid fats	Food products: margarine, shortening

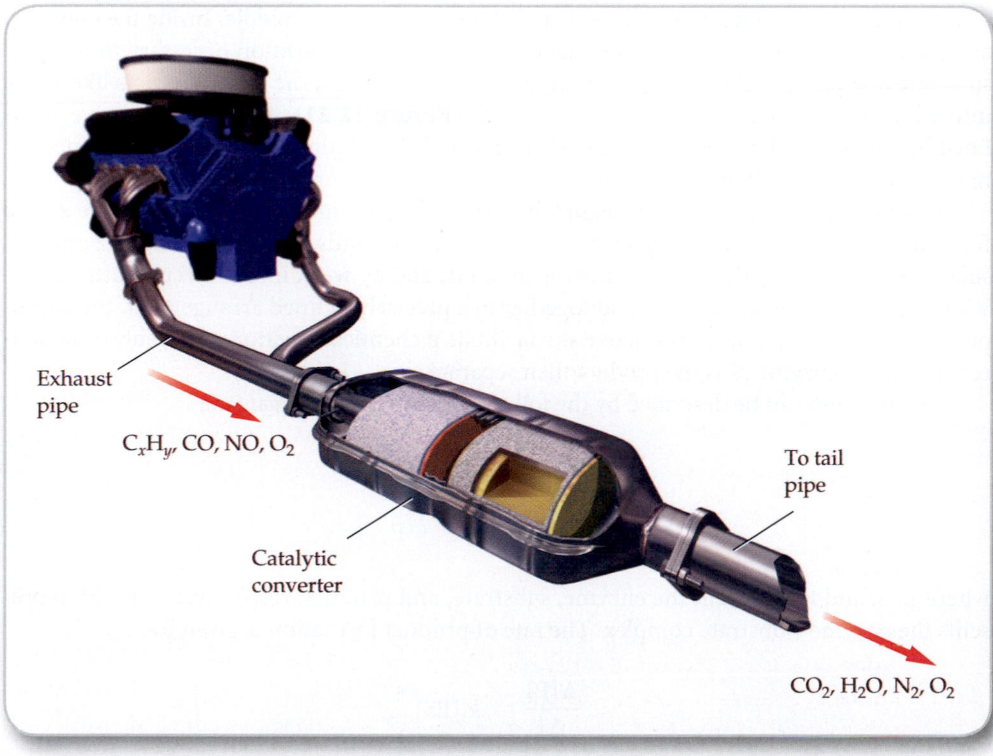

FIGURE 12.20

An automobile catalytic converter. The gases exhausted from an automobile engine pass through a catalytic converter where air pollutants such as unburned hydrocarbons (C_xH_y), CO, and NO are converted to CO_2, H_2O, N_2, and O_2. The catalytic converter contains beads that are impregnated with the heterogeneous catalyst.

Typical catalysts for these reactions are the so-called noble metals Pt, Pd, and Rh and the transition metal oxides V_2O_5, Cr_2O_3, and CuO. The surface of the catalyst is rendered ineffective, or poisoned, by the adsorption of lead, which is why automobiles with catalytic converters (all those built since 1975) use unleaded gasoline.

FYI HOW DO ENZYMES WORK?

Enzymes are large protein molecules that act as catalysts for biological reactions. They differ from simple inorganic catalysts in two important respects. First, enzymes have much larger and more complex structures than inorganic catalysts, with molecular weights ranging from about 10^4 to greater than 10^6 u. Second, enzymes are far more specific in their action than inorganic catalysts, often catalyzing only a single reaction of a single compound, called the enzyme's *substrate*. For example, the enzyme *amylase* found in human digestive systems is able to catalyze the breakdown of starch to yield glucose but has no effect on cellulose, even though starch and cellulose are structurally similar. Thus, humans can digest potatoes (starch) but not grass (cellulose).

$$\text{Starch} + H_2O \xrightarrow{\text{Amylase}} \text{Many glucose molecules}$$

$$\text{Cellulose} + H_2O \xrightarrow{\text{Amylase}} \text{No reaction}$$

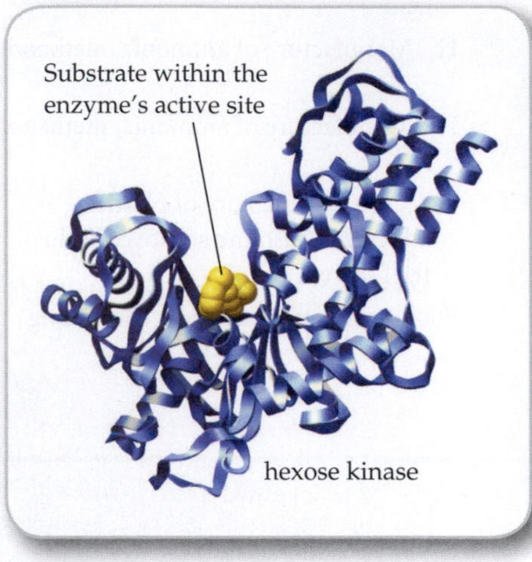

FIGURE 12.21

An enzyme's active site. An enzyme is a large, three-dimensional molecule containing a crevice with an active site. Only a substrate whose shape and structure are complementary to those of the active site can fit into the enzyme. This computer-generated structure depicts the active site of the enzyme hexose kinase.

The catalytic activity of an enzyme is measured by its *turnover number*, which is defined as the number of substrate molecules acted on by one molecule of enzyme per second. Most enzymes have turnover numbers in the range 1–20,000, but some have much higher values. Carbonic anhydrase, which catalyzes the reaction of CO_2 with water to yield HCO_3^- ion, acts on *1,000,000* substrate molecules per second.

How do enzymes work? According to the *lock-and-key model*, an enzyme is pictured as a large, irregularly shaped molecule with a cleft, or crevice, in its middle. Inside the crevice is an *active site*, a small region with the shape and chemical composition necessary to bind the substrate and catalyze the appropriate reaction. In other words, the active site acts like a lock into which only a specific key (substrate) can fit (**Figure 12.21**). An enzyme's active site is lined by various acidic, basic, and neutral amino acid side chains, all properly positioned for maximum interaction with the substrate.

Enzyme-catalyzed reactions begin when the substrate migrates into the active site to form an *enzyme–substrate complex*. Often, no covalent bonds are formed; the enzyme and substrate are held together only by hydrogen bonds and by weak intermolecular attractions. With enzyme and substrate thus held together in a precisely defined arrangement, the appropriately positioned atoms in the active site facilitate a chemical reaction of the substrate molecule, and the enzyme plus the product then separate.

The reaction can be described by the following simplified mechanism:

$$E + S \underset{k_{-1}}{\overset{k_1}{\rightleftharpoons}} ES$$

$$ES \xrightarrow{k_2} E + P$$

where E, S, and P represent the enzyme, substrate, and product, respectively, and ES represents the enzyme–substrate complex. The rate of product formation is given by

$$\frac{\Delta[P]}{\Delta t} = k_2[ES]$$

At low concentrations of the substrate, the concentration of the enzyme–substrate complex is proportional to the concentration of the substrate, and the rate of product formation is therefore first order in S. At high concentrations of the substrate, however, the enzyme becomes saturated with the substrate—that is, all the enzyme is in the form of the enzyme–substrate complex. At that point, the reaction rate reaches a maximum value and becomes independent of the concentration of the substrate (zeroth order in S) because only substrate bound to the enzyme can react. The dependence of the rate on substrate concentration is shown in **Figure 12.22**.

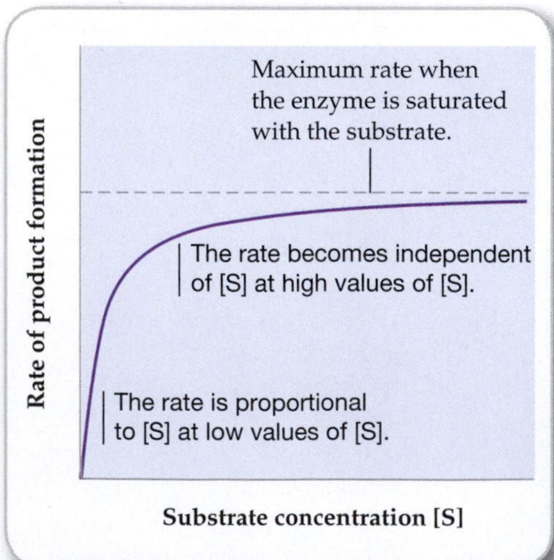

FIGURE 12.22

Rate of product formation as a function of substrate concentration for an enzyme-catalyzed reaction.

PROBLEM 12.23 Given the preceding mechanism for an enzyme-catalyzed reaction, sketch a potential energy profile for the reaction and label the appropriate parts of your graph with E, S, P, ES, and the activation energy E_a.

SUMMARY

Chemical kinetics is the area of chemistry concerned with reaction rates. A **reaction rate** is defined as the increase in the concentration of a product or the decrease in the concentration of a reactant, per unit time. It can be expressed as the average rate during a given time interval, the **instantaneous rate** at a particular time, or the **initial rate** at the beginning of the reaction.

Reaction rates depend on reactant concentrations, temperature, and the presence of catalysts. The concentration dependence is given by the **rate law**, rate $= k[A]^m[B]^n$, where k is the **rate constant**, m and n specify the **reaction order** with respect to reactants A and B, and $m + n$ is the overall reaction order. The values of m and n must be determined by experiment; they can't be deduced from the stoichiometry of the overall reaction.

The **integrated rate law** is a concentration–time equation that lets us calculate concentrations at any time t or the time required for an initial concentration to reach any particular value. For a **first-order reaction**, the integrated rate law is $\ln [A]_t = -kt + \ln [A]_0$. A graph of ln [A] versus time is a straight line with a slope equal to $-k$. For a **second-order reaction**, the integrated rate law is $1/[A]_t = kt + 1/[A]_0$. A graph of 1/[A] versus time is linear with a slope equal to k. For a **zeroth-order reaction**, the integrated rate law is $[A]_t = -kt + [A]_0$, and a graph of [A] versus time is linear with a slope equal to $-k$. The **half-life** ($t_{1/2}$) of a reaction is the time required for the reactant concentration to drop to one-half its initial value.

Radioactive decay is a first-order process, whose rate is proportional to the number of radioactive nuclei N in a sample times the **decay constant** k. The number N_t of radioactive nuclei remaining at time t can be calculated from the integrated rate law $\ln (N_t/N_0) = -kt$, where N_0 is the number of radioactive nuclei originally present.

The temperature dependence of rate constants is described by the **Arrhenius equation**, $k = Ae^{-E_a/RT}$, where A is the **frequency factor** and E_a is the **activation energy**. The value of E_a can be determined from the slope of a linear plot of ln k versus $1/T$, and it can be interpreted as the height of the potential energy barrier between the reactants and products. The configuration of atoms at the top of the barrier is called the **transition state**. According to **collision theory**, the rate constant is given by $k = pZe^{-E_a/RT}$, where p is a **steric factor** (the fraction of collisions in which the molecules have the proper orientation for reaction), Z is the collision frequency, and $e^{-E_a/RT}$ is the fraction of collisions with energy equal to or greater than E_a.

A **reaction mechanism** is the sequence of **elementary reactions**, or **elementary steps**, that defines the pathway from reactants to products. Elementary reactions are classified as **unimolecular**, **bimolecular**, or **termolecular**, depending on whether one, two, or three reactant molecules (or atoms) are involved. The rate law for an elementary reaction follows directly from its **molecularity**: rate $= k[A]$ for a unimolecular reaction, and rate $= k[A]^2$ or rate $= k[A][B]$ for a bimolecular reaction. The observed rate law for a multistep, overall reaction depends on the sequence of elementary steps and their relative rates. The slowest step is called the **rate-determining step**. A chemical species that is formed in one elementary step and consumed in a subsequent step is called a **reaction intermediate**. An acceptable mechanism must meet two criteria: (1) The elementary steps must sum to give the overall reaction, and (2) the mechanism must be consistent with the observed rate law.

A **catalyst** is a substance that increases the rate of a reaction without being consumed in the reaction. It functions by making available an alternative reaction pathway that has a lower activation energy. A **homogeneous catalyst** is present in the same phase as the reactants, whereas a **heterogeneous catalyst** is present in a different phase. **Enzymes** are large protein molecules that act as catalysts for biological reactions.

KEY WORDS

activation energy (E_a) *468*
Arrhenius equation *471*
bimolecular reaction *476*
catalyst *484*
chemical kinetics *442*
collision theory *468*
decay constant *461*
elementary reaction *474*
elementary step *474*
enzyme *490*
first-order reaction *454*
frequency factor *471*
half-life *458*
heterogeneous catalyst *487*
homogeneous catalyst *487*
initial rate *446*
instantaneous rate *446*
integrated rate law *455*
molecularity *476*
rate constant *447*
rate-determining step *479*
rate law *447*
reaction intermediate *475*
reaction mechanism *474*
reaction order *448*
reaction rate *444*
second-order reaction *463*
steric factor (p) *470*
termolecular reaction *476*
transition state *468*
unimolecular reaction *476*
zeroth-order reaction *467*

KEY EQUATIONS

- Generalized rate law for the reaction $a\,A + b\,B \longrightarrow$ Products **(Section 12.2)**

$$\text{Rate} = -\frac{\Delta[A]}{\Delta t} = k[A]^m[B]^n$$

where k is the rate constant, m is the order of the reaction in A, and n is the order of the reaction in B.

- Two forms of the integrated rate law for a first-order reaction **(Section 12.4)**

$$\ln \frac{[A]_t}{[A]_0} = -kt \qquad \ln [A]_t = -kt + \ln [A]_0$$

where $[A]_t$ is the concentration of A at time t and $[A]_0$ is the initial concentration of A.

- The relationship between the half-life ($t_{1/2}$) and the rate constant for a first-order reaction **(Section 12.5)**

$$t_{1/2} = \frac{0.693}{k}$$

- The relationship between the radioactive decay rate and the number N of radioactive nuclei **(Section 12.6)**

$$\text{Decay rate} = -\frac{\Delta N}{\Delta t} = kN$$

- The integrated rate law for radioactive decay **(Section 12.6)**

$$\ln \left(\frac{N_t}{N_0}\right) = -kt$$

- The integrated rate law and the half-life for a second order reaction **(Section 12.7)**

$$\frac{1}{[A]_t} = kt + \frac{1}{[A]_0} \qquad t_{1/2} = \frac{1}{k[A]_0}$$

- The integrated rate law for a zeroth-order reaction **(Section 12.8)**

$$[A]_t = -kt + [A]_0$$

- The rate constant predicted by collision theory: the Arrhenius equation **(Section 12.9)**

$$k = pZe^{-E_a/RT} = Ae^{-E_a/RT}$$

where p is the steric factor, Z is the collision frequency, E_a is the activation energy, and A (= pZ) is the frequency factor

- Two logarithmic forms of the Arrhenius equation **(Section 12.10)**

$$\ln k = \ln A - \frac{E_a}{RT} \qquad \ln \left(\frac{k_2}{k_1}\right) = \left(\frac{-E_a}{R}\right)\left(\frac{1}{T_2} - \frac{1}{T_1}\right)$$

where k_1 and k_2 are rate constants at temperatures T_1 and T_2

LEARNING OUTCOMES

- Determine the average and instantaneous reaction rate from a concentration versus time plot. **(Section 12.1)** ***Problems 34, 36, 37***
- Relate reaction orders and the rate law. **(Section 12.2)** ***Problems 3, 30, 42–47, 106, 115***
- Use the method of initial rates to find the rate law and the value of the rate constant. **(Section 12.3)** ***Problems 4, 6, 30, 31, 48–51, 112, 113, 115, 119, 128, 129, 131***
- Use the integrated rate law to determine the half-life and the concentrations remaining at various times for a first-order reaction. **(Sections 12.4, 12.5)** ***Problems 7–10, 52–55, 109, 114, 124, 126, 130, 134***
- Relate the half-life for radioactive decay to the decay constant and the amount of radioisotope remaining. **(Section 12.6)** ***Problems 11–13, 66–69, 121–123, 132***
- Determine the reaction order and the rate constant graphically. **(Section 12.7)** ***Problems 15, 60, 61, 104***
- Use the integrated rate law to determine the half-life and the concentrations remaining at various times for a second-order reaction. **(Section 12.7)** ***Problems 15, 56–59, 63, 126, 128, 131***
- Understand the collision theory model. **(Section 12.9)** ***Problems 16, 29, 32, 33, 76, 77, 84, 85, 107, 108, 111, 116***
- Use the Arrhenius equation to relate rate constants, temperatures, and the activation energy. **(Section 12.10)** ***Problems 17, 78–83, 105, 117, 124, 128, 129, 131, 133, 135, 136***
- Identify the molecularity of the elementary steps and the intermediates in a reaction mechanism. **(Section 12.11)** ***Problems 18, 29, 30, 88–93***
- Propose a mechanism consistent with the rate law for a multistep reaction, and derive a rate law from a reaction mechanism. **(Section 12.13)** ***Problems 20, 21, 30, 31, 92–97, 106, 115, 119***
- Identify catalysts and intermediates in the mechanism for a catalyzed reaction. **(Section 12.14)** ***Problems 22, 31, 100–103, 106, 115***

CONCEPTUAL PROBLEMS

Problems 12.1–12.23 appear within the chapter.

12.24 The following reaction is first order in A (red spheres) and first order in B (blue spheres):

$$A + B \longrightarrow \text{Products} \qquad \text{Rate} = k[A][B]$$

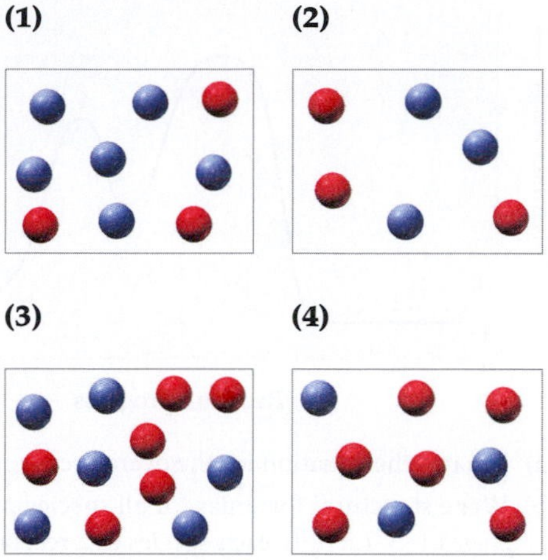

(a) What are the relative rates of this reaction in vessels (1)–(4)? Each vessel has the same volume.

(b) What are the relative values of the rate constant k for cases (1)–(4)?

12.25 Consider the first-order decomposition of A molecules (red spheres) in three vessels of equal volume.

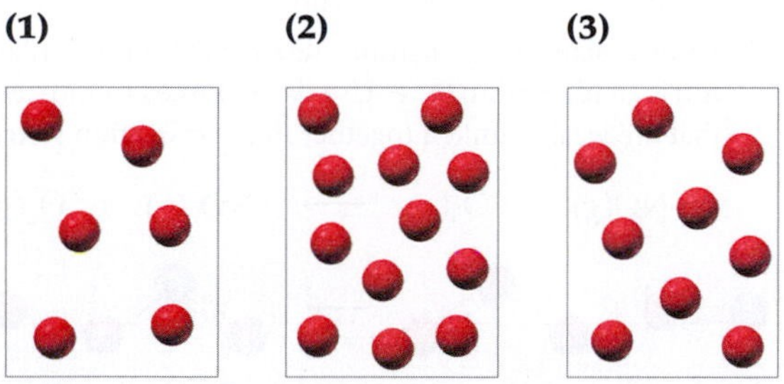

(a) What are the relative rates of decomposition in vessels (1)–(3)?

(b) What are the relative half-lives of the reactions in vessels (1)–(3)?

(c) How will the rates and half-lives be affected if the volume of each vessel is decreased by a factor of 2?

12.26 Consider the first-order reaction $A \rightarrow B$ in which A molecules (red spheres) are converted to B molecules (blue spheres).

(a) Given the following pictures at $t = 0$ min and $t = 1$ min, draw pictures that show the number of A and B molecules present at $t = 2$ min and $t = 3$ min.

(b) What is the half-life of the reaction?

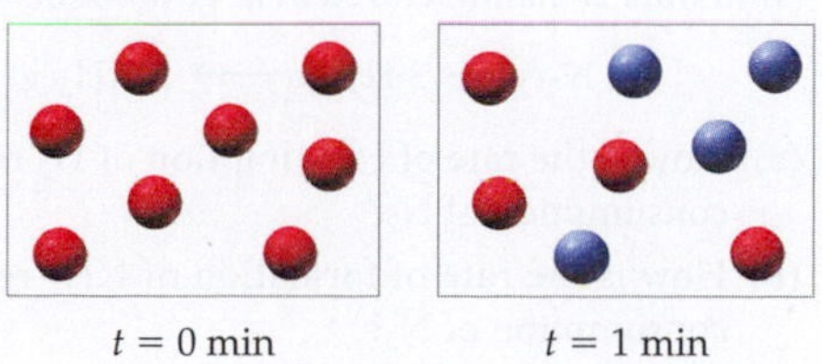

12.27 The following pictures represent the progress of the reaction $A \rightarrow B$ in which A molecules (red spheres) are converted to B molecules (blue spheres).

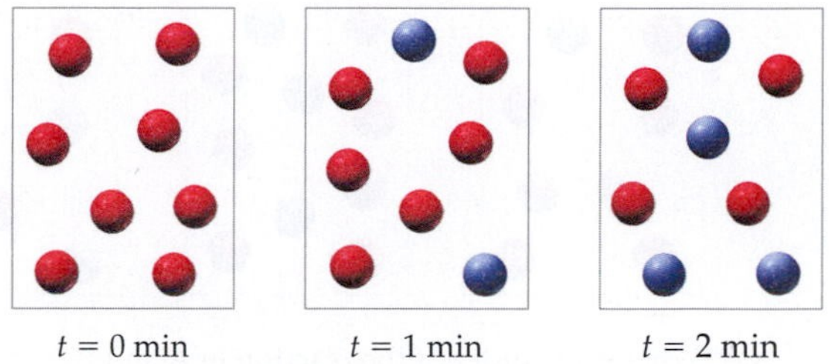

(a) What is the order of the reaction?

(b) Draw a picture that shows the number of A and B molecules present at $t = 3$ min.

(c) Suppose that each sphere represents 6.0×10^{21} molecules and that the volume of the container is 1.0 L. What is the rate constant for the reaction in the usual units?

12.28 The following pictures represent the progress of a reaction in which two A molecules combine to give a more complex molecule A_2, $2\,A \rightarrow A_2$.

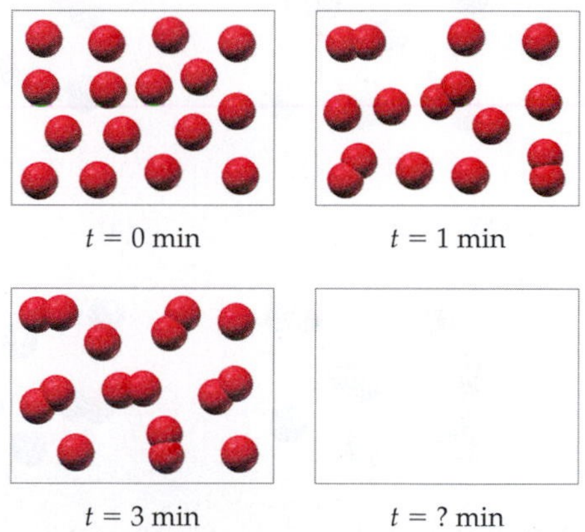

(a) Is the reaction first order or second order in A?

(b) What is the rate law?

(c) Draw an appropriate picture in the last box, and specify the time.

12.29 What is the molecularity of each of the following elementary reactions?

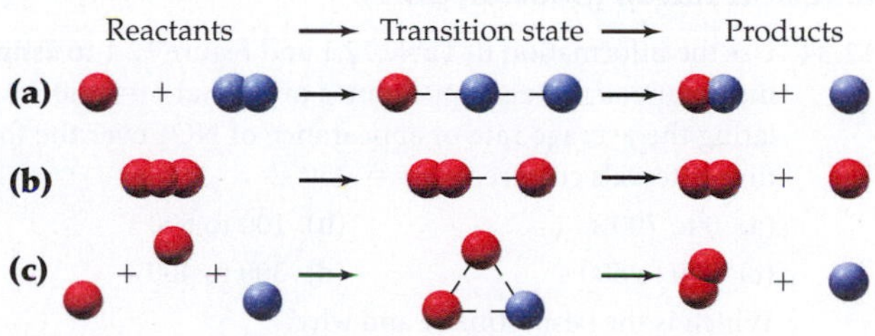

12.30 The relative rates of the reaction A + B → AB in vessels (1)–(3) are 4:4:1. Red spheres represent A molecules, and blue spheres represent B molecules.

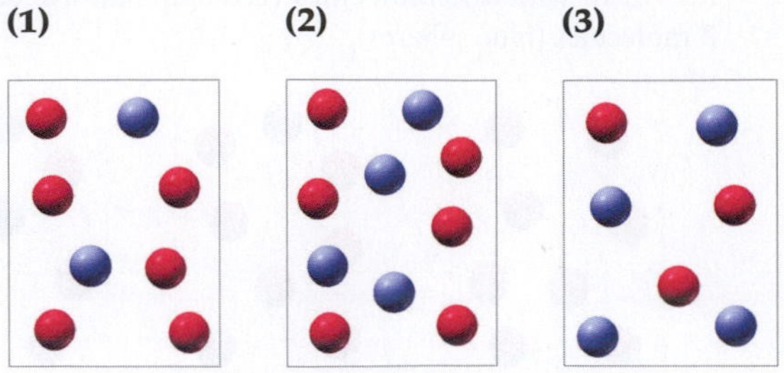

(a) What is the order of the reaction in A and B?

(b) Write the rate law.

(c) Write a mechanism that agrees with the rate law.

(d) Identify all intermediates in your mechanism.

12.31 The relative rates of the reaction AB + B → A + B_2 in vessels (1)–(4) are 2:1:1:4. Red spheres represent A, blue spheres represent B, and green spheres represent C.

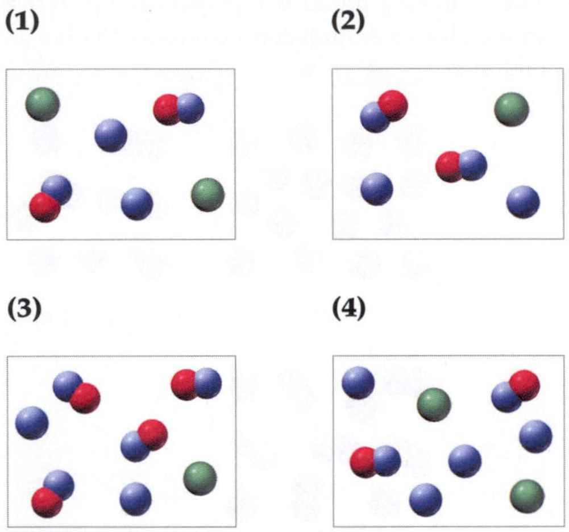

(a) Write the rate law.

(b) Write a mechanism that agrees with the rate law.

(c) Identify all intermediates and catalysts in your mechanism.

12.32 Consider a reaction that occurs by the following mechanism:

$$A + BC \longrightarrow AC + B$$
$$AC + D \longrightarrow A + CD$$

The potential energy profile for this reaction is as follows:

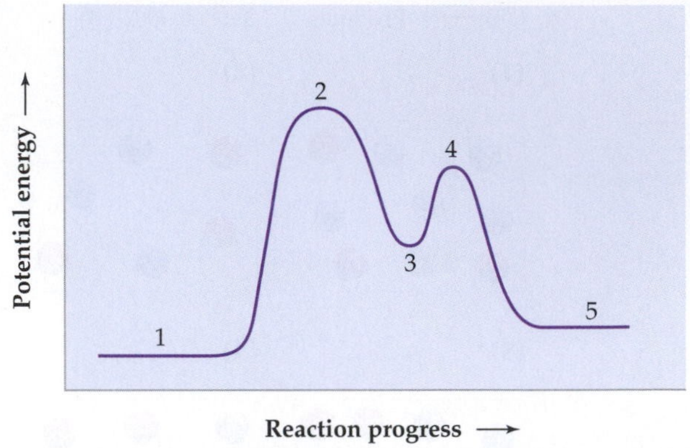

(a) What is the equation for the overall reaction?

(b) Write structural formulas for all species present at reaction stages 1–5. Identify each species as a reactant, product, catalyst, intermediate, or transition state.

(c) Which of the two steps in the mechanism is the rate-determining step? Write the rate law for the overall reaction.

(d) Is the reaction endothermic or exothermic? Add labels to the diagram that show the values of the energy of reaction ΔE and the activation energy E_a for the overall reaction.

(e) Does the reaction rate increase or decrease when the temperature increases? Explain.

12.33 Draw a plausible transition state for the bimolecular reaction of nitric oxide with ozone. Use dashed lines to indicate the atoms that are weakly linked together in the transition state.

$$NO(g) + O_3(g) \longrightarrow NO_2(g) + O_2(g)$$

SECTION PROBLEMS

Reaction Rates (Section 12.1)

12.34 Use the information in Table 12.1 and Figure 12.1 to estimate the instantaneous rate of appearance of NO_2 at $t = 350$ s by calculating the average rate of appearance of NO_2 over the following time intervals centered on $t = 350$ s:

(a) 0 to 700 s (b) 100 to 600 s

(c) 200 to 500 s (d) 300 to 400 s

Which is the best estimate and why?

12.35 Concentration–time data for the decomposition of nitrogen dioxide are given in Worked Example 12.12. Estimate the instantaneous rate of disappearance of NO_2 at $t = 250$ s by calculating the average rate of disappearance of NO_2 over the following time intervals centered on $t = 250$ s:

(a) 0 to 500 s (b) 100 to 400 s (c) 200 to 300 s

Which is the best estimate and why?

12.36 From the plot of concentration–time data in Figure 12.1, estimate:

(a) The instantaneous rate of decomposition of N_2O_5 at $t = 200$ s

(b) The initial rate of decomposition of N_2O_5

12.37 From a plot of the concentration–time data in Worked Example 12.12, estimate:

(a) The instantaneous rate of decomposition of NO_2 at $t = 100$ s

(b) The initial rate of decomposition of NO_2

12.38 Ammonia is manufactured in large amounts by the reaction

$$N_2(g) + 3\,H_2(g) \longrightarrow 2\,NH_3(g)$$

(a) How is the rate of consumption of H_2 related to the rate of consumption of N_2?

(b) How is the rate of formation of NH_3 related to the rate of consumption of N_2?

12.39 In the first step of the Ostwald process for the synthesis of nitric acid, ammonia is converted to nitric oxide by the high-temperature reaction

$$4\,NH_3(g) + 5\,O_2(g) \longrightarrow 4\,NO(g) + 6\,H_2O(g)$$

(a) How is the rate of consumption of O_2 related to the rate of consumption of NH_3?

(b) How are the rates of formation of NO and H_2O related to the rate of consumption of NH_3?

12.40 Chlorite is reduced by bromide in acidic solution according to the following balanced equation:

$$ClO_2^-(aq) + 4\,Br^-(aq) + 4\,H^+(aq) \longrightarrow Cl^-(aq) + 2\,Br_2(aq) + 2\,H_2O(l)$$

(a) If $\Delta[Br_2]/\Delta t = 4.8 \times 10^{-6}$ M/s, what is the value of $\Delta[ClO_2^-]/\Delta t$ during the same time interval?

(b) What is the average rate of consumption of Br^- during the same time interval?

12.41 The oxidation of 2-butanone ($CH_3COC_2H_5$) by the cerium(IV) ion in aqueous solution to form acetic acid (CH_3CO_2H) occurs according to the following balanced equation:

$$CH_3COC_2H_5(aq) + 6\,Ce^{4+}(aq) + 3\,H_2O(l) \longrightarrow 2\,CH_3CO_2H(aq) + 6\,Ce^{3+}(aq) + 6\,H^+(aq)$$

(a) If acetic acid appears at an average rate of 5.0×10^{-8} M/s, what is $\Delta[H^+]/\Delta t$ during the same time interval?

(b) What is the average rate of consumption of Ce^{4+} during the same time interval?

Rate Laws (Sections 12.2, 12.3)

12.42 The gas-phase reaction of hydrogen and iodine monochloride is first order in H_2 and first order in ICl. What is the rate law, and what are the units of the rate constant?

$$H_2(g) + 2\,ICl(g) \longrightarrow 2\,HCl(g) + I_2(g)$$

12.43 The reaction $2\,NO(g) + 2\,H_2(g) \longrightarrow N_2(g) + 2\,H_2O(g)$ is first order in H_2 and second order in NO. Write the rate law, and specify the units of the rate constant.

12.44 Bromomethane is converted to methanol in an alkaline solution. The reaction is first order in each reactant.

$$CH_3Br(aq) + OH^-(aq) \longrightarrow CH_3OH(aq) + Br^-(aq)$$

(a) Write the rate law.

(b) How does the reaction rate change if the OH^- concentration is decreased by a factor of 5?

(c) What is the change in rate if the concentrations of both reactants are doubled?

12.45 The oxidation of Br^- by BrO_3^- in acidic solution is described by the equation

$$5\,Br^-(aq) + BrO_3^-(aq) + 6\,H^+(aq) \longrightarrow 3\,Br_2(aq) + 3\,H_2O(l)$$

The reaction is first order in Br^-, first order in BrO_3^-, and second order in H^+.

(a) Write the rate law.

(b) What is the overall reaction order?

(c) How does the reaction rate change if the H^+ concentration is tripled?

(d) What is the change in rate if the concentrations of both Br^- and BrO_3^- are halved?

12.46 Oxidation of bis(bipyridine)copper(I) ion by molecular oxygen is described by the equation:

$$Cu(C_{10}H_8N_2)_2^+(aq) + O_2(aq) \longrightarrow \text{products}$$

The reaction is first order in oxygen and second order in $Cu(C_{10}H_8N_2)_2^+$.

(a) Write the rate law.

(b) What is the overall reaction order?

(c) How does the reaction rate change if the concentration of $Cu(C_{10}H_8N_2)_2^+$ is decreased by a factor of 4?

12.47 In aqueous solution the oxidation of nitric oxide occurs according to the equation:

$$4\,NO(aq) + O_2(aq) + 2\,H_2O(l) \longrightarrow 4\,HNO_2(aq)$$

The reaction is second order in nitric oxide and first order in oxygen.

(a) Write the rate law.

(b) What is the overall reaction order?

(c) How does the reaction rate change when the concentration of oxygen is doubled?

(d) How does the reaction rate change when the concentration of nitric oxide is doubled and the concentration of oxygen is halved?

12.48 At a certain temperature, initial rate data for the decomposition of gaseous N_2O_5 are as follows:

Experiment	Initial $[N_2O_5]$	Initial Rate of Decomposition of N_2O_5 (M/s)
1	0.014	2.4×10^{-5}
2	0.040	6.8×10^{-5}
3	0.065	1.1×10^{-4}

(a) What is the rate law?

(b) What is the value of the rate constant, including the proper units?

(c) What is the initial rate of decomposition of N_2O_5 when its initial concentration is 0.030 M?

12.49 The initial rates listed in the following table were determined for the reaction

$$2\,NO(g) + Cl_2(g) \longrightarrow 2\,NOCl(g)$$

Experiment	Initial [NO]	Initial $[Cl_2]$	Initial Rate of Consumption of Cl_2 (M/s)
1	0.13	0.20	1.0×10^{-2}
2	0.26	0.20	4.0×10^{-2}
3	0.13	0.10	5.0×10^{-3}

(a) What is the rate law?

(b) What is the value of the rate constant, including the proper units?

(c) What is the initial rate when the initial concentrations of both reactants are 0.12 M?

12.50 The initial rates listed in the following table were determined for the reaction

$$A + 2\,B + C \longrightarrow \text{Products}$$

Experiment	Initial [A]	Initial [B]	Initial [C]	Initial Rate of Consumption of A (M/s)
1	0.10	0.12	0.050	6.0×10^{-3}
2	0.10	0.24	0.075	1.2×10^{-2}
3	0.10	0.36	0.050	1.8×10^{-2}
4	0.20	0.36	0.050	7.2×10^{-2}

(a) What is the rate law?

(b) What is the value of the rate constant, including the proper units?

(c) What is the initial rate when the initial concentrations of all three reactants are 0.020 M?

12.51 The initial rates listed in the following table were determined for the reaction

$$A + B + 2\,C \longrightarrow \text{Products}$$

Experiment	Initial [A]	Initial [B]	Initial [C]	Initial Rate of Consumption of A (M/s)
1	0.20	0.16	0.12	1.0×10^{-4}
2	0.50	0.08	0.12	2.5×10^{-4}
3	0.20	0.16	0.48	4.0×10^{-4}
4	0.50	0.16	0.36	7.5×10^{-4}

(a) What is the rate law?

(b) What is the value of the rate constant, including the proper units?

(c) What is the initial rate when the initial concentrations of all three reactants are 0.10 M?

Integrated Rate Law; Half-Life (Sections 12.4, 12.5, 12.7, 12.8)

12.52 At 500 °C, cyclopropane (C_3H_6) rearranges to propene ($CH_3{-}CH{=}CH_2$). The reaction is first order, and the rate constant is $6.7 \times 10^{-4}\ s^{-1}$. If the initial concentration of C_3H_6 is 0.0500 M:

(a) What is the molarity of C_3H_6 after 30 min?

(b) How many minutes does it take for the C_3H_6 concentration to drop to 0.0100 M?

(c) How many minutes does it take for 25% of the C_3H_6 to react?

12.53 The rearrangement of methyl isonitrile (CH_3NC) to acetonitrile (CH_3CN) is a first-order reaction and has a rate constant of $5.11 \times 10^{-5}\ s^{-1}$ at 472 K.

$$CH_3{-}N{\equiv}C \longrightarrow CH_3{-}C{\equiv}N$$

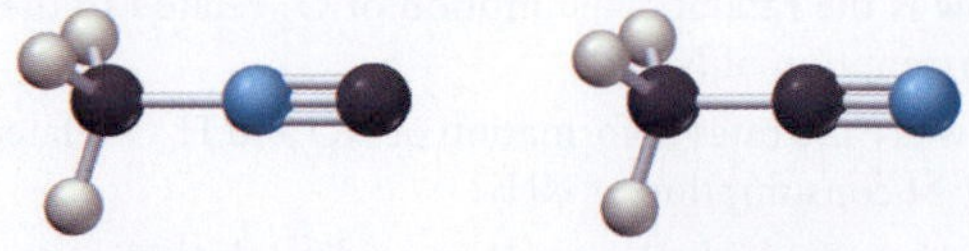

Methyl isonitrile   Acetonitrile

If the initial concentration of CH_3NC is 0.0340 M:

(a) What is the molarity of CH_3NC after 2.00 h?

(b) How many minutes does it take for the CH_3NC concentration to drop to 0.0300 M?

(c) How many minutes does it take for 20% of the CH_3NC to react?

12.54 What is the half-life (in minutes) of the reaction in Problem 12.52? How many minutes will it take for the concentration of cyclopropane to drop to 6.25% of its initial value?

12.55 What is the half-life (in hours) of the reaction in Problem 12.53? How many hours will it take for the concentration of methyl isonitrile to drop to 12.5% of its initial value?

12.56 Butadiene (C_4H_6) reacts with itself to form a dimer with the formula C_8H_{12}. The reaction is second order in C_4H_6. Assume that the rate constant at a particular temperature is $4.0 \times 10^{-2}\ M^{-1}\,s^{-1}$ and the initial concentration of C_4H_6 is 0.0200 M.

(a) What is its molarity after a reaction time of 1.00 h?

(b) What is the time (in hours) when the C_4H_6 concentration reaches a value of 0.0020 M?

12.57 Hydrogen iodide decomposes slowly to H_2 and I_2 at 600 K. The reaction is second order in HI and the rate constant is $9.7 \times 10^{-6}\ M^{-1}\,s^{-1}$. Assume the initial concentration of HI is 0.100 M.

(a) What is its molarity after a reaction time of 6.00 days?

(b) What is the time (in days) when the HI concentration reaches a value of 0.085 M?

12.58 What is the half-life (in minutes) of the reaction in Problem 12.56 when the initial C_4H_6 concentration is 0.0200 M? How many minutes does it take for the concentration of C_4H_6 to drop from 0.0100 M to 0.0050 M?

12.59 What is the half-life (in days) of the reaction in Problem 12.57 when the initial HI concentration is 0.100 M? How many days does it take for the concentration of HI to drop from 0.200 M to 0.100 M?

12.60 At elevated temperatures, nitrous oxide decomposes according to the equation

$$2\,N_2O(g) \longrightarrow 2\,N_2(g) + O_2(g)$$

Given the following data, plot the appropriate graphs to determine whether the reaction is zeroth, first, or second order. What is the value of the rate constant for the consumption of N_2O?

Time (min)	0	60	90	300	600
$[N_2O]$	0.250	0.228	0.216	0.128	0.0630

12.61 Nitrosyl bromide decomposes at 10 °C:

$$2\ NOBr(g) \longrightarrow 2\ NO(g) + Br_2(g)$$

Use the following kinetic data to determine the order of the reaction and the value of the rate constant for consumption of NOBr.

Time (s)	0	10	40	120	320
[NOBr]	0.0390	0.0315	0.0175	0.007 84	0.003 76

12.62 At 25 °C, the half-life of a certain first-order reaction is 248 s. What is the value of the rate constant at this temperature?

12.63 *trans*-cycloheptene, (C_7H_{12}), a strained cyclic hydrocarbon, converts to *cis*-cycloheptene at low temperatures. This molecular rearrangement is a second-order process with a rate constant of 0.030 $M^{-1}\,s^{-1}$ at 60 °C. If the initial concentration of *trans*-cycloheptene is 0.035 M:

(a) What is the concentration of *trans*-cycloheptene after a reaction time of 1600 s?

(b) At what time will the concentration drop to one-twentieth of its initial value?

(c) What is the half-life of *trans*-cycloheptene at an initial concentration of 0.075 M?

12.64 A reaction of the type $A \rightarrow B + C$ has a rate constant $k = 3.6 \times 10^{-5}$ M/s.

(a) What is the order of the reaction?

(b) What is the molarity of A after a reaction time of 30.0 min if the initial concentration of A is 0.096 M?

(c) What is the half-life (in minutes) of the reaction in part **(b)**?

12.65 Consider the following concentration–time data for the decomposition reaction $AB \rightarrow A + B$.

Time (min)	0	20	40	120	220
[AB]	0.206	0.186	0.181	0.117	0.036

(a) Determine the order of the reaction and the value of the rate constant.

(b) What is the molarity of AB after a reaction time of 192 min?

(c) What is the time (in minutes) when the AB concentration reaches a value of 0.0250 M?

Radioactive Decay Rates (Section 12.6)

12.66 The half-life of indium-111, a radioisotope used in studying the distribution of white blood cells, is $t_{1/2} = 2.805$ days. What is the decay constant of ^{111}In?

12.67 The decay constant of plutonium-239, a waste product from nuclear reactors, is 2.88×10^{-5} year^{-1}. What is the half-life of ^{239}Pu?

12.68 The decay constant of ^{35}S is 7.95×10^{-3} day^{-1}. What percentage of an ^{35}S sample remains after 185 days?

12.69 Plutonium-239 has a decay constant of 2.88×10^{-5} year^{-1}. What percentage of a ^{239}Pu sample remains after 1000 years? After 25,000 years? After 100,000 years?

12.70 Polonium-209, an α emitter, has a half-life of 102 years. How many α particles are emitted in 1.0 s from a 1.0 ng sample of ^{209}Po?

12.71 Chlorine-36 is a β emitter, with a half-life of 3.0×10^5 years. How many β particles are emitted in 1.0 min from a 5.0 mg sample of ^{36}Cl?

12.72 A 1.0 mg sample of ^{79}Se decays initially at a rate of 1.5×10^5 disintegrations/s. What is the half-life of ^{79}Se in years?

12.73 What is the half-life (in years) of ^{44}Ti if a 1.0 ng sample decays initially at a rate of 4.8×10^3 disintegrations/s?

12.74 A sample of ^{37}Ar undergoes 8540 disintegrations/min initially but undergoes 6990 disintegrations/min after 10.0 days. What is the half-life of ^{37}Ar in days?

12.75 A sample of ^{28}Mg decays initially at a rate of 53,500 disintegrations/min, but the decay rate falls to 10,980 disintegrations/min after 48.0 hours. What is the half-life of ^{28}Mg in hours?

The Arrhenius Equation (Sections 12.9, 12.10)

12.76 Why don't all collisions between reactant molecules lead to a chemical reaction?

12.77 Two reactions have the same activation energy, but their rates at the same temperature differ by a factor of 10. Explain.

12.78 Rate constants for the reaction $2\ N_2O_5(g) \rightarrow 4\ NO_2(g) + O_2(g)$ exhibit the following temperature dependence:

Temperature (°C)	k (s^{-1})	Temperature (°C)	k (s^{-1})
25	3.7×10^{-5}	55	1.7×10^{-3}
45	5.1×10^{-4}	65	5.2×10^{-3}

Plot an appropriate graph of the data, and determine the activation energy for this reaction in kJ/mol.

12.79 The following rate constants describe the thermal decomposition of nitrogen dioxide:

$$2\ NO_2(g) \longrightarrow 2\ NO(g) + O_2(g)$$

Temperature (°C)	k ($M^{-1}\,s^{-1}$)	Temperature (°C)	k ($M^{-1}\,s^{-1}$)
330	0.77	378	4.1
354	1.8	383	4.7

Plot an appropriate graph of the data, and calculate the value of E_a for this reaction in kJ/mol.

12.80 Rate constants for the reaction $NO_2(g) + CO(g) \rightarrow NO(g) + CO_2(g)$ are 1.3 $M^{-1}\,s^{-1}$ at 700 K and 23.0 $M^{-1}\,s^{-1}$ at 800 K.

(a) What is the value of the activation energy in kJ/mol?

(b) What is the rate constant at 750 K?

12.81 A certain first-order reaction has a rate constant of $1.0 \times 10^{-3}\ s^{-1}$ at 25 °C.

(a) If the reaction rate doubles when the temperature is increased to 35 °C, what is the activation energy for this reaction in kJ/mol?

(b) What is the E_a (in kJ/mol) if the same temperature change causes the rate to triple?

12.82 Reaction of the anti-cancer drug cisplatin, $Pt(NH_3)_2Cl_2$, with water is described by the equation

$$Pt(NH_3)_2Cl_2(aq) + H_2O(l) \longrightarrow Pt(NH_3)_2(H_2O)Cl^{+}(aq) + Cl^{-}(aq)$$

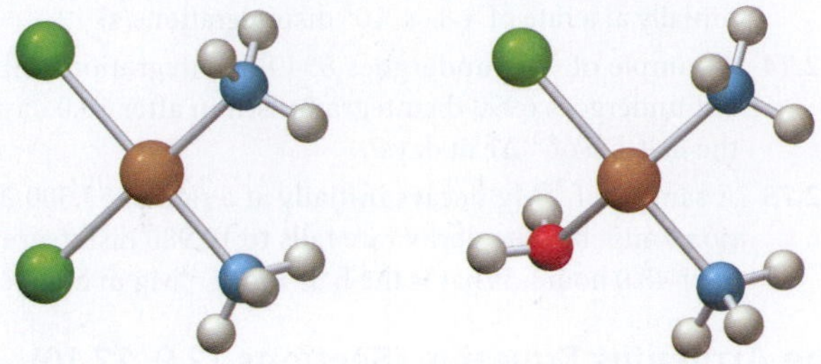

The rate of this reaction increases by a factor of 15 on raising the temperature from 25 °C to 50 °C. What is the value of the activation energy in kJ/mol?

12.83 The widely used solvent ethyl acetate undergoes the following reaction in basic solution:

$$CH_3CO_2C_2H_5(aq) + OH^{-}(aq) \longrightarrow CH_3CO_2^{-}(aq) + C_2H_5OH(aq)$$

The rate of this reaction increases by a factor of 6.37 on raising the temperature from 15 °C to 45 °C. Calculate the value of the activation energy in kJ/mol.

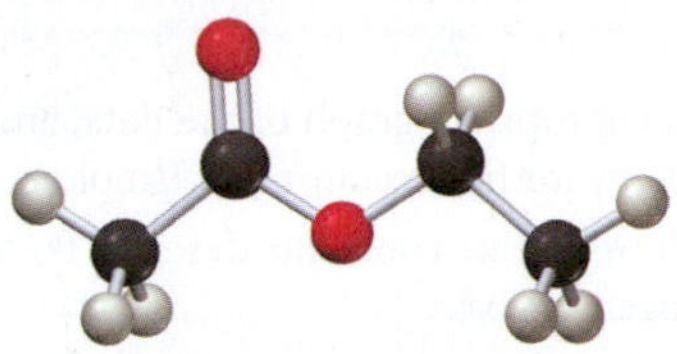

Ethyl acetate

12.84 The values of E_a = 183 kJ/mol and ΔE = 9 kJ/mol have been measured for the reaction

$$2\,HI(g) \longrightarrow H_2(g) + I_2(g)$$

Sketch a potential energy profile for this reaction that shows the potential energy of reactants, products, and the transition state. Include labels that define E_a and ΔE.

12.85 The values of E_a = 248 kJ/mol and ΔE = 41 kJ/mol have been measured for the reaction

$$H_2(g) + CO_2(g) \longrightarrow H_2O(g) + CO(g)$$

(a) Sketch a potential energy profile for this reaction that shows the potential energy of reactants, products, and the transition state. Include labels that define E_a and ΔE.

(b) Considering the geometry of the reactants and products, suggest a plausible structure for the transition state.

Reaction Mechanisms (Sections 12.11–12.13)

12.86 What is the relationship between the coefficients in a balanced chemical equation for an overall reaction and the exponents in the rate law?

12.87 What distinguishes the rate-determining step from the other steps in a reaction mechanism? How does the rate-determining step affect the observed rate law?

12.88 Consider the following mechanism for the reaction of hydrogen and iodine monochloride:

Step 1. $H_2(g) + ICl(g) \longrightarrow HI(g) + HCl(g)$

Step 2. $HI(g) + ICl(g) \longrightarrow I_2(g) + HCl(g)$

(a) Write the equation for the overall reaction.

(b) Identify any reaction intermediates.

(c) What is the molecularity of each elementary step?

12.89 The following mechanism has been proposed for the reaction of nitric oxide and chlorine:

Step 1. $NO(g) + Cl_2(g) \longrightarrow NOCl_2(g)$

Step 2. $NOCl_2(g) + NO(g) \longrightarrow 2\,NOCl(g)$

(a) What is the overall reaction?

(b) Identify any reaction intermediates.

(c) What is the molecularity of each elementary step?

12.90 Give the molecularity and the rate law for each of the following elementary reactions:

(a) $O_3(g) + Cl(g) \longrightarrow O_2(g) + ClO(g)$

(b) $NO_2(g) \longrightarrow NO(g) + O(g)$

(c) $ClO(g) + O(g) \longrightarrow Cl(g) + O_2(g)$

(d) $Cl(g) + Cl(g) + N_2(g) \longrightarrow Cl_2(g) + N_2(g)$

12.91 Identify the molecularity and write the rate law for each of the following elementary reactions:

(a) $I_2(g) \longrightarrow 2\,I(g)$

(b) $2\,NO(g) + Br_2(g) \longrightarrow 2\,NOBr(g)$

(c) $N_2O_5(g) \longrightarrow NO_2(g) + NO_3(g)$

12.92 The thermal decomposition of nitryl chloride, NO_2Cl, is believed to occur by the following mechanism:

$$NO_2Cl(g) \xrightarrow{k_1} NO_2(g) + Cl(g)$$
$$Cl(g) + NO_2Cl(g) \xrightarrow{k_2} NO_2(g) + Cl_2(g)$$

(a) What is the overall reaction?

(b) What is the molecularity of each of the elementary steps?

(c) What rate law is predicted by this mechanism if the first step is rate-determining?

12.93 The substitution reactions of molybdenum hexacarbonyl, $Mo(CO)_6$, with a variety of other molecules, L, are believed to occur by the following mechanism:

$$Mo(CO)_6 \xrightarrow{k_1} Mo(CO)_5 + CO$$
$$Mo(CO)_5 + L \xrightarrow{k_2} Mo(CO)_5L$$

(a) What is the overall reaction?

(b) What is the molecularity of each of the elementary steps?

(c) Write the rate law, assuming that the first step is rate-determining.

12.94 The reaction $2\,NO_2(g) + F_2(g) \rightarrow 2\,NO_2F(g)$ has a second-order rate law, rate = $k[NO_2][F_2]$. Suggest a mechanism that is consistent with this rate law.

12.95 The decomposition of ozone in the upper atmosphere is facilitated by NO. The overall reaction and the rate law are

$$O_3(g) + O(g) \longrightarrow 2\,O_2(g) \qquad \text{Rate} = k[O_3][NO]$$

Write a mechanism that is consistent with the rate law.

12.96 A proposed mechanism for the oxidation of nitric oxide to nitrogen dioxide was described in Problem 12.21. Another possible mechanism for this reaction is

$$2\,NO(g) \underset{k_{-1}}{\overset{k_1}{\rightleftharpoons}} N_2O_2(g) \quad \text{Faster, reversible}$$

$$N_2O_2(g) + O_2(g) \xrightarrow{k_2} 2\,NO_2(g) \quad \text{Slower, rate-determining}$$

(a) Write a balanced equation for the overall reaction.

(b) Show that this mechanism is consistent with the experimental rate law, rate $= -\Delta[NO]/\Delta t = k[NO]^2[O_2]$.

(c) Relate the rate constant k to the rate constants for the elementary reactions.

12.97 The following mechanism has been proposed for the decomposition of dinitrogen pentoxide, which has the experimental rate law, rate $= -\Delta[N_2O_5]/\Delta t = k[N_2O_5]$:

$$N_2O_5(g) \underset{k_{-1}}{\overset{k_1}{\rightleftharpoons}} NO_2(g) + NO_3(g) \quad \textbf{Fast, reversible}$$

$$NO_2(g) + NO_3(g) \xrightarrow{k_2} NO(g) + NO_2(g) + O_2(g) \quad \textbf{Slow, rate-determining}$$

$$NO(g) + NO_3(g) \xrightarrow{k_3} 2\,NO_2(g) \quad \textbf{Fast}$$

Note that the first step must be multiplied by 2 to obtain the overall reaction.

(a) Write a balanced equation for the overall reaction.

(b) Identify all reaction intermediates.

(c) Show that the proposed mechanism is consistent with the experimental rate law.

(d) Relate the rate constant k to the rate constants for the elementary reactions.

Catalysis (Sections 12.14, 12.15)

12.98 Comment on the following statement: "A catalyst increases the rate of a reaction, but it is not consumed because it does not participate in the reaction."

12.99 Why doesn't a catalyst appear in the overall chemical equation for a reaction?

12.100 In the upper atmosphere, chlorofluorocarbons such as $CFCl_3$ absorb sunlight and subsequently fragment, yielding Cl atoms. The Cl atoms then destroy ozone by the following mechanism:

$$Cl(g) + O_3(g) \longrightarrow ClO(g) + O_2(g)$$

$$ClO(g) + O(g) \longrightarrow Cl(g) + O_2(g)$$

(a) Write the chemical equation for the overall reaction.

(b) What is the role of the Cl atoms in this reaction?

(c) Is ClO a catalyst or a reaction intermediate?

(d) What distinguishes a catalyst from an intermediate?

12.101 Sulfur dioxide is oxidized to sulfur trioxide in the following sequence of reactions:

$$2\,SO_2(g) + 2\,NO_2(g) \longrightarrow 2\,SO_3(g) + 2\,NO(g)$$

$$2\,NO(g) + O_2(g) \longrightarrow 2\,NO_2(g)$$

(a) Write the chemical equation for the overall reaction.

(b) Identify any molecule that acts as a catalyst or intermediate in this reaction.

12.102 Consider the following mechanism for the decomposition of nitramide (NH_2NO_2) in aqueous solution:

$$NH_2NO_2(aq) + OH^-(aq) \longrightarrow NHNO_2^-(aq) + H_2O(l)$$

$$NHNO_2^-(aq) \longrightarrow N_2O(g) + OH^-(aq)$$

(a) Write the chemical equation for the overall reaction.

(b) Identify the catalyst and the reaction intermediate.

(c) How will the rate of the overall reaction be affected if HCl is added to the solution?

12.103 In Problem 12.95, you wrote a mechanism for the nitric oxide-facilitated decomposition of ozone. Does your mechanism involve a catalyst or a reaction intermediate? Explain.

CHAPTER PROBLEMS

12.104 You wish to determine the reaction order and rate constant for the following thermal decomposition reaction:

$$2\,AB_2 \longrightarrow A_2 + 2\,B_2$$

(a) What data would you collect?

(b) How would you use these data to determine whether the reaction is zeroth order, first order, or second order?

(c) Describe how you would determine the value of the rate constant.

12.105 You wish to determine the activation energy for the following first-order reaction:

$$A \longrightarrow B + C$$

(a) What data would you collect?

(b) How would you use these data to determine the activation energy?

12.106 The rate of the reaction $A + B_2 \rightarrow AB + B$ is directly proportional to the concentration of B_2, independent of the concentration of A, and directly proportional to the concentration of a substance C.

(a) What is the rate law?

(b) Write a mechanism that agrees with the experimental facts.

(c) What is the role of C in this reaction? Why doesn't C appear in the chemical equation for the overall reaction?

12.107 Consider three reactions with different values of E_a and ΔE:

Reaction 1. $E_a = 20$ kJ/mol; $\Delta E = -60$ kJ/mol

Reaction 2. $E_a = 10$ kJ/mol; $\Delta E = -20$ kJ/mol

Reaction 3. $E_a = 40$ kJ/mol; $\Delta E = +15$ kJ/mol

(a) Sketch a potential energy profile for each reaction that shows the potential energy of reactants, products, and the transition state. Include labels that define E_a and ΔE.

(b) Assuming that all three reactions are carried out at the same temperature and that all three have the same frequency factor A, which reaction is the fastest and which is the slowest?

(c) Which reaction is the most endothermic, and which is the most exothermic?

12.108 Consider the potential energy profile in Figure 12.18 for the iodide ion-catalyzed decomposition of H_2O_2. What point on the profile represents the potential energy of the transition state for the first step in the reaction? What point represents the potential energy of the transition state for the second step? What point represents the potential energy of the intermediate products $H_2O(l) + IO^-(aq)$?

12.109 The decomposition of N_2O_5 is a first-order reaction. At 25 °C, it takes 5.2 h for the concentration to drop from 0.120 M to 0.060 M. How many hours does it take for the concentration to drop from 0.030 M to 0.015 M? From 0.480 M to 0.015 M?

12.110 Consider the reaction $H_2(g) + I_2(g) \longrightarrow 2\,HI(g)$. The reaction of a fixed amount of H_2 and I_2 is studied in a cylinder fitted with a movable piston. Indicate the effect of each of the following changes on the rate of the reaction:

(a) An increase in temperature at constant volume

(b) An increase in volume at constant temperature

(c) The addition of a catalyst

(d) The addition of argon (an inert gas) at constant volume

12.111 When the temperature of a gas is raised by 10 °C, the collision frequency increases by only about 2%, but the reaction rate increases by 100% (a factor of 2) or more. Explain.

12.112 The initial rates listed in the following table were measured in methanol solution for the reaction

$$C_2H_4Br_2 + 3\,I^- \longrightarrow C_2H_4 + 2\,Br^- + I_3^-$$

Experiment	Initial $[C_2H_4Br_2]$	Initial $[I^-]$	Initial Rate of Formation of I_3^- (M/s)
1	0.127	0.102	6.45×10^{-5}
2	0.343	0.102	1.74×10^{-4}
3	0.203	0.125	1.26×10^{-4}

(a) What is the rate law?

(b) What is the value of the rate constant?

(c) What is the initial rate when the concentrations of both reactants are 0.150 M?

12.113 Trimethylamine and chlorine dioxide react in water in an electron transfer reaction to form the trimethylamine cation and chlorite anion:

$$(CH_3)_3N(aq) + ClO_2(aq) \longrightarrow (CH_3)_3N^+(aq) + ClO_2^-(aq)$$

Initial rate data obtained at 23 °C are listed in the following table:

Experiment	$[(CH_3)_3N]$	$[ClO_2]$	Initial Rate of Formation of ClO_2^- (M/s)
1	3.25×10^{-3}	4.60×10^{-3}	0.90
2	6.50×10^{-3}	2.30×10^{-3}	0.90
3	1.30×10^{-2}	2.30×10^{-3}	1.79
4	2.60×10^{-2}	9.20×10^{-3}	14.4

(a) What is the rate law, including the value of the rate constant?

(b) What would be the initial rate in an experiment with initial concentrations $[(CH_3)_3N] = 4.2 \times 10^{-2}$ M and $[ClO_2] = 3.4 \times 10^{-2}$ M?

12.114 The light-stimulated conversion of 11-*cis*-retinal to 11-*trans*-retinal is central to the vision process in humans. This reaction also occurs (more slowly) in the absence of light. At 80.0 °C in heptane solution, the reaction is first order with a rate constant of 1.02×10^{-5}/s.

(a) What is the molarity of 11-*cis*-retinal after 6.00 h if its initial concentration is 3.50×10^{-3} M?

(b) How many minutes does it take for 25% of the 11-*cis*-retinal to react?

(c) How many hours does it take for the concentration of 11-*trans*-retinal to reach 3.15×10^{-3} M?

12.115 Concentration–time data for the conversion of A and B to D are listed in the following table:

Experiment	Time (s)	[A]	[B]	[C]	[D]
1	0	5.00	2.00	1.00	0.00
	60	4.80	1.90	1.00	0.10
2	0	10.00	2.00	1.00	0.00
	60	9.60	1.80	1.00	0.20
3	0	5.00	4.00	1.00	0.00
	60	4.80	3.90	1.00	0.10
4	0	5.00	2.00	2.00	0.00
	60	4.60	1.80	2.00	0.20

(a) Write a balanced equation for the reaction.

(b) What is the reaction order with respect to A, B, and C? What is the overall reaction order?

(c) What is the rate law?

(d) Is a catalyst involved in this reaction? Explain.

(e) Suggest a mechanism that is consistent with the data.

(f) Calculate the rate constant for the formation of D.

12.116 What fraction of the molecules in a gas at 300 K collide with an energy equal to or greater than E_a when E_a equals 50 kJ/mol? What is the value of this fraction when E_a is 100 kJ/mol?

12.117 If the rate of a reaction increases by a factor of 2.5 when the temperature is raised from 20 °C to 30 °C, what is the value of the activation energy in kJ/mol? By what factor does the rate of this reaction increase when the temperature is raised from 120 °C to 130 °C?

12.118 Beginning with the integrated rate law, derive a general equation for the half-life of a zeroth-order reaction of the type $A \longrightarrow$ Products. How does the length of each half-life compare with the length of the previous one? Make the same comparison for first-order and second-order reactions.

12.119 Consider the following concentration–time data for the reaction of iodide ion and hypochlorite ion (OCl^-). The products are chloride ion and hypoiodite ion (OI^-).

Experiment	Time (s)	$[I^-]$	$[OCl^-]$	$[OH^-]$
1	0	2.40×10^{-4}	1.60×10^{-4}	1.00
	10	2.17×10^{-4}	1.37×10^{-4}	1.00
2	0	1.20×10^{-4}	1.60×10^{-4}	1.00
	10	1.08×10^{-4}	1.48×10^{-4}	1.00
3	0	2.40×10^{-4}	4.00×10^{-5}	1.00
	10	2.34×10^{-4}	3.40×10^{-5}	1.00
4	0	1.20×10^{-4}	1.60×10^{-4}	2.00
	10	1.14×10^{-4}	1.54×10^{-4}	2.00

(a) Write a balanced equation for the reaction.

(b) Determine the rate law, and calculate the value of the rate constant.

(c) Does the reaction occur by a single-step mechanism? Explain.

(d) Propose a mechanism that is consistent with the rate law, and express the rate constant in terms of the rate constants for the elementary steps in your mechanism. (Hint: Transfer of an H^+ ion between H_2O and OCl^- is a rapid reversible reaction.)

12.120 Consider the reversible, first-order interconversion of two molecules A and B:

$$A \underset{k_r}{\overset{k_f}{\rightleftharpoons}} B$$

where $k_f = 3.0 \times 10^{-3}\ s^{-1}$ is the rate constant for the forward reaction and $k_r = 1.0 \times 10^{-3}\ s^{-1}$ is the rate constant for the reverse reaction. We'll see in **Chapter 13** that a reaction does not go to completion but instead reaches a state of equilibrium with comparable concentrations of reactants and products if the rate constants k_f and k_r have comparable values.

(a) What are the rate laws for the forward and reverse reactions?

(b) Draw a qualitative graph that shows how the rates of the forward and reverse reactions vary with time.

(c) What are the relative concentrations of B and A when the rates of the forward and reverse reactions become equal?

12.121 Radioactive decay exhibits a first-order rate law, rate $= -\Delta N/\Delta t = kN$, where N denotes the number of radioactive nuclei present at time t. The half-life of strontium-90, a dangerous nuclear fission product, is 29 years.

(a) What fraction of the strontium-90 remains after three half-lives?

(b) What is the value of the decay constant for strontium-90?

(c) How many years are required for 99% of the strontium-90 to disappear?

12.122 The age of any remains from a once-living organism can be determined by *radiocarbon dating*, a procedure that works by determining the concentration of radioactive ^{14}C in the remains. All living organisms contain an equilibrium concentration of radioactive ^{14}C that gives rise to an average of 15.3 nuclear decay events per minute per gram of carbon. At death, however, no additional ^{14}C is taken in, so the concentration slowly drops as radioactive decay occurs. What is the age of a bone fragment from an archaeological dig if the bone shows an average of 2.3 radioactive events per minute per gram of carbon? For ^{14}C, $t_{1/2} = 5715$ years.

12.123 Potassium ion, K^+, is present in most foods and is an essential nutrient in the human body. Potassium-40, however, which has a natural abundance of 0.0117%, is radioactive with $t_{1/2} = 1.25 \times 10^9$ years. What is the decay constant of ^{40}K? How many $^{40}K^+$ ions are present in 1.00 g of KCl? How many disintegrations/s does 1.00 g of KCl undergo?

12.124 Assume that you are studying the first-order conversion of a reactant X to products in a reaction vessel with a constant volume of 1.000 L. At 1:00 p.m., you start the reaction at 25 °C with 1.000 mol of X. At 2:00 p.m., you find that 0.600 mol of X remains, and you immediately increase the temperature of the reaction mixture to 35 °C. At 3:00 p.m., you discover that 0.200 mol of X is still present. You want to finish the reaction by 4:00 p.m. but need to continue it until only 0.010 mol of X remains, so you decide to increase the temperature once again. What is the minimum temperature required to convert all but 0.010 mol of X to products by 4:00 p.m.?

12.125 The half-life for the first-order decomposition of N_2O_4 is 1.3×10^{-5} s.

$$N_2O_4(g) \longrightarrow 2\ NO_2(g)$$

If N_2O_4 is introduced into an evacuated flask at a pressure of 17.0 mm Hg, how many seconds are required for the pressure of NO_2 to reach 1.3 mm Hg?

12.126 Some reactions are so rapid that they are said to be diffusion-controlled; that is, the reactants react as quickly as they can collide. An example is the neutralization of H_3O^+ by OH^-, which has a second-order rate constant of $1.3 \times 10^{11}\ M^{-1}\ s^{-1}$ at 25 °C.

(a) If equal volumes of 2.0 M HCl and 2.0 M NaOH are mixed instantaneously, how much time is required for 99.999% of the acid to be neutralized?

(b) Under normal laboratory conditions, would you expect the rate of the acid–base neutralization to be limited by the rate of the reaction or by the speed of mixing?

12.127 The reaction $2\ NO(g) + O_2(g) \rightarrow 2\ NO_2(g)$ has the third-order rate law: rate $= k[NO]^2[O_2]$, where $k = 25\ M^{-2}\ s^{-1}$. Under the condition that $[NO] = 2\ [O_2]$, the integrated rate law is

$$\frac{1}{[O_2]^2} = 8\ kt + \frac{1}{([O_2]_0)^2}$$

What are the concentrations of NO, O_2, and NO_2 after 100.0 s if the initial concentrations are $[NO] = 0.0200$ M and $[O_2] = 0.0100$ M?

12.128 Consider the following data for the gas-phase decomposition of NO_2:

$$2\ NO_2(g) \longrightarrow 2\ NO(g) + O_2(g)$$

Temperature (K)	Initial $[NO_2]$	Initial Rate of Decomposition of NO_2 (M/s)
600	0.0010	5.4×10^{-7}
600	0.0020	2.2×10^{-6}
700	0.0020	5.2×10^{-5}

If 0.0050 mol of NO_2 is introduced into a 1.0 L flask and allowed to decompose at 650 K, how many seconds does it take for the NO_2 concentration to drop to 0.0010 M?

12.129 Use the following initial rate data to determine the activation energy (in kJ/mol) for the reaction $A + B \rightarrow C$:

Experiment	Temperature (K)	Initial [A]	Initial [B]	Initial Rate of Formation of C (M/s)
1	700	0.20	0.10	1.8×10^{-5}
2	700	0.40	0.10	3.6×10^{-5}
3	700	0.10	0.20	3.6×10^{-5}
4	600	0.50	0.50	4.3×10^{-5}

12.130 The reaction $A \rightarrow C$ is first order in the reactant A and is known to go to completion. The product C is colored and absorbs light strongly at 550 nm, while the reactant and intermediates are colorless. A solution of A was prepared, and the absorbance of C at 550 nm was measured as a function of time. (Note that the absorbance of C is directly proportional to its concentration.) Use the following data to determine the half-life of the reaction:

Time (s)	Absorbance
0	0.000
10	0.444
20	0.724
100	1.188
200	1.200
500	1.200

12.131 The following experimental data were obtained in a study of the reaction $2\,HI(g) \rightarrow H_2(g) + I_2(g)$. Predict the concentration of HI that would give a rate of 1.0×10^{-5} M/s at 650 K.

Experiment	Temperature (K)	Initial [HI]	Initial Rate (M/s)
1	700	0.10	1.8×10^{-5}
2	700	0.30	1.6×10^{-4}
3	800	0.20	3.9×10^{-3}
4	650	?	1.0×10^{-5}

12.132 A proposed nuclear theory suggests that the relative abundances of the uranium isotopes ^{235}U and ^{238}U were approximately equal at the time they were formed. Today, the observed ratio of these isotopes $^{235}U/^{238}U$ is 7.25×10^{-3}. Given that the half-lives for radioactive decay are 7.04×10^{8} y for ^{235}U and 4.47×10^{9} y for ^{238}U, calculate the age of the elements.

12.133 *Mycobacterium avium*, a human pathogen responsible for respiratory infections, is sometimes found in hot tubs. *M. avium* can be inactivated by many disinfectants including chlorine, bromine, chlorine dioxide, and ozone. For inactivation by chlorine dioxide, the following rate constants were obtained. What is the activation energy (in kJ/mol) for the inactivation of *M. avium* by chlorine dioxide?

Temperature (°C)	k [L/(mg · min)]
5	0.267
30	3.45

12.134 The half-life of a typical peptide bond (the C–N bond in a protein backbone) in neutral aqueous solution is about 500 years. When a protease enzyme acts on a peptide bond, the bond's half-life is about 0.010 s. Assuming that these half-lives correspond to first-order reactions, by what factor does the enzyme increase the rate of the peptide bond breaking reaction?

12.135 Poly(ethylene terephthalate) is a synthetic plastic used for fibers, bottles and packaging. This material can be broken down for recycling by treatment with water at elevated temperatures (230–270 °C) and pressures according to the following equation:

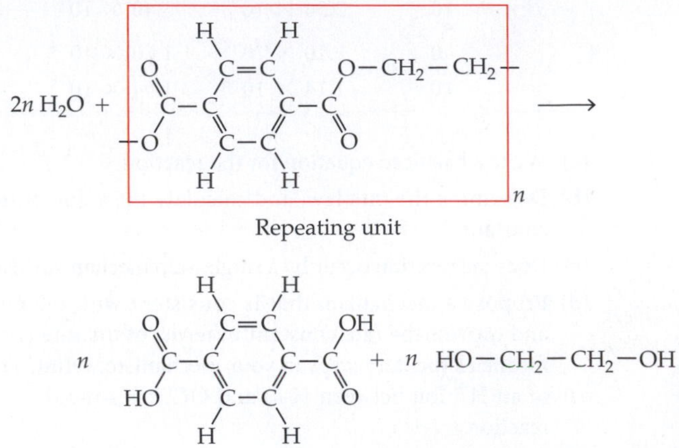

The reaction was carried out at different temperatures, and the following rate constants were obtained.

Temperature (°C)	$k\,(M^{-3/2}\,s^{-1})$
235	2.95×10^{-5}
250	5.77×10^{-5}
265	1.44×10^{-4}

Plot an appropriate graph of the data, and calculate the activation energy for this reaction in kJ/mol.

12.136 Polytetrafluoroethylene (Teflon) decomposes when heated above 500 °C. Rate constants for the decomposition are $2.60 \times 10^{-4}\,s^{-1}$ at 530 °C and $9.45 \times 10^{-3}\,s^{-1}$ at 620 °C.

(a) What is the activation energy in kJ/mol?

(b) What is the half life of this reaction at 580 °C?

MULTICONCEPT PROBLEMS

12.137 Values of E_a = 6.3 kJ/mol and $A = 6.0 \times 10^8\ M^{-1}\ s^{-1}$ have been measured for the bimolecular reaction:

$$NO(g) + F_2(g) \longrightarrow NOF(g) + F(g)$$

(a) Calculate the rate constant at 25 °C.

(b) The product of the reaction is nitrosyl fluoride. Its formula is usually written as NOF, but its structure is actually ONF. Is the ONF molecule linear or bent?

(c) Draw a plausible transition state for the reaction. Use dashed lines to indicate the atoms that are weakly linked together in the transition state.

(d) Why does the reaction have such a low activation energy?

12.138 A 1.50 L sample of gaseous HI having a density of 0.0101 g/cm^3 is heated at 410 °C. As time passes, the HI decomposes to gaseous H_2 and I_2. The rate law is $-\Delta[HI]/\Delta t = k[HI]^2$, where $k = 0.031/(M \cdot min)$ at 410 °C.

(a) What is the initial rate of production of I_2 in molecules/min?

(b) What is the partial pressure of H_2 after a reaction time of 8.00 h?

12.139 The rate constant for the decomposition of gaseous NO_2 to NO and O_2 is 4.7/(M · s) at 383 °C. Consider the decomposition of a sample of pure NO_2 having an initial pressure of 746 mm Hg in a 5.00 L reaction vessel at 383 °C.

(a) What is the order of the reaction?

(b) What is the initial rate of formation of O_2 in g/(L · s)?

(c) What is the mass of O_2 in the vessel after a reaction time of 1.00 min?

12.140 The rate constant for the first-order decomposition of gaseous N_2O_5 to NO_2 and O_2 is $1.7 \times 10^{-3}\ s^{-1}$ at 55 °C.

(a) If 2.70 g of gaseous N_2O_5 is introduced into an evacuated 2.00 L container maintained at a constant temperature of 55 °C, what is the total pressure in the container after a reaction time of 13.0 minutes?

(b) Use the data in Appendix B to calculate the initial rate at which the reaction mixture absorbs heat (in J/s). You may assume that the heat of the reaction is independent of temperature.

(c) What is the total amount of heat absorbed (in kilojoules) after a reaction time of 10.0 min?

12.141 For the thermal decomposition of nitrous oxide, $2\ N_2O(g) \rightarrow 2\ N_2(g) + O_2(g)$, values of the parameters in the Arrhenius equation are $A = 4.2 \times 10^9\ s^{-1}$ and E_a = 222 kJ/mol. If a stream of N_2O is passed through a tube 25 mm in diameter and 20 cm long at a flow rate of 0.75 L/min, at what temperature should the tube be maintained to have a partial pressure of 1.0 mm of O_2 in the exit gas? Assume that the total pressure of the gas in the tube is 1.50 atm.

12.142 A 0.500 L reaction vessel equipped with a movable piston is filled completely with a 3.00% aqueous solution of hydrogen peroxide. The H_2O_2 decomposes to water and O_2 gas in a first-order reaction that has a half-life of 10.7 h. As the reaction proceeds, the gas formed pushes the piston against a constant external atmospheric pressure of 738 mm Hg. Calculate the *PV* work done (in joules) after a reaction time of 4.02 h. (You may assume that the density of the solution is 1.00 g/mL and that the temperature of the system is maintained at 20 °C.)

12.143 At 791 K and relatively low pressures, the gas-phase decomposition of acetaldehyde (CH_3CHO) is second order in acetaldehyde.

$$CH_3CHO(g) \longrightarrow CH_4(g) + CO(g)$$

The total pressure of a particular reaction mixture was found to vary as follows:

Time (s)	0	75	148	308	605
Total pressure (atm)	0.500	0.583	0.641	0.724	0.808

(a) Use the pressure data to determine the value of the rate constant in units of atm^{-1} s^{-1}.

(b) What is the rate constant in the usual units of $M^{-1}\ s^{-1}$?

(c) If the volume of the reaction mixture is 1.00 L, what is the total amount of heat liberated (in joules) after a reaction time of 605 s?

CHAPTER 13

Chemical Equilibrium: The Extent of Chemical Reactions

When running for a touchdown, the muscles of football players need extra oxygen. The transfer of O_2 from the lungs to the blood and then to the muscle tissues depends on chemical equilibria involving hemoglobin, an oxygen-carrying protein.

At the beginning of Chapter 12, we raised three key questions about chemical reactions: What happens? How fast and by what mechanism does it happen? To what extent does it happen? The answer to the first question is given by the stoichiometry of the balanced chemical equation, and the answer to the second question is given by the kinetics of the reaction. In this chapter, we'll look at the answer to the third question: How far does a reaction proceed toward completion before it reaches a state of **chemical equilibrium**—a state in which the concentrations of reactants and products no longer change?

▶ **Chemical equilibrium** The state reached when the concentrations of reactants and products remain constant over time

We've already touched on the concept of equilibrium in connection with our study of the evaporation of liquids (Section 11.5). When a liquid evaporates in a closed container, it soon gives rise to a constant vapor pressure because of a dynamic equilibrium in which the number of molecules leaving the liquid equals the number returning from the vapor. Chemical reactions behave similarly. They can occur in both forward and reverse directions, and when the rates of the forward and reverse reactions become equal, the concentrations of reactants and products remain constant. At that point, the system is in a state of dynamic chemical equilibrium.

Chemical equilibria are important in numerous biological and environmental processes. For example, equilibria involving O_2 molecules and the protein hemoglobin in red blood cells play a crucial role in the transport and delivery of oxygen from our lungs to cells throughout our body. Similar equilibria involving CO molecules and hemoglobin account for the toxicity of carbon monoxide.

▲ In a closed container, liquid bromine and its vapor are in a dynamic equilibrium.

A mixture of reactants and products in the equilibrium state is called an **equilibrium mixture**. In this chapter, we'll address a number of important questions about the composition of equilibrium mixtures: What is the relationship between the concentrations of reactants and products in an equilibrium mixture? How can we determine equilibrium concentrations

CONTENTS

from initial concentrations? What factors can we exploit to alter the composition of an equilibrium mixture? This last question is particularly important when choosing conditions for the synthesis of industrial chemicals such as hydrogen, ammonia, and methanol (CH_3OH).

▲ Hong Kong on a smoggy day and on a clear day. The brown color of the smog is due primarily to NO_2 produced in automobile exhausts.

13.1 THE STATE OF CHEMICAL EQUILIBRIUM

In previous chapters, we've generally assumed that chemical reactions result in complete conversion of reactants to products. Many reactions, however, do not go to completion. Take, for example, the decomposition of the colorless gas dinitrogen tetroxide (N_2O_4) to the dark brown gas nitrogen dioxide (NO_2).

$$\underset{\text{Colorless}}{N_2O_4(g)} \rightleftharpoons \underset{\text{Brown}}{2\,NO_2(g)}$$

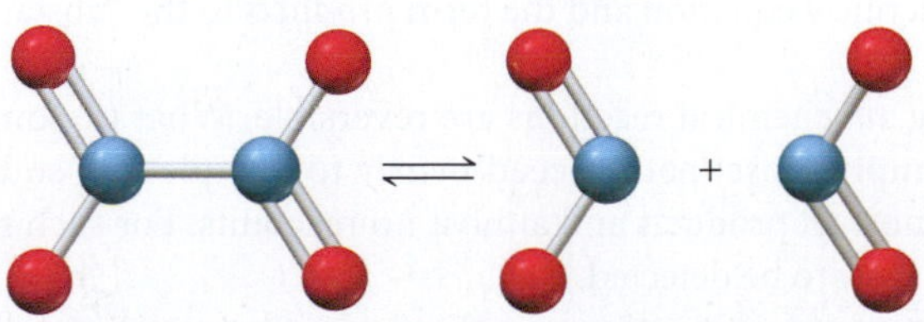

Figure 13.1 shows the results of two experiments at 25 °C that illustrate the interconversion of N_2O_4 and NO_2. In the first experiment (**Figure 13.1a**), we begin with N_2O_4 at an initial concentration of 0.0400 M. The formation of NO_2 is indicated by the appearance of a brown color, and its concentration can be monitored by measuring the intensity of the color with a spectrophotometer. According to the balanced equation, 2.0 mol of NO_2 forms for each mole of N_2O_4 that disappears, so the concentration of N_2O_4 at any time equals the initial concentration of N_2O_4 minus half the concentration of NO_2. As time passes, the concentration of

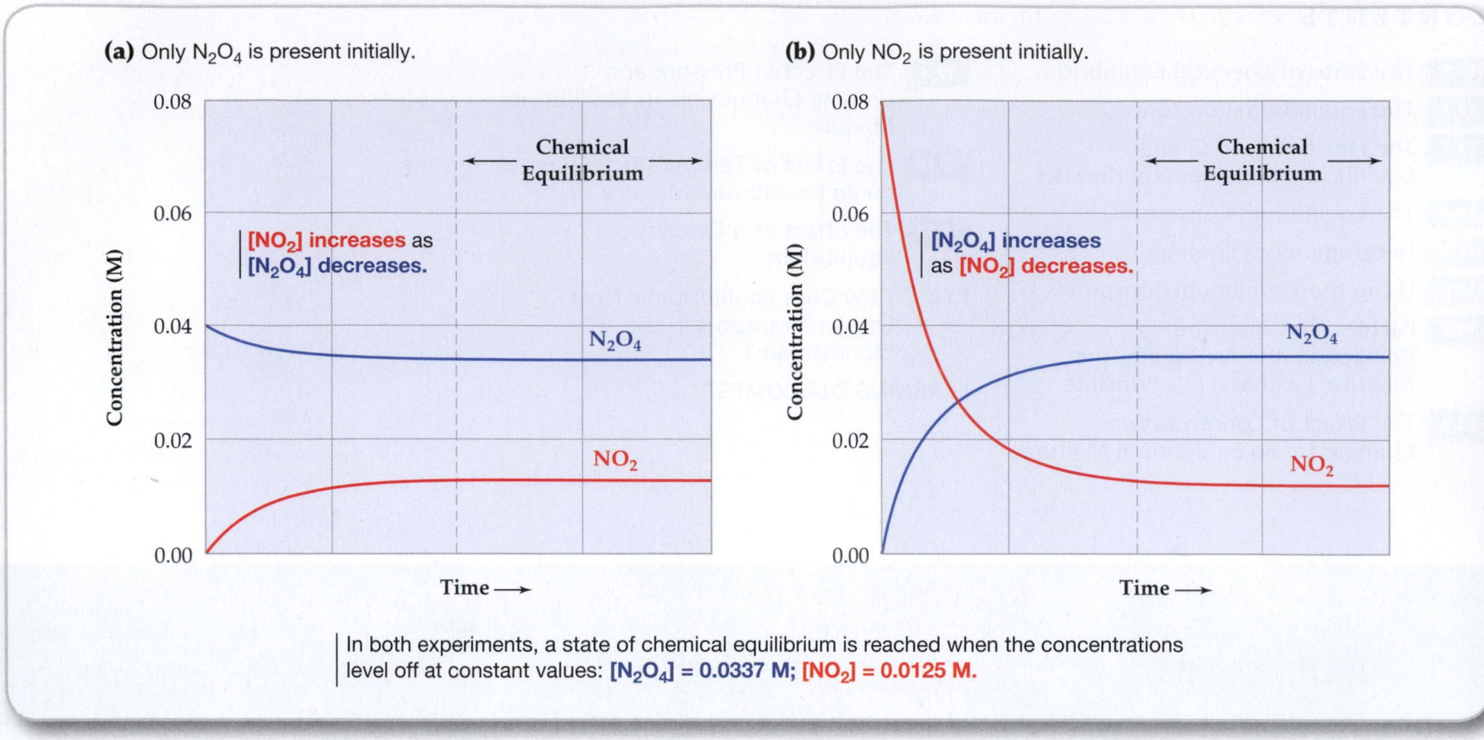

FIGURE 13.1

Change in the concentrations of N_2O_4 and NO_2 with time in two experiments at 25 °C.

N_2O_4 decreases and the concentration of NO_2 increases until both concentrations level off at constant, equilibrium values: $[N_2O_4] = 0.0337$ M; $[NO_2] = 0.0125$ M.

In the second experiment, shown in **Figure 13.1b**, we begin with NO_2 as the sole reactant at a concentration of 0.0800 M. The conversion of NO_2 to N_2O_4 proceeds until the concentrations level off at the same values as obtained in the first experiment. Taken together, the two experiments demonstrate that the interconversion of N_2O_4 and NO_2 is reversible and that the same equilibrium state is reached starting from either substance.

$$N_2O_4(g) \rightleftharpoons 2\,NO_2(g) \qquad \text{Reaction occurs in both directions}$$

To indicate that the reaction can proceed in both forward and reverse directions, we write the balanced equation with two arrows, one pointing from reactants to products and the other pointing from products to reactants. (The terms "reactants" and "products" could be confusing in this context because the products of the forward reaction are the reactants in the reverse reaction. To avoid confusion, we'll restrict the term *reactants* to the substances on the left side of the chemical equation and the term *products* to the substances on the right side of the equation.)

Strictly speaking, *all* chemical reactions are reversible. What we sometimes call irreversible reactions are simply those that proceed *nearly* to completion, so that the equilibrium mixture contains almost all products and almost no reactants. For such reactions, the reverse reaction is often too slow to be detected.

The reason why chemical reactions reach an equilibrium state follows from chemical kinetics (Chapter 12). Consider again the interconversion of N_2O_4 and NO_2. The rates of the forward reaction ($N_2O_4 \rightarrow 2\,NO_2$) and the reverse reaction ($N_2O_4 \leftarrow 2\,NO_2$) are given by the following **rate laws**:

REMEMBER...

A reaction rate is given by the **rate law** and ordinarily depends on the concentrations of at least some of the reacting species, usually increasing with increasing concentration and decreasing with decreasing concentration. (Section 12.2)

$$\text{Rate forward} = k_f[N_2O_4] \qquad \text{Rate reverse} = k_r[NO_2]^2$$

The rate of the forward reaction decreases as the concentration of the reactant N_2O_4 decreases, while the rate of the reverse reaction increases as the concentration of the product NO_2 increases. Eventually, the decreasing rate of the forward reaction and the increasing rate of the reverse reaction become equal. At that point, there are no further changes in concentrations, not because the reactions stop but because N_2O_4 and NO_2 both disappear as fast as they're formed. Thus, chemical equilibrium is a dynamic state in which forward and reverse reactions continue at equal rates so that there is no *net* conversion of reactants to products **(Figure 13.2)**.

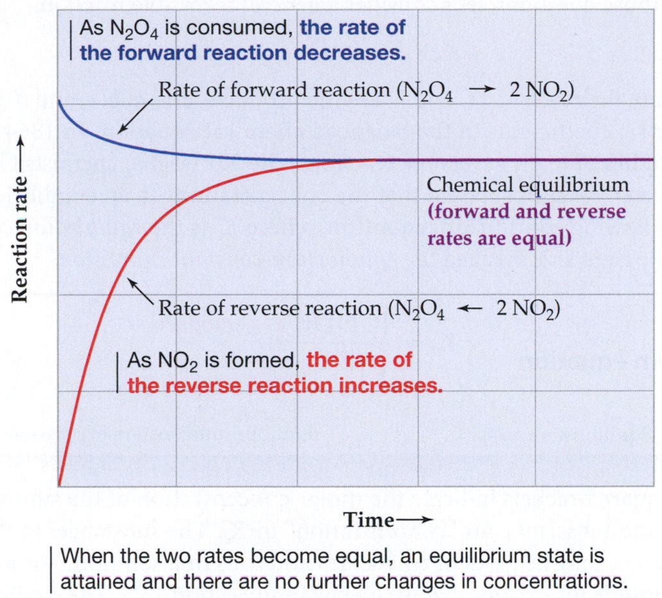

▲ Another example of a dynamic equilibrium: If the rate at which people move from the first floor to the second equals the rate at which people move from the second floor to the first, the number of people on each floor remains constant and the two populations are in dynamic equilibrium.

FIGURE 13.2

Rates of the forward and reverse reactions for the decomposition of N_2O_4 to NO_2

13.2 THE EQUILIBRIUM CONSTANT K_c

In the introduction to this chapter, we raised the question: What is the relationship between the concentrations of reactants and products in an equilibrium mixture? **Table 13.1** lists concentration data for the experiments in Figure 13.1 along with data for three additional experiments. In Experiments 1 and 2, the equilibrium mixtures have identical compositions because the initial concentration of N_2O_4 in Experiment 1 is half the initial concentration of NO_2 in Experiment 2; that is, the total number of N and O atoms is the same in both experiments. In Experiments 3–5, different initial concentrations of N_2O_4 and/or NO_2 give different equilibrium concentrations. In all the experiments, however, the equilibrium concentrations are related. The last column of Table 13.1 shows that, at equilibrium, the expression $[NO_2]^2/[N_2O_4]$ has a constant value of approximately 4.64×10^{-3} M.

TABLE 13.1 Concentration Data at 25 °C for the Reaction $N_2O_4(g) \rightleftharpoons 2\ NO_2(g)$

	Initial Concentrations (M)		Equilibrium Concentrations (M)		Equilibrium Constant Expression
Experiment	$[N_2O_4]$	$[NO_2]$	$[N_2O_4]$	$[NO_2]$	$[NO_2]^2/[N_2O_4]$
1	0.0400	0.0000	0.0337	0.0125	4.64×10^{-3}
2	0.0000	0.0800	0.0337	0.0125	4.64×10^{-3}
3	0.0600	0.0000	0.0522	0.0156	4.66×10^{-3}
4	0.0000	0.0600	0.0246	0.0107	4.65×10^{-3}
5	0.0200	0.0600	0.0429	0.0141	4.63×10^{-3}

The expression $[NO_2]^2/[N_2O_4]$ appears to be related to the balanced equation for the reaction $N_2O_4(g) \rightleftharpoons 2\,NO_2(g)$ in that the concentration of the product is in the numerator, raised to the power of its coefficient in the balanced equation, and the concentration of the reactant is in the denominator raised to the power of its coefficient in the balanced equation. Is there an analogous expression with a constant value for every chemical reaction? If so, how is the form of that expression related to the balanced equation for the reaction?

To answer those questions, let's consider a general reversible reaction:

$$a\,A + b\,B \rightleftharpoons c\,C + d\,D$$

where A and B are the reactants, C and D are the products, and *a*, *b*, *c*, and *d* are their respective stoichiometric coefficients in the balanced chemical equation. In 1864 on the basis of experimental studies of many reversible reactions, the Norwegian chemists Cato Maximilian Guldberg and Peter Waage proposed that the concentrations in an equilibrium mixture are related by the following **equilibrium equation**, where K_c is the *equilibrium constant* and the expression on the right side is called the *equilibrium-constant expression.*

▶ **Equilibrium equation**

$$K_c = \frac{[C]^c[D]^d}{[A]^a[B]^b} \begin{matrix} \longleftarrow \text{Products} \\ \longleftarrow \text{Reactants} \end{matrix}$$

Equilibrium constant (K_c) — Equilibrium-constant expression (the fraction)

LOOKING AHEAD ...

In Section 13.5, we'll see why **pure solids** and **pure liquids** are omitted from the equilibrium-constant expression.

As usual, square brackets indicate the molar concentration of the substance within the brackets, hence the subscript c for "concentration" in K_c. The substances in the equilibrium-constant expression may be gases or molecules and ions in a solution, but may not be **pure solids** or **pure liquids** for reasons that we'll explain in Section 13.5. The equilibrium equation is also known as the *law of mass action* because in the early days of chemistry, concentration was called "active mass."

The **equilibrium constant** K_c is the number obtained by multiplying the equilibrium concentrations of all the products and dividing by the product of the equilibrium concentrations of all the reactants, with the concentration of each substance raised to the power of its coefficient in the balanced chemical equation. No matter what the individual equilibrium concentrations may be in a particular experiment, *the equilibrium constant for a reaction at a particular temperature always has the same value.* Thus, the equilibrium equation for the decomposition reaction $N_2O_4 \rightleftharpoons 2\,NO_2$ is

$$K_c = \frac{[NO_2]^2}{[N_2O_4]} = 4.64 \times 10^{-3} \quad \text{at } 25\,°\text{C}$$

where the equilibrium-constant expression is $[NO_2]^2/[N_2O_4]$ and the equilibrium constant K_c has a value of 4.64×10^{-3} at 25 °C (Table 13.1).

REMEMBER...

The **thermodynamic standard state** is the set of conditions under which thermodynamic measurements are reported: 1 M concentration for each solute in solution, 1 atm pressure for each gas, and a specified temperature, usually 25 °C. (Section 8.6)

Values of K_c are generally reported without units because the concentrations in the equilibrium-constant expression are considered to be concentration ratios in which the molarity of each substance is divided by its molarity (1 M) in the **thermodynamic standard state** (Section 8.6). Because the units cancel, the concentration ratios and the values of K_c are dimensionless. For Experiment 1 in Table 13.1, for example,

$$K_c = \frac{[NO_2]^2}{[N_2O_4]} = \frac{\left(\dfrac{0.0125\,\cancel{M}}{1\,\cancel{M}}\right)^2}{\left(\dfrac{0.0337\,\cancel{M}}{1\,\cancel{M}}\right)} = 4.64 \times 10^{-3} \quad \text{at } 25\,°\text{C}$$

Equilibrium constants are temperature-dependent, so the temperature must be given when citing a value of K_c. For example, K_c for the decomposition of N_2O_4 increases from 4.64×10^{-3} at 25 °C to 1.53 at 127 °C.

The form of the equilibrium-constant expression and the numerical value of the equilibrium constant depend on the form of the balanced chemical equation. Look again at the chemical equation and the equilibrium equation for a general reaction:

$$a\text{A} + b\text{B} \rightleftharpoons c\text{C} + d\text{D} \qquad K_c = \frac{[\text{C}]^c[\text{D}]^d}{[\text{A}]^a[\text{B}]^b}$$

If we write the chemical equation in the reverse direction, the new equilibrium-constant expression is the reciprocal of the original expression, and the new equilibrium constant K_c' is the reciprocal of the original equilibrium constant K_c. (The prime distinguishes K_c' from K_c.)

$$c\text{C} + d\text{D} \rightleftharpoons a\text{A} + b\text{B} \qquad K_c' = \frac{[\text{A}]^a[\text{B}]^b}{[\text{C}]^c[\text{D}]^d} = \frac{1}{K_c}$$

Because the equilibrium constants K_c and K_c' have different numerical values, it's important to specify the form of the balanced chemical equation when quoting the value of an equilibrium constant.

Worked Example 13.1

Writing Equilibrium Equations for Gas-Phase Reactions

Write the equilibrium equation for each of the following reactions:

(a) $N_2(g) + 3\,H_2(g) \rightleftharpoons 2\,NH_3(g)$

(b) $2\,NH_3(g) \rightleftharpoons N_2(g) + 3\,H_2(g)$

STRATEGY

Put the concentrations of the reaction products in the numerator of the equilibrium-constant expression and the concentrations of the reactants in the denominator. Then raise the concentration of each substance to the power of its coefficient in the balanced chemical equation.

SOLUTION

(a) $$K_c = \frac{[NH_3]^2}{[N_2][H_2]^3}$$

(Annotations: exponent 2 — Coefficient of NH_3; exponent 3 — Coefficient of H_2)

(b) $$K_c' = \frac{[N_2][H_2]^3}{[NH_3]^2} \qquad K_c' = \frac{1}{K_c}$$

(Annotations: exponent 3 — Coefficient of H_2; exponent 2 — Coefficient of NH_3)

Because the balanced equation in part **(b)** is the reverse of that in part **(a)**, the equilibrium-constant expression in part **(b)** is the reciprocal of the expression in part **(a)** and the equilibrium constant K_c' is the reciprocal of K_c.

Worked Example 13.2

Calculating the Equilibrium Constant K_c

The following concentrations were measured for an equilibrium mixture at 500 K: $[N_2] = 3.0 \times 10^{-2}$ M; $[H_2] = 3.7 \times 10^{-2}$ M; $[NH_3] = 1.6 \times 10^{-2}$ M. Calculate the equilibrium constant at 500 K for each of the reactions in Worked Example 13.1.

STRATEGY

To calculate the value of the equilibrium constant, substitute the equilibrium concentrations into the equilibrium equation.

continued on next page

SOLUTION

(a) $$K_c = \frac{[NH_3]^2}{[N_2][H_2]^3} = \frac{(1.6 \times 10^{-2})^2}{(3.0 \times 10^{-2})(3.7 \times 10^{-2})^3} = 1.7 \times 10^2$$

(b) $$K_c' = \frac{[N_2][H_2]^3}{[NH_3]^2} = \frac{(3.0 \times 10^{-2})(3.7 \times 10^{-2})^3}{(1.6 \times 10^{-2})^2} = 5.9 \times 10^{-3}$$

Note that K_c' is the reciprocal of K_c. That is,

$$5.9 \times 10^{-3} = \frac{1}{1.7 \times 10^2}$$

▲ MTBE is a component of some gasolines but is being phased out because it has been found to be an unsafe contaminant of groundwater.

Worked Example 13.3

Writing Equilibrium Equations for Solution Reactions

Methyl *tert*-butyl ether (MTBE), a substance once used as a gasoline additive but now being phased out because of safety concerns, can be synthesized by heating methanol and *tert*-butyl alcohol with a catalytic amount of sulfuric acid:

$$\underset{\text{Methanol}}{CH_3OH(soln)} + \underset{\textit{tert}\text{-Butyl alcohol}}{C_4H_9OH(soln)} \underset{\text{catalyst}}{\overset{H^+}{\rightleftharpoons}} \underset{\text{Methyl }\textit{tert}\text{-butyl ether}}{C_4H_9OCH_3(soln)} + H_2O(soln)$$

In this equation, *soln* denotes a largely organic solution that also contains water. Write the equilibrium-constant expression for the reaction.

STRATEGY

The rules for writing the equilibrium-constant expression apply to reactions in liquid solutions as well as to gas-phase reactions (reactions in gaseous mixtures). Put the concentrations of the products in the numerator and the concentrations of the reactants in the denominator. No exponents are needed because all the coefficients in the balanced chemical equation equal 1.

SOLUTION

$$K_c = \frac{[C_4H_9OCH_3]\,[H_2O]}{[CH_3OH]\,[C_4H_9OH]} \begin{matrix} \leftarrow \text{Products} \\ \leftarrow \text{Reactants} \end{matrix}$$

PROBLEM 13.1 The oxidation of sulfur dioxide to give sulfur trioxide is an important step in the industrial process for the synthesis of sulfuric acid. Write the equilibrium equation for each of the following reactions:

(a) $2\,SO_2(g) + O_2(g) \rightleftharpoons 2\,SO_3(g)$
(b) $2\,SO_3(g) \rightleftharpoons 2\,SO_2(g) + O_2(g)$

PROBLEM 13.2 The following equilibrium concentrations were measured at 800 K: $[SO_2] = 3.0 \times 10^{-3}$ M; $[O_2] = 3.5 \times 10^{-3}$ M; $[SO_3] = 5.0 \times 10^{-2}$ M. Calculate the equilibrium constant at 800 K for each of the reactions in Problem 13.1.

PROBLEM 13.3 Lactic acid, which builds up in muscle tissue during strenuous exercise, is partially dissociated in aqueous solution:

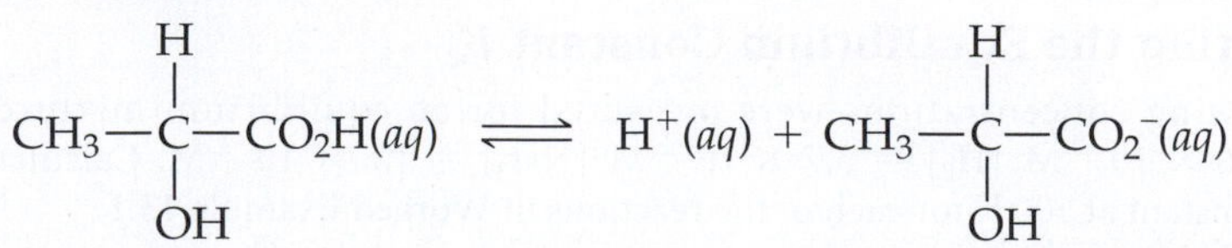

Lactic acid — Lactate ion

(a) Write the equilibrium-constant expression for K_c.
(b) What is the value of K_c if the extent of dissociation in 0.100 M lactic acid is 3.65% at 25 °C?

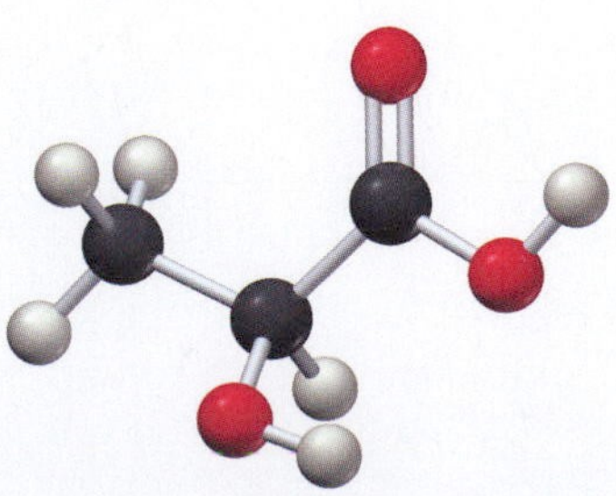
Lactic acid

WORKED CONCEPTUAL EXAMPLE 13.4

Judging Whether a Mixture Is at Equilibrium

The following pictures represent mixtures of A molecules (red spheres) and B molecules (blue spheres), which interconvert according to the equation A $\rightleftharpoons$ B. If mixture (1) is at equilibrium, which of the other mixtures are also at equilibrium? Explain.

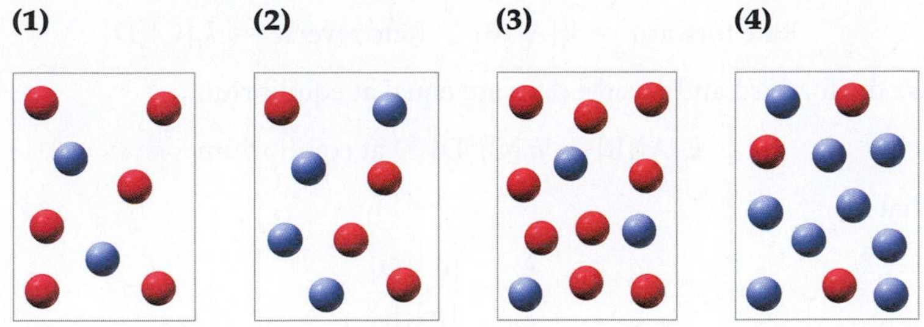

STRATEGY

The equilibrium constant for the reaction is given by $K_c = [B]/[A]$, where the concentrations are equilibrium concentrations in units of mol/L. Since the equilibrium-constant expression has the same number of concentration terms in the numerator and denominator, the volume cancels and $K_c =$ (moles of B)/(moles of A). Because the number of moles is directly proportional to the number of molecules, $K_c =$ (molecules of B)/(molecules of A) in the equilibrium mixture, mixture (1). To determine whether the other mixtures are at equilibrium, count the number of molecules and compare the B/A ratio in mixtures (2)–(4) with the B/A ratio in the equilibrium mixture.

SOLUTION

For mixture (1), $K_c = [B]/[A] = 2/6 = 1/3$.
For mixture (2), $[B]/[A] = 4/4 = 1 \neq K_c$.
For mixture (3), $[B]/[A] = 3/9 = 1/3 = K_c$.
For mixture (4), $[B]/[A] = 9/3 = 3 \neq K_c$.

Mixture (3) is at equilibrium, but mixtures (2) and (4) are not at equilibrium because their equilibrium-constant expressions $[B]/[A]$ do not equal K_c.

CONCEPTUAL PROBLEM 13.4 The following pictures represent mixtures that contain A atoms (red), B atoms (blue), and AB and B_2 molecules, which interconvert according to the equation $A + B_2 \rightleftharpoons AB + B$. If mixture (1) is at equilibrium, which of the other mixtures are also at equilibrium? Explain.

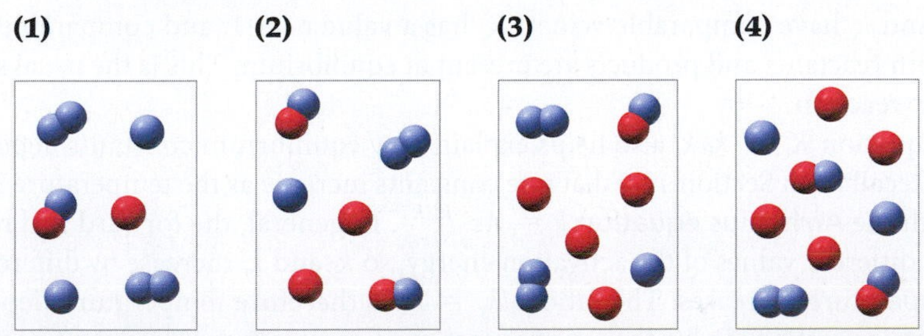

13.3 THE LINK BETWEEN CHEMICAL EQUILIBRIUM AND CHEMICAL KINETICS

In Section 13.1, we emphasized that the equilibrium state is a dynamic one in which reactant and product concentrations remain constant, not because the reaction stops but because the rates of the forward and reverse reactions are equal. Now that we have defined the equilibrium constant K_c, we can explore the relationship between K_c and the rate constants for the forward and reverse reactions. To derive that relationship, let's consider the general, reversible reaction

$$A + B \rightleftharpoons C + D$$

and assume that the forward and reverse reactions occur in a single bimolecular step; that is, they are **elementary reactions** (Section 12.11). We can then write the following rate laws:

REMEMBER...

Because an **elementary reaction** describes an individual molecular event, its rate law follows directly from its stoichiometry. (Sections 12.11 and 12.12)

$$\text{Rate forward} = k_f[A][B] \qquad \text{Rate reverse} = k_r[C][D]$$

Because the forward and reverse rates are equal at equilibrium,

$$k_f[A][B] = k_r[C][D] \quad \text{at equilibrium}$$

it follows that

$$\frac{k_r}{k_f} = \frac{[C][D]}{[A][B]}$$

The right side of this last equation is the equilibrium-constant expression for the forward reaction, which equals the **equilibrium constant K_c** since the reaction mixture is at equilibrium.

$$K_c = \frac{[C][D]}{[A][B]}$$

Therefore, the equilibrium constant is simply the ratio of the rate constants for the forward and reverse reactions:

▶ **Equilibrium constant** $\quad K_c = \dfrac{k_f}{k_r}$

In deriving this equation for K_c, we have assumed a single-step mechanism. For a multi-step mechanism, each step has a characteristic rate constant ratio, k_f/k_r. When equilibrium is reached, each step in the mechanism must be at equilibrium, and K_c for the overall reaction is equal to the product of the rate constant ratios for the individual steps.

The equation relating K_c to k_f and k_r provides a fundamental link between chemical equilibrium and chemical kinetics: The relative values of the rate constants for the forward and reverse reactions determine the composition of the equilibrium mixture. When k_f is much larger than k_r, K_c is very large and the reaction goes almost to completion. Such a reaction is sometimes said to be irreversible because the reverse reaction is often too slow to be detected. When k_f and k_r have comparable values, K_c has a value near 1, and comparable concentrations of both reactants and products are present at equilibrium. This is the usual situation for a reversible reaction.

The equation $K_c = k_f/k_r$ also helps explain why equilibrium constants depend on temperature. Recall from Section 12.9 that rate constants increase as the temperature increases, in accord with the **Arrhenius equation** $k = Ae^{-E_a/RT}$. In general, the forward and reverse reactions have different values of the activation energy, so k_f and k_r increase by different amounts as the temperature increases. The ratio $k_f/k_r = K_c$ is therefore temperature-dependent. For an exothermic reaction, E_a for the reverse reaction is greater than E_a for the forward reaction (Figure 12.12, page 469). Consequently, as the temperature increases, k_r increases by more than k_f increases, and so $K_c = k_f/k_r$ for an exothermic reaction decreases as the temperature increases. Conversely, K_c for an endothermic reaction increases as the temperature increases.

REMEMBER...

Because the fraction of molecular collisions with sufficient energy for reaction is given by $e^{-E_a/RT}$, the **Arrhenius equation** indicates that the rate constant decreases as E_a increases and increases as T increases. (Section 12.9)

Worked Example 13.5

Exploring the Link between Equilibrium and Kinetics

The equilibrium constant K_c for the reaction of hydrogen with iodine is 57.0 at 700 K, and the reaction is endothermic ($\Delta E = 9$ kJ).

$$H_2(g) + I_2(g) \underset{k_r}{\overset{k_f}{\rightleftharpoons}} 2\,HI(g) \qquad K_c = 57.0 \text{ at } 700\text{ K}$$

(a) Is the rate constant k_f for the formation of HI larger or smaller than the rate constant k_r for the decomposition of HI?
(b) The value of k_r at 700 K is $1.16 \times 10^{-3}\ M^{-1}\ s^{-1}$. What is the value of k_f at the same temperature?
(c) How are the values of k_f, k_r, and K_c affected by an increase in temperature?

STRATEGY

To answer these questions, make use of the relationship $K_c = k_f/k_r$. Also, recall that the temperature dependence of a rate constant increases with increasing value of the **activation energy** (Section 12.10).

REMEMBER...

The greater the **activation energy**, the steeper the slope of an Arrhenius plot (a graph of ln k versus $1/T$) and the greater the increase in k for a given increase in T. (Section 12.10)

SOLUTION

(a) Because $K_c = k_f/k_r = 57.0$, the rate constant for the formation of HI (forward reaction) is larger than the rate constant for the decomposition of HI (reverse reaction) by a factor of 57.0.
(b) Because $K_c = k_f/k_r$,

$$k_f = (K_c)(k_r) = (57.0)(1.16 \times 10^{-3}\ M^{-1}\ s^{-1}) = 6.61 \times 10^{-2}\ M^{-1}\ s^{-1}$$

(c) Because the reaction is endothermic, E_a for the forward reaction is greater than E_a for the reverse reaction. Consequently, as the temperature increases, k_f increases by more than k_r increases, and therefore $K_c = k_f/k_r$ increases.

PROBLEM 13.5 Nitric oxide emitted from the engines of supersonic aircraft can contribute to the destruction of stratospheric ozone:

$$NO(g) + O_3(g) \underset{k_r}{\overset{k_f}{\rightleftharpoons}} NO_2(g) + O_2(g)$$

This reaction is highly exothermic ($\Delta E = -201$ kJ), and its equilibrium constant K_c is 3.4×10^{34} at 300 K.
(a) Which rate constant is larger, k_f or k_r?
(b) The value of k_f at 300 K is $8.5 \times 10^{6}\ M^{-1}\ s^{-1}$. What is the value of k_r at the same temperature?
(c) A typical temperature in the stratosphere is 230 K. Do the values of k_f, k_r, and K_c increase or decrease when the temperature is lowered from 300 K to 230 K?

▲ The F/A Hornet Strike Fighter breaking the sound barrier. Nitric oxide emissions from supersonic aircraft can contribute to destruction of the ozone layer.

13.4 THE EQUILIBRIUM CONSTANT K_p

Because gas pressures are easily measured, equilibrium equations for gas-phase reactions are often written using **partial pressures** (Section 9.5) rather than molar concentrations. For example, the equilibrium equation for the decomposition of N_2O_4 can be written as

$$K_p = \frac{(P_{NO_2})^2}{P_{N_2O_4}} \qquad \text{for the reaction } N_2O_4(g) \rightleftharpoons 2\,NO_2(g)$$

REMEMBER...

The **partial pressure** of a gas in a mixture is the pressure the gas would exert if it were the only one present. Its partial pressure is independent of the partial pressures of the other gases in the mixture. (Section 9.5)

where $P_{N_2O_4}$ and P_{NO_2} are the partial pressures (in atmospheres) of reactants and products at equilibrium, and the subscript p on K reminds us that the **equilibrium constant K_p** is defined using partial pressures. As for K_c, values of K_p are dimensionless because the partial pressures in the equilibrium equation are actually ratios of partial pressures in atmospheres to the standard-state partial pressure of 1 atm. Thus, the units cancel. Note that the equilibrium equations for K_p and K_c have the same form except that the expression for K_p contains partial pressures instead of molar concentrations:

$$K_p = \frac{(P_{NO_2})^2}{P_{N_2O_4}} \qquad K_c = \frac{[NO_2]^2}{[N_2O_4]}$$

For the general gas-phase reaction $a\,A + b\,B \rightleftharpoons c\,C + d\,D$, K_p is given by

▶ **Equilibrium constant** $$K_p = \frac{(P_C)^c(P_D)^d}{(P_A)^a(P_B)^b}$$

K_p and K_c are related because the pressure of each component in a mixture of ideal gases is directly proportional to its molar concentration. For component A, for example,

$$P_A V = n_A RT$$

so

$$P_A = \frac{n_A}{V} RT = [A]RT$$

Similarly, $P_B = [B]RT$, $P_C = [C]RT$, and $P_D = [D]RT$. The equilibrium equation for K_p is therefore given by

$$K_p = \frac{(P_C)^c(P_D)^d}{(P_A)^a(P_B)^b} = \frac{([C]RT)^c([D]RT)^d}{([A]RT)^a([B]RT)^b} = \frac{[C]^c[D]^d}{[A]^a[B]^b} \times (RT)^{(c+d)-(a+b)}$$

Because the first term on the right side equals K_c, the values of K_p and K_c are related by the equation

▶ **Equilibrium constant** $K_p = K_c(RT)^{\Delta n}$ for the reaction $aA + bB \rightleftharpoons cC + dD$

Here, R is the gas constant, 0.082 06 (L · atm)/(K · mol), T is the absolute temperature, and $\Delta n = (c + d) - (a + b)$ is the number of moles of gaseous products minus the number of moles of gaseous reactants.

For the decomposition of 1 mol of N_2O_4 to 2 mol of NO_2, $\Delta n = 2 - 1 = 1$, and $K_p = K_c(RT)$:

$$N_2O_4(g) \rightleftharpoons 2\,NO_2(g) \quad K_p = K_c(RT)$$

For the reaction of 1 mol of hydrogen with 1 mol of iodine to give 2 mol of hydrogen iodide, $\Delta n = 2 - (1 + 1) = 0$, and $K_p = K_c(RT)^0 = K_c$:

$$H_2(g) + I_2(g) \rightleftharpoons 2\,HI(g) \quad K_p = K_c$$

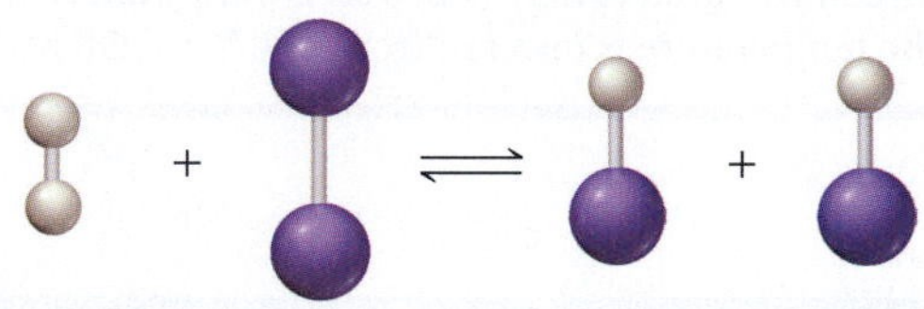

In general, K_p equals K_c only if the same number of moles of gases appear on both sides of the balanced chemical equation so that $\Delta n = 0$.

Worked Example 13.6

Determining the Equilibrium Constant K_p

Methane (CH_4) reacts with hydrogen sulfide to yield H_2 and carbon disulfide, a solvent used in manufacturing rayon and cellophane:

$$CH_4(g) + 2\,H_2S(g) \rightleftharpoons CS_2(g) + 4\,H_2(g)$$

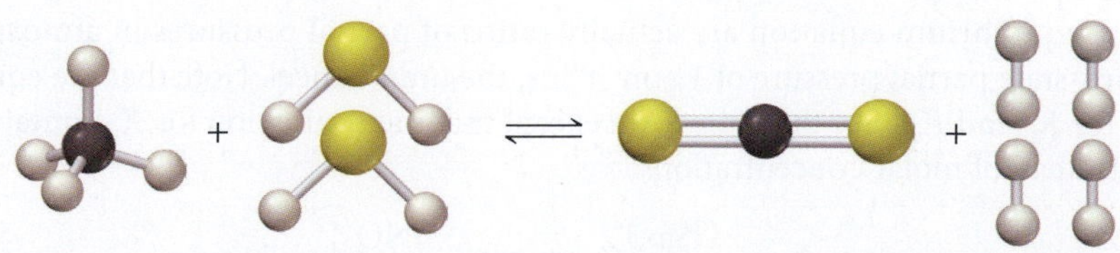

What is the value of K_p at 1000 K if the partial pressures in an equilibrium mixture at 1000 K are 0.20 atm of CH_4, 0.25 atm of H_2S, 0.52 atm of CS_2, and 0.10 atm of H_2?

STRATEGY

Write the equilibrium equation by setting K_p equal to the equilibrium-constant expression using partial pressures. Put the partial pressures of products in the numerator and the partial pressures of reactants in the denominator, with the pressure of each substance raised to the power of its coefficient in the balanced chemical equation. Then substitute the partial pressures into the equilibrium equation and solve for K_p.

SOLUTION

Coefficient of H_2

$$K_p = \frac{(P_{CS_2})(P_{H_2})^4}{(P_{CH_4})(P_{H_2S})^2}$$

Coefficient of H_2S

$$K_p = \frac{(P_{CS_2})(P_{H_2})^4}{(P_{CH_4})(P_{H_2S})^2} = \frac{(0.52)(0.10)^4}{(0.20)(0.25)^2} = 4.2 \times 10^{-3}$$

Note that the partial pressures must be in units of atmospheres (not mm Hg) because the standard-state partial pressure for gases is 1 atm.

Worked Example 13.7

Relating the Equilibrium Constants K_p and K_c

Hydrogen is produced industrially by the steam–hydrocarbon re-forming process. The reaction that takes place in the first step of this process is

$$H_2O(g) + CH_4(g) \rightleftharpoons CO(g) + 3\,H_2(g)$$

(a) If $K_c = 3.8 \times 10^{-3}$ at 1000 K, what is the value of K_p at the same temperature?
(b) If $K_p = 6.1 \times 10^4$ at 1125 °C, what is the value of K_c at 1125 °C?

STRATEGY

To calculate K_p from K_c, or vice versa, use the equation $K_p = K_c(RT)^{\Delta n}$, where R must be in units of $(L \cdot atm)/(K \cdot mol)$, T is the temperature in kelvin, and Δn is the number of moles of gaseous products minus the number of moles of gaseous reactants.

SOLUTION

(a) For this reaction, $\Delta n = (1 + 3) - (1 + 1) = 2$. Therefore,

$$K_p = K_c(RT)^{\Delta n} = K_c(RT)^2 = (3.8 \times 10^{-3})[(0.082\,06)(1000)]^2 = 26$$

(b) Solving the equation $K_p = K_c(RT)^2$ for K_c gives

$$K_c = \frac{K_p}{(RT)^2} = \frac{6.1 \times 10^4}{[(0.082\,06)(1398)]^2} = 4.6$$

Note that the temperature in these equations is the absolute temperature; 1125 °C corresponds to 1125 + 273 = 1398 K.

PROBLEM 13.6 In the industrial synthesis of hydrogen, mixtures of CO and H_2 are enriched in H_2 by allowing the CO to react with steam. The chemical equation for this so-called water-gas shift reaction is

$$CO(g) + H_2O(g) \rightleftharpoons CO_2(g) + H_2(g)$$

What is the value of K_p at 700 K if the partial pressures in an equilibrium mixture at 700 K are 1.31 atm of CO, 10.0 atm of H_2O, 6.12 atm of CO_2, and 20.3 atm of H_2?

continued on next page

PROBLEM 13.7 Nitric oxide reacts with oxygen to give nitrogen dioxide, an important reaction in the Ostwald process for the industrial synthesis of nitric acid:

$$2\,NO(g) + O_2(g) \rightleftharpoons 2\,NO_2(g)$$

If $K_c = 6.9 \times 10^5$ at 227 °C, what is the value of K_p at this temperature? If $K_p = 1.3 \times 10^{-2}$ at 1000 K, what is the value of K_c at 1000 K?

13.5 HETEROGENEOUS EQUILIBRIA

▲ The manufacture of cement begins with the thermal decomposition of limestone, $CaCO_3$, in large kilns.

Thus far we've been discussing **homogeneous equilibria**, in which all reactants and products are in the same phase, usually either gaseous or solution. **Heterogeneous equilibria**, by contrast, are those in which reactants and products are present in more than one phase. Take, for example, the thermal decomposition of solid calcium carbonate, a reaction used in manufacturing cement:

$$\underset{\text{Limestone}}{CaCO_3(s)} \rightleftharpoons \underset{\text{Lime}}{CaO(s)} + CO_2(g)$$

When the reaction is carried out in a closed container, three phases are present at equilibrium: solid calcium carbonate, solid calcium oxide, and gaseous carbon dioxide. If we were to write the usual equilibrium equation for the reaction, including all the reactants and products, we would have

$$K_c = \frac{[CaO][CO_2]}{[CaCO_3]}$$

Once again, though, the concentrations in the equilibrium equation are ratios of actual concentrations to concentrations in the standard state. Because the standard state of a pure solid is the pure solid itself, the concentration ratio for a pure solid (and also for a pure liquid) is equal to 1. Thus, $[CaO] = 1$ and $[CaCO_3] = 1$, and so these concentrations can be omitted from the equilibrium equation:

$$K_c = \frac{[CaO][CO_2]}{[CaCO_3]} = \frac{(1)[CO_2]}{(1)} = [CO_2]$$

The analogous equilibrium equation in terms of pressure is $K_p = P_{CO_2}$, where P_{CO_2} is the equilibrium pressure of CO_2 in atmospheres:

$$K_c = [CO_2] \qquad K_p = P_{CO_2}$$

As a general rule, the concentrations of pure solids and pure liquids are not included when writing an equilibrium equation. We include only the concentrations of gases and the concentrations of solutes in solutions because only those concentrations can be varied.

To establish equilibrium between solid $CaCO_3$, solid CaO, and gaseous CO_2, all three components must be present. It follows from the equations $K_c = [CO_2]$ and $K_p = P_{CO_2}$, however, that the concentration and pressure of CO_2 at equilibrium are constant, independent of how much solid CaO and $CaCO_3$ is present (**Figure 13.3**). If the temperature is changed, however, the concentration and pressure of CO_2 will also change because the values of K_c and K_p depend on temperature.

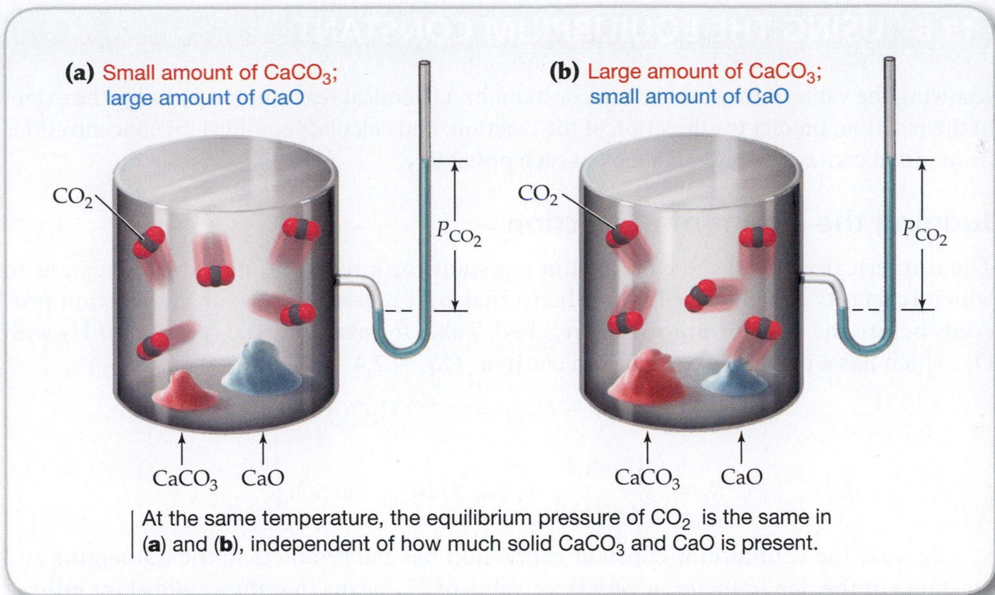

FIGURE 13.3

Thermal decomposition of calcium carbonate: $CaCO_3(s) \rightleftharpoons CaO(s) + CO_2(g)$.

Worked Example 13.8

Writing Equilibrium Equations for Heterogeneous Equilibria

Write the equilibrium equation for each of the following reactions:

(a) $CO_2(g) + C(s) \rightleftharpoons 2\,CO(g)$ **(b)** $Hg(l) + Hg^{2+}(aq) \rightleftharpoons Hg_2^{\ 2+}(aq)$

STRATEGY

Write the usual equilibrium-constant expressions but omit the pure solid carbon in part **(a)** and the pure liquid mercury in part **(b)** because the ratio of their concentrations to their concentrations in the standard state is equal to 1.

SOLUTION

(a) $K_c = \dfrac{[CO]^2}{[CO_2]}$

Alternatively, because CO and CO_2 are gases, the equilibrium equation can be written using partial pressures:

$$K_p = \frac{(P_{CO})^2}{P_{CO_2}}$$

The relationship between K_p and K_c is $K_p = K_c(RT)^{\Delta n} = K_c(RT)$, because $\Delta n = 2 - 1 = 1$.

(b) $K_c = \dfrac{[Hg_2^{\ 2+}]}{[Hg^{2+}]}$

In this case, it's not appropriate to write an expression for K_p because none of the reactants or products is a gas.

▶ **PROBLEM 13.8** For each of the following reactions, write the equilibrium-constant expression for K_c. Where appropriate, also write the equilibrium-constant expression for K_p and relate K_p to K_c.

(a) $2\,Fe(s) + 3\,H_2O(g) \rightleftharpoons Fe_2O_3(s) + 3\,H_2(g)$

(b) $2\,H_2O(l) \rightleftharpoons 2\,H_2(g) + O_2(g)$

(c) $SiCl_4(g) + 2\,H_2(g) \rightleftharpoons Si(s) + 4\,HCl(g)$

(d) $Hg_2^{\ 2+}(aq) + 2\,Cl^-(aq) \rightleftharpoons Hg_2Cl_2(s)$

13.6 USING THE EQUILIBRIUM CONSTANT

Knowing the value of the equilibrium constant for a chemical reaction lets us judge the extent of the reaction, predict the direction of the reaction, and calculate equilibrium concentrations from initial concentrations. Let's look at each possibility.

Judging the Extent of a Reaction

The numerical value of the equilibrium constant for a reaction indicates the extent to which reactants are converted to products; that is, it measures how far the reaction proceeds before the equilibrium state is reached. Take, for example, the reaction of H_2 with O_2, which has a very large equilibrium constant ($K_c = 2.4 \times 10^{47}$ at 500 K):

$$2\,H_2(g) + O_2(g) \rightleftharpoons 2\,H_2O(g)$$

$$K_c = \frac{[H_2O]^2}{[H_2]^2[O_2]} = 2.4 \times 10^{47} \quad \text{at 500 K}$$

Because the equilibrium-constant expression has the products in the numerator and reactants in the denominator, a very large value of K_c means that the equilibrium ratio of products to reactants is very large. In other words, the reaction proceeds nearly to completion. For example, if stoichiometric amounts of H_2 and O_2 are allowed to react and $[H_2O] = 0.10$ M at equilibrium, then the concentrations of H_2 and O_2 that remain at equilibrium are negligibly small: $[H_2] = 4.4 \times 10^{-17}$ M and $[O_2] = 2.2 \times 10^{-17}$ M. (Try substituting these concentrations into the equilibrium equation to show that they satisfy the equation.)

By contrast, if a reaction has a very small value of K_c, the equilibrium ratio of products to reactants is very small and the reaction proceeds hardly at all before equilibrium is reached. For example, the reverse of the reaction of H_2 with O_2 gives the same equilibrium mixture as obtained from the forward reaction ($[H_2] = 4.4 \times 10^{-17}$ M, $[O_2] = 2.2 \times 10^{-17}$ M, $[H_2O] = 0.10$ M). The reverse reaction does not occur to any appreciable extent, however, because its equilibrium constant is so small: $K_c' = 1/K_c = 1/(2.4 \times 10^{47}) = 4.2 \times 10^{-48}$.

$$2\,H_2O(g) \rightleftharpoons 2\,H_2(g) + O_2(g)$$

$$K_c' = \frac{[H_2]^2[O_2]}{[H_2O]^2} = 4.2 \times 10^{-48} \quad \text{at 500 K}$$

If a reaction has an intermediate value of K_c—say, a value in the range of 10^3 to 10^{-3}—then appreciable concentrations of both reactants and products are present in the equilibrium mixture. The reaction of hydrogen with iodine, for example, has $K_c = 57.0$ at 700 K:

$$H_2(g) + I_2(g) \rightleftharpoons 2\,HI(g)$$

$$K_c = \frac{[HI]^2}{[H_2][I_2]} = 57.0 \quad \text{at 700 K}$$

If the equilibrium concentrations of H_2 and I_2 are both 0.010 M, then the concentration of HI at equilibrium is 0.075 M:

$$[HI]^2 = K_c[H_2][I_2]$$

$$[HI] = \sqrt{K_c[H_2][I_2]} = \sqrt{(57.0)(0.010)(0.010)} = 0.075\ \text{M}$$

Thus, the concentrations of both reactants and products—0.010 M and 0.075 M—are appreciable.

(Note that pressing the $\sqrt{x}$ key on a calculator gives a positive number. Remember, though, that the square root of a positive number can be positive or negative. Of the two roots for the concentration of HI, ± 0.075 M, we choose the positive one because the concentration of a chemical substance is always a positive quantity.)

The gas-phase decomposition of N_2O_4 to NO_2 is another reaction with a value of K_c that is neither large nor small: $K_c = 4.64 \times 10^{-3}$ at 25 °C. Accordingly, equilibrium mixtures contain appreciable concentrations of both N_2O_4 and NO_2, as shown previously in Table 13.1.

We can make the following generalizations concerning the composition of equilibrium mixtures:

- If $K_c > 10^3$, products predominate over reactants. If K_c is very large, the reaction proceeds nearly to completion.
- If $K_c < 10^{-3}$, reactants predominate over products. If K_c is very small, the reaction proceeds hardly at all.
- If K_c is in the range 10^{-3} to 10^3, appreciable concentrations of both reactants and products are present.

These points are illustrated in **Figure 13.4**.

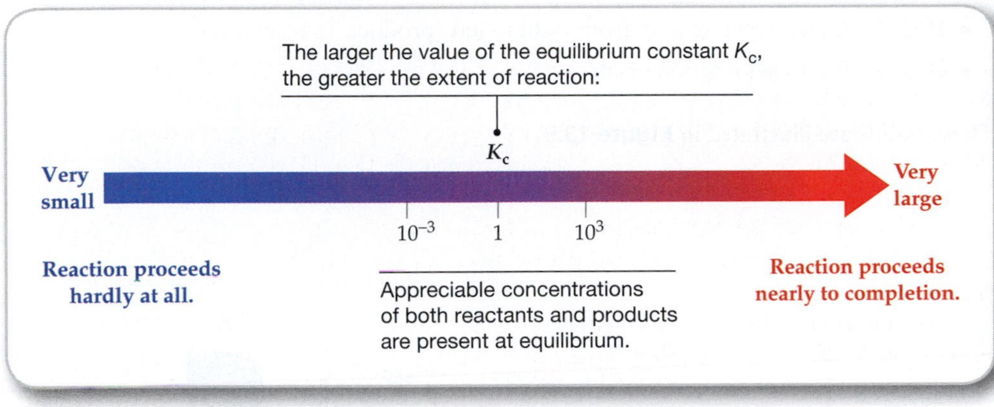

FIGURE 13.4

Judging the extent of a reaction.

PROBLEM 13.9 The value of K_c for the dissociation reaction $H_2(g) \rightleftharpoons 2\,H(g)$ is 1.2×10^{-42} at 500 K. Does the equilibrium mixture contain mainly H_2 molecules or H atoms? Explain in light of the electron configuration of hydrogen.

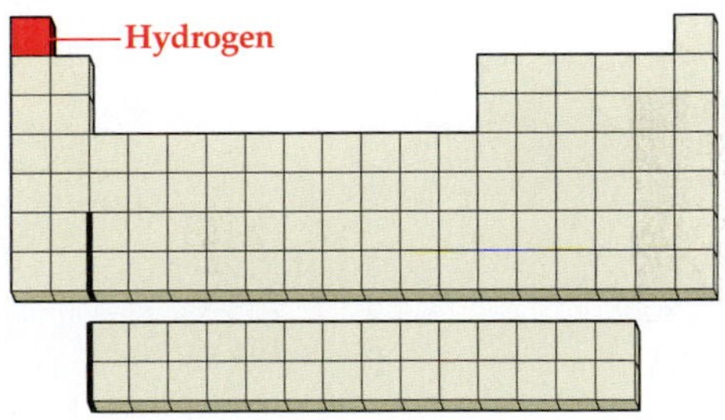

Predicting the Direction of a Reaction

Let's look again at the gaseous reaction of hydrogen with iodine:

$$H_2(g) + I_2(g) \rightleftharpoons 2\,HI(g) \qquad K_c = 57.0 \text{ at } 700 \text{ K}$$

Suppose that we have a mixture of $H_2(g)$, $I_2(g)$, and $HI(g)$ at 700 K with concentrations $[H_2]_t = 0.10$ M, $[I_2]_t = 0.20$ M, and $[HI]_t = 0.40$ M. (The subscript t on the concentration symbols means that the concentrations were measured at some arbitrary time t, not necessarily at equilibrium.) If we substitute these concentrations into the equilibrium-constant expression, we obtain a value called the *reaction quotient* Q_c.

$$Q_c = \frac{[HI]_t^2}{[H_2]_t[I_2]_t} = \frac{(0.40)^2}{(0.10)(0.20)} = 8.0$$

For the general reaction $a\,A + b\,B \rightleftharpoons c\,C + d\,D$, the **reaction quotient Q_c** is given by

▶ **Reaction quotient** $$Q_c = \frac{[C]_t^c[D]_t^d}{[A]_t^a[B]_t^b}$$

The reaction quotient Q_c is defined in the same way as the equilibrium constant K_c except that the concentrations in Q_c are not necessarily equilibrium values.

For the reaction of hydrogen with iodine, the numerical value of Q_c (8.0) does not equal K_c (57.0), so the mixture of $H_2(g)$, $I_2(g)$, and $HI(g)$ is not at equilibrium. As time passes, though, reaction will occur, changing the concentrations and thus changing the value of Q_c in the direction of K_c. After a sufficiently long time, an equilibrium state will be reached and $Q_c = K_c$.

The reaction quotient Q_c is useful because it lets us predict the direction of reaction by comparing the values of Q_c and K_c. If Q_c is less than K_c, movement toward equilibrium increases Q_c by converting reactants to products (that is, net reaction proceeds from left to right). If Q_c is greater than K_c, movement toward equilibrium decreases Q_c by converting products to reactants (that is, net reaction proceeds from right to left). If Q_c equals K_c, the reaction mixture is already at equilibrium, and no net reaction occurs.

Thus, we can make the following generalizations concerning the direction of the reaction:

- If $Q_c < K_c$, net reaction goes from left to right (reactants to products).
- If $Q_c > K_c$, net reaction goes from right to left (products to reactants).
- If $Q_c = K_c$, no net reaction occurs.

These points are illustrated in **Figure 13.5**.

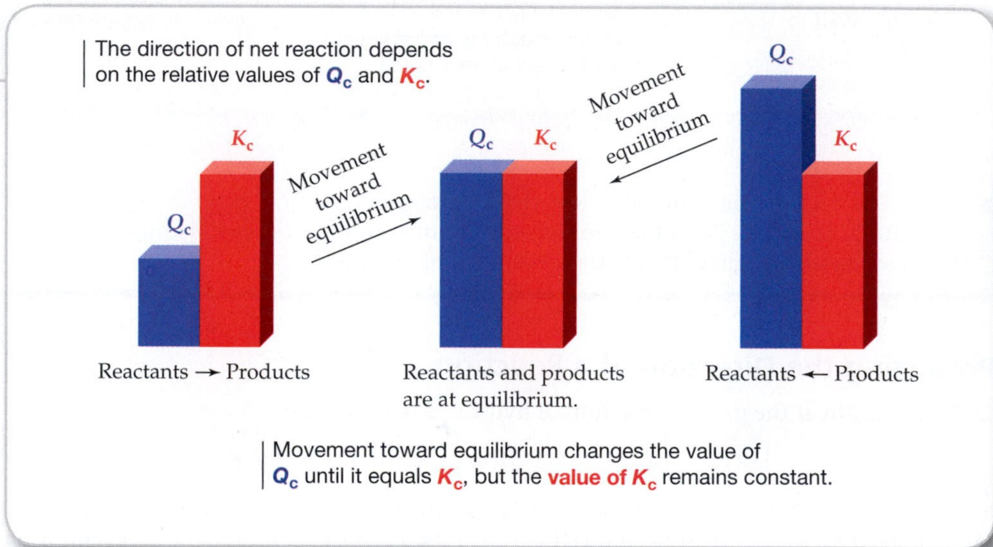

FIGURE 13.5

Predicting the direction of reaction.

Worked Example 13.9

Predicting the Direction of Reaction

A mixture of 1.57 mol of N_2, 1.92 mol of H_2, and 8.13 mol of NH_3 is introduced into a 20.0 L reaction vessel at 500 K. At this temperature, the equilibrium constant K_c for the reaction $N_2(g) + 3\,H_2(g) \rightleftharpoons 2\,NH_3(g)$ is 1.7×10^2. Is the reaction mixture at equilibrium? If not, what is the direction of the net reaction?

STRATEGY

To determine whether the reaction mixture is at equilibrium, we need to calculate the value of the reaction quotient Q_c and then compare it with the equilibrium constant K_c. If the mixture is not at equilibrium, the relative values of Q_c and K_c tell us the direction of the net reaction. Because we are given amounts in moles, we must first convert moles to molar concentrations before substituting into the expression for Q_c.

SOLUTION

The initial concentration of N_2 is (1.57 mol)/(20.0 L) = 0.0785 M. Similarly, $[H_2]$ = 0.0960 M and $[NH_3]$ = 0.406 M. Substituting these concentrations into the equilibrium-constant expression gives

$$Q_c = \frac{[NH_3]_t^{\,2}}{[N_2]_t[H_2]_t^{\,3}} = \frac{(0.406)^2}{(0.0785)(0.0960)^3} = 2.37 \times 10^3$$

Because Q_c does not equal K_c (1.7×10^2), the reaction mixture is not at equilibrium. Because Q_c is greater than K_c, net reaction will proceed from right to left, decreasing the concentration of NH_3 and increasing the concentrations of N_2 and H_2 until $Q_c = K_c = 1.7 \times 10^2$.

✓ BALLPARK CHECK

Approximate initial concentrations can be calculated by dividing rounded values of the number of moles of each substance by the volume; $[N_2] \approx$ (1.6 mol)/(20 L) $\approx$ 0.08 M, $[H_2] \approx$ (2 mol)/(20 L) $\approx$ 0.1 M, and $[NH_3] \approx$ (8 mol)/(20 L) $\approx$ 0.4 M. Substituting these concentrations into the expression for Q_c gives a ballpark estimate of Q_c:

$$Q_c = \frac{[NH_3]_t^{\,2}}{[N_2]_t[H_2]_t^{\,3}} = \frac{(0.4)^2}{(0.08)(0.1)^3} = 2 \times 10^3$$

You can calculate this value without a calculator because it equals $(16 \times 10^{-2})/(8 \times 10^{-5})$. The ballpark estimate of Q_c, like the more exact value (2.37×10^3), exceeds K_c, so the reaction mixture is not at equilibrium.

PROBLEM 13.10 The equilibrium constant K_c for the reaction $2\,NO(g) + O_2(g) \rightleftharpoons 2\,NO_2(g)$ is 6.9×10^5 at 500 K. A 5.0 L reaction vessel at 500 K was filled with 0.060 mol of NO, 1.0 mol of O_2, and 0.80 mol of NO_2.

(a) Is the reaction mixture at equilibrium? If not, in which direction does the net reaction proceed?

(b) What is the direction of the net reaction if the initial amounts are 5.0×10^{-3} mol of NO, 0.20 mol of O_2, and 4.0 mol of NO_2?

CONCEPTUAL PROBLEM 13.11 The reaction $A_2 + B_2 \rightleftharpoons 2\,AB$ has an equilibrium constant K_c = 4. The following pictures represent reaction mixtures that contain A_2 molecules (red), B_2 molecules (blue), and AB molecules:

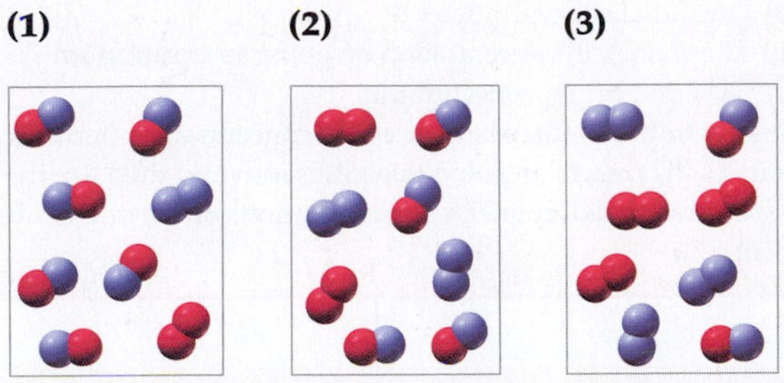

(a) Which reaction mixture is at equilibrium?

(b) For those reaction mixtures that are not at equilibrium, will the net reaction go in the forward or reverse direction to reach equilibrium?

Calculating Equilibrium Concentrations

If the equilibrium constant and all the equilibrium concentrations but one are known, the unknown concentration can be calculated directly from the equilibrium equation. To illustrate, let's consider the following problem: What is the concentration of NO in an equilibrium mixture of gaseous NO, O_2, and NO_2 at 500 K that contains 1.0×10^{-3} M O_2 and 5.0×10^{-2} M NO_2? At this temperature, the equilibrium constant K_c for the reaction $2\,NO(g) + O_2(g) \rightleftharpoons 2\,NO_2(g)$ is 6.9×10^5.

In this problem, K_c and all the equilibrium concentrations except one are known and we're asked to calculate the unknown equilibrium concentration. First, we write the equilibrium equation for the reaction and solve for the unknown concentration:

$$K_c = \frac{[NO_2]^2}{[NO]^2[O_2]} \qquad [NO] = \sqrt{\frac{[NO_2]^2}{[O_2]K_c}}$$

Then we substitute the known values of K_c, $[O_2]$, and $[NO_2]$ into the expression for [NO]. Taking the positive square root because the concentration of NO is a positive quantity, we obtain:

$$[NO] = \sqrt{\frac{(5.0 \times 10^{-2})^2}{(1.0 \times 10^{-3})\,(6.9 \times 10^5)}} = \sqrt{3.6 \times 10^{-6}} = 1.9 \times 10^{-3}\,\text{M}$$

To be sure that we haven't made any errors, it's a good idea to check the result by substituting it into the equilibrium equation:

$$K_c = 6.9 \times 10^5 = \frac{[NO_2]^2}{[NO]^2[O_2]} = \frac{(5.0 \times 10^{-2})^2}{(1.9 \times 10^{-3})^2(1.0 \times 10^{-3})} = 6.9 \times 10^5$$

How can we determine equilibrium concentrations from initial concentrations if we don't know any of the equilibrium concentrations? To solve this kind of problem, follow the series of steps summarized in **Figure 13.6** and illustrated in Worked Examples 13.10 and 13.11. The same approach can be used to calculate equilibrium partial pressures from initial partial pressures and K_p, as shown in Worked Example 13.12.

FIGURE 13.6

Steps to follow in calculating equilibrium concentrations from initial concentrations.

Step 1. Write the balanced equation for the reaction.

↓

Step 2. Under the balanced equation, make a table that lists for each substance involved in the reaction:
- **(a)** The initial concentration
- **(b)** The change in concentration on going to equilibrium
- **(c)** The equilibrium concentration

In constructing the table, define x as the concentration (mol/L) of one of the substances that reacts on going to equilibrium and then use the stoichiometry of the reaction to determine the concentrations of the other substances in terms of x.

↓

Step 3. Substitute the equilibrium concentrations into the equilibrium equation for the reaction and solve for x. If you must solve a quadratic equation, choose the mathematical solution that makes chemical sense.

↓

Step 4. Calculate the equilibrium concentrations from the calculated value of x.

↓

Step 5. Check your results by substituting them into the equilibrium equation.

Worked Example 13.10

Calculating Equilibrium Concentrations from Equal Initial Concentrations

The equilibrium constant K_c for the reaction of H_2 with I_2 is 57.0 at 700 K:

$$H_2(g) + I_2(g) \rightleftharpoons 2\,HI(g) \qquad K_c = 57.0 \text{ at } 700 \text{ K}$$

If 1.00 mol of H_2 is allowed to react with 1.00 mol of I_2 in a 10.0 L reaction vessel at 700 K, what are the concentrations of H_2, I_2, and HI at equilibrium? What is the composition of the equilibrium mixture in moles?

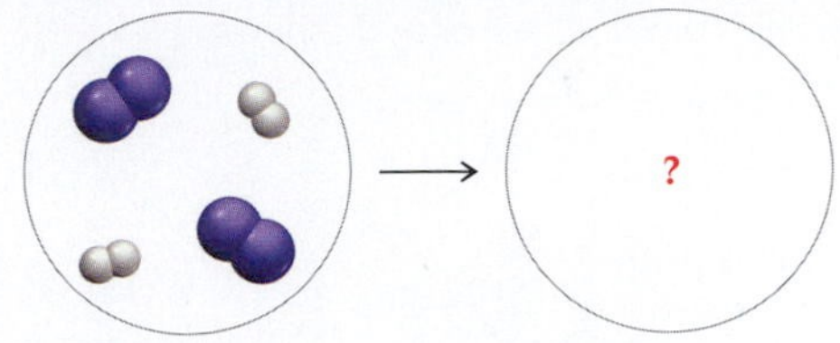

Initial state Equilibrium state

STRATEGY

We need to calculate equilibrium concentrations from initial concentrations, so we use the method outlined in Figure 13.6.

SOLUTION

Step 1 The balanced equation is given: $H_2(g) + I_2(g) \rightleftharpoons 2\,HI(g)$.

Step 2 The initial concentrations are $[H_2] = [I_2] = (1.00 \text{ mol})/(10.0 \text{ L}) = 0.100$ M. For convenience, define an unknown, x, as the concentration (mol/L) of H_2 that reacts. According to the balanced equation for the reaction, x mol/L of H_2 reacts with x mol/L of I_2 to give $2x$ mol/L of HI. This reduces the initial concentrations of H_2 and I_2 from 0.100 mol/L to $(0.100 - x)$ mol/L at equilibrium. Let's summarize these results in a table under the balanced equation:

	$H_2(g)$	+	$I_2(g)$	$\rightleftharpoons$	$2\,HI(g)$
Initial concentration (M)	0.100		0.100		0
Change (M)	$-x$		$-x$		$+2x$
Equilibrium concentration (M)	$(0.100 - x)$		$(0.100 - x)$		$2x$

Step 3 Substitute the equilibrium concentrations into the equilibrium equation for the reaction:

$$K_c = 57.0 = \frac{[HI]^2}{[H_2][I_2]} = \frac{(2x)^2}{(0.100 - x)(0.100 - x)} = \left(\frac{2x}{0.100 - x}\right)^2$$

Because the right side of this equation is a perfect square, we can take the square root of both sides:

$$\sqrt{57.0} = \pm 7.55 = \frac{2x}{0.100 - x}$$

Solving for x, we obtain two solutions. The equation with the positive square root of 57.0 gives

$$+7.55(0.100 - x) = 2x$$
$$0.755 = 2x + 7.55x$$
$$x = \frac{0.755}{9.55} = 0.0791 \text{ M}$$

The equation with the negative square root of 57.0 gives

$$-7.55(0.100 - x) = 2x$$
$$-0.755 = 2x - 7.55x$$
$$x = \frac{-0.755}{-5.55} = 0.136 \text{ M}$$

Because the initial concentrations of H_2 and I_2 are 0.100 M, x can't exceed 0.100 M. Therefore, discard $x = 0.136$ M as chemically unreasonable and choose the first solution, $x = 0.0791$ M.

Step 4 Calculate the equilibrium concentrations from the calculated value of x:

$$[H_2] = [I_2] = 0.100 - x = 0.100 - 0.0791 = 0.021 \text{ M}$$
$$[HI] = 2x = (2)(0.0791) = 0.158 \text{ M}$$

continued on next page

Step 5 Check the results by substituting them into the equilibrium equation:

$$K_c = 57.0 = \frac{[HI]^2}{[H_2][I_2]} = \frac{(0.158)^2}{(0.021)(0.021)} = 57$$

The number of moles of each substance in the equilibrium mixture is obtained by multiplying each concentration by the volume of the reaction vessel:

$$\text{Moles of } H_2 = \text{Moles of } I_2 = (0.021 \text{ mol/L})(10.0 \text{ L}) = 0.21 \text{ mol}$$
$$\text{Moles of HI} = (0.158 \text{ mol/L})(10.0 \text{ L}) = 1.58 \text{ mol}$$

Worked Example 13.11

Calculating Equilibrium Concentrations from Unequal Initial Concentrations

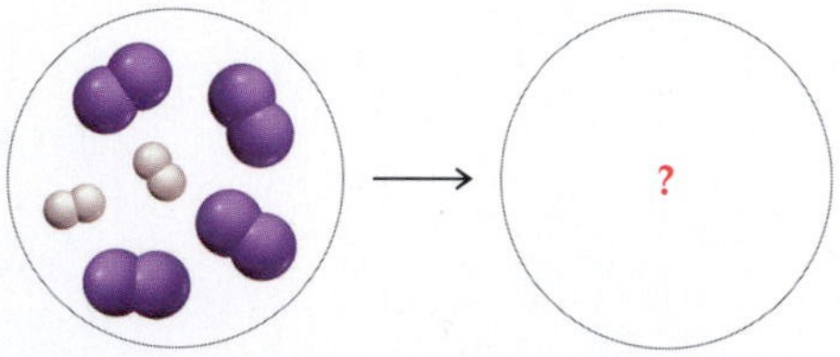

Initial state Equilibrium state

Calculate the equilibrium concentrations of H_2, I_2, and HI at 700 K if the initial concentrations are $[H_2] = 0.100$ M and $[I_2] = 0.200$ M. The equilibrium constant K_c for the reaction $H_2(g) + I_2(g) \rightleftharpoons 2\,HI(g)$ is 57.0 at 700 K.

STRATEGY

This problem is similar to Worked Example 13.10 except that the initial concentrations of H_2 and I_2 are unequal. Again, we follow the steps in Figure 13.6.

SOLUTION

Step 1 The balanced equation is $H_2(g) + I_2(g) \rightleftharpoons 2\,HI(g)$.

Step 2 Again, define x as the concentration of H_2 that reacts. Set up a table of concentrations under the balanced equation:

	$H_2(g)$	+	$I_2(g)$	$\rightleftharpoons$	$2\,HI(g)$
Initial concentration (M)	0.100		0.200		0
Change (M)	$-x$		$-x$		$+2x$
Equilibrium concentration (M)	$(0.100 - x)$		$(0.200 - x)$		$2x$

Step 3 Substitute the equilibrium concentrations into the equilibrium equation:

$$K_c = 57.0 = \frac{[HI]^2}{[H_2][I_2]} = \frac{(2x)^2}{(0.100 - x)(0.200 - x)}$$

Because the right side of this equation is not a perfect square, we must put the equation into the standard quadratic form, $ax^2 + bx + c = 0$, and then solve for x using the quadratic formula (Appendix A.4):

$$x = \frac{-b \pm \sqrt{b^2 - 4ac}}{2a}$$

Rearranging the equilibrium equation gives

$$(57.0)(0.0200 - 0.300x + x^2) = 4x^2$$

or

$$53.0x^2 - 17.1x + 1.14 = 0$$

Substituting the values of a, b, and c into the quadratic formula gives two solutions:

$$x = \frac{17.1 \pm \sqrt{(-17.1)^2 - 4(53.0)(1.14)}}{2(53.0)} = \frac{17.1 \pm 7.1}{106} = 0.228 \quad \text{and} \quad 0.0943$$

Now decide which of the two solutions is physically possible. Discard the solution that uses the positive square root ($x = 0.228$) because the H_2 concentration can't change by more than its initial value (0.100 M). Therefore, choose the solution that uses the negative square root ($x = 0.0943$).

Step 4 Calculate the equilibrium concentrations from the calculated value of x:

$$[H_2] = 0.100 - x = 0.100 - 0.0943 = 0.006 \text{ M}$$
$$[I_2] = 0.200 - x = 0.200 - 0.0943 = 0.106 \text{ M}$$
$$[HI] = 2x = (2)(0.0943) = 0.189 \text{ M}$$

Step 5 Check the results by substituting them into the equilibrium equation:

$$K_c = 57.0 = \frac{[HI]^2}{[H_2][I_2]} = \frac{(0.189)^2}{(0.006)(0.106)} = 56.2$$

The calculated value of K_c (56.2), which should be rounded to one significant figure (6×10^1), agrees with the value given in the problem (57.0).

Worked Example 13.12

Calculating Equilibrium Partial Pressures from Initial Partial Pressures

One reaction that occurs in producing steel from iron ore is the reduction of iron(II) oxide by carbon monoxide to give iron metal and carbon dioxide. The equilibrium constant K_p for the reaction at 1000 K is 0.259.

$$FeO(s) + CO(g) \rightleftharpoons Fe(s) + CO_2(g) \qquad K_p = 0.259 \text{ at } 1000 \text{ K}$$

What are the equilibrium partial pressures of CO and CO_2 at 1000 K if the initial partial pressures are $P_{CO} = 1.000$ atm and $P_{CO_2} = 0.500$ atm?

▲ The steel used in making automobiles is produced by the reaction of iron ore with carbon monoxide.

STRATEGY

We can calculate equilibrium partial pressures from initial partial pressures and K_p in the same way that we calculate equilibrium concentrations from initial concentrations and K_c. Follow the steps in Figure 13.6, but substitute partial pressures for concentrations.

SOLUTION

Step 1 The balanced equation is $FeO(s) + CO(g) \rightleftharpoons Fe(s) + CO_2(g)$.

Step 2 Define x as the partial pressure of CO that reacts. Set up a table of partial pressures of the gases under the balanced equation:

	$FeO(s)$ + $CO(g)$	$\rightleftharpoons$	$Fe(s)$ + $CO_2(g)$
Initial pressure (atm)	1.000		0.500
Change (atm)	$-x$		$+x$
Equilibrium pressure (atm)	$(1.000 - x)$		$(0.500 + x)$

Step 3 Substitute the equilibrium partial pressures into the equilibrium equation for K_p:

$$K_p = 0.259 = \frac{P_{CO_2}}{P_{CO}} = \frac{0.500 + x}{1.000 - x}$$

As usual for a heterogeneous equilibrium, we omit the pure solids from the equilibrium equation. Rearranging the equilibrium equation and solving for x gives

$$0.259 - 0.259x = 0.500 + x$$

$$x = \frac{-0.241}{1.259} = -0.191$$

Step 4 Calculate the equilibrium partial pressures from the calculated value of x:

$$P_{CO} = 1.000 - x = 1.000 - (-0.191) = 1.191 \text{ atm}$$

$$P_{CO_2} = 0.500 + x = 0.500 + (-0.191) = 0.309 \text{ atm}$$

Step 5 Check the results by substituting them into the equilibrium equation:

$$K_p = 0.259 = \frac{P_{CO_2}}{P_{CO}} = \frac{0.309}{1.191} = 0.259$$

A negative value for x means that the reaction goes from products to reactants to reach equilibrium. This makes sense because the initial reaction quotient, $Q_p = 0.500/1.000 = 0.500$, is greater than the equilibrium constant, $K_p = 0.259$. When $Q_p > K_p$, the net reaction always goes from products to reactants (right to left).

PROBLEM 13.12 In Problem 13.9, we found that an equilibrium mixture of H_2 molecules and H atoms at 500 K contains mainly H_2 molecules because the equilibrium constant for the dissociation reaction $H_2(g) \rightleftharpoons 2\,H(g)$ is very small ($K_c = 1.2 \times 10^{-42}$).

(a) What is the molar concentration of H atoms at equilibrium if the equilibrium concentration of H_2 is 0.10 M?

(b) How many H atoms and H_2 molecules are present at equilibrium in 1.0 L of 0.10 M H_2 at 500 K?

PROBLEM 13.13 The H_2/CO ratio in mixtures of carbon monoxide and hydrogen (called synthesis gas) is increased by the water-gas shift reaction $CO(g) + H_2O(g) \rightleftharpoons CO_2(g) + H_2(g)$, which has an equilibrium constant $K_c = 4.24$ at 800 K. Calculate the equilibrium concentrations of CO_2, H_2, CO, and H_2O at 800 K if only CO and H_2O are present initially at concentrations of 0.150 M.

PROBLEM 13.14 Calculate the equilibrium concentrations of N_2O_4 and NO_2 at 25 °C in a vessel that contains an initial N_2O_4 concentration of 0.0500 M. The equilibrium constant K_c for the reaction $N_2O_4(g) \rightleftharpoons 2\,NO_2(g)$ is 4.64×10^{-3} at 25 °C.

PROBLEM 13.15 Calculate the equilibrium concentrations at 25 °C for the reaction in Problem 13.14 if the initial concentrations are $[N_2O_4] = 0.0200$ M and $[NO_2] = 0.0300$ M.

PROBLEM 13.16 The equilibrium constant K_p for the reaction $C(s) + H_2O(g) \rightleftharpoons CO(g) + H_2(g)$ is 2.44 at 1000 K. What are the equilibrium partial pressures of H_2O, CO, and H_2 if the initial partial pressures are $P_{H_2O} = 1.20$ atm, $P_{CO} = 1.00$ atm, and $P_{H_2} = 1.40$ atm?

13.7 FACTORS THAT ALTER THE COMPOSITION OF AN EQUILIBRIUM MIXTURE: LE CHÂTELIER'S PRINCIPLE

One of the main goals of chemical synthesis is to maximize the conversion of reactants to products while minimizing the expenditure of energy. This objective is achieved easily if the reaction goes nearly to completion at mild temperature and pressure. If the reaction gives an equilibrium mixture that is rich in reactants and poor in products, however, then the experimental conditions must be adjusted. For example, in the Haber process for the synthesis of ammonia from N_2 and H_2 (**Figure 13.7**), the choice of experimental conditions is of real economic importance. Annual worldwide production of ammonia is about 120 million metric tons, primarily for use as fertilizer.

The following factors can be exploited to alter the composition of an equilibrium mixture:

- The concentration of reactants or products can be changed.
- The pressure and volume can be changed.
- The temperature can be changed.

LOOKING AHEAD ...

We'll see in Section 13.11 that a **catalyst** increases the rates of both the forward and the reverse reactions but does not change the composition of the equilibrium mixture.

A possible fourth factor, addition of a catalyst, increases only the rate at which equilibrium is reached. As we'll see in Section 13.11, a **catalyst** does not affect the equilibrium concentrations.

The qualitative effect of the listed changes on the composition of an equilibrium mixture can be predicted using a principle first described by the French chemist Henri-Louis Le Châtelier (pronounced Li Sha–tell–**yea**):

Le Châtelier's Principle If a stress is applied to a reaction mixture at equilibrium, net reaction occurs in the direction that relieves the stress.

The word "stress" in this context means a change in concentration, pressure, volume, or temperature that disturbs the original equilibrium. Reaction then occurs to change the composition of the mixture until a new state of equilibrium is reached. The direction that the

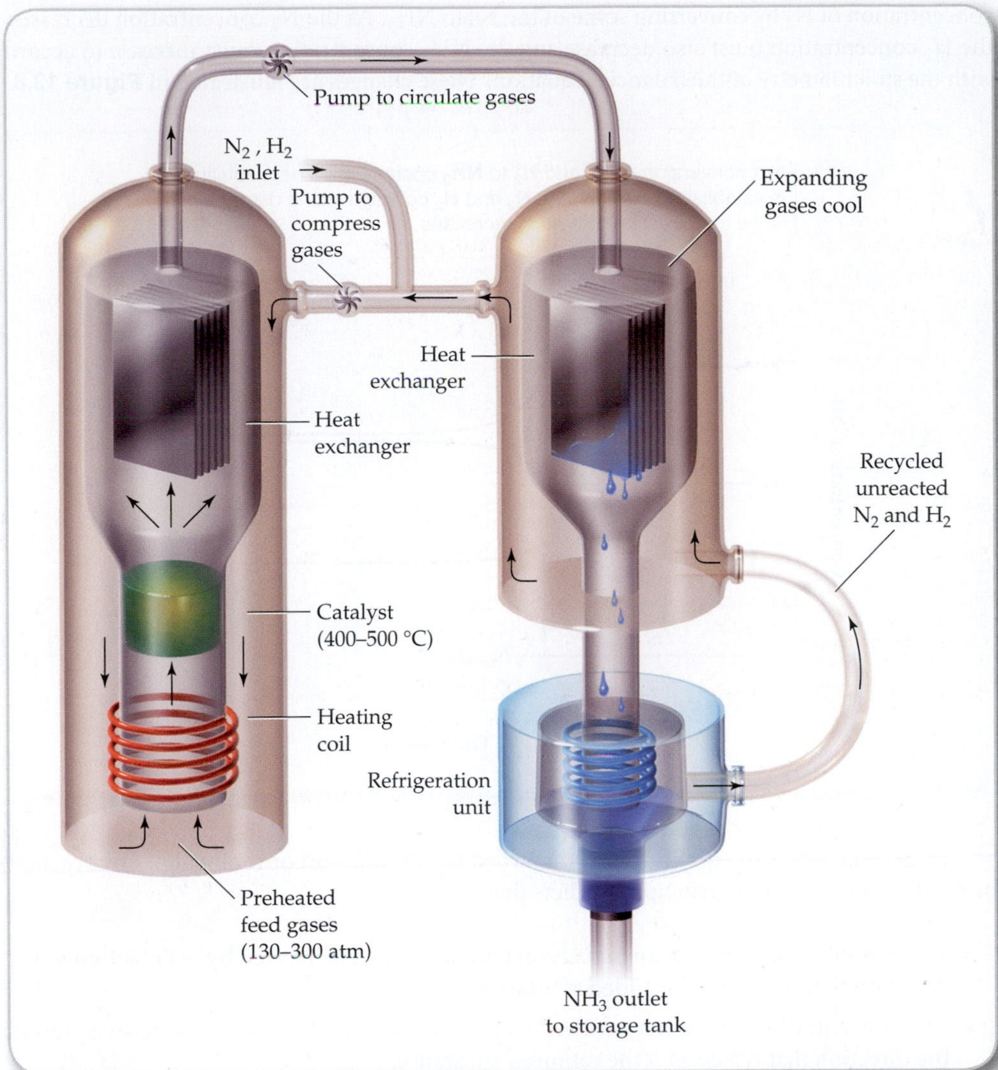

FIGURE 13.7

Representation of the Haber process for the industrial production of ammonia. A mixture of gaseous N_2 and H_2 at 130–300 atm pressure is passed over a catalyst at 400–500°C, and ammonia is produced by the reaction $N_2(g) + 3\,H_2(g) \rightleftharpoons 2\,NH_3(g)$. The NH_3 in the gaseous mixture of reactants and products is liquefied, and the unreacted N_2 and H_2 are recycled.

reaction takes (reactants to products or vice versa) is the one that reduces the stress. In the next three sections, we'll look at the different kinds of stress that can change the composition of an equilibrium mixture.

13.8 THE EFFECT OF CONCENTRATION CHANGES ON AN EQUILIBRIUM MIXTURE

Let's consider the equilibrium that occurs in the Haber process for the synthesis of ammonia:

$$N_2(g) + 3\,H_2(g) \rightleftharpoons 2\,NH_3(g) \qquad K_c = 0.291 \text{ at } 700\text{ K}$$

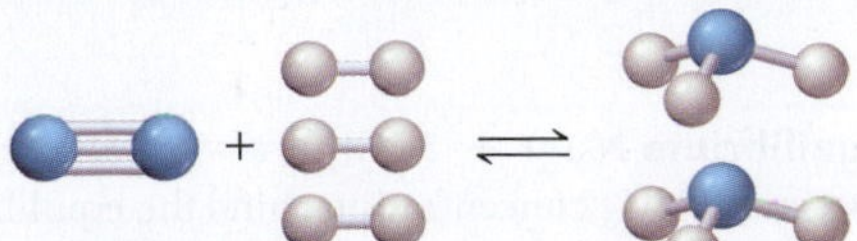

Suppose that we have an equilibrium mixture of 0.50 M N_2, 3.00 M H_2, and 1.98 M NH_3 at 700 K and that we disturb the equilibrium by increasing the N_2 concentration to 1.50 M. Le Châtelier's principle tells us that reaction will occur to relieve the stress of the increased concentration of N_2 by converting some of the N_2 to NH_3. As the N_2 concentration decreases, the H_2 concentration must also decrease and the NH_3 concentration must increase in accord with the stoichiometry of the balanced equation. These changes are illustrated in **Figure 13.8**.

FIGURE 13.8

Changes in concentrations when N_2 is added to an equilibrium mixture of N_2, H_2, and NH_3.

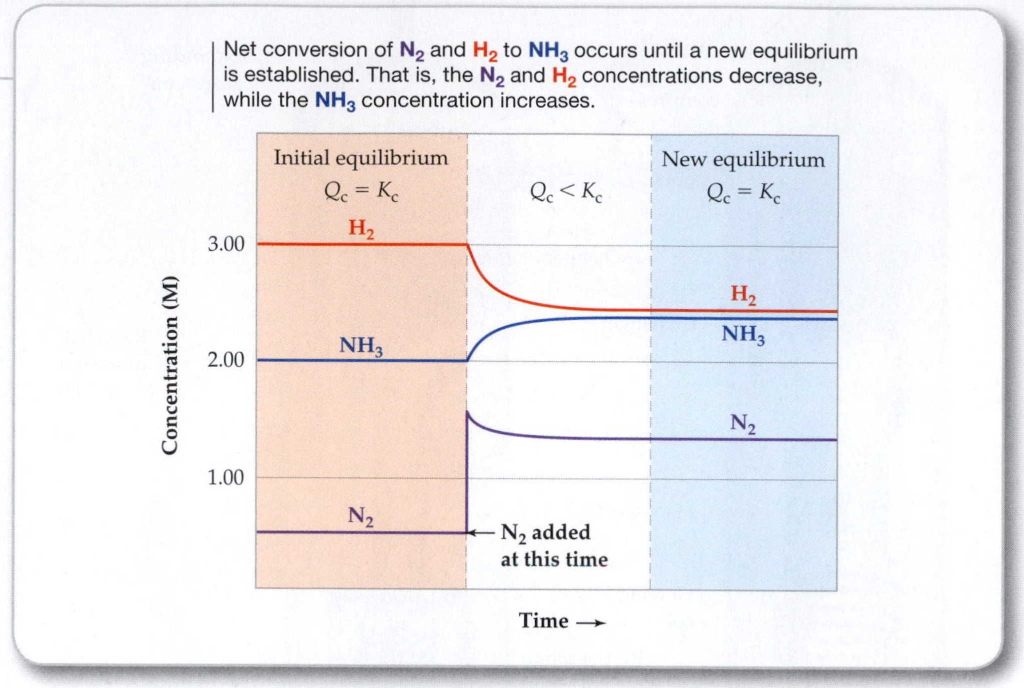

In general, when an equilibrium is disturbed by the addition or removal of any reactant or product, Le Châtelier's principle predicts that

- The concentration stress of an *added* reactant or product is relieved by net reaction in the direction that *consumes* the added substance.
- The concentration stress of a *removed* reactant or product is relieved by net reaction in the direction that *replenishes* the removed substance.

If these rules are applied to the equilibrium $N_2(g) + 3\,H_2(g) \rightleftharpoons 2\,NH_3(g)$, then the yield of ammonia is increased by an increase in the N_2 or H_2 concentration or by a decrease in the NH_3 concentration (**Figure 13.9**). In the industrial production of ammonia, the concentration of gaseous NH_3 is decreased by liquefying the ammonia (bp -33 °C) as it's formed, and so more ammonia is produced.

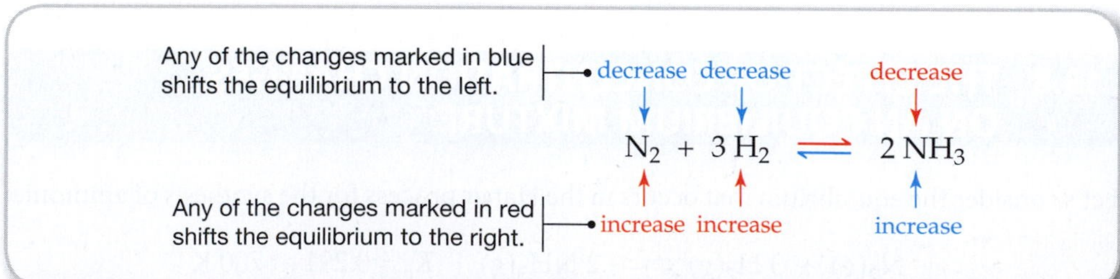

FIGURE 13.9

Effect of concentration changes on the equilibrium $N_2(g) + 3\,H_2(g) \rightleftharpoons 2\,NH_3(g)$. An increase in the N_2 or H_2 concentration or a decrease in the NH_3 concentration shifts the equilibrium from left to right. A decrease in the N_2 or H_2 concentration or an increase in the NH_3 concentration shifts the equilibrium from right to left.

Le Châtelier's principle is a handy rule for predicting changes in the composition of an equilibrium mixture, but it doesn't explain *why* those changes occur. To see why Le Châtelier's principle works, let's look again at the reaction quotient Q_c. For the initial equilibrium mixture of 0.50 M N_2, 3.00 M H_2, and 1.98 M NH_3 at 700 K, Q_c equals the equilibrium constant K_c (0.291) because the system is at equilibrium:

$$Q_c = \frac{[NH_3]_t^2}{[N_2]_t[H_2]_t^3} = \frac{(1.98)^2}{(0.50)(3.00)^3} = 0.29 = K_c$$

When we disturb the equilibrium by increasing the N_2 concentration to 1.50 M, the denominator of the equilibrium-constant expression increases and Q_c decreases to a value less than K_c:

$$Q_c = \frac{[NH_3]_t^2}{[N_2]_t[H_2]_t^3} = \frac{(1.98)^2}{(1.50)(3.00)^3} = 0.0968 < K_c$$

For the system to move to a new state of equilibrium, Q_c must increase; that is, the numerator of the equilibrium-constant expression must increase and the denominator must decrease. This implies the net conversion of N_2 and H_2 to NH_3, just as predicted by Le Châtelier's principle. When the new equilibrium is established (Figure 13.8), the concentrations are 1.31 M N_2, 2.43 M H_2, and 2.36 M NH_3, and Q_c again equals K_c:

$$Q_c = \frac{[NH_3]_t^2}{[N_2]_t[H_2]_t^3} = \frac{(2.36)^2}{(1.31)(2.43)^3} = 0.296 = K_c$$

As another example of how a change in concentration affects an equilibrium, let's consider the reaction in aqueous solution of iron(III) ions and thiocyanate (SCN^-) ions to give an equilibrium mixture that contains the Fe–N bonded red complex ion $FeNCS^{2+}$:

$$\underset{\text{Pale yellow}}{Fe^{3+}(aq)} + \underset{\text{Colorless}}{SCN^-(aq)} \rightleftharpoons \underset{\text{Red}}{FeNCS^{2+}(aq)}$$

Shifts in the position of this equilibrium can be detected by observing how the color of the solution changes when we add various reagents (**Figure 13.10**). If we add aqueous $FeCl_3$, the red color gets darker, as predicted by Le Châtelier's principle. The concentration stress of added Fe^{3+} is relieved by net reaction from left to right, which consumes some of the Fe^{3+} and increases the concentration of $FeNCS^{2+}$. (Note that the Cl^- ions are not involved in the reaction.) Similarly, if we add aqueous KSCN, the stress of added SCN^- shifts the equilibrium from left to right and again the red color gets darker.

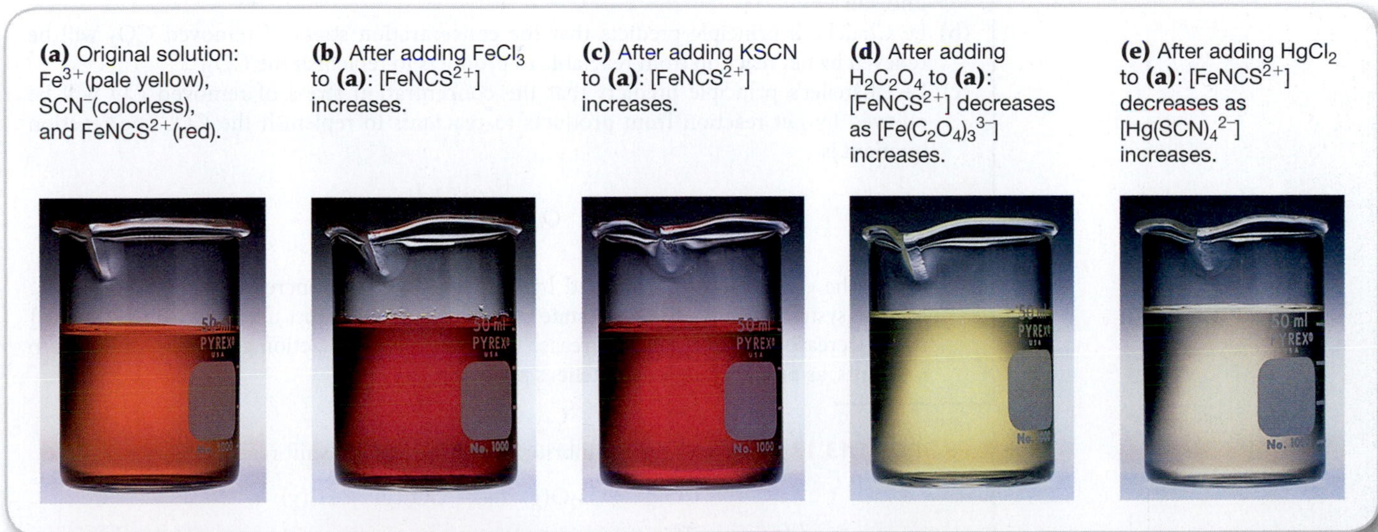

FIGURE 13.10

Color changes produced by adding various reagents to an equilibrium mixture of Fe^{3+} (pale yellow), SCN^- (colorless), and $FeNCS^{2+}$ (red).

The equilibrium can be shifted in the opposite direction by adding reagents that remove Fe^{3+} or SCN^- ions. For example, oxalic acid ($H_2C_2O_4$), a poisonous substance present in the leaves of plants such as rhubarb, reacts with Fe^{3+} to form the stable, yellow complex ion $Fe(C_2O_4)_3^{3-}$, thus decreasing the concentration of free $Fe^{3+}(aq)$. In accord with Le Châtelier's principle, the concentration stress of removed Fe^{3+} is relieved by the dissociation of $FeNCS^{2+}$

▲ The extremely sour leaves of rhubarb contain toxins such as oxalic acid, but the stalks and roots are nutritious.

to replenish the Fe^{3+} ions. Because the concentration of $FeNCS^{2+}$ decreases, the red color disappears.

$$3\,H_2C_2O_4(aq) + Fe^{3+}(aq) \longrightarrow Fe(C_2O_4)_3{}^{3-}(aq) + 6\,H^+(aq)$$
$$FeNCS^{2+}(aq) \longrightarrow Fe^{3+}(aq) + SCN^-(aq)$$

Addition of aqueous $HgCl_2$ also eliminates the red color because $HgCl_2$ reacts with SCN^- ions to form the stable, colorless Hg–S bonded complex ion $Hg(SCN)_4{}^{2-}$. Removal of free $SCN^-(aq)$ results in dissociation of the red $FeNCS^{2+}$ ions so as to replenish the SCN^- ions.

$$HgCl_2(aq) + 4\,SCN^-(aq) \longrightarrow Hg(SCN)_4{}^{2-}(aq) + 2\,Cl^-(aq)$$
$$FeNCS^{2+}(aq) \longrightarrow Fe^{3+}(aq) + SCN^-(aq)$$

Worked Example 13.13

Applying Le Châtelier's Principle to Concentration Changes

The reaction of iron(III) oxide with carbon monoxide occurs in a blast furnace when iron ore is reduced to iron metal:

$$Fe_2O_3(s) + 3\,CO(g) \rightleftharpoons 2\,Fe(l) + 3\,CO_2(g)$$

Use Le Châtelier's principle to predict the direction of the net reaction when an equilibrium mixture is disturbed by:

(a) Adding Fe_2O_3
(b) Removing CO_2
(c) Removing CO; also account for the change using the reaction quotient Q_c.

STRATEGY

To predict the direction of net reaction, recall that a concentration stress is relieved by reaction in the direction that consumes an added substance or replenishes a removed substance. This rule does not apply to pure solids or pure liquids because their concentrations have a constant value equal to 1.

SOLUTION

(a) Because Fe_2O_3 is a pure solid, its concentration is equal to 1 and doesn't change when more Fe_2O_3 is added. Therefore, there is no concentration stress and the original equilibrium is undisturbed.

(b) Le Châtelier's principle predicts that the concentration stress of removed CO_2 will be relieved by net reaction from reactants to products to replenish the CO_2.

(c) Le Châtelier's principle predicts that the concentration stress of removed CO will be relieved by net reaction from products to reactants to replenish the CO. The reaction quotient is

$$Q_c = \frac{[CO_2]_t{}^3}{[CO]_t{}^3}$$

When the equilibrium is disturbed by reducing [CO], Q_c increases, so that $Q_c > K_c$. For the system to move to a new state of equilibrium, Q_c must decrease—that is, $[CO_2]$ must decrease and [CO] must increase. Therefore, the net reaction goes from products to reactants, as predicted by Le Châtelier's principle.

▶ **PROBLEM 13.17** Consider the equilibrium for the water-gas shift reaction:

$$CO(g) + H_2O(g) \rightleftharpoons CO_2(g) + H_2(g)$$

Use Le Châtelier's principle to predict how the concentration of H_2 will change when the equilibrium is disturbed by:

(a) Adding CO
(b) Adding CO_2
(c) Removing H_2O
(d) Removing CO_2; also account for the change using the reaction quotient Q_c

13.9 THE EFFECT OF PRESSURE AND VOLUME CHANGES ON AN EQUILIBRIUM MIXTURE

To illustrate how an equilibrium mixture is affected by a change in pressure as a result of a change in the volume, let's return to the Haber synthesis of ammonia. The balanced equation for the reaction has 4 mol of gas on the reactant side of the equation and 2 mol on the product side:

$$N_2(g) + 3\,H_2(g) \rightleftharpoons 2\,NH_3(g) \quad K_c = 0.291 \text{ at } 700\text{ K}$$

What happens to the composition of the equilibrium mixture if we increase the pressure by decreasing the volume? (Recall from Sections 9.2 and 9.3 that the pressure of an ideal gas is inversely proportional to the volume at constant temperature and constant number of moles of gas; $P = nRT/V$.) According to Le Châtelier's principle, net reaction will occur in the direction that relieves the stress of the increased pressure, which means that the number of moles of gas must decrease. Therefore, we predict that the net reaction will proceed from left to right because the forward reaction converts 4 mol of gaseous reactants to 2 mol of gaseous products.

In general, Le Châtelier's principle predicts that

- An *increase* in pressure by reducing the volume will bring about net reaction in the direction that *decreases* the number of moles of gas.
- A *decrease* in pressure by expanding the volume will bring about net reaction in the direction that *increases* the number of moles of gas.

To see why Le Châtelier's principle works for pressure (volume) changes, let's look again at the reaction quotient for the equilibrium mixture of 0.50 M N_2, 3.00 M H_2, and 1.98 M NH_3 at 700 K:

$$Q_c = \frac{[NH_3]_t^2}{[N_2]_t[H_2]_t^3} = \frac{(1.98)^2}{(0.50)(3.00)^3} = 0.29 = K_c$$

If we disturb the equilibrium by reducing the volume by a factor of 2, we not only double the total pressure, we also double the partial pressure and thus the molar concentration of each reactant and product (because molarity $= n/V = P/RT$). Because the balanced equation has more moles of gaseous reactants than gaseous products, the increase in the denominator of the equilibrium-constant expression is greater than the increase in the numerator and the new value of Q_c is less than the equilibrium constant K_c:

$$Q_c = \frac{[NH_3]_t^2}{[N_2]_t[H_2]_t^3} = \frac{(3.96)^2}{(1.00)(6.00)^3} = 0.0726 < K_c$$

For the system to move to a new state of equilibrium, Q_c must increase, which means that the net reaction must go from reactants to products, as predicted by Le Châtelier's principle (**Figure 13.11**). In practice, the yield of ammonia in the Haber process is increased by running the reaction at high pressure, typically 130–300 atm.

$$N_2(g) + 3\,H_2(g) \rightleftharpoons 2\,NH_3(g).$$

The composition of an equilibrium mixture is unaffected by a change in pressure if the reaction involves no change in the number of moles of gas. For example, the reaction of hydrogen with gaseous iodine has 2 mol of gas on both sides of the balanced equation:

$$H_2(g) + I_2(g) \rightleftharpoons 2\,HI(g)$$

If we double the pressure by halving the volume, the numerator and denominator of the reaction quotient change by the same factor and Q_c remains unchanged:

$$Q_c = \frac{[HI]_t^2}{[H_2]_t[I_2]_t}$$

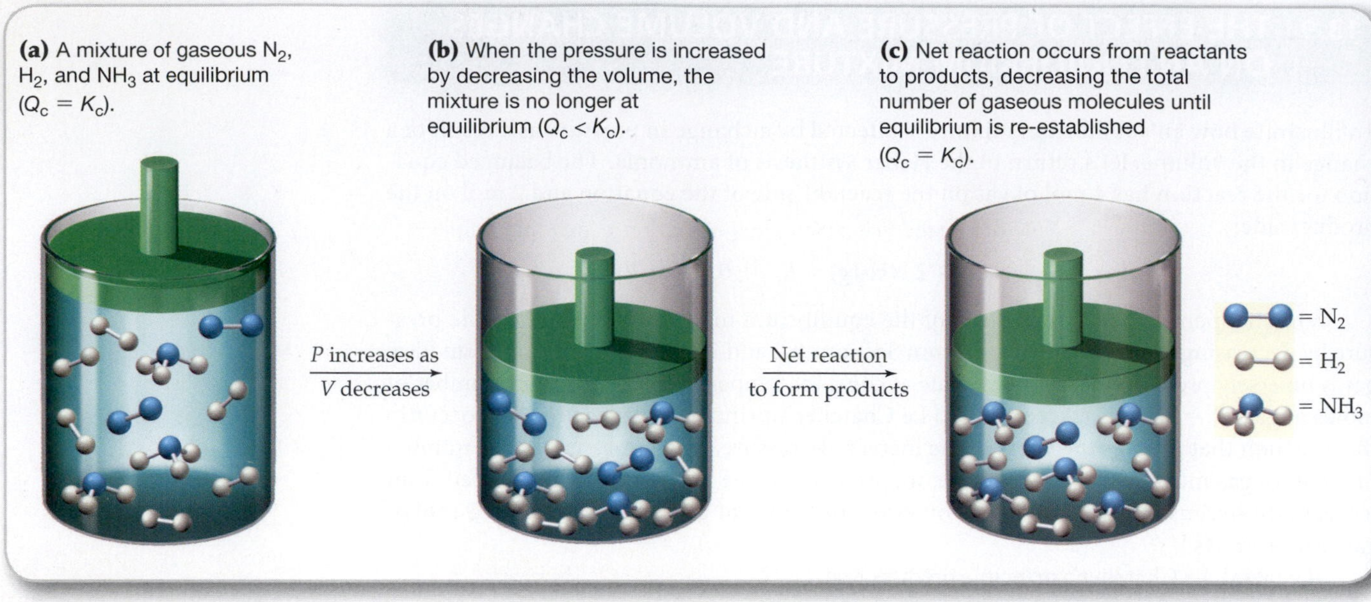

FIGURE 13.11

Qualitative effect of pressure and volume on the equilibrium $N_2(g) + 3\,H_2(g) \rightleftharpoons 2\,NH_3(g)$.

In applying Le Châtelier's principle to a heterogeneous equilibrium, the effect of pressure changes on solids and liquids can be ignored because the volume (and concentration) of a solid or a liquid is nearly independent of pressure. Take, for example, the high-temperature reaction of carbon with steam, the first step in converting coal to gaseous fuels:

$$C(s) + H_2O(g) \rightleftharpoons CO(g) + H_2(g)$$

Ignoring the carbon because it's a solid, we predict that a decrease in volume (increase in pressure) will shift the equilibrium from products to reactants because the reverse reaction decreases the amount of gas from 2 mol to 1 mol.

Throughout this section, we've been careful to limit the application of Le Châtelier's principle to pressure changes that result from a change in *volume*. What happens, though, if we keep the volume constant but increase the total pressure by adding a gas that is not involved in the reaction—say, an inert gas such as argon? In that case, the equilibrium remains undisturbed because adding an inert gas at constant volume does not change the partial pressures or the molar concentrations of the substances involved in the reaction. Only if the added gas is a reactant or product does the reaction quotient change.

Worked Example 13.14

Applying Le Châtelier's Principle to Pressure and Volume Changes

Does the number of moles of reaction products increase, decrease, or remain the same when each of the following equilibria is subjected to a decrease in pressure by increasing the volume?

(a) $PCl_5(g) \rightleftharpoons PCl_3(g) + Cl_2(g)$
(b) $CaO(s) + CO_2(g) \rightleftharpoons CaCO_3(s)$
(c) $3\,Fe(s) + 4\,H_2O(g) \rightleftharpoons Fe_3O_4(s) + 4\,H_2(g)$

STRATEGY

According to Le Châtelier's principle, the stress of a decrease in pressure is relieved by net reaction in the direction that increases the number of moles of gas.

SOLUTION

(a) Because the forward reaction converts 1 mol of gas to 2 mol of gas, net reaction will go from reactants to products, thus increasing the number of moles of PCl_3 and Cl_2.

(b) Because there is 1 mol of gas on the reactant side of the balanced equation and none on the product side, the stress of a decrease in pressure is relieved by net reaction from products to reactants. The number of moles of $CaCO_3$ therefore decreases.

(c) Because there are 4 mol of gas on both sides of the balanced equation, the composition of the equilibrium mixture is unaffected by a change in pressure. The number of moles of Fe_3O_4 and H_2 remains the same.

PROBLEM 13.18 Does the number of moles of products increase, decrease, or remain the same when each of the following equilibria is subjected to an increase in pressure by decreasing the volume?

(a) $CO(g) + H_2O(g) \rightleftharpoons CO_2(g) + H_2(g)$
(b) $2\,CO(g) \rightleftharpoons C(s) + CO_2(g)$
(c) $N_2O_4(g) \rightleftharpoons 2\,NO_2(g)$

CONCEPTUAL PROBLEM 13.19 The following picture represents the equilibrium mixture for the gas-phase reaction $A_2 \rightleftharpoons 2\,A$:

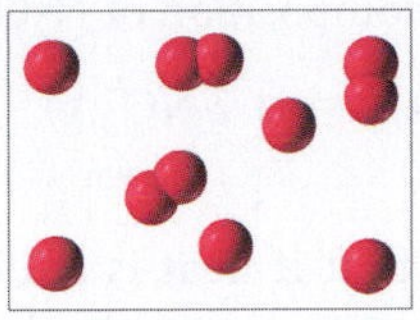

Draw a picture that shows how the concentrations change when the pressure is increased by reducing the volume.

13.10 THE EFFECT OF TEMPERATURE CHANGES ON AN EQUILIBRIUM MIXTURE

When an equilibrium is disturbed by a change in concentration, pressure, or volume, the composition of the equilibrium mixture changes because the reaction quotient Q_c no longer equals the equilibrium constant K_c. As long as the temperature remains constant, however, concentration, pressure, or volume changes don't change the value of the equilibrium constant.

What happens, though, to an equilibrium mixture when the temperature changes? A *change in temperature nearly always changes the value of the equilibrium constant.* For the Haber synthesis of ammonia, which is an exothermic reaction, the equilibrium constant K_c decreases by a factor of 10^{11} over the temperature range 300–1000 K (**Figure 13.12**).

$$N_2(g) + 3\,H_2(g) \rightleftharpoons 2\,NH_3(g) + 92.2\ \text{kJ} \qquad \Delta H^\circ = -92.2\ \text{kJ}$$

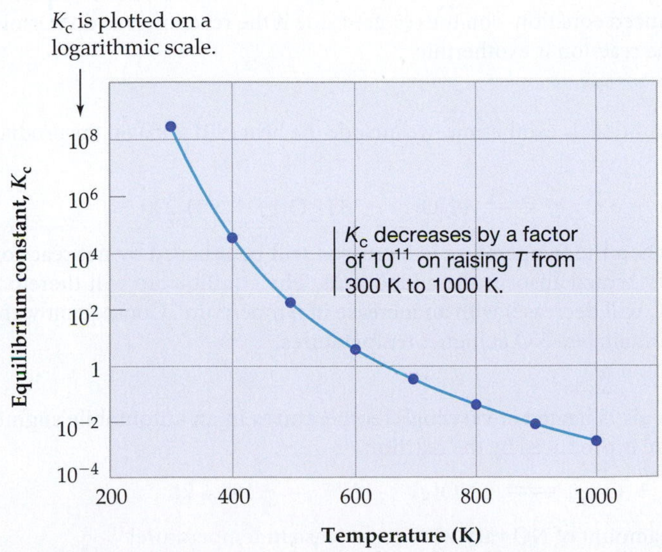

Temp (K)	K_c
300	2.6×10^8
400	3.9×10^4
500	1.7×10^2
600	4.2
700	2.9×10^{-1}
800	3.9×10^{-2}
900	8.1×10^{-3}
1000	2.3×10^{-3}

FIGURE 13.12

Temperature dependence of the equilibrium constant for the reaction $N_2(g) + 3\,H_2(g) \rightleftharpoons 2\,NH_3(g)$

At low temperatures, the equilibrium mixture is rich in NH_3 because K_c is large. At high temperatures, the equilibrium shifts in the direction of N_2 and H_2.

$N_2O_4(g) \rightleftharpoons 2\,NO_2(g);\ \Delta H° > 0$

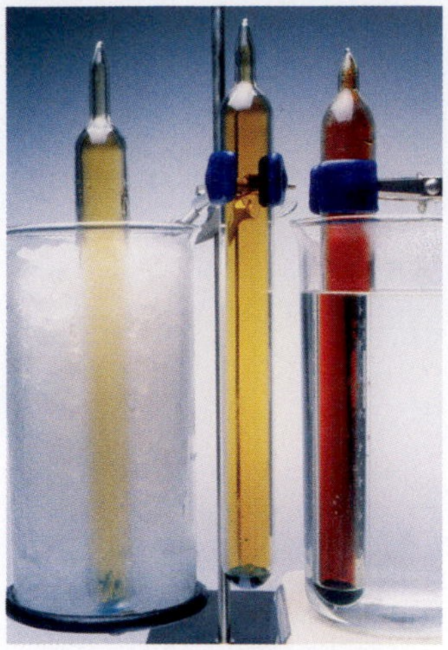

K_c increases as

T increases

The darker **brown** color of the sample at the highest temperature indicates that the equilibrium $N_2O_4(g) \rightleftharpoons 2\,NO_2(g)$ shifts from reactants to products with increasing temperature, as expected for an endothermic reaction.

FIGURE 13.13

Sample tubes containing an equilibrium mixture of N_2O_4 and NO_2 immersed in ice water (left), at room temperature (center), and immersed in hot water (right).

In general, the temperature dependence of an equilibrium constant depends on the sign of $\Delta H°$ for the reaction.

- The equilibrium constant for an exothermic reaction (negative $\Delta H°$) decreases as the temperature increases.
- The equilibrium constant for an endothermic reaction (positive $\Delta H°$) increases as the temperature increases.

You can predict the way in which K_c depends on temperature by using Le Châtelier's principle. Take the endothermic decomposition of N_2O_4, for example:

$$\underset{\text{Colorless}}{N_2O_4(g)} + 55.3\text{ kJ} \rightleftharpoons \underset{\text{Brown}}{2\,NO_2(g)} \qquad \Delta H° = +55.3\text{ kJ}$$

Le Châtelier's principle says that if heat is added to an equilibrium mixture, thus increasing its temperature, net reaction occurs in the direction that relieves the stress of the added heat. For an endothermic reaction, such as the decomposition of N_2O_4, heat is absorbed by reaction in the forward direction. The equilibrium therefore shifts to the product side at the higher temperature, which means that K_c increases with increasing temperature.

Because N_2O_4 is colorless and NO_2 has a brown color, the effect of temperature on the N_2O_4–NO_2 equilibrium is readily apparent from the color of the mixture (**Figure 13.13**). For an exothermic reaction, such as the Haber synthesis of NH_3, heat is absorbed by net reaction in the reverse direction, so K_c decreases with increasing temperature.

Worked Example 13.15

Applying Le Châtelier's Principle to Temperature Changes

In the first step of the Ostwald process for the synthesis of nitric acid, ammonia is oxidized to nitric oxide by the reaction

$$4\,NH_3(g) + 5\,O_2(g) \rightleftharpoons 4\,NO(g) + 6\,H_2O(g) \qquad \Delta H° = -901.2\text{ kJ}$$

How does the equilibrium amount of NO vary with an increase in temperature?

STRATEGY

Le Châtelier's principle predicts that the stress of added heat when the temperature is increased will be relieved by net reaction in the direction that absorbs the heat. It's helpful to include the heat in the balanced equation—on the reactant side if the reaction is endothermic, or on the product side if the reaction is exothermic.

SOLUTION

Because the oxidation of ammonia is exothermic, we include the heat (901.2 kJ) on the product side:

$$4\,NH_3(g) + 5\,O_2(g) \rightleftharpoons 4\,NO(g) + 6\,H_2O(g) + 901.2\text{ kJ}$$

The stress of added heat when the temperature is increased will be relieved by net reaction from products to reactants, which absorbs the added heat. The equilibrium will therefore shift to the reactant side (K_c will decrease) with an increase in temperature. Consequently, the equilibrium mixture will contain less NO at higher temperatures.

PROBLEM 13.20 When air is heated at very high temperatures in an automobile engine, the air pollutant nitric oxide is produced by the reaction

$$N_2(g) + O_2(g) \rightleftharpoons 2\,NO(g) \qquad \Delta H° = +182.6\text{ kJ}$$

How does the equilibrium amount of NO vary with an increase in temperature?

PROBLEM 13.21 Ethyl acetate, a solvent used in many fingernail-polish removers, is made by the reaction of acetic acid with ethanol:

$$\underset{\text{Acetic acid}}{CH_3CO_2H(soln)} + \underset{\text{Ethanol}}{CH_3CH_2OH(soln)} \rightleftharpoons \underset{\text{Ethyl acetate}}{CH_3CO_2CH_2CH_3(soln)} + H_2O(soln) \quad \Delta H° = -2.9\text{ kJ}$$

Does the amount of ethyl acetate in an equilibrium mixture increase or decrease when the temperature is increased? How does K_c change when the temperature is decreased? Justify your answers using Le Châtelier's principle.

▲ Fingernail polish can be removed by dissolving it in ethyl acetate.

WORKED CONCEPTUAL EXAMPLE 13.16

Applying Le Châtelier's Principle to Temperature Changes

The following pictures represent the composition of the equilibrium mixture at 600 K and 650 K for the combination of two A molecules, $2\text{ A}(g) \rightleftharpoons A_2(g)$:

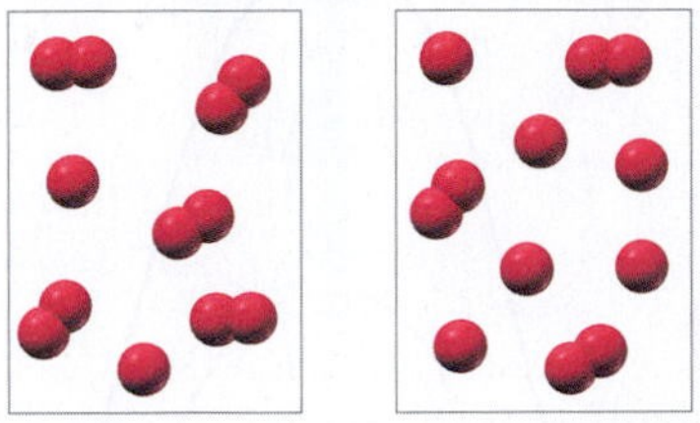

$T = 600$ K $\qquad$ $T = 650$ K

Is the reaction endothermic or exothermic? Explain using Le Châtelier's principle.

STRATEGY

We can determine the direction of net reaction on raising the temperature by counting the number of A and A_2 molecules at each temperature. According to Le Châtelier's principle, if the net reaction converts reactants to products on raising the temperature, heat is on the reactant side of the chemical equation and the reaction is endothermic. Conversely, if the net reaction converts products to reactants on raising the temperature, heat is on the product side and the reaction is exothermic.

SOLUTION

Two A and five A_2 molecules are present at 600 K, and six A and three A_2 molecules are present at 650 K. On raising the temperature, the net reaction converts products to reactants so heat is on the product side of the chemical equation:

$$2\text{ A}(g) \rightleftharpoons A_2(g) + \text{heat}$$

The reaction is therefore exothermic, as expected for a reaction in which a chemical bond is formed. Note that Le Châtelier's principle predicts that net reaction will occur in the direction that uses up the added heat.

CONCEPTUAL PROBLEM 13.22 The following pictures represent the composition of the equilibrium mixture for the reaction $A(g) + B(s) \rightleftharpoons AB(g)$ at 400 K and 500 K:

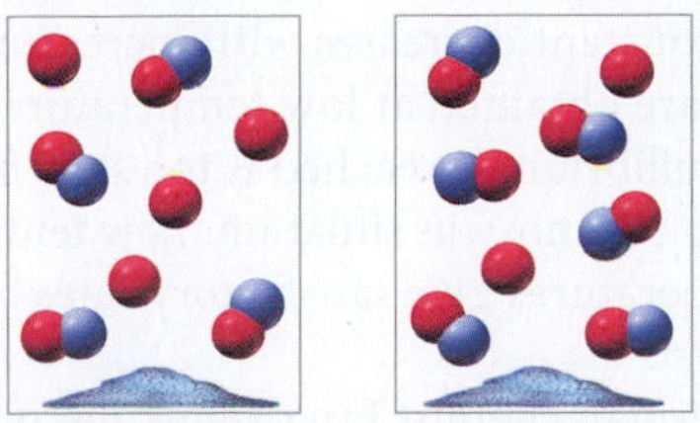

$T = 400$ K $\qquad$ $T = 500$ K

Is the reaction endothermic or exothermic? Explain using Le Châtelier's principle.

13.11 THE EFFECT OF A CATALYST ON EQUILIBRIUM

REMEMBER...

A **catalyst** is a substance that increases the rate of a transformation without being consumed in the process. (Section 12.14)

Recall from Section 12.14 that a **catalyst** increases the rate of a chemical reaction by making available a new, lower-energy pathway for the conversion of reactants to products. Because the forward and reverse reactions pass through the same transition state, a catalyst lowers the activation energy for the forward and reverse reactions by exactly the same amount (**Figure 13.14**). As a result, the rates of the forward and reverse reactions increase by the same factor and $K_c = k_f/k_r$ is unchanged.

FIGURE 13.14

Potential energy profiles for a reaction whose activation energy is lowered by the presence of a catalyst.

The activation energy for the **catalyzed pathway** (red curve) is lower than that for the **uncatalyzed pathway** (blue curve) by an amount ΔE_a.

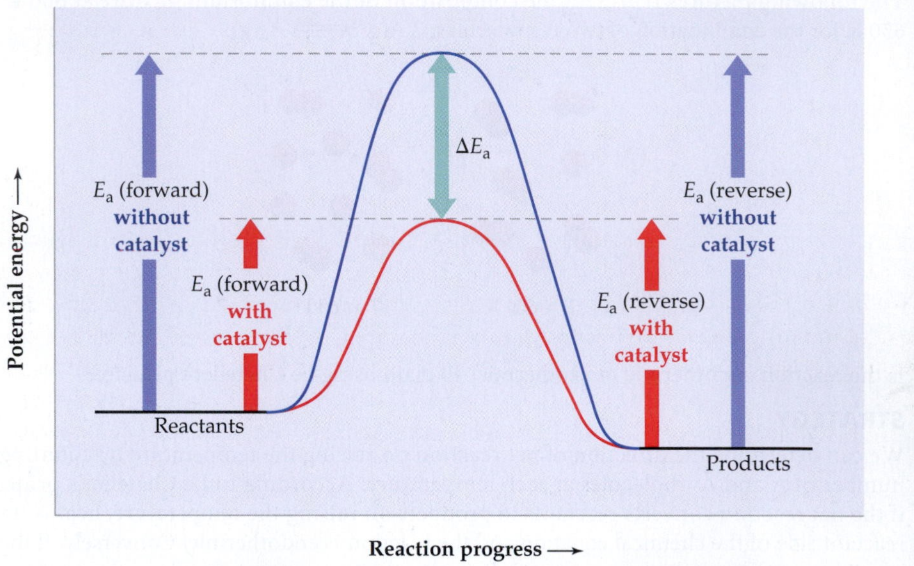

A catalyst lowers the activation energy barrier for the forward and reverse reactions by the same amount. The catalyst therefore accelerates the forward and reverse reactions by the same factor, and the composition of the equilibrium mixture is unchanged.

If a reaction mixture is at equilibrium in the absence of a catalyst (that is, the forward and reverse rates are equal), it will still be at equilibrium after a catalyst is added because the forward and reverse rates, though faster, remain equal. If a reaction mixture is not at equilibrium, a catalyst accelerates the rate at which equilibrium is reached, but it does not affect the composition of the equilibrium mixture. Because a catalyst has no effect on the equilibrium concentrations, it does not appear in the balanced chemical equation or in the equilibrium-constant expression.

Even though a catalyst doesn't change the position of an equilibrium, it can nevertheless significantly influence the choice of optimum conditions for a reaction. Look again at the Haber synthesis of ammonia. Because the reaction $N_2(g) + 3\,H_2(g) \rightleftharpoons 2\,NH_3(g)$ is exothermic, its equilibrium constant decreases with increasing temperature (Figure 13.12) and optimum yields of NH_3 are obtained at low temperatures. At those low temperatures, however, the rate at which equilibrium is reached is too slow for the reaction to be practical. We thus have what appears to be a no-win situation: Low temperatures give good yields but slow rates, whereas high temperatures give satisfactory rates but poor yields. The answer to the dilemma is to find a catalyst.

In the early 1900s, the German chemist Fritz Haber discovered that a catalyst consisting of iron mixed with certain metal oxides causes the reaction to occur at a satisfactory rate at temperatures where the equilibrium concentration of NH_3 is reasonably favorable. The yield of NH_3 can be improved further by running the reaction at high pressures. Typical reaction conditions for the industrial synthesis of ammonia are 400–500 °C and 130–300 atm.

PROBLEM 13.23 A platinum catalyst is used in automobile catalytic converters to hasten the oxidation of carbon monoxide:

$$2\,CO(g) + O_2(g) \xrightleftharpoons{Pt} 2\,CO_2(g) \qquad \Delta H° = -566\text{ kJ}$$

Suppose that you have a reaction vessel containing an equilibrium mixture of $CO(g)$, $O_2(g)$, and $CO_2(g)$. Will the amount of CO in the equilibrium mixture increase, decrease, or remain the same after each of the following changes?
(a) A platinum catalyst is added.
(b) The temperature is increased.
(c) The pressure is increased by decreasing the volume.
(d) The pressure is increased by adding argon gas.
(e) The pressure is increased by adding O_2 gas.

FYI HOW DOES EQUILIBRIUM AFFECT OXYGEN TRANSPORT IN THE BLOODSTREAM?

Humans, like all animals, need oxygen. The oxygen comes from breathing: About 500 mL of air is drawn into the lungs of an average person with each breath. When the freshly inspired air travels through the bronchial passages and enters the approximately 150 million alveolar sacs of the lungs, it picks up moisture and mixes with air remaining from the previous breath. As it mixes, the concentrations of both water vapor and carbon dioxide increase. These gas concentrations are measured by their partial pressures, with the partial pressure of oxygen in the lungs usually around 100 mm Hg (**Table 13.2**). Oxygen then diffuses through the delicate walls of the lung alveoli and into arterial blood, which transports it to all body tissues.

Only about 3% of the oxygen in blood is dissolved; the rest is chemically bound to *hemoglobin* molecules (Hb), large proteins that contain *heme* groups embedded in them. Each hemoglobin molecule contains four heme groups, and each heme group contains an iron atom that can bind to one O_2 molecule. Thus, a single hemoglobin molecule can bind four molecules of oxygen.

TABLE 13.2 Partial Pressure of Oxygen in Human Lungs and Blood at Sea Level

Source	P_{O_2} (mm Hg)
Dry air	159
Alveolar air	100
Arterial blood	95
Venous blood	40

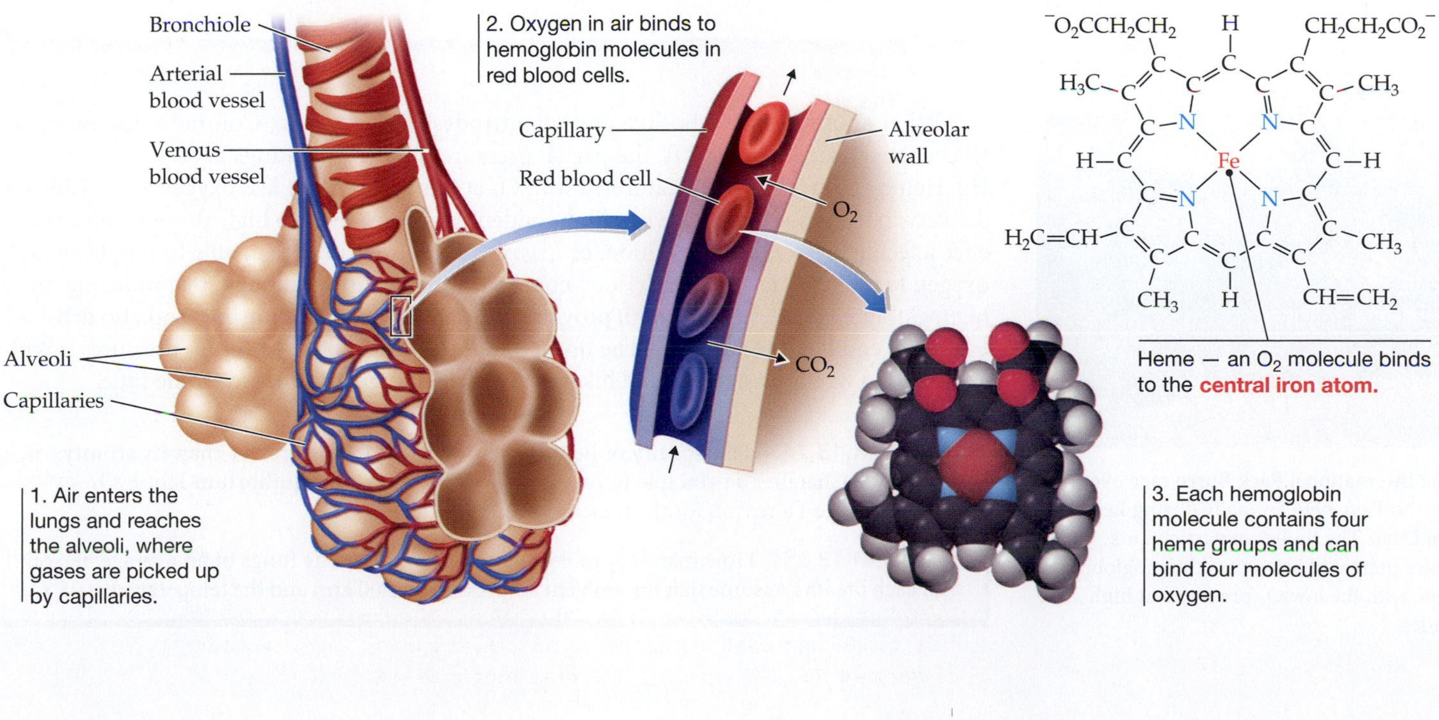

The entire system of oxygen transport and delivery in the body depends on the pickup and release of O_2 by hemoglobin according to the following series of equilibria:

$$Hb + O_2 \rightleftharpoons Hb(O_2)$$
$$Hb(O_2) + O_2 \rightleftharpoons Hb(O_2)_2$$
$$Hb(O_2)_2 + O_2 \rightleftharpoons Hb(O_2)_3$$
$$Hb(O_2)_3 + O_2 \rightleftharpoons Hb(O_2)_4$$

The positions of the different equilibria depend on the partial pressures of O_2 (P_{O_2}) in the various tissues. In hard-working, oxygen-starved muscles, where P_{O_2} is low, oxygen is released from hemoglobin as the equilibria shift toward the left, according to Le Châtelier's principle. In the lungs, where P_{O_2} is high, oxygen is absorbed by hemoglobin as the equilibria shift toward the right.

The amount of oxygen carried by hemoglobin at any given value of P_{O_2} is usually expressed as a percent saturation and can be found from the curve shown in **Figure 13.15**. The saturation is 97.5% in the lungs, where P_{O_2} = 100 mm Hg, meaning that each hemoglobin molecule is carrying close to its maximum possible amount of 4 O_2 molecules. When P_{O_2} = 26 mm Hg, however, the saturation drops to 50%.

FIGURE 13.15

An oxygen-carrying curve for hemoglobin. The percent saturation of the oxygen-binding sites on hemoglobin depends on the partial pressure of oxygen (P_{O_2}).

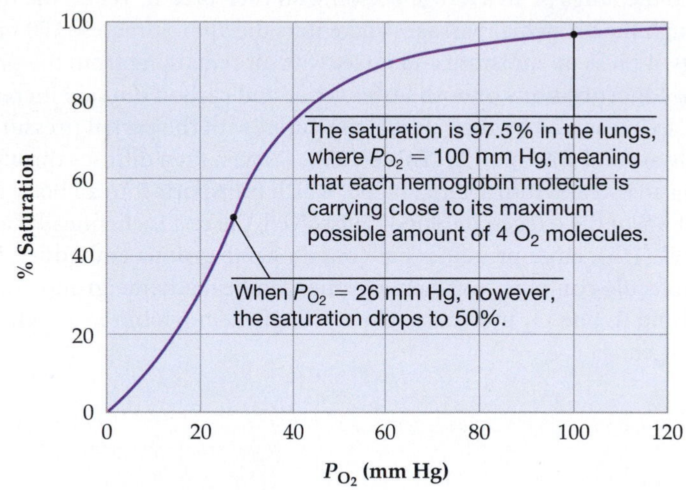

▲ The International Pack Burro race over Mosquito Pass, held annually during Leadville Boom Days. The bodies of participants produce increased amounts of hemoglobin to cope with the low O_2 pressures at high altitudes.

What about people who live at high altitudes? In Leadville, Colorado, for example, where the altitude is 10,156 ft, the partial pressure of O_2 in the lungs is only about 68 mm Hg. Hemoglobin is only 90% saturated with O_2 at this pressure, so less oxygen is available for delivery to the tissues. People who climb suddenly from sea level to high altitude thus experience a feeling of oxygen deprivation, or *hypoxia*, as their bodies are unable to supply enough oxygen to their tissues. The body soon copes with the situation, though, by producing more hemoglobin molecules, which both provide more capacity for O_2 transport and also drive the Hb + O_2 equilibria to the right. The time required to adapt to the lower O_2 pressures is typically days to weeks, so athletes and hikers must train at high altitudes for some time.

PROBLEM 13.24 The affinity of hemoglobin (Hb) for CO is greater than its affinity for O_2. Use Le Châtelier's principle to predict how CO affects the equilibrium $Hb + O_2 \rightleftharpoons Hb(O_2)$. Suggest a reason for the toxicity of CO.

PROBLEM 13.25 How many O_2 molecules are drawn into the lungs of an average person with each breath? Assume that the ambient air pressure is 1.00 atm and the temperature is 25°C.

SUMMARY

Chemical equilibrium is a dynamic state in which the concentrations of reactants and products remain constant because the rates of the forward and reverse reactions are equal. For the general reaction $a\,A + b\,B \rightleftharpoons c\,C + d\,D$, concentrations in the **equilibrium mixture** are related by the **equilibrium equation**:

$$K_c = \frac{[C]^c[D]^d}{[A]^a[B]^b}$$

The quotient on the right side of the equation is called the *equilibrium-constant expression*. The **equilibrium constant K_c** is the number obtained when equilibrium concentrations (in mol/L) are substituted into the equilibrium-constant expression. The value of K_c varies with temperature and depends on the form of the balanced chemical equation. The equilibrium constant for a single-step reaction equals the ratio of the rate constants for the forward and reverse reactions: $K_c = k_f/k_r$.

The **equilibrium constant K_p** can be used for gas-phase reactions. It is defined in the same way as K_c except that the equilibrium-constant expression contains partial pressures (in atmospheres) instead of molar concentrations. The constants K_p and K_c are related by the equation $K_p = K_c(RT)^{\Delta n}$, where $\Delta n = (c + d) - (a + b)$ is the number of moles of gaseous products minus the number of moles of gaseous reactants.

Homogeneous equilibria are those in which all reactants and products are in a single phase; **heterogeneous equilibria** are those in which reactants and products are present in more than one phase. The equilibrium equation for a heterogeneous equilibrium does not include concentrations of pure solids or pure liquids.

The value of the equilibrium constant for a reaction makes it possible to judge the extent of reaction, predict the direction of reaction, and calculate equilibrium concentrations (or partial pressures) from initial concentrations (or partial pressures). The farther the reaction proceeds toward completion, the larger the value of K_c. The direction of a reaction not at equilibrium depends on the relative values of K_c and the **reaction quotient Q_c**, which is defined in the same way as K_c except that the concentrations in the equilibrium-constant expression are not necessarily equilibrium concentrations. If $Q_c < K_c$, the net reaction goes from reactants to products to attain equilibrium; if $Q_c > K_c$, the net reaction goes from products to reactants; if $Q_c = K_c$, the system is at equilibrium.

The composition of an equilibrium mixture can be altered by changes in concentration, pressure (volume), or temperature. The qualitative effect of these changes is predicted by **Le Châtelier's principle**, which says that if a stress is applied to a reaction mixture at equilibrium, net reaction occurs in the direction that relieves the stress. Temperature changes affect equilibrium concentrations because K_c is temperature-dependent. As the temperature increases, K_c for an exothermic reaction decreases and K_c for an endothermic reaction increases. A catalyst increases the rate at which chemical equilibrium is reached, but it does not affect the equilibrium constant or the equilibrium concentrations.

KEY WORDS

chemical equilibrium *504*
equilibrium constant K_c *508*
equilibrium constant K_p *513*
equilibrium equation *508*
equilibrium mixture *504*
heterogeneous equilibria *516*
homogeneous equilibria *516*
Le Châtelier's principle *526*
reaction quotient Q_c *519*

KEY EQUATIONS

- Equilibrium equation for the reaction $a\,A + b\,B \rightleftharpoons c\,C + d\,D$ in terms of molar concentrations **(Section 13.2)**

$$K_c = \frac{[C]^c[D]^d}{[A]^a[B]^b}$$

- Relationship between the equilibrium constant K_c and the rate constants for the forward and reverse reactions **(Section 13.3)**

$$K_c = k_f/k_r$$

- Equilibrium equation for the gas phase reaction $a\,A + b\,B \rightleftharpoons c\,C + d\,D$ in terms of partial pressures **(Section 13.4)**

$$K_p = \frac{(P_C)^c(P_D)^d}{(P_A)^a(P_B)^b}$$

- Relationship between the equilibrium constants K_p and K_c **(Section 13.4)**

$$K_p = K_c(RT)^{\Delta n}$$

where $R = 0.082\,06\ (\text{L}\cdot\text{atm})/(\text{K}\cdot\text{mol})$ and $\Delta n = (c + d) - (a + b)$ is the number of moles of gaseous products minus the number of moles of gaseous reactants

- Reaction quotient Q_c for the reaction $a\,A + b\,B \rightleftharpoons c\,C + d\,D$ **(Section 13.6)**

$$Q_c = \frac{[C]_t^{\,c}[D]_t^{\,d}}{[A]_t^{\,a}[B]_t^{\,b}}$$

where the subscript t denotes concentrations at an arbitrary time t, not necessarily at equilibrium

LEARNING OUTCOMES

- Write the equilibrium-constant expression for gas-phase and solution reactions, and calculate the value of the equilibrium constant K_c. **(Section 13.2)** ***Problems 1, 2, 3, 38, 39, 44, 46, 122***
- Relate the equilibrium constant K_c to the rate constants for the forward and reverse reactions. **(Section 13.3)** ***Problems 5, 35, 58–63, 109***
- Write the equilibrium-constant expression for gas-phase reactions in terms of partial pressures, calculate the value of the equilibrium constant K_p, and relate K_p to K_c. **(Section 13.4)** ***Problems 6–8, 40–43, 45, 47, 52–57, 104, 106, 108, 118, 121, 122***
- Write the equilibrium equation for a heterogeneous equilibrium. **(Section 13.5)** ***Problems 8, 54–57, 100, 107, 115***
- Use the equilibrium constant to judge the extent of reaction. **(Section 13.6)** ***Problems 9, 66–69***
- Compare the equilibrium constant and the reaction quotient Q_c to predict the direction of reaction. **(Section 13.6)** ***Problems 10, 11, 28, 70–72, 74, 75, 104***
- Calculate equilibrium concentrations from K_c and initial concentrations. **(Section 13.6)** ***Problems 13–15, 75, 78–80, 82, 83, 98–100, 104, 105, 108, 115, 119, 122, 123, 126***
- Calculate equilibrium partial pressures from K_p and initial partial pressures. **(Section 13.6)** ***Problems 16, 74, 84, 85, 106, 108, 111, 118, 121***
- Apply Le Châtelier's principle to predict the effect on an equilibrium mixture of changes in concentration, pressure, volume, and temperature. **(Sections 13.7–13.10)** ***Problems 17–23, 29, 30, 32, 34, 36, 37, 86–97, 101–103, 106, 115***
- Describe the effect of a catalyst on an equilibrium mixture. **(Section 13.11)** ***Problems 23, 36, 93–95, 97, 101, 102, 109***

CONCEPTUAL PROBLEMS

Problems 13.1–13.25 appear within the chapter.

13.26 Consider the interconversion of A molecules (red spheres) and B molecules (blue spheres) according to the reaction $A \rightleftharpoons B$. Each of the following series of pictures represents a separate experiment in which time increases from left to right:

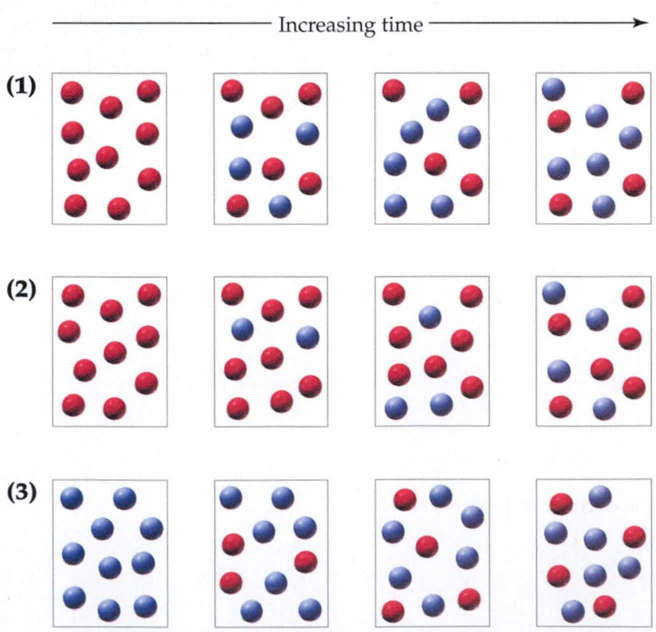

(a) Which of the experiments has resulted in an equilibrium state?

(b) What is the value of the equilibrium constant K_c for the reaction $A \rightleftharpoons B$?

(c) Explain why you can calculate K_c without knowing the volume of the reaction vessel.

13.27 The following pictures represent the equilibrium state for three different reactions of the type $A_2 + X_2 \rightleftharpoons 2\,AX$ (X = B, C, or D):

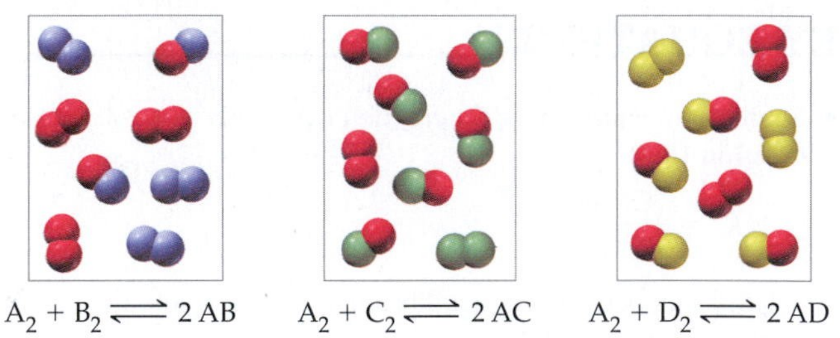

(a) Which reaction has the largest equilibrium constant?

(b) Which reaction has the smallest equilibrium constant?

13.28 The reaction $A_2 + B \rightleftharpoons A + AB$ has an equilibrium constant $K_c = 2$. The following pictures represent reaction mixtures that contain A atoms (red), B atoms (blue), and A_2 and AB molecules:

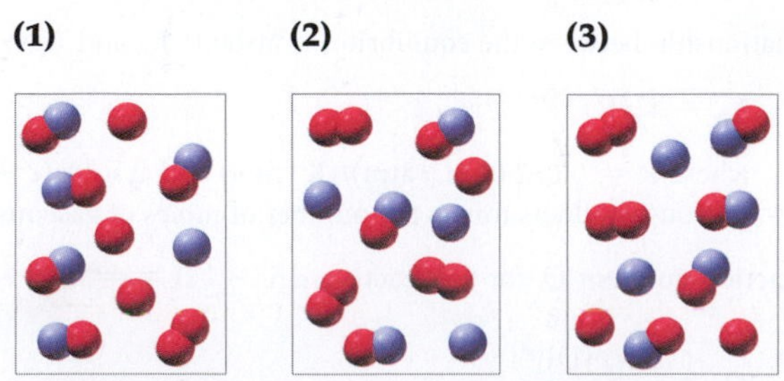

(a) Which reaction mixture is at equilibrium?

(b) For those mixtures that are not at equilibrium, will the reaction go in the forward or reverse direction to reach equilibrium? Explain.

13.29 The following pictures represent the initial state and the equilibrium state for the reaction of A_2 molecules (red) with B atoms (blue) to give AB molecules:

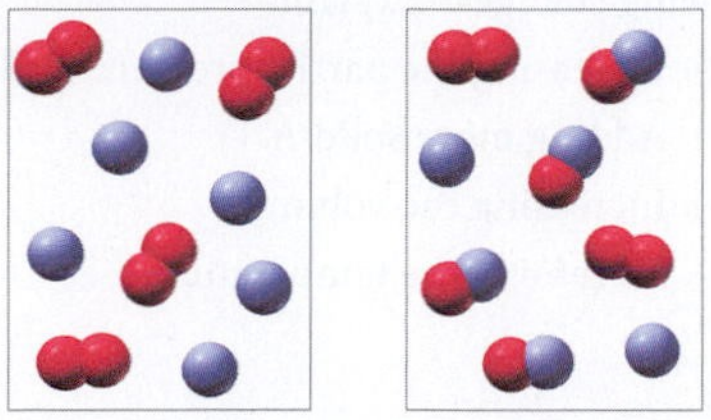

Initial state   Equilibrium state

(a) Write a balanced chemical equation for the reaction.

(b) If the volume of the equilibrium mixture is decreased, will the number of AB molecules increase, decrease, or remain the same? Explain.

13.30 Consider the reaction $A + B \rightleftharpoons AB$. The vessel on the right contains an equilibrium mixture of A molecules (red spheres), B molecules (blue spheres), and AB molecules. If the stopcock is opened and the contents of the two vessels are allowed to mix, will the reaction go in the forward or reverse direction? Explain.

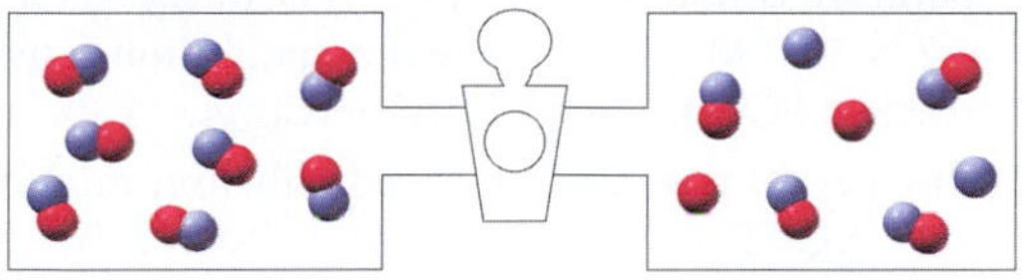

13.31 The following pictures represent the composition of the equilibrium mixture for the reaction $A + B \rightleftharpoons AB$ at 300 K and at 400 K:

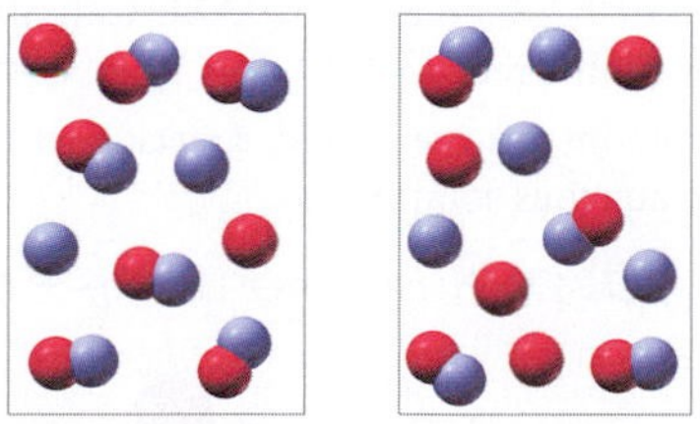

$T = 300$ K $T = 400$ K

Is the reaction exothermic or endothermic? Explain using Le Châtelier's principle.

13.32 The following pictures represent equilibrium mixtures at 325 K and 350 K for a reaction involving A atoms (red), B atoms (blue), and AB molecules:

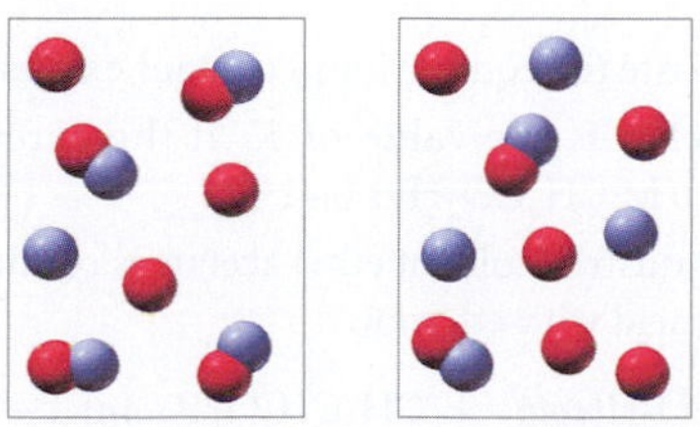

$T = 325$ K $T = 350$ K

(a) Write a balanced equation for the reaction that occurs on raising the temperature.

(b) Is the reaction exothermic or endothermic? Explain using Le Châtelier's principle.

(c) If the volume of the container is increased, will the number of A atoms increase, decrease, or remain the same? Explain.

13.33 The following picture represents an equilibrium mixture of solid $BaCO_3$, solid BaO, and gaseous CO_2 obtained as a result of the endothermic decomposition of $BaCO_3$:

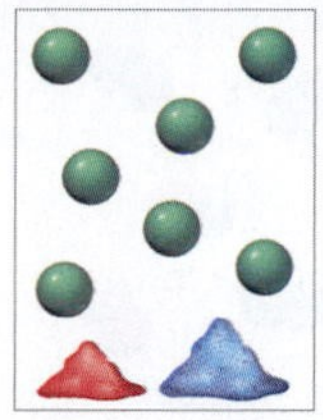

(a) Draw a picture that represents the equilibrium mixture after addition of four more CO_2 molecules.

(b) Draw a picture that represents the equilibrium mixture at a higher temperature.

13.34 The following picture represents the composition of the equilibrium mixture for the endothermic reaction $A_2 \rightleftharpoons 2\,A$ at 500 K:

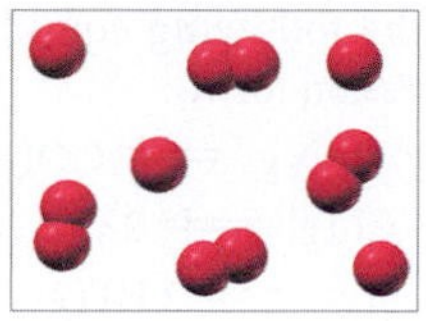

Draw a picture that represents the equilibrium mixture after each of the following changes:

(a) Adding a catalyst

(b) Increasing the volume

(c) Decreasing the temperature

13.35 The following picture represents the equilibrium state for the reaction $2\,AB \rightleftharpoons A_2 + B_2$:

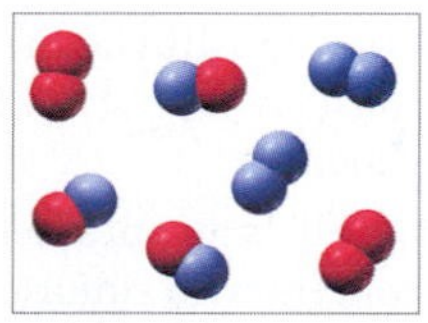

Which rate constant is larger, k_f or k_r? Explain.

13.36 The following pictures represent the initial and equilibrium states for the exothermic decomposition of gaseous A molecules (red) to give gaseous B molecules (blue):

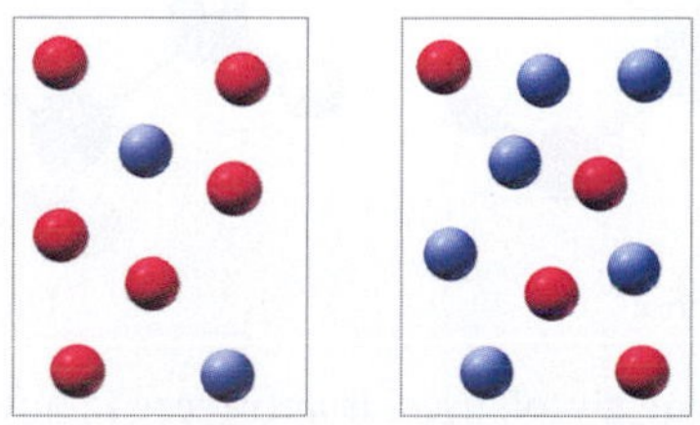

Initial state   Equilibrium state

(a) Write a balanced chemical equation for the reaction.

(b) Will the number of A molecules in the equilibrium mixture increase, decrease, or remain the same after each of the following changes? Explain.

(1) Increasing the temperature

(2) Decreasing the volume

(3) Increasing the pressure by adding an inert gas

(4) Adding a catalyst

13.37 The following pictures represent the initial and equilibrium states for the exothermic reaction of solid A (red) with gaseous B_2 (blue) to give gaseous AB:

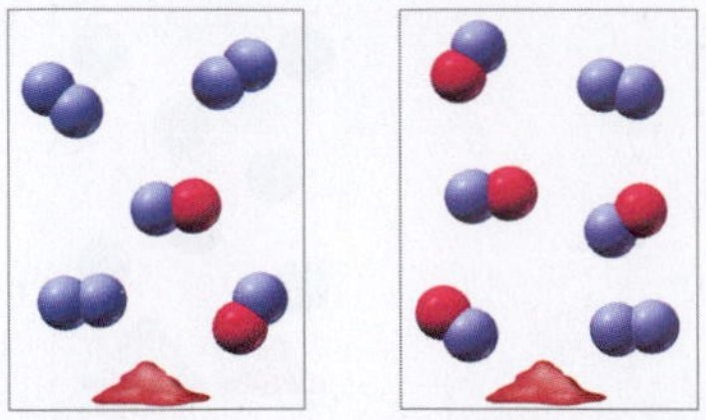

Initial state   Equilibrium state

(a) Write a balanced chemical equation for the reaction.

(b) Will the number of AB molecules in the equilibrium mixture increase, decrease, or remain the same after each of the following changes? Explain.

(1) Increasing the partial pressure of B_2

(2) Adding more solid A

(3) Increasing the volume

(4) Increasing the temperature

SECTION PROBLEMS

Equilibrium-Constant Expressions and Equilibrium Constants (Sections 13.1–13.2, 13.4–13.5)

13.38 For each of the following equilibria, write the equilibrium-constant expression for K_c:

(a) $CH_4(g) + H_2O(g) \rightleftharpoons CO(g) + 3\,H_2(g)$

(b) $3\,F_2(g) + Cl_2(g) \rightleftharpoons 2\,ClF_3(g)$

(c) $H_2(g) + F_2(g) \rightleftharpoons 2\,HF(g)$

13.39 For each of the following equilibria, write the equilibrium-constant expression for K_c:

(a) $2\,C_2H_4(g) + O_2(g) \rightleftharpoons 2\,CH_3CHO(g)$

(b) $2\,NO(g) \rightleftharpoons N_2(g) + O_2(g)$

(c) $4\,NH_3(g) + 5\,O_2(g) \rightleftharpoons 4\,NO(g) + 6\,H_2O(g)$

13.40 For each of the equilibria in Problem 13.38, write the equilibrium-constant expression for K_p and give the equation that relates K_p and K_c.

13.41 For each of the equilibria in Problem 13.39, write the equilibrium-constant expression for K_p and give the equation that relates K_p and K_c.

13.42 Diethyl ether, used as an anesthetic, is synthesized by heating ethanol with concentrated sulfuric acid. Write the equilibrium-constant expression for K_c.

$$2\,C_2H_5OH(soln) \rightleftharpoons C_2H_5OC_2H_5(soln) + H_2O(soln)$$

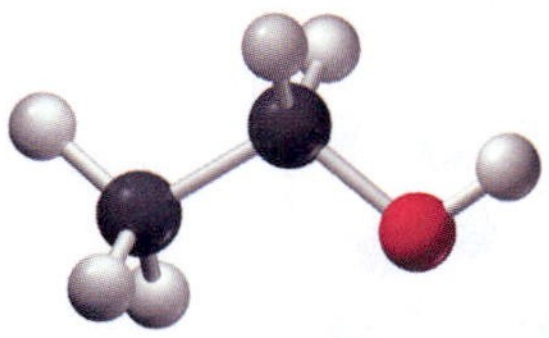

Ethanol

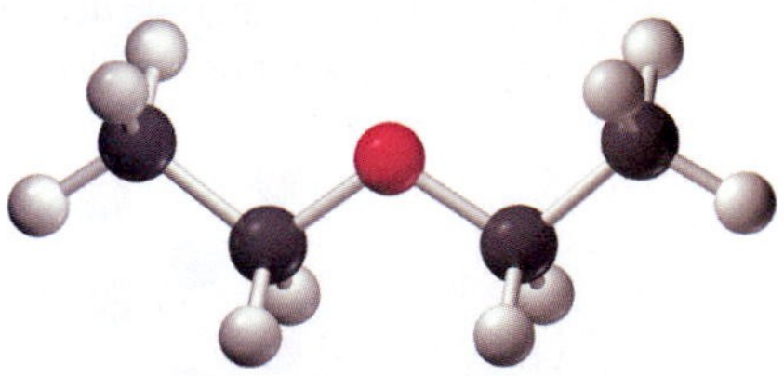

Diethyl ether

13.43 Ethylene glycol, used as antifreeze in automobile radiators, is manufactured by the hydration of ethylene oxide. Write the equilibrium-constant expression for K_c.

$$\overset{\large O}{\overset{/\ \backslash}{H_2C{-}CH_2}}(soln) + H_2O(soln) \rightleftharpoons HOCH_2CH_2OH(soln)$$

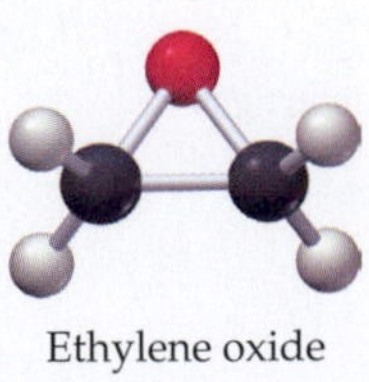

Ethylene oxide

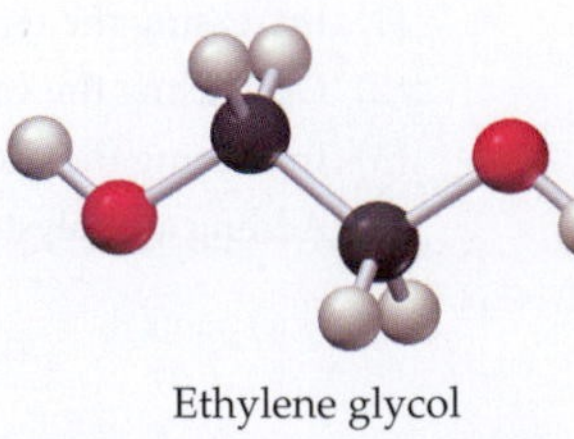

Ethylene glycol

13.44 If $K_c = 7.5 \times 10^{-9}$ at 1000 K for the reaction $N_2(g) + O_2(g) \rightleftharpoons 2\,NO(g)$, what is K_c at 1000 K for the reaction $2\,NO(g) \rightleftharpoons N_2(g) + O_2(g)$?

13.45 At 400 K, $K_p = 50.2$ for the reaction $N_2O_4(g) \rightleftharpoons 2\,NO_2(g)$. What is K_p at 400 K for the reaction $2\,NO_2(g) \rightleftharpoons N_2O_4(g)$?

13.46 An equilibrium mixture of PCl_5, PCl_3, and Cl_2 at a certain temperature contains 8.3×10^{-3} M PCl_5, 1.5×10^{-2} M PCl_3, and 3.2×10^{-2} M Cl_2. Calculate the equilibrium constant K_c for the reaction $PCl_5(g) \rightleftharpoons PCl_3(g) + Cl_2(g)$.

13.47 The partial pressures in an equilibrium mixture of NO, Cl_2, and NOCl at 500 K are as follows: $P_{NO} = 0.240$ atm; $P_{Cl_2} = 0.608$ atm; $P_{NOCl} = 1.35$ atm. What is K_p at 500 K for the reaction $2\,NO(g) + Cl_2(g) \rightleftharpoons 2\,NOCl(g)$?

13.48 A sample of HI (9.30×10^{-3} mol) was placed in an empty 2.00 L container at 1000 K. After equilibrium was reached, the concentration of I_2 was 6.29×10^{-4} M. Calculate the value of K_c at 1000 K for the reaction $H_2(g) + I_2(g) \rightleftharpoons 2\,HI(g)$.

13.49 Vinegar contains acetic acid, a weak acid that is partially dissociated in aqueous solution:

$$CH_3CO_2H(aq) \rightleftharpoons H^+(aq) + CH_3CO_2^-(aq)$$

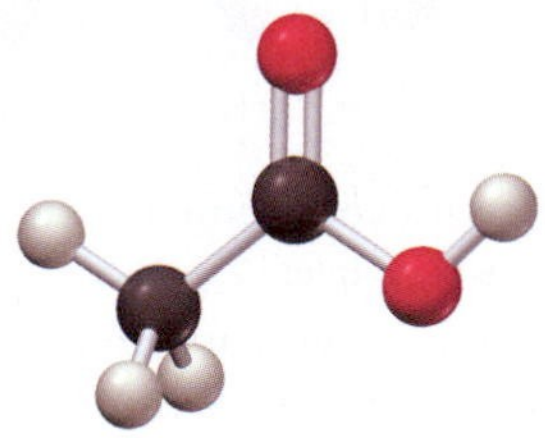

Acetic acid

(a) Write the equilibrium-constant expression for K_c.

(b) What is the value of K_c if the extent of dissociation in 1.0 M CH_3CO_2H is 0.42%?

13.50 The industrial solvent ethyl acetate is produced by the reaction of acetic acid with ethanol:

$$\underset{\text{Acetic acid}}{CH_3CO_2H(soln)} + \underset{\text{Ethanol}}{CH_3CH_2OH(soln)} \rightleftharpoons \underset{\text{Ethyl acetate}}{CH_3CO_2CH_2CH_3(soln)} + H_2O(soln)$$

(a) Write the equilibrium-constant expression for K_c.

(b) A solution prepared by mixing 1.00 mol of acetic acid and 1.00 mol of ethanol contains 0.65 mol of ethyl acetate at equilibrium. Calculate the value of K_c. Explain why you can calculate K_c without knowing the volume of the solution.

13.51 A characteristic reaction of ethyl acetate is hydrolysis, the reverse of the reaction in Problem 13.50. Write the equilibrium equation for the hydrolysis of ethyl acetate, and use the data in Problem 13.50 to calculate K_c for the hydrolysis reaction.

13.52 At 298 K, K_c is 2.2×10^5 for the reaction $F(g) + O_2(g) \rightleftharpoons O_2F(g)$. What is the value of K_p at this temperature?

13.53 At 298 K, K_p is 1.6×10^{-6} for the reaction $2\,NOCl(g) \rightleftharpoons 2\,NO(g) + Cl_2(g)$. What is the value of K_c at this temperature?

13.54 The vapor pressure of water at 25 °C is 0.0313 atm. Calculate the values of K_p and K_c at 25 °C for the equilibrium $H_2O(l) \rightleftharpoons H_2O(g)$.

13.55 Naphthalene, a white solid used to make mothballs, has a vapor pressure of 0.10 mm Hg at 27 °C. Calculate the values of K_p and K_c at 27 °C for the equilibrium $C_{10}H_8(s) \rightleftharpoons C_{10}H_8(g)$.

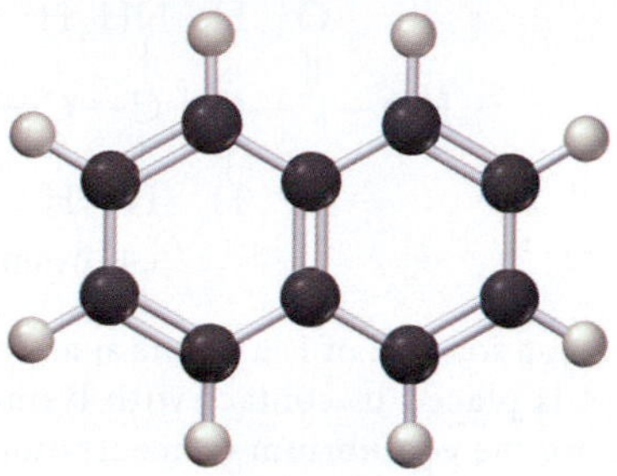

Naphthalene

13.56 For each of the following equilibria, write the equilibrium-constant expression for K_c. Where appropriate, also write the equilibrium-constant expression for K_p.

(a) $Fe_2O_3(s) + 3\,CO(g) \rightleftharpoons 2\,Fe(l) + 3\,CO_2(g)$
(b) $4\,Fe(s) + 3\,O_2(g) \rightleftharpoons 2\,Fe_2O_3(s)$
(c) $BaSO_4(s) \rightleftharpoons BaO(s) + SO_3(g)$
(d) $BaSO_4(s) \rightleftharpoons Ba^{2+}(aq) + SO_4^{2-}(aq)$

13.57 For each of the following equilibria, write the equilibrium-constant expression for K_c. Where appropriate, also write the equilibrium-constant expression for K_p.

(a) $WO_3(s) + 3\,H_2(g) \rightleftharpoons W(s) + 3\,H_2O(g)$
(b) $Ag^+(aq) + Cl^-(aq) \rightleftharpoons AgCl(s)$
(c) $2\,FeCl_3(s) + 3\,H_2O(g) \rightleftharpoons Fe_2O_3(s) + 6\,HCl(g)$
(d) $MgCO_3(s) \rightleftharpoons MgO(s) + CO_2(g)$

Chemical Equilibrium and Chemical Kinetics (Section 13.3)

13.58 Consider a general, single-step reaction of the type $A + B \rightleftharpoons C$. Show that the equilibrium constant is equal to the ratio of the rate constants for the forward and reverse reactions, $K_c = k_f/k_r$.

13.59 Which of the following relative values of k_f and k_r results in an equilibrium mixture that contains large amounts of reactants and small amounts of products?

(a) $k_f > k_r$
(b) $k_f = k_r$
(c) $k_f < k_r$

13.60 Consider the gas-phase hydration of hexafluoroacetone, $(CF_3)_2CO$:

$$(CF_3)_2CO(g) + H_2O(g) \underset{k_r}{\overset{k_f}{\rightleftharpoons}} (CF_3)_2C(OH)_2(g)$$

At 76 °C, the forward and reverse rate constants are $k_f = 0.13\ M^{-1}\,s^{-1}$ and $k_r = 6.2 \times 10^{-4}\ s^{-1}$. What is the value of the equilibrium constant K_c?

13.61 Consider the reaction of chloromethane with OH^- in aqueous solution:

$$CH_3Cl(aq) + OH^-(aq) \underset{k_r}{\overset{k_f}{\rightleftharpoons}} CH_3OH(aq) + Cl^-(aq)$$

At 25 °C, the rate constant for the forward reaction is $6 \times 10^{-6}\ M^{-1}\,s^{-1}$, and the equilibrium constant K_c is 1×10^{16}. Calculate the rate constant for the reverse reaction at 25 °C.

13.62 In automobile catalytic converters, the air pollutant nitric oxide is converted to nitrogen and oxygen. Listed in the table are forward and reverse rate constants for the reaction $2\,NO(g) \rightleftharpoons N_2(g) + O_2(g)$.

Temperature (K)	k_f ($M^{-1}\,s^{-1}$)	k_r ($M^{-1}\,s^{-1}$)
1400	0.29	1.1×10^{-6}
1500	1.3	1.4×10^{-5}

Is the reaction endothermic or exothermic? Explain in terms of kinetics.

13.63 Forward and reverse rate constants for the reaction $CO_2(g) + N_2(g) \rightleftharpoons CO(g) + N_2O(g)$ exhibit the following temperature dependence:

Temperature (K)	k_f ($M^{-1}\,s^{-1}$)	k_r ($M^{-1}\,s^{-1}$)
1200	9.1×10^{-11}	1.5×10^5
1300	2.7×10^{-9}	2.6×10^5

Is the reaction endothermic or exothermic? Explain in terms of kinetics.

13.64 For the reaction $A_2 + 2\,B \rightleftharpoons 2\,AB$, the rate of the forward reaction is 18 M/s and the rate of the reverse reaction is 12 M/s. The reaction is not at equilibrium. Will the reaction proceed in the forward or reverse direction to attain equilibrium?

13.65 For the reaction $2\,A_3 + B_2 \rightleftharpoons 2\,A_3B$, the rate of the forward reaction is 0.35 M/s and the rate of the reverse reaction is 0.65 M/s. The reaction is not at equilibrium. Will the reaction proceed in the forward or reverse direction to attain equilibrium?

Using the Equilibrium Constant (Section 13.6)

13.66 When the following reactions come to equilibrium, does the equilibrium mixture contain mostly reactants or mostly products?

(a) $H_2(g) + S(s) \rightleftharpoons H_2S(g)$; $K_c = 7.8 \times 10^5$
(b) $N_2(g) + 2\,H_2(g) \rightleftharpoons N_2H_4(g)$; $K_c = 7.4 \times 10^{-26}$

13.67 Which of the following reactions yield appreciable equilibrium concentrations of both reactants and products?

(a) $2\,Cu(s) + O_2(g) \rightleftharpoons 2\,CuO(s)$; $K_c = 4 \times 10^{45}$
(b) $H_3PO_4(aq) \rightleftharpoons H^+(aq) + H_2PO_4^-(aq)$; $K_c = 7.5 \times 10^{-3}$
(c) $2\,HBr(g) \rightleftharpoons H_2(g) + Br_2(g)$; $K_c = 2 \times 10^{-19}$

13.68 When wine spoils, ethanol is oxidized to acetic acid as O_2 from the air reacts with the wine:

$$\underset{\text{Ethanol}}{CH_3CH_2OH(aq)} + O_2(aq) \rightleftharpoons \underset{\text{Acetic acid}}{CH_3CO_2H(aq)} + H_2O(l)$$

The value of K_c for this reaction at 25 °C is 1.2×10^{82}. Will much ethanol remain when the reaction has reached equilibrium? Explain.

13.69 The value of K_c for the reaction $3\,O_2(g) \rightleftharpoons 2\,O_3(g)$ is 1.7×10^{-56} at 25 °C. Do you expect pure air at 25 °C to contain much O_3 (ozone) when O_2 and O_3 are in equilibrium? If the equilibrium concentration of O_2 in air at 25 °C is 8×10^{-3} M, what is the equilibrium concentration of O_3?

13.70 At 1400 K, $K_c = 2.5 \times 10^{-3}$ for the reaction $CH_4(g) + 2\,H_2S(g) \rightleftharpoons CS_2(g) + 4\,H_2(g)$. A 10.0 L reaction vessel at 1400 K contains 2.0 mol of CH_4, 3.0 mol of CS_2, 3.0 mol of H_2, and 4.0 mol of H_2S. Is the reaction mixture at equilibrium? If not, in which direction does the reaction proceed to reach equilibrium?

13.71 The first step in the industrial synthesis of hydrogen is the reaction of steam and methane to give synthesis gas, a mixture of carbon monoxide and hydrogen:

$$H_2O(g) + CH_4(g) \rightleftharpoons CO(g) + 3\,H_2(g) \quad K_c = 4.7 \text{ at } 1400 \text{ K}$$

A mixture of reactants and products at 1400 K contains 0.035 M H_2O, 0.050 M CH_4, 0.15 M CO, and 0.20 M H_2. In which direction does the reaction proceed to reach equilibrium?

13.72 Phosphine (PH_3) decomposes at elevated temperatures, yielding gaseous P_2 and H_2:

$$2\,PH_3(g) \rightleftharpoons P_2(g) + 3\,H_2(g) \quad K_p = 398 \text{ at } 873 \text{ K}$$

If the initial partial pressures are $P_{PH_3} = 0.0260$ atm, $P_{P_2} = 0.871$ atm, $P_{H_2} = 0.517$ atm, calculate Q_p and determine the direction of reaction to attain equilibrium.

13.73 When a mixture of PH_3, P_2, and H_2 comes to equilibrium at 873 K according to the reaction in Problem 13.72, $P_{P_2} = 0.412$ atm and $P_{H_2} = 0.822$ atm. What is P_{PH_3}?

13.74 Gaseous indium dihydride is formed from the elements at elevated temperature:

$$In(g) + H_2(g) \rightleftharpoons InH_2(g) \quad K_p = 1.48 \text{ at } 973 \text{ K}$$

Partial pressures measured in a reaction vessel are: $P_{In} = 0.0600$ atm, $P_{H_2} = 0.0350$ atm, $P_{InH_2} = 0.0760$ atm.

(a) Calculate Q_p, and determine the direction of reaction to attain equilibrium.

(b) Determine the equilibrium partial pressures of all the gases.

13.75 The following reaction has $K_c = 0.145$ at 298 K in carbon tetrachloride solution:

$$2\,BrCl(soln) \rightleftharpoons Br_2(soln) + Cl_2(soln)$$

If the concentrations are $[BrCl] = 0.050$ M, $[Br_2] = 0.035$ M, and $[Cl_2] = 0.030$ M,

(a) Calculate Q_c, and determine the direction of reaction to attain equilibrium.

(b) Determine the equilibrium concentrations of BrCl, Br_2, and Cl_2.

13.76 An equilibrium mixture of N_2, H_2, and NH_3 at 700 K contains 0.036 M N_2 and 0.15 M H_2. At this temperature, K_c for the reaction $N_2(g) + 3\,H_2(g) \rightleftharpoons 2\,NH_3(g)$ is 0.29. What is the concentration of NH_3?

13.77 An equilibrium mixture of O_2, SO_2, and SO_3 contains equal concentrations of SO_2 and SO_3. Calculate the concentration of O_2 if $K_c = 2.7 \times 10^2$ for the reaction $2\,SO_2(g) + O_2(g) \rightleftharpoons 2\,SO_3(g)$.

13.78 The air pollutant NO is produced in automobile engines from the high-temperature reaction $N_2(g) + O_2(g) \rightleftharpoons 2\,NO(g)$; $K_c = 1.7 \times 10^{-3}$ at 2300 K. If the initial concentrations of N_2 and O_2 at 2300 K are both 1.40 M, what are the concentrations of NO, N_2, and O_2 when the reaction mixture reaches equilibrium?

13.79 Recalculate the equilibrium concentrations in Problem 13.78 if the initial concentrations are 2.24 M N_2 and 0.56 M O_2. (This N_2/O_2 concentration ratio is the ratio in air.)

13.80 The interconversion of the amino acids L-α-lysine and L-β-lysine which has $K_c = 7.20$ at 333 K is catalyzed by the enzyme lysine 2,3-aminomutase.

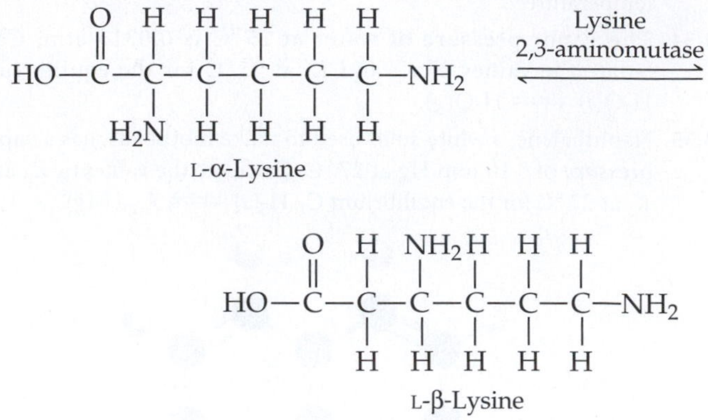

At 333 K, a solution of L-α-lysine at a concentration of 3.00×10^{-3} M is placed in contact with lysine 2,3-aminomutase. What are the equilibrium concentrations of L-α-lysine and L-β-lysine?

13.81 Chlorine monoxide and dichlorine dioxide are involved in the catalytic destruction of stratospheric ozone. They are related by the reaction

$$2\,ClO(g) \rightleftharpoons Cl_2O_2(g)$$

which has $K_c = 4.96 \times 10^{11}$ at 253 K. What is [ClO] in an equilibrium mixture that has $[Cl_2O_2] = 6.00 \times 10^{-6}$ M?

13.82 The value of K_c for the reaction of acetic acid with ethanol is 3.4 at 25 °C:

$$\underset{\text{Acetic acid}}{CH_3CO_2H(soln)} + \underset{\text{Ethanol}}{CH_3CH_2OH(soln)} \rightleftharpoons \underset{\text{Ethyl acetate}}{CH_3CO_2CH_2CH_3(soln)} + H_2O(soln) \quad K_c = 3.4$$

(a) How many moles of ethyl acetate are present in an equilibrium mixture that contains 4.0 mol of acetic acid, 6.0 mol of ethanol, and 12.0 mol of water at 25 °C?

(b) Calculate the number of moles of all reactants and products in an equilibrium mixture prepared by mixing 1.00 mol of acetic acid and 10.00 mol of ethanol.

13.83 In a basic aqueous solution, chloromethane undergoes a substitution reaction in which Cl^- is replaced by OH^-:

$$CH_3Cl(aq) + OH^-(aq) \rightleftharpoons CH_3OH(aq) + Cl^-(aq)$$

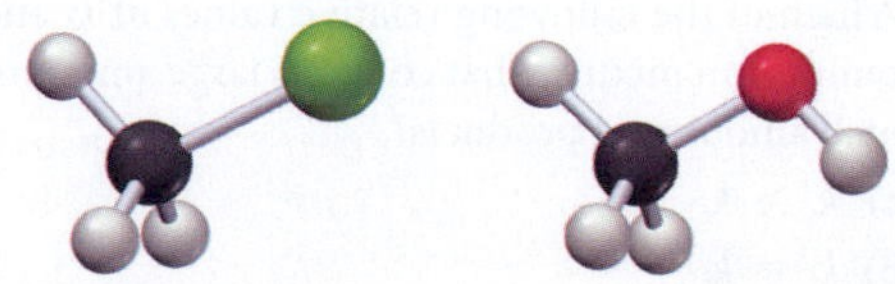

Chloromethane Methanol

The equilibrium constant K_c is 1×10^{16}. Calculate the equilibrium concentrations of CH_3Cl, CH_3OH, OH^-, and Cl^- in a solution prepared by mixing equal volumes of 0.1 M CH_3Cl and 0.2 M NaOH. (Hint: In defining x, assume that the reaction goes 100% to completion, and then take account of a small amount of the reverse reaction.)

13.84 At 700 K, $K_p = 0.140$ for the reaction $ClF_3(g) \rightleftharpoons ClF(g) + F_2(g)$. Calculate the equilibrium partial pressures of ClF_3, ClF, and F_2 if only ClF_3 is present initially, at a partial pressure of 1.47 atm.

13.85 The reaction of iron(III) oxide with carbon monoxide is important in making steel. At 1000 K, $K_p = 19.9$ for the reaction $Fe_2O_3(s) + 3\ CO(g) \rightleftharpoons 2\ Fe(s) + 3\ CO_2(g)$. What are the equilibrium partial pressures of CO and CO_2 if CO is the only gas present initially, at a partial pressure of 0.978 atm?

Le Châtelier's Principle (Sections 13.7–13.11)

13.86 Consider the following equilibrium: $Ag^+(aq) + Cl^-(aq) \rightleftharpoons AgCl(s)$. Use Le Châtelier's principle to predict how the amount of solid silver chloride will change when the equilibrium is disturbed by:
(a) Adding NaCl
(b) Adding $AgNO_3$
(c) Adding NH_3, which reacts with Ag^+ to form the complex ion $Ag(NH_3)_2^+$
(d) Removing Cl^-; also account for the change using the reaction quotient Q_c.

13.87 Will the concentration of NO_2 increase, decrease, or remain the same when the equilibrium $NO_2Cl(g) + NO(g) \rightleftharpoons NOCl(g) + NO_2(g)$ is disturbed by the following changes?
(a) Adding NOCl
(b) Adding NO
(c) Removing NO
(d) Adding NO_2Cl; also account for the change using the reaction quotient Q_c.

13.88 When each of the following equilibria is disturbed by increasing the pressure as a result of decreasing the volume, does the number of moles of reaction products increase, decrease, or remain the same?
(a) $2\ CO_2(g) \rightleftharpoons 2\ CO(g) + O_2(g)$
(b) $N_2(g) + O_2(g) \rightleftharpoons 2\ NO(g)$
(c) $Si(s) + 2\ Cl_2(g) \rightleftharpoons SiCl_4(g)$

13.89 For each of the following equilibria, use Le Châtelier's principle to predict the direction of reaction when the volume is increased.
(a) $C(s) + H_2O(g) \rightleftharpoons CO(g) + H_2(g)$
(b) $2\ H_2(g) + O_2(g) \rightleftharpoons 2\ H_2O(g)$
(c) $2\ Fe(s) + 3\ H_2O(g) \rightleftharpoons Fe_2O_3(s) + 3\ H_2(g)$

13.90 For the water-gas shift reaction $CO(g) + H_2O(g) \rightleftharpoons CO_2(g) + H_2(g)$, $\Delta H° = -41.2$ kJ, does the amount of H_2 in an equilibrium mixture increase or decrease when the temperature is increased? How does K_c change when the temperature is decreased? Justify your answers using Le Châtelier's principle.

13.91 The value of $\Delta H°$ for the reaction $3\ O_2(g) \rightleftharpoons 2\ O_3(g)$ is +285 kJ. Does the equilibrium constant for this reaction increase or decrease when the temperature increases? Justify your answer using Le Châtelier's principle.

13.92 Consider the exothermic reaction $CoCl_4^{2-}(aq) + 6\ H_2O(l) \rightleftharpoons Co(H_2O)_6^{2+}(aq) + 4\ Cl^-(aq)$, which interconverts the blue $CoCl_4^{2-}$ ion and the pink $Co(H_2O)_6^{2+}$ ion. Will the equilibrium concentration of $CoCl_4^{2-}$ increase or decrease when the following changes occur?
(a) HCl is added
(b) $Co(NO_3)_2$ is added
(c) The solution is diluted with water
(d) The temperature is increased

13.93 Consider the endothermic reaction $Fe^{3+}(aq) + Cl^-(aq) \rightleftharpoons FeCl^{2+}(aq)$. Use Le Châtelier's principle to predict how the equilibrium concentration of the complex ion $FeCl^{2+}$ will change when:
(a) $Fe(NO_3)_3$ is added
(b) Cl^- is precipitated as AgCl by addition of $AgNO_3$
(c) The temperature is increased
(d) A catalyst is added

13.94 Methanol (CH_3OH) is manufactured by the reaction of carbon monoxide with hydrogen in the presence of a $Cu/ZnO/Al_2O_3$ catalyst:

$$CO(g) + 2\ H_2(g) \xrightleftharpoons[\text{catalyst}]{Cu/ZnO/Al_2O_3} CH_3OH(g) \qquad \Delta H° = -91\ \text{kJ}$$

Does the amount of methanol increase, decrease, or remain the same when an equilibrium mixture of reactants and products is subjected to the following changes?
(a) The temperature is increased
(b) The volume is decreased
(c) Helium is added
(d) CO is added
(e) The catalyst is removed

13.95 In the gas phase at 400 °C, isopropyl alcohol (rubbing alcohol) decomposes to acetone, an important industrial solvent:

$$\underset{\text{Isopropyl alcohol}}{(CH_3)_2CHOH(g)} \rightleftharpoons \underset{\text{Acetone}}{(CH_3)_2CO(g)} + H_2(g) \qquad \Delta H° = +57.3\ \text{kJ}$$

Does the amount of acetone increase, decrease, or remain the same when an equilibrium mixture of reactants and products is subjected to the following changes?
(a) The temperature is increased
(b) The volume is increased
(c) Argon is added
(d) H_2 is added
(e) A catalyst is added

13.96 The following reaction is important in gold mining:

$$4\ Au(s) + 8\ CN^-(aq) + O_2(g) + 2\ H_2O(l) \rightleftharpoons 4\ Au(CN)_2^-(aq) + 4\ OH^-(aq)$$

For a reaction mixture at equilibrium, in which direction would the reaction go to re-establish equilibrium after each of the following changes?
(a) Adding gold
(b) Increasing the hydroxide concentration
(c) Increasing the partial pressure of oxygen
(d) Adding $Fe^{3+}(aq)$, which reacts with cyanide to form $Fe(CN)_6^{3-}(aq)$

13.97 The following endothermic reaction is catalyzed by iridium at 700 K:

$$CaO(s) + CH_4(g) + 2\ H_2O(g) \rightleftharpoons CaCO_3(s) + 4\ H_2(g)$$

If the following changes are made to an equilibrium mixture at 700 K, would the amount of $CaCO_3$ increase, decrease, or remain the same?
(a) Increasing the temperature
(b) Adding calcium oxide
(c) Removing methane (CH_4)
(d) Increasing the volume
(e) Adding iridium

CHAPTER PROBLEMS

13.98 The equilibrium concentrations in a gas mixture at a particular temperature are 0.13 M H_2, 0.70 M I_2, and 2.1 M HI. What equilibrium concentrations are obtained at the same temperature when 0.20 mol of HI is injected into an empty 500.0 mL container?

13.99 A 5.00 L reaction vessel is filled with 1.00 mol of H_2, 1.00 mol of I_2, and 2.50 mol of HI. Calculate the equilibrium concentrations of H_2, I_2, and HI at 500 K. The equilibrium constant K_c at 500 K for the reaction $H_2(g) + I_2(g) \rightleftharpoons 2\,HI(g)$ is 129.

13.100 At 1000 K, the value of K_c for the reaction $C(s) + H_2O(g) \rightleftharpoons CO(g) + H_2(g)$ is 3.0×10^{-2}. Calculate the equilibrium concentrations of H_2O, CO_2, and H_2 in a reaction mixture obtained by heating 6.00 mol of steam and an excess of solid carbon in a 5.00 L container. What is the molar composition of the equilibrium mixture?

13.101 The equilibrium constant K_p for the reaction $PCl_5(g) \rightleftharpoons PCl_3(g) + Cl_2(g)$ is 3.81×10^2 at 600 K and 2.69×10^3 at 700 K.

(a) Is the reaction endothermic or exothermic?

(b) How are the equilibrium amounts of reactants and products affected by (i) an increase in volume, (ii) addition of an inert gas, and (iii) addition of a catalyst?

13.102 Consider the following gas-phase reaction: $2\,A(g) + B(g) \rightleftharpoons C(g) + D(g)$. An equilibrium mixture of reactants and products is subjected to the following changes:

(a) A decrease in volume

(b) An increase in temperature

(c) Addition of reactants

(d) Addition of a catalyst

(e) Addition of an inert gas

Which of these changes affect the composition of the equilibrium mixture, but leave the value of the equilibrium constant K_c unchanged? Which of the changes affect the value of K_c? Which affect neither the composition of the equilibrium mixture nor the value of K_c?

13.103 Baking soda (sodium bicarbonate) decomposes when it is heated:

$$2\,NaHCO_3(s) \rightleftharpoons Na_2CO_3(s) + CO_2(g) + H_2O(g) \qquad \Delta H^\circ = +136\text{ kJ}$$

Consider an equilibrium mixture of reactants and products in a closed container. How does the number of moles of CO_2 change when the mixture is disturbed by the following:

(a) Adding solid $NaHCO_3$

(b) Adding water vapor

(c) Decreasing the volume of the container

(d) Increasing the temperature

13.104 When 1.000 mol of PCl_5 is introduced into a 5.000 L container at 500 K, 78.50% of the PCl_5 dissociates to give an equilibrium mixture of PCl_5, PCl_3, and Cl_2:

$$PCl_5(g) \rightleftharpoons PCl_3(g) + Cl_2(g)$$

(a) Calculate the values of K_c and K_p.

(b) If the initial concentrations in a particular mixture of reactants and products are $[PCl_5] = 0.500$ M, $[PCl_3] = 0.150$ M, and $[Cl_2] = 0.600$ M, in which direction does the reaction proceed to reach equilibrium? What are the concentrations when the mixture reaches equilibrium?

13.105 Heavy water, D_2O ($D = {}^2H$), finds use as a neutron moderator in nuclear reactors. In a mixture with ordinary water, exchange of isotopes occurs according to the following equation:

$$H_2O + D_2O \rightleftharpoons 2\,HDO \qquad K_c = 3.86 \text{ at } 298\text{ K}$$

When 1.00 mol of H_2O is combined with 1.00 mol of D_2O, what are the equilibrium amounts of H_2O, D_2O, and HDO (in moles) at 298 K? Assume the density of the mixture is constant at 1.05 g/cm^3.

13.106 Refining petroleum involves cracking large hydrocarbon molecules into smaller, more volatile pieces. A simple example of hydrocarbon cracking is the gas-phase thermal decomposition of butane to give ethane and ethylene:

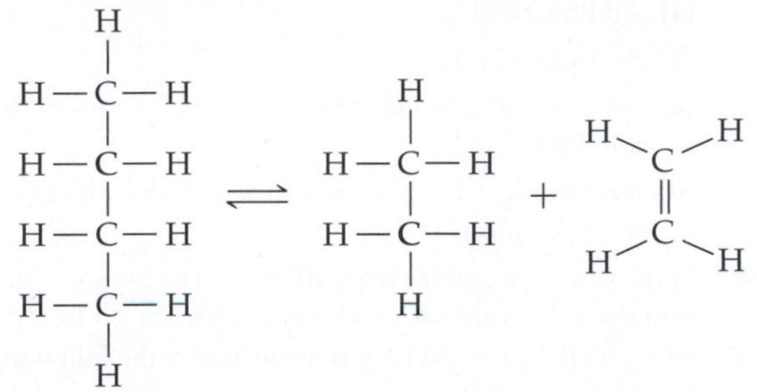

Butane, C_4H_{10} — Ethane, C_2H_6 — Ethylene, C_2H_4

(a) Write the equilibrium-constant expressions for K_p and K_c.

(b) The value of K_p at 500 °C is 12. What is the value of K_c?

(c) A sample of butane having a pressure of 50 atm is heated at 500 °C in a closed container at constant volume. When equilibrium is reached, what percentage of the butane has been converted to ethane and ethylene? What is the total pressure at equilibrium?

(d) How would the percent conversion in part **(c)** be affected by a decrease in volume.

13.107 Consider the reaction $C(s) + CO_2(g) \rightleftharpoons 2\,CO(g)$. When 1.50 mol of CO_2 and an excess of solid carbon are heated in a 20.0 L container at 1100 K, the equilibrium concentration of CO is 7.00×10^{-2} M.

(a) What is the equilibrium concentration of CO_2?

(b) What is the value of the equilibrium constant K_c at 1100 K?

13.108 The equilibrium constant K_p for the gas-phase thermal decomposition of *tert*-butyl chloride is 3.45 at 500 K:

$$(CH_3)_3CCl(g) \rightleftharpoons (CH_3)_2C{=}CH_2(g) + HCl(g)$$

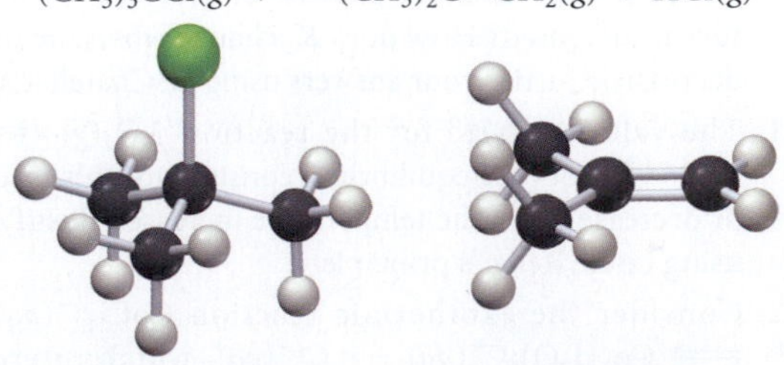

tert-Butyl chloride — Isobutylene

(a) Calculate the value of K_c at 500 K.

(b) Calculate the molar concentrations of reactants and products in an equilibrium mixture obtained by heating 1.00 mol of *tert*-butyl chloride in a 5.00 L vessel at 500 K.

(c) A mixture of isobutylene (0.400 atm partial pressure at 500 K) and HCl (0.600 atm partial pressure at 500 K) is allowed to reach equilibrium at 500 K. What are the equilibrium partial pressures of *tert*-butyl chloride, isobutylene, and HCl?

13.109 As shown in Figure 13.14, a catalyst lowers the activation energy for the forward and reverse reactions by the same amount, ΔE_a.

(a) Apply the Arrhenius equation, $k = Ae^{-E_a/RT}$, to the forward and reverse reactions, and show that a catalyst increases the rates of both reactions by the same factor.

(b) Use the relation between the equilibrium constant and the forward and reverse rate constants, $K_c = k_f/k_r$, to show that a catalyst does not affect the value of the equilibrium constant.

13.110 Given the Arrhenius equation, $k = Ae^{-E_a/RT}$, and the relation between the equilibrium constant and the forward and reverse rate constants, $K_c = k_f/k_r$, explain why K_c for an exothermic reaction decreases with increasing temperature.

13.111 At 1000 K, $K_p = 2.1 \times 10^6$ and $\Delta H° = -107.7$ kJ for the reaction $H_2(g) + Br_2(g) \rightleftharpoons 2\,HBr(g)$.

(a) A 0.974 mol quantity of Br_2 is added to a 1.00 L reaction vessel that contains 1.22 mol of H_2 gas at 1000 K. What are the partial pressures of H_2, Br_2, and HBr at equilibrium?

(b) For the equilibrium in part **(a)**, each of the following changes will increase the equilibrium partial pressure of HBr. Choose the change that will cause the greatest increase in the pressure of HBr, and explain your choice.

(i) Adding 0.10 mol of H_2
(ii) Adding 0.10 mol of Br_2
(iii) Decreasing the temperature to 700 K.

13.112 Consider the gas-phase decomposition of NOBr:

$$2\,NOBr(g) \rightleftharpoons 2\,NO(g) + Br_2(g)$$

(a) When 0.0200 mol of NOBr is added to an empty 1.00 L flask and the decomposition reaction is allowed to reach equilibrium at 300 K, the total pressure in the flask is 0.588 atm. What is the equilibrium constant K_c for this reaction at 300 K?

(b) What is the value of K_p for this reaction at 300 K?

13.113 At 100 °C, $K_c = 4.72$ for the reaction $2\,NO_2(g) \rightleftharpoons N_2O_4(g)$. An empty 10.0 L flask is filled with 4.60 g of NO_2 at 100 °C. What is the total pressure in the flask at equilibrium?

13.114 Halogen lamps are ordinary tungsten filament lamps in which the lamp bulb contains a small amount of a halogen (often bromine). At the high temperatures of the lamp, the halogens dissociate and exist as single atoms.

(a) In an ordinary tungsten lamp, the hot tungsten filament is constantly evaporating, and the tungsten condenses on the relatively cool walls of the bulb. In a Br-containing halogen lamp, the tungsten reacts with the Br atoms to give gaseous WBr_4:

$$W(s) + 4\,Br(g) \rightleftharpoons WBr_4(g)$$

At the walls of the lamp, where the temperature is about 900 K, this reaction has an equilibrium constant K_p of about 100. If the equilibrium pressure of Br(g) is 0.010 atm, what is the equilibrium pressure of $WBr_4(g)$ near the walls of the bulb?

(b) Near the tungsten filament, where the temperature is about 2800 K, the reaction in part **(a)** has a K_p value of about 5.0. Is the reaction exothermic or endothermic?

(c) When the $WBr_4(g)$ diffuses back toward the filament, it decomposes, depositing tungsten back onto the filament. Show quantitatively that the pressure of WBr_4 from part **(a)** will cause the reaction in part **(a)** to go in reverse direction at 2800 K. [The pressure of Br(g) is still 0.010 atm.] Thus, tungsten is continually recycled from the walls of the bulb back to the filament, allowing the bulb to last longer and burn brighter.

13.115 The decomposition of solid ammonium carbamate, $(NH_4)(NH_2CO_2)$, to gaseous ammonia and carbon dioxide is an endothermic reaction.

$$(NH_4)(NH_2CO_2)(s) \rightleftharpoons 2\,NH_3(g) + CO_2(g)$$

(a) When solid $(NH_4)(NH_2CO_2)$ is introduced into an evacuated flask at 25 °C, the total pressure of gas at equilibrium is 0.116 atm. What is the value of K_p at 25 °C?

(b) Given that the decomposition reaction is at equilibrium, how would the following changes affect the total quantity of NH_3 in the flask once equilibrium is re-established?

(i) Adding CO_2
(ii) Adding $(NH_4)(NH_2CO_2)$
(iii) Removing CO_2
(iv) Increasing the total volume
(v) Adding neon
(vi) Increasing the temperature

13.116 At 25 °C, $K_c = 216$ for the reaction $2\,NO_2(g) \rightleftharpoons N_2O_4(g)$. A 1.00 L flask containing a mixture of NO_2 and N_2O_4 at 25 °C has a total pressure of 1.50 atm. What is the partial pressure of each gas?

13.117 At 500 °C, F_2 gas is stable and does not dissociate, but at 840 °C, some dissociation occurs: $F_2(g) \rightleftharpoons 2\,F(g)$. A flask filled with 0.600 atm of F_2 at 500 °C was heated to 840 °C, and the pressure at equilibrium was measured to be 0.984 atm. What is the equilibrium constant K_p for the dissociation of F_2 gas at 840 °C?

13.118 The reaction $NO(g) + NO_2(g) \rightleftharpoons N_2O_3(g)$ takes place in the atmosphere with $K_c = 13$ at 298 K. A gas mixture is prepared with 2.0 mol NO and 3.0 mol NO_2 and an initial total pressure of 1.65 atm.

(a) What are the equilibrium partial pressures of NO, NO_2, and N_2O_3 at 298 K?

(b) What is the volume of the container?

13.119 Phosgene ($COCl_2$) is a toxic gas that damages the lungs. At 360 °C, $K_c = 8.4 \times 10^{-4}$ for the decomposition of phosgene:

$$COCl_2(g) \rightleftharpoons CO(g) + Cl_2(g)$$

If an empty 50.0 L container is charged with 1.00 mol of phosgene at 360 °C, what is the total pressure in the container after the system comes to equilibrium?

13.120 The equilibrium constant K_c for the reaction $N_2(g) + 3\,H_2(g) \rightleftharpoons 2\,NH_3(g)$ is 4.20 at 600 K. When a quantity of gaseous NH_3 was placed in a 1.00 L reaction vessel at 600 K and the reaction was allowed to reach equilibrium, the vessel was found to contain 0.200 mol of N_2. How many moles of NH_3 were placed in the vessel?

13.121 At 45 °C, $K_c = 0.619$ for the reaction $N_2O_4(g) \rightleftharpoons 2\,NO_2(g)$. If 46.0 g of N_2O_4 is introduced into an empty 2.00 L container, what are the partial pressures of NO_2 and N_2O_4 after equilibrium has been achieved at 45 °C?

13.122 When 9.25 g of ClF_3 was introduced into an empty 2.00 L container at 700.0 K, 19.8% of the ClF_3 decomposed to give an equilibrium mixture of ClF_3, ClF, and F_2.

$$ClF_3(g) \rightleftharpoons ClF(g) + F_2(g)$$

(a) What is the value of the equilibrium constant K_c at 700.0 K?

(b) What is the value of the equilibrium constant K_p at 700.0 K?

(c) In a separate experiment, 39.4 g of ClF_3 was introduced into an empty 2.00 L container at 700.0 K. What are the concentrations of ClF_3, ClF, and F_2 when the mixture reaches equilibrium?

13.123 The following reaction in aqueous solution is catalyzed by the enzyme aspartase and has $K_c = 6.95 \times 10^{-3}$ at 37 °C:

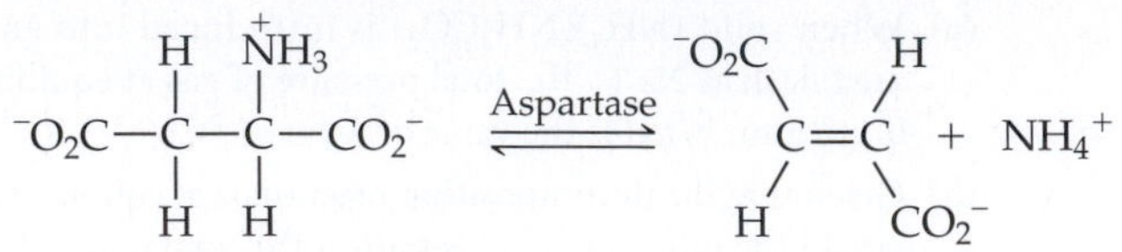

L-Aspartate   Fumarate

If the initial concentration of L-aspartate is 8.32×10^{-3} M, what are the equilibrium concentrations of L-aspartate, fumarate, and ammonium ion at 37 °C?

13.124 The reaction of fumarate (Problem 13.123) with water to form L-malate is catalyzed by the enzyme fumarase; $K_c = 3.3$ at 37 °C.

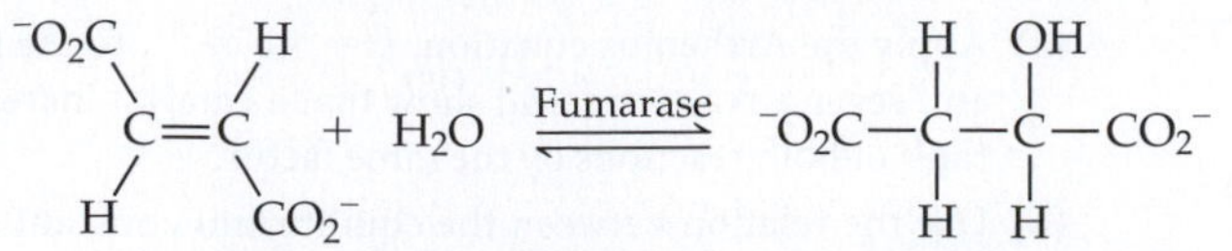

Fumarate   L-Malate

When a reaction mixture with [fumarate] = 1.56×10^{-3} M and [L-malate] = 2.27×10^{-3} M comes to equilibrium in the presence of fumarase at 37 °C, what are the equilibrium concentrations of fumarate and L-malate? (Water can be omitted from the equilibrium equation because its concentration in dilute solutions is essentially the same as that in pure water.)

MULTICONCEPT PROBLEMS

13.125 The F–F bond in F_2 is relatively weak because the lone pairs of electrons on one F atom repel the lone pairs on the other F atom; $K_p = 7.83$ at 1500 K for the reaction $F_2(g) \rightleftharpoons 2\,F(g)$.

(a) If the equilibrium partial pressure of F_2 molecules at 1500 K is 0.200 atm, what is the equilibrium partial pressure of F atoms in atm?

(b) What fraction of the F_2 molecules dissociate at 1500 K?

(c) Why is the F–F bond in F_2 weaker than the Cl–Cl bond in Cl_2?

13.126 When 0.500 mol of N_2O_4 is placed in a 4.00 L reaction vessel and heated at 400 K, 79.3% of the N_2O_4 decomposes to NO_2.

(a) Calculate K_c and K_p at 400 K for the reaction $N_2O_4(g) \rightleftharpoons 2\,NO_2(g)$.

(b) Draw an electron-dot structure for NO_2, and rationalize the structure of N_2O_4.

13.127 The equilibrium constant K_c for the gas-phase thermal decomposition of cyclopropane to propene is 1.0×10^5 at 500 K:

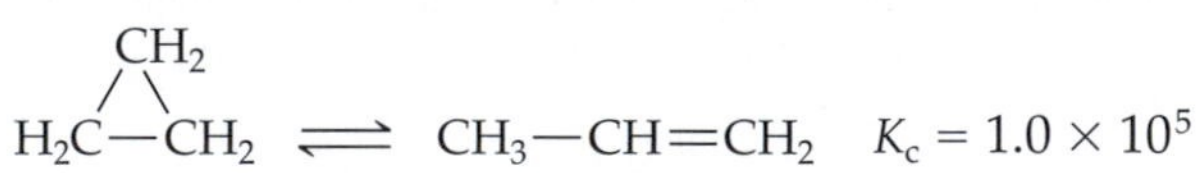

Cyclopropane   Propene

(a) What is the value of K_p at 500 K?

(b) What is the equilibrium partial pressure of cyclopropane at 500 K when the partial pressure of propene is 5.0 atm?

(c) Can you alter the ratio of the two concentrations at equilibrium by adding cyclopropane or by decreasing the volume of the container? Explain.

(d) Which has the larger rate constant, the forward reaction or the reverse reaction?

(e) Why is cyclopropane so reactive? (Hint: Consider the hybrid orbitals used by the C atoms.)

13.128 Acetic acid tends to form dimers, $(CH_3CO_2H)_2$, because of hydrogen bonding:

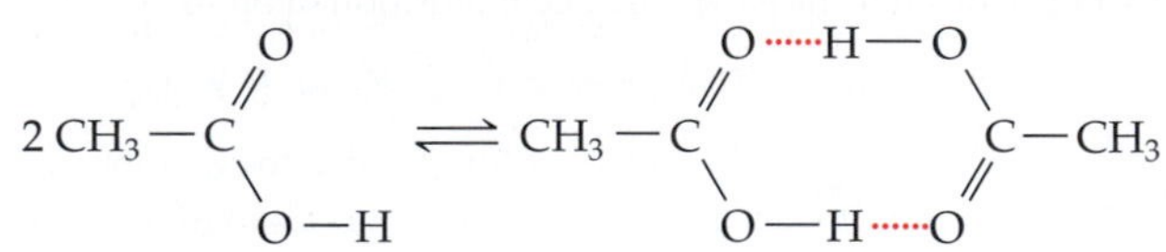

Monomer   Dimer

The equilibrium constant K_c for this reaction is 1.51×10^2 in benzene solution, but only 3.7×10^{-2} in water solution.

(a) Calculate the ratio of dimers to monomers for 0.100 M acetic acid in benzene.

(b) Calculate the ratio of dimers to monomers for 0.100 M acetic acid in water.

(c) Why is K_c for the water solution so much smaller than K_c for the benzene solution?

13.129 A 125.4 g quantity of water and an equal molar amount of carbon monoxide were placed in an empty 10.0 L vessel, and the mixture was heated to 700 K. At equilibrium, the partial pressure of CO was 9.80 atm. The reaction is

$$CO(g) + H_2O(g) \rightleftharpoons CO_2(g) + H_2(g)$$

(a) What is the value of K_p at 700 K?

(b) An additional 31.4 g of water was added to the reaction vessel, and a new state of equilibrium was achieved. What are the equilibrium partial pressures of each gas in the mixture? What is the concentration of H_2 in molecules/cm^3?

13.130 A 79.2 g chunk of dry ice (solid CO_2) and 30.0 g of graphite (carbon) were placed in an empty 5.00 L container, and the mixture was heated to achieve equilibrium. The reaction is

$$CO_2(g) + C(s) \rightleftharpoons 2\,CO(g)$$

(a) What is the value of K_p at 1000 K if the gas density at 1000 K is 16.3 g/L?

(b) What is the value of K_p at 1100 K if the gas density at 1100 K is 16.9 g/L?

(c) Is the reaction exothermic or endothermic? Explain.

13.131 The amount of carbon dioxide in a gaseous mixture of CO_2 and CO can be determined by passing the gas into an aqueous solution that contains an excess of $Ba(OH)_2$. The CO_2 reacts, yielding a precipitate of $BaCO_3$, but the CO does not react. This method was used to analyze the equilibrium composition of the gas obtained when 1.77 g of CO_2 reacted with 2.0 g of graphite in a 1.000 L container at 1100 K. The analysis yielded 3.41 g of $BaCO_3$. Use these data to calculate K_p at 1100 K for the reaction

$$CO_2(g) + C(s) \rightleftharpoons 2\,CO(g)$$

13.132 A 14.58 g quantity of N_2O_4 was placed in a 1.000 L reaction vessel at 400 K. The N_2O_4 decomposed to an equilibrium mixture of N_2O_4 and NO_2 that had a total pressure of 9.15 atm.

(a) What is the value of K_c for the reaction $N_2O_4(g) \rightleftharpoons 2\,NO_2(g)$ at 400 K?

(b) How much heat (in kilojoules) was absorbed when the N_2O_4 decomposed to give the equilibrium mixture? (Standard heats of formation may be found in Appendix B.)

13.133 Consider the sublimation of mothballs at 27 °C in a room having dimensions 8.0 ft × 10.0 ft × 8.0 ft. Assume that the mothballs are pure solid naphthalene (density 1.16 g/cm^3) and that they are spheres with a diameter of 12.0 mm. The equilibrium constant K_c for the sublimation of naphthalene is 5.40×10^{-6} at 27 °C.

$$C_{10}H_8(s) \rightleftharpoons C_{10}H_8(g)$$

(a) When excess mothballs are present, how many gaseous naphthalene molecules are in the room at equilibrium?

(b) How many mothballs are required to saturate the room with gaseous naphthalene?

13.134 Ozone is unstable with respect to decomposition to ordinary oxygen:

$$2\,O_3(g) \rightleftharpoons 3\,O_2(g) \qquad K_p = 1.3 \times 10^{57}$$

How many O_3 molecules are present at equilibrium in 10 million cubic meters of air at 25 °C and 720 mm Hg pressure?

13.135 The equilibrium constant for the dimerization of acetic acid in benzene solution is 1.51×10^2 at 25 °C (see Problem 13.128).

$$2\,CH_3CO_2H \rightleftharpoons (CH_3CO_2H)_2 \quad K_c = 1.51 \times 10^2 \text{ at } 25\ °C$$

(a) What are the equilibrium concentrations of monomer and dimer at 25 °C in a solution prepared by dissolving 0.0300 mol of pure acetic acid in enough benzene to make 250.0 mL of solution?

(b) What is the osmotic pressure of the solution at 25 °C?

13.136 For the decomposition reaction $PCl_5(g) \rightleftharpoons PCl_3(g) + Cl_2(g)$, $K_p = 381$ at 600 K and $K_c = 46.9$ at 700 K.

(a) Is the reaction endothermic or exothermic? Explain. Does your answer agree with what you would predict based on bond energies?

(b) If 1.25 g of PCl_5 is introduced into an evacuated 0.500 L flask at 700 K and the decomposition reaction is allowed to reach equilibrium, what percent of the PCl_5 will decompose and what will be the total pressure in the flask?

(c) Write electron-dot structures for PCl_5 and PCl_3, and indicate whether these molecules have a dipole moment. Explain.

13.137 Propanol (PrOH) and methyl methacrylate (MMA) associate in solution by an intermolecular force, forming an adduct represented as PrOH • MMA. The equilibrium constant for the association reaction is $K_c = 0.701$ at 298 K.

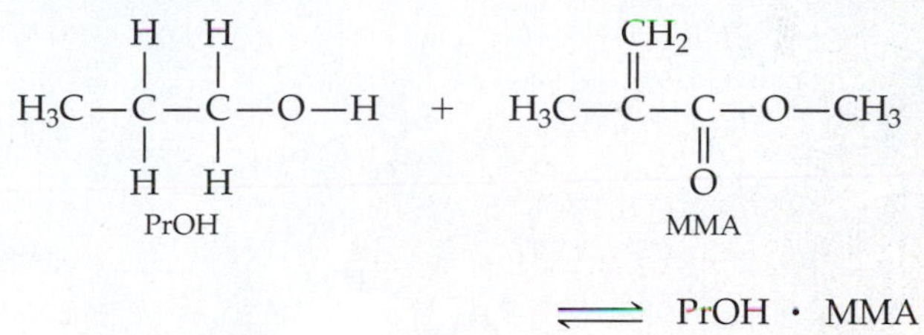

(a) What is the predominant intermolecular force accounting for the interaction between PrOH and MMA?

(b) Draw a plausible structure for the PrOH • MMA adduct. Use • • • to signify an intermolecular interaction.

(c) If the initial concentrations are [PrOH] = 0.100 M and [MMA] = 0.0500 M, what are the equilibrium concentrations of PrOH, MMA, and PrOH • MMA?

CHAPTER 14

Aqueous Equilibria: Acids and Bases

Leshan Buddha in the Sichuan province of China, the world's tallest Buddha, needs yet another facelift. To find out why, see the *FYI* at the end of this chapter.

In the previous chapter, we discussed the principles of chemical equilibria. We'll now apply those principles to aqueous solutions of acids and bases, perhaps the most familiar of all chemical compounds. Acetic acid in vinegar, citric acid in lemons and other citrus fruits, magnesium hydroxide in commercial antacids, and ammonia in household cleaning products are among the acids and bases that we encounter in everyday life. Hydrochloric acid of approximately 0.15 M concentration is the acid in gastric juice; it is essential to digestion and is secreted by the lining of our stomachs in quantities of 1.2–1.5 L per day.

The characteristic properties of acids and bases have been known for centuries. Acids react with metals such as iron and zinc to yield H_2 gas, and they change the color of the plant dye *litmus* from blue to red. By contrast, bases feel slippery, and they change the color of litmus from red to blue. When acids and bases are mixed in the right proportion, the characteristic acidic and basic properties disappear and new substances known as *salts* are obtained.

Before examining the quantitative aspects of acid–base equilibria—that is, how to calculate equilibrium concentrations in solutions of acids and bases—we need to ask what is it that makes an acid an acid and a base a base. We first raised those questions in Section 7.5, and we'll now take a closer look at some of the concepts that chemists have developed to describe the chemical behavior of acids and bases. An enormous amount of chemistry can be understood in terms of acid–base reactions, perhaps the most important reaction type in all of chemistry.

14.1 ACID–BASE CONCEPTS: THE BRØNSTED–LOWRY THEORY

Thus far, we've been using the Arrhenius theory of acids and bases (Section 7.5). According to Arrhenius, acids are substances that dissociate in water to produce hydrogen ions (H^+) and bases are substances that dissociate in water to yield hydroxide ions (OH^-). Thus, HCl and H_2SO_4 are acids, and NaOH and $Ba(OH)_2$ are bases.

CONTENTS

▶ **A generalized Arrhenius acid** $HA(aq) \rightleftharpoons H^+(aq) + A^-(aq)$

▶ **A generalized Arrhenius base** $MOH(aq) \rightleftharpoons M^+(aq) + OH^-(aq)$

The Arrhenius theory accounts for the properties of many common acids and bases, but it has important limitations. For one thing, the Arrhenius theory is restricted to aqueous solutions; for another, it doesn't account for the basicity of substances like ammonia (NH_3) that don't contain OH groups. In 1923, a more general theory of acids and bases was proposed independently by the Danish chemist Johannes Brønsted and the English chemist Thomas Lowry. According to the **Brønsted–Lowry theory**, an acid is any substance (molecule or ion) that can transfer a proton (H^+ ion) to another substance and a base is any substance that can accept a proton. In short, acids are proton donors, bases are proton acceptors, and acid–base reactions are proton-transfer reactions:

▶ **Brønsted–Lowry Acid** A substance that can transfer H^+

▶ **Brønsted–Lowry Base** A substance that can accept H^+

$$\underset{\substack{H^+\text{ donor}\\ \text{Acid}}}{HA} + \underset{\substack{H^+\text{ acceptor}\\ \text{Base}}}{B} \rightleftharpoons \underset{\substack{H^+\text{ donor}\\ \text{Acid}}}{BH^+} + \underset{\substack{H^+\text{ acceptor}\\ \text{Base}}}{A^-}$$

Conjugate acid–base pairs

It follows from this equation that the products of a Brønsted–Lowry acid–base reaction, BH^+ and A^-, are themselves acids and bases. The species BH^+ produced when the base B accepts a proton from HA can itself donate a proton back to A^-, meaning that it is a Brønsted–Lowry acid. Similarly, the species A^- produced when HA loses a proton can itself accept a proton back from BH^+, meaning that it is a Brønsted–Lowry base. Chemical species whose formulas differ only by one proton are said to be **conjugate acid–base pairs**. Thus, A^- is the **conjugate base** of the acid HA, and HA is the **conjugate acid** of the base A^-. Similarly, B is the conjugate base of the acid BH^+, and BH^+ is the conjugate acid of the base B.

To see what's going on in an acid–base reaction, keep your eye on the proton. For example, when a Brønsted–Lowry acid HA dissolves in water, it reacts reversibly with water in an *acid-dissociation equilibrium*. The acid transfers a proton to the solvent, which acts as a base (a proton acceptor). The products are the trigonal pyramidal **hydronium ion, H_3O^+** (the conjugate acid of H_2O), and A^- (the conjugate base of HA):

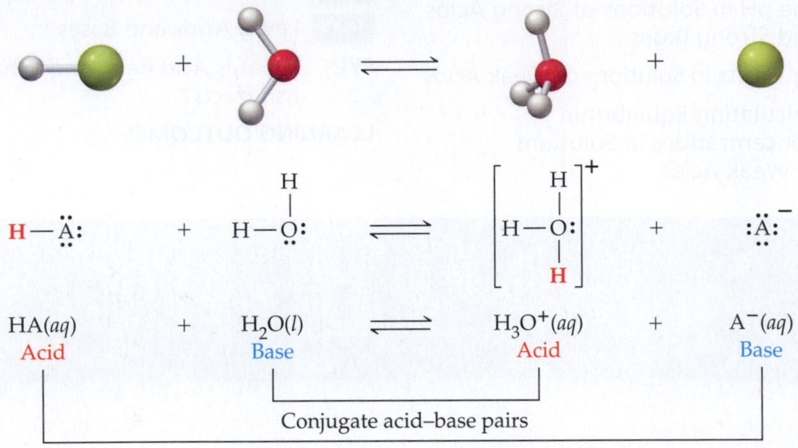

In the reverse reaction, H_3O^+ acts as the proton donor (acid) and A^- acts as the proton acceptor (base). Typical examples of Brønsted–Lowry acids include not only electrically neutral molecules, such as HCl, HNO_3, and HF, but also cations and anions of salts that contain transferable protons, such as NH_4^+, HSO_4^-, and HCO_3^-.

When a Brønsted–Lowry base such as NH_3 dissolves in water, it accepts a proton from the solvent, which acts as an acid. The products are the hydroxide ion, OH^- (the conjugate base of water), and the ammonium ion, NH_4^+ (the conjugate acid of NH_3). In the reverse reaction, NH_4^+ acts as the proton donor and OH^- acts as the proton acceptor:

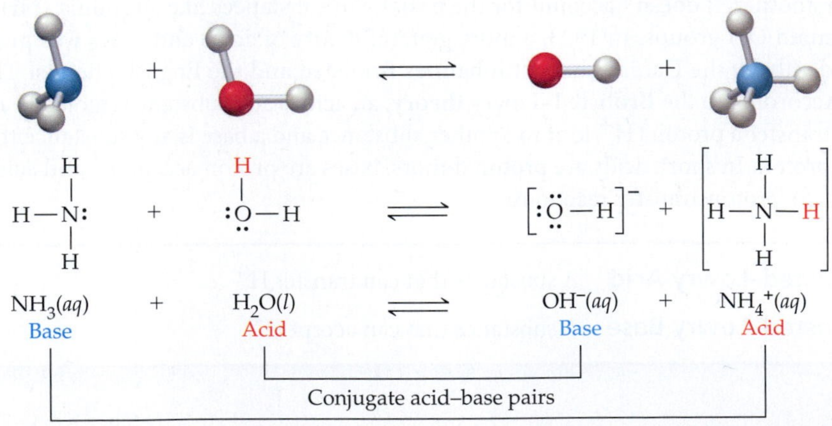

For a molecule or ion to accept a proton, it must have at least one unshared pair of electrons that it can use for bonding to the proton. As shown by the following electron-dot structures, all Brønsted–Lowry bases have one or more lone pairs of electrons:

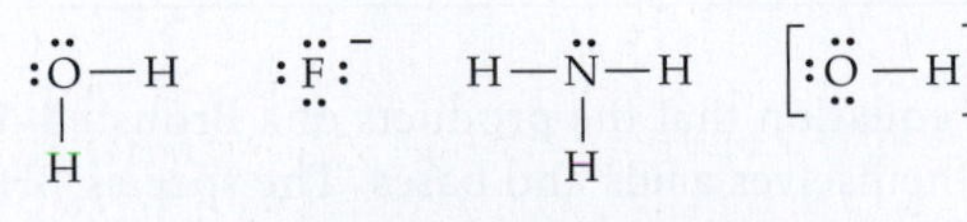

Some Brønsted–Lowry bases

Note that the hydronium ion, H_3O^+, can be regarded as the simplest hydrate of the proton, $[H(H_2O)]^+$. It can associate through hydrogen bonding with additional water molecules to give higher hydrates with the general formula $[H(H_2O)_n]^+$ ($n = 2, 3,$ or 4), such as $H_5O_2^+$, $H_7O_3^+$, and $H_9O_4^+$. It's likely that acidic aqueous solutions contain a distribution of $[H(H_2O)_n]^+$ ions having different values of n. In this book, though, we'll use the symbols $H^+(aq)$ and $H_3O^+(aq)$ to mean the same thing—namely, a proton hydrated by an unspecified number of water molecules. Ordinarily, we'll use H_3O^+ in acid–base reactions to emphasize the proton-transfer character of those reactions.

▲ H_3O^+—the hydronium ion, or hydrated H^+.

Worked Example 14.1

Explaining Acidity with the Arrhenius and Brønsted–Lowry Theories

Account for the acidic properties of nitrous acid (HNO_2) using the Arrhenius theory and the Brønsted–Lowry theory, and identify the conjugate base of HNO_2.

Nitrous acid

STRATEGY

To account for the acidity of a substance, consider how it can produce H^+ ions in water (Arrhenius theory) and how it can act as a proton donor (Brønsted–Lowry theory).

SOLUTION

HNO_2 is an Arrhenius acid because it dissociates in water to produce H^+ ions:

$$HNO_2(aq) \rightleftharpoons H^+(aq) + NO_2^-(aq)$$

Nitrous acid is a Brønsted–Lowry acid because it acts as a proton donor when it dissociates, transferring a proton to water to give the hydronium ion, H_3O^+:

$$\underset{\text{Acid}}{HNO_2(aq)} + \underset{\text{Base}}{H_2O(l)} \rightleftharpoons \underset{\text{Acid}}{H_3O^+(aq)} + \underset{\text{Base}}{NO_2^-(aq)}$$

Conjugate acid–base pairs

The conjugate base of HNO_2 is NO_2^-, the species that remains after HNO_2 has lost a proton.

PROBLEM 14.1 Write a balanced equation for the dissociation of each of the following Brønsted–Lowry acids in water:

(a) H_2SO_4 (b) HSO_4^- (c) H_3O^+ (d) NH_4^+

What is the conjugate base of each acid?

PROBLEM 14.2 What is the conjugate acid of each of the following Brønsted–Lowry bases?

(a) HCO_3^- (b) CO_3^{2-} (c) OH^- (d) $H_2PO_4^-$

WORKED CONCEPTUAL EXAMPLE 14.2

Identifying Brønsted–Lowry Acids, Bases, and Conjugate Acid–Base Pairs

For the following reaction in aqueous solution, identify the Brønsted–Lowry acids, bases, and conjugate acid–base pairs:

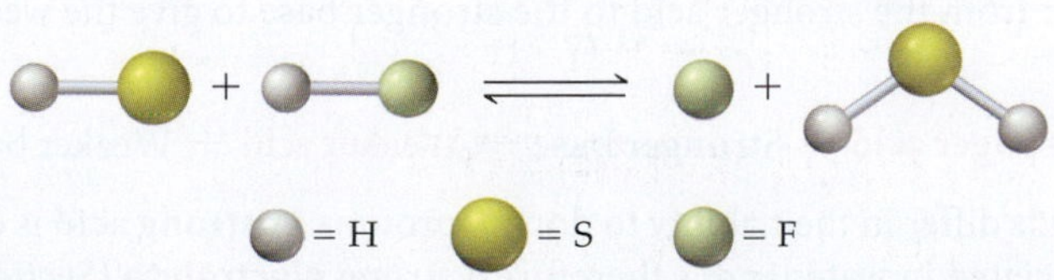

continued on next page

STRATEGY

The simplest approach is to identify the conjugate acid–base pairs, the species whose formulas differ by just one proton.

SOLUTION

The second reactant is HF, and the first product is its conjugate base F^-. The second product is H_2S, and the first reactant is its conjugate base HS^-. Therefore, the Brønsted–Lowry acids, bases, and conjugate acid–base pairs are as follows:

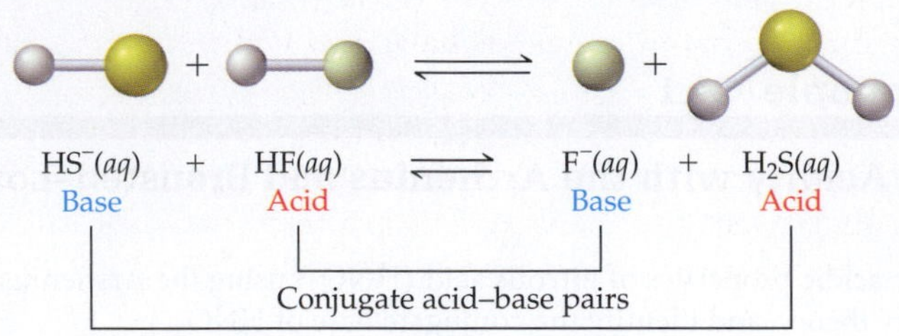

CONCEPTUAL PROBLEM 14.3 For the following reaction in aqueous solution, identify the Brønsted–Lowry acids, bases, and conjugate acid–base pairs:

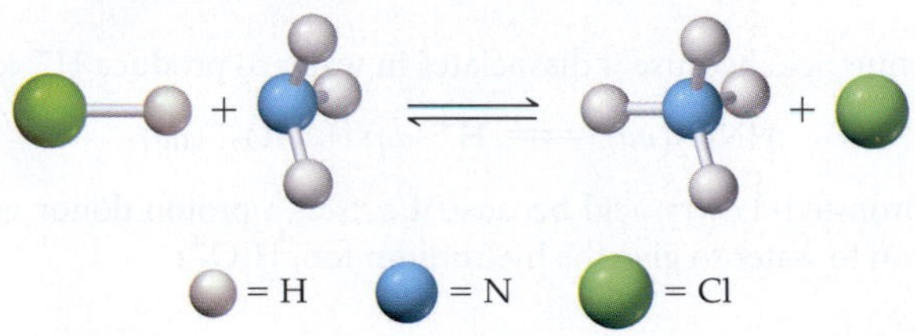

PROBLEM 14.4 Some of the following ions have been detected experimentally: $H_9O_4^+$, $H_{13}O_5^+$, $H_{19}O_9^+$, $H_{25}O_{11}^+$, $H_{43}O_{21}^+$.

(a) Which should be considered as hydrates of the proton?

(b) For the ions that are hydrates, tell how many water molecules are present in addition to the proton.

14.2 ACID STRENGTH AND BASE STRENGTH

A helpful way of viewing an acid-dissociation equilibrium is to realize that the two bases, H_2O and A^-, are competing for protons:

$$\underset{\text{Acid}}{HA(aq)} + \underset{\text{Base}}{H_2O(l)} \rightleftharpoons \underset{\text{Acid}}{H_3O^+(aq)} + \underset{\text{Base}}{A^-(aq)}$$

If H_2O is a stronger base (a stronger proton acceptor) than A^-, the H_2O molecules will get the protons and the solution will contain mainly H_3O^+ and A^-. If A^- is a stronger base than H_2O, the A^- ions will get the protons and the solution will contain mainly HA and H_2O. When beginning with equal concentrations of reactants and products, *the proton is always transferred to the stronger base.* This means that the direction of reaction to reach equilibrium is proton transfer from the stronger acid to the stronger base to give the weaker acid and the weaker base:

$$\text{Stronger acid} + \text{Stronger base} \rightarrow \text{Weaker acid} + \text{Weaker base}$$

REMEMBER...

Substances that dissolve in water to produce solutions that conduct electricity are called **electrolytes**. Molecular substances that dissociate into ions to a large extent are strong electrolytes, while those that dissociate to only a small extent are weak electrolytes. (Section 7.1)

Different acids differ in their ability to donate protons. A **strong acid** is one that is almost completely dissociated in water and is therefore a strong **electrolyte** (Section 7.1). Thus, the acid-dissociation equilibrium of a strong acid lies nearly 100% to the right, and the solution contains almost entirely H_3O^+ and A^- ions with only a negligible amount of undissociated HA molecules.

> ▶ **Strong acid** $HA(aq) + H_2O(l) \xrightarrow{\sim 100\%} H_3O^+(aq) + A^-(aq)$

Typical strong acids are perchloric acid ($HClO_4$), hydrochloric acid (HCl), hydrobromic acid (HBr), hydroiodic acid (HI), nitric acid (HNO_3), and sulfuric acid (H_2SO_4). It follows from this definition that strong acids have very weak conjugate bases. The anions ClO_4^-, Cl^-, Br^-, I^-, NO_3^-, and HSO_4^- have only a negligible tendency to combine with a proton in aqueous solution, and they are therefore much weaker bases than H_2O.

A **weak acid** is one that is only partially dissociated in water and is thus a weak electrolyte. Only a small fraction of the weak acid molecules transfer a proton to water, and the solution therefore contains mainly undissociated HA molecules along with small amounts of H_3O^+ and the conjugate base A^-.

> ▶ **Weak Acid** $HA(aq) + H_2O(l) \rightleftharpoons H_3O^+(aq) + A^-(aq)$

Typical weak acids are nitrous acid (HNO_2), hydrofluoric acid (HF), and acetic acid (CH_3CO_2H). In the case of very weak acids, such as NH_3, OH^-, and H_2, the acid has practically no tendency to transfer a proton to water and the acid-dissociation equilibrium lies essentially 100% to the left. It follows from this definition that very weak acids have strong conjugate bases. For example, the NH_2^-, O^{2-}, and H^- ions are essentially 100% protonated in aqueous solution and are much stronger bases than H_2O.

The equilibrium concentrations of HA, H_3O^+, and A^- for strong acids, weak acids, and very weak acids are represented in **Figure 14.1**, and the inverse relationship between the strength of an acid and the strength of its conjugate base is illustrated in **Table 14.1**.

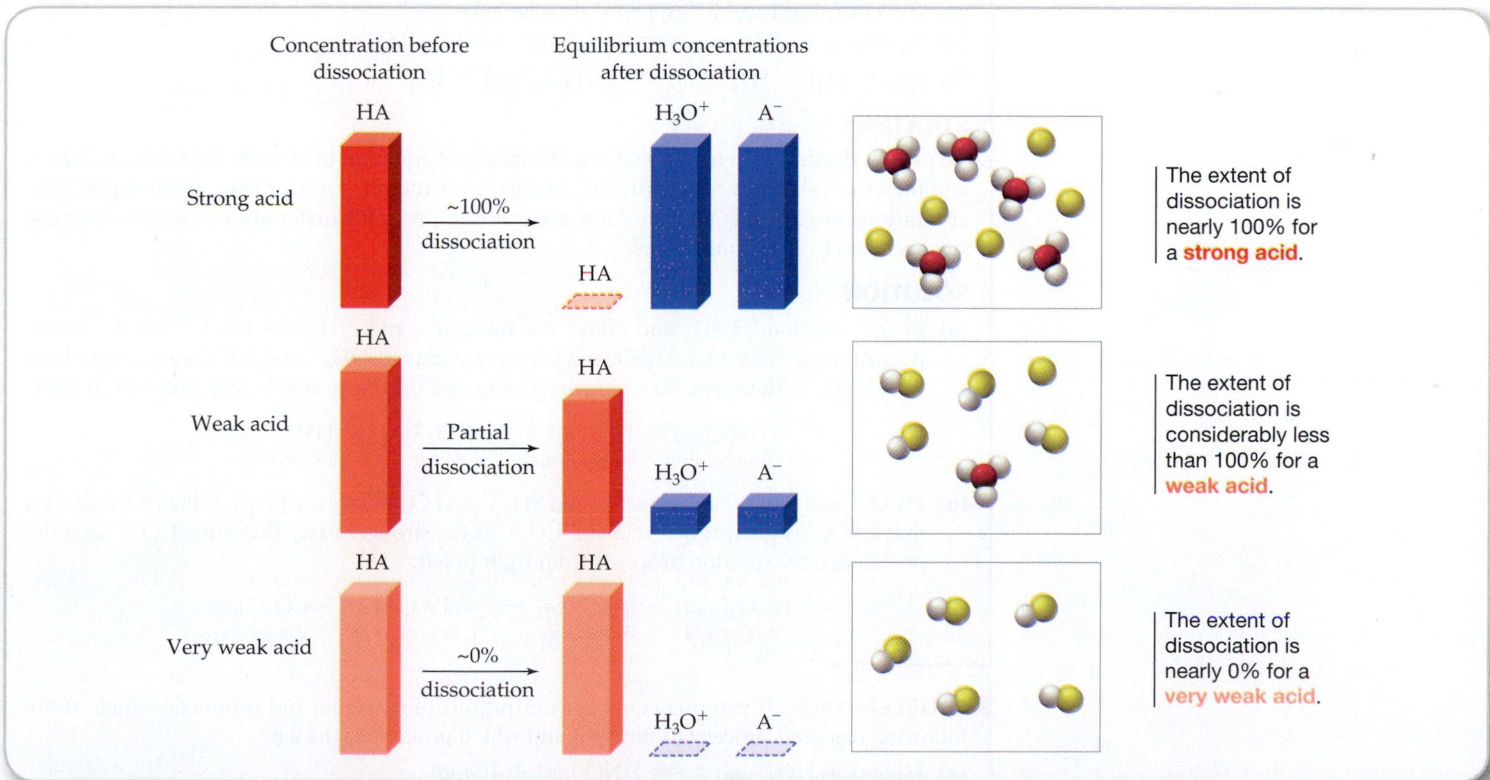

FIGURE 14.1

Dissociation of HA involves H^+ transfer to H_2O, yielding H_3O^+ and A^-.

TABLE 14.1 Relative Strengths of Conjugate Acid–Base Pairs

	Acid, HA		Base, A^-		
Stronger acid	$HClO_4$	**Strong acids:** 100% dissociated in aqueous solution.	ClO_4^-	**Very weak bases:** Negligible tendency to be protonated in aqueous solution.	Weaker base
	HCl		Cl^-		
	H_2SO_4		HSO_4^-		
	HNO_3		NO_3^-		
	H_3O^+		H_2O		
	HSO_4^-	**Weak acids:** Exist in solution as a mixture of HA, A^-, and H_3O^+.	SO_4^{2-}	**Weak bases:** Moderate tendency to be protonated in aqueous solution.	
	H_3PO_4		$H_2PO_4^-$		
	HNO_2		NO_2^-		
	HF		F^-		
	CH_3CO_2H		$CH_3CO_2^-$		
	H_2CO_3		HCO_3^-		
	H_2S		HS^-		
	NH_4^+		NH_3		
	HCN		CN^-		
	HCO_3^-		CO_3^{2-}		
	H_2O		OH^-		
	NH_3	**Very weak acids:** Negligible tendency to dissociate.	NH_2^-	**Strong bases:** 100% protonated in aqueous solution.	
	OH^-		O^{2-}		
Weaker acid	H_2		H^-		Stronger base

Worked Example 14.3

Predicting the Direction of Acid–Base Reactions

If you mix equal concentrations of reactants and products, which of the following reactions proceed to the right and which proceed to the left?

(a) $H_2SO_4(aq) + NH_3(aq) \rightleftharpoons NH_4^+(aq) + HSO_4^-(aq)$

(b) $HCO_3^-(aq) + SO_4^{2-}(aq) \rightleftharpoons HSO_4^-(aq) + CO_3^{2-}(aq)$

STRATEGY

To predict the direction of reaction, use the balanced equation to identify the acids and bases, and then use Table 14.1 to identify the stronger acid and the stronger base. When equal concentrations of reactants and products are present, proton transfer always occurs from the stronger acid to the stronger base.

SOLUTION

(a) In this reaction, H_2SO_4 and NH_4^+ are the acids, and NH_3 and HSO_4^- are the bases. According to Table 14.1, H_2SO_4 is a stronger acid than NH_4^+ and NH_3 is a stronger base than HSO_4^-. Therefore, NH_3 gets the proton and the reaction proceeds from left to right.

$$\underset{\text{Stronger acid}}{H_2SO_4(aq)} + \underset{\text{Stronger base}}{NH_3(aq)} \xrightarrow{100\%} \underset{\text{Weaker acid}}{NH_4^+(aq)} + \underset{\text{Weaker base}}{HSO_4^-(aq)}$$

(b) HCO_3^- and HSO_4^- are the acids, and SO_4^{2-} and CO_3^{2-} are the bases. Table 14.1 indicates that HSO_4^- is the stronger acid and CO_3^{2-} is the stronger base. Therefore, CO_3^{2-} gets the proton and the reaction proceeds from right to left.

$$\underset{\text{Weaker acid}}{HCO_3^-(aq)} + \underset{\text{Weaker base}}{SO_4^{2-}(aq)} \longleftarrow \underset{\text{Stronger acid}}{HSO_4^-(aq)} + \underset{\text{Stronger base}}{CO_3^{2-}(aq)}$$

PROBLEM 14.5 If you mix equal concentrations of reactants and products, which of the following reactions proceed to the right and which proceed to the left?

(a) $HF(aq) + NO_3^-(aq) \rightleftharpoons HNO_3(aq) + F^-(aq)$

(b) $NH_4^+(aq) + CO_3^{2-}(aq) \rightleftharpoons HCO_3^-(aq) + NH_3(aq)$

CONCEPTUAL PROBLEM 14.6 The following pictures represent aqueous solutions of two acids HA (A = X or Y); water molecules have been omitted for clarity.

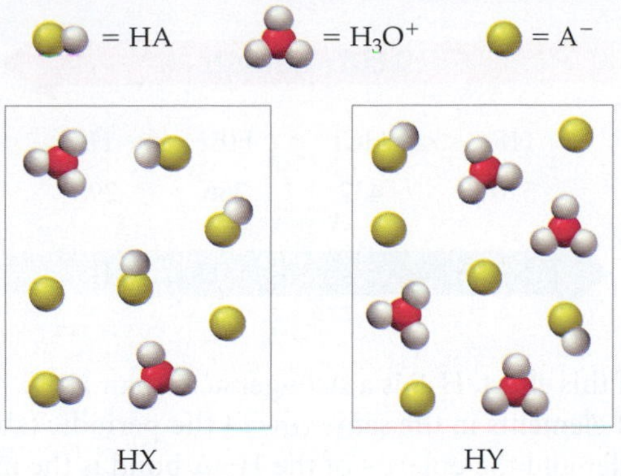

(a) Which is the stronger acid, HX or HY?
(b) Which is the stronger base, X^- or Y^-?
(c) If you mix equal concentrations of reactants and products, will the following reaction proceed to the right or to the left?

$$HX + Y^- \rightleftharpoons HY + X^-$$

14.3 FACTORS THAT AFFECT ACID STRENGTH

Why is one acid stronger than another? Although a complete analysis of the factors that determine the strength of an acid is complex, the extent of dissociation of a binary acid HA is often determined by the strength and polarity of the H–A bond. The **strength of the H – A bond**, as we saw in Section 8.10, is the enthalpy required to dissociate HA into an H atom and an A atom. The polarity of the H–A bond increases with an increase in the electronegativity of A and is related to the ease of electron transfer from an H atom to an A atom to give an H^+ cation and an A^- anion. In general, the weaker and more polar the H–A bond, the stronger the acid.

REMEMBER...
The **strength of a bond** is measured by the bond dissociation energy, which for an H–A bond equals $\Delta H°$ for the reaction $HA(g) \longrightarrow H(g) + A(g)$. (Section 8.10)

Let's look first at the hydrohalic acids HF, HCl, HBr, and HI. **Electrostatic potential maps** (Section 4.4) show that all these molecules are polar, with the halogen atom being electron rich (red) and the H atom being electron poor (blue).

REMEMBER...
Electrostatic potential maps use color to portray the calculated electron distribution in a molecule. Electron-rich regions are red, and electron-poor regions are blue. Intermediate regions may be yellow, orange, or green. (Section 4.4)

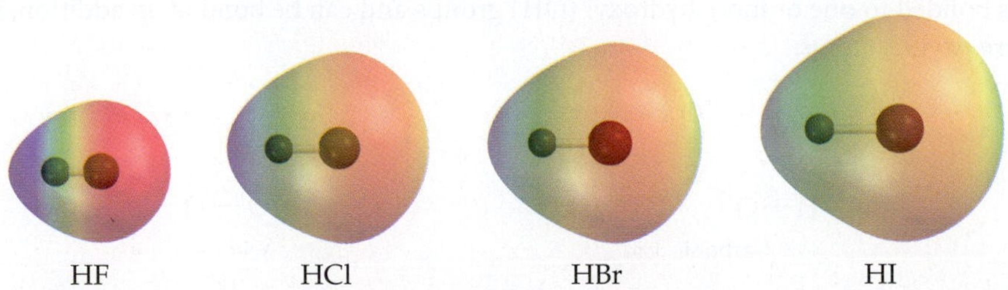

The variation in polarity in this series, however, is much less important than the variation in bond strength, which decreases markedly from 570 kJ/mol for HF to 298 kJ/mol for HI.

In general, for binary acids of elements in the same group of the periodic table, the H–A bond strength is the most important determinant of acidity. The H–A bond strength generally decreases down a group because overlap between the 1*s* orbital of the H atom and the larger, more diffuse *p* orbital of the A atom decreases with increasing size of the element A, and so acid strength increases. For HA (A = F, Cl, Br, or I), for example, the size of A increases from F to I, so bond strength decreases and acidity increases from HF to HI. Hydrofluoric acid is a weak acid, whereas HCl, HBr, and HI are strong acids.

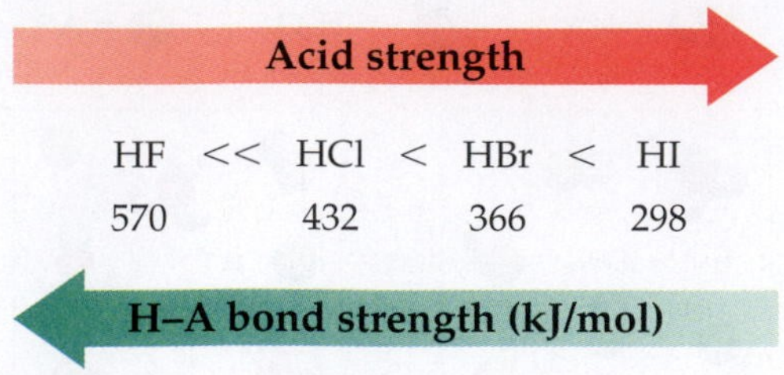

As a further example of this effect, H_2S is a stronger acid than H_2O.

For binary acids of elements in the same row of the periodic table, changes in the H—A bond strength are smaller and the polarity of the H—A bond is the most important determinant of acid strength. The strengths of binary acids of the second-row elements, for example, increase as the electronegativity of A increases:

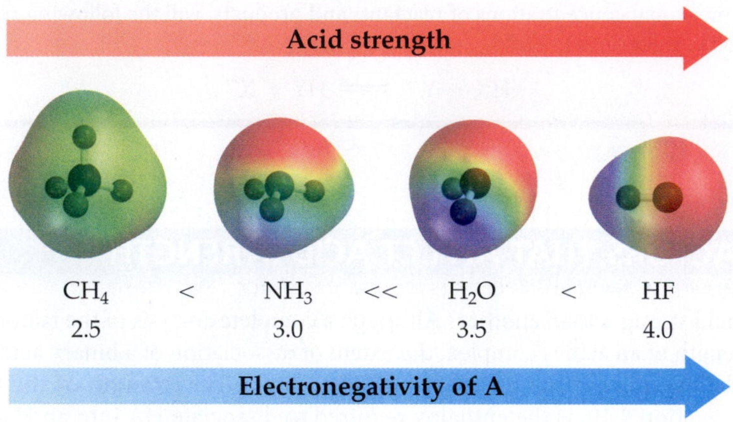

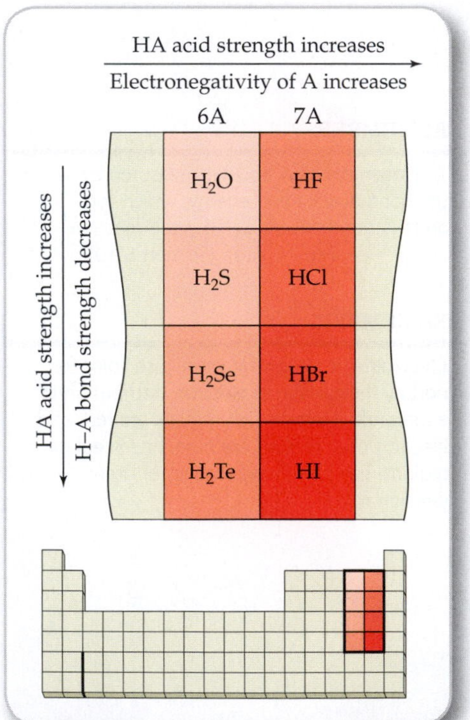

FIGURE 14.2

Periodic trends in the strength of binary acids. The acid strength of a binary acid HA increases from left to right in the periodic table, with increasing electronegativity of A, and from top to bottom, with decreasing H—A bond strength.

As the electrostatic potential maps show, the C–H bond is relatively nonpolar and methane has no tendency to dissociate in water into H_3O^+ and CH_3^- ions. The N–H bond is more polar, but dissociation of NH_3 into H_3O^+ and NH_2^- ions is still negligibly small. Water and hydrofluoric acid, however, are increasingly stronger acids. Periodic trends in the strength of binary acids are summarized in **Figure 14.2**.

Oxoacids, such as H_2CO_3, HNO_3, H_2SO_4, and HClO, have the general formula H_nYO_m, where Y is a nonmetallic atom, such as C, N, S, or Cl, and *n* and *m* are integers. The atom Y is always bonded to one or more hydroxyl (OH) groups and can be bonded, in addition, to one or more oxygen atoms:

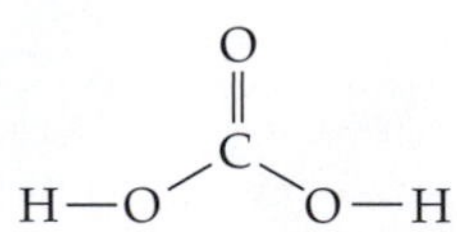

Carbonic acid

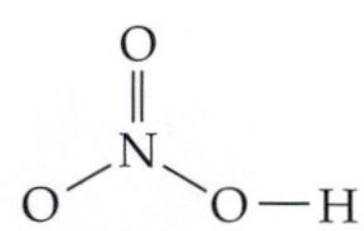

Nitric acid

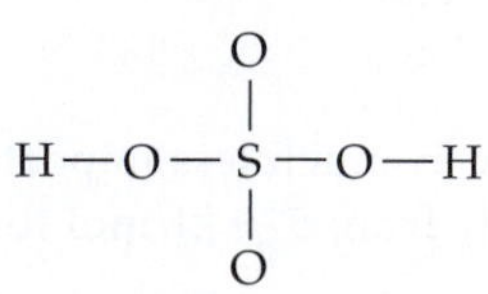

Sulfuric acid

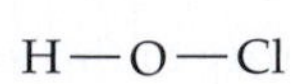

Hypochlorous acid

Because the dissociation of an oxoacid requires breaking an O—H bond, any factor that weakens the O—H bond or increases its polarity increases the strength of the acid.

Two such factors are the electronegativity of Y and the oxidation number of Y in the general reaction

$$-\overset{|}{\underset{|}{\mathrm{Y}}}-\mathrm{O}-\mathrm{H} + \mathrm{H_2O} \rightleftharpoons \mathrm{H_3O^+} + -\overset{|}{\underset{|}{\mathrm{Y}}}-\mathrm{O}^-$$

- For oxoacids that contain the same number of OH groups and the same number of O atoms, acid strength increases as the electronegativity of Y increases. For example, the acid strength of the hypohalous acids HOY (Y = Cl, Br, or I) increases as the electronegativity of the halogen increases:

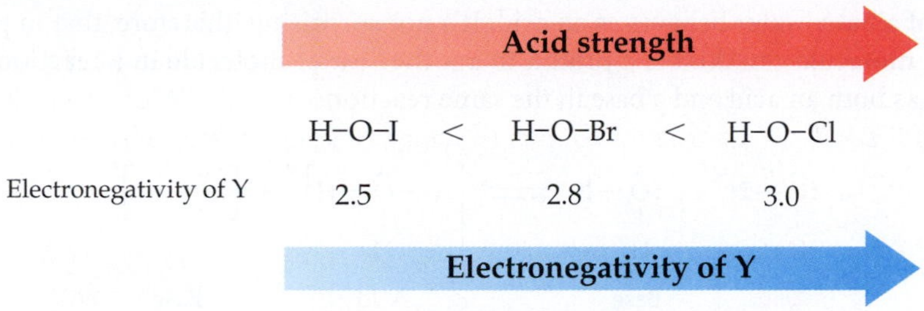

As the halogen becomes more electronegative, an increasing amount of electron density shifts from the O—H bond toward the halogen atom, thus weakening the O–H bond and increasing its polarity. As a result, the proton is more easily transferred to a solvent water molecule and so the acid strength increases.

- For oxoacids that contain the same atom Y but different numbers of oxygen atoms, acid strength increases as the oxidation number of Y increases. The oxidation number of Y increases, in turn, as the number of oxygen atoms increases. This effect is illustrated by the oxoacids of chlorine:

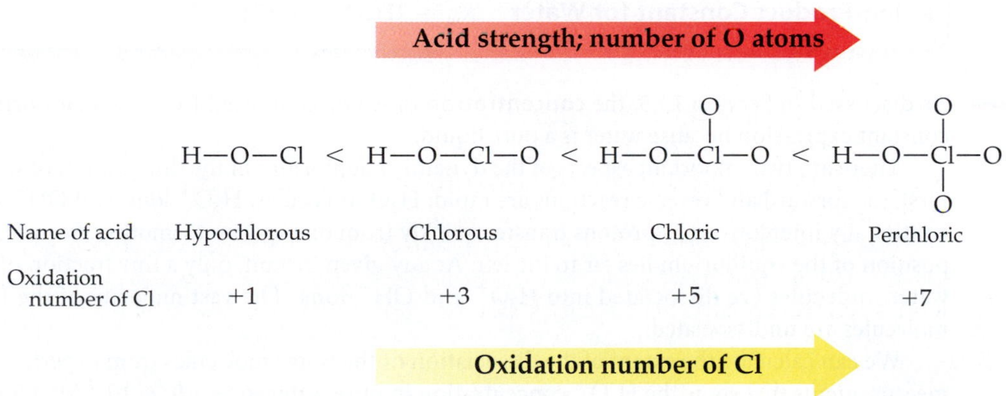

As the number of O atoms in $HClO_m$ increases, an increasing amount of electron density shifts from the Cl atom toward the more electronegative O atoms. The amount of positive charge on the Cl atom therefore increases as its oxidation number increases. The increased positive charge on the Cl atom in turn attracts an increasing amount of electron density from the O—H bond, thus weakening the O–H bond and increasing its polarity. As a result, the proton is more easily transferred to a solvent water molecule.

Another factor that affects the acid strength of oxoacids is the relative stability of the corresponding oxoanions. The ClO_m^- anion becomes more stable as the number of O atoms increases in the series $ClO^- < ClO_2^- < ClO_3^- < ClO_4^-$ because a larger number of electronegative O atoms can better accommodate the anion's negative charge. As the stability of the anion increases, the corresponding acid has a greater tendency to dissociate. The increase in acid strength with increasing number of O atoms is further illustrated by the oxoacids of sulfur: H_2SO_4 is a stronger acid than H_2SO_3.

PROBLEM 14.7 Identify the stronger acid in each of the following pairs:
(a) H_2S or H_2Se
(b) HI or H_2Te
(c) HNO_2 or HNO_3
(d) H_2SO_3 or H_2SeO_3

14.4 DISSOCIATION OF WATER

In Section 14.1, we saw that a water molecule can act both as a proton donor (an acid) and as a proton acceptor (a base). In the presence of an acid, water behaves as a base, whereas in the presence of a base, water behaves as an acid. It's not surprising, therefore, that in pure water one water molecule can donate a proton to another water molecule in a reaction in which water acts as both an acid and a base in the same reaction:

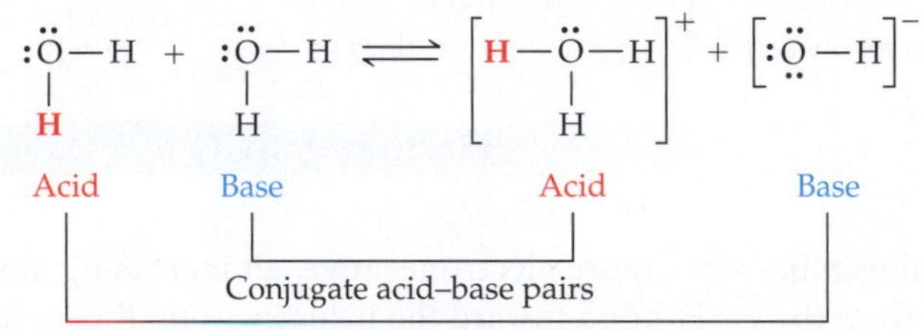

Called the *dissociation of water*, this reaction is characterized by the equilibrium equation $K_w = [H_3O^+][OH^-]$, where the equilibrium constant K_w is called the *ion-product constant for water*.

▶ **Dissociation of Water** $2\,H_2O(l) \rightleftharpoons H_3O^+(aq) + OH^-(aq)$

▶ **Ion-Product Constant for Water** $K_w = [H_3O^+][OH^-]$

REMEMBER...

The **concentration** of a pure liquid or a pure solid is omitted from the equilibrium-constant expression because the ratio of its actual concentration to its concentration in the thermodynamic standard state is equal to 1. (Section 13.5)

As discussed in Section 13.5, the **concentration** of water is omitted from the equilibrium-constant expression because water is a pure liquid.

There are two important aspects of the dynamic equilibrium in the dissociation of water. First, the forward and reverse reactions are rapid: H_2O molecules, H_3O^+ ions, and OH^- ions continually interconvert as protons transfer quickly from one species to another. Second, the position of the equilibrium lies far to the left: At any given instant, only a tiny fraction of the water molecules are dissociated into H_3O^+ and OH^- ions. The vast majority of the H_2O molecules are undissociated.

We can calculate the extent of the dissociation of the water molecules from experimental measurements that show the H_3O^+ concentration in pure water to be 1.0×10^{-7} M at 25 °C:

$$[H_3O^+] = 1.0 \times 10^{-7}\ \text{M} \quad \text{at } 25\ ^\circ\text{C}$$

Since the dissociation reaction of water produces equal concentrations of H_3O^+ and OH^- ions, the OH^- concentration in pure water is also 1.0×10^{-7} M at 25 °C:

$$[H_3O^+] = [OH^-] = 1.0 \times 10^{-7}\ \text{M} \quad \text{at } 25\ ^\circ\text{C}$$

Furthermore, we know that the molar concentration of pure water, calculated from its density and molar mass, is 55.4 M at 25 °C:

$$[H_2O] = \left(\frac{997\ \text{g}}{\text{L}}\right)\left(\frac{1\ \text{mol}}{18.0\ \text{g}}\right) = 55.4\ \text{mol/L} \quad \text{at } 25\ ^\circ\text{C}$$

From these facts, we conclude that the ratio of dissociated to undissociated water molecules is about 2 in 10^9, a very small number indeed:

$$\frac{[H_2O]_{\text{dissociated}}}{[H_2O]_{\text{undissociated}}} = \frac{1.0 \times 10^{-7}\ \text{M}}{55.4\ \text{M}} = 1.8 \times 10^{-9} \qquad \text{about 2 in } 10^9$$

In addition, we can calculate that the numerical value of K_w at 25 °C is 1.0×10^{-14}:

$$K_w = [H_3O^+][OH^-] = (1.0 \times 10^{-7})(1.0 \times 10^{-7})$$
$$= 1.0 \times 10^{-14} \quad \text{at } 25\ °C$$

In very dilute solutions, the water is almost a pure liquid and the product of the H_3O^+ and OH^- concentrations is unaffected by the presence of solutes. This is not true in more concentrated solutions, but we'll neglect that complication and assume that the product of the H_3O^+ and OH^- concentrations is always 1.0×10^{-14} at 25 °C in any aqueous solution.

We can distinguish acidic, neutral, and basic aqueous solutions by the relative values of the H_3O^+ and OH^- concentrations:

Acidic: $[H_3O^+] > [OH^-]$

Neutral: $[H_3O^+] = [OH^-]$

Basic: $[H_3O^+] < [OH^-]$

At 25 °C, $[H_3O^+] > 1.0 \times 10^{-7}$ M in an acidic solution, $[H_3O^+] = [OH^-] = 1.0 \times 10^{-7}$ M in a neutral solution, and $[H_3O^+] < 1.0 \times 10^{-7}$ M in a basic solution (**Figure 14.3**). If one of the concentrations, $[H_3O^+]$ or $[OH^+]$, is known, the other is readily calculated:

Since $[H_3O^+][OH^-] = K_w = 1.0 \times 10^{-14}$

then $[H_3O^+] = \dfrac{1.0 \times 10^{-14}}{[OH^-]}$ and $[OH^-] = \dfrac{1.0 \times 10^{-14}}{[H_3O^+]}$

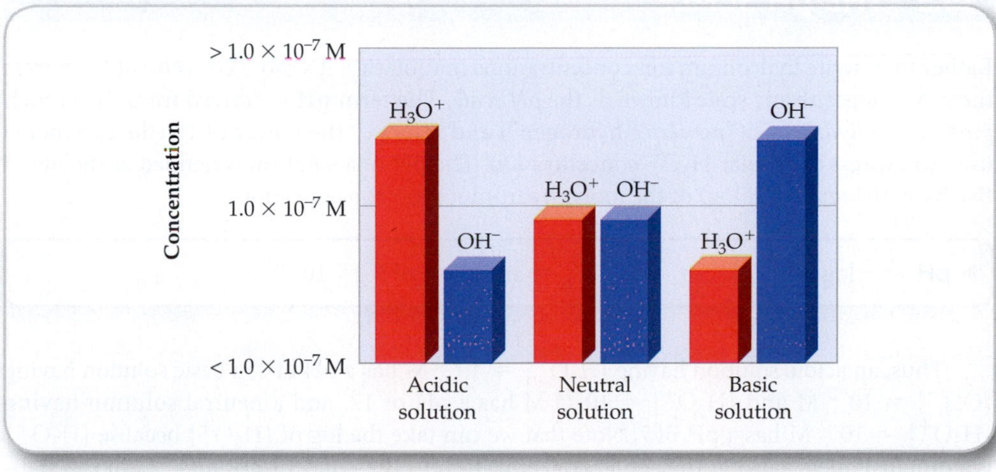

FIGURE 14.3

Values of the H_3O^+ and OH^- concentrations at 25 °C in acidic, neutral, and basic solutions.

In the previous discussion, we were careful to emphasize that the value of $K_w = 1.0 \times 10^{-14}$ applies only at 25 °C. That is because K_w, like all equilibrium constants, is affected by temperature and the H_3O^+ and OH^- concentrations in neutral aqueous solutions at temperatures other than 25 °C deviate from 1.0×10^{-7} M (see Problem 14.10). Unless otherwise indicated, we'll always assume a temperature of 25 °C.

Worked Example 14.4

Calculating $[OH^-]$ from K_w and $[H_3O^+]$

The concentration of H_3O^+ ions in a sample of lemon juice is 2.5×10^{-3} M. Calculate the concentration of OH^- ions, and classify the solution as acidic, neutral, or basic.

STRATEGY

When $[H_3O^+]$ is known, the OH^- concentration can be found from the expression $[OH^-] = K_w/[H_3O^+]$.

continued on next page

SOLUTION

$$[\text{OH}^-] = \frac{K_w}{[\text{H}_3\text{O}^+]}$$

$$= \frac{1.0 \times 10^{-14}}{2.5 \times 10^{-3}} = 4.0 \times 10^{-12}\ \text{M}$$

Because $[H_3O^+] > [OH^-]$, the solution is acidic.

✓ BALLPARK CHECK

Because the product of the H_3O^+ and OH^- concentrations must equal 10^{-14}, and because the H_3O^+ concentration is in the range 10^{-3} M to 10^{-2} M, the OH^- concentration must be in the range 10^{-11} M to 10^{-12} M. The ballpark check and the solution agree.

PROBLEM 14.8 The concentration of H_3O^+ ions in the runoff from a coal mine is 1.4×10^{-4} M. Calculate the concentration of OH^- ions, and classify the solution as acidic, neutral, or basic.

PROBLEM 14.9 The concentration of OH^- in a sample of seawater is 2.0×10^{-6} M. Calculate the concentration of H_3O^+ ions, and classify the solution as acidic, neutral, or basic.

PROBLEM 14.10 At 50 °C the value of K_w is 5.5×10^{-14}. What are the concentrations of H_3O^+ and OH^- in a neutral solution at 50 °C?

14.5 THE pH SCALE

Rather than write hydronium ion concentrations in molarity, it's more convenient to express them on a logarithmic scale known as the *pH scale*. The term **pH** is derived from the French *puissance d'hydrogène* ("power of hydrogen") and refers to the power of 10 (the exponent) used to express the molar H_3O^+ concentration. The pH of a solution is defined as the negative base-10 logarithm (log) of the molar hydronium ion concentration:

▶ **pH** $= -\log[\text{H}_3\text{O}^+]$ or $[\text{H}_3\text{O}^+] = \text{antilog}(-\text{pH}) = 10^{-\text{pH}}$

Thus, an acidic solution having $[H_3O^+] = 10^{-2}$ M has a pH of 2, a basic solution having $[OH^-] = 10^{-2}$ M and $[H_3O^+] = 10^{-12}$ M has a pH of 12, and a neutral solution having $[H_3O^+] = 10^{-7}$ M has a pH of 7. Note that we can take the log of $[H_3O^+]$ because $[H_3O^+]$ is a dimensionless ratio of the actual concentration to the concentration (1 M) in the standard state. Although much less frequently used than pH, a pOH can be defined in the same way as pH and used to express the molar OH^- concentration. Just as $\text{pH} = -\log[H_3O^+]$, so $\text{pOH} = -\log[OH^-]$. It follows from the equation $[H_3O^+][OH^-] = 1.0 \times 10^{-14}$ that $\text{pH} + \text{pOH} = 14.00$.

If you use a calculator to find the pH from the H_3O^+ concentration, your answer will have more decimal places than the proper number of significant figures. For example, the pH of the lemon juice in Worked Example 14.4 ($[H_3O^+] = 2.5 \times 10^{-3}$ M) is found on a calculator to be

$$\text{pH} = -\log(2.5 \times 10^{-3}) = 2.602\,06$$

This result should be rounded to pH 2.60 (two significant figures) because $[H_3O^+]$ has only two significant figures. Note that the only significant figures in a logarithm are the digits to the right of the decimal point; the number to the left of the decimal point is an exact number related to the integral power of 10 in the exponential expression for $[H_3O^+]$:

$$\text{pH} = -\log(\underbrace{2.5}_{\text{2 significant figures (2 SF's)}} \times \underbrace{10^{-3}}_{\text{Exact number}}) = -\log 10^{-3} - \log 2.5 = \underbrace{3}_{\text{Exact number}} - \underbrace{0.40}_{\text{2 SF's}} = \underbrace{2}_{\text{Exact number}}.\underbrace{60}_{\text{2 SF's}}$$

Because the pH scale is logarithmic, the pH changes by 1 unit when $[H_3O^+]$ changes by a factor of 10, by 2 units when $[H_3O^+]$ changes by a factor of 100, and by 6 units when $[H_3O^+]$ changes by a factor of 1,000,000. To appreciate the extent to which the pH scale is a compression of the $[H_3O^+]$ scale, compare the amounts of 12 M HCl required to change the pH of the water in a backyard swimming pool: Only about 100 mL of 12 M HCl is needed to change the pH from 7 to 6, but a 10,000 L truckload of 12 M HCl is needed to change the pH from 7 to 1.

The pH scale and pH values for some common substances are shown in **Figure 14.4**. Because the pH is the *negative* log of $[H_3O^+]$, the pH decreases as $[H_3O^+]$ increases. Thus, when $[H_3O^+]$ increases from 10^{-7} M to 10^{-6} M, the pH decreases from 7 to 6. At 25 °C, acidic solutions have pH less than 7 and basic solutions have pH greater than 7.

Acidic solution:	$pH < 7$	at 25 °C
Neutral solution:	$pH = 7$	at 25 °C
Basic solution:	$pH > 7$	at 25 °C

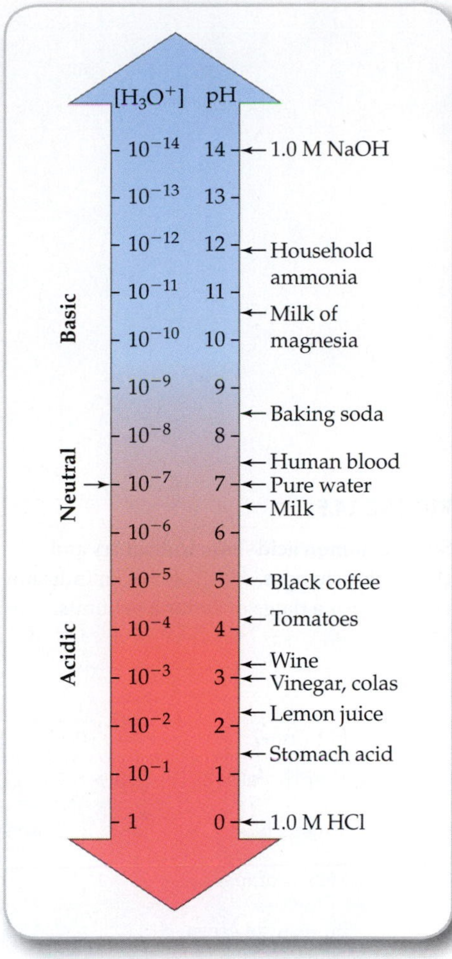

FIGURE 14.4

The pH scale and pH values for some common substances at 25 °C.

Worked Example 14.5

Calculating the pH from the H_3O^+ Concentration

Calculate the pH of an aqueous ammonia solution that has an OH^- concentration of 1.9×10^{-3} M.

STRATEGY

First, calculate the H_3O^+ concentration from the OH^- concentration, and then take the negative logarithm of $[H_3O^+]$ to convert to pH.

SOLUTION

$$[H_3O^+] = \frac{K_w}{[OH^-]} = \frac{1.0 \times 10^{-14}}{1.9 \times 10^{-3}} = 5.3 \times 10^{-12}\ M$$

$$pH = -\log [H_3O^+] = -\log (5.3 \times 10^{-12}) = 11.28$$

The pH is quoted to two significant figures (.28) because $[H_3O^+]$ is known to two significant figures (5.3).

✔ BALLPARK CHECK

Because $[OH^-]$ is between 10^{-3} M and 10^{-2} M, $[H_3O^+]$ is between 10^{-11} M and 10^{-12} M. Therefore, the pH is between 11 and 12, in agreement with the solution.

Worked Example 14.6

Calculating the H_3O^+ Concentration from the pH

Acid rain is a matter of serious concern because most species of fish die in waters having a pH lower than 4.5–5.0. (See the *FYI* at the end of this chapter.) Calculate the H_3O^+ concentration in a lake that has a pH of 4.5.

STRATEGY

Calculate the H_3O^+ concentration by taking the antilogarithm of the negative of the pH.

SOLUTION

$$[H_3O^+] = \text{antilog}(-pH) = 10^{-pH} = 10^{-4.5} = 3 \times 10^{-5}\ M$$

$[H_3O^+]$ is reported to only one significant figure because the pH has only one digit beyond the decimal point. (If you need help in finding the antilog of a number, see Appendix A.2.)

✔ BALLPARK CHECK

Because a pH of 4.5 is between 4 and 5, $[H_3O^+]$ is between 10^{-4} M and 10^{-5} M, in agreement with the solution.

PROBLEM 14.11 Calculate the pH of each of the following solutions:
(a) A sample of seawater that has an OH^- concentration of 1.58×10^{-6} M
(b) A sample of acid rain that has an H_3O^+ concentration of 6.0×10^{-5} M

PROBLEM 14.12 Calculate the concentrations of H_3O^+ and OH^- in each of the following solutions:
(a) Human blood (pH 7.40) **(b)** A cola beverage (pH 2.8)

14.6 MEASURING pH

The approximate pH of a solution can be determined by using an **acid–base indicator**, a substance that changes color in a specific pH range **(Figure 14.5)**. Indicators (abbreviated HIn) exhibit pH-dependent color changes because they are weak acids and have different colors in their acid (HIn) and conjugate base (In^-) forms:

$$\underset{\text{Color A}}{HIn(aq)} + H_2O(l) \rightleftharpoons H_3O^+(aq) + \underset{\text{Color B}}{In^-(aq)}$$

Bromthymol blue, for example, changes color in the pH range 6.0–7.6, from yellow in its acid form to blue in its base form.

FIGURE 14.5

Some common acid–base indicators and their color changes. The color of an indicator changes over a range of about 2 pH units.

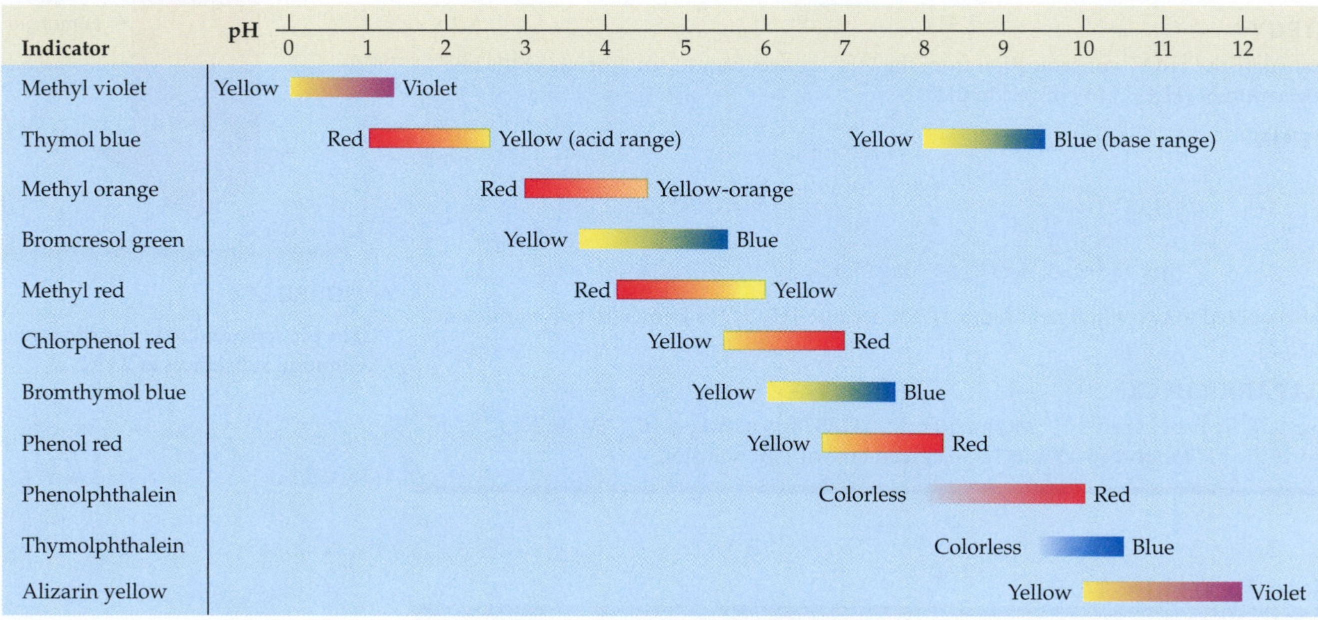

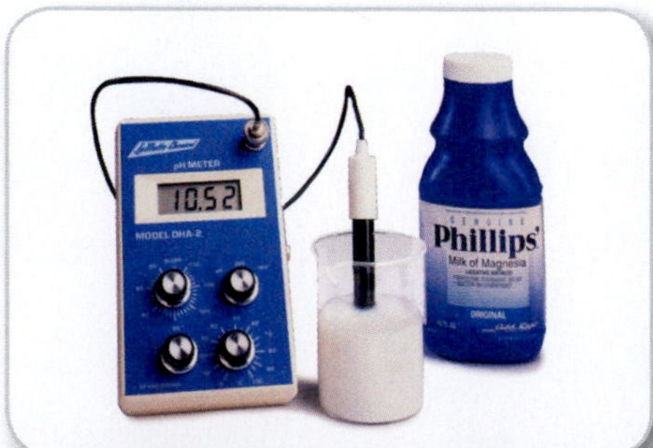

FIGURE 14.6

A pH meter with its electrical probe dipping into milk of magnesia. An accurate value of the pH (10.52) is shown on the meter.

Especially convenient for making approximate pH measurements is a commercially available mixture of indicators known as *universal indicator* that exhibits various colors depending on the pH (see Figure 14.9 on page 581). More accurate pH values can be determined with an electronic instrument called a *pH meter* **(Figure 14.6)**, a device that measures the pH-dependent electrical potential of the test solution. We'll learn more about pH meters in Section 18.7.

14.7 THE pH IN SOLUTIONS OF STRONG ACIDS AND STRONG BASES

How do we calculate the pH in solutions of strong acids and strong bases? The commonly encountered strong acids listed in Table 14.1 include three *monoprotic acids* ($HClO_4$, HCl, and HNO_3), which contain a single dissociable proton, and one *diprotic acid* (H_2SO_4), which has two dissociable protons. Because strong monoprotic acids are nearly 100% dissociated in aqueous solution, the H_3O^+ and A^- concentrations are equal to the initial concentration of the acid and the concentration of undissociated HA molecules is essentially zero.

$$HA(aq) + H_2O(l) \xrightarrow{100\%} H_3O^+(aq) + A^-(aq)$$

The pH of a solution of a strong monoprotic acid is easily calculated from the H_3O^+ concentration, as shown in Worked Example 14.7. Calculation of the pH of an H_2SO_4 solution is more complicated because essentially 100% of the H_2SO_4 molecules dissociate to give H_3O^+ and HSO_4^- ions but much less than 100% of the resulting HSO_4^- ions dissociate to give H_3O^+ and SO_4^{2-} ions. We'll have more to say about **diprotic acids** in Section 14.11.

LOOKING AHEAD ...

In Section 14.11 we'll see why a **diprotic acid** H_2A loses its first proton more easily than it loses its second proton.

The most familiar examples of strong bases are alkali metal hydroxides, MOH, such as NaOH (*caustic soda*) and KOH (*caustic potash*). These compounds are water-soluble ionic solids that exist in aqueous solution as alkali metal cations (M^+) and OH^- anions:

$$MOH(s) \xrightarrow{H_2O} MOH(aq) \xrightarrow{100\%} M^+(aq) + OH^-(aq)$$

Thus, 0.10 M NaOH contains 0.10 M Na^+ and 0.10 M OH^-, and the pH is readily calculated from the OH^- concentration, as shown in Worked Example 14.8a.

The alkaline earth metal hydroxides $M(OH)_2$ (M = Mg, Ca, Sr, or Ba) are also strong bases (~100% dissociated), but they give lower OH^- concentrations because they are less soluble. Their solubility at room temperature varies from 38 g/L for the relatively soluble $Ba(OH)_2$ to ~10^{-2} g/L for the relatively insoluble $Mg(OH)_2$. Aqueous suspensions of $Mg(OH)_2$, called *milk of magnesia*, are used as an antacid. The most common and least expensive alkaline earth hydroxide is $Ca(OH)_2$, which is used in making mortars and cements. $Ca(OH)_2$ is called *slaked lime* because it is made by treating *lime* (CaO) with water. Aqueous solutions of $Ca(OH)_2$, which has a solubility of only ~1.3 g/L, are known as *limewater*.

▲ Lime is spread on lawns and gardens to raise the pH of acidic soils.

Alkaline earth oxides, such as CaO, are even stronger bases than the corresponding hydroxides because the oxide ion (O^{2-}) is a stronger base than OH^- (Table 14.1). In fact, the O^{2-} ion can't exist in aqueous solutions because it is immediately and completely protonated by water, yielding OH^- ions:

$$O^{2-}(aq) + H_2O(l) \xrightarrow{100\%} OH^-(aq) + OH^-(aq)$$

Thus, when lime is dissolved in water, it gives 2 OH^- ions per CaO formula unit:

$$CaO(s) + H_2O(l) \longrightarrow Ca^{2+}(aq) + 2\,OH^-(aq)$$

Lime is the world's most important strong base. Annual worldwide production is approximately 280 million metric tons for use in steelmaking, water purification, and chemical manufacture. Lime is made by the decomposition of limestone, $CaCO_3$, at temperatures of 800–1000 °C:

$$CaCO_3(s) \xrightarrow{\text{Heat}} CaO(s) + CO_2(g)$$

Worked Example 14.7

Calculating the pH of a Strong Acid Solution

Calculate the pH of a 0.025 M HNO_3 solution.

STRATEGY

Because nitric acid is a strong acid, it is almost completely dissociated in aqueous solution. Therefore, the $[H_3O^+]$ equals the initial concentration of the HNO_3, and the pH equals the negative log of the $[H_3O^+]$.

SOLUTION

$$HNO_3(aq) + H_2O(l) \xrightarrow{100\%} H_3O^+(aq) + NO_3^-(aq)$$

$$pH = -\log[H_3O^+] = -\log(2.5 \times 10^{-2}) = 1.60$$

✔ BALLPARK CHECK

Because the $[H_3O^+]$ is between 10^{-1} M and 10^{-2} M, the pH is between 1 and 2, in agreement with the solution.

Worked Example 14.8

Calculating the pH of a Strong Base Solution

Calculate the pH of each of the following solutions:

(a) A 0.10 M solution of NaOH

(b) A 0.0050 M solution of slaked lime [$Ca(OH)_2$]

(c) A solution prepared by dissolving 0.28 g of lime (CaO) in enough water to make 1.00 L of limewater [$Ca(OH)_2(aq)$]

STRATEGY

Because NaOH and $Ca(OH)_2$ are strong bases, they are essentially 100% dissociated and their $[OH^-]$ is directly related to their initial concentrations. To calculate the $[OH^-]$ in a solution prepared by dissolving CaO, we must first do a mass-to-mole conversion and then use the balanced equation for the reaction of CaO with water to find the number of moles of OH^- in the solution. In each case, $[H_3O^+] = K_w/[OH^-]$ and $pH = -\log [H_3O^+]$.

SOLUTION

(a) Because NaOH is a strong base, $[OH^-] = 0.10\ M$, $[H_3O^+] = 1.0 \times 10^{-13}\ M$, and $pH = 13.00$.

$$[H_3O^+] = \frac{K_w}{[OH^-]} = \frac{1.0 \times 10^{-14}}{0.10} = 1.0 \times 10^{-13}\ M$$

$$pH = -\log (1.0 \times 10^{-13}) = 13.00$$

(b) Because slaked lime is a strong base, it provides 2 OH^- ions per $Ca(OH)_2$ formula unit. Therefore, $[OH^-] = 2(0.0050\ M) = 0.010\ M$, $[H_3O^+] = 1.0 \times 10^{-12}\ M$, and $pH = 12.00$:

$$[H_3O^+] = \frac{K_w}{[OH^-]} = \frac{1.0 \times 10^{-14}}{0.010} = 1.0 \times 10^{-12}\ M$$

$$pH = -\log (1.0 \times 10^{-12}) = 12.00$$

(c) First calculate the number of moles of CaO dissolved from the given mass of CaO and its molar mass (56.1 g/mol):

$$\text{Moles of CaO} = 0.28\ \text{g CaO} \times \frac{1\ \text{mol CaO}}{56.1\ \text{g CaO}} = 0.0050\ \text{mol CaO}$$

Protonation of the O^{2-} ion produces 2 mol of OH^- per mol of CaO dissolved:

$$CaO(s) + H_2O(l) \longrightarrow Ca^{2+}(aq) + 2\ OH^-(aq)$$

$$\text{Moles of } OH^- \text{ produced} = 2(0.0050\ \text{mol}) = 0.010\ \text{mol}$$

Since the solution volume is 1.00 L,

$$[OH^-] = \frac{0.010\ \text{mol}}{1.00\ \text{L}} = 0.010\ M$$

The $[OH^-]$ happens to be identical to that in part **(b)**. Therefore, $pH = 12.00$.

PROBLEM 14.13 Calculate the pH of the following solutions:

(a) 0.050 M $HClO_4$

(b) 6.0 M HCl

(c) 4.0 M KOH

(d) 0.010 M $Ba(OH)_2$

PROBLEM 14.14 Calculate the pH of a solution prepared by dissolving 0.25 g of BaO in enough water to make 0.500 L of solution.

14.8 EQUILIBRIA IN SOLUTIONS OF WEAK ACIDS

Now that we've discussed ion concentrations and the pH in solutions of strong acids and strong bases, we can examine the slightly more complicated case of weak acids. It's important to realize that a weak acid is not the same thing as a dilute solution of a strong acid. Whereas a strong acid is 100% dissociated in aqueous solution, a weak acid is only partially dissociated. It might therefore happen that the H_3O^+ concentration from complete dissociation of a dilute strong acid is the same as that from partial dissociation of a more concentrated weak acid.

Like the equilibrium reactions discussed in Chapter 13, the dissociation of a weak acid in water is characterized by an equilibrium equation. For the general reaction

$$HA(aq) + H_2O(l) \rightleftharpoons H_3O^+(aq) + A^-(aq)$$

the equilibrium constant is called the **acid-dissociation constant**, K_a.

▶ **Acid-dissociation constant** $K_a = \frac{[H_3O^+][A^-]}{[HA]}$

The right side of the equilibrium equation has the same form as the equilibrium-constant expressions in Chapter 13, and the equilibrium constant K_a can be regarded as a special kind of K_c. Note that water has been omitted from the equilibrium equation because its concentration in dilute solutions is essentially the same as that in pure water (55.4 M) and pure liquids are always omitted from equilibrium equations (Section 13.5).

Values of K_a and pK_a for some typical weak acids are listed in **Table 14.2**. Just as the pH is defined as $-\log\,[H^+]$, so the pK_a of an acid is defined as $-\log K_a$. Note that the pK_a decreases as K_a increases. As indicated by the equilibrium equation, the larger the value of K_a, the stronger the acid. Thus, methanol ($K_a = 2.9 \times 10^{-16}$; p$K_a$ = 15.54) is the weakest of the acids listed in Table 14.2, and nitrous acid ($K_a = 4.5 \times 10^{-4}$; pK_a = 3.35) is the strongest of the weak acids. Strong acids, such as HCl, have K_a values that are much greater than 1 and pK_a values that are negative. A more complete list of K_a values for weak acids is given in Appendix C.

TABLE 14.2 Acid-Dissociation Constants at 25 °C

	Acid	Molecular Formula	Structural Formula*	K_a	pK_a†
Stronger acid	Hydrochloric	HCl	H—Cl	2×10^{6}	−6.3
	Nitrous	HNO_2	H—O—N=O	4.5×10^{-4}	3.35
	Hydrofluoric	HF	H—F	3.5×10^{-4}	3.46
	Formic	HCO_2H	H—C(=O)—O—H	1.8×10^{-4}	3.74
	Ascorbic (vitamin C)	$C_6H_8O_6$	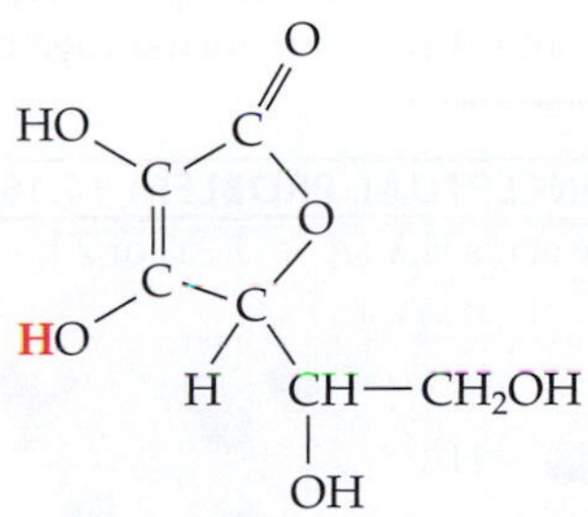	8.0×10^{-5}	4.10
	Acetic	CH_3CO_2H	CH_3—C(=O)—O—H	1.8×10^{-5}	4.74
	Hypochlorous	HOCl	H—O—Cl	3.5×10^{-8}	7.46
Weaker acid	Hydrocyanic	HCN	H—C≡N	4.9×10^{-10}	9.31
	Methanol	CH_3OH	CH_3—O—H	2.9×10^{-16}	15.54

* The proton that is transferred to water when the acid dissociates is shown in red.
† p$K_a = -\log K_a$.

Numerical values of acid-dissociation constants are determined from pH measurements, as shown in Worked Example 14.9.

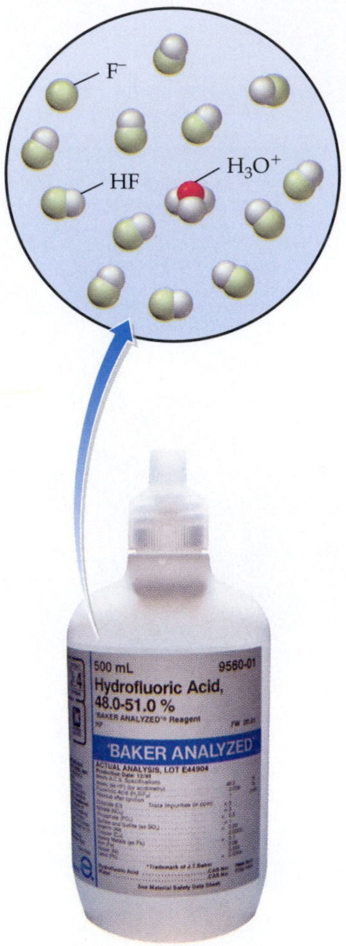

▲ Of the hydrohalic acids HF, HCl, HBr, and HI, hydrofluoric acid is the only weak acid. Which of these acids has a positive value of pK_a?

Worked Example 14.9

Calculating K_a and pK_a for a Weak Acid from the pH of the Solution

The pH of 0.250 M HF is 2.036. What are the values of K_a and pK_a for hydrofluoric acid?

STRATEGY

First, write the balanced equation for the dissociation equilibrium. Then, define x as the concentration of HF that dissociates and make the usual table under the balanced equation (Steps 1 and 2 of Figure 13.6 on page 522). Because x equals the H_3O^+ concentration, its value can be calculated from the pH. Finally, substitute the equilibrium concentrations into the equilibrium equation to obtain the value of K_a and take the negative log of K_a to obtain the pK_a.

SOLUTION

	$HF(aq)$ +	$H_2O(l) \rightleftharpoons$	$H_3O^+(aq)$ +	$F^-(aq)$
Initial concentration (M)	0.250		~0*	0
Change (M)	$-x$		$+x$	$+x$
Equilibrium concentration (M)	$(0.250 - x)$		x	x

*A very small concentration of H_3O^+ is present initially because of the dissociation of water.

We can calculate the value of x from the pH:

$$x = [H_3O^+] = \text{antilog}(-\text{pH}) = 10^{-\text{pH}} = 10^{-2.036} = 9.20 \times 10^{-3}\ \text{M}$$

The other equilibrium concentrations are

$$[F^-] = x = 9.20 \times 10^{-3}\ \text{M}$$
$$[HF] = 0.250 - x = 0.250 - 0.009\,20 = 0.241\ \text{M}$$

Substituting these concentrations into the equilibrium equation gives the value of K_a:

$$K_a = \frac{[H_3O^+][F^-]}{[HF]} = \frac{(x)(x)}{(0.250 - x)} = \frac{(9.20 \times 10^{-3})(9.20 \times 10^{-3})}{0.241} = 3.51 \times 10^{-4}$$

$$pK_a = -\log K_a = -\log(3.51 \times 10^{-4}) = 3.455$$

✓ BALLPARK CHECK

Because the pH is about 2, $[H_3O^+]$ and $[F^-]$ are about 10^{-2} M and [HF] is about 0.25 M (0.250 M − 10^{-2} M). The value of K_a is therefore about $(10^{-2})(10^{-2})/0.25$, or 4×10^{-4}, and the pK_a is between 3 and 4. The ballpark check and the solution agree.

PROBLEM 14.15 The pH of 0.10 M HOCl is 4.23. Calculate K_a and pK_a for hypochlorous acid, and check your answers against the values given in Table 14.2.

CONCEPTUAL PROBLEM 14.16 The following pictures represent aqueous solutions of three acids HA (A = X, Y, or Z); water molecules have been omitted for clarity:

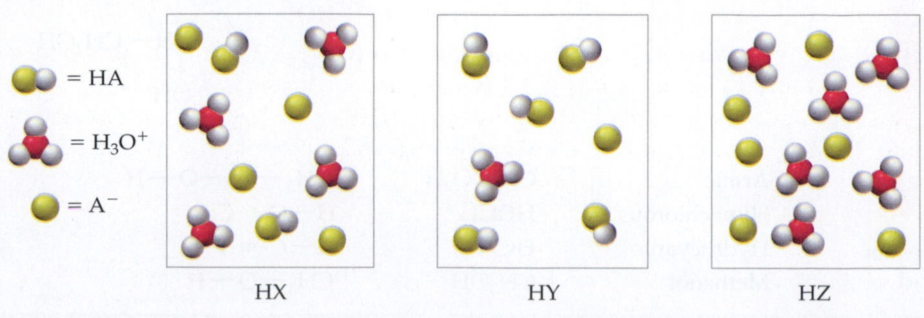

(a) Arrange the three acids in order of increasing value of K_a, assuming that all three have the same concentration.
(b) Which acid, if any, is a strong acid?
(c) Which solution has the highest pH, and which has the lowest pH?

14.9 CALCULATING EQUILIBRIUM CONCENTRATIONS IN SOLUTIONS OF WEAK ACIDS

Once the K_a value for a weak acid has been measured, it can be used to calculate equilibrium concentrations and the pH in a solution of the acid. We'll illustrate the approach to such a problem by calculating the concentrations of all species present (H_3O^+, CN^-, HCN, and OH^-) and the pH in a 0.10 M HCN solution. The approach we'll take is quite general and will be useful on numerous later occasions.

The key to solving acid–base equilibrium problems is to think about the chemistry—that is, to consider the possible proton-transfer reactions that can take place between Brønsted–Lowry acids and bases.

Step 1 Let's begin by listing the species present initially before any dissociation and by identifying them as acids or bases. Since water can behave either as an acid or a base, our list of species present initially is

HCN	H_2O
Acid	Acid or base

Step 2 Because we have two acids (HCN and H_2O) and just one base (H_2O), two proton-transfer reactions are possible:

$$HCN(aq) + H_2O(l) \rightleftharpoons H_3O^+(aq) + CN^-(aq) \qquad K_a = 4.9 \times 10^{-10}$$
$$H_2O(l) + H_2O(l) \rightleftharpoons H_3O^+(aq) + OH^-(aq) \qquad K_w = 1.0 \times 10^{-14}$$

The value of K_a for HCN comes from Table 14.2.

Step 3 The proton-transfer reaction that proceeds farther to the right—the one that has the larger equilibrium constant—is called the **principal reaction**. Any other proton-transfer reactions are called **subsidiary reactions**. Since K_a for HCN is more than 10,000 times greater than K_w, the principal reaction in this case is dissociation of HCN and dissociation of water is a subsidiary reaction.

Although the principal reaction and the subsidiary reaction both produce H_3O^+ ions, there is only one H_3O^+ concentration in the solution, which must simultaneously satisfy the equilibrium equations for both reactions. To make life simple, we'll assume that essentially all the H_3O^+ comes from the principal reaction:

$$[H_3O^+]\ (\text{total}) = [H_3O^+]\ (\text{from principal reaction}) + [H_3O^+]\ (\text{from subsidiary reaction})$$
$$\approx [H_3O^+]\ (\text{from principal reaction})$$

In other words, we'll assume that the equilibrium concentration of H_3O^+ is established entirely by the dissociation of the stronger acid, HCN, while dissociation of the weaker acid, H_2O, makes a negligible contribution.

$$[H_3O^+]\ (\text{total}) \approx [H_3O^+]\ (\text{from HCN})$$

Step 4 Next, we express the concentrations of the species involved in the principal reaction in terms of the concentration of HCN that dissociates—say, x mol/L. According to the balanced equation for the dissociation of HCN, if x mol/L of HCN dissociates, then x mol/L of H_3O^+ and x mol/L of CN^- are formed and the initial concentration of HCN before dissociation (0.10 mol/L in our example) is reduced to $(0.10 - x)$ mol/L at equilibrium. Let's summarize these considerations in the usual table under the principal reaction:

Principal Reaction:	$HCN(aq)$ + $H_2O(l)$ ⇌	$H_3O^+(aq)$ +	$CN^-(aq)$
Initial concentration (M)	0.10	~0	0
Change (M)	$-x$	$+x$	$+x$
Equilibrium concentration (M)	$0.10 - x$	x	x

Step 5 Substituting the equilibrium concentrations into the equilibrium equation for the principal reaction gives

$$K_a = 4.9 \times 10^{-10} = \frac{[H_3O^+][CN^-]}{[HCN]} = \frac{(x)(x)}{(0.10 - x)}$$

Because K_a is very small, the principal reaction will not proceed very far to the right and x will be negligibly small compared to 0.10. Therefore, we can make the approximation that $(0.10 - x) \approx 0.10$, which greatly simplifies the solution:

$$4.9 \times 10^{-10} = \frac{(x)(x)}{(0.10 - x)} = \frac{x^2}{0.10}$$

$$x^2 = 4.9 \times 10^{-11}$$

$$x = 7.0 \times 10^{-6}$$

Step 6 Next, we use the calculated value of x to obtain the equilibrium concentration of all species involved in the principal reaction:

$$[H_3O^+] = [CN^-] = x = 7.0 \times 10^{-6}\ \text{M}$$

$$[HCN] = 0.10 - x = 0.10 - (7.0 \times 10^{-6}) = 0.10\ \text{M}$$

Note that our simplifying approximation, $0.10 - x \approx 0.10$, is valid because x is only 7.0×10^{-6} and the initial [HCN] is known to only the second decimal place (0.10). In general, x can be neglected if the value of the initial concentration doesn't change when x is subtracted from it. In this example, x could not be neglected if it were greater than or equal to 0.01. *It's important to check the validity of the simplifying approximation in every problem* because x is not always negligible compared to the initial concentration of the acid. Worked Example 14.10 illustrates such a case.

Step 7 The species involved in the principal reaction are present in "big" concentrations. The species involved in the subsidiary reaction(s) are present in smaller concentrations that can be calculated from equilibrium equations for the subsidiary reaction(s) and the big concentrations already determined. In the present problem, only the OH^- concentration remains to be calculated. It is determined from the subsidiary equilibrium equation, $[H_3O^+][OH^-] = K_w$, and the H_3O^+ concentration (7.0×10^{-6} M) already calculated from the principal reaction:

$$[OH^-] = \frac{K_w}{[H_3O^+]} = \frac{1.0 \times 10^{-14}}{7.0 \times 10^{-6}} = 1.4 \times 10^{-9}\ \text{M}$$

Note that $[OH^-]$ is 5000 times smaller than $[H_3O^+]$.

At this point, we can check the initial assumption that essentially all the H_3O^+ comes from the principal reaction. Because the dissociation of water gives one H_3O^+ ion for each OH^- ion and because water is the only source of OH^-, $[H_3O^+]$ from the dissociation of water is equal to $[OH^-]$, which we just calculated to be 1.4×10^{-9} M. This value is negligible compared with $[H_3O^+]$ from the dissociation of HCN (7.0×10^{-6} M).

$$[H_3O^+]\ (\text{total}) = [H_3O^+]\ (\text{from HCN}) + [H_3O^+]\ (\text{from } H_2O)$$

$$= (7.0 \times 10^{-6}\ \text{M}) + (1.4 \times 10^{-9}\ \text{M}) = 7.0 \times 10^{-6}\ \text{M}$$

Step 8 Finally, we can calculate the pH:

$$\text{pH} = -\log(\text{total}\ [H_3O^+]) = -\log(7.0 \times 10^{-6}) = 5.15$$

Figure 14.7 summarizes the steps followed in solving this problem. This same systematic approach is applied to all acid–base equilibrium problems in this chapter and in Chapter 15.

Step 1. List the species present before dissociation and identify them as Brønsted–Lowry acids or bases.

↓

Step 2. Write balanced equations for all possible proton-transfer reactions.

↓

Step 3. Identify the principal reaction—the reaction that has the largest equilibrium constant.

↓

Step 4. Make a table that lists the following values for each of the species involved in the principal reaction:
- **(a)** The initial concentration
- **(b)** The change in concentration on proceeding to equilibrium
- **(c)** The equilibrium concentration

In constructing this table, define x as the concentration (mol/L) of the acid that dissociates.

↓

Step 5. Substitute the equilibrium concentrations into the equilibrium equation for the principal reaction, and solve for x.

↓

Step 6. Calculate the "big" concentrations—the concentrations of the species involved in the principal reaction.

↓

Step 7. Use the big concentrations and the equilibrium equations for the subsidiary reactions to calculate the small concentrations—the concentrations of the species involved in the subsidiary equilibria.

↓

Step 8. Calculate the pH = $-\log [H_3O^+]$.

FIGURE 14.7

Steps to follow in solving problems involving weak acids.

Worked Example 14.10

Calculating the pH and the Equilibrium Concentrations in a Solution of a Weak Acid

Calculate the pH and the concentrations of all species present (H_3O^+, F^-, HF, and OH^-) in 0.050 M HF.

STRATEGY

Follow the eight-step sequence outlined in Figure 14.7.

SOLUTION

Step 1 The species present initially are

HF	H_2O
Acid	Acid or base

Step 2 The possible proton-transfer reactions are

$$HF(aq) + H_2O(l) \rightleftharpoons H_3O^+(aq) + F^-(aq) \qquad K_a = 3.5 \times 10^{-4}$$
$$H_2O(l) + H_2O(l) \rightleftharpoons H_3O^+(aq) + OH^-(aq) \qquad K_w = 1.0 \times 10^{-14}$$

Step 3 Because $K_a \gg K_w$, the principal reaction is the dissociation of HF.

continued on next page

Step 4

Principal Reaction:	$HF(aq)$	+ $H_2O(l)$ $\rightleftharpoons$	$H_3O^+(aq)$	+ $F^-(aq)$
Initial concentration (M)	0.050		~0	0
Change (M)	$-x$		$+x$	$+x$
Equilibrium concentration (M)	$0.050 - x$		x	x

Step 5 Substituting the equilibrium concentrations into the equilibrium equation for the principal reaction gives

$$K_a = 3.5 \times 10^{-4} = \frac{[H_3O^+][F^-]}{[HF]} = \frac{(x)(x)}{(0.050 - x)}$$

Making the usual approximation that x is negligible compared with the initial concentration of the acid, we assume that $(0.050 - x) \approx 0.050$ and then solve for an approximate value of x:

$$x^2 \approx (3.5 \times 10^{-4})(0.050)$$
$$x \approx 4.2 \times 10^{-3}$$

Since the initial concentration of HF (0.050 M) is known to the third decimal place, x is negligible compared with the initial [HF] only if x is less than 0.001 M. Our approximate value of x (0.0042 M) is not negligible compared with 0.050 M because 0.050 M − 0.0042 M = 0.046 M. Therefore, our approximation, $0.050 - x \approx 0.050$, is invalid, and we must solve the quadratic equation without making approximations:

$$3.5 \times 10^{-4} = \frac{x^2}{(0.050 - x)}$$
$$x^2 + (3.5 \times 10^{-4})x - (1.75 \times 10^{-5}) = 0$$

We use the standard quadratic formula (Appendix A.4):

$$x = \frac{-b \pm \sqrt{b^2 - 4ac}}{2a}$$
$$= \frac{-(3.5 \times 10^{-4}) \pm \sqrt{(3.5 \times 10^{-4})^2 - 4(1)(-1.75 \times 10^{-5})}}{2(1)}$$
$$= \frac{-(3.5 \times 10^{-4}) \pm (8.37 \times 10^{-3})}{2}$$
$$= +4.0 \times 10^{-3} \quad \text{or} \quad -4.4 \times 10^{-3}$$

Of the two solutions for x, only the positive value has physical meaning, since x is the H_3O^+ concentration. Therefore,

$$x = 4.0 \times 10^{-3}$$

Note that whether we must solve the quadratic equation depends on both the size of x and the number of significant figures in the initial concentration.

Step 6 The big concentrations are

$$[H_3O^+] = [F^-] = x = 4.0 \times 10^{-3}\ M$$
$$[HF] = (0.050 - x) = (0.050 - 0.0040) = 0.046\ M$$

Step 7 The small concentration, $[OH^-]$, is obtained from the subsidiary equilibrium, the dissociation of water:

$$[OH^-] = \frac{K_w}{[H_3O^+]} = \frac{1.0 \times 10^{-14}}{4.0 \times 10^{-3}} = 2.5 \times 10^{-12}\ M$$

Step 8 $pH = -\log[H_3O^+] = -\log(4.0 \times 10^{-3}) = 2.40$

✓ BALLPARK CHECK

Arithmetic errors in solving quadratic equations are common, so it's a good idea to check that the value of x obtained from the quadratic equation is reasonable. If the approximate value of x (4.2×10^{-3}) is fairly small compared to the initial concentration of the acid (0.050 M), as is the case in this problem, then the value of x obtained from the quadratic equation (4.0×10^{-3}) should be fairly close to the approximate value of x. The approximate and more exact values of x agree.

PROBLEM 14.17 Acetic acid, CH_3CO_2H, is the solute that gives vinegar its characteristic odor and sour taste. Calculate the pH and the concentrations of all species present (H_3O^+, $CH_3CO_2^-$, CH_3CO_2H, and OH^-) in:
(a) 1.00 M CH_3CO_2H **(b)** 0.0100 M CH_3CO_2H

PROBLEM 14.18 A vitamin C tablet containing 250 mg of ascorbic acid ($C_6H_8O_6$; $K_a = 8.0 \times 10^{-5}$) is dissolved in a 250 mL glass of water. What is the pH of the solution?

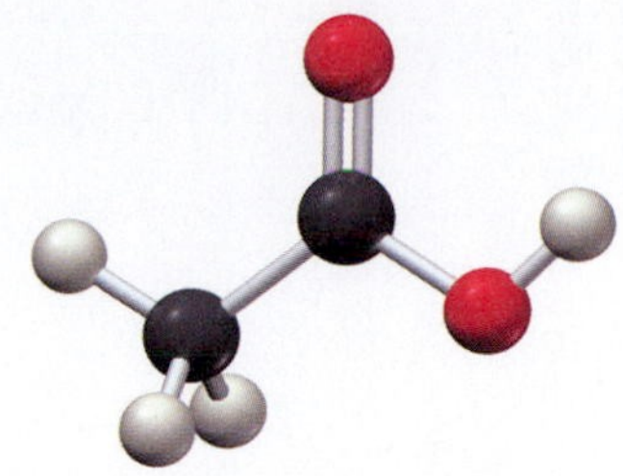

Acetic acid

14.10 PERCENT DISSOCIATION IN SOLUTIONS OF WEAK ACIDS

In addition to K_a, another useful measure of the strength of a weak acid is the **percent dissociation**, defined as the concentration of the acid that dissociates divided by the initial concentration of the acid times 100%:

$$\textbf{Percent dissociation} = \frac{[HA]_{dissociated}}{[HA]_{initial}} \times 100\%$$

Take, for example, the 1.00 M acetic acid solution in Problem 14.17a. If you solved that problem correctly, you found that 1.00 M CH_3CO_2H has $[H_3O^+] = 4.2 \times 10^{-3}$ M. Because $[H_3O^+]$ equals the concentration of CH_3CO_2H that dissociates, the percent dissociation in 1.00 M CH_3CO_2H is 0.42%:

$$\text{Percent dissociation} = \frac{[CH_3CO_2H]_{dissociated}}{[CH_3CO_2H]_{initial}} \times 100\%$$

$$= \frac{4.2 \times 10^{-3}\ \text{M}}{1.00\ \text{M}} \times 100\% = 0.42\%$$

In general, the percent dissociation depends on the acid and increases with increasing value of K_a. For a given weak acid, the percent dissociation increases with decreasing concentration, as shown in **Figure 14.8**. The 0.0100 M CH_3CO_2H solution in Problem 14.17b, for example, has $[H_3O^+] = 4.2 \times 10^{-4}$ M, and the percent dissociation is 4.2%:

$$\text{Percent dissociation} = \frac{[CH_3CO_2H]_{dissociated}}{[CH_3CO_2H]_{initial}} \times 100\%$$

$$= \frac{4.2 \times 10^{-4}\ \text{M}}{0.0100\ \text{M}} \times 100\% = 4.2\%$$

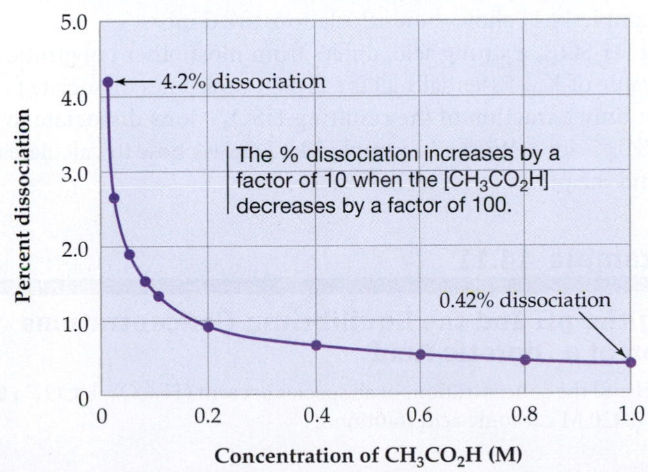

FIGURE 14.8
The percent dissociation of acetic acid increases as the concentration of the acid decreases.

PROBLEM 14.19 Calculate the percent dissociation of HF ($K_a = 3.5 \times 10^{-4}$) in:
(a) 0.050 M HF **(b)** 0.50 M HF

14.11 EQUILIBRIA IN SOLUTIONS OF POLYPROTIC ACIDS

Acids that contain more than one dissociable proton are called **polyprotic acids**. Polyprotic acids dissociate in a stepwise manner, and each dissociation step is characterized by its own acid-dissociation constant, K_{a1}, K_{a2}, and so forth. For example, carbonic acid (H_2CO_3), the diprotic acid that forms when gaseous carbon dioxide dissolves in water, is important in maintaining a constant pH in human blood. It undergoes the following dissociation reactions:

$$H_2CO_3(aq) + H_2O(l) \rightleftharpoons H_3O^+(aq) + HCO_3^-(aq) \qquad K_{a1} = \frac{[H_3O^+][HCO_3^-]}{[H_2CO_3]} = 4.3 \times 10^{-7}$$

$$HCO_3^-(aq) + H_2O(l) \rightleftharpoons H_3O^+(aq) + CO_3^{2-}(aq) \qquad K_{a2} = \frac{[H_3O^+][CO_3^{2-}]}{[HCO_3^-]} = 5.6 \times 10^{-11}$$

▲ Carbonated beverages contain polyprotic acids such as carbonic acid and phosphoric acid.

As shown in **Table 14.3**, the values of stepwise dissociation constants of polyprotic acids decrease, typically by a factor of 10^4 to 10^6, in the order $K_{a1} > K_{a2} > K_{a3}$. Because of electrostatic forces, it's more difficult to remove a positively charged proton from a negative ion such as HCO_3^- than from an uncharged molecule such as H_2CO_3, so $K_{a2} < K_{a1}$. In the case of triprotic acids (such as H_3PO_4), it's more difficult to remove H^+ from an anion with a double negative charge (such as HPO_4^{2-}) than from an anion with a single negative charge (such as $H_2PO_4^-$), so $K_{a3} < K_{a2}$.

TABLE 14.3 Stepwise Dissociation Constants for Polyprotic Acids at 25 °C

Name	Formula	K_{a1}	K_{a2}	K_{a3}
Carbonic acid	H_2CO_3	4.3×10^{-7}	5.6×10^{-11}	
Hydrogen sulfide*	H_2S	1.0×10^{-7}	$\sim 10^{-19}$	
Oxalic acid	$H_2C_2O_4$	5.9×10^{-2}	6.4×10^{-5}	
Phosphoric acid	H_3PO_4	7.5×10^{-3}	6.2×10^{-8}	4.8×10^{-13}
Sulfuric acid	H_2SO_4	Very large	1.2×10^{-2}	
Sulfurous acid	H_2SO_3	1.5×10^{-2}	6.3×10^{-8}	

*Because of its very small size, K_{a2} for H_2S is difficult to measure and its value is uncertain.

Polyprotic acid solutions contain a mixture of acids—H_2A, HA^-, and H_2O in the case of a diprotic acid. Because H_2A is by far the strongest acid, the principal reaction is the dissociation of H_2A and essentially all the H_3O^+ in the solution comes from the first dissociation step. Worked Example 14.11 shows how calculations are done.

Sulfuric acid (H_2SO_4), a strong acid, differs from most other polyprotic acids because it has a very large value of K_{a1}. Essentially all the H_2SO_4 molecules dissociate to give H_3O^+ and HSO_4^- ions, but only a fraction of the resulting HSO_4^- ions dissociate to give additional H_3O^+ ions and SO_4^{2-} ions. Worked Example 14.12 shows how to calculate the equilibrium concentrations and the pH.

Worked Example 14.11

Calculating the pH and the Equilibrium Concentrations in a Solution of a Diprotic Acid

Calculate the pH and the concentrations of all species present (H_2CO_3, HCO_3^-, CO_3^{2-}, H_3O^+, and OH^-) in a 0.020 M carbonic acid solution.

Carbonic acid

STRATEGY

Use the eight-step procedure summarized in Figure 14.7. The values of K_{a1} and K_{a2} may be found in Table 14.3.

SOLUTION

Steps 1–3 The species present initially are H_2CO_3 (acid) and H_2O (acid or base). Because $K_{a1} \gg K_w$, the principal reaction is the dissociation of H_2CO_3.

Step 4

Principal Reaction	$H_2CO_3(aq)$ + $H_2O(l)$ $\rightleftharpoons$	$H_3O^+(aq)$ +	$HCO_3^-(aq)$
Initial concentration (M)	0.020	~0	0
Change (M)	$-x$	$+x$	$+x$
Equilibrium concentration (M)	$0.020 - x$	x	x

Step 5 Substituting the equilibrium concentrations into the equilibrium equation for the principal reaction gives

$$K_{a1} = 4.3 \times 10^{-7} = \frac{[H_3O^+][HCO_3^-]}{[H_2CO_3]} = \frac{(x)(x)}{(0.020 - x)}$$

Assuming that $(0.020 - x) \approx 0.020$,

$$x^2 = (4.3 \times 10^{-7})(0.020)$$

$$x = 9.3 \times 10^{-5} \quad \text{Approximation } (0.020 - x) \approx 0.020 \text{ is justified.}$$

Step 6 The big concentrations are

$$[H_3O^+] = [HCO_3^-] = x = 9.3 \times 10^{-5}\ \text{M}$$

$$[H_2CO_3] = 0.020 - x = 0.020 - 0.000\,093 = 0.020\ \text{M}$$

Step 7 The small concentrations are obtained from the subsidiary equilibria—(1) dissociation of HCO_3^- and (2) dissociation of water—and from the big concentrations already determined:

$$(1)\ HCO_3^-(aq) + H_2O(l) \rightleftharpoons H_3O^+(aq) + CO_3^{2-}(aq)$$

$$K_{a2} = 5.6 \times 10^{-11} = \frac{[H_3O^+][CO_3^{2-}]}{[HCO_3^-]} = \frac{(9.3 \times 10^{-5})[CO_3^{2-}]}{9.3 \times 10^{-5}}$$

$$[CO_3^{2-}] = K_{a2} = 5.6 \times 10^{-11}\ \text{M}$$

(In general, for a solution of a weak diprotic acid that has a very small value of K_{a2}, $[A^{2-}] = K_{a2}$.)

$$(2)\ [OH^-] = \frac{K_w}{[H_3O^+]} = \frac{1.0 \times 10^{-14}}{9.3 \times 10^{-5}} = 1.1 \times 10^{-10}\ \text{M}$$

The second dissociation of H_2CO_3 produces a negligible amount of H_3O^+ compared with the H_3O^+ obtained from the first dissociation. Of the 9.3×10^{-5} mol/L of HCO_3^- produced by the first dissociation, only 5.6×10^{-11} mol/L dissociates to form H_3O^+ and CO_3^{2-}.

Step 8 $\text{pH} = -\log [H_3O^+] = -\log (9.3 \times 10^{-5}) = 4.03$

✓ BALLPARK CHECK

Because x can be neglected compared with 0.020 (Step 5), $[H_3O^+]$ equals the square root of K_{a1} times the initial concentration of the acid. In this problem, $[H_3O^+]$ is the square root of approximately $(4 \times 10^{-7})(2 \times 10^{-2})$, or about 10^{-4} M. Therefore, the pH $\approx$ 4, in agreement with the solution.

Worked Example 14.12

Calculating the pH and the Equilibrium Concentrations in a Sulfuric Acid Solution

Calculate the pH and the concentrations of all species present (HSO_4^-, SO_4^{2-}, H_3O^+, and OH^-) in 0.10 M H_2SO_4.

Sulfuric acid

STRATEGY

Use the eight-step procedure summarized in Figure 14.7.

continued on next page

▲ The brightness of the planet Venus is due in part to thick, highly reflective clouds in its upper atmosphere. These clouds consist of sulfur dioxide and droplets of sulfuric acid.

SOLUTION

Steps 1–3 Because the strong acid H_2SO_4 is essentially 100% dissociated to give H_3O^+ and HSO_4^- (K_{a1} is very large), the species present initially are H_3O^+ (acid), HSO_4^- (acid), and H_2O (acid or base). Although H_3O^+ is the strongest acid present, proton transfer from H_3O^+ to H_2O to give H_2O and H_3O^+ is not considered to be the principal reaction because it does not change any of the concentrations. The next strongest acid present is HSO_4^- ($K_{a2} > K_w$), so we consider the principal reaction to be dissociation of HSO_4^-.

Step 4

Principal Reaction	$HSO_4^-(aq)$ + $H_2O(l)$ ⇌	$H_3O^+(aq)$ +	$SO_4^{2-}(aq)$
Initial concentration (M)	0.10	0.10	0
Change (M)	$-x$	$+x$	$+x$
Equilibrium concentration (M)	$0.10 - x$	$0.10 + x$	x

Note that the second dissociation step takes place in the presence of 0.10 M H_3O^+ from the first dissociation step.

Step 5 Substituting the equilibrium concentrations into the equilibrium equation for the principal reaction gives

$$K_{a2} = 1.2 \times 10^{-2} = \frac{[H_3O^+][SO_4^{2-}]}{[HSO_4^-]} = \frac{(0.10 + x)(x)}{(0.10 - x)}$$

Neglecting x compared with 0.10 and solving this equation would give $x = K_{a2} = 0.012$, which is not negligible compared with 0.10. Therefore, we use the quadratic equation to obtain the value of x:

$$0.0012 - 0.012x = 0.10x + x^2$$

$$x^2 + 0.112x - 0.0012 = 0$$

$$x = \frac{-b \pm \sqrt{b^2 - 4ac}}{2a} = \frac{-0.112 \pm \sqrt{(0.112)^2 - 4(1)(-0.0012)}}{2(1)}$$

$$= \frac{-0.112 \pm 0.132}{2}$$

$$= +0.010 \quad \text{or} \quad -0.122$$

Because x is the SO_4^{2-} concentration, it must be positive. Therefore,

$$x = 0.010$$

Step 6 The big concentrations are

$$[SO_4^{2-}] = x = 0.010\text{ M}$$

$$[HSO_4^-] = 0.10 - x = 0.10 - 0.010 = 0.09\text{ M}$$

$$[H_3O^+] = 0.10 + x = 0.10 + 0.010 = 0.11\text{ M}$$

Step 7 The small concentration, $[OH^-]$, is obtained from the subsidiary equilibrium, dissociation of water:

$$[OH^-] = \frac{K_w}{[H_3O^+]} = \frac{1.0 \times 10^{-14}}{0.11} = 9.1 \times 10^{-14}\text{ M}$$

Step 8 $\text{pH} = -\log [H_3O^+] = -\log 0.11 = 0.96$

✓ BALLPARK CHECK

The easiest check in this case is to substitute the big concentrations obtained in Step 6 into the equilibrium equation for the second dissociation step and show that the equilibrium-constant expression equals K_{a2}:

$$\frac{[H_3O^+][SO_4^{2-}]}{[HSO_4^-]} = \frac{(0.11)(0.010)}{0.09} = 0.012 = K_{a2}$$

Since $[H_3O^+]$ is approximately 10^{-1} M, the pH should be about 1, in agreement with the detailed solution.

PROBLEM 14.20 Calculate the pH and the concentrations of all species present in 0.10 M H_2SO_3. Values of K_a are in Table 14.3.

PROBLEM 14.21 Like sulfuric acid, selenic acid (H_2SeO_4) is a diprotic acid that has a very large value of K_{a1}. Calculate the pH and the concentrations of all species present in 0.50 M H_2SeO_4 ($K_{a2} = 1.2 \times 10^{-2}$).

14.12 EQUILIBRIA IN SOLUTIONS OF WEAK BASES

Weak bases, such as ammonia, accept a proton from water to give the conjugate acid of the base and OH^- ions:

$$NH_3(aq) + H_2O(l) \rightleftharpoons NH_4^+(aq) + OH^-(aq)$$

The equilibrium reaction of any base B with water is characterized by an equilibrium equation similar in form to that for the dissociation of a weak acid. For the general reaction

$$B(aq) + H_2O(l) \rightleftharpoons BH^+(aq) + OH^-(aq)$$

the equilibrium constant is called the **base-dissociation constant**, K_b, and is a quantitative measure of the strength of the base:

▶ **Base-dissociation constant** $$K_b = \frac{[BH^+][OH^-]}{[B]}$$

As usual, $[H_2O]$ is omitted from the equilibrium-constant expression. **Table 14.4** lists some typical weak bases and gives their K_b values at 25 °C. (The term *base-protonation constant* might be a more descriptive name for K_b, but the term *base-dissociation constant* is still widely used.)

TABLE 14.4 K_b Values for Some Weak Bases and K_a Values for Their Conjugate Acids at 25 °C

Base	Formula, B	K_b	Conjugate Acid, BH^+	K_a
Ammonia	NH_3	1.8×10^{-5}	NH_4^+	5.6×10^{-10}
Aniline	$C_6H_5NH_2$	4.3×10^{-10}	$C_6H_5NH_3^+$	2.3×10^{-5}
Dimethylamine	$(CH_3)_2NH$	5.4×10^{-4}	$(CH_3)_2NH_2^+$	1.9×10^{-11}
Hydrazine	N_2H_4	8.9×10^{-7}	$N_2H_5^+$	1.1×10^{-8}
Hydroxylamine	NH_2OH	9.1×10^{-9}	NH_3OH^+	1.1×10^{-6}
Methylamine	CH_3NH_2	3.7×10^{-4}	$CH_3NH_3^+$	2.7×10^{-11}

Many weak bases are organic compounds called *amines*, derivatives of ammonia in which one or more hydrogen atoms are replaced by an organic, carbon-based group, such as a methyl group (CH_3). Methylamine (CH_3NH_2), for example, is an organic amine responsible for the odor of rotting fish.

Amines are organic compounds in which one, two, or three substituents, such as a methyl group, are attached to a basic nitrogen atom.

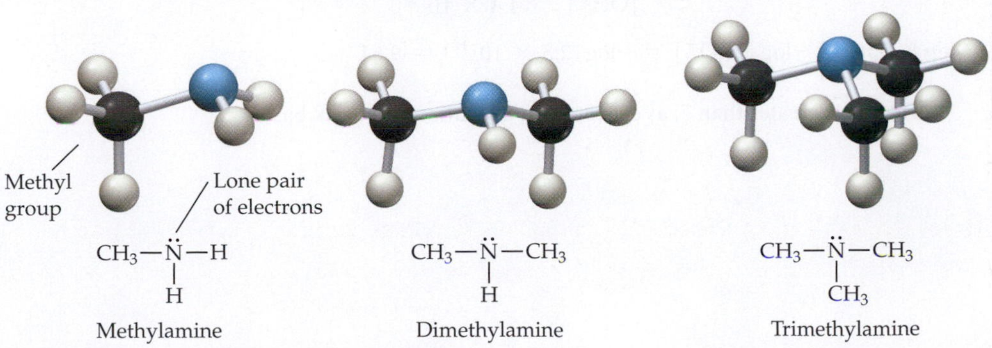

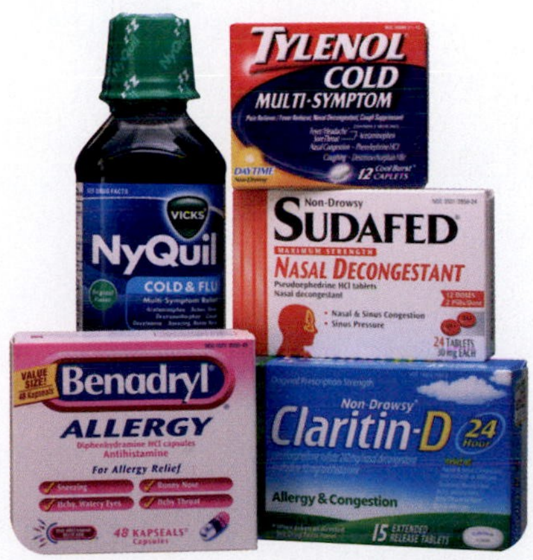

▲ Many over-the-counter drugs contain salts formed from amines and hydrochloric acid.

The basicity of an amine is due to the lone pair of electrons on the nitrogen atom, which can be used for bonding to a proton.

Equilibria in solutions of weak bases are treated by the same procedure used for solving problems involving weak acids. Worked Example 14.13 illustrates the procedure.

Worked Example 14.13

Calculating the pH and the Equilibrium Concentrations in a Solution of a Weak Base

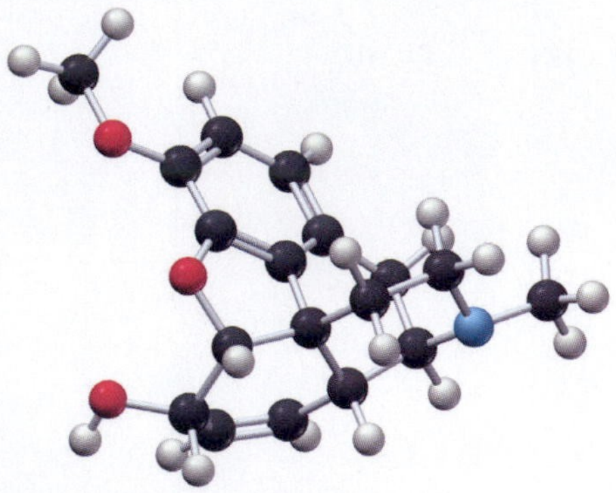

Codeine

Codeine ($C_{18}H_{21}NO_3$), a drug used in painkillers and cough medicines, is a naturally occurring amine that has $K_b = 1.6 \times 10^{-6}$. Calculate the pH and the concentrations of all species present in a 0.0012 M solution of codeine.

STRATEGY

Use the procedure outlined in Figure 14.7.

SOLUTION

Step 1 Let's use Cod as an abbreviation for codeine and $CodH^+$ for its conjugate acid. The species present initially are Cod (base) and H_2O (acid or base).

Step 2 There are two possible proton-transfer reactions:

$$\text{Cod}(aq) + H_2O(l) \rightleftharpoons \text{CodH}^+(aq) + OH^-(aq) \qquad K_b = 1.6 \times 10^{-6}$$

$$H_2O(l) + H_2O(l) \rightleftharpoons H_3O^+(aq) + OH^-(aq) \qquad K_w = 1.0 \times 10^{-14}$$

Step 3 Since Cod is a much stronger base than H_2O ($K_b \gg K_w$), the principal reaction involves the protonation of codeine.

Step 4

Principal Reaction	**Cod(*aq*) + H₂O(*l*)**	**⇌ CodH⁺(*aq*)**	**+ OH⁻(*aq*)**
Initial concentration (M)	0.0012	0	~0
Change (M)	$-x$	$+x$	$+x$
Equilibrium concentration (M)	$0.0012 - x$	x	x

Step 5 The value of x is obtained from the equilibrium equation:

$$K_b = 1.6 \times 10^{-6} = \frac{[\text{CodH}^+][OH^-]}{[\text{Cod}]} = \frac{(x)(x)}{(0.0012 - x)}$$

Assuming that $(0.0012 - x) \approx 0.0012$,

$$x^2 = (1.6 \times 10^{-6})(0.0012)$$

$$x = 4.4 \times 10^{-5} \quad \text{Approximation } (0.0012 - x) \approx 0.0012 \text{ is justified.}$$

Step 6 The big concentrations are

$$[\text{CodH}^+] = [OH^-] = x = 4.4 \times 10^{-5}\ \text{M}$$

$$[\text{Cod}] = 0.0012 - x = 0.0012 - 0.000\,044 = 0.0012\ \text{M}$$

Step 7 The small concentration is obtained from the subsidiary equilibrium, the dissociation of water:

$$[H_3O^+] = \frac{K_w}{[OH^-]} = \frac{1.0 \times 10^{-14}}{4.4 \times 10^{-5}} = 2.3 \times 10^{-10}\ \text{M}$$

Step 8 $\text{pH} = -\log[H_3O^+] = -\log(2.3 \times 10^{-10}) = 9.64$

The pH is greater than 7, as expected for a solution of a weak base.

✓ BALLPARK CHECK

Because x is negligible compared with the initial concentration of codeine (Step 5), $[OH^-]$ equals the square root of the product of K_b and the initial concentration of codeine. Therefore, $[OH^-]$ equals the square root of approximately $(16 \times 10^{-7})(1 \times 10^{-3})$, or about 4×10^{-5}. Since the $[OH^-]$ is between 10^{-5} M and 10^{-4} M, the $[H_3O^+]$ is between 10^{-9} M and 10^{-10} M, and so the pH is between 9 and 10. The ballpark check and the solution agree.

PROBLEM 14.22 Calculate the pH and the concentrations of all species present in 0.40 M NH_3 ($K_b = 1.8 \times 10^{-5}$).

PROBLEM 14.23 Strychnine ($C_{21}H_{22}N_2O_2$), a deadly poison used for killing rodents, is a weak base having $K_b = 1.8 \times 10^{-6}$. Calculate the pH of a saturated solution of strychnine (16 mg/100 mL).

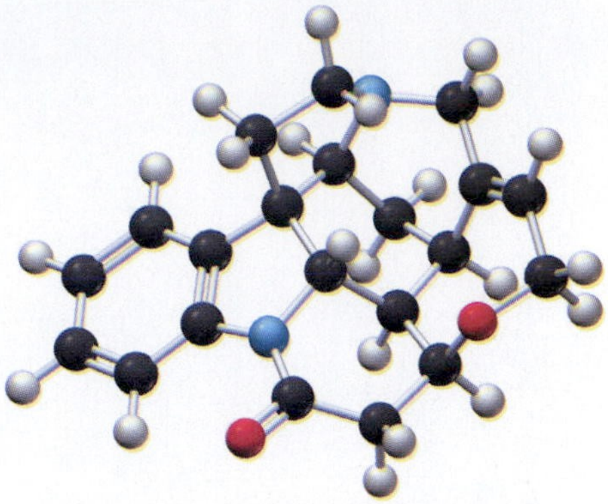

Strychnine

14.13 RELATION BETWEEN K_a AND K_b

We've seen in previous sections that the strength of an acid is indicated by its value of K_a and the strength of a base is given by its value of K_b. For a conjugate acid–base pair, the two equilibrium constants are related in a simple way that makes it possible to calculate either one from the other. To illustrate, let's consider the conjugate acid–base pair NH_4^+ and NH_3, where K_a refers to proton transfer from the acid NH_4^+ to water and K_b refers to proton transfer from water to the base NH_3. The sum of the two reactions is simply the dissociation of water:

$$\cancel{NH_4^+(aq)} + H_2O(l) \rightleftharpoons H_3O^+(aq) + \cancel{NH_3(aq)} \qquad K_a = \frac{[H_3O^+][NH_3]}{[NH_4^+]} = 5.6 \times 10^{-10}$$

$$\cancel{NH_3(aq)} + H_2O(l) \rightleftharpoons \cancel{NH_4^+(aq)} + OH^-(aq) \qquad K_b = \frac{[NH_4^+][OH^-]}{[NH_3]} = 1.8 \times 10^{-5}$$

$$\text{Net: } 2\,H_2O(l) \rightleftharpoons H_3O^+(aq) + OH^-(aq) \qquad K_w = [H_3O^+][OH^-] = 1.0 \times 10^{-14}$$

The equilibrium constant for the net reaction equals the product of the equilibrium constants for the reactions added:

$$K_a \times K_b = \frac{[H_3O^+]\cancel{[NH_3]}}{\cancel{[NH_4^+]}} \times \frac{\cancel{[NH_4^+]}[OH^-]}{\cancel{[NH_3]}} = [H_3O^+][OH^-] = K_w$$

$$= (5.6 \times 10^{-10})(1.8 \times 10^{-5}) = 1.0 \times 10^{-14}$$

What we've shown in this particular case is true in general. *Whenever chemical equations for two (or more) reactions are added to get the equation for a net reaction, the equilibrium constant for the net reaction equals the product of the equilibrium constants for the individual reactions*:

$$K_{net} = K_1 \times K_2 \times \ldots$$

For any conjugate acid–base pair, the product of the acid-dissociation constant for the acid and the base-dissociation constant for the base always equals the ion-product constant for water:

$$K_a \times K_b = K_w$$

As the strength of an acid increases (larger K_a), the strength of its conjugate base decreases (smaller K_b) because the product $K_a \times K_b$ remains constant at 1.0×10^{-14}. This inverse relationship between the strength of an acid and the strength of its conjugate base was illustrated qualitatively in Table 14.1. Taking the negative base-10 logarithm of both sides of the equation $K_a \times K_b = K_w$ gives another useful relationship:

$$pK_a + pK_b = pK_w$$

where $pK_a = -\log K_a$, $pK_b = -\log K_b$, and $pK_w = -\log K_w = 14.00$ at 25 °C.

Compilations of equilibrium constants, such as Appendix C, generally list either K_a or K_b, but not both, because K_a is easily calculated from K_b and vice versa:

$$K_a = \frac{K_w}{K_b} \quad \text{and} \quad K_b = \frac{K_w}{K_a}$$

Similarly, pK_a can be calculated from pK_b and vice versa:

$$pK_a = 14.00 - pK_b \quad \text{and} \quad pK_b = 14.00 - pK_a \quad \text{at 25 °C}$$

Worked Example 14.14

Relating K_a, K_b, pK_a, and pK_b

(a) K_b for trimethylamine is 6.5×10^{-5}. Calculate K_a for the trimethylammonium ion, $(CH_3)_3NH^+$.
(b) K_a for HCN is 4.9×10^{-10}. Calculate K_b for CN^-.
(c) Pyridine (C_5H_5N), an organic solvent, has pK_b = 8.74. What is the value of pK_a for the pyridinium ion, $C_5H_5NH^+$?

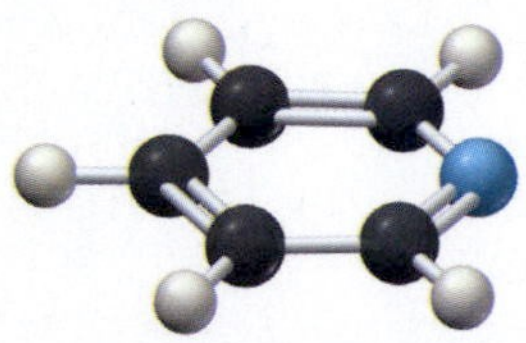

Pyridine

STRATEGY

To calculate K_a from K_b (or vice versa), use the equation $K_a = K_w/K_b$ or $K_b = K_w/K_a$. To calculate pK_a from pK_b, use the equation pK_a = 14.00 − pK_b.

SOLUTION

(a) K_a for $(CH_3)_3NH^+$ is the equilibrium constant for the acid-dissociation reaction

$$(CH_3)_3NH^+(aq) + H_2O(l) \rightleftharpoons H_3O^+(aq) + (CH_3)_3N(aq)$$

Because $K_a = K_w/K_b$, we can find K_a for $(CH_3)_3NH^+$ from K_b for its conjugate base $(CH_3)_3N$:

$$K_a = \frac{K_w}{K_b} = \frac{1.0 \times 10^{-14}}{6.5 \times 10^{-5}} = 1.5 \times 10^{-10}$$

(b) K_b for CN^- is the equilibrium constant for the base-protonation reaction

$$CN^-(aq) + H_2O(l) \rightleftharpoons HCN(aq) + OH^-(aq)$$

Because $K_b = K_w/K_a$, we can find K_b for CN^- from K_a for its conjugate acid HCN:

$$K_b = \frac{K_w}{K_a} = \frac{1.0 \times 10^{-14}}{4.9 \times 10^{-10}} = 2.0 \times 10^{-5}$$

(c) We can find pK_a for $C_5H_5NH^+$ from pK_b for C_5H_5N:

$$pK_a = 14.00 - pK_b = 14.00 - 8.74 = 5.26$$

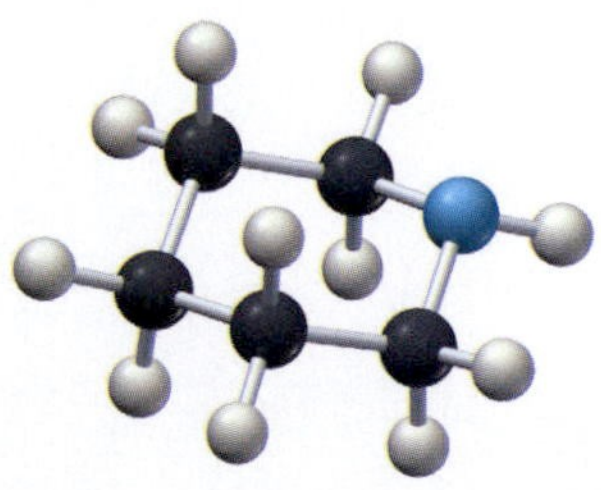

Piperidine

PROBLEM 14.24

(a) Piperidine ($C_5H_{11}N$) is an amine found in black pepper. Find K_b for piperidine in Appendix C, and then calculate K_a for the $C_5H_{11}NH^+$ cation.
(b) Find K_a for HOCl in Appendix C, and then calculate K_b for OCl^-.
(c) The value of pK_a for formic acid (HCO_2H) is 3.74. What is the value of pK_b for the formate ion (HCO_2^-)?

14.14 ACID–BASE PROPERTIES OF SALTS

When an acid neutralizes a base (Section 7.5), an ionic compound called a *salt* is formed. Salt solutions can be neutral, acidic, or basic, depending on the acid–base properties of the constituent cations and anions (**Figure 14.9**). As a general rule, salts formed by the reaction of a strong acid with a strong base give neutral solutions, salts formed by the reaction of a strong acid with a weak base give acidic solutions, and salts formed by the reaction of a weak acid with a strong base give basic solutions. It's as if the influence of the stronger partner dominates:

Strong acid + **Strong base** ⟶ Neutral solution
Strong acid + Weak base ⟶ **Acidic** solution
Weak acid + **Strong base** ⟶ **Basic** solution

FIGURE 14.9

Neutral, acidic, and basic salt solutions (0.10 M): NaCl, NH_4Cl, $AlCl_3$, NaCN, and $(NH_4)_2CO_3$. A few drops of universal indicator have been added to each solution. Universal indicator is red at pH 3, yellow at pH 5, green at pH 7, and blue at pH ≥ 10.

Salts That Yield Neutral Solutions

Salts such as NaCl that are derived from a strong base (NaOH) and a strong acid (HCl) yield neutral solutions because neither the cation nor the anion reacts appreciably with water to produce H_3O^+ or OH^- ions. As the conjugate base of a strong acid, Cl^- has no tendency to make the solution basic by picking up a proton from water. As the cation of a strong base, the hydrated Na^+ ion has no appreciable tendency to make the solution acidic by transferring a proton to a solvent water molecule.

The following ions do not react appreciably with water to produce either H_3O^+ or OH^- ions:

- **Cations from strong bases:**
  Alkali metal cations of group 1A (Li^+, Na^+, K^+)
  Alkaline earth cations of group 2A (Mg^{2+}, Ca^{2+}, Sr^{2+}, Ba^{2+}), except for Be^{2+}
- **Anions from strong monoprotic acids:**

  Cl^-, Br^-, I^-, NO_3^-, and ClO_4^-

  Salts that contain only these ions give neutral solutions in pure water (pH = 7).

Salts That Yield Acidic Solutions

Salts such as NH_4Cl that are derived from a weak base (NH_3) and a strong acid (HCl) produce acidic solutions. In such a case, the anion is neither an acid nor a base, but the cation is a weak acid:

$$NH_4^+(aq) + H_2O(l) \rightleftharpoons H_3O^+(aq) + NH_3(aq)$$

Related ammonium salts derived from amines, such as $[CH_3NH_3]Cl$, $[(CH_3)_2NH_2]Cl$, and $[(CH_3)_3NH]Cl$, also give acidic solutions because they too have cations with at least one dissociable proton. The pH of a solution that contains an acidic cation can be calculated by the standard procedure outlined in Figure 14.7. For a 0.10 M NH_4Cl solution, the pH is 5.12. Although the reaction of a cation or anion of a salt with water to produce H_3O^+ or OH^- ions is sometimes called a *salt hydrolysis reaction*, there is no fundamental difference between a salt hydrolysis reaction and any other Brønsted–Lowry acid–base reaction.

Another type of acidic cation is the hydrated cation of a small, highly charged metal ion, such as Al^{3+}. In aqueous solution, the Al^{3+} ion bonds to six water molecules to give the hydrated cation $Al(H_2O)_6{}^{3+}$. As shown in **Figure 14.10**, the negative (oxygen) end of each dipolar water molecule bonds to the positive metal cation and the six water molecules are located at the vertices of a regular octahedron.

FIGURE 14.10

Octahedral structure of the $Al(H_2O)_6{}^{3+}$ cation.

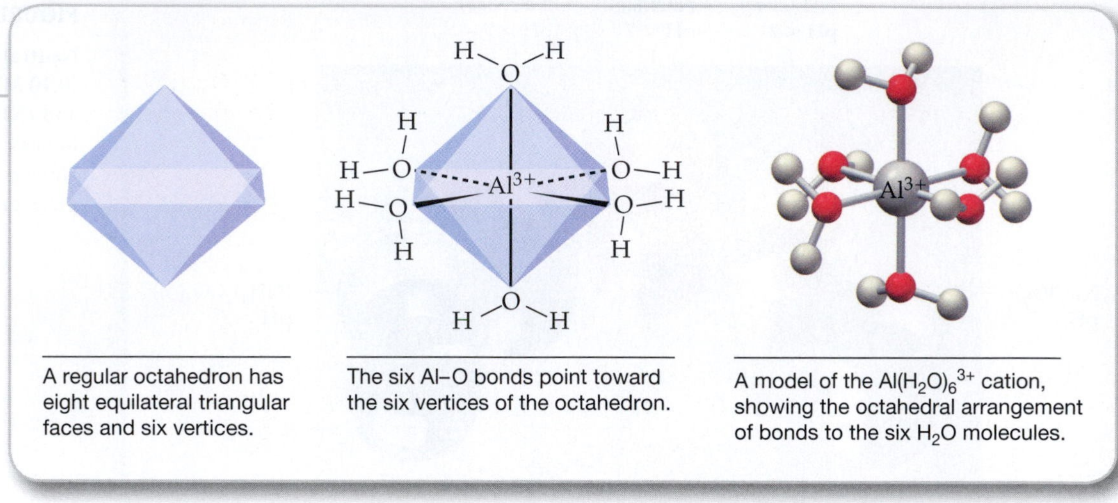

All metal ions exist in aqueous solution as hydrated cations, but the acidity of hydrated cations varies greatly depending on the charge and size of the unhydrated metal ion. Because of the high (3+) charge on the Al^{3+} ion, electrons in the O—H bonds of the bound water molecules are attracted toward the Al^{3+} ion. The attraction is strong because the Al^{3+} ion is small and the electrons in the O—H bonds are relatively close to the center of positive charge. As a result, electron density shifts from the O—H bonds toward the Al^{3+} ion, thus weakening the O—H bonds and increasing their polarity, which in turn eases the transfer of a proton to a solvent water molecule:

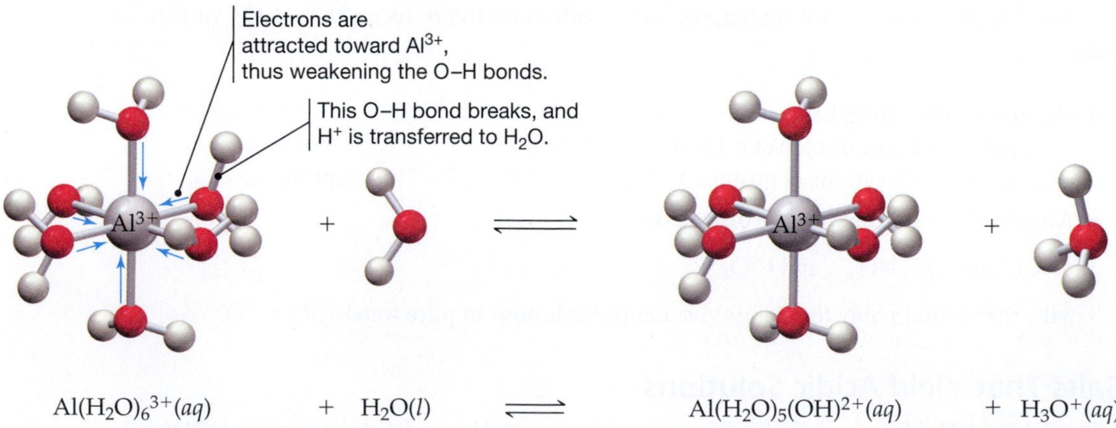

$$Al(H_2O)_6{}^{3+}(aq) + H_2O(l) \rightleftharpoons Al(H_2O)_5(OH)^{2+}(aq) + H_3O^+(aq)$$

The acid-dissociation constant for $Al(H_2O)_6{}^{3+}$, $K_a = 1.4 \times 10^{-5}$, is much larger than $K_w = 1.0 \times 10^{-14}$, which means that the water molecules in the hydrated cation are much stronger proton donors than are free solvent water molecules. In fact, the acid strength of $Al(H_2O)_6{}^{3+}$ is comparable to that of acetic acid, which has $K_a = 1.8 \times 10^{-5}$. In general, the acidity of hydrated main-group cations increases from left to right in the periodic table as the metal ion charge increases and the metal ion size decreases

($Li^+ < Be^{2+}$; $Na^+ < Mg^{2+} < Al^{3+}$). Transition metal cations, such as Zn^{2+}, Cr^{3+}, and Fe^{3+}, also give acidic solutions; their K_a values are listed in Table C.2 of Appendix C.

Worked Example 14.15

Calculating the pH of an Acidic Salt Solution

Calculate the pH of a 0.10 M solution of $AlCl_3$; K_a for $Al(H_2O)_6^{3+}$ is 1.4×10^{-5}.

STRATEGY

Because this problem is similar to others done earlier, we'll abbreviate the procedure in Figure 14.7.

SOLUTION

Steps 1–4 The species present initially are $Al(H_2O)_6^{3+}$ (acid), Cl^- (inert), and H_2O (acid or base). Because $Al(H_2O)_6^{3+}$ is a much stronger acid than water ($K_a \gg K_w$), the principal reaction is dissociation of $Al(H_2O)_6^{3+}$:

Principal Reaction	$Al(H_2O)_6^{3+}(aq) + H_2O(l) \rightleftharpoons$	$H_3O^+(aq)$	$+\ Al(H_2O)_5(OH)^{2+}(aq)$
Equilibrium concentration (M)	$0.10 - x$	x	x

Step 5 The value of x is obtained from the equilibrium equation:

$$K_a = 1.4 \times 10^{-5} = \frac{[H_3O^+][Al(H_2O)_5(OH)^{2+}]}{[Al(H_2O)_6^{3+}]} = \frac{(x)(x)}{(0.10 - x)} \approx \frac{x^2}{0.10}$$

$$x = [H_3O^+] = 1.2 \times 10^{-3}\ M$$

Step 8 $pH = -\log(1.2 \times 10^{-3}) = 2.92$

Thus, $Al(H_2O)_6^{3+}$ is a much stronger acid than NH_4^+, which agrees with the colors of the indicator in Figure 14.9.

PROBLEM 14.25 Predict whether the following salt solutions are neutral or acidic, and calculate the pH of each:

(a) 0.25 M NH_4Br

(b) 0.40 M $ZnCl_2$; K_a for $Zn(H_2O)_6^{2+}$ is 2.5×10^{-10}.

Salts That Yield Basic Solutions

Salts such as NaCN that are derived from a strong base (NaOH) and a weak acid (HCN) yield basic solutions. In this case, the cation is neither an acid nor a base but the anion is a weak base:

$$CN^-(aq) + H_2O(l) \rightleftharpoons HCN(aq) + OH^-(aq)$$

Other anions that exhibit basic properties are listed in Table 14.1 and include NO_2^-, F^-, $CH_3CO_2^-$, and CO_3^{2-}. The pH of a basic salt solution can be calculated by the standard procedure, as shown in Worked Example 14.16.

Worked Example 14.16

Calculating the pH of a Basic Salt Solution

Calculate the pH of a 0.10 M solution of NaCN; K_a for HCN is 4.9×10^{-10}.

STRATEGY

Use the procedure summarized in Figure 14.7.

continued on next page

SOLUTION

Step 1 The species present initially are Na^+ (inert), CN^- (base), and H_2O (acid or base).

Step 2 There are two possible proton-transfer reactions:

$$CN^-(aq) + H_2O(l) \rightleftharpoons HCN(aq) + OH^-(aq) \quad K_b$$
$$H_2O(l) + H_2O(l) \rightleftharpoons H_3O^+(aq) + OH^-(aq) \quad K_w$$

Step 3 As shown in Worked Example 14.14b, $K_b = K_w/(K_a \text{ for HCN}) = 2.0 \times 10^{-5}$. Because $K_b \gg K_w$, CN^- is a stronger base than H_2O and the principal reaction is proton transfer from H_2O to CN^-.

Step 4

Principal Reaction	$CN^-(aq)$	+	$H_2O(l)$	$\rightleftharpoons$	$HCN(aq)$	+	$OH^-(aq)$
Equilibrium concentration (M)	$0.10 - x$				x		x

Step 5 The value of x is obtained from the equilibrium equation:

$$K_b = 2.0 \times 10^{-5} = \frac{[HCN][OH^-]}{[CN^-]} = \frac{(x)(x)}{(0.10 - x)} \approx \frac{x^2}{0.10}$$

$$x = [OH^-] = 1.4 \times 10^{-3}\ M$$

Step 7 $$[H_3O^+] = \frac{K_w}{[OH^-]} = \frac{1.0 \times 10^{-14}}{1.4 \times 10^{-3}} = 7.1 \times 10^{-12}$$

Step 8 $pH = -\log(7.1 \times 10^{-12}) = 11.15$

The solution is basic, which agrees with the blue color of the indicator in Figure 14.9.

PROBLEM 14.26 Calculate the pH of 0.20 M $NaNO_2$; K_a for HNO_2 is 4.6×10^{-4}.

Salts That Contain Acidic Cations and Basic Anions

Finally, let's look at a salt such as $(NH_4)_2CO_3$ in which both the cation and the anion can undergo proton-transfer reactions. Because NH_4^+ is a weak acid and CO_3^{2-} is a weak base, the pH of an $(NH_4)_2CO_3$ solution depends on the relative acid strength of the cation and base strength of the anion:

$$NH_4^+(aq) + H_2O(l) \rightleftharpoons H_3O^+(aq) + NH_3(aq) \qquad \text{Acid strength } (K_a)$$
$$CO_3^{2-}(aq) + H_2O(l) \rightleftharpoons HCO_3^-(aq) + OH^-(aq) \qquad \text{Base strength } (K_b)$$

We can distinguish three possible cases:

- $K_a > K_b$: If K_a for the cation is greater than K_b for the anion, the solution will contain an excess of H_3O^+ ions (pH < 7).
- $K_a < K_b$: If K_a for the cation is less than K_b for the anion, the solution will contain an excess of OH^- ions (pH > 7).
- $K_a \approx K_b$: If K_a for the cation and K_b for the anion are comparable, the solution will contain approximately equal concentrations of H_3O^+ and OH^- ions (pH ≈ 7).

To determine whether an $(NH_4)_2CO_3$ solution is acidic, basic, or neutral, let's work out the values of K_a for NH_4^+ and K_b for CO_3^{2-}:

$$K_a \text{ for } NH_4^+ = \frac{K_w}{K_b \text{ for } NH_3} = \frac{1.0 \times 10^{-14}}{1.8 \times 10^{-5}} = 5.6 \times 10^{-10}$$

$$K_b \text{ for } CO_3^{2-} = \frac{K_w}{K_a \text{ for } HCO_3^-} = \frac{K_w}{K_{a2} \text{ for } H_2CO_3} = \frac{1.0 \times 10^{-14}}{5.6 \times 10^{-11}} = 1.8 \times 10^{-4}$$

Because $K_a < K_b$, the solution is basic (pH > 7), in accord with the blue color of the indicator in Figure 14.9.

A summary of the acid–base properties of salts is given in **Table 14.5**.

TABLE 14.5 Acid–Base Properties of Salts

Type of Salt	Examples	Ions That React with Water	pH of Solution
Cation from strong base; anion from strong acid	NaCl, KNO_3, BaI_2	None	~7
Cation from weak base; anion from strong acid	NH_4Cl, NH_4NO_3, $[(CH_3)_3NH]Cl$	Cation	<7
Small, highly charged cation; anion from strong acid	$AlCl_3$, $Cr(NO_3)_3$, $Fe(ClO_4)_3$	Hydrated cation	<7
Cation from strong base; anion from weak acid	NaCN, KF, Na_2CO_3	Anion	>7
Cation from weak base; anion from weak acid	NH_4CN, NH_4F, $(NH_4)_2CO_3$	Cation and anion	<7 if $K_a > K_b$ >7 if $K_a < K_b$ ~ 7 if $K_a \approx K_b$

PROBLEM 14.27 Calculate K_a for the cation and K_b for the anion in an aqueous NH_4CN solution. Is the solution acidic, basic, or neutral?

PROBLEM 14.28 Classify each of the following salt solutions as acidic, basic, or neutral:
(a) KBr **(b)** $NaNO_2$ **(c)** NH_4Br **(d)** $ZnCl_2$ **(e)** NH_4F

14.15 LEWIS ACIDS AND BASES

In 1923, the same year in which Brønsted and Lowry defined acids and bases in terms of their proton donor/acceptor properties, the American chemist G. N. Lewis proposed an even more general concept of acids and bases. Lewis noticed that when a base accepts a proton, it does so by sharing a lone pair of electrons with the proton to form a new covalent bond. Using ammonia as an example, the reaction can be written in the following format, in which the curved arrow represents the donation of the nitrogen lone pair to form a bond with H^+:

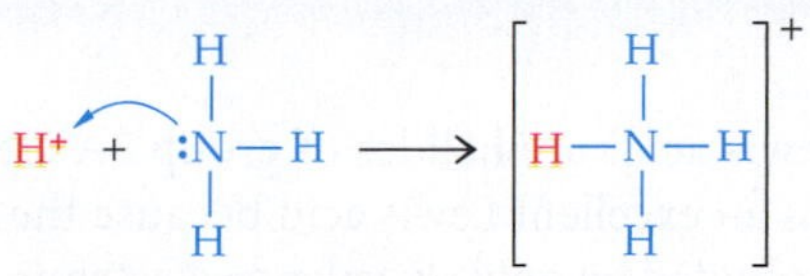

In this reaction, the proton behaves as an electron-pair acceptor and the ammonia molecule behaves as an electron-pair donor. Consequently, the Lewis definition of acids and bases states that a *Lewis acid* is an electron-pair acceptor and a *Lewis base* is an electron-pair donor.

- **Lewis acid** An electron-pair acceptor
- **Lewis base** An electron-pair donor

Because all proton acceptors have an unshared pair of electrons and all electron-pair donors can accept a proton, the Lewis and the Brønsted–Lowry definitions of a base are simply different ways of looking at the same property. All Lewis bases are Brønsted–Lowry bases, and all Brønsted–Lowry bases are Lewis bases. The Lewis definition of an acid, however, is considerably more general than the Brønsted–Lowry definition. Lewis acids include not only H^+ but also other cations and neutral molecules having vacant valence orbitals that can accept a share in a pair of electrons donated by a Lewis base.

Common examples of cationic Lewis acids are metal ions, such as Al^{3+} and Cu^{2+}. Hydration of the Al^{3+} ion, for example, is a Lewis acid–base reaction in which each of six H_2O molecules donates a pair of electrons to Al^{3+} to form the hydrated cation $Al(H_2O)_6{}^{3+}$:

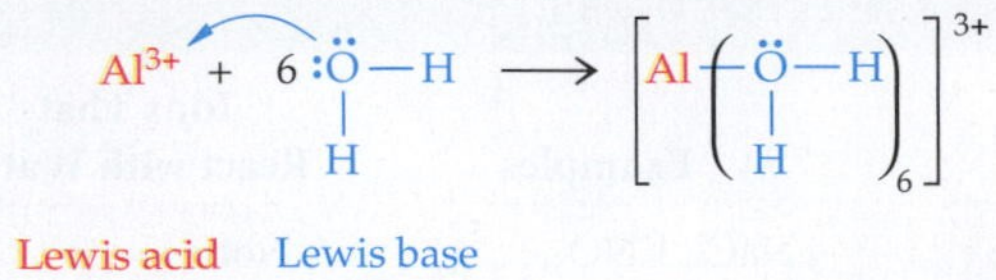

Similarly, the reaction of Cu^{2+} ion with ammonia is a Lewis acid–base reaction in which each of four NH_3 molecules donates a pair of electrons to Cu^{2+} to form the deep blue $Cu(NH_3)_4{}^{2+}$ ion (**Figure 14.11**).

$$\underset{\text{Lewis acid}}{Cu^{2+}} + \underset{\text{Lewis base}}{4\,{:}NH_3} \longrightarrow Cu(NH_3)_4{}^{2+}$$

FIGURE 14.11

The addition of aqueous ammonia to a solution of the Cu^{2+}(*aq*) ion is a Lewis acid–base reaction.

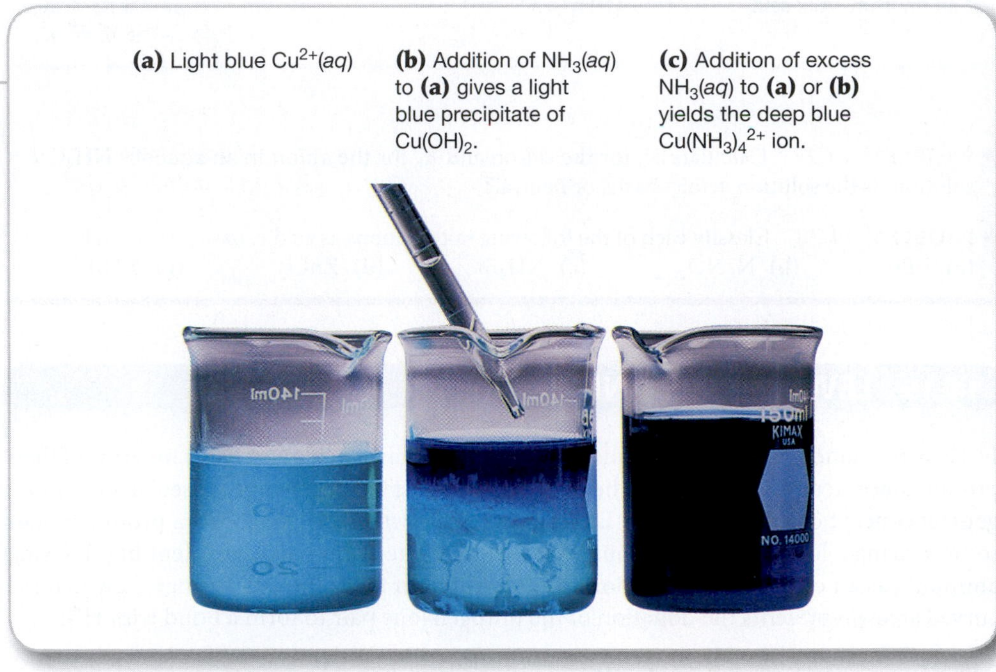

Examples of neutral Lewis acids are halides of group 3A elements, such as BF_3. Boron trifluoride, a colorless gas, is an excellent Lewis acid because the boron atom in the trigonal planar BF_3 molecule is surrounded by only six valence electrons (**Figure 14.12**). The boron atom uses three sp^2 hybrid orbitals to bond to the three F atoms and has a vacant $2p$ valence orbital that can accept a share in a pair of electrons from a Lewis base, such as NH_3. The Lewis acid and base sites are evident in electrostatic potential maps, which show the electron-poor B atom (blue) and the electron-rich N atom (red). In the product, called an *acid–base adduct*, the boron atom has acquired a stable octet of electrons.

Additional examples of neutral Lewis acids are oxides of nonmetals, such as CO_2, SO_2, and SO_3. The reaction of SO_3 with water, for example, can be viewed as a Lewis acid–base reaction in which SO_3 accepts a share in a lone pair of electrons from a water molecule:

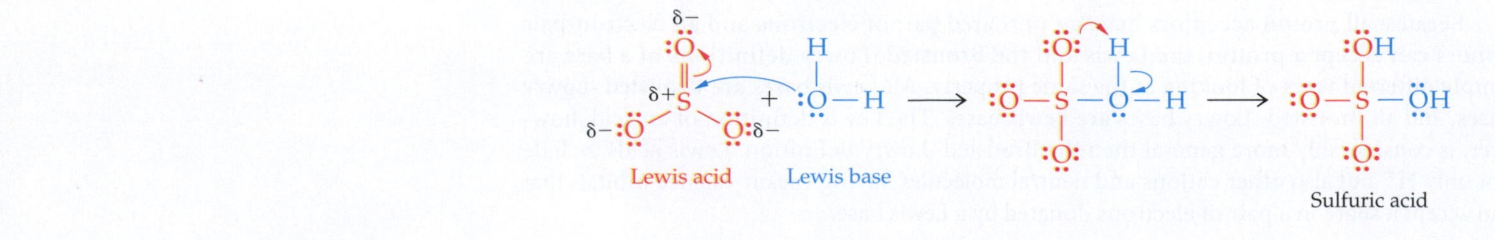

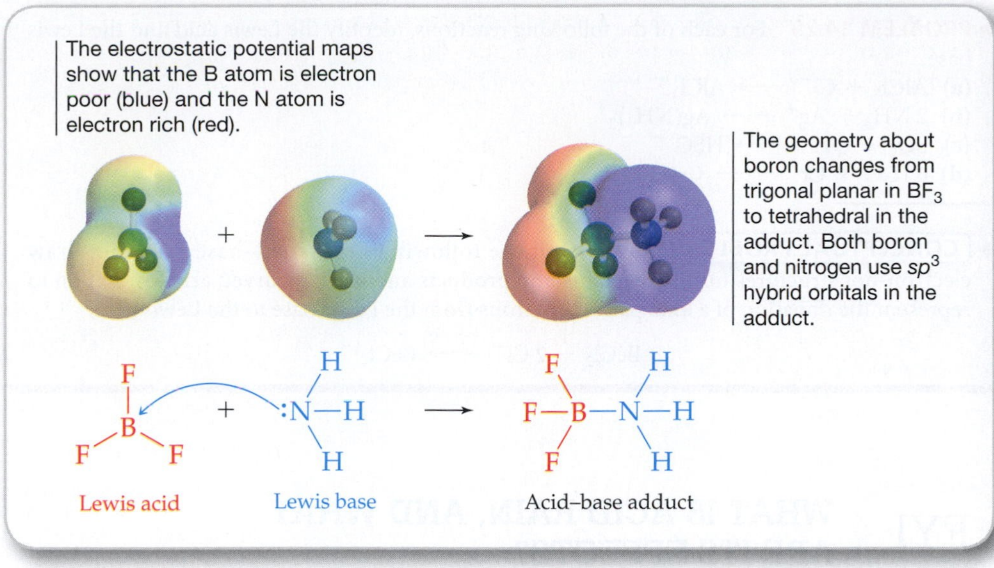

FIGURE 14.12

The reaction of the Lewis acid BF_3 with the Lewis base NH_3.

Because the sulfur–oxygen bonds are polar, with a partial positive charge ($\delta+$) on the less electronegative S atom, the S atom attracts an electron pair from H_2O. Formation of a bond from the water O atom to the S atom in the first step is helped along by the shift of a shared pair of electrons to oxygen. In the second step, a proton shifts from one oxygen atom to another, yielding sulfuric acid (H_2SO_4).

Worked Example 14.17

Identifying Lewis Acids and Bases

For each of the following reactions, identify the Lewis acid and the Lewis base.

(a) $CO_2 + OH^- \longrightarrow HCO_3^-$
(b) $B(OH)_3 + OH^- \longrightarrow B(OH)_4^-$
(c) $6\ CN^- + Fe^{3+} \longrightarrow Fe(CN)_6^{3-}$

STRATEGY

To identify the Lewis acid and the Lewis base, determine which molecule or ion can accept an electron pair (the Lewis acid) and which can donate an electron pair (the Lewis base).

SOLUTION

(a) The carbon atom of O=C=O bears a partial positive charge ($\delta+$) because carbon is less electronegative than oxygen. Therefore, the carbon atom attracts an electron pair from OH^-. Formation of a covalent bond from OH^- to CO_2 is helped along by a shift of a shared electron pair to oxygen:

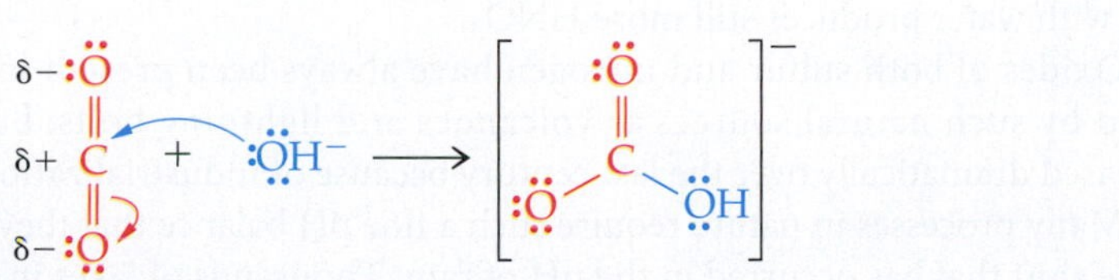

The Lewis acid (electron-pair acceptor) is CO_2; the Lewis base (electron-pair donor) is OH^-.

(b) The Lewis acid is boric acid, $B(OH)_3$, a weak acid and mild antiseptic used in eyewash. The boron atom in the trigonal planar $B(OH)_3$, molecule has a vacant $2p$ valence orbital and completes its octet by accepting a share in a pair of electrons from the Lewis base, OH^-.

(c) The Lewis acid is Fe^{3+}, and the Lewis base is CN^-. Each of the six :C≡N:$^-$ ions bonds to the Fe^{3+} ion by donating a lone pair of electrons on the C atom.

PROBLEM 14.29 For each of the following reactions, identify the Lewis acid and the Lewis base.

(a) $AlCl_3 + Cl^- \longrightarrow AlCl_4^-$
(b) $2\,NH_3 + Ag^+ \longrightarrow Ag(NH_3)_2^+$
(c) $SO_2 + OH^- \longrightarrow HSO_3^-$
(d) $6\,H_2O + Cr^{3+} \longrightarrow Cr(H_2O)_6^{3+}$

CONCEPTUAL PROBLEM 14.30 For the following Lewis acid–base reaction, draw electron-dot structures for the reactants and products and use the curved arrow notation to represent the donation of a lone pair of electrons from the Lewis base to the Lewis acid.

$$BeCl_2 + 2\,Cl^- \longrightarrow BeCl_4^{2-}$$

FYI WHAT IS ACID RAIN, AND WHAT ARE ITS EFFECTS?

▲ Plants that burn sulfur-containing coal and oil release large quantities of sulfur oxides into the atmosphere, ultimately leading to acid rain.

The problem of acid rain is one of the more important environmental issues of recent times. Both the causes and the effects of acid rain are well understood. The problem is what to do about it.

As the water that has evaporated from oceans and lakes condenses into raindrops, it dissolves small quantities of gases from the atmosphere. Under normal conditions, rain is slightly acidic, with a pH close to 5.6, because of dissolved CO_2. In recent decades, however, the acidity of rainwater in many industrialized areas of the world has increased by a factor of over 100, to a pH between 3 and 3.5.

The primary cause of acid rain is industrial and automotive pollution. Each year in industrialized countries, large power plants and smelters that burn sulfur-containing fossil fuels pour millions of tons of sulfur dioxide (SO_2) gas into the atmosphere, where some is oxidized by air to produce sulfur trioxide (SO_3). Sulfur oxides then dissolve in rain to form dilute sulfurous acid and sulfuric acid:

$$SO_2(g) + H_2O(l) \longrightarrow H_2SO_3(aq) \qquad \text{Sulfurous acid}$$
$$SO_3(g) + H_2O(l) \longrightarrow H_2SO_4(aq) \qquad \text{Sulfuric acid}$$

Nitrogen oxides, produced by the high-temperature reaction of N_2 with O_2 in coal-burning plants and in automobile engines, further contribute to the problem. Nitrogen dioxide (NO_2) reacts with water to form dilute nitric acid (HNO_3) and nitric oxide (NO):

$$3\,NO_2(g) + H_2O(l) \longrightarrow 2\,HNO_3(aq) + NO(g)$$

The NO is subsequently oxidized to NO_2 by atmospheric oxygen, and reaction of the resulting NO_2 with water produces still more HNO_3.

Oxides of both sulfur and nitrogen have always been present in the atmosphere, produced by such natural sources as volcanoes and lightning bolts, but their amounts have increased dramatically over the last century because of industrialization.

Many processes in nature require such a fine pH balance that they are dramatically upset by the shift that has occurred in the pH of rain. Thousands of lakes in the Adirondack region of upper New York state and in southeastern Canada have become so acidic that all fish life has disappeared. Massive tree die-offs have occurred throughout central and eastern Europe as acid rain has lowered the pH of the soil and leached nutrients from leaves. Countless marble statues have been slowly dissolved away as their calcium carbonate has been attacked by acid rain.

$$CaCO_3(s) + 2\,H^+(aq) \longrightarrow Ca^{2+}(aq) + H_2O(l) + CO_2(g)$$

Acid rain also contributes to the death of fish by leaching Al^{3+} from rocks into lakes. Though the lakes are acidic, the gills of fish are basic because fish excrete nitrogenous waste as NH_3 from their gills. The Al^{3+} then precipitates as $Al(OH)_3$ and clogs the gills:

$$NH_3(aq) + H_2O(l) \rightleftharpoons NH_4^+(aq) + OH^-(aq)$$
$$Al^{3+}(aq) + 3\,OH^-(aq) \longrightarrow Al(OH)_3(s)$$

Fortunately, acidic emissions from automobiles and power plants have been greatly reduced in recent years. Nitrogen oxide emissions have been lowered by equipping automobiles with catalytic converters (Section 12.15), which catalyze the decomposition of nitrogen oxides to N_2 and O_2. Sulfur dioxide emissions from power plants have been reduced by scrubbing combustion products before they are emitted from plant smokestacks. In this process, an aqueous suspension of lime (CaO) is added to the combustion chamber and the stack. The lime reacts with SO_2 to give calcium sulfite ($CaSO_3$):

$$CaO(s) + SO_2(g) \longrightarrow CaSO_3(s)$$

In some scrubbers, the economically worthless $CaSO_3$ is oxidized to $CaSO_4 \cdot 2\,H_2O$ (gypsum) which is marketable for use in the building industry.

The 1990 Amendments to the Clean Air Act established the Acid Rain Program administered by the Environmental Protection Agency. One of the goals of this program was an approximately 50% reduction in annual emissions of SO_2 from 1980 levels of 18.9 million tons to 8.95 million tons by 2010. The success of this market-based trading program has exceeded expectations, with SO_2 emissions in 2009 falling to just 5.7 million tons.

Another approach to combating acid rain is *coal gasification*, a promising technology for producing electric power with nearly zero harmful emissions. In a coal gasifier, coal is treated with steam and a controlled amount of air or oxygen at high temperature and pressure, producing a gaseous mixture of hydrogen and carbon monoxide along with methane and carbon dioxide. Combustion of this mixture, a synthetic fuel, is used to drive gas and steam turbines, which in turn generate electricity. Because sulfur and other impurities can be removed from the gas stream prior to combustion, air pollution is minimized. Moreover, compared to conventional coal-fired power plants, coal gasification plants would contribute less to global warming because carbon dioxide could be more easily captured for underground storage. Coal gasification plants are already operating in countries with no reserves of natural gas or petroleum, and this technology looks to be a promising option for future electric power plants.

Unfortunately, it takes many years for lakes and streams to recover from the damage done by acid rain, even after sulfur and nitrogen oxide emissions have been reduced and the pH of rainwater has returned to normal levels. One approach to accelerating the process and mitigating the effects of continued oxide emissions employs lime (CaO) and limestone ($CaCO_3$) to neutralize the acidity:

$$CaO(s) + 2\,H^+(aq) \longrightarrow Ca^{2+}(aq) + H_2O(l)$$
$$CaCO_3(s) + 2\,H^+(aq) \longrightarrow Ca^{2+}(aq) + H_2O(l) + CO_2(g)$$

This approach, called liming, has been widely used in Norway and Sweden but only infrequently in the United States. It is an expensive, short-term remediation procedure, but it does allow fish to survive in lakes and streams while longer-term solutions to the acid-rain problem are being pursued.

▲ This stone sculpture, the Leshan Buddha, is being slowly dissolved by reaction of calcium carbonate with acid rain.

PROBLEM 14.31 The reaction of lime (CaO) with SO_2 in the scrubber of a power plant is a Lewis acid–base reaction. Explain.

PROBLEM 14.32 What is the pH of 1.00 L of rainwater that has dissolved 5.47 mg of NO_2? Assume that all of the NO_2 has reacted with water to give nitric acid.

SUMMARY

According to the Arrhenius theory, acids (HA) are substances that dissociate in water to produce $H^+(aq)$ and bases (MOH) are substances that dissociate to yield $OH^-(aq)$. The more general **Brønsted–Lowry theory** defines an acid as a proton donor, a base as a proton acceptor, and an acid–base reaction as a proton-transfer reaction. Examples of Brønsted–Lowry acids are HCl, NH_4^+, and HSO_4^-; examples of Brønsted–Lowry bases are OH^-, F^-, and NH_3. The H_3O^+ ion, a hydrated proton, is called the hydronium ion.

A strong acid HA is nearly 100% dissociated, whereas a weak acid HA is only partially dissociated, existing as an equilibrium mixture of HA, H_3O^+, and A^-:

$$HA(aq) + H_2O(l) \rightleftharpoons H_3O^+(aq) + A^-(aq)$$

The strength of an acid (HA) and the strength of its **conjugate base** (A^-) are inversely related.

The acid strength of a binary acid HA increases down a group of the periodic table with decreasing strength of the H–A bond and increases from left to right across a row with increasing polarity of the H–A bond. The acid strength of an oxoacid, H_nYO_m (Y = C, N, S, Cl), increases with increasing electronegativity and increasing oxidation number of the central atom Y.

Water, which can act both as an acid and as a base, undergoes the dissociation reaction $H_2O + H_2O \rightleftharpoons H_3O^+ + OH^-$. In pure water at 25 °C, $[H_3O^+] = [OH^-] = 1.0 \times 10^{-7}$ M. The **ion-product constant for water, K_w,** is given by $K_w = [H_3O^+][OH^-] = 1.0 \times 10^{-14}$ at 25 °C. The acidity of an aqueous solution is expressed on the **pH** scale, where $pH = -\log [H_3O^+]$. At 25 °C, acidic solutions have pH < 7, basic solutions have pH > 7, and neutral solutions have pH = 7. The pH of a solution can be determined using an **acid–base indicator** or a pH meter. The extent of dissociation of a weak acid HA is measured by its **acid-dissociation constant, K_a**:

$$K_a = \frac{[H_3O^+][A^-]}{[HA]}$$

Polyprotic acids contain more than one dissociable proton and dissociate in a stepwise manner. Because the stepwise dissociation constants decrease in the order $K_{a1} \gg K_{a2} \gg K_{a3}$, nearly all the H_3O^+ in a polyprotic acid solution comes from the first dissociation step.

The extent of dissociation of a weak base B is measured by its **base-dissociation constant, K_b**:

$$B(aq) + H_2O(l) \rightleftharpoons BH^+(aq) + OH^-(aq) \quad K_b = \frac{[BH^+][OH^-]}{[B]}$$

Examples of weak bases are NH_3 and derivatives of NH_3 called *amines*. For any conjugate acid–base pair, (K_a for the acid) × (K_b for the base) = K_w.

Aqueous solutions of salts can be neutral, acidic, or basic, depending on the acid–base properties of the constituent ions. Group 1A and 2A cations (except Be^{2+}) and anions that are conjugate bases of strong acids, such as Cl^-, do not react appreciably with water to produce H_3O^+ or OH^- ions. Cations that are conjugate acids of weak bases, such as NH_4^+, and hydrated cations of small, highly charged metal ions, such as Al^{3+}, yield acidic solutions. Anions that are conjugate bases of weak acids, such as CN^-, yield basic solutions.

A **Lewis acid** is an electron-pair acceptor, and a **Lewis base** is an electron-pair donor. Lewis acids include not only H^+ but also other cations and neutral molecules that can accept a share in a pair of electrons from a Lewis base. Examples of Lewis acids are Al^{3+}, Cu^{2+}, BF_3, SO_3, and CO_2.

KEY WORDS

acid–base indicator *564*
acid-dissociation constant (K_a) *567*
Arrhenius acid *551*
Arrhenius base *551*
base-dissociation constant (K_b) *577*
Brønsted–Lowry acid *551*
Brønsted–Lowry base *551*
Brønsted–Lowry theory *551*
conjugate acid *552*
conjugate acid–base pair *552*
conjugate base *552*
dissociation of water *560*
hydronium ion (H_3O^+) *552*
ion-product constant for water (K_w) *560*
Lewis acid *585*
Lewis base *585*
percent dissociation *573*
pH *562*
polyprotic acid *574*
principal reaction *569*
strong acid *554*
subsidiary reaction *569*
weak acid *555*

KEY EQUATIONS

- Ion-product constant for water **(Section 14.4)**

$$K_w = [H_3O^+][OH^-] = 1.0 \times 10^{-14} \text{ at } 25\ °C$$

- Definition of pH **(Section 14.5)**

$$pH = -\log [H_3O^+] \qquad [H_3O^+] = 10^{-pH}$$

- Acid-dissociation constant for a weak acid HA **(Section 14.8)**

$$K_a = \frac{[H_3O^+][A^-]}{[HA]}$$

$$pK_a = -\log K_a$$

- Percent dissociation of a weak acid **(Section 14.10)**

$$\text{Percent dissociation} = \frac{[HA]_{\text{dissociated}}}{[HA]_{\text{initial}}} \times 100\%$$

- Base-dissociation, or base-protonation, constant for a weak base B **(Section 14.12)**

$$K_b = \frac{[BH^+][OH^-]}{[B]}$$

$$pK_b = -\log K_b$$

- Relation between K_a and K_b for a conjugate acid–base pair **(Section 14.13)**

$$K_a \times K_b = K_w \qquad pK_a + pK_b = pK_w$$

LEARNING OUTCOMES

- Identify Brønsted–Lowry acids, bases, and conjugate acid–base pairs. **(Section 14.1)** ***Problems 1–3, 33, 41, 44–49, 96, 97, 108***
- Predict the direction of an acid–base reaction based on the acid and base strengths of the conjugate acid–base pairs. **(Section 14.2)** ***Problems 5, 6, 52, 53***
- Predict the relative strengths of binary acids HA based on the size and electronegativity of the atom A, and predict the relative strengths of oxoacids H_nYO_m based on the electronegativity and oxidation number of the atom Y. **(Section 14.3)** ***Problems 7, 40, 54–59***
- Use $K_w = [H_3O^+][OH^-]$ to calculate $[OH^-]$ from $[H_3O^+]$ or $[H_3O^+]$ from $[OH^-]$. **(Section 14.4)** ***Problems 8, 9, 60, 61***
- Calculate the pH of a solution from $[H_3O^+]$ or $[OH^-]$, and calculate $[H_3O^+]$ or $[OH^-]$ from the pH. **(Section 14.5)** ***Problems 11, 12, 64–67***
- Calculate the pH of a solution of a strong acid or a strong base. **(Section 14.7)** ***Problems 13, 14, 32, 70–73***
- Write the net ionic equation and the equilibrium equation for dissociation of a weak acid, and calculate the dissociation constant K_a and the pK_a from the pH of the solution. **(Section 14.8)** ***Problems 15, 42, 76, 77, 114, 115, 126, 131***
- Given the value of K_a, calculate the pH, the equilibrium concentrations of all species present, and the percent dissociation of the acid in a solution of a weak acid. **(Sections 14.9, 14.10)** ***Problems 17–19, 78–81, 119, 120, 128, 130, 131, 133, 138***
- Given the stepwise dissociation constants, calculate the pH and the equilibrium concentrations of all species present in a solution of a polyprotic acid. **(Section 14.11)** ***Problems 20, 21, 84–87, 121–123, 132***
- Given the base-dissociation constant K_b, calculate the pH and the equilibrium concentrations of all species present in a solution of a weak base. **(Section 14.12)** ***Problems 22, 23, 92, 93***
- Relate K_a, K_b, pK_a, and pK_b for a conjugate acid–base pair. **(Section 14.13)** ***Problems 24, 94, 95, 113, 116***
- Predict whether a salt solution is acidic, basic, or neutral, and calculate the pH of the solution. **(Section 14.14)** ***Problems 25–28, 98–101, 117, 118, 124, 125, 127, 137***
- Identify the Lewis acid and the Lewis base in a Lewis acid–base reaction. **(Section 14.15)** ***Problems 29, 31, 41, 42, 102, 103, 106***

CONCEPTUAL PROBLEMS

Problems 14.1–14.32 appear within the chapter.

14.33 For each of the following reactions, identify the Brønsted–Lowry acids and bases:

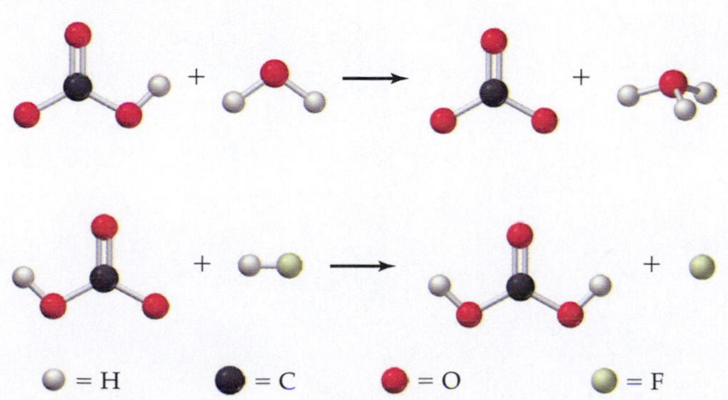

14.34 The following pictures represent aqueous solutions of three acids HA (A = X, Y, or Z); water molecules have been omitted for clarity:

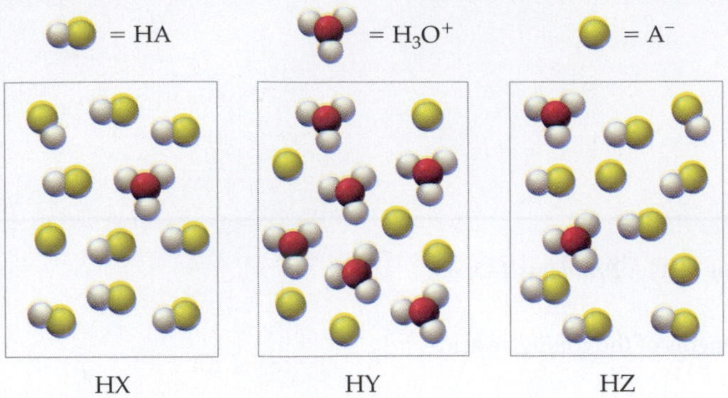

(a) What is the conjugate base of each acid?

(b) Arrange the three acids in order of increasing acid strength.

(c) Which acid, if any, is a strong acid?

(d) Which acid has the smallest value of K_a?

(e) What is the percent dissociation in the solution of HZ?

14.35 Which of the following pictures represents a solution of a weak diprotic acid H_2A? (Water molecules are omitted for clarity.) Which pictures represent an impossible situation? Explain.

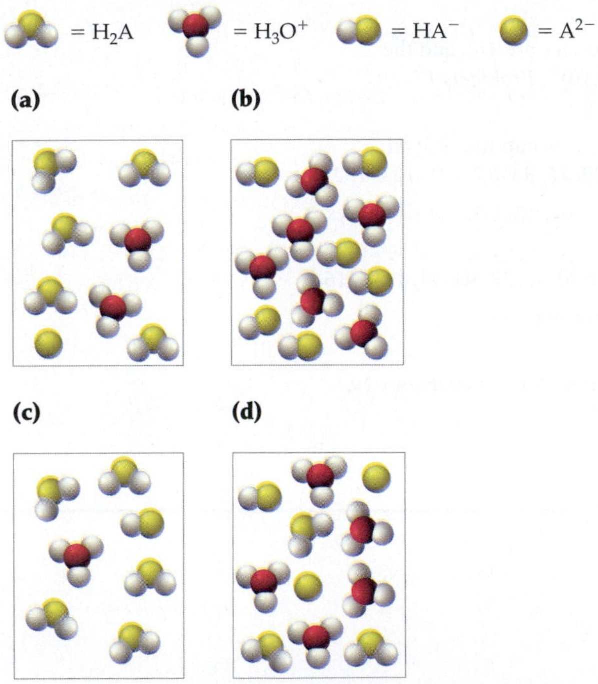

14.36 Which of the following pictures best represents an aqueous solution of sulfuric acid? Explain. (Water molecules have been omitted for clarity.)

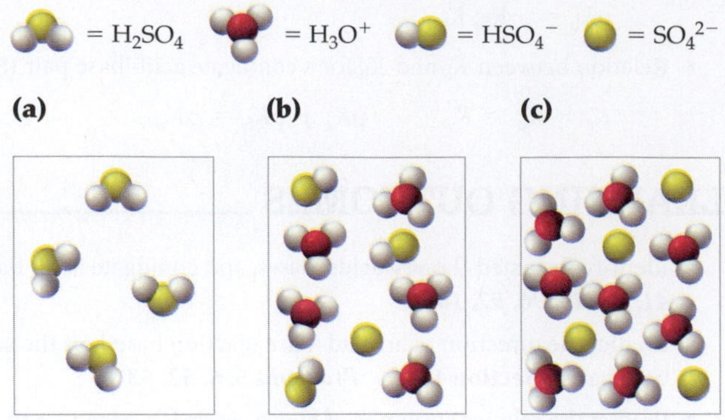

14.37 The following pictures represent aqueous solutions of three acids HA (A = X, Y, or Z); water molecules have been omitted for clarity:

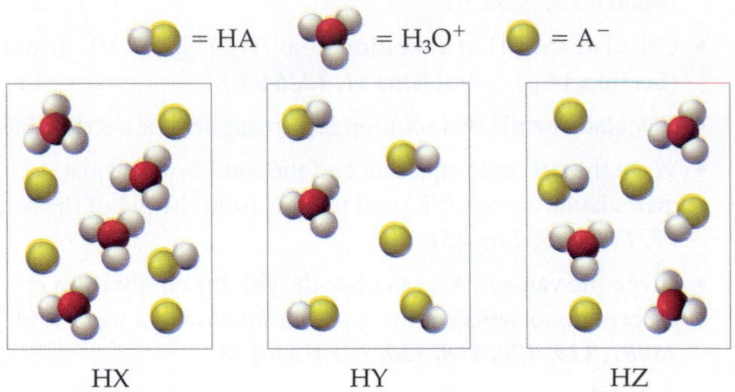

(a) Which conjugate base ($A^- = X^-, Y^-$, or Z^-) has the largest value of K_b?

(b) Which A^- ion is the weakest base?

14.38 The following pictures represent solutions of three salts NaA ($A^- = X^-, Y^-$, or Z^-); water molecules and Na^+ ions have been omitted for clarity:

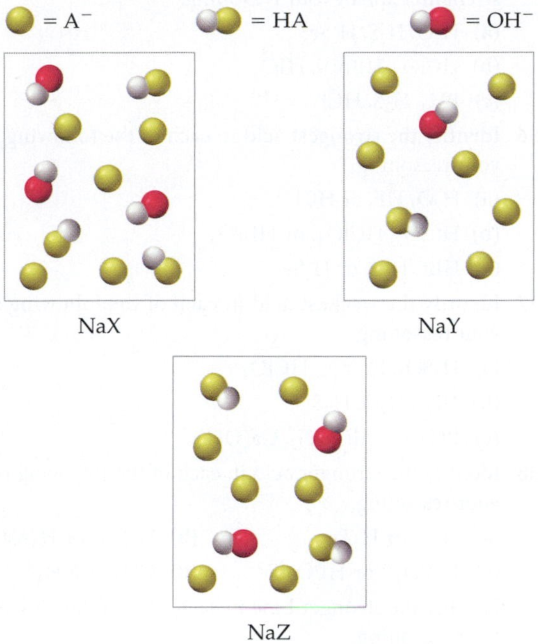

(a) Arrange the three A^- anions in order of increasing base strength.

(b) Which A^- anion has the strongest conjugate acid?

(c) Which A^- anion has the smallest value of pK_b?

(d) Why, within each box, is the number of HA molecules and OH^- anions the same?

14.39 The following picture represents the hydrated metal cation $M(H_2O)_6{}^{n+}$, where $n = 1, 2$, or 3.

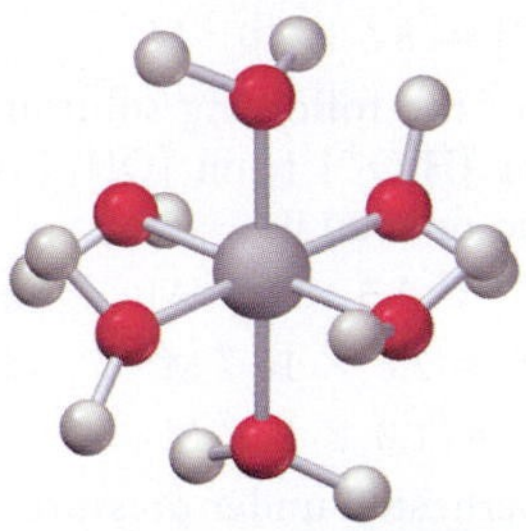

(a) Write a balanced equation for the reaction of $M(H_2O)_6{}^{n+}$ with water, and write the equilibrium equation for the reaction.

(b) Does the equilibrium constant increase, decrease, or remain the same as the value of n increases? Explain.

(c) Which $M(H_2O)_6{}^{n+}$ ion ($n = 1, 2$, or 3) is the strongest acid, and which has the strongest conjugate base?

14.40 Locate sulfur, selenium, chlorine, and bromine in the periodic table:

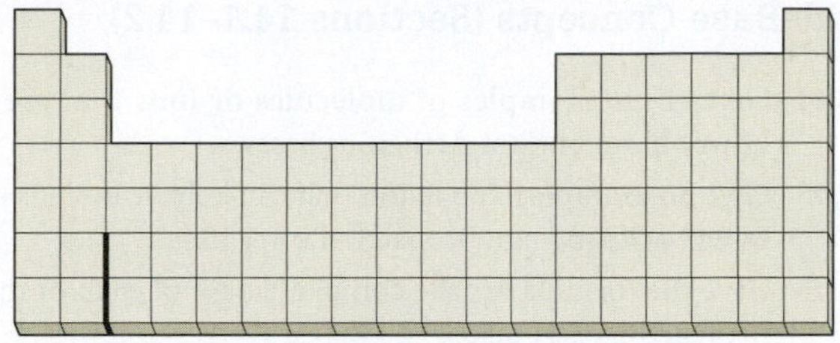

(a) Which binary acid (H_2S, H_2Se, HCl, or HBr) is the strongest? Which is the weakest? Explain.

(b) Which oxoacid (H_2SO_3, H_2SeO_3, $HClO_3$, or $HBrO_3$) is the strongest? Which is the weakest? Explain.

14.41 Look at the electron-dot structures of the following molecules and ions:

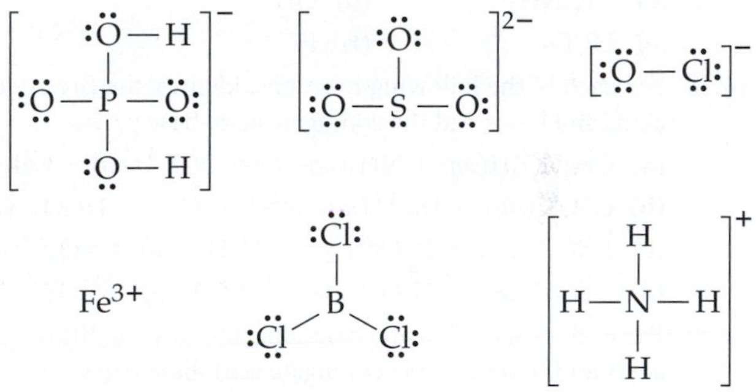

(a) Which of these molecules and ions can behave as a Brønsted–Lowry acid? Which can behave as a Brønsted–Lowry base?

(b) Which can behave as a Lewis acid? Which can behave as a Lewis base?

14.42 Boric acid (H_3BO_3) is a weak monoprotic acid that yields H_3O^+ ions in water. H_3BO_3 might behave either as a Brønsted–Lowry acid or as a Lewis acid, though it is, in fact, a Lewis acid.

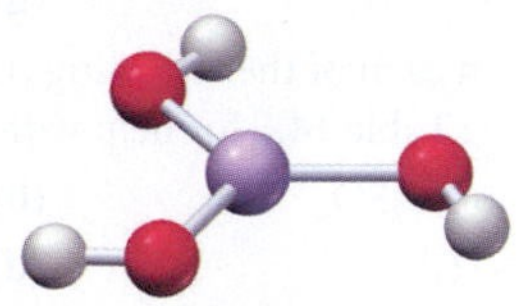

Boric acid

(a) Write a balanced equation for the reaction with water in which H_3BO_3 behaves as a Brønsted–Lowry acid.

(b) Write a balanced equation for the reaction with water in which H_3BO_3 behaves as a Lewis acid. Hint: One of the reaction products contains a tetrahedral boron atom.

14.43 The reaction of $PCl_4{}^+$ with Cl^- is a Lewis acid–base reaction. Draw electron-dot structures for the reactants and products, and use the curved arrow notation (Section 14.15) to represent the donation of a lone pair of electrons from the Lewis base to the Lewis acid.

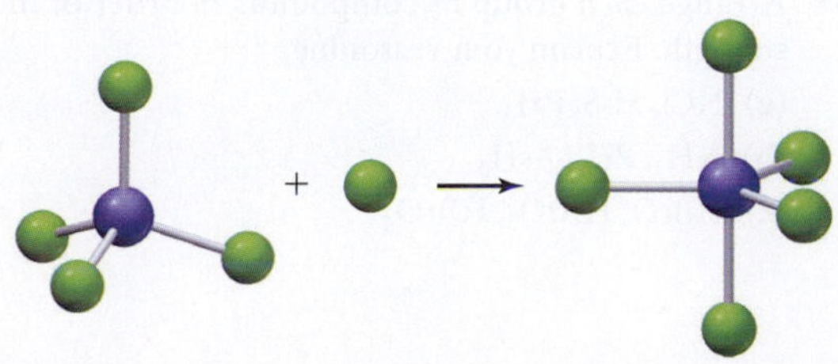

SECTION PROBLEMS

Acid–Base Concepts (Sections 14.1–14.2)

14.44 Give three examples of molecules or ions that are Brønsted–Lowry bases but not Arrhenius bases.

14.45 Give an example of an anion that can behave both as a Brønsted–Lowry acid and as a Brønsted–Lowry base.

14.46 Give the formula for the conjugate base of each of the following Brønsted–Lowry acids:

(a) HSO_4^- (b) H_2SO_3
(c) $H_2PO_4^-$ (d) NH_4^+
(e) H_2O (f) NH_3

14.47 Give the formula for the conjugate acid of each of the following Brønsted–Lowry bases:

(a) SO_3^{2-} (b) H_2O
(c) CH_3NH_2 (d) OH^-
(e) HCO_3^- (f) H^-

14.48 For each of the following reactions, identify the Brønsted–Lowry acids and bases and the conjugate acid–base pairs:

(a) $CH_3CO_2H(aq) + NH_3(aq) \rightleftharpoons NH_4^+(aq) + CH_3CO_2^-(aq)$
(b) $CO_3^{2-}(aq) + H_3O^+(aq) \rightleftharpoons H_2O(l) + HCO_3^-(aq)$
(c) $HSO_3^-(aq) + H_2O(l) \rightleftharpoons H_3O^+(aq) + SO_3^{2-}(aq)$
(d) $HSO_3^-(aq) + H_2O(l) \rightleftharpoons H_2SO_3(aq) + OH^-(aq)$

14.49 For each of the following reactions, identify the Brønsted–Lowry acids and bases and the conjugate acid–base pairs:

(a) $CN^-(aq) + H_2O(l) \rightleftharpoons OH^-(aq) + HCN(aq)$
(b) $H_2PO_4^-(aq) + H_2O(l) \rightleftharpoons H_3O^+(aq) + HPO_4^{2-}(aq)$
(c) $HPO_4^{2-}(aq) + H_2O(l) \rightleftharpoons OH^-(aq) + H_2PO_4^-(aq)$
(d) $NH_4^+(aq) + NO_2^-(aq) \rightleftharpoons HNO_2(aq) + NH_3(aq)$

14.50 Which of the following species behave as strong acids or as strong bases in aqueous solution? See Table 14.1 to check your answers.

(a) HNO_2 (b) HNO_3 (c) NH_4^+ (d) Cl^-
(e) H^- (f) O^{2-} (g) H_2SO_4

14.51 Which acid in each of the following pairs has the stronger conjugate base? See Table 14.1 for help with parts **(c)** and **(d)**.

(a) H_2CO_3 or H_2SO_4 (b) HCl or HF
(c) HF or NH_4^+ (d) HCN or HSO_4^-

14.52 Of the conjugate acid-base pairs HF/F^-, HCl/Cl^-, and HCN/CN^-, complete the following equation with the pair that gives an equilibrium constant $K_c > 1$.

$$\_\_\_\_ + H_2PO_4^- \rightleftharpoons \_\_\_\_ + H_3PO_4$$

14.53 Of the conjugate acid-base pairs HSO_4^-/SO_4^{2-}, NH_4^+/NH_3, and HNO_3/NO_3^-, complete the following equation with the pair that gives an equilibrium constant $K_c > 1$.

$$\_\_\_\_ + H_2S \rightleftharpoons \_\_\_\_ + HS^-$$

Factors That Affect Acid Strength (Section 14.3)

14.54 Arrange each group of compounds in order of increasing acid strength. Explain your reasoning.

(a) HCl, H_2S, PH_3
(b) NH_3, PH_3, AsH_3
(c) HBrO, $HBrO_3$, $HBrO_4$

14.55 Arrange each group of compounds in order of decreasing acid strength. Explain your reasoning.

(a) H_2O, H_2S, H_2Se
(b) $HClO_3$, $HBrO_3$, HIO_3
(c) PH_3, H_2S, HCl

14.56 Identify the strongest acid in each of the following sets. Explain your reasoning.

(a) H_2O, HF, or HCl
(b) $HClO_2$, $HClO_3$, or $HBrO_3$
(c) HBr, H_2S, or H_2Se

14.57 Identify the weakest acid in each of the following sets. Explain your reasoning.

(a) H_2SO_3, $HClO_3$, $HClO_4$
(b) NH_3, H_2O, H_2S
(c) $B(OH)_3$, $Al(OH)_3$, $Ga(OH)_3$

14.58 Identify the stronger acid in each of the following pairs. Explain your reasoning.

(a) H_2Se or H_2Te (b) H_3PO_4 or H_3AsO_4
(c) $H_2PO_4^-$ or HPO_4^{2-} (d) CH_4 or NH_4^+

14.59 Identify the stronger base in each of the following pairs. Explain your reasoning.

(a) ClO_2^- or ClO_3^- (b) HSO_4^- or $HSeO_4^-$
(c) HS^- or OH^- (d) HS^- or Br^-

Dissociation of Water; pH (Sections 14.4–14.6)

14.60 For each of the following solutions, calculate $[OH^-]$ from $[H_3O^+]$, or $[H_3O^+]$ from $[OH^-]$. Classify each solution as acidic, basic, or neutral.

(a) $[H_3O^+] = 3.4 \times 10^{-9}$ M (b) $[OH^-] = 0.010$ M
(c) $[OH^-] = 1.0 \times 10^{-10}$ M (d) $[H_3O^+] = 1.0 \times 10^{-7}$ M
(e) $[H_3O^+] = 8.6 \times 10^{-5}$ M

14.61 For each of the following solutions, calculate $[OH^-]$ from $[H_3O^+]$, or $[H_3O^+]$ from $[OH^-]$. Classify each solution as acidic, basic, or neutral.

(a) $[H_3O^+] = 2.5 \times 10^{-4}$ M (b) $[H_3O^+] = 2.0$ M
(c) $[OH^-] = 5.6 \times 10^{-9}$ M (d) $[OH^-] = 1.5 \times 10^{-3}$ M
(e) $[OH^-] = 1.0 \times 10^{-7}$ M

14.62 Water superheated under pressure to 200 °C and 750 atm has $K_w = 1.5 \times 10^{-11}$. What is $[H_3O^+]$ and $[OH^-]$ at 200 °C? Is the water acidic, basic, or neutral?

14.63 Water at 500 °C and 250 atm is a supercritical fluid. Under these conditions, K_w is approximately 1.7×10^{-19}. Estimate $[H_3O^+]$ and $[OH^-]$ at 500 °C. Is the water acidic, basic, or neutral?

14.64 Calculate the pH to the correct number of significant figures for solutions with the following concentrations of H_3O^+ or OH^-:

(a) $[H_3O^+] = 2.0 \times 10^{-5}$ M (b) $[OH^-] = 4 \times 10^{-3}$ M
(c) $[H_3O^+] = 3.56 \times 10^{-9}$ M (d) $[H_3O^+] = 10^{-3}$ M
(e) $[OH^-] = 12$ M

14.65 What is the pH to the correct number of significant figures for solutions with the following concentrations of H_3O^+ or OH^-?

(a) $[OH^-] = 7.6 \times 10^{-3}$ M (b) $[H_3O^+] = 10^{-8}$ M
(c) $[H_3O^+] = 5.0$ M (d) $[OH^-] = 1.0 \times 10^{-7}$ M
(e) $[H_3O^+] = 2.18 \times 10^{-10}$ M

14.66 Calculate the H_3O^+ concentration to the correct number of significant figures for solutions with the following pH values:

(a) 4.1 (b) 10.82 (c) 0.00
(d) 14.25 (e) −1.0 (f) 5.238

14.67 What is the H_3O^+ concentration to the correct number of significant figures for solutions with the following pH values?

(a) 9.0 (b) 7.00 (c) −0.3
(d) 15.18 (e) 2.63 (f) 10.756

14.68 Which of the indicators thymol blue, alizarin yellow, chlorphenol red, or methyl orange would be most appropriate to detect a pH change from:

(a) 7 to 5? (b) 8 to 10? (c) 3 to 5?

14.69 Which of the indicators methyl violet, bromcresol green, phenol red, or thymolphthalein, would be most appropriate to detect a pH change from:

(a) 4 to 6? (b) 8 to 10? (c) 2 to 0?

Strong Acids and Strong Bases (Section 14.7)

14.70 A solution of NaOH has a pH of 10.50. How many grams of CaO should be dissolved in sufficient water to make 1.00 L of a solution having the same pH?

14.71 A solution of KOH has a pH of 10.00. How many grams of SrO should be dissolved in sufficient water to make 2.00 L of a solution having the same pH?

14.72 Calculate the pH of solutions prepared by:

(a) Dissolving 4.8 g of lithium hydroxide in water to give 250 mL of solution

(b) Dissolving 0.93 g of hydrogen chloride in water to give 0.40 L of solution

(c) Diluting 50.0 mL of 0.10 M HCl to a volume of 1.00 L

(d) Mixing 100.0 mL of 2.0×10^{-3} M HCl and 400.0 mL of 1.0×10^{-3} M $HClO_4$ (Assume that volumes are additive.)

14.73 Calculate the pH of solutions prepared by:

(a) Dissolving 0.20 g of sodium oxide in water to give 100.0 mL of solution

(b) Dissolving 1.26 g of pure nitric acid in water to give 0.500 L of solution

(c) Diluting 40.0 mL of 0.075 M $Ba(OH)_2$ to a volume of 300.0 mL

(d) Mixing equal volumes of 0.20 M HCl and 0.50 M HNO_3 (Assume that volumes are additive.)

Weak Acids (Sections 14.8–14.10)

14.74 Look up the values of K_a in Appendix C for C_6H_5OH, HNO_3, CH_3CO_2H, and HOCl, and arrange these acids in order of:

(a) Increasing acid strength

(b) Decreasing percent dissociation

Also estimate $[H_3O^+]$ in a 1.0 M solution of each acid.

14.75 Look up the values of K_a in Appendix C for HCO_2H, HCN, $HClO_4$, and HOBr, and arrange these acids in order of:

(a) Increasing acid strength

(b) Decreasing percent dissociation

Also estimate $[H_3O^+]$ in a 1.0 M solution of each acid.

14.76 The pH of 0.040 M hypobromous acid (HOBr) is 5.05. Set up the equilibrium equation for the dissociation of HOBr, and calculate the value of the acid-dissociation constant.

14.77 Lactic acid ($C_3H_6O_3$), which occurs in sour milk and foods such as sauerkraut, is a weak monoprotic acid. The pH of a 0.10 M solution of lactic acid is 2.43. What are the values of K_a and pK_a for lactic acid?

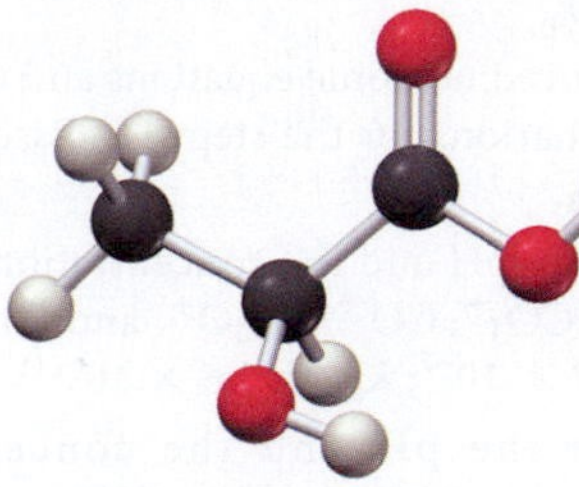

Lactic acid

14.78 Acrylic acid ($C_3H_4O_2$) is used in the manufacture of paints and plastics. The pK_a of acrylic acid is 4.25.

(a) Calculate the pH and the concentrations of all species (H_3O^+, $C_3H_3O_2^-$, $C_3H_4O_2$, and OH^-) in 0.150 M acrylic acid.

(b) Calculate the percent dissociation in 0.0500 M acrylic acid.

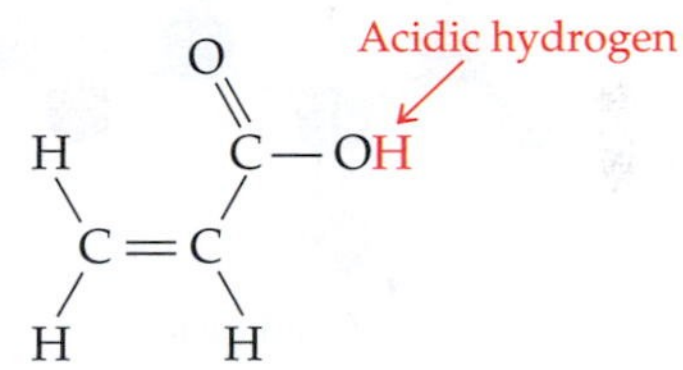

Acrylic acid

14.79 Hippuric acid ($C_9H_9NO_3$), found in horse urine, has pK_a = 3.62.

(a) Calculate the pH and the concentrations of all species (H_3O^+, $C_9H_8NO_3^-$, $C_9H_9NO_3$, and OH^-) in 0.100 M hippuric acid.

(b) Calculate the percent dissociation in 0.0750 M hippuric acid.

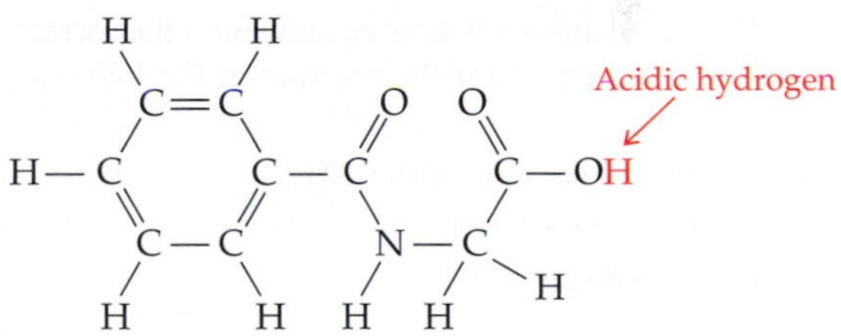

Hippuric acid

14.80 Calculate the pH and the percent dissociation in 1.5 M HNO_2 ($K_a = 4.5 \times 10^{-4}$).

14.81 A typical aspirin tablet contains 324 mg of aspirin (acetylsalicylic acid), a monoprotic acid having $K_a = 3.0 \times 10^{-4}$. If you dissolve two aspirin tablets in a 300 mL glass of water, what is the pH of the solution and the percent dissociation?

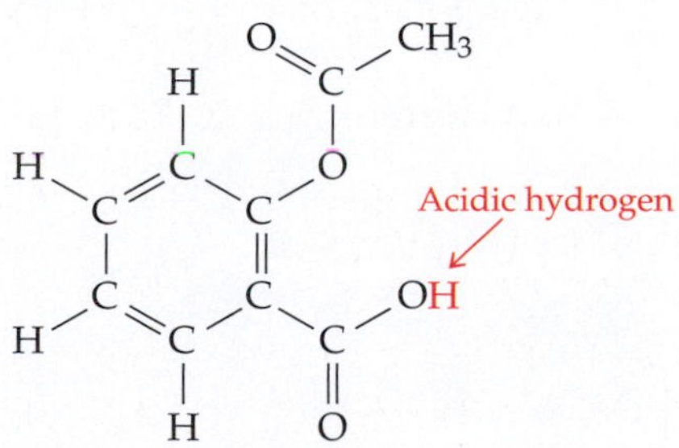

Acetylsalicylic acid (Aspirin)

Polyprotic Acids (Section 14.11)

14.82 Write balanced net ionic equations and the corresponding equilibrium equations for the stepwise dissociation of the diprotic acid H_2SeO_4.

14.83 Write balanced net ionic equations and the corresponding equilibrium equations for the stepwise dissociation of the triprotic acid H_3PO_4.

14.84 Calculate the pH and the concentrations of all species present (H_2CO_3, HCO_3^-, CO_3^{2-}, H_3O^+, and OH^-) in 0.010 M H_2CO_3 ($K_{a1} = 4.3 \times 10^{-7}$; $K_{a2} = 5.6 \times 10^{-11}$).

14.85 Calculate the pH and the concentrations of H_2SO_3, HSO_3^-, SO_3^{2-}, H_3O^+, and OH^- in 0.025 M H_2SO_3 ($K_{a1} = 1.5 \times 10^{-2}$; $K_{a2} = 6.3 \times 10^{-8}$).

14.86 Oxalic acid ($H_2C_2O_4$) is a diprotic acid that occurs in plants such as rhubarb and spinach. Calculate the pH and the concentration of $C_2O_4^{2-}$ ions in 0.20 M $H_2C_2O_4$ ($K_{a1} = 5.9 \times 10^{-2}$; $K_{a2} = 6.4 \times 10^{-5}$).

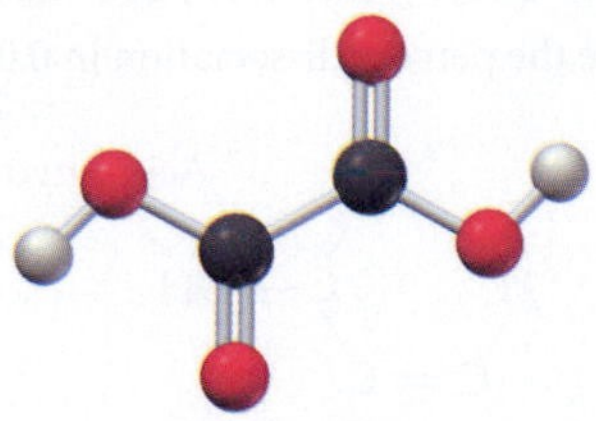

Oxalic acid

14.87 Calculate the concentrations of H_3O^+ and SO_4^{2-} in a solution prepared by mixing equal volumes of 0.2 M HCl and 0.6 M H_2SO_4 (K_{a2} for H_2SO_4 is 1.2×10^{-2}).

Weak Bases; Relation between K_a and K_b (Sections 14.12–14.13)

14.88 Write a balanced net ionic equation and the corresponding equilibrium equation for the reaction of the following weak bases with water:

(a) Dimethylamine, $(CH_3)_2NH$

(b) Aniline, $C_6H_5NH_2$

(c) Cyanide ion, CN^-

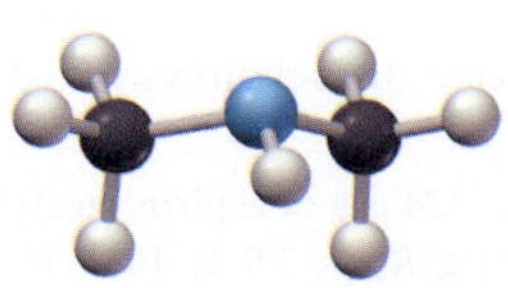

Dimethylamine

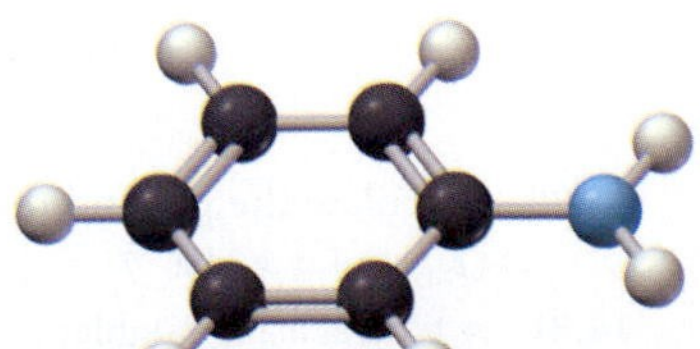

Aniline

14.89 Write a balanced net ionic equation and the corresponding equilibrium equation for the reaction of the following weak bases with water:

(a) Pyridine, C_5H_5N

(b) Ethylamine, $C_2H_5NH_2$

(c) Acetate ion, $CH_3CO_2^-$

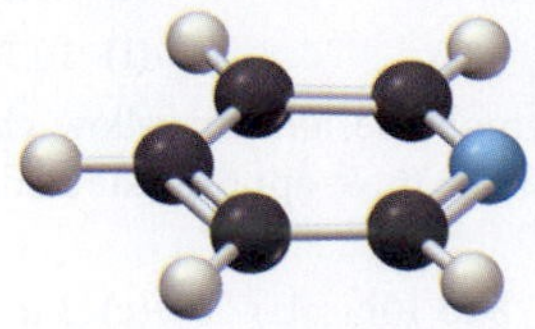

Pyridine

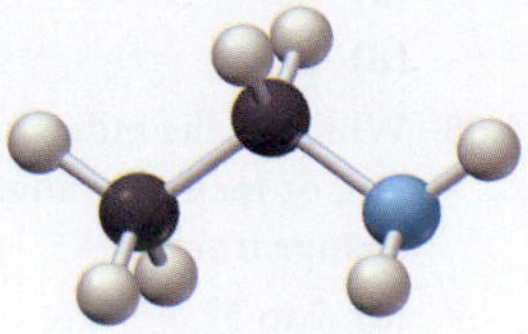

Ethylamine

14.90 Morphine ($C_{17}H_{19}NO_3$), a narcotic used in painkillers, is a weak organic base. If the pH of a 7.0×10^{-4} M solution of morphine is 9.50, what are the values of K_b and pK_b?

14.91 A 1.00×10^{-3} M solution of quinine, a drug used in treating malaria, has a pH of 9.75. What are the values of K_b and pK_b?

14.92 Oxycodone ($C_{18}H_{21}NO_4$), a narcotic analgesic, is a weak base with $pK_b = 5.47$. Calculate the pH and the concentrations of all species present ($C_{18}H_{21}NO_4$, $HC_{18}H_{21}NO_4^+$, H_3O^+, and OH^-) in a 0.002 50 M oxycodone solution.

14.93 Morpholine (C_4H_9NO) is a weak organic base with $pK_b = 5.68$. Calculate the pH and the concentrations of all species present (C_4H_9NO, $HC_4H_9NO^+$, H_3O^+, and OH^-) in a 0.0100 M morpholine solution.

14.94 Using values of K_b in Appendix C, calculate values of K_a for each of the following ions:

(a) Propylammonium ion, $C_3H_7NH_3^+$

(b) Hydroxylammonium ion, NH_3OH^+

(c) Anilinium ion, $C_6H_5NH_3^+$

(d) Pyridinium ion, $C_5H_5NH^+$

14.95 Using values of K_a in Appendix C, calculate values of K_b for each of the following ions:

(a) Fluoride ion, F^-

(b) Hypobromite ion, OBr^-

(c) Hydrogen sulfide ion, HS^-

(d) Sulfide ion, S^{2-}

Acid–Base Properties of Salts (Section 14.14)

14.96 Write a balanced net ionic equation for the reaction of each of the following ions with water. In each case, identify the Brønsted–Lowry acids and bases and the conjugate acid–base pairs.

(a) $CH_3NH_3^+$ **(b)** $Cr(H_2O)_6^{3+}$

(c) $CH_3CO_2^-$ **(d)** PO_4^{3-}

14.97 Write a balanced net ionic equation for the principal reaction in solutions of each of the following salts. In each case, identify the Brønsted–Lowry acids and bases and the conjugate acid–base pairs.

(a) Na_2CO_3 **(b)** NH_4NO_3

(c) NaCl **(d)** $ZnCl_2$

14.98 Classify each of the following ions according to whether they react with water to give a neutral, acidic, or basic solution:
(a) F^- (b) Br^- (c) NH_4^+
(d) $K(H_2O)_6^+$ (e) SO_3^{2-} (f) $Cr(H_2O)_6^{3+}$

14.99 Classify each of the following salt solutions as neutral, acidic, or basic. See Appendix C for values of equilibrium constants.
(a) $Fe(NO_3)_3$ (b) $Ba(NO_3)_2$ (c) NaOCl
(d) NH_4I (e) NH_4NO_2 (f) $(CH_3NH_3)Cl$

14.100 Calculate the concentrations of all species present and the pH in 0.10 M solutions of the following substances. See Appendix C for values of equilibrium constants.
(a) Ethylammonium nitrate, $(C_2H_5NH_3)NO_3$
(b) Sodium acetate, $Na(CH_3CO_2)$
(c) Sodium nitrate, $NaNO_3$

14.101 Calculate the pH and the percent dissociation of the hydrated cation in 0.020 M solutions of the following substances. See Appendix C for values of equilibrium constants.
(a) $Fe(NO_3)_2$ (b) $Fe(NO_3)_3$

Lewis Acids and Bases (Section 14.15)

14.102 For each of the following reactions, identify the Lewis acid and the Lewis base:
(a) $SiF_4 + 2\,F^- \longrightarrow SiF_6^{2-}$
(b) $4\,NH_3 + Zn^{2+} \longrightarrow Zn(NH_3)_4^{2+}$
(c) $2\,Cl^- + HgCl_2 \longrightarrow HgCl_4^{2-}$
(d) $CO_2 + H_2O \longrightarrow H_2CO_3$

14.103 For each of the following reactions, identify the Lewis acid and the Lewis base:
(a) $2\,Cl^- + BeCl_2 \longrightarrow BeCl_4^{2-}$
(b) $Mg^{2+} + 6\,H_2O \longrightarrow Mg(H_2O)_6^{2+}$
(c) $SO_3 + OH^- \longrightarrow HSO_4^-$
(d) $F^- + BF_3 \longrightarrow BF_4^-$

14.104 For each of the Lewis acid–base reactions in Problem 14.102, draw electron-dot structures for the reactants and products, and use the curved arrow notation (Section 14.15) to represent the donation of a lone pair of electrons from the Lewis base to the Lewis acid.

14.105 For each of the Lewis acid–base reactions in Problem 14.103, draw electron-dot structures for the reactants and products, and use the curved arrow notation (Section 14.15) to represent the donation of a lone pair of electrons from the Lewis base to the Lewis acid.

14.106 Classify each of the following as a Lewis acid or a Lewis base:
(a) CN^- (b) H^+ (c) H_2O
(d) Fe^{3+} (e) OH^- (f) CO_2
(g) $P(CH_3)_3$ (h) $B(CH_3)_3$

14.107 Which would you expect to be the stronger Lewis acid in each of the following pairs? Explain.
(a) BF_3 or BH_3 (b) SO_2 or SO_3
(c) Sn^{2+} or Sn^{4+} (d) CH_3^+ or CH_4

CHAPTER PROBLEMS

14.108 Aqueous solutions of hydrogen sulfide contain H_2S, HS^-, S^{2-}, H_3O^+, OH^-, and H_2O in varying concentrations. Which of these species can act only as an acid? Which can act only as a base? Which can act both as an acid and as a base?

14.109 The hydronium ion H_3O^+ is the strongest acid that can exist in aqueous solution because stronger acids dissociate by transferring a proton to water. What is the strongest base that can exist in aqueous solution?

14.110 Baking powder contains baking soda ($NaHCO_3$) and an acidic substance such as sodium alum, $NaAl(SO_4)_2 \cdot 12\,H_2O$. These components react in an aqueous medium to produce CO_2 gas, which "raises" the dough. Write a balanced net ionic equation for the reaction.

14.111 Arrange the following substances in order of increasing $[H_3O^+]$ for a 0.10 M solution of each:
(a) $Zn(NO_3)_2$ (b) Na_2O (c) NaOCl
(d) $NaClO_4$ (e) $HClO_4$

14.112 At 0 °C, the density of liquid water is 0.9998 g/mL and the value of K_w is 1.14×10^{-15}. What fraction of the molecules in liquid water are dissociated at 0 °C? What is the percent dissociation at 0 °C? What is the pH of a neutral solution at 0 °C?

14.113 Use the conjugate acid–base pair HCN and CN^- to derive the relationship between K_a and K_b.

14.114 The pH of 0.050 M gallic acid, an acid found in tea leaves, is 2.86. Calculate K_a and pK_a for gallic acid.

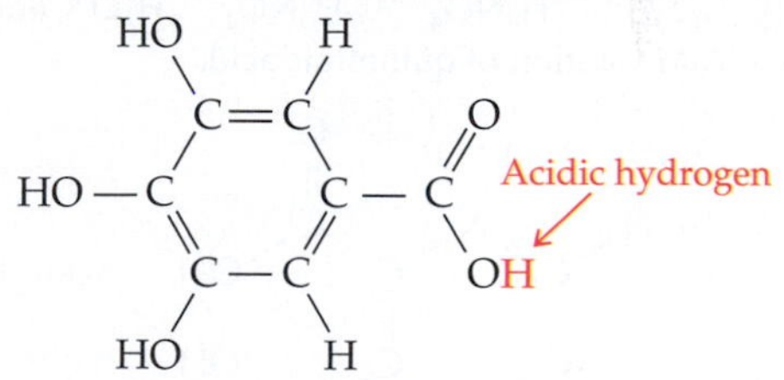

Gallic acid

14.115 The pH of 0.040 M pyruvic acid, a compound involved in metabolic pathways, is 1.96. Calculate K_a and pK_a for pyruvic acid.

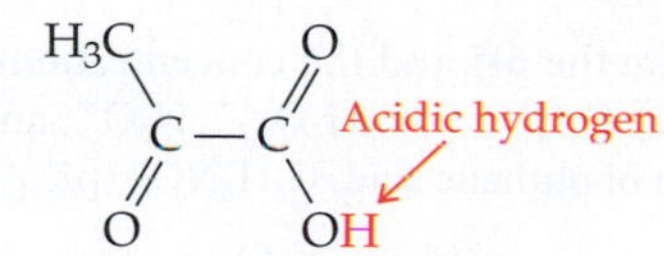

Pyruvic acid

14.116 Nicotine ($C_{10}H_{14}N_2$) can accept two protons because it has two basic N atoms ($K_{b1} = 1.0 \times 10^{-6}$; $K_{b2} = 1.3 \times 10^{-11}$). Calculate the values of K_a for the conjugate acids $C_{10}H_{14}N_2H^+$ and $C_{10}H_{14}N_2H_2^{2+}$.

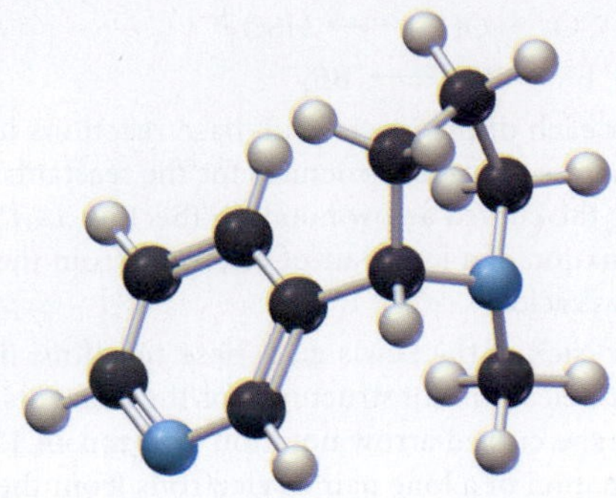

Nicotine

14.117 Sodium benzoate ($C_6H_5CO_2Na$) is used as a food preservative. Calculate the pH and the concentrations of all species present (Na^+, $C_6H_5CO_2^-$, $C_6H_5CO_2H$, H_3O^+, and OH^-) in 0.050 M sodium benzoate; K_a for benzoic acid ($C_6H_5CO_2H$) is 6.5×10^{-5}.

14.118 The hydrated cation $M(H_2O)_6^{3+}$ has $K_a = 10^{-4}$, and the acid HA has $K_a = 10^{-5}$. Identify the principal reaction in an aqueous solution of each of the following salts, and classify each solution as acidic, basic, or neutral:

(a) NaA **(b)** $M(NO_3)_3$ **(c)** $NaNO_3$ **(d)** MA_3

14.119 Calculate the pH and the concentrations of all species present (H_3O^+, F^-, HF, Cl^-, and OH^-) in a solution that contains 0.10 M HF ($K_a = 3.5 \times 10^{-4}$) and 0.10 M HCl.

14.120 Calculate the pH and the concentrations of all species present (H_3O^+, OH^-, HIO_3, and IO_3^-) in 0.0500 M HIO_3. K_a for HIO_3 is 1.7×10^{-1}.

14.121 Quinolinic acid, $C_7H_5NO_4$ ($pK_{a1} = 2.43$; $pK_{a2} = 4.78$), has been implicated in the progression of Alzheimer's disease. Calculate the pH and the concentrations of all species present ($C_7H_5NO_4$, $C_7H_4NO_4^-$, $C_7H_3NO_4^{2-}$, H_3O^+, and OH^-) in a 0.050 M solution of quinolinic acid.

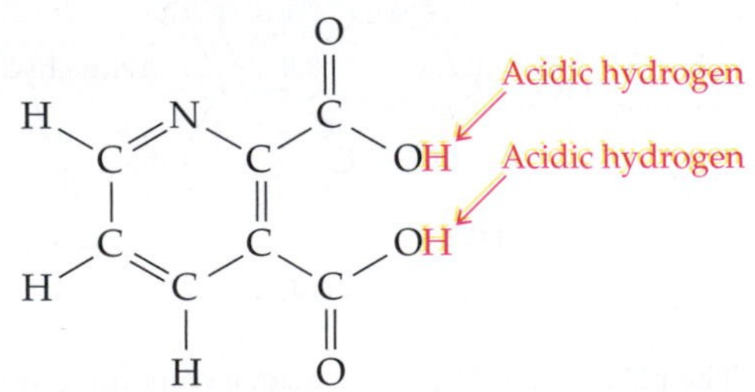

Quinolinic acid

14.122 Calculate the pH and the concentrations of all species present ($C_8H_6O_4$, $C_8H_5O_4^-$, $C_8H_4O_4^{2-}$, H_3O^+, and OH^-) in a 0.0250 M solution of phthalic acid, $C_8H_6NO_4$ ($pK_{a1} = 2.89$; $pK_{a2} = 5.51$).

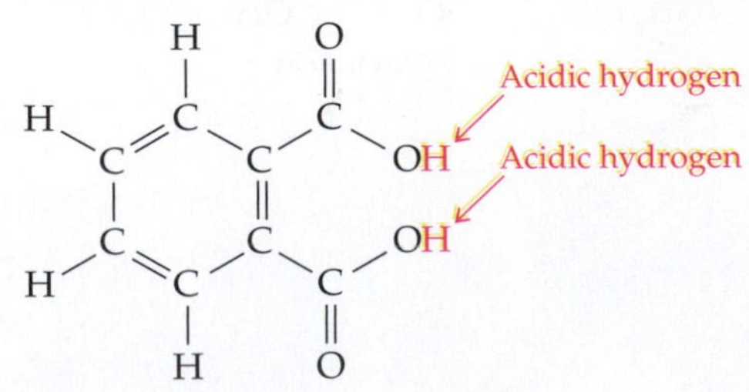

Phthalic acid

14.123 Sulfur dioxide is quite soluble in water:

$$SO_2(g) + H_2O(l) \rightleftharpoons H_2SO_3(aq) \quad K = 1.33$$

The H_2SO_3 produced is a weak diprotic acid ($K_{a1} = 1.5 \times 10^{-2}$; $K_{a2} = 6.3 \times 10^{-8}$). Calculate the pH and the concentrations of H_2SO_3, HSO_3^-, and SO_3^{2-} in a solution prepared by continuously bubbling SO_2 at a pressure of 1.00 atm into pure water.

14.124 Classify each of the following salt solutions as neutral, acidic, or basic. See Appendix C for values of equilibrium constants.

(a) NH_4F **(b)** $(NH_4)_2SO_3$

14.125 Calculate the pH and the percent dissociation of the hydrated cation in the following solutions. See Appendix C for the value of the equilibrium constant.

(a) 0.010 M $Cr(NO_3)_3$

(b) 0.0050 M $Cr(NO_3)_3$

(c) 0.0010 M $Cr(NO_3)_3$

14.126 Beginning with the equilibrium equation for the dissociation of a weak acid HA, show that the percent dissociation varies directly as the square root of K_a and inversely as the square root of the initial concentration of HA when the concentration of HA that dissociates is negligible compared with its initial concentration.

14.127 Calculate the pH and the concentrations of all species present in 0.25 M solutions of each of the salts in Problem 14.124. (Hint: The principal reaction is proton transfer from the cation to the anion.)

14.128 For a solution of two weak acids with comparable values of K_a, there is no single principal reaction. The two acid-dissociation equilibrium equations must therefore be solved simultaneously. Calculate the pH in a solution that is 0.10 M in acetic acid (CH_3CO_2H, $K_a = 1.8 \times 10^{-5}$) and 0.10 M in benzoic acid ($C_6H_5CO_2H$, $K_a = 6.5 \times 10^{-5}$). (Hint: Let $x = [CH_3CO_2H]$ that dissociates and $y = [C_6H_5CO_2H]$ that dissociates; then $[H_3O^+] = x + y$.)

14.129 What is the pH and the principal source of H_3O^+ ions in 1.0×10^{-10} M HCl? (Hint: The pH of an acid solution can't exceed 7.) What is the pH of 1.0×10^{-7} M HCl?

14.130 When NO_2 is bubbled into water, it is completely converted to HNO_3 and HNO_2:

$$2\,NO_2(g) + H_2O(l) \longrightarrow HNO_3(aq) + HNO_2(aq)$$

Calculate the pH and the concentrations of all species present (H_3O^+, OH^-, HNO_2, NO_2^-, and NO_3^-) in a solution prepared by dissolving 0.0500 mol of NO_2 in 1.00 L of water. K_a for HNO_2 is 4.5×10^{-4}.

14.131 Acid and base behavior can be observed in solvents other than water. One commonly used solvent is dimethyl sulfoxide (DMSO), which can be treated as a monoprotic acid "HSol." Just as water can behave either as an acid or a base, so HSol can behave either as a Brønsted–Lowry acid or base.

(a) The equilibrium constant for self-dissociation of HSol (call it K_{HSol}) is 1×10^{-35}. Write the chemical equation for the self-dissociation reaction and the corresponding equilibrium equation. (Hint: The equilibrium equation is analogous to the equilibrium equation for K_w in the case of water.)

(b) The weak acid HCN has an acid-dissociation constant $K_a = 1.3 \times 10^{-13}$ in the solvent HSol. If 0.010 mol of NaCN is dissolved in 1.00 L of HSol, what is the equilibrium concentration of H_2Sol^+?

MULTICONCEPT PROBLEMS

14.132 A 7.0 mass % solution of H_3PO_4 in water has a density of 1.0353 g/mL. Calculate the pH and the molar concentrations of all species present (H_3PO_4, $H_2PO_4^-$, HPO_4^{2-}, PO_4^{3-}, H_3O^+, and OH^-) in the solution. Values of equilibrium constants are listed in Appendix C.

14.133 In the case of very weak acids, $[H_3O^+]$ from the dissociation of water is significant compared with $[H_3O^+]$ from the dissociation of the weak acid. The sugar substitute saccharin ($C_7H_5NO_3S$), for example, is a very weak acid having $K_a = 2.1 \times 10^{-12}$ and a solubility in water of 348 mg/100 mL. Calculate $[H_3O^+]$ in a saturated solution of saccharin. (Hint: Equilibrium equations for the dissociation of saccharin and water must be solved simultaneously.)

14.134 In aqueous solution, sodium acetate behaves as a strong electrolyte, yielding Na^+ cations and $CH_3CO_2^-$ anions. A particular solution of sodium acetate has a pH of 9.07 and a density of 1.0085 g/mL. What is the molality of this solution, and what is its freezing point?

14.135 During a certain time period, 4.0 million tons of SO_2 was released into the atmosphere and was subsequently oxidized to SO_3. As explained in the *FYI*, the acid rain produced when the SO_3 dissolves in water can damage marble statues:

$$CaCO_3(s) + H_2SO_4(aq) \longrightarrow CaSO_4(aq) + CO_2(g) + H_2O(l)$$

(a) How many 500 pound marble statues could be damaged by the acid rain? (Assume that the statues are pure $CaCO_3$ and that a statue is damaged when 3.0% of its mass is dissolved.)

(b) How many liters of CO_2 gas at 20 °C and 735 mm Hg is produced as a byproduct?

(c) The cation in aqueous H_2SO_4 is trigonal pyramidal rather than trigonal planar. Explain.

14.136 Neutralization reactions involving either a strong acid or a strong base go essentially to completion, and therefore we must take such neutralizations into account before calculating concentrations in mixtures of acids and bases. Consider a mixture of 3.28 g of Na_3PO_4 and 300.0 mL of 0.180 M HCl. Write balanced net ionic equations for the neutralization reactions, and calculate the pH of the solution.

14.137 We've said that alkali metal cations do not react appreciably with water to produce H_3O^+ ions, but in fact, all cations are acidic to some extent. The most acidic alkali metal cation is the smallest one, Li^+, which has $K_a = 2.5 \times 10^{-14}$ for the reaction

$$Li(H_2O)_4^+(aq) + H_2O(l) \rightleftharpoons H_3O^+(aq) + Li(H_2O)_3(OH)(aq)$$

This reaction and the dissociation of water must be considered simultaneously in calculating the pH of Li^+ solutions, which nevertheless have pH $\approx$ 7. Check this by calculating the pH of 0.10 M LiCl.

14.138 A 1.000 L sample of HF gas at 20.0 °C and 0.601 atm pressure was dissolved in enough water to make 50.0 mL of hydrofluoric acid.

(a) What is the pH of the solution?

(b) To what volume must you dilute the solution to triple the percent dissociation. Why can't you solve this problem by using the result obtained in Problem 14.126?

14.139 A 200.0 mL sample of 0.350 M acetic acid (CH_3CO_2H) was allowed to react with 2.000 L of gaseous ammonia at 25 °C and a pressure of 650.8 mm Hg. Assuming no change in the volume of the solution, calculate the pH and the equilibrium concentrations of all species present (CH_3CO_2H, $CH_3CO_2^-$, NH_3, NH_4^+, H_3O^+, and OH^-). Values of equilibrium constants are listed in Appendix C.

14.140 You may have been told not to mix bleach and ammonia. The reason is that bleach (sodium hypochlorite) reacts with ammonia to produce toxic chloramines, such as NH_2Cl. For example, in basic solution:

$$OCl^-(aq) + NH_3(aq) \longrightarrow OH^-(aq) + NH_2Cl(aq)$$

(a) The following initial rate data for this reaction were obtained in basic solution at 25 °C:

pH	Initial $[OCl^-]$	Initial $[NH_3]$	Initial Rate (M/s)
12	0.001	0.01	0.017
12	0.002	0.01	0.033
12	0.002	0.03	0.100
13	0.002	0.03	0.010

What is the rate law for the reaction? What is the numerical value of the rate constant *k*, including the correct units?

(b) The following mechanism has been proposed for this reaction in basic solution:

$$H_2O + OCl^- \rightleftharpoons HOCl + OH^- \quad \text{Fast, equilibrium constant } K_1$$

$$HOCl + NH_3 \longrightarrow H_2O + NH_2Cl \quad \text{Slow, rate constant } k_2$$

Assuming that the first step is in equilibrium and the second step is rate-determining, calculate the value of the rate constant k_2 for the second step. K_a for HOCl is 3.5×10^{-8}.

CHAPTER

15 Applications of Aqueous Equilibria

Crystals of fluorapatite, $Ca_5(PO_4)_3F$. Fluoride ions help to prevent tooth decay by partially converting the hydroxyapatite, $Ca_5(PO_4)_3OH$, of tooth enamel to fluorapatite, thus making the enamel more resistant to attack by acids, as explained in the *FYI* at the end of this chapter.

Aqueous equilibria play a crucial role in many environmental and biological processes. For example, the pH of many lakes and streams must remain near 5.5 for plant and aquatic life to flourish. The pH of human blood is carefully controlled at a value of 7.4 by equilibria involving primarily the conjugate acid–base pair H_2CO_3 and HCO_3^-. At lower pH values, the affinity of hemoglobin in red blood cells for O_2 molecules decreases and the blood carries less oxygen to the tissues.

We began our study of the principles of chemical equilibria in Chapter 13 and learned how to write equilibrium equations and calculate the concentrations of reactants and products in an equilibrium mixture. In Chapter 14, we applied those principles to equilibria in aqueous solutions of a weak acid or a weak base. In this chapter, we'll continue our study by examining some applications of aqueous equilibria. First we'll see how to calculate the pH of *mixtures* of acids and bases. Then we'll look at the dissolution and precipitation of slightly soluble ionic compounds and the factors that affect solubility. Aqueous equilibria involving the dissolution and precipitation of ionic compounds are important in a great many natural processes, from tooth decay to the formation of coral reefs and limestone caves.

15.1 NEUTRALIZATION REACTIONS

We've seen on numerous occasions that the neutralization reaction of an acid with a base produces water and a salt. But to what extent does a neutralization reaction go to completion? We must answer that question before we can make pH calculations on mixtures of acids and bases. We'll look at four types of neutralization reactions: (1) strong acid–strong base, (2) weak acid–strong base, (3) strong acid–weak base, and (4) weak acid–weak base.

Strong Acid–Strong Base

As an example of a strong acid–strong base reaction, let's consider the reaction of hydrochloric acid with aqueous sodium hydroxide to give water and an aqueous solution of sodium chloride:

$$HCl(aq) + NaOH(aq) \longrightarrow H_2O(l) + NaCl(aq)$$

CONTENTS

Because HCl(*aq*), NaOH(*aq*), and NaCl(*aq*) are all completely dissociated, the net ionic equation for the neutralization reaction is

$$H_3O^+(aq) + OH^-(aq) \rightleftharpoons 2\,H_2O(l)$$

If we mix equal numbers of moles of HCl(*aq*) and NaOH(*aq*), the concentrations of H_3O^+ and OH^- remaining in the NaCl solution after neutralization will be the same as those in pure water, $[H_3O^+] = [OH^-] = 1.0 \times 10^{-7}$ M. In other words, the reaction of HCl with NaOH proceeds far to the right.

We come to the same conclusion by looking at the equilibrium constant for the reaction. Because the neutralization reaction of any strong acid with a strong base is the reverse of the dissociation of water, its equilibrium constant, K_n ("n" for neutralization), is the reciprocal of the ion-product constant for water, $K_n = 1/K_w$:

$$H_3O^+(aq) + OH^-(aq) \rightleftharpoons 2\,H_2O(l)$$

$$K_n = \frac{1}{[H_3O^+][OH^-]} = \frac{1}{K_w} = \frac{1}{1.0 \times 10^{-14}} = 1.0 \times 10^{14}$$

The value of K_n (1.0×10^{14}) for a strong acid–strong base reaction is a very large number, which means that the neutralization reaction proceeds essentially 100% to completion. After neutralization of equal molar amounts of acid and base, the solution contains a salt derived from a strong base and a strong acid. Because neither the cation nor the anion of the salt has acidic or basic properties, the pH is 7 (Section 14.14).

Weak Acid–Strong Base

Because a weak acid HA is largely undissociated, the net ionic equation for the neutralization reaction of a weak acid with a strong base involves proton transfer from HA to the strong base, OH^-:

$$HA(aq) + OH^-(aq) \rightleftharpoons H_2O(l) + A^-(aq)$$

Acetic acid (CH_3CO_2H), for example, reacts with aqueous NaOH to give water and aqueous sodium acetate (CH_3CO_2Na):

$$CH_3CO_2H(aq) + OH^-(aq) \rightleftharpoons H_2O(l) + CH_3CO_2^-(aq)$$

Weak acid–strong base neutralization
$CH_3CO_2H(aq) + OH^-(aq) \rightleftharpoons H_2O(l) + CH_3CO_2^-(aq)$

▲ When NaOH(*aq*) is added to $CH_3CO_2H(aq)$ containing the acid–base indicator phenolphthalein, the color of the indicator changes from colorless to pink in the pH range 8.2–9.8 because neutralization of the acetic acid gives a weakly basic solution.

Na^+ ions do not appear in the net ionic equation because both NaOH and CH_3CO_2Na are completely dissociated.

To obtain the equilibrium constant K_n for the neutralization of acetic acid, we multiply equilibrium constants for reactions that add to give the net ionic equation for the neutralization reaction. Because CH_3CO_2H is on the left side of the equation and $CH_3CO_2^-$ is on the right side, one of the reactions needed is the dissociation of CH_3CO_2H. Because H_2O is on the right side of the equation and OH^- is on the left side, the other reaction needed is the reverse of the dissociation of H_2O. Note that H_3O^+ and one H_2O molecule cancel when the two equations are added:

$$CH_3CO_2H(aq) + \cancel{H_2O(l)} \rightleftharpoons \cancel{H_3O^+(aq)} + CH_3CO_2^-(aq) \qquad K_a = 1.8 \times 10^{-5}$$

$$\cancel{H_3O^+(aq)} + OH^-(aq) \rightleftharpoons 2\,H_2O(l) \qquad 1/K_w = 1.0 \times 10^{14}$$

$$\text{Net: } CH_3CO_2H(aq) + OH^-(aq) \rightleftharpoons H_2O(l) + CH_3CO_2^-(aq) \qquad K_n = (K_a)(1/K_w) = (1.8 \times 10^{-5})(1.0 \times 10^{14}) = 1.8 \times 10^9$$

As we saw in Section 14.13, the equilibrium constant for the net reaction equals the product of the equilibrium constants for the reactions added. Therefore, we multiply K_a for CH_3CO_2H by the reciprocal of K_w to get K_n for the neutralization reaction. (We use $1/K_w$ because the H_2O dissociation reaction is written in the reverse direction.) The resulting large value of K_n (1.8×10^9) means that the neutralization reaction proceeds nearly 100% to completion.

As a general rule, the neutralization of any weak acid with a strong base will go 100% to completion because OH^- has a great affinity for protons. After neutralization of equal molar amounts of CH_3CO_2H and NaOH, the solution contains Na^+, which has no acidic or basic properties, and $CH_3CO_2^-$, which is a weak base. Therefore, the pH is greater than 7 (Section 14.14).

Strong acid–weak base neutralization
$H_3O^+(aq) + NH_3(aq) \rightleftharpoons H_2O(l) + NH_4^+(aq)$

▲ When hydrochloric acid is added to aqueous ammonia containing the acid–base indicator methyl red, the color of the indicator changes from yellow to red in the pH range 6.0–4.2 because the neutralization of the NH_3 gives a weakly acidic solution.

Strong Acid–Weak Base

A strong acid HA is completely dissociated into H_3O^+ and A^- ions, and its neutralization reaction with a weak base therefore involves proton transfer from H_3O^+ to the weak base B:

$$H_3O^+(aq) + B(aq) \rightleftharpoons H_2O(l) + BH^+(aq)$$

For example, the net ionic equation for the neutralization of hydrochloric acid with aqueous ammonia is

$$H_3O^+(aq) + NH_3(aq) \rightleftharpoons H_2O(l) + NH_4^+(aq)$$

As in the weak acid–strong base case, we can obtain the equilibrium constant for the neutralization reaction by multiplying equilibrium constants for reactions that add to give the net ionic equation:

$$NH_3(aq) + \cancel{H_2O(l)} \rightleftharpoons NH_4^+(aq) + \cancel{OH^-(aq)} \qquad K_b = 1.8 \times 10^{-5}$$

$$H_3O^+(aq) + \cancel{OH^-(aq)} \rightleftharpoons 2\,H_2O(l) \qquad 1/K_w = 1.0 \times 10^{14}$$

$$\text{Net: } H_3O^+(aq) + NH_3(aq) \rightleftharpoons H_2O(l) + NH_4^+(aq) \qquad K_n = (K_b)(1/K_w) = (1.8 \times 10^{-5})(1.0 \times 10^{14}) = 1.8 \times 10^9$$

Again, the neutralization reaction proceeds nearly 100% to the right because its equilibrium constant K_n is a very large number (1.8×10^9). (It's purely coincidental that the neutralization reactions of CH_3CO_2H with NaOH and of HCl with NH_3 have the same value of K_n. The two K_n values are the same because K_a for CH_3CO_2H happens to have the same value as K_b for NH_3.)

The neutralization of any weak base with a strong acid generally goes 100% to completion because H_3O^+ is a powerful proton donor. After the neutralization of equal molar amounts of NH_3 and HCl, the solution contains NH_4^+, which is a weak acid, and Cl^-, which has no acidic or basic properties. Therefore, the pH is less than 7 (Section 14.14).

Weak Acid–Weak Base

Both a weak acid HA and a weak base B are largely undissociated, and the neutralization reaction between them therefore involves proton transfer from the weak acid to the weak base. For example, the net ionic equation for the neutralization of acetic acid with aqueous ammonia is

$$CH_3CO_2H(aq) + NH_3(aq) \rightleftharpoons NH_4^+(aq) + CH_3CO_2^-(aq)$$

The equilibrium constant K_n can be obtained by multiplying equilibrium constants for (1) the acid dissociation of acetic acid, (2) the base protonation of ammonia, and (3) the reverse of the dissociation of water:

$$CH_3CO_2H(aq) + \cancel{H_2O(l)} \rightleftharpoons \cancel{H_3O^+(aq)} + CH_3CO_2^-(aq) \qquad K_a = 1.8 \times 10^{-5}$$

$$NH_3(aq) + \cancel{H_2O(l)} \rightleftharpoons NH_4^+(aq) + \cancel{OH^-(aq)} \qquad K_b = 1.8 \times 10^{-5}$$

$$\cancel{H_3O^+(aq)} + \cancel{OH^-(aq)} \rightleftharpoons \cancel{2\,H_2O(l)} \qquad 1/K_w = 1.0 \times 10^{14}$$

$$\text{Net:}\quad CH_3CO_2H(aq) + NH_3(aq) \rightleftharpoons NH_4^+(aq) + CH_3CO_2^-(aq) \qquad K_n = (K_a)(K_b)(1/K_w)$$

$$K_n = (K_a)(K_b)(1/K_w) = (1.8 \times 10^{-5})(1.8 \times 10^{-5})(1.0 \times 10^{14}) = 3.2 \times 10^4$$

The value of K_n in this case is smaller than it is for the preceding three cases, so the neutralization does not proceed as far toward completion.

In general, weak acid–weak base neutralizations have less tendency to proceed to completion than neutralizations involving strong acids or strong bases. The neutralization of HCN with aqueous ammonia, for example, has a value of K_n less than one, which means that the reaction proceeds less than halfway to completion:

$$HCN(aq) + NH_3(aq) \rightleftharpoons NH_4^+(aq) + CN^-(aq) \qquad K_n = 0.88$$

Worked Example 15.1

Writing an Equation for a Neutralization Reaction, and Estimating the pH of the Resulting Solution

Write a balanced net ionic equation for the neutralization of equal molar amounts of nitric acid and the weak base methylamine (CH_3NH_2). Indicate whether the pH after neutralization is greater than, equal to, or less than 7.

STRATEGY

The formulas that should appear in the net ionic equation depend on whether the acid and base are strong (completely dissociated) or weak (largely undissociated). The pH after neutralization depends on the acid–base properties of the cation and anion in the resulting salt solution (Section 14.14).

SOLUTION

Because HNO_3 is a strong acid and CH_3NH_2 is a weak base, the neutralization proceeds to completion and the net ionic equation is

$$H_3O^+(aq) + CH_3NH_2(aq) \longrightarrow H_2O(l) + CH_3NH_3^+(aq)$$

In this reaction methylamine uses the lone pair of electrons on its N atom to accept a proton from H_3O^+:

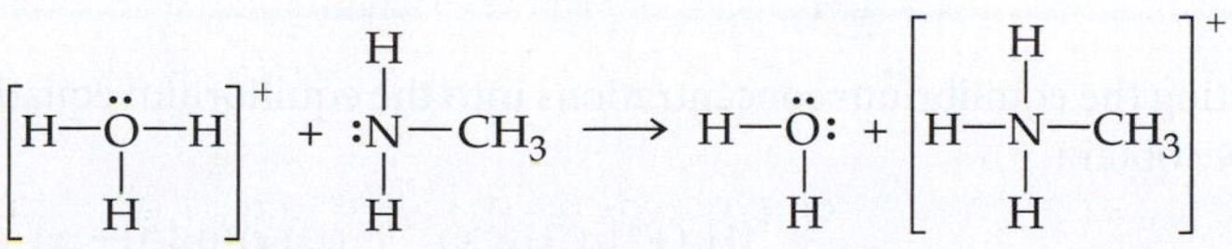

After neutralization, the solution contains $CH_3NH_3^+$ (a weak acid) and NO_3^-, which has no acidic or basic properties. Therefore, the pH is less than 7.

PROBLEM 15.1 Write balanced net ionic equations for the neutralization of equal molar amounts of the following acids and bases. Indicate whether the pH after neutralization is greater than, equal to, or less than 7. Values of K_a and K_b are listed in Appendix C.

(a) HNO_2 and KOH **(b)** HBr and NH_3 **(c)** KOH and $HClO_4$

continued on next page

PROBLEM 15.2 Write balanced net ionic equations for the neutralization of the following acids and bases, calculate the value of K_n for each neutralization reaction, and arrange the reactions in order of increasing tendency to proceed to completion. Values of K_a and K_b are listed in Appendix C.

(a) HF and NaOH **(b)** HCl and KOH **(c)** HF and NH_3

15.2 THE COMMON-ION EFFECT

A solution of a weak acid and its conjugate base is an important acid–base mixture because such mixtures regulate the pH in biological systems. To illustrate pH calculations for weak acid–conjugate base mixtures, let's calculate the pH of a solution prepared by dissolving 0.10 mol of acetic acid and 0.10 mol of sodium acetate in water and then diluting the solution to a volume of 1.00 L. This problem, like those discussed in Chapter 14, can be solved by thinking about the chemistry involved. First, identify the acid–base properties of the various species in solution, and then consider the possible proton-transfer reactions these species can undergo. We'll follow the procedure outlined in Figure 14.7 on page 571.

Step 1 Because acetic acid is largely undissociated in aqueous solution and the salt sodium acetate is essentially 100% dissociated, the species present initially are

CH_3CO_2H	Na^+	$CH_3CO_2^-$	H_2O
Acid	Inert	Base	Acid or base

Steps 2–3 Because we have two acids and two bases, there are four possible proton-transfer reactions. We know, however, that acetic acid is a stronger acid than water and that the principal reaction therefore involves proton transfer from CH_3CO_2H to either $CH_3CO_2^-$ or H_2O:

$$CH_3CO_2H(aq) + CH_3CO_2^-(aq) \rightleftharpoons CH_3CO_2^-(aq) + CH_3CO_2H(aq) \qquad K = 1$$

$$CH_3CO_2H(aq) + H_2O(l) \rightleftharpoons H_3O^+(aq) + CH_3CO_2^-(aq) \qquad K_a = 1.8 \times 10^{-5}$$

Although the first of these reactions has the larger equilibrium constant, we can't consider it to be the principal reaction because the reactants and products are identical. Proton transfer from acetic acid to its conjugate base is constantly occurring, but that reaction doesn't change any concentrations and therefore can't be used to calculate equilibrium concentrations. Consequently, we consider the dissociation of acetic acid to be the principal reaction.

Step 4 Now we can set up a table of concentrations for the species involved in the principal reaction. As in Section 14.9, we define x as the concentration of acid that dissociates—here acetic acid—but we need to remember that the acetate ions come from two sources: 0.10 mol/L of acetate comes from the sodium acetate present initially, and x mol/L comes from the dissociation of acetic acid.

Principal Reaction	$CH_3CO_2H(aq)$ + $H_2O(l)$ ⇌	$H_3O^+(aq)$ +	$CH_3CO_2^-(aq)$
Initial concentration (M)	0.10	~0	0.10
Change (M)	$-x$	$+x$	$+x$
Equilibrium concentration (M)	$0.10 - x$	x	$0.10 + x$

Step 5 Substituting the equilibrium concentrations into the equilibrium equation for the principal reaction, we obtain

$$K_a = 1.8 \times 10^{-5} = \frac{[H_3O^+][CH_3CO_2^-]}{[CH_3CO_2H]} = \frac{(x)(0.10 + x)}{(0.10 - x)}$$

Because K_a is small, x is small compared to 0.10 and we can make the approximation that $(0.10 + x) \approx (0.10 - x) \approx 0.10$, which simplifies the solution of the equation:

$$1.8 \times 10^{-5} = \frac{(x)(0.10 + x)}{(0.10 - x)} \approx \frac{(x)(0.10)}{0.10}$$

$$x = [H_3O^+] = 1.8 \times 10^{-5}\ M$$

Step 6 $pH = -\log(1.8 \times 10^{-5}) = 4.74$

It's interesting to compare the $[H_3O^+]$ in a 0.10 M acetic acid–0.10 M sodium acetate solution with the $[H_3O^+]$ in a 0.10 M solution of pure acetic acid. The principal reaction is the same in both cases, but in 0.10 M acetic acid all the acetate ions come from the dissociation of acetic acid:

Principal Reaction	$CH_3CO_2H(aq)$	+	$H_2O(l)$	$\rightleftharpoons$	$H_3O^+(aq)$	+	$CH_3CO_2^-(aq)$
Equilibrium concentration (M)	$0.10 - x$				x		x

The calculated $[H_3O^+]$ of 0.10 M acetic acid is 1.3×10^{-3} M, and the pH is 2.89 versus a pH of 4.74 for the acetic acid–sodium acetate solution. The difference in pH is illustrated in **Figure 15.1**.

FIGURE 15.1

A demonstration of the common-ion effect. The 0.10 M acetic acid–0.10 M sodium acetate solution on the left has a lower H_3O^+ concentration and a higher pH than the 0.10 M acetic acid solution on the right.

The decrease in $[H_3O^+]$ on adding acetate ions to an acetic acid solution is an example of the **common-ion effect**, the shift in an equilibrium on adding a substance that provides more of an ion already involved in the equilibrium. Thus, added acetate ions shift the acetic acid-dissociation equilibrium to the left, as shown in **Figure 15.2**.

$$CH_3CO_2H(aq) + H_2O(l) \rightleftharpoons H_3O^+(aq) + CH_3CO_2^-(aq)$$

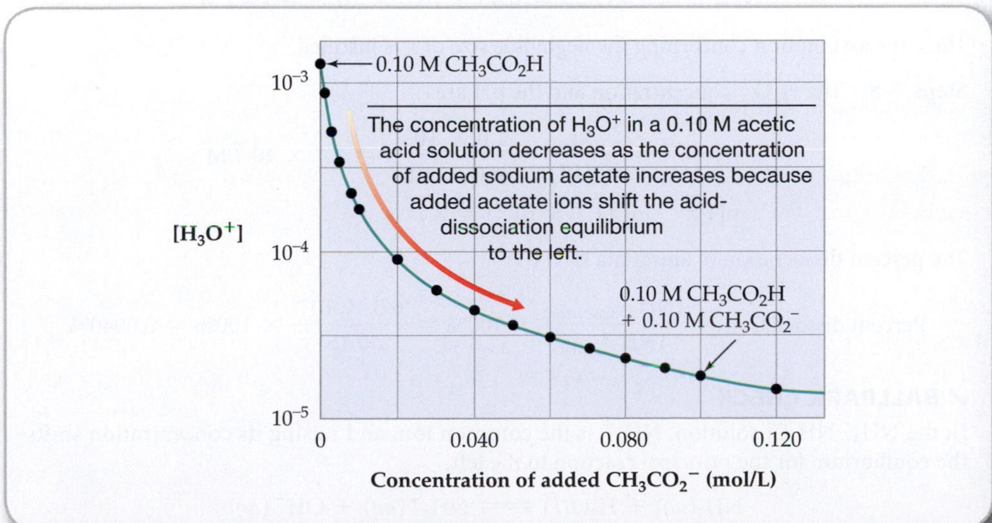

FIGURE 15.2

The common-ion effect. Note that $[H_3O^+]$ is plotted on a logarithmic scale.

REMEMBER...

Le Châtelier's principle says that if a concentration, pressure, or temperature stress is applied to a reaction mixture at equilibrium, net reaction occurs in the direction that relieves the stress. Net reaction continues until a new state of equilibrium is achieved. (Section 13.7)

The common-ion effect is just another example of **Le Châtelier's principle** (Section 13.7), in which the stress on the equilibrium of raising one of the product concentrations is relieved by shifting the equilibrium to the reactant side. Another case is discussed in Worked Example 15.2.

Worked Example 15.2

Calculating the Effect of a Common Ion on Concentrations, pH, and Percent Dissociation

Calculate the concentrations of all species present, the pH, and the percent dissociation of ammonia in a solution that is 0.15 M in NH_3 ($K_b = 1.8 \times 10^{-5}$) and 0.45 M in NH_4Cl. For comparison, the pH of a 0.15 M NH_3 solution that contains no NH_4Cl is 11.21 and the percent dissociation is 1.1%.

STRATEGY

The solution contains a weak base (NH_3) and its conjugate acid (NH_4^+), the common ion. Because this problem is similar to the acetic acid–sodium acetate problem, we'll abbreviate the procedure shown in Figure 14.7.

SOLUTION

Steps 1–3 The principal reaction is proton transfer to NH_3 from H_2O:

$$NH_3(aq) + H_2O(l) \rightleftharpoons NH_4^+(aq) + OH^-(aq)$$

Step 4 Since NH_4^+ ions come both from the NH_4Cl present initially (0.45 M) and from the reaction of NH_3 with H_2O, the concentrations of the species involved in the principal reaction are as follows:

Principal Reaction	$NH_3(aq)$ + $H_2O(l)$ $\rightleftharpoons$	$NH_4^+(aq)$ +	$OH^-(aq)$
Initial concentration (M)	0.15	0.45	~0
Change (M)	$-x$	$+x$	$+x$
Equilibrium concentration (M)	$0.15 - x$	$0.45 + x$	x

Steps 5–6 The equilibrium equation for the principal reaction is

$$K_b = 1.8 \times 10^{-5} = \frac{[NH_4^+][OH^-]}{[NH_3]} = \frac{(0.45 + x)(x)}{0.15 - x} \approx \frac{(0.45)(x)}{0.15}$$

We assume x is negligible compared to 0.45 and 0.15 because (1) the equilibrium constant K_b is small and (2) the equilibrium is shifted to the left by the common-ion effect. Therefore,

$$x = [OH^-] = \frac{(1.8 \times 10^{-5})(0.15)}{0.45} = 6.0 \times 10^{-6}\ M$$

$$[NH_3] = 0.15 - x = 0.15 - (6.0 \times 10^{-6}) = 0.15\ M$$

$$[NH_4^+] = 0.45 + x = 0.45 + (6.0 \times 10^{-6}) = 0.45\ M$$

Thus, the assumption concerning the negligible size of x is justified.

Steps 7–8 The H_3O^+ concentration and the pH are

$$[H_3O^+] = \frac{K_w}{[OH^-]} = \frac{1.0 \times 10^{-14}}{6.0 \times 10^{-6}} = 1.7 \times 10^{-9}\ M$$

$$pH = -\log(1.7 \times 10^{-9}) = 8.77$$

The percent dissociation of ammonia is

$$\text{Percent dissociation} = \frac{[NH_3]_{\text{dissociated}}}{[NH_3]_{\text{initial}}} \times 100\% = \frac{6.0 \times 10^{-6}}{0.15} \times 100\% = 0.0040\%$$

✔ BALLPARK CHECK

In the NH_3–NH_4Cl solution, NH_4^+ is the common ion, and raising its concentration shifts the equilibrium for the principal reaction to the left.

$$NH_3(aq) + H_2O(l) \rightleftharpoons NH_4^+(aq) + OH^-(aq)$$

Consequently, the OH^- concentration and the percent dissociation should decrease. As expected, the calculated values of the pH (8.77) and the percent dissociation (0.0040%) are less than the pH (11.21) and the percent dissociation (1.1%) in the 0.15 M NH_3 solution that contains no NH_4^+ ions.

PROBLEM 15.3 Calculate the concentrations of all species present, the pH, and the percent dissociation of HCN ($K_a = 4.9 \times 10^{-10}$) in a solution that is 0.025 M in HCN and 0.010 M in NaCN.

PROBLEM 15.4 Calculate the pH in a solution prepared by dissolving 0.10 mol of solid NH_4Cl in 0.500 L of 0.40 M NH_3. Assume that there is no volume change.

WORKED CONCEPTUAL EXAMPLE 15.3

Determining the Effect of a Common Ion on pH and Percent Dissociation

The following pictures represent solutions of a weak acid HA that may also contain the sodium salt NaA. Which solution has the highest pH? Which has the largest percent dissociation of HA? (Na^+ ions and solvent water molecules have been omitted for clarity.)

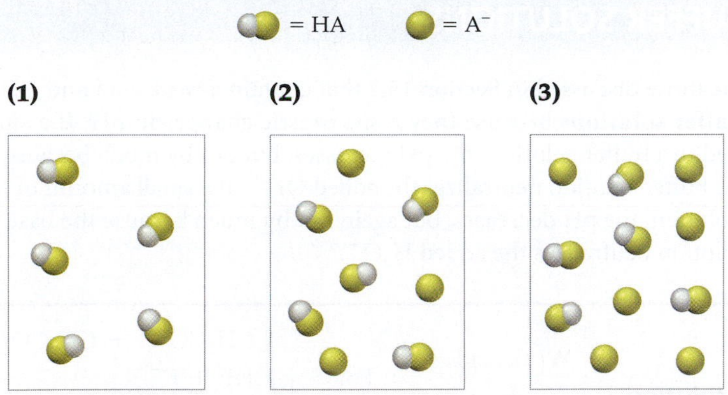

STRATEGY

The pH and the percent dissociation of HA are determined by the extent of the reaction $HA(aq) + H_2O(l) \rightleftharpoons H_3O^+(aq) + A^-(aq)$. The dissociation equilibrium shifts to the left on adding more of the common ion A^-, thus decreasing the H_3O^+ concentration (increasing the pH) and decreasing the percent dissociation. To answer the questions posed, simply count the number of A^- ions.

SOLUTION

All three solutions contain the same number of HA molecules, but different numbers of A^- ions—none for solution 1, three for solution 2, and six for solution 3. The dissociation equilibrium lies farthest to the left for solution 3, and therefore solution 3 has the lowest H_3O^+ concentration and the highest pH. For solution 1, no common ion is present to suppress the dissociation of HA, and therefore solution 1 has the largest percent dissociation.

CONCEPTUAL PROBLEM 15.5 The following pictures represent solutions of a weak base B that may also contain the chloride salt BH^+Cl^-. Which solution has the lowest pH? Which has the largest percent dissociation of B? (Cl^- ions and solvent water molecules have been omitted for clarity.)

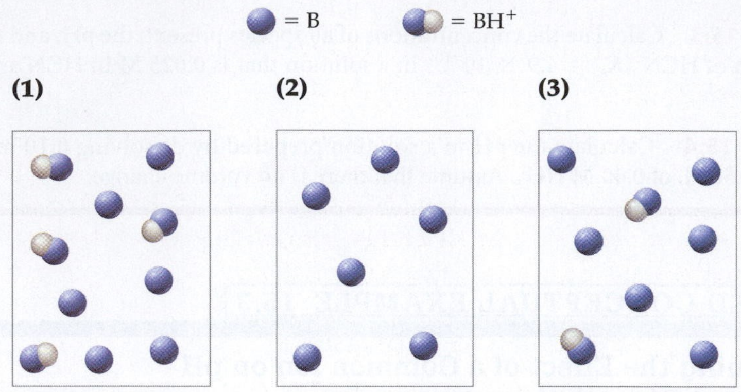

15.3 BUFFER SOLUTIONS

Solutions like those discussed in Section 15.2 that contain a weak acid and its conjugate base are called **buffer solutions** because they resist drastic changes in pH. If a small amount of OH^- is added to a buffer solution, the pH increases, but not by much because the acid component of the buffer solution neutralizes the added OH^-. If a small amount of H_3O^+ is added to a buffer solution, the pH decreases, but again not by much because the base component of the buffer solution neutralizes the added H_3O^+.

▶ **Buffer solution**

Weak acid + Conjugate base

For example:
- $CH_3CO_2H + CH_3CO_2^-$
- $HF + F^-$
- $NH_4^+ + NH_3$
- $H_2PO_4^- + HPO_4^{2-}$

Buffer solutions are very important in biological systems. Blood, for example, is a buffer solution that can soak up the acids and bases produced in biological reactions. The pH of human blood is carefully controlled at a value very close to 7.4 by conjugate acid–base pairs, primarily H_2CO_3 and its conjugate base HCO_3^-. The oxygen-carrying ability of blood depends on control of the pH to within 0.1 pH unit.

To see how a buffer solution works, let's return to the 0.10 M acetic acid–0.10 M sodium acetate solution discussed in Section 15.2. The principal reaction and the equilibrium concentrations for the solution are

Principal Reaction	$CH_3CO_2H(aq)$ + $H_2O(l)$ ⇌	$H_3O^+(aq)$ +	$CH_3CO_2^-(aq)$
Equilibrium concentration (M)	$0.10 - x$	x	$0.10 + x$

If we solve the equilibrium equation for $[H_3O^+]$, we obtain

$$K_a = \frac{[H_3O^+][CH_3CO_2^-]}{[CH_3CO_2H]}$$

$$[H_3O^+] = K_a \frac{[CH_3CO_2H]}{[CH_3CO_2^-]}$$

Thus, the H_3O^+ concentration in a buffer solution has a value close to the value of K_a for the weak acid but differs by a factor equal to the concentration ratio [weak acid]/[conjugate base]. In the 0.10 M acetic acid–0.10 M sodium acetate solution, where the concentration ratio is 1.0, $[H_3O^+]$ equals K_a:

$$[H_3O^+] = K_a\frac{(0.10 - x)}{(0.10 + x)} = K_a\left(\frac{0.10}{0.10}\right) = K_a = 1.8 \times 10^{-5}\ \text{M}$$

$$\text{pH} = \text{p}K_a = -\log(1.8 \times 10^{-5}) = 4.74$$

Note that in calculating this result we have set the *equilibrium* concentrations, $(0.10 - x)$ and $(0.10 + x)$, equal to the *initial* concentrations, 0.10, because x is negligible compared with the initial concentrations. For commonly used buffer solutions, K_a is small and the initial concentrations are relatively large. As a result, x is generally negligible compared with the initial concentrations, and we can use initial concentrations in the calculations.

Addition of OH^- to a Buffer Solution

Now let's consider what happens when we add H_3O^+ or OH^- to a buffer solution. First, suppose that we add 0.01 mol of solid NaOH to 1.00 L of the 0.10 M acetic acid–0.10 M sodium acetate solution. Because neutralization reactions involving strong acids or strong bases go essentially 100% to completion (Section 15.1), we must take account of neutralization before calculating $[H_3O^+]$. Initially, we have (1.00 L)(0.10 mol/L) = 0.10 mol of acetic acid and an equal amount of acetate ion. When we add 0.01 mol of NaOH, the neutralization reaction will alter the numbers of moles:

Neutralization Reaction	$CH_3CO_2H(aq)$ +	$OH^-(aq)$ $\xrightarrow{100\%}$	$H_2O(l)$ +	$CH_3CO_2^-(aq)$
Before reaction (mol)	0.10	0.01		0.10
Change (mol)	−0.01	−0.01		+0.01
After reaction (mol)	0.09	~0		0.11

If we assume that the solution volume remains constant at 1.00 L, the concentrations of the buffer components after neutralization are

$$[CH_3CO_2H] = \frac{0.09\ \text{mol}}{1.00\ \text{L}} = 0.09\ \text{M}$$

$$[CH_3CO_2^-] = \frac{0.11\ \text{mol}}{1.00\ \text{L}} = 0.11\ \text{M}$$

Substituting these concentrations into the expression for $[H_3O^+]$, we can then calculate the pH:

$$[H_3O^+] = K_a\frac{[CH_3CO_2H]}{[CH_3CO_2^-]}$$

$$= (1.8 \times 10^{-5})\left(\frac{0.09}{0.11}\right) = 1.5 \times 10^{-5}\ \text{M}$$

$$\text{pH} = 4.82$$

Adding 0.01 mol of NaOH changes $[H_3O^+]$ by only a small amount because the concentration ratio [weak acid]/[conjugate base] changes by only a small amount, from 1.0 to 9/11 (**Figure 15.3a**). The corresponding change in pH, from 4.74 to 4.82, is only 0.08 pH unit.

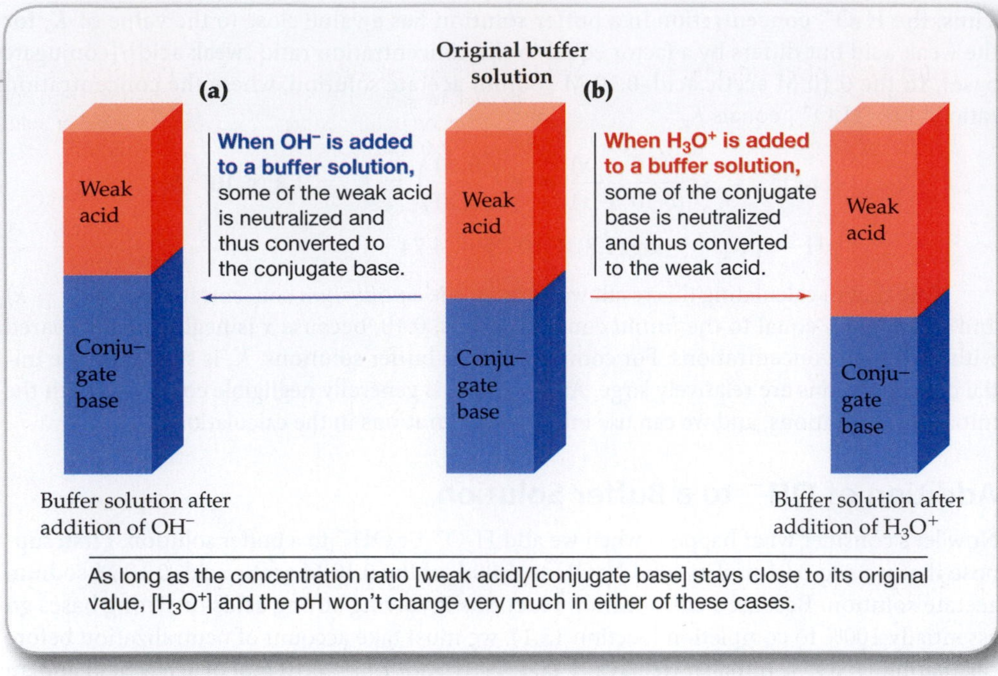

FIGURE 15.3

The addition of (a) OH^- and (b) H_3O^+ to a buffer solution.

Addition of H_3O^+ to a Buffer Solution

Now suppose that we add 0.01 mol of HCl to 1.00 L of the 0.10 M acetic acid–0.10 M sodium acetate buffer solution. The added strong acid will convert 0.01 mol of acetate ions to 0.01 mol of acetic acid because of the neutralization reaction

$$H_3O^+(aq) + CH_3CO_2^-(aq) \xrightarrow{100\%} H_2O(l) + CH_3CO_2H(aq)$$

The concentrations after neutralization will be $[CH_3CO_2H] = 0.11$ M and $[CH_3CO_2^-] = 0.09$ M, and the pH of the solution will be 4.66:

$$[H_3O^+] = K_a \frac{[CH_3CO_2H]}{[CH_3CO_2^-]}$$

$$= (1.8 \times 10^{-5})\left(\frac{0.11}{0.09}\right) = 2.2 \times 10^{-5}\ \text{M}$$

$$\text{pH} = 4.66$$

Again, the change in pH, from 4.74 to 4.66, is small because the concentration ratio [weak acid]/[conjugate base] remains close to its original value (**Figure 15.3b**).

Buffer Capacity

To appreciate the ability of a buffer solution to maintain a nearly constant pH, let's contrast the behavior of the 0.10 M acetic acid–0.10 M sodium acetate buffer with that of a 1.8×10^{-5} M HCl solution. This very dilute solution of a strong acid has the same pH (4.74) as the buffer solution, but it doesn't have the capacity to soak up added acid or base. For example, if we add 0.01 mol of solid NaOH to 1.00 L of 1.8×10^{-5} M HCl, a negligible amount of the OH^- (1.8×10^{-5} mol) is neutralized and the concentration of OH^- after neutralization is 0.01 mol/1.00 L = 0.01 M. As a result, the pH rises from 4.74 to 12.0:

$$[H_3O^+] = \frac{K_w}{[OH^-]} = \frac{1.0 \times 10^{-14}}{0.01} = 1 \times 10^{-12}\ \text{M}$$

$$\text{pH} = 12.0$$

The abilities of the HCl solution and the buffer solution to absorb added base are contrasted in **Figure 15.4**.

(a) 1.00 L of 1.8×10^{-5} M HCl ($pH = 4.74$)

(b) The solution from **(a)** turns yellow ($pH > 5.4$) after addition of only a few drops of 0.10 M NaOH.

(c) 1.00 L of a 0.10 M acetic acid–0.10 M sodium acetate buffer solution ($pH = 4.74$)

(d) The solution from **(c)** is still red ($pH < 5.4$) after addition of 100 mL of 0.10 M NaOH.

FIGURE 15.4

The abilities of a strong acid solution and a buffer solution to absorb added base. The color of each solution is due to the presence of a few drops of methyl red, an acid–base indicator that is red at pH less than about 5.4 and yellow at pH greater than about 5.4.

We sometimes talk about the buffering ability of a solution using the term **buffer capacity** as a measure of the amount of acid or base that the solution can absorb without a significant change in pH. Buffer capacity is also a measure of how little the pH changes with the addition of a given amount of acid or base. Buffer capacity depends on how many moles of weak acid and conjugate base are present. For equal volumes of solution, the more concentrated the solution, the greater the buffer capacity. For solutions having the same concentration, the greater the volume, the greater the buffer capacity.

CONCEPTUAL PROBLEM 15.6 The following pictures represent solutions that contain a weak acid HA and/or its sodium salt NaA. (Na^+ ions and solvent water molecules have been omitted for clarity.)

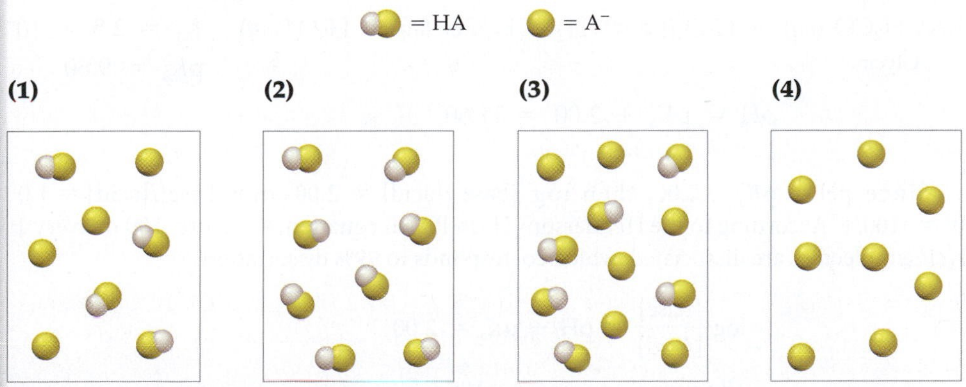

(a) Which of the solutions are buffer solutions?
(b) Which solution has the greatest buffer capacity?

PROBLEM 15.7 Calculate the pH of 0.100 L of a buffer solution that is 0.25 M in HF and 0.50 M in NaF. What is the change in pH on addition of the following?
(a) 0.002 mol of HNO_3
(b) 0.004 mol of KOH

PROBLEM 15.8 Calculate the change in pH when 0.002 mol of HNO_3 is added to 0.100 L of a buffer solution that is 0.050 M in HF and 0.100 M in NaF. Does this solution have more or less buffer capacity than the one in Problem 15.7?

15.4 THE HENDERSON–HASSELBALCH EQUATION

We saw in Section 15.3 that the H_3O^+ concentration in a buffer solution depends on the dissociation constant of the weak acid and on the concentration ratio [weak acid]/[conjugate base]:

$$[H_3O^+] = K_a \frac{[\text{Acid}]}{[\text{Base}]}$$

This equation can be rewritten in logarithmic form by taking the negative base-10 logarithm of both sides:

$$\text{pH} = -\log [H_3O^+] = -\log\left(K_a \frac{[\text{Acid}]}{[\text{Base}]}\right) = -\log K_a - \log \frac{[\text{Acid}]}{[\text{Base}]}$$

Because $\text{p}K_a = -\log K_a$ and

$$-\log \frac{[\text{Acid}]}{[\text{Base}]} = \log \frac{[\text{Base}]}{[\text{Acid}]}$$

we obtain an expression called the **Henderson–Hasselbalch equation**:

▶ **Henderson–Hasselbalch Equation** $\text{pH} = \text{p}K_a + \log \frac{[\text{Base}]}{[\text{Acid}]}$

The Henderson–Hasselbalch equation says that the pH of a buffer solution has a value close to the $\text{p}K_a$ of the weak acid, differing only by the amount log [base]/[acid]. When [base]/[acid] = 1, then log [base]/[acid] = 0 and the pH equals the $\text{p}K_a$.

The real importance of the Henderson–Hasselbalch equation, particularly in biochemistry, is that it tells us how the pH affects the percent dissociation of a weak acid. Suppose, for example, that you have a solution containing the amino acid glycine, one of the molecules from which proteins are made, and that the pH of the solution is 2.00 pH units greater than the $\text{p}K_a$ of glycine:

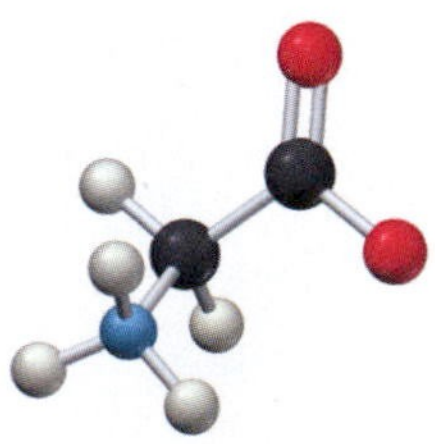
Glycine

$$H_3\overset{+}{N}CH_2\overset{O}{\overset{\|}{C}}O^-(aq) + H_2O(l) \rightleftharpoons H_2NCH_2\overset{O}{\overset{\|}{C}}O^-(aq) + H_3O^+(aq) \quad K_a = 2.5 \times 10^{-10}$$

Glycine

$$\text{p}K_a = 9.60$$

$$\text{pH} = \text{p}K_a + 2.00 = 11.60$$

Since $\text{pH} = \text{p}K_a + 2.00$, then log [base]/[acid] = 2.00 and [base]/[acid] = $1.0 \times 10^2 = 100/1$. According to the Henderson–Hasselbalch equation, therefore, 100 of every 101 glycine molecules are dissociated, which corresponds to 99% dissociation:

$$\log \frac{[\text{Base}]}{[\text{Acid}]} = \text{pH} - \text{p}K_a = 2.00$$

$$\frac{[\text{Base}]}{[\text{Acid}]} = 1.0 \times 10^2 = \frac{100}{1} \quad \text{99\% dissociation}$$

The Henderson–Hasselbalch equation thus gives the following relationships:

$$\text{At pH} = \text{p}K_a + 2.00 \quad \frac{[\text{Base}]}{[\text{Acid}]} = 1.0 \times 10^2 = \frac{100}{1} \quad \text{99\% dissociation}$$

$$\text{At pH} = \text{p}K_a + 1.00 \quad \frac{[\text{Base}]}{[\text{Acid}]} = 1.0 \times 10^1 = \frac{10}{1} \quad \text{91\% dissociation}$$

$$\text{At pH} = \text{p}K_a + 0.00 \quad \frac{[\text{Base}]}{[\text{Acid}]} = 1.0 \times 10^0 = \frac{1}{1} \quad \text{50\% dissociation}$$

$$\text{At pH} = pK_a - 1.00 \qquad \frac{[\text{Base}]}{[\text{Acid}]} = 1.0 \times 10^{-1} = \frac{1}{10} \qquad 9\% \text{ dissociation}$$

$$\text{At pH} = pK_a - 2.00 \qquad \frac{[\text{Base}]}{[\text{Acid}]} = 1.0 \times 10^{-2} = \frac{1}{100} \qquad 1\% \text{ dissociation}$$

The Henderson–Hasselbalch equation also tells us how to prepare a buffer solution with a given pH. The general idea is to select a weak acid whose pK_a is close to the desired pH and then adjust the [base]/[acid] ratio to the value specified by the Henderson–Hasselbalch equation. For example, to prepare a buffer having a pH near 7, we might use the $H_2PO_4^-$–HPO_4^{2-} conjugate acid–base pair because the pK_a for $H_2PO_4^-$ is $-\log(6.2 \times 10^{-8}) = 7.21$. Similarly, a mixture of NH_4Cl and NH_3 would be a good choice for a buffer having a pH near 9 because the pK_a for NH_4^+ is $-\log(5.6 \times 10^{-10}) = 9.25$. As a rule of thumb, the pK_a of the weak acid component of a buffer should be within ± 1 pH unit of the desired pH.

Because buffer solutions are widely used in the laboratory and in medicine, prepackaged buffers having a variety of precisely known pH values are commercially available (**Figure 15.5**). The manufacturer prepares these buffers by choosing a buffer system having an appropriate pK_a value and then adjusting the amounts of the ingredients so that the [base]/[acid] ratio has the proper value.

The pH of a buffer solution does not depend on the volume of the solution because a change in solution volume changes the concentrations of the acid and base by the same amount. Thus, the [base]/[acid] ratio and the pH remain unchanged. As a result, the volume of water used to prepare a buffer solution is not critical and you can dilute a buffer without changing its pH. The pH depends only on pK_a and on the relative molar amounts of the weak acid and the conjugate base.

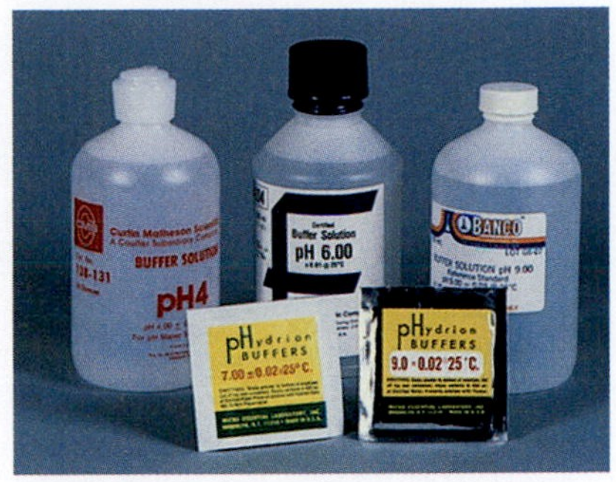

FIGURE 15.5

Prepackaged buffer solutions of known pH and solid ingredients for preparing buffer solutions of known pH.

Worked Example 15.4

Using the Henderson–Hasselbalch Equation

(a) Use the Henderson–Hasselbalch equation to calculate the pH of a buffer solution that is 0.45 M in NH_4Cl and 0.15 M in NH_3.

(b) How would you prepare an NH_4Cl–NH_3 buffer that has a pH of 9.00?

STRATEGY

(a) We've already solved this problem by another method in Worked Example 15.2. Now that we've discussed equilibria in buffer solutions, though, we can use the Henderson–Hasselbalch equation as a shortcut. Since NH_4^+ is the weak acid in an NH_4^+–NH_3 buffer solution, we need to find the pK_a for NH_4^+ from the tabulated K_b value for NH_3 (Appendix C). Then, substitute $[NH_3]$, $[NH_4^+]$, and the pK_a value into the Henderson–Hasselbalch equation to find the pH.

(b) Use the Henderson–Hasselbalch equation to calculate the $[NH_3]/[NH_4^+]$ ratio from the desired pH and the pK_a value for NH_4^+.

SOLUTION

(a) Since K_b for NH_3 is 1.8×10^{-5}, K_a for NH_4^+ is 5.6×10^{-10} and $pK_a = 9.25$.

$$K_a = \frac{K_w}{K_b} = \frac{1.0 \times 10^{-14}}{1.8 \times 10^{-5}} = 5.6 \times 10^{-10}$$

$$pK_a = -\log K_a = -\log(5.6 \times 10^{-10}) = 9.25$$

Since $[\text{base}] = [NH_3] = 0.15$ M and $[\text{acid}] = [NH_4^+] = 0.45$ M,

$$\text{pH} = pK_a + \log\frac{[\text{Base}]}{[\text{Acid}]} = 9.25 + \log\left(\frac{0.15}{0.45}\right) = 9.25 - 0.48 = 8.77$$

The pH of the buffer solution is 8.77.

continued on next page

(b) Rearrange the Henderson–Hasselbalch equation to obtain an expression for the relative amounts of NH_3 and NH_4^+ in a solution having pH = 9.00:

$$\log \frac{[\text{Base}]}{[\text{Acid}]} = \text{pH} - \text{p}K_a = 9.00 - 9.25 = -0.25$$

Therefore,

$$\frac{[NH_3]}{[NH_4^+]} = \text{antilog}(-0.25) = 10^{-0.25} = 0.56$$

The solution must contain 0.56 mol of NH_3 for every 1.00 mol of NH_4Cl, but the volume of the solution isn't critical. One way of preparing the buffer would be to combine 1.00 mol of NH_4Cl (53.5 g) with 0.56 mol of NH_3 (say, 560 mL of 1.00 M NH_3).

✔ BALLPARK CHECK

A common error in using the Henderson–Hasselbalch equation is to invert the [base]/[acid] ratio, so it's wise to check that your answer makes chemical sense. If the concentrations of the acid and its conjugate base are equal, the pH will equal the pK_a. If the acid predominates, the pH will be less than the pK_a, and if the conjugate base predominates, the pH will be greater than the pK_a. In part **(a)**, [acid] = $[NH_4^+]$ is greater than [base] = $[NH_3]$ and so the calculated pH (8.77) should be less than the pK_a (9.25). In part **(b)**, the desired pH is less than the pK_a so the buffer should contain more moles of acid than base, in agreement with the solution.

Worked Example 15.5

Using the Henderson–Hasselbalch Equation to Determine the Effectiveness of a Buffer Solution

What $[NH_3]/[NH_4^+]$ ratio is required for a buffer solution that has pH = 7.00? Why is a mixture of NH_3 and NH_4Cl a poor choice for a buffer having pH = 7.00?

STRATEGY

We can calculate the required $[NH_3]/[NH_4^+]$ ratio from the Henderson–Hasselbalch equation.

SOLUTION

$$\log \frac{[NH_3]}{[NH_4^+]} = \text{pH} - \text{p}K_a = 7.00 - 9.25 = -2.25$$

$$\frac{[NH_3]}{[NH_4^+]} = \text{antilog}(-2.25) = 10^{-2.25} = 5.6 \times 10^{-3}$$

For a typical value of $[NH_4^+]$—say, 1.0 M—the NH_3 concentration would have to be very small (0.0056 M). Such a solution is a poor buffer because it has little capacity to absorb added acid. Also, because the $[NH_3]/[NH_4^+]$ ratio is far from 1.0, addition of a small amount of H_3O^+ or OH^- will result in a large change in the pH.

✔ BALLPARK CHECK

Because the desired pH is much less than the pK_a value, the buffer must contain much more weak acid than conjugate base, in agreement with the solution.

▶ PROBLEM 15.9 Use the Henderson–Hasselbalch equation to calculate the pH of a buffer solution prepared by mixing equal volumes of 0.20 M $NaHCO_3$ and 0.10 M Na_2CO_3. (K_a values are given in Appendix C.)

▶ PROBLEM 15.10 How would you prepare an $NaHCO_3$–Na_2CO_3 buffer solution that has pH = 10.40?

▶ PROBLEM 15.11 Suppose you are performing an experiment that requires a constant pH of 7.50. Suggest an appropriate buffer system based on the K_a values in Appendix C.

PROBLEM 15.12 The pK_a of the amino acid serine is 9.15. At what pH is serine:
(a) 66% dissociated?
(b) 5% dissociated?

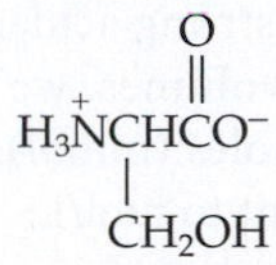

Serine
pK_a = 9.15

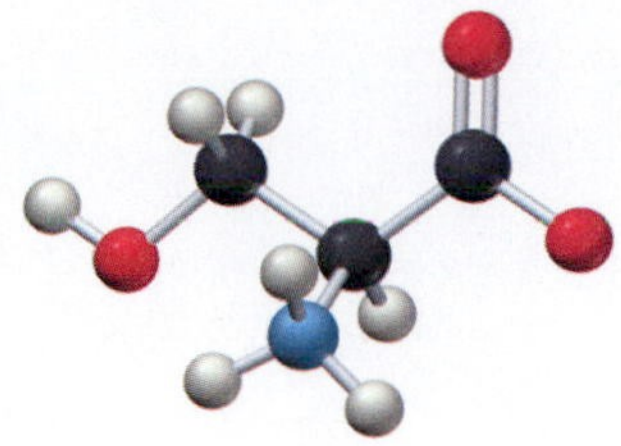

Serine

15.5 pH TITRATION CURVES

Let's continue our study of equilibria in mixtures of acids and bases by looking at the solutions obtained in the course of an acid–base titration. In a typical acid–base **titration** (Section 6.11), a solution containing a known concentration of base (or acid) is added slowly from a buret to a second solution containing an unknown concentration of acid (or base). The progress of the titration is monitored, either by using a pH meter (**Figure 15.6a**) or by observing the color of a suitable acid–base indicator. With a pH meter, you can record data to produce a **pH titration curve**, a plot of the pH of the solution as a function of the volume of added titrant (**Figure 15.6b**).

REMEMBER...
Titration is a procedure for determining the concentration of a solution by allowing a carefully measured volume to react with a standard solution of another substance, whose concentration is known. (Section 6.11)

(a) A pH titration in which 0.100 M NaOH is added slowly from a buret to an HCl solution of unknown concentration. The pH of the solution is measured with a pH meter and is recorded as a function of the volume of NaOH added.

(b) The pH titration curve for titration of 40.0 mL of 0.100 M HCl with 0.100 M NaOH.

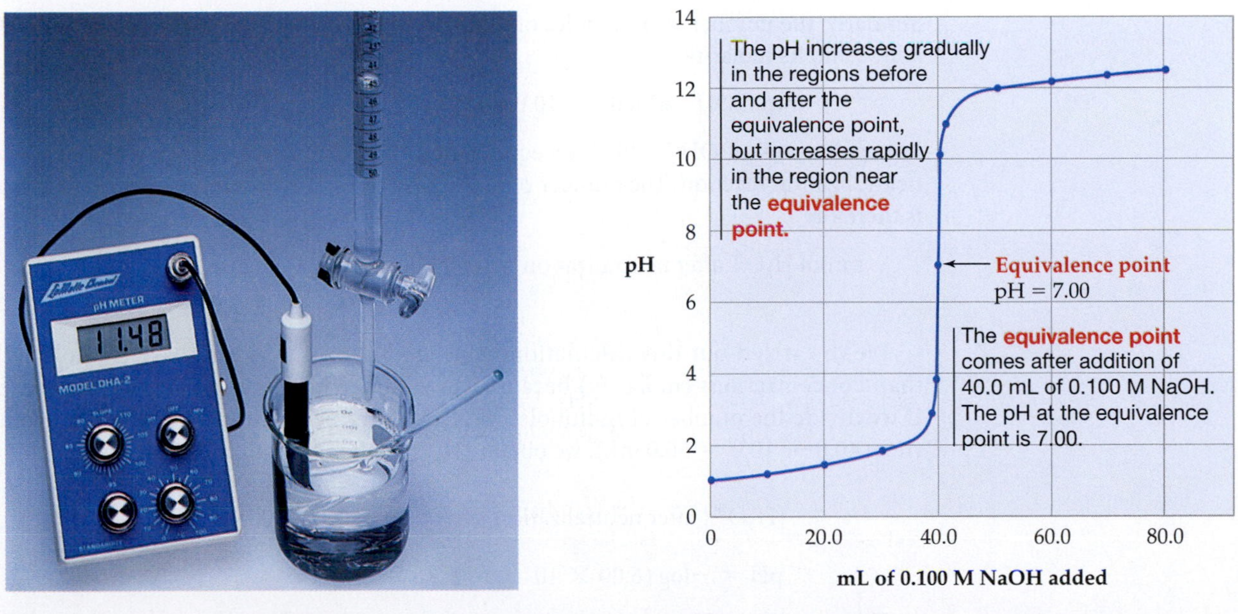

FIGURE 15.6
A strong acid–strong base titration.

Why study titration curves? The shape of a pH titration curve makes it possible to identify the **equivalence point** in a titration, the point at which stoichiometrically equivalent quantities of acid and base have been mixed together. Knowing the shape of the titration curve is also useful in selecting a suitable indicator to signal the equivalence point. We'll explore both of these points later.

We can calculate pH titration curves using the principles of aqueous solution equilibria discussed in Chapter 14. To understand why titration curves have certain characteristic shapes, let's calculate these curves for four common types of titration: (1) strong acid–strong base, (2) weak acid–strong base, (3) weak base–strong acid, and (4) polyprotic acid–strong base. Because titrations typically use milliliter volumes, we'll express solution volumes in milliliters (mL) and amounts of solute in millimoles (mmol). Molar concentration can thus be expressed in mmol/mL, a unit that is equivalent to mol/L:

$$\text{Molarity} = \frac{\text{mmol of solute}}{\text{mL of solution}} = \frac{10^{-3}\text{ mol of solute}}{10^{-3}\text{ L of solution}} = \frac{\text{mol of solute}}{\text{L of solution}}$$

15.6 STRONG ACID–STRONG BASE TITRATIONS

As an example of a strong acid–strong base titration, let's consider the titration of 40.0 mL of 0.100 M HCl with 0.100 M NaOH. We'll calculate the pH at selected points in the course of the titration to illustrate the procedures we use to calculate the entire curve.

1. **Before addition of any NaOH.** Since HCl is a strong acid, the initial concentration of H_3O^+ is 0.100 M and the pH is 1.00. (We've rounded the value of the pH to two significant figures.)
2. **Before the equivalence point.** Let's calculate the pH after addition of 10.0 mL of 0.100 M NaOH. The added OH^- ions will decrease $[H_3O^+]$ because of the neutralization reaction

$$H_3O^+(aq) + OH^-(aq) \xrightarrow{100\%} 2\,H_2O(l)$$

The number of millimoles of H_3O^+ present initially is the product of the initial volume of HCl and its molarity:

$$\text{mmol } H_3O^+ \text{ initial} = (40.0\text{ mL})(0.100\text{ mmol/mL}) = 4.00\text{ mmol}$$

Similarly, the number of millimoles of OH^- added is the product of the volume of NaOH added and its molarity:

$$\text{mmol } OH^- \text{ added} = (10.0\text{ mL})(0.100\text{ mmol/mL}) = 1.00\text{ mmol}$$

For each mmol of OH^- added, an equal amount of H_3O^+ will disappear because of the neutralization reaction. The number of millimoles of H_3O^+ remaining after neutralization is therefore

$$\begin{aligned}\text{mmol } H_3O^+ \text{ after neutralization} &= \text{mmol } H_3O^+{}_{\text{initial}} - \text{mmol } OH^-{}_{\text{added}}\\ &= 4.00\text{ mmol} - 1.00\text{ mmol} = 3.00\text{ mmol}\end{aligned}$$

We've carried out this calculation using *amounts* of acid and base (mmol) rather than concentrations (molarity) because the volume changes as the titration proceeds. If we divide the number of millimoles of H_3O^+ after neutralization by the total volume (now 40.0 + 10.0 = 50.0 mL), we obtain $[H_3O^+]$ after neutralization:

$$[H_3O^+] \text{ after neutralization} = \frac{3.00\text{ mmol}}{50.0\text{ mL}} = 6.00 \times 10^{-2}\text{ M}$$

$$\text{pH} = -\log(6.00 \times 10^{-2}) = 1.22$$

This same procedure can be used to calculate the pH at other points prior to the equivalence point, giving the results summarized in the top part of **Table 15.1**.

3. **At the equivalence point.** After addition of 40.0 mL of 0.100 M NaOH, we have added (40.0 mL)(0.100 mmol/mL) = 4.00 mmol of NaOH, which is just enough OH^- to neutralize all the 4.00 mmol of HCl present initially. This is the equivalence point of the titration, and the pH is 7.00 because the solution contains only water and NaCl, a salt derived from a strong base and a strong acid.
4. **Beyond the equivalence point.** After addition of 60.0 mL of 0.100 M NaOH, we have added (60.0 mL)(0.100 mmol/mL) = 6.00 mmol of NaOH, which is more than enough

TABLE 15.1 Sample Results for pH Calculations at Various Points in the Titration of 40.0 mL of 0.100 M HCl with 0.100 M NaOH

mL NaOH Added	mmol OH^- Added	mmol H_3O^+ after Neutralization	Total Volume (mL)		$[H_3O^+]$ after Neutralization	pH
Before the equivalence point:						
0.0	0.0	4.00	40.0		1.00×10^{-1}	1.00
10.0	1.00	3.00	50.0		6.00×10^{-2}	1.22
20.0	2.00	2.00	60.0		3.33×10^{-2}	1.48
30.0	3.00	1.00	70.0		1.43×10^{-2}	1.84
39.0	3.90	0.10	79.0		1.27×10^{-3}	2.90
39.9	3.99	0.01	79.9		1.3×10^{-4}	3.9
At the equivalence point:						
40.0	4.00	0.00	80.0		1.0×10^{-7}	7.00
Beyond the equivalence point:						
		mmol OH^- after Neutralization		$[OH^-]$ after Neutralization		
40.1	4.01	0.01	80.1	1.2×10^{-4}	8.3×10^{-11}	10.1
41.0	4.10	0.10	81.0	1.2×10^{-3}	8.3×10^{-12}	11.08
50.0	5.00	1.00	90.0	1.11×10^{-2}	9.0×10^{-13}	12.05
60.0	6.00	2.00	100.0	2.00×10^{-2}	5.0×10^{-13}	12.30
70.0	7.00	3.00	110.0	2.73×10^{-2}	3.7×10^{-13}	12.43
80.0	8.00	4.00	120.0	3.33×10^{-2}	3.0×10^{-13}	12.52

to neutralize the 4.00 mmol of HCl present initially. Consequently, an excess of OH^- (6.00 − 4.00 = 2.00 mmol) is present. Since the total volume is now 40.0 + 60.0 = 100.0 mL, the concentration of OH^- is

$$[OH^-] \text{ after neutralization} = \frac{2.00 \text{ mmol}}{100.0 \text{ mL}} = 2.00 \times 10^{-2} \text{ M}$$

The H_3O^+ concentration and the pH are

$$[H_3O^+] = \frac{K_w}{[OH^-]} = \frac{1.0 \times 10^{-14}}{2.00 \times 10^{-2}} = 5.0 \times 10^{-13} \text{ M}$$

$$\text{pH} = -\log(5.0 \times 10^{-13}) = 12.30$$

Sample results for pH calculations at other places beyond the equivalence point are also included in the bottom part of Table 15.1.

Plotting the pH data in Table 15.1 as a function of milliliters of NaOH added gives the pH titration curve shown previously in Figure 15.6b (page 615). This curve exhibits a gradual increase in pH in the regions before and after the equivalence point but a very sharp increase in pH in the region near the equivalence point. Thus, when the volume of added NaOH increases from 39.9 to 40.1 mL (0.2 mL is only about 4 drops), the pH increases from 3.9 to 10.1 (Table 15.1). This very sharp increase in pH in the region of the equivalence point is characteristic of the titration curve for any strong acid–strong base reaction, a feature that allows us to identify the equivalence point when the concentration of the acid is unknown.

The pH curve for the titration of a strong base with a strong acid is similar except that the initial pH is high and then decreases as acid is added (**Figure 15.7**).

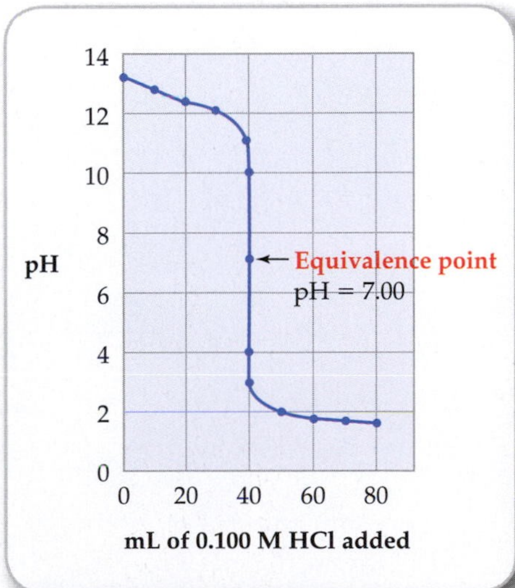

FIGURE 15.7

A strong base–strong acid titration curve. The curve shown is for the titration of 40.0 mL of 0.100 M NaOH with 0.100 M HCl.

PROBLEM 15.13 A 40.0 mL volume of 0.100 M HCl is titrated with 0.100 M NaOH. Calculate the pH of the solution after addition of **(a)** 35.0 mL and **(b)** 45.0 mL of base. Are your results consistent with the pH data in Table 15.1?

PROBLEM 15.14 A 40.0 mL volume of 0.100 M NaOH is titrated with 0.0500 M HCl. Calculate the pH after addition of the following volumes of acid:
(a) 60.0 mL **(b)** 80.2 mL **(c)** 100.0 mL

15.7 WEAK ACID–STRONG BASE TITRATIONS

As an example of a weak acid–strong base titration, let's consider the titration of 40.0 mL of 0.100 M acetic acid ($K_a = 1.8 \times 10^{-5}$; $pK_a = 4.74$) with 0.100 M NaOH. Calculating the pH at selected points along the titration curve is straightforward because we've already met all the equilibrium problems that arise.

REMEMBER...
The step-by-step procedure for solving equilibrium problems involving **weak acids** is outlined in Figure 14.7. (Section 14.9)

1. **Before addition of any NaOH.** The equilibrium problem at this point is the familiar one of calculating the pH of a solution of a **weak acid**. The calculated pH of 0.100 M acetic acid is 2.89.
2. **Before the equivalence point.** Since acetic acid is largely undissociated and NaOH is completely dissociated, the neutralization reaction is

$$CH_3CO_2H(aq) + OH^-(aq) \xrightarrow{100\%} H_2O(l) + CH_3CO_2^-(aq)$$

After addition of 10.0 mL of 0.100 M NaOH, we have added (10.0 mL)(0.100 mmol/mL) = 1.00 mmol of NaOH, which is enough OH^- to neutralize 1.00 mmol of the 4.00 mmol of CH_3CO_2H present initially. Neutralization gives a buffer solution that contains 1.00 mmol of $CH_3CO_2^-$ and 4.00 − 1.00 = 3.00 mmol of CH_3CO_2H. Consequently, the [base]/[acid] ratio is 1.00/3.00 and the Henderson–Hasselbalch equation gives a pH of 4.26.

$$pH = pK_a + \log\frac{[\text{Base}]}{[\text{Acid}]} = 4.74 + \log\frac{1.00}{3.00} = 4.26$$

This same procedure can be used to calculate the pH at other points prior to the equivalence point. Some results are summarized in the top part of **Table 15.2**.

TABLE 15.2 Sample Results for pH Calculations at Various Points in the Titration of 40.0 mL of 0.100 M Acetic Acid with 0.100 M NaOH

mL NaOH Added	mmol OH^- Added	mmol $CH_3CO_2^-$ after Neutralization	mmol CH_3CO_2H after Neutralization	pH
Before the equivalence point:				
0.0	0.0	0.0	4.00	2.89
10.0	1.00	1.00	3.00	4.26
20.0	2.00	2.00	2.00	4.74
30.0	3.00	3.00	1.00	5.22
At the equivalence point:				
40.0	4.00	4.00	0.00	8.72

Beyond the equivalence point:

mL NaOH Added	mmol OH^- Added	mmol OH^- after Neutralization	Total Volume (mL)	$[OH^-]$ after Neutralization	$[H_3O^+]$ after Neutralization	pH
50.0	5.00	1.00	90.0	1.11×10^{-2}	9.0×10^{-13}	12.05
60.0	6.00	2.00	100.0	2.00×10^{-2}	5.0×10^{-13}	12.30

3. **Halfway to the equivalence point.** After addition of 20.0 mL of 0.100 M NaOH, we have added 2.00 mmol of NaOH, which is enough OH^- to neutralize exactly half the 4.00 mmol of CH_3CO_2H present initially. Neutralization gives a buffer solution that contains 2.00 mmol of $CH_3CO_2^-$ and $4.00 - 2.00 = 2.00$ mmol of CH_3CO_2H. Consequently, the [base]/[acid] ratio is 1.00 and pH = pK_a:

$$\text{pH} = \text{p}K_a + \log\frac{[\text{Base}]}{[\text{Acid}]} = \text{p}K_a = 4.74$$

Halfway to the equivalence point is a special point in the titration of any weak acid with a strong base because at that point the pH always equals the pK_a of the weak acid.

4. **At the equivalence point.** The equivalence point is reached after adding 40.0 mL of 0.100 M NaOH (4.00 mmol), which is just enough OH^- to neutralize all the 4.00 mmol of $CH_3\,CO_2H$ present initially. After neutralization, the solution contains 0.0500 M $CH_3CO_2^-$:

$$[\text{Na}^+] = [\text{CH}_3\text{CO}_2^-] = \frac{4.00\text{ mmol}}{40.0\text{ mL} + 40.0\text{ mL}} = 0.0500\text{ M}$$

Because Na^+ is neither an acid nor a base and $CH_3CO_2^-$ is a weak base, we have a **basic salt solution** (Section 14.14), whose pH can be calculated as 8.72 by the method outlined in Worked Example 14.16 on page 583. For a weak monoprotic acid–strong base titration, the pH at the equivalence point is always greater than 7 because the anion of the weak acid is a base.

REMEMBER...

Acetate ion ($CH_3CO_2^-$) gives a **basic salt solution** because it can accept a proton from water, yielding OH^-: $CH_3CO_2^-(aq) + H_2O(l) \rightleftharpoons CH_3CO_2H(aq) + OH^-(aq)$. (Section 14.14)

5. **After the equivalence point.** After addition of 50.0 mL of 0.100 M NaOH, we have added 5.00 mmol of NaOH, which is more than enough OH^- to neutralize the 4.00 mmol of CH_3CO_2H present initially. The total volume is $40.0 + 50.0 = 90.0$ mL, and the concentrations after neutralization are

$$[\text{CH}_3\text{CO}_2^-] = \frac{4.00\text{ mmol}}{90.0\text{ mL}} = 0.0444\text{ M}$$

$$[\text{OH}^-] = \frac{5.00\text{ mmol} - 4.00\text{ mmol}}{90.0\text{ mL}} = 0.0111\text{ M}$$

The principal reaction is the same as that at the equivalence point:

$$\text{CH}_3\text{CO}_2^-(aq) + \text{H}_2\text{O}(l) \rightleftharpoons \text{CH}_3\text{CO}_2\text{H}(aq) + \text{OH}^-(aq)$$

In this case, however, $[OH^-]$ from the principal reaction is negligible compared with $[OH^-]$ from the excess NaOH. The hydronium ion concentration and the pH can be calculated from the excess $[OH^-]$:

$$[\text{H}_3\text{O}^+] = \frac{K_w}{[\text{OH}^-]} = \frac{1.0 \times 10^{-14}}{0.0111} = 9.0 \times 10^{-13}\text{ M}$$

$$\text{pH} = 12.05$$

In general, $[OH^-]$ from the reaction of the anion of a weak acid with water is negligible beyond the equivalence point, and the pH is determined by the concentration of OH^- from the excess NaOH.

The results of pH calculations for the titration of 0.100 M CH_3CO_2H with 0.100 M NaOH are plotted in **Figure 15.8**. Comparison of the titration curves for the weak acid–strong base titration and the strong acid–strong base case shows several significant differences:

- The initial rise in pH is greater for the titration of the weak acid than for the strong acid, but the curve then becomes more level in the region midway to the equivalence point. Both effects are due to the buffering action of the weak acid–conjugate base mixture. The curve has its minimum slope exactly halfway to the equivalence point, where the buffering action is maximized and the pH = pK_a for the weak acid.
- The increase in pH in the region near the equivalence point is smaller in the weak acid case than in the strong acid case. The weaker the acid, the smaller the increase in the pH, as illustrated in **Figure 15.9**.
- The pH at the equivalence point is greater than 7 in the weak acid titration because the anion of a weak acid is a base.

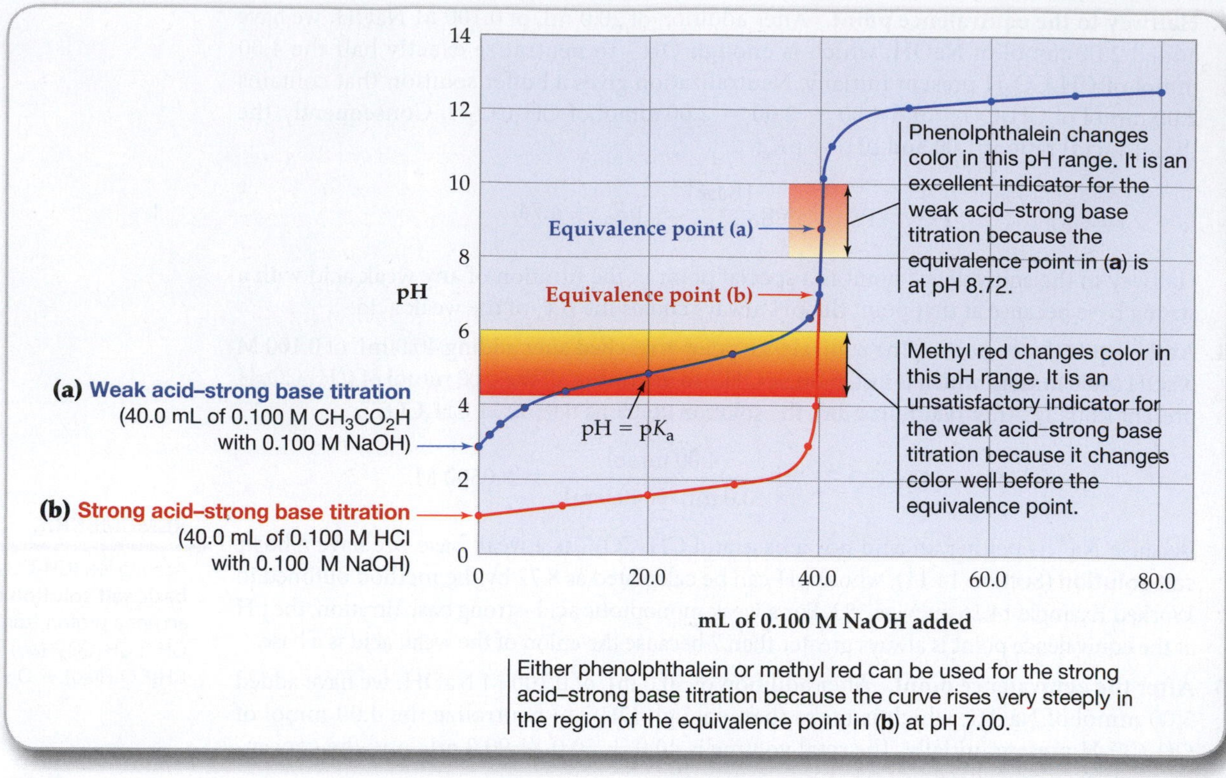

FIGURE 15.8

(a) A weak acid–strong base titration curve (blue) compared with (b) a strong acid–strong base curve (red). The pH ranges in which the acid–base indicators phenolphthalein and methyl red change color are indicated.

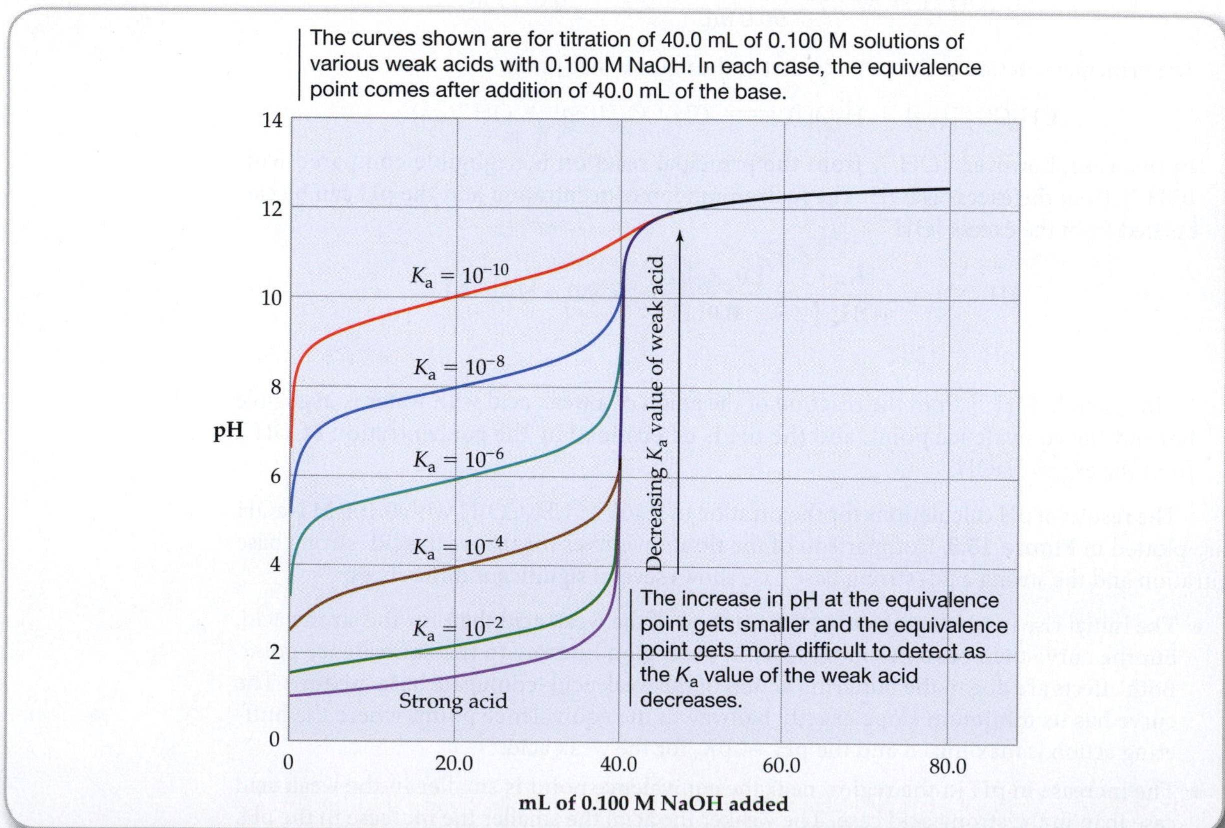

FIGURE 15.9

Changes in weak acid–strong base pH titration curves as the strength of the weak acid (K_a value) changes.

Beyond the equivalence point, the curves for the weak acid–strong base and strong acid–strong base titrations are identical because the pH in both cases is determined by the concentration of OH^- from the excess NaOH.

Figure 15.8 shows how knowing the shape of a pH titration curve makes it possible to select a suitable acid–base indicator to signal the equivalence point of a titration. Phenolphthalein is an excellent indicator for the CH_3CO_2H–NaOH titration because the pH at the equivalence point (8.72) falls within the pH range (8.2–9.8) in which phenolphthalein changes color. Methyl red, however, is an unacceptable indicator for this titration because it changes color in a pH range (4.2–6.0) well before the equivalence point. Anyone who tried to determine the acetic acid content in a solution of unknown concentration would badly underestimate the amount of acid present if methyl red were used as the indicator.

Either phenolphthalein or methyl red is a suitable indicator for a strong acid–strong base titration. The increase in pH in the region of the equivalence point is so steep that any indicator changing color in the pH range 4–10 can be used without making a significant error in locating the equivalence point.

CONCEPTUAL PROBLEM 15.15 The following pictures represent solutions at various points in the titration of a weak acid HA with aqueous NaOH. (Na^+ ions and solvent water molecules have been omitted for clarity.)

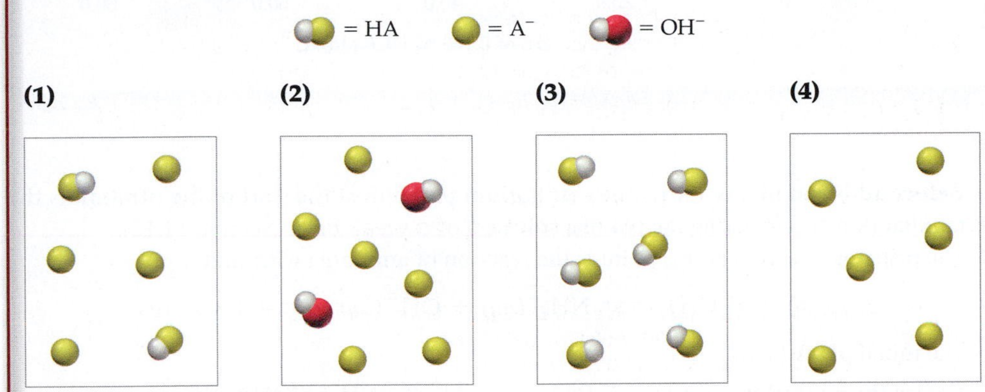

Which picture corresponds to each of the following points in the titration?
(a) Before the addition of any NaOH
(b) Before the equivalence point
(c) At the equivalence point
(d) After the equivalence point

PROBLEM 15.16 Consider the titration of 100.0 mL of 0.016 M HOCl ($K_a = 3.5 \times 10^{-8}$) with 0.0400 M NaOH. How many milliliters of 0.0400 M NaOH are required to reach the equivalence point? Calculate the pH:
(a) After the addition of 10.0 mL of 0.0400 M NaOH
(b) Halfway to the equivalence point
(c) At the equivalence point

PROBLEM 15.17 The following acid–base indicators change color in the indicated pH ranges: bromthymol blue (6.0–7.6), thymolphthalein (9.4–10.6), and alizarin yellow (10.1–12.0). Which indicator is best for the titration in Problem 15.16? Which indicator is unacceptable? Explain.

15.8 WEAK BASE–STRONG ACID TITRATIONS

Figure 15.10 shows the pH titration curve for a typical weak base–strong acid titration, the titration of 40.0 mL of 0.100 M NH_3 with 0.100 M HCl. The pH calculations are simply outlined to save space; you should verify the results yourself.

FIGURE 15.10

A weak base–strong acid titration curve.

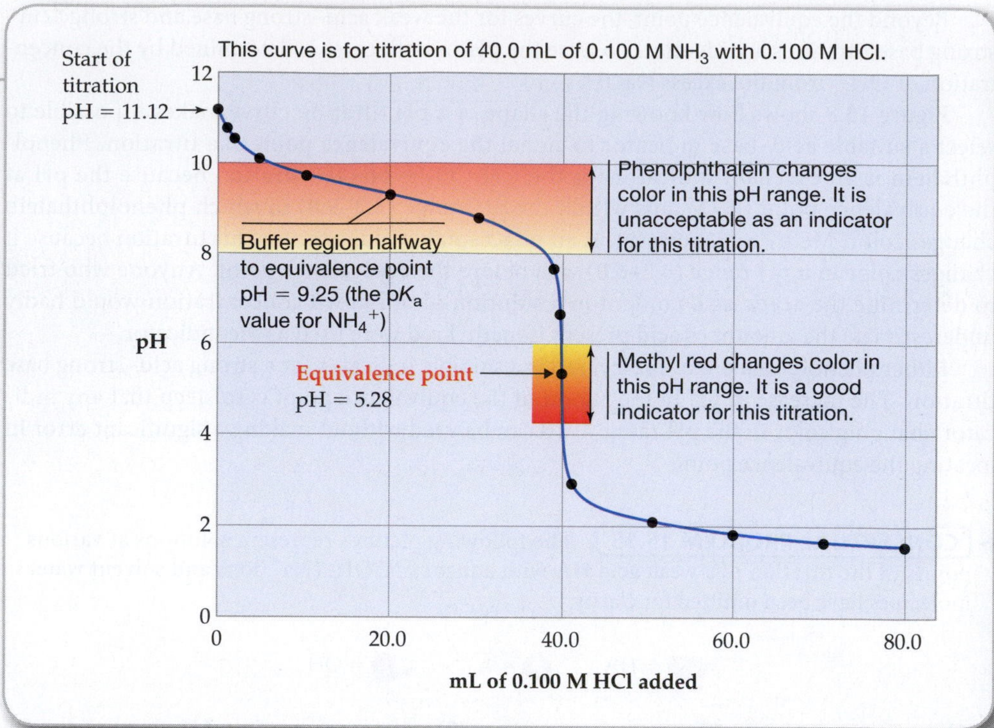

REMEMBER...

Equilibria in **solutions of weak bases** are treated by the same procedure used for solving problems involving weak acids. (Section 14.12)

1. **Before addition of any HCl.** The equilibrium problem at the start of the titration is the familiar one of calculating the pH of a **solution of a weak base** (Section 14.12). The principal reaction at this point is the reaction of ammonia with water:

$$NH_3(aq) + H_2O(l) \rightleftharpoons NH_4^+(aq) + OH^-(aq) \quad K_b = 1.8 \times 10^{-5}$$

The initial pH is 11.12.

2. **Before the equivalence point.** As HCl is added to the NH_3 solution, NH_3 is converted to NH_4^+ because of the neutralization reaction

$$NH_3(aq) + H_3O^+(aq) \xrightarrow{100\%} NH_4^+(aq) + H_2O(l)$$

The neutralization reaction goes to completion, but the amount of H_3O^+ added before the equivalence point is insufficient to convert all the NH_3 to NH_4^+. We therefore have an NH_4^+–NH_3 buffer solution, which accounts for the leveling of the titration curve in the buffer region between the start of the titration and the equivalence point. The pH at any specific point can be calculated from the Henderson–Hasselbalch equation. After addition of 20.0 mL of 0.100 M HCl (halfway to the equivalence point), $[NH_3] = [NH_4^+]$ and the pH equals pK_a for NH_4^+ (9.25).

3. **At the equivalence point.** The equivalence point is reached after adding 40.0 mL of 0.100 M HCl (4.00 mmol). At this point, the 4.00 mmol of NH_3 present initially has been converted to 4.00 mmol of NH_4^+; $[NH_4^+] = (4.00\ \text{mmol})/(80.0\ \text{mL}) = 0.0500\ \text{M}$. Since NH_4^+ is a weak acid and Cl^- is neither an acid nor a base, we have an acidic salt solution (Section 14.14). The principal reaction is

$$NH_4^+(aq) + H_2O(l) \rightleftharpoons H_3O^+(aq) + NH_3(aq) \quad K_a = K_w/K_b = 5.6 \times 10^{-10}$$

The pH at the equivalence point is 5.28.

4. **Beyond the equivalence point.** In this region, all the NH_3 has been converted to NH_4^+, and excess H_3O^+ is present from the excess HCl. Because the acid dissociation of NH_4^+ produces a negligible $[H_3O^+]$ compared with $[H_3O^+]$ from the excess HCl, the pH can be

calculated directly from the concentration of the excess HCl. For example, after addition of 60.0 mL of 0.100 M HCl:

$$[H_3O^+] = \frac{6.00 \text{ mmol} - 4.00 \text{ mmol}}{100.0 \text{ mL}} = 0.0200 \text{ M}$$

$$\text{pH} = 1.70$$

15.9 POLYPROTIC ACID–STRONG BASE TITRATIONS

As a final example of an acid–base titration, let's consider the gradual addition of NaOH to the protonated form of the amino acid alanine (H_2A^+), a substance that behaves as a diprotic acid. Amino acids are both acidic and basic and can be protonated by strong acids such as HCl, yielding salts such as $H_2A^+Cl^-$. The protonated form of the amino acid has two dissociable protons and can react with two molar amounts of OH^- to give first the neutral form and then the anionic form:

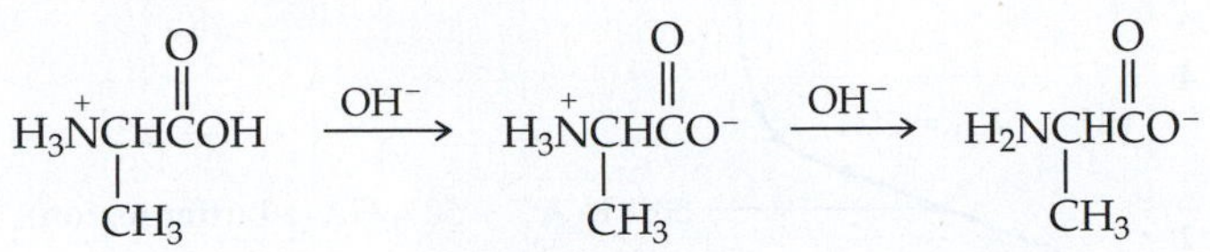

Protonated form (H_2A^+) of alanine — Neutral form (HA) of alanine — Anionic form (A^-) of alanine

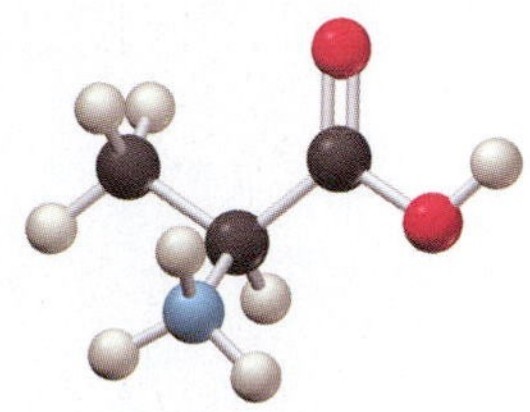

Alanine cation (H_2A^+)

The proton of the $-CO_2H$ group is more acidic than the proton of the $-NH_3^+$ group and is neutralized first. The neutral form of alanine (HA) that results has a plus charge on the $-NH_3^+$ group and a minus charge on the $-CO_2^-$ group but is electrically neutral overall. In a second step, the proton of the $-NH_3^+$ group is neutralized, yielding the anionic form of alanine (A^-). The dissociation equilibria and their K_a values are

$$H_2A^+(aq) + H_2O(l) \rightleftharpoons H_3O^+(aq) + HA(aq) \quad K_{a1} = 4.6 \times 10^{-3}; pK_{a1} = 2.34$$

$$HA(aq) + H_2O(l) \rightleftharpoons H_3O^+(aq) + A^-(aq) \quad K_{a2} = 2.0 \times 10^{-10}; pK_{a2} = 6.69$$

Figure 15.11 shows the pH titration curve for the addition of solid NaOH to 1.00 L of a 1.00 M solution of H_2A^+ (1.00 mol). Because K_{a1} and K_{a2} are separated by several powers of 10, the titration curve exhibits two well-defined equivalence points and two buffer regions.

Calculations for Diprotic Acid–Strong Base Titrations

As a review of many of the acid–base problems we've encountered in this chapter and in Chapter 14, let's calculate the pH at several points on the curve for titration of the protonated form of alanine, H_2A^+, with NaOH. We've met all these equilibrium problems previously, except for the situation at the first equivalence point. To simplify the calculations, we'll assume that the added base is solid NaOH so that we can neglect volume changes in the course of the titration.

1. **Before addition of any NaOH.** The equilibrium problem at the start of the titration is the familiar one of calculating the pH of a diprotic acid (Section 14.11). **The principal reaction is the dissociation of H_2A^+**, and $[H_3O^+]$ can be calculated from the equilibrium equation

$$K_{a1} = 4.6 \times 10^{-3} = \frac{[H_3O^+][HA]}{[H_2A^+]} = \frac{(x)(x)}{1.00 - x}$$

Solving the quadratic equation gives $[H_3O^+] = 0.066$ M and pH $= 1.18$.

2. **Halfway to the first equivalence point.** As NaOH is added, H_2A^+ is converted to HA because of the neutralization reaction

$$H_2A^+(aq) + OH^-(aq) \xrightarrow{100\%} H_2O(l) + HA(aq)$$

REMEMBER...

Because K_{a1} for H_2A^+ is much greater than K_{a2}, **the principal reaction is the dissociation of H_2A^+** and essentially all the H_3O^+ in the solution comes from the first dissociation step. (Section 14.11)

FIGURE 15.11

A diprotic acid–strong base titration curve. The change in the pH of 1.00 L of a 1.00 M solution of the protonated form of alanine (H_2A^+) on the addition of solid NaOH.

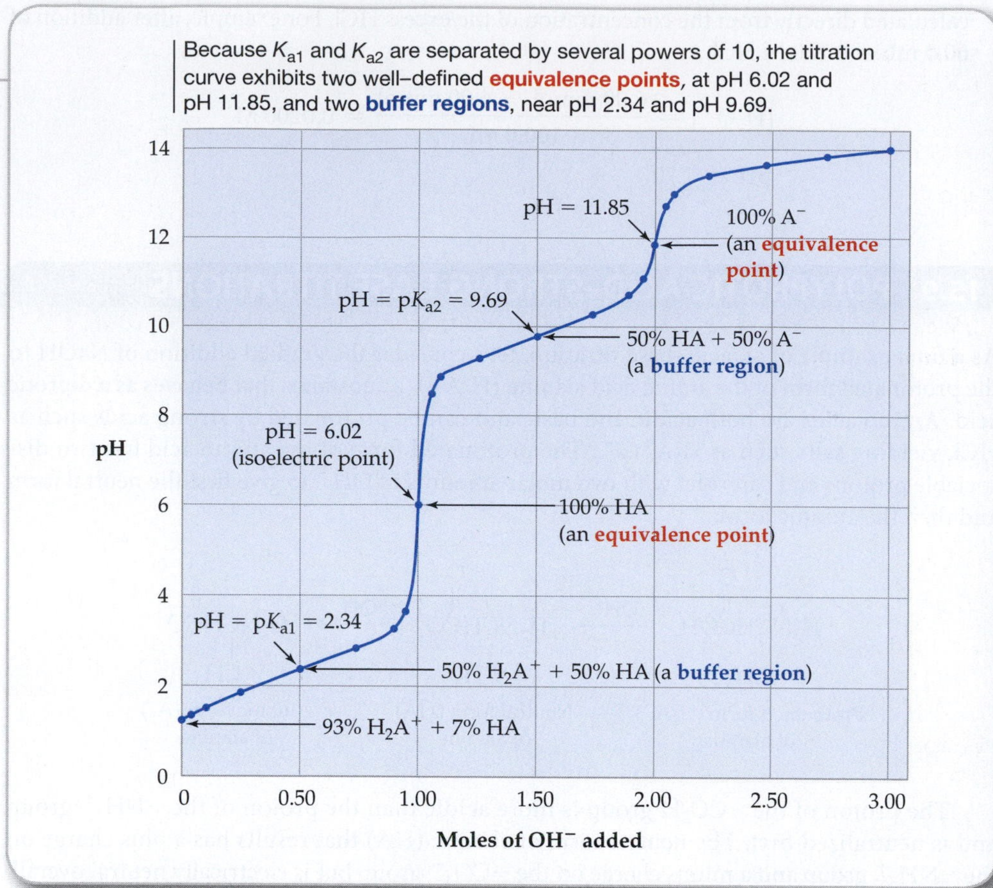

Halfway to the first equivalence point, we have an H_2A^+–HA buffer solution with $[H_2A^+] = [HA]$. The Henderson–Hasselbalch equation gives $pH = pK_{a1} = 2.34$.

3. **At the first equivalence point.** At this point, we have added just enough NaOH to convert all the H_2A^+ to HA. The principal reaction at the first equivalence point is proton transfer between HA molecules:

$$2\,HA(aq) \rightleftharpoons H_2A^+(aq) + A^-(aq) \qquad K = K_{a2}/K_{a1} = 4.3 \times 10^{-8}$$

It can be shown (see Problem 15.140) that the pH at this point equals the average of the two pK_a values:

$$\blacktriangleright \text{ pH at first equivalence point} = \frac{pK_{a1} + pK_{a2}}{2}$$

Since H_2A^+ has $pK_{a1} = 2.34$ and $pK_{a2} = 9.69$, the pH at the first equivalence point is $(2.34 + 9.69)/2 = 6.02$. The same situation holds for most polyprotic acids: The pH at the first equivalence point equals the average of pK_{a1} and pK_{a2}.

For an amino acid, the pH value, $(pK_{a1} + pK_{a2})/2$, is called the *isoelectric point* (Figure 15.11). At that point, the concentration of the neutral HA is at a maximum, and the very small concentrations of H_2A^+ and A^- are equal. Biochemists use isoelectric points to separate mixtures of amino acids and proteins.

4. **Halfway between the first and second equivalence points.** At this point, half the HA has been converted to A^- because of the neutralization reaction

$$HA(aq) + OH^-(aq) \xrightarrow{100\%} H_2O(l) + A^-(aq)$$

We thus have an HA–A^- buffer solution with $[HA] = [A^-]$. Therefore, $pH = pK_{a2} = 9.69$.

5. At the second equivalence point. At this point, we have added enough NaOH to convert all the HA to A^- and we have a 1.00 M solution of a **basic salt** (Section 14.14). The principal reaction is

$$A^-(aq) + H_2O(l) \rightleftharpoons HA(aq) + OH^-(aq)$$

and its equilibrium constant is

$$K_b = \frac{K_w}{K_a \text{ for HA}} = \frac{K_w}{K_{a2}} = \frac{1.0 \times 10^{-14}}{2.0 \times 10^{-10}} = 5.0 \times 10^{-5}$$

We can obtain $[OH^-]$ from the equilibrium equation for the principal reaction and then calculate $[H_3O^+]$ and the pH in the usual way:

$$K_b = 5.0 \times 10^{-5} = \frac{[HA][OH^-]}{[A^-]} = \frac{(x)(x)}{1.00 - x} \qquad x = OH^- = 7.1 \times 10^{-3}\text{ M}$$

$$[H_3O^+] = \frac{K_w}{[OH^-]} = \frac{1.0 \times 10^{-14}}{7.1 \times 10^{-3}} = 1.4 \times 10^{-12}\text{ M}$$

$$\text{pH} = 11.85$$

Since the initial solution of H_2A^+ contained 1.00 mol of H_2A^+, the amount of NaOH required to reach the second equivalence point is 2.00 mol. Beyond the second equivalence point, the pH is determined by $[OH^-]$ from the excess NaOH.

REMEMBER...

A **basic salt** contains the anion of a weak acid, so the anion is a proton acceptor. (Section 14.14)

Worked Example 15.6

Calculating the pH for a Diprotic Acid–Strong Base Titration

Assume that you are titrating 30.0 mL of a 0.0600 M solution of the protonated form of the amino acid methionine (H_2A^+) with 0.0900 M NaOH. Calculate the pH after addition of the following volumes of 0.0900 M NaOH:

(a) 20.0 mL
(b) 30.0 mL
(c) 35.0 mL

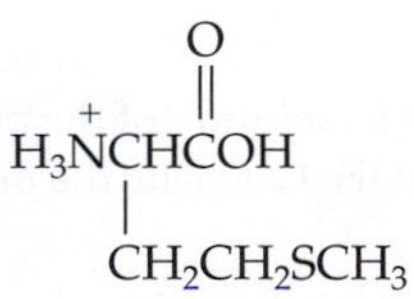

Methionine cation (H_2A^+)
$K_{a1} = 5.2 \times 10^{-3}$
$K_{a2} = 6.2 \times 10^{-10}$

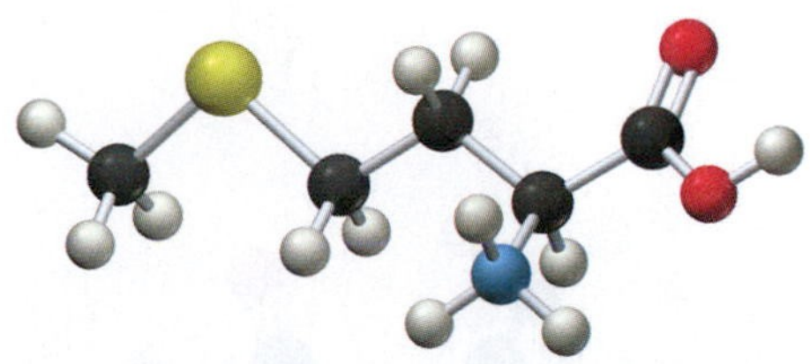

Methionine cation

STRATEGY

First calculate the number of millimoles of H_2A^+ present initially and the number of millimoles of NaOH added. Then you can tell which species remain after neutralization and therefore what type of equilibrium problem you must solve.

SOLUTION

(a) The number of millimoles of H_2A^+ present initially and the number of millimoles NaOH added are

$$\text{mmol } H_2A^+ \text{ initial} = (30.0\text{ mL})(0.0600\text{ mmol/mL}) = 1.80\text{ mmol}$$

$$\text{mmol NaOH added} = (20.0\text{ mL})(0.0900\text{ mmol/mL}) = 1.80\text{ mmol}$$

The added base is just enough to reach the first equivalence point, converting all the H_2A^+ to HA because of the neutralization reaction

$$H_2A^+(aq) + OH^-(aq) \xrightarrow{100\%} H_2O(l) + HA(aq)$$

Therefore, the pH equals the average of pK_{a1} and pK_{a2}:

$$\text{pH} = \frac{pK_{a1} + pK_{a2}}{2} = \frac{[-\log(5.2 \times 10^{-3})] + [-\log(6.2 \times 10^{-10})]}{2}$$

$$= \frac{2.28 + 9.21}{2} = 5.74$$

continued on next page

(b) The amount of NaOH added to the 1.80 mmol of H_2A^+ present initially is (30.0 mL) $\times$ (0.0900 mmol/mL) = 2.70 mmol. The added base is enough to convert all of the H_2A^+ to 1.80 mmol of HA and then convert half of the resulting HA to 0.90 mmol of A^- because of the neutralization reaction

$$HA(aq) + OH^-(aq) \xrightarrow{100\%} H_2O(l) + A^-(aq)$$

After neutralization, we have an HA–A^- buffer solution that contains equal amounts of HA and A^- (0.90 mmol), and so the pH equals pK_{a2} of the methionine cation H_2A^+.

$$pH = pK_{a2} = -\log(6.2 \times 10^{-10}) = 9.21$$

(c) The amount of NaOH added to the 1.80 mmol of H_2A^+ present initially is (35.0 mL) $\times$ (0.0900 mmol/mL) = 3.15 mmol. The added base is enough to convert all of the H_2A^+ to 1.80 mmol of HA and then convert 3.15 − 1.80 = 1.35 mmol of the resulting HA to 1.35 mmol of A^-. After neutralization, we have an HA–A^- buffer solution that contains 0.45 mmol of HA and 1.35 mmol of A^-. We can use the Henderson–Hasselbalch equation to calculate the pH:

$$pH = pK_{a2} + \log\frac{[\text{Base}]}{[\text{Acid}]} = 9.21 + \log\frac{1.35}{0.45} = 9.69$$

✔ BALLPARK CHECK

Because the conjugate base predominates in part **(c)**, the pH should be greater than pK_{a2}, in agreement with the solution. Also note that the pH increases as NaOH is added [from 5.74 in part **(a)** to 9.21 in part **(b)** to 9.69 in part **(c)**].

PROBLEM 15.18 Assume that 40.0 mL of 0.0800 M H_2SO_3 ($K_{a1} = 1.5 \times 10^{-2}$, $K_{a2} = 6.3 \times 10^{-8}$) is titrated with 0.160 M NaOH. Calculate the pH after addition of the following volumes of 0.160 M NaOH:
(a) 20.0 mL
(b) 30.0 mL
(c) 35.0 mL

PROBLEM 15.19 Assume that 40.0 mL of a 0.0250 M solution of the protonated form of the amino acid valine (H_2A^+) is titrated with 0.100 M NaOH. Calculate the pH after addition of the following volumes of 0.100 M NaOH:
(a) 10.0 mL
(b) 15.0 mL
(c) 20.0 mL

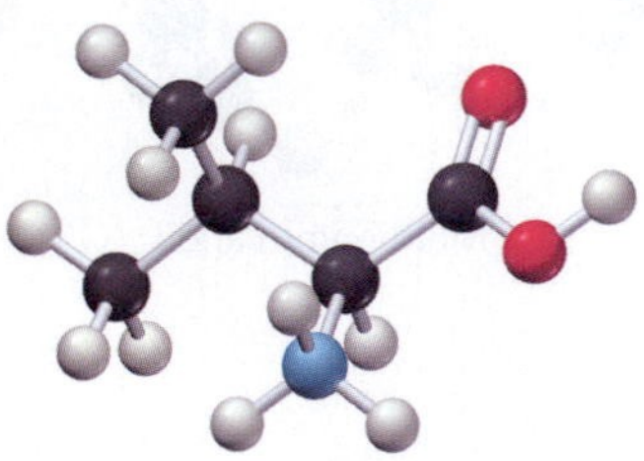

Valine cation

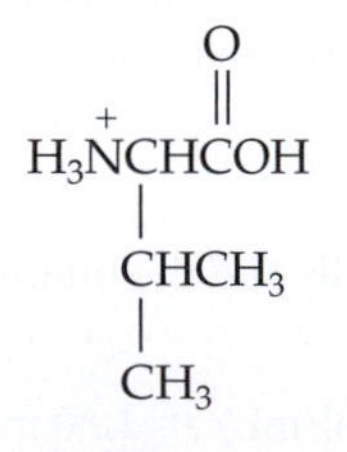

Valine cation (H_2A^+)
$K_{a1} = 4.8 \times 10^{-3}$
$K_{a2} = 2.4 \times 10^{-10}$

▲ Most kidney stones consist of insoluble calcium salts, such as calcium oxalate.

15.10 SOLUBILITY EQUILIBRIA FOR IONIC COMPOUNDS

Many biological and environmental processes involve the dissolution or precipitation of a sparingly soluble ionic compound. Tooth decay, for example, begins when tooth enamel, composed of the mineral hydroxyapatite, $Ca_5(PO_4)_3OH$, dissolves on reaction with organic acids produced by the bacterial decomposition of foods rich in sugar. Kidney stones form when moderately insoluble calcium salts, such as calcium oxalate, CaC_2O_4, precipitate slowly over a long period of time. To understand the quantitative aspects of such solubility and precipitation phenomena, we must examine the principles of solubility equilibria.

Let's consider the solubility equilibrium in a saturated solution of calcium fluoride in contact with an excess of solid calcium fluoride. Like most sparingly soluble ionic solutes, calcium fluoride is a strong electrolyte in water and exists in the aqueous phase as dissociated hydrated ions, $Ca^{2+}(aq)$ and $F^{-}(aq)$. At equilibrium, the ion concentrations remain constant because the rate at which solid CaF_2 dissolves to give $Ca^{2+}(aq)$ and $F^{-}(aq)$ exactly equals the rate at which the ions crystallize to form solid CaF_2:

$$CaF_2(s) \rightleftharpoons Ca^{2+}(aq) + 2\,F^{-}(aq)$$

The equilibrium equation for the dissolution reaction is

$$K_{sp} = [Ca^{2+}][F^{-}]^2$$

where the equilibrium constant K_{sp} is called the *solubility product constant*, or simply the **solubility product**. As usual for a heterogeneous equilibrium, **the concentration of the solid**, CaF_2, is omitted from the equilibrium equation (Section 13.5).

REMEMBER...

The concentration of a pure solid or a pure liquid is omitted from the equilibrium-constant expression because the ratio of its actual concentration to its concentration in the thermodynamic standard state is equal to 1. (Section 13.5)

For the general solubility equilibrium

$$M_mX_x(s) \rightleftharpoons m\,M^{n+}(aq) + x\,X^{y-}(aq)$$

the equilibrium-constant expression for K_{sp} is

▶ **Solubility product** $K_{sp} = [M^{n+}]^m[X^{y-}]^x$

Thus, K_{sp} always equals the product of the equilibrium concentrations of all the ions on the right side of the chemical equation, with the concentration of each ion raised to the power of its coefficient in the balanced equation.

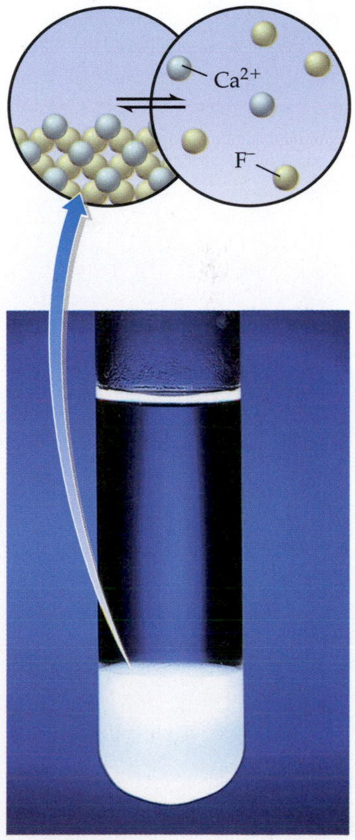

▲ A saturated solution of calcium fluoride in contact with solid CaF_2 contains constant equilibrium concentrations of $Ca^{2+}(aq)$ and $F^{-}(aq)$ because at equilibrium the ions crystallize at the same rate as the solid dissolves.

Worked Example 15.7

Writing Equilibrium-Constant Expressions for K_{sp}

Write the equilibrium-constant expression for the solubility product of silver chromate, Ag_2CrO_4.

STRATEGY AND SOLUTION

First write the balanced equation for the solubility equilibrium:

$$Ag_2CrO_4(s) \rightleftharpoons 2\,Ag^{+}(aq) + CrO_4^{2-}(aq)$$

The exponents in the equilibrium-constant expression for K_{sp} are the coefficients in the balanced equation. Therefore,

$$K_{sp} = [Ag^{+}]^2[CrO_4^{2-}]$$

▶ **PROBLEM 15.20** Write the equilibrium-constant expression for K_{sp} of:
(a) AgCl **(b)** PbI_2 **(c)** $Ca_3(PO_4)_2$ **(d)** $Cr(OH)_3$

15.11 MEASURING K_{sp} AND CALCULATING SOLUBILITY FROM K_{sp}

The numerical value of a solubility product K_{sp} is measured by experiment. For example, we could determine K_{sp} for CaF_2 by adding an excess of solid CaF_2 to water, stirring the mixture to give a saturated solution of CaF_2, and then measuring the concentrations of Ca^{2+} and F^{-} in the saturated solution. To make sure that the concentrations had reached constant equilibrium values, we would want to stir the mixture for an additional period of time and then repeat the measurements. Suppose that we found $[Ca^{2+}] = 2.0 \times 10^{-4}$ M and $[F^{-}] = 4.1 \times 10^{-4}$ M. (The value of $[F^{-}]$ is twice the value of $[Ca^{2+}]$ because each mole of

CaF_2 that dissolves yields 1 mol of Ca^{2+} ions and 2 mol of F^- ions.) We could then calculate K_{sp} for CaF_2:

$$K_{sp} = [Ca^{2+}][F^-]^2 = (2.0 \times 10^{-4})(4.1 \times 10^{-4})^2 = 3.4 \times 10^{-11}$$

Another way to measure K_{sp} for CaF_2 is to approach the equilibrium from the opposite direction—that is, by mixing sources of Ca^{2+} and F^- ions to give a precipitate of solid CaF_2 and a saturated solution of CaF_2. Suppose, for example, that we mix solutions of $CaCl_2$ and NaF, allow time for equilibrium to be reached, and then measure $[Ca^{2+}] = 3.5 \times 10^{-5}$ M and $[F^-] = 1.0 \times 10^{-3}$ M. Within experimental error, these ion concentrations yield the same value of K_{sp}:

$$K_{sp} = [Ca^{2+}][F^-]^2 = (3.5 \times 10^{-5})(1.0 \times 10^{-3})^2 = 3.5 \times 10^{-11}$$

The value of K_{sp} is unaffected by the presence of other ions in solution, such as Na^+ from NaF and Cl^- from $CaCl_2$, as long as the solution is very dilute. As ion concentrations increase, K_{sp} values are somewhat modified because of electrostatic interactions between ions, but we'll ignore that complication here.

If the saturated solution is prepared by a method other than the dissolution of CaF_2 in pure water, there are no separate restrictions on $[Ca^{2+}]$ and $[F^-]$; the only restriction on the ion concentrations is that the value of the equilibrium-constant expression $[Ca^{2+}][F^-]^2$ must equal the K_{sp}. That condition is satisfied by an infinite number of combinations of $[Ca^{2+}]$ and $[F^-]$, and therefore we can prepare many different solutions that are saturated with respect to CaF_2. For example, if $[F^-]$ is 1.0×10^{-2} M, then $[Ca^{2+}]$ must be 3.5×10^{-7} M:

$$[Ca^{2+}] = \frac{K_{sp}}{[F^-]^2} = \frac{3.5 \times 10^{-11}}{(1.0 \times 10^{-2})^2} = 3.5 \times 10^{-7}$$

Selected values of K_{sp} for various ionic compounds at 25 °C are listed in **Table 15.3**, and additional values can be found in Appendix C. Like all equilibrium constants, values of K_{sp} depend on temperature (Section 13.10).

TABLE 15.3 K_{sp} **Values for Some Ionic Compounds at 25 °C**

Name	Formula	K_{sp}
Aluminum hydroxide	$Al(OH)_3$	1.9×10^{-33}
Barium carbonate	$BaCO_3$	2.6×10^{-9}
Calcium carbonate	$CaCO_3$	5.0×10^{-9}
Calcium fluoride	CaF_2	3.5×10^{-11}
Lead(II) chloride	$PbCl_2$	1.2×10^{-5}
Lead(II) chromate	$PbCrO_4$	2.8×10^{-13}
Silver chloride	AgCl	1.8×10^{-10}
Silver sulfate	Ag_2SO_4	1.2×10^{-5}

Once the K_{sp} value for a compound has been measured, you can use it to calculate the solubility of the compound—the amount of compound that dissolves per unit volume of saturated solution. Because of two complications, however, calculated solubilities are often approximate. First, K_{sp} values can be difficult to measure, and values listed in different sources might differ by as much as a factor of 10 or more. Second, calculated solubilities can be less than observed solubilities because of side reactions. For example, dissolution of $PbCl_2$ gives both Pb^{2+} and $PbCl^+$ because of some ion association between Pb^{2+} and Cl^- ions:

$$(1)\ PbCl_2(s) \rightleftharpoons Pb^{2+}(aq) + 2\,Cl^-(aq)$$
$$(2)\ Pb^{2+}(aq) + Cl^-(aq) \rightleftharpoons PbCl^+(aq)$$

In this book, we will calculate approximate solubilities assuming that ionic solutes are completely dissociated (reaction 1). In the case of $PbCl_2$, ignoring the second equilibrium gives a calculated solubility that is too low by a factor of about 2.

Worked Example 15.8

Calculating K_{sp} from Ion Concentrations

A particular saturated solution of silver chromate, Ag_2CrO_4, has $[Ag^+] = 5.0 \times 10^{-5}$ M and $[CrO_4^{2-}] = 4.4 \times 10^{-4}$ M. What is the value of K_{sp} for Ag_2CrO_4?

STRATEGY

Substituting the equilibrium concentrations into the expression for K_{sp} of Ag_2CrO_4 (Worked Example 15.7) gives the value of K_{sp}.

SOLUTION

$$K_{sp} = [Ag^+]^2[CrO_4^{2-}] = (5.0 \times 10^{-5})^2(4.4 \times 10^{-4}) = 1.1 \times 10^{-12}$$

✔ BALLPARK CHECK

K_{sp} is approximately $(5)^2(4) = 100$ times $(10^{-5})^2(10^{-4})$. So K_{sp} is about 10^{-12}, in agreement with the solution.

▲ Addition of aqueous K_2CrO_4 to aqueous $AgNO_3$ gives a red precipitate of Ag_2CrO_4 and a saturated solution of Ag_2CrO_4.

Worked Example 15.9

Calculating K_{sp} from Solubility

A saturated solution of Ag_2CrO_4 prepared by dissolving solid Ag_2CrO_4 in water has $[CrO_4^{2-}] = 6.5 \times 10^{-5}$ M. Calculate K_{sp} for Ag_2CrO_4.

STRATEGY

Because both the Ag^+ and CrO_4^{2-} ions come from the dissolution of solid Ag_2CrO_4, $[Ag^+]$ must be twice $[CrO_4^{2-}]$.

$$Ag_2CrO_4(s) \rightleftharpoons 2\,Ag^+(aq) + CrO_4^{2-}(aq)$$

Substituting $[Ag^+]$ and $[CrO_4^{2-}]$ into the equilibrium equation gives the value of K_{sp}.

SOLUTION

$$[Ag^+] = (2)[CrO_4^{2-}] = (2)(6.5 \times 10^{-5}) = 1.3 \times 10^{-4}\ M$$

$$K_{sp} = [Ag^+]^2[CrO_4^{2-}] = (1.3 \times 10^{-4})^2(6.5 \times 10^{-5}) = 1.1 \times 10^{-12}$$

Worked Example 15.10

Calculating Solubility from K_{sp}

Calculate the solubility of MgF_2 in water at 25 °C in units of:
(a) Moles per liter **(b)** Grams per liter

STRATEGY

(a) Write the balanced equation for the solubility equilibrium assuming the complete dissociation of MgF_2, and look up K_{sp} for MgF_2 in Appendix C. If we define x as the number of moles per liter of MgF_2 that dissolves, then the saturated solution contains x mol/L of Mg^{2+} and $2x$ mol/L of F^-. Substituting these equilibrium concentrations into the expression for K_{sp} and solving for x gives the molar solubility.

(b) To convert the solubility from units of moles per liter to units of grams per liter, multiply the molar solubility of MgF_2 by its molar mass (62.3 g/mol).

SOLUTION

(a) It's helpful to summarize the equilibrium concentrations under the balanced equation:

Solubility Equilibrium	$MgF_2(s) \rightleftharpoons$	$Mg^{2+}(aq)$	$+\ 2\,F^-(aq)$
Equilibrium concentration (M)		x	$2x$

continued on next page

Substituting the equilibrium concentrations into the expression for K_{sp} gives

$$K_{sp} = 7.4 \times 10^{-11} = [Mg^{2+}][F^-]^2 = (x)(2x)^2$$
$$4x^3 = 7.4 \times 10^{-11}$$
$$x^3 = 1.8 \times 10^{-11}$$
$$x = [Mg^{2+}] = \text{Molar solubility} = 2.6 \times 10^{-4}\ \text{mol/L}.$$

Thus, the molar solubility of MgF_2 in water at 25 °C is 2.6×10^{-4} M.

Note that the number 2 appears twice in the expression $(x)(2x)^2$. The exponent 2 is required because of the equilibrium equation, $K_{sp} = [Mg^{2+}][F^-]^2$. The coefficient 2 in $2x$ is required because each mole of MgF_2 that dissolves gives 2 mol of $F^-(aq)$.

(b) $\text{Solubility (in g/L)} = \dfrac{2.6 \times 10^{-4}\ \text{mol}}{\text{L}} \times \dfrac{62.3\ \text{g}}{\text{mol}} = 1.6 \times 10^{-2}\ \text{g/L}$

✔ BALLPARK CHECK

(a) The molar solubility of MgF_2 is the cube root of $K_{sp}/4$, which equals the cube root of approximately 20×10^{-12}. Because $(2)^3 = 8$, $(3)^3 = 27$, and $(10^{-4})^3 = 10^{-12}$, the molar solubility is between 2×10^{-4} M and 3×10^{-4} M. The ballpark check and the solution agree.

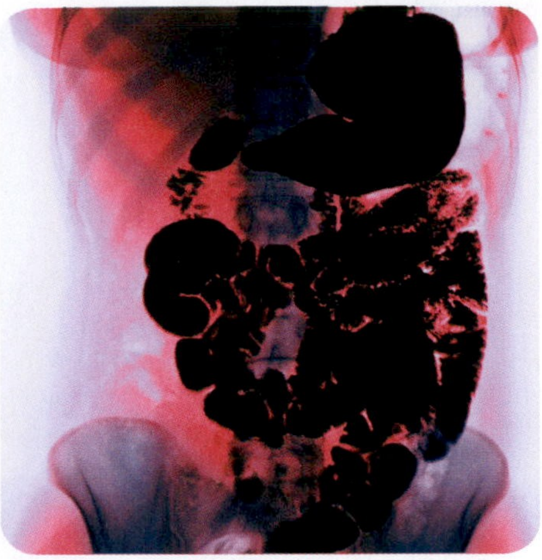

▲ The contrast in this X-ray photograph of the stomach and intestines is due to barium sulfate.

PROBLEM 15.21 A saturated solution of $Ca_3(PO_4)_2$ has $[Ca^{2+}] = 2.01 \times 10^{-8}$ M and $[PO_4^{3-}] = 1.6 \times 10^{-5}$ M. Calculate K_{sp} for $Ca_3(PO_4)_2$.

PROBLEM 15.22 Prior to having an X-ray exam of the upper gastrointestinal tract, a patient drinks an aqueous suspension of solid $BaSO_4$. (Scattering of X rays by barium greatly enhances the quality of the photograph.) Although Ba^{2+} is toxic, ingestion of $BaSO_4$ is safe because it is quite insoluble. If a saturated solution prepared by dissolving solid $BaSO_4$ in water has $[Ba^{2+}] = 1.05 \times 10^{-5}$ M, what is the value of K_{sp} for $BaSO_4$?

PROBLEM 15.23 Which has the greater molar solubility: AgCl with $K_{sp} = 1.8 \times 10^{-10}$ or Ag_2CrO_4 with $K_{sp} = 1.1 \times 10^{-12}$? Which has the greater solubility in grams per liter?

CONCEPTUAL PROBLEM 15.24 The following pictures represent saturated solutions of three silver salts: AgX, AgY, and AgZ. (Other ions and solvent water molecules have been omitted for clarity.)

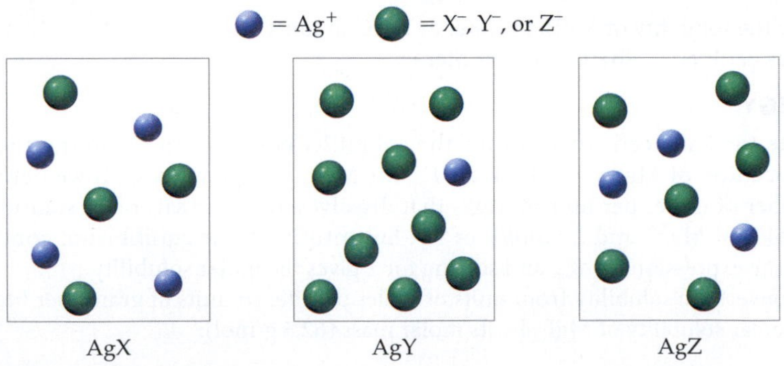

(a) Which salt has the largest value of K_{sp}?
(b) Which salt has the smallest value of K_{sp}?

15.12 FACTORS THAT AFFECT SOLUBILITY

Solubility and the Common-Ion Effect

We've already discussed the common-ion effect in connection with the dissociation of weak acids and bases (Section 15.2). To see how a common ion affects the position of a solubility equilibrium, let's look again at the solubility of MgF_2:

$$MgF_2(s) \rightleftharpoons Mg^{2+}(aq) + 2\,F^-(aq)$$

In Worked Example 15.10, we found that the molar solubility of MgF_2 in pure water at 25 °C is 2.6×10^{-4} M. Thus,

$$[Mg^{2+}] = 2.6 \times 10^{-4}\,M \qquad [F^-] = 2[Mg^{2+}] = 5.2 \times 10^{-4}\,M$$

When MgF_2 dissolves in a solution that contains a common ion from another source—say, F^- from NaF—the position of the solubility equilibrium is shifted to the left by the common-ion effect. If $[F^-]$ is larger than 5.2×10^{-4} M, then $[Mg^{2+}]$ must be correspondingly smaller than 2.6×10^{-4} M to maintain the equilibrium-constant expression $[Mg^{2+}][F^-]^2$ at a constant value of $K_{sp} = 7.4 \times 10^{-11}$. A smaller value of $[Mg^{2+}]$ means that MgF_2 is less soluble in a sodium fluoride solution than it is in pure water. Similarly, the presence of Mg^{2+} from another source—say, $MgCl_2$—shifts the solubility equilibrium to the left and decreases the solubility of MgF_2.

This must decrease . . . | . . . if this increases . . . | . . . because this is constant.

$$[Mg^{2+}][F^-]^2 = K_{sp} = 7.4 \times 10^{-11}$$

If this increases . . . | . . . this must decrease . . . | . . . because this is constant.

In general, the solubility of a slightly soluble ionic compound is decreased by the presence of a common ion in the solution, as predicted by Le Châtelier's principle (Section 13.7). This effect is illustrated in **Figure 15.12**. Quantitative aspects of the common-ion effect are explored in Worked Example 15.11.

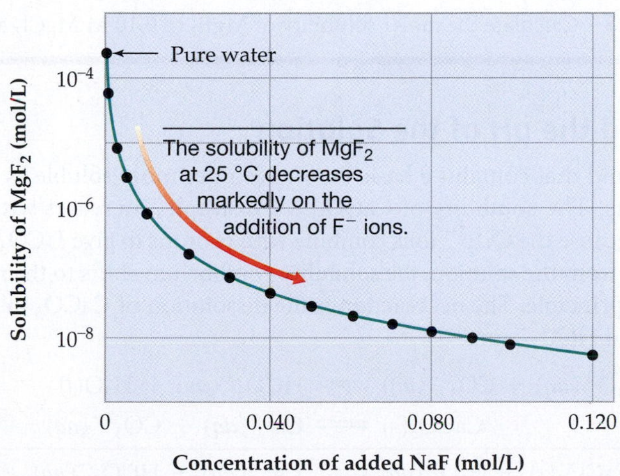

FIGURE 15.12

The common-ion effect. The calculated solubility of MgF_2 is plotted on a logarithmic scale.

Worked Example 15.11

Calculating Solubility in a Solution That Contains a Common Ion

Calculate the molar solubility of MgF_2 in 0.10 M NaF at 25 °C.

STRATEGY

Once again, we define x as the molar solubility of MgF_2. The dissolution of x mol/L of MgF_2 provides x mol/L of Mg^{2+} and $2x$ mol/L of F^-, but the total concentration of F^- is $(0.10 + 2x)$ mol/L because the solution already contains 0.10 mol/L of F^- from the completely dissociated NaF. Substituting the equilibrium concentrations into the expression for K_{sp} and solving for x gives the molar solubility.

SOLUTION

Again, we summarize the equilibrium concentrations under the balanced equation:

Solubility Equilibrium	$MgF_2(s) \rightleftharpoons$	$Mg^{2+}(aq)$	$+ \; 2\,F^-(aq)$
Equilibrium concentration (M)		x	$0.10 + 2x$

Substituting the equilibrium concentrations into the expression for K_{sp} gives

$$K_{sp} = 7.4 \times 10^{-11} = [Mg^{2+}][F^-]^2 = (x)(0.10 + 2x)^2$$

Because K_{sp} is small, $2x$ will be small compared to 0.10 and we can make the approximation that $(0.10 + 2x) \approx 0.10$. Therefore,

$$7.4 \times 10^{-11} = (x)(0.10 + 2x)^2 \approx (x)(0.10)^2$$

$$x = [Mg^{2+}] = \text{Molar solubility} = \frac{7.4 \times 10^{-11}}{(0.10)^2} = 7.4 \times 10^{-9}\text{ M}$$

In accord with Le Châtelier's principle, the calculated solubility of MgF_2 in 0.10 M NaF is less than that in pure water (by a factor of about 35,000!). (See Figure 15.12.)

✔ BALLPARK CHECK

We can check our results by substituting the calculated equilibrium concentrations into the expression for K_{sp}:

$$K_{sp} = 7.4 \times 10^{-11} = [Mg^{2+}][F^-]^2 = (7.4 \times 10^{-9})(0.10)^2 = 7.4 \times 10^{-11}$$

PROBLEM 15.25 Calculate the molar solubility of MgF_2 in 0.10 M $MgCl_2$ at 25 °C.

Solubility and the pH of the Solution

An ionic compound that contains a basic anion becomes more soluble as the acidity of the solution increases. The solubility of $CaCO_3$, for example, increases with decreasing pH (**Figure 15.13**) because the CO_3^{2-} ions combine with protons to give HCO_3^- ions. As CO_3^{2-} ions are removed from the solution, the solubility equilibrium shifts to the right, as predicted by Le Châtelier's principle. The net reaction is the dissolution of $CaCO_3$ in acidic solution to give Ca^{2+} ions and HCO_3^- ions:

$$H_3O^+(aq) + CO_3^{2-}(aq) \rightleftharpoons HCO_3^-(aq) + H_2O(l)$$

$$CaCO_3(s) \rightleftharpoons Ca^{2+}(aq) + CO_3^{2-}(aq)$$

$$\text{Net:}\quad CaCO_3(s) + H_3O^+(aq) \rightleftharpoons Ca^{2+}(aq) + HCO_3^-(aq) + H_2O(l)$$

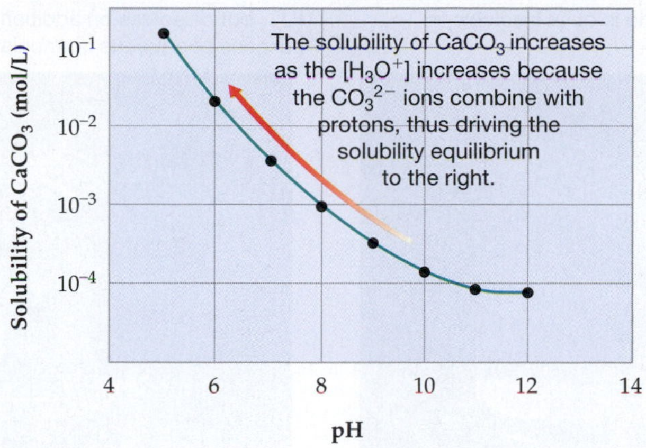

FIGURE 15.13

Plot of the solubility of $CaCO_3$ at 25 °C versus the pH of the solution. The solubility is plotted on a logarithmic scale.

Other salts that contain basic anions, such as CN^-, PO_4^{3-}, S^{2-}, or F^-, behave similarly. By contrast, pH has no effect on the solubility of salts that contain anions of strong acids (Cl^-, Br^-, I^-, NO_3^-, and ClO_4^-) because these anions are not protonated by H_3O^+.

The effect of pH on the solubility of $CaCO_3$ has important environmental consequences. For instance, the formation of limestone caves, such as Mammoth Cave in Kentucky, is due to the slow dissolution of limestone ($CaCO_3$) in the slightly acidic natural water of underground streams, and the formation of stalactites and stalagmites in these caves is due to the slow precipitation of $CaCO_3$ from dripping water. Marble, another form of $CaCO_3$, also dissolves in acid, which accounts for the deterioration of marble monuments on exposure to acid rain (*FYI*, pages 588–589).

The effect of pH on solubility is also important in understanding how fluoride ion reduces tooth decay. When tooth enamel, mainly hydroxyapatite, $Ca_5(PO_4)_3OH$, comes in contact with F^- ions in drinking water or fluoride-containing toothpaste, OH^- ions in $Ca_5(PO_4)_3OH$ are replaced by F^- ions, giving the mineral fluorapatite, $Ca_5(PO_4)_3F$. Because F^- is a much weaker base than OH^-, $Ca_5(PO_4)_3F$ is much more resistant than $Ca_5(PO_4)_3OH$ to dissolving in acids. (See the *FYI* at the end of this chapter.)

▲ These downward-growing, icicle-shaped structures, called stalactites, and the upward growing columns, called stalagmites, are formed in limestone caves by the slow precipitation of calcium carbonate from dripping water.

PROBLEM 15.26 Which of the following compounds are more soluble in acidic solution than in pure water?
(a) AgCN **(b)** PbI_2 **(c)** $Al(OH)_3$ **(d)** ZnS

Solubility and the Formation of Complex Ions

The solubility of an ionic compound increases dramatically if the solution contains a **Lewis base** that can form a **coordinate covalent bond** to the metal cation (Section 4.6). Silver chloride, for example, is quite insoluble in water and in acid, but it dissolves in an excess of aqueous ammonia, forming the *complex ion* $Ag(NH_3)_2^+$ **(Figure 15.14)**. A **complex ion** is an ion that contains a metal cation bonded to one or more small molecules or ions, such as NH_3, CN^-, or OH^-. In accord with Le Châtelier's principle, ammonia shifts the solubility equilibrium to the right by tying up the Ag^+ ion in the form of the complex ion:

$$AgCl(s) \rightleftharpoons Ag^+(aq) + Cl^-(aq)$$
$$Ag^+(aq) + 2\,NH_3(aq) \rightleftharpoons Ag(NH_3)_2^+(aq)$$

REMEMBER...

A **coordinate covalent bond** forms when one atom donates two electrons (a lone pair) to another atom that has a vacant orbital. (Section 4.6)
The electron-pair donor is a **Lewis base**. (Section 14.15)

LOOKING AHEAD ...

We'll see many more examples of **complex ions** in Chapter 20.

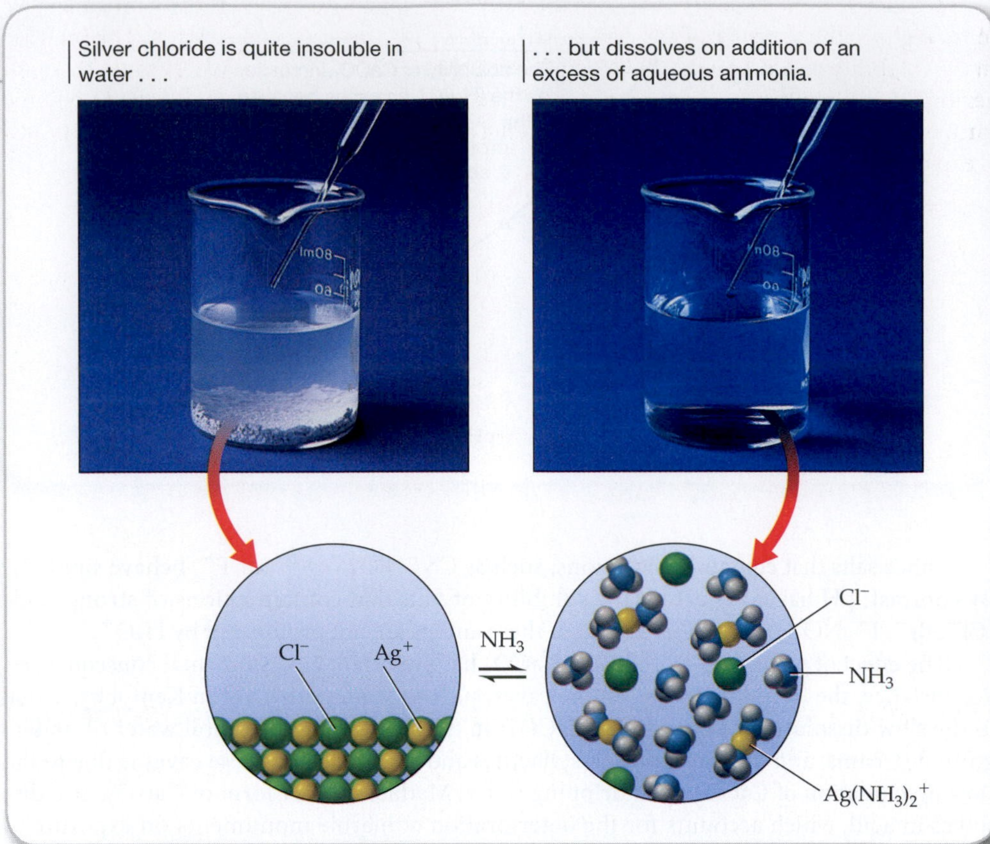

FIGURE 15.14

The reaction of silver chloride with aqueous ammonia.

The formation of a complex ion is a stepwise process, and each step has its own characteristic equilibrium constant. For the formation of $Ag(NH_3)_2^+$, the reactions are

$$Ag^+(aq) + NH_3(aq) \rightleftharpoons Ag(NH_3)^+(aq) \qquad K_1 = 2.1 \times 10^3$$
$$Ag(NH_3)^+(aq) + NH_3(aq) \rightleftharpoons Ag(NH_3)_2^+(aq) \qquad K_2 = 8.1 \times 10^3$$
$$\text{Net: } Ag^+(aq) + 2\,NH_3(aq) \rightleftharpoons Ag(NH_3)_2^+(aq) \qquad K_f = 1.7 \times 10^7$$

$$K_1 = \frac{[Ag(NH_3)^+]}{[Ag^+][NH_3]} = 2.1 \times 10^3 \qquad K_2 = \frac{[Ag(NH_3)_2^+]}{[Ag(NH_3)^+][NH_3]} = 8.1 \times 10^3$$

$$K_f = K_1K_2 = \frac{[Ag(NH_3)_2^+]}{[Ag^+][NH_3]^2} = 1.7 \times 10^7 \quad \text{at 25 °C}$$

The stability of a complex ion is measured by its **formation constant K_f** (or *stability constant*), the equilibrium constant for the formation of the complex ion from the hydrated metal cation. The large value of K_f for $Ag(NH_3)_2^+$ means that this complex ion is quite stable, and nearly all the Ag^+ ion in an aqueous ammonia solution is therefore present in the form of $Ag(NH_3)_2^+$ (see Worked Example 15.12).

The net reaction for dissolution of AgCl in aqueous ammonia is the sum of the equations for the dissolution of AgCl in water and the reaction of $Ag^+(aq)$ with $NH_3(aq)$ to give $Ag(NH_3)_2^+$:

$$AgCl(s) \rightleftharpoons Ag^+(aq) + Cl^-(aq) \qquad K_{sp} = 1.8 \times 10^{-10}$$
$$Ag^+(aq) + 2\,NH_3(aq) \rightleftharpoons Ag(NH_3)_2^+(aq) \qquad K_f = 1.7 \times 10^7$$
$$\text{Net: } AgCl(s) + 2\,NH_3(aq) \rightleftharpoons Ag(NH_3)_2^+(aq) + Cl^-(aq) \qquad K = 3.1 \times 10^{-3}$$

The equilibrium constant K for the net reaction is the product of the equilibrium constants for the reactions added:

$$K = \frac{[Ag(NH_3)_2^+][Cl^-]}{[NH_3]^2} = (K_{sp})(K_f) = (1.8 \times 10^{-10})(1.7 \times 10^7) = 3.1 \times 10^{-3}$$

Because K is much larger than K_{sp}, the solubility equilibrium for AgCl lies much farther to the right in the presence of ammonia than it does in the absence of ammonia. The increase in the solubility of AgCl on the addition of ammonia is shown graphically in **Figure 15.15**. In general, the solubility of an ionic compound increases when the metal cation is tied up in the form of a complex ion. The quantitative effect of complex formation on the solubility of AgCl is explored in Worked Example 15.13.

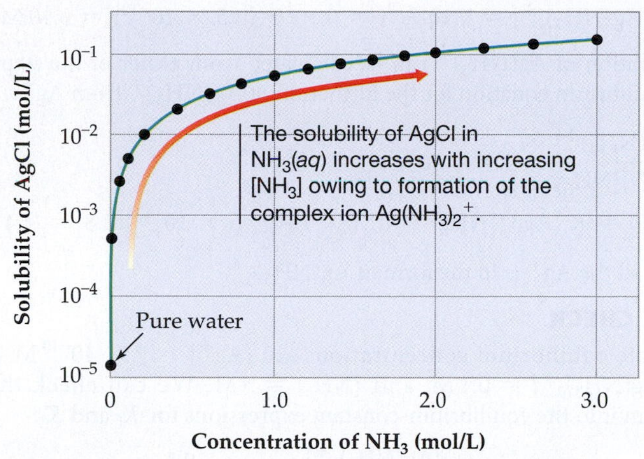

FIGURE 15.15

Plot of the solubility of AgCl in aqueous ammonia at 25 °C versus the concentration of ammonia. The solubility is plotted on a logarithmic scale.

Worked Example 15.12

Calculating the Concentrations of Complex Ions

What are the concentrations of Ag^+, $Ag(NH_3)^+$, and $Ag(NH_3)_2^+$ in a solution prepared by adding 0.10 mol of $AgNO_3$ to 1.0 L of 3.0 M NH_3? $K_f = 1.7 \times 10^7$, $K_1 = 2.1 \times 10^3$, and $K_2 = 8.1 \times 10^3$.

STRATEGY

Because K_1, K_2, and K_f for $Ag(NH_3)_2^+$ are all large numbers, nearly all the Ag^+ from $AgNO_3$ will be converted to $Ag(NH_3)_2^+$:

$$Ag^+(aq) + 2\,NH_3(aq) \rightleftharpoons Ag(NH_3)_2^+(aq) \qquad K_f = 1.7 \times 10^7$$

To calculate the concentrations, it's convenient to imagine that 100% of the Ag^+ is converted to $Ag(NH_3)_2^+$, followed by a tiny amount of back-reaction [dissociation of $Ag(NH_3)_2^+$] to give a small equilibrium concentration of Ag^+.

SOLUTION

The conversion of 0.10 mol/L of Ag^+ to $Ag(NH_3)_2^+$ consumes 0.20 mol/L of NH_3. Assuming 100% conversion to $Ag(NH_3)_2^+$, the following concentrations are obtained:

$$[Ag^+] = 0\ M$$
$$[Ag(NH_3)_2^+] = 0.10\ M$$
$$[NH_3] = 3.0 - 0.20 = 2.8\ M$$

The dissociation of x mol/L of $Ag(NH_3)_2^+$ in the back-reaction produces x mol/L of Ag^+ and $2x$ mol/L of NH_3. Therefore, the equilibrium concentrations (in mol/L) are

$$[Ag(NH_3)_2^+] = 0.10 - x$$
$$[Ag^+] = x$$
$$[NH_3] = 2.8 + 2x$$

Let's summarize our reasoning in a table under the balanced equation:

	$Ag^+(aq)$	$+\ 2\,NH_3(aq)$	$\rightleftharpoons Ag(NH_3)_2^+(aq)$
Initial concentration (M)	0.10	3.0	0
After 100% reaction (M)	0	2.8	0.10
Equilibrium concentration (M)	x	$2.8 + 2x$	$0.10 - x$

continued on next page

Substituting the equilibrium concentrations into the expression for K_f and making the approximation that x is negligible compared to 0.10 (and to 2.8) gives

$$K_1 = 1.7 \times 10^7 = \frac{[Ag(NH_3)_2{}^+]}{[Ag^+][NH_3]^2} = \frac{0.10 - x}{(x)(2.8 + 2x)^2} = \frac{0.10}{(x)(2.8)^2}$$

$$[Ag^+] = x = \frac{0.10}{(1.7 \times 10^7)(2.8)^2} = 7.5 \times 10^{-10}\ M$$

$$[Ag(NH_3)_2{}^+] = 0.10 - x = 0.10 - (7.5 \times 10^{-10}) = 0.10\ M$$

The concentration of $Ag(NH_3)^+$ can be calculated from either of the stepwise equilibria. Let's use the equilibrium equation for the formation of $Ag(NH_3)^+$ from Ag^+:

$$K_1 = \frac{[Ag(NH_3)^+]}{[Ag^+][NH_3]} = 2.1 \times 10^3$$

$$[Ag(NH_3)^+] = K_1[Ag^+][NH_3] = (2.1 \times 10^3)(7.5 \times 10^{-10})(2.8) = 4.4 \times 10^{-6}\ M$$

Thus, nearly all the Ag^+ is in the form of $Ag(NH_3)_2{}^+$.

✔ BALLPARK CHECK

The approximate equilibrium concentrations are $[Ag^+] = 7 \times 10^{-10}$ M, $[Ag(NH_3)^+] = 4 \times 10^{-6}$ M, $[Ag(NH_3)_2{}^+] = 0.1$ M, and $[NH_3] = 3$ M. We can check these results by substituting them into the equilibrium-constant expressions for K_f and K_1:

$$K_f = 1.7 \times 10^7 = \frac{[Ag(NH_3)_2{}^+]}{[Ag^+][NH_3]^2} \approx \frac{0.1}{(7 \times 10^{-10})(3)^2} = 2 \times 10^7$$

$$K_1 = 2.1 \times 10^3 = \frac{[Ag(NH_3)^+]}{[Ag^+][NH_3]} \approx \frac{4 \times 10^{-6}}{(7 \times 10^{-10})(3)} = 2 \times 10^3$$

The ballpark estimates of K_f and K_1 agree with the experimental values.

Worked Example 15.13

Exploring the Effect of Complex Formation on Solubility

Calculate the molar solubility of AgCl at 25 °C in:

(a) Pure water **(b)** 3.0 M NH_3

STRATEGY

Write the balanced equation for the dissolution reaction, and define x as the number of moles per liter of AgCl that dissolves. Then express the equilibrium concentrations in terms of x, and substitute them into the appropriate equilibrium equation. Solving for x gives the molar solubility.

SOLUTION

(a) In pure water, the solubility equilibrium is

$$AgCl(s) \rightleftharpoons Ag^+(aq) + Cl^-(aq)$$

Substituting the equilibrium concentrations (x mol/L) into the expression for K_{sp} gives

$$K_{sp} = 1.8 \times 10^{-10} = [Ag^+][Cl^-] = (x)(x)$$

$$x = \text{Molar solubility} = \sqrt{1.8 \times 10^{-10}} = 1.3 \times 10^{-5}\ M$$

(b) The balanced equation for the dissolution of AgCl in aqueous NH_3 is

$$AgCl(s) + 2\,NH_3(aq) \rightleftharpoons Ag(NH_3)_2{}^+(aq) + Cl^-(aq) \quad K = 3.1 \times 10^{-3}$$

If we define x as the number of moles per liter of AgCl that dissolves, then the saturated solution contains x mol/L of $Ag(NH_3)_2{}^+$, x mol/L of Cl^-, and $(3.0 - 2x)$ mol/L of NH_3. (We're assuming that essentially all the Ag^+ is in the form of $Ag(NH_3)_2{}^+$, as proved in Worked Example 15.12.) Substituting the equilibrium concentrations into the equilibrium equation gives

$$K = 3.1 \times 10^{-3} = \frac{[Ag(NH_3)_2{}^+][Cl^-]}{[NH_3]^2} = \frac{(x)(x)}{(3.0 - 2x)^2}$$

Taking the square root of both sides, we obtain

$$5.6 \times 10^{-2} = \frac{x}{3.0 - 2x}$$

$$x = (5.6 \times 10^{-2})(3.0 - 2x) = 0.17 - 0.11x$$

$$x = \frac{0.17}{1.11} = 0.15 \text{ M}$$

The molar solubility of AgCl in 3.0 M NH_3 is 0.15 M. Thus, AgCl is much more soluble in aqueous NH_3 than in pure water, as shown in Figure 15.15.

✔ BALLPARK CHECK

Check the calculated equilibrium concentrations by substituting them into the appropriate equilibrium equation. For part **(b)**, for example, $[Ag(NH_3)_2^+] = [Cl^-] = x = 0.15\text{ M}$ and $[NH_3] = 3.0 - 2x = 2.7\text{ M}$. Since $[NH_3]$ is 10% less than 3 M, $K = [Ag(NH_3)_2^+][Cl^-]/[NH_3]^2$ is about 20% greater than $(0.15/3)^2 = (0.05)^2 = 2.5 \times 10^{-3}$. The ballpark estimate and the experimental value of $K = 3.1 \times 10^{-3}$ agree.

PROBLEM 15.27 In an excess of $NH_3(aq)$, Cu^{2+} ion forms a deep blue complex ion, $Cu(NH_3)_4^{2+}$, which has a formation constant $K_f = 5.6 \times 10^{11}$. Calculate the concentration of Cu^{2+} in a solution prepared by adding 5.0×10^{-3} mol of $CuSO_4$ to 0.500 L of 0.40 M NH_3.

PROBLEM 15.28 Silver bromide dissolves in aqueous sodium thiosulfate, $Na_2S_2O_3$, yielding the complex ion $Ag(S_2O_3)_2^{3-}$:

$$AgBr(s) + 2\,S_2O_3^{2-}(aq) \rightleftharpoons Ag(S_2O_3)_2^{3-}(aq) + Br^-(aq) \quad K = ?$$

Calculate the equilibrium constant K for the dissolution reaction, and calculate the molar solubility of AgBr in 0.10 M $Na_2S_2O_3$. $K_{sp} = 5.4 \times 10^{-13}$ for AgBr and $K_f = 4.7 \times 10^{13}$ for $Ag(S_2O_3)_2^{3-}$.

Solubility and Amphoterism

Certain metal hydroxides, such as aluminum hydroxide (**Figure 15.16**), are soluble both in strongly acidic and in strongly basic solutions:

$$\text{In acid:} \quad Al(OH)_3(s) + 3\,H_3O^+(aq) \rightleftharpoons Al^{3+}(aq) + 6\,H_2O(l)$$

$$\text{In base:} \quad Al(OH)_3(s) + OH^-(aq) \rightleftharpoons Al(OH)_4^-(aq)$$

Such hydroxides are said to be **amphoteric** (am-fo-**tare**-ic), a term that comes from the Greek word *amphoteros,* meaning "in both ways."

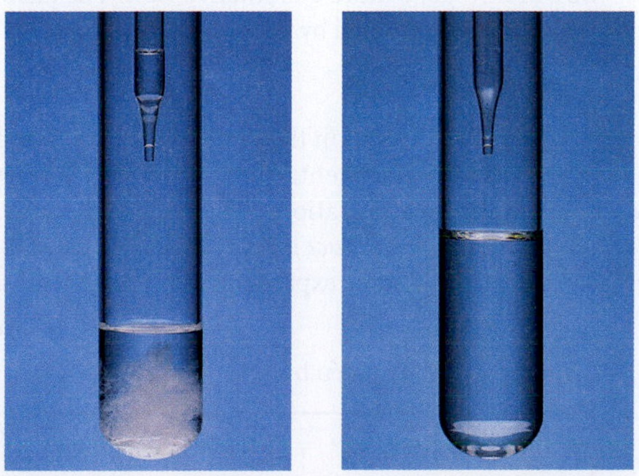

FIGURE 15.16

The amphoteric behavior of $Al(OH)_3$.

The dissolution of $Al(OH)_3$ in excess base is just a special case of the effect of complex-ion formation on solubility: $Al(OH)_3$ dissolves because excess OH^- ions convert it to the soluble complex ion $Al(OH)_4^-$ (aluminate ion). The effect of pH on the solubility of $Al(OH)_3$ is shown in **Figure 15.17**.

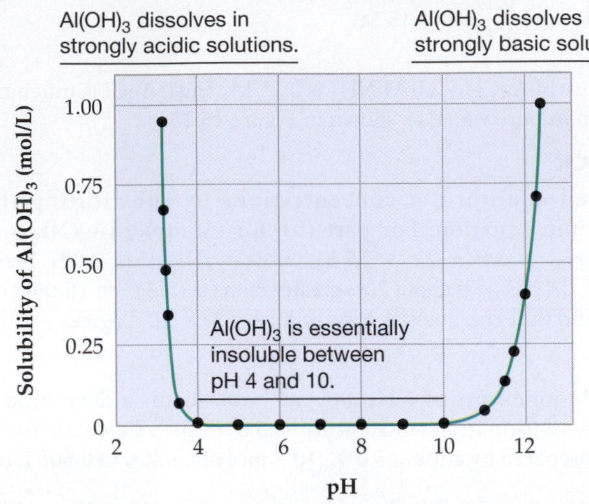

FIGURE 15.17

A plot of solubility versus pH shows that $Al(OH)_3$ is an amphoteric hydroxide.

Other examples of amphoteric hydroxides include $Zn(OH)_2$, $Cr(OH)_3$, $Sn(OH)_2$, and $Pb(OH)_2$, which react with excess OH^- ions to form the soluble complex ions $Zn(OH)_4^{2-}$ (zincate ion), $Cr(OH)_4^-$ (chromite ion), $Sn(OH)_3^-$ (stannite ion), and $Pb(OH)_3^-$ (plumbite ion), respectively. By contrast, basic hydroxides, such as $Mn(OH)_2$, $Fe(OH)_2$, and $Fe(OH)_3$, dissolve in strong acid but not in strong base.

15.13 PRECIPITATION OF IONIC COMPOUNDS

A common problem in chemistry is to decide whether a precipitate of an ionic compound will form when solutions that contain the constituent ions are mixed. For example, will CaF_2 precipitate on mixing solutions of $CaCl_2$ and NaF? In other words, will the dissolution reaction proceed in the reverse direction, from right to left?

$$CaF_2(s) \rightleftharpoons Ca^{2+}(aq) + 2\,F^-(aq)$$

We touched on this question briefly in Section 7.4 when we looked at solubility guidelines, but we can now get a more quantitative view. The answer depends on the value of the **ion product (IP)**, a number defined by the expression

$$IP = [Ca^{2+}]_t[F^-]_t^2$$

The IP is defined in the same way as K_{sp}, except that the concentrations in the expression for IP are initial concentrations—that is, arbitrary concentrations at time *t*, not necessarily equilibrium concentrations. Thus, the IP is actually a **reaction quotient Q_c** (Section 13.6), but the term *ion product* is more descriptive because, as usual, solid CaF_2 is omitted from the equilibrium-constant expression. For the general solubility equilibrium

REMEMBER...

The **reaction quotient Q_c** is defined in the same way as the equilibrium constant K_c except that the concentrations in the equilibrium-constant expression are not necessarily equilibrium values. (Section 13.6)

$$M_mX_x(s) \rightleftharpoons m\,M^{n+}(aq) + x\,X^{y-}(aq)$$

the ion product is given by

▶ **Ion product** $IP = [M^{n+}]_t^m[X^{y-}]_t^x$

If the value of IP is greater than K_{sp}, the solution is supersaturated with respect to CaF_2—a nonequilibrium situation. In that case, CaF_2 will precipitate, thus reducing the ion concentrations until IP equals K_{sp}. At that point, solubility equilibrium is reached and the solution is saturated.

In general, we only need to calculate the value of IP and then compare it with K_{sp} to decide whether an ionic compound will precipitate. Three cases arise:

1. If IP $> K_{sp}$, the solution is supersaturated and precipitation will occur.
2. If IP $= K_{sp}$, the solution is saturated and equilibrium exists already.
3. If IP $< K_{sp}$, the solution is unsaturated and precipitation will not occur.

Worked Example 15.14

Deciding Whether a Precipitate Will Form When Solutions Are Mixed

Will a precipitate form when 0.150 L of 0.10 M $Pb(NO_3)_2$ and 0.100 L of 0.20 M NaCl are mixed?

STRATEGY

Since ionic compounds are strong electrolytes, $Pb(NO_3)_2$ and NaCl exist in solution as separate cations and anions. Use the solubility guidelines discussed in Section 7.4 to decide which ions might form a precipitate, and calculate their concentrations after mixing. Then calculate the IP for the possible precipitate and compare it with the value of K_{sp}.

SOLUTION

After the two solutions are mixed, the combined solution contains Pb^{2+}, NO_3^-, Na^+, and Cl^- ions and has a volume of 0.150 L + 0.100 L = 0.250 L. Because sodium salts and nitrate salts are soluble in water, the only compound that might precipitate is $PbCl_2$, which has $K_{sp} = 1.2 \times 10^{-5}$ (Appendix C). To calculate the value of IP for $PbCl_2$, first calculate the number of moles of Pb^{2+} and Cl^- in the combined solution:

$$\text{Moles } Pb^{2+} = (0.150 \text{ L})(0.10 \text{ mol/L}) = 1.5 \times 10^{-2} \text{ mol}$$
$$\text{Moles } Cl^- = (0.100 \text{ L})(0.20 \text{ mol/L}) = 2.0 \times 10^{-2} \text{ mol}$$

Then convert moles to molar concentrations:

$$[Pb^{2+}] = \frac{1.5 \times 10^{-2} \text{ mol}}{0.250 \text{ L}} = 6.0 \times 10^{-2} \text{ M}$$

$$[Cl^-] = \frac{2.0 \times 10^{-2} \text{ mol}}{0.250 \text{ L}} = 8.0 \times 10^{-2} \text{ M}$$

The ion product is

$$\text{IP} = [Pb^{2+}]_t[Cl^-]_t^2 = (6.0 \times 10^{-2})(8.0 \times 10^{-2})^2 = 3.8 \times 10^{-4}$$

Since $K_{sp} = 1.2 \times 10^{-5}$, IP is greater than K_{sp} and $PbCl_2$ will precipitate.

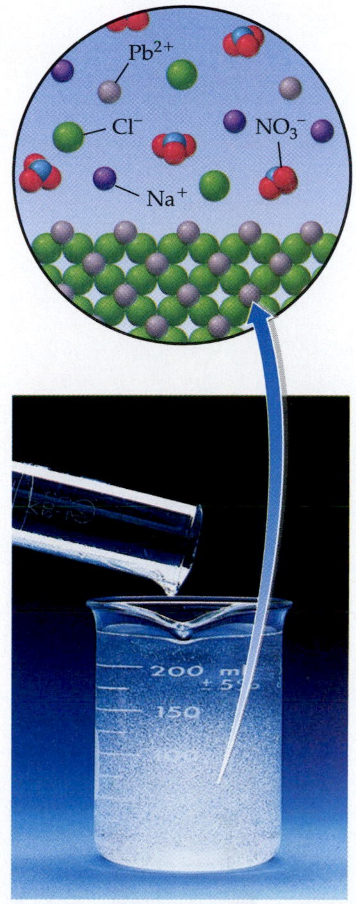

▲ When 0.150 L of 0.10 M $Pb(NO_3)_2$ and 0.100 L of 0.20 M NaCl are mixed, a white precipitate of $PbCl_2$ forms because the ion product is greater than K_{sp}.

PROBLEM 15.29 Will a precipitate form on mixing equal volumes of the following solutions?
(a) 3.0×10^{-3} M $BaCl_2$ and 2.0×10^{-3} M Na_2CO_3
(b) 1.0×10^{-5} M $Ba(NO_3)_2$ and 4.0×10^{-5} M Na_2CO_3

PROBLEM 15.30 Will a precipitate form on mixing 25 mL of 1.0×10^{-3} M $MnSO_4$, 25 mL of 1.0×10^{-3} M $FeSO_4$, and 200 mL of a buffer solution that is 0.20 M in NH_4Cl and 0.20 M in NH_3? Values of K_{sp} can be found in Appendix C.

15.14 SEPARATION OF IONS BY SELECTIVE PRECIPITATION

A convenient method for separating a mixture of ions is to add a solution that will precipitate some of the ions but not others. The anions SO_4^{2-} and Cl^-, for example, can be separated by addition of a solution of $Ba(NO_3)_2$. Insoluble $BaSO_4$ precipitates, but Cl^- remains in solution because $BaCl_2$ is soluble. Similarly, the cations Ag^+ and Zn^{2+} can be separated by addition of dilute HCl. Silver chloride, AgCl, precipitates, but Zn^{2+} stays in solution because $ZnCl_2$ is soluble.

In Section 15.15, we'll see that mixtures of metal cations, M^{2+}, can be separated into two groups by the selective precipitation of metal sulfides, MS. For example, Pb^{2+}, Cu^{2+}, and Hg^{2+}, which form very insoluble sulfides, can be separated from Mn^{2+}, Fe^{2+}, Co^{2+}, Ni^{2+}, and Zn^{2+}, which form more soluble sulfides. The separation is carried out in an acidic solution and makes use of the following solubility equilibrium:

$$MS(s) + 2\,H_3O^+(aq) \rightleftharpoons M^{2+}(aq) + H_2S(aq) + 2\,H_2O(l)$$

The equilibrium constant for this reaction, called the *solubility product in acid*, is given the symbol K_{spa}:

▶ **Solubility product in acid** $K_{spa} = \dfrac{[M^{2+}][H_2S]}{[H_3O^+]^2}$

The separation depends on adjusting the H_3O^+ concentration so that the reaction quotient Q_c exceeds K_{spa} for the very insoluble sulfides but not for the more soluble ones (**Table 15.4**).

TABLE 15.4 Solubility Products in Acid (K_{spa}) at 25 °C for Metal Sulfides

Metal Sulfide, MS	K_{spa}	Metal Sulfide, MS	K_{spa}
MnS	3×10^7	ZnS	3×10^{-2}
FeS	6×10^2	PbS	3×10^{-7}
CoS	3	CuS	6×10^{-16}
NiS	8×10^{-1}	HgS	2×10^{-32}

We use K_{spa} for metal sulfides rather than K_{sp} for two reasons. First, the ion separations are carried out in acidic solution, so use of K_{spa} is more convenient. Second, because the value of K_{a2} for H_2S ($\sim 10^{-19}$) is very small, S^{2-}, like O^{2-}, is highly basic and is not an important species in aqueous solutions. The principal sulfide-containing species in aqueous solutions are H_2S in acidic solutions and HS^- in basic solutions.

In a typical experiment, the M^{2+} concentrations are about 0.01 M, and the H_3O^+ concentration is adjusted to about 0.3 M by adding HCl. The solution is then saturated with H_2S gas, which gives an H_2S concentration of about 0.10 M. Substituting these concentrations into the equilibrium-constant expression, we find that the reaction quotient Q_c is 1×10^{-2}:

$$Q_c = \frac{[M^{2+}]_t[H_2S]_t}{[H_3O^+]_t^{\,2}} = \frac{(0.01)(0.10)}{(0.3)^2} = 1 \times 10^{-2}$$

This value of Q_c exceeds K_{spa} for PbS, CuS, and HgS (Table 15.4) but does not exceed K_{spa} for MnS, FeS, CoS, NiS, or ZnS. As a result, PbS, CuS, and HgS precipitate under these acidic conditions, but Mn^{2+}, Fe^{2+}, Co^{2+}, Ni^{2+}, and Zn^{2+} remain in solution.

▲ Adding H_2S to an acidic solution of Hg^{2+} and Ni^{2+} precipitates Hg^{2+} as black HgS but leaves green Ni^{2+} in solution.

PROBLEM 15.31 Determine whether Cd^{2+} can be separated from Zn^{2+} by bubbling H_2S through a 0.3 M HCl solution that contains 0.005 M Cd^{2+} and 0.005 M Zn^{2+}. (K_{spa} for CdS is 8×10^{-7}.)

15.15 QUALITATIVE ANALYSIS

Qualitative analysis is a procedure for identifying the ions present in an unknown solution. The ions are identified by specific chemical tests, but because one ion can interfere with the test for another, the ions must first be separated. In the traditional scheme of analysis for metal cations, some 20 cations are separated initially into five groups by selective precipitation (**Figure 15.18**).

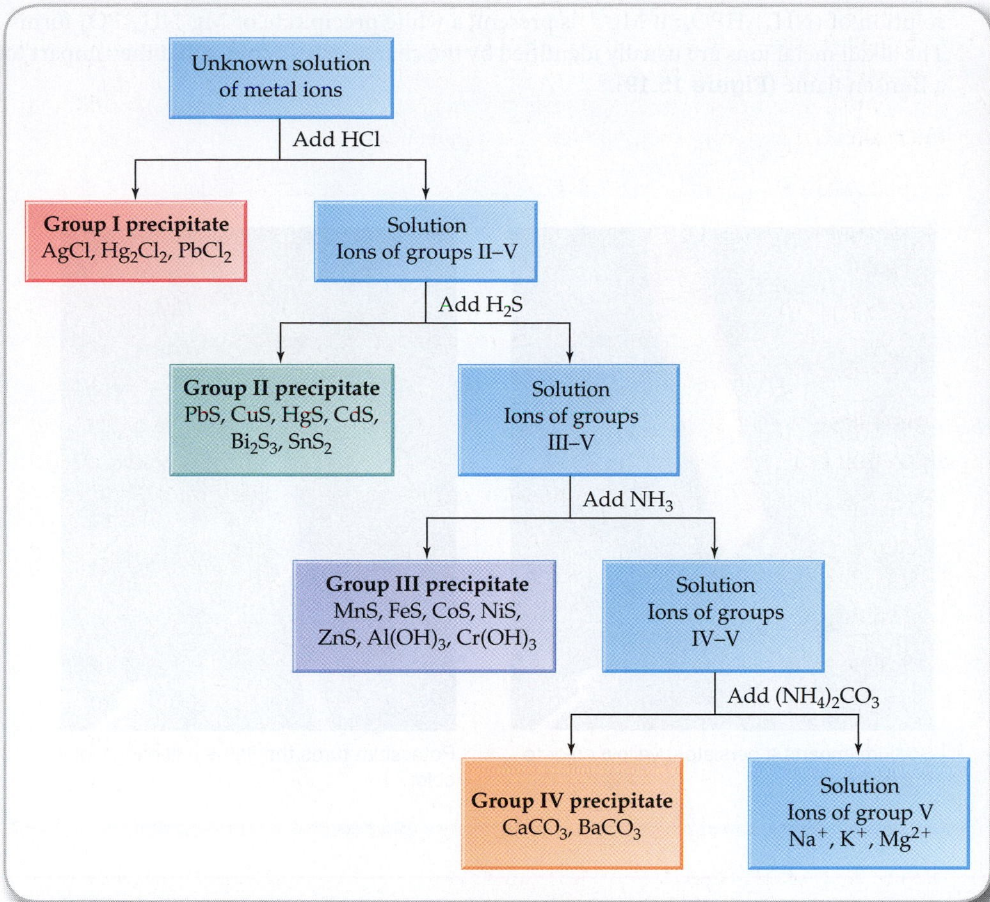

FIGURE 15.18

Flowchart for the separation of metal cations in qualitative analysis.

- **Group I: Ag^+, Hg_2^{2+}, and Pb^{2+}.** When aqueous HCl is added to the unknown solution, the cations of group I precipitate as insoluble chlorides: AgCl, Hg_2Cl_2, and $PbCl_2$. The cations of groups II–V, which form soluble chlorides, remain in solution. A small amount of the Pb^{2+} also remains in solution because $PbCl_2$ is slightly soluble.
- **Group II: Pb^{2+}, Cu^{2+}, Hg^{2+}, Cd^{2+}, Bi^{3+}, and Sn^{4+}.** After the insoluble chlorides have been removed, the solution is treated with H_2S to precipitate the cations of group II as insoluble sulfides: PbS, CuS, HgS, CdS, Bi_2S_3, and SnS_2. Because the solution is strongly acidic at this point ($[H_3O^+] \approx 0.3$ M), only the most insoluble sulfides precipitate. The acid-insoluble sulfides are then removed from the solution.
- **Group III: Mn^{2+}, Fe^{2+}, Co^{2+}, Ni^{2+}, Zn^{2+}, Al^{3+}, and Cr^{3+}.** At this point, aqueous NH_3 is added, neutralizing the acidic solution and giving an $NH_4^+-NH_3$ buffer that is slightly basic (pH ≈ 8). The decrease in $[H_3O^+]$ shifts the metal sulfide solubility equilibrium to the left, thus precipitating the 2+ cations of group III as insoluble sulfides: MnS, FeS, CoS, NiS, and ZnS. The 3+ cations precipitate from the basic solution, not as sulfides but as insoluble hydroxides: $Al(OH)_3$ and $Cr(OH)_3$.
- **Group IV: Ca^{2+} and Ba^{2+}.** After the base-insoluble sulfides and the insoluble hydroxides have been removed, the solution is treated with $(NH_4)_2CO_3$ to precipitate the cations of group IV as insoluble carbonates: $CaCO_3$ and $BaCO_3$. Magnesium carbonate does not precipitate at this point because $[CO_3^{2-}]$ in the $NH_4^+-NH_3$ buffer is maintained at a low value.
- **Group V: Na^+, K^+, and Mg^{2+}.** The only ions remaining in solution at this point are those whose chlorides, sulfides, and carbonates are soluble under the conditions of the previous reactions. Magnesium ion is separated and identified by the addition of a

solution of $(NH_4)_2HPO_4$; if Mg^{2+} is present, a white precipitate of $Mg(NH_4)PO_4$ forms. The alkali metal ions are usually identified by the characteristic colors that they impart to a Bunsen flame (**Figure 15.19**).

FIGURE 15.19

Flame tests for sodium and potassium.

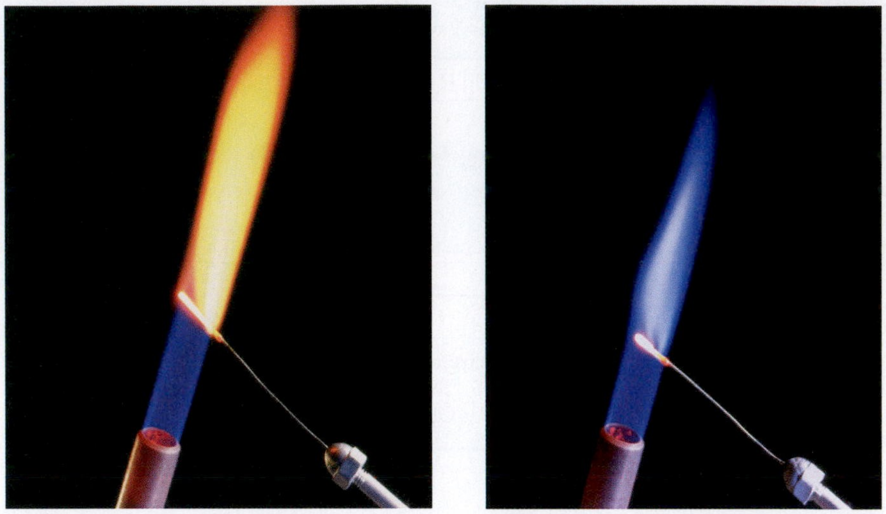

Sodium imparts a persistent yellow color to the flame.

Potassium turns the flame a fleeting violet color.

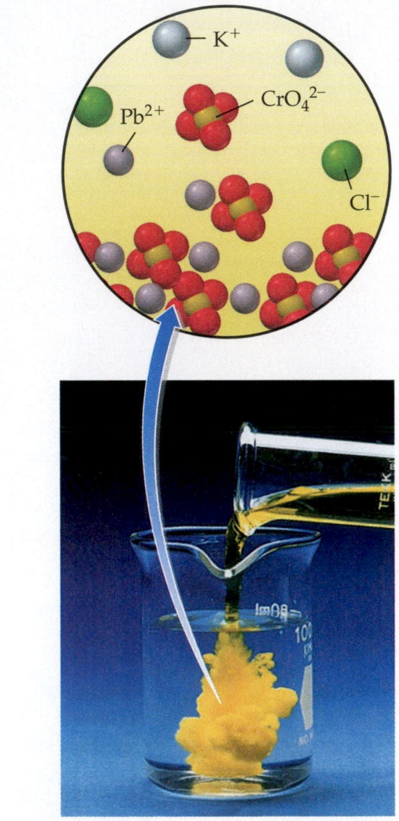

▲ When aqueous potassium chromate is added to a solution that contains Pb^{2+}, a yellow precipitate of $PbCrO_4$ forms.

Once the cations have been separated into groups, further separations and specific tests are carried out to determine the presence or absence of the ions in each group. In group I, for example, lead can be separated from silver and mercury by treating the precipitate with hot water. The more soluble $PbCl_2$ dissolves, but the less soluble AgCl and Hg_2Cl_2 do not. To test for Pb^{2+}, the solid chlorides are removed, and the solution is treated with a solution of K_2CrO_4. If Pb^{2+} is present, a yellow precipitate of $PbCrO_4$ forms.

Detailed procedures for separating and identifying all the ions can be found in general chemistry laboratory manuals. Although modern methods of metal-ion analysis employ sophisticated analytical instruments, qualitative analysis is still included in many general chemistry laboratory courses because it is an excellent vehicle for developing laboratory skills and for learning about acid–base, solubility, and complex-ion equilibria.

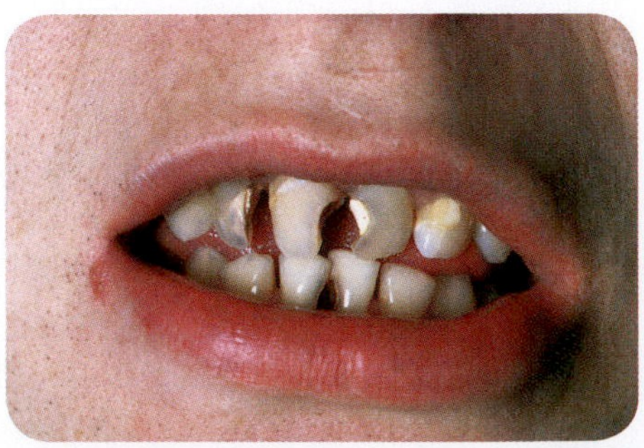

▲ Better take care of your teeth. A result of poor dental hygiene.

FYI HOW DOES FLUORIDE ION HELP TO PREVENT DENTAL CAVITIES?

Tooth enamel, the hardest substance in the human body, comes under constant chemical attack, both from the foods we eat and from bacterial infections. Made of the mineral hydroxyapatite, enamel forms the hard outer layer of the tooth where it protects the tooth from abrasion. Although it stands up well to the mechanical process of chewing, tooth enamel is nevertheless susceptible to erosion by acids. This erosion eventually leads to tooth decay and to the subsequent cavities that affect an estimated 75% of schoolchildren and virtually all adults.

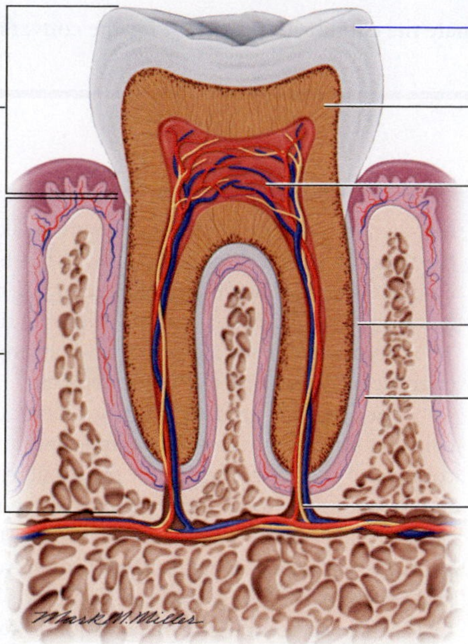

Like calcium phosphate, $Ca_3(PO_4)_2$, hydroxyapatite, $Ca_5(PO_4)_3(OH)$ is a very insoluble compound, with $K_{sp} = 2.3 \times 10^{-59}$. Despite its small K_{sp}, hydroxyapatite can be demineralized, or dissolved, by reaction with acid:

$$Ca_5(PO_4)_3(OH)(s) + 4\,H^+(aq) \underset{\text{Remineralization}}{\overset{\text{Demineralization}}{\rightleftharpoons}} 5\,Ca^{2+}(aq) + 3\,HPO_4^{\,2-}(aq) + H_2O(l)$$

Saliva, which contains relatively high concentrations Ca^{2+}, $PO_4^{\,3-}$, and $HPO_4^{\,2-}$, gives some protection to enamel, and some remineralization, or precipitation, of enamel from the ions in saliva can also take place. Thus, small erosions heal naturally before becoming cavities.

Unfortunately, this dynamic, balanced system in the mouth is overwhelmed by the presence of large amounts of acid. Both acidic beverages like sodas and bacterial metabolism of the sugars and starches left on teeth after eating are common sources of acid. When the initial decay can't be fully reversed by remineralization, small erosions, called dental caries (or, more commonly, "cavities"), reach and penetrate the softer dentin layer beneath the enamel causing considerable pain when they reach nerves in the underlying pulp.

The simplest and most common way to protect teeth from cavities is to brush them regularly, which physically removes the food that can feed bacteria. In addition, teeth can be protected by chemical treatment with a dilute solution of fluoride ion, which makes the enamel more resistant to attack by acid. Fluoride functions both by increasing the rate at which enamel remineralizes and by causing the partial conversion of hydroxyapatite to fluorapatite through exchange of F^- for OH^- in healthy enamel.

$$Ca_5(PO_4)_3(OH)(s) + F^-(aq) \rightleftharpoons Ca_5(PO_4)_3(F)(s) + OH^-(aq)$$

The K_{sp} of fluorapatite is about 10 times smaller than that of hydroxyapatite:

$$Ca_5(PO_4)_3(OH)(s) \rightleftharpoons 5\,Ca^{2+}(aq) + 3\,PO_4^{\,3-}(aq) + OH^-(aq) \qquad K_{sp} = 2.3 \times 10^{-59}$$

$$Ca_5(PO_4)_3(F)(s) \rightleftharpoons 5\,Ca^{2+}(aq) + 3\,PO_4^{\,3-}(aq) + F^-(aq) \qquad K_{sp} = 3.2 \times 10^{-60}$$

Thus, fluorapatite is less soluble than hydroxyapatite, and substitution of F^- for OH^- in tooth enamel is favored when the F^- concentration exceeds the OH^- concentration. In addition, the fluorapatite is less susceptible to dissolution by acid because F^- is a weaker base than OH^-.

Many communities now fluoridate their water supplies; just 1 mg/L (1.0 ppm) of fluoride ion in drinking water is sufficient to cause a 20–40% reduction in cavities among children. In regions where the water is not fluoridated, similar results can be obtained by brushing with toothpastes that contain a fluoride compound such as NaF, SnF_2, or Na_2PO_3F. Periodic treatments in a dentist's office with fluoride-containing gels or rinses also help to strengthen tooth enamel.

PROBLEM 15.32 Express the concentration of 1.0 ppm fluoride in units of molarity (assume that the solution density is 1.0 g/mL).

PROBLEM 15.33 Estimate the equilibrium constant for the conversion of hydroxyapatite to fluorapatite.

SUMMARY

Neutralization reactions involving a strong acid and/or a strong base have very large equilibrium constants (K_n) and proceed nearly 100% to completion. Weak acid–weak base neutralizations do not go to completion.

The **common-ion effect** is the shift in the position of an equilibrium that occurs when a substance is added that provides more of an ion already involved in the equilibrium. An example is the decrease in the percent dissociation of a weak acid on the addition of its conjugate base.

A solution of a weak acid and its conjugate base is called a **buffer solution** because it resists drastic changes in pH. **Buffer capacity**, the ability of a buffer solution to absorb small amounts of added H_3O^+ or OH^- without a significant change in pH, increases with increasing amounts of weak acid and conjugate base. The pH of a buffer solution has a value close to the pK_a ($-\log K_a$) of the weak acid and can be calculated from the **Henderson–Hasselbalch equation**:

$$\text{pH} = \text{p}K_a + \log \frac{[\text{Conjugate base}]}{[\text{Weak acid}]}$$

A **pH titration curve** is a plot of the pH of a solution as a function of the volume of base (or acid) added in the course of an acid–base titration. For a strong acid–strong base titration, the titration curve exhibits a sharp change in pH in the region of the **equivalence point**, the point at which stoichiometrically equivalent amounts of acid and base have been mixed together. For weak acid–strong base and weak base–strong acid titrations, the titration curves display a relatively flat buffer region midway to the equivalence point, a smaller change in pH in the region of the equivalence point, and a pH at the equivalence point that is not equal to 7.00. The titration curve for a diprotic acid–strong base titration exhibits two equivalence points and two buffer regions.

The **solubility product (K_{sp})** for an ionic compound is the equilibrium constant for dissolution of the compound in water. The solubility of the compound and K_{sp} are related by the equilibrium equation for the dissolution reaction. The solubility of an ionic compound is (1) suppressed by the presence of a common ion in the solution, (2) increased by decreasing the pH if the compound contains a basic anion, such as OH^-, S^{2-}, or CO_3^{2-}, and (3) increased by the presence of a Lewis base, such as NH_3, CN^-, or OH^-, that can bond to the metal cation to form a **complex ion**. The stability of a complex ion is measured by its **formation constant (K_f)**.

When solutions of soluble ionic compounds are mixed, an insoluble compound will precipitate if the **ion product (IP)** for the insoluble compound exceeds its K_{sp}. The IP is defined in the same way as K_{sp}, except that the concentrations in the expression for IP are not necessarily equilibrium concentrations. Certain metal cations can be separated by the selective precipitation of metal sulfides. Selective precipitation is important in **qualitative analysis**, a procedure for identifying the ions present in an unknown solution.

KEY WORDS

KEY EQUATIONS

- Henderson–Hasselbalch equation for calculating the pH of a buffer solution **(Section 15.4)**

$$\text{pH} = \text{p}K_a + \log \frac{[\text{Conjugate base}]}{[\text{Weak acid}]}$$

- pH at the first equivalence point in a diprotic acid–strong base titration **(Section 15.9)**

$$\text{pH at first equivalence point} = \frac{\text{p}K_{a1} + \text{p}K_{a2}}{2}$$

- Solubility product K_{sp} for the dissolution reaction $M_mX_x(s) \rightleftharpoons m\,M^{n+}(aq) + x\,X^{y-}(aq)$ **(Section 15.10)**

$$K_{sp} = [M^{n+}]^m[X^{y-}]^x$$

- Ion product for the reaction $M_mX_x(s) \rightleftharpoons m\,M^{n+}(aq) + x\,X^{y-}(aq)$ **(Section 15.13)**

$$\text{IP} = [M^{n+}]_t^{\,m}[X^{y-}]_t^{\,x}$$

where the subscript *t* denotes concentrations at an arbitrary time *t*, not necessarily at equilibrium.

- Solubility product in acid for the reaction $MS(s) + 2\,H_3O^+(aq) \rightleftharpoons M^{2+}(aq) + H_2S(aq) + 2\,H_2O(l)$ **(Section 15.14)**

$$K_{spa} = \frac{[M^{2+}][H_2S]}{[H_3O^+]^2}$$

LEARNING OUTCOMES

- Determine whether a neutralization reaction goes essentially 100% to completion, calculate the value of K_n, and determine whether the pH after neutralization is greater than, equal to, or less than 7. **(Section 15.1)** ***Problems 1, 2, 44–49***
- Calculate the effect of a common ion on concentrations, pH, and percent dissociation in a solution of a weak acid. **(Section 15.2)** ***Problems 3–5, 52–55, 58, 59***
- Calculate the pH of a buffer solution and the change in pH on addition of a strong acid or a strong base. **(Section 15.3)** ***Problems 7, 8, 64–67, 128***
- Use the Henderson–Hasselbalch equation to calculate the pH of a buffer solution and to prepare a buffer solution that has a given pH. **(Section 15.4)** ***Problems 9–11, 68–73, 132–134, 136***
- Understand the shape of the titration curves for various acid–base titrations. **(Sections 15.6–15.9)** ***Problems 42, 43, 76, 77, 84–87, 123, 124, 131, 142***
- Calculate the pH at various points in a strong acid–strong base titration. **(Section 15.6)** ***Problems 13, 14, 76–79, 86, 87***
- Calculate the pH at various points in a weak acid–strong base titration or a weak base–strong acid titration. **(Sections 15.7–15.8)** ***Problems 16, 80, 81, 86, 87, 124***
- Calculate the pH at various points in a diprotic acid–strong base titration. **(Sections 15.9)** ***Problems 18, 19, 82, 83***
- Write the equilibrium-constant expression for dissolution of an ionic compound, and calculate the value of its K_{sp}. **(Section 15.10–15.11)** ***Problems 20–22, 88–95, 126***
- Calculate ion concentrations and the solubility of an ionic compound from its K_{sp}. **(Section 15.11)** ***Problems 23, 25, 90, 91, 96, 97, 110, 137, 138***
- Understand the factors that affect the solubility of an ionic compound: (1) common-ion effect, (2) pH of the solution, (3) formation of complex ions, and (4) amphoterism. **(Section 15.12)** ***Problems 26, 98, 99, 102–105, 108, 109, 125***
- Calculate solubility in a solution that contains a common ion. **(Section 15.12)** ***Problems 25, 100, 101***
- Use the formation constant K_f to calculate ion concentrations in a solution that contains a complex ion, and calculate the solubility of an ionic compound when the cation can form a complex ion. **(Section 15.12)** ***Problems 27, 28, 106, 107, 110, 111, 138, 139***
- Calculate the ion product IP for an ionic compound, and determine whether a precipitate will form when various solutions are mixed. **(Section 15.13)** ***Problems 29, 30, 41, 112–115, 146***
- Understand selective precipitation and its use in qualitative analysis. **(Sections 15.14–15.15)** ***Problems 31, 116–122, 129, 130, 146***

CONCEPTUAL PROBLEMS

Problems 15.1–15.33 appear within the chapter.

15.34 The following pictures represent solutions that contain one or more of the compounds H_2A, NaHA, and Na_2A, where H_2A is a weak diprotic acid. (Na^+ ions and solvent water molecules have been omitted for clarity.)

(a) Which of the solutions are buffer solutions?

(b) Which solution has the greatest buffer capacity?

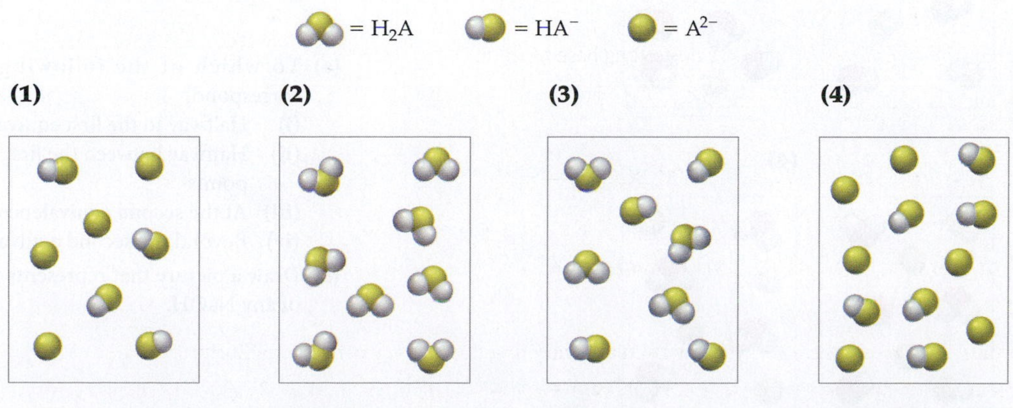

15.35 The following pictures represent solutions that contain a weak acid HA (pK_a = 6.0) and its sodium salt NaA. (Na^+ ions and solvent water molecules have been omitted for clarity.)

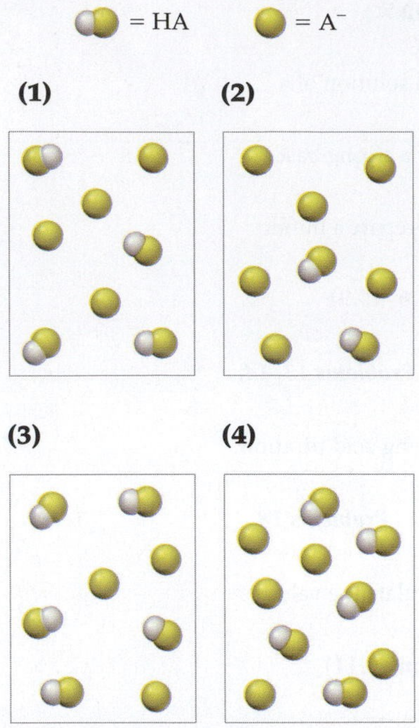

(a) Which solution has the highest pH? Which has the lowest pH?

(b) Draw a picture that represents the equilibrium state of solution (1) after the addition of two H_3O^+ ions.

(c) Draw a picture that represents the equilibrium state of solution (1) after the addition of two OH^- ions.

15.36 The strong acid HA is mixed with an equal molar amount of aqueous NaOH. Which of the following pictures represents the equilibrium state of the solution? (Na^+ ions and solvent water molecules have been omitted for clarity.)

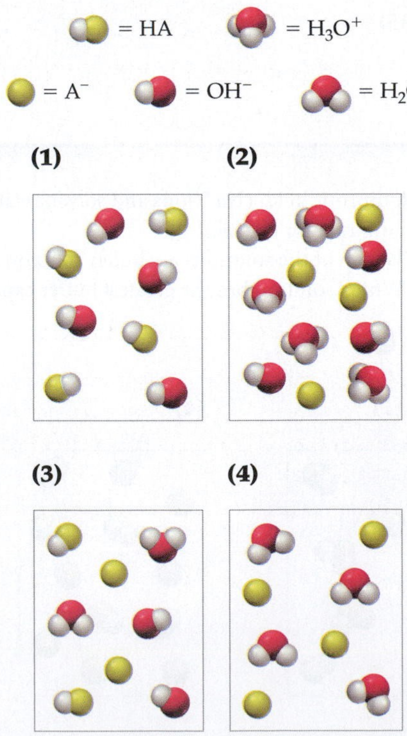

15.37 The following pictures represent solutions at various stages in the titration of a weak diprotic acid H_2A with aqueous NaOH. (Na^+ ions and water molecules have been omitted for clarity.)

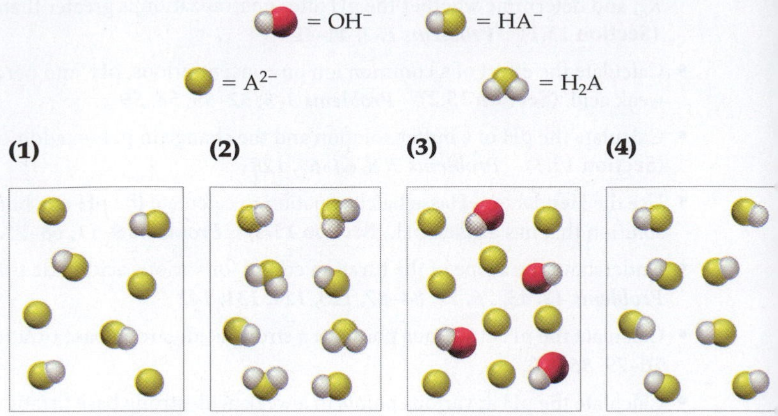

(a) To which of the following stages do solutions 1–4 correspond?

(i) Halfway to the first equivalence point
(ii) At the first equivalence point
(iii) Halfway between the first and second equivalence points
(iv) Beyond the second equivalence point

(b) Which solution has the highest pH? Which has the lowest pH?

15.38 The following pictures represent solutions at various stages in the titration of sulfuric acid H_2A ($A^{2-} = SO_4^{\ 2-}$) with aqueous NaOH. (Na^+ ions and water molecules have been omitted for clarity.)

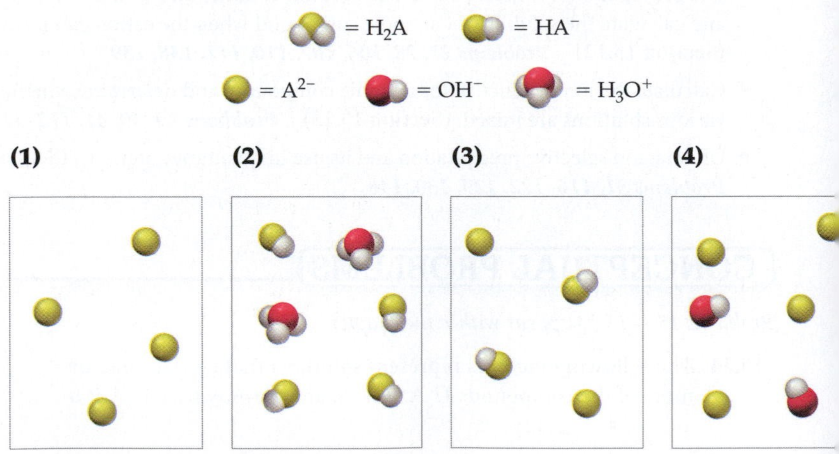

(a) To which of the following stages do solutions 1–4 correspond?

(i) Halfway to the first equivalence point
(ii) Halfway between the first and second equivalence points
(iii) At the second equivalence point
(iv) Beyond the second equivalence point

(b) Draw a picture that represents the solution prior to addition of any NaOH.

15.39 The following pictures represent solutions at various stages in the titration of a weak base B with aqueous HCl. (Cl^- ions and solvent water molecules have been omitted for clarity.)

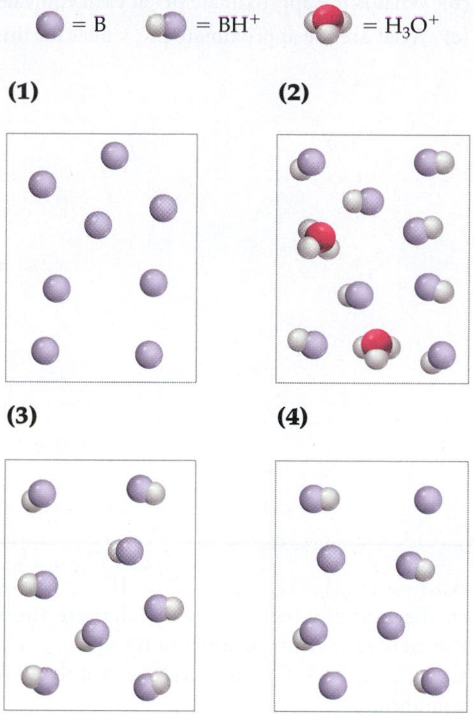

(a) To which of the following stages do solutions 1–4 correspond?

- **(i)** The initial solution before addition of any HCl
- **(ii)** Halfway to the equivalence point
- **(iii)** At the equivalence point
- **(iv)** Beyond the equivalence point

(b) Is the pH at the equivalence point more or less than 7?

15.40 The following pictures represent solutions of AgCl, which also may contain ions other than Ag^+ and Cl^- that are not shown. If solution 1 is a saturated solution of AgCl, classify solutions 2–4 as unsaturated, saturated, or supersaturated.

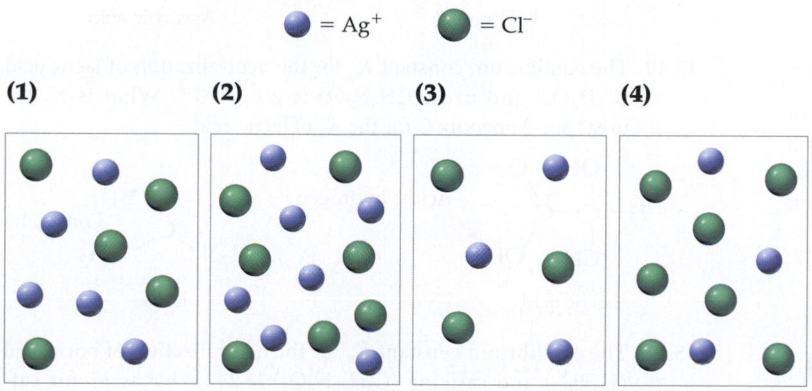

15.41 The following pictures represent solutions of Ag_2CrO_4, which also may contain ions other than Ag^+ and CrO_4^{2-} that are not shown. Solution 1 is in equilibrium with solid Ag_2CrO_4. Will a precipitate of solid Ag_2CrO_4 form in solutions 2–4? Explain.

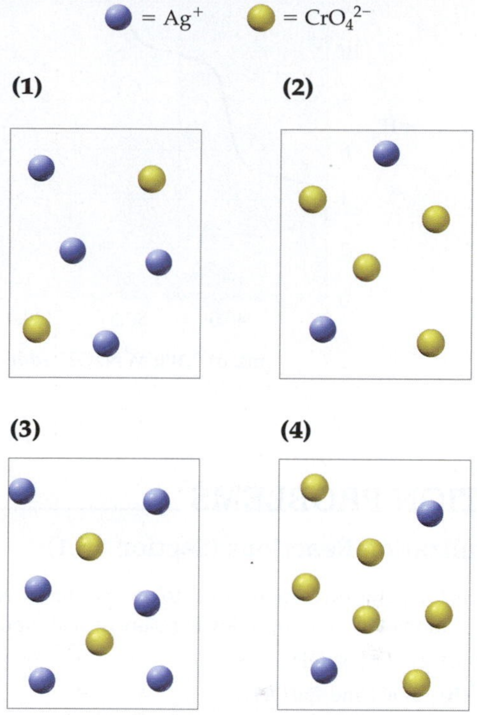

15.42 The following plot shows two pH titration curves, each representing the titration of 50.0 mL of 0.100 M acid with 0.100 M NaOH:

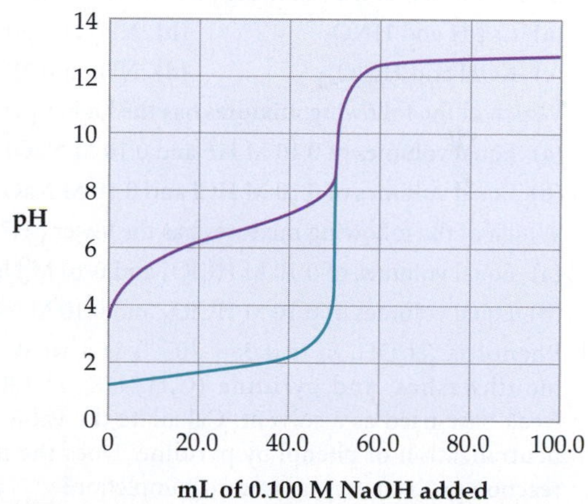

(a) Which of the two curves represents the titration of a strong acid? Which represents a weak acid?

(b) What is the approximate pH at the equivalence point for each of the acids?

(c) What is the approximate pK_a of the weak acid?

15.43 The following pH titration curve applies to the titration of 40.0 mL of a 0.100 M solution of a diprotic acid with 0.100 M NaOH:

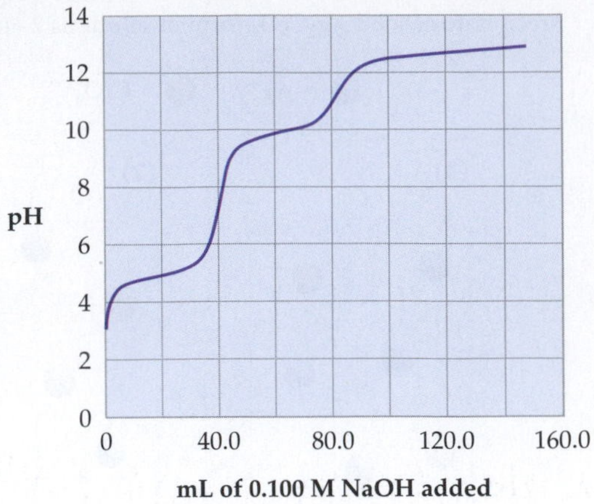

(a) How many equivalence points are present, and how many milliliters of base are required to reach each equivalence point?

(b) What is the approximate pH at each equivalence point?

(c) What are the approximate pK_a values for this acid?

SECTION PROBLEMS

Neutralization Reactions (Section 15.1)

15.44 Is the pH greater than, equal to, or less than 7 after the neutralization of each of the following pairs of acids and bases?

(a) HI and NaOH

(b) HOCl and $Ba(OH)_2$

(c) HNO_3 and aniline ($C_6H_5NH_2$)

(d) Benzoic acid ($C_6H_5CO_2H$) and KOH

15.45 Is the pH greater than, equal to, or less than 7 after the neutralization of each of the following pairs of acids and bases?

(a) CsOH and HNO_2 **(b)** NH_3 and HBr

(c) KOH and $HClO_4$ **(d)** NH_3 and HOBr

15.46 Which of the following mixtures has the higher pH?

(a) Equal volumes of 0.10 M HF and 0.10 M NaOH

(b) Equal volumes of 0.10 M HCl and 0.10 M NaOH

15.47 Which of the following mixtures has the lower pH?

(a) Equal volumes of 0.10 M $HClO_4$ and 0.10 M NaOH

(b) Equal volumes of 0.10 M $HClO_4$ and 0.10 M NH_3

15.48 Phenol (C_6H_5OH, $K_a = 1.3 \times 10^{-10}$) is a weak acid used in mouthwashes, and pyridine (C_5H_5N, $K_b = 1.8 \times 10^{-9}$) is a weak base used as a solvent. Calculate the value of K_n for the neutralization of phenol by pyridine. Does the neutralization reaction proceed very far toward completion?

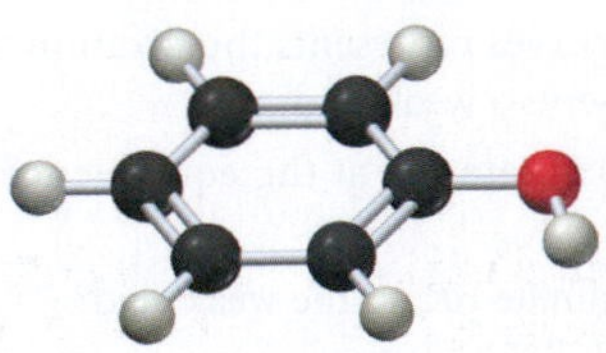

Phenol

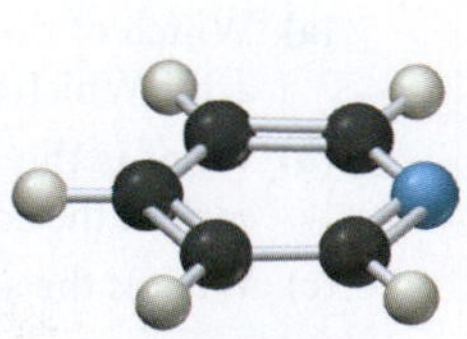

Pyridine

15.49 Aniline ($C_6H_5NH_2$, $K_b = 4.3 \times 10^{-10}$) is a weak base used in the manufacture of dyes. Calculate the value of K_n for the neutralization of aniline by vitamin C (ascorbic acid, $C_6H_8O_6$, $K_a = 8.0 \times 10^{-5}$). Does much aniline remain at equilibrium?

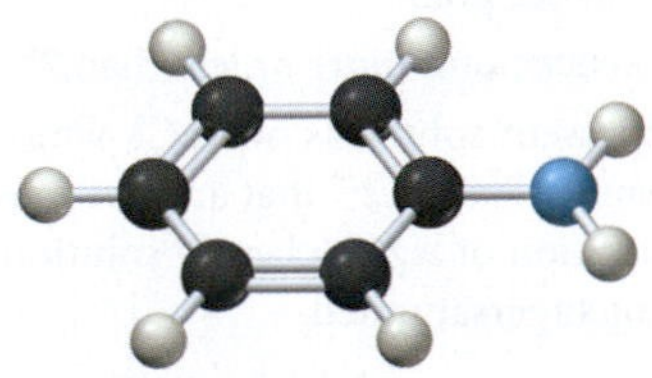

Aniline

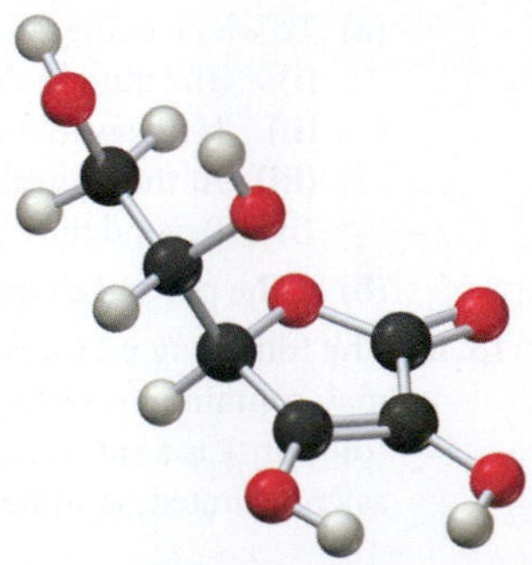

Ascorbic acid

15.50 The equilibrium constant K_n for the neutralization of lactic acid ($C_3H_6O_3$) and urea (CH_4N_2O) is 2.1×10^{-4}. What is K_b for urea? See Appendix C for the K_a of lactic acid.

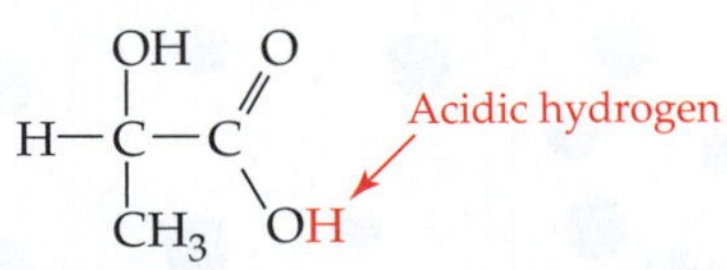

Lactic acid

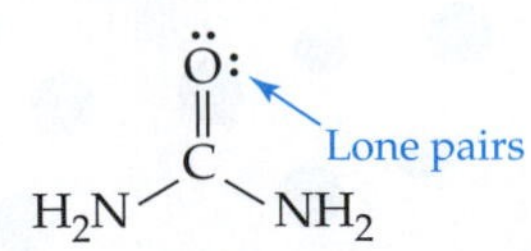

Urea

15.51 The equilibrium constant K_n for the neutralization of boric acid (H_3BO_3) and caffeine ($C_8H_{10}N_4O_2$) is 24. What is K_b for caffeine? See Appendix C for the K_a of boric acid.

The Common-Ion Effect (Section 15.2)

15.52 Does the pH increase, decrease, or remain the same on the addition of each of the following?

(a) LiF to an HF solution

(b) KI to an HI solution

(c) NH_4Cl to an NH_3 solution

15.53 Does the pH increase, decrease, or remain the same on the addition of each of the following?

(a) NH_4NO_3 to an NH_3 solution

(b) Na_2CO_3 to an $NaHCO_3$ solution

(c) $NaClO_4$ to an NaOH solution

15.54 Calculate the pH of a solution that is 0.25 M in HF and 0.10 M in NaF.

15.55 Calculate the pH of a solution prepared by mixing equal volumes of 0.20 M methylamine (CH_3NH_2, $K_b = 3.7 \times 10^{-4}$) and 0.60 M CH_3NH_3Cl.

15.56 The pH of a solution of HN_3 ($K_a = 1.9 \times 10^{-5}$) and NaN_3 is 4.86. What is the molarity of NaN_3 if the molarity of HN_3 is 0.016 M?

15.57 The pH of a solution of NH_3 and NH_4Br is 8.90. What is the molarity of NH_4Br if the molarity of NH_3 is 0.016 M?

15.58 Calculate the percent dissociation of 0.10 M hydrazoic acid (HN_3, $K_a = 1.9 \times 10^{-5}$). Recalculate the percent dissociation of 0.10 M HN_3 in the presence of 0.10 M HCl, and explain the change.

15.59 Calculate the pH of 100 mL of 0.30 M NH_3 before and after the addition of 4.0 g of NH_4NO_3, and account for the change. Assume that the volume remains constant.

Buffer Solutions (Sections 15.3–15.4)

15.60 Which of the following gives a buffer solution when equal volumes of the two solutions are mixed?

(a) 0.10 M HF and 0.10 M NaF

(b) 0.10 M HF and 0.10 M NaOH

(c) 0.20 M HF and 0.10 M NaOH

(d) 0.10 M HCl and 0.20 M NaF

15.61 Which of the following gives a buffer solution when equal volumes of the two solutions are mixed?

(a) 0.10 M NH_3 and 0.10 M HCl

(b) 0.20 M NH_3 and 0.10 M HCl

(c) 0.10 M NH_4Cl and 0.10 M NH_3

(d) 0.20 M NH_4Cl and 0.10 M NaOH

15.62 Which of the following solutions has the greater buffer capacity: (a) 100 mL of 0.30 M HNO_2–0.30 M $NaNO_2$ or (b) 100 mL of 0.10 M HNO_2–0.10 M $NaNO_2$? Explain.

15.63 Which of the following solutions has the greater buffer capacity: (a) 50 mL of 0.20 M NH_4Br–0.30 M NH_3 or (b) 50 mL of 0.40 M NH_4Br–0.60 M NH_3? Explain.

15.64 Calculate the pH of a buffer solution that is 0.20 M in HCN and 0.12 M in NaCN. Will the pH change if the solution is diluted by a factor of 2? Explain.

15.65 Calculate the pH of a buffer solution prepared by dissolving 4.2 g of $NaHCO_3$ and 5.3 g of Na_2CO_3 in 0.20 L of water. Will the pH change if the solution volume is increased by a factor of 10? Explain.

15.66 Calculate the pH of 0.250 L of a 0.36 M formic acid–0.30 M sodium formate buffer before and after the addition of **(a)** 0.0050 mol of NaOH and **(b)** 0.0050 mol of HCl. Assume that the volume remains constant.

15.67 Calculate the pH of 0.375 L of a 0.18 M acetic acid–0.29 M sodium acetate buffer before and after the addition of **(a)** 0.0060 mol of KOH and **(b)** 0.0060 mol of HBr. Assume that the volume remains constant.

15.68 Use the Henderson–Hasselbalch equation to calculate the pH of a buffer solution that is 0.25 M in formic acid (HCO_2H) and 0.50 M in sodium formate (HCO_2Na).

15.69 Use the Henderson–Hasselbalch equation to calculate the ratio of H_2CO_3 to HCO_3^- in blood having a pH of 7.40.

15.70 In what volume ratio should you mix 1.0 M solutions of NH_4Cl and NH_3 to produce a buffer solution having pH = 9.80?

15.71 Give a recipe for preparing a CH_3CO_2H–CH_3CO_2Na buffer solution that has pH = 4.44.

15.72 You need a buffer solution that has pH = 7.00. Which of the following buffer systems should you choose? Explain.

(a) H_3PO_4 and $H_2PO_4^-$

(b) $H_2PO_4^-$ and HPO_4^{2-}

(c) HPO_4^{2-} and PO_4^{3-}

15.73 Which of the following conjugate acid–base pairs should you choose to prepare a buffer solution that has pH = 4.50? Explain.

(a) HSO_4^- and SO_4^{2-}

(b) HOCl and OCl^-

(c) $C_6H_5CO_2H$ and $C_6H_5CO_2^-$

15.74 What is the K_a of the amino acid leucine if it is 34.5% dissociated at pH = 9.46?

15.75 What is the K_a of the amino acid serine if it is 86.3% dissociated at pH = 9.95?

pH Titration Curves (Sections 15.5–15.9)

15.76 Consider the titration of 60.0 mL of 0.150 M HNO_3 with 0.450 M NaOH.

(a) How many millimoles of HNO_3 are present at the start of the titration?

(b) How many milliliters of NaOH are required to reach the equivalence point?

(c) What is the pH at the equivalence point?

(d) Sketch the general shape of the pH titration curve.

15.77 Make a rough plot of pH versus milliliters of acid added for the titration of 50.0 mL of 1.0 M NaOH with 1.0 M HCl. Indicate the pH at the following points, and tell how many milliliters of acid are required to reach the equivalence point.

(a) At the start of the titration

(b) At the equivalence point

(c) After the addition of a large excess of acid

15.78 Consider the titration of 25.0 mL of 0.125 M HCl with 0.100 M KOH. Calculate the pH after the addition of each of the following volumes of base:

(a) 3.0 mL (b) 20 mL (c) 65 mL

15.79 Consider the titration of 50.0 mL of 0.116 M NaOH with 0.0750 M HCl. Calculate the pH after the addition of each of the following volumes of acid:

(a) 5.0 mL (b) 50 mL (c) 0.10 L

15.80 Consider the titration of 40.0 mL of 0.250 M HF with 0.200 M NaOH. How many milliliters of base are required to reach the equivalence point? Calculate the pH at each of the following points:

(a) After the addition of 10.0 mL of base
(b) Halfway to the equivalence point
(c) At the equivalence point
(d) After the addition of 80.0 mL of base

15.81 A 100.0 mL sample of 0.100 M methylamine (CH_3NH_2, $K_b = 3.7 \times 10^{-4}$) is titrated with 0.250 M HNO_3. Calculate the pH after the addition of each of the following volumes of acid:

(a) 0.0 mL (b) 20.0 mL (c) 40.0 mL (d) 60.0 mL

15.82 Consider the titration of 50.0 mL of a 0.100 M solution of the protonated form of the amino acid alanine (H_2A^+; $K_{a1} = 4.6 \times 10^{-3}$, $K_{a2} = 2.0 \times 10^{-10}$) with 0.100 M NaOH. Calculate the pH after the addition of each of the following volumes of base:

(a) 10.0 mL (b) 25.0 mL (c) 50.0 mL (d) 75.0 mL
(e) 100.0 mL

15.83 Consider the titration of 25.0 mL of 0.0200 M H_2CO_3 with 0.0250 M KOH. Calculate the pH after the addition of each of the following volumes of base:

(a) 10.0 mL (b) 20.0 mL (c) 30.0 mL (d) 40.0 mL
(e) 50.0 mL

15.84 The equivalence point was reached in titrations of three unknown acids at pH 9.16 (acid A), 8.88 (acid B), and 8.19 (acid C).

(a) Which is the strongest acid?
(b) Which is the weakest acid?

15.85 The equivalence point was reached in titrations of three unknown bases at pH 5.53 (base A), 4.11 (base B), and 6.00 (base C).

(a) Which is the strongest base?
(b) Which is the weakest base?

15.86 What is the pH at the equivalence point for the titration of 0.10 M solutions of the following acids and bases, and which of the indicators in Figure 14.5 would be suitable for each titration?

(a) HNO_2 and NaOH
(b) HI and NaOH
(c) CH_3NH_2 (methylamine) and HCl

15.87 What is the pH at the equivalence point for the titration of 0.20 M solutions of the following acids and bases? Which of the indicators in Figure 14.5 would be suitable for each titration?

(a) $C_5H_{11}N$ (piperidine) and HNO_3
(b) $NaHSO_3$ and NaOH
(c) $Ba(OH)_2$ and HBr

Solubility Equilibria (Sections 15.10–15.11)

15.88 For each of the following compounds, write a balanced net ionic equation for the dissolution of the compound in water, and write the equilibrium-constant expression for K_{sp}:

(a) Ag_2CO_3 (b) $PbCrO_4$ (c) $Al(OH)_3$ (d) Hg_2Cl_2

15.89 For each of the following, write the equilibrium-constant expression for K_{sp}:

(a) $Ca(OH)_2$ (b) Ag_3PO_4 (c) $BaCO_3$
(d) $Ca_5(PO_4)_3OH$

15.90 A particular saturated solution of PbI_2 has $[Pb^{2+}] = 5.0 \times 10^{-3}$ M and $[I^-] = 1.3 \times 10^{-3}$ M.

(a) What is the value of K_{sp} for PbI_2?
(b) What is $[I^-]$ in a saturated solution of PbI_2 that has $[Pb^{2+}] = 2.5 \times 10^{-4}$ M?
(c) What is $[Pb^{2+}]$ in a saturated solution that has $[I^-] = 2.5 \times 10^{-4}$ M?

15.91 A particular saturated solution of $Ca_3(PO_4)_2$ has $[Ca^{2+}] = [PO_4^{3-}] = 2.9 \times 10^{-7}$ M.

(a) What is the value of K_{sp} for $Ca_3(PO_4)_2$?
(b) What is $[Ca^{2+}]$ in a saturated solution of $Ca_3(PO_4)_2$ that has $[PO_4^{3-}] = 0.010$ M?
(c) What is $[PO_4^{3-}]$ in a saturated solution that has $[Ca^{2+}] = 0.010$ M?

15.92 If a saturated solution prepared by dissolving Ag_2CO_3 in water has $[Ag^+] = 2.56 \times 10^{-4}$ M, what is the value of K_{sp} for Ag_2CO_3?

15.93 If a saturated aqueous solution of the shock-sensitive compound lead(II) azide, $Pb(N_3)_2$, has $[Pb^{2+}] = 8.5 \times 10^{-4}$ M, what is the value of K_{sp} for $Pb(N_3)_2$?

15.94 Use the following solubility data to calculate a value of K_{sp} for each compound:

(a) SrF_2; 1.03×10^{-3} M
(b) CuI; 1.05×10^{-6} M
(c) MgC_2O_4; 0.094 g/L
(d) $Zn(CN)_2$; 4.95×10^{-4} g/L

15.95 Use the following solubility data to calculate a value of K_{sp} for each compound:

(a) $CdCO_3$; 1.0×10^{-6} M
(b) $Ca(OH)_2$; 1.06×10^{-2} M
(c) $PbBr_2$; 4.34 g/L
(d) $BaCrO_4$; 2.8×10^{-3} g/L

15.96 Use the values of K_{sp} in Appendix C to calculate the molar solubility of the following compounds:

(a) $BaCrO_4$ (b) $Mg(OH)_2$ (c) Ag_2SO_3

15.97 Use the values of K_{sp} in Appendix C to calculate the solubility of the following compounds (in g/L):

(a) Ag_2CO_3 (b) CuBr (c) $Cu_3(PO_4)_2$

Factors That Affect Solubility (Section 15.12)

15.98 Use Le Châtelier's principle to explain the following changes in the solubility of Ag_2CO_3 in water:

(a) Decrease on addition of $AgNO_3$
(b) Increase on addition of HNO_3
(c) Decrease on addition of Na_2CO_3
(d) Increase on addition of NH_3

15.99 Use Le Châtelier's principle to predict whether the solubility of BaF_2 will increase, decrease, or remain the same on addition of each of the following substances:

(a) HCl (b) KF (c) $NaNO_3$
(d) $Ba(NO_3)_2$

15.100 Calculate the molar solubility of $PbCrO_4$ in:

(a) Pure water (b) 1.0×10^{-3} M K_2CrO_4

15.101 Calculate the molar solubility of SrF_2 in:

(a) 0.010 M $Sr(NO_3)_2$
(b) 0.010 M NaF

15.102 Which of the following compounds are more soluble in acidic solution than in pure water? Write a balanced net ionic equation for each dissolution reaction.

(a) $AgBr$ (b) $CaCO_3$
(c) $Ni(OH)_2$ (d) $Ca_3(PO_4)_2$

15.103 Which of the following compounds are more soluble in acidic solution than in pure water? Write a balanced net ionic equation for each dissolution reaction.

(a) MnS (b) $Fe(OH)_3$
(c) $AgCl$ (d) $BaCO_3$

15.104 Is the solubility of $Zn(OH)_2$ increased, decreased, or unchanged on addition of each of the following substances? Write a balanced net ionic equation for each dissolution reaction. (See Appendix C.6 for formulas of complex ions.)

(a) HCl (b) KOH (c) $NaCN$

15.105 Is the solubility of $Fe(OH)_3$ increased, decreased, or unchanged on addition of each of the following substances? Write a balanced net ionic equation for each dissolution reaction. (See Appendix C.6 for formulas of complex ions.)

(a) $HBr(aq)$ (b) $NaOH(aq)$ (c) $KCN(aq)$

15.106 Silver ion reacts with excess CN^- to form a colorless complex ion, $Ag(CN)_2^-$, which has a formation constant $K_f = 3.0 \times 10^{20}$. Calculate the concentration of Ag^+ in a solution prepared by mixing equal volumes of 2.0×10^{-3} M $AgNO_3$ and 0.20 M NaCN.

15.107 Dissolution of 5.0×10^{-3} mol of $Cr(OH)_3$ in 1.0 L of 1.0 M NaOH gives a solution of the complex ion $Cr(OH)_4^-$ ($K_f = 8 \times 10^{29}$). What fraction of the chromium in such a solution is present as uncomplexed Cr^{3+}?

15.108 Write a balanced net ionic equation for each of the following dissolution reactions, and use the appropriate K_{sp} and K_f values in Appendix C to calculate the equilibrium constant for each.

(a) AgI in aqueous NaCN to form $Ag(CN)_2^-$
(b) $Al(OH)_3$ in aqueous NaOH to form $Al(OH)_4^-$
(c) $Zn(OH)_2$ in aqueous NH_3 to form $Zn(NH_3)_4^{2+}$

15.109 Write a balanced net ionic equation for each of the following dissolution reactions, and use the appropriate K_{sp} and K_f values in Appendix C to calculate the equilibrium constant for each.

(a) $Zn(OH)_2$ in aqueous NaOH to form $Zn(OH)_4^{2-}$
(b) $Cu(OH)_2$ in aqueous NH_3 to form $Cu(NH_3)_4^{2+}$
(c) AgBr in aqueous NH_3 to form $Ag(NH_3)_2^+$

15.110 Calculate the molar solubility of AgI in:

(a) Pure water
(b) 0.10 M NaCN; K_f for $Ag(CN)_2^-$ is 3.0×10^{20}

15.111 Calculate the molar solubility of $Cr(OH)_3$ in 0.50 M NaOH; K_f for $Cr(OH)_4^-$ is 8×10^{29}.

Precipitation; Qualitative Analysis (Sections 15.13–15.15)

15.112 What compound, if any, will precipitate when 80 mL of 1.0×10^{-5} M $Ba(OH)_2$ is added to 20 mL of 1.0×10^{-5} M $Fe_2(SO_4)_3$?

15.113 "Hard" water contains alkaline earth cations such as Ca^{2+}, which reacts with CO_3^{2-} to form insoluble deposits of $CaCO_3$. Will a precipitate of $CaCO_3$ form if a 250 mL sample of hard water having $[Ca^{2+}] = 8.0 \times 10^{-4}$ M is treated with the following?

(a) 0.10 mL of 2.0×10^{-3} M Na_2CO_3
(b) 10 mg of solid Na_2CO_3

15.114 The pH of a sample of hard water (Problem 15.113) having $[Mg^{2+}] = 2.5 \times 10^{-4}$ M is adjusted to pH 10.80. Will $Mg(OH)_2$ precipitate?

15.115 In qualitative analysis, Al^{3+} and Mg^{2+} are separated in an NH_4^+–NH_3 buffer having pH ≈ 8. Assuming cation concentrations of 0.010 M, show why $Al(OH)_3$ precipitates but $Mg(OH)_2$ does not.

15.116 Can Fe^{2+} be separated from Sn^{2+} by bubbling H_2S through a 0.3 M HCl solution that contains 0.01 M Fe^{2+} and 0.01 M Sn^{2+}? A saturated solution of H_2S has $[H_2S] \approx 0.10$ M. Values of K_{spa} are 6×10^2 for FeS and 1×10^{-5} for SnS.

15.117 Can Co^{2+} be separated from Zn^{2+} by bubbling H_2S through a 0.3 M HCl solution that contains 0.01 M Co^{2+} and 0.01 M Zn^{2+}? A saturated solution of H_2S has $[H_2S] \approx 0.10$ M. Values of K_{spa} are 3 for CoS and 3×10^{-2} for ZnS.

15.118 Will FeS precipitate in a solution that is 0.10 M in $Fe(NO_3)_2$, 0.4 M in HCl, and 0.10 M in H_2S? Will FeS precipitate if the pH of the solution is adjusted to pH 8 with an NH_4^+–NH_3 buffer? $K_{spa} = 6 \times 10^2$ for FeS.

15.119 Will CoS precipitate in a solution that is 0.10 M in $Co(NO_3)_2$, 0.5 M in HCl, and 0.10 M in H_2S? Will CoS precipitate if the pH of the solution is adjusted to pH 8 with an NH_4^+–NH_3 buffer? $K_{spa} = 3$ for CoS.

15.120 Using the qualitative analysis flowchart in Figure 15.18, tell how you could separate the following pairs of ions:

(a) Ag^+ and Cu^{2+} (b) Na^+ and Ca^{2+}
(c) Mg^{2+} and Mn^{2+} (d) K^+ and Cr^{3+}

15.121 Give a method for separating the following pairs of ions by the addition of no more than two substances:

(a) Hg_2^{2+} and Co^{2+} (b) Na^+ and Mg^{2+}
(c) Fe^{2+} and Hg^{2+} (d) Ba^{2+} and Pb^{2+}

CHAPTER PROBLEMS

15.122 Assume that you have three white solids: NaCl, KCl, and $MgCl_2$. What tests could you do to tell which is which?

15.123 On the same graph, sketch pH titration curves for the titration of (1) a strong acid with a strong base and (2) a weak acid with a strong base. How do the two curves differ with respect to the following?

(a) The initial pH
(b) The pH in the region between the start of the titration and the equivalence point
(c) The pH at the equivalence point
(d) The pH beyond the equivalence point
(e) The volume of base required to reach the equivalence point

15.124 Consider the titration of 50.0 mL of 0.010 M HA ($K_a = 1.0 \times 10^{-4}$) with 0.010 M NaOH.

(a) Sketch the pH titration curve, and label the equivalence point.

(b) How many milliliters of 0.010 M NaOH are required to reach the equivalence point?

(c) Is the pH at the equivalence point greater than, equal to, or less than 7?

(d) What is the pH exactly halfway to the equivalence point?

15.125 Consider saturated solutions of the slightly soluble salts AgBr and $BaCO_3$.

(a) Is the solubility of AgBr increased, decreased, or unaffected by the addition of each of the following substances?

(i) HBr **(ii)** HNO_3
(iii) $AgNO_3$ **(iv)** NH_3

(b) Is the solubility of $BaCO_3$ increased, decreased, or unaffected by the addition of each of the following substances?

(i) HNO_3 **(ii)** $Ba(NO_3)_2$
(iii) Na_2CO_3 **(iv)** CH_3CO_2H

15.126 A saturated solution of $Mg(OH)_2$ in water has pH = 10.35. Calculate K_{sp} for $Mg(OH)_2$.

15.127 Consider a buffer solution that contains equal concentrations of $H_2PO_4^-$ and HPO_4^{2-}. Will the pH increase, decrease, or remain the same when each of the following substances is added?

(a) Na_2HPO_4 **(b)** HBr **(c)** KOH
(d) KI **(e)** H_3PO_4 **(f)** Na_3PO_4

15.128 Calculate the concentrations of NH_4^+ and NH_3 and the pH in a solution prepared by mixing 20.0 g of NaOH and 0.500 L of 1.5 M NH_4Cl. Assume that the volume remains constant.

15.129 In qualitative analysis, Ag^+, Hg_2^{2+}, and Pb^{2+} are separated from other cations by the addition of HCl. Calculate the concentration of Cl^- required to just begin the precipitation of **(a)** AgCl, **(b)** Hg_2Cl_2, **(c)** $PbCl_2$ in a solution having metal-ion concentrations of 0.030 M. What fraction of the Pb^{2+} remains in solution when the Ag^+ just begins to precipitate?

15.130 Calculate the molar solubility of MnS in a 0.30 M NH_4Cl–0.50 M NH_3 buffer solution that is saturated with H_2S ($[H_2S] \approx 0.10$ M). What is the solubility of MnS (in g/L)? (K_{spa} for MnS is 3×10^7.)

15.131 The titration of 0.02500 L of a diprotic acid solution with 0.1000 M NaOH requires 34.72 mL of titrant to reach the second equivalence point. The pH is 3.95 at the first equivalence point and 9.27 at the second equivalence point. If the acid solution contained 0.2015 g of the acid, what is the molar mass, pK_{a1}, and pK_{a2} of the acid?

15.132 What is the pH of a solution prepared by mixing 60.0 mL of 1.00 M H_3PO_4 and 1.00 L of 0.100 M LiOH?

15.133 A student dilutes a solution of 0.100 mol of H_3PO_4 and 0.150 mol of NaOH with water to a volume of 1.00 L but then realizes that the distilled water used was in fact dilute hydrochloric acid.

(a) If the pH of the final solution is 6.73, what are the concentrations of H_3PO_4, $H_2PO_4^-$, and HPO_4^{2-}?

(b) How many moles of HCl were inadvertently added to the original solution?

15.134 Given a buret filled with 0.10 M acetic acid and a second buret filled with 0.15 M sodium acetate, how many milliliters of each solution should be mixed together to produce 20.0 mL of a solution with a pH of 4.85?

15.135 The acidity of lemon juice is derived primarily from citric acid (H_3Cit), a triprotic acid. What are the concentrations of H_3Cit, H_2Cit^-, $HCit^{2-}$, and Cit^{3-} in a sample of lemon juice that has a pH of 2.37 and a total concentration of the four citrate-containing species of 0.350 M?

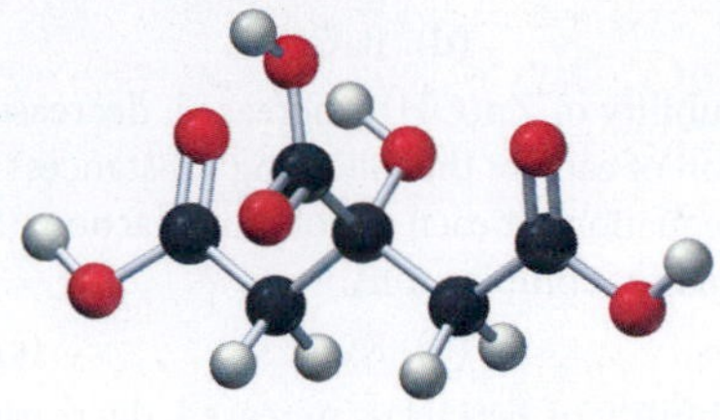

Citric acid

15.136 A 100.0 mL sample of a solution that is 0.100 M in HCl and 0.100 M in HCN is titrated with 0.100 M NaOH. Calculate the pH after the addition of the following volumes of NaOH:

(a) 0.0 mL **(b)** 75.0 mL
(c) 100.0 mL **(d)** 125.0 mL

15.137 A 0.0100 mol sample of solid $Cd(OH)_2$ ($K_{sp} = 5.3 \times 10^{-15}$) in 100.0 mL of water is titrated with 0.100 M HNO_3.

(a) What is the molar solubility of $Cd(OH)_2$ in pure water? What is the pH of the solution before the addition of any HNO_3?

(b) What is the pH of the solution after the addition of 90.0 mL of 0.100 M HNO_3?

(c) How many milliliters of 0.100 M HNO_3 must be added to completely neutralize the $Cd(OH)_2$?

15.138 Zinc hydroxide, $Zn(OH)_2$ ($K_{sp} = 4.1 \times 10^{-17}$), is nearly insoluble in water but is more soluble in strong base because Zn^{2+} forms the soluble complex ion $Zn(OH)_4^{2-}$ ($K_f = 3 \times 10^{15}$).

(a) What is the molar solubility of $Zn(OH)_2$ in pure water? (You may ignore OH^- from the self-dissociation of water.)

(b) What is the pH of the solution in part **(a)**?

(c) What is the molar solubility of $Zn(OH)_2$ in 0.10 M NaOH?

15.139 Citric acid (H_3Cit) can be used as a household cleaning agent to dissolve rust stains. The rust, represented as $Fe(OH)_3$, dissolves because the citrate ion forms a soluble complex with Fe^{3+}:

$$Fe(OH)_3(s) + H_3Cit(aq) \rightleftharpoons Fe(Cit)(aq) + 3\,H_2O(l)$$

(a) Using the equilibrium constants in Appendix C and $K_f = 6.3 \times 10^{11}$ for Fe(Cit), calculate the equilibrium constant K for the reaction.

(b) Calculate the molar solubility of $Fe(OH)_3$ in a 0.500 M solution of H_3Cit.

MULTICONCEPT PROBLEMS

15.140 When a typical diprotic acid H_2A ($K_{a1} = 10^{-4}$; $K_{a2} = 10^{-10}$) is titrated with NaOH, the principal A-containing species at the first equivalence point is HA^-.

(a) By considering all four proton-transfer reactions that can occur in an aqueous solution of HA^-, show that the principal reaction is $2\ HA^- \rightleftharpoons H_2A + A^{2-}$.

(b) Assuming that this is the principal reaction, show that the pH at the first equivalence point equals the average of pK_{a1} and pK_{a2}.

(c) How many A^{2-} ions are present in 50.0 mL of 1.0 M NaHA?

15.141 Ethylenediamine ($NH_2CH_2CH_2NH_2$, abbreviated en) is an organic base that can accept two protons:

$$en(aq) + H_2O(l) \rightleftharpoons enH^+(aq) + OH^-(aq) \qquad K_{b1} = 5.2 \times 10^{-4}$$

$$enH^+(aq) + H_2O(l) \rightleftharpoons enH_2^{\ 2+}(aq) + OH^-(aq) \qquad K_{b2} = 3.7 \times 10^{-7}$$

(a) Consider the titration of 30.0 mL of 0.100 M ethylenediamine with 0.100 M HCl. Calculate the pH after the addition of the following volumes of acid, and construct a qualitative plot of pH versus milliliters of HCl added:

(i) 0.0 mL (ii) 15.0 mL (iii) 30.0 mL
(iv) 45.0 mL (v) 60.0 mL (vi) 75.0 mL

(b) Draw the structure of ethylenediamine, and explain why it can accept two protons.

(c) What hybrid orbitals do the N atoms use for bonding?

15.142 A 40.0 mL sample of a mixture of HCl and H_3PO_4 was titrated with 0.100 M NaOH. The first equivalence point was reached after 88.0 mL of base, and the second equivalence point was reached after 126.4 mL of base.

(a) What is the concentration of H_3O^+ at the first equivalence point?

(b) What are the initial concentrations of HCl and H_3PO_4 in the mixture?

(c) What percent of the HCl is neutralized at the first equivalence point?

(d) What is the pH of the mixture before the addition of any base?

(e) Sketch the pH titration curve, and label the buffer regions and equivalence points.

(f) What indicators would you select to signal the equivalence points?

15.143 A 1.000 L sample of HCl gas at 25 °C and 732.0 mm Hg was absorbed completely in an aqueous solution that contained 6.954 g of Na_2CO_3 and 250.0 g of water.

(a) What is the pH of the solution?

(b) What is the freezing point of the solution?

(c) What is the vapor pressure of the solution? (The vapor pressure of pure water at 25 °C is 23.76 mm Hg.)

15.144 A saturated solution of an ionic salt MX exhibits an osmotic pressure of 74.4 mm Hg at 25 °C. Assuming that MX is completely dissociated in solution, what is the value of its K_{sp}?

15.145 Consider the reaction that occurs on mixing 50.0 mL of 0.560 M $NaHCO_3$ and 50.0 mL of 0.400 M NaOH at 25 °C.

(a) Write a balanced net ionic equation for the reaction.

(b) What is the pH of the resulting solution?

(c) How much heat (in joules) is liberated by the reaction? (Standard heats of formation are given in Appendix B.)

(d) What is the final temperature of the solution to the nearest 0.1 °C? You may assume that all the heat liberated is absorbed by the solution, the mass of the solution is 100.0 g, and its specific heat is 4.18 J/(g · °C).

15.146 In qualitative analysis, Ca^{2+} and Ba^{2+} are separated from Na^+, K^+, and Mg^{2+} by adding aqueous $(NH_4)_2CO_3$ to a solution that also contains aqueous NH_3 (Figure 15.18). Assume that the concentrations after mixing are 0.080 M $(NH_4)_2CO_3$ and 0.16 M NH_3.

(a) List all the Brønsted–Lowry acids and bases present initially, and identify the principal reaction.

(b) Calculate the pH and the concentrations of all species present in the solution.

(c) In order for the human eye to detect the appearance of a precipitate, a very large number of ions must come together to form solid particles. For this and other reasons, the ion product must often exceed K_{sp} by a factor of about 10^3 before a precipitate can be detected in a typical qualitative analysis experiment. Taking this fact into account, show quantitatively that the $CO_3^{\ 2-}$ concentration is large enough to give observable precipitation of $CaCO_3$ and $BaCO_3$, but not $MgCO_3$. Assume that the metal-ion concentrations are 0.010 M.

(d) Show quantitatively which of the Mg^{2+}, Ca^{2+}, and Ba^{2+} ions, if any, should give an observable precipitate of the metal hydroxide.

(e) Could the separation of Ca^{2+} and Ba^{2+} from Mg^{2+} be accomplished using 0.08 M Na_2CO_3 in place of 0.080 M $(NH_4)_2CO_3$? Show quantitatively why or why not.

15.147 A railroad tank car derails and spills 36 tons of concentrated sulfuric acid. The acid is 98.0 mass % H_2SO_4 and has a density of 1.836 g/mL.

(a) What is the molarity of the acid?

(b) How many kilograms of sodium carbonate are needed to completely neutralize the acid?

(c) How many liters of carbon dioxide at 18 °C and 745 mm Hg are produced as a byproduct?

15.148 Some progressive hair coloring products marketed to men, such as Grecian Formula 16, contain lead acetate, $Pb(CH_3CO_2)_2$. As the coloring solution is rubbed on the hair, the Pb^{2+} ions react with the sulfur atoms in hair proteins to give lead(II) sulfide (PbS), which is black. A typical coloring solution contains 0.3 mass % $Pb(CH_3CO_2)_2$, and about 2 mL of the solution is used per application.

(a) Assuming that 30% of the $Pb(CH_3CO_2)_2$ is converted to PbS, how many milligrams of PbS are formed per application of the coloring solution?

(b) Suppose the hair is washed with shampoo and water that has pH = 5.50. How many washings would be required to remove 50% of the black color? Assume that 3 gal of water is used per washing and that the water becomes saturated with PbS.

(c) Does the calculated number of washings look reasonable, given that frequent application of the coloring solution is recommended? What process(es) in addition to dissolution might contribute to the loss of color?

CHAPTER

16 Thermodynamics: Entropy, Free Energy, and Equilibrium

If the second law of thermodynamics states that all spontaneous processes lead to increased randomness, how can living things grow and evolve, creating ever more complex and highly ordered structures? The answer is in the *FYI* at the end of this chapter.

What factors determine the direction and extent of a chemical reaction? Some reactions, such as the combustion of hydrocarbon fuels, go almost to completion. Others, such as the combination of gold and oxygen, occur hardly at all. Still others—for example, the industrial synthesis of ammonia from N_2 and H_2 at 400–500 °C—result in an equilibrium mixture that contains appreciable amounts of both reactants and products.

As we saw in Section 13.6, the extent of any particular reaction is described by the value of its equilibrium constant *K*: A value of *K* much larger than 1 indicates that the reaction goes far toward completion, and a value of *K* much smaller than 1 means that the reaction does not proceed very far before reaching an equilibrium state. But what determines the value of the equilibrium constant, and can we predict its value without measuring the concentrations or partial pressures of the reactants and products? Put another way, what fundamental properties of nature determine the direction and extent of a particular chemical reaction? For answers to these questions, we turn to **thermodynamics**, the area of science that deals with the interconversion of heat and other forms of energy.

16.1 SPONTANEOUS PROCESSES

In Section 8.12 we defined a **spontaneous process** as one that proceeds on its own without any external influence. The reverse of a spontaneous process is always nonspontaneous and takes place only in the presence of some continuous external influence. Take, for example, the expansion of a gas into a vacuum. When the stopcock in the apparatus shown in **Figure 16.1** is opened, the gas in bulb A expands spontaneously into the evacuated bulb B until the gas pressure in the two bulbs is the same. The reverse process, migration of all the gas molecules into one bulb, does not occur spontaneously. To compress a gas from a larger to a smaller volume, we would have to push on the gas with a piston.

CONTENTS

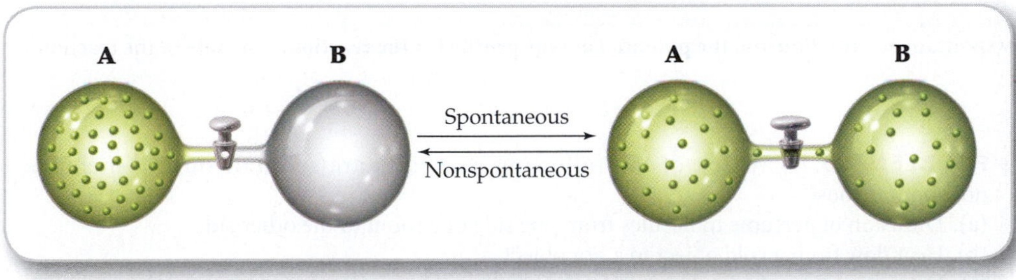

FIGURE 16.1

The expansion of a gas into a vacuum. When the stopcock is opened, the gas in bulb A expands spontaneously into evacuated bulb B to fill all the available volume. The reverse process, compression of the gas into a single bulb, is nonspontaneous.

As a second example, consider the combination of hydrogen and oxygen in the presence of a platinum catalyst:

$$2\,H_2(g) + O_2(g) \xrightarrow{\text{Catalyst}} 2\,H_2O(l)$$

The forward reaction occurs spontaneously, but the reverse reaction, decomposition of water into its elements, does not occur no matter how long we wait. We'll see in Section 17.12 that we can force the reverse reaction to occur by **electrolysis**, but that reverse process is nonspontaneous and requires a continuous input of electrical energy.

In general, whether the forward or reverse reaction is spontaneous depends on the temperature, pressure, and composition of the reaction mixture. Consider, for example, the Haber synthesis of ammonia:

$$N_2(g) + 3\,H_2(g) \xrightleftharpoons{\text{Catalyst}} 2\,NH_3(g)$$

A mixture of gaseous N_2, H_2, and NH_3, each at a partial pressure of 1 atm, reacts spontaneously at 300 K to convert some of the N_2 and H_2 to NH_3. We can predict the direction of spontaneous reaction from the relative values of the equilibrium constant K and the **reaction quotient Q** (Section 13.6). Since $K_p = 4.4 \times 10^5$ at 300 K and $Q_p = 1$ for partial pressures of 1 atm, the reaction will proceed in the forward direction because Q_p is less than K_p. Under these conditions, the reverse reaction is nonspontaneous. At 700 K, however, $K_p = 8.8 \times 10^{-5}$ and the reverse reaction is spontaneous because Q_p is greater than K_p.

LOOKING AHEAD ...

As we'll see in Section 17.12, **electrolysis** is the process of using an electric current to bring about a chemical change.

REMEMBER...

The **reaction quotient Q** is defined in the same way as the equilibrium constant K except that the concentrations or partial pressures in the equilibrium-constant expression are not necessarily equilibrium values. (Section 13.6)

A spontaneous reaction always moves a reaction mixture toward equilibrium. By contrast, a nonspontaneous reaction moves the composition of a mixture away from the equilibrium composition. Remember, though, that the word spontaneous doesn't necessarily mean "fast." A spontaneous reaction can be either fast or slow—for example, the gradual rusting of iron metal is a slow spontaneous reaction. Thermodynamics tells us where a reaction is headed, but it says nothing about how long it takes to get there. As discussed in Section 12.9, the rate at which equilibrium is achieved depends on kinetics, especially on the height of the **activation energy** barrier between the reactants and products (**Figure 16.2**).

REMEMBER...

The **activation energy E_a**, the energy difference between the transition state and the reactants, is the minimum collision energy needed for reaction. (Section 12.9)

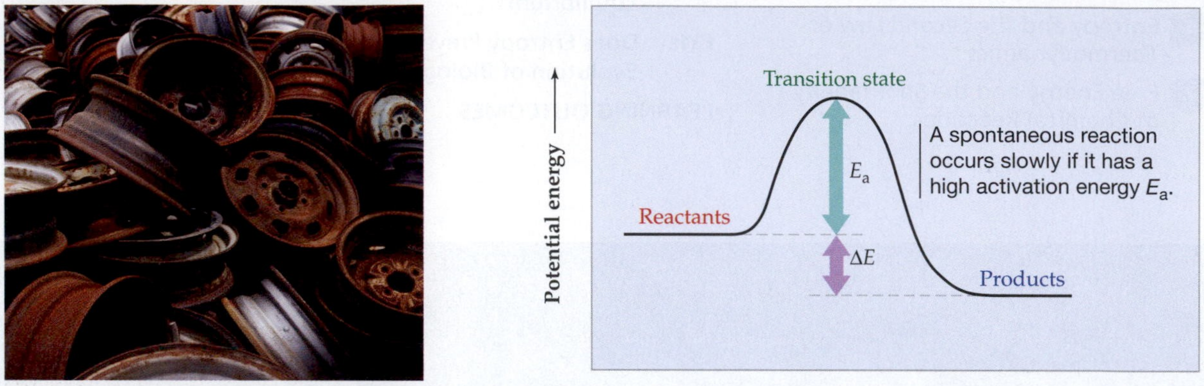

The rusting of these wheel rims is a slow but spontaneous reaction.

FIGURE 16.2

A familiar example of a slow spontaneous reaction and the potential energy profile for the reaction. The rate of the reaction depends on kinetics.

PROBLEM 16.1 Which of the following processes are spontaneous? Which are nonspontaneous?

(a) Diffusion of perfume molecules from one side of a room to the other side

(b) Heat flow from a cold object to a hot object

(c) Decomposition of rust ($Fe_2O_3 \cdot H_2O$) to iron metal, oxygen, and water

(d) Decomposition of solid $CaCO_3$ to solid CaO and gaseous CO_2 at 25 °C and 1 atm pressure ($K_p = 1.4 \times 10^{-23}$)

16.2 ENTHALPY, ENTROPY, AND SPONTANEOUS PROCESSES: A BRIEF REVIEW

▲ The combustion of natural gas (mainly CH_4) in air is a spontaneous, exothermic reaction.

Let's look more closely at spontaneous processes and at the thermodynamic factors that cause them to occur. We saw in Chapter 8 that most spontaneous chemical reactions are accompanied by the conversion of potential energy to heat. For example, when methane burns in air, the potential energy stored in the chemical bonds of CH_4 and O_2 is partly converted to heat, which flows from the system (reactants plus products) to the surroundings:

$$CH_4(g) + 2\,O_2(g) \longrightarrow CO_2(g) + 2\,H_2O(l) \qquad \Delta H^\circ = -890.3 \text{ kJ}$$

Because heat is lost by the system, the reaction is exothermic and the standard enthalpy of reaction is negative ($\Delta H^\circ = -890.3$ kJ). The total energy is conserved, so all the energy lost by the system shows up as heat gained by the surroundings.

Because spontaneous reactions so often give off heat, the nineteenth-century French chemist Marcellin Berthelot proposed that spontaneous chemical or physical changes are *always* exothermic. But Berthelot's proposal can't be correct. Ice, for example, spontaneously absorbs heat from the surroundings and melts at temperatures above 0 °C. Similarly, liquid water absorbs heat and spontaneously boils at temperatures above 100 °C. As further examples, gaseous N_2O_4 absorbs heat when it decomposes to NO_2 at 400 K, and table salt absorbs heat when it dissolves in water at room temperature:

$$H_2O(s) \longrightarrow H_2O(l) \qquad \Delta H_{fusion} = +6.01 \text{ kJ}$$
$$H_2O(l) \longrightarrow H_2O(g) \qquad \Delta H_{vap} = +40.7 \text{ kJ}$$
$$N_2O_4(g) \longrightarrow 2\,NO_2(g) \qquad \Delta H° = +55.3 \text{ kJ}$$
$$NaCl(s) \longrightarrow Na^+(aq) + Cl^-(aq) \qquad \Delta H° = +3.88 \text{ kJ}$$

All these processes are endothermic, yet all are spontaneous. In all cases, the system moves spontaneously to a state of *higher* enthalpy by absorbing heat from the surroundings.

Because some spontaneous reactions are exothermic and others are endothermic, enthalpy alone can't account for the direction of spontaneous change; a second factor must be involved. This second determinant of spontaneous change is nature's tendency to move to a condition of maximum randomness (Section 8.12).

Molecular randomness is called **entropy** and is denoted by the symbol *S*. Entropy is a **state function** (Section 8.2), and the entropy change ΔS for a process thus depends only on the initial and final states of the system:

$$\Delta S = S_{final} - S_{initial}$$

When the randomness of a system increases, ΔS has a positive value; when randomness decreases, ΔS is negative.

REMEMBER...

A **state function** is a function or property whose value depends only on the present state (condition) of the system, not on the path used to arrive at that condition. Pressure, volume, temperature, enthalpy, and entropy are state functions. (Section 8.2)

The randomness of a system comes about because the particles in the system (atoms, ions, and molecules) are in incessant motion, moving about in the accessible volume, colliding with each other and continually exchanging energy. Randomness—and thus entropy—is a probability concept, related to the number of ways that a particular state of a system can be achieved. A particular state of a macroscopic system, characterized by its temperature, pressure, volume, and number of particles, can be achieved in a vast number of ways in which the fluctuating positions and energies of the individual particles differ but the volume and total energy are constant.

We'll examine the relationship between entropy and probability in the next section, but first let's take a qualitative look at the four spontaneous endothermic processes mentioned previously (melting of ice, boiling of liquid water, decomposition of N_2O_4, and dissolving of NaCl in water). Each of these processes involves an increase in the randomness of the system. When ice melts, for example, randomness increases because the highly ordered crystalline arrangement of tightly held water molecules collapses and the molecules become free to move about in the liquid. When liquid water vaporizes, randomness further increases because the molecules can now move independently in the much larger volume of the gas. In general, processes that convert a solid to a liquid or a liquid to a gas involve an increase in randomness and thus an increase in entropy (**Figure 16.3**).

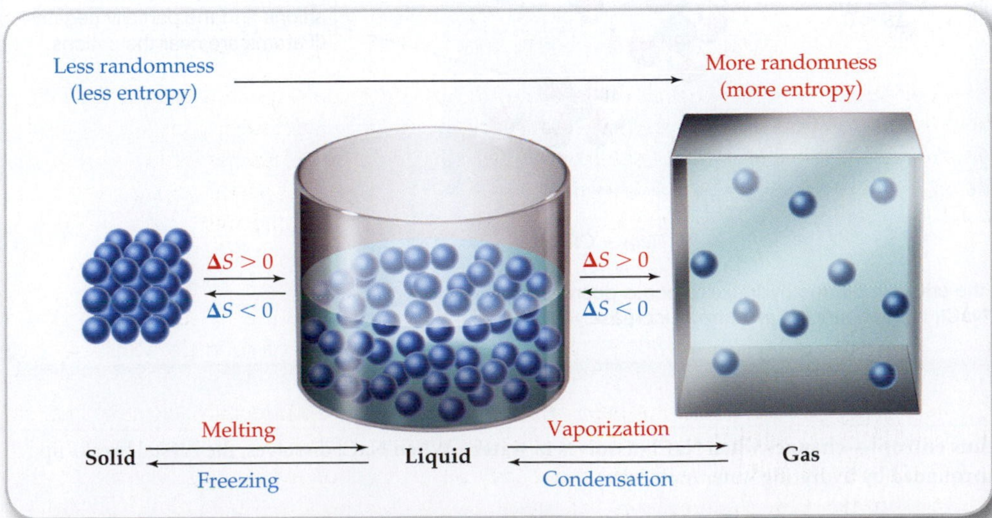

FIGURE 16.3

How molecular randomness—and thus entropy—changes when solids, liquids, and gases interconvert.

The decomposition of N_2O_4 ($O_2N{-}NO_2$) is accompanied by an increase in randomness because breaking the N–N bond allows the two gaseous NO_2 fragments to move independently. Whenever a molecule breaks into two or more pieces, the amount of molecular randomness increases. More generally, randomness—and thus entropy—increases whenever a reaction results in an increase in the number of gaseous particles (**Figure 16.4**).

FIGURE 16.4

How molecular randomness—and thus entropy—changes when the number of gaseous particles changes.

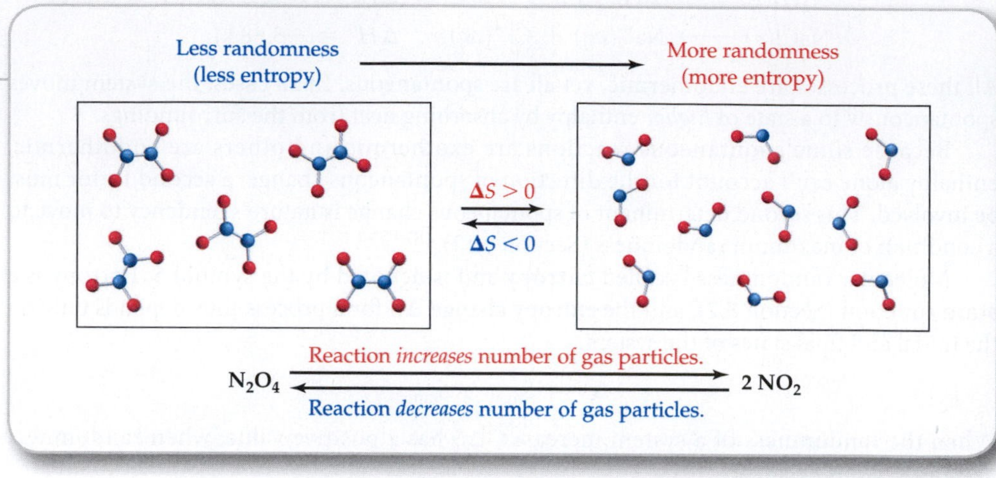

REMEMBER...

Hydrated ions are surrounded and stabilized by an ordered shell of solvent water molecules. The stabilization results from ion–dipole attractions. (Section 11.2)

The entropy change on dissolving sodium chloride in water occurs because the crystal structure of solid NaCl is disrupted and the Na^+ and Cl^- ions become **hydrated** (Section 11.2). Disruption of the crystal increases randomness because the Na^+ and Cl^- ions are tightly held in the solid but are free to move about in the liquid. The hydration process, however, *decreases* randomness because the polar, hydrating water molecules adopt an orderly arrangement about the Na^+ and Cl^- ions. It turns out that the overall dissolution process for NaCl results in a net increase in randomness, and ΔS is thus positive (**Figure 16.5**). This is

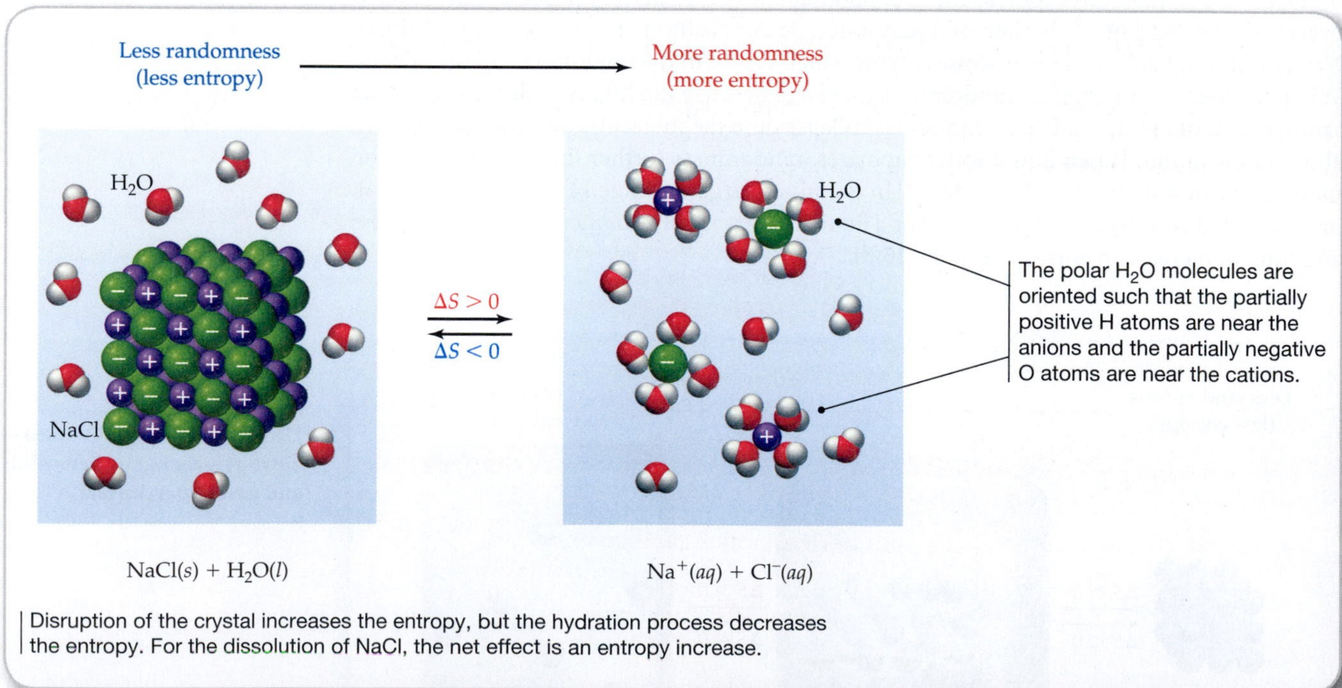

FIGURE 16.5

How molecular randomness—and thus entropy—changes when NaCl dissolves in water. When NaCl dissolves, the crystal breaks up and the Na^+ and Cl^- ions become surrounded by hydrating water molecules.

usually the case for the dissolution of molecular solids, such as $HgCl_2$, and salts that contain $+1$ cations and -1 anions. For salts such as $CaSO_4$, which contain more highly charged ions, the hydrating water molecules are more strongly attached to the ions and the dissolution process often results in a net decrease in entropy. The following dissolution reactions illustrate the point:

$$HgCl_2(s) \longrightarrow HgCl_2(aq) \qquad \Delta S = +9\ \text{J/(K}\cdot\text{mol)}$$
$$NaCl(s) \longrightarrow Na^+(aq) + Cl^-(aq) \qquad \Delta S = +43\ \text{J/(K}\cdot\text{mol)}$$
$$CaSO_4(s) \longrightarrow Ca^{2+}(aq) + SO_4^{2-}(aq) \qquad \Delta S = -140\ \text{J/(K}\cdot\text{mol)}$$

Worked Example 16.1

Predicting the Sign of ΔS

Predict the sign of ΔS in the system for each of the following processes:

(a) $CO_2(s) \longrightarrow CO_2(g)$ (sublimation of dry ice)
(b) $CaSO_4(s) \longrightarrow CaO(s) + SO_3(g)$
(c) $N_2(g) + 3\,H_2(g) \longrightarrow 2\,NH_3(g)$
(d) $I_2(s) \longrightarrow I_2(aq)$ (dissolution of iodine in water)

STRATEGY

To predict the sign of ΔS, look to see whether the process involves a phase change, a change in the number of gaseous molecules, or the dissolution (or precipitation) of a solid. Entropy generally increases for phase transitions that convert a solid to a liquid or a liquid to a gas, for reactions that increase the number of gaseous molecules, and for the dissolution of molecular solids or salts with $+1$ cations and -1 anions.

SOLUTION

(a) The molecules in a gas are free to move about randomly, whereas the molecules in a solid are tightly held in a highly ordered arrangement. Therefore, randomness increases when a solid sublimes and ΔS is positive.

(b) One mole of gaseous molecules appears on the product side of the equation and none appears on the reactant side. Because the reaction increases the number of gaseous molecules, the entropy change is positive.

(c) The entropy change is negative because the reaction decreases the number of gaseous molecules from 4 mol to 2 mol. Fewer particles can move independently after reaction than before.

(d) Iodine molecules are electrically neutral and form a molecular solid. The dissolution process destroys the order of the crystal and enables the iodine molecules to move about randomly in the liquid. Therefore, ΔS is positive.

PROBLEM 16.2 Predict the sign of ΔS in the system for each of the following processes:

(a) $H_2O(g) \longrightarrow H_2O(l)$ (formation of rain droplets)
(b) $I_2(g) \longrightarrow 2\,I(g)$
(c) $CaCO_3(s) \longrightarrow CaO(s) + CO_2(g)$
(d) $Ag^+(aq) + Br^-(aq) \longrightarrow AgBr(s)$
(e) Deposition of frost on a cold morning

WORKED CONCEPTUAL EXAMPLE 16.2

Predicting the Sign of ΔS for a Gas-Phase Reaction

Consider the gas-phase reaction of A_2 molecules (red) with B atoms (blue):

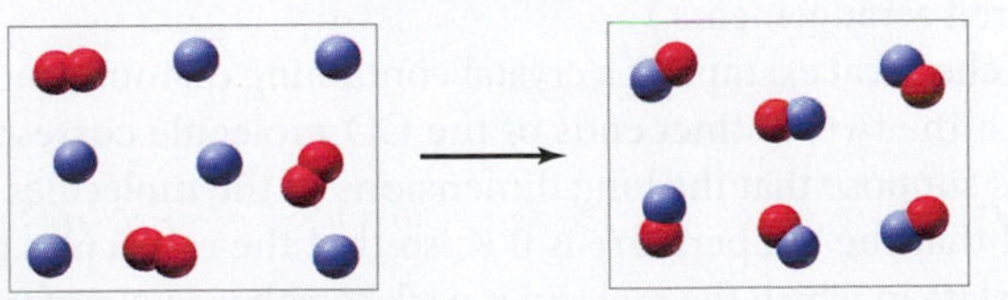

(a) Write a balanced equation for the reaction.
(b) Predict the sign of ΔS for the reaction.

continued on next page

STRATEGY

To determine the stoichiometry of the reaction, count the number of reactant A_2 molecules and B atoms and the number of product AB molecules. To predict the sign of the entropy change, see if the reaction increases or decreases the number of gaseous particles.

SOLUTION

(a) In this reaction, 3 A_2 molecules and 6 B atoms are consumed and 6 AB molecules are formed ($3\,A_2 + 6\,B \rightarrow 6\,AB$). Dividing by 3 to reduce the coefficients to their smallest whole number values gives the balanced equation $A_2(g) + 2\,B(g) \rightarrow 2\,AB(g)$.

(b) Because the reaction decreases the number of gaseous particles from 3 mol to 2 mol, the entropy change is negative.

CONCEPTUAL PROBLEM 16.3 Consider the gas-phase reaction of AB_3 and A_2 molecules:

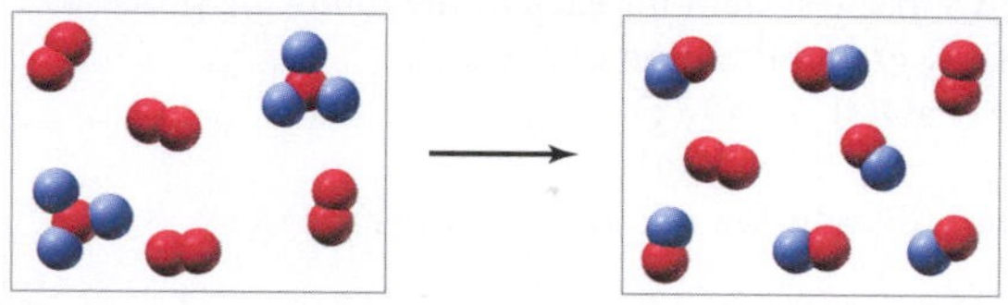

(a) Write a balanced equation for the reaction.

(b) What is the sign of the entropy change for the reaction?

16.3 ENTROPY AND PROBABILITY

▲ Shaking a box that contains 20 quarters gives a random arrangement of heads and tails.

Why do systems tend to move spontaneously to a state of maximum randomness? The answer is that a random arrangement of particles is more probable than an ordered arrangement because a random arrangement can be achieved in more ways. To begin with a simple example, suppose that you shake a box containing 20 identical coins and then count the number of heads (H) and tails (T). It's very unlikely that all 20 coins will come up heads; that is, a perfectly ordered arrangement of 20 heads (or 20 tails) is much less probable than a random mixture of heads and tails.

The probabilities of the ordered and random arrangements are proportional to the number of ways that the arrangements can be achieved. The perfectly ordered arrangement of 20 heads can be achieved in only one way because it consists of a single configuration. In how many ways, though, can a random arrangement be achieved? If there were just two coins in the box, each of them could come up in one of two ways (H or T), and the two together could come up in $2 \times 2 = 2^2 = 4$ ways (HH, HT, TH, or TT). Three coins could come up in $2 \times 2 \times 2 = 2^3 = 8$ ways (HHH, THH, HTH, HHT, HTT, THT, TTH, or TTT), and so on. For the case of 20 coins, the number of possible arrangements is $2^{20} = 1{,}048{,}576$.

Because the ordered arrangement of 20 heads (or 20 tails) can be achieved in only one way and a random mixture of heads and tails can be achieved in $2^{20} - 2 \approx 2^{20}$ ways, a random arrangement is 2^{20} times more probable than a perfectly ordered arrangement. If you begin with an ordered arrangement of 20 heads and shake the box, the system will move to a state with a random mixture of heads and tails because that state is more probable. (Note that the state with a random arrangement of coins includes all possible arrangements except the two perfectly ordered arrangements.)

An analogous chemical example is a crystal containing diatomic molecules such as carbon monoxide in which the two distinct ends of the CO molecule correspond to the heads and tails of a coin. Let's suppose that the long dimensions of the molecules are oriented vertically (**Figure 16.6**) and that the temperature is 0 K, so that the molecules are locked into a fixed arrangement. The state in which the molecules pack together in a perfectly ordered "heads-up" arrangement (**Figure 16.6a**) can be achieved in only one way, whereas the state in which the molecules are arranged randomly with respect to the vertical direction can be achieved in many ways—2^{20} ways for a hypothetical crystal containing 20 CO molecules (**Figure 16.6b**). Therefore, a structure in which the molecules are arranged randomly is 2^{20} times more probable than the perfectly ordered heads-up structure.

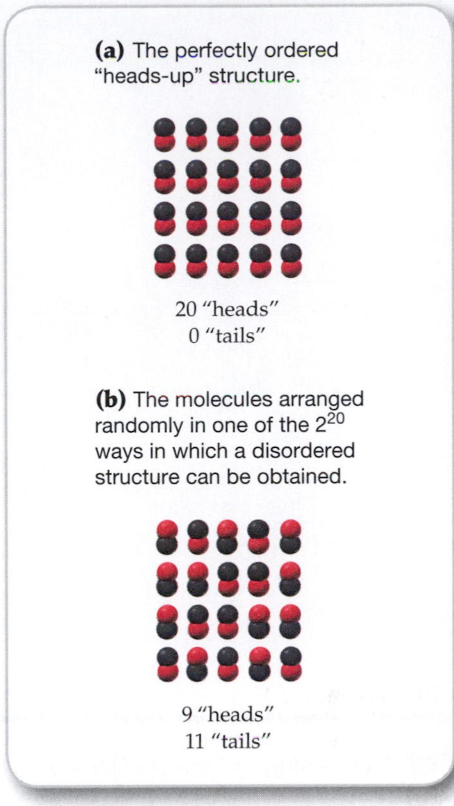

FIGURE 16.6

A hypothetical crystal containing 20 CO molecules.

Boltzmann's Equation

The Austrian physicist Ludwig Boltzmann proposed that the entropy of a particular state is related to the number of ways that the state can be achieved, according to the equation

$$S = k \ln W$$

where S is the entropy of the state, $\ln W$ is the natural logarithm of the number of ways that the state can be achieved, and k, now known as *Boltzmann's constant*, is a universal constant equal to the gas constant R divided by Avogadro's number ($k = R/N_A = 1.38 \times 10^{-23}$ J/K). Because a logarithm is dimensionless, the Boltzmann equation implies that entropy has the same units as the constant k, joules per kelvin.

Now let's apply Boltzmann's equation to our hypothetical crystal containing 20 CO molecules. Because a perfectly ordered state can be achieved in only one way ($W = 1$ in the Boltzmann equation) and because $\ln 1 = 0$, the entropy of the perfectly ordered state is zero:

$$\begin{aligned} S &= k \ln W = k \ln 1 \\ &= 0 \end{aligned}$$

The more probable state in which the molecules are arranged randomly can be achieved in 2^{20} ways and thus has a higher entropy:

$$\begin{aligned} S &= k \ln W = k \ln 2^{20} \\ &= (1.38 \times 10^{-23}\ \text{J/K})(20)(\ln 2) \\ &= 1.91 \times 10^{-22}\ \text{J/K} \end{aligned}$$

where we have made use of the relation $\ln x^a = a \ln x$ (Appendix A.2).

If our crystal contained 1 mol of CO molecules, the entropy of the perfectly ordered state (6.02×10^{23} C atoms up) would still be zero, but the entropy of the state with a random arrangement of CO molecules would be much higher because Avogadro's number of molecules can be arranged randomly in a huge number of ways ($W = 2^{N_A} = 2^{6.02 \times 10^{23}}$).

According to Boltzmann's equation, the entropy of the state with a random arrangement of CO molecules is

$$S = k \ln W = k \ln 2^{N_A} = kN_A \ln 2$$

Because $k = R/N_A$,

$$\begin{aligned} S &= R \ln 2 = (8.314\ \text{J/K})(0.693) \\ &= 5.76\ \text{J/K} \end{aligned}$$

Based on experimental measurements, the entropy of 1 mol of solid carbon monoxide near 0 K is about 5 J/K, indicating that the CO molecules adopt a nearly random arrangement. Entropy associated with a random arrangement of molecules in space is sometimes called positional, or configurational, entropy.

The nearly random arrangement of CO molecules in crystalline carbon monoxide is unusual but can be understood in terms of molecular structure. Because CO molecules have a **dipole moment** of only 0.11 D, intermolecular **dipole–dipole forces** are unusually weak (Section 10.2), and the molecules therefore have little preference for a slightly lower energy, completely ordered arrangement. By contrast, HCl, with a larger dipole moment of 1.11 D, forms an ordered crystalline solid, and so the entropy of 1 mol of solid HCl at 0 K is 0 J/K.

REMEMBER...

The **dipole moment** (μ) is a measure of the net polarity of a molecule and is defined as $\mu = Q \times r$, where Q is the magnitude of the charge at either end of the molecular dipole and r is the distance between the charges. (Section 10.1)

Dipole–dipole forces result from electrical interactions among neighboring polar molecules. (Section 10.2)

Expansion of an Ideal Gas

Boltzmann's equation also explains why a gas expands into a vacuum. If the two bulbs in Figure 16.1 have equal volumes, each molecule has one chance in two of being in bulb A (heads, in our coin example) and one chance in two of being in bulb B (tails) when the stopcock is opened. It's exceedingly unlikely that all the molecules in 1 mol of gas will be in bulb A because that state can be achieved in only one way. The state in which Avogadro's number of molecules are randomly distributed between bulbs A and B can be achieved in $2^{6.02 \times 10^{23}}$ ways, and the entropy of that state is therefore higher than the entropy of the ordered state by the now familiar amount, $R \ln 2 = 5.76$ J/K. Thus, a gas expands spontaneously because the state of greater volume is more probable.

We can derive a general equation for the entropy change that occurs on the expansion of an ideal gas at constant temperature by considering the distribution of N molecules among B hypothetical boxes, or cells, each having an equal volume v.

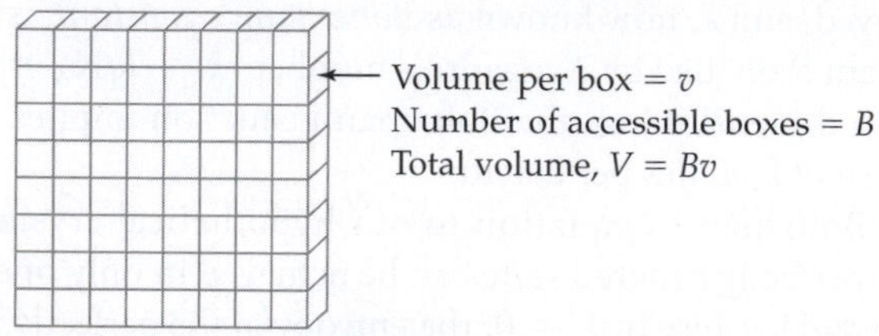

REMEMBER...

According to the kinetic–molecular theory, the **kinetic energy** of 1 mol of an **ideal gas** equals $3RT/2$ and is independent of pressure and volume. (Section 9.6)

Since the **energy of an ideal gas** depends only on the temperature (Section 9.6), ΔE for the expansion of an ideal gas at constant temperature is zero. To calculate the entropy change $\Delta S = S_{\text{final}} - S_{\text{initial}}$ using the Boltzmann equation, we have only to find the number of ways N molecules can be distributed among the B boxes.

A single molecule can go into any one of the boxes and can thus be assigned to B boxes in B ways. Two molecules can occupy the boxes in $B \times B = B^2$ ways, three molecules can fill the boxes in $B \times B \times B = B^3$ ways, and so on. The number of ways that N molecules can occupy B boxes is $W = B^N$.

Now suppose that the initial volume comprises B_{initial} boxes and the final volume consists of B_{final} boxes:

$$V_{\text{initial}} = B_{\text{initial}}v \quad \text{and} \quad V_{\text{final}} = B_{\text{final}}v$$

Then the probabilities of the initial and final states—that is, the number of ways they can be achieved—are

$$W_{\text{initial}} = (B_{\text{initial}})^N \quad \text{and} \quad W_{\text{final}} = (B_{\text{final}})^N$$

According to the Boltzmann equation, the entropy change due to a change in volume is

$$\Delta S = S_{\text{final}} - S_{\text{initial}} = k \ln W_{\text{final}} - k \ln W_{\text{initial}} = k \ln \frac{W_{\text{final}}}{W_{\text{initial}}}$$

$$= k \ln \left(\frac{B_{\text{final}}}{B_{\text{initial}}}\right)^N = kN \ln \frac{B_{\text{final}}}{B_{\text{initial}}}$$

Because $V_{\text{initial}} = B_{\text{initial}}v$ and $V_{\text{final}} = B_{\text{final}}v$

$$\Delta S = kN \ln \left(\frac{V_{\text{final}}/v}{V_{\text{initial}}/v}\right) = kN \ln \frac{V_{\text{final}}}{V_{\text{initial}}}$$

Finally, because $k = R/N_A$ and the number of particles equals the number of moles of gas times Avogadro's number ($N = nN_A$), then

$$kN = \left(\frac{R}{N_A}\right)(nN_A) = nR$$

and so the entropy change for expansion (or compression) of n moles of an ideal gas at constant temperature is

$$\Delta S = nR \ln \frac{V_{\text{final}}}{V_{\text{initial}}}$$

For a twofold expansion of 1 mol of an ideal gas at constant temperature, $\Delta S = R \ln 2$, the same result as we obtained previously.

Because the pressure and volume of an ideal gas are related inversely ($P = nRT/V$), we can also write

$$\Delta S = nR \ln \frac{P_{\text{initial}}}{P_{\text{final}}}$$

Thus, the entropy of a gas *increases* when its pressure *decreases* at constant temperature, and the entropy *decreases* when its pressure *increases*. Common sense tells us that the more we squeeze the gas, the less space the gas molecules have and so randomness decreases.

PROBLEM 16.4 Which state has the higher entropy? Explain in terms of probability.

(a) A perfectly ordered crystal of solid nitrous oxide (N≡N–O) or a disordered crystal in which the molecules are oriented randomly
(b) Quartz glass (Section 10.10) or a quartz crystal
(c) 1 mol of N_2 gas at STP or 1 mol of N_2 gas at 273 K in a volume of 11.2 L
(d) 1 mol of N_2 gas at STP or 1 mol of N_2 gas at 273 K and 0.25 atm

16.4 ENTROPY AND TEMPERATURE

Thus far we've seen that entropy is associated with the orientation and distribution of molecules in space. Disordered crystals have higher entropy than ordered crystals, and expanded gases have higher entropy than compressed gases.

Entropy is also associated with molecular motion. As the temperature of a substance increases, random molecular motion increases and there is a corresponding increase in the average kinetic energy of the molecules. But not all the molecules have the same energy. As we saw in Section 9.6, there is a distribution of molecular speeds in a gas, a distribution that broadens and shifts to higher speeds with increasing temperature (Figure 9.12, page 337). In solids, liquids, and gases, the total energy of a substance can be distributed among the individual molecules in a number of ways that increases as the total energy increases. According to Boltzmann's equation, the more ways that the energy can be distributed, the greater the randomness of the state and the higher its entropy. Therefore, the entropy of a substance increases with increasing temperature (**Figure 16.7**).

FIGURE 16.7

A substance at a higher temperature has greater entropy than the same substance at a lower temperature.

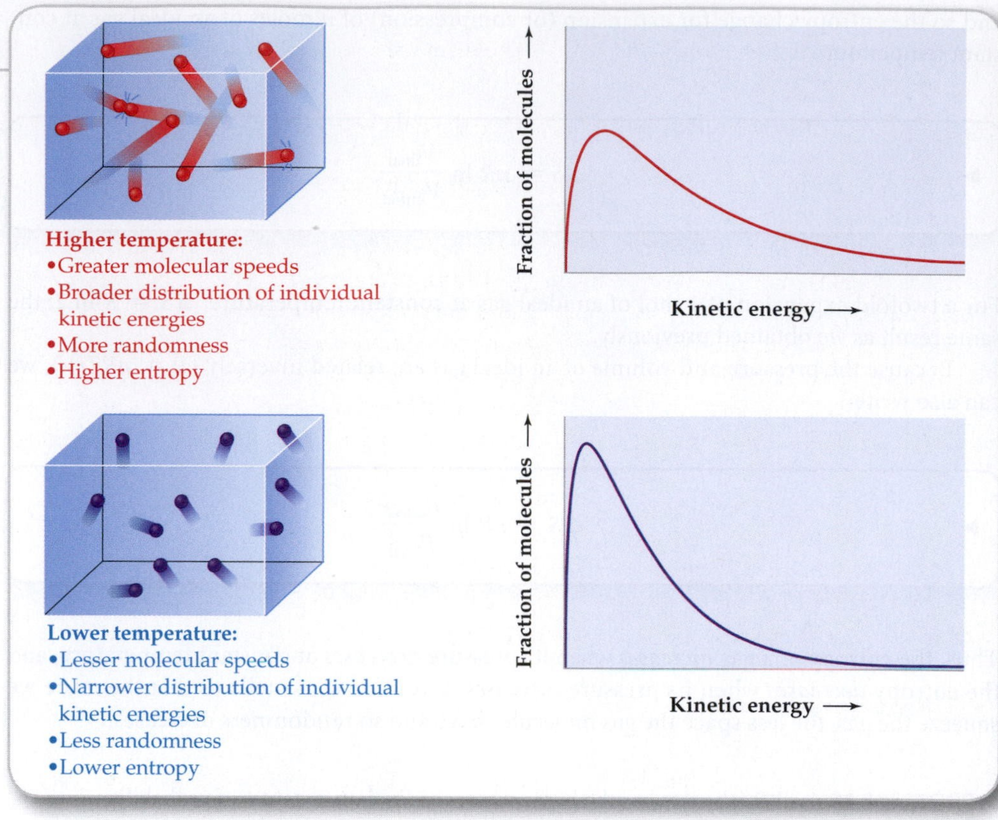

A typical plot of entropy versus temperature is shown in **Figure 16.8**. At absolute zero, every substance is a solid whose particles are tightly held in a crystalline structure. If there is no residual orientational disorder, like that in carbon monoxide (Figure 16.6b), the entropy of the substance at 0 K will be zero, a general result summarized in the **third law of thermodynamics**:

FIGURE 16.8

Entropy versus temperature. The entropy of a pure substance, equal to zero at 0 K, shows a steady increase with rising temperature, punctuated by discontinuous jumps in entropy at the temperatures of the phase transitions.

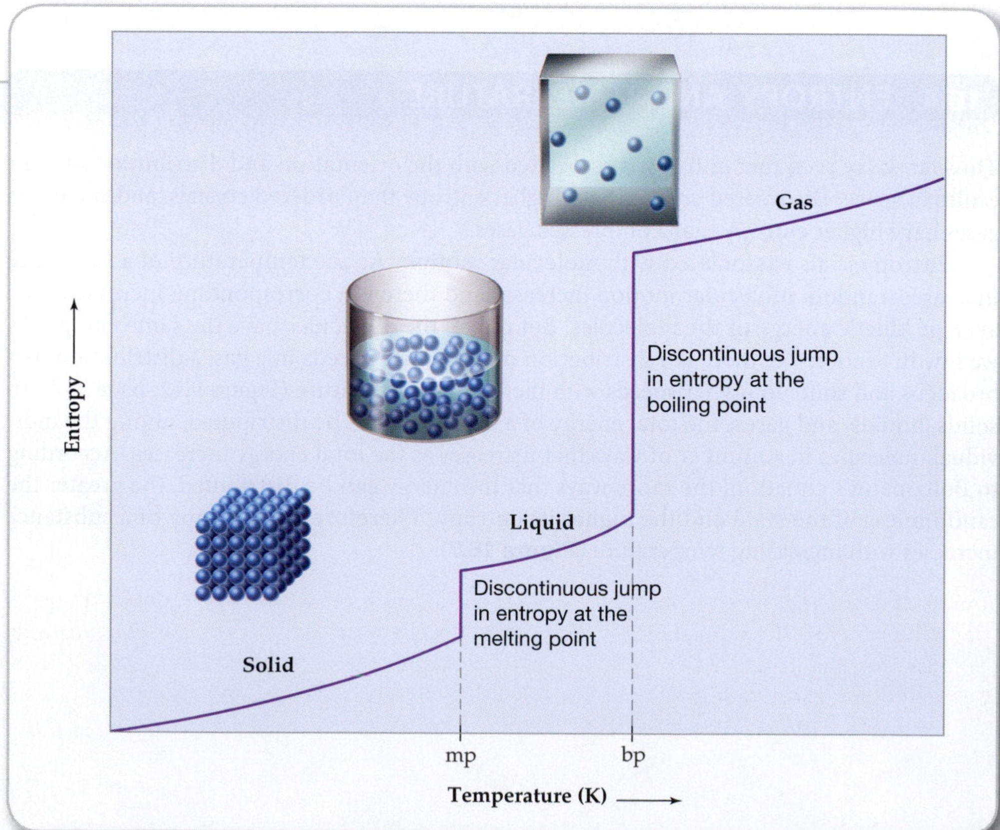

▶ **Third law of thermodynamics** The entropy of a perfectly ordered crystalline substance at 0 K is zero.

(The first law of thermodynamics was discussed in Section 8.1. We'll review the first law and discuss the second law in Section 16.6.)

As the temperature of a solid is raised, the added energy increases the vibrational motion of the molecules about their equilibrium positions in the crystal. The number of ways in which the vibrational energy can be distributed increases with rising temperature, and the entropy of the solid thus increases steadily as the temperature increases.

At the melting point, there is a discontinuous jump in entropy because there are many more ways of arranging the molecules in the liquid than in the solid. Furthermore, the molecules in the liquid can undergo translational and rotational as well as vibrational motion (**Figure 16.9**), and so there are many more ways of distributing the total energy in the liquid. An even greater jump in entropy is observed at the boiling point because molecules in the gas are free to occupy a much larger volume. Between the melting point and the boiling point, the entropy of a liquid increases steadily as molecular motion increases and the number of ways of distributing the total energy among the individual molecules increases. For the same reason, the entropy of a gas rises steadily as its temperature increases.

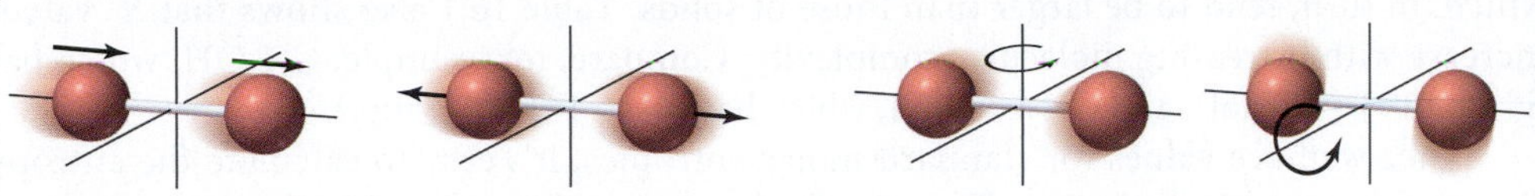

FIGURE 16.9

Types of molecular motion, illustrated for a diatomic molecule. Translational motion is motion of the center of mass. Vibrational motion is the stretching and compression of chemical bonds. Rotational motion is rotation of the molecule about fixed coordinate axes.

16.5 STANDARD MOLAR ENTROPIES AND STANDARD ENTROPIES OF REACTION

We won't describe how the entropy of a substance is determined, except to note that two approaches are available: (1) calculations based on Boltzmann's equation and (2) experimental measurements of **heat capacities** (Section 8.5) down to very low temperatures. Suffice it to say that **standard molar entropies**, denoted by $S°$, are known for many substances.

REMEMBER...

The molar **heat capacity** is the amount of heat needed to raise the temperature of 1 mol of a substance by 1 °C. (Section 8.5)

▶ **Standard molar entropy, $S°$** The entropy of 1 mol of the pure substance at 1 atm pressure and a specified temperature, usually 25 °C.

Values of $S°$ for some common substances at 25 °C are listed in **Table 16.1**, and additional values are given in Appendix B. Note that the units of $S°$ are *joules* (not kilojoules) per kelvin mole [J/(K · mol)]. Standard molar entropies are often called *absolute entropies* because they are measured with respect to an absolute reference point—the entropy of the perfectly ordered crystalline substance at 0 K [$S° = 0$ J/(K · mol) at $T = 0$ K].

TABLE 16.1 Standard Molar Entropies for Some Common Substances at 25 °C

Substance	Formula	$S°$ [J/(K·mol)]	Substance	Formula	$S°$ [J/(K·mol)]
Gases			**Liquids**		
Acetylene	C_2H_2	200.8	Acetic acid	CH_3CO_2H	160
Ammonia	NH_3	192.3	Ethanol	CH_3CH_2OH	161
Carbon dioxide	CO_2	213.6	Methanol	CH_3OH	127
Carbon monoxide	CO	197.6	Water	H_2O	69.9
Ethylene	C_2H_4	219.5	**Solids**		
Hydrogen	H_2	130.6	Calcium carbonate	$CaCO_3$	91.7
Methane	CH_4	186.2	Calcium oxide	CaO	38.1
Nitrogen	N_2	191.5	Diamond	C	2.4
Nitrogen dioxide	NO_2	240.0	Graphite	C	5.7
Dinitrogen tetroxide	N_2O_4	304.3	Iron	Fe	27.3
Oxygen	O_2	205.0	Iron(III) oxide	Fe_2O_3	87.4

Standard molar entropies make it possible to compare the entropies of different substances under the same conditions of temperature and pressure. It's apparent from Table 16.1, for example, that the entropies of gaseous substances tend to be larger than those of liquids, which, in turn, tend to be larger than those of solids. Table 16.1 also shows that $S°$ values increase with increasing molecular complexity. Compare, for example, CH_3OH, which has $S° = 127$ J/(K·mol), and CH_3CH_2OH, which has $S° = 161$ J/(K·mol).

Once we have values for standard molar entropies, it's easy to calculate the entropy change for a chemical reaction. The **standard entropy of reaction ($\Delta S°$)** can be obtained simply by subtracting the standard molar entropies of all the reactants from the standard molar entropies of all the products:

▶ **Standard entropy of reaction** $\Delta S° = S°(\text{products}) - S°(\text{reactants})$

Because $S°$ values are quoted on a per-mole basis, the $S°$ value for each substance must be multiplied by the stoichiometric coefficient of that substance in the balanced chemical equation. Thus, for the general reaction

$$a\text{A} + b\text{B} \longrightarrow c\text{C} + d\text{D}$$

the standard entropy of reaction is

▶ $$\Delta S° = [c\,S°(\text{C}) + d\,S°(\text{D})] - [a\,S°(\text{A}) + b\,S°(\text{B})]$$

In this equation, the units of the coefficients are moles, the units of $S°$ are J/(K·mol), and the units of $\Delta S°$ are J/K.

As an example, let's calculate the standard entropy change for the reaction

$$N_2O_4(g) \longrightarrow 2\,NO_2(g)$$

Using the appropriate $S°$ values from Table 16.1, we find that $\Delta S° = 175.7$ J/K:

$$\begin{aligned}\Delta S° &= 2\,S°(NO_2) - S°(N_2O_4)\\ &= (2\text{ mol})\left(240.0\,\frac{\text{J}}{\text{K}\cdot\text{mol}}\right) - (1\text{ mol})\left(304.3\,\frac{\text{J}}{\text{K}\cdot\text{mol}}\right)\\ &= 175.7\text{ J/K}\end{aligned}$$

Although the standard molar entropy of N_2O_4 is larger than that of NO_2, as expected for a more complex molecule, $\Delta S°$ for the reaction is positive because 1 mol of N_2O_4 is converted to 2 mol of NO_2. As noted in Figure 16.4, we expect an increase in entropy whenever a molecule breaks into two or more pieces.

Worked Example 16.3

Calculating the Standard Entropy of Reaction

Calculate the standard entropy of reaction at 25 °C for the Haber synthesis of ammonia:

$$N_2(g) + 3\,H_2(g) \longrightarrow 2\,NH_3(g)$$

STRATEGY

To calculate $\Delta S°$ for the reaction, subtract the standard molar entropies of all the reactants from the standard molar entropies of all the products. Look up the $S°$ values in Table 16.1 or Appendix B, and remember to multiply the $S°$ value for each substance by its coefficient in the balanced chemical equation.

SOLUTION

$$\Delta S° = 2\,S°(NH_3) - [S°(N_2) + 3\,S°(H_2)]$$

$$= (2\text{ mol})\left(192.3\,\frac{\text{J}}{\text{K}\cdot\text{mol}}\right) - \left[(1\text{ mol})\left(191.5\,\frac{\text{J}}{\text{K}\cdot\text{mol}}\right) + (3\text{ mol})\left(130.6\,\frac{\text{J}}{\text{K}\cdot\text{mol}}\right)\right]$$

$$= -198.7\text{ J/K}$$

✔ BALLPARK CHECK

As predicted in Worked Example 16.1c, $\Delta S°$ should be negative because the reaction decreases the number of gaseous molecules from 4 mol to 2 mol.

PROBLEM 16.5 Calculate the standard entropy of reaction at 25 °C for the decomposition of calcium carbonate:

$$CaCO_3(s) \longrightarrow CaO(s) + CO_2(g)$$

16.6 ENTROPY AND THE SECOND LAW OF THERMODYNAMICS

We've seen thus far that molecular systems tend to move spontaneously toward a state of minimum enthalpy and maximum entropy. In any particular reaction, though, the enthalpy of the system can either increase or decrease. Similarly, the entropy of the system can either increase or decrease. How, then, can we decide whether a reaction will occur spontaneously? In Section 8.13, we said that it is the value of the *free-energy change*, ΔG, that is the criterion for spontaneity, where $\Delta G = \Delta H - T\Delta S$. If $\Delta G < 0$, the reaction is spontaneous; if $\Delta G > 0$, the reaction is nonspontaneous; and if $\Delta G = 0$, the reaction is at equilibrium. In this section and the next, we'll see how that conclusion was reached. Let's begin by looking at the first and second laws of thermodynamics:

- **First law of thermodynamics** In any process, spontaneous or nonspontaneous, the total energy of a system and its surroundings is constant.
- **Second law of thermodynamics** In any *spontaneous* process, the total entropy of a system and its surroundings always increases.

The first law is simply a statement of the conservation of energy (Section 8.1). It says that energy (or enthalpy) can flow between a system and its surroundings but the total energy of the system plus the surroundings always remains constant. In an exothermic reaction, the system loses enthalpy to the surroundings; in an endothermic reaction, the system gains enthalpy from the surroundings. Because energy is conserved in all chemical processes, spontaneous and nonspontaneous, the first law helps us keep track of energy flow between the system and the surroundings but it doesn't tell us whether a particular reaction will be spontaneous or nonspontaneous.

The second law, however, provides a clear-cut criterion of spontaneity. It says that the direction of spontaneous change is always determined by the sign of the total entropy change:

$$\Delta S_{total} = \Delta S_{system} + \Delta S_{surroundings}$$

Specifically,

If $\Delta S_{total} > 0$, the reaction is spontaneous.

If $\Delta S_{total} < 0$, the reaction is nonspontaneous.

If $\Delta S_{total} = 0$, the reaction mixture is at equilibrium.

All reactions proceed spontaneously in the direction that increases the entropy of the system plus surroundings. A reaction that is nonspontaneous in the forward direction is spontaneous in the reverse direction because ΔS_{total} for the reverse reaction equals $-\Delta S_{total}$ for the forward reaction. If ΔS_{total} is zero, the reaction doesn't go spontaneously in either direction, and so the reaction mixture is at equilibrium.

To determine the value of ΔS_{total}, we need values for the entropy changes in the system and the surroundings. The entropy change in the system, ΔS_{sys}, is just the entropy of reaction, which can be calculated from standard molar entropies (Table 16.1), as described in Section 16.5. For a reaction that occurs at constant pressure, the entropy change in the surroundings is directly proportional to the enthalpy change for the reaction (ΔH) and inversely proportional to the kelvin temperature (T) of the surroundings, according to the equation

$$\Delta S_{surr} = \frac{-\Delta H}{T}$$

Although we won't derive this equation to calculate ΔS_{surr}, we can nevertheless justify its form. To see why ΔS_{surr} is proportional to $-\Delta H$, recall that for an exothermic reaction ($\Delta H < 0$), the system loses heat to the surroundings **(Figure 16.10a)**. As a result, the random motion of the molecules in the surroundings increases and the entropy of the surroundings also increases ($\Delta S_{surr} > 0$). Conversely, for an endothermic reaction ($\Delta H > 0$), the system gains heat from the surroundings **(Figure 16.10b)** and the entropy of the surroundings therefore decreases ($\Delta S_{surr} < 0$). Because ΔS_{surr} is positive when ΔH is negative and vice versa, ΔS_{surr} is proportional to $-\Delta H$:

$$\Delta S_{surr} \propto -\Delta H$$

FIGURE 16.10

The relationship between the enthalpy change in the system (ΔH) and the entropy change in the surroundings (ΔS_{surr}).

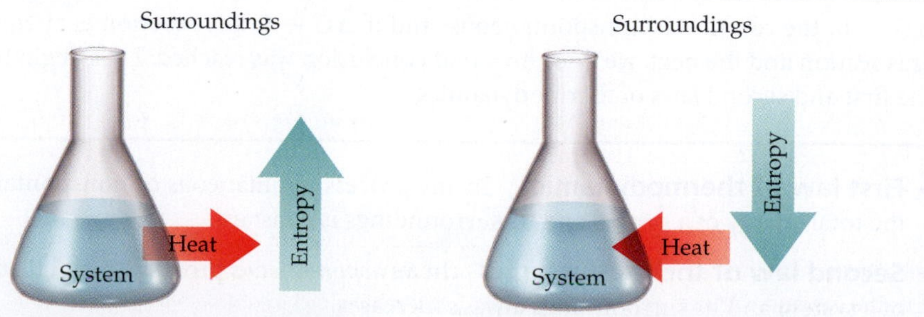

(a) When an exothermic reaction occurs in the system ($\Delta H < 0$), the surroundings gains heat and its entropy increases ($\Delta S_{surr} > 0$).

(b) When an endothermic reaction occurs in the system ($\Delta H > 0$), the surroundings loses heat and its entropy decreases ($\Delta S_{surr} < 0$).

The reason why ΔS_{surr} is inversely proportional to the absolute temperature T is more subtle. We can think of the surroundings as an infinitely large constant-temperature bath to which heat can be added without changing its temperature. If the surroundings has a low temperature, it has only a small amount of randomness, in which case the addition of a given quantity of heat results in a substantial increase in the amount of randomness (a relatively large value of ΔS_{surr}). If the surroundings has a high temperature, it already has a large amount of randomness, and the addition of the same quantity of heat produces only a marginal increase in the amount of randomness (a relatively small value of ΔS_{surr}). Thus, ΔS_{surr} varies inversely with temperature:

$$\Delta S_{surr} \propto \frac{1}{T}$$

Adding heat to the surroundings is somewhat analogous to tossing a rock into a lake. If the lake exhibits little motion (calm, smooth surface) before the rock is thrown, the rock's impact produces considerable motion, evident in a circular pattern of waves. If the lake already exhibits appreciable motion (rough, choppy surface), the additional motion produced when the rock hits the water is hardly noticeable.

▲ Adding heat to cold surroundings is analogous to tossing a rock into calm waters. Both processes produce a considerable increase in the amount of motion and thus a relatively large increase in entropy.

▲ Adding heat to hot surroundings is analogous to tossing a rock into rough waters. Both processes produce a relatively small increase in the amount of motion and thus a relatively small increase in entropy.

Worked Example 16.4

Determining Whether a Reaction Is Spontaneous

Consider the oxidation of iron metal:

$$4\ Fe(s) + 3\ O_2(g) \longrightarrow 2\ Fe_2O_3(s)$$

By determining the sign of ΔS_{total}, show whether the reaction is spontaneous at 25 °C.

STRATEGY

To determine the sign of $\Delta S_{total} = \Delta S_{sys} + \Delta S_{surr}$, we need to calculate the values of ΔS_{sys} and ΔS_{surr}. The entropy change in the system equals the standard entropy of reaction and can be calculated using the standard molar entropies in Table 16.1. To obtain $\Delta S_{surr} = -\Delta H°/T$, calculate $\Delta H°$ for the reaction from standard enthalpies of formation (Section 8.9).

SOLUTION

$$\Delta S_{sys} = \Delta S° = 2\ S°(Fe_2O_3) - [4\ S°(Fe) + 3\ S°(O_2)]$$

$$= (2\ \text{mol})\left(87.4\ \frac{\text{J}}{\text{K}\cdot\text{mol}}\right) - \left[(4\ \text{mol})\left(27.3\ \frac{\text{J}}{\text{K}\cdot\text{mol}}\right) + (3\ \text{mol})\left(205.0\ \frac{\text{J}}{\text{K}\cdot\text{mol}}\right)\right]$$

$$= -549.5\ \text{J/K}$$

$$\Delta H° = 2\ \Delta H°_f(Fe_2O_3) - [4\ \Delta H°_f(Fe) + 3\ \Delta H°_f(O_2)]$$

Because $\Delta H°_f = 0$ for elements and $\Delta H°_f = -824.2$ kJ/mol for Fe_2O_3 (Appendix B), $\Delta H°$ for the reaction is

$$\Delta H° = 2\ \Delta H°_f(Fe_2O_3) = (2\ \text{mol})(-824.2\ \text{kJ/mol}) = -1648.4\ \text{kJ}$$

Therefore, at 25 °C = 298.15 K,

$$\Delta S_{surr} = \frac{-\Delta H}{T} = \frac{-(-1{,}648{,}400\ \text{J})}{298.15\ \text{K}} = 5529\ \text{J/K}$$

$$\Delta S_{total} = \Delta S_{sys} + \Delta S_{surr} = -549.5\ \text{J/K} + 5529\ \text{J/K} = 4980\ \text{J/K}$$

Because the total entropy change is positive, the reaction is spontaneous under standard-state conditions at 25 °C.

✔ BALLPARK CHECK

Since the reaction consumes 3 mol of gas, ΔS_{sys} is negative. Because the oxidation of iron metal is highly exothermic, $\Delta S_{surr} = -\Delta H°/T$ is positive and very large. The value of ΔS_{surr} is greater than the absolute value of ΔS_{sys}, and so ΔS_{total} is positive, in agreement with the solution.

PROBLEM 16.6 By determining the sign of ΔS_{total}, show whether the decomposition of calcium carbonate is spontaneous under standard-state conditions at 25 °C.

$$CaCO_3(s) \longrightarrow CaO(s) + CO_2(g)$$

16.7 FREE ENERGY AND THE SPONTANEITY OF CHEMICAL REACTIONS

Chemists are generally more interested in the system (the reaction mixture) than the surroundings, and it's therefore convenient to restate the second law in terms of the thermodynamic properties of the system, without regard to the surroundings. For this purpose, we use the thermodynamic property called **free energy**, denoted by G in honor of J. Willard Gibbs (1839–1903), the American mathematical physicist who laid the foundations of chemical thermodynamics. The free energy G of a system is defined as

▶ **Free energy** $G = H - TS$

As you might expect from its name, free energy has units of energy (J or kJ).

Because enthalpy, entropy, and temperature are state functions, free energy is also a state function and the change in free energy (ΔG) for a process is independent of path. As we saw in Section 8.13, ΔG for a reaction at constant temperature equals the change in enthalpy minus the product of temperature times the change in entropy:

▶ **Free energy change** $\Delta G = \Delta H - T\Delta S$

To see what this equation for free-energy change has to do with spontaneity, let's return to the relationship

$$\Delta S_{total} = \Delta S_{sys} + \Delta S_{surr} = \Delta S + \Delta S_{surr}$$

where we have now dropped the subscript *sys*. (It's generally understood that symbols without a subscript refer to the system, not the surroundings.) Since $\Delta S_{surr} = -\Delta H/T$, where ΔH is the heat gained by the system at constant pressure, we can also write

$$\Delta S_{total} = \Delta S - \frac{\Delta H}{T}$$

Multiplying both sides by $-T$ gives

$$-T\Delta S_{total} = \Delta H - T\Delta S$$

The right side of this equation is just ΔG, the change in the free energy of the system at constant temperature and pressure. Therefore,

$$-T\Delta S_{total} = \Delta G$$

Note that ΔG and ΔS_{total} have opposite signs because the absolute temperature T is always positive.

According to the second law of thermodynamics, a reaction is spontaneous if ΔS_{total} is positive, nonspontaneous if ΔS_{total} is negative, and at equilibrium if ΔS_{total} is zero. Because $-T\Delta S_{total} = \Delta G$, and because ΔG and ΔS_{total} have opposite signs, we can restate the thermodynamic criterion for the spontaneity of a reaction carried out at constant temperature and pressure in the following way:

If $\Delta G < 0$, the reaction is spontaneous.
If $\Delta G > 0$, the reaction is nonspontaneous.
If $\Delta G = 0$, the reaction mixture is at equilibrium.

In other words, *in any spontaneous process at constant temperature and pressure, the free energy of the system always decreases.*

The relationship between the signs of the enthalpy and entropy contributions to ΔG in the free-energy equation $\Delta G = \Delta H - T\Delta S$ is summarized in **Table 16.2**. When ΔH is negative and ΔS is positive, ΔG is negative at all temperatures; and when ΔH is positive and ΔS is negative, ΔG is positive at all temperatures. Thus, exothermic reactions with a positive entropy change are always spontaneous, and endothermic reactions with a negative entropy change are always nonspontaneous, no matter what the temperature.

TABLE 16.2 Signs of Enthalpy, Entropy, and Free-Energy Changes and Reaction Spontaneity for a Reaction at Constant Temperature and Pressure

ΔH	ΔS	$\Delta G = \Delta H - T\Delta S$	Reaction Spontaneity	Example
−	+	−	Spontaneous at all temperatures	$2\,NO_2(g) \longrightarrow N_2(g) + 2\,O_2(g)$
+	−	+	Nonspontaneous at all temperatures	$3\,O_2(g) \longrightarrow 2\,O_3(g)$
−	−	− or +	Spontaneous at low temperatures where ΔH outweighs $T\Delta S$ Nonspontaneous at high temperatures where $T\Delta S$ outweighs ΔH	$N_2(g) + 3\,H_2(g) \longrightarrow 2\,NH_3(g)$
+	+	− or +	Spontaneous at high temperatures where $T\Delta S$ outweighs ΔH Nonspontaneous at low temperatures where ΔH outweighs $T\Delta S$	$2\,HgO(s) \longrightarrow 2\,Hg(l) + O_2(g)$

If, however, ΔH and ΔS are either both negative or both positive, the sign of ΔG (and therefore the spontaneity of the reaction) depends on the temperature. As discussed in Section 8.13, the temperature T in the equation $\Delta G = \Delta H - T\Delta S$ acts as a weighting factor that determines the relative importance of the enthalpy and entropy contributions to ΔG. If ΔH and ΔS are both negative, the reaction will be spontaneous only if the absolute value of ΔH is larger than the absolute value of $T\Delta S$. This is most likely at low temperatures, where the weighting factor T in $T\Delta S$ is small. If ΔH and ΔS are both positive, the reaction will be spontaneous only if $T\Delta S$ is larger than ΔH, which is most likely at high temperatures. We've already seen how these considerations apply to **phase changes** (Section 10.4).

REMEMBER...

The **melting** of ice, for example, is spontaneous above 0 °C where $T\Delta S$ outweighs ΔH and ΔG is negative but is nonspontaneous below 0 °C where ΔH outweighs $T\Delta S$ and ΔG is positive. (Section 10.4)

When ΔH and ΔS are either both negative or both positive, we can estimate the temperature at which a nonspontaneous reaction becomes spontaneous, or a spontaneous reaction becomes nonspontaneous, because that's the temperature at which the reaction is at equilibrium. At equilibrium, $\Delta G = \Delta H - T\Delta S = 0$ and so the crossover temperature at which a reaction changes between spontaneous and nonspontaneous is given by $T = \Delta H/\Delta S$.

$$\Delta G = \Delta H - T\Delta S = 0 \quad \text{At equilibrium}$$

$$T = \frac{\Delta H}{\Delta S} \quad \text{Crossover temperature}$$

This equation provides only an estimate, not an exact value, of the crossover temperature because the values of ΔH and ΔS are somewhat dependent on temperature. In general, the enthalpies and entropies of both reactants and products increase with increasing temperature, but the increases for the products tend to cancel the increases for the reactants. As a result, values of ΔH and ΔS for a reaction are relatively independent of temperature, at least over a small temperature range.

Worked Example 16.5

Determining the Temperature at Which a Reaction Becomes Spontaneous

Iron metal can be produced by reducing iron(III) oxide with hydrogen:

$$Fe_2O_3(s) + 3\,H_2(g) \longrightarrow 2\,Fe(s) + 3\,H_2O(g) \quad \Delta H^\circ = +98.8\text{ kJ};\ \Delta S^\circ = +141.5\text{ J/K}$$

(a) Is this reaction spontaneous under standard-state conditions at 25 °C?
(b) At what temperature will the reaction become spontaneous?

STRATEGY

To determine whether the reaction is spontaneous at 25 °C, we need to determine the sign of $\Delta G = \Delta H - T\Delta S$. To find the crossover temperature at which the reaction becomes spontaneous, we use the equation $T = \Delta H/\Delta S$.

SOLUTION

(a) At 25 °C (298 K), ΔG for the reaction is

$$\begin{aligned}\Delta G = \Delta H - T\Delta S &= (98.8\text{ kJ}) - (298\text{ K})(0.1415\text{ kJ/K})\\ &= (98.8\text{ kJ}) - (42.2\text{ kJ})\\ &= 56.6\text{ kJ}\end{aligned}$$

Because the positive ΔH term is larger than the positive $T\Delta S$ term, ΔG is positive and the reaction is nonspontaneous at 298 K.

(b) At sufficiently high temperatures, $T\Delta S$ becomes larger than ΔH, ΔG becomes negative, and the reaction becomes spontaneous. We can estimate the crossover temperature at which ΔG changes from positive to negative by substituting the values of ΔH and ΔS into the equation $T = \Delta H/\Delta S$:

$$T = \frac{\Delta H}{\Delta S} = \frac{98.8\text{ kJ}}{0.1415\text{ kJ/K}} = 698\text{ K}$$

Because this calculation assumes the values of ΔH and ΔS are unchanged on going from the standard-state temperature of 298 K to 698 K, the calculated value of T is only an estimate.

✔ BALLPARK CHECK

(a) Let's use rounded values of ΔH (100 kJ), T (300 K), and ΔS (0.14 kJ/K) to estimate the relative values of ΔH and $T\Delta S$. Because $T\Delta S = 300\text{ K} \times 0.14\text{ kJ/K} = 42\text{ kJ}$ is smaller than ΔH (100 kJ), $\Delta G = \Delta H - T\Delta S$ is positive and the reaction is nonspontaneous at 300 K. The ballpark check agrees with the solution.

(b) The temperature at which the reaction becomes spontaneous is approximately $(100\text{ kJ})/(0.14\text{ kJ/K}) \approx 700\text{ K}$, in agreement with the solution.

PROBLEM 16.7 Consider the decomposition of gaseous N_2O_4:

$$N_2O_4(g) \longrightarrow 2\,NO_2(g) \quad \Delta H^\circ = +55.3\text{ kJ};\ \Delta S^\circ = +175.7\text{ J/K}$$

(a) Is this reaction spontaneous under standard-state conditions at 25 °C?
(b) Estimate the temperature at which the reaction becomes spontaneous.

PROBLEM 16.8 The following data apply to the vaporization of mercury: $\Delta H_{vap} = 59.11$ kJ/mol; $\Delta S_{vap} = 93.9$ J/(K · mol).
(a) Does mercury boil at 325 °C and 1 atm pressure?
(b) What is the normal boiling point of mercury?

CONCEPTUAL PROBLEM 16.9 What are the signs (+, −, or 0) of ΔH, ΔS, and ΔG for the following spontaneous reaction of A atoms (red) and B atoms (blue)?

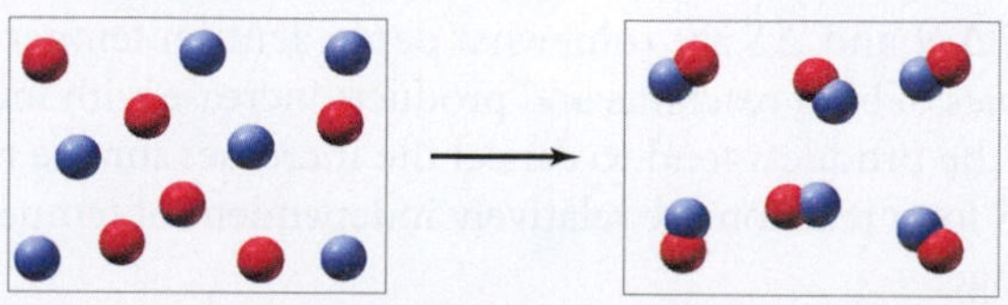

16.8 STANDARD FREE-ENERGY CHANGES FOR REACTIONS

The free energy of a substance, like its enthalpy and entropy, depends on its temperature, pressure, physical state (solid, liquid, or gas), and concentration (in the case of solutions). As a result, free-energy changes for chemical reactions must be compared under a well-defined set of standard-state conditions, as discussed in Section 8.6.

▶ **Standard-state conditions**
- Solids, liquids, and gases in pure form at 1 atm pressure
- Solutes at 1 M concentration
- A specified temperature, usually 25 °C

The **standard free-energy change ($\Delta G°$)** for a reaction is the change in free energy that occurs when reactants in their standard states are converted to products in their standard states. As with $\Delta H°$ (Section 8.9), the value of $\Delta G°$ is an **extensive property** (Section 1.3) that refers to the number of moles indicated in the balanced chemical equation. For example, $\Delta G°$ at 25 °C for the reaction

REMEMBER...

Extensive properties, like mass and volume, have values that depend on the sample size. **Intensive properties**, like temperature and melting point, have values that do not depend on the sample size. (Section 1.3)

$$2\,Na(s) + 2\,H_2O(l) \longrightarrow H_2(g) + 2\,Na^+(aq) + 2\,OH^-(aq)$$

is the change in free energy that occurs when 2 mol of solid sodium reacts completely with 2 mol of liquid water to give 1 mol of hydrogen gas at 1 atm pressure, along with an aqueous solution that contains 2 mol of Na^+ ions and 2 mol of OH^- ions at 1 M concentrations, with all reactants and products at a temperature of 25 °C. For this reaction, $\Delta G° = -364$ kJ.

Because the free-energy change for any process at constant temperature and pressure is $\Delta G = \Delta H - T\Delta S$, we can calculate the standard free-energy change $\Delta G°$ for a reaction from the standard enthalpy change $\Delta H°$ and the standard entropy change $\Delta S°$. Consider again the Haber synthesis of ammonia:

$$N_2(g) + 3\,H_2(g) \longrightarrow 2\,NH_3(g) \quad \Delta H° = -92.2\text{ kJ}; \Delta S° = -198.7\text{ J/K}$$

The standard free-energy change at 25 °C (298 K) is

$$\begin{aligned}\Delta G° = \Delta H° - T\Delta S° &= (-92.2 \times 10^3\text{ J}) - (298\text{ K})(-198.7\text{ J/K}) \\ &= (-92.2 \times 10^3\text{ J}) - (-59.2 \times 10^3\text{ J}) \\ &= -33.0 \times 10^3\text{ J} \quad \text{or} \quad -33.0\text{ kJ}\end{aligned}$$

Because the negative $\Delta H°$ term is larger than the negative $T\Delta S°$ term at 25 °C, $\Delta G°$ is negative and the reaction is spontaneous under standard-state conditions.

It's important to note that a standard free-energy change applies to a hypothetical process rather than an actual process. In the hypothetical process, separate reactants in their standard states are completely converted to separate products in their standard states. In an actual process, however, reactants and products are mixed together and the reaction may not go to completion.

Take the Haber synthesis, for example. The hypothetical process is

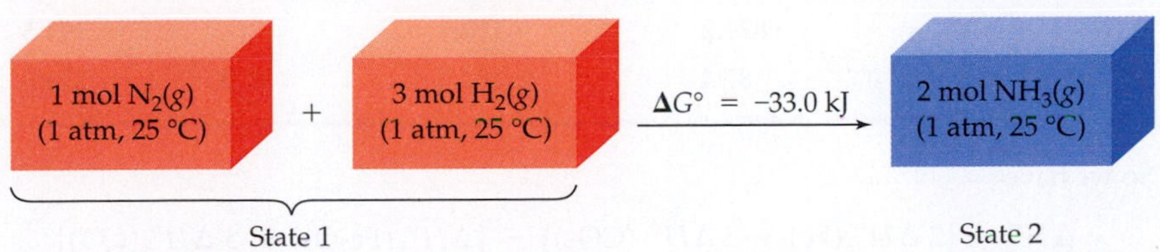

where $\Delta G° = -33.0$ kJ is the change in the free energy of the system on going from state 1 to state 2. In an actual synthesis of ammonia, however, the reactants N_2 and H_2 are not separate but are mixed together. Moreover, the reaction doesn't go to completion; it reaches an equilibrium state in which both reactants and products are present together.

How, then, should we think about the meaning of $\Delta G°$ in the context of an actual reaction? One way would be to suppose that we have a mixture of N_2, H_2, and NH_3, with each substance present at a partial pressure of 1 atm. Suppose further that the mixture behaves as an ideal gas so that the free energy of each component in the mixture is the same as the free energy of the pure substance. Finally, suppose that the number of moles of each component—say *x*, *y*, and *z*—is very large so that the partial pressures don't change appreciably when 1 mol of N_2 and 3 mol of H_2 are converted to 2 mol of NH_3. In other words, we are imagining the following real process:

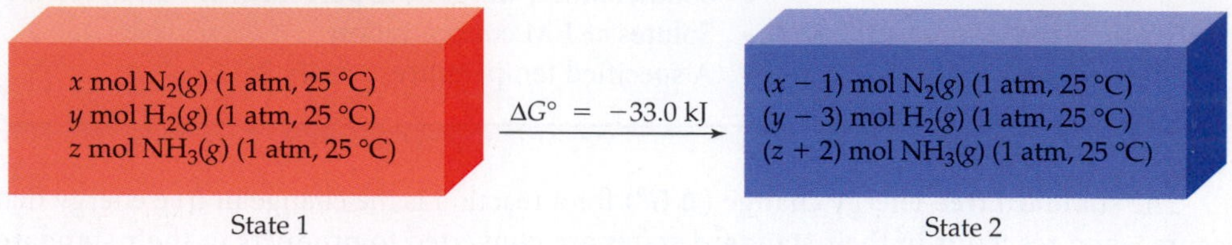

The free-energy change for this process is the standard free-energy change $\Delta G°$ because each reactant and product is present at 1 atm pressure. If $\Delta G°$ is negative, the reaction will proceed spontaneously to give more products. If $\Delta G°$ is positive, the reaction will proceed in the reverse direction to give more reactants. As always, the value of $\Delta G°$ indicates only whether the reaction is spontaneous and provides no information about its rate.

Worked Example 16.6

Calculating $\Delta G°$ for a Reaction from $\Delta H°$ and $\Delta S°$

Iron metal is produced commercially by reducing iron(III) oxide in iron ore with carbon monoxide:

$$Fe_2O_3(s) + 3\,CO(g) \longrightarrow 2\,Fe(s) + 3\,CO_2(g)$$

(a) Calculate the standard free-energy change for this reaction at 25 °C.
(b) Is the reaction spontaneous under standard-state conditions at 25 °C?
(c) Does the reverse reaction become spontaneous at higher temperatures? Explain.

STRATEGY

(a) We can calculate the standard free-energy change from the relation $\Delta G° = \Delta H° - T\Delta S°$, but first we must find $\Delta H°$ and $\Delta S°$ from standard enthalpies of formation ($\Delta H°_f$) and standard molar entropies ($S°$).
(b) The reaction is spontaneous under standard-state conditions if $\Delta G°$ is negative.
(c) The spontaneity of the reaction at higher temperatures depends on the signs and magnitudes of $\Delta H°$ and $\Delta S°$.

SOLUTION

(a) The following values of $\Delta H°_f$ and $S°$ are found in Appendix B:

	$Fe_2O_3(s)$	$CO(g)$	$Fe(s)$	$CO_2(g)$
$\Delta H°_f$ (kJ/mol)	−824.2	−110.5	0	−393.5
$S°$ [J/(K · mol)]	87.4	197.6	27.3	213.6

So we have

$$\begin{aligned}\Delta H° &= [2\,\Delta H°_f(Fe) + 3\,\Delta H°_f(CO_2)] - [\Delta H°_f(Fe_2O_3) + 3\,\Delta H°_f(CO)]\\ &= [(2\text{ mol})(0\text{ kJ/mol}) + (3\text{ mol})(-393.5\text{ kJ/mol})]\\ &\qquad - [(1\text{ mol})(-824.2\text{ kJ/mol}) + (3\text{ mol})(-110.5\text{ kJ/mol})]\\ \Delta H° &= -24.8\text{ kJ}\end{aligned}$$

and

$$\Delta S^\circ = [2\,S^\circ(\text{Fe}) + 3\,S^\circ(\text{CO}_2)] - [S^\circ(\text{Fe}_2\text{O}_3) + 3\,S^\circ(\text{CO})]$$

$$= \left[(2\text{ mol})\left(27.3\,\frac{\text{J}}{\text{K}\cdot\text{mol}}\right) + (3\text{ mol})\left(213.6\,\frac{\text{J}}{\text{K}\cdot\text{mol}}\right)\right]$$

$$-\left[(1\text{ mol})\left(87.4\,\frac{\text{J}}{\text{K}\cdot\text{mol}}\right) + (3\text{ mol})\left(197.6\,\frac{\text{J}}{\text{K}\cdot\text{mol}}\right)\right]$$

$$\Delta S^\circ = +15.2\text{ J/K} \quad \text{or} \quad 0.0152\text{ kJ/K}$$

Therefore,

$$\Delta G^\circ = \Delta H^\circ - T\Delta S^\circ$$

$$= (-24.8\text{ kJ}) - (298\text{ K})(0.0152\text{ kJ/K})$$

$$\Delta G^\circ = -29.3\text{ kJ}$$

(b) Because ΔG° is negative, the reaction is spontaneous at 25 °C. This means that a mixture of $Fe_2O_3(s)$, $CO(g)$, $Fe(s)$, and $CO_2(g)$, with each gas at a partial pressure of 1 atm, will react at 25 °C to produce more iron metal.

(c) Because ΔH° is negative and ΔS° is positive, ΔG° will be negative at all temperatures. The forward reaction is therefore spontaneous at all temperatures, and the reverse reaction does not become spontaneous at higher temperatures.

PROBLEM 16.10 Consider the thermal decomposition of calcium carbonate:

$$CaCO_3(s) \longrightarrow CaO(s) + CO_2(g)$$

(a) Use the data in Appendix B to calculate the standard free-energy change for this reaction at 25 °C.

(b) Will a mixture of solid $CaCO_3$, solid CaO, and gaseous CO_2 at 1 atm pressure react spontaneously at 25 °C to produce more CaO and CO_2?

(c) Assuming that ΔH° and ΔS° are independent of temperature, estimate the temperature at which the reaction becomes spontaneous.

CONCEPTUAL PROBLEM 16.11 Consider the following endothermic decomposition of AB_2 molecules:

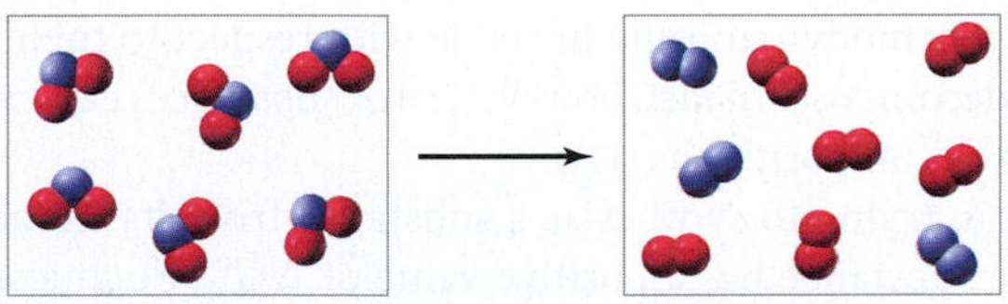

(a) What is the sign (+, −, or 0) of ΔS° for the reaction?

(b) Is the reaction more likely to be spontaneous at high temperatures or at low temperatures? Explain.

16.9 STANDARD FREE ENERGIES OF FORMATION

The **standard free energy of formation** (ΔG°_f) of a substance is the free-energy change for the formation of 1 mol of the substance in its standard state from the most stable form of its constituent elements in their standard states. For example, we found in Section 16.8 that the standard free-energy change ΔG° for the synthesis of 2 mol of NH_3 from its constituent elements is −33.0 kJ:

$$N_2(g) + 3\,H_2(g) \longrightarrow 2\,NH_3(g) \qquad \Delta G^\circ = -33.0\text{ kJ}$$

Therefore, ΔG°_f for ammonia is −33.0 kJ/2 mol, or −16.5 kJ/mol.

Values of ΔG°_f at 25 °C for some common substances are listed in **Table 16.3**, and additional values are given in Appendix B. Note that ΔG°_f for an element in its most stable form at 25 °C is defined to be zero. Thus, solid graphite has $\Delta G^\circ_f = 0$ kJ/mol, but diamond, a less stable form of solid carbon at 25 °C, has $\Delta G^\circ_f = 2.9$ kJ/mol.

TABLE 16.3 Standard Free Energies of Formation for Some Common Substances at 25 °C

Substance	Formula	ΔG_f° (kJ/mol)	Substance	Formula	ΔG_f° (kJ/mol)
Gases			**Liquids**		
Acetylene	C_2H_2	209.9	Acetic acid	CH_3CO_2H	−390
Ammonia	NH_3	−16.5	Ethanol	CH_3CH_2OH	−174.9
Carbon dioxide	CO_2	−394.4	Methanol	CH_3OH	−166.6
Carbon monoxide	CO	−137.2	Water	H_2O	−237.2
Ethylene	C_2H_4	68.1	**Solids**		
Hydrogen	H_2	0	Calcium carbonate	$CaCO_3$	−1129.1
Methane	CH_4	−50.8	Calcium oxide	CaO	−603.3
Nitrogen	N_2	0	Diamond	C	2.9
Nitrogen dioxide	NO_2	51.3	Graphite	C	0
Dinitrogen tetroxide	N_2O_4	99.8	Iron(III) oxide	Fe_2O_3	−742.2

As with standard enthalpies of formation, ΔH°_f, a zero value of ΔG°_f for elements in their most stable form establishes a thermochemical "sea level," or reference point, with respect to which the standard free energies of other substances are measured. We can't measure the absolute value of a substance's free energy (as we can the entropy), but that's not a problem because we are interested only in free-energy *differences* between reactants and products.

The standard free energy of formation of a substance measures its thermodynamic stability with respect to its constituent elements. Substances that have a negative value of ΔG°_f, such as carbon dioxide and water, are stable and do not decompose to their constituent elements under standard-state conditions. Substances that have a positive value of ΔG°_f, such as ethylene and nitrogen dioxide, are thermodynamically unstable with respect to their constituent elements and can, in principle, decompose. In fact, however, such substances can exist for long periods of time if the rate of their decomposition is slow.

There's no point in trying to synthesize a substance from its elements under standard-state conditions if the substance has a positive value of ΔG°_f. Such a substance would have to be prepared at other temperatures and/or pressures, or it would have to be made from alternative starting materials using a reaction that has a negative free-energy change. Thus, a knowledge of thermodynamics can save considerable time in chemical synthesis.

In Section 16.8, we calculated standard free-energy changes for reactions from the equation $\Delta G^\circ = \Delta H^\circ - T\Delta S^\circ$, using tabulated values of ΔH°_f and S° to find ΔH° and ΔS°. Alternatively, we can calculate ΔG° more directly by subtracting the standard free energies of formation of all the reactants from the standard free energies of formation of all the products.

▶ **Standard free-energy change** $\Delta G^\circ = \Delta G^\circ_f(\text{products}) - \Delta G^\circ_f(\text{reactants})$

For the general reaction

$$a\text{A} + b\text{B} \longrightarrow c\text{C} + d\text{D}$$

the standard free-energy change is

▶ $$\Delta G^\circ = [c\,\Delta G^\circ_f(\text{C}) + d\,\Delta G^\circ_f(\text{D})] - [a\,\Delta G^\circ_f(\text{A}) + b\,\Delta G^\circ_f(\text{B})]$$

To illustrate, let's calculate the standard free-energy change for the reaction in Worked Example 16.6—the reduction of iron(III) oxide by carbon monoxide:

$$Fe_2O_3(s) + 3\,CO(g) \longrightarrow 2\,Fe(s) + 3\,CO_2(g)$$

Using the $\Delta G°_f$ values in Table 16.3, we obtain

$$\begin{aligned}\Delta G° &= [2\,\Delta G°_f(Fe) + 3\,\Delta G°_f(CO_2)] - [\Delta G°_f(Fe_2O_3) + 3\,\Delta G°_f(CO)]\\ &= [(2\text{ mol})(0\text{ kJ/mol}) + (3\text{ mol})(-394.4\text{ kJ/mol})]\\ &\qquad - [(1\text{ mol})(-742.2\text{ kJ/mol}) + (3\text{ mol})(-137.2\text{ kJ/mol})]\\ \Delta G° &= -29.4\text{ kJ}\end{aligned}$$

This result agrees well with the value of −29.3 kJ calculated from $\Delta G° = \Delta H° - T\Delta S°$ in Worked Example 16.6.

Worked Example 16.7

Calculating $\Delta G°$ for a Reaction from $\Delta G°_f$ Values

(a) Calculate the standard free-energy change for the oxidation of ammonia to give nitric oxide (NO) and water. Is it worth trying to find a catalyst for this reaction under standard-state conditions at 25 °C?

$$4\,NH_3(g) + 5\,O_2(g) \longrightarrow 4\,NO(g) + 6\,H_2O(l)$$

(b) Is it worth trying to find a catalyst for the synthesis of NO from gaseous N_2 and O_2 under standard-state conditions at 25 °C?

STRATEGY

We can calculate $\Delta G°$ most easily from tabulated standard free energies of formation (Appendix B). It's worth trying to find a catalyst for a reaction only if the reaction has a negative free-energy change.

SOLUTION

(a)

$$\begin{aligned}\Delta G° &= [4\,\Delta G°_f(NO) + 6\,\Delta G°_f(H_2O)] - [4\,\Delta G°_f(NH_3) + 5\,\Delta G°_f(O_2)]\\ &= [(4\text{ mol})(87.6\text{ kJ/mol}) + (6\text{ mol})(-237.2\text{ kJ/mol})]\\ &\qquad - [(4\text{ mol})(-16.5\text{ kJ/mol}) + (5\text{ mol})(0\text{ kJ/mol})]\\ \Delta G° &= -1006.8\text{ kJ}\end{aligned}$$

It is worth looking for a catalyst because the negative value of $\Delta G°$ indicates that the reaction is spontaneous under standard-state conditions. (This reaction is the first step in the Ostwald process for the production of nitric acid. In industry, the reaction is carried out using a platinum–rhodium catalyst.)

(b) It's not worth looking for a catalyst for the reaction $N_2(g) + O_2(g) \rightarrow 2\,NO(g)$ because the standard free energy of formation of NO is positive ($\Delta G°_f = 87.6$ kJ/mol). This means that NO is unstable with respect to decomposition to N_2 and O_2 under standard-state conditions at 25 °C. A catalyst could only increase the rate of its decomposition. A catalyst can't affect the composition of the equilibrium mixture (Section 13.11), and so it can't reverse the direction of the reaction.

PROBLEM 16.12

(a) Using values of $\Delta G°_f$ in Appendix B, calculate the standard free-energy change for the reaction of calcium carbide (CaC_2) with water. Might this reaction be used for synthesis of acetylene ($HC{\equiv}CH$, or C_2H_2)?

$$CaC_2(s) + 2\,H_2O(l) \longrightarrow C_2H_2(g) + Ca(OH)_2(s)$$

(b) Is it possible to synthesize acetylene from solid graphite and gaseous H_2 at 25 °C and 1 atm pressure?

16.10 FREE-ENERGY CHANGES FOR REACTIONS UNDER NONSTANDARD-STATE CONDITIONS

The sign of the standard free-energy change $\Delta G°$ tells us the direction of spontaneous reaction when both reactants and products are present at standard-state conditions. In actual reactions, however, the composition of the reaction mixture seldom corresponds to standard-state pressures and concentrations. Moreover, the partial pressures and concentrations change as a reaction proceeds. How, then, do we calculate the free-energy change ΔG for a reaction when the reactants and products are present at nonstandard-state pressures and concentrations?

The answer is given by the relation

$$\Delta G = \Delta G° + RT \ln Q$$

where ΔG is the free-energy change under nonstandard-state conditions, $\Delta G°$ is the free-energy change under standard-state conditions, R is the gas constant, T is the temperature in kelvin, and Q is the reaction quotient (Q_p for reactions involving gases because the standard state for gases is 1 atm pressure, or Q_c for reactions involving solutes in solution because the standard state for solutes is 1 M concentration). Recall from Section 13.6 that the reaction quotient Q_c is an expression having the same form as the equilibrium-constant expression K_c except that the concentrations do not necessarily have equilibrium values. Similarly, Q_p has the same form as K_p except that the partial pressures have arbitrary values. For example, for the Haber synthesis of ammonia:

$$N_2(g) + 3\,H_2(g) \rightleftharpoons 2\,NH_3(g) \quad Q_p = \frac{(P_{NH_3})^2}{(P_{N_2})(P_{H_2})^3}$$

For reactions that involve both gases and solutes in solution, the reaction quotient Q contains partial pressures of gases in atmospheres and molar concentrations of solutes.

We won't derive the equation for ΔG under nonstandard-state conditions. It's hardly surprising, however, that ΔG and Q should turn out to be related because both predict the direction of a reaction. Worked Example 16.8 shows how to use this equation.

Worked Example 16.8

Calculating ΔG for a Reaction under Nonstandard-State Conditions

Calculate the free-energy change for ammonia synthesis at 25 °C (298 K) given the following sets of partial pressures:

(a) 1.0 atm N_2, 3.0 atm H_2, 0.020 atm NH_3

(b) 0.010 atm N_2, 0.030 atm H_2, 2.0 atm NH_3

$$N_2(g) + 3\,H_2(g) \rightleftharpoons 2\,NH_3(g) \quad \Delta G° = -33.0 \text{ kJ}$$

STRATEGY

We can calculate ΔG from the relation $\Delta G = \Delta G° + RT \ln Q$, where Q is Q_p for the reaction $N_2(g) + 3\,H_2(g) \rightleftharpoons 2\,NH_3(g)$.

SOLUTION

(a) The value of Q_p is

$$Q_p = \frac{(P_{NH_3})^2}{(P_{N_2})(P_{H_2})^3} = \frac{(0.020)^2}{(1.0)(3.0)^3} = 1.5 \times 10^{-5}$$

Substituting this value of Q_p into the equation for ΔG gives

$$\begin{aligned}\Delta G &= \Delta G° + RT \ln Q_p \\ &= (-33.0 \times 10^3 \text{ J/mol}) + [8.314 \text{ J/(K}\cdot\text{mol)}](298 \text{ K})(\ln 1.5 \times 10^{-5}) \\ &= (-33.0 \times 10^3 \text{ J/mol}) + (-27.5 \times 10^3 \text{ J/mol}) \\ \Delta G &= -60.5 \text{ kJ/mol}\end{aligned}$$

To maintain consistent units in this calculation, we have expressed ΔG and $\Delta G°$ in units of J/mol because R has units of J/(K · mol). The phrase *per mole* in this context means per molar amounts of reactants and products indicated by the coefficients in the balanced equation. Thus, the free-energy change is −60.5 kJ when 1 mol of N_2 and 3 mol of H_2 are converted to 2 mol of NH_3 under the specified conditions.

ΔG is more negative than $\Delta G°$ because Q_p is less than 1 and ln Q_p is therefore a negative number. Thus, the reaction has a greater thermodynamic tendency to occur under the cited conditions than it does under standard-state conditions. When each reactant and product is present at a partial pressure of 1 atm, $Q_p = 1$, $\ln Q_p = 0$, and $\Delta G = \Delta G°$.

(b) The value of Q_p is

$$Q_p = \frac{(P_{NH_3})^2}{(P_{N_2})(P_{H_2})^3} = \frac{(2.0)^2}{(0.010)(0.030)^3} = 1.5 \times 10^7$$

The corresponding value of ΔG is

$$\begin{aligned}\Delta G &= \Delta G° + RT \ln Q_p \\ &= (-33.0 \times 10^3 \text{ J/mol}) + [8.314 \text{ J/(K·mol)}](298 \text{ K})(\ln 1.5 \times 10^7) \\ &= (-33.0 \times 10^3 \text{ J/mol}) + (40.9 \times 10^3 \text{ J/mol}) \\ \Delta G &= 7.9 \text{ kJ/mol}\end{aligned}$$

Because Q_p is large enough to give a positive value for ΔG, the reaction is nonspontaneous in the forward direction but spontaneous in the reverse direction. Thus, as we saw already in Section 13.6, the direction in which a reaction proceeds spontaneously depends on the composition of the reaction mixture.

✔ BALLPARK CHECK

(a) Based on Le Châtelier's principle (Section 13.7), we expect that the reaction will have a greater tendency to occur under the cited conditions than under standard-state conditions because one of the reactant partial pressures is greater than 1 atm and the product partial pressure is less than 1 atm. We therefore predict that Q_p will be less than 1 and ΔG will be more negative than $\Delta G°$, in agreement with the solution.

(b) In this case, the reaction mixture is rich in the product and poor in the reactants. Therefore, Q_p is expected to be greater than 1 and ΔG should be more positive than $\Delta G°$, in agreement with the solution.

▶ PROBLEM 16.13 Calculate ΔG for the formation of ethylene (C_2H_4) from carbon and hydrogen at 25 °C when the partial pressures are 100 atm H_2 and 0.10 atm C_2H_4.

$$2\,C(s) + 2\,H_2(g) \longrightarrow C_2H_4(g) \qquad \Delta G° = 68.1 \text{ kJ}$$

Is the reaction spontaneous in the forward or the reverse direction?

▶ CONCEPTUAL PROBLEM 16.14 Consider the following gas-phase reaction of A_2 (red) and B_2 (blue) molecules:

$$A_2 + B_2 \rightleftharpoons 2\,AB \qquad \Delta G° = 15 \text{ kJ}$$

(a) Which of the following reaction mixtures has the largest ΔG of reaction? Which has the smallest?

(1) **(2)** **(3)**

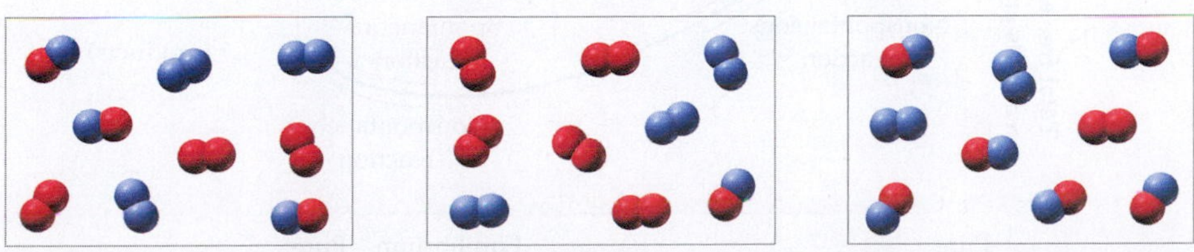

(b) If the partial pressure of each reactant and product in reaction mixture 1 is equal to 1 atm, what is the value of ΔG for the reaction in mixture 1?

16.11 FREE ENERGY AND CHEMICAL EQUILIBRIUM

Now that we've seen how ΔG for a reaction depends on the composition of a reaction mixture, we can understand how the total free energy of a reaction mixture changes as the reaction progresses toward equilibrium. Look again at the expression for calculating ΔG:

$$\Delta G = \Delta G^\circ + RT \ln Q$$

If the reaction mixture contains mainly reactants and almost no products, Q will be much less than 1 and $RT \ln Q$ will be a very large negative number (minus infinity when $Q = 0$). Consequently, no matter what the value of ΔG° (positive or negative), the negative $RT \ln Q$ term will dominate the ΔG° term and ΔG will be negative. This means that the forward reaction is always spontaneous when the concentration of products is very small. Conversely, if the reaction mixture contains mainly products and almost no reactants, Q will be much greater than 1 and $RT \ln Q$ will be a very large positive number (plus infinity when no reactants are present). Consequently, the positive $RT \ln Q$ term will dominate the ΔG° term and ΔG will be positive. Thus, the reverse reaction is always spontaneous when the concentration of reactants is very small. These conditions are summarized as follows:

- When the reaction mixture is mostly reactants,

  $$Q \ll 1 \qquad RT \ln Q \ll 0 \qquad \Delta G < 0$$

  the total free energy decreases as the reaction proceeds spontaneously in the forward direction.
- When the reaction mixture is mostly products,

  $$Q \gg 1 \qquad RT \ln Q \gg 0 \qquad \Delta G > 0$$

  the total free energy decreases as the reaction proceeds spontaneously in the reverse direction.

Figure 16.11 shows how the total free energy of a reaction mixture changes as the reaction progresses. Because the free energy decreases as pure reactants form products and also decreases as pure products form reactants, the free-energy curve must go through a

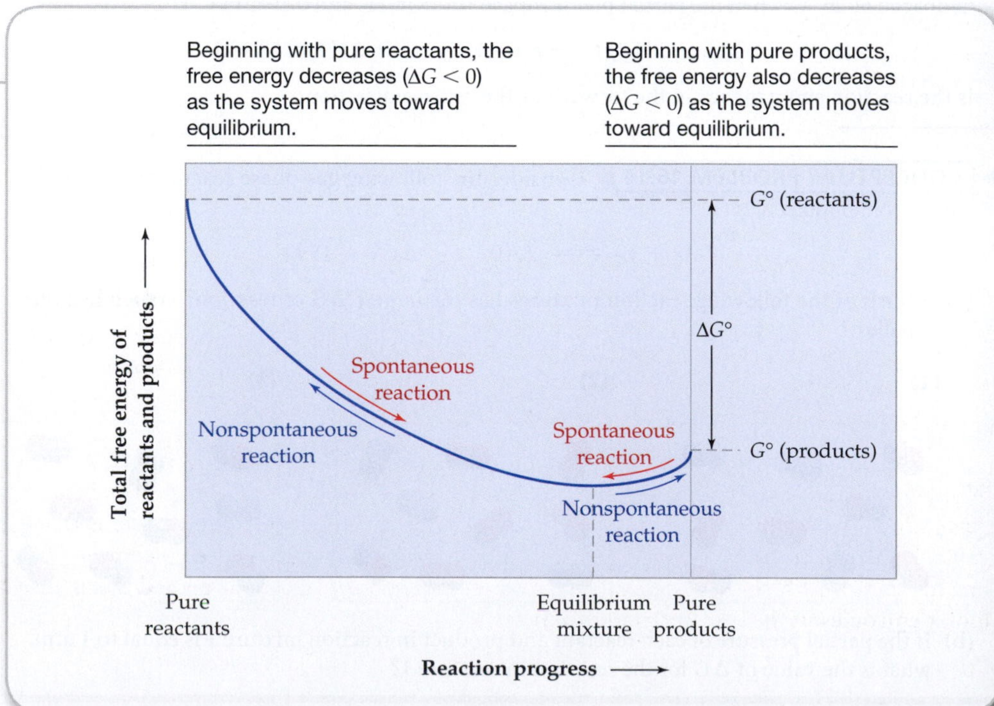

FIGURE 16.11

The total free energy of a reaction mixture as a function of the progress of the reaction. The graph is drawn assuming that the pure reactants and pure products are in their standard states and that ΔG° for the reaction is negative, so the equilibrium composition is rich in products.

minimum somewhere between pure reactants and pure products. At that minimum free-energy composition, the system is at equilibrium because the conversion of either reactants to products or products to reactants would increase the free energy. The equilibrium composition persists indefinitely unless the system is disturbed by an external influence.

The sign of ΔG for the reaction is the same as the sign of the slope of the free-energy curve (Figure 16.11). To the left of the equilibrium composition, ΔG and the slope of the curve are negative and the free energy decreases as reactants are converted to products. To the right of the equilibrium composition, ΔG and the slope of the curve are positive. Exactly at the equilibrium composition, ΔG and the slope of the curve are zero and no net reaction occurs.

We can now derive a relationship between free energy and the equilibrium constant. At equilibrium, ΔG for a reaction is zero and the reaction quotient Q equals the equilibrium constant K. Substituting $\Delta G = 0$ and $Q = K$ into the equation

$$\Delta G = \Delta G^\circ + RT \ln Q$$

gives $$0 = \Delta G^\circ + RT \ln K$$ or

$$\Delta G^\circ = -RT \ln K$$

In this equation, K is K_p for reactions involving gases and K_c for reactions involving solutes in solution. For reactions involving both gases and solutes in solution, the equilibrium-constant expression contains partial pressures of gases and molar concentrations of solutes.

The equation $\Delta G^\circ = -RT \ln K$ is one of the most important relationships in chemical thermodynamics because it allows us to calculate the equilibrium constant for a reaction from the standard free-energy change, or vice versa. This relationship is especially useful when K is difficult to measure. Consider a reaction so slow that it takes more than an experimenter's lifetime to reach equilibrium or a reaction that goes essentially to completion, so that the equilibrium concentrations of the reactants are extremely small and hard to measure. We can't measure K directly in such cases, but we can calculate its value from ΔG°.

The relationship between ΔG° and the equilibrium constant K is summarized in **Table 16.4**. A reaction with a negative value of ΔG° has an equilibrium constant greater than 1, which corresponds to a minimum in the free-energy curve of Figure 16.11 at a composition rich in products. Conversely, a reaction that has a positive value of ΔG° has an equilibrium constant less than 1 and a minimum in the free-energy curve at a composition rich in reactants. Try redrawing Figure 16.11 for the case where $\Delta G^\circ > 0$.

TABLE 16.4 Relationship Between the Standard Free-Energy Change and the Equilibrium Constant for a Reaction: $\Delta G^\circ = -RT \ln K$

ΔG°	$\ln K$	K	Comment
$\Delta G^\circ < 0$	$\ln K > 0$	$K > 1$	The equilibrium mixture is mainly products.
$\Delta G^\circ > 0$	$\ln K < 0$	$K < 1$	The equilibrium mixture is mainly reactants.
$\Delta G^\circ = 0$	$\ln K = 0$	$K = 1$	The equilibrium mixture contains comparable amounts of reactants and products.

We have now answered the fundamental question posed at the beginning of this chapter: What determines the value of the equilibrium constant—that is, what properties of nature determine the direction and extent of a particular chemical reaction? The answer is that the value of the equilibrium constant is determined by the *standard* free-energy change (ΔG°) for the reaction, which depends, in turn, on the standard heats of formation and the standard molar entropies of the reactants and products.

Worked Example 16.9

Calculating an Equilibrium Constant from $\Delta G°$ for a Reaction

Methanol (CH_3OH), an important alcohol used in the manufacture of adhesives, fibers, and plastics, is synthesized industrially by the reaction

$$CO(g) + 2\,H_2(g) \rightleftharpoons CH_3OH(g)$$

Use the thermodynamic data in Appendix B to calculate the equilibrium constant for this reaction at 25 °C.

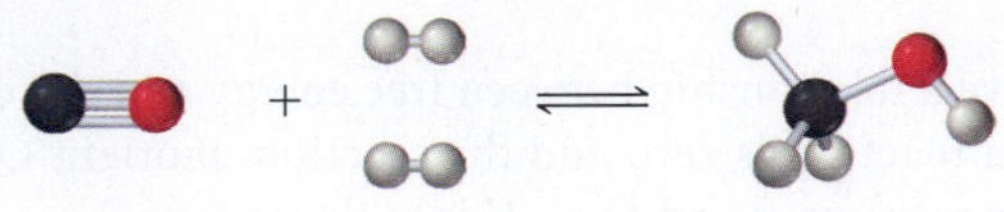

STRATEGY

First calculate $\Delta G°$ for the reaction from the tabulated values of $\Delta G°_f$ for the reactants and products. Then use the equation $\Delta G° = -RT \ln K$ to find the value of the equilibrium constant.

SOLUTION

$$\Delta G° = \Delta G°_f(CH_3OH) - [\Delta G°_f(CO) + 2\,\Delta G°_f(H_2)]$$
$$= (1\text{ mol})(-162.3\text{ kJ/mol}) - [(1\text{ mol})(-137.2\text{ kJ/mol}) + (2\text{ mol})(0\text{ kJ/mol})]$$
$$\Delta G° = -25.1\text{ kJ}$$

Solving the equation $\Delta G° = -RT \ln K$

for ln K gives $\ln K = \dfrac{-\Delta G°}{RT} = \dfrac{-(-25.1 \times 10^3\text{ J/mol})}{[8.314\text{ J/(K}\cdot\text{mol)}](298\text{ K})} = 10.1$

Therefore, $K = K_p = \text{antiln } 10.1 = e^{10.1} = 2 \times 10^4$

The equilibrium constant obtained by this procedure is K_p because the reactants and products are gases and their standard states are defined in terms of pressure. If we want the value of K_c, we must calculate it from the relation $K_p = K_c(RT)^{\Delta n}$ (Section 13.4), where R must be expressed in the proper units $[R = 0.082\,06\text{ (L}\cdot\text{atm)/(K}\cdot\text{mol)}]$.

✔ BALLPARK CHECK

$\Delta G°$ is negative, so the equilibrium mixture should be mainly products ($K > 1$), in agreement with the solution.

▲ Mercury has an appreciable vapor pressure at room temperature, and its handling requires adequate ventilation.

Worked Example 16.10

Calculating a Vapor Pressure from $\Delta G°$

The value of $\Delta G°_f$ at 25 °C for gaseous mercury is 31.85 kJ/mol. What is the vapor pressure of mercury at 25 °C?

STRATEGY

The vapor pressure (in atm) equals K_p for the reaction

$$Hg(l) \rightleftharpoons Hg(g) \qquad K_p = P_{Hg}$$

$Hg(l)$ is omitted from the equilibrium-constant expression because it is a pure liquid. Because the standard state for elemental mercury is the pure liquid, $\Delta G°_f = 0$ for $Hg(l)$ and $\Delta G°$ for the vaporization reaction simply equals $\Delta G°_f$ for $Hg(g)$(31.85 kJ/mol). We can calculate K_p from the equation $\Delta G° = -RT \ln K_p$, as in Worked Example 16.9.

SOLUTION

$$\ln K_p = \frac{-\Delta G°}{RT} = \frac{-(31.85 \times 10^3\ \text{J/mol})}{[8.314\ \text{J/(K} \cdot \text{mol)}](298\ \text{K})} = -12.86$$

$$K_p = \text{antiln}\,(-12.86) = e^{-12.86} = 2.6 \times 10^{-6}$$

Since K_p is defined in units of atmospheres, the vapor pressure of mercury at 25 °C is 2.6×10^{-6} atm (0.0020 mm Hg). Because the vapor pressure is appreciable and mercury is toxic in the lungs, mercury should not be handled without adequate ventilation.

✔ BALLPARK CHECK

$\Delta G°$ is positive, so the vaporization reaction should not proceed very far before reaching equilibrium. Thus, K_p should be less than 1, in agreement with the solution.

Worked Example 16.11

Calculating $\Delta G°$ from an Equilibrium Constant

At 25 °C, K_{sp} for $PbCrO_4$ is 2.8×10^{-13}. Calculate the standard free-energy change at 25 °C for the reaction $PbCrO_4(s) \rightleftharpoons Pb^{2+}(aq) + CrO_4^{2-}(aq)$.

STRATEGY AND SOLUTION

We can calculate $\Delta G°$ directly from the equilibrium constant K_{sp}:

$$\Delta G° = -RT \ln K_{sp} = -[8.314\ \text{J/(K} \cdot \text{mol)}](298\ \text{K})(\ln 2.8 \times 10^{-13})$$
$$= 71.6 \times 10^3\ \text{J/mol} = 71.6\ \text{kJ/mol}$$

✔ BALLPARK CHECK

The K_{sp} value is much less than 1, so $\Delta G°$ should be a large positive number, in agreement with the solution.

PROBLEM 16.15 Given the data in Appendix B, calculate K_p at 25 °C for the reaction

$$CaCO_3(s) \rightleftharpoons CaO(s) + CO_2(g).$$

PROBLEM 16.16 Use the data in Appendix B to calculate the vapor pressure of water at 25 °C.

PROBLEM 16.17 At 25 °C, K_w for the dissociation of water is 1.0×10^{-14}. Calculate $\Delta G°$ for the reaction $2\ H_2O(l) \rightleftharpoons H_3O^+(aq) + OH^-(aq)$.

FYI DOES ENTROPY PREVENT THE EVOLUTION OF BIOLOGICAL COMPLEXITY?

The idea of entropy has intrigued thinkers for more than a century. Poets, philosophers, physicists, and biologists have all struggled to understand the consequences of entropy.

Poets and philosophers have spoken of entropy as "time's arrow," a metaphor that arises out of the second law of thermodynamics. According to the second law, all spontaneously occurring processes are accompanied by an increase in the entropy of the universe. At some far distant time, when all matter and energy is randomly distributed and there is no available energy left, the universe as we know it must end.

Physicists speak of entropy as giving a directionality to time. There is a symmetry to basic physical laws that makes them equally valid when the signs of the quantities are reversed. Thus, the attraction between a positively charged proton and a negatively charged electron,

▲ At some distant time, when all is disorder, the universe as we know it must end.

for example, is exactly the same as the attraction between a negatively charged proton (a so-called *antiproton*) and a positively charged electron (a *positron*). Time, however, cannot be reversed because of entropy. Any process that takes place spontaneously over time must increase the entropy of the universe. The reverse process, which would have to go backward over time, can't occur because it would decrease the entropy of the universe. Thus, there is a one-way nature to time that is not shared by other quantities.

Biologists, too, have been intrigued by entropy and its consequences. Their problem is that if the randomness of the universe is always increasing, how is it possible for enormously complex and increasingly sophisticated life forms to evolve? After all, the more complex the organism, the less the amount of randomness and the *lower* the entropy. The answer to the biologists' question again arises from the second law of thermodynamics: All spontaneous processes increase the randomness of the universe—that is, the total randomness of both system and surroundings. It's perfectly possible, however, for the randomness of any *system* to decrease spontaneously as long as the randomness of the *surroundings* increases by an even greater amount.

The energy used to power living organisms comes from sunlight, caused ultimately by nuclear reactions in the Sun. Photosynthetic cells in plants use the Sun's energy to make glucose, which is then used by animals as their primary source of energy. The energy an animal obtains from glucose is then used to build and organize complex molecules, resulting in a decrease in entropy for the animal. At the same time, however, the entropy of the surroundings increases as the animal releases small, simple waste products such as CO_2 and H_2O. Furthermore, heat is released by the animal, further increasing the entropy of the surroundings. Thus, an organism pays for its decrease in entropy by increasing the entropy of the rest of the universe.

PROBLEM 16.18 Consider the growth of a human adult from a single cell. Does this process violate the second law of thermodynamics? Explain.

PROBLEM 16.19 If you watched a movie run backward, would you expect to see violations of the second law? Explain.

SUMMARY

Thermodynamics deals with the interconversion of heat and other forms of energy and allows us to predict the direction and extent of chemical reactions and other spontaneous processes. A **spontaneous process** proceeds on its own without any external influence. All spontaneous reactions move toward equilibrium.

Entropy, denoted by *S*, is a state function that measures molecular randomness. The entropy of a system (reactants plus products) increases (ΔS is positive) for the following processes: phase transitions that convert a solid to a liquid or a liquid to a gas, reactions that increase the number of gaseous particles, dissolution of molecular solids and certain salts in water, raising the temperature of a substance, and the expansion of a gas at constant temperature. For a change in the volume of *n* mol of an ideal gas at constant temperature, $\Delta S = nR \ln (V_{\text{final}}/V_{\text{initial}})$.

A state of a system with a random distribution of molecules and molecular energies (a state of high entropy) can be achieved in more ways (W) than an ordered state and is therefore more probable. The entropy of a state can be calculated from Boltzmann's equation, $S = k \ln W$.

According to the **third law of thermodynamics**, the entropy of a pure, perfectly ordered crystalline substance at 0 K is zero.

The **standard molar entropy** ($S°$) of a substance is the absolute entropy of 1 mol of the pure substance at 1 atm pressure and a specified temperature, usually 25 °C. The **standard entropy of reaction** ($\Delta S°$) can be calculated from the relation $\Delta S° = S°(\text{products}) - S°(\text{reactants})$.

The **first law of thermodynamics** states that in any process, the total energy of a system and its surroundings remains constant. The **second law of thermodynamics** says that in any *spontaneous* process, the total entropy of a system and its surroundings ($\Delta S_{\text{total}} = \Delta S_{\text{sys}} + \Delta S_{\text{surr}}$) always increases. A chemical reaction is spontaneous if $\Delta S_{\text{total}} > 0$, nonspontaneous if $\Delta S_{\text{total}} < 0$, and at equilibrium if $\Delta S_{\text{total}} = 0$. Reactions that are nonspontaneous in the forward direction are spontaneous in the reverse direction. For a reaction at constant pressure, $\Delta S_{\text{surr}} = -\Delta H/T$, and ΔS_{sys} is ΔS for the reaction.

Free energy, $G = H - TS$, is a state function that indicates whether a reaction is spontaneous or nonspontaneous. A reaction at constant temperature and pressure is spontaneous if $\Delta G < 0$, nonspontaneous if $\Delta G > 0$, and at equilibrium if $\Delta G = 0$. In the equation $\Delta G = \Delta H - T\Delta S$, temperature is a weighting factor that determines the relative importance of the enthalpy and entropy contributions to ΔG.

The **standard free-energy change** ($\Delta G°$) for a reaction is the change in free energy that occurs when reactants in their standard states are completely converted to products in their standard states. The **standard free energy of formation** ($\Delta G°_f$) of a substance is the free-energy change for formation of 1 mol of the substance in its standard state from the most stable form of the constituent elements in their standard states. Substances with a negative value of $\Delta G°_f$ are thermodynamically stable with respect to the constituent elements. We can calculate $\Delta G°$ for a reaction in either of two ways: (1) $\Delta G° = \Delta G°_f(\text{products}) - \Delta G°_f(\text{reactants})$ or (2) $\Delta G° = \Delta H° - T\Delta S°$.

The free-energy change, ΔG, for a reaction under nonstandard-state conditions is given by $\Delta G = \Delta G° + RT \ln Q$, where Q is the reaction quotient. At equilibrium, $\Delta G = 0$ and $Q = K$. As a result, $\Delta G° = -RT \ln K$, which allows us to calculate the equilibrium constant from $\Delta G°$ and vice versa.

KEY WORDS

KEY EQUATIONS

- Boltzmann equation for calculating the entropy S of a state of a system from the number of ways W that the state can be achieved **(Section 16.3)**

  $S = k \ln W$ where the Boltzmann constant $k = 1.38 \times 10^{-23}$ J/K

- Entropy change for the expansion (or compression) of n moles of an ideal gas at constant temperature **(Section 16.3)**

$$\Delta S = nR \ln \frac{V_{\text{final}}}{V_{\text{initial}}} \qquad \Delta S = nR \ln \frac{P_{\text{initial}}}{P_{\text{final}}}$$

- Standard entropy change for a reaction **(Section 16.5)**

  $\Delta S° = S°(\text{products}) - S°(\text{reactants})$ where $S°$ is the standard molar entropy

- Standard entropy change for the reaction $a\text{A} + b\text{B} \rightarrow c\text{C} + d\text{D}$ **(Section 16.5)**

$$\Delta S° = [c\,S°(\text{C}) + d\,S°(\text{D})] - [a\,S°(\text{A}) + b\,S°(\text{B})]$$

- Total entropy change for a reaction **(Section 16.6)**

$$\Delta S_{\text{total}} = \Delta S_{\text{system}} + \Delta S_{\text{surroundings}}$$

  where ΔS_{system} is the entropy of reaction and $\Delta S_{\text{surroundings}}$ is the entropy change in the surroundings

- Entropy change in the surroundings for a reaction at constant pressure **(Section 16.6)**

$$\Delta S_{\text{surroundings}} = -\Delta H/T$$

  where ΔH is the enthalpy of reaction and T is the kelvin temperature of the surroundings

- Definition of free energy **(Section 16.7)**

$$G = H - TS$$

- Free energy change for a reaction at constant temperature and pressure **(Section 16.7)**

$$\Delta G = \Delta H - T\Delta S$$

- Standard free energy change for a reaction **(Section 16.9)**

  $\Delta G° = \Delta G°_f(\text{products}) - \Delta G°_f(\text{reactants})$ where $\Delta G°_f$ is the standard free energy of formation

- Standard free energy change for the reaction $a\text{A} + b\text{B} \longrightarrow c\text{C} + d\text{D}$ **(Section 16.9)**

$$\Delta G^\circ = [c\,\Delta G^\circ_f(\text{C}) + d\,\Delta G^\circ_f(\text{D})] - [a\,\Delta G^\circ_f(\text{A}) + b\,\Delta G^\circ_f(\text{B})]$$

- Free energy change for a reaction under nonstandard-state conditions **(Section 16.10)**

$$\Delta G = \Delta G^\circ + RT \ln Q \quad \text{where } Q \text{ is the reaction quotient}$$

- Relationship between ΔG° and the equilibrium constant for a reaction **(Section 16.11)**

$$\Delta G^\circ = -RT \ln K$$

LEARNING OUTCOMES

- Define a spontaneous process, and classify various physical processes and chemical reactions as spontaneous or nonspontaneous. **(Section 16.1)** ***Problems 1, 30–33, 107***
- Define entropy, and predict the sign of ΔS for various physical processes and chemical reactions. **(Section 16.2)** ***Problems 2, 3, 9, 11, 20–25, 34, 36, 38, 39***
- Use Boltzmann's equation to calculate the entropy of a system. **(Section 16.3)** ***Problems 40–45***
- Calculate ΔS for the expansion or compression of an ideal gas at constant temperature. **(Section 16.3)** ***Problems 46–49, 106***
- State the third law of thermodynamics, and explain how and why the entropy of a substance changes with increasing temperature. **(Section 16.4)** ***Problems 46–47, 110***
- Use standard molar entropies (S°) to calculate the standard entropy of reaction (ΔS°). **(Section 16.5)** ***Problems 52–55, 120, 126, 128, 129, 132, 134–138, 140***
- Understand and apply the second law of thermodynamics. **(Section 16.6)** ***Problems 20, 56–59, 71, 108***
- Calculate values of ΔS_{sys}, ΔS_{surr}, and ΔS_{total} for a reaction, and use these values to determine whether the reaction is spontaneous. **(Section 16.6)** ***Problems 60–67***
- State and apply the free-energy criterion for reaction spontaneity, and determine the temperature, if any, at which a reaction changes between spontaneous and nonspontaneous. **(Section 16.7)** ***Problems 7, 8, 68–75, 86, 87, 112, 115, 116, 119, 123, 129, 139***
- Define the standard free-energy change (ΔG°) for a reaction, and calculate ΔG° from ΔH° and ΔS°. **(Section 16.8)** ***Problems 10, 76, 78, 79, 115, 116, 119, 121, 123, 124, 128–130, 134–138, 140***
- Define the standard free energy of formation (ΔG°_f) of a substance, and calculate ΔG° for a reaction from values of ΔG°_f. **(Section 16.9)** ***Problems 12, 76, 80–89, 100, 139***
- Calculate ΔG for a reaction under nonstandard-state conditions. **(Section 16.10)** ***Problems 13, 90–95, 111, 115, 124, 136***
- Understand the relationship between ΔG° and the equilibrium constant for a reaction, and be able to calculate one from the other. **(Section 16.11)** ***Problems 15–17, 28, 29, 96–105, 111, 115, 118, 121–124, 128, 130, 132, 134, 135, 137–140***

CONCEPTUAL PROBLEMS

Problems 16.1–16.19 appear within the chapter.

16.20 Ideal gases A (red spheres) and B (blue spheres) occupy two separate bulbs. The contents of both bulbs constitute the initial state of an isolated system. Consider the process that occurs when the stopcock is opened.

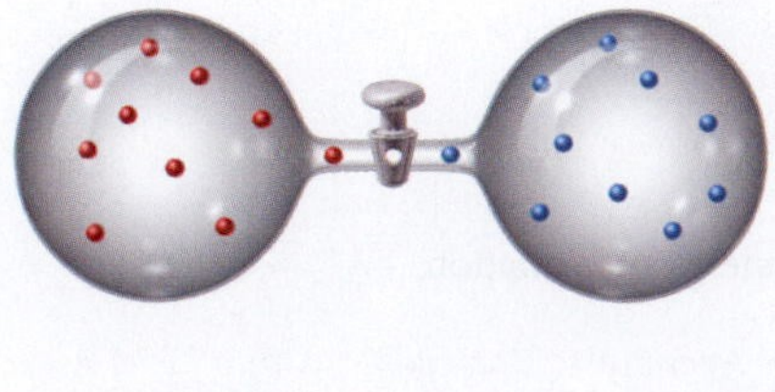

(a) Sketch the final (equilibrium) state of the system.

(b) What are the signs (+, −, or 0) of ΔH, ΔS, and ΔG for this process? Explain.

(c) How does this process illustrate the second law of thermodynamics?

(d) Is the reverse process spontaneous or nonspontaneous? Explain.

16.21 What are the signs (+, −, or 0) of ΔH, ΔS, and ΔG for the spontaneous sublimation of a crystalline solid? Explain.

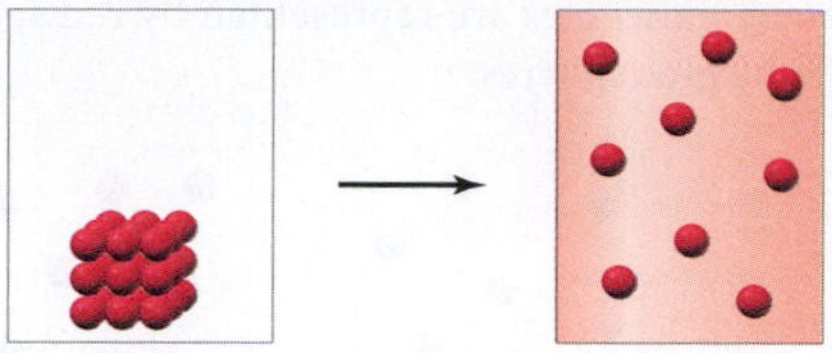

16.22 What are the signs (+, −, or 0) of ΔH, ΔS, and ΔG for the spontaneous condensation of a vapor to a liquid? Explain.

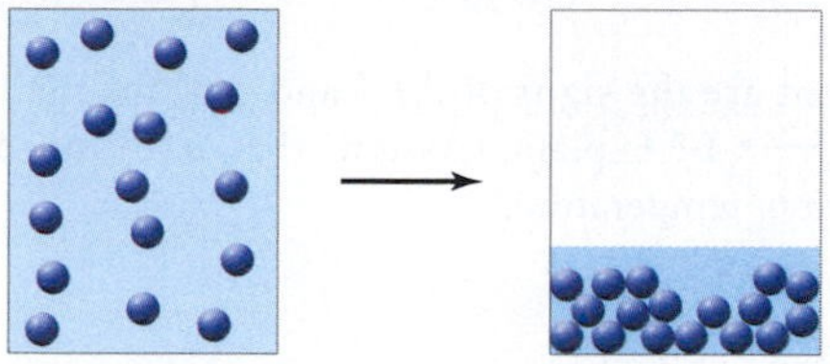

16.23 An ideal gas is compressed at constant temperature. What are the signs (+, −, or 0) of ΔH, ΔS, and ΔG for the process? Explain.

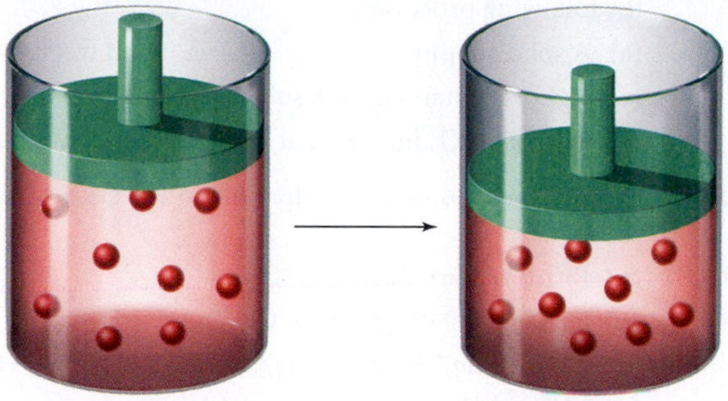

16.24 Consider the following spontaneous reaction of A_2 molecules (red) and B_2 molecules (blue):

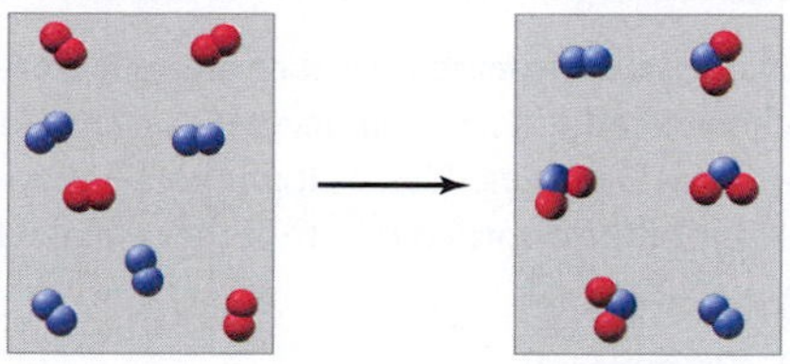

(a) Write a balanced equation for the reaction.

(b) What are the signs (+, −, or 0) of ΔH, ΔS, and ΔG for the reaction? Explain.

16.25 Consider the dissociation reaction $A_2(g) \rightleftharpoons 2\,A(g)$. The following pictures represent two possible initial states and the equilibrium state of the system:

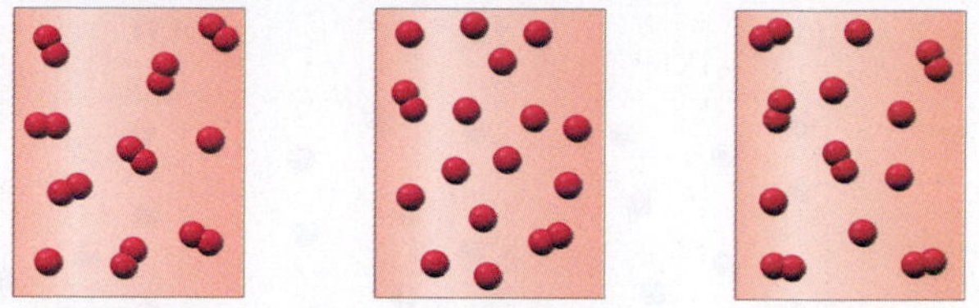

Initial state 1   Initial state 2   Equilibrium state

(a) Is the reaction quotient Q_p for initial state 1 greater than, less than, or equal to the equilibrium constant K_p? Is Q_p for initial state 2 greater than, less than, or equal to K_p?

(b) What are the signs (+, −, or 0) of ΔH, ΔS, and ΔG when the system goes from initial state 1 to the equilibrium state? Explain. Is this a spontaneous process?

(c) What are the signs (+, −, or 0) of ΔH, ΔS, and ΔG when the system goes from initial state 2 to the equilibrium state? Explain. Is this a spontaneous process?

(d) Relate each of the pictures to the graph in Figure 16.11.

16.26 Consider again the dissociation reaction $A_2(g) \rightleftharpoons 2\,A(g)$ (Problem 16.25).

(a) What are the signs (+, −, or 0) of the standard enthalpy change, $\Delta H°$, and the standard entropy change, $\Delta S°$, for the forward reaction?

(b) Distinguish between the meaning of $\Delta S°$ for the dissociation reaction and ΔS for the process in which the system goes from initial state 1 to the equilibrium state (pictured in Problem 16.25).

(c) What can you say about the sign of $\Delta G°$ for the dissociation reaction? How does $\Delta G°$ depend on temperature? Will $\Delta G°$ increase, decrease, or remain the same if the temperature increases?

(d) Will the equilibrium constant K_p increase, decrease, or remain the same if the temperature increases? How will the picture for the equilibrium state (Problem 16.25) change if the temperature increases?

(e) What is the value of ΔG for the dissociation reaction when the system is at equilibrium?

16.27 Consider the following graph of the total free energy of reactants and products versus reaction progress for a general reaction, Reactants $\longrightarrow$ Products:

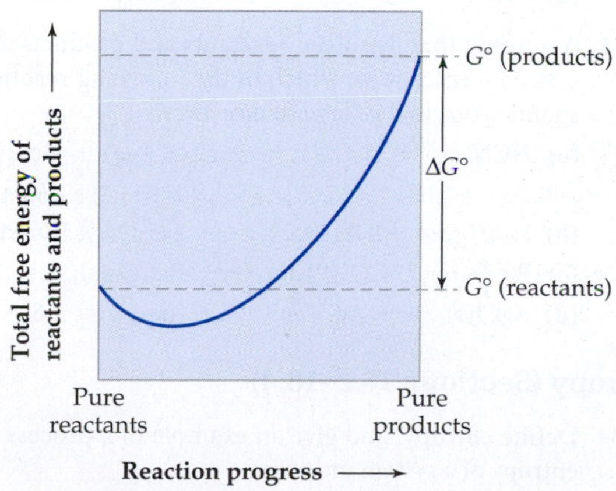

(a) Account for the shape of the curve, and identify the point at which $\Delta G = 0$. What is the significance of that point?

(b) Why is the minimum in the plot on the left side of the graph?

16.28 The following pictures represent equilibrium mixtures for the interconversion of A molecules (red) and X, Y, or Z molecules (blue):

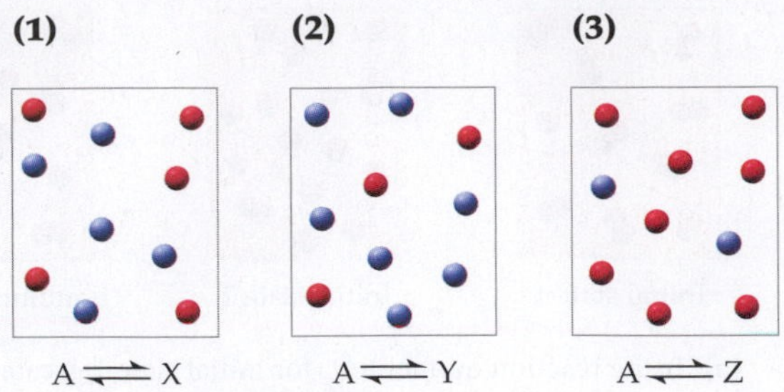

What is the sign of $\Delta G°$ for each of the three reactions?

16.29 The following pictures represent the composition of the equilibrium mixture at 25 °C and 45 °C for the reaction $A \rightleftharpoons B$, where A molecules are represented by red spheres and B molecules by blue spheres:

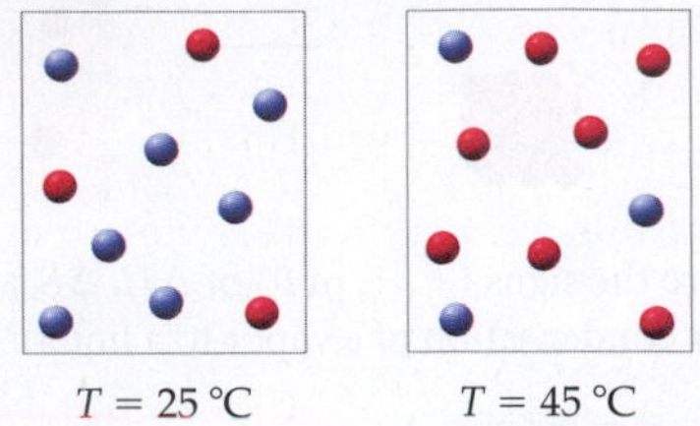

What are the signs of $\Delta H°$ and $\Delta S°$ for the forward reaction $A \longrightarrow B$? Explain. (Assume that $\Delta H°$ and $\Delta S°$ are independent of temperature.)

SECTION PROBLEMS

Spontaneous Processes (Section 16.1)

16.30 Which of the following processes are spontaneous, and which are nonspontaneous?

(a) Freezing of water at 2 °C

(b) Corrosion of iron metal

(c) Expansion of a gas to fill the available volume

(d) Separation of an unsaturated aqueous solution of potassium chloride into solid KCl and liquid water

16.31 Tell whether the following processes are spontaneous or nonspontaneous:

(a) Dissolving sugar in hot coffee

(b) Decomposition of NaCl to solid sodium and gaseous chlorine at 25 °C and 1 atm pressure

(c) Uniform mixing of bromine vapor and nitrogen gas

(d) Boiling of gasoline at 25 °C and 1 atm pressure

16.32 Assuming that gaseous reactants and products are present at 1 atm partial pressure, which of the following reactions are spontaneous in the forward direction?

(a) $N_2(g) + 2\,H_2(g) \longrightarrow N_2H_4(l);\ K_p = 7 \times 10^{-27}$

(b) $2\,Mg(s) + O_2(g) \longrightarrow 2\,MgO(s);\ K_p = 2 \times 10^{198}$

(c) $MgCO_3(s) \longrightarrow MgO(s) + CO_2(g);\ K_p = 9 \times 10^{-10}$

(d) $2\,CO(g) + O_2(g) \longrightarrow 2\,CO_2(g);\ K_p = 1 \times 10^{90}$

16.33 Assuming that dissolved reactants and products are present at 1 M concentrations, which of the following reactions are nonspontaneous in the forward direction?

(a) $HCN(aq) + H_2O(l) \longrightarrow H_3O^+(aq) + CN^-(aq);$ $K = 4.9 \times 10^{-10}$

(b) $H_3O^+(aq) + OH^-(aq) \longrightarrow 2\,H_2O(l);\ K = 1.0 \times 10^{14}$

(c) $Ba^{2+}(aq) + CO_3^{\ 2-}(aq) \longrightarrow BaCO_3(s);\ K = 3.8 \times 10^{8}$

(d) $AgCl(s) \longrightarrow Ag^+(aq) + Cl^-(aq);\ K = 1.8 \times 10^{-10}$

Entropy (Sections 16.2–16.4)

16.34 Define entropy, and give an example of a process in which the entropy of a system increases.

16.35 Comment on the following statement: Exothermic reactions are spontaneous, but endothermic reactions are nonspontaneous.

16.36 Predict the sign of the entropy change in the system for each of the following processes:

(a) A solid sublimes

(b) A liquid freezes

(c) AgI precipitates from a solution containing Ag^+ and I^- ions

(d) Gaseous CO_2 bubbles out of a carbonated beverage

16.37 Predict the sign of ΔS in the system for each of the following reactions:

(a) $PCl_5(s) \longrightarrow PCl_3(l) + Cl_2(g)$

(b) $CH_4(g) + 2\,O_2(g) \longrightarrow CO_2(g) + 2\,H_2O(l)$

(c) $2\,H_3O^+(aq) + CO_3^{\ 2-}(aq) \longrightarrow CO_2(g) + 3\,H_2O(l)$

(d) $Mg(s) + Cl_2(g) \longrightarrow MgCl_2(s)$

16.38 Predict the sign of ΔS for each process in Problem 16.30.

16.39 Predict the sign of ΔS for each process in Problem 16.31.

16.40 Consider a disordered crystal of monodeuteriomethane in which each tetrahedral CH_3D molecule is oriented randomly in one of four possible ways. Use Boltzmann's equation to calculate the entropy of the disordered state of the crystal if the crystal contains:

(a) 12 molecules

(b) 120 molecules

(c) 1 mol of molecules

What is the entropy of the crystal if the C–D bond of each of the CH_3D molecules points in the same direction?

16.41 Consider the distribution of ideal gas molecules among three bulbs (A, B, and C) of equal volume. For each of the following states, determine the number of ways (W) that the state can be achieved, and use Boltzmann's equation to calculate the entropy of the state:

(a) 2 molecules in bulb A

(b) 2 molecules randomly distributed among bulbs A, B, and C

(c) 3 molecules in bulb A

(d) 3 molecules randomly distributed among bulbs A, B, and C

(e) 1 mol of molecules in bulb A

(f) 1 mol of molecules randomly distributed among bulbs A, B, and C

What is ΔS on going from state **(e)** to state **(f)**? Compare your result with ΔS calculated from the equation $\Delta S = nR \ln (V_{final}/V_{initial})$.

16.42 What is the entropy of 100 molecules in a system of 1000 boxes?

16.43 What is the entropy of 100 molecules in a system of 10,000 boxes?

16.44 By what factor does the entropy increase for a collection of 1000 molecules moved from 1.00×10^6 boxes to 1.00×10^7 boxes? For a move from 1.00×10^{16} to 1.00×10^{17} boxes? (Express your answers to three significant figures.)

16.45 If the entropy of a collection of molecules in 50,000 boxes is 3.73×10^{-20} J/K, how many molecules are there?

16.46 Which state in each of the following pairs has the higher entropy per mole of substance?

(a) H_2 at 25 °C in a volume of 10 L or H_2 at 25 °C in a volume of 50 L

(b) O_2 at 25 °C and 1 atm or O_2 at 25 °C and 10 atm

(c) H_2 at 25 °C and 1 atm or H_2 at 100 °C and 1 atm

(d) CO_2 at STP or CO_2 at 100 °C and 0.1 atm

16.47 Which state in each of the following pairs has the higher entropy per mole of substance?

(a) Ice at −40 °C or ice at 0 °C

(b) N_2 at STP or N_2 at 0 °C and 10 atm

(c) N_2 at STP or N_2 at 0 °C in a volume of 50 L

(d) Water vapor at 150 °C and 1 atm or water vapor at 100 °C and 2 atm

14.48 What is the entropy change when the volume of 1.6 g of O_2 increases from 2.5 L to 3.5 L at a constant temperature of 75 °C? Assume that O_2 behaves as an ideal gas.

14.49 What is the value of ΔS when 2.4 g of CH_4 is compressed from 30.0 L to 20.0 L at a constant temperature of 100 °C? Assume that CH_4 behaves as an ideal gas.

Standard Molar Entropies and Standard Entropies of Reaction (Section 16.5)

16.50 Which substance in each of the following pairs would you expect to have the higher standard molar entropy? Explain.

(a) $C_2H_2(g)$ or $C_2H_6(g)$ **(b)** $CO_2(g)$ or $CO(g)$

(c) $I_2(s)$ or $I_2(g)$ **(d)** $CH_3OH(g)$ or $CH_3OH(l)$

16.51 Which substance in each of the following pairs would you expect to have the higher standard molar entropy? Explain.

(a) $NO(g)$ or $NO_2(g)$ **(b)** $CH_3CO_2H(l)$ or $HCO_2H(l)$

(c) $Br_2(l)$ or $Br_2(s)$ **(d)** $S(s)$ or $SO_3(g)$

16.52 Use the standard molar entropies in Appendix B to calculate the standard entropy of reaction for the oxidation of carbon monoxide to carbon dioxide:

$$2\,CO(g) + O_2(g) \longrightarrow 2\,CO_2(g)$$

16.53 Use the standard molar entropies in Appendix B to calculate the standard entropy of reaction for the oxidation of graphite to carbon dioxide:

$$C(s) + O_2(g) \longrightarrow 2\,CO_2(g)$$

16.54 Use the standard molar entropies in Appendix B to calculate $\Delta S°$ at 25 °C for each of the following reactions. Account for the sign of the entropy change in each case.

(a) $2\,H_2O_2(l) \longrightarrow 2\,H_2O(l) + O_2(g)$

(b) $2\,Na(s) + Cl_2(g) \longrightarrow 2\,NaCl(s)$

(c) $2\,O_3(g) \longrightarrow 3\,O_2(g)$

(d) $4\,Al(s) + 3\,O_2(g) \longrightarrow 2\,Al_2O_3(s)$

16.55 Use the $S°$ values in Appendix B to calculate $\Delta S°$ at 25 °C for each of the following reactions. Suggest a reason for the sign of $\Delta S°$ in each case.

(a) $2\,S(s) + 3\,O_2(g) \longrightarrow 2\,SO_3(g)$

(b) $SO_3(g) + H_2O(l) \longrightarrow H_2SO_4(aq)$

(c) $AgCl(s) \longrightarrow Ag^+(aq) + Cl^-(aq)$

(d) $NH_4NO_3(s) \longrightarrow N_2O(g) + 2\,H_2O(g)$

Entropy and the Second Law of Thermodynamics (Section 16.6)

16.56 State the second law of thermodynamics.

16.57 An isolated system is one that exchanges neither matter nor energy with the surroundings. What is the entropy criterion for spontaneous change in an isolated system? Give an example of a spontaneous process in an isolated system.

16.58 Give an equation that relates the entropy change in the surroundings to the enthalpy change in the system. What is the sign of ΔS_{surr} for the following?

(a) An exothermic reaction

(b) An endothermic reaction

16.59 When heat is added to the surroundings, the entropy of the surroundings increases. How does ΔS_{surr} depend on the temperature of the surroundings? Explain.

16.60 Reduction of mercury(II) oxide with zinc gives metallic mercury:

$$HgO(s) + Zn(s) \longrightarrow ZnO(s) + Hg(l)$$

(a) If $\Delta H° = -259.7$ kJ/mol and $\Delta S° = +7.8$ J/K, what is ΔS_{total} for this reaction? Is the reaction spontaneous under standard-state conditions at 25 °C?

(b) Estimate at what temperature, if any, the reaction will become nonspontaneous.

16.61 Elemental sulfur is formed by the reaction of zinc sulfide with oxygen:

$$2\,ZnS(s) + O_2(g) \longrightarrow 2\,ZnO(s) + 2\,S(s)$$

(a) If $\Delta H° = -289.0$ kJ/mol and $\Delta S° = -169.4$ J/K, what is ΔS_{total} for this reaction? Is the reaction spontaneous under standard-state conditions at 25 °C?

(b) At what temperature, if any, will the reaction become nonspontaneous?

16.62 In lightning storms, oxygen is converted to ozone:

$$3\,O_2(g) \longrightarrow 2\,O_3(g)$$

By determining the sign of ΔS_{total}, show whether the reaction is spontaneous at 25 °C.

16.63 Sulfur dioxide emitted from coal-fired power plants is oxidized to sulfur trioxide in the atmosphere:

$$2\,SO_2(g) + O_2(g) \longrightarrow 2\,SO_3(g)$$

By determining the sign of ΔS_{total}, show whether the reaction is spontaneous at 25 °C.

16.64 Elemental mercury can be produced from its oxide:

$$2\,HgO(s) \longrightarrow 2\,Hg(l) + O_2(g)$$

(a) Use data in Appendix B to calculate ΔS_{sys}, ΔS_{surr}, and ΔS_{total} for this reaction. Is the reaction spontaneous under standard-state conditions at 25 °C?

(b) Estimate the temperature at which the reaction will become spontaneous.

16.55 Phosphorus pentachloride forms from phosphorus trichloride and chlorine:

$$PCl_3(g) + Cl_2(g) \longrightarrow PCl_5(g)$$

(a) Use data in Appendix B to calculate ΔS_{sys}, ΔS_{surr}, and ΔS_{total} for this reaction. Is the reaction spontaneous under standard-state conditions at 25 °C?

(b) Estimate the temperature at which the reaction will become nonspontaneous.

16.66 For the vaporization of benzene, $\Delta H_{vap} = 30.7$ kJ/mol and $\Delta S_{vap} = 87.0$ J/(K·mol). Calculate ΔS_{surr} and ΔS_{total} at:

(a) 70 °C **(b)** 80 °C **(c)** 90 °C

Does benzene boil at 70 °C and 1 atm pressure? Calculate the normal boiling point of benzene.

16.67 For the melting of sodium chloride, $\Delta H_{fusion} = 28.16$ kJ/mol and $\Delta S_{fusion} = 26.22$ J/(K·mol). Calculate ΔS_{surr} and ΔS_{total} at:

(a) 1050 K **(b)** 1074 K **(c)** 1100 K

Does NaCl melt at 1100 K? Calculate the melting point of NaCl.

Free Energy (Section 16.7)

16.68 Describe how the signs of ΔH and ΔS determine whether a reaction is spontaneous or nonspontaneous at constant temperature and pressure.

16.69 What determines the direction of a spontaneous reaction when ΔH and ΔS are both positive or both negative? Explain.

16.70 The melting point of benzene is 5.5 °C. Predict the signs of ΔH, ΔS, and ΔG for melting of benzene at:

(a) 0 °C **(b)** 15 °C

16.71 Consider a twofold expansion of 1 mol of an ideal gas at 25 °C in the isolated system shown in Figure 16.1 on page 655.

(a) What are the values of ΔH, ΔS, and ΔG for the process?

(b) How does this process illustrate the second law of thermodynamics?

16.72 Given the data in Problem 16.66, calculate ΔG for the vaporization of benzene at:

(a) 70 °C **(b)** 80 °C **(c)** 90 °C

Predict whether benzene will boil at each of these temperatures and 1 atm pressure.

16.73 Given the data in Problem 16.67, calculate ΔG for the melting of sodium chloride at:

(a) 1050 K **(b)** 1074 K **(c)** 1100 K

Predict whether NaCl will melt at each of these temperatures and 1 atm pressure.

16.74 Calculate the melting point of benzoic acid ($C_6H_5CO_2H$), given the following data: $\Delta H_{fusion} = 18.02$ kJ/mol and $\Delta S_{fusion} = 45.56$ J/(K·mol).

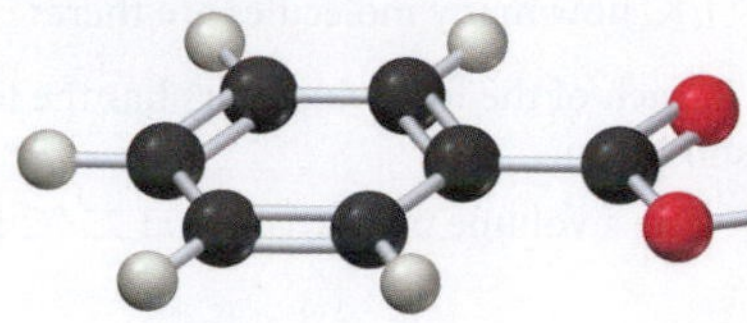

Benzoic acid

16.75 Calculate the enthalpy of fusion of naphthalene ($C_{10}H_8$), given that its melting point is 128 °C and its entropy of fusion is 47.7 J/(K·mol).

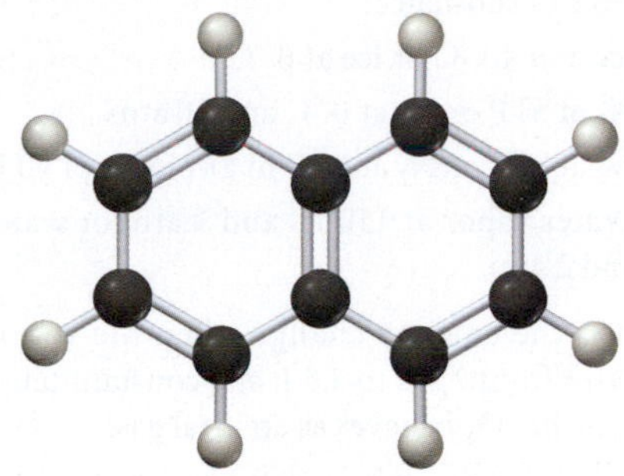

Naphthalene

Standard Free-Energy Changes and Standard Free Energies of Formation (Sections 16.8–16.9)

16.76 Define **(a)** the standard free-energy change, $\Delta G°$, for a reaction and **(b)** the standard free energy of formation, $\Delta G°_f$, of a substance.

16.77 What is meant by the standard state of a substance?

16.78 Use the data in Appendix B to calculate $\Delta H°$ and $\Delta S°$ for each of the following reactions. From the values of $\Delta H°$ and $\Delta S°$, calculate $\Delta G°$ at 25 °C and predict whether each reaction is spontaneous under standard-state conditions.

(a) $N_2(g) + 2\,O_2(g) \longrightarrow 2\,NO_2(g)$

(b) $2\,KClO_3(s) \longrightarrow 2\,KCl(s) + 3\,O_2(g)$

(c) $CH_3CH_2OH(l) + O_2(g) \longrightarrow CH_3CO_2H(l) + H_2O(l)$

16.79 Use the data in Appendix B to calculate $\Delta H°$ and $\Delta S°$ for each of the following reactions. From the values of $\Delta H°$ and $\Delta S°$, calculate $\Delta G°$ at 25 °C and predict whether each reaction is spontaneous under standard-state conditions.

(a) $2\,SO_2(g) + O_2(g) \longrightarrow 2\,SO_3(g)$

(b) $N_2(g) + 2\,H_2(g) \longrightarrow N_2H_4(l)$

(c) $CH_3OH(l) + O_2(g) \longrightarrow HCO_2H(l) + H_2O(l)$

16.80 Use the standard free energies of formation in Appendix B to calculate ΔG° at 25 °C for each reaction in Problem 16.78.

16.81 Use the standard free energies of formation in Appendix B to calculate ΔG° at 25 °C for each reaction in Problem 16.79.

16.82 Use the data in Appendix B to tell which of the following compounds are thermodynamically stable with respect to their constituent elements at 25 °C:

(a) $BaCO_3(s)$ **(b)** $HBr(g)$
(c) $N_2O(g)$ **(d)** $C_2H_4(g)$

16.83 Use the data in Appendix B to decide whether synthesis of the following compounds from their constituent elements is thermodynamically feasible at 25 °C:

(a) $C_6H_6(l)$ **(b)** $NO(g)$
(c) $PH_3(g)$ **(d)** $FeO(s)$

16.84 Use the values of ΔG°_f in Appendix B to calculate the standard free-energy change for the synthesis of dichloroethane from ethylene and chlorine.

$$C_2H_4(g) + Cl_2(g) \longrightarrow CH_2ClCH_2Cl(l)$$

Is it possible to synthesize dichloroethane from gaseous C_2H_4 and Cl_2, each at 25 °C and 1 atm pressure?

16.85 Use the values of ΔG°_f in Appendix B to calculate the standard free-energy change for the conversion of ammonia to hydrazine:

$$2\,NH_3(g) \longrightarrow H_2(g) + N_2H_4(l)$$

Is it worth trying to find a catalyst for this reaction under standard-state conditions at 25 °C?

16.86 Ethanol is manufactured in industry by the hydration of ethylene:

$$CH_2{=}CH_2(g) + H_2O(l) \longrightarrow CH_3CH_2OH(l)$$

Ethylene

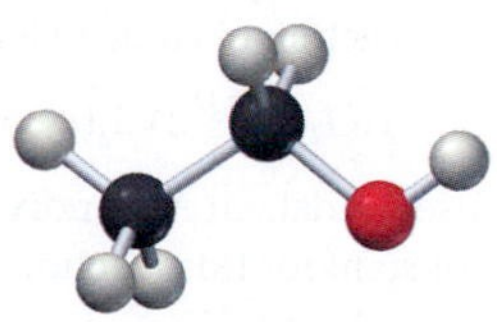

Ethanol

Using the data in Appendix B, calculate ΔG° and show that this reaction is spontaneous at 25 °C. Why does this reaction become nonspontaneous at higher temperatures? Estimate the temperature at which the reaction becomes nonspontaneous.

16.87 Sulfur dioxide in the effluent gases from coal-burning electric power plants is one of the principal causes of acid rain. One method for reducing SO_2 emissions involves partial conversion of SO_2 to H_2S, followed by catalytic conversion of the H_2S and the remaining SO_2 to elemental sulfur:

$$2\,H_2S(g) + SO_2(g) \longrightarrow 3\,S(s) + 2\,H_2O(g)$$

Using the data in Appendix B, calculate ΔG° and show that this reaction is spontaneous at 25 °C. Why does this reaction become nonspontaneous at high temperatures? Estimate the temperature at which the reaction becomes nonspontaneous.

16.88 Consider the conversion of acetylene to benzene:

$$3\,C_2H_2(g) \longrightarrow C_6H_6(l)$$

Acetylene

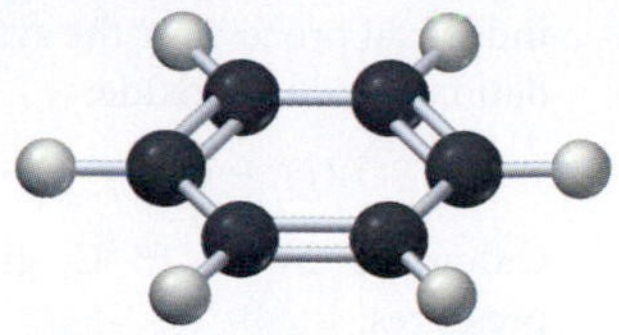

Benzene

Is a catalyst for this reaction possible? Is it possible to synthesize benzene from graphite and gaseous H_2 at 25 °C and 1 atm pressure?

16.89 Consider the conversion of 1,2-dichloroethane to vinyl chloride, the starting material for manufacturing poly(vinyl chloride) (PVC) plastics:

$$CH_2ClCH_2Cl(l) \longrightarrow CH_2{=}CHCl(g) + HCl(g)$$

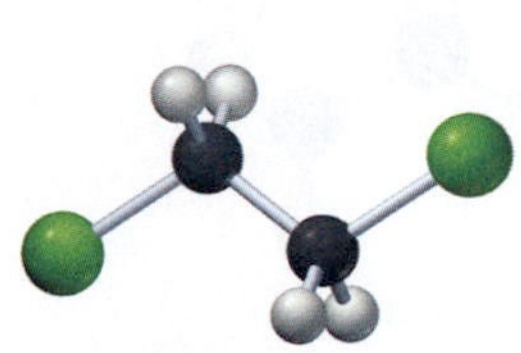

1,2-Dichloroethane

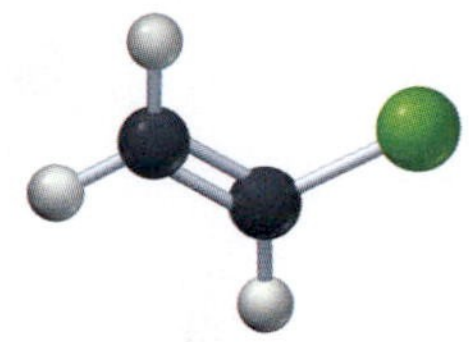

Vinyl chloride

Is this reaction spontaneous under standard-state conditions? Would it help to carry out the reaction in the presence of base to remove HCl? Explain. Is it possible to synthesize vinyl chloride from graphite, gaseous H_2, and gaseous Cl_2 at 25 °C and 1 atm pressure?

Free Energy, Composition, and Chemical Equilibrium (Sections 16.10–16.11)

16.90 What is the relationship between the free-energy change under nonstandard-state conditions, ΔG, the free-energy change under standard-state conditions, ΔG°, and the reaction quotient, Q?

16.91 Compare the values of ΔG and ΔG° when:

(a) $Q < 1$ **(b)** $Q = 1$ **(c)** $Q > 1$

Does the thermodynamic tendency for the reaction to occur increase or decrease as Q increases?

16.92 Use the data in Appendix B to calculate ΔG for the decomposition of nitrosyl chloride at 25 °C when the partial pressures are 2.00 atm of NOCl, 1.00×10^{-3} atm of NO, and 1.00×10^{-3} atm of Cl_2.

$$2\,NO(g) + Cl_2(g) \longrightarrow 2\,NOCl(g)$$

Is the reaction spontaneous in the forward or the reverse direction under these conditions?

16.93 What is ΔG for the formation of solid uranium hexafluoride from uranium and fluorine at 25 °C when the partial pressure of F_2 is 0.045 atm? The standard free energy of formation of $UF_6(s)$ is −2068 kJ/mol.

$$U(s) + 3\,F_2(g) \longrightarrow UF_6(s)$$

Is the reaction spontaneous in the forward or the reverse direction under these conditions?

16.94 Sulfuric acid is produced in larger amounts by weight than any other chemical. It is used in manufacturing fertilizers, oil refining, and hundreds of other processes. An intermediate step in the industrial process for the synthesis of H_2SO_4 is the catalytic oxidation of sulfur dioxide:

$$2\,SO_2(g) + O_2(g) \longrightarrow 2\,SO_3(g) \quad \Delta G° = -141.8\text{ kJ}$$

Calculate ΔG at 25 °C, given the following sets of partial pressures:

(a) 100 atm SO_2, 100 atm O_2, 1.0 atm SO_3

(b) 2.0 atm SO_2, 1.0 atm O_2, 10 atm SO_3

(c) Each reactant and product at a partial pressure of 1.0 atm

16.95 Urea (NH_2CONH_2), an important nitrogen fertilizer, is produced industrially by the reaction

$$2\,NH_3(g) + CO_2(g) \longrightarrow NH_2CONH_2(aq) + H_2O(l)$$

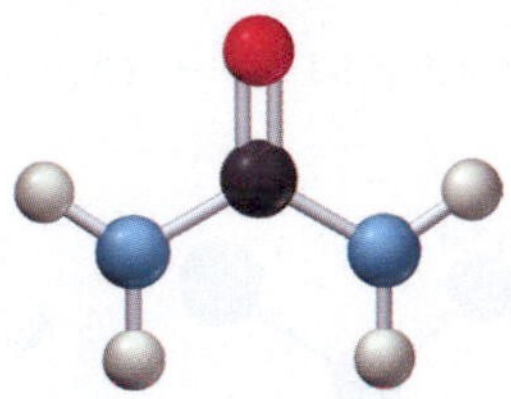

Urea

Given that $\Delta G° = -13.6$ kJ, calculate ΔG at 25 °C for the following sets of conditions:

(a) 10 atm NH_3, 10 atm CO_2, 1.0 M NH_2CONH_2

(b) 0.10 atm NH_3, 0.10 atm CO_2, 1.0 M NH_2CONH_2

Is the reaction spontaneous for the conditions in part **(a)** and/or part **(b)**?

16.96 What is the relationship between the standard free-energy change, $\Delta G°$, for a reaction and the equilibrium constant, K? What is the sign of $\Delta G°$ when:

(a) $K > 1$ **(b)** $K = 1$ **(c)** $K < 1$

16.97 Do you expect a large or small value of the equilibrium constant for a reaction with the following values of $\Delta G°$?

(a) $\Delta G°$ is positive. **(b)** $\Delta G°$ is negative.

16.98 Calculate the equilibrium constant K_p at 25 °C for the reaction in Problem 16.94.

16.99 Calculate the equilibrium constant at 25 °C for the reaction in Problem 16.95.

16.100 Given values of $\Delta G°_f$ at 25 °C for liquid ethanol (−174.9 kJ/mol) and gaseous ethanol (−167.9 kJ/mol), calculate the vapor pressure of ethanol at 25 °C.

16.101 At 25 °C, K_a for acid dissociation of aspirin ($C_9H_8O_4$) is 3.0×10^{-4}. Calculate $\Delta G°$ for the reaction $C_9H_8O_4(aq) + H_2O(l) \rightleftharpoons H_3O^+(aq) + C_9H_7O_4^-(aq)$

16.102 If $\Delta G°_f$ for gaseous bromine is 3.14 kJ/mol at 25 °C, what is the vapor pressure of bromine at 25 °C?

16.103 Calculate the equilibrium partial pressure of iodine vapor above solid iodine at 25 °C if $\Delta G°_f$ for gaseous iodine is 19.4 kJ/mol at 25 °C.

16.104 Ethylene oxide, C_2H_4O, is used to make antifreeze (ethylene glycol, $HOCH_2CH_2OH$). It is produced industrially by the catalyzed air oxidation of ethylene:

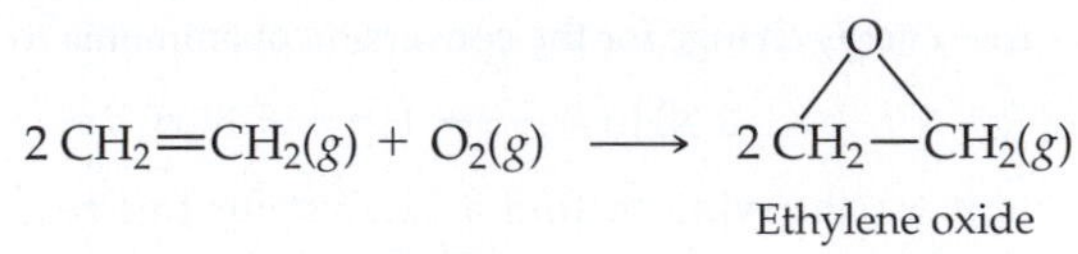

Use the data in Appendix B to calculate $\Delta G°$ and K_p for this reaction at 25 °C.

16.105 The first step in the commercial production of titanium metal is the reaction of rutile (TiO_2) with chlorine and graphite:

$$TiO_2(s) + 2\,Cl_2(g) + 2\,C(s) \longrightarrow TiCl_4(l) + 2\,CO(g)$$

Use the data in Appendix B to calculate $\Delta G°$ and the equilibrium constant for this reaction at 25 °C.

CHAPTER PROBLEMS

16.106 What is the entropy change when 1.32 g of propane (C_3H_8) at 0.100 atm pressure is compressed by a factor of five at a constant temperature of 20 °C? Assume that propane behaves as an ideal gas.

16.107 Indicate whether the following processes are spontaneous or nonspontaneous:

(a) Heat transfer from a block of ice to a room maintained at 25 °C

(b) Evaporation of water from an open beaker

(c) Conversion of iron(III) oxide to iron metal and oxygen

(d) Uphill motion of an automobile

16.108 Do you agree with the following statements? If not, explain.

(a) Spontaneous reactions are always fast.

(b) In any spontaneous process, the entropy of the system always increases.

(c) An endothermic reaction is always nonspontaneous.

(d) A reaction that is nonspontaneous in the forward direction is always spontaneous in the reverse direction.

16.109 When rolling a pair of dice, there are two ways to get a point total of 3 (1 + 2; 2 + 1) but only one way to get a point total of 2 (1 + 1). How many ways are there of getting point totals of 4–12? What is the most probable point total?

16.110 Make a rough, qualitative plot of the standard molar entropy versus temperature for methane from 0 K to 298 K. Incorporate the following data into your plot: mp = −182 °C; bp = −164 °C; $S°$ = 186.2 J/(K · mol) at 25 °C.

16.111 The standard free-energy change at 25 °C for the dissociation of water is 79.9 kJ:

$$2\,H_2O(l) \rightleftharpoons H_3O^+(aq) + OH^-(aq) \qquad \Delta G° = 79.9\text{ kJ}$$

For each of the following sets of concentrations, calculate ΔG at 25 °C and indicate whether the reaction is spontaneous in the forward or reverse direction:

(a) $[H_3O^+] = [OH^-] = 1.0$ M

(b) $[H_3O^+] = [OH^-] = 1.0 \times 10^{-7}$ M

(c) $[H_3O^+] = 1.0 \times 10^{-7}$ M, $[OH^-] = 1.0 \times 10^{-10}$ M

Are your results consistent with Le Châtelier's principle? Use the thermodynamic data to calculate the equilibrium constant for the reaction.

16.112 Calculate the normal boiling point of ethanol (CH_3CH_2OH), given that its enthalpy of vaporization is 38.6 kJ/mol and its entropy of vaporization is 110 J/(K · mol).

16.113 Chloroform ($CHCl_3$) has a normal boiling point of 61 °C and an enthalpy of vaporization of 29.24 kJ/mol. What are its values of ΔG_{vap} and ΔS_{vap} at 61 °C?

16.114 The entropy change for a certain nonspontaneous reaction at 50 °C is 104 J/K.
(a) Is the reaction endothermic or exothermic?
(b) What is the minimum value of ΔH (in kJ) for the reaction?

16.115 Ammonium nitrate is dangerous because it decomposes (sometimes explosively) when heated:

$$NH_4NO_3(s) \longrightarrow N_2O(g) + 2\,H_2O(g)$$

(a) Using the data in Appendix B, show that this reaction is spontaneous at 25 °C.
(b) How does $\Delta G°$ for the reaction change when the temperature is raised?
(c) Calculate the equilibrium constant K_p at 25 °C.
(d) Calculate ΔG for the reaction when the partial pressure of each gas is 30 atm.

16.116 Use the data in Appendix B to calculate $\Delta H°$, $\Delta S°$, and $\Delta G°$ at 25 °C for each of the following reactions:
(a) $2\,Mg(s) + O_2(g) \longrightarrow 2\,MgO(s)$
(b) $MgCO_3(s) \longrightarrow MgO(s) + CO_2(g)$
(c) $Fe_2O_3(s) + 2\,Al(s) \longrightarrow Al_2O_3(s) + 2\,Fe(s)$
(d) $2\,NaHCO_3(s) \longrightarrow Na_2CO_3(s) + CO_2(g) + H_2O(g)$
Are these reactions spontaneous or nonspontaneous at 25 °C and 1 atm pressure? How does $\Delta G°$ change when the temperature is raised?

16.117 *Trouton's rule* says that the ratio of the molar heat of vaporization of a liquid to its normal boiling point (in kelvin) is approximately the same for all liquids: $\Delta H_{vap}/T_{bp} \approx 88$ J/(K · mol).
(a) Check the reliability of Trouton's rule for the liquids listed in the following table.
(b) Explain why liquids tend to have the same value of $\Delta H_{vap}/T_{bp}$.
(c) Which of the liquids in the table deviate(s) from Trouton's rule? Explain.

Liquid	bp (°C)	ΔH_{vap} (kJ/mol)
Ammonia	−33.3	23.4
Benzene	80.1	30.8
Carbon tetrachloride	76.8	29.8
Chloroform	61.1	29.2
Mercury	356.6	59.11

16.118 Ammonium hydrogen sulfide, a stink bomb ingredient, decomposes to ammonia and hydrogen sulfide.

$$NH_4HS(s) \longrightarrow NH_3(g) + H_2S(g)$$

Calculate the standard free energy of formation of $NH_4HS(s)$ at 25 °C if the total pressure resulting from NH_4HS placed in an evacuated container is 0.658 atm at 25 °C.

16.119 Nickel tetracarbonyl, a volatile liquid used to purify nickel, forms when carbon monoxide contacts nickel. For $Ni(CO)_4(l)$, $\Delta H°_f = -633.0$ kJ/mol and $S° = 313.4$ J/(K · mol).

$$Ni(s) + 4\,CO(g) \longrightarrow Ni(CO)_4(l)$$

(a) Calculate $\Delta H°$, $\Delta S°$, and $\Delta G°$ for the reaction under standard-state conditions at 25 °C.
(b) Estimate the temperature at which the reaction becomes nonspontaneous.
(c) Calculate $\Delta G°_f$ for $Ni(CO)_4(l)$.

16.120 Just as we can define a standard enthalpy of formation ($\Delta H°_f$) and a standard free energy of formation ($\Delta G°_f$), we can define an analogous standard entropy of formation ($\Delta S°_f$) as being the entropy change for the formation of a substance in its standard state from its constituent elements in their standard states. Use the standard molar entropies given in Appendix B to calculate $\Delta S°_f$ for the following substances:
(a) Benzene, $C_6H_6(l)$ **(b)** $CaSO_4(s)$
(c) Ethanol, $CH_3CH_2OH(l)$
Check your answers by calculating $\Delta S°_f$ from the values given in Appendix B for $\Delta H°_f$ and $\Delta G°_f$.

16.121 Use the data in Appendix B to calculate the equilibrium pressure of CO_2 in a closed 1 L vessel that contains each of the following samples:
(a) 15 g of $MgCO_3$ and 1.0 g of MgO at 25 °C
(b) 15 g of $MgCO_3$ and 1.0 g of MgO at 280 °C
(c) 30 g of $MgCO_3$ and 1.0 g of MgO at 280 °C
Assume that $\Delta H°$ and $\Delta S°$ are independent of temperature.

16.122 The equilibrium constant K_b for dissociation of aqueous ammonia is 1.710×10^{-5} at 20 °C and 1.892×10^{-5} at 50 °C. What are the values of $\Delta H°$ and $\Delta S°$ for the reaction?

$$NH_3(aq) + H_2O(l) \rightleftharpoons NH_4^+(aq) + OH^-(aq)$$

16.123 Consider the Haber synthesis of gaseous NH_3 ($\Delta H°_f = -46.1$ kJ/mol; $\Delta G°_f = -16.5$ kJ/mol):

$$N_2(g) + 3\,H_2(g) \longrightarrow 2\,NH_3(g)$$

(a) Use only these data to calculate $\Delta H°$ and $\Delta S°$ for the reaction at 25 °C.
(b) Account for the sign of $\Delta S°$.
(c) Is the reaction spontaneous under standard-state conditions at 25 °C? Explain.
(d) What are the equilibrium constants K_p and K_c for the reaction at 350 K? Assume that $\Delta H°$ and $\Delta S°$ are independent of temperature.

16.124 Consider the dissolution of AgBr in water at 25 °C:

$$AgBr(s) \rightleftharpoons Ag^+(aq) + Br^-(aq)$$

(a) Use the standard heats of formation and standard molar entropies in Appendix B to calculate $\Delta G°$ for the reaction at 25 °C.
(b) Calculate K_{sp} for AgBr at 25 °C.
(c) Calculate ΔG for the dissolution of AgBr at 25 °C when $[Ag^+] = [Br^-] = 1.00 \times 10^{-5}$ M. Is your result consistent with the relative values of Q and K_{sp}?

16.125 The temperature dependence of the equilibrium constant is given by the equation

$$\ln K = \frac{-\Delta H^\circ}{R}\left(\frac{1}{T}\right) + \frac{\Delta S^\circ}{R}$$

where ΔH° and ΔS° are assumed to be independent of temperature.

(a) Derive this equation from equations given in this chapter.

(b) Explain how this equation can be used to determine experimental values of ΔH° and ΔS° from values of K at several different temperatures.

(c) Use this equation to predict the sign of ΔH° for a reaction whose equilibrium constant increases with increasing temperature. Is the reaction endothermic or exothermic? Is your prediction in accord with Le Châtelier's principle?

16.126 The normal boiling point of bromine is 58.8 °C, and the standard entropies of the liquid and vapor are $S^\circ[Br_2(l)] = 152.2$ J/(K · mol); $S^\circ[Br_2(g)] = 245.4$ J/(K · mol). At what temperature does bromine have a vapor pressure of 227 mm Hg?

16.127 The molar solubility of lead iodide is 1.45×10^{-3} M at 20 °C and 6.85×10^{-3} M at 80 °C. What are the values of ΔH° and ΔS° for dissolution of PbI_2?

$$PbI_2(s) \longrightarrow Pb^{2+}(aq) + 2\,I^-(aq)$$

Assume that ΔH° and ΔS° are independent of temperature.

16.128 Use the data in Appendix B to calculate the equilibrium constant K for the following reaction at 80 °C:

$$2\,Br^-(aq) + Cl_2(g) \longrightarrow Br_2(l) + 2\,Cl^-(aq)$$

Assume that ΔH° and ΔS° are independent of temperature.

16.129 Use the data from Appendix B to determine the normal boiling point of carbon disulfide (CS_2).

16.130 A humidity sensor consists of a cardboard square that is colored blue in dry weather and red in humid weather. The color change is due to the reaction:

$$\underset{\text{Blue}}{CoCl_2(s)} + 6\,H_2O(g) \rightleftharpoons \underset{\text{Red}}{[Co(H_2O)_6]Cl_2(s)}$$

For this reaction at 25 °C, $\Delta H^\circ = -352$ kJ/mol and $\Delta S^\circ = -899$ J/(K · mol). Assuming that ΔH° and ΔS° are independent of temperature, what is the vapor pressure of water (in mm Hg) at equilibrium for the above reaction at 35 °C on a hot summer day?

16.131 The following reaction, sometimes used in the laboratory to generate small quantities of oxygen gas, has $\Delta G^\circ = -224.4$ kJ/mol at 25 °C:

$$2\,KClO_3(s) \longrightarrow 2\,KCl(s) + 3\,O_2(g)$$

Use the following additional data at 25 °C to calculate the standard molar entropy S° of O_2 at 25 °C: $\Delta H^\circ_f(KClO_3) = -397.7$ kJ/mol, $\Delta H^\circ_f(KCl) = -436.5$ kJ/mol, $S^\circ(KClO_3) = 143.1$ J/(K · mol), and $S^\circ(KCl) = 82.6$ J/(K · mol).

16.132 Consider the equilibrium $N_2O_4(g) \rightleftharpoons 2\,NO_2(g)$.

(a) Use the thermodynamic data in Appendix B to determine the temperature at which an equilibrium mixture with a total pressure of 1.00 atm will contain twice as much NO_2 as N_2O_4. Assume that ΔH° and ΔS° are independent of temperature.

(b) At what temperature will an equilibrium mixture with a total pressure of 1.00 atm contain equal amounts of NO_2 and N_2O_4?

MULTICONCEPT PROBLEMS

16.133 Sorbitol ($C_6H_{14}O_6$), a substance used as a sweetener in foods, is prepared by the reaction of glucose with hydrogen in the presence of a catalyst:

$$C_6H_{12}O_6(aq) + H_2(g) \longrightarrow C_6H_{14}O_6(aq)$$

Which of the following quantities are affected by the catalyst?

(a) Rate of the forward reaction

(b) Rate of the reverse reaction

(c) Spontaneity of the reaction

(d) ΔH°

(e) ΔS°

(f) ΔG°

(g) The equilibrium constant

(h) Time required to reach equilibrium

16.134 A mixture of 14.0 g of N_2 and 3.024 g of H_2 in a 5.00 L container is heated to 400 °C. Use the data in Appendix B to calculate the molar concentrations of N_2, H_2, and NH_3 at equilibrium. Assume that ΔH° and ΔS° are independent of temperature, and remember that the standard state of a gas is defined in terms of pressure.

$$N_2(g) + 3\,H_2(g) \rightleftharpoons 2\,NH_3(g)$$

16.135 One step in the commercial synthesis of sulfuric acid is the catalytic oxidation of sulfur dioxide:

$$2\,SO_2(g) + O_2(g) \rightleftharpoons 2\,SO_3(g)$$

(a) A mixture of 192 g of SO_2, 48.0 g of O_2, and a V_2O_5 catalyst is heated to 800 K in a 15.0 L vessel. Use the data in Appendix B to calculate the partial pressures of SO_3, SO_2, and O_2 at equilibrium. Assume that ΔH° and ΔS° are independent of temperature.

(b) Does the percent yield of SO_3 increase or decrease on raising the temperature from 800 K to 1000 K? Explain.

(c) Does the total pressure increase or decrease on raising the temperature from 800 K to 1000 K? Calculate the total pressure (in atm) at 1000 K.

16.136 The lead storage battery uses the reaction

$$Pb(s) + PbO_2(s) + 2\,H^+(aq) + 2\,HSO_4^-(aq) \longrightarrow 2\,PbSO_4(s) + 2\,H_2O(l)$$

(a) Use the data in Appendix B to calculate $\Delta G°$ for this reaction.

(b) Calculate ΔG for this reaction on a cold winter's day (10 °F) in a battery that has run down to the point where the sulfuric acid concentration is only 0.100 M.

16.137 What is the molar solubility of $CaCO_3$ at 50 °C in a solution prepared by dissolving 1.000 L of CO_2 gas (at 20 °C and 731 mm Hg) and 3.335 g of solid $Ca(OH)_2$ in enough water to make 500.0 mL of solution at 50 °C? Is the solubility of $CaCO_3$ at 50 °C larger or smaller than at 25 °C? Explain. You may assume that $\Delta H°$ and $\Delta S°$ are independent of temperature.

16.138 A 1.00 L volume of gaseous ammonia at 25.0 °C and 744 mm Hg was dissolved in enough water to make 500.0 mL of aqueous ammonia at 2.0 °C. What is K_b for NH_3 at 2.0 °C, and what is the pH of the solution? Assume that $\Delta H°$ and $\Delta S°$ are independent of temperature.

16.139 Consider the unbalanced equation: $I_2(s) \longrightarrow I^-(aq) + IO_3^-(aq)$.

(a) Balance the equation for this reaction in basic solution.

(b) Use the data in Appendix B and $\Delta G°_f$ for $IO_3^-(aq) = -128.0$ kJ/mol to calculate $\Delta G°$ for the reaction at 25 °C.

(c) Is the reaction spontaneous or nonspontaneous under standard-state conditions?

(d) What pH is required for the reaction to be at equilibrium at 25 °C when $[I^-] = 0.10$ M and $[IO_3^-] = 0.50$ M?

16.140 A mixture of NO_2 and N_2O_4, each at an initial partial pressure of 1.00 atm and a temperature of 100 °C, is allowed to react.

(a) Use the data in Appendix B to calculate the partial pressure of each gas at equilibrium. Assume that $\Delta H°$ and $\Delta S°$ are independent of temperature.

$$N_2O_4(g) \rightleftharpoons 2\,NO_2(g)$$

(b) What is the shape of the N_2O_4 molecule, and what hybrid orbitals do the N atoms use in bonding?

CHAPTER 17

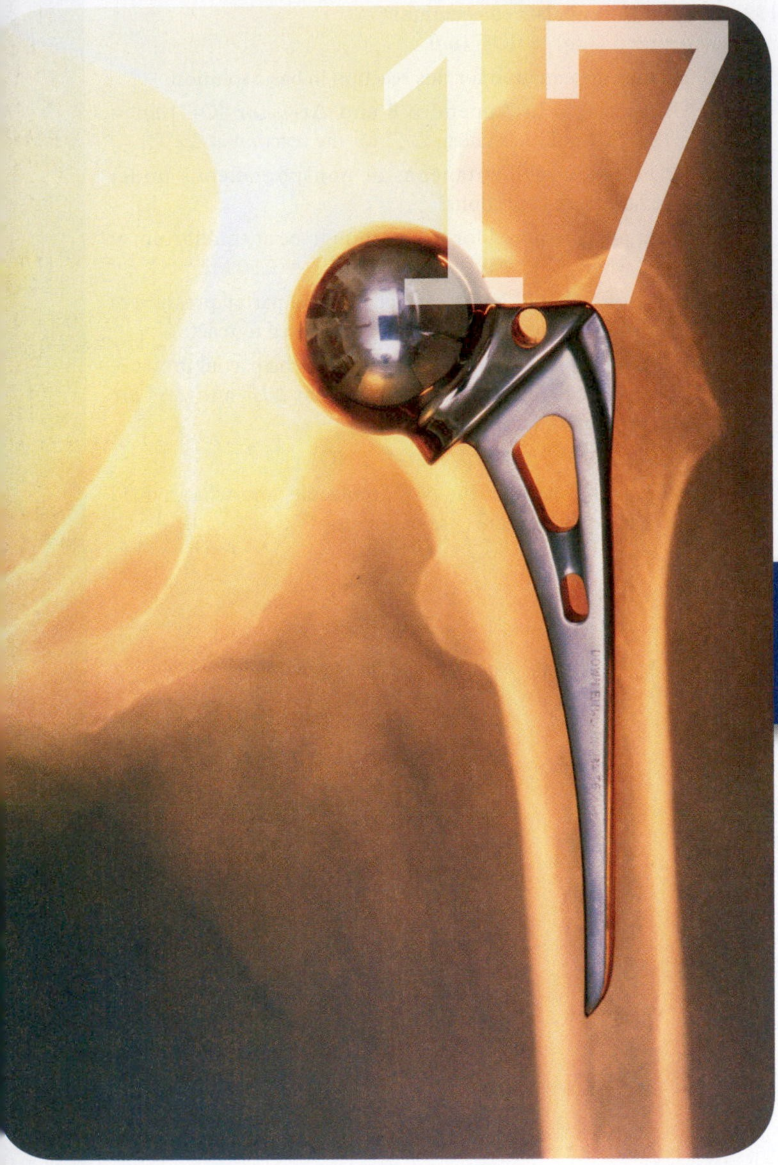

How is electrochemistry applied to orthopedic implants such as this artificial hip joint? See the *FYI* at the end of this chapter.

Electrochemistry

Batteries are everywhere. They provide the electric current to start our automobiles and to power a host of products such as digital watches and cameras, electronic calculators and laptop computers, heart pacemakers, radios, and cell phones. In hybrid and all-electric vehicles, batteries power an electric motor that supplements or replaces a gasoline engine. A battery is an *electrochemical cell*, a device for converting chemical energy to electrical energy. A battery takes the energy released by a spontaneous chemical reaction and uses it to produce electricity.

Electrochemistry, the area of chemistry concerned with the interconversion of chemical and electrical energy, is enormously important in modern science and technology not only because of batteries but also because it makes possible the manufacture of essential industrial chemicals and materials. Sodium hydroxide, for example, which is used in the manufacture of paper, textiles, soaps, and detergents, is produced by passing an electric current through an aqueous solution of sodium chloride. Chlorine, essential to the manufacture of plastics such as poly(vinyl chloride) (PVC), is obtained in the same process. Aluminum metal is also produced in an electrochemical process, as is pure copper for use in electrical wiring.

In this chapter, we'll look at the principles involved in the design and operation of electrochemical cells. In addition, we'll explore some important connections between electrochemistry and thermodynamics.

17.1 GALVANIC CELLS

Electrochemical cells are of two types: **galvanic cells** (also called *voltaic cells* and commonly known as batteries) and **electrolytic cells**. The names "galvanic" and "voltaic" honor the Italian scientists Luigi Galvani (1737–1798) and Alessandro Volta (1745–1827), who

CONTENTS

conducted pioneering work in the field of electrochemistry. In a galvanic cell, a spontaneous chemical reaction generates an electric current. In an electrolytic cell, an electric current drives a nonspontaneous reaction. The two types are therefore the reverse of each other. We'll take up galvanic cells in this section and will examine electrolytic cells later. First, though, let's review some of the basics of oxidation–reduction, or redox, reactions, first described in Section 7.6.

If you immerse a strip of zinc metal in an aqueous solution of copper sulfate, you find that a dark-colored solid deposits on the surface of the zinc and that the blue color characteristic of the Cu^{2+} ion slowly disappears from the solution (**Figure 17.1**). Chemical analysis shows that the dark-colored deposit consists of tiny particles of copper metal and that the solution now contains zinc ions. Therefore, the reaction is

$$\mathrm{Zn}(s) + \mathrm{Cu}^{2+}(aq) \longrightarrow \mathrm{Zn}^{2+}(aq) + \mathrm{Cu}(s)$$

This is a redox reaction in which Zn is oxidized to Zn^{2+} and Cu^{2+} is reduced to Cu. Recall that an *oxidation* is a loss of electrons (an increase in oxidation number) and a *reduction* is a gain of electrons (a decrease in oxidation number).

As discussed in Section 7.9, we can represent the oxidation and reduction aspects of the reaction by separating the overall process into *half-reactions*, one representing the oxidation reaction and the other representing the reduction:

Oxidation half-reaction: $\mathrm{Zn}(s) \longrightarrow \mathrm{Zn}^{2+}(aq) + 2\,e^-$

Reduction half-reaction: $\mathrm{Cu}^{2+}(aq) + 2\,e^- \longrightarrow \mathrm{Cu}(s)$

We say that Cu^{2+} is the *oxidizing agent* because, in gaining electrons from Zn, it causes the oxidation of Zn to Zn^{2+}. Similarly, we say that Zn is the *reducing agent* because, in losing electrons to Cu^{2+}, it causes the reduction of Cu^{2+} to Cu.

If the reaction is carried out as shown in Figure 17.1, electrons are transferred directly from Zn to Cu^{2+} and the enthalpy of reaction is lost to the surroundings as heat. If, however,

FIGURE 17.1

The redox reaction of zinc metal with aqueous Cu^{2+} ions.

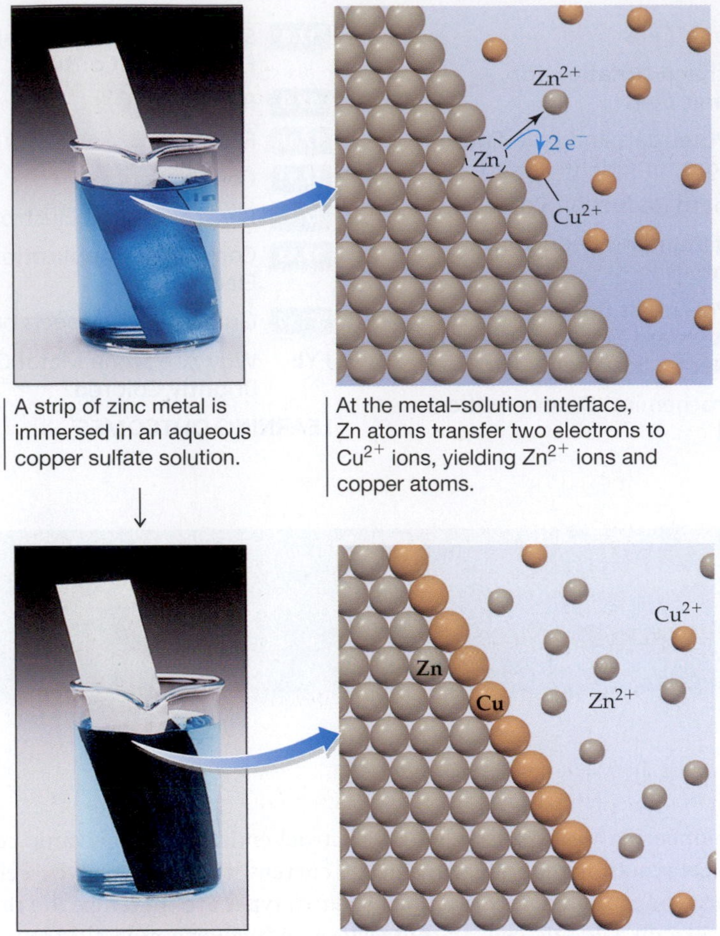

A strip of zinc metal is immersed in an aqueous copper sulfate solution.

At the metal–solution interface, Zn atoms transfer two electrons to Cu^{2+} ions, yielding Zn^{2+} ions and copper atoms.

As time passes, a dark deposit of copper metal appears on the zinc, and the blue color of $Cu^{2+}(aq)$ fades from the solution.

The surface of the zinc becomes covered with copper atoms, and Zn^{2+} ions replace Cu^{2+} ions in the solution.

the reaction is carried out using the electrochemical cell depicted in **Figure 17.2,** then some of the chemical energy released by the reaction is converted to electrical energy, which can be used to light a lightbulb or run an electric motor.

The apparatus shown in Figure 17.2 is a type of galvanic cell called a *Daniell cell*, after John Frederick Daniell, the English chemist who first constructed it in 1836. It consists of two separate *half-cells*, a beaker containing a strip of zinc that dips into an aqueous solution of zinc sulfate and a second beaker containing a strip of copper that dips into aqueous copper sulfate. The strips of zinc and copper are called **electrodes** and are connected by an electrically conducting wire. In addition, the two solutions are connected by a **salt bridge**, a U-shaped tube that contains a gel permeated with a solution of an inert electrolyte, such as Na_2SO_4. The ions of the inert electrolyte do not react with the other ions in the solutions, and they are not oxidized or reduced at the electrodes.

The reaction that occurs in the Daniell cell is the same one that occurs when Zn reacts directly with Cu^{2+}, but now, because the Zn metal and Cu^{2+} ions are in separate compartments, the electrons are transferred from Zn to Cu^{2+} through the wire. Consequently, the oxidation and reduction half-reactions occur at separate electrodes and an electric current flows through the wire. Electrons are not transferred through the solution because the metal wire is a much better conductor of electrons than is water. In fact, free electrons react rapidly with water and are therefore unstable in aqueous solutions.

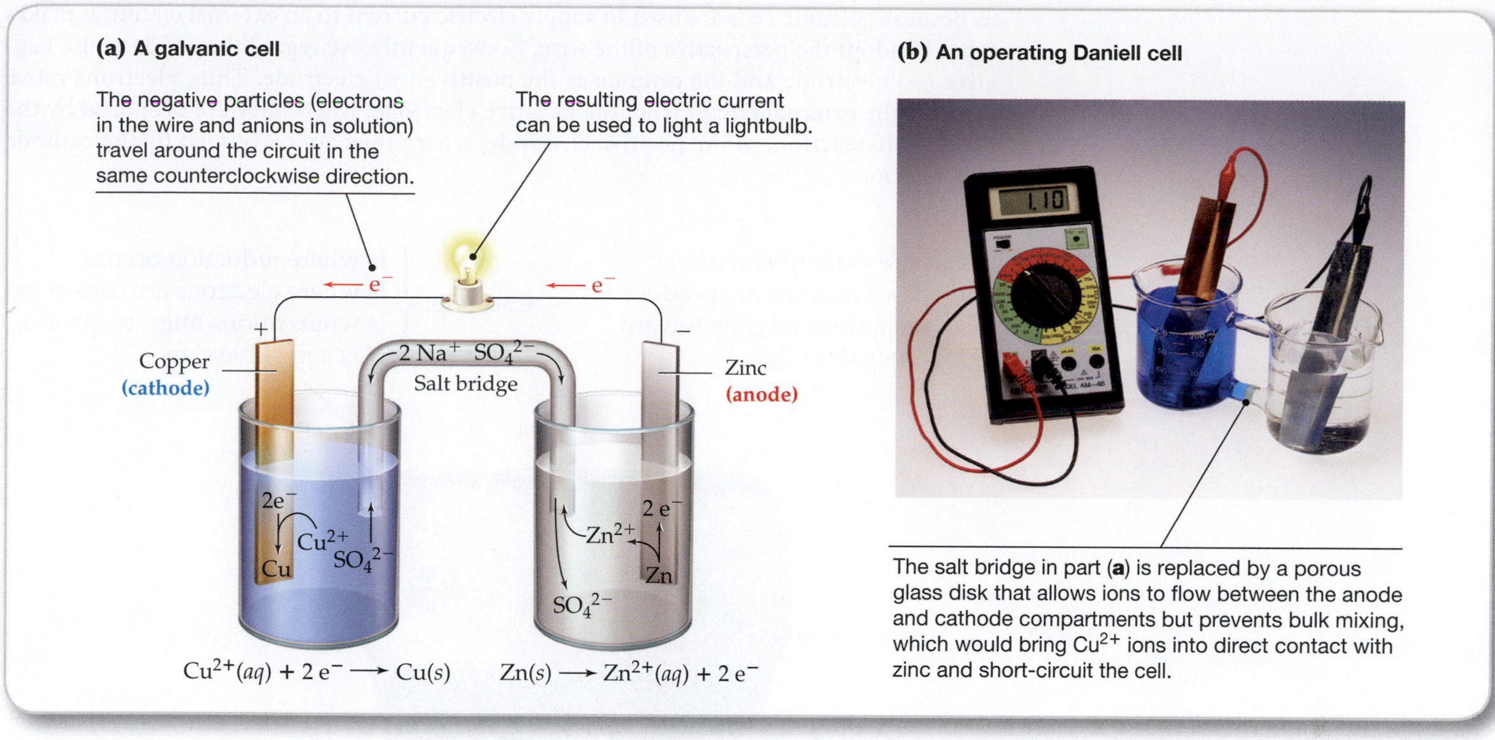

FIGURE 17.2

A Daniell cell, an example of a galvanic cell. **(a)** A Daniell cell uses the oxidation of zinc metal to Zn^{2+} ions and the reduction of Cu^{2+} ions to copper metal. **(b)** In this operating Daniell cell, the lightbulb in part **(a)** is replaced with a digital voltmeter (more about that in Section 17.3).

The electrode at which oxidation takes place is called the **anode** (the zinc strip in this example), and the electrode at which reduction takes place is called the **cathode** (the copper strip). The anode and cathode half-reactions must add to give the overall cell reaction:

$$\begin{array}{ll} \text{Anode (oxidation) half-reaction:} & Zn(s) \longrightarrow Zn^{2+}(aq) + 2\,e^- \\ \text{Cathode (reduction) half-reaction:} & Cu^{2+}(aq) + 2\,e^- \longrightarrow Cu(s) \\ \hline \text{Overall cell reaction:} & Zn(s) + Cu^{2+}(aq) \longrightarrow Zn^{2+}(aq) + Cu(s) \end{array}$$

The salt bridge is necessary to complete the electrical circuit. Without it, the solution in the anode compartment would become positively charged as Zn^{2+} ions appeared in it, and the solution in the cathode compartment would become negatively charged as Cu^{2+} ions were removed from it. Because of the charge imbalance, the electrode reactions would quickly come to a halt and electron flow through the wire would cease.

With the salt bridge in place, electrical neutrality is maintained in both compartments by a flow of ions. Anions (in this case SO_4^{2-}) flow through the salt bridge from the cathode compartment to the anode compartment, and cations migrate through the salt bridge from the anode compartment to the cathode compartment. For the cell shown in Figure 17.2, Na^+ ions move out of the salt bridge into the cathode compartment and Zn^{2+} ions move into the salt bridge from the anode compartment. The anode and cathode get their names from the direction of ion flow between the two compartments: *An*ions move toward the *an*ode, and *cat*ions move toward the *cat*hode.

The electrodes of commercial galvanic cells are generally labeled with plus (+) and minus (−) signs, although the magnitude of the actual charge on the electrodes is infinitesimally small and the sign of the charge associated with each electrode depends on the point of view. From the perspective of the wire, the anode looks negative because a stream of negatively charged electrons comes from it. From the perspective of the solution, however, the anode looks positive because a stream of positively charged Zn^{2+} ions move from it. Because galvanic cells are used to supply electric current to an external circuit, it makes sense to adopt the perspective of the wire. Consequently, we regard the anode as the negative (−) electrode and the cathode as the positive (+) electrode. Thus, electrons move through the external circuit from the negative electrode, where they are produced by the anode half-reaction, to the positive electrode, where they are consumed by the cathode half-reaction.

Anode:
- Is where oxidation occurs
- Is where electrons are produced
- Is what anions migrate toward
- Has a negative sign

Cathode:
- Is where reduction occurs
- Is where electrons are consumed
- Is what cations migrate toward
- Has a positive sign

Why do anions move toward the anode? Shouldn't the negative ions be repelled by the negative charge of the anode? The answer is that the anode's negative charge is shielded by the surrounding Zn^{2+} cations, which enter the solution from the surface of the electrode when zinc is oxidized. From the perspective of the anions, the solution around the anode looks positive, and so the anions move toward the anode to neutralize the positive charge of the Zn^{2+} cations (**Figure 17.3**).

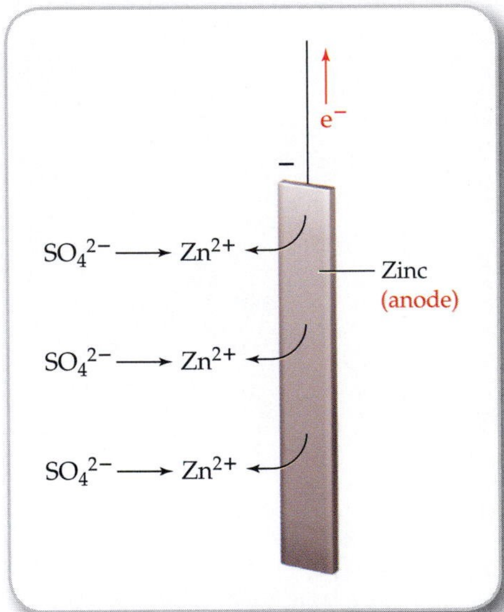

FIGURE 17.3

Why do negative ions move toward the negative electrode?

In some galvanic cells, one (or both) of the half-cell reactions occurs at an inert electrode. Take, for instance, a cell that uses the reaction

$$Fe(s) + 2\,Fe^{3+}(aq) \longrightarrow 3\,Fe^{2+}(aq)$$

In this reaction, iron metal is oxidized to iron(II) ions and iron(III) ions are reduced to iron(II) ions. Therefore, the cell half-reactions are

$$\begin{array}{lrcl} \text{Anode (oxidation):} & Fe(s) & \longrightarrow & Fe^{2+}(aq) + 2\,e^- \\ \text{Cathode (reduction):} & 2 \times [Fe^{3+}(aq) + e^- & \longrightarrow & Fe^{2+}(aq)] \\ \hline \text{Overall cell reaction:} & Fe(s) + 2\,Fe^{3+}(aq) & \longrightarrow & 3\,Fe^{2+}(aq) \end{array}$$

Notice that the cathode half-reaction is multiplied by a factor of 2 so that the two half-reactions will add to give the overall cell reaction. Whenever half-reactions are added, the electrons must cancel. No electrons can appear in the overall reaction because all the electrons lost by the reducing agent are gained by the oxidizing agent.

A possible experimental setup for this cell is shown in **Figure 17.4**. The anode compartment consists of an iron metal electrode dipping into an aqueous solution of $Fe(NO_3)_2$. Note, though, that any inert electrolyte can be used to carry the current in the anode compartment; Fe^{2+} does not need to be present initially because it's not a reactant in the anode half-reaction.

Since Fe^{3+} is a reactant in the cathode half-reaction, $Fe(NO_3)_3$ is a good electrolyte for the cathode compartment. The cathode can be any electrical conductor that doesn't react with the ions in the solution. A platinum wire is a common inert electrode, but iron metal

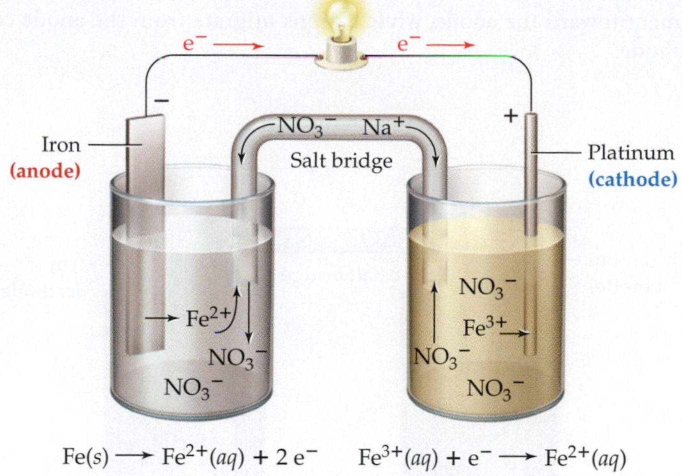

FIGURE 17.4

A galvanic cell consisting of an Fe/Fe^{2+} half-cell and an Fe^{3+}/Fe^{2+} half-cell. The cathode reaction takes place at an inert platinum electrode.

can't be used because it would react directly with the Fe^{3+} ions, thus short-circuiting the cell. The salt bridge contains $NaNO_3$, but any inert electrolyte would do. Electrons flow through the wire from the iron anode (−) to the platinum cathode (+). Anions move from the cathode compartment toward the anode while cations migrate from the anode compartment toward the cathode.

Worked Example 17.1

Designing a Galvanic Cell

Design a galvanic cell that uses the redox reaction

$$2\,Cr(s) + 3\,Sn^{2+}(aq) \longrightarrow 2\,Cr^{3+}(aq) + 3\,Sn(s)$$

Identify the anode and cathode half-reactions, and sketch the experimental setup. Label the anode and cathode, indicate the direction of electron and ion flow, and identify the sign of each electrode.

STRATEGY

First, separate the overall cell reaction into anode (oxidation) and cathode (reduction) half-reactions. Then, set up two half-cells that use these half-reactions and connect the half-cells with a conducting wire and a salt bridge.

SOLUTION

The cell half-reactions involve oxidation of chromium metal to chromium(III) ions at the anode and reduction of tin(II) ions to tin metal at the cathode:

Anode (oxidation):	$2 \times [Cr(s) \longrightarrow Cr^{3+}(aq) + 3\,e^-]$
Cathode (reduction):	$3 \times [Sn^{2+}(aq) + 2\,e^- \longrightarrow Sn(s)]$
Overall cell reaction:	$2\,Cr(s) + 3\,Sn^{2+}(aq) \longrightarrow 2\,Cr^{3+}(aq) + 3\,Sn(s)$

The anode half-reaction is multiplied by a factor of 2 and the cathode half-reaction is multiplied by a factor of 3 so that the electrons will cancel when the two half-reactions are added to give the overall cell reaction.

A possible experimental setup is shown below. The anode compartment consists of a chromium metal electrode dipping into an aqueous solution of $Cr(NO_3)_3$, and the cathode compartment consists of a tin metal electrode dipping into an aqueous solution of $Sn(NO_3)_2$.

continued on next page

The salt bridge contains $NaNO_3$, but any inert electrolyte can be used. Electrons flow through the wire from the chromium anode (−) to the tin cathode (+). Anions move from the cathode compartment toward the anode, while cations migrate from the anode compartment toward the cathode.

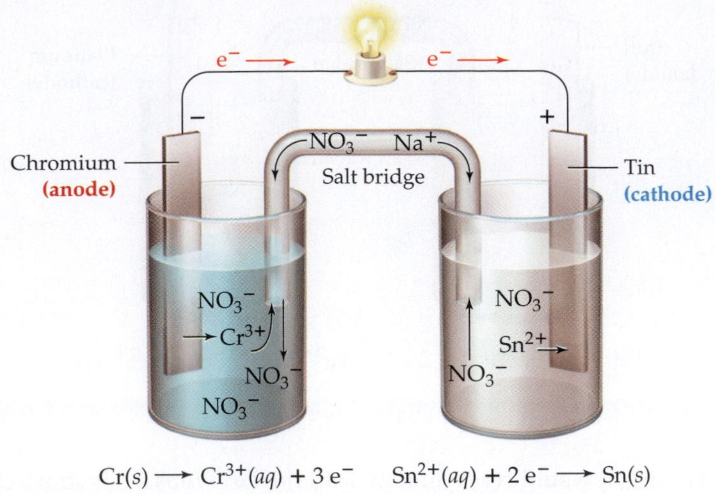

PROBLEM 17.1 Describe a galvanic cell that uses the reaction

$$2\,Ag^+(aq) + Ni(s) \longrightarrow 2\,Ag(s) + Ni^{2+}(aq)$$

Identify the anode and cathode half-reactions, and sketch the experimental setup. Label the anode and cathode, indicate the direction of electron and ion flow, and identify the sign of each electrode.

17.2 SHORTHAND NOTATION FOR GALVANIC CELLS

Rather than describing a galvanic cell in words, it's convenient to use a shorthand notation for representing the cell. For the Daniell cell in Figure 17.2, which uses the reaction

$$Zn(s) + Cu^{2+}(aq) \longrightarrow Zn^{2+}(aq) + Cu(s)$$

we can write the following expression:

$$Zn(s)\,|\,Zn^{2+}(aq)\,||\,Cu^{2+}(aq)\,|\,Cu(s)$$

In this notation, a single vertical line (|) represents a phase boundary, such as that between a solid electrode and an aqueous solution, and the double vertical line (||) denotes a salt bridge. The shorthand for the anode half-cell is always written on the left of the salt-bridge symbol, followed on the right by the shorthand for the cathode half-cell. The electrodes are written on the extreme left (anode) and the extreme right (cathode), and the reactants in each half-cell are always written first, followed by the products. With these arbitrary conventions, electrons move through the external circuit from left to right (from anode to cathode). Reading the shorthand thus suggests the overall cell reaction: Zn is oxidized to Zn^{2+}, and Cu^{2+} is reduced to Cu.

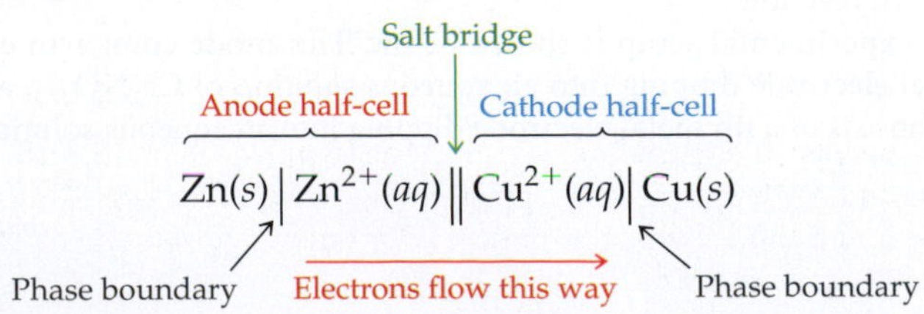

For the galvanic cell in Figure 17.4, based on the reaction

$$Fe(s) + 2\,Fe^{3+}(aq) \longrightarrow 3\,Fe^{2+}(aq)$$

the shorthand notation is

$$Fe(s)\,|\,Fe^{2+}(aq)\,||\,Fe^{3+}(aq), Fe^{2+}(aq)\,|\,Pt(s)$$

The shorthand for the cathode half-cell includes both reactant (Fe^{3+}) and product (Fe^{2+}) as well as the electrode (Pt). The two ions $Fe^{3+}(aq)$ and $Fe^{2+}(aq)$ are separated by a comma rather than a vertical line because they are in the same phase.

The notation for a cell involving a gas has an additional vertical line because an additional phase is present. Thus, the notation

$$Cu(s)\,|\,Cu^{2+}(aq)\,||\,Cl_2(g)\,|\,Cl^-(aq)\,|\,C(s)$$

specifies a cell in which copper is oxidized to Cu^{2+} at a copper anode and Cl_2 gas is reduced to Cl^- at an inert graphite (carbon) cathode. The cell reaction is

$$Cu(s) + Cl_2(g) \longrightarrow Cu^{2+}(aq) + 2\,Cl^-(aq)$$

A more detailed notation would include ion concentrations and gas pressures, as in the following:

$$Cu(s)\,|\,Cu^{2+}(1.0\text{ M})\,||\,Cl_2(1\text{ atm})\,|\,Cl^-(1.0\text{ M})\,|\,C(s)$$

Although the anode half-cell always appears on the left in the shorthand notation, its location in a cell drawing is arbitrary. This means that you can't infer which electrode is the anode and which is the cathode from the location of the electrodes in a cell drawing. You must identify the electrodes based on whether each electrode half-reaction is an oxidation or a reduction.

Worked Example 17.2

Interpreting the Shorthand Notation for a Galvanic Cell

Given the following shorthand notation

$$Pt(s)\,|\,Sn^{2+}(aq), Sn^{4+}(aq)\,||\,Ag^+(aq)\,|\,Ag(s)$$

write a balanced equation for the cell reaction, and give a brief description of the cell.

STRATEGY

We can obtain the cell half-reactions simply by reading the shorthand notation. To find the balanced equation for the cell reaction, add the two half-reactions after multiplying each by an appropriate factor so that the electrons will cancel. The shorthand notation specifies the anode on the extreme left, the cathode on the extreme right, and the reactants in the half-cell compartments.

SOLUTION

Because the anode always appears at the left in the shorthand notation, the anode (oxidation) half-reaction is

$$Sn^{2+}(aq) \longrightarrow Sn^{4+}(aq) + 2\,e^-$$

The platinum electrode is inert and serves only to conduct electrons. The cathode (reduction) half-reaction is

$$2 \times [Ag^+(aq) + e^- \longrightarrow Ag(s)]$$

The cathode half-reaction is multiplied by a factor of 2 so that the electrons will cancel when the two half-reactions are summed to give the cell reaction:

$$Sn^{2+}(aq) + 2\,Ag^+(aq) \longrightarrow Sn^{4+}(aq) + 2\,Ag(s)$$

The cell consists of a platinum wire anode dipping into an Sn^{2+} solution—say, $Sn(NO_3)_2(aq)$—and a silver cathode dipping into an Ag^+ solution—say, $AgNO_3(aq)$. As usual, the anode and cathode half-cells must be connected by a wire and a salt bridge containing inert ions.

PROBLEM 17.2 Write a balanced equation for the overall cell reaction, and give a brief description of a galvanic cell represented by the following shorthand notation:

$$Pb(s)\,|\,Pb^{2+}(aq)\,||\,Br_2(l)\,|\,Br^-(aq)\,|\,Pt(s)$$

PROBLEM 17.3 Write the shorthand notation for a galvanic cell that uses the reaction

$$2\,Fe(s) + Cr_2O_7^{\,2-}(aq) + 14\,H^+(aq) \longrightarrow 2\,Fe^{3+}(aq) + 2\,Cr^{3+}(aq) + 7\,H_2O(l)$$

CONCEPTUAL PROBLEM 17.4 Consider the following galvanic cell:

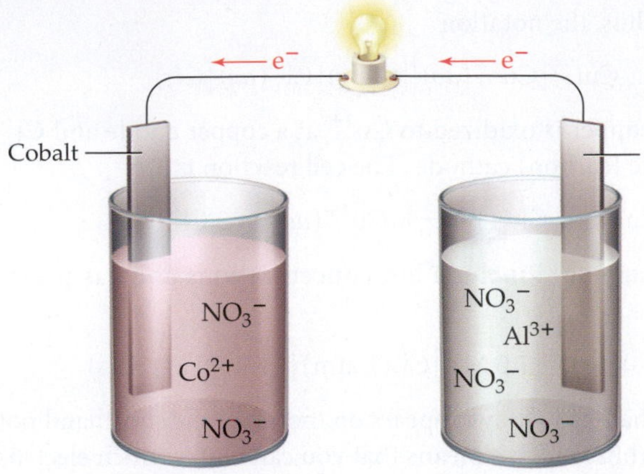

(a) Complete the drawing by adding any components essential for a functioning cell.
(b) Label the anode and cathode, and indicate the direction of ion flow.
(c) Write a balanced equation for the cell reaction.
(d) Write the shorthand notation for the cell.

17.3 CELL POTENTIALS AND FREE-ENERGY CHANGES FOR CELL REACTIONS

Let's return to the Daniell cell shown in Figure 17.2 to find an electrical measure of the tendency for a cell reaction to occur. Electrons move through the external circuit from the zinc anode to the copper cathode because they have lower energy when on copper than on zinc. The force that pushes the negatively charged electrons away from the anode (− electrode) and pulls them toward the cathode (+ electrode) is an electrical potential called the **electromotive force (emf)**, also known as the **cell potential** (*E*) or the **cell voltage**. The SI unit of electrical potential is the volt (V), and the potential of a galvanic cell is defined as a positive quantity.

The relationship between the volt and the SI units of energy (joule, J) and electric charge (coulomb, C) is given by the equation

$$1\,J = 1\,C \times 1\,V$$

REMEMBER...
The **ampere** (1 A = 1 C/s) is the SI unit of electric current. (Section 0.2)

where 1 C is the amount of charge transferred when a current of 1 **ampere (A)** flows for 1 second (s). When 1 C of charge moves between two electrodes that differ in electrical potential by 1 V, 1 J of energy is released by the cell and can be used to do electrical work.

The relationship between the watt (W), the ampere, and the volt is given by

$$1\,W = \frac{1\,J}{s} = \frac{1\,C \times 1\,V}{s} = 1\,A \times 1\,V$$

where 1 W is 1 J/s and 1 A is 1 C/s. Thus, the current passing through a 75 W household lightbulb at a household voltage of 110 V is about 0.7 A, which means that the electric charge of the electrons passing through the bulb in 1 s is 0.7 C.

A cell potential is measured with an electronic instrument called a *voltmeter* (Figure 17.2b), which is designed to give a positive reading when the + and − terminals of the voltmeter are connected to the + (cathode) and − (anode) electrodes of the cell, respectively.

Thus, the voltmeter–cell connections required to get a positive reading on the voltmeter indicate which electrode is the anode and which is the cathode.

We've now seen two quantitative measures of the tendency for a chemical reaction to occur: the cell potential E, an electrochemical quantity, and the **free-energy change ΔG**, a thermochemical quantity (Section 16.7). The values of ΔG and E are directly proportional and are related by the equation

REMEMBER...

If the **free-energy change ΔG** for a reaction is negative, the reaction is spontaneous. The greater the negative value of ΔG, the greater the tendency for the reaction to occur. (Section 16.7)

$$\Delta G = -nFE$$

where n is the number of moles of electrons transferred in the reaction and F is the **faraday** (or *Faraday constant*), the electric charge on 1 mol of electrons (96,485 C/mol e^-). In our calculations, we'll round the value of F to three significant figures:

$$F = 96{,}500 \text{ C/mol } e^-$$

The faraday is named in honor of Michael Faraday (1791–1867), the nineteenth-century English scientist who laid the foundations for our current understanding of electricity.

Two features of the equation $\Delta G = -nFE$ are worth noting: the units and the minus sign. When we multiply the charge transferred (nF) in coulombs by the cell potential (E) in volts, we obtain an energy (ΔG) in joules, in accord with the relationship $1 \text{ J} = 1 \text{ C} \times 1 \text{ V}$. The minus sign is required because E and ΔG have opposite signs: The spontaneous reaction in a galvanic cell has a positive cell potential but a negative free-energy change (Section 16.7).

Later in this chapter, we'll see that cell potentials, like free-energy changes, depend on the composition of the reaction mixture. The **standard cell potential** $E°$ is the cell potential when both reactants and products are in their standard states—solutes at 1 M concentrations, gases at a partial pressure of 1 atm, solids and liquids in pure form, with all at a specified temperature, usually 25 °C. For example, $E°$ for the reaction

$$Zn(s) + Cu^{2+}(aq) \longrightarrow Zn^{2+}(aq) + Cu(s) \qquad E° = 1.10 \text{ V}$$

is the cell potential measured at 25 °C for a cell that has pure Zn and Cu metal electrodes and 1 M concentrations of Zn^{2+} and Cu^{2+}.

The standard free-energy change and the standard cell potential are related by the equation

$$\Delta G° = -nFE°$$

Because $\Delta G°$ and $E°$ are directly proportional, a voltmeter can be regarded as a "free-energy meter." When a voltmeter measures $E°$, it also indirectly measures $\Delta G°$.

Worked Example 17.3

Calculating a Standard Free-Energy Change from a Standard Cell Potential

The standard cell potential at 25 °C is 0.90 V for the reaction

$$2\,Al(s) + 3\,Zn^{2+}(aq) \longrightarrow 2\,Al^{3+}(aq) + 3\,Zn(s)$$

Calculate the standard free-energy change for this reaction at 25 °C.

STRATEGY

To calculate $\Delta G°$, we use the equation $\Delta G° = -nFE°$, where n can be inferred from the balanced chemical equation, F is 96,500 C/mol e^-, and $E°$ is given.

SOLUTION

When two moles of Al are oxidized to 2 moles of Al^{3+} ions, six moles of electrons are transferred to three moles of Zn^{2+} ions, which are reduced to three moles of Zn. Therefore, $n = 6$ mol e^- for this reaction, and the standard free-energy change is

$$\Delta G° = -nFE° = -(6 \text{ mol } e^-)\left(\frac{96{,}500 \text{ C}}{\text{mol } e^-}\right)(0.90 \text{ V})\left(\frac{1 \text{ J}}{1 \text{ C} \cdot \text{V}}\right)$$

$$= -5.2 \times 10^5 \text{ J} = -520 \text{ kJ}$$

continued on next page

✔ BALLPARK CHECK

F is a bit less than 10^5 C/mol e$^-$ and $E°$ is 10% less than 1 V, so $\Delta G° = -nFE°$ is approximately 10% less than $-(6 \text{ mol e}^-)(10^5 \text{ C/mol e}^-)(1 \text{ V}) = -6 \times 10^5$ J, or about -600 kJ. Ten percent less than -600 kJ is -540 kJ, in agreement with the solution. Note that $\Delta G°$ is negative, as expected for a spontaneous reaction.

PROBLEM 17.5 The standard cell potential at 25 °C is 1.21 V for the reaction

$$Cr_2O_7^{2-}(aq) + 3\,Sn^{2+}(aq) + 14\,H^+(aq) \longrightarrow 2\,Cr^{3+}(aq) + 3\,Sn^{4+}(aq) + 7\,H_2O(l)$$

What is the standard free-energy change for this reaction at 25 °C?

17.4 STANDARD REDUCTION POTENTIALS

The standard potential of any galvanic cell is the sum of the standard half-cell potentials for oxidation at the anode and reduction at the cathode:

$$E°_{\text{cell}} = E°_{\text{ox}} + E°_{\text{red}}$$

Consider, for example, a cell in which H_2 gas is oxidized to H^+ ions at the anode and Cu^{2+} ions are reduced to copper metal at the cathode (**Figure 17.5**):

Anode (oxidation):	$H_2(g) \longrightarrow 2\,H^+(aq) + 2\,e^-$
Cathode (reduction):	$Cu^{2+}(aq) + 2\,e^- \longrightarrow Cu(s)$
Overall cell reaction:	$H_2(g) + Cu^{2+}(aq) \longrightarrow 2\,H^+(aq) + Cu(s)$

[In this chapter we represent the hydrated proton as $H^+(aq)$ rather than $H_3O^+(aq)$ because we're interested here in electron transfer, not proton transfer as in Chapter 14.] The standard potential for this cell, 0.34 V at 25 °C, measures the sum of the reaction tendencies of the oxidation and reduction half-reactions:

$$E°_{\text{cell}} = E°_{\text{ox}} + E°_{\text{red}} = E°_{H_2 \rightarrow H^+} + E°_{Cu^{2+} \rightarrow Cu} = 0.34 \text{ V}$$

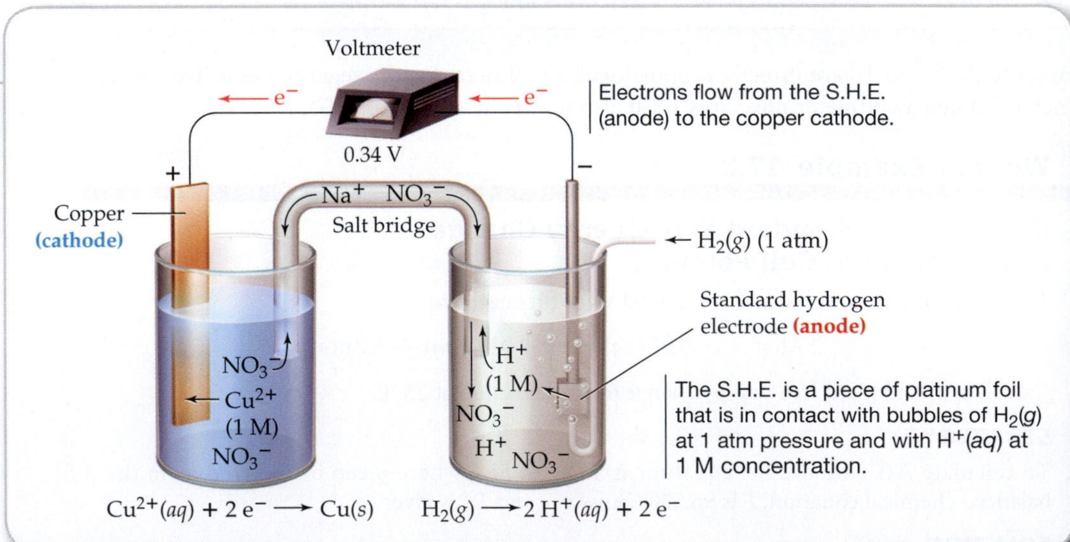

FIGURE 17.5

A galvanic cell consisting of a Cu^{2+}(1 M)/Cu half-cell and a standard hydrogen electrode (S.H.E.). The measured standard cell potential at 25 °C is 0.34 V.

If we could determine $E°$ values for individual half-reactions, we could combine those values to obtain $E°$ values for a host of cell reactions. Unfortunately, it's not possible to measure the potential of a single electrode; we can measure only a potential *difference* by placing a voltmeter between *two* electrodes. Nevertheless, we can develop a set of standard half-cell potentials by choosing an arbitrary standard half-cell as a reference point, assigning it an arbitrary potential, and then expressing the potential of all other half-cells relative to the reference half-cell. Recall that this same approach was used in Section 8.9 for determining **standard enthalpies of formation, $\Delta H°_f$.**

REMEMBER...

The **standard enthalpy of formation, $\Delta H°_f$,** of a substance is expressed relative to an arbitrary reference of $\Delta H°_f = 0$ kJ/mol for its constituent elements in their standard states. (Section 8.9)

To define an electrochemical "sea level," chemists have chosen a reference half-cell called the **standard hydrogen electrode (S.H.E.)**, shown in Figure 17.5. It consists of a platinum electrode in contact with H_2 gas and aqueous H^+ ions at standard-state conditions [1 atm $H_2(g)$, 1 M $H^+(aq)$, 25 °C]. The corresponding half-reaction, written in either direction, is assigned an arbitrary potential of exactly 0 V:

$$2\,H^+(aq, 1\,M) + 2\,e^- \longrightarrow H_2(g, 1\,atm) \qquad E° = 0\,V$$
$$H_2(g, 1\,atm) \longrightarrow 2\,H^+(aq, 1\,M) + 2\,e^- \qquad E° = 0\,V$$

With this choice of standard reference electrode, the entire potential of the cell

$$Pt(s)\,|\,H_2(1\,atm)\,|\,H^+(1\,M)\,||\,Cu^{2+}(1\,M)\,|\,Cu(s)$$

can be attributed to the Cu^{2+}/Cu half-cell:

$$\underset{0.34\,V}{\underset{\uparrow}{E°_{cell}}} = \underset{0\,V}{\underset{\uparrow}{E°_{H_2 \rightarrow H^+}}} + \underset{0.34\,V}{\underset{\uparrow}{E°_{Cu^{2+} \rightarrow Cu}}} = 0.34\,V$$

Because the Cu^{2+}/Cu half-reaction is a reduction, the corresponding half-cell potential, $E° = 0.34$ V, is called a **standard reduction potential**:

$$Cu^{2+}(aq) + 2\,e^- \longrightarrow Cu(s) \qquad \text{Standard reduction potential: } E° = 0.34\,V$$

In a cell in which this half-reaction occurs in the opposite direction, the corresponding half-cell potential has the same magnitude but opposite sign:

$$Cu(s) \longrightarrow Cu^{2+}(aq) + 2\,e^- \qquad E° = -0.34\,V$$

Whenever the direction of a half-reaction is reversed, the sign of $E°$ must also be reversed. Thus, the standard potential for an oxidation half-reaction is the negative of the standard reduction potential.

We can determine standard potentials for other half-cells simply by constructing galvanic cells in which each half-cell of interest is paired up with the S.H.E. For example, to find the potential of a half-cell consisting of a zinc electrode dipping into a 1 M Zn^{2+} solution, we would build the cell shown in **Figure 17.6**. The voltmeter–electrode connections required to get a positive reading on the voltmeter (0.76 V) tell us that the zinc electrode is the anode and the S.H.E. is the cathode. Therefore, the half-cell reactions involve oxidation of Zn and

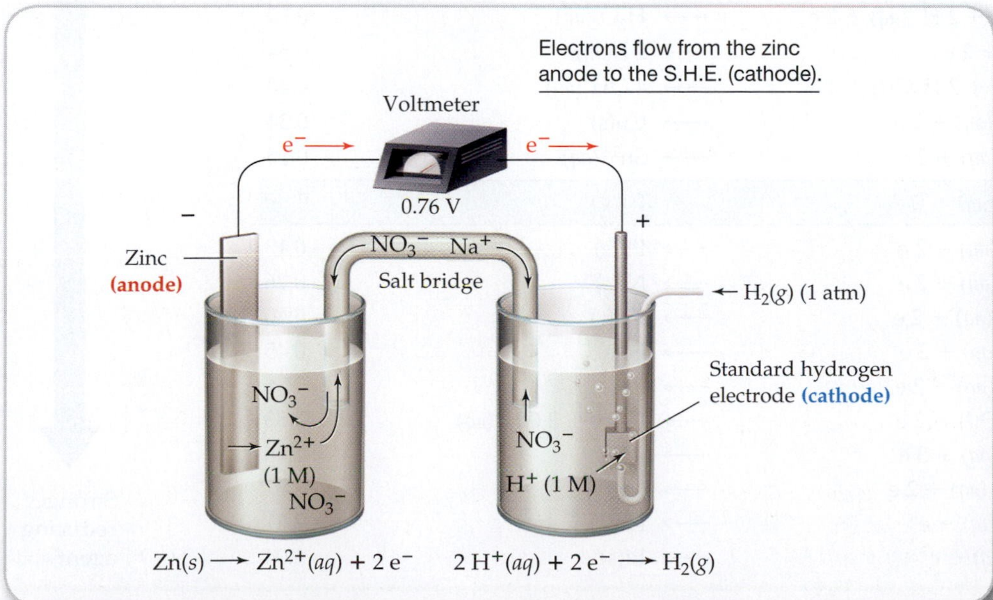

FIGURE 17.6

A galvanic cell consisting of a Zn/Zn^{2+} (1 M) half-cell and a standard hydrogen electrode. The measured standard cell potential at 25 °C is 0.76 V.

reduction of H^+. Alternatively, we could identify the direction of the half-reactions by noting that the H^+ concentration decreases as the cell reaction progresses.

Anode (oxidation):	$Zn(s) \longrightarrow Zn^{2+}(aq) + 2\,e^-$	$E° = ?$
Cathode (reduction):	$2\,H^+(aq) + 2\,e^- \longrightarrow H_2(g)$	$E° = 0\text{ V}$
Overall cell reaction:	$Zn(s) + 2\,H^+(aq) \longrightarrow Zn^{2+}(aq) + H_2(g)$	$E° = 0.76\text{ V}$

Since the anode and cathode half-cell potentials must sum to give the overall cell potential, the $E°$ value for oxidation of Zn to Zn^{2+} must be 0.76 V and the standard *reduction* potential for the Zn^{2+}/Zn half-cell is therefore −0.76 V.

We could also have calculated the standard reduction potential for the Zn^{2+}/Zn half-cell from the observed standard potential of the Daniell cell (1.10 V) and the standard reduction potential for the Cu^{2+}/Cu half-cell (0.34 V):

Anode (oxidation):	$Zn(s) \longrightarrow Zn^{2+}(aq) + 2\,e^-$	$E° = ?$
Cathode (reduction):	$Cu^{2+}(aq) + 2\,e^- \longrightarrow Cu(s)$	$E° = 0.34\text{ V}$
Overall cell reaction:	$Zn(s) + Cu^{2+}(aq) \longrightarrow Zn^{2+}(aq) + Cu(s)$	$E° = 1.10\text{ V}$

Since 1.10 V = 0.76 V + 0.34 V, we would again conclude that the $E°$ value for oxidation of Zn to Zn^{2+} is 0.76 V and the standard reduction potential for the Zn^{2+}/Zn half-cell is therefore −0.76 V.

From experiments of the sort just described, hundreds of half-cell potentials have been determined. A short list is presented in **Table 17.1**, and a more complete tabulation is

TABLE 17.1 Standard Reduction Potentials at 25 °C

	Reduction Half-Reaction		$E°$ (V)	
Stronger oxidizing agent	$F_2(g) + 2\,e^-$	$\longrightarrow 2\,F^-(aq)$	2.87	Weaker reducing agent
	$H_2O_2(aq) + 2\,H^+(aq) + 2\,e^-$	$\longrightarrow 2\,H_2O(l)$	1.78	
	$MnO_4^-(aq) + 8\,H^+(aq) + 5\,e^-$	$\longrightarrow Mn^{2+}(aq) + 4\,H_2O(l)$	1.51	
	$Cl_2(g) + 2\,e^-$	$\longrightarrow 2\,Cl^-(aq)$	1.36	
	$Cr_2O_7^{2-}(aq) + 14\,H^+(aq) + 6\,e^-$	$\longrightarrow 2\,Cr^{3+}(aq) + 7\,H_2O(l)$	1.36	
	$O_2(g) + 4\,H^+(aq) + 4\,e^-$	$\longrightarrow 2\,H_2O(l)$	1.23	
	$Br_2(aq) + 2\,e^-$	$\longrightarrow 2\,Br^-(aq)$	1.09	
	$Ag^+(aq) + e^-$	$\longrightarrow Ag(s)$	0.80	
	$Fe^{3+}(aq) + e^-$	$\longrightarrow Fe^{2+}(aq)$	0.77	
	$O_2(g) + 2\,H^+(aq) + 2\,e^-$	$\longrightarrow H_2O_2(aq)$	0.70	
	$I_2(s) + 2\,e^-$	$\longrightarrow 2\,I^-(aq)$	0.54	
	$O_2(g) + 2\,H_2O(l) + 4\,e^-$	$\longrightarrow 4\,OH^-(aq)$	0.40	
	$Cu^{2+}(aq) + 2\,e^-$	$\longrightarrow Cu(s)$	0.34	
	$Sn^{4+}(aq) + 2\,e^-$	$\longrightarrow Sn^{2+}(aq)$	0.15	
	$2\,H^+(aq) + 2\,e^-$	$\longrightarrow H_2(g)$	0	
	$Pb^{2+}(aq) + 2\,e^-$	$\longrightarrow Pb(s)$	−0.13	
	$Ni^{2+}(aq) + 2\,e^-$	$\longrightarrow Ni(s)$	−0.26	
	$Cd^{2+}(aq) + 2\,e^-$	$\longrightarrow Cd(s)$	−0.40	
	$Fe^{2+}(aq) + 2\,e^-$	$\longrightarrow Fe(s)$	−0.45	
	$Zn^{2+}(aq) + 2\,e^-$	$\longrightarrow Zn(s)$	−0.76	
	$2\,H_2O(l) + 2\,e^-$	$\longrightarrow H_2(g) + 2\,OH^-(aq)$	−0.83	
	$Al^{3+}(aq) + 3\,e^-$	$\longrightarrow Al(s)$	−1.66	
Weaker oxidizing agent	$Mg^{2+}(aq) + 2\,e^-$	$\longrightarrow Mg(s)$	−2.37	Stronger reducing agent
	$Na^+(aq) + e^-$	$\longrightarrow Na(s)$	−2.71	
	$Li^+(aq) + e^-$	$\longrightarrow Li(s)$	−3.04	

given in Appendix D. The following conventions are observed when constructing a table of half-cell potentials:

- **The half-reactions are written as reductions** rather than as oxidations. This means that oxidizing agents and electrons are on the left side of each half-reaction and reducing agents are on the right side.
- **The listed half-cell potentials are standard reduction potentials**, also known as **standard electrode potentials.**
- **The half-reactions are listed in order of decreasing standard reduction potential**, meaning a decreasing tendency to occur in the forward direction and an increasing tendency to occur in the reverse direction. Consequently, the strongest oxidizing agents are located in the upper left of the table (F_2, H_2O_2, MnO_4^-, and so forth), and the strongest reducing agents are found in the lower right of the table (Li, Na, Mg, and so forth).

By choosing $E° = 0$ V for the standard hydrogen electrode, we obtain standard reduction potentials that range from about +3 V to −3 V.

Note that the strongest oxidizing agents in Table 17.1 tend to be either elements on the right side of the periodic table that have large electron affinities, such as F_2 and Cl_2, or oxoanions that have a metal atom in a high oxidation state, such as MnO_4^- and $Cr_2O_7^{2-}$. Metal atoms in high oxidation states have a strong affinity for electrons and are relatively easily reduced to lower oxidation states. Note also that the strongest reducing agents in Table 17.1 are the main-group metals on the left side of the periodic table that have relatively low ionization energies and are therefore relatively easily oxidized.

Finally, note how the ordering of the half-reactions in Table 17.1 corresponds to their ordering in the activity series in Table 7.5 (page 255). The more active metals at the top of the activity series have the more positive oxidation potentials and therefore the more negative standard reduction potentials.

PROBLEM 17.6 The standard potential for the following galvanic cell is 0.92 V:

$$Al(s)\,|\,Al^{3+}(aq)\,||\,Cr^{3+}(aq)\,|\,Cr(s)$$

Look up the standard reduction potential for the Al^{3+}/Al half-cell in Table 17.1, and calculate the standard reduction potential for the Cr^{3+}/Cr half-cell.

17.5 USING STANDARD REDUCTION POTENTIALS

A table of standard reduction potentials summarizes an enormous amount of chemical information in a very small space. It enables us to arrange oxidizing or reducing agents in order of increasing strength, and it permits us to predict the spontaneity or nonspontaneity of thousands of redox reactions. Suppose, for example, that the table contains just 100 half-reactions. Each reduction half-reaction can be paired with any one of the remaining 99 oxidation half-reactions to give a total of $100 \times 99 = 9900$ cell reactions. By calculating the $E°$ values for these cell reactions, we would find that half of them are spontaneous and the other half are nonspontaneous. (Can you see why?)

To illustrate how to use the tabulated $E°$ values, let's calculate $E°$ for the oxidation of $Zn(s)$ by $Ag^+(aq)$:

$$2\,Ag^+(aq) + Zn(s) \longrightarrow 2\,Ag(s) + Zn^{2+}(aq)$$

First, find the relevant half-reactions in Table 17.1 and write them in the appropriate direction for reduction of Ag^+ and oxidation of Zn. Next, add the half-reactions to get the overall reaction. Before adding, though, multiply the Ag^+/Ag half-reaction by a factor of 2 so that the electrons will cancel:

Reduction:	$2 \times [Ag^+(aq) + e^- \longrightarrow Ag(s)]$	$E° = 0.80\ V$
Oxidation:	$Zn(s) \longrightarrow Zn^{2+}(aq) + 2\,e^-$	$E° = -(-0.76\ V)$
Overall reaction:	$2\,Ag^+(aq) + Zn(s) \longrightarrow 2\,Ag(s) + Zn^{2+}(aq)$	$E° = 1.56\ V$

Then, tabulate the $E°$ values for the half-reactions, remembering that $E°$ for oxidation of zinc is the negative of the standard reduction potential (−0.76 V). Do not multiply the $E°$ value for reduction of Ag^+ (0.80 V) by a factor of 2, however, because an electrical potential does not depend on how much reaction occurs.

The reason why $E°$ values are independent of the amount of reaction can be understood by looking at the equation $\Delta G° = -nFE°$. Free energy is an **extensive property** (Section 1.3) because it depends on the amount of substance. If twice as much Ag^+ is reduced, the free-energy change, $\Delta G°$, doubles. The number of electrons transferred, n, also doubles, however, so the ratio $E° = -\Delta G°/nF$ is constant. Electrical potential is therefore an **intensive property**, which does not depend on the amount of substance.

REMEMBER...

Extensive properties, like mass and volume, have values that depend on the sample size. **Intensive properties**, like temperature and melting point, have values that do not depend on the sample size. (Section 1.3)

The $E°$ value for the overall reaction of Zn(s) with $Ag^+(aq)$ is the sum of the $E°$ values for the two half-reactions: 0.80 V + 0.76 V = 1.56 V. Because $E°$ is positive (and $\Delta G°$ is negative), oxidation of zinc by Ag^+ is a spontaneous reaction under standard-state conditions. Just as Ag^+ can oxidize Zn, it's evident from Table 17.1 that Ag^+ can oxidize any reducing agent that lies below it in the table (Fe^{2+}, H_2O_2, Cu, and so forth). Summing $E°$ for the Ag^+/Ag reduction (0.80 V) and $-E°$ for any half-reaction that lies below the Ag^+/Ag half-reaction always gives a positive $E°$ for the overall reaction.

In general, an oxidizing agent can oxidize any reducing agent that lies below it in the table but can't oxidize a reducing agent that appears above it in the table. Thus, Ag^+ can't oxidize Br^-, H_2O, Cr^{3+}, and so forth because $E°$ for the overall reaction is negative. Simply by glancing at the locations of the oxidizing and reducing agents in the table, we can predict whether a reaction is spontaneous or nonspontaneous.

Worked Example 17.4

Arranging Oxidizing and Reducing Agents in Order of Increasing Strength

(a) Arrange the following oxidizing agents in order of increasing strength under standard-state conditions: $Br_2(aq)$, $Fe^{3+}(aq)$, $Cr_2O_7^{2-}(aq)$.

(b) Arrange the following reducing agents in order of increasing strength under standard-state conditions: Al(s), Na(s), Zn(s).

STRATEGY

Pick out the half-reactions in Table 17.1 that involve the given oxidizing or reducing agents, and list them, along with their $E°$ values, in the order in which they occur in the table. The strength of an oxidizing agent increases as the $E°$ value increases, and the strength of a reducing agent increases as the $E°$ value decreases.

SOLUTION

(a) List the half-reactions that involve Br_2, Fe^{3+}, and $Cr_2O_7^{2-}$ in the order in which they occur in Table 17.1:

$$Cr_2O_7^{2-}(aq) + 14\,H^+(aq) + 6\,e^- \longrightarrow 2\,Cr^{3+}(aq) + 7\,H_2O(l) \qquad E° = 1.36\ V$$
$$Br_2(aq) + 2\,e^- \longrightarrow 2\,Br^-(aq) \qquad E° = 1.09\ V$$
$$Fe^{3+}(aq) + e^- \longrightarrow Fe^{2+}(aq) \qquad E° = 0.77\ V$$

You can see that $Cr_2O_7^{2-}$ has the greatest tendency to be reduced (largest $E°$) and Fe^{3+} has the least tendency to be reduced (smallest $E°$). The species that has the greatest tendency to be reduced is the strongest oxidizing agent, so oxidizing strength increases in the order $Fe^{3+} < Br_2 < Cr_2O_7^{2-}$. As a shortcut, simply note that the strength of the oxidizing agents, listed on the left side of Table 17.1, increases on moving up in the table.

(b) List the half-reactions that involve Al(s), Na(s), and Zn(s) in the order in which they occur in Table 17.1:

$$Zn^{2+}(aq) + 2\,e^- \longrightarrow Zn(s) \qquad E° = -0.76\text{ V}$$
$$Al^{3+}(aq) + 3\,e^- \longrightarrow Al(s) \qquad E° = -1.66\text{ V}$$
$$Na^{+}(aq) + e^- \longrightarrow Na(s) \qquad E° = -2.71\text{ V}$$

The last half-reaction has the least tendency to occur in the forward direction (most negative $E°$) and the greatest tendency to occur in the reverse direction. Therefore, Na is the strongest reducing agent, and reducing strength increases in the order Zn < Al < Na. As a shortcut, note that the strength of the reducing agents, listed on the right side of Table 17.1, increases on moving down in the table.

Worked Example 17.5

Predicting Whether a Redox Reaction Is Spontaneous

Predict from Table 17.1 whether $Pb^{2+}(aq)$ can oxidize Al(s) or Cu(s) under standard-state conditions. Calculate $E°$ for each reaction at 25 °C.

STRATEGY

To predict whether a redox reaction is spontaneous, remember that an oxidizing agent can oxidize any reducing agent that lies below it in the table but can't oxidize one that lies above it. To calculate $E°$ for a redox reaction, sum the $E°$ values for the reduction and oxidation half-reactions.

SOLUTION

$Pb^{2+}(aq)$ is above Al(s) in the table but below Cu(s). Therefore, $Pb^{2+}(aq)$ can oxidize Al(s) but can't oxidize Cu(s). To confirm these predictions, calculate $E°$ values for the overall reactions.

For the oxidation of Al by Pb^{2+}, $E°$ is positive (1.53 V) and the reaction is therefore spontaneous:

$$3 \times [Pb^{2+}(aq) + 2\,e^- \longrightarrow Pb(s)] \qquad E° = -0.13\text{ V}$$
$$2 \times [Al(s) \longrightarrow Al^{3+}(aq) + 3\,e^-] \qquad E° = 1.66\text{ V}$$
$$3\,Pb^{2+}(aq) + 2\,Al(s) \longrightarrow 3\,Pb(s) + 2\,Al^{3+}(aq) \qquad E° = 1.53\text{ V}$$

Note that we have multiplied the Pb^{2+}/Pb half-reaction by a factor of 3 and the Al/Al^{3+} half-reaction by a factor of 2 so that the electrons will cancel, but we do not multiply the $E°$ values by these factors because electrical potential is an intensive property.

For the oxidation of Cu by Pb^{2+}, $E°$ is negative (−0.47 V) and the reaction is therefore nonspontaneous:

$$Pb^{2+}(aq) + 2\,e^- \longrightarrow Pb(s) \qquad E° = -0.13\text{ V}$$
$$Cu(s) \longrightarrow Cu^{2+}(aq) + 2\,e^- \qquad E° = -0.34\text{ V}$$
$$Pb^{2+}(aq) + Cu(s) \longrightarrow Pb(s) + Cu^{2+}(aq) \qquad E° = -0.47\text{ V}$$

PROBLEM 17.7 Which is the stronger oxidizing agent, $Cl_2(g)$ or $Ag^{+}(aq)$? Which is the stronger reducing agent, Fe(s) or Mg(s)?

PROBLEM 17.8 Predict from Table 17.1 whether each of the following reactions can occur under standard-state conditions:

(a) $2\,Fe^{3+}(aq) + 2\,I^-(aq) \longrightarrow 2\,Fe^{2+}(aq) + I_2(s)$
(b) $3\,Ni(s) + 2\,Al^{3+}(aq) \longrightarrow 3\,Ni^{2+}(aq) + 2\,Al(s)$

Confirm your predictions by calculating the value of $E°$ for each reaction. Which reaction(s) can occur in the reverse direction under standard-state conditions?

CONCEPTUAL PROBLEM 17.9 Consider the following table of standard reduction potentials:

Reduction Half-Reaction	$E°$ (V)
$A^{3+} + 2\,e^- \longrightarrow A^+$	1.47
$B^{2+} + 2\,e^- \longrightarrow B$	0.60
$C^{2+} + 2\,e^- \longrightarrow C$	−0.21
$D^+ + e^- \longrightarrow D$	−1.38

(a) Which substance is the strongest reducing agent? Which is the strongest oxidizing agent?
(b) Which substances can be oxidized by B^{2+}? Which can be reduced by C?
(c) Write a balanced equation for the overall cell reaction that delivers the highest voltage, and calculate $E°$ for the reaction.

17.6 CELL POTENTIALS UNDER NONSTANDARD-STATE CONDITIONS: THE NERNST EQUATION

Cell potentials, like free-energy changes, depend on temperature and on the composition of the reaction mixture—that is, on the concentrations of solutes and the partial pressures of gases. This dependence can be derived from the equation

$$\Delta G = \Delta G° + RT \ln Q$$

Recall from Section 16.10 that ΔG is the free-energy change for a reaction under nonstandard-state conditions, $\Delta G°$ is the free-energy change under standard-state conditions, and **Q is the reaction quotient**. Since $\Delta G = -nFE$ and $\Delta G° = -nFE°$, we can rewrite the equation for ΔG in the form

REMEMBER... The **reaction quotient Q** is an expression having the same form as the equilibrium constant expression *K* except that the concentrations and partial pressures do not necessarily have equilibrium values. (Section 16.10)

$$-nFE = -nFE° + RT \ln Q$$

Dividing by $-nF$, we obtain the **Nernst equation**, named after Walther Nernst (1864–1941), the German chemist who first derived it:

▶ **Nernst equation** $$E = E° - \frac{RT}{nF} \ln Q \quad \text{or} \quad E = E° - \frac{2.303\,RT}{nF} \log Q$$

Because of an intimate connection between the cell voltage and pH (Section 17.7), we will write the Nernst equation in terms of base-10 logarithms: log Q rather than ln Q. At 25 °C, 2.303 RT/F has a value of 0.0592 V, and therefore

$$E = E° - \frac{0.0592\ \text{V}}{n} \log Q \qquad \text{in volts, at 25 °C}$$

In actual galvanic cells, the concentrations and partial pressures of reactants and products seldom have standard-state values, and the values change as the cell reaction proceeds. The Nernst equation is useful because it enables us to calculate cell potentials under nonstandard-state conditions, as shown in Worked Example 17.6.

Worked Example 17.6

Using the Nernst Equation to Calculate the Cell Potential under Nonstandard-State Conditions

Consider a galvanic cell that uses the reaction

$$Zn(s) + 2\,H^+(aq) \longrightarrow Zn^{2+}(aq) + H_2(g)$$

Calculate the cell potential at 25 °C when $[H^+] = 1.0$ M, $[Zn^{2+}] = 0.0010$ M, and $P_{H_2} = 0.10$ atm.

STRATEGY

We can calculate the standard cell potential $E°$ from the standard reduction potentials in Table 17.1 and then use the Nernst equation to find the cell potential E under the cited conditions.

SOLUTION

The standard cell potential is

$$E° = E°_{Zn \rightarrow Zn^{2+}} + E°_{H^+ \rightarrow H_2} = -(-0.76\text{ V}) + 0\text{ V} = 0.76\text{ V}$$

The cell potential at 25 °C under nonstandard-state conditions is given by the Nernst equation:

$$E = E° - \frac{0.0592\text{ V}}{n}\log Q = E° - \left(\frac{0.0592\text{ V}}{n}\right)\left(\log\frac{[Zn^{2+}](P_{H_2})}{[H^+]^2}\right)$$

where the reaction quotient contains both molar concentrations of solutes and the partial pressure of a gas (in atm). As usual, zinc has been omitted from the reaction quotient because it is a pure solid. For this reaction, 2 mol of electrons are transferred, so $n = 2$. Substituting into the Nernst equation the appropriate values of $E°$, n, $[H^+]$, $[Zn^{2+}]$, and P_{H_2} gives

$$E = (0.76\text{ V}) - \left(\frac{0.0592\text{ V}}{2}\right)\left(\log\frac{(0.0010)(0.10)}{(1.0)^2}\right) = (0.76\text{ V}) - \left(\frac{0.0592\text{ V}}{2}\right)(-4.0)$$

$$= 0.76\text{ V} + 0.12\text{ V} = 0.88\text{ V} \quad \text{at } 25\text{ °C}$$

✔ BALLPARK CHECK

We expect that the reaction will have a greater tendency to occur under the cited conditions than under standard-state conditions because the product concentrations are lower than standard-state values. We therefore predict that the cell potential E will be greater than the standard cell potential $E°$, in agreement with the solution.

PROBLEM 17.10 Consider a galvanic cell that uses the reaction

$$Cu(s) + 2\,Fe^{3+}(aq) \longrightarrow Cu^{2+}(aq) + 2\,Fe^{2+}(aq)$$

What is the potential of a cell at 25 °C that has the following ion concentrations?

$$[Fe^{3+}] = 1.0 \times 10^{-4}\text{ M} \qquad [Cu^{2+}] = 0.25\text{ M} \qquad [Fe^{2+}] = 0.20\text{ M}$$

WORKED CONCEPTUAL EXAMPLE 17.7

Exploring How Changes in Concentrations Affect Cell Voltage

Consider the following galvanic cell:

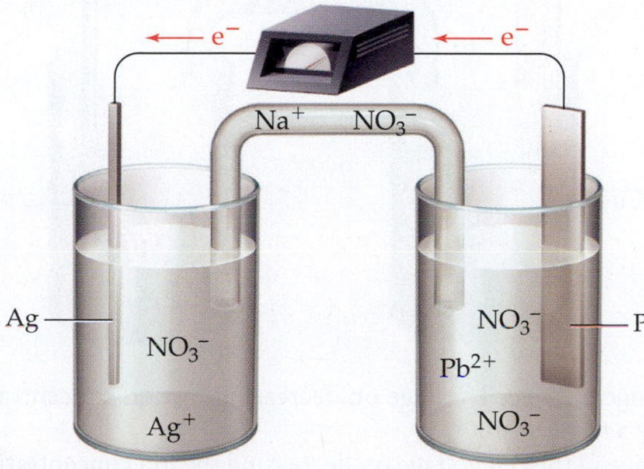

(a) What is the change in the cell voltage on increasing the ion concentrations in the anode compartment by a factor of 10?

(b) What is the change in the cell voltage on increasing the ion concentrations in the cathode compartment by a factor of 10?

continued on next page

STRATEGY

The direction of electron flow in the picture indicates that lead is the anode and silver is the cathode. Therefore, the cell reaction is

$$Pb(s) + 2\,Ag^+(aq) \longrightarrow Pb^{2+}(aq) + 2\,Ag(s)$$

The cell potential at 25 °C is given by the Nernst equation, where $n = 2$ and $Q = [Pb^{2+}]/[Ag^+]^2$:

$$E = E° - \frac{0.0592\text{ V}}{n}\log Q = E° - \left(\frac{0.0592\text{ V}}{2}\right)\left(\log\frac{[Pb^{2+}]}{[Ag^+]^2}\right)$$

The change in E on changing the ion concentrations will be determined by the change in the log term in the Nernst equation.

SOLUTION

(a) Pb^{2+} is in the anode compartment, and Ag^+ is in the cathode compartment. Suppose that the original concentrations of Pb^{2+} and Ag^+ are 1 M, so that $E = E°$. Increasing $[Pb^{2+}]$ to 10 M gives

$$E = E° - \left(\frac{0.0592\text{ V}}{2}\right)\left(\log\frac{(10)}{(1)^2}\right)$$

Because log 10 = 1.0, $E = E° - 0.03$ V. Thus, increasing the Pb^{2+} concentration by a factor of 10 decreases the cell voltage by 0.03 V.

(b) Increasing $[Ag^+]$ to 10 M gives

$$E = E° - \left(\frac{0.0592\text{ V}}{2}\right)\left(\log\frac{(1)}{(10)^2}\right)$$

Because $\log (10)^{-2} = -2.0$, $E = E° + 0.06$ V. Thus, increasing the Ag^+ concentration by a factor of 10 increases the cell voltage by 0.06 V.

✔ BALLPARK CHECK

We expect that the reaction will have a lesser tendency to occur when the product ion concentration, $[Pb^{2+}]$, is increased and a greater tendency to occur when the reactant ion concentration, $[Ag^+]$, is increased. Therefore, the cell voltage E will decrease when $[Pb^{2+}]$ is increased and will increase when $[Ag^+]$ is increased. The ballpark check and the solution agree.

CONCEPTUAL PROBLEM 17.11 Consider the following galvanic cell:

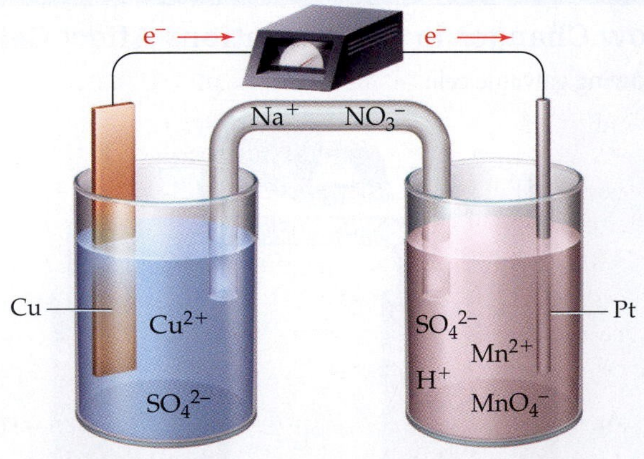

(a) What is the change in the cell voltage on decreasing the ion concentrations in the anode compartment by a factor of 100?

(b) What is the change in the cell voltage on decreasing the ion concentrations in the cathode compartment by a factor of 100?

17.7 ELECTROCHEMICAL DETERMINATION OF pH

The electrochemical determination of pH using a pH meter is a particularly important application of the Nernst equation. Consider, for example, a cell with a hydrogen electrode as the anode and a second reference electrode as the cathode:

$$\text{Pt}\,|\,\text{H}_2(1\text{ atm})\,|\,\text{H}^+(?\text{ M})\,||\,\text{Reference cathode}$$

The hydrogen electrode consists of a platinum wire that is in contact with H_2 at 1 atm and dips into a solution of unknown pH. The potential of this cell is

$$E_{\text{cell}} = E_{\text{H}_2 \rightarrow \text{H}^+} + E_{\text{ref}}$$

We can calculate the potential for the hydrogen electrode half-reaction

$$\text{H}_2(g) \longrightarrow 2\,\text{H}^+(aq) + 2\,\text{e}^-$$

by applying the Nernst equation to this half-reaction:

$$E_{\text{H}_2 \rightarrow \text{H}^+} = (E°_{\text{H}_2 \rightarrow \text{H}^+}) - \left(\frac{0.0592\text{ V}}{n}\right)\left(\log \frac{[\text{H}^+]^2}{P_{\text{H}_2}}\right)$$

Since $E° = 0\text{ V}$ for the standard hydrogen electrode, $n = 2$, and $P_{H_2} = 1$ atm, we can rewrite this equation as

$$E_{\text{H}_2 \rightarrow \text{H}^+} = -\left(\frac{0.0592\text{ V}}{n}\right)(\log\,[\text{H}^+]^2)$$

Further, because $\log\,[H^+]^2 = 2\log\,[H^+]$ and $-\log\,[H^+] = \text{pH}$, the half-cell potential for the hydrogen electrode is directly proportional to the pH:

$$\begin{aligned} E_{\text{H}_2 \rightarrow \text{H}^+} &= -\left(\frac{0.0592\text{ V}}{2}\right)(2)(\log\,[\text{H}^+]) = (0.0592\text{ V})(-\log\,[\text{H}^+]) \\ &= (0.0592\text{ V})(\text{pH}) \end{aligned}$$

The overall cell potential is

$$E_{\text{cell}} = (0.0592\text{ V})(\text{pH}) + E_{\text{ref}}$$

and the pH is therefore a linear function of the cell potential:

$$\text{pH} = \frac{E_{\text{cell}} - E_{\text{ref}}}{0.0592\text{ V}}$$

Thus, a higher cell potential indicates a higher pH, meaning that we can measure the pH of a solution simply by measuring E_{cell}.

In actual pH measurements, a *glass electrode* replaces the cumbersome hydrogen electrode and a *calomel electrode* is used as the reference. A glass electrode consists of a silver wire coated with silver chloride that dips into a reference solution of dilute hydrochloric acid (**Figure 17.7**). The hydrochloric acid is separated from the test solution of unknown pH by a thin glass membrane. For rugged applications, the exterior of the electrode is made of an epoxy resin, which protects the glass tip. A calomel electrode consists of mercury(I) chloride (Hg_2Cl_2, commonly called calomel) in contact with liquid mercury and aqueous KCl. The cell half-reactions are

$$\begin{aligned} &2 \times [\text{Ag}(s) + \text{Cl}^-(aq) \longrightarrow \text{AgCl}(s) + \text{e}^-] && E° = -0.22\text{ V} \\ &\text{Hg}_2\text{Cl}_2(s) + 2\,\text{e}^- \longrightarrow 2\,\text{Hg}(l) + 2\,\text{Cl}^-(aq) && E° = 0.28\text{ V} \end{aligned}$$

The overall cell potential, E_{cell}, depends not only on the potentials of these two half-reactions but also on the boundary potential that develops across the thin glass membrane separating the reference HCl solution from the test solution. Because the boundary potential

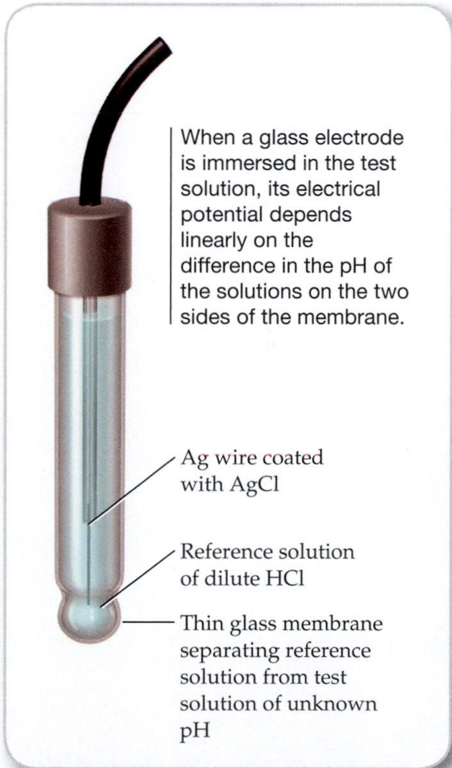

FIGURE 17.7

A glass electrode consists of a silver wire coated with silver chloride that dips into a reference solution of dilute hydrochloric acid.

depends linearly on the difference in the pH of the solutions on the two sides of the membrane, the pH of the test solution can be determined by measuring E_{cell}. The cell potential is measured with a pH meter, a voltage-measuring device that electronically converts E_{cell} to pH and displays the result in pH units.

Worked Example 17.8

Using the Nernst Equation in the Electrochemical Determination of pH

The following cell has a potential of 0.55 V at 25 °C:

$$Pt(s)\,|\,H_2(1\ atm)\,|\,H^+(?\ M)\,||\,Cl^-(1\ M)\,|\,Hg_2Cl_2(s)\,|\,Hg(l)$$

What is the pH of the solution in the anode compartment?

STRATEGY

First, read the shorthand notation to obtain the cell reaction. Then, calculate the half-cell potential for the hydrogen electrode from the observed cell potential and the half-cell potential for the calomel reference electrode. Finally, apply the Nernst equation to find the pH.

SOLUTION

The cell reaction is

$$H_2(g) + Hg_2Cl_2(s) \longrightarrow 2\,H^+(aq) + 2\,Hg(l) + 2\,Cl^-(aq)$$

and the cell potential is

$$E_{cell} = E_{H_2 \rightarrow H^+} + E_{Hg_2Cl_2 \rightarrow Hg} = 0.55\ V$$

Because the reference electrode is the standard calomel electrode, which has $E = E^\circ = 0.28$ V (Appendix D), the half-cell potential for the hydrogen electrode is 0.27 V:

$$E_{H_2 \rightarrow H^+} = E_{cell} - E_{Hg_2Cl_2 \rightarrow Hg} = 0.55\ V - 0.28\ V = 0.27\ V$$

We can then apply the Nernst equation to the half-reaction $H_2(g) \rightarrow 2\,H^+(aq) + 2\,e^-$:

$$E_{H_2 \rightarrow H^+} = (E^\circ_{H_2 \rightarrow H^+}) - \left(\frac{0.0592\ V}{n}\right)\left(\log \frac{[H^+]^2}{P_{H_2}}\right)$$

Substituting in the values of E, E°, n, and P_{H_2} gives

$$0.27\ V = (0\ V) - \left(\frac{0.0592\ V}{2}\right)\left(\log \frac{[H^+]^2}{1}\right) = (0.0592\ V)(pH)$$

Therefore, the pH is

$$pH = \frac{0.27\ V}{0.0592\ V} = 4.6$$

PROBLEM 17.12 What is the pH of the solution in the anode compartment of the following cell if the measured cell potential at 25 °C is 0.28 V?

$$Pt(s)\,|\,H_2(1\ atm)\,|\,H^+(?\ M)\,||\,Pb^{2+}(1\ M)\,|\,Pb(s)$$

17.8 STANDARD CELL POTENTIALS AND EQUILIBRIUM CONSTANTS

We saw in Section 17.3 that the standard free-energy change for a reaction is related to the standard cell potential by the equation

$$\Delta G^\circ = -nFE^\circ$$

In addition, we showed in Section 16.11 that the standard free-energy change is also related to the equilibrium constant for the reaction:

$$\Delta G^\circ = -RT \ln K$$

REMEMBER...

The equation $\Delta G^\circ = -RT \ln K$ follows directly from the equation $\Delta G = \Delta G^\circ + RT \ln Q$ because, at equilibrium, $\Delta G = 0$ and $Q = K$. (Section 16.11)

Combining these two equations, we obtain

$$-nFE^\circ = -RT \ln K$$

or

$$E^\circ = \frac{RT}{nF} \ln K = \frac{2.303\,RT}{nF} \log K$$

Since 2.303 RT/F has a value of 0.0592 V at 25 °C, we can rewrite this equation in the following simplified form, which relates the standard cell potential and the equilibrium constant for any redox reaction:

$$E^\circ = \frac{0.0592\text{ V}}{n} \log K \quad \text{in volts, at 25 °C}$$

Very small concentrations are difficult to measure, so the determination of an equilibrium constant from concentration measurements is not feasible when K is either very large or very small. Standard cell potentials, however, are relatively easy to measure. Consequently, the most common use of this equation is in calculating equilibrium constants from standard cell potentials.

As an example, let's calculate the value of K for the reaction in the Daniell cell:

$$Zn(s) + Cu^{2+}(aq) \longrightarrow Zn^{2+}(aq) + Cu(s) \qquad E^\circ = 1.10\text{ V}$$

Solving for log K in the equation that relates E° and log K and substituting in the appropriate values of E° and n, we obtain

$$E^\circ = \frac{0.0592\text{ V}}{n} \log K$$

$$\log K = \frac{nE^\circ}{0.0592\text{ V}} = \frac{(2)(1.10\text{ V})}{0.0592\text{ V}} = 37.2$$

$$K = \text{antilog } 37.2 = 10^{37.2} = 2 \times 10^{37} \quad \text{at 25 °C}$$

Because $K = [Zn^{2+}]/[Cu^{2+}]$ is a very large number, the reaction goes essentially to completion. When $[Zn^{2+}] = 1$ M, for example, $[Cu^{2+}]$ is less than 10^{-37} M.

The preceding calculation shows that even a relatively small value of E° (+1.10 V) corresponds to a huge value of K (2×10^{37}). A positive value of E° corresponds to a positive value of log K and therefore $K > 1$, and a negative value of E° corresponds to a negative value of log K and therefore $K < 1$. Because the standard reduction potentials in Table 17.1 span a range of about 6 V, E° for a redox reaction can range from +6 V for reaction of the strongest oxidizing agent with the strongest reducing agent to −6 V for reaction of the weakest oxidizing agent with the weakest reducing agent. However, E° values outside the range +3 V to −3 V are uncommon. For the case of $n = 2$, the correspondence between the values of E° and K is indicated in **Figure 17.8**. Equilibrium constants for redox reactions tend to be either very large or very small compared to equilibrium constants for acid–base reactions, which are in the range of 10^{14} to 10^{-14}. Thus, redox reactions typically go either essentially to completion or almost not at all.

FIGURE 17.8

The relationship between the equilibrium constant K for a redox reaction with $n = 2$ and the standard cell potential $E°$. Note that K is plotted on a logarithmic scale.

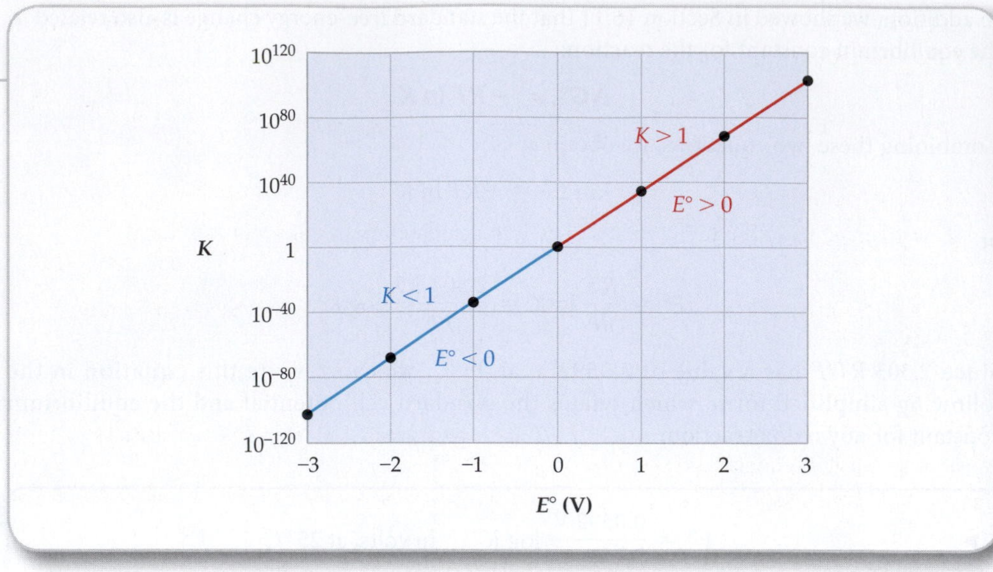

In previous chapters, we discussed two different ways to determine the value of an equilibrium constant K: from concentration data (Section 13.2) and from thermochemical data (Section 16.11). In this section, we've added a third way: from electrochemical data. The following are the key relationships needed for each approach:

1. K from concentration data for solutes: $K = \dfrac{[C]^c[D]^d}{[A]^a[B]^b}$
2. K from thermochemical data: $\Delta G° = -RT \ln K$; $\quad \ln K = \dfrac{-\Delta G°}{RT}$
3. K from electrochemical data: $E° = \dfrac{RT}{nF} \ln K$; $\quad \ln K = \dfrac{nFE°}{RT}$

Worked Example 17.9

Using Standard Reduction Potentials to Calculate an Equilibrium Constant

Use the standard reduction potentials in Table 17.1 to calculate the equilibrium constant at 25 °C for the reaction

$$6\,Br^-(aq) + Cr_2O_7^{2-}(aq) + 14\,H^+(aq) \longrightarrow 3\,Br_2(aq) + 2\,Cr^{3+}(aq) + 7\,H_2O(l)$$

STRATEGY

Calculate $E°$ for the reaction from standard reduction potentials, as in Worked Example 17.5. Then use the equation $\log K = nE°/0.0592$ V to determine the equilibrium constant.

SOLUTION

Find the relevant half-reactions in Table 17.1, and write them in the proper direction for oxidation of Br^- and reduction of $Cr_2O_7^{2-}$. Before adding the half-reactions to get the overall reaction, multiply the Br^-/Br_2 half-reaction by a factor of 3 so that the electrons will cancel:

$$\begin{array}{rlr} 3 \times [2\,Br^-(aq) & \longrightarrow Br_2(aq) + 2\,e^-] & E° = -1.09\text{ V} \\ Cr_2O_7^{2-}(aq) + 14\,H^+(aq) + 6\,e^- & \longrightarrow 2\,Cr^{3+}(aq) + 7\,H_2O(l) & E° = 1.36\text{ V} \\ \hline 6\,Br^-(aq) + Cr_2O_7^{2-}(aq) + 14\,H^+(aq) & \longrightarrow 3\,Br_2(aq) + 2\,Cr^{3+}(aq) + 7\,H_2O(l) & E° = 0.27\text{ V} \end{array}$$

Note that $E°$ for the Br^-/Br_2 oxidation is the negative of the tabulated standard reduction potential (1.09 V), and remember that we don't multiply this $E°$ value by a factor of 3 because electrical potential is an intensive property. The $E°$ value for the overall reaction is the sum of the $E°$ values for the half-reactions: $-1.09\ \text{V} + 1.36\ \text{V} = 0.27\ \text{V}$. To calculate the equilibrium constant, use the relation between log K and $nE°$, with $n = 6$:

$$\log K = \frac{nE°}{0.0592\ \text{V}} = \frac{(6)(0.27\ \text{V})}{0.0592\ \text{V}} = 27 \qquad K = 1 \times 10^{27} \quad \text{at } 25\ °\text{C}$$

✔ BALLPARK CHECK

$E°$ is positive, so K should be greater than 1, in agreement with the solution.

PROBLEM 17.13 Use the data in Table 17.1 to calculate the equilibrium constant at 25 °C for the reaction

$$4\ Fe^{2+}(aq) + O_2(g) + 4\ H^+(aq) \longrightarrow 4\ Fe^{3+}(aq) + 2\ H_2O(l)$$

PROBLEM 17.14 What is the value of $E°$ for a redox reaction involving the transfer of 2 mol of electrons if its equilibrium constant is 1.8×10^{-5}?

17.9 BATTERIES

By far the most important practical application of galvanic cells is their use as *batteries*. In multicell batteries, such as those in automobiles, the individual galvanic cells are linked in series, with the anode of each cell connected to the cathode of the adjacent cell. The voltage provided by the battery is the sum of the individual cell voltages.

The features required in a battery depend on the application. In general, though, a commercially successful battery should be compact, lightweight, physically rugged, and inexpensive, and it must provide a stable source of power for relatively long periods of time. Battery design is an active area of research that requires considerable ingenuity as well as a solid understanding of electrochemistry. Let's look at several of the most common types of commercial batteries.

Lead Storage Battery

The *lead storage battery* is perhaps the most familiar of all galvanic cells because it has been used as a reliable source of power for starting automobiles for nearly a century. A typical 12 V battery consists of six cells connected in series, each cell providing a potential of about 2 V. The cell design is illustrated in **Figure 17.9**. The anode, a series of lead grids packed with spongy lead, and the cathode, a second series of grids packed with lead dioxide, dip into the electrolyte, an aqueous solution of sulfuric acid (38% by weight). When the cell is discharging (providing current) the electrode half-reactions and the overall cell reaction are:

Anode: $Pb(s) + HSO_4^-(aq) \longrightarrow PbSO_4(s) + H^+(aq) + 2\,e^- \qquad E° = 0.296\ \text{V}$

Cathode: $PbO_2(s) + 3\,H^+(aq) + HSO_4^-(aq) + 2\,e^- \longrightarrow PbSO_4(s) + 2\,H_2O(l) \qquad E° = 1.628\ \text{V}$

Overall: $Pb(s) + PbO_2(s) + 2\,H^+(aq) + 2\,HSO_4^-(aq) \longrightarrow 2\,PbSO_4(s) + 2\,H_2O(l) \qquad E° = 1.924\ \text{V}$

Note that these equations contain HSO_4^- ions rather than SO_4^{2-} because SO_4^{2-} is protonated in strongly acidic solutions.

Lead is oxidized to lead sulfate at the anode, and lead dioxide is reduced to lead sulfate at the cathode. The cell doesn't need to have separate anode and cathode compartments because the oxidizing and reducing agents are both solids (PbO_2 and Pb) that are kept from coming in contact by the presence of insulating spacers between the grids.

Because the reaction product, solid $PbSO_4$, adheres to the surface of the electrodes, a run-down lead storage battery can be recharged by using an external source of direct current to drive the cell reaction in the reverse, nonspontaneous direction. In an automobile, the battery is continuously recharged by a device called an *alternator*, which is driven by the engine.

FIGURE 17.9

A lead storage battery and a cutaway view of one cell.

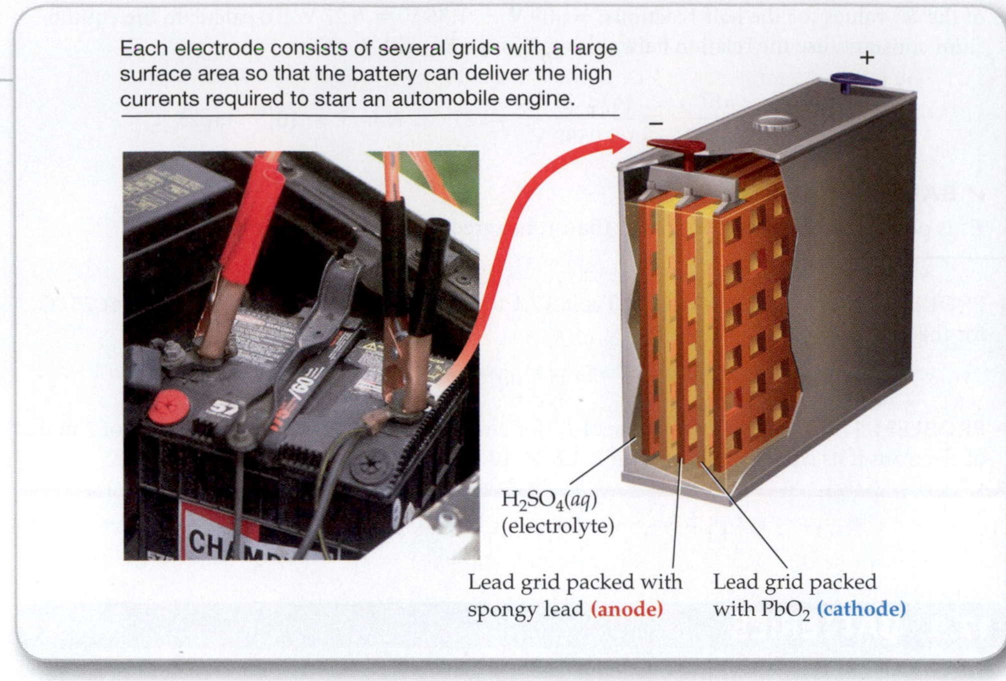

A lead storage battery typically provides good service for several years, but eventually the spongy $PbSO_4$ deposits on the electrodes turn into a hard, crystalline form that can't be converted back to Pb and PbO_2. Then it's no longer possible to recharge the battery and it must be replaced. Because lead is extremely toxic, replaced batteries should be recycled, as is mandated by most states.

Dry-Cell Batteries

The *dry cell*, or *Leclanché cell*, was patented in 1866 by the Frenchman Georges Leclanché. Used for many years as a power source for flashlights and portable radios, it consists of a zinc metal can, which serves as the anode, and an inert graphite rod surrounded by a paste of solid manganese dioxide and carbon black, which functions as the cathode (**Figure 17.10**). Surrounding the MnO_2-containing paste is the electrolyte, a moist paste of ammonium chloride and zinc chloride in starch. A dry cell is not completely dry but gets its name because the electrolyte is a viscous, aqueous paste rather than a liquid solution. The electrode reactions, which are rather complicated, can be represented in simplified form by the following equations:

Anode: $Zn(s) \longrightarrow Zn^{2+}(aq) + 2\,e^-$

Cathode: $2\,MnO_2(s) + 2\,NH_4^+(aq) + 2\,e^- \longrightarrow Mn_2O_3(s) + 2\,NH_3(aq) + H_2O(l)$

FIGURE 17.10

Leclanché dry cell and a cutaway view.

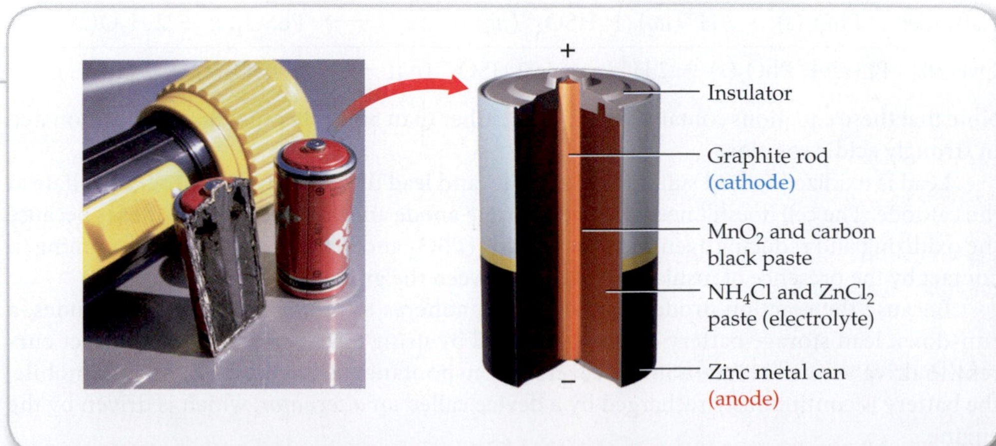

The Leclanché cell is largely of historical interest and has been displaced by the *alkaline dry cell*, a modified version in which the acidic NH_4Cl electrolyte of the Leclanché cell is replaced by a basic electrolyte, either NaOH or KOH. As in the Leclanché cell, the electrode reactions involve oxidation of zinc and reduction of manganese dioxide, but the oxidation product is zinc oxide, as is appropriate to the basic conditions:

$$\text{Anode:} \quad Zn(s) + 2\,OH^-(aq) \longrightarrow ZnO(s) + H_2O(l) + 2\,e^-$$
$$\text{Cathode:} \quad 2\,MnO_2(s) + H_2O(l) + 2\,e^- \longrightarrow Mn_2O_3(s) + 2\,OH^-(aq)$$

Corrosion of the zinc anode is a significant side reaction under acidic conditions because zinc reacts with $H^+(aq)$ to give $Zn^{2+}(aq)$ and $H_2(g)$. Under basic conditions, however, the cell has a longer life because zinc corrodes more slowly. The alkaline cell also produces higher power and a more stable current and voltage because of more efficient ion transport in the alkaline electrolyte. The voltage of an alkaline cell is about 1.5 V.

Nickel–Cadmium Batteries

Nickel-cadmium, or *"ni-cad," batteries* are popular for use in calculators and portable power tools because, unlike most other dry-cell batteries, they are rechargeable. The anode of a ni-cad battery is cadmium metal, and the cathode is the nickel(III) compound NiO(OH) supported on nickel metal. The electrode reactions are

$$\text{Anode:} \quad Cd(s) + 2\,OH^-(aq) \longrightarrow Cd(OH)_2(s) + 2\,e^-$$
$$\text{Cathode:} \quad NiO(OH)(s) + H_2O(l) + e^- \longrightarrow Ni(OH)_2(s) + OH^-(aq)$$

Ni–cad batteries can be recharged hundreds of times because the solid products of the electrode reactions adhere to the surface of the electrodes.

Nickel–Metal Hydride Batteries

Because cadmium is an expensive, toxic, heavy metal, ni–cad batteries have been replaced in many applications by lighter, more environmentally friendly *nickel–metal hydride*, or *NiMH, batteries.* An NiMH battery delivers about the same voltage (1.2 V) as a ni-cad battery but has about twice the energy density, the amount of energy stored per unit mass. Though both batteries have the same cathode, they have different anodes. In an NiMH battery, the anode is a special metal alloy, such as $LaNi_5$, that is capable of absorbing and releasing large amounts of hydrogen at ordinary temperatures and pressures. When the cell is discharging, the electrode and overall cell reactions can be written as

$$\text{Anode:} \quad MH_{ab}(s) + OH^-(aq) \longrightarrow M(s) + H_2O(l) + e^-$$
$$\text{Cathode:} \quad NiO(OH)(s) + H_2O(l) + e^- \longrightarrow Ni(OH)_2(s) + OH^-(aq)$$
$$\text{Overall:} \quad MH_{ab}(s) + NiO(OH)(s) \longrightarrow M(s) + Ni(OH)_2(s)$$

where M represents the hydrogen-absorbing metal alloy and H_{ab} denotes the absorbed hydrogen. Thus, the metal hydride is oxidized at the anode, NiO(OH) is reduced at the cathode, and the overall cell reaction transfers hydrogen from the anode to the cathode with no change in the KOH electrolyte.

▲ Rechargeable nickel–metal hydride batteries

NiMH batteries are used in a host of consumer electronic products and also in hybrid gas–electric automobiles, which are powered by both a gasoline engine and an electric motor. Battery packs consisting of many NiMH batteries supply energy to the electric motor and also store energy that is captured on braking. When the brakes are applied, the electric motor acts as a generator and recharges the batteries. Some of the car's kinetic energy, which would otherwise be dissipated as heat, is converted to electrical energy, thus reducing use of the gasoline engine and increasing gas mileage.

Lithium and Lithium-Ion Batteries

Lithium and lithium-ion batteries are even more popular for portable electronic products than nickel-based batteries because of their light weight and high voltage (3–4 V). Lithium has a higher E° value for oxidation and a lower atomic weight than any other metal; only 6.94 g of lithium is needed to provide 1 mol of electrons.

Lithium batteries use a lithium metal anode and an electrolyte consisting of a lithium salt, such as $LiClO_4$, in an organic solvent. Most lithium batteries have a manganese dioxide cathode, which can absorb a variable number of Li^+ ions into its solid-state structure. The electrode reactions involve oxidation of lithium at the anode and reduction of MnO_2 at the cathode:

$$\text{Anode:} \quad x\,\text{Li}(s) \longrightarrow x\,\text{Li}^+(soln) + x\,\text{e}^-$$
$$\text{Cathode:} \quad \text{MnO}_2(s) + x\,\text{Li}^+(soln) + x\,\text{e}^- \longrightarrow \text{Li}_x\text{MnO}_2(s)$$

Li^+ ions migrate through the electrolyte from the anode to the cathode, while electrons move from the anode to the cathode through the external circuit. Lithium batteries are used in watches, calculators, and other small consumer devices.

The distinction between a lithium battery and a lithium-ion battery lies in the nature of the anode. Whereas a lithium battery uses a lithium metal anode, a lithium-ion battery uses a graphite anode that has lithium atoms inserted between its layers of carbon atoms. This so-called lithiated graphite, written as Li_xC_6, contains no lithium metal as such, which is why a lithium-ion battery isn't called a lithium battery. The usual cathode in a lithium-ion battery is CoO_2, another metal oxide that can incorporate Li^+ ions into its structure. The electrolyte is either a solution of a lithium salt in an organic solvent or a solid-state polymer electrolyte that can transport Li^+ ions.

Because an uncharged lithium-ion battery has a graphite anode and an $LiCoO_2$ cathode, the electrode reactions on discharging the battery can be written as

$$\text{Anode:} \quad \text{Li}_x\text{C}_6(s) \longrightarrow x\,\text{Li}^+(soln) + 6\,\text{C}(s) + x\,\text{e}^-$$
$$\text{Cathode:} \quad \text{Li}_{1-x}\text{CoO}_2(s) + x\,\text{Li}^+(soln) + x\,\text{e}^- \longrightarrow \text{LiCoO}_2(s)$$

When the battery is charged initially or recharged subsequently, the electrode reactions occur in the reverse direction.

Lithium-ion batteries are used in cell phones, laptop computers, digital cameras, power tools, and electric cars. Recent concerns about the possible overheating and explosion of lithium ion batteries are being addressed by the development of alternative electrode materials, such as lithium titanate (Li_2TiO_3) and lithium iron phosphate ($LiFePO_4$), which have greater thermal stability than $LiCoO_2$.

▲ This Tesla Motors roadster, a zero-emission electric vehicle, is powered by 6831 lithium-ion batteries. It has a top speed of 125 mph, a 0-to-60 mph acceleration time of 3.9 s, and a 245-mile range.

PROBLEM 17.15 Write a balanced equation for the overall cell reaction when each of the following batteries is producing current:
(a) Leclanché dry cell
(b) Alkaline dry cell
(c) Nickel–cadmium battery
(d) Lithium battery
(e) Lithium-ion battery

17.10 FUEL CELLS

A **fuel cell** is a galvanic cell in which one of the reactants is a fuel such as hydrogen or methanol. Traditional hydrocarbon fuels, such as methane or propane, can't be used directly but must first be converted to hydrogen by the steam–hydrocarbon reforming process we'll describe in Section 18.2. A fuel cell differs from an ordinary battery in that the reactants are not contained within the cell but instead are continuously supplied from an external reservoir.

LOOKING AHEAD ...

We'll see in Section 18.2 that steam reacts with hydrocarbons, such as CH_4, at high temperatures in the presence of a catalyst, yielding H_2 and carbon monoxide.

Perhaps the best-known example is the hydrogen–oxygen fuel cell that is used as a source of electric power in space vehicles (**Figure 17.11**). This cell contains porous carbon electrodes impregnated with metallic catalysts and an electrolyte consisting of hot, aqueous KOH. The fuel, gaseous H_2, and the oxidizing agent, gaseous O_2, don't react directly but instead flow into separate cell compartments where H_2 is oxidized at the anode and O_2 is reduced at the cathode. The overall cell reaction is simply the conversion of hydrogen and oxygen to water:

$$\begin{array}{ll} \text{Anode:} & 2\,H_2(g) + 4\,OH^-(aq) \longrightarrow 4\,H_2O(l) + 4\,e^- \\ \text{Cathode:} & O_2(g) + 2\,H_2O(l) + 4\,e^- \longrightarrow 4\,OH^-(aq) \\ \hline \text{Overall:} & 2\,H_2(g) + O_2(g) \longrightarrow 2\,H_2O(l) \end{array}$$

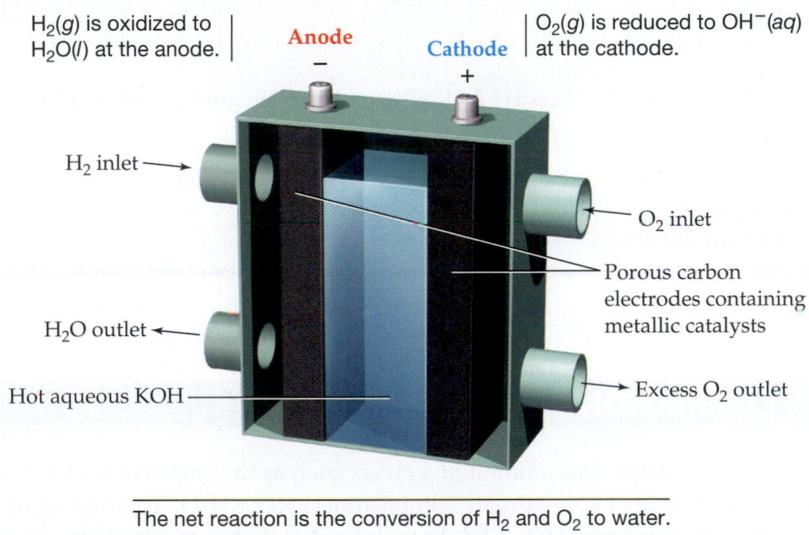

FIGURE 17.11

A hydrogen–oxygen fuel cell.

In fuel cells designed to power environmentally clean electric vehicles, the aqueous KOH electrolyte is replaced by a special, polymer membrane that conducts protons but not electrons. The electrode and overall cell reactions in these so-called *proton-exchange membrane (PEM) fuel cells* are

$$\begin{array}{ll} \text{Anode:} & 2\,H_2(g) \longrightarrow 4\,H^+(aq) + 4\,e^- \\ \text{Cathode:} & O_2(g) + 4\,H^+(aq) + 4\,e^- \longrightarrow 2\,H_2O(l) \\ \hline \text{Overall:} & 2\,H_2(g) + O_2(g) \longrightarrow 2\,H_2O(l) \end{array}$$

Protons pass through the membrane from the anode to the cathode, while electrons move through the external circuit from the anode to the cathode, driving an electric motor that powers the vehicle. Because the only reaction product is water, the vehicle itself doesn't produce environmental pollutants or CO_2, although CO_2 is produced in the steam–hydrocarbon reforming process used to obtain hydrogen.

Fuel-cell-powered demonstration vehicles have been in existence for several years: Buses have been tested in several North American and European cities, the first fuel-cell-powered truck has been delivered to the U.S. Army, and General Motors has recently unveiled a fuel-cell concept car that accelerates from 0 to 60 mph in 10 s and has a range of 300 mi. Despite these advances and vigorous current research efforts, mass marketing of commercial fuel-cell vehicles that are competitive with internal combustion vehicles is at least a decade away. Among the developments needed for commercialization are improvements in the performance and a decrease in the cost of proton-exchange membranes, development of less expensive catalysts for the electrode reactions, discovery of new and safer methods for onboard storage of hydrogen, and development of a hydrogen-fuel infrastructure.

Consumer applications of fuel cells might also become available in small electronic products, such as cell phones and laptop computers. The cell proposed for use in these products is the *direct methanol fuel cell* (DMFC). It is similar to the PEM fuel cell but uses aqueous methanol (CH_3OH) as the fuel, rather than gaseous hydrogen:

Anode:	$2\,CH_3OH(aq) + 2\,H_2O(l) \longrightarrow 2\,CO_2(g) + 12\,H^+(aq) + 12\,e^-$
Cathode:	$3\,O_2(g) + 12\,H^+(aq) + 12\,e^- \longrightarrow 6\,H_2O(l)$
Overall:	$2\,CH_3OH(aq) + 3\,O_2(g) \longrightarrow 2\,CO_2(g) + 4\,H_2O(l)$

The DMFC is lighter and has a higher energy density than conventional batteries. Moreover, methanol is more available, safer, and easier to store than hydrogen.

Fuel cells are also likely to find use as power generators for hospitals, hotels, and apartment buildings. Indeed, for some time now, the Tokyo Electric Power Company in Japan has been operating an 11 megawatt fuel-cell power plant capable of supplying electricity to about 4000 households.

PROBLEM 17.16 In what ways are fuel cells and batteries similar, and in what ways are they different?

PROBLEM 17.17 What is the standard potential for the cell reaction in
(a) the hydrogen–oxygen fuel cell used in space vehicles?
(b) the PEM fuel cell used in electric automobiles?

17.11 CORROSION

▲ This steel railroad trestle collapsed during a tornado due to corrosion of the bolts that connected the structure to its concrete footings.

Corrosion is the oxidative deterioration of a metal, such as the conversion of iron to rust, a hydrated iron(III) oxide of approximate composition $Fe_2O_3 \cdot H_2O$. The rusting of iron has enormous economic consequences. It has been estimated that as much as one-fourth of the steel produced in the United States goes to replace steel structures and products that have been destroyed by corrosion.

To prevent corrosion, we first have to understand how it occurs. One important fact is that the rusting of iron requires both oxygen and water; it doesn't occur in oxygen-free water or in dry air. Another clue is the observation that rusting involves pitting of the metal surface, but the rust is deposited at a location physically separated from the pits. This suggests that rust does not form by direct reaction of iron and oxygen but rather by an electrochemical process in which iron is oxidized in one region of the surface and oxygen is reduced in another region.

A possible mechanism for rusting, consistent with the known facts, is illustrated in **Figure 17.12**. The surface of the iron and a droplet of surface water constitute a tiny galvanic cell in which different regions of the surface act as anode and cathode while the aqueous phase serves as the electrolyte. Iron is oxidized more readily in some regions (anode regions) than in others (cathode regions) because the composition of the metal is somewhat inhomogeneous and the surface is irregular. Factors such as impurities, phase boundaries, and mechanical stress may influence the ease of oxidation in a particular region of the surface.

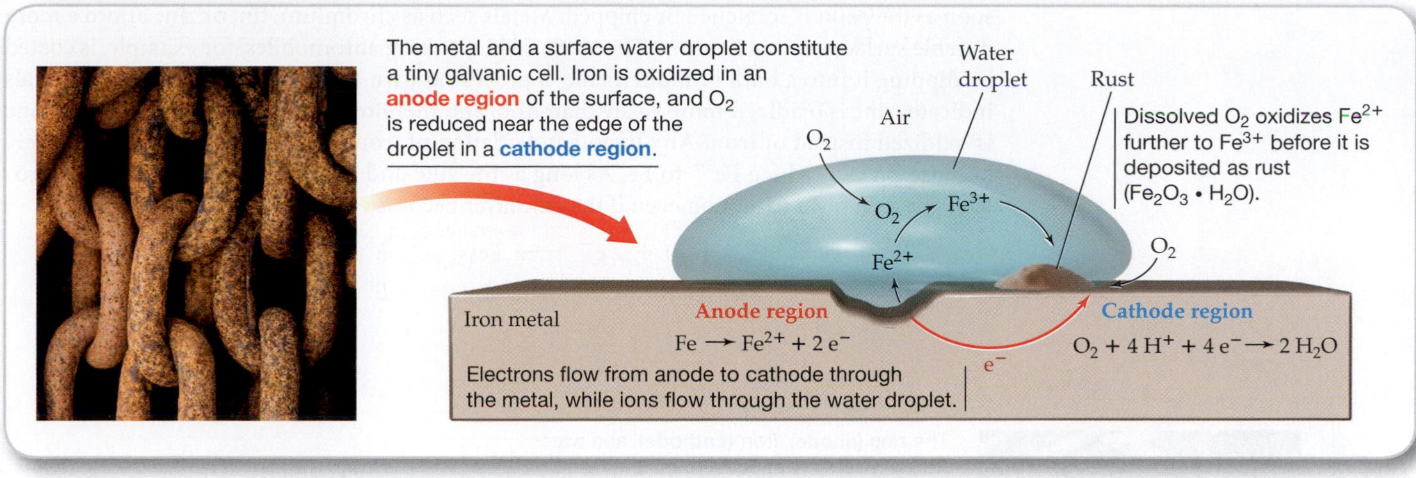

FIGURE 17.12

Why does iron corrode?

At an anode region, iron is oxidized to Fe^{2+} ions,

$$Fe(s) \longrightarrow Fe^{2+}(aq) + 2\,e^- \qquad E^\circ = 0.45\text{ V}$$

while at a cathode region, oxygen is reduced to water:

$$O_2(g) + 4\,H^+(aq) + 4\,e^- \longrightarrow 2\,H_2O(l) \qquad E^\circ = 1.23\text{ V}$$

The actual potential for the reduction half-reaction is less than the standard potential (1.23 V) because the water droplet is not 1 M in H^+ ions. (In fact, the water is only slightly acidic because the main source of H^+ ions is the reaction of water with dissolved atmospheric carbon dioxide.) Even at pH 7, however, the potential for the reduction half-reaction is 0.81 V, which means that the cell potential is highly positive, indicative of a spontaneous reaction.

The electrons required for reduction of O_2 at the cathode region are supplied by a current that flows through the metal from the more easily oxidized anode region (Figure 17.12). The electrical circuit is completed by migration of ions in the water droplet. When Fe^{2+} ions migrate away from the pitted anode region, they come in contact with O_2 dissolved in the surface portion of the water droplet and are further oxidized to Fe^{3+} ions:

$$4\,Fe^{2+}(aq) + O_2(g) + 4\,H^+(aq) \longrightarrow 4\,Fe^{3+}(aq) + 2\,H_2O(l)$$

Iron(III) forms a very insoluble hydrated oxide even in moderately acidic solutions, and so the iron(III) is deposited as the familiar red-brown material that we call rust:

$$2\,Fe^{3+}(aq) + 4\,H_2O(l) \longrightarrow \underset{\text{Rust}}{Fe_2O_3 \cdot H_2O(s)} + 6\,H^+(aq)$$

An electrochemical mechanism for corrosion also explains nicely why automobiles rust more rapidly in places where road salt is used to melt snow and ice. Dissolved salts in the water droplet greatly increase the conductivity of the electrolyte, thus accelerating the pace of corrosion.

A glance at a table of standard reduction potentials indicates that the O_2/H_2O half-reaction lies above the M^{n+}/M half-reaction for nearly all metals, so O_2 can oxidize all metals except a few, such as gold and platinum. Aluminum, for example, has $E^\circ = -1.66$ V for the Al^{3+}/Al half-reaction and is oxidized more readily than iron. In other words, the corrosion of aluminum products such as aircraft and automobile parts, window frames, cooking utensils, and soda cans should be a serious problem. Fortunately, it isn't, because oxidation of aluminum gives a very hard, almost impenetrable film of Al_2O_3 that adheres to the surface of the metal and protects it from further contact and reaction with oxygen. Other metals such as magnesium, chromium, titanium, and zinc form similar protective oxide coatings. In the case of iron, however, rust is too porous to shield the underlying metal from further oxidation.

▲ Actor Jerry Seinfeld's titanium bicycle doesn't corrode because a hard, impenetrable layer of TiO_2 adheres to the surface and protects the metal from further oxidation.

Prevention of Corrosion

Corrosion of iron can be prevented, or at least minimized, by shielding the metal surface from oxygen and moisture. A coat of paint is effective for a while, but rust begins to form as soon as the paint is scratched or chipped. Metals such as chromium, tin, or zinc afford a more durable surface coating for iron. The steel used in making automobiles, for example, is coated by dipping it into a bath of molten zinc, a process known as **galvanizing**. As the potentials indicate, zinc is oxidized more easily than iron, and therefore, when the metal corrodes, zinc is oxidized instead of iron. Any incipient oxidation of iron would be reversed immediately because Zn can reduce Fe^{2+} to Fe. As long as the zinc and iron are in contact, the zinc protects the iron from oxidation even if the zinc layer becomes scratched (**Figure 17.13**).

$$Fe^{2+}(aq) + 2\,e^- \longrightarrow Fe(s) \qquad E^\circ = -0.45\text{ V}$$
$$Zn^{2+}(aq) + 2\,e^- \longrightarrow Zn(s) \qquad E^\circ = -0.76\text{ V}$$

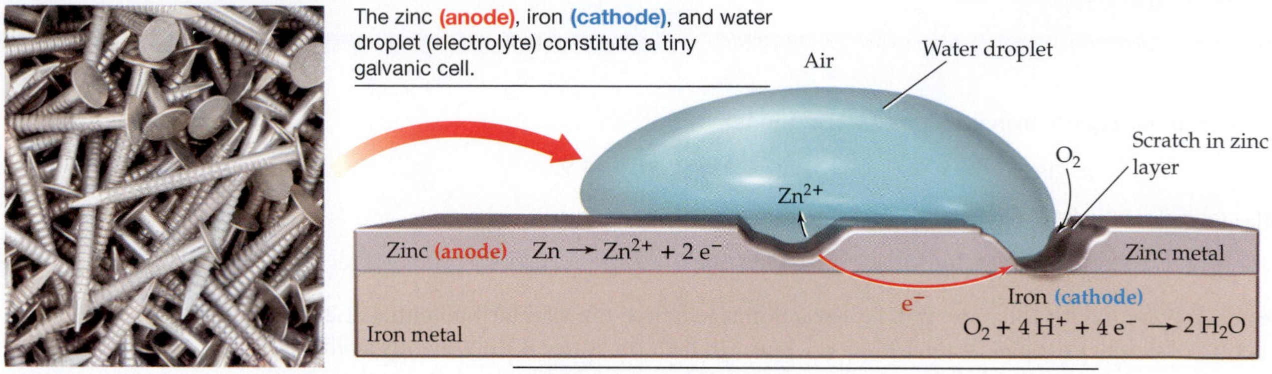

FIGURE 17.13

Galvanized iron. A layer of zinc protects iron from oxidation, even when the zinc layer becomes scratched.

▲ A ship's steel propeller and housing with an attached strip of zinc that serves as a sacrificial anode. The zinc corrodes instead of the iron.

The technique of protecting a metal from corrosion by connecting it to a second metal that is more easily oxidized is called **cathodic protection**. It's unnecessary to cover the entire surface of the metal with a second metal, as in galvanizing iron. All that's required is electrical contact with the second metal. An underground steel pipeline, for example, can be protected by connecting it through an insulated wire to a stake of magnesium, which acts as a *sacrificial anode* and corrodes instead of the iron. In effect, the arrangement is a galvanic cell in which the easily oxidized magnesium acts as the anode, the pipeline behaves as the cathode, and moist soil is the electrolyte. The cell half-reactions are

$$\text{Anode:} \quad Mg(s) \longrightarrow Mg^{2+}(aq) + 2\,e^- \qquad E^\circ = 2.37\text{ V}$$
$$\text{Cathode:} \quad O_2(g) + 4\,H^+(aq) + 4\,e^- \longrightarrow 2\,H_2O(l) \qquad E^\circ = 1.23\text{ V}$$

For large steel structures such as pipelines, storage tanks, bridges, and ships, cathodic protection is the best defense against premature rusting.

PROBLEM 17.18 Magnesium is often attached to the steel hulls of ships to protect the steel from rusting. Write balanced equations for the corrosion reactions that occur **(a)** in the presence of Mg and **(b)** in the absence of Mg.

17.12 ELECTROLYSIS AND ELECTROLYTIC CELLS

Thus far, we've been concerned only with galvanic cells—electrochemical cells in which a spontaneous redox reaction produces an electric current. A second important kind of electrochemical cell is the electrolytic cell, in which an electric current is used to drive a nonspontaneous reaction. Thus, the processes occurring in galvanic and electrolytic cells are the reverse of each other: A galvanic cell converts chemical energy to electrical energy when a reaction with a positive value of E (and a negative value of ΔG) proceeds toward equilibrium; an electrolytic cell converts electrical energy to chemical energy when an electric current drives a reaction with a negative value of E (and a positive value of ΔG) in a direction away from equilibrium.

The process of using an electric current to bring about chemical change is called **electrolysis**. The opposite signs of E and $\Delta G = -nFE$ for the two kinds of cells are summarized in **Table 17.2**, along with the situation for a reaction that has reached equilibrium—a dead battery!

TABLE 17.2 Relationship between Cell Potentials E and Free-Energy Changes ΔG

Reaction Type	E	ΔG	Cell Type
Spontaneous	+	−	Galvanic
Nonspontaneous	−	+	Electrolytic
Equilibrium	0	0	Dead battery

Electrolysis of Molten Sodium Chloride

An electrolytic cell has two electrodes that dip into an electrolyte and are connected to a battery or some other source of direct electric current. A cell for electrolysis of molten sodium chloride, for example, is illustrated in **Figure 17.14**. The battery serves as an electron pump, pushing electrons into one electrode and pulling them out of the other. The negative electrode attracts Na^+ cations, which combine with the electrons supplied by the battery and are thereby reduced to liquid sodium metal. Similarly, the positive electrode attracts Cl^- anions, which replenish the electrons removed by the battery and are thereby oxidized to chlorine gas. The electrode reactions and overall cell reaction are

$$\begin{array}{ll} \text{Anode (oxidation):} & 2\,Cl^-(l) \longrightarrow Cl_2(g) + 2\,e^- \\ \text{Cathode (reduction):} & 2\,Na^+(l) + 2\,e^- \longrightarrow 2\,Na(l) \\ \hline \text{Overall cell reaction:} & 2\,Na^+(l) + 2\,Cl^-(l) \longrightarrow 2\,Na(l) + Cl_2(g) \end{array}$$

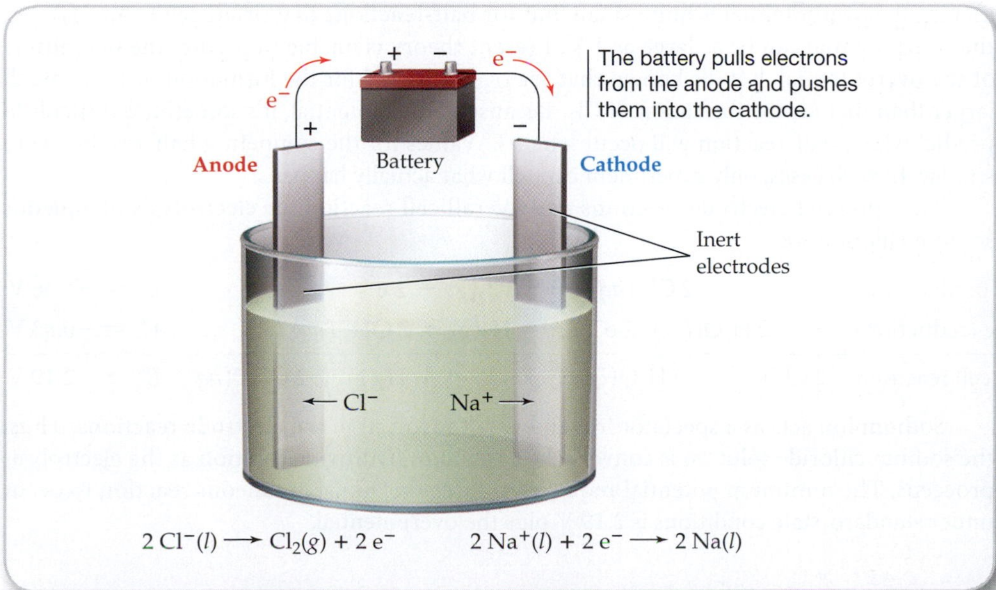

FIGURE 17.14

Electrolysis of molten sodium chloride. Chloride ions are oxidized to Cl_2 gas at the anode, and Na^+ ions are reduced to sodium metal at the cathode.

As in a galvanic cell, the anode is the electrode where oxidation takes place and the cathode is the electrode where reduction takes place. The signs of the electrodes, however, are opposite for galvanic and electrolytic cells. In a galvanic cell, the anode is considered to be negative because it supplies electrons to the external circuit, but in an electrolytic cell, the anode is considered to be positive because electrons are pulled out of it by the battery. The sign of each electrode in the electrolytic cell is the same as the sign of the battery electrode to which it is attached.

Electrolysis of Aqueous Sodium Chloride

When an aqueous salt solution is electrolyzed, the electrode reactions may differ from those for electrolysis of the molten salt because water may be involved. In the electrolysis of aqueous sodium chloride, for example, the cathode half-reaction might be either the reduction of Na^+ to sodium metal, as in the case of molten sodium chloride, or the reduction of water to hydrogen gas:

$$Na^+(aq) + e^- \longrightarrow Na(s) \qquad E° = -2.71\ V$$
$$2\,H_2O(l) + 2\,e^- \longrightarrow H_2(g) + 2\,OH^-(aq) \qquad E° = -0.83\ V$$

Because the standard potential is much less negative for the reduction of water than for the reduction of Na^+, water is reduced preferentially and bubbles of hydrogen gas are produced at the cathode.

The anode half-reaction might be either oxidation of Cl^- to Cl_2 gas, as in the case of molten sodium chloride, or oxidation of water to oxygen gas:

$$2\,Cl^-(aq) \longrightarrow Cl_2(g) + 2\,e^- \qquad E° = -1.36\ V$$
$$2\,H_2O(l) \longrightarrow O_2(g) + 4\,H^+(aq) + 4\,e^- \qquad E° = -1.23\ V$$

Based on the $E°$ values, we might expect a slight preference for oxidation of water in a solution having 1 M ion concentrations. For a neutral solution ($[H^+] = 10^{-7}$ M), the preference for water oxidation will be even greater because its oxidation potential at pH 7 is −0.81 V. The observed product at the anode, however, is Cl_2, not O_2, because of a phenomenon called *overpotential*.

Experiments indicate that the applied voltage required for an electrolysis is always greater than the voltage calculated from standard reduction potentials. The additional voltage required is the **overpotential**, also called the *overvoltage*. The overpotential is needed because the rate of electron transfer at the electrode–solution interface for one or both of the cell half-reactions is often slow, thus limiting the amount of current that passes through an electrolytic cell. For electrode half-reactions involving solution or deposition of metals, the overpotential is quite small, but for half-reactions that produce O_2 (or H_2) gas, the overpotential can be as large as 1 V. Present theory is unable to predict the magnitude of the overpotential, but it's known that the overpotential for the formation of O_2 is much larger than that for the formation of Cl_2. Because of overpotential, it's sometimes difficult to predict which half-reaction will occur when $E°$ values for the competing half-reactions are similar. In such cases, only experiment can tell what actually happens.

The observed electrode reactions and overall cell reaction for electrolysis of aqueous sodium chloride are

Anode (oxidation):	$2\,Cl^-(aq) \longrightarrow Cl_2(g) + 2\,e^-$	$E° = -1.36\ V$
Cathode (reduction):	$2\,H_2O(l) + 2\,e^- \longrightarrow H_2(g) + 2\,OH^-(aq)$	$E° = -0.83\ V$
Overall cell reaction:	$2\,Cl^-(aq) + 2\,H_2O(l) \longrightarrow Cl_2(g) + H_2(g) + 2\,OH^-(aq)$	$E° = -2.19\ V$

Sodium ion acts as a spectator ion and is not involved in the electrode reactions. Thus, the sodium chloride solution is converted to a sodium hydroxide solution as the electrolysis proceeds. The minimum potential required to force this nonspontaneous reaction to occur under standard-state conditions is 2.19 V plus the overpotential.

Electrolysis of Water

The electrolysis of any aqueous solution requires the presence of an electrolyte to carry the current in the solution. If the ions of the electrolyte are less easily oxidized and reduced than water, however, then water will react at both electrodes. Take, for instance, the electrolysis of an aqueous solution of the inert electrolyte Na_2SO_4. Water is oxidized at the anode in preference to SO_4^{2-} ions and is reduced at the cathode in preference to Na^+ ions. The electrode and overall cell reactions are

Anode (oxidation):	$2\,H_2O(l) \longrightarrow O_2(g) + 4\,H^+(aq) + 4\,e^-$
Cathode (reduction):	$4\,H_2O(l) + 4\,e^- \longrightarrow 2\,H_2(g) + 4\,OH^-(aq)$
Overall cell reaction:	$6\,H_2O(l) \longrightarrow 2\,H_2(g) + O_2(g) + 4\,H^+(aq) + 4\,OH^-(aq)$

If the anode and cathode solutions are mixed, the H^+ and OH^- ions react to form water:

$$4\,H^+(aq) + 4\,OH^-(aq) \longrightarrow 4\,H_2O(l)$$

The net electrolysis reaction is therefore the decomposition of water, a process sometimes used in the laboratory to produce small amounts of pure H_2 and O_2:

$$2\,H_2O(l) \longrightarrow 2\,H_2(g) + O_2(g)$$

CONCEPTUAL PROBLEM 17.19 Metallic potassium was first prepared by Humphrey Davy in 1807 by electrolysis of molten potassium hydroxide:

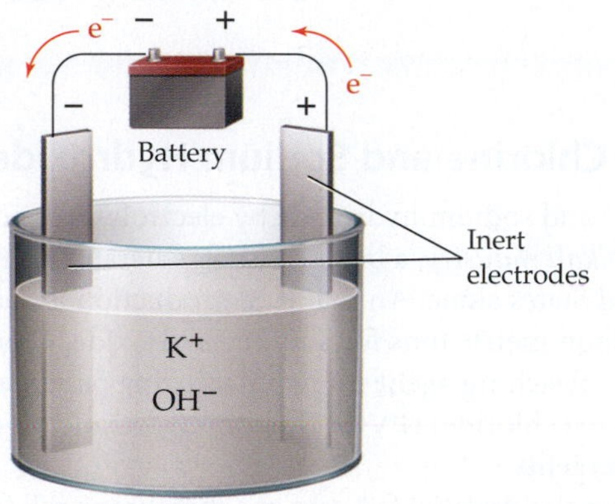

(a) Label the anode and cathode, and show the direction of ion flow.
(b) Write balanced equations for the anode, cathode, and overall cell reactions.

PROBLEM 17.20 Predict the half-cell reactions that occur when aqueous solutions of the following salts are electrolyzed in a cell with inert electrodes. What is the overall cell reaction in each case?

(a) LiCl **(b)** $CuSO_4$ **(c)** K_2SO_4

17.13 COMMERCIAL APPLICATIONS OF ELECTROLYSIS

Electrolysis is used in the manufacture of many important chemicals and in numerous processes for purification and electroplating of metals. Let's look at some examples.

Manufacture of Sodium

Sodium metal is produced commercially in a *Downs cell* by electrolysis of a molten mixture of sodium chloride and calcium chloride (**Figure 17.15**). The presence of $CaCl_2$ allows the cell to operate at a lower temperature because the melting point of the NaCl–$CaCl_2$ mixture (about 580 °C) is depressed well below that of pure NaCl (801 °C). The liquid sodium produced at the cylindrical steel cathode is less dense than the molten salt and thus floats to the top part of the cell, where it is drawn off into a suitable container. Chlorine gas forms at the graphite anode, which is separated from the cathode by an iron screen to keep the highly reactive sodium and chlorine away from each other. Because the Downs process requires high currents (typically 25,000–40,000 A), plants for producing sodium are located near sources of inexpensive hydroelectric power, such as Niagara Falls, New York.

FIGURE 17.15

Cross-sectional view of a Downs cell for commercial production of sodium metal by electrolysis of molten sodium chloride. The cell design keeps the sodium and chlorine apart from each other and out of contact with air.

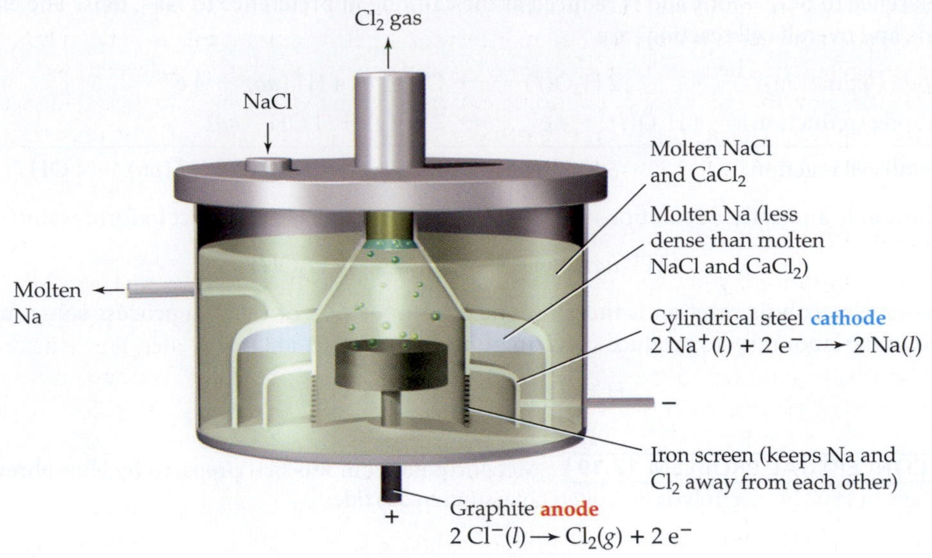

Manufacture of Chlorine and Sodium Hydroxide

Production of chlorine and sodium hydroxide by electrolysis of aqueous sodium chloride is the basis of the *chlor-alkali industry*, a business that generates annual sales of approximately \$4 billion in the United States alone. Annual U.S. production is about 10 million metric tons for chlorine and 8 million metric tons for sodium hydroxide. Chlorine is used in water and sewage treatment, as a bleaching agent in manufacturing paper, and in the manufacture of plastics such as poly(vinyl chloride) (PVC). Sodium hydroxide is employed in making paper, textiles, soaps, and detergents.

Figure 17.16 shows the essential features of a membrane cell for commercial production of chlorine and sodium hydroxide. A saturated aqueous solution of sodium chloride (*brine*) flows into the anode compartment, where Cl^- is oxidized to Cl_2 gas, and water enters the

FIGURE 17.16

A membrane cell for electrolytic production of Cl_2 and NaOH. Chloride ion is oxidized to Cl_2 gas at the anode, and water is converted to H_2 gas and OH^- ions at the cathode.

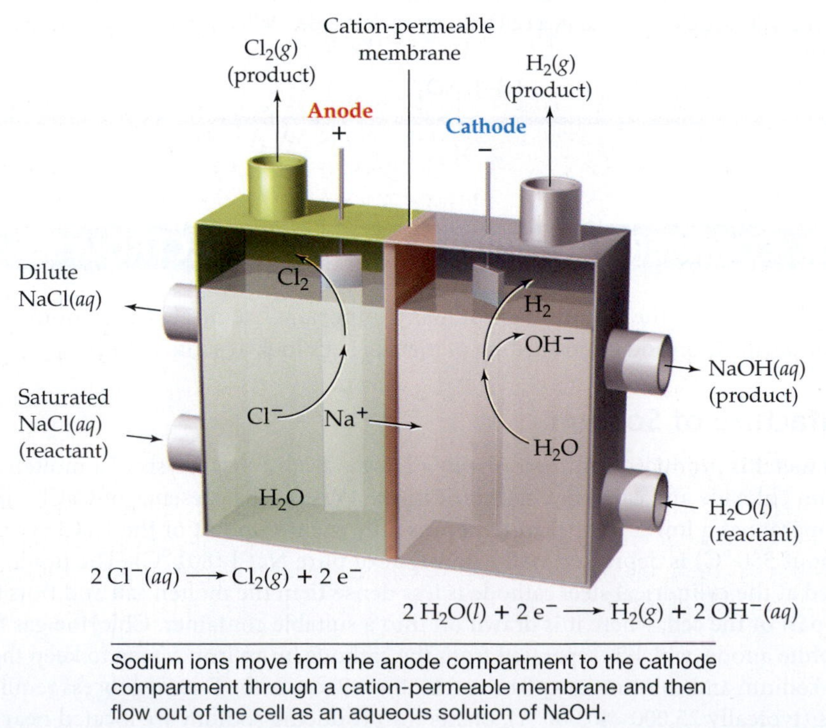

cathode compartment, where it is converted to H_2 gas and OH^- ions. Between the anode and cathode compartments is a special polymer membrane that is permeable to cations but not to anions or water. The membrane keeps the Cl_2 and OH^- ions apart but allows a current of Na^+ ions to flow into the cathode compartment, thus carrying the current in the solution and maintaining electrical neutrality in both compartments. The Na^+ and OH^- ions flow out of the cathode compartment as an aqueous solution of NaOH.

Manufacture of Aluminum

Although aluminum is the third most abundant element in the Earth's crust at 8.23% by mass, it remained a rare and expensive metal until 1886, when a 22-year-old American, Charles Martin Hall, and a 23-year-old Frenchman, Paul Héroult, independently devised a practical process for the electrolytic production of aluminum. Still used today, the **Hall–Héroult process** involves electrolysis of a molten mixture of aluminum oxide (Al_2O_3) and cryolite (Na_3AlF_6) at about 1000 °C in a cell with graphite electrodes **(Figure 17.17)**. Electrolysis of pure Al_2O_3 is impractical because it melts at a very high temperature (2045 °C), and electrolysis of aqueous Al^{3+} solutions is not feasible because water is reduced in preference to Al^{3+} ions. Thus, the use of cryolite as a solvent for Al_2O_3 is the key to the success of the Hall–Héroult process.

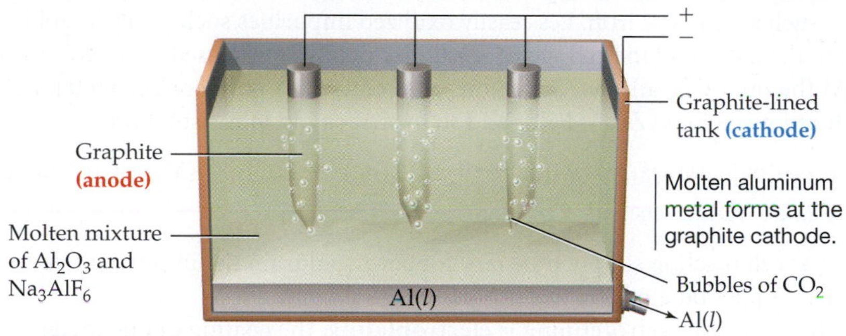

FIGURE 17.17

An electrolytic cell for production of aluminum by the Hall–Héroult process.

The electrode reactions are still not fully understood, but they probably involve complex anions of the type $AlF_xO_y^{+3-x-2y}$, formed by the reaction of Al_2O_3 and Na_3AlF_6. The complex anions are reduced at the cathode to molten aluminum metal and are oxidized at the anode to O_2 gas, which reacts with the graphite anodes to give CO_2 gas. As a result, the anodes are chewed up rapidly and must be replaced frequently. The cell operates at a low voltage (5–6 V) but with very high currents (up to 250,000 A) because 1 mol of electrons produces only 9.0 g of aluminum. Electrolytic production of aluminum is the largest single consumer of electricity in the United States, making recycling of aluminum products highly desirable. As with the manufacture of sodium, aluminum production generally takes place near hydroelectric plants, such as those along the Columbia River in Washington.

▲ Alcoa aluminum plant along the Columbia River at Wenatchee, Washington.

Electrorefining and Electroplating

The purification of a metal by means of electrolysis is called **electrorefining**. For example, impure copper obtained from ores is converted to pure copper in an electrolytic cell that has impure copper as the anode and pure copper as the cathode **(Figure 17.18)**. The electrolyte is an aqueous solution of copper sulfate.

Alternating slabs of impure copper and pure copper serve as the electrodes in electrolytic cells for the refining of copper.

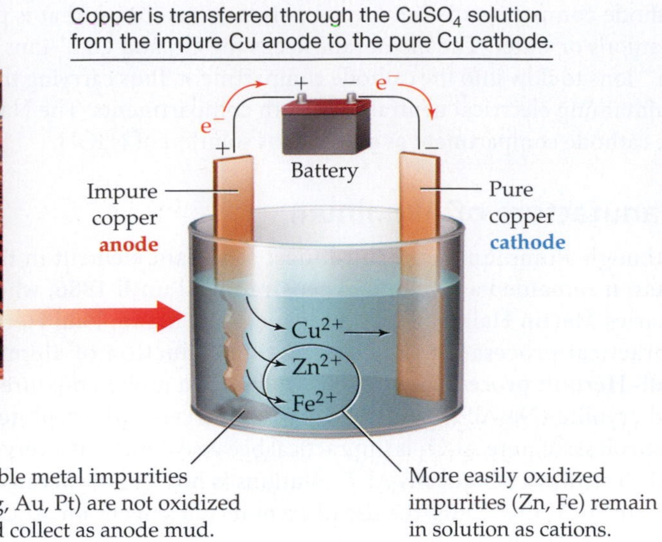

FIGURE 17.18

Electrorefining of copper metal.

At the impure Cu anode, copper is oxidized along with more easily oxidized metallic impurities such as zinc and iron. Less easily oxidized impurities such as silver, gold, and platinum fall to the bottom of the cell as *anode mud*, which is reprocessed to recover the precious metals. At the pure Cu cathode, Cu^{2+} ions are reduced to pure copper metal, but the less easily reduced metal ions (Zn^{2+}, Fe^{2+}, and so forth) remain in the solution.

Anode (oxidation): $M(s) \longrightarrow M^{2+}(aq) + 2\,e^-$ $\quad$ (M = Cu, Zn, Fe)

Cathode (reduction): $Cu^{2+}(aq) + 2\,e^- \longrightarrow Cu(s)$

Thus, the net cell reaction simply transfers copper metal from the impure anode to the pure cathode. The copper obtained by this process is 99.95% pure.

Closely related to electrorefining is **electroplating**, the coating of one metal on the surface of another using electrolysis. For example, steel motorcycle parts are often plated with chromium to protect them from corrosion, and silver-plating is commonly used to make items of fine table service. The object to be plated is carefully cleaned and then set up as the cathode of an electrolytic cell that contains a solution of ions of the metal to be deposited.

▲ Silver-plated objects such as this teapot are made by electroplating pure silver metal onto another metal such as stainless steel, copper, or nickel.

PROBLEM 17.21 Sketch an electrolytic cell suitable for electroplating a silver spoon. Describe the electrodes and the electrolyte, label the anode and cathode, and indicate the direction of electron and ion flow. Write balanced equations for the anode and cathode half-reactions. What is the overall cell reaction?

17.14 QUANTITATIVE ASPECTS OF ELECTROLYSIS

In the 1830s, Michael Faraday showed that the amount of substance produced at an electrode by electrolysis depends on the quantity of charge passed through the cell. For example, passing 1 mol of electrons through a Downs cell yields 1 mol (23.0 g) of sodium at the cathode:

$$\underset{\text{1 mol}}{Na^+(l)} + \underset{\text{1 mol}}{e^-} \longrightarrow \underset{\text{1 mol (23.0 g)}}{Na(l)}$$

Similarly, passing 1 mol of electrons in the Hall–Héroult process produces 1/3 mol (9.0 g) of aluminum, because 3 mol of electrons are required to reduce 1 mol of Al^{3+} to aluminum metal:

$$\underset{\text{1/3 mol}}{Al^{3+}(l)} + \underset{\text{1 mol}}{3\,e^-} \longrightarrow \underset{\text{1/3 mol (9.0 g)}}{Al(l)}$$

In general, the amount of product formed in an electrode reaction follows directly from the stoichiometry of the reaction and the molar mass of the product.

To find out how many moles of electrons pass through a cell in a particular experiment, we need to measure the electric current and the time that the current flows. The number of coulombs of charge passed through the cell equals the product of the current in amperes (coulombs per second) and the time in seconds:

$$\text{Charge (C)} = \text{Current (A)} \times \text{Time (s)}$$

Because the charge on 1 mol of electrons is 96,500 C (Section 17.3), the number of moles of electrons passed through the cell is

$$\text{Moles of e}^- = \text{Charge (C)} \times \frac{1 \text{ mol e}^-}{96{,}500 \text{ C}}$$

The sequence of conversions in **Figure 17.19** is used to calculate the mass or volume of product produced by passing a known current through a cell for a fixed period of time. The key is to think of the electrons as a reactant in a balanced chemical equation and then to proceed as with any other stoichiometry problem. Worked Example 17.10 illustrates the calculations. Alternatively, we can calculate the current (or time) required to produce a given amount of product by working through the sequence in Figure 17.19 in the reverse direction, as shown in Worked Example 17.11.

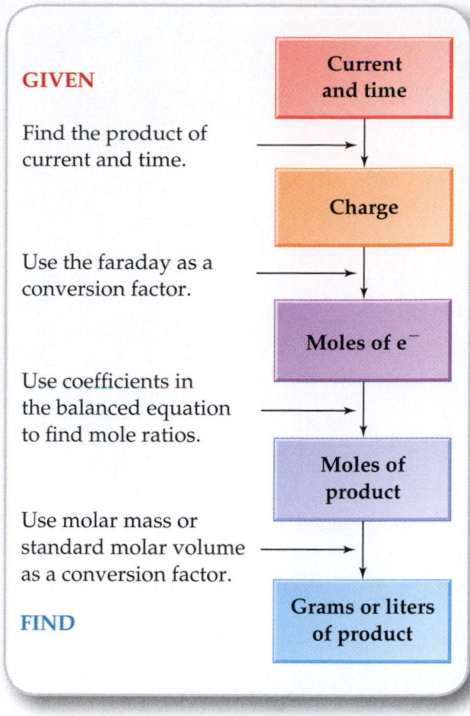

FIGURE 17.19

Sequence of conversions used to calculate the amount of product produced by passing a current through an electrolytic cell for a fixed period of time.

Worked Example 17.10

Calculating the Amount of Product Produced by Electrolysis

A constant current of 30.0 A is passed through an aqueous solution of NaCl for 1.00 h. How many grams of NaOH and how many liters of Cl_2 gas at STP are produced?

STRATEGY

To convert the current and time to grams or liters of product, carry out the sequence of conversions in Figure 17.19.

SOLUTION

Because electrons can be thought of as a reactant in the electrolysis process, the first step is to calculate the charge and the number of moles of electrons passed through the cell:

$$\text{Charge} = \left(30.0 \frac{\text{C}}{\text{s}}\right)(1.00 \text{ h})\left(\frac{60 \text{ min}}{\text{h}}\right)\left(\frac{60 \text{ s}}{\text{min}}\right) = 1.08 \times 10^5 \text{ C}$$

$$\text{Moles of e}^- = (1.08 \times 10^5 \text{ C})\left(\frac{1 \text{ mol e}^-}{96{,}500 \text{ C}}\right) = 1.12 \text{ mol e}^-$$

The cathode reaction yields 2 mol of OH^- per 2 mol of electrons (Section 17.12), so 1.12 mol of NaOH will be obtained:

$$2\,H_2O(l) + 2\,e^- \longrightarrow H_2(g) + 2\,OH^-(aq)$$

$$\text{Moles of NaOH} = (1.12 \text{ mol e}^-)\left(\frac{2 \text{ mol NaOH}}{2 \text{ mol e}^-}\right) = 1.12 \text{ mol Na OH}$$

Converting the number of moles of NaOH to grams gives 44.8 g of NaOH:

$$\text{Grams of NaOH} = (1.12 \text{ mol NaOH})\left(\frac{40.0 \text{ g NaOH}}{\text{mol NaOH}}\right) = 44.8 \text{ g NaOH}$$

The anode reaction gives 1 mol of Cl_2 per 2 mol of electrons, so 0.560 mol of Cl_2 will be obtained:

$$2\,Cl^-(aq) \longrightarrow Cl_2(g) + 2\,e^-$$

$$\text{Moles of } Cl_2 = (1.12 \text{ mol e}^-)\left(\frac{1 \text{ mol } Cl_2}{2 \text{ mol e}^-}\right) = 0.560 \text{ mol } Cl_2$$

continued on next page

Since 1 mol of an ideal gas occupies 22.4 L at STP, the volume of Cl_2 obtained is 12.5 L:

$$\text{Liters of } Cl_2 = (0.560 \text{ mol } Cl_2)\left(\frac{22.4 \text{ L } Cl_2}{\text{mol } Cl_2}\right) = 12.5 \text{ L } Cl_2$$

As a shortcut, the entire sequence of conversions can be carried out in one step. For example, the volume of Cl_2 produced at the anode is

$$\left(30.0 \frac{\text{C}}{\text{s}}\right)(1.00 \text{ h})\left(\frac{3600 \text{ s}}{\text{h}}\right)\left(\frac{1 \text{ mol e}^-}{96{,}500 \text{ C}}\right)\left(\frac{1 \text{ mol } Cl_2}{2 \text{ mol e}^-}\right)\left(\frac{22.4 \text{ L } Cl_2}{\text{mol } Cl_2}\right) = 12.5 \text{ L } Cl_2$$

✔ BALLPARK CHECK

Since approximately 1 mol of electrons is passed through the cell and the electrode reactions yield 1 mol of NaOH and 0.5 mol of Cl_2 per mole of electrons, 1 mol of NaOH (~40 g) and 0.5 mol of Cl_2 (~11 L at STP) will be formed. The ballpark check and the solution agree.

Worked Example 17.11

Calculating the Current Required to Produce a Given Amount of Product in a Given Time

How many amperes must be passed through a Downs cell to produce sodium metal at a rate of 30.0 kg/h?

STRATEGY

Proceed through a sequence of conversions similar to that in Figure 17.19 and Worked Example 17.10, but in reverse order.

SOLUTION

Because the molar mass of sodium is 23.0 g/mol, the number of moles of sodium produced per hour is

$$\text{Moles of Na} = (30.0 \text{ kg Na})\left(\frac{1000 \text{ g}}{1 \text{ kg}}\right)\left(\frac{1 \text{ mol Na}}{23.0 \text{ g Na}}\right) = 1.30 \times 10^3 \text{ mol Na}$$

To produce each mole of sodium, 1 mol of electrons must be passed through the cell:

$$Na^+(l) + e^- \longrightarrow Na(l)$$

Therefore, the charge passed per hour is

$$\text{Charge} = (1.30 \times 10^3 \text{ mol Na})\left(\frac{1 \text{ mol e}^-}{1 \text{ mol Na}}\right)\left(\frac{96{,}500 \text{ C}}{\text{mol e}^-}\right) = 1.25 \times 10^8 \text{ C}$$

Since there are 3600 s in 1 h, the current required is

$$\text{Current} = \frac{1.25 \times 10^8 \text{ C}}{3600 \text{ s}} = 3.47 \times 10^4 \text{ C/s} = 34{,}700 \text{ A}$$

PROBLEM 17.22 How many kilograms of aluminum can be produced in 8.00 h by passing a constant current of 1.00×10^5 A through a molten mixture of aluminum oxide and cryolite?

PROBLEM 17.23 A layer of silver is electroplated on a coffee server using a constant current of 0.100 A. How much time is required to deposit 3.00 g of silver?

FYI WHY ARE SOME METAL OBJECTS BRIGHTLY COLORED?

If you're like many people, you may have used a brightly colored aluminum sports bottle. These bottles are lightweight, inexpensive, and impart no taste or odor to their contents. Or perhaps you know someone who has had a total joint replacement. If so, the titanium screws used to implant the artificial joint may have been colored because color coding of the components of orthopedic implants increases efficiency and accuracy during surgery. Given that aluminum and titanium are silvery metals, how do aluminum and titanium objects become brightly colored?

▲ Anodized aluminum sports bottles.

Aluminum, titanium, and several other metals can be colored by an electrochemical process called *anodizing*. Unlike electroplating, in which a metal ion in the electrolyte is *reduced* and the metal is coated onto the surface of the cathode, anodizing *oxidizes* a metal anode to yield a metal oxide coating. In the oxidation of aluminum, for instance, the electrode reactions are

Cathode (reduction):	$6\,H^+(aq) + 6\,e^- \longrightarrow 3\,H_2(g)$
Anode (oxidation):	$2\,Al(s) + 3\,H_2O(l) \longrightarrow Al_2O_3(s) + 6\,H^+(aq) + 6\,e^-$
Overall cell reaction:	$2\,Al(s) + 3\,H_2O(l) \longrightarrow Al_2O_3(s) + 3\,H_2(g)$

The thickness of the aluminum oxide coating that forms on the anode can be controlled by varying the current flow during the electrolysis. Typically, the coating is about 0.01 mm thick, which corresponds to about 4×10^4 atomic layers of Al_2O_3 on top of the underlying aluminum metal. The porosity of the coating can also be controlled by varying the electrochemical conditions, and it's this porosity that makes coloring possible. When an organic dye is added to the electrolyte, dye molecules soak into the spongy surface coating as it forms and become trapped when the surface hardens. A broad range of dyes is available, so anodized aluminum objects like water bottles, bicycle parts, and key chains come in a wide variety of colors.

Titanium anodizing proceeds much like that of aluminum, but the resultant coat of TiO_2 is much thinner (10^{-4} mm) than the corresponding coat of Al_2O_3 (10^{-2} mm). Furthermore, the iridescent colors of anodized titanium result not from the absorption of organic dyes but from the interference of light as it is reflected by the anodized surface. When a beam of white light strikes the anodized surface, part of the light is reflected from the outer TiO_2, while part penetrates through the semitransparent TiO_2 and is reflected from the inner metal. If the two reflections of a particular wavelength are out of phase, they interfere destructively and that wavelength is canceled from the reflected light. As a result, the light that remains is colored. Similar effects are responsible for the diffraction of X-rays (see Figure 10.18 on page 376).

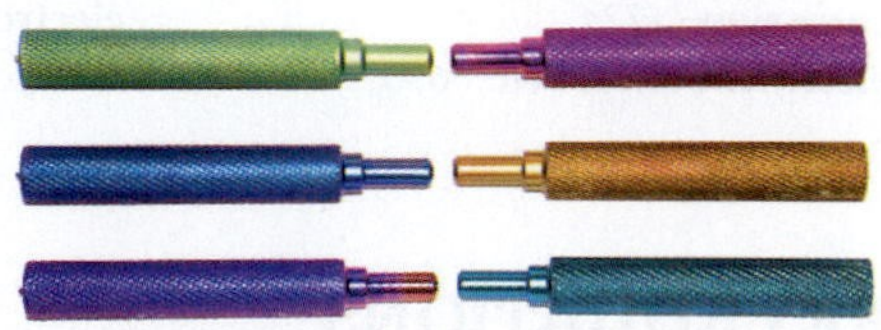

▲ Anodized titanium medical implant components such as bone screws are color coded to assist the surgeon during surgery.

Anodized titanium is particularly useful for biomedical applications because bone cells adhere to the microscopically rough surface of the TiO_2 coating about 10 times better than they do to the smooth surface of bare titanium metal, thus hastening the integration of the implant into the body. Moreover, postoperative infections, which occur in up to 4% of total joint replacements, can be minimized by incorporating antibiotics into the rough surface of the TiO_2 coating. Locating antibiotics at the site of the implant is more effective than administering them either orally or intravenously.

PROBLEM 17.24 What is the overall cell reaction and cell potential for anodizing titanium if the anode half-reaction is

$$Ti(s) + 2\,H_2O(l) \longrightarrow TiO_2(s) + 4\,H^+(aq) + 4\,e^- \qquad E° = +1.066\text{ V}$$

PROBLEM 17.25 How many minutes are required to produce a 0.0100 mm thick coating of Al_2O_3 (density 3.97 g/cm^3) on a square piece of aluminum metal 10.0 cm on an edge if the current passed through the piece is 0.600 A?

SUMMARY

Electrochemistry is the area of chemistry concerned with the interconversion of chemical and electrical energy. Chemical energy is converted to electrical energy in a **galvanic cell**, a device in which a spontaneous redox reaction is used to produce an electric current. Electrical energy is converted to chemical energy in an **electrolytic cell**, a cell in which an electric current drives a nonspontaneous reaction. It's convenient to separate cell reactions into half-reactions because oxidation and reduction occur at separate **electrodes**. The electrode at which oxidation occurs is called the **anode**, and the electrode at which reduction occurs is called the **cathode**.

The **cell potential *E*** (also called the **cell voltage** or **electromotive force**) is an electrical measure of the cell reaction's tendency to occur. Cell potentials depend on temperature, ion concentrations, and gas pressures. The **standard cell potential *E*°** is the cell potential when reactants and products are in their standard states. Cell potentials are related to free-energy changes by the equations $\Delta G = -nFE$ and $\Delta G^\circ = -nFE^\circ$, where n is the number of moles of electrons transferred in the cell reaction and $F = 96{,}500\ \text{C/mol e}^-$ is the **faraday**, the charge on 1 mol of electrons.

The **standard reduction potential** for a half-reaction is defined relative to an arbitrary value of 0 V for the **standard hydrogen electrode (S.H.E.)**:

$$2\,H^+(aq, 1\ M) + 2\,e^- \longrightarrow H_2(g, 1\ \text{atm}) \qquad E^\circ = 0\ \text{V}$$

Tables of standard reduction potentials—also called **standard electrode potentials**—are used to arrange oxidizing and reducing agents in order of increasing strength, to calculate E° values for cell reactions, and to decide whether a particular redox reaction is spontaneous.

Cell potentials under nonstandard-state conditions can be calculated using the **Nernst equation**,

$$E = E^\circ - \frac{0.0592\ \text{V}}{n} \log Q \qquad \text{in volts, at } 25^\circ\ \text{C}$$

where Q is the reaction quotient. The equilibrium constant K and the standard cell potential E° are related by the equation

$$E^\circ = \frac{0.0592\ \text{V}}{n} \log K \qquad \text{in volts, at } 25\ ^\circ\text{C}$$

A **battery** consists of one or more galvanic cells. A **fuel cell** differs from a battery in that the reactants are continuously supplied to the cell.

Corrosion of iron (rusting) is an electrochemical process in which iron is oxidized in an anode region of the metal surface and oxygen is reduced in a cathode region. Corrosion can be prevented by covering iron with another metal, such as zinc in the process called **galvanizing**, or simply by putting the iron in electrical contact with a second metal that is more easily oxidized, a process called **cathodic protection**.

Electrolysis, the process of using an electric current to bring about chemical change, is employed to produce sodium, chlorine, sodium hydroxide, and aluminum (**Hall–Héroult process**) and is used in **electrorefining** and **electroplating**.

The product obtained at an electrode depends on the reduction potentials and **overpotential**. The amount of product obtained is related to the number of moles of electrons passed through the cell, which depends on the current and the time that the current flows.

KEY WORDS

anode *699*
cathode *699*
cathodic protection *726*
cell potential (*E*) *704*
cell voltage *704*
corrosion *724*
electrochemical cell *696*
electrochemistry *696*
electrode *698*
electrolysis *727*
electrolytic cell *696*
electromotive force (emf) *704*
electroplating *732*
electrorefining *731*
faraday *705*
fuel cell *723*
galvanic cell *696*
galvanizing *726*
Hall–Héroult process *731*
Nernst equation *712*
overpotential *728*
salt bridge *698*
standard cell potential (*E*°) *705*
standard electrode potential *709*
standard hydrogen electrode (S.H.E.) *707*
standard reduction potential *707*

KEY EQUATIONS

- Relationships between the free-energy change and the cell potential for a redox reaction **(Section 17.3)**

$$\Delta G = -nFE \quad \text{and} \quad \Delta G^\circ = -nFE^\circ$$

where n is the number of moles of electrons transferred in the cell reaction and $F = 96{,}500\ \text{C/mol e}^-$ is the faraday

- The Nernst equation, which relates the cell potential under nonstandard-state conditions E to the standard cell potential E° and the reaction quotient Q **(Section 17.6)**

$$E = E^\circ - \frac{0.0592\ \text{V}}{n} \log Q \qquad \text{in volts, at } 25\ ^\circ\text{C}$$

- Relationship between the standard cell potential and the equilibrium constant **(Section 17.8)**

$$E^\circ = \frac{0.0592\ \text{V}}{n} \log K \qquad \text{in volts, at } 25\ ^\circ\text{C}$$

LEARNING OUTCOMES

- Draw and interpret the sketch a galvanic cell. Label the anode and cathode, indicate the direction of electron and ion flow, and write balanced equations for the electrode and overall cell reactions. (Section 17.1) ***Problems 1, 4, 26, 27, 38, 39, 44–47***
- Write and interpret the shorthand notation for a galvanic cell. (Section 17.2) ***Problems 2–4, 27, 28, 40–47, 119, 122, 126***
- Relate the cell potential for a redox reaction and its free-energy change. (Section 17.3) ***Problems 5, 50, 51, 54, 55, 62–65, 112, 114, 116, 131, 132, 136, 139, 140***
- Use standard reduction potentials to arrange oxidizing and reducing agents in order of increasing strength. (Section 17.5) ***Problems 7, 9, 35, 58–61, 112, 118***
- Use standard reduction potentials to calculate $E°$ for a cell reaction and to predict whether the reaction is spontaneous. (Section 17.5) ***Problems 8, 9, 35, 62–71, 110–112, 118, 124, 125***
- Use the Nernst equation to relate the cell potential and concentrations under nonstandard-state conditions. (Section 17.6) ***Problems 10, 11, 28, 32, 33, 72–79, 112–114, 120, 123, 124, 126, 128, 129, 133, 134, 136, 141***
- Use the Nernst equation to calculate the pH of a solution. (Section 17.7) ***Problems 12, 80, 81***
- Relate standard potentials and the equilibrium constant for a redox reaction. (Section 17.8) ***Problems 13, 14, 82–89, 112, 114, 122, 126, 127, 131, 136***
- Describe batteries and fuel cells, and calculate values of $E°$, $\Delta G°$, and K for the cell reaction. (Sections 17.9, 17.10) ***Problems 16, 17, 34, 90–93, 137, 138***
- Describe how corrosion occurs and methods used to prevent it. (Section 17.11) ***Problems 18, 94–97***
- Draw and interpret the sketch of an electrolytic cell. Label the anode and cathode, indicate the direction of electron and ion flow, and write balanced equations for the electrode and overall cell reactions. (Sections 17.12, 17.13) ***Problems 19, 21, 29–31, 98, 99***
- Predict the products formed and the electrode and overall cell reactions when an aqueous solution of an ionic compound is electrolyzed. (Sections 17.12, 17.13) ***Problems 20, 100–103***
- Relate the current, time, and amount of product produced in an electrolytic or galvanic cell. (Section 17.14) ***Problems 22, 23, 104–109, 114, 115, 119, 130, 137, 138***

CONCEPTUAL PROBLEMS

Problems 17.1–17.25 appear within the chapter.

17.26 The following picture of a galvanic cell has lead and zinc electrodes:

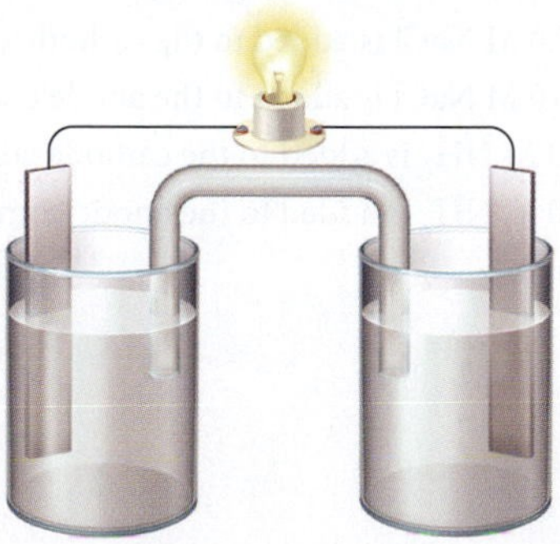

(a) Label the electrodes, and identify the ions present in the solutions.

(b) Label the anode and cathode.

(c) Indicate the direction of electron flow in the wire and ion flow in the solutions.

(d) Tell what electrolyte could be used in the salt bridge, and indicate the direction of ion flow.

(e) Write balanced equations for the electrode and overall cell reactions.

17.27 Consider the following galvanic cell:

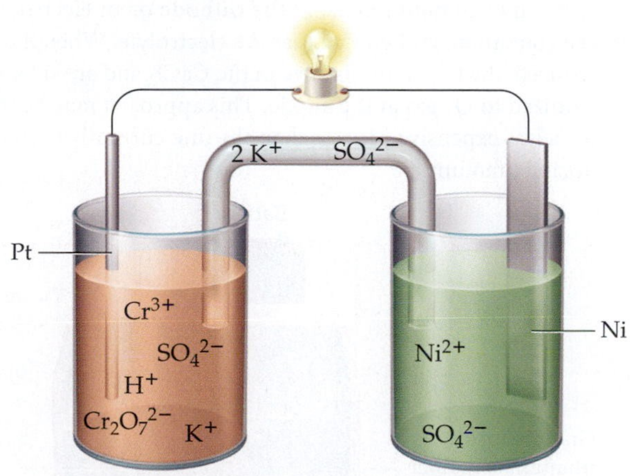

(a) Identify the anode and cathode.

(b) Write a balanced equation for the cell reaction.

(c) Write the shorthand notation for the cell.

17.28 Consider the following galvanic cells:

(1) $Cu(s)|Cu^{2+}(1\ M)||Fe^{3+}(1\ M), Fe^{2+}(1\ M)|Pt(s)$

(2) $Cu(s)|Cu^{2+}(1\ M)||Fe^{3+}(1\ M), Fe^{2+}(5\ M)|Pt(s)$

(3) $Cu(s)|Cu^{2+}(0.1\ M)||Fe^{3+}(0.1\ M), Fe^{2+}(0.1\ M)|Pt(s)$

(a) Write a balanced equation for each cell reaction.

(b) Sketch each cell. Label the anode and cathode, and indicate the direction of electron and ion flow.

(c) Which of the three cells has the largest cell potential? Which has the smallest cell potential? Explain.

17.29 Sketch a cell with inert electrodes suitable for electrolysis of aqueous $CuBr_2$.

(a) Label the anode and cathode.

(b) Indicate the direction of electron and ion flow.

(c) Write balanced equations for the anode, cathode, and overall cell reactions.

17.30 Consider the following electrochemical cell:

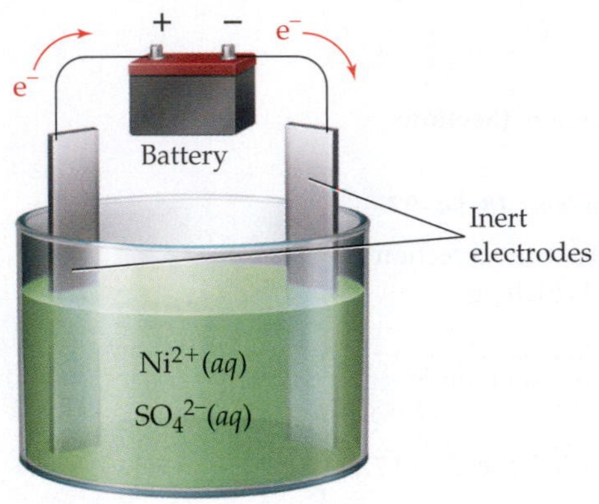

(a) Is the cell a galvanic or an electrolytic cell? Explain.

(b) Label the anode and cathode, and show the direction of ion flow.

(c) Write balanced equations for the anode, cathode, and overall cell reactions.

17.31 It has recently been reported that porous pellets of TiO_2 can be reduced to titanium metal at the cathode of an electrochemical cell containing molten $CaCl_2$ as the electrolyte. When the TiO_2 is reduced, the O^{2-} ions dissolve in the $CaCl_2$ and are subsequently oxidized to O_2 gas at the anode. This approach may be the basis for a less expensive process than the one currently used for producing titanium.

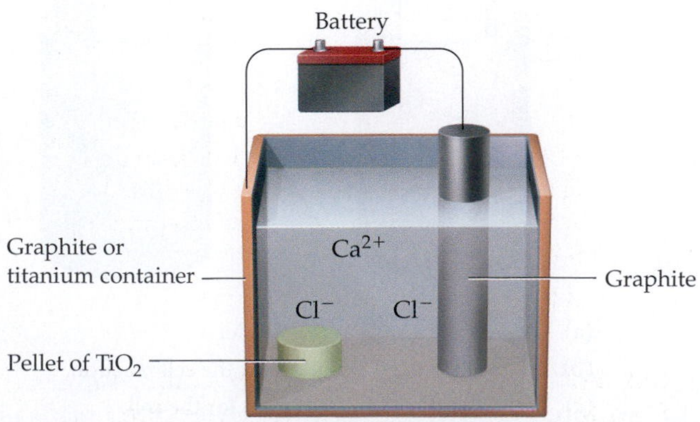

(a) Label the anode and cathode, and indicate the signs of the electrodes.

(b) Indicate the direction of electron and ion flow.

(c) Write balanced equations for the anode, cathode, and overall cell reactions.

17.32 Consider a Daniell cell with 1.0 M ion concentrations:

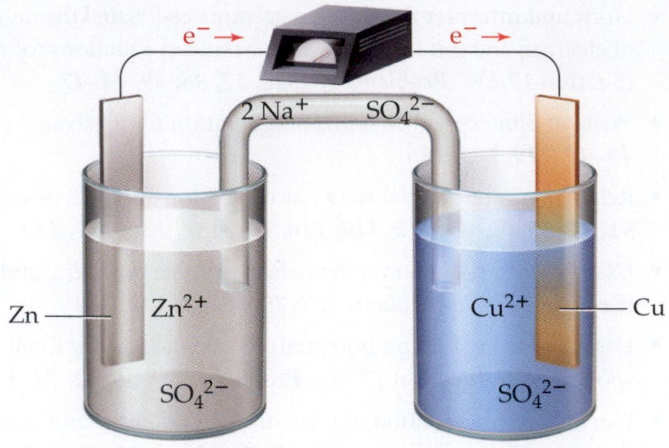

Does the cell voltage increase, decrease, or remain the same when each of the following changes is made? Explain.

(a) 5.0 M $CuSO_4$ is added to the cathode compartment.

(b) 5.0 M H_2SO_4 is added to the cathode compartment.

(c) 5.0 M $Zn(NO_3)_2$ is added to the anode compartment.

(d) 1.0 M $Zn(NO_3)_2$ is added to the anode compartment.

17.33 Consider the following galvanic cell with 0.10 M concentrations:

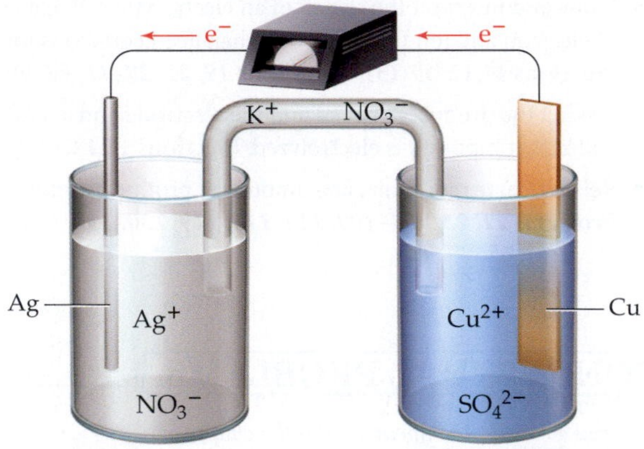

Does the cell voltage increase, decrease, or remain the same when each of the following changes is made? Explain.

(a) 0.10 M NaCl is added to the cathode compartment.

(b) 0.10 M NaCl is added to the anode compartment.

(c) 1.0 M NH_3 is added to the cathode compartment.

(d) 1.0 M NH_3 is added to the anode compartment.

17.34 Consider a PEM fuel cell, consisting of a sandwich of two porous graphite electrodes and the proton-exchange membrane, with separate compartments for entrance of H_2 and O_2:

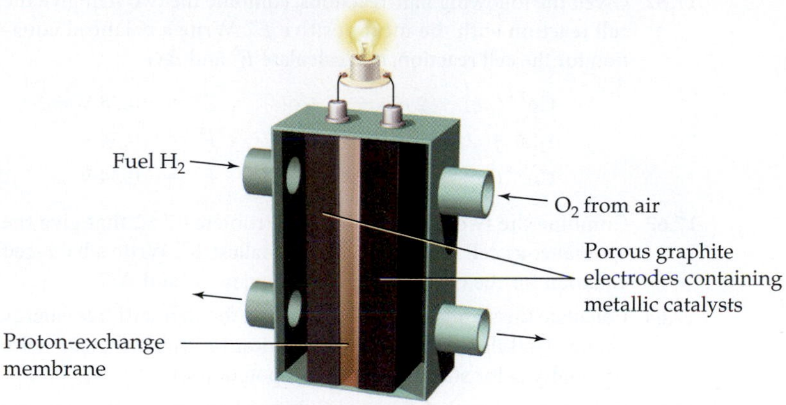

(a) Label the anode and cathode, and indicate the signs of the electrodes.
(b) Indicate the direction of electron and ion flow.
(c) Identify the substances that exit from each cell compartment.
(d) Write balanced equations for the anode, cathode, and overall cell reactions, and calculate the standard cell potential.

17.35 Consider the following table of standard reduction potentials:

Reduction Half-Reaction	$E°$ (V)
$A^+ + e^- \longrightarrow A$	0.80
$B^{2+} + 2\,e^- \longrightarrow B$	0.38
$C_2 + 2\,e^- \longrightarrow 2\,C^-$	0.17
$D^{3+} + 3\,e^- \longrightarrow D$	−1.36

(a) Which substance is the strongest oxidizing agent? Which is the strongest reducing agent?
(b) Which substances can be oxidized by B^{2+}? Which can be reduced by D?
(c) Write a balanced equation for the overall cell reaction that delivers a voltage of 1.53 V under standard-state conditions.

SECTION PROBLEMS

Galvanic Cells (Sections 17.1–17.2)

17.36 Why is the cathode of a galvanic cell considered to be the positive electrode?

17.37 What is the function of a salt bridge in a galvanic cell?

17.38 Describe galvanic cells that use the following reactions. In each case, write the anode and cathode half-reactions and sketch the experimental setup. Label the anode and cathode, identify the sign of each electrode, and indicate the direction of electron and ion flow.

(a) $Cd(s) + Sn^{2+}(aq) \longrightarrow Cd^{2+}(aq) + Sn(s)$
(b) $2\,Al(s) + 3\,Cd^{2+}(aq) \longrightarrow 2\,Al^{3+}(aq) + 3\,Cd(s)$
(c) $Cr_2O_7^{\;2-}(aq) + 6\,Fe^{2+}(aq) + 14\,H^+(aq) \longrightarrow 2\,Cr^{3+}(aq) + 6\,Fe^{3+}(aq) + 7\,H_2O(l)$

17.39 Describe galvanic cells that use the following reactions. In each case, write the anode and cathode half-reactions and sketch the experimental setup. Label the anode and cathode, identify the sign of each electrode, and indicate the direction of electron and ion flow.

(a) $3\,Cu^{2+}(aq) + 2\,Cr(s) \longrightarrow 3\,Cu(s) + 2\,Cr^{3+}(aq)$
(b) $Pb(s) + 2\,H^+(aq) \longrightarrow Pb^{2+}(aq) + H_2(g)$
(c) $Cl_2(g) + Sn^{2+}(aq) \longrightarrow Sn^{4+}(aq) + 2\,Cl^-(aq)$

17.40 Write the standard shorthand notation for each cell in Problem 17.38.

17.41 Write the standard shorthand notation for each cell in Problem 17.39.

17.42 Write a balanced equation for the overall cell reaction in the following galvanic cell, and tell why inert electrodes are required at the anode and cathode.

$$Pt(s)\,|\,Br^-(aq)\,|\,Br_2(l)\,||\,Cl_2(g)\,|\,Cl^-(aq)\,|\,Pt(s)$$

17.43 Write the standard shorthand notation for a galvanic cell that uses the following cell reaction. Include inert electrodes if necessary.

$$Fe(s) + I_2(s) \longrightarrow Fe^{2+}(aq) + 2\,I^-(aq)$$

17.44 An H_2/H^+ half-cell (anode) and an Ag^+/Ag half-cell (cathode) are connected by a wire and a salt bridge.

(a) Sketch the cell, indicating the direction of electron and ion flow.
(b) Write balanced equations for the electrode and overall cell reactions.
(c) Give the shorthand notation for the cell.

17.45 A galvanic cell is constructed from a Zn/Zn^{2+} half-cell (anode) and a Cl_2/Cl^- half-cell (cathode).

(a) Sketch the cell, indicating the direction of electron and ion flow.
(b) Write balanced equations for the electrode and overall cell reactions.
(c) Give the shorthand notation for the cell.

17.46 Write balanced equations for the electrode and overall cell reactions in the following galvanic cells. Sketch each cell, labeling the anode and cathode and showing the direction of electron and ion flow.

(a) $Co(s)\,|\,Co^{2+}(aq)\,||\,Cu^{2+}(aq)\,|\,Cu(s)$
(b) $Fe(s)\,|\,Fe^{2+}(aq)\,||\,O_2(g)\,|\,H^+(aq), H_2O(l)\,|\,Pt(s)$

17.47 Write balanced equations for the electrode and overall cell reactions in the following galvanic cells. Sketch each cell, labeling the anode and cathode and showing the direction of electron and ion flow.

(a) $Mn(s)\,|\,Mn^{2+}(aq)\,||\,Pb^{2+}(aq)\,|\,Pb(s)$
(b) $Pt(s)\,|\,H_2(g)\,|\,H^+(aq)\,||\,Cl^-(aq)\,|\,AgCl(s)\,|\,Ag(s)$

Cell Potentials and Free-Energy Changes; Standard Reduction Potentials (Sections 17.3–17.5)

17.48 What conditions must be met for a cell potential E to qualify as a standard cell potential $E°$?

17.49 How are standard reduction potentials defined?

17.50 The silver oxide–zinc battery used in watches delivers a voltage of 1.60 V. Calculate the free-energy change (in kilojoules) for the cell reaction

$$Zn(s) + Ag_2O(s) \longrightarrow ZnO(s) + 2\,Ag(s)$$

17.51 The standard cell potential for a lead storage battery is 1.924 V. Calculate $\Delta G°$ (in kilojoules) for the cell reaction

$$Pb(s) + PbO_2(s) + 2\,H^+(aq) + 2\,HSO_4^-(aq) \longrightarrow 2\,PbSO_4(s) + 2\,H_2O(l)$$

17.52 What is the value of x for the following reaction if $E° = 1.43$ V and $\Delta G° = -414$ kJ?

$$A + B^{x+} \longrightarrow A^{x+} + B$$

17.53 What are the values of x and y for the following reaction if $E° = 0.91$ V and $\Delta G° = -527$ kJ?

$$2\,A^{x+} + 3\,B \longrightarrow 2\,A + 3\,B^{y+}$$

17.54 Use the standard free energies of formation in Appendix B to calculate the standard cell potential for the reaction in the hydrogen–oxygen fuel cell:

$$2\,H_2(g) + O_2(g) \longrightarrow 2\,H_2O(l)$$

17.55 Consider a fuel cell that uses the reaction

$$CH_4(g) + 2\,O_2(g) \longrightarrow CO_2(g) + 2\,H_2O(l)$$

Given the standard free energies of formation in Appendix B, what is the value of $E°$ for the cell reaction?

17.56 The standard potential for the following galvanic cell is 0.40 V:

$$Zn(s)\,|\,Zn^{2+}(aq)\,||\,Eu^{3+}(aq), Eu^{2+}(aq)\,|\,Pt(s)$$

(Europium, Eu, is one of the lanthanide elements.) Use the data in Table 17.1 to calculate the standard reduction potential for the Eu^{3+}/Eu^{2+} half-cell.

17.57 The following reaction has an $E°$ value of 0.27 V:

$$Cu^{2+}(aq) + 2\,Ag(s) + 2\,Br^-(aq) \longrightarrow Cu(s) + 2\,AgBr(s)$$

Use the data in Table 17.1 to calculate the standard reduction potential for the half-reaction

$$AgBr(s) + e^- \longrightarrow Ag(s) + Br^-(aq)$$

17.58 Arrange the following oxidizing agents in order of increasing strength under standard-state conditions: $Br_2(aq)$, $MnO_4^-(aq)$, $Sn^{4+}(aq)$.

17.59 List the following reducing agents in order of increasing strength under standard-state conditions: $Al(s)$, $Pb(s)$, $Fe(s)$.

17.60 Consider the following substances: $I_2(s)$, $Fe^{2+}(aq)$, $Cr_2O_7^{2-}(aq)$. Which is the strongest oxidizing agent? Which is the weakest oxidizing agent?

17.61 Consider the following substances: $Fe^{2+}(aq)$, $Sn^{2+}(aq)$, $I^-(aq)$. Identify the strongest reducing agent and the weakest reducing agent.

17.62 Given the following half-reactions, combine the two that give the cell reaction with the most positive $E°$. Write a balanced equation for the cell reaction, and calculate $E°$ and $\Delta G°$.

$$Co^{2+}(aq) + 2\,e^- \longrightarrow Co(s) \qquad E° = -0.28\text{ V}$$
$$I_2(s) + 2\,e^- \longrightarrow 2\,I^-(aq) \qquad E° = 0.54\text{ V}$$
$$Cu^{2+}(aq) + 2\,e^- \longrightarrow Cu(s) \qquad E° = 0.34\text{ V}$$

17.63 Combine the two half-reactions in Problem 17.62 that give the spontaneous cell reaction with the smallest $E°$. Write a balanced equation for the cell reaction, and calculate $E°$ and $\Delta G°$.

17.64 Calculate the standard cell potential and the standard free-energy change (in kilojoules) for each reaction in Problem 17.38. (See Appendix D for standard reduction potentials.)

17.65 Calculate $E°$ and $\Delta G°$ (in kilojoules) for the cell reactions in Problem 17.39. (See Appendix D for standard reduction potentials.)

17.66 Calculate $E°$ for each of the following reactions, and tell which are spontaneous under standard-state conditions:

(a) $2\,Fe^{2+}(aq) + Pb^{2+}(aq) \longrightarrow 2\,Fe^{3+}(aq) + Pb(s)$

(b) $Mg(s) + Ni^{2+}(aq) \longrightarrow Mg^{2+}(aq) + Ni(s)$

17.67 Calculate $E°$ for each of the following reactions, and tell which are spontaneous under standard-state conditions:

(a) $5\,Ag^+(aq) + Mn^{2+}(aq) + 4\,H_2O(l) \longrightarrow 5\,Ag(s) + MnO_4^-(aq) + 8\,H^+(aq)$

(b) $2\,H_2O_2(aq) \longrightarrow O_2(g) + 2\,H_2O(l)$

17.68 Use the data in Appendix D to predict whether the following reactions can occur under standard-state conditions:

(a) Oxidation of $Sn^{2+}(aq)$ by $Br_2(aq)$

(b) Reduction of $Ni^{2+}(aq)$ by $Sn^{2+}(aq)$

(c) Oxidation of $Ag(s)$ by $Pb^{2+}(aq)$

(d) Reduction of $I_2(s)$ by $H_2SO_3(aq)$

17.69 Use the data in Appendix D to predict whether the following reactions can occur under standard-state conditions:

(a) Reduction of $Pb^{2+}(aq)$ by $Ni(s)$

(b) Oxidation of $Au^+(aq)$ by $Mn^{2+}(aq)$

(c) Reduction of $I_2(s)$ by $Mn(s)$

(d) Oxidation of $Fe^{2+}(aq)$ by $Br_2(aq)$

17.70 What reaction can occur, if any, when the following experiments are carried out under standard-state conditions?

(a) Oxygen gas is bubbled through an acidic solution of $Cr(NO_3)_3$.

(b) A strip of lead is dipped into an aqueous solution of $AgNO_3$.

(c) Chlorine gas is bubbled through aqueous $H_2C_2O_4$.

(d) A nickel wire is dipped into an aqueous solution of HClO.

17.71 What reaction can occur, if any, when the following experiments are carried out under standard-state conditions?

(a) A strip of zinc is dipped into an aqueous solution of $Pb(NO_3)_2$.

(b) An acidic solution of $FeSO_4$ is exposed to oxygen.

(c) A silver wire is immersed in an aqueous solution of $NiCl_2$.

(d) Hydrogen gas is bubbled through aqueous $Cd(NO_3)_2$.

The Nernst Equation (Sections 17.6–17.7)

17.72 Consider a galvanic cell that uses the reaction

$$2\,Ag^+(aq) + Sn(s) \longrightarrow 2\,Ag(s) + Sn^{2+}(aq)$$

Calculate the potential at 25 °C for a cell that has the following ion concentrations: $[Ag^+] = 0.010$ M, $[Sn^{2+}] = 0.020$ M.

17.73 Consider a galvanic cell based on the reaction

$$2\,Fe^{2+}(aq) + Cl_2(g) \longrightarrow 2\,Fe^{3+}(aq) + 2\,Cl^-(aq)$$

Calculate the cell potential at 25 °C when $[Fe^{2+}] = 1.0$ M, $[Fe^{3+}] = 1.0 \times 10^{-3}$ M, $[Cl^-] = 3.0 \times 10^{-3}$ M, and $P_{Cl_2} =$ 0.50 atm.

17.74 What is the cell potential at 25 °C for the following galvanic cell?

$$Pb(s)\,|\,Pb^{2+}(1.0\text{ M})\,||\,Cu^{2+}(1.0 \times 10^{-4}\text{ M})\,|\,Cu(s)$$

If the Pb^{2+} concentration is maintained at 1.0 M, what is the Cu^{2+} concentration when the cell potential drops to zero?

17.75 A galvanic cell has an iron electrode in contact with 0.10 M $FeSO_4$ and a copper electrode in contact with a $CuSO_4$ solution. If the measured cell potential at 25 °C is 0.67 V, what is the concentration of Cu^{2+} in the $CuSO_4$ solution?

17.76 What is the Zn^{2+}: Cu^{2+} concentration ratio in the following cell at 25 °C if the measured cell potential is 1.07 V?

$$Zn(s)\,|\,Zn^{2+}(aq)\,||\,Cu^{2+}(aq)\,|\,Cu(s)$$

17.77 What is the Fe^{2+}: Sn^{2+} concentration ratio in the following cell at 25 °C if the measured cell potential is 0.35 V?

$$Fe(s)\,|\,Fe^{2+}(aq)\,||\,Sn^{2+}(aq)\,|\,Sn(s)$$

17.78 The Nernst equation applies to both cell reactions and half-reactions. For the conditions specified, calculate the potential for the following half-reactions at 25 °C:

(a) $I_2(s) + 2\,e^- \longrightarrow 2\,I^-(aq)$; $[I^-] = 0.020$ M

(b) $Fe^{3+}(aq) + e^- \longrightarrow Fe^{2+}(aq)$; $[Fe^{3+}] = [Fe^{2+}] = 0.10$ M

(c) $Sn^{2+}(aq) \longrightarrow Sn^{4+}(aq) + 2\,e^-$; $[Sn^{2+}] = 1.0 \times 10^{-3}$ M, $[Sn^{4+}] = 0.40$ M

(d) $2\,Cr^{3+}(aq) + 7\,H_2O(l) \longrightarrow Cr_2O_7^{\,2-}(aq) + 14\,H^+(aq) + 6\,e^-$; $[Cr^{3+}] = [Cr_2O_7^{\,2-}] = 1.0$ M, $[H^+] = 0.010$ M

17.79 What is the reduction potential at 25 °C for the hydrogen electrode in each of the following solutions? The half-reaction is

$$2\,H^+(aq) + 2\,e^- \longrightarrow H_2(g, 1\text{ atm})$$

(a) 1.0 M HCl

(b) A solution having pH 4.00

(c) Pure water

(d) 1.0 M NaOH

17.80 The following cell has a potential of 0.27 V at 25 °C:

$$Pt(s)\,|\,H_2(1\text{ atm})\,|\,H^+(?\text{ M})\,||\,Ni^{2+}(1\text{ M})\,|\,Ni(s)$$

What is the pH of the solution in the anode compartment?

17.81 What is the pH of the solution in the cathode compartment of the following cell if the measured cell potential at 25 °C is 0.58 V?

$$Zn(s)\,|\,Zn^{2+}(1\text{ M})\,||\,H^+(?\text{ M})\,|\,H_2(1\text{ atm})\,|\,Pt(s)$$

Standard Cell Potentials and Equilibrium Constants (Section 17.8)

17.82 Beginning with the equations that relate $E°$, $\Delta G°$, and K, show that $\Delta G°$ is negative and $K > 1$ for a reaction that has a positive value of $E°$.

17.83 If a reaction has an equilibrium constant $K < 1$, is $E°$ positive or negative? What is the value of K when $E° = 0$ V?

17.84 Use the data in Table 17.1 to calculate the equilibrium constant at 25 °C for the reaction

$$Ni(s) + 2\,Ag^+(aq) \longrightarrow Ni^{2+}(aq) + 2\,Ag(s)$$

17.85 From standard reduction potentials, calculate the equilibrium constant at 25 °C for the reaction

$$2\,MnO_4^{\,-}(aq) + 10\,Cl^-(aq) + 16\,H^+(aq) \longrightarrow 2\,Mn^{2+}(aq) + 5\,Cl_2(g) + 8\,H_2O(l)$$

17.86 Calculate the equilibrium constant at 25 °C for each reaction in Problem 17.38.

17.87 Calculate the equilibrium constant at 25 °C for each reaction in Problem 17.39.

17.88 Calculate the equilibrium constant at 25 °C for the reaction

$$Hg_2^{\,2+}(aq) \longrightarrow Hg(l) + Hg^{2+}(aq)$$

See Appendix D for standard reduction potentials.

17.89 Use standard reduction potentials to calculate the equilibrium constant at 25 °C for decomposition of hydrogen peroxide:

$$2\,H_2O_2(l) \longrightarrow 2\,H_2O(l) + O_2(g)$$

Batteries; Fuel Cells; Corrosion (Sections 17.9–17.11)

17.90 For a lead storage battery:

(a) Sketch one cell that shows the anode, cathode, electrolyte, direction of electron and ion flow, and sign of the electrodes.

(b) Write the anode, cathode, and overall cell reactions.

(c) Calculate the equilibrium constant for the cell reaction ($E° = 1.924$ V).

(d) What is the cell voltage when the cell reaction reaches equilibrium?

17.91 A mercury battery uses the following electrode half-reactions:

$$HgO(s) + H_2O(l) + 2\,e^- \longrightarrow Hg(l) + 2\,OH^-(aq) \quad E° = 0.098\text{ V}$$

$$ZnO(s) + H_2O(l) + 2\,e^- \longrightarrow Zn(s) + 2\,OH^-(aq) \quad E° = -1.260\text{ V}$$

(a) Write a balanced equation for the overall cell reaction.

(b) Calculate $\Delta G°$ (in kilojoules) and K at 25 °C for the cell reaction.

(c) What is the effect on the cell voltage of a tenfold change in the concentration of KOH in the electrolyte? Explain.

17.92 Calculate the values of $E°$, $\Delta G°$ (in kilojoules), and K at 25 °C for the cell reaction in a hydrogen–oxygen fuel cell: $2\,H_2(g) + O_2(g) \longrightarrow 2\,H_2O(l)$. What is the cell voltage at 25 °C if the partial pressure of each gas is 25 atm?

17.93 Use the thermodynamic data in Appendix B to calculate the standard cell potential and the equilibrium constant at 25 °C for the cell reaction in a direct methanol fuel cell:

$$2\,CH_3OH(l) + 3\,O_2(g) \longrightarrow 2\,CO_2(g) + 4\,H_2O(l)$$

17.94 What is rust? What causes it to form? What can be done to prevent its formation?

17.95 The standard oxidation potential for the reaction $Cr(s) \longrightarrow Cr^{3+}(aq) + 3\,e^-$ is 0.74 V. Despite the large, positive oxidation potential, chromium is sometimes used as a protective coating on steel. Why doesn't the chromium corrode?

17.96 What is meant by cathodic protection? Which of the following metals can offer cathodic protection to iron?

Zn, Ni, Al, Sn

17.97 What is a sacrificial anode? Give an example.

Electrolysis (Sections 17.12–17.14)

17.98 Magnesium metal is produced by the electrolysis of molten magnesium chloride using inert electrodes.

(a) Sketch the cell, label the anode and cathode, indicate the sign of the electrodes, and show the direction of electron and ion flow.

(b) Write balanced equations for the anode, cathode, and overall cell reactions.

17.99 **(a)** Sketch a cell with inert electrodes suitable for the electrolysis of an aqueous solution of sulfuric acid. Label the anode and cathode, and indicate the direction of electron and ion flow. Identify the positive and negative electrodes.

(b) Write balanced equations for the anode, cathode, and overall cell reactions.

17.100 List the anode and cathode half-reactions that might occur when an aqueous solution of $MgCl_2$ is electrolyzed in a cell having inert electrodes. Predict which half-reactions will occur, and justify your answers.

17.101 What products should be formed when the following reactants are electrolyzed in a cell having inert electrodes? Account for any differences.

(a) Molten KCl **(b)** Aqueous KCl

17.102 Predict the anode, cathode, and overall cell reactions when an aqueous solution of each of the following salts is electrolyzed in a cell having inert electrodes:

(a) NaBr **(b)** $CuCl_2$ **(c)** LiOH

17.103 Predict the anode, cathode, and overall cell reactions when an aqueous solution of each of the following salts is electrolyzed in a cell having inert electrodes:

(a) Ag_2SO_4 **(b)** $Ca(OH)_2$ **(c)** KI

17.104 How many grams of silver will be obtained when an aqueous silver nitrate solution is electrolyzed for 20.0 min with a constant current of 2.40 A?

17.105 A constant current of 100.0 A is passed through an electrolytic cell having an impure copper anode, a pure copper cathode, and an aqueous $CuSO_4$ electrolyte. How many kilograms of copper are refined by transfer from the anode to the cathode in a 24.0 h period?

17.106 How many hours are required to produce 1.00×10^3 kg of sodium by the electrolysis of molten NaCl with a constant current of 3.00×10^4 A? How many liters of Cl_2 at STP will be obtained as a byproduct?

17.107 What constant current (in amperes) is required to produce aluminum by the Hall–Héroult process at a rate of 40.0 kg/h?

17.108 Electrolysis of a metal nitrate solution $M(NO_3)_2(aq)$ for 325 min with a constant current of 20.0 A gives 111 g of the metal. Identify the metal ion M^{2+}.

17.109 What is the metal ion in a metal nitrate solution $M(NO_3)_3(aq)$ if 90.52 g of metal was recovered from a 4.00 h electrolysis at a constant current of 35.0 A?

CHAPTER PROBLEMS

17.110 Consider a galvanic cell that uses the following half-reactions:

$$MnO_4^-(aq) + 8\,H^+(aq) + 5\,e^- \longrightarrow Mn^{2+}(aq) + 4\,H_2O(l)$$
$$Sn^{4+}(aq) + 2\,e^- \longrightarrow Sn^{2+}(aq)$$

(a) Write a balanced equation for the overall cell reaction.

(b) What is the oxidizing agent, and what is the reducing agent?

(c) Calculate the standard cell potential.

17.111 Given the following half-reactions and $E°$ values,

$$Mn^{3+}(aq) + e^- \longrightarrow Mn^{2+}(aq) \qquad E° = 1.54\text{ V}$$
$$MnO_2(s) + 4\,H^+(aq) + e^- \longrightarrow Mn^{3+}(aq) + 2\,H_2O(l) \qquad E° = 0.95\text{ V}$$

write a balanced equation for the formation of Mn^{2+} and MnO_2 from Mn^{3+}, and calculate the value of $E°$ for this reaction. Is the reaction spontaneous under standard-state conditions?

17.112 Consider the following half-reactions and $E°$ values:

$$Ag^+(aq) + e^- \longrightarrow Ag(s) \qquad E° = 0.80\text{ V}$$
$$Cu^{2+}(aq) + 2\,e^- \longrightarrow Cu(s) \qquad E° = 0.34\text{ V}$$
$$Pb^{2+}(aq) + 2\,e^- \longrightarrow Pb(s) \qquad E° = -0.13\text{ V}$$

(a) Which of these metals or ions is the strongest oxidizing agent? Which is the strongest reducing agent?

(b) The half-reactions can be used to construct three different galvanic cells. Tell which cell delivers the highest voltage, identify the anode and cathode, and tell the direction of electron and ion flow.

(c) Write the cell reaction for part **(b)**, and calculate the values of $E°$, $\Delta G°$ (in kilojoules), and K for this reaction at 25 °C.

(d) Calculate the voltage for the cell in part **(b)** if both ion concentrations are 0.010 M.

17.113 Standard reduction potentials for the Pb^{2+}/Pb and Cd^{2+}/Cd half-reactions are −0.13 V and −0.40 V, respectively. At what relative concentrations of Pb^{2+} and Cd^{2+} will these half-reactions have the same reduction potential?

17.114 Consider a galvanic cell that uses the following half-reactions:

$$2\,H^+(aq) + 2\,e^- \longrightarrow H_2(g)$$
$$Al^{3+}(aq) + 3\,e^- \longrightarrow Al(s)$$

(a) What materials are used for the electrodes? Identify the anode and cathode, and indicate the direction of electron and ion flow.

(b) Write a balanced equation for the cell reaction, and calculate the standard cell potential.

(c) Calculate the cell potential at 25 °C if the ion concentrations are 0.10 M and the partial pressure of H_2 is 10.0 atm.

(d) Calculate $\Delta G°$ (in kilojoules) and K for the cell reaction at 25 °C.

(e) Calculate the mass change (in grams) of the aluminum electrode after the cell has supplied a constant current of 10.0 A for 25.0 min.

17.115 When the nickel-zinc battery, used in digital cameras, is recharged, the following cell reaction occurs:

$$2\,Ni(OH)_2(s) + Zn(OH)_2(s) \longrightarrow 2\,Ni(OH)_3(s) + Zn(s)$$

(a) How many grams of zinc are formed when 3.35×10^{-2} g of $Ni(OH)_2$ are consumed?

(b) How many minutes are required to fully recharge a dead battery that contains 6.17×10^{-2} g of Zn with a constant current of 0.100 A?

17.116 Chlorine can be prepared in the laboratory by the reaction of hydrochloric acid and potassium permanganate.

(a) Use data in Appendix D to write a balanced equation for the reaction. The reduction product is Mn^{2+}.

(b) Calculate $E°$ and $\Delta G°$ for the reaction.

(c) How many liters of Cl_2 at 1.0 atm and 25 °C will result from the reaction of 179 g $KMnO_4$ with an excess of HCl?

17.117 The following cell reactions occur spontaneously:

$$B + A^+ \longrightarrow B^+ + A$$
$$C + A^+ \longrightarrow C^+ + A$$
$$B + C^+ \longrightarrow B^+ + C$$

(a) Arrange the following reduction half-reactions in order of decreasing tendency to occur: $A^+ + e^- \rightarrow A$, $B^+ + e^- \rightarrow B$, and $C^+ + e^- \rightarrow C$.

(b) Which of these substances (A, A^+, B, B^+, C, C^+) is the strongest oxidizing agent? Which is the strongest reducing agent?

(c) Which of the three cell reactions delivers the highest voltage?

17.118 Consider the following substances: Fe(s), $PbO_2(s)$, $H^+(aq)$, Al(s), Ag(s), $Cr_2O_7^{2-}(aq)$.

(a) Look at the $E°$ values in Appendix D, and classify each substance as an oxidizing agent or a reducing agent.

(b) Which is the strongest oxidizing agent? Which is the weakest oxidizing agent?

(c) Which is the strongest reducing agent? Which is the weakest reducing agent?

(d) Which substances can be oxidized by $Cu^{2+}(aq)$? Which can be reduced by $H_2O_2(aq)$?

17.119 The sodium-sulfur battery has molybdenum electrodes with anode and cathode compartments separated by β-alumina, a ceramic through which sodium ions can pass. Because the battery operates at temperatures above 300 °C, all the reactants and products are present in a molten solution. The cell voltage is about 2.0 V.

(a) What is the cell reaction if the shorthand notation is $Mo(s)\,|\,Na(soln), Na^+(soln)\,||\,S(soln), S^{2-}(soln)\,|\,Mo(s)$?

(b) How many kilograms of sodium are consumed when a 25 kW sodium-sulfur battery produces current for 32 minutes?

17.120 When suspected drunk drivers are tested with a Breathalyzer, the alcohol (ethanol) in the exhaled breath is oxidized to acetic acid with an acidic solution of potassium dichromate:

$$\underset{\text{Ethanol}}{3\,CH_3CH_2OH(aq)} + 2\,Cr_2O_7^{2-}(aq) + 16\,H^+(aq) \longrightarrow \underset{\text{Acetic acid}}{3\,CH_3CO_2H(aq)} + 4\,Cr^{3+}(aq) + 11H_2O(l)$$

The color of the solution changes because some of the orange $Cr_2O_7^{2-}$ is converted to the green Cr^{3+}. The Breathalyzer measures the color change and produces a meter reading calibrated in blood alcohol content.

(a) What is $E°$ for the reaction if the standard half-cell potential for the reduction of acetic acid to ethanol is 0.058 V?

(b) What is the value of E for the reaction when the concentrations of ethanol, acetic acid, $Cr_2O_7^{2-}$, and Cr^{3+} are 1.0 M and the pH is 4.00?

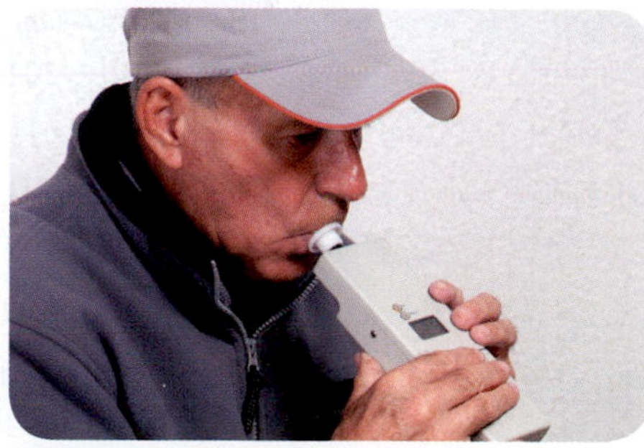

17.121 Consider the addition of the following half-reactions:

(1) $Fe^{3+}(aq) + 3\,e^- \longrightarrow Fe(s)$	$E°_1 = -0.04$ V
(2) $Fe(s) \longrightarrow Fe^{2+}(aq) + 2\,e^-$	$E°_2 = 0.45$ V
(3) $Fe^{3+}(aq) + e^- \longrightarrow Fe^{2+}(aq)$	$E°_3 = ?$

Because half-reactions (1) and (2) contain a different number of electrons, the net reaction (3) is another half-reaction, and $E°_3$ can't be obtained simply by adding $E°_1$ and $E°_2$. The free-energy changes, however, are additive because G is a state function:

$$\Delta G°_3 = \Delta G°_1 + \Delta G°_2$$

(a) Starting with the relationship between $\Delta G°$ and $E°$, derive a general equation that relates the $E°$ values for half-reactions (1), (2), and (3).

(b) Calculate the value of $E°_3$ for the Fe^{3+}/Fe^{2+} half-reaction.

(c) Explain why the $E°$ values would be additive ($E°_3 = E°_1 + E°_2$) if reaction (3) were an overall cell reaction rather than a half-reaction.

17.122 The following galvanic cell has a potential of 0.578 V at 25 °C:

$$Ag(s)\,|\,AgCl(s)\,|\,Cl^-(1.0\,M)\,||\,Ag^+(1.0\,M)\,|\,Ag(s)$$

Use this information to calculate K_{sp} for AgCl at 25 °C.

17.123 A galvanic cell has a silver electrode in contact with 0.050 M $AgNO_3$ and a copper electrode in contact with 1.0 M $Cu(NO_3)_2$.

(a) Write a balanced equation for the cell reaction, and calculate the cell potential at 25 °C.

(b) Excess NaBr(*aq*) is added to the $AgNO_3$ solution to precipitate AgBr. What is the cell potential at 25 °C after the precipitation of AgBr if the concentration of excess Br^- is 1.0 M? Write a balanced equation for the cell reaction under these conditions. (K_{sp} for AgBr at 25 °C is 5.4×10^{-13}.)

(c) Use the result in part **(b)** to calculate the standard reduction potential $E°$ for the half-reaction

$$AgBr(s) + e^- \longrightarrow Ag(s) + Br^-(aq).$$

17.124 At one time on Earth, iron was present mostly as iron(II). Later, once plants had produced a significant quantity of oxygen in the atmosphere, the iron became oxidized to iron(III). Show that $Fe^{2+}(aq)$ can be spontaneously oxidized to $Fe^{3+}(aq)$ by $O_2(g)$ at 25 °C assuming the following reasonable environmental conditions: $[Fe^{2+}] = [Fe^{3+}] = 1 \times 10^{-7}$ M; pH = 7.0; $P_{O_2} = 160$ mm Hg.

17.125 Given the following standard reduction potentials at 25 °C, **(a)** balance the equation for the reaction of H_2MoO_4 with elemental arsenic in acidic solution to give Mo^{3+} and H_3AsO_4, and **(b)** calculate $E°$ for this reaction.

Half-Reaction	$E°$ (V)
$H_3AsO_4(aq) + 2\,H^+(aq) + 2\,e^- \longrightarrow H_3AsO_3(aq) + H_2O(l)$	+0.560
$H_3AsO_3(aq) + 3\,H^+(aq) + 3\,e^- \longrightarrow As(s) + 3\,H_2O(l)$	+0.240
$H_2MoO_4(aq) + 2\,H^+(aq) + 2\,e^- \longrightarrow MoO_2(s) + 2\,H_2O(l)$	+0.646
$MoO_2(s) + 4\,H^+(aq) + e^- \longrightarrow Mo^{3+}(aq) + 2\,H_2O(l)$	−0.008

17.126 The following galvanic cell has a potential of 1.214 V at 25 °C:

$$Hg(l)\,|\,Hg_2Br_2(s)\,|\,Br^-(0.10\text{ M})\,||\,MnO_4^-(0.10\text{ M}), Mn^{2+}(0.10\text{ M}), H^+(0.10\text{ M})\,|\,Pt(s)$$

Calculate the value of K_{sp} for Hg_2Br_2 at 25 °C.

17.127 For the following half-reaction, $E° = 1.103$ V:

$$Cu^{2+}(aq) + 2\,CN^-(aq) + e^- \longrightarrow Cu(CN)_2^-(aq)$$

Calculate the formation constant K_f for $Cu(CN)_2^-$.

17.128 Accidentally chewing on a stray fragment of aluminum foil can cause a sharp tooth pain if the aluminum comes in contact with an amalgam filling. The filling, an alloy of silver, tin, and mercury, acts as the cathode of a tiny galvanic cell, the aluminum behaves as the anode, and saliva serves as the electrolyte. When the aluminum and the filling come in contact, an electric current passes from the aluminum to the filling, which is sensed by a nerve in the tooth. Aluminum is oxidized at the anode, and O_2 gas is reduced to water at the cathode.

(a) Write balanced equations for the anode, cathode, and overall cell reactions.

(b) Write the Nernst equation in a form that applies at body temperature (37 °C).

(c) Calculate the cell voltage at 37 °C. You may assume that $[Al^{3+}] = 1.0 \times 10^{-9}$ M, $P_{O_2} = 0.20$ atm, and that saliva has a pH of 7.0. Also assume that the $E°$ values in Appendix D apply at 37 °C.

17.129 Copper reduces dilute nitric acid to nitric oxide (NO) but reduces concentrated nitric acid to nitrogen dioxide (NO_2):

$$\text{(1)}\quad 3\,Cu(s) + 2\,NO_3^-(aq) + 8\,H^+(aq) \longrightarrow 3\,Cu^{2+}(aq) + 2\,NO(g) + 4\,H_2O(l) \qquad E° = 0.62\text{ V}$$

$$\text{(2)}\quad Cu(s) + 2\,NO_3^-(aq) + 4\,H^+(aq) \longrightarrow Cu^{2+}(aq) + 2\,NO_2(g) + 2\,H_2O(l) \qquad E° = 0.45\text{ V}$$

Assuming that $[Cu^{2+}] = 0.10$ M and that the partial pressures of NO and NO_2 are 1.0×10^{-3} atm, calculate the potential (E) for reactions (1) and (2) at 25 °C and show which reaction has the greater thermodynamic tendency to occur when the concentration of HNO_3 is

(a) 1.0 M

(b) 10.0 M

(c) At what HNO_3 concentration do reactions (1) and (2) have the same value of E?

MULTICONCEPT PROBLEMS

17.130 Adiponitrile, a key intermediate in the manufacture of nylon, is made industrially by an electrolytic process that reduces acrylonitrile:

Anode (oxidation): $2\,H_2O \longrightarrow O_2 + 4\,H^+ + 4\,e^-$

Cathode (reduction):

$$\underset{\text{Acrylonitrile}}{2\,CH_2{=}CHCN} + 2\,H^+ + 2\,e^- \longrightarrow \underset{\text{Adiponitrile}}{NC(CH_2)_4CN}$$

(a) Write a balanced equation for the overall cell reaction.

(b) How many kilograms of adiponitrile are produced in 10.0 h in a cell that has a constant current of 3.00×10^3 A?

(c) How many liters of O_2 at 740 mm Hg and 25 °C are produced as a byproduct?

17.131 The reaction of MnO_4^- with oxalic acid ($H_2C_2O_4$) in acidic solution, yielding Mn^{2+} and CO_2 gas, is widely used to determine the concentration of permanganate solutions:

(a) Write a balanced net ionic equation for the reaction.

(b) Use the data in Appendix D to calculate $E°$ for the reaction.

(c) Show that the reaction goes to completion by calculating the values of $\Delta G°$ and K at 25 °C.

(d) A 1.200 g sample of sodium oxalate ($Na_2C_2O_4$) is dissolved in dilute H_2SO_4 and then titrated with a $KMnO_4$ solution. If 32.50 mL of the $KMnO_4$ solution is required to reach the equivalence point, what is the molarity of the $KMnO_4$ solution?

17.132 Calculate the standard reduction potential for $Ba^{2+}(aq) + 2\,e^- \rightarrow Ba(s)$ given that $\Delta G° = 16.7$ kJ for the reaction $Ba^{2+}(aq) + 2\,Cl^-(aq) \rightarrow BaCl_2(s)$. Use any data needed from Appendixes B and D.

17.133 A concentration cell has the same half-reactions at the anode and cathode, but a voltage results from different concentrations in the two electrode compartments.

(a) What is x in the concentration cell $Cu(s)\,|\,Cu^{2+}(x\text{ M})\,||\,Cu^{2+}(0.10\text{ M})\,|\,Cu(s)$ if the measured cell potential is 0.0965 V?

(b) A similar cell has 0.10 M Cu^{2+} in both compartments. When a stoichiometric amount of ethylenediamine ($NH_2CH_2CH_2NH_2$) is added to one compartment, the measured cell potential is 0.179 V. Calculate the formation constant K_f for the complex ion $Cu(NH_2CH_2CH_2NH_2)_2^{2+}$. Assume there is no volume change.

17.134 Consider the redox titration (Section 7.10) of 120.0 mL of 0.100 M $FeSO_4$ with 0.120 M $K_2Cr_2O_7$ at 25 °C, assuming that the pH of the solution is maintained at 2.00 with a suitable buffer. The solution is in contact with a platinum electrode and constitutes one half-cell of an electrochemical cell. The other half-cell is a standard hydrogen electrode. The two half-cells are connected with a wire and a salt bridge, and the progress of the titration is monitored by measuring the cell potential with a voltmeter.

(a) Write a balanced net ionic equation for the titration reaction, assuming that the products are Fe^{3+} and Cr^{3+}.

(b) What is the cell potential at the equivalence point?

17.135 Consider the reaction that occurs in the hydrogen–oxygen fuel cell:

$$2\,H_2(g) + O_2(g) \longrightarrow 2\,H_2O(l)$$

(a) Use the thermodynamic data in Appendix B to calculate the values of $\Delta G°$ and $E°$ at 95 °C, assuming that $\Delta H°$ and $\Delta S°$ are independent of temperature.

(b) Calculate the cell voltage at 95 °C when the partial pressures of H_2 and O_2 are 25 atm.

17.136 Consider a galvanic cell that utilizes the following half-reactions:

Anode: $Zn(s) + H_2O(l) \longrightarrow ZnO(s) + 2\,H^+(aq) + 2\,e^-$

Cathode: $Ag^+(aq) + e^- \longrightarrow Ag(s)$

(a) Write a balanced equation for the cell reaction, and use the thermodynamic data in Appendix B to calculate the values of $\Delta H°$, $\Delta S°$, and $\Delta G°$ for the reaction.

(b) What are the values of $E°$ and the equilibrium constant K for the cell reaction at 25 °C?

(c) What happens to the cell voltage if aqueous ammonia is added to the cathode compartment? Calculate the cell voltage assuming that the solution in the cathode compartment was prepared by mixing 50.0 mL of 0.100 M $AgNO_3$ and 50.0 mL of 4.00 M NH_3.

(d) Will AgCl precipitate if 10.0 mL of 0.200 M NaCl is added to the solution in part **(c)**? Will AgBr precipitate if 10.0 mL of 0.200 M KBr is added to the resulting solution?

17.137 The nickel–iron battery has an iron anode, an NiO(OH) cathode, and a KOH electrolyte. This battery uses the following half-reactions and has an $E°$ value of 1.37 V at 25 °C:

$$Fe(s) + 2\,OH^-(aq) \longrightarrow Fe(OH)_2(s) + 2\,e^-$$
$$NiO(OH)(s) + H_2O(l) + e^- \longrightarrow Ni(OH)_2(s) + OH^-(aq)$$

(a) Write a balanced equation for the cell reaction.

(b) Calculate $\Delta G°$ (in kilojoules) and the equilibrium constant K for the cell reaction at 25 °C.

(c) What is the cell voltage at 25 °C when the concentration of KOH in the electrolyte is 5.0 M?

(d) How many grams of $Fe(OH)_2$ are formed at the anode when the battery produces a constant current of 0.250 A for 40.0 min? How many water molecules are consumed in the process?

17.138 Experimental solid-oxide fuel cells that use butane (C_4H_{10}) as the fuel have been reported recently. These cells contain composite metal/metal oxide electrodes and a solid metal oxide electrolyte. The cell half-reactions are

Anode: $C_4H_{10}(g) + 13\,O^{2-}(s) \longrightarrow 4\,CO_2(g) + 5\,H_2O(l) + 26\,e^-$

Cathode: $O_2(g) + 4\,e^- \longrightarrow 2\,O^{2-}(s)$

(a) Write a balanced equation for the cell reaction.

(b) Use the thermodynamic data in Appendix B to calculate the values of $E°$ and the equilibrium constant K for the cell reaction at 25 °C. Will $E°$ and K increase, decrease, or remain the same on raising the temperature?

(c) How many grams of butane are required to produce a constant current of 10.5 A for 8.00 h? How many liters of gaseous butane at 20 °C and 815 mm Hg pressure are required?

17.139 The half-reactions that occur in ordinary alkaline batteries can be written as

Cathode: $MnO_2(s) + H_2O(l) + e^- \longrightarrow MnO(OH)(s) + OH^-(aq)$

Anode: $Zn(s) + 2\,OH^-(aq) \longrightarrow Zn(OH)_2(s) + 2\,e^-$

In 1999, researchers in Israel reported a new type of alkaline battery, called a "super-iron" battery. This battery uses the same anode reaction as an ordinary alkaline battery but involves the reduction of FeO_4^{2-} ion (from K_2FeO_4) to solid $Fe(OH)_3$ at the cathode.

(a) Use the following standard reduction potential and any data from Appendixes C and D to calculate the standard cell potential expected for an ordinary alkaline battery:

$$MnO(OH)(s) + H_2O(l) + e^- \longrightarrow Mn(OH)_2(s) + OH^-(aq) \qquad E° = -0.380\text{ V}$$

(b) Write a balanced equation for the cathode half-reaction in a super-iron battery. The half-reaction occurs in a basic environment.

(c) A super-iron battery should last longer than an ordinary alkaline battery of the same size and weight because its cathode can provide more charge per unit mass. Quantitatively compare the number of coulombs of charge released by the reduction of 10.0 g of K_2FeO_4 to $Fe(OH)_3$ with the number of coulombs of charge released by the reduction of 10.0 g of MnO_2 to MnO(OH).

17.140 Gold metal is extracted from its ore by treating the crushed rock with an aerated cyanide solution. The unbalanced equation for the reaction is

$$Au(s) + CN^-(aq) + O_2(g) \longrightarrow Au(CN)_2^-(aq)$$

(a) Balance the equation for this reaction in basic solution.

(b) Use any of the following data at 25 °C to calculate $\Delta G°$ for this reaction at 25 °C: K_f for $Au(CN)_2^- = 6.2 \times 10^{38}$, K_a for HCN $= 4.9 \times 10^{-10}$, and standard reduction potentials are

$$O_2(g) + 4\,H^+(aq) + 4\,e^- \longrightarrow 2\,H_2O(l) \qquad E° = 1.229\text{ V}$$
$$Au^{3+}(aq) + 3\,e^- \longrightarrow Au(s) \qquad E° = 1.498\text{ V}$$
$$Au^{3+}(aq) + 2\,e^- \longrightarrow Au^+(aq) \qquad E° = 1.401\text{ V}$$

17.141 Consider the redox titration of 100.0 mL of a solution of 0.010 M Fe^{2+} in 1.50 M H_2SO_4 with a 0.010 M solution of $KMnO_4$, yielding Fe^{3+} and Mn^{2+}. The titration is carried out in an electrochemical cell equipped with a platinum electrode and a calomel reference electrode consisting of an Hg_2Cl_2/Hg electrode in contact with a saturated KCl solution having $[Cl^-] = 2.9$ M. Using any data in Appendixes C and D, calculate the cell potential after addition of **(a)** 5.0 mL, **(b)** 10.0 mL, **(c)** 19.0 mL, and **(d)** 21.0 mL of the $KMnO_4$ solution.

CHAPTER 18

Hydrogen, Oxygen, and Water

Scott Niedermayer, the captain of the 2010 Canadian Olympic hockey team, in his Honda FCX hydrogen fuel-cell car. Hydrogen may be the fuel of the future.

Hydrogen is the most abundant element in the universe. It combines with many transition elements and with every main-group element except the noble gases and perhaps indium and thallium. Hydrogen forms more compounds than any other element. Industrially, large amounts of elemental hydrogen are produced for use in the synthesis of important chemicals such as ammonia and methanol. Because its only combustion product is water vapor, hydrogen is being considered as an environmentally clean fuel of the future. (See the *FYI* at the end of this chapter.)

Oxygen is the most abundant element on the Earth's surface. It is essential for respiration and is the oxidizing agent in the energy-generating combustion processes that maintain our industrialized civilization. Approximately 29 million metric tons of oxygen is produced annually in the United States, largely for use in making steel.

When chemically combined, hydrogen and oxygen yield water, the most familiar of all chemical compounds. Water is essential to life, and humans can survive for only a few days without it. In our bodies, water is the solvent for thousands of biochemical reactions, the medium for transport of nutrients to our cells and waste products away from our cells, and the substance that helps to regulate body temperature.

In this chapter, we'll take a detailed look at the chemistry of hydrogen and oxygen, and we'll discuss some of the properties of water, the most important solvent in both chemistry and biology.

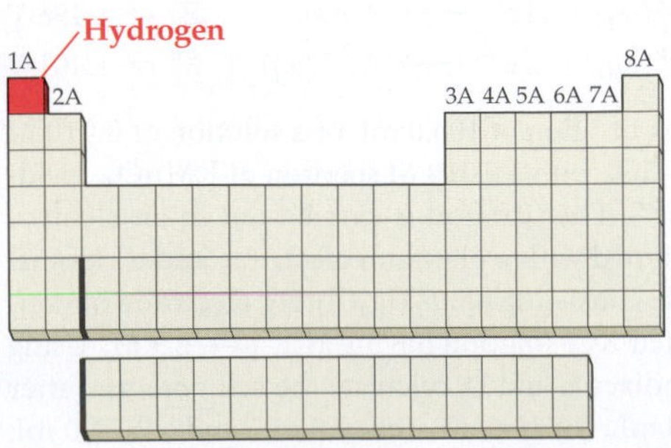

18.1 HYDROGEN

Henry Cavendish (1731–1810), an English chemist and physicist, was the first person to isolate hydrogen in pure form. Cavendish showed that the action of acids on metals, such as zinc, iron, and tin, produces a flammable gas that can be distinguished from other gases by its unusually low density:

$$2\,H^+(aq) + Zn(s) \longrightarrow H_2(g) + Zn^{2+}(aq)$$

CONTENTS

The French chemist Lavoisier called the gas "hydrogen," meaning "water former," because it combines with oxygen to produce water.

At ordinary temperatures and pressures, hydrogen is a colorless, odorless, and tasteless gas comprised of diatomic H_2 molecules. Because H_2 is a nonpolar molecule that contains only two electrons, intermolecular forces are extremely weak (Section 10.2). As a result, hydrogen has a very low melting point (−259.2 °C) and a very low boiling point (−252.8 °C). The bonding forces within the H_2 molecule are exceptionally strong, however, because of the small size of H atoms and the good overlap of their $1s$ orbitals: The H—H bond dissociation energy is 436 kJ/mol, greater than that for any other single bond between two atoms of the same element (Section 4.3):

$$H_2(g) \longrightarrow 2\,H(g) \qquad D = 436\ \text{kJ/mol}$$

Even at 2000 K, only 1 of every 2500 H_2 molecules is dissociated into H atoms at 1 atm pressure.

Hydrogen is thought to account for approximately 75% of the mass of the universe. Our Sun and other stars, for instance, are composed mainly of hydrogen, which serves as their nuclear fuel. On Earth, though, hydrogen is rarely found in uncombined form because the Earth's gravity is too weak to hold such a light molecule. The Earth's atmosphere contains only 0.53 ppm of H_2 by volume. In the Earth's crust and oceans, hydrogen is the ninth most abundant element on a mass basis (0.9 mass %) and the third most abundant on an atom basis (15.4 atom %). Hydrogen is found in water, petroleum, proteins, carbohydrates, fats, and many millions of other compounds.

As mentioned in Section 1.8, there are three **isotopes** of hydrogen: *protium*, or ordinary hydrogen (^1_1H); *deuterium*, or heavy hydrogen (^2_1H or D); and *tritium* (^3_1H or T), a radioactive isotope of hydrogen. Nearly all (99.985%) the atoms in naturally occurring hydrogen are protium. The terrestrial abundance of deuterium is only 0.015 atom %, and tritium is present only in trace amounts ($\sim 10^{-16}$ atom %). Although the masses of protium, deuterium, and tritium differ, their properties are similar because chemical behavior is determined primarily by electronic structure and all three isotopes have the same electron configuration ($1s^1$).

PROBLEM 18.1 Hydrogen is used to inflate weather balloons because it is much less dense than air. Calculate the density of gaseous H_2 at 25 °C and 1 atm pressure. Compare your result with the density of dry air under the same conditions (1.185×10^{-3} g/cm^3).

▲ The Lagoon Nebula in the constellation Sagittarius. These interstellar gas clouds consist largely of atomic hydrogen, the most abundant element in the universe. The gas is heated by radiation from nearby stars. Can you explain its characteristic red glow? (Recall Section 2.2.)

REMEMBER...

Atoms with identical atomic numbers (*Z*) but different mass numbers (*A*) are called **isotopes**. Isotopes of an element have the same number of protons but a different number of neutrons. (Section 1.8)

PROBLEM 18.2 The most abundant elements by mass in the body of a healthy human adult are oxygen (61.4%), carbon (22.9%), hydrogen (10.0%), and nitrogen (2.6%).
(a) Calculate the mass percent D if all the hydrogen atoms in a human were deuterium atoms.
(b) How much weight would a 150-pound person gain if all ^1H atoms were replaced by ^2H atoms?

▲ Electrolysis of water gives H_2 gas at one electrode and O_2 gas at the other electrode. Which gas is at which electrode?

18.2 PREPARATION AND USES OF HYDROGEN

Hydrogen of greater than 99.95% purity is made by the electrolysis of water. However, that process requires a large amount of energy—286 kJ per mole of H_2 produced—and thus is not economical for large-scale production.

$$2\,H_2O(l) \longrightarrow 2\,H_2(g) + O_2(g) \qquad \Delta H^\circ = +572\text{ kJ}$$

Small amounts of hydrogen are conveniently prepared in the laboratory by the reaction of dilute acid with an active metal such as zinc:

$$Zn(s) + 2\,H^+(aq) \longrightarrow H_2(g) + Zn^{2+}(aq)$$

A typical apparatus for generating and collecting hydrogen is shown in **Figure 18.1**.

FIGURE 18.1

Preparation of hydrogen by the reaction of zinc metal with dilute acid.

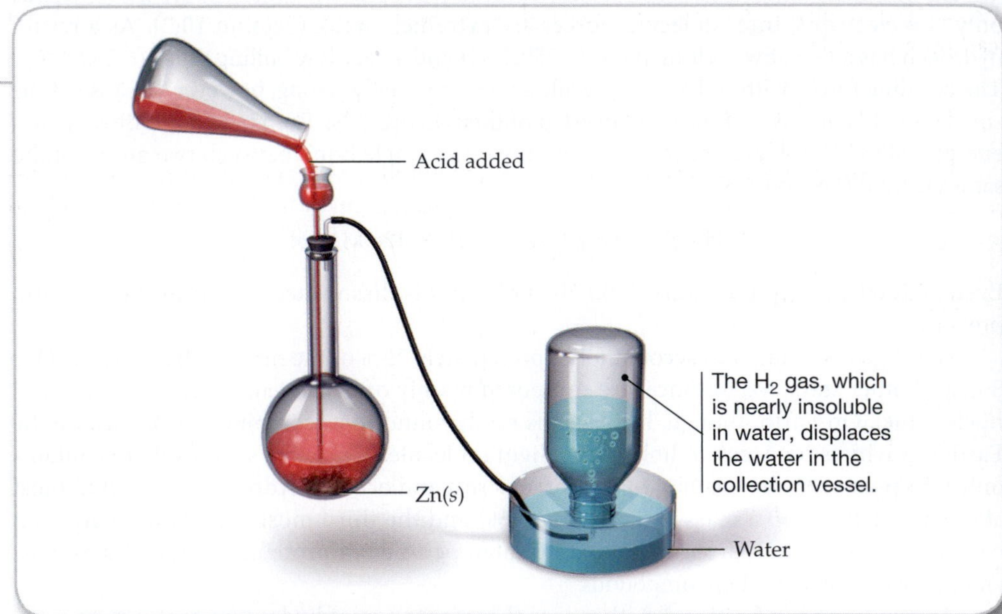

Because water is the cheapest and most readily available source of hydrogen, all large-scale, industrial methods for producing hydrogen use an inexpensive reducing agent such as hot iron, carbon, or methane (natural gas) to extract the oxygen from steam:

$$4\,H_2O(g) + 3\,Fe(s) \xrightarrow{\text{Heat}} Fe_3O_4(s) + 4\,H_2(g) \qquad \Delta H^\circ = -151\text{ kJ}$$

$$H_2O(g) + C(s) \xrightarrow{1000\,^\circ C} CO(g) + H_2(g) \qquad \Delta H^\circ = +131\text{ kJ}$$

At present, the most important industrial method for producing hydrogen is the three-step, **steam–hydrocarbon re-forming process**. The first step in the process is the conversion of steam and methane to a mixture of carbon monoxide and hydrogen known as *synthesis gas*, so-called because it can be used as the starting material for the synthesis of liquid fuels. The reaction requires high temperature, moderately high pressure, and a nickel catalyst:

$$H_2O(g) + CH_4(g) \xrightarrow[\text{Ni catalyst}]{1100\,^\circ C} CO(g) + 3\,H_2(g) \qquad \Delta H^\circ = +206\text{ kJ}$$

In the second step, the synthesis gas and additional steam are passed over a metal oxide catalyst at about 400 °C. Under these conditions, the carbon monoxide component of the synthesis gas and the steam are converted to carbon dioxide and more hydrogen. This reaction of CO with H_2O is called the **water-gas shift reaction** because it shifts the composition of synthesis gas by removing the toxic carbon monoxide and producing more of the economically important hydrogen:

$$CO(g) + H_2O(g) \xrightarrow[\text{Catalyst}]{400\,^\circ C} CO_2(g) + H_2(g) \qquad \Delta H^\circ = -41\text{ kJ}$$

Finally, the unwanted carbon dioxide is removed in a third step by passing the H_2/CO_2 mixture through a basic aqueous solution. This treatment converts the carbon dioxide to carbonate ion, which remains in the aqueous phase:

$$CO_2(g) + 2\,OH^-(aq) \longrightarrow CO_3^{2-}(aq) + H_2O(l)$$

Approximately 95% of the H_2 produced in industry is synthesized and consumed on the same site in industrial plants that manufacture other chemicals. The largest single consumer of hydrogen is the Haber process for synthesizing ammonia (Sections 13.7–13.11):

$$N_2(g) + 3\,H_2(g) \rightleftharpoons 2\,NH_3(g)$$

Large amounts of hydrogen are also used for the synthesis of methanol, CH_3OH, from carbon monoxide:

$$CO(g) + 2\,H_2(g) \xrightarrow[\text{Cu/ZnO/Al}_2\text{O}_3\text{ catalyst}]{250\,^\circ C} CH_3OH(g)$$

Methanol is an industrial solvent and a starting material for the manufacture of formaldehyde, CH_2O, used in making plastics. Annual U.S. production of methanol, about 1.3 billion gallons, consumes about 490,000 metric tons of hydrogen.

Worked Example 18.1

Writing Equations for the Preparation of Hydrogen

Write a balanced equation for the production of synthesis gas from butane, $C_4H_{10}(g)$.

STRATEGY

Synthesis gas is a mixture of H_2 and CO. It is produced when hydrocarbons, such as methane or butane, react with steam. Balance the equation.

SOLUTION

$$C_4H_{10}(g) + 4\,H_2O(g) \longrightarrow 9\,H_2(g) + 4\,CO(g)$$

PROBLEM 18.3 Write a balanced equation for the production of synthesis gas from each of the following:
(a) graphite **(b)** propane, $C_3H_8(g)$

18.3 REACTIVITY OF HYDROGEN

The hydrogen atom is the simplest of all atoms, containing only a single 1*s* electron and a single proton. In most versions of the periodic table, hydrogen is located in group 1A above the alkali metals because they too have just one valence electron. Alternatively, hydrogen could be placed in group 7A above the halogens because, like the halogens, it is just one electron short of a noble gas configuration (He). Thus, hydrogen has properties similar to those of both alkali metals and halogens. A hydrogen atom can lose an electron to form a hydrogen cation, H^+, or it can gain an electron to yield a hydride anion, H^-:

$$\text{Alkali-metal-like reaction:} \quad H(g) \longrightarrow H^+(g) + e^- \qquad E_i = +1312 \text{ kJ/mol}$$
$$\text{Halogen-like reaction:} \quad H(g) + e^- \longrightarrow H^-(g) \qquad E_{ea} = -73 \text{ kJ/mol}$$

Because the amount of energy needed to ionize a hydrogen atom is so large ($E_i = 1312$ kJ/mol), hydrogen doesn't completely transfer its valence electron in chemical reactions. Instead, it shares this electron with a nonmetallic element to give a covalent compound such as CH_4, NH_3, H_2O, or HF. In this regard, hydrogen differs markedly from the alkali metals, which have much smaller ionization energies, ranging from 520 kJ/mol for Li to 376 kJ/mol for Cs, and form ionic compounds with nonmetals.

Complete ionization of a hydrogen atom is possible in the gas phase, but in liquids and solids the bare proton is too reactive to exist by itself. Instead, it bonds to a molecule that has a lone pair of electrons. In water, for example, the proton bonds to an H_2O molecule to give a hydronium ion, $H_3O^+(aq)$.

Although adding an electron to hydrogen ($E_{ea} = -73$ kJ/mol) releases less energy than adding an electron to the halogens ($E_{ea} = -295$ to -349 kJ/mol), hydrogen will accept an electron from an active metal to give an ionic hydride, such as NaH or CaH_2. In this regard, the behavior of hydrogen parallels that of the halogens, which form ionic halides such as NaCl and $CaCl_2$.

At room temperature, H_2 is relatively unreactive because of its strong H—H bond, although it does react with F_2 to give HF and with Cl_2 in the presence of light to give HCl. Reactions of H_2 with O_2, N_2, or C, however, require high temperatures, the presence of a catalyst, or both. Catalysts such as metallic iron, nickel, palladium, or platinum facilitate the dissociation of H_2 into highly reactive H atoms (Section 12.15). The reaction of hydrogen and oxygen is highly exothermic, and gas mixtures that contain as little as 4% hydrogen by volume in air are highly flammable and potentially explosive.

$$2\,H_2(g) + O_2(g) \longrightarrow 2\,H_2O(l) \qquad \Delta H^\circ = -572 \text{ kJ}$$

▶ Explosive burning of the hydrogen-filled dirigible Hindenburg during landing at Lakehurst, New Jersey, on May 6, 1937, killed 36 of the 97 persons aboard. The Challenger space shuttle disaster was also caused by an explosion of hydrogen and oxygen.

PROBLEM 18.4 Write the chemical formula of a compound that contains hydrogen and each of the following elements, and tell which compounds are covalent and which are ionic.
(a) Silicon **(b)** Potassium **(c)** Selenium

18.4 BINARY HYDRIDES

The **binary hydrides** are compounds that contain hydrogen and just one other element. Formulas and melting points of the simplest hydrides of the main-group elements are listed in **Figure 18.2**. Binary hydrides can be classified as ionic, covalent, or metallic.

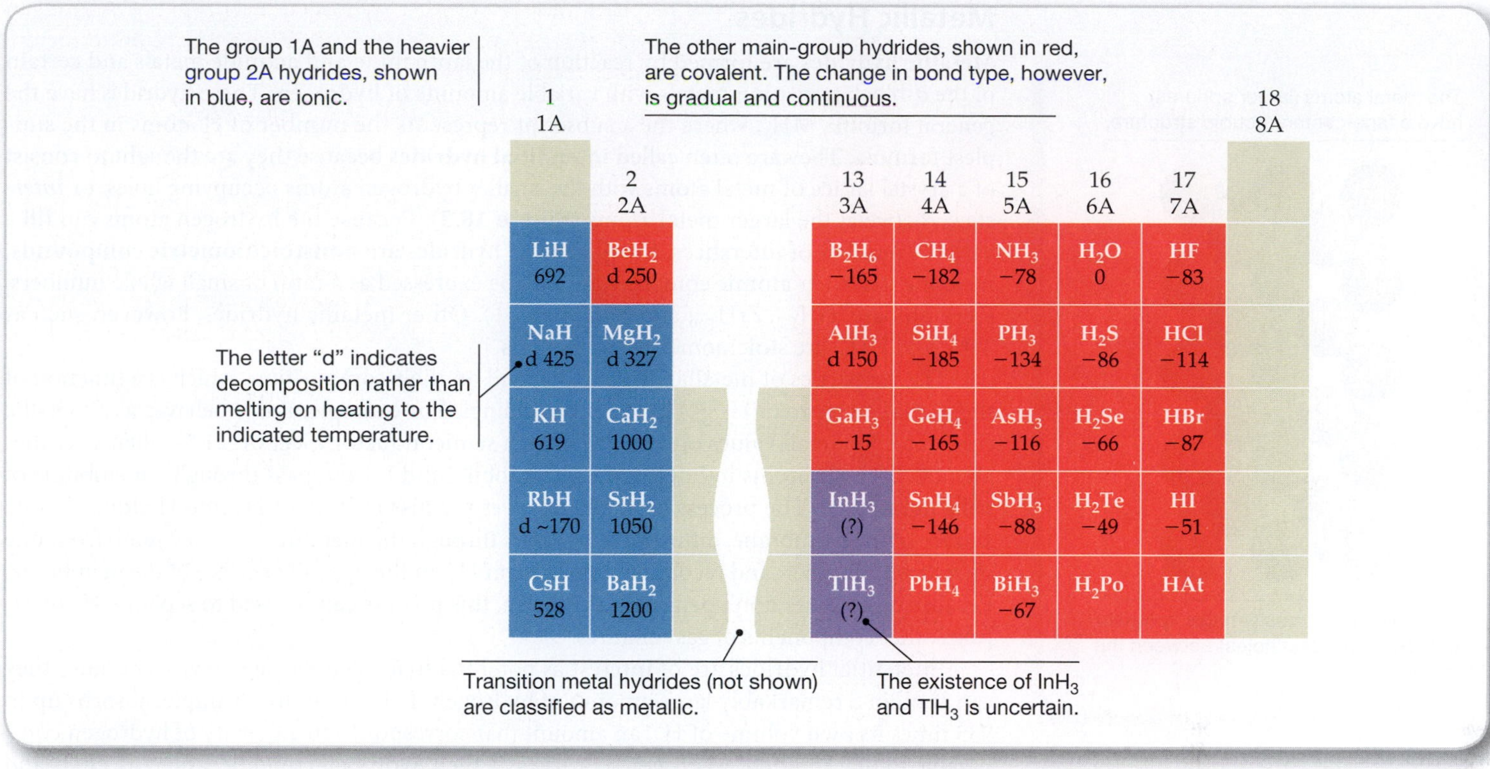

FIGURE 18.2

Formulas and melting points (°C) of the simplest hydrides of the main-group elements.

Ionic Hydrides

Ionic hydrides are saltlike, high-melting, white, crystalline compounds formed by the alkali metals and the heavier alkaline earth metals Ca, Sr, and Ba. They can be prepared by direct reaction of the elements at about 400 °C:

$$2\,Na(l) + H_2(g) \longrightarrow 2\,NaH(s) \qquad \Delta H° = -112.6\text{ kJ}$$
$$Ca(s) + H_2(g) \longrightarrow CaH_2(s) \qquad \Delta H° = -181.5\text{ kJ}$$

The alkali metal hydrides contain alkali metal cations and H^- anions in a face-centered cubic crystal structure like that of sodium chloride (Section 10.9).

The H^- anion is a good proton acceptor (Brønsted–Lowry base). Consequently, ionic hydrides react with water to give H_2 gas and OH^- ions:

$$CaH_2(s) + 2\,H_2O(l) \longrightarrow 2\,H_2(g) + Ca^{2+}(aq) + 2\,OH^-(aq)$$

This reaction of an ionic hydride with water is a redox reaction, as well as an acid–base reaction, because the hydride ion reduces the water (+1 oxidation state for H) to H_2 (0 oxidation state). In turn, the hydride ion (−1 oxidation state for H) is oxidized to H_2. In general, ionic hydrides are good reducing agents.

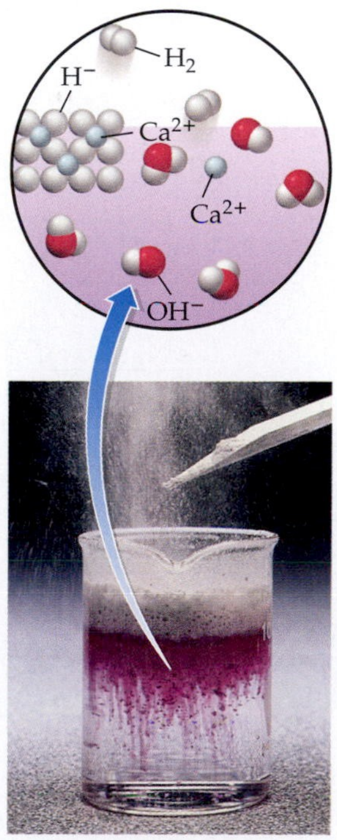

▲ Calcium hydride reacts with water to give bubbles of H_2 gas and OH^- ions. The red color is due to added phenolphthalein, which turns from colorless to red in a basic solution.

Covalent Hydrides

Covalent hydrides, as their name implies, are compounds in which hydrogen is attached to another element by a covalent bond. The most common examples are hydrides of nonmetallic elements, such as diborane (B_2H_6), methane (CH_4), ammonia (NH_3), water (H_2O), and the hydrogen halides (HX; X = F, Cl, Br, or I). Only the simplest covalent hydrides are listed in Figure 18.2, though more complex examples, such as hydrogen peroxide (H_2O_2) and hydrazine (N_2H_4), are also known. Because most covalent hydrides consist of discrete, small molecules that have relatively weak intermolecular forces, they are gases or volatile liquids at ordinary temperatures.

Metallic Hydrides

Metallic hydrides are formed by reaction of the lanthanide and actinide metals and certain of the *d*-block transition metals with variable amounts of hydrogen. These hydrides have the general formula MH_x, where the *x* subscript represents the number of H atoms in the simplest formula. They are often called **interstitial hydrides** because they are thought to consist of a crystal lattice of metal atoms with the smaller hydrogen atoms occupying holes, or *interstices*, between the larger metal atoms (**Figure 18.3**). Because the hydrogen atoms can fill a variable number of interstices, many metallic hydrides are **nonstoichiometric compounds**, meaning that their atomic composition can't be expressed as a ratio of small whole numbers. Examples are $TiH_{1.7}$, $ZrH_{1.9}$, and PdH_x ($x < 1$). Other metallic hydrides, however, such as TiH_2 and UH_3, are stoichiometric compounds.

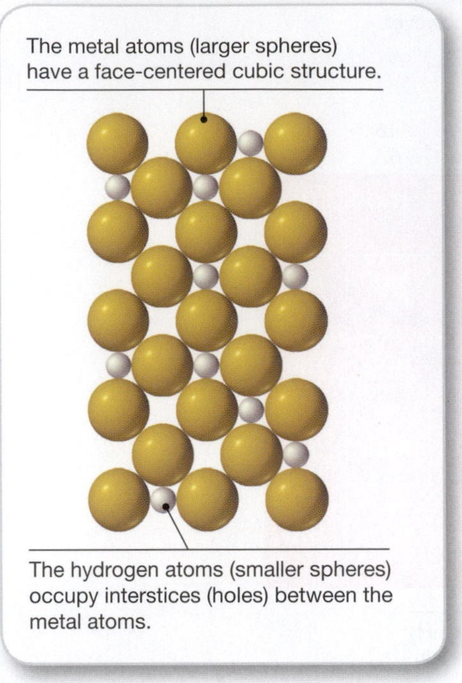

FIGURE 18.3

One plane of the structure of an interstitial metallic hydride.

The properties of metallic hydrides depend on their composition, which is a function of the partial pressure of H_2 gas in the surroundings. For example, PdH_x behaves as a metallic conductor for small values of *x* but becomes a semiconductor (Section 21.5) when *x* reaches about 0.5. The H atoms in PdH_x are highly mobile, and H_2 can pass through a membrane of palladium metal. The process probably involves the dissociation of H_2 into H atoms on one surface of the membrane, diffusion of H atoms through the membrane as they jump from one interstice to another, and recombination to form H_2 on the opposite surface of the membrane. Because other gases don't penetrate palladium, this process can be used to separate H_2 or D_2 from other components of gas mixtures.

Interstitial hydrides are of interest as potential hydrogen-storage devices because they can contain a remarkably large amount of hydrogen. Palladium, for example, absorbs up to 935 times its own volume of H_2, an amount that corresponds to a density of hydrogen comparable to that in liquid hydrogen. For use as a fuel, hydrogen could be stored as PdH_x and then liberated when needed simply by heating the PdH_x.

Favored at higher temperature $\quad Pd(s) + \frac{x}{2} H_2(g) \rightleftharpoons PdH_x(s) \quad$ Favored at lower temperature

This approach, however, may not be economical compared to other methods of storing and transporting hydrogen because of the high cost and weight of palladium.

Worked Example 18.2

Writing Equations for Reactions of Ionic Hydrides with Water

Write a balanced net ionic equation for the reaction of each of the following hydrides with water:

(a) Lithium hydride **(b)** Barium hydride

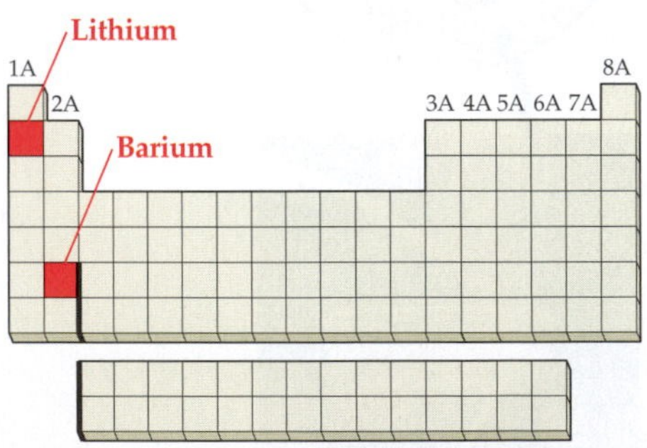

STRATEGY

Because lithium is a group 1A metal and barium is a group 2A metal, their hydrides have the formulas LiH and BaH_2, respectively. Both are ionic hydrides and therefore react with water to give H_2 gas and OH^- ions. The equations are balanced by trial and error or by the half-reaction method (Section 7.9).

SOLUTION

(a) $LiH(s) + H_2O(l) \longrightarrow H_2(g) + Li^+(aq) + OH^-(aq)$

Oxidation numbers: H in LiH: -1; H in H_2O: $+1$; H in H_2: 0

The H^- ion reduces water to H_2 gas, and in the process H^- is oxidized to H_2 gas.

(b) $BaH_2(s) + 2\,H_2O(l) \longrightarrow 2\,H_2(g) + Ba^{2+}(aq) + 2\,OH^-(aq)$

Oxidation numbers: H in BaH_2: -1; H in H_2O: $+1$; H in H_2: 0

PROBLEM 18.5 Write a balanced net ionic equation for the reaction of each of the following hydrides with water:

(a) Strontium hydride **(b)** Potassium hydride

WORKED CONCEPTUAL EXAMPLE 18.3

Determining Formulas and Properties of Binary Hydrides

The following pictures represent binary hydrides AH_x, where A = K, Ti, C, or F. Ivory spheres represent H atoms or ions, and purple spheres represent atoms or ions of the element A.

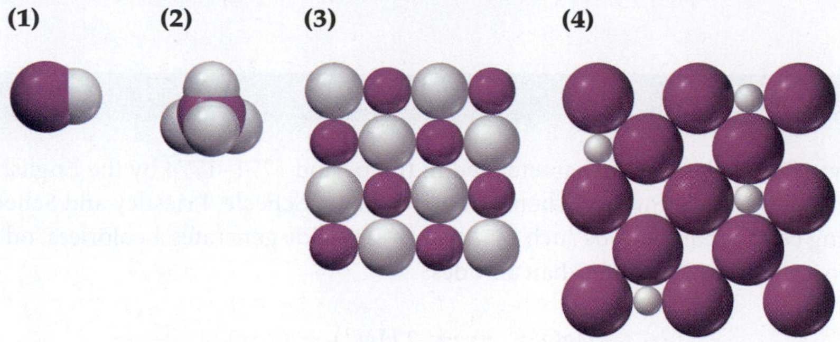

(a) Write the formula of each hydride, and classify each as ionic, covalent, or interstitial.
(b) Which hydride has the lowest melting point? Explain.
(c) Which hydride reacts with water at 25 °C to give H_2 gas?

STRATEGY

(a) The location of each element in the periodic table tells us the formula and the type of hydride it forms. Alkali metals form ionic hydrides and have the formula AH. Some transition elements form interstitial hydrides and often have nonstoichiometric formulas AH_x. Nonmetals form covalent hydrides and have formulas that depend on their periodic group.
(b) Covalent hydrides have low melting points because they have only relatively weak intermolecular forces.
(c) Ionic hydrides react with water to give H_2 gas.

SOLUTION

(a) Pictures (1) and (2) represent discrete AH and AH_4 molecules, respectively, and are therefore covalent hydrides HF and CH_4 of the nonmetallic elements F and C. Picture (3) shows a slice of the face-centered-cubic crystal structure of a solid ionic hydride with formula AH. Thus, hydride (3) is the ionic hydride KH. Picture (4) represents a slice of the structure of a nonstoichiometric interstitial hydride, which must be TiH_x.
(b) The lowest melting point belongs to one of the covalent hydrides, HF or CH_4. Polar HF molecules have dipole–dipole forces, hydrogen-bonding forces, and dispersion forces, while nonpolar CH_4 molecules have only dispersion forces. Thus, CH_4 has the lowest melting point.
(c) KH, the only ionic hydride, reacts with water at 25 °C to give H_2 gas:

$$KH(s) + H_2O(l) \longrightarrow H_2(g) + K^+(aq) + OH^-(aq).$$

CONCEPTUAL PROBLEM 18.6 The following pictures represent binary hydrides AH_x, where A = Br, Li, P, or Zr. Ivory spheres represent H atoms or ions, and purple spheres represent atoms or ions of the element A.

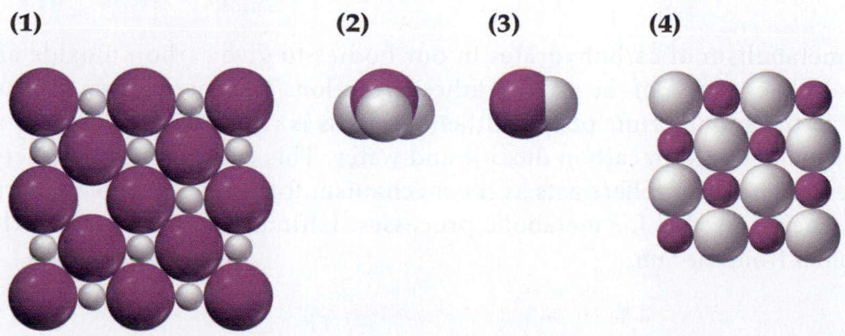

(a) Write the formula of each hydride, and classify each as ionic, covalent, or interstitial.
(b) Which hydrides are likely to be solids at 25 °C, and which are likely to be gases? Explain.
(c) Which hydride reacts with water at 25 °C to give a basic solution?

PROBLEM 18.7 If palladium metal (density 12.0 g/cm^3) dissolves 935 times its own volume of H_2 at STP, what is the value of x in the formula PdH_x? What is the density of hydrogen in PdH_x? What is the molarity of H atoms in PdH_x? Assume that the volume of palladium is unchanged when the H atoms go into the interstices.

18.5 OXYGEN

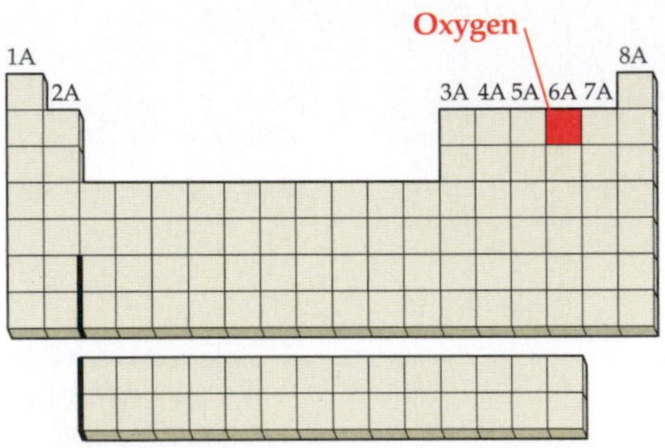

Oxygen was first isolated and characterized in the period 1771–1774 by the English chemist Joseph Priestley and the Swedish chemist Karl Wilhelm Scheele. Priestley and Scheele found that heating certain compounds such as mercury(II) oxide generates a colorless, odorless gas that supports combustion better than air does:

$$2\,HgO(s) \xrightarrow{\text{Heat}} 2\,Hg(l) + O_2(g)$$

Priestley called the gas "dephlogisticated air," but Lavoisier soon recognized it as an element and named it "oxygen," meaning "acid-former."

Gaseous O_2 condenses at −183 °C to form a pale blue liquid and freezes at −219 °C to give a pale blue solid. In all three phases—gas, liquid, and solid—O_2 is paramagnetic, as illustrated previously in Figure 5.11. The bond length in O_2 is 121 pm, appreciably shorter than the O—O single bond in H_2O_2 (148 pm), and the bond dissociation energy of O_2 (498 kJ/mol) is intermediate between that for the single bond in F_2 (159 kJ/mol) and the triple bond in N_2 (945 kJ/mol). These properties are consistent with the presence of a **double bond in O_2** (Section 5.6).

REMEMBER...
Molecular orbital theory predicts the presence of both paramagnetism and a **double bond** in the O_2 molecule. (Section 5.6)

Oxygen is the most abundant element on the surface of our planet and is crucial to human life. It's in the air we breathe, the water we drink, and the food we eat. It's the oxidizing agent in the metabolic "burning" of foods, and it's an important component of biological molecules: Approximately one-fourth of the atoms in living organisms are oxygen.

▲ Liquid oxygen has a pale blue color.

On a mass basis, oxygen constitutes 23% of the atmosphere (21% by volume), 46% of the lithosphere (the Earth's crust), and more than 85% of the hydrosphere. In the atmosphere, oxygen is found primarily as O_2, sometimes called *dioxygen*. The oxygen in the hydrosphere is in the form of H_2O, but enough dissolved O_2 is typically present within the water to maintain aquatic life. In the lithosphere, oxygen is combined with other elements in crustal rocks composed of silicates, carbonates, oxides, and other oxygen-containing minerals.

The amount of oxygen in the atmosphere remains fairly constant at about 1.18×10^{18} kg because the combustion and respiration processes that remove O_2 are balanced by photosynthesis, the complex process in which green plants use solar energy to produce O_2 and glucose from carbon dioxide and water:

$$6\,CO_2 + 6\,H_2O \xrightarrow{h\nu} 6\,O_2 + \underset{\text{Glucose}}{C_6H_{12}O_6}$$

The metabolism of carbohydrates in our bodies to give carbon dioxide and water is essentially the reverse of the photosynthesis reaction. The energy from the Sun that is absorbed in the endothermic photosynthetic process is released when organic matter is burned or metabolized to carbon dioxide and water. This cycling of oxygen between the atmosphere and the biosphere acts as the mechanism for converting solar energy to the chemical energy needed for metabolic processes. Ultimately, therefore, nearly all our energy comes from the Sun.

18.6 PREPARATION AND USES OF OXYGEN

Small amounts of O_2 can be prepared in the laboratory by the thermal decomposition of an oxoacid salt, such as potassium chlorate, $KClO_3$:

$$2\,KClO_3(s) \xrightarrow[MnO_2\text{ catalyst}]{\text{Heat}} 2\,KCl(s) + 3\,O_2(g)$$

Because oxygen is relatively insoluble in water, it can be collected by water displacement using the same apparatus employed to collect hydrogen (Figure 18.1). Oxygen is seldom prepared in the laboratory, though, because it's commercially available as a compressed gas in high-pressure steel cylinders.

Oxygen is produced on an industrial scale, along with nitrogen and argon, by the fractional distillation of liquefied air. When liquid air warms in a suitable distilling column, the more volatile components—nitrogen (bp −196 °C) and argon (bp −186 °C)—can be removed as gases from the top of the column. The less volatile oxygen (bp −183 °C) remains as a liquid at the bottom. Annual production of oxygen in the United States is approximately 29 million metric tons; only sulfuric acid and nitrogen are produced in greater quantities.

More than two-thirds of the oxygen produced industrially is used in making **steel**. Among its other uses, oxygen is employed in sewage treatment to destroy malodorous compounds and in paper bleaching to oxidize compounds that impart unwanted colors. In the oxyacetylene torch, the highly exothermic reaction between O_2 and acetylene provides the high temperatures (>3000 °C) needed for cutting and welding metals:

$$\underset{\text{Acetylene}}{2\,HC{\equiv}CH(g)} + 5\,O_2(g) \longrightarrow 4\,CO_2(g) + 2\,H_2O(g) \qquad \Delta H° = -2512\text{ kJ}$$

In all its applications, O_2 serves as an inexpensive and readily available oxidizing agent.

▲ Crude iron is converted to steel by oxidizing impurities with O_2 gas.

LOOKING AHEAD...

We'll see in Section 21.3 that the impure iron obtained by reduction of iron ore is purified and converted to **steel** by exposing the hot, molten metal to a jet of pure oxygen.

18.7 REACTIVITY OF OXYGEN

We can anticipate the reactivity of oxygen from the electron configuration of an oxygen atom ($1s^2\,2s^2\,2p^4$) and its high electronegativity (3.5). With six valence electrons, oxygen is just two electrons short of the octet configuration of neon, the next noble gas. Oxygen can therefore achieve an octet configuration either by accepting two electrons from an active metal or by gaining a share in two additional electrons through covalent bonding. Thus, oxygen reacts with active metals, such as lithium and magnesium, to give *ionic oxides*:

$$4\,Li(s) + O_2(g) \longrightarrow 2\,Li_2O(s)$$
$$2\,Mg(s) + O_2(g) \longrightarrow 2\,MgO(s)$$

On the other hand, with nonmetals, such as hydrogen, carbon, sulfur, and phosphorus, oxygen forms *covalent oxides*:

$$2\,H_2(g) + O_2(g) \longrightarrow 2\,H_2O(l)$$
$$C(s) + O_2(g) \longrightarrow CO_2(g)$$
$$S_8(s) + 8\,O_2(g) \longrightarrow 8\,SO_2(g)$$
$$P_4(s) + 5\,O_2(g) \longrightarrow P_4O_{10}(s)$$

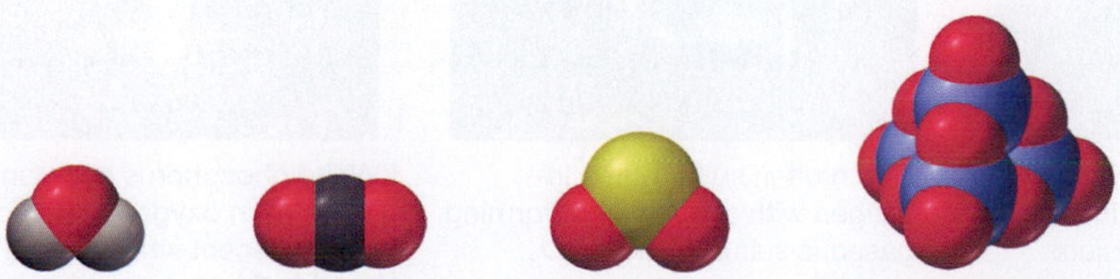

Water Carbon dioxide Sulfur dioxide Tetraphosphorus decoxide

In covalent compounds, oxygen generally achieves an octet configuration either by forming two single bonds, as in H_2O, or one double bond, as in CO_2. Oxygen often forms a double bond to small atoms such as carbon and nitrogen because there is good **π overlap** between the relatively compact $2p$ orbitals of second-row atoms (Section 5.4). With larger atoms such as silicon, however, there is less efficient π overlap, and double bond formation is therefore less common (**Figure 18.4**).

REMEMBER ...

Sigma (σ) bonds arise from head-on, σ overlap of orbitals. Pi (π) bonds arise from parallel, sideways, **π overlap** of orbitals. (Section 5.4)

FIGURE 18.4

Pi overlap between the p orbitals of oxygen and other atoms. Why does oxygen form a double bond with a small atom like carbon but a single bond with a larger atom like silicon?

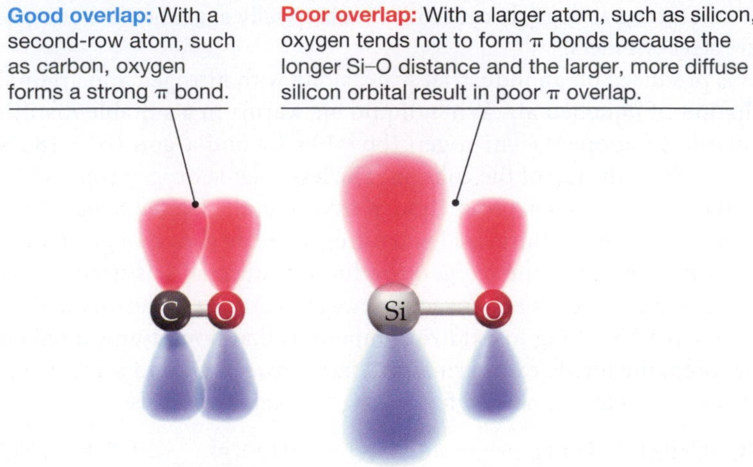

Oxygen reacts directly with all the elements in the periodic table except the noble gases and a few inactive metals, such as platinum and gold. Fortunately, most of these reactions are slow at room temperature; otherwise, cars would rust faster and many elements would spontaneously burst into flame in air. At higher temperatures, however, oxygen is extremely reactive. **Figure 18.5** illustrates reactions of oxygen with magnesium, sulfur, and white phosphorus. The brilliant white light emitted by burning magnesium makes this element useful in fireworks.

FIGURE 18.5

Some reactions of oxygen.

Magnesium metal burns in oxygen with a bright white flame, producing a white smoke of solid magnesium oxide, MgO.

Hot, molten sulfur burns in oxygen with a blue flame, forming gaseous sulfur dioxide, SO_2.

White phosphorus spontaneously inflames in oxygen, yielding an incandescent white smoke of solid P_4O_{10}.

18.8 OXIDES

Binary oxygen-containing compounds can be classified on the basis of oxygen's oxidation state. Binary compounds with oxygen in the –2 oxidation state are called **oxides**, compounds with oxygen in the –1 oxidation state are **peroxides**, and compounds with oxygen in the −1/2 oxidation state are **superoxides**. We'll look first at oxides and then consider peroxides and superoxides in Section 18.9.

Oxides can be categorized as basic, acidic, or amphoteric, as shown in **Figure 18.6**. Amphoteric oxides, like the corresponding **amphoteric hydroxides** described in Section 15.12, exhibit both acidic and basic properties.

REMEMBER...

Some hydroxides, such as $Al(OH)_3$, $Cr(OH)_3$, $Zn(OH)_2$, $Sn(OH)_2$, and $Pb(OH)_2$, are said to be **amphoteric** because they are soluble in both strongly acidic and strongly basic solutions. (Section 15.12)

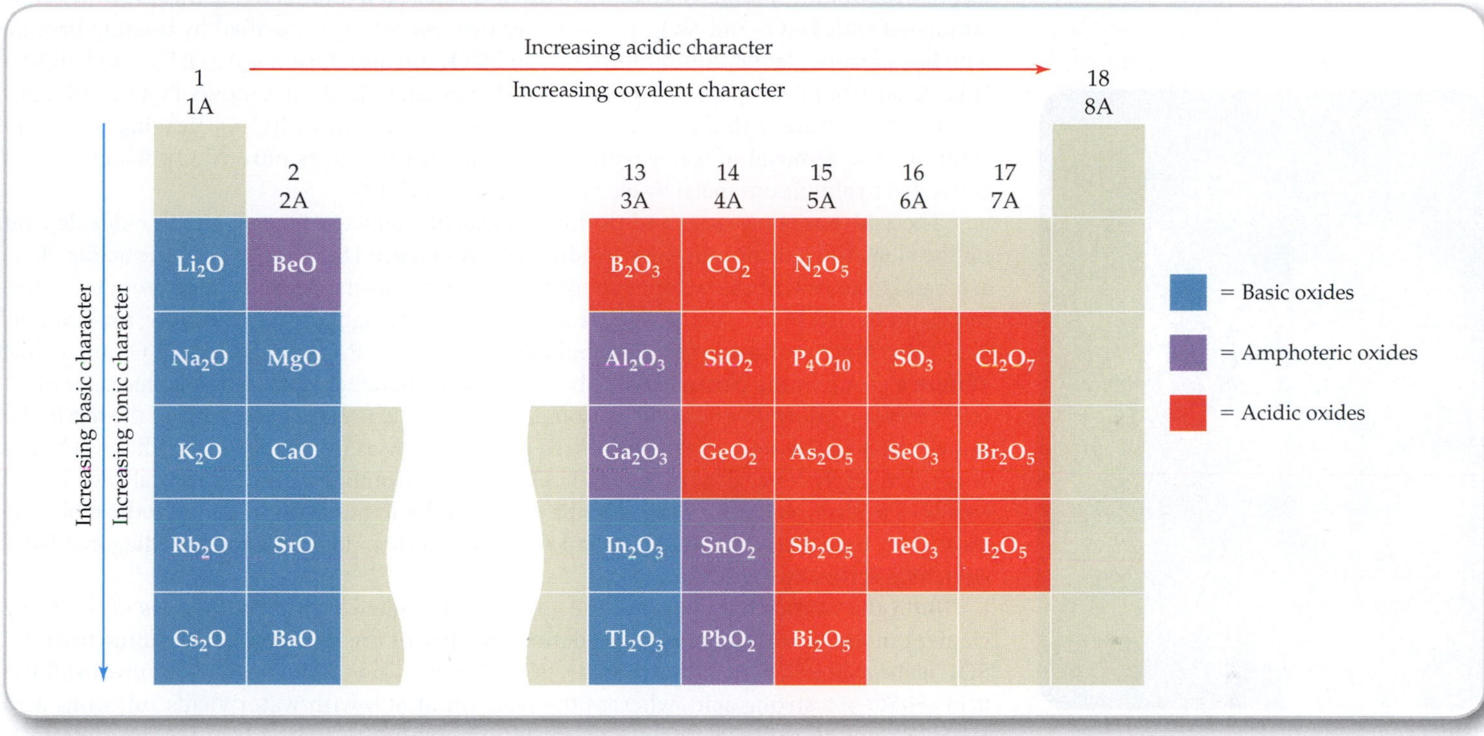

FIGURE 18.6

Formulas, acid–base properties, and the covalent–ionic character of the oxides of main-group elements in their highest oxidation states.

Basic oxides, also called base anhydrides, are ionic and are formed by metals on the left side of the periodic table. Water-soluble basic oxides, such as Na_2O, dissolve by reacting with water to produce OH^- ions:

$$Na_2O(s) + H_2O(l) \longrightarrow 2\,Na^+(aq) + 2\,OH^-(aq)$$

Water-insoluble basic oxides, such as MgO, can dissolve in strong acids because H^+ ions from the acid combine with the O^{2-} ion to produce water:

$$MgO(s) + 2\,H^+(aq) \longrightarrow Mg^{2+}(aq) + H_2O(l)$$

Acidic oxides, also called acid anhydrides, are covalent and are formed by the nonmetals on the right side of the periodic table. Water-soluble acidic oxides, such as N_2O_5, dissolve by reacting with water to produce aqueous H^+ ions:

$$N_2O_5(s) + H_2O(l) \longrightarrow 2\,H^+(aq) + 2\,NO_3^-(aq)$$

Water-insoluble acidic oxides, such as SiO_2, can dissolve in strong bases:

$$SiO_2(s) + 2\,OH^-(aq) \longrightarrow SiO_3^{2-}(aq) + H_2O(l)$$

Amphoteric oxides can behave both as bases and as acids. For example, Al_2O_3 is insoluble in water, but it dissolves both in strong acids and in strong bases. Al_2O_3 behaves as a base when it reacts with acids, giving the Al^{3+} ion, but it behaves as an acid when it reacts with bases, yielding the aluminate ion, $Al(OH)_4^-$.

Basic behavior: $Al_2O_3(s) + 6\,H^+(aq) \longrightarrow 2\,Al^{3+}(aq) + 3\,H_2O(l)$

Acidic behavior: $Al_2O_3(s) + 2\,OH^-(aq) + 3\,H_2O(l) \longrightarrow 2\,Al(OH)_4^-(aq)$

The elements that form amphoteric oxides have intermediate electronegativities, and the bonds in their oxides have intermediate ionic–covalent character.

The dissolution of Al_2O_3 in base is important in the production of aluminum metal from its ore. Aluminum is mined as bauxite ($Al_2O_3 \cdot x\,H_2O$), a hydrated oxide that is always contaminated with Fe_2O_3 and SiO_2. In the **Bayer process**, Al_2O_3 is purified by treating bauxite with hot aqueous NaOH. Aluminum oxide and SiO_2 dissolve, forming $Al(OH)_4^-$ and SiO_3^{2-} (silicate ion), but Fe_2O_3 remains undissolved. Subsequent filtration removes Fe_2O_3, and treatment of the filtrate with a weak acid (CO_2 in air) precipitates $Al(OH)_3$, leaving SiO_3^{2-} in solution. The removal of water from $Al(OH)_3$ by heating gives pure Al_2O_3, which is then converted to aluminum metal by electrolysis (Section 17.13).

The acid–base properties and the ionic–covalent character of an element's oxide depend on the element's position in the periodic table. As Figure 18.6 shows, both the acidic character and the covalent character of an oxide increase across the periodic table, from the active metals on the left to the electronegative nonmetals on the right. In the third row, for example, Na_2O and MgO are basic, Al_2O_3 is amphoteric, and SiO_2, P_4O_{10}, SO_3, and Cl_2O_7 are acidic. Within a group in the periodic table, both the basic character and the ionic character of an oxide increase going down the table, from the more electronegative elements at the top to the less electronegative ones at the bottom. In group 3A, for example, B_2O_3 is acidic, Al_2O_3 and Ga_2O_3 are amphoteric, and In_2O_3 and Tl_2O_3 are basic. Combining the horizontal and vertical trends in acidity, we find the most acidic oxides in the upper right of the periodic table, the most basic oxides in the lower left, and the amphoteric oxides in a roughly diagonal band stretching across the middle.

▲ Telephone messages are transmitted by light passing through optical fibers made of SiO_2.

Both the acidic character and the covalent character of different oxides of the same element increase with increasing oxidation number of the element. Thus, sulfur trioxide, SO_3, is more acidic than sulfur dioxide, SO_2. The reaction of SO_3 with water gives sulfuric acid (H_2SO_4), a strong acid, whereas the reaction of SO_2 with water yields sulfurous acid (H_2SO_3), a weak acid. The oxides of chromium exhibit the same trend. Chromium(VI) oxide (CrO_3) is acidic, chromium(III) oxide (Cr_2O_3) is amphoteric, and chromium(II) oxide (CrO) is basic.

As the bonding in oxides gradually changes from ionic to covalent, a corresponding change in structure occurs, from extended three-dimensional structures to discrete molecular structures. This results in a change in physical properties, as illustrated by the melting points in **Figure 18.7**. MgO, an ionic oxide with the face-centered cubic NaCl crystal structure, has a very high melting point because of its high **lattice energy** (Section 3.10). Al_2O_3 and SiO_2 also have **extended three-dimensional structures** (Section 10.10), but they exhibit increasingly covalent character and have increasingly lower melting points. The trend continues to P_4O_{10}, SO_3, and Cl_2O_7, which are molecular substances with strong covalent bonds within each molecule but only weak intermolecular forces between molecules. Thus, P_4O_{10} is a volatile, relatively low-melting solid, and SO_3 and Cl_2O_7 are volatile liquids.

REMEMBER...

The **lattice energy** is the amount of energy that must be supplied to break up an ionic crystal into its individual gaseous ions, so it is the measure of the strength of the crystal's ionic bonds. (Section 3.10)

REMEMBER...

SiO_2 has a covalent network structure in which each Si atom is bonded to four O atoms and each O atom is bonded to two Si atoms to give an **extended three-dimensional array**. The entire crystal can be regarded as one giant molecule. (Section 10.10)

The uses of oxides are determined by their properties. Because of their high thermal stability, mechanical strength, and electrical resistance, MgO and Al_2O_3 are used as high-temperature electrical insulators in products such as electrical heaters and automobile spark plugs. SiO_2 is the main component of the optical fibers used for communications. The acidic oxides of the nonmetals are important as precursors to industrial acids, such as HNO_3, H_3PO_4, and H_2SO_4.

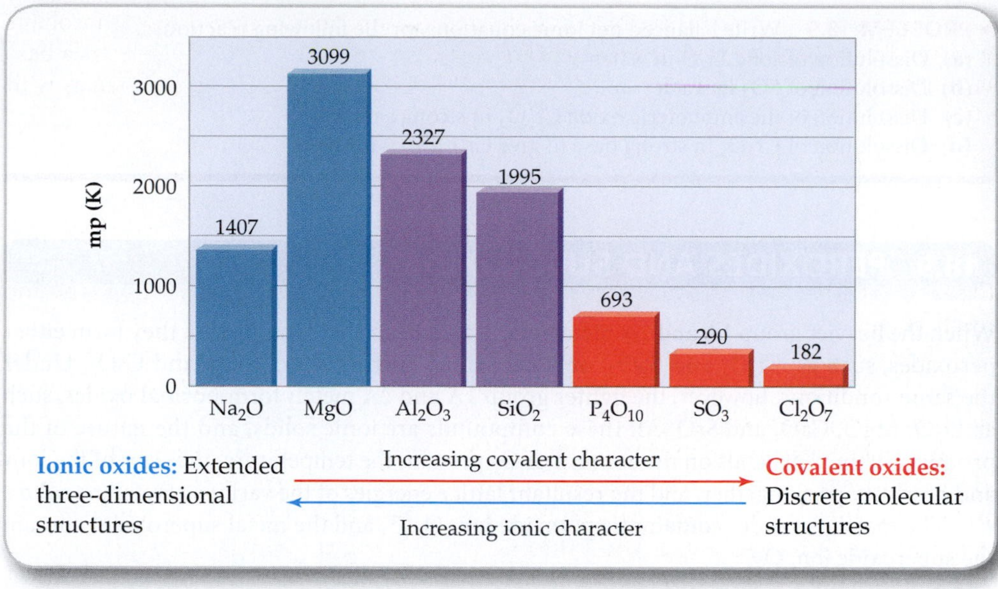

FIGURE 18.7

Melting points (in kelvin) of oxides of the third-period elements in their highest oxidation states.

WORKED CONCEPTUAL EXAMPLE 18.4

Determining Formulas and Properties of Oxides

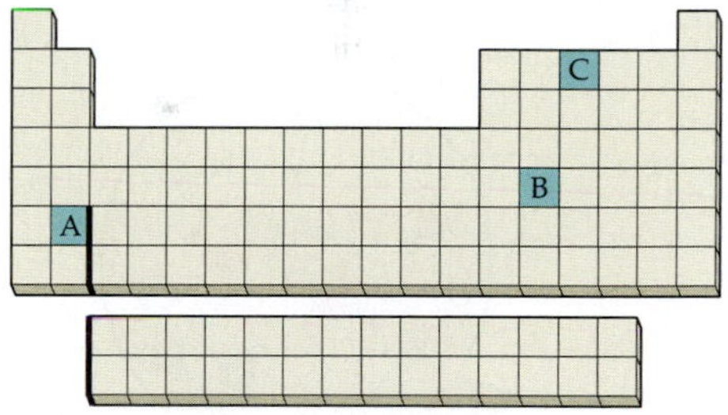

Look at the location of elements A, B, and C in the periodic table.

(a) Write the formula of the oxide that has each of these elements in its highest oxidation state.

(b) Which oxide is the most ionic, and which is the most covalent?

(c) Classify each oxide as basic, acidic, or amphoteric.

STRATEGY

(a) The oxidation number of a main-group element in its highest oxidation state and the formula of its oxide are determined by the element's location in the periodic table. For example, an element M in group 1A has an oxidation number of +1, and its oxide has formula M_2O because the oxidation number of oxygen in an oxide is −2 and the oxidation numbers of all the atoms in a neutral compound must sum to zero.

(b) Recall that ionic character decreases and covalent character increases in the periodic table from left to right and from bottom to top.

(c) Recall that basic character decreases and acidic character increases in the periodic table from left to right and from bottom to top.

SOLUTION

(a) Elements A, B, and C can be identified as Ba, Sn, and N, respectively. Because Ba, Sn, and N are in groups 2A, 4A, and 5A, respectively, the formulas of their highest-oxidation-state oxides are BaO, SnO_2, and N_2O_5.

(b) BaO is the most ionic because Ba is in the lower left region of the periodic table, and N_2O_5 is the most covalent because N is in the upper right region.

(c) BaO is basic, N_2O_5 is acidic, and SnO_2 is amphoteric. Note that Sn is one of five main-group elements that form an amphoteric oxide (Figure 18.6).

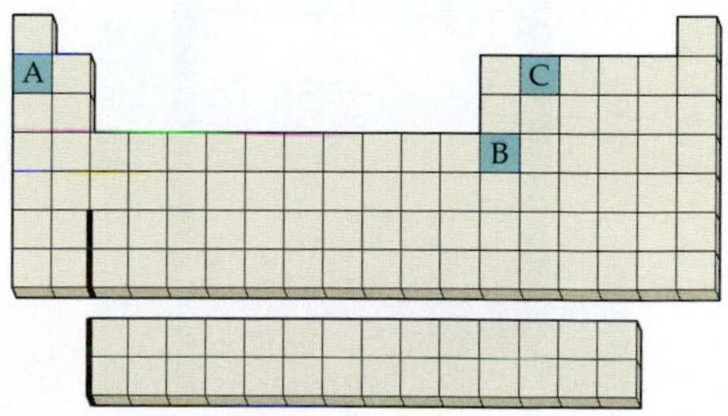

CONCEPTUAL PROBLEM 18.8 Look at the location of elements A, B, and C in the periodic table.

(a) Write the formula of the oxide that has each of these elements in its highest oxidation state.

(b) Which oxide is the most ionic, and which is the most covalent?

(c) Which oxide is the most acidic, and which is the most basic?

(d) Which oxide can react with both $H^+(aq)$ and $OH^-(aq)$?

PROBLEM 18.9 Write balanced net ionic equations for the following reactions:
(a) Dissolution of solid Li_2O in water
(b) Dissolution of SO_3 in water
(c) Dissolution of the amphoteric oxide Cr_2O_3 in strong acid
(d) Dissolution of Cr_2O_3 in strong base to give $Cr(OH)_4^-$ ions

18.9 PEROXIDES AND SUPEROXIDES

When the heavier group 1A and 2A metals are heated in an excess of oxygen, they form either peroxides, such as Na_2O_2 and BaO_2, or superoxides, such as KO_2, RbO_2, and CsO_2. Under the same conditions, however, the lighter group 1A and 2A metals form normal oxides, such as Li_2O, MgO, CaO, and SrO. All these compounds are ionic solids, and the nature of the product obtained depends on the amount of O_2 present, the temperature, the sizes of the ions and how they pack together, and the resultant lattice energies of the various crystalline solids.

The metal peroxides contain the peroxide ion, O_2^{2-}, and the metal superoxides contain the superoxide ion, O_2^-:

REMEMBER...
The population of the **MOs of O_2** in order of increasing energy is $\sigma_{2s}^2\ \sigma^*{}_{2s}^2\ \sigma_{2p}^2\ \pi_{2p}^4\ \pi^*{}_{2p}^2$. Thus, O_2 has 8 bonding electrons, 4 antibonding electrons, and a bond order of 2. (Section 5.6)

Peroxide ion　　　Superoxide ion

The peroxide ion is diamagnetic and has an O–O single bond length of 149 pm. The superoxide ion, with one unpaired electron, is paramagnetic and has an O–O bond length of 128 pm, intermediate between the bond length of the peroxide ion (O_2^{2-}) and the O=O double bond length of 121 pm in O_2. This suggests a bond order between 1 and 2 for O_2^-.

The trend in bond lengths and the magnetic properties of O_2, O_2^-, and O_2^{2-} are all nicely explained by the **molecular orbital (MO) theory** discussed in Section 5.6. Recall that the occupied MOs of highest energy in O_2 are the two degenerate antibonding $\pi^*{}_{2p}$ orbitals. Each of these orbitals contains one electron, and the two electrons have the same spin, in accord with Hund's rule (Section 2.11). Thus, the O_2 molecule is paramagnetic and has a bond order of 2. The O_2^- ion has one additional electron (a total of 3) in the $\pi^*{}_{2p}$ orbitals, and the O_2^{2-} ion has two additional electrons (a total of 4) in the $\pi^*{}_{2p}$ orbitals. As the electron population of the antibonding $\pi^*{}_{2p}$ orbitals increases, the bond order decreases and the bond length increases (**Table 18.1**).

TABLE 18.1 Bond and Magnetic Properties of Diatomic Oxygen Species

Species	Number of $\pi^*{}_{2p}$ Electrons	Number of Unpaired Electrons	Bond Order	Bond Length (pm)	Magnetic Properties
O_2	2	2	2	121	Paramagnetic
O_2^-	3	1	1.5	128	Paramagnetic
O_2^{2-}	4	0	1	149	Diamagnetic

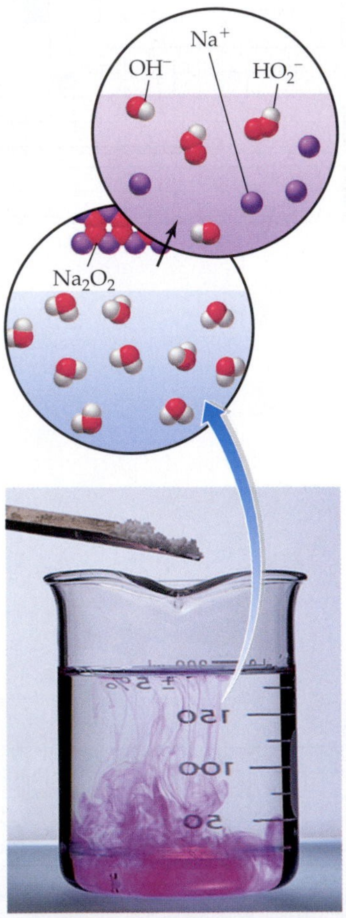

▲ Dissolution of sodium peroxide in water that contains phenolphthalein gives a red color due to formation of OH^- ions.

Like the oxide ion, the peroxide ion is a basic anion. When a metal peroxide dissolves in water, the O_2^{2-} ion picks up a proton from water, forming HO_2^- and OH^- ions. The dissolution of Na_2O_2 is typical:

$$Na_2O_2(s) + H_2O(l) \longrightarrow 2\,Na^+(aq) + HO_2^-(aq) + OH^-(aq)$$

In the presence of a strong acid, the O_2^{2-} ion combines with two protons, yielding hydrogen peroxide, H_2O_2. For example, aqueous solutions of H_2O_2 are conveniently prepared in the laboratory by the reaction of barium peroxide with a stoichiometric amount of sulfuric acid:

$$BaO_2(s) + H_2SO_4(aq) \longrightarrow BaSO_4(s) + H_2O_2(aq)$$

The byproduct, barium sulfate, is insoluble and easily removed by filtration.

When metal superoxides, such as KO_2, dissolve in water, they decompose with evolution of oxygen:

$$2\,KO_2(s) + H_2O(l) \longrightarrow O_2(g) + 2\,K^+(aq) + HO_2^-(aq) + OH^-(aq)$$

Oxidation numbers: KO_2: $-1/2$; O_2: 0; HO_2^-: -1

This decomposition is a redox reaction in which the oxygen in KO_2 is simultaneously oxidized from the $-1/2$ oxidation state in O_2^- to the 0 oxidation state in O_2 and reduced from the $-1/2$ oxidation state in O_2^- to the -1 oxidation state in HO_2^-. Such a reaction, in which a substance is both oxidized and reduced, is called a **disproportionation** reaction.

In the human body, the superoxide ion is a highly reactive and destructive species. Because it has one unpaired electron, it rapidly captures an electron from a bond in some other molecule, resulting in the breaking of that bond. The breaking of bonds in proteins, DNA, and other biologically important molecules can result in cancer, heart disease, and inflammatory disorders. Though the O_2^- ion is produced in small amounts as a byproduct of breathing, fortunately most of it is destroyed by enzymes. For more discussion of the role in the body of species having unpaired electrons—the so-called free radicals—see the *FYI* at the end of Chapter 4.

Worked Example 18.5

Identifying Oxides, Peroxides, and Superoxides

What is the oxidation number of oxygen in each of the following compounds? Tell whether the compound is an oxide, a peroxide, or a superoxide.

(a) KO_2 **(b)** BaO_2 **(c)** SiO_2

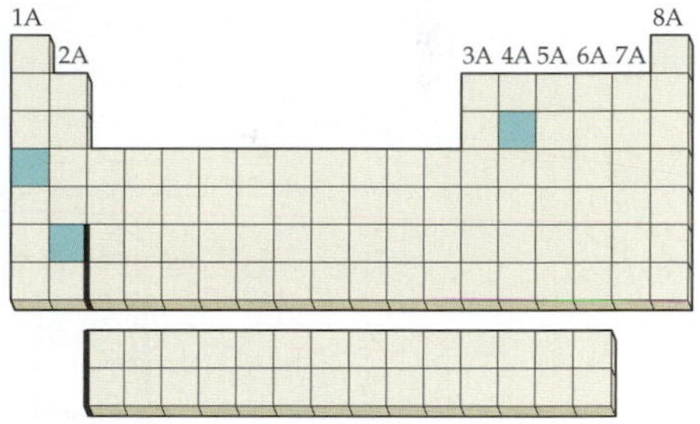

STRATEGY

First, determine the oxidation number of the other element, which is generally equal to its periodic group number. Then assign an oxidation number to oxygen so that the sum of the oxidation numbers is zero.

SOLUTION

(a) Because K in group 1A has an oxidation number of +1, the oxidation number of oxygen in KO_2 must be $-1/2$. Thus, KO_2 is a superoxide.

(b) Because Ba in group 2A has an oxidation number of +2, the oxidation number of oxygen in BaO_2 must be -1. Thus, BaO_2 is a peroxide.

(c) Because Si in group 4A has an oxidation number of +4, the oxidation number of oxygen in SiO_2 must be -2. Thus, SiO_2 is an oxide.

PROBLEM 18.10 What is the oxidation number of oxygen in each of the following compounds? Tell whether the compound is an oxide, a peroxide, or a superoxide.

(a) Rb_2O_2 **(b)** CaO **(c)** CsO_2 **(d)** SrO_2 **(e)** CO_2

PROBLEM 18.11 Write a balanced net ionic equation for the reaction of water with each of the oxygen compounds listed in Problem 18.10.

CONCEPTUAL PROBLEM 18.12 Draw a molecular orbital energy-level diagram for O_2^-, including the MOs derived from the oxygen 2*s* and 2*p* orbitals. Show the electron population of the MOs, and verify that O_2^- is paramagnetic and has a bond order of 1.5.

18.10 HYDROGEN PEROXIDE

Hydrogen peroxide (HOOH) is sold in drugstores as a 3% aqueous solution for domestic use and is marketed as a 30% aqueous solution for industrial and laboratory use. Because of its oxidizing properties, hydrogen peroxide is used as a mild antiseptic and as a bleach for textiles, paper pulp, and hair. In the chemical industry, hydrogen peroxide is a starting material for the synthesis of other peroxide compounds, some of which are used as catalysts in the manufacture of plastics.

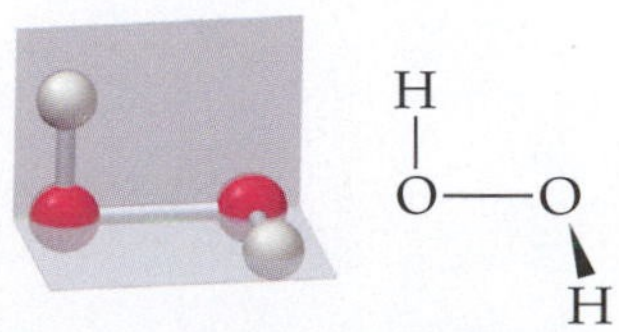

▲ The two H–O–O planes in the H_2O_2 molecule are nearly perpendicular to each other.

REMEMBER...

A **hydrogen bond** is an attractive interaction between an H atom bonded to a very electronegative atom (O, N, or F) and an unshared electron pair on another electronegative atom. Hydrogen bonding is responsible for the relatively high boiling points of H_2O, NH_3, and HF. (Section 10.2)

Pure hydrogen peroxide is an almost colorless, syrupy liquid that freezes at $-0.4\ °C$ and boils at an estimated 150 °C. The exact boiling point is not known with certainty, however, because pure H_2O_2 explodes when heated. The relatively high estimated boiling point indicates strong **hydrogen bonding** between H_2O_2 molecules in the liquid (Section 10.2). In aqueous solutions, hydrogen peroxide behaves as a weak acid, partially dissociating to give H_3O^+ and HO_2^- ions.

Hydrogen peroxide is both a strong oxidizing agent and a reducing agent:

$$H_2O_2(aq) + 2\,H^+(aq) + 2\,e^- \longrightarrow 2\,H_2O(l) \qquad E° = 1.78\ V$$

$$H_2O_2(aq) \longrightarrow O_2(g) + 2\,H^+(aq) + 2\,e^- \qquad E° = -0.70\ V$$

When hydrogen peroxide acts as an oxidizing agent, oxygen is reduced from the -1 oxidation state in H_2O_2 to the -2 oxidation state in H_2O (or OH^-). For example, hydrogen peroxide oxidizes Br^- to Br_2 in acidic solution:

$$H_2O_2(aq) + 2\,H^+(aq) + 2\,Br^-(aq) \longrightarrow 2\,H_2O(l) + Br_2(aq)$$

Oxidation numbers: -1 (O in H_2O_2), -1 (Br in Br^-), -2 (O in H_2O), 0 (Br in Br_2)

When hydrogen peroxide acts as a reducing agent, oxygen is oxidized from the -1 oxidation state in H_2O_2 to the 0 oxidation state in O_2. A typical example is the reaction of hydrogen peroxide with permanganate ion in acidic solution:

$$5\,H_2O_2(aq) + 2\,MnO_4^-(aq) + 6\,H^+(aq) \longrightarrow 5\,O_2(g) + 2\,Mn^{2+}(aq) + 8\,H_2O(l)$$

-1 (O in H_2O_2), $+7$ (Mn in MnO_4^-), 0 (O in O_2), $+2$ (Mn in Mn^{2+})

The reduction of manganese in this reaction from the $+7$ oxidation state in MnO_4^- to the $+2$ oxidation state in Mn^{2+} is accompanied by a beautiful color change (**Figure 18.8**).

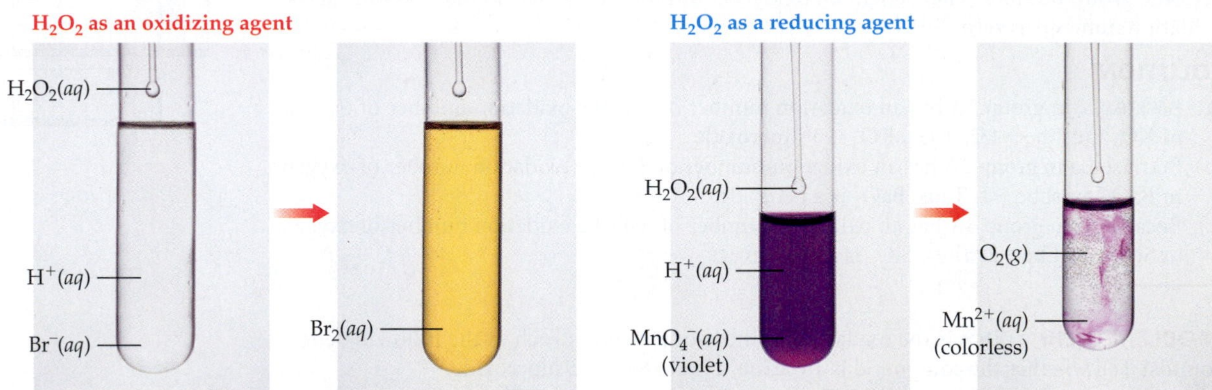

Addition of a colorless aqueous solution of H_2O_2 to a colorless acidic solution of NaBr produces a yellow-orange color due to formation of bromine, Br_2.

Addition of aqueous H_2O_2 to a violet acidic solution of $KMnO_4$ produces bubbles of O_2 gas and decolorizes the solution as the violet MnO_4^- ion is converted to the nearly colorless Mn^{2+} ion.

FIGURE 18.8

Some reactions of hydrogen peroxide.

Because it is both an oxidizing agent and a reducing agent, H_2O_2 can oxidize and reduce itself. Thus, hydrogen peroxide is unstable and undergoes disproportionation to water and oxygen:

$$2\,H_2O_2(aq) \longrightarrow 2\,H_2O(l) + O_2(g) \qquad E° = 1.08\ V$$

Oxidation numbers: -1 (O in H_2O_2), -2 (O in H_2O), 0 (O in O_2)

In the absence of a catalyst, the disproportionation reaction is too slow to be observed at room temperature. The rapid, exothermic, and potentially explosive decomposition of hydrogen peroxide is initiated, however, by heat and by a broad range of catalysts, including transition metal ions, certain anions (such as I^-), metal surfaces, blood (**Figure 18.9**), and even tiny particles of dust. Because decomposition is accelerated by light, hydrogen peroxide is stored in dark bottles. It is best handled in dilute aqueous solutions; concentrated solutions and the pure liquid are extremely hazardous materials.

The disproportionation of H_2O_2

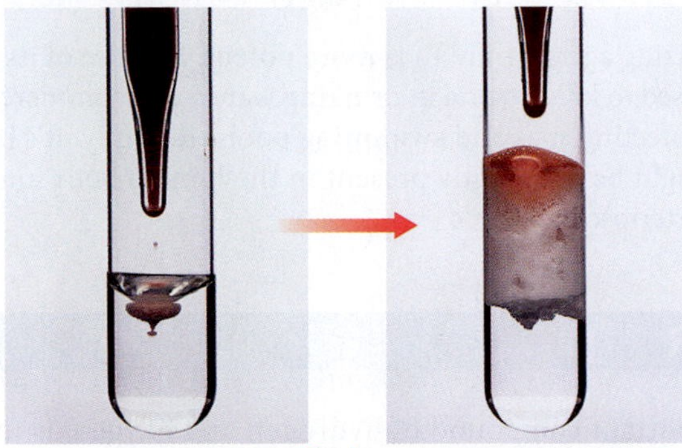

$$2\,H_2O_2(aq) \xrightarrow{\text{Catalase}} 2\,H_2O(l) + O_2(g)$$

FIGURE 18.9

Catalytic decomposition of hydrogen peroxide. When a few drops of blood are added to aqueous hydrogen peroxide (left), the hydrogen peroxide decomposes rapidly, evolving bubbles of oxygen that produce a thick foam (right). The reaction is catalyzed by *catalase*, an enzyme present in blood.

PROBLEM 18.13 Draw an electron-dot structure for H_2O_2. Is your structure consistent with its O–O bond length of 148 pm? (Look at Table 18.1 to see how bond length and bond order are related.)

PROBLEM 18.14 The discoloration and restoration of old oil paintings involves some interesting chemistry. On exposure to polluted air containing H_2S, white lead carbonate pigments are converted to PbS, a black solid. Hydrogen peroxide has been used to restore the original white color. Write a balanced equation for the reaction, which involves the oxidation of black PbS to white $PbSO_4$.

18.11 OZONE

Oxygen exists in two allotropes: ordinary dioxygen (O_2) and ozone (O_3). Ozone is a toxic, pale blue gas with a characteristic sharp, penetrating odor that you can detect at concentrations as low as 0.01 ppm. You may have noticed the odor of ozone after a severe electrical storm. Ozone is produced when an electric discharge passes through O_2, providing the energy needed to bring about the following endothermic reaction:

$$3\,O_2(g) \xrightarrow[\text{discharge}]{\text{Electric}} 2\,O_3(g) \qquad \Delta H° = +285 \text{ kJ}$$

Ozone is a pollutant in the lower atmosphere, but in the upper atmosphere it acts as a filter to protect us from harmful ultraviolet solar radiation (Section 9.9).

Although ozone can be prepared in the laboratory by passing O_2 through a strong electric field, it is unstable and decomposes exothermically to O_2. The decomposition of the dilute gas is slow, but the concentrated gas, liquid ozone (bp −111 °C), or solid ozone (mp −193 °C) can decompose explosively.

As discussed in Section 4.9, two **resonance structures** are required to explain the structure of the ozone molecule because the two O–O bonds have equal lengths:

REMEMBER...

Neither electron dot **resonance structure** on its own can account for the two equal O–O bond lengths in O_3. The actual electronic structure is a resonance hybrid (average) of the two resonance structures. (Section 4.9)

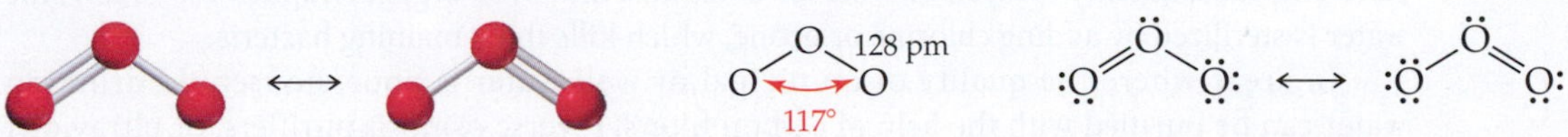

REMEMBER...

According to the **valence-shell electron-pair repulsion (VSEPR) model**, electrons in bonds and in lone pairs are treated as charge clouds that repel one another and stay as far apart as possible, thus causing molecules to assume specific shapes. (Section 5.1)

The bent structure is in accord with the **VSEPR model** described in Section 5.1, which predicts a bond angle near 120° for a triatomic molecule having a central atom surrounded by three charge clouds. The π electrons are shared by the three oxygen atoms, giving a net bond order of 1.5 between each pair of oxygen atoms (1 σ bond + 0.5 π bond). This bonding description is in agreement with the O–O bond length of 128 pm, intermediate between the lengths of an O–O single bond (148 pm) and an O=O double bond (121 pm).

Ozone is an extremely powerful oxidizing agent:

$$O_3(g) + 2\,H^+(aq) + 2\,e^- \longrightarrow O_2(g) + H_2O(l) \qquad E° = 2.08\text{ V}$$

Of the common oxidizing agents, only F_2 is more potent. Because of its oxidizing properties, ozone is sometimes used to kill bacteria in drinking water, and commercially available ozonators are used for disinfecting spas and swimming pools. Recently, it's been found that small amounts of ozone might be transiently present in the human body and may be involved in the development of arteriosclerosis.

18.12 WATER

Water, the most important compound of hydrogen and oxygen, is the most familiar and abundant compound on Earth. Nearly three-fourths of the Earth's surface is covered with water, and an estimated $1.35 \times 10^{18}\ m^3$ of water is present in the oceans. (It's interesting to note that the volume of the oceans in milliliters, 1.35×10^{24} mL, is roughly twice Avogadro's number.) Water accounts for nearly two-thirds of the mass of the adult human body and 93% of the mass of the human embryo in the first month.

Because water is essential to life as we know it, the recent exploration of the planet Mars has focused on a search for water. In early 2004, one of the Mars rovers discovered rocks that contain a high concentration of sulfate salts, as well as bromide and chloride salts, evidence that these rocks were formed by evaporation of seawater. The composition and surface features of the rocks point to a time when water was abundant on the surface of Mars.

On the Earth, approximately 97.3% of our vast supply of water is in the oceans. Most of the rest is in the form of polar ice caps and glaciers (2.0%) and underground fresh water (0.6%). Freshwater lakes and rivers account for less than 0.01% of the total, yet they still contain an enormous amount of water ($1.26 \times 10^{14}\ m^3$).

Seawater is unfit for drinking or agriculture because each kilogram contains about 35 g of dissolved salts. The most abundant salt in seawater is sodium chloride, but more than 60 different elements are present in small amounts. **Table 18.2** lists the ions that account for more than 99% of the mass of the dissolved salts. Although the oceans represent an almost unlimited source of chemicals, ion concentrations are so low that recovery costs are high. Only three substances are obtained from seawater commercially: sodium chloride, magnesium, and bromine.

TABLE 18.2 Major Ionic Constituents of Seawater

Ion	g/kg of Seawater
Cl^-	19.0
Na^+	10.5
SO_4^{2-}	2.65
Mg^{2+}	1.35
Ca^{2+}	0.40
K^+	0.38
HCO_3^-	0.14
Br^-	0.065

Water for use in homes, agriculture, and industry is generally obtained from freshwater lakes, rivers, or underground aquifers. The water you drink must be purified to remove solid particles, colloidal material, bacteria, and other harmful impurities. Important steps in a typical purification process include preliminary filtration, sedimentation, sand filtration, aeration, and sterilization (**Figure 18.10**).

The sedimentation, or settling, of suspended matter takes place in large settling tanks and is accelerated by the addition of lime, CaO, and aluminum sulfate, $Al_2(SO_4)_3$. The lime makes the water slightly basic, which precipitates the added Al^{3+} ions as aluminum hydroxide:

$$CaO(s) + H_2O(l) \longrightarrow Ca^{2+}(aq) + 2\,OH^-(aq)$$
$$Al^{3+}(aq) + 3\,OH^-(aq) \longrightarrow Al(OH)_3(s)$$

As the gelatinous precipitate of aluminum hydroxide slowly settles, it carries with it suspended solids, colloidal material, and most of the bacteria. The water is then filtered through a bed of sand and subsequently sprayed into the air to oxidize dissolved organic impurities. Finally, the water is sterilized by adding chlorine or ozone, which kills the remaining bacteria.

In areas where the quality of municipal or well water is poor, household drinking water can be purified with the help of carbon filters, reverse osmosis purifiers, or ultraviolet radiation. Reverse osmosis uses a pressure greater than the osmotic pressure of the impure water to force pure water through a semipermeable membrane (Section 11.9).

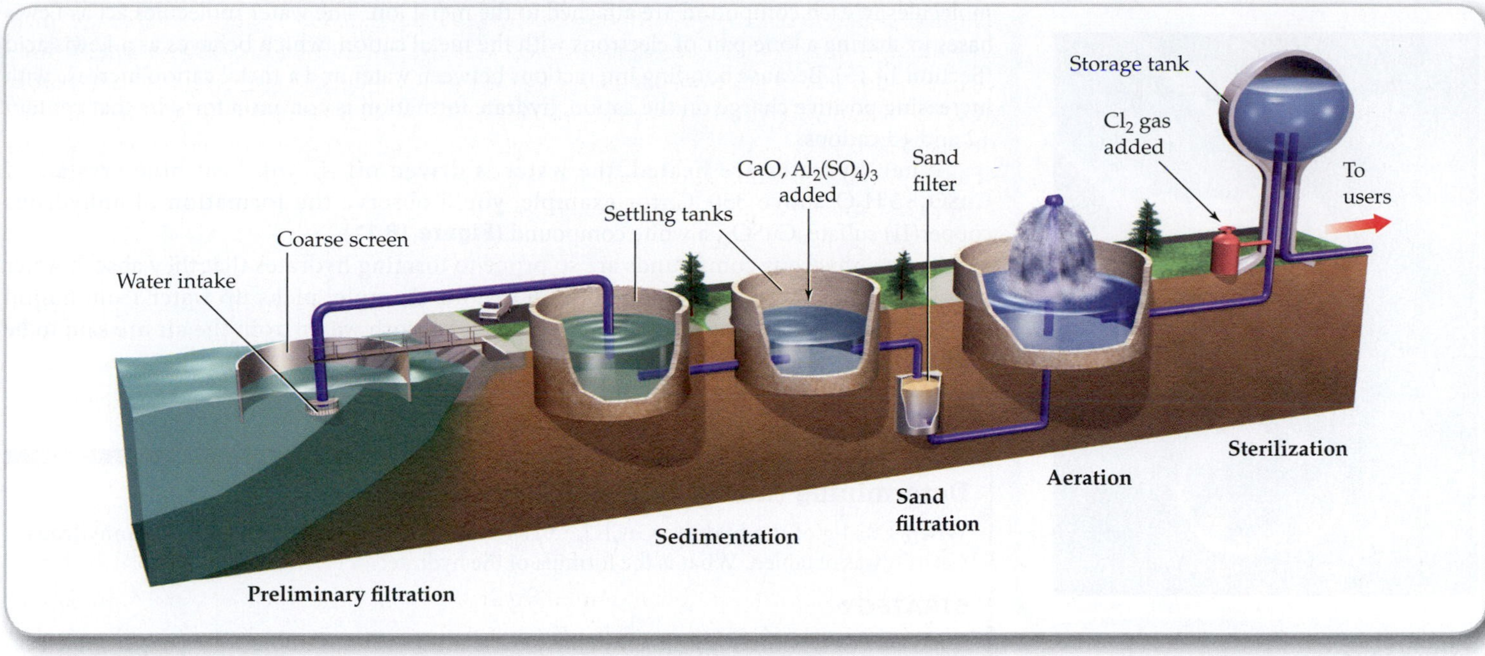

FIGURE 18.10

Purification of drinking water.

Water that contains appreciable concentrations of doubly charged cations such as Ca^{2+}, Mg^{2+}, and Fe^{2+} is called *hard water*. These cations combine with organic anions in soaps to give the undesirable, insoluble precipitates (soap scum) often observed in sinks and bathtubs. They also form the unwanted metal carbonate precipitates, known as *boiler scale*, that deposit in boilers, hot water heaters, and teakettles, thus reducing the efficiency of heat transfer.

You can understand the formation of boiler scale in terms of **Le Châtelier's principle** (Section 13.7). When hard water containing HCO_3^- anions is heated, the equilibrium for the decomposition of HCO_3^- to CO_2 and CO_3^{2-} shifts to the right as CO_2 gas escapes from the solution. The resulting CO_3^{2-} ions then combine with cations, such as Ca^{2+}, to form insoluble metal carbonates.

$$2\,HCO_3^-(aq) \xrightleftharpoons{\text{Heat}} CO_2(g) + H_2O(l) + CO_3^{2-}(aq)$$

$$Ca^{2+}(aq) + CO_3^{2-}(aq) \longrightarrow CaCO_3(s)$$

REMEMBER...

Le Châtelier's principle says that the concentration stress of a removed product—here $CO_2(g)$—is relieved by net reaction in the direction that replenishes the removed substance. (Section 13.7)

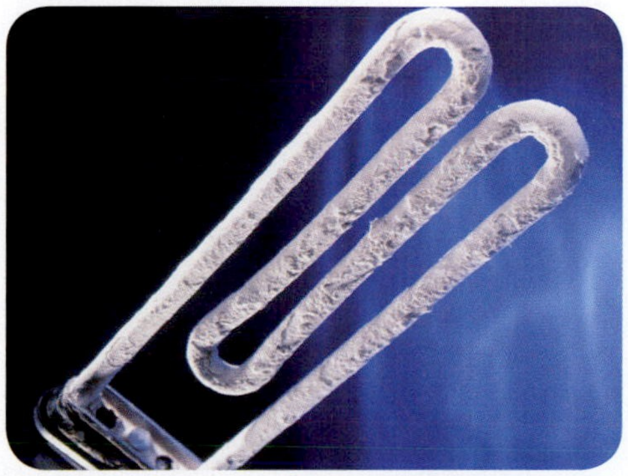

▲ Boiler scale, consisting of insoluble metal carbonates, such as $CaCO_3$ and $MgCO_3$, forms when hard water is heated.

Hard water can be softened by **ion exchange**, a process in which the Ca^{2+} and Mg^{2+} ions are replaced by Na^+. When hard water is passed through a resin that has ionic $SO_3^-Na^+$ groups attached to it, the more highly charged Ca^{2+} and Mg^{2+} cations bond to the negative SO_3^- groups more strongly than Na^+ does and the freed Na^+ ions move into the water. The ion-exchange process can be represented by the following equation, where RSO_3^- represents the ionic groups on the resin:

$$2\,RSO_3^-Na^+(s) + \underset{\text{Hard water}}{Ca^{2+}(aq)} \longrightarrow (RSO_3^-)_2Ca^{2+}(s) + \underset{\text{Soft water}}{2\,Na^+(aq)}$$

Ion-exchange resins make convenient household water softeners because the Na^+ ion form of the resin is easily regenerated by treating the used resin with a concentrated solution of NaCl.

18.13 HYDRATES

Solid compounds that contain water molecules are called **hydrates**. Magnesium perchlorate hexahydrate, $Mg(ClO_4)_2 \cdot 6\,H_2O$, and aluminum chloride hexahydrate, $AlCl_3 \cdot 6\,H_2O$, are examples. Because the structures of hydrates are sometimes complex or unknown, a dot is used in the formula of a hydrate to specify the composition without indicating how the water is bound. If the structure is known, a more informative formula can be given. The formulas $[Mg(H_2O)_6](ClO_4)_2$ and $[Al(H_2O)_6]Cl_3$, for instance, indicate that the six water

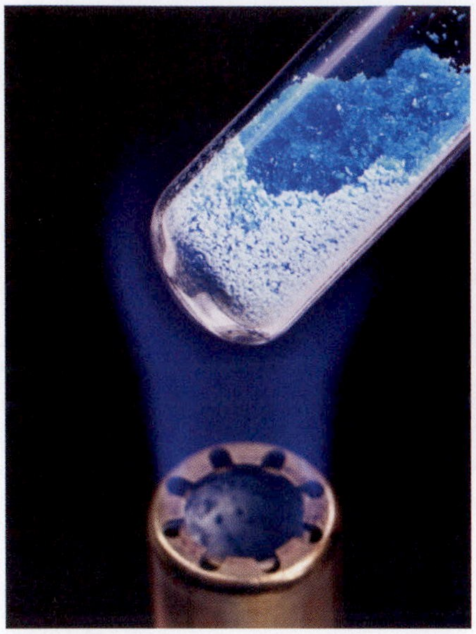

FIGURE 18.11

Blue crystals of $CuSO_4 \cdot 5\,H_2O$ are converted to white anhydrous $CuSO_4$ when the sample is heated with a Bunsen flame.

molecules in each compound are attached to the metal ion. The water molecules act as Lewis bases in sharing a lone pair of electrons with the metal cation, which behaves as a Lewis acid (Section 14.15). Because bonding interactions between water and a metal cation increase with increasing positive charge on the cation, hydrate formation is common for salts that contain +2 and +3 cations.

When hydrates are heated, the water is driven off. If you heat blue crystals of $CuSO_4 \cdot 5\,H_2O$ above 350 °C, for example, you'll observe the formation of anhydrous copper(II) sulfate, $CuSO_4$, a white compound (**Figure 18.11**).

Some anhydrous compounds are so prone to forming hydrates that they absorb water from the atmosphere. Anhydrous $Mg(ClO_4)_2$, for example, picks up water from humid air, yielding $Mg(ClO_4)_2 \cdot 6\,H_2O$. Compounds that absorb water from the air are said to be **hygroscopic** and are often useful as drying agents.

Worked Example 18.6

Determining the Formula of a Hydrate

When 1.823 g of the hydrate $CaCrO_4 \cdot x\,H_2O$ was heated at 200 °C, 1.479 g of anhydrous $CaCrO_4$ was obtained. What is the formula of the hydrate?

STRATEGY

The value of x in the formula of the hydrate is the number of moles of H_2O per mole of $CaCrO_4$, so we need to use the masses and the molar masses of H_2O (18.02 g/mol) and $CaCrO_4$ (156.1 g/mol) to calculate their mole ratio.

SOLUTION

$$\text{Grams of } H_2O = 1.823\text{ g} - 1.479\text{ g} = 0.344\text{ g } H_2O$$

$$\text{Moles of } H_2O = (0.344\text{ g } H_2O)\left(\frac{1\text{ mol } H_2O}{18.02\text{ g } H_2O}\right) = 1.91 \times 10^{-2}\text{ mol } H_2O$$

$$\text{Moles of } CaCrO_4 = (1.479\text{ g } CaCrO_4)\left(\frac{1\text{ mol } CaCrO_4}{156.1\text{ g } CaCrO_4}\right) = 9.475 \times 10^{-3}\text{ mol } CaCrO_4$$

$$x = \frac{\text{Moles of } H_2O}{\text{Moles of } CaCrO_4} = \frac{1.91 \times 10^{-2}\text{ mol}}{9.475 \times 10^{-3}\text{ mol}} = 2.02$$

Since $x = 2$ within experimental uncertainty, the formula of the hydrate is $CaCrO_4 \cdot 2\,H_2O$.

✔ BALLPARK CHECK

If the hydrate contained 1 mol of H_2O per mole of $CaCrO_4$, the mass ratio of H_2O to $CaCrO_4$ would be 18 g/mol divided by 156 g/mol, or about 0.1. The actual mass ratio of H_2O to $CaCrO_4$ in the hydrate is approximately $(1.8 - 1.5)\text{ g}/1.5\text{ g} = 0.2$, so the hydrate must contain 2 mol of H_2O per mole of $CaCrO_4$. The ballpark check agrees with the solution.

PROBLEM 18.15 A 3.10 g sample of anhydrous $NiSO_4$ was exposed to moist air. If 5.62 g of a hydrate of nickel sulfate was obtained, what is the formula of the hydrate?

FYI WHAT ROLE FOR HYDROGEN IN OUR ENERGY FUTURE?

Hydrogen is an enormously attractive fuel because it's environmentally clean, giving only water as a combustion product. If hydrogen is burned in air, small amounts of nitrogen oxides can be produced because of the high-temperature combination of nitrogen and oxygen, but the combustion products are free of CO, CO_2, SO_2, unburned hydrocarbons, and other environmental pollutants that result from the combustion of petroleum fuels. In addition, the amount of heat liberated when hydrogen burns is 242 kJ/mol (121 kJ/g), more than twice that of gasoline, oil, or natural gas on a mass basis.

$$H_2(g) + 1/2\, O_2(g) \longrightarrow H_2O(g) \qquad \Delta H^\circ = -242 \text{ kJ}$$

As a result, some people envision what they call a "hydrogen economy" in which our energy needs are met by gaseous, liquid, and solid hydrogen. For heating homes, gaseous hydrogen could be conveyed through underground pipes much like natural gas, while liquid hydrogen could be shipped by truck or by rail in large vacuum-insulated tanks. Automobiles might be powered by compressed hydrogen gas, liquid hydrogen, or "solid hydrogen"—hydrogen stored in the form of solid metal hydrides, porous solids with high surface areas, or solid hydrogen storage compounds such as NH_3BH_3. Prototype cars have already been built with their engines modified to run on hydrogen. Current work centers on the development of electric vehicles powered by hydrogen fuel cells (Section 17.10) and on the discovery of new solid materials for onboard hydrogen storage.

◀ This Boeing demonstrator airplane is powered by a hydrogen fuel cell. It has flown for about 20 minutes at an altitude of 3,300 feet.

What is keeping us from reaching a hydrogen economy? Before a hydrogen economy can become a reality, cheaper ways of producing hydrogen must be found. Because hydrogen is not a naturally occurring energy source like coal, oil, or natural gas, energy must first be expended to produce the hydrogen before it can be used. Current research therefore focuses on finding cheaper methods for extracting hydrogen from its compounds.

At present, hydrogen is produced mainly by the steam–hydrocarbon re-forming process (Section 18.2). This method can contribute to global warming because it produces CO_2 as a byproduct. It may be possible, however, to capture the CO_2 and sequester it in depleted gas wells or deep saline aquifers, thus avoiding addition of CO_2 to the atmosphere.

Where inexpensive electric power is available—for example off-peak hydropower or nuclear power—hydrogen can be produced by electrolysis of water. Another approach is to use solar energy to "split" water into H_2 and O_2. The economic feasibility of this scheme, however, depends on the development of catalysts that absorb sunlight and then use the solar energy to reduce water to hydrogen.

Liquid hydrogen has been used as a fuel in the U.S. space program for many years. Hydrogen powered the Saturn V rocket that carried the first astronauts to the moon, and it fueled the rocket engines of the space shuttle. It is, however, an extremely dangerous substance, whose storage and distribution presents substantial hazards at present.

PROBLEM 18.16 Hydrogen is a gas at ordinary temperatures. Explain how it can be stored as a solid.

PROBLEM 18.17 The space shuttle fuel tank contained 1.45×10^6 L of liquid hydrogen, which has a density of 0.088 kg/L. How much heat (in kilojoules) was liberated when the hydrogen burned in an excess of oxygen? How many kilograms of oxygen were needed to oxidize the hydrogen?

SUMMARY

Hydrogen, the most abundant element in the universe, has three isotopes: protium ($^{1}_{1}H$), deuterium ($^{2}_{1}H$), and tritium ($^{3}_{1}H$). A hydrogen atom, which has the electron configuration $1s^1$, can lose its electron, forming a hydrogen cation (H^+), or it can gain an electron, yielding a hydride anion (H^-). At ordinary temperatures, hydrogen exists as diatomic H_2 molecules, which are thermally stable and unreactive because of the strong H—H bond.

Hydrogen for industrial purposes is produced by the **steam–hydrocarbon re-forming process** and is used in the synthesis of ammonia and methanol. In the laboratory, hydrogen is prepared by the reaction of dilute acid with an active metal, such as zinc.

Hydrogen forms three types of **binary hydrides**. Active metals give **ionic hydrides**, such as LiH and CaH_2; nonmetals give **covalent hydrides**, such as NH_3, H_2O, and HF; and transition metals give **metallic**, or **interstitial**, **hydrides**, such as PdH_x. Interstitial hydrides are often **nonstoichiometric compounds**.

Oxygen is the most abundant element in the Earth's crust. Dioxygen (O_2) can be prepared in the laboratory by the thermal decomposition of $KClO_3$. Oxygen is manufactured by the fractional distillation of liquefied air and is used in making steel. The O_2 molecule is paramagnetic and has an O=O double bond. Ozone (O_3), an allotrope of oxygen, is a powerful oxidizing agent.

Oxygen forms ionic **oxides**, such as Li_2O and MgO, with active metals, and covalent oxides, such as P_4O_{10} and SO_3, with nonmetals. Oxides can also be classified according to their acid–base properties. Basic oxides are ionic, and acidic oxides are covalent. Amphoteric oxides, such as Al_2O_3, exhibit both acidic and basic properties.

Metal **peroxides**, such as Na_2O_2, are ionic compounds that contain the ${O_2}^{2-}$ anion and have oxygen in the −1 oxidation state. Metal **superoxides**, such as KO_2, contain the ${O_2}^{-}$ anion and have oxygen in the −1/2 oxidation state. Hydrogen peroxide (H_2O_2), a strong oxidizing agent and also a reducing agent, is unstable with respect to **disproportionation** to H_2O and O_2. A disproportionation reaction is one in which an element in a substance is simultaneously oxidized and reduced.

Water is the most abundant compound on Earth. Seawater, which accounts for 97.3% of the world's water supply, contains 3.5 mass % of dissolved salts. Purification of drinking water involves preliminary filtration, sedimentation, sand filtration, aeration, and sterilization. Hard water, which contains appreciable concentrations of doubly charged cations such as Ca^{2+}, Mg^{2+}, and Fe^{2+}, can be softened by **ion exchange**. Solid compounds that contain water are known as **hydrates**.

KEY WORDS

Bayer process *758*
binary hydride *750*
covalent hydride *751*
disproportionation *761*
hydrate *765*
hygroscopic *766*
interstitial hydride *752*
ion exchange *765*
ionic hydride *751*
metallic hydride *752*
nonstoichiometric compound *752*
oxide *757*
peroxide *757*
steam–hydrocarbon re-forming process *748*
superoxide *757*
water-gas shift reaction *749*

LEARNING OUTCOMES

- Describe how elemental hydrogen is prepared. (Section 18.2) ***Problems 3, 30–35, 94***
- Classify a binary hydride as ionic, covalent, or interstitial. (Sections 18.3, 18.4) ***Problems 4, 6, 18, 19, 36, 37***
- Describe the structures of binary hydrides and their physical and chemical properties. (Section 18.4) ***Problems 5–7, 19, 24, 26, 38, 39, 42–45***
- Describe the preparation and properties of elemental oxygen. (Sections 18.5–18.7) ***Problems 50, 52–54, 56, 57***
- Classify the oxide of an element as ionic or covalent, and as acidic, basic, or amphoteric, and describe how these properties vary with the position of the element in the periodic table. (Section 18.8) ***Problems 8, 9, 21–23, 62–67***
- Write balanced net ionic equations for the reactions of oxides. (Section 18.8) ***Problems 9, 68–71, 105***
- Determine whether a binary oxygen compound is an oxide, peroxide, or superoxide, and describe its properties and reactions. (Sections 18.8, 18.9) ***Problems 10–12, 20, 27, 73, 75, 101, 112***
- Write balanced net ionic equations for reactions of hydrogen peroxide. (Section 18.10) ***Problems 14, 78, 79***
- Describe a hydrate and use experimental data to determine its formula. (Section 18.13) ***Problems 15, 88–93, 108, 110***

CONCEPTUAL PROBLEMS

Problems 18.1–18.17 appear within the chapter.

18.18 In the following pictures of binary hydrides, ivory spheres represent H atoms or ions, and purple spheres represent atoms or ions of the other element.

(1) (2) (3) (4)

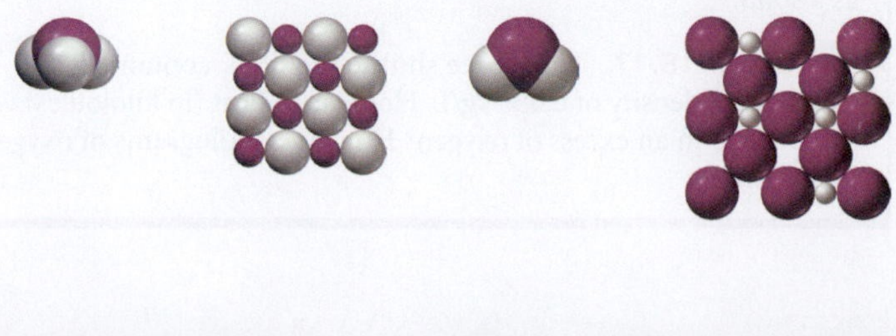

(a) Identify each binary hydride as ionic, covalent, or interstitial.

(b) What is the oxidation state of hydrogen in compounds (1), (2), and (3)? What is the oxidation state of the other element?

18.19 Look at the location of elements A, B, C, and D in the following periodic table:

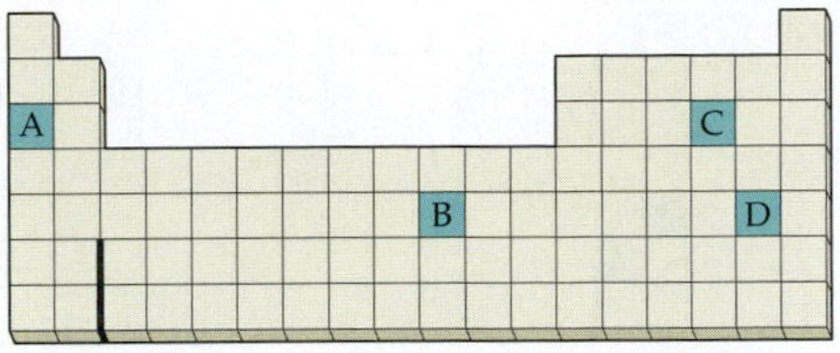

(a) Write the formula of the simplest binary hydride of each element.

(b) Classify each binary hydride as ionic, covalent, or interstitial.

(c) Which of these hydrides are molecular? Which are solids with an infinitely extended three-dimensional crystal structure?

(d) What are the oxidation states of hydrogen and the other element in the hydrides of A, C, and D?

18.20 The following molecular orbital energy-level diagram shows the energies and electron occupancies of the MOs derived from the atomic $2p$ orbitals for the anion in an oxygen-containing binary compound of potassium.

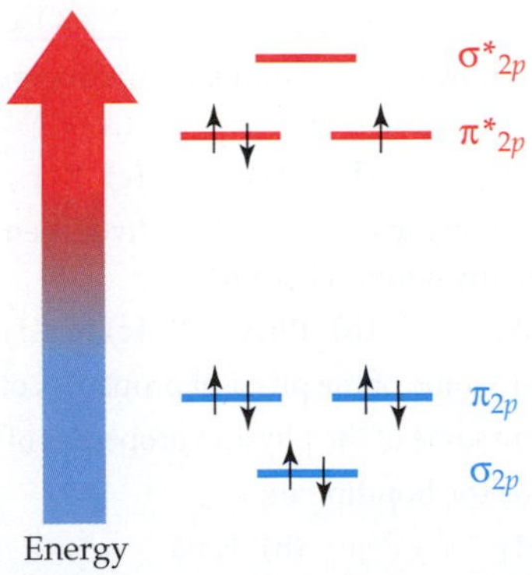

(a) Is the compound a peroxide or a superoxide? Write its chemical formula.

(b) Is the compound attracted or repelled by magnetic fields?

(c) Compare the oxygen–oxygen bond energy and bond length in this compound with the corresponding properties of the O_2 molecule.

(d) When this compound is treated with water, is the resulting aqueous solution acidic or basic? Explain.

18.21 In the following pictures of oxides, red spheres represent O atoms or ions, and green spheres represent atoms or ions of a second- or third-row element in its highest oxidation state.

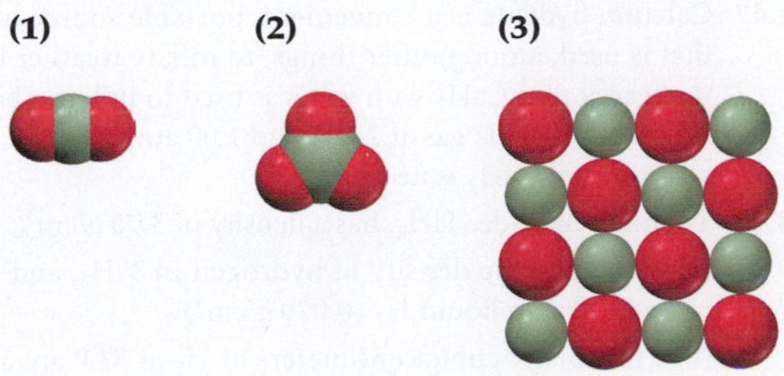

(a) What is the oxidation state of oxygen in each oxide? What is the oxidation state of the other element?

(b) Identify each oxide as ionic or covalent.

(c) Identify each oxide as acidic or basic.

(d) What is the identity of the other element in (1) and (2)?

18.22 In the following pictures of oxides, red spheres represent O atoms or ions, and green spheres represent atoms or ions of a first- or second-row element in its highest oxidation state.

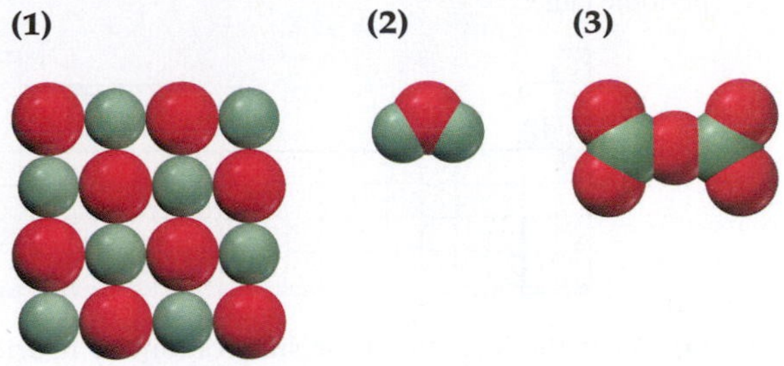

(a) What is the oxidation state of oxygen in each oxide? What is the oxidation state of the other element?

(b) Which of these oxides is (are) molecular, and which has (have) an infinitely extended three-dimensional structure?

(c) Which of these oxides is (are) likely to be a gas or a liquid, and which is (are) likely to be a high-melting solid?

(d) Identify the other element in (2) and (3).

18.23 Look at the location of elements A, B, C, and D in the following periodic table:

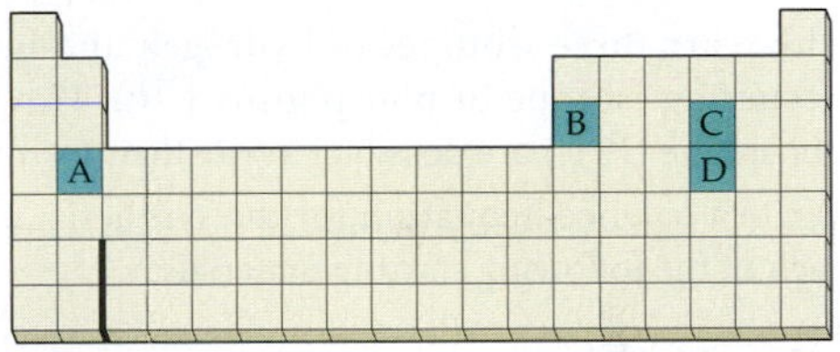

(a) Write the formula of the oxide that has each of these elements in its highest oxidation state.

(b) Classify each oxide as basic, acidic, or amphoteric.

(c) Which oxide is the most ionic? Which is the most covalent?

(d) Which of these oxides are molecular? Which are solids with an infinitely extended three-dimensional crystal structure?

(e) Which of these oxides has the highest melting point? Which has the lowest melting point?

18.24 The following pictures represent structures of the hydrides of four second-row elements.

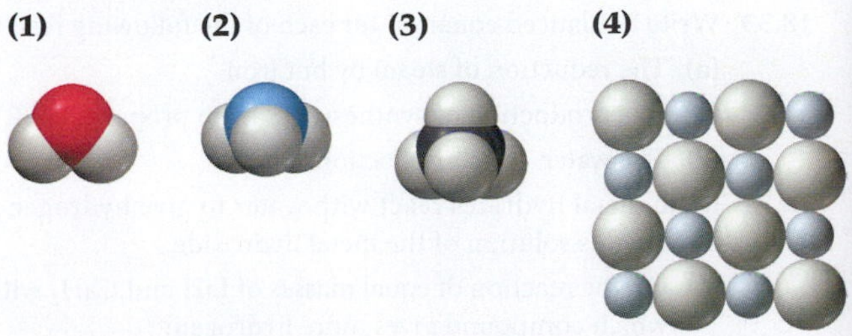

(a) Which compound has the highest melting point?

(b) Which compound has the lowest boiling point?

(c) Which compounds yield H_2 gas when they are mixed together?

18.25 The following pictures represent the structures of oxides of carbon and sulfur. Which has the stronger bonds? Explain.

(1)

(2)

18.26 Look at the location of elements A, B, C, and D in the following periodic table:

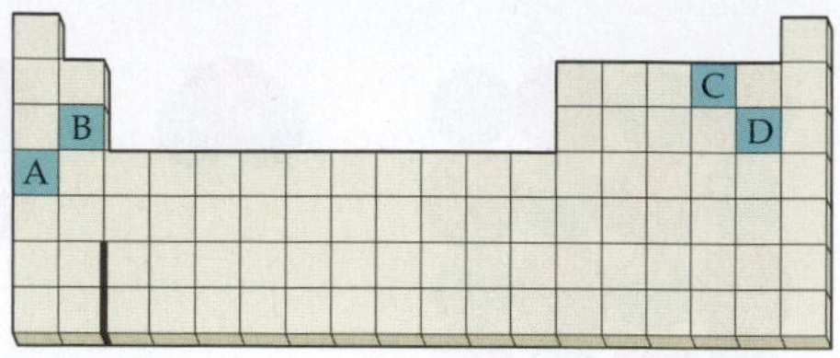

(a) Write the formula of the simplest binary hydride of each element.

(b) Which hydride has the lowest boiling point?

(c) Which hydrides react with water to give H_2 gas? Write a balanced net ionic equation for each reaction.

(d) Which hydrides react with water to give an acidic solution, and which give a basic solution?

18.27 Look at the location of elements A and B in the following periodic table:

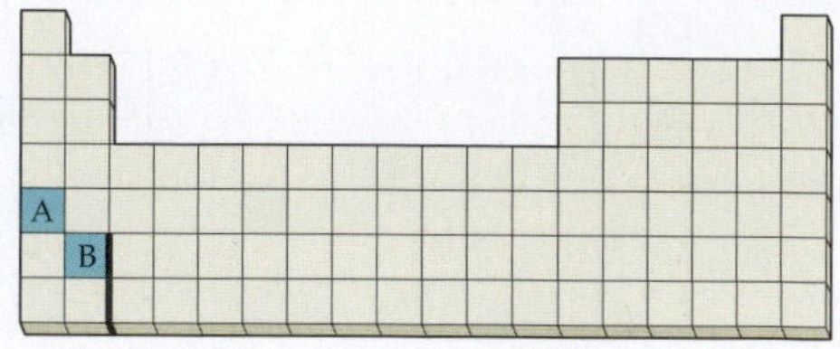

(a) Write the formula of the oxide, peroxide, and superoxide of elements A and B.

(b) Which of the six compounds in part **(a)** is (are) paramagnetic?

(c) Write a balanced equation for the reaction of the oxide, peroxide, and superoxide of element B with water.

SECTION PROBLEMS

Chemistry of Hydrogen (Sections 18.1–18.4)

18.28 There are three isotopes of hydrogen and three naturally occurring isotopes of oxygen (^{16}O, ^{17}O, and ^{18}O). How many kinds of water are possible? Write their formulas.

18.29 There are three isotopes of hydrogen and just one naturally occurring isotope of phosphorus (^{31}P). How many kinds of phosphine (PH_3) are possible? Write their formulas.

18.30 Write a balanced equation for the synthesis of hydrogen using each of the following starting materials.

(a) Zn **(b)** C

(c) CH_4 **(d)** H_2O

18.31 Complete and balance the equation for each of the following reactions:

(a) $Fe(s) + H^+(aq) \longrightarrow$

(b) $Ca(s) + H_2O(l) \longrightarrow$

(c) $Al(s) + H^+(aq) \longrightarrow$

(d) $C_2H_6(g) + H_2O(g) \xrightarrow[\text{Catalyst}]{\text{Heat}}$

18.32 What is the most important method for the industrial production of hydrogen? Write balanced equations for the reactions involved.

18.33 Write a balanced equation for each of the following reactions:

(a) The reduction of steam by hot iron

(b) The production of synthesis gas from propane, C_3H_8

(c) The water-gas shift reaction

18.34 Ionic metal hydrides react with water to give hydrogen gas and an aqueous solution of the metal hydroxide.

(a) On the reaction of equal masses of LiH and CaH_2 with water, which compound gives more hydrogen?

(b) How many kilograms of CaH_2 are needed to fill a 100 L tank with compressed H_2 gas at 150 atm pressure and 25 °C?

18.35 The hydrogen-filled dirigible Hindenburg had a volume of 1.99×10^8 L. If the hydrogen used was produced by the reaction of carbon with steam, how many kilograms of carbon would have been needed to produce enough hydrogen to fill the dirigible at 20 °C and 740 mm pressure?

$$C(s) + H_2O(g) \longrightarrow CO(g) + H_2(g)$$

18.36 In the following compounds, is hydrogen present as H^+, H^-, or a covalently bound H atom?

(a) MgH_2 **(b)** PH_3 **(c)** KH **(d)** HBr

18.37 In the following compounds, is hydrogen present as H^+, H^-, or a covalently bound H atom?

(a) H_2Se **(b)** RbH **(c)** CaH_2 **(d)** GeH_4

18.38 Compare some of the physical properties of H_2S, NaH, and PdH_x.

18.39 Compare some of the physical properties of $TiH_{1.7}$, HCl, and CaH_2.

18.40 Describe the bonding in:

(a) CH_4 **(b)** NaH

18.41 Describe the bonding in:

(a) CaH_2 **(b)** NH_3

18.42 Predict the molecular structure of:

(a) H_2Se **(b)** AsH_3 **(c)** SiH_4

18.43 Describe the molecular structure of:

(a) GeH_4 **(b)** H_2S **(c)** NH_3

18.44 What is a nonstoichiometric compound? Give an example, and account for its lack of stoichiometry in terms of structure.

18.45 Explain why the hydrogen atoms in interstitial hydrides are mobile.

18.46 How many grams of barium hydride must be treated with water to obtain 4.36 L of hydrogen at 20 °C and 0.975 atm pressure?

18.47 Calcium hydride is a convenient, portable source of hydrogen that is used, among other things, to inflate weather balloons. If the reaction of CaH_2 with water is used to inflate a balloon with 2.0×10^5 L of H_2 gas at 25 °C and 1.00 atm pressure, how many kilograms of CaH_2 is needed?

18.48 Titanium hydride, TiH_2, has a density of 3.75 g/cm^3.

(a) Calculate the density of hydrogen in TiH_2, and compare it with that in liquid H_2 (0.070 g/cm^3).

(b) How many cubic centimeters of H_2 at STP are absorbed in making 1.00 cm^3 of TiH_2?

18.49 The density of LiH is 0.78 g/cm³.

(a) Calculate the density of hydrogen in LiH and the number of H atoms in 5.0 cm³ of LiH.

(b) How many liters of H_2 at 20 °C and 740 mm Hg are required to make 5.0 cm³ of LiH?

Chemistry of Oxygen (Sections 18.5–18.11)

18.50 How is O_2 prepared (a) in industry and (b) in the laboratory? Write balanced equations for the reactions involved.

18.51 In what forms is oxygen commonly found in nature?

18.52 How many milliliters of O_2 gas at 25 °C and 1.00 atm pressure can be obtained by the thermal decomposition of 0.200 g of $KClO_3$? The balanced equation for the reaction is $2\ KClO_3(s) \longrightarrow 2\ KCl(s) + 3\ O_2(g)$.

18.53 How many milliliters of O_2 gas at 25 °C and 1.00 atm pressure are obtained from the thermal decomposition of 0.200 g of $KMnO_4$? The balanced equation for the reaction is $2\ KMnO_4(s) \longrightarrow K_2MnO_4(s) + MnO_2(s) + O_2(g)$.

18.54 In the presence of a catalyst, hydrogen peroxide decomposes to water and O_2. How many liters of O_2 gas at 25 °C and 0.985 atm pressure can be obtained by the catalytic decomposition of 20.4 g of hydrogen peroxide?

18.55 In the oxyacetylene torch, how many grams of acetylene and how many liters of O_2 at STP are needed to generate 1000 kJ of heat?

18.56 Write a balanced equation for the reaction of an excess of O_2 with each of the following elements:

(a) Li (b) P (c) Al (d) Si

18.57 Write a balanced equation for the reaction of an excess of O_2 with each of the following elements:

(a) Ca (b) C (c) As (d) B

18.58 Draw some electron-dot structures for O_2, and explain why they are inconsistent with both the paramagnetism of O_2 and its O=O double bond.

18.59 Use molecular orbital theory to account for the paramagnetism of O_2 and its O=O double bond.

18.60 If an element forms an acidic oxide, is it more likely to form an ionic or covalent hydride?

18.61 If an element forms an ionic hydride, is it more likely to form an acidic or basic oxide?

18.62 Arrange the following oxides in order of increasing covalent character: B_2O_3, BeO, CO_2, Li_2O, N_2O_5.

18.63 Arrange the following oxides in order of increasing ionic character: SiO_2, K_2O, P_4O_{10}, Ga_2O_3, GeO_2.

18.64 Arrange the following oxides in order of increasing basic character: Al_2O_3, Cs_2O, K_2O, N_2O_5.

18.65 Arrange the following oxides in order of increasing acidic character: BaO, Cl_2O_7, SO_3, SnO_2.

18.66 Which is more acidic?

(a) Cr_2O_3 or CrO_3 (b) N_2O_5 or N_2O_3

(c) SO_2 or SO_3

18.67 Which is more basic?

(a) CrO or Cr_2O_3 (b) SnO_2 or SnO

(c) As_2O_3 or As_2O_5

18.68 Write a balanced net ionic equation for the reaction of each of the following oxides with water:

(a) Cl_2O_7 (b) K_2O (c) SO_3

18.69 Write a balanced net ionic equation for the reaction of each of the following oxides with water:

(a) BaO (b) Cs_2O (c) N_2O_5

18.70 Write a balanced net ionic equation for the reaction of the amphoteric oxide ZnO with:

(a) Hydrochloric acid

(b) Aqueous sodium hydroxide; the product is $Zn(OH)_4{}^{2-}$

18.71 Write a balanced net ionic equation for the reaction of the amphoteric oxide Ga_2O_3 with:

(a) Aqueous sulfuric acid

(b) Aqueous potassium hydroxide; the product is $Ga(OH)_4{}^-$

18.72 Distinguish between a peroxide and a superoxide, and give an example of each.

18.73 Classify each of the following compounds as an oxide, a peroxide, or a superoxide:

(a) Na_2O_2 (b) GeO_2 (c) RbO_2 (d) BaO_2

18.74 What products are formed when the following metals are burned in an excess of air?

(a) Ba (b) Ca (c) Cs (d) Li (e) Na

18.75 Write balanced net ionic equations for the reaction of water with:

(a) BaO_2 (b) CsO_2

18.76 Draw MO energy-level diagrams for O_2, $O_2{}^-$, and $O_2{}^{2-}$, including only MOs derived from the oxygen $2p$ atomic orbitals. Show the electron population of the MOs. (See Section 5.6.)

(a) Why does the O–O bond length increase in the series O_2, $O_2{}^-$, $O_2{}^{2-}$?

(b) Why is $O_2{}^-$ paramagnetic, whereas $O_2{}^{2-}$ is diamagnetic?

18.77 Draw an MO energy-level diagram for $O_2{}^+$, including only MOs derived from the oxygen $2p$ atomic orbitals. Show the electron population of the MOs.

(a) Predict the bond order in $O_2{}^+$, and tell whether the O–O bond should be longer or shorter than that in O_2.

(b) Is $O_2{}^+$ paramagnetic or diamagnetic?

18.78 Write a balanced net ionic equation for each of the following reactions:

(a) Oxidation by H_2O_2 of I^- to I_2 in acidic solution

(b) Reduction by H_2O_2 of $Cr_2O_7{}^{2-}$ to Cr^{3+} in acidic solution

18.79 Write a balanced net ionic equation for each of the following reactions:

(a) Oxidation by H_2O_2 of Fe^{2+} to Fe^{3+} in acidic solution

(b) Reduction by H_2O_2 of $IO_4{}^-$ to $IO_3{}^-$ in basic solution

18.80 Give a description of the electronic structure of ozone that is consistent with the two O–O bond lengths being equal.

18.81 Explain how resonance and the VSEPR model help to account for the molecular structure of ozone.

18.82 How is ozone made in the laboratory?

18.83 How many kilojoules must be supplied to convert 10.0 g of O_2 to O_3?

Water and Hydrates (Sections 18.12–18.13)

18.84 If seawater contains 3.5 mass % of dissolved salts, how many kilograms of salts are present in 1.0 mi³ of seawater? (1 mi = 1609 m; density of seawater = 1.025 g/cm³)

18.85 How many kilograms of magnesium are present in a cubic meter of seawater? Assume the Mg^{2+} ion concentration listed in Table 18.2 and a density for seawater of 1.025 g/cm³.

18.86 Why does boiler scale form when hard water is heated?

18.87 What is an ion exchange resin, and how can it be used to soften hard water?

18.88 Give an example of a hydrate, and indicate how the water is bound.

18.89 Describe the structure of the cation in $[Fe(H_2O)_6](NO_3)_3$.

18.90 What is the mass percent of water in plaster of paris, $CaSO_4 \cdot 1/2\ H_2O$?

18.91 Calculate the mass percent of water in $CuSO_4 \cdot 5\ H_2O$.

18.92 When 3.44 g of the mineral gypsum, $CaSO_4 \cdot x\,H_2O$, is heated to 128 °C, 2.90 g of $CaSO_4 \cdot 1/2\ H_2O$ is obtained. What is the value of *x* in the formula of gypsum?

18.93 Anhydrous, hygroscopic, blue $CoCl_2$ forms red-violet $CoCl_2 \cdot x\,H_2O$ on exposure to moist air. If the color change is accompanied by an 83.0% increase in mass, what is the formula of the hydrate?

CHAPTER PROBLEMS

18.94 To prepare H_2 from water, would you allow water to react with an oxidizing agent or a reducing agent? Which of the following metals could be used in the reaction? (Hint: Recall the activity series, Section 7.8.)

(a) Ag (b) Al

(c) Au (d) Ca

18.95 The volume of the oceans is $1.35 \times 10^{18}\ m^3$.

(a) How many kilograms of deuterium are present in the oceans if the abundance of deuterium is 0.0156 atom %? (Neglect the presence of dissolved substances, and assume that the density of water is 1.00 g/cm^3.)

(b) Do the same calculation for tritium, assuming that the abundance of tritium is approximately 10^{-16} atom %.

18.96 How many liters of H_2 at STP are required for the hydrogenation of 2.7 kg of 1,3-butadiene?

$$\underset{\text{1,3-Butadiene}}{H_2C{=}CH{-}CH{=}CH_2(g)} + 2\,H_2(g) \longrightarrow \underset{\text{Butane}}{CH_3{-}CH_2{-}CH_2{-}CH_3(g)}$$

18.97 Give the chemical formula for each of the following compounds:

(a) Calcium hydroxide

(b) Chromium(III) oxide

(c) Rubidium superoxide

(d) Sodium peroxide

(e) Barium hydride

(f) Hydrogen selenide

18.98 Name each of the following compounds:

(a) B_2O_3 (b) H_2O_2

(c) SrH_2 (d) CsO_2

(e) $HClO_4$ (f) BaO_2

18.99 Which of the following molecules have similar properties, and which have quite different properties?

(a) $H_2{}^{16}O$ (b) $D_2{}^{17}O$

(c) $H_2{}^{16}O_2$ (d) $H_2{}^{17}O$

(e) $D_2{}^{17}O_2$

18.100 Three isotopes of oxygen exist (^{16}O, ^{17}O, and ^{18}O).

(a) How many kinds of dioxygen (O_2) molecules are possible? Write their formulas.

(b) How many kinds of ozone (O_3) molecules are possible? Write their formulas.

18.101 What is the oxidation number of oxygen in each of the following compounds?

(a) Al_2O_3 (b) SrO_2

(c) SnO_2 (d) CsO_2

18.102 Write a balanced equation for a reaction in which each of the following acts as an oxidizing agent:

(a) O_2 (b) O_3

(c) H_2O_2 (d) H_2

(e) H_2O

18.103 How many liters of seawater (density 1.025 g/cm^3) must be processed to obtain 2.0 million kg of bromine? Assume the Br^- ion concentration listed in Table 18.2 and a recovery rate of 20%.

18.104 Use the standard heats of formation in Appendix B to calculate $\Delta H°$ (in kilojoules) for each of the following reactions:

(a) $CO(g) + 2\,H_2(g) \longrightarrow CH_3OH(l)$

(b) $CO(g) + H_2O(g) \longrightarrow CO_2(g) + H_2(g)$

(c) $2\,KClO_3(s) \longrightarrow 2\,KCl(s) + 3\,O_2(g)$

(d) $6\,CO_2(g) + 6\,H_2O(l) \longrightarrow 6\,O_2(g) + C_6H_{12}O_6(s)$

18.105 One model of an acid–base reaction involves proton transfer from the acid to a solvent water molecule to give an H_3O^+ ion. Write balanced chemical equations that show how aqueous solutions of acidic oxides such as SO_2 and SO_3 can yield H_3O^+ ions. Which of these two oxides gives a higher concentration of H_3O^+ ions? Explain.

18.106 Oxygen reacts with cyclohexane (C_6H_{12}) in the presence of a catalyst to form cyclohexanol, an intermediate in nylon production:

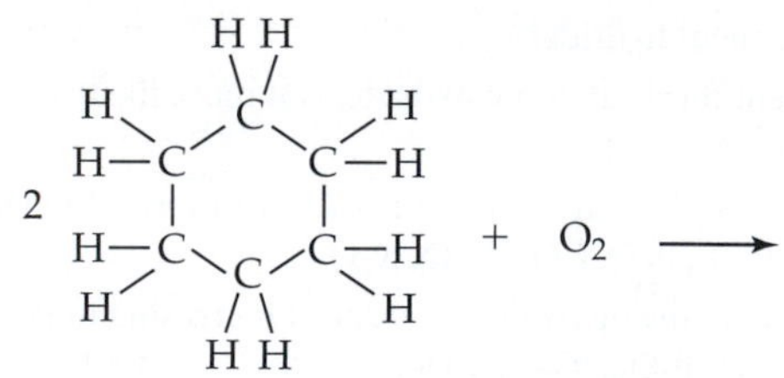

Cyclohexane

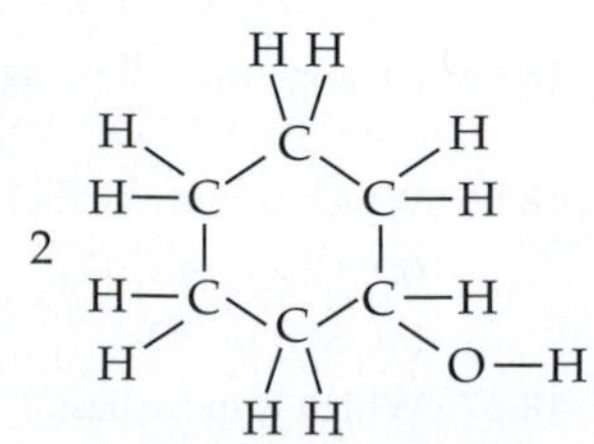

Cyclohexanol

How many liters of oxygen at 25.0 °C and 1.00 atm are required to completely react with 555 g of cyclohexane?

18.107 Samarium(III) oxide (Sm_2O_3), used as a neutron absorber in nuclear power reactors, is formed by the reaction of samarium with oxygen at 200 °C. How many liters of oxygen at 200 °C and 1.00 atm are required to completely react with 712 g of Sm?

18.108 In addition to hydrated salts, a type of hydrate exists in which a hydrogen-bonded, ice-like, network of water molecules surrounds atoms or small molecules like xenon or methane. If the so-called methane hydrate has the elemental composition 13.06% H, 10.04% C, and 76.90% O, what is the value of x in the formula $(CH_4)_x(H_2O)_{23}$?

18.109 Anions such as sulfate can be removed from water by ion exchange using a resin that can be represented as $RN(CH_3)_3{}^+Cl^-$. Write a balanced equation for the exchange of sulfate by chloride.

18.110 A boiler with water that contained high concentrations of Ca^{2+} and $SO_4{}^{2-}$—but no $CO_3{}^{2-}$ or $HCO_3{}^-$—developed a boiler scale. Elemental analysis of the scale showed 2.34% H and 55.76% O. Assuming the ions in the boiler scale are Ca^{2+} and $SO_4{}^{2-}$ and that the scale is a hydrate, what is its formula?

18.111 Ammonia–borane ($H_3N{-}BH_3$) has been proposed as a hydrogen-storage material. What is the mass percent hydrogen in ammonia–borane?

18.112 The binary compound formed by the reaction of sodium and oxygen at 300 °C and 5 atm is 58.19% O. What is the formula of the compound? Is it an oxide, peroxide, or superoxide?

18.113 Hydrogen peroxide reacts with sulfuric acid to generate H_2SO_5, a potent oxidizing agent. Write a balanced redox equation for the oxidation of Fe^{2+} to Fe^{3+} by aqueous H_2SO_5 if the byproducts are sulfate ion and water.

18.114 At 175 °C in a 0.500 L vessel, 3.22 g of a gaseous binary compound of hydrogen and germanium exerted a pressure of 1.048 atm. When the temperature was raised to 250 °C, the compound decomposed to its elements. The vessel was then cooled to 175 °C, and the pressure was found to be 4.192 atm. Assuming that the pressure after decomposition is only from H_2, determine the formula of the germanium hydride.

18.115 The sulfur–iodine cycle, an alternate process for hydrogen production, has the following three steps:

$$I_2(g) + SO_2(g) + 2\,H_2O(g) \longrightarrow 2\,HI(g) + H_2SO_4(g)$$
$$2\,H_2SO_4(g) \longrightarrow 2\,SO_2(g) + 2\,H_2O(g) + O_2(g)$$
$$2\,HI(g) \longrightarrow I_2(s) + H_2(g)$$

The first two steps are carried out at high temperatures but the third step is spontaneous only at lower temperatures.

(a) Estimate the temperature below which the third step is spontaneous. (Use data from Appendix B, and assume the I_2 is a solid under the reaction conditions.)

(b) Write a balanced equation for the overall process.

MULTICONCEPT PROBLEMS

18.116 How much heat (in joules) is liberated when 1.000 L of ozone at 20 °C and 63.6 mm Hg decomposes to ordinary oxygen, O_2?

18.117 Sodium hydride, which has the NaCl crystal structure, has a density of 1.39 g/cm^3. If the ionic radius of Na^+ is 102 pm, what is the ionic radius of H^- in NaH?

18.118 A 1.84 g sample of an alkaline earth metal hydride was treated with an excess of dilute hydrochloric acid, and the resulting gas was collected in a 1.000 L container at 20 °C. The measured pressure of the gas was 750.0 mm Hg. Identify the alkaline earth metal, and write the formula for the metal hydride.

18.119 A 250.0 mL gaseous sample of a sulfur oxide at 77 °C and 720.0 mm Hg pressure was allowed to react completely with an excess of water. Prior to reaction, the density of the gas was determined to be 2.64 g/L.

(a) What is the formula for the sulfur oxide?

(b) How much heat (in kilojoules) is released when the gas reacts with the water?

(c) How many milliliters of 0.160 M NaOH is needed to neutralize the aqueous solution?

18.120 The putrid gas phosphine, PH_3, reacts with oxygen to form phosphoric acid, H_3PO_4, a low-melting solid.

(a) Write a balanced equation for the reaction. (Phosphoric acid is the only product.)

(b) The product of the reaction of 0.646 g of phosphine with excess oxygen is dissolved in water, and the solution is diluted to a volume of 0.250 L. What is the pH of the solution?

18.121 What volume of PH_3 at STP combined with excess oxygen will make 2.00 L of aqueous H_3PO_4 with pH = 1.95? (See Problem 18.120.)

18.122 At 25 °C and 740 mm Hg, electrolysis of water for 10.0 min with a constant current of 5.00 A yielded 352.5 mL of hydrogen. If the water volume was 2.0 L and no hydrogen escaped from the system, calculate the Henry's Law constant for hydrogen.

18.123 The volume of oxygen collected in the electrolysis in Problem 18.122 was 131.7 mL. Using the information in Problem 18.122, calculate the Henry's Law constant for oxygen.

18.124 A 300.0 mL sample of ordinary water was allowed to react with 5.4 g of N_2O_5, and an excess of zinc metal was then added. The resulting gas was collected in a 500.0 mL container at 25 °C.

(a) What is the partial pressure (in mm Hg) of HD in the container?

(b) How many HD molecules are present in the container?

(c) How many D_2 molecules are present in the container?

18.125 Sodium amalgam is an alloy of sodium and mercury. The percent sodium in the alloy can be determined by treating the amalgam with an excess of hydrochloric acid and collecting the liberated H_2 gas.

(a) When a 5.26 g sample of amalgam was treated with 250.0 mL of 0.2000 M HCl and the liberated H_2 was collected in a 500.0 mL container at 22 °C, the gas had a pressure of 434 mm Hg. What is the mass percent of sodium in the amalgam?

(b) After the reaction in part **(a)**, a 50.00 mL portion of the solution was titrated with 0.1000 M NaOH. How many milliliters of the NaOH solution is needed to neutralize the excess HCl?

18.126 Direct measurement of the normal boiling point of hydrogen peroxide is not possible because pure H_2O_2 explodes on heating. The boiling point can be estimated, however, from vapor-pressure data. Use the vapor pressures of 75.0 mm Hg at 89.0 °C and 319.2 mm Hg at 125.0 °C to estimate the boiling point. Compare your result with the value given in Section 18.10.

18.127 Use the data in this chapter to calculate the number of Mg^{2+} ions and Br^- ions in the Earth's oceans. How many metric tons of $MgBr_2$ could be made from these ions assuming 100% recovery? (The density of seawater is 1.025 g/cm^3.)

CHAPTER

The Main-Group Elements

Because the electrical conductivity of the main-group element selenium increases by a factor of 1000 when it is exposed to light, selenium is used in making laser printers. See the *FYI* at the end of this chapter.

The main-group elements are the 50 elements that occupy groups 1A–8A of the periodic table. They are subdivided into the *s*-block elements of groups 1A and 2A, with valence electron configuration ns^1 or ns^2, and the *p*-block elements of groups 3A–8A, with valence configurations $ns^2\,np^{1-6}$. Main-group elements are important because of their high natural abundance and their presence in commercially valuable chemicals. Eight of the 10 most abundant elements in the Earth's crust and all 10 of the most abundant elements in the human body are main-group elements (**Figure 19.1**). In addition, the 10 most widely used industrial chemicals contain only main-group elements (**Table 19.1**).

In Sections 3.11–3.14, we surveyed the alkali and alkaline earth metals (groups 1A and 2A), the halogens (group 7A), and the noble gases (group 8A). In Chapter 18, we looked at the chemistry of

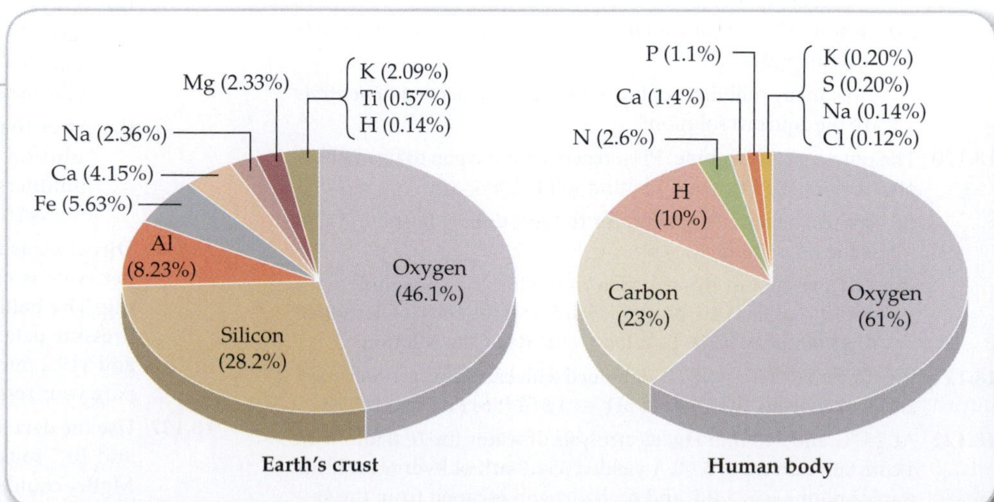

FIGURE 19.1

The 10 most abundant elements by mass in the Earth's crust and in the human body. All are main-group elements except iron and titanium.

CONTENTS

TABLE 19.1 The Top 10 Chemicals (2010 U.S. Production)*

Chemical	Millions of Metric Tons	Principal Uses
Sulfuric acid (H_2SO_4)	32.5	Fertilizers, chemicals, oil refining
Ethylene ($CH_2{=}CH_2$)	24.0	Plastics, antifreeze
Lime (CaO)	21.3	Steelmaking, chemicals, water treatment
Propylene ($CH_3CH{=}CH_2$)	14.1	Plastics, fibers, solvents
Ammonia (NH_3)	9.7	Fertilizers, nitric acid
Chlorine (Cl_2)	9.7	Chemicals, plastics, water treatment
Phosphoric acid (H_3PO_4)	9.4	Fertilizers, detergents
Ethylene dichloride ($ClCH_2CH_2Cl$)	8.8	Plastics, insecticides, solvent
Sodium hydroxide (NaOH)	7.5	Chemicals, textiles, soaps
Nitric acid (HNO_3)	6.3	Fertilizers, explosives, plastics, dyes

*2006 data for CaO

hydrogen and oxygen. Now we'll examine the remaining main-group elements, paying special attention to boron and aluminum (group 3A), carbon and silicon (group 4A), nitrogen and phosphorus (group 5A), and sulfur (group 6A).

19.1 A REVIEW OF GENERAL PROPERTIES AND PERIODIC TRENDS

Let's begin our survey of the main-group elements by reviewing the periodic trends that make it possible to classify these elements as metals, nonmetals, or semimetals. **Figure 19.2** shows the main group regions of the periodic table, with metals to the left of the heavy stairstep line, nonmetals to the right of the line, and semimetals—elements with intermediate properties—along the line. The elements usually classified as semimetals are boron (group 3A), silicon and germanium (group 4A), arsenic and antimony (group 5A), tellurium (group 6A), and astatine (group 7A).

FIGURE 19.2

Periodic trends in the properties of the main-group elements.

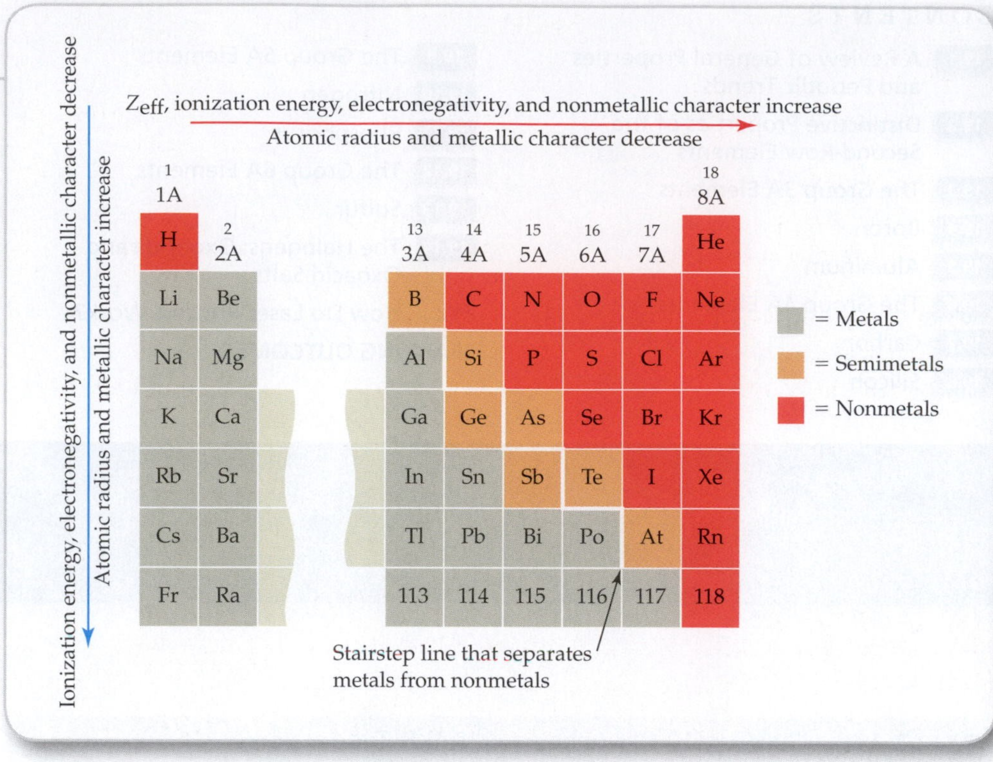

REMEMBER...

The nuclear charge actually felt by an electron, called the **effective nuclear charge Z_{eff}**, is less than the actual nuclear charge because of electron shielding. The valence electrons are less well shielded by other valence electrons than by the inner-shell electrons. (Section 2.11)

From left to right across the periodic table, the **effective nuclear charge Z_{eff}** increases because each additional valence electron does not completely shield the additional nuclear charge (Section 2.11). As a result, the atom's electrons are more strongly attracted to the nucleus, ionization energy generally increases, atomic radius decreases, and electronegativity increases, as shown in Figure 19.2. The elements on the left side of the table tend to form cations by losing electrons, and those on the right tend to form anions by gaining electrons. Thus, metallic character decreases and nonmetallic character increases across the table from left to right. In the third row, for example, sodium, magnesium, and aluminum are metals, silicon is a semimetal, and phosphorus, sulfur, and chlorine are nonmetals.

From the top to the bottom of a group in the periodic table, additional shells of electrons are occupied and atomic radius therefore increases. Because the valence electrons are farther from the nucleus, though, ionization energy and electronegativity generally decrease. As a result, metallic character increases and nonmetallic character decreases down a group from top to bottom. In group 4A, for example, carbon is a nonmetal, silicon and germanium are semimetals, and tin and lead are metals. The horizontal and vertical periodic trends combine to locate the element with the most metallic character (francium) in the lower left of the periodic table, the element with the most nonmetallic character (fluorine) in the upper right, and the semimetals along the diagonal stairstep that stretches across the middle.

REMEMBER...

Binary hydrides are compounds that contain hydrogen and just one other element. The ions in ionic hydrides are held together by strong ionic bonds, whereas the molecules in covalent hydrides are held together by relatively weak intermolecular forces. (Section 18.4)

In earlier chapters, we saw how the metallic or nonmetallic character of an element affects its chemistry. Metals tend to form ionic compounds with nonmetals, whereas nonmetals tend to form covalent, molecular compounds with one another. Thus, **binary metallic hydrides**, such as NaH and CaH_2, are ionic solids with high melting points, whereas **binary nonmetallic hydrides**, such as CH_4, NH_3, H_2O, and HF, are covalent, molecular compounds that exist at room temperature as gases or volatile liquids (Section 18.4).

Oxides exhibit similar trends. In the third row, for example, Na_2O and MgO are typical high-melting, ionic solids, whereas P_4O_{10}, SO_3, and Cl_2O_7 are volatile, covalent, molecular compounds (Section 18.8). The metallic or nonmetallic character of an oxide also affects its acid–base properties. Na_2O and MgO are basic, for example, whereas P_4O_{10}, SO_3, and Cl_2O_7 are acidic. **Table 19.2** summarizes some of the properties that distinguish metallic and nonmetallic elements.

TABLE 19.2 Properties of Metallic and Nonmetallic Elements

Metals	Nonmetals
All are solids at 25 °C except Hg, which is a liquid	Eleven are gases at 25 °C, one is a liquid (Br), and five are solids (C, P, S, Se, and I)
Most have a silvery shine	Most lack a metallic luster
Malleable and ductile	Nonmalleable and brittle
Good conductors of heat and electricity	Poor conductors of heat and electricity, except graphite
Relatively low ionization energies	Relatively high ionization energies
Relatively low electronegativities	Relatively high electronegativities
Lose electrons to form cations	Gain electrons to form anions; share electrons to form oxoanions
Hydrides are ionic (or interstitial)	Hydrides are covalent and molecular
Oxides are ionic and basic	Oxides are covalent, molecular, and acidic

Worked Example 19.1

Using the Periodic Table to Predict Metallic Character

Use the periodic table to predict which element in each of the following pairs has more metallic character:

(a) Ga or As **(b)** P or Bi **(c)** Sb or S **(d)** Sn or Ba

STRATEGY

Because metallic character increases from right to left and from top to bottom in the periodic table, look to see which of the pair of elements lies farther toward the lower left of the table.

SOLUTION

(a) Ga and As are in the same row of the periodic table, but Ga (group 3A) lies to the left of As (group 5A). Therefore, Ga is more metallic.
(b) Bi lies below P in group 5A and is therefore more metallic.
(c) Sb (group 5A) has more metallic character because it lies below and to the left of S (group 6A).
(d) Ba (group 2A) is more metallic because it lies below and to the left of Sn (group 4A).

PROBLEM 19.1 Predict which element in each of the following pairs has more nonmetallic character:
(a) B or Al **(b)** Ge or Br **(c)** In or Se **(d)** Cl or Te

19.2 DISTINCTIVE PROPERTIES OF THE SECOND-ROW ELEMENTS

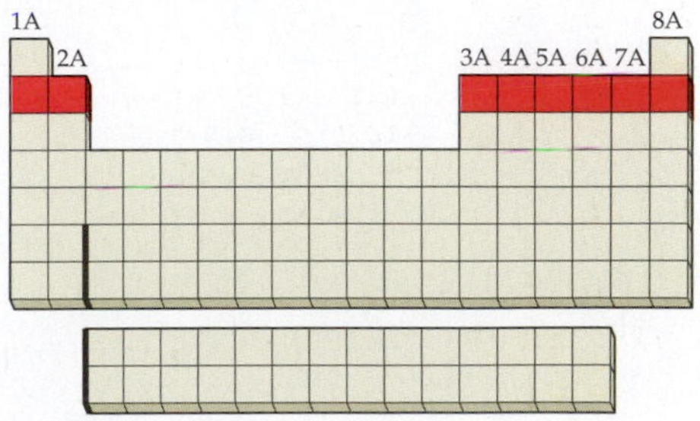

The properties of elements in the second row of the periodic table differ markedly from those of heavier elements in the same periodic group. The second-row atoms have especially small sizes and especially high electronegativities. In group 5A, for example, the electronegativity of N is 3.0, whereas the electronegativities of P, As, Sb, and Bi are all in the range 2.1–1.9. Figure 4.4 (page 143) shows the discontinuity in electronegativity that distinguishes the second-row elements from other elements of the same periodic group.

The small sizes and high electronegativities of the second-row elements accentuate their nonmetallic behavior. Thus, BeO is amphoteric but the oxides of the other group 2A elements are basic. Boron differs from the metallic elements of group 3A in forming mainly covalent, molecular compounds. For example, BF_3 (bp −100 °C) is a gaseous, molecular halide, but

REMEMBER...

A **hydrogen bond** is an attractive interaction between an H atom bonded to a very electronegative atom (O, N, or F) and an unshared electron pair on another electronegative atom. (Section 10.2)

REMEMBER...

A **weak acid** HA is only partially dissociated in water to give H_3O^+ and A^- ions, whereas a strong acid is essentially 100% dissociated. (Section 14.2)

AlF_3 (mp 1290 °C) is a typical high-melting, ionic solid. Furthermore, **hydrogen bonding** interactions are generally restricted to compounds of the highly electronegative second-row elements N, O, and F (Section 10.2). Recall also that HF contrasts with HCl, HBr, and HI in being the only **weak hydrohalic acid** (Section 14.2).

Because of the small sizes of their atoms, the second-row elements generally form a maximum of four covalent bonds. By contrast, the larger atoms of the third-row elements can accommodate more than four nearest neighbors and can therefore form more than four bonds. Thus, nitrogen forms only NCl_3, but phosphorus forms both PCl_3 and PCl_5:

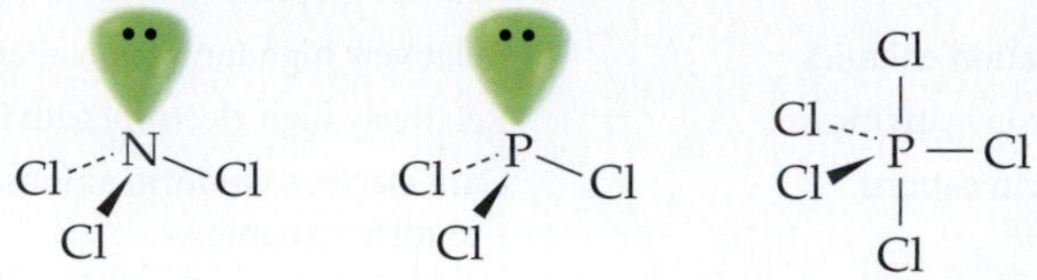

A further consequence of the small size of the second-row atoms C, N, and O is their ability to form multiple bonds by π overlap of the 2*p* orbitals. By contrast, the 3*p* orbitals of the corresponding third-row atoms Si, P, and S are more diffuse, and the longer bond distances for these larger atoms result in poor π overlap.

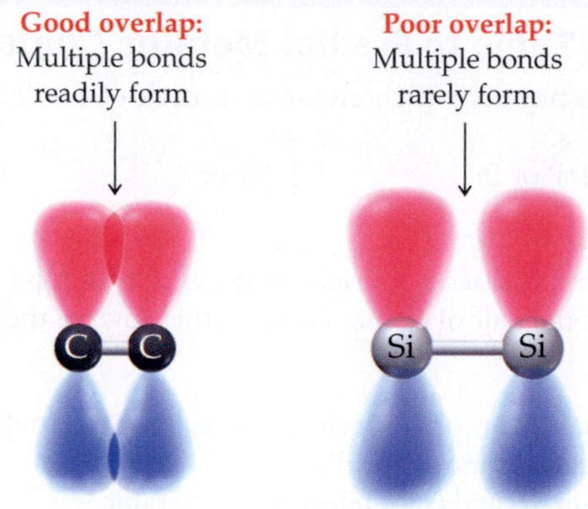

As a result of this poor overlap, π bonds involving *p* orbitals are rare for elements of the third and higher rows. Although compounds with C=C and C≡C multiple bonds are common, molecules with Si=Si and Si≡Si multiple bonds are uncommon and have been synthesized only recently. In group 5A, elemental nitrogen contains triply bonded N_2 molecules, whereas white phosphorus contains tetrahedral P_4 molecules in which each P atom forms three single bonds rather than one triple bond. Similarly, O_2 contains an O=O double bond, whereas elemental sulfur contains crown-shaped S_8 rings in which each S atom forms two single bonds rather than one double bond.

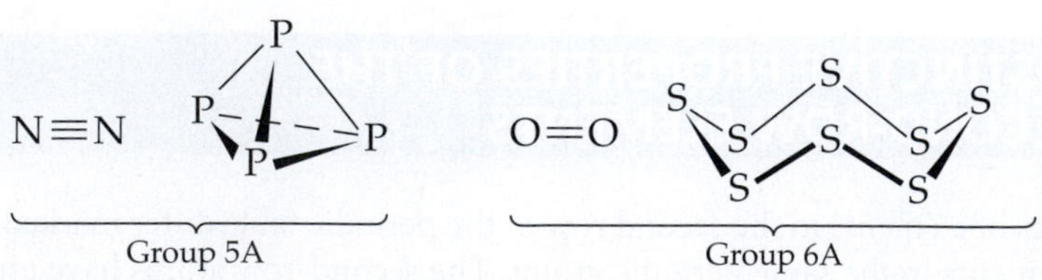

Worked Example 19.2

Exploring Differences between Second- and Third-Row Elements

Account for the following observations:

(a) CO_2 is a gaseous molecular substance, whereas SiO_2 is a covalent network solid in which SiO_4 tetrahedra are linked to four neighboring SiO_4 tetrahedra by shared oxygen atoms.

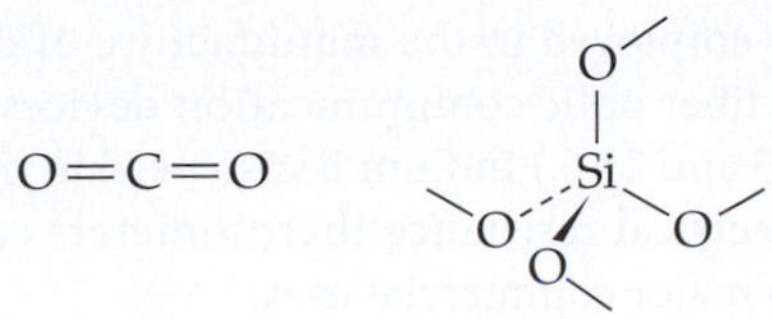

Carbon dioxide

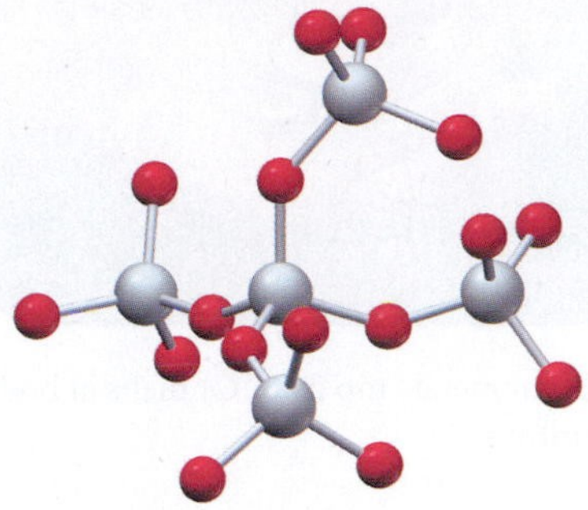

A portion of the structure of SiO_2

(b) Glass made of SiO_2 is attacked by hydrofluoric acid with the formation of SiF_6^{2-} anions. The analogous CF_6^{2-} anion does not exist.

STRATEGY

To account for these differences, remember that

(a) second-row atoms are smaller and form stronger multiple bonds than third-row atoms and

(b) third-row atoms can form more than four bonds because of their larger size.

SOLUTION

(a) Because of its small size and good π overlap with other small atoms, carbon forms strong double bonds with two oxygens to give discrete CO_2 molecules. Because the larger Si atom does not have good π overlap with other atoms, it uses its four valence electrons to form four single bonds rather than two double bonds.

(b) The larger silicon atom can bond to six F^- ions, whereas the smaller carbon atom can form a maximum of only four bonds.

PROBLEM 19.2

(a) Draw electron-dot structures for HNO_3 and H_3PO_4, and suggest a reason for the difference in the formulas of these acids.

(b) Sulfur forms SF_6, but oxygen bonds to a maximum of two F atoms, yielding OF_2. Explain.

CONCEPTUAL PROBLEM 19.3 The organic solvent acetone has the molecular formula $(CH_3)_2CO$. The silicon analogue, a thermally stable lubricant, is a polymer, $[(CH_3)_2SiO]_n$. Account for the difference in structure:

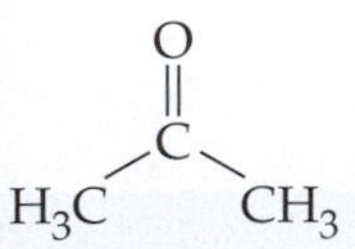

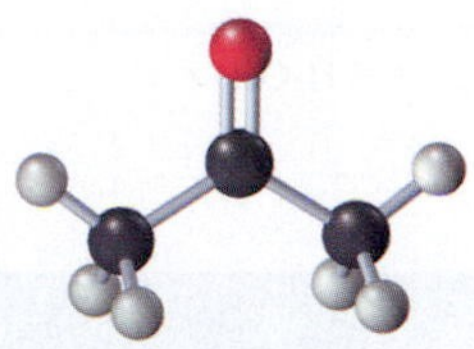

Acetone

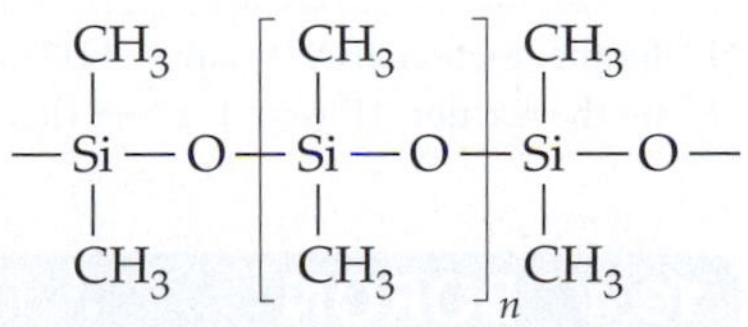

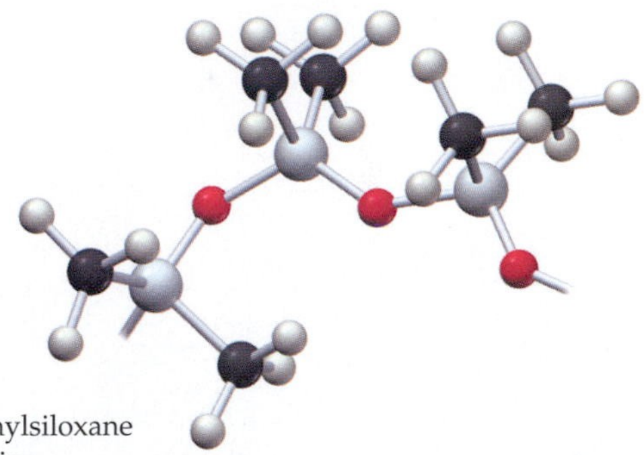

A portion of a dimethylsiloxane polymer chain

19.3 THE GROUP 3A ELEMENTS

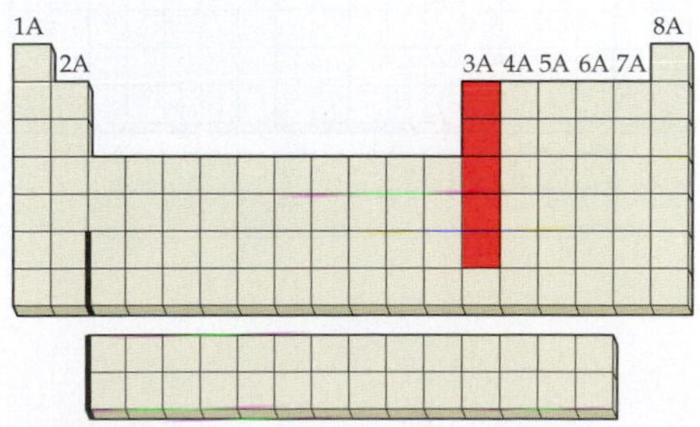

The elements of group 3A—boron, aluminum, gallium, indium, and thallium—are the first of the ***p*-block elements** (Section 2.13) and have the valence electron configuration $ns^2\ np^1$. (The electron configuration and chemical properties of element 113, and also elements 114–118, are as yet unknown because only a few atoms of these recently discovered, radioactive elements have been synthesized.)

With the exception of boron, which behaves as a semimetal, the group 3A elements are silvery metals and good conductors of electricity. Aluminum is the most important element of the group because of its use in making aircraft, electrical transmission lines, cooking utensils, soda cans, and numerous other commercial products. Boron is of interest because of its ability to form unusual kinds of chemical bonds. We'll take up boron and aluminum in the next two sections.

Gallium is remarkable for its unusually low melting point (29.8 °C) and unusually large liquid range (29.8–2204 °C). Its most important use is in making gallium arsenide (GaAs),

REMEMBER...

The periodic table can be divided into regions according to the subshell being filled. The group 1A and 2A elements are the *s*-block elements, and the group 3A–8A elements are the ***p*-block elements**. (Section 2.13)

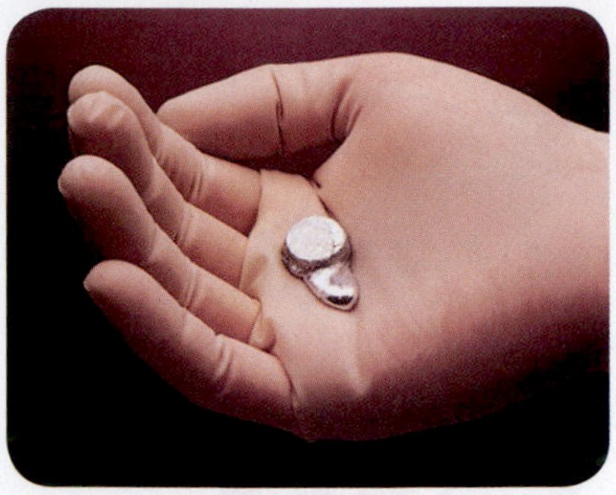

▲ Gallium metal (mp 29.8 °C) melts at body temperature.

a semiconductor material employed in the manufacture of diode lasers for laser printers, compact disc players, and fiber optic communication devices. (We'll say more about semiconductors in Sections 21.5 and 21.6.) Indium is also used in making semiconductor devices, such as transistors and electrical resistance thermometers called *thermistors*. Thallium is extremely toxic and has no major commercial uses.

Despite some irregularities, the properties of the group 3A elements are generally consistent with increasing metallic character down the group (**Table 19.3**). Boron has a much smaller atomic radius and a higher electronegativity than the other elements of the group; it therefore shares its valence electrons in covalent bonds rather than transferring them to another element. With a valence electron configuration of $ns^2\ np^1$, the group 3A elements have a primary oxidation state of +3. In addition, the heavier elements exhibit a +1 state, which is uncommon for gallium and indium but is the most stable oxidation state for thallium.

TABLE 19.3 Properties of the Group 3A Elements

Property	Boron	Aluminum	Gallium	Indium	Thallium
Valence electron configuration	$2s^2\ 2p^1$	$3s^2\ 3p^1$	$4s^2\ 4p^1$	$5s^2\ 5p^1$	$6s^2\ 6p^1$
Melting point (°C)	2075	660	29.8	157	304
Boiling point (°C)	4000	2519	2204	2072	1473
Density (g/cm^3)	2.34	2.70	5.91	7.31	11.8
Abundance in Earth's crust (mass %)	0.0010	8.23	0.0019	0.000 02	0.000 08
Common oxidation states	+3	+3	+3	+3	+3, +1
Atomic radius (pm)	83	143	135	167	170
M^{3+} ionic radius (pm)		54	62	80	89
First ionization energy (kJ/mol)	801	578	579	558	589
Electronegativity	2.0	1.5	1.6	1.7	1.8
Redox potential, $E°$ (V) for $M^{3+}(aq) + 3\ e^- \rightarrow M(s)$	−0.87*	−1.66	−0.55	−0.34	−0.34†

*$E°$ for the reaction $B(OH)_3(aq) + 3\ H^+(aq) + 3\ e^- \rightarrow B(s) + 3\ H_2O(l)$
†$E°$ for the reaction $Tl^+(aq) + e^- \rightarrow Tl(s)$

19.4 BORON

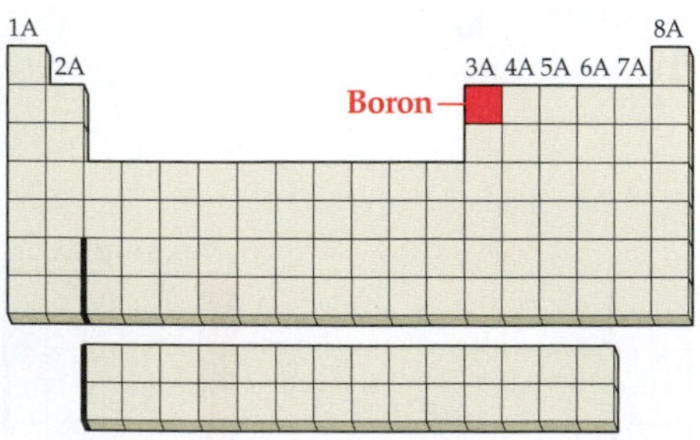

Boron is a relatively rare element, accounting for only 0.0010% of the Earth's crust by mass. Nevertheless, boron is readily available because it occurs in concentrated deposits of borate minerals such as borax, $Na_2B_4O_7 \cdot 10\ H_2O$.

◀ The borate minerals in this open-pit borax mine near Boron, California are believed to have been formed by the evaporation of water from hot springs that were once present.

High-purity, crystalline boron is best obtained by the reaction of boron tribromide and hydrogen on a heated tantalum filament at high temperatures:

$$2\,BBr_3(g) + 3\,H_2(g) \xrightarrow[1200\ ^\circ C]{\text{Ta wire}} 2\,B(s) + 6\,HBr(g)$$

Crystalline boron is a strong, hard, high-melting substance (mp 2075 °C) that is chemically inert at room temperature except for reaction with fluorine. These properties make boron fibers a desirable component in high-strength **composite materials** used in making sports equipment and military aircraft. Unlike Al, Ga, In, and Tl, which are metallic conductors, boron is a semiconductor.

LOOKING AHEAD...

We'll see in Section 21.9 that a **composite material** is a ceramic material that is mixed with another component such as a metal, organic polymer, or another ceramic.

Boron Compounds

Boron Halides The boron halides are highly reactive, volatile, covalent compounds that consist of trigonal planar BX_3 molecules. In their most important reactions, the boron halides behave as Lewis acids. For example, BF_3 reacts with ammonia to give the Lewis acid–base adduct $F_3B{-}NH_3$ (Section 14.15); it reacts with metal fluorides, yielding salts that contain the tetrahedral BF_4^- anion; and it acts as a Lewis acid catalyst in many industrially important organic reactions. In all these reactions, the boron atom uses its vacant $2p$ orbital to accept a share in a pair of electrons from a Lewis base.

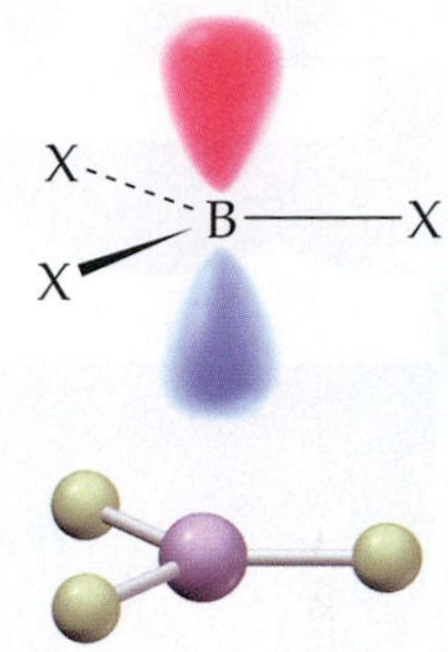

▲ The boron halides are Lewis acids because the boron atom has a vacant $2p$ orbital.

Boron Hydrides The boron hydrides, or **boranes**, are volatile, molecular compounds with formulas B_nH_m. The simplest is diborane (B_2H_6), the dimer of the unstable BH_3.

Boranes are of interest because of their unusual structures and bonding. The diborane molecule has a structure in which two BH_2 groups are connected by two bridging H atoms. The geometry about the B atoms is roughly tetrahedral, and the bridging B–H bonds are significantly longer than the terminal B–H bonds, 133 pm versus 119 pm. The structure differs from that of ethane (C_2H_6) and is unusual because hydrogen normally forms only one bond.

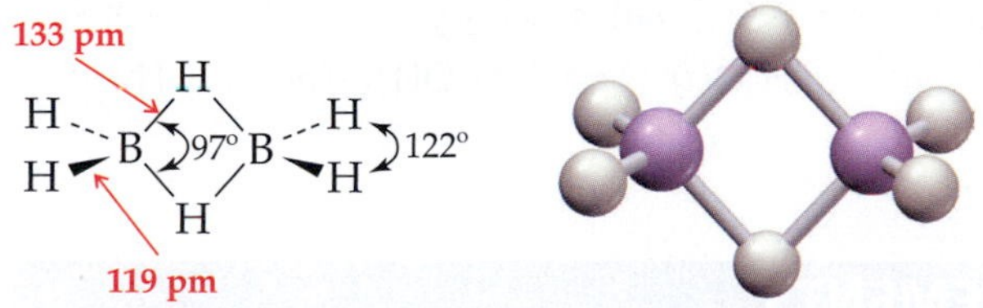

Diborane

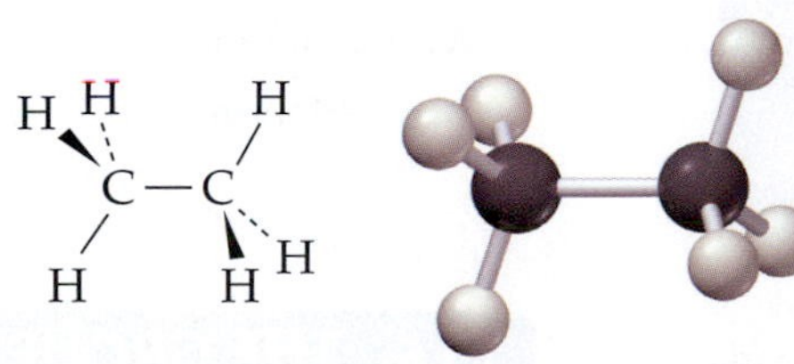

Ethane

If each line in the structural formula of diborane represented an ordinary 2-electron covalent bond between 2 atoms (a *two-center, two-electron bond*, or 2c–2e bond), there would be 8 pairs, or 16 valence electrons. Diborane, however, has a total of only 12 valence electrons—3 electrons from each boron and 1 from each hydrogen. Thus, diborane is said to be *electron-deficient*: It doesn't have enough electrons to form a 2c–2e bond between each pair of bonded atoms.

Because the geometry about the B atoms is roughly tetrahedral, we can assume that each boron uses sp^3 hybrid orbitals to bond to the four neighboring H atoms. The four terminal B–H bonds are assumed to be ordinary 2c–2e bonds, formed by the overlap of a boron sp^3 hybrid orbital and a hydrogen $1s$ orbital, thereby using 4 of the 6 pairs of electrons. Each of the 2 remaining pairs of electrons forms a **three-center, two-electron bond** (3c–2e bond), which joins each bridging H atom to *both* B atoms. Each electron pair occupies a three-center orbital formed by the interaction of three atomic orbitals—one sp^3 hybrid orbital from each B atom and the $1s$ orbital on one bridging H atom (**Figure 19.3**). Because the 2 electrons in the B–H–B bridge are spread out over 3 atoms, the electron density between adjacent atoms is less than in an ordinary 2c–2e bond. The bridging B–H bonds are therefore weaker and correspondingly longer than the terminal B–H bonds.

Each of the two 3-center bonds (one shown darker than the other) is formed by the interaction of an sp^3 hybrid orbital on each B atom and the $1s$ orbital on one bridging H atom.

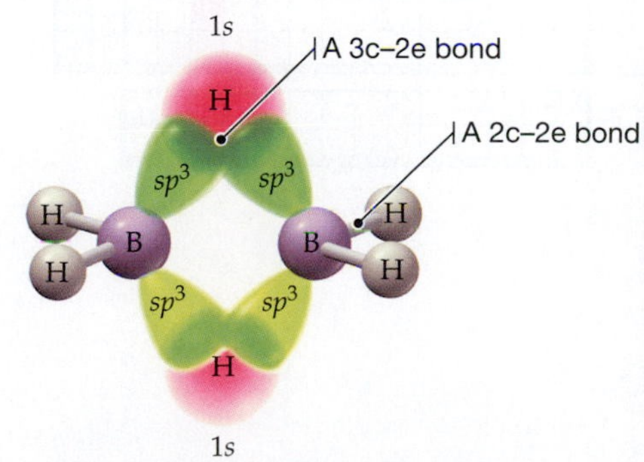

Each 3-center bond contains one pair of electrons and accounts for the bonding in one B–H–B bridge.

FIGURE 19.3

Three-center bonding orbitals in diborane.

PROBLEM 19.4 Why doesn't diborane have an ethane-like structure?

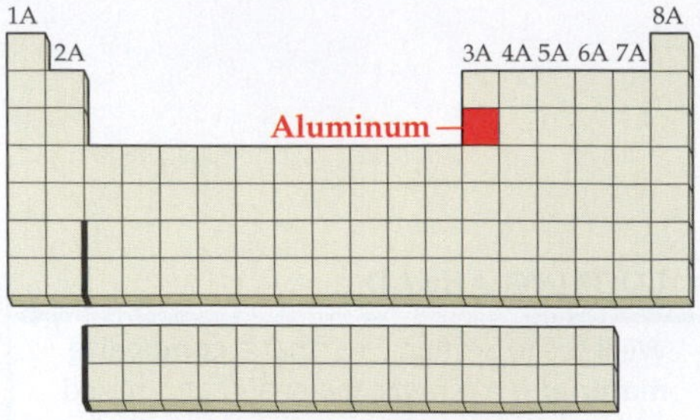

19.5 ALUMINUM

Aluminum, the most abundant metal in the Earth's crust at 8.23%, takes its name from alum, $KAl(SO_4)_2 \cdot 12\,H_2O$, a salt that has been used medicinally since Roman times. In spite of its abundance, the metal nevertheless proved difficult to isolate in pure form. It was such a precious substance in the mid-nineteenth century, in fact, that aluminum cutlery was sometimes used for elegant dinners and the Washington Monument was originally capped by a pyramid of pure aluminum. Not until 1886 did an economical manufacturing process become available.

Aluminum occurs in many common minerals and clays, as well as in gemstones. Ruby and sapphire are both impure forms of Al_2O_3 and receive their color from the presence of small amounts of other elements (Cr in ruby, Fe and Ti in sapphire). Most aluminum is currently obtained from bauxite, $Al_2O_3 \cdot x\,H_2O$. The Al_2O_3 is purified by the Bayer process, which exploits the amphoterism of Al_2O_3 to separate it from Fe_2O_3 and SiO_2 impurities (Section 18.8). Aluminum metal is then obtained by the Hall–Héroult process, the electrolysis of a molten mixture of Al_2O_3 and cryolite (Section 17.13).

Like gallium and indium, aluminum is a reducing agent that undergoes redox reactions by losing all three valence electrons to yield Al^{3+} ions. For example, it reacts with the halogens to yield colorless halides, AlX_3, with oxygen to yield Al_2O_3, and with nitrogen to yield a nitride, AlN:

$$2\,Al(s) + 3\,X_2(g, l, \text{or } s) \longrightarrow 2\,AlX_3(s) \qquad \text{where X = F, Cl, Br, or I}$$
$$4\,Al(s) + 3\,O_2(g) \longrightarrow 2\,Al_2O_3(s)$$
$$2\,Al(s) + N_2(g) \longrightarrow 2\,AlN(s)$$

Aluminum reacts with both acidic and basic solutions to give H_2 gas and either Al^{3+} ions in acidic solutions or aluminate ions, $Al(OH)_4^-$, in basic solutions.

$$\text{Acidic solution:} \quad 2\,Al(s) + 6\,H^+(aq) \longrightarrow 2\,Al^{3+}(aq) + 3\,H_2(g)$$
$$\text{Basic solution:} \quad 2\,Al(s) + 2\,OH^-(aq) + 6\,H_2O(l) \longrightarrow 2\,Al(OH)_4^-(aq) + 3\,H_2(g)$$

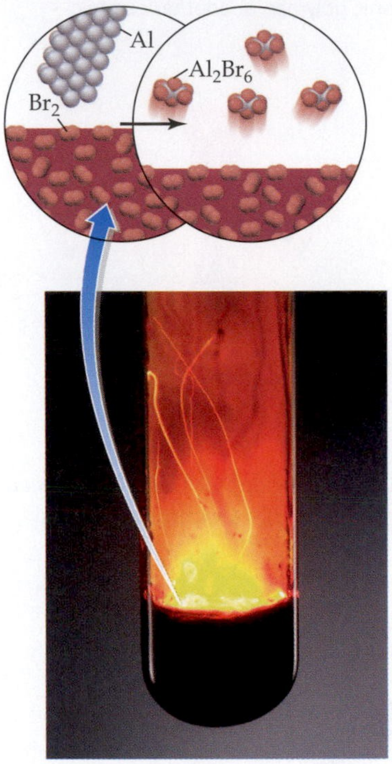

▲ Aluminum metal reacts with liquid bromine in a spectacular display of sparks, yielding the dimer Al_2Br_6.

19.6 THE GROUP 4A ELEMENTS

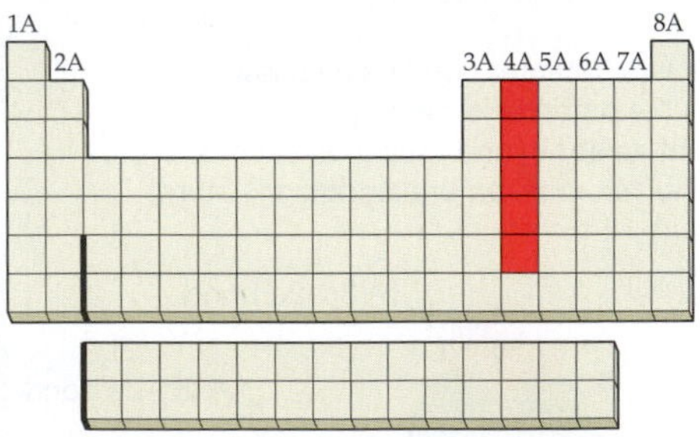

The group 4A elements—carbon, silicon, germanium, tin, and lead—are especially important both in industry and in living organisms. Carbon is present in all plants and animals, accounts for 23% of the mass of the human body, and is an essential constituent of the molecules on which life is based. Silicon is equally important in the mineral world: It is present in numerous silicate minerals and is the second most abundant element in the Earth's crust. Both silicon and germanium are used in making modern solid-state electronic devices. Tin and lead have been known and used since ancient times.

The group 4A elements exemplify the increase in metallic character down a group in the periodic table: Carbon is a nonmetal, silicon and germanium are semimetals, and tin and lead are metals. The usual periodic trends in atomic size, ionization energy, and electronegativity are evident in the data of **Table 19.4**.

The physical properties of the heavier group 4A elements nicely illustrate the gradual transition from semimetallic to metallic character. Germanium is a relatively high-melting, brittle semiconductor that has the same crystal structure as diamond and silicon. Tin exists in two allotropic forms: the usual silvery white metallic form called *white tin* and a brittle, semiconducting form with the diamond structure called *gray tin*. White tin is the stable form at room temperature, but when kept for long periods of time below the transition temperature of 13 °C, it slowly crumbles to gray tin, a phenomenon known as tin disease:

$$\text{White tin} \xrightleftharpoons{13\ ^\circ\text{C}} \text{Gray tin}$$

Both white tin and lead are soft, malleable, low-melting metals. Only the metallic form occurs for lead.

TABLE 19.4 Properties of the Group 4A Elements

Property	Carbon	Silicon	Germanium	Tin	Lead
Valence electron configuration	$2s^2\,2p^2$	$3s^2\,3p^2$	$4s^2\,4p^2$	$5s^2\,5p^2$	$6s^2\,6p^2$
Melting point (°C)	~8700*	1414	938	232†	327
Boiling point (°C)		3265	2833	2602	1749
Density (g/cm^3)	3.51*	2.33	5.32	7.29†	11.3
Abundance in Earth's crust (mass %)	0.020	28.2	0.0005	0.0002	0.0013
Common oxidation states	+2, +4	+4	+4	+2, +4	+2, +4
Atomic radius (pm)	77	118	122	140	175
First ionization energy (kJ/mol)	1086	787	762	709	716
Electronegativity	2.5	1.8	1.8	1.8	1.9
Redox potential, $E°$ (V) for $M^{2+}(aq) + 2\,e^- \rightarrow M(s)$				−0.14	−0.13

*Diamond; mp at 6–10 million atm.

†White Sn

Germanium

Tin

Lead

Because the group 4A elements have the valence electron configuration $ns^2\,np^2$, they adopt an oxidation state of +4, as in CCl_4, $SiCl_4$, $GeCl_4$, $SnCl_4$, and $PbCl_4$. These compounds are volatile, molecular liquids in which the group 4A atom uses tetrahedral sp^3 hybrid orbitals to form covalent bonds to the Cl atoms. The +2 oxidation state occurs for tin and lead and is the most stable oxidation state for lead. Both $Sn^{2+}(aq)$ and $Pb^{2+}(aq)$ are common solution species, but there are no simple $M^{4+}(aq)$ ions for any of the group 4A elements. Instead, M(IV) species exist in solution as covalently bonded complex ions—for example, $SiF_6{}^{2-}$, $GeCl_6{}^{2-}$, $Sn(OH)_6{}^{2-}$, and $Pb(OH)_6{}^{2-}$. In general, the +4 oxidation-state compounds are covalent, and the compounds with tin and lead in the +2 oxidation state are largely ionic.

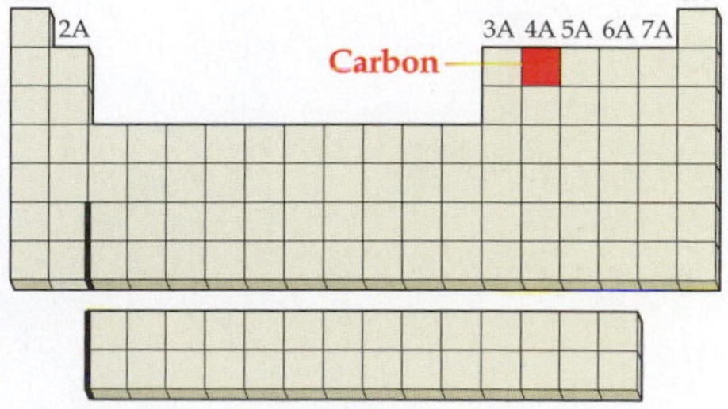

19.7 CARBON

Carbon, although the second most abundant element in living organisms, accounts for only 0.02% of the mass of the Earth's crust. It is present in carbonate minerals, such as limestone ($CaCO_3$), and in fossil fuels, such as coal, petroleum, and natural gas. In uncombined form, carbon is found as diamond and graphite.

Recall from Section 10.10 that diamond has a covalent network structure in which each C atom uses sp^3 hybrid orbitals to form a tetrahedral array of σ bonds, with bond lengths of 154 pm. (See Figure 10.26b, page 385.) The interlocking, three-dimensional network of strong bonds makes diamond an extremely hard substance and gives it the highest melting point for an element—about 9000 K at a pressure of 6–10 million atm. At 1 atm pressure, diamond doesn't melt at any temperature. Naturally occurring diamonds are formed over several billion years at the high temperatures and pressures that exist some 200 km below the Earth's surface. Subsequently, the diamonds are brought near the surface by volcanic activity. Because the valence electrons in diamond are localized in the σ bonds, they are not free to move when an electrical potential is applied and diamond is therefore an electrical insulator.

Graphite has a two-dimensional sheetlike structure (Figure 10.26a, page 385) in which each C atom uses sp^2 hybrid orbitals to form trigonal planar σ bonds to three neighboring C atoms. In addition, each C atom uses its remaining p orbital, which is perpendicular to the plane of the sheet, to form a π bond. Because each C atom must share its π bond with its three neighbors, the π electrons are delocalized and are free to move in the plane of the sheet. As a result, the electrical conductivity of graphite in a direction parallel to the sheets is about 10^{20} times greater than the conductivity of diamond, which makes graphite useful as an electrode material.

The carbon sheets in graphite are separated by a distance of 335 pm and are held together by only London dispersion forces. Atmospheric gases can be absorbed between the sheets, thus enabling the sheets to easily slide over one another. As a result, graphite has a slippery feel and can be used as a lubricant.

Closely related to graphite is *graphene*, a remarkable two-dimensional array of hexagonally arranged carbon atoms that is just one atom thick. First prepared in 2004 by pulling apart layers of graphite with sticky tape—the so-called Scotch tape method—graphene is extremely strong and flexible, and is a superb conductor of electricity. Possible future applications of graphene include use in new, smaller and faster computer chips, thin coatings for solar cells and LCD displays, and components of strong composite materials. For their work on the preparation and properties of graphene, Andre Geim and Konstantin Novoselov were awarded the 2010 Nobel Prize in Physics.

Still another crystalline allotrope of carbon, called *fullerene*, was discovered in 1985 by Robert F. Curl, Jr., Harold Kroto, and Richard E. Smalley, winners of the 1996 Nobel Prize in Chemistry. This allotrope consists of nearly spherical C_{60} molecules with the shape of a soccer ball (Figure 10.27a, page 386). Fullerene can be prepared in relatively large amounts by electrically heating a graphite rod in an atmosphere of helium. Unlike diamond, graphite, and graphene, C_{60} is a molecular substance and is therefore soluble in nonpolar organic solvents.

Related to C_{60} are other carbon clusters, such as the egg-shaped molecule C_{70} and the tube-shaped molecules called *carbon nanotubes*. These ultrasmall tubes are stronger than steel. Some are metallic and conduct electricity 10 times better than copper. Others are semiconductors and might eventually appear in some electronic devices.

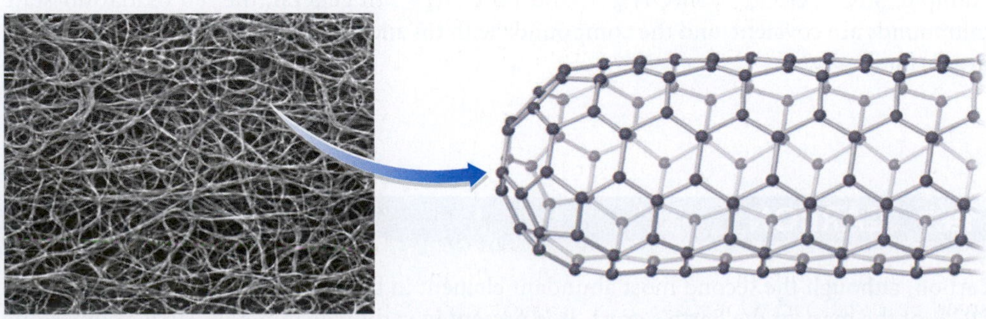

▲ A carbon nanotube has walls made of graphite sheets rolled into a cylinder and hemispherical tips made of a portion of a fullerene molecule.

Carbon Compounds

Carbon forms more than 40 million known compounds, most of which are classified as *organic*; only CO, CO_2, $CaCO_3$, HCN, CaC_2, and a handful of others are considered to be *inorganic*. The distinction is a historical one rather than a scientific one, though, as discussed in more detail in Chapter 23. For the present, we'll look only at some simple inorganic compounds of carbon.

Oxides of Carbon The most important oxides of carbon are carbon monoxide (CO) and carbon dioxide (CO_2). Carbon monoxide is a colorless, odorless, toxic gas that forms when carbon or hydrocarbon fuels are burned in a limited supply of oxygen. In an excess of oxygen, CO burns to give CO_2:

$$2\,C(s) + O_2(g) \longrightarrow 2\,CO(g) \qquad \Delta H° = -221\text{ kJ}$$
$$2\,CO(g) + O_2(g) \longrightarrow 2\,CO_2(g) \qquad \Delta H° = -566\text{ kJ}$$

The high toxicity of CO results from its ability to bond strongly to the iron(II) atom of hemoglobin, the oxygen-carrying protein in red blood cells. Because hemoglobin has a greater affinity for CO than for O_2 by a factor of 200, even small concentrations of CO in the blood can convert a substantial fraction of the O_2-bonded hemoglobin, called *oxyhemoglobin*, to the CO-bonded form, called *carboxyhemoglobin*, thus impairing the ability of hemoglobin to carry O_2 to the tissues:

$$\underset{\text{Oxyhemoglobin}}{Hb{-}O_2} + CO \rightleftharpoons \underset{\text{Carboxyhemoglobin}}{Hb{-}CO} + O_2$$

▲ Known as "the silent killer," carbon monoxide is the cause of 500 accidental poisoning deaths in the United States each year. Household carbon monoxide detectors are designed to monitor CO concentrations over time and sound an alarm before life-threatening levels are reached.

A CO concentration in air of only 200 ppm can produce symptoms such as headache, dizziness, and nausea, and a concentration of 1000 ppm can cause death within 4 hours. One hazard of cigarette smoking is chronic exposure to low levels of CO. Because CO reduces the blood's ability to carry O_2, the heart must work harder to supply O_2 to the tissues, thus increasing the risk of heart attack.

Carbon dioxide, a colorless, odorless, nonpoisonous gas, is produced when fuels are burned in an excess of oxygen and is an end product of food metabolism in humans and animals. Commercially, carbon dioxide is obtained as a byproduct of the yeast-catalyzed fermentation of sugar in the manufacture of alcoholic beverages:

$$\underset{\text{Glucose}}{C_6H_{12}O_6(aq)} \xrightarrow{\text{Yeast}} \underset{\text{Ethanol}}{2\,CH_3CH_2OH(aq)} + 2\,CO_2(g)$$

It is produced in the laboratory when metal carbonates are treated with strong acids:

$$Na_2CO_3(s) + 2\,H^+(aq) \longrightarrow 2\,Na^+(aq) + CO_2(g) + H_2O(l)$$

Carbon dioxide is of environmental concern not only because of global warming but also because approximately one-third of the CO_2 produced by human activities winds up in the Earth's oceans, lowering their pH and dissolving $CaCO_3$, with adverse effects on marine life, such as coral and mollusks, that require $CaCO_3$ to grow:

$$CaCO_3(s) + H^+(aq) \rightleftharpoons Ca^{2+}(aq) + HCO_3^-(aq)$$

CO_2 is used in fire extinguishers because it is nonflammable and is about 1.5 times more dense than air. It therefore settles over a small fire like a blanket, separating the fire from its source of oxygen. Solid CO_2 (dry ice), which sublimes at −78 °C, is used primarily as a refrigerant.

Above 31 °C and 73 atm pressure, carbon dioxide becomes a **supercritical fluid**, a phase that is neither a true liquid nor a true gas (Section 10.11). Because supercritical carbon dioxide ($scCO_2$) is nontoxic, nonflammable, and easily recovered, it has been touted as a "green" solvent, a replacement for environmentally unfriendly organic solvents such as chlorinated hydrocarbons. $scCO_2$ is already used as a solvent in the extraction of caffeine from coffee and tea, and it is likely to replace organic solvents and water in the manufacture of computer microchips. The largest potential market for $scCO_2$ is as a replacement for chlorinated organic solvents in the dry cleaning of clothes. Biological applications of $scCO_2$ are also on the horizon, for example, as a medium for coating drug particles with polymers in the production of time-released drugs.

REMEMBER...

The critical point in a phase diagram is the point at which the liquid/gas boundary line abruptly ends. Beyond that point the liquid and gas phases blend into each other to give a **supercritical fluid**. (Section 10.11)

Carbonates Carbonic acid, H_2CO_3, forms two series of salts: carbonates, which contain the trigonal planar CO_3^{2-} ion, and hydrogen carbonates (bicarbonates), which contain the HCO_3^- ion. Several million tons of soda ash, Na_2CO_3, are used annually to make glass, and $Na_2CO_3 \cdot 10\,H_2O$, known as washing soda, is used to launder textiles. The carbonate ion removes cations such as Ca^{2+} and Mg^{2+} from hard water, and it acts as a base to give OH^- ions, which help remove grease from fabrics:

$$Ca^{2+}(aq) + CO_3^{2-}(aq) \longrightarrow CaCO_3(s)$$
$$CO_3^{2-}(aq) + H_2O(l) \rightleftharpoons HCO_3^-(aq) + OH^-(aq)$$

Sodium hydrogen carbonate, $NaHCO_3$, is called baking soda because it reacts with acidic substances in food to yield bubbles of CO_2 gas that cause dough to rise:

$$NaHCO_3(s) + H^+(aq) \longrightarrow Na^+(aq) + CO_2(g) + H_2O(l)$$

Hydrogen Cyanide and Cyanides Hydrogen cyanide is a highly toxic, volatile substance (bp 26 °C) produced when metal cyanide solutions are acidified:

$$CN^-(aq) + H^+(aq) \longrightarrow HCN(aq)$$

Aqueous solutions of HCN, known as hydrocyanic acid, are very weakly acidic ($K_a = 4.9 \times 10^{-10}$).

The cyanide ion, CN^-, acts as a Lewis base, bonding to transition metals through the lone pair of electrons on its carbon atom. In fact, the toxicity of HCN and other cyanides is due to the strong bonding of CN^- to iron(III) in cytochrome oxidase, an enzyme involved in the oxidation of food molecules. With CN^- attached to the iron, the enzyme is unable to function. Cellular energy production thus comes to a halt, and rapid death follows.

▲ A cyanide leaching pool for extracting gold operated by the Zolotaya Zvezda mining company in Khakassia, Russia.

The bonding of CN^- to gold and silver is exploited in the extraction of these metals from their ores. The crushed rock containing small amounts of the precious metals is treated with an aerated cyanide solution, and the metals are then recovered from their $M(CN)_2^-$ complex ions by reduction with zinc. For gold, the reactions are

$$4\,Au(s) + 8\,CN^-(aq) + O_2(g) + 2\,H_2O(l) \longrightarrow 4\,Au(CN)_2^-(aq) + 4\,OH^-(aq)$$
$$2\,Au(CN)_2^-(aq) + Zn(s) \longrightarrow 2\,Au(s) + Zn(CN)_4^{2-}(aq)$$

Because of the toxicity of cyanides, this method is now banned in several U.S. states and foreign countries.

Carbides Carbon forms a number of binary inorganic compounds called **carbides**, in which the carbon atom has a negative oxidation state. Examples include ionic carbides of active metals such as CaC_2 and Al_4C_3, interstitial carbides of transition metals such as Fe_3C, and covalent network carbides such as SiC, the industrial abrasive called carborundum. Calcium carbide is a high-melting, colorless solid that has an NaCl type of structure with Ca^{2+} ions in place of Na^+ and C_2^{2-} ions in place of Cl^-. It is prepared by heating lime (CaO) and coke (C) at high temperatures and is used to prepare acetylene (C_2H_2) for oxyacetylene welding:

$$CaO(s) + 3\,C(s) \xrightarrow{2200\ °C} CaC_2(s) + CO(g)$$
$$CaC_2(s) + 2\,H_2O(l) \longrightarrow C_2H_2(g) + Ca(OH)_2(s)$$

▶ **PROBLEM 19.5** The equilibrium between oxyhemoglobin and carboxyhemoglobin suggests an approach to treating mild cases of carbon monoxide poisoning. Explain.

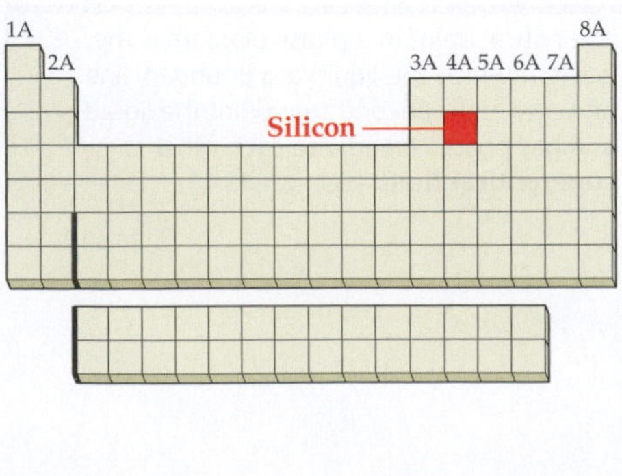

19.8 SILICON

Silicon is a hard, gray, semiconducting solid that melts at 1414 °C. It crystallizes in a diamondlike structure but does not form a graphitelike allotrope because of the relatively poor overlap of silicon π orbitals. In nature, silicon is generally found combined with oxygen in

SiO_2 and in various silicate minerals. It is obtained in elemental form by the reduction of silica sand (SiO_2) with coke (C) in an electric furnace:

$$SiO_2(l) + 2\,C(s) \xrightarrow{\text{Heat}} Si(l) + 2\,CO(g)$$

The silicon used for making solid-state semiconductor devices such as transistors, computer chips, and solar cells must be ultrapure, with impurities at a level of less than $10^{-7}\%$ (1 ppb). For electronic applications, silicon is purified by converting it to $SiCl_4$, a volatile liquid (bp 58 °C) that can be separated from impurities by fractional distillation (Section 11.10) and then converted back to elemental silicon by reduction with hydrogen:

$$Si(s) + 2\,Cl_2(g) \longrightarrow SiCl_4(l)$$
$$SiCl_4(g) + 2\,H_2(g) \xrightarrow{\text{Heat}} Si(s) + 4\,HCl(g)$$

The silicon is purified further by a process called **zone refining** (**Figure 19.4a**), in which a heater melts a narrow zone of a silicon rod. Because the impurities are more soluble in the liquid phase than in the solid, they concentrate in the molten zone. As the heater sweeps slowly down the rod, ultrapure silicon crystallizes at the trailing edge of the molten zone and the impurities are dragged to the rod's lower end. **Figure 19.4b** shows some samples of ultrapure silicon.

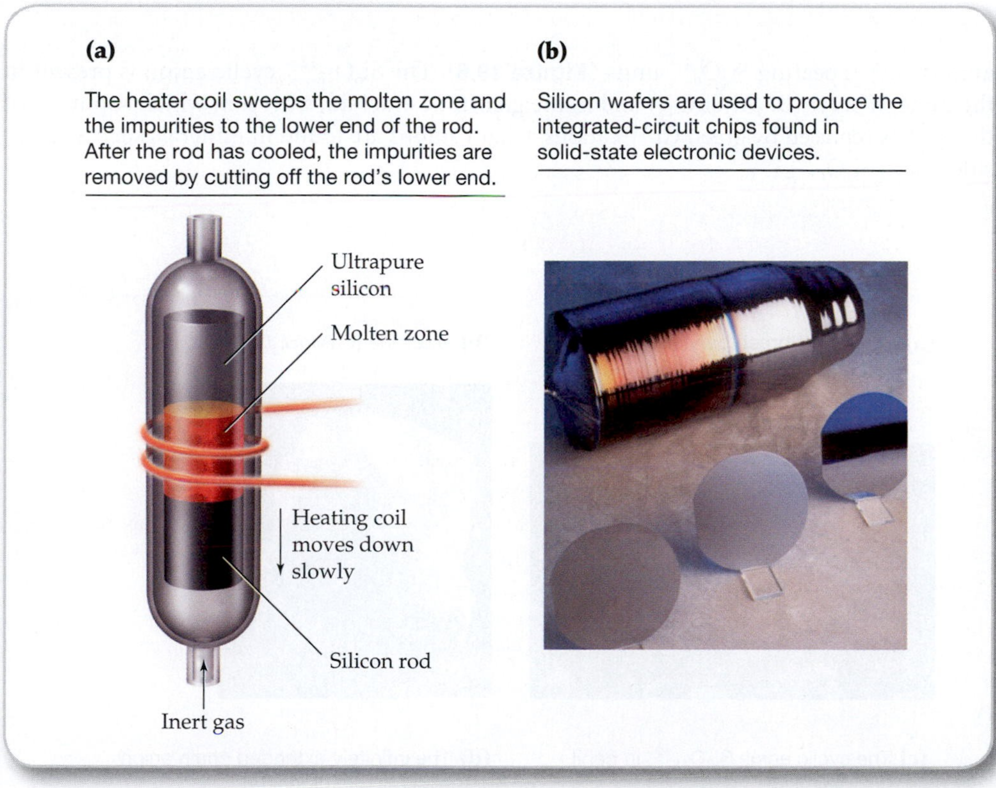

FIGURE 19.4

Ultrapure silicon for solid-state semiconductor devices. **(a)** Purification of silicon by zone refining. **(b)** A rod of ultrapure silicon and silicon wafers cut from the rod.

Silicates Approximately 90% of the Earth's crust consists of **silicates**, ionic compounds that contain silicon oxoanions along with cations such as Na^+, K^+, Mg^{2+}, or Ca^{2+} to balance the negative charge of the anions. As shown in **Figure 19.5**, the basic structural building block in silicates is the SiO_4 tetrahedron, a unit that occurs as the simple orthosilicate ion ($SiO_4{}^{4-}$) in the mineral zircon, $ZrSiO_4$. (Zircon is not to be confused with cubic zirconia, the cubic form of ZrO_2.) If two SiO_4 tetrahedra share a common O atom, the disilicate anion $Si_2O_7{}^{6-}$, found in $Sc_2Si_2O_7$, results.

Simple anions such as $SiO_4{}^{4-}$ and $Si_2O_7{}^{6-}$ are relatively rare in silicate minerals. More common are larger anions in which two or more O atoms bridge between Si atoms to give rings, chains, layers, and extended three-dimensional structures. The sharing of two O atoms per SiO_4 tetrahedron gives either cyclic anions, such as $Si_6O_{18}{}^{12-}$, or infinitely extended chain

▲ The mineral zircon ($ZrSiO_4$) is a relatively inexpensive gemstone.

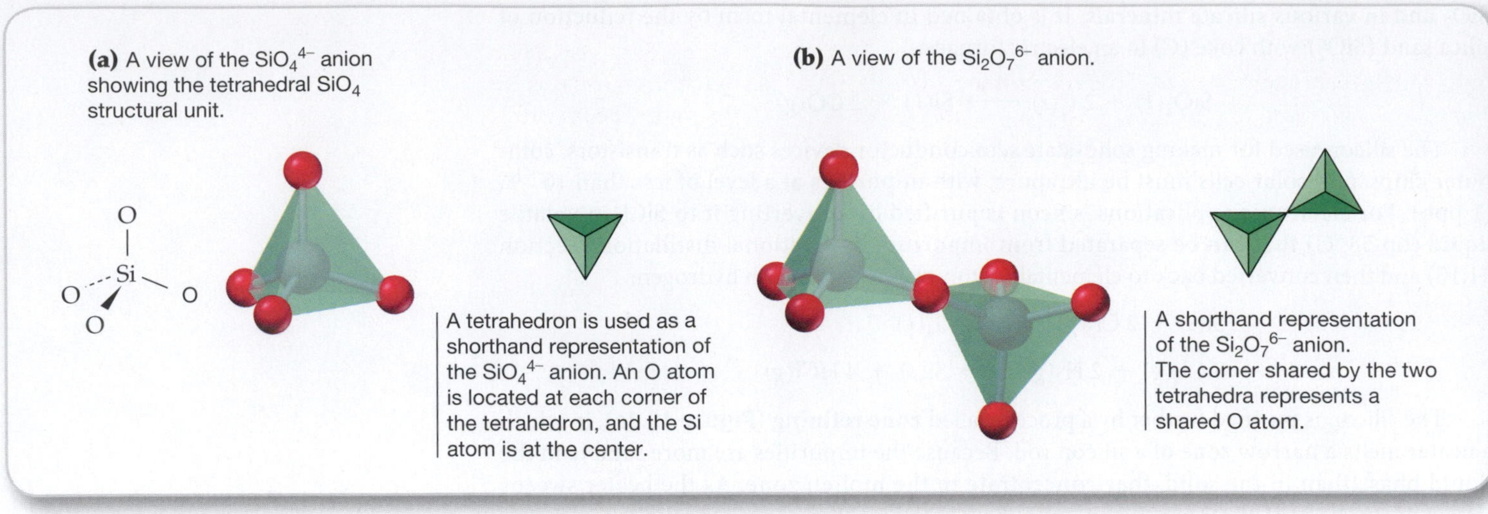

FIGURE 19.5

The SiO_4^{4-} and $Si_2O_7^{6-}$ anions and their shorthand representations.

anions with repeating $Si_2O_6^{4-}$ units (**Figure 19.6**). The $Si_6O_{18}^{12-}$ cyclic anion is present in the mineral beryl ($Be_3Al_2Si_6O_{18}$) and in the gemstone emerald, a beryl in which about 2% of the Al^{3+} is replaced by green Cr^{3+} cations. Chain anions are found in minerals such as diopside, $CaMgSi_2O_6$.

FIGURE 19.6

Samples of the silicate minerals emerald and diopside and the structures of their anions.

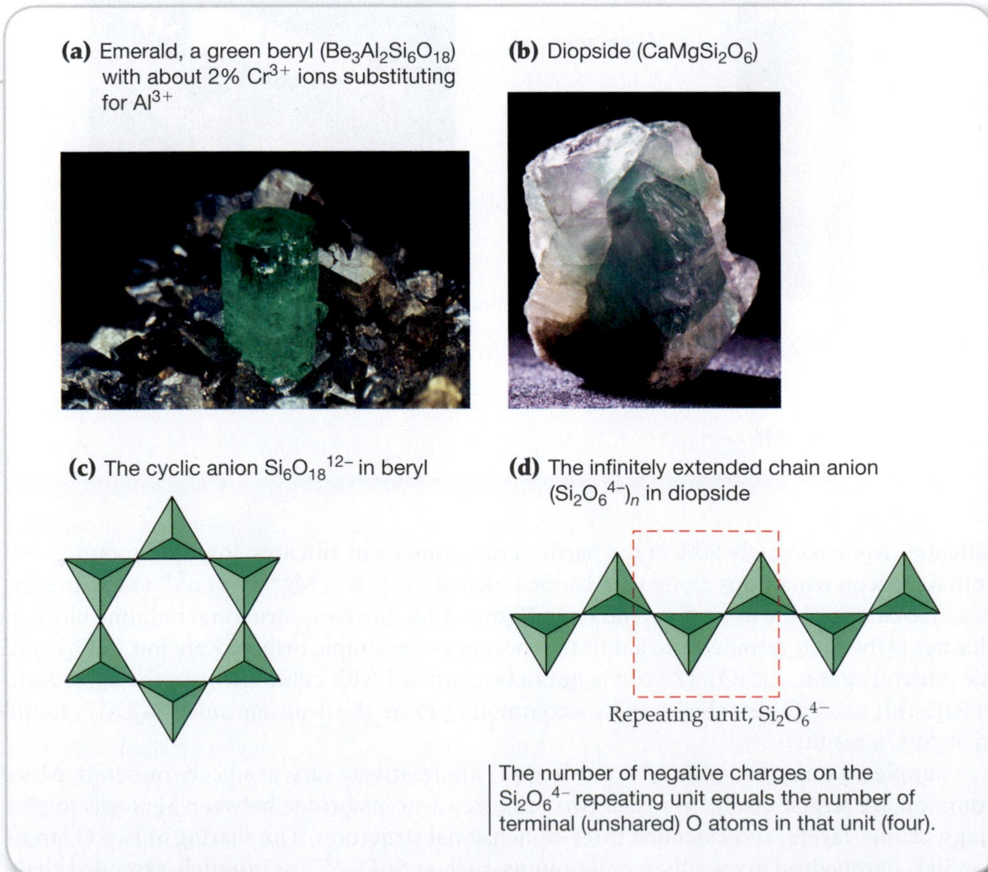

As shown in **Figure 19.7**, additional sharing of O atoms gives the double-stranded chain anions $(Si_4O_{11}{}^{6-})_n$ found in asbestos minerals such as tremolite, $Ca_2Mg_5(Si_4O_{11})_2(OH)_2$, and the infinitely extended two-dimensional layer anions $(Si_4O_{10}{}^{4-})_n$ found in clay minerals, micas, and talc, $Mg_3(OH)_2(Si_4O_{10})$. Asbestos is fibrous, as shown in the figure, because the ionic bonds between the silicate chain anions and the Ca^{2+} and Mg^{2+} cations that lie between the chains and hold them together are relatively weak and easily broken. Mica is sheetlike because the ionic bonds between the two-dimensional layer anions and the interposed metal cations are much weaker than the Si–O covalent bonds within the layer anions.

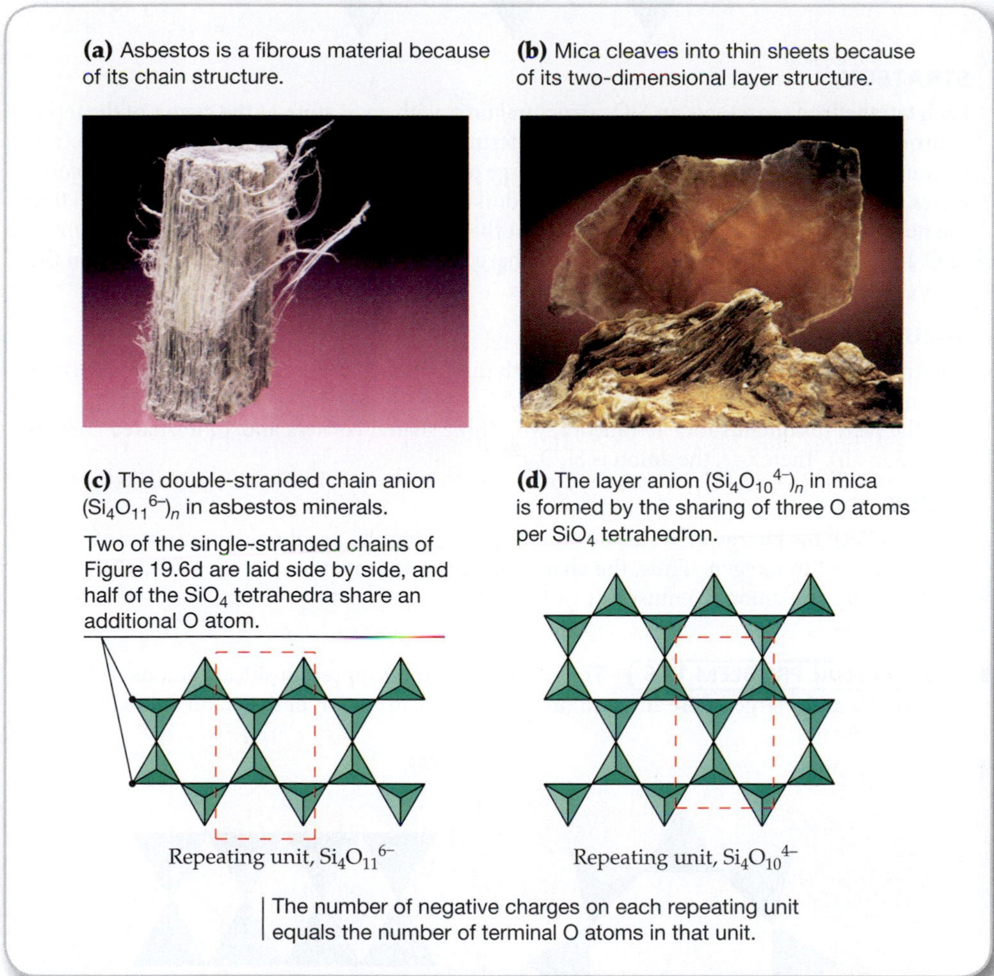

FIGURE 19.7

Samples of asbestos and mica, and the structures of their anions.

If the layer anions of Figure 19.7d are stacked on top of one another and the terminal O atoms are shared, an infinitely extended three-dimensional structure is obtained in which all four O atoms of each SiO_4 tetrahedron are shared between two Si atoms, resulting in *silica* (SiO_2). The mineral quartz is one of many crystalline forms of SiO_2.

Partial substitution of the Si^{4+} of SiO_2 with Al^{3+} gives **aluminosilicates** called *feldspars*, the most abundant of all minerals. An example is orthoclase, $KAlSi_3O_8$, which has a three-dimensional structure like that of SiO_2. The $(AlSi_3O_8{}^{-})_n$ framework consists of SiO_4 and AlO_4 tetrahedra that share all four of their corners with neighboring tetrahedra. The K^+ cation balances the negative charge. In the aluminosilicates known as *zeolites*, the SiO_4 and AlO_4 tetrahedra are joined together in an open structure that has a three-dimensional network of cavities linked by channels. Because only small molecules can enter these channels, zeolites act as molecular sieves for separating small molecules from larger ones. They are also used as catalysts in many industrial processes, including the manufacture of gasoline.

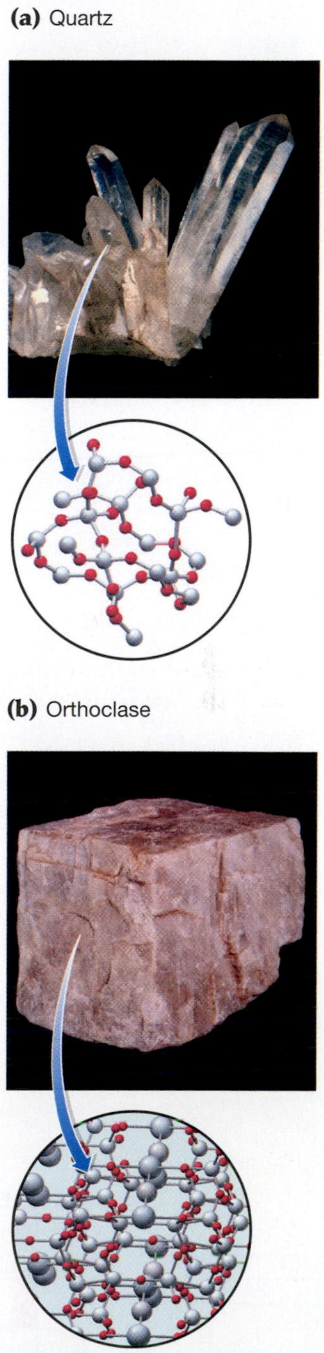

▲ (a) Pure crystalline quartz, one form of SiO_2, is colorless. (b) Orthoclase, $KAlSi_3O_8$, has a structure similar to that of SiO_2.

WORKED CONCEPTUAL EXAMPLE 19.3

Interpreting Representations of Silicate Anions

The following pictures represent silicate anions. What are the formula and charge of each anion?

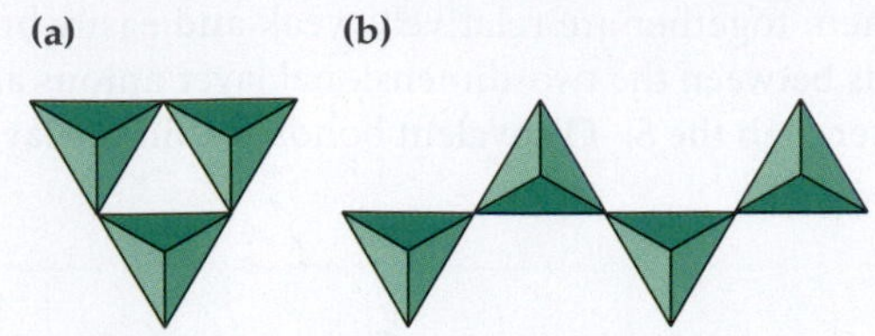

STRATEGY

Each tetrahedron represents an SiO_4 structural unit, with an Si atom at the center of the tetrahedron and an O atom at each corner. Each terminal (unshared) O atom needs one electron to complete its octet and therefore has a charge of -1. Each bridging (shared) O atom completes its octet by forming bonds to two Si atoms and is therefore electrically neutral. To find the number of Si atoms in the formula, count the number of tetrahedra. To find the number of O atoms, count the number of corners (shared and unshared). To find the charge on the anion, count the number of unshared corners.

SOLUTION

(a) The picture contains three tetrahedra, with three shared corners and six unshared corners (nine in all). Therefore, the anion is $Si_3O_9^{6-}$.

(b) The picture contains four tetrahedra, with three shared corners and 10 unshared corners (13 in all). Therefore, the anion is $Si_4O_{13}^{10-}$.

CHECK

We can check the charge on a silicate anion by assigning the usual oxidation states of $+4$ to silicon and -2 to oxygen. Thus, the charge on anion **(a)** must be $(3)(4) + (9)(-2) = -6$, and the charge on anion **(b)** must be $(4)(4) + (13)(-2) = -10$.

CONCEPTUAL PROBLEM 19.6 The following pictures represent silicate anions. What are the formula and charge of the anion in **(a)** and the repeating unit of the chain anion in **(b)**?

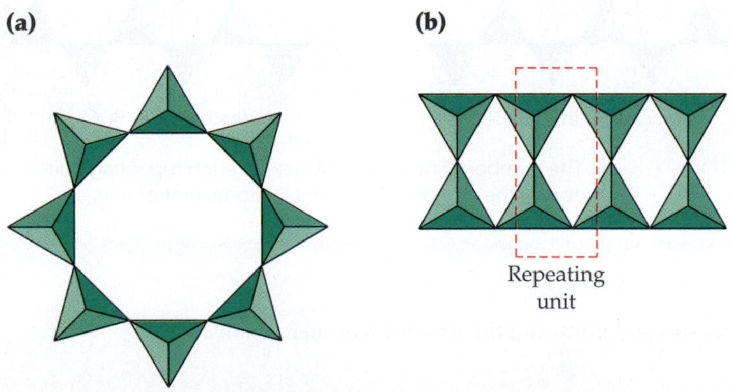

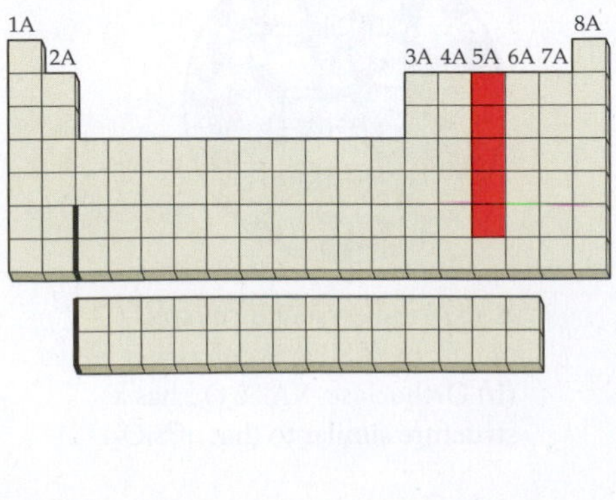

19.9 THE GROUP 5A ELEMENTS

The group 5A elements are nitrogen, phosphorus, arsenic, antimony, and bismuth. As shown in **Table 19.5**, these elements exhibit the expected trends of increasing atomic size, decreasing ionization energy, and decreasing electronegativity down the group from N to Bi. Accordingly, metallic character increases in the same order: N and P are typical nonmetals, As and Sb are semimetals, and Bi is a metal. Thus, nitrogen is a gaseous substance comprised

TABLE 19.5 Properties of the Group 5A Elements

Property	Nitrogen	Phosphorus	Arsenic	Antimony	Bismuth
Valence electron configuration	$2s^2\,2p^3$	$3s^2\,3p^3$	$4s^2\,4p^3$	$5s^2\,5p^3$	$6s^2\,6p^3$
Melting point (°C)	−210	44*	614†	631	271
Boiling point (°C)	−196	280		1587	1564
Atomic radius (pm)	75	110	120	140	150
First ionization energy (kJ/mol)	1402	1012	944	831	703
Electronegativity	3.0	2.1	2.0	1.9	1.9

*White phosphorus

†Sublimes

of N_2 molecules, but bismuth is a silvery solid having an extended three-dimensional structure. The increasing metallic character of the heavier elements is also evident in the acid–base properties of their oxides: Most nitrogen and phosphorus oxides are acidic, arsenic and antimony oxides are amphoteric, and Bi_2O_3 is basic.

Arsenic

Antimony

Bismuth

The valence electron configuration of the group 5A elements is $ns^2\,np^3$. They exhibit a maximum oxidation state of +5 in compounds such as HNO_3 and PF_5, in which they share all 5 valence electrons with a more electronegative element. They show a minimum oxidation state of −3 in compounds such as NH_3 and PH_3, where they share 3 valence electrons with a less electronegative element. The −3 state also occurs in ionic compounds such as Li_3N and Mg_3N_2, which contain the N^{3-} anion.

Nitrogen and phosphorus are unusual in that they exhibit all oxidation states between −3 and +5. For arsenic and antimony, the most important oxidation states are +3, as in $AsCl_3$, As_2O_3, and H_3AsO_3, and +5, as in AsF_5, As_2O_5, and H_3AsO_4. The +5 state becomes increasingly less stable from As to Sb to Bi. Another indication of increasing metallic character down the group is the existence of Sb^{3+} and Bi^{3+} cations in salts such as $Sb_2(SO_4)_3$ and $Bi(NO_3)_3$. By contrast, no simple cations are found in compounds of N or P.

As, Sb, and Bi are found in sulfide ores and are used to make various metal alloys. Arsenic is also used to make pesticides and semiconductors, such as GaAs. Bismuth compounds are present in some pharmaceuticals, such as Pepto-Bismol.

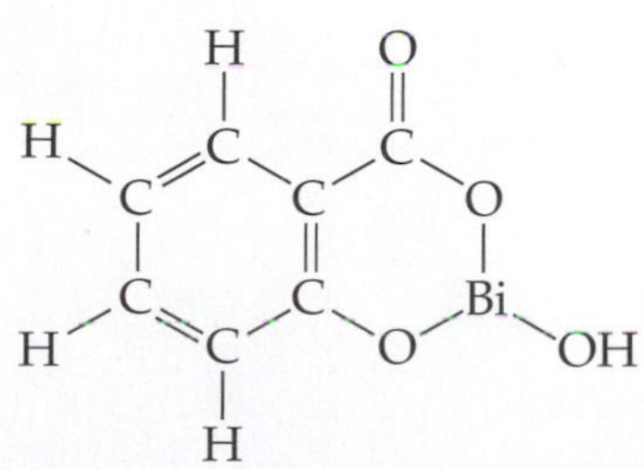

▲ Pepto-Bismol (bismuth subsalicylate, $C_7H_5BiO_4$)

19.10 NITROGEN

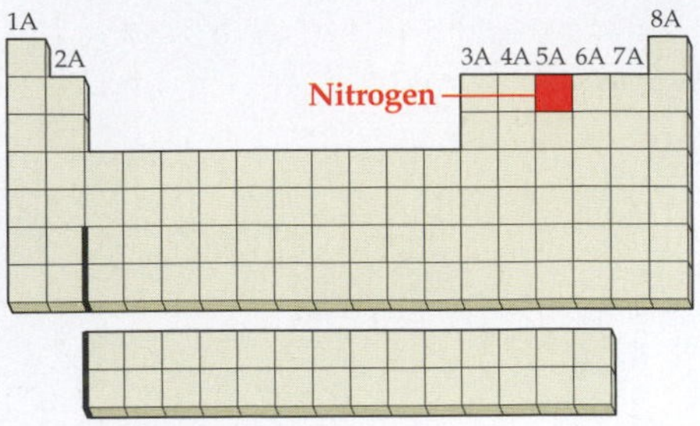

Elemental nitrogen is a colorless, odorless, tasteless gas that makes up 78% of the Earth's atmosphere by volume. Because nitrogen (bp −196 °C) is the most volatile component of liquid air, it is readily separated from the less volatile oxygen (bp −183 °C) and argon (bp −186 °C) by fractional distillation. Nitrogen gas is used as a protective inert atmosphere in manufacturing processes, and the liquid is used as a refrigerant. By far the most important use of nitrogen, however, is in the Haber process for the manufacture of ammonia, used in nitrogen fertilizers (Section 13.7).

Under most conditions, the N_2 molecule is unreactive because a large amount of energy is required to break its strong nitrogen–nitrogen triple bond:

$$:\text{N}{\equiv}\text{N}: \longrightarrow 2\,:\dot{\text{N}}\cdot \qquad \Delta H^\circ = 945\text{ kJ}$$

As a result, reactions involving N_2 often have a high activation energy and/or an unfavorable equilibrium constant. For example, the equilibrium constant for formation of nitric oxide from N_2 and O_2 is 2.0×10^{-31} at 25 °C:

$$\text{N}_2(g) + \text{O}_2(g) \rightleftharpoons 2\,\text{NO}(g) \qquad \Delta H^\circ = 182.6\text{ kJ}; \qquad K_c = 2.0 \times 10^{-31} \text{ at } 25\ ^\circ\text{C}$$

REMEMBER...

The temperature dependence of the equilibrium constant *K* depends on the sign of ΔH° for the reaction. **For an endothermic reaction ($\Delta H^\circ > 0$), *K* increases as *T* increases.** (Section 13.10)

At higher temperatures, however, this reaction does occur because it is endothermic and the **equilibrium constant increases as the temperature increases** (Section 13.10). Indeed, the high-temperature formation of NO from air in automobile engines is a major source of air pollution. Atmospheric N_2 and O_2 also react to form NO during electrical storms, where lightning discharges provide the energy required for the highly endothermic reaction.

It's convenient to classify nitrogen compounds by oxidation state, as shown in **Table 19.6**.

TABLE 19.6 Oxidation States of Nitrogen and Representative Compounds

Oxidation State	Compound	Formula	Electron-Dot Structure
−3	Ammonia	NH_3	H—N̈—H, with H below N
−2	Hydrazine	N_2H_4	H—N̈—N̈—H, with H below each N
−1	Hydroxylamine	NH_2OH	H—N̈—Ö—H, with H below N
+1	Nitrous oxide	N_2O	:N≡N—Ö:
+2	Nitric oxide	NO	:Ṅ=Ö:
+3	Nitrous acid	HNO_2	H—Ö—N̈=Ö:
+4	Nitrogen dioxide	NO_2	:Ö—Ṅ=Ö:
+5	Nitric acid	HNO_3	H—Ö—N=Ö:, with :Ö: below N

Nitrogen Compounds

Among the more important compounds of nitrogen are ammonia, hydrazine, oxides, and oxoacids.

Ammonia Ammonia and its synthesis by the Haber process serve as the gateway to nitrogen chemistry because ammonia is the starting material for the industrial synthesis of other important nitrogen compounds, such as nitric acid. Used in agriculture as a fertilizer, ammonia is the most commercially important compound of nitrogen.

Ammonia

Ammonia is a colorless, pungent-smelling gas, consisting of polar, trigonal pyramidal NH_3 molecules that have a lone pair of electrons on the N atom. Because of hydrogen bonding (Section 10.2), gaseous NH_3 is extremely soluble in water and is easily condensed to liquid NH_3, which boils at −33 °C. Neutralization of aqueous ammonia with acids yields ammonium salts, which resemble alkali metal salts in their solubility.

Hydrazine Hydrazine (H_2NNH_2) can be regarded as a derivative of NH_3 in which one H atom is replaced by an *amino* (NH_2) group. It can be prepared by the reaction of ammonia with a basic solution of sodium hypochlorite (NaOCl):

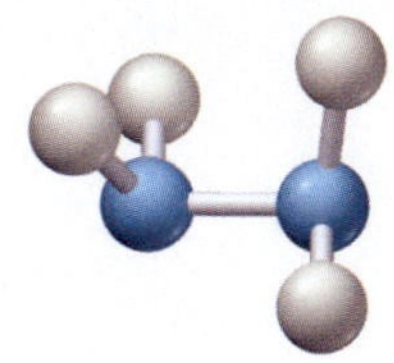
Hydrazine

$$2\,NH_3(aq) + OCl^-(aq) \longrightarrow N_2H_4(aq) + H_2O(l) + Cl^-(aq)$$

You have perhaps heard that household cleaners should never be mixed because exothermic reactions may occur or dangerous products may form. The formation of hydrazine on mixing household ammonia and hypochlorite-containing chlorine bleaches is a case in point.

Pure hydrazine is a poisonous, colorless liquid that smells like ammonia, freezes at 2 °C, and boils at 114 °C. It is explosive when mixed with air or other oxidizing agents and is used as a rocket fuel. Hydrazine can be handled safely in aqueous solutions, where it behaves as a weak base ($K_b = 8.9 \times 10^{-7}$) and a versatile reducing agent. It reduces Fe^{3+} to Fe^{2+}, I_2 to I^-, and Ag^+ to metallic Ag, for example.

Oxides of Nitrogen Nitrogen forms a large number of oxides, but we'll discuss only three: nitrous oxide (dinitrogen monoxide, N_2O), nitric oxide (nitrogen monoxide, NO), and nitrogen dioxide (NO_2). We discussed N_2O_5 and N_2O_4 in Chapters 12 and 13.

Nitrous oxide (N_2O) is a colorless, sweet-smelling gas obtained when molten ammonium nitrate is heated gently at about 270 °C. Strong heating can cause an explosion.

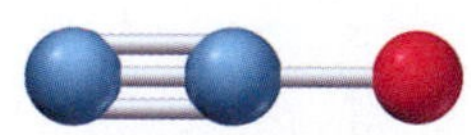
Nitrous oxide

$$NH_4NO_3(l) \xrightarrow{\text{Heat}} N_2O(g) + 2\,H_2O(g)$$

Known as "laughing gas" because small doses are mildly intoxicating, nitrous oxide is used as a dental anesthetic and as a propellant for dispensing whipped cream and cooking sprays.

Nitric oxide (NO) is a colorless gas, produced in the laboratory when copper metal is treated with dilute nitric acid:

Nitric oxide

$$3\,Cu(s) + 2\,NO_3^-(aq) + 8\,H^+(aq) \longrightarrow 3\,Cu^{2+}(aq) + 2\,NO(g) + 4\,H_2O(l)$$

Nitric oxide is important in many biological processes where it helps transmit messages between nerve cells and kills harmful bacteria. It also helps to protect the heart from insufficient oxygen levels by dilating blood vessels. Very recently, it has been learned that the drug nitroglycerin, long used to relieve the pain of angina, is a source of NO, which in turn dilates clogged arteries. Nitric oxide also plays a role in controlling penile erections; the drug Viagra works by triggering the release of NO, thus increasing blood flow to the penis.

Nitrogen dioxide (NO_2) is the highly toxic, reddish brown gas that forms rapidly when nitric oxide is exposed to air (**Figure 19.8**):

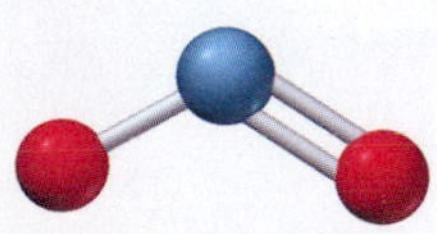
Nitrogen dioxide

$$2\,NO(g) + O_2(g) \longrightarrow 2\,NO_2(g)$$

NO_2 is also produced when copper reacts with concentrated nitric acid:

$$Cu(s) + 2\,NO_3^-(aq) + 4\,H^+(aq) \longrightarrow Cu^{2+}(aq) + 2\,NO_2(g) + 2\,H_2O(l)$$

Because NO_2 has an odd number of electrons (23), it is **paramagnetic** (Section 5.6). It tends to dimerize, forming colorless, diamagnetic N_2O_4, in which the unpaired electrons of two NO_2 molecules pair up to give an N–N bond:

REMEMBER...

Substances with unpaired electrons are attracted by magnetic fields and are said to be **paramagnetic.** (Section 5.6)

$$\underset{\text{Brown}}{O_2N\cdot + \cdot NO_2} \rightleftharpoons \underset{\text{Colorless}}{O_2N{-}NO_2} \qquad \Delta H^\circ = -55.3\text{ kJ}$$

FIGURE 19.8

Nitric oxide (NO) and nitrogen dioxide (NO_2).

Nitric oxide is a colorless gas.

NO turns brown on contact with air because it is rapidly oxidized to NO_2 .

Copper reacts with concentrated HNO_3, yielding noxious, red-brown fumes of NO_2. The blue color of the solution is due to Cu^{2+} ions.

In the gas phase, NO_2 and N_2O_4 are present in equilibrium. Because dimer formation is exothermic, N_2O_4 predominates at lower temperatures and NO_2 predominates at higher temperatures, as was illustrated in Figure 13.13 (page 534).

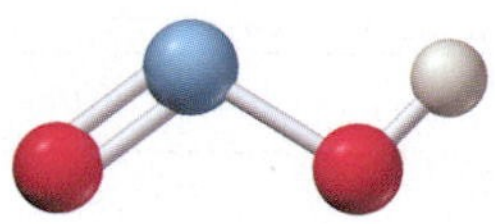

Nitrous acid

Nitrous Acid Nitrogen dioxide reacts with water to yield a mixture of nitrous acid (HNO_2) and nitric acid (HNO_3):

$$2\,NO_2(g) + H_2O(l) \longrightarrow HNO_2(aq) + H^+(aq) + NO_3^-(aq)$$

This is a disproportionation reaction in which nitrogen goes from the +4 oxidation state in NO_2 to the +3 state in HNO_2 and the +5 state in HNO_3. Nitrous acid is a weak acid ($K_a = 4.5 \times 10^{-4}$) that tends to disproportionate to nitric oxide and nitric acid. It has not been isolated as a pure compound:

$$3\,HNO_2(aq) \rightleftharpoons 2\,NO(g) + H^+(aq) + NO_3^-(aq) + H_2O(l)$$

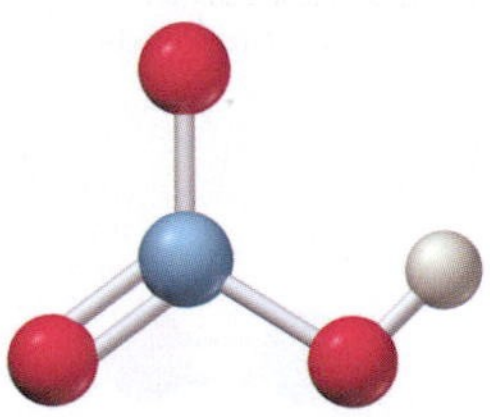

Nitric acid

Nitric Acid Nitric acid, one of the most important inorganic acids, is used mainly to make ammonium nitrate for fertilizers, but it is also used to manufacture explosives, plastics, and dyes. Annual U.S. production of HNO_3 is approximately 6.3 million metric tons.

Nitric acid is produced industrially by the multistep **Ostwald process**, which involves (1) air oxidation of ammonia to nitric oxide at about 850 °C over a platinum–rhodium catalyst, (2) rapid oxidation of the nitric oxide to nitrogen dioxide, and (3) disproportionation of NO_2 in water:

$$(1)\quad 4\,NH_3(g) + 5\,O_2(g) \xrightarrow[\text{Pt/Rh catalyst}]{850\,^\circ\text{C}} 4\,NO(g) + 6\,H_2O(g)$$

$$(2)\quad 2\,NO(g) + O_2(g) \longrightarrow 2\,NO_2(g)$$

$$(3)\quad 3\,NO_2(g) + H_2O(l) \longrightarrow 2\,HNO_3(aq) + NO(g)$$

Distillation of the resulting aqueous HNO_3 removes some of the water and gives concentrated (15 M) nitric acid, an HNO_3–H_2O mixture that is 68.5% HNO_3 by mass.

Further removal of water is required to obtain pure nitric acid, a colorless liquid that boils at 83 °C. In the laboratory, concentrated nitric acid often has a yellow-brown color due to the presence of a small amount of NO_2 produced by a slight amount of decomposition:

$$4\,HNO_3(aq) \longrightarrow 4\,NO_2(aq) + O_2(g) + 2\,H_2O(l)$$

▲ Why does freshly prepared concentrated nitric acid (left) turn yellow-brown on standing (right)?

Nitric acid is a strong acid and is essentially 100% dissociated in water. It's also a strong oxidizing agent, as indicated by large, positive $E°$ values for reduction to lower oxidation states:

$$NO_3^-(aq) + 2\,H^+(aq) + e^- \longrightarrow NO_2(g) + H_2O(l) \qquad E° = 0.79\ \text{V}$$

$$NO_3^-(aq) + 4\,H^+(aq) + 3\,e^- \longrightarrow NO(g) + 2\,H_2O(l) \qquad E° = 0.96\ \text{V}$$

Thus, nitric acid is a stronger oxidizing agent than $H^+(aq)$ and can oxidize relatively inactive metals like copper and silver that are not oxidized by aqueous HCl. The reduction product of HNO_3 in a particular reaction depends on the nature of the reducing agent and the reaction conditions. We've already seen, for instance, that copper reduces dilute HNO_3 to NO, but it reduces concentrated HNO_3 to NO_2.

An even more potent oxidizing agent than HNO_3 is *aqua regia*, a mixture of concentrated HCl and concentrated HNO_3 in a 3:1 ratio by volume. Aqua regia can oxidize even inactive metals like gold, which do not react with either HCl or HNO_3 separately:

$$Au(s) + 3\,NO_3^-(aq) + 6\,H^+(aq) + 4\,Cl^-(aq) \longrightarrow AuCl_4^-(aq) + 3\,NO_2(g) + 3\,H_2O(l)$$

The NO_3^- ion serves as the oxidizing agent, and Cl^- facilitates the reaction by converting the Au(III) oxidation product to the $AuCl_4^-$ complex ion.

PROBLEM 19.7 For the molecules in Table 19.6 (page 792) that must be described as resonance hybrids, draw electron-dot resonance structures, assign formal charges, and indicate which resonance structure(s) is (are) most important. In addition, indicate which of the molecules in Table 19.6 are paramagnetic.

19.11 PHOSPHORUS

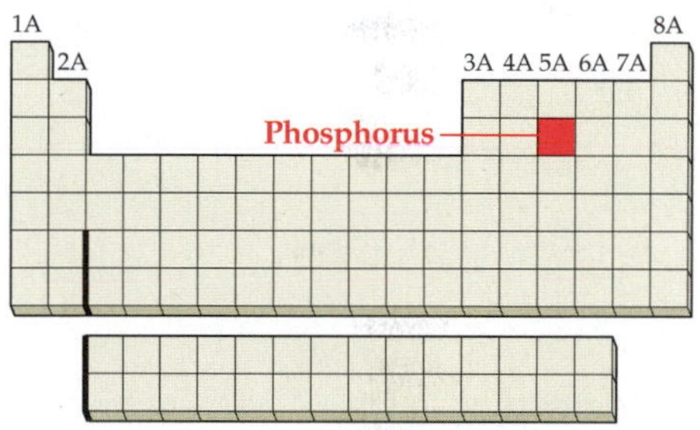

Phosphorus is the most abundant element of group 5A, accounting for 0.10% of the mass of the Earth's crust. It is found in phosphate rock, which is mostly calcium phosphate, $Ca_3(PO_4)_2$, and in fluorapatite, $Ca_5(PO_4)_3F$. The apatites are phosphate minerals with the formula $3\,Ca_3(PO_4)_2 \cdot CaX_2$, where X^- is usually F^- or OH^-. Phosphorus is also important in living systems and is the sixth most abundant element in the human body (Figure 19.1). Our bones are mostly hydroxyapatite, $Ca_5(PO_4)_3OH$, along with the fibrous protein collagen, and tooth enamel is almost pure hydroxyapatite. Phosphate groups are also an integral part of the nucleic acids DNA and RNA, the molecules that pass genetic information from generation to generation, and phospholipids, the phosphate-containing molecules that are major components of cell membranes.

Elemental phosphorus is produced industrially by heating phosphate rock, coke, and silica sand at about 1500 °C in an electric furnace. The reaction can be represented by the simplified equation

$$2\,Ca_3(PO_4)_2(s) + 10\,C(s) + 6\,SiO_2(s) \longrightarrow P_4(s) + 10\,CO(g) + 6\,CaSiO_3(l)$$

Phosphorus is used to make phosphoric acid, one of the top 10 industrial chemicals (Table 19.1).

Phosphorus exists as two common allotropes: white phosphorus and red phosphorus (**Figure 19.9**). White phosphorus, the form produced in the industrial synthesis, is a toxic, waxy, white solid that contains discrete tetrahedral P_4 molecules. Red phosphorus, by contrast, is essentially nontoxic and has a polymeric structure.

As expected for a molecular solid that contains small, nonpolar molecules, white phosphorus has a low melting point (44 °C) and is soluble in nonpolar solvents. It is highly reactive, bursting into flames when exposed to air, and is thus stored under water. When white phosphorus is heated in the absence of air at about 300 °C, it is converted to the more stable red form. Consistent with its polymeric structure, red phosphorus is higher melting (mp 579 °C), less soluble, and less reactive than white phosphorus, and it does not ignite on contact with air (Figure 19.9).

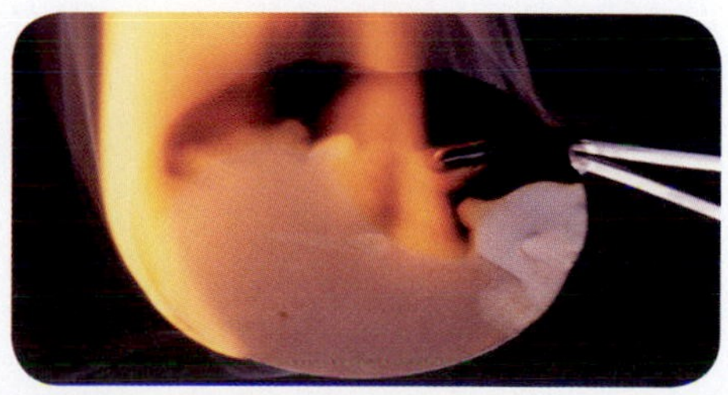

▲ A piece of filter paper coated with white phosphorus bursts into flame in air.

The high reactivity of white phosphorus is due to an unusual bonding that produces considerable strain in the tetrahedral P_4 molecules. If each P atom uses three $3p$ orbitals to form its three P—P bonds, all the bond angles should be 90°. The geometry of P_4, however,

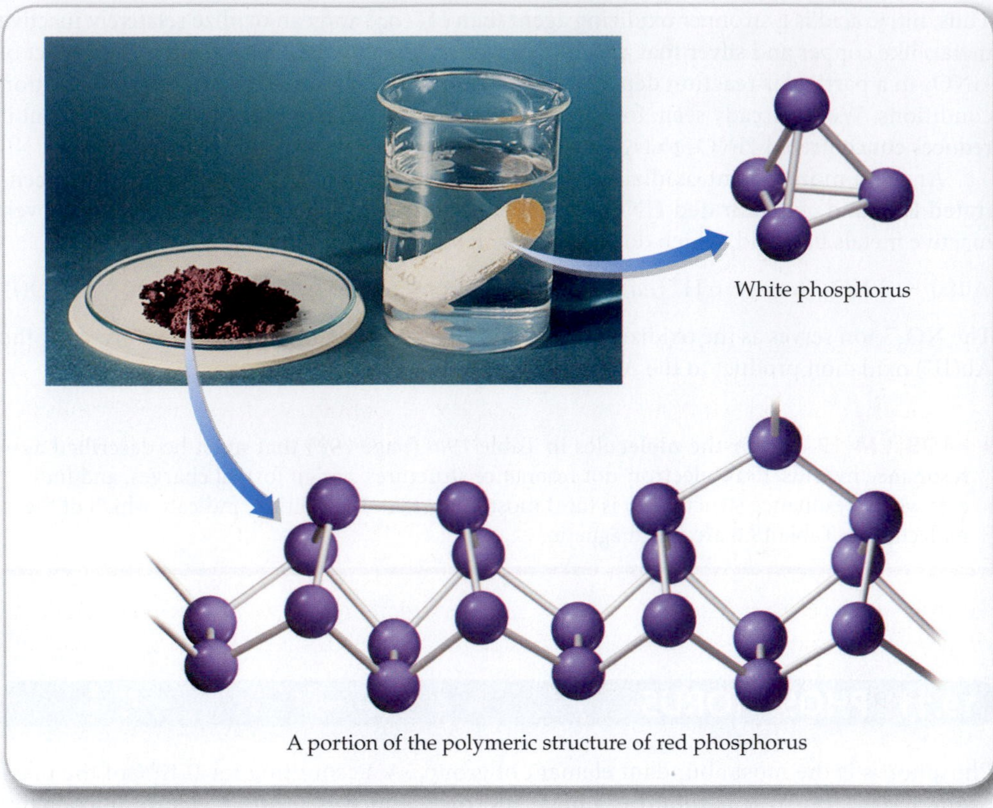

FIGURE 19.9

Red phosphorus (left) and white phosphorus stored under water (right) because it reacts with oxygen in air.

requires that all the bonds have 60° angles, which means that the *p* orbitals can't overlap in a head-on fashion. As a result, the P—P bonds are "bent," relatively weak, and highly reactive (**Figure 19.10**).

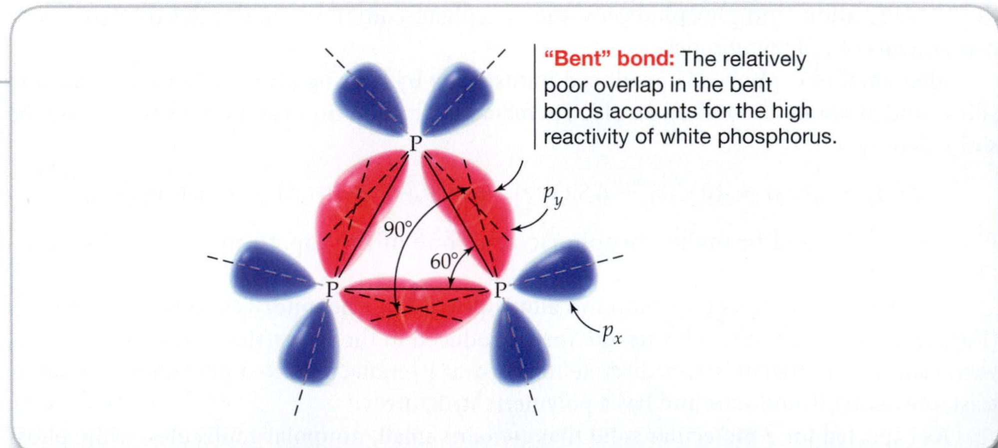

FIGURE 19.10

One equilateral triangular face of a tetrahedral P_4 molecule, showing the 60° bond angles and the 90° angles between the *p* orbitals.

Phosphorus Compounds

Like nitrogen, phosphorus forms compounds in all oxidation states between −3 and +5, but the +3 state, as in PCl_3, P_4O_6, and H_3PO_3, and the +5 state, as in PCl_5, P_4O_{10}, and H_3PO_4, are the most common. Compared to nitrogen, phosphorus is more likely to be found in a positive oxidation state because of its lower electronegativity (Table 19.5, page 791).

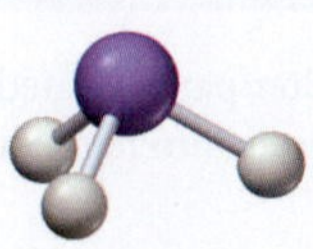

Phosphine

Phosphine Phosphine (PH_3), a colorless, extremely poisonous gas, is the simplest hydride of phosphorus. Like NH_3, phosphine has a trigonal pyramidal structure and has the group 5A atom in the −3 oxidation state. Unlike NH_3, however, its aqueous solutions are neutral, indicating that PH_3 is a poor proton acceptor. In accord with the low electronegativity of phosphorus, phosphine is easily oxidized, burning in air to form phosphoric acid:

$$PH_3(g) + 2\,O_2(g) \longrightarrow H_3PO_4(l)$$

Phosphorus Halides Phosphorus reacts with all the halogens, forming phosphorus(III) halides, PX_3, or phosphorus(V) halides, PX_5 (X = F, Cl, Br, or I), depending on the relative amounts of the reactants:

$$\text{Limited amount of } X_2\text{:}\quad P_4 + 6\,X_2 \longrightarrow 4\,PX_3$$
$$\text{Excess amount of } X_2\text{:}\quad P_4 + 10\,X_2 \longrightarrow 4\,PX_5$$

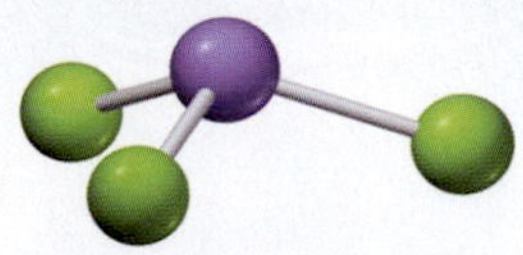

Phosphorus trichloride

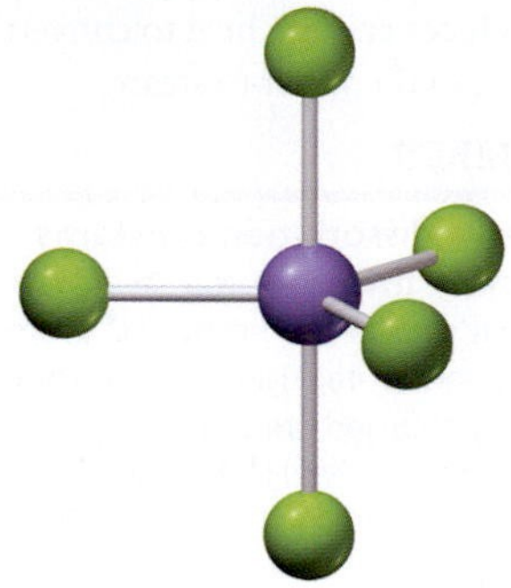

Phosphorus pentachloride

At room temperature, all of these halides are gases, volatile liquids, or low-melting solids. For example, phosphorus trichloride is a colorless liquid that boils at 76 °C, and phosphorus pentachloride is an off-white solid that melts at 167 °C. Both fume on contact with moist air because of a reaction with water that breaks the P–Cl bonds, converting PCl_3 to phosphorous acid (H_3PO_3), and PCl_5 to phosphoric acid (H_3PO_4):

$$PCl_3(l) + 3\,H_2O(l) \longrightarrow H_3PO_3(aq) + 3\,HCl(aq)$$
$$PCl_5(s) + 4\,H_2O(l) \longrightarrow H_3PO_4(aq) + 5\,HCl(aq)$$

Note the difference between the words *phosphorus* and *phosphorous.* Phosphorus, with a -us ending, is the name of the element, but phosphorous, with an -ous ending, is the name of a specific acid compound. Many people confuse the two.

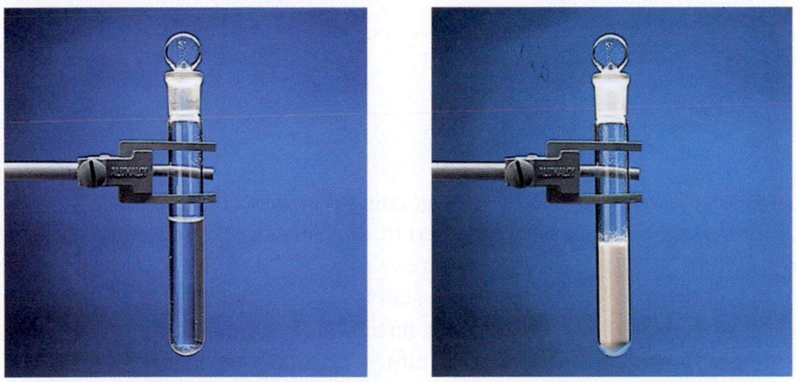

▲ Samples of PCl_3 (left) and PCl_5 (right).

Oxides and Oxoacids of Phosphorus When phosphorus burns in air or oxygen, it yields either tetraphosphorus hexoxide (P_4O_6, mp 24 °C) or tetraphosphorus decoxide (P_4O_{10}, mp 420 °C), depending on the amount of oxygen present:

$$\text{Limited amount of } O_2\text{:}\quad P_4(s) + 3\,O_2(g) \longrightarrow P_4O_6(s)$$
$$\text{Excess amount of } O_2\text{:}\quad P_4(s) + 5\,O_2(g) \longrightarrow P_4O_{10}(s)$$

Both oxides are molecular compounds and have structures with a tetrahedral array of P atoms, as in white phosphorus. One O atom bridges each of the six edges of the P_4 tetrahedron, and an additional, terminal O atom is bonded to each P atom in P_4O_{10}.

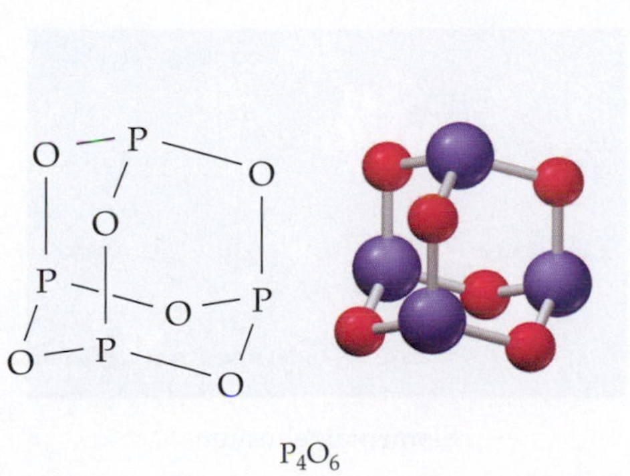

P_4O_6

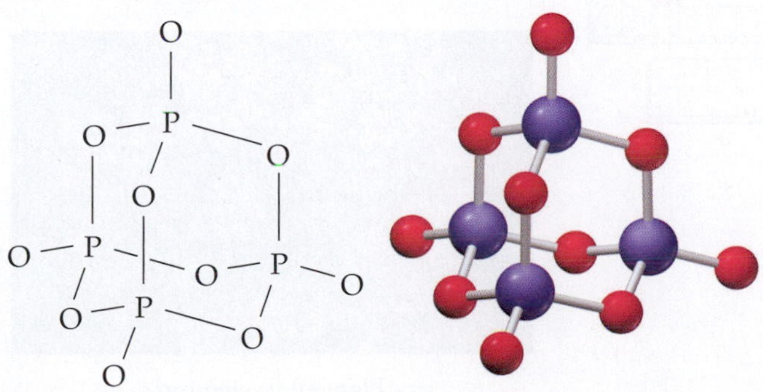

P_4O_{10}

▲ The reaction of P_4O_{10} with water produces enough heat to convert some of the water to steam.

REMEMBER...

Successive dissociation constants decrease in the order $K_{a1} > K_{a2} > K_{a3}$ because it's easier to remove H^+ from a neutral molecule (H_3PO_4) than from a negatively charged anion ($H_2PO_4^-$). Similarly, it's easier to remove H^+ from $H_2PO_4^-$ than from a doubly charged anion (HPO_4^{2-}). (Section 14.11)

Both P_4O_6 and P_4O_{10} are acidic oxides and react with water to form aqueous solutions of phosphorous acid and phosphoric acid, respectively:

$$P_4O_6(s) + 6\,H_2O(l) \longrightarrow 4\,H_3PO_3(aq)$$
$$P_4O_{10}(s) + 6\,H_2O(l) \longrightarrow 4\,H_3PO_4(aq)$$

Because P_4O_{10} has a great affinity for water, it is widely used as a drying agent for gases and organic solvents.

Phosphorous acid (H_3PO_3) is a weak diprotic acid because only two of its three H atoms are bonded to oxygen. The H atom bonded directly to phosphorus is not acidic because phosphorus and hydrogen have the same electronegativity and so the P—H bond is nonpolar. In phosphoric acid, however, all three hydrogens are attached to oxygen, and thus phosphoric acid is a weak triprotic acid. The geometry about the P atom in both molecules is tetrahedral, as expected. Note that the **successive dissociation constants decrease** by a factor of about 10^5 (Section 14.11).

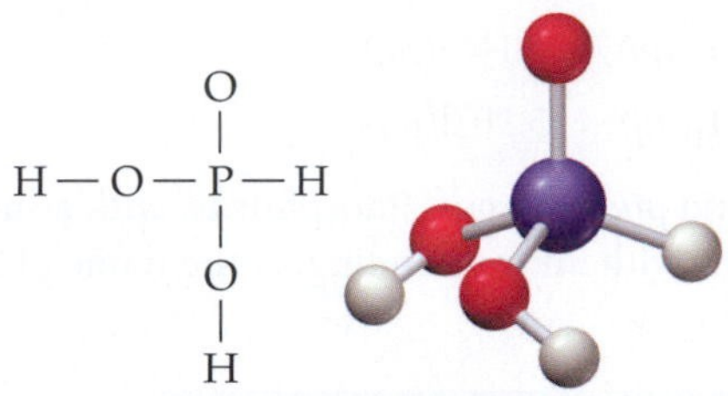

Phosphorous acid, H_3PO_3
$K_{a1} = 1.0 \times 10^{-2}$
$K_{a2} = 2.6 \times 10^{-7}$

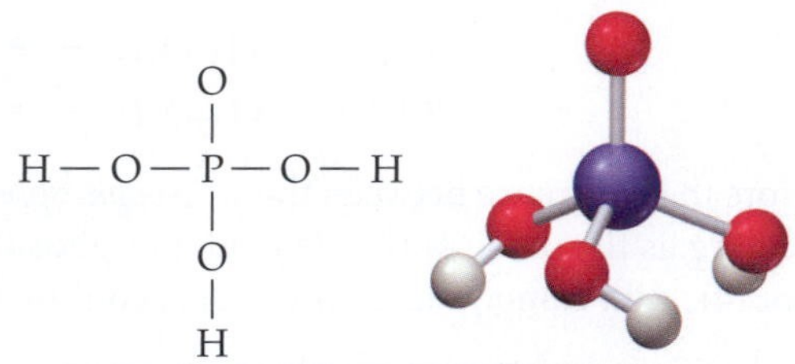

Phosphoric acid, H_3PO_4
$K_{a1} = 7.5 \times 10^{-3}$
$K_{a2} = 6.2 \times 10^{-8}$
$K_{a3} = 4.8 \times 10^{-13}$

Pure phosphoric acid is a low-melting, colorless, crystalline solid (mp 42 °C), but the commercially available phosphoric acid used in the laboratory is a syrupy, aqueous solution containing about 82% H_3PO_4 by mass. For use as a food additive—for example, as the tart ingredient in soft drinks—pure phosphoric acid is made by burning molten phosphorus in a mixture of air and steam. For use in making fertilizers, an impure form of phosphoric acid is produced by treating phosphate rock with sulfuric acid:

$$Ca_3(PO_4)_2(s) + 3\,H_2SO_4(aq) \longrightarrow 2\,H_3PO_4(aq) + 3\,CaSO_4(aq)$$

19.12 THE GROUP 6A ELEMENTS

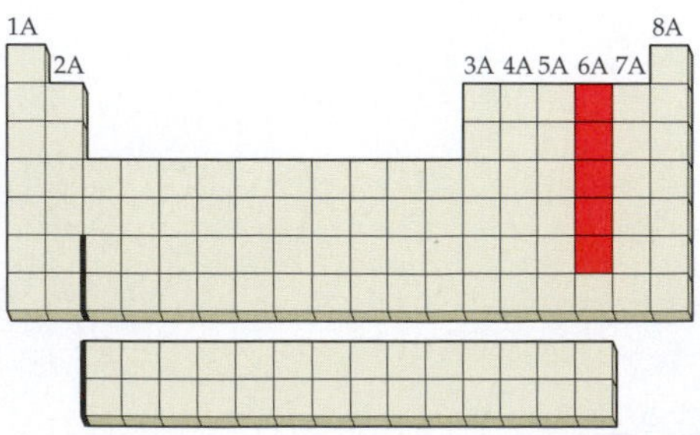

The group 6A elements are oxygen, sulfur, selenium, tellurium, and polonium. As shown in **Table 19.7**, their properties exhibit the usual periodic trends. Both oxygen and sulfur are typical nonmetals. Selenium and tellurium are primarily nonmetallic in character, though the most stable allotrope of selenium, gray selenium, is a lustrous semiconducting solid.

Elemental selenium

Elemental tellurium

TABLE 19.7 Properties of the Group 6A Elements

Property	Oxygen	Sulfur	Selenium	Tellurium	Polonium
Valence electron configuration	$2s^2\,2p^4$	$3s^2\,3p^4$	$4s^2\,4p^4$	$5s^2\,5p^4$	$6s^2\,6p^4$
Melting point (°C)	−219	113*	221†	450	254
Boiling point (°C)	−183	445	685	988	962
Atomic radius (pm)	66	104	116	143	167
X^{2-} ionic radius (pm)	140	184	198	221	
First ionization energy (kJ/mol)	1314	1000	941	869	812
Electron affinity (kJ/mol)	−141	−200	−195	−190	−183
Electronegativity	3.5	2.5	2.4	2.1	2.0
Redox potential, $E°$ (V) for $X + 2\,H^+ + 2\,e^- \longrightarrow H_2X$	1.23	0.14	−0.40	−0.79	

*Rhombic S

†Gray Se

Tellurium is also a semiconductor and is usually classified as a semimetal. Polonium, a radioactive element that occurs in trace amounts in uranium ores, is a silvery white metal.

With valence electron configuration $ns^2\,np^4$, the group 6A elements are just two electrons short of an octet configuration and the −2 oxidation state is therefore common. The stability of the −2 state decreases, however, with increasing metallic character, as indicated by the $E°$ values in Table 19.7. Thus, oxygen is a powerful oxidizing agent, but $E°$ values for reduction of Se and Te are negative, which means that H_2Se and H_2Te are reducing agents. Because S, Se, and Te are much less electronegative than oxygen, they are commonly found in positive oxidation states, especially +4, as in SF_4, SO_2, and H_2SO_3, and +6, as in SF_6, SO_3, and H_2SO_4.

Commercial uses of Se, Te, and Po are limited, though selenium is used to make red colored glass and is used in photocopiers and laser printers (see the *FYI* at the end of this chapter). Tellurium is used in alloys to improve their machinability, and polonium (^{210}Po) has been used as a heat source in space equipment and as a source of alpha particles in scientific research.

▲ The color of the red glass in this traffic signal is due to cadmium selenide, CdSe.

19.13 SULFUR

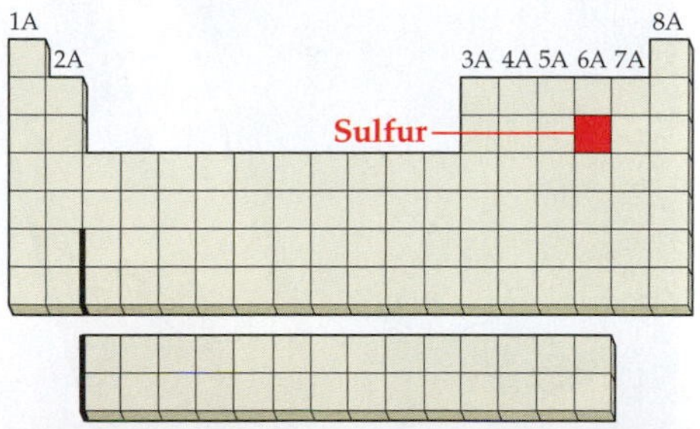

Sulfur occurs in elemental form in large underground deposits and is present in numerous minerals such as pyrite (FeS_2, which contains the S_2^{2-} ion), galena (PbS), cinnabar (HgS), and gypsum ($CaSO_4 \cdot 2\,H_2O$). Sulfur is also present in natural gas as H_2S and in crude oil as organic sulfur compounds. In plants and animals, sulfur occurs in various proteins, and it is one of the 10 most abundant elements in the human body (Figure 19.1).

Elemental sulfur is obtained from underground deposits and is recovered from natural gas and crude oil. In the United States, 88% of the sulfur produced is used to manufacture sulfuric acid.

Sulfur exists in many allotropic forms, but the most stable at 25 °C is rhombic sulfur, a yellow crystalline solid (mp 113 °C) that contains crown-shaped S_8 rings:

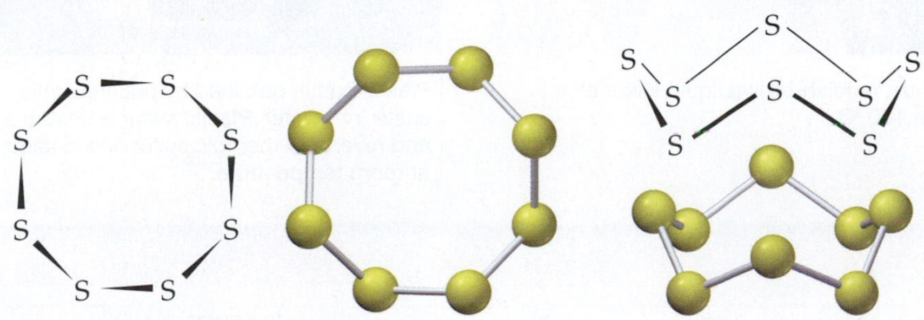

Top view   Side view

▲ Pyrite (FeS_2) is often called fool's gold because of its golden yellow color. It contains the disulfide ion (S_2^{2-}), the sulfur analogue of the peroxide ion (O_2^{2-}).

▲ A sample of rhombic sulfur, the most stable allotrope of sulfur.

Above 95 °C, rhombic sulfur is less stable than monoclinic sulfur (mp ~119 °C), an allotrope in which the cyclic S_8 molecules pack differently in the crystal. The phase transition from rhombic to monoclinic sulfur is very slow, however, and rhombic sulfur simply melts at 113 °C when heated at an ordinary rate.

As shown in **Figure 19.11**, molten sulfur exhibits some striking changes when its temperature is increased. Just above its melting point, sulfur is a fluid, straw-colored liquid, but between 160 °C and 195 °C its color becomes dark reddish brown and its viscosity increases by a factor of more than 10,000. At still higher temperatures, the liquid becomes more fluid again and then boils at 445 °C. If the hot liquid is cooled rapidly by pouring it into water, the sulfur forms an amorphous, rubbery material called *plastic sulfur*.

The dramatic increase in the viscosity of molten sulfur at 160–195 °C is due to the opening of the S_8 rings, yielding S_8 chains that subsequently form long polymers with more than 200,000 S atoms in the chain:

$$S_8 \text{ rings} \xrightarrow{\text{Heat}} \cdot S{-}S_6{-}S\cdot \text{ chains}$$

$$\cdot S{-}S_6{-}S\cdot \text{ chains} + S_8 \text{ rings} \longrightarrow \cdot S{-}S_{14}{-}S\cdot \text{ chains} \longrightarrow S_n \text{ chains}$$

$$n > 200{,}000$$

Whereas the small S_8 rings easily slide over one another in the liquid, the long polymer chains become entangled, thus accounting for the increase in viscosity. Above 200 °C, the polymer chains begin to fragment into smaller pieces, and the viscosity therefore decreases. On rapid cooling, the chains are temporarily frozen in a disordered, tangled arrangement, which accounts for the elastic properties of *plastic sulfur*.

Fluid, straw-colored liquid sulfur at about 120 °C.

Viscous, reddish-brown liquid sulfur at about 180 °C.

Plastic sulfur, obtained by pouring liquid sulfur into water. Plastic sulfur is unstable and reverts to rhombic sulfur on standing at room temperature.

FIGURE 19.11

Effect of temperature on the properties of sulfur.

Sulfur Compounds

Hydrogen Sulfide Hydrogen sulfide is a colorless gas (bp −60 °C) with the strong, foul odor we associate with rotten eggs, in which it occurs because of the bacterial decomposition of sulfur-containing proteins. Hydrogen sulfide is extremely toxic, causing headaches and nausea at concentrations of 10 ppm and sudden paralysis and death at 100 ppm. On initial exposure, the odor of H_2S can be detected at about 0.02 ppm, but unfortunately the gas tends to dull the sense of smell. It is thus an extremely insidious poison, even more dangerous than HCN.

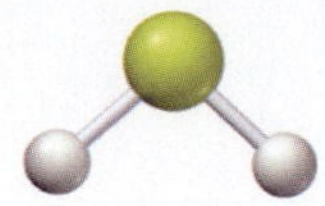
Hydrogen sulfide

In the laboratory, H_2S can be prepared by treating iron(II) sulfide with dilute sulfuric acid:

$$FeS(s) + 2\,H^+(aq) \longrightarrow H_2S(g) + Fe^{2+}(aq)$$

For use in qualitative analysis (Section 15.15), H_2S is usually generated in solution by the hydrolysis of thioacetamide:

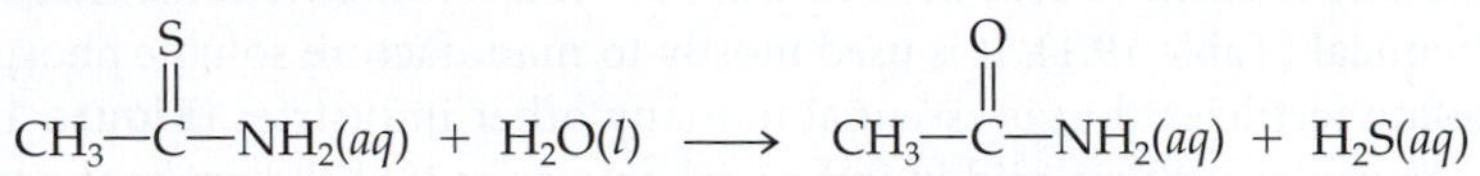
Thioacetamide Acetamide

Hydrogen sulfide is a very weak diprotic acid ($K_{a1} = 1.0 \times 10^{-7}$; $K_{a2} \approx 10^{-19}$) and a mild reducing agent. In reactions with mild oxidizing agents, it is oxidized to a milky white suspension of elemental sulfur:

$$H_2S(aq) + 2\,Fe^{3+}(aq) \longrightarrow S(s) + 2\,Fe^{2+}(aq) + 2\,H^+(aq)$$

Oxides and Oxoacids of Sulfur Sulfur dioxide (SO_2) and sulfur trioxide (SO_3) are the most important of the various oxides of sulfur. Sulfur dioxide, a colorless, toxic gas (bp −10 °C) with a pungent, choking odor, is formed when sulfur burns in air:

$$S(s) + O_2(g) \longrightarrow SO_2(g)$$

Sulfur dioxide

Sulfur dioxide is slowly oxidized in the atmosphere to SO_3, which dissolves in rainwater to give sulfuric acid. The burning of sulfur-containing fuels is thus a major cause of acid rain (Section 9.9). In the laboratory, SO_2 is conveniently prepared by treating sodium sulfite with dilute acid:

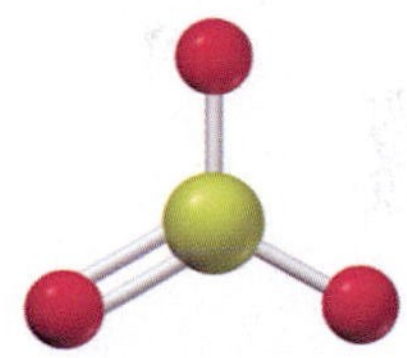
Sulfur trioxide

$$Na_2SO_3(s) + 2\,H^+(aq) \longrightarrow SO_2(g) + H_2O(l) + 2\,Na^+(aq)$$

Because SO_2 is toxic to microorganisms, it is used for sterilizing wine and dried fruit.

Both SO_2 and SO_3 are acidic, though aqueous solutions of SO_2 contain mainly dissolved SO_2 and little, if any, sulfurous acid (H_2SO_3):

$$SO_2(aq) + H_2O(l) \rightleftharpoons H_2SO_3(aq)$$

(The longer left-pointing arrow means that the equilibrium lies far to the left.) Nevertheless, it's convenient to write the acidic species as H_2SO_3, a weak diprotic acid. Sulfurous acid has never been isolated in pure form.

Sulfuric acid (H_2SO_4), the world's most important industrial chemical, is manufactured by the **contact process**, a three-step reaction sequence in which (1) sulfur burns in air to give SO_2, (2) SO_2 is oxidized to SO_3 in the presence of a vanadium(V) oxide catalyst, and (3) SO_3 reacts with water to give H_2SO_4:

(1) $S(s) + O_2(g) \longrightarrow SO_2(g)$

(2) $2SO_2(g) + O_2(g) \xrightarrow[V_2O_5\text{ catalyst}]{\text{Heat}} 2\,SO_3(g)$

(3) $SO_3(g) + H_2O\text{ (in conc }H_2SO_4) \longrightarrow H_2SO_4\text{ (in conc }H_2SO_4)$

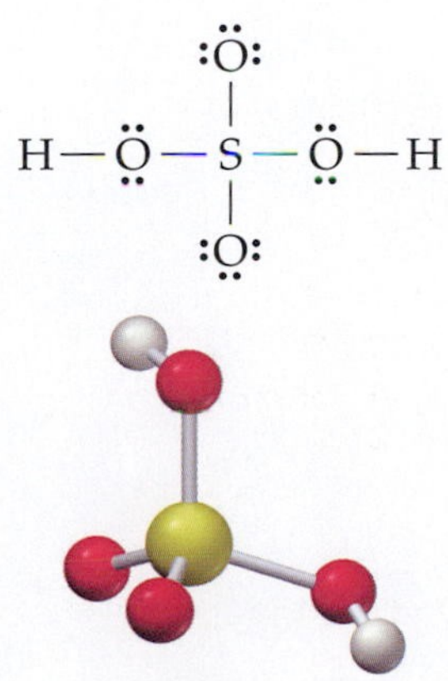
Sulfuric acid

In the third step, the SO_3 is absorbed in concentrated sulfuric acid rather than in water because the dissolution of SO_3 in water is slow. Water is then added to achieve the desired concentration. Commercial concentrated sulfuric acid is 98% H_2SO_4 by mass (18 M H_2SO_4). Anhydrous (100%) H_2SO_4 is a viscous, colorless liquid that freezes at 10.3 °C and boils above 300 °C. As predicted by the VSEPR model, the H_2SO_4 molecule is tetrahedral (Section 5.1).

Sulfuric acid is a strong acid for the dissociation of its first proton and has $K_{a2} = 1.2 \times 10^{-2}$ for the dissociation of its second proton. As a diprotic acid, it forms two series of salts: hydrogen sulfates, such as $NaHSO_4$, and sulfates, such as Na_2SO_4.

The oxidizing properties of sulfuric acid depend on its concentration and temperature. In dilute solutions at room temperature, H_2SO_4 behaves like HCl, oxidizing metals that stand above hydrogen in the activity series (Table 7.5, page 255):

$$Fe(s) + 2\,H^+(aq) \longrightarrow Fe^{2+}(aq) + H_2(g)$$

Hot, concentrated H_2SO_4 is a stronger oxidizing agent than the dilute, cold acid and can oxidize metals like copper, which are not oxidized by $H^+(aq)$. In the process, H_2SO_4 is reduced to SO_2:

$$Cu(s) + 2\,H_2SO_4(l) \longrightarrow Cu^{2+}(aq) + SO_4^{2-}(aq) + SO_2(g) + 2\,H_2O(l)$$

U.S. production of sulfuric acid in 2010 was 32.5 million metric tons, exceeding that of any other chemical (Table 19.1). It is used mostly to manufacture soluble phosphate and ammonium sulfate fertilizers but is essential to many other industries (**Figure 19.12**). So widespread is the use of sulfuric acid in industrial countries that the amount produced is sometimes regarded as an indicator of economic activity.

FIGURE 19.12

Uses of sulfuric acid in the United States.

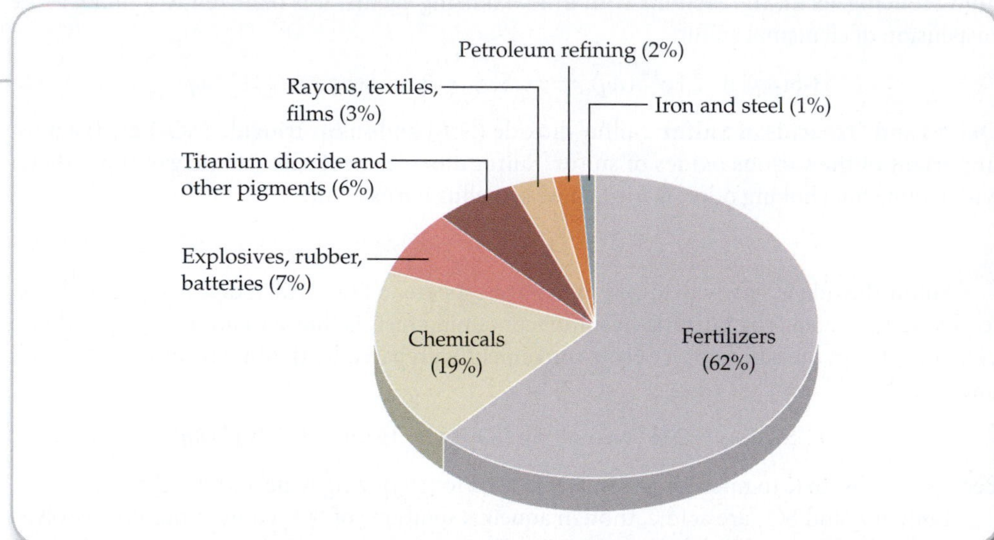

CONCEPTUAL PROBLEM 19.8 Consider the following sulfur-containing oxoanions:

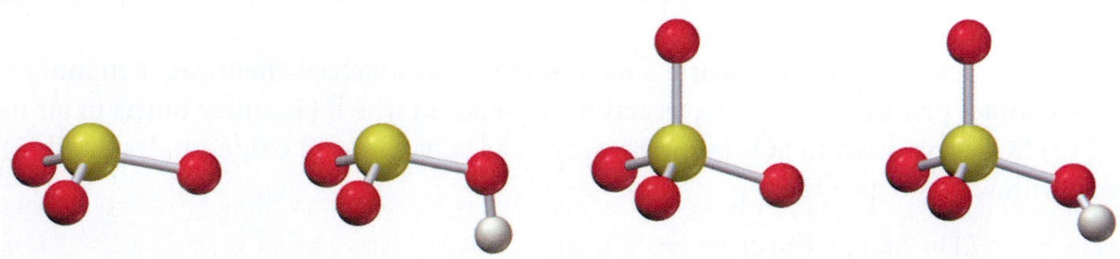

(a) Write the formula of each oxoanion, including its charge.
(b) Which oxoanion is the strongest acid?
(c) Which is the strongest base?
(d) Which is the weakest base?

PROBLEM 19.9 Write electron-dot structures for each of the following molecules, and use VSEPR theory to predict the structure of each:

(a) H_2S (b) SO_2 (c) SO_3

19.14 THE HALOGENS: OXOACIDS AND OXOACID SALTS

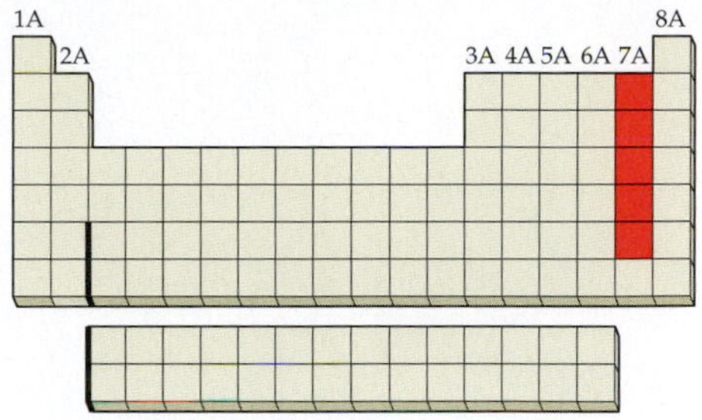

The halogens (group 7A) have valence electron configuration $ns^2\,np^5$ and are the most electronegative group of elements in the periodic table. We saw in Section 3.13 that halogen atoms tend to achieve an octet configuration by gaining one electron in reactions with metals, thus forming ionic compounds such as NaCl. They also share one electron with nonmetals to give molecular compounds such as HCl, BCl_3, PF_5, and SF_6. In all these compounds, the halogen is in the -1 oxidation state.

Among the most important compounds of halogens in positive oxidation states are the oxoacids of Cl, Br, and I (**Table 19.8**) and the corresponding oxoacid salts. In these compounds, the halogen shares its valence electrons with oxygen, a more electronegative element. (Electronegativities are O, 3.5; Cl, 3.0; Br, 2.8; I, 2.5.) The general formula for a halogen oxoacid is HXO_n, and the oxidation state of the halogen is $+1$, $+3$, $+5$, or $+7$, depending on the value of n.

TABLE 19.8 Oxoacids of the Halogens

Oxidation State	Generic Name (formula)	Chlorine	Bromine	Iodine
+1	Hypohalous acid (HXO)	HClO	HBrO	HIO
+3	Halous acid (HXO_2)	$HClO_2$	—	—
+5	Halic acid (HXO_3)	$HClO_3$	$HBrO_3$	HIO_3
+7	Perhalic acid (HXO_4)	$HClO_4$	$HBrO_4$	HIO_4, H_5IO_6

Only four of the acids listed in Table 19.8 have been isolated in pure form: perchloric acid ($HClO_4$), iodic acid (HIO_3), and the two periodic acids, metaperiodic acid (HIO_4) and paraperiodic acid (H_5IO_6). The others are stable only in aqueous solution or in the form of their salts. Bromous acid ($HBrO_2$) is known only as an unstable reaction intermediate, and iodous acid (HIO_2) is unknown.

The acid strength of the halogen oxoacids increases with the **increasing oxidation state of the halogen** (Section 14.3). For example, acid strength increases from HClO, a weak acid ($K_a = 3.5 \times 10^{-8}$), to $HClO_4$, a very strong acid ($K_a \gg 1$). The acidic proton is bonded to oxygen, not to the halogen, even though we usually write the molecular formula of these acids as HXO_n. All the halogen oxoacids and their salts are strong oxidizing agents.

REMEMBER...

The **higher the oxidation state of the halogen** in HXO_n, the greater the shift of electron density from the O–H bond toward the halogen, thus weakening the O–H bond, increasing its polarity, and facilitating proton transfer from HXO_n to a solvent water molecule. (Section 14.3)

A hypohalous acid is formed when Cl_2, Br_2, or I_2 dissolves in cold water:

$$X_2(g, l, \text{ or } s) + H_2O(l) \rightleftharpoons HOX(aq) + H^+(aq) + X^-(aq)$$

In this reaction, the halogen disproportionates, going to the $+1$ oxidation state in HOX and the -1 state in X^-. The equilibrium lies to the left but is shifted to the right in basic solution:

$$X_2(g, l, \text{ or } s) + 2\,OH^-(aq) \rightleftharpoons OX^-(aq) + X^-(aq) + H_2O(l)$$

Large amounts of aqueous sodium hypochlorite (NaOCl) are produced in the chlor-alkali industry (Section 17.13) when the Cl_2 gas and aqueous NaOH from the electrolysis of aqueous NaCl are allowed to mix. Aqueous NaOCl is a strong oxidizing agent and is sold in a 5% solution as chlorine bleach.

Further disproportionation of OCl^- to ClO_3^- and Cl^- is slow at room temperature but becomes fast at higher temperatures. Thus, when Cl_2 gas reacts with hot aqueous NaOH, it gives a solution that contains sodium chlorate ($NaClO_3$) rather than NaOCl:

$$3\,Cl_2(g) + 6\,OH^-(aq) \longrightarrow ClO_3^-(aq) + 5\,Cl^-(aq) + 3\,H_2O(l)$$

Chlorate salts are used as weed killers and as strong oxidizing agents. Potassium chlorate, for example, is an oxidizer in matches, fireworks, and explosives.

Sodium perchlorate ($NaClO_4$) is produced commercially by the electrolytic oxidation of aqueous sodium chlorate and is converted to perchloric acid by reaction with concentrated HCl:

$$ClO_3^-(aq) + H_2O(l) \xrightarrow{\text{Electrolysis}} ClO_4^-(aq) + H_2(g)$$
$$NaClO_4(s) + HCl(aq) \longrightarrow HClO_4(aq) + NaCl(s)$$

The $HClO_4$ is then concentrated by distillation at reduced pressure.

Pure, anhydrous perchloric acid is a colorless, shock-sensitive liquid that explodes on heating. It is a powerful and dangerous oxidizing agent, violently oxidizing organic matter and rapidly oxidizing even silver and gold. Perchlorate salts are also strong oxidants, and they too must be handled with caution. Ammonium perchlorate (NH_4ClO_4), in fact, was the oxidizer in the solid booster rockets used to propel the space shuttle.

Perchlorate ion has been detected in drinking water in 35 states at levels of at least 4 ppb. It is also present in dairy milk and human breast milk. It's not yet clear how much of the ClO_4^- comes from rocket fuel and how much derives from natural sources. Nevertheless, relatively high levels of ClO_4^- in the environment are of concern because this ion inhibits the uptake of iodide ion by the thyroid gland, which in turn may lower the level of thyroid hormone in the body and result in neurological damage.

Iodine differs from the other halogens because it forms more than one perhalic acid. Paraperiodic acid (H_5IO_6) is obtained as white crystals (mp 128 °C) when periodic acid solutions are evaporated. When heated to 100 °C at reduced pressure, these crystals lose water and are converted to metaperiodic acid (HIO_4):

$$H_5IO_6(s) \xrightarrow[\text{12 mm Hg}]{100\ °\text{C}} HIO_4(s) + 2\,H_2O(g)$$

Metaperiodic acid is a strong monoprotic acid, whereas paraperiodic acid is a weak polyprotic acid ($K_{a1} = 5.1 \times 10^{-4}$, $K_{a2} = 4.9 \times 10^{-9}$). It has an octahedral structure in which a central iodine atom is bonded to one O atom and five OH groups:

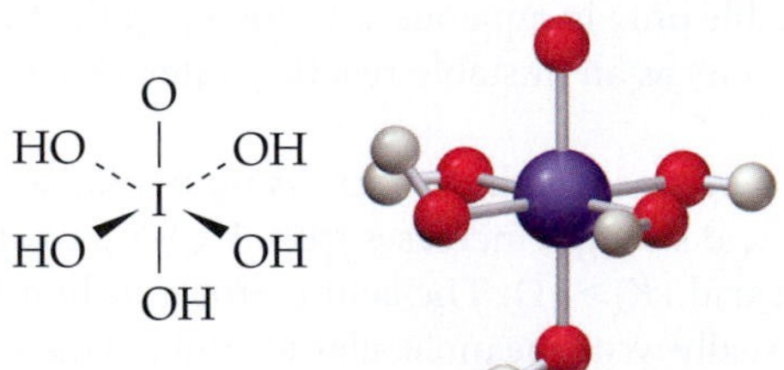

Paraperiodic acid, H_5IO_6

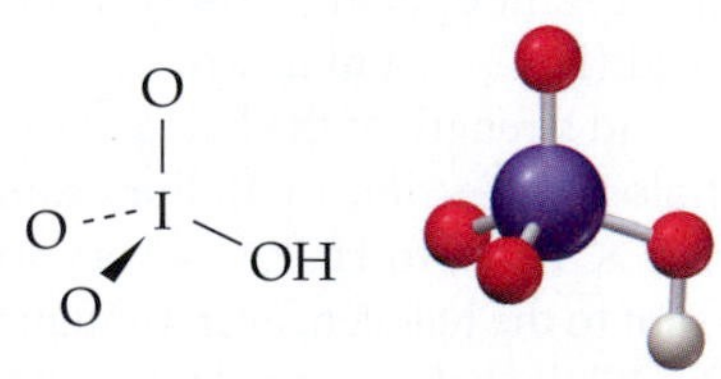

Metaperiodic acid, HIO_4

Chlorine and bromine do not form perhalic acids of the type H_5XO_6 because their smaller sizes favor a tetrahedral structure over an octahedral one.

FYI HOW DO LASER PRINTERS WORK?

Laser printers are the descendants of the plain-paper photocopy machine; the prototype laser printer was built from one. A half century ago, the photocopier revolutionized the way offices handle paper by making possible the creation of numerous, identical copies of a document. Today, the laser printers that sit on our desktops make possible the rapid and economical production of original, computer-generated documents.

Photocopiers and laser printers take advantage of an unusual property of selenium, the group 6A element below sulfur in the periodic table. Selenium is a *photoconductor*, a substance that is a poor electrical conductor when dark but whose conductivity increases (by a factor of 1000) when exposed to light. When the light is removed, the conductivity again drops.

As illustrated in **Figure 19.13**, the image-forming process in a laser printer begins when a selenium-coated drum, kept in the dark, is given a large, uniform negative charge. As the drum rotates, it is exposed to computer-generated flashes of infrared laser light ($\lambda = 780$ nm) directed by lenses and mirrors. Areas on the drum hit by the precisely focused laser light become conducting and lose most of their negative charge. These charge-depleted areas correspond to the dark parts of the document. Those areas that correspond to the light parts of the document remain nonconducting and retain their large negative charge. Thus, an image of the document is formed on the drum as an array of slightly negative (largely neutralized) areas amidst the large area of undepleted negative charges.

Following its formation, the image is developed by exposing the drum to negatively charged dry ink particles (*toner*), which are repelled from the strongly negative areas of the drum and collect in the largely neutralized areas. The developed image is then transferred to paper by passing a sheet of paper between the drum and a positively charged development electrode, which induces the negatively charged toner particles to jump from the drum to the paper.

▲ A selenium-coated drum used in laser printers.

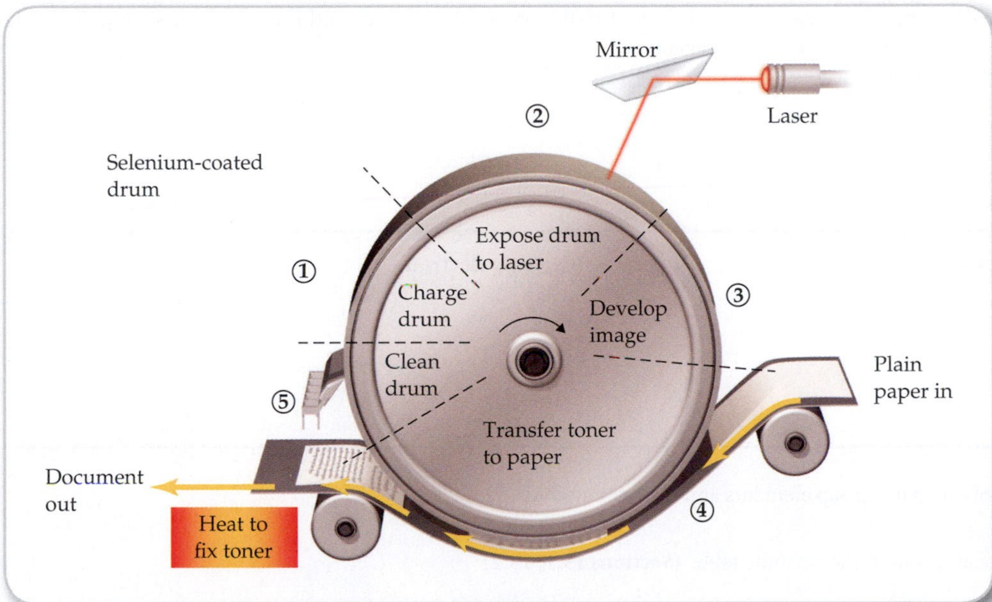

FIGURE 19.13

The laser printer process.
① A selenium-coated rotating drum is given a large, uniform negative charge.
② The drum is then exposed to a scanning laser beam, which depletes the charge in the exposed areas.
③ Negatively charged toner particles applied to the drum avoid the areas with large negative charges and adhere to the charge-depleted areas.
④ The image is transferred from the drum to a sheet of paper. Heating then fixes the image.
⑤ The drum is flooded with light and cleaned to ready it for printing the next page.

At this point, the toner could easily be brushed off the paper; you may have done this if you've ever opened a copier or printer to remove a paper jam. The toner particles, made of a low-melting plastic resin mixed with carbon black, are affixed to the paper by the fuser, a hot roller at about 170 °C that melts and bonds the toner to the paper. The document then rolls out of the machine, and the drum is restored to its original condition by flooding it with light to remove all remaining charges and gently scraping off any bits of excess toner.

The laser printing process has been adapted for printing color documents. To obtain a color image, four toner colors are needed to generate the entire spectrum of colors: cyan (blue-green), magenta (red-purple), yellow, and black. Color printers work like monochrome printers, except they repeat the printing process four times, once for each color. Careful alignment of the colors with one another is required to obtain a crisp image.

PROBLEM 19.10 What would be the effect of coating the laser printer drum with copper instead of selenium? Explain.

SUMMARY

The main-group elements are the *s*-block elements of groups 1A and 2A and the *p*-block elements of groups 3A–8A. From left to right across the periodic table, ionization energy, electronegativity, and nonmetallic character generally increase, while atomic radius and metallic character decrease. From top to bottom of a group in the periodic table, ionization energy, electronegativity, and nonmetallic character generally decrease, while atomic radius and metallic character increase. The second-row elements form strong multiple bonds but are generally unable to form more than four bonds because of the small size of their atoms.

The group 3A elements—B, Al, Ga, In, and Tl—are metals except for boron, which is a semimetal. Boron is a semiconductor and forms molecular compounds. **Boranes**, such as diborane (B_2H_6), are electron-deficient molecules that contain **three-center, two-electron bonds** (B—H—B).

The group 4A elements—C, Si, Ge, Sn, and Pb—exhibit the usual increase in metallic character down the group. They often adopt an oxidation state of +4, but the +2 state becomes increasingly more stable from Ge to Sn to Pb. In elemental form, carbon exists as diamond, graphite, graphene, and fullerene.

Silicon, the second most abundant element in the Earth's crust, is obtained by reducing silica sand (SiO_2) with coke. It is purified for use in the semiconductor industry by **zone refining**. In the **silicates**, SiO_4 tetrahedra share common O atoms to give silicon oxoanions with ring, chain, layer, and extended three-dimensional structures. In **aluminosilicates**, such as $KAlSi_3O_8$, Al^{3+} replaces some of the Si^{4+}.

Molecular nitrogen (N_2) is unreactive because of its strong N≡N triple bond. Nitrogen exhibits all oxidation states between −3 and +5. Nitric acid is manufactured by the **Ostwald process**.

Phosphorus, the most abundant group 5A element, exists in two common allotropic forms—white phosphorus, which contains highly reactive tetrahedral P_4 molecules, and red phosphorus, which is polymeric. The most common oxidation states of P are −3, as in phosphine (PH_3); +3, as in PCl_3, P_4O_6, and H_3PO_3; and +5, as in PCl_5, P_4O_{10}, and H_3PO_4.

Sulfur is obtained from underground deposits and is recovered from natural gas and crude oil. The properties of sulfur change dramatically on heating as the S_8 rings of rhombic sulfur open and polymerize to give long chains, which then fragment at higher temperatures. The most common oxidation states of S are −2, as in H_2S; +4, as in SO_2 and H_2SO_3; and +6, as in SO_3 and H_2SO_4. Sulfuric acid, the world's most important industrial chemical, is manufactured by the **contact process**.

Chlorine, bromine, and iodine form a series of oxoacids: hypohalous acid (HXO), halous acid (HXO_2 for X = Cl), halic acid (HXO_3), and perhalic acid (HXO_4). Acid strength increases as the oxidation state of the halogen increases from +1 to +7. Iodine forms two perhalic acids, HIO_4 and H_5IO_6. Halogen oxoacids and their salts are strong oxidizing agents.

KEY WORDS

aluminosilicate *789*
boranes *781*
carbide *786*
contact process *801*
Ostwald process *794*
silicate *787*
three-center, two-electron bond *781*
zone refining *787*

LEARNING OUTCOMES

- Use the periodic table to predict the properties of the main-group elements and their compounds. **(Section 19.1)** ***Problems 1, 20–35, 108, 109***
- Use the properties of main-group elements to locate them in the periodic table. **(Sections 19.1, 19.2)** ***Problems 11–14***
- Account for differences in the structure and properties of second-row elements (and their compounds) and the structure and properties of third- and higher-row elements. **(Section 19.2)** ***Problems 2, 3, 15, 16, 36–39, 43***
- Describe the structure, bonding, and properties of molecules and ions in main-group compounds. **(Sections 19.6–19.11, 19.14)** ***Problems 9, 50, 51, 71–73, 75, 96, 97, 120, 123, 130***
- Identify a main-group element given a distinguishing physical or chemical property. **(Sections 19.3–19.13)** ***Problems 46–49, 66, 67, 80, 81, 127***
- Describe the structure and bonding in diborane. **(Section 19.4)** ***Problems 4, 44, 45, 124***
- Describe the structure and properties of the allotropes of carbon. **(Section 19.7)** ***Problems 52–55, 110***
- Use the shorthand notation for a silicate anion to represent its structure and interpret the notation to find the formula and charge of a silicate anion. **(Section 19.7)** ***Problems 6, 19, 62, 63***
- Write electron-dot structures for molecules and ions in main-group compounds. **(Sections 19.3–19.14)** ***Problems 7, 17, 18, 75, 97***
- Relate compound names, compound formulas, and oxidation states in compounds of the main-group elements. **(Sections 19.3–19.14)** ***Problems 41, 56, 68, 69, 84, 85, 92–95***
- Account for the properties of the group 5A elements in terms of structure and bonding. **(Sections 19.10, 19.11)** ***Problems 70, 74, 76, 128***
- Write balanced equations for reactions of the main-group elements and their compounds. **(Sections 19.4, 19.7, 19.10–19.14)** ***Problems 78, 87–89, 100–102, 106, 107, 115, 116***

CONCEPTUAL PROBLEMS

Problems 19.1–19.10 appear within the chapter.

19.11 Locate each of the following groups of elements on the periodic table:

(a) Main-group elements
(b) *s*-Block elements
(c) *p*-Block elements
(d) Main-group metals
(e) Nonmetals
(f) Semimetals

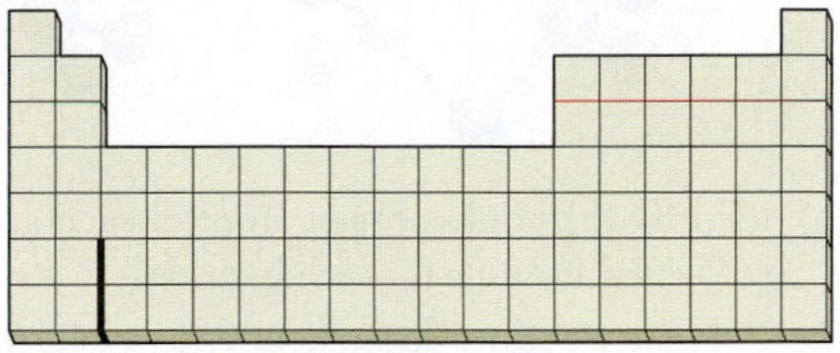

19.12 Locate each of the following elements on the periodic table:

(a) Element with the lowest ionization energy
(b) Most electronegative element
(c) Group 4A element with the largest atomic radius
(d) Group 6A element with the smallest atomic radius
(e) Group 3A element that is a semiconductor
(f) Group 5A element that forms the strongest π bonds

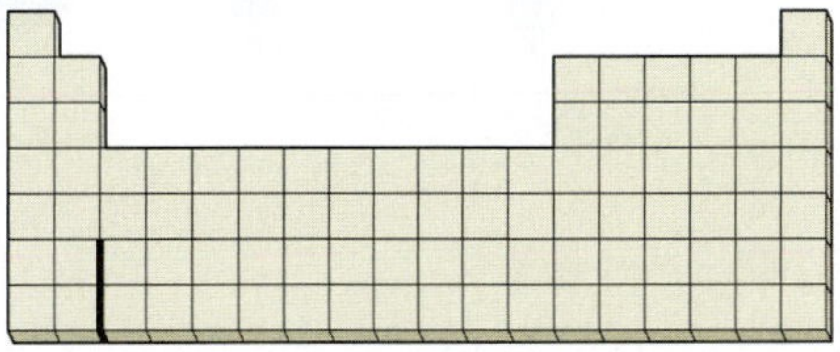

19.13 Locate the following elements on the periodic table:

(a) Elements that are gases at room temperature (25 °C)
(b) A main-group element that is a liquid at 25 °C
(c) Nonmetals that are solids at 25 °C
(d) Elements that exist as diatomic molecules at 25 °C

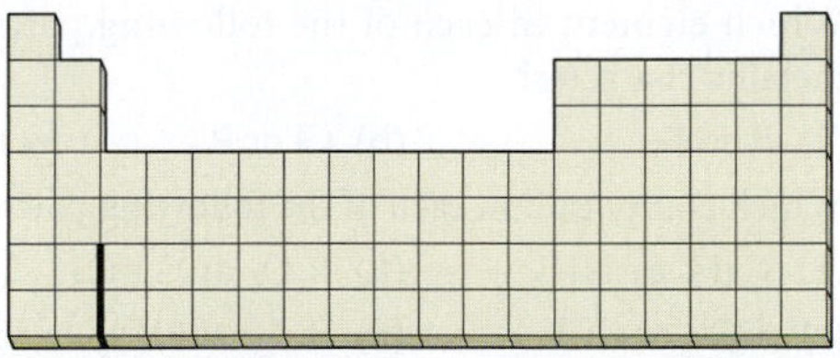

19.14 Locate the following elements on the periodic table, and write the formula of a compound that justifies each of your answers:

(a) Two nonmetals that can form more than four bonds
(b) Two nonmetals that form a maximum of four bonds
(c) Two nonmetals that form oxides that are gases at 25 °C
(d) A nonmetal that forms an oxide that is a solid at 25 °C

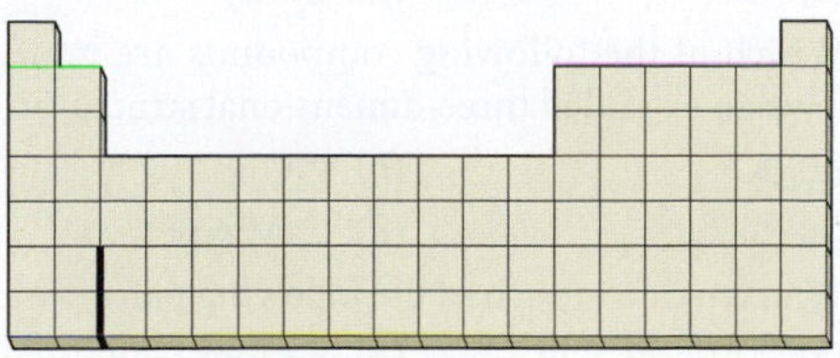

19.15 Consider the six second- and third-row elements in groups 5A–7A of the periodic table:

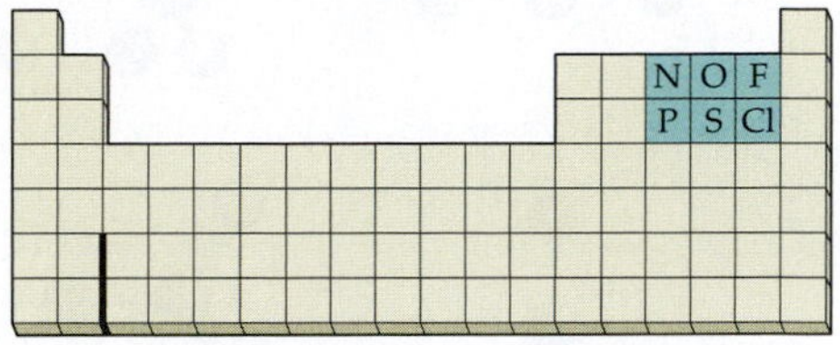

Possible molecular structures for common allotropes of these elements are shown below:

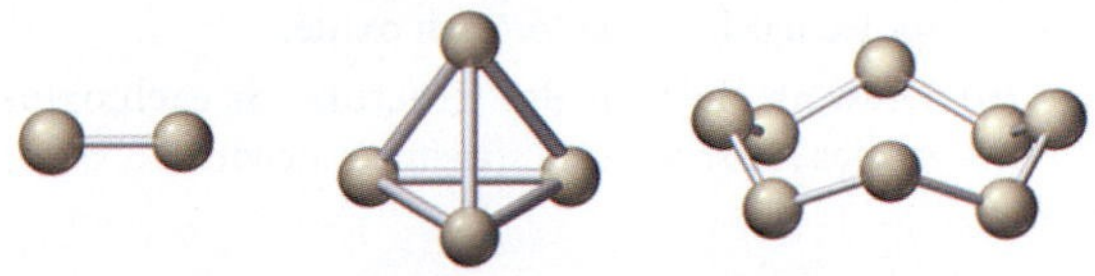

(a) What is the molecular structure of each of the six elements?
(b) Using electron-dot structures, explain why each element has its particular molecular structure.
(c) Explain why nitrogen and phosphorus have different molecular structures and why oxygen and sulfur have different molecular structures, but fluorine and chlorine have the same molecular structure.

19.16 Consider the six second- and third-row elements in groups 4A–6A of the periodic table:

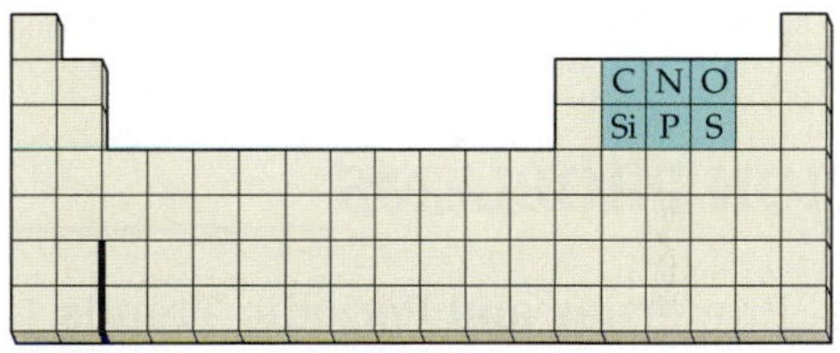

Possible structures for the binary fluorides of each of these elements in its highest oxidation state are shown below.

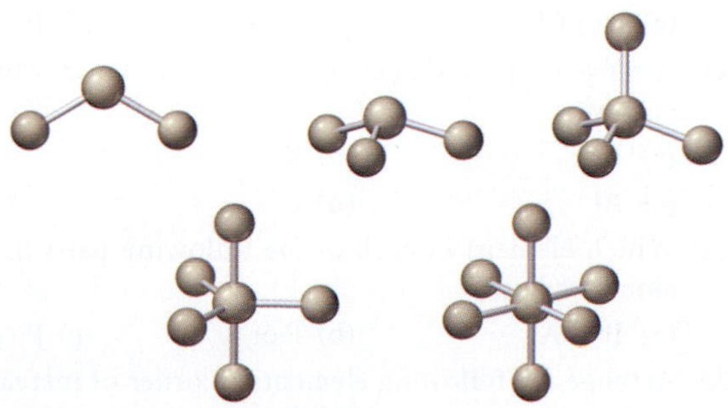

(a) Identify the nonfluorine atom in each case, and write the molecular formula of each fluoride.
(b) Explain why the fluorides of nitrogen and phosphorus have different molecular structures, but the fluorides of carbon and silicon have the same molecular structure.

19.17 The following models represent the structures of binary oxides of second- and third-row elements in their highest oxidation states:

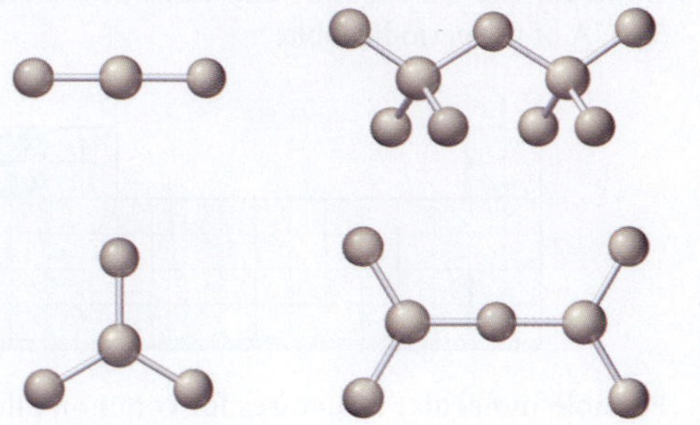

(a) Identify the non-oxygen atom in each case, and write the molecular formula for each oxide.

(b) Draw an electron-dot structure for each oxide. For which oxides are resonance structures needed?

19.18 The following models represent the structures of binary hydrides of second-row elements:

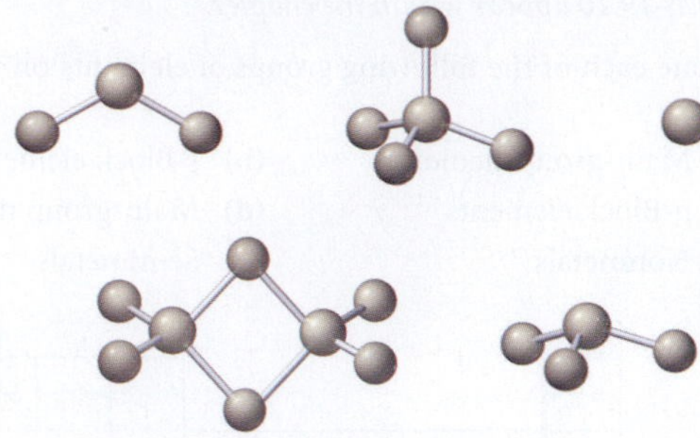

(a) Identify the nonhydrogen atom in each case, and write the molecular formula for each hydride.

(b) Draw an electron-dot structure for each hydride. For which hydride is there a problem in drawing the structure? Explain.

19.19 The following pictures represent various silicate anions. Write the formula and charge of each anion.

(a) **(b)** **(c)**

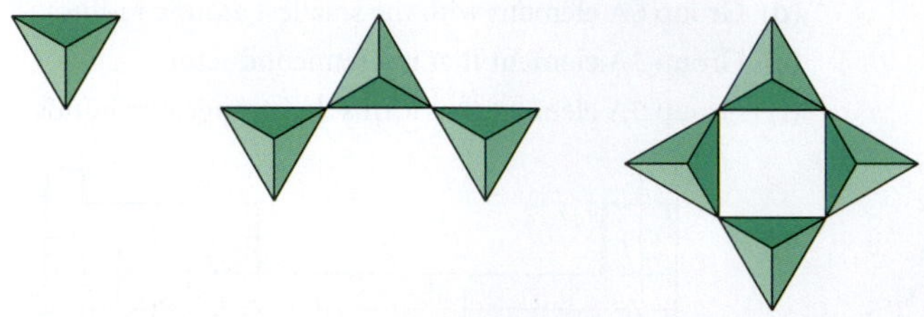

SECTION PROBLEMS

General Properties and Periodic Trends (Sections 19.1, 19.2)

19.20 Which element in each of the following pairs has the higher ionization energy?

(a) S or Cl (b) Si or Ge (c) In or O

19.21 Arrange the following elements in order of increasing ionization energy:

(a) P (b) K
(c) Al (d) F

19.22 Which element in each of the following pairs has the larger atomic radius?

(a) B or Al (b) P or S (c) Pb or Br

19.23 Arrange the following elements in order of increasing atomic radius:

(a) As (b) O
(c) Sn (d) S

19.24 Which element in each of the following pairs has the higher electronegativity?

(a) Te or I (b) N or P (c) In or F

19.25 Arrange the following elements in order of increasing electronegativity:

(a) N (b) Ge
(c) O (d) P

19.26 Which element in each of the following pairs has more metallic character?

(a) Si or Sn (b) Ge or Se (c) Bi or I

19.27 Which element in each of the following pairs has more nonmetallic character?

(a) S or Te (b) Cl or P (c) Bi or Br

19.28 Which compound in each of the following pairs is more ionic?

(a) CaH_2 or NH_3 (b) P_4O_6 or Ga_2O_3
(c) $SiCl_4$ or KCl (d) BCl_3 or $AlCl_3$

19.29 Which compound in each of the following pairs is more covalent?

(a) PCl_3 or AlF_3 (b) CaO or NO
(c) NH_3 or KH (d) SnO_2 or SiO_2

19.30 Which of the following compounds are molecular, and which have an extended three-dimensional structure?

(a) B_2H_6 (b) $KAlSi_3O_8$
(c) SO_3 (d) $GeCl_4$

19.31 Which of the following compounds are molecular, and which have an extended three-dimensional structure?

(a) KF (b) P_4O_{10}
(c) $SiCl_4$ (d) $CaMgSi_2O_6$

19.32 Which oxide in each of the following pairs is more acidic?

(a) Al_2O_3 or P_4O_{10} (b) B_2O_3 or Ga_2O_3
(c) SO_2 or SnO_2 (d) As_2O_3 or N_2O_3

19.33 Which oxide in each of the following pairs is more basic?
(a) SO_2 or SnO_2 (b) In_2O_3 or Ga_2O_3
(c) Al_2O_3 or N_2O_5 (d) BaO or MgO

19.34 Consider the elements C, Se, B, Sn, Cl. Identify the element on this list that:
(a) Has the largest atomic radius
(b) Is the most electronegative
(c) Is the best electrical conductor
(d) Has a maximum oxidation state of +6
(e) Forms a hydride with the empirical formula XH_3

19.35 Consider the elements N, Si, Al, S, F. Identify which of these elements:
(a) Has the highest ionization energy
(b) Has the most metallic character
(c) Forms the strongest π bonds
(d) Is a semiconductor
(e) Forms a 2− anion

19.36 BF_3 reacts with F^- to give BF_4^-, but AlF_3 reacts with F^- to give AlF_6^{3-}. Explain.

19.37 $GeCl_4$ reacts with Cl^- to give $GeCl_6^{2-}$, but CCl_4 does not react with excess Cl^-. Explain.

19.38 At ordinary temperatures, sulfur exists as S_8 but oxygen exists as O_2. Explain.

19.39 Elemental nitrogen exists as N_2, but white phosphorus exists as P_4. Explain.

The Group 3A Elements (Sections 19.3–19.5)

19.40 What is the most common oxidation state for each of the group 3A elements?

19.41 What is the oxidation state of the group 3A element in each of the following compounds?
(a) $NaBF_4$ (b) $GaCl_3$
(c) TlCl (d) B_2H_6

19.42 List three ways in which the properties of boron differ from those of the other group 3A elements.

19.43 Explain why the properties of boron differ so markedly from the properties of the other group 3A elements.

19.44 Tell what is meant by:
(a) An electron-deficient molecule
(b) A three-center, two-electron bond
Illustrate each definition with an example.

19.45 Describe the structure of diborane (B_2H_6), and explain why the bridging B—H bonds are longer than the terminal B—H bonds.

19.46 Identify the group 3A element that best fits each of the following descriptions:
(a) Is the most abundant element of the group
(b) Is stable in the +1 oxidation state
(c) Is a semiconductor
(d) Forms a molecular fluoride

19.47 Identify the group 3A element that best fits each of the following descriptions:
(a) Has an unusually low melting point
(b) Is the most electronegative
(c) Is extremely toxic
(d) Forms an acidic oxide

The Group 4A Elements (Sections 19.6–19.8)

19.48 Identify the group 4A element that best fits each of the following descriptions:
(a) Prefers the +2 oxidation state
(b) Forms the strongest π bonds
(c) Is the second most abundant element in the Earth's crust
(d) Forms the most acidic oxide

19.49 Select the group 4A element that best fits each of the following descriptions:
(a) Forms the most basic oxide
(b) Is the least dense semimetal
(c) Is the second most abundant element in the human body
(d) Is the most electronegative

19.50 Describe the shape of each of the following molecules or ions, and tell which hybrid orbitals are used by the central atom:
(a) $GeBr_4$ (b) CO_2
(c) CO_3^{2-} (d) $SnCl_3^-$

19.51 What is the shape of each of the following molecules or ions, and which hybrid orbitals are used by the central atom?
(a) SiO_4^{4-} (b) CCl_4
(c) $SnCl_2$ (d) HCN

19.52 List three properties of diamond, and account for them in terms of structure and bonding.

19.53 Describe the structure and bonding in graphite, and explain why graphite is a good lubricant and a good electrical conductor.

19.54 What is graphene, and how does it differ from graphite?

19.55 Fullerene is soluble in nonpolar solvents, but graphite isn't. Explain.

19.56 Give the name and formula of a compound in which carbon exhibits an oxidation state of:
(a) +4 (b) +2 (c) −4

19.57 Give an example of an ionic carbide. What is the oxidation state of carbon in this substance?

19.58 List three commercial uses for carbon dioxide, and relate each use to one of carbon dioxide's properties.

19.59 Why are CO and CN^- so toxic to humans?

19.60 Describe the preparation of silicon from silica sand, and tell how silicon is purified for use in semiconductor devices. Write balanced equations for all reactions.

19.61 How do the structures and properties of elemental silicon and germanium differ from those of tin and lead?

19.62 Using the shorthand notation of Figure 19.5 (page 788), draw the structure of the silicate anion in:
(a) K_4SiO_4 (b) $Ag_{10}Si_4O_{13}$
What is the relationship between the charge on the anion and the number of terminal O atoms?

19.63 Using the shorthand notation of Figure 19.5 (page 788), draw the structure of the cyclic silicate anion in which four SiO_4 tetrahedra share O atoms to form an eight-membered ring of alternating Si and O atoms. Give the formula and charge of the anion.

19.64 The silicate anion in the mineral kinoite is a chain of three SiO_4 tetrahedra that share corners with adjacent tetrahedra. The mineral also contains Ca^{2+} ions, Cu^{2+} ions, and water molecules in a 1:1:1 ratio.
(a) Give the formula and charge of the silicate anion.
(b) Give the complete formula for the mineral.

19.65 Suggest a plausible structure for the silicate anion in each of the following minerals:
(a) Spodumene, $LiAlSi_2O_6$
(b) Wollastonite, $Ca_3Si_3O_9$
(c) Thortveitite, $Sc_2Si_2O_7$
(d) Albite, $NaAlSi_3O_8$

The Group 5A Elements (Sections 19.9–19.11)

19.66 Identify the group 5A element(s) that best fits each of the following descriptions:
(a) Makes up part of bones and teeth
(b) Forms stable salts containing M^{3+} ions
(c) Is the most abundant element in the atmosphere
(d) Forms a basic oxide

19.67 Identify the group 5A element that best fits each of the following descriptions:
(a) Forms strong π bonds
(b) Is a metal
(c) Is the most abundant group 5A element in the Earth's crust
(d) Forms oxides with the group 5A element in the +1, +2, and +4 oxidation states

19.68 Give the chemical formula for each of the following compounds, and indicate the oxidation state of the group 5A element:
(a) Nitrous oxide
(b) Hydrazine
(c) Calcium phosphide
(d) Phosphorous acid
(e) Arsenic acid

19.69 Give the chemical formula for each of the following compounds, and indicate the oxidation state of the group 5A element:
(a) Nitric oxide
(b) Nitrous acid
(c) Phosphine
(d) Tetraphosphorus decoxide
(e) Phosphoric acid

19.70 Draw an electron-dot structure for N_2, and explain why this molecule is so unreactive.

19.71 Draw electron-dot structures for:
(a) Nitrous oxide
(b) Nitric oxide
(c) Nitrogen dioxide
Predict the geometrical structure of each, and indicate which are expected to be paramagnetic.

19.72 Predict the geometrical structure of each of the following molecules or ions:
(a) NO_2^- (b) PH_3
(c) PF_5 (d) PCl_4^+

19.73 Predict the geometrical structure of each of the following molecules or ions:
(a) PCl_6^- (b) N_2O
(c) H_3PO_3 (d) NO_3^-

19.74 Describe the structures of the white and red allotropes of phosphorus, and explain why white phosphorus is so reactive.

19.75 Draw the structure of each of the following molecules:
(a) Tetraphosphorus hexoxide
(b) Tetraphosphorus decoxide
(c) Phosphorous acid
(d) Phosphoric acid

19.76 Account for each of the following observations:
(a) Phosphorous acid is a diprotic acid.
(b) Nitrogen doesn't exist as a four-atom molecule like P_4.

19.77 Account for each of the following observations:
(a) Nitric acid is a strong oxidizing agent, but phosphoric acid is not.
(b) Phosphorus, arsenic, and antimony form trichlorides and pentachlorides, but nitrogen forms only NCl_3.

19.78 Write a balanced equation to account for each of the following observations:
(a) Nitric oxide turns brown when exposed to air.
(b) Nitric acid turns yellow-brown on standing.
(c) Silver dissolves in dilute HNO_3, yielding a colorless gas.
(d) Hydrazine reduces iodine to I^- and in the process is oxidized to N_2 gas.

19.79 Describe the process used for the industrial production of the following chemicals:
(a) Nitrogen
(b) Ammonia
(c) Nitric acid
(d) Phosphoric acid
Write balanced equations for all chemical reactions.

The Group 6A Elements (Sections 19.12, 19.13)

19.80 Identify the group 6A element that best fits each of the following descriptions:
(a) Is the most electronegative
(b) Is a semimetal
(c) Is radioactive
(d) Is the most abundant element in the Earth's crust

19.81 Identify the group 6A element that best fits each of the following descriptions:
(a) Is a metal
(b) Is the most abundant element in the human body
(c) Is the strongest oxidizing agent
(d) Has the most negative electron affinity

19.82 Describe the structure of the sulfur molecules in:
(a) Rhombic sulfur
(b) Monoclinic sulfur
(c) Plastic sulfur
(d) Liquid sulfur above 160 °C

19.83 The viscosity of liquid sulfur increases sharply at about 160 °C and then decreases again above 200 °C. Explain.

19.84 Give the name and formula of two compounds in which sulfur exhibits an oxidation state of:
(a) −2 (b) +4 (c) +6

19.85 What is the oxidation state of sulfur in each of the following compounds?
(a) HgS (b) $Ca(HSO_4)_2$ (c) H_2SO_3
(d) FeS_2 (e) SF_4

19.86 Describe the contact process for the manufacture of sulfuric acid, and write balanced equations for all reactions.

19.87 Describe a convenient laboratory method for preparing each of the following compounds, and write balanced equations for all reactions:

(a) Sulfur dioxide
(b) Hydrogen sulfide
(c) Sodium hydrogen sulfate

19.88 Write a balanced net ionic equation for each of the following reactions:

(a) $Zn(s)$ + dilute $H_2SO_4(aq) \longrightarrow ?$
(b) $BaSO_3(s) + HCl(aq) \longrightarrow ?$
(c) $Cu(s)$ + hot, conc. $H_2SO_4(l) \longrightarrow ?$
(d) $H_2S(aq) + I_2(aq) \longrightarrow ?$

19.89 Write a balanced net ionic equation for each of the following reactions:

(a) $ZnS(s) + HCl(aq) \longrightarrow ?$
(b) $H_2S(aq) + Fe(NO_3)_3(aq) \longrightarrow ?$
(c) $Fe(s)$ + dilute $H_2SO_4(aq) \longrightarrow ?$
(d) $BaO(s) + H_2SO_4(aq) \longrightarrow ?$

19.90 Account for each of the following observations:

(a) H_2SO_4 is a stronger acid than H_2SO_3.
(b) SF_4 exists, but OF_4 does not.
(c) The S_8 ring is nonplanar.

19.91 Account for each of the following observations:

(a) Oxygen is more electronegative than sulfur.
(b) Sulfur forms long S_n chains, but oxygen does not.
(c) The SO_3 molecule is trigonal planar, but the SO_3^{2-} ion is trigonal pyramidal.

Halogen Oxoacids and Oxoacid Salts (Section 19.14)

19.92 Write the formula for each of the following compounds, and indicate the oxidation state of the halogen:

(a) Bromic acid
(b) Hypoiodous acid
(c) Sodium chlorite
(d) Potassium metaperiodate

19.93 Write the formula for each of the following compounds, and indicate the oxidation state of the halogen:

(a) Potassium hypobromite
(b) Paraperiodic acid
(c) Sodium bromate
(d) Chlorous acid

19.94 Name each of the following compounds:

(a) HIO_3 (b) $HClO_2$
(c) $NaOBr$ (d) $LiClO_4$

19.95 Name each of the following compounds:

(a) $KClO_2$ (b) HIO_4
(c) $HOBr$ (d) $NaBrO_3$

19.96 Write an electron-dot structure for each of the following molecules or ions, and predict the geometrical structure:

(a) HIO_3 (b) ClO_2^-
(c) $HOCl$ (d) IO_6^{5-}

19.97 Write an electron-dot structure for each of the following molecules or ions, and predict the geometrical structure:

(a) BrO_4^- (b) ClO_3^-
(c) HIO_4 (d) $HOBr$

19.98 Explain why acid strength increases in the order $HClO < HClO_2 < HClO_3 < HClO_4$.

19.99 Explain why acid strength increases in the order $HIO < HBrO < HClO$.

19.100 Write a balanced net ionic equation for each of the following reactions:

(a) $Br_2(l)$ + cold $NaOH(aq) \longrightarrow ?$
(b) $Cl_2(g)$ + cold $H_2O(l) \longrightarrow ?$
(c) $Cl_2(g)$ + hot $NaOH(aq) \longrightarrow ?$

19.101 Write a balanced equation for the reaction of potassium chlorate and sucrose. The products are $KCl(s)$, $CO_2(g)$, and $H_2O(g)$.

CHAPTER PROBLEMS

19.102 Write a balanced net ionic equation for the reaction between magnesium and hot, concentrated sulfuric acid, a reaction similar to that between hot, concentrated sulfuric acid and copper.

19.103 What compound in Table 19.6 is used as a rocket fuel?

19.104 Does the silicate hedenbergite, $CaFeSi_2O_6$, contain single-stranded or double-stranded silicate chains? (Draw comparisons with Figures 19.6 and 19.7.)

19.105 An iron object exposed to rain is eventually damaged by rusting, but a similarly placed aluminum object is visibly unchanged. Explain.

19.106 Hydrazine can be used to remove small amounts of dissolved oxygen from the water used in boilers. Write a balanced equation for the reaction if the products are nitrogen and water.

19.107 Iodine forms the acid anhydride I_2O_5. Write a balanced equation for the reaction of this anhydride with water, and name the acid that is formed.

19.108 Which compound in each of the following pairs has the higher melting point?

(a) $LiCl$ or PCl_3
(b) CO_2 or SiO_2
(c) P_4O_{10} or NO_2

19.109 Which element in each of the following pairs is the better electrical conductor?

(a) B or Ga
(b) In or S
(c) Pb or P

19.110 Draw the structure of graphene.

19.111 Compare and contrast the properties of ammonia and phosphine.

19.112 Identify as many as you can of the four most abundant elements in the Earth's crust and in the human body without consulting Figure 19.1.

19.113 Identify as many of the 10 most important industrial chemicals as you can without consulting Table 19.1.

19.114 Which of the group 4A elements have allotropes with the diamond structure? Which have metallic allotropes? How does the variation in the structure of the group 4A elements illustrate how metallic character varies down a periodic group?

19.115 Write a balanced chemical equation for a laboratory preparation of each of the following compounds:

(a) NH_3 **(b)** CO_2 **(c)** C_2H_2 (acetylene)
(d) N_2O **(e)** NO_2

19.116 Write balanced equations for the reactions of **(a)** H_3PO_4 and **(b)** $B(OH)_3$ with water. Classify each acid as a Brønsted–Lowry acid or a Lewis acid.

19.117 What oxoanion is used for these purposes?

(a) Oxidizing agent in space shuttle booster rockets
(b) Oxidizing agent in chlorine bleach
(c) Oxidizing agent in matches and fireworks

19.118 Account for each of the following observations:

(a) Diamond is extremely hard and high melting, whereas graphite is very soft and high melting.
(b) Chlorine does not form a perhalic acid, H_5ClO_6.

19.119 So-called fuming sulfuric acid is formed when sulfur trioxide dissolves in anhydrous sulfuric acid to form $H_2S_2O_7$. Propose a structure for $H_2S_2O_7$, which contains an S–O–S linkage.

19.120 Chlorine reacts with molten sulfur to yield disulfur dichloride, a yellowish-red liquid. Propose a structure for disulfur dichloride.

19.121 The organ pipes in unheated northern European churches are often observed to be pitted and crumbling to powder in places. Suggest an explanation, given that the pipes are made from tin.

19.122 Of ammonia, hydrazine, and hydroxylamine, which reacts to the greatest extent with the weak acid HNO_2? Consult Appendix C for equilibrium constants.

19.123 Describe the relationship between the geometrical structures of white phosphorus, tetraphosphorus hexoxide, and tetraphosphorus decoxide.

19.124 Suggest a structure for the mixed aluminum–boron hydride $AlBH_6$.

19.125 Suggest a benefit and a limitation to using gallium as the material in a thermometer.

19.126 A sample of P_4 burned in excess oxygen and formed a phosphorus oxide. The phosphorus oxide was dissolved carefully in enough water to make 1.00 L of solution with pH = 1.93. What is the identity of the phosphorus oxide and the oxoacid that was formed, and how much P_4, in grams, was burned? See Appendix C for acid dissociation constants.

19.127 Give one example from main group chemistry that illustrates each of the following descriptions:

(a) Covalent network solid
(b) Disproportionation reaction
(c) Paramagnetic oxide
(d) Polar molecule that violates the octet rule
(e) Lewis acid
(f) Amphoteric oxide
(g) Semiconductor
(h) Strong oxidizing agent
(i) Allotropes

19.128 Could the strain in the P_4 molecule be reduced by using sp^3 hybrid orbitals in bonding instead of pure *p* orbitals? Explain.

19.129 Carbon is an essential element in the molecules on which life is based. Would silicon be equally satisfactory? Explain.

MULTICONCEPT PROBLEMS

19.130 An important physiological reaction of nitric oxide (NO) is its interaction with the superoxide ion (O_2^-) to form the peroxynitrite ion ($ONOO^-$).

(a) Write electron-dot structures for NO, O_2^-, and $ONOO^-$, and predict the O–N–O bond angle in $ONOO^-$.
(b) The bond length in NO (115 pm) is intermediate between the length of an NO triple bond and an NO double bond. Account for the bond length and the paramagnetism of NO using molecular orbital theory.

19.131 Consider phosphorous acid, a polyprotic acid with formula H_3PO_3.

(a) Draw two plausible structures for H_3PO_3. For each one, predict the shape of the pH titration curve for the titration of the H_3PO_3 ($K_{a1} = 1.0 \times 10^{-2}$) with aqueous NaOH.
(b) For the structure with the H atoms in two different environments, calculate the pH at the first and second equivalence points, assuming that 30.00 mL of 0.1240 M H_3PO_3 ($K_{a2} = 2.6 \times 10^{-7}$) is titrated with 0.1000 M NaOH.

19.132 We've said that the +1 oxidation state is uncommon for indium but is the most stable state for thallium. Verify this statement by calculating $E°$ and $\Delta G°$ (in kilojoules) for the disproportionation reaction

$$3\,M^+(aq) \longrightarrow M^{3+}(aq) + 2\,M(s) \qquad M = \text{In or Tl}$$

Is disproportionation a spontaneous reaction for In^+ and/or Tl^+? Standard reduction potentials for the relevant half-reactions are

$$In^{3+}(aq) + 2\,e^- \longrightarrow In^+(aq) \qquad E° = -0.44\text{ V}$$
$$In^+(aq) + e^- \longrightarrow In(s) \qquad E° = -0.14\text{ V}$$
$$Tl^{3+}(aq) + 2\,e^- \longrightarrow Tl^+(aq) \qquad E° = +1.25\text{ V}$$
$$Tl^+(aq) + e^- \longrightarrow Tl(s) \qquad E° = -0.34\text{ V}$$

19.133 Terrorists often use ammonium nitrate fertilizer as an ingredient in car bombs. When ammonium nitrate explodes, it decomposes to gaseous nitrogen, oxygen, and water vapor. The force of the explosion results from the sudden production of a huge volume of hot gas.

(a) Write a balanced equation for the reaction.

(b) What volume of gas (in liters) is produced from the explosion of 1.80 m^3 of solid NH_4NO_3? Assume that the gas has a temperature of 500 °C and a pressure of 1.00 atm. The density of NH_4NO_3 is 1.725 g/cm^3.

(c) Use the thermodynamic data in Appendix B to calculate the amount of heat (in kilojoules) released in the reaction.

19.134 It has been claimed that NH_4NO_3 fertilizer can be rendered unexplodable (see Problem 19.133) by adding compounds such as diammonium hydrogen phosphate, $(NH_4)_2HPO_4$. Analysis of such a desensitized sample of NH_4NO_3 showed the mass % nitrogen to be 33.81%.

(a) Assuming that the mixture contains only NH_4NO_3 and $(NH_4)_2HPO_4$, what is the mass percent of each of these two components?

(b) A 0.965 g sample of the mixture was dissolved in enough water to make 50.0 mL of solution. What is the pH of the solution? (Hint: The strongest base present is HPO_4^{2-}.)

19.135 A 5.00 g quantity of white phosphorus was burned in an excess of oxygen, and the product was dissolved in enough water to make 250.0 mL of solution.

(a) Write balanced equations for the reactions.

(b) What is the pH of the solution?

(c) When the solution was treated with an excess of aqueous $Ca(NO_3)_2$, a white precipitate was obtained. Write a balanced equation for the reaction, and calculate the mass of the precipitate in grams.

(d) The precipitate in part **(c)** was removed, and the solution that remained was treated with an excess of zinc, yielding a colorless gas that was collected at 20 °C and 742 mm Hg. Identify the gas, and determine its volume.

19.136 A 500.0 mL sample of an equilibrium mixture of gaseous N_2O_4 and NO_2 at 25 °C and 753 mm Hg pressure was allowed to react with enough water to make 250.0 mL of solution at 25 °C. You may assume that all the dissolved N_2O_4 is converted to NO_2, which disproportionates in water, yielding a solution of nitrous acid and nitric acid. Assume further that the disproportionation reaction goes to completion and that none of the nitrous acid disproportionates. The equilibrium constant K_p for the reaction $N_2O_4(g) \rightleftharpoons 2\,NO_2(g)$ is 0.113 at 25 °C. K_a for HNO_2 is 4.5×10^{-4} at 25 °C.

(a) Write a balanced equation for the disproportionation reaction.

(b) What is the molar concentration of NO_2^-, and what is the pH of the solution?

(c) What is the osmotic pressure of the solution in atmospheres?

(d) How many grams of lime (CaO) would be needed to neutralize the solution?

CHAPTER

20 Transition Elements and Coordination Chemistry

How do transition elements help to convert atmospheric nitrogen to the nitrogen compounds needed for plant growth? The answer is in the *FYI* at the end of this chapter.

The *transition elements* occupy the central part of the periodic table, bridging the gap between the active *s*-block metals of groups 1A and 2A on the left and the *p*-block metals, semimetals, and nonmetals of groups 3A–8A on the right (**Figure 20.1**). Because the *d* subshells are being filled in this region of the periodic table, the transition elements are also called the *d-block elements*.

Each *d* subshell consists of five orbitals and can accommodate 10 electrons, so each transition series consists of 10 elements. The first series extends from scandium through zinc and includes many familiar metals, such as chromium, iron, and copper. The second series runs from yttrium through cadmium, and the third series runs from lutecium through mercury. In addition, there is a fourth transition series made up of the unstable, radioactive elements lawrencium through the recently discovered element 112, copernicium (Cn).

Tucked into the periodic table between barium (atomic number 56) and lutecium (atomic number 71) are the *lanthanides*. In this series of 14 metallic elements, the seven 4*f* orbitals are progressively filled, as shown in Figure 2.17 (page 84). Following radium (atomic number 88) is a second series of 14 elements, the *actinides*, in which the 5*f* subshell is progressively filled. The lanthanides and actinides together comprise the *f-block elements*, or *inner transition elements*.

The transition metals iron and copper have been known since antiquity and have played an important role in the development of civilization. Iron, the main constituent of steel, is still important as a structural material. Worldwide production of steel was 1.41 billion metric tons in 2010. In newer technologies, other transition elements are useful. For example, the strong, lightweight metal titanium is a major component in modern jet aircraft and is used in manufacturing artificial joints and dental implants. Transition metals are also used as heterogeneous catalysts in automobile catalytic converters and in the industrial synthesis of essential chemicals such as sulfuric acid, nitric acid, and ammonia.

The role of the transition elements in living systems is equally important. Iron is present in biomolecules such as hemoglobin, which transports oxygen from our lungs to other

CONTENTS

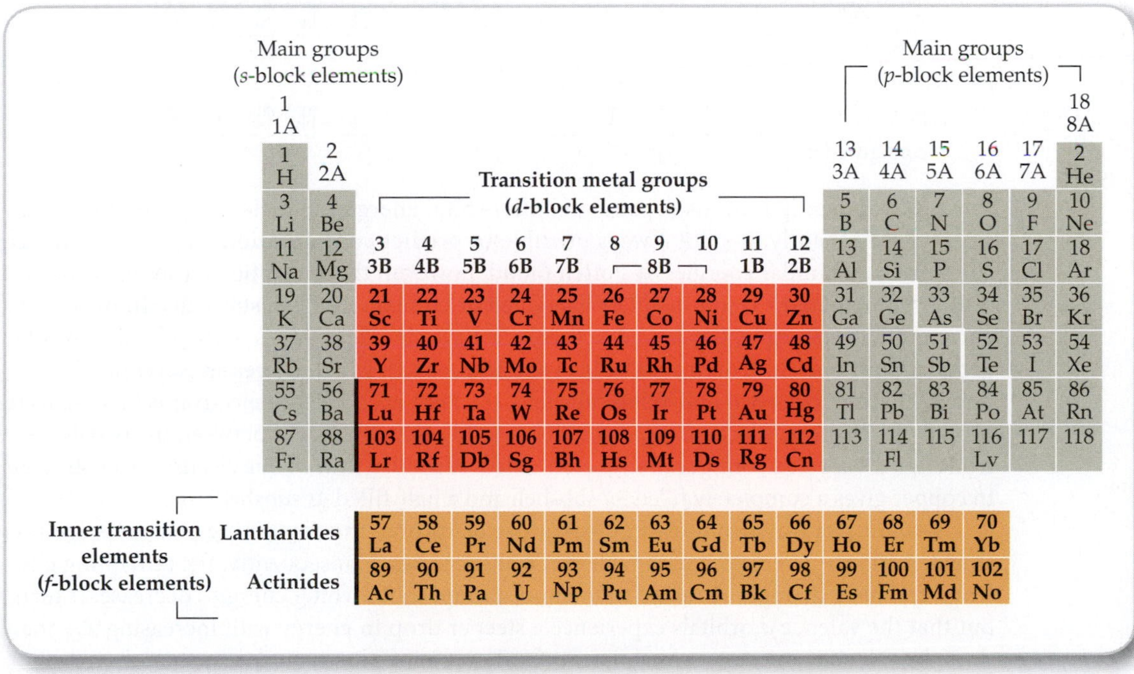

FIGURE 20.1

The transition elements are located in the central region of the periodic table.

parts of the body. Cobalt is an essential component of vitamin B_{12}. Nickel, copper, and zinc are vital constituents of many enzymes, the large protein molecules that catalyze biochemical reactions.

In this chapter, we'll look at the properties and chemical behavior of transition metal compounds, paying special attention to *coordination compounds*, in which a central metal ion or atom—usually a transition metal—is attached to a group of surrounding molecules or ions by coordinate covalent bonds (Section 4.6).

20.1 ELECTRON CONFIGURATIONS

Look at the electron configurations of potassium and calcium, the *s*-block elements immediately preceding the first transition series. These atoms have 4*s* valence electrons, but no *d* electrons:

$$\textbf{K}\ (Z = 19)\text{: [Ar] } 3d^0\, 4s^1 \qquad \textbf{Ca}\ (Z = 20)\text{: [Ar] } 3d^0\, 4s^2$$

The filling of the 3*d* subshell begins at atomic number 21 (scandium) and continues until the subshell is completely filled at atomic number 30 (zinc):

$$\textbf{Sc}\ (Z = 21)\text{: [Ar] } 3d^1\, 4s^2 \longrightarrow \textbf{Zn}\ (Z = 30)\text{: [Ar] } 3d^{10}\, 4s^2$$

(It's convenient to list the valence electrons in order of principal quantum number rather than in order of orbital filling because the 4*s* electrons are lost first in forming ions.)

For main-group elements, the valence electrons are generally considered to be those in the outermost shell because they are the ones that are involved in chemical bonding. For transition elements, however, both the $(n - 1)d$ and the *ns* electrons are involved in bonding and are considered valence electrons.

REMEMBER...

According to **Hund's rule**, if two or more orbitals with the same energy are available, one electron goes into each until all are half-filled and the electrons in the half-filled orbitals all have the same spin. (Section 2.11)

The filling of the 3*d* subshell generally proceeds according to **Hund's rule** with one electron adding to each of the five 3*d* orbitals before a second electron adds to any one of them (Section 2.11). There are just two exceptions to the expected regular filling pattern, chromium and copper:

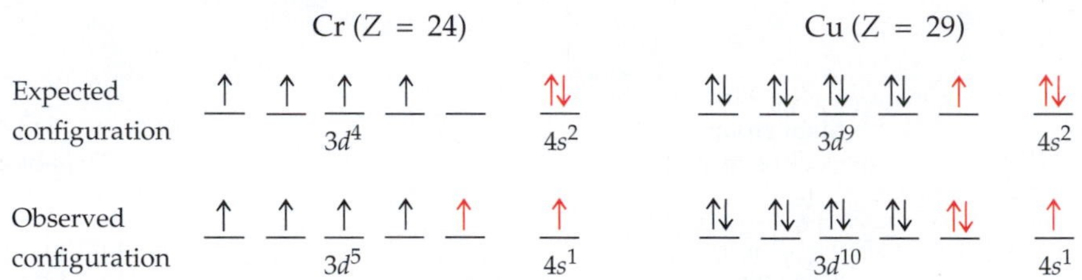

Electron configurations depend on both orbital energies and electron–electron repulsions. Consequently, it's not always possible to predict configurations when two valence subshells have similar energies. It's often found, however, that exceptions from the expected orbital filling pattern result in either half-filled or completely filled subshells. In the case of chromium, for example, the 3*d* and 4*s* subshells have similar energies. It's evidently advantageous to shift one electron from the 4*s* to the 3*d* subshell, which decreases electron–electron repulsions and gives two half-filled subshells. Because each valence electron is in a separate orbital, the electron–electron repulsion that would otherwise occur between the two 4*s* electrons in the expected configuration is eliminated. A similar shift of one electron from 4*s* to 3*d* in copper gives a completely filled 3*d* subshell and a half-filled 4*s* subshell.

When a neutral atom loses one or more electrons, the remaining electrons are less shielded and the effective nuclear charge (Z_{eff}) increases. Consequently, the remaining electrons are more strongly attracted to the nucleus and their orbital energies decrease. It turns out that the valence *d* orbitals experience a steeper drop in energy with increasing Z_{eff} than does the valence *s* orbital, making the *d* orbitals in cations lower in energy than the *s* orbital. As a result, the valence *s* orbital is vacant and all the valence electrons occupy the *d* orbitals in transition metal cations. Iron, for example, which forms 2+ and 3+ cations, has the following valence electron configurations:

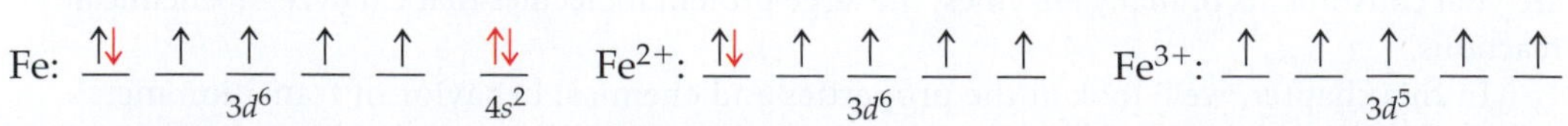

In neutral molecules and complex anions, the metal atom usually has a positive oxidation state. It therefore has a partial positive charge and a higher Z_{eff} than that of the neutral atom. As a result, the valence *d* orbitals are again lower in energy than the valence *s* orbital, and so all the metal's valence electrons occupy the *d* orbitals. The metal atom in both VCl_4 and MnO_4^{2-}, for example, has the valence configuration $3d^1$ rather than $4s^1$. Electron configurations and other properties for atoms and common ions of first-series transition elements are summarized in **Table 20.1**.

TABLE 20.1 Selected Properties of First-Series Transition Elements

Group:	3B	4B	5B	6B	7B	8B			1B	2B
Element:	**Sc**	**Ti**	**V**	**Cr**	**Mn**	**Fe**	**Co**	**Ni**	**Cu**	**Zn**
Valence electron configuration										
M atom	$3d^1 4s^2$	$3d^2 4s^2$	$3d^3 4s^2$	$3d^5 4s^1$	$3d^5 4s^2$	$3d^6 4s^2$	$3d^7 4s^2$	$3d^8 4s^2$	$3d^{10} 4s^1$	$3d^{10} 4s^2$
M^{2+} ion		$3d^2$	$3d^3$	$3d^4$	$3d^5$	$3d^6$	$3d^7$	$3d^8$	$3d^9$	$3d^{10}$
M^{3+} ion	$3d^0$	$3d^1$	$3d^2$	$3d^3$	$3d^4$	$3d^5$	$3d^6$			
Elec. conductivity*	3	4	8	11	1	17	24	24	96	27
Melting point (°C)	1541	1668	1910	1907	1246	1538	1495	1455	1085	420
Boiling point (°C)	2836	3287	3407	2671	2061	2861	2927	2913	2562	907
Density (g/cm^3)	2.99	4.51	6.0	7.15	7.3	7.87	8.86	8.90	8.96	7.13
Atomic radius (pm)	162	147	134	128	127	126	125	124	128	134
E_i (kJ/mol)†										
First	633	659	651	653	717	762	760	737	745	906
Second	1235	1310	1410	1591	1509	1562	1648	1753	1958	1733
Third	2389	2653	2828	2987	3248	2957	3232	3395	3555	3833

*Electrical conductivity relative to an arbitrary value of 100 for silver
†Ionization energy

Historically, transition elements are placed in groups of the periodic table designated 1B–8B because their valence electron configurations are similar to those of analogous elements in the main groups 1A–8A. Thus, copper in group 1B ([Ar] $3d^{10} 4s^1$) and zinc in group 2B ([Ar] $3d^{10} 4s^2$) have valence electron configurations similar to those of potassium in group 1A ([Ar] $4s^1$) and calcium in group 2A ([Ar] $4s^2$). Similarly, scandium in group 3B ([Ar] $3d^1 4s^2$) through iron in group 8B ([Ar] $3d^6 4s^2$) have the same number of valence electrons as the *p*-block elements aluminum in group 3A ([Ne] $3s^2 3p^1$) through argon in group 8A ([Ne] $3s^2 3p^6$). Cobalt ([Ar] $3d^7 4s^2$) and nickel ([Ar] $3d^8 4s^2$) are also assigned to group 8B although there are no main-group elements with 9 or 10 valence electrons.

Worked Example 20.1

Writing Electron Configurations of Transition Metals

Write the electron configuration of the metal in each of the following atoms or ions:
(a) Ni **(b)** Cr^{3+} **(c)** FeO_4^{2-} (ferrate ion)

STRATEGY

In neutral atoms of the first transition series, the 4*s* orbital is usually filled with 2 electrons and the remaining electrons occupy the 3*d* orbitals. In transition metal ions, all the valence electrons occupy the *d* orbitals. For polyatomic ions, first determine the oxidation number of the transition metal, and then assign the valence electrons to the *d* orbitals as you would if the metal were a simple ion.

continued on next page

SOLUTION

(a) Nickel ($Z = 28$) has a total of 28 electrons, including the argon core of 18. Two of the 10 valence electrons occupy the 4*s* orbital, and the remaining eight are assigned to the 3*d* orbitals. The electron configuration is therefore [Ar] $3d^8\,4s^2$.

(b) A neutral Cr atom ($Z = 24$) has a total of 24 electrons; a Cr^{3+} ion has $24 - 3 = 21$ electrons. Because all the valence electrons occupy the 3*d* orbitals, the electron configuration of Cr^{3+} is [Ar] $3d^3$.

(c) The oxidation number of each of the four oxygens in FeO_4^{2-} is -2 and the overall charge on the oxoanion is -2, so the oxidation number of the iron must be $+6$. An iron(VI) atom has a total of 20 electrons, 6 less than a neutral iron atom ($Z = 26$). Because the valence electrons occupy the 3*d* orbitals, the electron configuration of Fe(VI) is [Ar] $3d^2$.

PROBLEM 20.1 Write the electron configuration of the metal in each of the following:
(a) V **(b)** Co^{2+} **(c)** MnO_2 **(d)** $CuCl_4^{2-}$

CONCEPTUAL PROBLEM 20.2 On the periodic table below, locate the transition metal atom or ion with the following electron configurations. Identify each atom or ion.

(a) An atom: [Ar] $3d^5\,4s^2$
(b) A 2+ ion: [Ar] $3d^8$
(c) An atom: [Kr] $4d^{10}\,5s^1$
(d) A 3+ ion: [Kr] $4d^3$

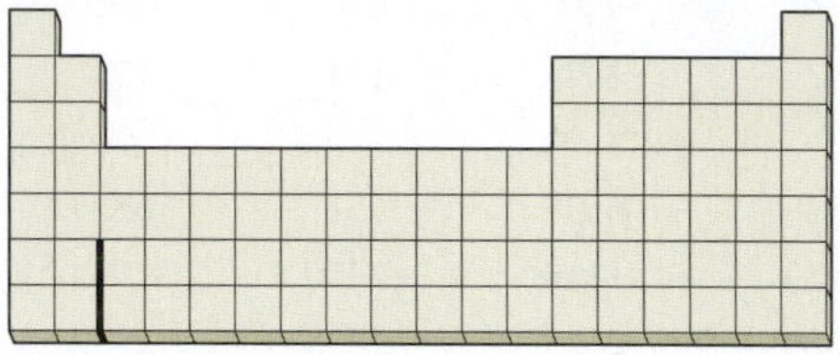

20.2 PROPERTIES OF TRANSITION ELEMENTS

Let's look at some trends in the properties of the transition elements shown in Table 20.1 and try to understand them in terms of electron configurations.

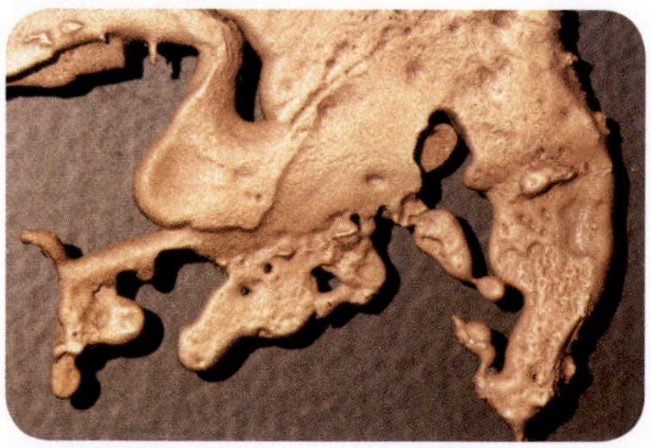

▲ Native copper.

LOOKING AHEAD ...

In **Section 21.4**, we'll see that the properties of metals can be understood in terms of band theory, a version of molecular orbital theory.

Metallic Properties

All the transition elements are metals. Like the metals of groups 1A and 2A, the transition metals are malleable, ductile, lustrous, and good conductors of heat and electricity. Silver has the highest electrical conductivity of any element at room temperature, with copper a close second. The transition metals are harder, have higher melting and boiling points, and are more dense than the group 1A and 2A metals, largely because the sharing of *d*, as well as *s*, electrons results in stronger bonding.

From left to right across the first transition series, melting points increase from 1541 °C for Sc to a maximum of 1910 °C for V in group 5B, then decrease to 420 °C for Zn (Table 20.1, **Figure 20.2**). The second- and third-series elements exhibit a similar maximum in melting point, but at group 6B rather than 5B: 2623 °C for Mo and 3422 °C for W, the metal with the highest melting point. The melting points increase as the number of unpaired *d* electrons available for bonding increases and then decrease as the *d* electrons pair up and become less available for bonding. Zinc ($3d^{10}\,4s^2$), in which all the *d* and *s* electrons are paired, has a relatively low melting point (420 °C) and mercury ($4f^{14}\,5d^{10}\,6s^2$) is a liquid at room temperature (mp -39 °C). We'll look at bonding in metals in more detail in **Section 21.4**.

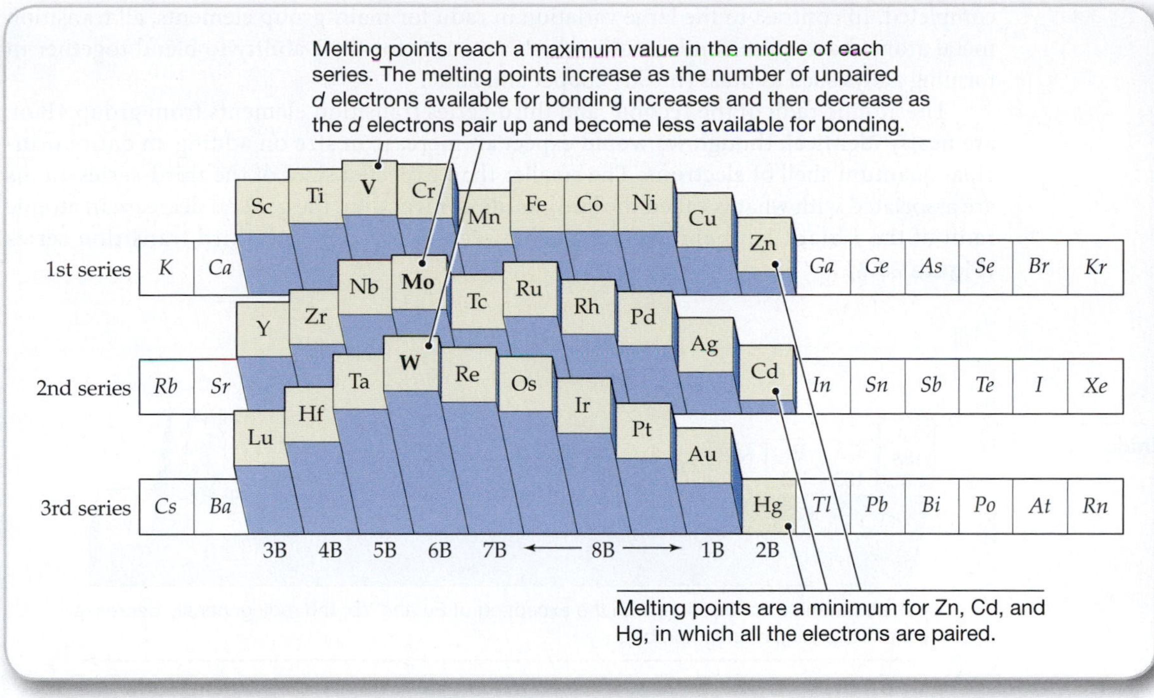

FIGURE 20.2

Relative melting points of the transition elements.

Atomic Radii and Densities

Atomic radii are given in **Figure 20.3**. From left to right across a transition series, the atomic radii decrease, at first markedly and then more gradually after group 6B. Toward the end of each series, the radii increase again. The decrease in radii with increasing atomic number occurs because the added *d* electrons only partially **shield** the added nuclear charge (Section 2.14). As a result, the effective nuclear charge Z_{eff} increases. With increasing Z_{eff}, the electrons are more strongly attracted to the nucleus and atomic radii decrease. The upturn in radii toward the end of each series is probably due to more effective shielding and increasing electron–electron repulsion as double occupation of the *d* orbitals is

REMEMBER...

The valence electrons are only weakly **shielded** by electrons in the same subshell, which are at the same distance from the nucleus. (Section 2.14)

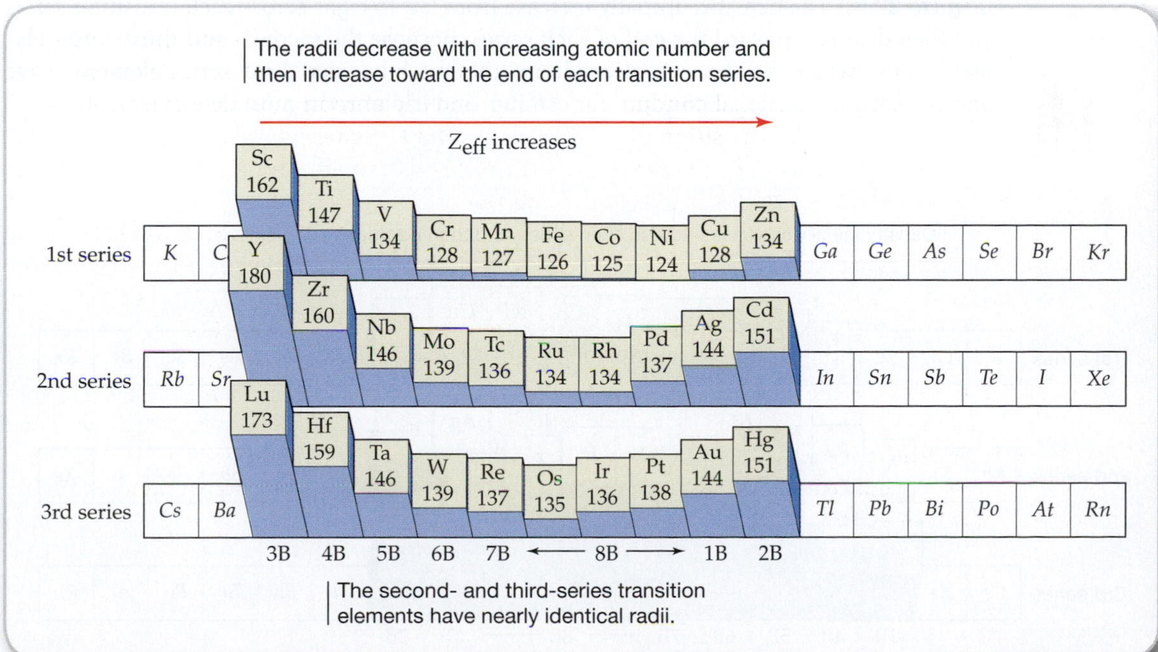

FIGURE 20.3

Atomic radii (in pm) of the transition elements.

completed. In contrast to the large variation in radii for main-group elements, all transition metal atoms have quite similar radii, which accounts for their ability to blend together in forming alloys such as brass (mostly copper and zinc).

The atomic radii of the second- and third-series transition elements from group 4B on are nearly identical, though we would expect an increase in size on adding an entire principal quantum shell of electrons. The smaller than expected sizes of the third-series atoms are associated with what is called the **lanthanide contraction**, the general decrease in atomic radii of the *f*-block lanthanide elements between the second and third transition series (**Figure 20.4**).

FIGURE 20.4

Atomic radii (in pm) of the lanthanide elements.

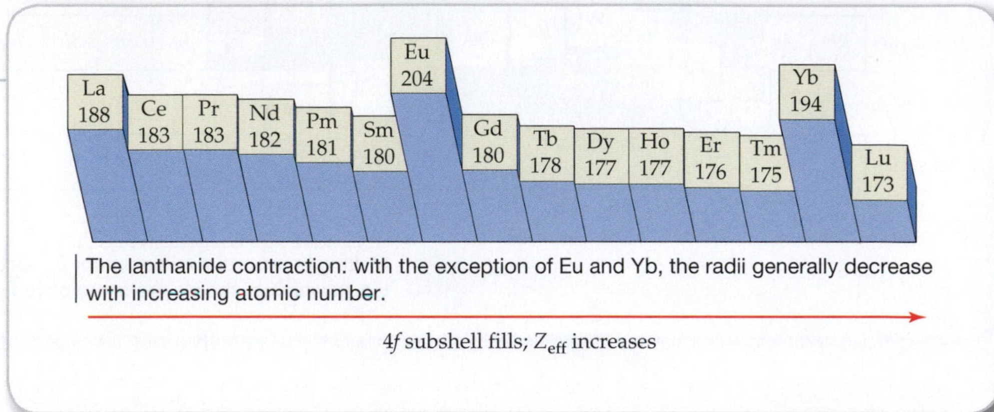

The lanthanide contraction is due to the increase in effective nuclear charge with increasing atomic number as the 4*f* subshell is filled. By the end of the lanthanides, the size *decrease* due to a larger Z_{eff} almost exactly compensates for the expected size *increase* due to an added quantum shell of electrons. Consequently, atoms of the third transition series have radii very similar to those of the second transition series (Figure 20.3). The irregularly large atomic radii in Figure 20.4 for Eu and Yb are related to their half-filled and completely filled 4*f* subshells, [Xe] $4f^7\,6s^2$ for Eu and [Xe] $4f^{14}\,6s^2$ for Yb, but a full explanation of their anomalous radii is complicated.

The densities of the transition metals are inversely related to their atomic radii (**Figure 20.5**). The densities initially increase from left to right across each transition series and then decrease toward the end of each series. Because the second- and third-series elements have nearly the same atomic volume, the much heavier third-series elements have unusually high densities: 22.6 g/cm^3 for osmium and iridium, the most dense elements.

FIGURE 20.5

Relative densities of the transition metals.

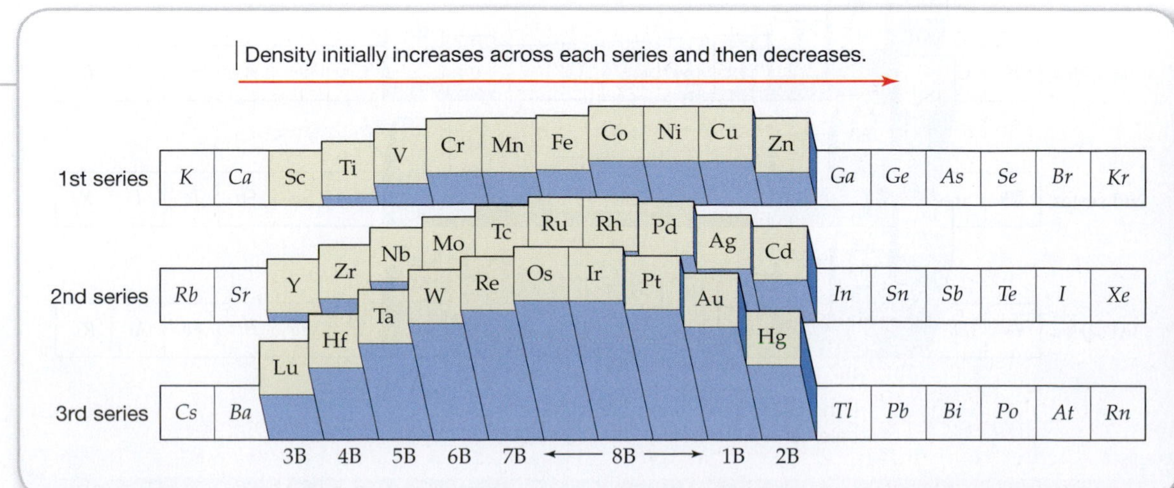

Ionization Energies and Oxidation Potentials

Ionization energies generally increase from left to right across each transition series, though there are some irregularities, as indicated in Table 20.1 for the atoms of the first transition series. The general trend correlates with an increase in effective nuclear charge and a decrease in atomic radius.

Table 20.2 lists standard potentials $E°$ for oxidation of first-series transition metals. Note that these potentials are the negative of the corresponding standard reduction potentials that were given in Table 17.1 on page 708. Except for copper, all the $E°$ values are positive, which means that the solid metal is oxidized to its aqueous cation more readily than H_2 gas is oxidized to $H^+(aq)$.

$$M(s) \longrightarrow M^{2+}(aq) + 2\,e^- \qquad E° > 0\text{ V} \qquad \text{Product is } M^{3+} \text{ for } M = Sc$$
$$H_2(g) \longrightarrow 2\,H^+(aq) + 2\,e^- \qquad E° = 0\text{ V}$$

In other words, the first-series metals, except for copper, are stronger reducing agents than H_2 gas and can therefore be oxidized by the H^+ ion in acids like HCl that lack an oxidizing anion:

$$M(s) + 2\,H^+(aq) \longrightarrow M^{2+}(aq) + H_2(g) \qquad E° > 0\text{ V (except for } M = Cu)$$

Oxidation of copper requires a stronger oxidizing agent, such as HNO_3.

TABLE 20.2 Standard Potentials for Oxidation of First-Series Transition Metals

Oxidation Half-Reaction	$E°$ (V)	Oxidation Half-Reaction	$E°$ (V)
$Sc(s) \longrightarrow Sc^{3+}(aq) + 3\,e^-$	2.08	$Fe(s) \longrightarrow Fe^{2+}(aq) + 2\,e^-$	0.45
$Ti(s) \longrightarrow Ti^{2+}(aq) + 2\,e^-$	1.63	$Co(s) \longrightarrow Co^{2+}(aq) + 2e^-$	0.28
$V(s) \longrightarrow V^{2+}(aq) + 2\,e^-$	1.18	$Ni(s) \longrightarrow Ni^{2+}(aq) + 2\,e^-$	0.26
$Cr(s) \longrightarrow Cr^{2+}(aq) + 2\,e^-$	0.91	$Cu(s) \longrightarrow Cu^{2+}(aq) + 2\,e^-$	−0.34
$Mn(s) \longrightarrow Mn^{2+}(aq) + 2\,e^-$	1.18	$Zn(s) \longrightarrow Zn^{2+}(aq) + 2\,e^-$	0.76

The standard potential for the oxidation of a metal is a composite property that depends on $\Delta G°$ for sublimation of the metal, the ionization energies of the metal atom, and $\Delta G°$ for hydration of the metal ion:

$$M(s) \xrightarrow{\Delta G°_{subl}} M(g) \xrightarrow{E_i(-2e^-)} M^{2+}(g) \xrightarrow{\Delta G°_{hydr}} M^{2+}(aq)$$

Nevertheless, the general trend in the $E°$ values shown in Table 20.2 correlates with the general trend in the ionization energies in Table 20.1. The ease of oxidation of the metal decreases as the ionization energies increase across the transition series from Sc to Zn. (Only Mn and Zn deviate from the trend of decreasing $E°$ values.) Thus, the so-called *early transition metals*, those on the left side of the *d* block (Sc through Mn), are oxidized most easily and are the strongest reducing agents.

20.3 OXIDATION STATES OF TRANSITION ELEMENTS

The transition elements differ from most main-group metals in that they exhibit a variety of oxidation states. Sodium, magnesium, and aluminum, for example, have a single oxidation state equal to their periodic group number (Na^+, Mg^{2+}, and Al^{3+}), but the transition elements frequently have oxidation states less than their group number. For example, manganese in group 7B shows oxidation states of +2 in $Mn^{2+}(aq)$, +3 in $Mn(OH)_3(s)$, +4 in $MnO_2(s)$,

+6 in $MnO_4^{2-}(aq)$ (manganate ion), and +7 in $MnO_4^{-}(aq)$ (permanganate ion). **Figure 20.6** summarizes the common oxidation states for elements of the first transition series, with the most frequently encountered ones indicated in red.

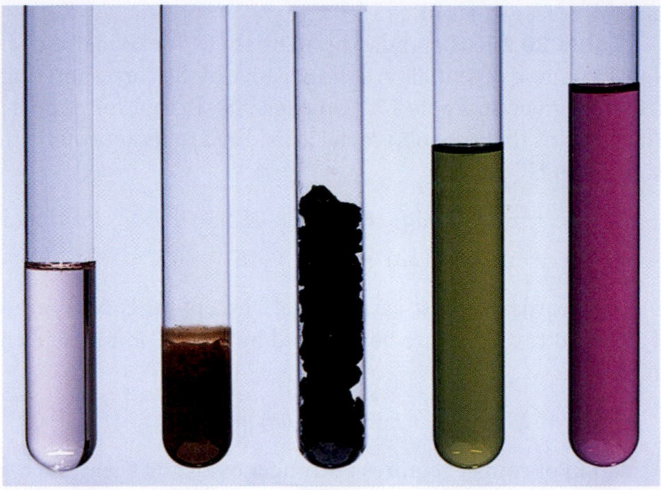

Manganese species:	$Mn^{2+}(aq)$	$Mn(OH)_3(s)$	$MnO_2(s)$	$MnO_4^{2-}(aq)$	$MnO_4^{-}(aq)$
Oxidation state of manganese:	+2	+3	+4	+6	+7

▶ Manganese has different oxidation states and different colors in these ions and solid compounds.

All the first-series transition elements except scandium form a 2+ cation, corresponding to loss of the two 4*s* valence electrons. Because the 3*d* and 4*s* orbitals have similar energies, loss of a 3*d* electron is also possible, yielding 3+ cations such as $V^{3+}(aq)$, $Cr^{3+}(aq)$, and $Fe^{3+}(aq)$. Additional energy is required to remove the third electron, but this is more than compensated for by the larger (more negative) $\Delta G°$ of hydration of the more highly charged 3+ cation. Still higher oxidation states result from loss or sharing of additional *d* electrons. In their highest oxidation states, the transition elements are combined with the most electronegative elements, F and O. For example, vanadium forms $VF_5(l)$ and $V_2O_5(s)$, chromium forms CrO_4^{2-} (chromate ion) and $Cr_2O_7^{2-}$ (dichromate ion), and manganese forms MnO_4^{-}.

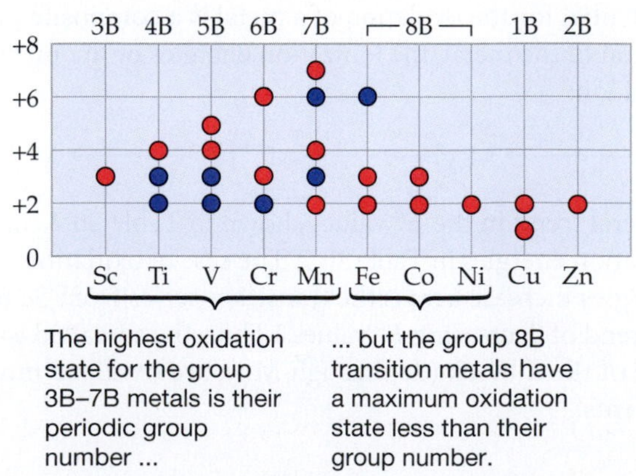

FIGURE 20.6

Common oxidation states for first-series transition elements. The states encountered most frequently are shown in red. Most transition elements have more than one common oxidation state.

Note in Figure 20.6 that the highest oxidation state for the group 3B–7B metals is the group number, corresponding to loss or sharing of all the valence *s* and *d* electrons. For the group 8B transition metals, though, loss or sharing of all the valence electrons is energetically prohibitive because of the increasing value of Z_{eff}. Consequently, only lower oxidation states are accessible for these transition metals—for example, +6 in FeO_4^{2-} and +3 in Co^{3+}. Even these species have a great tendency to be reduced to still lower oxidation states. For example, the aqueous Co^{3+} ion is unstable because it oxidizes water to O_2 gas and is thereby reduced to Co^{2+}:

$$4\,Co^{3+}(aq) + 2\,H_2O(l) \longrightarrow 4\,Co^{2+}(aq) + O_2(g) + 4\,H^+(aq) \qquad E^\circ = +0.58\text{ V}$$

In general, ions that have the transition metal in a high oxidation state tend to be good oxidizing agents—for example, $Cr_2O_7^{2-}$, MnO_4^-, and FeO_4^{2-}. Conversely, early transition metal ions with the metal in a low oxidation state are good reducing agents—for example, V^{2+} and Cr^{2+}. Divalent ions of the later transition metals on the right side of the *d* block, such as Co^{2+}, Ni^{2+}, Cu^{2+}, and Zn^{2+}, are poor reducing agents because of the larger value of Z_{eff}. In fact, zinc has only one stable oxidation state (+2).

CONCEPTUAL PROBLEM 20.3 How does the effective nuclear charge, Z_{eff}, vary from left to right across the first transition series?

Sc	Ti	V	Cr	Mn	Fe	Co	Ni	Cu	Zn

Based on the variation in Z_{eff},

(a) Which M^{2+} ion (M = Ti–Zn) should be the strongest reducing agent? Which should be the weakest?

(b) Which oxoanion (VO_4^{3-}, CrO_4^{2-}, MnO_4^{2-}, or FeO_4^{2-}) should be the strongest oxidizing agent? Which should be the weakest?

20.4 CHEMISTRY OF SELECTED TRANSITION ELEMENTS

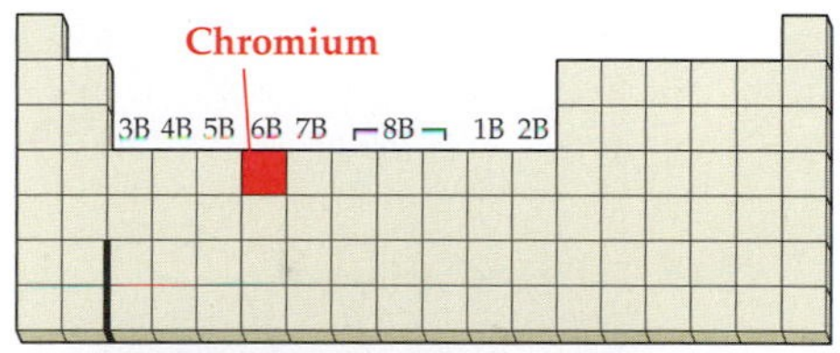

Experimental work in transition metal chemistry is particularly enjoyable because most transition metal compounds have brilliant colors. In this section, we'll look at the chemistry of some representative elements commonly encountered in the laboratory.

Chromium

Chromium, which gets its name from the Greek word for color (*chroma*), is obtained from the ore chromite, a mixed metal oxide with the formula $FeO \cdot Cr_2O_3$, or $FeCr_2O_4$. Reduction of chromite with carbon gives the alloy ferrochrome, which is used to make stainless steel, a hard, corrosion-resistant steel that contains up to 30% chromium.

$$FeCr_2O_4(s) + 4\,C(s) \longrightarrow \underbrace{Fe(s) + 2\,Cr(s)}_{\text{Ferrochrome}} + 4\,CO(g)$$

Pure chromium is obtained by reducing chromium(III) oxide with aluminum:

$$Cr_2O_3(s) + 2\,Al(s) \longrightarrow 2\,Cr(s) + Al_2O_3(s)$$

In addition to its use in making steels, chromium is widely used to electroplate metallic objects with an attractive, protective coating (Section 17.13). Chromium is hard and lustrous, takes a high polish, and resists corrosion because an invisible, microscopic film of chromium(III) oxide shields the surface from further oxidation.

The aqueous solution chemistry of chromium can be systematized according to its oxidation states and the species that exist under acidic and basic conditions (**Table 20.3**). The common oxidation states are +2, +3, and +6, with the +3 state the most stable.

▲ This sculpture, *Herakles in Ithaka* by Jason Seley, was made from chromium-plated automobile bumpers.

TABLE 20.3 Chromium Species in Common Oxidation States

	Oxidation State				
	+2		+3		+6
Acidic solution:	$Cr^{2+}(aq)$ Chromium(II) ion (Chromous ion) Blue	$\xrightarrow{+0.41\ V}$	$Cr^{3+}(aq)$ Chromium(III) ion (Chromic ion) Violet	$\xleftarrow{+1.36\ V}$	$Cr_2O_7^{2-}(aq)$ Dichromate ion Orange
Basic solution:	$Cr(OH)_2(s)$ Light blue		$Cr(OH)_3(s)$ Pale green $Cr(OH)_4^-(aq)$ Chromite ion Deep green	$\xleftarrow{-0.13\ V}$	$CrO_4^{2-}(aq)$ Chromate ion Yellow

Chromium metal reacts with aqueous acids in the absence of oxygen to give $Cr^{2+}(aq)$, the beautiful blue chromium(II) (chromous) ion, in which Cr^{2+} is bound to six water molecules, $Cr(H_2O)_6^{2+}$ (**Figure 20.7a**):

$$Cr(s) + 2\,H^+(aq) \longrightarrow Cr^{2+}(aq) + H_2(g) \qquad E° = +0.91\ V$$

In air, the reaction of chromium metal with acids yields chromium(III) because chromium(II) is rapidly oxidized by atmospheric oxygen:

$$4\,Cr^{2+}(aq) + O_2(g) + 4\,H^+(aq) \longrightarrow 4\,Cr^{3+}(aq) + 2\,H_2O(l) \qquad E° = +1.64\ V$$

Although the $Cr(H_2O)_6^{3+}$ ion is violet, chromium(III) solutions are often green because anions replace some of the bound water molecules to give green complex ions such as $Cr(H_2O)_5Cl^{2+}$ and $Cr(H_2O)_4Cl_2^+$ (**Figure 20.7b**).

FIGURE 20.7

Reaction of chromium metal with aqueous acids.

In basic solution, chromium(III) precipitates as chromium(III) hydroxide, a pale green solid that dissolves both in acid and in excess base (**Figure 20.8**):

$$\text{In acid:} \quad Cr(OH)_3(s) + 3\,H_3O^+(aq) \rightleftharpoons Cr^{3+}(aq) + 6\,H_2O(l)$$

$$\text{In excess base:} \quad Cr(OH)_3(s) + OH^-(aq) \rightleftharpoons Cr(OH)_4^-(aq)$$

Recall from Section 15.12 that this behavior is typical of amphoteric metal hydroxides.

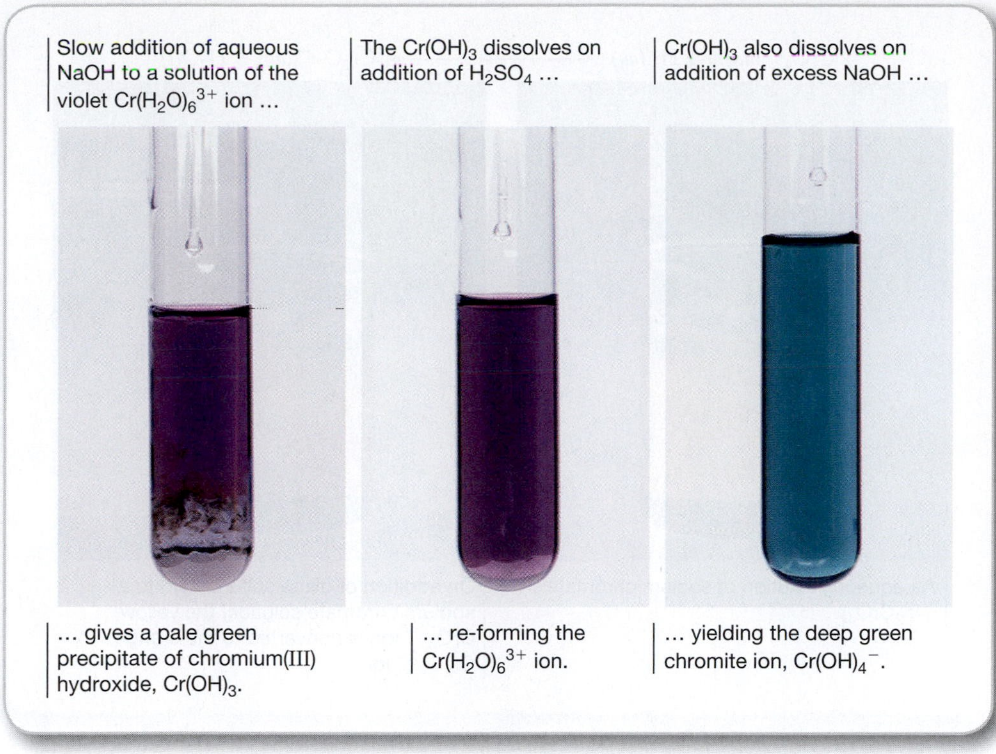

FIGURE 20.8

Amphoteric behavior of chromium(III) hydroxide.

In contrast to the amphoteric $Cr(OH)_3$, chromium(II) hydroxide is a typical basic hydroxide. It dissolves in acid but not in excess base. Conversely, the chromium(VI) compound, $CrO_2(OH)_2$, is a strong acid (chromic acid, H_2CrO_4). Recall from Section 14.3 that **acid strength** increases with increasing polarity of the O–H bonds, which increases, in turn, with increasing oxidation number of the chromium atom.

+2	+3	+6
$Cr(OH)_2$	$Cr(OH)_3$	$CrO_2(OH)_2$
Basic	Amphoteric	Acidic

Increasing strength as a proton donor ⟶

REMEMBER...

Acid strength increases with increasing oxidation number of the atom to which an OH group is attached. Electron density shifts from the OH group to the attached atom, thus weakening the O–H bond, increasing its polarity, and facilitating proton transfer to a solvent water molecule. (Section 14.3)

In the +6 oxidation state, the most important solution species are the yellow chromate ion (CrO_4^{2-}) and the orange dichromate ion ($Cr_2O_7^{2-}$). These ions are interconverted by the rapid equilibrium reaction

$$2\,CrO_4^{2-}(aq) + 2\,H^+(aq) \rightleftharpoons Cr_2O_7^{2-}(aq) + H_2O(l) \qquad K \approx 10^{14}$$

Because the equilibrium constant is about 10^{14}, CrO_4^{2-} ions predominate in basic solutions and $Cr_2O_7^{2-}$ ions predominate in acidic solutions (**Figure 20.9**). The CrO_4^{2-} ion is tetrahedral, and the $Cr_2O_7^{2-}$ ion has a structure in which two CrO_4^{2-} ions share a common oxygen atom.

The $Cr_2O_7^{2-}$ ion is a powerful oxidizing agent in acidic solution and is widely used as an oxidant in analytical chemistry.

$$Cr_2O_7^{2-}(aq) + 14\,H^+(aq) + 6\,e^- \longrightarrow 2\,Cr^{3+}(aq) + 7\,H_2O(l) \quad E^\circ = +1.36\text{ V}$$

In basic solution, where CrO_4^{2-} is the predominant species, chromium(VI) is a much weaker oxidizing agent:

$$CrO_4^{2-}(aq) + 4\,H_2O(l) + 3\,e^- \longrightarrow Cr(OH)_3(s) + 5\,OH^-(aq) \quad E^\circ = -0.13\text{ V}$$

Chromium(VI) compounds must be handled with care because they are toxic and are known to cause cancer.

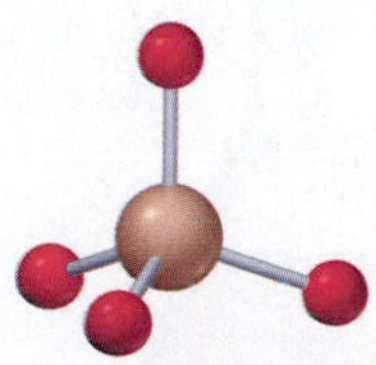

Chromate ion, CrO_4^{2-}

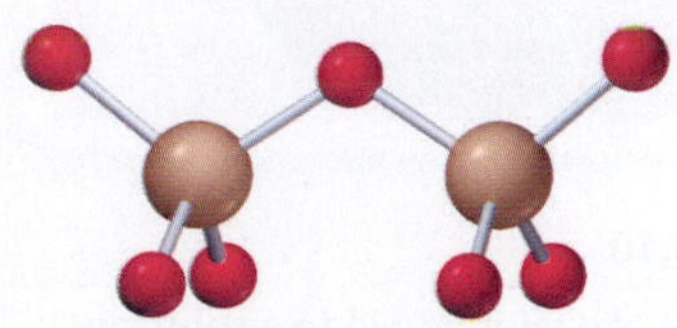

Dichromate ion, $Cr_2O_7^{2-}$

FIGURE 20.9

Conversion of the yellow $CrO_4^{2-}(aq)$ ion to the orange $Cr_2O_7^{2-}(aq)$ ion.

$$2\,CrO_4^{2-}(aq) + 2\,H^+(aq) \longrightarrow Cr_2O_7^{2-}(aq) + H_2O(l)$$

An aqueous solution of sodium chromate (Na_2CrO_4).

On addition of dilute sulfuric acid to a sodium chromate solution, the yellow CrO_4^{2-} ion is converted to the orange $Cr_2O_7^{2-}$ ion.

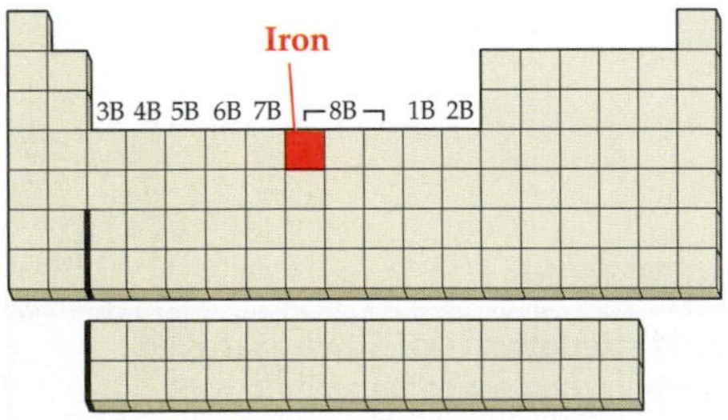

Iron

Iron, the fourth most abundant element in the Earth's crust at 5.6% by mass, is immensely important both in human civilization and in living systems. Because iron is relatively soft and easily corroded, it is combined with carbon and other metals, such as vanadium, chromium, and manganese, to make steel alloys that are harder and less reactive than pure iron. Iron is obtained from its most important ores, hematite (Fe_2O_3) and magnetite (Fe_3O_4), by reduction with coke in a blast furnace (see Section 21.3). In living systems, iron is an essential constituent of numerous biomolecules. The body of a healthy human adult contains about 4 g of iron, 65% of which is present in the oxygen-carrying protein hemoglobin (Chapter 13 *FYI*).

The most common oxidation states of iron are +2 (ferrous) and +3 (ferric). When iron metal reacts in the absence of air with an acid such as HCl, which lacks an oxidizing anion, the product is the light green iron(II) ion, $Fe(H_2O)_6^{2+}$:

$$Fe(s) + 2\,H^+(aq) \longrightarrow Fe^{2+}(aq) + H_2(g) \quad E° = +0.45\text{ V}$$

The oxidation stops at the iron(II) stage because the standard potential for oxidation of the iron(II) ion is negative:

$$Fe^{2+}(aq) \longrightarrow Fe^{3+}(aq) + e^- \quad E° = -0.77\text{ V}$$

In air, the iron(II) ion is slowly oxidized by atmospheric oxygen to the iron(III) ion, $Fe(H_2O)_6^{3+}$:

$$4\,Fe^{2+}(aq) + O_2(g) + 4\,H^+(aq) \longrightarrow 4\,Fe^{3+}(aq) + 2\,H_2O(l) \quad E° = +0.46\text{ V}$$

When iron is treated with an acid that has an oxidizing anion—for example, dilute nitric acid—the metal is oxidized directly to iron(III):

$$Fe(s) + NO_3^-(aq) + 4\,H^+(aq) \longrightarrow Fe^{3+}(aq) + NO(g) + 2\,H_2O(l) \quad E° = +1.00\text{ V}$$

The addition of a base to iron(III) solutions precipitates the gelatinous, red-brown hydrous oxide, $Fe_2O_3 \cdot x\,H_2O$, usually written as $Fe(OH)_3$ **(Figure 20.10)**:

$$Fe^{3+}(aq) + 3\,OH^-(aq) \longrightarrow Fe(OH)_3(s)$$

$$Fe^{3+}(aq) + 3\,OH^-(aq) \longrightarrow Fe(OH)_3(s)$$

FIGURE 20.10

When dilute NaOH is added to a solution of iron(III) sulfate, $Fe_2(SO_4)_3$, a gelatinous, red-brown precipitate of $Fe(OH)_3$ forms.

Because $Fe(OH)_3$ is very insoluble ($K_{sp} = 2.6 \times 10^{-39}$), it forms as soon as the pH rises above pH 2. The red-brown rust stains that you've seen in sinks and bathtubs are due to air oxidation of $Fe^{2+}(aq)$ in hard water followed by deposition of hydrous iron(III) oxide. Unlike $Cr(OH)_3$, $Fe(OH)_3$ is not appreciably amphoteric. It dissolves in acid, but not in excess base.

Copper

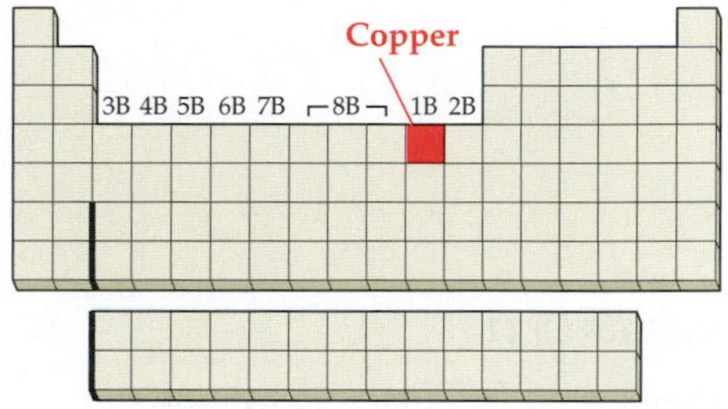

Copper, a reddish metal, is a relatively rare element, accounting for only 0.0068% of the Earth's crust by mass. Like the other group 1B elements silver and gold, copper is found in nature in the elemental state. Copper's most important ores are sulfides, such as chalcopyrite, $CuFeS_2$. In a multistep process, copper sulfides are concentrated, separated from iron, and converted to molten copper(I) sulfide, which is then reduced to elemental copper by blowing air through the hot liquid:

$$Cu_2S(l) + O_2(g) \longrightarrow 2\,Cu(l) + SO_2(g)$$

The product, containing about 99% copper, is purified by electrolysis (Section 17.13).

Because of its high electrical conductivity and negative oxidation potential, copper is widely used to make electrical wiring and corrosion-resistant water pipes. Copper is also used in coins and is combined with other metals to make alloys such as brass (mostly copper and zinc) and bronze (mostly copper and tin). Because of the relatively high cost of copper, the U.S. copper penny is only 2.5% copper (copper-plated zinc) and has not been made of pure copper since 1837. It's interesting to note that it costs more than a penny to make a penny (about 2.4 cents), and Canada's recent decision to eliminate the penny is expected to save about \$11 million per year.

Though less reactive than other first-series transition metals, copper is oxidized on prolonged exposure to O_2, CO_2, and water in moist air, forming basic copper(II) carbonate, $Cu_2(OH)_2CO_3$. Subsequent reaction with dilute H_2SO_4 in acid rain then forms $Cu_2(OH)_2SO_4$, the green patina seen on bronze monuments.

▲ Why do statues made of copper or bronze turn green with age?

$$2\,Cu(s) + O_2(g) + CO_2(g) + H_2O(g) \longrightarrow Cu_2(OH)_2CO_3(s)$$
$$Cu_2(OH)_2CO_3(s) + H_2SO_4(aq) \longrightarrow Cu_2(OH)_2SO_4(s) + CO_2(g) + H_2O(l)$$

In its compounds, copper exists in two common oxidation states, +1 (cuprous) and +2 (cupric). Because $E°$ for the Cu^+/Cu^{2+} half-reaction is less negative than that for the Cu/Cu^+ half-reaction, any oxidizing agent strong enough to oxidize copper to the copper(I) ion is also able to oxidize the copper(I) ion to the copper(II) ion.

$$Cu(s) \longrightarrow Cu^+(aq) + e^- \qquad E° = -0.52\ V$$
$$Cu^+(aq) \longrightarrow Cu^{2+}(aq) + e^- \qquad E° = -0.15\ V$$

Dilute nitric acid, for example, oxidizes copper to the +2 oxidation state:

$$3\,Cu(s) + 2\,NO_3^-(aq) + 8\,H^+(aq) \longrightarrow 3\,Cu^{2+}(aq) + 2\,NO(g) + 4\,H_2O(l) \qquad E° = +0.62\ V$$

It follows from the $E°$ values that $Cu^+(aq)$ can disproportionate, oxidizing and reducing itself:

$$Cu^+(aq) + e^- \longrightarrow Cu(s) \qquad E° = +0.52\ V$$
$$Cu^+(aq) \longrightarrow Cu^{2+}(aq) + e^- \qquad E° = -0.15\ V$$
$$\overline{2\,Cu^+(aq) \longrightarrow Cu(s) + Cu^{2+}(aq) \qquad E° = +0.37\ V}$$

The positive $E°$ value for the disproportionation corresponds to a large equilibrium constant, indicating that the reaction proceeds far toward completion:

$$2\,Cu^+(aq) \rightleftharpoons Cu(s) + Cu^{2+}(aq) \qquad K_c = 1.8 \times 10^6$$

Thus, the copper(I) ion is not an important species in aqueous solution, though copper(I) does exist in solid compounds such as CuCl. In the presence of Cl^- ions, the disproportionation equilibrium is reversed because precipitation of the insoluble, white copper(I) chloride drives the following reaction to the right:

$$Cu(s) + Cu^{2+}(aq) + 2\,Cl^-(aq) \longrightarrow 2\,CuCl(s)$$

The more common +2 oxidation state is found in the blue aqueous ion, $Cu(H_2O)_6^{2+}$, and in numerous solid compounds and complex ions. The addition of base (aqueous ammonia) to a solution of a copper(II) salt gives a blue precipitate of copper(II) hydroxide, which dissolves in excess aqueous ammonia, yielding the dark blue complex ion $Cu(NH_3)_4^{2+}$ (**Figure 20.11**):

$$Cu^{2+}(aq) + 2\,OH^-(aq) \longrightarrow Cu(OH)_2(s)$$
$$Cu(OH)_2(s) + 4\,NH_3(aq) \longrightarrow Cu(NH_3)_4^{2+}(aq) + 2\,OH^-(aq)$$

FIGURE 20.11

Reactions of $Cu^{2+}(aq)$ with aqueous ammonia.

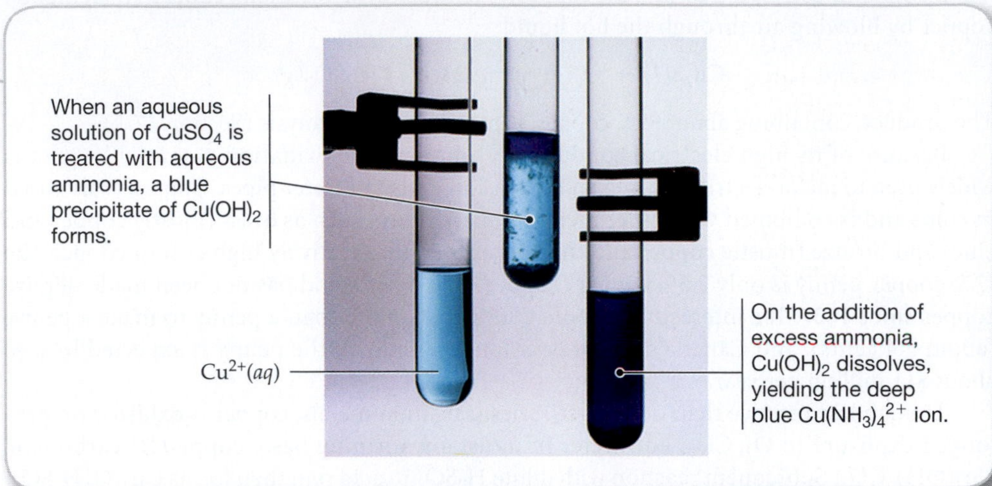

PROBLEM 20.4 What product is formed when dilute H_2SO_4, a nonoxidizing acid, is added to each of the following ions or compounds?

(a) CrO_4^{2-} (b) $Cr(OH)_3$
(c) $Cr(OH)_2$ (d) $Fe(OH)_2$
(e) $Cu(OH)_2$

PROBLEM 20.5 What product is formed when excess aqueous NaOH is added to a solution of each of the following ions?

(a) Cr^{2+} (b) Cr^{3+}
(c) $Cr_2O_7^{2-}$ (d) Fe^{2+}
(e) Fe^{3+}

20.5 COORDINATION COMPOUNDS

A **coordination compound** is a compound in which a central metal ion or atom is attached to a group of surrounding molecules or ions by coordinate covalent bonds. A good example is the anticancer drug cisplatin, $Pt(NH_3)_2Cl_2$, in which two NH_3 molecules and two Cl^- ions use lone pairs of electrons to bond to the platinum(II) ion:

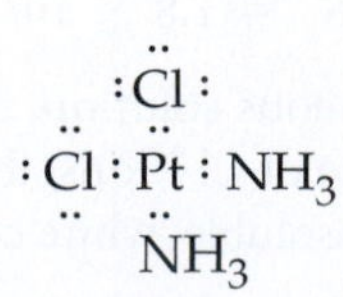

Cisplatin

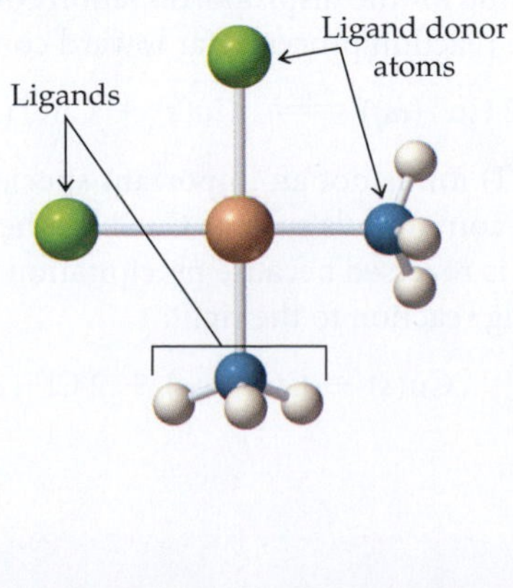

The molecules or ions that surround the central metal ion in a coordination compound are called **ligands**, and the atoms that are attached directly to the metal are called **ligand donor atoms**. In cisplatin, for example, the ligands are NH_3 and Cl^-, and the ligand donor atoms are N and Cl. The formation of a coordination compound is a **Lewis acid–base reaction** in which the ligands act as Lewis bases, or electron-pair donors, and the central metal ion behaves as a Lewis acid, an electron-pair acceptor (Section 14.15).

REMEMBER...

In a **Lewis acid–base reaction**, a coordinate covalent bond is formed when a Lewis acid accepts a share in a lone pair of electrons from a Lewis base, which acts as an electron-pair donor. (Section 14.15)

Not all coordination compounds are neutral molecules like $Pt(NH_3)_2Cl_2$. Many are salts, such as $[Ni(NH_3)_6]Cl_2$ and $K_3[Fe(CN)_6]$, which contain a complex cation or anion along with enough ions of opposite charge to give a compound that is electrically neutral overall. To emphasize that the complex ion is a discrete structural unit, it is always enclosed in brackets in the formula of a salt. Thus, $[Ni(NH_3)_6]Cl_2$ contains $[Ni(NH_3)_6]^{2+}$ cations and Cl^- anions. The term **metal complex**, or simply complex, refers both to neutral molecules, such as $Pt(NH_3)_2Cl_2$, and to complex ions, such as $[Ni(NH_3)_6]^{2+}$ and $[Fe(CN)_6]^{3-}$.

The number of ligand donor atoms that surround a central metal ion in a complex is called the **coordination number** of the metal. Thus, platinum(II) has a coordination number of 4 in $Pt(NH_3)_2Cl_2$, and iron(III) has a coordination number of 6 in $[Fe(CN)_6]^{3-}$. The most common coordination numbers are 4 and 6, but others are well known (**Table 20.4**). The coordination number of a metal ion in a particular complex depends on the metal ion's size, charge, and electron configuration, and on the size and shape of the ligands.

TABLE 20.4 Examples of Complexes with Various Coordination Numbers

Coordination Number	Complex
2	$[Ag(NH_3)_2]^+$, $[CuCl_2]^-$
3	$[HgI_3]^-$
4	$[Zn(NH_3)_4]^{2+}$, $[Ni(CN)_4]^{2-}$
5	$[Ni(CN)_5]^{3-}$, $Fe(CO)_5$
6	$[Cr(H_2O)_6]^{3+}$, $[Fe(CN)_6]^{3-}$
7	$[ZrF_7]^{3-}$
8	$[Mo(CN)_8]^{4-}$

Metal complexes have characteristic shapes that depend on the metal ion's coordination number. Two-coordinate complexes, such as $[Ag(NH_3)_2]^+$, are linear. Four-coordinate complexes are either tetrahedral or square planar; for example, $[Zn(NH_3)_4]^{2+}$ is tetrahedral, and $[Ni(CN)_4]^{2-}$ is square planar. Nearly all six-coordinate complexes are octahedral. The more common coordination geometries are illustrated in **Figure 20.12**. Coordination geometries were first deduced by the Swiss chemist Alfred Werner, who was awarded the 1913 Nobel Prize in Chemistry for his pioneering studies.

FIGURE 20.12

The arrangement of ligand donor atoms (L) in ML_n complexes with coordination numbers 2, 4, and 6.

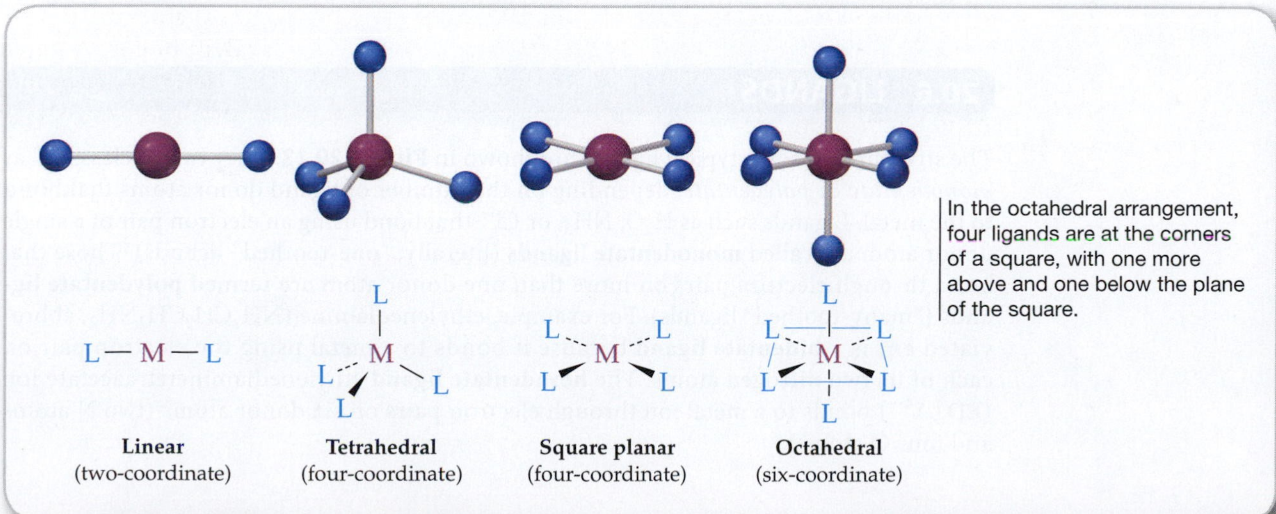

The charge on a metal complex equals the charge on the metal ion (its oxidation state) plus the sum of the charges on the ligands. Thus, if we know the charge on each ligand and the charge on the complex, we can easily find the oxidation state of the metal.

Worked Example 20.2

Determining the Formula of a Metal Complex

A cobalt(III) ion forms a complex with four ammonia molecules and two chloride ions. What is the formula of the complex, including its charge?

STRATEGY

The charge on the complex is the sum of the charges on the Co^{3+} ion, the four electrically neutral NH_3 ligands, and the two Cl^- ligands:

$$\begin{array}{ccccccc} Co^{3+} & & 4\,NH_3 & & 2\,Cl^- & & \\ +3 & + & (4)(0) & + & (2)(-1) & = & +1 \end{array}$$

SOLUTION

The formula of the complex is $[Co(NH_3)_4Cl_2]^+$.

Worked Example 20.3

Determining the Oxidation State of the Metal in a Coordination Compound

What is the oxidation state of platinum in the coordination compound $K[Pt(NH_3)Cl_5]$?

STRATEGY

Because the compound is electrically neutral overall and contains one K^+ cation per complex anion, the anion must be $[Pt(NH_3)Cl_5]^-$. Since ammonia is neutral and chloride has a charge of −1, the sum of the oxidation numbers is $+1 + n + 0 + (5)(-1) = 0$, where n is the oxidation number of Pt:

$$\begin{array}{cccccccl} K^+ & & Pt^{n+} & & NH_3 & & 5\,Cl^- & \\ +1 & + & n & + & 0 & + & (5)(-1) = 0; & \text{so, } n = +4 \end{array}$$

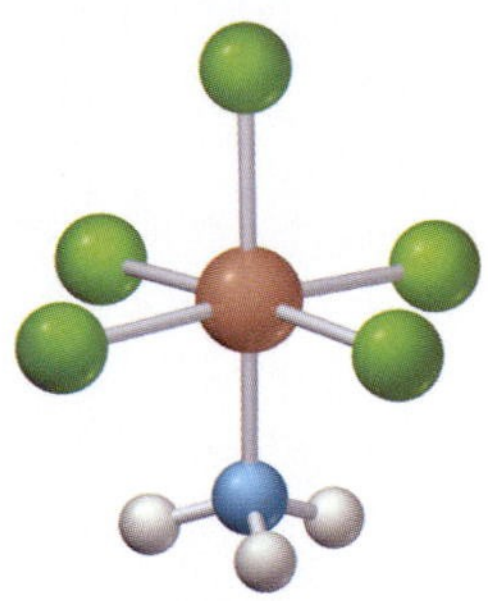

$[Pt(NH_3)Cl_5]^-$

SOLUTION

The oxidation state of platinum is +4.

PROBLEM 20.6 What is the formula of the chromium(III) complex that contains two ammonia and four thiocyanate (SCN^-) ligands?

PROBLEM 20.7 What is the oxidation state of iron in $Na_4[Fe(CN)_6]$?

20.6 LIGANDS

The structures of some typical ligands are shown in **Figure 20.13**. They can be classified as *monodentate* or *polydentate*, depending on the number of ligand donor atoms that bond to the metal. Ligands such as H_2O, NH_3, or Cl^- that bond using an electron pair of a single donor atom are called **monodentate ligands** (literally, "one-toothed" ligands). Those that bond through electron pairs on more than one donor atom are termed **polydentate ligands** ("many-toothed" ligands). For example, ethylenediamine ($NH_2CH_2CH_2NH_2$, abbreviated en) is a **bidentate ligand** because it bonds to a metal using the electron pair on each of its two nitrogen atoms. The **hexadentate ligand** ethylenediaminetetraacetate ion ($EDTA^{4-}$) bonds to a metal ion through electron pairs on six donor atoms (two N atoms and four O atoms).

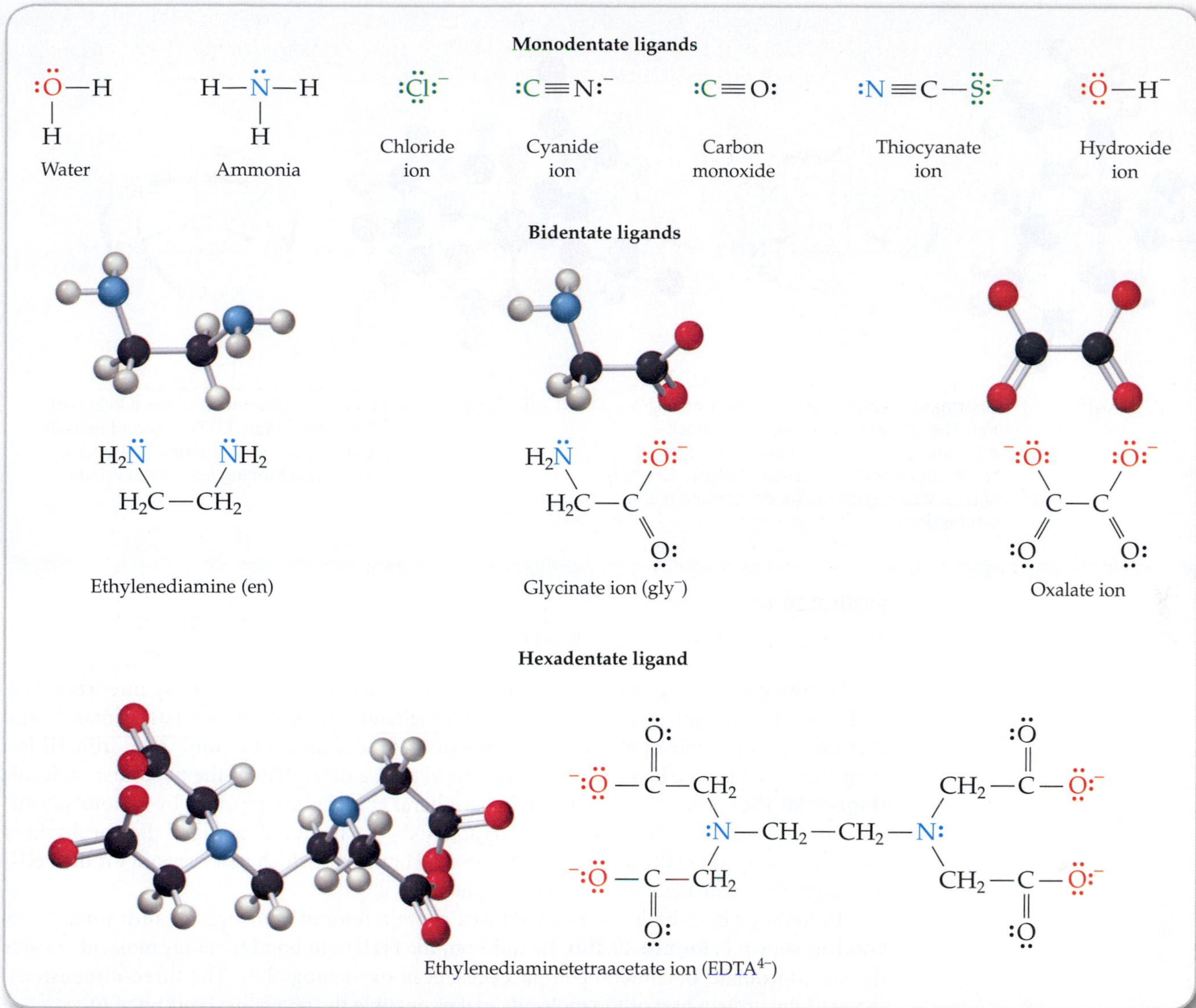

FIGURE 20.13

Structures of some common ligands. Ligand donor atoms are in color. The thiocyanate ion can bond to a metal through either the S atom or the N atom.

Polydentate ligands are known as **chelating agents** (pronounced **key**-late-ing, from the Greek word *chele*, meaning "claw") because their multipoint attachment to a metal ion resembles the grasping of an object by the claws of a crab. For example, ethylenediamine holds a cobalt(III) ion with two claws, its two nitrogen donor atoms **(Figure 20.14)**. The resulting five-membered ring consisting of the Co(III) ion, two N atoms, and two C atoms of the ligand is called a **chelate ring**. A complex such as $[Co(en)_3]^{3+}$ or $[Co(EDTA)]^-$ that contains one or more chelate rings is known as a metal **chelate**.

Because $EDTA^{4-}$ bonds to a metal ion through six donor atoms, it forms especially stable complexes and is often used to hold metal ions in solution. For example, in the treatment of lead poisoning, $EDTA^{4-}$ bonds to Pb^{2+}, which is then excreted by the kidneys as the soluble chelate $[Pb(EDTA)]^{2-}$. $EDTA^{4-}$ is commonly added to food products such as commercial salad dressings to complex any metal cations that might be present in trace amounts. The free metal ions might otherwise catalyze the oxidation of oils, thus causing the dressing to become rancid.

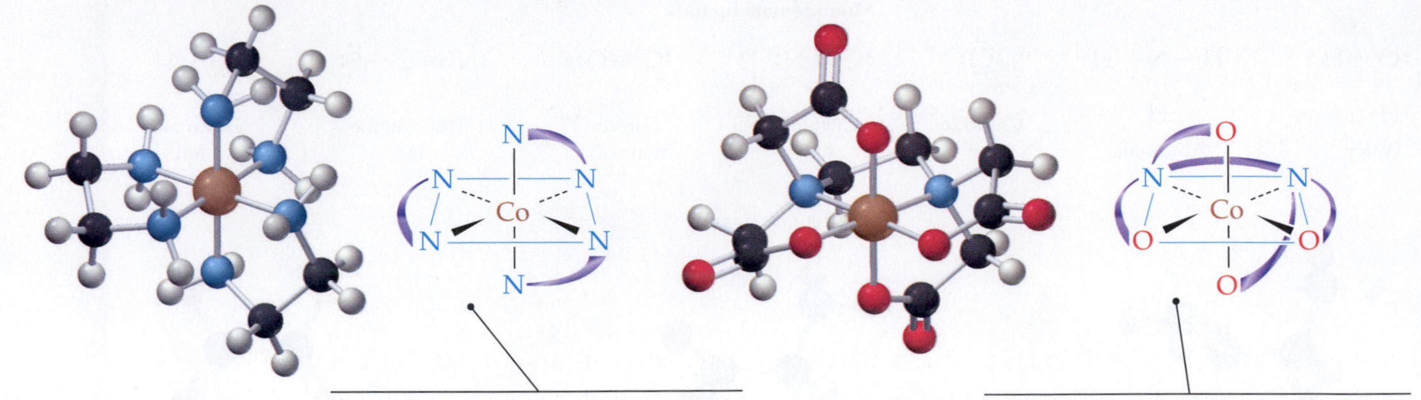

$[Co(en)_3]^{3+}$

Shorthand representation of the $[Co(en)_3]^{3+}$ ion. The complex contains three cobalt–ethylenediamine chelate rings. The symbol N⌒N represents a bidentate $NH_2CH_2CH_2NH_2$ ligand, which spans adjacent corners of the octahedron.

$[Co(EDTA)]^-$

Shorthand representation of the $[Co(EDTA)]^-$ ion. The hexadentate $EDTA^{4-}$ ligand uses its two N atoms and four O atoms to bond to the metal, thus forming five chelate rings.

FIGURE 20.14

The structures of $[Co(en)_3]^{3+}$ and $[Co(EDTA)]^-$.

Naturally occurring chelating ligands are essential constituents of many important biomolecules. For example, the heme group in hemoglobin contains a planar, tetradentate ligand that uses the lone pair of electrons on each of its four N atoms to bond to an iron(II) ion (Chapter 13 *FYI*). The ligand in heme is a *porphyrin*, a derivative of the porphine molecule **(Figure 20.15a)** in which the porphine's peripheral H atoms are replaced by various substituent groups ($-CH_3$, $-CH{=}CH_2$, and $-CH_2CH_2CO_2^-$ in heme). Bonding of the porphyrin to the Fe(II) ion involves the prior loss of the two NH protons, which makes room for the Fe(II) to occupy the cavity between the four N atoms.

In hemoglobin, the heme is linked to a protein (globin) through an additional Fe–N bond, as shown in **Figure 20.15b**. In addition, the Fe(II) can bond to an O_2 molecule to give the six-coordinate, octahedral complex present in oxyhemoglobin. The three-dimensional shape of the protein part of the molecule makes possible the reversible binding of O_2.

FIGURE 20.15

Chelating ligands in biomolecules. (a) The structure of the porphine molecule. The porphyrins are derivatives of porphine in which some of the peripheral H atoms are replaced by various substituent groups. **(b)** Schematic of the planar heme group, the attached protein chain, and the bound O_2 molecule in oxyhemoglobin.

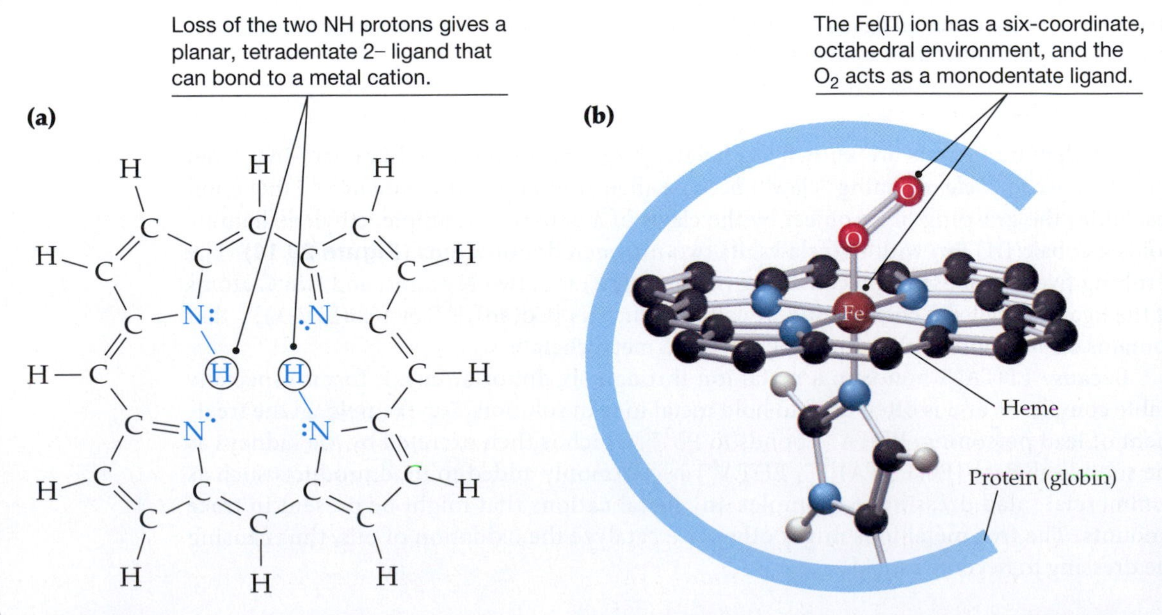

PROBLEM 20.8 Chromium forms the metal chelate $Cr(gly)_3$, where gly^- is the glycinate ligand shown in Figure 20.13.

(a) Using the abbreviation N⌒O for the glycinate ligand, draw a possible structure for $Cr(gly)_3$.
(b) Identify the Lewis acid, Lewis base, ligand donor atoms, and chelate rings.
(c) What is the coordination number, coordination geometry, and oxidation state of chromium?

20.7 NAMING COORDINATION COMPOUNDS

In the early days of coordination chemistry, coordination compounds were named after their discoverer or according to their color. Now, we use systematic names that specify the number of ligands of each particular type, the metal, and its oxidation state. Before listing the rules used to name coordination compounds, let's consider a few examples that will illustrate how to apply the rules:

$K_3[Fe(CN)_6]$ Potassium hexacyanoferrate(III)

- Potassium: Cation
- hexacyano: 6 CN^- ligands
- ferrate(III): Metal in +3 oxidation state
- hexacyanoferrate(III): Anion

$Pt(NH_3)_2Cl_2$ Diamminedichloroplatinum(II)

- Diammine: 2 NH_3 ligands
- dichloro: 2 Cl^- ligands
- platinum(II): Metal in +2 oxidation state

$[Cr(H_2O)_4Cl_2]Cl$ Tetraaquadichlorochromium(III) chloride

- Tetraaqua: 4 H_2O ligands
- dichloro: 2 Cl^- ligands
- chromium(III): Metal in +3 oxidation state
- Tetraaquadichlorochromium(III): Cation
- chloride: Anion

The following list summarizes the nomenclature rules recommended by the International Union of Pure and Applied Chemistry:

1. If the compound is a salt, name the cation first and then the anion, just as in naming simple salts. For example, $K_3[Fe(CN)_6]$ is potassium hexacyanoferrate(III).
2. In naming a complex ion or a neutral complex, name the ligands first and then the metal. The names of anionic ligands end in *-o*. As shown in **Table 20.5**, they are usually obtained by changing the anion endings *-ide* to *-o* and *-ate* to *-ato*. Neutral ligands are specified by their usual names, except for H_2O, NH_3, and CO, which are called aqua, ammine (note spelling), and carbonyl, respectively. The name of a complex is one word, with no space between the various ligand names and no space between the names of the last ligand and the metal.
3. If the complex contains more than one ligand of a particular type, indicate the number with the appropriate Greek prefix: *di-*, *tri-*, *tetra-*, *penta-*, *hexa-*, and so forth. The ligands are listed in alphabetical order, and the prefixes are ignored in determining the order. Thus, tetra*a*qua precedes dichloro in the name for $[Cr(H_2O)_4Cl_2]Cl$: tetraaquadichlorochromium(III) chloride.
4. If the name of a ligand itself contains a Greek prefix—for example, ethylene*di*amine—put the ligand name in parentheses and use one of the following alternative prefixes to specify the number of ligands: *bis-* (2), *tris-* (3), *tetrakis-* (4), and so forth. Thus, the name of $[Co(en)_3]Cl_3$ is tris(ethylenediamine)cobalt(III) chloride.
5. Use a Roman numeral in parentheses immediately following the name of the metal to indicate the metal's oxidation state. As shown by the preceding examples, there is no space between the name of the metal and the parenthesis.

▲ A sample of tris(ethylenediamine)cobalt(III) chloride, $[Co(en)_3]Cl_3$.

TABLE 20.5 Names of Some Common Ligands

Anionic Ligand	Ligand Name	Neutral Ligand	Ligand Name
Bromide, Br^-	Bromo	Ammonia, NH_3	Ammine
Carbonate, CO_3^{2-}	Carbonato	Water, H_2O	Aqua
Chloride, Cl^-	Chloro	Carbon monoxide, CO	Carbonyl
Cyanide, CN^-	Cyano	Ethylenediamine, en	Ethylenediamine
Fluoride, F^-	Fluoro		
Glycinate, gly^-	Glycinato		
Hydroxide, OH^-	Hydroxo		
Oxalate, $C_2O_4^{2-}$	Oxalato		
Thiocyanate, SCN^-	Thiocyanato* Isothiocyanato†		

* Ligand donor atom is S
† Ligand donor atom is N

6. In naming the metal, use the ending *-ate* if the metal is in an anionic complex. Thus, $[Fe(CN)_6]^{3-}$ is the hexacyanoferrate(III) anion. There are no simple rules for going from the name of the metal to the name of the metallate anion, partly because some of the anions have Latin names. Some common examples are given in **Table 20.6**.

TABLE 20.6 Names of Some Common Metallate Anions

Metal	Anion Name	Metal	Anion Name
Aluminum	Aluminate	Iron	Ferrate
Chromium	Chromate	Manganese	Manganate
Cobalt	Cobaltate	Nickel	Nickelate
Copper	Cuprate	Platinum	Platinate
Gold	Aurate	Zinc	Zincate

The rules for naming coordination compounds make it possible to go from a formula to the systematic name or from a systematic name to the appropriate formula. Worked Examples 20.4 and 20.5 provide some practice.

Worked Example 20.4

Determining the Name of a Coordination Compound from Its Formula

Name each of the following:

(a) $[Co(NH_3)_6]Cl_3$, prepared in 1798 by B. M. Tassaert and generally considered to be the first coordination compound

(b) $[Rh(NH_3)_5I]I_2$, a yellow compound obtained by heating $[Rh(NH_3)_5(H_2O)]I_3$ at 100 °C

(c) $Fe(CO)_5$, a highly toxic, volatile liquid

(d) $[Fe(C_2O_4)_3]^{3-}$, the ion formed when Fe_2O_3 rust stains are dissolved in oxalic acid ($H_2C_2O_4$)

STRATEGY

First determine the oxidation state of the metal, as in Worked Example 20.3. Then apply the six rules above to name the compound or ion.

SOLUTION

(a) Because the chloride ion has a charge of -1 and ammonia is neutral, the oxidation state of cobalt is $+3$. Use the prefix *hexa-* to indicate that the cation contains six NH_3 ligands, and use a Roman numeral III in parentheses to indicate the oxidation state of cobalt. The name of $[Co(NH_3)_6]Cl_3$ is hexaamminecobalt(III) chloride.

(b) Because the iodide ion has a charge of -1, the complex cation is $[Rh(NH_3)_5I]^{2+}$ and the rhodium has an oxidation state of $+3$. List the *a*mmine ligands before the *i*odo ligand, and use the prefix *penta-* to indicate the presence of five NH_3 ligands. The name of $[Rh(NH_3)_5I]I_2$ is pentaammineiodorhodium(III) iodide.

(c) Because the carbonyl ligand is neutral, the oxidation state of iron is zero. The systematic name of $Fe(CO)_5$ is pentacarbonyliron(0), but the common name iron pentacarbonyl is often used.

(d) Because each oxalate ligand ($C_2O_4^{2-}$) has a charge of -2 and because $[Fe(C_2O_4)_3]^{3-}$ has an overall charge of -3, iron must have an oxidation state of $+3$. Use the name ferrate(III) for the metal because the complex is an anion. The name of $[Fe(C_2O_4)_3]^{3-}$ is the trioxalatoferrate(III) ion.

▲ The compound $K_3[Fe(C_2O_4)_3] \cdot 3\ H_2O$ contains the trioxalatoferrate(III) anion.

Worked Example 20.5

Determining the Formula of a Coordination Compound from Its Name

Write the formula for each of the following:

(a) Potassium tetracyanonickelate(II)

(b) Aquachlorobis(ethylenediamine)cobalt(III) chloride

(c) Sodium hexafluoroaluminate

(d) Diamminesilver(I) ion

STRATEGY

To find the formula and charge of the complex cation or anion, note the number of ligands of each type, their charge, and the oxidation state of the metal in the name of the compound or ion. If the name refers to an electrically neutral compound, balance the charge of the complex cation or anion with the appropriate number of ions of opposite charge.

SOLUTION

(a) The bonding of four CN^- ligands to Ni^{2+} gives an $[Ni(CN)_4]^{2-}$ anion, which must be balanced by two K^+ cations. The compound's formula is therefore $K_2[Ni(CN)_4]$. This compound is obtained when excess KCN is added to a solution of a nickel(II) salt.

(b) Because the complex cation contains one H_2O, one Cl^-, and two neutral en ligands, and because the metal is Co^{3+}, the cation is $[Co(en)_2(H_2O)Cl]^{2+}$. The $2+$ charge of the cation must be balanced by two Cl^- anions, so the formula of the compound is $[Co(en)_2(H_2O)Cl]Cl_2$. The cation is the first product formed when $[Co(en)_2Cl_2]^+$ reacts with water.

(c) Sodium hexafluoroaluminate, also called cryolite, is used in the electrolytic production of aluminum metal (Section 17.13) and is a coordination compound of a main-group element. The oxidation state of aluminum is omitted from the name because aluminum has only one oxidation state ($+3$). Because F^- has a charge of -1, the anion is $[AlF_6]^{3-}$. The charge of the anion must be balanced by three Na^+ cations; therefore, the formula for the compound is $Na_3[AlF_6]$.

(d) Diamminesilver(I) ion, formed when silver chloride dissolves in an excess of aqueous ammonia, has the formula $[Ag(NH_3)_2]^+$.

▶ **PROBLEM 20.9** Name each of the following:

(a) $[Cu(NH_3)_4]SO_4$, a deep blue compound obtained when $CuSO_4$ is treated with an excess of ammonia

(b) $Na[Cr(OH)_4]$, the compound formed when $Cr(OH)_3$ is dissolved in an excess of aqueous NaOH

(c) $Co(gly)_3$, a complex that contains the anion of the amino acid glycine

(d) $[Fe(H_2O)_5(NCS)]^{2+}$, the red complex ion formed in a qualitative analysis test for iron

PROBLEM 20.10 Write the formula for each of the following:

(a) Tetraamminezinc(II) nitrate, the compound formed when zinc nitrate is treated with an excess of ammonia

(b) Tetracarbonylnickel(0), the first metal carbonyl (prepared in 1888) and an important compound in the industrial refining of nickel metal

(c) Potassium amminetrichloroplatinate(II), a compound that contains a square planar anion

(d) The dicyanoaurate(I) ion, an ion important in the extraction of gold from its ores

20.8 ISOMERS

One of the more interesting aspects of coordination chemistry is the existence of **isomers**, compounds that have the same formula but a different arrangement of their constituent atoms. Because their atoms are arranged differently, isomers are different compounds with different chemical reactivity and different physical properties such as color, solubility, and melting point. **Figure 20.16** shows a scheme for classifying some of the kinds of isomers in coordination chemistry. As we'll see in Chapter 23, isomers are also important in organic chemistry.

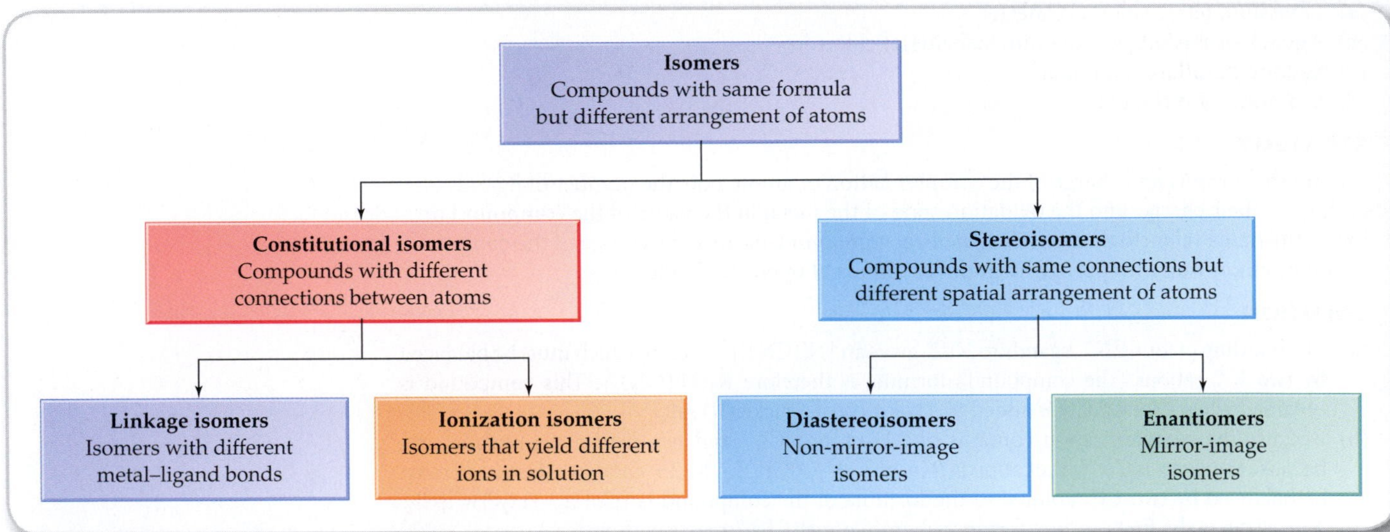

FIGURE 20.16

Classification scheme for the kinds of isomers in coordination chemistry. See the text for examples.

Constitutional Isomers

Isomers that have different connections among their constituent atoms are called **constitutional isomers**. Of the various types of constitutional isomers, we'll discuss just two: *linkage isomers* and *ionization isomers*.

Linkage isomers arise when a ligand can bond to a metal through either of two different donor atoms. For example, the nitrite (NO_2^-) ion forms two different pentaamminecobalt(III) complexes: a yellow *nitro* complex that contains a Co–N bond, $[Co(NH_3)_5(NO_2)]^{2+}$, and a red *nitrito* complex that contains a Co–O bond, $[Co(NH_3)_5(ONO)]^{2+}$ (**Figure 20.17**). The ligand in the nitrito complex is written as ONO to emphasize that it's linked to the cobalt through an oxygen atom. The thiocyanate (SCN^-) ion is another ligand that gives linkage isomers because it can bond to a metal through either the sulfur atom to give a thiocyanato complex or the nitrogen atom to give an isothiocyanato complex.

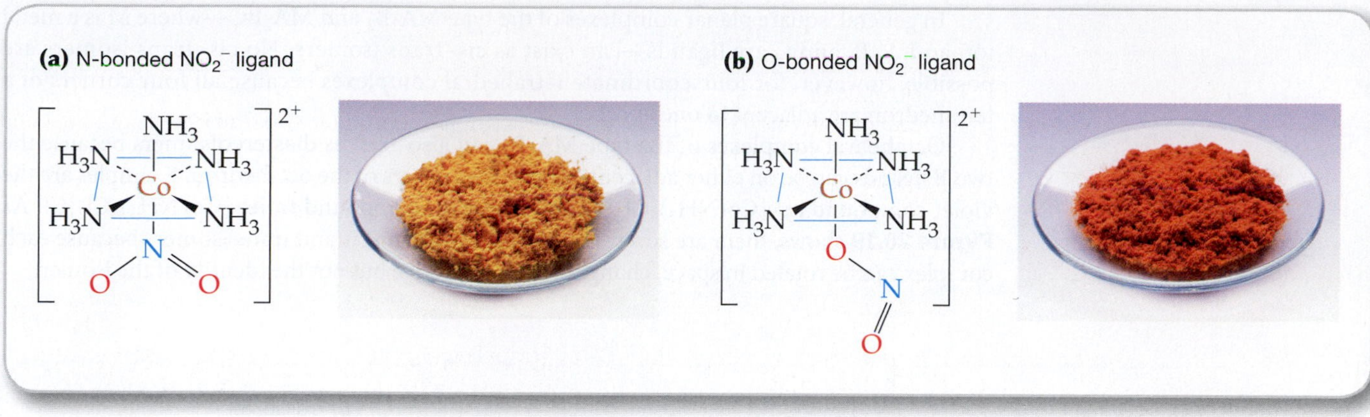

FIGURE 20.17

Linkage isomers. Structures and samples of **(a)** the nitro complex $[Co(NH_3)_5(NO_2)]^{2+}$ and **(b)** the nitrito complex $[Co(NH_3)_5(ONO)]^{2+}$.

Ionization isomers differ in the anion that is bonded to the metal ion. An example is the pair $[Co(NH_3)_5Br]SO_4$, a violet compound that has a Co—Br bond and a free sulfate anion, and $[Co(NH_3)_5SO_4]Br$, a red compound that has a Co—sulfate bond and a free bromide ion. Ionization isomers get their name because they yield different ions in solution.

Stereoisomers

Isomers that have the same connections among atoms but a different arrangement of the atoms in space are called **stereoisomers**. In coordination chemistry, there are two kinds of stereoisomers: *diastereoisomers* and *enantiomers*. We'll discuss diastereoisomers in this section and enantiomers in Section 20.9.

Diastereoisomers, also called *geometric isomers*, are non-mirror-image stereoisomers. For example, in the square planar complex $Pt(NH_3)_2Cl_2$, the two Pt—Cl bonds can be oriented either adjacent at a 90° angle or opposite at a 180° angle, as shown in **Figure 20.18**. The isomer in which identical ligands occupy adjacent corners of the square is called the **cis isomer**, and the one in which identical ligands are across from each other is called the **trans isomer**. (The Latin word *cis* means "next to"; *trans* means "across.")

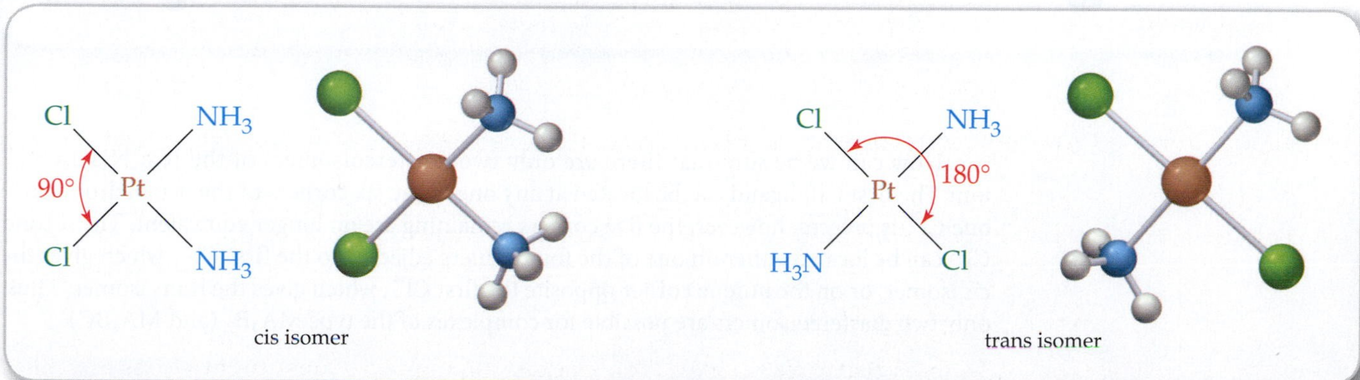

FIGURE 20.18

Diastereoisomers of the square planar complex $Pt(NH_3)_2Cl_2$. The two compounds have the same connections among atoms but different arrangements of the atoms in space.

Cis and trans isomers are different compounds with different properties. Thus, *cis*-$Pt(NH_3)_2Cl_2$ is a polar molecule and is more soluble in water than *trans*-$Pt(NH_3)_2Cl_2$. The trans isomer is nonpolar because the two Pt—Cl and the two Pt—NH_3 bond dipoles point in opposite directions and therefore cancel. It's also interesting that *cis*-$Pt(NH_3)_2Cl_2$ (cisplatin) is a widely prescribed anticancer drug, whereas the trans isomer is inactive as an antitumor agent.

In general, square planar complexes of the type MA_2B_2 and MA_2BC—where M is a metal ion and A, B, and C are ligands—can exist as cis–trans isomers. No cis–trans isomers are possible, however, for four-coordinate tetrahedral complexes because all four corners of a tetrahedron are adjacent to one another.

Octahedral complexes of the type MA_4B_2 can also exist as diastereoisomers because the two B ligands can be on either adjacent or opposite corners of the octahedron. Examples are the violet compound *cis*-$[Co(NH_3)_4Cl_2]Cl$ and the green compound *trans*-$[Co(NH_3)_4Cl_2]Cl$. As **Figure 20.19** shows, there are several ways of drawing the cis and trans isomers because each complex can be rotated in space, changing the perspective but not the identity of the isomer.

FIGURE 20.19

Diastereoisomers of the $[Co(NH_3)_4Cl_2]^+$ ion.

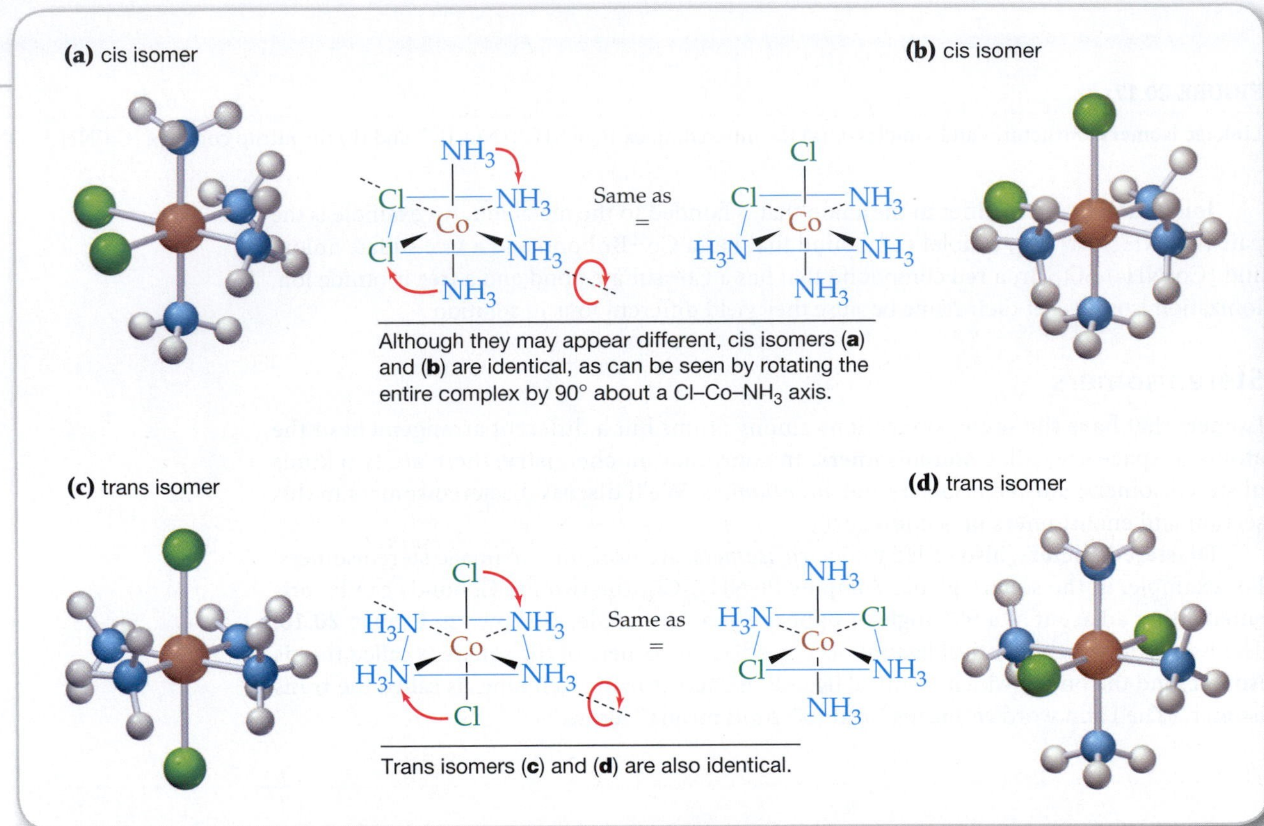

How can we be sure that there are only two diastereoisomers of the $[Co(NH_3)_4Cl_2]^+$ ion? The first Cl^- ligand can be located at any one of the six corners of the octahedron. Once one Cl^- is present, however, the five corners remaining are no longer equivalent. The second Cl^- can be located either on one of the four corners adjacent to the first Cl^-, which gives the cis isomer, or on the unique corner opposite the first Cl^-, which gives the trans isomer. Thus, only two diastereoisomers are possible for complexes of the type MA_4B_2 (and MA_4BC).

Worked Example 20.6

Identifying Diastereoisomers for Square Planar and Octahedral Complexes

Platinum(II) forms square planar complexes, and platinum(IV) gives octahedral complexes. How many diastereoisomers are possible for each of the following complexes? Draw their structures.

(a) $[Pt(NH_3)_3Cl]^+$ **(b)** $[Pt(NH_3)Cl_5]^-$

(c) $Pt(NH_3)_2Cl(NO_2)$ **(d)** $[Pt(NH_3)_4ClBr]^{2+}$

STRATEGY

Recall that cis and trans isomers are possible for square planar complexes of the type MA_2B_2 and MA_2BC and for octahedral complexes of the type MA_4B_2 and MA_4BC. Cis and trans isomers are not possible when only one ligand differs from the others, as in complexes of the type MA_3B and MA_5B.

SOLUTION

(a) No isomers are possible for $[Pt(NH_3)_3Cl]^+$, a square planar complex of the type MA_3B.

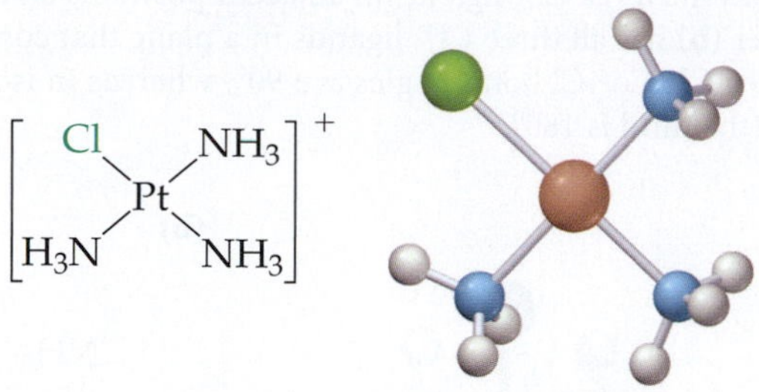

(b) No isomers are possible for $[Pt(NH_3)Cl_5]^-$, an octahedral complex of the type MA_5B.

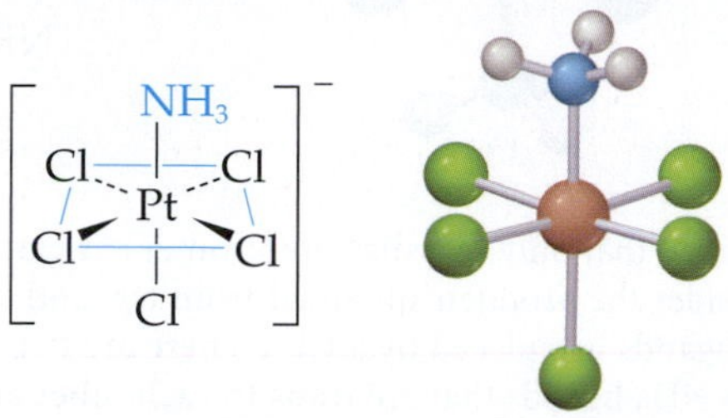

(c) Cis and trans isomers are possible for $Pt(NH_3)_2Cl(NO_2)$, a square planar complex of the type MA_2BC. The Cl^- and NO_2^- ligands can be on either adjacent or opposite corners of the square.

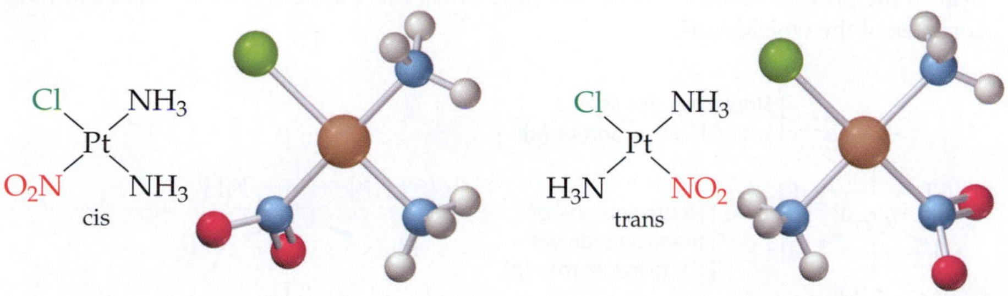

(d) Cis and trans isomers are possible for $[Pt(NH_3)_4ClBr]^{2+}$, an octahedral complex of the type MA_4BC. The Cl^- and Br^- ligands can be on either adjacent or opposite corners of the octahedron.

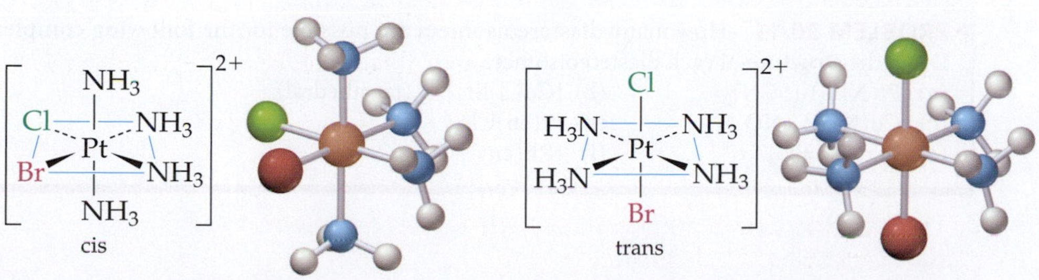

Worked Example 20.7

Drawing the Structures of Diastereoisomers

Draw the structures of all possible diastereoisomers of triamminetrichlorocobalt(III), $Co(NH_3)_3Cl_3$.

STRATEGY AND SOLUTION

Two diastereoisomers are possible for an octahedral complex of the type MA_3B_3. Isomer **(a)** shown below has the three Cl^- ligands in adjacent positions on one triangular face of the octahedron; isomer **(b)** has all three Cl^- ligands in a plane that contains the Co(III) ion. In isomer **(a)**, all three Cl–Co–Cl bond angles are 90°, whereas in isomer **(b)** two Cl–Co–Cl angles are 90° and the third is 180°.

(a)

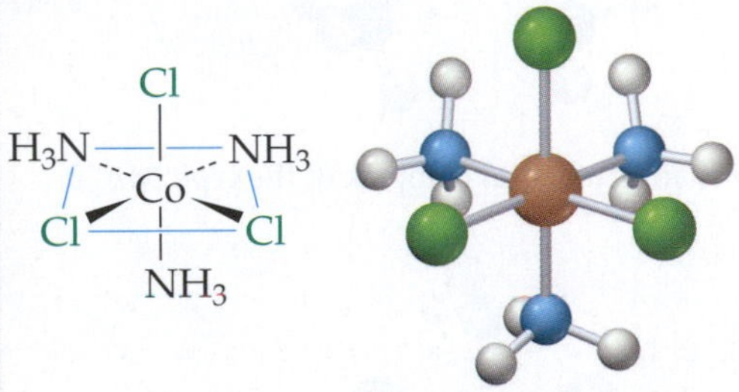

(b)

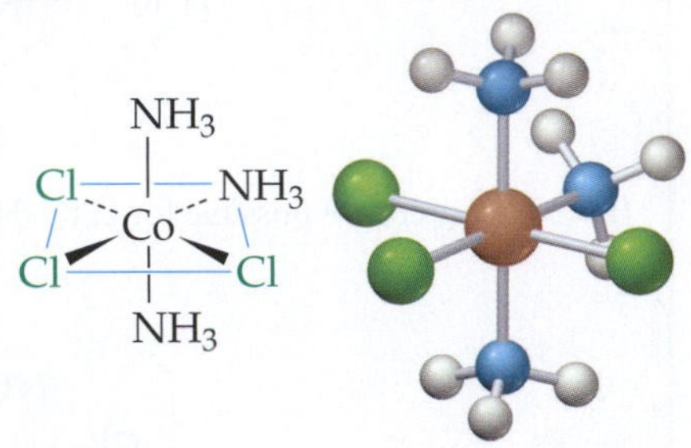

To convince yourself that only two diastereoisomers of $Co(NH_3)_3Cl_3$ exist, look back at Figure 20.19 and consider the products obtained from *cis*- and *trans*-$[Co(NH_3)_4Cl_2]^+$ when one of the four NH_3 ligands is replaced by a Cl^-. There are two kinds of NH_3 ligands in *cis*-$[Co(NH_3)_4Cl_2]^+$: two NH_3 ligands that are trans to each other and two NH_3 ligands that are trans to a Cl^-. Replacing an NH_3 trans to another NH_3 with Cl^- gives $Co(NH_3)_3Cl_3$ isomer **(a)**, whereas replacing an NH_3 trans to a Cl^- with another Cl^- gives isomer **(b)**. All four NH_3 ligands in *trans*-$[Co(NH_3)_4Cl_2]^+$ are equivalent, and replacing any one of them with a Cl^- gives isomer **(b)**. [If you rotate the resulting $Co(NH_3)_3Cl_3$ complex, you can see that it is identical to the previous drawing of isomer **(b)**.] Thus, there are only two diastereoisomers for a complex of the type MA_3B_3.

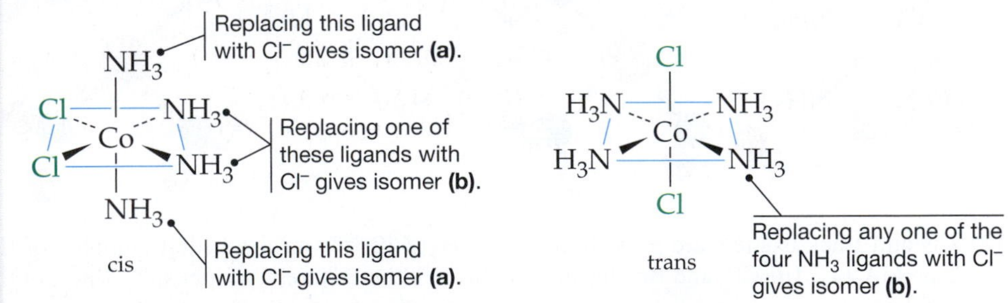

PROBLEM 20.11 How many diastereoisomers are possible for the following complexes? Draw the structure of each diastereoisomer.

(a) $Pt(NH_3)_2(SCN)_2$
(b) $[CoCl_2Br_2]^{2-}$ (tetrahedral)
(c) $Co(NH_3)_3(NO_2)_3$
(d) $Pt(en)Cl_2$
(e) $[Cr(en)_2Br_2]^+$
(f) $[Rh(en)_3]^{3+}$

WORKED CONCEPTUAL EXAMPLE 20.8

Identifying cis and trans Isomers

Consider the following isomers of the $[Co(NH_3)_4(H_2O)Cl]^{2+}$ ion:

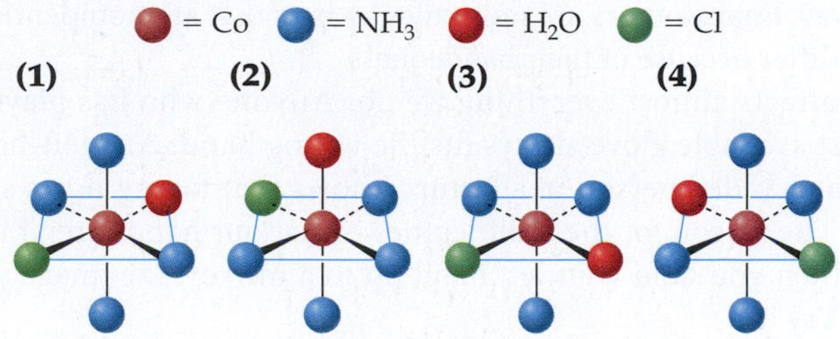

(a) Label the isomers as cis or trans.
(b) Which isomers are identical, and which are different?

STRATEGY

Look at the location of the H_2O and Cl^- ligands. Isomers that have H_2O and Cl^- on adjacent corners of the octahedron are cis, whereas those that have H_2O and Cl^- on opposite corners are trans. Because only two isomers are possible for complexes of the type MA_4BC, all the cis isomers are identical and all the trans isomers are identical.

SOLUTION

(a) Isomers (1) and (4) have H_2O and Cl^- on opposite corners of the octahedron and are therefore trans isomers. Isomers (2) and (3) have H_2O and Cl^- on adjacent corners of the octahedron and are therefore cis isomers.

(b) Isomers (1) and (4) are identical, as can be seen by rotating (1) counterclockwise by 90° about the vertical (z) axis:

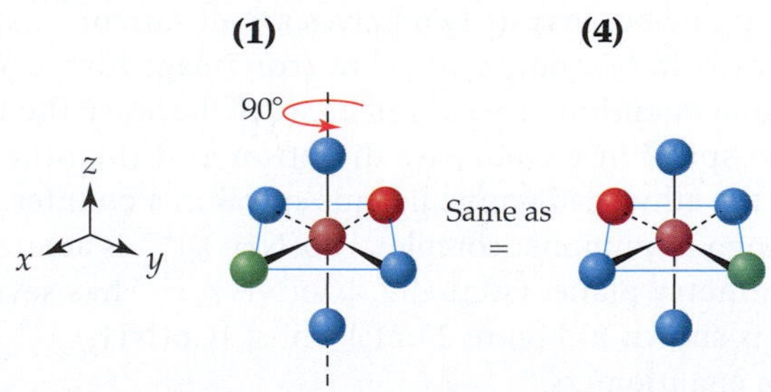

Isomers (2) and (3) differ from (1) and (4) and are identical, as can be seen by rotating (2) clockwise by 90° about the x axis and then counterclockwise by 90° about the y axis:

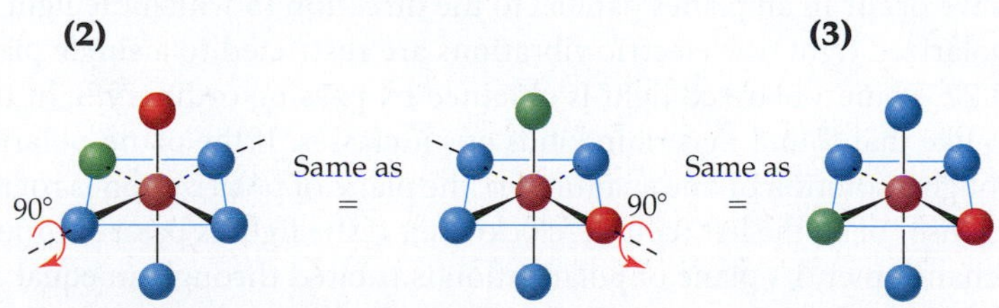

CONCEPTUAL PROBLEM 20.12 Which of the following $[Co(en)(NH_3)_3Cl]^{2+}$ isomers are identical, and which are different?

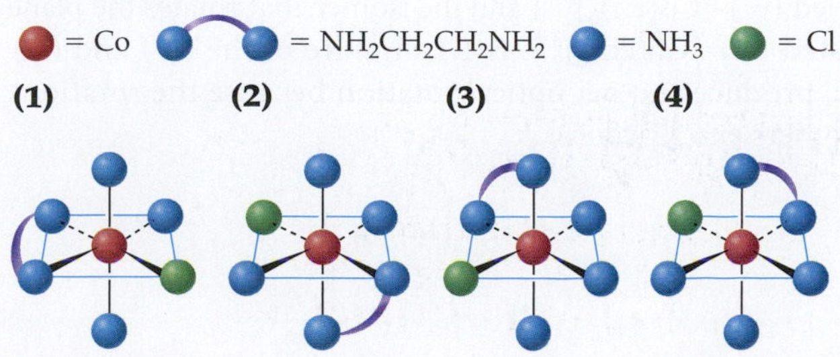

20.9 ENANTIOMERS AND MOLECULAR HANDEDNESS

Diastereoisomers are relatively easy to distinguish because the various bonds in the cis and trans isomers point in different directions. *Enantiomers*, however, are stereoisomers that differ in a more subtle way. **Enantiomers** are molecules or ions that are nonidentical mirror images of each other and differ because of their handedness.

Handedness affects almost everything we do. Anyone who has played much softball knows that the last available glove always fits the wrong hand. Any left-handed person sitting next to a right-handed person in a lecture knows that taking notes sometimes means bumping elbows. The reason for these difficulties is that our hands aren't identical—they're mirror images. When you hold your left hand up to a mirror, the image you see looks like your right hand. (Try it.)

Not all objects are handed. There's no such thing as a right-handed tennis ball or a left-handed coffee mug. When a tennis ball or a coffee mug is held up to a mirror, the image reflected is identical to the ball or mug itself. Objects that do have a handedness to them are said to be **chiral** (pronounced **ky**-ral, from the Greek *cheir*, meaning "hand"), and objects that lack handedness are said to be nonchiral, or **achiral**.

Why is it that some objects are chiral but others aren't? In general, an object is not chiral if, like the coffee mug, it has a **symmetry plane** cutting through its middle so that one half of the object is a mirror image of the other half. If you were to cut the mug in half, one half of the mug would be the mirror image of the other half. A hand, however, has no symmetry plane and is therefore chiral. If you were to cut a hand in two, one "half" of the hand would not be a mirror image of the other half (**Figure 20.20**).

Symmetry plane

An achiral object like the coffee mug has a symmetry plane passing through it, making the two halves mirror images.

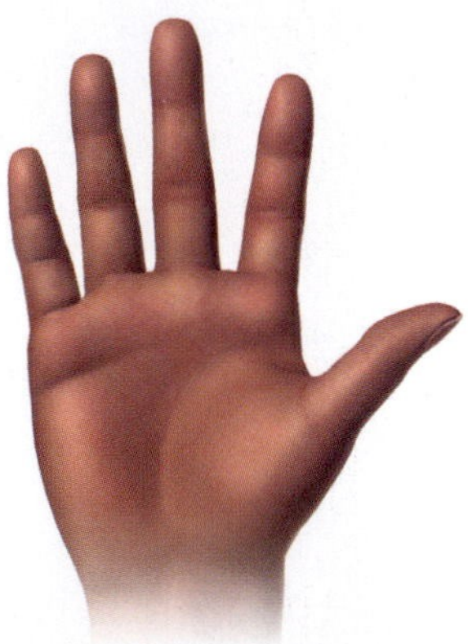

A chiral object like the hand has no symmetry plane because the two halves of the hand are not mirror images.

FIGURE 20.20

The meaning of a symmetry plane.

Just as certain objects like a hand are chiral, certain molecules and ions are also chiral. Consider, for example, the tris(ethylenediamine)cobalt(III) ion, $[Co(en)_3]^{3+}$, an ion with the shape of a three-bladed propeller (**Figure 20.21**). The $[Co(en)_3]^{3+}$ cation has no symmetry plane because its two halves aren't mirror images. Thus, $[Co(en)_3]^{3+}$ is chiral and can exist in two nonidentical mirror-image forms. When viewed as drawn in Figure 20.21, one enantiomer is "right-handed" because the three ethylenediamine ligands appear to spiral in a clockwise direction and the other enantiomer is "left-handed" because the ethylenediamine ligands spiral in a counterclockwise direction. By contrast, the analogous ammonia complex $[Co(NH_3)_6]^{3+}$ is achiral because, like a coffee mug, it has a symmetry plane. (Actually, $[Co(NH_3)_6]^{3+}$ has several symmetry planes, though only one is shown in Figure 20.21.) Thus, $[Co(NH_3)_6]^{3+}$ exists in a single form and does not have enantiomers.

Enantiomers have identical properties except for their reactions with other chiral substances and their effect on *plane-polarized light*. In ordinary light, the electric vibrations of the light wave occur in all planes parallel to the direction in which the light is traveling, but in plane-polarized light, the electric vibrations are restricted to a single plane, as shown in **Figure 20.22**. Plane-polarized light is obtained by passing ordinary light through a polarizing filter, like that found in certain kinds of sunglasses. If the plane-polarized light is then passed through a solution of one enantiomer, the plane of polarization is rotated, either to the right (clockwise) or to the left (counterclockwise). If the light is passed through a solution of the other enantiomer, its plane of polarization is rotated through an equal angle, but in the opposite direction. Enantiomers are sometimes called *optical isomers* because of their effect on plane-polarized light.

Enantiomers are labeled (+) or (−), depending on the direction of rotation of the plane of polarization. For example, the isomer of $[Co(en)_3]^{3+}$ that rotates the plane of polarization to the right is labeled (+)-$[Co(en)_3]^{3+}$, and the isomer that rotates the plane of polarization to the left is designated (−)-$[Co(en)_3]^{3+}$. A 1:1 mixture of the (+) and (−) isomers, called a **racemic mixture**, produces no net optical rotation because the rotations produced by the individual enantiomers exactly cancel.

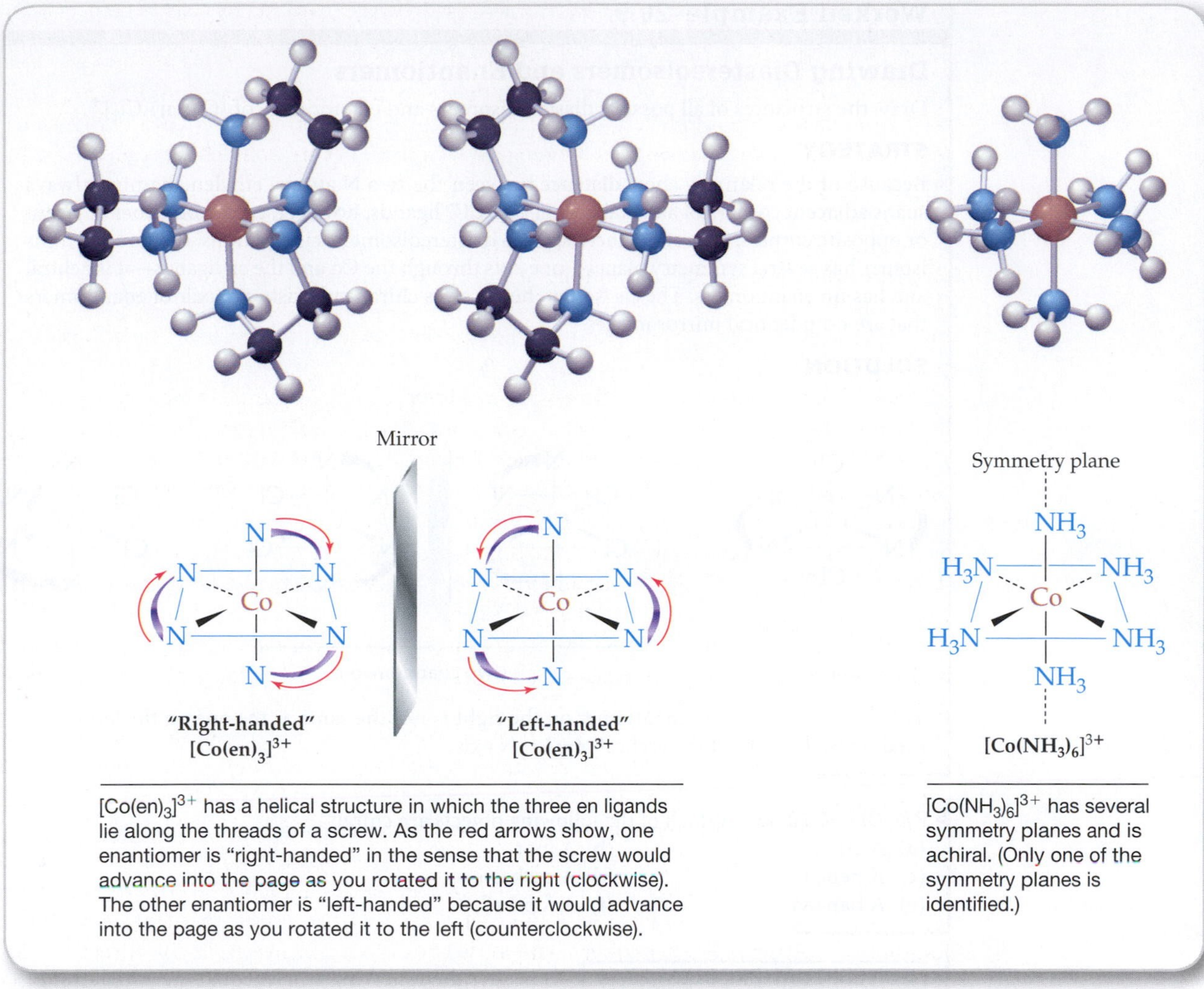

FIGURE 20.21

The structures of the chiral $[Co(en)_3]^{3+}$ enantiomers and the achiral $[Co(NH_3)_6]^{3+}$ ion.

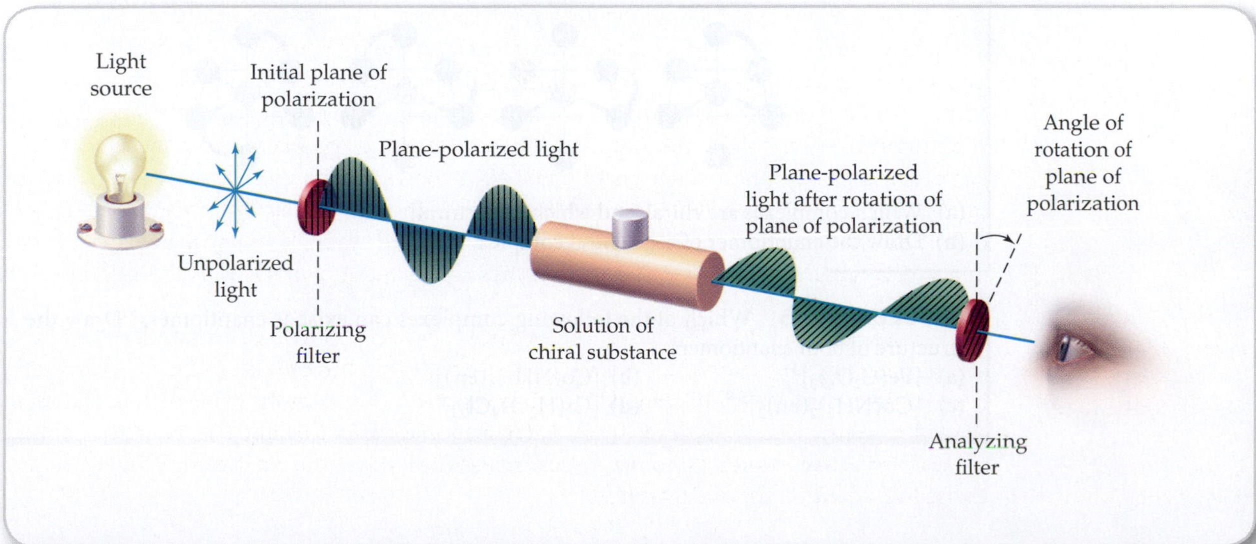

FIGURE 20.22

The essential features of a polarimeter. The polarimeter measures the angle through which the plane of plane-polarized light is rotated when the light is passed through a solution of a chiral substance.

Worked Example 20.9

Drawing Diastereoisomers and Enantiomers

Draw the structures of all possible diastereoisomers and enantiomers of $[Co(en)_2Cl_2]^+$.

STRATEGY

Because of the relatively short distance between the two N atoms, ethylenediamine always spans adjacent corners of an octahedron. The Cl^- ligands, however, can be on either adjacent or opposite corners. Therefore, there are two diastereoisomers, cis and trans. Because the trans isomer has several symmetry planes—one cuts through the Co and the en ligands—it is achiral and has no enantiomers. The cis isomer, however, is chiral and exists as a pair of enantiomers that are nonidentical mirror images.

SOLUTION

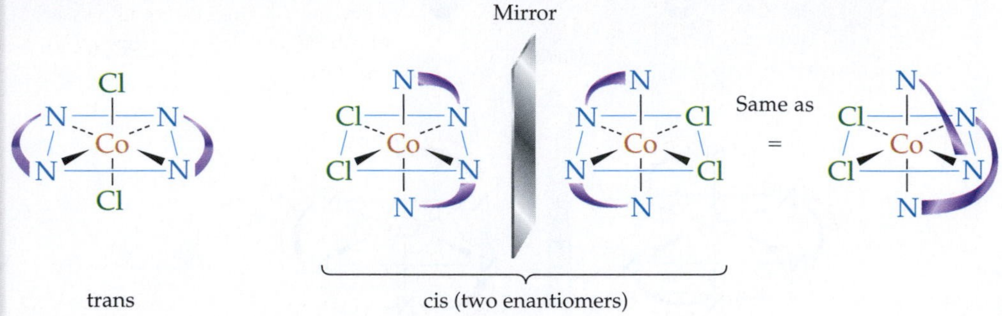

You can see that the cis enantiomer on the right is not the same as the one on the left if you rotate it by 180° about the vertical N–Co–N axis.

PROBLEM 20.13 Which of the following objects are chiral?

(a) A chair (b) A foot
(c) A pencil (d) A corkscrew
(e) A banana (f) A football

CONCEPTUAL PROBLEM 20.14 Consider the following ethylenediamine complexes of rhodium:

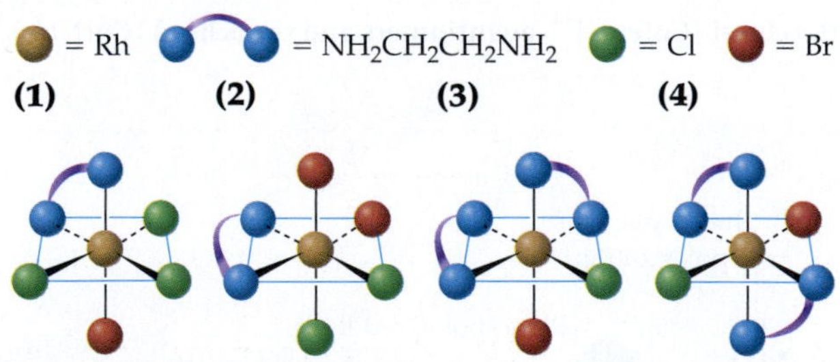

(a) Which complexes are chiral, and which are achiral?
(b) Draw the enantiomer of each chiral complex.

PROBLEM 20.15 Which of the following complexes can exist as enantiomers? Draw the structure of both enantiomers.

(a) $[Fe(C_2O_4)_3]^{3-}$ (b) $[Co(NH_3)_4(en)]^{3+}$
(c) $[Co(NH_3)_2(en)_2]^{3+}$ (d) $[Cr(H_2O)_4Cl_2]^+$

20.10 COLOR OF TRANSITION METAL COMPLEXES

Most transition metal complexes have beautiful colors that depend on the identity of the metal and the ligands. The color of an aqua complex, for example, depends on the metal: $[Co(H_2O)_6]^{2+}$ is pink, $[Ni(H_2O)_6]^{2+}$ is green, $[Cu(H_2O)_6]^{2+}$ is blue, but $[Zn(H_2O)_6]^{2+}$ is colorless (**Figure 20.23**). If we keep the metal constant but vary the ligand, the color also changes. For example, $[Ni(H_2O)_6]^{2+}$ is green, $[Ni(NH_3)_6]^{2+}$ is blue, and $[Ni(en)_3]^{2+}$ is violet (**Figure 20.24**).

Different metals, but the same ligand

FIGURE 20.23

Aqueous solutions of the nitrate salts of cobalt(II), nickel(II), copper(II), and zinc(II).

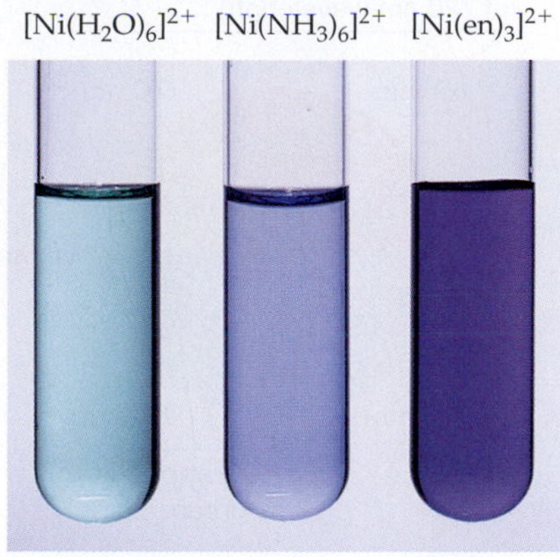

The same metal, but different ligands

FIGURE 20.24

Aqueous solutions that contain $[Ni(H_2O)_6]^{2+}$, $[Ni(NH_3)_6]^{2+}$, and $[Ni(en)_3]^{2+}$.

How can we account for the color of transition metal complexes? Let's begin by recalling that white light consists of a continuous spectrum of wavelengths corresponding to different colors (Section 2.1). When white light strikes a colored substance, some wavelengths are transmitted while others are absorbed. Just as atoms can absorb light by undergoing electronic transitions between atomic energy levels, thereby giving rise to **atomic spectra**, so a metal complex can absorb light by undergoing an electronic transition from its lowest (ground) energy state to a higher (excited) energy state.

REMEMBER...

Atomic line spectra result when an electron moves between two states having different energy. (Section 2.9)

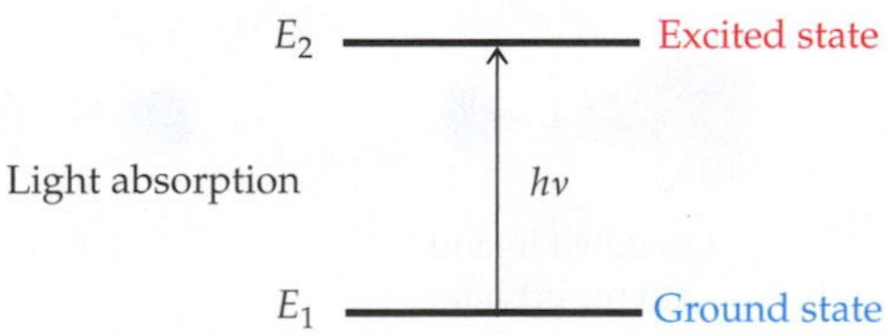

The wavelength λ of the light absorbed by a metal complex depends on the **energy separation $\Delta E = E_2 - E_1$ between the two states**, as given by the equation $\Delta E = h\nu = hc/\lambda$ (Section 2.3), where h is Planck's constant, ν is the frequency of the light, and c is the speed of light:

$$\Delta E = E_2 - E_1 = h\nu = \frac{hc}{\lambda} \quad \text{or} \quad \lambda = \frac{hc}{\Delta E}$$

REMEMBER...

To cause an electronic transition between two energy states, the energy of the photon must match **the energy difference between the two states**. (Section 2.3)

The measure of the amount of light absorbed by a substance is called the *absorbance*, and a plot of absorbance versus wavelength is called an **absorption spectrum**. For example, the absorption spectrum of the red-violet $[Ti(H_2O)_6]^{3+}$ ion has a broad absorption band at about 500 nm, a wavelength in the blue-green part of the visible spectrum (**Figure 20.25**). In general, the color that we see is complementary to the color absorbed (**Figure 20.26**).

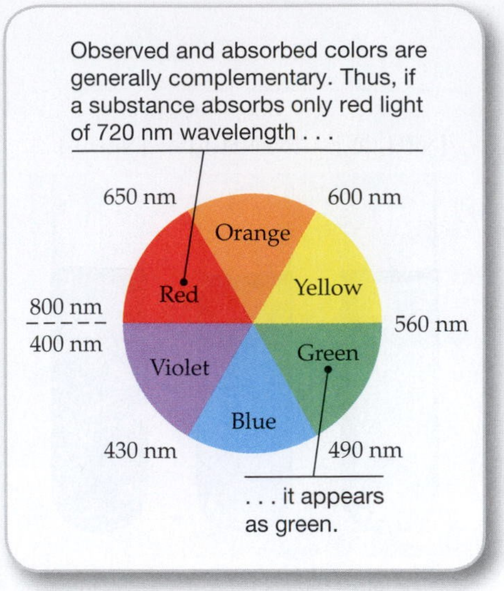

FIGURE 20.26

An artist's color wheel. The observed color of a substance is complementary to the color of the light absorbed and appears on the opposite side of the wheel.

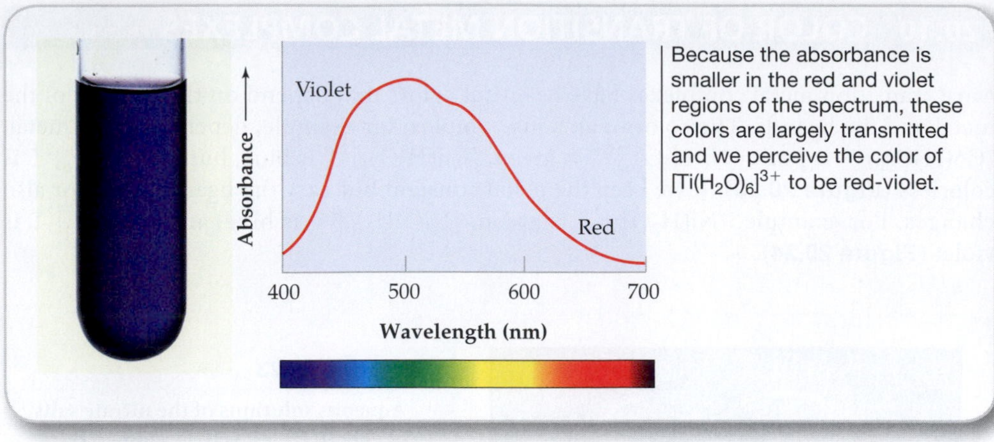

FIGURE 20.25

A solution that contains the $[Ti(H_2O)_6]^{3+}$ ion and its visible absorption spectrum.

PROBLEM 20.16 Consider a complex ion that has just one absorption band at 625 nm.

(a) What is the most likely color of this ion?

(b) What is the energy difference ΔE (in joules) between the ground and excited states for this ion?

20.11 BONDING IN COMPLEXES: VALENCE BOND THEORY

REMEMBER...

The **valence bond theory** describes a covalent bond as a pair of electrons of opposite spin that are shared between two atoms as a result of orbital overlap. (Section 5.2)

According to the **valence bond theory** (Section 5.2), the bonding in metal complexes arises when a filled ligand orbital containing a pair of electrons overlaps a vacant hybrid orbital on the metal ion to give a coordinate covalent bond:

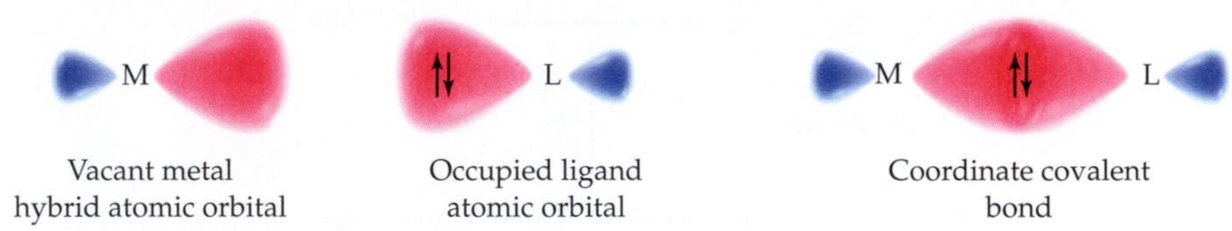

REMEMBER...

Hybrid orbitals are formed by a mathematical combination of atomic orbital wave functions. Combination of one *s* and one *p* orbital gives a set of two equivalent *sp* hybrids that are oriented at 180° to each other. The combination of one *s* and three *p* orbitals gives a set of four equivalent sp^3 hybrids that are oriented at 109.5° to each other and point toward the corners of a regular tetrahedron. (Sections 5.3, 5.4)

The **hybrid orbitals** that a metal ion uses to accept a share in the ligand electrons are those that point in the directions of the ligands. In linear complexes the metal uses *sp* hybrids, while in tetrahedral complexes the metal uses sp^3 hybrids. Remember that geometry and hybridization go together. Once you know the geometry of a complex, you automatically know which hybrid orbitals the metal ion uses.

In octahedral complexes, the metal ion uses valence *d* orbitals in addition to *s* and *p* orbitals to form six bonds to the ligands. Hybridization occurs by a combination of two *d* orbitals, one *s* orbital, and three *p* orbitals, resulting in either six d^2sp^3 hybrids or six sp^3d^2 hybrids. As we'll see later, the difference between d^2sp^3 and sp^3d^2 hybrids lies in the principal quantum number of the *d* orbitals. Both sets of hybrids, however, point toward the six corners of a regular octahedron, as shown in **Figure 20.27**. All six orbitals within each set are equivalent, and the angle between any two adjacent orbitals is 90°.

FIGURE 20.27

The octahedral geometry of the six d^2sp^3 and sp^3d^2 hybrid orbitals.

In square planar complexes, the metal ion uses a set of four equivalent dsp^2 hybrid orbitals formed by the hybridization of one *d* orbital, one *s* orbital, and two *p* orbitals. These hybrids point toward the four corners of a square, as shown in **Figure 20.28**.

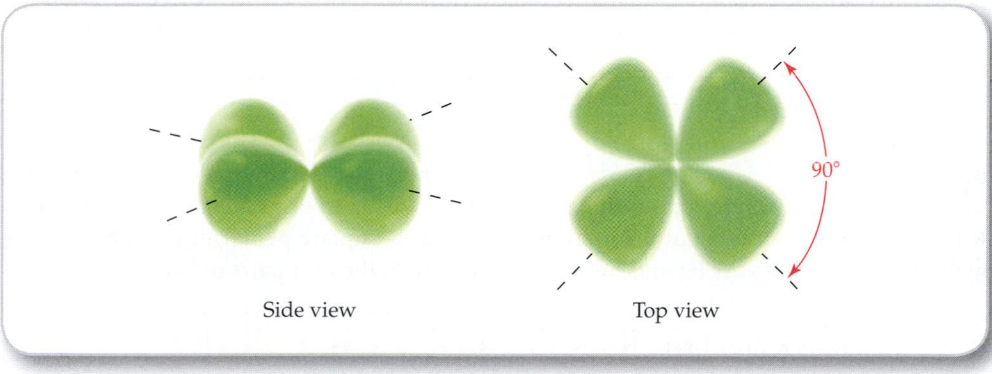

FIGURE 20.28

The square planar geometry of the four dsp^2 hybrid orbitals.

The relationship between the geometry of a complex and the hybrid orbitals used by the metal ion is summarized in **Table 20.7**.

TABLE 20.7 Hybrid Orbitals for Common Coordination Geometries

Coordination Number	Geometry	Hybrid Orbitals	Example
2	Linear	*sp*	$[Ag(NH_3)_2]^+$
4	Tetrahedral	sp^3	$[CoCl_4]^{2-}$
4	Square planar	dsp^2	$[Ni(CN)_4]^{2-}$
6	Octahedral	d^2sp^3 or sp^3d^2	$[Cr(H_2O)_6]^{3+}$, $[Co(H_2O)_6]^{2+}$

To illustrate the relationship between geometry and hybridization, let's consider the tetrahedral complex $[CoCl_4]^{2-}$. A free Co^{2+} ion has the electron configuration [Ar] $3d^7$ and the following orbital diagram:

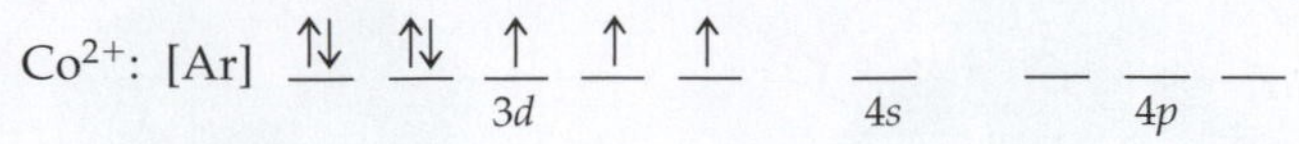

Because the geometry of $[CoCl_4]^{2-}$ is tetrahedral, the hybrid orbitals that Co^{2+} uses to accept a share in the four pairs of ligand electrons must be sp^3 hybrids formed from the vacant $4s$ and $4p$ orbitals. The following orbital diagram represents the bonding in the complex, showing the hybridization of the metal orbitals and the four pairs of ligand electrons, now shared in the bonds between the metal and the ligands:

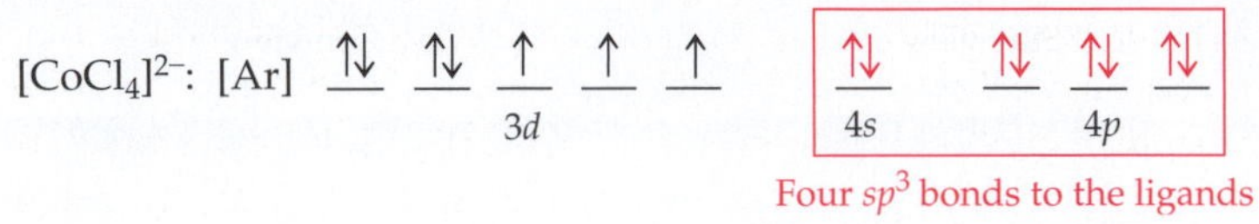

REMEMBER...

The O_2 molecule has two unpaired π^*_{2p} electrons and is **paramagnetic**, whereas N_2 and F_2 have no unpaired electrons and are diamagnetic. (Section 5.6)

In accord with this description, $[CoCl_4]^{2-}$ has three unpaired electrons and is paramagnetic. Recall from Section 5.6 that **paramagnetic** substances contain unpaired electrons and are attracted by magnetic fields, whereas diamagnetic substances contain only paired electrons and are weakly repelled by magnetic fields. The number of unpaired electrons in a transition metal complex can be determined by quantitatively measuring the force exerted on the complex by a magnetic field.

As an example of a square planar complex, consider $[Ni(CN)_4]^{2-}$. A free Ni^{2+} ion has eight $3d$ electrons, two of which are unpaired in accord with Hund's rule (Section 2.11):

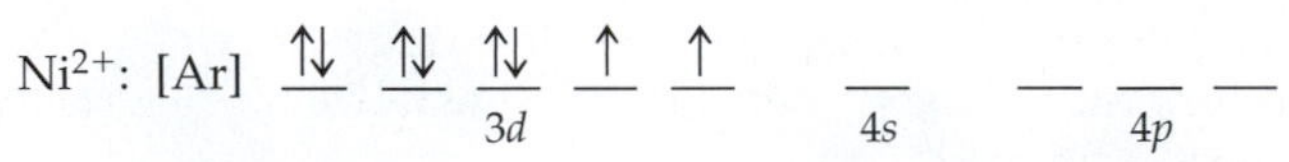

Because $[Ni(CN)_4]^{2-}$ is square planar, the metal ion must use dsp^2 hybrids. By pairing up the two unpaired d electrons in one d orbital, we obtain a vacant $3d$ orbital that can be hybridized with the $4s$ orbital and two of the $4p$ orbitals to give the square planar dsp^2 hybrids. These hybrids form bonds to the ligands by accepting a share in the four pairs of ligand electrons:

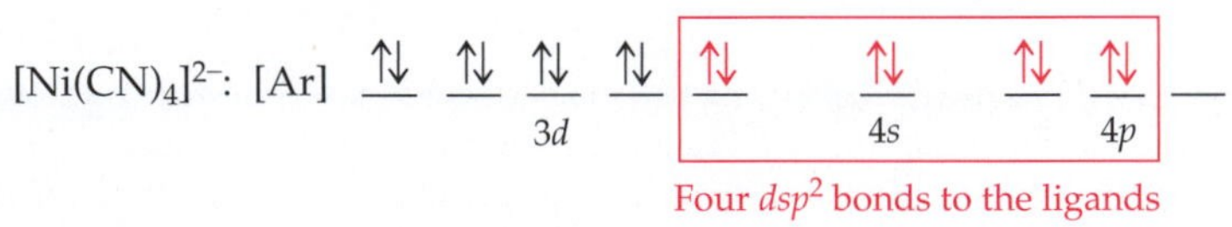

In agreement with this description, $[Ni(CN)_4]^{2-}$ is diamagnetic.

In octahedral complexes, the metal ion uses either sp^3d^2 or d^2sp^3 hybrid orbitals. To see the difference between these two kinds of hybrids, let's consider the cobalt(III) complexes $[CoF_6]^{3-}$ and $[Co(CN)_6]^{3-}$. A free Co^{3+} ion has six $3d$ electrons, four of which are unpaired:

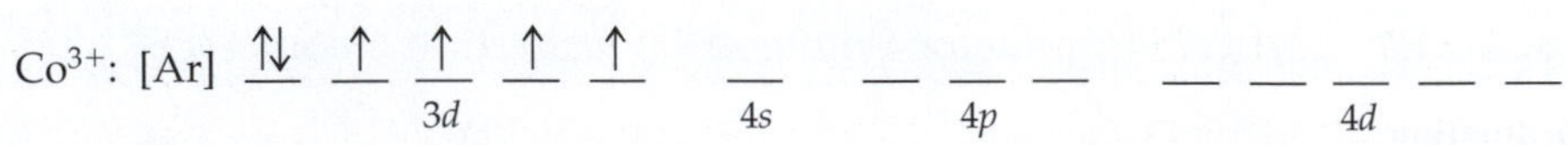

Magnetic measurements indicate that $[CoF_6]^{3-}$ is paramagnetic and contains four unpaired electrons. Evidently, none of the $3d$ orbitals is available to accept a share in the ligand electrons because each is already at least partially occupied. Consequently, the octahedral hybrids that Co^{3+} uses are formed from the vacant $4s$, $4p$, and $4d$ orbitals. These orbitals, called sp^3d^2 hybrids, share in the six pairs of ligand electrons, as shown in the following orbital diagram:

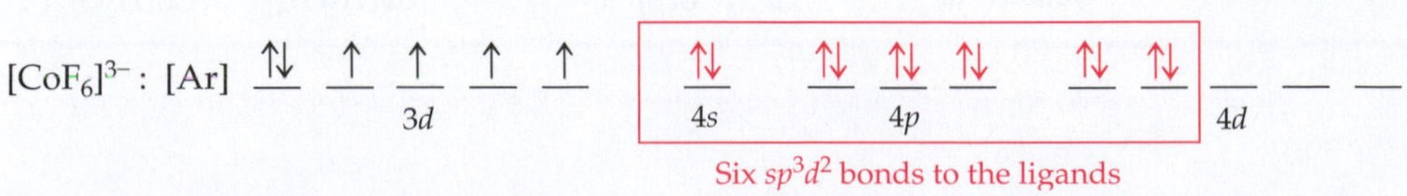

In contrast to $[CoF_6]^{3-}$, magnetic measurements indicate that $[Co(CN)_6]^{3-}$ is diamagnetic. All six $3d$ electrons are therefore paired and occupy just three of the five $3d$ orbitals.

That leaves two vacant 3*d* orbitals, which combine with the vacant 4*s* and 4*p* orbitals to give a set of six octahedral hybrid orbitals called d^2sp^3 hybrids. The d^2sp^3 hybrids form bonds to the ligands by accepting a share in the six pairs of ligand electrons:

$[Co(CN)_6]^{3-}$: [Ar] ↑↓ ↑↓ ↑↓ [↑↓ ↑↓ (3*d*) ↑↓ (4*s*) ↑↓ ↑↓ ↑↓ (4*p*)] __ __ __ __ __ (4*d*)

Six d^2sp^3 bonds to the ligands

The difference between d^2sp^3 and sp^3d^2 hybrids lies in the principal quantum number of the *d* orbitals. In d^2sp^3 hybrids, the principal quantum number of the *d* orbitals is one less than the principal quantum number of the *s* and *p* orbitals. In sp^3d^2 hybrids, the *s*, *p*, and *d* orbitals have the same principal quantum number. To determine which set of hybrids is used in any given complex, we must know the magnetic properties of the complex.

Complexes of metals like Co^{3+} that exhibit more than one spin state are classified as *high-spin* or *low-spin*. A **high-spin complex**, such as $[CoF_6]^{3-}$, is one in which the *d* electrons are arranged according to Hund's rule to give the maximum number of unpaired electrons. A **low-spin complex**, such as $[Co(CN)_6]^{3-}$, is one in which the *d* electrons are paired up to give a maximum number of doubly occupied *d* orbitals and a minimum number of unpaired electrons.

Worked Example 20.10

Describing the Bonding in Complexes in Terms of Valence Bond Hybrid Orbitals

Give a valence bond description of the bonding in $[V(NH_3)_6]^{3+}$. Include orbital diagrams for the free metal ion and the metal ion in the complex. Tell which hybrid orbitals the metal ion uses and the number of unpaired electrons present.

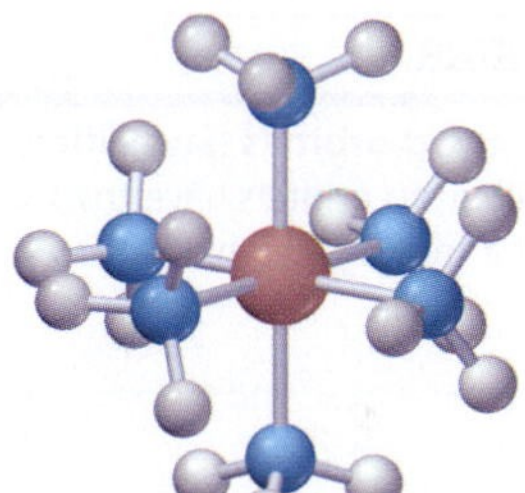

$[V(NH_3)_6]^{3+}$

STRATEGY AND SOLUTION

The free V^{3+} ion has the electron configuration [Ar] $3d^2$ and the orbital diagram

V^{3+}: [Ar] ↑ ↑ __ __ __ (3*d*) __ (4*s*) __ __ __ (4*p*)

Because $[V(NH_3)_6]^{3+}$ is octahedral, the V^{3+} ion must use either d^2sp^3 or sp^3d^2 hybrid orbitals in accepting a share in six pairs of electrons from the six NH_3 ligands. The preferred hybrids are d^2sp^3 because several 3*d* orbitals are vacant and d^2sp^3 hybrids have lower energy than sp^3d^2 hybrids (because the 3*d* orbitals have lower energy than the 4*d* orbitals). Thus, $[V(NH_3)_6]^{3+}$ has the following orbital diagram:

$[V(NH_3)_6]^{3+}$: [Ar] ↑ ↑ __ [↑↓ ↑↓ (3*d*) ↑↓ (4*s*) ↑↓ ↑↓ ↑↓ (4*p*)]

Six d^2sp^3 bonds to the ligands

The complex has two unpaired electrons and is therefore paramagnetic.

PROBLEM 20.17 Give a valence bond description of the bonding in each of the following complexes. Include orbital diagrams for the free metal ion and the metal ion in the complex. Tell which hybrid orbitals the metal ion uses and the number of unpaired electrons in each complex.

(a) $[Fe(CN)_6]^{3-}$ (low-spin) **(b)** $[Co(H_2O)_6]^{2+}$ (high-spin)

(c) $[VCl_4]^-$ (tetrahedral) **(d)** $[PtCl_4]^{2-}$ (square planar)

20.12 CRYSTAL FIELD THEORY

FIGURE 20.29

Crystal field model of the octahedral $[TiF_6]^{3-}$ complex. The metal ion and ligands are regarded as charged particles held together by electrostatic attraction. The ligands lie along the $\pm x$, $\pm y$, and $\pm z$ directions.

Valence bond theory helps us visualize the bonding in complexes, but it doesn't account for their color or explain why some complexes are high-spin whereas others are low-spin. To explain these properties, we turn to the **crystal field theory**, a model that views the bonding in complexes as arising from electrostatic interactions and considers the effect of the ligand charges on the energies of the metal ion *d* orbitals. This model was first applied to transition metal ions in ionic crystals—hence the name crystal field theory—but it also applies to metal complexes where the "crystal field" is the electric field due to the charges or dipoles of the ligands.

Octahedral Complexes

Let's first consider an octahedral complex such as $[TiF_6]^{3-}$ (**Figure 20.29**). According to the crystal field theory, the bonding is ionic and involves electrostatic attraction between the positively charged Ti^{3+} ion and the negatively charged F^- ligands. Of course, the F^- ligands repel one another, which is why they adopt the geometry (octahedral) that locates them as far apart from one another as possible. Because the metal–ligand attractions are greater than the ligand–ligand repulsions, the complex is more stable than the separated ions, thus accounting for the bonding.

Note the differences between crystal field theory and valence bond theory. In crystal field theory, there are no covalent bonds, no shared electrons, and no hybrid orbitals—just electrostatic interactions within an array of ions. In complexes that contain neutral dipolar ligands, such as H_2O or NH_3, the electrostatic interactions are of the ion–dipole type (Section 10.2). For example, in $[Ti(H_2O)_6]^{3+}$, the Ti^{3+} ion attracts the negative end of the water dipoles.

REMEMBER... Although the ***d* orbitals** have different shapes, all five *d* orbitals have the same energy in a free metal ion. (Section 2.7)

To explain why complexes are colored, we need to look at the effect of the ligand charges on the energies of the *d* orbitals. Recall that four of the ***d* orbitals** are shaped like a cloverleaf, while the fifth (d_{z^2}) is shaped like a dumbbell inside a donut (Section 2.7). **Figure 20.30** shows the spatial orientation of the *d* orbitals with respect to an octahedral array of charged ligands located along the *x*, *y*, and *z* coordinate axes.

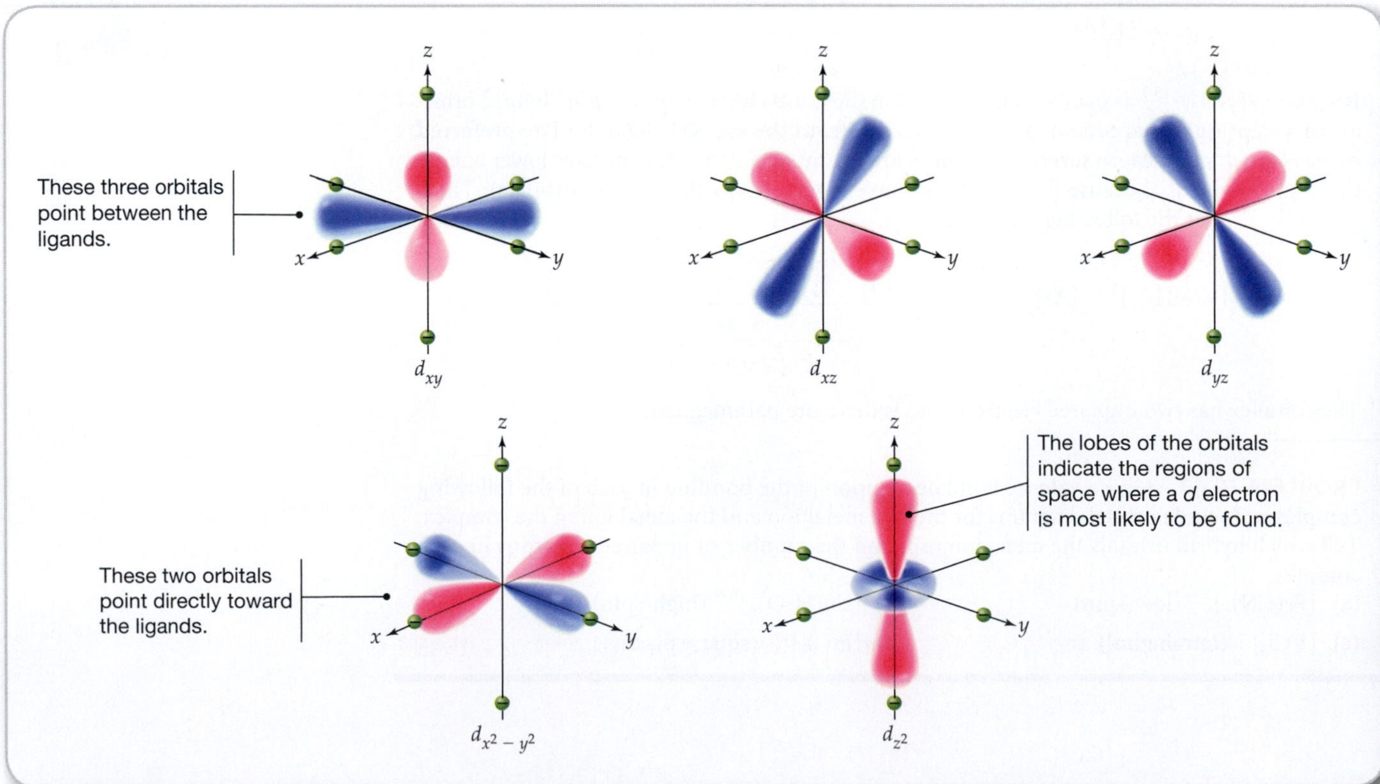

FIGURE 20.30

The shapes of the five *d* orbitals and their orientation with respect to an octahedral array of charged ligands.

Because the *d* electrons are negatively charged, they are repelled by the negatively charged ligands. Thus, their orbital energies are higher in the complex than in the free metal ion. But not all the *d* orbitals are raised in energy by the same amount. As shown in **Figure 20.31**, the d_{z^2} and $d_{x^2-y^2}$ orbitals, which point directly at the ligands, are raised in energy to a greater extent than the d_{xy}, d_{xz}, and d_{yz} orbitals, which point between the ligands. This energy splitting between the two sets of *d* orbitals is called the **crystal field splitting** and is represented by the Greek letter Δ.

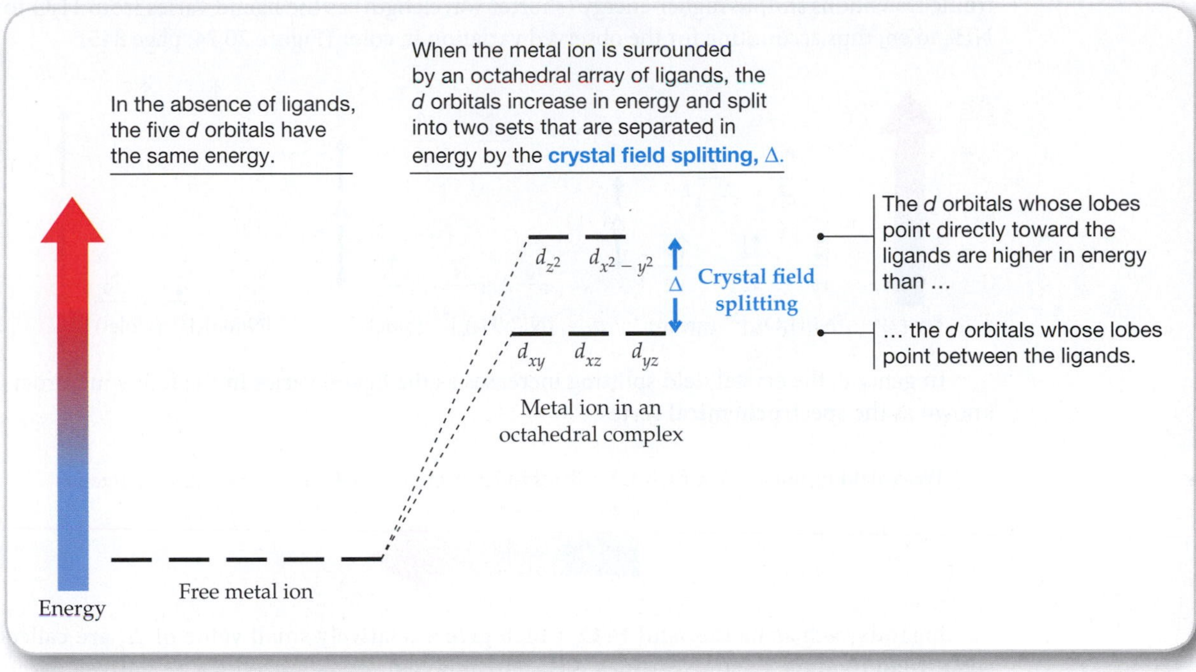

FIGURE 20.31
A *d*-orbital energy-level diagram for a free metal ion and a metal ion in an octahedral complex.

In general, the crystal field splitting energy Δ corresponds to wavelengths of light in the visible region of the spectrum, and the colors of complexes can therefore be attributed to electronic transitions between the lower- and higher-energy sets of *d* orbitals. Consider, for example, $[Ti(H_2O)_6]^{3+}$, a complex that contains a single *d* electron (Ti^{3+} has the electron configuration [Ar] $3d^1$). In the ground-energy state, the *d* electron occupies one of the lower-energy orbitals—*xy*, *xz*, or *yz* (from now on we'll denote the *d* orbitals by their subscripts). When $[Ti(H_2O)_6]^{3+}$ absorbs blue-green light with a wavelength of about 500 nm, the absorbed energy promotes the *d* electron to one of the higher-energy orbitals, z^2 or $x^2 - y^2$:

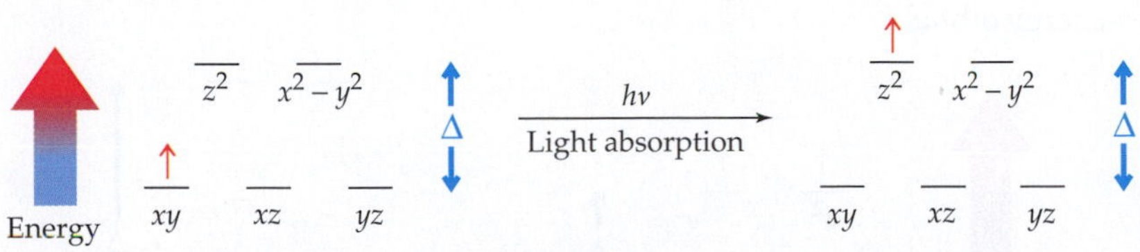

We can calculate the value of Δ from the wavelength of the absorbed light:

$$\Delta = h\nu = \frac{hc}{\lambda}$$

For $[Ti(H_2O)_6]^{3+}$, which absorbs at about 500 nm, the value of Δ is given by

$$\Delta = \frac{hc}{\lambda} = \frac{(6.626 \times 10^{-34}\ J \cdot s)(3.00 \times 10^8\ m/s)}{500 \times 10^{-9}\ m} = 4.0 \times 10^{-19}\ J$$

This is the energy needed to excite a single $[Ti(H_2O)_6]^{3+}$ ion. To express Δ on a per-mole basis, we multiply by Avogadro's number:

$$\Delta = \left(4.0 \times 10^{-19}\,\frac{\text{J}}{\text{ion}}\right)\left(6.02 \times 10^{23}\,\frac{\text{ions}}{\text{mol}}\right) = 2.4 \times 10^5\ \text{J/mol} = 240\ \text{kJ/mol}$$

The absorption spectra of different complexes indicate that the size of the crystal field splitting depends on the nature of the ligands. For example, Δ for Ni^{2+} ([Ar] $3d^8$) complexes increases as the ligand varies from H_2O to NH_3 to ethylenediamine (en). Accordingly, the electronic transitions shift to higher energy (shorter wavelength) as the ligand varies from H_2O to NH_3 to en, thus accounting for the observed variation in color (Figure 20.24; page 845):

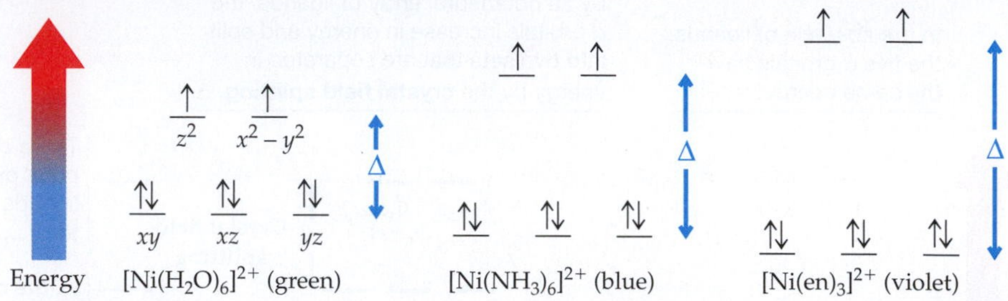

In general, the crystal field splitting increases as the ligand varies in the following order, known as the **spectrochemical series**:

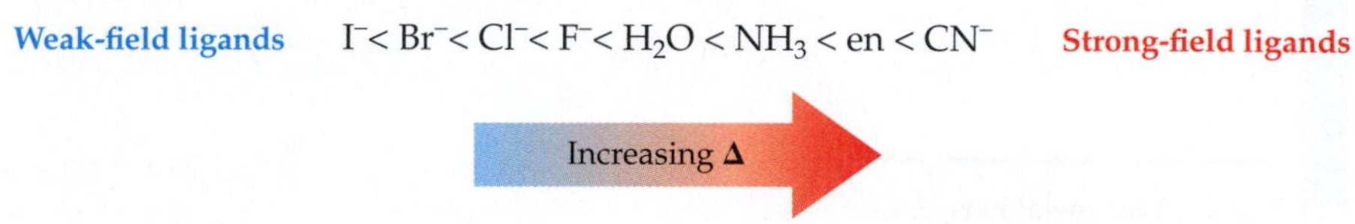

Ligands such as halides and H_2O, which give a relatively small value of Δ, are called **weak-field ligands**. Ligands such as NH_3, en, and CN^-, which produce a relatively large value of Δ, are known as **strong-field ligands**. Different metal ions have different values of Δ, which explains why their complexes with the same ligand have different colors (Figure 20.23; page 845). Because d^0 ions such as Ti^{4+}, d^{10} ions such as Zn^{2+}, and main-group ions don't have partially filled *d* subshells, they can't undergo *d–d* electronic transitions and most of their compounds are therefore colorless.

Crystal field theory accounts for the magnetic properties of complexes as well as for their color. It explains, for example, why complexes with weak-field ligands, such as $[CoF_6]^{3-}$, are high-spin, whereas related complexes with strong-field ligands, such as $[Co(CN)_6]^{3-}$, are low-spin. In $[CoF_6]^{3-}$, the six *d* electrons of Co^{3+} ([Ar] $3d^6$) occupy both the higher- and lower-energy *d* orbitals, whereas in $[Co(CN)_6]^{3-}$, all six *d* electrons are spin-paired in the lower-energy orbitals:

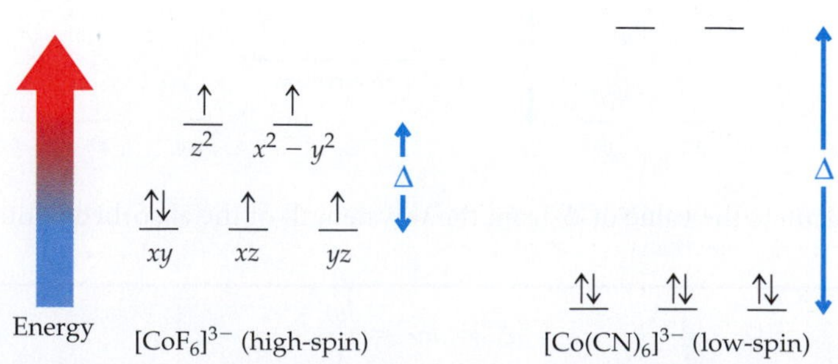

What determines which of the two spin states has the lower energy? In general, when an electron moves from a z^2 or $x^2 - y^2$ orbital to one of the lower-energy orbitals, the orbital energy decreases by Δ. But because of electron–electron repulsion, it costs energy to put the electron into an orbital that already contains another electron. The energy required is called the spin-pairing energy *P*. If Δ is greater than *P*, as it is for $[Co(CN)_6]^{3-}$, then the low-spin arrangement has lower energy. If Δ is less than *P*, as it is for $[CoF_6]^{3-}$, the high-spin arrangement has lower energy. Thus, the observed spin state depends on the relative values of Δ and *P*. In general, strong-field ligands give low-spin complexes, and weak-field ligands give high-spin complexes.

A choice between the high-spin and low-spin electron configurations arises only for complexes of metal ions with four to seven *d* electrons, so-called d^4–d^7 complexes. For d^1–d^3 and d^8–d^{10} complexes, only one ground-state electron configuration is possible. In d^1–d^3 complexes, all the electrons occupy the lower-energy *d* orbitals, independent of the value of Δ. In d^8–d^{10} complexes, the lower-energy set of *d* orbitals is filled with three pairs of electrons, while the higher-energy set contains two, three, or four electrons, again independent of the value of Δ.

Worked Example 20.11

Drawing Crystal Field Energy-Level Diagrams for Octahedral Complexes

Draw a crystal field orbital energy-level diagram, and predict the number of unpaired electrons for each of the following complexes:

(a) $[Cr(en)_3]^{3+}$ **(b)** $[Mn(CN)_6]^{3-}$ **(c)** $[Co(H_2O)_6]^{2+}$

STRATEGY

All three complexes are octahedral, so the energy-level diagrams will show three lower-energy and two higher-energy *d* orbitals. For d^1–d^3 and d^8–d^{10} complexes, the electrons occupy the orbitals in accord with Hund's rule so as to give the maximum number of unpaired electrons. For d^4–d^7 complexes, the orbital occupancy and number of unpaired electrons depend on the position of the ligand in the spectrochemical series.

SOLUTION

(a) Cr^{3+} ([Ar] $3d^3$) has three unpaired electrons. In the complex, they occupy the lower-energy set of *d* orbitals as shown below.

(b) Mn^{3+} ([Ar] $3d^4$) can have a high-spin or a low-spin configuration. Because CN^- is a strong-field ligand, all four *d* electrons go into the lower-energy *d* orbitals. The complex is low-spin, with two unpaired electrons.

(c) Co^{2+} ([Ar] $3d^7$) has a high-spin configuration with three unpaired electrons because H_2O is a weak-field ligand.

In the following orbital energy-level diagrams, the relative values of the crystal field splitting Δ agree with the positions of the ligands in the spectrochemical series (H_2O < en < CN^-):

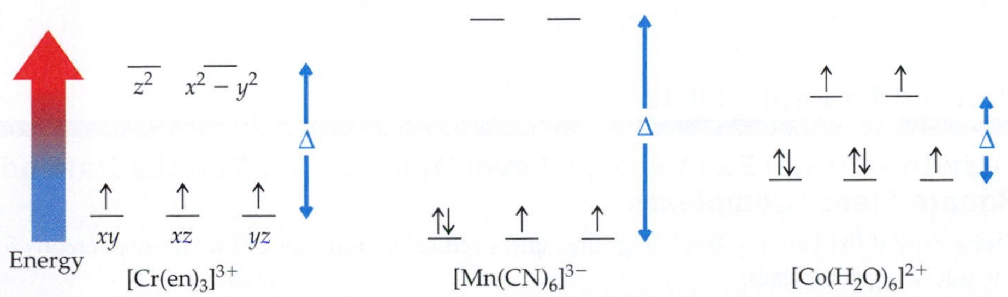

PROBLEM 20.18 Draw a crystal field *d*-orbital energy-level diagram, and predict the number of unpaired electrons for each of the following complexes:

(a) $[Fe(H_2O)_6]^{2+}$ **(b)** $[Fe(CN)_6]^{4-}$

(c) $[VF_6]^{3-}$ **(d)** $[Ni(en)_3]^{2+}$

Tetrahedral and Square Planar Complexes

Different geometric arrangements of the ligands give different energy splittings for the *d* orbitals. **Figure 20.32** shows *d*-orbital energy-level diagrams for tetrahedral and square planar complexes.

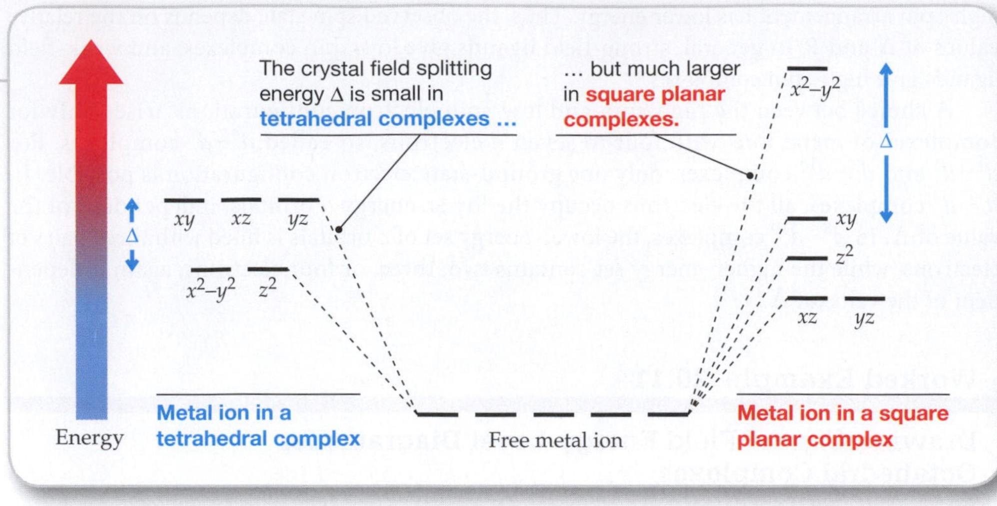

FIGURE 20.32

Energies of the *d* orbitals in tetrahedral and square planar complexes relative to their energy in the free metal ion.

The splitting pattern in tetrahedral complexes is just the opposite of that in octahedral complexes; that is, the *xy*, *xz*, and *yz* orbitals have higher energy than the $x^2 - y^2$ and z^2 orbitals. (As with octahedral complexes, the energy ordering follows from the relative orientation of the orbital lobes and the ligands, but we won't derive the result.) Because none of the orbitals points directly at the ligands in tetrahedral geometry and because there are only four ligands instead of six, the crystal field splitting in tetrahedral complexes is only about half of that in octahedral complexes. Consequently, Δ is almost always smaller than the spin-pairing energy *P*, and nearly all tetrahedral complexes are high-spin.

Square planar complexes look like octahedral ones (Figure 20.30) except that the two trans ligands along the *z* axis are missing. In square planar complexes, the $x^2 - y^2$ orbital is high in energy (Figure 20.32) because it points directly at all four ligands, which lie along the *x* and *y* axes. The splitting pattern is more complicated than for octahedral and tetrahedral complexes, but the main point to remember is that a large energy gap exists between the $x^2 - y^2$ orbital and the four lower-energy orbitals. Square planar geometry is most common for metal ions with electron configuration d^8 because this configuration favors low-spin complexes in which all four lower-energy orbitals are filled and the higher-energy $x^2 - y^2$ orbital is vacant. Common examples are $[Ni(CN)_4]^{2-}$, $[PdCl_4]^{2-}$, and $Pt(NH_3)_2Cl_2$.

Worked Example 20.12

Drawing Crystal Field Energy-Level Diagrams for Tetrahedral and Square Planar Complexes

Draw crystal field energy-level diagrams, and predict the number of unpaired electrons for the following complexes:

(a) $[FeCl_4]^-$ (tetrahedral) **(b)** $[PtCl_4]^{2-}$ (square planar)

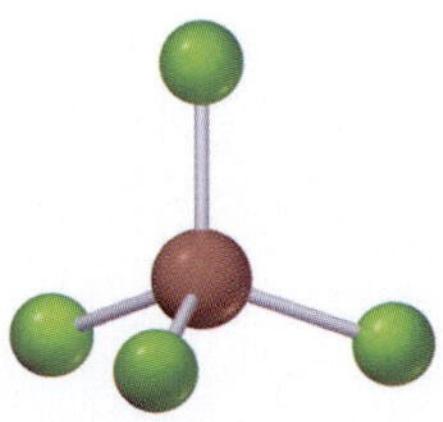

$[FeCl_4]^-$

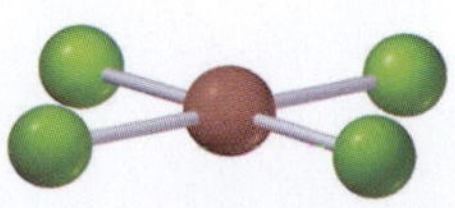

$[PtCl_4]^{2-}$

STRATEGY

Begin with the energy-level diagrams in Figure 20.32, and remember that nearly all tetrahedral complexes are high-spin (small Δ) while square planar complexes have a large energy gap (large Δ) between the $x^2 - y^2$ orbital and the four lower-energy orbitals.

SOLUTION

(a) The five *d* electrons of Fe^{3+} ([Ar] $3d^5$) are distributed between the higher- and lower-energy orbitals as shown below. Because Δ is small in tetrahedral complexes, $[FeCl_4]^-$ is high-spin with five unpaired electrons.

(b) Pt^{2+} ([Xe] $4f^{14}\,5d^8$) has eight *d* electrons. Because Δ is large in square planar complexes, all the electrons occupy the four lower-energy *d* orbitals. There are no unpaired electrons, and the complex is diamagnetic.

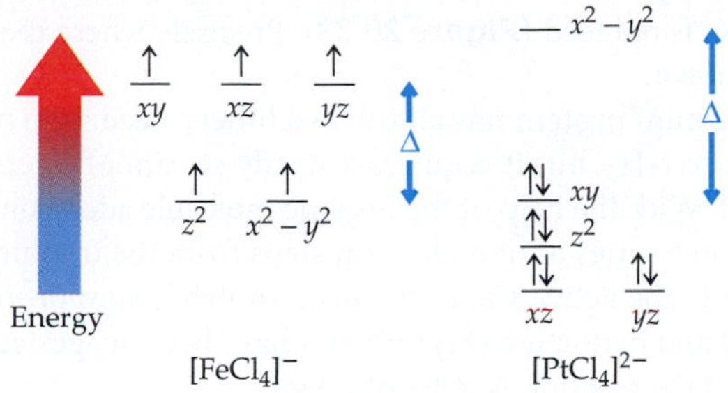

PROBLEM 20.19 Draw a crystal field energy-level diagram, and predict the number of unpaired electrons for the following complexes:
(a) $[NiCl_4]^{2-}$ (tetrahedral) **(b)** $[Ni(CN)_4]^{2-}$ (square planar)

FYI HOW DO LIVING THINGS ACQUIRE NITROGEN?

As a constituent of DNA, RNA, and proteins, nitrogen is an essential element for every life form. Although elemental nitrogen (N_2) makes up some 78% of our atmosphere, it is not directly usable by plants and animals because its extremely strong N≡N triple bond prevents easy conversion to nitrogen compounds that living organisms can use. Elemental nitrogen does engage in some chemical reactions, but overcoming the high activation energy barrier for breaking its triple bond requires an electric discharge, elevated temperature and pressure, or a metal catalyst.

Converting elemental nitrogen into simple nitrogen compounds that plants and animals can use is referred to as "fixing nitrogen." Examples of compounds made by fixing nitrogen include ammonia, nitrites, nitrates, and organic compounds such as urea (H_2NCONH_2).

Nitrogen is fixed industrially by the Haber process (Section 13.7) in which elemental N_2 and H_2 are converted to ammonia at 400–500 °C and 130–300 atm pressure in the presence of an iron-containing catalyst. Produced on an immense scale—about 131 million metric tons per year worldwide—ammonia is used in cleaning products and chemical manufacture, but its most important use is as a fertilizer and as a starting material for synthesis of other nitrogen fertilizers such as urea. Global food production today is estimated to be double what it would be without the ammonia produced by the Haber process.

▲ Ammonia is injected directly into a field from tanks containing liquid ammonia under pressure.

In addition to the Haber process, an equally important source of ammonia for agriculture is biological nitrogen fixation, a natural process in which certain bacteria—such as those of the genus *Rhizobium* found in the root nodules of leguminous plants like peas, beans, and clover—fix atmospheric N_2. Plants harboring these bacteria will grow in nitrogen-depleted soil and can be plowed under to serve as natural fertilizer for another crop. Whereas the Haber synthesis of ammonia requires high temperature and pressure, biological nitrogen fixation occurs under ambient conditions.

How do *Rhizobia* accomplish nitrogen fixation? The bacteria contain an enzyme complex called a *nitrogenase*, which catalyzes the reduction of N_2 to NH_3:

$$N_2 + 6\,H^+ + 6\,e^- \longrightarrow 2\,NH_3$$

Nitrogenases contain iron along with another transition metal such as molybdenum or vanadium. The well-studied molybdenum nitrogenase consists of two proteins: an iron protein

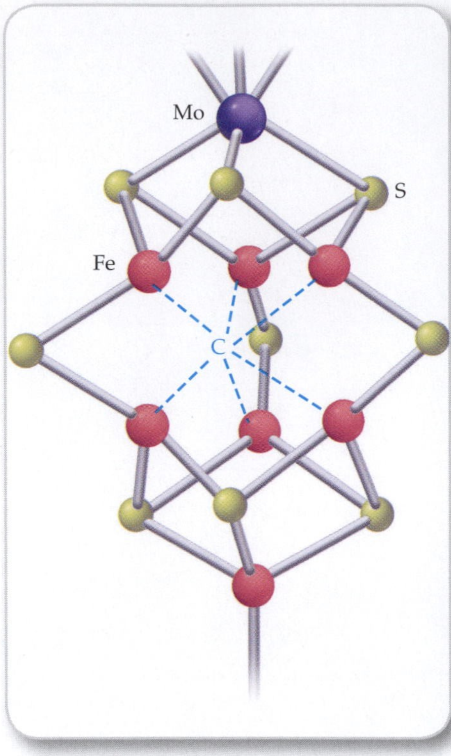

FIGURE 20.33

The Fe_7MoS_9 active site of nitrogenase as determined by X-ray crystallography (Section 10.7).

containing a single Fe_4S_4 cluster with a tetrahedral iron environment and an iron–molybdenum protein containing Fe_7MoS_9 clusters with a tetrahedral environment for the iron and an octahedral environment for the molybdenum. The Mo atom is surrounded by three sulfur ligands and three other ligand donor atoms, and the Fe_7MoS_9 cluster is believed to be the active site where the N_2 is reduced (**Figure 20.33**). Precisely where the N_2 molecule binds in the active site is not known.

The iron–molybdenum protein has all the machinery needed to break the strong N≡N bond of N_2 and produce NH_3, but it requires a steady stream of electrons—six electrons to break each N≡N bond. With the help of the organic molecule adenosine triphosphate (ATP), electrons are pumped in a series of one-electron steps from the iron protein through a chain of iron–sulfur clusters to the active site of the iron–molybdenum protein. Compounds such as diimide (HN=NH) and hydrazine ($H_2N–NH_2$) have been suggested as intermediates, but the exact mechanism of the reaction is still unknown.

PROBLEM 20.20 Assume that hydrogenation of N_2 occurs in the following steps:

$$N_2(g) + H_2(g) \longrightarrow HN{=}NH(g) \qquad \Delta G°_1 = ?$$
$$HN{=}NH(g) + H_2(g) \longrightarrow H_2N{-}NH_2(g) \qquad \Delta G°_2 = ?$$
$$H_2N{-}NH_2(g) + H_2(g) \longrightarrow 2\,NH_3(g) \qquad \Delta G°_3 = ?$$

Use the data in Appendix B to calculate $\Delta G°$ for each step ($\Delta G°_f$ for HN=NH(g) is +243.8 kJ/mol). Which step is the most favorable under standard-state conditions?

PROBLEM 20.21 Draw the two diastereoisomers of diimide, HN=NH.

SUMMARY

Transition elements, or *d-block elements*, are the metallic elements in the central part of the periodic table. Most of the neutral atoms have a valence electron configuration $(n-1)d^{1-10}\,ns^2$, and the cations have a configuration $(n-1)d^{0-10}$. Transition metals exhibit a variety of oxidation states. Ions with the metal in a high oxidation state tend to be good oxidizing agents ($Cr_2O_7^{2-}$, MnO_4^-), while ions with an early transition metal in a low oxidation state are good reducing agents (V^{2+}, Cr^{2+}).

Coordination compounds (metal complexes) are compounds in which a central metal ion is attached to a group of surrounding **ligands** by coordinate covalent bonds. Ligands can be **monodentate** or **polydentate**, depending on the number of **ligand donor atoms** attached to the metal. Polydentate ligands are also called **chelating agents**. They form complexes known as metal **chelates** that contain rings of atoms known as **chelate rings**.

The number of ligand donor atoms bonded to a metal is called the **coordination number** of the metal. Common coordination numbers and geometries are 2 (linear), 4 (tetrahedral or square planar), and 6 (octahedral). Systematic names for complexes specify the number of ligands of each particular type, the metal, and its oxidation state.

Many complexes exist as **isomers**, compounds that have the same formula but a different arrangement of the constituent atoms. **Constitutional isomers**, such as **linkage isomers** and **ionization isomers**, have different connections between their constituent atoms. **Stereoisomers** have the same connections but a different spatial arrangement of the atoms. **Diastereoisomers** are non-mirror-image stereoisomers, such as the **cis** and **trans isomers** of square planar MA_2B_2 and octahedral MA_4B_2 complexes. **Enantiomers** are mirror-image stereoisomers, such as "right-handed" and "left-handed" $[Co(en)_3]^{3+}$. One enantiomer rotates the plane of plane-polarized light to the right, and the other rotates the plane through an equal angle but in the opposite direction. A 1:1 mixture of the two enantiomers is called a **racemic mixture**. Molecules that have handedness are said to be **chiral**. All other molecules are **achiral**.

Valence bond theory describes the bonding in complexes in terms of two-electron, coordinate covalent bonds resulting from the overlap of filled ligand orbitals with vacant metal hybrid orbitals that point in the direction of the ligands: *sp* (linear), sp^3 (tetrahedral), dsp^2 (square planar), and d^2sp^3 or sp^3d^2 (octahedral).

Crystal field theory assumes that the metal–ligand bonding is entirely ionic. Because of electrostatic repulsions between the *d* electrons and the ligands, the *d* orbitals are raised in energy and are differentiated by an energy separation called the **crystal field splitting**, Δ. In octahedral complexes, the d_{z^2} and $d_{x^2-y^2}$ orbitals have higher energy than the d_{xy}, d_{xz}, and d_{yz} orbitals. Tetrahedral and square planar complexes exhibit different splitting patterns. The colors of complexes are due to electronic transitions from one set of *d* orbitals to another, and the transition energies depend on the position of the ligand in the **spectrochemical series**. **Weak-field ligands** give small Δ values, and **strong-field ligands** give large Δ values. Crystal field theory accounts for the magnetic properties of complexes in terms of the relative values of Δ and the spin-pairing energy *P*. Small Δ values favor **high-spin complexes**, and large Δ values favor **low-spin complexes**.

KEY WORDS

absorption spectrum *845*
achiral *842*
bidentate ligand *830*
chelate *831*
chelate ring *831*
chelating agent *831*
chiral *842*
cis isomer *837*
constitutional isomer *836*
coordination compound *828*
coordination number *829*
crystal field splitting (Δ) *851*
crystal field theory *850*
diastereoisomer *837*
enantiomer *842*
hexadentate ligand *830*
high-spin complex *849*
ionization isomer *837*
isomer *836*
lanthanide contraction *820*
ligand *829*
ligand donor atom *829*
linkage isomer *836*
low-spin complex *849*
metal complex *829*
monodentate ligand *830*
polydentate ligand *830*
racemic mixture *842*
spectrochemical series *852*
stereoisomer *837*
strong-field ligand *852*
symmetry plane *842*
trans isomer *837*
weak-field ligand *852*

KEY EQUATION

Relationship between the crystal field splitting energy Δ and the wavelength of light λ absorbed in a *d–d* electronic transition (Section 20.12)

$$\Delta = h\nu = \frac{hc}{\lambda}$$

where h is Planck's constant, ν is the frequency of the light, and c is the speed of light.

LEARNING OUTCOMES

- Relate the electron configuration of a transition metal atom or ion and its location in the periodic table. (Section 20.1) ***Problems 1, 2, 22, 23, 32–35***
- Account for the properties of transition metals in terms of their location in the periodic table. (Section 20.2) ***Problems 24, 36–43***
- Know the oxidation states of first-series transition metals, and relate the oxidizing and reducing strength of a transition metal ion to its oxidation state and location in the periodic table. (Section 20.3) ***Problems 46–55***
- Predict the products and write balanced equations for the reactions of first-series transition metals, their ions, and their compounds. (Sections 20.2, 20.4) ***Problems 44, 45, 56–59, 61–67, 116, 117, 141***
- Determine the formula of a coordination complex, identify the oxidation state and coordination number of the metal atom, and draw the structure of the complex. (Sections 20.5, 20.6) ***Problems 6, 7, 27, 71–79, 136, 137***
- Determine the systematic name of a coordination compound from its formula, and write the formula of a coordination compound from its name. (Section 20.7) ***Problems 9, 10, 26, 80–85, 136, 137***
- Identify and draw the structures of isomers of coordination complexes—constitutional isomers, diastereoisomers, and enantiomers. (Sections 20.8, 20.9) ***Problems 11, 12, 14, 15, 28, 29, 86–94, 119, 127–130, 134, 135, 139, 144, 145***
- Relate the color of a metal complex, the wavelength of light it absorbs, and the energy difference between its ground and excited state. (Section 20.10) ***Problems 16, 96, 97, 126, 131, 138***
- Describe the bonding in a metal complex in terms of valence bond theory, including orbital diagrams, the hybrid orbitals used by the metal atom, and the number of unpaired electrons. (Section 20.11) ***Problems 17, 98–101, 121, 142***
- Draw a crystal field *d*-orbital energy-level diagram for a metal complex, and predict (or account for) its color and magnetic properties in terms of crystal field theory. (Section 20.12) ***Problems 18, 19, 30, 102, 104–115, 120–126, 132, 133, 136–139, 141–143, 145, 147***

CONCEPTUAL PROBLEMS

Problems 20.1–20.21 appear within the chapter.

20.22 Locate on the periodic table the transition elements with the following electron configurations. Identify each element.

(a) [Ar] $3d^7\,4s^2$ **(b)** [Ar] $3d^5\,4s^1$
(c) [Kr] $4d^2\,5s^2$ **(d)** [Xe] $4f^3\,6s^2$

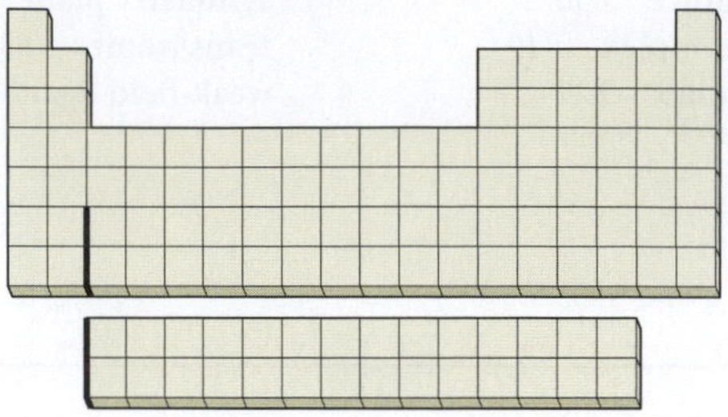

20.23 Look at the location in the periodic table of elements A, B, C, and D.

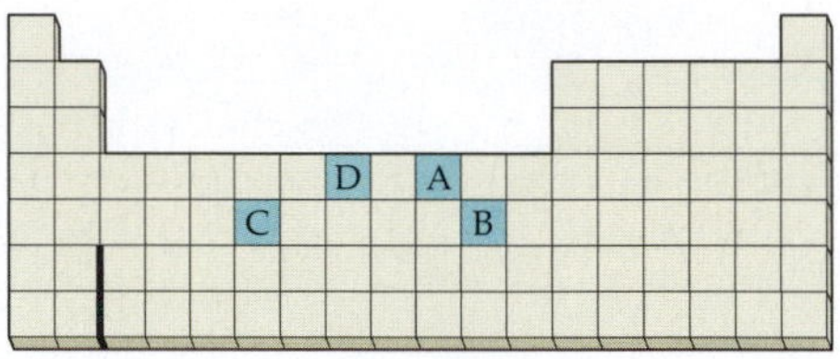

What is the electron configuration of the transition metal in each of the following ions?

(a) A^{2+} **(b)** B^{+} **(c)** C^{3+} **(d)** $DO_4^{\,2-}$

20.24 What is the general trend in the following properties from left to right across the first transition series (Sc to Zn)? Explain each trend.

(a) Atomic radius **(b)** Density
(c) Ionization energy **(d)** Standard oxidation potential

Sc	Ti	V	Cr	Mn	Fe	Co	Ni	Cu	Zn

20.25 Classify the following ligands as monodentate, bidentate, or tridentate. Which can form chelate rings?

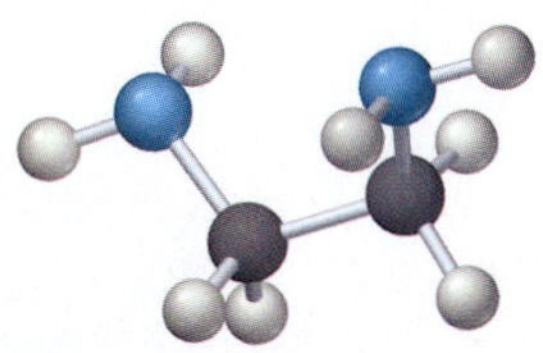

(a) $NH_2CH_2CH_2NH_2$

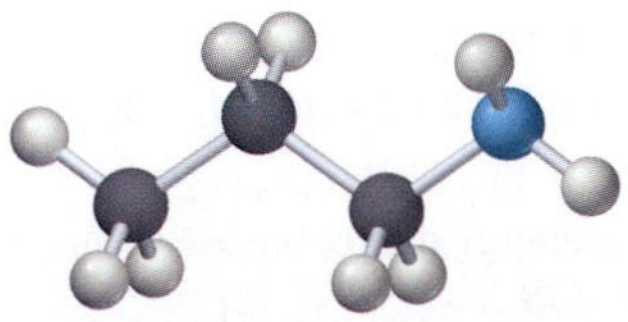

(b) $CH_3CH_2CH_2NH_2$

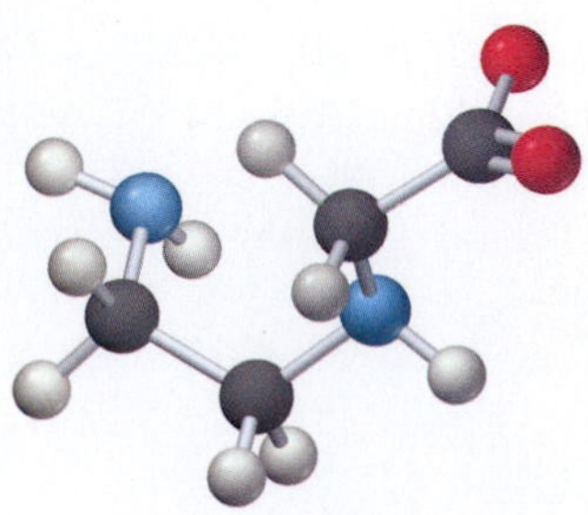

(c) $NH_2CH_2CH_2NHCH_2CO_2^-$

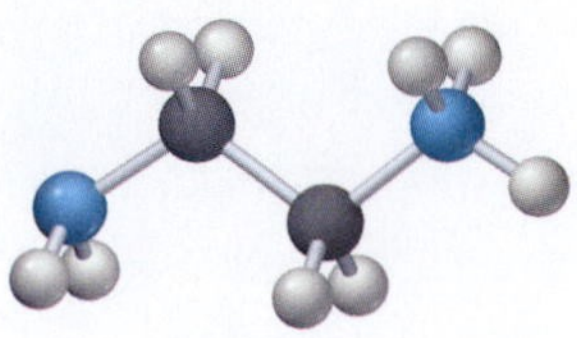

(d) $NH_2CH_2CH_2NH_3^+$

20.26 What is the systematic name for each of the following molecules or ions? Include *cis* or *trans* prefixes for diastereoisomers. Platinum is Pt(II) in square planar complexes and Pt(IV) in octahedral complexes.

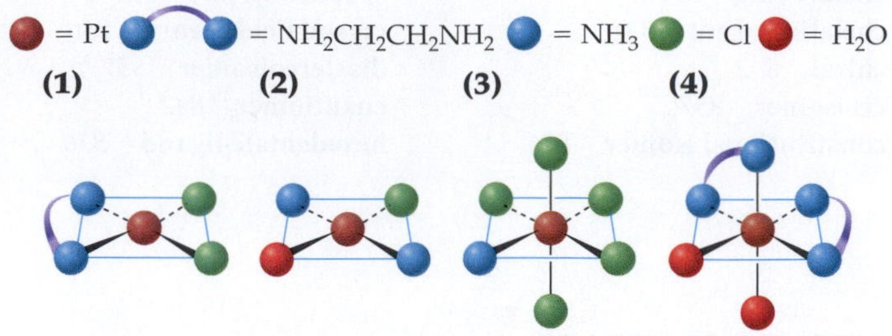

20.27 Draw the structure of the following complexes. What are the oxidation state, coordination number, and coordination geometry of the metal in each?

(a) $Na[Au(CN)_2]$ **(b)** $[Co(NH_3)_5Br]SO_4$
(c) $Pt(en)Cl_2$ **(d)** $(NH_4)_2[PtCl_2(C_2O_4)_2]$

20.28 Consider the following isomers of $[Cr(NH_3)_2Cl_4]^-$

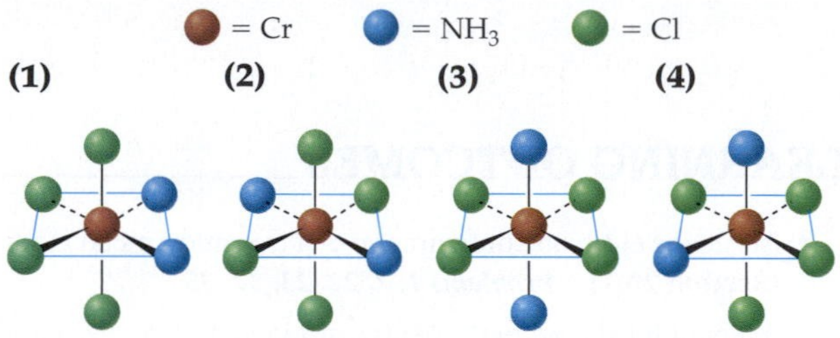

(a) Label the isomers as cis or trans.
(b) Which isomers are identical, and which are different?
(c) Do any of these isomers exist as enantiomers? Explain.

20.29 Consider the following ethylenediamine complexes:

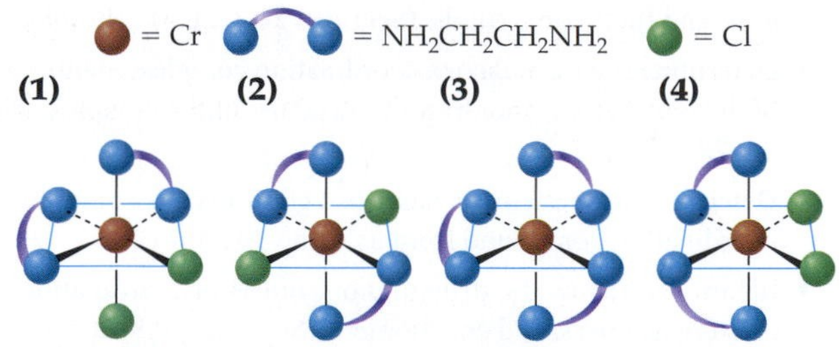

(a) Which complexes are chiral, and which are achiral?
(b) Draw the enantiomer of each chiral complex.
(c) Which, if any, of the chiral complexes are enantiomers of one another?

20.30 Predict the crystal field energy-level diagram for a square pyramidal ML_5 complex that has two ligands along the $\pm x$ and $\pm y$ axes but only one ligand along the z axis. Your diagram should be intermediate between those for an octahedral ML_6 complex and a square planar ML_4 complex.

20.31 Imagine two complexes, one tetrahedral and one square planar, in which the central atom is bonded to four different ligands (shown here in four different colors). Is either complex chiral? Explain.

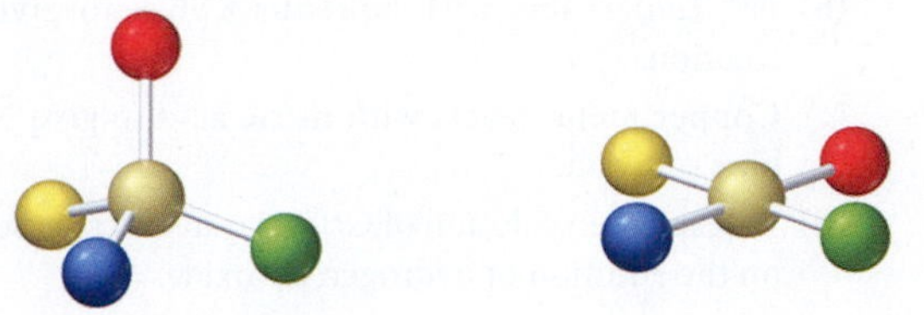

SECTION PROBLEMS

Electron Configurations and Properties of Transition Elements (Sections 20.1, 20.2)

20.32 Use the periodic table to give the electron configuration for each of the following atoms and ions:
(a) Cr (b) Zr (c) Co^{2+}
(d) Fe^{3+} (e) Mo^{3+} (f) Cr(VI) in CrO_4^{2-}

20.33 Specify the electron configuration for each of the following atoms and ions:
(a) Mn (b) Cd (c) Cr^{3+}
(d) Ag^{+} (e) Rh^{3+} (f) Mn(VI) in MnO_4^{2-}

20.34 Predict the number of unpaired electrons for each of the following:
(a) Cu^{2+} (b) Ti^{2+}
(c) Zn^{2+} (d) Cr^{3+}

20.35 Predict the number of unpaired electrons for each of the following:
(a) Sc^{3+} (b) Co^{2+}
(c) Mn^{3+} (d) Cr^{2+}

20.36 Titanium, used to make jet aircraft engines, is much harder than potassium or calcium. Explain.

20.37 Molybdenum (mp 2623 °C) has a higher melting point than yttrium (mp 1522 °C) or cadmium (321 °C). Explain.

20.38 Briefly account for each of the following observations:
(a) Atomic radii decrease in the order Sc > Ti > V.
(b) Densities increase in the order Ti < V < Cr.

20.39 Arrange the following atoms in order of decreasing atomic radius, and account for the trend:
(a) Cr (b) Ti
(c) Mn (d) V

20.40 What is the lanthanide contraction, and why does it occur?

20.41 The atomic radii of zirconium and hafnium are nearly identical. Explain.

20.42 Calculate the sum of the first two ionization energies for the first-series transition elements, and account for the general trend.

20.43 What is the general trend in standard potentials for the oxidation of first-series transition metals from Sc to Zn? What is the reason for the trend?

20.44 Write a balanced net ionic equation for the reaction of each of the following metals with hydrochloric acid in the absence of air. If no reaction occurs, indicate N.R.
(a) Cr (b) Zn
(c) Cu (d) Fe

20.45 Write a balanced net ionic equation for the reaction of each of the following metals with dilute sulfuric acid in the absence of air. If no reaction occurs, indicate N.R.
(a) Mn (b) Ag
(c) Sc (d) Ni

Oxidation States (Section 20.3)

20.46 Which of the following metals have more than one oxidation state?
(a) Zn (b) Mn
(c) Sr (d) Cu

20.47 Which of the following metals have only one oxidation state?
(a) V (b) Al
(c) Co (d) Sc

20.48 What is the highest oxidation state for each of the elements from Sc to Zn?

20.49 The highest oxidation state for the early transition metals Sc, Ti, V, Cr, and Mn is the periodic group number. The highest oxidation state for the later transition elements Fe, Co, and Ni is less than the periodic group number. Explain.

20.50 Which is the stronger oxidizing agent, Cr^{2+} or Cu^{2+}? Explain.

20.51 Which is the stronger reducing agent, Ti^{2+} or Ni^{2+}? Explain.

20.52 Do you expect a compound with vanadium in the +2 oxidation state to be an oxidizing or a reducing agent? Explain.

20.53 Will a compound that contains a Co^{3+} ion be an oxidizing agent or a reducing agent? Explain.

20.54 Arrange the following substances in order of increasing strength as an oxidizing agent, and account for the trend:
(a) Mn^{2+} (b) MnO_2 (c) MnO_4^{-}

20.55 Arrange the following ions in order of increasing strength as a reducing agent, and account for the trend:
(a) Cr^{2+} (b) Cr^{3+} (c) $Cr_2O_7^{2-}$

Chemistry of Selected Transition Elements (Section 20.4)

20.56 Write a balanced equation for the industrial production of:
(a) Chromium metal from chromium(III) oxide
(b) Copper metal from copper(I) sulfide

20.57 Write a balanced net ionic equation for the reaction of nitric acid with:
(a) Iron metal
(b) Copper metal
(c) Chromium metal

20.58 Choose the chromium hydroxide that is amphoteric, and write a balanced net ionic equation for its reaction with hydroxide ion. What is the oxidation state of the chromium and the color of the product?
(a) $CrO_2(OH)_2$
(b) $Cr(OH)_3$
(c) $Cr(OH)_2$

20.59 Explain how $Cr(OH)_3$ can act both as an acid and as a base.

20.60 Which of the following compounds is amphoteric?
(a) $Cr(OH)_2$
(b) $Fe(OH)_2$
(c) $Cr(OH)_3$
(d) $Fe(OH)_3$

20.61 Which of the following ions disproportionates in aqueous solution? Write a balanced net ionic equation for the reaction.
(a) Cr^{3+} (b) Fe^{3+}
(c) Cu^{+} (d) Cu^{2+}

20.62 How might you separate the following pairs of ions by the addition of a single reagent? Include formulas for the major products of the reactions.
(a) Fe^{3+} and Na^{+}
(b) Cr^{3+} and Fe^{3+}
(c) Fe^{3+} and Cu^{2+}

20.63 Give a method for separating the following pairs of ions by the addition of a single reagent. Include formulas for the major products of the reactions.
(a) K^{+} and Cu^{2+}
(b) Cu^{2+} and Cr^{3+}
(c) Fe^{3+} and Al^{3+}

20.64 Complete and balance the net ionic equation for each of the following reactions in acidic solution:
(a) $Cr_2O_7^{2-}(aq) + Fe^{2+}(aq) \longrightarrow ?$
(b) $Fe^{2+}(aq) + O_2(g) \longrightarrow ?$
(c) $Cu_2O(s) + H^{+}(aq) \longrightarrow ?$
(d) $Fe(s) + H^{+}(aq) \longrightarrow ?$

20.65 Complete and balance the net ionic equation for each of the following reactions in acidic solution:
(a) $Cr^{2+}(aq) + Cr_2O_7^{2-}(aq) \longrightarrow ?$
(b) $Cu(s) + \text{conc } HNO_3(aq) \longrightarrow ?$
(c) $Cu^{2+}(aq) + \text{excess } NH_3(aq) \longrightarrow ?$
(d) $Cr(OH)_4^{-}(aq) + \text{excess } H^{+}(aq) \longrightarrow ?$

20.66 Write a balanced net ionic equation for each of the following reactions:
(a) A CrO_4^{2-} solution turns from yellow to orange upon the addition of acid.
(b) $Fe^{3+}(aq)$ reacts with aqueous KSCN to give a deep red solution.
(c) Copper metal reacts with nitric acid to give NO gas and a blue solution.
(d) A deep green solution of $Cr(OH)_3$ in excess base turns yellow on the addition of hydrogen peroxide.

20.67 Write a balanced net ionic equation for each of the following reactions:
(a) A $CuSO_4$ solution becomes dark blue when excess ammonia is added.
(b) A solution of $Na_2Cr_2O_7$ turns from orange to yellow on the addition of base.
(c) When base is added to a solution of $Fe(NO_3)_3$, a red-brown precipitate forms.
(d) The dissolution of CuS in hot HNO_3 gives NO gas, a blue solution, and a yellow solid.

Coordination Compounds; Ligands (Sections 20.5, 20.6)

20.68 Forming $[Ni(en)_3]^{2+}$ from Ni^{2+} and ethylenediamine is a Lewis acid–base reaction. Explain.

20.69 Identify the Lewis acid and the Lewis base in the reaction of oxalate ions ($C_2O_4^{2-}$) with Fe^{3+} to give $[Fe(C_2O_4)_3]^{3-}$.

20.70 Give an example of a coordination compound in which the metal exhibits a coordination number of:
(a) 2 (b) 4 (c) 6

20.71 What is the coordination number of the metal in each of the following complexes?
(a) $AgCl_2^{-}$
(b) $[Cr(H_2O)_5Cl]^{2+}$
(c) $[Co(NCS)_4]^{2-}$
(d) $[ZrF_8]^{4-}$
(e) $Co(NH_3)_3(NO_2)_3$
(f) $[Fe(EDTA)(H_2O)]^{-}$

20.72 What is the oxidation state of the metal in each of the complexes in Problem 20.71?

20.73 Identify the oxidation state of the metal in each of the following complexes:
(a) $[Ni(CN)_5]^{3-}$
(b) $Ni(CO)_4$
(c) $[Co(en)_2(H_2O)Br]^{2+}$
(d) $[Cu(H_2O)_2(C_2O_4)_2]^{2-}$

20.74 What is the formula of a complex that has each of the following geometries?
(a) Tetrahedral (b) Linear
(c) Octahedral (d) Square planar

20.75 What is the formula, including the charge, for each of the following complexes?

(a) An iridium(III) complex with three ammonia and three chloride ligands

(b) A chromium(III) complex with two water and two oxalate ligands

(c) A platinum(IV) complex with two ethylenediamine and two thiocyanate ligands

20.76 Draw the structure of the iron oxalate complex $[Fe(C_2O_4)_3]^{3-}$. Describe the coordination geometry, and identify any chelate rings. What are the coordination number and the oxidation number of the iron?

20.77 Draw the structure of the platinum ethylenediamine complex $[Pt(en)_2]^{2+}$. Describe the coordination geometry, and identify any chelate rings. What are the coordination number and the oxidation number of the platinum?

20.78 Identify the oxidation state of the metal in each of the following compounds:

(a) $Co(NH_3)_3(NO_2)_3$

(b) $[Ag(NH_3)_2]NO_3$

(c) $K_3[Cr(C_2O_4)_2Cl_2]$

(d) $Cs[CuCl_2]$

20.79 What is the oxidation state of the metal in each of the following compounds?

(a) $(NH_4)_3[RhCl_6]$

(b) $[Cr(NH_3)_4(SCN)_2]Br$

(c) $[Cu(en)_2]SO_4$

(d) $Na_2[Mn(EDTA)]$

Naming Coordination Compounds (Section 20.7)

20.80 What is the systematic name for each of the following ions?

(a) $[MnCl_4]^{2-}$

(b) $[Ni(NH_3)_6]^{2+}$

(c) $[Co(CO_3)_3]^{3-}$

(d) $[Pt(en)_2(SCN)_2]^{2+}$

20.81 Assign a systematic name to each of the following ions:

(a) $[AuCl_4]^-$

(b) $[Fe(CN)_6]^{4-}$

(c) $[Fe(H_2O)_5NCS]^{2+}$

(d) $[Cr(NH_3)_2(C_2O_4)_2]^-$

20.82 What is the systematic name for each of the following coordination compounds?

(a) $Cs[FeCl_4]$

(b) $[V(H_2O)_6](NO_3)_3$

(c) $[Co(NH_3)_4Br_2]Br$

(d) $Cu(gly)_2$

20.83 What is the systematic name for each of the following compounds?

(a) $[Cu(NH_3)_4]SO_4$

(b) $Cr(CO)_6$

(c) $K_3[Fe(C_2O_4)_3]$

(d) $[Co(en)_2(NH_3)CN]Cl_2$

20.84 Write the formula for each of the following compounds:

(a) Tetraammineplatinum(II) chloride

(b) Sodium hexacyanoferrate(III)

(c) Tris(ethylenediamine)platinum(IV) sulfate

(d) Triamminetrithiocyanatorhodium(III)

20.85 Write the formula for each of the following compounds:

(a) Diamminesilver(I) nitrate

(b) Potassium diaquadioxalatocobaltate(III)

(c) Hexacarbonylmolybdenum(0)

(d) Diamminebis(ethylenediamine)chromium(III) chloride

Isomers (Sections 20.8, 20.9)

20.86 Draw all constitutional isomers of $[Ru(NH_3)_5(NO_2)]Cl$. Label the isomers as linkage isomers or ionization isomers.

20.87 There are six possible isomers for a square planar palladium(II) complex that contains two Cl^- and two SCN^- ligands. Sketch the structures of all six, and label them according to the isomer classification scheme in Figure 20.16.

20.88 Which of the following complexes can exist as diastereoisomers?

(a) $[Cr(NH_3)_2Cl_4]^-$

(b) $[Co(NH_3)_5Br]^{2+}$

(c) $[FeCl_2(NCS)_2]^{2-}$ (tetrahedral)

(d) $[PtCl_2Br_2]^{2-}$ (square planar)

20.89 Tell how many diastereoisomers are possible for each of the following complexes, and draw their structures:

(a) $Pt(NH_3)_2(CN)_2$

(b) $[Co(en)(SCN)_4]^-$

(c) $[Cr(H_2O)_4Cl_2]^+$

(d) $Ru(NH_3)_3I_3$

20.90 Which of the following complexes are chiral?

(a) $Pt(en)Cl_2$

(b) *cis*-$[Co(NH_3)_4Br_2]^+$

(c) *cis*-$[Cr(en)_2(H_2O)_2]^{3+}$

(d) $[Cr(C_2O_4)_3]^{3-}$

20.91 Which of the following complexes can exist as enantiomers? Draw their structures.

(a) $[Cr(en)_3]^{3+}$

(b) *cis*-$[Co(en)_2(NH_3)Cl]^{2+}$

(c) *trans*-$[Co(en)_2(NH_3)Cl]^{2+}$

(d) $[Pt(NH_3)_3Cl_3]^+$

20.92 Draw all possible diastereoisomers of $[Cr(C_2O_4)_2(H_2O)_2]^-$. Which can exist as a pair of enantiomers?

20.93 Draw the three possible diastereoisomers of the triethylenetetramine complex $[Co(trien)Cl_2]^+$. Abbreviate the flexible tetradentate trien ligand $H_2NCH_2CH_2NHCH_2CH_2NHCH_2CH_2NH_2$ as N⌒N⌒N⌒N. Which of the isomers can exist as a pair of enantiomers?

20.94 How does plane-polarized light differ from ordinary light? Draw the structure of a chromium complex that rotates the plane of plane-polarized light.

20.95 What is a racemic mixture? Does it affect plane-polarized light? Explain.

Color of Complexes; Valence Bond and Crystal Field Theories (Sections 20.10–20.12)

20.96 What is an absorption spectrum? If the absorption spectrum of a complex has just one band at 455 nm, what is the color of the complex?

20.97 A red complex has just one absorption band in the visible region of the spectrum. Predict the approximate wavelength of this band.

20.98 Give a valence bond description of the bonding in each of the following complexes. Include orbital diagrams for the free metal ion and the metal ion in the complex. Indicate which hybrid orbitals the metal ion uses for bonding, and specify the number of unpaired electrons.

(a) $[Ti(H_2O)_6]^{3+}$
(b) $[NiBr_4]^{2-}$ (tetrahedral)
(c) $[Fe(CN)_6]^{3-}$ (low-spin)
(d) $[MnCl_6]^{3-}$ (high-spin)

20.99 For each of the following complexes, describe the bonding using valence bond theory. Include orbital diagrams for the free metal ion and the metal ion in the complex. Indicate which hybrid orbitals the metal ion uses for bonding, and specify the number of unpaired electrons.

(a) $[AuCl_4]^-$ (square planar)
(b) $[Ag(NH_3)_2]^+$
(c) $[Fe(H_2O)_6]^{2+}$ (high-spin)
(d) $[Fe(CN)_6]^{4-}$ (low-spin)

20.100 There are two possible $[M(OH)_4]^-$ complexes of first-series transition metals that have three unpaired electrons.

(a) What is the oxidation state and the identity of M in these complexes?
(b) Using orbital diagrams, give a valence bond description of the bonding in each complex.
(c) Based on common oxidation states of first-series transition metals (Figure 20.6), which $[M(OH)_4]^-$ complex is more likely to exist?

20.101 Two first-series transition metals have three unpaired electrons in complex ions of the type $[MCl_4]^{2-}$.

(a) What is the oxidation state and the identity of M in these ions?
(b) Draw valence bond orbital diagrams for the two possible ions.
(c) Based on common oxidation states of first-series transition metals (Figure 20.6), which ion is more likely to exist?

20.102 Draw a crystal field energy-level diagram for the 3*d* orbitals of titanium in $[Ti(H_2O)_6]^{3+}$. Indicate the crystal field splitting, and explain why $[Ti(H_2O)_6]^{3+}$ is colored.

20.103 Use a sketch to explain why the d_{xy} and $d_{x^2-y^2}$ orbitals have different energies in an octahedral complex. Which of the two orbitals has higher energy?

20.104 The $[Ti(NCS)_6]^{3-}$ ion exhibits a single absorption band at 544 nm. Calculate the crystal field splitting energy Δ in kJ/mol. Is NCS^- a stronger or weaker field ligand than water? Predict the color of $[Ti(NCS)_6]^{3-}$.

20.105 The $[Cr(H_2O)_6]^{3+}$ ion is violet, and $[Cr(CN)_6]^{3-}$ is yellow. Explain this difference using crystal field theory. Use the colors to order H_2O and CN^- in the spectrochemical series.

20.106 For each of the following complexes, draw a crystal field energy-level diagram, assign the electrons to orbitals, and predict the number of unpaired electrons:

(a) $[CrF_6]^{3-}$
(b) $[V(H_2O)_6]^{3+}$
(c) $[Fe(CN)_6]^{3-}$

20.107 Draw a crystal field energy-level diagram, assign the electrons to orbitals, and predict the number of unpaired electrons for each of the following:

(a) $[Cu(en)_3]^{2+}$
(b) $[FeF_6]^{3-}$
(c) $[Co(en)_3]^{3+}$ (low-spin)

20.108 The $Ni^{2+}(aq)$ cation is green, but $Zn^{2+}(aq)$ is colorless. Explain.

20.109 The $Cr^{3+}(aq)$ cation is violet, but $Y^{3+}(aq)$ is colorless. Explain.

20.110 Weak-field ligands tend to give high-spin complexes, but strong-field ligands tend to give low-spin complexes. Explain.

20.111 Explain why nearly all tetrahedral complexes are high-spin.

20.112 Draw a crystal field energy-level diagram for a square planar complex, and explain why square planar geometry is especially common for d^8 complexes.

20.113 For each of the following complexes, draw a crystal field energy-level diagram, assign the electrons to orbitals, and predict the number of unpaired electrons:

(a) $[Pt(NH_3)_4]^{2+}$ (square planar)
(b) $[MnCl_4]^{2-}$ (tetrahedral)
(c) $[Co(NCS)_4]^{2-}$ (tetrahedral)
(d) $[Cu(en)_2]^{2+}$ (square planar)

CHAPTER PROBLEMS

20.114 Which of the following complexes are paramagnetic?

(a) $[Mn(CN)_6]^{3-}$ **(b)** $[Zn(NH_3)_4]^{2+}$
(c) $[Fe(CN)_6]^{4-}$ **(d)** $[FeF_6]^{4-}$

20.115 Which of the following complexes are diamagnetic?

(a) $[Ni(H_2O)_6]^{2+}$ **(b)** $[Co(CN)_6]^{3-}$
(c) $[HgI_4]^{2-}$ **(d)** $[Cu(NH_3)_4]^{2+}$

20.116 For each of the following reactions in acidic solution, predict the products and write a balanced net ionic equation:

(a) $Co^{3+}(aq) + H_2O(l) \longrightarrow ?$
(b) $Cr^{2+}(aq) + O_2(g) \longrightarrow ?$
(c) $Cu(s) + Cr_2O_7^{2-}(aq) \longrightarrow ?$
(d) $CrO_4^{2-}(aq) + H^+(aq) \longrightarrow ?$

20.117 Write a balanced net ionic equation for each of the following reactions in acidic, basic, or neutral solution:

(a) $MnO_4^-(aq) + C_2O_4^{2-}(aq) \xrightarrow{\text{Acidic}} Mn^{2+}(aq) + CO_2(g)$

(b) $Cr_2O_7^{2-}(aq) + Ti^{3+}(aq) \xrightarrow{\text{Acidic}} Cr^{3+}(aq) + TiO^{2+}(aq)$

(c) $MnO_4^-(aq) + SO_3^{2-}(aq) \xrightarrow{\text{Basic}} MnO_4^{2-}(aq) + SO_4^{2-}(aq)$

(d) $Fe(OH)_2(s) + O_2(g) \xrightarrow{\text{Neutral}} Fe(OH)_3(s)$

20.118 What role does $EDTA^{4-}$ play as a trace additive to mayonnaise? Would the glycinate ion ($H_2NCH_2CO_2^-$) be an effective substitute for $EDTA^{4-}$?

20.119 Tris(2-aminoethyl)amine, abbreviated tren, is the tetradentate ligand $N(CH_2CH_2NH_2)_3$. Using $N\frown N$ to represent each of the three $NCH_2CH_2NH_2$ segments of the ligand, sketch all possible isomers of the octahedral complex $[Co(tren)BrCl]^+$.

20.120 Nickel(II) complexes with the formula NiX_2L_2, where X^- is Cl^- or N-bonded NCS^- and L is the monodentate triphenylphosphine ligand $P(C_6H_5)_3$, can be square planar or tetrahedral.

(a) Draw crystal field energy-level diagrams for a square planar and a tetrahedral nickel(II) complex, and show the population of the orbitals.

(b) If $NiCl_2L_2$ is paramagnetic and $Ni(NCS)_2L_2$ is diamagnetic, which of the two complexes is tetrahedral and which is square planar?

(c) Draw possible structures for each of the NiX_2L_2 complexes, and tell which ones have a dipole moment.

20.121 Describe the bonding in $[Mn(CN)_6]^{3-}$, using both crystal field theory and valence bond theory. Include the appropriate crystal field *d*-orbital energy-level diagram and the valence bond orbital diagram. Which model allows you to predict the number of unpaired electrons? How many do you expect?

20.122 The complex $[FeCl_6]^{3-}$ is more paramagnetic than $[Fe(CN)_6]^{3-}$. Explain.

20.123 Although Cl^- is a weak-field ligand and CN^- is a strong-field ligand, $[CrCl_6]^{3-}$ and $[Cr(CN)_6]^{3-}$ exhibit approximately the same amount of paramagnetism. Explain.

20.124 In octahedral complexes, the choice between high-spin and low-spin electron configurations arises only for d^4–d^7 complexes. Explain.

20.125 Draw a crystal field energy-level diagram, and predict the number of unpaired electrons for each of the following:

(a) $[Mn(H_2O)_6]^{2+}$

(b) $Pt(NH_3)_2Cl_2$

(c) $[FeO_4]^{2-}$

(d) $[Ru(NH_3)_6]^{2+}$ (low-spin)

20.126 Explain why $[CoCl_4]^{2-}$ (blue) and $[Co(H_2O)_6]^{2+}$ (pink) have different colors. Which complex has its absorption bands at longer wavelengths?

20.127 Draw the structure of all isomers of the octahedral complex $[NbX_2Cl_4]^-$ ($X^- = NCS^-$), and identify those that are linkage isomers.

20.128 The glycinate anion, $gly^- = NH_2CH_2CO_2^-$, bonds to metal ions through the N atom and one of the O atoms. Using $N\frown O$ to represent gly^-, sketch the structures of the four stereoisomers of $Co(gly)_3$.

20.129 There are three coordination compounds with the empirical formula $Co(NH_3)_3(NO_2)_3$ in which all the nitrite ions are bonded through the N atom. Two isomers have the same molar mass but different nonzero dipole moments. The third compound is a salt with singly charged ions and a molar mass twice that of the other compounds.

(a) Draw the structures of the isomeric compounds.

(b) What is the chemical formula of the third compound?

20.130 Draw the structures of all possible diastereoisomers of an octahedral complex with the formula $MA_2B_2C_2$. Which of the diastereoisomers, if any, can exist as enantiomers?

20.131 Look at the colors of the isomeric complexes in Figure 20.17 on page 837, and predict which is the stronger field ligand, nitro ($-NO_2$) or nitrito ($-ONO$)? Explain.

20.132 The amount of paramagnetism for a first-series transition metal complex is related approximately to its spin-only magnetic moment. The spin-only value of the magnetic moment in units of Bohr magnetons (BM) is given by $\sqrt{n(n+2)}$, where n is the number of unpaired electrons. Calculate the spin-only value of the magnetic moment for the 2+ ions of the first-series transition metals (except Sc) in octahedral complexes with **(a)** weak-field ligands and **(b)** strong-field ligands. For which electron configurations can the magnetic moment distinguish between high-spin and low-spin electron configurations?

20.133 Predict the crystal field energy-level diagram for a linear ML_2 complex that has two ligands along the $\pm z$ axis:

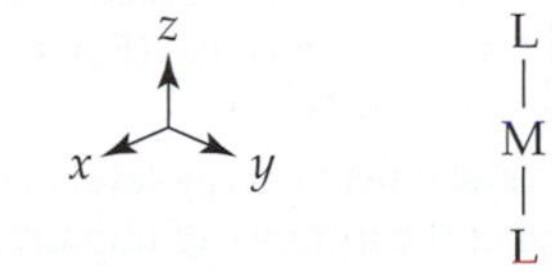

20.134 The reaction of the octahedral complex $Co(NH_3)_3(NO_2)_3$ with HCl yields a complex $[Co(NH_3)_3(H_2O)Cl_2]^+$ in which the two chloride ligands are trans to one another.

(a) Draw the two possible stereoisomers of the starting material $[Co(NH_3)_3(NO_2)_3]$. (All three NO_2^- ligands are bonded to Co through the N atom.)

(b) Assuming that the NH_3 groups remain in place, which of the two starting isomers could give rise to the observed product?

20.135 Consider the octahedral complex $[Co(en)(dien)Cl]^{2+}$, where dien $= H_2NCH_2CH_2NHCH_2CH_2NH_2$, which can be abbreviated $N\frown N\frown N$.

(a) The dien (diethylenetriamine) ligand is a tridentate ligand. Explain what is meant by "tridentate" and why dien can act as a tridentate ligand.

(b) Draw all possible stereoisomers of $[Co(en)(dien)Cl]^{2+}$ (dien is a flexible ligand). Which stereoisomers are chiral, and which are achiral?

20.136 For each of the following, (i) give the systematic name of the compound and specify the oxidation state of the transition metal, (ii) draw a crystal field energy-level diagram and assign the *d* electrons to orbitals, (iii) indicate whether the complex is high-spin or low-spin (for d^4–d^7 complexes), and (iv) specify the number of unpaired electrons.

(a) $(NH_4)[Cr(H_2O)_6](SO_4)_2$

(b) $Mo(CO)_6$ (CO is a strong-field ligand)

(c) $[Ni(NH_3)_4(H_2O)_2](NO_3)_2$

(d) $K_4[Os(CN)_6]$

(e) $[Pt(NH_3)_4](ClO_4)_2$

(f) $Na_2[Fe(CO)_4]$

20.137 The drug Nipride, $Na_2[Fe(CN)_5NO]$, is an inorganic complex used as a source of NO to lower blood pressure during surgery.

(a) The nitrosyl ligand in this complex is believed to be NO^+ rather than neutral NO. What is the oxidation state of iron, and what is the systematic name for $Na_2[Fe(CN)_5NO]$?

(b) Draw a crystal field energy-level diagram for $[Fe(CN)_5NO]^{2-}$, assign the electrons to orbitals, and predict the number of unpaired electrons.

20.138 Based on the colors of their Cr(III) complexes, arrange the following ligands in a spectrochemical series in order of increasing value of the crystal field splitting Δ: $acac^-$ (a bidentate ligand), $CH_3CO_2^-$ (acetate), Cl^-, H_2O, NH_3, and urea. The colors of the Cr(III) complexes are red for $Cr(acac)_3$, violet for $[Cr(H_2O)_6]^{3+}$, green for $[CrCl_2(H_2O)_4]^+$, green for $[Cr(urea)_6]^{3+}$, yellow for $[Cr(NH_3)_6]^{3+}$, and blue-violet for $Cr(CH_3CO_2)_3(H_2O)_3$.

MULTICONCEPT PROBLEMS

20.139 Spinach contains a lot of iron but is not a good source of dietary iron because nearly all the iron is tied up in the oxalate complex $[Fe(C_2O_4)_3]^{3-}$.

(a) The formation constant K_f for $[Fe(C_2O_4)_3]^{3-}$ is 3.3×10^{20}. Calculate the equilibrium concentration of free Fe^{3+} in a 0.100 M solution of $[Fe(C_2O_4)_3]^{3-}$. (Ignore any acid–base reactions.)

(b) Under the acidic conditions in the stomach, the Fe^{3+} concentration should be greater because of the reaction

$$[Fe(C_2O_4)_3]^{3-}(aq) + 6\,H_3O^+(aq) \rightleftharpoons Fe^{3+}(aq) + 3\,H_2C_2O_4\,(aq) + 6\,H_2O(l)$$

Show, however, that this reaction is nonspontaneous under standard-state conditions. (For $H_2C_2O_4$, $K_{a1} = 5.9 \times 10^{-2}$ and $K_{a2} = 6.4 \times 10^{-5}$.)

(c) Draw a crystal field energy-level diagram for $[Fe(C_2O_4)_3]^{3-}$, and predict the number of unpaired electrons. ($C_2O_4^{2-}$ is a weak-field bidentate ligand.)

(d) Draw the structure of $[Fe(C_2O_4)_3]^{3-}$. Is the complex chiral or achiral?

20.140 Formation constants for the ammonia and ethylenediamine complexes of nickel(II) indicate that $Ni(en)_3^{2+}$ is much more stable than $Ni(NH_3)_6^{2+}$:

$$(1)\ Ni(H_2O)_6^{2+}(aq) + 6\,NH_3(aq) \rightleftharpoons Ni(NH_3)_6^{2+}(aq) + 6\,H_2O(l) \qquad K_f = 2.0 \times 10^8$$

$$(2)\ Ni(H_2O)_6^{2+}(aq) + 3\,en(aq) \rightleftharpoons Ni(en)_3^{2+}(aq) + 6\,H_2O(l) \qquad K_f = 4 \times 10^{17}$$

The enthalpy changes for the two reactions, $\Delta H°_1$ and $\Delta H°_2$, should be about the same because both complexes have six Ni–N bonds.

(a) Which of the two reactions should have the larger entropy change, $\Delta S°$? Explain.

(b) Account for the greater stability of $Ni(en)_3^{2+}$ in terms of the relative values of $\Delta S°$ for the two reactions.

(c) Assuming that $\Delta H°_2 - \Delta H°_1$ is zero, calculate the value of $\Delta S°_2 - \Delta S°_1$.

20.141 The percent iron in iron ore can be determined by dissolving the ore in acid, then reducing the iron to Fe^{2+}, and finally titrating the Fe^{2+} with aqueous $KMnO_4$. The reaction products are Fe^{3+} and Mn^{2+}.

(a) Write a balanced net ionic equation for the titration reaction.

(b) Use the $E°$ values in Appendix D to calculate $\Delta G°$ (in kilojoules) and the equilibrium constant for the reaction.

(c) Draw a crystal field energy-level diagram for the reactants and products, MnO_4^-, $[Fe(H_2O)_6]^{2+}$, $[Fe(H_2O)_6]^{3+}$, and $[Mn(H_2O)_6]^{2+}$, and predict the number of unpaired electrons for each.

(d) Does the paramagnetism of the solution increase or decrease as the reaction proceeds? Explain.

(e) What is the mass % Fe in the iron ore if titration of the Fe^{2+} from a 1.265 g sample of ore requires 34.83 mL of 0.051 32 M $KMnO_4$ to reach the equivalence point?

20.142 The complete reaction of 2.60 g of chromium metal with 50.00 mL of 1.200 M H_2SO_4 in the absence of air gave a blue solution and a colorless gas that was collected at 25 °C and a pressure of 735 mm Hg.

(a) Write a balanced net ionic equation for the reaction.

(b) How many liters of gas were produced?

(c) What is the pH of the solution?

(d) Describe the bonding in the blue-colored ion, using both the crystal field theory and the valence bond theory. Include the appropriate crystal field *d*-orbital energy-level diagram and the valence bond orbital diagram. Identify the hybrid orbitals used in the valence bond description.

(e) When an excess of KCN is added to the solution, the color changes and the paramagnetism of the solution decreases. Explain.

20.143 In acidic aqueous solution, the complex *trans*-$[Co(en)_2Cl_2]^+$ undergoes the following substitution reaction:

$$trans\text{-}[Co(en)_2Cl_2]^+(aq) + H_2O(l) \longrightarrow trans\text{-}[Co(en)_2(H_2O)Cl]^{2+}(aq) + Cl^-(aq)$$

The reaction is first order in *trans*-$[Co(en)_2Cl_2]^+$, and the rate constant at 25 °C is $3.2 \times 10^{-5}\ s^{-1}$.

(a) What is the half-life of the reaction in hours?

(b) If the initial concentration of *trans*-$[Co(en)_2Cl_2]^+$ is 0.138 M, what is its molarity after a reaction time of 16.5 h?

(c) Devise a possible reaction mechanism with a unimolecular rate-determining step.

(d) Is the reaction product chiral or achiral? Explain.

(e) Draw a crystal field energy-level diagram for *trans*-$[Co(en)_2Cl_2]^+$ that takes account of the fact that Cl^- is a weaker-field ligand than ethylenediamine.

20.144 Chromium forms three isomeric compounds A, B, and C with percent composition 19.52% Cr, 39.91% Cl, and 40.57% H_2O. When a sample of each compound was dissolved in water and aqueous $AgNO_3$ was added, a precipitate of AgCl formed immediately. A 0.225 g sample of compound A gave 0.363 g of AgCl, 0.263 g of B gave 0.283 g of AgCl, and 0.358 g of C gave 0.193 g of AgCl. One of the three compounds is violet, while the other two are green. In all three, chromium has coordination number 6.

(a) What are the empirical formulas of A, B, and C?

(b) What are the probable structural formulas of A, B, and C? Draw the structure of the cation in each compound. Which cation can exist as diastereoisomers?

(c) Which of the three compounds is likely to be the violet one? Explain.

(d) What are the approximate freezing points of 0.25 *m* solutions of A, B, and C, assuming complete dissociation?

20.145 Cobalt(III) trifluoroacetylacetonate, $Co(tfac)_3$, is a six-coordinate, octahedral metal chelate in which three planar, bidentate tfac ligands are attached to a central Co atom:

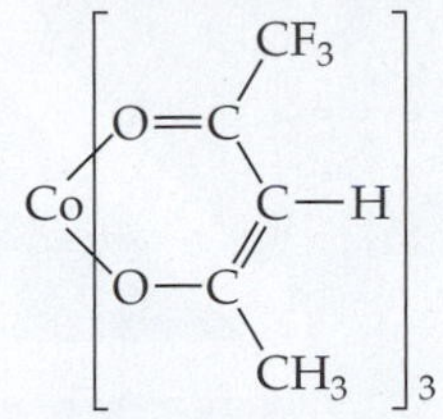

(a) Draw all possible diastereoisomers and enantiomers of $Co(tfac)_3$.

(b) Diastereoisomers A and B have dipole moments of 6.5 D and 3.8 D, respectively. Which of your diastereoisomers is A and which is B?

(c) The isomerization reaction A → B in chloroform solution has first-order rate constants of $0.0889\ h^{-1}$ at 66.1 °C and $0.0870\ min^{-1}$ at 99.2 °C. What is the activation energy for the reaction?

(d) Draw a crystal field energy-level diagram for $Co(tfac)_3$, and predict its magnetic properties. (In this complex, tfac is a strong-field ligand.)

20.146 Consider the following reaction, and assume that its equilibrium constant is 1.00×10^{14}:

$$2\,CrO_4^{2-}(aq) + 2\,H^+(aq) \rightleftharpoons Cr_2O_7^{2-}(aq) + H_2O(l)$$

(a) Write the equilibrium equation for the reaction, and explain why CrO_4^{2-} ions predominate in basic solutions and $Cr_2O_7^{2-}$ ions predominate in acidic solutions.

(b) Calculate the CrO_4^{2-} and $Cr_2O_7^{2-}$ concentrations in a solution that has a total chromium concentration of 0.100 M and a pH of 4.000.

(c) What are the CrO_4^{2-} and $Cr_2O_7^{2-}$ concentrations if the pH is 2.000?

20.147 An alternative to cyanide leaching of gold ores is leaching with thiocyanate ion, which forms a square planar gold(III) complex, $[Au(SCN)_4]^-$.

(a) If the formation constant for $[Au(SCN)_4]^-$ is $K_f = 10^{37}$, what is the equilibrium concentration of Au^{3+} in a 0.050 M solution of $[Au(SCN)_4]^-$?

(b) Draw a crystal field energy-level diagram for $[Au(SCN)_4]^-$, and predict the number of unpaired electrons.

CHAPTER

21

Metals and Solid-State Materials

Why do these cadmium selenide nanoparticles have different colors? The answer is in the *FYI* at the end of this chapter.

The materials available for making tools and weapons, houses and skyscrapers, computers and lasers have had a profound effect on the development of human civilization. Indeed, archaeologists organize early human history in terms of materials—the Stone Age, in which only natural materials such as wood and stone were available; the Bronze Age, in which implements were made of copper alloyed with tin; and the Iron Age, in which ornaments, weapons, and tools were made of iron.

Copper and iron are still of enormous importance. Copper is used to make electrical wiring, and iron is the main constituent of steel. Today, metals unknown in ancient times, such as aluminum and titanium, play a leading role in modern technology. These metals are widely used, for instance, in the aircraft industry, because of their low densities and high resistance to corrosion.

Modern technology is made possible by a host of *solid-state materials* such as *semiconductors*, *superconductors*, *advanced ceramics*, *composites*, and *nanoparticles*. Semiconductors are used in the miniature solid-state electronic devices found in computers. Superconductors are used to make the powerful electromagnets found in the magnetic resonance imaging (MRI) instruments employed in medical diagnosis. Advanced ceramics and composites have numerous engineering, electronic, and biomedical applications and are likely to be among the more important materials in future technologies. Nanoparticles, used in sunscreens, cosmetics, and many other products, are being investigated as a possible weapon in the war on cancer.

In this chapter, we'll look at both metals and solid-state materials. We'll examine the natural sources of the metallic elements, the methods used to obtain metals from their ores, and the models used to describe the bonding in metals. We'll also look at the structure, bonding, properties, and applications of semiconductors, superconductors, ceramics, composites, and nanoparticles.

CONTENTS

◀ The MRI instruments used in hospitals contain electromagnets made with superconductors.

21.1 SOURCES OF THE METALLIC ELEMENTS

Most metals occur in nature as **minerals**, the crystalline, inorganic constituents of the rocks that make up the Earth's crust. **Silicates** and **aluminosilicates** (Section 19.8) are the most abundant minerals, but they are difficult to concentrate and reduce and are therefore generally unimportant as commercial sources of metals. More important are oxides and sulfides, such as hematite (Fe_2O_3), rutile (TiO_2), and cinnabar (HgS) (**Figure 21.1**), which yield iron,

REMEMBER...

Silicates contain oxoanions in which one or more O atoms bridge Si atoms to give rings, chains, layers, and extended three-dimensional structures. Partial substitution of Si^{4+} with Al^{3+} gives **aluminosilicates**. (Section 19.8)

FIGURE 21.1

Samples of hematite (Fe_2O_3), rutile (TiO_2), and cinnabar (HgS).

titanium, and mercury, respectively. Mineral deposits from which metals can be produced economically are called **ores** (**Table 21.1**).

TABLE 21.1 Principal Ores of Some Important Metals

Metal	Ore	Formula	Location of Important Deposits
Aluminum	Bauxite	$Al_2O_3 \cdot x\,H_2O$	Australia, Guinea, Jamaica
Chromium	Chromite	$FeCr_2O_4$	South Africa, Zimbabwe, Russia
Copper	Chalcopyrite	$CuFeS_2$	U.S., Chile, Canada
Iron	Hematite	Fe_2O_3	Australia, Ukraine, U.S.
Lead	Galena	PbS	U.S., Australia, Canada
Manganese	Pyrolusite	MnO_2	Russia, Gabon, South Africa
Mercury	Cinnabar	HgS	Spain, Peru, U.S.
Tin	Cassiterite	SnO_2	Malaysia, Bolivia
Titanium	Rutile	TiO_2	Australia, Sierra Leone
	Ilmenite	$FeTiO_3$	Australia, South Africa, Canada
Zinc	Sphalerite	ZnS	U.S., Canada, Australia

The chemical compositions of the most common ores correlate with the locations of the metals in the periodic table (**Figure 21.2**). The early transition metals on the left side of the *d* block generally occur as oxides, and the more electronegative, late transition metals on the right side of the *d* block occur as sulfides. This pattern makes sense because the less electronegative metals tend to form ionic compounds by losing electrons to highly electronegative nonmetals such as oxygen. By contrast, the more electronegative metals tend to form compounds with more covalent character by bonding to the less electronegative nonmetals such as sulfur.

We might also expect to find oxide ores for the *s*-block metals and sulfide ores for the more electronegative *p*-block metals. In fact, sulfide ores are common for the *p*-block metals, except for Al and Sn, but oxides of the *s*-block metals are strongly basic and far too reactive to exist in an environment that contains acidic oxides such as CO_2 and SiO_2. Consequently, *s*-block metals are found in nature as carbonates, as silicates, and, in the case of Na and K, as chlorides. Only gold (Au) and the platinum-group metals (Ru, Os, Rh, Ir, Pd, and Pt) are sufficiently unreactive to occur commonly in uncombined form as the free metals.

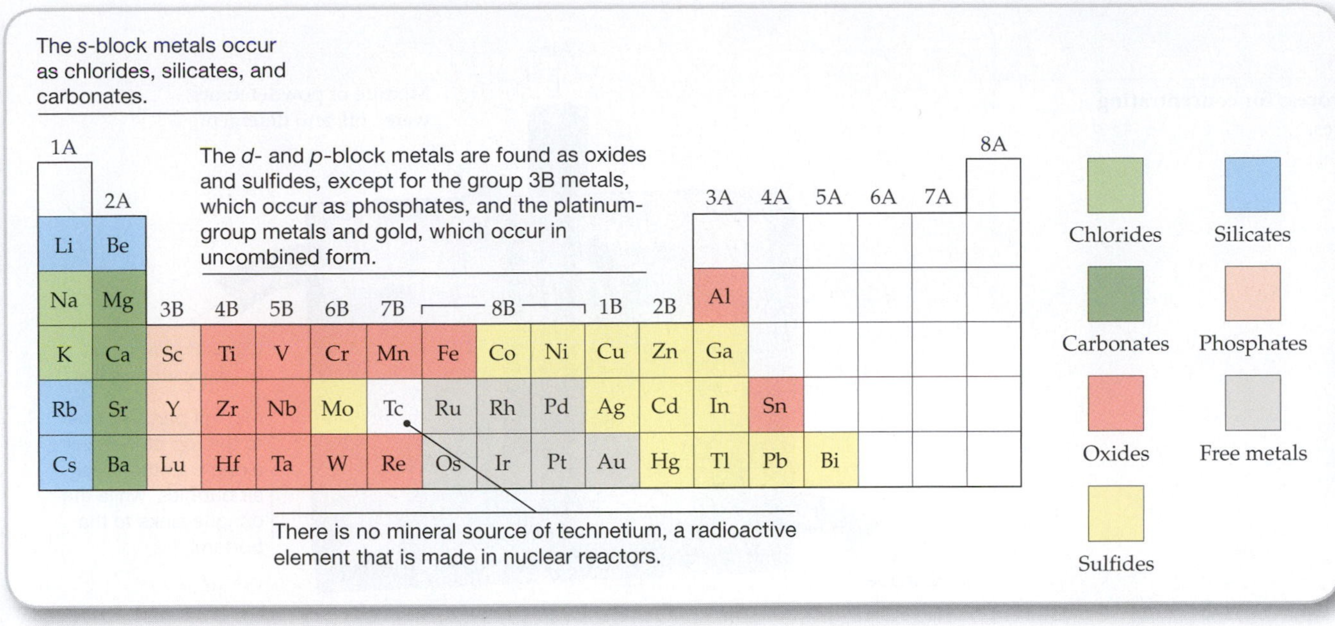

FIGURE 21.2

Primary mineral sources of metals.

PROBLEM 21.1 In view of the 3− charge on the PO_4^{3-} ion, explain why the group 3B transition metals are found in nature as phosphates.

21.2 METALLURGY

An ore is a complex mixture of a metal-containing mineral and economically worthless material called **gangue** (pronounced "gang"), consisting of sand, clay, and other impurities. The extraction of a metal from its ore requires several steps: (1) concentration of the ore and, if necessary, chemical treatment prior to reduction; (2) reduction of the mineral to the free metal; and (3) refining or purification of the metal. These processes are a part of **metallurgy**, the science and technology of extracting metals from their ores. Another aspect of metallurgy is the making of **alloys**, solid metallic solutions composed of two or more elements. Bronze, a mixture of copper with tin, and different steels, mixtures of iron with carbon, manganese, chromium, vanadium, or tungsten, are examples.

Concentration and Chemical Treatment of Ores

Ores are concentrated by separating the mineral from the gangue. The mineral and the gangue have different properties, which are exploited in various separation methods. Density differences, for example, are important in panning for gold, a procedure in which prospectors flush water over gold-bearing earth in a pan. The less dense gangue is washed away, and the more dense gold particles remain at the bottom of the pan. Differences in magnetic properties are used to concentrate the iron ore magnetite (Fe_3O_4). The Fe_3O_4 is strongly attracted by magnets, but the gangue is unaffected.

Metal sulfide ores are concentrated by **flotation**, a process that exploits differences in the ability of water and oil to wet the surfaces of the mineral and the gangue. A powdered ore such as chalcopyrite ($CuFeS_2$) is mixed with water, oil, and a detergent, and the mixture is vigorously agitated in a tank by blowing air through the liquid (**Figure 21.3**). The gangue, which contains ionic silicates, is moistened by the polar water molecules and sinks to the bottom of the tank. The mineral particles, which contain the less polar metal sulfide, are coated by the nonpolar oil molecules and become attached to the soapy air bubbles created by the detergent. The metal sulfide particles are thus carried to the surface in the soapy froth, which is skimmed off at the top of the tank.

▲ In panning for gold, prospectors use density differences to separate gold from the gangue.

FIGURE 21.3

The flotation process for concentrating metal sulfide ores.

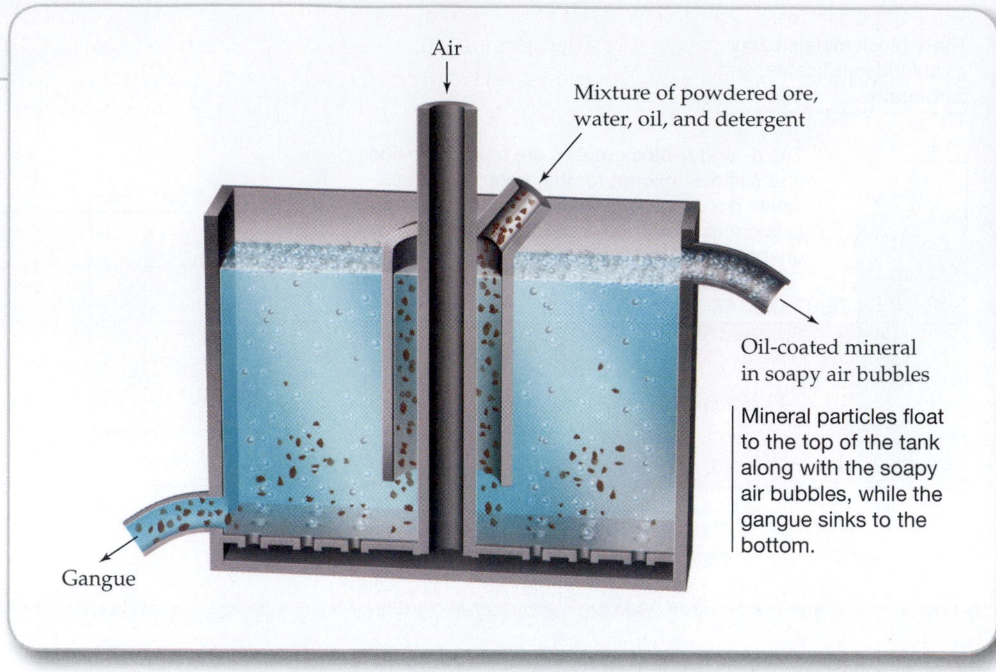

Sometimes, ores are concentrated by chemical treatment. In the *Bayer process*, for instance, the Al_2O_3 in bauxite ($Al_2O_3 \cdot x\,H_2O$) is separated from Fe_2O_3 impurities by treating the ore with hot aqueous NaOH. The amphoteric Al_2O_3 dissolves as the aluminate ion, $Al(OH)_4^-$, but the basic Fe_2O_3 does not:

$$Al_2O_3(s) + 2\,OH^-(aq) + 3\,H_2O(l) \longrightarrow 2\,Al(OH)_4^-(aq)$$

The Bayer process was described in more detail in Section 18.8.

Chemical treatment is also used to convert minerals to compounds that are more easily reduced to the metal. For example, sulfide minerals, such as sphalerite (ZnS), are converted to oxides by **roasting**, a process that involves heating the mineral in air:

$$2\,ZnS(s) + 3\,O_2(g) \xrightarrow{\text{Heat}} 2\,ZnO(s) + 2\,SO_2(g)$$

Formerly, the sulfur dioxide byproduct was a source of acid rain because it is oxidized in the atmosphere to SO_3, which reacts with water vapor to yield sulfuric acid. In modern roasting facilities, however, the SO_2 is converted to sulfuric acid rather than being vented to the atmosphere.

Reduction

Once an ore has been concentrated, it is reduced to the free metal, either by chemical reduction or by electrolysis (**Table 21.2**). The method used depends on the activity of the metal, as measured by its standard reduction potential (Table 17.1). The most active metals have the most negative standard reduction potentials and are the most difficult to reduce; the least active metals have the most positive standard reduction potentials and are the easiest to reduce.

TABLE 21.2 Reduction Methods for Producing Some Common Metals

	Metal	Reduction Method
Least active	Au, Pt	None; found in nature as the free metal
↓	Cu, Ag, Hg	Roasting of the metal sulfide
↓	V, Cr, Mn, Fe, Ni, Zn, W, Pb	Reduction of the metal oxide with carbon, hydrogen, or a more active metal
↓	Al	Electrolysis of molten Al_2O_3 in cryolite
Most active	Li, Na, Mg	Electrolysis of the molten metal chloride

Gold and platinum are so inactive that they are usually found in nature in uncombined form, but copper and silver, which are slightly more active, are found in both combined and uncombined form. Copper, silver, and mercury commonly occur in sulfide ores that are easily reduced by roasting. Cinnabar (HgS), for example, yields elemental mercury when the ore is heated at 600 °C in a stream of air:

$$HgS(s) + O_2(g) \xrightarrow{600\,°C} Hg(g) + SO_2(g)$$

Although it may seem strange that heating a substance in oxygen reduces it to the metal, both oxidation and reduction of HgS occur in this reaction: Sulfide ions are oxidized and mercury(II) is reduced.

More active metals, such as chromium, zinc, and tungsten, are obtained by reducing their oxides with a chemical reducing agent such as carbon, hydrogen, or a more active metal (Na, Mg, or Al). Pure chromium, for example, is produced by reducing Cr_2O_3 with aluminum (Section 20.4). Zinc, used in the automobile industry for galvanizing steel (Section 17.11), is obtained by reducing ZnO with coke, a form of carbon produced by heating coal in the absence of air:

$$ZnO(s) + C(s) \xrightarrow{\text{Heat}} Zn(g) + CO(g)$$

Although carbon is the cheapest available reducing agent, it is unsatisfactory for reducing oxides of metals like tungsten that form very stable carbides. (Tungsten carbide, WC, is an extremely hard material used to make high-speed cutting tools.) The preferred method for producing tungsten reduces tungsten(VI) oxide with hydrogen:

$$WO_3(s) + 3\,H_2(g) \xrightarrow{850\,°C} W(s) + 3\,H_2O(g)$$

Because of its high strength, high melting point (3422 °C), low volatility, and high efficiency for converting electrical energy into light, tungsten is used to make filaments for electric lightbulbs (**Figure 21.4**). Although tungsten has the highest boiling point of any element (5555 °C), the tungsten slowly evaporates from the hot filament and condenses as the black spot of tungsten metal usually visible on the inside surface of a burned-out bulb.

There are no chemical reducing agents strong enough to reduce compounds of the most active metals, so these metals are produced by electrolytic reduction (Section 17.13). Lithium, sodium, and magnesium, for example, are obtained by the electrolysis of their molten chlorides. Aluminum is manufactured by the electrolysis of purified Al_2O_3 in molten cryolite (Na_3AlF_6).

FIGURE 21.4

The incandescent lightbulb. The bulb emits white light when an electric current passes through the tungsten wire filament, heating it to a high temperature.

Refining

For most applications, the metals obtained from processing and reducing ores need to be purified. The methods used include distillation, chemical purification, and electrorefining. Zinc (bp 907 °C), for example, is volatile enough to be separated from cadmium, lead, and other impurities by distillation. In this way, the purity of the zinc obtained from reduction of ZnO is increased from 99% to 99.99%.

Nickel, used as a catalyst and as a battery material, is purified by the **Mond process**, a chemical method in which the volatile compound nickel tetracarbonyl, $Ni(CO)_4$ (bp 42 °C), is formed and subsequently decomposed. Carbon monoxide is passed over impure nickel at about 150 °C and 20 atm pressure, forming $Ni(CO)_4$ and leaving metal impurities behind. The $Ni(CO)_4$ is then decomposed at higher temperatures (about 230 °C) on pellets of pure nickel. The process works because the equilibrium shifts to the left with increasing temperature:

$$Ni(s) + 4\,CO(g) \underset{\text{Higher temp.}}{\overset{\text{Lower temp.}}{\rightleftharpoons}} Ni(CO)_4(g) \qquad \Delta H° = -160.8\text{ kJ}; \Delta S° = -410\text{ J/K}$$

A similar strategy is employed to purify zirconium, which is used to make the cladding (metallic exterior) of the uranium oxide fuel rods in nuclear reactors. The crude metal is heated at about 200 °C with a small amount of iodine in an evacuated container to form the volatile ZrI_4. The ZrI_4 is then decomposed to pure zirconium by letting the vapor come in contact with an electrically heated tungsten or zirconium filament at about 1300 °C:

$$Zr(s) + 2\,I_2(g) \underset{1300\,°C}{\overset{200\,°C}{\rightleftharpoons}} ZrI_4(g)$$

▲ Zirconium is used as cladding for uranium dioxide fuel rods in nuclear reactors.

Copper from the reduction of ores must be purified for use in making electrical wiring because impurities increase its electrical resistance. The method used is electrorefining, an electrolytic process in which copper is oxidized to Cu^{2+} at an impure copper anode and Cu^{2+} from an aqueous copper sulfate solution is reduced to copper at a pure copper cathode. The process was described in Section 17.13.

PROBLEM 21.2 Write a balanced equation for the production of each of the following metals:

(a) Chromium by reducing Cr_2O_3 with aluminum
(b) Copper by roasting Cu_2S
(c) Lead by reducing PbO with coke
(d) Potassium by the electrolysis of molten KCl

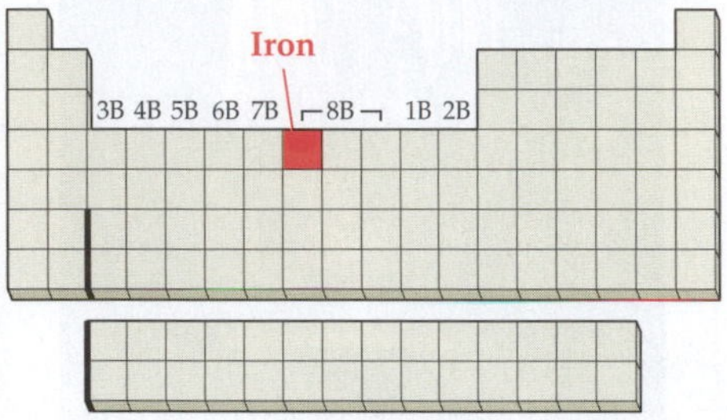

21.3 IRON AND STEEL

The metallurgy of iron is of special technological importance because iron is the major constituent of steel, the most widely used of all metals. Worldwide steel production amounts to 1.4 billion metric tons per year. Iron is produced by carbon monoxide reduction of iron ore, usually hematite (Fe_2O_3), in a huge reactor called a *blast furnace* (**Figure 21.5**). Modern blast furnaces are as large as 60 m in height and 14 m in diameter. They are designed for continuous operation and produce up to 9000 metric tons of iron per day.

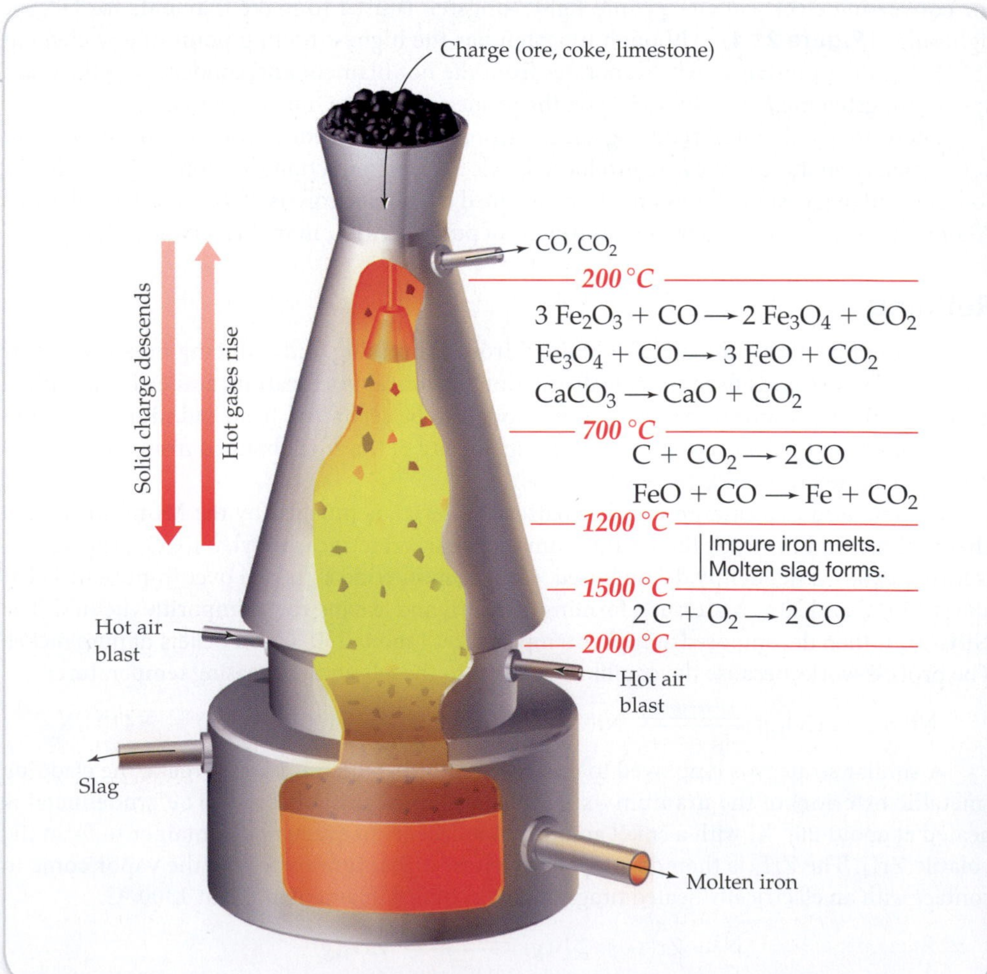

FIGURE 21.5

A diagram of a blast furnace for the reduction of iron ore. Note the approximate temperatures and the chemical reactions that occur in the various regions of the furnace.

A charge of iron ore, coke, and limestone ($CaCO_3$) is introduced at the top of the furnace, and a blast of hot air is sent in at the bottom, burning coke and yielding carbon monoxide at a temperature of about 2000 °C:

$$2\,C(s) + O_2(g) \longrightarrow 2\,CO(g) \qquad \Delta H° = -221.0\text{ kJ}$$

As the charge descends and the hot carbon monoxide rises, a complex series of high-temperature reactions occur in the various regions of the furnace, as shown in Figure 21.5. The key overall reaction is the reduction of Fe_2O_3 to iron metal (mp 1538 °C), which is obtained as an impure liquid at the bottom of the furnace:

$$Fe_2O_3(s) + 3\,CO(g) \longrightarrow 2\,Fe(l) + 3\,CO_2(g)$$

The purpose of the limestone is to remove the gangue from the iron ore. At the high temperatures of the furnace, the limestone decomposes yielding lime (CaO), a basic oxide that reacts with SiO_2 and other acidic oxides present in the gangue. The product, called **slag**, is a molten material consisting mainly of calcium silicate:

$$CaCO_3(s) \longrightarrow CaO(s) + CO_2(g)$$

$$\underset{\text{Lime}}{CaO(s)} + \underset{\text{Sand}}{SiO_2(s)} \longrightarrow \underset{\text{Slag}}{CaSiO_3(l)}$$

Because the slag is less dense than the molten iron, it floats on the surface of the iron, thus allowing the iron and the slag to be removed from the bottom of the furnace through separate taps.

The iron obtained from a blast furnace is a brittle material called *cast iron*, or *pig iron*. It contains about 4% elemental carbon and smaller amounts of other impurities such as elemental silicon, phosphorus, sulfur, and manganese, which are formed from their compounds in the reducing atmosphere of the furnace. The most important of several methods for purifying the iron and converting it to steel is the **basic oxygen process**. Molten iron from the blast furnace is exposed to a jet of pure oxygen gas for about 20 min in a furnace that is lined with basic oxides such as CaO. The impurities in the iron are oxidized, and the acidic oxides that form react with the basic CaO to yield a molten slag that can be poured off. Phosphorus, for example, is oxidized to P_4O_{10}, which then reacts with CaO to give molten calcium phosphate:

$$P_4(l) + 5\,O_2(g) \longrightarrow P_4O_{10}(l)$$

$$\underset{\text{Basic oxide}}{6\,CaO(s)} + \underset{\text{Acidic oxide}}{P_4O_{10}(l)} \longrightarrow \underset{\text{Slag}}{2\,Ca_3(PO_4)_2(l)}$$

Manganese also passes into the slag because its oxide is basic and reacts with added SiO_2, yielding molten manganese silicate:

$$2\,Mn(l) + O_2(g) \longrightarrow 2\,MnO(s)$$

$$\underset{\text{Basic oxide}}{MnO(s)} + \underset{\text{Acidic oxide}}{SiO_2(s)} \longrightarrow \underset{\text{Slag}}{MnSiO_3(l)}$$

The basic oxygen process produces steels that contain about 1% carbon but only very small amounts of phosphorus and sulfur. Usually, the composition of the liquid steel is monitored by chemical analysis, and the amounts of oxygen and impure iron used are adjusted to achieve the desired concentrations of carbon and other impurities. The hardness, strength, and malleability of the steel depend on its chemical composition, on the rate at which the liquid steel is cooled, and on subsequent heat treatment of the solid. The mechanical and chemical properties of a steel can also be altered by adding other metals. Stainless steel, for example, is a corrosion-resistant iron alloy that contains up to 30% chromium along with smaller amounts of nickel.

▲ Using the basic oxygen process to make steel.

PROBLEM 21.3 The formation of $CaSiO_3$ slag in a blast furnace can be considered an acid–base reaction. Explain.

21.4 BONDING IN METALS

Thus far, we've discussed the sources, production, and properties of some important metals. Some properties, such as hardness and melting point, vary considerably among metals, but other properties are characteristic of metals in general. For instance, all metals can be drawn into wires (ductility) or beaten into sheets (malleability) without breaking into pieces like glass or an ionic crystal. Furthermore, all metals have a high thermal and electrical conductivity. When you touch a metal, it feels cold because the metal efficiently conducts heat away from your hand, and when you connect a metal wire to the terminals of a battery, it conducts an electric current.

To understand those properties, we need to look at the bonding in metals. We'll consider two theoretical models that are commonly used: the *electron-sea model* and the *molecular orbital theory*.

▶ Because of its ductility, aluminum can be drawn into the wires used in electric power lines.

Electron-Sea Model of Metals

If you try to draw an electron-dot structure for a metal, you'll quickly realize that there aren't enough valence electrons available to form an electron-pair bond between every pair of adjacent atoms. Sodium, for example, which has just one valence electron per atom ($3s^1$), crystallizes in a **body-centered cubic** structure in which each Na atom is surrounded by eight nearest neighbors (Section 10.8). Consequently, the valence electrons can't be localized in a bond between any particular pair of atoms. Instead, they are delocalized and belong to the crystal as a whole.

REMEMBER...

A **body-centered cubic** unit cell has one atom at each of the eight corners of the cell and an additional atom at the center. Thus, each atom is surrounded by eight others. (Section 10.8)

In the **electron-sea model**, a metal crystal is viewed as a three-dimensional array of metal cations immersed in a sea of delocalized electrons that are free to move throughout the crystal (**Figure 21.6**). The continuum of delocalized, mobile valence electrons acts as an electrostatic glue that holds the metal cations together.

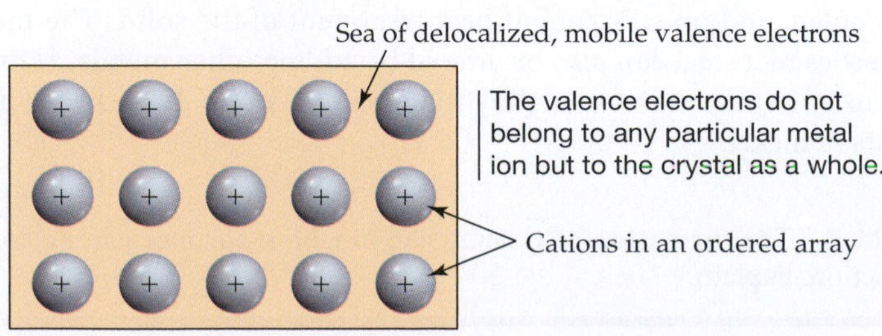

FIGURE 21.6

A two-dimensional representation of the electron-sea model of a metal. An ordered array of cations is immersed in a continuous distribution of delocalized, mobile valence electrons.

The electron-sea model affords a simple qualitative explanation for the electrical and thermal conductivity of metals. Because the electrons are mobile, they are free to move away from a negative electrode and toward a positive electrode when a metal is subjected to an electrical potential. The mobile electrons can conduct heat by carrying kinetic energy from one part of the crystal to another. Metals are malleable and ductile because the delocalized bonding extends in all directions; that is, it is not confined to oriented bond directions, as in covalent network solids like diamond or SiO_2. When a metallic crystal is deformed, no localized bonds are broken. Instead, the electron sea simply adjusts to the new distribution of cations, and the energy of the deformed structure is similar to that of the original. Thus, the energy required to deform a metal like sodium is relatively small. The energy required to deform a transition metal like iron is greater because iron has more valence electrons ($4s^2\,3d^6$), and the electrostatic glue is therefore more dense.

Molecular Orbital Theory for Metals

A more detailed understanding of the bonding in metals is provided by the molecular orbital theory, a model that is a logical extension of the **molecular orbital** description of small molecules discussed in Sections 5.5–5.7. Recall that in the H_2 molecule the 1*s* orbitals of the two H atoms overlap to give a σ bonding MO and a higher-energy σ^* antibonding MO. The bonding in the gaseous Na_2 molecule is similar: The 3*s* orbitals of the two Na atoms combine to give a σ and a σ^* MO. Because each Na atom has just one 3*s* valence electron, the lower-energy bonding orbital is filled and the higher-energy antibonding orbital is empty:

REMEMBER...

A **molecular orbital (MO)** is a wave function whose square gives the probability of finding an electron within a given region of space in a molecule. MOs are formed by mathematically combining atomic orbitals on different atoms. (Sections 5.5–5.7)

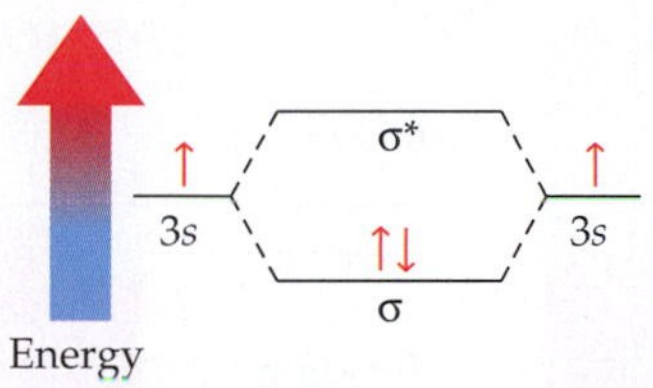

Now consider what happens if we bring together an increasingly larger number of Na atoms to build up a crystal of sodium metal. The key idea to remember from Section 5.5 is that the number of molecular orbitals formed is the same as the number of atomic orbitals combined. Thus, there will be three MOs for a triatomic Na_3 molecule, four MOs for Na_4, and so on. A cubic crystal of sodium metal, 1.5 mm on an edge, contains about 10^{20} Na atoms and therefore has about 10^{20} MOs, each of which is delocalized over all the atoms in the crystal.

As shown in **Figure 21.7**, the difference in energy between successive MOs in an Na_n molecule decreases as the number of Na atoms increases so that the MOs merge into an almost continuous band of energy levels for large values of *n*. Consequently, MO theory for

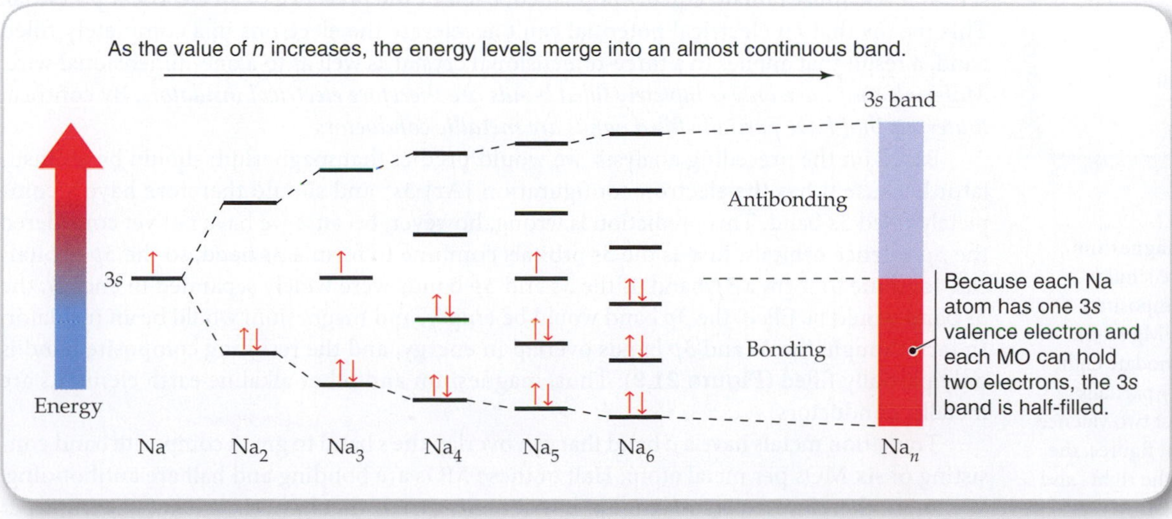

FIGURE 21.7

Molecular orbital energy levels for Na_n molecules. A crystal of sodium metal can be regarded as a giant Na_n molecule, where *n* has a value of about 10^{20}. In this and subsequent figures, the red color denotes the filled portion of a band.

metals is often called **band theory**. The bottom half of the band consists of bonding MOs and is filled, whereas the top half of the band consists of antibonding MOs and is empty.

How does band theory account for the electrical conductivity of metals? Because each of the MOs in a metal has a definite energy (Figure 21.7), each electron in a metal has a specific kinetic energy and a specific velocity. These values depend on the particular MO energy level and increase from the bottom of a band to the top. For a one-dimensional metal wire, electrons traveling in opposite directions at the same speed have the same kinetic energy. Thus, the energy levels within a band occur in **degenerate** pairs; one set of energy levels applies to electrons moving to the right, and the other set applies to electrons moving to the left.

REMEMBER...

Degenerate energy levels have the same energy. (Section 2.11)

In the absence of an electrical potential, the two sets of levels are equally populated. That is, for each electron moving to the right, another electron moves to the left with exactly the same speed (**Figure 21.8**). As a result, there is no net electric current in either direction. In the presence of an electrical potential, however, those electrons moving to the right (toward the positive terminal of a battery) are accelerated, those moving to the left (toward the negative terminal) are slowed down, and those moving to the left with very slow speeds undergo a change of direction. Thus, the number of electrons moving to the right is now greater than the number moving to the left and there is a net electric current.

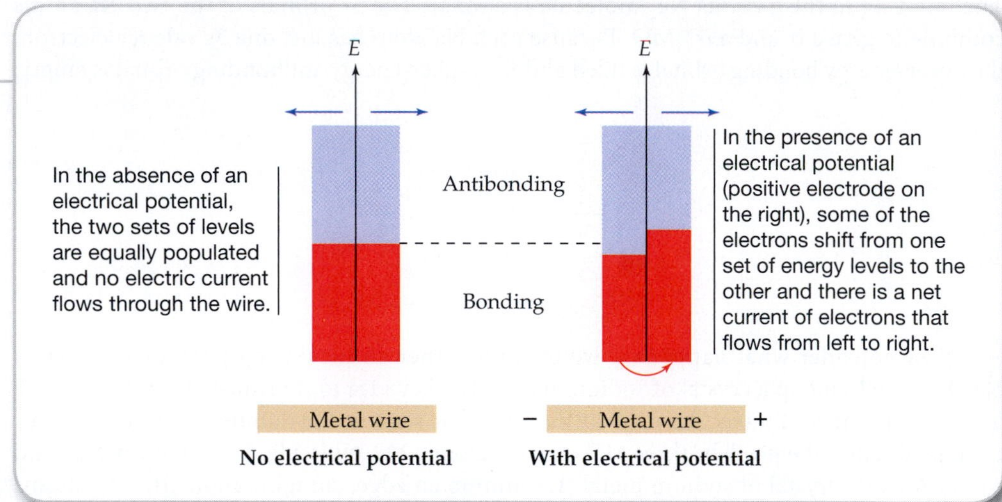

FIGURE 21.8

Half-filled 3s band of MO energy levels for a one-dimensional sodium metal wire. The direction of electron motion for the two degenerate sets of energy levels is indicated by the horizontal arrows.

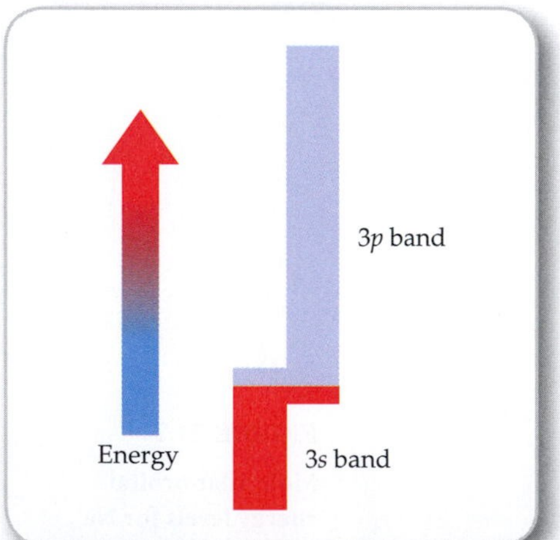

FIGURE 21.9

The composite 3s–3p band in magnesium metal. The 3s and 3p bands have similar energies and overlap to give a composite band consisting of four MOs per Mg atom. The composite band can accommodate eight electrons per Mg atom but is only partially filled, since each Mg atom has just two valence electrons. (In this and subsequent figures, the separate sets of energy levels for the right- and left-moving electrons aren't shown.)

Figure 21.8 shows that an electrical potential can shift electrons from one set of energy levels to the other only if the band is partially filled. If the band is completely filled, there are no available vacant energy levels to which electrons can be excited, and therefore the two sets of levels must remain equally populated, even in the presence of an electrical potential. This means that an electrical potential can't accelerate the electrons in a completely filled band, a result that applies to a three-dimensional crystal as well as to a one-dimensional wire. *Materials that have only completely filled bands are therefore electrical insulators.* By contrast, *materials that have partially filled bands are metallic conductors.*

Based on the preceding analysis, we would predict that magnesium should be an insulator because it has the electron configuration [Ar] $3s^2$ and should therefore have a completely filled 3*s* band. This prediction is wrong, however, because we have not yet considered the 3*p* valence orbitals. Just as the 3*s* orbitals combine to form a 3*s* band, so the 3*p* orbitals can combine to form a 3*p* band. If the 3*s* and 3*p* bands were widely separated in energy, the 3*s* band would be filled, the 3*p* band would be empty, and magnesium would be an insulator. In fact, though, the 3*s* and 3*p* bands overlap in energy, and the resulting composite band is only partially filled (**Figure 21.9**). Thus, magnesium and other alkaline earth elements are metallic conductors.

Transition metals have a *d* band that can overlap the *s* band to give a composite band consisting of six MOs per metal atom. Half of these MOs are bonding and half are antibonding. We might therefore expect maximum bonding for metals that have six valence electrons per metal atom because six electrons will just fill the bonding MOs and leave the antibonding MOs empty. In accord with this picture, maximum bonding near group 6B causes the melting points of the transition metals to also be at a maximum near group 6B (Section 20.2).

Worked Example 21.1

Using Band Theory to Account for Melting Points

The melting points of chromium and zinc are 1907 °C and 420 °C, respectively. Use band theory to account for the difference.

STRATEGY

The electron configurations are [Ar] $3d^5\ 4s^1$ for Cr and [Ar] $3d^{10}\ 4s^2$ for Zn. Assume that the $3d$ and $4s$ bands overlap. The composite band, which can accommodate 12 valence electrons per metal atom, will be half-filled for Cr and completely filled for Zn. The melting points will depend on the occupancy of the bonding and antibonding MOs.

SOLUTION

Strong bonding and a consequent high melting point are expected for Cr because all the bonding MOs are occupied and all the antibonding MOs are empty. Weak bonding and a low melting point are expected for Zn because both the bonding and the antibonding MOs are occupied. (The fact that Zn is a metal suggests that the $4p$ orbitals also contribute to the composite band.)

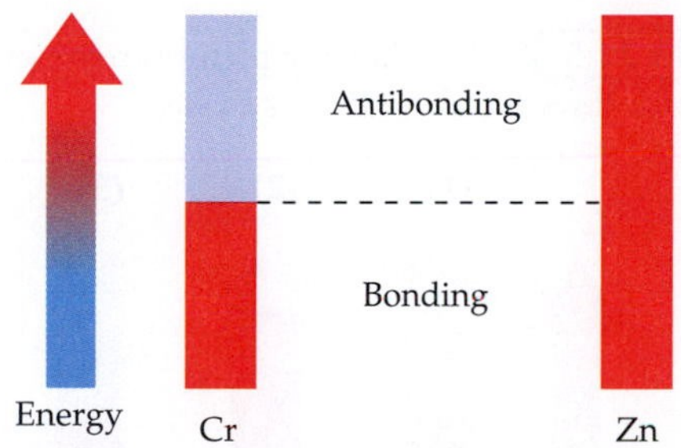

WORKED CONCEPTUAL EXAMPLE 21.2

Using Band Theory to Account for Physical Properties

The following pictures represent the electron population of the composite s–d band for three metals—Ag, Mo, and Y:

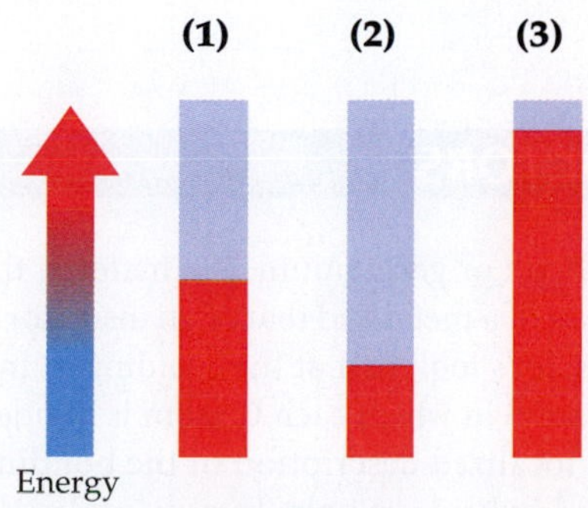

(a) Which picture corresponds to which metal? Explain.
(b) Which metal has the highest melting point, and which has the lowest? Explain.
(c) Molybdenum is very hard, whereas silver is relatively soft. Explain.

continued on next page

STRATEGY

As second-series transition metals, Ag, Mo, and Y have 5*s* and 4*d* valence electrons. All the valence electrons will occupy a composite *s–d* band, which can accommodate 12 electrons per metal atom. To identify each metal, count the number of 5*s* and 4*d* valence electrons and compare that number with the population of the bands in pictures (1), (2), and (3) above. The melting point and hardness of a metal is expected to increase as the difference between the number of bonding and antibonding electrons increases.

SOLUTION

(a) Ag in group 1B has 11 valence electrons. Thus, its *s–d* band is eleven-twelfths filled, which corresponds to picture (3). Mo in group 6B has six valence electrons. Its *s–d* band is exactly half-filled, which corresponds to picture (1). Y in group 3B has three valence electrons. Its *s–d* band is one-fourth filled, which corresponds to picture (2).

(b) Mo has the highest melting point because its bonding MOs are completely filled and its antibonding MOs are completely vacant. Ag has the lowest melting point because it has six bonding electrons and five antibonding electrons per Ag atom, an excess of only one bonding electron per Ag atom.

(c) Mo is very hard because it has an excess of six bonding electrons per Mo atom. Ag is very soft because it has an excess of only one bonding electron per Ag atom. Both hardness and melting point increase as metal–metal bonding increases.

PROBLEM 21.4 Mercury metal is a liquid at room temperature. Use band theory to suggest a reason for its low melting point (–39 °C).

CONCEPTUAL PROBLEM 21.5 The following pictures represent the electron population of the composite *s–d* band for three metals—Hf, Pt, and Re:

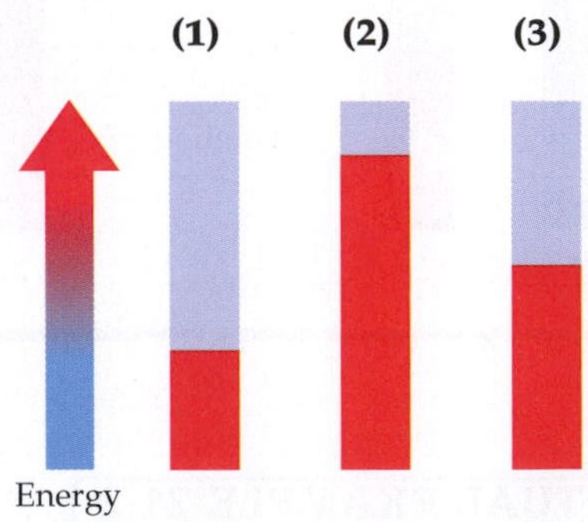

(a) Which picture corresponds to which metal? Explain.
(b) Which metal has the highest melting point, and which is the hardest? Explain.
(c) Which metal has the lowest melting point, and which is the softest? Explain.

21.5 SEMICONDUCTORS

A **semiconductor**, such as silicon or germanium, is a material that has an electrical conductivity intermediate between that of a metal and that of an insulator. To understand the electrical properties of semiconductors, let's look first at the bonding in insulators. Take diamond, for example, a covalent network solid in which each C atom is bonded tetrahedrally to four other C atoms (Figure 10.26). In a localized description of the bonding, C–C electron-pair bonds result from the overlap of sp^3 hybrid orbitals. In a delocalized description, the 2*s* and 2*p* valence orbitals of all the C atoms combine to give bands of bonding and antibonding MOs—a total of four MOs per C atom. As is generally the case for insulators, the bonding MOs, called the **valence band**, and the higher-energy antibonding MOs, called the **conduction band**, are separated in energy by a large **band gap**. The band gap in diamond is about 520 kJ/mol.

Each of the two bands in diamond can accommodate four electrons per C atom. Because carbon has just four valence electrons ($2s^2\,2p^2$), the valence band is completely filled and the conduction band is completely empty. Diamond is therefore an electrical insulator because there are no vacant MOs in the valence band to which electrons can be excited by an electrical potential and because population of the vacant MOs of the conduction band is prevented by the large band gap. By contrast, metallic conductors have no energy gap between the highest occupied and lowest unoccupied MOs. This fundamental difference between the energy levels of metals and insulators is illustrated in **Figure 21.10**.

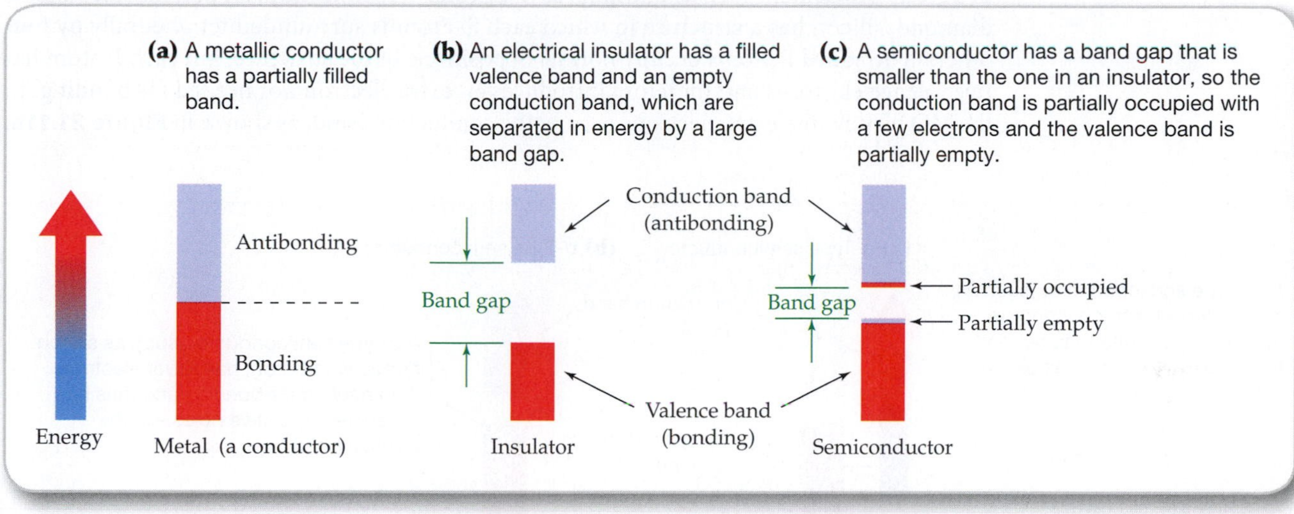

FIGURE 21.10

Bands of MO energy levels for (a) a metallic conductor, (b) an electrical insulator, and (c) a semiconductor. Electrical conductivity in metals and semiconductors results from the presence of partially filled bands.

The MOs of a semiconductor are similar to those of an insulator, but the band gap in a semiconductor is smaller (Figure 21.10). As a result, a few electrons have enough thermal energy to jump the gap and occupy the higher-energy conduction band. The conduction band is thus partially filled, and the valence band is partially empty because it now contains a few unoccupied MOs. When an electrical potential is applied to a semiconductor, it conducts a small amount of current because the potential can accelerate the electrons in the partially filled bands. **Table 21.3** shows how the electrical properties of the group 4A elements vary with the size of the band gap.

TABLE 21.3 Band Gaps for the Group 4A Elements

Element*	Band Gap (kJ/mol)	Type of Material
C (diamond)	520	Insulator
Si	107	Semiconductor
Ge	65	Semiconductor
Sn (gray tin)	8	Semiconductor
Sn (white tin)	0	Metal
Pb	0	Metal

*Si, Ge, and gray Sn have the same structure as diamond.

The electrical conductivity of a semiconductor increases with increasing temperature because the number of electrons with sufficient thermal energy to occupy the conduction band increases as the temperature rises. At higher temperatures, there are more charge carriers (electrons) in the conduction band and more vacancies in the valence band. By contrast, the electrical conductivity of a metal decreases with increasing temperature. At higher temperatures, the metal cations undergo increased vibrational motion about their lattice sites and vibration of the cations disrupts the flow of electrons through the crystal. Thus, the temperature dependence of the electrical conductivity is the best criterion for distinguishing a metal from a semiconductor: As the temperature increases, the conductivity of a metal decreases, whereas the conductivity of a semiconductor increases.

The conductivity of a semiconductor can be greatly increased by adding small (ppm) amounts of certain impurities, a process called **doping**. Consider, for example, the addition of a group 5A element such as phosphorus to a group 4A semiconductor such as silicon. Like diamond, silicon has a structure in which each Si atom is surrounded tetrahedrally by four others. The added P atoms occupy normal Si positions in the structure, but each P atom has five valence electrons and therefore introduces an extra electron not needed for bonding. In the MO picture, the extra electrons occupy the conduction band, as shown in **Figure 21.11a**.

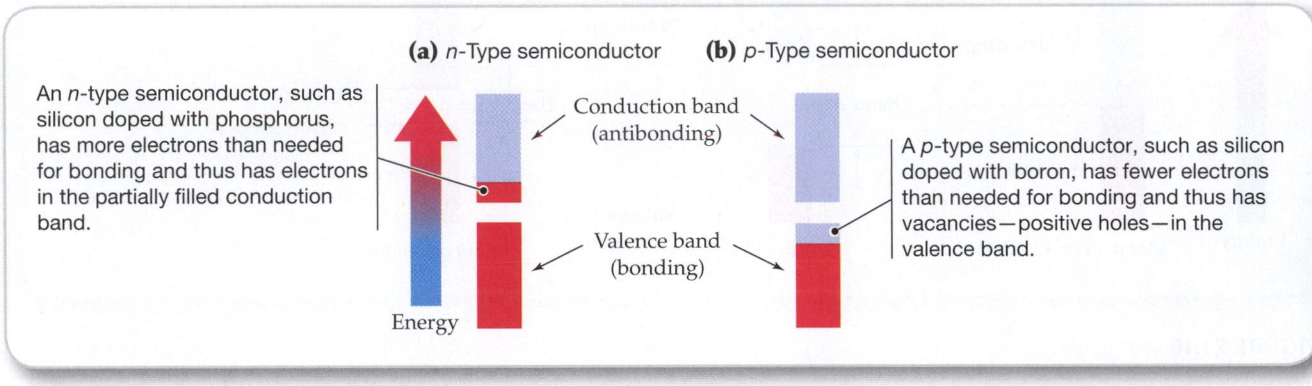

FIGURE 21.11

MO energy levels for doped semiconductors.

The number of electrons in the conduction band of the doped silicon is much greater than the number in pure silicon, and the conductivity of the doped semiconductor is therefore correspondingly higher. When just one of every 1 million Si atoms is replaced by P, the number of electrons in the conduction band increases from $\sim 10^{10}/cm^3$ to $\sim 10^{17}/cm^3$ and the conductivity increases by a factor of $\sim 10^7$. Because the charge carriers are electrons, which are *negatively* charged, the silicon doped with a group 5A element is called an ***n*-type semiconductor**.

Now let's consider a semiconductor in which silicon is doped with a group 3A element such as boron. Each B atom has just three valence electrons and therefore does not have enough electrons to form bonds to its four Si neighbors. In the MO picture, the bonding MOs of the valence band are only partially filled, as shown in **Figure 21.11b**.

The vacancies in the valence band can be thought of as positive holes in a filled band. When the electrons in the partially filled valence band move under the influence of an applied potential, the positive holes move in the opposite direction. Because the charge carriers can be regarded as the *positive* holes, silicon doped with a group 3A element is called a ***p*-type semiconductor**. In a localized picture, a positive hole is a missing electron in a B–Si electron-pair bond. When an electron from an adjacent Si–Si bond moves into the hole, the hole moves in the opposite direction.

Worked Example 21.3

Identifying the Type of a Doped Semiconductor

Consider a crystal of germanium that has been doped with a small amount of aluminum. Is the doped crystal an *n*-type or *p*-type semiconductor? Compare the conductivity of the doped crystal with that of pure germanium.

STRATEGY

Consider the location of germanium and aluminum in the periodic table and the number of valence electrons in the doped crystal relative to the number in pure germanium. Doped semiconductors with more electrons than the pure semiconductor are *n*-type, and those with fewer electrons are *p*-type.

SOLUTION

Germanium, like silicon, is a group 4A semiconductor, and aluminum, like boron, is a group 3A element. The doped germanium is therefore a *p*-type semiconductor because each Al atom has one less valence electron than needed for bonding to the four neighboring Ge atoms. (Like silicon, germanium has the diamond structure.) The valence band is thus partially filled, which accounts for the electrical conductivity. The conductivity is greater than that of pure germanium because the doped germanium has many more positive holes in the valence band. That is, it has more vacant MOs available to which electrons can be excited by an electrical potential.

PROBLEM 21.6 Is germanium doped with arsenic an *n*-type or a *p*-type semiconductor? Why is its conductivity greater than that of pure germanium?

CONCEPTUAL PROBLEM 21.7 The following pictures show the electron population of the bands of MO energy levels for four materials—diamond, silicon, silicon doped with aluminum, and white tin:

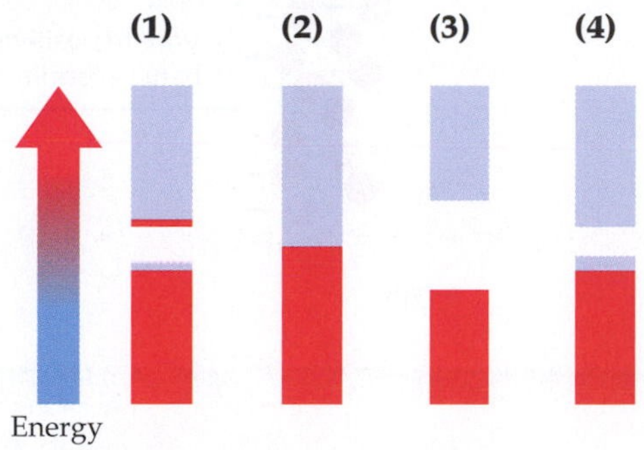

(a) Which picture corresponds to which material?
(b) Arrange the four materials in order of increasing electrical conductivity. Explain.

21.6 SEMICONDUCTOR APPLICATIONS

Doped semiconductors are essential components in the modern solid-state electronic devices found in numerous commercial products. These devices include

- *diodes* that convert alternating current to direct current.
- *light-emitting diodes (LEDs)* that serve as light sources in traffic signals, vehicle brake lights, digital clocks, and hospital pulse oximeters that monitor the oxygen saturation of hemoglobin in blood.
- *diode lasers* that produce the bright, red light of laser pointers and supermarket bar-code scanners.
- *photovoltaic (solar) cells* that convert sunlight into electricity.
- *transistors* that control and amplify electrical signals in the integrated circuits of computers, cell phones, and many other consumer electronic products.

All these devices are made from *n*- and *p*-type semiconductors.

Diodes

A diode is a device that permits electric current to flow in one direction but is highly resistant to current flow in the opposite direction. It consists of a *p*-type semiconductor in contact with an *n*-type semiconductor to give a *p*–*n* junction (**Figure 21.12**).

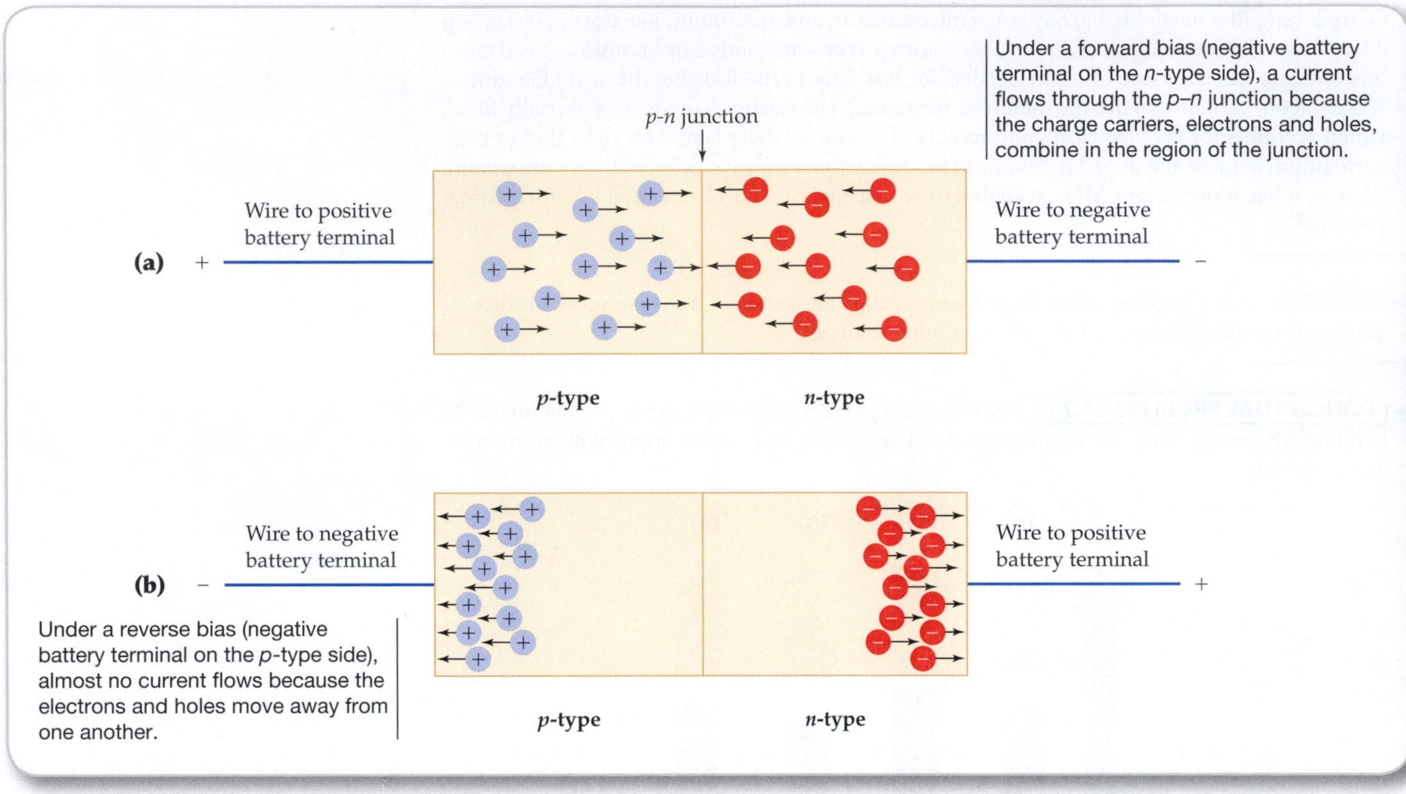

FIGURE 21.12

A semiconductor diode.

If the *n*-type semiconductor is connected to the negative terminal of a battery and the *p*-type semiconductor is connected to the positive terminal (**Figure 21.12a**), electrons in the conduction band of the *n*-type semiconductor are repelled by the negative terminal and are attracted to the positive terminal. Consequently, they move into the region of the *p*–*n* junction from the *n*-side to the *p*-side where they fall into vacancies (positive holes) in the valence band of the *p*-type semiconductor. At the same time, the positive holes move in the opposite direction, from the *p*-side to the *n*-side where they combine with electrons in the conduction band of the *n*-type semiconductor. The motion of the electrons and holes constitutes an electric current that persists as long as the device is connected to the battery because the battery continues to pump electrons into the *n*-side and holes into the *p*-side. When the device is connected to the battery in this way, the *p*–*n* junction is said to be under a *forward bias*.

If the connections to the battery are reversed (*reverse bias*, **Figure 21.12b**), the charge carriers move in the reverse directions: Negative electrons move toward the positive battery terminal, and positive holes move toward the negative terminal. Because the charge carriers move away from the *p*–*n* junction, almost no electric current can flow through the junction; current flows only when the junction is under a forward bias. A *p*–*n* junction that is part of a circuit and subjected to an alternating potential acts as a *rectifier*, allowing current to flow in only one direction, thereby converting alternating current to direct current.

Light-Emitting Diodes (LEDs)

Energy is released as light when electrons and holes combine in the *p*–*n* junction of a diode under a forward bias because of an electronic transition from the conduction band to the valence band. A schematic of a light-emitting diode, or LED, is shown in **Figure 21.13**, along with the corresponding band theory energy-level diagram.

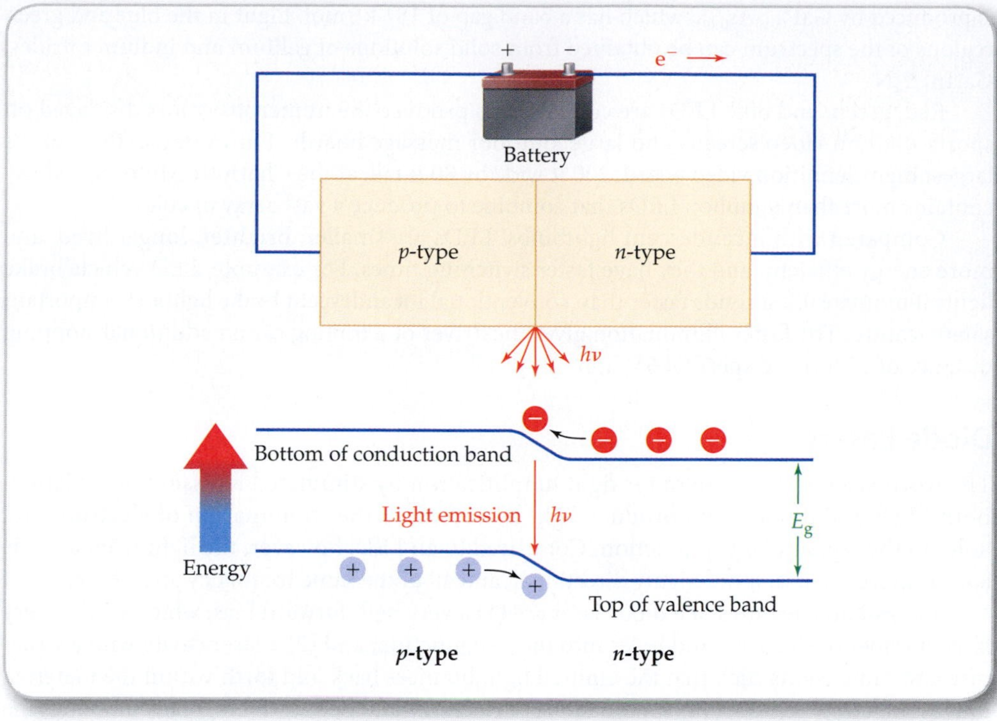

FIGURE 21.13

A schematic of a light-emitting diode (LED), along with the corresponding band theory energy-level diagram.

In the energy-level diagram, the valence and conduction bands have different energies on the *n*- and *p*-type sides because migration of a few electrons from the *n*-side to the *p*-side in the absence of an external voltage gives the *n*-side a small positive charge and the *p*-side a small negative charge. This charge difference lowers the energy of the bands on the *n*-side, raises the energy of the bands on the *p*-side, and prevents current flow in the absence of an external voltage. In the presence of an external voltage, the energy required for electrons to climb the energy hill in the region of the *p–n* junction is provided by the forward bias from the battery. The energy of the light emitted by an LED is roughly equal to the band gap energy, E_g, and is related to the wavelength of the light by the equation

$$E_g = h\nu = hc/\lambda$$

The semiconductors commonly used for making LEDs are 1:1 compounds of group 3A and 5A elements, such as GaAs, GaP, AlAs, and InP. These so-called 3–5 semiconductors have the same diamond structure and the same number of valence electrons as the elemental semiconductors Si and Ge. In GaAs, for example, each Ga atom is surrounded tetrahedrally by four As atoms and each As atom is surrounded tetrahedrally by four Ga atoms. The average number of valence electrons per atom is four—three from Ga and five from As—so the valence band is nearly filled and the conduction band is nearly empty, as is the case for pure Si and Ge. *n*-Type GaAs can be made by doping with an element having one more valence electron than As (for example, Se), and *p*-type GaAs can be made by doping with an element having one less valence electron than Ga (for example, Zn).

In one method for manufacturing LEDs, thin films of *n*-type and *p*-type semiconductors are layered onto a transparent substrate by a process called *chemical vapor deposition*. For example, a GaAs layer can be deposited by bringing highly purified vapors of trimethylgallium, $Ga(CH_3)_3$, and arsine, AsH_3, into contact with the hot substrate at about 1000 °C.

$$Ga(CH_3)_3(g) + AsH_3(g) \xrightarrow{1000\,°C} GaAs(s) + 3\,CH_4(g)$$

Because mixtures of different 3–5 semiconductors can form a continuous series of solid solutions, the band gap and thus the color of the light emitted by an LED can be tuned by varying the composition of the mixture. For example, GaAs and GaP form solid solutions having composition GaP_xAs_{1-x} $(0 \le x \le 1)$. The band gaps of these solutions vary from ~135 kJ/mol, corresponding to infrared light, to ~222 kJ/mol, corresponding to green light, as the value of x increases from 0 to 1. The familiar red light emitted by many commercial LEDs

▲ LED vehicle brake lights.

is produced by $GaP_{0.40}As_{0.60}$, which has a band gap of 181 kJ/mol. Light in the blue and green regions of the spectrum can be obtained from solid solutions of gallium and indium nitrides, $Ga_xIn_{1-x}N$.

Red, green, and blue LEDs are combined to produce the numerous colors displayed on sports stadium video screens and large, outdoor message boards. For example, the world's largest high-definition video board, 200 ft wide by 80 ft tall, at the Charlotte Motor Speedway contains more than 9 million LEDs that combine to produce a vast array of colors.

Compared with incandescent lightbulbs, LEDs are smaller, brighter, longer lived, and more energy efficient, and they have faster switching times. For example, LED vehicle brake lights illuminate 0.2 seconds faster than conventional incandescent brake lights, an important safety feature. The faster illumination gives the driver of a trailing car an additional stopping distance of 19 feet at a speed of 65 mph.

Diode Lasers

The word *laser* is an acronym for *l*ight *a*mplification by *s*timulated *e*mission of *r*adiation. Both LEDs and diode lasers produce light as a result of the combination of electrons and holes in the region of a *p–n* junction. Compared to an LED, however, the light from a diode laser is more intense, more highly directional, and all of the same frequency and phase.

The essential features of a diode laser are (1) a very high forward bias, which sends a very large number of electrons and holes into the *p–n* junction, and (2) a laser cavity with a structure and dimensions such that the emitted light bounces back and forth within the material, thus stimulating a cascade of combinations of electrons and holes, which greatly amplifies the amount of light produced. Diode lasers are widely used in laser pointers and printers, bar-code readers, compact disc players, and fiber optic systems for data transmission.

Photovoltaic (Solar) Cells

Like an LED, a photovoltaic cell contains a *p–n* junction, but the two devices involve opposite processes. Whereas an LED converts electrical energy to light, a photovoltaic, or solar, cell converts light to electricity. When light with an energy greater than or equal to the band gap shines on a *p–n* junction, electrons are excited from the valence band of the *p*-type semiconductor into the conduction band of the *n*-type semiconductor (**Figure 21.14**). If the *p–n* junction is made part of an electrical circuit, electrons will flow through the junction from the *p*-side to the *n*-side because of the negative (downhill) slope of the conduction band in

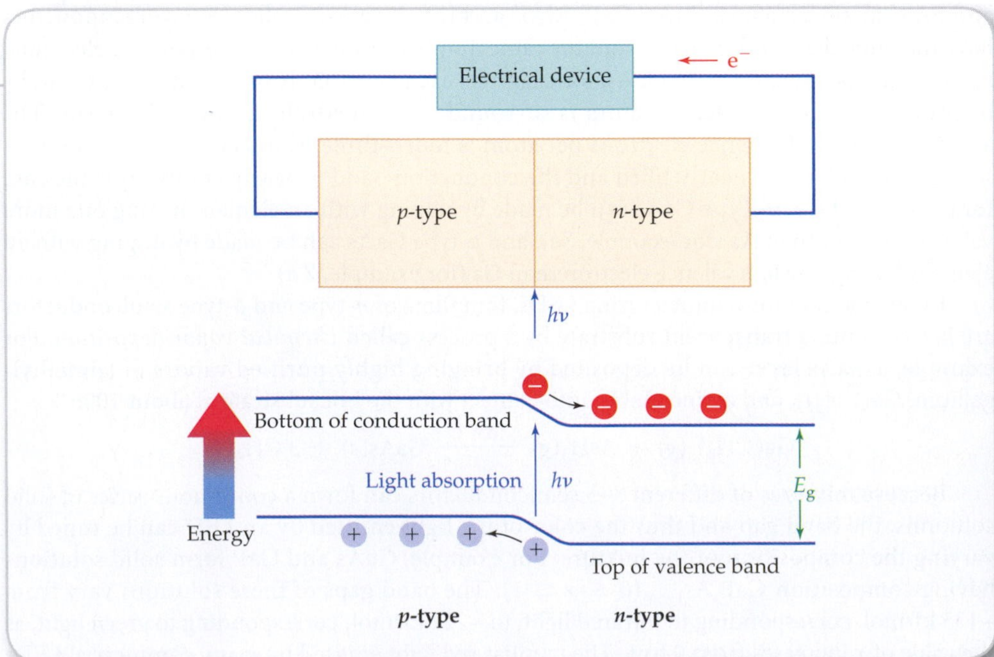

FIGURE 21.14

A schematic of a photovoltaic (solar) cell, along with the corresponding band theory energy-level diagram.

the region of the junction. Positive holes left behind in the valence band move in the opposite direction. The resulting current can be used to charge a battery or power an electrical device, such as a calculator or a lightbulb.

Silicon, which has a band gap of 107 kJ/mol, is commonly used to make solar cells; it absorbs sunlight with a wavelength less than about 1100 nm (ultraviolet, visible, and some of the infrared). Because the efficiency of light-to-electricity conversion in presently available solar cells is only about 20%, current research focuses on increasing the efficiency of these cells and reducing their cost.

Transistors

The transistor, invented in 1947 at Bell Laboratories, is an essential component in radios, television sets, computers, cell phones, and a host of other electronic products. Transistors consist of *n–p–n* or *p–n–p* junctions that control and amplify electrical signals in modern integrated circuits. An amazing number of these extremely small devices can be packed into a small space, thus decreasing the size and increasing the speed of electrical equipment. For example, computer microprocessors now contain tens of billions of transistors on a silicon chip with a surface area of about 2 cm^2 and are able to execute more than 100 billion instructions per second.

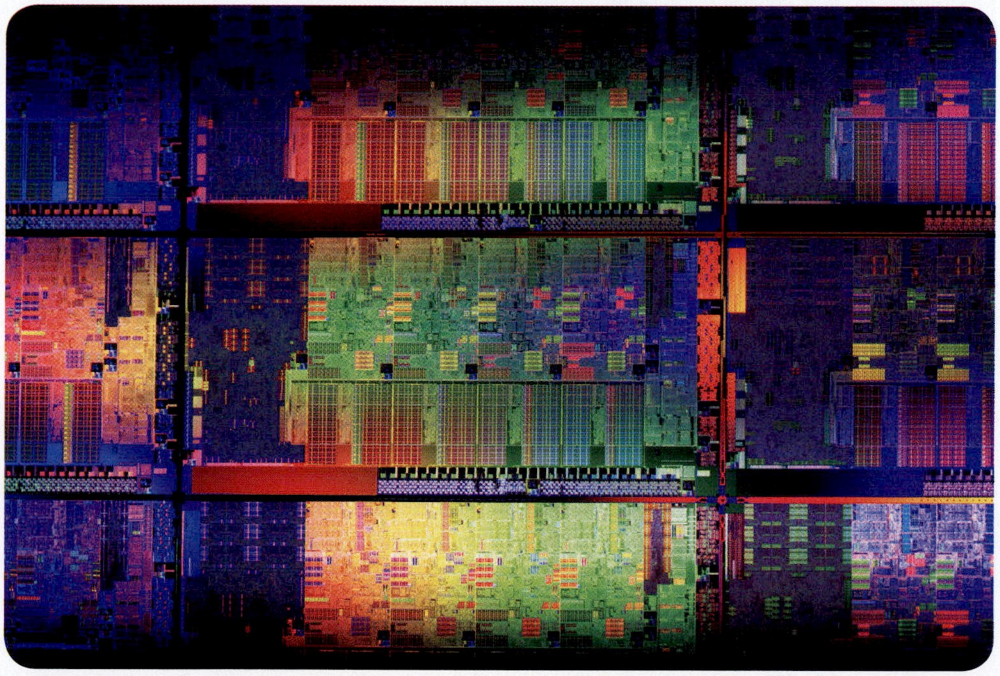

◀ An Intel Core vPro Processor.

PROBLEM 21.8 What is the wavelength of the light emitted by a GaP LED with a band gap of 222 kJ/mol?

21.7 SUPERCONDUCTORS

The discovery of high-temperature superconductors is one of the more exciting scientific developments in the last 25 years. It has stimulated an enormous amount of research in chemistry, physics, and materials science that could some day lead to a world of superfast computers, magnetically levitated trains, and power lines that carry electric current without loss of energy.

A **superconductor** is a material that loses all electrical resistance below a characteristic temperature called the **superconducting transition temperature,** T_c. This phenomenon was discovered in 1911 by the Dutch physicist Heike Kamerlingh Onnes, who found that mercury abruptly loses its electrical resistance when it is cooled with liquid helium to 4.2 K (**Figure 21.15**). Below its T_c, a superconductor becomes a perfect conductor and an electric current, once started, flows indefinitely without loss of energy.

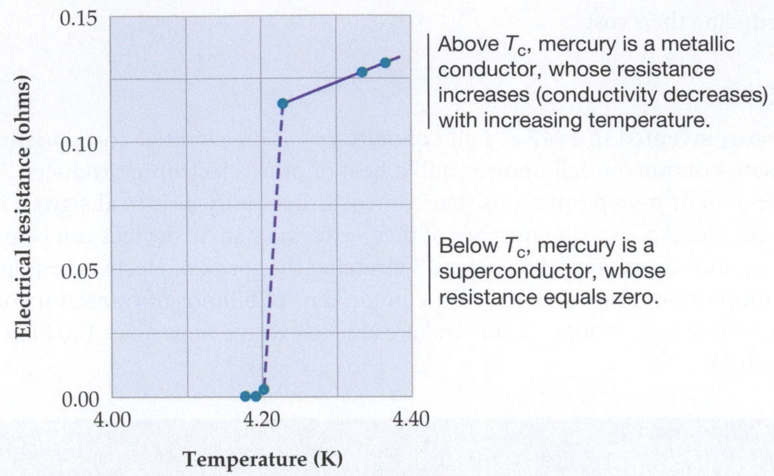

FIGURE 21.15

The electrical resistance of mercury falls to zero at its superconducting transition temperature, $T_c = 4.2$ K.

Ever since the initial discovery in 1911, scientists have been searching for materials that superconduct at higher temperatures, and more than 6000 superconductors are now known. Until 1986, however, the record value of T_c was only 23.2 K, for the compound Nb_3Ge. The situation changed dramatically in 1986, when K. Alex Müller and J. Georg Bednorz of the IBM Zürich Research Laboratory reported a T_c of 35 K for the nonstoichiometric barium lanthanum copper oxide $Ba_xLa_{2-x}CuO_4$, where x has a value of about 0.1. Soon thereafter, scientists found even higher values of T_c for other copper-containing oxides: 90 K for $YBa_2Cu_3O_7$, 125 K for $Tl_2Ba_2Ca_2Cu_3O_{10}$, 133 K for $HgBa_2Ca_2Cu_3O_{8+x}$, and 138 K for $Hg_{0.8}Tl_{0.2}Ba_2Ca_2Cu_3O_{8.33}$, the recordholder as of this writing. High values of T_c for these compounds were completely unexpected because most metal oxides—nonmetallic inorganic solids called *ceramics*—are electrical insulators. Within just one year of discovering the first ceramic superconductor, Müller and Bednorz were awarded the 1987 Nobel Prize in Physics.

The crystal structure of $YBa_2Cu_3O_7$, the so-called 1-2-3 compound (1 yttrium, 2 bariums, and 3 coppers), is illustrated in **Figure 21.16**, and one unit cell of the structure is shown in **Figure 21.17**. The crystal contains parallel planes of Y, Ba, and Cu atoms. Two-thirds of the

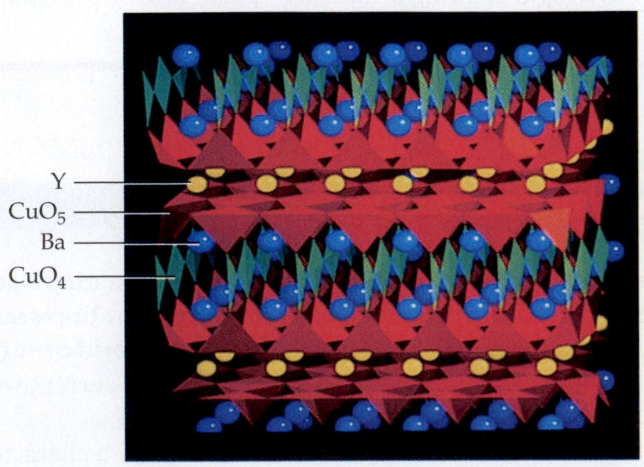

FIGURE 21.16

A computer graphics representation of the crystal structure of $YBa_2Cu_3O_7$.

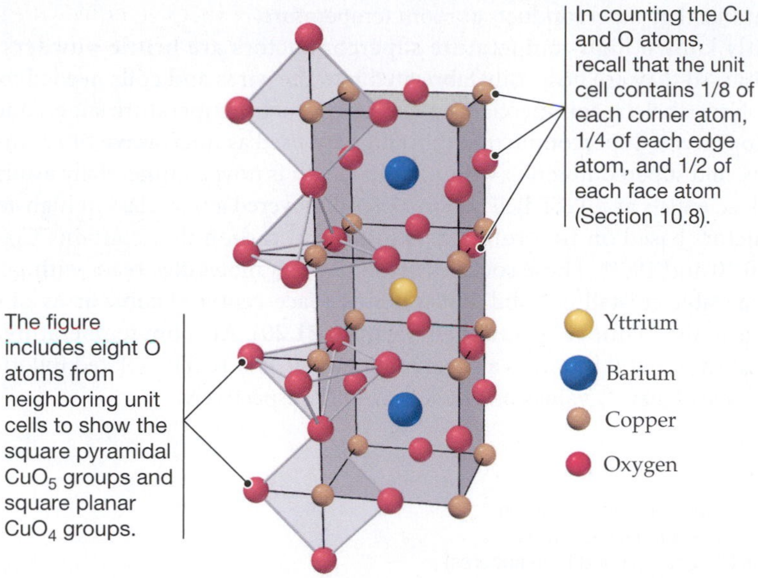

FIGURE 21.17

One unit cell of the crystal structure of $YBa_2Cu_3O_7$. The unit cell contains one Y atom, two Ba atoms, three Cu atoms, and seven O atoms.

Cu atoms are surrounded by a square pyramid of five O atoms, some of which are shared with neighboring CuO_5 groups to give two-dimensional layers of square pyramids. The remaining Cu atoms are surrounded by a square of four O atoms, two of which are shared with neighboring CuO_4 squares to give chains of CuO_4 groups. It's interesting to note that the Cu atoms have a fractional oxidation number of +2.33, based on the usual oxidation numbers of +3 for Y, +2 for Ba, and −2 for O. Both the infinitely extended layers of Cu and O atoms and the fractional oxidation number of Cu appear to play a role in the current flow, but a generally accepted theory of superconductivity in ceramic superconductors is not yet available. This is a field where experiment is far ahead of theory.

One of the most dramatic properties of a superconductor is its ability to levitate a magnet. When a superconductor is cooled below its T_c and a magnet is lowered toward it, the superconductor and the magnet repel each other, and the magnet hovers above the superconductor as though suspended in midair (**Figure 21.18**).

FIGURE 21.18

Levitation of a magnet above a pellet of $YBa_2Cu_3O_7$ cooled to 77 K with liquid nitrogen. $YBa_2Cu_3O_7$ becomes a superconductor at approximately 90 K.

The force responsible for levitation arises in the following way: When the magnet moves toward the superconductor, it induces a supercurrent in the surface of the superconductor that continues to flow even after the magnet stops moving. The supercurrent, in turn, induces a magnetic field in the superconductor that exactly cancels the field from the magnet. Thus, the net magnetic field within the bulk of the superconductor is zero, a phenomenon called the *Meissner effect*. Outside the superconductor, however, the magnetic fields due to the magnet and the supercurrent repel each other, just as the north poles of two bar magnets do. The magnet therefore experiences an upward magnetic force as well as the usual downward gravitational force, and it remains suspended above the superconductor at the point where the two forces are equal. Potential applications of the Meissner effect include high-speed, magnetically levitated trains, such as the one now operating in Shanghai, China (**Figure 21.19**).

FIGURE 21.19

A Maglev train is now in commercial operation in Shanghai, China, running at 430 km/h.

Other common applications of superconductors exist. For example, powerful superconducting magnets are essential components in the magnetic resonance imaging (MRI) instruments widely used in medical diagnosis. Superconductors are also used to make the magnets that bend the path of charged particles in high-energy particle accelerators. These applications, however, use conventional superconductors ($T_c \leq 20$ K) that are cooled to 4.2 K with liquid helium, an expensive substance that requires sophisticated cryogenic (cooling) equipment. Much of the excitement surrounding the high-temperature superconductors arises because their T_c values are above the boiling point of liquid nitrogen (bp 77 K), an abundant refrigerant that is cheaper than milk. Of course, the search goes on for materials with still higher values of T_c. For applications such as long-distance electric power transmission, the goal is a material that superconducts at room temperature.

Presently known high-temperature superconductors are brittle powders with high melting points, so they are not easily fabricated into the wires and coils needed for electrical equipment. Nevertheless, commercial applications of high-temperature superconductors are beginning to emerge. Superconducting thin films are used as microwave filters in cell-phone base stations, and superconducting wire 1 km in length is now commercially available.

REMEMBER...

Carbon exists in more than 40 known structural forms, or allotropes, most of which are amorphous. Three of the most common crystalline forms are graphite, diamond, and **fullerene**. (Sections 10.10 and 19.7)

In 1991, scientists at AT&T Bell Laboratories discovered a new class of high-temperature superconductors based on **fullerene**, the allotrope of carbon that contains C_{60} molecules (Sections 10.10 and 19.7). These soccer ball-shaped C_{60} molecules react with potassium to give K_3C_{60}, a stable crystalline solid that contains a face-centered cubic array of $C_{60}{}^{3-}$ ions, with K^+ ions in the cavities between them (**Figure 21.20**). At room temperature, K_3C_{60} is a metallic conductor, but it becomes a superconductor at 19.5 K. The related fullerides Rb_3C_{60} and Cs_3C_{60} have higher T_c values of 29.5 K and 40 K, respectively.

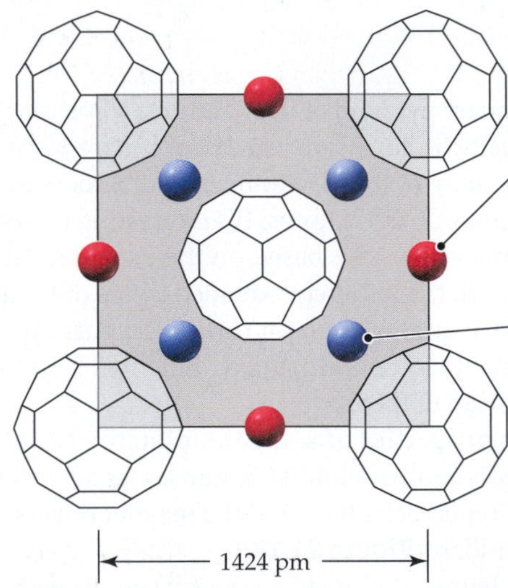

FIGURE 21.20

A portion of one unit cell of the face-centered cubic structure of K_3C_{60} viewed perpendicular to a cube face.

Still another class of high-temperature superconductors has been discovered very recently. In February of 2008 scientists at the Tokyo Institute of Technology reported that a fluorine-doped lanthanum oxide iron arsenide, $LaO_{1-x}F_xFeAs$ ($x = 0.11$), superconducts at 26 K. Subsequent replacement of lanthanum by other lanthanide elements has raised the value of T_c to a current maximum of 55 K for the fluorine-doped samarium compound.

PROBLEM 21.9 Show that one unit cell of $YBa_2Cu_3O_7$ (Figure 21.17) contains three Cu atoms and seven O atoms.

21.8 CERAMICS

Ceramics are inorganic, nonmetallic, nonmolecular solids, including both crystalline materials such as quartz (SiO_2) and amorphous materials such as glasses. Known since ancient times, traditional silicate ceramics, such as pottery and porcelain, are made by heating aluminosilicate clays to high temperatures. Modern, so-called **advanced ceramics**—materials that have high-tech engineering, electronic, and biomedical applications—include *oxide ceramics*, such as alumina (Al_2O_3), and *nonoxide ceramics*, such as silicon carbide (SiC) and silicon nitride (Si_3N_4). Additional examples are listed in **Table 21.4**, which compares the properties of ceramics with those of aluminum and steel. Note that oxide ceramics are named by replacing the *-um* ending of the element name with an *-a*; thus, BeO is beryllia, ZrO_2 is zirconia, and so forth.

TABLE 21.4 Properties of Some Ceramic and Metallic Materials

Material	Melting Point (°C)	Density (g/cm^3)	Elastic Modulus (GPa)*	Hardness (Mohs scale)†
Oxide ceramics				
Alumina, Al_2O_3	2054	3.99	380	9
Beryllia, BeO	2578	3.01	370	8
Zirconia, ZrO_2	2710	5.68	210	8
Nonoxide ceramics				
Boron carbide, B_4C	2350	2.50	280	9
Silicon carbide, SiC	2830	3.16	400	9
Silicon nitride, Si_3N_4	1900	3.17	310	9
Metals				
Aluminum	660	2.70	70	3
Plain carbon steel	1515	7.86	205	5

*The elastic modulus, measured in units of pressure (1 gigapascal = 1 GPa = 10^9 Pa) indicates the stiffness of a material when it is subjected to a load. The larger the value, the stiffer the material.

†Numbers on the Mohs hardness scale range from 1 for talc, a very soft material, to 10 for diamond, the hardest known natural substance.

In many respects, the properties of ceramics are superior to those of metals: Ceramics have higher melting points and are stiffer, harder, and more resistant to wear and corrosion. Moreover, they maintain much of their strength at high temperatures, where metals either melt or corrode because of oxidation. Silicon nitride and silicon carbide, for example, are stable to oxidation in air up to 1400–1500 °C, and oxide ceramics don't react with oxygen because they are already fully oxidized. Because ceramics are less dense than steel, they are attractive lightweight, high-temperature materials for replacing metal components in aircraft, space vehicles, and automotive engines.

▲ A silicon nitride rotor for use in gas-turbine engines.

Unfortunately, ceramics are brittle, as anyone who has dropped a coffee cup well knows. The brittleness, hardness, stiffness, and high melting points of ceramics are all due to strong chemical bonding. Take silicon carbide, for example, a covalent network solid that crystallizes in the diamond structure (**Figure 21.21**). Each Si atom is bonded tetrahedrally to four C atoms, and each C atom is bonded tetrahedrally to four Si atoms. The strong, highly directional covalent bonds ($D_{Si-C} = 435$ kJ/mol) prevent the planes of atoms from sliding over one another when the solid is subjected to the stress of a load or an impact. As a result, the solid can't deform to relieve the stress. It maintains its shape up to a point, but then the bonds give way suddenly, and the material fails catastrophically when the stress exceeds a certain threshold value. Oxide ceramics, in which the bonding is largely ionic, behave similarly. By contrast, metals are able to deform under stress because their planes of metal cations can slide easily in the electron sea (Section 21.4). As a result, metals dent but ceramics shatter.

Ceramic processing, the series of steps that leads from raw material to the finished ceramic object, is important in determining the strength and the resistance to fracture of the product. Processing often begins with a fine powder, which is combined with an organic binder, shaped, compacted, and finally *sintered* at temperatures of 1300–2000 °C. **Sintering**, which occurs below the melting point, is a process in which the particles of the powder are

FIGURE 21.21

One unit cell of the cubic form of silicon carbide, SiC. The crystal can't deform under stress because the bonds are strong and highly directional.

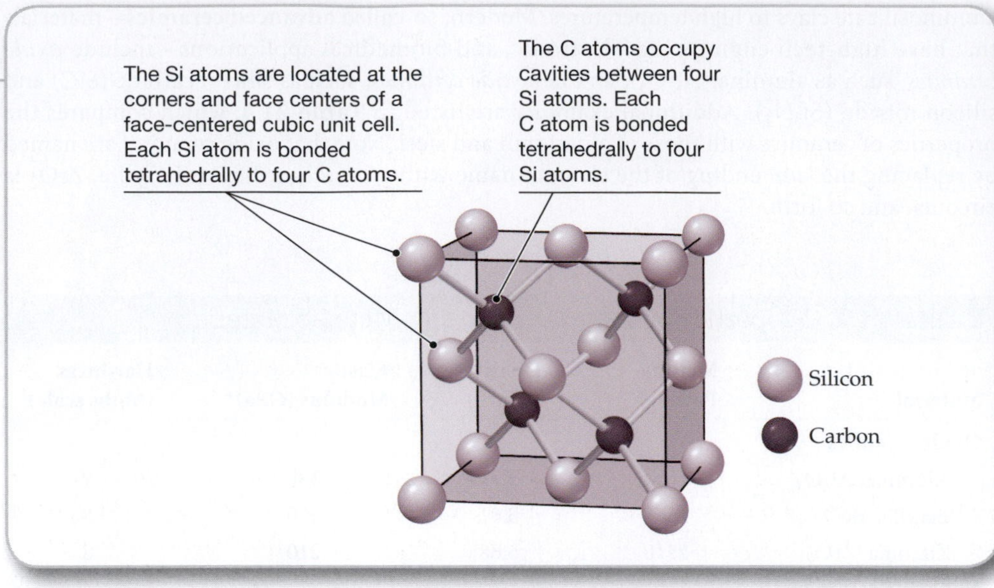

welded together without completely melting. During sintering, the crystal grains grow larger and the density of the material increases as the void spaces between particles disappear. Unfortunately, impurities and remaining voids can lead to microscopic cracks that cause the material to fail under stress. It is therefore important to minimize impurities and voids by beginning with high-purity, fine powders that can be tightly compacted prior to sintering.

One approach to making such powders is the **sol–gel method**, in which a metal oxide powder is synthesized from a metal alkoxide, a compound derived from a metal and an alcohol. In the synthesis of titania (TiO_2) from titanium ethoxide, $Ti(OCH_2CH_3)_4$, for example, the $Ti(OCH_2CH_3)_4$ starting material is made by the reaction of titanium(IV) chloride with ethanol and ammonia in a benzene solution:

$$\underset{}{TiCl_4} + \underset{\text{Ethanol}}{4\,HOCH_2CH_3} + 4\,NH_3(g) \xrightarrow{\text{Benzene}} \underset{\text{Titanium ethoxide}}{Ti(OCH_2CH_3)_4} + 4\,NH_4Cl(s)$$

Pure $Ti(OCH_2CH_3)_4$ is then dissolved in an appropriate organic solvent, and water is added to bring about a *hydrolysis* reaction:

$$Ti(OCH_2CH_3)_4 + 4\,H_2O \xrightarrow{\text{Solution}} Ti(OH)_4(s) + 4\,HOCH_2CH_3$$

In this reaction, Ti–OCH_2CH_3 bonds are broken, Ti–OH bonds are formed, and ethanol is regenerated. The $Ti(OH)_4$ forms in the solution as a colloidal dispersion called a *sol*, consisting of extremely fine particles having a diameter of only 0.001–0.1 μm. Subsequent reactions eliminate water molecules and form oxygen bridges between Ti atoms:

$$(HO)_3Ti{-}O{-}H + H{-}O{-}Ti(OH)_3 \longrightarrow (HO)_3Ti{-}O{-}Ti(OH)_3 + H_2O$$

Because all the OH groups can undergo this reaction, the particles of the sol link together through a three-dimensional network of oxygen bridges and the sol is converted to a more rigid, gelatinlike material called a *gel*. The remaining water and solvent are then removed by heating the gel, and TiO_2 is obtained as a fine powder consisting of high-purity particles with a diameter less than 1 μm (**Figure 21.22**).

Oxide ceramics have many important uses. Alumina, for example, is the material of choice for making spark-plug insulators because of its high electrical resistance, high strength, high thermal stability, and chemical inertness. Because alumina is nontoxic and essentially inert in biological systems, it is used in constructing dental crowns and the heads of artificial hips. Alumina is also used as a substrate material for electronic circuit boards. High purity silica (SiO_2) fibers were used to make the lightweight, heat-resistant ceramic tiles that protected the space shuttle on re-entry into the atmosphere. The shuttle was covered with more than 20,000 tiles with a density less than that of Styrofoam and able to withstand temperatures up to 1600 °C.

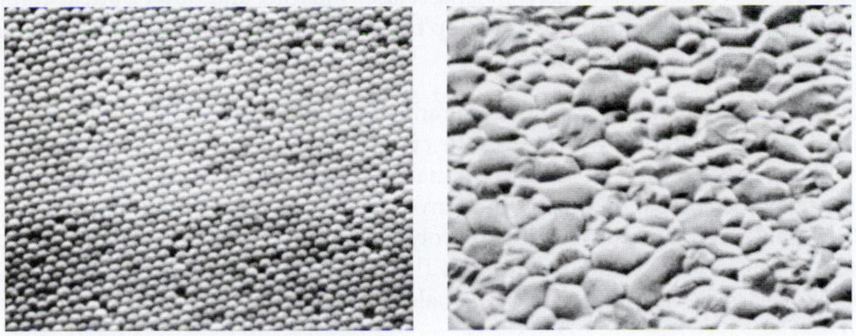

FIGURE 21.22

Electron micrographs of a titania powder (left) and the dense ceramic produced by sintering the powder (right). The powder, consisting of tightly packed particles of TiO_2 with a diameter less than 1 μm, was made by the sol–gel method.

In the military, ceramics are widely used because they are lightweight and are strong enough to shatter and deflect bullets and other projectiles. For example, body armor, once made of steel (density, 7.86 g/cm^3), is now made of light ceramic materials such as boron carbide (density, 2.50 g/cm^3). Use of lightweight ceramic armor panels in place of steel in combat vehicles and military aircraft has increased maneuverability and decreased fuel consumption.

Worked Example 21.4

Synthesizing the 1-2-3 Superconductor by Hydrolysis of a Mixture of Metal Ethoxides

The 1-2-3 ceramic superconductor $YBa_2Cu_3O_7$ has been synthesized by the sol–gel method from a stoichiometric mixture of yttrium ethoxide, barium ethoxide, and copper(II) ethoxide in an appropriate organic solvent. The oxide product, before being heated in oxygen, has the formula $YBa_2Cu_3O_{6.5}$. Write a balanced equation for the hydrolysis of the stoichiometric mixture of metal ethoxides.

STRATEGY

First, determine the chemical formulas of the metal ethoxides, and then write a balanced equation for the reaction of the mixture of metal ethoxides with water.

SOLUTION

The ethoxide ligand is the anion of ethanol, $HOCH_2CH_3$, and has a charge of −1. Because yttrium is a group 3B element and has an oxidation number of +3, the formula of yttrium ethoxide must be $Y(OCH_2CH_3)_3$. Similarly, because both barium (in group 2A) and copper(II) have an oxidation number of +2, the formulas of barium ethoxide and copper ethoxide must be $Ba(OCH_2CH_3)_2$ and $Cu(OCH_2CH_3)_2$, respectively.

Hydrolysis of $Y(OCH_2CH_3)_3$, which breaks the Y–OCH_2CH_3 bonds and forms Y-OH bonds, requires one H_2O molecule for each ethoxide ligand:

$$Y(OCH_2CH_3)_3 + 3\,H_2O \longrightarrow Y(OH)_3 + 3\,HOCH_2CH_3$$

Similarly, hydrolysis of $Ba(OCH_2CH_3)_2$ and $Cu(OCH_2CH_3)_2$ yields $Ba(OH)_2$ and $Cu(OH)_2$, respectively:

$$Ba(OCH_2CH_3)_2 + 2\,H_2O \longrightarrow Ba(OH)_2 + 2\,HOCH_2CH_3$$
$$Cu(OCH_2CH_3)_2 + 2\,H_2O \longrightarrow Cu(OH)_2 + 2\,HOCH_2CH_3$$

Because the three metal ethoxides are present together in a 1:2:3 ratio in the synthesis of the superconductor, we can write the product of the hydrolysis as $Y(OH)_3 \cdot 2\,Ba(OH)_2 \cdot 3\,Cu(OH)_2$, or $YBa_2Cu_3(OH)_{13}$. Thus, the hydrolysis reaction requires 13 H_2O molecules for 13 OCH_2CH_3 ligands, and the balanced equation is

$$Y(OCH_2CH_3)_3 + 2\,Ba(OCH_2CH_3)_2 + 3\,Cu(OCH_2CH_3)_2 + 13\,H_2O \longrightarrow YBa_2Cu_3(OH)_{13} + 13\,HOCH_2CH_3$$

Subsequent heating of $YBa_2Cu_3(OH)_{13}$ removes water, converting the mixed-metal hydroxide to the oxide $YBa_2Cu_3O_{6.5}$, which is then oxidized to $YBa_2Cu_3O_7$ by heating in O_2 gas.

PROBLEM 21.10 Silica glasses used in lenses, laser mirrors, and other optical components can be made by the sol–gel method. One step in the process is the hydrolysis of $Si(OCH_3)_4$. Write a balanced equation for the reaction.

PROBLEM 21.11 Crystals of the oxide ceramic barium titanate, $BaTiO_3$, have an unsymmetrical arrangement of ions, which gives the crystals an electric dipole moment. Such materials are called ferroelectrics and are used to make various electronic devices. Barium titanate can be made by the sol–gel method, which involves hydrolysis of a mixture of metal alkoxides. Write a balanced equation for the hydrolysis of a 1:1 mixture of barium isopropoxide and titanium isopropoxide, and explain how the resulting sol is converted to $BaTiO_3$. (The isopropoxide ligand is the anion of isopropyl alcohol, $HOCH(CH_3)_2$, also known as rubbing alcohol.)

21.9 COMPOSITES

In the previous section, we saw that ceramics are brittle and prone to fracture. They can be strengthened and toughened, though, by mixing the ceramic powder prior to sintering with fibers of a second material, such as carbon, boron, or silicon carbide. The resulting hybrid material, called a **ceramic composite**, combines the advantageous properties of both components. An example is the composite consisting of fine grains of alumina reinforced with *whiskers* of silicon carbide. Whiskers are tiny, fiber-shaped particles, about 0.5 μm in diameter and 50 μm long, that are very strong because they are single crystals. Silicon carbide-reinforced alumina possesses high strength and high shock resistance, even at high temperatures, and has therefore been used to make high-speed cutting tools for machining very hard steels.

How do fibers and whiskers increase the strength and fracture toughness of a composite material? First, fibers have great strength along the fiber axis because most of the chemical bonds are aligned in that direction. Second, there are several ways in which fibers can prevent microscopic cracks from propagating to the point that they lead to the fracture of the material. The fibers can deflect cracks, thus preventing them from moving cleanly in one direction, and they can bridge cracks, thus holding the two sides of a crack together.

Silicon carbide-reinforced alumina is a composite in which both the fibers and the surrounding matrix are ceramics. There are other composites, however, in which the two phases are different types of materials. Examples are **ceramic–metal composites**, or **cermets**, such as aluminum metal reinforced with boron fiber, and **ceramic–polymer composites**, such as boron/epoxy and carbon/epoxy. (Epoxy is a resin consisting of long-chain organic molecules.) These materials are popular for aerospace and military applications because of their high

▶ The skin of the B-2 advanced technology aircraft is a strong, lightweight composite material that contains carbon fibers.

strength-to-weight ratios. Boron-reinforced aluminum, for example, was used as a lightweight structural material in the space shuttle, and boron/epoxy and carbon/epoxy skins are used on military aircraft. Increased use of composite materials in commercial aircraft could result in weight savings of 20–30% and corresponding savings in fuel, which accounts for as much as 40% of the cost of operating an airline. Carbon/epoxy composites are also used to make strong, lightweight sports equipment, such as golf clubs, tennis racquets, fishing rods, and bicycles.

Ceramic fibers used in composites are usually made by high-temperature methods. Carbon (graphite) fiber, for example, can be made by the thermal decomposition of fibers of polyacrylonitrile, a long-chain organic molecule also used to make the textile Orlon:

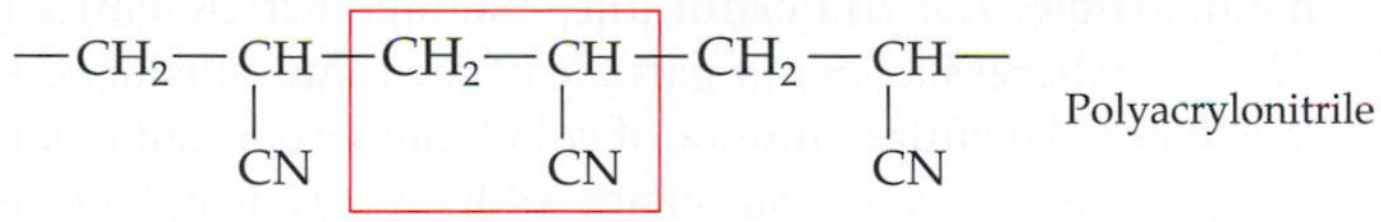

Repeating unit

In the final step of the multistep process, the carbon in the fiber is converted to graphite by heating at 1400–2500 °C. Similarly, silicon carbide fiber can be made by heating fibers that contain long-chain molecules with alternating silicon and carbon atoms:

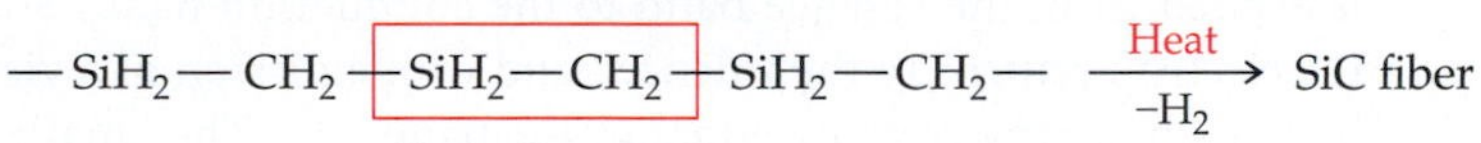

Repeating unit

PROBLEM 21.12 Classify each of these composites as ceramic–ceramic, ceramic–metal, or ceramic–polymer:

(a) Cobalt/tungsten carbide
(b) Silicon carbide/zirconia
(c) Boron nitride/epoxy
(d) Boron carbide/titanium

FYI WILL THE NEXT BIG THING BE REALLY SMALL?

Imagine a world of new lightweight replacements for metals, synthetic scaffolds on which bones can be re-grown, and drugs that target and kill cancer cells with a minimum of side effects. Those are but a few of the developments that might emerge from *nanotechnology*, one of the hottest research areas in science today.

Nanotechnology is the study and production of materials and structures that have at least one dimension between 1 nm and 100 nm, where one nanometer is one billionth of a meter ($1\ nm = 10^{-9}\ m$). Nanotechnology is an explosively growing, multidisciplinary enterprise, spanning the fields of chemistry, physics, biology, medicine, materials science, environmental science, and engineering.

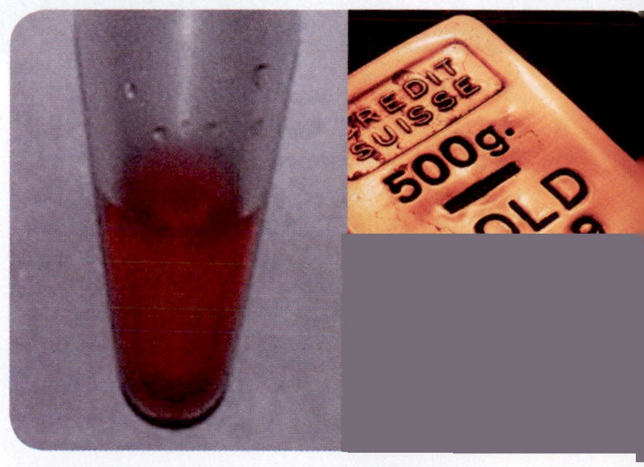

▲ Wine-colored, 15 nm gold nanoparticles in solution (left) and a bar of bulk gold (right).

To appreciate the extremely small size of nanomaterials, it's helpful to note that 1 nm is the width of four H atoms laid side by side and about 50,000 times smaller than the diameter of a human hair. Because of their small size, nanoparticles have unique properties. They tend to have lower melting points, different colors, and greater reactivity than bulk materials. For instance, gold nanoparticles are red and chemically reactive, whereas the bulk metal is yellow and inert. The lower melting points and greater reactivity of nanoparticles result because a substantial percentage of the atoms in a nanoparticle are on the surface or on edges and are therefore bound less tightly than those within the bulk solid; for a 5 nm particle, about 30% of the atoms are on the surface.

▲ The color of these semiconductor nanoparticles, called quantum dots, varies from red to violet as the size of the nanoparticles decreases.

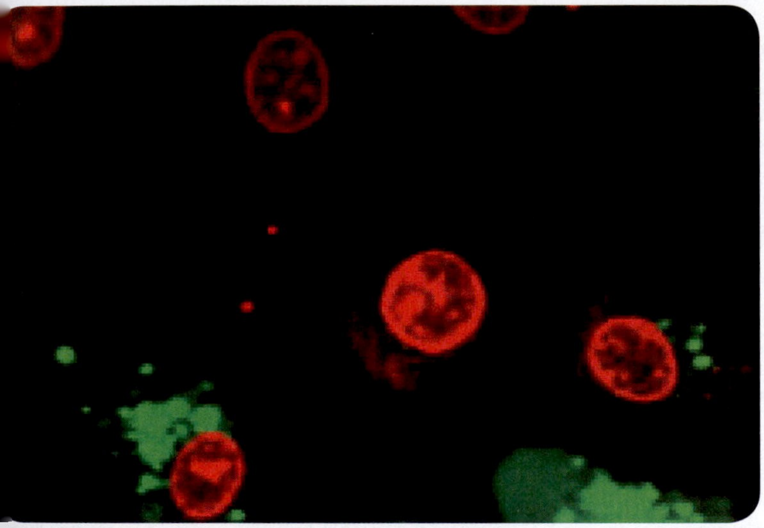

▲ Upon excitation with light, nanoparticles that contain a ruthenium chelate generate a highly reactive form of oxygen that is lethal to rat cancer cells. Dead cells are shown in red, live cells in green.

Because nanoparticles have dimensions intermediate between those of individual atoms and bulk solids, the spacing of their electronic energy levels is intermediate between the widely separated energy levels of individual atoms and the nearly continuous band of energy levels in a bulk metal. Moreover, the energy spacing depends on the size of the nanoparticle.

For example, when a so-called *quantum dot*, a semiconductor nanoparticle that is about 1–10 nm in diameter, is irradiated with ultraviolet light, it emits visible light having a wavelength that depends on the size of the nanoparticle. A 3 nm cadmium selenide particle emits green light at 520 nm, whereas a 5.5 nm particle of the same substance emits red light at 620 nm. An entire rainbow of colors can be emitted by differently sized nanoparticles of a single substance, with the wavelength of the emitted light decreasing from the red to the violet part of the spectrum as the size of the particle decreases.

Color, as we've seen in Sections 2.9 and 20.10, is due to electronic transitions. When a quantum dot is irradiated with ultraviolet light, an electron is excited from the valence band to the conduction band. Subsequently, the electron returns to the valence band with emission of a visible photon having an energy roughly equal to the band gap. The smaller the nanoparticle, the larger the band gap and the greater the shift in the color of the emitted light from the red to the violet.

Because of their distinctive optical properties, quantum dots may have numerous applications. For example, the surface of a quantum dot can be chemically modified by attaching an organic molecule that targets and bonds to a specific biomolecule, such as DNA or a protein. When a quantum dot bonded to a biomolecule is irradiated with ultraviolet light, it serves as a fluorescent probe, or label, for that particular biomolecule, with the color of the emitted light dependent on the size of the quantum dot. Thus, irradiation with a single light source of a mixture of quantum dots having different sizes and different surface modifications allows simultaneous detection, tracking, and imaging in color of a number of different biomolecules within a cell. Several types of nanoparticles, including quantum dots, are being investigated as potential weapons in the fight against cancer. For example, nanoparticles can be modified by attaching antibodies that target and bond to receptors found on the outside of cancer cell membranes but have little affinity for the membranes of healthy cells. Use of nanoparticles in imaging cancer cells may allow cancer to be detected at its earliest and most curable stages.

Although human cancer applications are still in the future, low-tech applications of nanoparticles exist already. For example, metal oxide nanoparticles are widely used in sunscreens and cosmetics, stain-resistant fabrics, dirt-repellent paints, and coatings for self-cleaning windows. Carbon nanotubes are used as reinforcing rods and toughening elements in structural composite materials, and gold nanoparticles are effective as highly selective oxidation catalysts.

PROBLEM 21.13 You wish to track the motions of two proteins within a cell. To protein A you attach a quantum dot that emits yellow light, and to protein B, a quantum dot of the same substance that emits blue light. Which quantum dot is larger? Explain.

PROBLEM 21.14 Calculate the percentage of the atoms that are on the surface of a cubic nanoparticle if the diameter of the atoms is 250 pm and the edge length of the particle is
(a) 5.0 nm **(b)** 10 nm

SUMMARY

Most metals occur in nature as **minerals**—the silicates, carbonates, oxides, sulfides, phosphates, and chlorides that make up the rocks of the Earth's crust. The type of mineral in which a metal is found depends on the metal's location in the periodic table. Minerals from which metals can be produced economically are called **ores**.

Metallurgy, the science and technology of extracting metals from their ores, involves three steps: (1) concentration and chemical treatment to separate the mineral from impurities called **gangue**; (2) reduction of the mineral to the free metal; and (3) refining or purification of the metal. The first step uses processes such as **flotation** and **roasting**. The reduction method (roasting, chemical reduction, or electrolysis) depends on the activity of the metal. The refining step involves distillation, chemical purification (for example, the **Mond process** for nickel), or electrorefining.

Iron is produced in a blast furnace by reducing iron oxide ores with CO. Added limestone ($CaCO_3$) decomposes to CaO, which reacts with SiO_2 and other acidic oxide impurities to yield a molten **slag** (mainly $CaSiO_3$) that is separated from the molten iron at the bottom of the furnace. The iron is converted to steel by the **basic oxygen process**, which removes impurities such as elemental Si, P, S, and most of the C.

Two bonding models are used for metals. The **electron-sea model** pictures a metal as an array of metal cations immersed in a sea of delocalized, mobile valence electrons that act as an electrostatic glue. In the molecular orbital theory for metals, also called **band theory**, the delocalized valence electrons occupy a vast number of MO energy levels that are so closely spaced that they merge into an almost continuous band. Both theories account for properties such as malleability, ductility, and high thermal and electrical conductivity, but band theory better explains how the number of valence electrons affects properties such as melting point and hardness.

Band theory also accounts for the electrical properties of metals, insulators, and semiconductors. Materials with partially filled bands are metallic conductors, and materials with only completely filled bands are electrical insulators. In insulators, the bonding MOs, called the **valence band**, and the antibonding MOs, called the **conduction band**, are separated in energy by a large **band gap**. In **semiconductors**, the band gap is smaller, and a few electrons have enough thermal energy to occupy the conduction band. The resulting partially filled valence and conduction bands give rise to a small conductivity. The conductivity can be increased by **doping**—adding a group 5A impurity to a group 4A element, which gives an ***n*-type semiconductor**, or adding a group 3A impurity to a group 4A element, which gives a ***p*-type semiconductor**. Doped semiconductors are essential components in modern solid-state electronic devices, such as diodes, light-emitting diodes (LEDs), diode lasers, photovoltaic (solar) cells, and transistors.

A **superconductor** is a material that loses all electrical resistance below a characteristic temperature called the **superconducting transition temperature, T_c**. An example is $YBa_2Cu_3O_7$ ($T_c = 90$ K). Below T_c, a superconductor can levitate a magnet, a consequence of the Meissner effect.

Ceramics are inorganic, nonmetallic, nonmolecular solids. Modern **advanced ceramics** include oxide ceramics such as Al_2O_3 and $YBa_2Cu_3O_7$ and nonoxide ceramics such as SiC and Si_3N_4. Ceramics are generally higher-melting, lighter, stiffer, harder, and more resistant to wear and corrosion than metals. One approach to ceramic processing is the **sol–gel method**, involving hydrolysis of a metal alkoxide. The strength and fracture toughness of ceramics can be increased by making **ceramic composites**, hybrid materials such as Al_2O_3/SiC in which fine grains of Al_2O_3 are reinforced by whiskers of SiC. Other types of composites include **ceramic–metal composites** and **ceramic–polymer composites**.

KEY WORDS

advanced ceramic *889*
alloy *869*
band gap *878*
band theory *876*
basic oxygen process *873*
ceramic *888*
ceramic composite *892*
ceramic–metal composite (cermet) *892*
ceramic–polymer composite *892*
conduction band *878*
doping *880*
electron-sea model *874*
flotation *869*
gangue *869*
metallurgy *869*
mineral *867*
Mond process *871*
***n*-type semiconductor** *880*
ore *868*
***p*-type semiconductor** *880*
roasting *870*
semiconductor *878*
sintering *889*
slag *873*
sol–gel method *890*
superconducting transition temperature, T_c *885*
superconductor *885*
valence band *878*

LEARNING OUTCOMES

- Relate the formulas of metal-containing minerals and the location of the metals in the periodic table. (Section 21.1) ***Problems 1, 15, 22–27, 114***
- Describe the concentration, reduction, and refining aspects of metallurgical processes, and write balanced equations for the reactions that occur when a metal is produced from its ores. (Section 21.2) ***Problems 2, 16, 30, 31, 34, 35***
- Describe the commercial production and purification iron metal and write balanced equations for the reactions that occur in these processes. (Section 21.3) ***Problems 3, 42–49, 121***
- Describe the electron-sea model of metals, and explain how it accounts for their properties. (Section 21.4) ***Problems 53–57, 116, 117***
- Describe the molecular orbital theory for metals (band theory), draw MO energy-level diagrams for metals, and use band theory to account for their properties. (Section 21.4) ***Problems 4, 5, 7, 17, 18, 20, 58–65, 80, 81, 116–118***
- Describe the bonding in semiconductors and account for their properties in terms of band theory. Classify doped semiconductors as *n*-type or *p*-type. (Section 21.5) ***Problems 6, 7, 17, 19, 20, 68–77, 80, 81, 119, 123, 127–129, 134***
- Describe some semiconductor applications. Relate the wavelength of light emitted by an LED or a diode laser to the band gap in its band theory energy-level diagram. (Section 21.6) ***Problems 8, 21, 82–87, 113***
- Describe the synthesis and properties of ceramics, and write balanced equations for the reactions that occur in the sol–gel synthesis method. (Section 21.8) ***Problems 10, 11, 92–105, 122, 124***
- Classify composites as ceramic–ceramic, ceramic–metal, or ceramic–polymer, and account for some of their properties. (Section 21.9) ***Problems 12, 108, 109***

CONCEPTUAL PROBLEMS

Problems 21.1–21.14 appear within the chapter.

21.15 Look at the location of elements A, B, C, and D in the following periodic table:

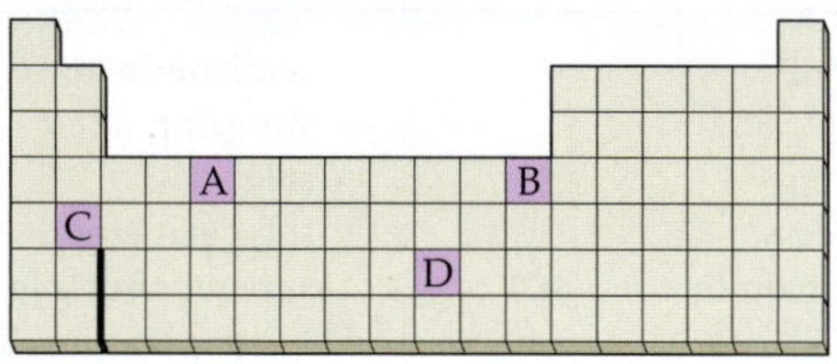

Without looking at Figure 21.2, predict whether these elements are likely to be found in nature as carbonates, oxides, sulfides, or in uncombined form. Explain.

21.16 The methods for extracting metals from their ores include (i) roasting a metal sulfide, (ii) chemical reduction of a metal oxide, and (iii) electrolysis. The preferred method depends on the $E°$ value for the reduction half-reaction $M^{n+}(aq) + n\,e^- \longrightarrow M(s)$.

(a) Which method is appropriate for the metals with the most negative $E°$ values?

(b) Which method is appropriate for the metals with the most positive $E°$ values?

(c) Which method is appropriate for metals A, B, C, and D in the following periodic table?

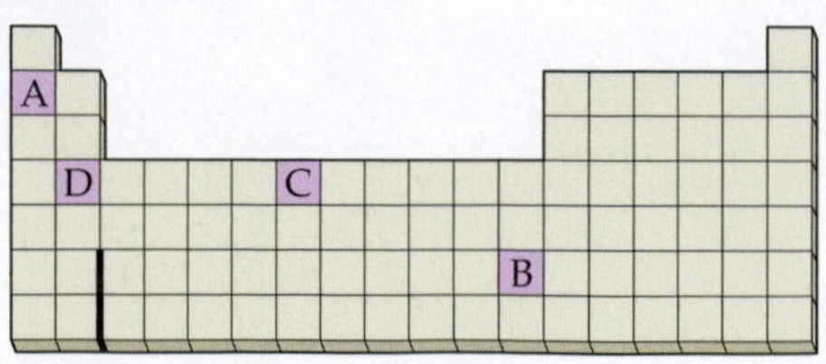

21.17 The following pictures show the electron populations of the bands of MO energy levels for four different materials:

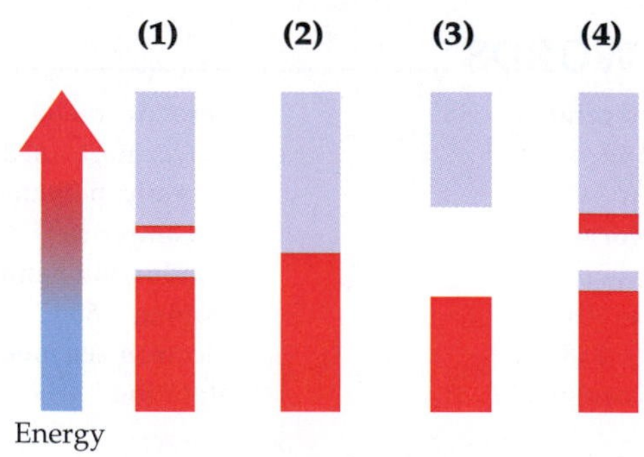

(a) Classify each material as an insulator, a semiconductor, or a metal.

(b) Arrange the four materials in order of increasing electrical conductivity. Explain.

(c) Tell whether the conductivity of each material increases or decreases when the temperature increases.

21.18 The following pictures show the electron populations of the composite *s–d* bands for three different transition metals:

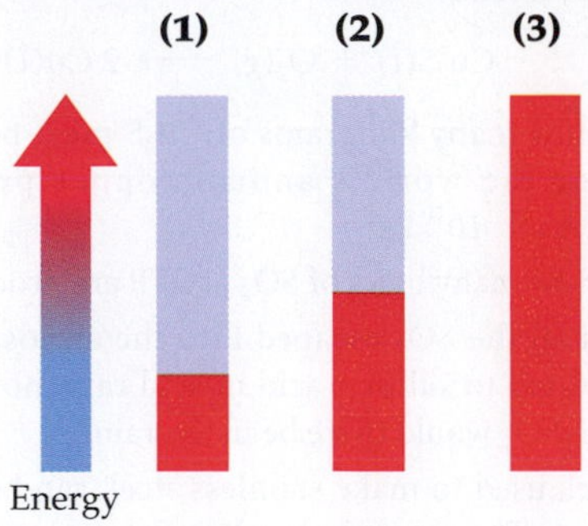

(a) Which metal has the highest melting point? Explain.

(b) Which metal has the lowest melting point? Explain.

(c) Arrange the metals in order of increasing hardness. Explain.

21.19 The following picture represents the electron population of the bands of MO energy levels for elemental silicon:

(a) Identify the valence band, conduction band, and band gap.

(b) In a drawing, show how the electron population changes when the silicon is doped with gallium.

(c) In a drawing, show how the electron population changes when the silicon is doped with arsenic.

(d) Compare the electrical conductivity of the doped silicon samples with that of pure silicon. Account for the differences.

21.20 Sketch the electron populations of the bands of MO energy levels for elemental carbon (diamond), silicon, germanium, gray tin, and white tin. (Band gap data are given in Table 21.3.) Your sketches should show how the populations of the different bands vary with a change in the group 4A element.

21.21 Pictures (1) and (2) are energy-level diagrams for two different LEDs. One LED emits red light, and the other emits blue light. Which one emits red, and which blue? Explain.

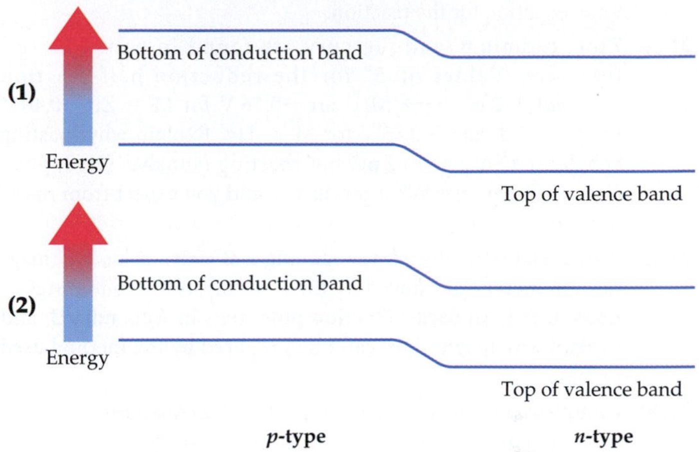

SECTION PROBLEMS

Sources of the Metallic Elements (Section 21.1)

21.22 List three metals that are found in nature as oxide ores.

21.23 List three metals that are found in nature in uncombined form.

21.24 Locate the following metals in the periodic table, and predict whether they are likely to be found in nature as oxides, sulfides, or in uncombined form:

(a) Copper **(b)** Zirconium
(c) Palladium **(d)** Bismuth

21.25 Locate the following metals in the periodic table, and predict whether they are likely to be found in nature as oxides, sulfides, or in uncombined form:

(a) Vanadium **(b)** Silver
(c) Rhodium **(d)** Hafnium

21.26 Explain why early transition metals occur in nature as oxides but late transition metals occur as sulfides.

21.27 Explain why iron occurs in nature as an oxide but calcium occurs as a carbonate.

21.28 Name minerals that contain the following compounds:

(a) Fe_2O_3 **(b)** PbS
(c) TiO_2 **(d)** $CuFeS_2$

21.29 Write chemical formulas for the following minerals:

(a) Cinnabar **(b)** Bauxite
(c) Sphalerite **(d)** Chromite

Metallurgy (Section 21.2)

21.30 Describe the flotation process for concentrating a metal sulfide ore, and explain why this process would not work well for concentrating a metal oxide ore.

21.31 Gallium is a minor constituent in bauxite, $Al_2O_3 \cdot x\,H_2O$, and, like aluminum, forms an amphoteric oxide, Ga_2O_3. Explain how the Bayer process makes it possible to separate Ga_2O_3 from Fe_2O_3, which is also present in bauxite. Write a balanced net ionic equation for the reaction.

21.32 Zinc, cadmium, and mercury are found in nature as sulfide ores. Values of $E°$ for the reduction half-reaction $M^{2+}(aq) + 2\,e^- \longrightarrow M(s)$ are −0.76 V for M = Zn, −0.40 V for M = Cd, and +0.85 V for M = Hg. Explain why roasting sphalerite (ZnS) gives ZnO but roasting cinnabar (HgS) gives elemental mercury. What product would you expect from roasting CdS?

21.33 Zinc metal is produced by reducing ZnO with coke, and magnesium metal is produced by the electrolysis of molten $MgCl_2$. Look at the standard reduction potentials in Appendix D, and explain why magnesium can't be prepared by the method used for zinc.

21.34 Complete and balance each of the following equations:

(a) $V_2O_5(s) + Ca(s) \longrightarrow$ **(b)** $PbS(s) + O_2(g) \longrightarrow$

(c) $MoO_3(s) + H_2(g) \longrightarrow$ **(d)** $MnO_2(s) + Al(s) \longrightarrow$

(e) $MgCl_2(l) \xrightarrow{\text{Electrolysis}}$

21.35 Complete and balance each of the following equations:

(a) $Fe_2O_3(s) + H_2(g) \longrightarrow$ **(b)** $Cr_2O_3(s) + Al(s) \longrightarrow$

(c) $Ag_2S(s) + O_2(g) \longrightarrow$ **(d)** $TiCl_4(g) + Mg(l) \longrightarrow$

(e) $LiCl(l) \xrightarrow{\text{Electrolysis}}$

21.36 Use the thermodynamic data in Appendix B to calculate $\Delta H°$ and $\Delta G°$ for the roasting of sphalerite, ZnS:

$$2\,ZnS(s) + 3\,O_2(g) \longrightarrow 2\,ZnO(s) + 2\,SO_2(g)$$

Explain why $\Delta H°$ and $\Delta G°$ have different values, and account for the sign of $(\Delta H° - \Delta G°)$.

21.37 Use the thermodynamic data in Appendix B to calculate $\Delta H°$ and $\Delta G°$ for the roasting of cinnabar, HgS:

$$HgS(s) + O_2(g) \longrightarrow Hg(l) + SO_2(g)$$

Explain why $\Delta H°$ and $\Delta G°$ have different values, and account for the sign of $(\Delta H° - \Delta G°)$.

21.38 Ferrochrome, an iron–chromium alloy used to make stainless steel, is produced by reducing chromite ($FeCr_2O_4$) with coke:

$$FeCr_2O_4(s) + 4\,C(s) \longrightarrow \underbrace{Fe(s) + 2\,Cr(s)}_{\text{Ferrochrome}} + 4\,CO(g)$$

(a) How many kilograms of chromium can be obtained by the reaction of 236 kg of chromite with an excess of coke?

(b) How many liters of carbon monoxide at 25 °C and 740 mm Hg are obtained as a byproduct?

21.39 In one step of the industrial process for producing copper from chalcopyrite ($CuFeS_2$), molten copper(I) sulfide is reduced with a blast of hot air:

$$Cu_2S(l) + O_2(g) \longrightarrow 2\,Cu(l) + SO_2(g)$$

(a) How many kilograms of Cu_2S must be reduced to account for the world's annual copper production of about 1.4×10^{10} kg?

(b) How many liters of SO_2 at STP are produced as a byproduct?

(c) If all the SO_2 escaped into the atmosphere and were converted to sulfuric acid in acid rain, how many kilograms of H_2SO_4 would there be in the rain?

21.40 Nickel, used to make stainless steel, can be purified by electrorefining. The electrolysis cell has an impure nickel anode, a pure nickel cathode, and an aqueous solution of nickel sulfate as the electrolyte. How many kilograms of nickel can be refined in 8.00 h if the current passed through the cell is held constant at 52.5 A?

21.41 Pure copper for use in electrical wiring is obtained by electrorefining. How many hours are required to transfer 7.50 kg of copper from an impure copper anode to a pure copper cathode if the current passed through the electrolysis cell is held constant at 40.0 A?

Iron and Steel (Section 21.3)

21.42 Write a balanced equation for the overall reaction that occurs in a blast furnace when iron ore is reduced to iron metal. Identify the oxidizing agent and the reducing agent.

21.43 What is the role of coke in the commercial process for producing iron? Write balanced equations for the relevant chemical reactions.

21.44 What is slag, and what is its role in the commercial process for producing iron?

21.45 Why is limestone added to the blast furnace in the commercial process for producing iron? Write balanced equations for the relevant chemical reactions.

21.46 Briefly describe the basic oxygen process. Write balanced equations for the relevant reactions.

21.47 Why does slag form in the basic oxygen process? Write balanced equations for the relevant reactions.

21.48 When iron ore is reduced in a blast furnace, some of the SiO_2 impurity is also reduced by reaction with carbon to give elemental silicon and carbon monoxide. The silicon is subsequently reoxidized in the basic oxygen process, and the resulting SiO_2 reacts with CaO, yielding slag, which is then separated from the molten steel. Write balanced equations for the three reactions involving SiO_2.

21.49 In a blast furnace, some of the $CaSO_4$ impurity in iron ore is reduced by carbon, yielding elemental sulfur and carbon monoxide. The sulfur is subsequently oxidized in the basic oxygen process, and the product reacts with CaO to give a molten slag. Write balanced equations for the reactions.

21.50 Could tungsten be produced in a blast furnace by the reduction of WO_3? Explain.

21.51 Could aluminum be produced in a blast furnace by the reduction of Al_2O_3? Explain.

Bonding in Metals (Section 21.4)

21.52 Potassium metal crystallizes in a body-centered cubic structure. Draw one unit cell, and try to draw an electron-dot structure for bonding of the central K atom to its nearest-neighbor K atoms. What is the problem?

21.53 Describe the electron-sea model of the bonding in cesium metal. Cesium has a body-centered cubic structure.

21.54 How does the electron-sea model account for the malleability and ductility of metals?

21.55 How does the electron-sea model account for the electrical and thermal conductivity of metals?

21.56 Cesium metal is very soft, and tungsten metal is very hard. Explain the difference using the electron-sea model.

21.57 Sodium melts at 98 °C, and magnesium melts at 650 °C. Account for the higher melting point of magnesium using the electron-sea model.

21.58 Why is the molecular orbital theory for metals called band theory?

21.59 Draw an MO energy-level diagram that shows the population of the 4*s* band for potassium metal.

21.60 How does band theory account for the electrical conductivity of metals?

21.61 Materials with partially filled bands are metallic conductors, and materials with only completely filled bands are electrical insulators. Explain why the population of the bands affects the conductivity.

21.62 Draw an MO energy-level diagram for beryllium metal, and show the population of the MOs for the following two cases:

(a) The 2*s* and 2*p* bands are well separated in energy.

(b) The 2*s* and 2*p* bands overlap in energy.

Which diagram agrees with the fact that beryllium has a high electrical conductivity? Explain.

21.63 Draw an MO energy-level diagram for calcium metal, and show the population of the MOs for the following two cases:

(a) The 4*s* and 3*d* bands are well separated in energy.

(b) The 4*s* and 3*d* bands overlap in energy.

Which diagram agrees with the fact that calcium has a high electrical conductivity? Explain.

21.64 The melting points for the second-series transition elements increase from 1522 °C for yttrium to 2623 °C for molybdenum and then decrease to 321 °C for cadmium. Account for the trend using band theory.

21.65 Copper has a Mohs hardness value of 3, and iron has a Mohs hardness value of 5. Use band theory to explain why copper is softer than iron.

Semiconductors and Semiconductor Applications (Sections 21.5, 21.6)

21.66 Define a semiconductor, and give three examples.

21.67 Tell what is meant by each of the following terms:

(a) Valence band **(b)** Conduction band

(c) Band gap **(d)** Doping

21.68 Draw the bands of MO energy levels and the electron population for:

(a) A semiconductor **(b)** An electrical insulator

Explain why a semiconductor has the higher electrical conductivity.

21.69 Draw the bands of MO energy levels and the electron population for:

(a) A semiconductor **(b)** A metallic conductor

Explain why a semiconductor has the lower electrical conductivity.

21.70 How does the electrical conductivity of a semiconductor change as the size of the band gap increases? Explain.

21.71 How does the electrical conductivity of a semiconductor change as the temperature increases? Explain.

21.72 Explain what an *n*-type semiconductor is, and give an example. Draw an MO energy-level diagram, and show the population of the valence band and the conduction band for an *n*-type semiconductor.

21.73 Explain what a *p*-type semiconductor is, and give an example. Draw an MO energy-level diagram, and show the population of the valence band and the conduction band for a *p*-type semiconductor.

21.74 Explain why germanium doped with phosphorus has a higher electrical conductivity than pure germanium.

21.75 Explain why silicon doped with gallium has a higher electrical conductivity than pure silicon.

21.76 Classify the following semiconductors as *n*-type or *p*-type:

(a) Si doped with In

(b) Ge doped with Sb

(c) Gray Sn doped with As

21.77 Classify the following semiconductors as *n*-type or *p*-type:

(a) Ge doped with As

(b) Ge doped with B

(c) Si doped with Sb

21.78 The so-called 2–6 semiconductors are 1:1 compounds of group 2B and 6A elements. Write a balanced equation for the synthesis of the 2–6 semiconductor CdSe by chemical vapor deposition from the molecular precursors $Cd(CH_3)_2(g)$ and $H_2Se(g)$.

21.79 Write a balanced equation for the synthesis of the 2–6 semiconductor ZnTe by chemical vapor deposition from dimethylzinc, $Zn(CH_3)_2$, and diethyltellurium, $Te(CH_2CH_3)_2$, assuming that the other product is gaseous propane ($CH_3CH_2CH_3$).

21.80 Arrange the following materials in order of increasing electrical conductivity:

(a) Cu **(b)** Al_2O_3

(c) Fe **(d)** Pure Ge

(e) Ge doped with In

21.81 Arrange the following materials in order of increasing electrical conductivity:

(a) Pure gray Sn **(b)** Gray Sn doped with Sb

(c) NaCl **(d)** Ag

(e) Pure Si

21.82 Explain how a diode converts alternating current to direct current.

21.83 An LED produces light only when the *n*- and *p*-type semiconductors are connected to the proper terminals of a battery. Explain why no light is produced if you reverse the battery connections.

21.84 What are the main differences between an LED and a photovoltaic cell?

21.85 In what ways are LEDs and diode lasers similar, and in what ways are they different?

21.86 What is the wavelength and the color of the light emitted by a GaP_xAs_{1-x} LED with a band gap of 193 kJ/mol? (See Figure 20.26, page 846, for the relationship between wavelength and color.)

21.87 What band gap is needed for an LED to produce blue light with a wavelength of 470 nm?

Superconductors (Section 21.7)

21.88 What are the two most striking properties of a superconductor?

21.89 $YBa_2Cu_3O_7$ is a superconductor below its T_c of 90 K and a metallic conductor above 90 K. Make a rough plot of electrical resistance versus temperature for $YBa_2Cu_3O_7$.

21.90 What is the coordination environment of the K^+ ions in the fullerene-based superconductor K_3C_{60}.

21.91 Look at Figure 21.17, and identify the coordination numbers of the Cu, Y, and Ba atoms.

Ceramics and Composites (Sections 21.8, 21.9)

21.92 What is a ceramic, and what properties distinguish a ceramic from a metal?

21.93 Contrast the bonding in ceramics with the bonding in metals.

21.94 Why are ceramics more wear-resistant than metals?

21.95 Why are oxide ceramics more corrosion-resistant than metals?

21.96 Silicon nitride (Si_3N_4), a high-temperature ceramic useful for making engine components, is a covalent network solid in which each Si atom is bonded to four N atoms and each N atom is bonded to three Si atoms. Explain why silicon nitride is more brittle than a metal like copper.

21.97 Magnesia (MgO), used as an insulator for electrical heating devices, has a face-centered cubic structure like that of NaCl. Draw one unit cell of the structure of MgO, and explain why MgO is more brittle than magnesium metal.

21.98 What is ceramic processing?

21.99 Describe what happens when a ceramic powder is sintered.

21.100 Zirconia (ZrO_2), an unusually tough oxide ceramic, has been used to make very sharp table knives. Write a balanced equation for the hydrolysis of zirconium isopropoxide in the sol-gel method for making zirconia powders. The isopropoxide ligand is the anion of isopropyl alcohol, $HOCH(CH_3)_2$.

21.101 Zinc oxide is a semiconducting ceramic used to make *varistors* (variable resistors). Write a balanced equation for the hydrolysis of zinc ethoxide in the sol–gel method for making ZnO powders.

21.102 Describe the reactions that occur when an $Si(OH)_4$ sol becomes a gel. What is the formula of the ceramic obtained when the gel is dried and sintered?

21.103 Describe the reactions that occur when a $Y(OH)_3$ sol becomes a gel. What is the chemical formula of the ceramic obtained when the gel is dried and sintered?

21.104 Some ceramics can be synthesized easily from molecular or ionic precursors. Titanium diboride (TiB_2) powder forms from the decomposition of titanium(III) borohydride dissolved in an organic solvent at 140 °C. Write a balanced equation for this reaction if the other products are diborane (B_2H_6) and molecular hydrogen. (The borohydride ion is BH_4^-.)

21.105 GeS_2 is a nonoxide ceramic accessible by the sol-gel method. Write a balanced equation for the preparation of GeS_2 from tetraethoxygermane, $Ge(OCH_2CH_3)_4$, and H_2S in an organic solvent.

21.106 Silicon nitride powder can be made by the reaction of silicon tetrachloride vapor with gaseous ammonia. The byproduct is gaseous hydrogen chloride. Write a balanced equation for the reaction.

21.107 Boron, which is used to make composites, is deposited on a tungsten wire when the wire is heated electrically in the presence of boron trichloride vapor and gaseous hydrogen. Write a balanced equation for the reaction.

21.108 Explain why graphite/epoxy composites are good materials for making tennis rackets and golf clubs.

21.109 Explain why silicon carbide-reinforced alumina is stronger and tougher than pure alumina.

CHAPTER PROBLEMS

21.110 Estimate the temperature at which the Mond reaction for the formation of $Ni(CO)_4$ becomes nonspontaneous. (See Section 21.2 for thermodynamic data.)

21.111 What is the standard free-energy change in kJ per mole of metal for decomposition of M_2O_3 (M = Fe or Al) to the metal and O_2 gas? Can these metals be obtained by decomposition of their oxides under standard-state conditions? (See Appendix B for thermodynamic data.)

21.112 To separate europium (Eu) from a mixture of lanthanides (symbolized by Ln) an aqueous solution of $Ln(NO_3)_3$ is treated with zinc amalgam, an alloy of zinc and mercury. Of all the Ln^{3+} ions, only Eu^{3+} is reduced by Zn (to Eu^{2+}). The Eu^{2+} ions can then be precipitated by addition of sulfate ions. Write balanced net ionic equations for the reduction and precipitation reactions.

21.113 Wide band gap semiconductors have a band gap between 2 and 7 electron volts (eV), where 1 eV = 96.485 kJ/mol. The wide band gap semiconductor GaN, used to construct the laser in Blu-ray DVD players, has a band gap of 3.44 eV. The material in the laser, $Ga_xIn_{1-x}N$, has some indium substituted for gallium.

(a) What wavelength of light (in nm) would GaN emit, based on its band gap?

(b) If the light from the device is blue, does partial substitution of indium for gallium increase or decrease the band gap of $Ga_xIn_{1-x}N$ compared to GaN?

21.114 Imagine a planet with an atmosphere that contains O_2 and SO_2 but no CO_2. Give the chemical composition of the minerals you would expect to find for the alkaline earth metals on such a planet.

21.115 Superconductors with values of T_c above 77 K are of special interest. What's so special about 77 K?

21.116 What properties of metals are better explained by band theory than by the electron-sea model?

21.117 Tungsten is hard and has a very high melting point (3422 °C), and gold is soft and has a relatively low melting point (1064 °C). Are these facts in better agreement with the electron-sea model or the MO model (band theory)? Explain.

21.118 Explain why the enthalpy of vaporization of vanadium (460 kJ/mol) is much larger than that of zinc (114 kJ/mol).

21.119 Classify each of the following materials as a metallic conductor, an *n*-type semiconductor, a *p*-type semiconductor, or an electrical insulator:

(a) MgO **(b)** Si doped with Sb
(c) White tin **(d)** Ge doped with Ga
(e) Stainless steel

21.120 Gallium arsenide, a material used to manufacture laser printers and compact disc players, has a band gap of 130 kJ/mol. Is GaAs a metallic conductor, a semiconductor, or an electrical insulator? With what group 4A element is GaAs isoelectronic? (Isoelectronic substances have the same number of electrons.)

21.121 A 3.4×10^3 kg batch of cast iron contains 0.45% by mass of phosphorus as an impurity. In the conversion of cast iron to steel by the basic oxygen process, the phosphorus is oxidized to P_4O_{10}, which then reacts with CaO and is removed as slag.

(a) Write balanced equations for the oxidation of P_4 and for the formation of slag.

(b) How many kilograms of CaO are required to react with all the P_4O_{10}?

21.122 The $YBa_2Cu_3O_7$ superconductor can be synthesized by the sol–gel method from a stoichiometric mixture of metal ethoxides followed by heating in oxygen. How many grams of $Y(OCH_2CH_3)_3$ and how many grams of $Ba(OCH_2CH_3)_2$ are required to react with 75.4 g of $Cu(OCH_2CH_3)_2$ and an excess of water? Assuming a 100% yield, how many grams of $YBa_2Cu_3O_7$ are obtained?

21.123 Does the conductivity increase, decrease, or remain the same when

(a) *n*-type InP is doped with Cd? Explain.

(b) *p*-type InP is doped with Se? Explain.

21.124 Mullite, $3\,Al_2O_3 \cdot 2\,SiO_2$, a high-temperature ceramic being considered for use in engines, can be made by the sol–gel method.

(a) Write a balanced equation for formation of the sol by hydrolysis of a stoichiometric mixture of aluminum ethoxide and tetraethoxysilane, $Si(OCH_2CH_3)_4$.

(b) Describe the reactions that convert the sol to a gel.

(c) What additional steps are required to convert the gel to the ceramic product?

21.125 The glass in photochromic sunglasses is called a nanocomposite because it contains tiny silver halide particles with a diameter of less than 10 nm. This glass darkens and becomes less transparent outdoors in sunlight but returns to its original transparency indoors, where there is less light. How might these reversible changes occur?

21.126 Small molecules with C=C double bonds, called monomers, can join with one another to form long chain molecules called polymers. Thus, acrylonitrile, $H_2C{=}CHCN$, polymerizes under appropriate conditions to give polyacrylonitrile, a common starting material for producing the carbon fibers used in composites.

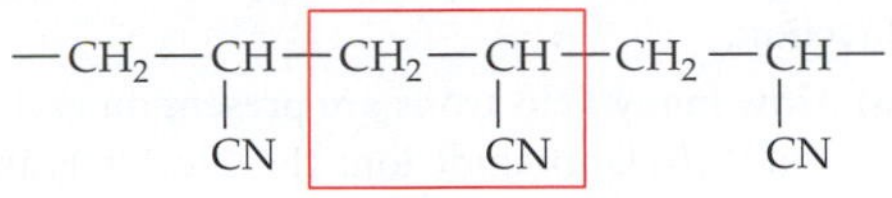

Repeating unit

Polyacrylonitrile

(a) Write electron-dot structures for acrylonitrile and polyacrylonitrile, and show how rearranging the electrons can lead to formation of the polymer.

(b) Use the bond dissociation energies in Table 7.1 to calculate ΔH per $H_2C{=}CHCN$ unit for the conversion of acrylonitrile to polyacrylonitrile. Is the reaction endothermic or exothermic?

21.127 Zinc selenide (ZnSe) doped with Ga has some Ga atoms in place of Zn atoms and is an *n*-type semiconductor. Draw an MO energy-level diagram for doped ZnSe, show the population of the bands, and explain why Ga substitution gives an *n*-type semiconductor.

21.128 Consider a hypothetical material consisting of an infinite one-dimensional array of hydride ions (H^-).

(a) Show how a band of MOs could be formed from individual atomic orbitals, and explain why this material should be an electrical insulator.

(b) If some of the hydride ions in the above material were replaced with neutral hydrogen atoms, the resulting doped material should be a better electrical conductor. Draw an MO energy-level diagram for the doped material, show the population of the band, and explain why the doped material can conduct electricity.

(c) Would you describe the doped material as *n*-type or *p*-type? Explain.

21.129 The small size of cellular telephones results from the use of sophisticated gallium arsenide (GaAs) semiconductors instead of ordinary silicon-based semiconductors.

(a) Draw an MO energy-level diagram for GaAs, and show the population of the bands.

(b) Explain why the conductivity of GaAs increases as the temperature increases.

(c) Does the conductivity of GaAs increase or decrease when pure GaAs is doped with (i) Zn or (ii) S? Explain.

MULTICONCEPT PROBLEMS

21.130 An 8.894 g block of aluminum was pressed into a thin square of foil with 36.5 cm edge lengths.

(a) If the density of Al is 2.699 g/cm^3, how thick is the foil in cm?

(b) How many unit cells thick is the foil? Aluminum crystallizes in a face-centered cubic structure and has an atomic radius of 143 pm.

21.131 Europium(II) oxide is a semiconductor with a band gap of 108 kJ/mol. Below 69 K, it is also ferromagnetic, meaning all the unpaired electrons on europium are aligned in the same direction.

(a) How many *f* electrons are present on each europium ion in EuO? (In lanthanide ions the 4*f* orbitals are lower in energy than the 6*s* orbitals.)

(b) All the *f* electrons are unpaired in EuO. Predict the spin-only magnetic moment of EuO, using the formula in Problem 20.132 on page 863.

21.132 A 0.3249 g sample of stainless steel was analyzed for iron by dissolving the sample in sulfuric acid and titrating the Fe^{2+} in the resulting solution with 0.018 54 M $K_2Cr_2O_7$. If 38.89 mL of the $K_2Cr_2O_7$ solution was required to reach the equivalence point, what is the mass percentage of iron in the steel?

21.133 Red light with a wavelength of 660 nm from a 3.0 mW diode laser shines on a solar cell.

(a) How many photons per second are emitted by the laser? (1 W = 1 J/s)

(b) How much current (in amperes) flows in the circuit of the solar cell if all the photons are absorbed by the cell and each photon produces one electron?

21.134 The mineral wustite is a nonstoichiometric iron oxide with the empirical formula Fe_xO, where *x* is a number slightly less than 1. Wustite can be regarded as an FeO in which some of the Fe sites are vacant. It has a density of 5.75 g/cm^3, a cubic unit cell with an edge length of 431 pm, and a face-centered cubic arrangement of oxygen atoms.

(a) What is the value of *x* in the formula Fe_xO?

(b) Based on the formula in part **(a)**, what is the average oxidation state of Fe?

(c) Each Fe atom in wustite is in either the +2 or the +3 oxidation state. What percent of the Fe atoms are in the +3 oxidation state?

(d) Using X rays with a wavelength of 70.93 pm, at what angle would third-order diffraction be observed from the planes of atoms that coincide with the faces of the unit cells? Third-order diffraction means that the value of *n* in the Bragg equation is equal to 3.

(e) Wustite is a semiconducting iron(II) oxide in which some of the Fe^{2+} has been replaced by Fe^{3+}. Should it be described as an *n*-type or a *p*-type semiconductor? Explain.

21.135 At high temperatures, coke reduces silica impurities in iron ore to elemental silicon:

$$SiO_2(s) + 2\,C(s) \longrightarrow Si(s) + 2\,CO(g)$$

(a) Use the data in Appendix B to calculate the values of $\Delta H°$, $\Delta S°$, and $\Delta G°$ for this reaction at 25 °C.

(b) Is the reaction endothermic or exothermic?

(c) Account for the sign of the entropy change.

(d) Is the reaction spontaneous at 25 °C and 1 atm pressure of CO?

(e) Assuming that $\Delta H°$ and $\Delta S°$ do not depend on temperature, estimate the temperature at which the reaction becomes spontaneous at 1 atm pressure of CO.

21.136 The Mond process for purifying nickel involves the reaction of impure nickel with carbon monoxide at about 150 °C to give nickel tetracarbonyl. The nickel tetracarbonyl then decomposes to pure nickel at about 230 °C:

$$Ni(s) + 4\,CO(g) \rightleftharpoons Ni(CO)_4(g) \qquad \Delta H° = -160.8\text{ kJ};\ \Delta S° = -410\text{ J/K}$$

The values of $\Delta H°$ and $\Delta S°$ apply at 25 °C, but they are relatively independent of temperature and can be used at 150 °C and 230 °C.

(a) Calculate $\Delta G°$ and the equilibrium constant K_p at 150 °C.

(b) Calculate $\Delta G°$ and the equilibrium constant K_p at 230 °C.

(c) Why does the reaction have a large negative value for $\Delta S°$? Show that the change in $\Delta G°$ with increasing temperature is consistent with a negative value of $\Delta S°$.

(d) Show that the change in K_p with increasing temperature is consistent with a negative value of $\Delta H°$.

21.137 Figure 21.5 indicates that carbon reacts with carbon dioxide to form carbon monoxide at high temperatures:

$$C(s) + CO_2(g) \longrightarrow 2\,CO(g)$$

(a) Use the thermodynamic data in Appendix B to calculate the total pressure and the molar concentrations of CO and CO_2 at 500 °C in a 50.00 L vessel that initially contains 100.0 g of CO_2 and an excess of solid carbon. Assume that the reaction mixture has reached equilibrium and that $\Delta H°$ and $\Delta S°$ are independent of temperature.

(b) Calculate the total pressure and the equilibrium concentrations at 1000 °C.

(c) Use $\Delta H°$ and $\Delta S°$ to explain why the concentrations increase or decrease on raising the temperature.

21.138 Zinc chromite ($ZnCr_2O_4$), once used to make magnetic tape for cassette recorders, can be prepared by the thermal decomposition of $(NH_4)_2Zn(CrO_4)_2$. The other reaction products are N_2 and water vapor.

(a) Write a balanced equation for the reaction.

(b) How many grams of $ZnCr_2O_4$ can be prepared from 10.36 g of $(NH_4)_2Zn(CrO_4)_2$.

(c) How many liters of gaseous byproducts are obtained at 292 °C and 745 mm pressure?

(d) $ZnCr_2O_4$ has the cubic spinel structure consisting of a cubic closest-packed (face-centered cubic) array of O^{2-} ions with Zn^{2+} ions in tetrahedral holes surrounded by four O^{2-} ions and Cr^{3+} ions in octahedral holes surrounded by six O^{2-} ions. The tetrahedral and octahedral holes are like those in K_3C_{60} (see Figure 21.20). What fraction of each of the two types of holes is occupied?

(e) Draw a crystal field *d*-orbital energy-level diagram for the Cr^{3+} ion in $ZnCr_2O_4$. Would you expect this compound to be colored? Explain. Does the Zn^{2+} ion contribute to the color? Explain.

21.139 The alkali metal fulleride superconductors M_3C_{60} have a cubic closest-packed (face-centered cubic) arrangement of nearly spherical $C_{60}{}^{3-}$ anions with M^+ cations in the holes between the larger $C_{60}{}^{3-}$ ions. The holes are of two types: octahedral holes, which are surrounded octahedrally by six $C_{60}{}^{3-}$ ions, and tetrahedral holes, which are surrounded tetrahedrally by four $C_{60}{}^{3-}$ ions.

(a) Sketch the three-dimensional structure of one unit cell.

(b) How many $C_{60}{}^{3-}$ ions, octahedral holes, and tetrahedral holes are present per unit cell?

(c) Specify fractional coordinates for all the octahedral and tetrahedral holes. (Fractional coordinates are fractions of the unit cell edge lengths. For example, a hole at the center of the cell has fractional coordinates $\frac{1}{2}, \frac{1}{2}, \frac{1}{2}$.)

(d) The radius of a $C_{60}{}^{3-}$ ion is about 500 pm. Assuming that the $C_{60}{}^{3-}$ ions are in contact along the face diagonals of the unit cell, calculate the radii of the octahedral and tetrahedral holes.

(e) The ionic radii of Na^+, K^+, and Rb^+ are 102, 138, and 152 pm, respectively. Which of these ions will fit into the octahedral and tetrahedral holes? Which ions will fit only if the framework of $C_{60}{}^{3-}$ ions expands?

21.140 A single-walled carbon nanotube can be regarded as a single sheet of graphite rolled into a cylinder (Figure 10.27b, page 386). Calculate the number of C atoms in a single-walled nanotube having a length of 1.0 mm and a diameter of 1.08 nm. The C–C bond length in graphite is 141.5 pm.

CHAPTER

22 Nuclear Chemistry

A look at the core of a modern TRIGA nuclear reactor. How long do you think nuclear reactors have been on Earth? The answer, given in the *FYI* at the end of this chapter, may surprise you.

We've seen repeatedly in previous chapters that only the bonds between atoms change when a chemical reaction takes place, while the identities of the atoms themselves remain unchanged. Yet we also saw in Sections 1.10 and 1.11 that atoms *can* change their identities when *nuclear* reactions occur. You may recall that five kinds of nuclear processes commonly take place, as summarized in **Table 22.1**.

You may also recall from Section 12.6 that radioactive decay is kinetically a **first-order process**, whose rate is proportional to the number of radioactive nuclei N in a sample times the first-order rate constant k. That is, the decay rate $= k \times N$. Like all first-order processes, radioactive decay is characterized by a **half-life, $t_{1/2}$**, the time required for the number of radioactive nuclei in a sample to decay to half its initial value. The half-lives of different **radioisotopes** vary enormously. Some, such as uranium-238, decay at a barely perceptible rate over billions of years, while others, such as carbon-21, decay within microseconds.

Let's now look further into the nature of nuclear decay, focusing particularly on the energy changes that occur during nuclear transformations and on the uses and effects of radioactivity.

REMEMBER...

A **first-order process** is one whose rate depends on the concentration of only one substance raised to the first power. (Section 12.4)

REMEMBER...

The **half-life, $t_{1/2}$,** of a nuclear decay is the time required for the number of radioactive nuclei in the sample to drop to one-half its initial value and is related to the rate constant k by the equation $t_{1/2} = 0.693/k$. (Section 12.6)

REMEMBER...

A **radioisotope** is an isotope that undergoes spontaneous radioactive decay. (Section 1.11)

TABLE 22.1 A Summary of Nuclear Decay Processes

Process	Symbol	Change in Atomic Number	Change in Mass Number	Change in Neutron Number
Alpha emission	${}^{4}_{2}He$ or α	−2	−4	−2
Beta emission	${}^{0}_{-1}e$ or β^-	+1	0	−1
Gamma emission	${}^{0}_{0}\gamma$ or γ	0	0	0
Positron emission	${}^{0}_{1}e$ or β^+	−1	0	+1
Electron capture	E. C.	−1	0	+1

CONTENTS

22.1 NUCLEAR STABILITY

Why do some nuclei undergo spontaneous radioactive decay while others do not? Why, for instance, does a carbon-*14* nucleus, with 6 protons and *8* neutrons, spontaneously emit a β particle, whereas a carbon-*13* nucleus, with 6 protons and *7* neutrons, is nonradioactive?

The answer has to do with the ratio of neutrons to protons in the nucleus and with the forces holding the nucleus together. To see the effect of the neutron/proton ratio on nuclear stability, look at the grid in **Figure 22.1**. Numbers on the side of the grid give the number of neutrons in different nuclei, and numbers along the bottom give the number of protons.

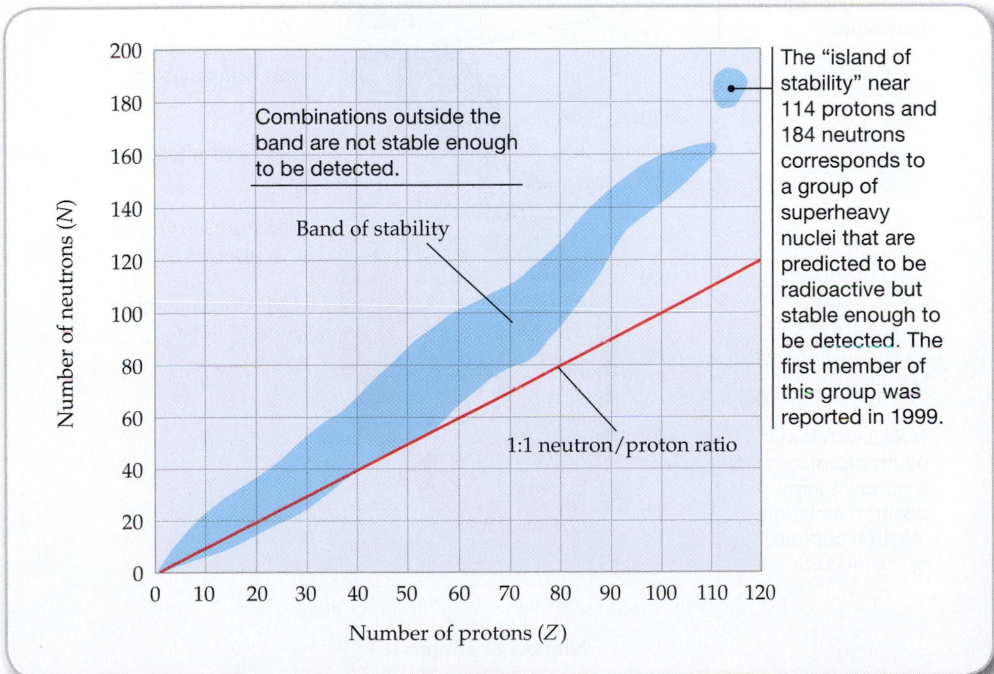

FIGURE 22.1

The band of nuclear stability. The band indicates various neutron/proton combinations that give rise to nuclei that are either nonradioactive or that are radioactive but decay slowly enough to exist for a measurable time.

The first 92 elements are naturally occurring, while the remainder consists of the artificially produced **transuranium elements**. (Actually, only 90 of the first 92 elements occur naturally. Technetium and promethium do not occur naturally because all their isotopes are radioactive and have very short lifetimes. Francium and astatine occur on Earth only in very tiny amounts.)

When the more than 3600 known isotopes are plotted on the neutron/proton grid in Figure 22.1, they fall in a curved band sometimes called the *band of nuclear stability*. Even within the band, only 264 of the isotopes are nonradioactive. The others decay spontaneously, although their rates of decay vary enormously. On either side of the band is a so-called sea of instability representing the large number of unstable neutron–proton combinations that have never been detected. Particularly interesting is the island of stability predicted to exist for a few superheavy isotopes near 114 protons (flerovium, Fl) and 184 neutrons. The first members of this group—^{287}Fl, ^{288}Fl, and ^{289}Fl—were prepared in 1999 and do indeed seem to be stable enough to live for several seconds before they decay.

Several generalizations can be made about nuclear stability:

- Every element in the periodic table has at least one radioactive isotope.
- Hydrogen is the only element whose most abundant isotope ($^{1}_{1}H$) contains more protons (1) than neutrons (0).
- The ratio of neutrons to protons gradually increases, giving a curved appearance to the band of stability.
- All isotopes heavier than bismuth-209 are radioactive, even though they may decay slowly and be stable enough to occur naturally.

A close-up look at a segment of the band of nuclear stability (**Figure 22.2**) shows the interesting trend that radioactive nuclei with higher neutron/proton ratios (top side of the band)

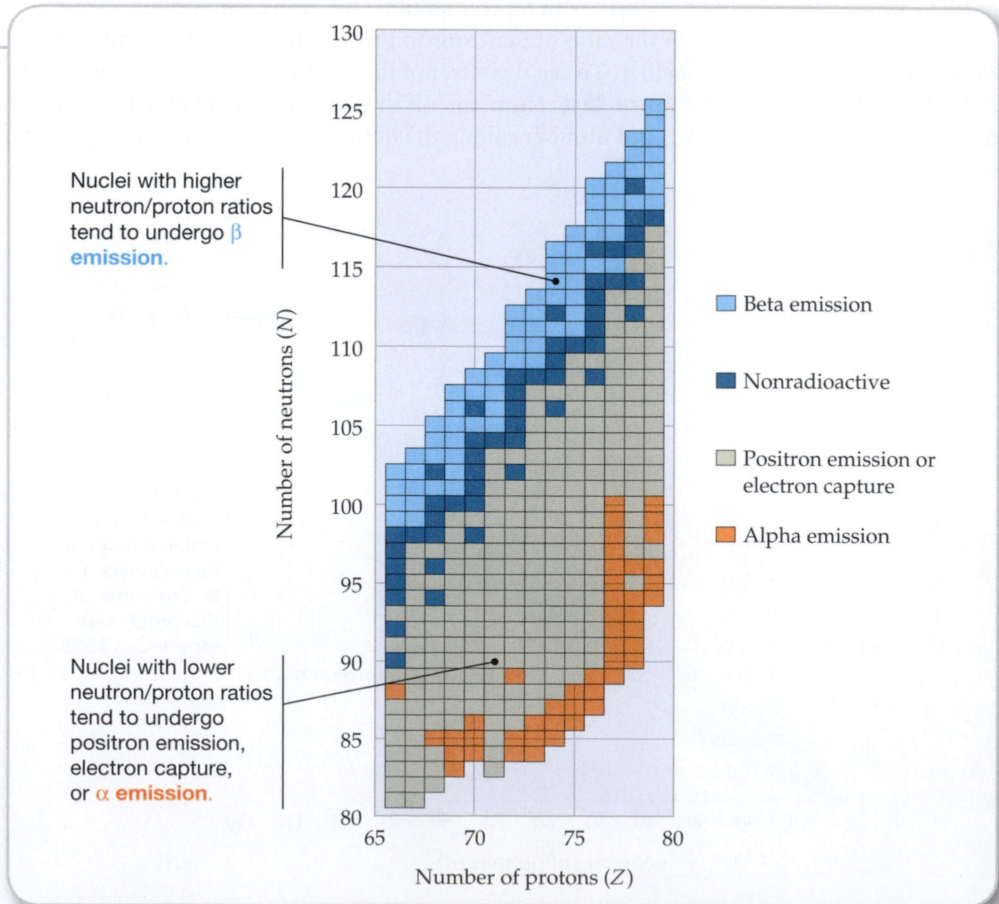

FIGURE 22.2

A close-up look at the band of nuclear stability. This look at the region from $Z = 66$ (dysprosium) through $Z = 79$ (gold) shows the types of radioactive processes that various radioisotopes undergo.

tend to emit β particles while nuclei with lower neutron/proton ratios (bottom side of the band) tend to undergo nuclear decay by positron emission, electron capture, or α emission.

The trends shown in Figure 22.2 makes sense if you think about it. The nuclei on the top side of the band are neutron-rich and therefore undergo a process—β emission—that *decreases* the neutron/proton ratio by converting a neutron into a proton. The nuclei on the bottom side of the band, by contrast, are neutron-poor and therefore undergo processes that *increase* the neutron/proton ratio. Take a minute to convince yourself that α emission does, in fact, increase the neutron/proton ratio for heavy nuclei in which $n > p$.

This process decreases the neutron/proton ratio:	Beta emission:	Neutron $\longrightarrow$ Proton + β^-
These processes increase the neutron/proton ratio:	Positron emission:	Proton $\longrightarrow$ Neutron + β^+
	Electron capture:	Proton + Electron $\longrightarrow$ Neutron
	Alpha emission:	${}^{A}_{Z}X \longrightarrow {}^{A-4}_{Z-2}Y + {}^{4}_{2}He$

PROBLEM 22.1

(a) Of the two isotopes ^{173}Au and ^{199}Au, one decays by β emission and one decays by α emission. Which is which?

(b) Of the two isotopes ^{196}Pb and ^{206}Pb, one is nonradioactive and one decays by positron emission. Which is which?

22.2 ENERGY CHANGES DURING NUCLEAR REACTIONS

We saw in Figure 22.1 that the band of nuclear stability obtained by plotting the number of protons versus the number of neutrons for observable nuclei is slightly curved. Lighter elements have an approximately 1:1 neutron/proton ratio, but heavier elements need a substantially higher ratio of neutrons to protons for stability. Such observations suggest that neutrons function as a kind of nuclear glue to overcome the proton–proton repulsions that would otherwise cause a nucleus to fly apart. The more protons there are in the nucleus, the more glue is needed.

In principle, it should be possible to measure the strength of the forces holding a nucleus together by measuring the amount of energy necessary to tear the nucleus apart into its component protons and neutrons. That is, the amount of energy that must be *added* to tear the nucleus apart is numerically equal to the amount of energy that is *released* when a stable nucleus forms from its constituent particles. The two processes are simply the reverse of one another. With a helium-4 nucleus, for instance, the energy change associated with combining two neutrons and two protons to form ${}^{4}_{2}He$ is a direct measure of helium-4 nuclear stability.

$$2\,{}^{1}_{1}H + 2\,{}^{1}_{0}n \longrightarrow {}^{4}_{2}He \qquad \Delta E = ?$$

Unfortunately, there is a problem with actually carrying out the measurement, because temperatures rivaling those in the interior of the Sun (10^7 K) are necessary for nuclei to form. That is, the **activation energy** required to force the elementary particles close enough for reaction is enormous. Nevertheless, the energy change for the process can be calculated using the now-famous equation $\Delta E = \Delta mc^2$, which was proposed by Albert Einstein in 1905 as part of his special theory of relativity and which relates the energy change ΔE of a process to a corresponding mass change Δm. (Recall that the symbol c represents the speed of light; its value is $c = 299\,792\,458$ m/s.)

REMEMBER...

Activation energy is the potential-energy barrier that must be surmounted before reactants can be converted to products. The height of the barrier controls the rate of the reaction. (Section 12.9)

Using the helium-4 nucleus for an example calculation, the mass of the two neutrons and two protons that combine to form the nucleus is 4.031 88 **unified atomic mass units (u)**:

REMEMBER...

Atomic masses are given using the **unified atomic mass unit (u)**, defined as 1/12 the mass of a ^{12}C atom. In practice, the unit is not often specified and atomic masses are written as dimensionless. (Section 1.9)

Mass of 2 neutrons	= (2)(1.008 67)	= 2.017 34
Mass of 2 protons	= (2)(1.007 28)	= 2.014 56
Total mass of 2 n + 2 p		= 4.031 90

The mass of the helium-4 *nucleus* can be found by subtracting the mass of two electrons from the experimentally measured mass of a helium-4 *atom* (4.002 60).

$$\begin{array}{lcl} \text{Mass of helium-4 atom} & & = \ \ 4.002\,60 \\ -\text{Mass of 2 electrons} & = -(2)(5.486 \times 10^{-4}) & = -0.001\,10 \\ \hline \text{Mass of helium-4 nucleus} & & = \ \ 4.001\,50 \end{array}$$

Subtracting the mass of the helium nucleus (4.001 50) from the combined mass of its constituent particles (4.031 90) shows a difference of 0.030 40. That is, 0.030 40 u (or 0.030 40 g/mol) is lost when two protons and two neutrons combine to form a helium-4 nucleus:

$$\begin{array}{lcl} \text{Mass of 2 n + 2 p} & = & 4.031\,90 \\ -\text{Mass of } ^4\text{He nucleus} & = & -4.001\,50 \\ \hline \text{Mass difference} & = & 0.030\,40 \text{ u} \end{array}$$

$$\text{Mass difference in grams} = \left(0.030\,40\ \frac{\text{u}}{\text{atom}}\right)\left(1.6605 \times 10^{-24}\ \frac{\text{g}}{\text{u}}\right) = 5.048 \times 10^{-26}\ \text{g/atom}$$

$$\text{Mass difference in g/mol} = \left(5.048 \times 10^{-26}\ \frac{\text{g}}{\text{atom}}\right)\left(6.022 \times 10^{23}\ \frac{\text{atom}}{\text{mol}}\right) = 3.040 \times 10^{-2}\ \frac{\text{g}}{\text{mol}}$$

The loss in mass that occurs when protons and neutrons combine to form a nucleus is called the **mass defect** of the nucleus. This lost mass is converted into energy that is released during the nuclear reaction and is thus a direct measure of the **binding energy** holding the nucleus together. The larger the binding energy, the more stable the nucleus. For a helium-4 nucleus, the binding energy is

$$\Delta E = \Delta mc^2 = (5.048 \times 10^{-26}\ \text{g})\left(10^{-3}\ \frac{\text{kg}}{\text{g}}\right)\left(3.00 \times 10^8\ \frac{\text{m}}{\text{s}}\right)^2$$

$$= 4.54 \times 10^{-12}\left(\frac{\text{kg} \cdot \text{m}^2}{\text{s}^2}\right) = 4.54 \times 10^{-12}\ \text{J}$$

The binding energy can be expressed on a kJ/mol basis by converting joules to kilojoules and multiplying by Avogadro's number:

$$\text{Binding energy} = \left(4.54 \times 10^{-12}\ \frac{\text{J}}{\text{atom}}\right)\left(\frac{1\ \text{kJ}}{1000\ \text{J}}\right)\left(6.022 \times 10^{23}\ \frac{\text{atom}}{\text{mol}}\right)$$

$$= 2.73 \times 10^9\ \frac{\text{kJ}}{\text{mol}}$$

The binding energy of a helium-4 nucleus is 4.54×10^{-12} J, or 2.73×10^9 kJ/mol. In other words, 4.54×10^{-12} J is released when a helium-4 nucleus is formed from 2 protons and 2 neutrons, and 4.54×10^{-12} J must be supplied to disintegrate a helium-4 nucleus into isolated protons and neutrons. This amount of energy is more than *10 million times* the energy change associated with a typical chemical process.

REMEMBER...

A **nucleon** is the general term for a nuclear particle, either proton or neutron. (Section 1.11)

To make comparisons among different nuclei easier, binding energies are usually expressed on a per-**nucleon** basis using *electron volts* (*eV*) as the energy unit, where 1 eV = 1.60×10^{-19} J and 1 million electron volts (1 MeV) = 1.60×10^{-13} J. Thus, the helium-4 binding energy is 7.09 MeV/nucleon:

$$\text{Helium-4 binding energy} = \left(4.54 \times 10^{-12}\ \frac{\text{J}}{\text{nucleus}}\right)\left(\frac{1\ \text{Mev}}{1.60 \times 10^{-13}\ \text{J}}\right)\left(\frac{1\ \text{nucleus}}{4\ \text{nucleons}}\right)$$

$$= 7.09\ \text{MeV/nucleon}$$

A plot of binding energy per nucleon for the most stable isotope of each element is shown in **Figure 22.3**. Since a larger binding energy per nucleon corresponds to greater stability, the most stable nuclei are at the top of the curve and the least stable nuclei are at the bottom. Iron-56, with a binding energy of 8.79 MeV/nucleon, is the most stable isotope known.

The idea that mass and energy are interconvertible is potentially disturbing because it seems to overthrow two of the fundamental principles on which chemistry is based—the

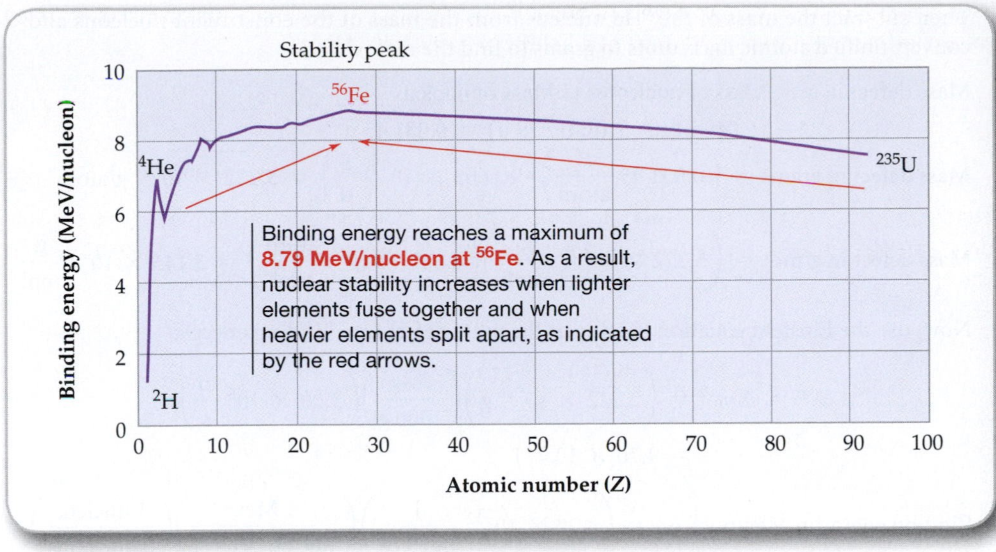

FIGURE 22.3

The binding energy per nucleon for the most stable isotope of each naturally occurring element.

law of mass conservation and the **law of energy conservation**. In fact, what the mass–energy interconversion means is that neither mass nor energy is conserved independently; rather, it is the *combination* of mass and energy that is conserved. Every time we do a reaction, whether nuclear *or* chemical, mass and energy are interconverted, but the combination of the two is conserved. For the energy change involved in a typical chemical reaction, however, the effect is so small that the mass change can't be detected by even the best analytical balance. Worked Example 22.2 illustrates such a mass–energy calculation for a chemical reaction.

REMEMBER...

The **law of mass conservation** says that mass is neither created nor destroyed in chemical reactions. (Section 1.4)

REMEMBER...

The **law of energy conservation** says that energy can neither be created nor destroyed but only converted from one form to another. (Section 8.1)

Worked Example 22.1

Calculating a Mass Defect and a Binding Energy

Helium-6 is a radioactive isotope with $t_{1/2} = 0.807$ s. Calculate the mass defect for the formation of a ^{6}He nucleus in both grams and g/mol, and calculate the binding energy in both MeV/nucleon and kJ/mol. Is a ^{6}He nucleus more stable or less stable than a ^{4}He nucleus? The mass of a ^{6}He atom is 6.018 89 u.

STRATEGY

Find the mass defect by subtracting the mass of the ^{6}He nucleus from the mass of its constituent nucleons, and use the Einstein equation to find the ^{6}He binding energy. Then use Avogadro's number ($6.022 \times 10^{23}\ \text{mol}^{-1}$) to find the per-mole value.

SOLUTION

First, calculate the total mass of the nucleons (4 n + 2 p):

$$\begin{aligned}
\text{Mass of 4 neutrons} &= (4)(1.008\,67) = 4.034\,68 \\
\text{Mass of 2 protons} &= (2)(1.007\,28) = 2.014\,56 \\
\hline
\text{Mass of 4 n + 2 p} &= 6.049\,24
\end{aligned}$$

Next, calculate the mass of a ^{6}He nucleus by subtracting the mass of two electrons from the mass of a ^{6}He atom:

$$\begin{aligned}
\text{Mass of helium-6 atom} &= 6.018\,89 \\
-\text{Mass of 2 electrons} = -(2)(5.486 \times 10^{-4}) &= -0.001\,10 \\
\hline
\text{Mass of helium-6 nucleus} &= 6.017\,79
\end{aligned}$$

continued on next page

Then subtract the mass of the ^6He nucleus from the mass of the constituent nucleons and convert unified atomic mass units to grams to find the mass defect:

$$\text{Mass defect in u} = \text{Mass of nucleons} - \text{Mass of nucleus}$$
$$= (6.049\ 24\ \text{u}) - (6.017\ 79\ \text{u}) = 0.031\ 45\ \text{u}$$

$$\text{Mass defect in grams} = \left(0.031\ 45\ \frac{\text{u}}{\text{atom}}\right)\left(1.6605 \times 10^{-24}\ \frac{\text{g}}{\text{u}}\right) = 5.222 \times 10^{-26}\ \text{g/atom}$$

$$\text{Mass defect in g/mol} = \left(5.222 \times 10^{-26}\ \frac{\text{g}}{\text{atom}}\right)\left(6.022 \times 10^{23}\ \frac{\text{atom}}{\text{mol}}\right) = 3.145 \times 10^{-2}\ \frac{\text{g}}{\text{mol}}$$

Now, use the Einstein equation to convert the mass defect into binding energy:

$$\Delta E = \Delta mc^2 = \left(5.222 \times 10^{-26}\ \text{g}\right)\left(\frac{1\ \text{kg}}{1000\ \text{g}}\right)\left(3.00 \times 10^8\ \frac{\text{m}}{\text{s}}\right)^2$$
$$= 4.70 \times 10^{-12}\ \text{J}$$

$$\text{Binding energy in Mev/nucleon} = \left(4.70 \times 10^{-12}\ \frac{\text{J}}{\text{nucleus}}\right)\left(\frac{1\ \text{Mev}}{1.60 \times 10^{-13}\ \text{J}}\right)\left(\frac{1\ \text{nucleus}}{6\ \text{nucleons}}\right)$$
$$= 4.90\ \text{MeV/nucleon}$$

$$\text{Binding energy in kJ/mol} = \left(4.70 \times 10^{-12}\ \frac{\text{J}}{\text{atom}}\right)\left(\frac{1\ \text{kJ}}{1000\ \text{J}}\right)\left(6.022 \times 10^{23}\ \frac{\text{atom}}{\text{mol}}\right)$$
$$= 2.83 \times 10^9\ \frac{\text{kJ}}{\text{mol}}$$

The binding energy of a radioactive ^6He nucleus is 4.90 MeV/nucleon, making ^6He less stable than ^4He, whose binding energy is 7.09 MeV/nucleon.

Worked Example 22.2

Calculating the Mass Defect in a Chemical Reaction

What is the mass change in g/mol when hydrogen atoms (H) combine to form hydrogen molecules (H_2)?

$$2\ \text{H} \longrightarrow \text{H}_2 \qquad \Delta E = -436\ \text{kJ/mol}$$

STRATEGY

The problem asks us to calculate a mass defect Δm when the energy change ΔE is known. To do this, we have to rearrange the Einstein equation to solve for mass, remembering that $1\ \text{J} = 1\ \text{kg} \cdot \text{m}^2/\text{s}^2$.

SOLUTION

$$\Delta m = \frac{\Delta E}{c^2} = \frac{\left(-436\ \frac{\text{kJ}}{\text{mol}}\right)\left(\frac{1\ \text{kg} \cdot \text{m}^2/\text{s}^2}{\text{J}}\right)\left(\frac{1000\ \text{J}}{1\ \text{kJ}}\right)\left(\frac{1000\ \text{g}}{1\ \text{kg}}\right)}{\left(3.00 \times 10^8\ \frac{\text{m}}{\text{s}}\right)^2} = -4.84 \times 10^{-9}\ \frac{\text{g}}{\text{mol}}$$

The loss in mass accompanying formation of H_2 molecules from their constituent H atoms is 4.84×10^{-9} g/mol, much too small an amount to be detectable by any balance and far smaller than the mass defect of a stable nucleus, which we calculated to be approximately 10^{-2} g/mol.

PROBLEM 22.2 Calculate the mass defect for the formation of an oxygen-16 nucleus in both grams and g/mol, and calculate the binding energy in both MeV/nucleon and kJ/mol. The mass of an ^{16}O atom is 15.994 91 u.

PROBLEM 22.3 What is the mass change in g/mol for the reaction of sodium metal with chlorine gas (Cl_2) to give sodium chloride?

$$2\,Na(s) + Cl_2(g) \longrightarrow 2\,NaCl(s) \qquad \Delta E = -820\text{ kJ}$$

22.3 NUCLEAR FISSION AND FUSION

A careful look at the plot of atomic number versus binding energy per nucleon in Figure 22.3 leads to some interesting and important conclusions. The fact that binding energy per nucleon begins at a relatively low value for ${}^{2}_{1}H$, reaches a maximum for ${}^{56}_{26}Fe$, and then gradually tails off implies that both lighter and heavier nuclei are less stable than midweight nuclei near iron-56. Very heavy nuclei can therefore gain stability and release energy if they fragment to yield midweight nuclei, while very light nuclei can gain stability and release energy if they fuse together. The two resultant processes—**fission** for the fragmenting of heavy nuclei and **fusion** for the joining together of light nuclei—have changed the world since their discovery in the late 1930s and early 1940s.

Nuclear Fission

Certain nuclei—uranium-233, uranium-235, and plutonium-239, for example—do more than undergo simple radioactive decay; they break into fragments when struck by neutrons. As illustrated in **Figure 22.4**, an incoming neutron causes the nucleus to split into two smaller pieces of roughly similar size.

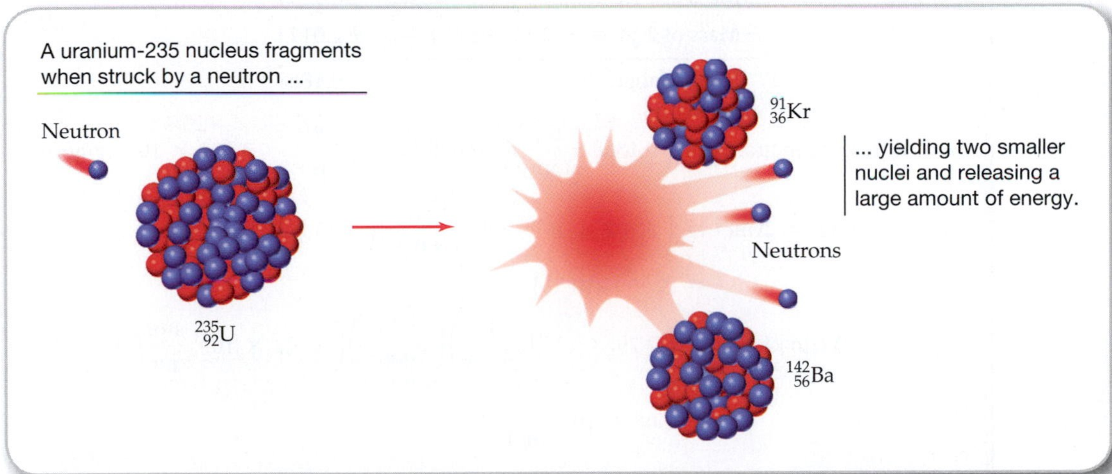

FIGURE 22.4
A representation of nuclear fission.

The fission of a given nucleus does not occur in exactly the same way each time, and nearly 800 different fission products have been identified from uranium-235. One of the more frequently occurring pathways generates barium-142 and krypton-91, along with two additional neutrons plus the one neutron that initiated the fission:

$$^{1}_{0}n + {}^{235}_{92}U \longrightarrow {}^{142}_{56}Ba + {}^{91}_{36}Kr + 3\,{}^{1}_{0}n$$

The three neutrons released by fission of a ^{235}U nucleus can induce up to three more fissions yielding as many as nine neutrons, which can induce up to nine more fissions yielding 27 neutrons, and so on indefinitely. The result is a **chain reaction** that continues to occur even if the external supply of bombarding neutrons is cut off. If the sample size is small, many of the neutrons escape before initiating additional fission events, and the chain reaction soon stops. If there is a sufficient amount of ^{235}U, though—an amount called the **critical mass**—enough neutrons remain for the chain reaction to become self-sustaining. Under high-pressure

▲ An enormous amount of energy is released in the explosion that accompanies an uncontrolled nuclear chain reaction.

conditions that confine the ^{235}U to a small volume, the chain reaction may even occur so rapidly that a nuclear explosion results. For ^{235}U, the critical mass is about 56 kg, although the amount can be reduced to 15 kg by placing a coating of ^{238}U around the ^{235}U to reflect back some of the escaping neutrons.

The amount of energy released during nuclear fission can be calculated as in Worked Example 22.3 by finding the accompanying mass change and then using the Einstein mass–energy relationship discussed in the previous section. (When calculating the mass change for a nuclear reaction, it's simplest to use the masses of the *atoms* corresponding to the relevant nuclei, rather than the masses of the nuclei themselves, because the number of electrons is the same in both reactants and products and thus cancels from the calculation.)

Worked Example 22.3

Calculating the Energy Released in a Nuclear Reaction

How much energy in both joules and kJ/mol is released by the fission of uranium-235 to form barium-142 and krypton-91? The atomic masses are ^{235}U (235.0439 u), ^{142}Ba (141.9164 u), ^{91}Kr (90.9234 u), and $^{1}_{0}n$ (1.008 67 u).

$$^{1}_{0}n + {}^{235}_{92}U \longrightarrow {}^{142}_{56}Ba + {}^{91}_{36}Kr + 3\,{}^{1}_{0}n$$

STRATEGY

Calculate the mass change by subtracting the masses of the products from the mass of the ^{235}U reactant, and then use the Einstein equation to convert mass to energy.

SOLUTION

$$\begin{array}{lcr} \text{Mass of } {}^{235}U & = & 235.0439 \\ -\text{Mass of } {}^{142}Ba & = & -141.9164 \\ -\text{Mass of } {}^{91}Kr & = & -90.9234 \\ -\text{Mass of } 2\,{}^{1}_{0}n = -(2)(1.008\ 67) & = & -2.0173 \\ \hline \text{Mass change:} & = & 0.1868 \end{array}$$

$$\text{Mass difference in grams} = \left(0.1868\ \frac{\text{u}}{\text{atom}}\right)\left(1.6605 \times 10^{-24}\ \frac{\text{g}}{\text{u}}\right) = 3.102 \times 10^{-25}\ \text{g/atom}$$

$$\Delta E = \Delta mc^2 = \left(3.102 \times 10^{-25}\ \text{g}\right)\left(\frac{1\ \text{kg}}{1000\ \text{g}}\right)\left(3.00 \times 10^{8}\ \frac{\text{m}}{\text{s}}\right)^2$$
$$= 2.79 \times 10^{-11}\ \text{J}$$

$$\Delta E \text{ in kJ/mol} = \left(2.79 \times 10^{-11}\ \frac{\text{J}}{\text{atom}}\right)\left(\frac{1\ \text{kJ}}{1000\ \text{J}}\right)\left(6.022 \times 10^{23}\ \frac{\text{atom}}{\text{mol}}\right)$$
$$= 1.68 \times 10^{10}\ \frac{\text{kJ}}{\text{mol}}$$

Nuclear fission of a ^{235}U atom releases 2.79×10^{-11} J, or 1.68×10^{10} kJ/mol.

PROBLEM 22.4 An alternative pathway for the nuclear fission of ^{235}U produces tellurium-137 and zirconium-97. How much energy in kJ/mol is released in this fission pathway?

$$^{1}_{0}n + {}^{235}_{92}U \rightarrow {}^{137}_{52}Te + {}^{97}_{40}Zr + 2\,{}^{1}_{0}n$$

The masses are ^{235}U (235.0439 u), ^{137}Te (136.9254 u), ^{97}Zr (96.9110 u), and $^{1}_{0}n$ (1.008 67 u).

Nuclear Reactors

The same fission process that leads to a nuclear explosion under uncontrolled conditions can be used to generate electric power when carried out in a controlled manner in a nuclear reactor (**Figure 22.5**). The principle behind a nuclear reactor is straightforward: Uranium fuel is placed in a pressurized containment vessel surrounded by water, and *control rods* made of boron and cadmium are added. The water slows the neutrons released during fission so their escape is more difficult, and the control rods absorb and thus regulate the flow of neutrons when raised and lowered. As a result, the fission is maintained at a barely self-sustainable rate to prevent overheating. Energy from the controlled fission then heats a circulating coolant, which in turn produces steam to drive a turbine and produce electricity.

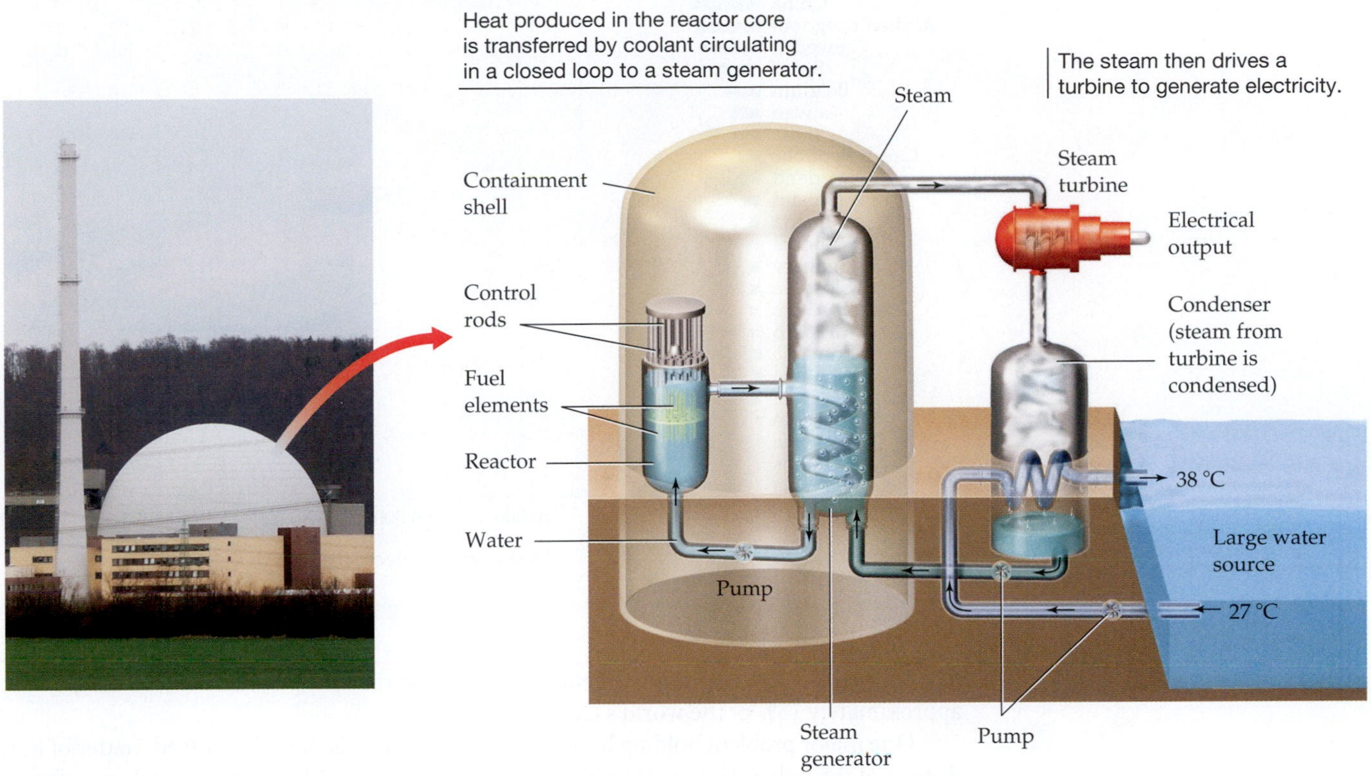

FIGURE 22.5

A nuclear power plant.

Naturally occurring uranium is a mixture of two major isotopes. The nonfissionable ^{238}U isotope has a natural abundance of 99.3%, while the fissionable ^{235}U isotope is present only to the extent of 0.7%. In a typical nuclear reactor, the fuel is made of compressed pellets of UO_2 that have been isotopically enriched to a 3% concentration of ^{235}U and then encased in zirconium rods. The rods are placed in a pressure vessel filled with water, which acts as a moderator to slow the neutrons so they can be captured more readily.

No nuclear explosion can occur in a reactor because the amount and concentration of fissionable fuel is too low and because the fuel is not confined by pressure into a small volume. In a worst-case accident, however, uncontrolled fission can lead to enormous overheating that can melt the reactor and surrounding containment vessel, thereby releasing large amounts of radioactivity to the environment. This is exactly what happened in the Fukushima Daiichi nuclear disaster in Japan on March 11, 2011, when an earthquake and tsunami struck the plant. Newer reactors get around this problem by adopting what are called *passive safety* designs that automatically slow runaway reactors without any electronic feedback or operator input, but the Fukushima reactors were too old to have this technology.

Thirty-one countries around the world now obtain some of their electricity from nuclear energy (**Figure 22.6**). The United States produces more nuclear power than any other country—approximately 101,000 megawatts per year from 104 commercial generators. France produces the highest percentage of its electricity from nuclear reactors—75% from 58 plants. Worldwide, 432 nuclear plants were in operation in mid-2011, with an additional

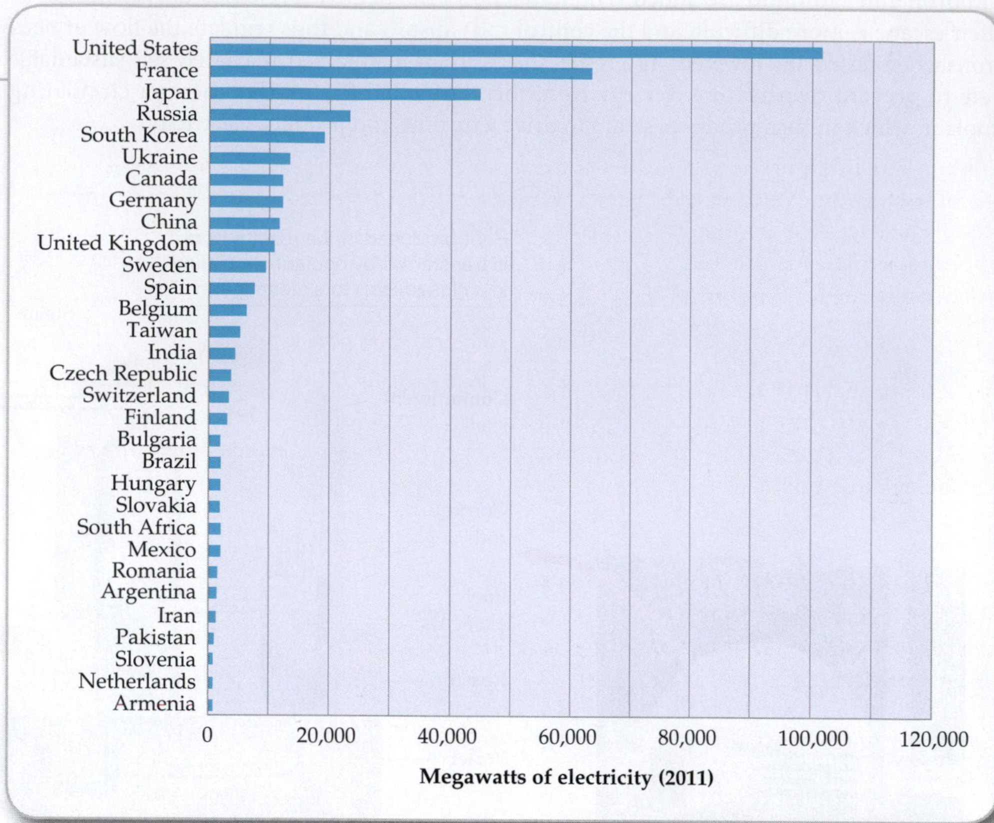

FIGURE 22.6

Megawatts of electricity generated by nuclear power in 2011.

65 under construction and more being planned. The total output was 369,000 megawatts, or approximately 15% of the world's electrical power.

One major problem holding back future development is the yet unsolved matter of how to dispose of the radioactive wastes generated by the plants. It will take at least 600 years for waste strontium-90 to decay to safe levels, and at least 20,000 years for plutonium-239 to decay.

Nuclear Fusion

Just as heavy nuclei such as ^{235}U release energy when they undergo *fission*, very light nuclei such as the isotopes of hydrogen release enormous amounts of energy when they undergo *fusion*. In fact, the fusion of hydrogen nuclei into helium is what powers our Sun and other stars. Among the processes thought to occur in the Sun are those in the following sequence leading to helium-4:

$$^{1}_{1}H + ^{1}_{1}H \longrightarrow ^{2}_{1}H + ^{0}_{1}e$$
$$^{1}_{1}H + ^{2}_{1}H \longrightarrow ^{3}_{2}He$$
$$^{3}_{2}He + ^{3}_{2}He \longrightarrow ^{4}_{2}He + 2\,^{1}_{1}H$$
$$^{3}_{2}He + ^{1}_{1}H \longrightarrow ^{4}_{2}He + ^{0}_{1}e$$

The main appeal of nuclear fusion as a power source is that the hydrogen isotopes used as fuel are cheap and plentiful and that the fusion products are nonradioactive and nonpolluting. The technical problems that must be solved before achieving a practical and controllable fusion method are staggering, however. Not the least of the problems is that a temperature of approximately 40 *million* kelvin is needed to initiate the fusion process.

PROBLEM 22.5 Calculate the amount of energy released in kJ/mol for the fusion reaction of ^{1}H and ^{2}H atoms to yield a ^{3}He atom:

$$^{1}_{1}H + ^{2}_{1}H \longrightarrow ^{3}_{2}He$$

The atomic masses are ^{1}H (1.007 83 u), ^{2}H (2.014 10 u), and ^{3}He (3.016 03 u).

22.4 NUCLEAR TRANSMUTATION

Fewer than 300 of the more than 3600 known isotopes occur naturally. The remainder have been made by **nuclear transmutation**, the change of one element into another. Such transmutation is often brought about by bombardment of an atom with a high-energy particle such as a proton, neutron, or α particle. In the ensuing collision between particle and atom, an unstable nucleus is momentarily created, a nuclear change occurs, and a different element is produced. The first nuclear transmutation was accomplished in 1917 by Ernest Rutherford, who bombarded ^{14}N nuclei with α particles and found that ^{17}O was produced:

$$^{14}_{7}N + ^{4}_{2}He \longrightarrow ^{17}_{8}O + ^{1}_{1}H$$

Other nuclear transmutations can lead to the synthesis of entirely new elements never before seen on Earth. In fact, all the transuranium elements—those elements with atomic numbers greater than 92—have been produced by transmutation. Plutonium, for example, can be made by bombarding uranium-238 with α particles:

$$^{238}_{92}U + ^{4}_{2}He \longrightarrow ^{241}_{94}Pu + ^{1}_{0}n$$

The plutonium-241 that results from U-238 bombardment is itself radioactive with a half-life of 14.3 years, decaying by β emission to yield americium-241. (If that name sounds familiar, it's because americium is used commercially in making smoke detectors.) Americium-241 is also radioactive, decaying by α emission with a half-life of 433 years.

$$^{241}_{94}Pu \longrightarrow ^{241}_{95}Am + ^{0}_{-1}e$$
$$^{241}_{95}Am \longrightarrow ^{237}_{93}Np + ^{4}_{2}He$$

Still other nuclear transmutations are carried out using neutrons, protons, or even whole atoms for bombardment. The cobalt-60 used in radiation therapy for cancer patients can be prepared by neutron bombardment of iron-58. Iron-58 first absorbs a neutron to yield iron-59, the iron-59 undergoes β decay to yield cobalt-59, and the cobalt-59 then absorbs a second neutron to yield cobalt-60:

$$^{58}_{26}Fe + ^{1}_{0}n \longrightarrow ^{59}_{26}Fe$$
$$^{59}_{26}Fe \longrightarrow ^{59}_{27}Co + ^{0}_{-1}e$$
$$^{59}_{27}Co + ^{1}_{0}n \longrightarrow ^{60}_{27}Co$$

The overall change can be written as

$$^{58}_{26}Fe + 2\,^{1}_{0}n \longrightarrow ^{60}_{27}Co + ^{0}_{-1}e$$

▲ The Fermi National Accelerator Laboratory has a particle accelerator 4 mi in circumference that is able to accelerate protons to energies of 1 trillion eV.

Worked Example 22.4

Balancing a Nuclear Transmutation Equation

The element berkelium was first prepared at the University of California at Berkeley in 1949 by α bombardment of $^{241}_{95}Am$. Two neutrons are also produced during the reaction. What isotope of berkelium results from this transmutation? Write a balanced nuclear equation.

STRATEGY AND SOLUTION

According to the periodic table, berkelium has $Z = 97$. Since the sum of the reactant mass numbers is $241 + 4 = 245$ and 2 neutrons are produced, the berkelium isotope must have a mass number of $245 - 2 = 243$.

$$^{241}_{95}Am + ^{4}_{2}He \longrightarrow ^{243}_{97}Bk + 2\,^{1}_{0}n$$

PROBLEM 22.6 Write a balanced nuclear equation for the reaction of argon-40 with a proton:

$$^{40}_{18}Ar + ^{1}_{1}H \longrightarrow \text{?} + ^{1}_{0}n$$

continued on next page

PROBLEM 22.7 Write a balanced nuclear equation for the reaction of uranium-238 with a deuteron ($^{2}_{1}H$):

$$^{238}_{92}U + ^{2}_{1}H \longrightarrow ? + 2\,^{1}_{0}n$$

22.5 DETECTING AND MEASURING RADIOACTIVITY

Radioactive emissions are invisible. We can't see, hear, smell, touch, or taste them, no matter how high the dose. We can, however, measure and detect radioactivity using a suitable instrument, such as a Geiger counter (**Figure 22.7**), and the effects of that radiation on living organisms can be readily observed.

FIGURE 22.7

A Geiger counter.

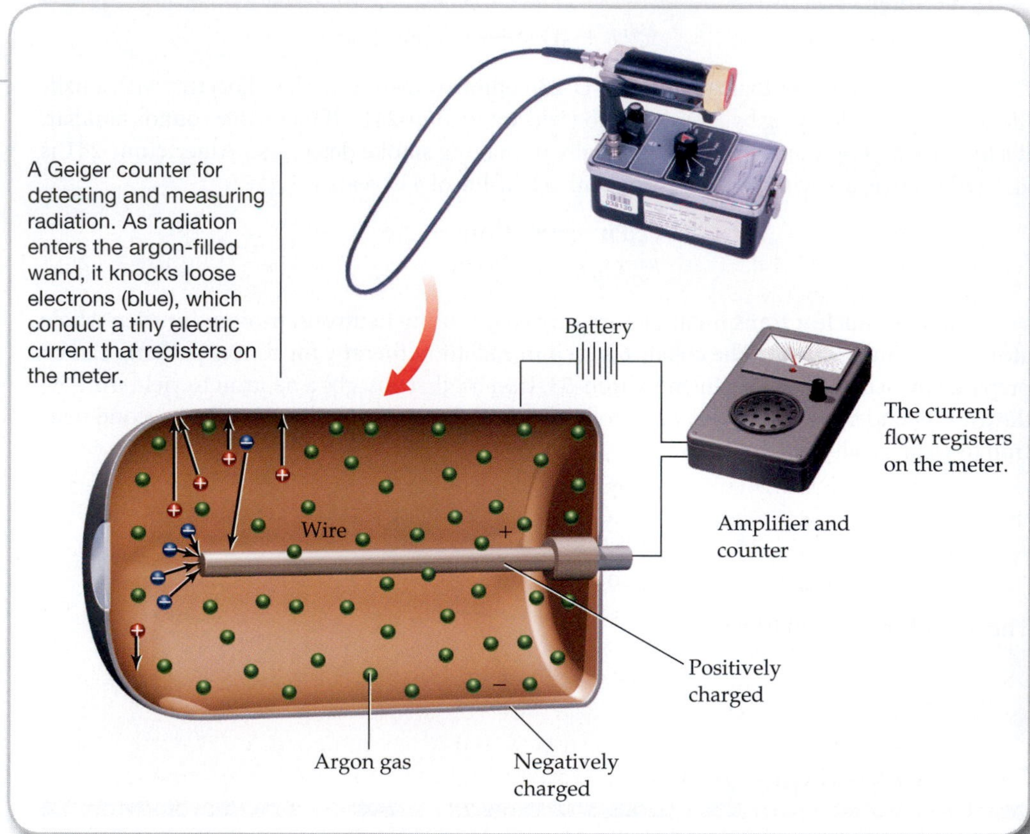

Radiation intensity is expressed in different ways, depending on what is being measured. Some units measure the number of nuclear decay events, while others measure the amount of exposure to radiation or the biological consequences of radiation (**Table 22.2**).

TABLE 22.2 Units for Measuring Radiation

Unit	Quantity Measured	Description
Becquerel (Bq)	Decay events	1 Bq = 1 disintegration/s
Curie (Ci)	Decay events	1 Ci = 3.7×10^{10} disintegrations/s
Gray (Gy)	Energy absorbed per kilogram of tissue	1 Gy = 1 J/kg tissue
Rad	Energy absorbed per kilogram of tissue	1 rad = 0.01 Gy
Sievert (Sv)	Tissue damage	1 Sv = 1 J/kg
Rem	Tissue damage	1 rem = 0.01 Sv

- **The *becquerel* (Bq)** is the SI unit for measuring the number of nuclear disintegrations occurring per second in a sample: 1 Bq = 1 disintegration/s. The *curie* (Ci) and *millicurie* (mCi) also measure disintegrations per unit time, but they are far larger units than the becquerel and are more often used, particularly in medicine and biochemistry. One curie is the decay rate of 1 g of radium, equal to 3.7×10^{10} Bq:

$$1 \text{ Bq} = 1 \text{ disintegration/s}$$
$$1 \text{ Ci} = 3.7 \times 10^{10} \text{ Bq} = 3.7 \times 10^{10} \text{ disintegrations/s}$$

As an example, a 1.5 mCi sample of tritium is equal to 5.6×10^7 Bq, meaning that it undergoes 5.6×10^7 disintegrations/s.

$$(1.5 \text{ mCi})\left(10^{-3}\,\frac{\text{Ci}}{\text{mCi}}\right)\left(3.7 \times 10^{10}\,\frac{\text{Bq}}{\text{Ci}}\right) = 5.6 \times 10^7 \text{ Bq}$$

- **The *gray* (Gy)** is the SI unit for measuring the amount of energy absorbed per kilogram of tissue exposed to a radiation source: 1 Gy = 1 J/kg. The *rad* (*radiation absorbed dose*) also measures tissue exposure and is more often used in medicine.

$$1 \text{ Gy} = 1 \text{ J/kg} \qquad 1 \text{ rad} = 0.01 \text{ Gy}$$

- **The *sievert* (Sv)** is the SI unit that measures the amount of tissue damage caused by radiation. It takes into account not just the energy absorbed per kilogram of tissue but also the different biological effects of different kinds of radiation. For example, 1 Gy of α radiation causes 20 times more tissue damage than 1 Gy of γ rays, but 1 Sv of α radiation and 1 Sv of γ rays cause the same amount of damage. The *rem* (*roentgen equivalent for man*) is an analogous non-SI unit that is more frequently used in medicine.

$$1 \text{ rem} = 0.01 \text{ Sv}$$

The effects of radiation on the human body vary with the energy and kind of radiation, as well as with the length of exposure and whether the radiation is from an external or internal source. When coming from an external source, X rays and γ radiation are more harmful than α and β particles because they penetrate clothing and skin. When coming from an internal source, however, α and β particles are particularly dangerous because all their energy is given up to surrounding tissue. Alpha emitters, although relatively harmless externally, are especially hazardous internally and are almost never used in medical applications.

Because of their relatively large mass, α particles move slowly—up to only one-tenth the speed of light—and can be stopped by a few sheets of paper or by the top layer of skin. Beta particles, because they are much lighter, move at up to nine-tenths the speed of light and have about 100 times the penetrating power of α particles. A block of wood or heavy protective clothing is necessary to stop β radiation, which would otherwise penetrate and burn the skin.

Gamma rays and X rays move at the speed of light and have about 1000 times the penetrating power of α particles. A lead block several inches thick is needed to stop γ and X radiation, which could otherwise penetrate and damage the body's internal organs. Some properties of different kinds of radiation are summarized in **Table 22.3**.

TABLE 22.3 Some Properties of Radiation

Type of Radiation	Energy Range	Penetrating Distance in Water*
α	3–9 MeV	0.02–0.04 mm
β	0–3 MeV	0–4 mm
X	100 eV–10 keV	0.01–1 cm
γ	10 keV–10 MeV	1–20 cm

*Distances at which one-half of the radiation has been stopped.

The biological effects of different radiation doses are given in **Table 22.4**. Although the effects sound fearful, the average radiation dose received annually by most people is only about 120 mrem. About 70% of this radiation comes from natural sources such as rocks and cosmic rays—energetic particles coming from interstellar space. The remaining 30% comes from medical procedures such as X rays. The amount due to emissions from nuclear power plants and to fallout from atmospheric testing of nuclear weapons in the 1950s is barely detectable.

TABLE 22.4 Biological Effects of Short-Term Radiation on Humans

Dose (rem)	Biological Effects
0–25	No detectable effects
25–100	Temporary decrease in white blood cell count
100–200	Nausea, vomiting, longer-term decrease in white blood cells
200–300	Vomiting, diarrhea, loss of appetite, listlessness
300–600	Vomiting, diarrhea, hemorrhaging, eventual death in some cases
Above 600	Eventual death in nearly all cases

22.6 SOME APPLICATIONS OF NUCLEAR CHEMISTRY

Nuclear chemistry has many applications in daily life, in numerous scientific fields, and in medicine. Here are a few:

Dating with Radioisotopes

▲ Radiocarbon dating places the age of this bog mummy at 2200 years BP. The abbreviation BP means "Before Present," with 1950 taken as the reference point.

Biblical scrolls are found in a cave near the Dead Sea. Are they authentic? A mummified body is discovered in a peat bog near Ramten, Denmark. How old is it? The burned bones of a man are dug up near Lubbock, Texas. How long have humans lived in the area? These and many other questions can be answered by archaeologists using a technique called *radiocarbon dating*. (The Dead Sea Scrolls are 1900 years old and authentic, the mummy is 2200 years old, and the human remains found in Texas are 9900 years old.)

Radiocarbon dating of archaeological artifacts depends on the slow and constant production of radioactive carbon-14 in the upper atmosphere by neutron bombardment of nitrogen atoms. (The neutrons come from the bombardment of other atoms by cosmic rays.)

$$^{14}_{7}\text{N} + ^{1}_{0}\text{n} \longrightarrow ^{14}_{6}\text{C} + ^{1}_{1}\text{H}$$

Carbon-14 atoms produced in the upper atmosphere combine with oxygen to yield $^{14}CO_2$, which slowly diffuses into the lower atmosphere, where it mixes with ordinary $^{12}CO_2$ and is taken up by plants during photosynthesis. When these plants are eaten, carbon-14 enters the food chain and is ultimately distributed throughout all living organisms.

As long as a plant or animal is living, a dynamic equilibrium exists in which an organism excretes or exhales the same ratio of ^{14}C to ^{12}C that it takes in. As a result, the ratio of ^{14}C to ^{12}C in the living organism remains constant and is the same as that in the atmosphere—about 1 part in 10^{12}. When the plant or animal dies, however, it no longer takes in more ^{14}C and the $^{14}C/^{12}C$ ratio in the organism slowly decreases as ^{14}C undergoes radioactive decay by β emission, with $t_{1/2} = 5715$ years.

$$^{14}_{6}\text{C} \longrightarrow ^{14}_{7}\text{N} + ^{\ 0}_{-1}\text{e}$$

At 5715 years (one ^{14}C half-life) after the death of the organism, the $^{14}C/^{12}C$ ratio has decreased by a factor of 2; at 11,430 years after death, the $^{14}C/^{12}C$ ratio has decreased by a factor of 4; and so on. By measuring the present $^{14}C/^{12}C$ ratio in the traces of any once-living organism, archaeologists can determine how long ago the organism died. The results are usually given as the number of years BP, meaning "Before Present," with the year 1950 used as the reference point.

Human or animal hair from well-preserved remains, charcoal or wood fragments from once-living trees, and cotton or linen from once-living plants are all useful sources for radiocarbon dating. The technique becomes less accurate as samples get older and the amount of ^{14}C they contain diminishes, but artifacts with an age of 1000–20,000 years can be dated with reasonable accuracy. The outer limit of the technique is about 60,000 years.

Just as radiocarbon measurements allow the dating of once-living organisms, similar measurements on other radioisotopes make possible the dating of rocks. Uranium-238, for example, has a half-life of 4.47×10^9 years and decays through a series of events to yield lead-206. The age of a uranium-containing rock can therefore be determined by measuring the $^{238}U/^{206}Pb$ ratio. Similarly, potassium-40 has a half-life of 1.25×10^9 years and decays through electron capture and positron emission to yield argon-40. (Both processes yield the same product.)

$$^{40}_{19}K + ^{0}_{-1}e \longrightarrow ^{40}_{18}Ar$$
$$^{40}_{19}K \longrightarrow ^{40}_{18}Ar + ^{0}_{1}e$$

The age of a rock can be found by crushing a sample, measuring the amount of ^{40}Ar gas that escapes, and comparing the amount of ^{40}Ar with the amount of ^{40}K remaining in the sample. It is through techniques such as these that the age of the Earth has been estimated at 4.54 billion years.

Worked Example 22.5

Using Radiocarbon Dating

Radiocarbon measurements made in 1988 on the Shroud of Turin, a religious artifact thought by some to be the burial shroud of Christ, showed a ^{14}C decay rate of 14.2 disintegrations/min per gram of carbon. What age BP is implied by this result if currently living organisms decay at the rate of 15.3 disintegrations/min per gram of carbon? The half-life of ^{14}C is 5715 years.

STRATEGY

For a first-order process like radioactive decay, in which rate $= kN_t$, the ratio of the decay rate at any time t to the decay rate at time $t = 0$ is the same as the ratio of N_t to N_0:

$$\frac{\text{Decay rate at time } t}{\text{Decay rate at time } t = 0} = \frac{kN_t}{kN_0} = \frac{N_t}{N_0}$$

To date the shroud, we need to calculate the time t that corresponds to the observed decay rate. This can be done by solving for t in the integrated rate law, as explained in Section 12.6.

$$\ln\left(\frac{N_t}{N_0}\right) = (-\ln 2)\left(\frac{t}{t_{1/2}}\right)$$

SOLUTION

Substituting the proper values into the equation gives

$$\ln\left(\frac{14.2}{15.3}\right) = -0.693\left(\frac{t}{5715 \text{ years}}\right) \quad \text{or} \quad -0.0746 = -0.693\left(\frac{t}{5715 \text{ years}}\right)$$

$$\text{so} \quad t = \frac{(-0.0746)(5715 \text{ years})}{-0.693} = 615 \text{ years}$$

An age of 615 years in 1988 corresponds to an age of 615 − 38 = 577 years in 1950, or 577 BP. This result implies that the Shroud of Turin comes from medieval times.

PROBLEM 22.8 Charcoal found in the Lascaux cave in France, site of many prehistoric cave paintings, was observed in 1950 to decay at a rate of 2.4 disintegrations/min per gram of carbon. What is the age of the charcoal BP if currently living organisms decay at the rate of 15.3 disintegrations/min per gram of carbon? The half-life of ^{14}C is 5715 years.

▲ What is the age of this painting from the Lascaux cave in France?

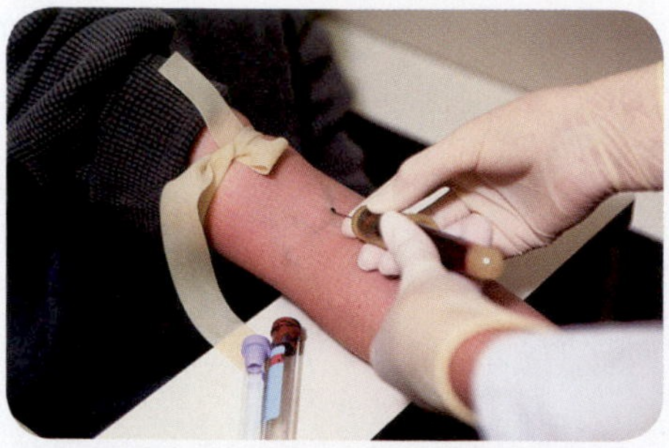

▲ A person's blood volume can be found by injecting a small amount of radioactive chromium-51 and measuring the dilution factor.

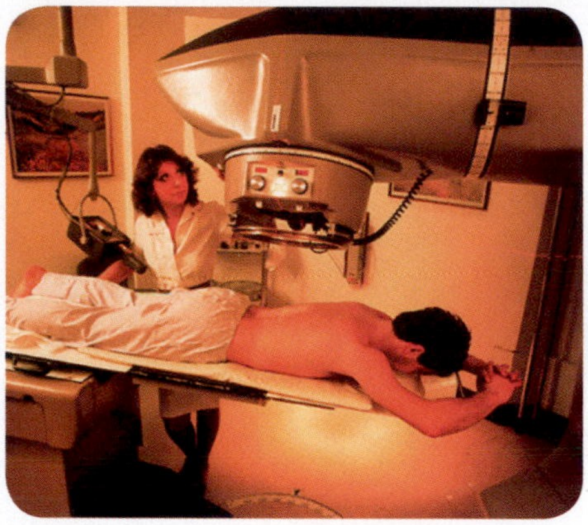

▲ Cancerous tumors can be treated by irradiation with γ rays from this cobalt-60 source.

▲ Technetium-99*m*, a short-lived radioisotope used for brain scans, is obtained from the decay of ^{99}Mo, which is itself prepared by neutron bombardment of ^{98}Mo and then stored in this "technetium cow" in the form of MoO_4^{2-}. Small amounts of ^{99m}Tc are removed by passing a saline solution through the cylinder.

Medical Uses of Radioactivity

The origins of nuclear medicine date to 1901, when the French physician Henri Danlos first used radium in the treatment of a tuberculous skin lesion. Since that time, uses of radioactivity have become a crucial part of modern medical care, both diagnostic and therapeutic. Current nuclear techniques can be grouped into three classes: (1) in vivo procedures, (2) therapeutic procedures, and (3) imaging procedures.

In Vivo Procedures In vivo studies—those that take place *inside* the body—are carried out to assess the functioning of a particular organ or body system. A radiopharmaceutical agent is administered, and its path in the body—whether it is absorbed, excreted, diluted, or concentrated—is determined by analysis of blood or urine samples.

An example of the many in vivo procedures using radioactive agents is the determination of whole-blood volume by injecting a known quantity of red blood cells labeled with radioactive chromium-51. After a suitable interval to allow the labeled cells to be distributed evenly throughout the body, a blood sample is taken, the amount of dilution of the ^{51}Cr is measured, and the blood volume is calculated.

Therapeutic Procedures Therapeutic procedures—those in which radiation is used to kill diseased tissue—can involve either external or internal sources of radiation. External radiation therapy for the treatment of cancer is often carried out with γ rays from a cobalt-60 source. The highly radioactive source is shielded by a thick lead container and has a small opening directed toward the site of the tumor. By focusing the radiation beam on the tumor and rotating the patient's body, the tumor receives the full exposure while the exposure of surrounding parts of the body is minimized. Nevertheless, sufficient exposure occurs so that most patients suffer some effects of radiation sickness.

Internal radiation therapy is a much more selective technique than external therapy. In the treatment of thyroid disease, for example, iodine-131, a powerful β emitter known to localize in the target tissue, is administered internally. Because β particles penetrate no farther than several millimeters, the localized ^{131}I produces a high radiation dose that destroys only the surrounding diseased tissue.

The half-lives of some medically useful radioisotopes are listed in **Table 22.5**. As you might expect if you think about it, the radioisotopes used internally in medical applications have fairly short half-lives so that they decay rapidly and don't cause long-term health hazards.

TABLE 22.5 Half-Lives of Some Useful Radioisotopes

Radioisotope	Symbol	Radiation	Half-life	Use
Fluorine-18	$^{18}_{9}F$	β^+	110 minutes	PET scans
Phosphorus-32	$^{32}_{15}P$	β^-	14.28 days	Leukemia therapy
Cobalt-60	$^{60}_{27}Co$	β^-, γ	5.27 years	Cancer therapy
Technetium-99*m**	$^{99m}_{43}Tc$	γ	6.01 hours	Brain scans
Iodine-123	$^{123}_{53}I$	γ	13.2 hours	Thyroid therapy

* The *m* in technetium-99*m* stands for *metastable*, meaning that it undergoes γ emission but does not change its mass number or atomic number.

Imaging Procedures Imaging procedures give diagnostic information about the health of body organs by analyzing the distribution pattern of radioisotopes introduced into the body. A radiopharmaceutical agent that is known to concentrate in a specific tissue or organ is injected into the body, and its distribution pattern is monitored by external radiation detectors. Depending on the disease and the organ, a diseased organ might concentrate more of the radiopharmaceutical than a normal organ and thus show up as a radioactive "hot" spot against a "cold" background. Alternatively, the diseased organ might concentrate less of the radiopharmaceutical than a normal organ and thus show up as a cold spot on a hot background.

The radioisotope most widely used today is technetium-99*m*, whose short half-life of 6.01 hours minimizes a patient's exposure to harmful effects. Bone scans using Tc-99*m*, such as that shown in **Figure 22.8**, are an important tool in the diagnosis of stress fractures, cancer, and other conditions.

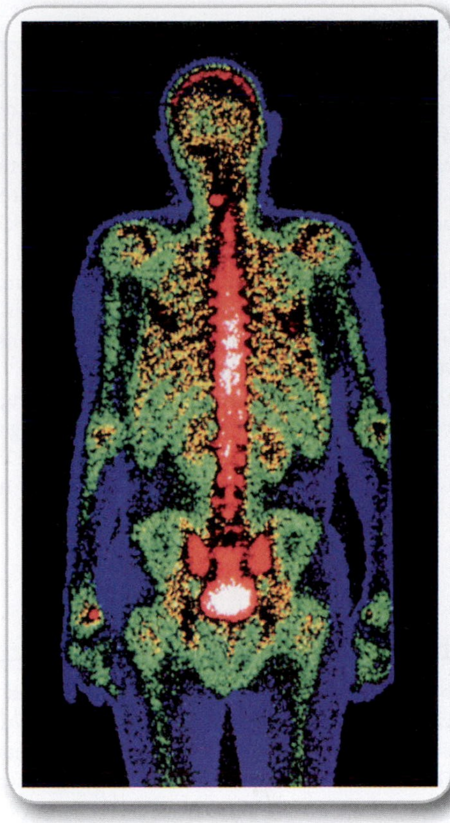

FIGURE 22.8

A bone scan using radioactive technetium-99*m*.

FYI NATURALLY OCCURRING NUCLEAR REACTORS

Nuclear reactors aren't easy to build. A critical mass of fissionable fuel has to be obtained, typically uranium-235, enriched from its 0.7% natural abundance to about 3%. Purified control rods have to be made, neutron moderators selected, and many safety precautions taken. The first man-made nuclear reactor, in fact, was constructed in Chicago, Illinois, only 70 years ago, in 1942. Nevertheless, and as hard as it is to believe, the world's true first nuclear reactor is located in Gabon, Africa, and is approximately *2 billion* years old. Sixteen of these natural reactors are now known, collectively called the Oklo Fossil Reactors.

The reactors were discovered in early 1972 when chemical analysis of uranium ore samples from the Oklo mine in Gabon showed abnormally low amounts of ^{235}U. Although naturally occurring uranium typically contains 0.72% of the fissionable ^{235}U isotope, some of the samples from the Oklo mine contained as little as 0.44% of ^{235}U. Furthermore, the abundances of several other isotopes in the ore, including ^{142}Nd and ^{99}Ru, were also dramatically different from typical values. All these differences closely resemble those found in the spent fuel from modern nuclear reactors, leading investigators to conclude that spontaneous and self-sustaining nuclear chain reactions must have occurred at the mine site nearly 2 billion years ago.

The Oklo nuclear reactors are thought to have formed when unusually rich uranium deposits were flooded by groundwater, which acted as a moderator to slow the neutrons released by fission of ^{235}U and prevent their escape, thereby allowing a chain reaction to begin. As soon as the chain reaction started, though, the heat released rapidly boiled away the groundwater, which stopped the process and shut down the reactor until it cooled. Groundwater then returned, and the chain reaction restarted. Isotope measurements in nearby rocks indicate that the reactors cycled on and off every 2–3 hours for several hundred thousand years until the concentration of ^{235}U in the ore became too low to sustain a chain reaction. Up to 2 tons of ^{235}U underwent fission.

Could another natural reactor form today? When the Oklo reactors formed 2 billion years ago, the abundance of the fissionable ^{235}U isotope in naturally occurring uranium was about 3%, the same percentage as that used today in nuclear power plants. Today, however, the natural abundance of ^{235}U is only about 0.7% because of nuclear decay over the past few billion years. Because a chain-reaction is no longer self-sustaining at the present 0.7% level of ^{235}U, the conditions needed by natural reactors are no longer present on Earth.

▲ Believe it or not, naturally occurring nuclear chain reactions took place 2 billion years ago at this site in west equatorial Africa near present-day Oklo, Gabon.

PROBLEM 22.9 What were the conditions that allowed natural nuclear reactors to form 2 billion years ago, and why can't those reactors form today?

SUMMARY

Nuclear chemistry is the study of the properties and reactions of atomic nuclei. Nuclear reactions differ from chemical reactions in that they involve a change in an atom's nucleus, often producing a different element. The rate of radioactive decay is characterized kinetically by a first-order decay constant and by a half-life, $t_{1/2}$, the time required for the number of radioactive nuclei in a sample to drop to half its initial value.

The stability of a given nucleus is related to its neutron/proton ratio. The strength of the forces involved can be measured by calculating an atom's **mass defect**—the difference in mass between a given nucleus and the total mass of its constituent nucleons (protons and neutrons). Applying the Einstein equation $\Delta E = \Delta mc^2$ then allows calculation of the nuclear **binding energy**. The energy change accompanying a nuclear reaction is far greater than that accompanying any chemical reaction.

Certain heavy nuclei, such as uranium-235, undergo **nuclear fission** when struck by neutrons, breaking apart into fragment nuclei and releasing enormous amounts of energy. Light nuclei, such as the isotopes of hydrogen, undergo **nuclear fusion** when heated to sufficiently high temperatures, forming heavier nuclei and releasing energy.

Of the more than 3600 known isotopes, most have been made by **nuclear transmutation**, the change of one element into another. Such transmutation is often brought about by bombardment of an atom with a high-energy particle such as a proton, neutron, or α particle.

Radiation intensity is expressed in different ways, according to what property is being measured. The becquerel (Bq) and the curie (Ci) measure the number of radioactive disintegrations per second in a sample. The gray (Gy) and the rad measure the amount of radiation absorbed per kilogram of tissue. The sievert (Sv) and the rem measure the amount of tissue damage caused by radiation. Radiation effects become noticeable with a human exposure of 25 rem and become lethal at an exposure above 600 rem.

KEY WORDS

binding energy *908*
chain reaction *911*
critical mass *911*
fission *911*
fusion *911*
mass defect *908*
nuclear transmutation *915*
transuranium element *906*

LEARNING OUTCOMES

- Write nuclear equations for the five nuclear decay processes. **(Section 22.1)** ***Problems 1, 10–13, 37, 45–47***
- Calculate the mass defect and binding energy of an atom. **(Section 22.2)** ***Problems 2, 3, 14–21, 38, 43***
- Calculate the amount of energy released in a nuclear reaction. **(Section 22.3)** ***Problems 4, 5, 22–27, 34, 36, 40, 44, 46***
- Write balanced equations for nuclear transmutation reactions. **(Section 22.4)** ***Problems 6, 7, 28–33, 42***
- Calculate the age of an object using radiocarbon dating. **(Section 22.6)** ***Problems 8, 39, 41, 48***

SECTION PROBLEMS

Problems 22.1–22.9 appear within the chapter.

Nuclear Stability (Section 22.1)

22.10 Of the two isotopes of tungsten, ^{160}W and ^{185}W, one decays by β emission and one decays by α emission. Which does which? Explain.

22.11 Of the two isotopes of iodine, ^{136}I and ^{122}I, one decays by β emission and one decays by positron emission. Which does which? Explain.

22.12 Why do nuclei that are neutron-rich emit β particles?

22.13 Why do nuclei that are neutron-poor emit α particles or positrons or undergo electron capture?

Energy Changes during Nuclear Reactions (Section 22.2)

22.14 Why does a given nucleus have less mass than the sum of its constituent protons and neutrons?

22.15 In an endothermic chemical reaction, do the products have more mass, less mass, or the same mass as the reactants? Explain.

22.16 What is the wavelength (in nm) of γ rays whose energy is 1.50 MeV?

22.17 What is the frequency (in Hz) of X rays whose energy is 6.82 keV?

22.18 Calculate the mass defect (in g/mol) for the following nuclei:

(a) ^{52}Fe (atomic mass = 51.948 11 u)

(b) ^{92}Mo (atomic mass = 91.906 81 u)

22.19 Calculate the mass defect (in g/mol) for the following nuclei:
(a) ^{32}S (atomic mass = 31.972 07 u)
(b) ^{40}Ca (atomic mass = 39.962 59 u)

22.20 Calculate the binding energy (in MeV/nucleon) for the following nuclei:
(a) ^{58}Ni (atomic mass = 57.935 35 u)
(b) ^{84}Kr (atomic mass = 83.911 51 u)

22.21 Calculate the binding energy (in MeV/nucleon) for the following nuclei:
(a) ^{63}Cu (atomic mass = 62.939 60 u)
(b) ^{84}Sr (atomic mass = 83.913 43 u)

22.22 What is the energy change ΔE (in kJ/mol) when an α particle is emitted from ^{174}Ir? The atomic mass of ^{174}Ir is 173.966 66 u, the atomic mass of ^{170}Re is 169.958 04 u, and the atomic mass of a ^{4}He atom is 4.002 60 u.

$$^{174}_{77}Ir \longrightarrow ^{170}_{75}Re + ^{4}_{2}He \qquad \Delta E = ?$$

22.23 Magnesium-28 is a β emitter that decays to aluminum-28. How much energy is released in kJ/mol? The atomic mass of ^{28}Mg is 27.983 88 u, and the atomic mass of ^{28}Al is 27.981 91 u.

22.24 What is the mass change in grams accompanying the formation of NH_3 from H_2 and N_2?

$$N_2(g) + 3\,H_2(g) \longrightarrow 2\,NH_3(g) \qquad \Delta H° = -92.2\text{ kJ}$$

22.25 What is the mass change in grams accompanying the formation of CO and H_2 in the water-gas reaction?

$$C(s) + H_2O(g) \longrightarrow CO(g) + H_2(g) \qquad \Delta H° = +131\text{ kJ}$$

22.26 A positron has the same mass as an electron (9.109×10^{-31} kg) but an opposite charge. When the two particles encounter each other, annihilation occurs and only γ rays are produced. How much energy (in kJ/mol) is produced?

22.27 How much energy is released (in kJ) in the fusion reaction of ^{2}H to yield 1 mol of ^{3}He? The atomic mass of ^{2}H is 2.0141 u, and the atomic mass of ^{3}He is 3.0160 u.

$$2\,^{2}_{1}H \longrightarrow ^{3}_{2}He + ^{1}_{0}n$$

Nuclear Transmutation (Section 22.4)

22.28 Give the products of the following nuclear reactions:
(a) $^{109}_{47}Ag + ^{4}_{2}He \longrightarrow ?$
(b) $^{10}_{5}B + ^{4}_{2}He \longrightarrow ? + ^{1}_{0}n$

22.29 Balance the following equations for the nuclear fission of ^{235}U:
(a) $^{235}_{92}U \longrightarrow ^{160}_{62}Sm + ^{72}_{30}Zn + ?\,^{1}_{0}n$
(b) $^{235}_{92}U \longrightarrow ^{87}_{35}Br + ? + 2\,^{1}_{0}n$

22.30 Element 109 ($^{266}_{109}Mt$) was prepared in 1982 by bombardment of ^{209}Bi atoms with ^{58}Fe atoms. Identify the other product that must have formed, and write a balanced nuclear equation assuming no other products were formed.

22.31 Molybdenum-99 is formed by neutron bombardment of a naturally occurring isotope. If one neutron is absorbed and no by-products are formed, what is the starting isotope?

22.32 Californium-246 is formed by bombardment of uranium-238 atoms. If four neutrons are formed as by-products, what particle is used for the bombardment?

22.33 Balance the following transmutation reactions:
(a) $^{246}_{96}Cm + ^{12}_{6}C \longrightarrow ? + 4\,^{1}_{0}n$
(b) $^{253}_{99}Es + ? \longrightarrow ^{256}_{101}Md + ^{1}_{0}n$
(c) $^{250}_{98}Cf + ^{11}_{5}B \longrightarrow ? + 4\,^{1}_{0}n$

CHAPTER PROBLEMS

22.34 The electronic systems on the New Horizons spacecraft, launched on January 19, 2006 and scheduled to reach Pluto on July 14, 2015, are powered by electricity generated by heat. The heat comes from the radioactive decay of ^{238}Pu in the 11 kg of $^{238}PuO_2$ fuel onboard. The generator provided 240 W when the spacecraft was launched. If the power output is directly proportional to the amount of ^{238}Pu in the generator, what will the power output be when the spacecraft reaches Pluto? The half-life of ^{238}Pu is 87.7 y.

22.35 To treat a brain tumor with gamma knife radiosurgery, the patient's head is positioned within a hemispherical dome covered by 201 individual ^{60}Co sources whose narrow beams are directed inward toward the tumor target point. The tumor receives a very high dose of radiation because all the beams converge on it, while any irradiated healthy tissue receives only the radiation of a single beam. For a prescribed dose of 1800 rad, how long should the radiation treatment go if 2.2×10^{11} disintegrations of ^{60}Co are required to give a dose of 1.0 rad? Assume that all the sources are directed at the tumor with an activity of 30 Ci each.

22.36 Thorium-232 decays by a 10-step series of nuclear reactions, ultimately yielding lead-208, along with 6 α particles and 4 β particles. How much energy (in kJ/mol) is released during the overall process? The relevant masses are: ^{232}Th = 232.038 054 u; ^{208}Pb = 207.976 627 u; electron = 0.000 548 6 u, ^{4}He = 4.002 603 u.

22.37 Fraud in science is rare but does happen occasionally. In 1999, the creation of three superheavy elements (one new) was claimed when ^{208}Pb was bombarded with ^{86}Kr. The claim was subsequently found to be fraudulent and was withdrawn. Identify the isotopes X, Y, and Z that were claimed.

$$^{208}_{82}Pb + ^{86}_{36}Kr \xrightarrow{-n} X \xrightarrow{-\alpha} Y \xrightarrow{-\alpha} Z$$

22.38 Calculate the mass defect (in g/mol) and the binding energy (in MeV/nucleon) for the following nuclei. Which of the two is more stable?

(a) ^{50}Cr (atomic mass = 49.946 05 u)

(b) ^{64}Zn (atomic mass = 63.929 15 u)

22.39 What is the age BP of a bone fragment that shows an average of 2.9 disintegrations/min per gram of carbon in 2005? The carbon in living organisms undergoes an average of 15.3 disintegrations/min per gram, and the half-life of ^{14}C is 5715 years. (See Section 12.6.)

22.40 How much energy (in kJ/mol) is released in the fusion reaction of ^{2}H with ^{3}He?

$$^{2}_{1}H + ^{3}_{2}He \longrightarrow ^{4}_{2}He + ^{1}_{1}H$$

The relevant masses are ^{2}H (2.0141 u), ^{3}He (3.0160 u), ^{4}He (4.0026 u), and ^{1}H (1.0078 u).

22.41 What is the age of a rock whose $^{40}Ar/^{40}K$ ratio is 1.15? The half-life of ^{40}K is 1.28×10^9 years.

22.42 The most abundant isotope of uranium, ^{238}U, does not undergo fission. In a *breeder reactor*, however, a ^{238}U atom captures a neutron and emits two β particles to make a fissionable isotope of plutonium, which can then be used as fuel in a nuclear reactor. Write a balanced nuclear equation.

22.43 It has been estimated that 3.9×10^{23} kJ/s is radiated into space by the Sun. What is the rate of the Sun's mass loss in kg/s?

22.44 In a cancer treatment called boron neutron-capture therapy, a compound containing boron-10 is injected into a patient where it selectively binds to cancer cells. Irradiating the affected area with neutrons then induces the following reaction:

$$^{10}B + ^{1}n \longrightarrow ^{4}He + ^{7}Li + \gamma$$

The α radiation kills the cancer cells, leaving the surrounding tissue unharmed. The reactants in this nuclear process have essentially no kinetic energy, but the products have a total kinetic energy of 2.31 MeV. What is the energy of the γ photon released? Relevant masses are: ^{4}He (4.002 603 u), ^{7}Li (7.016 004 u), ^{10}B (10.012 937 u), e^- (0.000 548 6 u), $^{1}_{0}n$ (1.008 665 u).

22.45 Neptunium-237 decays by a series of steps to bismuth-209. How many α and β particles are produced by this decay process?

22.46 The radioactive isotope ^{100}Tc decays to form the stable isotope ^{100}Mo.

(a) There are two possible pathways for this decay. Write balanced equations for both.

(b) Only one of the pathways is observed. Calculate the energy released by both pathways, and explain why only one is observed. Relevant masses are: ^{100}Tc (99.907 657 u), ^{100}Mo (99.907 48 u), e^- (0.000 548 6 u).

22.47 The radioisotope ^{226}Ac can decay by any of three different nuclear processes: α emission, β emission, or electron capture.

(a) Write a balanced nuclear equation for the decay of ^{226}Ac by each decay mode.

(b) For the decay of ^{226}Ac by all processes combined, the first-order rate constant is $k = 0.556\ d^{-1}$. How many days are required for 80.0% of a sample of ^{226}Ac to decay?

MULTICONCEPT PROBLEMS

22.48 A small sample of wood from an archeological site in Clovis, New Mexico, was burned in O_2, and the CO_2 produced was bubbled through a solution of $Ba(OH)_2$ to produce a precipitate of $BaCO_3$. When the $BaCO_3$ was collected by filtration, a 1.000 g sample was found to have a radioactivity of 4.0×10^{-3} Bq. The half-life of ^{14}C is 5715 y, and living organisms have a radioactivity due to ^{14}C of 15.3 disintegrations/min per gram of carbon. Assuming that the analysis was carried out in 1960, what is the age BP of the Clovis site?

22.49 Polonium-210, a naturally occurring radioisotope, is an α emitter, with $t_{1/2}$ = 138 d. Assume that a sample of ^{210}Po with a mass of 0.700 mg was placed in a 250.0 mL flask, which was evacuated, sealed, and allowed to sit undisturbed. What would the pressure be inside the flask (in mm Hg) at 20 °C after 365 days if all the α particles emitted had become helium atoms?

22.50 A blood-volume determination was carried out on a patient by injection with 20.0 mL of blood that had been radioactively labeled with Cr-51 to an activity of 4.10 μCi/mL. After a brief period to allow for mixing in the body, blood was drawn from the patient for analysis. Unfortunately, a mix-up in the laboratory prevented an immediate analysis, and it was not until 17.0 days later that a measurement on the blood was made. The radiation level was then determined to be 0.009 35 μCi/mL. If ^{51}Cr has $t_{1/2} = 27.7$ days, what is the volume of blood in the patient?

22.51 Imagine that you have a 0.007 50 M aqueous $MgCl_2$ solution, prepared so that it contains a small amount of radioactive ^{28}Mg. The half-life of ^{28}Mg is 20.91 h, and the initial activity of the $MgCl_2$ solution is 0.112 μCi/mL. Assume that 20.00 mL of this $MgCl_2$ solution is added to 15.00 mL of 0.012 50 M aqueous Na_2CO_3 solution and that the resultant precipitate is then removed by filtration to yield a clear filtrate. After a long break to go for a run, you find that the activity of the filtrate measured 2.40 h after beginning the experiment is 0.029 μCi/mL. What are the molar concentrations of Mg^{2+} and CO_3^{2-} in the filtrate, and what is the solubility product constant of $MgCO_3$?

CHAPTER 23

Organic and Biological Chemistry

Real flamingos owe their color to organic chemicals in their diet. Plastic flamingoes owe their color to organic chemicals in their dyes.

If the ultimate goal of chemistry is to understand the world around us on a molecular level, then a knowledge of **biochemistry**—the chemistry of living organisms—is a central part of that goal. Biochemistry, in turn, is a branch of *organic chemistry*, a term originally used to mean the study of compounds from living organisms while *inorganic chemistry* was used for the study of compounds from nonliving sources. Today, however, we know that there are no fundamental differences between organic and inorganic compounds; the same principles apply to both. The only common characteristic of compounds from living sources is that all contain the element carbon. Thus, **organic chemistry** is now defined as the study of carbon compounds.

But why is carbon special, and why do chemists still treat organic chemistry as a separate branch of science? The answers to these questions involve the ability of carbon atoms to bond together, forming long chains and rings. Of all the elements, only carbon is able to form such an immense array of compounds, from methane, with one carbon atom, to deoxyribonucleic acid (DNA), with tens of billions of carbon atoms. In fact, more than 67 million organic compounds have been made in laboratories around the world, and living organisms contain additional millions.

23.1 ORGANIC MOLECULES AND THEIR STRUCTURES: ALKANES

Why are there so many organic compounds? The answer is that a relatively small number of atoms can bond together in a great many ways. Take molecules that contain only carbon and hydrogen—**hydrocarbons**—and have only single bonds. Such compounds belong to the family of organic molecules called **alkanes**. Because carbon atoms have four outer-shell electrons and form four covalent bonds (Section 4.6), the only possible one-carbon alkane is methane, CH_4. Similarly, the only possible two-carbon alkane is ethane, C_2H_6, and the only possible three-carbon alkane is propane, C_3H_8.

CONTENTS

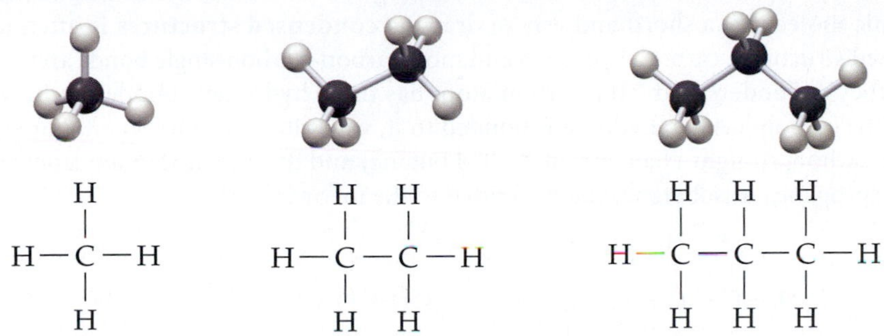

Methane, CH_4 — Ethane, C_2H_6 — Propane, C_3H_8

When larger numbers of carbons combine with hydrogen, however, more than one structure can result. There are two four-carbon alkanes with the formula C_4H_{10}, for instance. In one compound, the four carbons are in a row, while in the other they have a branched arrangement. Similarly, there are three alkanes with the formula C_5H_{12} and even more possibilities for larger alkanes. Compounds with all their carbons connected in a row are called **straight-chain alkanes**, and those with a branching connection of carbons are called **branched-chain alkanes**.

C_4H_{10}

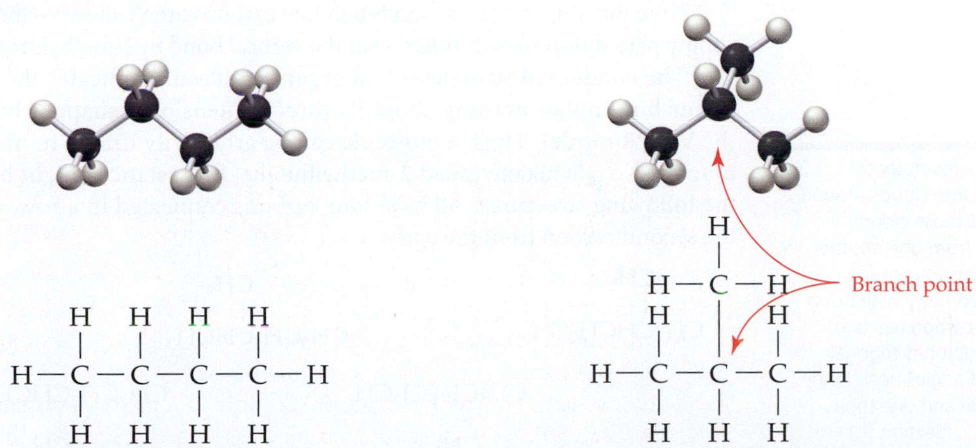

Butane (straight chain) — 2-Methylpropane (branched chain)

C_5H_{12}

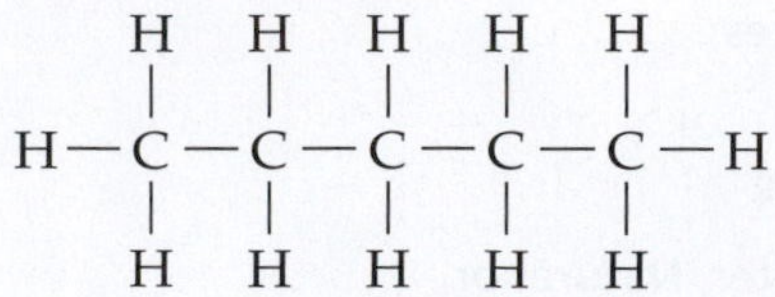

Pentane (straight chain)

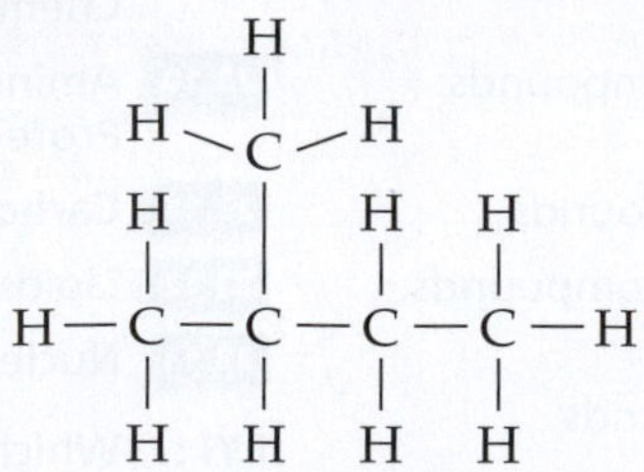

2-Methylbutane (branched chain)

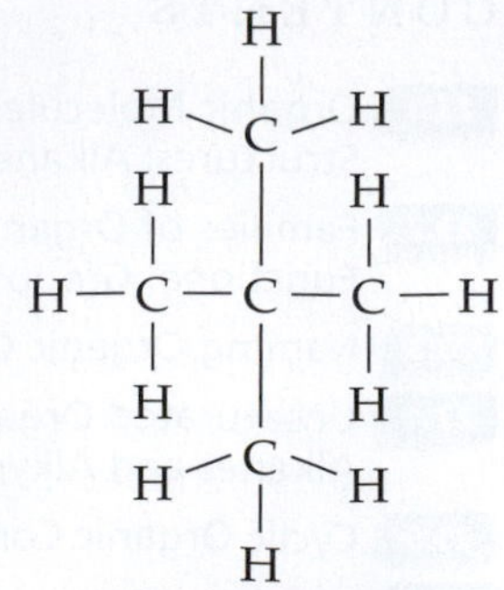

2,2-Dimethylpropane (branched chain)

Compounds like the two different C_4H_{10} molecules and the three different C_5H_{12} molecules, which have the same molecular formula but different chemical structures, are called **isomers**. The number of possible alkane isomers grows rapidly as the number of carbon atoms increases, from five isomers for C_6H_{14} to more than 60 *trillion* possible isomers for $C_{40}H_{82}$! As noted in Section 20.8, different isomers are different chemical compounds. They have different structures, different chemical properties, and different physical properties, such as melting point and boiling point.

REMEMBER...

Isomers are compounds that have the same formula but differ in the way their atoms are arranged. (Section 20.8)

Because it's both awkward and time-consuming to draw all the bonds and all the atoms in organic molecules, a shorthand way of drawing **condensed structures** is often used. In condensed structures, carbon–hydrogen and most carbon–carbon single bonds aren't shown; rather, they're "understood." If a carbon atom has three hydrogens bonded to it, we write CH_3; if the carbon has two hydrogens bonded to it, we write CH_2; and so on. For example, the four-carbon, straight-chain alkane (called butane) and its branched-chain isomer (called 2-methylpropane, or isobutane) can be written in the following way:

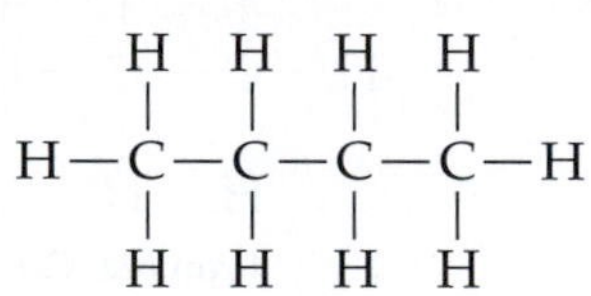

= $CH_3CH_2CH_2CH_3$ **Butane**

In condensed structures, most single bonds are omitted.

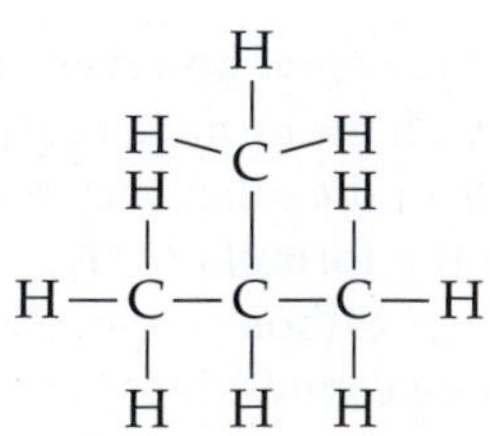

=

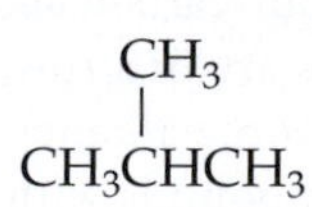

2-Methylpropane (Isobutane)

Note that the horizontal bonds between carbons aren't shown—the CH_3 and CH_2 units are simply placed next to each other—but the vertical bond in 2-methylpropane is shown for clarity.

The condensed structure of an organic molecule indicates the connections among its atoms but implies nothing about its three-dimensional shape, which can be predicted by the **VSEPR model**. Thus, a molecule can be arbitrarily drawn in many different ways. The branched-chain alkane called 2-methylbutane, for instance, might be represented by any of the following structures. All have four carbons connected in a row, with a $-CH_3$ branch on the second carbon from the end.

REMEMBER...

The **VSEPR model** predicts geometry by counting the number of charge clouds around an atom and assuming that those clouds orient in space as far away from one another as possible. When carbon has four single bonds, it has tetrahedral geometry with bond angles near 109.5°. When carbon has two single and one double bond, it has trigonal planar geometry with bond angles near 120°. When carbon has one single and one triple bond, it has linear geometry. (Section 5.1)

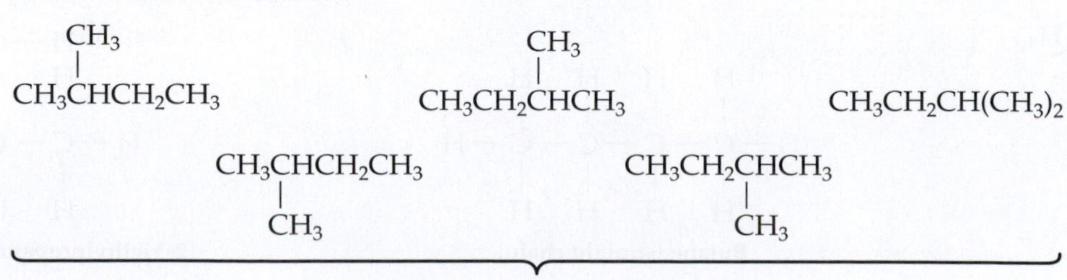

Some representations of 2-methylbutane

In fact, 2-methylbutane has no one single shape because rotation occurs around carbon–carbon single bonds. The two parts of a molecule joined by a C–C single bond are free to spin around the bond, giving rise to an infinite number of possible three-dimensional structures, or *conformations*. Thus, a large sample of 2-methylbutane contains a great many molecules that are constantly changing their shape. At any given instant, though, most of the molecules have an extended, zigzag shape, which is slightly more stable than other possibilities. The same is true for other alkanes. Remember when drawing three-dimensional structures that normal lines represent bonds in the plane of the page, wedges represent bonds coming out of the page toward the viewer, and dashed lines represent bonds receding into the page away from the viewer (Section 5.1).

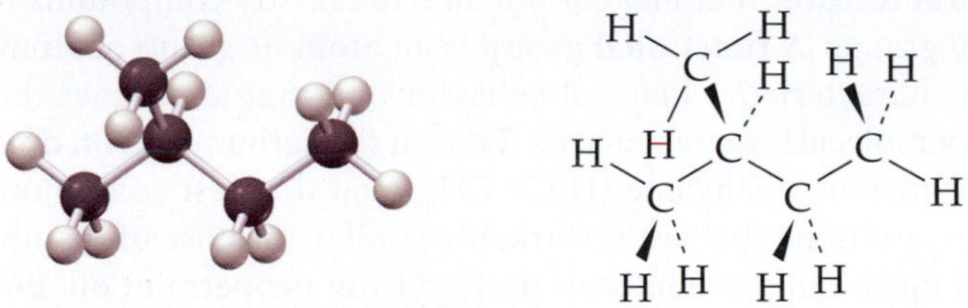

2-Methylbutane

Worked Example 23.1

Identifying Isomeric Compounds

The following condensed structures have the same formula, C_8H_{18}. Which of them represent the same molecule?

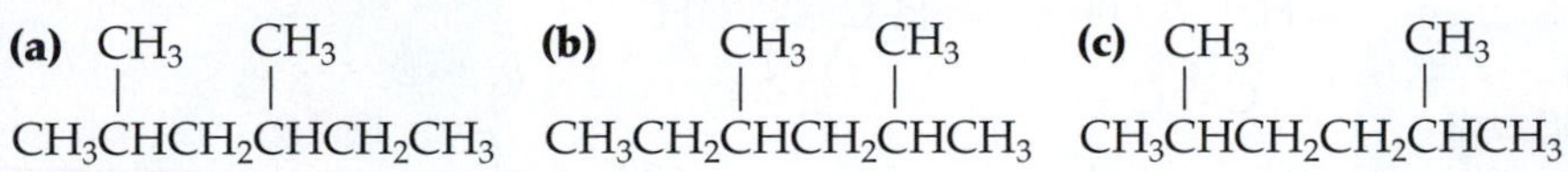

STRATEGY

Pay attention to the connections between atoms. Don't get confused by the apparent differences caused by writing a structure right-to-left versus left-to-right.

SOLUTION

Structure **(a)** has a straight chain of six carbons with $-CH_3$ branches on the second and fourth carbons from the end. Structure **(b)** also has a straight chain of six carbons with $-CH_3$ branches on the second and fourth carbons from the end and is therefore identical to **(a)**. The only difference between **(a)** and **(b)** is that one is written "forward" and one is written "backward." Structure **(c)** has a straight chain of six carbons with $-CH_3$ branches on the second and *fifth* carbons from the end, so it is an isomer of **(a)** and **(b)**.

PROBLEM 23.1 Draw the straight-chain alkane with the formula C_7H_{16} as a condensed structure.

PROBLEM 23.2 Draw the five alkane isomers with the formula C_6H_{14} as condensed structures.

PROBLEM 23.3 Which of the following structures are identical?

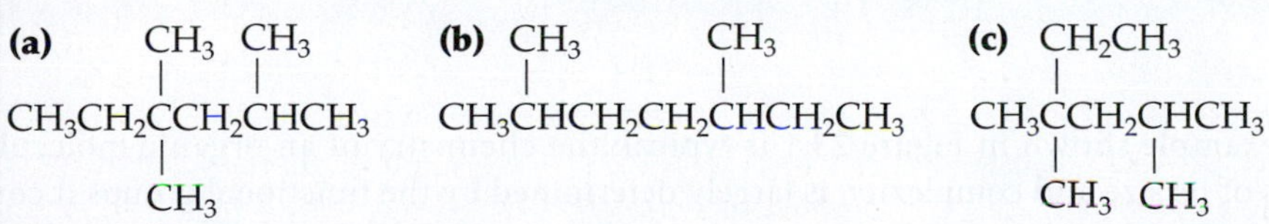

CONCEPTUAL PROBLEM 23.4 Give the formula of the following molecular model, and convert the model into a condensed structure:

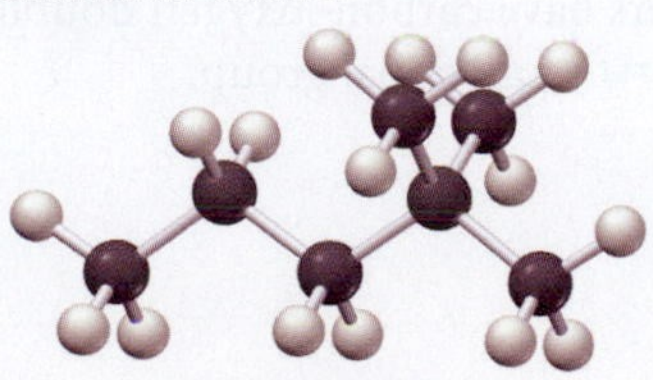

23.2 FAMILIES OF ORGANIC COMPOUNDS: FUNCTIONAL GROUPS

Chemists have learned through experience that organic compounds can be classified into families according to their structural features and that the members of a given family often have similar chemical reactivity. Instead of 45 million compounds with random chemical behavior, there are a small number of families of compounds whose behavior is reasonably predictable.

The structural features that make it possible to classify compounds into families are called *functional groups*. A **functional group** is an atom or group of atoms within a molecule that has a characteristic chemical behavior and that undergoes the same kinds of reactions in every molecule where it occurs. Look at the carbon–carbon **double-bond** functional group, for instance. Ethylene ($H_2C{=}CH_2$), the simplest compound with a double bond, undergoes reactions that are remarkably similar to those of menthene ($C_{10}H_{18}$), a much larger and more complex molecule derived from peppermint oil. Both, for example, react with Br_2 to give products in which a Br atom has added to each of the double-bond carbons (**Figure 23.1**).

REMEMBER...

A **double bond** is formed by sharing two pairs, or four electrons, between atoms. (Section 4.6)

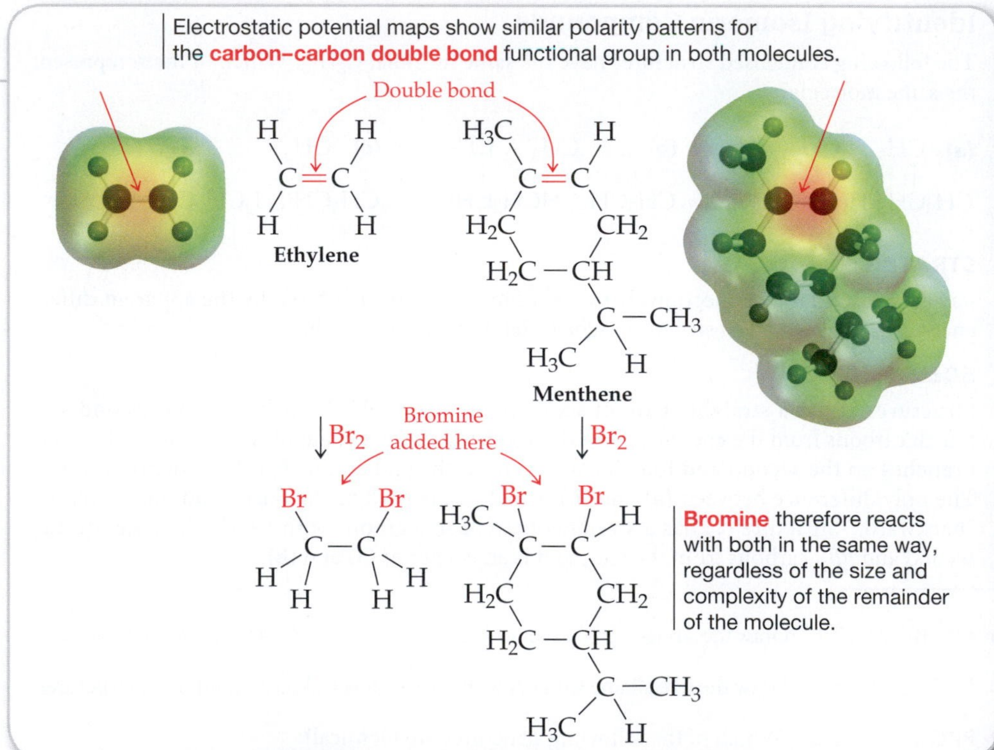

FIGURE 23.1

Reactions of the carbon–carbon double bond in ethylene and menthene with bromine.

The example shown in Figure 23.1 is typical: the chemistry of an organic molecule, regardless of its size and complexity, is largely determined by the functional groups it contains. **Table 23.1** lists some of the most common functional groups and gives examples of their occurrence. Some functional groups, such as alkenes, alkynes, and aromatic rings, have only carbon–carbon double or triple bonds; others contain single bonds to halogen, oxygen, or nitrogen atoms; and still others have carbon–oxygen double bonds. Most biological molecules, in particular, contain a C=O functional group.

TABLE 23.1 Some Families of Organic Compounds

Family Name	Functional Group Structure	Simple Example	Name	Name Ending
Alkane	(contains only C—H and C—C single bonds)	CH_3CH_3	Ethane	*-ane*
Alkene	$\rangle C{=}C\langle$	$H_2C{=}CH_2$	Ethene (Ethylene)	*-ene*
Alkyne	$-C{\equiv}C-$	$H-C{\equiv}C-H$	Ethyne (Acetylene)	*-yne*
Arene (aromatic)	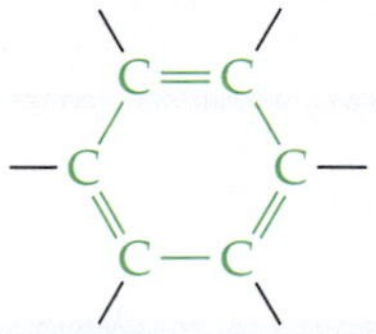	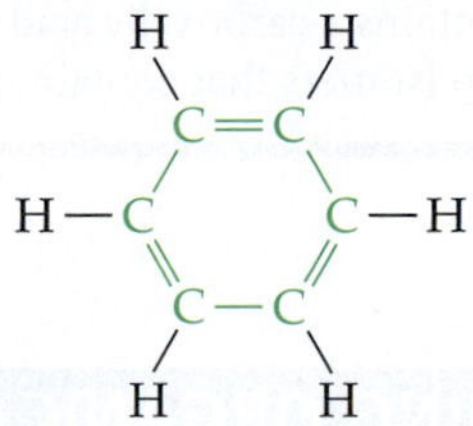	Benzene	None
Alcohol	$-\overset{\vert}{\underset{\vert}{C}}-O-H$	CH_3OH	Methanol	*-ol*
Ether	$-\overset{\vert}{\underset{\vert}{C}}-O-\overset{\vert}{\underset{\vert}{C}}-$	CH_3OCH_3	Dimethyl ether	*ether*
Amine	$-\overset{\vert}{\underset{\vert}{C}}-\overset{\vert}{N}-$	CH_3NH_2	Methylamine	*-amine*
Aldehyde	$-\overset{\vert}{\underset{\vert}{C}}-\overset{O}{\overset{\Vert}{C}}-H$	$CH_3\overset{O}{\overset{\Vert}{C}}H$	Ethanal (Acetaldehyde)	*-al*
Ketone	$-\overset{\vert}{\underset{\vert}{C}}-\overset{O}{\overset{\Vert}{C}}-\overset{\vert}{\underset{\vert}{C}}-$	$CH_3\overset{O}{\overset{\Vert}{C}}CH_3$	Propanone (Acetone)	*-one*
Carboxylic acid	$-\overset{\vert}{\underset{\vert}{C}}-\overset{O}{\overset{\Vert}{C}}-O-H$	$CH_3\overset{O}{\overset{\Vert}{C}}OH$	Ethanoic acid (Acetic acid)	*-oic acid*
Ester	$-\overset{\vert}{\underset{\vert}{C}}-\overset{O}{\overset{\Vert}{C}}-O-\overset{\vert}{\underset{\vert}{C}}-$	$CH_3\overset{O}{\overset{\Vert}{C}}OCH_3$	Methyl ethanoate (Methyl acetate)	*-oate*
Amide	$-\overset{\vert}{\underset{\vert}{C}}-\overset{O}{\overset{\Vert}{C}}-\overset{\vert}{N}-$	$CH_3\overset{O}{\overset{\Vert}{C}}NH_2$	Ethanamide (Acetamide)	*-amide*

The bonds whose connections aren't specified are assumed to be attached to carbon or hydrogen atoms in the rest of the molecule.

PROBLEM 23.5 Locate and identify the functional groups in the following molecules:

(a) Lactic acid, from sour milk

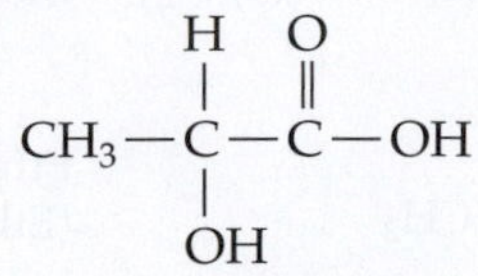

(b) Styrene, used to make polystyrene

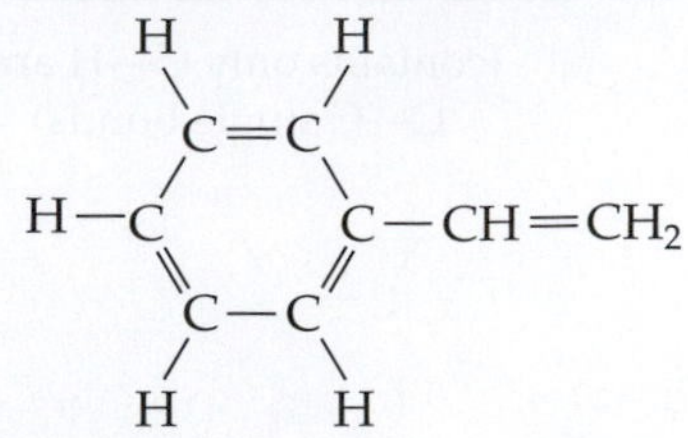

PROBLEM 23.6 Propose structures for molecules that fit the following descriptions:

(a) C_2H_4O; contains an aldehyde functional group

(b) $C_3H_6O_2$, contains a carboxylic acid functional group

(c) $C_4H_8O_2$: two isomers that contain ester groups

23.3 NAMING ORGANIC COMPOUNDS

Because there are so many organic compounds, it's particularly important to have a systematic way of assigning a unique and informative name to each. The system generally used for naming organic compounds is that devised by the International Union of Pure and Applied Chemistry, abbreviated IUPAC. In the IUPAC system, a chemical name has three parts: prefix, parent, and suffix. The parent name tells how many carbon atoms are present in the longest continuous chain, the suffix identifies the family the molecule belongs to, and the prefix (if needed) specifies the various substituent groups attached to the parent chain:

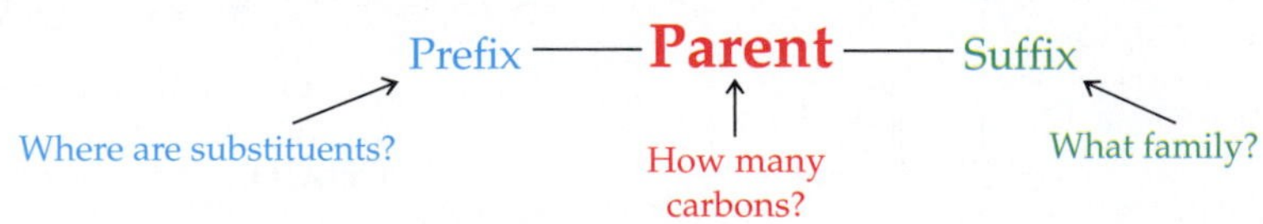

Let's use alkanes to illustrate the naming procedure. Straight-chain alkanes are named by counting the number of carbon atoms in the longest continuous chain and adding the family suffix *-ane*. With the exception of the first four compounds—methane, ethane, propane, and butane—whose names have historical origins, the alkanes are named from Greek numbers according to how many carbons they contain. Thus, *pent*ane is the five-carbon alkane, *hex*ane is the six-carbon alkane, *hept*ane is the seven-carbon alkane, and so on, as shown in **Table 23.2**.

TABLE 23.2 Names of Straight-Chain Alkanes

Number of Carbons	Structure	Name	Number of Carbons	Structure	Name
1	CH_4	*Meth*ane	6	$CH_3CH_2CH_2CH_2CH_2CH_3$	*Hex*ane
2	CH_3CH_3	*Eth*ane	7	$CH_3CH_2CH_2CH_2CH_2CH_2CH_3$	*Hept*ane
3	$CH_3CH_2CH_3$	*Prop*ane	8	$CH_3CH_2CH_2CH_2CH_2CH_2CH_2CH_3$	*Oct*ane
4	$CH_3CH_2CH_2CH_3$	*But*ane	9	$CH_3CH_2CH_2CH_2CH_2CH_2CH_2CH_2CH_3$	*Non*ane
5	$CH_3CH_2CH_2CH_2CH_3$	*Pent*ane	10	$CH_3CH_2CH_2CH_2CH_2CH_2CH_2CH_2CH_2CH_3$	*Dec*ane

Branched-chain alkanes are named by following four steps:

Step 1. Name the parent chain. Find the longest continuous chain of carbons in the molecule, and use the name of that chain as the parent. The longest chain may not always be obvious from the manner of writing; you may have to "turn corners" to find it.

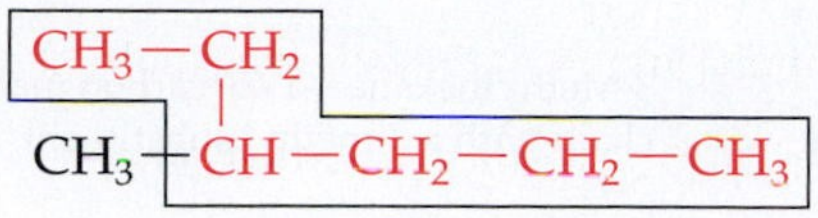

Named as hexane, not as a pentane, because the longest chain has six carbons.

If you prefer, you can redraw the structure so that the longest chain is on one line.

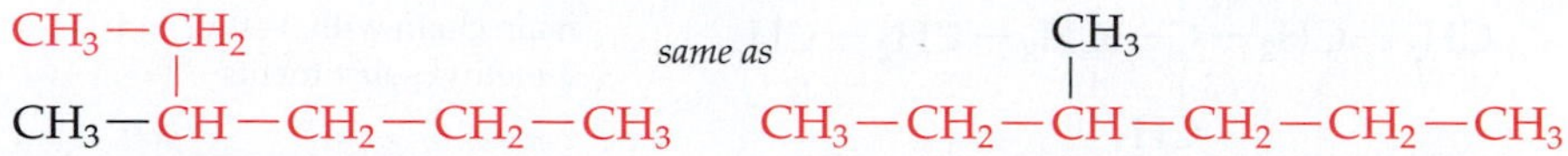

Step 2. Number the carbon atoms in the main chain. Beginning at the end nearer the first branch point, number the carbon atoms in the parent chain.

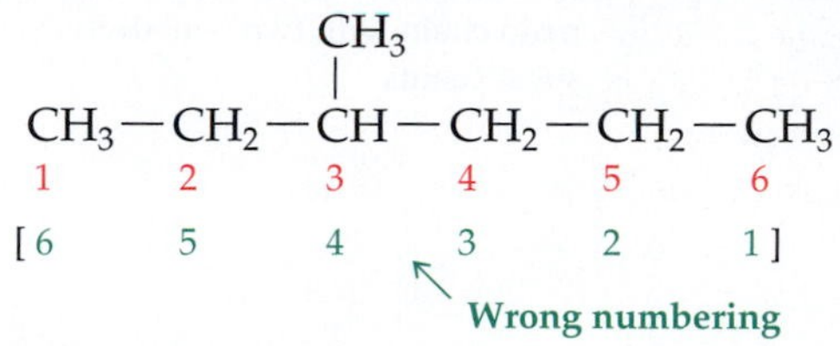

The first (and only) branch occurs at the third carbon, C3, if we start numbering from the left, but would occur at C4 if we started from the right by mistake.

Step 3. Identify and number the branching substituent. Assign a number to each branching substituent on the parent chain according to its point of attachment.

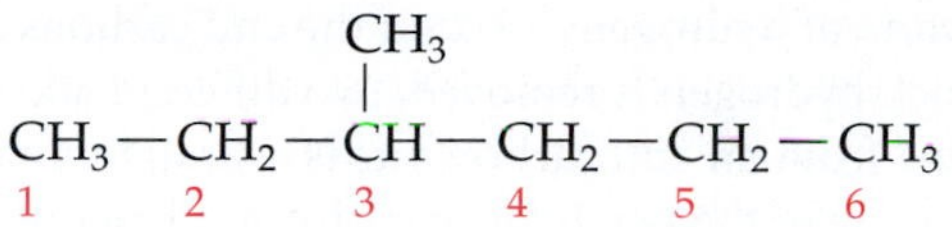

The main chain is a hexane. There is a $-CH_3$ substituent group connected to C3 of the chain.

If there are two substituent groups on the same carbon, assign the same number to both. There must always be as many numbers in the name as there are substituents.

A substituent derived from an alkane by the removal of a hydrogen atom is called an alkyl group.

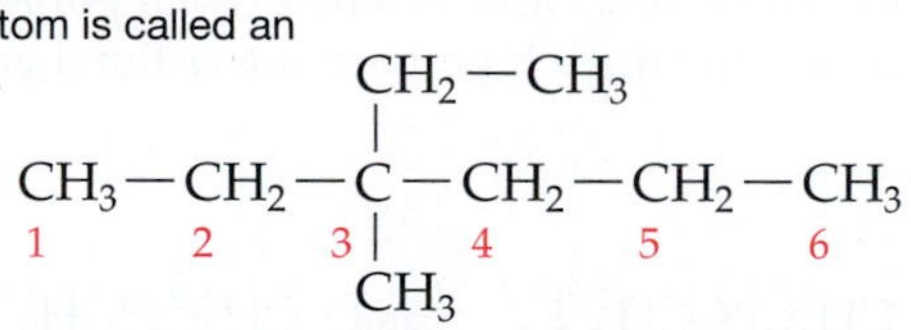

The main chain is hexane. There are two substituents, a $-CH_3$ and a $-CH_2CH_3$, both connected to C3 of the chain.

The $-CH_3$ and $-CH_2CH_3$ substituents that branch off the main chain are called **alkyl groups**. You can think of an alkyl group as the part of an alkane that remains when a hydrogen is removed. For example, removal of a hydrogen from methane, CH_4, gives the *methyl group*, $-CH_3$, and removal of a hydrogen from ethane, CH_3CH_3, gives the *ethyl group*, $-CH_2CH_3$. Alkyl groups are named by replacing the *-ane* ending of the parent alkane with an *-yl* ending.

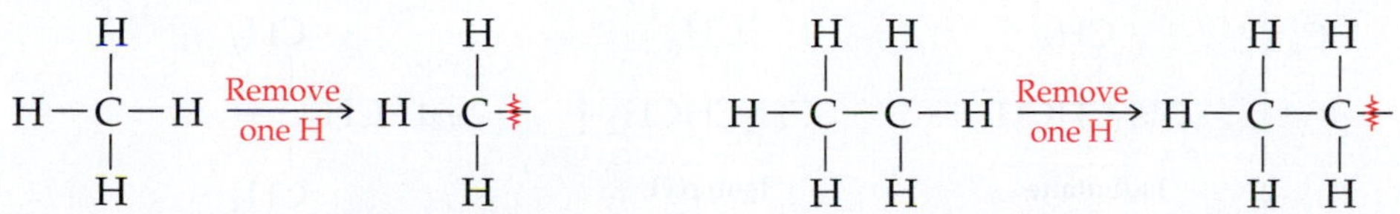

Methane **A methyl group** **Ethane** **An ethyl group**

Step 4. Write the name as a single word. Use hyphens to separate the different prefixes, and use commas to separate numbers if there is more than one. If two or more different substituent groups are present, list them in alphabetical order. If two or more identical substituents are present, use a numerical prefix *di-*, *tri-*, *tetra-*, and so forth, but don't use these numerical prefixes for alphabetizing. That is, a prefix like "dimethyl" is listed alphabetically under "m" rather than under "d." Look at the following examples to see how names are written:

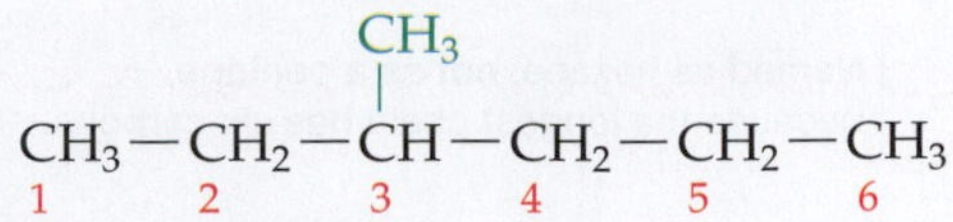

3-Methylhexane—a six-carbon main chain with a 3-methyl substituent

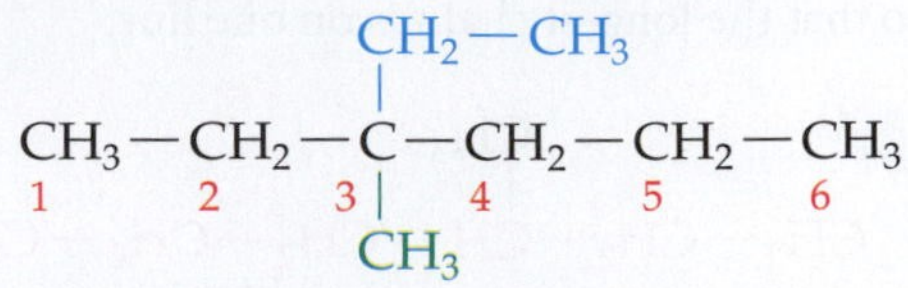

3-Ethyl-3-methylhexane—a six-carbon main chain with 3-ethyl and 3-methyl substituents

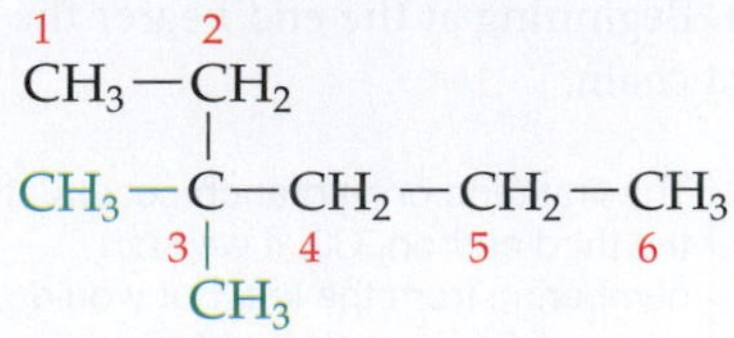

3,3-Dimethylhexane—a six-carbon main chain with two 3-methyl substituents

More About Alkyl Groups

It doesn't matter which hydrogen is removed from CH_4 to form a methyl group or which hydrogen is removed from CH_3CH_3 to form an ethyl group because all the hydrogen atoms in both molecules are equivalent. The eight hydrogens in propane, $CH_3CH_2CH_3$, however, are not all equivalent. Propane has two sorts of hydrogens—six on the end carbons and two on the middle carbon. Depending on which hydrogen is removed, two different alkyl groups result. Removing one of the six hydrogens from an end carbon yields a straight-chain alkyl group called simply *propyl*, and removing one of the two hydrogens from the middle carbon yields a branched-chain alkyl group called *isopropyl*.

Similarly, there are four different butyl groups. Two (butyl and *sec*-butyl) are derived from straight-chain butane, and two (isobutyl and *tert*-butyl) are derived from branched-chain isobutane. The prefixes *sec*- (for *secondary*) and *tert*- (for *tertiary*) refer to the number of other carbon atoms attached to the branching carbon. The branch point in a *sec*-butyl group has two carbons attached to it, and the branch point in a *tert*-butyl group has three carbons attached.

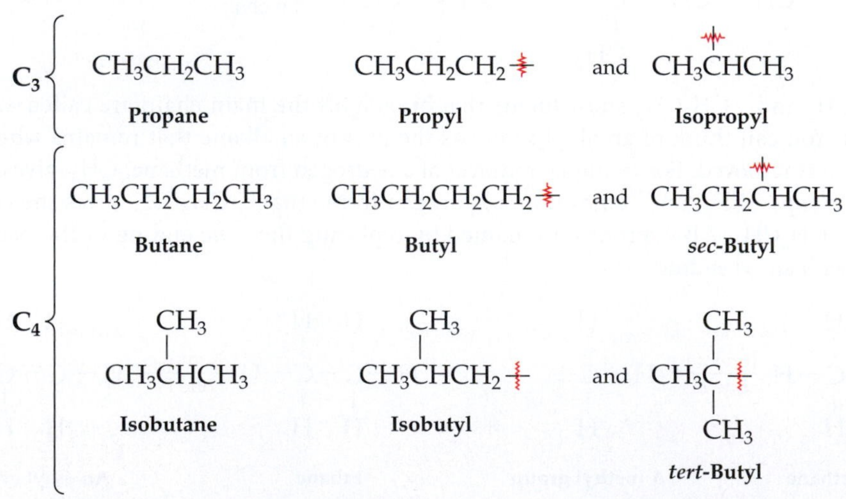

Keep in mind that alkyl groups themselves are not stable compounds and that the "removal" of a hydrogen from an alkane is just a useful way of looking at things, not a chemical reaction. Alkyl groups are simply parts of molecules that help us to name compounds.

Worked Example 23.2

Converting a Name into a Chemical Structure

Draw the structure of 3-isopropyl-2-methylhexane.

STRATEGY AND SOLUTION

First, look at the parent name (hexane) and draw its carbon structure:

C—C—C—C—C—C **Hexane**

Next, find the substituents (3-isopropyl and 2-methyl) and place them on the proper carbons:

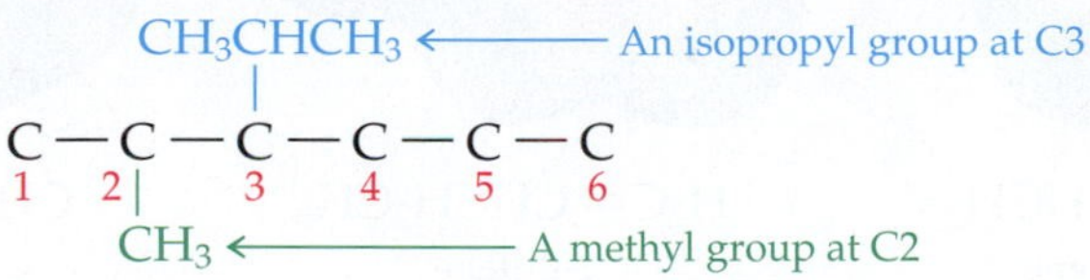

Finally, add hydrogens to complete the structure:

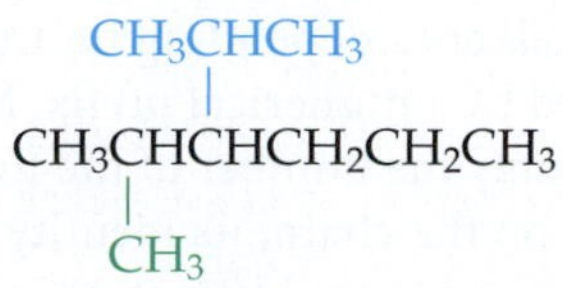

3-Isopropyl-2-methylhexane

PROBLEM 23.7 What are the IUPAC names of the following alkanes?

(a) The three isomers of C_5H_{12}

(b)

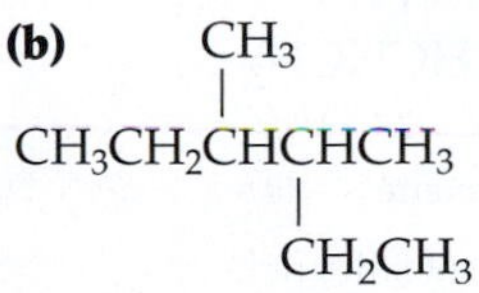

(c)

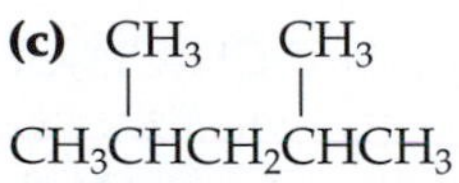

(d)

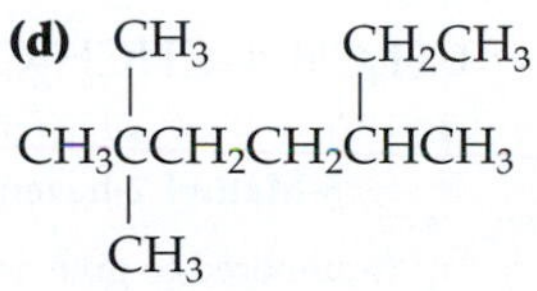

PROBLEM 23.8 Draw condensed structures corresponding to the following IUPAC names:

(a) 3,4-Dimethylnonane
(b) 3-Ethyl-4,4-dimethylheptane
(c) 2,2-Dimethyl-4-propyloctane
(d) 2,2,4-Trimethylpentane

CONCEPTUAL PROBLEM 23.9 What is the IUPAC name of the following alkane?

23.4 UNSATURATED ORGANIC COMPOUNDS: ALKENES AND ALKYNES

In contrast to alkanes, which have only single bonds, alkenes and alkynes have multiple bonds. **Alkenes** are hydrocarbons that contain a carbon–carbon double bond, and **alkynes** are hydrocarbons that contain a carbon–carbon triple bond. Both groups of compounds are said to be **unsaturated**, meaning that they have fewer hydrogens per carbon than the related alkanes, which have the maximum possible number of hydrogens and are thus **saturated**. Ethylene ($H_2C{=}CH_2$), for example, is unsaturated and has the formula C_2H_4, whereas ethane (CH_3CH_3) is saturated and has the formula C_2H_6.

Alkenes are named by counting the longest chain of carbons that contains the double bond and adding the suffix *-ene*. Thus, ethylene, the simplest alkene, is followed by propene, butene, pentene, hexene, and so on. Note that ethylene should properly be called *ethene*, but the name ethylene has been used for so long that it is universally accepted. Similarly, the name *propylene* is often used for propene.

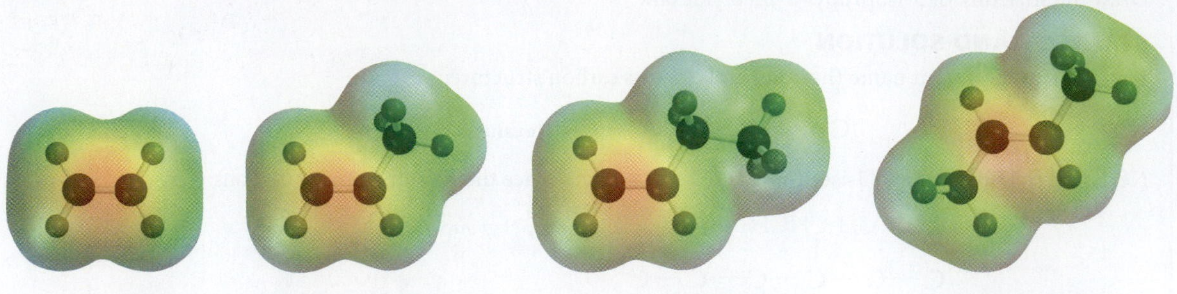

$H_2C{=}CH_2$ **Ethylene (ethene)** — $H_2C{=}CHCH_3$ **Propylene (propene)** — $H_2C{=}CHCH_2CH_3$ **1-Butene** — $CH_3CH{=}CHCH_3$ **2-Butene**

Isomers are possible for butene and higher alkenes depending on the position of the double bond in the chain, which must be specified by a numerical prefix. Numbering starts from the chain end nearer the double bond, and only the number of the first of the double-bond carbons is cited. If a substituent is present on the chain, its identity is noted and the position of its attachment is given. If the double bond is equidistant from both ends of the chain, numbering starts from the end nearer the substituent.

$CH_3CH{=}CHCH_2CH(CH_3)CH_3$ (carbons numbered 1 2 3 4 5 6)

5-Methyl-2-hexene

(numbered to give double bond the lower number)

$CH_3CH_2CH{=}CHCH(CH_3)CH_3$ (carbons numbered 6 5 4 3 2 1)

2-Methyl-3-hexene

(numbered to give substituent the lower number when the double bond is equidistant from both ends)

In addition to the alkene isomers that occur because of double-bond *position*, alkene isomers can also occur because of double-bond *geometry*. For instance, there are two geometrical isomers, or **cis–trans isomers**, of 2-butene, which differ in their geometry about the double bond. The cis isomer has its two $-CH_3$ groups on the same side of the double bond, and the trans isomer has its two $-CH_3$ groups on opposite sides.

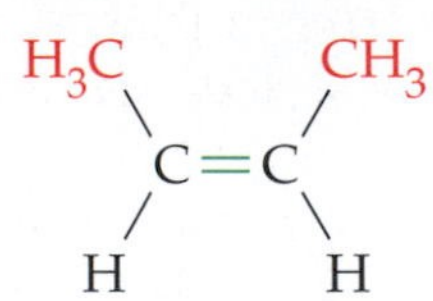

***cis*-2-Butene**
(methyl groups on the same side)

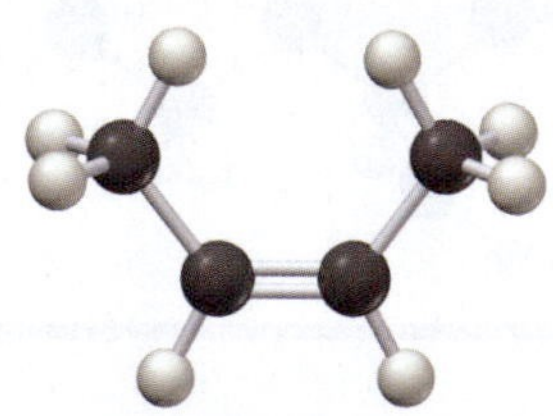

(Top view)

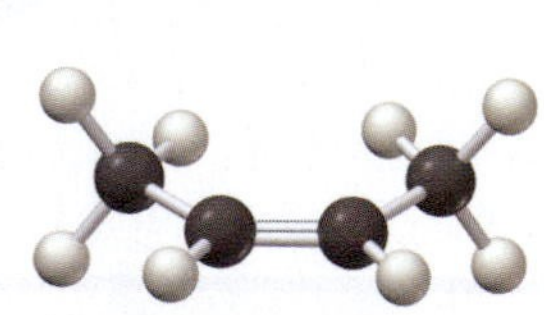

(Side view)

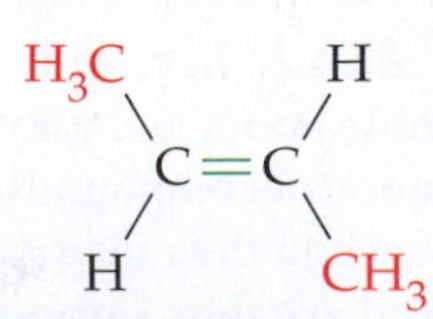

***trans*-2-Butene**
(methyl groups on opposite sides)

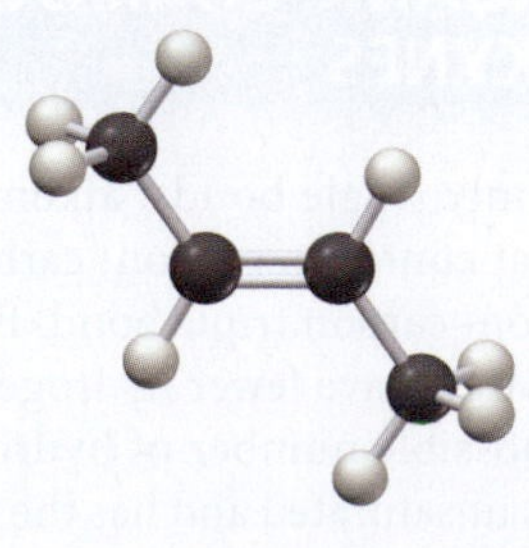

(Top view)

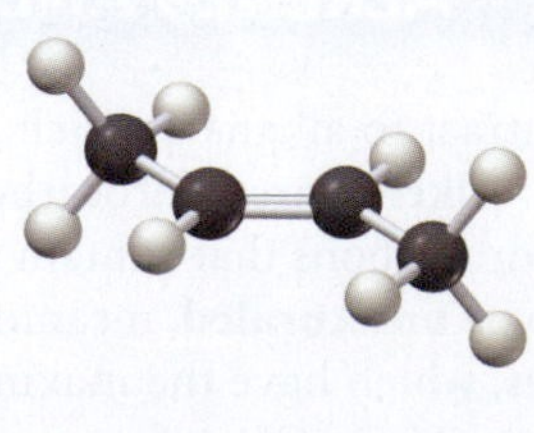

(Side view)

Cis–trans isomerism in alkenes arises because of the **electronic structure of the double bond**, which makes bond rotation energetically unfavorable. Were rotation to occur, it would break the π part of the double bond by disrupting the sideways overlap of two parallel *p* orbitals (**Figure 23.2**). In fact, an energy input of 240 kJ/mol is needed to cause bond rotation.

REMEMBER...

The **electronic structure of a carbon–carbon double bond** consists of two parts: head-on (σ) overlap of sp^2 hybridized orbitals and sideways (π) overlap of unhybridized *p* orbitals. (Section 5.4)

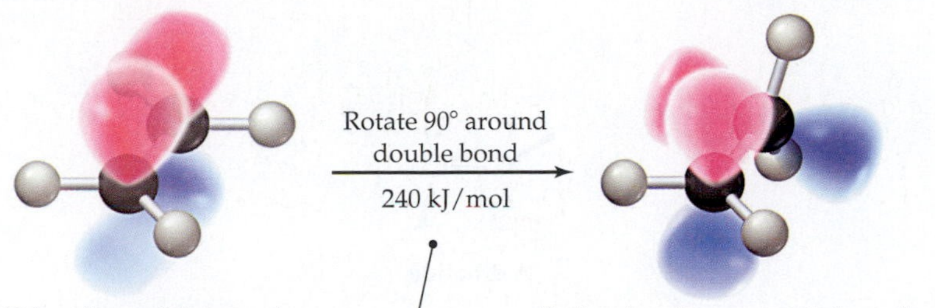

FIGURE 23.2

Rotation around a carbon–carbon double bond.

Cis–trans alkene isomerism is also common in biochemistry. For example, our visual system is based on the light-induced cis–trans isomerization in our eyes of a compound called metarhodopsin II. In the absence of light, the isomerization would take approximately 1100 years to occur, but in the presence of light, it occurs within 2×10^{-13} seconds! The isomerization results in a change in molecular shape, which causes a nerve impulse to be sent to the brain where it is perceived as vision.

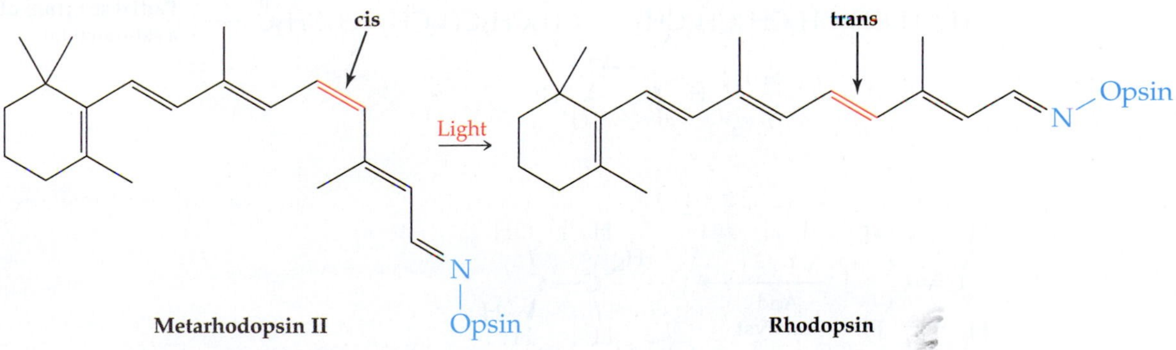

Alkynes are similar in many respects to alkenes and are named using the suffix *-yne*. The simplest alkyne, HC≡CH, is often called by its alternative name *acetylene* rather than by its systematic name *ethyne*.

HC≡CH	$CH_3C{\equiv}CH$	$\overset{1}{C}H_3\overset{2}{C}{\equiv}\overset{3}{C}\overset{4}{C}H_3$	$\overset{4}{C}H_3\overset{3}{C}H_2\overset{2}{C}{\equiv}\overset{1}{C}H$
Ethyne (Acetylene)	**Propyne**	**2-Butyne**	**1-Butyne**

As with alkenes, isomers are also possible for butyne and higher alkynes, depending on the position of the triple bond in the chain. Unlike the situation with alkenes, however, no cis–trans isomers are possible for alkynes because of their linear geometry.

Chemically, the most common transformations of alkenes and alkynes are **addition reactions**, in which a reagent adds to the multiple bond of the unsaturated reactant to give a saturated product. Alkenes can thus be converted into alkanes by addition of hydrogen, into dihalides by addition of Cl_2 or Br_2, and into alcohols by addition of water.

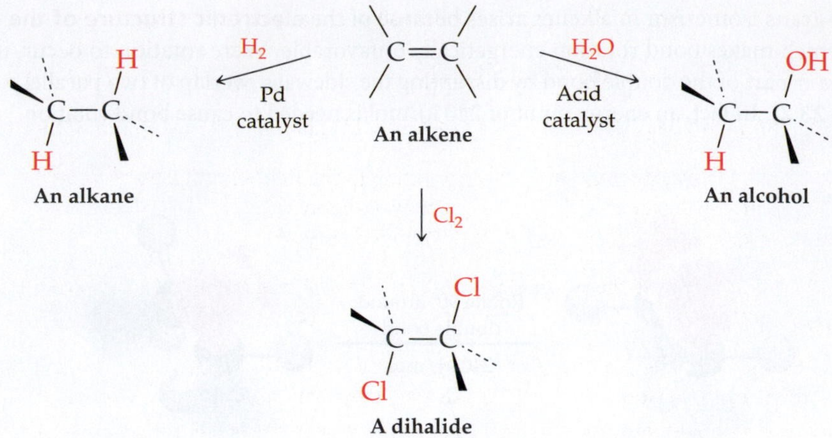

Addition of hydrogen in the presence of a metal catalyst, called **hydrogenation**, is used commercially to convert unsaturated vegetable oils to the saturated fats used in margarine and cooking fats. Addition of water, called hydration, is used commercially to convert ethylene into ethyl alcohol.

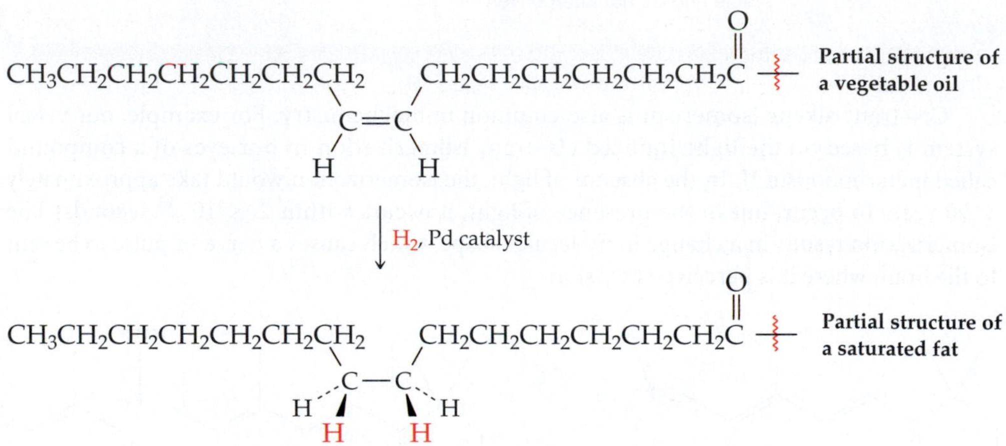

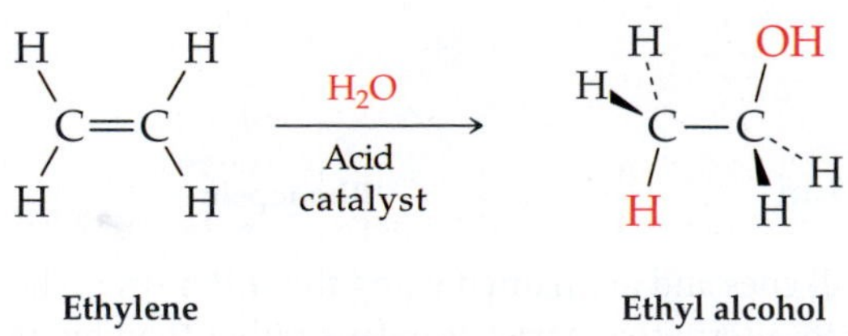

Worked Example 23.3

Converting a Name into a Structure

Draw the structure of *cis*-3-heptene.

STRATEGY

The name 3-heptene indicates that the molecule has seven carbons (*hept-*) and has a double bond between carbons 3 and 4. The prefix *cis-* indicates that the two alkyl groups attached to the double-bond carbons lie on the same side of the double bond.

SOLUTION

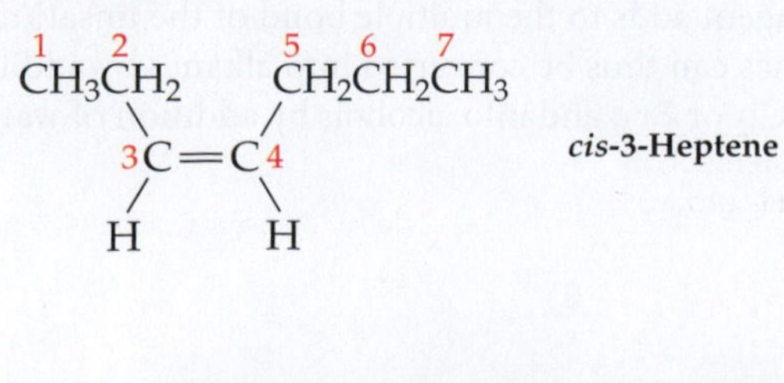

PROBLEM 23.10 Give IUPAC names for the following alkenes and alkynes:

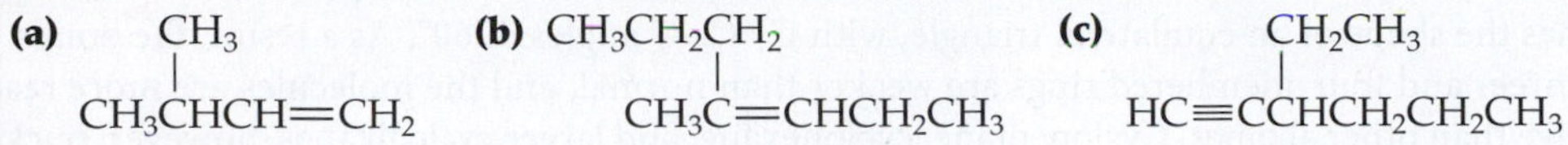

PROBLEM 23.11 Draw structures corresponding to the following IUPAC names:

(a) 2,2-Dimethyl-3-hexene (b) 4-Isopropyl-2-heptyne
(c) *trans*-3-Heptene (d) *cis*-2-Methyl-3-hexene

PROBLEM 23.12 Show the products of the reaction of 2-butene with the following:

(a) H_2, Pd catalyst (b) Br_2 (c) H_2O, H_2SO_4 catalyst

PROBLEM 23.13 The reaction of 2-pentene with H_2O in the presence of an acid catalyst yields a mixture of two isomeric alcohol products. Draw their structures.

23.5 CYCLIC ORGANIC COMPOUNDS

The molecules we've been dealing with thus far have been open-chain, or *acyclic*, compounds. *Cyclic* compounds, which contain *rings* of carbon atoms, are also well known and are widespread throughout nature. Compounds of all ring sizes from 3 through 30 carbons and beyond have been prepared.

The simplest cyclic compounds are the **cycloalkanes** which, like their open-chain counterparts, contain only C–C and C–H single bonds. The compounds having three carbons (cyclopropane), four carbons (cyclobutane), five carbons (cyclopentane), and six carbons (cyclohexane) are shown in **Figure 23.3**.

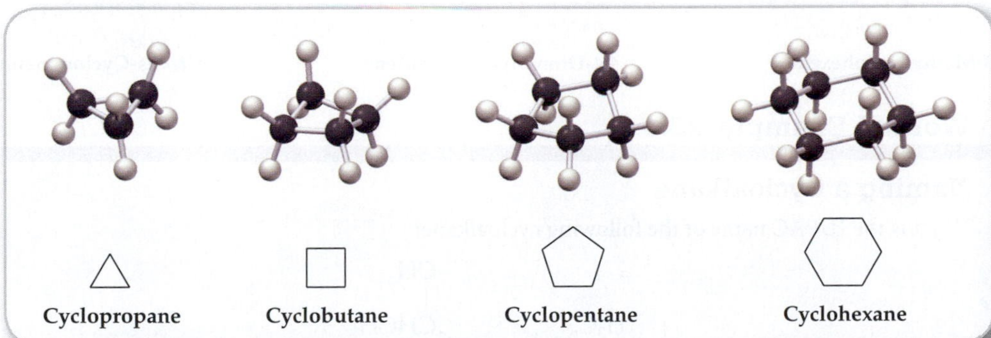

FIGURE 23.3

Structures of some simple cycloalkanes. The rings are shown as polygons.

Even condensed structures are awkward for cyclic molecules, so a streamlined way of drawing structures is often used in which cycloalkanes are represented by polygons. A triangle represents cyclopropane, a square represents cyclobutane, and so on. Carbon and hydrogen atoms aren't shown explicitly in these structures. A carbon atom is simply understood to be at every junction of lines, and the proper number of hydrogen atoms needed to give each carbon four bonds is supplied mentally. Methylcyclohexane, for instance, looks like this:

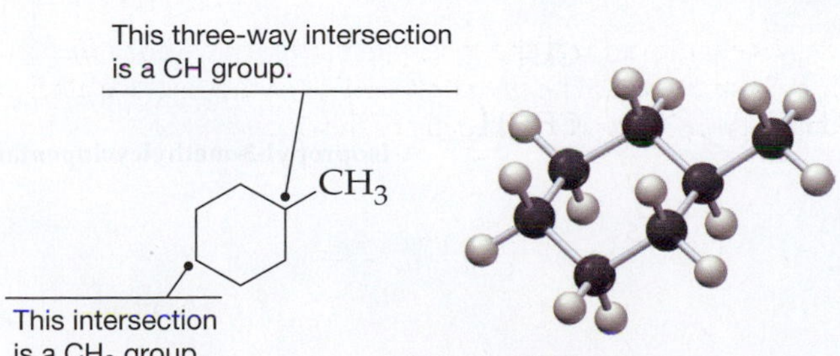

Methylcyclohexane

As you might imagine, the C–C–C bond angles in cyclopropane and cyclobutane are considerably distorted from the ideal 109.5° tetrahedral value. Cyclopropane, for example, has the shape of an equilateral triangle, with C–C–C angles of 60°. As a result, the bonds in three- and four-membered rings are weaker than normal, and the molecules are more reactive than other alkanes. Cyclopentane, cyclohexane, and larger cycloalkanes, however, pucker into shapes that allow bond angles to be near their normal tetrahedral value, as shown by the models in Figure 23.3.

Substituted cycloalkanes are named using the cycloalkane as the parent name and identifying the positions on the ring where substituents are attached. Start numbering at the group that has alphabetical priority, and proceed around the ring in the direction that gives the second substituent the lower possible number. For example

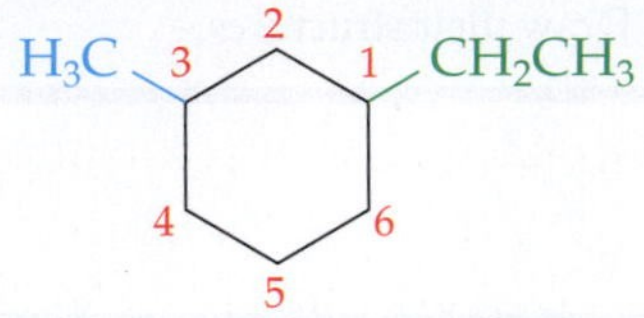

1-Ethyl-3-methylcyclohexane

Not 1-methyl-3-ethylcyclohexane
or 1-ethyl-5-methylcyclohexane
or 1-methyl-5-ethylcyclohexane

Unsaturated cyclic compounds are also well known, particularly cycloalkenes. Unlike their open-chain counterparts, however, cis–trans isomerism does not normally occur because a trans double bond is not geometrically possible within rings having fewer than eight carbons. When naming cycloalkenes, numbering always starts with the two double bond carbons and proceeds around the ring in a direction that gives the first substituent as low a number as possible.

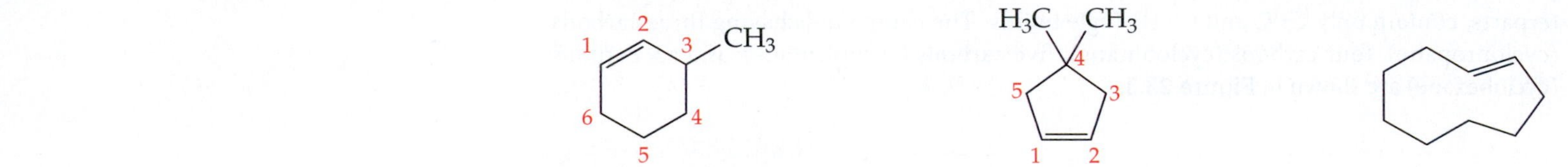

3-Methylcyclohexene **4,4-Dimethylcyclopentene** ***trans*-Cyclononene**

Worked Example 23.4

Naming a Cycloalkane

What is the IUPAC name of the following cycloalkane?

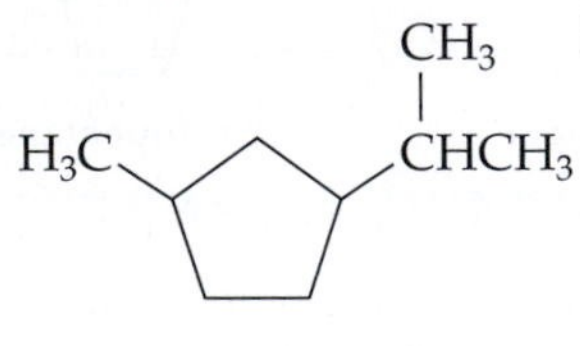

STRATEGY

First, identify the parent cycloalkane (cyclopentane) and the two substituents (a methyl group and an isopropyl group). Then, number the ring beginning at the group having alphabetical priority (isopropyl rather than methyl) and proceed in a direction that gives the second group the lower possible number.

SOLUTION

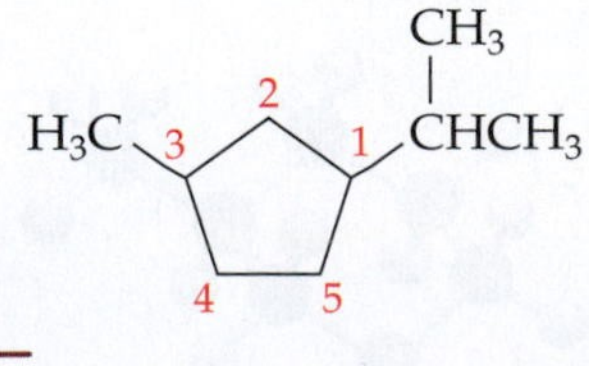

1-Isopropyl-3-methylcyclopentane

PROBLEM 23.14 Give IUPAC names for the following cycloalkanes:

(a)

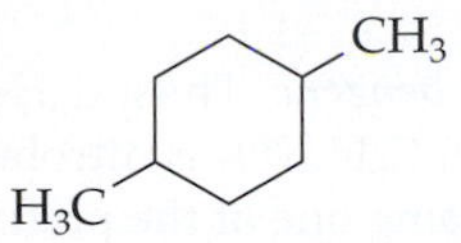

(b)

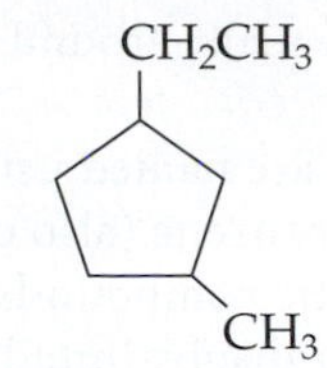

(c)

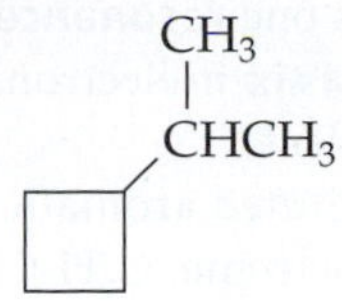

PROBLEM 23.15 Draw structures corresponding to the following IUPAC names. Use polygons for the rings.
(a) 1,1-Dimethylcyclobutane
(b) 1-*tert*-Butyl-2-methylcyclopentane
(c) 1,3,5-Trimethylcycloheptane
(d) 4-Methyl-1,1-diethylcyclohexane

23.6 AROMATIC COMPOUNDS

A second major class of cyclic compounds in addition to cycloalkanes are the so-called aromatic compounds. In the early days of organic chemistry, the word *aromatic* was used to describe certain fragrant substances found in fruits, trees, and other natural sources. Chemists soon realized, however, that substances grouped as aromatic behaved in a chemically different manner from most other organic compounds. Today, the term **aromatic** refers to the class of compounds that can be represented as having a six-membered ring with alternating single and double bonds. Benzene is the simplest aromatic compound, but aspirin, the steroid sex hormone estradiol, and many important biological molecules and pharmaceutical agents also contain aromatic rings.

Benzene

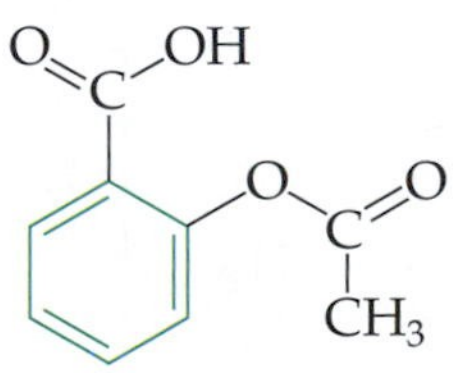

Aspirin

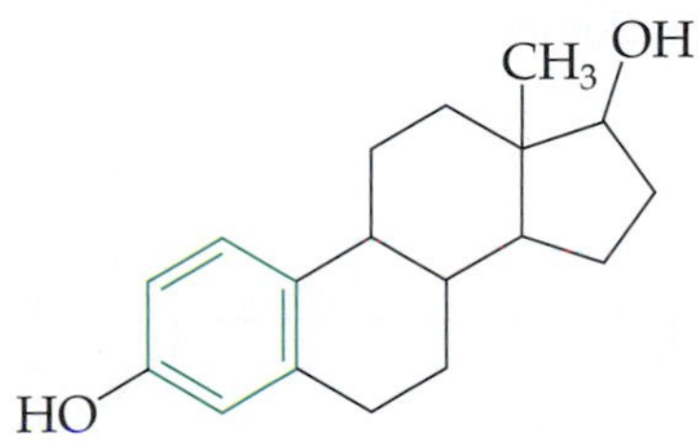

Estradiol

▲ Benzaldehyde, a close structural relative of benzene, is an aromatic compound responsible for the odor of cherries.

Benzene is a flat, symmetrical molecule that is usually drawn as a six-membered ring with three double bonds. The problem with this representation, however, is shown in **Figure 23.4a** and the symmetrical electrostatic potential map in **Figure 23.4b**. Each of the six carbons in benzene is sp^2-hybridized and has a p orbital perpendicular to the ring. When these p orbitals overlap to form π bonds, there are two possibilities, shown in **Figure 23.4c**.

(a) An orbital picture shows that all six carbons have a *p* orbital.

(b) An electrostatic potential map shows that all six C–C bonds are equivalent.

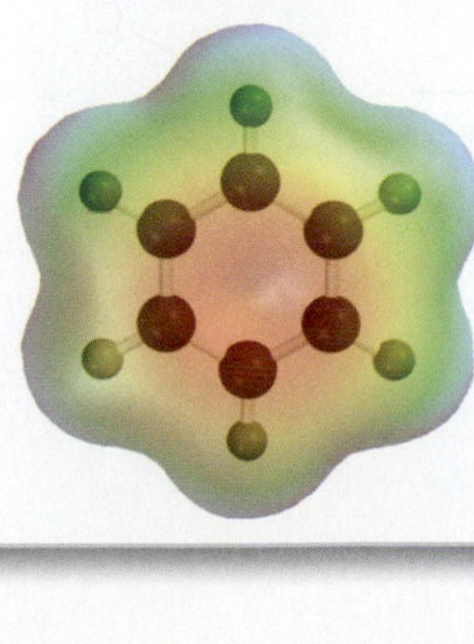

(c) Benzene has two equivalent resonance structures, which differ only in the positions in which the double bonds are drawn.

FIGURE 23.4 **Some representations of benzene.**

REMEMBER...

Different **resonance forms** of a substance differ only in the placement of bonding and nonbonding electrons. The connections between atoms and the relative positions of the atoms remain the same. (Section 4.9)

Neither of the two equivalent structures in Figure 23.4c is correct by itself. Rather, each represents one **resonance form** of the true benzene structure, which is a hybrid of the two forms. The six π electrons are spread equally around the entire ring, and all six C–C bonds are equivalent.

Substituted aromatic compounds are named using the suffix *-benzene*. Thus, C_6H_5Br is bromobenzene, $C_6H_5CH_3$ is methylbenzene (also called *toluene*), $C_6H_5NO_2$ is nitrobenzene, and so on. Disubstituted aromatic compounds are named using one of the prefixes *ortho-*, *meta-*, or *para-*. An *ortho-* or *o*-disubstituted benzene has its two substituents in a 1,2 relationship on the ring; a *meta-* or *m*-disubstituted benzene has its two substituents in a 1,3 relationship; and a *para-* or *p*-disubstituted benzene has its substituents in a 1,4 relationship. When the benzene ring itself is a substituent, the name *phenyl* (pronounced **fen**-nil) is used.

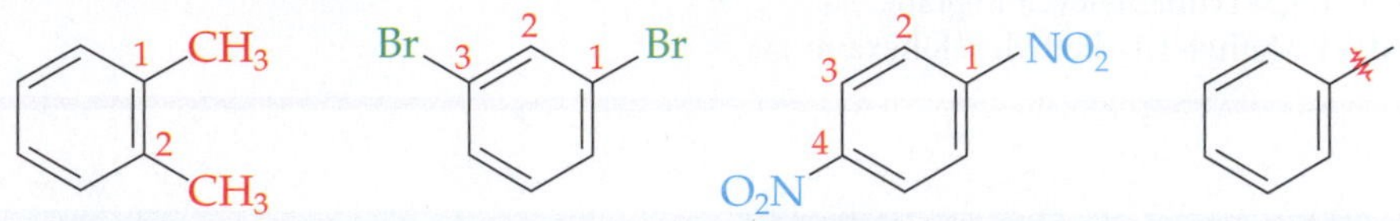

ortho-**Dimethylbenzene** *meta*-**Dibromobenzene** *para*-**Dinitrobenzene** **A phenyl group**

Unlike alkenes, which undergo addition reactions, aromatic compounds typically undergo **substitution reactions** to give products that retain the aromatic ring. That is, a group –X substitutes for one of the hydrogen atoms on the aromatic ring without changing the ring itself. It doesn't matter which of the six ring hydrogens in benzene is replaced, because all six are equivalent.

Substitution of a nitro group ($-NO_2$) for a ring hydrogen occurs when benzene reacts with nitric acid in the presence of sulfuric acid as catalyst. The reaction is a key step in the synthesis of explosives such as TNT (trinitrotoluene) and many important pharmaceutical agents. Nitrobenzene itself is a starting material for preparing many of the brightly colored dyes used in clothing.

▲ The dyes used to add the bright colors to clothing are made by a process that begins with substitution of a nitro group onto benzene.

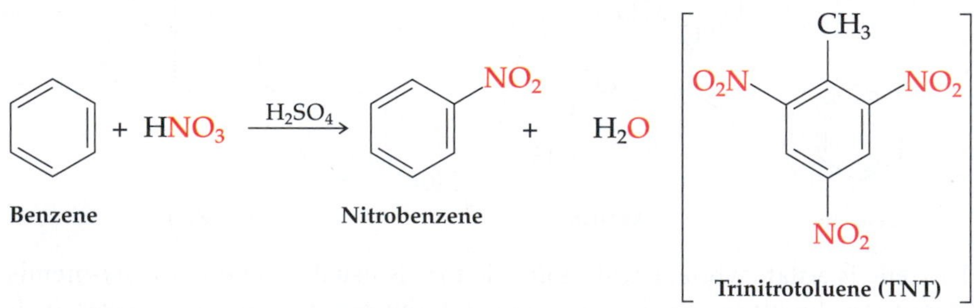

Substitution of a bromine or chlorine for a ring hydrogen occurs when benzene reacts with Br_2 or Cl_2 in the presence of $FeBr_3$ or $FeCl_3$ as catalyst. The reaction is used in the synthesis of many pharmaceutical agents, including the antiallergy medication loratadine, marketed as Claritin.

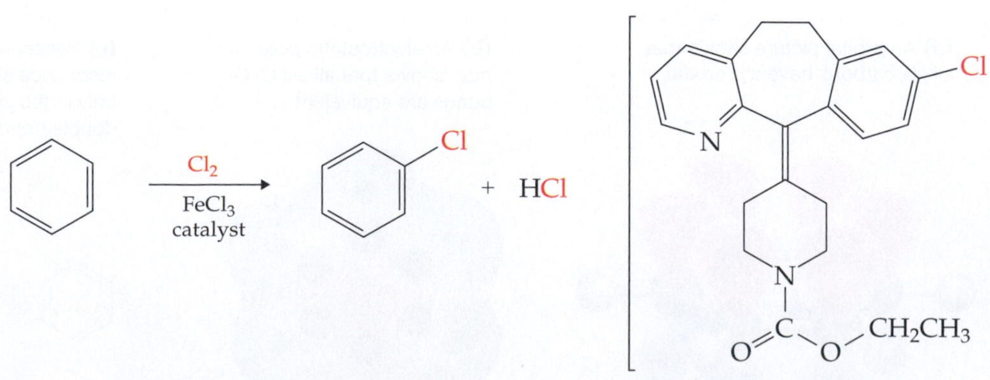

Loratadine

PROBLEM 23.16 Draw structures corresponding to the following names:
(a) *o*-Dibromobenzene
(b) *p*-Chloronitrobenzene
(c) *m*-Diethylbenzene

PROBLEM 23.17 Write the products from reaction of the following reagents with *p*-dimethylbenzene (also called *p*-xylene):
(a) Br_2, $FeBr_3$
(b) HNO_3, H_2SO_4

PROBLEM 23.18 Reaction of $Br_2/FeBr_3$ with toluene (methylbenzene) might lead to a mixture of *three* isomeric substitution products. Show the structure of each.

23.7 ALCOHOLS, ETHERS, AND AMINES

Alcohols, ethers, and amines are compounds that have a carbon singly bonded to an electronegative atom—either oxygen or nitrogen.

Alcohols

Alcohols can be viewed either as derivatives of water, in which one of the hydrogens is replaced by an organic substituent, or as derivatives of alkanes, in which one of the hydrogens is replaced by a *hydroxyl group* (−OH).

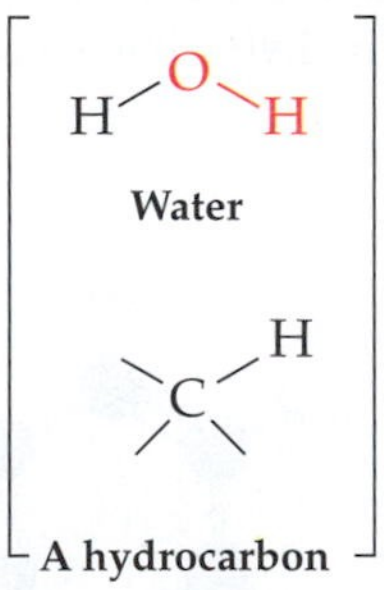

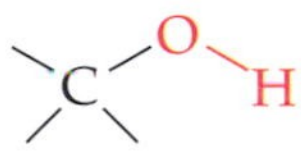

for example: CH_3CH_2OH

Ethanol

Like water, alcohols form **hydrogen bonds**, which affect many of their chemical and physical properties. Alcohols are generally higher boiling than alkanes of similar size, and simple alcohols are often soluble in water because of their ability to form hydrogen bonds (**Figure 23.5**).

REMEMBER...
A **hydrogen bond** is an attractive interaction between a hydrogen atom bonded to an electronegative atom (O, N, or F) and an unshared electron pair on another electronegative atom. (Section 10.2)

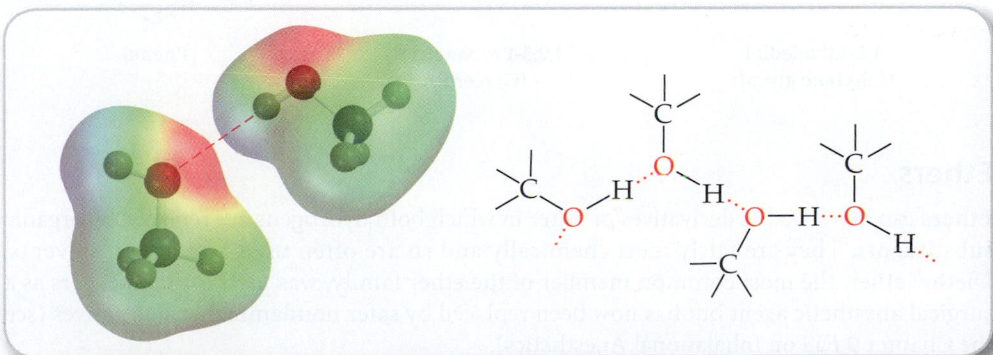

FIGURE 23.5
Hydrogen bonding in alcohols. Because of their hydrogen bonds, alcohols are relatively high boiling and are often soluble in water.

Alcohols are named by identifying the point of attachment of the $-OH$ group to the hydrocarbon chain and using the suffix *-ol* to replace the terminal *-e* in the alkane name. Numbering of the chain begins at the end nearer the $-OH$ group. For example,

$CH_3CH_2CH_2OH$ (carbons numbered 3 2 1)
1-Propanol

$CH_3CH(OH)CH_3$ (carbons numbered 1 2 3)
2-Propanol

$CH_3C(OH)(CH_3)CH_3$ (carbons numbered 1 2 3)
2-Methyl-2-propanol

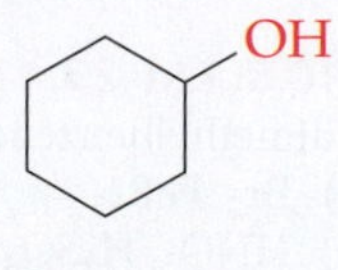

Cyclohexanol

CH_3OH
Methanol

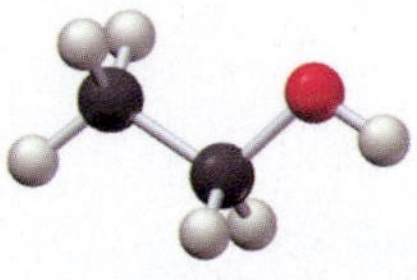

CH_3CH_2OH
Ethanol

Alcohols are among the most important and commonly encountered of all organic chemicals. Methanol (CH_3OH), the simplest member of the family, was once known as *wood alcohol* because it was prepared by heating wood in the absence of air. Methanol is toxic to humans, causing blindness in low doses (15 mL) and death in larger amounts (100–200 mL), but it is nevertheless an important industrial starting material for preparing formaldehyde (CH_2O), acetic acid (CH_3CO_2H), and other chemicals.

Ethanol (CH_3CH_2OH) is one of the oldest known pure organic chemicals. Its production by fermentation of grains and sugars goes back at least 8000 years in the Middle East and perhaps as many as 9000 years in China. Sometimes called *grain alcohol*, ethanol is the "alcohol" present in all wines (10–14%), beers (3–10%), and distilled liquors (35–95%). Fermentation is carried out by adding yeast to an aqueous sugar solution and allowing enzymes in the yeast to break down carbohydrates into ethanol and CO_2.

Still other important alcohols include 2-propanol (isopropyl alcohol), 1,2-ethanediol (ethylene glycol), 1,2,3-propanetriol (glycerol), and the aromatic compound phenol. Isopropyl alcohol is commonly called rubbing alcohol and is used as a disinfectant; ethylene glycol is the principal constituent of automobile antifreeze; glycerol is used as a moisturizing agent in many foods and cosmetics; and phenol is used for preparing nylon, epoxy adhesives, and heat-setting resins.

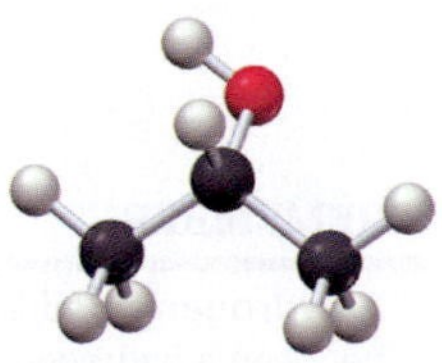

$CH_3CH(OH)CH_3$
2-Propanol
(Isopropyl alcohol)

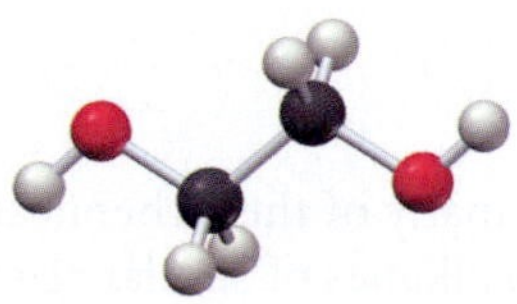

$HOCH_2CH_2OH$
1,2-Ethanediol
(Ethylene glycol)

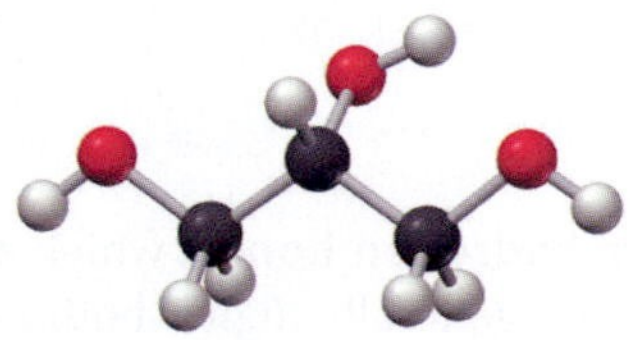

$HOCH_2CH(OH)CH_2OH$
1,2,3-Propanetriol
(Glycerol)

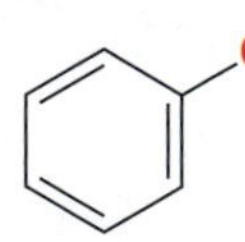

Phenol

Ethers

Ethers can be viewed as derivatives of water in which both hydrogens are replaced by organic substituents. They are fairly inert chemically and so are often used as reaction solvents. Diethyl ether, the most common member of the ether family, was used for many years as a surgical anesthetic agent but has now been replaced by safer nonflammable alternatives (see the Chapter 9 *FYI* on Inhalational Anesthetics).

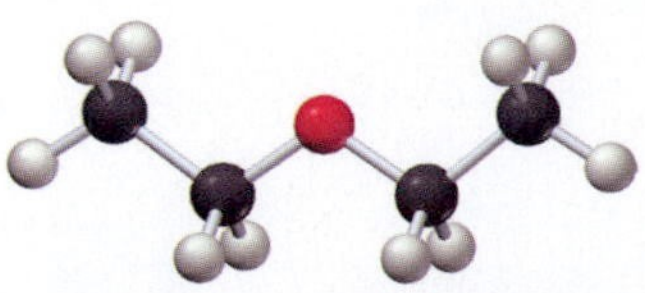

$CH_3CH_2OCH_2CH_3$
Diethyl ether

Amines

Amines are organic derivatives of ammonia in the same way that alcohols and ethers are organic derivatives of water. That is, one or more of the ammonia hydrogens is replaced in amines by an organic substituent. As the following examples indicate, the suffix *-amine* is used in naming these compounds:

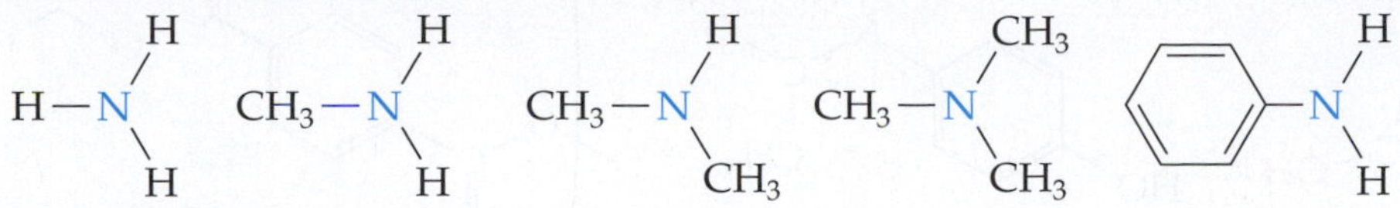

Ammonia — Methylamine — Dimethylamine — Trimethylamine — Benzeneamine (Aniline)

▲ The characteristic aroma of ripe fish is due to methylamine, CH_3NH_2.

Like ammonia, amines are bases because they can use the lone pair of electrons on nitrogen to accept H^+ from an acid and give ammonium salts (Section 14.12). Because they're ionic, ammonium salts are much more soluble in water than neutral amines are. Thus, a water-insoluble amine such as triethylamine dissolves readily in water when converted to its ammonium salt by reaction with HCl.

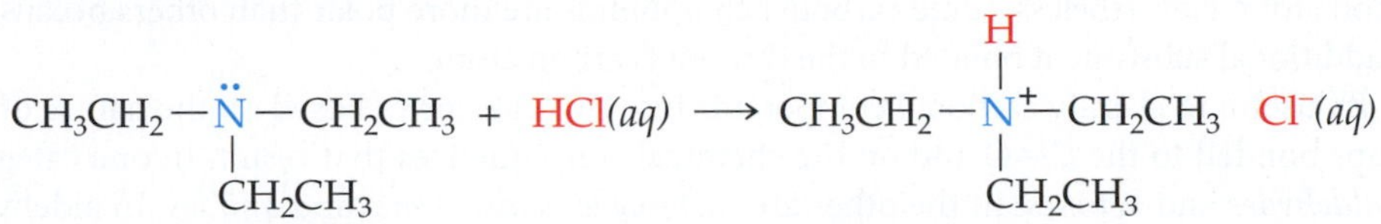

Triethylamine (water-insoluble) — **Triethylammonium chloride (water-soluble)**

This increase in water solubility on conversion of an amine to its protonated salt has enormous practical consequences in drug delivery. Many important amine-containing drugs, such as morphine (an analgesic, or painkiller) and tetracycline (an antibiotic), are insoluble in aqueous body fluids and are thus difficult to deliver to the appropriate site within the body. Converting these drugs to their ammonium salts, however, increases their water solubility to the point where delivery through the bloodstream becomes possible.

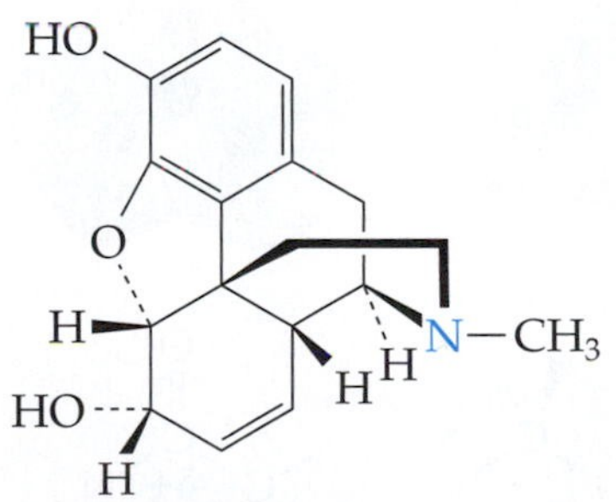

Morphine (analgesic)

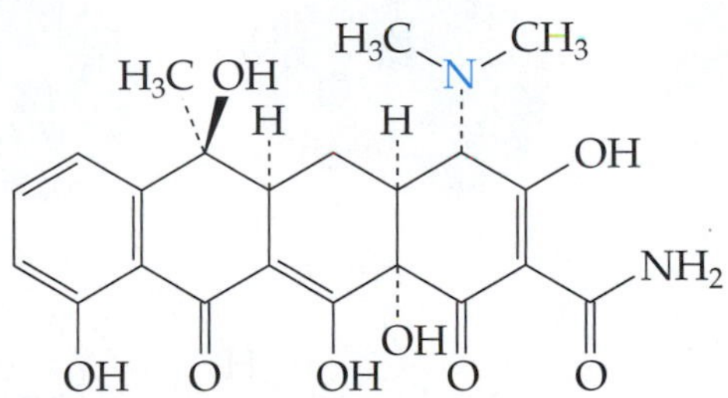

Tetracycline (antibiotic)

PROBLEM 23.19 Write the structures of the ammonium salts produced by reaction of the following amines with HCl:

(a)

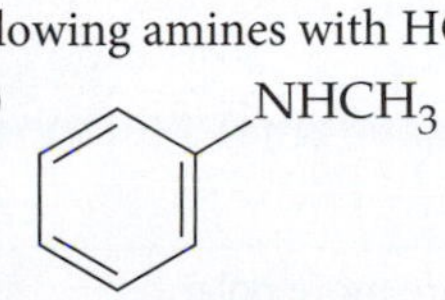

(b) $CH_3CH_2CH_2NH_2$

23.8 CARBONYL COMPOUNDS

Look back at the functional groups listed in Table 23.1 and you'll see that many of them have a carbon–oxygen double bond (C=O), called a **carbonyl group** (car-bo-**neel**). Carbonyl-containing compounds are everywhere. Carbohydrates, fats, proteins, and nucleic acids all contain carbonyl groups; most pharmaceutical agents contain carbonyl groups; and many of the synthetic polymers we encounter in everyday life contain carbonyl groups.

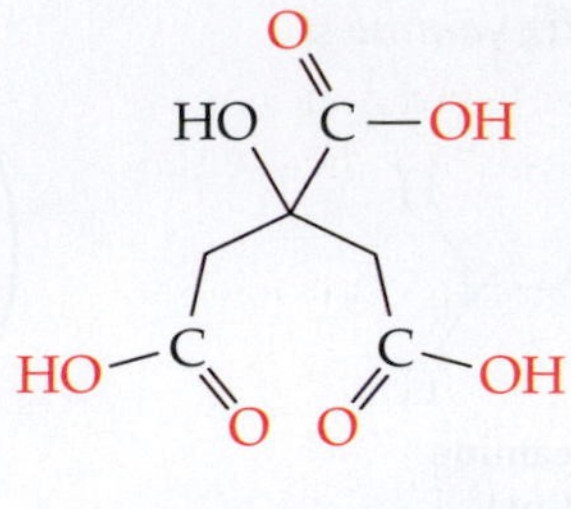

Citric acid
(a carboxylic acid)

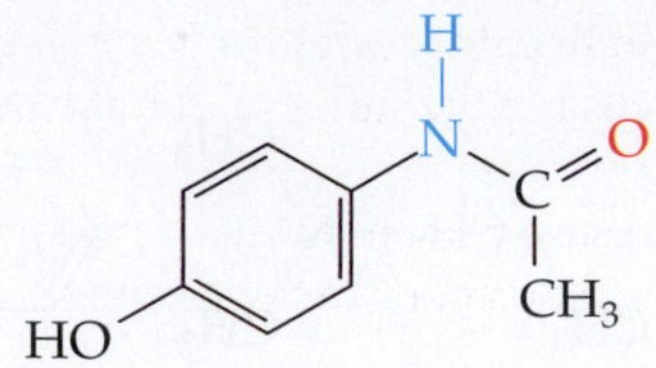

Acetaminophen
(an amide)

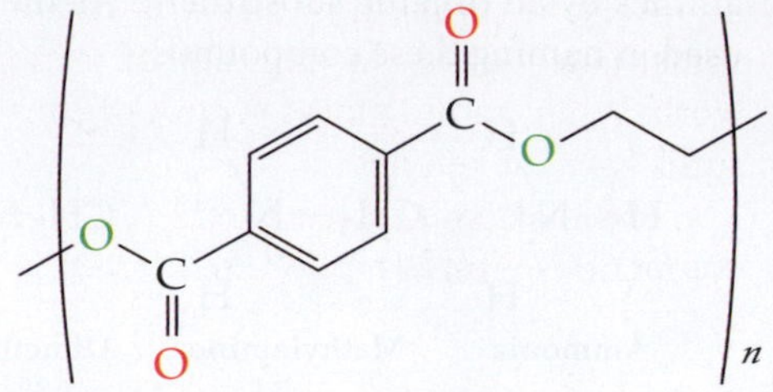

Dacron
(a polyester)

As shown by the electrostatic potential maps in **Figure 23.6**, the C=O bond in carbonyl compounds is polar because the electronegative oxygen atom attracts electrons from the carbon atom. Nevertheless, some carbonyl compounds are more polar than others because of the additional substituent bonded to the carbonyl carbon atom.

It's useful to classify carbonyl compounds into two categories based on the nature of the groups bonded to the C=O and on the chemical consequences that result. In one category are *aldehydes* and *ketones*. In the other are *carboxylic acids*, *esters*, and *amides*. In aldehydes and ketones, the carbonyl carbon is bonded to atoms (H and C) that are not strongly electronegative and thus contribute no additional polarity to the molecule. In carboxylic acids, esters, and amides, however, the carbonyl carbon is bonded to an atom (O or N) that *is* strongly electronegative, giving these compounds even greater polarity and greater chemical reactivity.

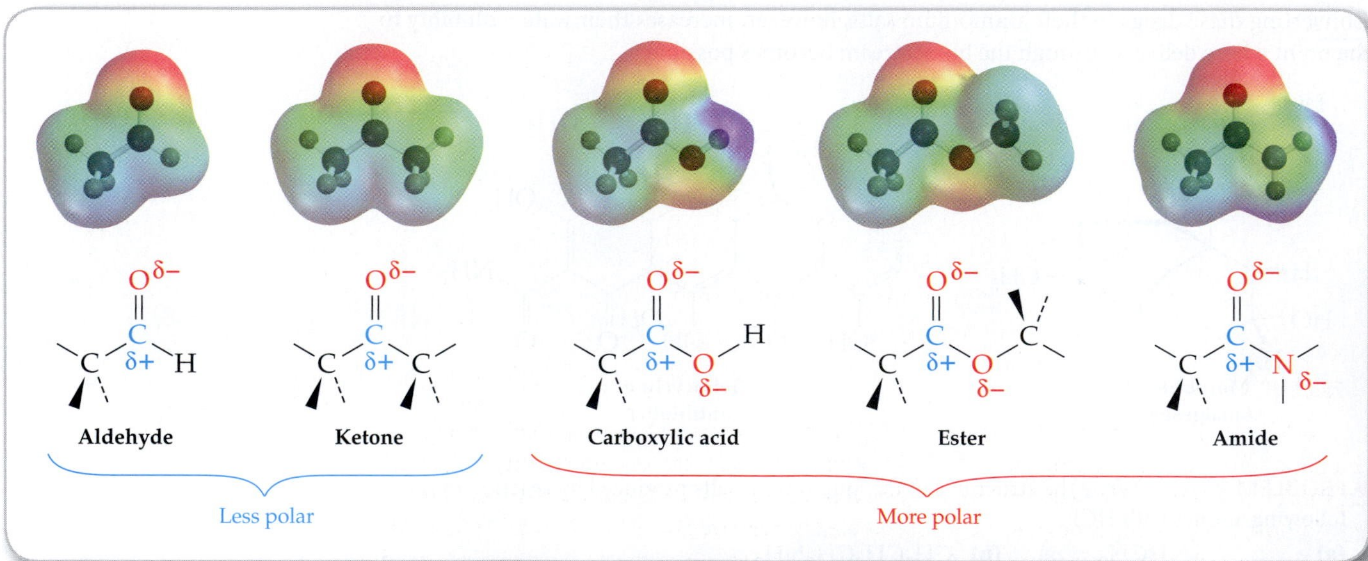

FIGURE 23.6

Kinds of carbonyl compounds. Aldehydes and ketones are less polar, while carboxylic acids, esters, and amides are more polar.

Aldehydes and Ketones

Aldehydes, which have a hydrogen atom bonded to the carbonyl group, and **ketones,** which have two carbon atoms bonded to the carbonyl group, are used throughout chemistry and biology. For example, an aqueous solution of formaldehyde (properly named *methanal*) is used under the name *formalin* as a biological sterilant and preservative. Formaldehyde is also used in the chemical industry as a starting material for the manufacture of the plastics Bakelite and melamine and as a component of the adhesives used to bind plywood. Note that formaldehyde differs from other aldehydes in having two hydrogens attached to the carbonyl group rather than one.

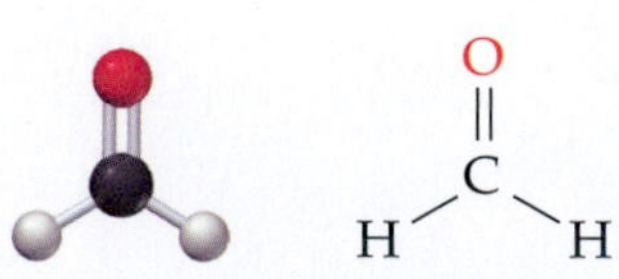

Formaldehyde

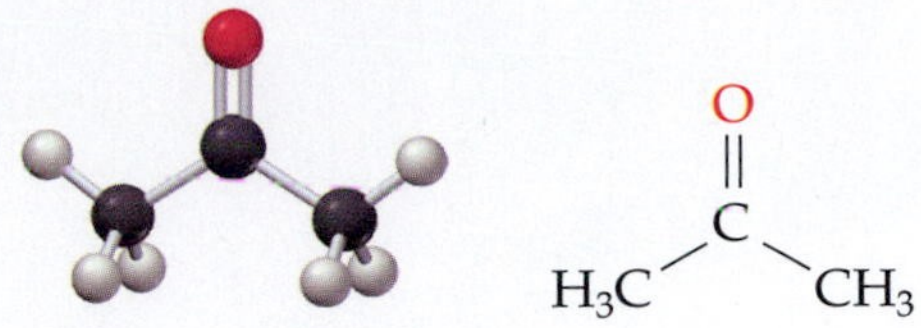

Acetone

Acetone (properly named *propanone*) is perhaps the most widely used of all organic solvents. You might have seen cans of acetone sold in paint stores for general-purpose cleanup work. When naming these groups of compounds, use the suffix *-al* for aldehydes and use the suffix *-one* for ketones.

Aldehyde and ketone functional groups are also present in many biologically important compounds. Testosterone and many other steroid hormones are examples.

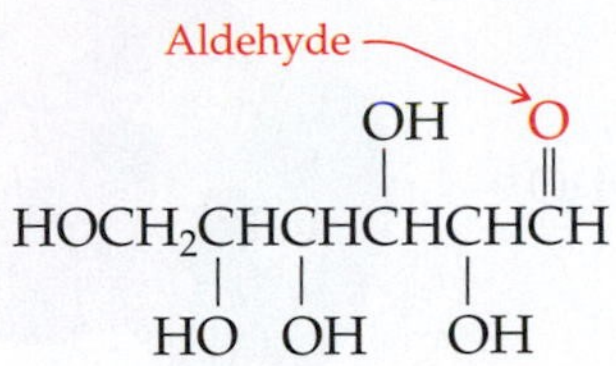

Glucose—a pentahydroxyhexanal

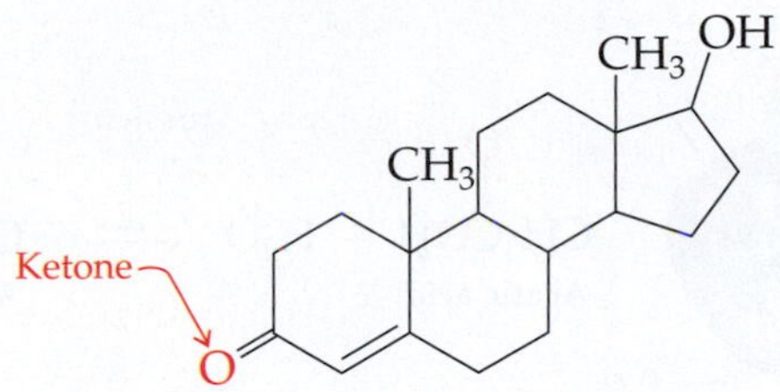

Testosterone—a steroid hormone

Carboxylic acids, esters, and amides differ from aldehydes and ketones in that their carbonyl groups are bonded to strongly electronegative atoms (O or N). All three families undergo carbonyl-group substitution reactions, in which a group we can represent as —Y substitutes for the —OH, —OC, or —N group of the carbonyl reactant.

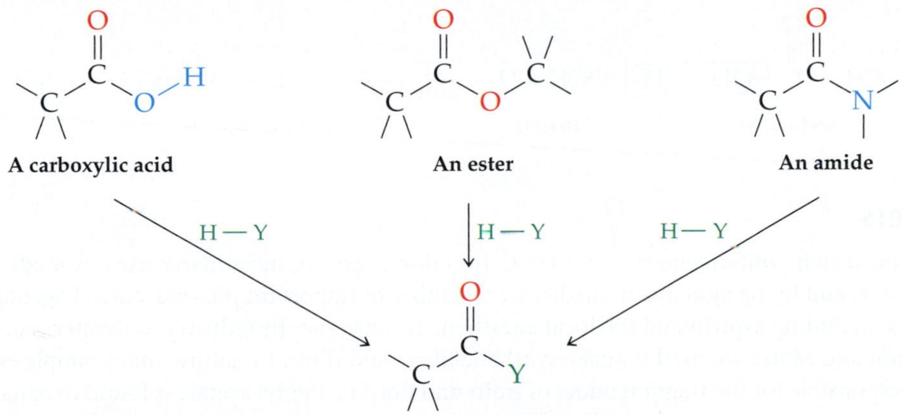

A carbonyl-group substitution reaction

Carboxylic Acids

Carboxylic acids, which contain the $-\overset{\overset{\displaystyle O}{\|}}{C}-OH$ functional group, occur widely throughout the plant and animal kingdoms. Acetic acid (ethanoic acid), for instance, is the principal organic constituent of vinegar, and butanoic acid is responsible for the odor of rancid butter. In addition, long-chain carboxylic acids such as stearic acid are constituents of all animal fats and vegetable oils. Although many carboxylic acids have common names—*acetic* acid instead of *ethanoic* acid, for instance—systematic names are derived by replacing the final *-e* of the corresponding alkane with *-oic acid.*

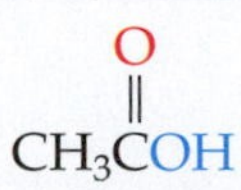

Acetic acid
(Ethanoic acid)

Butanoic acid

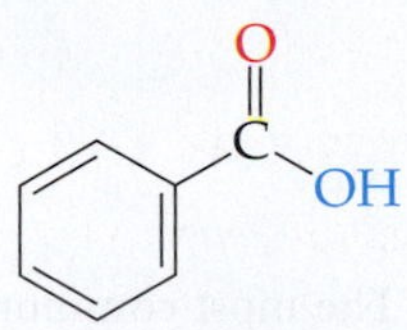

Benzoic acid

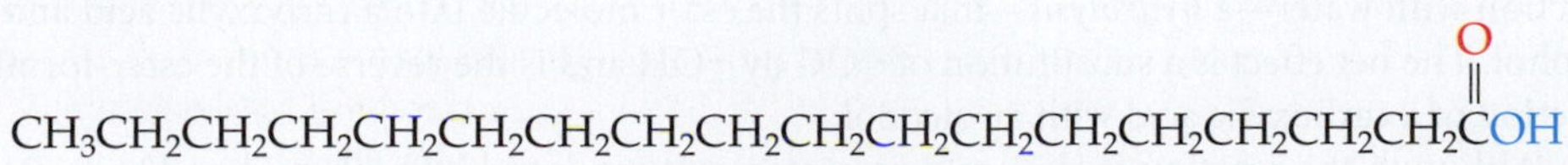

Stearic acid
(Octadecanoic acid)

As their name implies, carboxylic acids are *acidic*—they dissociate slightly in aqueous solution to give H_3O^+ and a **carboxylate anion**. Carboxylic acids are much weaker than inorganic acids like HCl or H_2SO_4, however. The K_a of acetic acid, for example, is 1.8×10^{-5} ($pK_a = 4.74$), meaning that only about 1% of acetic acid molecules dissociate in a 1.0 M aqueous solution. Note in the following electrostatic potential map of acetic acid that the acidic —OH hydrogen is positively polarized (blue).

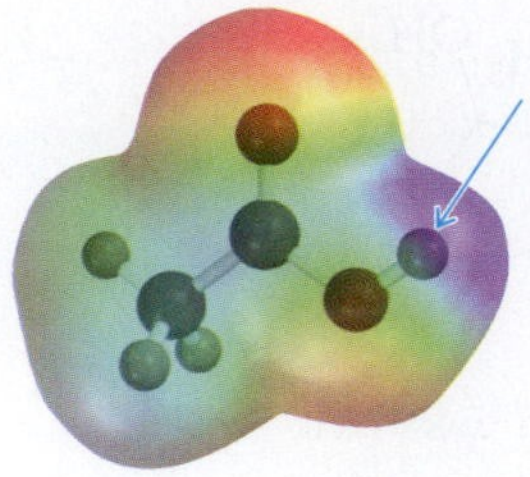

$$CH_3\overset{\overset{\large O}{\|}}{C}OH + H_2O \rightleftharpoons CH_3\overset{\overset{\large O}{\|}}{C}O^- + H_3O^+$$

Acetic acid **Acetate ion**

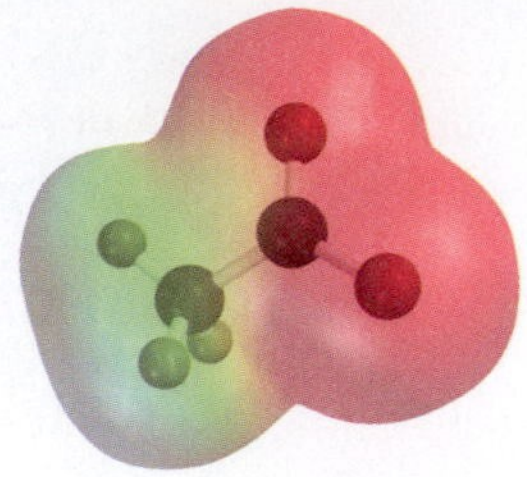

One of the most useful chemical transformations of carboxylic acids is their acid-catalyzed reaction with an alcohol to yield an ester. Acetic acid, for example, reacts with ethanol in the presence of H_2SO_4 to yield ethyl acetate, a widely used solvent. The reaction is a typical carbonyl-group substitution, with $-OCH_2CH_3$ from the alcohol replacing —OH from the acid.

$$CH_3-\overset{\overset{\large O}{\|}}{C}-\boxed{OH + H}-OCH_2CH_3 \xrightarrow[\text{catalyst}]{H^+} CH_3-\overset{\overset{\large O}{\|}}{C}-OCH_2CH_3 + H_2O$$

Acetic acid **Ethanol** **Ethyl acetate**

Esters

Esters, which contain the $-\overset{\overset{\large O}{\|}}{C}-O-C$ functional group, have many uses in medicine, industry, and living systems. In medicine, a number of important pharmaceutical agents are esters, including aspirin and the local anesthetic benzocaine. In industry, polyesters such as Dacron and Mylar are used to make synthetic fibers and films. In nature, many simple esters are responsible for the fragrant odors of fruits and flowers. Pentyl acetate is found in bananas, for instance, and octyl acetate is found in oranges.

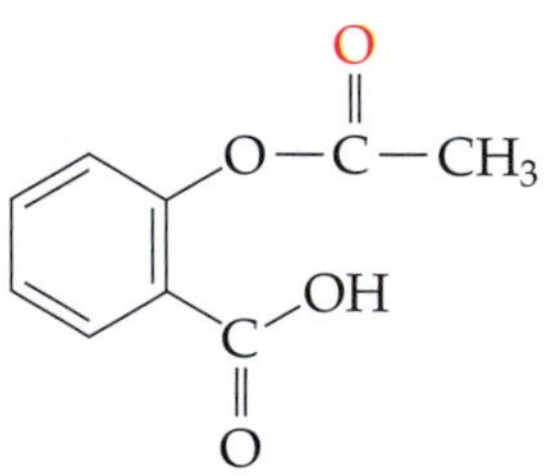

Aspirin

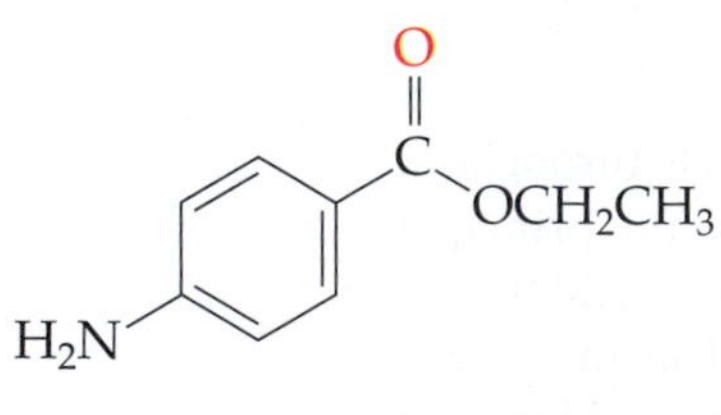

Benzocaine

$$CH_3\overset{\overset{\large O}{\|}}{C}OCH_2CH_2CH_2CH_2CH_3$$

Pentyl acetate

▲ The odor of bananas is due to pentyl acetate, a simple ester.

The most common reaction of esters is their conversion by a carbonyl-group substitution reaction into carboxylic acids. Both in the laboratory and in the body, esters undergo a reaction with water—a *hydrolysis*—that splits the ester molecule into a carboxylic acid and an alcohol. The net effect is a substitution of —OC by —OH and is the reverse of the ester-forming reaction of a carboxylic acid with an alcohol.

Although the reaction is slow in pure water, it is catalyzed by both acid and base. Base-catalyzed ester hydrolysis is often called *saponification*, from the Latin word *sapo* meaning "soap." Soap, in fact, is a mixture of sodium salts of long-chain carboxylic acids produced by hydrolysis of the naturally occurring esters in animal fat.

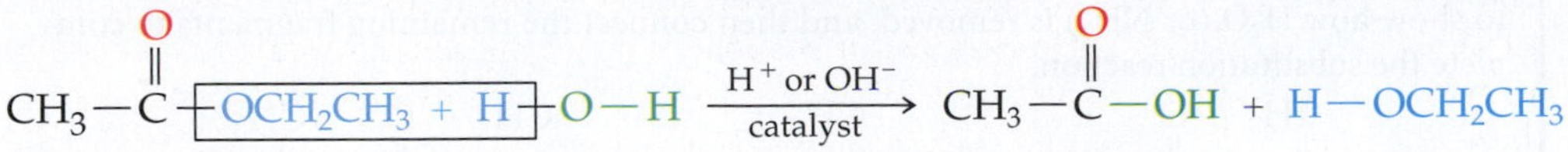

Ethyl acetate Acetic acid Ethanol

Because esters are derived from carboxylic acids and alcohols, they are named by first identifying the alcohol-related part and then the acid-related part, using the *-ate* ending. Ethyl acetate, for example, is the ester derived from ethanol and acetic acid.

Amides

Amides are compounds with the $-\overset{\overset{\displaystyle O}{\|}}{C}-N$ functional group. Without amides, there would be no life because the amide bond between nitrogen and a carbonyl-group carbon is the fundamental link used by organisms for forming proteins. In addition, some synthetic polymers such as nylon contain amide groups, and important pharmaceutical agents such as acetaminophen, the aspirin substitute found in Tylenol and Excedrin, are amides.

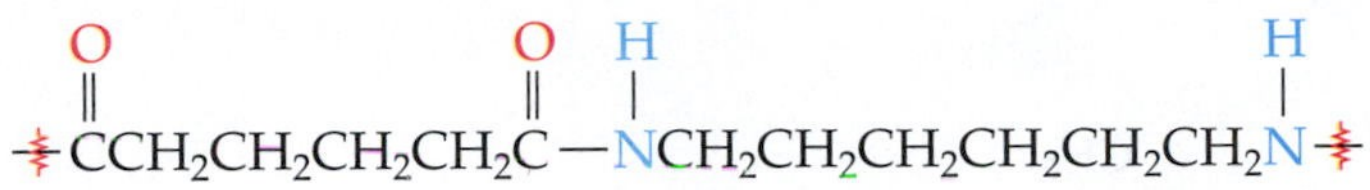

Repeating unit of nylon 66

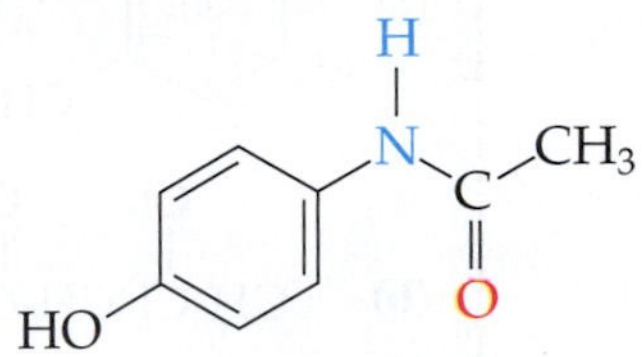

Acetaminophen

Unlike amines, which also contain nitrogen (Section 23.7), amides are neutral rather than basic. Amides don't act as proton acceptors and don't form ammonium salts when treated with acid. The neighboring carbonyl group causes the unshared pair of electrons on nitrogen to be held tightly, thus preventing the electrons from bonding to H^+.

Amides undergo an acid- or base-catalyzed hydrolysis reaction with water in the same way that esters do. Just as an ester yields a carboxylic acid and an alcohol, an amide yields a carboxylic acid and an amine (or ammonia). The net effect is a substitution of –N by –OH. This hydrolysis of amides is the key process that occurs in the stomach during digestion of proteins.

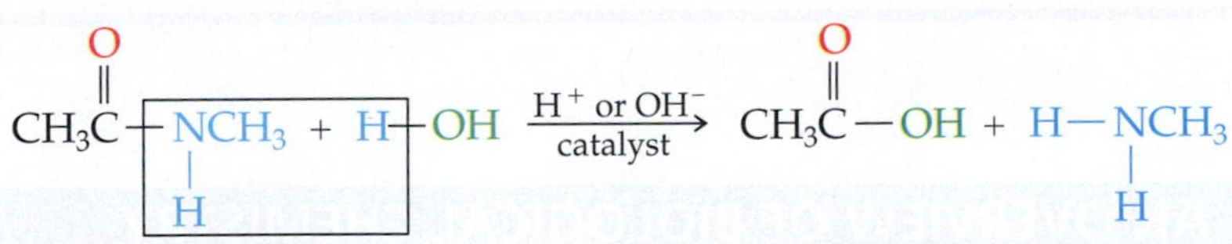

N-Methylacetamide Acetic acid Methylamine

Worked Example 23.5

Predicting the Products of Organic Reactions

Write the products of the following reactions:

(a) 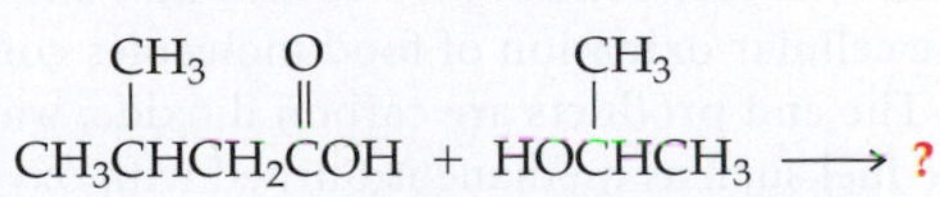

(b) 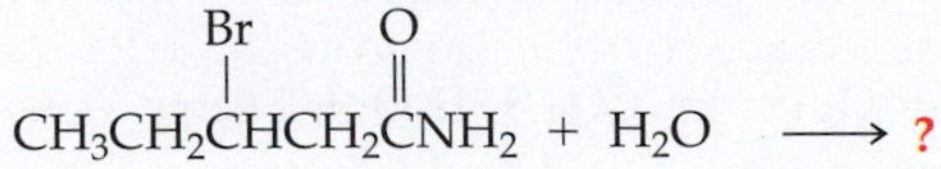

continued on next page

STRATEGY AND SOLUTION

The reaction of a carboxylic acid with an alcohol yields an ester plus water, and the reaction of an amide with water yields a carboxylic acid and an amine (or ammonia). Write the reactants to show how H_2O (or NH_3) is removed, and then connect the remaining fragments to complete the substitution reaction.

(a) 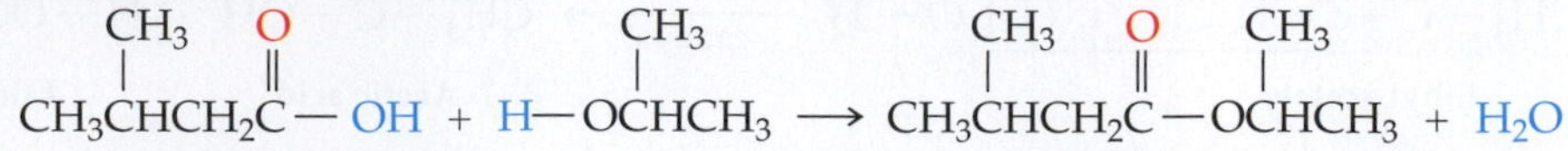

(b) 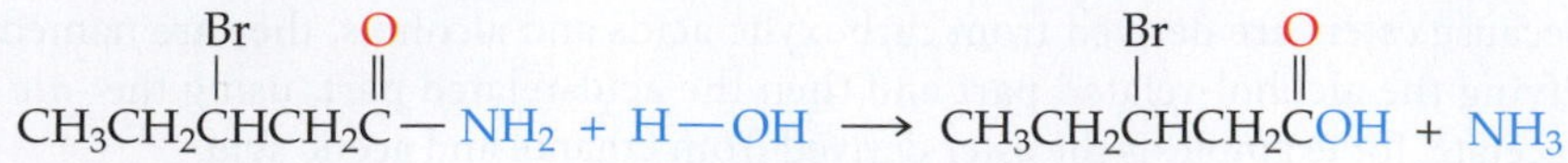

PROBLEM 23.20 Draw structures corresponding to the following names:
(a) 4-Methylpentanoic acid (b) Isopropyl benzoate (c) *N*-Ethylpropanamide

PROBLEM 23.21 Write the products of the following reactions:

(a) 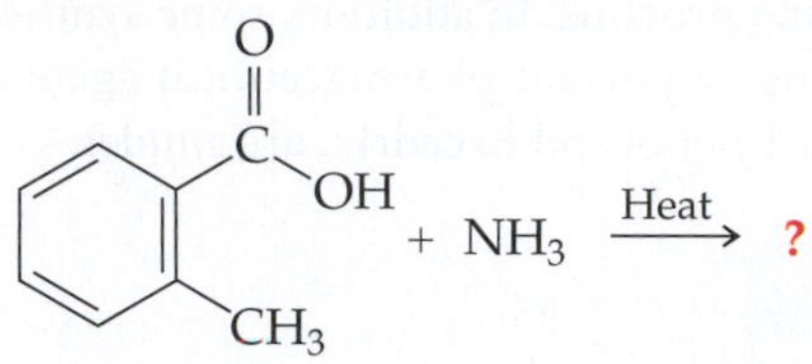

(b) 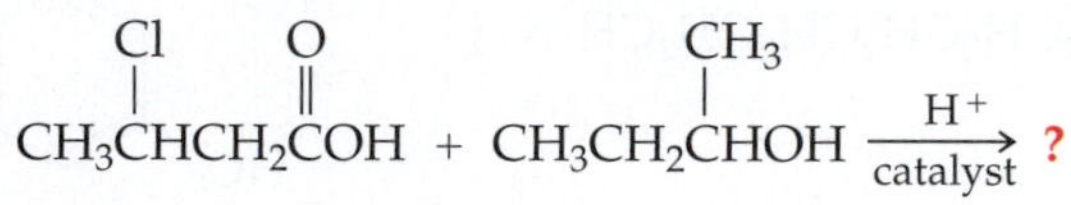

CONCEPTUAL PROBLEM 23.22 Draw the structure of the ester you would obtain by acid-catalyzed reaction of the following carboxylic acid with 2-propanol:

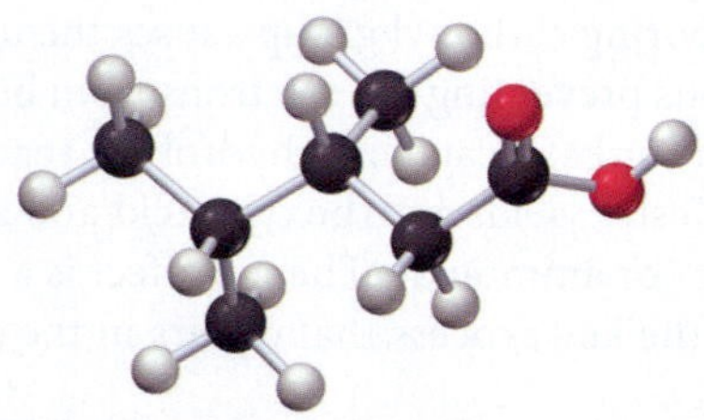

23.9 AN OVERVIEW OF BIOLOGICAL CHEMISTRY

Now that we've looked at some of the fundamental families of organic compounds, let's see their relevance to biological chemistry. We'll begin with a quick overview of biological energy and then look briefly at the main classes of biological molecules.

All living organisms do mechanical work. Microorganisms engulf food, plants bend toward the sun, and animals move about. Organisms also do chemical work in synthesizing the biomolecules needed for growth and repair. In animals, it is the energy extracted from food and released in the interconnected reactions of *metabolism* that allows work to be done. Animals are powered by the cellular oxidation of food molecules containing mainly carbon, hydrogen, and oxygen. The end products are carbon dioxide, water, and energy, just as they are when an organic fuel such as methane is burned with oxygen in a furnace.

$$\text{C, H, O (food molecules)} + O_2 \longrightarrow CO_2 + H_2O + \text{Energy}$$

The many organic reactions that take place in the cells of living organisms are collectively called **metabolism**. Those reaction sequences that break down larger molecules into smaller ones are known as **catabolism**, while the sequences that synthesize larger molecules from smaller ones are known as **anabolism**. Catabolic reactions usually release energy, and anabolic reactions generally absorb energy. The overall picture of catabolism and energy production can be roughly divided into the four stages shown in **Figure 23.7**.

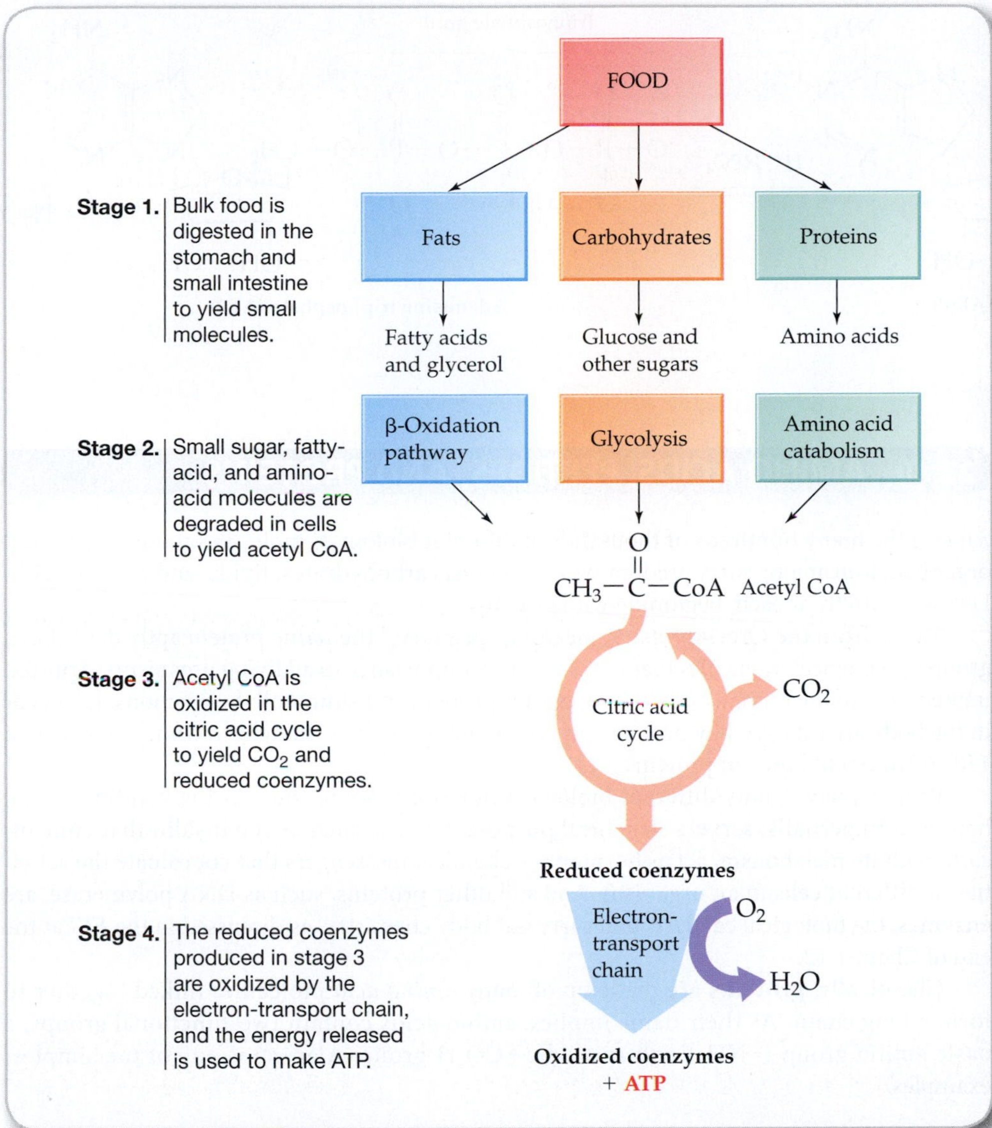

FIGURE 23.7

The four stages of food catabolism and the production of biochemical energy.

The first stage of catabolism, commonly called *digestion*, takes place in the stomach and small intestine when bulk food is broken down into small molecules such as simple sugars, long-chain carboxylic acids called *fatty acids*, and amino acids. In stage 2, these small molecules are further degraded to yield two-carbon acetyl groups $\left(CH_3\overset{|}{C}{=}O\right)$ attached to the large carrier molecule *coenzyme A*. The resultant compound, *acetyl coenzyme A* (*acetyl CoA*), is an intermediate in the breakdown of all the main classes of food molecules.

Acetyl groups are oxidized in the third stage of catabolism, the *citric acid cycle*, to yield CO_2 and water. This stage releases a great deal of energy that is used in stage 4, the *electron-transport chain*, to make molecules of adenosine triphosphate (ATP) by the endothermic reaction of adenosine diphosphate (ADP) with hydrogen phosphate ion, $HPO_4^{\,2-}$. ATP, the final product of food catabolism, plays a pivotal role in the production of biological energy. As a crucial molecule in many metabolic reactions, ATP has been called the "energy currency of the living cell." Catabolic reactions "pay off" in ATP by synthesizing it from ADP, while anabolic reactions "spend" ATP by transferring a phosphate group to another molecule, thereby regenerating ADP. The entire process of energy production thus revolves around the ATP $\rightleftharpoons$ ADP interconversion.

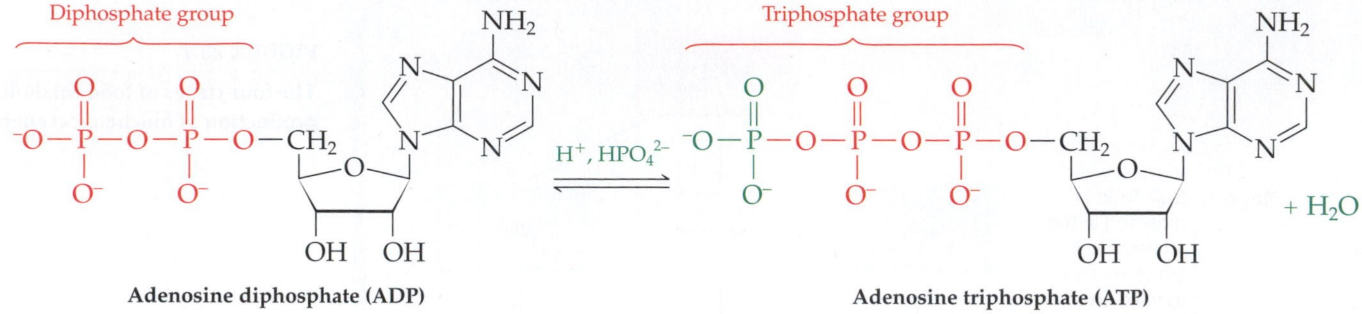

▲ Bird feathers are made largely of the protein keratin.

23.10 AMINO ACIDS, PEPTIDES, AND PROTEINS

Among the many hundreds of thousands of different biological molecules found in a typical organism, four major sorts predominate: proteins, carbohydrates, lipids, and nucleic acids. Let's look briefly at each, beginning with proteins.

Taken from the Greek *proteios*, meaning "primary," the name *protein* aptly describes a group of biological molecules that are of primary importance to all living organisms. Approximately 50% of the human body's dry weight is protein, and almost all the reactions that occur in the body are catalyzed by proteins. In fact, a human body is thought to contain more than *150,000* different kinds of proteins.

Proteins have many different biological functions. Some, such as the keratin in skin, hair, and fingernails, serve a structural purpose. Others, such as the insulin that controls carbohydrate metabolism, act as hormones—chemical messengers that coordinate the activities of different cells in an organism. And still other proteins, such as DNA polymerase, are **enzymes**, the biological catalysts that carry out body chemistry, as discussed in the *FYI* at the end of Chapter 12.

Chemically, **proteins** are made up of many *amino acid* molecules linked together to form a long chain. As their name implies, amino acids contain two functional groups, a basic amino group ($-NH_2$) and an acidic $-CO_2H$ group. Alanine is one of the simplest examples.

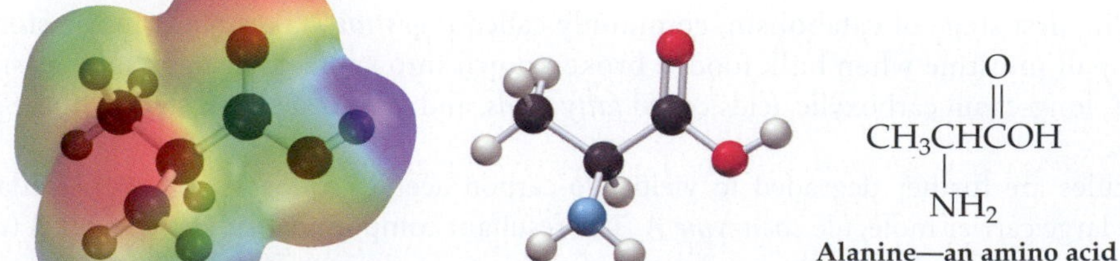

Two or more amino acids can link together by forming amide bonds (Section 23.8), usually called **peptide bonds**, between the $-NH_2$ group of one and the $-CO_2H$ group of the other. A *dipeptide* results when two amino acids link together by one amide bond, a *tripeptide* results when three amino acids link together with two peptide bonds, and so on. Short chains of up to 50 amino acids are usually called **peptides**, while the terms *polypeptide* and *protein* are generally used for longer chains.

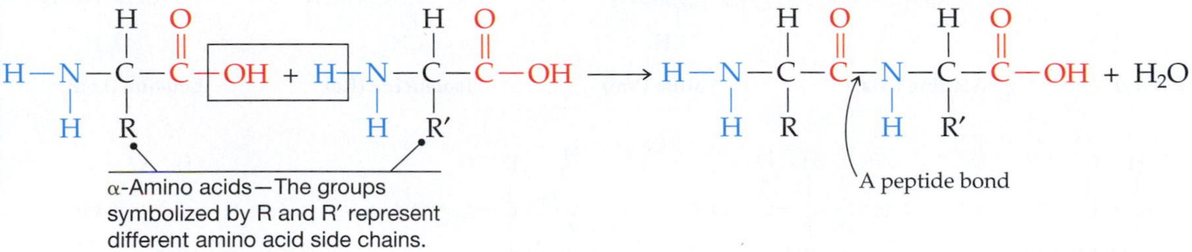

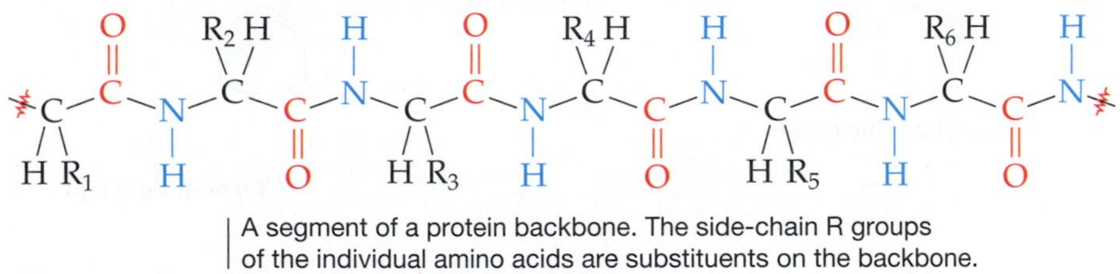

A polypeptide

Twenty different amino acids are commonly found in proteins, as shown in **Figure 23.8**. For convenience, each amino acid is referred to by a three-letter shorthand code, such as Ala (alanine), Gly (glycine), Pro (proline), and so on. All 20 are called **alpha- (α-) amino acids** because the amine nitrogen atom in each is connected to the carbon atom *alpha to* (next to) the carboxylic acid group. Nineteen of the 20 have an $-NH_2$ amino group, and one (proline) has an $-NH-$ amino group as part of a ring.

The 20 amino acids differ in the nature of the group attached to the α carbon. Called the *side chain*, this group can be symbolized in a general way by the letter **R**. Our bodies can synthesize only 11 of the 20 amino acids. The remaining 9, highlighted in Figure 23.8, are called *essential amino acids* because they must be obtained from the diet.

▲ Corn is particularly low in lysine, one of the essential amino acids.

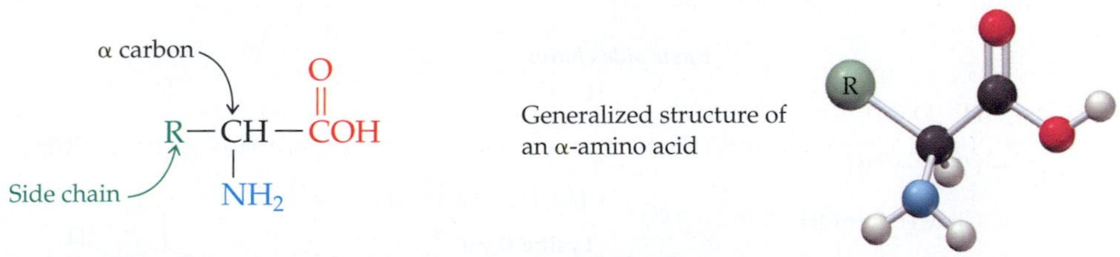

Generalized structure of an α-amino acid

The 20 common amino acids are classified as *neutral, basic,* or *acidic,* depending on the structure of their side chains. Fifteen of the 20 have neutral side chains. Two (aspartic acid and glutamic acid) have an additional carboxylic acid group in their side chains and are classified as acidic amino acids. Three (lysine, arginine, and histidine) have an additional amine function in their side chains and are classified as basic amino acids. The 15 neutral amino acids can be further divided into those with nonpolar side chains and those with polar functional groups such as amide or hydroxyl groups. Nonpolar side chains are often described as *hydrophobic* (water fearing) because they are not attracted to water, while polar side chains are described as *hydrophilic* (water loving) because they *are* attracted to water.

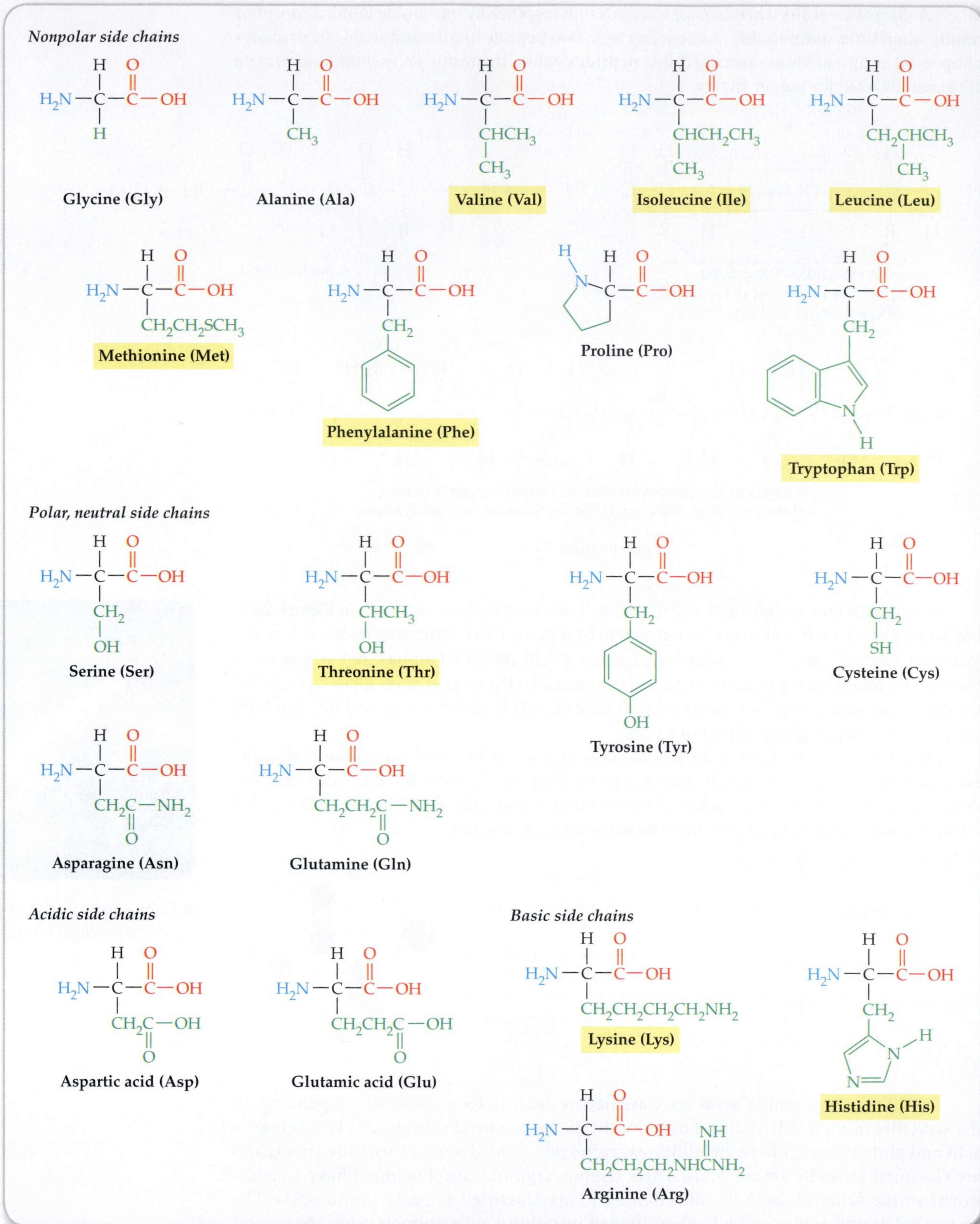

FIGURE 23.8

Structures of the 20 α-amino acids found in proteins. Fifteen of the 20 have neutral side chains, 2 have acidic side chains, and 3 have basic side chains. The names of the 9 essential amino acids are highlighted.

Because amino acids can be assembled in any order, depending on which $-CO_2H$ group forms an amide bond with which $-NH_2$ group, the number of possible isomeric peptides increases rapidly as the number of amino acids increases. There are six ways in which three different amino acids can be joined, more than 40,000 ways in which the eight amino acids present in the blood pressure-regulating hormone angiotensin II can be joined (**Figure 23.9**), and a staggering number of ways in which the *1800* amino acids in myosin, the major component of muscle filaments, can be arranged.

FIGURE 23.9

The structure of angiotensin II, an octapeptide present in blood plasma.

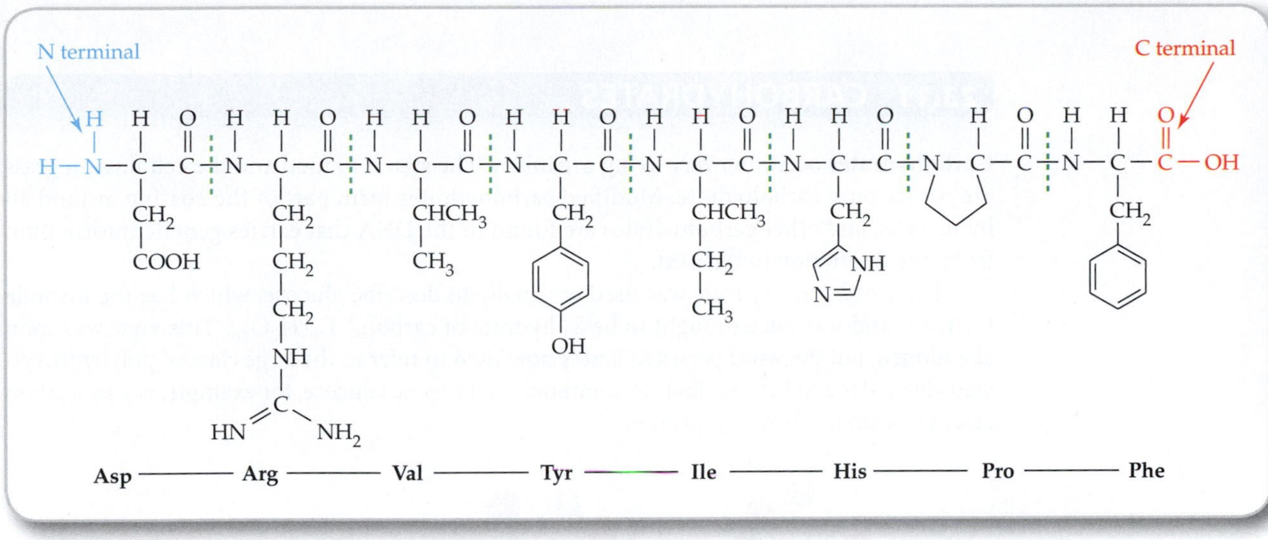

No matter how long the chain, all noncyclic proteins have an *N-terminal amino acid* with a free $-NH_2$ on one end and a *C-terminal amino acid* with a free $-CO_2H$ on the other end. By convention, a protein is written with the free $-NH_2$ on the left and the free $-CO_2H$ on the right, and its name is indicated using the three-letter abbreviations listed in Figure 23.8.

Worked Example 23.6

Drawing a Dipeptide Structure

Draw the structure of the dipeptide Ala-Ser.

STRATEGY

First, look up the names and structures of the two amino acids, Ala (alanine) and Ser (serine). Since alanine is N-terminal and serine is C-terminal, Ala-Ser must have an amide bond between the alanine $-CO_2H$ and the serine $-NH_2$.

SOLUTION

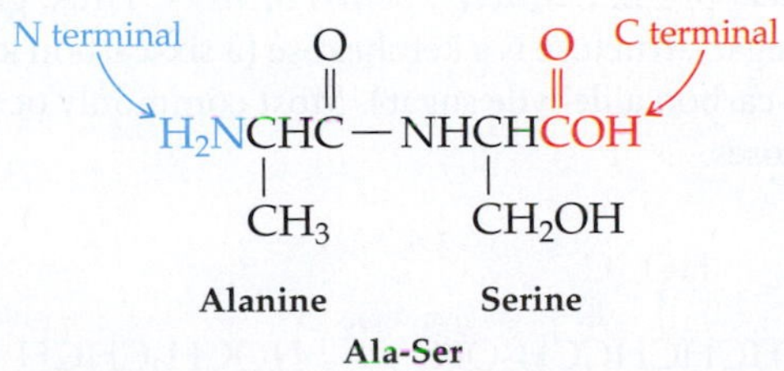

PROBLEM 23.23 Which common amino acids contain an aromatic (benzene-like) ring? Which contain sulfur? Which are alcohols? Which have alkyl-group side chains?

PROBLEM 23.24 Use the three-letter shorthand notations to name the two isomeric dipeptides that can be made from valine and cysteine. Draw both structures.

PROBLEM 23.25 Name the six tripeptides that contain valine, tyrosine, and glycine.

23.11 CARBOHYDRATES

Carbohydrates occur in every living organism. The starch in food and the cellulose in grass are nearly pure carbohydrate. Modified carbohydrates form part of the coating around all living cells, and other carbohydrates are found in the DNA that carries genetic information from one generation to the next.

The word *carbohydrate* was used originally to describe glucose, which has the formula $C_6H_{12}O_6$ and was once thought to be a "hydrate of carbon," $C_6(H_2O)_6$. This view was soon abandoned, but the word persisted and is now used to refer to the large class of polyhydroxylated aldehydes and ketones that we commonly call *sugars*. Glucose, for example, is a six-carbon aldehyde with five hydroxyl groups.

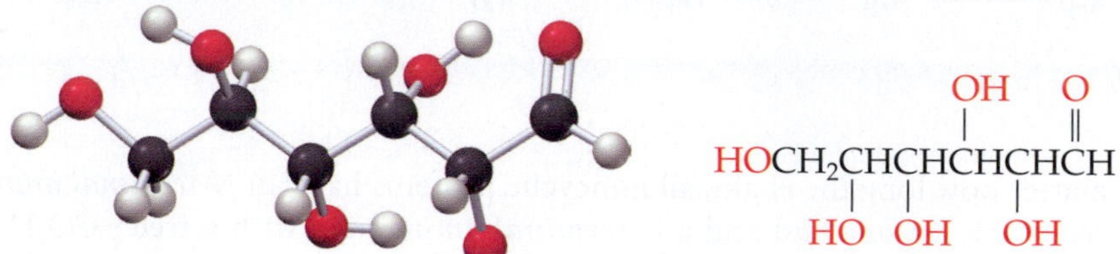

Glucose—a pentahydroxy aldehyde

Carbohydrates are classified as either *simple* or *complex*. Simple sugars, or **monosaccharides**, are carbohydrates such as glucose and fructose that can't be broken down into smaller molecules by hydrolysis with aqueous acid. Complex carbohydrates, or **polysaccharides**, are compounds such as cellulose and starch that are made of many simple sugars linked together and *can* be broken down by hydrolysis.

Monosaccharides

Monosaccharides are further classified as either aldoses or ketoses. An *aldose* contains an aldehyde carbonyl group; a *ketose* contains a ketone carbonyl group (Section 23.8). The *-ose* suffix indicates a sugar, and the number of carbon atoms in the sugar is specified by using the appropriate numerical prefix *tri-*, *tetr-*, *pent-*, or *hex-*. Thus, glucose is an aldohexose (a six-carbon aldehyde sugar), fructose is a ketohexose (a six-carbon ketone sugar), and ribose is an aldopentose (a five-carbon aldehyde sugar). Most commonly occurring sugars are either aldopentoses or aldohexoses.

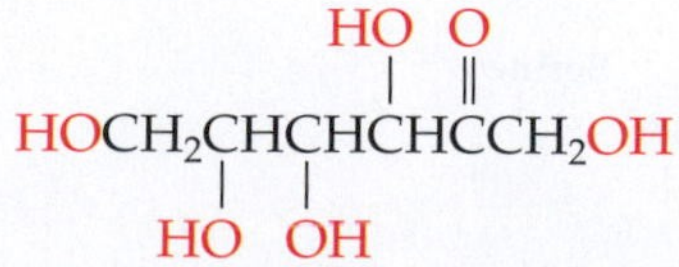

Fructose—a ketohexose

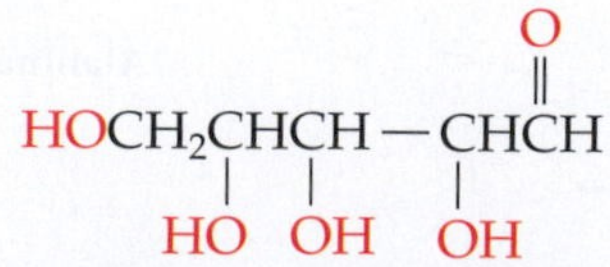

Ribose—an aldopentose

Glucose and other monosaccharides are often shown for convenience as having open-chain structures. They actually exist, however, primarily as cyclic molecules in which an –OH group near one end of the chain adds to the carbonyl group at or near the other end of the chain to form a ring. In glucose, ring formation occurs between the –OH group on C5 and the C=O group at C1 (**Figure 23.10**).

FIGURE 23.10

The cyclic α and β forms of glucose.

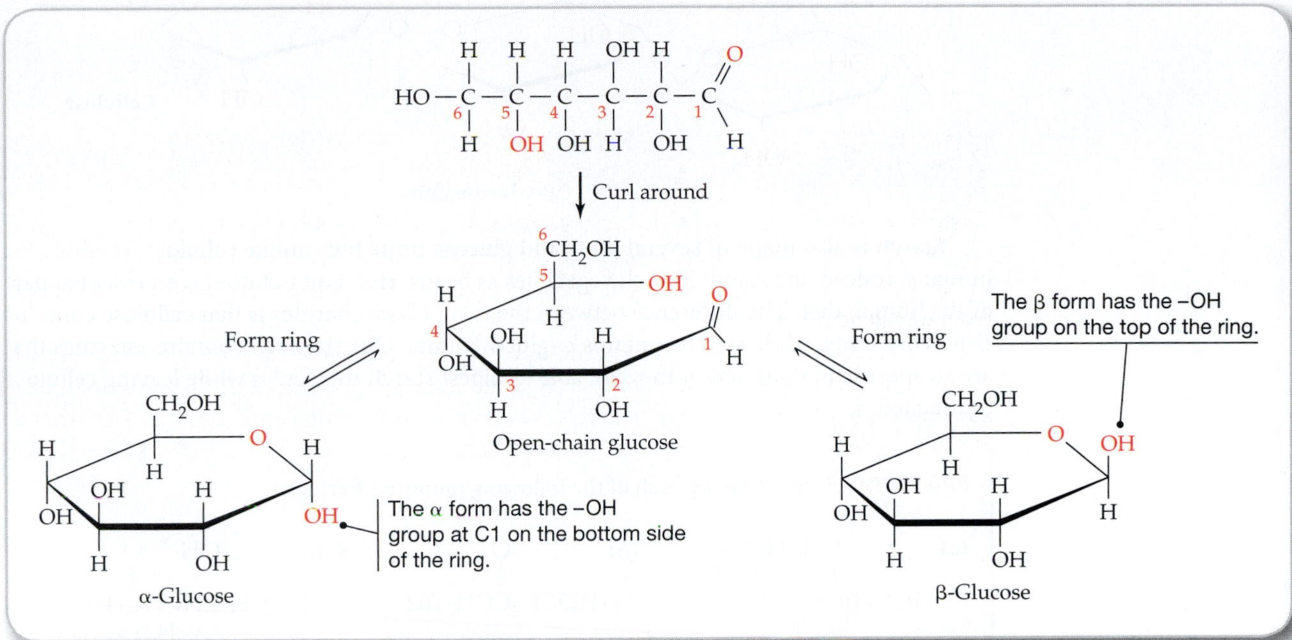

Two cyclic forms of glucose can result from ring formation, depending on whether the newly formed –OH group at C1 is on the bottom or top side of the ring. The ordinary crystalline glucose you might take from a bottle is entirely the cyclic α form, in which the C1 –OH group is on the bottom side of the ring. At equilibrium in water solution, however, all three forms are present in the proportion 0.02% open-chain form, 36% α form, and 64% β form.

Polysaccharides

Sucrose, or plain table sugar, is probably the most common pure organic chemical in the world. Although sucrose is found in many plants, sugar beets (20% by mass) and sugar cane (15% by mass) are the most common sources. Chemically, sucrose is a *disaccharide* composed of one molecule of glucose and one molecule of fructose joined together. The 1:1 mixture of glucose and fructose that results from hydrolysis of sucrose, often called *invert sugar*, is commonly used as a food additive.

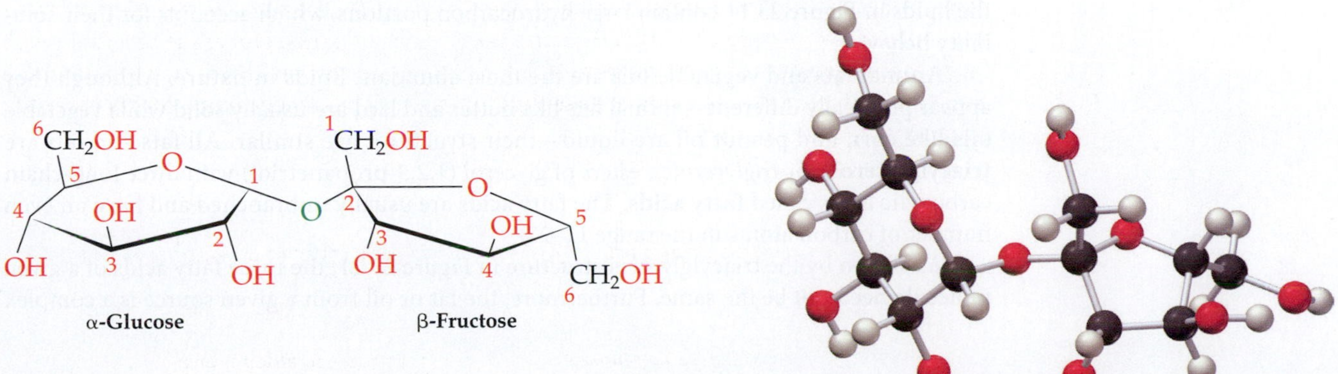

Sucrose

Cellulose, the fibrous substance that forms the structural material in grasses, leaves, and stems, is a polysaccharide composed of several thousand β-glucose molecules joined together to form an immense chain.

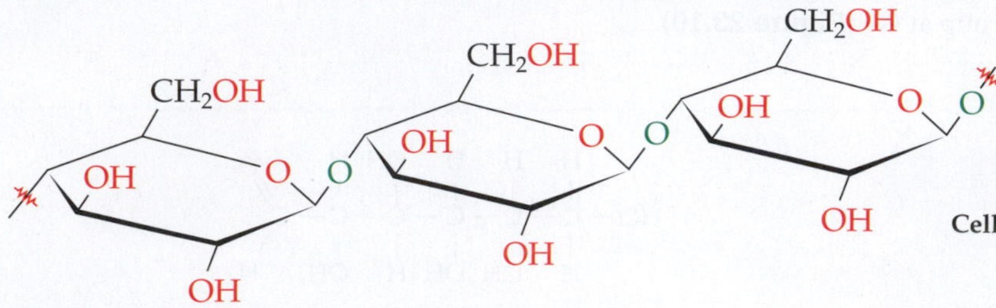

Cellulose

β-Glucose units

Starch is also made of several thousand glucose units but, unlike cellulose, is edible for humans. Indeed, the starch in such vegetables as beans, rice, and potatoes is an essential part of the human diet. The difference between the two polysaccharides is that cellulose contains β-glucose units, while starch contains α-glucose units. Our stomach contains enzymes that are so specific in their action they are able to digest starch molecules while leaving cellulose untouched.

PROBLEM 23.26 Classify each of the following monosaccharides:

(a) **(b)** **(c)**

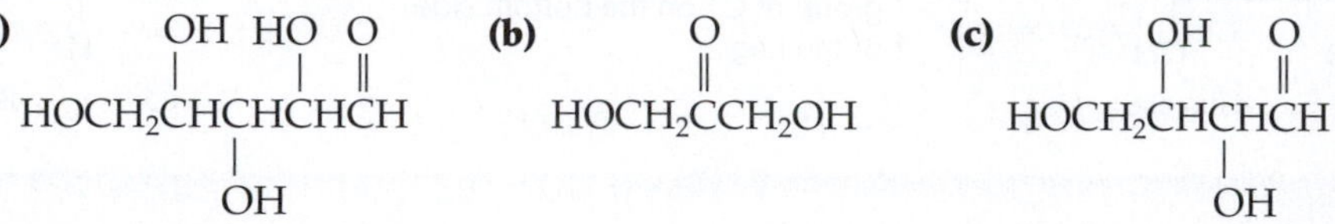

23.12 LIPIDS

Lipids are less well known to most people than are proteins or carbohydrates, yet they are just as essential to life. Lipids have many important biological functions, serving as sources of fuel, as protective coatings around many plants and insects, and as components of the membranes that enclose every living cell.

Chemically, a **lipid** is a naturally occurring organic molecule that dissolves in a nonpolar organic solvent when a sample of plant or animal tissue is crushed or ground. Because they're defined by solubility, a physical property, rather than by chemical structure, it's not surprising that there are a great many different kinds of lipids (**Figure 23.11**). Note that all the lipids in Figure 23.11 contain large hydrocarbon portions, which accounts for their solubility behavior.

Animal fats and vegetable oils are the most abundant lipids in nature. Although they appear physically different—animal fats like butter and lard are usually solid while vegetable oils like corn and peanut oil are liquid—their structures are similar. All fats and oils are **triacylglycerols**, or *triglycerides*, esters of glycerol (1,2,3-propanetriol) with three long-chain carboxylic acids called **fatty acids**. The fatty acids are usually unbranched and have an even number of carbon atoms in the range 12–22.

As shown by the triacylglycerol structure in Figure 23.11, the three fatty acids of a given molecule need not be the same. Furthermore, the fat or oil from a given source is a complex

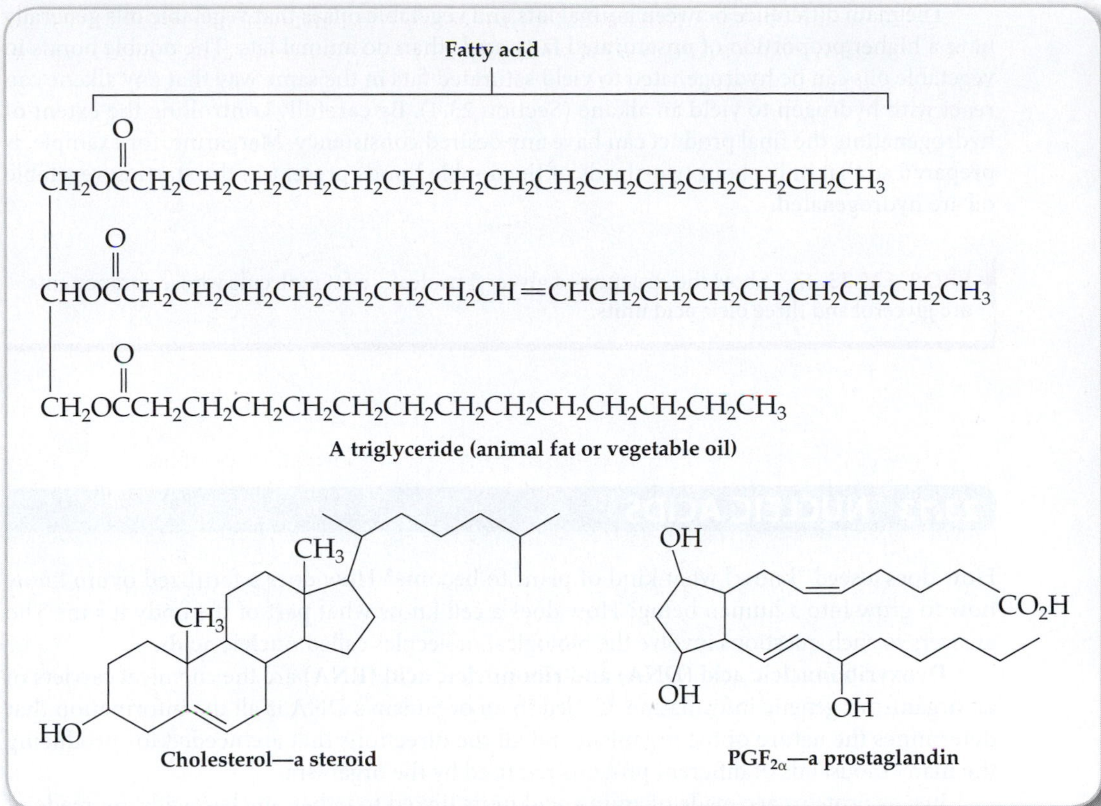

FIGURE 23.11

Structures of some representative lipids. All are isolated from plant and animal tissue by extraction with nonpolar organic solvents, and all have large hydrocarbon portions.

mixture of many different triacylglycerols. **Table 23.3** shows the structures of some commonly occurring fatty acids.

About 40 different fatty acids occur naturally. Palmitic acid (C_{16}) and stearic acid (C_{18}) are the most abundant saturated acids; oleic and linoleic acids (both C_{18}) are the most abundant unsaturated ones. Oleic acid is monounsaturated because it has only one double bond, but linoleic and linolenic acids are *polyunsaturated fatty acids* because they have more than one carbon–carbon double bond. For reasons that are not yet clear, a diet rich in saturated fats leads to a higher level of blood cholesterol and consequent higher risk of heart attack than a diet rich in unsaturated fats.

TABLE 23.3 Structures of Some Common Fatty Acids

Name	No. of Carbons	No. of Double Bonds	Structure
Saturated			
Myristic	14	0	$CH_3(CH_2)_{12}CO_2H$
Palmitic	16	0	$CH_3(CH_2)_{14}CO_2H$
Stearic	18	0	$CH_3(CH_2)_{16}CO_2H$
Unsaturated			
Oleic	18	1	$CH_3(CH_2)_7CH{=}CH(CH_2)_7CO_2H$ (cis)
Linoleic	18	2	$CH_3(CH_2)_4CH{=}CHCH_2CH{=}CH(CH_2)_7CO_2H$ (all cis)
Linolenic	18	3	$CH_3CH_2CH{=}CHCH_2CH{=}CHCH_2CH{=}CH(CH_2)_7CO_2H$ (all cis)

The main difference between animal fats and vegetable oils is that vegetable oils generally have a higher proportion of unsaturated fatty acids than do animal fats. The double bonds in vegetable oils can be hydrogenated to yield saturated fats in the same way that any alkene can react with hydrogen to yield an alkane (Section 23.4). By carefully controlling the extent of hydrogenation, the final product can have any desired consistency. Margarine, for example, is prepared so that only about two-thirds of the double bonds present in the starting vegetable oil are hydrogenated.

PROBLEM 23.27 Show the structure of glyceryl trioleate, a fat molecule whose components are glycerol and three oleic acid units.

23.13 NUCLEIC ACIDS

How does a seed "know" what kind of plant to become? How does a fertilized ovum know how to grow into a human being? How does a cell know what part of the body it's in? The answers to such questions involve the biological molecules called **nucleic acids**.

Deoxyribonucleic acid (DNA) and **ribonucleic acid (RNA)** are the chemical carriers of an organism's genetic information. Coded in an organism's DNA is all the information that determines the nature of the organism and all the directions that are needed for producing the many thousands of different proteins required by the organism.

Just as proteins are made of amino acid units linked together, nucleic acids are made of **nucleotide** units linked together in a long chain. Each nucleotide is composed of a **nucleoside** plus phosphoric acid, H_3PO_4, and each nucleoside is composed of an aldopentose sugar plus an amine base.

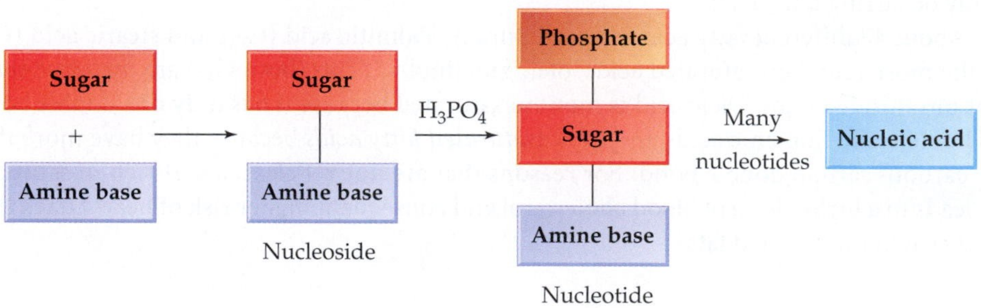

The sugar component in RNA is ribose, and the sugar in DNA is 2-deoxyribose, where "2-deoxy" means that oxygen is missing from C2 of ribose.

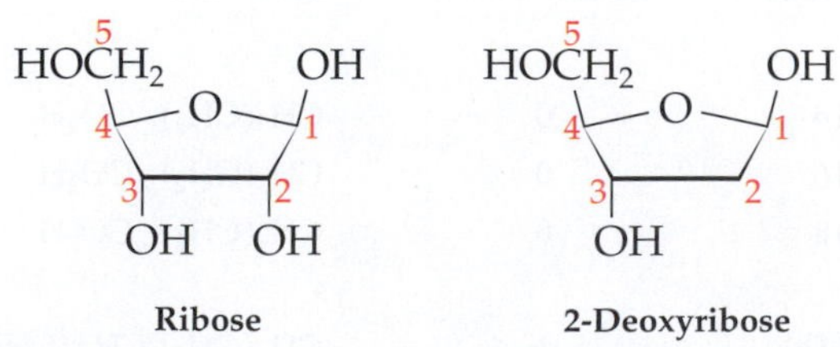

Four different cyclic amine bases occur in DNA: adenine, guanine, cytosine, and thymine. Adenine, guanine, and cytosine also occur in RNA, but thymine is replaced in RNA by a related base called uracil.

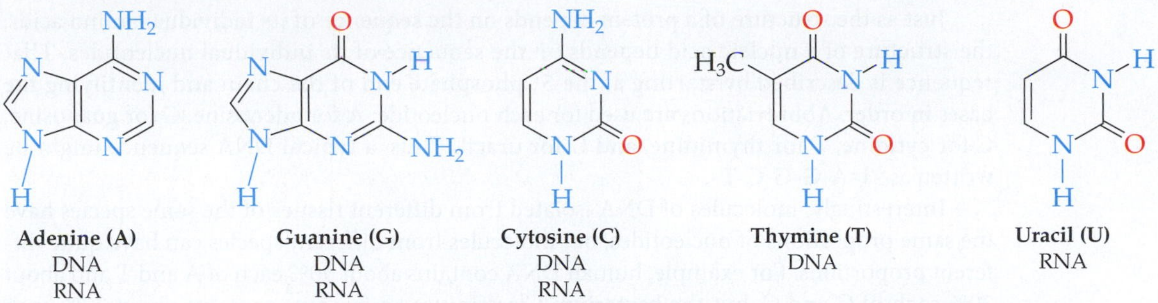

In both DNA and RNA, the cyclic amine base is bonded to C1′ of the sugar, and the phosphoric acid is bonded to the C5′ sugar position. Thus, nucleosides and nucleotides have the general structures shown in **Figure 23.12**. (Numbers with a prime superscript refer to positions on the sugar component of a nucleotide, and numbers without a prime refer to positions on the cyclic amine base.)

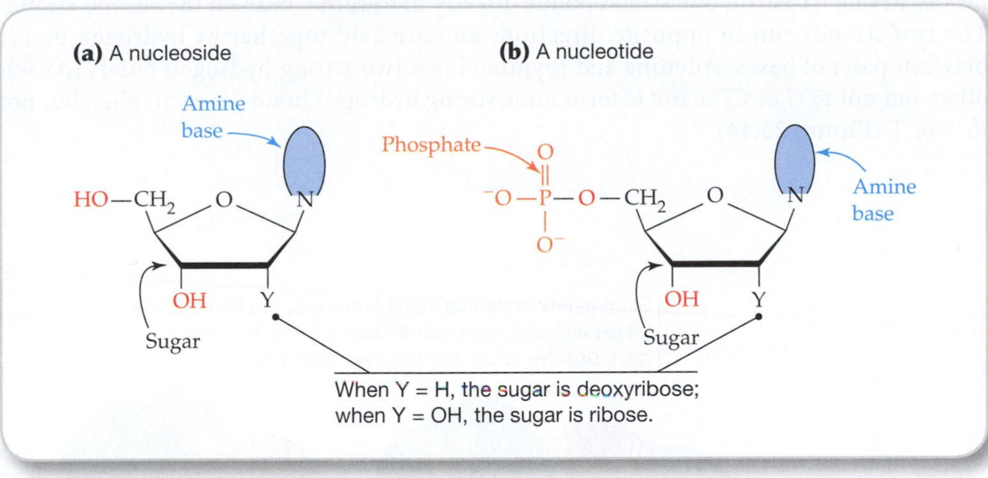

FIGURE 23.12

General structures of (a) a nucleoside and (b) a nucleotide.

Nucleotides join together in nucleic acids by forming a bond between the phosphate group at the 5′ position of one nucleotide and the hydroxyl group on the sugar component at the 3′ position of another nucleotide (**Figure 23.13**).

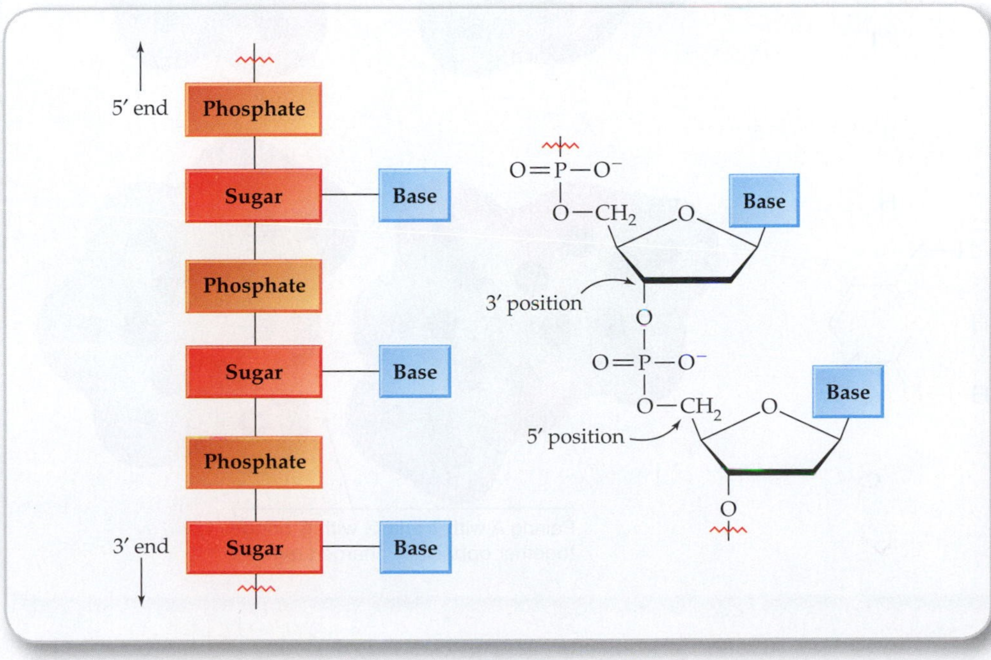

FIGURE 23.13

Generalized structure of a nucleic acid.

Just as the structure of a protein depends on the sequence of its individual amino acids, the structure of a nucleic acid depends on the sequence of its individual nucleotides. That sequence is described by starting at the 5′ phosphate end of the chain and identifying the bases in order. Abbreviations are used for each nucleotide: A for adenosine, G for guanosine, C for cytidine, T for thymidine, and U for uracil. Thus, a typical DNA sequence might be written as -T-A-G-G-C-T-.

Interestingly, molecules of DNA isolated from different tissues of the same species have the same proportions of nucleotides, but molecules from different species can have quite different proportions. For example, human DNA contains about 30% each of A and T and about 20% each of G and C, but the bacterium *Clostridium perfringens* contains about 37% each of A and T and only 13% each of G and C. Note that in both cases, the bases occur in pairs. Adenine and thymine are usually present in equal amounts, as are guanine and cytosine. Why should this be?

According to the **Watson–Crick model**, DNA consists of two polynucleotide strands coiled around each other in a *double helix* like the handrails on a spiral staircase. The sugar–phosphate backbone is on the outside of the helix, and the amine bases are on the inside, so that a base on one strand points directly in toward a base on the second strand. The two strands run in opposite directions and are held together by hydrogen bonds between pairs of bases. Adenine and thymine form two strong hydrogen bonds to each other, but not to G or C; G and C form three strong hydrogen bonds to each other, but not to A or T (**Figure 23.14**).

FIGURE 23.14

Hydrogen bonding between base pairs in the DNA double helix.

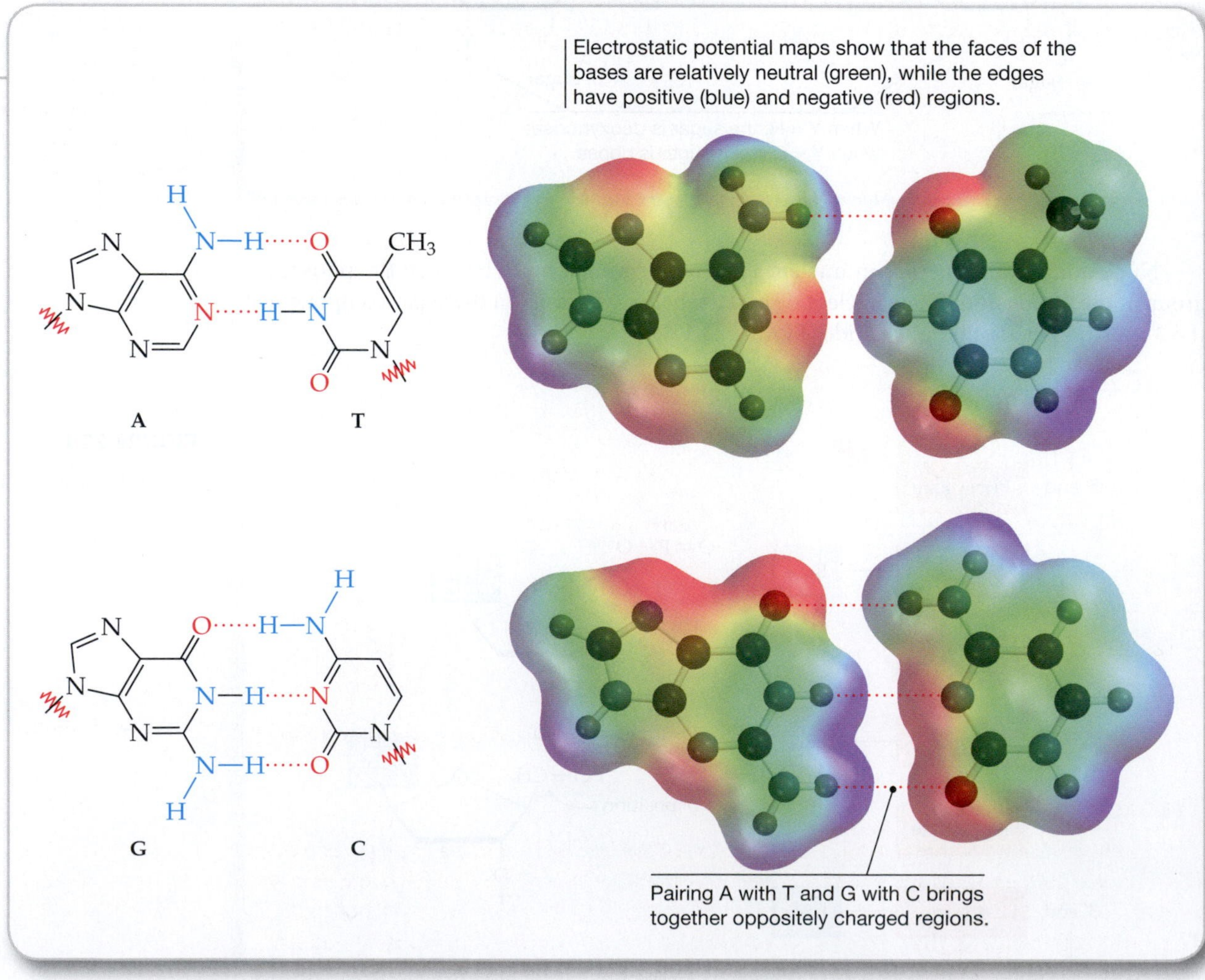

The two strands of the DNA double helix aren't identical; rather, they're complementary. Whenever a G base occurs in one strand, a C base occurs opposite it in the other strand because of hydrogen bonding. When an A base occurs in one strand, a T base occurs in the other strand. This complementary pairing of bases explains why A and T are always found in equal amounts, as are G and C. **Figure 23.15** shows how the two complementary strands coil into the double helix.

The DNA of higher organisms, both plant and animal, is found primarily in the nucleus of cells in the form of threadlike strands that are coated with proteins and wound into complex assemblies called *chromosomes*. Each chromosome contains several thousand *genes*, where a gene is a segment of a DNA chain that contains the instructions necessary to make a specific protein. By decoding the right genes at the right time, an organism uses genetic information to synthesize the thousands of proteins needed for living. DNA thus acts as the storage medium for an organism's genetic information, which RNA then reads, decodes, and uses to make proteins.

Three main processes take place in the transfer and use of genetic information. **Replication** is the process by which identical copies of DNA are made, forming additional molecules and preserving genetic information for passing on to offspring. **Transcription** is the process by which information in the DNA is transferred to and decoded by RNA. **Translation** is the process by which RNA uses the information to build proteins.

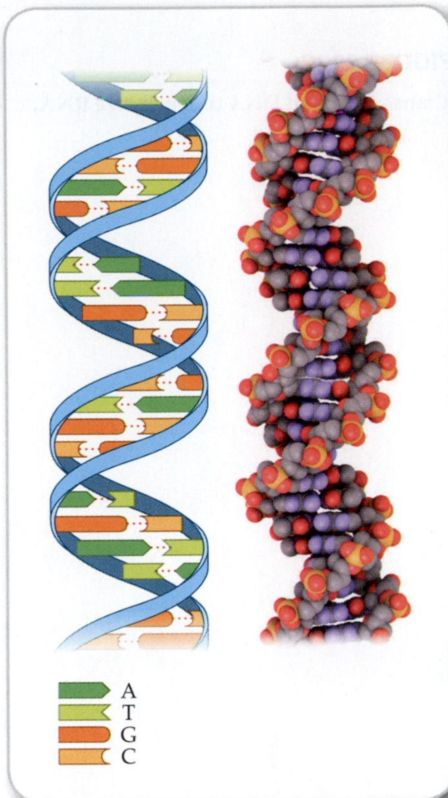

FIGURE 23.15

The DNA double helix. The coil of the sugar–phosphate backbone is visible on the outside of the DNA double helix, while the hydrogen-bonded pairs of amine bases lie flat on the inside.

Replication

DNA replication begins with a partial unwinding of the double helix. As the DNA strands separate and bases are exposed, new nucleotides line up on each strand in a complementary manner, A to T and C to G, and two new strands begin to grow. Each new strand is complementary to its old template strand, so two new, identical DNA double helixes are produced (**Figure 23.16**).

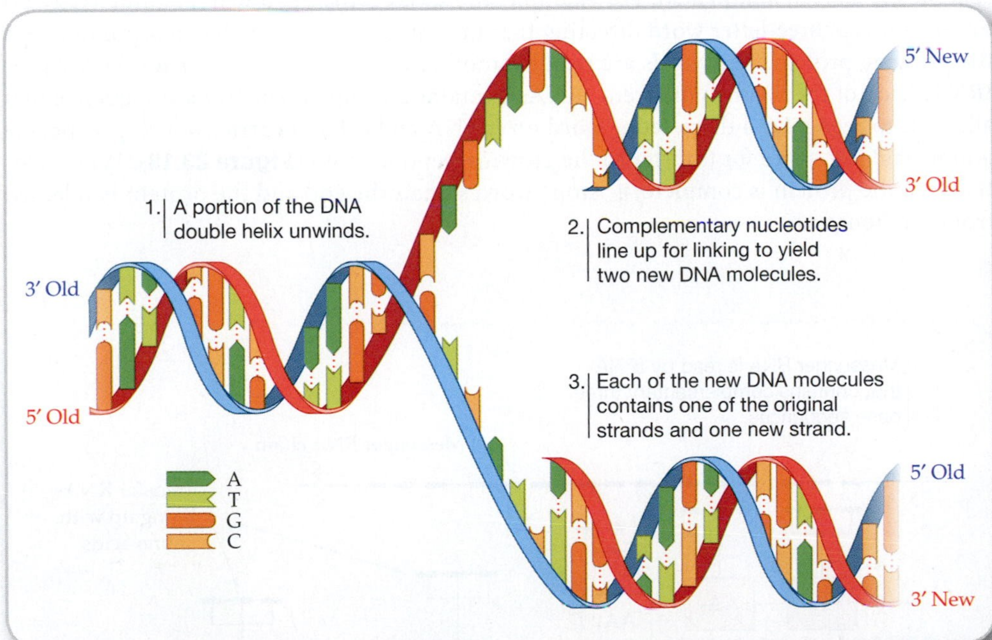

FIGURE 23.16

DNA replication.

Transcription

The genetic instructions contained in DNA are transcribed into RNA when a small portion of the DNA double helix unwinds and one of the two DNA strands acts as a template for complementary *ribonucleotides* to line up, a process similar to that of DNA replication (**Figure 23.17**). The only difference is that uracil (U) rather than thymine lines up opposite adenine. Once completed, the RNA molecule separates from the DNA template, and the DNA rewinds to its stable double-helix conformation.

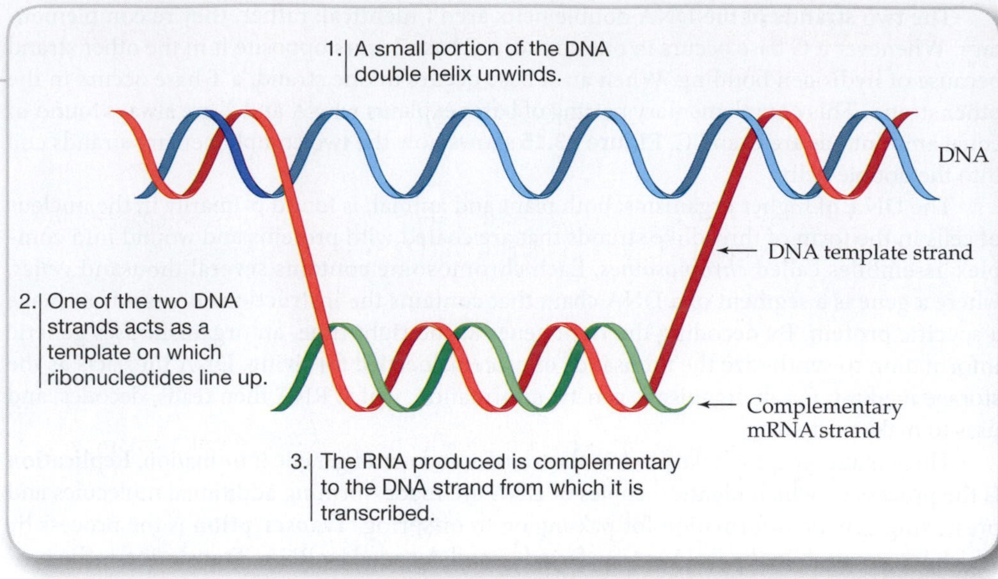

FIGURE 23.17

Transcription of DNA to synthesize RNA.

Translation

Protein biosynthesis is directed by a special kind of RNA called *messenger RNA*, or *mRNA*, and takes place on knobby protuberances within a cell called *ribosomes*. The specific ribonucleotide sequence in mRNA acts like a long coded sentence to specify the order in which different amino acid units are to be joined.

Each "word" along the mRNA chain consists of a series of three ribonucleotides that is specific for a given amino acid. For example, the series cytosine–uracil–guanine (C-U-G) on mRNA is a three-letter word directing that the amino acid leucine be incorporated into the growing protein. The words are read by another kind of RNA called *transfer RNA*, or *tRNA*. Each of the 60 or so different tRNAs contains a complementary base sequence that allows it to recognize a three-letter word on mRNA and act as a carrier to bring a specific amino acid into place for transfer to the growing peptide chain (**Figure 23.18**). When synthesis of the protein is complete, a "stop" word signals the end and the protein is released from the ribosome.

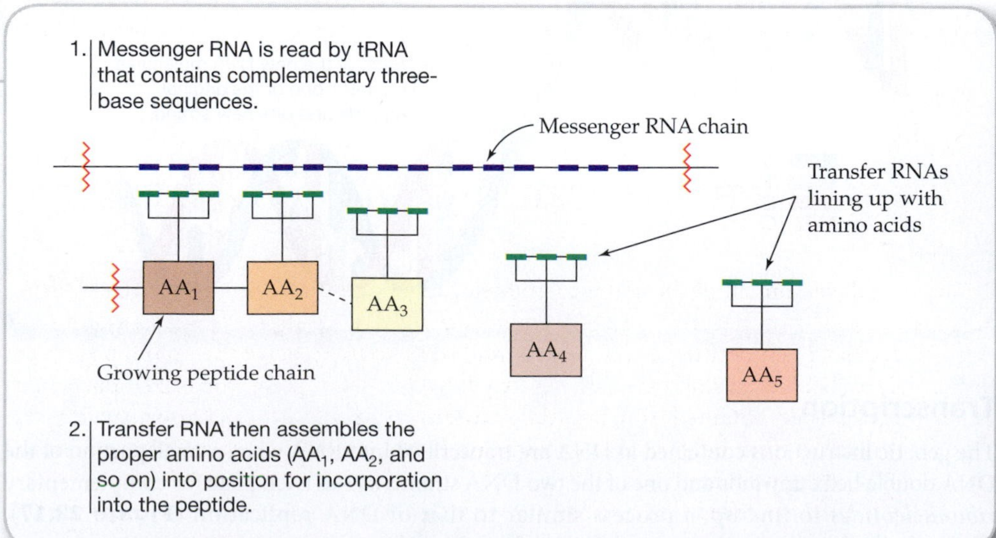

FIGURE 23.18

Protein biosynthesis.

Worked Example 23.7

DNA Complementarity

What sequence of bases on one strand of DNA is complementary to the sequence G-C-A-T-T-A-T on another strand?

STRATEGY

Because A and G form complementary pairs with T and C, respectively, go through the given sequence replacing A with T, G with C, T with A, and C with G.

SOLUTION

Original: G-C-A-T-T-A-T
Complement: C-G-T-A-A-T-A

PROBLEM 23.28 What sequence of bases on one strand of DNA is complementary to the following sequence on another strand?

G-G-C-C-C-G-T-A-A-T

PROBLEM 23.29 What RNA sequence is complementary to the following DNA sequence?

C-G-T-G-A-T-T-A-C-A

PROBLEM 23.30 From what DNA sequence was the following RNA sequence transcribed?

U-G-C-A-U-C-G-A-G-U

CONCEPTUAL PROBLEM 23.31 Identify the following bases, and tell whether each is found in DNA, RNA, or both.

(a) (b)

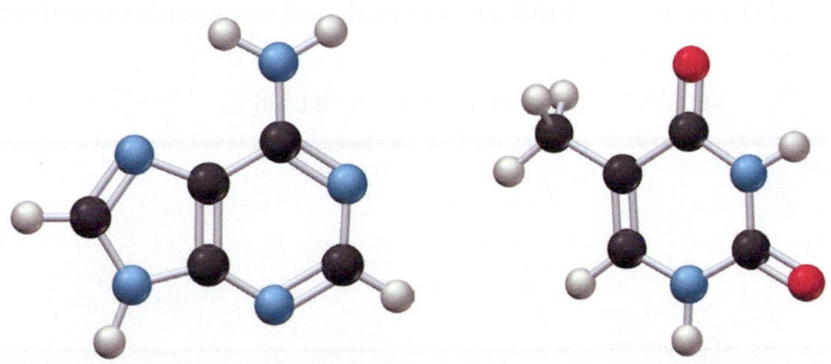

FYI WHICH IS BETTER, NATURAL OR SYNTHETIC?

▲ Whether from the laboratory or from food, the vitamin C is the same.

Prior to the development of the chemical industry in the late 19th and early 20th centuries, only substances from natural sources were available for treating our diseases, dying our clothes, cleansing and perfuming our bodies, and so forth. Extracts of the opium poppy, for instance, have been used since the 17th century for the relief of pain. The prized purple dye called *Tyrian purple,* obtained from a Middle Eastern mollusk, has been known since antiquity. Oils distilled from bergamot, sweet bay, rose, and lavender have been employed for centuries in making perfume.

Many of these so-called *natural products* were first used without any knowledge of their chemical composition. As organic chemistry developed, though, chemists learned how to work out the structures of the compounds in natural products. The disease-curing properties of limes and other citrus fruits, for example, were known for centuries, but the chemical structure of vitamin C, the active ingredient, was not determined until 1933. Today there is a revival of interest in folk remedies, and a large effort is being made to identify medicinally important chemical compounds found in plants.

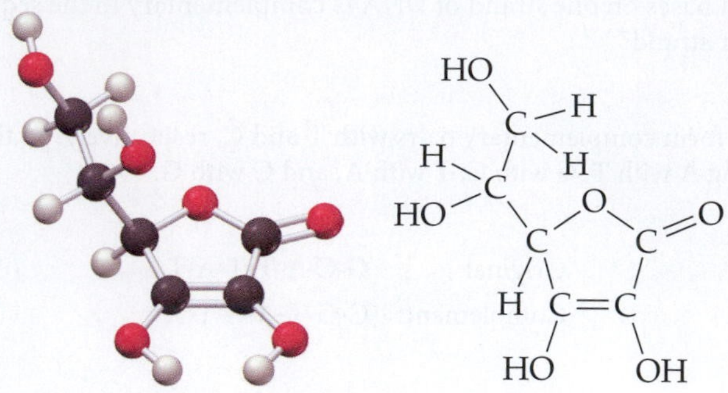

Vitamin C
(ascorbic acid)

Once a structure is known, organic chemists try to synthesize the compound in the laboratory. If the starting materials are inexpensive and the synthesis process is simple enough, it may become more economical to manufacture a compound than to isolate it from a plant or bacterium. In the case of vitamin C, a complete synthesis was achieved in 1933, and it is now much cheaper to synthesize it starting from glucose than to extract it from citrus or other natural sources. Worldwide, more than 110,000 metric tons of vitamin C are synthesized each year.

But is the "synthetic" vitamin C as good as the "natural" one? Some people still demand vitamins only from natural sources, assuming that natural is somehow better. Although eating an orange is probably better than taking a tablet, the difference lies in the many other substances present in the orange. The vitamin C itself is exactly the same, just as the NaCl produced by reacting sodium and chlorine in the laboratory is exactly the same as the NaCl found in the ocean. Natural and synthetic compounds are identical in all ways; neither is better than the other.

PROBLEM 23.32 Identify the functional groups in vitamin C.

SUMMARY

Organic chemistry is the study of carbon compounds. The more than 45 million known organic compounds can be organized into families according to the **functional groups** they contain.

The simplest compounds are the **alkanes**, which contain only carbon and hydrogen (**hydrocarbons**) and have only single bonds. **Straight-chain alkanes** have all their carbons connected in a row, **branched-chain alkanes** have a branched connection of atoms in their chain, and **cycloalkanes** have a ring of carbon atoms. Straight-chain alkanes are named in the IUPAC system by adding the family ending *-ane* to the Greek number that tells how many carbon atoms are present. Branched-chain alkanes are named by identifying the longest continuous chain of carbon atoms and then identifying what **alkyl groups** are present as branches.

Alkenes are hydrocarbons that contain a carbon–carbon double bond, and **alkynes** are hydrocarbons that contain a carbon–carbon triple bond. **Cis–trans isomers** are possible for substituted alkenes because of the lack of rotation about the carbon–carbon double bond. The cis isomer has two substituents on the same side of the double bond, and the trans isomer has two substituents on opposite sides. The most important transformations of alkenes and alkynes are **addition reactions**, in which a substance adds to the multiple bond to yield a saturated product.

Aromatic compounds, which are often represented as having a six-membered ring with three double bonds, usually undergo **substitution reactions** in which a group substitutes for one of the hydrogen atoms on the aromatic ring. **Alcohols** and **ethers** are derivatives of water in which one or both of the hydrogens are replaced by an organic substituent. Similarly, **amines** are derivatives of ammonia in which one or more of the ammonia hydrogens are replaced by an organic substituent. Amines are bases and can be protonated by acids to yield ammonium salts.

Compounds that contain a **carbonyl group**, C=O, can be classified into two categories. In **aldehydes** and **ketones**, the carbonyl-group carbon is bonded to atoms (H and C) that don't attract electrons strongly. In **carboxylic acids**, **esters**, and **amides**, the carbonyl-group carbon is bonded to an electronegative atom (O or N) that *does* attract electrons strongly. As a result, these three families of compounds undergo **substitution reactions**, in which a group –Y substitutes for the –OH, –OC, or –N group of the carbonyl reactant.

Proteins are large biomolecules consisting of **α-amino acids** linked together by amide, or **peptide bonds**. Twenty amino acids are commonly found in proteins. **Carbohydrates** are polyhydroxy aldehydes and ketones. Simple carbohydrates, or **monosaccharides**, can't be hydrolyzed to smaller molecules; complex carbohydrates such as starch and cellulose contain many simple sugars linked together.

Lipids are the naturally occurring organic molecules that dissolve in a nonpolar solvent. Animal fats and vegetable oils are **triacylglycerols**—esters of glycerol with three long-chain **fatty acids**. The fatty acids are unbranched, have an even number of carbon atoms, and may be either saturated or unsaturated.

Deoxyribonucleic acid (DNA) and **ribonucleic acid (RNA)** are the chemical carriers of an organism's genetic information. They are made up of **nucleotides**, linked together to form a long chain. Each nucleotide consists of a cyclic amine base linked to C1 of a sugar, with the sugar in turn linked to phosphoric acid. The sugar component in RNA is ribose; the sugar in DNA is 2-deoxyribose. The bases in DNA are adenine (A), guanine (G), cytosine (C), and thymine (T); the bases in RNA are adenine, guanine, cytosine, and uracil (U). Molecules of DNA consist of two complementary nucleotide strands held together by hydrogen bonds and coiled into a double helix.

Three processes take place in the transfer of genetic information. **Replication** is the process by which identical copies of DNA are made. **Transcription** is the process by which messenger RNA is produced. **Translation** is the process by which mRNA directs protein synthesis.

KEY WORDS

addition reaction *937*
alcohol *943*
aldehyde *946*
alkane *926*
alkene *935*
alkyl group *933*
alkyne *935*
alpha- (α-) amino acid *953*
amide *949*
amine *944*
anabolism *951*
aromatic *941*
biochemistry *926*
branched-chain alkane *927*
carbohydrate *956*
carbonyl group *948*
carboxylate anion *948*
carboxylic acid *947*
catabolism *951*
cis–trans isomers *936*
condensed structure *928*
cycloalkane *939*
deoxyribonucleic acid (DNA) *960*
enzyme *952*
ester *948*
ether *944*
fatty acid *958*
functional group *930*
hydrocarbon *926*
hydrogenation *938*
ketone *946*
lipid *958*
metabolism *951*
monosaccharide *956*
nucleic acid *960*
nucleoside *960*
nucleotide *960*
organic chemistry *926*
peptide *953*
peptide bond *952*
polysaccharide *956*
protein *952*
replication *963*
ribonucleic acid (RNA) *960*
saturated *935*
straight-chain alkane *927*
substitution reaction *942*
transcription *963*
translation *963*
triacylglycerol *958*
unsaturated *935*
Watson–Crick model *962*

LEARNING OUTCOMES

- Draw organic molecules as condensed structures. (Section 23.1) ***Problems 1–4, 8, 11, 33***
- Identify and name the functional groups in a molecule. (Section 23.2) ***Problems 5, 6, 34, 40–45, 106***
- Write and interpret the IUPAC name of an alkane. (Section 23.3) ***Problems 7–9, 35, 46–52, 98, 99***
- Write and interpret the IUPAC name of an alkene or alkyne, and describe a typical addition reaction. (Section 23.4) ***Problems 10–13, 35, 55–62***
- Write IUPAC names for cycloalkanes, and draw their structures. (Section 23.5) ***Problems 14, 15, 53***
- Write IUPAC names for aromatic compounds, and describe a typical aromatic substitution reaction. (Section 23.6) ***Problems 16–18, 63***
- Draw and name alcohols and amines. (Section 23.7) ***Problems 64, 65***
- Identify the major classes of carbonyl compounds and the typical reactions of each. (Section 23.8) ***Problems 20–22, 66–73, 100, 105***
- Identify amino acids, show how they join together to form peptides, and tell how to name small peptides using the three-letter shorthand notation. (Section 23.10) ***Problems 23–25, 36, 38, 74–79, 102***
- Classify a monosaccharide as an aldose or ketose, and draw a typical open-chain structure. (Section 23.11) ***Problems 26, 80–83***
- Recognize triacylglycerols, and draw a typical structure. (Section 23.12) ***Problems 27, 84–89, 107***
- Identify the bases found in nucleic acids, write the DNA sequence that is complementary to a specified strand, and write the RNA sequence transcribed from it. (Section 23.13) ***Problems 28–31, 90–97, 104***

CONCEPTUAL PROBLEMS

Problems 23.1–23.32 appear within the chapter.

23.33 Convert the following model into a condensed structure, and draw the structures of two isomeric compounds.

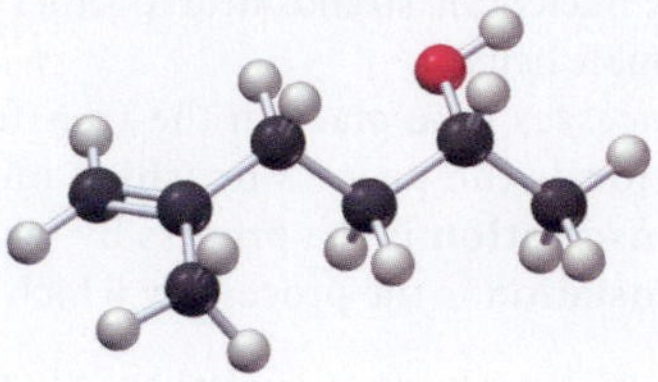

23.34 Identify the functional groups in each of the following compounds:

(a)

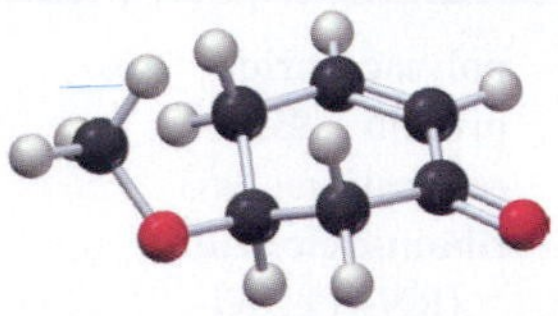

(b)

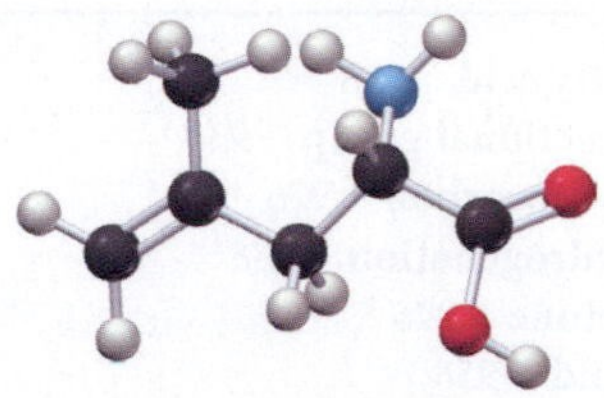

23.35 Give systematic names for the following compounds:

(a)

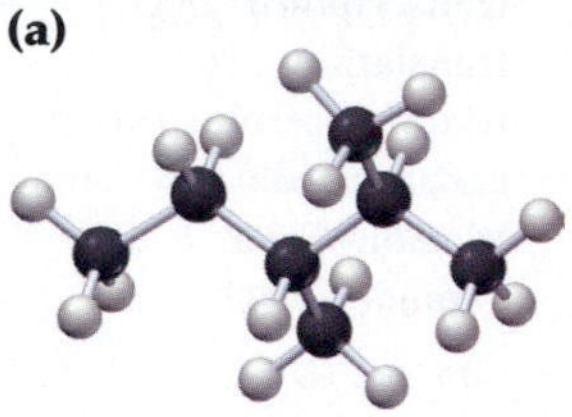

(b)

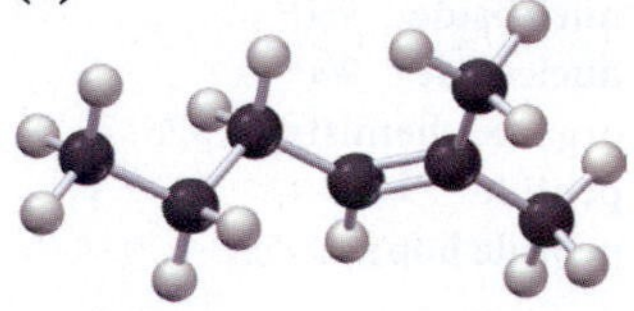

23.36 Identify the following amino acids (yellow = S):

(a)

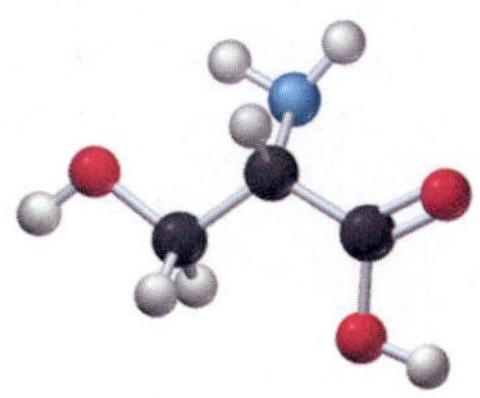

(b)

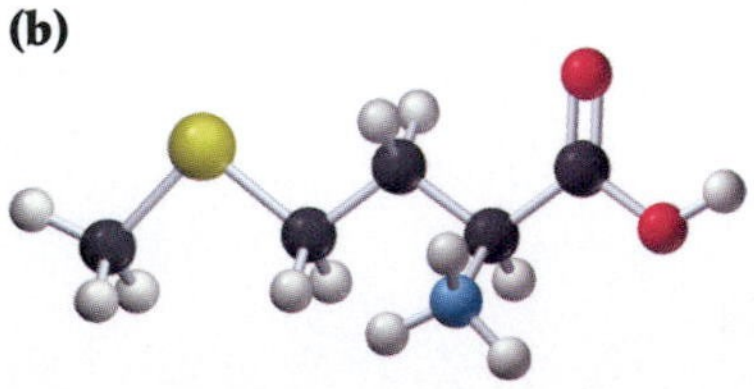

23.37 Draw three resonance forms for naphthalene, showing the positions of the double bonds.

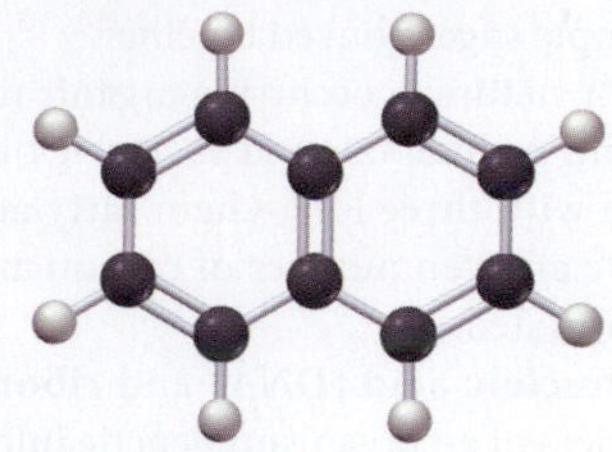

23.38 Identify the following dipeptide:

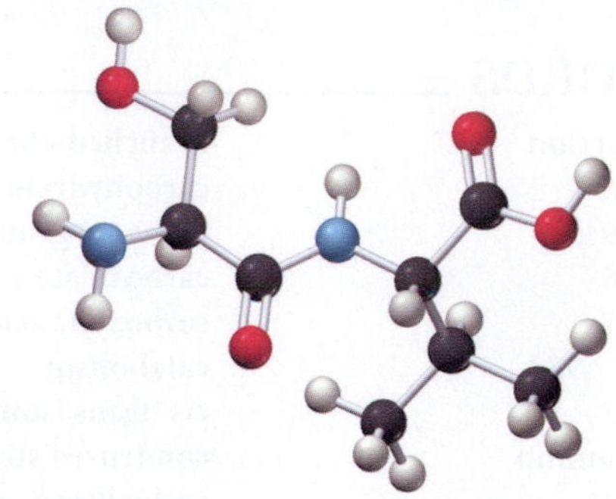

23.39 Identify the following amine bases found in nucleic acids:

(a)

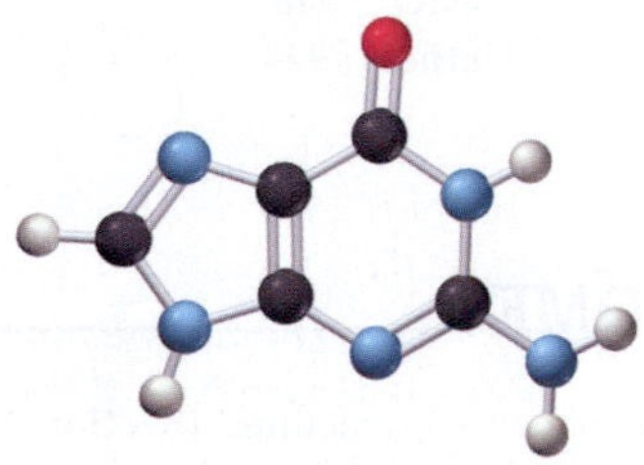

(b)

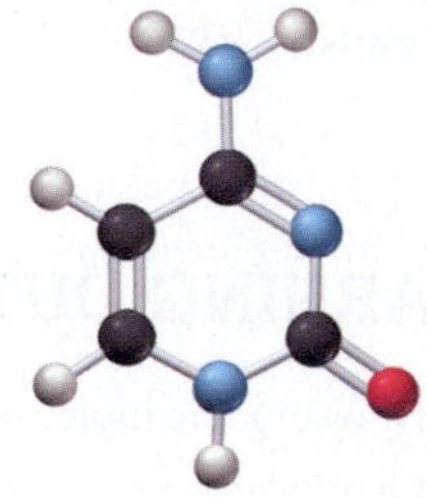

SECTION PROBLEMS

Functional Groups and Isomers (Sections 23.2–23.3)

23.40 What are functional groups, and why are they important?

23.41 Describe the structure of the functional group in each of the following families:

(a) Alkene **(b)** Alcohol

(c) Ester **(d)** Amine

23.42 Propose structures for molecules that meet the following descriptions:

(a) A ketone with the formula $C_5H_{10}O$

(b) An ester with the formula $C_6H_{12}O_2$

(c) A compound with formula $C_2H_5NO_2$ that is both an amine and a carboxylic acid

23.43 Write structures for each of the following molecular formulas. You may have to use rings and/or multiple bonds in some instances.

(a) C_2H_7N **(b)** C_4H_8

(c) C_2H_4O **(d)** CH_2O_2

23.44 Draw the structures of the three isomers with the formula C_3H_8O.

23.45 Identify the functional groups in cocaine.

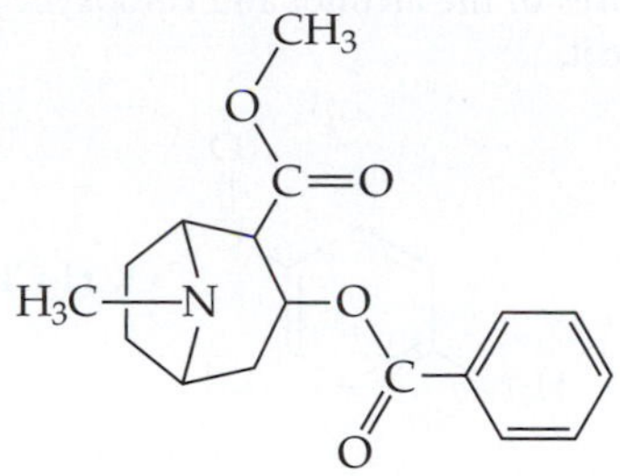

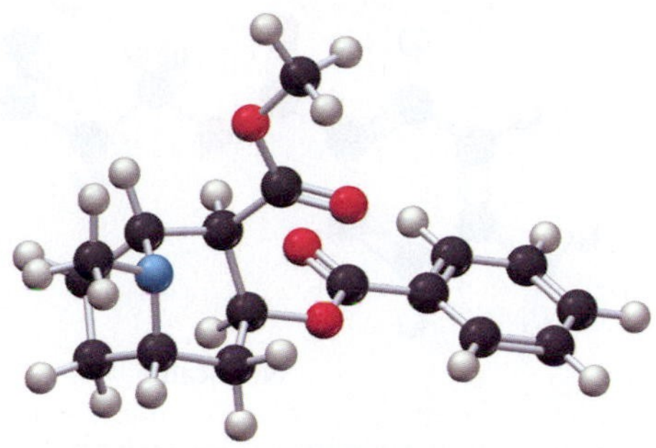

Cocaine

Alkanes and Their Names (Sections 23.1, 23.3)

23.46 What is the difference between a straight-chain alkane and a branched-chain alkane?

23.47 What is the difference between an alkane and an alkyl group?

23.48 If someone reported the preparation of a compound with the formula C_3H_9, most chemists would be skeptical. Why?

23.49 What is wrong with each of the following structures?

(a) $CH_3{=}CHCH_2CH_2OH$

(b) $CH_3CH_2CH{=}\overset{O}{\overset{\|}{C}}CH_3$

(c) $CH_3CH_2C{\equiv}CH_2CH_3$

23.50 What are the IUPAC names of the following alkanes?

(a) $CH_3CH_2CH_2CH_2\overset{CH_2CH_3}{\overset{|}{C}}H\underset{CH_3}{\underset{|}{C}}HCH_2CH_3$

(b) $CH_3CH_2CH_2\overset{CH_3CHCH_3}{\overset{|}{C}}HCH_2\underset{CH_3}{\underset{|}{C}}HCH_3$

(c) $CH_3\overset{CH_3}{\overset{|}{\underset{CH_3}{\underset{|}{C}}}}CH_2CH_2CH_2\overset{CH_3}{\overset{|}{C}}HCH_3$

(d) $CH_3CH_2CH_2CH_2\overset{CH_2CH_2CH_2CH_3}{\overset{|}{\underset{CH_2CH_3}{\underset{|}{C}}}}CH_3$

23.51 The following compound, known commonly as isooctane, is used as a reference substance for determining the octane rating of gasoline. What is the IUPAC name of isooctane?

$$CH_3\overset{CH_3}{\overset{|}{\underset{CH_3}{\underset{|}{C}}}}CH_2\overset{CH_3}{\overset{|}{C}}HCH_3$$

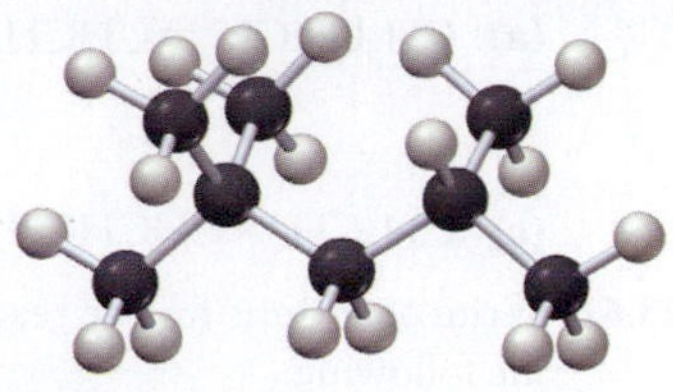

Isooctane

23.52 Write condensed structures for each of the following compounds:

(a) 3-Ethylhexane

(b) 2,2,3-Trimethylpentane

(c) 3-Ethyl-3,4-dimethylheptane

(d) 5-Isopropyl-2-methyloctane

23.53 Draw structures corresponding to the following IUPAC names:

(a) Cyclooctane

(b) 1,1-Dimethylcyclopentane

(c) 1,2,3,4-Tetramethylcyclobutane

(d) 4-Ethyl-1,1-dimethylcyclohexane

Alkenes, Alkynes, and Aromatic Compounds (Sections 23.4, 23.6)

23.54 Not all compounds that smell nice are called "aromatic," and not all compounds called "aromatic" smell nice. Explain.

23.55 What is meant by the term *addition reaction*?

23.56 Write structural formulas for compounds that meet the following descriptions:

(a) An alkene with five carbons

(b) An alkyne with four carbons

(c) A substituted aromatic compound with eight carbons

23.57 How many dienes (compounds with two double bonds) are there with the formula C_5H_8? Draw structures of as many as you can.

23.58 Give IUPAC names for the following compounds:

(a) $CH_3\overset{CH_3}{\overset{|}{C}}HCH{=}CHCH_3$ **(b)** $CH_3CH_2\overset{CH{=}CH_2}{\overset{|}{C}}HCH_3$

(c)

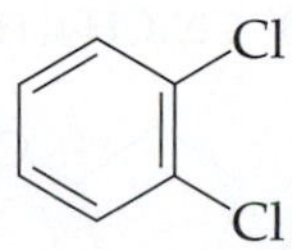

(d) $CH_3CH{=}\overset{CH_3}{\overset{|}{C}}CH_3$

(e) $CH_3CH_2C{\equiv}CCH_2CH_2\overset{CH_3}{\overset{|}{C}}HCH_3$

23.59 Draw structures corresponding to the following IUPAC names:

(a) *cis*-2-Hexene **(b)** 2-Methyl-3-hexene

(c) 2-Methyl-1,3-butadiene

23.60 Which of the following compounds are capable of cis–trans isomerism?

(a) 1-Hexene **(b)** 2-Hexene **(c)** 3-Hexene

23.61 Which of the following compounds are capable of cis–trans isomerism?

(a) $CH_3CH(CH_3)CH{=}CHCH_3$ (b) $CH_3CH_2CH(CH{=}CH_2)CH_3$

(c) $CH_3CH{=}CHCH(Cl)CH_2CH_3$

23.62 Write equations for the reaction of 2,3-dimethyl-2-butene with the following:

(a) H_2 and Pd catalyst
(b) Br_2
(c) H_2O and H_2SO_4 catalyst

23.63 Write equations for the reaction of *p*-dichlorobenzene with the following:

(a) Br_2 and $FeBr_3$ catalyst
(b) HNO_3 and H_2SO_4 catalyst
(c) Cl_2 and $FeCl_3$ catalyst

Alcohols, Ethers, Amines, and Carbonyl Compounds (Sections 23.7–23.8)

23.64 Draw structures corresponding to the following names:

(a) 2,4-Dimethyl-2-pentanol
(b) 2,2-Dimethylcyclohexanol
(c) 5,5-Diethyl-1-heptanol
(d) 3-Ethyl-3-hexanol

23.65 Draw structures corresponding to the following names:

(a) Propylamine
(b) Diethylamine
(c) *N*-Methylpropylamine

23.66 What is the structural difference between an aldehyde and a ketone?

23.67 What general kind of reaction do carboxylic acids, esters, and amides undergo?

23.68 Identify the kinds of carbonyl groups in the following molecules (aldehyde, amide, ester, or ketone):

(a) 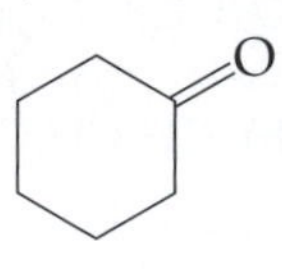

(b) $CH_3CH_2CH_2CHO$

(c) $CH_3CH(CH_3)CH_2COCH_3$

(d) 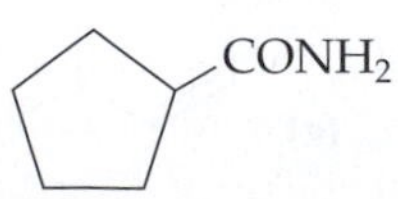

(e) $CH_3CH(CH_3)CH_2COOCH_3$

23.69 Draw and name compounds that meet the following descriptions:

(a) Three different amides with the formula $C_5H_{11}NO$
(b) Three different esters with the formula $C_6H_{12}O_2$

23.70 Draw structures corresponding to the following IUPAC names:

(a) Methyl pentanoate
(b) Isopropyl 2-methylbutanoate
(c) Cyclohexyl acetate

23.71 Draw structures corresponding to the following IUPAC names:

(a) 3-Methylpentanamide
(b) *N*-Phenylacetamide
(c) *N*-Ethyl-*N*-methylbenzamide

23.72 Novocaine, a local anesthetic, has the following structure. Identify the functional groups present in novocaine, and show the structures of the alcohol and carboxylic acid you would use to prepare it.

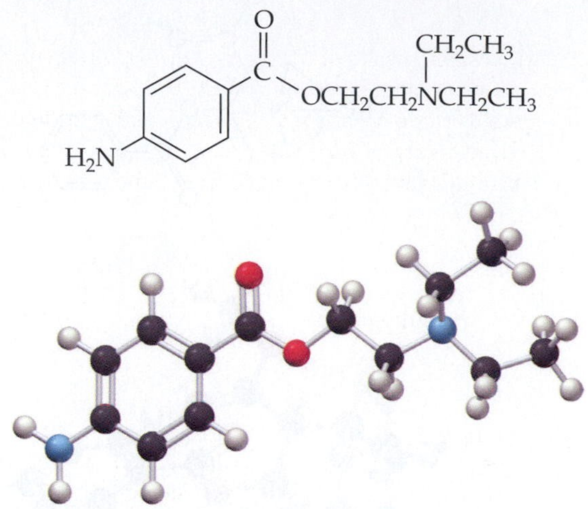

Novocaine

23.73 Ordinary soap is a mixture of the sodium or potassium salts of long-chain carboxylic acids that arise from base-catalyzed hydrolysis (saponification) of animal fat. Draw the structures of soap molecules produced in the following reaction:

$$\begin{array}{l} CH_2OC(O)(CH_2)_{14}CH_3 \\ | \\ CHOC(O)(CH_2)_7CH{=}CH(CH_2)_7CH_3 \ + \ 3\ KOH \longrightarrow \\ | \\ CH_2OC(O)(CH_2)_{16}CH_3 \end{array}$$

A fat

Amino Acids, Peptides, and Proteins (Section 23.10)

23.74 What amino acids do the following abbreviations stand for?

(a) Ser (b) Thr (c) Pro
(d) Phe (e) Cys

23.75 Name and draw the structures of amino acids that fit the following descriptions:

(a) Contains an isopropyl group
(b) Contains an alcohol group
(c) Contains a thiol (—SH) group
(d) Contains an aromatic ring

23.76 Identify the amino acids present in the following hexapeptide:

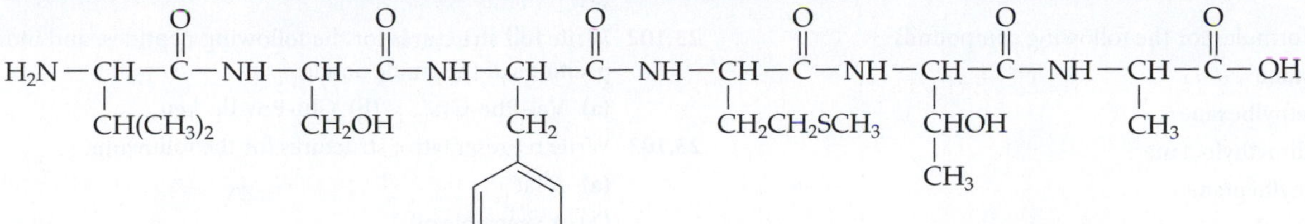

23.77 *Aspartame*, marketed for use as a nonnutritive sweetener under such trade names as Equal, NutraSweet, and Canderel, is the methyl ester of a simple dipeptide. Identify the two amino acids present in aspartame, and show all the products of digestion, assuming that both amide and ester bonds are hydrolyzed in the stomach.

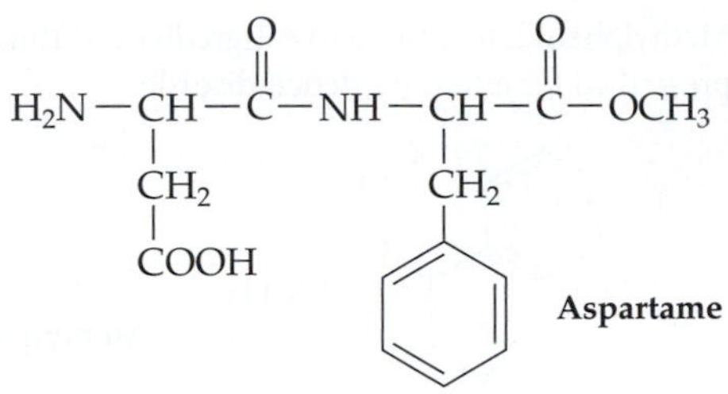

Aspartame

23.78 Use the three-letter abbreviations to name all tripeptides containing methionine, isoleucine, and lysine.

23.79 How many tetrapeptides containing alanine, serine, leucine, and glutamic acid do you think there are? Use the three-letter abbreviations to name three.

Carbohydrates (Section 23.11)

23.80 What is the structural difference between an aldose and a ketose?

23.81 Classify each of the following carbohydrates by indicating the nature of its carbonyl group and the number of carbon atoms present. For example, glucose is an aldohexose.

(a) 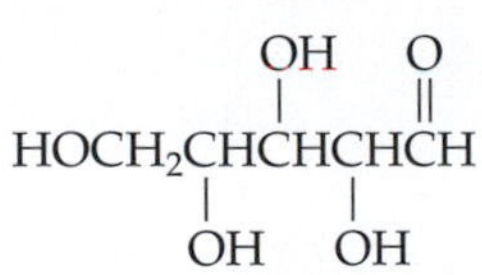

(b) 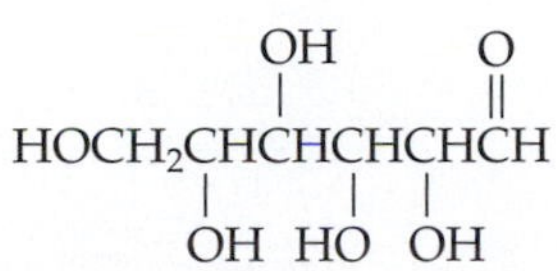

(c) 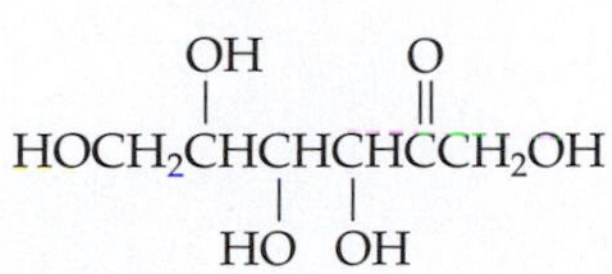

23.82 Write the open-chain structure of a ketotetrose.

23.83 Write the open-chain structure of a four-carbon deoxy sugar.

Lipids (Section 23.12)

23.84 What is a fatty acid?

23.85 What does it mean to say that fats and oils are triacylglycerols?

23.86 Draw the structure of glycerol myristate, a fat made from glycerol and three myristic acid molecules (see Table 23.3).

23.87 Spermaceti, a fragrant substance isolated from sperm whales, was a common ingredient in cosmetics until its use was banned in 1976 to protect the whales from extinction. Chemically, spermaceti is cetyl palmitate, the ester of palmitic acid (see Table 23.3) with cetyl alcohol (the straight-chain C_{16} alcohol). Show the structure of spermaceti.

23.88 There are two isomeric fat molecules whose components are glycerol, one palmitic acid, and two stearic acids (see Table 23.3). Draw the structures of both, and explain how they differ.

23.89 Draw the structures of all products you would obtain by reaction of the following lipid with aqueous KOH. What are the names of the products?

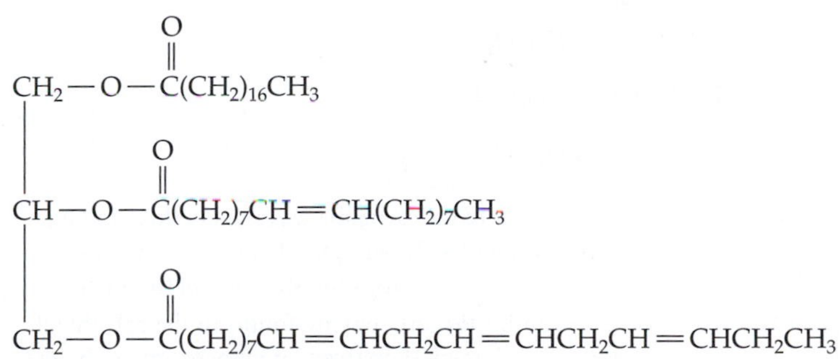

Nucleic Acids (Section 23.13)

23.90 What is a nucleotide, and what three kinds of components does it contain?

23.91 What are the names of the sugars in DNA and RNA, and how do they differ?

23.92 Show by drawing structures how the phosphate and sugar components of a nucleic acid are joined.

23.93 Show by drawing structures how the sugar and amine base components of a nucleic acid are joined.

23.94 If the sequence T-A-C-C-G-A appeared on one strand of DNA, what sequence would appear opposite it on the other strand?

23.95 What sequence would appear on the mRNA molecule transcribed from the DNA in Problem 23.94?

23.96 Human insulin is composed of two peptide chains. One chain contains 21 amino acids, and the other contains 30 amino acids. How many nucleotides are present in the DNA to code for each chain?

23.97 The DNA from sea urchins contains about 32% A and about 18% G. What percentages of T and C would you expect in sea urchin DNA? Explain.

CHAPTER PROBLEMS

23.98 Draw structural formulas for the following compounds:

(a) 2-Methylheptane

(b) 4-Ethyl-2-methylhexane

(c) 4-Ethyl-3,4-dimethyloctane

(d) 2,4,4-Trimethylheptane

(e) 1,1-Dimethylcyclopentane

(f) 4-Isopropyl-3-methylheptane

23.99 Give IUPAC names for the following alkanes:

(a) 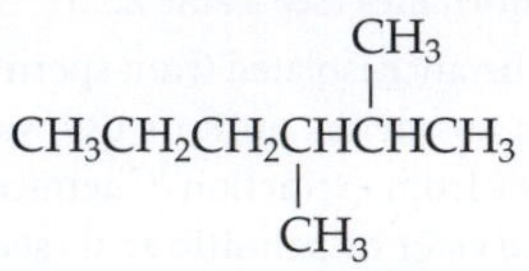

(b) 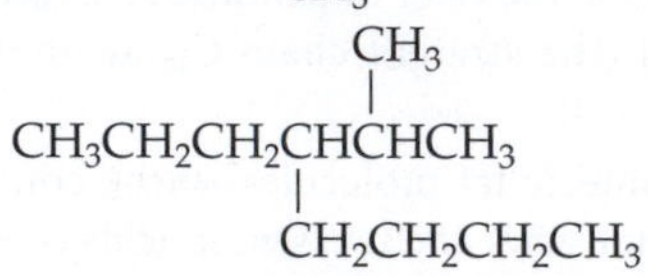

(c) 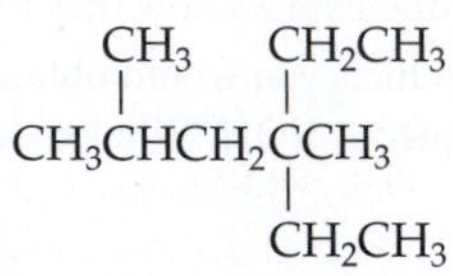

(d) 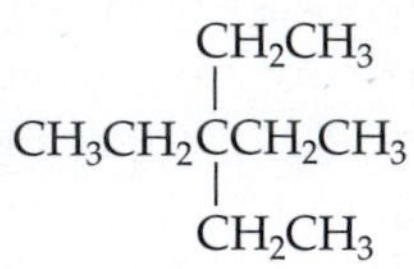

23.100 One of the constituents of the carnauba wax used in floor and furniture polish is an ester of a C_{32} straight-chain alcohol with a C_{20} straight-chain carboxylic acid. Draw the structure of this ester.

23.101 Cytochrome *c* is an important enzyme found in the cells of all aerobic organisms. Elemental analysis of cytochrome *c* shows that it contains 0.43% iron. What is the minimum molecular weight of this enzyme?

23.102 Write full structures for the following peptides, and indicate the positions of the amide bonds:

(a) Val-Phe-Cys **(b)** Glu-Pro-Ile-Leu

23.103 Write representative structures for the following:

(a) A fat

(b) A vegetable oil

(c) An aldotetrose

23.104 What DNA sequence is complementary to the following sequence?

A-G-T-T-C-A-T-C-G

23.105 Jojoba wax, used in candles and cosmetics, is partially composed of the ester of stearic acid and a straight-chain C_{22} alcohol. Draw the structure of this ester.

23.106 Methylphenidate is the active ingredient in Ritalin, a medication prescribed for attention-deficit disorder.

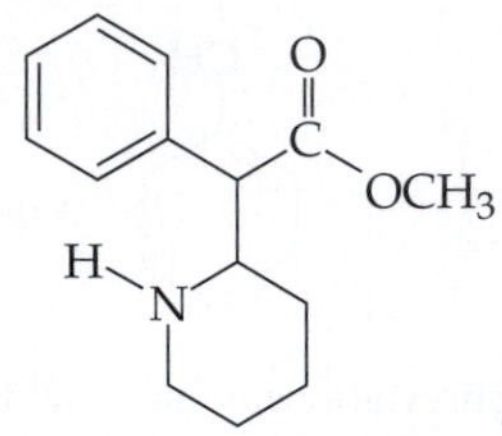

Methylphenidate

(a) What functional groups are present in methylphenidate?

(b) Draw the products you would obtain by treating methylphenidate with aqueous base.

23.107 Elaidic acid, a component of so-called trans fats, is the trans isomer of oleic acid (Table 23.3). Draw both isomers using condensed structures.

MULTICONCEPT PROBLEMS

23.108 Fumaric acid is an organic substance widely used as a food additive. Its elemental composition is 41.4% C, 3.5% H, and 55.1% O. A solution made by dissolving 0.1500 g of fumaric acid in water and diluting to a volume of 100.0 mL gave rise to an osmotic pressure of 240.3 mm Hg at 298 K. On titration of a sample weighing 0.573 g, 94.1 mL of 0.105 M NaOH was required to reach an equivalence point. Fumaric acid reacts with 1 mol of H_2 to give a hydrogenation product.

(a) What is the empirical formula of fumaric acid?

(b) What is the molecular weight of fumaric acid?

(c) Draw three possible structures for fumaric acid.

(d) If fumaric acid contains a trans double bond, which of your structures is correct?

23.109 The protonated form of a neutral amino acid such as alanine is a diprotic acid, H_2A^+, with two acid dissociation constants, one for the $-NH_3^+$ group and one for the $-CO_2H$ group.

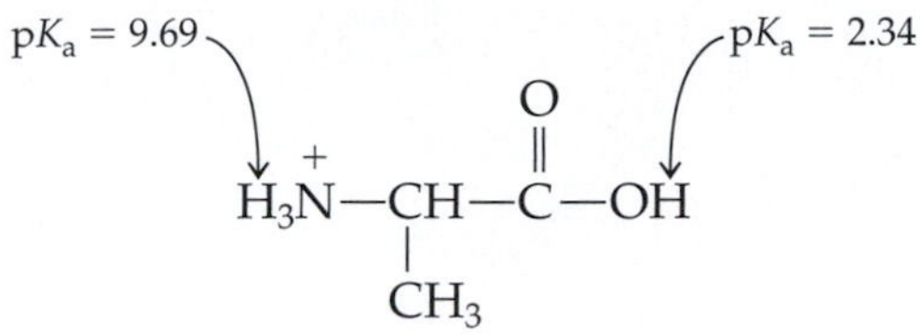

Alanine (protonated)

(a) Which group is more acidic, the $-NH_3^+$ or the $-CO_2H$?

(b) What percentage of each form—protonated (H_2A^+), neutral (HA), and deprotonated (A^-)—is present in aqueous solution at pH = 4.00?

(c) What percentage of each form—protonated (H_2A^+), neutral (HA), and deprotonated (A^-)—is present in aqueous solution at pH = 8.50?

(d) At what pH is the neutral form present in maximum amount?

23.110 The relative amount of unsaturation in a fat or oil is expressed as an *iodine number*. Olive oil, for instance, is highly unsaturated and has an iodine number of 172, while butter is much less unsaturated and has an iodine number of 37. Defined as the number of grams of I_2 absorbed per 100 grams of fat, the iodine number is based on the fact that the carbon–carbon double bonds in fats and oils undergo an addition reaction with I_2. The larger the number of double bonds, the larger the amount of I_2 that reacts.

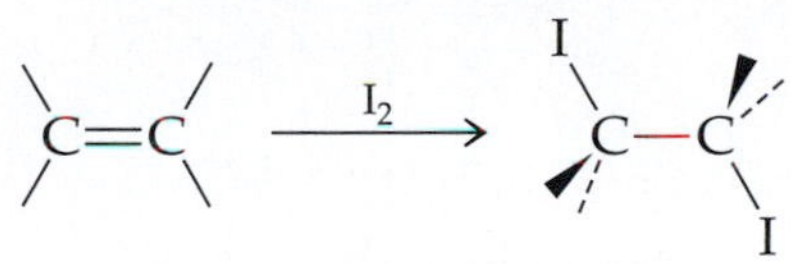

To determine an iodine number, a known amount of fat is treated with a known amount of I_2. When the addition reaction is complete, the amount of excess I_2 remaining is determined by titration with $Na_2S_2O_3$ according to the equation

$$2\,Na_2S_2O_3(aq) + I_2(aq) \longrightarrow Na_2S_4O_6(aq) + 2\,NaI(aq)$$

Knowing both the amount of I_2 originally added and the amount remaining after reaction, the iodine number can be calculated.

Assume that 0.500 g of human milk fat is allowed to react with 25.0 mL of 0.200 M I_2 solution and that 81.99 mL of 0.100 M $Na_2S_2O_3$ is required for complete reaction with the excess I_2.

(a) What amount (in grams) of I_2 was added initially?

(b) How many grams of I_2 reacted with the milk fat, and how many grams were in excess?

(c) What is the iodine number of human milk fat?

(d) Assuming a molecular weight of 800, how many double bonds does an average molecule of milk fat contain?

APPENDIX A

Mathematical Operations

A.1 SCIENTIFIC NOTATION

The numbers that you encounter in chemistry are often either very large or very small. For example, there are about 33,000,000,000,000,000,000,000 H_2O molecules in 1.0 mL of water, and the distance between the H and O atoms in an H_2O molecule is 0.000 000 000 095 7 m. These quantities are more conveniently written in scientific notation as 3.3×10^{22} molecules and 9.57×10^{-11} m, respectively. In scientific notation, numbers are written in the exponential format $A \times 10^n$, where A is a number between 1 and 10, and the exponent n is a positive or negative integer.

How do you convert a number from ordinary notation to scientific notation? If the number is greater than or equal to 10, shift the decimal point to the *left* by n places until you obtain a number between 1 and 10. Then, multiply the result by 10^n. For example, the number 8137.6 is written in scientific notation as 8.1376×10^3:

$$8137.6 = 8.1376 \times 10^3$$

Number of places decimal point was shifted to the left

Shift decimal point to the left by 3 places to get a number between 1 and 10.

When you shift the decimal point to the left by three places, you are in effect dividing the number by $10 \times 10 \times 10 = 1000 = 10^3$. Therefore, you must multiply the result by 10^3 so that the value of the number is unchanged.

To convert a number less than 1 to scientific notation, shift the decimal point to the *right* by n places until you obtain a number between 1 and 10. Then, multiply the result by 10^{-n}. For example, the number 0.012 is written in scientific notation as 1.2×10^{-2}:

$$0.012 = 1.2 \times 10^{-2}$$

Number of places decimal point was shifted to the right

Shift decimal point to the right by 2 places to get a number between 1 and 10.

When you shift the decimal point to the right by two places, you are in effect multiplying the number by $10 \times 10 = 100 = 10^2$. Therefore, you must multiply the result by 10^{-2} so that the value of the number is unchanged. ($10^2 \times 10^{-2} = 10^0 = 1$.)

The following table gives some additional examples. To convert from scientific notation to ordinary notation, simply reverse the preceding process. Thus, to write the number 5.84×10^4 in ordinary notation, drop the factor of 10^4 and move the decimal point by 4 places to the *right* ($5.84 \times 10^4 = 58{,}400$). To write the number 3.5×10^{-1} in ordinary notation, drop the factor of 10^{-1} and move the decimal point by 1 place to the *left* ($3.5 \times 10^{-1} = 0.35$). Note that you don't need scientific notation for numbers between 1 and 10 because $10^0 = 1$.

Number	Scientific Notation
58,400	5.84×10^4
0.35	3.5×10^{-1}
7.296	$7.296 \times 10^0 = 7.296$

Addition and Subtraction

To add or subtract two numbers expressed in scientific notation, both numbers must have the same exponent. Thus, to add 7.16×10^3 and 1.32×10^2, first write the latter number as 0.132×10^3 and then add:

$$\begin{array}{r} 7.16 \times 10^3 \\ +0.132 \times 10^3 \\ \hline 7.29 \times 10^3 \end{array}$$

The answer has three significant figures. (Significant figures are discussed in Section 0.9.) Alternatively, you can write the first number as 71.6×10^2 and then add:

$$\begin{array}{r} 71.6 \times 10^2 \\ +1.32 \times 10^2 \\ \hline 72.9 \times 10^2 \end{array} = 7.29 \times 10^3$$

Multiplication and Division

To multiply two numbers expressed in scientific notation, multiply the factors in front of the powers of 10 and then add the exponents:

$$(A \times 10^n)(B \times 10^m) = AB \times 10^{n+m}$$

For example,

$$(2.5 \times 10^4)(4.7 \times 10^7) = (2.5)(4.7) \times 10^{4+7} = 12 \times 10^{11} = 1.2 \times 10^{12}$$
$$(3.46 \times 10^5)(2.2 \times 10^{-2}) = (3.46)(2.2) \times 10^{5+(-2)} = 7.6 \times 10^3$$

Both answers have two significant figures.

To divide two numbers expressed in scientific notation, divide the factors in front of the powers of 10 and then subtract the exponent in the denominator from the exponent in the numerator:

$$\frac{A \times 10^n}{B \times 10^m} = \frac{A}{B} \times 10^{n-m}$$

For example,

$$\frac{3 \times 10^6}{7.2 \times 10^2} = \frac{3}{7.2} \times 10^{6-2} = 0.4 \times 10^4 = 4 \times 10^3 \quad \text{(1 significant figure)}$$

$$\frac{7.50 \times 10^{-5}}{2.5 \times 10^{-7}} = \frac{7.50}{2.5} \times 10^{-5-(-7)} = 3.0 \times 10^2 \quad \text{(2 significant figures)}$$

Powers and Roots

To raise a number $A \times 10^n$ to a power m, raise the factor A to the power m and then multiply the exponent n by the power m:

$$(A \times 10^n)^m = A^m \times 10^{n \times m}$$

For example, 3.6×10^2 raised to the 3rd power is 4.7×10^7:

$$(3.6 \times 10^2)^3 = (3.6)^3 \times 10^{2 \times 3} = 47 \times 10^6 = 4.7 \times 10^7 \quad \text{(2 significant figures)}$$

To take the mth root of a number $A \times 10^n$, raise the number to the power $1/m$. That is, raise factor A to the power $1/m$ and then divide the exponent n by the root m:

$$\sqrt[m]{A \times 10^n} = (A \times 10^n)^{1/m} = A^{1/m} \times 10^{n/m}$$

For example, the square root of 9.0×10^8 is 3.0×10^4:

$$\sqrt[2]{9.0 \times 10^8} = (9.0 \times 10^8)^{1/2} = (9.0)^{1/2} \times 10^{8/2} = 3.0 \times 10^4 \quad \text{(2 significant figures)}$$

Because the exponent in the answer (n/m) is an integer, we must sometimes rewrite the original number by shifting the decimal point so that the exponent n is an integral multiple of the root m. For example, to take the cube root of 6.4×10^{10}, we first rewrite this number as 64×10^9 so that the exponent (9) is an integral multiple of the root 3:

$$\sqrt[3]{6.4 \times 10^{10}} = \sqrt[3]{64 \times 10^9} = (64)^{1/3} \times 10^{9/3} = 4.0 \times 10^3$$

Scientific Notation and Electronic Calculators

With a scientific calculator you can carry out calculations in scientific notation. You should consult the instruction manual for your particular calculator to learn how to enter and manipulate numbers expressed in an exponential format. On most calculators, you enter the number $A \times 10^n$ by (i) entering the number A, (ii) pressing a key labeled EXP or EE, and (iii) entering the exponent n. If the exponent is negative, you press a key labeled +/− before entering the value of n. (Note that you do not enter the number 10.) The calculator displays the number $A \times 10^n$ with the number A on the left followed by some space and then the exponent n. For example,

4.625×10^2 is displayed as 4.625 02

To add, subtract, multiply, or divide exponential numbers, use the same sequence of keystrokes as you would in working with ordinary numbers. When you add or subtract on a calculator, the numbers need not have the same exponent; the calculator automatically takes account of the different exponents. Remember, though, that the calculator often gives more digits in the answer than the allowed number of significant figures. It's sometimes helpful to outline the calculation on paper, as in the preceding examples, in order to keep track of the number of significant figures.

Most calculators have x^2 and $\sqrt{x}$ keys for squaring a number and finding its square root. Just enter the number and press the appropriate key. You probably have a y^x (or a^x) key for raising a number to a power. To raise 4.625×10^2 to the 3rd power, for example, use the following keystrokes: (i) enter the number 4.625×10^2 in the usual way, (ii) press the y^x key, (iii) enter the power 3, and (iv) press the = key. The result is displayed as 9.8931641 07, but it must be rounded to 4 significant figures. Therefore, $(4.625 \times 10^2)^3 = 9.893 \times 10^7$.

To take the mth root of a number, raise the number to the power $1/m$. For example, to take the 5th root of 4.52×10^{11}, use the following keystrokes: (i) enter the number 4.52×10^{11}, (ii) press the y^x key, (iii) enter the number 5 (for the 5th root), (iv) press the $1/x$ key (to convert the 5th root to the power 1/5), and (v) press the = key. The result is

$$\sqrt[5]{4.52 \times 10^{11}} = (4.52 \times 10^{11})^{1/5} = 2.14 \times 10^2$$

The calculator is able to handle the nonintegral exponent 11/5, and there is therefore no need to enter the number as 45.2×10^{10} so that the exponent is an integral multiple of the root 5.

PROBLEM A.1 Perform the following calculations, expressing the result in scientific notation with the correct number of significant figures. (You don't need a calculator for these.)

(a) $(1.50 \times 10^4) + (5.04 \times 10^3)$
(b) $(2.5 \times 10^{-2}) - (5.0 \times 10^{-3})$
(c) $(4.0 \times 10^4)^2$
(d) $\sqrt[3]{8 \times 10^{12}}$
(e) $\sqrt{2.5 \times 10^5}$

ANSWERS:
(a) 2.00×10^4 **(b)** 2.0×10^{-2} **(c)** 1.6×10^9
(d) 2×10^4 **(e)** $\pm 5.0 \times 10^2$

continued on next page

PROBLEM A.2 Perform the following calculations, expressing the result in scientific notation with the correct number of significant figures. (Use a calculator for these.)

(a) $(9.72 \times 10^{-1}) + (3.4823 \times 10^{2})$ **(b)** $(3.772 \times 10^{3}) - (2.891 \times 10^{4})$

(c) $(7.62 \times 10^{-3})^4$ **(d)** $\sqrt[3]{8.2 \times 10^7}$

(e) $\sqrt[5]{3.47 \times 10^{-12}}$

ANSWERS:

(a) 3.4920×10^2 **(b)** -2.514×10^4 **(c)** 3.37×10^{-9}
(d) 4.3×10^2 **(e)** 5.11×10^{-3}

A.2 LOGARITHMS

Common Logarithms

Any positive number x can be written as 10 raised to some power z, that is, $x = 10^z$. The exponent z is called the *common*, or *base 10, logarithm* of the number x and is denoted $\log_{10} x$, or simply $\log x$:

$$x = 10^z \qquad \log x = z$$

For example, 100 can be written as 10^2, and log 100 is therefore equal to 2:

$$100 = 10^2 \qquad \log 100 = 2$$

Similarly,

$$10 = 10^1 \qquad \log 10 = 1$$
$$1 = 10^0 \qquad \log 1 = 0$$
$$0.1 = 10^{-1} \qquad \log 0.1 = -1$$

In general, the logarithm of a number x is the power z to which 10 must be raised to equal the number x.

As **Figure A.1** shows, the logarithm of a number greater than 1 is positive, the logarithm of 1 is zero, and the logarithm of a positive number less than 1 is negative. The logarithm of a *negative* number is undefined because 10 raised to any power is always positive ($x = 10^z > 0$).

FIGURE A.1

Values of log x for values of x in the range 0.1 to 10.

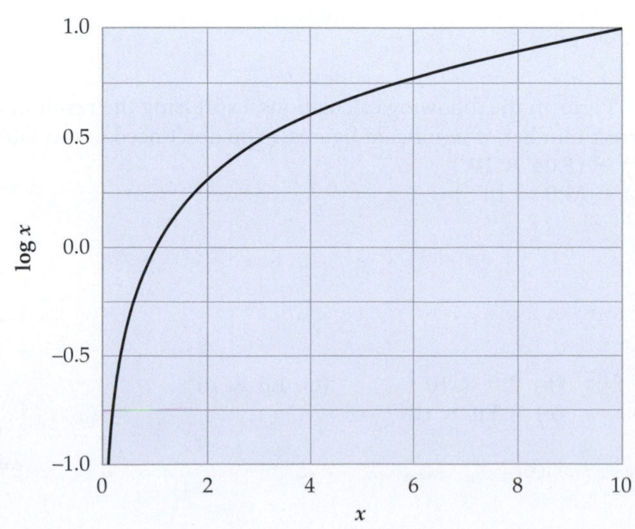

You can use a calculator to find the logarithm of a number that is not an integral power of 10. For example, to find the logarithm of 61.2, simply enter 61.2, and press the LOG key. The logarithm should be between 1 and 2 because 61.2 is between 10^1 and 10^2. The calculator gives a value of 1.786751422, which must be rounded to 1.787 because 61.2 has three significant figures.

Significant Figures and Common Logarithms

The only significant figures in a logarithm are the digits to the right of the decimal point; the number to the left of the decimal point is an exact number related to the integral power of 10 in the exponential expression for the number whose logarithm is to be found. Thus, the logarithm of 61.2, which has three significant figures, can be written as follows:

$$\log \underbrace{61.2}_{\text{3 SF's}} = \log(\underbrace{6.12}_{\text{3 SF's}} \times \underbrace{10^1}_{\text{Exact number}}) = \log 6.12 + \log 10^1 = 0.\underbrace{787}_{\text{3 SF's}} + \underbrace{1}_{\text{Exact number}} = \underbrace{1}_{\text{Exact number}}.\underbrace{787}_{\text{3 SF's}}$$

The digit (1) to the left of the decimal point in the logarithm (1.787) is an exact number and is not a significant figure; it merely indicates the location of the decimal point in the number 61.2. There are only three significant figures in the logarithm (7, 8, 7) because 61.2 has only three significant figures. Similarly, $\log 61 = 1.79$ (2 significant figures), and $\log(6 \times 10^1) = 1.8$ (1 significant figure).

Antilogarithms

The antilogarithm, denoted antilog, is the inverse of the common logarithm. If z is the logarithm of x, then x is the antilogarithm of z. But since x can be written as 10^z, the antilogarithm of z is 10^z:

$$\text{If } z = \log x \qquad \text{then} \qquad x = \text{antilog } z = 10^z$$

In other words, the antilog of a number is 10 raised to a power equal to that number. For example, the antilog of 2 is $10^2 = 100$, and the antilog of 3.71 is $10^{3.71}$.

To find the value of antilog 3.71, use your calculator. If you have a 10^x key, enter 3.71 and press the 10^x key. If you have a y^x key, use the following keystrokes: (i) enter 10, (ii) press the y^x key, (iii) enter the exponent 3.71, and (iv) press the = key. If you have an INV (inverse) key, enter 3.71, press the INV key, and then press the LOG key. The calculator gives antilog $3.71 = 5.12861384 \times 10^3$, which must be rounded to 5.1×10^3 (2 significant figures) because the logarithm (3.71) has just two significant figures, the two digits to the right of the decimal point.

Natural Logarithms

The number $e = 2.718\,28\ldots$, like $\pi = 3.141\,59\ldots$, turns up in many scientific problems. It is therefore convenient to define a logarithm based on e, just as we defined a logarithm based on 10. Just as a number x can be written as 10^z, it can also be written as e^u. The exponent u is called the *natural*, or *base* e, *logarithm* of the number x and is denoted $\log_e x$, or more commonly, $\ln x$:

$$x = e^u \qquad \ln x = u$$

The natural logarithm of a number x is the power u to which e must be raised to equal the number x. For example, the number 10.0 can be written as $e^{2.303}$, and therefore the natural logarithm of 10.0 equals 2.303:

$$10.0 = e^{2.303} = (2.718\,28\ldots)^{2.303} \qquad \ln 10.0 = 2.303 \qquad \text{(3 significant figures)}$$

To find the natural logarithm of a number on your calculator, simply enter the number, and press the LN key.

The natural antilogarithm, denoted antiln, is the inverse of the natural logarithm. If u is the natural logarithm of x, then x $(= e^u)$ is the natural antilogarithm of u:

$$\text{If } u = \ln x \quad \text{then} \quad x = \text{antiln } u = e^u$$

In other words, the natural antilogarithm of a number is e raised to a power equal to that number. For example, the natural antilogarithm of 3.71 is $e^{3.71}$, which equals 41:

$$\text{antiln } 3.71 = e^{3.71} = 41 \quad \text{(2 significant figures)}$$

Your calculator probably has an INV (inverse) key or an e^x key. To find the natural antilogarithm of a number—say, 3.71—enter 3.71, press the INV key, and then press the LN key. Alternatively, you can enter 3.71, and press the e^x key.

Some Mathematical Properties of Logarithms

Because logarithms are exponents, the algebraic properties of exponents can be used to derive the following useful relationships involving logarithms:

1. The logarithm (either common or natural) of a product xy equals the sum of the logarithm of x and the logarithm of y:

$$\log xy = \log x + \log y \qquad \ln xy = \ln x + \ln y$$

2. The logarithm of a quotient x/y equals the difference between the logarithm of x and the logarithm of y:

$$\log \frac{x}{y} = \log x - \log y \qquad \ln \frac{x}{y} = \ln x - \ln y$$

It follows from these relationships that

$$\log \frac{y}{x} = -\log \frac{x}{y} \qquad \ln \frac{y}{x} = -\ln \frac{x}{y}$$

Because $\log 1 = \ln 1 = 0$, it also follows that

$$\log \frac{1}{x} = -\log x \qquad \ln \frac{1}{x} = -\ln x$$

3. The logarithm of x raised to a power a equals a times the logarithm of x:

$$\log x^a = a \log x \qquad \ln x^a = a \ln x$$

Similarly,

$$\log x^{1/a} = \frac{1}{a} \log x \qquad \ln x^{1/a} = \frac{1}{a} \ln x$$

where

$$x^{1/a} = \sqrt[a]{x}$$

What is the numerical relationship between the common logarithm and the natural logarithm? To derive it, we begin with the definitions of $\log x$ and $\ln x$:

$$\log x = z \quad \text{where} \quad x = 10^z$$

$$\ln x = u \quad \text{where} \quad x = e^u$$

We then write $\ln x$ in terms of 10^z and make use of the property that $\ln x^a = a \ln x$:

$$\ln x = \ln 10^z = z \ln 10$$

Because $z = \log x$ and $\ln 10.0 = 2.303$, we find that the natural logarithm is 2.303 times the common logarithm:

$$\ln x = 2.303 \log x$$

Since the natural and common logarithms differ by a factor of only 2.303, the same rule can be used to find the number of significant figures in both: The only digits that are significant figures in both natural and common logarithms are those to the right of the decimal point.

PROBLEM A.3 Use a calculator to evaluate the following expressions and round each result to the correct number of significant figures:

(a) log 705 **(b)** ln (3.4×10^{-6})
(c) antilog (−2.56) **(d)** antiln 8.1

ANSWERS:

(a) 2.848 **(b)** −12.59
(c) 2.8×10^{-3} **(d)** 3×10^{3}

A.3 STRAIGHT-LINE GRAPHS AND LINEAR EQUATIONS

The results of a scientific experiment are often summarized in the form of a graph. Consider an experiment in which some property *y* is measured as a function of some variable *x*. (A real example would be measurement of the volume of a gas as a function of its temperature, but we'll use *y* and *x* to keep the discussion general.) Suppose that we obtain the following experimental data:

x	y
−1	−5
1	1
3	7
5	13

The graph in **Figure A.2** shows values of *x*, called the independent variable, along the horizontal axis and values of *y*, the dependent variable, along the vertical axis. Each pair of experimental values of *x* and *y* is represented by a point on the graph. For this particular experiment, the four data points lie on a straight line.

The equation of a straight line can be written as

$$y = mx + b$$

where *m* is the slope of the line and *b* is the intercept, the value of *y* at the point where the line crosses the *y* axis—that is, the value of *y* when $x = 0$. The slope of the line is the change in $y\,(\Delta y)$ for a given change in $x\,(\Delta x)$:

$$m = \text{slope} = \frac{\Delta y}{\Delta x}$$

The right-triangle in Figure A.2 shows that *y* changes from 4 to 13 when *x* changes from 2 to 5. Therefore, the slope of the line is 3:

$$m = \text{slope} = \frac{\Delta y}{\Delta x} = \frac{13 - 4}{5 - 2} = \frac{9}{3} = 3$$

The graph shows a *y* intercept of −2 ($b = -2$), and the equation of the line is therefore

$$y = 3x - 2$$

An equation of the form $y = mx + b$ is called a *linear equation* because values of *x* and *y* that satisfy such an equation are the coordinates of points that lie on a straight line. We also say that *y* is a *linear function* of *x* and, when $b = 0$, that *y* is *directly proportional* to *x*. In our example, the rate of change of *y* is 3 times that of *x*.

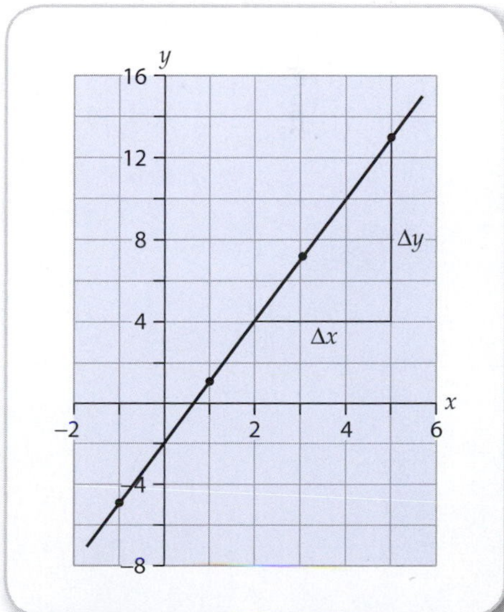

FIGURE A.2

A straight-line *y* versus *x* plot of the data in the table.

A.4 QUADRATIC EQUATIONS

A quadratic equation is an equation that can be written in the form

$$ax^2 + bx + c = 0$$

where *a*, *b*, and *c* are constants. The equation contains only powers of *x* and is called quadratic because the highest power of *x* is 2. The solutions to a quadratic equation (values of *x* that satisfy the equation) are given by the *quadratic formula*:

$$x = \frac{-b \pm \sqrt{b^2 - 4ac}}{2a}$$

The $\pm$ indicates that there are two solutions, one given by the + sign and the other given by the − sign.

As an example, let's solve the equation

$$x^2 = \frac{2 - 6x}{3}$$

First, we put the equation into the form $ax^2 + bx + c = 0$ by multiplying it by 3 and moving $2 - 6x$ to the left side. The result is

$$3x^2 + 6x - 2 = 0$$

Then we apply the quadratic formula with $a = 3$, $b = 6$, and $c = -2$:

$$x = \frac{-6 \pm \sqrt{(6)^2 - 4(3)(-2)}}{2(3)}$$

$$= \frac{-6 \pm \sqrt{36 + 24}}{6} = \frac{-6 \pm \sqrt{60}}{6} = \frac{-6 \pm 7.746}{6}$$

The two solutions are

$$x = \frac{-6 + 7.746}{6} = \frac{1.746}{6} = 0.291 \quad \text{and} \quad x = \frac{-6 - 7.746}{6} = \frac{-13.746}{6} = -2.291$$

A.5 CALCULUS DERIVATIONS OF INTEGRATED RATE LAWS

Zeroth-Order Reactions (Section 12.8)

The rate law for a zeroth-order reaction is

$$\text{Rate} = -\frac{\Delta[A]}{\Delta t} = k$$

In differential form, we can write this equation as

$$-\frac{d[A]}{dt} = k$$

$$\text{or} \quad d[A] = -kdt$$

Integrating from $t = 0$ to $t = t$ gives

$$\int_{[A]_0}^{[A]_t} d[A] = -k\int_0^t dt$$

$$[A]_t - [A]_0 = -kt$$

$$\text{or} \quad [A]_t = -kt + [A]_0$$

First-Order Reactions (Section 12.4)

The rate law for a first-order reaction is

$$\text{Rate} = -\frac{\Delta[A]}{\Delta t} = k[A]$$

In differential form, we can write this equation as

$$-\frac{d[A]}{dt} = k[A]$$

$$\text{or} \quad \frac{d[A]}{[A]} = -kdt$$

Integrating from $t = 0$ to $t = t$ gives

$$\int_{[A]_0}^{[A]_t} \frac{d[A]}{[A]} = -k \int_0^t dt$$

$$\ln [A]_t - \ln [A]_0 = -kt$$

$$\text{or} \quad \ln \frac{[A]_t}{[A]_0} = -kt$$

Second-Order Reactions (Section 12.7)

The rate law for a second-order reaction is

$$\text{Rate} = -\frac{\Delta[A]}{\Delta t} = k[A]^2$$

In differential form, we can write this equation as

$$-\frac{d[A]}{dt} = k[A]^2$$

$$\text{or} \quad \frac{d[A]}{[A]^2} = -kdt$$

Integrating from $t = 0$ to $t = t$ gives

$$\int_{[A]_0}^{[A]_t} \frac{d[A]}{[A]^2} = -k \int_0^t dt$$

$$-\frac{1}{[A]_t} + \frac{1}{[A]_0} = -kt$$

$$\text{or} \quad \frac{1}{[A]_t} = kt + \frac{1}{[A]_0}$$

APPENDIX B

Thermodynamic Properties at 25 °C

TABLE B.1 Inorganic Substances

Substance and State	ΔH°_f (kJ/mol)	ΔG°_f (kJ/mol)	S° [J/(K·mol)]
Aluminum			
$Al(s)$	0	0	28.3
$Al(g)$	330.0	289.4	164.5
$AlCl_3(s)$	−704.2	−628.8	109.3
$Al_2O_3(s)$	−1676	−1582	50.9
Barium			
$Ba(s)$	0	0	62.5
$Ba(g)$	180.0	146.0	170.1
$Ba^{2+}(aq)$	−537.6	−560.8	9.6
$BaCl_2(s)$	−855.0	−806.7	123.7
$BaO(s)$	−548.0	−520.3	72.1
$BaCO_3(s)$	−1213.0	−1134.4	112.1
$BaSO_4(s)$	−1473.2	−1362.2	132.2
Beryllium			
$Be(s)$	0	0	9.5
$BeO(s)$	−609.4	−580.1	13.8
$Be(OH)_2(s)$	−902.5	−815.0	45.5
Boron			
$B(s)$	0	0	5.9
$BF_3(g)$	−1136.0	−1119.4	254.3
$BCl_3(g)$	−403.8	−388.7	290.0
$B_2H_6(g)$	36.4	87.6	232.0
$B_2O_3(s)$	−1273.5	−1194.3	54.0
$H_3BO_3(s)$	−1094.3	−968.9	90.0
Bromine			
$Br(g)$	111.9	82.4	174.9
$Br^-(aq)$	−121.5	−104.0	82.4
$Br_2(l)$	0	0	152.2
$Br_2(g)$	30.9	3.14	245.4
$HBr(g)$	−36.3	−53.4	198.6
Cadmium			
$Cd(s)$	0	0	51.8
$Cd(g)$	111.8	77.3	167.6
$Cd^{2+}(aq)$	−75.9	−77.6	−73.2
$CdCl_2(s)$	−391.5	−343.9	115.3
$CdO(s)$	−258.4	−228.7	54.8
$CdS(s)$	−161.9	−156.5	64.9
$CdSO_4(s)$	−933.3	−822.7	123.0

Substance and State	ΔH°_f (kJ/mol)	ΔG°_f (kJ/mol)	S° [J/(K·mol)]
Calcium			
$Ca(s)$	0	0	41.4
$Ca(g)$	177.8	144.0	154.8
$Ca^{2+}(aq)$	−542.8	−553.6	−53.1
$CaF_2(s)$	−1228.0	−1175.6	68.5
$CaCl_2(s)$	−795.4	−748.8	108.4
$CaH_2(s)$	−181.5	−142.5	41.4
$CaC_2(s)$	−59.8	−64.8	70.0
$CaO(s)$	−634.9	−603.3	38.1
$Ca(OH)_2(s)$	−985.2	−897.5	83.4
$CaCO_3(s)$	−1207.6	−1129.1	91.7
$CaSO_4(s)$	−1434.1	−1321.9	107
$Ca_3(PO_4)_2(s)$	−4120.8	−3884.7	236.0
Carbon			
$C(s$, graphite)	0	0	5.7
$C(s$, diamond)	1.9	2.9	2.4
$C(g)$	716.7	671.3	158.0
$CO(g)$	−110.5	−137.2	197.6
$CO_2(g)$	−393.5	−394.4	213.6
$CO_2(aq)$	−413.8	−386.0	117.6
$CO_3^{2-}(aq)$	−677.1	−527.8	−56.9
$HCO_3^-(aq)$	−692.0	−586.8	91.2
$H_2CO_3(aq)$	−699.7	−623.2	187.4
$HCN(l)$	108.9	125.0	112.8
$HCN(g)$	135.1	124.7	201.7
$CS_2(l)$	89.0	64.6	151.3
$CS_2(g)$	116.7	67.1	237.7
$COCl_2(g)$	−219.1	−204.9	283.4
Cesium			
$Cs(s)$	0	0	85.2
$Cs(g)$	76.5	49.6	175.6
$Cs^+(aq)$	−258.3	−292.0	133.1
$CsF(s)$	−553.5	−525.5	92.8
$CsCl(s)$	−443.0	−414.5	101.2
$CsBr(s)$	−405.8	−391.4	113.1
$CsI(s)$	−346.6	−340.6	123.1
Chlorine			
$Cl(g)$	121.3	105.3	165.1
$Cl^-(aq)$	−167.2	−131.3	56.5

TABLE B.1 Inorganic Substances (*continued*)

Substance and State	$\Delta H°_f$ (kJ/mol)	$\Delta G°_f$ (kJ/mol)	$S°$ [J/(K · mol)]
$Cl_2(g)$	0	0	223.0
$HCl(g)$	−92.3	−95.3	186.8
$HCl(aq)$	−167.2	−131.2	56.5
$ClO_2(g)$	102.5	120.5	256.7
$Cl_2O(g)$	80.3	97.9	266.1
Chromium			
$Cr(s)$	0	0	23.8
$Cr(g)$	396.6	351.8	174.4
$Cr_2O_3(s)$	−1140	−1058	81.2
Cobalt			
$Co(s)$	0	0	30.0
$Co(g)$	424.7	380.3	179.4
$CoO(s)$	−237.9	−214.2	53.0
Copper			
$Cu(s)$	0	0	33.1
$Cu(g)$	337.4	297.7	166.3
$Cu^{2+}(aq)$	64.8	65.5	−99.6
$CuCl(s)$	−137.2	−119.9	86.2
$CuCl_2(s)$	−220.1	−175.7	108.1
$CuO(s)$	−157.3	−129.7	42.6
$Cu_2O(s)$	−168.6	−146.0	93.1
$CuS(s)$	−53.1	−53.6	66.5
$Cu_2S(s)$	−79.5	−86.2	120.9
$CuSO_4(s)$	−771.4	−662.2	109.2
Fluorine			
$F(g)$	79.4	62.3	158.7
$F^-(aq)$	−332.6	−278.8	−13.8
$F_2(g)$	0	0	202.7
$HF(g)$	−273.3	−275.4	173.7
Hydrogen			
$H(g)$	218.0	203.3	114.6
$H^+(aq)$	0	0	0
$H_2(g)$	0	0	130.6
$OH^-(aq)$	−230.0	−157.3	−10.8
$H_2O(l)$	−285.8	−237.2	69.9
$H_2O(g)$	−241.8	−228.6	188.7
$H_2O_2(l)$	−187.8	−120.4	110
$H_2O_2(g)$	−136.3	−105.6	232.6
$H_2O_2(aq)$	−191.2	−134.1	144
Iodine			
$I(g)$	106.8	70.3	180.7
$I^-(aq)$	−55.2	−51.6	111
$I_2(s)$	0	0	116.1
$I_2(g)$	62.4	19.4	260.6
$HI(g)$	26.5	1.7	206.5
Iron			
$Fe(s)$	0	0	27.3
$Fe(g)$	416.3	370.3	180.5
$FeCl_2(s)$	−341.8	−302.3	118.0
$FeCl_3(s)$	−399.5	−334.0	142.3
$FeO(s)$	−272	−255	61
$Fe_2O_3(s)$	−824.2	−742.2	87.4
$Fe_3O_4(s)$	−1118	−1015	146
$FeS_2(s)$	−178.2	−166.9	52.9
Lead			
$Pb(s)$	0	0	64.8
$Pb(g)$	195.2	162.2	175.3
$PbCl_2(s)$	−359.4	−314.1	136.0
$PbBr_2(s)$	−278.7	−261.9	161.5
$PbO(s)$	−217.3	−187.9	68.7
$PbO_2(s)$	−277	−217.4	68.6
$PbS(s)$	−100	−98.7	91.2
$PbCO_3(s)$	−699.1	−625.5	131.0
$PbSO_4(s)$	−919.9	−813.2	148.6
Lithium			
$Li(s)$	0	0	29.1
$Li(g)$	159.3	126.6	138.7
$Li^+(aq)$	−278.5	−293.3	13
$LiF(s)$	−616.0	−587.7	35.7
$LiCl(s)$	−408.6	−384.4	59.3
$LiBr(s)$	−351.2	−342.0	74.3
$LiI(s)$	−270.4	−270.3	86.8
$Li_2O(s)$	−597.9	−561.2	37.6
$LiOH(s)$	−487.5	−441.5	42.8
Magnesium			
$Mg(s)$	0	0	32.7
$Mg(g)$	147.1	112.5	148.6
$MgCl_2(s)$	−641.6	−591.8	89.6
$MgO(s)$	−601.7	−569.4	26.9
$MgCO_3(s)$	−1096	−1012	65.7
$MgSO_4(s)$	−1284.9	−1170.6	91.6
Manganese			
$Mn(s)$	0	0	32.0
$Mn(g)$	280.7	238.5	173.6
$MnO(s)$	−385.2	−362.9	59.7
$MnO_2(s)$	−520.0	−465.1	53.1
Mercury			
$Hg(l)$	0	0	76.0
$Hg(g)$	61.32	31.85	174.8
$Hg^{2+}(aq)$	171.1	164.4	−32.2

TABLE B.1 Inorganic Substances (*continued*)

Substance and State	$\Delta H°_f$ (kJ/mol)	$\Delta G°_f$ (kJ/mol)	$S°$ [J/(K · mol)]
$Hg_2^{2+}(aq)$	172.4	153.5	84.5
$HgCl_2(s)$	−224.3	−178.6	146.0
$Hg_2Cl_2(s)$	−265.4	−210.7	191.6
$HgO(s)$	−90.8	−58.6	70.3
$HgS(s)$	−58.2	−50.6	82.4
Nickel			
$Ni(s)$	0	0	29.9
$Ni(g)$	429.7	384.5	182.1
$NiCl_2(s)$	−305.3	−259.1	97.7
$NiO(s)$	−240	−212	38.0
$NiS(s)$	−82.0	−79.5	53.0
Nitrogen			
$N(g)$	472.7	455.6	153.2
$N_2(g)$	0	0	191.5
$NH_3(g)$	−46.1	−16.5	192.3
$NH_3(aq)$	−80.3	−26.6	111
$NH_4^+(aq)$	−132.5	−79.4	113
$N_2H_4(l)$	50.6	149.2	121.2
$N_2H_4(g)$	95.4	159.3	238.4
$NO(g)$	91.3	87.6	210.7
$NO_2(g)$	33.2	51.3	240.0
$N_2O(g)$	82.0	104.2	219.7
$N_2O_4(g)$	11.1	99.8	304.3
$N_2O_5(g)$	13.3	117.1	355.6
$NOCl(g)$	51.7	66.1	261.6
$NO_2Cl(g)$	12.6	54.4	272.1
$HNO_3(l)$	−174.1	−80.8	155.6
$HNO_3(g)$	−133.9	−73.5	266.8
$HNO_2(aq)$	−119	−50.6	136
$HNO_3(aq)$	−207.4	−111.3	146.4
$NO_3^-(aq)$	−207.4	−111.3	146.4
$NH_4Cl(s)$	−314.4	−202.9	94.6
$NH_4NO_3(s)$	−365.6	−184.0	151.1
Oxygen			
$O(g)$	249.2	231.7	160.9
$O_2(g)$	0	0	205.0
$O_3(g)$	143	163	238.8
Phosphorus			
P(*s*, white)	0	0	41.1
P(*s*, red)	−18	−12	22.8
$P_4(g)$	58.9	24.5	279.9
$PH_3(g)$	5.4	13.5	210.1
$PCl_3(l)$	−319.7	−272.3	217.1
$PCl_3(g)$	−287.0	−267.8	311.7
$PCl_5(s)$	−443.5		

Substance and State	$\Delta H°_f$ (kJ/mol)	$\Delta G°_f$ (kJ/mol)	$S°$ [J/(K · mol)]
$PCl_5(g)$	−374.9	−305.0	364.5
$P_4O_{10}(s)$	−2984	−2698	228.9
$PO_4^{3-}(aq)$	−1277.4	−1018.7	−220.5
$HPO_4^{2-}(aq)$	−1292.1	−1089.2	−33.5
$H_2PO_4^-(aq)$	−1296.3	−1130.2	90.4
$H_3PO_4(s)$	−1284.4	−1124.3	110.5
Potassium			
$K(s)$	0	0	64.7
$K(g)$	89.2	60.6	160.2
$K^+(aq)$	−252.4	−283.3	102.5
$KF(s)$	−567.3	−537.8	66.6
$KCl(s)$	−436.5	−408.5	82.6
$KBr(s)$	−393.8	−380.7	95.9
$KI(s)$	−327.9	−324.9	106.3
$K_2O(s)$	−361.5		
$K_2O_2(s)$	−494.1	−425.1	102.1
$KO_2(s)$	−284.9	−239.4	116.7
$KOH(s)$	−424.6	−379.4	81.2
$KOH(aq)$	−482.4	−440.5	91.6
$KClO_3(s)$	−397.7	−296.3	143.1
$KClO_4(s)$	−432.8	−303.1	151.0
$KNO_3(s)$	−494.6	−394.9	133.1
Rubidium			
$Rb(s)$	0	0	76.8
$Rb(g)$	80.9	53.1	170.0
$Rb^+(aq)$	−251.2	−284.0	121.5
$RbF(s)$	−557.7		
$RbCl(s)$	−435.4	−407.8	95.9
$RbBr(s)$	−394.6	−381.8	110.0
$RbI(s)$	−333.8	−328.9	118.4
Selenium			
Se(*s*, black)	0	0	42.44
$H_2Se(g)$	29.7	15.9	219.0
Silicon			
$Si(s)$	0	0	18.8
$Si(g)$	450.0	405.5	167.9
$SiF_4(g)$	−1615.0	−1572.8	282.7
$SiCl_4(l)$	−687.0	−619.8	239.7
SiO_2(*s*, quartz)	−910.7	−856.3	41.5
Silver			
$Ag(s)$	0	0	42.6
$Ag(g)$	284.9	246.0	173.0
$Ag^+(aq)$	105.6	77.1	72.7
$AgF(s)$	−204.6		
$AgCl(s)$	−127.1	−109.8	96.2

TABLE B.1 Inorganic Substances (*continued*)

Substance and State	ΔH°_f (kJ/mol)	ΔG°_f (kJ/mol)	S° [J/(K · mol)]
$AgBr(s)$	−100.4	−96.9	107.1
$AgI(s)$	−61.8	−66.2	115.5
$Ag_2O(s)$	−31.1	−11.2	121.3
$Ag_2S(s)$	−32.6	−40.7	144.0
$AgNO_3(s)$	−124.4	−33.4	140.9
Sodium			
$Na(s)$	0	0	51.2
$Na(g)$	107.3	76.8	153.6
$Na^+(aq)$	−240.1	−261.9	59.0
$NaF(s)$	−576.6	−546.3	51.1
$NaCl(s)$	−411.2	−384.2	72.1
$NaBr(s)$	−361.1	−349.0	86.8
$NaI(s)$	−287.8	−286.1	98.5
$NaH(s)$	−56.3	−33.5	40.0
$Na_2O(s)$	−414.2	−375.5	75.1
$Na_2O_2(s)$	−510.9	−447.7	95.0
$NaO_2(s)$	−260.2	−218.4	115.9
$NaOH(s)$	−425.6	−379.5	64.5
$NaOH(aq)$	−470.1	−419.2	48.2
$Na_2CO_3(s)$	−1130.7	−1044.5	135.0
$NaHCO_3(s)$	−950.8	−851.0	102
$NaNO_3(s)$	−467.9	−367.0	116.5
$NaNO_3(aq)$	−447.5	−373.2	205.4
$Na_2SO_4(s)$	−1387.1	−1270.2	149.6
Sulfur			
$S(s$, rhombic)	0	0	31.8
$S(s$, monoclinic)	0.3		
$S(g)$	277.2	236.7	167.7
$S_2(g)$	128.6	79.7	228.2
$H_2S(g)$	−20.6	−33.6	205.7
$H_2S(aq)$	−39.7	−27.9	121
$HS^-(aq)$	−17.6	12.1	62.8
$SO_2(g)$	−296.8	−300.2	248.1
$SO_3(g)$	−395.7	−371.1	256.6
$H_2SO_4(l)$	−814.0	−690.1	156.9
$H_2SO_4(aq)$	−909.3	−744.6	20
$HSO_4^-(aq)$	−887.3	−756.0	132
$SO_4^{2-}(aq)$	−909.3	−744.6	20
Tin			
$Sn(s$, white)	0	0	51.2
$Sn(s$, gray)	−2.1	0.1	44.1
$Sn(g)$	301.2	266.2	168.4
$SnCl_4(l)$	−511.3	−440.1	258.6
$SnCl_4(g)$	−471.5	−432.2	365.8
$SnO(s)$	−280.7	−251.9	57.4
$SnO_2(s)$	−577.6	−515.8	49.0
Titanium			
$Ti(s)$	0	0	30.6
$Ti(g)$	473.0	428.4	180.2
$TiCl_4(l)$	−804.2	−737.2	252.3
$TiCl_4(g)$	−763.2	−726.3	353.2
$TiO_2(s)$	−944.0	−888.8	50.6
Tungsten			
$W(s)$	0	0	32.6
$W(g)$	849.4	807.1	174.0
$WO_3(s)$	−842.9	−764.0	75.9
Zinc			
$Zn(s)$	0	0	41.6
$Zn(g)$	130.4	94.8	160.9
$Zn^{2+}(aq)$	−153.9	−147.1	−112.1
$ZnCl_2(s)$	−415.1	−369.4	111.5
$ZnO(s)$	−350.5	−320.5	43.7
$ZnS(s)$	−206.0	−201.3	57.7
$ZnSO_4(s)$	−982.8	−871.5	110.5

TABLE B.2 Organic Substances

Substance and State	Formula	ΔH°_f (kJ/mol)	ΔG°_f (kJ/mol)	S° [J/(K · mol)]
Acetaldehyde(*g*)	CH_3CHO	−166.2	−133.0	263.8
Acetic acid(*l*)	CH_3CO_2H	−484.5	−390	160
Acetylene(*g*)	C_2H_2	227.4	209.9	200.8
Benzene(*l*)	C_6H_6	49.1	124.5	173.4
Butane(*g*)	C_4H_{10}	−126	−17	310
Carbon tetrachloride(*l*)	CCl_4	−135.4	−65.3	216.4
Dichloroethane(*l*)	CH_2ClCH_2Cl	−165.2	−79.6	208.5
Ethane(*g*)	C_2H_6	−84.0	−32.0	229.1
Ethanol(*l*)	C_2H_5OH	−277.7	−174.9	161
Ethanol(*g*)	C_2H_5OH	−234.8	−167.9	281.5
Ethylene(*g*)	C_2H_4	52.3	68.1	219.5
Ethylene oxide(*g*)	C_2H_4O	−52.6	−13.1	242.4
Formaldehyde(*g*)	HCHO	−108.6	−102.5	218.8
Formic acid(*l*)	HCO_2H	−424.7	−361.4	129.0
Glucose(*s*)	$C_6H_{12}O_6$	−1273.3	−910	209.2
Methane(*g*)	CH_4	−74.8	−50.8	186.2
Methanol(*l*)	CH_3OH	−239.2	−166.6	127
Methanol(*g*)	CH_3OH	−201.0	−162.3	239.8
Propane(*g*)	C_3H_8	−103.8	−23.4	270.2
Vinyl chloride(*g*)	$CH_2{=}CHCl$	35	51.9	263.9

APPENDIX C

Equilibrium Constants at 25 °C

TABLE C.1 Acid-Dissociation Constants at 25 °C

Acid	Formula	K_{a1}	K_{a2}	K_{a3}
Acetic	CH_3CO_2H	1.8×10^{-5}		
Acetylsalicylic	$C_9H_8O_4$	3.0×10^{-4}		
Arsenic	H_3AsO_4	5.6×10^{-3}	1.7×10^{-7}	4.0×10^{-12}
Arsenious	H_3AsO_3	6×10^{-10}		
Ascorbic	$C_6H_8O_6$	8.0×10^{-5}		
Benzoic	$C_6H_5CO_2H$	6.5×10^{-5}		
Boric	H_3BO_3	5.8×10^{-10}		
Carbonic	H_2CO_3	4.3×10^{-7}	5.6×10^{-11}	
Chloroacetic	CH_2ClCO_2H	1.4×10^{-3}		
Citric	$C_6H_8O_7$	7.1×10^{-4}	1.7×10^{-5}	4.1×10^{-7}
Formic	HCO_2H	1.8×10^{-4}		
Hydrazoic	HN_3	1.9×10^{-5}		
Hydrocyanic	HCN	4.9×10^{-10}		
Hydrofluoric	HF	3.5×10^{-4}		
Hydrogen peroxide	H_2O_2	2.4×10^{-12}		
Hydrosulfuric	H_2S	1.0×10^{-7}	$\sim 10^{-19}$	
Hypobromous	$HOBr$	2.0×10^{-9}		
Hypochlorous	$HOCl$	3.5×10^{-8}		
Hypoiodous	HOI	2.3×10^{-11}		
Iodic	HIO_3	1.7×10^{-1}		
Lactic	$HC_3H_5O_3$	1.4×10^{-4}		
Nitrous	HNO_2	4.5×10^{-4}		
Oxalic	$H_2C_2O_4$	5.9×10^{-2}	6.4×10^{-5}	
Phenol	C_6H_5OH	1.3×10^{-10}		
Phosphoric	H_3PO_4	7.5×10^{-3}	6.2×10^{-8}	4.8×10^{-13}
Phosphorous	H_3PO_3	1.0×10^{-2}	2.6×10^{-7}	
Saccharin	$C_7H_5NO_3S$	2.1×10^{-12}		
Selenic	H_2SeO_4	Very large	1.2×10^{-2}	
Selenious	H_2SeO_3	3.5×10^{-2}	5×10^{-8}	
Sulfuric	H_2SO_4	Very large	1.2×10^{-2}	
Sulfurous	H_2SO_3	1.5×10^{-2}	6.3×10^{-8}	
Tartaric	$C_4H_6O_6$	1.0×10^{-3}	4.6×10^{-5}	
Water	H_2O	1.8×10^{-16}		

TABLE C.2 Acid-Dissociation Constants at 25 °C for Hydrated Metal Cations

Cation	K_a	Cation	K_a
$Fe^{2+}(aq)$	3.2×10^{-10}	$Be^{2+}(aq)$	3×10^{-7}
$Co^{2+}(aq)$	1.3×10^{-9}	$Al^{3+}(aq)$	1.4×10^{-5}
$Ni^{2+}(aq)$	2.5×10^{-11}	$Cr^{3+}(aq)$	1.6×10^{-4}
$Zn^{2+}(aq)$	2.5×10^{-10}	$Fe^{3+}(aq)$	6.3×10^{-3}

Note: As an example, K_a for $Fe^{2+}(aq)$ is the equilibrium constant for the reaction

$$Fe(H_2O)_6{}^{2+}(aq) + H_2O(l) \rightleftharpoons H_3O^+(aq) + Fe(H_2O)_5(OH)^+(aq)$$

TABLE C.3 Base-Dissociation Constants at 25 °C

Base	Formula	K_b
Ammonia	NH_3	1.8×10^{-5}
Aniline	$C_6H_5NH_2$	4.3×10^{-10}
Codeine	$C_{18}H_{21}NO_3$	1.6×10^{-6}
Dimethylamine	$(CH_3)_2NH$	5.4×10^{-4}
Ethylamine	$C_2H_5NH_2$	6.4×10^{-4}
Hydrazine	N_2H_4	8.9×10^{-7}
Hydroxylamine	NH_2OH	9.1×10^{-9}
Methylamine	CH_3NH_2	3.7×10^{-4}
Morphine	$C_{17}H_{19}NO_3$	1.6×10^{-6}
Piperidine	$C_5H_{11}N$	1.3×10^{-3}
Propylamine	$C_3H_7NH_2$	5.1×10^{-4}
Pyridine	C_5H_5N	1.8×10^{-9}
Strychnine	$C_{21}H_{22}N_2O_2$	1.8×10^{-6}
Trimethylamine	$(CH_3)_3N$	6.5×10^{-5}

TABLE C.4 Solubility Product Constants at 25 °C

Compound	Formula	K_{sp}
Aluminum hydroxide	$Al(OH)_3$	1.9×10^{-33}
Barium carbonate	$BaCO_3$	2.6×10^{-9}
Barium chromate	$BaCrO_4$	1.2×10^{-10}
Barium fluoride	BaF_2	1.8×10^{-7}
Barium hydroxide	$Ba(OH)_2$	5.0×10^{-3}
Barium sulfate	$BaSO_4$	1.1×10^{-10}
Cadmium carbonate	$CdCO_3$	1.0×10^{-12}
Cadmium hydroxide	$Cd(OH)_2$	5.3×10^{-15}
Calcium carbonate	$CaCO_3$	5.0×10^{-9}
Calcium fluoride	CaF_2	3.5×10^{-11}
Calcium hydroxide	$Ca(OH)_2$	4.7×10^{-6}
Calcium phosphate	$Ca_3(PO_4)_2$	2.1×10^{-33}
Calcium sulfate	$CaSO_4$	7.1×10^{-5}
Chromium(III) hydroxide	$Cr(OH)_3$	6.7×10^{-31}
Cobalt(II) hydroxide	$Co(OH)_2$	5.9×10^{-15}
Copper(I) bromide	$CuBr$	6.3×10^{-9}
Copper(I) chloride	$CuCl$	1.7×10^{-7}
Copper(II) carbonate	$CuCO_3$	2.5×10^{-10}
Copper(II) hydroxide	$Cu(OH)_2$	1.6×10^{-19}
Copper (II) phosphate	$Cu_3(PO_4)_2$	1.4×10^{-37}
Iron(II) hydroxide	$Fe(OH)_2$	4.9×10^{-17}
Iron(III) hydroxide	$Fe(OH)_3$	2.6×10^{-39}
Lead(II) bromide	$PbBr_2$	6.6×10^{-6}
Lead(II) chloride	$PbCl_2$	1.2×10^{-5}
Lead(II) chromate	$PbCrO_4$	2.8×10^{-13}
Lead(II) iodide	PbI_2	8.5×10^{-9}
Lead(II) sulfate	$PbSO_4$	1.8×10^{-8}
Magnesium carbonate	$MgCO_3$	6.8×10^{-6}
Magnesium fluoride	MgF_2	7.4×10^{-11}
Magnesium hydroxide	$Mg(OH)_2$	5.6×10^{-12}
Manganese(II) carbonate	$MnCO_3$	2.2×10^{-11}
Manganese(II) hydroxide	$Mn(OH)_2$	2.1×10^{-13}
Mercury(I) bromide	Hg_2Br_2	6.4×10^{-23}
Mercury(I) chloride	Hg_2Cl_2	1.4×10^{-18}
Mercury(I) iodide	Hg_2I_2	5.3×10^{-29}
Mercury(II) hydroxide	$Hg(OH)_2$	3.1×10^{-26}
Nickel(II) hydroxide	$Ni(OH)_2$	5.5×10^{-16}
Silver bromide	$AgBr$	5.4×10^{-13}
Silver carbonate	Ag_2CO_3	8.4×10^{-12}
Silver chloride	$AgCl$	1.8×10^{-10}
Silver chromate	Ag_2CrO_4	1.1×10^{-12}
Silver cyanide	$AgCN$	6.0×10^{-17}
Silver iodide	AgI	8.5×10^{-17}
Silver sulfate	Ag_2SO_4	1.2×10^{-5}
Silver sulfite	Ag_2SO_3	1.5×10^{-14}
Strontium carbonate	$SrCO_3$	5.6×10^{-10}
Tin(II) hydroxide	$Sn(OH)_2$	5.4×10^{-27}
Zinc carbonate	$ZnCO_3$	1.2×10^{-10}
Zinc hydroxide	$Zn(OH)_2$	4.1×10^{-17}

TABLE C.5 Solubility Products in Acid (K_{spa}) at 25 °C

Compound	Formula	K_{spa}
Cadmium sulfide	CdS	8×10^{-7}
Cobalt(II) sulfide	CoS	3
Copper(II) sulfide	CuS	6×10^{-16}
Iron(II) sulfide	FeS	6×10^{2}
Lead(II) sulfide	PbS	3×10^{-7}
Manganese(II) sulfide	MnS	3×10^{7}
Mercury(II) sulfide	HgS	2×10^{-32}
Nickel(II) sulfide	NiS	8×10^{-1}
Silver sulfide	Ag_2S	6×10^{-30}
Tin(II) sulfide	SnS	1×10^{-5}
Zinc sulfide	ZnS	3×10^{-2}

Note: K_{spa} for MS is the equilibrium constant for the reaction

$$MS(s) + 2\,H_3O^+(aq) \rightleftharpoons M^{2+}(aq) + H_2S(aq) + 2\,H_2O(l)$$

TABLE C.6 Formation Constants for Complex Ions at 25 °C

Complex Ion	K_f	Complex Ion	K_f
$Ag(CN)_2^-$	3.0×10^{20}	$Ga(OH)_4^-$	3×10^{39}
$Ag(NH_3)_2^+$	1.7×10^{7}	$Ni(CN)_4^{2-}$	1.7×10^{30}
$Ag(S_2O_3)_2^{3-}$	4.7×10^{13}	$Ni(NH_3)_6^{2+}$	2.0×10^{8}
$Al(OH)_4^-$	3×10^{33}	$Ni(en)_3^{2+}$	4×10^{17}
$Be(OH)_4^{2-}$	4×10^{18}	$Pb(OH)_3^-$	8×10^{13}
$Cr(OH)_4^-$	8×10^{29}	$Sn(OH)_3^-$	3×10^{25}
$Cu(NH_3)_4^{2+}$	5.6×10^{11}	$Zn(CN)_4^{2-}$	4.7×10^{19}
$Fe(CN)_6^{4-}$	3×10^{35}	$Zn(NH_3)_4^{2+}$	7.8×10^{8}
$Fe(CN)_6^{3-}$	4×10^{43}	$Zn(OH)_4^{2-}$	3×10^{15}

APPENDIX D

Standard Reduction Potentials at 25 °C

Half-Reaction	$E°$ (V)
$F_2(g) + 2\,e^- \longrightarrow 2\,F^-(aq)$	2.87
$O_3(g) + 2\,H^+(aq) + 2\,e^- \longrightarrow O_2(g) + H_2O(l)$	2.08
$S_2O_8^{2-}(aq) + 2\,e^- \longrightarrow 2\,SO_4^{2-}(aq)$	2.01
$Co^{3+}(aq) + e^- \longrightarrow Co^{2+}(aq)$	1.81
$H_2O_2(aq) + 2\,H^+(aq) + 2\,e^- \longrightarrow 2\,H_2O(l)$	1.78
$Ce^{4+}(aq) + e^- \longrightarrow Ce^{3+}(aq)$	1.72
$MnO_4^-(aq) + 4\,H^+(aq) + 3\,e^- \longrightarrow MnO_2(s) + 2\,H_2O(l)$	1.68
$PbO_2(s) + 3\,H^+(aq) + HSO_4^-(aq) + 2\,e^- \longrightarrow PbSO_4(s) + 2\,H_2O(l)$	1.628
$2\,HClO(aq) + 2\,H^+(aq) + 2\,e^- \longrightarrow Cl_2(g) + 2\,H_2O(l)$	1.61
$Mn^{3+}(aq) + e^- \longrightarrow Mn^{2+}(aq)$	1.54
$MnO_4^-(aq) + 8\,H^+(aq) + 5\,e^- \longrightarrow Mn^{2+}(aq) + 4\,H_2O(l)$	1.51
$2\,BrO_3^-(aq) + 12\,H^+(aq) + 10\,e^- \longrightarrow Br_2(l) + 6\,H_2O(l)$	1.48
$ClO_3^-(aq) + 6\,H^+(aq) + 5\,e^- \longrightarrow 1/2\,Cl_2(g) + 3\,H_2O(l)$	1.47
$Au^{3+}(aq) + 2\,e^- \longrightarrow Au^+(aq)$	1.40
$Cl_2(g) + 2\,e^- \longrightarrow 2\,Cl^-(aq)$	1.36
$Cr_2O_7^{2-}(aq) + 14\,H^+(aq) + 6\,e^- \longrightarrow 2\,Cr^{3+}(aq) + 7\,H_2O(l)$	1.36
$O_2(g) + 4\,H^+(aq) + 4\,e^- \longrightarrow 2\,H_2O(l)$	1.23
$MnO_2(s) + 4\,H^+(aq) + 2\,e^- \longrightarrow Mn^{2+}(aq) + 2\,H_2O(l)$	1.22
$2\,IO_3^-(aq) + 12\,H^+(aq) + 10\,e^- \longrightarrow I_2(s) + 6\,H_2O(l)$	1.20
$Br_2(aq) + 2\,e^- \longrightarrow 2\,Br^-(aq)$	1.09
$HNO_2(aq) + H^+(aq) + e^- \longrightarrow NO(g) + H_2O(l)$	0.98
$NO_3^-(aq) + 4\,H^+(aq) + 3\,e^- \longrightarrow NO(g) + 2\,H_2O(l)$	0.96
$2\,Hg^{2+}(aq) + 2\,e^- \longrightarrow Hg_2^{2+}(aq)$	0.92
$HO_2^-(aq) + H_2O(l) + 2\,e^- \longrightarrow 3\,OH^-(aq)$	0.88
$Hg^{2+}(aq) + 2\,e^- \longrightarrow Hg(l)$	0.85
$ClO^-(aq) + H_2O(l) + 2\,e^- \longrightarrow Cl^-(aq) + 2\,OH^-(aq)$	0.84
$Ag^+(aq) + e^- \longrightarrow Ag(s)$	0.80
$Hg_2^{2+}(aq) + 2\,e^- \longrightarrow 2\,Hg(l)$	0.80
$NO_3^-(aq) + 2\,H^+(aq) + e^- \longrightarrow NO_2(g) + H_2O(l)$	0.79
$Fe^{3+}(aq) + e^- \longrightarrow Fe^{2+}(aq)$	0.77
$O_2(g) + 2\,H^+(aq) + 2\,e^- \longrightarrow H_2O_2(aq)$	0.70
$MnO_4^-(aq) + e^- \longrightarrow MnO_4^{2-}(aq)$	0.56
$H_3AsO_4(aq) + 2\,H^+(aq) + 2\,e^- \longrightarrow H_3AsO_3(aq) + H_2O(l)$	0.56
$I_2(s) + 2\,e^- \longrightarrow 2\,I^-(aq)$	0.54
$Cu^+(aq) + e^- \longrightarrow Cu(s)$	0.52
$H_2SO_3(aq) + 4\,H^+(aq) + 4\,e^- \longrightarrow S(s) + 3\,H_2O(l)$	0.45
$O_2(g) + 2\,H_2O(l) + 4\,e^- \longrightarrow 4\,OH^-(aq)$	0.40
$Cu^{2+}(aq) + 2\,e^- \longrightarrow Cu(s)$	0.34
$BiO^+(aq) + 2\,H^+(aq) + 3\,e^- \longrightarrow Bi(s) + H_2O(l)$	0.32
$Hg_2Cl_2(s) + 2\,e^- \longrightarrow 2\,Hg(l) + 2\,Cl^-(aq)$	0.28

(continued)

Half-Reaction	$E°$ **(V)**
$AgCl(s) + e^- \longrightarrow Ag(s) + Cl^-(aq)$	0.22
$SO_4^{2-}(aq) + 4\,H^+(aq) + 2\,e^- \longrightarrow H_2SO_3(aq) + H_2O(l)$	0.17
$Cu^{2+}(aq) + e^- \longrightarrow Cu^+(aq)$	0.15
$Sn^{4+}(aq) + 2\,e^- \longrightarrow Sn^{2+}(aq)$	0.15
$S(s) + 2\,H^+(aq) + 2\,e^- \longrightarrow H_2S(aq)$	0.14
$AgBr(s) + e^- \longrightarrow Ag(s) + Br^-(aq)$	0.07
$2\,H^+(aq) + 2\,e^- \longrightarrow H_2(g)$	0
$Fe^{3+}(aq) + 3\,e^- \longrightarrow Fe(s)$	−0.04
$Pb^{2+}(aq) + 2\,e^- \longrightarrow Pb(s)$	−0.13
$CrO_4^{2-}(aq) + 4\,H_2O(l) + 3\,e^- \longrightarrow Cr(OH)_3(s) + 5\,OH^-(aq)$	−0.13
$Sn^{2+}(aq) + 2\,e^- \longrightarrow Sn(s)$	−0.14
$AgI(s) + e^- \longrightarrow Ag(s) + I^-(aq)$	−0.15
$Ni^{2+}(aq) + 2\,e^- \longrightarrow Ni(s)$	−0.26
$Co^{2+}(aq) + 2\,e^- \longrightarrow Co(s)$	−0.28
$PbSO_4(s) + H^+(aq) + 2\,e^- \longrightarrow Pb(s) + HSO_4^-(aq)$	−0.296
$Tl^+(aq) + e^- \longrightarrow Tl(s)$	−0.34
$Se(s) + 2\,H^+(aq) + 2\,e^- \longrightarrow H_2Se(aq)$	−0.40
$Cd^{2+}(aq) + 2\,e^- \longrightarrow Cd(s)$	−0.40
$Cr^{3+}(aq) + e^- \longrightarrow Cr^{2+}(aq)$	−0.41
$Fe^{2+}(aq) + 2\,e^- \longrightarrow Fe(s)$	−0.45
$2\,CO_2(g) + 2\,H^+(aq) + 2\,e^- \longrightarrow H_2C_2O_4(aq)$	−0.49
$Ga^{3+}(aq) + 3\,e^- \longrightarrow Ga(s)$	−0.55
$Cr^{3+}(aq) + 3\,e^- \longrightarrow Cr(s)$	−0.74
$Zn^{2+}(aq) + 2\,e^- \longrightarrow Zn(s)$	−0.76
$2\,H_2O(l) + 2\,e^- \longrightarrow H_2(g) + 2\,OH^-(aq)$	−0.83
$Cr^{2+}(aq) + 2\,e^- \longrightarrow Cr(s)$	−0.91
$Mn^{2+}(aq) + 2\,e^- \longrightarrow Mn(s)$	−1.18
$Al^{3+}(aq) + 3\,e^- \longrightarrow Al(s)$	−1.66
$Mg^{2+}(aq) + 2\,e^- \longrightarrow Mg(s)$	−2.37
$Na^+(aq) + e^- \longrightarrow Na(s)$	−2.71
$Ca^{2+}(aq) + 2\,e^- \longrightarrow Ca(s)$	−2.87
$Ba^{2+}(aq) + 2\,e^- \longrightarrow Ba(s)$	−2.91
$K^+(aq) + e^- \longrightarrow K(s)$	−2.93
$Li^+(aq) + e^- \longrightarrow Li(s)$	−3.04

APPENDIX E

Properties of Water

Normal melting point	0 °C = 273.15 K
Normal boiling point	100 °C = 373.15 K
Heat of fusion	6.01 kJ/mol at 0 °C
Heat of vaporization	44.94 kJ/mol at 0 °C
	44.02 kJ/mol at 25 °C
	40.67 kJ/mol at 100 °C
Specific heat	4.179 J/(g · °C) at 25 °C
Ion-product constant, K_w	1.15×10^{-15} at 0 °C
	1.01×10^{-14} at 25 °C
	5.43×10^{-13} at 100 °C

TABLE E.1 Vapor Pressure of Water at Various Temperatures

Temp (°C)	P_{vap} (mm Hg)	Temp (°C)	P_{vap} (mm Hg)
0	4.58	60	149.4
10	9.21	70	233.7
20	17.5	80	355.1
30	31.8	90	525.9
40	55.3	100	760.0
50	92.5	105	906.0

Answers to Selected Problems

CHAPTER 0

0.1 (a) 3.72×10^{-10} m; (b) 1.5×10^{11} m **0.2** (a) microgram; (b) decimeter; (c) picosecond; (d) kiloampere; (e) millimole **0.3** 37.0 °C; 310.2 K **0.4** (a) 195 K; (b) 316 °F; (c) 215 °F **0.5** 2.212 g/cm^3 **0.6** 6.32 mL **0.7** (a) 2300 kJ; (b) 6.3 h **0.8** The results are both precise and accurate. **0.9** (a) 5 significant figures; (b) 6 significant figures; (c) 1, 2, 3, or 4 significant figures; (d) 3 significant figures; (e) 18 students is an exact number; (f) 1 significant figure; (g) 4 significant figures; (h) 3 or 4 significant figures **0.10** (a) good precision, poor accuracy; (b) good precision, good accuracy; (c) poor precision, poor accuracy **0.11** (a) 3.774 L; (b) 255 K; (c) 55.27 kg; (d) 906.40 kJ **0.12** (a) 24.612 g; (b) 1.26×10^3 g/L; (c) 41.1 mL **0.13** 32.6 °C (3 significant figures) **0.14** (a) 1947 °F; (b) 6×10^{-11} cm^3 **0.15** 8.88 g; 0.313 ounces **0.16** 2.52 cm^3; 4.45×10^{23} C atoms **0.17** mass of dry rhubarb leaves = 21.0 lb; mass of undried rhubarb leaves = 420 lb ***Conceptual Problems*** **0.18** (a) red; (b) green **0.20** The 5 mL graduated cylinder will give more accurate measurements. ***Section Problems*** **0.22** Mass measures the amount of matter in an object, whereas weight measures the pull of gravity on an object by the earth or other celestial body. **0.24** (a) kilogram, kg; (b) meter, m; (c) kelvin, K; (d) cubic meter, m^3; (e) joule, (kg · m^2)/s^2; (f) kg/m^3 or g/cm^3 **0.26** A Celsius degree is larger than a Fahrenheit degree by a factor of 9/5. **0.28** The volume of a cubic decimeter (dm^3) and a liter (L) are the same. **0.30** (a) and (b) **0.32** cL is centiliter (10^{-2} L) **0.34** (a) 6.02×10^1 km; (b) 46 μs; (c) 200,098 g **0.36** 1×10^9 pg/mg; 3.5×10^4 pg/35 ng **0.38** (a) 5 pm = 5×10^{-10} cm = 5×10^{-3} nm; (b) 8.5 cm^3 = 8.5×10^{-6} m^3 = 8.5×10^3 mm^3; (c) 65.2 mg = 0.0652 g = 6.52×10^{10} pg **0.40** (a) 6 significant figures; (b) 6 significant figures; (c) 4 significant figures; (d) 3 significant figures; (e) 2, 3, 4, or 5 significant figures; (f) 5 significant figures **0.42** 3.6665×10^6 m^3. **0.44** (a) 4.5332×10^2 mg; (b) 4.21×10^{-5}mL; (c) 6.67×10^5 g **0.46** (a) 3.567×10^4 m (4 significant figures); 35,670.1 m (6 significant figures); (b) 69 g (2 significant figures); 68.5 g (3 significant figures); (c) 5.00×10^3 cm (3 significant figures); (d) 2.3099×10^{-4} kg (5 significant figures) **0.48** (a) 10.0; (b) 26; (c) 0.039; (d) 5526; (e) 87.6; (f) 13 **0.50** 11.394 mi/h **0.52** 103.8 °F; 72.0 °F **0.54** 3422 °C; 3695 K **0.56** (a) 1.021 °E/ °C; (b) 0.5675 °E/ °F; (c) H_2O melting point = 119.8 °E; H_2O boiling point = 222.0 °E; (d) 157.6 °E; (e) Because the outside temperature is 50.0 °F, I would wear a sweater or light jacket. **0.58** 2500 kg **0.60** 0.18 cm^3; 162,000 cm^3 **0.62** 11 g/cm^3 **0.64** 169 kcal **0.66** (a) 110 g; (b) 443.2 m; (c) 7.6181×10^{12} m^2 **0.68** (a) 43, 560 ft^3; (b) 3.92×10^8 acre-ft **0.70** (a) 2000 mg/L; (b) 2000 μg/mL; (c) 2 g/L; (d) 2000 ng/μL; (e) 10 g **0.72** 0.61 cm/shake ***Chapter Problems*** **0.74** 2.33 g/cm^3 **0.76** mp = 801 °C = 1474 °F; bp = 1413 °C = 2575 °F **0.78** 75.85 mL **0.80** (a) 0.958 61 g/mL; (b) 4047 m^2; (c) 1400 kg; (d) 140 kg; (e) 792 Cal from fat **0.82** (a) 200 kisses; (b) 3.3 mL; (c) 26 Cal/kiss; (d) 51% **0.84** The Celsius and Fahrenheit scales "cross" at −40 °C (−40 °F). **0.86** 5 tablets **0.88** 34.1 °C **0.90** 45.9 g **0.92** (a) 5.904 g/cm^3; (b) 355 °G

CHAPTER 1

1.1 (a) Cd; (b) Sb; (c) Am **1.2** (a) silver; (b) rhodium; (c) rhenium; (d) cesium; (e) argon; (f) arsenic **1.3** (a) Ti, metal; (b) Te, semimetal; (c) Se, nonmetal; (d) Sc, metal; (e) At, semimetal; (f) Ar, nonmetal **1.4** copper (Cu), silver (Ag), and gold (Au) **1.5** 3/2 **1.6** 2×10^4 Au atoms **1.7** 40 times **1.8** 34 p, 34 e$^-$, 41 n **1.9** ${}^{35}_{17}Cl$ has 18 n; ${}^{37}_{17}Cl$ has 20 n **1.10** ${}^{109}_{47}Ag$ **1.11** 63.55 **1.12** 2.04×10^{22} Cu atoms **1.13** (a) 72.04 g Ti; (b) 7.75 g Na; (c) 614.8 g U **1.14** (a) 0.2405 mol Ti; (b) 1.2670 mol Na; (c) 6.205 mol U **1.15** (a) ${}^{106}_{44}Ru \rightarrow {}^{0}_{-1}e + {}^{106}_{45}Rh$; (b) ${}^{189}_{83}Bi \rightarrow {}^{4}_{2}He + {}^{185}_{81}Tl$; (c) ${}^{204}_{84}Po + {}^{0}_{-1}e \rightarrow {}^{204}_{83}Bi$ **1.16** ${}^{4}_{2}He$ **1.17** ${}^{148}_{69}Tm$ decays to ${}^{148}_{68}Er$ by either positron emission or electron capture. **1.18** H and He **1.19** (a) ${}^{1}_{1}H + {}^{1}_{1}H \rightarrow {}^{2}_{1}H + {}^{0}_{1}n$ (b) ${}^{1}_{1}H + {}^{2}_{1}H \rightarrow {}^{3}_{2}He$

Conceptual Problems

1.20

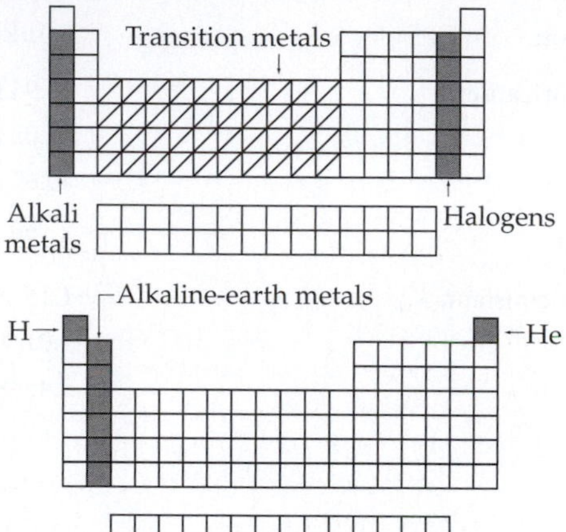

1.22 red—gas; blue—42; green—lithium, sodium, potassium or rubidium are possible answers **1.24** Drawing (a) represents a collection of SO_2 units. **1.26** Figures (b) and (d) **1.28** Figures (b) and (c) represent isotopes of the same element. Figure (a) represents a different element. ***Section Problems*** **1.30** 118 elements are presently known. About 90 elements occur naturally. **1.32** There are 18 groups in the periodic table. They are labeled as follows: 1A, 2A, 3B, 4B, 5B, 6B, 7B, 8B (3 groups), 1B, 2B, 3A, 4A, 5A, 6A, 7A, 8A.

1.34

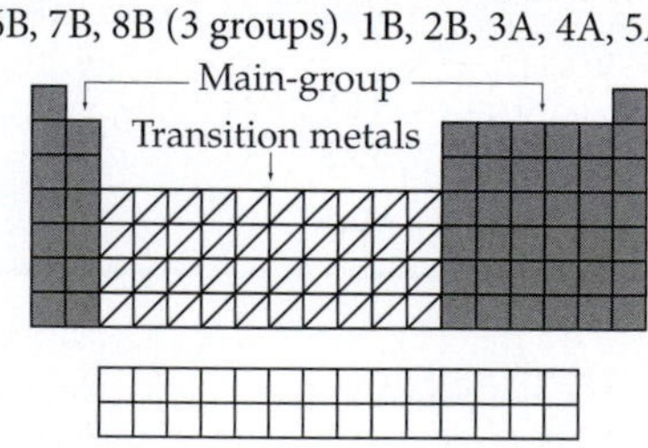

1.36

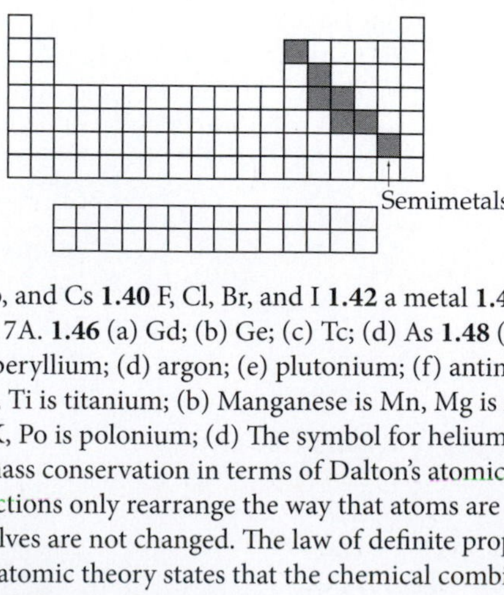

1.38 Li, Na, K, Rb, and Cs **1.40** F, Cl, Br, and I **1.42** a metal **1.44** All match in groups 2A and 7A. **1.46** (a) Gd; (b) Ge; (c) Tc; (d) As **1.48** (a) tellurium; (b) rhenium; (c) beryllium; (d) argon; (e) plutonium; (f) antimony **1.50** (a) Tin is Sn, Ti is titanium; (b) Manganese is Mn, Mg is magnesium; (c) Potassium is K, Po is polonium; (d) The symbol for helium is He. **1.52** The law of mass conservation in terms of Dalton's atomic theory states that chemical reactions only rearrange the way that atoms are combined; the atoms themselves are not changed. The law of definite proportions in terms of Dalton's atomic theory states that the chemical combination of elements to make different substances occurs when atoms join together in small, whole-number ratios. **1.54** 3.7 g

1.56.
$$\frac{\text{C : H mass ratio in benzene}}{\text{C : H mass ratio in ethane}} = \frac{12}{4.00} = \frac{3}{1}$$
$$\frac{\text{C : H mass ratio in benzene}}{\text{C : H mass ratio in ethylene}} = \frac{12}{6.0} = \frac{2}{1}$$
$$\frac{\text{C : H mass ratio in ethylene}}{\text{C : H mass ratio in ethane}} = \frac{6.0}{4.00} = \frac{3}{2}$$

1.58 (a) benzene, CH; ethane, CH_3; ethylene, CH_2 (b) These ratios are consistent with their modern formulas. **1.60** x g **1.62** $5.27 \times 10^{-4}\,x$ g **1.64** 1 Zn:1 S **1.66** The "other" compound is not methane
$$\frac{\text{C : H mass ratio in "other"}}{\text{C : H mass ratio in methane}} = \frac{4}{3}$$

1.68 The atomic number is equal to the number of protons. The mass number is equal to the sum of the number of protons and the number of neutrons. **1.70** The subscript giving the atomic number of an atom is often left off of an isotope symbol because one can readily look up the atomic number in the periodic table. **1.72** ^{63}Cu **1.74** (a) carbon, C; (b) argon, Ar; (c) vanadium, V **1.76** (a) $^{220}_{26}Rn$; (b) $^{210}_{84}Po$; (c) $^{197}_{79}Au$ **1.78** (a) 7 p, 7 e^-, 8 n; (b) 27 p, 27 e^- 33 n; (c) 53 p, 53 e^-, 78 n; (d) 58 p, 58 e^-, 84 n **1.80** (a) $^{24}_{12}Mg$, magnesium; (b) $^{58}_{28}Ni$, nickel; (c) $^{104}_{46}Pd$, palladium; (d) $^{183}_{74}W$, tungsten **1.82** $^{12}_{5}C$; $^{33}_{35}Br$; $^{11}_{5}Bo$ **1.84** 10.8 **1.86** 25.982 for ^{26}Mg; **1.88** Positron emission is the conversion of a proton in the nucleus into a neutron plus an ejected positron. Electron capture is the process in which a proton in the nucleus captures an inner-shell electron and is thereby converted into a neutron. **1.90** In beta emission a neutron is converted to a proton and the atomic number increases. In positron emission a proton is converted to a neutron and the atomic number decreases. **1.92** (a) $^{126}_{50}Sn \rightarrow {}^{0}_{-1}e + {}^{126}_{51}Sb$; (b) $^{210}_{88}Ra \rightarrow {}^{4}_{2}He + {}^{206}_{86}Rn$; (c) $^{77}_{37}Rb \rightarrow {}^{0}_{1}e + {}^{77}_{36}Kr$; (d) $^{76}_{36}Kr + {}^{0}_{-1}e \rightarrow {}^{76}_{35}Br$ **1.94** (a) $^{0}_{1}e$; (b) $^{4}_{2}He$; (c) $^{0}_{-1}e$ **1.96** (a) $^{162}_{75}Re \rightarrow {}^{158}_{73}Ta + {}^{4}_{2}He$; (b) $^{138}_{62}Sm + {}^{0}_{-1}e \rightarrow {}^{138}_{61}Pm$; (c) $^{188}_{74}W \rightarrow {}^{188}_{75}Re + {}^{0}_{-1}e$; (d) $^{165}_{73}Ta \rightarrow {}^{165}_{72}Hf + {}^{0}_{1}e$ **1.98** $^{237}_{93}Np$, $^{233}_{91}Pa$, $^{233}_{92}U$, $^{229}_{90}Th$, $^{225}_{88}Ra$, $^{225}_{89}Ac$, $^{221}_{87}Fr$, $^{217}_{85}At$, $^{213}_{83}Bi$, $^{213}_{84}Po$, $^{209}_{82}Pb$, $^{209}_{83}Bi$ **1.100** 6 α, 4 β ***Chapter Problems*** **1.102** 72.6 **1.104** For NH_3, 0.505 g H; For N_2H_4, 0.337 g H **1.106** (a) I; (b) Kr **1.108** 12.0005 amu **1.110** 151.165 **1.112** $^{100}_{43}Tc \rightarrow {}^{0}_{1}e + {}^{100}_{42}Mo$ (positron emission); $^{100}_{43}Tc + {}^{0}_{-1}e \rightarrow {}^{100}_{42}Mo$ (electron capture) **1.114** (a) 5.53×10^{-17} C/blorvek; (b) 2.77×10^{-17} C/blorvek

CHAPTER 2

2.1 gamma ray, 8.43×10^{18} Hz; radar wave, 2.91×10^{9} Hz **2.2** 2.93 m; 3.14×10^{-10} m **2.3** (b) has the higher frequency. (b) represents the more intense beam of light. (b) represents blue light. (a) represents red light. **2.4** 397.0 nm **2.5** 1875 nm **2.6** 820.4 nm **2.7** 1310 kJ/mol **2.8** IR, 77.2 kJ/mol; UV, 479 kJ/mol; X ray, 2.18×10^{4} kJ/mol **2.9** 390 nm photons have E = 307 kJ/mol, which is not enough energy to eject electrons from zinc. **2.10** 487 kJ/mol, ultraviolet **2.11** 2.34×10^{-38} m **2.12**

n	l	m_l	Orbital	No. of Orbitals
5	0	0	5s	1
	1	−1, 0, +1	5p	3
	2	−2, −1, 0, +1, +2	5d	5
	3	−3, −2, −1, 0, +1, +2, +3	5f	7
	4	−4, −3, −2, −1, 0, +1, +2, +3, +4	5g	9

There are 25 possible orbitals in the fifth shell. **2.13** (a) 2p; (b) 4f; (c) 3d **2.14** (a) $n = 3, l = 0, m_l = 0$; (b) $n = 2, l = 1, m_l = -1, 0, +1$; (c) $n = 4, l = 2, m_l = -2, -1, 0, +1, +2$ **2.15** four nodal planes. **2.16** $n = 4$ and $l = 2$ **2.17** 1.31×10^{3} kJ/mol **2.18** Cr, Cu, Nb, Mo, Ru, Rh, Pd, Ag, La, Ce, Gd, Pt, Au, Ac, Th, Pa, U, Np, Cm, Ds, Rg **2.19** (a) Ti, $1s^2 2s^2 2p^6 3s^2 3p^6 4s^2 3d^2$ or [Ar] $4s^2 3d^2$;

$$[\text{Ar}]\ \underset{4s}{\underline{\uparrow\downarrow}}\quad \underline{\uparrow}\ \underline{\uparrow}\ \underset{3d}{\underline{\ \ }}\ \underline{\ \ }\ \underline{\ \ }$$

(b) Zn, $1s^2 2s^2 2p^6 3s^2 3p^6 4s^2 3d^{10}$ or [Ar] $4s^2 3d^{10}$;

$$[\text{Ar}]\ \underset{4s}{\underline{\uparrow\downarrow}}\quad \underline{\uparrow\downarrow}\ \underline{\uparrow\downarrow}\ \underset{3d}{\underline{\uparrow\downarrow}}\ \underline{\uparrow\downarrow}\ \underline{\uparrow\downarrow}$$

(c) Sn, $1s^2 2s^2 2p^6 3s^2 3p^6 4s^2 3d^{10} 4p^6 5s^2 4d^{10} 5p^2$ or [Kr] $5s^2 4d^{10} 5p^2$;

$$[\text{Kr}]\ \underset{5s}{\underline{\uparrow\downarrow}}\quad \underline{\uparrow\downarrow}\ \underline{\uparrow\downarrow}\ \underset{4d}{\underline{\uparrow\downarrow}}\ \underline{\uparrow\downarrow}\ \underline{\uparrow\downarrow}\quad \underline{\uparrow}\ \underset{5p}{\underline{\uparrow}}\ \underline{\ \ }$$

(d) Pb, [Xe] $6s^2 4f^{14} 5d^{10} 6p^2$ **2.20** Na^+, $1s^2 2s^2 2p^6$; Cl^-, $1s^2 2s^2 2p^6 3s^2 3p^6$ **2.21** Ni **2.22** (a) Ba; (b) Hf; (c) Sn; (d) Lu **2.23** Excited mercury atoms in a fluorescent bulb emit photons, some in the visible but most in the ultraviolet region. Visible photons contribute to light we can see; ultraviolet photons are invisible to our eyes. To utilize this ultraviolet energy, fluorescent bulbs are coated on the inside with a phosphor that absorbs ultraviolet photons and re-emits the energy as visible light. ***Conceptual Problems*** **2.24**

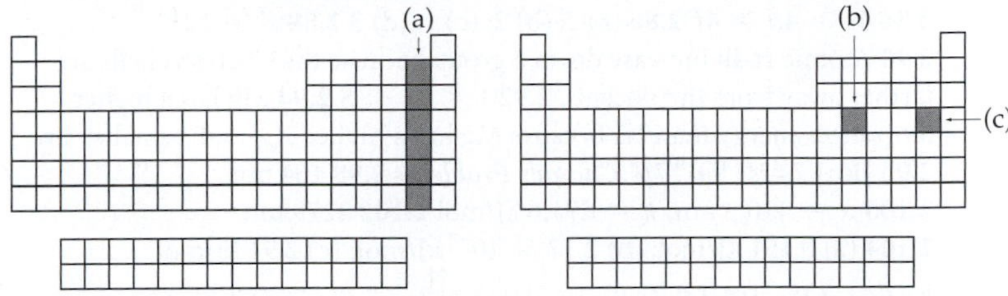

2.26 The green element, molybdenum, has an anomalous electron configuration. Its predicted configuration is [Ar] $5s^2 4d^4$. Its anomalous configuration is [Ar] $5s^1 4d^5$ because of the resulting half-filled d-orbitals. **2.28** Ga **2.30** Sr (215 pm) > Ca (197 pm) > Br (114 pm) ***Section Problems*** **2.32** Violet has the higher frequency and energy. Red has the higher wavelength. **2.34** visible; completely within this range; ultraviolet and infrared; partially in this range. **2.36** 5.5×10^{-8} m **2.38** (a) ν = 99.5 MHz, $E = 3.97 \times 10^{-5}$ kJ/mol, ν = 1150 kHz, $E = 4.589 \times 10^{-7}$ kJ/mol. The FM radio wave (99.5 MHz) has the higher energy. (b) $\lambda = 3.44 \times 10^{-9}$ m, $E = 3.48 \times 10^{4}$ kJ/mol, $\lambda = 6.71 \times 10^{-2}$ m, $E = 1.78 \times 10^{-3}$ kJ/mol. The X ray ($\lambda = 3.44 \times 10^{-9}$ m) has the higher energy. **2.40** (a) 36.4 cm; (b) 34.3 cm **2.42** (a) 1320 nm, near IR; (b) 0.149 m, radio wave; (c) 65.4 nm, UV **2.44** (a) 779 nm, 2.55×10^{-19} J; (b) 649 nm, 3.06×10^{-19} J; (c) 405 nm, 4.91×10^{-19} J **2.46** For $n = 3$, E = 182.3 kJ/mol; For $n = 4$, E = 246.1 kJ/mol; For $n = 5$, E = 275.6 kJ/mol **2.48** 1.09×10^{15} Hz **2.50** 2.45×10^{-12} m, γ ray **2.52** 1.06×10^{-34} m **2.54** 9.14×10^{-24} m/s **2.56** 8×10^{-31} m **2.58** n is the principal quantum number. The size and energy level of an orbital depend on n. l is the angular-momentum quantum number. l defines the three-dimensional shape of an orbital. m_l is the magnetic quantum number. m_l defines the spatial orientation of an orbital. m_s is the spin quantum number. m_s indicates the spin of the electron and can have either of two values, +1/2 or −1/2. **2.60** The probability of finding the electron drops off rapidly as distance from the nucleus increases, although it never drops to zero, even at large distances. As a result, there is no definite boundary or size for an orbital. However, we usually imagine the boundary surface of an orbital enclosing the volume where an electron spends 95% of its time. **2.62** Part of the electron–nucleus attraction is canceled by the electron–electron repulsion, an effect we describe by saying that the electrons are shielded from the nucleus by the other electrons. The net nuclear charge actually felt by an electron is called the effective nuclear charge, Z_{eff}, and is often substantially lower than the actual nuclear charge, Z_{actual}. $Z_{eff} = Z_{actual}$ − electron shielding **2.64** (a) 4s: $n = 4$; $l = 0$; $m_l = 0$; $m_s = \pm 1/2$; (b) 3p: $n = 3$; $l = 1$; $m_l = -1, 0, +1$; $m_s = \pm 1/2$; (c) 5f: $n = 5$; $l = 3$; $m_l = -3, -2, -1, 0, +1, +2, +3$; $m_s = \pm 1/2$; (d) 5d: $n = 5$; $l = 2$; $m_l = -2, -1, 0, +1, +2$; $m_s = \pm 1/2$ **2.66** (a) not allowed because for $l = 0$, $m_l = 0$ only; (b) allowed; (c) not allowed because for $n = 4$, l = 0, 1, 2, or 3 only **2.68** The maximum number of electrons will occur when the 5g orbital is filled: [Rn] $7s^2 5f^{14} 6d^{10} 7p^6 8s^2 5g^{18}$ = 138 electrons **2.70** E = 328.1 kJ/mol **2.72** 363 kJ/mol **2.74** The principal quantum number *n* increases by 1 from one period to the next. As the principal quantum number increases, the number of orbitals in a shell increases. The progression of elements parallels the number of electrons in a particular shell. **2.76** (a) 5*d*; (b) 4*s*; (c) 6*s* **2.78** (a) 3*d*; (b) 4*p*; (c) 6*d*; (d) 6*s* **2.80** (a) $1s^22s^22p^63s^23p^64s^23d^2$; (b) $1s^22s^22p^63s^23p^64s^23d^{10}4p^65s^24d^6$; (c) $1s^22s^22p^63s^23p^64s^23d^{10}4p^65s^24d^{10}5p^2$; (d) $1s^2\,2s^22p^63s^23p^64s^23d^{10}4p^65s^2$; (e) $1s^22s^22p^63s^23p^64s^23d^{10}4p^4$

2.82 (a) Rb, [Kr] ↑ (5*s*)

(b) W, [Xe] ↑↓ (6*s*) ↑↓ ↑↓ ↑↓ ↑↓ ↑↓ ↑↓ ↑↓ (4*f*) ↑ ↑ ↑ ↑ __ (5*d*)

(c) Ge, [Ar] ↑↓ (4*s*) ↑↓ ↑↓ ↑↓ ↑↓ ↑↓ (3*d*) ↑ ↑ __ (4*p*)

(d) Zr, [Kr] ↑↓ (5*s*) ↑ ↑ __ __ __ (4*d*)

2.84 4*s* > 4*d* > 4*f* **2.86** (a) 2; (b) 2; (c) 1; (d) 3 **2.88** $Z = 121$
2.90 Atomic radii increase down a group because the electron shells are farther away from the nucleus. **2.92** F < O < S **2.94** Mg has a higher ionization energy than Na because Mg has a higher Z_{eff} and a smaller size. **2.96** [Rn] $7s^25f^{14}6d^{10}7p^4$ ***Chapter Problems*** **2.98** 164 nm **2.100** $\lambda = 410.2$ nm; $E = 291.6$ kJ/mol **2.102** 2279 nm **2.104** (a) 0.151 kJ/mol; (b) 2.17×10^{-8} kJ/mol; (c) 2.91 kJ/mol

2.106 (a) Ra, [Rn] $7s^2$ [Rn] ↑↓ (7*s*)

(b) Sc, [Ar] $4s^2\,3d^1$ [Ar] ↑↓ (4*s*) ↑ __ __ __ __ (3*d*)

(c) Lr, [Rn] $7s^2\,5f^{14}\,6d^1$

[Rn] ↑↓ (7*s*) ↑↓ ↑↓ ↑↓ ↑↓ ↑↓ ↑↓ ↑↓ (5*f*) ↑ __ __ __ __ (6*d*)

(d) B, [He] $2s^2\,2p^1$ [He] ↑↓ (2*s*) ↑ __ __ (2*p*)

(e) Te, [Kr] $5s^2\,4d^{10}\,5p^4$

[Kr] ↑↓ (5*s*) ↑↓ ↑↓ ↑↓ ↑↓ ↑↓ (4*d*) ↑↓ ↑ ↑ (5*p*)

2.108 637 nm

2.110 (a) Sr, [Kr] ↑↓ (5*s*)

(b) Cd, [Kr] ↑↓ (5*s*) ↑↓ ↑↓ ↑↓ ↑↓ ↑↓ (4*d*)

(c) Ti, [Ar] ↑↓ (4*s*) ↑ ↑ __ __ __ (3*d*)

(d) Se, [Ar] ↑↓ (4*s*) ↑↓ ↑↓ ↑↓ ↑↓ ↑↓ (3*d*) ↑↓ ↑ ↑ (4*p*)

2.112 (a) 1.0×10^{-10} m; (b) 1.2×10^6 kJ/mol; (c) X rays **2.114** K, $Z_{eff} = 2.26$; Kr, $Z_{eff} = 4.06$ **1.116** 8.3×10^{28} photons **1.118** 940 kJ/mol **1.120**

$$\Delta E = \frac{Z^2e^2}{2a_0}\left[\frac{1}{{n_1}^2} - \frac{1}{{n_2}^2}\right]$$

This equation shows that ΔE is proportional to

$$\left[\frac{1}{{n_1}^2} - \frac{1}{{n_2}^2}\right]$$

where n_1 and n_2 are integers with $n_2 > n_1$. This is similar to the Balmer–Rydberg equation where $1/\lambda$ or ν is proportional to

$$\left[\frac{1}{m^2} - \frac{1}{n^2}\right]$$

where *m* and *n* are integers with $n > m$. **2.122** (a) 3*d*, $n = 3, l = 2$; (b) 2*p*, $n = 2, l = 1, m_l = -1, 0, +1$; 3*p*, $n = 3, l = 1, m_l = -1, 0, +1$; 3*d*, $n = 3, l = 2, m_l = -2, -1, 0, +1, +2$; (c) N, $1s^22s^2\,2p^3$ so the 3*s*, 3*p*, and 3*d* orbitals are empty; (d) C, $1s^22s^22p^2$ so the 1*s* and 2*s* orbitals are filled; (e) Be, $1s^22s^2$ so the 2*s* orbital contains the outermost electrons; (f) 2*p* and 3*p* (↑ ↑ __) and 3*d* (↑ ↑ __ __ __)

Multiconcept Problems **2.124** (a) 1.09×10^{15} s^{-1}; (b) 1.8 nm **2.126** (a) 5*f* subshell: $n = 5, l = 3, m_l = -3, -2, -1, 0, +1, +2, +3$; 3*d* subshell: $n = 3, l = 2, m_l = -2, -1, 0, +1, +2$; (b) 1282 nm; (c) 146 kJ/mol

CHAPTER 3

3.1 (a) ionic; (b) molecular; (c) molecular; (d) ionic **3.2** (a) ionic; (b) molecular **3.3** (a) cesium fluoride; (b) potassium oxide; (c) copper(II) oxide; (d) barium sulfide; (e) beryllium bromide **3.4** (a) VCl_3; (b) MnO_2; (c) CuS; (d) Al_2O_3 **3.5** red—potassium sulfide, K_2S; green—strontium iodide, SrI_2; blue—gallium oxide, Ga_2O_3 **3.6** (a) calcium hypochlorite; (b) silver thiosulfate; (c) sodium dihydrogen phosphate; (d) tin(II) nitrate; (e) lead(IV) acetate; (f) ammonium sulfate **3.7** (a) Li_3PO_4; (b) $Mg(HSO_4)_2$; (c) $Mn(NO_3)_2$; (d) $Cr_2(SO_4)_3$ **3.8** Drawing 1 $CaCl_2$; Drawing 2 LiBr and $NaNO_2$ **3.9** [Rn]; (b) [Kr]; (c) [Ar]; (d) [Ne] **3.10** Zn^{2+} **3.11** (a) O^{2-}; (b) S; (c) Fe; (d) H^- **3.12** K^+, $r = 133$ pm; Cl^-, $r = 184$ pm; K, $r = 227$ pm **3.13** (a) Br; (b) S; (c) Se; (d) Ne **3.14** (a) Be; (b) Ga **3.15** (b) Cl has the highest E_{i1} and smallest E_{i4}. **3.16** Al < Kr < Ca **3.17** Cr can accept an electron into a 4*s* orbital. Both Mn and Fe accept the added electron into a 3*d* orbital that contains an electron, but Mn has a lower value of Z_{eff}. **3.18** The least favorable E_{ea} is for Kr. The most favorable E_{ea} is for Ge. **3.19** (a) [Kr]; (b) [Xe]; (c) [Ar]-like configuration (note that Ga^{3+} has ten 3*d* electrons in addition to the two 3*s* and six 3*p* electrons); (d) [Ne] **3.20** gain 2 electrons **3.21** −562 kJ/mol **3.22** (a) KCl; (b) CaF_2; (c) CaO **3.23** (b) MgO has the larger lattice energy. **3.24** (a) O^{2-}; (b) ${O_2}^{2-}$; (c) ${O_2}^-$ **3.25** (a) $2\,Cs(s) + 2\,H_2O(l) \rightarrow 2\,Cs^+(aq) + 2\,OH^-(aq) + H_2(g)$; (b) $Rb(s) + O_2(g) \rightarrow RbO_2(s)$; (c) $2\,K(s) + 2\,NH_3(g) \rightarrow 2\,KNH_2(s) + H_2(g)$; **3.26** (a) $Be(s) + Br_2(l) \rightarrow BeBr_2(s)$; (b) $Sr(s) + 2\,H_2O(l) \rightarrow Sr(OH)_2(aq) + H_2(g)$; (c) $2\,Mg(s) + O_2(g) \rightarrow 2\,MgO(s)$ **3.27** MgS(*s*); −2 **3.28** evaporation of seawater; mining deposits of *halite*, or *rock salt* ***Conceptual Problems*** **3.30** (a) ionic compound; (b) covalent compound **3.32**

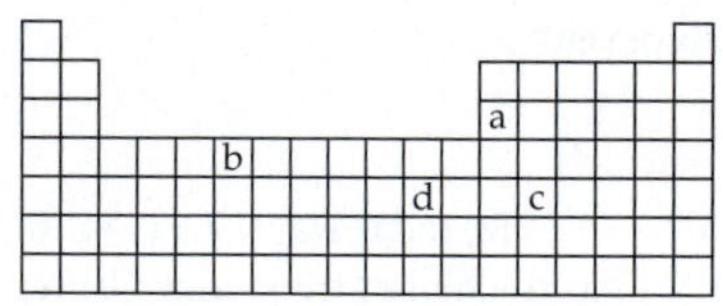

(a) Al^{3+}; (b) Cr^{3+}; (c) Sn^{2+}; (d) Ag^+ **3.34** (a) I_2; (b) Na; (c) NaCl; (d) Cl_2 **3.36** Green, CBr_4; Blue, SrF_2; Red, PbS or PbS_2 ***Section Problems*** **3.38** (a) KCl; (b) $SnBr_2$; (c) CaO; (d) $BaCl_2$; (e) AlH_3 **3.40** (a) barium ion; (b) cesium ion; (c) vanadium(III) ion; (d) hydrogen carbonate ion; (e) ammonium ion; (f) nickel(II) ion; (g) nitrite ion; (h) chlorite ion; (i) manganese(II) ion; (j) perchlorate ion **3.42** (a) $CaBr_2$; (b) $CaSO_4$; (c) $Al_2(SO_4)_3$ **3.44** (a) $CaCl_2$; (b) CaO; (c) CaS **3.46** (a) ${SO_3}^{2-}$; (b) ${PO_4}^{3-}$; (c) Zr^{4+}; (d) ${CrO_4}^{2-}$; (e) ${CH_3CO_2}^-$; (f) ${S_2O_3}^{2-}$ **3.48** (a) Na_2SO_4; (b) $Ba_3(PO_4)_2$; (c) $Ga_2(SO_4)_3$ **3.50** A covalent bond results when two atoms share several (usually two) of their electrons. An ionic bond results from a complete transfer of one or more electrons from one atom to another. The C—H bonds in methane (CH_4) are covalent bonds. The bond in NaCl (Na^+Cl^-) is an ionic bond. **3.52** (a) 4 p, 2 e$^-$, (b) 37 p, 36 e$^-$; (c) 34 p, 36 e$^-$; (d) 79 p, 76 e$^-$; **3.54** (a) La^{3+}, [Xe]; (b) Ag^+, [Kr] $4d^{10}$; (c) Sn^{2+}, [Kr] $5s^2\,4d^{10}$ **3.56** Ca^{2+}, [Ar]; Ti^{2+}, [Ar] $3d^2$ **3.58** Cr^{2+}; Fe^{2+} **3.60** Largest E_{i1} in group 8A because of the largest values of Z_{eff}; smallest E_{i1} in group 1A because of the smallest values of Z_{eff}. **3.62** (a) Ca; (b) Ca **3.64** Ar has the highest E_{i2} and the lowest E_{i7}. **3.66** boron **3.68** (a) lowest: K; highest: Li; (b) lowest: B; highest: Cl; (c) lowest: Ca; highest: Cl **3.70** They have the same magnitude but opposite signs. **3.72** Na^+ has a more negative electron affinity than either Na or Cl. **3.74** because of the positive Z_{eff}. **3.76** The electron–electron repulsion is large and Z_{eff} is low. **3.78** $MgCl_2$ > LiCl > KCl > KBr **3.80** +195 kJ/mol

3.82 −325 kJ/mol **3.84** 808 kJ/mol **3.86** −537 kJ/mol **3.88** −176 kJ/mol
3.90

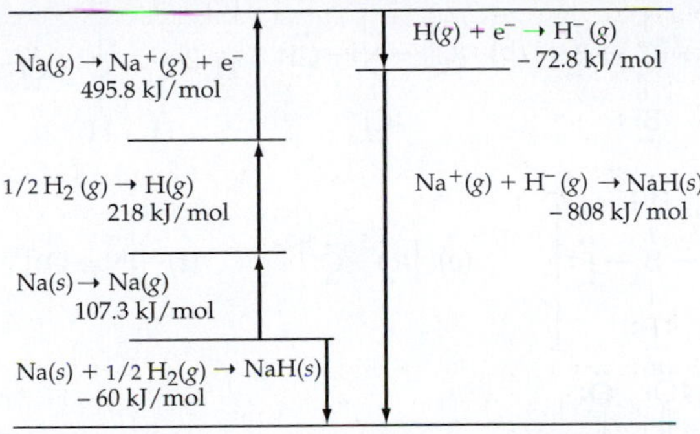

3.92 (a) solid; (b) likely to react with Na, NaAt **3.94** Down each group, the valence electrons are farther from the nucleus and more easily removed. **3.96** (a) $2\,K(s) + 2\,H_2O(l) \rightarrow 2\,K^+(aq) + 2\,OH^-(aq) + H_2(g)$; (b) $2\,K(s) + 2\,NH_3(g) \rightarrow 2\,KNH_2(s) + H_2(g)$; (c) $2\,K(s) + Br_2(l) \rightarrow 2\,KBr(s)$; (d) $K(s) + O_2(g) \rightarrow KO_2(s)$ **3.98** (a) $Cl_2(g) + H_2(g) \rightarrow 2\,HCl(g)$; (b) $Cl_2(g) + Ar(g) \rightarrow$ N. R.; (c) $Cl_2(g) + 2\,Rb(s) \rightarrow 2\,RbCl(s)$ ***Chapter Problems*** **3.100** Cu^{2+} has fewer electrons and a larger effective nuclear charge. **3.102** MgF, −294 kJ/mol; MgF_2, −1114 kJ/mol; MgF_2 will form because the overall energy for the formation of MgF_2 is much more negative than for the formation of MgF. **3.104** Cl^+Na^-, +640 kJ/mol. Formation of Cl^+Na^- from its elements is not favored because the net energy change is positive whereas it is negative for the formation of Na^+Cl^-. **3.106** (a) $2\,Li(s) + 2\,H_2O(l) \rightarrow 2\,Li^+(aq) + 2\,OH^-(aq) + H_2(g)$; (b) $2\,Li(s) + 2\,NH_3(g) \rightarrow 2\,LiNH_2(s) + H_2(g)$; (c) $2\,Li(s) + Br_2(l) \rightarrow 2\,LiBr(s)$; (d) $6\,Li(s) + N_2(g) \rightarrow 2\,Li_3N(s)$; (e) $4\,Li(s) + O_2(g) \rightarrow 2\,Li_2O(s)$ **3.108** (a) $MgCl_2$, magnesium chloride; (b) CaO, calcium oxide; (c) Li_3N, lithium nitride; (d) Al_2O_3, aluminum oxide **3.110** When moving diagonally down and right on the periodic table, the increase in atomic radius caused by going to a larger shell is offset by a decrease caused by a higher Z_{eff}. **3.112** $E_{ea2} = +744$ kJ/mol. O^{2-} is not stable in the gas phase. It is stable in MgO because of the large lattice energy that results from the +2 and −2 charge of the ions and their small size. **3.114** (a) F; (b) Se^{2-} **3.116** 5295 kJ/mol ***Multiconcept Problems*** **3.118** (a) Fe: [Ar] $4s^2\,3d^6$; Fe^{2+}: [Ar] $3d^6$; Fe^{3+}: [Ar] $3d^5$; (b) $n = 3$, $l = 2$. (c) 40.6 nm; (d) The electron removed from Ru to go from Ru^{2+} to Ru^{3+} is a 4*d* electron. The electron with the higher principal quantum number, $n = 4$, is farther from the nucleus, less tightly held, and requires less energy to remove.

CHAPTER 4

4.1

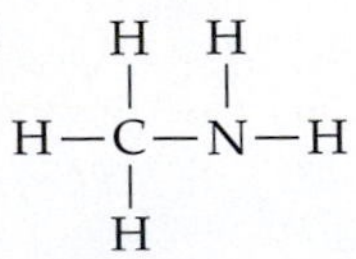

4.2 $C_5H_{11}NO_2S$ **4.3** $C_9H_{13}NO_3$ **4.4** (a) polar covalent; (b) ionic; (c) polar covalent; (d) polar covalent **4.5** $CCl_4 \sim ClO_2 < TiCl_3 < BaCl_2$
4.6 H is positively polarized (blue). O is negatively polarized (red). This is consistent with the electronegativity values for O (3.5) and H (2.1).
4.7 (a) nitrogen trichloride; (b) tetraphosphorus hexoxide; (c) disulfur difluoride; (d) selenium dioxide **4.8** (a) S_2Cl_2; (b) ICl; (c) NI_3
4.9 (a) phosphorus pentachloride; (b) dinitrogen monoxide
4.10 (a) H:S̈:H (b)

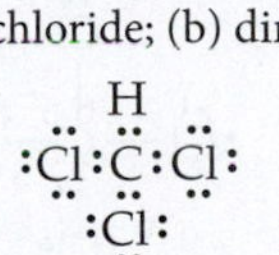

4.11

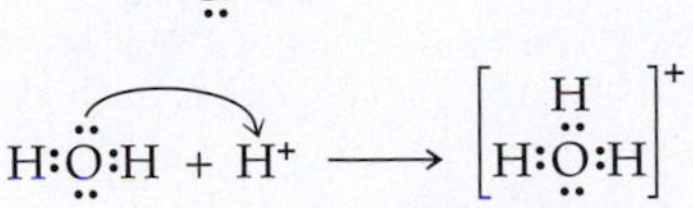

Hydronium ion

4.12

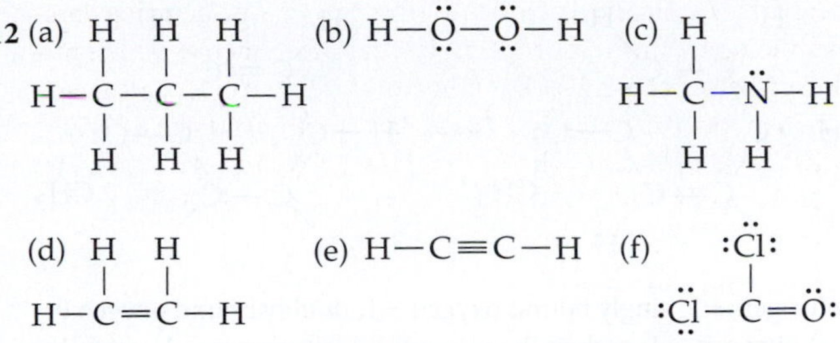

4.13

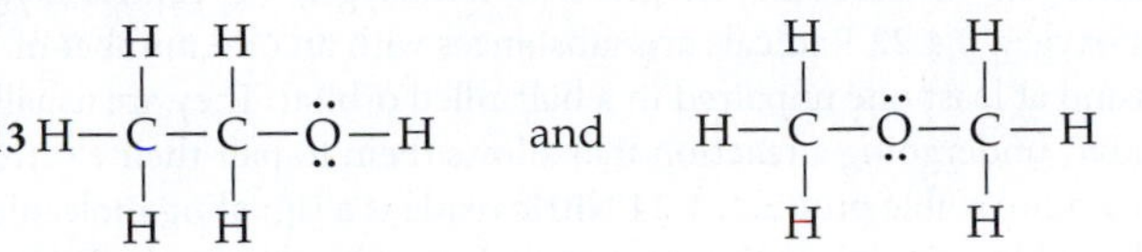

4.14 $C_4H_5N_3O$

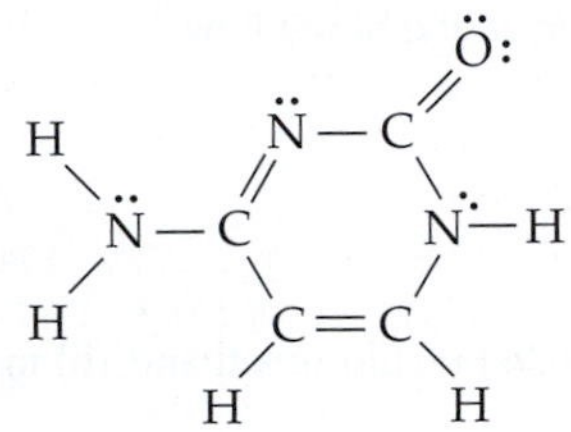

4.15 :C≡O:
4.16

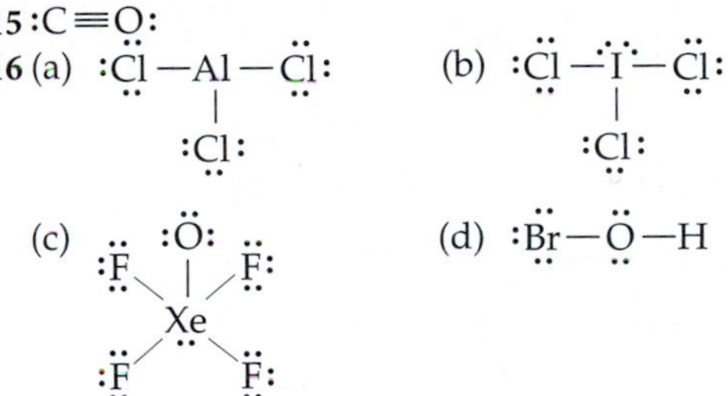

4.17

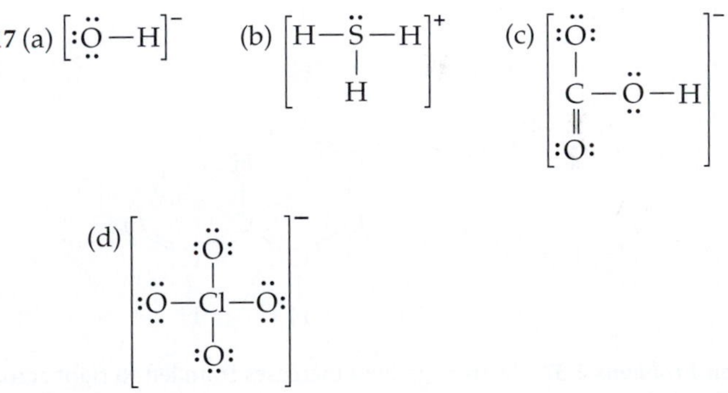

4.18 :N̈=N=Ö: ⟷ :N≡N—Ö̤:
4.19

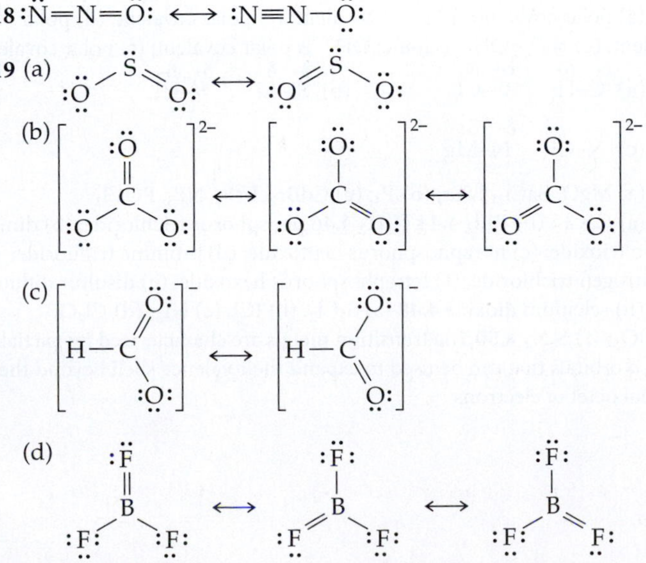

4.20

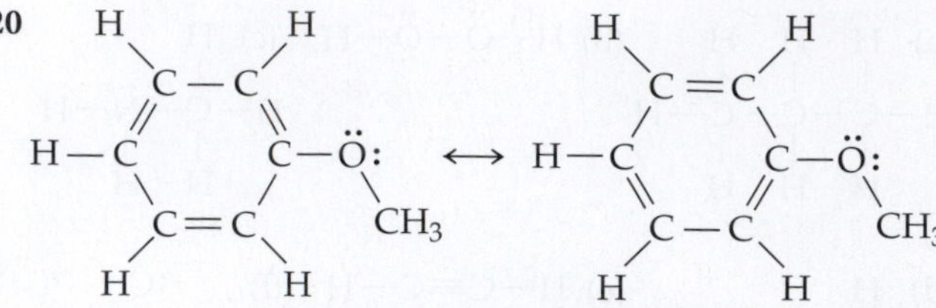

4.21 nitrogen +1; singly bound oxygen −1; doubly bound oxygen 0 **4.22** (a) nitrogen −1; carbon 0; oxygen 0 (b) left oxygen −1; central oxygen +1; right oxygen 0 **4.23** Radicals are substances with an odd number of electrons and at least one unpaired in a half-filled orbital. They are usually very reactive, undergoing a reaction that allows them to pair their electrons and form a more stable product. **4.24** Nitric oxide is a signaling molecule that is released by cells lining the interior surface of blood vessels. This release causes the muscles surrounding the vessels to relax, thereby dilating the blood vessels and increasing blood flow.

4.25

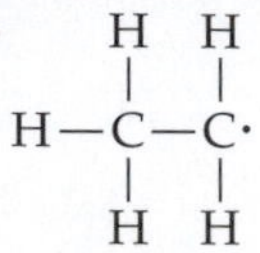

Conceptual Problems **4.26** (a) chloromethane; (b) methyllithium
4.28 $C_8H_9NO_2$;

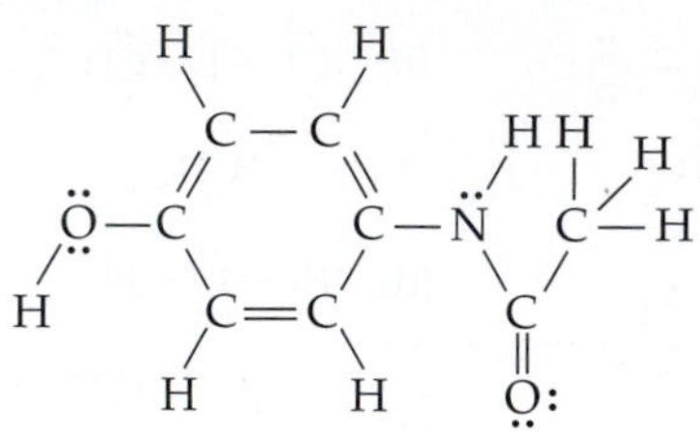

4.30

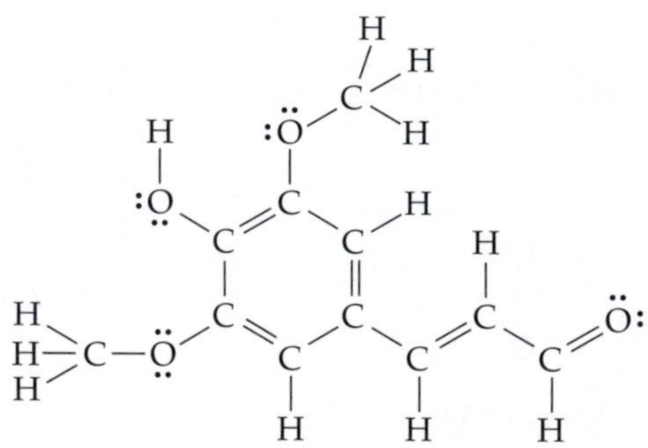

Section Problems **4.32** Electronegativity increases from left to right across a period and decreases down a group. **4.34** K < Li < Mg < Pb < C < Br **4.36** (a) polar covalent; (b) polar covalent; (c) polar covalent; (d) polar covalent; (e) $Na^+{-}OH^-$ is ionic; OH^- is polar covalent; (f) polar covalent.

4.38 (a) $\overset{\delta-}{C}$–$\overset{\delta+}{H}$ $\overset{\delta+}{C}$–$\overset{\delta-}{Cl}$ (b) $\overset{\delta-}{Si}$–$\overset{\delta+}{Li}$ $\overset{\delta+}{Si}$–$\overset{\delta-}{Cl}$

(c) N–Cl $\overset{\delta-}{N}$–$\overset{\delta+}{Mg}$

4.40 (a) MgO, $BaCl_2$, LiBr; (b) P_4; (c) $CdBr_2$, BrF_3, NF_3, $POCl_3$ **4.42** (a) $CaCl_2$; (b) PCl_3 **4.44** N_2H_2 **4.46** phosphorus trichloride; (b) dinitrogen trioxide; (c) tetraphosphorus heptoxide; (d) bromine trifluoride; (e) nitrogen trichloride; (f) tetraphosphorus hexoxide; (g) disulfur difluoride; (h) selenium dioxide **4.48** (a) S_2Cl_2; (b) ICl; (c) NI_3; (d) Cl_2O; (e) ClO_3; (f) S_4N_4 **4.50** The transition metals are characterized by partially filled *d* orbitals that can be used to expand their valence shell beyond the normal octet of electrons.

4.52

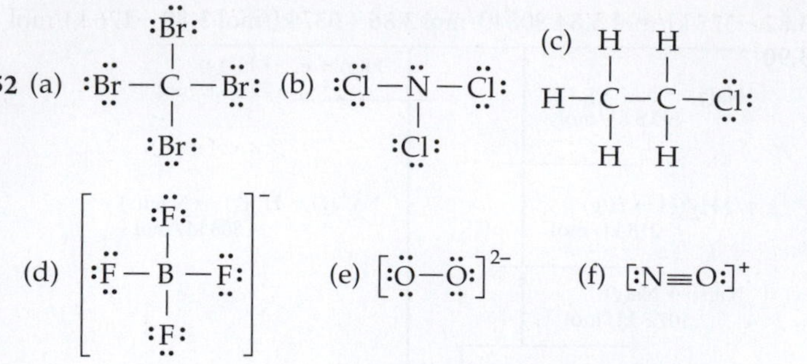

4.54

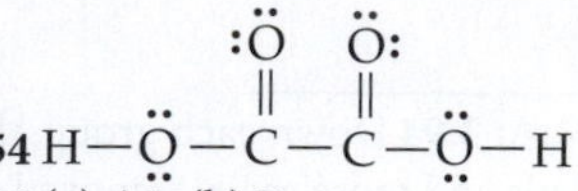

4.56 (a) Al; (b) P

4.58

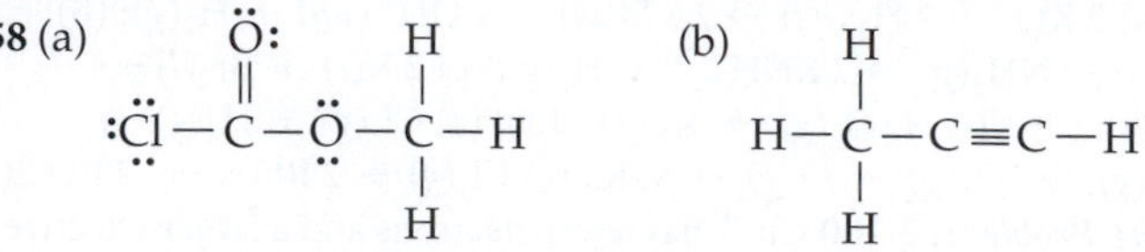

4.60

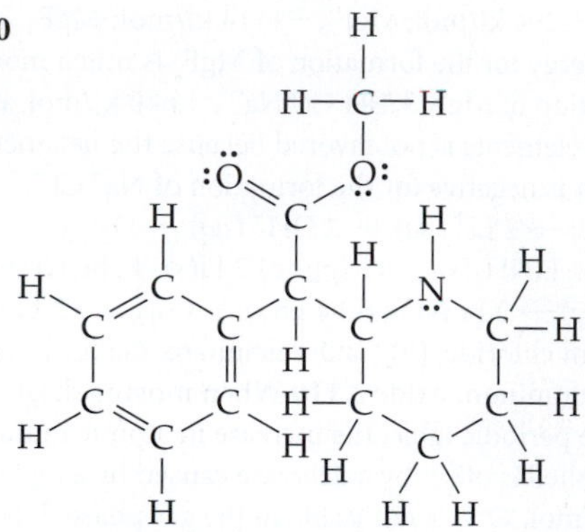

4.62

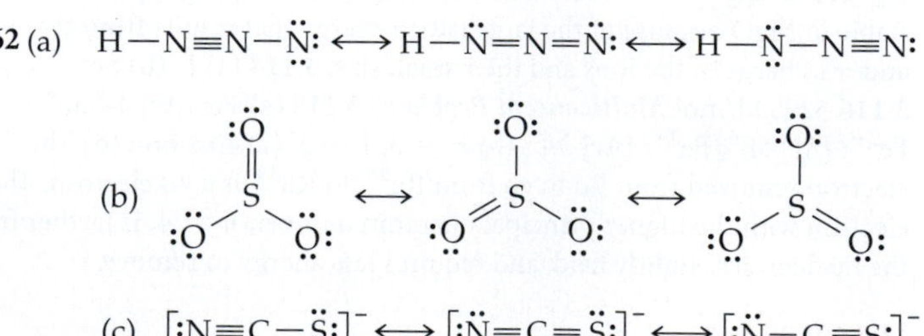

4.64 (a) yes; (b) yes; (c) yes; (d) yes
4.66 :C≡O: carbon −1; oxygen +1;
4.68

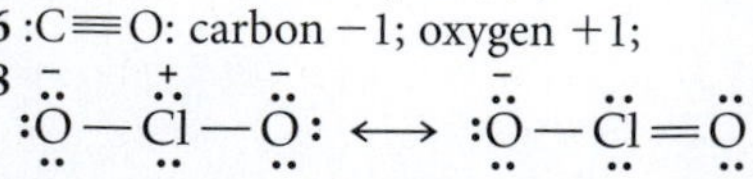

4.70

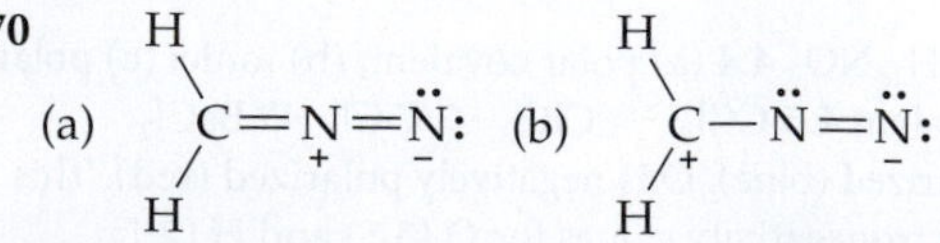

Structure (a) is more important because of the octet of electrons around carbon.

4.72

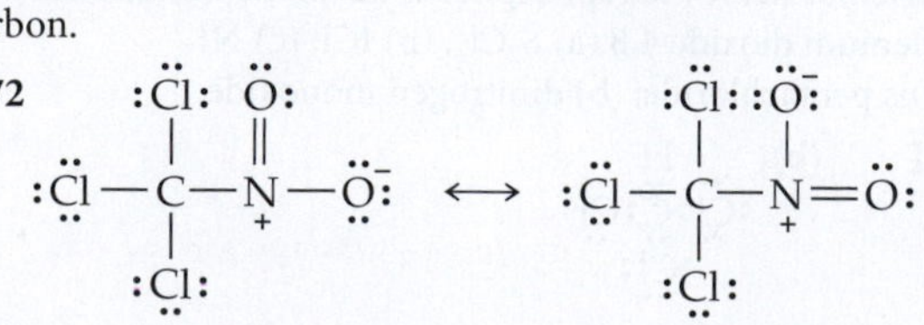

Chapter Problems **4.74** (a) reactants: boron 0; oxygen 0; product: boron −1; oxygen +1

4.76

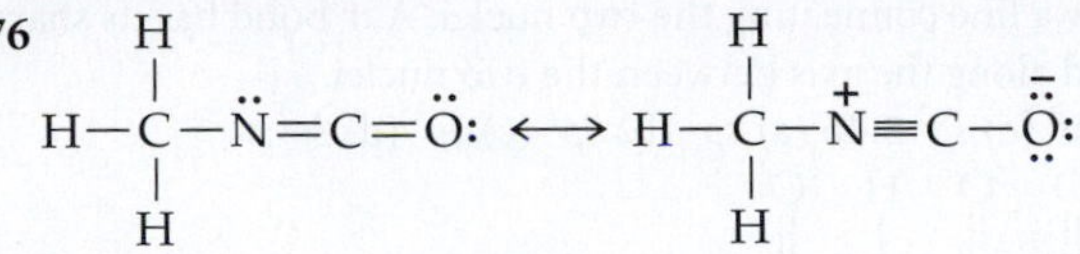

4.78 −109 kJ

4.80

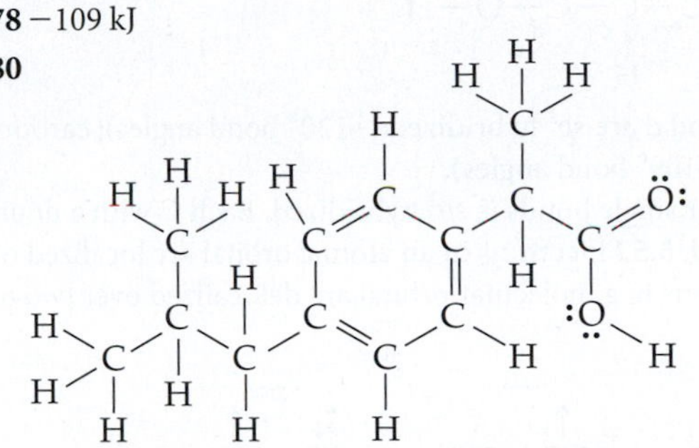

4.82 BF_3 is a resonance hybrid of the following structures:

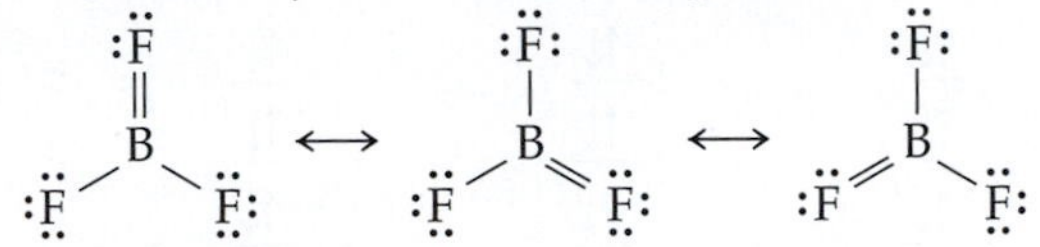

4.84 (a) magnesium fluoride; (b) phosphorus tribromide

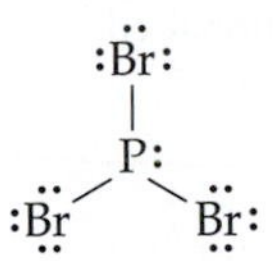

(c) Germanium tetrachoride

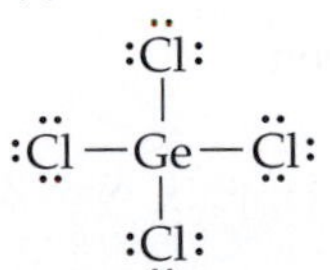

(d) Calcium oxide

Multiconcept Problems **4.86** :C̤̈l—S̤̈—S̤̈—C̤̈l:

4.88 [:Ö—H]⁻   :Ö—H

(b) The oxygen in OH has a half-filled 2*p* orbital that can accept the additional electron. $n = 2$ and $l = 1$; (c) When OH gains an additional electron, it achieves an octet configuration.

CHAPTER 5

5.1 (a) bent; (b) trigonal pyramidal; (c) linear; (d) octahedral;(e) square pyramidal; (f) tetrahedral; (g) tetrahedral; (h) tetrahedral; (i) square planar; (j) trigonal planar

5.2

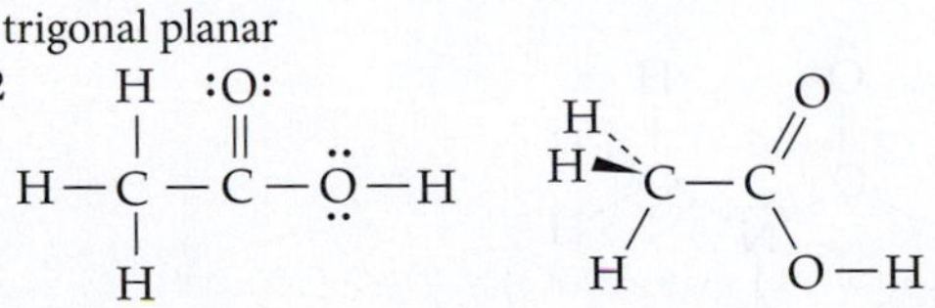

5.3

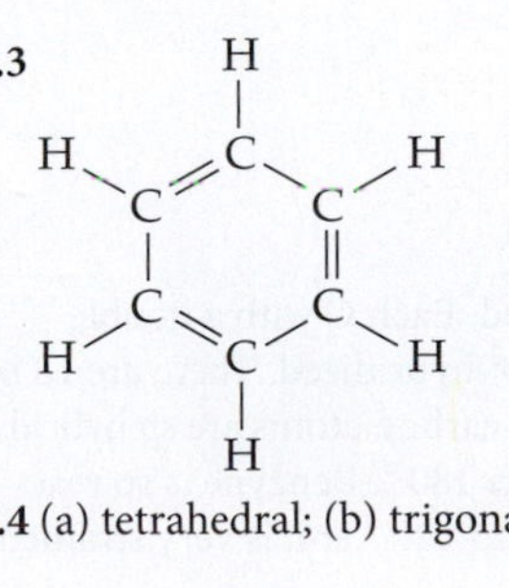

Benzene is a planar hexagon.

5.4 (a) tetrahedral; (b) trigonal bipyramidal

5.5

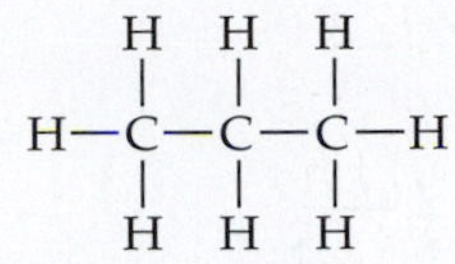

Each C is sp^3 hybridized. The C—C bond are formed by the overlap of one singly occupied sp^3 hybrid orbital from each C. The C—H bonds are formed by the overlap of one singly occupied sp^3orbital on C with a singly occupied H 1*s* orbital. **5.6** The carbon in formaldehyde is sp^2 hybridized.

5.7

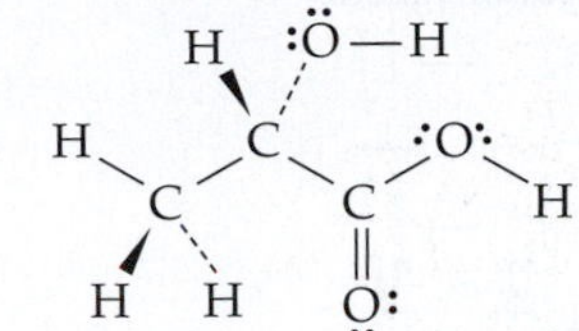

The two C's with four single bonds are sp^3 hybridized. The C with a double bond is sp^2 hybridized. The C—C bonds are formed by the overlap of one singly occupied sp^3 or sp^2 hybrid orbital from each C. The C—H bonds are formed by the overlap of one singly occupied sp^3 orbital on C with a singly occupied H 1*s* orbital. **5.8** In HCN the carbon is *sp* hybridized.

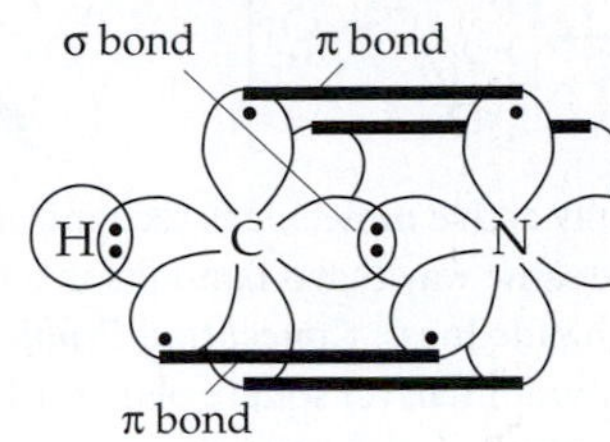

5.9 In CO_2 the carbon is *sp* hybridized.

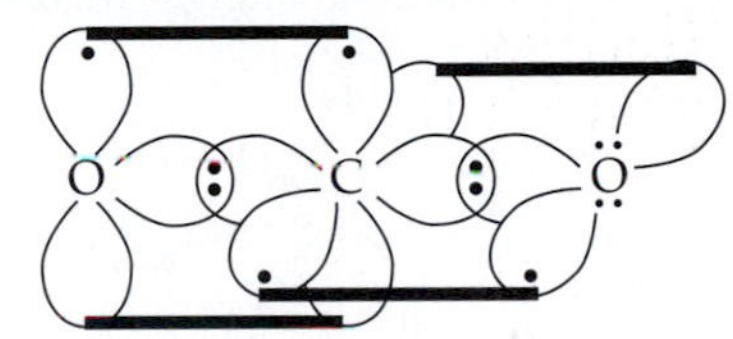

5.10 In Cl_2CO the carbon is sp^2 hybridized.

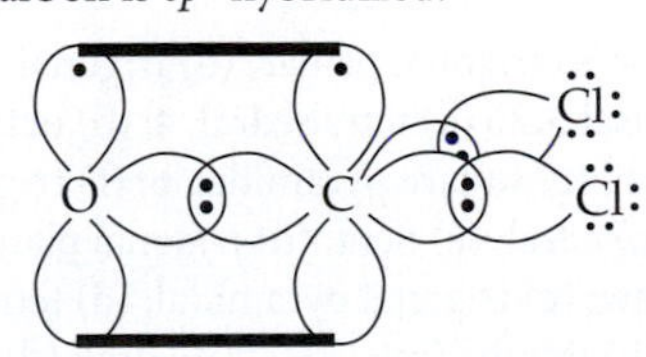

5.11 (a) sp^2 (b) *sp* (c) sp^3

5.12

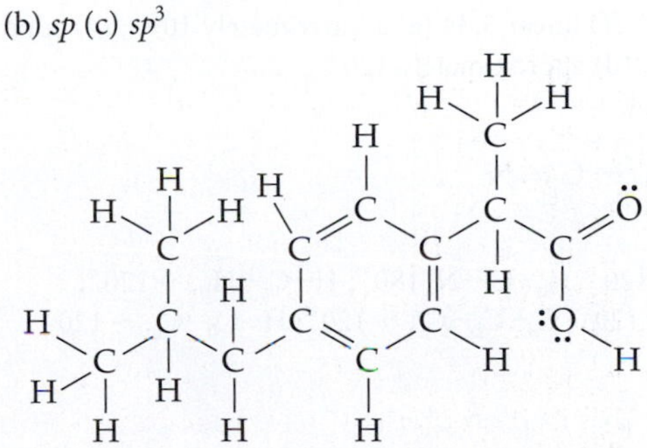

Each C with four single bonds is sp^3 hybridized with tetrahedral geometry around it. Each C with a double bond is sp^2 hybridized with trigonal planar geometry.

5.13 For He_2^+ σ^*_{1s} ↑ σ_{1s} ↑↓

He_2^+ should be stable with a bond order of 1/2.

5.14. For B_2

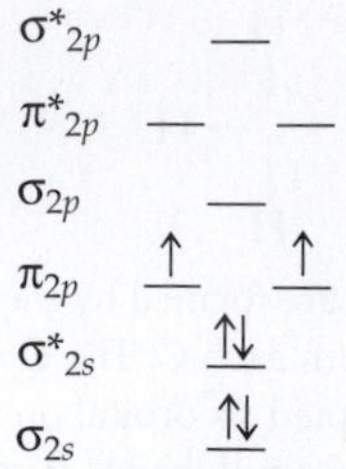

B_2 Bond order = 1; paramagnetic

For C_2

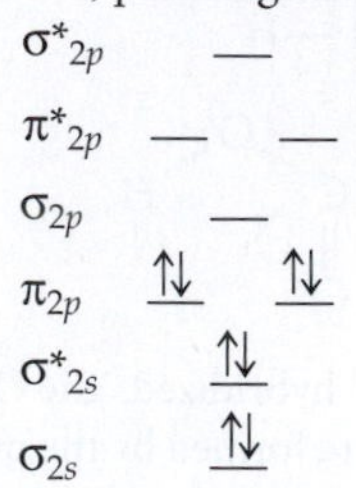

C_2 Bond order = 2; diamagnetic

5.15

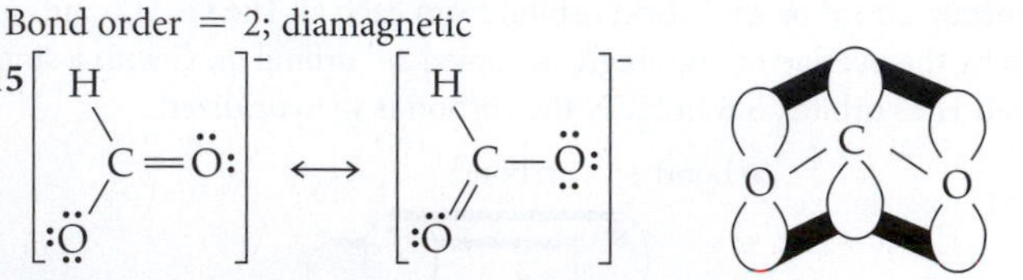

5.16 because a biologically active molecule is recognized and fits in a receptor molecule in the same way that a hand fits in a glove. **5.17** (a) has no handedness; (b) has handedness. ***Conceptual Problems*** **5.18** (a) square pyramidal; (b) trigonal pyramidal; (c) square planar; (d) trigonal planar **5.20** (c) does not have a tetrahedral central atom.

5.22

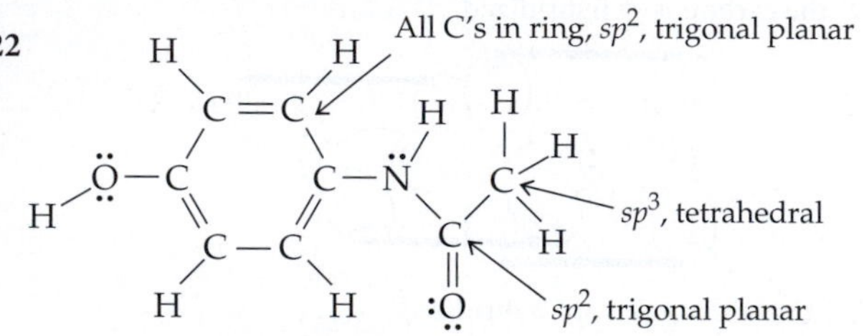

Section Problems **5.24** (a) trigonal planar; (b) trigonal bipyramidal; (c) linear; (d) octahedral **5.26** (a) tetrahedral, 4; (b) octahedral, 6; (c) bent, 3 or 4; (d) linear, 2 or 5; (e) square pyramidal, 6; (f) trigonal pyramidal, 4 **5.28** (a) bent; (b) tetrahedral; (c) bent; (d) trigonal planar **5.30** (a) trigonal bipyramidal; (b) seesaw; (c) trigonal pyramidal; (d) tetrahedral **5.32** (a) tetrahedral; (b) tetrahedral; (c) tetrahedral; (d) trigonal pyramidal; (e) tetrahedral; (f) linear **5.34** (a) approximately 109 °; (b) approximately 120 °; (c) 90 °; (d) approximately 120 °

5.36

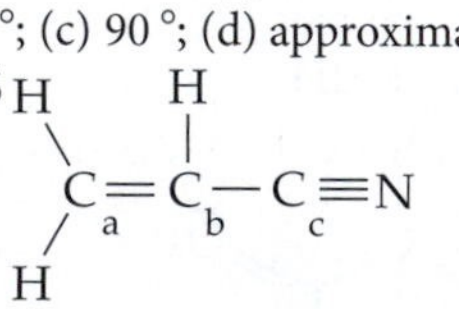

$H-C_a-H$, ~120 °, C_b-C_c-N, 180 °; $H-C_a-H_b$, ~120 °; C_a-C_b-H, ~120 °; $C_a-C_b-C_c$, ~120 °; $H-C_b-C_c$, ~120 °

5.38

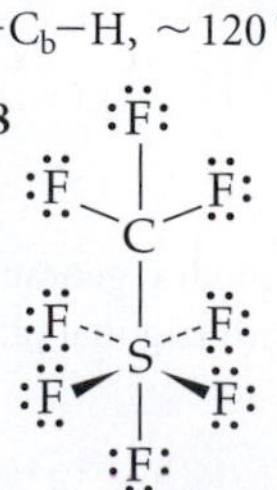

109.5 ° around carbon and 90 ° around S. **5.40** The geometry about each carbon is tetrahedral with a C–C–C bond angle of approximately 109 °. Because the geometry about each carbon is tetrahedral, the cyclohexane ring cannot be flat. **5.42** In a π bond, the shared electrons occupy a region above and below a line connecting the two nuclei. A σ bond has its shared electrons located along the axis between the two nuclei. **5.44** (a) sp (b) sp^2 (c) sp^3 **5.46** (a) sp^2 (b) sp^2 (c) sp^3 (d) sp^2

5.48

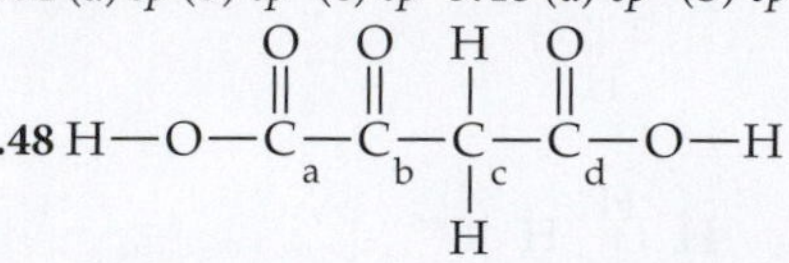

Carbon atoms a, b, and d are sp^2 hybridized (~120° bond angles); carbon c is sp^3 hybridized (~109° bond angles). **5.50** Each C with four single bonds is sp^2 hybridized. Each C with a double bond is sp^2 hybridzed. **5.52** Electrons in an atomic orbital are localized on a single atom. Electrons in a molecular orbital are delocalized over two or more atoms.

5.54

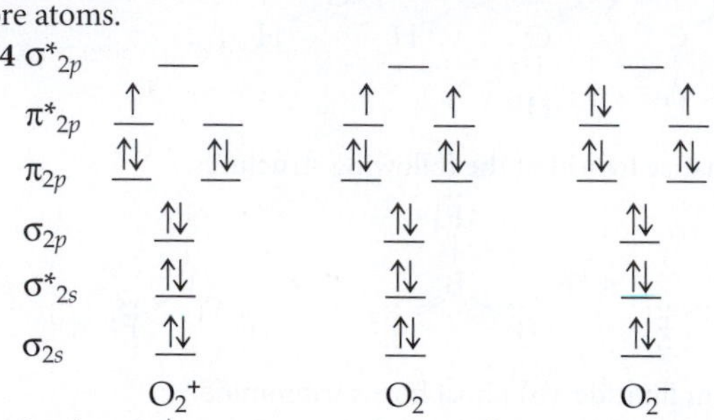

Bond order: O_2^+, 2.5; O_2, 2; O_2^-, 1.5. All are stable. All have unpaired electrons. **5.56** (a) C_2 bond order = 2; (b) Add one electron; (c) C_2^- bond order = 2.5 **5.58** (a) diamagnetic; (b) paramagnetic; (c) paramagnetic; (d) diamagnetic; (e) paramagnetic

5.60

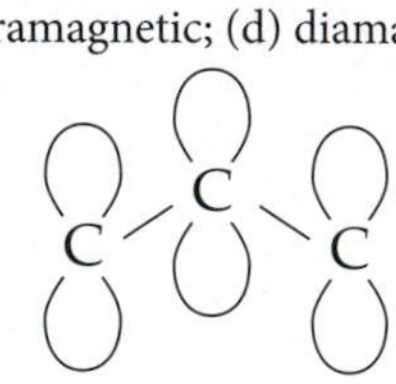

p orbitals in allyl cation

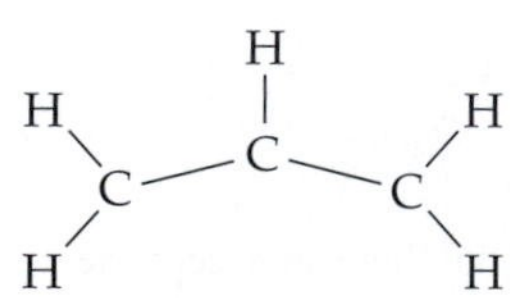

allyl cation showing only the σ bonds (each C is sp^2 hybridized)

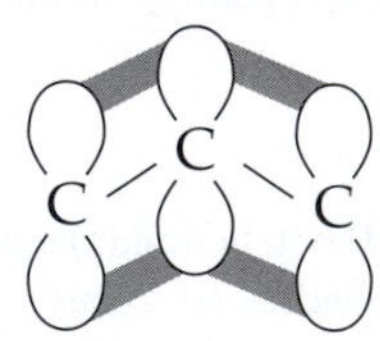

delocalized MO model for π bonding in the allyl cation

5.62

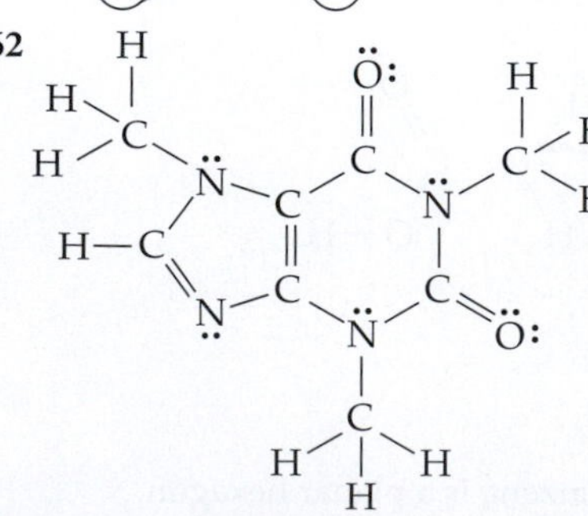

Each C with four single bonds is sp^3 hybridized. Each C with a double bond is sp^2 hybridized **5.64** Every carbon is sp^2 hybridized. There are 18 σ bonds and 5 π bonds. **5.66** The triply bonded carbon atoms are sp hybridized. The theoretical bond angle for C–C≡C is 180 °. Benzyne is so reactive because the C–C≡C bond angle is closer to 120 ° and is very strained.

5.68 C_2^{2-}

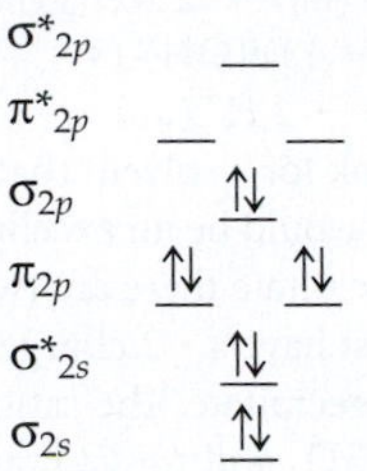

bond order = 3

5.70 (a)

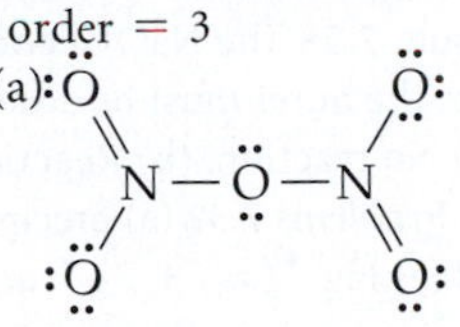

The molecule is bent about the central O and trigonal planar about each N. (b) The central O is sp^3 hybridized. The N's are sp^2 hybridized. (c)

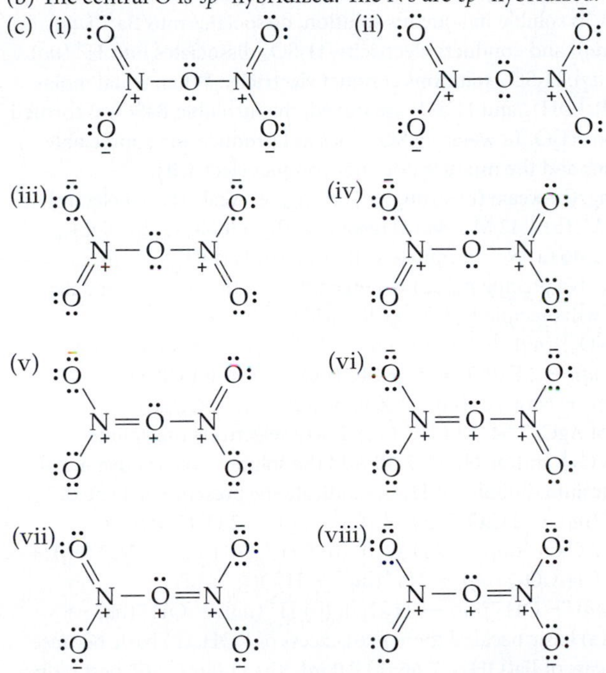

Structures (a)–(d) make more important contributions to the resonance hybrid because of only −1 and 0 formal charges on the oxygens.

5.72

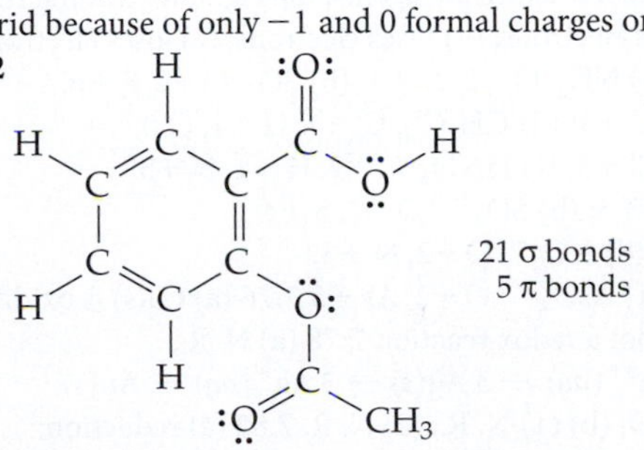

Each C with a double bond is sp^2 hybridized. The CH_3 carbon is sp^3 hybridized.

5.74 (a)

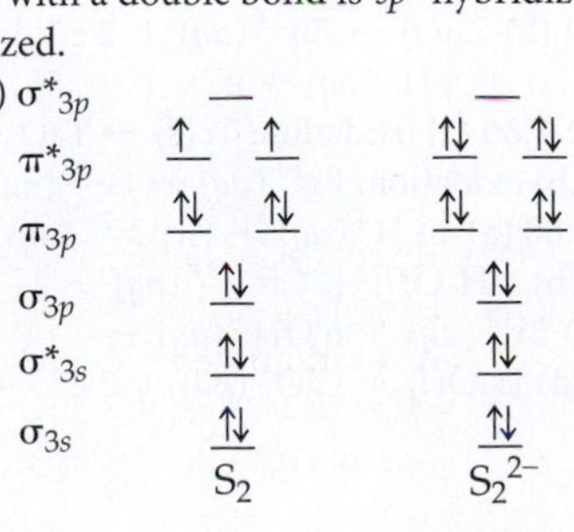

(b) paramagnetic; (c) 2; (d) The two added electrons go into the antibonding π^*_{3p} MOs, and the bond length in S_2^{2-} should be longer.

5.76 (a)

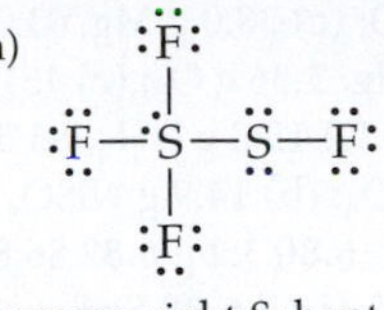

left S: seesaw; right S: bent

(b)

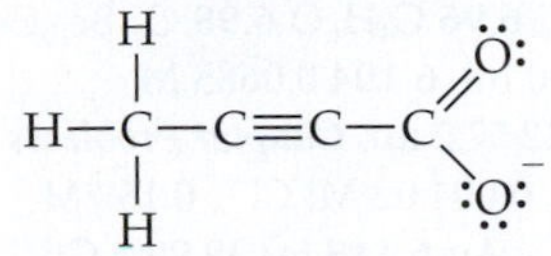

left C: tetrahedral; right C: trigonal planar; central two C's: linear.

5.78 (a) PF_3; (b) PCl_4^+; (c) PCl_6^-

5.80 (a)

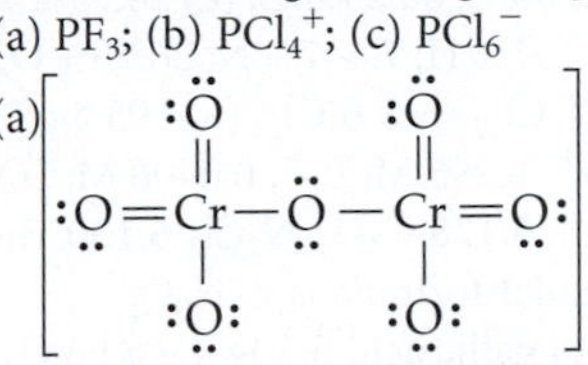

(b) Each Cr atom has 6 pairs of electrons around it. The likely geometry about each Cr atom is tetrahedral because each Cr has 4 charge clouds.

5.82 (a) Each carbon is sp^2 hybridized; (b) & (c);

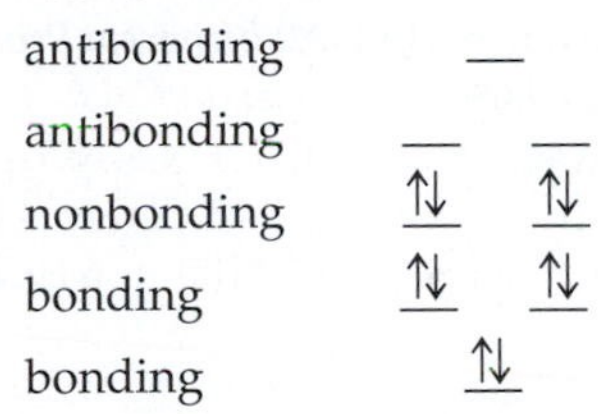

(d) The cyclooctatetraene dianion has only paired electrons and is diamagnetic.

CHAPTER 6

6.1 $2\ NaClO_3 \rightarrow 2\ NaCl + 3\ O_2$ **6.2** (a) $C_6H_{12}O_6 \rightarrow 2\ C_2H_6O + 2\ CO_2$; (b) $6\ CO_2 + 6\ H_2O \rightarrow C_6H_{12}O_6 + 6\ O_2$; (c) $4\ NH_3 + Cl_2 \rightarrow N_2H_4 + 2\ NH_4Cl$ **6.3** $3\ A_2 + 2\ B \rightarrow 2\ BA_3$ **6.4** (a) 159.7; (b) 98.1; (c) 192.1; (d) 334.4 **6.5** 2.77×10^{-3} mol; 1.67×10^{21} molecules **6.6** (a) 3.33 g $C_4H_6O_3$; (b) 5.87 g $C_9H_8O_4$; (c) 1.96 g CH_3CO_2H **6.7** 63% **6.8** 4220 g **6.9** Li_2O is the limiting reactant; 41 kg H_2O **6.10** 921 g CO_2 **6.11** (a) $A + B_2 \rightarrow AB_2$; A is the limiting reactant; (b) 1.0 mol of AB_2 **6.12** CH_4N; 39.9% C, 13.4% H, 46.6% N **6.13** $MgCO_3$ **6.14** 37.5% C, 4.21% H, 58.3% O **6.15** $C_{10}H_{20}O$ **6.16** $C_5H_{10}O_5$ **6.17** (a) B_2H_6; (b) $C_3H_6O_3$ **6.18** (a) 0.025 mol; (b) 1.62 mol **6.19** (a) 25.0 g; (b) 67.6 g **6.20** 690 mL **6.21** 1 g **6.22** 0.656 M **6.23** Dilute 6.94 mL of 18.0 M H_2SO_4 with enough water to make 250.0 mL of solution. **6.24** 10.0 mL **6.25** 5.47×10^{-2} M **6.26** 0.758 M **6.27** Because the two volumes are equal and the concentrations are proportional to the number of solute ions, $[OH^-] = 0.67$ M. **6.28** The assumptions that (i) the oil molecules are tiny cubes; (ii) the oil layer is one molecule thick; (iii) the molecular mass of 900 for the oil **6.29** 1.0×10^{24} molecules/mole ***Conceptual Problems*** **6.30** box (b) **6.32** 0.004 mol CO_2; 0.0025 mol H_2O **6.34** 309.36 **6.36** (a) $A_2 + 3\ B_2 \rightarrow 2\ AB_3$; B_2 is the limiting reactant; (b) 2/3 mol AB_3 ***Section Problems*** **6.38** Equation (b) is balanced, (a) is not balanced. **6.40** (a) $Mg + 2\ HNO_3 \rightarrow H_2 + Mg(NO_3)_2$; (b) $CaC_2 + 2\ H_2O \rightarrow Ca(OH)_2 + C_2H_2$; (c) $2\ S + 3\ O_2 \rightarrow 2\ SO_3$; (d) $UO_2 + 4\ HF \rightarrow UF_4 + 2\ H_2O$ **6.42** (a) $SiCl_4 + 2\ H_2O \rightarrow SiO_2 + 4\ HCl$; (b) $P_4O_{10} + 6\ H_2O \rightarrow 4\ H_3PO_4$; (c) $CaCN_2 + 3\ H_2O \rightarrow CaCO_3 + 2\ NH_3$; (d) $3\ NO_2 + H_2O \rightarrow 2\ HNO_3 + NO$ **6.44** (a) Hg_2Cl_2: 472.1; (b) $C_4H_8O_2$: 88.1; (c) CF_2Cl_2: 120.9 **6.46** (a) 558.7; (b) 444.5; (c) 321.8 **6.48** (a) 47.87 g; (b) 159.81 g; (c) 200.59 g; (d) 18.02 g **6.50** 0.867 mol **6.52** 119 **6.54** 1.97×10^{-3} mol $FeSO_4$; 1.19×10^{21} Fe(II) atoms

6.56 6.44×10^{-4} mol; 3.88×10^{20} molecules **6.58** Ne **6.60** 166.8 kg **6.62** (a) $2\ Fe_2O_3 + 3\ C \rightarrow 4\ Fe + 3\ CO_2$; (b) 4.93 mol C; (c) 59.2 g C **6.64** (a) $2\ Mg + O_2 \rightarrow 2\ MgO$; (b) 16.5 g O_2, 41.5 g MgO; (c) 38.0 g Mg, 63.0 g MgO **6.66** (a) $2\ HgO \rightarrow 2\ Hg + O_2$; (b) 42.1 g Hg, 3.36 g O_2; (c) 451 g HgO **6.68** AgCl **6.70** (a) 581 g I_2; (b) 1847 g HI **6.72** 15.8 g NH_3, 83.3 g N_2 left over **6.74** 5.22 g $C_2H_4Cl_2$ **6.76** (a) 12.0 g H_2SO_4; (b) 14.9 g $NiSO_4$ **6.78** 0.526 L CO_2, $CaCO_3$ is the limiting reactant. **6.80** 3.2 g **6.82** 86.8% **6.84** 20.0% C, 6.72% H, 46.6% N, 26.6% O **6.86** C_7H_7Cl **6.88** SnF_2 **6.90** C_7H_8 **6.92** 13,000 **6.94** Disilane is Si_2H_6 **6.96** C_4H_6O **6.98** $C_{12}Br_{10}O$ **6.100** (a) 0.0420 mol; (b) 0.12 mol **6.102** 160 mL **6.104** 0.0685 M **6.106** 0.958 M **6.108** 1.71 M **6.110** 15.5 g **6.112** 57.2 mL ***Chapter Problems*** **6.114** Na^+, 0.147 M; Ca^{2+}, 0.002 98 M; K^+, 0.004 02 M; Cl^-, 0.157 M **6.116** (a) 5.1×10^{-11} mol/L; (b) 1.3×10^{13} g Au **6.118** (a) 39.99% C; 6.713% H; 53.27% O; (b) 2.055% H; 32.70% S; 65.25% O; (c) 24.75% K; 34.77% Mn; 40.51% O; (d) 45.89% C; 2.751% H; 7.647% N; 26.20% O; 17.51% S **6.120** (a) $6\ WCl_6 + 8\ Bi \rightarrow W_6Cl_{12} + 8\ BiCl_3$; (b) 105.4 g Bi; (c) 146 g W_6Cl_{12} **6.122** $C_{10}H_{10}Fe$ **6.124** Na^+, 0.295 M; Li^+, 0.0406 M; SO_4^{2-}, 0.0590 M; PO_4^{3-}, 0.0725 M **6.126** 18.1 lb **6.128** $C_3H_{12}N_2Cl_2$ **6.130** The empirical formula is $C_7H_6O_3$; the molecular formula is $C_7H_6O_3$. **6.132** 0.63 g of benzoic acid and 0.37 g of gallic acid **6.134** 4.4 g Fe_2O_3; 5.6 g FeO **6.136** $C_6H_{12}O_6 + 6\ O_2 \rightarrow 6\ CO_2 + 6\ H_2O$; 97.2 g CO_2, 56.1 L CO_2 **6.138** (a) 79.91% Cu, 20.1% S; (b) Cu_2S; (c) 4.2×10^{22} Cu^+ ions/cm^3 **6.140** The mass ratio of NH_4NO_3 to $(NH_4)_2HPO_4$ in the mixture is 2 to 1. ***Multiconcept Problems*** **6.142** (a) 80; Br; (b) 64; Cu **6.144** (a) 64.0% Cl; (b) Ca; (c) $Ca(s) + Cl_2(g) \rightarrow CaCl_2(s)$: $CaCl_2(aq) + 2\ AgNO_3(aq) \rightarrow 2\ AgCl(s) + Ca(NO_3)_2(aq)$; (d) 0.47 g of Cl_2 unreacted **6.146** (a) (i) $M_2O_3(s) + 3\ C(s) + 3\ Cl_2(g) \rightarrow 2\ MCl_3(l) + 3\ CO(g)$; (ii) $2\ MCl_3(l) + 3\ H_2(g) \rightarrow 2\ M(s) + 6\ HCl(g)$; (b) 10.8; B; (c) 0.265 g

CHAPTER 7

7.1 0.675 M **7.2** A_2Y is the strongest electrolyte because it is completely dissociated into ions. A_2X is the weakest electrolyte because it is the least dissociated. **7.3** (a) precipitation; (b) redox; (c) acid–base neutralization **7.4** (a) $2\ Ag^+(aq) + CrO_4^{2-}(aq) \rightarrow Ag_2CrO_4(s)$; (b) $2\ H^+(aq) + MgCO_3(s) \rightarrow H_2O(l) + CO_2(g) + Mg^{2+}(aq)$; (c) $Hg^{2+}(aq) + 2\ I^-(aq) \rightarrow HgI_2(s)$ **7.5** (a) $CdCO_3$, insoluble; (b) MgO, insoluble; (c) Na_2S, soluble; (d) $PbSO_4$, insoluble; (e) $(NH_4)_3PO_4$, soluble; (f) $HgCl_2$, soluble **7.6** (a) $Ni^{2+}(aq) + S^{2-}(aq) \rightarrow NiS(s)$; (b) $Pb^{2+}(aq) + CrO_4^{2-}(aq) \rightarrow PbCrO_4(s)$; (c) $Ag^+(aq) + Br^-(aq) \rightarrow AgBr(s)$; (d) $Zn^{2+}(aq) + CO_3^{2-}(aq) \rightarrow ZnCO_3(s)$ **7.7** $3\ CaCl_2(aq) + 2\ Na_3PO_4(aq) \rightarrow Ca_3(PO_4)_2(s) + 6\ NaCl(aq)$; $3\ Ca^{2+}(aq) + 2\ PO_4^{3-}(aq) \rightarrow Ca_3(PO_4)_2(s)$ **7.8** The precipitate is either $Mg_3(PO_4)_2$ or $Zn_3(PO_4)_2$. **7.9** (a) periodic acid; (b) bromous acid; (c) chromic acid **7.10** (a) H_3PO_3; (b) H_2Se **7.11** (a) $2\ Cs^+(aq) + 2\ OH^-(aq) + 2\ H^+(aq) + SO_4^{2-}(aq) \rightarrow 2\ Cs^+(aq) + SO_4^{2-}(aq) + 2\ H_2O(l)$; $H^+(aq) + OH^-(aq) \rightarrow H_2O(l)$; (b) $Ca^{2+}(aq) + 2\ OH^-(aq) + 2\ CH_3CO_2H(aq) \rightarrow Ca^{2+}(aq) + 2\ CH_3CO_2^-(aq) + 2\ H_2O(l)$; $CH_3CO_2H(aq) + OH^-(aq) \rightarrow CH_3CO_2^-(aq) + H_2O(l)$ **7.12** HY is the strongest acid; HX is the weakest acid. **7.13** (a) Cl −1, Sn +4; (b) O −2, Cr +6; (c) O −2, Cl −1, V +5; (d) O −2, V +3; (e) O −2, H +1, N +5; (f) O −2, S +6, Fe +2 **7.14** $2\ Cu^{2+}(aq) + 4\ I^-(aq) \rightarrow 2\ CuI(s) + I_2(aq)$; Cu^{2+} + 2; I^- − 1; CuI: Cu +1, I −1; I_2: 0; oxidizing agent, Cu^{2+}; reducing agent, I^- **7.15** (a) C is oxidized. C is the reducing agent. The Sn in SnO_2 is reduced. SnO_2 is the oxidizing agent. (b) Sn^{2+} is oxidized. Sn^{2+} is the reducing agent. Fe^{3+} is reduced. Fe^{3+} is the oxidizing agent. (c) The N in NH_3 is oxidized. NH_3 is the reducing agent. Each O in O_2 is reduced. O_2 is the oxidizing agent. **7.16** (a) N. R.; (b) N. R. **7.17** Because B will reduce A^+, B is above A in the activity series. Because B will not reduce C^+, C is above B in the activity series. Therefore C must be above A in the activity series and C will reduce A^+. **7.18** A > D > B > C **7.19** (a) $MnO_4^-(aq) \rightarrow MnO_2(s)$; $IO_3^-(aq) \rightarrow IO_4^-(aq)$; (b) $NO_3^-(aq) \rightarrow NO_2(g)$; $SO_2(aq) \rightarrow SO_4^{2-}(aq)$ **7.20** $2\ NO_3^-(aq) + 8\ H^+(aq) + 3\ Cu(s) \rightarrow 3\ Cu^{2+}(aq) + 2\ NO(g) + 4\ H_2O(l)$ **7.21** $4\ Fe(OH)_2(s) + 2\ H_2O(l) + O_2(g) \rightarrow 4\ Fe(OH)_3(s)$ **7.22** 1.98 M **7.23** $Pb(s) + PbO_2(s) + 2\ HSO_4^-(aq) + 2\ H^+(aq) \rightarrow 2\ PbSO_4(s) + 2\ H_2O(l)$ **7.24** For a green process look for a solvent that is safe, non-toxic, non-polluting, and renewable. H_2O would be an excellent green solvent. ***Conceptual Problems*** **7.26** In the precipitate there are two cations (blue) for each anion (red), so the anion must have a −2 charge and the cation a +1 charge for charge neutrality of the precipitate. The cation must be Ag^+ because all Na^+ salts are soluble. Ag_2CrO_4 and Ag_2CO_3 are insoluble and consistent with the observed result. **7.28** The NaOCl concentration is 0.040 M. 67% of the I^- solution from the buret must be added to the flask to react with all of the OCl^-. **7.30** (a) No reaction. (b) Reaction would occur. (c) Reaction would occur. ***Section Problems*** **7.32** (a) precipitation; (b) redox; (c) acid–base neutralization **7.34** (a) $Hg^{2+}(aq) + 2\ I^-(aq) \rightarrow HgI_2(s)$; (b) $2\ HgO(s) \xrightarrow{\text{Heat}} 2\ Hg(l) + O_2(g)$; (c) $H_3PO_4(aq) + 3\ OH^-(aq) \rightarrow PO_4^{3-}(aq) + 3\ H_2O(l)$ **7.36** (a) bright; (b) dark; (c) dim **7.38** $Ba(OH)_2$ is soluble in aqueous solution, dissociates into $Ba^{2+}(aq)$ and $2\ OH^-(aq)$, and conducts electricity. H_2SO_4 dissociates into $H^+(aq)$ and $HSO_4^-(aq)$. H_2SO_4 solutions conduct electricity. When equal molar solutions of $Ba(OH)_2$ and H_2SO_4 are mixed, the insoluble $BaSO_4$ is formed along with two H_2O. In water, $BaSO_4$ does not produce any appreciable amount of ions and the mixture does not conduct electricity. **7.40** (a) strong; (b) weak; (c) strong; (d) strong; (e) weak; (f) nonelectrolyte **7.42** (a) 2.25 M; (b) 1.42 M **7.44** (a) insoluble; (b) soluble; (c) insoluble; (d) insoluble **7.46** (a) No precipitate will form; (b) $Fe(OH)_2(s)$ will precipitate; (c) No precipitate; (d) No precipitate. **7.48** (a) no precipitate; (b) $BaSO_4(s)$ will precipitate; (c) $AgCl(s)$ will precipitate. **7.50** (a) $Pb(NO_3)_2(aq) + Na_2SO_4(aq) \rightarrow PbSO_4(s) + 2\ NaNO_3(aq)$; (b) $3\ MgCl_2(aq) + 2\ K_3PO_4(aq) \rightarrow Mg_3(PO_4)_2(s) + 6\ KCl(aq)$; (c) $ZnSO_4(aq) + Na_2CrO_4(aq) \rightarrow ZnCrO_4(s) + Na_2SO_4(aq)$ **7.52** 0.645 g of AgCl **7.54** Add HCl(*aq*); it will selectively precipitate AgCl(*s*). **7.56** Cs^+ and/or NH_4^+ **7.58** Add the solution to an active metal, such as magnesium. Bubbles of H_2 gas indicate the presence of an acid. **7.60** (a) $2\ H^+(aq) + 2\ ClO_4^-(aq) + Ca^{2+}(aq) + 2\ OH^-(aq) \rightarrow Ca^{2+}(aq) + 2\ ClO_4^-(aq) + 2\ H_2O(l)$; (b) $CH_3CO_2H(aq) + Na^+(aq) + OH^-(aq) \rightarrow CH_3CO_2^-(aq) + Na^+(aq) + H_2O(l)$ **7.62** (a) $H^+(aq) + OH^-(aq) \rightarrow H_2O(l)$; (b) $H^+(aq) + OH^-(aq) \rightarrow H_2O(l)$ **7.64** (a) basic because there is an excess of KOH; (b) basic because there is an excess of $Ba(OH)_2$. **7.66** (a) 9.0 mL; (b) 11.0 mL **7.68** best reducing agents, the bottom left; best oxidizing agents, top right; (excluding the noble gases) **7.70** (a) gains electrons; (b) loses electrons; (c) loses electrons; (d) gains electrons **7.72** (a) NO_2: O −2, N +4; (b) SO_3: O −2, S +6; (c) $COCl_2$: O −2, Cl −1, C +4; (d) CH_2Cl_2: Cl −1, H +1, C 0; (e) $KClO_3$: O −2, K +1, Cl +5; (f) HNO_3: O −2, H +1, N +5 **7.74** (a) ClO_3^-: O −2, Cl +5; (b) SO_3^{2-}: O −2, S +4; (c) $C_2O_4^{2-}$: O −2, C +3; (d) NO_2^-: O −2, N +3; (e) BrO^-: O −2, Br +1; (f) AsO_4^{3-}: O −2, As +5 **7.76** (a) Ca(*s*) is oxidized; $Sn^{2+}(aq)$ is reduced; (b) not a redox reaction **7.78** (a) N. R.; (b) N. R.; (c) N. R.; (d) $Au^{3+}(aq) + 3\ Ag(s) \rightarrow 3\ Ag^+(aq) + Au(s)$ **7.80** (a) A > B > C > D; (b) (1) N. R.; (2) N. R. **7.82** (a) reduction; (b) oxidation; (c) oxidation; (d) reduction **7.84** (a) $3\ e^- + 4\ H^+(aq) + NO_3^-(aq) \rightarrow NO(g) + 2\ H_2O(l)$; (b) $Zn(s) \rightarrow Zn^{2+}(aq) + 2\ e^-$; (c) $Ti^{3+}(aq) + 2\ H_2O(l) \rightarrow TiO_2(s) + 4\ H^+(aq) + e^-$; (d) $Sn^{4+}(aq) + 2\ e^- \rightarrow Sn^{2+}(aq)$ **7.86** (a) oxidation: $Te(s) \rightarrow TeO_2(s)$; reduction: $NO_3^-(aq) \rightarrow NO(g)$; (b) oxidation: $Fe^{2+}(aq) \rightarrow Fe^{3+}(aq)$; reduction: $H_2O_2(aq) \rightarrow H_2O(l)$ **7.88** (a) $14\ H^+(aq) + Cr_2O_7^{2-}(aq) + 6\ e^- \rightarrow 2\ Cr^{3+}(aq) + 7\ H_2O(l)$; (b) $4\ H_2O(l) + CrO_4^{2-}(aq) + 3\ e^- \rightarrow Cr(OH)_4^-(aq) + 4\ OH^-(aq)$; (c) $Bi^{3+}(aq) + 6\ OH^-(aq) \rightarrow BiO_3^-(aq) + 3\ H_2O(l) + 2\ e^-$; (d) $H_2O(l) + ClO^-(aq) + 2\ e^- \rightarrow Cl^-(aq) + 2\ OH^-(aq)$

7.90 (a) $H_2O(l) + 2\,MnO_4^-(aq) + 3\,IO_3^-(aq) \rightarrow 2\,MnO_2(s) + 3\,IO_4^-(aq) + 2\,OH^-(aq)$; (b) $2\,Cu(OH)_2(s) + N_2H_4(aq) \rightarrow 2\,Cu(s) + 4\,H_2O(l) + N_2(g)$; (c) $3\,Fe(OH)_2(s) + 4\,H_2O(l) + CrO_4^{2-}(aq) \rightarrow 3\,Fe(OH)_3(s) + Cr(OH)_4^-(aq) + OH^-(aq)$; (d) $ClO_4^-(aq) + 2\,H_2O_2(aq) \rightarrow ClO_2^-(aq) + 2\,H_2O(l) + 2\,O_2(g)$ **7.92** (a) $Zn(s) + 2\,VO^{2+}(aq) + 4\,H^+(aq) \rightarrow Zn^{2+}(aq) + 2\,V^{3+}(aq) + 2\,H_2O(l)$; (b) $2\,H^+(aq) + Ag(s) + NO_3^-(aq) \rightarrow Ag^+(aq) + NO_2(g) + H_2O(l)$; (c) $3\,Mg(s) + 16\,H^+(aq) + 2\,VO_4^{3-}(aq) \rightarrow 3\,Mg^{2+}(aq) + 2\,V^{2+}(aq) + 8\,H_2O(l)$; (d) $6\,H^+(aq) + IO_3^-(aq) + 8\,I^-(aq) \rightarrow 3\,I_3^-(aq) + 3\,H_2O(l)$ **7.94** 0.670 g I_2 **7.96** 1.130 M **7.98** 0.134 M **7.100** 80.32% **7.102** 0.101% ***Chapter Problems*** **7.104** (a) $4[Fe(CN)_6]^{3-}(aq) + N_2H_4(aq) + 4\,OH^-(aq) \rightarrow 4[Fe(CN)_6]^{4-}(aq) + N_2(g) + 4\,H_2O(l)$; (b) $SeO_3^{2-}(aq) + Cl_2(g) + 2\,OH^-(aq) \rightarrow SeO_4^{2-}(aq) + 2\,Cl^-(aq) + H_2O(l)$; (c) $2\,Co^{2+}(aq) + H_2O(l) + HO_2^-(aq) + 3\,OH^-(aq) \rightarrow 2\,Co(OH)_3(s)$ **7.106** (a) C_2H_6: H +1, C −3; (b) $Na_2B_4O_7$: O −2, Na +1, B +3; (c) Mg_2SiO_4: O −2, Mg +2, Si +4 **7.108** (a) C > A > D > B; (b) (1) The reaction, $A^+ + C \rightarrow A + C^+$, will occur; (2) $A^+ + B \rightarrow A + B^+$, will not occur. **7.110** 0.4450% Cu **7.112** $K_{sp} = [Mg^{2+}][F^-]^2 = 7.0 \times 10^{-11}$ **7.114** (a) $Hg_2^{2+}(aq) + 2\,Cl^-(aq) \rightarrow Hg_2Cl_2(s)$; (b) $Pb^{2+}(aq) + SO_4^{2-}(aq) \rightarrow PbSO_4(s)$; (c) $Ca^{2+}(aq) + CO_3^{2-}(aq) \rightarrow CaCO_3(s)$; (d) $Ba^{2+}(aq) + SO_4^{2-}(aq) \rightarrow BaSO_4(s)$ **7.116** (a) $2\,Mn(OH)_2(s) + H_2O_2(aq) \rightarrow 2\,Mn(OH)_3(s)$; (b) $4\,H^+(aq) + 3\,MnO_4^{2-}(aq) \rightarrow MnO_2(s) + 2\,MnO_4^-(aq) + 2\,H_2O(l)$; (c) $8\,I^-(aq) + IO_3^-(aq) + 6\,H^+(aq) \rightarrow 3\,I_3^-(aq) + 3\,H_2O(l)$; (d) $2\,H_2O(l) + 2\,P(s) + 3\,PO_4^{3-}(aq) + OH^-(aq) \rightarrow 5\,HPO_3^{2-}(aq)$ **7.118** (a) $20\,H^+(aq) + S_4O_6^{2-}(aq) + 6\,Al(s) \rightarrow 4\,H_2S(aq) + 6\,Al^{3+}(aq) + 6\,H_2O(l)$; (b) $14\,H^+(aq) + 6\,S_2O_3^{2-}(aq) + Cr_2O_7^{2-}(aq) \rightarrow 3\,S_4O_6^{2-}(aq) + 2\,Cr^{3+}(aq) + 7\,H_2O(l)$; (c) $18\,H_2O(l) + 14\,ClO_3^-(aq) + 3\,As_2S_3(s) \rightarrow 14\,Cl^-(aq) + 6\,H_2AsO_4^-(aq) + 9\,HSO_4^-(aq) + 15\,H^+(aq)$; (d) $3\,H_2O(l) + 7\,IO_3^-(aq) + 6\,Re(s) \rightarrow 7\,I^-(aq) + 6\,ReO_4^-(aq) + 6\,H^+(aq)$; (e) $26\,H^+(aq) + 30\,HSO_4^-(aq) + As_4(s) + 10\,Pb_3O_4(s) \rightarrow 4H_2AsO_4^-(aq) + 30\,PbSO_4(s) + 24\,H_2O(l)$; (f) $3\,HNO_2(aq) \rightarrow NO_3^-(aq) + 2\,NO(g) + H_2O(l) + H^+(aq)$ **7.120** 3.06 g Cu_2O; 7.44 g CuO **7.122** Al **7.124** (a) $2\,H^+ + 2\,Cl^- + H_2O_2 \rightarrow Cl_2 + 2\,H_2O$, H_2O_2 reduced; (b) $6\,H^+ + 2\,MnO_4^- + 5\,H_2O_2 \rightarrow 2\,Mn^{2+} + 5\,O_2 + 8\,H_2O$, H_2O_2 oxidized; (c) $H_2O_2 + Cl_2 \rightarrow 2\,H^+ + 2\,Cl^- + O_2$, H_2O_2 oxidized. **7.126** 63.68% ***Multiconcept Problems*** **7.128** 6.5 g $Ba(OH)_2$; 3.5 g NaOH **7.130** (a) $BaSO_4(s)$ and $AgCl(s)$; (b) 150 g $BaCl_2$, 50 g NaCl, 50 g KNO_3 **7.132** (a) $14\,H^+(aq) + Cr_2O_7^{2-}(aq) + 6\,Cr^{2+}(aq) \rightarrow 8\,Cr^{3+}(aq) + 7\,H_2O(l)$; (b) K^+, 0.0833 M; NO_3^-, 0.617 M; H^+, 0.183 M; $Cr_2O_7^{2-}$, 0.0250 M; Cr^{3+}, 0.133 M **7.134** (a) (1) $3\,Cu(s) + 8\,H^+(aq) + 2\,NO_3^-(aq) \rightarrow 3\,Cu^{2+}(aq) + 2\,NO(g) + 4\,H_2O(l)$; (2) $2\,Cu^{2+}(aq) + 2\,SCN^-(aq) + H_2O(l) + HSO_3^-(aq) \rightarrow 2\,CuSCN(s) + HSO_4^-(aq) + 2\,H^+(aq)$; (3) $10\,Cu^+(aq) + 12\,H^+(aq) + 2\,IO_3^-(aq) \rightarrow 10\,Cu^{2+}(aq) + I_2(aq) + 6\,H_2O(l)$; (4) $I_2(aq) + 2\,S_2O_3^{2-}(aq) \rightarrow 2\,I^-(aq) + S_4O_6^{2-}(aq)$; (5) $2\,ZnNH_4PO_4 \rightarrow Zn_2P_2O_7 + H_2O + 2\,NH_3$; (b) 77.1% Cu; (c) 19.5% Zn **7.136** (a) $5\,H_3MO_3(aq) + 2\,MnO_4^-(aq) + 6\,H^+(aq) \rightarrow 5\,H_3MO_4(aq) + 2\,Mn^{2+}(aq) + 3\,H_2O(l)$; (b) 1.34×10^{-3} mol M_2O_3; 2.68×10^{-3} mol M (c) As.

CHAPTER 8

8.1 (a) and (b) are state functions; (c) is not. **8.2** $+1.9 \times 10^4$ J flows into the system **8.3** −0.25 kJ; The expanding system loses work energy and does work on the surroundings. **8.4** (a) $P\Delta V$ is negative for this reaction because the system volume is decreased at constant pressure. (b) ΔH is negative. Its value is slightly more negative than ΔE. **8.5** $q = -32$ kJ **8.6** 0.130 J/(g · °C) **8.7** -1.1×10^2 kJ **8.8** $w = 0.57$ kJ; $\Delta E = -120$ kJ **8.9** −45.2 kJ **8.10** (a) 780 kJ evolved; (b) 1.24 kJ absorbed **8.11** 1.000×10^3 kJ **8.12** −202 kJ **8.13** (a) $A + 2\,B \rightarrow D$; $\Delta H° = -150$ kJ; (b) red arrow: step 1; green arrow: step 2; blue arrow: overall reaction; (c) top energy level represents A + 2 B, middle energy level represents C + B, bottom energy level represents D **8.14**

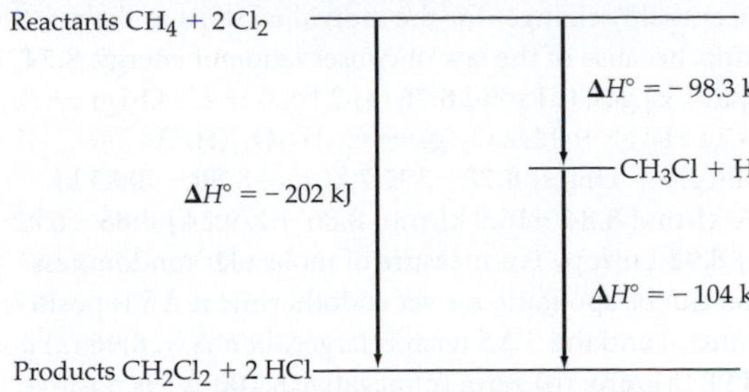

8.15 −901.2 kJ **8.16** +2803 kJ **8.17** +78 kJ **8.18** −81 kJ **8.19** −2635.5 kJ/mol; −45.35 kJ/g; −26.3 kJ/mL **8.20** $\Delta S°$ is negative because the reaction decreases the number of moles of gaseous molecules. **8.21** The reaction proceeds from a solid and a gas (reactants) to all gas (product). Randomness increases, so $\Delta S°$ is positive. **8.22** (a) spontaneous; (b) nonspontaneous **8.23** $\Delta G° = -32.9$ kJ; reaction is spontaneous; $T = 190$ °C **8.24** (a) $2\,A_2 + B_2 \rightarrow 2\,A_2B$; (b) ΔH is negative; ΔS is negative; (c) low temperatures only **8.25** $C_2H_6O + 3\,O_2 \rightarrow 2\,CO_2 + 3\,H_2O$; $2\,C_{19}H_{38}O_2 + 55\,O_2 \rightarrow 38\,CO_2 + 38\,H_2O$ **8.26** Because the standard heat of formation of $CO_2(g)$ (−393.5 kJ/mol), is more negative than that of $H_2O(g)$ (−241.8 kJ/mol), formation of CO_2 releases more heat than formation of H_2O. According to the balanced equations, combustion of ethanol yields a 2:3 ratio of CO_2 to H_2O, whereas combustion of biodiesel yields a 1:1 ratio. Thus biodiesel has a more favorable (more negative) combustion enthalpy per gram ***Conceptual Problems*** **8.28** (a) yes; $w < 0$; (b) yes; $\Delta H < 0$; exothermic

8.30

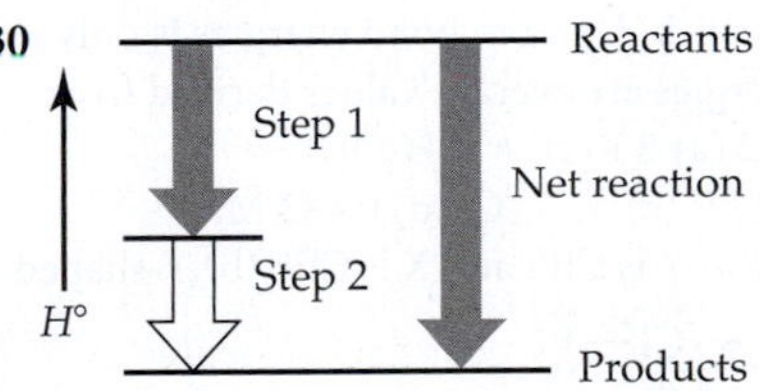

8.32 The volume decreases from 5 L to 3 L.

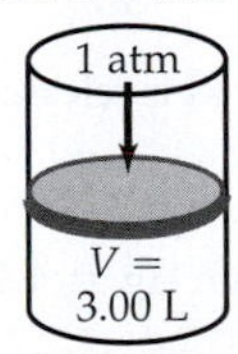

8.34 $\Delta G < 0$; $\Delta H > 0$; $\Delta S > 0$ ***Section Problems*** **8.36** Heat is the energy transferred from one object to another as the result of a temperature difference between them. Temperature is a measure of the kinetic energy of molecular motion. Energy is the capacity to do work or supply heat. Work is defined as the distance moved times the force that opposes the motion ($w = d \times F$). Kinetic energy is the energy of motion. Potential energy is stored energy. **8.38** Car: 7.1×10^5 J; truck: 6.7×10^5 J **8.40** −70 J. The energy change is negative. **8.42** (a) 35 min; (b) 150 min **8.44** (a) −593 J; (b) 26.4 cm **8.46** Heat capacity is the amount of heat required to raise the temperature of a substance a given amount. Specific heat is the amount of heat necessary to raise the temperature of exactly 1 g of a substance by exactly 1 °C. **8.48** 1.23 J/(g · °C) **8.50** −83.7 kJ **8.52** $\Delta H = -56$ kJ/mol; same temperature increase because NaOH is still the limiting reactant. **8.54** $\Delta E = q_v$ is the heat of a reaction at constant volume. $\Delta H = q_p$ is the heat of a reaction at constant pressure. **8.56** ΔH and ΔE are nearly equal when there are no gases involved in a chemical reaction, or, if gases are involved, $\Delta V = 0$. **8.58** -0.30 kJ **8.60** 45.4 kJ **8.62** $\Delta E = -314$ kJ **8.64** 25.5 kJ **8.66** 131 kJ **8.68** 0.388 kJ is evolved; exothermic. **8.70** The standard state of an element is its most stable form at 1 atm and 25 °C. **8.72** The overall enthalpy change for a reaction is equal to the sum of the enthalpy changes for the individual steps in the reaction. Hess's Law works because of the law of conservation of energy. **8.74** (a) gas; (b) liquid; (c) gas; (d) solid **8.76** (a) $2\ Fe(s) + 3/2\ O_2(g) \rightarrow Fe_2O_3(s)$; (b) $12\ C(s) + 11\ H_2(g) + 11/2\ O_2(g) \rightarrow C_{12}H_{22}O_{11}(s)$; (c) $U(s) + 3\ F_2(g) \rightarrow UF_6(s)$ **8.78** -395.7 kJ/mol **8.80** -909.3 kJ **8.82** $= +104$ kJ/mol **8.84** -16.9 kJ/mol **8.86** $+179.2$ kJ **8.88** -6 kJ **8.90** -2645 kJ **8.92** Entropy is a measure of molecular randomness. **8.94** A reaction can be spontaneous yet endothermic if ΔS is positive (more randomness) and the $T\Delta S$ term is larger than ΔH. **8.96** (a) positive; (b) negative **8.98** (a) zero; (b) zero; (c) negative **8.100** ΔS is positive. The reaction increases the total number of molecules. **8.102** (a) spontaneous; exothermic. (b) nonspontaneous; exothermic. (c) spontaneous; endothermic. (d) nonspontaneous; endothermic. **8.104** 570 K **8.106** (a) spontaneous at all temperatures; (b) has a crossover temperature; (c) has a crossover temperature; (d) nonspontaneous at all temperatures **8.108** 31.6 J/(K · mol) ***Chapter Problems*** **8.110** -468 kJ **8.112** -171.5 kJ **8.114** 279 K **8.116** (a) $+34$ kJ; (b) -451 kJ; (c) -87 kJ **8.118** (a) $2\ C_8H_{18}(l) + 25\ O_2(g) \rightarrow 16\ CO_2(g) + 18\ H_2O(l)$; (b) -259 kJ/mol **8.120** (a) $\Delta G = -T\Delta S_{total}$; (b) $\Delta S_{surr} = -9399$ J/(K · mol) **8.122** $\Delta H° = +201.9$ kJ **8.124** 311 g **8.126** -56 kJ **8.128** -1198.4 kJ ***Multiconcept Problems*** **8.130** (a)

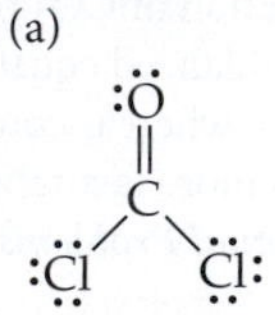

(b) -183 kJ/mol. The calculation of $\Delta H°_f$ from bond energies is only an estimate because the bond energies are average values derived from many different compounds. **8.132** (a) $2\ K(s) + 2\ H_2O(l) \rightarrow 2\ KOH(aq) + H_2(g)$; (b) -393.2 kJ; (c) 47.7 °C; (d) 0.483 M; 174 mL of 0.554 M H_2SO_4 **8.134** (a) Y is ClF_3 and X is ClF; (b) T-shaped

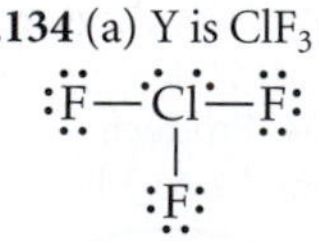

(c) -139.2 kJ/mol ClF_3; (d) 55.9 kJ is released.

CHAPTER 9

9.1 1.00 atm = 14.7 psi; 1.00 mm Hg = 1.93×10^{-2} psi **9.2** 10.3 m **9.3** 0.650 atm **9.4** 1000 mm Hg

9.5 (a) (b)

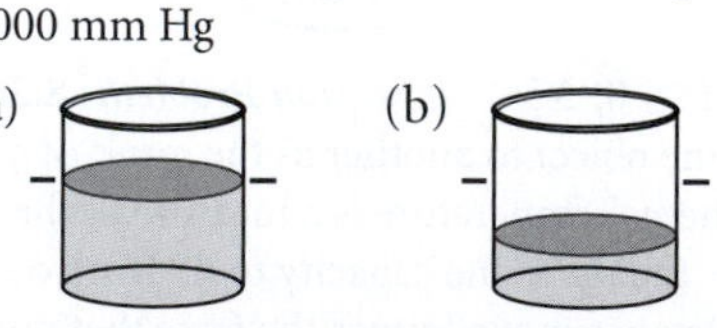

9.6 4.461×10^3 mol; 7.155×10^4 g **9.7** 5.0 atm **9.8** 267 mol **9.9** 28 °C **9.10** (a) The volume should increase by about 10%.

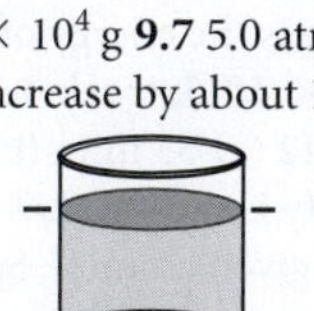

(b) the volume should decrease by half;

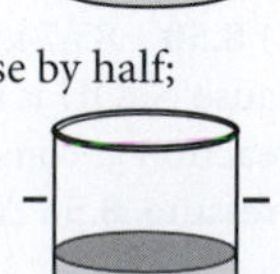

(c) the volume is unchanged.

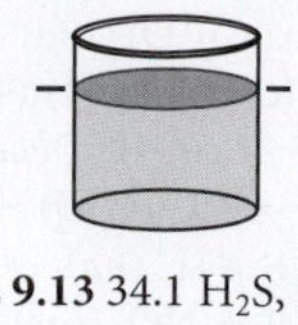

9.11 14.8 g; 7.55 L **9.12** 190 L **9.13** 34.1 H_2S, hydrogen sulfide **9.14** $X_{H_2} = 0.7281$; $X_{N_2} = 0.2554$; $X_{NH_3} = 0.0165$ **9.15** $P_{total} = 25.27$ atm; $P_{H_2} = 18.4$ atm; $P_{N_2} = 6.45$ atm; $P_{NH_3} = 0.417$ atm **9.16** 0.0280 atm **9.17** $P_{red} = 300$ mm Hg; $P_{yellow} = 100$ mm Hg; $P_{green} = 200$ mm Hg **9.18** at 37 °C, 525 m/s; at -25 °C, 470 m/s **9.19** -187.0 °C **9.20** (a) O_2, 1.62; (b) C_2H_2, 1.04 **9.21** $^{20}Ne(1.05) > {}^{21}Ne(1.02) > {}^{22}Ne(1.00)$ **9.22** ideal gas law: 20.5 atm; van der Waals equation: 20.3 atm **9.23** 3.8×10^{-5} m **9.24** 2.0% **9.25** (a) 5.9 mm Hg; (b) 0.41 g ***Conceptual Problems*** **9.26** The picture on the right will be the same as that on the left, apart from random scrambling of the He and Ar atoms. **9.28** (a) the volume will increase by a factor of 1.5; (b) the volume will decrease by a factor of 2; (c) there is no change in volume.

(a) (b) (c)

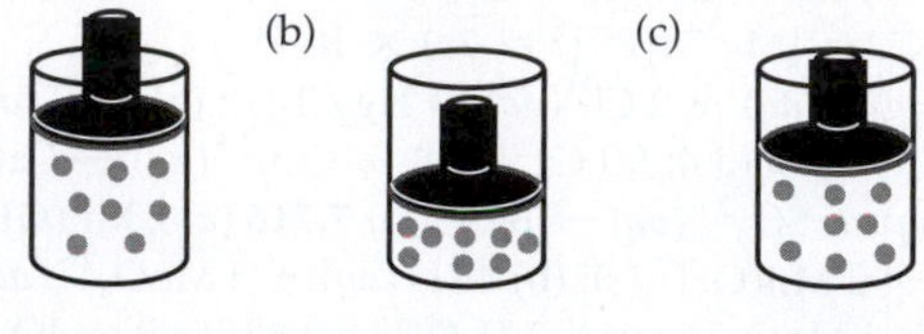

9.30 (c). **9.32** The gas pressure in the bulb in mm Hg is equal to the difference in the height of the Hg in the two arms of the manometer. **9.34** (a) yellow; (b) 36 ***Section Problems*** **9.36** Temperature is a measure of the average kinetic energy of gas particles. **9.38** 0.632 atm; 6.40×10^4 Pa **9.40** 930 mm Hg **9.42** 1.046×10^5 Pa **9.44** 28.96 **9.46** (a) P would triple; (b) P would be 1/3 the initial pressure. (c) P would increase by 1.8 times; (d) P would be 0.17 times the initial pressure. **9.48** They all contain the same number of gas molecules. **9.50** 7210 L; 51.5 L **9.52** 2.1×10^4 mm Hg **9.54** 1×10^{-17} mm Hg **9.56** 1.23×10^4 g **9.58** ice **9.60** Weigh the containers. The heavier container contains O_2. **9.62** 1.5×10^4 g O_2 **9.64** (a) 0.716 g/L; (b) 1.96 g/L; (c) 1.43 g/L **9.66** 34.0 **9.68** 0.5469 L **9.70** (a) 9.44 L; (b) 6.05 g Zn **9.72** (a) 380 g; (b) 5.4 days **9.74** $P_{N_2} = 0.7808$ atm; $P_{O_2} = 0.2095$ atm; $P_{Ar} = 0.0093$ atm; $P_{CO_2} = 0.000\ 38$ atm **9.76** $P_{O_2} = 0.970$ atm; $P_{CO_2} = 0.007\ 11$ atm; **9.78** $X_{HCl} = 0.026$; $X_{H_2} = 0.094$; $X_{Ne} = 0.88$ **9.80** (a) 1.68 atm; (b) 0.219 atm. **9.82** $P_{H_2} = 723$ mm Hg; 3.36 g Mg **9.84** See list in text Section 9.6. **9.86** Heat is the energy transferred from one object to another as the result of a temperature difference between them. Temperature is a measure of the kinetic energy of molecular motion. **9.88** $u = 443$ m/s **9.90** For H_2, $u = 1360$ m/s. For He, $u = 2010$ m/s **9.92** 17.2 **9.94** Relative rates of diffusion are HCl (1.05) > F_2(1.02) > Ar (1.00). **9.96** -272.83 °C ***Chapter Problems*** **9.98** (a) ozone depletion; (b) acid rain; (c) global warming; (d) air pollution **9.100** Relative rates of diffusion are $^{35}Cl_2(1.03) > {}^{35}Cl^{37}Cl\ (1.01) > {}^{37}Cl_2(1.00)$. **9.102** 1.1 L **9.104** (a) ideal gas law: 1.79 atm; (b) van der Waals equation: 1.74 atm **9.106** (a) Kr < O_2; (b) O_2 < Kr; (c) Kr < O_2; (d) Both are the same. **9.108** (a) 372 mm Hg; (b) 222 mm Hg; (c) 46.5 mm Hg **9.110** 504.3 g **9.112** (a) Bulb A contains $CO_2(g)$ and $N_2(g)$; B contains $CO_2(g)$, $N_2(g)$, and $H_2O(s)$; (b) $n = 0.0013$ mol H_2O; (c) A contains $N_2(g)$; B contains $N_2(g)$ and $H_2O(s)$; C contains $N_2(g)$ and $CO_2(s)$; (d) 0.010 92 mol N_2; (e) 0.0181 mol CO_2 **9.114** ideal gas law: $P = 59.1$ atm at 0 °C; 70.0 atm at 50 °C; 80.8 atm at 100 °C. van der Waals equation: $P = 36.5$ atm at 0 °C; 48.5 atm at 50 °C; 60.5 atm at 100 °C **9.116** 816 atm **9.118** (a) 0.901 mol; (b) 1.44 atm; (c) $P_{CH_4} = 1.32$ atm; $P_{C_2H_6} = 0.12$ atm; (d) Heat liberated $= -843$ kJ **9.120** (a) 196 g/mol; XeO_4; (b) $P_{Xe} = 0.313$ atm; $P_{O_2} = 0.626$ atm; $P_{total} = 0.939$ atm **9.122** 1.43 atm **9.124** 12.7% **9.126** $X_{NO_2} = 0.380$;

$X_{N_2O_4} = 0.620$ ***Multiconcept Problems*** **9.128** (a) 3.75 g; (b) 310 °C; (c) 1.02 atm **9.130** (a) $2\ C_8H_{18}(l) + 25\ O_2(g) \rightarrow 16\ CO_2(g) + 18\ H_2O(g)$; (b) 1.1×10^{11} kg; (c) 5.7×10^{13} L; (d) 59.5 mol, 1.33×10^3 L **9.132** (a) 0.0290 mol; (b) 0.0100 mol A; A = H_2O; (c) 0.0120 mol B; B = CO_2; (d) 0.001 00 mol C; C = O_2; 0.006 00 mol D; 28.0 g/mol; D = N_2; (e) $4\ C_3H_5N_3O_9(l) \rightarrow 10\ H_2O(g) + 12\ CO_2(g) + O_2(g) + 6\ N_2(g)$

CHAPTER 10

10.1 41%; HF has more ionic character than HCl. **10.2** (a) SF_6 is symmetrical (octahedral) and has no dipole moment; (b) $H_2C{=}CH_2$ is symmetrical; no dipole moment.

(c)

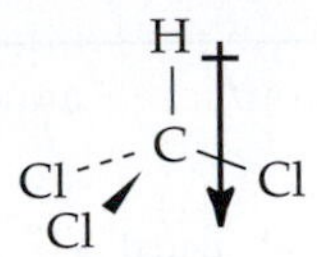

The C–Cl bonds in $CHCl_3$ are polar covalent bonds, and the molecule is polar.

(d)

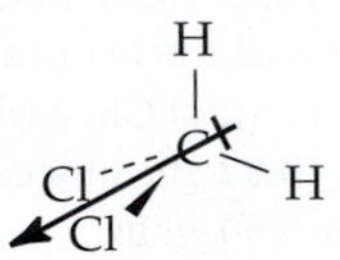

The C–Cl bonds in CH_2Cl_2 are polar covalent bonds, and the molecule is polar.

10.3

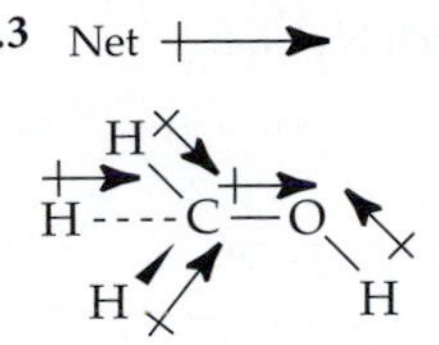

10.4 The N atom is electron rich (red) because of its high electronegativity. The C and H atoms are electron poor (blue) because they are less electronegative. **10.5** (a) HNO_3; (b) HNO_3; (c) Ar **10.6** H_2S, dipole–dipole, dispersion; CH_3OH, hydrogen bonding, dipole–dipole, dispersion; C_2H_6, dispersion; Ar, dispersion; Ar < C_2H_6 < H_2S < CH_3OH **10.7** (a) positive; (b) negative; (c) positive **10.8** 334 K **10.9** 47 °C **10.10** 31.4 kJ/mol **10.11** (a) 2 atoms; (b) 4 atoms **10.12** 167 pm **10.13** 9.31 g/cm^3 **10.14** There are several possibilities. Here's one.

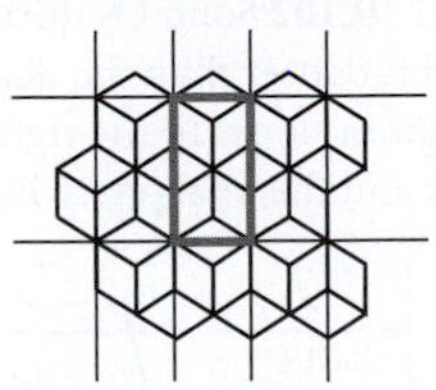

10.15 For CuCl: 4 minuses, 4 pluses. For $BaCl_2$: 8 pluses, 8 minuses **10.16** (a) 1 Re atom, 3 O atoms; (b) ReO_3; (c) +6; (d) linear; (e) octahedral **10.17** The triple point pressure of 5.11 atm **10.18** (a) $CO_2(s) \rightarrow CO_2(g)$; (b) $CO_2(l) \rightarrow CO_2(g)$; (c) $CO_2(g) \rightarrow CO_2(l) \rightarrow$ supercritical CO_2 **10.19** (a)

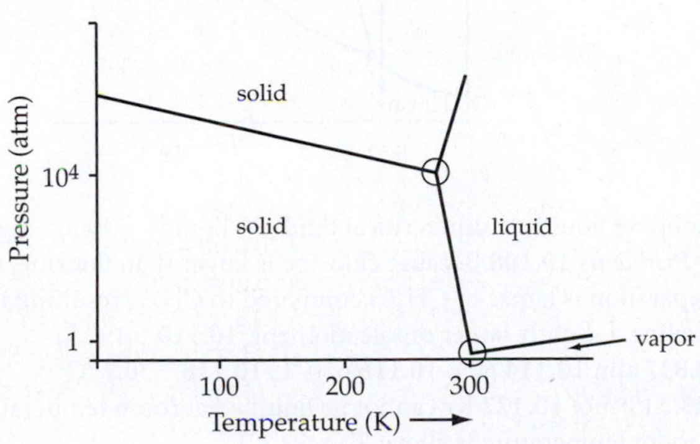

(b) two; (c) Increasing the pressure favors the liquid phase, giving the solid/liquid boundary a negative slope. At 1 atm pressure the liquid phase is more dense than the solid phase. **10.20** As the name implies, the constituent particles in an ionic liquid are cations and anions rather than molecules. **10.21** In ionic liquids the cation has an irregular shape and one or both of the ions are large and bulky to disperse charges over a large volume. Both factors minimize the crystal lattice energy, making the solid less stable and favoring the liquid. ***Conceptual Problems*** **10.22** The electronegative O atoms are electron rich (red), while the rest of the molecule is electron poor (blue). **10.24** (a) cubic closest-packed; (b) 4 S^{2-}; 4 Zn^{2+} **10.26** (a) normal boiling point ≈ 300 K, normal melting point ≈ 180 K; (b) (i) solid, (ii) gas, (iii) supercritical fluid **10.28** Here are two possibilities.

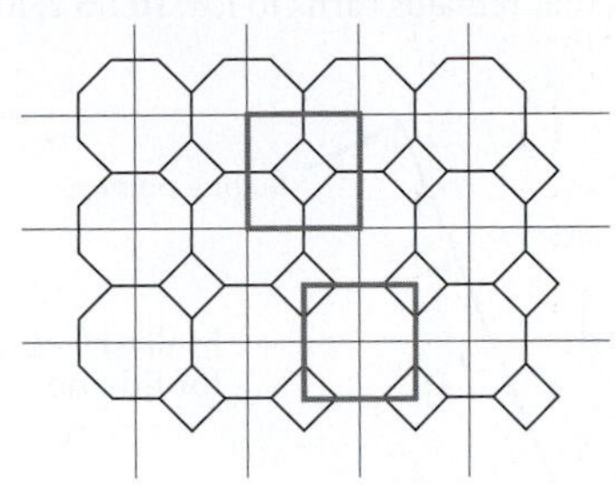

10.30.

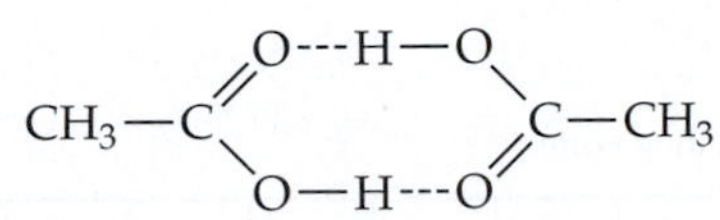

Section Problems **10.32** If a molecule has polar covalent bonds, the molecular shape (and location of lone pairs of electrons) determines whether the bond dipoles cancel and thus whether the molecule has a dipole moment. **10.34** (a) Dipole–dipole forces; dispersion forces are also present; (b) dispersion forces; (c) dispersion forces; (d) dipole–dipole forces and hydrogen bonding; dispersion forces are also present. **10.36** For CH_3OH and CH_4, dispersion forces are small. CH_3OH can hydrogen bond; CH_4 cannot. This accounts for the large difference in boiling points. For 1-decanol and decane, dispersion forces are comparable and relatively large along the C–H chain. 1-decanol can hydrogen bond; decane cannot. This accounts for the 57 °C higher boiling point for 1-decanol.

10.38

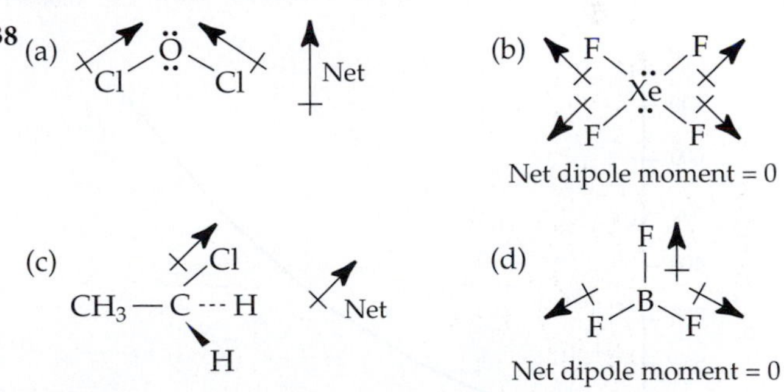

10.40 5.05%

10.42

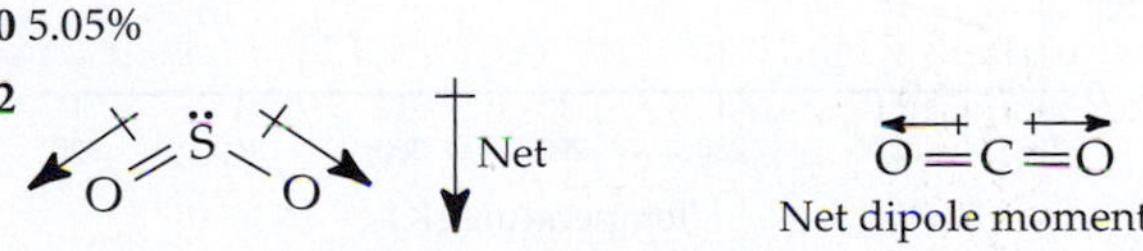

SO_2 is bent and the individual bond dipole moments add to give a net dipole moment. CO_2 is linear and the individual bond dipole moments cancel.

10.44

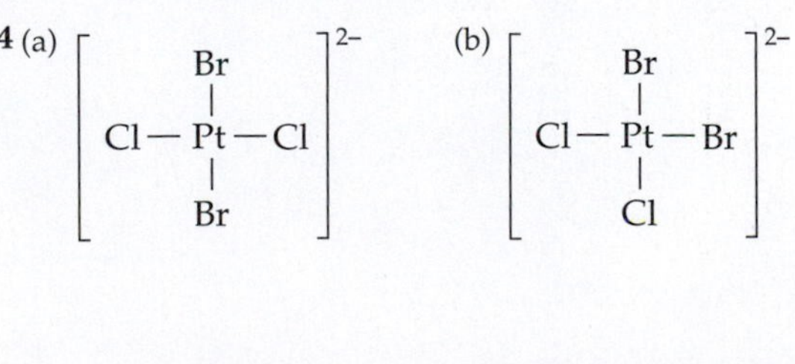

10.46

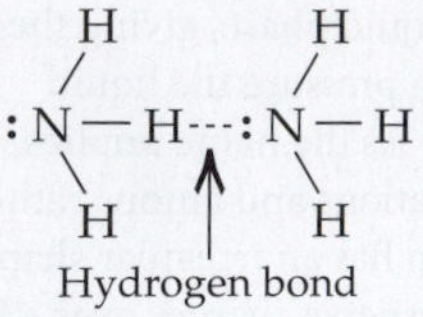

10.48 surface tension **10.50** ΔH_{vap} is usually larger than ΔH_{fusion} because ΔH_{vap} is the heat required to overcome all intermolecular forces. **10.52** (a) $Hg(l) \rightarrow Hg(g)$; (b) no change of state, Hg remains a liquid; (c) $Hg(g) \rightarrow Hg(l) \rightarrow Hg(s)$ **10.54** As the pressure is lowered, more of the liquid H_2O is converted to H_2O vapor. This conversion is an endothermic process and the temperature decreases. The decrease in pressure and temperature takes the system across the liquid/solid boundary in the phase diagram so the H_2O that remains turns to ice. **10.56** 2.40 kJ **10.58** 3.73 kJ

10.60

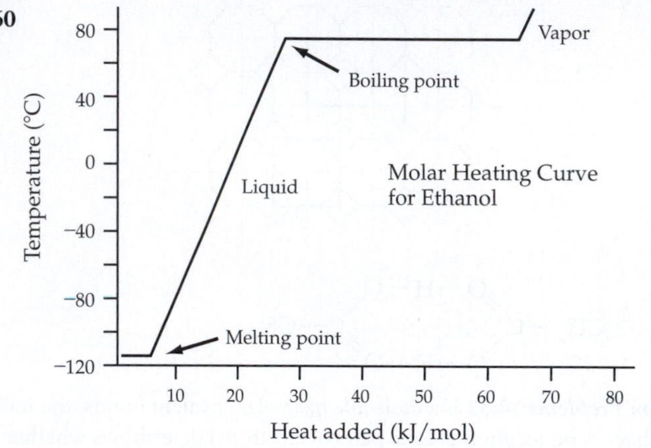

10.62 88.2 J/(K · mol) **10.64** 28.0 kJ/mol **10.66** 294 mm Hg
10.68

T(K)	P_{vap}(mm Hg)	ln P_{vap}	1/*T*
263	80.1	4.383	0.003 802
273	133.6	4.8949	0.003 663
283	213.3	5.3627	0.003 534
293	329.6	5.7979	0.003 413
303	495.4	6.2054	0.003 300
313	724.4	6.5853	0.003 195

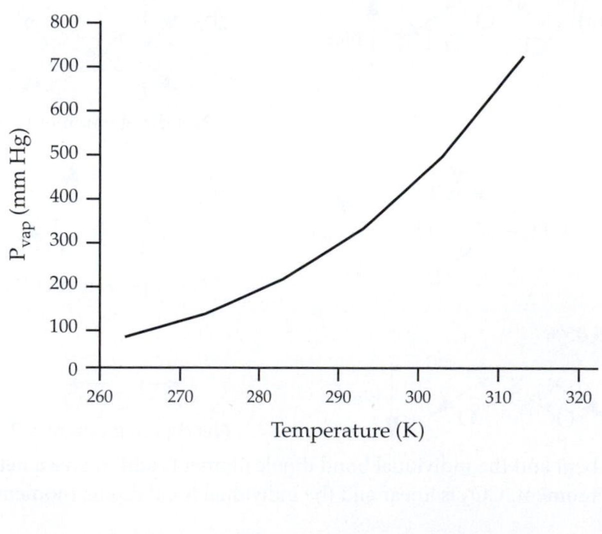

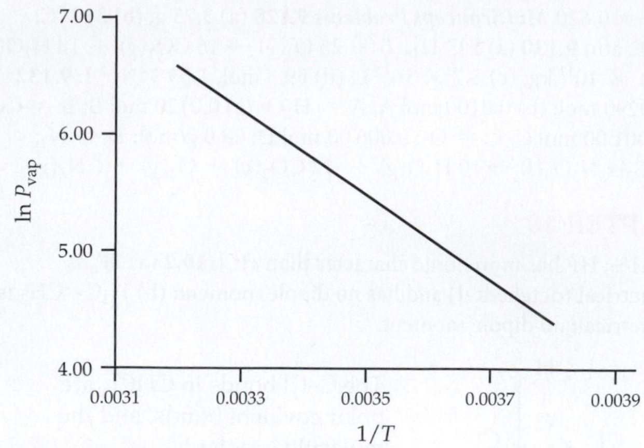

10.70 ΔH_{vap} = 30.1 kJ/mol **10.72** 30.1 kJ/mol. The calculated ΔH_{vap} and that obtained from the plot in Problem 10.68 are the same. **10.74** molecular solid, CO_2, I_2; metallic solid, any metallic element; covalent network solid, diamond; ionic solid, NaCl **10.76** (a) rubber; (b) Na_3PO_4; (c) CBr_4; (d) quartz; (e) Au **10.78** covalent network **10.80** 201 pm **10.82** The unit cell is the smallest repeating unit in a crystal. **10.84** 128 pm; 8.90 g/cm^3 **10.86** 404.9 pm **10.88** 137 pm **10.90** face-centered cubic **10.92** Six Na^+ ions touch each H^- ion, and six H^- ions touch each Na^+ ion. **10.94** 244 pm **10.96** (a) gas; (b) liquid; (c) solid

10.98.

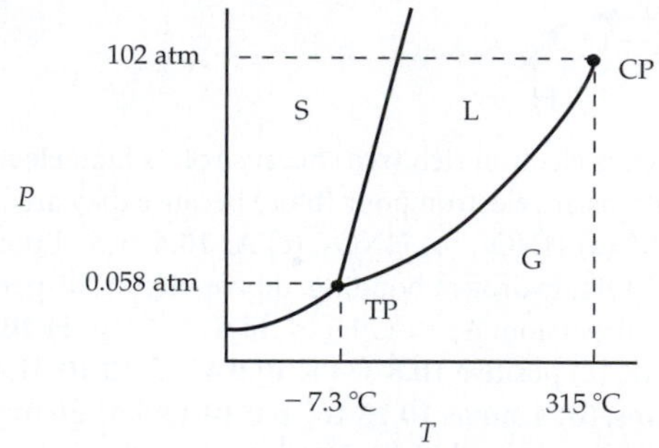

10.100 (a) $Br_2(s)$; (b) $Br_2(l)$ **10.102** Solid O_2 does not melt when pressure is applied because the solid is denser than the liquid, and the solid/liquid boundary in the phase diagram slopes to the right. **10.104** The starting phase is benzene as a solid, and the final phase is benzene as a gas.

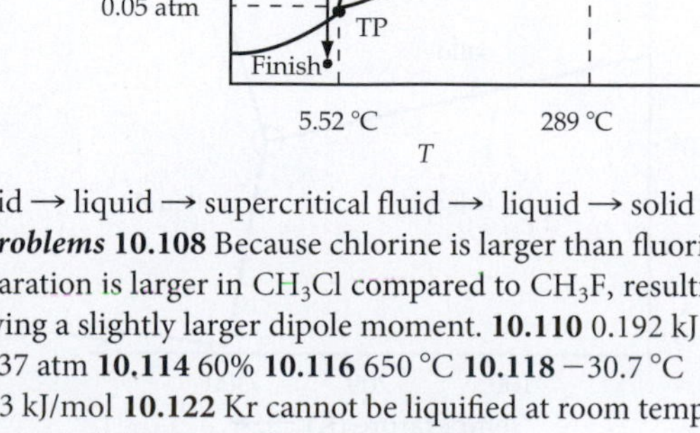

10.106 solid → liquid → supercritical fluid → liquid → solid → gas ***Chapter Problems*** **10.108** Because chlorine is larger than fluorine, the charge separation is larger in CH_3Cl compared to CH_3F, resulting in CH_3Cl having a slightly larger dipole moment. **10.110** 0.192 kJ **10.112** 0.837 atm **10.114** 60% **10.116** 650 °C **10.118** −30.7 °C **10.120** 23.3 kJ/mol **10.122** Kr cannot be liquified at room temperature because room temperature is above T_c(−63 °C).

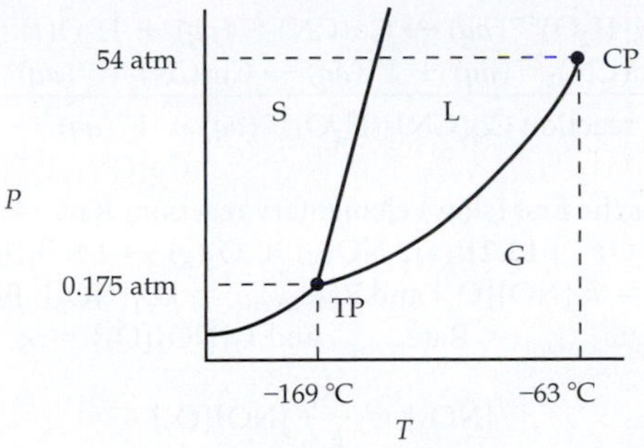

10.124 68% **10.126** 6.01×10^{23} atoms/mol **10.128** (a) 556 pm; (b) 2.26 g/cm^3 **10.130** Al_2O_3, ionic (greater lattice energy than NaCl because of higher ion charges); F_2, dispersion; H_2O, H–bonding, dipole–dipole; Br_2, dispersion (larger and more polarizable than F_2), ICl, dipole–dipole, NaCl, ionic; $F_2 < Br_2 < ICl < H_2O < NaCl < Al_2O_3$
10.132 (a)

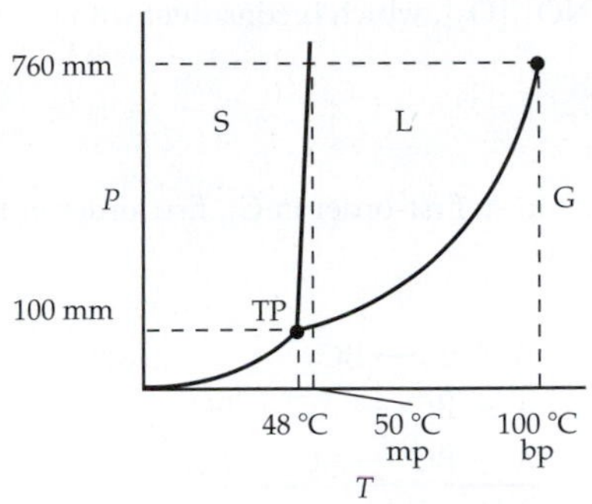

(b) (i) solid, (ii) gas, (iii) liquid, (iv) liquid, (v) solid ***Multiconcept Problems*** **10.134** (a) 1.926 g; (b) Fe_3O_4; (c) 24 Fe atoms; 32 O atoms **10.136** (a) 231 pm; (b) K; (c) density of solid = 0.857 g/cm^3; density of vapor = 7.29×10^{-6} g/cm^3

CHAPTER 11

11.1 toluene < Br_2 < KBr **11.2** (a) Na^+ because the Na^+ ion is smaller than the Cs^+ ion; (b) Ba^{2+} because of its higher charge. **11.3** 5.52 mass % **11.4** 4.6×10^{-5} g **11.5** 0.614 M **11.6** molality = 0.0249 m; $X_{C_{27}H_{46}O} = 2.96 \times 10^{-3}$ **11.7** 312 g **11.8** 0.251 M **11.9** 0.513 m **11.10** 0.621 m **11.11** 3.2×10^{-2} mol/(L · atm) **11.12** (a) 0.080 M; (b) 1.3×10^{-5} M **11.13** 98.6 mm Hg **11.14** 17.6 g **11.15** upper curve: pure solvent; lower curve: solution **11.16** (a) 27.1 mm Hg; (b) 46.6 mm Hg **11.17** (a) lower boiling point;
(b)

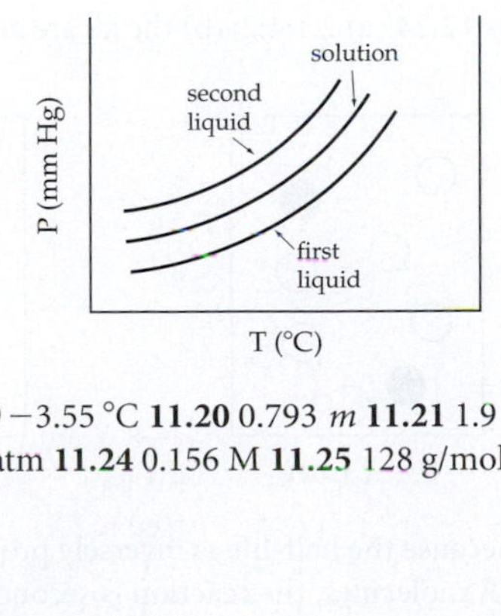

11.18 62.1 °C **11.19** −3.55 °C **11.20** 0.793 m **11.21** 1.9 **11.22** (a) 62 °C; (b) 2 m **11.23** 9.54 atm **11.24** 0.156 M **11.25** 128 g/mol **11.26** 342 g/mol **11.27** (a) and (c)

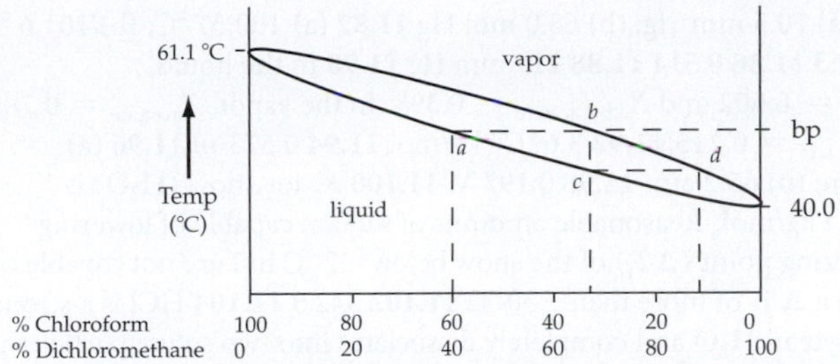

(b) ~50 °C; (d) 90% dichloromethane and 10% chloroform **11.28** Both solvent molecules and small solute particles can pass through a semipermeable dialysis membrane. Only large colloidal particles such as proteins can't pass through. Only solvent molecules can pass through a semipermeable membrane used for osmosis. ***Conceptual Problems***
11.30 (a) < (b) < (c)

11.32

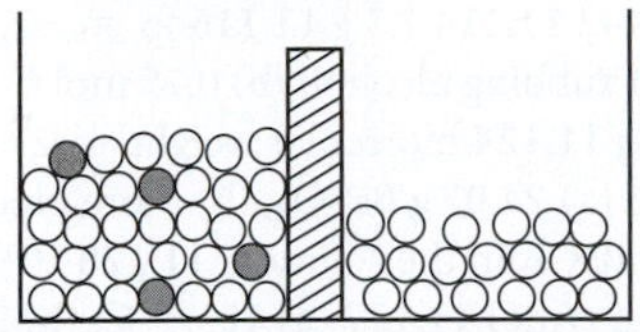

Assume that only the blue (open) spheres (solvent) can pass through the semipermeable membrane. There will be a net transfer of solvent from the right compartment (pure solvent) to the left compartment (solution) to achieve equilibrium. **11.34** (b) **11.36** (a) Solution B ($i = 3$): red line; Solution A ($i = 1$): blue line; (b) 15.5 °C; (c) Both the solutions are 0.50 m. ***Section Problems*** **11.38** The larger the surface area, the more solid-solvent interactions, and the more rapidly the solid will dissolve. Powdered NaCl has a much larger surface area than a large block of NaCl. **11.40** Energy is required to overcome intermolecular forces holding solute particles together in the crystal. For an ionic solid, this is the lattice energy. **11.42** Ethyl alcohol and water are both polar with small dispersion forces. They both can hydrogen bond, and are miscible. Pentyl alcohol is slightly polar and can hydrogen bond. It has, however, a relatively large dispersion force because of its size, which limits its water solubility. **11.44** 42.5 °C **11.46** (a) 0.049 ppb; (b) 1.9×10^{-9} mol/L **11.48** (a) Dissolve 0.150 mol of glucose in water; dilute to 1.00 L; (b) Dissolve 1.135 mol of KBr in 1.00 kg of H_2O; (c) Mix together 0.15 mol of CH_3OH with 0.85 mol of H_2O. **11.50** Dissolve 4.42×10^{-3} mol (0.540 g) of $C_7H_6O_2$ in enough $CHCl_3$ to make 165 mL of solution. **11.52** (a) 0.500 M KCl; (b) 1.75 M glucose **11.54** (a) 11.2 mass %; (b) 0.002 70 mass %; (c) 3.65 mass % **11.56** 0.20 ppm **11.58** (a) 0.196 m; (b) $X_{C_{10}H_{14}N_2} = 0.0145$; $X_{CH_2Cl_2} = 0.985$ **11.60** 1.81 M **11.62** 10.7 m **11.64** 3.7 g **11.66** (a) 0.0187; (b) 16.0%; (c) 1.06 m **11.68** (a) NH_4Cl; (b) CH_3CO_2Na **11.70** 0.068 M **11.72** 0.06 atm **11.74** 4.8×10^{-4} mol/L **11.76** NaCl is nonvolatile. Methyl alcohol is volatile. When NaCl is added to water, the vapor pressure of the solution is decreased, which means that the boiling point of the solution will increase. When methyl alcohol is added to water, the vapor pressure of the solution is increased, which means that the boiling point of the solution will decrease.

11.78

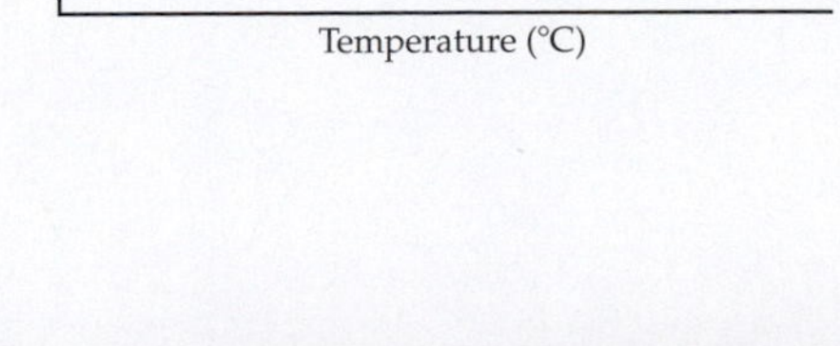

11.80 (a) 70.5 mm Hg; (b) 68.0 mm Hg **11.82** (a) 100.57 °C; (b) 101.6 °C **11.84** 2.3 **11.86** 0.514 **11.88** 219 mm Hg **11.90** In the liquid, $X_{acetone} = 0.602$ and $X_{ethyl\ acetate} = 0.398$. In the vapor, $X_{acetone} = 0.785$; $X_{ethyl\ acetate} = 0.215$ **11.92** 3.6 °C · kg/mol **11.94** 0.573 *m* **11.96** (a) 13.0 atm; (b) 65.2 atm **11.98** 0.197 M **11.100** K_f for snow (H_2O) is 1.86 °C · kg/mol. Reasonable amounts of salt are capable of lowering the freezing point (ΔT_f) of the snow below −2 °C but are not capable of causing a ΔT_f of more than −30 °C **11.102** 342.5 **11.104** HCl is a strong electrolyte in H_2O and completely dissociates into two solute particles per each HCl. HF is a weak electrolyte in H_2O. Only a few percent of the HF molecules dissociates into ions. **11.106** molecular weight = 538; molecular formula is $C_{40}H_{56}$ ***Chapter Problems*** **11.108** 2.60×10^3 g $C_2H_6O_2$ **11.110** When solid $CaCl_2$ is added to liquid water, the temperature rises because ΔH_{soln} for $CaCl_2$ is exothermic. When solid $CaCl_2$ is added to ice at 0 °C, some of the ice will melt (an endothermic process) and the temperature will fall because the $CaCl_2$ lowers the freezing point of an ice/water mixture. **11.112** 0.442 **11.114** 1.3 g **11.116** fp = −2.3 °C; bp = 100.63 °C **11.118** (a) 11.7 g of rubbing alcohol; (b) 0.75 mol C_3H_8O **11.120** 2.45% **11.122** 0.51 mm Hg **11.124** molecular weight = 271; molecular formula is $C_{18}H_{24}O_2$ **11.126** (a) 24.07 g $BaSO_4$; (b) More precipitate will form because of the excess $BaCl_2$ in the solution. **11.128** 10% LiCl; 41% H_2O; 49% CH_3OH **11.130** 334 g/mol **11.132** (a) 103.3 °C; (b) 21.3 mm Hg **11.134** (a) 1.31 *m*; (b) −4.87 °C; (c) 83% **11.136** 89.9 g of $C_{12}H_{22}O_{11}$; 10.2 g of NaCl ***Multiconcept Problems*** **11.138** (a) C_2H_3; (b) 108; (c) C_8H_{12} **11.140** 5.75 g LiCl; 3.15 g $CaCl_2$

CHAPTER 12

12.1 (a) 1.6×10^{-4} M/s; (b) 3.2×10^{-4} M/s **12.2** Rate of decomposition of $N_2O_5 = 2.2 \times 10^{-5}$ M/s; Rate of formation of $O_2 = 1.1 \times 10^{-5}$ M/s **12.3** Rate = $k[BrO_3^-][Br^-][H^+]^2$; 1st order in Br^-, 2nd order in H^+, 4th order overall; Rate = $k[H_2][I_2]$, 1st order in H_2, 1st order in I_2, 2nd order overall **12.4** (a) Rate = $k[H_2O_2][I^-]$; (b) $1.15 \times 10^{-2}/(M \cdot s)$; (c) 1.38×10^{-3} M/s **12.5** 1/s; 1/s; $1/(M^3 \cdot s)$; $1/(M \cdot s)$ **12.6** (a) Zeroth order in A; second order in B; second order overall; (b) rate = $k[B]^2$ **12.7** (a) 0.080 M; (b) 61 h **12.8** A plot of ln [cyclopropane] versus time is linear, indicating that the data fit the equation for a first-order reaction. $k = 6.6 \times 10^{-4}$/s (0.040/min) **12.9** (a) 11 h; (b) 0.019 M; (c) 22 h **12.10** (a) 5 min; (b)

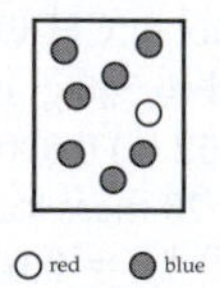

12.11 64.2 h **12.12** 0.0229 y^{-1} **12.13** 14.0% **12.14** 44.5 d **12.15** (a) A plot of 1/[HI] versus time is linear. The reaction is second order. (b) $0.0308/(M \cdot min)$; (c) 260 min; (d) 81.2 min **12.16** (a) 80 kJ/mol; (b) endothermic; (c)

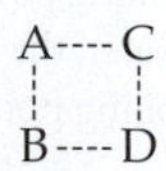

(d) The reaction rate increases as the temperature increases because at higher temperatures more collisions occur with an energy greater than the activation energy **12.17** (a) 104 kJ/mol; (b) 1.4×10^{-4}/s **12.18** (a) $2\,NO_2(g) + F_2(g) \rightarrow 2\,NO_2F(g)$; F(*g*) is a reaction intermediate. (b) each elementary reaction is bimolecular. **12.19** (a) Rate = $k[O_3][O]$; (b) Rate = $k[Br]^2[Ar]$; (c) Rate = $k[Co(CN)_5(H_2O)^{2-}]$

12.20

$$\begin{array}{l} Co(CN)_5(H_2O)^{2-}(aq) \rightarrow Co(CN)_5{}^{2-}(aq) + H_2O(l) \text{ (slow)} \\ Co(CN)_5{}^{2-}(aq) + I^-(aq) \rightarrow Co(CN)_5I^{3-}(aq) \text{ (fast)} \\ \hline \text{Overall reaction } Co(CN)_5(H_2O)^{2-}(aq) + I^-(aq) \rightarrow Co(CN)_5I^{3-}(aq) + H_2O(l) \end{array}$$

The rate law for the first (slow) elementary reaction: Rate = $k[Co(CN)_5(H_2O)^{2-}]$ **12.21** (a) $2\,NO(g) + O_2(g) \rightarrow 2\,NO_2(g)$; (b) $Rate_{forward} = k_1[NO][O_2]$ and $Rate_{reverse} = k_{-1}[NO_3]$. Because of the equilibrium, $Rate_{forward} = Rate_{reverse}$, and $k_1[NO][O_2] = k_{-1}[NO_3]$.

$$[NO_3] = \frac{k_1}{k_{-1}}[NO][O_2]$$

The rate law for the rate determining step is Rate = $k_2[NO_3][NO]$. Because 2 NO disappear in the overall reaction for every NO that reacts in the second step, the rate law for the overall reaction is Rate = $-\Delta[NO]/\Delta t = 2\,k_2[NO_3][NO]$. In this rate law substitute for $[NO_3]$.

Rate = $2\,k_2 \dfrac{k_1}{k_{-1}}[NO]^2[O_2]$, which is consistent with the experimental rate law; (c) $k = \dfrac{2\,k_2k_1}{k_{-1}}$

12.22 (a) zero-order in A, first-order in C_2, first-order in B. (b) Rate = $k[B][C_2]$; (c)

$$\begin{array}{ll} B + C_2 \rightarrow BC_2 & \text{(slow)} \\ A + BC_2 \rightarrow AC + BC & \\ A + BC \rightarrow AC + B & \\ \hline 2\,A + C_2 \rightarrow 2\,AC & \text{(overall)} \end{array}$$

(d) B is a catalyst. BC_2 and BC are intermediates

12.23

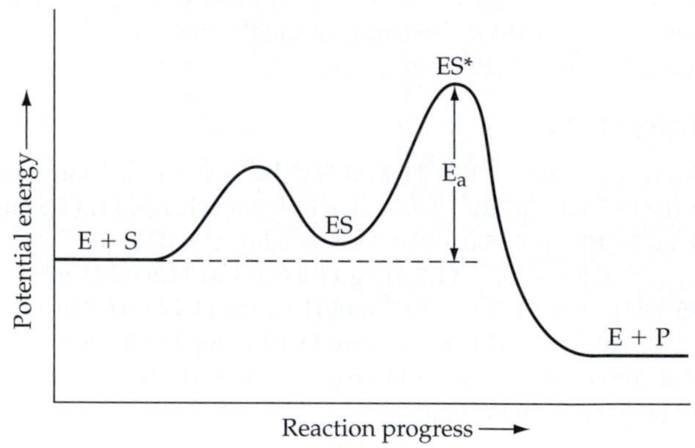

Conceptual Problems **12.24** (a) 2:1:4:2; (b) the *k*'s are all the same **12.26** (a)

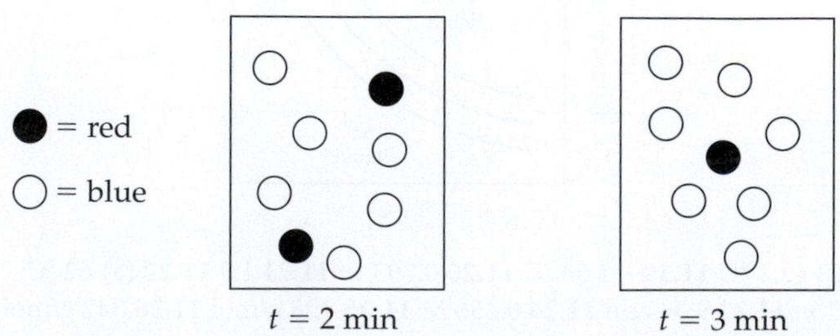

(b) 1 min **12.28** (a) Because the half-life is inversely proportional to the concentration of A molecules, the reaction is second order in A. (b) Rate = $k[A]^2$;

(c)

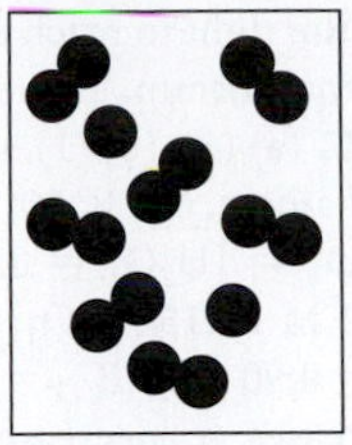

$t = 3 \text{ min} + 4 \text{ min} = 7 \text{ min}$

12.30 (a) zeroth-order in B; second-order in A. (b) Rate $= k[A]^2$.

(c) $2\,A \rightarrow A_2$ (slow)

$A_2 + B \rightarrow AB + A$ (d) A_2

$A + B \rightarrow AB$ (overall)

12.32 (a) $BC + D \rightarrow B + CD$; (b) 1. B—C + D (reactants), A (catalyst); 2. B—C—A (transition state), D (reactant); 3. A—C (intermediate), B (product), D (reactant); 4. A—C—D (transition state), B (product); 5. A (catalyst), C—D + B (products); (c) first step; Rate $= k[A][BC]$; (d) Endothermic (e) The rate increases because as the temperature increases the fraction of the collisions with an energy greater than the activation energy also increases. ***Section Problems*** **12.34** (a) 4.0×10^{-5} M/s; (b) 3.9×10^{-5} M/s; (c) 3.8×10^{-5} M/s; (d) 3.7×10^{-5} M/s Rate (d) because it is determined from measurements taken over the smallest time interval. **12.36** (a) 2.4×10^{-5} M/s; (b) 3.1×10^{-5} M/s **12.38** (a) 3 times faster; (b) 2 times faster. **12.40** (a) -2.4×10^{-6} M/s; (b) 9.6×10^{-6} M/s **12.42** Rate $= k[H_2][ICl]$; $1/(M \cdot s)$ **12.44** (a) Rate $= k[CH_3Br][OH^-]$; (b) The rate will decrease by a factor of 5; (c) The rate will increase by a factor of 4. **12.46** (a) Rate $= k[Cu(C_{10}H_8N_2)_2{}^+]^2[O_2]$; (b) Overall reaction order = 3; (c) The rate will decrease by a factor of 16. **12.48** (a) Rate $= k[N_2O_5]$; (b) 1.7×10^{-3}/s; 5.1×10^{-5} M/s **12.50** (a) Rate $= k[A]^2[B]$; (b) $5.0/(M^2 \cdot s)$; (c) 4.0×10^{-5} M/s **12.52** (a) 0.015 M; (b) 40 min; (c) 7.2 min **12.54** $t_{1/2} = 17$ min; $t = 69$ min **12.56** (a) 5.2×10^{-3} M; (b) 3.1 h **12.58** $t_{1/2} = 21$ min; $t = 42$ min **12.60** A plot of $\ln[N_2O]$ versus time is linear. The reaction is first order in N_2O. $k = 3.92 \times 10^{-5}$/s **12.62** 2.79×10^{-3}/s **12.64** (a) zeroth-order; (b) 0.031 M; (c) 22 min **12.66** 0.247 d^{-1} **12.68** 23.0% **12.70** 621 α particles **12.72** 1.1×10^6 y **12.74** 34.6 d **12.76** Very few collisions involve a collision energy greater than or equal to the activation energy, and only a fraction of those have the proper orientation for reaction. **12.78** 104 kJ/mol

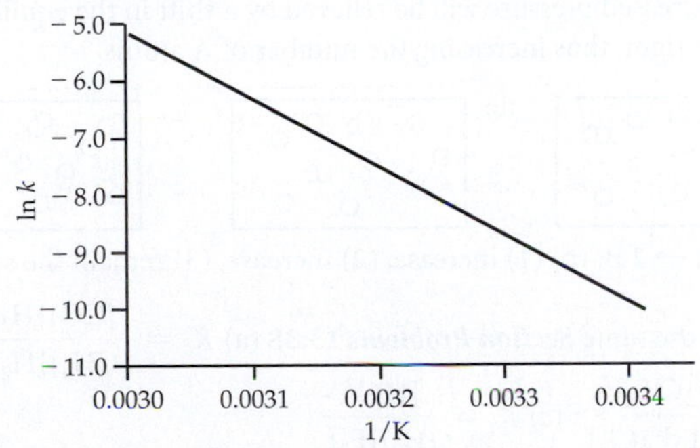

12.80 (a) 134 kJ/mol; (b) $6.0/(M \cdot s)$ **12.82** 87 kJ/mol

12.84

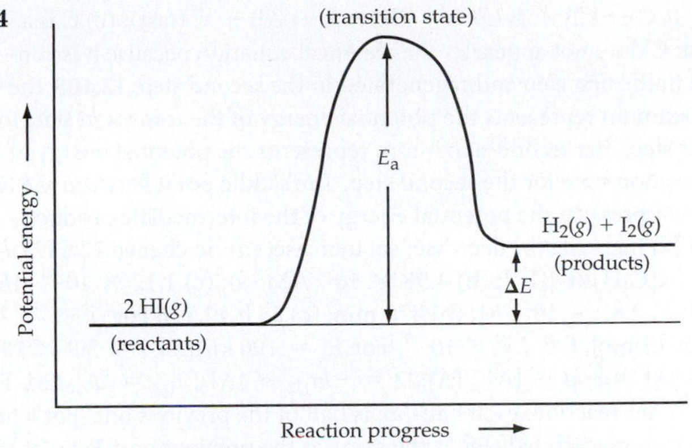

12.86 There is no relationship between the coefficients in a balanced chemical equation for an overall reaction and the exponents in the rate law unless the overall reaction occurs in a single elementary step, in which case the coefficients in the balanced equation are the exponents in the rate; law. **12.88** (a) $H_2(g) + 2\,ICl(g) \rightarrow I_2(g) + 2\,HCl(g)$; (b) HI(g); (c) Each elementary reaction is bimolecular. **12.90** (a) bimolecular, Rate $= k[O_3][Cl]$; (b) unimolecular, Rate $= k[NO_2]$; (c) bimolecular, Rate $= k[ClO][O]$; (d) termolecular, Rate $= k[Cl]^2[N_2]$

12.92 (a) $2\,NO_2Cl(g) \rightarrow 2\,NO_2(g) + Cl_2(g)$; (b) 1. unimolecular; 2. bimolecular; (c) Rate $= k[NO_2Cl]$ **12.94** $NO_2(g) + F_2(g) \rightarrow NO_2F(g) + F(g)$ (slow); $F(g) + NO_2(g) \rightarrow NO_2F(g)$(fast) **12.96** (a) $2\,NO(g) + O_2(g) \rightarrow 2\,NO_2(g)$; (b) $\text{Rate}_{forward} = k_1[NO]^2$ and $\text{Rate}_{reverse} = k_{-1}[N_2O_2]$ Because of the equilibrium, $\text{Rate}_{forward} = \text{Rate}_{reverse}$, and $k_1[NO]^2 = k_{-1}[N_2O_2]$.

$$[N_2O_2] = \frac{k_1}{k_{-1}}[NO]^2$$

The rate law for the rate determining step is Rate $= -\Delta[NO]/\Delta t = 2k_2[N_2O_2][O_2]$ because two NO molecules are consumed in the overall reaction for every $[N_2O_2]$ that reacts in the second step. In this rate law substitute for $[N_2O_2]$.

$$\text{Rate} = 2k_2\frac{k_1}{k_{-1}}[NO]^2[O_2] \quad (c)\ k = \frac{2k_2k_1}{k_{-1}}$$

12.98 A catalyst does participate in the reaction, but it is not consumed because it reacts in one step of the reaction and is regenerated in a subsequent step. **12.100** (a) $O_3(g) + O(g) \rightarrow 2\,O_2(g)$; (b) Cl acts as a catalyst; (c) ClO is a reaction intermediate; (d) A catalyst reacts in one step and is regenerated in a subsequent step. A reaction intermediate is produced in one step and consumed in another. **12.102** (a) $NH_2NO_2(aq) \rightarrow N_2O(g) + H_2O(l)$; (b) OH^- acts as a catalyst; $NHNO_2{}^-$ is a reaction intermediate; (c) The rate will decrease because added acid decreases the concentration of OH^-, which appears in the rate law since it is a catalyst. ***Chapter Problems*** **12.104** (a) Measure the change in the concentration of AB_2 as a function of time. (b) and (c) If a plot of $[AB_2]$ versus time is linear, the reaction is zeroth-order and $k = -\text{slope}$. If a plot of $\ln[AB_2]$ versus time is linear, the reaction is first-order and $k = -\text{slope}$. If a plot of $1/[AB_2]$ versus time is linear, the reaction is second-order and $k = \text{slope}$. **12.106** (a) Rate $= k[B_2][C]$;

(b) $B_2 + C \rightarrow CB + B$ (slow); $CB + A \rightarrow AB + C$ (fast); (c) C is a catalyst. C does not appear in the chemical equation because it is consumed in the first step and regenerated in the second step. **12.108** The first maximum represents the potential energy of the transition state for the first step. The second maximum represents the potential energy of the transition state for the second step. The saddle point between the two maxima represents the potential energy of the intermediate products. **12.110** (a) increase; (b) decrease; (c) increase; (d) no change **12.112** (a) Rate $= k[C_2H_4Br_2][I^-]$; (b) $4.98 \times 10^{-3}/(M \cdot s)$; (c) 1.12×10^{-4} M/s **12.114** (a) 2.81×10^{-3} M; (b) 470 min; (c) 63 h **12.116** For $E_a = 50$ kJ/mol, $f = 2.0 \times 10^{-9}$. For $E_a = 100$ kJ/mol, $f = 3.9 \times 10^{-18}$ **12.118** $[A] = -kt + [A]_0$; $[A]_0/2 = -kt_{1/2} + [A]_0$; $t_{1/2} = [A]_0/2\,k$. For a zeroth-order reaction, each half-life is half of the previous one. For a first-order reaction, each half-life is the same as the previous one. For a second-order reaction, each half-life is twice the previous one. **12.120** (a) $\text{Rate}_f = k_f[A]$ and $\text{Rate}_r = k_r[B]$;

(b)

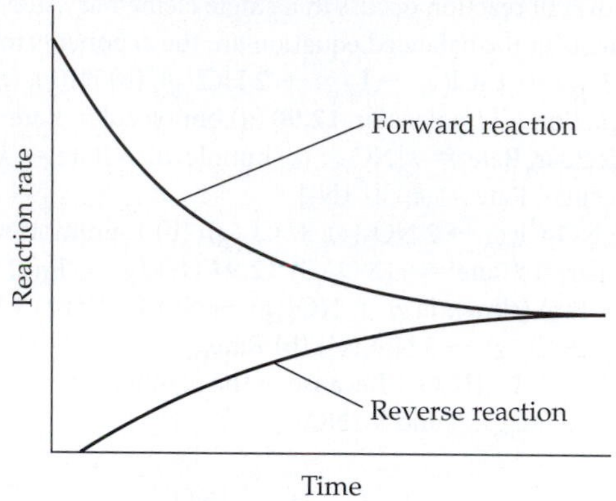

(c) $\dfrac{[B]}{[A]} = 3$ **12.122** 1.6×10^4 y **12.124** 49 °C **12.126** (a) 7.7×10^{-7} s; (b) the speed of mixing **12.128** 2.7×10^2 s **12.130** 15 s **12.132** 5.9×10^9 y **12.134** 1.6×10^{12} **12.136** (a) 238 kJ/mol; (b) 330 s ***Multiconcept Problems*** **12.138** (a) 8.7×10^{19} molecules/min; (b) 1.2 atm **12.140** (a) 0.71 atm; (b) 2.4 J/s; (c) 0.89 kJ **12.142** −122 J

CHAPTER 13

13.1 (a) $K_c = \dfrac{[SO_3]^2}{[SO_2]^2[O_2]}$; (b) $K_c = \dfrac{[SO_2]^2[O_2]}{[SO_3]^2}$

13.2 (a) 7.9×10^4; (b) 1.3×10^{-5}

13.3 (a) $K_c = \dfrac{[H^+][C_3H_5O_3^-]}{[C_3H_6O_3]}$; (b) 1.38×10^{-4}

13.4 Mixture (2) and (4) because $K_c = 1$, as for mixture (1). **13.5** (a) k_f; (b) $2.5 \times 10^{-28}\ M^{-1}\,s^{-1}$; (c) k_r decreases more than k_f decreases, and therefore K_c increases. **13.6** 9.48 **13.7** at 500 K: $K_p = 1.7 \times 10^4$; at 1000 K: $K_c = 1.1$

13.8 (a) $K_c = \dfrac{[H_2]^3}{[H_2O]^3}$, $K_p = \dfrac{(P_{H_2})^3}{(P_{H_2O})^3}$, $K_p = K_c$

(b) $K_c = [H_2]^2[O_2]$, $K_p = (P_{H_2})^2(P_{O_2})$, $K_p = K_c(RT)^3$

(c) $K_c = \dfrac{[HCl]^4}{[SiCl_4][H_2]^2}$, $K_p = \dfrac{(P_{HCl})^4}{(P_{SiCl_4})(P_{H_2})^2}$, $K_p = K_c(RT)$

(d) $K_c = \dfrac{1}{[Hg_2^{\ 2+}][Cl^-]^2}$

13.9 $K_c = 1.2 \times 10^{-42}$. Because K_c is very small, the equilibrium mixture contains mostly H_2 molecules. H is in periodic group 1A. A very small value of K_c is consistent with strong bonding between 2 H atoms, each with one valence electron. **13.10** (a) Because $Q_c < K_c$, the reaction is not at equilibrium. The reaction will proceed to the right to reach equilibrium. (b) Because $Q_c > K_c$, the reaction is not at equilibrium. The reaction will proceed to the left to reach equilibrium. **13.11** (a) (2); (b) (1), reverse; (3), forward **13.12** (a) 3.5×10^{-22} M; (b) 210 H atoms, 6.0×10^{22} H_2 molecules **13.13** $[CO_2] = [H_2] = 0.101$ M; $[CO] = [H_2O] = 0.049$ M **13.14** $[N_2O_4] = 0.0429$ M; $[NO_2] = 0.0141$ M **13.15** $[N_2O_4] = 0.0292$ M; $[NO_2] = 0.0116$ M **13.16** $P_{H_2O} = 0.90$ atm; $P_{CO} = 1.30$ atm; $P_{H_2} = 1.70$ atm **13.17** (a) increases; (b) decreases; (c) decreases; (d) increases; At equilibrium,

$$Q_c = K_c = \frac{[CO_2][H_2]}{[CO][H_2O]}$$

If some CO_2 is removed from the equilibrium mixture, the numerator in Q_c is decreased, which means that $Q_c < K_c$ and the reaction will shift to the right, increasing the H_2 concentration. **13.18** (a) remains the same; (b) increases; (c) decreases

13.19

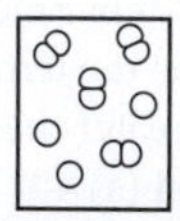

13.20 The equilibrium mixture will contain more NO, the higher the temperature. **13.21** As the temperature is increased the reaction shifts from right to left. The amount of ethyl acetate decreases. The product concentrations increase, and the reactant concentrations decrease. This corresponds to an increase in K_c. **13.22** There are more AB(g) molecules at the higher temperature. The equilibrium shifted to the right at the higher temperature, which means the reaction is endothermic. **13.23** (a) remains the same; (b) increases; (c) decreases; (d) remains the same; (e) decreases **13.24** The reaction will shift to the left. This will decrease the effectiveness of Hb for carrying O_2. **13.25** 2.58×10^{21} O_2 molecules ***Conceptual Problems*** **13.26** (a) (1) and (3); (b) $K_c = 1.5$; (c) Because the same number of molecules appear on both sides of the equation, the volume terms in K_c all cancel. Therefore, we can calculate K_c without including the volume. **13.28** (a) (3); (b) reaction (1): reverse direction; reaction (2): forward direction **13.30** When the stopcock is opened, the reaction will go in the reverse direction because there will be initially an excess of AB molecules. **13.32** (a) $AB \rightarrow A + B$; (b) The reaction is endothermic because a stress of added heat (higher temperature) shifts the $AB \rightarrow A + B$ equilibrium to the right. (c) If the volume is increased, the pressure is decreased. The stress of decreased pressure will be relieved by a shift in the equilibrium from left to right, thus increasing the number of A atoms.

13.34 (a)

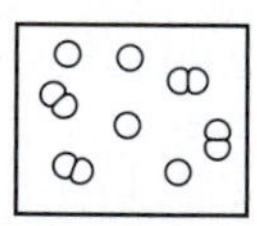

(b)

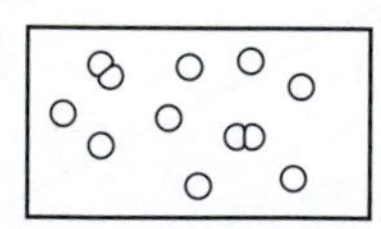

(c)

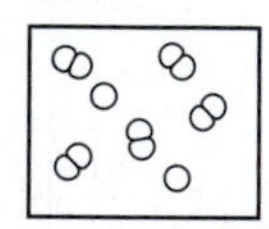

13.36 (a) $A \rightarrow 2\,B$; (b) (1) increase, (2) increase, (3) remain the same, (4) remain the same ***Section Problems*** **13.38** (a) $K_c = \dfrac{[CO][H_2]^3}{[CH_4][H_2O]}$;

(b) $K_c = \dfrac{[ClF_3]^2}{[F_2]^3[Cl_2]}$; (c) $K_c = \dfrac{[HF]^2}{[H_2][F_2]}$

13.40 (a) $K_p = \dfrac{(P_{CO})(P_{H_2})^3}{(P_{CH_4})(P_{H_2O})}$, $\Delta n = 2$ and $K_p = K_c(RT)^2$

(b) $K_p = \dfrac{(P_{ClF_3})^2}{(P_{F_2})^3(P_{Cl_2})}$, $\Delta n = -2$ and $K_p = K_c(RT)^{-2}$

(c) $K_p = \dfrac{(P_{HF})^2}{(P_{H_2})(P_{F_2})}$, $\Delta n = 0$ and $K_p = K_c$

13.42 $K_c = \dfrac{[C_2H_5OC_2H_5][H_2O]}{[C_2H_5OH]^2}$

13.44 1.3×10^8 **13.46** 0.058 **13.48** 29.0

13.50 (a) $K_c = \dfrac{[CH_3CO_2C_2H_5][H_2O]}{[CH_3CO_2H][C_2H_5OH]}$

(b) $K_c = 3.4$. Because there are the same number of molecules on both sides of the equation, the volume terms in K_c cancel. Therefore, we can calculate K_c without including the volume. **13.52** 9.0×10^3 **13.54** $K_p = 0.0313$; $K_c = 1.28 \times 10^{-3}$

13.56 (a) $K_c = \dfrac{[CO_2]^3}{[CO]^3}$, $K_p = \dfrac{(P_{CO_2})^3}{(P_{CO})^3}$; (b) $K_c = \dfrac{1}{[O_2]^3}$, $K_p = \dfrac{1}{(P_{O_2})^3}$

(c) $K_c = [SO_3]$, $K_p = P_{SO_3}$; (d) $K_c = [Ba^{2+}][SO_4^{\,2-}]$

13.58 $k_f[A][B] = k_r[C]$; $\dfrac{k_f}{k_r} = \dfrac{[C]}{[A][B]} = K_c$

13.60 210 **13.62** k_r increases more than k_f, this means that E_a (reverse) is greater than E_a (forward). The reaction is exothermic when E_a (reverse) $> E_a$ (forward). **13.64** forward direction **13.66** (a) mostly product; (b) mostly reactants **13.68** $K_c = 1.2 \times 10^{82}$ is very large. When equilibrium is reached, very little if any ethanol will remain because the reaction goes to completion. **13.70** The reaction is not at equilibrium because $Q_c > K_c$. The reaction will proceed in the reverse direction to attain equilibrium. **13.72** $Q_p = 178$; forward direction **13.74** (a) $Q_p = 36.2$; the reaction will proceed in the reverse direction; (b) $P_{In} = 0.1193$ atm; $P_{H_2} = 0.0943$ atm; $P_{InH_2} = 0.0167$ atm **13.76** 5.9×10^{-3} M **13.78** [NO] = 0.056 M; $[N_2] = [O_2] = 1.37$ M **13.80** [L-α-lysine] = 0.000 37 M; [L-β-lysine] = 0.002 63 M **13.82** (a) 6.8 moles; (b) 0.03 mol CH_3CO_2H; 9.03 mol C_2H_5OH **13.84** $P_{ClF} = P_{F_2} = 0.389$ atm; $P_{ClF_3} = 1.08$ atm **13.86** (a) increases; (b) increases; (c) decreases; (d) decreases. Disturbing the equilibrium by decreasing $[Cl^-]$ increases Q_c ($Q_c = 1/[Ag^+]_t[Cl^-]_t$) to a value greater than K_c. To reach a new state of equilibrium, Q_c must decrease, which means that the denominator must increase; that is, the reaction must go from right to left, thus decreasing the amount of solid AgCl. **13.88** (a) decreases; (b) remains the same; (c) increases **13.90** $[H_2]$ decreases when the temperature is increased. As the temperature is decreased, the reaction shifts to the right. $[CO_2]$ and $[H_2]$ increase, [CO] and $[H_2O]$ decrease, and K_c increases. **13.92** (a) increases; (b) increases; (c) decreases; (d) increases **13.94** (a) decreases; (b) increases; (c) no change; (d) increases **13.96** (a) no shift; (b) toward reactants; (c) toward products; (d) toward reactants ***Chapter Problems*** **13.98** $[H_2] = [I_2] = 0.045$ M; [HI] = 0.31 M **13.100** $[CO] = [H_2] = 0.18$ M; $[H_2O] = 1.02$ M **13.102** A decrease in volume (a) and the addition of reactants (c) will affect the composition of the equilibrium mixture, but leave the value of K_c unchanged. A change in temperature (b) affects the value of K_c. Addition of a catalyst (d) or an inert gas (e) affects neither the composition of the equilibrium mixture nor the value of K_c. **13.104** (a) $K_c = 0.573$; $K_p = 23.5$; (b) The reaction proceeds to the right to reach equilibrium. $[PCl_5] = 0.365$ M; $[PCl_3] = 0.285$ M; $[Cl_2] = 0.735$ M;

13.106 (a) $K_c = \dfrac{[C_2H_6][C_2H_4]}{[C_4H_{10}]}$ $K_p = \dfrac{(P_{C_2H_6})(P_{C_2H_4})}{P_{C_4H_{10}}}$

(b) 0.19; (c) 38%; $P_{total} = 69$ atm; (d) A decrease in volume would decrease the % conversion of C_4H_{10}. **13.108** (a) 0.0840; (b) $[(CH_3)_2C{=}CCH_2] = [HCl] = 0.094$ M; $[(CH_3)_3CCl] = 0.106$ M; (c) $P_{t\text{-butyl chloride}} = 0.055$ atm; $P_{isobutylene} = 0.345$ atm; $P_{HCl} = 0.545$ atm **13.110** The activation energy (E_a) is positive, and for an exothermic reaction, $E_{a,r} > E_{a,f}$.

$$K_c = \frac{k_f}{k_r} = \frac{A_f e^{-E_{a,f}/RT}}{A_r e^{-E_{a,r}/RT}} = \frac{A_f}{A_r} e^{(E_{a,r}-E_{a,f})/RT}$$

$(E_{a,r} - E_{a,f})$ is positive, so the exponent is always positive. As the temperature increases, the exponent, $(E_{a,r} - E_{a,f})/RT$, decreases and the value for K_c decreases as well. **13.112** (a) $K_c = 1.6 \times 10^{-3}$; (b) $K_p = 0.039$

13.114 (a) 1.0×10^{-6} atm; (b) exothermic; (c) At 2800 K, $Q_p = 100$, $Q_p > K_p$ so the reaction will go from products to reactants **13.116** $P_{NO_2} = 0.359$ atm; $P_{N_2O_4} = 1.14$ atm **13.118** (a) $P_{NO} = 0.46$ atm; $P_{NO_2} = 0.79$ atm; $P_{N_2O_3} = 0.20$ atm; (b) 74 L **13.120** 0.826 mol **13.122** (a) 0.002 44; (b) $K_p = 0.140$; (c) $[ClF_3] = 0.191$ M, $[ClF] = [F_2] = 0.0216$ M **13.124** [Fumarate] = 8.9×10^{-4} M; [L-Malate] = 2.94×10^{-3} M ***Multiconcept Problems*** **13.126** **(a)** $K_c = 1.51$; $K_p = 49.6$;

(b)

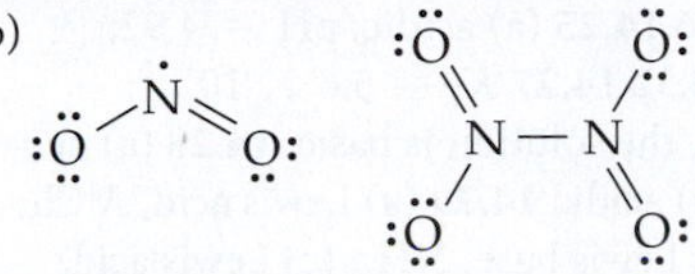

13.128 (a) 2.5; (b) 0.0034; (c) K_c for the water solution is so much smaller than K_c for the benzene solution because H_2O can hydrogen bond with acetic acid, thus preventing acetic acid dimer formation. Benzene cannot hydrogen bond with acetic acid. **13.130** (a) 1.47; (b) 10.3; (c) In agreement with Le Châtelier's principle, the reaction is endothermic because K_p increases with increasing temperature. **13.132** (a) 1.52; (b) 6.65 kJ **13.134** 6.0×10^2 O_3 molecules **13.136** (a) Because K_p is larger at the higher temperature, the reaction is endothermic. Because the reaction involves breaking two P–Cl bonds and forming just one Cl–Cl bond, it should be endothermic. (b) 99.97%; 1.38 atm

(c)

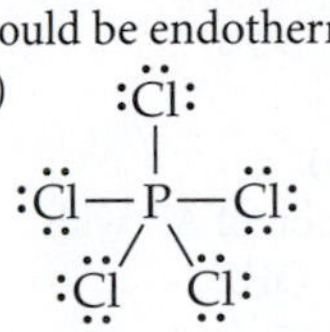

PCl_5 has no dipole moment because of a symmetrical distribution of Cl's around the central P.

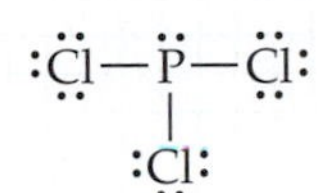

PCl_3 has a dipole moment because the geometry is trigonal pyramidal.

CHAPTER 14

14.1 (a) $H_2SO_4(aq) + H_2O(l) \rightleftharpoons H_3O^+(aq) + HSO_4^{\,-}(aq)$ conjugate base

(b) $HSO_4^{\,-}(aq) + H_2O(l) \rightleftharpoons H_3O^+(aq) + SO_4^{\,2-}(aq)$ conjugate base

(c) $H_3O^+(aq) + H_2O(l) \rightleftharpoons H_3O^+(aq) + H_2O(l)$ conjugate base

(d) $NH_4^{\,+}(aq) + H_2O(l) \rightleftharpoons H_3O^+(aq) + NH_3(aq)$ conjugate base

14.2 (a) H_2CO_3; (b) $HCO_3^{\,-}$; (c) H_2O; (d) H_3PO_4

14.3 $HCl(aq) + NH_3(aq) \rightleftharpoons NH_4^{\,+}(aq) + Cl^-(aq)$
Acid Base Acid Base
Conjugate acid–base pairs

14.4 (a) $H_9O_4^{\,+}$, $H_{19}O_9^{\,+}$, $H_{43}O_{21}^{\,+}$; (b) $H_9O_4^{\,+}$ ($4\,H_2O$), $H_{19}O_9^{\,+}$ ($9\,H_2O$), $H_{43}O_{21}^{\,+}$ ($21\,H_2O$) **14.5** (a) from right to left; (b) from left to right **14.6** (a) HY; (b) X^-; (c) to the left. **14.7** (a) H_2Se; (b) HI; (c) HNO_3; (d) H_2SO_3 **14.8** $[OH^-] = 7.1 \times 10^{-11}$ M, acidic. **14.9** $[H_3O^+] = 5.0 \times 10^{-9}$ M, basic. **14.10** 2.3×10^{-7} M **14.11** (a) 8.20; (b) 4.22 **14.12** (a) $[H_3O^+] = 4.0 \times 10^{-8}$ M; $[OH^-] = 2.5 \times 10^{-7}$ M; (b) $[H_3O^+] = 2 \times 10^{-3}$ M; $[OH^-] = 5 \times 10^{-12}$ M **14.13** (a) 1.30; (b) −0.78; (c) 14.60; (d) 12.30 **14.14** 11.81 **14.15** $K_a = 3.5 \times 10^{-8}$. This value of K_a agrees with the value in Table 14.2. **14.16** (a) HY < HX < HZ; (b) HZ; (c) HY has the highest pH; HX has the lowest pH **14.17** (a) pH = 2.38; $[H_3O^+] = [CH_3CO_2^{\,-}] = 4.2 \times 10^{-3}$ M; $[CH_3CO_2H] = 1.00$ M; $[OH^-] = 2.4 \times 10^{-12}$ M; (b) pH = 3.38;

$[H_3O^+] = [CH_3CO_2^-] = 4.2 \times 10^{-4}$ M; $[CH_3CO_2H] = 0.0096$ M; $[OH^-] = 2.4 \times 10^{-11}$ M **14.18** pH = 3.20 **14.19** (a) 8.0%; (b) 2.6% **14.20** $[H_3O^+] = [HSO_3^-] = 0.032$ M; $[H_2SO_3] = 0.07$ M; $[SO_3^{2-}] = 6.3 \times 10^{-8}$ M; $[OH^-] = 3.1 \times 10^{-13}$ M; pH = 1.49 **14.21** $[H_2SeO_4] = 0$ M; $[HSeO_4^-] = 0.49$ M; $[SeO_4^{2-}] = 0.011$ M; $[H_3O^+] = 0.51$ M; pH = 0.29; $[OH^-] = 2.0 \times 10^{-14}$ M **14.22** $[NH_4^+] = [OH^-] = 2.7 \times 10^{-3}$ M; $[NH_3] = 0.40$ M; $[H_3O^+] = 3.7 \times 10^{-12}$ M; pH = 11.43 **14.23** pH = 9.45 **14.24** (a) 7.7×10^{-12}; (b) 2.9×10^{-7} (c) 10.26 **14.25** (a) acidic, pH = 4.92; (b) acidic, pH = 5.00 **14.26** pH = 8.32 **14.27** $K_a = 5.6 \times 10^{-10}$; $K_b = 2.0 \times 10^{-5}$; Because $K_b > K_a$, the solution is basic. **14.28** (a) neutral; (b) basic; (c) acidic; (d) acidic; (e) acidic **14.29** (a) Lewis acid, $AlCl_3$; Lewis base, Cl^-; (b) Lewis acid, Ag^+; Lewis base, NH_3; (c) Lewis acid, SO_2; Lewis base, OH^-; (d) Lewis acid, Cr^{3+}; Lewis base, H_2O

14.30

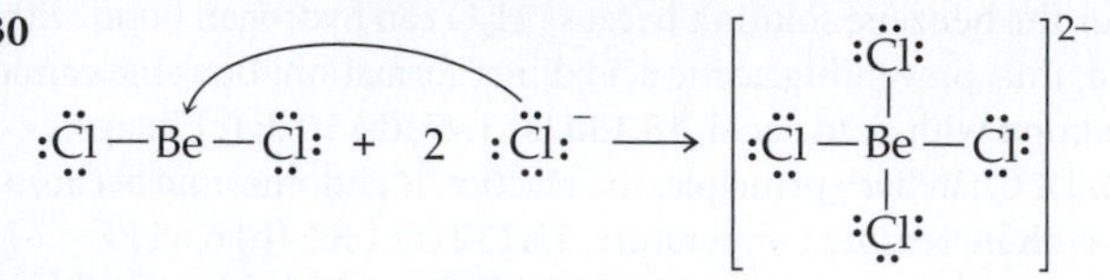

14.31 The O^{2-} from CaO is the Lewis base and SO_2 is the Lewis acid.

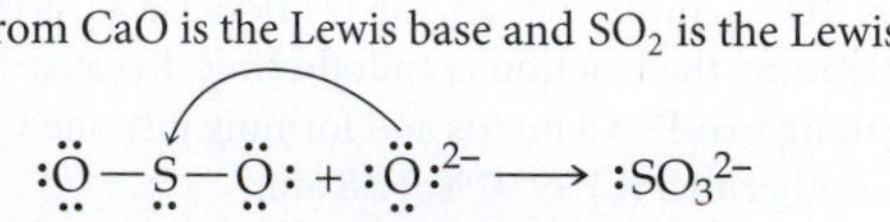

14.32 4.101 ***Conceptual Problems*** **14.34** (a) X^-, Y^-, Z^-; (b) HX < HZ < HY; (c) HY; (d) HX; (e) 20% **14.36** (b) **14.38** (a) $Y^- < Z^- < X^-$; (b) Y^-; (c) X^- (d) the reaction of A^- with water has a 1:1 stoichiometry: $A^- + H_2O \rightleftharpoons HA + OH^-$

14.40.

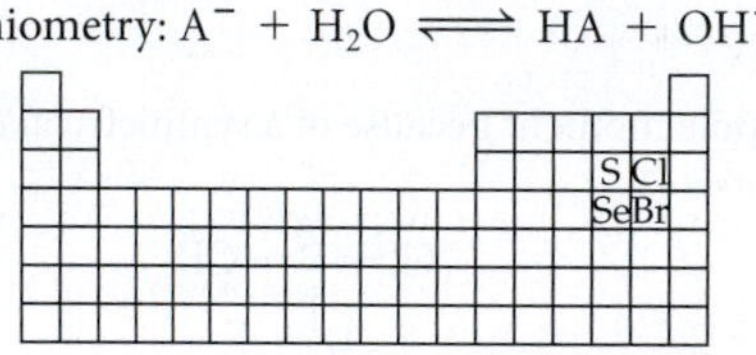

(a) H_2S, weakest; HBr, strongest. Acid strength for H_nX increases with increasing polarity of the H–X bond and with increasing size of X. (b) H_2SeO_3, weakest; $HClO_3$, strongest. Acid strength for H_nYO_3 increases with increasing electronegativity of Y. **14.42** (a) $H_3BO_3(aq) + H_2O(l) \rightleftharpoons H_3O^+(aq) + H_2BO_3^-(aq)$ (b) $H_3BO_3(aq) + 2\,H_2O(l) \rightleftharpoons H_3O^+(aq) + B(OH)_4^-(aq)$ ***Section Problems*** **14.44** NH_3, CN^-, and NO_2^- **14.46** (a) SO_4^{2-}; (b) HSO_3^-; (c) HPO_4^{2-}; (d) NH_3; (e) OH^-; (f) NH_2^-

14.48

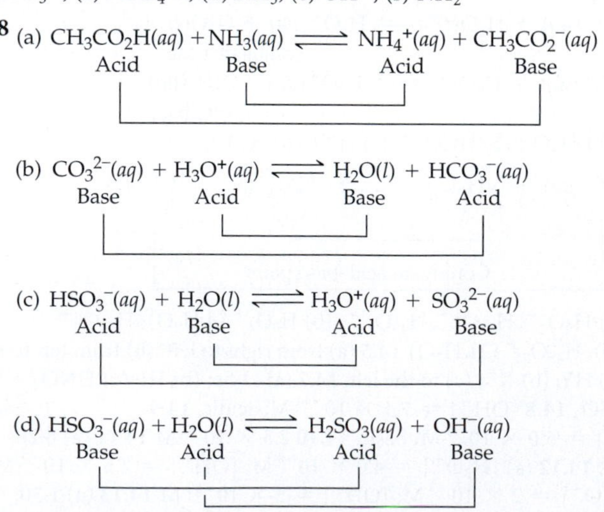

14.50 Strong acids: HNO_3 and H_2SO_4; Strong bases: H^- and O^{2-} **14.52** HCl, Cl^- **14.54** (a) $PH_3 < H_2S < HCl$; electronegativity increases from P to Cl; (b) $NH_3 < PH_3 < AsH_3$; X–H bond strength decreases from N to As (down a group); (c) $HBrO < HBrO_2 < HBrO_3$; acid strength increases with the number of O atoms **14.56** (a) HCl; The strength of a binary acid H_nA increases as A moves from left to right and from top to bottom in the periodic table. (b) $HClO_3$; The strength of an oxoacid increases with increasing electronegativity and increasing oxidation state of the central atom. (c) HBr; The strength of a binary acid H_nA increases as A moves from left to right and from top to bottom in the periodic table. **14.58** (a) H_2Te, weaker X–H bond; (b) H_3PO_4, P has higher electronegativity; (c) $H_2PO_4^-$, lower negative charge; (d) NH_4^+, higher positive charge and N is more electronegative than C **14.60** (a) $[OH^-] = 2.9 \times 10^{-6}$ M, basic; (b) $[H_3O^+] = 1.0 \times 10^{-12}$ M, basic; (c) $[H_3O^+] = 1.0 \times 10^{-4}$ M, acidic; (d) $[OH^-] = 1.0 \times 10^{-7}$ M, neutral; (e) $[OH^-] = 1.2 \times 10^{-10}$ M, acidic **14.62** $[H_3O^+] = [OH^-] = 3.9 \times 10^{-6}$M, neutral **14.64** (a) 4.70; (b) 11.6; (c) 8.449; (d) 3; (e) 15.08 **14.66** (a) 8×10^{-5} M; (b) 1.5×10^{-11} M; (c) 1.0 M; (d) 5.6×10^{-15} M, (e) 10 M; (f) 5.78×10^{-6} M **14.68** (a) chlorphenol red; (b) thymol blue; (c) methyl orange **14.70** 0.0089 g CaO **14.72** (a) 13.90; (b) 1.19; (c) 2.30; (d) 2.92 **14.74** (a) $C_6H_5OH < HOCl < CH_3CO_2H < HNO_3$; (b) $HNO_3 > CH_3CO_2H > HOCl > C_6H_5OH$; 1 M HNO_3, $[H_3O^+] = 1$ M; 1 M CH_3CO_2H, $[H_3O^+] = 4 \times 10^{-3}$ M; 1 M HOCl, $[H_3O^+] = 2 \times 10^{-4}$ M; 1 M C_6H_5OH, $[H_3O^+] = 1 \times 10^{-5}$ M **14.76** $K_a = 2.0 \times 10^{-9}$ **14.78** $[H_3O^+] = [C_3H_5O_2^-] = 0.0029$ M; $[HC_3H_5O_2] = 0.147$ M; pH = 2.54; $[OH^-] = 3.4 \times 10^{-12}$ M; (b) 3.3% **14.80** pH = 1.59; % dissociation = 1.7% **14.82** $H_2SeO_4(aq) + H_2O(l) \rightleftharpoons H_3O^+(aq) + HSeO_4^-(aq)$;

$$K_{a1} = \frac{[H_3O^+][HSeO_4^-]}{[H_2SeO_4]}$$

$HSeO_4^-(aq) + H_2O(l) \rightleftharpoons H_3O^+(aq) + SeO_4^{2-}(aq)$;

$$K_{a2} = \frac{[H_3O^+][SeO_4^{2-}]}{[HSeO_4^-]}$$

14.84 $[H_3O^+] = [HCO_3^-] = 6.6 \times 10^{-5}$ M; $[H_2CO_3] = 0.010$ M; $[CO_3^{2-}] = 5.6 \times 10^{-11}$ M; $[OH^-] = 1.5 \times 10^{-10}$ M; pH = 4.18 **14.86** pH = 1.08; $[C_2O_4^{2-}] = 6.4 \times 10^{-5}$ M **14.88** (a) $(CH_3)_2NH(aq) + H_2O(l) \rightleftharpoons (CH_3)_2NH_2^+(aq) + OH^-(aq)$;

$$K_b = \frac{[(CH_3)_2NH_2^+][OH^-]}{[(CH_3)_2NH]}$$

(b) $C_6H_5NH_2(aq) + H_2O(l) \rightleftharpoons C_6H_5NH_3^+(aq) + OH^-(aq)$;

$$K_b = \frac{[C_6H_5NH_3^+][OH^-]}{[C_6H_5NH_2]}$$

(c) $CN^-(aq) + H_2O(l) \rightleftharpoons HCN(aq) + OH^-(aq)$;

$$K_b = \frac{[HCN][OH^-]}{[CN^-]}$$

14.90 $K_b = 1 \times 10^{-6}$; $pK_b = 5.8$ **14.92** $[OH^-] = [HC_{18}H_{21}NO_4^+] = 9.0 \times 10^{-5}$ M; $[C_{18}H_{21}NO_4] = 0.002\,41$ M; $[H_3O^+] = 1.1 \times 10^{-10}$ M; pH = 9.96 **14.94** (a) 2.0×10^{-11}; (b) 1.1×10^{-6}; (c) 2.3×10^{-5}; (d) 5.6×10^{-6}

14.96 (a) $CH_3NH_3^+(aq) + H_2O(l) \rightleftharpoons H_3O^+(aq) + CH_3NH_2(aq)$
Acid Base Acid Base

(b) $Cr(H_2O)_6^{3+}(aq) + H_2O(l) \rightleftharpoons H_3O^+(aq) + Cr(H_2O)_5OH^{2+}(aq)$
Acid Base Acid Base

(c) $CH_3CO_2^-(aq) + H_2O(l) \rightleftharpoons OH^-(aq) + CH_3CO_2H(aq)$
Base Acid Base Acid

(d) $PO_4^{3-}(aq) + H_2O(l) \rightleftharpoons OH^-(aq) + HPO_4^{2-}(aq)$
Base Acid Base Acid

14.98 (a) basic; (b) neutral; (c) acidic; (d) neutral; (e) basic; (f) acidic **14.100** (a) $[H_3O^+] = [C_2H_5NH_2] = 1.2 \times 10^{-6}$ M; pH = 5.90; $[C_2H_5NH_3^+] = 0.10$ M; $[NO_3^-] = 0.10$ M; $[OH^-] = 8.0 \times 10^{-9}$ M; (b) $[CH_3CO_2H] = [OH^-] = 7.5 \times 10^{-6}$ M; $[CH_3CO_2^-] = 0.10$ M; $[Na^+] = 0.10$ M; $[H_3O^+] = 1.3 \times 10^{-9}$ M; pH = 8.89; (c) $[Na^+] = [NO_3^-] = 0.10$ M; $[H_3O^+] = [OH^-] = 1.0 \times 10^{-7}$ M; pH = 7.00 **14.102** (a) Lewis acid, SiF_4; Lewis base, F^-; (b) Lewis acid, Zn^{2+}; Lewis base, NH_3; (c) Lewis acid, $HgCl_2$; Lewis base, Cl^-; (d) Lewis acid, CO_2; Lewis base, H_2O

14.104

(a) $2\ :\ddot{F}:^- + SiF_4 \longrightarrow SiF_6^{2-}$

(b) $4\ \ddot{N}H_3 + Zn^{2+} \longrightarrow Zn(NH_3)_4^{2+}$

(c) $2\ :\ddot{Cl}:^- + HgCl_2 \longrightarrow HgCl_4^{2-}$

(d) $H_2\ddot{O}: + CO_2 \longrightarrow H_2CO_3$

14.106 (a) CN^-, Lewis base; (b) H^+, Lewis acid; (c) H_2O, Lewis base; (d) Fe^{3+}, Lewis acid; (e) OH^-, Lewis base; (f) CO_2, Lewis acid; (g) $P(CH_3)_3$, Lewis base; (h) $B(CH_3)_3$, Lewis acid ***Chapter Problems*** **14.108** H_2S, acid only. HS^-, acid and a base. S^{2-}, base only. H_2O, acid and a base. H_3O^+, acid only. OH^-, base only. **14.110** $HCO_3^-(aq) + Al(H_2O)_6^{3+}(aq) \rightarrow H_2O(l) + CO_2(g) + Al(H_2O)_5(OH)^{2+}(aq)$ **14.112** fraction dissociated = 6.09×10^{-10}; % dissociation = 6.09×10^{-8}% **14.114** $K_a = 3.9 \times 10^{-5}$; $pK_a = 4.41$ **14.116** $K_{a1} = 1.0 \times 10^{-8}$; $K_{a2} = 7.7 \times 10^{-4}$ **14.118** (a) $A^-(aq) + H_2O(l) \rightleftharpoons HA(aq) + OH^-(aq)$; basic; (b) $M(H_2O)_6^{3+}(aq) + H_2O(l) \rightleftharpoons H_3O^+(aq) + M(H_2O)_5(OH)^{2+}(aq)$; acidic; (c) $2\,H_2O(l) \rightleftharpoons H_3O^+(aq) + OH^-(aq)$; neutral; (d) $M(H_2O)_6^{3+}(aq) + A^-(aq) \rightleftharpoons HA(aq) + M(H_2O)_5(OH)^{2+}(aq)$; acidic because K_a for $M(H_2O)_6^{3+}$ (10^{-4}) is greater than K_b for A^- (10^{-9}). **14.120** $[H_3O^+] = 0.040$ M; pH = 1.39; $[HIO_3] = 0.010$ M; $[IO_3^-] = 0.040$ M; $[OH^-] = 2.5 \times 10^{-13}$ M **14.122** $[H_3O^+] = [HC_8H_4O_4^-] = 0.0051$ M; $[H_2C_8H_4O_4] = 0.020$ M; $[C_8H_4O_4^{2-}] = 3.1 \times 10^{-6}$ M; $[OH^-] = 2.0 \times 10^{-12}$ M; pH = 2.29 **14.124** (a) acidic; (b) basic

14.126 Fraction dissociated $= \dfrac{[HA]_{diss}}{[HA]_{initial}}$

For a weak acid, $[HA]_{diss} = [H_3O^+] = [A^-]$

$$K_a = \frac{[H_3O^+][A^-]}{[HA]} = \frac{[H_3O^+]^2}{[HA]};\ [H_3O^+] = \sqrt{K_a[HA]}$$

$$\text{Fraction dissociated} = \frac{[HA]_{diss}}{[HA]} = \frac{[H_3O^+]}{[HA]} = \frac{\sqrt{K_a[HA]}}{[HA]} = \sqrt{\frac{K_a}{[HA]}}$$

When the concentration of HA that dissociates is negligible compared with its initial concentration, the equilibrium concentration, [HA], equals the initial concentration, $[HA]_{initial}$. % dissociation $= \sqrt{\dfrac{K_a}{[HA]_{initial}}} \times 100\%$

14.128 pH = 2.54 **14.130** $[H_3O^+] = 0.0254$ M; pH = 1.59; $[OH^-] = 3.9 \times 10^{-13}$ M; $[NO_3^-] = 0.0250$ M; $[HNO_2] = 0.0246$ M; $[NO_2^-] = 4.3 \times 10^{-4}$ M ***Multiconcept Problems*** **14.132** $[H_3PO_4] = 0.67$ M; $[H_2PO_4^-] = 0.071$ M; $[HPO_4^{2-}] = 6.2 \times 10^{-8}$ M; $[PO_4^{3-}] = 4.2 \times 10^{-19}$ M; $[H_3O^+] = 0.071$ M; $[OH^-] = 1.4 \times 10^{-13}$ M; pH = 1.15 **14.134** 0.25 *m*, −0.93 °C **14.136** $H_3O^+(aq) + PO_4^{3-}(aq) \rightleftharpoons HPO_4^{2-}(aq) + H_2O(l)$; $H_3O^+(aq) + HPO_4^{2-}(aq) \rightleftharpoons H_2PO_4^-(aq) + H_2O(l)$; $H_3O^+(aq) + H_2PO_4^-(aq) \rightleftharpoons H_3PO_4(aq) + H_2O(l)$; pH = 2.02 **14.138** (a) pH = 1.88; (b) 0.48 L. The result in Problem 14.126 can't be used here because the concentration of HF that dissociates can't be neglected compared with the initial HF concentration.

14.140 (a) Rate $= k\dfrac{[OCl^-][NH_3]}{[OH^-]}$; $k = 17\ s^{-1}$;
(b) $k_2 = 5.9 \times 10^7\ M^{-1}\,s^{-1}$

CHAPTER 15

15.1 (a) $HNO_2(aq) + OH^-(aq) \rightleftharpoons NO_2^-(aq) + H_2O(l)$; NO_2^- (basic anion), pH > 7.00; (b) $H_3O^+(aq) + NH_3(aq) \rightleftharpoons NH_4^+(aq) + H_2O(l)$; NH_4^+(acidic cation), pH < 7.00; (c) $OH^-(aq) + H_3O^+(aq) \rightleftharpoons 2\,H_2O(l)$; pH = 7.00 **15.2** (a) $HF(aq) + OH^-(aq) \rightleftharpoons H_2O(l) + F^-(aq)$; $K_n = 3.5 \times 10^{10}$; (b) $H_3O^+(aq) + OH^-(aq) \rightleftharpoons 2\,H_2O(l)$; $K_n = 1.0 \times 10^{14}$; (c) $HF(aq) + NH_3(aq) \rightleftharpoons NH_4^+(aq) + F^-(aq)$; $K_n = 6.3 \times 10^5$ **15.3** $[H_3O^+] = 1.2 \times 10^{-9}$ M; pH = 8.91; $[OH^-] = 8.2 \times 10^{-6}$ M; $[Na^+] = [CN^-] = 0.010$ M; $[HCN] = 0.025$ M; % dissociation = 4.9×10^{-6}% **15.4** pH = 9.55 **15.5** Solution (2) has the largest percent dissociation. Solution (1) has the lowest pH. **15.6** (a) (1) and (3); (b) (3) **15.7** For the buffer, pH = 3.76; (a) pH = 3.71; (b) pH = 3.87 **15.8** This solution has less buffering capacity than the solution in Problem 15.7 because it contains less HF and F^- per 100 mL. Note that the change in pH is greater than that in Problem 15.7. **15.9** pH = 9.95 **15.10** Make the Na_2CO_3 concentration 1.4 times the concentration of $NaHCO_3$. **15.11** HOCl ($K_a = 3.5 \times 10^{-8}$) and NaOCl. **15.12** (a) 9.44; (b) 7.87 **15.13** (a) 2.17; (b) 11.77. The results are consistent with the pH data in Table 15.1 **15.14** (a) 12.00; (b) 4.08; (c) 2.15 **15.15** (a) (3); (b) (1); (c) (4); (d) (2) **15.16** 40 mL of 0.0400 M NaOH is required to reach the equivalence point. (a) 6.98; (b) 7.46; (c) 9.75 **15.17** Use thymolphthalein (pH 9.4–10.6). Bromthymol blue is unacceptable because it changes color halfway to the equivalence point. **15.18** (a) 4.51; (b) 7.20; (c) 7.68 **15.19** (a) 5.97; (b) 9.62; (c) 10.91 **15.20** (a) $K_{sp} = [Ag^+][Cl^-]$; (b) $K_{sp} = [Pb^{2+}][I^-]^2$; (c) $K_{sp} = [Ca^{2+}]^3[PO_4^{3-}]^2$; (d) $K_{sp} = [Cr^{3+}][OH^-]^3$ **15.21** 2.1×10^{-33} **15.22** 1.10×10^{-10} **15.23** (a) AgCl, solubility = 1.3×10^{-5} mol/L (0.0019 g/L); (b) Ag_2CrO_4, solubility = 6.5×10^{-5} mol/L (0.022 g/L); Ag_2CrO_4 has both the higher molar and gram solubility, despite its smaller value of K_{sp}. **15.24** (a) AgZ; (b) AgY **15.25** 1.4×10^{-5} M **15.26** AgCN, $Al(OH)_3$, and ZnS **15.27** 1.1×10^{-12} M **15.28** $K = 25.4$; 0.045 mol/L **15.29** (a) IP > K_{sp}; a precipitate of $BaCO_3$ will form. (b) IP < K_{sp}; no precipitate will form. **15.30** For $Mn(OH)_2$, IP < K_{sp}; no precipitate will form. For $Fe(OH)_2$, IP > K_{sp}; a precipitate of $Fe(OH)_2$ will form. **15.31** $Q_c > K_{spa}$ for CdS; CdS will precipitate. $Q_c < K_{spa}$ for ZnS; Zn^{2+} will remain in solution. **15.32** 5.3×10^{-5} M **15.33** 7.2 ***Conceptual Problems*** **15.34** (a) (1), (3), and (4); (b) (4) **15.36** (4) **15.38** (a) (1) corresponds to (iii); (2) to (i); (3) to (ii); and (4) to (iv)

(b) 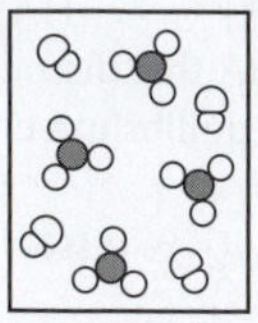

15.40 (2) is supersaturated; (3) is unsaturated; (4) is unsaturated. **15.42** (a) The lower curve represents the titration of a strong acid; the upper curve represents the titration of a weak acid. (b) pH = 7 for titration of the strong acid; pH = 10 for titration of the weak acid. (c) $pK_a \sim 6.3$. ***Section Problems*** **15.44** (a) pH = 7.00; (b) pH > 7.00; (c) pH < 7.00; (d) pH > 7.00. **15.46** Solution (a) **15.48** $K_n = 2.3 \times 10^{-5}$; K_n is small so the neutralization reaction does not proceed very far to completion. **15.50** 1.5×10^{-14} **15.52** (a) pH increases; (b) no change; (c) pH decreases **15.54** 3.06 **15.56** 0.022 M **15.58** For 0.10 M HN_3: % dissociation = 1.4. For 0.10 M HN_3 in 0.10 M HCl: % dissociation = 0.019%. The % dissociation is less because of the common ion (H_3O^+) effect. **15.60** Solutions (a), (c), and (d) **15.62** Solution (a) **15.64** pH = 9.09. The pH of a buffer solution will not change on dilution because the acid and base concentrations will change by the same amount and their ratio will remain the same. **15.66** pH = 3.67; (a) pH = 3.72; (b) pH = 3.61 **15.68** 4.04 **15.70** The volume of the 1.0 M NH_3 solution should be 3.5 times the volume of the 1.0 M NH_4Cl solution. **15.72** $H_2PO_4^-$ – HPO_4^{2-} because the pK_a for $H_2PO_4^-$ (7.21) is closest to 7.00. **15.74** 1.8×10^{-10} **15.76** (a) 9.00 mmol; (b) 20.0 mL; (c) 7.00;

(d) 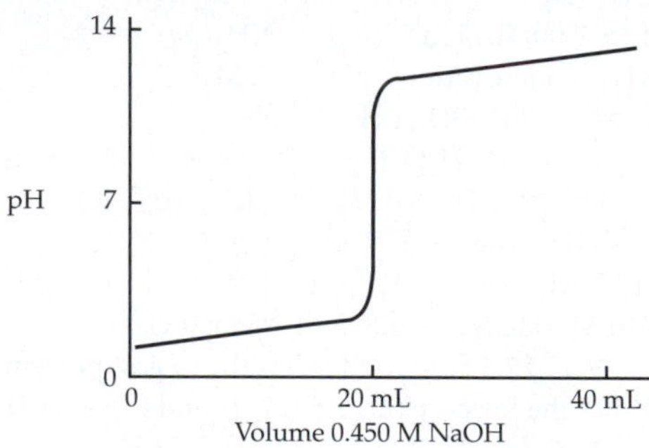

15.78 (a) 0.996; (b) 1.60; (c) 12.57 **15.80** 50.0 mL (a) 2.85; (b) 3.46; (c) 8.25; (d) 12.70 **15.82** (a) 1.74; (b) 2.34; (c) 6.02; (d) 9.70; (e) 11.11 **15.84** (a) C; (b) A **15.86** (a) pH = 8.04, phenol red; (b) pH = 7.00, bromthymol blue or phenol red (Any indicator that changes color in the pH range 4–10 is satisfactory for a strong acid–strong base titration.); (c) pH = 5.92, chlorphenol red **15.88** (a) $Ag_2CO_3(s) \rightleftharpoons 2\,Ag^+(aq) + CO_3^{2-}(aq)$; $K_{sp} = [Ag^+]^2[CO_3^{2-}]$; (b) $PbCrO_4(s) \rightleftharpoons Pb^{2+}(aq) + CrO_4^{2-}(aq)$; $K_{sp} = [Pb^{2+}][CrO_4^{2-}]$; (c) $Al(OH)_3(s) \rightleftharpoons Al^{3+}(aq) + 3\,OH^-(aq)$; $K_{sp} = [Al^{3+}][OH^-]^3$; (d) $Hg_2Cl_2(s) \rightleftharpoons Hg_2^{2+}(aq) + 2\,Cl^-(aq)$; $K_{sp} = [Hg_2^{2+}][Cl^-]^2$ **15.90** (a) 8.4×10^{-9}; (b) 5.8×10^{-3} M; (c) 0.13 M **15.92** 8.39×10^{-12} **15.94** (a) 4.37×10^{-9}; (b) 1.10×10^{-12}; (c) 7.0×10^{-7}; (d) 3.00×10^{-16} **15.96** (a) 1.1×10^{-5} M; (b) 1.1×10^{-4} M; (c) 1.6×10^{-5} M **15.98** (a) $AgNO_3$, source of Ag^+; equilibrium shifts left; (b) HNO_3, source of H_3O^+, removes CO_3^{2-}; equilibrium shifts right; (c) Na_2CO_3, source of CO_3^{2-}; equilibrium shifts left; (d) NH_3, forms $Ag(NH_3)_2^+$; removes Ag^+; equilibrium shifts right **15.100** (a) 5.3×10^{-7} M; (b) 2.8×10^{-10} M **15.102** (b), (c), and (d) are more soluble in acidic solution. (a) $AgBr(s) \rightleftharpoons Ag^+(aq) + Br^-(aq)$; (b) $CaCO_3(s) + H_3O^+(aq) \rightleftharpoons Ca^{2+}(aq) + HCO_3^-(aq) + H_2O(l)$; (c) $Ni(OH)_2(s) + 2\,H_3O^+(aq) \rightleftharpoons Ni^{2+}(aq) + 4\,H_2O(l)$; (d) $Ca_3(PO_4)_2(s) + 2\,H_3O^+(aq) \rightleftharpoons 3\,Ca^{2+}(aq) + 2\,HPO_4^{2-}(aq) + 2\,H_2O(l)$ **15.104** (a) more soluble, $Zn(OH)_2(s) + 2\,H_3O^+(aq) \rightleftharpoons Zn^{2+}(aq) + 4\,H_2O(l)$; (b) more soluble, $Zn(OH)_2(s) + 2\,OH^-(aq) \rightleftharpoons Zn(OH)_4^{2-}(aq)$; (c) more soluble, $Zn(OH)_2(s) + 4\,CN^-(aq) \rightleftharpoons Zn(CN)_4^{2-}(aq) + 2\,OH^-(aq)$ **15.106** 3.5×10^{-22} M **15.108** (a) $AgI(s) + 2\,CN^-(aq) \rightleftharpoons Ag(CN)_2^-(aq) + I^-(aq)$; $K = 2.6 \times 10^4$; (b) $Al(OH)_3(s) + OH^-(aq) \rightleftharpoons Al(OH)_4^-(aq)$; $K = 6$; (c) $Zn(OH)_2(s) + 4\,NH_3(aq) \rightleftharpoons Zn(NH_3)_4^{2+}(aq) + 2\,OH^-(aq)$; $K = 3.2 \times 10^{-8}$ **15.110** (a) 9.2×10^{-9} M; (b) 0.050 M **15.112** For $BaSO_4$, IP = 4.8×10^{-11}; IP < K_{sp}; $BaSO_4$ will not precipitate. For $Fe(OH)_3$, IP = 1.6×10^{-20}; IP > K_{sp}; $Fe(OH)_3(s)$ will precipitate. **15.114** IP = 9.6×10^{-11}; IP > K_{sp}; $Mg(OH)_2(s)$ will precipitate. **15.116** Yes. $Q_c = 1.1 \times 10^{-2}$. For FeS, $Q_c < K_{spa}$, and FeS will not precipitate. For SnS, $Q_c > K_{spa}$, and SnS will precipitate. **15.118** (i) $Q_c < K_{spa}$; FeS will not precipitate (ii) $Q_c > K_{spa}$; FeS(s) will precipitate. **15.120** (a) Add Cl^- to precipitate AgCl; (b) Add CO_3^{2-} to precipitate $CaCO_3$; (c) Add H_2S to precipitate MnS; (d) Add NH_3 and NH_4Cl to precipitate $Cr(OH)_3$. (Need buffer to control $[OH^-]$; excess OH^- produces the soluble $Cr(OH)_4^-$.) ***Chapter Problems*** **15.122** Prepare aqueous solutions of the three salts. Add a solution of $(NH_4)_2HPO_4$. If a white precipitate forms, the solution contains Mg^{2+}. Perform flame tests on the other two solutions. A yellow flame test indicates Na^+. A violet flame test indicates K^+. **15.124** (a)

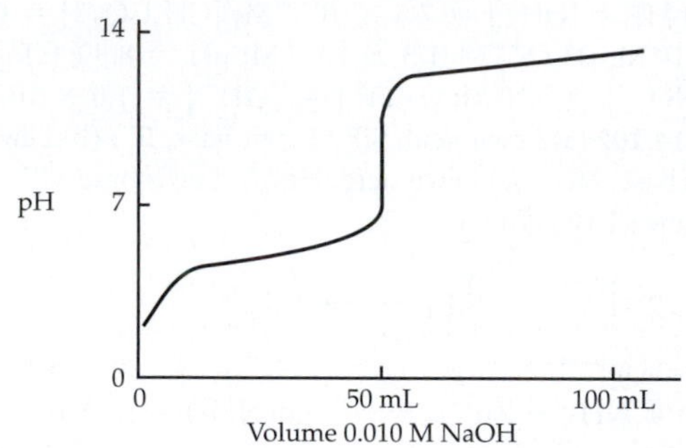

(b) 50 mL; (c) pH > 7.00; (d) pH = pK_a = 4.00 **15.126** 5.3×10^{-12} **15.128** $[NH_4^+]$ = 0.50 M; $[NH_3]$ = 1.0 M; pH = 9.55 **15.130** molar solubility = 3.5×10^{-11} M; solubility = 3×10^{-9} g/L **15.132** 7.51 **15.134** 10.7 mL of 0.10 M CH_3CO_2H and 9.3 mL of 0.15 M CH_3CO_2Na **15.136** (a) pH = 1.00; (b) pH = 1.84; (c) pH = 5.31; (d) pH = 8.83 **15.138** (a) 2.2×10^{-6} M; (b) 8.64; (c) 1.2×10^{-3} M ***Multiconcept Problems*** **15.140** (a)

$$HA^-(aq) + H_2O(l) \rightleftharpoons H_3O^+(aq) + A^{2-}(aq) \quad K_{a2} = 10^{-10}$$

$$HA^-(aq) + H_2O(l) \rightleftharpoons H_2A(aq) + OH^-(aq) \quad K_b = \frac{K_w}{K_{a1}} = 10^{-10}$$

$$2\,HA^-(aq) \rightleftharpoons H_2A(aq) + A^{2-}(aq) \quad K = \frac{K_{a2}}{K_{a1}} = 10^{-6}$$

$$2\,H_2O(l) \rightleftharpoons H_3O^+(aq) + OH^-(aq) \quad K_w = 1.0 \times 10^{-14}$$

The principal reaction is the one with the largest K, the third reaction.

(b) $K_{a1} = \dfrac{[H_3O^+][HA^-]}{[H_2A]}$ and $K_{a2} = \dfrac{[H_3O^+][A^{2-}]}{[HA^-]}$

$[H_3O^+] = \dfrac{K_{a1}[H_2A]}{[HA^-]}$ and $[H_3O^+] = \dfrac{K_{a2}[HA^-]}{[A^{2-}]}$

$\dfrac{K_{a1}[H_2A]}{[HA^-]} \times \dfrac{K_{a2}[HA^-]}{[A^{2-}]} = [H_3O^+]^2$; $\dfrac{K_{a1}K_{a2}[H_2A]}{[A^{2-}]} = [H_3O^+]^2$

Because the principal reaction is

$2\,HA^-(aq) \rightleftharpoons H_2A(aq) + A^{2-}(aq)$

$[H_2A] = [A^{2-}]$; $K_{a1}K_{a2} = [H_3O^+]^2$

$\log K_{a1} + \log K_{a2} = 2 \log [H_3O^+]$

$\dfrac{\log K_{a1} + \log K_{a2}}{2} = \log [H_3O^+]$;

$\dfrac{-\log K_{a1} + (-\log K_{a2})}{2} = -\log [H_3O^+]$; $\dfrac{pK_{a1} + pK_{a2}}{2} = \text{pH}$;

(c) 3×10^{19} A^{2-} ions **15.142** (a) 2.2×10^{-5} M; (b) [HCl] = 0.124 M; $[H_3PO_4]$ = 0.0960 M; (c) 100%; (d) 0.89;

(e)

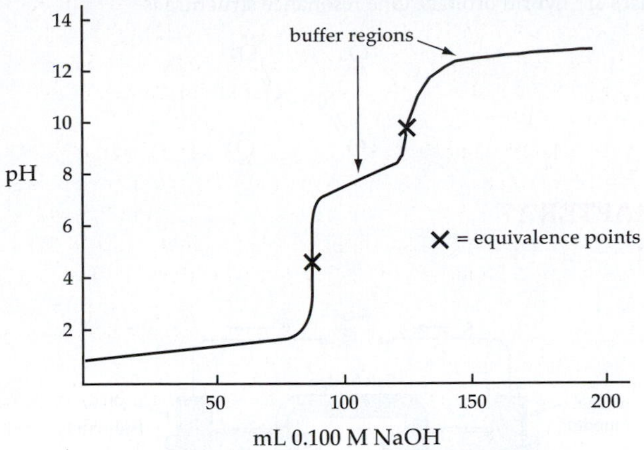

(f) Bromcresol green or methyl orange for the first equivalence point. Thymolphthalein for the second equivalence point. **15.144** 4.00×10^{-6} **15.146** (a)

NH_4^+	CO_3^{2-}	H_2O
acid	base	acid or base

$NH_4^+(aq) + CO_3^{2-}(aq) \rightleftharpoons HCO_3^-(aq) + NH_3(aq)$; (b) $[NH_4^+]$ = 0.095 M; $[NH_3]$ = 0.22 M; $[CO_3^{2-}]$ = 0.015 M; $[HCO_3^-]$ = 0.065 M; pH = 9.62; $[H_3O^+] = 2.4 \times 10^{-10}$ M; $[OH^-] = 4.2 \times 10^{-5}$ M; $[H_2CO_3] = 3.6 \times 10^{-4}$ M; (c) For MCO_3, IP = 1.5×10^{-4}; IP > 10^3 K_{sp} for $CaCO_3$ and $BaCO_3$, but IP < 10^3 K_{sp} for $MgCO_3$ so the $[CO_3^{2-}]$ is large enough to give observable precipitation of $CaCO_3$ and $BaCO_3$, but not $MgCO_3$. (d) For $M(OH)_2$, IP = 1.7×10^{-11}; IP < 10^3 K_{sp} for all three $M(OH)_2$. None precipitate. (e) In 0.08 M Na_2CO_3, $[OH^-] = 3.7 \times 10^{-3}$ M. Ca^{2+} and Ba^{2+} cannot be separated from Mg^{2+} using 0.08 M Na_2CO_3. Na_2CO_3 is more basic than $(NH_4)_2CO_3$, and $Mg(OH)_2$ would precipitate along with $CaCO_3$ and $BaCO_3$. **15.148** (a) 1.3 mg; (b) 1.4×10^2 washings; (c) The number of washings seems too high considering that frequent dye application is recommended. If the PbS is located mainly on the surface of the hair, solid particles of PbS can be lost by abrasion during shampooing.

CHAPTER 16

16.1 (a) spontaneous; (b), (c), and (d) nonspontaneous **16.2** (a) negative; (b) positive; (c) positive; (d) negative; (e) negative **16.3** (a) $A_2 + AB_3 \rightarrow 3\ AB$; (b) positive **16.4** (a) disordered N_2O (more randomness); (b) quartz glass (amorphous solid, more randomess); (c) 1 mole N_2 at STP (larger volume, more randomness); (d) 1 mole N_2 at 273 K and 0.25 atm (larger volume, more randomness) **16.5** +160.0 J/K **16.6** ΔS_{total} = −441 J/K. Because ΔS_{total} is negative, the reaction is not spontaneous under standard-state conditions at 25 °C. **16.7** (a) nonspontaneous at 25 °C; (b) 42 °C **16.8** (a) no; (b) 356 °C **16.9** $\Delta H < 0$; $\Delta S < 0$; $\Delta G < 0$ **16.10** (a) + 131.5 kJ; (b) Because $\Delta G > 0$, the reaction is nonspontaneous at 25 °C. (c) 847 °C **16.11** (a) $\Delta S°$ is positive; (b) at high temperature **16.12** (a) −148.4 kJ; (b) No because $\Delta G°_f(C_2H_2) > 0$. **16.13** ΔG = +39.6 kJ mol. Because $\Delta G > 0$, the reaction is spontaneous in the reverse direction. **16.14** (a) (3) has the largest ΔG; (2) has the smallest ΔG; (b) 15 kJ **16.15** 9×10^{-24} **16.16** 0.03 atm **16.17** 80 kJ **16.18** No, because the decrease in entropy of the growing person is more than compensated for by an increase in the entropy of the surroundings. **16.19** Yes. You might see an explosion run backwards, which is impossible because it would decrease the entropy of the universe.

Conceptual Problems **16.20** (a)

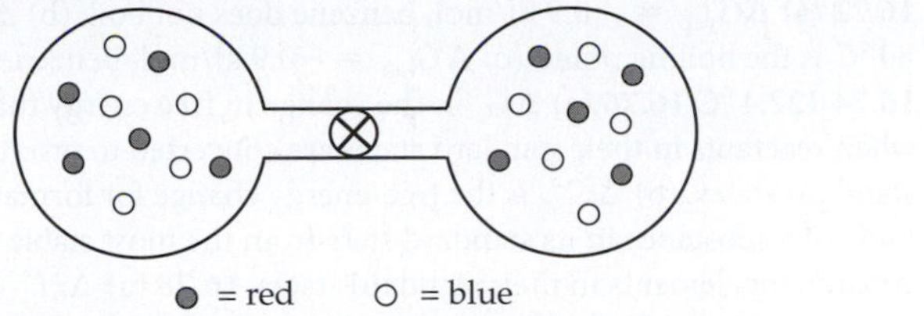

(b) $\Delta H = 0$ (no heat is gained or lost in the mixing of ideal gases); $\Delta S > 0$ (the mixture of the two gases has more randomness); $\Delta G < 0$ (the mixing of the two gases is spontaneous); (c) For an isolated system, $\Delta S_{surr} = 0$ and $\Delta S_{sys} = \Delta S_{Total} > 0$ for the spontaneous process; (d) $\Delta G > 0$ and the process is nonspontaneous. **16.22** $\Delta H < 0$ (heat is lost during condensation); $\Delta S < 0$ (liquid has less randomness than vapor); $\Delta G < 0$ (the reaction is spontaneous) **16.24** (a) $2\ A_2 + B_2 \rightarrow 2\ A_2B$; (b) $\Delta H < 0$ (because ΔS is negative, ΔH must also be negative in order for ΔG to be negative); $\Delta S < 0$ (the mixture becomes less random going from reactants [3 molecules] to products [2 molecules]); $\Delta G < 0$ (the reaction is spontaneous) **16.26** (a) $\Delta H° > 0$, $\Delta S° > 0$; (b) $\Delta S°$ is for the complete conversion of 1 mole of A_2 in its standard state to 2 moles of A in its standard state; (c) There is not enough information to say anything about the sign of $\Delta G°$. $\Delta G°$ decreases (becomes less positive or more negative) as the temperature increases; (d) K_p increases as the temperature increases. As the temperature increases there will be more A and less A_2. (e) $\Delta G = 0$ at equilibrium. **16.28** (1) $\Delta G° = 0$; (2) negative; (3) positive ***Section Problems*** **16.30** (a) and (d) nonspontaneous; (b) and (c) spontaneous **16.32** (b) and (d) spontaneous (because of the large positive K_p's) **16.34** Molecular randomness is called entropy. $H_2O(s) \rightarrow H_2O(l)$ at 25 °C. **16.36** (a) +; (b) −; (c) −; (d) + **16.38** (a) −; (b) −; (c) +; (d) − **16.40** (a) 2.30×10^{-22} J/K; (b) 2.30×10^{-21} J/K; (c) 11.5 J/K **16.42** 9.53×10^{-21} J/K **16.44** 1.17, 1.06 **16.46** (a) H_2 at 25 °C in 50 L; (b) O_2 at 25 °C, 1 atm; (c) H_2 at 100 °C, 1 atm; (d) CO_2 at 100 °C, 0.1 atm **16.48** 0.14 J/K **16.50** (a) $C_2H_6(g)$; (b) $CO_2(g)$; (c) $I_2(g)$; (d) $CH_3OH(g)$ **16.52** −173.0 J/K **16.54** (a) +125 J/K (+, because moles of gas increase); (b) −181.2 J/K (−, because moles of gas decrease); (c) +137.4 J/K (+, because moles of gas increase); (d) −626.4 J/K (−, because moles of gas decrease) **16.56** In any spontaneous process, the total entropy of a system and its surroundings always increases. **16.58** $\Delta S_{surr} = \frac{-\Delta H}{T}$; (a) positive; (b) negative. **16.60** (a) +879.3 J/K; (b) There is no temperature at which the reaction is not spontaneous. **16.62** ΔS_{total} = −1097 J/K; Because $\Delta S_{total} < 0$, the reaction is not spontaneous under standard-state conditions at 25 °C. **16.64** ΔS_{sys} = +216.4 J/K; ΔS_{surr} = −609.4 J/K; ΔS_{total} = −393.0 J/K; Because $\Delta S_{total} < 0$, the reaction is not spontaneous under standard-state conditions at 25 °C. (b) 839.2 K **16.66** (a) ΔS_{surr} = −89.5 J/(K·mol); ΔS_{total} = −2.5 J/(K·mol); (b) ΔS_{surr} = −87.0 J/(K·mol); ΔS_{total} = 0; (c) ΔS_{surr} = −84.6 J/(K·mol); ΔS_{total} = +2.4 J/(K·mol). Benzene does not boil at 70 °C (343 K) because ΔS_{total} is negative. The normal boiling point for benzene is 80 °C (353 K), where ΔS_{total} = 0. **16.68**

ΔH	ΔS	$\Delta G = \Delta H - T\Delta S$	Reaction Spontaneity
−	+	−	Spontaneous at all temperatures
−	−	− or +	Spontaneous at low temperatures where $\lvert\Delta H\rvert > \lvert T\Delta S\rvert$
			Nonspontaneous at high temperatures where $\lvert\Delta H\rvert < \lvert T\Delta S\rvert$
+	−	+	Nonspontaneous at all temperatures
+	+	− or +	Spontaneous at high temperatures where $T\Delta S > \Delta H$
			Nonspontaneous at low temperature where $T\Delta S < \Delta H$

16.70 (a) $\Delta H > 0, \Delta S > 0, \Delta G > 0$; (b) $\Delta H > 0, \Delta S > 0, \Delta G < 0$ **16.72** (a) $\Delta G_{vap} = +0.9$ kJ/mol, benzene does not boil; (b) $\Delta G_{vap} = 0$, 80 °C is the boiling point; (c) $\Delta G_{vap} = -0.9$ kJ/mol, benzene boils **16.74** 122.4 °C **16.76** (a) $\Delta G°$ is the change in free energy that occurs when reactants in their standard states are converted to products in their standard states. (b) $\Delta G°_f$ is the free-energy change for formation of one mole of a substance in its standard state from the most stable form of the constituent elements in their standard states. **16.78** (a) $\Delta H° = 66.4$ kJ; $\Delta S° = -121.5$ J/K; $\Delta G° = +102.6$ kJ. Because $\Delta G°$ is positive, the reaction is nonspontaneous under standard-state conditions at 25 °C. (b) $\Delta H° = -77.6$ kJ; $\Delta S° = 494.0$ J/K; $\Delta G° = -224.8$ kJ. Because $\Delta G°$ is negative, the reaction is spontaneous under standard-state conditions at 25 °C. (c) $\Delta H° = -492.6$ kJ; $\Delta S° = -136.1$ J/K; $\Delta G° = -452.0$ kJ; the reaction is spontaneous **16.80** (a) +102.6 kJ; (b) −224.4 kJ; (c) −452 kJ; **16.82** (a) stable; (b) stable; (c) unstable; (d) unstable **16.84** $\Delta G° = -147.7$ kJ. Because $\Delta G° < 0$, dichloroethane can be synthesized from gaseous C_2H_4 and Cl_2, each at 25 °C and 1 atm pressure. **16.86** $\Delta G° = -6.1$ kJ. The reaction becomes nonspontaneous at high temperatures because $\Delta S°$ is negative. The reaction becomes nonspontaneous at 72 °C. **16.88** $\Delta G° = -505.2$ kJ. Because $\Delta G°$ is negative, the reaction is possible. Look for a catalyst. Because $\Delta G°_f$ for benzene is positive (+124.5 kJ/mol), the synthesis of benzene from graphite and gaseous H_2 at 25 °C and 1 atm pressure is not possible. **16.90** $\Delta G = \Delta G° + RT \ln Q$ **16.92** $\Delta G = +11.8$ kJ/mol; reverse direction. **16.94** (a) −176.0 kJ; (b) −133.8 kJ; (c) −141.8 kJ **16.96** (a) If $K > 1$, $\Delta G°$ is negative. (b) If $K = 1$, $\Delta G° = 0$. (c) If $K < 1$, $\Delta G°$ is positive. **16.98** $K_p = 7.2 \times 10^{24}$ **16.100** 0.059 atm **16.102** 0.28 atm **16.104** 2.9×10^{28} ***Chapter Problems*** **16.106** −0.401 J/K **16.108** (a) Spontaneous does not mean fast, just possible. (b) For a spontaneous reaction, $\Delta S_{total} > 0$. ΔS_{sys} can be positive or negative. (c) An endothermic reaction can be spontaneous if $\Delta S_{sys} > 0$. (d) This statement is true because the sign of ΔG changes when the direction of a reaction is reversed.

16.110

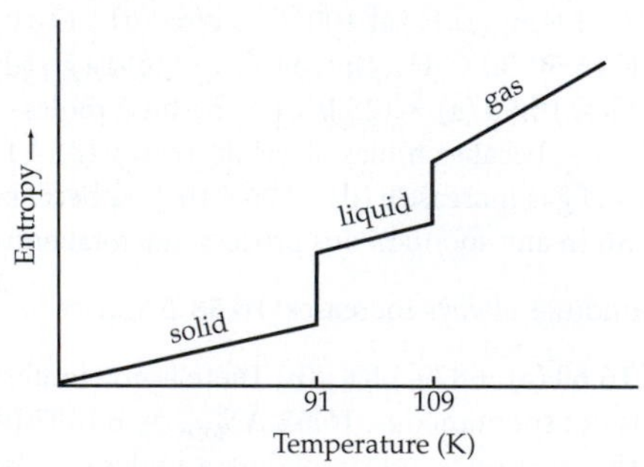

16.112 78 °C **16.114** (a) endothermic; (b) 33.6 kJ **16.116** (a) $\Delta H° = -1203.4$ kJ; $\Delta S° = -216.6$ J/K; $\Delta G° = -1138.8$ kJ; spontaneous. $\Delta G°$ becomes less negative as the temperature is raised. (b) $\Delta H° = +101$ kJ; $\Delta S° = 174.8$ J/K; $\Delta G° = +49$ kJ; not spontaneous. $\Delta G°$ becomes less positive as the temperature is raised. (c) $\Delta H° = -852$ kJ; $\Delta S° = -38.5$ J/K; $\Delta G° = -840$ kJ; spontaneous. $\Delta G°$ becomes less negative as the temperature is raised. (d) $\Delta H° = +135.6$ kJ; $\Delta S° = +333$ J/K; $\Delta G° = +36.4$ kJ; not spontaneous. $\Delta G°$ becomes less positive as the temperature is raised. **16.118** −55.6 kJ/mol **16.120** (a) −252.6 J/(K · mol); (b) −376 J/(K · mol); (c) −345 J/(K · mol) **16.122** $\Delta S° = -82.0$ J/K; $\Delta H° = +2.71$ kJ **16.124** (a) $\Delta G° = +70.2$ kJ; (b) $K_{sp} = 5 \times 10^{-13}$; (c) $\Delta G = 13.2$ kJ/mol. A positive value of ΔG means that the reverse reaction is spontaneous, which is consistent with the fact that $Q > K_{sp}$. **16.126** 26 °C **16.128** 1.3×10^7 **16.130** 5.7 mm Hg **16.132** (a) 46 °C; (b) 32 °C ***Multiconcept Problems*** **16.134** $[N_2] = 0.078$ M; $[H_2] = 0.235$ M; $[NH_3] = 0.043$ M **16.136** (a) −371.4 kJ; (b) −350.1 kJ **16.138** $K_b = 1.6 \times 10^{-5}$; pH = 11.05 **16.140** (a) $P_{N_2O_4} = 0.24$ atm; $P_{NO_2} = 2.52$ atm (b) N_2O_4 is planar with an N–N bond and trigonal plsnar geometry about the N atom. N uses sp^2 hybrid orbitals. One resonance structure is

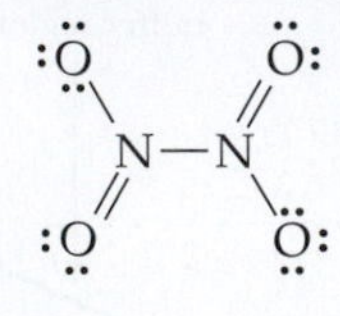

CHAPTER 17

17.1

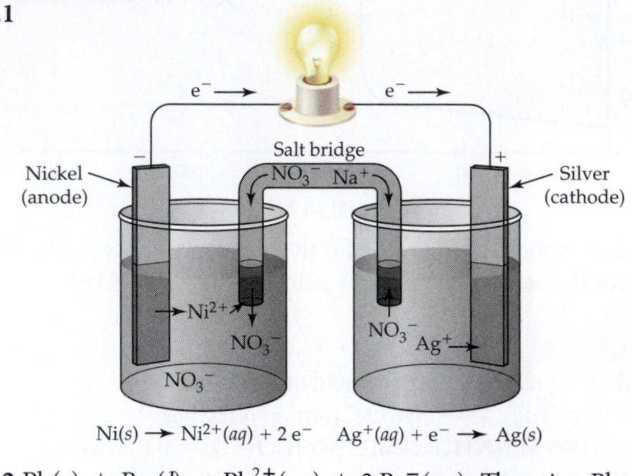

17.2 $Pb(s) + Br_2(l) \rightarrow Pb^{2+}(aq) + 2\,Br^-(aq)$. There is a Pb anode in an aqueous solution of Pb^{2+}. The cathode is a Pt wire that dips into a pool of liquid Br_2 and an aqueous solution that is saturated with Br_2. A salt bridge connects the anode and cathode compartment. The electrodes are connected through an external circuit.**17.3** $Fe(s)\,|\,Fe^{3+}(aq)\,||\,Cr_2O_7^{2-}(aq), Cr^{3+}(aq)\,|\,Pt(s)$ **17.4** (a) and (b)

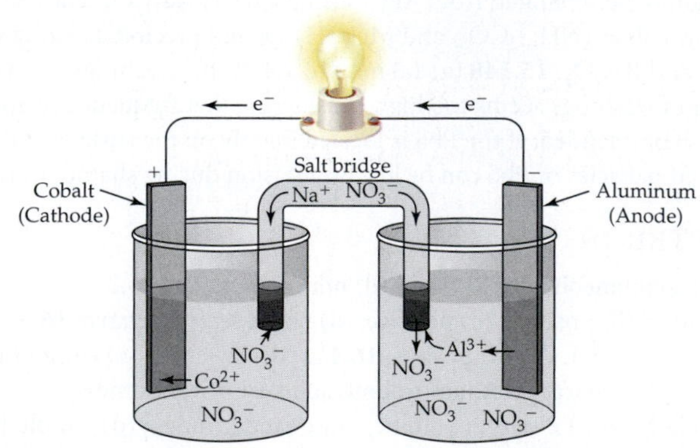

(c) $2\,Al(s) + 3\,Co^{2+}(aq) \rightarrow 2\,Al^{3+}(aq) + 3\,Co(s)$; (d) $Al(s)\,|\,Al^{3+}(aq)\,||\,Co^{2+}(aq)\,|\,Co(s)$ **17.5** 701 kJ **17.6** −0.74 V **17.7** Cl_2 is the stronger oxidizing agent. Mg is the stronger reducing agent. **17.8** (a) $E° = 0.23$ V; reaction can occur; (b) $E° = -1.40$ V; reaction cannot occur **17.9** (a) D is the strongest reducing agent. A^{3+} is the strongest oxidizing agent. (b) B^{2+} can oxidize C and D. C can reduce A^{3+} and B^{2+}. (c) $A^{3+} + 2\,D \rightarrow A^+ + 2\,D^+$; 2.85 V **17.10** 0.25 V **17.11** (a) +0.059 V; (b) −0.19 V **17.12** 6.9 **17.13** $K = 10^{31}$ **17.14** −0.140 V **17.15** (a) $Zn(s) + 2\,MnO_2(s) + 2\,NH_4^+(aq) \rightarrow Zn^{2+}(aq) + Mn_2O_3(s) + 2\,NH_3(aq) + H_2O(l)$; (b) $Zn(s) + 2\,MnO_2(s) \rightarrow ZnO(s) + Mn_2O_3(s)$; (c) $Cd(s) + 2\,NiO(OH)(s) + 2\,H_2O(l) \rightarrow Cd(OH)_2(s) + 2\,Ni(OH)_2(s)$; (d) $x\,Li(s) + MnO_2(s) \rightarrow Li_xMnO_2(s)$; (e) $Li_xC_6(s) + Li_{1-x}CoO_2(s) \rightarrow 6\,C(s) + LiCoO_2(s)$ **17.16** A fuel cell and a battery are both galvanic cells that convert chemical energy into electrical energy utilizing a spontaneous redox reacton. A fuel cell differs from a battery in that the reactants are not contained within the cell but instead are continuously supplied from an external reservoir.

17.17 (a) 1.23 V; (b) 1.23 V

17.18 (a)

$[Mg(s) \rightarrow Mg^{2+}(aq) + 2\,e^-] \times 2$

$O_2(g) + 4\,H^+(aq) + 4\,e^- \rightarrow 2\,H_2O(l)$

$2\,Mg(s) + O_2(g) + 4\,H^+(aq) \rightarrow 2\,Mg^{2+}(aq) + 2\,H_2O(l)$

(b)

$[Fe(s) \rightarrow Fe^{2+}(aq) + 2\,e^-] \times 4$

$[O_2(g) + 4\,H^+(aq) + 4\,e^- \rightarrow 2\,H_2O(l)] \times 2$

$4\,Fe^{2+}(aq) + O_2(g) + 4\,H^+(aq) \rightarrow 4\,Fe^{3+}(aq) + 2\,H_2O(l)$

$[2\,Fe^{3+}(aq) + 4\,H_2O(l) \rightarrow Fe_2O_3 \cdot H_2O(s) + 6\,H^+(aq)] \times 2$

$4\,Fe(s) + 3\,O_2(g) + 2\,H_2O(l) \rightarrow 2\,Fe_2O_3 \cdot H_2O(s)$

17.19 (a)

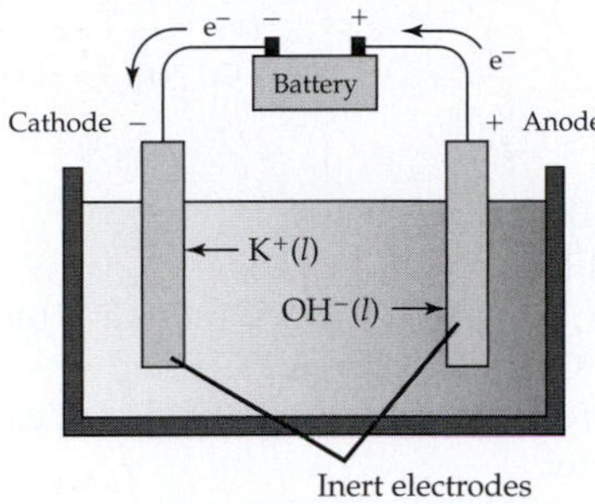

(b)	anode	$4\,OH^-(l) \rightarrow O_2(g) + 2\,H_2O(l) + 4\,e^-$
	cathode	$4\,K^+(l) + 4\,e^- \rightarrow 4\,K(l)$
	overall	$4\,K^+(l) + 4\,OH^-(l) \rightarrow 4\,K(l) + O_2(g) + 2\,H_2O(l)$

17.20 (a)

	anode	$2\,Cl^-(aq) \rightarrow Cl_2(g) + 2\,e^-$
	cathode	$2\,H_2O(l) + 2\,e^- \rightarrow H_2(g) + 2\,OH^-(aq)$
	overall	$2\,Cl^-(aq) + 2\,H_2O(l) \rightarrow Cl_2(g) + H_2(g) + 2\,OH^-(aq)$
(b)	anode	$2\,H_2O(l) \rightarrow O_2(g) + 4\,H^+(aq) + 4\,e^-$
	cathode	$2\,Cu^{2+}(aq) + 4\,e^- \rightarrow 2\,Cu(s)$
	overall	$2\,Cu^{2+}(aq) + 2\,H_2O(l) \rightarrow 2\,Cu(s) + O_2(g) + 4\,H^+(aq)$
(c)	anode	$2\,H_2O(l) \rightarrow O_2(g) + 4\,H^+(aq) + 4\,e^-$
	cathode	$4\,H_2O(l) + 4\,e^- \rightarrow 2\,H_2(g) + 4\,OH^-(aq)$
	overall	$2\,H_2O(l) \rightarrow 2\,H_2(g) + O_2(g)$

17.21

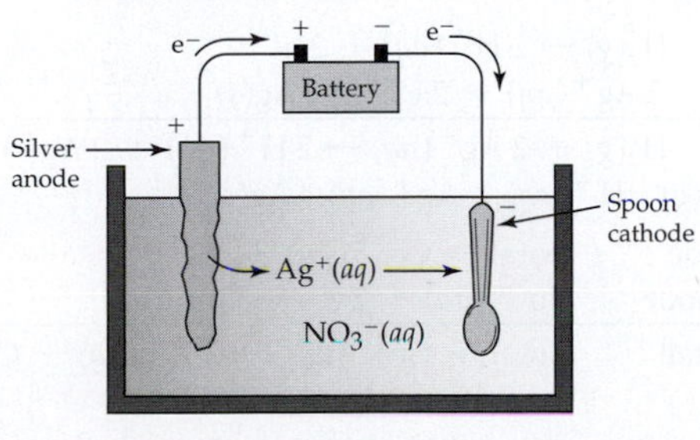

anode	$Ag(s) \rightarrow Ag^+(aq) + e^-$
cathode	$Ag^+(aq) + e^- \rightarrow Ag(s)$

The overall reaction is transfer of silver metal from the silver anode to the spoon. **17.22** 268 kg **17.23** 7.45 h

17.24 $Ti(s) + 2\,H_2O(l) \rightarrow TiO_2(s) + 2\,H_2(g) \quad E° = 1.066\ V$

17.25 62.6 min ***Conceptual Problems***

17.26 (a)–(d)

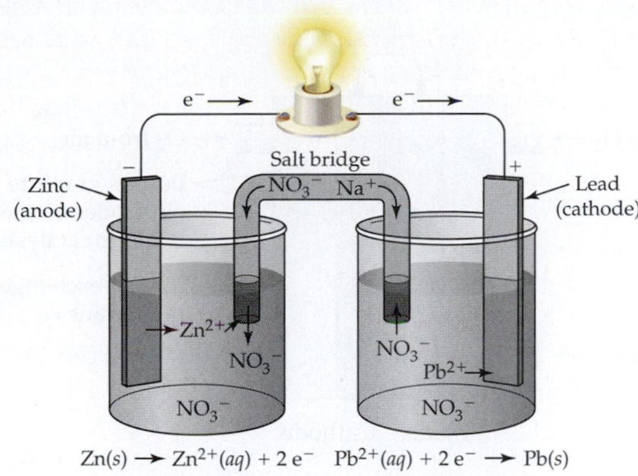

(e)	anode	$Zn(s) \rightarrow Zn^{2+}(aq) + 2\,e^-$
	cathode	$Pb^{2+}(aq) + 2\,e^- \rightarrow Pb(s)$
	overall	$Zn(s) + Pb^{2+}(aq) \rightarrow Zn^{2+}(aq) + Pb(s)$

17.28 (a) The three cell reactions are the same except for cation concentrations.

$Cu(s) + 2\,Fe^{3+}(aq) \rightarrow Cu^{2+}(aq) + 2\,Fe^{2+}(aq)$;

(b)

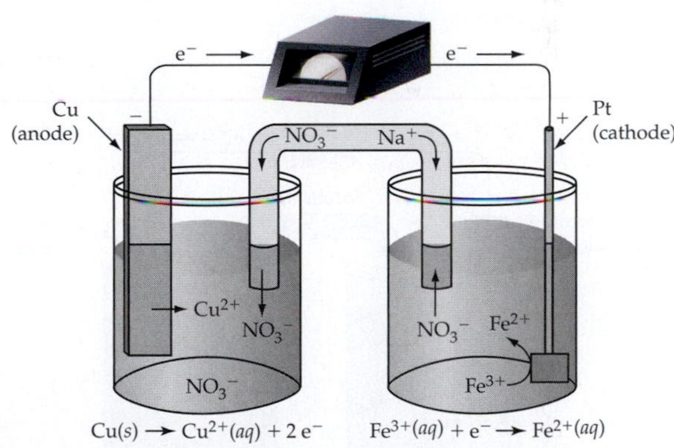

(c) (1) $E = 0.43$ V; (2) $E = 0.39$ V; (3) $E = 0.46$ V. Cell (3) has the largest potential, while cell (2) has the smallest as calculated from the Nernst equation.

17.30 (a) electrolytic

(b)

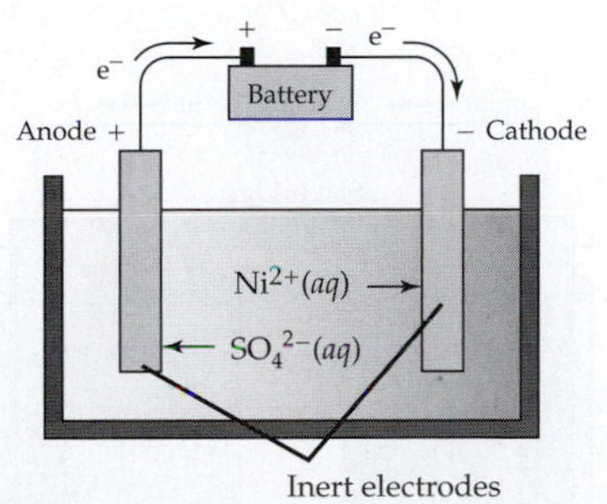

(c)	anode	$2\,H_2O(l) \rightarrow O_2(g) + 4\,H^+(aq) + 4\,e^-$
	cathode	$Ni^{2+}(aq) + 2\,e^- \rightarrow Ni(s)$
	overall	$2\,Ni^{2+}(aq) + 2\,H_2O(l) \rightarrow 2\,Ni(s) + O_2(g) + 4\,H^+(aq)$

17.32 (a) increase; (b) decrease; (c) decrease; (d) no change

17.34 (a), (b) & (c)

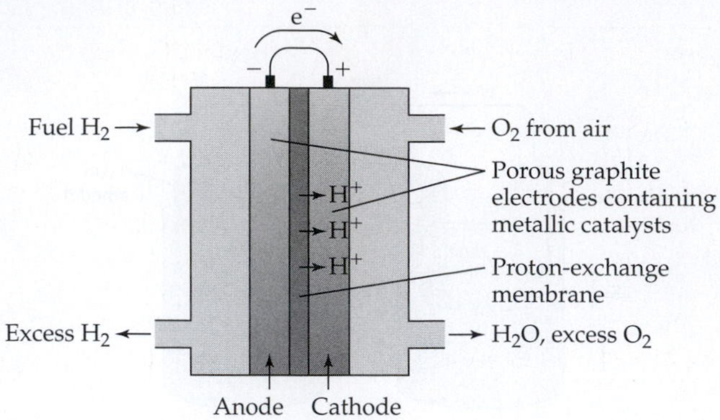

(d)

anode	$2\,H_2(g) \rightarrow 4\,H^+(aq) + 4\,e^-$	$E° = 0.00\ V$
cathode	$O_2(g) + 4\,H^+(aq) + 4\,e^- \rightarrow 2\,H_2O(l)$	$E° = 1.23\ V$
overall	$2\,H_2(g) + O_2(g) \rightarrow 2\,H_2O(l)$	$E° = 1.23\ V$

Section Problems

17.36 The cathode of a galvanic cell is considered to be the positive electrode because electrons flow through the external circuit toward the positive electrode (the cathode).

17.38

(a)

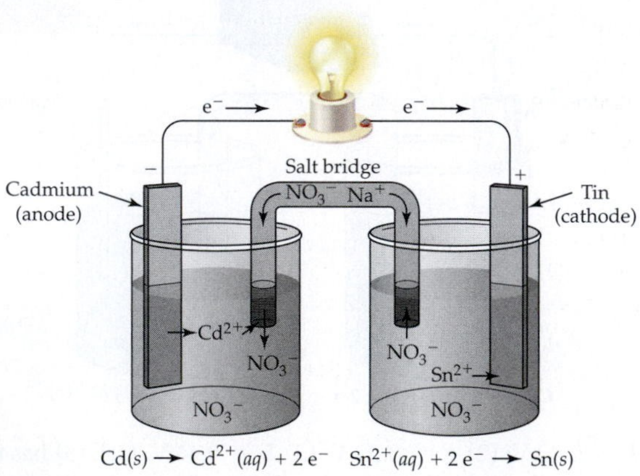

(b)

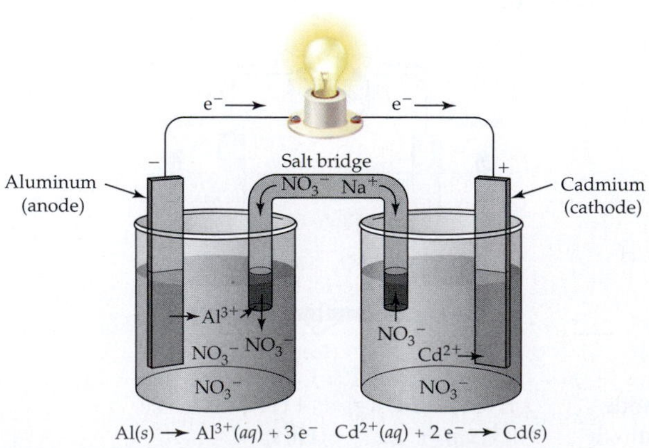

(c)

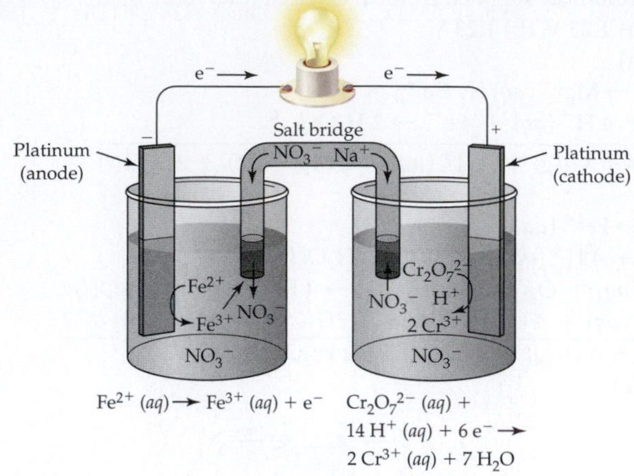

17.40

(a) $Cd(s)\,|\,Cd^{2+}(aq)\,||\,Sn^{2+}(aq)\,|\,Sn(s)$;

(b) $Al(s)\,|\,Al^{3+}(aq)\,||\,Cd^{2+}(aq)\,|\,Cd(s)$;

(c) $Pt(s)\,|\,Fe^{2+}(aq), Fe^{3+}(aq)\,||\,Cr_2O_7^{2-}(aq), Cr^{3+}(aq)\,|\,Pt(s)$

17.42 $2\,Br^-(aq) + Cl_2(g) \rightarrow Br_2(l) + 2\,Cl^-(aq)$

Inert electrodes are required because none of the reactants or products is an electrical conductor.

17.44 (a)

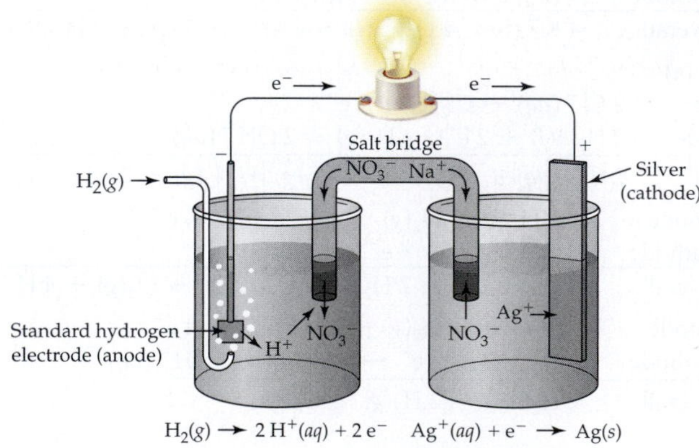

(b)	anode	$H_2(g) \rightarrow 2\,H^+(aq) + 2\,e^-$
	cathode	$2\,Ag^+(aq) + 2\,e^- \rightarrow 2\,Ag(s)$
	overall	$H_2(g) + 2\,Ag^+(aq) \rightarrow 2\,H^+(aq) + 2\,Ag(s)$

(c) $Pt(s)\,|\,H_2(g)\,|\,H^+(aq)\,||\,Ag^+(aq)\,|\,Ag(s)$

17.46 (a)	anode	$Co(s) \rightarrow Co^{2+}(aq) + 2\,e^-$
	cathode	$Cu^{2+}(aq) + 2\,e^- \rightarrow Cu(s)$
	overall	$Co(s) + Cu^{2+}(aq) \rightarrow Co^{2+}(aq) + Cu(s)$

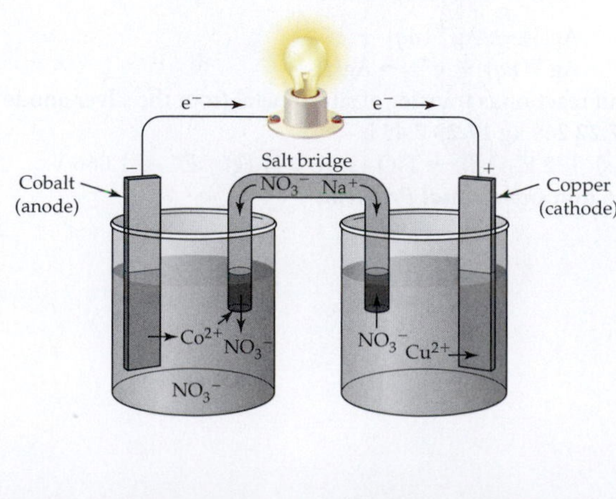

(b)

anode	$2\,Fe(s) \rightarrow 2\,Fe^{2+}(aq) + 4\,e^-$
cathode	$O_2(g) + 4\,H^+(aq) + 4\,e^- \rightarrow 2\,H_2O(l)$
overall	$2\,Fe(s) + O_2(g) + 4\,H^+(aq) \rightarrow 2\,Fe^{2+}(aq) + 2\,H_2O(l)$

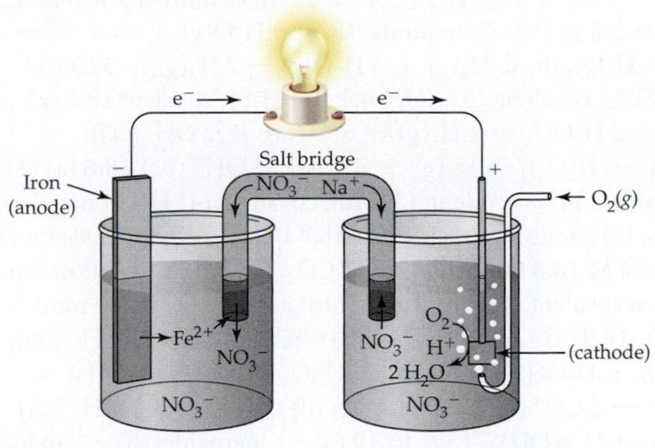

17.48 E is the standard cell potential ($E°$) when all reactants and products are in their standard states—solutes at 1 M concentrations, gases at a partial pressure of 1 atm, solids and liquids in pure form, all at 25 °C. **17.50** −309 kJ **17.52** $x = 3$ **17.54** +1.23 V **17.56** −0.36 V **17.58** $Sn^{4+}(aq) < Br_2(aq) < MnO_4^-(aq)$ **17.60** $Cr_2O_7^{2-}(aq)$ strongest; $Fe^{2+}(aq)$ weakest **17.62** $I_2(s) + Co(s) \rightarrow Co^{2+}(aq) + 2\,I^-(aq)$; $E° = 0.82$ V, $\Delta G° = -1.6 \times 10^2$ kJ **17.64** (a) $E° = 0.26$ V, $\Delta G° = -50$ kJ; (b) $E° = 1.26$ V, $\Delta G° = -730$ kJ; (c) $E° = 0.59$ V, $\Delta G° = -342$ kJ **17.66** (a) $E° = -0.90$ V, nonspontaneous; (b) $E° = 2.11$ V, spontaneous **17.68** (a) $E° = +0.94$ V, $Sn^{2+}(aq)$ can be oxidized by $Br_2(aq)$; (b) $E° = -0.41$ V, $Ni^{2+}(aq)$ cannot be reduced by $Sn^{2+}(aq)$; (c) $E° = -0.93$ V, Ag(s) cannot be oxidized by $Pb^{2+}(aq)$; (d) $E° = +0.37$ V, $I_2(s)$ can be reduced by H_2SO_3 **17.70** (a) no reaction; (b) $Pb(s) + 2\,Ag^+(aq) \rightarrow Pb^{2+}(aq) + 2\,Ag(s)$; (c) $Cl_2(g) + H_2C_2O_4(aq) \rightarrow 2\,Cl^-(aq) + 2\,CO_2(g) + 2\,H^+(aq)$; (d) $Ni(s) + 2\,HClO(aq) + 2\,H^+(aq) \rightarrow Ni^{2+}(aq) + Cl_2(g) + H_2O(l)$ **17.72** 0.87 V **17.74** $E = 0.35$ V, $[Cu^{2+}] = 1 \times 10^{-16}$ M **17.76** 10 **17.78** (a) 0.64 V; (b) 0.77 V; (c) −0.23 V; (d) −1.08 V **17.80** 9.0 **17.82** $\Delta G° = -nFE°$. Because n and F are always positive, $\Delta G°$ is negative when $E°$ is positive because of the negative sign in the equation.

$$E° = \frac{0.0592\ \text{V}}{n}\log K;\quad \log K = \frac{nE°}{0.0592\ \text{V}};\quad K = 10^{\frac{nE°}{0.0592}}$$

If $E°$ is positive, the exponent is positive (because n is positive), and K is greater than 1. **17.84** 6×10^{35} **17.86** (a) 6×10^8; (b) 10^{128}; (c) 10^{60} **17.88** 9×10^{-3}
17.90 (a)

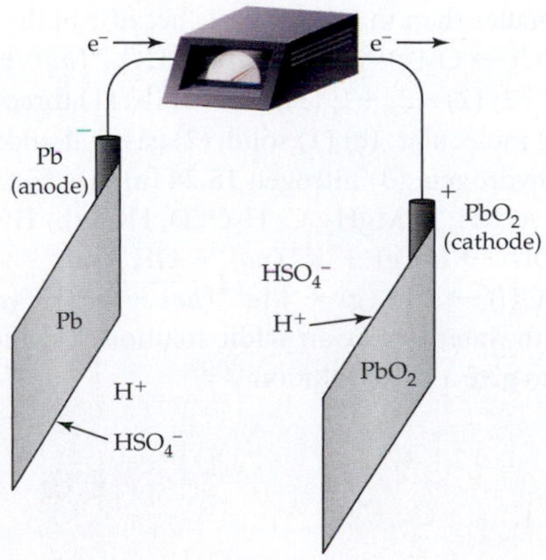

(b)

anode	$Pb(s) + HSO_4^-(aq) \rightarrow PbSO_4(s) + H^+(aq) + 2\,e^-$
cathode	$PbO_2(s) + 3\,H^+(aq) + HSO_4^-(aq) + 2\,e^- \rightarrow PbSO_4(s) + 2\,H_2O(l)$
overall	$Pb(s) + PbO_2(s) + 2\,H^+(aq) + 2\,HSO_4^-(aq) \rightarrow 2\,PbSO_4(s) + 2\,H_2O(l)$

(c) $K = 1 \times 10^{65}$; (d) cell voltage $= 0$ **17.92** $E° = 1.23$ V; $\Delta G° = -475$ kJ; $K = 1 \times 10^{83}$; $E = 1.29$ V **17.94** Rust is a hydrated form of iron(III) oxide ($Fe_2O_3 \cdot H_2O$). Rust forms from the oxidation of Fe in the presence of O_2 and H_2O. Rust can be prevented by coating Fe with Zn (galvanizing). **17.96** Cathodic protection is the attachment of a more easily oxidized metal to the metal you want to protect. This forces the metal you want to protect to be the cathode, hence the name, cathodic protection. Zn and Al can offer cathodic protection to Fe (Ni and Sn cannot).
17.98 (a)

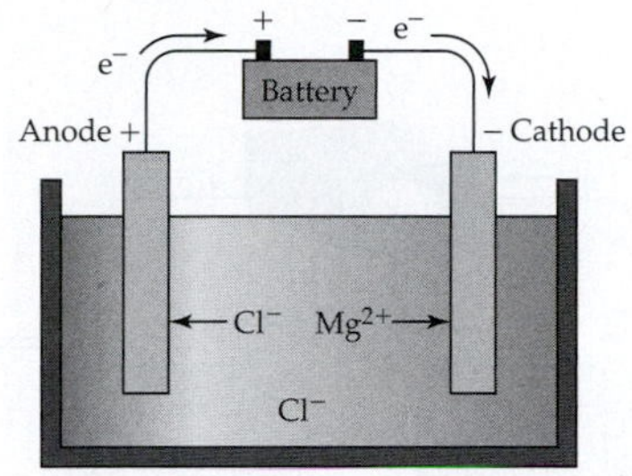

(b)

anode	$2\,Cl^-(l) \rightarrow Cl_2(g) + 2\,e^-$
cathode	$Mg^{2+}(l) + 2\,e^- \rightarrow Mg(l)$
overall	$Mg^{2+}(l) + 2\,Cl^-(l) \rightarrow Mg(l) + Cl_2(g)$

17.100 Possible anode reactions:
$2\,Cl^-(aq) \rightarrow Cl_2(g) + 2\,e^-$
$2\,H_2O(l) \rightarrow O_2(g) + 4\,H^+(aq) + 4\,e^-$
Possible cathode reactions:
$2\,H_2O(l) + 2\,e^- \rightarrow H_2(g) + 2\,OH^-(aq)$
$Mg^{2+}(aq) + 2\,e^- \rightarrow Mg(s)$
Actual reactions:
anode $2\,Cl^-(aq) \rightarrow Cl_2(g) + 2\,e^-$
cathode $2\,H_2O(l) + 2\,e^- \rightarrow H_2(g) + 2\,OH^-(aq)$
This anode reaction takes place instead of $2\,H_2O(l) \rightarrow O_2(g) + 4\,H^+(aq) + 4\,e^-$ because of a high overpotential for formation of gaseous O_2. This cathode reaction takes place instead of $Mg^{2+}(aq) + 2\,e^- \rightarrow Mg(s)$ because H_2O is easier to reduce than Mg^{2+}.
17.102 (a) NaBr

anode	$2\,Br^-(aq) \rightarrow Br_2(l) + 2\,e^-$
cathode	$2\,H_2O(l) + 2\,e^- \rightarrow H_2(g) + 2\,OH^-(aq)$
overall	$2\,H_2O(l) + 2\,Br^-(aq) \rightarrow Br_2(l) + H_2(g) + 2\,OH^-(aq)$

(b) $CuCl_2$

anode	$2\,Cl^-(aq) \rightarrow Cl_2(g) + 2\,e^-$
cathode	$Cu^{2+}(aq) + 2\,e^- \rightarrow Cu(s)$
overall	$Cu^{2+}(aq) + 2\,Cl^-(aq) \rightarrow Cu(s) + Cl_2(g)$

(c) LiOH

anode	$4\,OH^-(aq) \rightarrow O_2(g) + 2\,H_2O(l) + 4\,e^-$
cathode	$4\,H_2O(l) + 4\,e^- \rightarrow 2\,H_2(g) + 4\,OH^-(aq)$
overall	$2\,H_2O(l) \rightarrow O_2(g) + 2\,H_2(g)$

17.104 3.22 g **17.106** Time = 38.9 h; 4.87×10^5 L Cl_2 **17.108** Mn^{2+}
Chapter Problems **17.110** (a) $2\,MnO_4^-(aq) + 16\,H^+(aq) + 5\,Sn^{2+}(aq) \rightarrow 2\,Mn^{2+}(aq) + 5\,Sn^{4+}(aq) + 8\,H_2O(l)$; (b) MnO_4^- is the oxidizing agent; Sn^{2+} is the reducing agent; (c) 1.36 V
17.112 (a) Ag^+ is the strongest oxidizing agent; Pb is the strongest reducing agent.

(b)

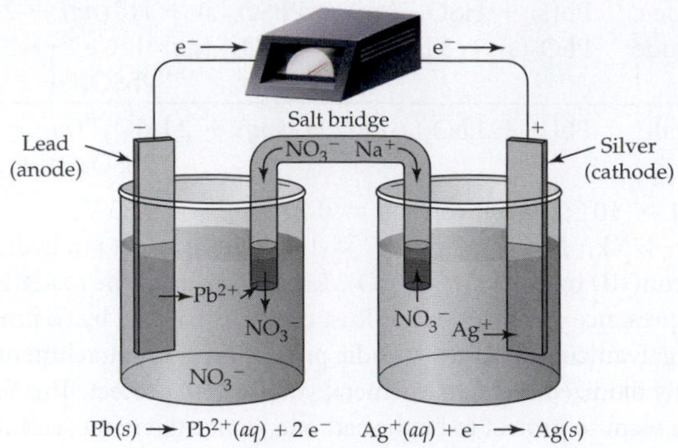

(c) $Pb(s) + 2\,Ag^+(aq) \rightarrow Pb^{2+}(aq) + 2\,Ag(s)$; $E° = 0.93$ V; (d) 0.87 V **17.114** (a)

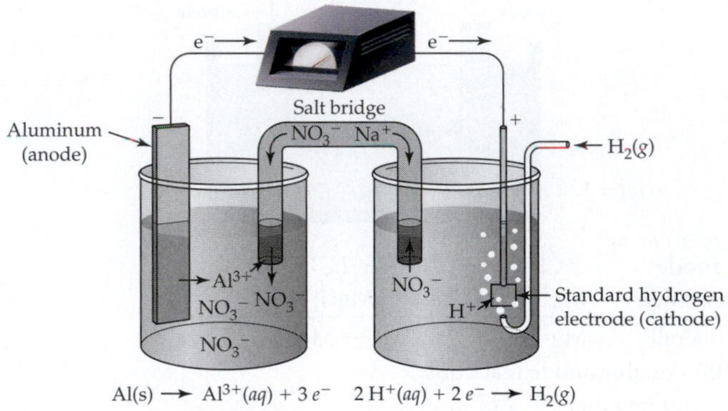

(b) $2\,Al(s) + 6\,H^+(aq) \rightarrow 2\,Al^{3+}(aq) + 3\,H_2(g)$; $E° = 1.66$ V; (c) 1.59 V; (d) $K = 10^{168}$; (e) 1.40 g
17.116 (a) $10\,Cl^-(aq) + 2\,MnO_4^-(aq) + 16\,H^+(aq) \rightarrow 5\,Cl_2(g) + 2\,Mn^{2+}(aq) + 8\,H_2O(l)$; (b) $E° = 0.15$ V, $\Delta G° = -1.4 \times 10^2$ kJ; (c) 69 L **17.118** (a) oxidizing agents: PbO_2, H^+, $Cr_2O_7^{2-}$; reducing agents: Al, Fe, Ag; (b) PbO_2 is the strongest oxidizing agent. H^+ is the weakest oxidizing agent. (c) Al is the strongest reducing agent. Ag is the weakest reducing agent. (d) oxidized by Cu^{2+}: Fe and Al; reduced by H_2O_2: PbO_2 and $Cr_2O_7^{2-}$ **17.120** (a) 1.30 V; (b) 0.98 V **17.122** 1.7×10^{-10} **17.124** $E° = 0.46$ V; $E = 0.04$ V. Because E is positive, the reaction is spontaneous. **17.126** 6×10^{-23}
17.128 (a)

anode	$4[Al(s) \rightarrow Al^{3+}(aq) + 3\,e^-]$
cathode	$3[O_2(g) + 4\,H^+(aq) + 4\,e^- \rightarrow 2\,H_2O(l)]$
overall	$4\,Al(s) + 3\,O_2(g) + 12\,H^+(aq) \rightarrow 4\,Al^{3+}(aq) + 6\,H_2O(l)$

(b) $E = E° - \dfrac{2.303RT}{nF} \log \dfrac{[Al^{3+}]^4}{(P_{O_2})^3[H^+]^{12}}$ where $T = 310$ K

(c) 2.63 V ***Multiconcept Problems*** **17.130** (a) $4\,CH_2{=}CHCN + 2\,H_2O \rightarrow 2\,NC(CH_2)_4CN + O_2$; (b) 60.5 kg; (c) 7030 L **17.132** $E° = -2.91$ V **17.134** (a) $Cr_2O_7^{2-}(aq) + 6\,Fe^{2+}(aq) + 14\,H^+(aq) \rightarrow 2\,Cr^{3+}(aq) + 6\,Fe^{3+}(aq) + 7\,H_2O(l)$; (b) 1.05 V **17.136** (a) $Zn(s) + 2\,Ag^+(aq) + H_2O(l) \rightarrow ZnO(s) + 2\,Ag(s) + 2\,H^+(aq)$; $\Delta H° = -275.9$ kJ; $\Delta S° = -128.0$ J/K; $\Delta G° = -237.8$ kJ; (b) 1.232 V; $K = 4 \times 10^{41}$; (c) decreases; 0.694 V; (d) $IP = 1.6 \times 10^{-11}$; $IP < K_{sp}$, AgCl will not precipitate. $IP = 1.6 \times 10^{-11}$; $IP > K_{sp}$, AgBr will precipitate. **17.138** (a) $2\,C_4H_{10}(g) + 13\,O_2(g) \rightarrow 8\,CO_2(g) + 10\,H_2O(l)$; (b) $E° = 1.09$ V; $K = 7 \times 10^{962}$; decrease because $\Delta H° < 0$; (c) 7.00 g; 2.69 L **17.140** (a) $4\,Au(s) + 8\,CN^-(aq) + O_2(g) + 2\,H_2O(l) \rightarrow 4\,Au(CN)_2^-(aq) + 4\,OH^-(aq)$; (b) −387.0 kJ

CHAPTER 18

18.1 $d_{H_2} = 8.24 \times 10^{-5}$ g/cm^3. Air is 14 times more dense than H_2. **18.2** (a) 18.2% D; (b) 15.0 pounds. **18.3** (a) $H_2O(g) + C(s) \xrightarrow{1000\,°C} CO(g) + H_2(g)$; (b) $C_3H_8(g) + 3\,H_2O(g) \rightarrow 7\,H_2(g) + 3\,CO(g)$ **18.4** (a) SiH_4, covalent; (b) KH, ionic; (c) H_2Se, covalent **18.5** (a) $SrH_2(s) + 2\,H_2O(l) \rightarrow 2\,H_2(g) + Sr^{2+}(aq) + 2\,OH^-(aq)$; (b) $KH(s) + H_2O(l) \rightarrow H_2(g) + K^+(aq) + OH^-(aq)$ **18.6** (a) (1) ZrH_x, interstitial, (2) PH_3, covalent, (3) HBr, covalent, (4) LiH, ionic; (b) (1) and (4) solids, (2) and (3) gases; (c) LiH **18.7** $PdH_{0.74}$; $d_H = 0.0841$ g/cm^3; $M_H = 83.4$ M **18.8** (a) Li_2O, Ga_2O_3, CO_2; (b) Li_2O is the most ionic. CO_2 is the most covalent. (c) CO_2 is the most acidic. Li_2O is the most basic. (d) Ga_2O_3 **18.9** (a) $Li_2O(s) + H_2O(l) \rightarrow 2\,Li^+(aq) + 2\,OH^-(aq)$; (b) $SO_3(l) + H_2O(l) \rightarrow H^+(aq) + HSO_4^-(aq)$; (c) $Cr_2O_3(s) + 6\,H^+(aq) \rightarrow 2\,Cr^{3+}(aq) + 3\,H_2O(l)$; (d) $Cr_2O_3(s) + 2\,OH^-(aq) + 3\,H_2O(l) \rightarrow 2\,Cr(OH)_4^-(aq)$ **18.10** (a) −1, peroxide; (b) −2, oxide; (c) −1/2, superoxide; (d) −1, peroxide; (e) −2, oxide **18.11** (a) $Rb_2O_2(s) + H_2O(l) \rightarrow 2\,Rb^+(aq) + HO_2^-(aq) + OH^-(aq)$; (b) $CaO(s) + H_2O(l) \rightarrow Ca^{2+}(aq) + 2\,OH^-(aq)$; (c) $2\,CsO_2(s) + H_2O(l) \rightarrow O_2(g) + 2\,Cs^+(aq) + HO_2^-(aq) + OH^-(aq)$; (d) $SrO_2(s) + H_2O(l) \rightarrow Sr^{2+}(aq) + HO_2^-(aq) + OH^-(aq)$; (e) $CO_2(g) + H_2O(l) \rightleftharpoons H^+(aq) + HCO_3^-(aq)$
18.12

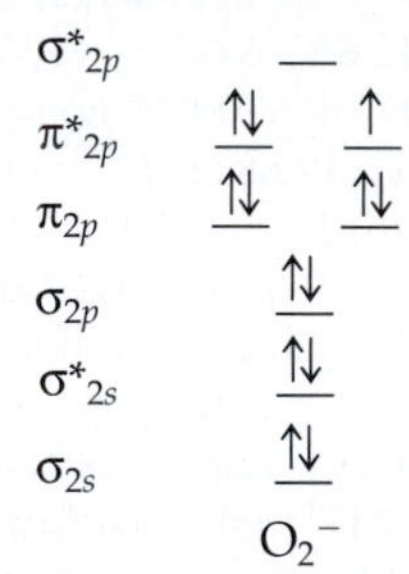

O_2^- is paramagnetic with one unpaired electron.
O_2^- bond order = (8−5)/2 = 1.5
18.13 H—Ö—Ö—H The electron dot structure indicates a single bond which is consistent with an O–O bond length of 148 pm.
18.14 $PbS(s) + 4\,H_2O_2(aq) \rightarrow PbSO_4(s) + 4\,H_2O(l)$ **18.15** $NiSO_4 \cdot 7\,H_2O$ **18.16** Hydrogen can be stored as a solid in the form of solid interstitial hydrides or in the recently discovered tube-shaped molecules called carbon nanotubes. **18.17** 1.5×10^{10} kJ; 1.0×10^6 kg O_2
Conceptual Problems **18.18** (a) (1) covalent, (2) ionic, (3) covalent, (4) interstitial; (b) (1) H, +1; other element, −3, (2) H, −1; other element, +1; (3) H, +1; other element, −2 **18.20** (a) a superoxide KO_2; (b) attracted; (c) The O–O bond length in O_2^- is longer and the bond energy is smaller than in O_2; (d) basic because of the reaction $2\,KO_2(s) + H_2O(l) \rightarrow O_2(g) + 2\,K^+(aq) + HO_2^-(aq) + OH^-(aq)$
18.22 (a) (1) −2, +2; (2) −2, +1; (3) −2, +5; (b) (1) three-dimensional; (2) molecular; (3) molecular; (c) (1) solid; (2) gas or liquid; (3) gas or liquid; (d) (2) hydrogen; (3) nitrogen **18.24** (a) (4); (b) (3); (c) (1) and (4) **18.26** (a) A, KH; B, MgH_2; C, H_2O; D, HCl; (b) HCl; (c) $KH(s) + H_2O(l) \rightarrow H_2(g) + K^+(aq) + OH^-(aq)$
$MgH_2(s) + 2\,H_2O(l) \rightarrow 2\,H_2(g) + Mg^{2+}(aq) + 2\,OH^-(aq)$;
(d) HCl reacts with water to give an acidic solution. KH and MgH_2 react with water to give a basic solution.

Section Problems **18.28** 18

$H_2^{16}O$	$H_2^{17}O$	$H_2^{18}O$
$D_2^{16}O$	$D_2^{17}O$	$D_2^{18}O$
$T_2^{16}O$	$T_2^{17}O$	$T_2^{18}O$
$HD^{16}O$	$HD^{17}O$	$HD^{18}O$
$HT^{16}O$	$HT^{17}O$	$HT^{18}O$
$DT^{16}O$	$DT^{17}O$	$DT^{18}O$

18.30 (a) $Zn(s) + 2\,H^+(aq) \rightarrow H_2(g) + Zn^{2+}(aq)$; (b) $H_2O(g) + C(s) \rightarrow CO(g) + H_2(g)$; (c) $H_2O(g) + CH_4(g) \rightarrow CO(g) + 3\,H_2(g)$; (d) electrolysis: $2\,H_2O(l) \rightarrow 2\,H_2(g) + O_2(g)$
18.32 The steam–hydrocarbon reforming process

$$CH_4(g) + H_2O(g) \xrightarrow[\text{Ni catalyst}]{1100\,°C} CO(g) + 3\,H_2(g)$$
$$CO(g) + H_2O(g) \xrightarrow{400\,°C} CO_2(g) + H_2(g)$$
$$CO_2(g) + 2\,OH^-(aq) \rightarrow CO_3^{2-}(aq) + H_2O(l)$$

18.34 (a) LiH; (b) 12.9 kg **18.36** (a) MgH_2, H^-; (b) PH_3, covalent; (c) KH, H^-; (d) HBr, covalent **18.38** H_2S: covalent hydride, gas, weak acid in H_2O; NaH: ionic hydride, solid (salt like), reacts with H_2O to produce H_2; PdH_x: metallic (interstitial) hydride, solid, stores hydrogen
18.40 (a) CH_4, covalent bonding; (b) NaH, ionic bonding
18.42 (a) H—S̈e—H bent; (b) H—Äs—H trigonal pyramidal; (with H bonded to As below)

(c) 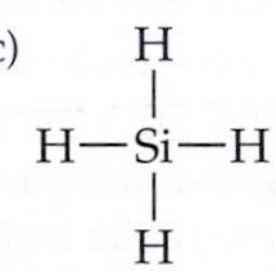 tetrahedral

18.44 A nonstoichiometric compound is a compound whose atomic composition cannot be expressed as a ratio of small whole numbers. An example is PdH_x. The lack of stoichiometry results from the hydrogen occupying holes in the solid state structure. **18.46** 12.3 g BaH_2
18.48 (a) $d_H = 0.15\ g/cm^3$; the density of H in TiH_2 is about 2.1 times the density of liquid H_2. (b) $1.7 \times 10^3\ cm^3$ **18.50** (a) O_2 is obtained in industry by the fractional distillation of liquid air; (b) In the laboratory, O_2 is prepared by the thermal decomposition of $KClO_3(s)$.
$2\,KClO_3(s) \xrightarrow[MnO_2]{\text{Heat}} 2\,KCl(s) + 3\,O_2(g)$ **18.52** 59.7 mL O_2
18.54 7.34 L **18.56** (a) $4\,Li(s) + O_2(g) \rightarrow 2\,Li_2O(s)$; (b) $P_4(s) + 5\,O_2(g) \rightarrow P_4O_{10}(s)$; (c) $4\,Al(s) + 3\,O_2(g) \rightarrow 2\,Al_2O_3(s)$; (d) $Si(s) + O_2(g) \rightarrow SiO_2(s)$ **18.58** :Ö::Ö: The electron dot structure shows an O=O double bond. It also shows all electrons paired. This is not consistent with the fact that O_2 is paramagnetic.
18.60 covalent hydride. **18.62** $Li_2O < BeO < B_2O_3 < CO_2 < N_2O_5$
18.64 $N_2O_5 < Al_2O_3 < K_2O < Cs_2O$ **18.66** (a) CrO_3; (b) N_2O_5; (c) SO_3
18.68 (a) $Cl_2O_7(l) + H_2O(l) \rightarrow 2\,H^+(aq) + 2\,ClO_4^-(aq)$; (b) $K_2O(s) + H_2O(l) \rightarrow 2\,K^+(aq) + 2\,OH^-(aq)$; (c) $SO_3(l) + H_2O(l) \rightarrow H^+(aq) + HSO_4^-(aq)$ **18.70** (a) $ZnO(s) + 2\,H^+(aq) \rightarrow Zn^{2+}(aq) + H_2O(l)$; (b) $ZnO(s) + 2\,OH^-(aq) + H_2O(l) \rightarrow Zn(OH)_4^{2-}(aq)$ **18.72** A peroxide has oxygen in the −1 oxidation state, for example, H_2O_2. A superoxide has oxygen in the −1/2 oxidation state, for example, KO_2.
18.74 (a) BaO_2; (b) CaO; (c) CsO_2; (d) Li_2O; (e) Na_2O_2
18.76

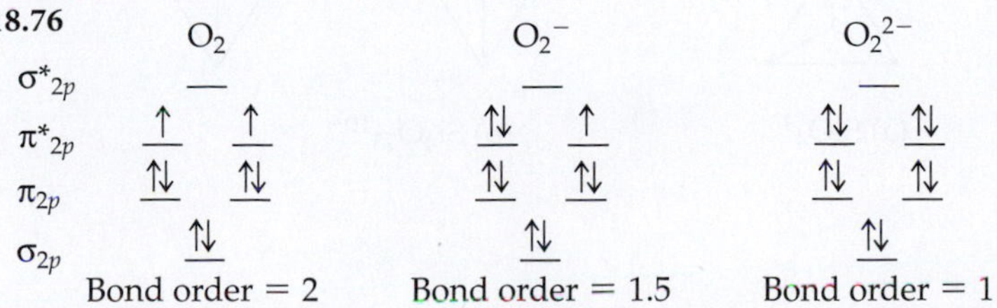

(a) The O—O bond length increases because the bond order decreases. The bond order decreases because of the increased occupancy of antibonding orbitals. (b) O_2^- has 1 unpaired electron and is paramagnetic. O_2^{2-} has no unpaired electrons and is diamagnetic. **18.78** (a) $H_2O_2(aq) + 2\,H^+(aq) + 2\,I^-(aq) \rightarrow I_2(aq) + 2\,H_2O(l)$; (b) $3\,H_2O_2(aq) + 8\,H^+(aq) + Cr_2O_7^{2-}(aq) \rightarrow 2\,Cr^{3+}(aq) + 3\,O_2(g) + 7\,H_2O(l)$
18.80

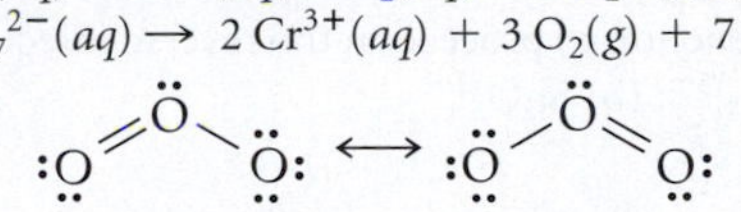

18.82 $3\,O_2(g) \xrightarrow[\text{discharge}]{\text{Electric}} 2\,O_3(g)$ **18.84** 1.5×10^{11} kg
18.86 The following equilibrium shifts to the right as heating expels gaseous CO_2 from the HCO_3^--containing hard water, yielding CO_3^{2-} ions, which precipitate Ca^{2+} as $CaCO_3$ (boiler scale):
$2\,HCO_3^-(aq) \rightleftharpoons CO_2(g) + H_2O(l) + CO_3^{2-}(aq)$
$Ca^{2+}(aq) + CO_3^{2-}(aq) \rightarrow CaCO_3(s)$
18.88 $AlCl_3 \cdot 6\,H_2O$

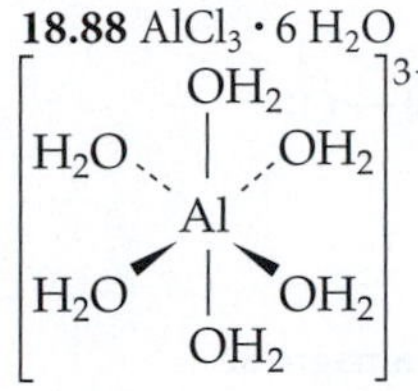

18.90 6.21% **18.92** The mineral gypsum $CaSO_4 \cdot 2\,H_2O$; $x = 2$

Chapter Problems

18.94 a reducing agent; Ca or Al **18.96** 2.2×10^3 L **18.98** (a) B_2O_3, diboron trioxide; (b) H_2O_2, hydrogen peroxide; (c) SrH_2, strontium hydride; (d) CsO_2, cesium superoxide; (e) $HClO_4$, perchloric acid (f) BaO_2, barium peroxide **18.100** (a) 6; $^{16}O_2$, $^{17}O_2$, $^{18}O_2$, $^{16}O^{17}O$, $^{16}O^{18}O$, $^{17}O^{18}O$; (b) 18

$^{16}O_3$	$^{16}O_2^{17}O$	$^{18}O_3$
$^{17}O_2^{16}O$	$^{17}O_3$	$^{16}O_2^{18}O$
$^{18}O_2^{16}O$	$^{18}O_2^{17}O$	$^{17}O_2^{18}O$
$^{16}O^{17}O^{16}O$	$^{17}O^{18}O^{17}O$	$^{16}O^{17}O^{18}O$
$^{16}O^{18}O^{16}O$	$^{17}O^{16}O^{17}O$	$^{18}O^{16}O^{18}O$
$^{17}O^{18}O^{16}O$	$^{18}O^{16}O^{17}O$	$^{18}O^{17}O^{18}O$

18.102 (a) $2\,H_2(g) + O_2(g) \rightarrow 2\,H_2O(l)$; (b) $O_3(g) + 2\,I^-(aq) + H_2O(l) \rightarrow O_2(g) + I_2(aq) + 2\,OH^-(aq)$; (c) $H_2O_2(aq) + 2\,H^+(aq) + 2\,Br^-(aq) \rightarrow 2\,H_2O(l) + Br_2(aq)$; (d) $2\,Na(l) + H_2(g) \rightarrow 2\,NaH(s)$; (e) $2\,Na(s) + 2\,H_2O(l) \rightarrow H_2(g) + 2\,Na^+(aq) + 2\,OH^-(aq)$
18.104 (a) −128.7 kJ; (b) −41.2 kJ; (c) −77.6 kJ; (d) 2802.5 kJ
18.106 80.7 L **18.108** $x = 4$ **18.110** $CaSO_4 \cdot 2\,H_2O$ **18.112** NaO_2 superoxide. **18.114** Ge_3H_8 ***Multiconcept Problems*** **18.116** 496 J
18.118 M = Sr; SrH_2 **18.120** (a) $PH_3(g) + 2\,O_2(g) \rightarrow H_3PO_4(s)$; (b) pH = 1.69 **18.122** 7.7×10^{-4}mol/(L · atm) **18.124** (a) 0.58 mm Hg; (b) 9.4×10^{18} HD molecules; (c) 7.3×10^{14} D_2 molecules **18.126** 150 °C

CHAPTER 19

19.1 (a) B; (b) Br; (c) Se; (d) Cl

19.2 (a)

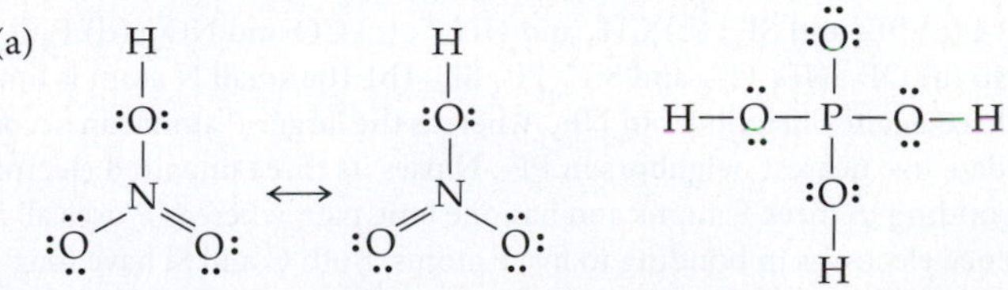

Nitrogen can form very strong $p\pi - p\pi$ bonds. Phosphorus forms weaker $p\pi - p\pi$ bonds, so it tends to form more single bonds. (b) The larger S atom can accommodate six bond pairs in its valence shell, but the smaller O atom is limited to two bond pairs and two lone pairs. **19.3** Carbon forms strong π bonds with oxygen. Silicon does not form strong π bonds with

oxygen, and what results are chains of alternating silicon and oxygen singly bonded to each other. **19.4** An ethane-like structure would require 14 valence electrons and diborane only has 12.
19.5 $Hb\text{–}O_2 + CO \rightleftharpoons Hb\text{–}CO + O_2$. Mild cases of carbon monoxide poisoning can be treated with O_2. Le Châtelier's principle says that adding a product (O_2) will cause the reaction to proceed in the reverse direction, back to $Hb\text{–}O_2$. **19.6** (a) $Si_8O_{24}^{16-}$; (b) $Si_2O_5^{2-}$
19.7 Nitrous oxide, N_2O

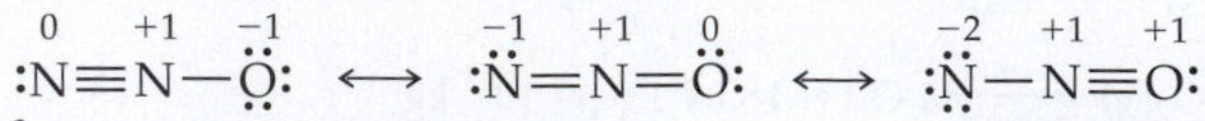

The first structure is most important.

Nitric oxide, NO

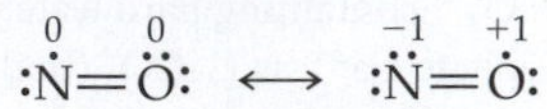

paramagnetic

The first structure is more important.

Nitrous acid, HNO_2

The first structure is more important.

Nitrogen dioxide, NO_2

paramagnetic

Both structures are of equal importance.

Nitric acid, HNO_3

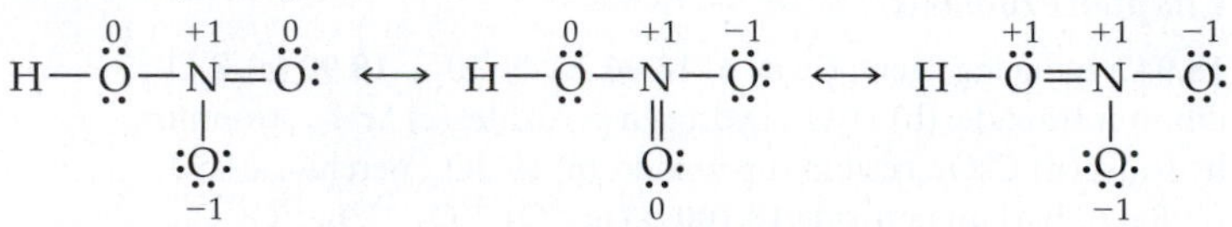

The first two structures are of equal importance. Both are more important than the third structure **19.8** (a) SO_3^{2-}, HSO_3^-, SO_4^{2-}, HSO_4^-; (b) HSO_4^-; (c) SO_3^{2-}; (d) HSO_4^-
19.9 (a) 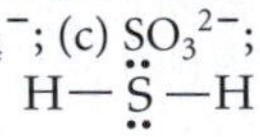 bent

(b) 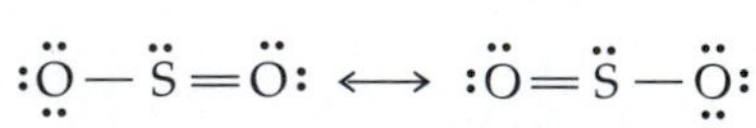 bent

(c) 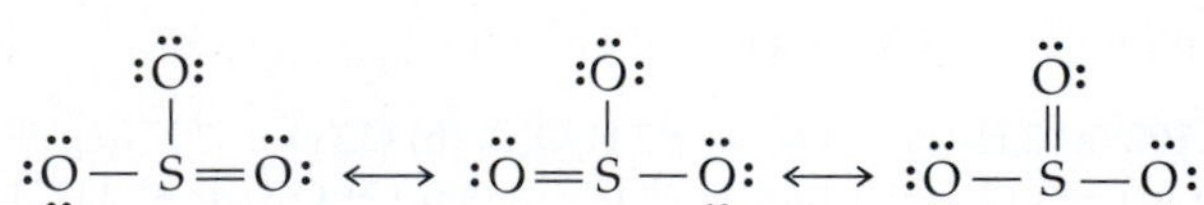

trigonal planar

19.10 Because copper is always conducting, a laser printer drum coated with copper would not hold any charge so no document image would adhere to the drum for printing.

Conceptual Problems

19.12

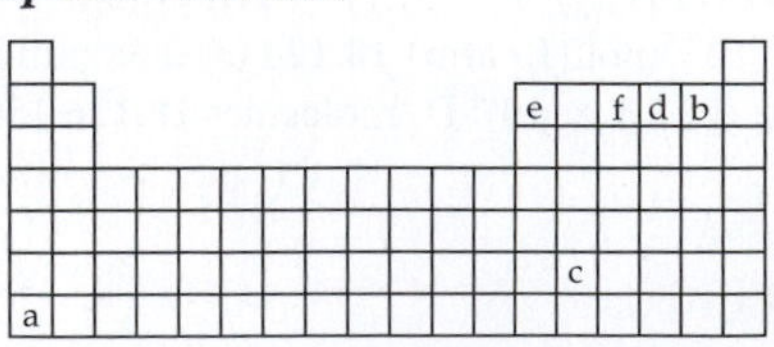

19.14 (a) PF_5 and SF_6; (b) CH_4 and NH_4^+; (c) CO and NO_2; (d) P_4O_{10}
19.16 (a) OF_2, NF_3, CF_4 and SiF_4, PF_5, SF_6; (b) The small N atom is limited to three nearest neighbors in NF_3, whereas the larger P atom can accommodate five nearest neighbors in PF_5. N uses its three unpaired electrons in bonding to three F atoms and has one lone pair, whereas P uses all five valence electrons in bonding to five F atoms. Both C and Si have four valence electrons and use sp^3 hybrid orbitals to bond to four F atoms.

19.18 (a) H_2O, CH_4, HF, B_2H_6, NH_3;
(b)

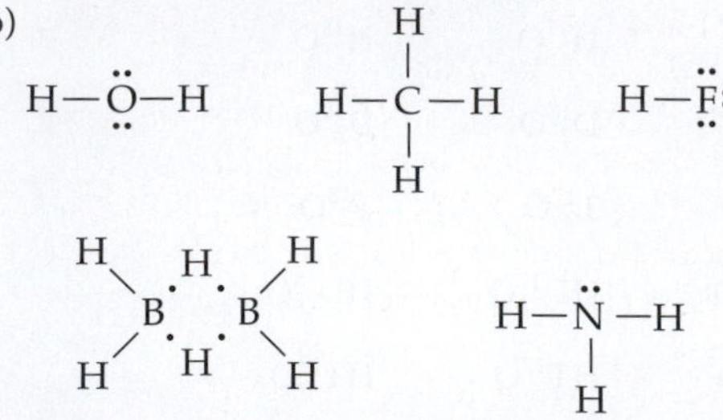

There is a problem in drawing an electron-dot structure for B_2H_6 because this molecule is electron deficient and has two three-center, two-electron bonds.

Section Problems

19.20 (a) Cl; (b) Si; (c) O **19.22** (a) Al; (b) P; (c) Pb **19.24** (a) I; (b) N; (c) F **19.26** (a) Sn; (b) Ge; (c) Bi **19.28** (a) CaH_2; (b) Ga_2O_3; (c) KCl; (d) $AlCl_3$ **19.30** Molecular: (a) B_2H_6; (c) SO_3; (d) $GeCl_4$. Extended three-dimensional structure: (b) $KAlSi_3O_8$ **19.32** (a) P_4O_{10}; (b) B_2O_3; (c) SO_2; (d) N_2O_3 **19.34** (a) Sn; (b) Cl; (c) Sn; (d) Se; (e) B **19.36** The smaller B atom can bond to a maximum of four nearest neighbors, whereas the larger Al atom can accommodate more than four nearest neighbors. **19.38** In O_2 a π bond is formed by 2*p* orbitals on each O. S does not form strong π bonds with its 3*p* orbitals, which leads to the S_8 ring structure with single bonds. **19.40** +3 for B, Al, Ga and In; +1 for Tl **19.42** Boron is a hard semiconductor with a high melting point. Boron forms only molecular compounds and does not form an aqueous B^{3+} ion. $B(OH)_3$ is an acid. **19.44** (a) An electron deficient molecule is a molecule that doesn't have enough electrons to form a two-center, two-electron bond between each pair of bonded atoms. B_2H_6 is an electron deficient molecule. (b) A three-center, two-electron bond has three atoms bonded together using just two electrons. The B–H–B bridging bond in B_2H_6 is a three-center two-electron bond. **19.46** (a) Al; (b) Tl; (c) B; (d) B **19.48** (a) Pb; (b) C; (c) Si; (d) C **19.50** (a) tetrahedral, sp^3 (b) linear, *sp*; (c) trigonal planar, sp^2; (d) trigonal pyramidal, sp^3 **19.52** Diamond is a very hard, high melting solid. It is an electrical insulator. Diamond has a covalent network structure in which each C atom uses sp^3 hybrid orbitals to form a tetrahedral array of σ bonds. The interlocking, three-dimensional network of strong bonds makes diamond the hardest natural substance with the highest melting point for an element. Because the valence electrons are localized in the σ bonds, diamond is an electrical insulator. **19.54** Graphene is a two-dimensional array of hexagonally arranged carbon atoms just one atom thick, essentially one layer of graphite. Graphene is extremely strong and flexible, and is a superb conductor of electricity. **19.56** (a) carbon tetrachloride, CCl_4; (b) carbon monoxide, CO; (c) methane, CH_4
19.58 (1) To provide the "bite" in soft drinks; $CO_2(aq) + H_2O(l) \rightleftharpoons H_2CO_3(aq)$. (2) CO_2 in fire extinguishers; CO_2 is nonflammable and 1.5 times more dense than air. (3) Refrigerant; dry ice, sublimes at −78 °C.
19.60 $\underset{\text{sand}}{SiO_2(l)} + 2\,C(s) \rightarrow Si(l) + 2\,CO(g)$ Purification of silicon for semiconductor devices: $Si(s) + 2\,Cl_2(g) \rightarrow SiCl_4(l)$; $SiCl_4$ is purified by distillation. $SiCl_4(g) + 2\,H_2(g) \xrightarrow{\text{Heat}} Si(s) + 4\,HCl(g)$; Si is purified by zone refining.
19.62

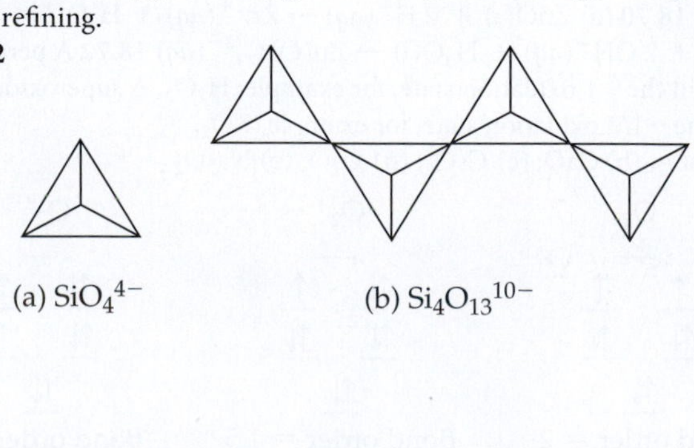

(a) SiO_4^{4-} (b) $Si_4O_{13}^{10-}$

The charge on the anion is equal to the number of terminal O atoms. **19.64** (a) $Si_3O_{10}{}^{8-}$; (b) $Ca_2Cu_2Si_3O_{10} \cdot 2\,H_2O$ **19.66** (a) P; (b) Sb and Bi; (c) N; (d) Bi **19.68** (a) N_2O, +1; (b) N_2H_4, −2; (c) Ca_3P_2, −3; (d) H_3PO_3, +3; (e) H_3AsO_4, +5 **19.70** :N≡N:; N_2 is unreactive because of the large amount of energy necessary to break the N≡N triple bond. **19.72** (a) bent; (b) trigonal pyramidal; (c) trigonal bipyramidal; (d) tetrahedral **19.74** White phosphorus contains tetrahedral P_4 molecules with 60° bond angles. Red phosphorus is polymeric. White phosphorus is reactive due to the considerable strain in the P_4 molecule.

19.76 (a) The structure for phosphorous acid is

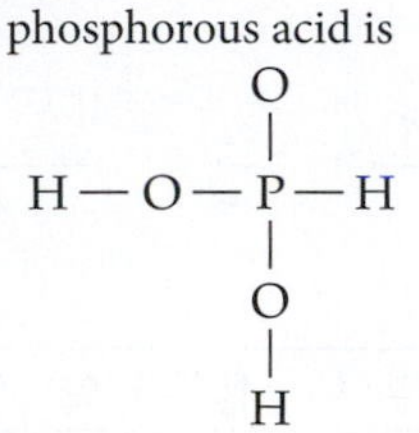

Only the two hydrogens bonded to oxygen are acidic. (b) Nitrogen forms strong π bonds, and in N_2 the nitrogen atoms are triple bonded to each other. Phosphorus does not form strong $p\pi$–$p\pi$ bonds, and so the P atoms are single bonded to each other in P_4.

19.78 (a) $2\,NO(g) + O_2(g) \rightarrow 2\,NO_2(g)$;
(b) $4\,HNO_3(aq) \rightarrow 4\,NO_2(aq) + O_2(g) + 2\,H_2O(l)$;
(c) $3\,Ag(s) + 4\,H^+(aq) + NO_3{}^-(aq) \rightarrow 3\,Ag^+(aq) + NO(g) + 2\,H_2O(l)$;
(d) $N_2H_4(aq) + 2\,I_2(aq) \rightarrow N_2(g) + 4\,H^+(aq) + 4\,I^-(aq)$
19.80 (a) O; (b) Te; (c) Po; (d) O **19.82** (a) crown-shaped S_8 rings; (b) S_8 rings pack differently; (c) disordered, tangled S_n chains; (d) long polymer chains (S_n, $n > 200{,}000$) **19.84** (a) hydrogen sulfide, H_2S; lead(II) sulfide, PbS; (b) sulfur dioxide, SO_2; sulfurous acid, H_2SO_3; (c) sulfur trioxide, SO_3; sulfur hexafluoride, SF_6 **19.86** a three-step reaction sequence in which (1) sulfur burns in air to give SO_2, (2) SO_2 is oxidized to SO_3 in the presence of a vanadium(V) oxide catalyst, and; (3) SO_3 reacts with water to give H_2SO_4.
(1) $S(s) + O_2(g) \rightarrow SO_2(g)$ (2) $2\,SO_2(g) + O_2(g) \xrightarrow[V_2O_5\text{ catalyst}]{\text{Heat}} 2\,SO_3(g)$
(3) $SO_3(g) + H_2O\ (\text{in conc } H_2SO_4) \rightarrow H_2SO_4(l)$
19.88 (a) $Zn(s) + 2\,H_3O^+(aq) \rightarrow Zn^{2+}(aq) + H_2(g) + 2\,H_2O(l)$;
(b) $BaSO_3(s) + 2\,H_3O^+(aq) \rightarrow H_2SO_3(aq) + Ba^{2+}(aq) + 2\,H_2O(l)$;
(c) $Cu(s) + 2\,H_2SO_4(l) \rightarrow Cu^{2+}(aq) + SO_4{}^{2-}(aq) + SO_2(g) + 2\,H_2O(l)$;
(d) $H_2S(aq) + I_2(aq) \rightarrow S(s) + 2\,H^+(aq) + 2\,I^-(aq)$
19.90 (a) Acid strength increases as the number of O atoms increases. (b) In comparison with S, O is much too electronegative to form compounds of O in the +4 oxidation state. Also, an S atom is large enough to accommodate four bond pairs and a lone pair in its valence shell, but an O atom is too small to do so. (c) Each S is sp^3 hybridized with two lone pairs of electrons. The bond angles are therefore 109.5°. A planar ring would require bond angles of 135°. **19.92** (a) $HBrO_3$, +5; (b) HIO, +1; (c) $NaClO_2$, +3; (d) $NaIO_4$, +7 **19.94** (a) iodic acid; (b) chlorous acid; (c) sodium hypobromite; (d) lithium perchlorate

19.96 (a) 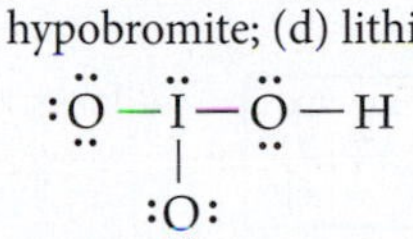 Trigonal pyramidal

(b) 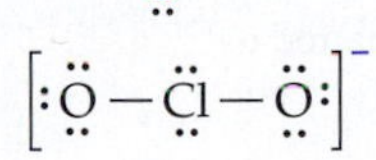 Bent

(c) 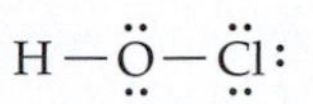 Bent

(d) 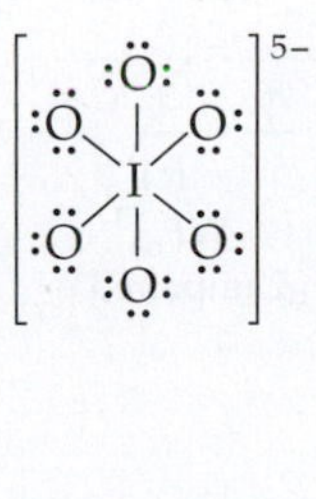 Octahedral

19.98 Oxygen atoms are highly electronegative. Increasing the number of oxygen atoms increases the polarity of the O–H bond and increases the acid strength. **19.100**
(a) $Br_2(l) + 2\,OH^-(aq) \rightarrow OBr^-(aq) + Br^-(aq) + H_2O(l)$;
(b) $Cl_2(g) + H_2O(l) \rightarrow HOCl(aq) + H^+(aq) + Cl^-(aq)$;
(c) $3\,Cl_2(g) + 6\,OH^-(aq) \rightarrow ClO_3{}^-(aq) + 5\,Cl^-(aq) + 3\,H_2O(l)$
Chapter Problems **19.102** $Mg(s) + 2\,H_2SO_4(l) \rightarrow Mg^{2+}(aq) + SO_4{}^{2-}(aq) + SO_2(g) + 2\,H_2O(l)$ **19.104** single-stranded silicate chains with an $Si_2O_6{}^{4-}$ repeating unit **19.106** $N_2H_4(aq) + O_2(aq) \rightarrow N_2(aq) + 2\,H_2O(l)$ **19.108** (a) LiCl; (b) SiO_2; (c) P_4O_{10} **19.110** A portion of the extended graphene

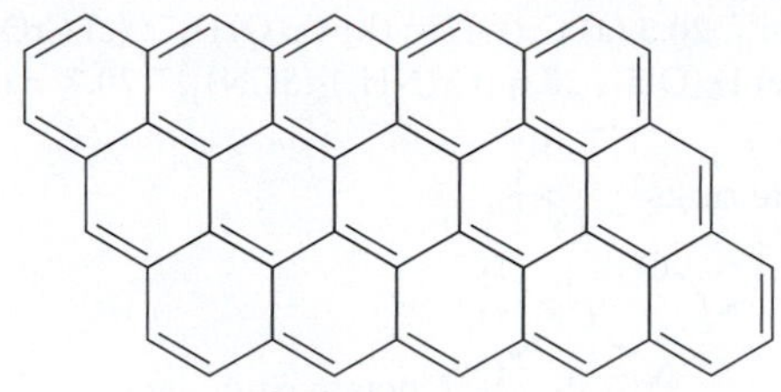

19.112 Earth's crust: O, Si, Al, Fe; Human body: O, C, H, N
19.114 C, Si, Ge and Sn have allotropes with the diamond structure. Sn and Pb have metallic allotropes. C (nonmetal), Si (semimetal), Ge (semimetal), Sn (semimetal and metal), Pb (metal)
19.116 (a) $H_3PO_4(aq) + H_2O(l) \rightleftharpoons H_3O^+(aq) + H_2PO_4{}^-(aq)$; H_3PO_4 is a Brønsted–Lowry acid. (b) $B(OH)_3(aq) + 2\,H_2O(l) \rightleftharpoons B(OH)_4{}^-(aq) + H_3O^+(aq)$; $B(OH)_3$ is a Lewis acid. **19.118** (a) In diamond each C is covalently bonded to four additional C atoms in a rigid three-dimensional network solid. Graphite is a two-dimensional covalent network solid of carbon sheets that can slide over each other. Both are high melting because melting requires the breaking of C–C bonds.
(b) Chlorine does not form perhalic acids of the type H_5XO_6 because its smaller size favors a tetrahedral structure over an octahedral one.
19.120 Cl–S–S–Cl **19.122** NH_3
19.124 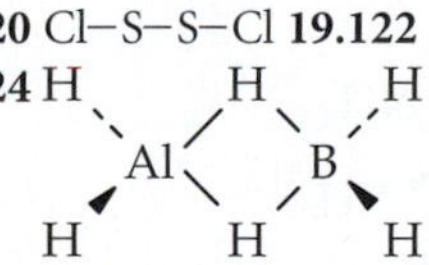

19.126 P_4O_{10}, H_3PO_4, 0.93 g P_4 **19.128** The angle required by P_4 is 60°. The strain would not be reduced by using sp^3 hybrid orbitals because their angle is ~109°.

Multiconcept Problems

19.130 (a)

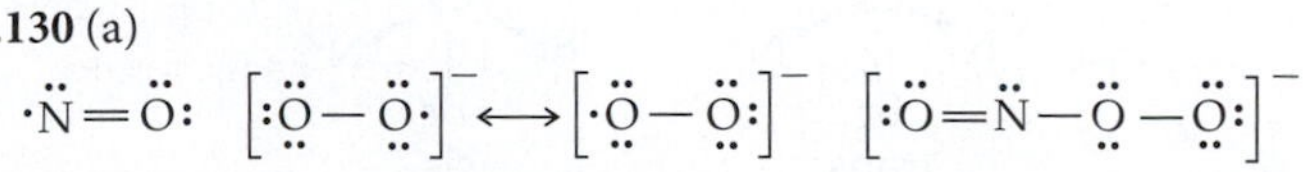

The O–N–O bond angle should be ~120°.
(b)

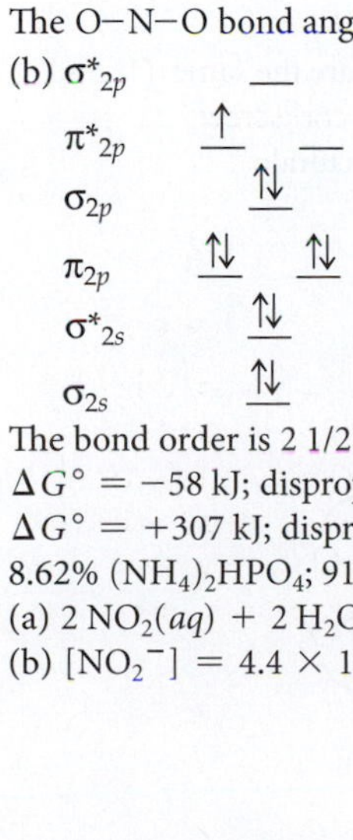

The bond order is 2 1/2 with one unpaired electron. **19.132** $E° = 0.30$ V; $\Delta G° = -58$ kJ; disproportionation of In^+ is spontaneous. $E° = -1.59$ V; $\Delta G° = +307$ kJ; disproportionation of Tl^+ is nonspontaneous. **19.134** (a) 8.62% $(NH_4)_2HPO_4$; 91.38%; NH_4NO_3; (b) 7.50 **19.136**
(a) $2\,NO_2(aq) + 2\,H_2O(l) \rightarrow HNO_2(aq) + H_3O^+(aq) + NO_3{}^-(aq)$;
(b) $[NO_2{}^-] = 4.4 \times 10^{-4}$ M; pH = 1.15; (c) 5.12 atm; (d) 0.976 g

CHAPTER 20

20.1 (a) [Ar] $3d^3\,4s^2$; (b) [Ar] $3d^7$; (c) [Ar] $3d^3$; (d) [Ar] $3d^9$ **20.2** (a) Mn; (b) Ni^{2+}; (c) Ag; (d) Mo^{3+}

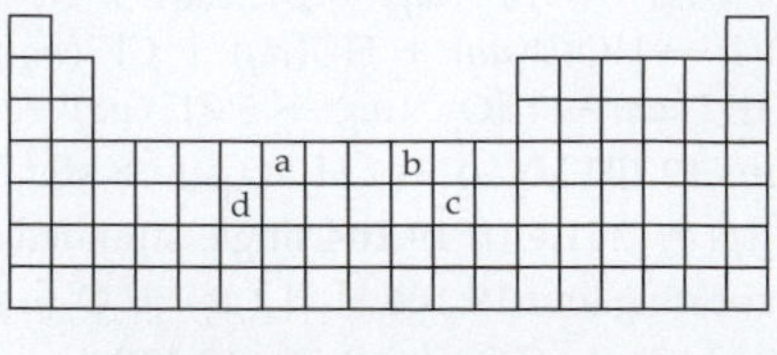

20.3 Z_{eff} increases from left to right. (a) Ti strongest, Zn weakest; (b) FeO_4^{2-} strongest, VO_4^{3-} weakest **20.4** (a) $Cr_2O_7^{2-}$; (b) Cr^{3+}; (c) Cr^{2+}; (d) Fe^{2+}; (e) Cu^{2+} **20.5** (a) $Cr(OH)_2$; (b) $Cr(OH)_4^-$; (c) CrO_4^{2-}; (d) $Fe(OH)_2$; (e) $Fe(OH)_3$ **20.6** $[Cr(NH_3)_2(SCN)_4]^-$ **20.7** +2 **20.8** (a)

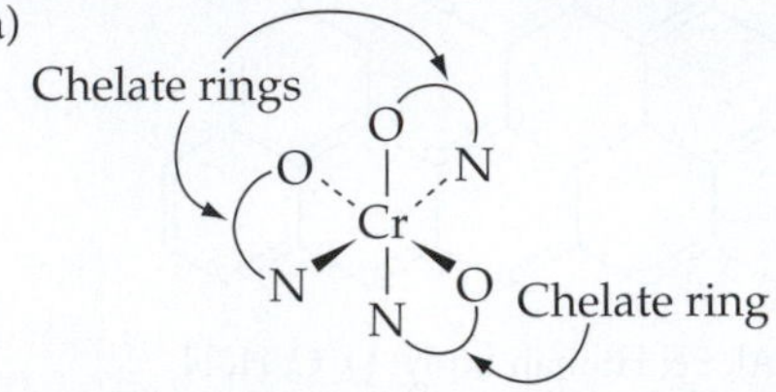

(b) Cr^{3+}, glycinate ligand, N and O; the chelate rings are identified in the drawing; (c) 6, octahedral, +3 **20.9** (a) tetraamminecopper(II) sulfate; (b) sodium tetrahydroxochromate(III); (c) triglycinatocobalt(III); (d) pentaaquaisothiocyanatoiron(III) ion **20.10** (a) $[Zn(NH_3)_4](NO_3)_2$; (b) $Ni(CO)_4$; (c) $K[Pt(NH_3)Cl_3]$; (d) $[Au(CN)_2]^-$ **20.11** (a) Two

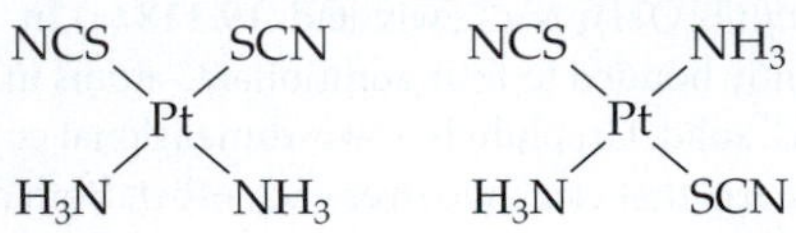

cis trans

(b) No isomers; (c) Two

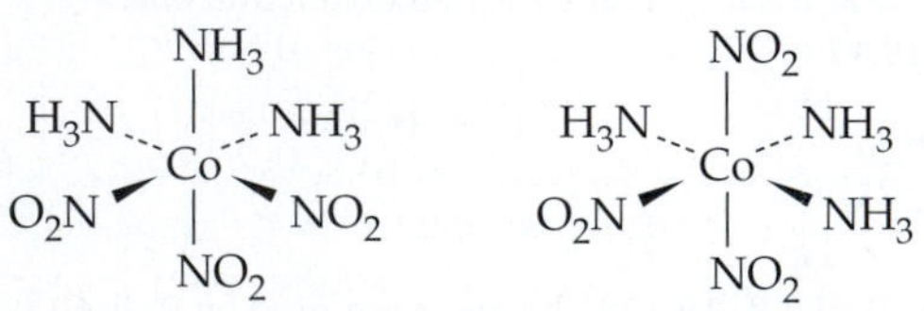

(d) No isomers; (e) Two

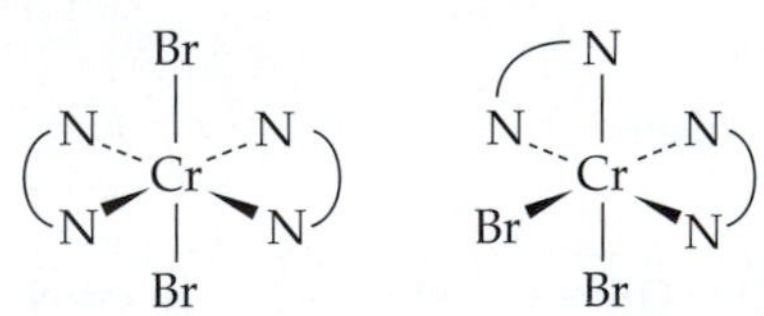

trans cis

(f) None **20.12** (1) and (2) are the same. (3) and (4) are the same. (1) and (2) are different from (3) and (4). **20.13** (b) foot; (d) corkscrew **20.14** (a) (2) and (3) are chiral, and (1) and (4) are achiral; (b)

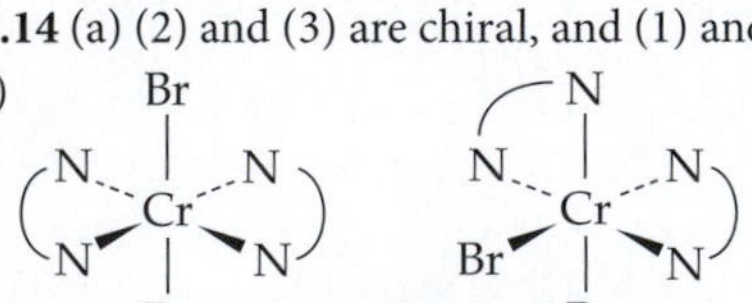

20.15 (a) $[Fe(C_2O_4)_3]^{3-}$

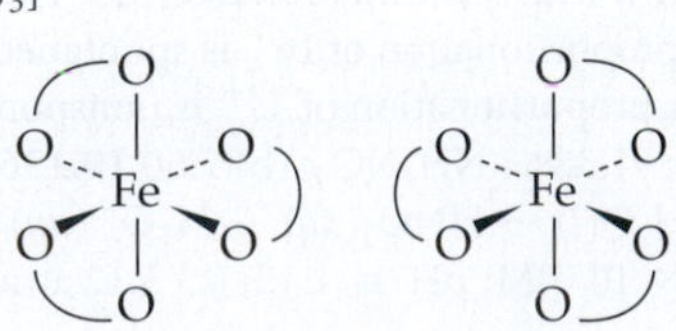

(c) $[Co(NH_3)_2(en)_2]^{3+}$

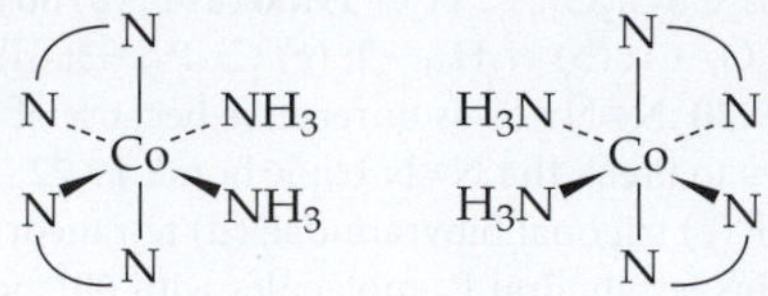

20.16 (a) blue; (b) 3.18×10^{-19} J
20.17
(a) Fe^{3+}

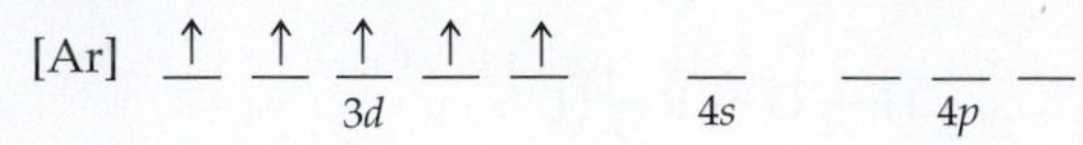

$[Fe(CN)_6]^{3-}$

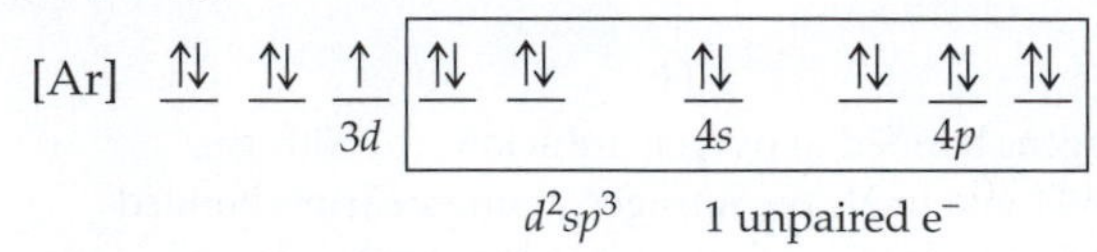

(b) Co^{2+}

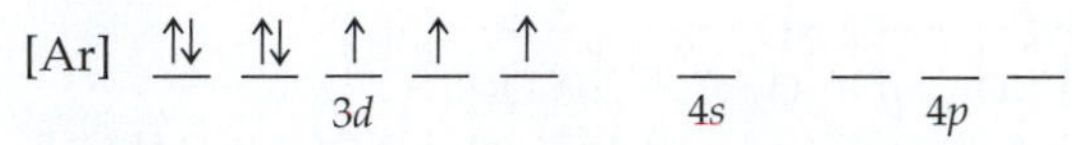

$[Co(H_2O)_6]^{2+}$

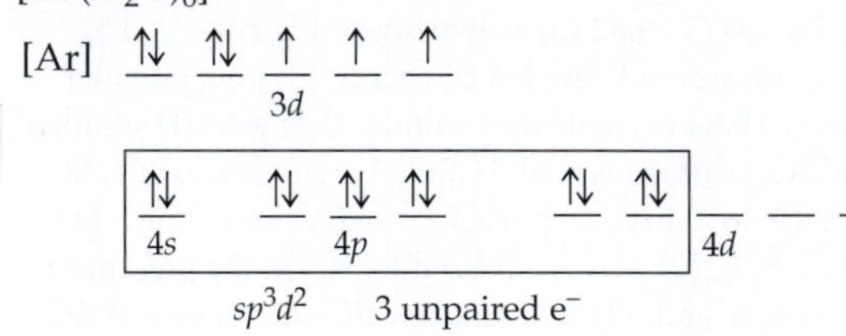

(c) V^{3+}

$[VCl_4]^-$

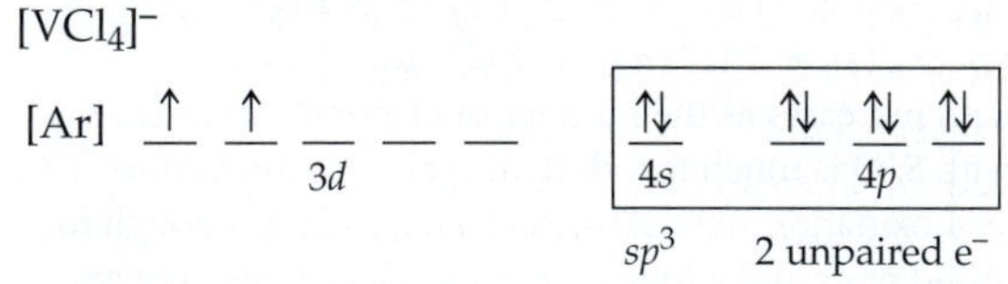

(d) Pt^{2+}

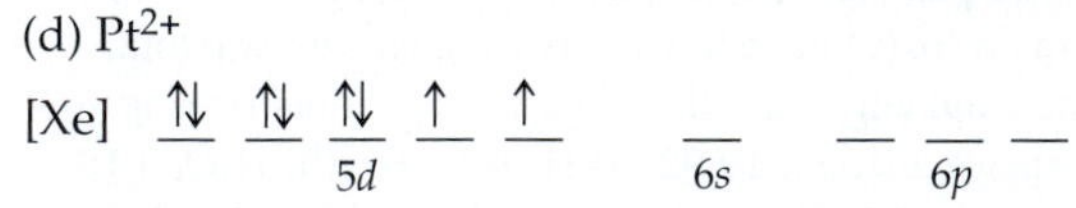

$[PtCl_4]^{2-}$

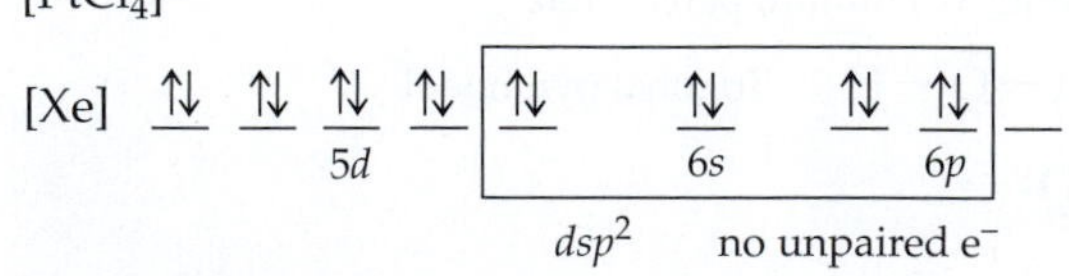

20.18.

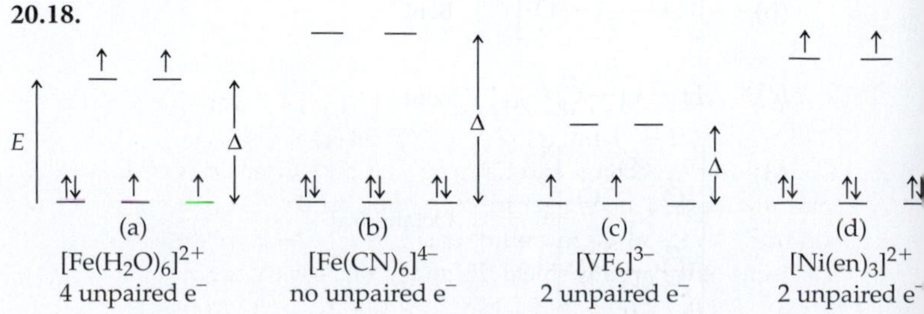

20.19 (a) $[NiCl_4]^{2-}$ (tetrahedral) (b) $[Ni(CN)_4]^{2-}$ (square planar)

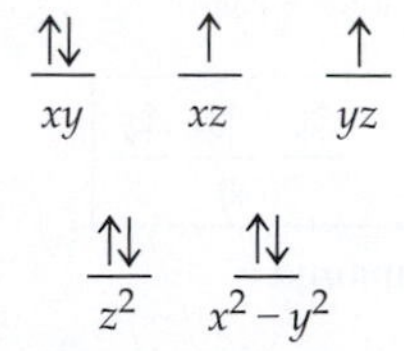

2 unpaired electrons

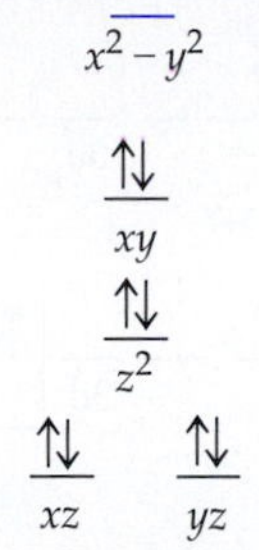

no unpaired electrons

20.20 $\Delta G°_1 = 243.8$ kJ; $\Delta G°_2 = -84.5$ kJ; $\Delta G°_3 = -192.3$ kJ; The third step is the most favorable
20.21

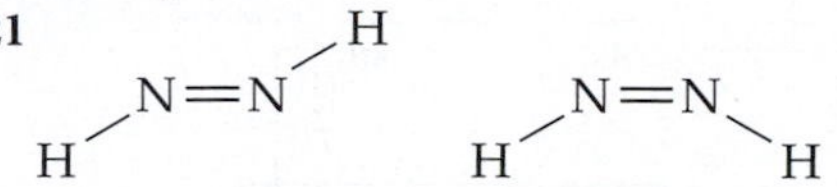

Conceptual Problems **20.22** (a) Co; (b) Cr; (c) Zr; (d) Pr

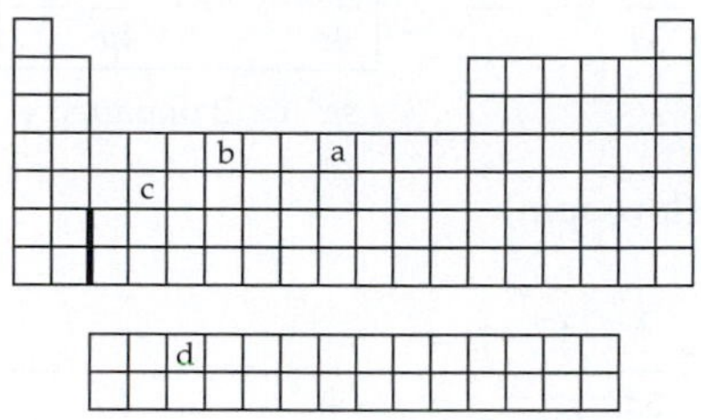

20.24 (a) The atomic radii decrease, at first markedly and then more gradually. Toward the end of the series, the radii increase again. The decrease in atomic radii is a result of an increase in Z_{eff}. The increase is due to electron–electron repulsions in doubly occupied *d* orbitals. (b) The densities of the transition metals are inversely related to their atomic radii. The densities initially increase from left to right and then decrease toward the end of the series. (c) Ionization energies generally increase from left to right across the series. The general trend correlates with an increase in Z_{eff} and a decrease in atomic radii. (d) The standard oxidation potentials generally decrease from left to right across the first transition series. This correlates with the general trend in ionization energies.
20.26 (1) dichloroethylenediamineplatinum(II); (2) *trans*-diammineaquachloroplatinum(II) ion; (3) amminepentachloroplatinum(IV) ion; (4) *cis*-diaquabis(ethylenediamine)platinum(IV) ion **20.28** (a) (1) cis; (2) trans; (3) trans; (4) cis; (b) (1) and (4) are the same. (2) and (3) are the same. (c) None of the isomers exist as enantiomers because their mirror images are identical.
20.30

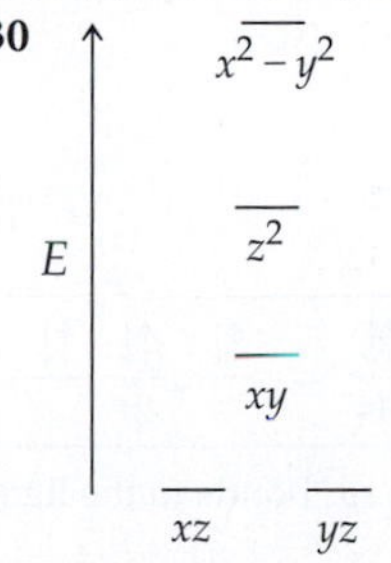

Section Problems **20.32** (a) [Ar] $3d^5\,4s^1$; (b) [Kr] $4d^2\,5s^2$; (c) [Ar] $3d^7$; (d) [Ar] $3d^5$; (e) [Kr] $4d^3$; (f) [Ar] $3d^0$ **20.34** (a) 1; (b) 2; (c) 0; (d) 3
20.36 Ti is harder than K and Ca largely because the sharing of *d*, as well as *s*, electrons results in stronger metallic bonding. **20.38** (a) The decrease in radii with increasing atomic number is expected because the added *d* electrons only partially shield the added nuclear charge. As a result, Z_{eff} increases. With increasing Z_{eff}, the electrons are more strongly attracted to the nucleus, and atomic size decreases. (b) The densities of the transition metals are inversely related to their atomic radii.
20.40 The lanthanide contraction is the general decrease in atomic radii of the *f*-block lanthanide elements. It is due to the increase in Z_{eff} as the 4*f* subshell is filled.
20.42

Element	$E_{i1} + E_{i2}$(kJ/mol)
Sc	1866
Ti	1969
V	2061
Cr	2224
Mn	2226
Fe	2324
Co	2408
Ni	2490
Cu	2703
Zn	2639

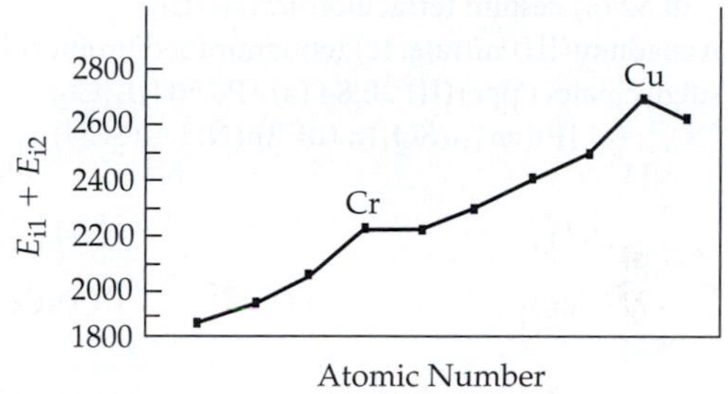

The general trend is due to the increase in Z_{eff} with increasing atomic number. Higher than expected values for the sum of the first two ionization energies are observed for Cr and Cu because of their anomalous electron configurations (Cr: $3d^5\,4s^1$; Cu: $3d^{10}\,4s^1$). An increasing Z_{eff} affects 3*d* orbitals more than the 4*s* orbital and the second ionization energy for an electron from the 3*d* orbital is higher than expected. **20.44** (a) $Cr(s) + 2\,H^+(aq) \rightarrow Cr^{2+}(aq) + H_2(g)$; (b) $Zn(s) + 2\,H^+(aq) \rightarrow Zn^{2+}(aq) + H_2(g)$; (c) N.R.; (d) $Fe(s) + 2\,H^+(aq) \rightarrow Fe^{2+}(aq) + H_2(g)$ **20.46** (b) Mn; (d) Cu
20.48 Sc(III), Ti(IV), V(V), Cr(VI), Mn(VII), Fe(VI), Co(III), Ni(II), Cu(II), Zn(II) **20.50** Cu^{2+} is a stronger oxidizing agent than Cr^{2+} because of a higher Z_{eff}. **20.52** a reducing agent, because early transition metal atoms have a relatively low effective nuclear charge and are easily oxidized to higher oxidation states. **20.54** $Mn^{2+} < MnO_2 < MnO_4^-$ because of increasing oxidation state of Mn. **20.56** (a) $Cr_2O_3(s) + 2\,Al(s) \rightarrow 2\,Cr(s) + Al_2O_3(s)$; (b) $Cu_2S(l) + O_2(g) \rightarrow 2\,Cu(l) + SO_2(g)$ **20.58** $Cr(OH)_3(s) + OH^-(aq) \rightarrow Cr(OH)_4^-(aq)$, + 3 oxidation state. deep green.
20.60 (c) $Cr(OH)_3$ **20.62** (a) Add excess KOH(*aq*) and Fe^{3+} will precipitate as $Fe(OH)_3(s)$. $Na^+(aq)$ will remain in solution. (b) Add excess NaOH(*aq*) and Fe^{3+} will precipitate as $Fe(OH)_3(s)$. $Cr(OH)_4^-(aq)$ will remain in solution. (c) Add excess $NH_3(aq)$ and Fe^{3+} will precipitate as $Fe(OH)_3(s)$. $Cu(NH_3)_4^{2+}(aq)$ will remain in solution.
20.64 (a) $Cr_2O_7^{2-}(aq) + 6\,Fe^{2+}(aq) + 14\,H^+(aq) \rightarrow 2\,Cr^{3+}(aq) + 6\,Fe^{3+}(aq) + 7\,H_2O(l)$; (b) $4\,Fe^{2+}(aq) + O_2(g) + 4\,H^+(aq) \rightarrow 4\,Fe^{3+}(aq) + 2\,H_2O(l)$; (c) $Cu_2O(s) + 2\,H^+(aq) \rightarrow Cu(s) + Cu^{2+}(aq) + H_2O(l)$; (d) $Fe(s) + 2\,H^+(aq) \rightarrow Fe^{2+}(aq) + H_2(g)$
20.66 (a) $2\,CrO_4^{2-}(aq) + 2\,H_3O^+(aq) \rightarrow Cr_2O_7^{2-}(aq) + 3\,H_2O(l)$; (b) $[Fe(H_2O)_6]^{3+}(aq) + SCN^-(aq) \rightarrow [Fe(H_2O)_5(SCN)]^{2+}(aq) + H_2O(l)$; (c) $3\,Cu(s) + 2\,NO_3^-(aq) + 8\,H^+(aq) \rightarrow 3\,Cu^{2+}(aq) + 2\,NO(g) + 4\,H_2O(l)$; (d) $Cr(OH)_3(s) + OH^-(aq) \rightarrow Cr(OH)_4^-(aq)$; $2\,Cr(OH)_4^-(aq) + 3\,HO_2^-(aq) \rightarrow 2\,CrO_4^{2-}(aq) + 5\,H_2O(l) + OH^-(aq)$ **20.68** Ni^{2+} accepts six pairs of electrons, two each from the three ethylenediamine ligands. Ni^{2+} is an electron pair acceptor, a Lewis acid. The two nitrogens in each ethylenediamine donate a pair of electrons to the Ni^{2+}. The ethylenediamine is an electron pair donor, a Lewis base. **20.70** (a) $[Ag(NH_3)_2]^+$; (b) $[Ni(CN)_4]^{2-}$; (c) $[Cr(H_2O)_6]^{3+}$

20.72 (a) +1; (b) +3; (c) +2; (d) +4; (e) +3 **20.74** (a) $Ni(CO)_4$; (b) $[Ag(NH_3)_2]^+$; (c) $[Fe(CN)_6]^{3-}$; (d) $[Ni(CN)_4]^{2-}$
20.76

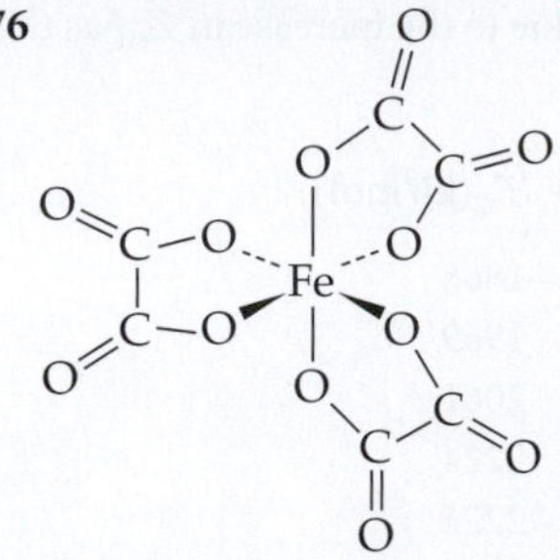

The iron is in the +3 oxidation state, and the coordination number is six. The geometry about the Fe is octahedral. There are three chelate rings, one formed by each oxalate ligand. **20.78** (a) +3; (b) +1; (c) +3; (d) +1 **20.80** (a) tetrachloromanganate(II); (b) hexaamminenickel(II); (c) tricarbonatocobaltate(III); (d) bis(ethylenediamine)dithiocyanatoplatinum(IV) **20.82** (a) cesium tetrachloroferrate(III); (b) hexaaquavanadium(III) nitrate; (c) tetraamminedibromocobalt(III) bromide; (d) diglycinatocopper(II) **20.84** (a) $[Pt(NH_3)_4]Cl_2$; (b) $Na_3[Fe(CN)_6]$; (c) $[Pt(en)_3](SO_4)_2$; (d) $Rh(NH_3)_3(SCN)_3$
20.86

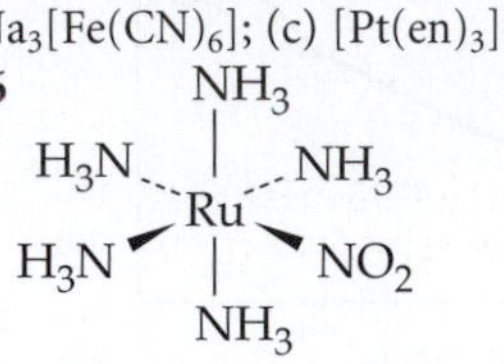

$[Ru(NH_3)_5(NO_2)]Cl$

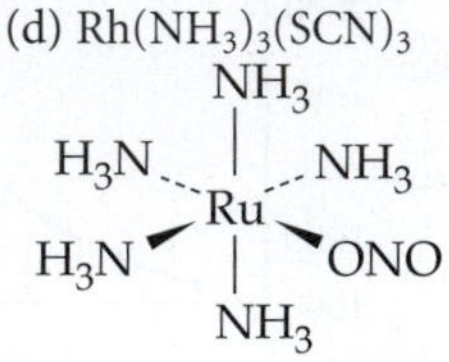

$[Ru(NH_3)_5(ONO)]Cl$

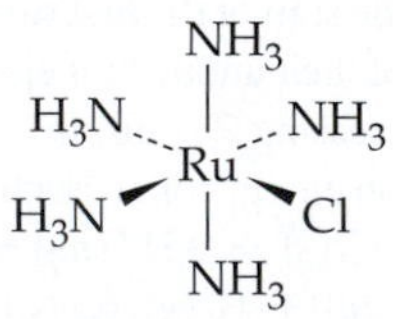

$[Ru(NH_3)_5Cl]NO_2$

$[Ru(NH_3)_5(NO_2)]Cl$ and $[Ru(NH_3)_5(ONO)]Cl$ are linkage isomers. $[Ru(NH_3)_5Cl]NO_2$ is an ionization isomer of both $[Ru(NH_3)_5(NO_2)]Cl$ and $[Ru(NH_3)_5(ONO)]Cl$. **20.88** (a) $[Cr(NH_3)_2Cl_4]^-$ can exist as cis and trans diastereoisomers. (d) $[PtCl_2Br_2]^{2-}$ (square planar) can exist as cis and trans diastereoisomers. **20.90** (c) cis-$[Cr(en)_2(H_2O)_2]^{3+}$; (d) $[Cr(C_2O_4)_3]^{3-}$
20.92

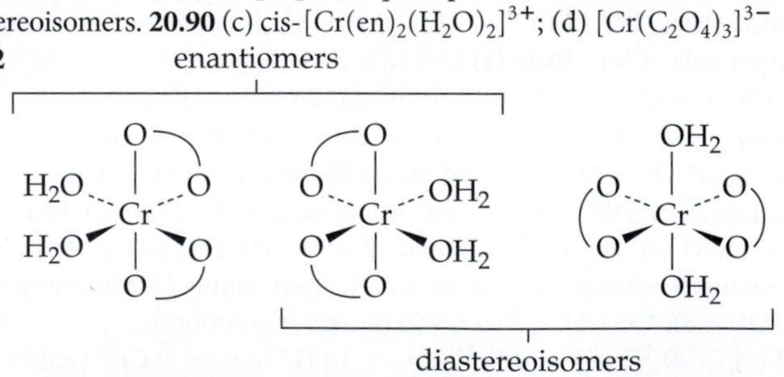

20.94 Plane-polarized light is light in which the electric vibrations of the light wave are restricted to a single plane. The following chromium complex can rotate the plane of plane-polarized light.

$[Cr(en)_3]^{3+}$

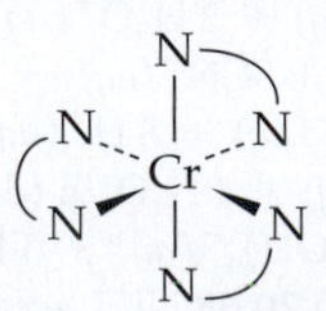

20.96 The measure of the amount of light absorbed by a substance is called the absorbance, and a graph of absorbance versus wavelength is called an absorption spectrum; orange

20.98 (a) $[Ti(H_2O)_6]^{3+}$
Ti^{3+}

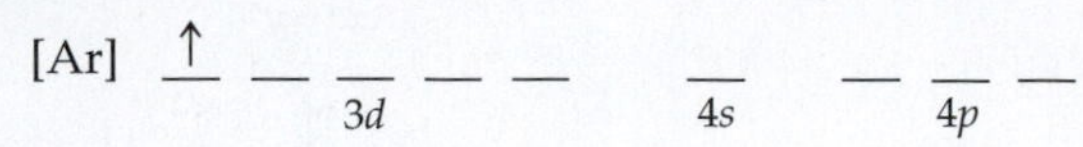

$[Ti(H_2O)_6]^{3+}$

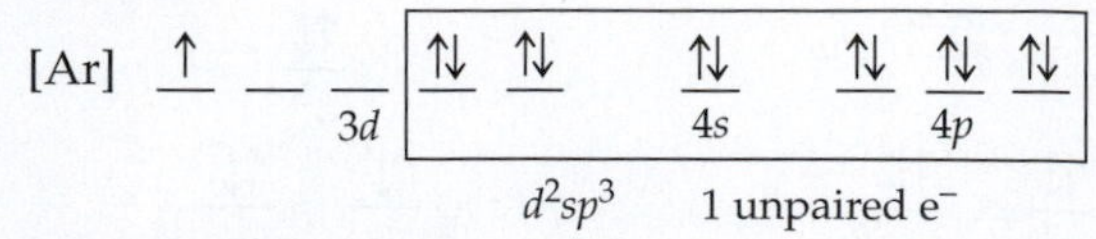

(b) $[NiBr_4]^{2-}$
Ni^{2+}

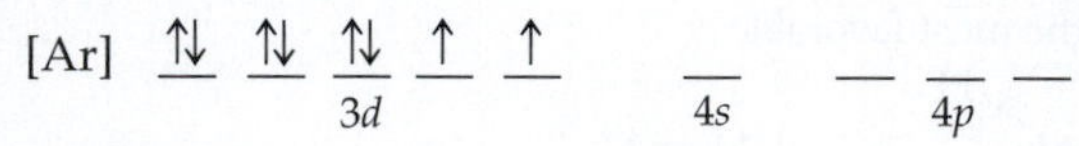

$[NiBr_4]^{2-}$

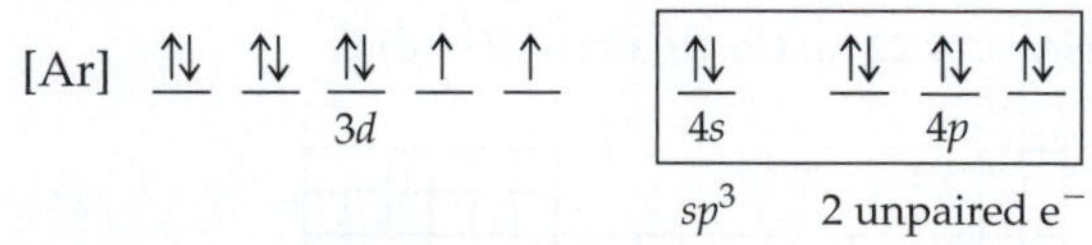

(c) $[Fe(CN)_6]^{3-}$ (low-spin)
Fe^{3+}

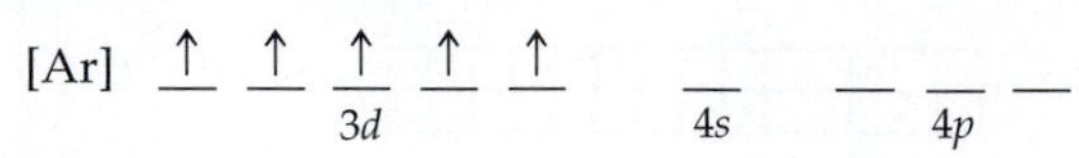

$[Fe(CN)_6]^{3-}$

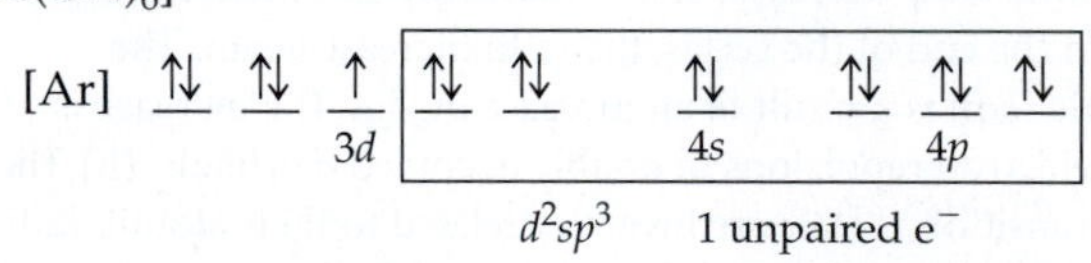

(d) $[MnCl_6]^{3-}$ (high-spin)
Mn^{3+}

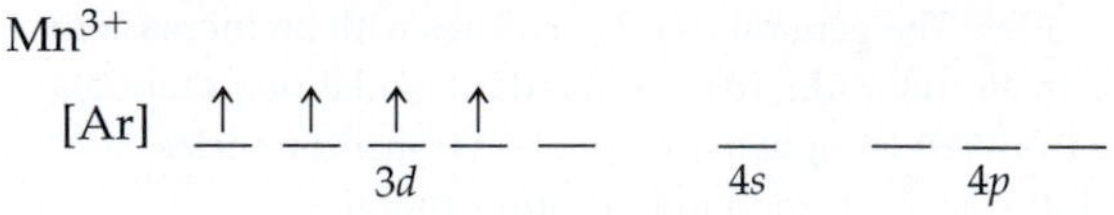

$[MnCl_6]^{3-}$

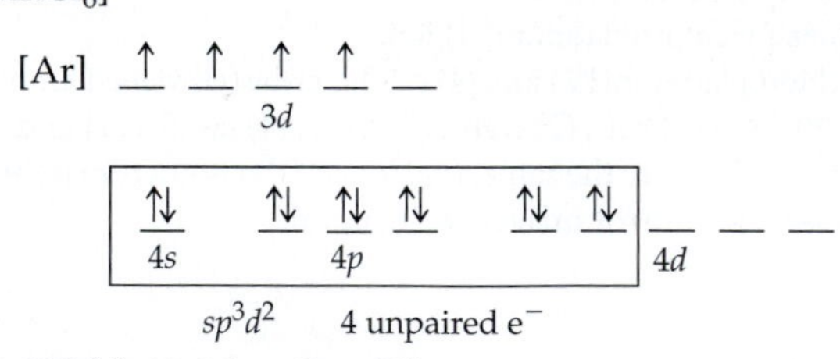

20.100 (a) +3, M = Cr or Ni;
(b)

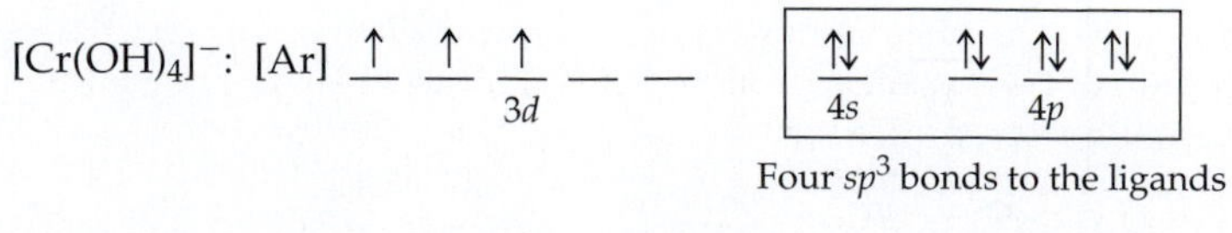

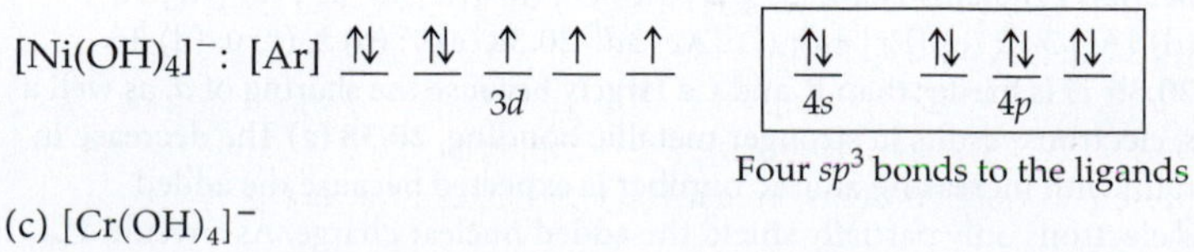

(c) $[Cr(OH)_4]^-$

20.102 $[Ti(H_2O)_6]^{3+}$, Ti^{3+}, $3d^1$

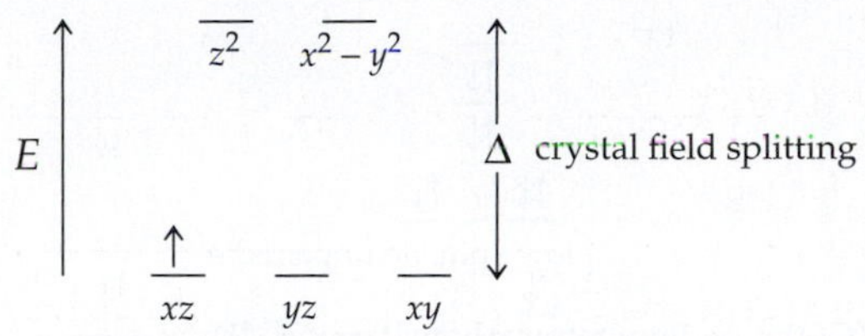

$[Ti(H_2O)_6]^{3+}$ is colored because it can absorb light in the visible region, exciting the electron to the higher-energy set of orbitals.
20.104 $\Delta = 220$ kJ/mol; NCS^- is a weaker-field ligand than H_2O; $[Ti(NCS)_6]^{3-}$ color should be red
20.106 (a) $[CrF_6]^{3-}$ (b) $[V(H_2O)_6]^{3+}$ (c) $[Fe(CN)_6]^{3-}$

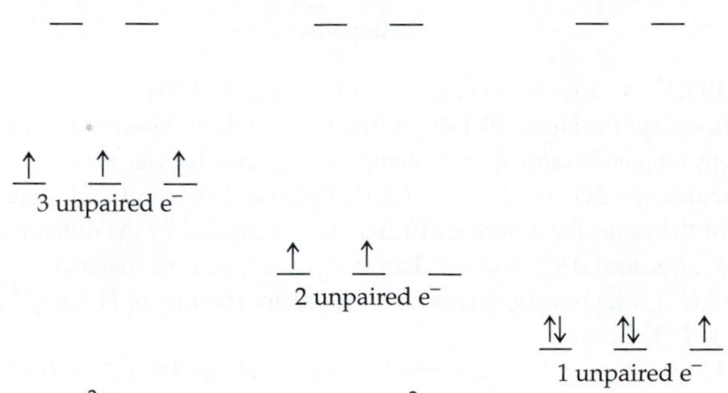

20.108 $Ni^{2+}(aq)$ $Zn^{2+}(aq)$

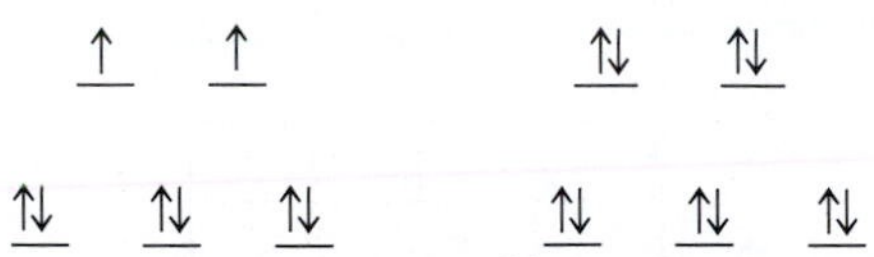

$Ni^{2+}(aq)$ is green because the Ni^{2+} ion can absorb light, which promotes electrons from the filled d orbitals to the higher energy half-filled d orbitals. $Zn^{2+}(aq)$ is colorless because the d orbitals are completely filled and no electrons can be promoted, so no light is absorbed. **20.110** Weak-field ligands produce a small Δ. Strong-field ligands produce a large Δ. For a metal complex with weak-field ligands, $\Delta < P$, where P is the pairing energy, and it is easier to place an electron in either d_{z^2} or $d_{x^2-y^2}$ than to pair up electrons; high-spin complexes result. For a metal complex with strong-field ligands, $\Delta > P$ and it is easier to pair up electrons than to place them in either d_{z^2} or $d_{x^2-y^2}$; low-spin complexes result.
20.112

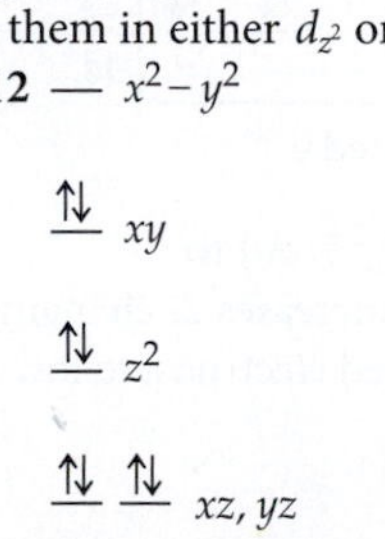

Square planar geometry is most common for metal ions with d^8 configurations because this configuration favors low-spin complexes in which all four lower energy d orbitals are filled and the higher energy $d_{x^2-y^2}$ orbital is vacant.
Chapter Problems **20.114** (a) $[Mn(CN)_6]^{3-}$; (d) $[FeF_6]^{4-}$
20.116 (a) $4\,Co^{3+}(aq) + 2\,H_2O(l) \rightarrow 4\,Co^{2+}(aq) + O_2(g) + 4\,H^+(aq)$; (b) $4\,Cr^{2+}(aq) + O_2(g) + 4\,H^+(aq) \rightarrow 4\,Cr^{3+}(aq) + 2\,H_2O(l)$; (c) $3\,Cu(s) + Cr_2O_7^{2-}(aq) + 14\,H^+(aq) \rightarrow 3\,Cu^{2+}(aq) + 2\,Cr^{3+}(aq) + 7\,H_2O(l)$; (d) $2\,CrO_4^{2-}(aq) + 2\,H^+(aq) \rightarrow Cr_2O_7^{2-}(aq) + H_2O(l)$
20.118 $EDTA^{4-}$ in mayonnaise will complex any metal cations that are present in trace amounts. Free metal ions can catalyze the oxidation of oils, causing the mayonnaise to become rancid. The bidentate ligand $H_2NCH_2CO_2^-$ will not bind to metal ions as strongly as does the hexadentate $EDTA^{4-}$ and so would not be an effective substitute for $EDTA^{4-}$.
20.120 (a)

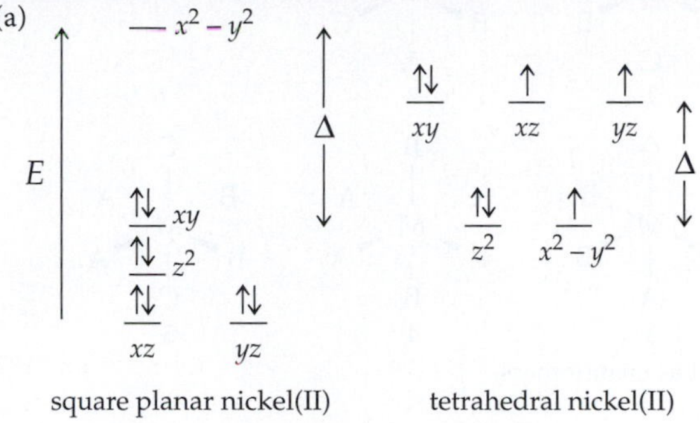

(b) $NiCl_2L_2$ is tetrahedral, $Ni(NCS)_2L_2$ is square planar.

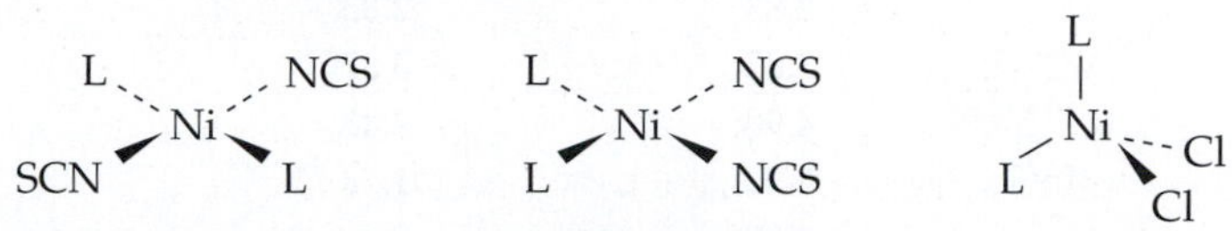

(c) Square planar cis-$Ni(NCS)_2L_2$ and tetrahedral $NiCl_2L_2$ have a dipole moment. **20.122** Cl^- is a weak-field ligand, whereas CN^- is a strong-field ligand. Δ for $[Fe(CN)_6]^{3-}$ is larger than the pairing energy P; Δ for $[FeCl_6]^{3-}$ is smaller than P. Fe^{3+} has a $3d^5$ electron configuration.

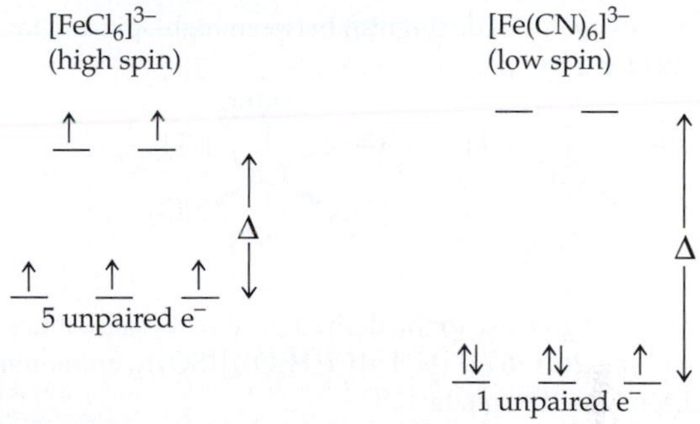

Because of the difference in Δ, $[FeCl_6]^{3-}$ is high-spin with five unpaired electrons, whereas $[Fe(CN)_6]^{3-}$ is low-spin with only one unpaired electron. **20.124** For d^1–d^3 and d^8–d^{10} complexes, only one ground-state electron configuration is possible. In d^1–d^3 complexes, all the electrons occupy the lower-energy d orbitals, independent of the value of Δ. In d^8–d^{10} complexes, the lower-energy set of d orbitals is filled with three pairs of electrons, while the higher-energy set contains two, three, or four electrons, again independent of the value of Δ. **20.126** $[CoCl_4]^{2-}$ is tetrahedral. $[Co(H_2O)_6]^{2+}$ is octahedral. Because $\Delta_{tet} < \Delta_{oct}$, these complexes have different colors. $[CoCl_4]^{2-}$ has absorption bands at longer wavelengths.
20.128

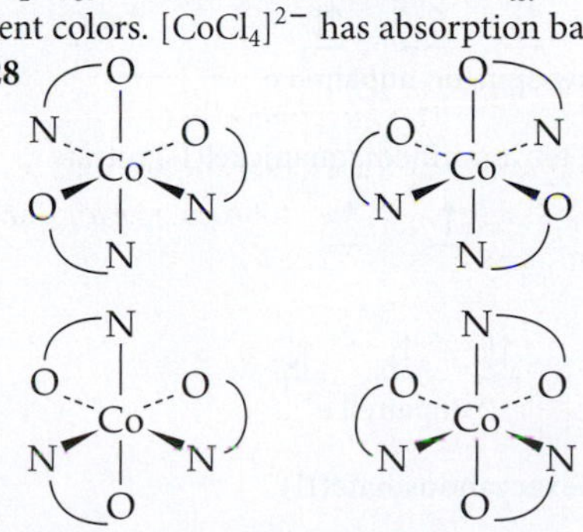

20.130

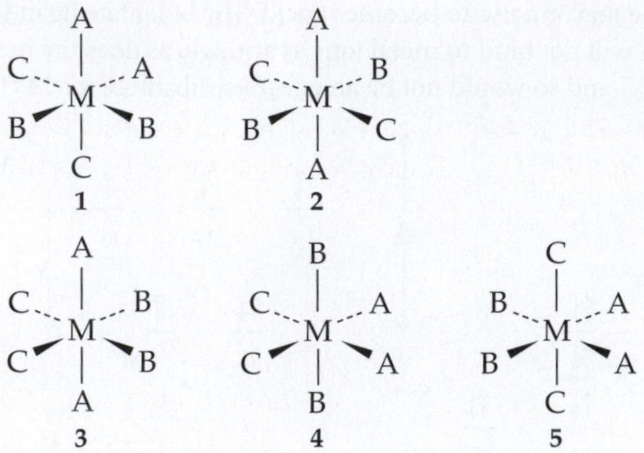

1 can exist as enantiomers.

20.132

	Weak-field ligands	Strong-field ligands
Ti^{2+}	2.83	2.83
V^{2+}	3.87	3.87
Cr^{2+}	4.90	2.83
Mn^{2+}	5.92	1.73
Fe^{2+}	4.90	0
Co^{2+}	3.87	1.73
Ni^{2+}	2.83	2.83
Cu^{2+}	1.73	1.73
Zn^{2+}	0	0

The magnetic moment can distinguish between high-spin and low-spin configurations for $d^4 - d^7$.

20.134 (a)

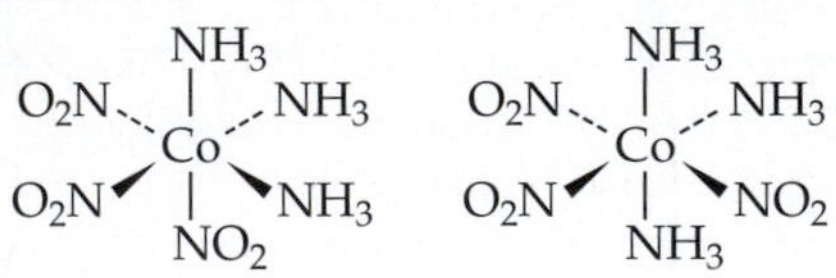

1 **2**

(b) Isomer **2** would give rise to the desired product because it has two trans NO_2 groups. **20.136** (a) $(NH_4)[Cr(H_2O)_6](SO_4)_2$, ammonium hexaaquachromium(III) sulfate

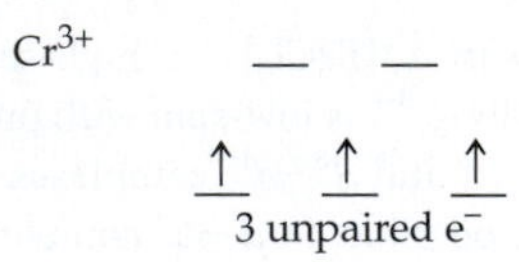

(b) $Mo(CO)_6$, hexacarbonylmolybdenum(0)

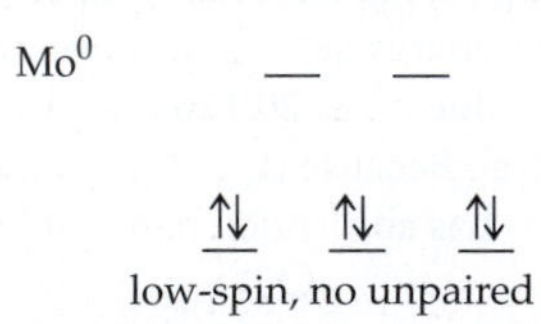

(c) $[Ni(NH_3)_4(H_2O)_2](NO_3)_2$, tetramminediaquanickel(II) nitrate

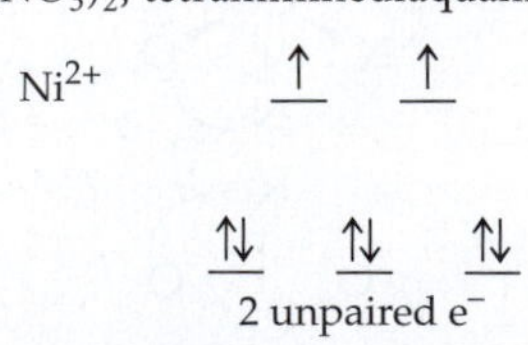

(d) $K_4[Os(CN)_6]$, potassium hexacyanoosmate(II)

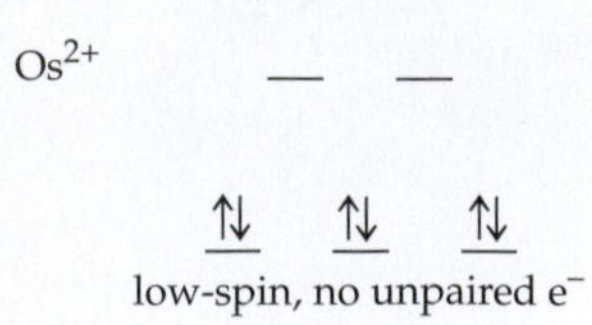

(e) $[Pt(NH_3)_4](ClO_4)_2$, tetraammineplatinum(II) perchlorate

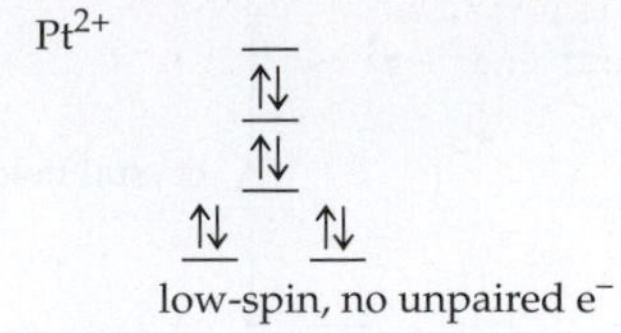

(f) $Na_2[Fe(CO)_4]$, sodium tetracarbonylferrate(–II)

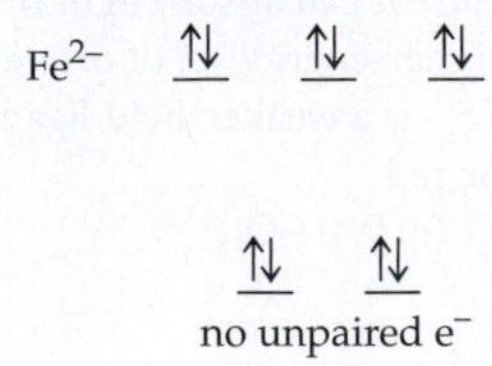

20.138 $Cl^- <$ urea $<$ acetate $< H_2O <$ acac $< NH_3$
Multiconcept Problems **20.140** (a) Reaction (2) should have the larger entropy change because three bidentate en ligands displace six water molecules. (b) $\Delta G° = \Delta H° - T\Delta S°$. Because $\Delta H°_1$ and $\Delta H°_2$ are almost the same, the difference in $\Delta G°$ is determined by the difference in $\Delta S°$. Because $\Delta S°_2$ is larger than $\Delta S°_1$, $\Delta G°_2$ is more negative than $\Delta G°_1$, which is consistent with the greater stability of $Ni(en)_3{}^{2+}$; (c) 180 J/(K · mol)
20.142 (a) $Cr(s) + 2\,H^+(aq) \rightarrow Cr^{2+}(aq) + H_2(g)$; (b) 1.26 L; (c) 2.23; (d)

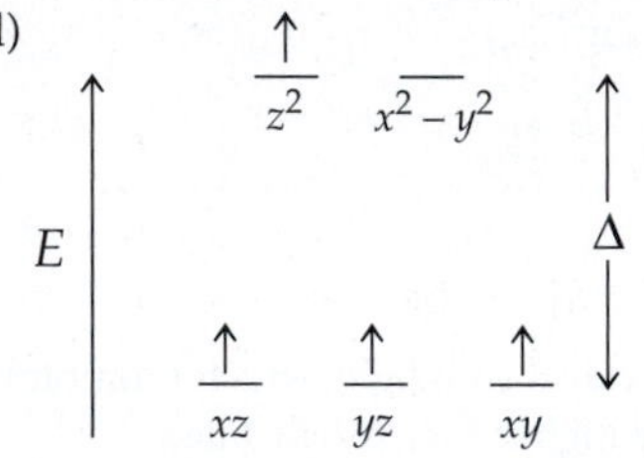

Valence bond orbital diagram

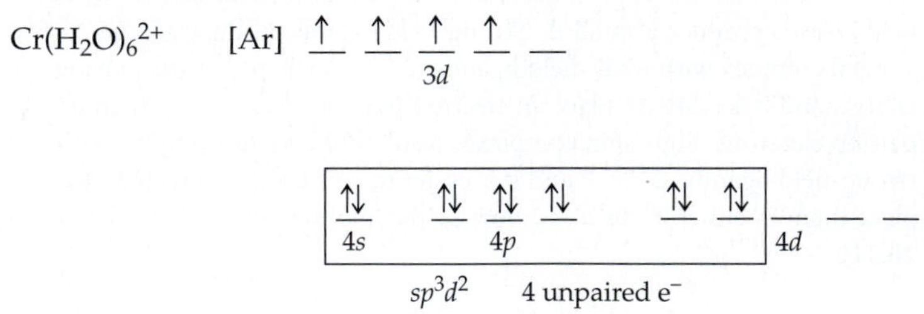

(e) The addition of excess KCN converts $Cr(H_2O)_6{}^{2+}(aq)$ to $Cr(CN)_6{}^{4-}(aq)$. CN^- is a strong field ligand and increases Δ changing the chromium complex from high spin, with 4 unpaired electrons, to low spin, with only 2 unpaired electrons.
20.144 (a) $CrCl_3(H_2O)_6$;
(b)

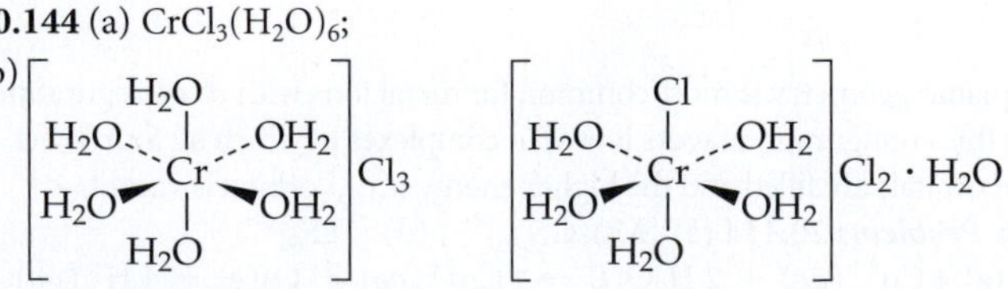

A **B**

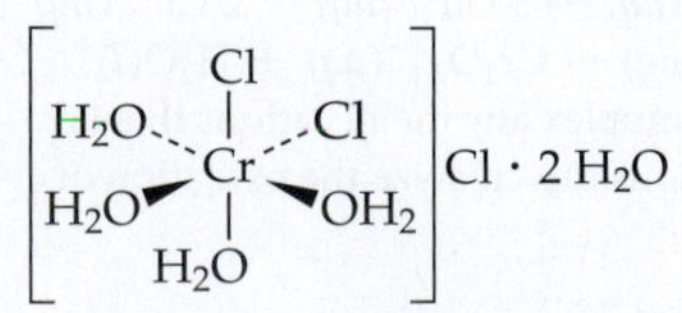

C

(c) H_2O is a stronger field ligand than Cl^-. Compound **A** is likely to be violet absorbing in the yellow. Compounds **B** and **C** have weaker field ligands and would appear blue or green absorbing in the orange or red, respectively. (d) **A**, −1.86 °C; **B**, −1.39 °C; **C**, −0.93 °C

20.146 (a) $K = \dfrac{[Cr_2O_7^{2-}]}{[CrO_4^{2-}]^2[H^+]^2} = 1.00 \times 10^{14}$

$[Cr_2O_7^{2-}]/[CrO_4^{2-}]^2 = 1.00 \times 10^{14}\,[H^+]^2$ In neutral solution, $[H^+] = 1.0 \times 10^{-7}$ and $[Cr_2O_7^{2-}]/[CrO_4^{2-}]^2 = 1$, so $[Cr_2O_7^{2-}]$ and $[CrO_4^{2-}]$ are comparable. In basic solution, $[H^+] < 1.0 \times 10^{-7}$ and $[Cr_2O_7^{2-}]/[CrO_4^{2-}]^2 < 1$, so $[CrO_4^{2-}]$ predominates. In acidic solution, $[H^+] > 1.0 \times 10^{-7}$ and $[Cr_2O_7^{2-}]/[CrO_4^{2-}]^2 > 1$, so $[Cr_2O_7^{2-}]$ predominates. (b) $[CrO_4^{2-}] = 2.23 \times 10^{-4}$ M, $[Cr_2O_7^{2-}] = 4.99 \times 10^{-2}$ M; (c) $[CrO_4^{2-}] = 2.24 \times 10^{-6}$ M, $[Cr_2O_7^{2-}] = 5.00 \times 10^{-2}$ M.

CHAPTER 21

21.1 The most common oxidation state for the 3B transition metals (Sc, Y, and La) is 3+. The 3+ oxidation state of the cations conveniently matches the 3− charge of the phosphate ion, resulting in a large lattice energy and corresponding insolubility for MPO_4 compounds. **21.2** (a) $Cr_2O_3(s) + 2\,Al(s) \rightarrow 2\,Cr(s) + Al_2O_3(s)$; (b) $Cu_2S(s) + O_2(g) \rightarrow 2\,Cu(s) + SO_2(g)$; (c) $PbO(s) + C(s) \rightarrow Pb(s) + CO(g)$; (d) $2\,K^+(l) + 2\,Cl^-(l) \xrightarrow{\text{electrolysis}} 2\,K(l) + Cl_2(g)$ **21.3** $CaO(s) + SiO_2(s) \rightarrow CaSiO_3(l)$ (slag). The O^{2-} in CaO behaves as a Lewis base and SiO_2 is the Lewis acid. **21.4** The electron configuration for Hg is [Xe] $4f^{14}\,5d^{10}\,6s^2$. Assuming the 5d and 6s bands overlap, the composite band can accommodate 12 valence electrons per metal atom. Weak bonding and a low melting point are expected for Hg because both the bonding and antibonding MOs are occupied. **21.5** (a) The composite *s–d* band can accommodate 12 valence electrons per metal atom. (1) is 1/4 filled, Hf [Xe] $6s^2\,4f^{14}\,5d^2$. (2) is 5/6 filled, Pt [Xe] $6s^2 4f^{14} 5d^8$. (3) is 7/12 filled, Re [Xe] $6s^2\,4f^{14}\,5d^5$. (b) Re has an excess of 5 bonding electrons and it has the highest melting point and is the hardest; (c) Pt has an excess of only 2 bonding electrons and it has the lowest melting point and is the softest **21.6** Ge doped with As is an *n*-type semiconductor because As has an additional valence electron. The extra electrons are in the conduction band. The number of electrons in the conduction band of the doped Ge is much higher than for pure Ge, and the conductivity of the doped semiconductor is higher. **21.7** (a) (1), silicon; (2), white tin; (3), diamond; (4), silicon doped with aluminum; (b) (3) < (1) < (4) < (2). Diamond (3) is an insulator with a large band gap. Silicon (1) is a semiconductor with a band gap smaller than diamond. Silicon doped with aluminum (4) is a *p*-type semiconductor that has vacancies (positive holes) in the valence band. White tin (2) has a partially filled *s–p* composite band and is a metallic conductor. **21.8** 539 nm
21.9

8 Cu at corners	8 × 1/8 = 1 Cu
8 Cu on edges	8 × 1/4 = 2 Cu
	Total = 3 Cu
12 O on edges	12 × 1/4 = 3 O
8 O on faces	8 × 1/2 = 4 O
	Total = 7 O

21.10 $Si(OCH_3)_4 + 4\,H_2O \rightarrow Si(OH)_4 + 4\,HOCH_3$
21.11 $Ba[OCH(CH_3)_2]_2 + Ti[OCH(CH_3)_2]_4 + 6\,H_2O \rightarrow BaTi(OH)_6(s) + 6\,HOCH(CH_3)_2$

$BaTi(OH)_6(s) \xrightarrow{\text{Heat}} BaTiO_3(s) + 3\,H_2O(g)$

21.12 (a) ceramic–metal; (b) ceramic–ceramic; (c) ceramic–polymer; (d) ceramic–metal **21.13** The smaller the particle, the larger the band gap and the greater the shift in the color of the emitted light from the red to the violet. The yellow quantum dot is larger because yellow is closer to the red than is blue. **21.14** (a) 27%; (b) 14%

Conceptual Problems **21.16** (a) electrolysis; (b) roasting a metal sulfide; (c) A = Li, electrolysis, B = Hg, roasting of the metal sulfide, C = Mn, reduction of the metal oxide, D = Ca, reduction of the metal oxide. **21.18** (a) (2), bonding MO's are filled. (b) (3), bonding and antibonding MO's are filled. (c) (3) < (1) < (2). Hardness increases with increasing MO bond order.
21.20

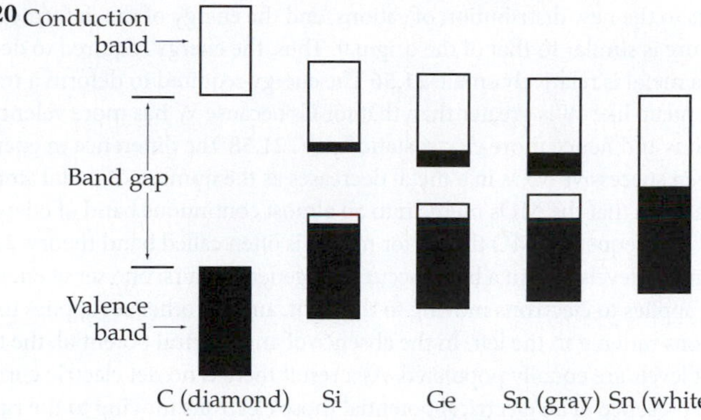

Section Problems **21.22** TiO_2, MnO_2, and Fe_2O_3 **21.24** (a) sulfide; (b) oxide; (c) uncombined; (d) sulfide **21.26** The less electronegative early transition metals tend to form ionic compounds by losing electrons to highly electronegative nonmetals such as oxygen. The more electronegative late transition metals tend to form compounds with more covalent character by bonding to the less electronegative nonmetals such as sulfur. **21.28** (a) hematite; (b) galena; (c) rutile; (d) chalcopyrite **21.30** The flotation process exploits the differences in the ability of water and oil to wet the surfaces of the mineral and the gangue. The gangue, which contains ionic silicates, is moistened by the polar water molecules and sinks to the bottom of the tank. The mineral particles, which contain the less polar metal sulfide, are coated by the oil and become attached to the soapy air bubbles created by the detergent. The metal sulfide particles are carried to the surface in the soapy froth, which is skimmed off at the top of the tank. This process would not work well for a metal oxide because it is too polar and will be wet by the water and sink with the gangue. **21.32** Because $E° < 0$ for Zn^{2+}, the reduction of Zn^{2+} is not favored. Because $E° > 0$ for Hg^{2+}, the reduction of Hg^{2+} is favored. The roasting of CdS should yield CdO because, like Zn^{2+}, $E° < 0$ for the reduction of Cd^{2+}. **21.34** (a) $V_2O_5(s) + 5\,Ca(s) \rightarrow 2\,V(s) + 5\,CaO(s)$; (b) $2\,PbS(s) + 3\,O_2(g) \rightarrow 2\,PbO(g) + 2\,SO_2(g)$; (c) $MoO_3(s) + 3\,H_2(g) \rightarrow Mo(s) + 3\,H_2O(g)$; (d) $3\,MnO_2(s) + 4\,Al(s) \rightarrow 3\,Mn(s) + 2\,Al_2O_3(s)$; (e) $MgCl_2(l) \xrightarrow{\text{Electrolysis}} Mg(l) + Cl_2(g)$ **21.36** $\Delta H° = -882.6$ kJ; $\Delta G° = -838.8$ kJ. $\Delta H°$ and $\Delta G°$ are different because of the entropy change associated with the reaction. The minus sign for $(\Delta H° - \Delta G°)$ indicates that the entropy is negative, which is consistent with a decrease in the number of moles of gas from 3 mol to 2 mol. **21.38** (a) 110 kg; (b) 1.06×10^5 L **21.40** 0.460 kg **21.42** $Fe_2O_3(s) + 3\,CO(g) \rightarrow 2\,Fe(l) + 3\,CO_2(g)$. Fe_2O_3 is the oxidizing agent. CO is the reducing agent. **21.44** Slag is a byproduct of iron production, consisting mainly of $CaSiO_3$. It is produced from the gangue in iron ore. **21.46** Molten iron from a blast furnace is exposed to a jet of pure oxygen gas for about 20 minutes. The impurities are oxidized to yield a molten slag that can be poured off. $P_4(l) + 5\,O_2(g) \rightarrow P_4O_{10}(l)$; $6\,CaO(s) + P_4O_{10}(l) \rightarrow 2\,Ca_3(PO_4)_2(l)$ (slag); $2\,Mn(l) + O_2(g) \rightarrow 2\,MnO(s)$; $MnO(s) + SiO_2(s) \rightarrow MnSiO_3(l)$ (slag) **21.48** $SiO_2(s) + 2\,C(s) \rightarrow Si(s) + 2\,CO(g)$; $Si(s) + O_2(g) \rightarrow SiO_2(s)$; $CaO(s) + SiO_2(s) \rightarrow CaSiO_3(l)$ (slag) **21.50** No. In a blast furnace tungsten carbide (WC) would be formed.
21.52

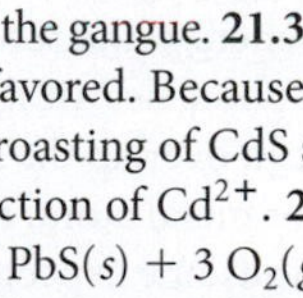

Each K has a single valence electron and has eight nearest neighbor K atoms. The valence electrons can't be localized in an electron-pair bond between any particular pair of K atoms. **21.54** Malleability and ductility of metals follow from the fact that the delocalized bonding extends in all directions. When a metallic crystal is deformed, no localized bonds are broken. Instead, the electron sea simply adjusts to the new distribution of cations, and the energy of the deformed structure is similar to that of the original. Thus, the energy required to deform a metal is relatively small. **21.56** The energy required to deform a transition metal like W is greater than that for Cs because W has more valence electrons and hence more electrostatic "glue". **21.58** The difference in energy between successive MOs in a metal decreases as the number of metal atoms increases so that the MOs merge into an almost continuous band of energy levels. Consequently, MO theory for metals is often called band theory. **21.60** The energy levels within a band occur in degenerate pairs; one set of energy levels applies to electrons moving to the right, and the other set applies to electrons moving to the left. In the absence of an electrical potential, the two sets of levels are equally populated. As a result there is no net electric current. In the presence of an electrical potential those electrons moving to the right are accelerated, those moving to the left are slowed down, and some change direction. Thus, the two sets of energy levels are now unequally populated. The number of electrons moving to the right is now greater than the number moving to the left, and so there is a net electric current.

21.62 (a) (b)

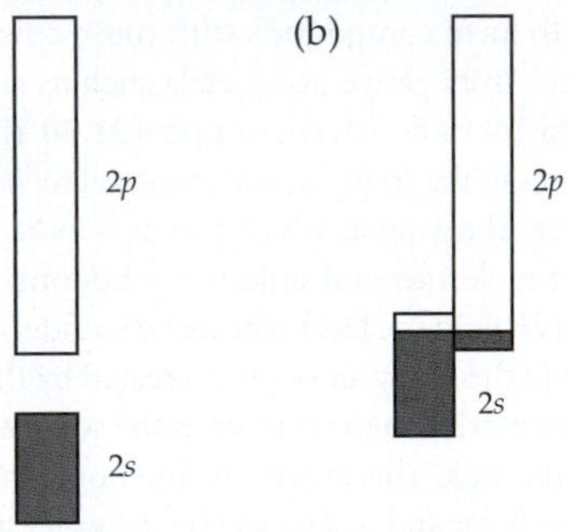

Diagram (b) shows the 2*s* and 2*p* bands overlapping in energy. The resulting composite band is only partially filled. Thus, Be is a good electrical conductor. **21.64** Transition metals have a *d* band that can overlap the *s* band to give a composite band consisting of six MOs per metal atom. Half of the MOs are bonding and half are antibonding, and thus one expects maximum bonding for metals that have six valence electrons per metal atom. Accordingly, the melting points of the transition metals go through a maximum at or near group 6B. **21.66** A semiconductor is a material that has an electrical conductivity intermediate between that of a metal and that of an insulator. Si, Ge, and Sn (gray) are semiconductors.

21.68 (a) (b)

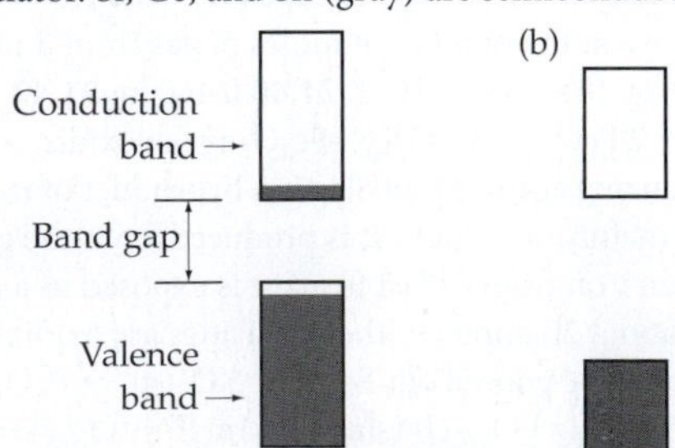

Semiconductor

Insulator

The MOs of a semiconductor are similar to those of an insulator, but the band gap in a semiconductor is smaller. As a result, a few electrons have enough energy to jump the gap and occupy the higher-energy, conduction band. The conduction band is thus partially filled, and the valence band is partially empty. When an electrical potential is applied to a semiconductor, it conducts a small amount of current because the potential can accelerate the electrons in the partially filled bands. **21.70** As the band gap increases, the number of electrons able to jump the gap and occupy the higher-energy conduction band decreases, and thus the conductivity decreases. **21.72** An *n*-type semiconductor is a semiconductor doped with a substance with more valence electrons than the semiconductor itself. Si doped with P is an example.

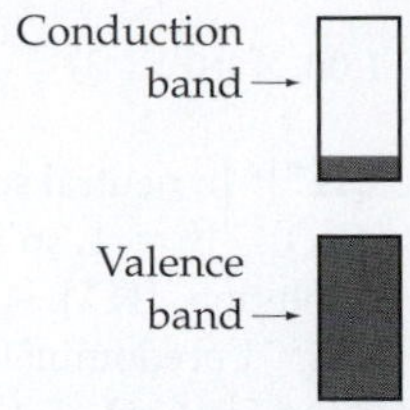

n-Type semiconductor

21.74 In the MO picture, the extra electrons occupy the conduction band. The number of electrons in the conduction band of the doped Ge is much greater than for pure Ge, and the conductivity of the doped semiconductor is correspondingly higher. **21.76** (a) *p*-type; (b) *n*-type; (c) *n*-type **21.78** $Cd(CH_3)_2(g) + H_2Se(g) \rightarrow CdSe(s) + 2\,CH_4(g)$ **21.80** $Al_2O_3 < Ge < $ Ge doped with In $< Fe < Cu$ **21.82** In a diode, current flows only when the junction is under a forward bias. A $p-n$ junction that is part of a circuit and subjected to an alternating potential acts as a rectifier, allowing current to flow in only one direction, thereby converting alternating current to direct current. **21.84** Both an LED and a photovoltaic cell contain $p-n$ junctions, but the two devices involve opposite processes. An LED converts electrical energy to light; a photovoltaic, or solar, cell converts light to electricity. **21.86** 620 nm, orange **21.88** (1) A superconductor is able to levitate a magnet. (2) In a superconductor, once an electric current is started, it flows indefinitely without loss of energy. A superconductor has no electrical resistance. **21.90** Some K^+ ions are surrounded octahedrally by six $C_{60}{}^{3-}$ ions; others are surrounded tetrahedrally by four $C_{60}{}^{3-}$ ions. **21.92** Ceramics are inorganic, nonmetallic, nonmolecular solids, including both crystalline and amorphous materials. Ceramics have higher melting points, and they are stiffer, harder, and more resistant to wear and corrosion than are metals. **21.94** Ceramics have higher melting points, and they are stiffer, harder, and more wear resistant than metals because they have stronger bonding. They maintain much of their strength at high temperatures, where metals either melt or corrode because of oxidation. **21.96** The brittleness of ceramics is due to strong chemical bonding. In silicon nitride each Si atom is bonded to four N atoms and each N atom is bonded to three Si atoms. The strong, highly directional covalent bonds prevent the planes of atoms from sliding over one another when the solid is subjected to a stress. As a result, the solid can't deform to relieve the stress. It maintains its shape up to a point, but then the bonds give way suddenly and the material fails catastrophically when the stress exceeds a certain threshold value. By contrast, metals are able to deform under stress because their planes of metal cations can slide easily in the electron sea. **21.98** Ceramic processing is the series of steps that leads from raw material to the finished ceramic object.
21.100 $Zr[OCH(CH_3)_2]_4 + 4\,H_2O \rightarrow Zr(OH)_4 + 4\,HOCH(CH_3)_2$
21.102 $(HO)_3Si{-}O{-}H + H{-}O{-}Si(OH)_3 \rightarrow$
$(HO)_3Si{-}O{-}Si(OH)_3 + H_2O$.
Further reactions of this sort give a three-dimensional network of Si–O–Si bridges. On heating, SiO_2 is obtained. **21.104** $2\,Ti(BH_4)_3(soln) \rightarrow 2\,TiB_2(s) + B_2H_6(g) + 9\,H_2(g)$
21.106 $3\,SiCl_4(g) + 4\,NH_3(g) \rightarrow Si_3N_4(s) + 12\,HCl(g)$
21.108 Graphite/epoxy composites are good materials for making tennis rackets and golf clubs because of their high strength-to-weight ratios.
Chapter Problems **21.110** 119 °C **21.112** $2\,Eu^{3+}(aq) + Zn(s) \rightarrow 2\,Eu^{2+}(aq) + Zn^{2+}(aq)$; $Eu^{2+}(aq) + SO_4{}^{2-}(aq) \rightarrow EuSO_4(s)$
21.114 The chemical composition of the alkaline earth minerals is that of metal sulfates and sulfites, MSO_4 and MSO_3. **21.116** Band theory better explains how the number of valence electrons affects properties such as melting point and hardness. **21.118** Transition metals have a *d* band that can overlap the *s* band to give a composite band consisting of six MOs per

metal atom. Half of the MOs are bonding and half are antibonding. Strong bonding and a high enthalpy of vaporization are expected for V because almost all of the bonding MOs are occupied and all of the antibonding MOs are empty. Weak bonding and a low enthalpy of vaporization are expected for Zn because both the bonding and the antibonding MOs are occupied. **21.120** With a band gap of 130 kJ/mol, GaAs is a semiconductor. Because Ge lies between Ga and As in the periodic table, GaAs is isoelectronic with Ge.

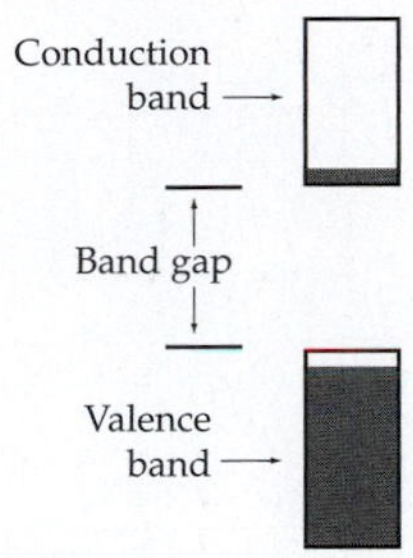

GaAs semiconductor

21.122 36.7 g $Y(OCH_2CH_3)_3$; 74.4 g $Ba(OCH_2CH_3)_2$; 109 g $YBa_2Cu_3O_7$ **21.124** (a) $6\ Al(OCH_2CH_3)_3 + 2\ Si(OCH_2CH_3)_4 + 26\ H_2O \rightarrow$

$\underbrace{6\ Al(OH)_3(s) + 2\ Si(OH)_4(s)}_{\text{sol}} + 26\ HOCH_2CH_3$;

(b) H_2O is eliminated from the sol through a series of reactions linking the sol particles together through a three-dimensional network of O bridges to form the gel. $(HO)_2Al{-}O{-}H + H{-}O{-}Si(OH)_3 \rightarrow (HO)_2Al{-}O{-}Si(OH)_3 + H_2O$; (c) The remaining H_2O and solvent are removed from the gel by heating to produce the ceramic, $3\ Al_2O_3 \cdot 2\ SiO_2$.

21.126

(a)

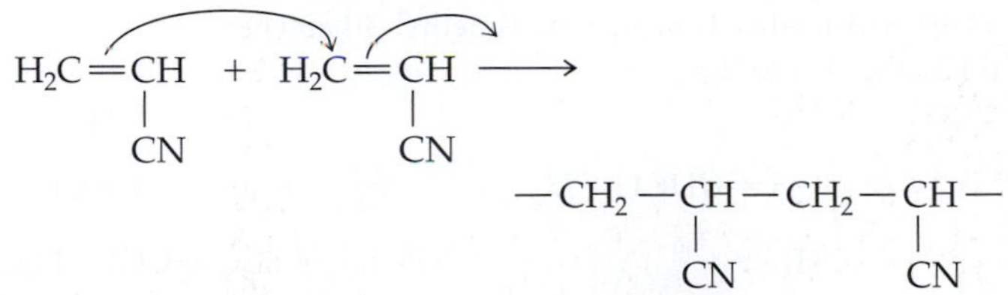

(b) −89 kJ/unit; exothermic

21.128 (a)

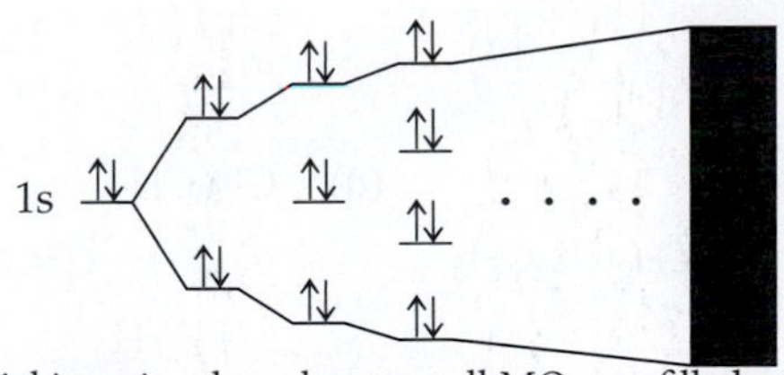

This material is an insulator because all MOs are filled, preventing the movement of electrons.

(b)

Neutral hydrogen atoms have only 1 valence electron, compared with 2 in H^-. Partially empty antibonding MOs will allow the movement of electrons, so the doped material will be a conductor. (c) The missing electrons in the doped material are positive "holes"; *p*-type *Multiconcept Problems* **21.130** (a) 2.47×10^{-3}cm; (b) 6.11×10^4 unit cells thick **21.132** 74.36% Fe **21.134** (a) 0.954; (b) 2.096 (c) 9.6%; (d) 14.3°; (e) The presence of Fe^{3+} in the semiconductor leads to missing electrons ("positive holes"). This type of doped material is a *p*-type semiconductor.

21.136 (a) $\Delta G° = +12.6$ kJ; $K_p = 0.028$; (b) $\Delta G° = +45.4$ kJ; $K_p = 1.9 \times 10^{-5}$; (c) $\Delta S°$ is large and negative because as the reaction proceeds in the forward direction, the number of moles of gas decreases from four to one. Because $\Delta S°$ is negative, $-T\Delta S°$ is positive, and as *T* increases, $\Delta G°$ becomes more positive because $\Delta G° = \Delta H° - T\Delta S°$; (d) The reaction is exothermic because $\Delta H°$ is negative. $Ni(s) + 4\ CO(g) \rightleftharpoons Ni(CO)_4(g) +$ heat. Heat is added as the temperature is raised and the reaction proceeds in the reverse direction to relieve this stress, as predicted by Le Châtelier's principle. As the reverse reaction proceeds, the partial pressure of CO increases and the partial pressure of $Ni(CO)_4$ decreases. K_p decreases as calculated because $K_p = P_{Ni(CO)4}/(P_{CO})^4$.
21.138 (a) $(NH_4)_2Zn(CrO_4)_2(s) \rightarrow ZnCr_2O_4(s) + N_2(g) + 4\ H_2O(g)$; (b) 7.251 g; (c) 7.35 L; (d) 1/8 of the tetrahedral holes and 1/2 of the octahedral holes are filled.

(e)

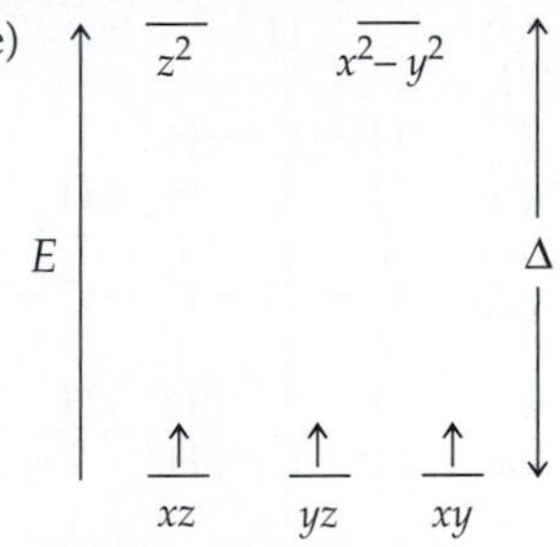

Octahedral Cr^{3+} has three unpaired electrons in the lower-energy *d* orbitals (*xy, xz, yz*). Cr^{3+} can absorb visible light to promote one of these *d* electrons to one of the higher-energy *d* orbitals making this compound colored. All of the *d* orbitals in Zn^{2+} are filled and no *d* electrons can be promoted; consequently the Zn^{2+} ion does not contribute to the color. **21.140** 1.3×10^8

CHAPTER 22

22.1 (a) ^{199}Au decays by beta emission. ^{173}Au decays by alpha emission; (b) ^{196}Pb decays by positron emission. ^{206}Pb is nonradioactive.
22.2 (a) Mass defect $= 2.2748 \times 10^{-25}$ g, 0.136 99 g/mol
Binding energy $=$ 8.00 MeV/nucleon $= 1.233 \times 10^{10}$kJ/mol
22.3 -9.11×10^{-9} g **22.4** 1.79×10^{10} kJ/mol **22.5** 5.31×10^8 kJ/mol
22.6 $^{40}_{18}Ar + ^{1}_{1}p \rightarrow ^{40}_{19}K + ^{1}_{0}n$ **22.7** $^{238}_{92}U + ^{2}_{1}H \rightarrow ^{238}_{93}Np + 2^{1}_{0}n$
22.8 1.53×10^4 y **22.9** Two billion years ago uranium deposits with the fissionable ^{235}U isotope at a natural abundance of about 3% were flooded with groundwater, which acted as a moderator to slow the neutrons released by fission of ^{235}U, allowing a nuclear chain reaction to take place. Today, however, the natural abundance of ^{235}U is only about 0.7% because of nuclear decay over the past 2 billion years. At that natural abundance, a chain reaction is no longer self sustaining.
Section Problems **22.10** ^{160}W is neutron poor and decays by alpha emission. ^{185}W is neutron rich and decays by beta emission.**22.12** Nuclei that are neutron-rich emit β particles to decrease the number of neutrons and increase the number of protons. **22.14** The lost mass is converted into the binding energy that is used to hold the nucleons together.
22.16 0.000 828 nm **22.18** (a) 0.480 59 g/mol; (b) 0.854 99 g/mol
22.20 (a) 8.76 MeV/nucleon; (b) 8.74 MeV/nucleon
22.22 5.42×10^8 kJ/mol **22.24** 1.02×10^{-9} g **22.26** 9.87×10^7 kJ/mol
22.28 (a) $^{113}_{49}In$; (b) $^{13}_{7}N$ **22.30** $^{209}_{83}Bi + ^{58}_{26}Fe \rightarrow ^{266}_{109}Mt + ^{1}_{0}n$
22.32 carbon-12
Chapter Problems **22.34** 220 W **22.36** 4.12×10^9 kJ/mol
22.38 (a) 0.467 00 g/mol; 8.72 MeV/nucleon; (b) 0.600 15 g/mol; 8.76 MeV/nucleon; ^{64}Zn is more stable. **22.40** 1.77×10^9 kJ/mol
22.42 $^{238}_{92}U$; $+ ^{1}_{0}n \rightarrow ^{239}_{94}Pu + 2^{\ 0}_{-1}e$ **22.44** 7.80×10^{-14} J **22.46** (a) $^{100}_{43}Tc + ^{\ 0}_{-1}e \rightarrow ^{100}_{42}Mo$ (positron emission); $^{100}_{43}Tc + ^{\ 0}_{-1}e \rightarrow ^{100}_{42}Mo$ (electron capture); (b) positron emission, -8.3×10^7 kJ/mol; electron capture, 1.6×10^7 kJ/mol. Only electron capture is observed because only this process involves a mass decrease and a release of energy.
Multiconcept Problems **22.48** 11,000 y **22.50** 5.73 L

CHAPTER 23

23.1

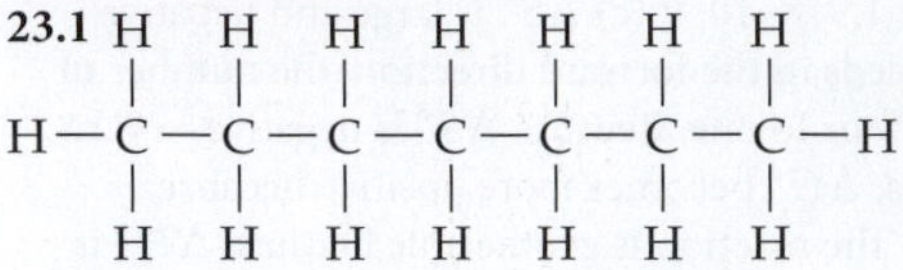

23.2

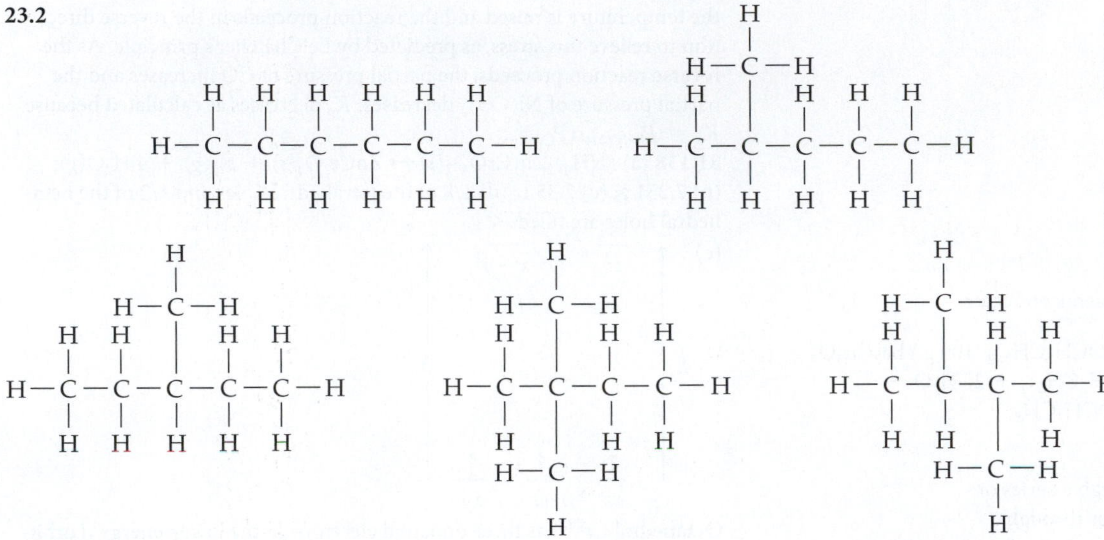

23.3 Structures (a) and (c) are identical.

23.4 C_7H_{16}

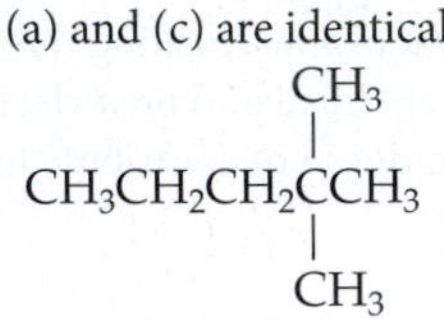

23.5

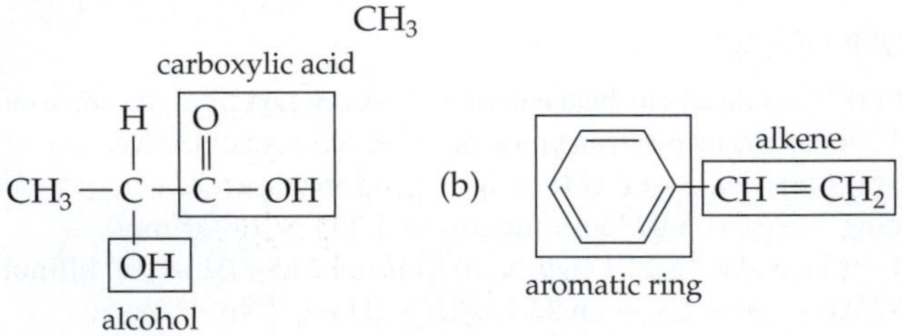

23.6 (a)

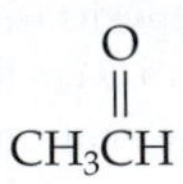

(b)

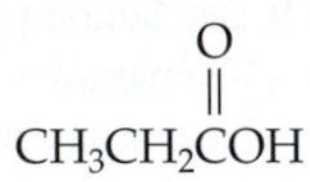

(c)

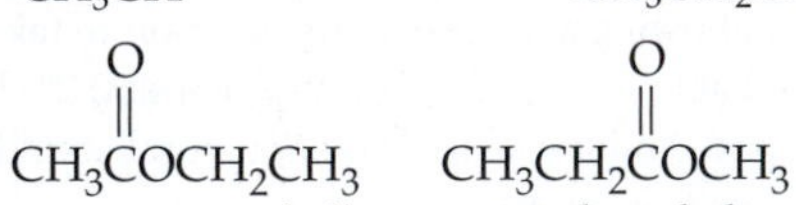

23.7 (a) pentane; 2-methylbutane; 2,2-dimethylpropane; (b) 3,4-dimethylhexane; (c) 2,4-dimethylpentane; (d) 2,2,5-trimethylheptane

23.8 (a)

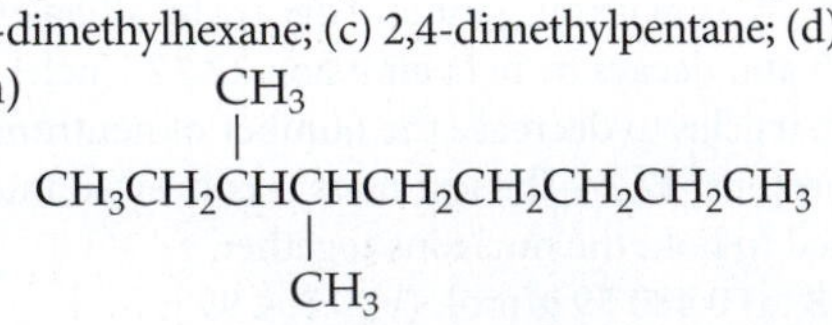

(b)

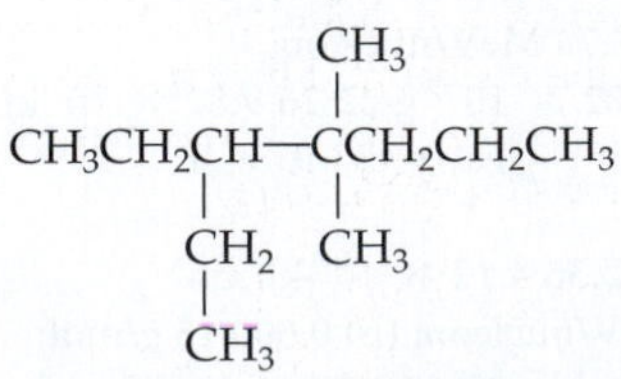

(c)

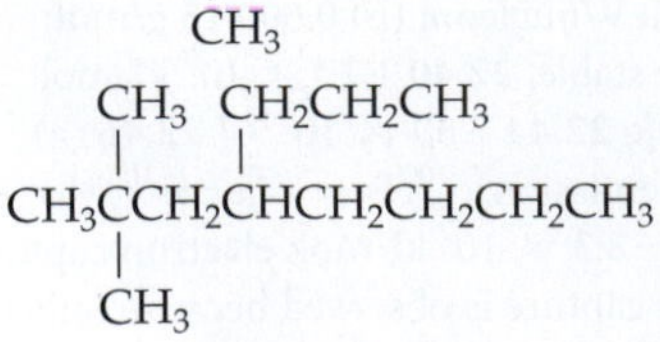

(d)

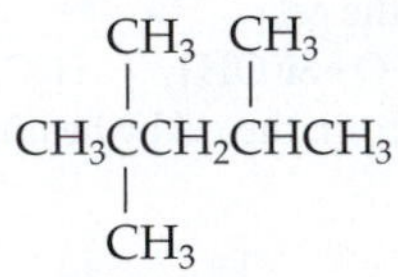

23.9 2,3-dimethylhexane

23.10 (a) 3-methyl-1-butene; (b) 4-methyl-3-heptene; (c) 3-ethyl-1-hexyne;

23.11

(a)

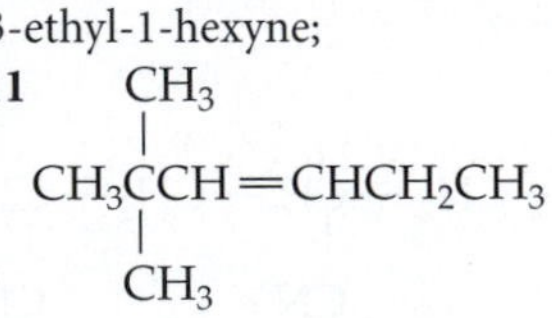

(b)

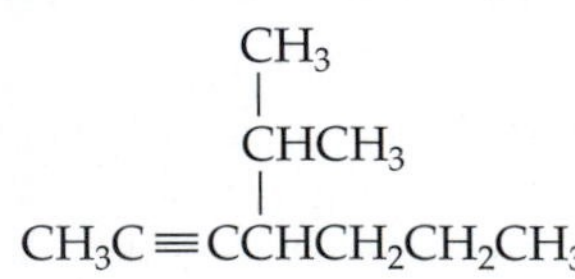

(c)

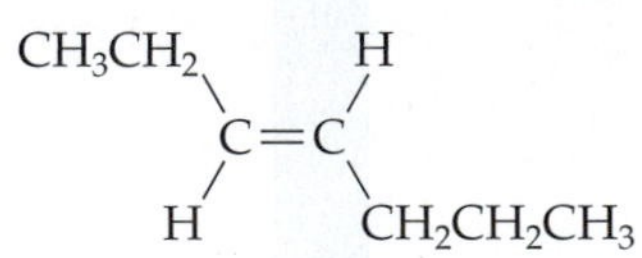

(d)

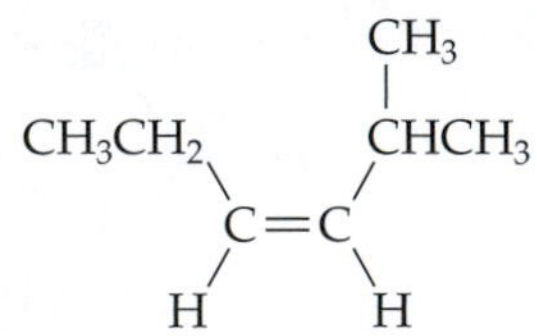

23.12 (a) $CH_3CH_2CH_2CH_3$ (b)

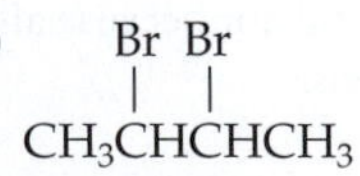

(c)

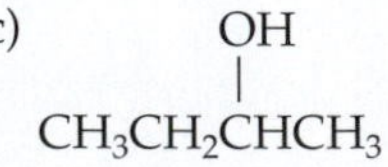

23.13

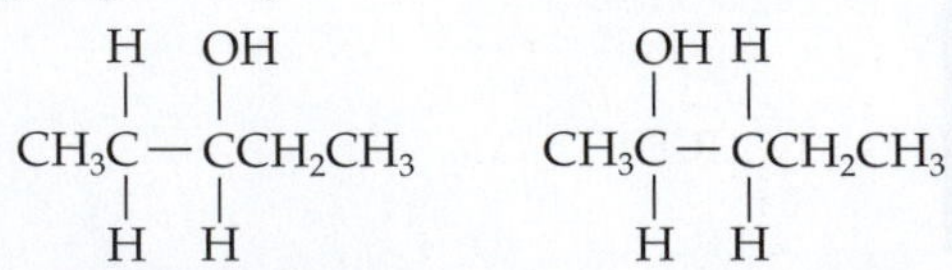

23.14 (a) 1,4-dimethylcyclohexane; (b) 1-ethyl-3-methylcyclopentane; (c) isopropylcyclobutane

23.15 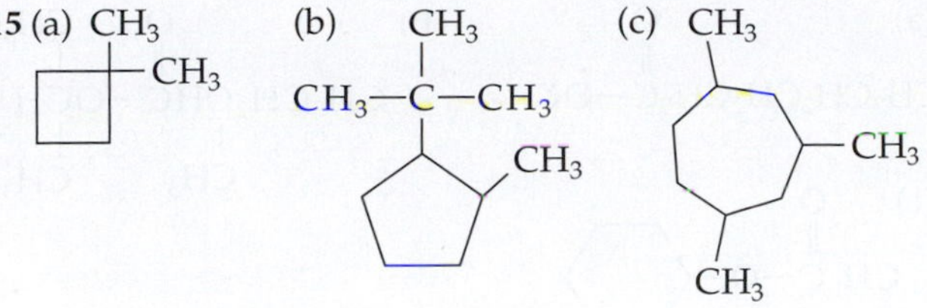

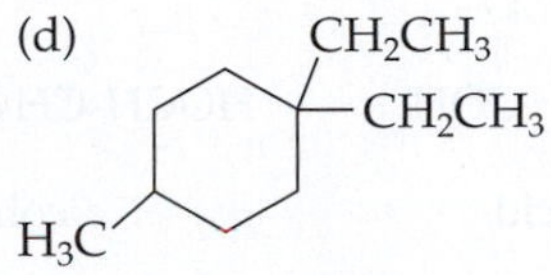

23.16 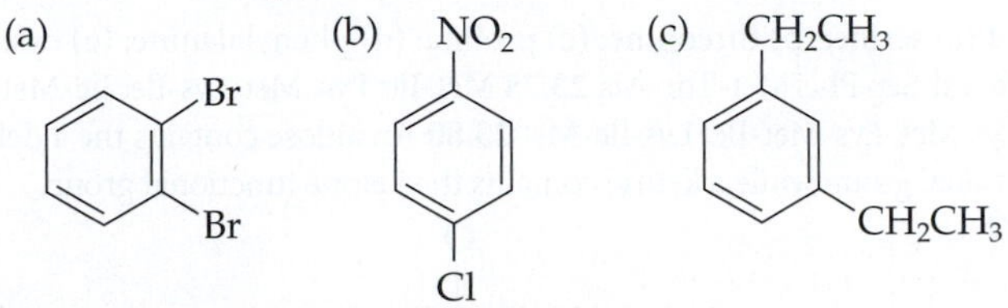

23.17 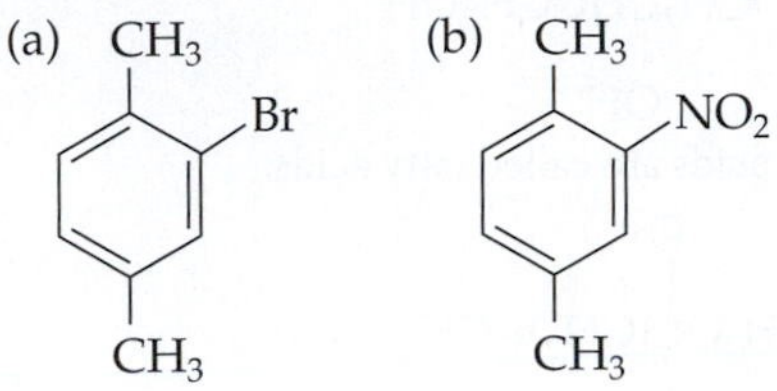

23.18 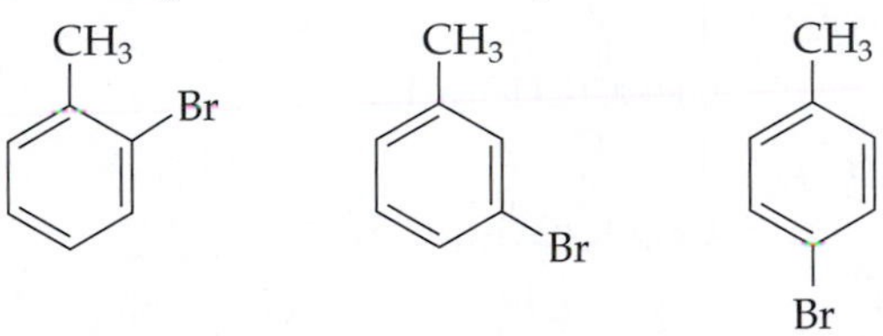

23.19 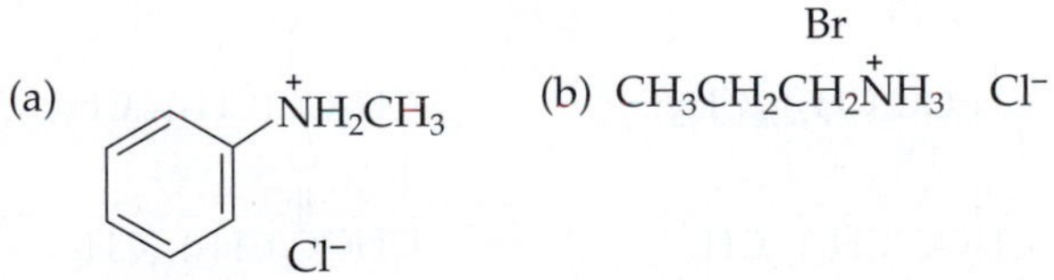

23.20 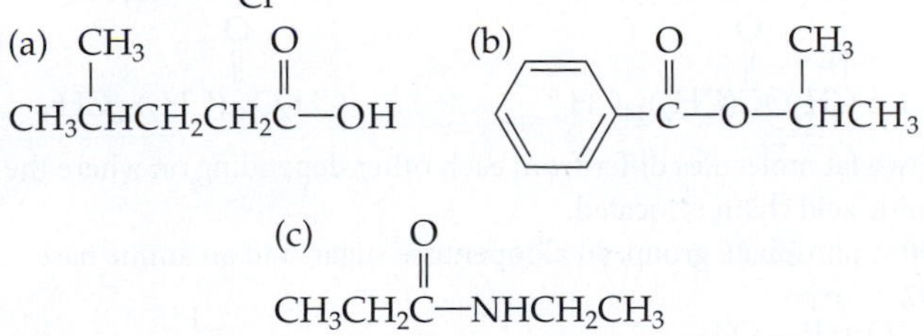

23.21 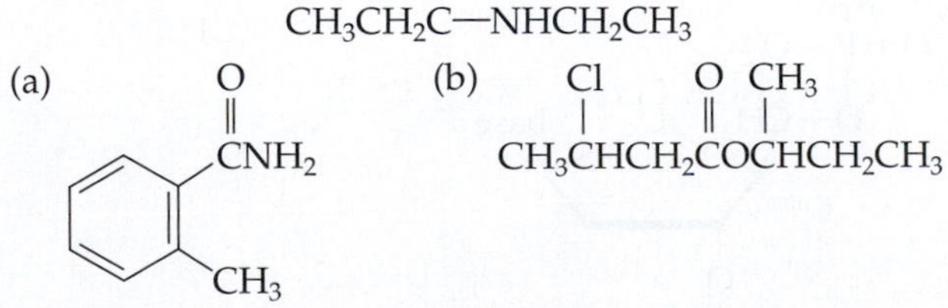

23.22 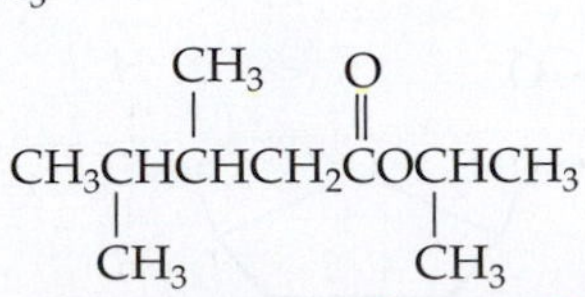

23.23 Amino acids that contain an aromatic ring: phenylalanine, tryptophan, tyrosine. Amino acids that contain sulfur: methionine, cysteine. Amino acids that are alcohols: serine, threonine, tyrosine. Amino acids that have alkyl-group side chains: alanine, valine, isoleucine, leucine.

23.24 Val-Cys Cys-Val

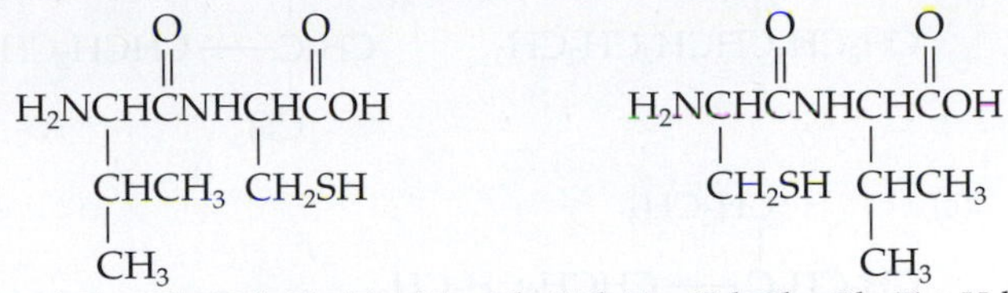

23.25 Val-Tyr-Gly; Val-Gly-Tyr; Tyr-Gly-Val; Tyr-Val-Gly; Gly-Tyr-Val; Gly-Val-Tyr **23.26** (a) aldopentose; (b) ketotriose; (c) aldotetrose

23.27

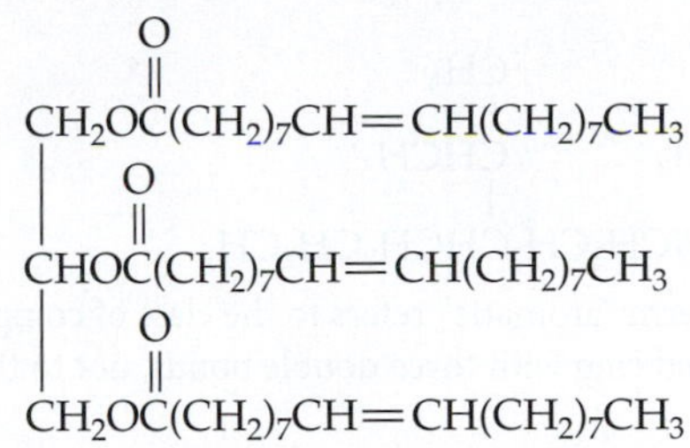

23.28 Original: G-G-C-C-C-G-T-A-A-T
Complement: C-C-G-G-G-C-A-T-T-A

23.29 C-G-T-G-A-T-T-A-C-A (DNA)
G-C-A-C-U-A-A-U-G-U (RNA)

23.30 U-G-C-A-U-C-G-A-G-U (RNA)
A-C-G-T-A-G-C-T-C-A (DNA)

23.31 (a) adenine (DNA, RNA); (b) thymine (DNA)

23.32

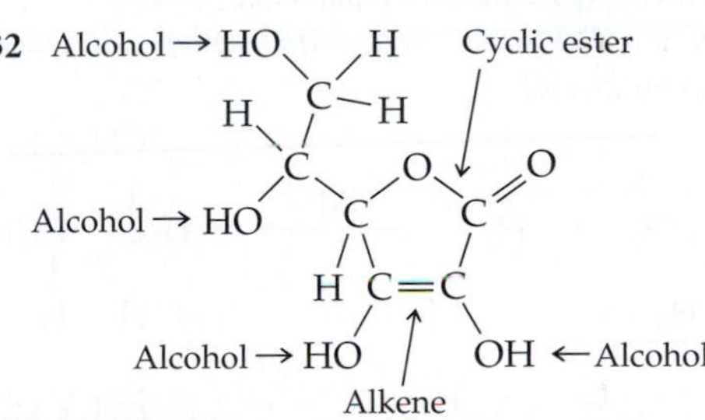

Conceptual Problems **23.34** (a) alkene, ketone, ether; (b) alkene, amine, carboxylic acid **23.36** (a) serine; (b) methionine **23.38** Ser-Val

Section Problems **23.40** A functional group is a part of a larger molecule and is composed of an atom or group of atoms that has a characteristic chemical behavior. They are important because their chemistry controls the chemistry in molecules that contain them.

23.42 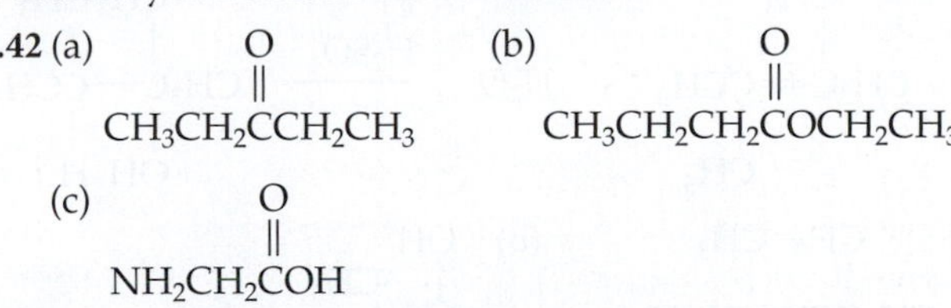

23.44 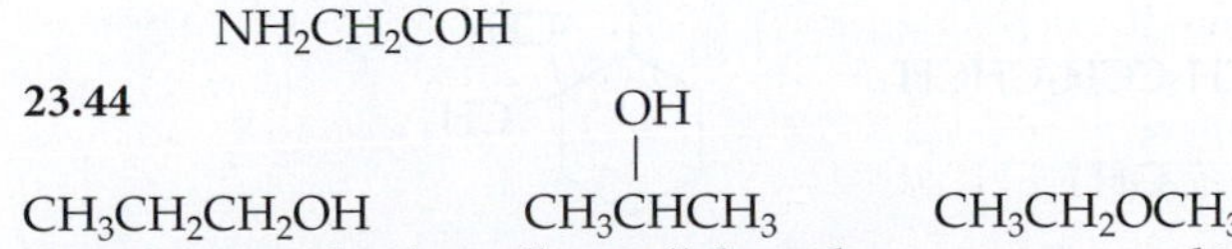

23.46 In a straight-chain alkane, all the carbons are connected in a row. In a branched-chain alkane, there are branching connections of carbons along the carbon chain. **23.48** C_3H_9 contains one more H than needed for an alkane. **23.50** (a) 4-ethyl-3-methyloctane; (b) 4-isopropyl-2-methylheptane; (c) 2,2,6-trimethylheptane; (d) 4-ethyl-4-methyloctane

23.52

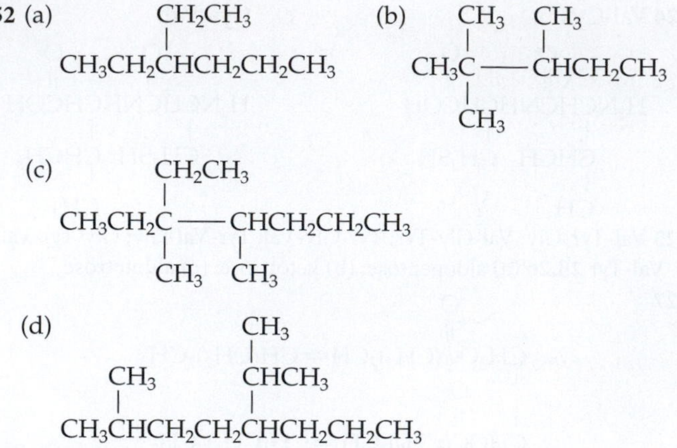

23.54 Today the term "aromatic" refers to the class of compounds containing a six-membered ring with three double bonds, not to the fragrance of a compound.

23.56 (a) $CH_3CH{=}CHCH_2CH_3$ (b) $HC{\equiv}CCH_2CH_3$; (c)

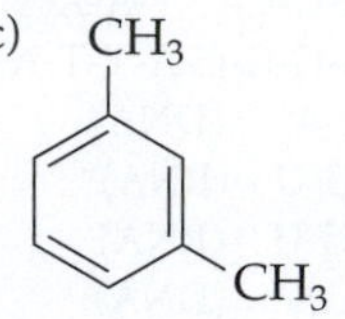

23.58 (a) 4-methyl-2-pentene; (b) 3-methyl-1-pentene; (c) 1,2-dichlorobenzene, or o-dichlorobenzene; (d) 2-methyl-2-butene; (e) 7-methyl-3-octyne **23.60** (b); (c)

23.62

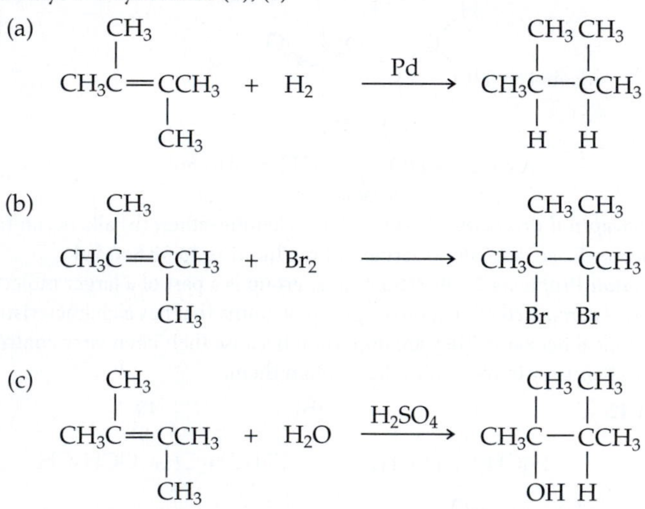

23.64

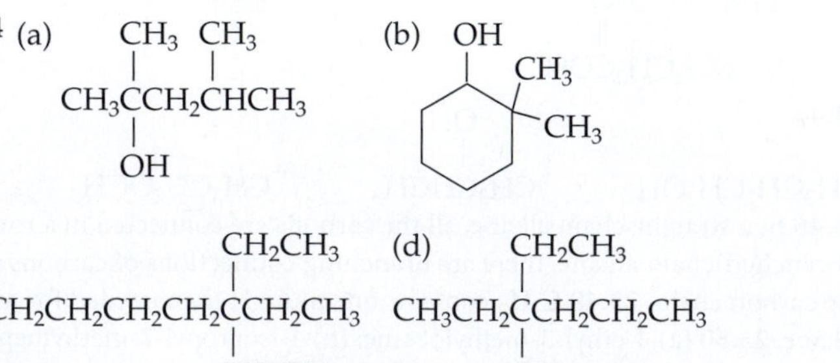

23.66 An aldehyde has a terminal carbonyl group. A ketone has the carbonyl group located between two carbon atoms. **23.68** (a) ketone; (b) aldehyde; (c) ketone; (d) amide; (e) ester

23.70

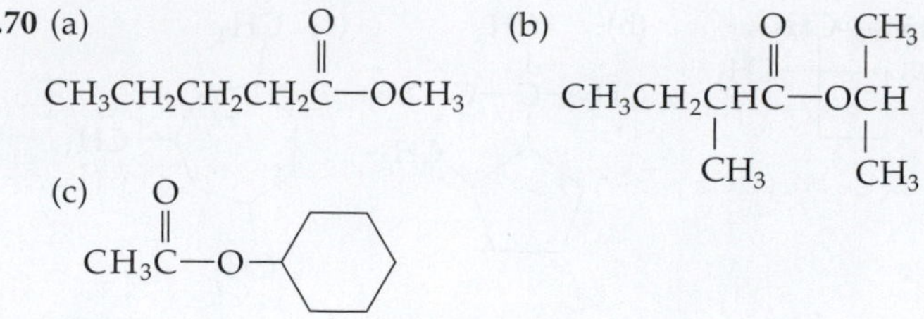

23.72 amine, aromatic ring, and ester

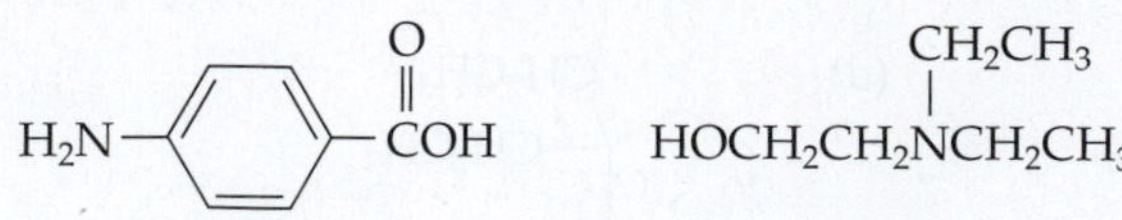

23.74 (a) serine; (b) threonine; (c) proline; (d) phenylalanine; (e) cysteine **23.76** Val-Ser-Phe-Met-Thr-Ala **23.78** Met-Ile-Lys, Met-Lys-Ile, Ile-Met-Lys, Ile-Lys-Met, Lys-Met-Ile, Lys-Ile-Met **23.80** An aldose contains the aldehyde functional group while a ketose contains the ketone functional group.

23.82

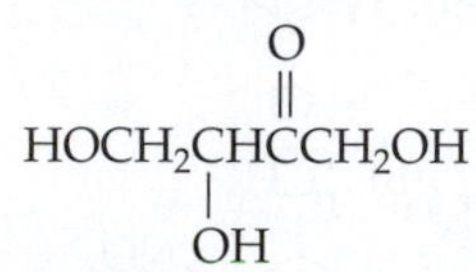

23.84 Long-chain carboxylic acids are called fatty acids.

23.86

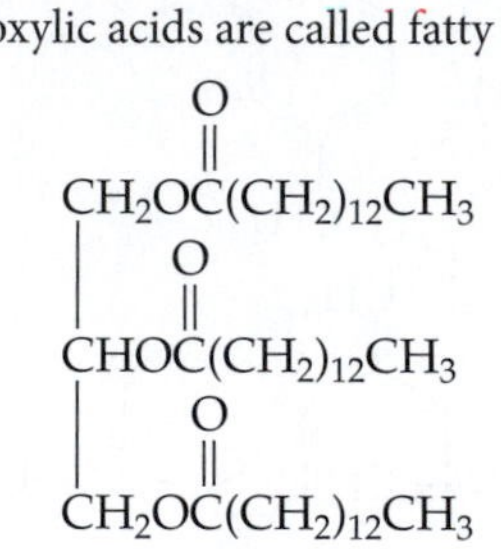

23.88

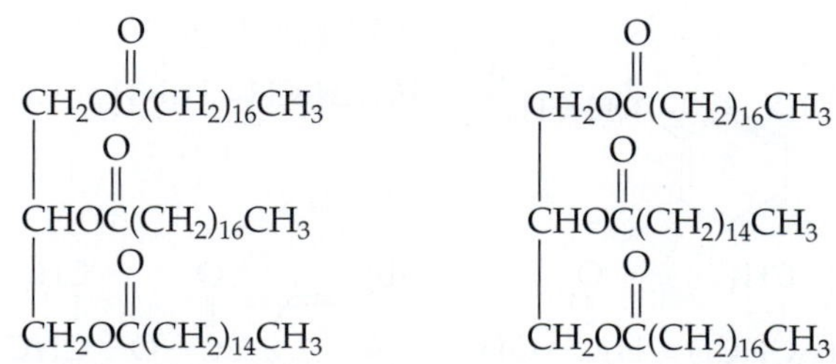

The two fat molecules differ from each other depending on where the palmitic acid chain is located.

23.90 A phosphate group, an aldopentose sugar, and an amine base.

23.92

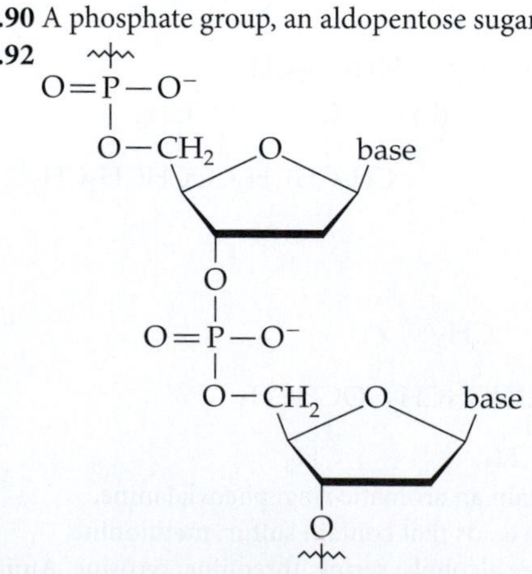

23.94 Original: T-A-C-C-G-A
Complement: A-T-G-G-C-T

23.96 It takes three nucleotides to code for a specific amino acid. In insulin the 21 amino acid chain would require 63 nucleotides and the 30 amino acid chain would require 90 nucleotides.

Chapter Problems
23.98
(a)
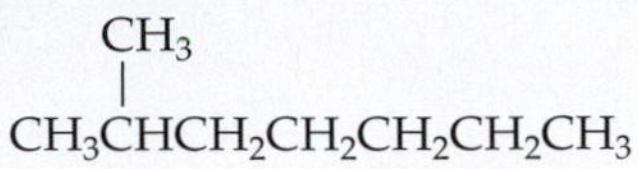
(b)
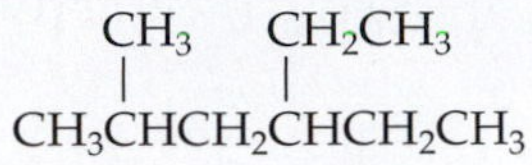
(c)
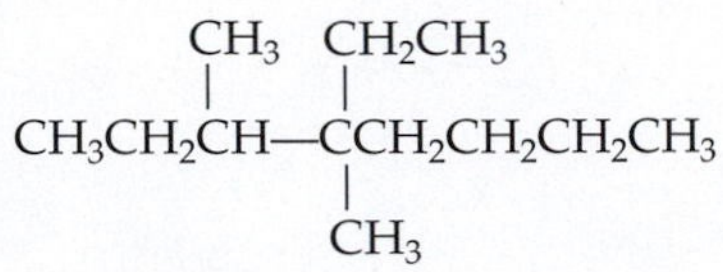
(d)
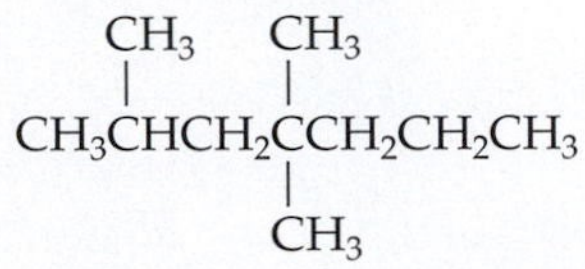
(e)
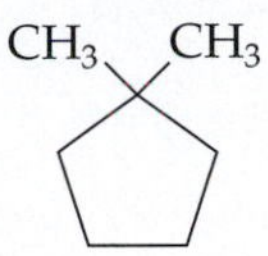
(f)
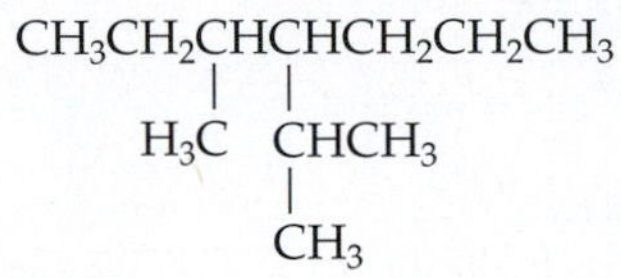

23.100
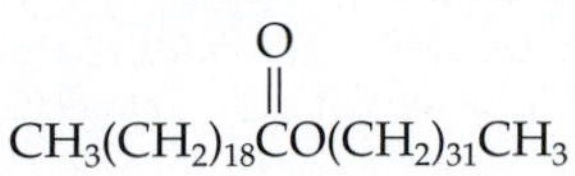

23.102 (a)
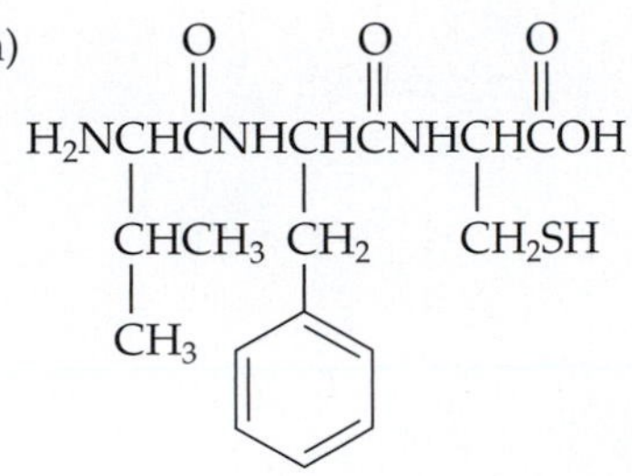
(b)
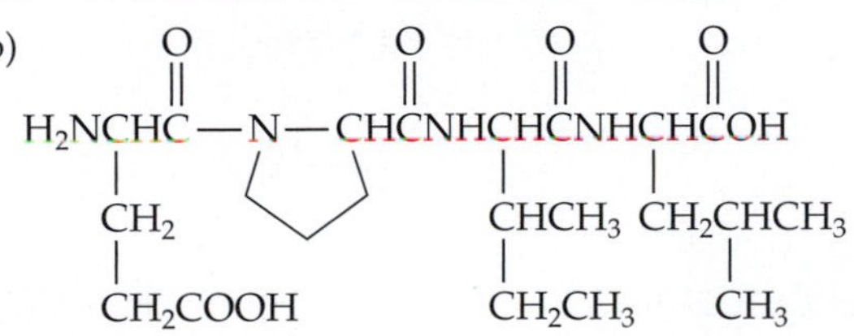

23.104 Original: A-G-T-T-C-A-T-C-G
Complement: T-C-A-A-G-T-A-G-C

23.106 (a) Amine, ester, and arene;
(b)
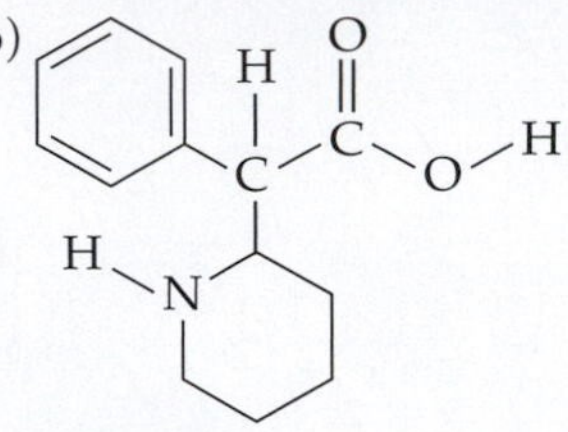
and CH_3OH

Multiconcept Problems **23.108** (a) CHO; (b) 116;
(c)
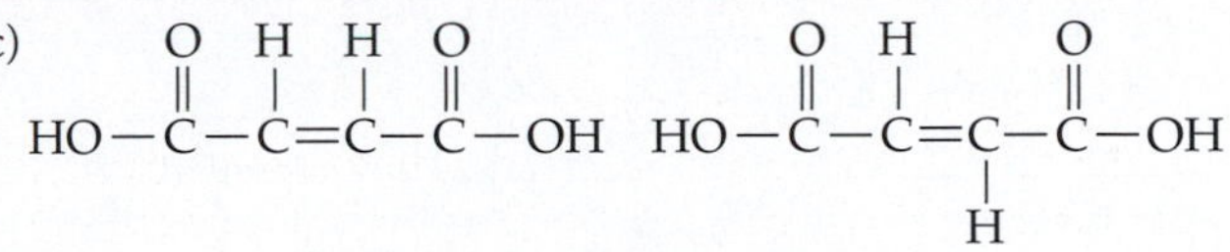
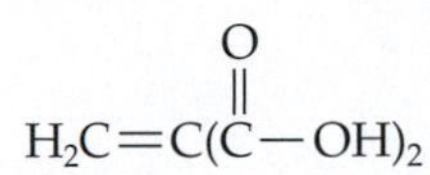
(d)
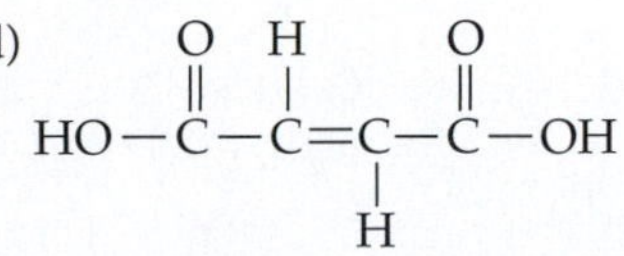

23.110 (a) 1.27 g I_2; (b) grams excess I_2 = 1.04 g, grams I_2 reacted = 0.23 g; (c) 46; (d) 1.4

GLOSSARY

Absorption spectrum a plot of the amount of light absorbed versus wavelength (*Section 20.10*)

Accuracy how close to the true value a given measurement is (*Section 0.9*)

Achiral lacking handedness (*Section 20.9*)

Acid a substance that can donate H^+ ions (*Section 7.5; Chapter 14*)

Acid–base indicator a substance that changes color in a specific pH range (*Section 14.6*)

Acid–base neutralization reaction a process in which an acid reacts with a base to yield water plus an ionic compound called a salt (*Sections 7.5, 15.1*)

Acid-dissociation constant (K_a) the equilibrium constant for the dissociation of an acid in water (*Section 14.8*)

Actinide one of the 14 inner transition metals following radium in the periodic table (*Section 1.2*)

Activation energy (E_a) the height of the energy barrier between reactants and products (*Section 12.9*)

Active site a small three-dimensional region of an enzyme with the specific shape necessary to bind the substrate and catalyze the appropriate reaction (*Chapter 12 FYI*)

Activity series a list of elements in order of their reducing ability in aqueous solution (*Section 7.8*)

Acyclic describes an open-chain compound; not part of a ring (*Section 23.5*)

Addition reaction an organic reaction in which a reagent adds to the multiple bond of the unsaturated reactant to yield a saturated product (*Section 23.4*)

Adenosine triphosphate (ATP) a molecule that is formed as the final result of food catabolism and plays a pivotal role in the production of biological energy (*Section 23.9*)

Advanced ceramic a ceramic material that has high-tech engineering, electronic, or biomedical applications (*Section 21.8*)

Alcohol an organic molecule that contains an –OH group (*Section 23.7*)

Aldehyde an organic molecule that contains one alkyl group and one hydrogen bonded to a C=O carbon (*Section 23.8*)

Alkali metal an element in group 1A of the periodic table (*Sections 1.3, 3.11*)

Alkaline-earth metal an element in group 2A of the periodic table (*Sections 1.3, 3.12*)

Alkane a compound that contains only carbon and hydrogen and has only single bonds (*Section 23.1*)

Alkene a hydrocarbon that has a carbon–carbon double bond (*Section 23.4*)

Alkyl group the part of an alkane that remains when a hydrogen is removed (*Section 23.3*)

Alkyne a hydrocarbon that has a carbon–carbon triple bond (*Section 23.4*)

Allotropes different structural forms of an element (*Section 10.10*)

Alloy a solid solution of two or more metals (*Section 21.2*)

Alpha (α) radiation a type of radioactive emission; a helium nucleus (*Sections 1.7, 1.11*)

Aluminosilicate a silicate mineral in which partial substitution of Si^{4+} by Al^{3+} has occurred (*Section 19.8*)

Amide an organic molecule that contains one alkyl group and one nitrogen bonded to a C=O carbon (*Section 23.8*)

Amine an organic derivative of ammonia (*Section 23.7*)

Amino acid a molecule that contains both a basic amine group ($-NH_2$) and an acidic carboxyl group ($-CO_2H$); the building block from which proteins are made (*Section 23.10*)

Amorphous solid a solid whose constituent particles are randomly arranged and have no ordered, long-range structure (*Section 10.6*)

Amphoteric exhibiting both acidic and basic properties (*Sections 15.12, 18.8*)

Amplitude a wave's height measured from the midpoint between peak and trough (*Section 2.1*)

Anabolism metabolic reaction sequences that put building blocks together to assemble larger molecules (*Section 23.9*)

Angular-momentum quantum number (l) a variable in the solutions to the Schrödinger wave equation that gives the three-dimensional shape of an orbital (*Section 2.6*)

Anion a negatively charged atom or group of atoms (*Section 3.1*)

Anode the electrode at which oxidation takes place (*Section 17.1*)

Anodizing the oxidation of a metal anode to yield a protective metal oxide coat (*Chapter 17 FYI*)

Antibonding molecular orbital a molecular orbital that is higher in energy than the atomic orbitals it is derived from (*Section 5.5*)

Aqueous solution a solution with water as solvent (*Chapter 7 Introduction*)

Aromatic compound the class of compounds related to benzene (*Section 23.6*)

Arrhenius acid a substance that provides H^+ ions when dissolved in water (*Sections 7.5, 14.1*)

Arrhenius base a substance that provides OH^- ions when dissolved in water (*Sections 7.5, 14.1*)

Arrhenius equation an equation relating reaction rate constant, temperature, and activation energy (*Section 12.9*)

Atmosphere (atm) a common unit of pressure measurement; standard atmospheric pressure at sea level is defined as exactly 760 mm Hg (*Section 9.1*)

Atom the smallest particle that retains the chemical properties of an element (*Chapter 1 Introduction, Section 1.5*)

Atomic mass the mass of a specific atom in unified atomic mass units (*Section 1.9*)

Atomic mass unit (amu) an older name for the unified atomic mass unit; 1/12th the mass of a $^{12}_{6}C$ atom (*Section 1.9*)

Atomic number (Z) the number of protons in an atom's nucleus (*Section 1.8*)

Atomic weight the weighted average mass of an element's naturally occurring atoms (*Section 1.9*)

Aufbau principle a set of rules that guides the electron filling order of orbitals in atoms (*Section 2.11*)

Avogadro's law The volume of a gas at a fixed pressure and temperature is proportional to its molar amount (*Section 9.2*).

Avogadro's number (N_A) the number of units in a mole; 6.022×10^{23} (*Section 1.9*)

Balanced said of a chemical equation in which the numbers and kinds of atoms are the same on both sides of the reaction arrow (*Section 6.2*) or of a nuclear equation in which the number of nuclear particles and charges is the same on both sides (*Section 1.11*).

Balmer–Rydberg equation an equation that accounts for all lines in the hydrogen spectrum (*Section 2.2*)

Band gap the energy difference between the bonding MOs in the valence band and the antibonding MOs in the conduction band of a semiconductor (*Section 21.5*)

Band theory the molecular orbital theory for metals (*Section 21.4*)

Base a substance that can accept H^+ ions (*Section 7.5; Chapter 14*)

Base-dissociation constant (K_b) the equilibrium constant for the reaction of a base with water (*Section 14.12*)

Basic oxygen process a method for purifying iron and converting it to steel (*Section 21.3*)

Battery *see* Galvanic cell (*Sections 17.1, 17.9*)

Bayer process purification of Al_2O_3 by treating bauxite with hot aqueous NaOH (*Section 18.8*)

Beta (β) radiation a type of radioactive emission consisting of electrons (*Section 1.11*)

Bidentate ligand a ligand that bonds to a metal using electron pairs on two donor atoms (*Section 20.6*)

Bimolecular reaction an elementary reaction that results from a collision between two reactant molecules (*Section 12.11*)

Binary compound a compound made up of only two elements (*Section 3.2*)

Binary hydride a compound that contains hydrogen and one other element (*Section 18.4*)

Binding energy the energy that holds nucleons together in the nucleus of an atom (*Section 22.1*)

Biochemistry the chemistry of living organisms (*Chapter 23 Introduction*)

Biofuel a fuel such as biodiesel, made from a renewable plant source (*Chapter 8 FYI*)

Blast furnace a huge reactor in which iron is produced by reduction of iron ore with carbon monoxide (*Section 21.3*)

Bleaching decolorizing a colored material (*Section 7.11*)

Body-centered cubic packing a packing arrangement of spheres into a body-centered cubic unit cell (*Section 10.8*)

Body-centered cubic unit cell a cubic unit cell with an atom at each of its eight corners and an additional atom in the center of the cube (*Section 10.8*)

Boiling point the temperature at which liquid and vapor coexist in equilibrium (*Section 10.4*)

Bond angle the angle at which two adjacent bonds intersect (*Section 5.1*)

Bond dissociation energy (D) the amount of energy necessary to break a chemical bond in an isolated molecule in the gaseous state (*Section 4.3*)

Bond length the minimum-energy distance between nuclei in a covalent bond (*Section 4.2*)

Bond order the number of electron pairs shared between two bonded atoms (*Section 4.6*)

Bonding electron pair a pair of valence electrons in a covalent bond (*Section 4.6*)

Bonding molecular orbital a molecular orbital that is lower in energy than the atomic orbitals it is derived from (*Section 5.5*)

Borane a compound of boron and hydrogen (*Section 19.4*)

Born–Haber cycle a pictorial way of viewing the energy changes in the various steps during formation of an ionic solid from its elements (*Section 3.9*)

Boyle's law The volume of a fixed amount of gas at a constant temperature varies inversely with its pressure (*Section 9.2*).

Bragg equation an equation used in X-ray crystallography for calculating the distance between atoms in a crystal (*Section 10.7*)

Branched-chain alkane an alkane with a branching connection of carbons (*Section 23.1*)

Brønsted–Lowry acid a substance that can transfer H^+ to a base in an acid–base reaction (*Section 14.1*)

Brønsted–Lowry base a substance that can accept H^+ from an acid in an acid–base reaction (*Section 14.1*)

Buffer capacity a measure of the amount of acid or base that a buffer can absorb without a significant change in pH (*Section 15.3*)

Buffer solution a solution of a weak acid and its conjugate base that resists drastic changes in pH (*Section 15.3*)

C-Terminal amino acid the amino acid with a free $-CO_2H$ group on the end of a protein chain (*Section 23.10*)

Carbide a carbon compound in which the carbon atom has a negative oxidation state (*Section 19.7*)

Carbohydrate a large class of organic molecules commonly called sugars and related to glucose (*Section 23.11*)

Carbonyl group the C=O group (*Section 23.8*)

Carboxylate anion the anion $-CO_2^-$ that results from deprotonation of a carboxylic acid (*Section 23.8*)

Carboxylic acid an organic molecule that contains the $-CO_2H$ group (*Section 23.8*)

Catabolism metabolic reaction sequences that break molecules apart (*Section 23.9*)

Catalyst a substance that increases the rate of a reaction without itself being consumed (*Section 12.14*)

Cathode the electrode at which reduction takes place (*Section 17.1*)

Cathode ray the visible glow emitted when an electric potential is applied across two electrodes in an evacuated chamber (*Section 1.6*)

Cathodic protection a technique for protecting a metal from corrosion by connecting it to a second metal that is more easily oxidized (*Section 17.11*)

Cation a positively charged atom or group of atoms (*Section 3.1*)

Cell potential (E) *see* Electromotive force (*Section 17.3*)

Cell voltage *see* Electromotive force (*Section 17.3*)

Celsius degree (°C) a common unit of temperature; 0 °C = 273.15 K (*Section 0.5*)

Centimeter (cm) a common unit of length; 1 cm = 0.01 m (*Section 0.4*)

Ceramic an inorganic, nonmetallic, nonmolecular solid (*Section 21.8*)

Ceramic composite a hybrid material made of two ceramics (*Section 21.9*)

Ceramic–metal composite a hybrid material made of a metal reinforced with a ceramic (*Section 21.9*)

Ceramic–polymer composite a hybrid material made of a polymer reinforced with a ceramic (*Section 21.9*)

Chain reaction a self-sustaining reaction whose product initiates further reaction (*Section 22.2*)

Change of state *see* Phase change (*Section 10.4*)

Charles's law The volume of a fixed amount of gas at a constant pressure varies directly with its absolute temperature (*Section 9.2*).

Chelate the cyclic complex formed by a metal atom and a polydentate ligand (*Section 20.6*)

Chelating agent a polydentate ligand (*Section 20.6*)

Chemical bond the force that holds atoms together in chemical compounds (*Section 3.1*)

Chemical compound a chemical substance composed of atoms of more than one element (*Section 1.4*)

Chemical energy potential energy stored in chemical bonds (*Section 8.1*)

Chemical equation a format for writing a chemical reaction, listing reactants on the left, products on the right, and an arrow between them (*Section 1.4*)

Chemical equilibrium the state reached when the concentrations of reactants and products remain constant in time (*Section 13.1*)

Chemical formula a format for listing the number and kind of constituent elements in a compound (*Section 1.4*)

Chemical kinetics the area of chemistry concerned with reaction rates and the sequence of steps by which reactions occur (*Chapter 12 Introduction*)

Chemical property a characteristic that results in a change in the chemical makeup of a sample (*Section 1.3*)

Chemical reaction the transformation of one substance into another (*Section 1.4*)

Chemistry the study of the composition, properties, and transformations of matter (*Chapter 0*)

Chiral having handedness (*Section 20.9*)

Chlor-alkali industry the commercial method of production for Cl_2 and NaOH by electrolysis of aqueous sodium chloride (*Section 17.13*)

Chromosome a threadlike strand of DNA in the nucleus of cells (*Section 23.13*)

Cis isomer the isomer of a metal complex or alkene in which identical ligands or groups are adjacent rather than opposite (*Sections 20.8, 23.4*)

Clausius–Clapeyron equation a mathematical relationship between vapor pressure and heat of vaporization for a substance (*Section 10.5*)

Coefficient a number placed before a formula in a chemical equation to indicate how many formula units are required to balance the equation (*Section 6.2*)

Colligative property a property that depends only on the amount of dissolved solute rather than on the chemical identity of the solute (*Section 11.5*)

Collision theory a model by which bimolecular reactions occur when two properly oriented reactant molecules come together in a sufficiently energetic collision (*Section 12.9*)

Colloid a homogeneous mixture containing particles with diameters in the range 2–500 nm (*Section 11.1*)

Combustion a chemical reaction that sustains a flame (*Sections 7.11, 8.5*)

Common-ion effect the shift in the position of an equilibrium on addition of a substance that provides an ion in common with one of the ions already involved in the equilibrium (*Section 15.2*)

Complex ion an ion that contains a metal cation bonded to one or more small molecules or ions (*Section 15.12*)

Condensation the change of a gas to a liquid (*Section 10.4*)

Condensed structure a shorthand method for drawing organic structures in which C–H and C–C single bonds are "understood" rather than shown (*Section 23.1*)

Conduction band the antibonding molecular orbitals in a semiconductor (*Section 21.5*)

Conjugate acid the species HA formed by addition of H^+ to a base A^- (*Section 14.1*)

Conjugate acid–base pair chemical species whose formulas differ only by one proton (*Section 14.1*)

Conjugate base the species A^- formed by loss of H^+ from an acid HA (*Section 14.1*)

Constitutional isomers isomers that have different connections among their constituent atoms (*Section 20.8*)

Contact process the commercial process for making sulfuric acid from sulfur (*Section 19.13*)

Conversion factor an expression that describes the relationship between different units (*Section 0.11*)

Coordinate covalent bond a bond formed when one atom donates two electrons to another atom that has a vacant valence orbital (*Section 4.6*)

Coordination compound a compound in which a central metal ion is attached to a group of surrounding molecules or ions by coordinate covalent bonds (*Section 20.5*)

Coordination number the number of nearest-neighbor atoms in a crystal (*Section 10.8*) or the number of ligand donor atoms that surround a central metal ion in a complex (*Section 20.5*)

Core electrons inner-shell electrons (*Section 3.5*)

Corrosion the oxidative deterioration of a metal, such as the conversion of iron to rust (*Sections 7.11, 17.11*)

Cosmic ray a stream of energetic particles, primarily protons, coming from interstellar space (*Section 22.4*)

Coulomb's law The force resulting from the interaction of two electric charges is equal to a constant *k* times the product of the charges divided by the square of the distance between them (*Section 3.10*).

Covalent bond a bond that occurs when two atoms share several (usually two) electrons (*Sections 3.1, 4.2*)

Covalent hydride a compound in which hydrogen is attached to another element by a covalent bond (*Section 18.4*)

Covalent network solid a solid whose atoms are linked together by covalent bonds into a giant three-dimensional array (*Sections 10.6, 10.10*)

Critical mass the amount of material necessary for a nuclear chain reaction to become self-sustaining (*Section 22.2*)

Critical point a combination of temperature and pressure beyond which a gas cannot be liquefied (*Section 10.11*)

Crystal field splitting the energy splitting between two sets of *d* orbitals in a metal complex (*Section 20.12*)

Crystal field theory a model that views the bonding in metal complexes as arising from electrostatic interactions and considers the effect of the ligand charges on the energies of the metal ion *d* orbitals (*Section 20.12*)

Crystalline solid a solid whose atoms, ions, or molecules have an ordered arrangement extending over a long range (*Section 10.6*)

Cubic centimeter (cm^3) a common unit of volume, equal in size to the milliliter; $1\ cm^3 = 10^{-6}\ m^3$ (*Section 0.6*)

Cubic closest-packed a packing arrangement of spheres into a face-centered cubic unit cell with three alternating layers (*Section 10.8*)

Cubic meter (m^3) the SI unit of volume (*Section 0.6*)

Cycloalkane an alkane that contains a ring of carbon atoms (*Section 23.5*)

***d*-Block element** a transition metal element in which *d* orbitals are filled (*Section 2.13; Chapter 20*)

Dalton (Da) used in biological chemistry as an alternative name for the unified atomic mass unit, u (*Section 1.9*)

Dalton's law of partial pressures The total pressure exerted by a mixture of gases in a container at constant *V* and *T* is equal to the sum of the pressures exerted by each individual gas in the container (*Section 9.5*).

de Broglie hypothesis an equation that relates mass, wavelength, and velocity, $m = h/\lambda v$ (*Section 2.4*)

Decay constant the first-order rate constant for radioactive decay (*Section 12.6*)

Degenerate having the same energy level (*Section 2.11*)

Density an intensive physical property that relates the mass of an object to its volume (*Section 0.7*)

Deoxyribonucleic acid (DNA) an immense biological molecule, made up of deoxyribonucleotide units and containing an organism's genetic information (*Section 23.13*)

Deposition the change of a gas directly to a solid (*Section 10.4*)

Diamagnetic a substance that has no unpaired electrons and is weakly repelled by a magnetic field (*Section 5.6*)

Diastereoisomers non-mirror-image stereoisomers (*Section 20.8*)

Diffraction scattering of a light beam by an object containing regularly spaced lines or points (*Section 10.7*)

Diffusion the mixing of different gases by random molecular motion with frequent collisions (*Section 9.7*)

Dimensional-analysis method a method of problem solving whereby problems are set up so that unwanted units cancel (*Section 0.11*)

Diode a semiconductor device that permits electric current to flow in one direction but is highly resistant to current flow in the opposite direction (*Section 21.6*)

Dipole a pair of separated electrical charges (*Section 10.1*)

Dipole–dipole force an intermolecular force resulting from electrical interactions among dipoles on neighboring molecules (*Section 10.2*)

Dipole moment (μ) the measure of net molecular polarity; $\mu = Q \times r$ (*Section 10.1*)

Diprotic acid an acid that has two dissociable protons (*Sections 7.5, 14.7*)

Disproportionation reaction a reaction in which a substance is both oxidized and reduced (*Section 18.9*)

Dissociate split apart to give ions when dissolved in water (*Section 7.1*)

Donor atom the atom attached directly to a metal in a coordination compound (*Section 20.6*)

Doping the addition of a small amount of an impurity to increase the conductivity of a semiconductor (*Section 21.5*)

Double bond a covalent bond formed by sharing four electrons between atoms (*Section 4.6*)

Effective nuclear charge (Z_{eff}) the net nuclear charge actually felt by an electron (*Sections 2.10, 2.14*)

Effusion the escape of gas molecules through a tiny hole in a membrane without molecular collisions (*Section 9.7*)

Electrochemical cell a device for interconverting chemical and electrical energy (*Section 17.1*)

Electrochemistry the area of chemistry concerned with the interconversion of chemical and electrical energy (*Chapter 17*)

Electrode a conductor through which electrical current enters or leaves a cell (*Section 17.1*)

Electrolysis the process of using an electric current to bring about chemical change (*Section 17.12*)

Electrolyte a substance that dissolves in water to produce ions (*Section 7.1*)

Electrolytic cell an electrochemical cell in which an electric current drives a nonspontaneous reaction (*Section 17.12*)

Electromagnetic radiation radiant energy (*Section 2.1*)

Electromagnetic spectrum the range of different kinds of electromagnetic radiation (*Section 2.1*)

Electromotive force (emf) the electrical potential that pushes electrons away from the anode and pulls them toward the cathode (*Section 17.3*)

Electron a negatively charged, fundamental atomic particle (*Section 1.6*)

Electron affinity (E_{ea}) the energy change that occurs when an electron is added to an isolated atom in the gaseous state (*Section 3.7*)

Electron capture a nuclear reaction in which a proton in the nucleus captures an inner-shell electron and is thereby converted into a neutron (*Section 1.11*)

Electron configuration a description of which orbitals in an atom are occupied by electrons (*Section 2.11*)

Electron-dot structure a representation of a molecule that shows valence electrons as dots; also called a Lewis structure (*Section 4.6*)

Electron-sea model a model that visualizes metals as a three-dimensional array of metal cations immersed in a sea of delocalized electrons that are free to move about (*Section 21.4*)

Electronegativity (EN) the ability of an atom in a molecule to attract the shared electrons in a covalent bond (*Section 4.4*)

Electroplating the coating of one metal on the surface of another using electrolysis (*Section 17.13*)

Electrorefining the purification of a metal by means of electrolysis (*Section 17.13*)

Element a fundamental substance that can't be chemically changed or broken down into anything simpler (*Sections 1.1, 1.4*)

Elemental analysis a technique for determining the identities and amounts of elements in a compound (*Section 6.7*)

Elementary reaction a single chemical step in a reaction mechanism (*Section 12.11*)

Elementary step *see* Elementary reaction (*Section 12.11*)

Empirical formula a formula that gives the ratios of atoms in a chemical compound (*Section 6.6*)

Enantiomers stereoisomers that are nonidentical mirror images of each other (*Section 20.9*)

Endothermic a reaction in which heat is absorbed and the temperature of the surroundings falls (*Section 8.7*)

Energy the capacity to do work or supply heat (*Sections 0.8, 8.1*)

Enthalpy (*H*) the quantity $E + PV$ (*Section 8.4*)

Enthalpy change (ΔH) the heat change in a reaction or process at constant pressure; $\Delta H = \Delta E + P\Delta V$ (*Section 8.4*)

Enthalpy of solution (ΔH_{soln}) the enthalpy change during formation of a solution (*Section 11.2*)

Entropy (*S*) the amount of molecular randomness in a system (*Sections 8.12, 16.3*)

Entropy of solution (ΔS_{soln}) the entropy change during formation of a solution (*Section 11.2*)

Enzyme a large protein that acts as a catalyst for a biological reaction (*Chapter 12 FYI; Section 23.10*)

Equilibrium constant (K_c) the constant in the equilibrium equation (*Section 13.2*)

Equilibrium constant (K_p) the equilibrium constant for reaction of gases, defined using partial pressures (*Section 13.4*)

Equilibrium equation an equation that relates the concentrations in an equilibrium mixture (*Section 13.2*)

Equilibrium mixture a mixture of reactants and products at equilibrium (*Section 13.1*)

Equivalence point the point in a titration at which stoichiometrically equivalent quantities of reactants have been mixed together (*Section 15.5*)

Ester an organic molecule that contains the $-CO_2R$ group (*Section 23.8*)

Ether an organic molecule that contains two alkyl groups bonded to the same oxygen atom (*Section 23.7*)

Exothermic a reaction in which heat is evolved and the temperature of the surroundings rises (*Section 8.7*)

Expansion work the work done in a chemical system as the result of a volume change (*Section 8.3*)

Extensive property a property whose value depends on the sample size (*Section 1.3*)

***f*-Block element** a lanthanide or actinide element, in which *f* orbitals are filled (*Section 2.13*)

Face-centered cubic unit cell a cubic unit cell with an atom at each of its eight corners and an additional atom on each of its six faces (*Section 10.8*)

Faraday the electrical charge on 1 mol of electrons (96,485 C/mol e^-) (*Section 17.3*)

Fatty acid a long-chain carboxylic acid found as a constituent of fats and oils (*Section 23.12*)

First law of thermodynamics The total internal energy of an isolated system is constant (*Sections 8.1, 8.2*).

First-order reaction a reaction whose rate depends on the concentration of a single reactant raised to the first power (*Section 12.4*)

Flotation a metallurgical process that exploits differences in the ability of water and oil to wet the surfaces of mineral and gangue (*Section 21.2*)

Formal charge an electron bookkeeping device that tells whether an atom in a molecule has gained or lost electrons compared to an isolated atom (*Section 4.10*)

Formation constant (K_f) the equilibrium constant for formation of a complex ion (*Section 15.12*)

Formula unit one unit (atom, ion, or molecule) corresponding to a given formula (*Section 6.2*)

Formula weight the sum of atomic weights of all atoms in one formula unit of a substance (*Section 6.3*)

Fractional distillation the separation of volatile liquids on the basis of boiling point (*Section 11.10*)

Free-energy change (ΔG) $\Delta G = \Delta H - T\Delta S$ (*Sections 8.13, 16.7*)

Free radical a compound that contains an odd number of electrons and therefore has an unpaired electron in a half-filled orbital (*Chapter 4 FYI*)

Freezing the change of a liquid to a solid (*Section 10.4*)

Frequency (ν) the number of wave maxima that pass by a fixed point per unit time (*Section 2.1*)

Frequency factor the parameter A (=pZ) in the Arrhenius equation (*Section 12.9*)

Fuel cell a galvanic cell in which one of the reactants is a traditional fuel such as methane or hydrogen (*Section 17.10*)

Fullerene a crystalline allotrope of carbon consisting of C_{60} molecules shaped like a soccer ball (*Section 19.7*)

Functional group a part of a larger molecule; composed of an atom or group of atoms that has characteristic chemical behavior (*Section 23.2*)

Fusion melting, or the change of a solid to a liquid (*Section 10.4*); also, the joining together of two nuclei in a nuclear reaction, accompanied by release of an enormous amount of energy (*Section 22.2*)

Galvanic cell an electrochemical cell in which a spontaneous chemical reaction generates an electric current (*Section 17.1*)

Galvanizing a process for protecting steel from corrosion by coating it with zinc (*Section 17.11*)

Gamma (γ) radiation a type of radioactive emission consisting of a stream of high-energy photons (*Section 1.11*)

Gangue the economically worthless material consisting of sand, clay, and other impurities that accompanies an ore (*Section 21.2*)

Gas constant (*R*) the constant in the ideal gas law $PV = nRT$ (*Section 9.3*)

Gas laws relationships among the variables *P*, *V*, *n*, and *T* for a gas sample (*Section 9.2*)

Gas-shift reaction the reaction of CO with H_2O to yield hydrogen (*Section 18.2*)

Gene a segment of a DNA chain that contains the instructions necessary to make a specific protein (*Section 23.13*)

Geometric isomers *see* Diastereoisomers (*Section 20.8*)

Gibbs free-energy change (ΔG) $\Delta G = \Delta H - T\Delta S$ (*Sections 8.13, 16.7*)

Graham's law The rate of effusion of a gas is inversely proportional to the square root of its molar mass (*Section 9.7*).

Gram (g) a common unit of mass; 1 g = 0.001 kg (*Section 0.3*)

Green chemistry a set of guidelines describing environmentally benign chemical reactions and practices (*Chapter 7 FYI*)

Ground-state electron configuration the lowest-energy electron configuration of an atom (*Section 2.11*)

Group a column of elements in the periodic table (*Section 1.2*)

Half-life ($t_{1/2}$) the time required for a reactant concentration to drop to one-half its initial value (*Section 12.5*)

Half-reaction the oxidation or reduction part of a redox reaction (*Sections 7.9, 17.1*)

Half-reaction method a method for balancing redox equations (*Section 7.9*)

Hall–Héroult process the commercial method for producing aluminum by electrolysis of a molten mixture of aluminum oxide and cryolite (*Section 17.13*)

Halogen an element in group 7A of the periodic table (*Sections 1.3, 3.13, 19.14*)

Halogenation the addition of halogen (Cl_2 or Br_2) to an alkene (*Section 23.4*)

Hard water water that contains appreciable concentrations of Ca^{2+}, Mg^{2+}, or Fe^{2+} cations (*Section 18.12*)

Heat the energy transferred from one object to another as the result of a temperature difference between them (*Section 8.1*)

Heat capacity (*C*) the amount of heat required to raise the temperature of an object or substance a given amount (*Section 8.5*)

Heat of combustion the amount of energy released on burning a substance (*Section 8.11*)

Heat of formation ($\Delta H°_f$) *see* Standard heat of formation (*Section 8.9*)

Heat of fusion (ΔH_{fusion}) the amount of heat required for melting a solid to a liquid (*Section 10.4*)

Heat of reaction (ΔH) the enthalpy change for a reaction (*Section 8.4*)

Heat of solution (ΔH_{soln}) the enthalpy change during formation of a solution (*Section 11.2*)

Heat of sublimation (ΔH_{subl}) the amount of heat required for sublimation of a solid to a gas (*Section 8.7*)

Heat of vaporization (ΔH_{vap}) the amount of heat required for vaporization of a liquid to a gas (*Section 10.4*)

Heisenberg uncertainty principle The position and the velocity of an electron can never both be known beyond a certain level of precision (*Section 2.5*).

Henderson–Hasselbalch equation an equation relating the pH of a solution to the pK_a of the weak acid; pH = pK_a + log ([base]/[acid]) (*Section 15.4*)

Henry's law The solubility of a gas in a liquid at a given temperature is directly proportional to the partial pressure of the gas over the solution (*Section 11.4*).

Hertz (Hz) a unit of frequency; $1\ Hz = 1\ s^{-1}$ (*Section 2.1*)

Hess's law The overall enthalpy change for a reaction is equal to the sum of the enthalpy changes for the individual steps in the reaction (*Section 8.8*).

Heterogeneous catalyst a catalyst that exists in a different phase than the reactants (*Section 12.15*)

Heterogeneous equilibria equilibria in which reactants and products are present in more than one phase (*Section 13.5*)

Heterogeneous mixture a mixture having regions with differing compositions (*Chapter 11 Introduction*)

Hexadentate ligand a ligand that bonds to a metal using six donor atoms (*Section 20.6*)

Hexagonal closest-packed a packing arrangement of spheres into a noncubic unit cell with two alternating layers (*Section 10.8*)

High-spin complex a metal complex in which the *d* electrons are arranged to give the maximum number of unpaired electrons (*Section 20.11*)

Homogeneous catalyst a catalyst that exists in the same phase as the reactants (*Section 12.15*)

Homogeneous equilibria equilibria in which all reactants and products are in a single phase, usually either gaseous or solution (*Section 13.5*)

Homogeneous mixture a mixture having a constant composition throughout (*Chapter 11 Introduction*)

Hund's rule If two or more degenerate orbitals are available, one electron goes in each until all are half full (*Section 2.11*).

Hybrid atomic orbital a wave function derived by combination of atomic wave functions (*Section 5.3*)

Hydrate a solid compound that contains water molecules (*Section 18.13*)

Hydration the addition of water to an alkene (*Section 23.4*)

Hydrocarbon a compound that contains only carbon and hydrogen (*Section 23.1*)

Hydrogen bond an attractive intermolecular force between a hydrogen atom bonded to an electronegative O, N, or F atom and an unshared electron pair on a nearby electronegative atom (*Section 10.2*)

Hydrogenation the addition of H_2 to an alkene to yield an alkane (*Section 23.4*)

Hydronium ion the protonated water molecule, H_3O^+ (*Sections 7.5, 14.1*)

Hygroscopic absorbing water from the air (*Section 18.13*)

Ideal gas a gas whose behavior exactly follows the ideal gas law (*Section 9.2*)

Ideal gas law a description of how the volume of a gas is affected by changes in pressure, temperature, and amount; $PV = nRT$ (*Section 9.3*)

Infrared radiation electromagnetic radiation with wavelengths in the range 750 nm to 0.1 mm (*Section 2.1*)

Initial rate the instantaneous rate at the beginning of a reaction (*Section 12.1*)

Inner transition metal element an element in the 14 groups shown separately at the bottom of the periodic table (*Section 1.2*)

Instantaneous rate the rate of a reaction at a particular time (*Section 12.1*)

Integrated rate law the integrated form of a rate law (*Section 12.4*)

Intensive property a property whose value does not depend on the sample size (*Section 1.3*)

Intermolecular force an attractive interaction between molecules (*Section 10.2*)

Internal energy (*E*) the sum of kinetic and potential energies for each particle in a system (*Sections 8.2, 8.4*)

International System of Units (SI) a group of seven fundamental units, along with others derived from them, used for all scientific measurements (*Section 0.2*)

Interstitial hydride a metallic hydride that consists of a crystal lattice of metal atoms with the smaller hydrogen atoms occupying holes between the larger metal atoms (*Section 18.4*)

Ion a charged atom or group of atoms (*Section 3.1*)

Ion–dipole force an intermolecular force resulting from electrical interactions between an ion and the partial charges on a polar molecule (*Section 10.2*)

Ion exchange a process for softening hard water in which the Ca^{2+} and Mg^{2+} ions are replaced by Na^+ (*Section 18.12*)

Ion product (IP) a number defined in the same way as K_{sp}, except that the concentrations in the expression for IP are not necessarily equilibrium values (*Section 15.13*)

Ion-product constant for water (K_w) $[H_3O^+] \times [OH^-] = 1.0 \times 10^{-14}$ (*Section 14.4*)

Ionic bond a bond that results from a transfer of one or more electrons between atoms (*Section 3.9*)

Ionic equation a chemical equation written so that ions are explicitly shown (*Section 7.3*)

Ionic hydride a saltlike, high-melting, white, crystalline compound formed by the alkali metals and the heavier alkaline earth metals (*Section 18.4*)

Ionic liquid a liquid whose constituent particles are ions rather than molecules (*Chapter 10 FYI*)

Ionic solid a solid whose constituent particles are ions ordered into a regular three-dimensional arrangement held together by ionic bonds (*Sections 3.9, 10.6*)

Ionization energy (E_i) the amount of energy necessary to remove the outermost electron from an isolated neutral atom in the gaseous state (*Section 3.5*)

Ionization isomers isomers that differ in the anion bonded to the metal ion (*Section 20.8*)

Isomers compounds that have the same formula but a different bonding arrangement of their constituent atoms (*Section 20.8*)

Isotopes atoms with identical atomic numbers but different mass numbers (*Section 1.8*)

joule (J) the SI unit of energy, equal to $1\ (kg \cdot m^2)/s^2$ (*Section 0.8*)

K_a acid-dissociation constant; the equilibrium constant for dissociation of an acid in water (*Section 14.8*)

K_b base-dissociation constant; the equilibrium constant for the reaction of a base with water (*Section 14.12*)

K_b molal boiling-point-elevation constant; the amount by which the boiling point of a solvent is raised by dissolved substances (*Section 11.7*)

K_c equilibrium constant; the constant in the equilibrium equation (*Section 13.2*)

K_f formation constant; the equilibrium constant for formation of a complex ion (*Section 15.12*)

K_f molal freezing-point-depression constant; the amount by which the melting point of a solvent is lowered by dissolved substances (*Section 11.7*)

K_p equilibrium constant; the equilibrium constant for reaction of gases, defined using partial pressures (*Section 13.4*)

K_{sp} solubility-product constant; the equilibrium constant for a dissolution reaction (*Section 15.10*)

K_{spa} solubility-product constant in acid; the equilibrium constant for a dissolution reaction in acid (*Section 15.14*)

K_w ion-product constant for water; $[H_3O^+] \times [OH^-] = 1.0 \times 10^{-14}$ (*Section 14.4*)

kelvin (K) the SI unit of temperature; $0\ K =$ absolute zero (*Section 0.5*)

Ketone an organic molecule that contains two alkyl groups bonded to a C=O carbon (*Section 23.8*)

kilogram (kg) the SI unit of mass; 1 kg = 2.205 U.S. lb (*Section 0.3*)

Kinetic energy (E_K) the energy of motion; $E_K = (1/2)mv^2$ (*Section 8.1*)

Kinetic–molecular theory a theory describing the quantitative behavior of gases (*Section 9.6*)

Lanthanide one of the 14 inner transition metals following barium in the periodic table (*Section 1.2*)

Lanthanide contraction the decrease in atomic radii across the *f*-block lanthanide elements (*Section 20.2*)

Lattice energy (*U*) the sum of the electrostatic interactions between ions in a solid that must be overcome to break a crystal into individual ions (*Section 3.10*)

Law of conservation of energy Energy can be neither created nor destroyed (*Section 8.1*).

Law of definite proportions Different samples of a pure chemical substance always contain the same proportion of elements by mass (*Section 1.4*).

Law of mass conservation Mass is neither created nor destroyed in chemical reactions (*Section 1.4*).

Law of multiple proportions When two elements combine in different ways to form different substances, the mass ratios are small, whole-number multiples of one another (*Section 1.5*).

LD_{50} the amount of a substance that is lethal to 50% of test animals (*Chapter 0 FYI*)

Le Châtelier's principle If a stress is applied to a reaction mixture at equilibrium, reaction occurs in the direction that relieves the stress (*Section 13.7*).

Lewis acid an electron-pair acceptor (*Section 14.15*)

Lewis base an electron-pair donor (*Section 14.15*)

Lewis structure an electron-dot structure, or representation of a molecule that shows valence electrons as dots (*Section 4.6*)

Ligand a molecule or ion that bonds to the central metal ion in a complex (*Sections 20.5, 20.6*)

Ligand donor atom an atom attached directly to the metal ion in a metal complex (*Section 20.5*)

Light-emitting diode (LED) a semiconductor device that converts electrical energy into light (*Section 21.6*)

Limiting reactant the reactant present in limiting amount that controls the extent to which a reaction occurs (*Section 6.5*)

Line spectrum the wavelengths of light emitted by an energetically excited atom (*Section 2.2*)

Linkage isomers isomers that arise when a ligand bonds to a metal through either of two different donor atoms (*Section 20.8*)

Lipid a naturally occurring organic molecule that dissolves in nonpolar organic solvents when a sample of plant or animal tissue is crushed or ground (*Section 23.12*)

Liter (L) a common unit of volume; $1\ L = 10^{-3}\ m^3$ (*Section 0.6*)

Lock-and-key model a model that pictures an enzyme as a large, irregularly shaped molecule with a cleft into which substrate can fit (*Chapter 12 FYI*)

London dispersion force an intermolecular force resulting from the presence of temporary dipoles in atoms or molecules (*Section 10.2*)

Lone pair electrons a pair of valence electrons not used for bonding (*Section 4.6*)

Low-spin complex a metal complex in which the *d* electrons are paired up to give a maximum number of doubly occupied *d* orbitals and a minimum number of unpaired electrons (*Section 20.11*)

Magnetic quantum number (m_l) a variable in the solutions to the Schrödinger wave equation that defines the spatial orientation of an orbital (*Section 2.6*)

Main-group element an element in the two groups on the left and the six groups on the right of the periodic table (*Section 1.2; Chapters 3, 19*)

Manometer a simple instrument for measuring gas pressure; similar in principle to the mercury barometer (*Section 9.1*)

Mass the amount of matter in an object (*Section 0.3*)

Mass defect the loss in mass that occurs when protons and neutrons combine to form a nucleus (*Section 22.1*)

Mass number (*A*) the total number of protons and neutrons in an atom (*Section 1.8*)

Mass percent a unit of concentration; the mass of one component divided by the total mass of the solution times 100% (*Section 11.3*)

Matter a term used to describe anything that has mass (*Section 0.3*)

Melting point the temperature at which solid and liquid coexist in equilibrium (*Section 10.4*)

Mesosphere the region of the atmosphere from 50–85 km above the earth's surface (*Section 9.9*)

Metabolism the sum of the many organic reactions that go on in cells (*Section 23.9*)

Metal complex *see* Coordination compound (*Section 20.5*)

Metal an element on the left side of the periodic table, bounded on the right by a zigzag line running from boron to astatine (*Section 1.3; Chapter 21*)

Metallic hydride a compound formed by reaction of lanthanide, actinide, or some *d*-block transition metals with variable amounts of hydrogen (*Section 18.4*)

Metallic solid a solid consisting of metal atoms, whose crystals have metallic properties such as electrical conductivity (*Section 10.6*)

Metalloid *see* Semimetal (*Section 1.3*)

Metallurgy the science and technology of extracting metals from their ores (*Section 21.2*)

Meter (m) the SI unit of length (*Section 0.4*)

Microgram (μg) a common unit of mass; $1\ \mu g = 0.001\ mg = 10^{-6}\ g$ (*Section 0.3*)

Micrometer (μm) a common unit of length; $1\ \mu m = 0.001\ mm = 10^{-6}\ m$ (*Section 0.4*)

Milligram (mg) a common unit of mass; $1\ mg = 0.001\ g = 10^{-6}\ kg$ (*Section 0.3*)

Milliliter (mL) a common unit of volume; $1\ mL = 1\ cm^3$ (*Section 0.6*)

Millimeter (mm) a common unit of length; $1\ mm = 0.001\ m$ (*Section 0.4*)

Millimeter of mercury (mm Hg) a common unit of pressure; the millimeter of mercury, also called a *torr*, is based on atmospheric pressure measurements using a mercury barometer (*Section 9.1*)

Mineral a crystalline, inorganic constituent of the rocks that make up the earth's crust (*Section 21.1*)

Miscible mutually soluble in all proportions (*Section 11.4*)

Mixture any combination of two or more pure substances blended together in some arbitrary proportion without chemically changing the individual substances themselves (*Chapter 11 Introduction*)

Molal boiling-point-elevation constant (K_b) the amount by which the boiling point of a solvent is raised by dissolved substances (*Section 11.7*)

Molal freezing-point-depression constant (K_f) the amount by which the melting point of a solvent is lowered by dissolved substances (*Section 11.7*)

Molality (*m*) a unit of concentration; the number of moles of solute per kilogram of solvent (mol/kg) (*Section 11.3*)

Molar heat capacity (C_m) the amount of heat necessary to raise the temperature of 1 mol of a substance 1°C (*Section 8.5*)

Molar mass the mass of 1 mol of substance; equal to the molecular or formula weight of the substance in grams (*Section 1.9*)

Molarity (M) a common unit of concentration; the number of moles of solute per liter of solution (*Sections 6.8, 11.3*)

Mole (mol) the SI unit for amount of substance; the quantity of a substance that contains as many molecules or formula units as there are atoms in exactly 12 g of carbon-12 (*Section 1.9*)

Mole fraction (*X*) a unit of concentration; the number of moles of a component divided by the total number of moles in the mixture (*Sections 9.5, 11.3*)

Molecular equation a chemical equation written using the complete formulas of reactants and products (*Section 7.3*)

Molecular formula a formula that tells the identity and numbers of atoms in a molecule (*Section 6.6*)

Molecular orbital a wave function whose square gives the probability of finding an electron within a given region of space in a molecule (*Section 5.5*)

Molecular orbital theory a quantum mechanical description of bonding in which electrons occupy molecular orbitals that belong to the entire molecule rather than to an individual atom (*Section 5.5*)

Molecular solid a solid whose constituent particles are molecules held together by intermolecular forces (*Section 10.6*)

Molecular weight the sum of atomic weights of the atoms in a molecule (*Section 6.3*)

Molecularity the number of molecules on the reactant side of the chemical equation for an elementary reaction (*Section 12.11*)

Molecule the unit of matter that results when two or more atoms are joined by covalent bonds (*Section 3.1, Chapter 4*)

Mond process a chemical method for purification of nickel from its ore (*Section 21.2*)

Monodentate ligand a ligand that bonds to a metal using the electron pair of a single donor atom (*Section 20.6*)

Monoprotic acid an acid that has a single dissociable proton (*Sections 7.5, 14.7*)

Monosaccharide a carbohydrate such as glucose that can't be broken down into smaller molecules by hydrolysis (*Section 23.11*)

N-Terminal amino acid the amino acid with a free $-NH_2$ group on the end of a protein chain (*Section 23.10*)

***n*-Type semiconductor** a semiconductor doped with an impurity that has more electrons than necessary for bonding (*Section 21.5*)

Nanometer (nm) a common unit of length; $1\ nm = 10^{-9}\ m$ (*Section 0.4*)

Nanotechnology the study and production of materials and structures that have at least one dimension between 1 nm and 100 nm (*Chapter 21 FYI*)

Nernst equation an equation for calculating cell potentials under non-standard-state conditions; $E = E° - (RT \ln Q)/(nF)$ (*Section 17.6*)

Net ionic equation a chemical equation written so that spectator ions are removed (*Section 7.3*)

Neutralization reaction *see* Acid–base neutralization reaction (*Sections 7.5, 15.1*)

Neutron a neutral, fundamental atomic particle in the nucleus of atoms (*Section 1.7*)

Newton (N) the SI unit for force (*Section 9.1*)

Noble gas an element in group 8A of the periodic table (*Sections 1.3, 3.14*)

Node a region where a wave has zero amplitude (*Section 2.7*)

Nonelectrolyte a substance that does not produce ions when dissolved in water (*Section 7.1*)

Nonmetal hydrogen plus elements on the right side of the periodic table, bounded on the left by a zigzag line running from boron to astatine (*Section 1.3*)

Nonspontaneous process a process that requires a continuous input of energy to proceed (*Section 8.12*)

Nonstoichiometric compound a compound whose atomic composition can't be expressed as a ratio of small whole numbers (*Section 18.4*)

Normal boiling point the temperature at which boiling occurs when there is exactly 1 atm of external pressure (*Sections 10.5, 10.11*)

Normal melting point the temperature at which melting occurs when there is exactly 1 atm of external pressure (*Section 10.11*)

Nuclear chemistry the study of the properties and reactions of atomic nuclei (*Section 1.10*)

Nuclear equation an equation for a nuclear reaction in which the sums of the nucleons are the same on both sides and the sums of the charges on the nuclei and any elementary particles are the same on both sides (*Section 1.10*)

Nuclear fission the fragmenting of heavy nuclei (*Section 22.2*)

Nuclear fusion the joining together of light nuclei (*Section 22.2*)

Nuclear reaction a reaction that changes an atomic nucleus (*Section 1.10*)

Nuclear transmutation the change of one element into another by a nuclear reaction (*Section 22.3*)

Nucleic acid a biological molecule made up of nucleotide units linked together to form a long chain (*Section 23.13*)

Nucleon a general term for both protons and neutrons (*Section 1.11*)

Nucleoside a constituent of nucleotides; composed of an aldopentose sugar plus an amine base (*Section 23.13*)

Nucleotide a building block from which nucleic acids are made; composed of a nucleoside plus phosphoric acid (*Section 23.13*)

Nucleus the central core of an atom consisting of protons and neutrons (*Section 1.7*)

Octet rule the statement that main-group elements tend to undergo reactions that leave them with eight valence electrons (*Section 3.8*)

Orbital a solution to the Schrödinger wave equation, describing a region of space where an electron is likely to be found (*Section 2.6*)

Ore a mineral deposit from which a metal can be produced economically (*Section 21.1*)

Organic chemistry the study of carbon compounds (*Chapter 23*)

Osmosis the passage of solvent through a membrane from the less concentrated side to the more concentrated side (*Section 11.8*)

Osmotic pressure the amount of pressure necessary to cause osmosis to stop (*Section 11.8*)

Ostwald process the commercial process for making nitric acid from ammonia (*Section 19.10*)

Overpotential the additional voltage required above that calculated for an electrolysis reaction (*Section 17.12*)

Oxidation the loss of one or more electrons by a substance (*Section 7.6*)

Oxidation number a value that measures whether an atom in a compound is neutral, electron-rich, or electron-poor compared to an isolated atom (*Section 7.6*)

Oxidation–reduction (redox) reaction a process in which one or more electrons are transferred between reaction partners (*Section 7.6*)

Oxide a binary compound with oxygen in the -2 oxidation state (*Section 18.8*)

Oxidizing agent a substance that causes an oxidation by accepting an electron (*Section 7.7*)

Oxoacid an acid that contains oxygen in addition to hydrogen and another element (*Section 7.5*)

Oxoanion an anion of an oxoacid (*Sections 3.2, 7.5*)

Ozone layer an atmospheric band stretching from about 20 to 40 km above the earth's surface (*Section 9.9*)

***p*-Block element** an element in groups 3A–8A, in which *p* orbitals are filled (*Section 2.13*)

***p*-Type semiconductor** a semiconductor doped with an impurity that has fewer electrons than necessary for bonding (*Section 21.5*)

Paramagnetic a substance that contains unpaired electrons and is attracted by a magnetic field (*Section 5.6*)

Parts per billion (ppb) a concentration unit for very dilute solutions; a concentration of 1 ppb means that each kilogram of solution contains 1 μg of solute (*Section 11.3*)

Parts per million (ppm) a concentration unit for very dilute solutions; a concentration of 1 ppm means that each kilogram of solution contains 1 mg of solute (*Section 11.3*)

Pascal (Pa) the SI unit for pressure (*Section 9.1*)

Pauli exclusion principle No two electrons in an atom can have the same four quantum numbers (*Section 2.8*).

Percent composition a list of elements present in a compound and the mass percent of each (*Section 6.6*)

Percent dissociation the concentration of the acid that dissociates divided by the initial concentration of the acid times 100% (*Section 14.10*)

Percent yield the amount of product actually formed in a reaction divided by the amount theoretically possible and multiplied by 100% (*Section 6.4*)

Period a row of elements in the periodic table (*Section 1.2*)

Periodic table a chart of the elements arranged by increasing atomic number so that elements in a given group have similar chemical properties (*Section 1.2*)

Peroxide a binary compound with oxygen in the -1 oxidation state (*Sections 18.8, 18.9*)

Petroleum a complex mixture of organic substances, primarily hydrocarbons (*Section 8.11*)

pH the negative base-10 logarithm of the molar hydronium ion concentration (*Section 14.5*)

pH titration curve a plot of the pH of a solution as a function of the volume of added base or acid (*Section 15.5*)

Phase (of matter) a state of matter, whether solid, liquid, or gas (*Section 10.4*)

Phase (of waves) the nonzero-amplitude parts of a wave, corresponding to different algebraic signs (*Section 2.7*)

Phase change a process in which the physical form but not the chemical identity of a substance changes (*Section 10.4*)

Phase diagram a plot showing the effects of pressure and temperature on the physical state of a substance (*Section 10.11*)

Photoelectric effect the ejection of electrons from a metal on exposure to radiant energy (*Section 2.3*)

Photon the smallest possible amount of radiant energy; a quantum (*Section 2.3*)

Photovoltaic (solar) cell a semiconductor device that converts light into electrical energy (*Section 21.6*)

Physical property a characteristic that can be determined without changing the chemical makeup of a sample (*Section 1.3*)

Pi (π) bond a covalent bond formed by sideways overlap of orbitals in which shared electrons occupy a region above and below a line connecting the two nuclei (*Section 5.4*)

Picometer (pm) a common unit of length; $1\ \text{pm} = 10^{-12}\ \text{m}$ (*Section 0.4*)

Planck's constant (*h*) $6.626 \times 10^{-34}\ \text{J} \cdot \text{s}$; a fundamental physical constant that relates energy and frequency, $E = h\nu$ (*Section 2.3*)

Planck's postulate an equation that relates energy, frequency, and wavelength of radiant energy, $E = h\nu = hc/\lambda$ (*Section 2.3*)

Polar covalent bond a bond in which the bonding electrons are attracted somewhat more strongly by one atom than by the other (*Section 4.4*)

Polarizability the ease with which a molecule's electron cloud can be distorted by a nearby electric field (*Section 10.2*)

Polyatomic ion a charged, covalently bonded group of atoms (*Section 3.1*)

Polydentate ligand a ligand that bonds to a metal through electron pairs on more than one donor atom (*Section 20.6*)

Polyprotic acid an acid that contains more than one dissociable proton (*Section 14.11*)

Polysaccharide a compound such as cellulose that is made of many simple sugars linked together (*Section 23.11*)

Positron emission a nuclear reaction that converts a proton into a neutron plus an ejected positron (*Section 1.11*)

Potential energy (E_P) energy that is stored, either in an object because of its position or in a molecule because of its chemical composition (*Section 8.1*)

Precipitation reaction a reaction in which an insoluble solid precipitate forms and drops out of solution (*Section 7.4*)

Precision how well a number of independent measurements agree with one another (*Section 0.9*)

Pressure a force exerted per unit area (*Section 9.1*)

Primitive-cubic unit cell a cubic unit cell with an atom at each of its eight corners (*Section 10.8*)

Principal quantum number (*n*) a variable in the solutions to the Schrödinger wave equation on which the size and energy level of an orbital primarily depends (*Section 2.6*)

Principal reaction the proton-transfer reaction that proceeds farther to the right when calculating equilibrium concentrations in solutions of weak acids (*Section 14.9*)

Property any characteristic that can be used to describe or identify matter (*Section 1.3*)

Protein a biological molecule made up of many amino acids linked together to form a long chain (*Section 23.10*)

Proton a positively charged, fundamental atomic particle in the nucleus of atoms (*Section 1.7*)

Qualitative analysis a procedure for identifying the ions present in an unknown solution (*Section 15.15*)

Quantized changing only in discrete amounts (*Section 2.3*)

Quantum the smallest possible amount of radiant energy (*Section 2.3*)

Quantum mechanical model a model of atomic structure that concentrates on an electron's wave-like properties (*Section 2.5*)

Quantum number a variable in the Schrödinger wave equation that describes the energy level and position in space where an electron is most likely to be found (*Section 2.6*)

Racemic mixture a 1:1 mixture of enantiomers (*Section 20.9*)

Radiant energy electromagnetic radiation (*Section 2.1*)

Radical a compound that contains an odd number of electrons and therefore has an unpaired electron in a half-filled orbital (*Chapter 4 FYI*)

Radioactivity the spontaneous emission of radiation accompanying a nuclear reaction (*Section 1.11*)

Radiocarbon dating a technique for dating archaeological artifacts by measuring the amount of ^{14}C in the sample (*Section 22.5*)

Radioisotope a radioactive isotope (*Section 1.11*)

Raoult's law The vapor pressure of a solution containing a nonvolatile solute is equal to the vapor pressure of pure solvent times the mole fraction of the solvent (*Section 11.6*).

Rate constant the proportionality constant in a rate law (*Section 12.2*)

Rate-determining step the slowest step in a reaction mechanism (*Section 12.13*)

Rate law an equation that tells how reaction rate depends on the concentration of each reactant (*Section 12.2*)

Reaction intermediate a species that is formed in one step of a reaction mechanism and consumed in a subsequent step (*Section 12.11*)

Reaction mechanism the sequence of molecular events that defines the pathway from reactants to products (*Section 12.11*)

Reaction order the value of the exponents of concentration terms in the rate law (*Section 12.2*)

Reaction quotient (Q_c) similar to the equilibrium constant K_c except that the concentrations in the equilibrium constant expression are not necessarily equilibrium values (*Section 13.6*)

Reaction rate the increase in the concentration of a product per unit time or the decrease in the concentration of a reactant per unit time (*Section 12.1*)

Redox reaction an oxidation–reduction reaction (*Section 7.6*)

Redox titration a procedure for determining the concentration of a redox agent (*Section 7.10*)

Reducing agent a substance that causes a reduction by donating an electron (*Section 7.7*)

Reduction the gain of one or more electrons by a substance (*Section 7.6*)

Replication in biochemistry, the process by which identical copies of DNA are made (*Section 23.13*)

Resonance hybrid an average of several valid electron-dot structures for a molecule (*Section 4.9*)

Respiration the process of breathing and using oxygen for biological redox reactions (*Section 7.11*)

Reverse osmosis the passage of solvent through a membrane from the more concentrated side to the less concentrated side (*Section 11.9*)

Ribonucleic acid (RNA) a biological polymer of ribonucleotide units that serves to transcribe the genetic information in DNA and uses that information to direct the synthesis of proteins (*Section 23.13*)

Roasting a metallurgical process that involves heating a mineral in air (*Section 21.2*)

Rounding off deleting digits to keep only the correct number of significant figures in a calculation (*Section 0.10*)

s-Block element an element in groups 1A or 2A, in which *s* orbitals are filled (*Section 2.13*)

Sacrificial anode an easily oxidized metal that corrodes instead of a less reactive metal to which it is connected (*Section 17.11*)

Salt an ionic compound formed in an acid–base neutralization reaction (*Section 7.5*)

Salt bridge a tube that contains a gel permeated with a solution of an inert electrolyte connecting the two sides of an electrochemical cell (*Section 17.1*)

Saponification the base-catalyzed hydrolysis of an ester to yield a carboxylic acid and an alcohol (*Section 23.8*)

Saturated hydrocarbon a hydrocarbon that contains only single bonds (*Section 23.4*)

Saturated solution a solution containing the maximum possible amount of dissolved solute at equilibrium (*Section 11.4*)

Schrödinger wave equation an equation describing the behavior of an electron in an atom (*Section 2.6*)

Scientific notation a system in which a large or small number is written as a number between 1 and 10 times a power of 10 (*Section 0.2*)

Second law of thermodynamics In any spontaneous process, the total entropy of a system and its surroundings always increases (*Section 16.6*).

Second-order reaction a reaction whose rate depends on the concentration of a single reactant raised to the second power or on the concentrations of two different reactants, each raised to the first power (*Section 12.7*)

Semiconductor a material that has an electrical conductivity intermediate between that of a metal and that of an insulator (*Section 21.5*)

Semimetal an element adjacent to the zigzag boundary between metals and nonmetals (*Section 1.3*)

Semipermeable membrane a membrane that allows passage of water or other small molecules but not the passage of large solute molecules or ions (*Section 11.8*)

Shell a grouping of orbitals according to principal quantum number (*Section 2.6*)

SI unit a system of units based on the metric system and used throughout the world for scientific measurements (*Section 0.2*)

Side chain the group attached to the α carbon of an amino acid (*Section 23.10*)

Sigma (σ) bond a covalent bond formed by head-on overlap of orbitals in which the shared electrons are centered about the axis between the two nuclei (*Section 5.2*)

Significant figures the total number of digits in a measurement (*Section 0.9*)

Silicate an ionic compound that contains silicon oxoanions along with cations, such as Na^+, K^+, Mg^{2+}, or Ca^{2+} (*Section 19.8*)

Simple cubic packing a packing arrangement of spheres into a primitive-cubic unit cell (*Section 10.8*)

Single bond a covalent bond formed by sharing two electrons between atoms (*Section 4.6*)

Sintering a process in which the particles of a powder are "welded" together without completely melting (*Section 21.8*)

Slag a by-product of iron production, consisting mainly of calcium silicate (*Section 21.3*)

Solubility the amount of a substance that dissolves in a given volume of solvent at a given temperature (*Sections 7.4, 11.4*)

Solubility product (K_{sp}) the equilibrium constant for a dissolution reaction (*Section 15.10*)

Solute the dissolved substance in a solution (*Sections 6.8, 11.1*)

Solution a homogeneous mixture containing particles the size of a typical ion or covalent molecule (*Section 11.1*)

Solvent the major component in a solution (*Section 11.1*)

Sol–gel method a method of preparing ceramics, involving synthesis of a metal oxide powder from a metal alkoxide (*Section 21.8*)

***sp* Hybrid orbital** a hybrid orbital formed by combination of one atomic *s* orbital with one *p* orbital (*Section 5.4*)

sp^2 Hybrid orbital a hybrid orbital formed by combination of one *s* and two *p* atomic orbitals (*Section 5.4*)

sp^3 Hybrid orbital a hybrid orbital formed by combination of one *s* and three *p* atomic orbitals (*Section 5.3*)

sp^3d Hybrid orbital a hybrid orbital formed by combination of one *s*, three *p*, and one *d* atomic orbitals (*Section 20.11*)

sp^3d^2 Hybrid orbital a hybrid orbital formed by combination of one *s*, three *p*, and two *d* atomic orbitals (*Section 20.11*)

Specific heat the amount of heat necessary to raise the temperature of 1 gram of a substance 1 °C (*Section 8.5*)

Spectator ion an ion that appears on both sides of the reaction arrow (*Section 7.3*)

Spectrochemical series an ordered list of ligands in which crystal field splitting increases (*Section 20.12*)

Spin quantum number (m_s) a variable that describes the spin of an electron, either $+1/2$ or $-1/2$ (*Section 2.8*)

Spontaneous process one that proceeds on its own without any continuous external influence (*Section 8.12*)

Standard cell potential ($E°$) the cell potential when both reactants and products are in their standard states (*Section 17.3*)

Standard electrode potential *see* Standard reduction potential (*Section 17.4*)

Standard enthalpy of reaction ($\Delta H°$) enthalpy change under standard-state conditions (*Section 8.5*)

Standard entropy of reaction ($\Delta S°$) the entropy change for a chemical reaction under standard-state conditions (*Section 16.5*)

Standard free-energy change ($\Delta G°$) the free-energy change that occurs when reactants in their standard states are converted to products in their standard states (*Section 16.8*)

Standard free energy of formation ($\Delta G°_f$) the free-energy change for formation of 1 mol of a substance in its standard state from the most stable form of the constituent elements in their standard states (*Section 16.9*)

Standard heat of formation ($\Delta H°_f$) the enthalpy change $\Delta H°_f$ for the hypothetical formation of 1 mol of a substance in its standard state from the most stable forms of its constituent elements in their standard states (*Section 8.9*)

Standard hydrogen electrode (S.H.E.) a reference half-cell consisting of a platinum electrode in contact with H_2 gas and aqueous H^+ ions at standard-state conditions (*Section 17.4*)

Standard molar entropy ($S°$) the entropy of 1 mol of a pure substance at 1 atm pressure and a specified temperature, usually 25 °C (*Section 16.5*)

Standard molar volume the volume of 1 mol of a gas at 0°C and 1 atm pressure; 22.414 L (*Section 9.2*)

Standard reduction potential the standard potential for a reduction half-cell (*Section 17.4*)

Standard temperature and pressure (STP) $T = 273.15$ K; $P = 1$ atm (*Section 9.3*)

State function a function or property whose value depends only on the present condition of the system, not on the path used to arrive at that condition (*Section 8.2*)

Steam–hydrocarbon re-forming process an important industrial method for producing hydrogen from methane (*Section 18.2*)

Stereoisomers isomers that have the same connections among atoms but have a different arrangement of the atoms in space (*Section 20.8*)

Steric factor the fraction of collisions with the proper orientation for converting reactants to products (*Section 12.9*)

Stoichiometry mole/mass relationships between reactants and products (*Section 6.3*)

Straight-chain alkane an alkane that has all its carbons connected in a row (*Section 23.1*)

Stratosphere the region of the atmosphere from 20–50 km above the earth's surface (*Section 9.9*)

Strong acid an acid that dissociates completely in water to give H^+ ions and is a strong electrolyte (*Sections 7.5, 14.2*)

Strong base a base that dissociates or reacts completely with water to give OH^- ions and is a strong electrolyte (*Sections 7.5, 14.2*)

Strong electrolyte a compound that dissociates completely into ions when dissolved in water (*Section 7.1*)

Strong-field ligand a ligand that has a large crystal field splitting (*Section 20.12*)

Structural formula a representation that shows the specific connections between atoms in a molecule (*Section 4.2*)

Sublimation the direct conversion of a solid to a vapor without going through a liquid state (*Sections 8.7, 10.4*)

Subshell a grouping of orbitals by angular-momentum quantum number (*Section 2.6*)

Subsidiary reaction any proton-transfer process other than the principal one in an acid–base reaction (*Section 14.9*)

Substitution reaction an organic reaction in which one group substitutes for another, particularly on aromatic rings (*Section 23.6*)

Substrate the compound acted on by an enzyme (*Chapter 12 FYI*)

Superconducting transition temperature (T_c) the temperature below which a superconductor loses all electrical resistance (*Section 21.7*)

Superconductor a material that loses all electrical resistance below a certain temperature (*Section 21.7*)

Supercritical fluid a state of matter beyond the critical point that is neither liquid nor gas (*Section 10.11*)

Superoxide a binary compound with oxygen in the –1/2 oxidation state (*Sections 18.8, 18.9*)

Supersaturated solution a solution containing a greater-than-equilibrium amount of solute (*Section 11.4*)

Surface tension the resistance of a liquid to spreading out and increasing its surface area (*Section 10.3*)

Suspension a homogeneous mixture containing particles greater than about 1000 nm in diameter that are visible with a low-power microscope (*Section 11.1*)

Symmetry plane a plane that cuts through an object so that one half of the object is a mirror image of the other half (*Section 20.9*)

Synthesis gas the mixture of carbon monoxide and hydrogen produced in the steam–hydrocarbon re-forming process (*Section 18.2*)

Temperature a measure of the kinetic energy of molecular motion (*Section 8.1*)

Termolecular reaction an elementary reaction that results from collisions between three reactant molecules (*Section 12.11*)

Theory a consistent explanation of known observations (*Section 0.1*)

Thermochemistry a study of the heat changes that take place during reactions (*Chapter 8 Introduction*)

Thermodynamic standard state conditions under which thermodynamic measurements are reported; 298.15 K (25°C), 1 atm pressure for each gas, 1 M concentration for solutions

(*Section 8.6*)

Thermodynamics the study of the interconversion of heat and other forms of energy (*Chapter 16*)

Thermosphere the region of the atmosphere from 85–120 km above the earth's surface (*Section 9.9*)

Third law of thermodynamics The entropy of a perfectly ordered crystalline substance at 0 K is zero (*Section 16.4*).

Three-center, two-electron bond a covalent bond in which three atoms share two electrons (*Section 19.4*)

Titration a procedure for determining the concentration of a solution (*Section 6.11*)

Trans isomer the isomer of a metal complex or alkene in which identical ligands or groups are opposite one another rather than adjacent (*Sections 20.8, 23.4*)

Transcription the process by which information in DNA is transferred to and decoded by RNA (*Section 23.13*)

Transistor a semiconductor device that controls and amplifies electrical signals (*Section 21.6*)

Transition metal element an element in the 10 groups in the middle of the periodic table (*Section 1.2; Chapter 20*)

Transition state the configuration of atoms at the maximum in the potential energy profile for a reaction (*Section 12.9*)

Translation the process by which RNA builds proteins (*Section 23.13*)

Transuranium elements the 26 artificially produced elements beyond uranium in the periodic table (*Section 22.1*)

Triacylglycerol a triester of glycerol (1,2,3-propanetriol) with three long-chain carboxylic acids (*Section 23.12*)

Triple bond a covalent bond formed by sharing six electrons between atoms (*Section 4.6*)

Triple point a unique combination of pressure and temperature at which gas, liquid, and solid phases coexist in equilibrium (*Section 10.11*)

Triprotic acid an acid that can provide three H^+ ions (*Section 7.5*)

Troposphere the region of the atmosphere from 0–20 km above the earth's surface (*Section 9.9*)

Turnover number the number of substrate molecules acted on by one molecule of enzyme per unit time (*Chapter 12 FYI*)

Ultraviolet radiation electromagnetic radiation with wavelengths in the range 10–400 nm (*Section 2.1*)

Unified atomic mass unit (*u*) a unit of mass equal to 1/12th the mass of a $^{12}_{6}C$ atom (*Section 1.9*)

Unimolecular reaction an elementary reaction that involves a single reactant molecule (*Section 12.11*)

Unit cell a small repeating unit that makes up a crystal (*Section 10.8*)

Unsaturated an organic molecule that contains a double or triple bond (*Section 23.4*)

Valence band the bonding molecular orbitals in a semiconductor (*Section 21.5*)

Valence bond theory a quantum mechanical description of bonding that pictures covalent bond formation as the overlap of two singly occupied atomic orbitals (*Section 5.2*)

Valence shell the outermost electron shell (*Section 2.13*)

Valence-shell electron-pair repulsion (VSEPR) model a model for predicting the approximate shape of a molecule (*Section 5.1*)

van der Waals equation a modification of the ideal gas law that introduces correction factors to account for the behavior of real gases (*Section 9.8*)

van der Waals forces an alternative name for intermolecular forces (*Section 10.2*)

van't Hoff factor (*i*) a measure of the extent of dissociation of a substance, used in interpreting colligative property measurements (*Section 11.6*)

Vapor pressure (P_{vap}) the partial pressure of a gas in equilibrium with liquid (*Section 10.5*)

Vaporization the change of a liquid to a gas (*Section 10.4*)

Viscosity the measure of a liquid's resistance to flow (*Section 10.3*)

Visible light electromagnetic radiation with wavelengths in the range 400–780 nm (*Section 2.1*)

Voltaic cell *see* Galvanic cell (*Section 17.1*)

Volume the amount of space occupied by an object (*Section 0.6*)

Water–gas shift reaction a reaction used in the industrial preparation of H_2 by reaction of CO with H_2O (*Section 18.2*)

Watson–Crick model a model of DNA structure, consisting of two polynucleotide strands coiled around each other in a double helix (*Section 23.13*)

Wave function a solution to the Schrödinger wave equation (*Section 2.6*)

Wavelength (λ) the length of a wave from one maximum to the next (*Section 2.1*)

Weak acid an acid that dissociates incompletely in water and is a weak electrolyte (*Sections 7.5, 14.2*)

Weak base a base that dissociates or reacts incompletely with water and is a weak electrolyte (*Sections 7.5, 14.2*)

Weak electrolyte a compound that dissociates incompletely when dissolved in water (*Section 7.1*)

Weak-field ligand a ligand that has a small crystal field splitting (*Section 20.12*)

Work (*w*) the distance (*d*) moved times the force (*F*) that opposes the motion (*Section 8.3*)

Work function (Φ) the amount of energy necessary to eject an electron from a metal (*Section 2.3*)

X ray electromagnetic radiation with wavelengths in the range 0.01–10 nm (*Section 2.1*)

X-ray crystallography a technique for determining molecular structure by analysis of the diffraction pattern produced when a crystal is struck by X rays (*Section 10.7*)

Yield the amount of product formed in a reaction (*Section 6.4*)

Zeroth-order reaction one whose rate remains constant, independent of reactant concentrations (*Section 12.8*)

Zone refining a purification technique in which a heater melts a narrow zone at the top of a rod of some material and then sweeps slowly down the rod bringing impurities with it (*Section 19.8*)

INDEX

Photo Credits

(T) = top; (B) = bottom; (C) = center; (L) = left; (R) = right

Chapter 0: **2**, Science and Society/SuperStock; **3**, Best View Stock/age fotostock; **5**, Tom Krueger/iStockphoto; **6 (TL)**, Mettler Toledo North America; **6 (TR)**, Ohaus Corporation; **6 (B)**, Dr. Tony Brain/Science Photo Library/Photo Researchers, Inc.; **8**, Richard Megna/Fundamental Photographs; **9 (L-R)**, McCracken Photographers/Pearson Education; **10**, **(T)** Richard Megna/Fundamental Photographs; **10 (B)**, Christian Wilkinson/Shutterstock; **11**, YuriSH/iStockphoto; **12 (T)**, Nikmd/Dreamstime; **12 (B)**, McCracken Photographers/Pearson Education; **14**, Texas Instruments Incorporated; **16**, REUTERS/David W Cerny; **17 (T)**, Susumu Nishinaga/Photo Researchers, Inc.; **17 (B)**, EPA/Uli Deck/Newscom; **19 (T)**, mathieukor/iStockphoto; **19 (B)**, Reuters/Stringer; **20**, Cephalon

Chapter 1: **26 (T)**, NASA; **26 (B)**, McCracken Photographers/Pearson Education; **29**, Sciencephotos/Alamy; **30 (T-B)**, Richard Megna/Fundamental Photographs; **31**, Pearson Education/McCracken Photographers; **32 (L-R)**, Richard Megna/Fundamental Photographs; **33 (L)**, McCracken Photographers/Pearson Education; **33 (R)**, Travel Pictures/Alamy; **34 (L-R)**, Richard Megna/Fundamental Photographs; **35**, Richard Megna/Fundamental Photographs; **36 (L-R)**, Richard Megna/Fundamental Photographs; **37 (T-B)**, Richard Megna/Fundamental Photographs; **38**, Richard Megna/Fundamental Photographs; **39 (L-R)**, Richard Megna/Fundamental Photographs; **41**, REUTERS/Kiyoshi Ota; **44**, Colin Monteath/age fotostock; **46**, Richad Megna/Fundamental Photographs; **47 (TL)**, Sebastien Burel/Fotolia; **47 (TR)**, NASA; **47 (BL)**, Nikolai Sorokin/Fotolia; **47 (BC)**, Dmitry Nikolaev/Fotolia; **47 (BR)**, gwimages/Fotolia; **49**, Thorsten Schier/Shutterstock; **52**, Steven Needell/Newscom; **53**, NASA, ESA, and the Hubble Heritage Team (STScI/AURA); **54**, NASA/JPL-Caltech/O. Krause (Steward Observatory)

Chapter 2: **60**, Pfeiffer; J/ARCO/age fotostock; **63 (T)**, Phillip Long/Getty Images; **63 (B)**, Natalia Bratslavsky/Shutterstock; **64 (L)**, GIPhotoStock/Photo Researchers, Inc.; **64 (R)**, Karin Lau/Shutterstock; **65 (L-R)**, Debra Crowe; **67**, Bill Corwin; **68**, Jens Johnson Photography; **69**, Nestor Noci/Shutterstock; **72**, Colorful High Speed Photographs/Getty Images; **76**, Richard Megna/Fundamental Photographs; **89**, Cultura Creative/Alamy

Chapter 3: **98**, Christophe Boisvieux/age fotostock/SuperStock; **99**, Zurijeta/Shutterstock; **100**, Richard Megna/Fundamental Photographs; **101**, usas/iStockphoto; **102**, REUTERS/Goran Tomasevic; **103**, Pearson Education/Eric Schrader; **104**, Richard Megna/Fundamental Photographs; **123**, Richard Megna/Fundamental Photographs; **124 (L-R)**, Richard Megna/Fundamental Photographs; **125 (L-R)**, Richard Megna/Fundamental Photographs; **126**, Richard Megna/Fundamental Photographs; **127**, Marc Tielemans/Alamy; **128**, Worachat Sodsri/Shutterstock

Chapter 4: **136**, .shock/Shutterstock; **137**, McCracken Photographers/Pearson Education; **160**, Dmitriy Shironosov/Shutterstock

Chapter 5: **168**, Andrea Seemann/Shutterstock; **187**, McCracken Photographers/Pearson Education; **191**, Jaimie D. Travis/iStockphoto; **193 (L)**, sunnyfrog/iStockphoto; **193 (R)**, pixelman/Shutterstock

Chapter 6: **200**, Christopher Berkey/EPA/Newscom; **202**, Blend Images/Shutterstock; **203**, Corbis; **204**, Associated Press/Anja Niedringhaus; **208**, moodboard/Alamy; **220 (L-R)**, Richard Megna/Fundamental Photographs; **222 (T)**, Paul Silverman/Fundamental Photographs; **222 (BL-BR)**, Richard Megna/Fundamental Photographs; **224**, Paul Silverman/Fundamental Photographs; **225 (L-R)**, Richard Megna/Fundamental Photographs; **226 (T)**, Library of Congress; **226 (B)**, Science Photo Library/Photo Researchers, Inc.

Chapter 7: **236**, Reuters/Petr Josek Snr; **237 (L-R)**, Richard Megna/Fundamental Photographs; **239**, McCracken Photographers/Pearson Education; **241**, Richard Megna/Fundamental Photographs; **242**, McCracken Photographers/Pearson Education; **245**, McCracken Photographers/Pearson Education; **248**, McCracken Photographers/Pearson Education; **252**, Gianni Dagli Orti/The Art Archive at Art Resource, NY; **254 (L-R)**, McCracken Photographers/Pearson Education; **255 (L-R)**, Kip Peticolas & Richard Megna/Fundamental Photographs; **257 (L-R)**, Richard Megna/Fundamental Photographs; **261 (T-B)**, McCracken Photographers/Pearson Education; **262**, Paul Silverman/Fundamental Photographs; **263**, Jeff Gentner/Stringer/Getty Images; **264**, ICCA/CEFIC

Chapter 8: **274**, imagebroker/Alamy; **278**, Dwight Smith/Fotolia; **279**, amygdala_imagery/iStockphoto; **281**, mtr/Shutterstock; **285 (T)**, John E. McMurry; **285 (B)**, McCracken Photographers/Pearson Education; **288 (T)**, Dex Image/Alamy; **288 (C)**, luchschen/Shutterstock; **288 (B)**, Mikko Pitkanen/Shutterstock; **289**, Richard Megna/Fundamental Photographs; **294**, Cephas Picture Library/Alamy; **295**, alle12/iStockphoto; **298**, Rob Belknap/iStockphoto; **299**, Rakoskerti/iStockphoto; **300 (T)**, World Pictures/Alamy; **300 (BL)**, Mikko Pitkanen/Shutterstock; **300 (BC)**, luchschen/Shutterstock; **300 (BR)**, Dex Image/Alamy; **303 (L-R)**, Vladone/iStockphoto; **306**, Teptong/Fotolia

Chapter 9: **316**, cZUMA Press/Newscom; **317**, Iakov Kalinin/Shutterstock; **318**, NASA; **320**, Pichugin Dmitry/Shutterstock; **322**, Leapingllamas; **324**, vate/iStockphoto; **328**, Paulo Resende/Shutterstock; **329**, fStop/Alamy; **332**, Eric Schrader/Pearson Education; **334**, Carlos Villoch - MagicSea.com/Alamy; **339**, U.S. Department of Energy/Photo Researchers, Inc.; **342 (T)**, Niilo Tippler/iStockphoto; **342 (B)**, Richard Megna/Fundamental Photographs; **343**, NASA; **344**, Charlotte Zeepvat

Chapter 10: **354**, BASF/Fundamental Photographs; **359**, David Taylor/Science Photo Library/Photo Researchers, Inc.; **363**, Bronwyn Photo/Shutterstock; **365 (T)**, Alexel Zaycev/iStockphoto; **365 (B)**, Harry Taylor/Dorling Kindersley; **367 (T)**, UpperCut Images/Alamy; **367 (B)**, Associated Press; **370**, Richard Megna/Fundamental Photographs; **372**, Richard Megna/Fundamental Photographs; **374 (L)**, Jonnysek/Fotolia; **374 (R)**, Ryan McVay/Getty Images; **375**, Pearson Education; **378**, Yenwen Lu/iStockphoto; **379**, amete/iStockphoto; **385**, General Electric Corporate Research & Development Center; **387 (T)**, Linda Whitwam/Dorling Kindersley; **387 (B)**, Richard Megna/Fundamental Photographs; **389**, Paul Silverman/Fundamental Photographs; **390**, BASF SE/Detlef W. Schmalow

Chapter 11: **400**, Galyna Andrushko/Fotolia; **401**, Judith Miller/Dorling Kindersley/Lyon and Turnbull Ltd.; **402**, Reuters/Sean Gardner; **404 (L-R)**, Richard Megna/Fundamental Photographs; **407**, dianacostescu/Fotolia; **412 (L-R)**, Charles D. Winters/Photo Researchers, Inc.; **414**, USC Catalina Hyperbaric Chamber; **427 (L)**, Janice Carr/Centers for Disease Control and Prevention; **427 (R)**, Custom Medical Stock Photography; **428**, Reuters/Nir Elias; **430**, GSPhotography/Shutterstock; **432**, Reuters/Suhaib Salem

Chapter 12: **442**, D. Hurst/Alamy; **443 (L)**, Richard Megna/Fundamental Photographs; **443 (R)**, HO/Associated Press; **444**, Adi; **449 (L-R)**, McCracken Photographers/Pearson Education; **453 (L-R)**, McCracken Photographers/Pearson Education; **458**, Richard Megna/Fundamental Photographs; **461**, Library of Congress; **462**, Viktor Gmyria/Fotolia; **468 (T-B)**, McCracken Photographers/Pearson Education; **476**, hornet72/Fotolia; **480**, DPAphotos/Newscom; **484**, Richard Megna/Fundamental Photographs; **485 (L-R)**, McCracken Photographers/Pearson Education

Chapter 13: **504 (T)**, Cal Sport Media/Newscom; **504 (B)**, Richard Megna/Fundamental Photographs; **505 (C)**, EPA/Newscom; **505 (B)**, Cintract Romain/age fotostock; **507**, Andrew Rubtsov/Alamy; **510**, Eric Schrader/Pearson Education; **513**, Ensign John Gay/US Navy; **516**, LinguisticDemographer; **525**, Rainer Plendl/Shutterstock; **529 (L-R)**, Richard Megna/Fundamental Photographs; **530**, gryth/Fotolia; **534**, Charles D. Winters/Photo Researchers, Inc.; **535**, Renn Sminkey/Pearson Science; **538**, Kevin Mastin

Chapter 14: **550**, Reuters/China Daily; **564**, Richard Megna/Fundamental Photographs; **565**, Dave Bevan/age fotostock; **568**, Richard Megna/Fundamental Photographs; **574**, Jeff Greenberg/Alamy; **576**, NASA; **577**, Pearson Education/Eric Schrader; **581**, Richard Megna/Fundamental Photographs; **586**, Richard Megna/Fundamental Photographs; **588**, Ben Osborne/Getty Images; **589**, Chi King, Used under a Creative Commons License. http://creativecommons.org/licenses/by/2.0/deed.en

Chapter 15: **600**, Rob Lavinsky/iRocks.com; **601**, Richard Megna/Fundamental Photographs; **602**, Richard Megna/Fundamental Photographs; **605**, Richard Megna/Fundamental Photographs; **611 (L-R)**, Richard Megna/Fundamental Photographs; **611**, Richard Megna/Fundamental Photographs; **613**, Donald Clegg and Roxy Wilson/Pearson Education; **615**, Richard Megna/Fundamental Photographs; **626**, Steve Sant/Alamy; **627**, Richard Megna/Fundamental Photographs; **629**, Richard Megna/Fundamental Photographs; **630**, Gustoimages/Photo Researchers, Inc.; **633**, Danita Delimont/Alamy; **634 (L-R)**, Richard Megna/Fundamental Photographs; **637 (L-R)**, **639**, Richard Megna/Fundamental Photographs; **640**, Richard Megna/Fundamental Photographs; **642 (TL-TR)**, Richard Megna/Fundamental Photographs; **642 (C)**, Richard Megna/Fundamental Photographs; **642 (B)**, Mediscan; **643**, Mark Miller/Photo Researchers, Inc.

Chapter 16: **654**, Cathy Keifer/Shutterstock; **656 (T)**, archives/iStockphoto; **656 (B)**, Tatiana Popova/Shutterstock; **660**, Paul Silverman/Fundamental Photographs; **669 (T)**, Vitaly Raduntsev/Fotolia; **669 (C)**, Taiga/Shutterstock; **682**, Richard Megna/Fundamental Photographs; **684**, NASA, ESA, and M. Livio and the Hubble 20th Anniversary Team (STScI)

Chapter 17: **696**, Steve Allen/Stone/Getty Images; **698 (T-B)**, Richard Megna/Fundamental Photographs; **699**, Richard Megna/Fundamental Photographs; **700**, Eric Schrader/Pearson Education; **720 (T)**, Liz Strenk/SuperStock; **720 (B)**, Tony Freeman/PhotoEdit; **721**, Photos.com; **722**, Drive Images/Newscom; **724**, Caitlin Mirra/Shutterstock; **725 (T)**, Tobias Helbig/iStockphoto; **725 (B)**, Lawrence Schwartzwald/SplashNews/Newscom; **726 (T)**, PiLensPhoto/Fotolia; **726 (B)**, Mar Photographics/Alamy; **731**, ALCOA; **732 (T)**, AP Photo/Axel Seidemann; **732 (B)**, David Huntley/Shutterstock; **735 (T)**, Radius Images/Alamy; **735 (B)**, Chis DeRuggiero/Anodic Process Tech.; **740**, Eric Schrader/Pearson Education; **742**, Freefly/Fotolia; **743**, SLOFotomedia/iStockphoto

Chapter 18: **746**, Honda Media/American Honda Motor Co.; **747**, National Optical Astronomy Observatories; **748**, Charles Winters/Photo Researchers, Inc.; **750**, SuperStock; **751**, Richard Megna/Fundamental Photographs; **754**, Richard Megna/Fundamental Photographs; **755**, Oleg - F/Shutterstock; **756 (L)**, McCracken Photographers/Pearson Education; **756 (C)**, Richard Megna/Fundamental Photographs; **756 (R)**, Tom Bochsler/Pearson Education; **758**, PeterPhoto123/Shutterstock; **760**, Richard Megna/Fundamental Photographs; **762 (L-R)**, Richard Megna/Fundamental Photographs; **763 (L-R)**, Richard Megna/Fundamental Photographs; **765**, Sheila Terry/Photo Researchers, Inc.; **766**, Richard Megna/Fundamental Photographs; **767**, Adam Brookes

Chapter 19: **774**, Steve Reisman; **780 (T)**, Richard Megna/Fundamental Photographs; **780 (B)**, U.S. Borax Inc.; **782**, Richard Megna/Fundamental Photographs; **783 (L)**, Richard Megna/Fundamental Photographs; **783 (C)**, Russell Lappa/Photo Researchers, Inc.; **783 (R)**, Charles D. Winters/Photo Researchers, Inc.; **784**, Michael J. Bronikowski; **785**, Martyn F. Chillmaid/Photo Researchers, Inc.; **786**, ITAR-TASS/Alexander Kolbasov/Newscom; **787 (T)**, Texas Instruments Incorporated; **787 (B)**, Harry Taylor/Dorling Kindersley; **788 (L)**, Marvin Dembinsky Photo Associates/Alamy; **788 (R)**, Sergey Lavrentev/Shutterstock; **789 (CL-CR)**, Charles D. Winters/Photo Researchers, Inc.; **789 (TR)**, Gontar/Shutterstock; **789 (BR)**, ephotocorp/Alamy;

791 (TL), Harry Taylor/Dorling Kindersley; **791 (BL)**, Russ Lappa/Science Source/Photo Researchers, Inc.; **791 (R)**, Paul Silverman/Fundamental Photographs; **794 (TL-TC)**, Donald Clegg and Roxy Wilson/Pearson Education; **794 (TR)**, Richard Megna/Fundamental Photographs; **794 (B)**, Kristen Brochmann/Fundamental Photographs; **795**, Richard Megna/Fundamental Photographs; **796**, Tom Bochsler/Pearson Education; **797 (L-R)**, Richard Megna/Fundamental Photographs; **798 (T)**, Richard Megna/Fundamental Photographs; **798 (BL)**, Charles D. Winters/Photo Researchers, Inc.; **798 (BR)**, Russ Lappa/Photo Researchers, Inc.; **799 (T)**, J.D. Dallet/age fotostock; **799 (B)**, Marvin Dembinsky Photo Associates/Alamy; **800 (T)**, lissart/Shutterstock; **800 (BL-BR)**, Richard Megna/Fundamental Photographs; **805**, Chris Howes/Wild Places Photography/Alamy

Chapter 20: **814**, Neil Emmerson/age fotostock; **818**, AlessandroZocc/Shutterstock; **822**, Richard Megna/Fundamental Photographs; **823**, Robert Barker, Cornell University Photography/Herbert F. Johnson Museum of Art; **824**, Donald Clegg and Roxy Wilson/Pearson Education; **825 (L-R)**, Richard Megna/Fundamental Photographs; **826 (TL-TR)**, Richard Megna/Fundamental Photographs; **826 (B)**, Richard Megna/Fundamental Photographs; **827**, Allison Cornford-Matheson/iStockphoto; **828**, Paul Silverman/Fundamental Photographs; **833**, Richard Megna/Fundamental Photographs; **835**, Richard Megna/Fundamental Photographs; **837 (L-R)**, Richard Megna/Fundamental Photographs; **845 (L-R)**, Richard Megna/Fundamental Photographs; **846**, Richard Megna/Fundamental Photographs; **855**, Howard Ande/age fotostock; **862**, Richard Megna/Fundamental Photographs; **863**, Richard Megna/Fundamental Photographs

Chapter 21: **866**, Associated Press/Nathan Denettek,CP; **867**, Philippe Psaila/SPL/Photo Researchers, Inc.; **868 (L)**, Linda Burgess/Dorling Kindersley; **868 (C-R)**, Paul Silverman/Fundamental Photographs; **869**, Jupiterimages/Photos.com; **871 (T)**, ThePropShoppe/iStockphoto; **871 (B)**, Erich Hartmann/Magnum Photos; **873**, Bethlehem Steel/AFP/Newscom; **874**, Peter Poulides/Getty Images; **884**, Sami Sarkis/Getty Images; **885**, Intel Corporation; **886**, Chemical Design Ltd./Science Photo Library/Photo Researchers, Inc.; **887 (T)**, Richard Megna/Fundamental Photographs; **887 (B)**, Associated Press; **889**, Dr. Hua-Tay Lin/Oak Ridge National Laboratory; **891 (L-R)**, Corning Incorporated; **892**, Northrop Grumman Corporation/Military Aircraft Systems Division; **893 (L)**, Dr Mathias Brust; **893 (R)**, Comstock Images/Photos.com; **894 (T)**, Michael S. Wong; **894 (B)**, Dr. Raoul Kopelman

Chapter 22: **904**, Reuters/Alessandro Bianchi; **912**, Los Alamos National Laboratory; **913**, Philip Lange/Shutterstock; **915**, Fermilab Visual Media Services; **916**, Ablestock.com/Thinkstock; **918**, C Gascoigne/Robert Harding; **919**, Peter Buckley/Pearson Education; **920 (T)**, Keith Brofsky/Getty Images; **920 (C)**, Martin Dohrn/Science Photo Library/Photo Researchers, Inc.; **920 (B)**, Australian Nuclear Science and Technology Organisation; **921 (T)**, Roger Tully/Getty Images; **921 (B)**, Nagra

Chapter 23: **926**, Robert Holmgren/Getty Images; **941**, Jill Chen/iStockphoto; **942**, Liveshot/Shutterstock; **945**, Grant Faith/Getty Images; **948**, Richard Weiss/Silver Burdett Ginn; **952**, witthaya/Fotolia; **953**, mycola/Shutterstock; **965**, Laura Clay-Ballard/iStockphoto

Useful Conversion Factors and Relationships

Length

SI unit: meter (m)

1 km = 10^3 m = 0.621 37 mi
1 mi = 5280 ft = 1760 yd = 1.6093 km
1 m = 10^2 cm = 1.0936 yd
1 in. = 2.54 cm (exactly)
1 cm = 0.393 70 in.
1 Å = 10^{-10} m = 100 pm

Mass

SI unit: kilogram (kg)

1 kg = 10^3 g = 2.2046 lb
1 lb = 16 oz = 453.59 g
1 oz = 28.35 g
1 ton = 2000 lb = 907.185 kg
1 metric ton = 10^3 kg = 1.102 tons
1 u = $1.660\ 54 \times 10^{-27}$ kg

Temperature

SI unit: kelvin (K)

0 K = −273.15 °C = −459.67 °F
K = °C + 273.15
$°C = \frac{5}{9}(°F - 32)$
$°F = \frac{9}{5}(°C) + 32$

Energy (derived)

SI unit: joule (J)

$1\ J = 1\ (kg \cdot m^2)/s^2 = 0.239\ 01$ cal
= 1 C × 1 V
1 cal = 4.184 J (exactly)
1 eV = $1.602\ 176 \times 10^{-19}$ J
1 MeV = $1.602\ 176 \times 10^{-13}$ J
1 kWh = 3.600×10^6 J
1 Btu = 1055 J

Pressure (derived)

SI unit: pascal (Pa)

$1\ Pa = 1\ N/m^2 = 1\ kg/(m \cdot s^2)$
1 atm = 101,325 Pa = 1.013 25 bar
= 760 mm Hg (torr)
= 14.70 lb/in.^{2}
1 bar = 10^5 Pa

Volume (derived)

SI unit: cubic meter (m^3)

$1\ L = 10^{-3}\ m^3 = 1\ dm^3 = 10^3\ cm^3$
= 1.0567 qt
1 gal = 4 qt = 3.7854 L
1 cm^3 = 1 mL
1 in.^{3} = 16.4 cm^3

Fundamental Constants

Atomic mass unit	1 u	$= 1.660\ 539 \times 10^{-27}$ kg
	1 g	$= 6.022\ 141 \times 10^{23}$ u
Avogadro's number	N_A	$= 6.022\ 141 \times 10^{23}$/mol
Boltzmann's constant	k	$= 1.380\ 649 \times 10^{-23}$ J/K
Electron charge	$-e$	$= -1.602\ 177 \times 10^{-19}$ C
Electron charge-to-mass ratio	$-e/m_e$	$= -1.758\ 820 \times 10^{11}$ C/kg
Electron mass	m_e	$= 5.485\ 798 \times 10^{-4}$ u
		$= 9.109\ 383 \times 10^{-31}$ kg
Elementary charge	e	$= 1.602\ 177 \times 10^{-19}$ C
Faraday's constant	F	$= 9.648\ 534 \times 10^4$ C/mol
Gas constant	R	$= 8.314\ 462$ J/(mol · K)
		$= 0.082\ 058\ 1$ (L · atm)/(mol · K)
Neutron mass	m_n	$= 1.008\ 665$ u
		$= 1.674\ 927 \times 10^{-27}$ kg
Pi	π	$= 3.141\ 592\ 6536$
Planck's constant	h	$= 6.626\ 069\ 57 \times 10^{-34}$ J · s
Proton mass	m_p	$= 1.007\ 276$ u
		$= 1.672\ 622 \times 10^{-27}$ kg
Rydberg constant	R_∞	$= 1.097\ 373 \times 10^7$/m
Speed of light	c	$= 2.997\ 924\ 58 \times 10^8$ m/s